Probabilistic Graphical Models

Adaptive Computation and Machine Learning

Thomas Dietterich, Editor

Christopher Bishop, David Heckerman, Michael Jordan, and Michael Kearns, Associate Editors

Probabilistic Graphical Models

Principles and Techniques

Daphne Koller

Nir Friedman

The MIT Press
Cambridge, Massachusetts
London, England

For information about special quantity discounts, please email special_sales@mitpress.mit.edu

This book was set by the authors in LaTeX2$_\epsilon$.
Printed and bound in the United States of America.

Library of Congress Cataloging-in-Publication Data

Koller, Daphne.
Probabilistic Graphical Models: Principles and Techniques / Daphne Koller and Nir Friedman.
p. cm. – (Adaptive computation and machine learning)
Includes bibliographical references and index.
ISBN 978-0-262-01319-2 (hardcover : alk. paper)
1. Graphical modeling (Statistics) 2. Bayesian statistical decision theory—Graphic methods. I. Koller, Daphne. II. Friedman, Nir.
QA279.5.K65 2010
519.5'420285–dc22
2009008615

10   9   8   7   6   5

To our families

> *my parents Dov and Ditza*
> *my husband Dan*
> *my daughters Natalie and Maya*
> D.K.

> *my parents Noga and Gad*
> *my wife Yael*
> *my children Roy and Lior*
> N.F.

As far as the laws of mathematics refer to reality, they are not certain, as far as they are certain, they do not refer to reality.

Albert Einstein, 1921

When we try to pick out anything by itself, we find that it is bound fast by a thousand invisible cords that cannot be broken, to everything in the universe.

John Muir, 1869

The actual science of logic is conversant at present only with things either certain, impossible, or entirely doubtful ... Therefore the true logic for this world is the calculus of probabilities, which takes account of the magnitude of the probability which is, or ought to be, in a reasonable man's mind.

James Clerk Maxwell, 1850

The theory of probabilities is at bottom nothing but common sense reduced to calculus; it enables us to appreciate with exactness that which accurate minds feel with a sort of instinct for which ofttimes they are unable to account.

Pierre Simon Laplace, 1819

Misunderstanding of probability may be the greatest of all impediments to scientific literacy.

Stephen Jay Gould

Contents

Acknowledgments

This book owes a considerable debt of gratitude to the many people who contributed to its creation, and to those who have influenced our work and our thinking over the years.

First and foremost, we want to thank our students, who, by asking the right questions, and forcing us to formulate clear and precise answers, were directly responsible for the inception of this book and for any clarity of presentation.

We have been fortunate to share the same mentors, who have had a significant impact on our development as researchers and as teachers: Joe Halpern, Stuart Russell. Much of our core views on probabilistic models have been influenced by Judea Pearl. Judea through his persuasive writing and vivid presentations inspired us, and many other researchers of our generation, to plunge into research in this field.

There are many people whose conversations with us have helped us in thinking through some of the more difficult concepts in the book: Nando de Freitas, Gal Elidan, Dan Geiger, Amir Globerson, Uri Lerner, Chris Meek, David Sontag, Yair Weiss, and Ramin Zabih. Others, in conversations and collaborations over the year, have also influenced our thinking and the presentation of the material: Pieter Abbeel, Jeff Bilmes, Craig Boutilier, Moises Goldszmidt, Carlos Guestrin, David Heckerman, Eric Horvitz, Tommi Jaakkola, Michael Jordan, Kevin Murphy, Andrew Ng, Ben Taskar, and Sebastian Thrun.

We especially want to acknowledge Gal Elidan for constant encouragement, valuable feedback, and logistic support at many critical junctions, throughout the long years of writing this book.

Over the course of the years of work on this book, many people have contributed to it by providing insights, engaging in enlightening discussions, and giving valuable feedback. It is impossible to individually acknowledge all of the people who made such contributions. However, we specifically wish to express our gratitude to those people who read large parts of the book and gave detailed feedback: Rahul Biswas, James Cussens, James Diebel, Yoni Donner, Tal El-Hay, Gal Elidan, Stanislav Funiak, Amir Globerson, Russ Greiner, Carlos Guestrin, Tim Heilman, Geremy Heitz, Maureen Hillenmeyer, Ariel Jaimovich, Tommy Kaplan, Jonathan Laserson, Ken Levine, Brian Milch, Kevin Murphy, Ben Packer, Ronald Parr, Dana Pe'er, and Christian Shelton.

We are deeply grateful to the following people, who contributed specific text and/or figures, mostly to the case studies and concept boxes without which this book would be far less interesting: Gal Elidan, to chapter 11, chapter 18, and chapter 19; Stephen Gould, to chapter 4 and chapter 13; Vladimir Jojic, to chapter 12; Jonathan Laserson, to chapter 19; Uri Lerner, to chapter 14; Andrew McCallum and Charles Sutton, to chapter 4; Brian Milch, to chapter 6; Kevin

Murphy, to chapter 15; and Benjamin Packer, to many of the exercises used throughout the book. In addition, we are very grateful to Amir Globerson, David Sontag and Yair Weiss whose insights on chapter 13 played a key role in the development of the material in that chapter.

Special thanks are due to Bob Prior at MIT Press who convinced us to go ahead with this project and was constantly supportive, enthusiastic and patient in the face of the recurring delays and missed deadlines. We thank Greg McNamee, our copy editor, and Mary Reilly, our artist, for their help in improving this book considerably. We thank Chris Manning, for allowing us to use his LaTeX macros for typesetting this book, and for providing useful advice on how to use them. And we thank Miles Davis for invaluable technical support.

We also wish to thank the many colleagues who used drafts of this book in teaching provided enthusiastic feedback that encouraged us to continue this project at times where it seemed unending. Sebastian Thrun deserves a special note of thanks, for forcing us to set a deadline for completion of this book and to stick to it.

We also want to thank the past and present members of the DAGS group at Stanford, and the Computational Biology group at the Hebrew University, many of whom also contributed ideas, insights, and useful comments. We specifically want to thank them for bearing with us while we devoted far too much of our time to working on this book.

Finally, noone deserves our thanks more than our long-suffering families — Natalie Anna Koller Avida, Maya Rika Koller Avida, and Dan Avida; Lior, Roy, and Yael Friedman — for their continued love, support, and patience, as they watched us work evenings and weekends to complete this book. We could never have done this without you.

List of Figures

List of Algorithms

List of Boxes

1 *Introduction*

1.1 Motivation

Most tasks require a person or an automated system to *reason*: to take the available information and reach conclusions, both about what might be true in the world and about how to act. For example, a doctor needs to take information about a patient — his symptoms, test results, personal characteristics (gender, weight) — and reach conclusions about what diseases he may have and what course of treatment to undertake. A mobile robot needs to synthesize data from its sonars, cameras, and other sensors to conclude where in the environment it is and how to move so as to reach its goal without hitting anything. A speech-recognition system needs to take a noisy acoustic signal and infer the words spoken that gave rise to it.

In this book, we describe a general framework that can be used to allow a computer system to answer questions of this type. In principle, one could write a special-purpose computer program for every domain one encounters and every type of question that one may wish to answer. The resulting system, although possibly quite successful at its particular task, is often very brittle: If our application changes, significant changes may be required to the program. Moreover, this general approach is quite limiting, in that it is hard to extract lessons from one successful solution and apply it to one which is very different.

declarative representation

model

We focus on a different approach, based on the concept of a *declarative representation*. In this approach, we construct, within the computer, a *model* of the system about which we would like to reason. This model encodes our knowledge of how the system works in a computer-readable form. This representation can be manipulated by various algorithms that can answer questions based on the model. For example, a model for medical diagnosis might represent our knowledge about different diseases and how they relate to a variety of symptoms and test results. A reasoning algorithm can take this model, as well as observations relating to a particular patient, and answer questions relating to the patient's diagnosis. **The key property of a declarative representation is the separation of knowledge and reasoning. The representation has its own clear semantics, separate from the algorithms that one can apply to it. Thus, we can develop a general suite of algorithms that apply any model within a broad class, whether in the domain of medical diagnosis or speech recognition. Conversely, we can improve our model for a specific application domain without having to modify our reasoning algorithms constantly.**

Declarative representations, or model-based methods, are a fundamental component in many fields, and models come in many flavors. Our focus in this book is on models for complex sys-

uncertainty

tems that involve a significant amount of *uncertainty*. Uncertainty appears to be an inescapable aspect of most real-world applications. It is a consequence of several factors. We are often uncertain about the true state of the system because our observations about it are partial: only some aspects of the world are observed; for example, the patient's true disease is often not directly observable, and his future prognosis is never observed. Our observations are also noisy — even those aspects that are observed are often observed with some error. The true state of the world is rarely determined with certainty by our limited observations, as most relationships are simply not deterministic, at least relative to our ability to model them. For example, there are few (if any) diseases where we have a clear, universally true relationship between the disease and its symptoms, and even fewer such relationships between the disease and its prognosis. Indeed, while it is not clear whether the universe (quantum mechanics aside) is deterministic when modeled at a sufficiently fine level of granularity, it is quite clear that it is not deterministic relative to our current understanding of it. To summarize, uncertainty arises because of limitations in our ability to observe the world, limitations in our ability to model it, and possibly even because of innate nondeterminism.

Because of this ubiquitous and fundamental uncertainty about the true state of world, we need to allow our reasoning system to consider different possibilities. One approach is simply to consider any state of the world that is possible. Unfortunately, it is only rarely the case that we can completely eliminate a state as being impossible given our observations. In our medical diagnosis example, there is usually a huge number of diseases that are *possible* given a particular set of observations. Most of them, however, are highly unlikely. If we simply list all of the possibilities, our answers will often be vacuous of meaningful content (e.g., "the patient can have any of the following 573 diseases"). **Thus, to obtain meaningful conclusions, we need to reason not just about what is possible, but also about what is *probable*.**

probability theory

The calculus of *probability theory* (see section 2.1) provides us with a formal framework for considering multiple possible outcomes and their likelihood. It defines a set of mutually exclusive and exhaustive possibilities, and associates each of them with a *probability* — a number between 0 and 1, so that the total probability of all possibilities is 1. This framework allows us to consider options that are unlikely, yet not impossible, without reducing our conclusions to content-free lists of every possibility.

Furthermore, one finds that probabilistic models are very liberating. Where in a more rigid formalism we might find it necessary to enumerate every possibility, here we can often sweep a multitude of annoying exceptions and special cases under the "probabilistic rug," by introducing outcomes that roughly correspond to "something unusual happens." In fact, as we discussed, this type of approximation is often inevitable, as we can only rarely (if ever) provide a deterministic specification of the behavior of a complex system. Probabilistic models allow us to make this fact explicit, and therefore often provide a model which is more faithful to reality.

1.2 Structured Probabilistic Models

This book describes a general-purpose framework for constructing and using probabilistic models of complex systems. We begin by providing some intuition for the principles underlying this framework, and for the models it encompasses. This section requires some knowledge of basic concepts in probability theory; a reader unfamiliar with these concepts might wish to read section 2.1 first.

Complex systems are characterized by the presence of multiple interrelated aspects, many of which relate to the reasoning task. For example, in our medical diagnosis application, there are multiple possible diseases that the patient might have, dozens or hundreds of symptoms and diagnostic tests, personal characteristics that often form predisposing factors for disease, and many more matters to consider. These domains can be characterized in terms of a set of *random variables*, where the value of each variable defines an important property of the world. For example, a particular disease, such as *Flu*, may be one variable in our domain, which takes on two values, for example, *present* or *absent*; a symptom, such as *Fever*, may be a variable in our domain, one that perhaps takes on continuous values. The set of possible variables and their values is an important design decision, and it depends strongly on the questions we may wish to answer about the domain.

Our task is to reason probabilistically about the values of one or more of the variables, possibly given observations about some others. In order to do so using principled probabilistic reasoning, we need to construct a *joint distribution* over the space of possible assignments to some set of random variables $\mathcal{X}$. This type of model allows us to answer a broad range of interesting queries. For example, we can make the observation that a variable X_i takes on the specific value x_i, and ask, in the resulting *posterior distribution*, what the probability distribution is over values of another variable X_j.

random variable

joint probability distribution

posterior distribution

Example 1.1

Consider a very simple medical diagnosis setting, where we focus on two diseases — flu and hayfever; these are not mutually exclusive, as a patient can have either, both, or none. Thus, we might have two binary-valued random variables, Flu and Hayfever. We also have a 4-valued random variable Season, which is correlated both with flu and hayfever. We may also have two symptoms, Congestion and Muscle Pain, each of which is also binary-valued. Overall, our probability space has $2 \times 2 \times 4 \times 2 \times 2 = 64$ values, corresponding to the possible assignments to these five variables. Given a joint distribution over this space, we can, for example, ask questions such as how likely the patient is to have the flu given that it is fall, and that she has sinus congestion but no muscle pain; as a probability expression, this query would be denoted

$$P(Flu = true \mid Season = fall, Congestion = true, Muscle\ Pain = false).$$

■

1.2.1 Probabilistic Graphical Models

Specifying a joint distribution over 64 possible values, as in example 1.1, already seems fairly daunting. When we consider the fact that a typical medical- diagnosis problem has dozens or even hundreds of relevant attributes, the problem appears completely intractable. This book describes the framework of probabilistic graphical models, which provides a mechanism for exploiting structure in complex distributions to describe them compactly, and in a way that allows them to be constructed and utilized effectively.

Probabilistic graphical models use a graph-based representation as the basis for compactly encoding a complex distribution over a high-dimensional space. In this graphical representation, illustrated in figure 1.1, the nodes (or ovals) correspond to the variables in our domain, and the edges correspond to direct probabilistic interactions between them. For example, figure 1.1a (top)

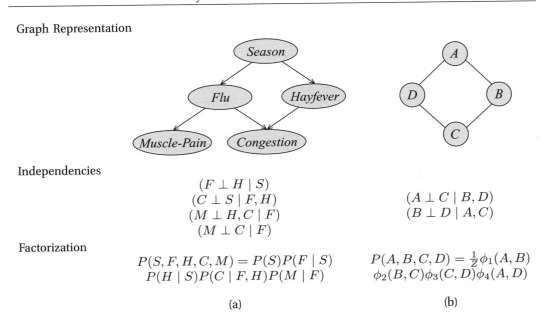

Figure 1.1 middle/bottom content:

Independencies

$(F \perp H \mid S)$
$(C \perp S \mid F, H)$
$(M \perp H, C \mid F)$
$(M \perp C \mid F)$

$(A \perp C \mid B, D)$
$(B \perp D \mid A, C)$

Factorization

$P(S, F, H, C, M) = P(S)P(F \mid S)$
$P(H \mid S)P(C \mid F, H)P(M \mid F)$

$P(A, B, C, D) = \frac{1}{Z}\phi_1(A, B)$
$\phi_2(B, C)\phi_3(C, D)\phi_4(A, D)$

(a)   (b)

Figure 1.1   Different perspectives on probabilistic graphical models: top — the graphical representation; middle — the independencies induced by the graph structure; bottom — the factorization induced by the graph structure. (a) A sample Bayesian network. (b) A sample Markov network.

illustrates one possible graph structure for our flu example. In this graph, we see that there is no direct interaction between *Muscle Pain* and *Season*, but both interact directly with *Flu*.

There is a dual perspective that one can use to interpret the structure of this graph. From one perspective, the graph is a compact representation of a set of *independencies* that hold in the distribution; these properties take the form X is independent of Y given Z, denoted $(X \perp Y \mid Z)$, for some subsets of variables X, Y, Z. For example, our "target" distribution P for the preceding example — the distribution encoding our beliefs about this particular situation — may satisfy the conditional independence (*Congestion* $\perp$ *Season* $\mid$ *Flu, Hayfever*). This statement asserts that

$$P(\text{*Congestion*} \mid \text{*Flu, Hayfever, Season*}) = P(\text{*Congestion*} \mid \text{*Flu, Hayfever*});$$

that is, if we are interested in the distribution over the patient having congestion, and we know whether he has the flu and whether he has hayfever, the season is no longer informative. Note that this assertion does *not* imply that *Season* is independent of *Congestion*; only that all of the information we may obtain from the season on the chances of having congestion we already obtain by knowing whether the patient has the flu and has hayfever. Figure 1.1a (middle) shows the set of independence assumptions associated with the graph in figure 1.1a (top).

factor

The other perspective is that the graph defines a skeleton for compactly representing a high-dimensional distribution: Rather than encode the probability of every possible assignment to all of the variables in our domain, we can "break up" the distribution into smaller *factors*, each over a much smaller space of possibilities. We can then define the overall joint distribution as a product of these factors. For example, figure 1.1(a-bottom) shows the factorization of the distribution associated with the graph in figure 1.1 (top). It asserts, for example, that the probability of the event "spring, no flu, hayfever, sinus congestion, muscle pain" can be obtained by multiplying five numbers: $P(Season = spring)$, $P(Flu = false \mid Season = spring)$, $P(Hayfever = true \mid Season = spring)$, $P(Congestion = true \mid Hayfever = true, Flu = false)$, and $P(Muscle\ Pain = true \mid Flu = false)$. This parameterization is significantly more compact, requiring only $3+4+4+4+2 = 17$ nonredundant parameters, as opposed to 63 nonredundant parameters for the original joint distribution (the 64th parameter is fully determined by the others, as the sum over all entries in the joint distribution must sum to 1). The graph structure defines the factorization of a distribution P associated with it — the set of factors and the variables that they encompass.

☞ **It turns out that these two perspectives — the graph as a representation of a set of independencies, and the graph as a skeleton for factorizing a distribution — are, in a deep sense, equivalent. The independence properties of the distribution are precisely what allow it to be represented compactly in a factorized form. Conversely, a particular factorization of the distribution guarantees that certain independencies hold.**

Bayesian network

Markov network

We describe two families of graphical representations of distributions. One, called *Bayesian networks*, uses a directed graph (where the edges have a source and a target), as shown in figure 1.1a (top). The second, called *Markov networks*, uses an undirected graph, as illustrated in figure 1.1b (top). It too can be viewed as defining a set of independence assertions (figure 1.1b [middle] or as encoding a compact factorization of the distribution (figure 1.1b [bottom]). Both representations provide the duality of independencies and factorization, but they differ in the set of independencies they can encode and in the factorization of the distribution that they induce.

1.2.2 Representation, Inference, Learning

The graphical language exploits structure that appears present in many distributions that we want to encode in practice: the property that variables tend to interact *directly* only with very few others. Distributions that exhibit this type of structure can generally be encoded naturally and compactly using a graphical model.

This framework has many advantages. First, it often allows the distribution to be written down tractably, even in cases where the explicit representation of the joint distribution is astronomically large. Importantly, the type of representation provided by this framework is *transparent*, in that a human expert can understand and evaluate its semantics and properties. This property is important for constructing models that provide an accurate reflection of our understanding of a domain. Models that are opaque can easily give rise to unexplained, and even undesirable, answers.

inference

Second, as we show, the same structure often also allows the distribution to be used effectively for *inference* — answering queries using the distribution as our model of the world. In particular, we provide algorithms for computing the posterior probability of some variables given evidence on others. For example, we might observe that it is spring and the patient has muscle pain, and we wish to know how likely he is to have the flu, a query that can formally be written as $P(Flu = true \mid Season = spring, Muscle\ Pain = true)$. These inference algorithms work directly on the graph structure and are generally orders of magnitude faster than manipulating the joint distribution explicitly.

Third, this framework facilitates the effective construction of these models, whether by a human expert or automatically, by *learning* from data a model that provides a good approximation to our past experience. For example, we may have a set of patient records from a doctor's office and wish to learn a probabilistic model encoding a distribution consistent with our aggregate experience. Probabilistic graphical models support a *data-driven approach* to model construction that is very effective in practice. In this approach, a human expert provides some rough guidelines on how to model a given domain. For example, the human usually specifies the attributes that the model should contain, often some of the main dependencies that it should encode, and perhaps other aspects. The details, however, are usually filled in automatically, by fitting the model to data. The models produced by this process are usually much better reflections of the domain than models that are purely hand-constructed. Moreover, they can sometimes reveal surprising connections between variables and provide novel insights about a domain.

data-driven
approach

 These three components — representation, inference, and learning — are critical components in constructing an intelligent system. We need a declarative representation that is a reasonable encoding of our world model. We need to be able to use this representation effectively to answer a broad range of questions that are of interest. And we need to be able to acquire this distribution, combining expert knowledge and accumulated data. Probabilistic graphical models are one of a small handful of frameworks that support all three capabilities for a broad range of problems.

1.3   Overview and Roadmap

1.3.1   Overview of Chapters

The framework of probabilistic graphical models is quite broad, and it encompasses both a variety of different types of models and a range of methods relating to them. This book describes several types of models. For each one, we describe the three fundamental cornerstones: representation, inference, and learning.

We begin in part I, by describing the most basic type of graphical models, which are the focus of most of the book. These models encode distributions over a fixed set $\mathcal{X}$ of random variables. We describe how graphs can be used to encode distributions over such spaces, and what the properties of such distributions are.

Specifically, in chapter 3, we describe the Bayesian network representation, based on directed graphs. We describe how a Bayesian network can encode a probability distribution. We also analyze the independence properties induced by the graph structure.

In chapter 4, we move to Markov networks, the other main category of probabilistic graphical models. Here also we describe the independencies defined by the graph and the induced factorization of the distribution. We also discuss the relationship between Markov networks and Bayesian networks, and briefly describe a framework that unifies both.

In chapter 5, we delve a little deeper into the representation of the parameters in probabilistic models, focusing mostly on Bayesian networks, whose parameterization is more constrained. We describe representations that capture some of the finer-grained structure of the distribution, and show that, here also, capturing structure can provide significant gains.

In chapter 6, we turn to formalisms that extend the basic framework of probabilistic graphical models to settings where the set of variables is no longer rigidly circumscribed in advance. One such setting is a *temporal* one, where we wish to model a system whose state evolves over time, requiring us to consider distributions over entire trajectories, We describe a compact representation — a *dynamic Bayesian network* — that allows us to represent structured systems that evolve over time. We then describe a family of extensions that introduce various forms of higher level structure into the framework of probabilistic graphical models. Specifically, we focus on domains containing *objects* (whether concrete or abstract), characterized by attributes, and related to each other in various ways. Such domains can include repeated structure, since different objects of the same type share the same probabilistic model. These languages provide a significant extension to the expressive power of the standard graphical models.

In chapter 7, we take a deeper look at models that include continuous variables. Specifically, we explore the properties of the multivariate Gaussian distribution and the representation of such distributions as both directed and undirected graphical models. Although the class of Gaussian distributions is a limited one and not suitable for all applications, it turns out to play a critical role even when dealing with distributions that are not Gaussian.

In chapter 8, we take a deeper, more technical look at probabilistic models, defining a general framework called the *exponential family*, that encompasses a broad range of distributions. This chapter provides some basic concepts and tools that will turn out to play an important role in later development.

We then turn, in part II, to a discussion of the inference task. In chapter 9, we describe the basic ideas underlying exact inference in probabilistic graphical models. We first analyze the fundamental difficulty of the exact inference task, separately from any particular inference algorithm we might develop. We then present two basic algorithms for exact inference — variable elimination and conditioning — both of which are equally applicable to both directed and undirected models. Both of these algorithms can be viewed as operating over the graph structure defined by the probabilistic model. They build on basic concepts, such as graph properties and dynamic programming algorithms, to provide efficient solutions to the inference task. We also provide an analysis of their computational cost in terms of the graph structure, and we discuss where exact inference is feasible.

In chapter 10, we describe an alternative view of exact inference, leading to a somewhat different algorithm. The benefit of this alternative algorithm is twofold. First, it uses dynamic programming to avoid repeated computations in settings where we wish to answer more than a single query using the same network. Second, it defines a natural algorithm that uses message passing on a graph structure; this algorithm forms the basis for approximate inference algorithms developed in later chapters.

Because exact inference is computationally intractable for many models of interest, we then proceed to describe approximate inference algorithms, which trade off accuracy with computational cost. We present two main classes of such algorithms. In chapter 11, we describe a class of methods that can be viewed from two very different perspectives: On one hand, they are direct generalizations of the graph-based message-passing approach developed for the case of exact inference in chapter 10. On the other hand, they can be viewed as solving an optimization problem: one where we approximate the distribution of interest using a simpler representation that allows for feasible inference. The equivalence of these views provides important insights and suggests a broad family of algorithms that one can apply to approximate inference.

In chapter 12, we describe a very different class of methods: *particle-based methods*, which approximate a complex joint distribution by considering samples from it (also known as particles). We describe several methods from this general family. These methods are generally based on core techniques from statistics, such as importance sampling and Markov-chain Monte Carlo methods. Once again, the connection to this general class of methods suggests multiple opportunities for new algorithms.

While the representation of probabilistic graphical models applies, to a great extent, to models including both discrete and continuous-valued random variables, inference in models involving continuous variables is significantly more challenging than the purely discrete case. In chapter 14, we consider the task of inference in continuous and *hybrid* (continuous/discrete) networks, and we discuss whether and how the exact and approximate inference methods developed in earlier chapters can be applied in this setting.

The representation that we discussed in chapter 6 allows a compact encoding of networks whose size can be unboundedly large. Such networks pose particular challenges to inference algorithms. In this chapter, we discuss some special-purpose methods that have been developed for the particular settings of networks that model dynamical systems.

We then turn, in part III, to the third of our main topics — learning probabilistic models from data. We begin in chapter 16 by reviewing some of the fundamental concepts underlying the general task of learning models from data. We then present the spectrum of learning problems that we address in this part of the book. These problems vary along two main axes: the extent to which we are given prior knowledge specifying the model, and whether the data from which we learn contain complete observations of all of the relevant variables. In contrast to the inference task, where the same algorithms apply equally to Bayesian networks and Markov networks, the learning task is quite different for these two classes of models. We begin with studying the learning task for Bayesian networks.

In chapter 17, we focus on the most basic learning task: learning parameters for a Bayesian network with a given structure, from fully observable data. Although this setting may appear somewhat restrictive, it turns out to form the basis for our entire development of Bayesian network learning. As we show, the factorization of the distribution, which was central both to representation and to inference, also plays a key role in making inference feasible.

We then move, in chapter 18, to the harder problem of learning both Bayesian network structure and the parameters, still from fully observed data. The learning algorithms we present trade off the accuracy with which the learned network represents the empirical distribution for the complexity of the resulting structure. As we show, the type of independence assumptions underlying the Bayesian network representation often hold, at least approximately, in real-world distributions. Thus, these learning algorithms often result in reasonably compact structures that capture much of the signal in the distribution.

In chapter 19, we address the Bayesian network learning task in a setting where we have access only to partial observations of the relevant variables (for example, when the available patient records have missing entries). This type of situation occurs often in real-world settings. Unfortunately, the resulting learning task is considerably harder, and the resulting algorithms are both more complex and less satisfactory in terms of their performance.

We conclude the discussion of learning in chapter 20 by considering the problem of learning Markov networks from data. It turns out that the learning tasks for Markov networks are significantly harder than the corresponding problem for Bayesian networks. We explain the difficulties and discuss the existing solutions.

Finally, in part IV, we turn to a different type of extension, where we consider the use of this framework for other forms of reasoning. Specifically, we consider cases where we can act, or intervene, in the world.

In chapter 21, we focus on the semantics of intervention and its relation to causality. We present the notion of a *causal model*, which allows us to answer not only queries of the form "if I observe X, what do I learn about Y," but also *intervention queries*, of the form "if I manipulate X, what effect does it have on Y."

We then turn to the task of *decision making* under uncertainty. Here, we must consider not only the distribution over different states of the world, but also the preferences of the agent regarding these outcomes. In chapter 22, we discuss the notion of *utility functions* and how they can encode an agent's preferences about complex situations involving multiple variables. As we show, the same ideas that we used to provide compact representations of probability distribution can also be used for utility functions.

In chapter 23, we describe a unified representation for decision making, called *influence diagrams*. Influence diagrams extend Bayesian networks by introducing actions and utilities. We present algorithms that use influence diagrams for making decisions that optimize the agent's expected utility. These algorithms utilize many of the same ideas that formed the basis for exact inference in Bayesian networks.

We conclude with a high-level synthesis of the techniques covered in this book, and with some guidance on how to use them in tackling a new problem.

1.3.2 Reader's Guide

As we mentioned, the topics described in this book relate to multiple fields, and techniques from other disciplines — probability theory, computer science, information theory, optimization, statistics, and more — are used in various places throughout it. While it is impossible to present all of the relevant material within the scope of this book, we have attempted to make the book somewhat self-contained by providing a very brief review of the key concepts from these related disciplines in chapter 2.

Some of this material, specifically the review of probability theory and of graph-related concepts, is very basic yet central to most of the development in this book. Readers who are less familiar with these topics may wish to read these sections carefully, and even knowledgeable readers may wish to briefly review them to gain familiarity with the notations used. Other background material, covering such topics as information theory, optimization, and algorithmic concepts, can be found in the appendix.

The chapters in the book are structured as follows. The main text in each chapter provides the detailed technical development of the key ideas. Beyond the main text, most chapters contain boxes that contain interesting material that augments these ideas. These boxes come in three types: *Skill boxes* describe "hands-on" tricks and techniques, which, while often heuristic in nature, are important for getting the basic algorithms described in the text to work in practice. *Case study boxes* describe empirical case studies relating to the techniques described in the text.

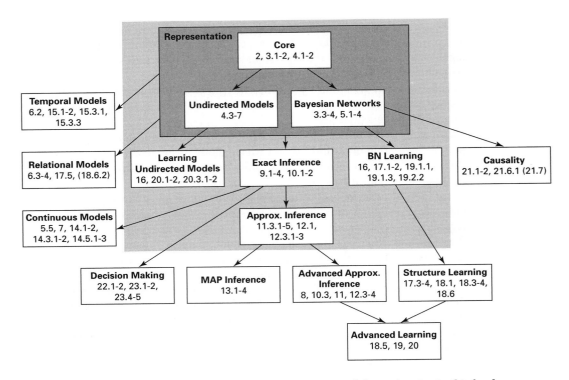

Figure 1.2   A reader's guide to the structure and dependencies in this book

These case studies include both empirical results on how the algorithms perform in practice and descriptions of applications of these algorithms to interesting domains, illustrating some of the issues encountered in practice. Finally, *concept boxes* present particular instantiations of the material described in the text, which have had significant impact in their own right.

This textbook is clearly too long to be used in its entirety in a one-semester class. Figure 1.2 tries to delineate some coherent subsets of the book that can be used for teaching and other purposes. The small, labeled boxes represent "units" of material on particular topics. Arrows between the boxes represent dependencies between these units. The first enclosing box (solid line) represents material that is fundamental to everything else, and that should be read by anyone using this book. One can then use the dependencies between the boxes to expand or reduce the depth of the coverage on any given topic. The material in the larger box (dashed line) forms a good basis for a one-semester (or even one-quarter) overview class. Some of the sections in the book are marked with an asterisk, denoting the fact that they contain more technically advanced material. In most cases, these sections are self-contained, and they can be skipped without harming the reader's ability to understand the rest of the text.

We have attempted in this book to present a synthesis of ideas, most of which have been developed over many years by multiple researchers. To avoid futile attempts to divide up the credit precisely, we have omitted all bibliographical references from the technical presentation in the chapters. Rather, each chapter ends with a section called "Relevant Literature," which describes the historical evolution of the material in the chapter, acknowledges the papers and books that developed the key concepts, and provides some additional readings on material relevant to the chapter. We encourage the reader who is interested in a topic to follow up on some of these additional readings, since there are many interesting developments that we could not cover in this book.

Finally, each chapter includes a set of exercises that explore in additional depth some of the material described in the text and present some extensions to it. The exercises are annotated with an asterisk for exercises that are somewhat more difficult, and with two asterisks for ones that are truly challenging.

Additional material related to this book, including slides and figures, solutions to some of the exercises, and errata, can be found online at `http://pgm.stanford.edu`.

1.3.3 Connection to Other Disciplines

The ideas we describe in this book are connected to many fields. From probability theory, we inherit the basic concept of a probability distribution, as well as many of the operations we can use to manipulate it. From computer science, we exploit the key idea of using a graph as a data structure, as well as a variety of algorithms for manipulating graphs and other data structures. These algorithmic ideas and the ability to manipulate probability distributions using discrete data structures are some of the key elements that make the probabilistic manipulations tractable. Decision theory extends these basic ideas to the task of decision making under uncertainty and provides the formal foundation for this task.

From computer science, and specifically from artificial intelligence, these models inherit the idea of using a declarative representation of the world to separate procedural reasoning from our domain knowledge. This idea is of key importance to the generality of this framework and its applicability to such a broad range of tasks.

Various ideas from other disciplines also arise in this field. Statistics plays an important role both in certain aspects of the representation and in some of the work on learning models from data. Optimization plays a role in providing algorithms both for approximate inference and for learning models from data. Bayesian networks first arose, albeit in a restricted way, in the setting of modeling genetic inheritance in human family trees; in fact, restricted version of some of the exact inference algorithms we discuss were first developed in this context. Similarly, undirected graphical models first arose in physics as a model for systems of electrons, and some of the basic concepts that underlie recent work on approximate inference developed from that setting.

Information theory plays a dual role in its interaction with this field. Information-theoretic concepts such as entropy and information arise naturally in various settings in this framework, such as evaluating the quality of a learned model. Thus, tools from this discipline are a key component in our analytic toolkit. On the other side, the recent successes in coding theory, based on the relationship between inference in probabilistic models and the task of decoding messages sent over a noisy channel, have led to a resurgence of work on approximate inference in graphical models. The resulting developments have revolutionized both the development of error-correcting codes and the theory and practice of approximate message-passing algorithms in graphical models.

1.3.3.1   What Have We Gained?

Although the framework we describe here shares common elements with a broad range of other topics, it has a coherent common core: the use of structure to allow a compact representation, effective reasoning, and feasible learning of general-purpose, factored, probabilistic models. These elements provide us with a general infrastructure for reasoning and learning about complex domains.

As we discussed earlier, by using a declarative representation, we essentially separate out the description of the model for the particular application, and the general-purpose algorithms used for inference and learning. Thus, this framework provides a general algorithmic toolkit that can be applied to many different domains.

Indeed, probabilistic graphical models have made a significant impact on a broad spectrum of real-world applications. For example, these models have been used for medical and fault diagnosis, for modeling human genetic inheritance of disease, for segmenting and denoising images, for decoding messages sent over a noisy channel, for revealing genetic regulatory processes, for robot localization and mapping, and more. Throughout this book, we will describe how probabilistic graphical models were used to address these applications and what issues arise in the application of these models in practice.

In addition to practical applications, these models provide a formal framework for a variety of fundamental problems. For example, the notion of conditional independence and its explicit graph-based representation provide a clear formal semantics for irrelevance of information. This framework also provides a general methodology for handling data fusion — we can introduce *sensor variables* that are noisy versions of the true measured quantity, and use Bayesian conditioning to combine the different measurements. The use of a probabilistic model allows us to provide a formal measure for model quality, in terms of a numerical fit of the model to observed data; this measure underlies much of our work on learning models from data. The temporal models we define provide a formal framework for defining a general trend toward persistence of state over time, in a way that does not raise inconsistencies when change does occur.

In general, part of the rich development in this field is due to the close and continuous interaction between theory and practice. In this field, unlike many others, the distance between theory and practice is quite small, and there is a constant flow of ideas and problems between them. Problems or ideas arise in practical applications and are analyzed and subsequently developed in more theoretical papers. Algorithms for which no theoretical analysis exists are tried out in practice, and the profile of where they succeed and fail often provides the basis for subsequent analysis. This rich synergy leads to a continuous and vibrant development, and it is a key factor in the success of this area.

1.4   Historical Notes

The foundations of probability theory go back to the sixteenth century, when Gerolamo Cardano began a formal analysis of games of chance, followed by additional key developments by Pierre de Fermat and Blaise Pascal in the seventeenth century. The initial development involved only discrete probability spaces, and the analysis methods were purely combinatorial. The foundations of modern probability theory, with its measure-theoretic underpinnings, were laid by Andrey Kolmogorov in the 1930s.

Particularly central to the topics of this book is the so-called *Bayes theorem*, shown in the eighteenth century by the Reverend Thomas Bayes (Bayes 1763). This theorem allows us to use a model that tells us the conditional probability of event a given event b (say, a symptom given a disease) in order to compute the contrapositive: the conditional probability of event b given event a (the disease given the symptom). This type of reasoning is central to the use of graphical models, and it explains the choice of the name *Bayesian network*.

The notion of representing the interactions between variables in a multidimensional distribution using a graph structure originates in several communities, with very different motivations. In the area of statistical physics, this idea can be traced back to Gibbs (1902), who used an undirected graph to represent the distribution over a system of interacting particles. In the area of genetics, this idea dates back to the work on path analysis of Sewal Wright (Wright 1921, 1934). Wright proposed the use of a directed graph to study inheritance in natural species. This idea, although largely rejected by statisticians at the time, was subsequently adopted by economists and social scientists (Wold 1954; Blalock, Jr. 1971). In the field of statistics, the idea of analyzing interactions between variables was first proposed by Bartlett (1935), in the study of *contingency tables*, also known as *log-linear models*. This idea became more accepted by the statistics community in the 1960s and 70s (Vorobev 1962; Goodman 1970; Haberman 1974).

In the field of computer science, probabilistic methods lie primarily in the realm of Artificial Intelligence (AI). The AI community first encountered these methods in the endeavor of building *expert systems*, computerized systems designed to perform difficult tasks, such as oil-well location or medical diagnosis, at an expert level. Researchers in this field quickly realized the need for methods that allow the integration of multiple pieces of evidence, and that provide support for making decisions under uncertainty. Some early systems (de Bombal et al. 1972; Gorry and Barnett 1968; Warner et al. 1961) used probabilistic methods, based on the very restricted *naive Bayes model*. This model restricts itself to a small set of possible hypotheses (e.g., diseases) and assumes that the different evidence variables (e.g., symptoms or test results) are independent given each hypothesis. These systems were surprisingly successful, performing (within their area of expertise) at a level comparable to or better than that of experts. For example, the system of de Bombal et al. (1972) averaged over 90 percent correct diagnoses of acute abdominal pain, whereas expert physicians were averaging around 65 percent.

Despite these successes, this approach fell into disfavor in the AI community, owing to a combination of several factors. One was the belief, prevalent at the time, that artificial intelligence should be based on similar methods to human intelligence, combined with a strong impression that people do not manipulate numbers when reasoning. A second issue was the belief that the strong independence assumptions made in the existing expert systems were fundamental to the approach. Thus, the lack of a flexible, scalable mechanism to represent interactions between variables in a distribution was a key factor in the rejection of the probabilistic framework.

The rejection of probabilistic methods was accompanied by the invention of a range of alternative formalisms for reasoning under uncertainty, and the construction of expert systems based on these formalisms (notably Prospector by Duda, Gaschnig, and Hart 1979 and Mycin by Buchanan and Shortliffe 1984). Most of these formalisms used the production rule framework, where each rule is augmented with some number(s) defining a measure of "confidence" in its validity. These frameworks largely lacked formal semantics, and many exhibited significant problems in key reasoning patterns. Other frameworks for handling uncertainty proposed at the time included fuzzy logic, possibility theory, and Dempster-Shafer belief functions. For a

expert systems

discussion of some of these alternative frameworks see Shafer and Pearl (1990); Horvitz et al. (1988); Halpern (2003).

The widespread acceptance of probabilistic methods began in the late 1980s, driven forward by two major factors. The first was a series of seminal theoretical developments. The most influential among these was the development of the Bayesian network framework by Judea Pearl and his colleagues in a series of paper that culminated in Pearl's highly influential textbook *Probabilistic Reasoning in Intelligent Systems* (Pearl 1988). In parallel, the key paper by S.L. Lauritzen and D.J. Spiegelhalter 1988 set forth the foundations for efficient reasoning using probabilistic graphical models. The second major factor was the construction of large-scale, highly successful expert systems based on this framework that avoided the unrealistically strong assumptions made by early probabilistic expert systems. The most visible of these applications was the Pathfinder expert system, constructed by Heckerman and colleagues (Heckerman et al. 1992; Heckerman and Nathwani 1992b), which used a Bayesian network for diagnosis of pathology samples.

At this time, although work on other approaches to uncertain reasoning continues, probabilistic methods in general, and probabilistic graphical models in particular, have gained almost universal acceptance in a wide range of communities. They are in common use in fields as diverse as medical diagnosis, fault diagnosis, analysis of genetic and genomic data, communication and coding, analysis of marketing data, speech recognition, natural language understanding, and many more. Several other books cover aspects of this growing area; examples include Pearl (1988); Lauritzen (1996); Jensen (1996); Castillo et al. (1997a); Jordan (1998); Cowell et al. (1999); Neapolitan (2003); Korb and Nicholson (2003). The Artificial Intelligence textbook of Russell and Norvig (2003) places this field within the broader endeavor of constructing an intelligent agent.

2 *Foundations*

In this chapter, we review some important background material regarding key concepts from probability theory, information theory, and graph theory. This material is included in a separate introductory chapter, since it forms the basis for much of the development in the remainder of the book. Other background material — such as discrete and continuous optimization, algorithmic complexity analysis, and basic algorithmic concepts — is more localized to particular topics in the book. Many of these concepts are presented in the appendix; others are presented in concept boxes in the appropriate places in the text. All of this material is intended to focus only on the minimal subset of ideas required to understand most of the discussion in the remainder of the book, rather than to provide a comprehensive overview of the field it surveys. We encourage the reader to explore additional sources for more details about these areas.

2.1 Probability Theory

The main focus of this book is on complex probability distributions. In this section we briefly review basic concepts from probability theory.

2.1.1 Probability Distributions

When we use the word "probability" in day-to-day life, we refer to a degree of confidence that an event of an uncertain nature will occur. For example, the weather report might say "there is a low probability of light rain in the afternoon." Probability theory deals with the formal foundations for discussing such estimates and the rules they should obey.

Before we discuss the representation of probability, we need to define what the events are to which we want to assign a probability. These events might be different outcomes of throwing a die, the outcome of a horse race, the weather configurations in California, or the possible failures of a piece of machinery.

2.1.1.1 Event Spaces

event

outcome space

Formally, we define *events* by assuming that there is an agreed upon *space* of possible outcomes, which we denote by Ω. For example, if we consider dice, we might set $\Omega = \{1, 2, 3, 4, 5, 6\}$. In the case of a horse race, the space might be all possible orders of arrivals at the finish line, a much larger space.

measurable event   In addition, we assume that there is a set of *measurable events* $\mathcal{S}$ to which we are willing to assign probabilities. Formally, each event $\alpha \in \mathcal{S}$ is a subset of Ω. In our die example, the event $\{6\}$ represents the case where the die shows 6, and the event $\{1, 3, 5\}$ represents the case of an odd outcome. In the horse-race example, we might consider the event "Lucky Strike wins," which contains all the outcomes in which the horse Lucky Strike is first.

Probability theory requires that the event space satisfy three basic properties:

- It contains the *empty event* $\emptyset$, and the *trivial event* Ω.
- It is closed under union. That is, if $\alpha, \beta \in \mathcal{S}$, then so is $\alpha \cup \beta$.
- It is closed under complementation. That is, if $\alpha \in \mathcal{S}$, then so is $\Omega - \alpha$.

The requirement that the event space is closed under union and complementation implies that it is also closed under other Boolean operations, such as intersection and set difference.

2.1.1.2   Probability Distributions

Definition 2.1   *A probability distribution P over $(\Omega, \mathcal{S})$ is a mapping from events in $\mathcal{S}$ to real values that satisfies*
probability   *the following conditions:*
distribution
- $P(\alpha) \geq 0$ *for all* $\alpha \in \mathcal{S}$.
- $P(\Omega) = 1$.
- *If* $\alpha, \beta \in \mathcal{S}$ *and* $\alpha \cap \beta = \emptyset$, *then* $P(\alpha \cup \beta) = P(\alpha) + P(\beta)$.   ∎

The first condition states that probabilities are not negative. The second states that the "trivial event," which allows all possible outcomes, has the maximal possible probability of 1. The third condition states that the probability that one of two mutually disjoint events will occur is the sum of the probabilities of each event. These two conditions imply many other conditions. Of particular interest are $P(\emptyset) = 0$, and $P(\alpha \cup \beta) = P(\alpha) + P(\beta) - P(\alpha \cap \beta)$.

2.1.1.3   Interpretations of Probability

Before we continue to discuss probability distributions, we need to consider the interpretations that we might assign to them. Intuitively, the probability $P(\alpha)$ of an event α quantifies the degree of confidence that α will occur. If $P(\alpha) = 1$, we are certain that one of the outcomes in α occurs, and if $P(\alpha) = 0$, we consider all of them impossible. Other probability values represent options that lie between these two extremes.

This description, however, does not provide an answer to what the numbers mean. There are two common interpretations for probabilities.

frequentist   The *frequentist* interpretation views probabilities as frequencies of events. More precisely, the
interpretation   probability of an event is the fraction of times the event occurs if we repeat the experiment indefinitely. For example, suppose we consider the outcome of a particular die roll. In this case, the statement $P(\alpha) = 0.3$, for $\alpha = \{1, 3, 5\}$, states that if we repeatedly roll this die and record the outcome, then the fraction of times the outcomes in α will occur is 0.3. More precisely, the limit of the sequence of fractions of outcomes in α in the first roll, the first two rolls, the first three rolls, ..., the first n rolls, ... is 0.3.

The frequentist interpretation gives probabilities a tangible semantics. When we discuss concrete physical systems (for example, dice, coin flips, and card games) we can envision how these frequencies are defined. It is also relatively straightforward to check that frequencies must satisfy the requirements of proper distributions.

The frequentist interpretation fails, however, when we consider events such as "It will rain tomorrow afternoon." Although the time span of "Tomorrow afternoon" is somewhat ill defined, we expect it to occur exactly once. It is not clear how we define the frequencies of such events. Several attempts have been made to define the probability for such an event by finding a *reference class* of similar events for which frequencies are well defined; however, none of them has proved entirely satisfactory. Thus, the frequentist approach does not provide a satisfactory interpretation for a statement such as "the probability of rain tomorrow afternoon is 0.3."

An alternative interpretation views probabilities as *subjective* degrees of belief. Under this interpretation, the statement $P(\alpha) = 0.3$ represents a subjective statement about one's own degree of belief that the event α will come about. Thus, the statement "the probability of rain tomorrow afternoon is 50 percent" tells us that in the opinion of the speaker, the chances of rain and no rain tomorrow afternoon are the same. Although tomorrow afternoon will occur only once, we can still have uncertainty about its outcome, and represent it using numbers (that is, probabilities).

This description still does not resolve what exactly it means to hold a particular degree of belief. What stops a person from stating that the probability that Bush will win the election is 0.6 and the probability that he will lose is 0.8? The source of the problem is that we need to explain how subjective degrees of beliefs (something that is internal to each one of us) are reflected in our actions.

This issue is a major concern in subjective probabilities. One possible way of attributing degrees of beliefs is by a betting game. Suppose you believe that $P(\alpha) = 0.8$. Then you would be willing to place a bet of \$1 against \$3. To see this, note that with probability 0.8 you gain a dollar, and with probability 0.2 you lose \$3, so on average this bet is a good deal with expected gain of 20 cents. In fact, you might be even tempted to place a bet of \$1 against \$4. Under this bet the average gain is 0, so you should not mind. However, you would not consider it worthwhile to place a bet \$1 against \$4 and 10 cents, since that would have negative expected gain. Thus, by finding which bets you are willing to place, we can assess your degrees of beliefs.

The key point of this mental game is the following. If you hold degrees of belief that do not satisfy the rule of probability, then by a clever construction we can find a series of bets that would result in a sure negative outcome for you. Thus, the argument goes, a rational person must hold degrees of belief that satisfy the rules of probability.[1]

In the remainder of the book we discuss probabilities, but we usually do not explicitly state their interpretation. Since both interpretations lead to the same mathematical rules, the technical definitions hold for both interpretations.

---

1. As stated, this argument assumes as that people's preferences are directly proportional to their expected earnings. For small amounts of money, this assumption is quite reasonable. We return to this topic in chapter 22.

(margin notes)
reference class

☞

subjective
interpretation

2.1.2   Basic Concepts in Probability

2.1.2.1   Conditional Probability

To use a concrete example, suppose we consider a distribution over a population of students taking a certain course. The space of outcomes is simply the set of all students in the population. Now, suppose that we want to reason about the students' intelligence and their final grade. We can define the event α to denote "all students with grade A," and the event β to denote "all students with high intelligence." Using our distribution, we can consider the probability of these events, as well as the probability of $\alpha \cap \beta$ (the set of intelligent students who got grade A). This, however, does not directly tell us how to update our beliefs given new evidence. Suppose we learn that a student has received the grade A; what does that tell us about her intelligence?

This kind of question arises every time we want to use distributions to reason about the real world. More precisely, after learning that an event α is true, how do we change our probability about β occurring? The answer is via the notion of *conditional probability*. Formally, the conditional probability of β given α is defined as

conditional
probability

$$P(\beta \mid \alpha) = \frac{P(\alpha \cap \beta)}{P(\alpha)} \qquad (2.1)$$

That is, the probability that β is true given that we know α is the relative proportion of outcomes satisfying β among these that satisfy α. (Note that the conditional probability is not defined when $P(\alpha) = 0$.)

The conditional probability given an event (say α) satisfies the properties of definition 2.1 (see exercise 2.4), and thus it is a probability distribution by its own right. Hence, we can think of the conditioning operation as taking one distribution and returning another over the same probability space.

2.1.2.2   Chain Rule and Bayes Rule

From the definition of the conditional distribution, we immediately see that

$$P(\alpha \cap \beta) = P(\alpha)P(\beta \mid \alpha). \qquad (2.2)$$

chain rule

This equality is known as the *chain rule* of conditional probabilities. More generally, if $\alpha_1, \ldots, \alpha_k$ are events, then we can write

$$P(\alpha_1 \cap \ldots \cap \alpha_k) = P(\alpha_1)P(\alpha_2 \mid \alpha_1) \cdots P(\alpha_k \mid \alpha_1 \cap \ldots \cap \alpha_{k-1}). \qquad (2.3)$$

In other words, we can express the probability of a combination of several events in terms of the probability of the first, the probability of the second given the first, and so on. It is important to notice that we can expand this expression using any order of events — the result will remain the same.

Bayes' rule

Another immediate consequence of the definition of conditional probability is *Bayes' rule*

$$P(\alpha \mid \beta) = \frac{P(\beta \mid \alpha)P(\alpha)}{P(\beta)}. \qquad (2.4)$$

A more general conditional version of Bayes' rule, where all our probabilities are conditioned on some background event γ, also holds:

$$P(\alpha \mid \beta \cap \gamma) = \frac{P(\beta \mid \alpha \cap \gamma)P(\alpha \mid \gamma)}{P(\beta \mid \gamma)}.$$

Bayes' rule is important in that it allows us to compute the conditional probability $P(\alpha \mid \beta)$ from the "inverse" conditional probability $P(\beta \mid \alpha)$.

Example 2.1

Consider the student population, and let Smart denote smart students and GradeA denote students who got grade A. Assume we believe (perhaps based on estimates from past statistics) that $P(GradeA \mid Smart) = 0.6$, and now we learn that a particular student received grade A. Can we estimate the probability that the student is smart? According to Bayes' rule, this depends on our **prior** *probability for students being smart (before we learn anything about them) and the prior probability of students receiving high grades. For example, suppose that $P(Smart) = 0.3$ and $P(GradeA) = 0.2$, then we have that $P(Smart \mid GradeA) = 0.6 * 0.3/0.2 = 0.9$. That is, an A grade strongly suggests that the student is smart. On the other hand, if the test was easier and high grades were more common, say, $P(GradeA) = 0.4$ then we would get that $P(Smart \mid GradeA) = 0.6 * 0.3/0.4 = 0.45$, which is much less conclusive about the student.* ∎

prior

Another classic example that shows the importance of this reasoning is in disease screening. To see this, consider the following hypothetical example (none of the mentioned figures are related to real statistics).

Example 2.2

Suppose that a tuberculosis (TB) skin test is 95 percent accurate. That is, if the patient is TB-infected, then the test will be positive with probability 0.95, and if the patient is not infected, then the test will be negative with probability 0.95. Now suppose that a person gets a positive test result. What is the probability that he is infected? Naive reasoning suggests that if the test result is wrong 5 percent of the time, then the probability that the subject is infected is 0.95. That is, 95 percent of subjects with positive results have TB.

If we consider the problem by applying Bayes' rule, we see that we need to consider the prior probability of TB infection, and the probability of getting positive test result. Suppose that 1 in 1000 of the subjects who get tested is infected. That is, $P(TB) = 0.001$. What is the probability of getting a positive test result? From our description, we see that $0.001 \cdot 0.95$ infected subjects get a positive result, and $0.999 \cdot 0.05$ uninfected subjects get a positive result. Thus, $P(Positive) = 0.0509$. Applying Bayes' rule, we get that $P(TB \mid Positive) = 0.001 \cdot 0.95/0.0509 \approx 0.0187$. Thus, although a subject with a positive test is much more probable to be TB-infected than is a random subject, fewer than 2 percent of these subjects are TB-infected. ∎

2.1.3 Random Variables and Joint Distributions

2.1.3.1 Motivation

Our discussion of probability distributions deals with events. Formally, we can consider any event from the set of measurable events. The description of events is in terms of sets of outcomes. In many cases, however, it would be more natural to consider *attributes* of the outcome. For example, if we consider a patient, we might consider attributes such as "age," "gender," and "smoking history" that are relevant for assigning probability over possible diseases and symptoms. We would like then consider events such as "age > 55, heavy smoking history, and suffers from repeated cough."

To use a concrete example, consider again a distribution over a population of students in a course. Suppose that we want to reason about the intelligence of students, their final grades, and so forth. We can use an event such as *GradeA* to denote the subset of students that received the grade A and use it in our formulation. However, this discussion becomes rather cumbersome if we also want to consider students with grade B, students with grade C, and so on. Instead, we would like to consider a way of directly referring to a student's grade in a clean, mathematical way.

random variable   The formal machinery for discussing attributes and their values in different outcomes are *random variables*. A random variable is a way of reporting an attribute of the outcome. For example, suppose we have a random variable *Grade* that reports the final grade of a student, then the statement $P(Grade = A)$ is another notation for $P(GradeA)$.

2.1.3.2   What Is a Random Variable?

Formally, a random variable, such as *Grade*, is defined by a function that associates with each outcome in Ω a value. For example, *Grade* is defined by a function f_{Grade} that maps each person in Ω to his or her grade (say, one of A, B, or C). The event *Grade* $= A$ is a shorthand for the event $\{\omega \in \Omega : f_{Grade}(\omega) = A\}$. In our example, we might also have a random variable *Intelligence* that (for simplicity) takes as values either "high" or "low." In this case, the event "*Intelligence = high*" refers, as can be expected, to the set of smart (high intelligence) students.

Random variables can take different sets of values. We can think of *categorical* (or *discrete*) random variables that take one of a few values, as in our two preceding examples. We can also talk about random variables that can take infinitely many values (for example, integer or real values), such as *Height* that denotes a student's height. We use $Val(X)$ to denote the set of values that a random variable X can take.

In most of the discussion in this book we examine either categorical random variables or random variables that take real values. We will usually use uppercase roman letters X, Y, Z to denote random variables. In discussing generic random variables, we often use a lowercase letter to refer to a value of a random variable. Thus, we use x to refer to a generic value of X. For example, in statements such as "$P(X = x) \geq 0$ for all $x \in Val(X)$." When we discuss categorical random variables, we use the notation $x^1, \ldots, x^k$, for $k = |Val(X)|$ (the number of elements in $Val(X)$), when we need to enumerate the specific values of X, for example, in statements such as

$$\sum_{i=1}^{k} P(X = x^i) = 1.$$

multinomial distribution   The distribution over such a variable is called a *multinomial*. In the case of a binary-valued random variable X, where $Val(X) = \{false, true\}$, we often use x^1 to denote the value *true* for X, and x^0 to denote the value *false*. The distribution of such a random variable is called a **Bernoulli distribution**   *Bernoulli distribution*.

We also use boldface type to denote sets of random variables. Thus, $\boldsymbol{X}$, $\boldsymbol{Y}$, or $\boldsymbol{Z}$ are typically used to denote a set of random variables, while $\boldsymbol{x}$, $\boldsymbol{y}$, $\boldsymbol{z}$ denote assignments of values to the variables in these sets. We extend the definition of $Val(\boldsymbol{X})$ to refer to sets of variables in the obvious way. Thus, $\boldsymbol{x}$ is always a member of $Val(\boldsymbol{X})$. For $\boldsymbol{Y} \subseteq \boldsymbol{X}$, we use $\boldsymbol{x}\langle \boldsymbol{Y} \rangle$ to refer to the assignment within $\boldsymbol{x}$ to the variables in $\boldsymbol{Y}$. For two assignments $\boldsymbol{x}$ (to $\boldsymbol{X}$) and $\boldsymbol{y}$ (to $\boldsymbol{Y}$), we say that $\boldsymbol{x} \sim \boldsymbol{y}$ if they agree on the variables in their intersection, that is, $\boldsymbol{x}\langle \boldsymbol{X} \cap \boldsymbol{Y} \rangle = \boldsymbol{y}\langle \boldsymbol{X} \cap \boldsymbol{Y} \rangle$.

In many cases, the notation $P(X = x)$ is redundant, since the fact that x is a value of X is already reported by our choice of letter. Thus, in many texts on probability, the identity of a random variable is not explicitly mentioned, but can be inferred through the notation used for its value. Thus, we use $P(x)$ as a shorthand for $P(X = x)$ when the identity of the random variable is clear from the context. Another shorthand notation is that $\sum_x$ refers to a sum over all possible values that X can take. Thus, the preceding statement will often appear as $\sum_x P(x) = 1$. Finally, another standard notation has to do with conjunction. Rather than write $P((X = x) \cap (Y = y))$, we write $P(X = x, Y = y)$, or just $P(x, y)$.

2.1.3.3 Marginal and Joint Distributions

Once we define a random variable X, we can consider the distribution over events that can be described using X. This distribution is often referred to as the *marginal distribution* over the random variable X. We denote this distribution by $P(X)$.

marginal distribution

Returning to our population example, consider the random variable *Intelligence*. The marginal distribution over *Intelligence* assigns probability to specific events such as $P(\textit{Intelligence} = \textit{high})$ and $P(\textit{Intelligence} = \textit{low})$, as well as to the trivial event $P(\textit{Intelligence} \in \{\textit{high}, \textit{low}\})$. Note that these probabilities are defined by the probability distribution over the original space. For concreteness, suppose that $P(\textit{Intelligence} = \textit{high}) = 0.3$, $P(\textit{Intelligence} = \textit{low}) = 0.7$.

If we consider the random variable *Grade*, we can also define a marginal distribution. This is a distribution over all events that can be described in terms of the *Grade* variable. In our example, we have that $P(\textit{Grade} = A) = 0.25$, $P(\textit{Grade} = B) = 0.37$, and $P(\textit{Grade} = C) = 0.38$.

It should be fairly obvious that the marginal distribution is a probability distribution satisfying the properties of definition 2.1. In fact, the only change is that we restrict our attention to the subsets of $\mathcal{S}$ that can be described with the random variable X.

In many situations, we are interested in questions that involve the values of several random variables. For example, we might be interested in the event "*Intelligence* = *high* and *Grade* = A." To discuss such events, we need to consider the *joint distribution* over these two random variables. In general, the joint distribution over a set $\mathcal{X} = \{X_1, \ldots, X_n\}$ of random variables is denoted by $P(X_1, \ldots, X_n)$ and is the distribution that assigns probabilities to events that are specified in terms of these random variables. We use ξ to refer to a full assignment to the variables in $\mathcal{X}$, that is, $\xi \in Val(\mathcal{X})$.

joint distribution

The joint distribution of two random variables has to be consistent with the marginal distribution, in that $P(x) = \sum_y P(x, y)$. This relationship is shown in figure 2.1, where we compute the marginal distribution over *Grade* by summing the probabilities along each row. Similarly, we find the marginal distribution over *Intelligence* by summing out along each column. The resulting sums are typically written in the row or column margins, whence the term "marginal distribution."

Suppose we have a joint distribution over the variables $\mathcal{X} = \{X_1, \ldots, X_n\}$. The most fine-grained events we can discuss using these variables are ones of the form "$X_1 = x_1$ and $X_2 = x_2$, …, and $X_n = x_n$" for a choice of values $x_1, \ldots, x_n$ for all the variables. Moreover,

		Intelligence		
		low	high	
	A	0.07	0.18	0.25
Grade	B	0.28	0.09	0.37
	C	0.35	0.03	0.38
		0.7	0.3	1

Figure 2.1   Example of a joint distribution $P(\textbf{\textit{Intelligence}}, \textbf{\textit{Grade}})$**:** Values of *Intelligence* (columns) and *Grade* (rows) with the associated marginal distribution on each variable.

any two such events must be either identical or disjoint, since they both assign values to all the variables in $\mathcal{X}$. In addition, any event defined using variables in $\mathcal{X}$ must be a union of a set of
canonical outcome space such events. Thus, we are effectively working in a *canonical outcome space*: a space where each outcome corresponds to a joint assignment to $X_1, \ldots, X_n$. More precisely, all our probability computations remain the same whether we consider the original outcome space (for example, all students), or the canonical space (for example, all combinations of intelligence and grade).
atomic outcome We use ξ to denote these *atomic outcomes*: those assigning a value to each variable in $\mathcal{X}$. For example, if we let $\mathcal{X} = \{Intelligence, Grade\}$, there are six atomic outcomes, shown in figure 2.1. The figure also shows one possible joint distribution over these six outcomes.

Based on this discussion, from now on we will not explicitly specify the set of outcomes and measurable events, and instead implicitly assume the canonical outcome space.

2.1.3.4   Conditional Probability

conditional distribution The notion of conditional probability extends to induced distributions over random variables. For example, we use the notation $P(Intelligence \mid Grade = A)$ to denote the *conditional distribution* over the events describable by *Intelligence* given the knowledge that the student's grade is A. Note that the conditional distribution over a random variable given an observation of the value of another one is not the same as the marginal distribution. In our example, $P(Intelligence = high) = 0.3$, and $P(Intelligence = high \mid Grade = A) = 0.18/0.25 = 0.72$. Thus, clearly $P(Intelligence \mid Grade = A)$ is different from the marginal distribution $P(Intelligence)$. The latter distribution represents our *prior* knowledge about students before learning anything else about a particular student, while the conditional distribution represents our more informed distribution after learning her grade.

We will often use the notation $P(X \mid Y)$ to represent a set of conditional probability distributions. Intuitively, for each value of Y, this object assigns a probability over values of X using the conditional probability. This notation allows us to write the shorthand version of the chain rule: $P(X, Y) = P(X)P(Y \mid X)$, which can be extended to multiple variables as

$$P(X_1, \ldots, X_k) = P(X_1)P(X_2 \mid X_1) \cdots P(X_k \mid X_1, \ldots, X_{k-1}). \tag{2.5}$$

Similarly, we can state Bayes' rule in terms of conditional probability distributions:

$$P(X \mid Y) = \frac{P(X)P(Y \mid X)}{P(Y)}. \tag{2.6}$$

2.1.4 Independence and Conditional Independence

2.1.4.1 Independence

As we mentioned, we usually expect $P(\alpha \mid \beta)$ to be different from $P(\alpha)$. That is, learning that β is true changes our probability over α. However, in some situations equality can occur, so that $P(\alpha \mid \beta) = P(\alpha)$. That is, learning that β occurs did not change our probability of α.

Definition 2.2

independent events

> *We say that an event α is* independent *of event β in P, denoted $P \models (\alpha \perp \beta)$, if $P(\alpha \mid \beta) = P(\alpha)$ or if $P(\beta) = 0$.* ∎

We can also provide an alternative definition for the concept of independence:

Proposition 2.1

> *A distribution P satisfies $(\alpha \perp \beta)$ if and only if $P(\alpha \cap \beta) = P(\alpha)P(\beta)$.*

PROOF Consider first the case where $P(\beta) = 0$; here, we also have $P(\alpha \cap \beta) = 0$, and so the equivalence immediately holds. When $P(\beta) \neq 0$, we can use the chain rule; we write $P(\alpha \cap \beta) = P(\alpha \mid \beta)P(\beta)$. Since α is independent of β, we have that $P(\alpha \mid \beta) = P(\alpha)$. Thus, $P(\alpha \cap \beta) = P(\alpha)P(\beta)$. Conversely, suppose that $P(\alpha \cap \beta) = P(\alpha)P(\beta)$. Then, by definition, we have that

$$P(\alpha \mid \beta) = \frac{P(\alpha \cap \beta)}{P(\beta)} = \frac{P(\alpha)P(\beta)}{P(\beta)} = P(\alpha).$$

∎

As an immediate consequence of this alternative definition, we see that independence is a symmetric notion. That is, $(\alpha \perp \beta)$ implies $(\beta \perp \alpha)$.

Example 2.3

> *For example, suppose that we toss two coins, and let α be the event "the first toss results in a head" and β the event "the second toss results in a head." It is not hard to convince ourselves that we expect that these two events to be independent. Learning that β is true would not change our probability of α. In this case, we see two different physical processes (that is, coin tosses) leading to the events, which makes it intuitive that the probabilities of the two are independent. In certain cases, the same process can lead to independent events. For example, consider the event α denoting "the die outcome is even" and the event β denoting "the die outcome is 1 or 2." It is easy to check that if the die is fair (each of the six possible outcomes has probability $\frac{1}{6}$), then these two events are independent.* ∎

2.1.4.2 Conditional Independence

 While independence is a useful property, it is not often that we encounter two independent events. A more common situation is when two events are independent given an additional event. For example, suppose we want to reason about the chance that our student is accepted to graduate studies at Stanford or MIT. Denote by *Stanford* the event "admitted to Stanford" and by *MIT* the event "admitted to MIT." In most reasonable distributions, these two events are not independent. If we learn that a student was admitted to Stanford, then our estimate of her probability of being accepted at MIT is now higher, since it is a sign that she is a promising student.

Now, suppose that both universities base their decisions only on the student's grade point average (GPA), and we know that our student has a GPA of A. In this case, we might argue that learning that the student was admitted to Stanford should not change the probability that she will be admitted to MIT: Her GPA already tells us the information relevant to her chances of admission to MIT, and finding out about her admission to Stanford does not change that. Formally, the statement is

$$P(\text{MIT} \mid \text{Stanford}, \text{GradeA}) = P(\text{MIT} \mid \text{GradeA}).$$

In this case, we say that *MIT* is *conditionally independent* of *Stanford* given *GradeA*.

Definition 2.3

conditional
independence

We say that an event α is conditionally independent *of event β given event γ in P, denoted $P \models (\alpha \perp \beta \mid \gamma)$, if $P(\alpha \mid \beta \cap \gamma) = P(\alpha \mid \gamma)$ or if $P(\beta \cap \gamma) = 0$.* ∎

It is easy to extend the arguments we have seen in the case of (unconditional) independencies to give an alternative definition.

Proposition 2.2

P satisfies $(\alpha \perp \beta \mid \gamma)$ if and only if $P(\alpha \cap \beta \mid \gamma) = P(\alpha \mid \gamma)P(\beta \mid \gamma)$.

2.1.4.3 Independence of Random Variables

Until now, we have focused on independence between events. Thus, we can say that two events, such as one toss landing heads and a second also landing heads, are independent. However, we would like to say that any pair of outcomes of the coin tosses is independent. To capture such statements, we can examine the generalization of independence to sets of random variables.

Definition 2.4

conditional
independence

observed variable

marginal
independence

Let X, Y, Z be sets of random variables. We say that X is conditionally independent *of Y given Z in a distribution P if P satisfies $(X = x \perp Y = y \mid Z = z)$ for all values $x \in \text{Val}(X)$, $y \in \text{Val}(Y)$, and $z \in \text{Val}(Z)$. The variables in the set Z are often said to be* observed. *If the set Z is empty, then instead of writing $(X \perp Y \mid \emptyset)$, we write $(X \perp Y)$ and say that X and Y are* marginally independent. ∎

Thus, an independence statement over random variables is a universal quantification over all possible values of the random variables.

The alternative characterization of conditional independence follows immediately:

Proposition 2.3

The distribution P satisfies $(X \perp Y \mid Z)$ if and only if $P(X, Y \mid Z) = P(X \mid Z)P(Y \mid Z)$.

Suppose we learn about a conditional independence. Can we conclude other independence properties that must hold in the distribution? We have already seen one such example:

symmetry

- **Symmetry**:

$$(X \perp Y \mid Z) \implies (Y \perp X \mid Z). \tag{2.7}$$

There are several other properties that hold for conditional independence, and that often provide a very clean method for proving important properties about distributions. Some key properties are:

decomposition

- **Decomposition**:

$$(X \perp Y, W \mid Z) \implies (X \perp Y \mid Z). \tag{2.8}$$

weak union

- **Weak union**:

$$(X \perp Y, W \mid Z) \implies (X \perp Y \mid Z, W). \tag{2.9}$$

contraction

- **Contraction**:

$$(X \perp W \mid Z, Y) \,\&\, (X \perp Y \mid Z) \implies (X \perp Y, W \mid Z). \tag{2.10}$$

An additional important property does not hold in general, but it does hold in an important subclass of distributions.

Definition 2.5

positive
distribution

A distribution P is said to be positive *if for all events $\alpha \in S$ such that $\alpha \neq \emptyset$, we have that $P(\alpha) > 0$.* ∎

For positive distributions, we also have the following property:

intersection

- **Intersection**: For positive distributions, and for mutually disjoint sets X, Y, Z, W:

$$(X \perp Y \mid Z, W) \,\&\, (X \perp W \mid Z, Y) \implies (X \perp Y, W \mid Z). \tag{2.11}$$

The proof of these properties is not difficult. For example, to prove Decomposition, assume that $(X \perp Y, W \mid Z)$ holds. Then, from the definition of conditional independence, we have that $P(X, Y, W \mid Z) = P(X \mid Z)P(Y, W \mid Z)$. Now, using basic rules of probability and arithmetic, we can show

$$
\begin{aligned}
P(X, Y \mid Z) &= \sum_w P(X, Y, w \mid Z) \\
&= \sum_w P(X \mid Z)P(Y, w \mid Z) \\
&= P(X \mid Z)\sum_w P(Y, w \mid Z) \\
&= P(X \mid Z)P(Y \mid Z).
\end{aligned}
$$

The only property we used here is called "reasoning by cases" (see exercise 2.6). We conclude that $(X \perp Y \mid Z)$.

2.1.5 Querying a Distribution

Our focus throughout this book is on using a joint probability distribution over multiple random variables to answer queries of interest.

2.1.5.1   Probability Queries

probability query

Perhaps the most common query type is the *probability query*. Such a query consists of two parts:

evidence

- The *evidence*: a subset E of random variables in the model, and an instantiation e to these variables;

query variables

- the *query variables*: a subset Y of random variables in the network.

Our task is to compute

$$P(Y \mid E = e),$$

posterior distribution

that is, the *posterior probability distribution* over the values y of Y, conditioned on the fact that $E = e$. This expression can also be viewed as the marginal over Y, in the distribution we obtain by conditioning on e.

2.1.5.2   MAP Queries

A second important type of task is that of finding a high-probability joint assignment to some subset of variables. The simplest variant of this type of task is the *MAP* query (also called *most probable explanation (MPE)*), whose aim is to find the *MAP assignment* — the most likely

MAP assignment

assignment to all of the (non-evidence) variables. More precisely, if we let $W = \mathcal{X} - E$, our task is to find the most likely assignment to the variables in W given the evidence $E = e$:

$$\text{MAP}(W \mid e) = \arg\max_{w} P(w, e), \tag{2.12}$$

where, in general, $\arg\max_x f(x)$ represents the value of x for which $f(x)$ is maximal. Note that there might be more than one assignment that has the highest posterior probability. In this case, we can either decide that the MAP task is to return the set of possible assignments, or to return an arbitrary member of that set.

It is important to understand the difference between MAP queries and probability queries. In a MAP query, we are finding the most likely *joint* assignment to W. To find the most likely assignment to a single variable A, we could simply compute $P(A \mid e)$ and then pick the most likely value. **However, the assignment where each variable individually picks its most likely value can be quite different from the most likely joint assignment to all variables simultaneously.** This phenomenon can occur even in the simplest case, where we have no evidence.

☞

Example 2.4

Consider a two node chain $A \to B$ where A and B are both binary-valued. Assume that:

$$\begin{array}{cc} a^0 & a^1 \\ \hline 0.4 & 0.6 \end{array} \qquad \begin{array}{c|cc} A & b^0 & b^1 \\ \hline a^0 & 0.1 & 0.9 \\ a^1 & 0.5 & 0.5 \end{array} \tag{2.13}$$

We can see that $P(a^1) > P(a^0)$, so that $\text{MAP}(A) = a^1$. However, $\text{MAP}(A, B) = (a^0, b^1)$: Both values of B have the same probability given a^1. Thus, the most likely assignment containing a^1 has probability $0.6 \times 0.5 = 0.3$. On the other hand, the distribution over values of B is more skewed given a^0, and the most likely assignment (a^0, b^1) has the probability $0.4 \times 0.9 = 0.36$. Thus, we have that $\arg\max_{a,b} P(a, b) \neq (\arg\max_a P(a), \arg\max_b P(b))$. ∎

2.1.5.3   Marginal MAP Queries

To motivate our second query type, let us return to the phenomenon demonstrated in example 2.4. Now, consider a medical diagnosis problem, where the most likely disease has multiple possible symptoms, each of which occurs with some probability, but not an overwhelming probability. On the other hand, a somewhat rarer disease might have only a few symptoms, each of which is very likely given the disease. As in our simple example, the MAP assignment to the data and the symptoms might be higher for the second disease than for the first one. The solution here is to look for the most likely assignment to the disease variable(s) only, rather than the most likely assignment to both the disease and symptom variables. This approach suggests the use of a more general query type. In the *marginal MAP* query, we have a subset of variables Y that forms our query. The task is to find the most likely assignment to the variables in Y given the evidence $E = e$:

marginal MAP

$$\mathrm{MAP}(Y \mid e) = \arg \max_{y} P(y \mid e).$$

If we let $Z = \mathcal{X} - Y - E$, the marginal MAP task is to compute:

$$\mathrm{MAP}(Y \mid e) = \arg \max_{Y} \sum_{Z} P(Y, Z \mid e).$$

Thus, marginal MAP queries contain both summations and maximizations; in a way, it contains elements of both a conditional probability query and a MAP query.

Note that example 2.4 shows that marginal MAP assignments are not monotonic: the most likely assignment $\mathrm{MAP}(Y_1 \mid e)$ might be completely different from the assignment to Y_1 in $\mathrm{MAP}(\{Y_1, Y_2\} \mid e)$. Thus, in particular, we cannot use a MAP query to give us the correct answer to a marginal MAP query.

2.1.6   Continuous Spaces

In the previous section, we focused on random variables that have a finite set of possible values. In many situations, we also want to reason about continuous quantities such as weight, height, duration, or cost that take real numbers in $\mathbb{R}$.

When dealing with probabilities over continuous random variables, we have to deal with some technical issues. For example, suppose that we want to reason about a random variable X that can take values in the range between 0 and 1. That is, $Val(X)$ is the interval $[0, 1]$. Moreover, assume that we want to assign each number in this range equal probability. What would be the probability of a number x? Clearly, since each x has the same probability, and there are infinite number of values, we must have that $P(X = x) = 0$. This problem appears even if we do not require uniform probability.

2.1.6.1   Probability Density Functions

How do we define probability over a continuous random variable? We say that a function $p : \mathbb{R} \mapsto \mathbb{R}$ is a *probability density function* or *(PDF)* for X if it is a nonnegative integrable

density function

function such that

$$\int_{Val(X)} p(x)dx = 1.$$

That is, the integral over the set of possible values of X is 1. The PDF defines a distribution for X as follows: for any x in our event space:

$$P(X \leq a) = \int_{-\infty}^{a} p(x)dx.$$

cumulative
distribution

The function P is the *cumulative distribution* for X. We can easily employ the rules of probability to see that by using the density function we can evaluate the probability of other events. For example,

$$P(a \leq X \leq b) = \int_{a}^{b} p(x)dx.$$

Intuitively, the value of a PDF $p(x)$ at a point x is the incremental amount that x adds to the cumulative distribution in the integration process. The higher the value of p at and around x, the more mass is added to the cumulative distribution as it passes x.

The simplest PDF is the uniform distribution.

Definition 2.6

uniform
distribution

A variable X has a uniform distribution *over $[a, b]$, denoted $X \sim \mathrm{Unif}[a, b]$ if it has the PDF*

$$p(x) = \begin{cases} \frac{1}{b-a} & b \geq x \geq a \\ 0 & \text{otherwise.} \end{cases}$$ ∎

Thus, the probability of any subinterval of $[a, b]$ is proportional its size relative to the size of $[a, b]$. Note that, if $b - a < 1$, then the density can be greater than 1. Although this looks unintuitive, this situation can occur even in a legal PDF, if the interval over which the value is greater than 1 is not too large. We have only to satisfy the constraint that the total area under the PDF is 1.

As a more complex example, consider the Gaussian distribution.

Definition 2.7

Gaussian
distribution

A random variable X has a Gaussian distribution *with mean μ and variance σ^2, denoted $X \sim \mathcal{N}\left(\mu; \sigma^2\right)$, if it has the PDF*

$$p(x) = \frac{1}{\sqrt{2\pi}\sigma} e^{-\frac{(x-\mu)^2}{2\sigma^2}}.$$

standard
Gaussian

A standard Gaussian *is one with mean 0 and variance 1.*   ∎

A Gaussian distribution has a bell-like curve, where the mean parameter μ controls the location of the peak, that is, the value for which the Gaussian gets its maximum value. The variance parameter σ^2 determines how peaked the Gaussian is: the smaller the variance, the

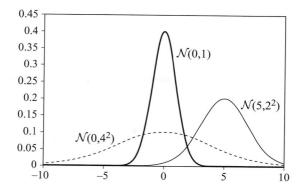

Figure 2.2   Example PDF of three Gaussian distributions

more peaked the Gaussian. Figure 2.2 shows the probability density function of a few different Gaussian distributions.

More technically, the probability density function is specified as an exponential, where the expression in the exponent corresponds to the square of the number of standard deviations σ that x is away from the mean μ. The probability of x decreases exponentially with the square of its deviation from the mean, as measured in units of its standard deviation.

2.1.6.2   Joint Density Functions

The discussion of density functions for a single variable naturally extends for joint distributions of continuous random variables.

Definition 2.8

joint density

Let P be a joint distribution over continuous random variables $X_1, \ldots, X_n$. A function $p(x_1, \ldots, x_n)$ is a joint density function *of $X_1, \ldots, X_n$ if*

- $p(x_1, \ldots, x_n) \geq 0$ *for all values $x_1, \ldots, x_n$ of $X_1, \ldots, X_n$.*
- *p is an integrable function.*
- *For any choice of $a_1, \ldots, a_n$, and $b_1, \ldots, b_n$,*

$$P(a_1 \leq X_1 \leq b_1, \ldots, a_n \leq X_n \leq b_n) = \int\limits_{a_1}^{b_1} \cdots \int\limits_{a_n}^{b_n} p(x_1, \ldots, x_n) dx_1 \ldots dx_n.$$

∎

Thus, a joint density specifies the probability of any joint event over the variables of interest.

Both the uniform distribution and the Gaussian distribution have natural extensions to the multivariate case. The definition of a multivariate uniform distribution is straightforward. We defer the definition of the multivariate Gaussian to section 7.1.

From the joint density we can derive the marginal density of any random variable by integrating out the other variables. Thus, for example, if $p(x, y)$ is the joint density of X and Y, then

$$p(x) = \int\limits_{-\infty}^{\infty} p(x,y)dy.$$

To see why this equality holds, note that the event $a \leq X \leq b$ is, by definition, equal to the event "$a \leq X \leq b$ and $-\infty \leq Y \leq \infty$." This rule is the direct analogue of marginalization for discrete variables. Note that, as with discrete probability distributions, we abuse notation a bit and use p to denote both the joint density of X and Y and the marginal density of X. In cases where the distinction is not clear, we use subscripts, so that p_X will be the marginal density, of X, and $p_{X,Y}$ the joint density.

2.1.6.3   Conditional Density Functions

As with discrete random variables, we want to be able to describe conditional distributions of continuous variables. Suppose, for example, we want to define $P(Y \mid X = x)$. Applying the definition of conditional distribution (equation (2.1)), we run into a problem, since $P(X = x) = 0$. Thus, the ratio of $P(Y, X = x)$ and $P(X = x)$ is undefined.

To avoid this problem, we might consider conditioning on the event $x - \epsilon \leq X \leq x + \epsilon$, which can have a positive probability. Now, the conditional probability is well defined. Thus, we might consider the limit of this quantity when $\epsilon \to 0$. We define

$$P(Y \mid x) = \lim_{\epsilon \to 0} P(Y \mid x - \epsilon \leq X \leq x + \epsilon).$$

When does this limit exist? If there is a continuous joint density function $p(x,y)$, then we can derive the form for this term. To do so, consider some event on Y, say $a \leq Y \leq b$. Recall that

$$P(a \leq Y \leq b \mid x - \epsilon \leq X \leq x + \epsilon) = \frac{P(a \leq Y \leq b, x - \epsilon \leq X \leq x + \epsilon)}{P(x - \epsilon \leq X \leq x + \epsilon)}$$

$$= \frac{\int_a^b \int_{x-\epsilon}^{x+\epsilon} p(x',y)dydx'}{\int_{x-\epsilon}^{x+\epsilon} p(x')dx'}.$$

When ϵ is sufficiently small, we can approximate

$$\int\limits_{x-\epsilon}^{x+\epsilon} p(x')dx' \approx 2\epsilon p(x).$$

Using a similar approximation for $p(x',y)$, we get

$$P(a \leq Y \leq b \mid x - \epsilon \leq X \leq x + \epsilon) \approx \frac{\int_a^b 2\epsilon p(x,y)dy}{2\epsilon p(x)}$$

$$= \int\limits_a^b \frac{p(x,y)}{p(x)}dy.$$

We conclude that $\frac{p(x,y)}{p(x)}$ is the density of $P(Y \mid X = x)$.

Definition 2.9

conditional
density function

Let $p(x, y)$ be the joint density of X and Y. The conditional density function of Y given X is defined as

$$p(y \mid x) = \frac{p(x, y)}{p(x)}$$

When $p(x) = 0$, the conditional density is undefined. ∎

The conditional density $p(y \mid x)$ characterizes the conditional distribution $P(Y \mid X = x)$ we defined earlier.

The properties of joint distributions and conditional distributions carry over to joint and conditional density functions. In particular, we have the chain rule

$$p(x, y) = p(x)p(y \mid x) \tag{2.14}$$

and Bayes' rule

$$p(x \mid y) = \frac{p(x)p(y \mid x)}{p(y)}. \tag{2.15}$$

As a general statement, whenever we discuss joint distributions of continuous random variables, we discuss properties with respect to the joint density function instead of the joint distribution, as we do in the case of discrete variables. Of particular interest is the notion of (conditional) independence of continuous random variables.

Definition 2.10

conditional
independence

Let $\boldsymbol{X}$, $\boldsymbol{Y}$, and $\boldsymbol{Z}$ be sets of continuous random variables with joint density $p(\boldsymbol{X}, \boldsymbol{Y}, \boldsymbol{Z})$. We say that $\boldsymbol{X}$ is conditionally independent of $\boldsymbol{Y}$ given $\boldsymbol{Z}$ if

$$p(\boldsymbol{x} \mid \boldsymbol{z}) = p(\boldsymbol{x} \mid \boldsymbol{y}, \boldsymbol{z}) \text{ for all } \boldsymbol{x}, \boldsymbol{y}, \boldsymbol{z} \text{ such that } p(\boldsymbol{z}) > 0.$$ ∎

2.1.7 Expectation and Variance

2.1.7.1 Expectation

expectation

Let X be a discrete random variable that takes numerical values; then the *expectation* of X under the distribution P is

$$\boldsymbol{E}_P[X] = \sum_x x \cdot P(x).$$

If X is a continuous variable, then we use the density function

$$\boldsymbol{E}_P[X] = \int x \cdot p(x)dx.$$

For example, if we consider X to be the outcome of rolling a fair die with probability $1/6$ for each outcome, then $\boldsymbol{E}[X] = 1 \cdot \frac{1}{6} + 2 \cdot \frac{1}{6} + \cdots + 6 \cdot \frac{1}{6} = 3.5$. On the other hand, if we consider a biased die where $P(X = 6) = 0.5$ and $P(X = x) = 0.1$ for $x < 6$, then $\boldsymbol{E}[X] = 1 \cdot 0.1 + \cdots + 5 \cdot 0.1 + \cdots + 6 \cdot 0.5 = 4.5$.

Often we are interested in expectations of a function of a random variable (or several random variables). Thus, we might consider extending the definition to consider the expectation of a functional term such as $X^2 + 0.5X$. Note, however, that any function g of a set of random variables $X_1, \ldots, X_k$ is essentially defining a new random variable Y: For any outcome $\omega \in \Omega$, we define the value of Y as $g(f_{X_1}(\omega), \ldots, f_{X_k}(\omega))$.

Based on this discussion, we often define new random variables by a functional term. For example $Y = X^2$, or $Y = e^X$. We can also consider functions that map values of one or more categorical random variables to numerical values. One such function that we use quite often is the *indicator function*, which we denote $\mathit{1}\{\boldsymbol{X} = \boldsymbol{x}\}$. This function takes value 1 when $\boldsymbol{X} = \boldsymbol{x}$, and 0 otherwise.

indicator function

In addition, we often consider expectations of functions of random variables without bothering to name the random variables they define. For example $\boldsymbol{E}_P[X + Y]$. Nonetheless, we should keep in mind that such a term does refer to an expectation of a random variable.

We now turn to examine properties of the expectation of a random variable.

First, as can be easily seen, the expectation of a random variable is a linear function in that random variable. Thus,

$$\boldsymbol{E}[a \cdot X + b] = a\boldsymbol{E}[X] + b.$$

A more complex situation is when we consider the expectation of a function of several random variables that have some joint behavior. An important property of expectation is that the expectation of a sum of two random variables is the sum of the expectations.

Proposition 2.4
$$\boldsymbol{E}[X + Y] = \boldsymbol{E}[X] + \boldsymbol{E}[Y].$$

linearity of
expectation

This property is called *linearity of expectation*. It is important to stress that this identity is true even when the variables are not independent. As we will see, this property is key in simplifying many seemingly complex problems.

Finally, what can we say about the expectation of a product of two random variables? In general, very little:

Example 2.5
Consider two random variables X and Y, each of which takes the value $+1$ with probability $1/2$, and the value -1 with probability $1/2$. If X and Y are independent, then $\boldsymbol{E}[X \cdot Y] = 0$. On the other hand, if X and Y are correlated in that they always take the same value, then $\boldsymbol{E}[X \cdot Y] = 1$.∎

However, when X and Y are independent, then, as in our example, we can compute the expectation simply as a product of their individual expectations:

Proposition 2.5
If X and Y are independent, then

$$\boldsymbol{E}[X \cdot Y] = \boldsymbol{E}[X] \cdot \boldsymbol{E}[Y].$$

conditional
expectation

We often also use the expectation given some evidence. The *conditional expectation* of X given $\boldsymbol{y}$ is

$$\boldsymbol{E}_P[X \mid \boldsymbol{y}] = \sum_x x \cdot P(x \mid \boldsymbol{y}).$$

2.1.7.2 Variance

variance

The expectation of X tells us the mean value of X. However, It does not indicate how far X deviates from this value. A measure of this deviation is the *variance* of X.

$$\mathbf{Var}_P[X] = \mathbf{E}_P\left[(X - \mathbf{E}_P[X])^2\right].$$

Thus, the variance is the expectation of the squared difference between X and its expected value. It gives us an indication of the spread of values of X around the expected value.

An alternative formulation of the variance is

$$\mathbf{Var}[X] = \mathbf{E}\left[X^2\right] - (\mathbf{E}[X])^2. \tag{2.16}$$

(see exercise 2.11).

Similar to the expectation, we can consider the expectation of a functions of random variables.

Proposition 2.6

If X and Y are independent, then

$$\mathbf{Var}[X + Y] = \mathbf{Var}[X] + \mathbf{Var}[Y].$$

It is straightforward to show that the variance scales as a quadratic function of X. In particular, we have:

$$\mathbf{Var}[a \cdot X + b] = a^2 \mathbf{Var}[X].$$

standard deviation

For this reason, we are often interested in the square root of the variance, which is called the *standard deviation* of the random variable. We define

$$\sigma_X = \sqrt{\mathbf{Var}[X]}.$$

The intuition is that it is improbable to encounter values of X that are farther than several standard deviations from the expected value of X. Thus, σ_X is a normalized measure of "distance" from the expected value of X.

As an example consider the Gaussian distribution of definition 2.7.

Proposition 2.7

Let X be a random variable with Gaussian distribution $N(\mu, \sigma^2)$, then $\mathbf{E}[X] = \mu$ and $\mathbf{Var}[X] = \sigma^2$.

Thus, the parameters of the Gaussian distribution specify the expectation and the variance of the distribution. As we can see from the form of the distribution, the density of values of X drops exponentially fast in the distance $\frac{x-\mu}{\sigma}$.

Not all distributions show such a rapid decline in the probability of outcomes that are distant from the expectation. However, even for arbitrary distributions, one can show that there is a decline.

Theorem 2.1

Chebyshev's inequality

(Chebyshev inequality):

$$P(|X - \mathbf{E}_P[X]| \geq t) \leq \frac{\mathbf{Var}_P[X]}{t^2}.$$

We can restate this inequality in terms of standard deviations: We write $t = k\sigma_X$ to get

$$P(|X - \boldsymbol{E}_P[X]| \geq k\sigma_X) \leq \frac{1}{k^2}.$$

Thus, for example, the probability of X being more than two standard deviations away from $\boldsymbol{E}[X]$ is less than $1/4$.

2.2 Graphs

Perhaps the most pervasive concept in this book is the representation of a probability distribution using a graph as a data structure. In this section, we survey some of the basic concepts in graph theory used in the book.

2.2.1 Nodes and Edges

A graph is a data structure $\mathcal{K}$ consisting of a set of nodes and a set of edges. Throughout most this book, we will assume that the set of nodes is $\mathcal{X} = \{X_1, \ldots, X_n\}$. A pair of nodes X_i, X_j

directed edge

undirected edge

can be connected by a *directed edge* $X_i \rightarrow X_j$ or an *undirected edge* $X_i{-}X_j$. Thus, the set of edges $\mathcal{E}$ is a set of pairs, where each pair is one of $X_i \rightarrow X_j$, $X_j \rightarrow X_i$, or $X_i{-}X_j$, for $X_i, X_j \in \mathcal{X}$, $i < j$. We assume throughout the book that, for each pair of nodes X_i, X_j, at most one type of edge exists; thus, we cannot have both $X_i \rightarrow X_j$ and $X_j \rightarrow X_i$, nor can we have $X_i \rightarrow X_j$ and $X_i{-}X_j$.[2] The notation $X_i \leftarrow X_j$ is equivalent to $X_j \rightarrow X_i$, and the notation $X_j{-}X_i$ is equivalent to $X_i{-}X_j$. We use $X_i \rightleftharpoons X_j$ to represent the case where X_i and X_j are connected via some edge, whether directed (in any direction) or undirected.

In many cases, we want to restrict attention to graphs that contain only edges of one kind

directed graph

or another. We say that a graph is *directed* if all edges are either $X_i \rightarrow X_j$ or $X_j \rightarrow X_i$. We usually denote directed graphs as $\mathcal{G}$. We say that a graph is *undirected* if all edges are $X_i{-}X_j$.

undirected graph

We denote undirected graphs as $\mathcal{H}$. We sometimes convert a general graph to an undirected graph by ignoring the directions on the edges.

Definition 2.11

graph's undirected version

Given a graph $\mathcal{K} = (\mathcal{X}, \mathcal{E})$, its undirected version *is a graph $\mathcal{H} = (\mathcal{X}, \mathcal{E}')$ where $\mathcal{E}' = \{X{-}Y : X \rightleftharpoons Y \in \mathcal{E}\}$.*   ■

child

Whenever we have that $X_i \rightarrow X_j \in \mathcal{E}$, we say that X_j is the *child* of X_i in $\mathcal{K}$, and that X_i is the *parent* of X_j in $\mathcal{K}$. When we have $X_i{-}X_j \in \mathcal{E}$, we say that X_i is a *neighbor* of

parent

X_j in $\mathcal{K}$ (and vice versa). We say that X and Y are adjacent whenever $X \rightleftharpoons Y \in \mathcal{E}$. We use

neighbor

Pa_X to denote the parents of X, Ch_X to denote its children, and Nb_X to denote its neighbors. We define the *boundary* of X, denoted $\mathrm{Boundary}_X$, to be $\mathrm{Pa}_X \cup \mathrm{Nb}_X$; for DAGs, this set is

boundary

simply X's parents, and for undirected graphs X's neighbors.[3] Figure 2.3 shows an example of a graph $\mathcal{K}$. There, we have that A is the only parent of C, and F, I are the children of C. The

degree

only neighbor of C is D, but its adjacent nodes are A, D, F, I. The *degree* of a node X is the number of edges in which it participates. Its *indegree* is the number of directed edges $Y \rightarrow X$.

indegree

The *degree* of a graph is the maximal degree of a node in the graph.

---

2. Note that our definition is somewhat restricted, in that it disallows cycles of length two, where $X_i \rightarrow X_j \rightarrow X_i$, and allows self-loops where $X_i \rightarrow X_i$.

3. When the graph is not clear from context, we often add the graph as an additional argument.

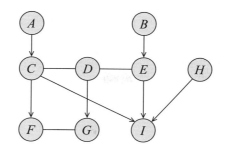

Figure 2.3   An example of a partially directed graph $\mathcal{K}$

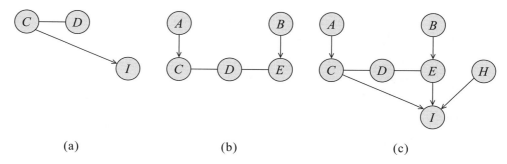

(a)   (b)   (c)

Figure 2.4   Induced graphs and their upward closure: (a) The induced subgraph $\mathcal{K}[C, D, I]$. (b) The upwardly closed subgraph $\mathcal{K}^+[C]$. (c) The upwardly closed subgraph $\mathcal{K}^+[C, D, I]$.

2.2.2   Subgraphs

In many cases, we want to consider only the part of the graph that is associated with a particular subset of the nodes.

Definition 2.12

induced subgraph

Let $\mathcal{K} = (\mathcal{X}, \mathcal{E})$, and let $\boldsymbol{X} \subset \mathcal{X}$. We define the induced subgraph $\mathcal{K}[\boldsymbol{X}]$ to be the graph $(\boldsymbol{X}, \mathcal{E}')$ where $\mathcal{E}'$ are all the edges $X \rightleftharpoons Y \in \mathcal{E}$ such that $X, Y \in \boldsymbol{X}$.   ∎

For example, figure 2.4a shows the induced subgraph $\mathcal{K}[C, D, I]$.

A type of subgraph that is often of particular interest is one that contains all possible edges.

Definition 2.13

complete subgraph

clique

A subgraph over $\boldsymbol{X}$ is complete if every two nodes in $\boldsymbol{X}$ are connected by some edge. The set $\boldsymbol{X}$ is often called a clique; we say that a clique $\boldsymbol{X}$ is maximal if for any superset of nodes $\boldsymbol{Y} \supset \boldsymbol{X}$, $\boldsymbol{Y}$ is not a clique.   ∎

Although the subset of nodes $\boldsymbol{X}$ can be arbitrary, we are often interested in sets of nodes that preserve certain aspects of the graph structure.

Definition 2.14

upward closure

We say that a subset of nodes $\boldsymbol{X} \in \mathcal{X}$ is upwardly closed in $\mathcal{K}$ if, for any $X \in \boldsymbol{X}$, we have that $\mathrm{Boundary}_X \subset \boldsymbol{X}$. We define the upward closure of $\boldsymbol{X}$ to be the minimal upwardly closed subset

Y that contains X. We define the upwardly closed subgraph of X, denoted $\mathcal{K}^+[X]$, to be the induced subgraph over Y, $\mathcal{K}[Y]$.   ∎

For example, the set A, B, C, D, E is the upward closure of the set $\{C\}$ in $\mathcal{K}$. The upwardly closed subgraph of $\{C\}$ is shown in figure 2.4b. The upwardly closed subgraph of $\{C, D, I\}$ is shown in figure 2.4c.

2.2.3   Paths and Trails

Using the basic notion of edges, we can define different types of longer-range connections in the graph.

Definition 2.15

path

We say that $X_1, \ldots, X_k$ form a path *in the graph $\mathcal{K} = (\mathcal{X}, \mathcal{E})$ if, for every $i = 1, \ldots, k - 1$, we have that either $X_i \rightarrow X_{i+1}$ or $X_i - X_{i+1}$. A path is* directed *if, for at least one i, we have $X_i \rightarrow X_{i+1}$.*   ∎

Definition 2.16

trail

We say that $X_1, \ldots, X_k$ form a trail *in the graph $\mathcal{K} = (\mathcal{X}, \mathcal{E})$ if, for every $i = 1, \ldots, k - 1$, we have that $X_i \rightleftharpoons X_{i+1}$.*   ∎

In the graph $\mathcal{K}$ of figure 2.3, A, C, D, E, I is a path, and hence also a trail. On the other hand, A, C, F, G, D is a trail, which is not a path.

Definition 2.17

connected graph

A graph is connected *if for every X_i, X_j there is a trail between X_i and X_j.*   ∎

We can now define longer-range relationships in the graph.

Definition 2.18

ancestor

descendant

We say that X is an ancestor *of Y in $\mathcal{K} = (\mathcal{X}, \mathcal{E})$, and that Y is a* descendant *of X, if there exists a directed path $X_1, \ldots, X_k$ with $X_1 = X$ and $X_k = Y$. We use $\mathrm{Descendants}_X$ to denote X's descendants, $\mathrm{Ancestors}_X$ to denote X's ancestors, and $\mathrm{NonDescendants}_X$ to denote the set of nodes in $\mathcal{X} - \mathrm{Descendants}_X$.*   ∎

In our example graph $\mathcal{K}$, we have that F, G, I are descendants of C. The ancestors of C are A, via the path A, C, and B, via the path B, E, D, C.

A final useful notion is that of an ordering of the nodes in a directed graph that is consistent with the directionality its edges.

Definition 2.19

topological
ordering

Let $\mathcal{G} = (\mathcal{X}, \mathcal{E})$ be a graph. An ordering of the nodes $X_1, \ldots, X_n$ is a topological ordering *relative to $\mathcal{K}$ if, whenever we have $X_i \rightarrow X_j \in \mathcal{E}$, then $i < j$.*   ∎

Appendix A.3.1 presents an algorithm for finding such a topological ordering.

2.2.4   Cycles and Loops

Note that, in general, we can have a cyclic path that leads from a node to itself, making that node its own descendant.

Definition 2.20

cycle

acyclic

A cycle *in $\mathcal{K}$ is a directed path $X_1, \ldots, X_k$ where $X_1 = X_k$. A graph is* acyclic *if it contains no cycles.* ∎

For most of this book, we will restrict attention to graphs that do not allow such cycles, since it is quite difficult to define a coherent probabilistic model over graphs with directed cycles.

DAG

A *directed acyclic graph (DAG)* is one of the central concepts in this book, as DAGs are the basic graphical representation that underlies Bayesian networks. For some of this book, we also use acyclic graphs that are partially directed. The graph $\mathcal{K}$ of figure 2.3 is acyclic. However, if we add the undirected edge A—E to $\mathcal{K}$, we have a path A, C, D, E, A from A to itself. Clearly, adding a directed edge $E \rightarrow A$ would also lead to a cycle. Note that prohibiting cycles does not imply that there is no *trail* from a node to itself. For example, $\mathcal{K}$ contains several trails: C, D, E, I, C as well as C, D, G, F, C.

PDAG

chain component

An acyclic graph containing both directed and undirected edges is called a *partially directed acyclic graph* or *PDAG*. The acyclicity requirement on a PDAG implies that the graph can be decomposed into a directed graph of *chain components*, where the nodes within each chain component are connected to each other only with undirected edges. The acyclicity of a PDAG guarantees us that we can order the components so that all edges point from lower-numbered components to higher-numbered ones.

Definition 2.21

Let $\mathcal{K}$ be a PDAG over $\mathcal{X}$. Let $\boldsymbol{K}_1, \ldots, \boldsymbol{K}_\ell$ be a disjoint partition of $\mathcal{X}$ such that:

- *the induced subgraph over $\boldsymbol{K}_i$ contains no directed edges;*
- *for any pair of nodes $X \in \boldsymbol{K}_i$ and $Y \in \boldsymbol{K}_j$ for $i < j$, an edge between X and Y can only be a directed edge $X \rightarrow Y$.*

chain component

Each component $\boldsymbol{K}_i$ is called a chain component. ∎

chain graph

Because of its chain structure, a PDAG is also called a *chain graph*.

Example 2.6

In the PDAG of figure 2.3, we have six chain components: $\{A\}, \{B\}, \{C, D, E\}, \{F, G\}, \{H\},$ and $\{I\}$. This ordering of the chain components is one of several possible legal orderings. ∎

Note that when the PDAG is an undirected graph, the entire graph forms a single chain component. Conversely, when the PDAG is a directed graph (and therefore acyclic), each node in the graph is its own chain component.

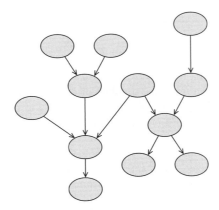

Figure 2.5   An example of a polytree

Different from a cycle is the notion of a loop:

Definition 2.22	*A loop in $\mathcal{K}$ is a trail $X_1, \ldots, X_k$ where $X_1 = X_k$. A graph is* singly connected *if it contains*
loop	*no loops. A node in a singly connected graph is called a* leaf *if it has exactly one adjacent node.*
singly connected	*A singly connected directed graph is also called a* polytree. *A singly connected undirected graph is*
leaf	*called a* forest; *if it is also connected, it is called a* tree. ■
polytree	

We can also define a notion of a forest, or of a tree, for directed graphs.

forest	*A directed graph is a* forest *if each node has at most one parent. A directed forest is a* tree *if it is*
tree	*also connected.* ■
Definition 2.23	

Note that polytrees are very different from trees. For example, figure 2.5 shows a graph that is a polytree but is not a tree, because several nodes have more than one parent. As we will discuss later in the book, loops in the graph increase the computational cost of various tasks.

We conclude this section with a final definition relating to loops in the graph. This definition will play an important role in evaluating the cost of reasoning using graph-based representations.

Definition 2.24	*Let $X_1 - X_2 - \cdots - X_k - X_1$ be a loop in the graph; a* chord *in the loop is an edge connecting*
chordal graph	X_i *and* X_j *for two nonconsecutive nodes* X_i, X_j. *An undirected graph* $\mathcal{H}$ *is said to be* chordal *if*
	any loop $X_1 - X_2 - \cdots - X_k - X_1$ *for* $k \geq 4$ *has a chord.* ■

Thus, for example, a loop $A - B - C - D - A$ (as in figure 1.1b) is nonchordal, but adding an edge $A - C$ would render it chordal. In other words, in a chordal graph, the longest "minimal loop" (one that has no shortcut) is a triangle. Thus, chordal graphs are often also called

triangulated   *triangulated.*
graph   We can extend the notion of chordal graphs to graphs that contain directed edges.

Definition 2.25	*A graph* $\mathcal{K}$ *is said to be* chordal *if its underlying undirected graph is chordal.* ■

2.3   Relevant Literature

Section 1.4 provides some history on the development of of probabilistic methods. There are many good textbooks about probability theory; see, for example, DeGroot (1989), Ross (1988) or Feller (1970). The distinction between the frequentist and subjective view of probability was a major issue during much of the late nineteenth and early twentieth centuries. Some references that touch on this discussion include Cox (2001) and Jaynes (2003) on the Bayesian side, and Feller (1970) on the frequentist side; these books also contain much useful general material about probabilistic reasoning.

Dawid (1979, 1980) was the first to propose the axiomatization of conditional independence properties, and he showed how they can help unify a variety of topics within probability and statistics. These axioms were studied in great detail by Pearl and colleagues, work that is presented in detail in Pearl (1988).

2.4   Exercises

Exercise 2.1

Prove the following properties using basic properties of definition 2.1:

a. $P(\emptyset) = 0$.
b. If $\alpha \subseteq \beta$, then $P(\alpha) \leq P(\beta)$.
c. $P(\alpha \cup \beta) = P(\alpha) + P(\beta) - P(\alpha \cap \beta)$.

Exercise 2.2

a. Show that for binary random variables X, Y, the event-level independence $(x^0 \perp y^0)$ implies random-variable independence $(X \perp Y)$.
b. Show a counterexample for nonbinary variables.
c. Is it the case that, for a binary-valued variable Z, we have that $(X \perp Y \mid z^0)$ implies $(X \perp Y \mid Z)$?

Exercise 2.3

Consider two events α and β such that $P(\alpha) = p_a$ and $P(\beta) = p_b$. Given only that knowledge, what is the maximum and minimum values of the probability of the events $\alpha \cap \beta$ and $\alpha \cup \beta$. Can you characterize the situations in which each of these extreme values occurs?

Exercise 2.4★

Let P be a distribution over $(\Omega, \mathcal{S})$, and let $a \in \mathcal{S}$ be an event such that $P(\alpha) > 0$. The conditional probability $P(\cdot \mid \alpha)$ assigns a value to each event in $\mathcal{S}$. Show that it satisfies the properties of definition 2.1.

Exercise 2.5

Let $\boldsymbol{X}, \boldsymbol{Y}, \boldsymbol{Z}$ be three disjoint subsets of variables such that $\mathcal{X} = \boldsymbol{X} \cup \boldsymbol{Y} \cup \boldsymbol{Z}$. Prove that $P \models (\boldsymbol{X} \perp \boldsymbol{Y} \mid \boldsymbol{Z})$ if and only if we can write P in the form:

$$P(\mathcal{X}) = \phi_1(\boldsymbol{X}, \boldsymbol{Z}) \phi_2(\boldsymbol{Y}, \boldsymbol{Z}).$$

Exercise 2.6

An often useful rule in dealing with probabilities is known as *reasoning by cases*. Let X, Y, and Z be random variables, then

$$P(X \mid Y) = \sum_z P(X, z \mid Y).$$

Prove this equality using the chain rule of probabilities and basic properties of (conditional) distributions.

Exercise 2.7★

In this exercise, you will prove the properties of conditional independence discussed in section 2.1.4.3.

   a. Prove that the weak union and contraction properties hold for any probability distribution P.

   b. Prove that the intersection property holds for any positive probability distribution P.

   c. Provide a counterexample to the intersection property in cases where the distribution P is not positive.

Exercise 2.8

   a. Show that for binary random variables X and Y, $(x^1 \perp y^1)$ implies $(X \perp Y)$.

   b. Provide a counterexample to this property for nonbinary variables.

   c. Is it the case that, for binary Z, $(X \perp Y \mid z^1)$ implies $(X \perp Y \mid Z)$?   Prove or provide a counterexample.

Exercise 2.9

Show how you can use breadth-first search to determine whether a graph $\mathcal{K}$ is cyclic.

Exercise 2.10★

In appendix A.3.1, we describe an algorithm for finding a topological ordering for a directed graph. Extend this algorithm to one that finds a topological ordering for the chain components in a PDAG. Your algorithm should construct both the chain components of the PDAG, as well as an ordering over them that satisfies the conditions of definition 2.21. Analyze the asymptotic complexity of your algorithm.

Exercise 2.11

Use the properties of expectation to show that we can rewrite the variance of a random variable X as

$$\textbf{\textit{Var}}[X] = \textbf{\textit{E}}\big[X^2\big] - (\textbf{\textit{E}}[X])^2 .$$

Exercise 2.12★

Prove the following property of expectations

Theorem 2.2

Markov inequality

(Markov inequality): *Let X be a random variable such that $P(X \geq 0) = 1$, then for any $t \geq 0$,*

$$P(X \geq t) \leq \frac{\textbf{\textit{E}}_P[X]}{t}.$$

You may assume in your proof that X is a discrete random variable with a finite number of values.

Exercise 2.13

Prove Chebyshev's inequality using the Markov inequality shown in exercise 2.12. (Hint: define a new random variable Y, so that the application of the Markov inequality with respect to this random variable gives the required result.)

Exercise 2.14★

Let $X \sim \mathcal{N}\left(\mu; \sigma^2\right)$, and define a new variable $Y = a \cdot X + b$. Show that $Y \sim \mathcal{N}\left(a \cdot \mu + b; a^2\sigma^2\right)$.

Exercise 2.15★

concave function

convex function

A function f is *concave* if for any $0 \leq \alpha \leq 1$ and any x and y, we have that $f(\alpha x + (1 - \alpha)y) \geq \alpha f(x) + (1 - \alpha)f(y)$. A function is *convex* if the opposite holds, that is, $f(\alpha x + (1 - \alpha)y) \leq \alpha f(x) + (1 - \alpha)f(y)$.

a. Prove that a continuous and differentiable function f is concave if and only if $f''(x) \leq 0$ for all x.

b. Show that $\log(x)$ is concave (over the positive real numbers).

Exercise 2.16★

Proposition 2.8

Jensen inequality

Jensen's inequality: Let f be a concave function and P a distribution over a random variable X. Then

$$\boldsymbol{E}_P[f(X)] \leq f(\boldsymbol{E}_P[X])$$

Use this inequality to show that:

a. $\boldsymbol{H}_P(X) \leq \log|Val(X)|$.

b. $\boldsymbol{H}_P(X) \geq 0$.

c. $\boldsymbol{D}(P\|Q) \geq 0$.

See appendix A.1 for the basic definitions.

Exercise 2.17

Show that, for any probability distribution $P(X)$, we have that

$$\boldsymbol{H}_P(X) = \log K - \boldsymbol{D}(P(X)\|P_u(X))$$

where $P_u(X)$ is the uniform distribution over $Val(X)$ and $K = |Val(X)|$.

Exercise 2.18★

Prove proposition A.3, and use it to show that $\boldsymbol{I}(X;Y) \geq 0$.

Exercise 2.19

conditional mutual information

As with entropies, we can define the notion of *conditional mutual information*

$$\boldsymbol{I}_P(X;Y \mid Z) = \boldsymbol{E}_P\left[\log \frac{P(X \mid Y, Z)}{P(X \mid Z)}\right].$$

Prove that:

a. $\boldsymbol{I}_P(X;Y \mid Z) = \boldsymbol{H}_P(X \mid Z) - \boldsymbol{H}_P(X, Y \mid Z)$.

chain rule of mutual information

b. The *chain rule of mutual information*:

$$\boldsymbol{I}_P(X;Y, Z) = \boldsymbol{I}_P(X;Y) + \boldsymbol{I}_P(X;Z \mid Y).$$

Exercise 2.20

Use the chain law of mutual information to prove that

$$\boldsymbol{I}_P(X;Y) \leq \boldsymbol{I}_P(X;Y, Z).$$

That is, the information that Y and Z together convey about X cannot be less than what Y alone conveys about X.

Exercise 2.21★

Consider a sequence of N independent samples from a binary random variable X whose distribution is $P(x^1) = p$, $P(x^0) = 1 - p$. As in appendix A.2, let S_N be the number of trials whose outcome is x^1. Show that

$$P(S_N = r) \approx \exp[-N \cdot \boldsymbol{D}((p, 1 - p) \| (r/N, 1 - r/N))].$$

Your proof should use Stirling's approximation to the factorial function:

$$m! \approx \frac{1}{2\pi m} m^m e^{-m}.$$

PART I

Representation

3   *The Bayesian Network Representation*

Our goal is to represent a joint distribution P over some set of random variables $\mathcal{X} = \{X_1, \ldots, X_n\}$. Even in the simplest case where these variables are binary-valued, a joint distribution requires the specification of $2^n - 1$ numbers — the probabilities of the 2^n different assignments of values $x_1, \ldots, x_n$. For all but the smallest n, **the explicit representation of the joint distribution is unmanageable from every perspective. Computationally, it is very expensive to manipulate and generally too large to store in memory. Cognitively, it is impossible to acquire so many numbers from a human expert; moreover, the numbers are very small and do not correspond to events that people can reasonably contemplate. Statistically, if we want to learn the distribution from data, we would need ridiculously large amounts of data to estimate this many parameters robustly. These problems were the main barrier to the adoption of probabilistic methods for expert systems until the development of the methodologies described in this book.**

In this chapter, we first show how independence properties in the distribution can be used to represent such high-dimensional distributions much more compactly. We then show how a combinatorial data structure — a directed acyclic graph — can provide us with a general-purpose modeling language for exploiting this type of structure in our representation.

3.1   Exploiting Independence Properties

The compact representations we explore in this chapter are based on two key ideas: the representation of independence properties of the distribution, and the use of an alternative parameterization that allows us to exploit these finer-grained independencies.

3.1.1   Independent Random Variables

To motivate our discussion, consider a simple setting where we know that each X_i represents the outcome of a toss of coin i. In this case, we typically assume that the different coin tosses are marginally independent (definition 2.4), so that our distribution P will satisfy $(X_i \perp X_j)$ for any i, j. More generally (strictly more generally — see exercise 3.1), we assume that the distribution satisfies $(\boldsymbol{X} \perp \boldsymbol{Y})$ for any disjoint subsets of the variables $\boldsymbol{X}$ and $\boldsymbol{Y}$. Therefore, we have that:

$$P(X_1, \ldots, X_n) = P(X_1)P(X_2) \cdots P(X_n).$$

If we use the standard parameterization of the joint distribution, this independence structure is obscured, and the representation of the distribution requires 2^n parameters. However, we can use a more natural set of parameters for specifying this distribution: If θ_i is the probability with which coin i lands heads, the joint distribution P can be specified using the n *parameters* $\theta_1, \ldots, \theta_n$. These parameters implicitly specify the 2^n probabilities in the joint distribution. For example, the probability that all of the coin tosses land heads is simply $\theta_1 \cdot \theta_2 \cdot \ldots \cdot \theta_n$. More generally, letting $\theta_{x_i} = \theta_i$ when $x_i = x_i^1$ and $\theta_{x_i} = 1 - \theta_i$ when $x_i = x_i^0$, we can define:

parameters

$$P(x_1, \ldots, x_n) = \prod_i \theta_{x_i}. \tag{3.1}$$

This representation is limited, and there are many distributions that we cannot capture by choosing values for $\theta_1, \ldots, \theta_n$. This fact is obvious not only from intuition, but also from a somewhat more formal perspective. The space of all joint distributions is a $2^n - 1$ dimensional subspace of $I\!\!R^{2^n}$ — the set $\{(p_1, \ldots, p_{2^n}) \in I\!\!R^{2^n} : p_1 + \ldots + p_{2^n} = 1\}$. On the other hand, the space of all joint distributions specified in a factorized way as in equation (3.1) is an n-dimensional manifold in $I\!\!R^{2^n}$.

independent parameters

A key concept here is the notion of *independent parameters* — parameters whose values are not determined by others. For example, when specifying an arbitrary multinomial distribution over a k dimensional space, we have $k - 1$ independent parameters: the last probability is fully determined by the first $k - 1$. In the case where we have an arbitrary joint distribution over n binary random variables, the number of independent parameters is $2^n - 1$. On the other hand, the number of independent parameters for distributions represented as n independent binomial coin tosses is n. Therefore, the two spaces of distributions cannot be the same. (While this argument might seem trivial in this simple case, it turns out to be an important tool for comparing the expressive power of different representations.)

As this simple example shows, certain families of distributions — in this case, the distributions generated by n independent random variables — permit an alternative parameterization that is substantially more compact than the naive representation as an explicit joint distribution. Of course, in most real-world applications, the random variables are not marginally independent. However, a generalization of this approach will be the basis for our solution.

3.1.2 The Conditional Parameterization

Let us begin with a simple example that illustrates the basic intuition. Consider the problem faced by a company trying to hire a recent college graduate. The company's goal is to hire intelligent employees, but there is no way to test intelligence directly. However, the company has access to the student's SAT scores, which are informative but not fully indicative. Thus, our probability space is induced by the two random variables *Intelligence* (I) and *SAT* (S). For simplicity, we assume that each of these takes two values: $Val(I) = \{i^1, i^0\}$, which represent the values high intelligence (i^1) and low intelligence (i^0); similarly $Val(S) = \{s^1, s^0\}$, which also represent the values *high* (score) and *low* (score), respectively.

Thus, our joint distribution in this case has four entries. For example, one possible joint distribution P would be

$$
\begin{array}{cc|c}
I & S & P(I,S) \\
\hline
i^0 & s^0 & 0.665 \\
i^0 & s^1 & 0.035 \\
i^1 & s^0 & 0.06 \\
i^1 & s^1 & 0.24.
\end{array}
\tag{3.2}
$$

There is, however, an alternative, and even more natural way of representing the same joint distribution. Using the chain rule of conditional probabilities (see equation (2.5)), we have that

$$P(I,S) = P(I)P(S \mid I).$$

Intuitively, we are representing the process in a way that is more compatible with causality. Various factors (genetics, upbringing, ...) first determined (stochastically) the student's intelligence. His performance on the SAT is determined (stochastically) by his intelligence. We note that the models we construct are not required to follow causal intuitions, but they often do. We return to this issue later on.

From a mathematical perspective, this equation leads to the following alternative way of representing the joint distribution. Instead of specifying the various joint entries $P(I,S)$, we would specify it in the form of $P(I)$ and $P(S \mid I)$. Thus, for example, we can represent the *prior* joint distribution of equation (3.2) using the following two tables, one representing the *prior distribution* over I and the other the *conditional probability distribution (CPD)* of S given I:

prior distribution

CPD

$$
\begin{array}{cc}
i^0 & i^1 \\
\hline
0.7 & 0.3
\end{array}
\qquad
\begin{array}{c||cc}
I & s^0 & s^1 \\
\hline
i^0 & 0.95 & 0.05 \\
i^1 & 0.2 & 0.8
\end{array}
\tag{3.3}
$$

The CPD $P(S \mid I)$ represents the probability that the student will succeed on his SATs in the two possible cases: the case where the student's intelligence is low, and the case where it is high. The CPD asserts that a student of low intelligence is extremely unlikely to get a high SAT score ($P(s^1 \mid i^0) = 0.05$); on the other hand, a student of high intelligence is likely, but far from certain, to get a high SAT score ($P(s^1 \mid i^1) = 0.8$).

It is instructive to consider how we could parameterize this alternative representation. Here, we are using three binomial distributions, one for $P(I)$, and two for $P(S \mid i^0)$ and $P(S \mid i^1)$. Hence, we can parameterize this representation using three independent parameters, say θ_{i^1}, $\theta_{s^1 \mid i^1}$, and $\theta_{s^1 \mid i^0}$. Our representation of the joint distribution as a four-outcome multinomial also required three parameters. Thus, although the conditional representation is more natural than the explicit representation of the joint, it is not more compact. However, as we will soon see, the conditional parameterization provides a basis for our compact representations of more complex distributions.

Although we will only define Bayesian networks formally in section 3.2.2, it is instructive to see how this example would be represented as one. The Bayesian network, as shown in figure 3.1a, would have a node for each of the two random variables I and S, with an edge from I to S representing the direction of the dependence in this model.

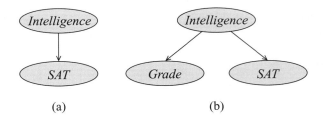

         (a)   (b)

Figure 3.1   **Simple Bayesian networks for the** student **example**

3.1.3   The Naive Bayes Model

We now describe perhaps the simplest example where a conditional parameterization is combined with conditional independence assumptions to produce a very compact representation of a high-dimensional probability distribution. Importantly, unlike the previous example of fully independent random variables, none of the variables in this distribution are (marginally) independent.

3.1.3.1   The Student Example

Elaborating our example, we now assume that the company also has access to the student's grade G in some course. In this case, our probability space is the joint distribution over the three relevant random variables I, S, and G. Assuming that I and S are as before, and that G takes on three values g^1, g^2, g^3, representing the grades A, B, and C, respectively, then the joint distribution has twelve entries.

Before we even consider the specific numerical aspects of our distribution P in this example, we can see that independence does not help us: for any reasonable P, there are no independencies that hold. The student's intelligence is clearly correlated both with his SAT score and with his grade. The SAT score and grade are also not independent: if we condition on the fact that the student received a high score on his SAT, the chances that he gets a high grade in his class are also likely to increase. Thus, we may assume that, for our particular distribution P, $P(g^1 \mid s^1) > P(g^1 \mid s^0)$.

However, it is quite plausible that our distribution P in this case satisfies a conditional independence property. If we know that the student has high intelligence, a high grade on the SAT no longer gives us information about the student's performance in the class. More formally:

$$P(g \mid i^1, s^1) = P(g \mid i^1).$$

More generally, we may well assume that

$$P \models (S \perp G \mid I). \tag{3.4}$$

Note that this independence statement holds only if we assume that the student's intelligence is the only reason why his grade and SAT score might be correlated. In other words, it assumes that there are no correlations due to other factors, such as the student's ability to take timed exams. These assumptions are also not "true" in any formal sense of the word, and they are often only approximations of our true beliefs. (See box 3.C for some further discussion.)

As in the case of marginal independence, conditional independence allows us to provide a compact specification of the joint distribution. Again, the compact representation is based on a very natural alternative parameterization. By simple probabilistic reasoning (as in equation (2.5)), we have that

$$P(I, S, G) = P(S, G \mid I)P(I).$$

But now, the conditional independence assumption of equation (3.4) implies that

$$P(S, G \mid I) = P(S \mid I)P(G \mid I).$$

Hence, we have that

$$P(I, S, G) = P(S \mid I)P(G \mid I)P(I). \tag{3.5}$$

Thus, we have factorized the joint distribution $P(I, S, G)$ as a product of three conditional probability distributions (CPDs). This factorization immediately leads us to the desired alternative parameterization. In order to specify fully a joint distribution satisfying equation (3.4), we need the following three CPDs: $P(I)$, $P(S \mid I)$, and $P(G \mid I)$. The first two might be the same as in equation (3.3). The latter might be

I	g^1	g^2	g^3
i^0	0.2	0.34	0.46
i^1	0.74	0.17	0.09

Together, these three CPDs fully specify the joint distribution (assuming the conditional independence of equation (3.4)). For example,

$$\begin{aligned} P(i^1, s^1, g^2) &= P(i^1)P(s^1 \mid i^1)P(g^2 \mid i^1) \\ &= 0.3 \cdot 0.8 \cdot 0.17 = 0.0408. \end{aligned}$$

Once again, we note that this probabilistic model would be represented using the Bayesian network shown in figure 3.1b.

In this case, the alternative parameterization is more compact than the joint. We now have three binomial distributions — $P(I)$, $P(S \mid i^1)$ and $P(S \mid i^0)$, and two three-valued multinomial distributions — $P(G \mid i^1)$ and $P(G \mid i^0)$. Each of the binomials requires one independent parameter, and each three-valued multinomial requires two independent parameters, for a total of seven. By contrast, our joint distribution has twelve entries, so that eleven independent parameters are required to specify an arbitrary joint distribution over these three variables.

It is important to note another advantage of this way of representing the joint: modularity. When we added the new variable G, the joint distribution changed entirely. Had we used the explicit representation of the joint, we would have had to write down twelve new numbers. In the factored representation, we could reuse our local probability models for the variables I and S, and specify only the probability model for G — the CPD $P(G \mid I)$. This property will turn out to be invaluable in modeling real-world systems.

3.1.3.2 The General Model

naive Bayes   This example is an instance of a much more general model commonly called the *naive Bayes*

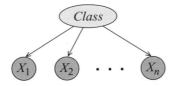

Figure 3.2   The Bayesian network graph for a naive Bayes model

model (also known as the *Idiot Bayes model*). The naive Bayes model assumes that instances fall into one of a number of mutually exclusive and exhaustive *classes*. Thus, we have a class variable C that takes on values in some set $\{c^1, \ldots, c^k\}$. In our example, the class variable is the student's intelligence I, and there are two classes of instances — students with high intelligence and students with low intelligence.

features   The model also includes some number of *features* $X_1, \ldots, X_n$ whose values are typically observed. The *naive Bayes assumption* is that the features are conditionally independent given the instance's class. In other words, within each class of instances, the different properties can be determined independently. More formally, we have that

$$(X_i \perp \boldsymbol{X}_{-i} \mid C) \quad \text{for all } i, \tag{3.6}$$

where $\boldsymbol{X}_{-i} = \{X_1, \ldots, X_n\} - \{X_i\}$. This model can be represented using the Bayesian network of figure 3.2. In this example, and later on in the book, we use a darker oval to represent variables that are always observed when the network is used.

factorization   Based on these independence assumptions, we can show that the model *factorizes* as:

$$P(C, X_1, \ldots, X_n) = P(C) \prod_{i=1}^{n} P(X_i \mid C). \tag{3.7}$$

(See exercise 3.2.) Thus, in this model, we can represent the joint distribution using a small set of factors: a prior distribution $P(C)$, specifying how likely an instance is to belong to different classes a priori, and a set of CPDs $P(X_j \mid C)$, one for each of the n finding variables. These factors can be encoded using a very small number of parameters. For example, if all of the variables are binary, the number of independent parameters required to specify the distribution is $2n + 1$ (see exercise 3.6). Thus, the number of parameters is linear in the number of variables, as opposed to exponential for the explicit representation of the joint.

Box 3.A — Concept: The Naive Bayes Model. *The naive Bayes model, despite the strong assumptions that it makes, is often used in practice, because of its simplicity and the small number of parameters required. The model is generally used for* classification *— deciding, based on the*
classification   *values of the evidence variables for a given instance, the class to which the instance is most likely to belong. We might also want to compute our confidence in this decision, that is, the extent to which our model favors one class c^1 over another c^2. Both queries can be addressed by the following ratio:*

$$\frac{P(C = c^1 \mid x_1, \ldots, x_n)}{P(C = c^2 \mid x_1, \ldots, x_n)} = \frac{P(C = c^1)}{P(C = c^2)} \prod_{i=1}^{n} \frac{P(x_i \mid C = c^1)}{P(x_i \mid C = c^2)}; \tag{3.8}$$

see exercise 3.2). This formula is very natural, since it computes the posterior probability ratio of c^1 versus c^2 as a product of their prior probability ratio (the first term), multiplied by a set of terms $\frac{P(x_i|C=c^1)}{P(x_i|C=c^2)}$ that measure the relative support of the finding x_i for the two classes.

medical diagnosis

This model was used in the early days of medical diagnosis, *where the different values of the class variable represented different diseases that the patient could have. The evidence variables represented different symptoms, test results, and the like. Note that the model makes several strong assumptions that are not generally true, specifically that the patient can have at most one disease, and that, given the patient's disease, the presence or absence of different symptoms, and the values of different tests, are all independent. This model was used for medical diagnosis because the small number of interpretable parameters made it easy to elicit from experts. For example, it is quite natural to ask of an expert physician what the probability is that a patient with pneumonia has high fever. Indeed, several early medical diagnosis systems were based on this technology, and some were shown to provide better diagnoses than those made by expert physicians.*

However, later experience showed that the strong assumptions underlying this model decrease its diagnostic accuracy. In particular, the model tends to overestimate the impact of certain evidence by "overcounting" it. For example, both hypertension (high blood pressure) and obesity are strong indicators of heart disease. However, because these two symptoms are themselves highly correlated, equation (3.8), which contains a multiplicative term for each of them, double-counts the evidence they provide about the disease. Indeed, some studies showed that the diagnostic performance of a naive Bayes model degraded as the number of features increased; this degradation was often traced to violations of the strong conditional independence assumption. This phenomenon led to the use of more complex Bayesian networks, with more realistic independence assumptions, for this application (see box 3.D).

Nevertheless, the naive Bayes model is still useful in a variety of applications, particularly in the context of models learned from data in domains with a large number of features and a relatively small number of instances, such as classifying documents into topics using the words in the documents as features; see box 17.E).

3.2 Bayesian Networks

Bayesian networks build on the same intuitions as the naive Bayes model by exploiting conditional independence properties of the distribution in order to allow a compact and natural representation. However, they are not restricted to representing distributions satisfying the strong independence assumptions implicit in the naive Bayes model. They allow us the flexibility to tailor our representation of the distribution to the independence properties that appear reasonable in the current setting.

The core of the Bayesian network representation is a directed acyclic graph (DAG) $\mathcal{G}$, whose nodes are the random variables in our domain and whose edges correspond, intuitively, to direct influence of one node on another. **This graph $\mathcal{G}$ can be viewed in two very different ways:**

☞
- **as a data structure that provides the skeleton for representing a joint distribution compactly in a factorized way;**

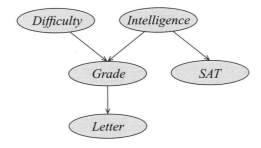

Figure 3.3 **The Bayesian Network graph for the** Student **example**

- **as a compact representation for a set of conditional independence assumptions about a distribution.**

As we will see, these two views are, in a strong sense, equivalent.

3.2.1   The Student **Example Revisited**

We begin our discussion with a simple toy example, which will accompany us, in various versions, throughout much of this book.

3.2.1.1   **The Model**

Consider our student from before, but now consider a slightly more complex scenario. The student's grade, in this case, depends not only on his intelligence but also on the difficulty of the course, represented by a random variable D whose domain is $Val(D) = \{easy, hard\}$. Our student asks his professor for a recommendation letter. The professor is absentminded and never remembers the names of her students. She can only look at his grade, and she writes her letter for him based on that information alone. The quality of her letter is a random variable L, whose domain is $Val(L) = \{strong, weak\}$. The actual quality of the letter depends stochastically on the grade. (It can vary depending on how stressed the professor is and the quality of the coffee she had that morning.)

We therefore have five random variables in this domain: the student's intelligence (I), the course difficulty (D), the grade (G), the student's SAT score (S), and the quality of the recommendation letter (L). All of the variables except G are binary-valued, and G is ternary-valued. Hence, the joint distribution has 48 entries.

As we saw in our simple illustrations of figure 3.1, a Bayesian network is represented using a directed graph whose nodes represent the random variables and whose edges represent direct influence of one variable on another. We can view the graph as encoding a generative sampling process executed by nature, where the value for each variable is selected by nature using a distribution that depends only on its parents. In other words, each variable is a stochastic function of its parents.

Based on this intuition, perhaps the most natural network structure for the distribution in this example is the one presented in figure 3.3. The edges encode our intuition about

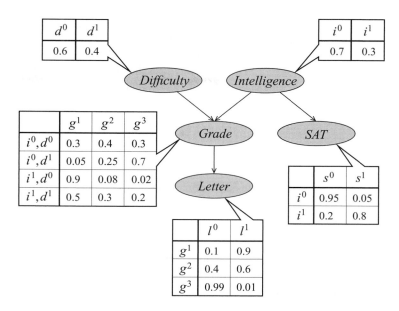

Figure 3.4 Student **Bayesian network** $\mathcal{B}^{student}$ **with CPDs**

the way the world works. The course difficulty and the student's intelligence are determined independently, and before any of the variables in the model. The student's grade depends on both of these factors. The student's SAT score depends only on his intelligence. The quality of the professor's recommendation letter depends (by assumption) only on the student's grade in the class. Intuitively, each variable in the model depends directly only on its parents in the network. We formalize this intuition later.

local probability model

The second component of the Bayesian network representation is a set of *local probability models* that represent the nature of the dependence of each variable on its parents. One such model, $P(I)$, represents the distribution in the population of intelligent versus less intelligent student. Another, $P(D)$, represents the distribution of difficult and easy classes. The distribution over the student's grade is a conditional distribution $P(G \mid I, D)$. It specifies the distribution over the student's grade, inasmuch as it depends on the student's intelligence and the course difficulty. Specifically, we would have a different distribution for each assignment of values i, d. For example, we might believe that a smart student in an easy class is 90 percent likely to get an A, 8 percent likely to get a B, and 2 percent likely to get a C. Conversely, a smart student in a hard class may only be 50 percent likely to get an A. In general, each variable X in the

CPD

model is associated with a *conditional probability distribution (CPD)* that specifies a distribution over the values of X given each possible joint assignment of values to its parents in the model. For a node with no parents, the CPD is conditioned on the empty set of variables. Hence, the CPD turns into a marginal distribution, such as $P(D)$ or $P(I)$. One possible choice of CPDs for this domain is shown in figure 3.4. The network structure together with its CPDs is a *Bayesian network* $\mathcal{B}$; we use $\mathcal{B}^{student}$ to refer to the Bayesian network for our student example.

How do we use this data structure to specify the joint distribution? Consider some particular state in this space, for example, i^1, d^0, g^2, s^1, l^0. Intuitively, the probability of this event can be computed from the probabilities of the basic events that comprise it: the probability that the student is intelligent; the probability that the course is easy; the probability that a smart student gets a B in an easy class; the probability that a smart student gets a high score on his SAT; and the probability that a student who got a B in the class gets a weak letter. The total probability of this state is:

$$P(i^1, d^0, g^2, s^1, l^0) = P(i^1)P(d^0)P(g^2 \mid i^1, d^0)P(s^1 \mid i^1)P(l^0 \mid g^2)$$
$$= 0.3 \cdot 0.6 \cdot 0.08 \cdot 0.8 \cdot 0.4 = 0.004608.$$

Clearly, we can use the same process for any state in the joint probability space. In general, we will have that

$$P(I, D, G, S, L) = P(I)P(D)P(G \mid I, D)P(S \mid I)P(L \mid G). \tag{3.9}$$

chain rule for
Bayesian
networks

This equation is our first example of the *chain rule for Bayesian networks* which we will define in a general setting in section 3.2.3.2.

3.2.1.2   Reasoning Patterns

A joint distribution $P_{\mathcal{B}}$ specifies (albeit implicitly) the probability $P_{\mathcal{B}}(\boldsymbol{Y} = \boldsymbol{y} \mid \boldsymbol{E} = \boldsymbol{e})$ of any event $\boldsymbol{y}$ given any observations $\boldsymbol{e}$, as discussed in section 2.1.3.3: We condition the joint distribution on the event $\boldsymbol{E} = \boldsymbol{e}$ by eliminating the entries in the joint inconsistent with our observation $\boldsymbol{e}$, and renormalizing the resulting entries to sum to 1; we compute the probability of the event $\boldsymbol{y}$ by summing the probabilities of all of the entries in the resulting posterior distribution that are consistent with $\boldsymbol{y}$. To illustrate this process, let us consider our $\mathcal{B}^{student}$ network and see how the probabilities of various events change as evidence is obtained.

Consider a particular student, George, about whom we would like to reason using our model. We might ask how likely George is to get a strong recommendation (l^1) from his professor in Econ101. Knowing nothing else about George or Econ101, this probability is about 50.2 percent. More precisely, let $P_{\mathcal{B}^{student}}$ be the joint distribution defined by the preceding BN; then we have that $P_{\mathcal{B}^{student}}(l^1) \approx 0.502$. We now find out that George is not so intelligent (i^0); the probability that he gets a strong letter from the professor of Econ101 goes down to around 38.9 percent; that is, $P_{\mathcal{B}^{student}}(l^1 \mid i^0) \approx 0.389$. We now further discover that Econ101 is an easy class (d^0). The probability that George gets a strong letter from the professor is now $P_{\mathcal{B}^{student}}(l^1 \mid i^0, d^0) \approx 0.513$. Queries such as these, where we predict the "downstream" effects of various factors (such as George's intelligence), are instances of *causal reasoning* or *prediction*.

causal reasoning

Now, consider a recruiter for Acme Consulting, trying to decide whether to hire George based on our previous model. A priori, the recruiter believes that George is 30 percent likely to be intelligent. He obtains George's grade record for a particular class Econ101 and sees that George received a C in the class (g^3). His probability that George has high intelligence goes down significantly, to about 7.9 percent; that is, $P_{\mathcal{B}^{student}}(i^1 \mid g^3) \approx 0.079$. We note that the probability that the class is a difficult one also goes up, from 40 percent to 62.9 percent.

Now, assume that the recruiter fortunately (for George) lost George's transcript, and has only the recommendation letter from George's professor in Econ101, which (not surprisingly) is

evidential
reasoning

weak. The probability that George has high intelligence still goes down, but only to 14 percent: $P_{\mathcal{B}^{student}}(i^1 \mid l^0) \approx 0.14$. Note that if the recruiter has both the grade and the letter, we have the same probability as if he had only the grade: $P_{\mathcal{B}^{student}}(i^1 \mid g^3, l^0) \approx 0.079$; we will revisit this issue. Queries such as this, where we reason from effects to causes, are instances of *evidential reasoning* or *explanation*.

Finally, George submits his SAT scores to the recruiter, and astonishingly, his SAT score is high. The probability that George has high intelligence goes up dramatically, from 7.9 percent to 57.8 percent: $P_{\mathcal{B}^{student}}(i^1 \mid g^3, s^1) \approx 0.578$. Intuitively, the reason that the high SAT score outweighs the poor grade is that students with low intelligence are extremely unlikely to get good scores on their SAT, whereas students with high intelligence can still get C's. However, smart students are much more likely to get C's in hard classes. Indeed, we see that the probability that Econ101 is a difficult class goes up from the 62.9 percent we saw before to around 76 percent.

This last pattern of reasoning is a particularly interesting one. The information about the SAT gave us information about the student's intelligence, which, in conjunction with the student's grade in the course, told us something about the difficulty of the course. In effect, we have one causal factor for the *Grade* variable — *Intelligence* — giving us information about another — *Difficulty*.

Let us examine this pattern in its pure form. As we said, $P_{\mathcal{B}^{student}}(i^1 \mid g^3) \approx 0.079$. On the other hand, if we now discover that Econ101 is a hard class, we have that $P_{\mathcal{B}^{student}}(i^1 \mid g^3, d^1) \approx 0.11$. In effect, we have provided at least a partial explanation for George's grade in Econ101. To take an even more striking example, if George gets a B in Econ 101, we have that $P_{\mathcal{B}^{student}}(i^1 \mid g^2) \approx 0.175$. On the other hand, if Econ101 is a hard class, we get $P_{\mathcal{B}^{student}}(i^1 \mid g^2, d^1) \approx 0.34$. In effect we have *explained away* the poor grade via the difficulty of the class. **Explaining away is an instance of a general reasoning pattern called *intercausal reasoning*, where different causes of the same effect can interact. This type of reasoning is a very common pattern in human reasoning.** For example, when we have fever and a sore throat, and are concerned about mononucleosis, we are greatly relieved to be told we have the flu. Clearly, having the flu does not prohibit us from having mononucleosis. Yet, having the flu provides an alternative explanation of our symptoms, thereby reducing substantially the probability of mononucleosis.

explaining away

intercausal
reasoning

This intuition of providing an alternative explanation for the evidence can be made very precise. As shown in exercise 3.3, if the flu deterministically causes the symptoms, the probability of mononucleosis goes down to its prior probability (the one prior to the observations of any symptoms). On the other hand, if the flu might occur without causing these symptoms, the probability of mononucleosis goes down, but it still remains somewhat higher than its base level. Explaining away, however, is not the only form of intercausal reasoning. The influence can go in any direction. Consider, for example, a situation where someone is found dead and may have been murdered. The probabilities that a suspect has motive and opportunity both go up. If we now discover that the suspect has motive, the probability that he has opportunity goes up. (See exercise 3.4.)

It is important to emphasize that, although our explanations used intuitive concepts such as cause and evidence, there is nothing mysterious about the probability computations we performed. They can be replicated simply by generating the joint distribution, as defined in equation (3.9), and computing the probabilities of the various events directly from that.

3.2.2   Basic Independencies in Bayesian Networks

As we discussed, a Bayesian network graph $\mathcal{G}$ can be viewed in two ways. In the previous section, we showed, by example, how it can be used as a skeleton data structure to which we can attach local probability models that together define a joint distribution. In this section, we provide a formal semantics for a Bayesian network, starting from the perspective that the graph encodes a set of conditional independence assumptions. We begin by understanding, intuitively, the basic conditional independence assumptions that we want a directed graph to encode. We then formalize these desired assumptions in a definition.

3.2.2.1   Independencies in the Student Example

In the Student example, we used the intuition that edges represent direct dependence. For example, we made intuitive statements such as "the professor's recommendation letter depends only on the student's grade in the class"; this statement was encoded in the graph by the fact that there are no direct edges into the L node except from G. This intuition, that "a node depends directly only on its parents," lies at the heart of the semantics of Bayesian networks.

We give formal semantics to this assertion using conditional independence statements. For example, the previous assertion can be stated formally as the assumption that L is conditionally independent of all other nodes in the network given its parent G:

$$(L \perp I, D, S \mid G). \tag{3.10}$$

In other words, once we know the student's grade, our beliefs about the quality of his recommendation letter are not influenced by information about any other variable. Similarly, to formalize our intuition that the student's SAT score depends only on his intelligence, we can say that S is conditionally independent of all other nodes in the network given its parent I:

$$(S \perp D, G, L \mid I). \tag{3.11}$$

Now, let us consider the G node. Following the pattern blindly, we may be tempted to assert that G is conditionally independent of all other variables in the network given its parents. However, this assumption is false both at an intuitive level and for the specific example distribution we used earlier. Assume, for example, that we condition on i^1, d^1; that is, we have a smart student in a difficult class. In this setting, is G independent of L? Clearly, the answer is no: if we observe l^1 (the student got a strong letter), then our probability in g^1 (the student received an A in the course) should go up; that is, we would expect

$$P(g^1 \mid i^1, d^1, l^1) > P(g^1 \mid i^1, d^1).$$

Indeed, if we examine our distribution, the latter probability is 0.5 (as specified in the CPD), whereas the former is a much higher 0.712.

Thus, we see that we do not expect a node to be conditionally independent of all other nodes given its parents. In particular, even given its parents, it can still depend on its descendants. Can it depend on other nodes? For example, do we expect G to depend on S given I and D? Intuitively, the answer is no. Once we know, say, that the student has high intelligence, his SAT score gives us no additional information that is relevant toward predicting his grade. Thus, we would want the property that:

$$(G \perp S \mid I, D). \tag{3.12}$$

It remains only to consider the variables I and D, which have no parents in the graph. Thus, in our search for independencies given a node's parents, we are now looking for marginal independencies. As the preceding discussion shows, in our distribution $P_{\mathcal{B}^{student}}$, I is not independent of its descendants G, L, or S. Indeed, the only nondescendant of I is D. Indeed, we assumed implicitly that *Intelligence* and *Difficulty* are independent. Thus, we expect that:

$$(I \perp D). \tag{3.13}$$

This analysis might seem somewhat surprising in light of our earlier examples, where learning something about the course difficulty drastically changed our beliefs about the student's intelligence. In that situation, however, we were reasoning in the presence of information about the student's grade. In other words, we were demonstrating the dependence of I and D *given* G. This phenomenon is a very important one, and we will return to it.

For the variable D, both I and S are nondescendants. Recall that, if $(I \perp D)$ then $(D \perp I)$. The variable S increases our beliefs in the student's intelligence, but knowing that the student is smart (or not) does not influence our beliefs in the difficulty of the course. Thus, we have that

$$(D \perp I, S). \tag{3.14}$$

We can see a pattern emerging. Our intuition tells us that the parents of a variable "shield" it from probabilistic influence that is causal in nature. In other words, once I know the value of the parents, no information relating directly or indirectly to its parents or other ancestors can influence my beliefs about it. However, information about its descendants *can* change my beliefs about it, via an evidential reasoning process.

3.2.2.2   Bayesian Network Semantics

We are now ready to provide the formal definition of the semantics of a Bayesian network structure. We would like the formal definition to match the intuitions developed in our example.

Definition 3.1

Bayesian network structure

local independencies

A Bayesian network structure $\mathcal{G}$ *is a directed acyclic graph whose nodes represent random variables* $X_1, \ldots, X_n$. *Let* $\mathrm{Pa}_{X_i}^{\mathcal{G}}$ *denote the parents of* X_i *in* $\mathcal{G}$, *and* $\mathrm{NonDescendants}_{X_i}$ *denote the variables in the graph that are not descendants of* X_i. *Then* $\mathcal{G}$ *encodes the following set of conditional independence assumptions, called the* local independencies, *and denoted by* $\mathcal{I}_\ell(\mathcal{G})$:

> *For each variable* X_i: $(X_i \perp \mathrm{NonDescendants}_{X_i} \mid \mathrm{Pa}_{X_i}^{\mathcal{G}})$.   ■

In other words, the local independencies state that each node X_i is conditionally independent of its nondescendants given its parents.

Returning to the Student network $G_{student}$, the local Markov independencies are precisely the ones dictated by our intuition, and specified in equation (3.10) – equation (3.14).

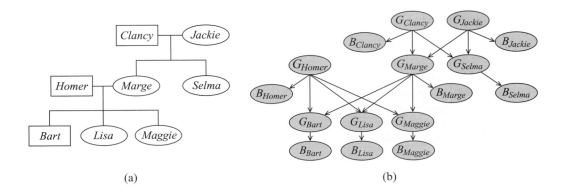

Figure 3.B.1 — Modeling Genetic Inheritance (a) A small family tree. (b) A simple BN for genetic inheritance in this domain. The G variables represent a person's genotype, and the B variables the result of a blood-type test.

Box 3.B — Case Study: The Genetics **Example.** *One of the very earliest uses of a Bayesian network model (long before the general framework was defined) is in the area of genetic pedigrees. In this setting, the local independencies are particularly intuitive. In this application, we want to model the transmission of a certain property, say blood type, from parent to child. The blood type of a person is an observable quantity that depends on her genetic makeup. Such properties are called* phenotypes. *The genetic makeup of a person is called* genotype.

To model this scenario properly, we need to introduce some background on genetics. The human genetic material consists of 22 pairs of autosomal *chromosomes and a pair of the* sex *chromosomes (X and Y). Each chromosome contains a set of genetic material, consisting (among other things) of genes that determine a person's properties. A region of the chromosome that is of interest is called a* locus; *a locus can have several variants, called* alleles.

For concreteness, we focus on autosomal chromosome pairs. In each autosomal pair, one chromosome is the paternal *chromosome, inherited from the father, and the other is the* maternal *chromosome, inherited from the mother. For genes in an autosomal pair, a person has two copies of the gene, one on each copy of the chromosome. Thus, one of the gene's alleles is inherited from the person's mother, and the other from the person's father. For example, the region containing the gene that encodes a person's blood type is a locus. This gene comes in three variants, or alleles: A, B, and O. Thus, a person's genotype is denoted by an ordered pair, such as $\langle A, B \rangle$; with three choices for each entry in the pair, there are 9 possible genotypes. The blood type phenotype is a function of both copies of the gene. For example, if the person has an A allele and an O allele, her observed blood type is "A." If she has two O alleles, her observed blood type is "O."*

To represent this domain, we would have, for each person, two variables: one representing the person's genotype, and the other her phenotype. We use the name $G(p)$ to represent person p's genotype, and $B(p)$ to represent her blood type.

In this example, the independence assumptions arise immediately from the biology. Since the blood type is a function of the genotype, once we know the genotype of a person, additional evidence about other members of the family will not provide new information about the blood type. Similarly, the process of genetic inheritance implies independence assumption. Once we know the genotype of both parents, we know what each of them can pass on to the offspring. Thus, learning new information about ancestors (or nondescendants) does not provide new information about the genotype of the offspring. These are precisely the local independencies in the resulting network structure, shown for a simple family tree in figure 3.B.1. The intuition here is clear; for example, Bart's blood type is correlated with that of his aunt Selma, but once we know Homer's and Marge's genotype, the two become independent.

To define the probabilistic model fully, we need to specify the CPDs. There are three types of CPDs in this model:

- *The penetrance model $P(B(c) \mid G(c))$, which describes the probability of different variants of a particular phenotype (say different blood types) given the person's genotype. In the case of the blood type, this CPD is a deterministic function, but in other cases, the dependence can be more complex.*

- *The transmission model $P(G(c) \mid G(p), G(m))$, where c is a person and p, m her father and mother, respectively. Each parent is equally likely to transmit either of his or her two alleles to the child.*

- *Genotype priors $P(G(c))$, used when person c has no parents in the pedigree. These are the general genotype frequencies within the population.*

Our discussion of blood type is simplified for several reasons. First, some phenotypes, such as late-onset diseases, are not a deterministic function of the genotype. Rather, an individual with a particular genotype might be more likely to have the disease than an individual with other genotypes. Second, the genetic makeup of an individual is defined by many genes. Some phenotypes might depend on multiple genes. In other settings, we might be interested in multiple phenotypes, which (naturally) implies a dependence on several genes. Finally, as we now discuss, the inheritance patterns of different genes are not independent of each other.

Recall that each of the person's autosomal chromosomes is inherited from one of her parents. However, each of the parents also has two copies of each autosomal chromosome. These two copies, within each parent, recombine to produce the chromosome that is transmitted to the child. Thus, the maternal chromosome inherited by Bart is a combination of the chromosomes inherited by his mother Marge from her mother Jackie and her father Clancy. The recombination process is stochastic, but only a handful recombination events take place within a chromosome in a single generation. Thus, if Bart inherited the allele for some locus from the chromosome his mother inherited from her mother Jackie, he is also much more likely to inherit Jackie's copy for a nearby locus. Thus, to construct an appropriate model for multilocus inheritance, we must take into consideration the probability of a recombination taking place between pairs of adjacent loci.

We can facilitate this modeling by introducing selector variables *that capture the inheritance pattern along the chromosome. In particular, for each locus ℓ and each child c, we have a variable $S(\ell, c, m)$ that takes the value 1 if the locus ℓ in c's maternal chromosome was inherited from c's maternal grandmother, and 2 if this locus was inherited from c's maternal grandfather. We have a similar selector variable $S(\ell, c, p)$ for c's paternal chromosome. We can now model correlations induced by low recombination frequency by correlating the variables $S(\ell, c, m)$ and $S(\ell', c, m)$ for adjacent loci ℓ, ℓ'.*

This type of model has been used extensively for many applications. In genetic counseling and prediction, one takes a phenotype with known loci and a set of observed phenotype and genotype data for some individuals in the pedigree to infer the genotype and phenotype for another person in the pedigree (say, a planned child). The genetic data can consist of direct measurements of the relevant disease loci (for some individuals) or measurements of nearby loci, which are correlated with the disease loci.

In linkage analysis, the task is a harder one: identifying the location of disease genes from pedigree data using some number of pedigrees where a large fraction of the individuals exhibit a disease phenotype. Here, the available data includes phenotype information for many individuals in the pedigree, as well as genotype information for loci whose location in the chromosome is known. Using the inheritance model, the researchers can evaluate the likelihood of these observations under different hypotheses about the location of the disease gene relative to the known loci. By repeated calculation of the probabilities in the network for different hypotheses, researchers can pinpoint the area that is "linked" to the disease. This much smaller region can then be used as the starting point for more detailed examination of genes in that area. This process is crucial, for it can allow the researchers to focus on a small area (for example, 1/10, 000 of the genome).

As we will see in later chapters, the ability to describe the genetic inheritance process using a sparse Bayesian network provides us the capability to use sophisticated inference algorithms that allow us to reason about large pedigrees and multiple loci. It also allows us to use algorithms for model learning to obtain a deeper understanding of the genetic inheritance process, such as recombination rates in different regions or penetrance probabilities for different diseases.

3.2.3   Graphs and Distributions

The formal semantics of a Bayesian network graph is as a set of independence assertions. On the other hand, our Student BN was a graph annotated with CPDs, which defined a joint distribution via the chain rule for Bayesian networks. In this section, we show that these two definitions are, in fact, equivalent. A distribution P satisfies the local independencies associated with a graph $\mathcal{G}$ if and only if P is representable as a set of CPDs associated with the graph $\mathcal{G}$. We begin by formalizing the basic concepts.

3.2.3.1   I-Maps

We first define the set of independencies associated with a distribution P.

Definition 3.2

independencies
in P

Let P be a distribution over $\mathcal{X}$. We define $\mathcal{I}(P)$ to be the set of independence assertions of the form $(\boldsymbol{X} \perp \boldsymbol{Y} \mid \boldsymbol{Z})$ that hold in P. ∎

We can now rewrite the statement that "P satisfies the local independencies associated with $\mathcal{G}$" simply as $\mathcal{I}_{\ell}(\mathcal{G}) \subseteq \mathcal{I}(P)$. In this case, we say that $\mathcal{G}$ is an *I-map* (independency map) for P. However, it is useful to define this concept more broadly, since different variants of it will be used throughout the book.

Definition 3.3

I-map

Let $\mathcal{K}$ be any graph object associated with a set of independencies $\mathcal{I}(\mathcal{K})$. We say that $\mathcal{K}$ is an I-map for a set of independencies $\mathcal{I}$ if $\mathcal{I}(\mathcal{K}) \subseteq \mathcal{I}$. ∎

We now say that $\mathcal{G}$ is an I-map for P if $\mathcal{G}$ is an I-map for $\mathcal{I}(P)$.

As we can see from the direction of the inclusion, for $\mathcal{G}$ to be an I-map of P, it is necessary that $\mathcal{G}$ does not mislead us regarding independencies in P: any independence that $\mathcal{G}$ asserts must also hold in P. Conversely, P may have additional independencies that are not reflected in $\mathcal{G}$.

Let us illustrate the concept of an I-map on a very simple example.

Example 3.1

Consider a joint probability space over two independent random variables X and Y. There are three possible graphs over these two nodes: $\mathcal{G}_\emptyset$, which is a disconnected pair $X \quad Y$; $\mathcal{G}_{X \to Y}$, which has the edge $X \to Y$; and $\mathcal{G}_{Y \to X}$, which contains $Y \to X$. The graph $\mathcal{G}_\emptyset$ encodes the assumption that $(X \perp Y)$. The latter two encode no independence assumptions.

Consider the following two distributions:

X	Y	$P(X,Y)$
x^0	y^0	0.08
x^0	y^1	0.32
x^1	y^0	0.12
x^1	y^1	0.48

X	Y	$P(X,Y)$
x^0	y^0	0.4
x^0	y^1	0.3
x^1	y^0	0.2
x^1	y^1	0.1

In the example on the left, X and Y are independent in P; for example, $P(x^1) = 0.48 + 0.12 = 0.6$, $P(y^1) = 0.8$, and $P(x^1, y^1) = 0.48 = 0.6 \cdot 0.8$. Thus, $(X \perp Y) \in \mathcal{I}(P)$, and we have that $\mathcal{G}_\emptyset$ is an I-map of P. In fact, all three graphs are I-maps of P: $\mathcal{I}_\ell(\mathcal{G}_{X \to Y})$ is empty, so that trivially P satisfies all the independencies in it (similarly for $\mathcal{G}_{Y \to X}$). In the example on the right, $(X \perp Y) \notin \mathcal{I}(P)$, so that $\mathcal{G}_\emptyset$ is not an I-map of P. Both other graphs are I-maps of P. ∎

3.2.3.2 I-Map to Factorization

A BN structure $\mathcal{G}$ encodes a set of conditional independence assumptions; every distribution for which $\mathcal{G}$ is an I-map must satisfy these assumptions. This property is the key to allowing the compact factorized representation that we saw in the Student example in section 3.2.1. The basic principle is the same as the one we used in the naive Bayes decomposition in section 3.1.3.

Consider any distribution P for which our Student BN $\mathcal{G}_{student}$ is an I-map. We will decompose the joint distribution and show that it factorizes into local probabilistic models, as in section 3.2.1. Consider the joint distribution $P(I, D, G, L, S)$; from the chain rule for probabilities (equation (2.5)), we can decompose this joint distribution in the following way:

$$P(I, D, G, L, S) = P(I)P(D \mid I)P(G \mid I, D)P(L \mid I, D, G)P(S \mid I, D, G, L). \qquad (3.15)$$

This transformation relies on no assumptions; it holds for any joint distribution P. However, it is also not very helpful, since the conditional probabilities in the factorization on the right-hand side are neither natural nor compact. For example, the last factor requires the specification of 24 conditional probabilities: $P(s^1 \mid i, d, g, l)$ for every assignment of values i, d, g, l.

This form, however, allows us to apply the conditional independence assumptions induced from the BN. Let us assume that $\mathcal{G}_{student}$ is an I-map for our distribution P. In particular, from equation (3.13), we have that $(D \perp I) \in \mathcal{I}(P)$. From that, we can conclude that $P(D \mid I) = P(D)$, allowing us to simplify the second factor on the right-hand side. Similarly, we know from equation (3.10) that $(L \perp I, D \mid G) \in \mathcal{I}(P)$. Hence, $P(L \mid I, D, G) = P(L \mid G)$, allowing us to simplify the third term. Using equation (3.11) in a similar way, we obtain that

$$P(I, D, G, L, S) = P(I)P(D)P(G \mid I, D)P(L \mid G)P(S \mid I). \tag{3.16}$$

This factorization is precisely the one we used in section 3.2.1.

This result tells us that any entry in the joint distribution can be computed as a product of factors, one for each variable. Each factor represents a conditional probability of the variable given its parents in the network. This factorization applies to *any* distribution P for which $G_{student}$ is an I-map.

We now state and prove this fundamental result more formally.

Definition 3.4

factorization

Let $\mathcal{G}$ be a BN graph over the variables $X_1, \ldots, X_n$. We say that a distribution P over the same space factorizes according to $\mathcal{G}$ if P can be expressed as a product

$$P(X_1, \ldots, X_n) = \prod_{i=1}^{n} P(X_i \mid \mathrm{Pa}_{X_i}^{\mathcal{G}}). \tag{3.17}$$

chain rule for Bayesian networks

CPD

This equation is called the chain rule for Bayesian networks. *The individual factors* $P(X_i \mid \mathrm{Pa}_{X_i}^{\mathcal{G}})$ *are called* conditional probability distributions (CPDs) *or local probabilistic models.* ∎

Definition 3.5

Bayesian network

A Bayesian network *is a pair* $\mathcal{B} = (\mathcal{G}, P)$ *where* P *factorizes over* $\mathcal{G}$, *and where* P *is specified as a set of CPDs associated with* $\mathcal{G}$'s *nodes. The distribution* P *is often annotated* $P_{\mathcal{B}}$. ∎

We can now prove that the phenomenon we observed for $G_{student}$ holds more generally.

Theorem 3.1

topological ordering

Let $\mathcal{G}$ *be a BN structure over a set of random variables* $\mathcal{X}$, *and let* P *be a joint distribution over the same space. If* $\mathcal{G}$ *is an I-map for* P, *then* P *factorizes according to* $\mathcal{G}$.

PROOF Assume, without loss of generality, that $X_1, \ldots, X_n$ is a *topological ordering* of the variables in $\mathcal{X}$ relative to $\mathcal{G}$ (see definition 2.19). As in our example, we first use the chain rule for probabilities:

$$P(X_1, \ldots, X_n) = \prod_{i=1}^{n} P(X_i \mid X_1, \ldots, X_{i-1}).$$

Now, consider one of the factors $P(X_i \mid X_1, \ldots, X_{i-1})$. As $\mathcal{G}$ is an I-map for P, we have that $(X_i \perp \mathrm{NonDescendants}_{X_i} \mid \mathrm{Pa}_{X_i}^{\mathcal{G}}) \in \mathcal{I}(P)$. By assumption, all of X_i's parents are in the set $X_1, \ldots, X_{i-1}$. Furthermore, none of X_i's descendants can possibly be in the set. Hence,

$$\{X_1, \ldots, X_{i-1}\} = \mathrm{Pa}_{X_i} \cup \boldsymbol{Z}$$

where $\boldsymbol{Z} \subseteq \mathrm{NonDescendants}_{X_i}$. From the local independencies for X_i and from the decomposition property (equation (2.8)) it follows that $(X_i \perp \boldsymbol{Z} \mid \mathrm{Pa}_{X_i})$. Hence, we have that

$$P(X_i \mid X_1, \ldots, X_{i-1}) = P(X_i \mid \mathrm{Pa}_{X_i}).$$

Applying this transformation to all of the factors in the chain rule decomposition, the result follows. ∎

Thus, the conditional independence assumptions implied by a BN structure $\mathcal{G}$ allow us to factorize a distribution P for which $\mathcal{G}$ is an I-map into small CPDs. Note that the proof is constructive, providing a precise algorithm for constructing the factorization given the distribution P and the graph $\mathcal{G}$.

The resulting factorized representation can be substantially more compact, particularly for sparse structures.

Example 3.2

In our Student *example, the number of independent parameters is fifteen: we have two binomial distributions $P(I)$ and $P(D)$, with one independent parameter each; we have four multinomial distributions over G — one for each assignment of values to I and D — each with two independent parameters; we have three binomial distributions over L, each with one independent parameter; and similarly two binomial distributions over S, each with an independent parameter. The specification of the full joint distribution would require $48 - 1 = 47$ independent parameters.* ∎

More generally, in a distribution over n binary random variables, the specification of the joint distribution requires $2^n - 1$ independent parameters. If the distribution factorizes according to a graph $\mathcal{G}$ where each node has at most k parents, the total number of independent parameters required is less than $n \cdot 2^k$ (see exercise 3.6). In many applications, we can assume a certain locality of influence between variables: although each variable is generally correlated with many of the others, it often depends *directly* on only a small number of other variables. Thus, in many cases, k will be very small, even though n is large. As a consequence, the number of parameters in the Bayesian network representation is typically exponentially smaller than the number of parameters of a joint distribution. This property is one of the main benefits of the Bayesian network representation.

3.2.3.3 Factorization to I-Map

Theorem 3.1 shows one direction of the fundamental connection between the conditional independencies encoded by the BN structure and the factorization of the distribution into local probability models: that the conditional independencies imply factorization. The converse also holds: factorization according to $\mathcal{G}$ implies the associated conditional independencies.

Theorem 3.2

Let $\mathcal{G}$ be a BN structure over a set of random variables $\mathcal{X}$ and let P be a joint distribution over the same space. If P factorizes according to $\mathcal{G}$, then $\mathcal{G}$ is an I-map for P.

We illustrate this theorem by example, leaving the proof as an exercise (exercise 3.9). Let P be some distribution that factorizes according to $\mathcal{G}_{student}$. We need to show that $\mathcal{I}_\ell(\mathcal{G}_{student})$ holds in P. Consider the independence assumption for the random variable S — $(S \perp D, G, L \mid I)$. To prove that it holds for P, we need to show that

$$P(S \mid I, D, G, L) = P(S \mid I).$$

By definition,

$$P(S \mid I, D, G, L) = \frac{P(S, I, D, G, L)}{P(I, D, G, L)}.$$

By the chain rule for BNs equation (3.16), the numerator is equal to $P(I)P(D)P(G \mid I, D)P(L \mid G)P(S \mid I)$. By the process of marginalizing over a joint distribution, we have that the denominator is:

$$
\begin{aligned}
P(I, D, G, L) &= \sum_S P(I, D, G, L, S) \\
&= \sum_S P(I)P(D)P(G \mid I, D)P(L \mid G)P(S \mid I) \\
&= P(I)P(D)P(G \mid I, D)P(L \mid G) \sum_S P(S \mid I) \\
&= P(I)P(D)P(G \mid I, D)P(L \mid G),
\end{aligned}
$$

where the last step is a consequence of the fact that $P(S \mid I)$ is a distribution over values of S, and therefore it sums to 1. We therefore have that

$$
\begin{aligned}
P(S \mid I, D, G, L) &= \frac{P(S, I, D, G, L)}{P(I, D, G, L)} \\
&= \frac{P(I)P(D)P(G \mid I, D)P(L \mid G)P(S \mid I)}{P(I)P(D)P(G \mid I, D)P(L \mid G)} \\
&= P(S \mid I).
\end{aligned}
$$

Box 3.C — Skill: Knowledge Engineering. *Our discussion of Bayesian network construction focuses on the process of going from a given distribution to a Bayesian network. Real life is not like that. We have a vague model of the world, and we need to crystallize it into a network structure and parameters. This task breaks down into several components, each of which can be quite subtle. Unfortunately, modeling mistakes can have significant consequences for the quality of the answers obtained from the network, or to the cost of using the network in practice.*

Picking variables *When we model a domain, there are many possible ways to describe the relevant entities and their attributes. Choosing which random variables to use in the model is often one of the hardest tasks, and this decision has implications throughout the model. A common problem is using ill-defined variables. For example, deciding to include the variable Fever to describe a patient in a medical domain seems fairly innocuous. However, does this random variable relate to the internal temperature of the patient? To the thermometer reading (if one is taken by the medical staff)? Does it refer to the temperature of the patient at a specific moment (for example, the time of admission to the hospital) or to occurrence of a fever over a prolonged period? Clearly, each of these might be a reasonable attribute to model, but the interaction of Fever with other variables depends on the specific interpretation we use.*

clarity test
As this example shows, we must be precise in defining the variables in the model. The clarity *test is a good way of evaluating whether they are sufficiently well defined. Assume that we are a million years after the events described in the domain; can an omniscient being, one who saw everything, determine the value of the variable? For example, consider a Weather variable with a value sunny. To be absolutely precise, we must define where we check the weather, at what time,*

and what fraction of the sky must be clear in order for it to be sunny. For a variable such as Heart-attack, we must specify how large the heart attack has to be, during what period of time it has to happen, and so on. By contrast, a variable such as Risk-of-heart-attack is meaningless, as even an omniscient being cannot evaluate whether a person had high risk or low risk, only whether the heart attack occurred or not. Introducing variables such as this confounds actual events and their probability. Note, however, that we can *use a notion of "risk group," as long as it is defined in terms of clearly specified attributes such as age or lifestyle.*

If we are not careful in our choice of variables, we will have a hard time making sure that evidence observed and conclusions made are coherent.

Generally speaking, we want our model to contain variables that we can potentially observe or that we may want to query. However, sometimes we want to put in a hidden variable *that is neither observed nor directly of interest. Why would we want to do that? Let us consider an example relating to a cholesterol test. Assume that, for the answers to be accurate, the subject has to have eaten nothing after 10:00 PM the previous evening. If the person eats (having no willpower), the results are consistently off. We do not really care about a Willpower variable, nor can we observe it. However, without it, all of the different cholesterol tests become correlated. To avoid graphs where all the tests are correlated, it is better to put in this additional hidden variable, rendering them conditionally independent given the true cholesterol level and the person's willpower.*

hidden variable

On the other hand, it is not necessary to add every variable that might be relevant. In our Student example, the student's SAT score may be affected by whether he goes out for drinks on the night before the exam. Is this variable important to represent? The probabilities already account for the fact that he may achieve a poor score despite being intelligent. It might not be worthwhile to include this variable if it cannot be observed.

It is also important to specify a reasonable domain of values for our variables. In particular, if our partition is not fine enough, conditional independence assumptions may be false. For example, we might want to construct a model where we have a person's cholesterol level, and two cholesterol tests that are conditionally independent given the person's true cholesterol level. We might choose to define the value normal to correspond to levels up to 200, and high to levels above 200. But it may be the case that both tests are more likely to fail if the person's cholesterol is marginal (200–240). In this case, the assumption of conditional independence given the value (high/normal) of the cholesterol test is false. It is only true if we add a marginal value.

Picking structure *As we saw, there are many structures that are consistent with the same set of independencies. One successful approach is to choose a structure that reflects the causal order and dependencies, so that causes are parents of the effect. Such structures tend to work well. Either because of some real locality of influence in the world, or because of the way people perceive the world, causal graphs tend to be sparser. It is important to stress that the causality is in the world, not in our inference process. For example, in an automobile insurance network, it is tempting to put Previous-accident as a parent of Good-driver, because that is how the insurance company thinks about the problem. This is not the causal order in the world, because being a bad driver causes previous (and future) accidents. In principle, there is nothing to prevent us from directing the edges in this way. However, a noncausal ordering often requires that we introduce many additional edges to account for induced dependencies (see section 3.4.1).*

One common approach to constructing a structure is a backward construction process. We begin with a variable of interest, say Lung-Cancer. We then try to elicit a prior probability for that variable. If our expert responds that this probability is not determinable, because it depends on other factors, that is a good indication that these other factors should be added as parents for that variable (and as variables into the network). For example, we might conclude using this process that Lung-Cancer really should have Smoking as a parent, and (perhaps not as obvious) that Smoking should have Gender and Age as a parent. This approach, called extending the conversation, *avoids probability estimates that result from an average over a heterogeneous population, and therefore leads to more precise probability estimates.*

When determining the structure, however, we must also keep in mind that approximations are inevitable. For many pairs of variables, we can construct a scenario where one depends on the other. For example, perhaps Difficulty depends on Intelligence, because the professor is more likely to make a class difficult if intelligent students are registered. In general, **there are many weak influences that we might choose to model, but if we put in all of them, the network can become very complex.** *Such networks are problematic from a representational perspective: they are hard to understand and hard to debug, and eliciting (or learning) parameters can get very difficult. Moreover, as reasoning in Bayesian networks depends strongly on their connectivity (see section 9.4), adding such edges can make the network too expensive to use.*

This final consideration may lead us, in fact, to make approximations that we know to be wrong. For example, in networks for fault or medical diagnosis, the correct approach is usually to model each possible fault as a separate random variable, allowing for multiple failures. However, such networks might be too complex to perform effective inference in certain settings, and so we may sometimes resort to a single fault approximation, *where we have a single random variable encoding the primary fault or disease.*

Picking probabilities   *One of the most challenging tasks in constructing a network manually is eliciting probabilities from people. This task is somewhat easier in the context of causal models, since the parameters tend to be natural and more interpretable. Nevertheless, people generally dislike committing to an exact estimate of probability.*

One approach is to elicit estimates qualitatively, using abstract terms such as "common," "rare," and "surprising," and then assign these to numbers using a predefined scale. This approach is fairly crude, and often can lead to misinterpretation. There are several approaches developed for assisting in eliciting probabilities from people. For example, one can visualize the probability of the event as an area (slice of a pie), or ask people how they would compare the probability in question to certain predefined lotteries. Nevertheless, probability elicitation is a long, difficult process, and one whose outcomes are not always reliable: the elicitation method can often influence the results, and asking the same question using different phrasing can often lead to significant differences in the answer. For example, studies show that people's estimates for an event such as "Death by disease" are significantly lower than their estimates for this event when it is broken down into different possibilities such as "Death from cancer," "Death from heart disease," and so on.

How important is it that we get our probability estimates exactly right? In some cases, small errors have very little effect. For example, changing a conditional probability of 0.7 to 0.75 generally does not have a significant effect. Other errors, however, can have a significant effect:

- **Zero probabilities**: *A common mistake is to assign a probability of zero to an event that is extremely unlikely, but not impossible. The problem is that* **one can never condition away a zero probability, no matter how much evidence we get. When an event is unlikely**

but not impossible, giving it probability zero is guaranteed to lead to irrecoverable errors. *For example, in one of the early versions of the the Pathfinder system (box 3.D), 10 percent of the misdiagnoses were due to zero probability estimates given by the expert to events that were unlikely but not impossible. As a general rule, very few things (except definitions) have probability zero, and we must be careful in assigning zeros.*

- **Orders of magnitude:** *Small differences in very low probability events can make a large difference to the network conclusions. Thus, a (conditional) probability of 10^{-4} is very different from 10^{-5}.*

- **Relative values:** *The qualitative behavior of the conclusions reached by the network — the value that has the highest probability — is fairly sensitive to the relative sizes of $P(x \mid y)$ for different values y of Pa_X. For example, it is important that the network encode correctly that the probability of having a high fever is greater when the patient has pneumonia than when he has the flu.*

sensitivity
analysis

A very useful tool for estimating network parameters is sensitivity analysis, *which allows us to determine the extent to which a given probability parameter affects the outcome. This process allows us to evaluate whether it is important to get a particular CPD entry right. It also helps us figure out which CPD entries are responsible for an answer to some query that does not match our intuitions.*

medical diagnosis

expert system

Pathfinder

Box 3.D — Case Study: Medical Diagnosis Systems. *One of the earliest applications of Bayesian networks was to the task of medical diagnosis. In the 1980s, a very active area of research was the construction of* expert systems — *computer-based systems that replace or assist an expert in performing a complex task. One such task that was tackled in several ways was medical diagnosis. This task, more than many others, required a treatment of uncertainty, due to the complex, nondeterministic relationships between findings and diseases. Thus, it formed the basis for experimentation with various formalisms for uncertain reasoning.*

The Pathfinder *expert system was designed by Heckerman and colleagues (Heckerman and Nathwani 1992a; Heckerman et al. 1992; Heckerman and Nathwani 1992b) to help a pathologist diagnose diseases in lymph nodes. Ultimately, the model contained more than sixty different diseases and around a hundred different features. It evolved through several versions, including some based on nonprobabilistic formalisms, and several that used variants of Bayesian networks. Its diagnostic ability was evaluated over real pathological cases and compared to the diagnoses of pathological experts.*

One of the first models used was a simple naive Bayes model, which was compared to the models based on alternative uncertainty formalisms, and judged to be superior in its diagnostic ability. It therefore formed the basis for subsequent development of the system.

The same evaluation pointed out important problems in the way in which parameters were elicited from the expert. First, it was shown that 10 percent of the cases were diagnosed incorrectly, because the correct disease was ruled out by a finding that was unlikely, but not impossible, to manifest in that disease. Second, in the original construction, the expert estimated the probabilities $P(\text{Finding} \mid \text{Disease})$ by fixing a single disease and evaluating the probabilities of all its findings.

It was found that the expert was more comfortable considering a single finding and evaluating its probability across all diseases. This approach allows the expert to compare the relative values of the same finding across multiple diseases, as described in box 3.C.

With these two lessons in mind, another version of Pathfinder — Pathfinder III — was constructed, still using the naive Bayes model. Finally, Pathfinder IV used a full Bayesian network, with a single disease hypothesis but with dependencies between the features. Pathfinder IV was constructed using a similarity network (see box 5.B), significantly reducing the number of parameters that must be elicited. Pathfinder IV, viewed as a Bayesian network, had a total of around 75,000 parameters, but the use of similarity networks allowed the model to be constructed with fewer than 14,000 distinct parameters. Overall, the structure of Pathfinder IV took about 35 hours to define, and the parameters 40 hours.

A comprehensive evaluation of the performance of the two models revealed some important insights. First, the Bayesian network performed as well or better on most cases than the naive Bayes model. In most of the cases where the Bayesian network performed better, the use of richer dependency models was a contributing factor. As expected, these models were useful because they address the strong conditional independence assumptions of the naive Bayes model, as described in box 3.A. Somewhat more surprising, they also helped in allowing the expert to condition the probabilities on relevant factors other than the disease, using the process of extending the conversation described in box 3.C, leading to more accurate elicited probabilities. Finally, the use of similarity networks led to more accurate models, for the smaller number of elicited parameters reduced irrelevant fluctuations in parameter values (due to expert inconsistency) that can lead to spurious dependencies.

Overall, the Bayesian network model agreed with the predictions of an expert pathologist in 50/53 cases, as compared with 47/53 cases for the naive Bayes model, with significant therapeutic implications. A later evaluation showed that the diagnostic accuracy of Pathfinder IV was at least as good as that of the expert used to design the system. When used with less expert pathologists, the system significantly improved the diagnostic accuracy of the physicians alone. Moreover, the system showed greater ability to identify important findings and to integrate these findings into a correct diagnosis.

Unfortunately, multiple reasons prevent the widespread adoption of Bayesian networks as an aid for medical diagnosis, including legal liability issues for misdiagnoses and incompatibility with the physicians' workflow. However, several such systems have been fielded, with significant success. Moreover, similar technology is being used successfully in a variety of other diagnosis applications (see box 23.C).

3.3   Independencies in Graphs

Dependencies and independencies are key properties of a distribution and are crucial for understanding its behavior. As we will see, independence properties are also important for answering queries: they can be exploited to reduce substantially the computation cost of inference. Therefore, it is important that our representations make these properties clearly visible both to a user and to algorithms that manipulate the BN data structure.

As we discussed, a graph structure $\mathcal{G}$ encodes a certain set of conditional independence assumptions $\mathcal{I}_\ell(\mathcal{G})$. Knowing only that a distribution P factorizes over $\mathcal{G}$, we can conclude that it satisfies $\mathcal{I}_\ell(\mathcal{G})$. An immediate question is whether there are other independencies that we can "read off" directly from $\mathcal{G}$. That is, are there other independencies that hold for *every* distribution P that factorizes over $\mathcal{G}$?

3.3.1 D-separation

Our aim in this section is to understand when we can *guarantee* that an independence ($\boldsymbol{X} \perp \boldsymbol{Y} \mid \boldsymbol{Z}$) holds in a distribution associated with a BN structure $\mathcal{G}$. To understand when a property is guaranteed to hold, it helps to consider its converse: "Can we imagine a case where it does not?" Thus, we focus our discussion on analyzing when it is *possible* that $\boldsymbol{X}$ can influence $\boldsymbol{Y}$ given $\boldsymbol{Z}$. If we construct an example where this influence occurs, then the converse property ($\boldsymbol{X} \perp \boldsymbol{Y} \mid \boldsymbol{Z}$) cannot hold for all of the distributions that factorize over $\mathcal{G}$, and hence the independence property ($\boldsymbol{X} \perp \boldsymbol{Y} \mid \boldsymbol{Z}$) cannot follow from $\mathcal{I}_\ell(\mathcal{G})$.

We therefore begin with an intuitive case analysis: Here, we try to understand when an observation regarding a variable X can possibly change our beliefs about Y, in the presence of evidence about the variables $\boldsymbol{Z}$. Although this analysis will be purely intuitive, we will show later that our conclusions are actually provably correct.

Direct connection We begin with the simple case, when X and Y are directly connected via an edge, say $X \rightarrow Y$. For any network structure $\mathcal{G}$ that contains the edge $X \rightarrow Y$, it is possible to construct a distribution where X and Y are correlated regardless of any evidence about any of the other variables in the network. In other words, if X and Y are directly connected, we can always get examples where they influence each other, regardless of $\boldsymbol{Z}$.

In particular, assume that $Val(X) = Val(Y)$; we can simply set $X = Y$. That, by itself, however, is not enough; if (given the evidence $\boldsymbol{Z}$) X deterministically takes some particular value, say 0, then X and Y both deterministically take that value, and are uncorrelated. We therefore set the network so that X is (for example) uniformly distributed, regardless of the values of any of its parents. This construction suffices to induce a correlation between X and Y, regardless of the evidence.

Indirect connection Now consider the more complicated case when X and Y are not directly connected, but there is a trail between them in the graph. We begin by considering the simplest such case: a three-node network, where X and Y are not directly connected, but where there is a trail between them via Z. It turns out that this simple case is the key to understanding the whole notion of indirect interaction in Bayesian networks.

There are four cases where X and Y are connected via Z, as shown in figure 3.5. The first two correspond to causal chains (in either direction), the third to a common cause, and the fourth to a common effect. We analyze each in turn.

Indirect causal effect (figure 3.5a). To gain intuition, let us return to the Student example, where we had a causal trail $I \rightarrow G \rightarrow L$. Let us begin with the case where G is not observed. Intuitively, if we observe that the student is intelligent, we are more inclined to believe that he gets an A, and therefore that his recommendation letter is strong. In other words, the probability of these latter events is higher conditioned on the observation that the student is intelligent.

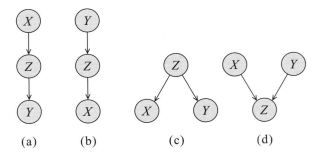

Figure 3.5 **The four possible two-edge trails from X to Y via Z:** (a) An indirect causal effect; (b) An indirect evidential effect; (c) A common cause; (d) A common effect.

In fact, we saw precisely this behavior in the distribution of figure 3.4. Thus, in this case, we believe that X can influence Y via Z.

Now assume that Z is observed, that is, $Z \in \mathbf{Z}$. As we saw in our analysis of the Student example, if we observe the student's grade, then (as we assumed) his intelligence no longer influences his letter. In fact, the local independencies for this network tell us that $(L \perp I \mid G)$. Thus, we conclude that X cannot influence Y via Z if Z is observed.

Indirect evidential effect (figure 3.5b). Returning to the Student example, we have a chain $I \to G \to L$. We have already seen that observing a strong recommendation letter for the student changes our beliefs in his intelligence. Conversely, once the grade is observed, the letter gives no additional information about the student's intelligence. Thus, our analysis in the case $Y \to Z \to X$ here is identical to the causal case: X can influence Y via Z, but only if Z is not observed. The similarity is not surprising, as dependence is a symmetrical notion. Specifically, if $(X \perp Y)$ does not hold, then $(Y \perp X)$ does not hold either.

Common cause (figure 3.5c). This case is one that we have analyzed extensively, both within the simple naive Bayes model of section 3.1.3 and within our Student example. Our example has the student's intelligence I as a parent of his grade G and his SAT score S. As we discussed, S and G are correlated in this model, in that observing (say) a high SAT score gives us information about a student's intelligence and hence helps us predict his grade. However, once we observe I, this correlation disappears, and S gives us no additional information about G. Once again, for this network, this conclusion follows from the local independence assumption for the node G (or for S). Thus, our conclusion here is identical to the previous two cases: X can influence Y via Z if and only if Z is not observed.

Common effect (figure 3.5d). In all of the three previous cases, we have seen a common pattern: X can influence Y via Z if and only if Z is not observed. Therefore, we might expect that this pattern is universal, and will continue through this last case. Somewhat surprisingly, this is not the case. Let us return to the Student example and consider I and D, which are parents of G. When G is not observed, we have that I and D are independent. In fact, this conclusion follows (once again) from the local independencies from the network. Thus, in this case, influence cannot "flow" along the trail $X \to Z \leftarrow Y$ if the intermediate node Z is not observed.

On the other hand, consider the behavior when Z is observed. In our discussion of the

Student example, we analyzed precisely this case, which we called intercausal reasoning; we showed, for example, that the probability that the student has high intelligence goes down dramatically when we observe that his grade is a C ($G = g^3$), but then goes up when we observe that the class is a difficult one $D = d^1$. Thus, in presence of the evidence $G = g^3$, we have that I and D are correlated.

Let us consider a variant of this last case. Assume that we do not observe the student's grade, but we do observe that he received a weak recommendation letter ($L = l^0$). Intuitively, the same phenomenon happens. The weak letter is an indicator that he received a low grade, and therefore it suffices to correlate I and D.

When influence can flow from X to Y via Z, we say that the trail $X \rightleftharpoons Z \rightleftharpoons Y$ is *active*. The results of our analysis for active two-edge trails are summarized thus:

- **Causal trail** $X \to Z \to Y$: active if and only if Z is not observed.
- **Evidential trail** $X \leftarrow Z \leftarrow Y$: active if and only if Z is not observed.
- **Common cause** $X \leftarrow Z \to Y$: active if and only if Z is not observed.
- **Common effect** $X \to Z \leftarrow Y$: active if and only if either Z or one of Z's descendants is observed.

v-structure

A structure where $X \to Z \leftarrow Y$ (as in figure 3.5d) is also called a *v-structure*.

It is useful to view probabilistic influence as a flow in the graph. Our analysis here tells us when influence from X can "flow" through Z to affect our beliefs about Y.

General Case Now consider the case of a longer trail $X_1 \rightleftharpoons \cdots \rightleftharpoons X_n$. Intuitively, for influence to "flow" from X_1 to X_n, it needs to flow through every single node on the trail. In other words, X_1 can influence X_n if every two-edge trail $X_{i-1} \rightleftharpoons X_i \rightleftharpoons X_{i+1}$ along the trail allows influence to flow.

We can summarize this intuition in the following definition:

Definition 3.6

observed variable

active trail

Let $\mathcal{G}$ be a BN structure, and $X_1 \rightleftharpoons \ldots \rightleftharpoons X_n$ a trail in $\mathcal{G}$. Let $\mathbf{Z}$ be a subset of observed variables. The trail $X_1 \rightleftharpoons \ldots \rightleftharpoons X_n$ is active *given* $\mathbf{Z}$ if

- *Whenever we have a v-structure* $X_{i-1} \to X_i \leftarrow X_{i+1}$, *then* X_i *or one of its descendants are in* $\mathbf{Z}$;

- *no other node along the trail is in* $\mathbf{Z}$. ∎

Note that if X_1 or X_n are in $\mathbf{Z}$ the trail is not active.

In our Student BN, we have that $D \to G \leftarrow I \to S$ is not an active trail for $\mathbf{Z} = \emptyset$, because the v-structure $D \to G \leftarrow I$ is not activated. That same trail is active when $\mathbf{Z} = \{L\}$, because observing the descendant of G activates the v-structure. On the other hand, when $\mathbf{Z} = \{L, I\}$, the trail is not active, because observing I blocks the trail $G \leftarrow I \to S$.

What about graphs where there is more than one trail between two nodes? Our flow intuition continues to carry through: one node can influence another if there is any trail along which influence can flow. Putting these intuitions together, we obtain the notion of *d-separation*, which provides us with a notion of separation between nodes in a directed graph (hence the term d-separation, for directed separation):

d-separation

Definition 3.7   *Let X, Y, Z be three sets of nodes in $\mathcal{G}$. We say that X and Y are d-separated given Z, denoted*
d-sep$_\mathcal{G}(X; Y \mid Z)$, if there is no active trail between any node $X \in X$ and $Y \in Y$ given Z.
We use $\mathcal{I}(\mathcal{G})$ to denote the set of independencies that correspond to d-separation:

$$\mathcal{I}(\mathcal{G}) = \{(X \perp Y \mid Z) \ : \ \text{d-sep}_\mathcal{G}(X; Y \mid Z)\}. \qquad \blacksquare$$

global Markov   This set is also called the set of *global Markov independencies*. The similarity between the nota-
independencies   tion $\mathcal{I}(\mathcal{G})$ and our notation $\mathcal{I}(P)$ is not coincidental: As we discuss later, the independencies
in $\mathcal{I}(\mathcal{G})$ are precisely those that are guaranteed to hold for every distribution over $\mathcal{G}$.

3.3.2   Soundness and Completeness

So far, our definition of d-separation has been based on our intuitions regarding flow of influence, and on our one example. As yet, we have no guarantee that this analysis is "correct."
Perhaps there is a distribution over the BN where X can influence Y despite the fact that all
trails between them are blocked.

soundness of   Hence, the first property we want to ensure for d-separation as a method for determining
d-separation   independence is *soundness*: if we find that two nodes X and Y are d-separated given some Z,
then we are guaranteed that they are, in fact, conditionally independent given Z.

Theorem 3.3   *If a distribution P factorizes according to $\mathcal{G}$, then $\mathcal{I}(\mathcal{G}) \subseteq \mathcal{I}(P)$.*

In other words, any independence reported by d-separation is satisfied by the underlying distribution. The proof of this theorem requires some additional machinery that we introduce in
chapter 4, so we defer the proof to that chapter (see section 4.5.1.1).

completeness of   A second desirable property is the complementary one — *completeness*: d-separation detects
d-separation   *all* possible independencies. More precisely, if we have that two variables X and Y are independent given Z, then they are d-separated. A careful examination of the completeness property
reveals that it is ill defined, inasmuch as it does not specify the distribution in which X and Y
are independent.

To formalize this property, we first define the following notion:

Definition 3.8   *A distribution P is faithful to $\mathcal{G}$ if, whenever $(X \perp Y \mid Z) \in \mathcal{I}(P)$, then d-sep$_\mathcal{G}(X; Y \mid Z)$. In*
faithful   *other words, any independence in P is reflected in the d-separation properties of the graph.* $\qquad \blacksquare$

We can now provide one candidate formalization of the completeness property is as follows:

- For any distribution P that factorizes over $\mathcal{G}$, we have that P is faithful to $\mathcal{G}$; that is, if X
  and Y are *not d-separated* given Z in $\mathcal{G}$, then X and Y are *dependent in all* distributions P
  that factorize over $\mathcal{G}$.

This property is the obvious converse to our notion of soundness: If true, the two together
would imply that, for any P that factorizes over $\mathcal{G}$, we have that $\mathcal{I}(P) = \mathcal{I}(\mathcal{G})$. Unfortunately,
this highly desirable property is easily shown to be false: Even if a distribution factorizes over
$\mathcal{G}$, it can still contain additional independencies that are not reflected in the structure.

Example 3.3

Consider a distribution P over two variables A and B, where A and B are independent. One possible I-map for P is the network $A \to B$. For example, we can set the CPD for B to be

	b^0	b^1
a^0	0.4	0.6
a^1	0.4	0.6

This example clearly violates the first candidate definition of completeness, because the graph $\mathcal{G}$ is an I-map for the distribution P, yet there are independencies that hold for this distribution but do not follow from d-separation. In fact, these are not independencies that we can hope to discover by examining the network structure. ∎

Thus, the completeness property does not hold for this candidate definition of completeness. We therefore adopt a weaker yet still useful definition:

- If $(X \perp Y \mid \boldsymbol{Z})$ in *all* distributions P that factorize over $\mathcal{G}$, then $d\text{-}sep_{\mathcal{G}}(X; Y \mid \boldsymbol{Z})$. And the contrapositive: If X and Y are *not d-separated* given $\boldsymbol{Z}$ in $\mathcal{G}$, then X and Y are *dependent in some* distribution P that factorizes over $\mathcal{G}$.

Using this definition, we can show:

Theorem 3.4

Let $\mathcal{G}$ be a BN structure. If X and Y are not d-separated given $\boldsymbol{Z}$ in $\mathcal{G}$, then X and Y are dependent given $\boldsymbol{Z}$ in some distribution P that factorizes over $\mathcal{G}$.

PROOF The proof constructs a distribution P that makes X and Y correlated. The construction is roughly as follows. As X and Y are not d-separated, there exists an active trail $U_1, \ldots, U_k$ between them. We define CPDs for the variables on the trail so as to make each pair U_i, U_{i+1} correlated; in the case of a v-structure $U_i \to U_{i+1} \leftarrow U_{i+2}$, we define the CPD of U_{i+1} so as to ensure correlation, and also define the CPDs of the path to some downstream evidence node, in a way that guarantees that the downstream evidence activates the correlation between U_i and U_{i+2}. All other CPDs in the graph are chosen to be uniform, and thus the construction guarantees that influence only flows along this single path, preventing cases where the influence of two (or more) paths cancel out. The details of the construction are quite technical and laborious, and we omit them. ∎

We can view the completeness result as telling us that our definition of $\mathcal{I}(\mathcal{G})$ is the maximal one. For any independence assertion that is not a consequence of d-separation in $\mathcal{G}$, we can always find a counterexample distribution P that factorizes over $\mathcal{G}$. In fact, this result can be strengthened significantly:

Theorem 3.5

For almost all distributions P that factorize over $\mathcal{G}$, that is, for all distributions except for a set of measure zero in the space of CPD parameterizations, we have that $\mathcal{I}(P) = \mathcal{I}(\mathcal{G})$.[1]

---

1. A set has measure zero if it is infinitesimally small relative to the overall space. For example, the set of all rationals has measure zero within the interval $[0, 1]$. A straight line has measure zero in the plane. This intuition is defined formally in the field of *measure theory*.

This result strengthens theorem 3.4 in two distinct ways: First, whereas theorem 3.4 shows that any dependency in the graph can be found in some distribution, this new result shows that there exists a single distribution that is faithful to the graph, that is, where all of the dependencies in the graph hold simultaneously. Second, not only does this property hold for a single distribution, but it also holds for almost all distributions that factorize over $\mathcal{G}$.

PROOF At a high level, the proof is based on the following argument: Each conditional independence assertion is a set of polynomial equalities over the space of CPD parameters (see exercise 3.13). A basic property of polynomials is that a polynomial is either identically zero or it is nonzero almost everywhere (its set of roots has measure zero). Theorem 3.4 implies that polynomials corresponding to assertions outside $\mathcal{I}(\mathcal{G})$ cannot be identically zero, because they have at least one counterexample. Thus, the set of distributions P, which exhibit any one of these "spurious" independence assertions, has measure zero. The set of distributions that do not satisfy $\mathcal{I}(P) = \mathcal{I}(\mathcal{G})$ is the union of these separate sets, one for each spurious independence assertion. The union of a finite number of sets of measure zero is a set of measure zero, proving the result. ∎

☞ **These results state that for almost all parameterizations P of the graph $\mathcal{G}$ (that is, for almost all possible choices of CPDs for the variables), the d-separation test precisely characterizes the independencies that hold for P.** In other words, even if we have a distribution P that satisfies more independencies than $\mathcal{I}(\mathcal{G})$, a slight perturbation of the CPDs of P will almost always eliminate these "extra" independencies. This guarantee seems to state that such independencies are always accidental, and we will never encounter them in practice. However, as we illustrate in example 3.7, there are cases where our CPDs have certain local structure that is not accidental, and that implies these additional independencies that are not detected by d-separation.

3.3.3 An Algorithm for d-Separation

The notion of d-separation allows us to infer independence properties of a distribution P that factorizes over $\mathcal{G}$ simply by examining the connectivity of $\mathcal{G}$. However, in order to be useful, we need to be able to determine d-separation effectively. Our definition gives us a constructive solution, but a very inefficient one: We can enumerate all trails between X and Y, and check each one to see whether it is active. The running time of this algorithm depends on the number of trails in the graph, which can be exponential in the size of the graph.

Fortunately, there is a much more efficient algorithm that requires only linear time in the size of the graph. The algorithm has two phases. We begin by traversing the graph bottom up, from the leaves to the roots, marking all nodes that are in Z or that have descendants in Z. Intuitively, these nodes will serve to enable v-structures. In the second phase, we traverse breadth-first from X to Y, stopping the traversal along a trail when we get to a blocked node. A node is blocked if: (a) it is the "middle" node in a v-structure and unmarked in phase I, or (b) is not such a node and is in Z. If our breadth-first search gets us from X to Y, then there is an active trail between them.

The precise algorithm is shown in algorithm 3.1. The first phase is straightforward. The second phase is more subtle. For efficiency, and to avoid infinite loops, the algorithm must keep track of all nodes that have been visited, so as to avoid visiting them again. However, in graphs

---

Algorithm 3.1 Algorithm for finding nodes reachable from X given Z via active trails

---

    Procedure Reachable (
        $\mathcal{G}$,   // Bayesian network graph
        X,   // Source variable
        $\boldsymbol{Z}$ // Observations
)

```
1        // Phase I: Insert all ancestors of Z into A
2        L ← Z     // Nodes to be visited
3        A ← ∅     // Ancestors of Z
4        while L ≠ ∅
5          Select some Y from L
6          L ← L − {Y}
7          if Y ∉ A then
8            L ← L ∪ Pa_Y     // Y's parents need to be visited
9            A ← A ∪ {Y}     // Y is ancestor of evidence
10
11       // Phase II: traverse active trails starting from X
12       L ← {(X, ↑)}     // (Node,direction) to be visited
13       V ← ∅     // (Node,direction) marked as visited
14       R ← ∅     // Nodes reachable via active trail
15       while L ≠ ∅
16         Select some (Y, d) from L
17         L ← L − {(Y, d)}
18         if (Y, d) ∉ V then
19           if Y ∉ Z then
20             R ← R ∪ {Y}     // Y is reachable
21           V ← V ∪ {(Y, d)}     // Mark (Y, d) as visited
22           if d = ↑ and Y ∉ Z then     // Trail up through Y active if Y not in Z
23             for each Z ∈ Pa_Y
24               L ← L ∪ {(Z, ↑)}     // Y's parents to be visited from bottom
25             for each Z ∈ Ch_Y
26               L ← L ∪ {(Z, ↓)}     // Y's children to be visited from top
27           else if d = ↓ then     // Trails down through Y
28             if Y ∉ Z then
29               // Downward trails to Y's children are active
30               for each Z ∈ Ch_Y
31                 L ← L ∪ {(Z, ↓)}     // Y's children to be visited from top
32             if Y ∈ A then     // v-structure trails are active
33               for each Z ∈ Pa_Y
34                 L ← L ∪ {(Z, ↑)}     // Y's parents to be visited from bottom
35       return R
```

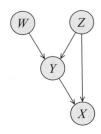

Figure 3.6   A simple example for the d-separation algorithm

with loops (multiple trails between a pair of nodes), an intermediate node Y might be involved in several trails, which may require different treatment within the algorithm:

Example 3.4   *Consider the Bayesian network of figure 3.6, where our task is to find all nodes reachable from X. Assume that Y is observed, that is, $Y \in \mathbf{Z}$. Assume that the algorithm first encounters Y via the direct edge $Y \rightarrow X$. Any extension of this trail is blocked by Y, and hence the algorithm stops the traversal along this trail. However, the trail $X \leftarrow Z \rightarrow Y \leftarrow W$ is not blocked by Y. Thus, when we encounter Y for the second time via the edge $Z \rightarrow Y$, we should not ignore it. Therefore, after the first visit to Y, we can mark it as visited for the purpose of trails coming in from children of Y, but not for the purpose of trails coming in from parents of Y.* ∎

In general, we see that, for each node Y, we must keep track separately of whether it has been visited from the top and whether it has been visited from the bottom. Only when both directions have been explored is the node no longer useful for discovering new active trails.

Based on this intuition, we can now show that the algorithm achieves the desired result:

Theorem 3.6   *The algorithm* Reachable$(\mathcal{G}, X, \mathbf{Z})$ *returns the set of all nodes reachable from X via trails that are active in $\mathcal{G}$ given $\mathbf{Z}$.*

The proof is left as an exercise (exercise 3.14).

3.3.4   I-Equivalence

The notion of $\mathcal{I}(\mathcal{G})$ specifies a set of conditional independence assertions that are associated with a graph. This notion allows us to abstract away the details of the graph structure, viewing it purely as a specification of independence properties. In particular, one important implication of this perspective is the observation that very different BN structures can actually be equivalent, in that they encode precisely the same set of conditional independence assertions. Consider, for example, the three networks in figure 3.5a,(b),(c). All three of them encode precisely the same independence assumptions: $(X \perp Y \mid Z)$.

Definition 3.9

I-equivalence

Two graph structures $\mathcal{K}_1$ and $\mathcal{K}_2$ over $\mathcal{X}$ are I-equivalent if $\mathcal{I}(\mathcal{K}_1) = \mathcal{I}(\mathcal{K}_2)$. The set of all graphs over $\mathcal{X}$ is partitioned into a set of mutually exclusive and exhaustive I-equivalence classes, which are the set of equivalence classes induced by the I-equivalence relation. ∎

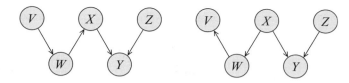

Figure 3.7   Skeletons and v-structures in a network. The two networks shown have the same skeleton and v-structures ($X \rightarrow Y \leftarrow Z$).

Note that the v-structure network in figure 3.5d induces a very different set of d-separation assertions, and hence it does not fall into the same I-equivalence class as the first three. Its I-equivalence class contains only that single network.

I-equivalence of two graphs immediately implies that any distribution P that can be factorized over one of these graphs can be factorized over the other. Furthermore, **there is no intrinsic property of P that would allow us to associate it with one graph rather than an equivalent one. This observation has important implications with respect to our ability to determine the directionality of influence.** In particular, although we can determine, for a distribution $P(X, Y)$, whether X and Y are correlated, there is nothing in the distribution that can help us determine whether the correct structure is $X \rightarrow Y$ or $Y \rightarrow X$. We return to this point when we discuss the causal interpretation of Bayesian networks in chapter 21.

The d-separation criterion allows us to test for I-equivalence using a very simple graph-based algorithm. We start by considering the trails in the networks.

Definition 3.10

skeleton

The skeleton *of a Bayesian network graph $\mathcal{G}$ over $\mathcal{X}$ is an undirected graph over $\mathcal{X}$ that contains an edge $\{X, Y\}$ for every edge (X, Y) in $\mathcal{G}$.* ∎

In the networks of figure 3.7, the networks (a) and (b) have the same skeleton.

If two networks have a common skeleton, then the set of trails between two variables X and Y is same in both networks. If they do not have a common skeleton, we can find a trail in one network that does not exist in the other and use this trail to find a counterexample for the equivalence of the two networks.

Ensuring that the two networks have the same trails is clearly not enough. For example, the networks in figure 3.5 all have the same skeleton. Yet, as the preceding discussion shows, the network of figure 3.5d is not equivalent to the networks of figure 3.5a–(c). The difference, is of course, the v-structure in figure 3.5d. Thus, it seems that if the two networks have the same skeleton and exactly the same set of v-structures, they are equivalent. Indeed, this property provides a sufficient condition for I-equivalence:

Theorem 3.7

Let $\mathcal{G}_1$ and $\mathcal{G}_2$ be two graphs over $\mathcal{X}$. If $\mathcal{G}_1$ and $\mathcal{G}_2$ have the same skeleton and the same set of v-structures then they are I-equivalent.

The proof is left as an exercise (see exercise 3.16).

Unfortunately, this characterization is not an equivalence: there are graphs that are I-equivalent but do not have the same set of v-structures. As a counterexample, consider *complete* graphs over a set of variables. Recall that a complete graph is one to which we cannot add additional arcs without causing cycles. Such graphs encode the empty set of conditional independence assertions. Thus, any two complete graphs are I-equivalent. Although they have the same skeleton, they invariably have different v-structures. Thus, by using the criterion on theorem 3.7, we can conclude (in certain cases) only that two networks are I-equivalent, but we cannot use it to guarantee that they are not.

We can provide a stronger condition that does correspond exactly to I-equivalence. Intuitively, the unique independence pattern that we want to associate with a v-structure $X \to Z \leftarrow Y$ is that X and Y are independent (conditionally on their parents), but dependent given Z. If there is a direct edge between X and Y, as there was in our example of the complete graph, the first part of this pattern is eliminated.

Definition 3.11

immorality

covering edge

A v-structure $X \to Z \leftarrow Y$ is an immorality *if there is no direct edge between X and Y. If there is such an edge, it is called a* covering edge *for the v-structure.* ∎

Note that not every v-structure is an immorality, so that two networks with the same immoralities do not necessarily have the same v-structures. For example, two different complete directed graphs always have the same immoralities (none) but different v-structures.

Theorem 3.8

Let $\mathcal{G}_1$ and $\mathcal{G}_2$ be two graphs over $\mathcal{X}$. Then $\mathcal{G}_1$ and $\mathcal{G}_2$ have the same skeleton and the same set of immoralities if and only if they are I-equivalent.

The proof of this (more difficult) result is also left as an exercise (see exercise 3.17).

We conclude with a final characterization of I-equivalence in terms of local operations on the graph structure.

Definition 3.12

covered edge

An edge $X \to Y$ in a graph $\mathcal{G}$ is said to be covered *if* $\mathrm{Pa}_Y^{\mathcal{G}} = \mathrm{Pa}_X^{\mathcal{G}} \cup \{X\}$. ∎

Theorem 3.9

Two graphs $\mathcal{G}$ and $\mathcal{G}'$ are I-equivalent if and only if there exists a sequence of networks $\mathcal{G} = \mathcal{G}_1, \ldots, \mathcal{G}_k = \mathcal{G}'$ that are all I-equivalent to $\mathcal{G}$ such that the only difference between $\mathcal{G}_i$ and $\mathcal{G}_{i+1}$ is a single reversal of a covered edge.

The proof of this theorem is left as an exercise (exercise 3.18).

3.4   From Distributions to Graphs

In the previous sections, we showed that, if P factorizes over $\mathcal{G}$, we can derive a rich set of independence assertions that hold for P by simply examining $\mathcal{G}$. This result immediately leads to the idea that we can use a graph as a way of revealing the structure in a distribution. In particular, we can test for independencies in P by constructing a graph $\mathcal{G}$ that represents P and testing d-separation in $\mathcal{G}$. As we will see, having a graph that reveals the structure in P has other important consequences, in terms of reducing the number of parameters required to specify or learn the distribution, and in terms of the complexity of performing inference on the network.

In this section, we examine the following question: Given a distribution P, to what extent can we construct a graph $\mathcal{G}$ whose independencies are a reasonable surrogate for the independencies in P? It is important to emphasize that we will never actually take a fully specified distribution P and construct a graph $\mathcal{G}$ for it: As we discussed, a full joint distribution is much too large to represent explicitly. However, answering this question is an important conceptual exercise, which will help us later on when we try to understand the process of constructing a Bayesian network that represents our model of the world, whether manually or by learning from data.

3.4.1 Minimal I-Maps

One approach to finding a graph that represents a distribution P is simply to take any graph that is an I-map for P. The problem with this naive approach is clear: As we saw in example 3.3, the complete graph is an I-map for any distribution, yet it does not reveal any of the independence structure in the distribution. However, examples such as this one are not very interesting. The graph that we used as an I-map is clearly and trivially unrepresentative of the distribution, in that there are edges that are obviously redundant. This intuition leads to the following definition, which we also define more broadly:

Definition 3.13

minimal I-map

A graph $\mathcal{K}$ is a minimal I-map *for a set of independencies $\mathcal{I}$ if it is an I-map for $\mathcal{I}$, and if the removal of even a single edge from $\mathcal{K}$ renders it not an I-map.* ∎

This notion of an I-map applies to multiple types of graphs, both Bayesian networks and other types of graphs that we will encounter later on. Moreover, because it refers to a set of independencies $\mathcal{I}$, it can be used to define an I-map for a distribution P, by taking $\mathcal{I} = \mathcal{I}(P)$, or to another graph $\mathcal{K}'$, by taking $\mathcal{I} = \mathcal{I}(\mathcal{K}')$.

Recall that definition 3.5 defines a Bayesian network to be a distribution P that factorizes over $\mathcal{G}$, thereby implying that $\mathcal{G}$ is an I-map for P. It is standard to restrict the definition even further, by requiring that $\mathcal{G}$ be a minimal I-map for P.

How do we obtain a minimal I-map for the set of independencies induced by a given distribution P? The proof of the factorization theorem (theorem 3.1) gives us a procedure, which is shown in algorithm 3.2. We assume we are given a predetermined *variable ordering*, say, $\{X_1, \ldots, X_n\}$. We now examine each variable X_i, $i = 1, \ldots, n$ in turn. For each X_i, we pick some minimal subset U of $\{X_1, \ldots, X_{i-1}\}$ to be X_i's parents in $\mathcal{G}$. More precisely, we require that U satisfy $(X_i \perp \{X_1, \ldots, X_{i-1}\} - U \mid U)$, and that no node can be removed from U without violating this property. We then set U to be the parents of X_i.

variable ordering

The proof of theorem 3.1 tells us that, if each node X_i is independent of $X_1, \ldots, X_{i-1}$ given its parents in $\mathcal{G}$, then P factorizes over $\mathcal{G}$. We can then conclude from theorem 3.2 that $\mathcal{G}$ is an I-map for P. By construction, $\mathcal{G}$ is minimal, so that $\mathcal{G}$ is a minimal I-map for P.

Note that our choice of U may not be unique. Consider, for example, a case where two variables A and B are logically equivalent, that is, our distribution P only gives positive probability to instantiations where A and B have the same value. Now, consider a node C that is correlated with A. Clearly, we can choose either A or B to be a parent of C, but having chosen the one, we cannot choose the other without violating minimality. Hence, the minimal parent set U in our construction is not necessarily unique. However, one can show that, if the distribution is positive (see definition 2.5), that is, if for any instantiation ξ to all the network variables $\mathcal{X}$ we have that $P(\xi) > 0$, then the choice of parent set, given an ordering, is unique. Under this assumption, algorithm 3.2 can produce all minimal I-maps for P: Let $\mathcal{G}$ be any

---

Algorithm 3.2 Procedure to build a minimal I-map given an ordering

   Procedure Build-Minimal-I-Map (
     $X_1, \ldots, X_n$ // an ordering of random variables in $\mathcal{X}$
     $\mathcal{I}$ // Set of independencies
)

1   Set $\mathcal{G}$ to an empty graph over $\mathcal{X}$
2   **for** $i = 1, \ldots, n$
3       $U \leftarrow \{X_1, \ldots, X_{i-1}\}$ // U is the current candidate for parents of X_i
4   **for** $U' \subseteq \{X_1, \ldots, X_{i-1}\}$
5   **if** $U' \subset U$ and $(X_i \perp \{X_1, \ldots, X_{i-1}\} - U' \mid U') \in \mathcal{I}$ **then**
6         $U \leftarrow U'$
7   // At this stage U is a minimal set satisfying $(X_i \perp$
         $\{X_1, \ldots, X_{i-1}\} - U \mid U)$
8   // Now set U to be the parents of X_i
9   **for** $X_j \in U$
10   Add $X_j \rightarrow X_i$ to $\mathcal{G}$
11   **return** $\mathcal{G}$

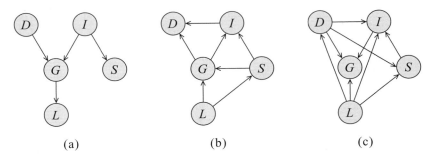

        (a)   (b)   (c)

Figure 3.8 **Three minimal I-maps for $P_{\mathcal{B}^{student}}$, induced by different orderings:** (a) D, I, S, G, L; (b) L, S, G, I, D; (C) L, D, S, I, G.

minimal I-map for P. If we give call Build-Minimal-I-Map with an ordering $\prec$ that is topological for $\mathcal{G}$, then, due to the uniqueness argument, the algorithm must return $\mathcal{G}$.

    At first glance, the minimal I-map seems to be a reasonable candidate for capturing the structure in the distribution: It seems that if $\mathcal{G}$ is a minimal I-map for a distribution P, then we should be able to "read off" all of the independencies in P directly from $\mathcal{G}$. Unfortunately, this intuition is false.

Example 3.5   *Consider the distribution $P_{\mathcal{B}^{student}}$, as defined in figure 3.4, and let us go through the process of constructing a minimal I-map for $P_{\mathcal{B}^{student}}$. We note that the graph $G_{student}$ precisely reflects the independencies in this distribution $P_{\mathcal{B}^{student}}$ (that is, $\mathcal{I}(P_{\mathcal{B}^{student}}) = \mathcal{I}(G_{student})$), so that we can use $G_{student}$ to determine which independencies hold in $P_{\mathcal{B}^{student}}$.*

    Our construction process starts with an arbitrary ordering on the nodes; we will go through this *process for three different orderings. Throughout this process, it is important to remember that we are testing independencies relative to the distribution $P_{\mathcal{B}^{student}}$. We can use G_{student} (figure 3.4) to guide our intuition about which independencies hold in $P_{\mathcal{B}^{student}}$, but we can always resort to testing these independencies in the joint distribution $P_{\mathcal{B}^{student}}$.*

The first ordering is a very natural one: D, I, S, G, L. We add one node at a time and see which of the possible edges from the preceding nodes are redundant. We start by adding D, then I. We can now remove the edge from D to I because this particular distribution satisfies $(I \perp D)$, so I is independent of D given its other parents (the empty set). Continuing on, we add S, but we can remove the edge from D to S because our distribution satisfies $(S \perp D \mid I)$. We then add G, but we can remove the edge from S to G, because the distribution satisfies $(G \perp S \mid I, D)$. Finally, we add L, but we can remove all edges from D, I, S. Thus, our final output is the graph in figure 3.8a, which is precisely our original network for this distribution.

Now, consider a somewhat less natural ordering: L, S, G, I, D. In this case, the resulting I-map is not quite as natural or as sparse. To see this, let us consider the sequence of steps. We start by adding L to the graph. Since it is the first variable in the ordering, it must be a root. Next, we consider S. The decision is whether to have L as a parent of S. Clearly, we need an edge from L to S, because the quality of the student's letter is correlated with his SAT score in this distribution, and S has no other parents that help render it independent of L. Formally, we have that $(S \perp L)$ does not *hold in the distribution. In the next iteration of the algorithm, we introduce G. Now, all possible subsets of $\{L, S\}$ are potential parents set for G. Clearly, G is dependent on L. Moreover, although G is independent of S given I, it is not independent of S given L. Hence, we must add the edge between S and G. Carrying out the procedure, we end up with the graph shown in figure 3.8b.*

Finally, consider the ordering: L, D, S, I, G. In this case, a similar analysis results in the graph shown in figure 3.8c, which is almost a complete graph, missing only the edge from S to G, which we can remove because G is independent of S given I. ∎

Note that the graphs in figure 3.8b,c really are minimal I-maps for this distribution. However, they fail to capture some or all of the independencies that hold in the distribution. Thus, they show that the fact that $\mathcal{G}$ is a minimal I-map for P is far from a guarantee that $\mathcal{G}$ captures the independence structure in P.

3.4.2 Perfect Maps

We aim to find a graph $\mathcal{G}$ that precisely captures the independencies in a given distribution P.

Definition 3.14

perfect map

We say that a graph $\mathcal{K}$ is a perfect map *(P-map) for a set of independencies $\mathcal{I}$ if we have that $\mathcal{I}(\mathcal{K}) = \mathcal{I}$. We say that $\mathcal{K}$ is a perfect map for P if $\mathcal{I}(\mathcal{K}) = \mathcal{I}(P)$.* ∎

If we obtain a graph $\mathcal{G}$ that is a P-map for a distribution P, then we can (by definition) read the independencies in P directly from $\mathcal{G}$. By construction, our original graph $G_{student}$ is a P-map for $P_{\mathcal{B}^{student}}$.

If our goal is to find a perfect map for a distribution, an immediate question is whether every distribution has a perfect map. Unfortunately, the answer is no, and for several reasons. The first type of counterexample involves regularity in the parameterization of the distribution that cannot be captured in the graph structure.

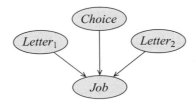

Figure 3.9   Network for the OneLetter **example**

Example 3.6   *Consider a joint distribution P over 3 random variables X,Y,Z such that:*

$$P(x, y, z) = \begin{cases} 1/12 & x \oplus y \oplus z = false \\ 1/6 & x \oplus y \oplus z = true \end{cases}$$

where $\oplus$ is the XOR (exclusive OR) function. A simple calculation shows that $(X \perp Y) \in \mathcal{I}(P)$, and that Z is not independent of X given Y or of Y given X. Hence, one minimal I-map for this distribution is the network $X \rightarrow Z \leftarrow Y$, using a deterministic XOR for the CPD of Z. However, this network is not a perfect map; a precisely analogous calculation shows that $(X \perp Z) \in \mathcal{I}(P)$, but this conclusion is not supported by a d-separation analysis.   ∎

Thus, we see that deterministic relationships can lead to distributions that do not have a P-map. Additional examples arise as a consequence of other regularities in the CPD.

Example 3.7   *Consider a slight elaboration of our* Student *example. During his academic career, our student George has taken both Econ101 and CS102. The professors of both classes have written him letters, but the recruiter at Acme Consulting asks for only a single recommendation. George's chance of getting the job depends on the quality of the letter he gives the recruiter. We thus have four random variables: $L1$ and $L2$, corresponding to the quality of the recommendation letters for Econ101 and CS102 respectively; C, whose value represents George's choice of which letter to use; and J, representing the event that George is hired by Acme Consulting.*

The obvious minimal I-map for this distribution is shown in figure 3.9. Is this a perfect map? Clearly, it does not reflect independencies that are not at the variable level. In particular, we have that $(L1 \perp J \mid C = 2)$. However, this limitation is not surprising; by definition, a BN structure makes independence assertions only at the level of variables. (We return to this issue in section 5.2.2.) However, our problems are not limited to these finer-grained independencies. Some thought reveals that, in our target distribution, we also have that $(L1 \perp L2 \mid C, J)$! This independence is not implied by d-separation, because the v-structure $L1 \rightarrow J \leftarrow L2$ is enabled. However, we can convince ourselves that the independence holds using reasoning by cases. If $C = 1$, then there is no dependence of J on $L2$. Intuitively, the edge from $L2$ to J disappears, eliminating the trail between $L1$ and $L2$, so that $L1$ and $L2$ are independent in this case. A symmetric analysis applies in the case that $C = 2$. Thus, in both cases, we have that $L1$ and $L2$ are independent. This independence assertion is not captured by our minimal I-map, which is therefore not a P-map.   ∎

A different class of examples is not based on structure within a CPD, but rather on symmetric variable-level independencies that are not naturally expressed within a Bayesian network.

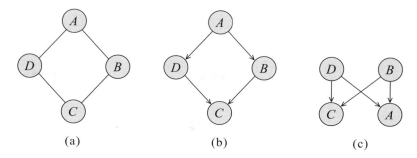

Figure 3.10 **Attempted Bayesian network models for the** Misconception **example:** (a) Study pairs over four students. (b) First attempt at a Bayesian network model. (c) Second attempt at a Bayesian network model.

A second class of distributions that do not have a perfect map are those for which the independence assumptions imposed by the structure of Bayesian networks is simply not appropriate.

Example 3.8

Consider a scenario where we have four students who get together in pairs to work on the homework for a class. For various reasons, only the following pairs meet: Alice and Bob; Bob and Charles; Charles and Debbie; and Debbie and Alice. (Alice and Charles just can't stand each other, and Bob and Debbie had a relationship that ended badly.) The study pairs are shown in figure 3.10a.

In this example, the professor accidentally misspoke in class, giving rise to a possible misconception among the students in the class. Each of the students in the class may subsequently have figured out the problem, perhaps by thinking about the issue or reading the textbook. In subsequent study pairs, he or she may transmit this newfound understanding to his or her study partners. We therefore have four binary random variables, representing whether the student has the misconception or not. We assume that for each $X \in \{A, B, C, D\}$, x^1 denotes the case where the student has the misconception, and x^0 denotes the case where he or she does not.

Because Alice and Charles never speak to each other directly, we have that A and C are conditionally independent given B and D. Similarly, B and D are conditionally independent given A and C. Can we represent this distribution (with these independence properties) using a BN? One attempt is shown in figure 3.10b. Indeed, it encodes the independence assumption that $(A \perp C \mid \{B, D\})$. However, it also implies that B and D are independent given only A, but dependent given both A and C. Hence, it fails to provide a perfect map for our target distribution. A second attempt, shown in figure 3.10c, is equally unsuccessful. It also implies that $(A \perp C \mid \{B, D\})$, but it also implies that B and D are marginally independent. It is clear that all other candidate BN structures are also flawed, so that this distribution does not have a perfect map. ∎

3.4.3 Finding Perfect Maps ★

Earlier we discussed an algorithm for finding minimal I-maps. We now consider an algorithm for finding a perfect map (P-map) of a distribution. Because the requirements from a P-map are stronger than the ones we require from an I-map, the algorithm will be more involved.

Throughout the discussion in this section, we assume that P has a P-map. In other words, there is an unknown DAG $\mathcal{G}^*$ that is P-map of P. Since $\mathcal{G}^*$ is a P-map, we will interchangeably refer to independencies in P and in $\mathcal{G}^*$ (since these are the same). We note that the algorithms we describe do fail when they are given a distribution that does not have a P-map. We discuss this issue in more detail later.

Thus, our goal is to identify $\mathcal{G}^*$ from P. One obvious difficulty that arises when we consider this goal is that $\mathcal{G}^*$ is, in general, not uniquely identifiable from P. A P-map of a distribution, if one exists, is generally not unique: As we saw, for example, in figure 3.5, multiple graphs can encode precisely the same independence assumptions. However, the P-map of a distribution *is* unique up to I-equivalence between networks. That is, a distribution P can have many P-maps, but all of them are I-equivalent.

If we require that a P-map construction algorithm return a single network, the output we get may be some arbitrary member of the I-equivalence class of $\mathcal{G}^*$. A more correct answer would be to return the entire equivalence class, thus avoiding an arbitrary commitment to a possibly incorrect structure. Of course, we do not want our algorithm to return a (possibly very large) set of distinct networks as output. Thus, one of our tasks in this section is to develop a compact representation of an entire equivalence class of DAGs. As we will see later in the book, this representation plays a useful role in other contexts as well.

This formulation of the problem points us toward a solution. Recall that, according to theorem 3.8, two DAGs are I-equivalent if they share the same skeleton and the same set of immoralities. Thus, we can construct the I-equivalence class for $\mathcal{G}^*$ by determining its skeleton and its immoralities from the independence properties of the given distribution P. We then use both of these components to build a representation of the equivalence class.

3.4.3.1　Identifying the Undirected Skeleton

At this stage we want to construct an undirected graph S that contains an edge $X{-}Y$ if X and Y are adjacent in $\mathcal{G}^*$; that is, if either $X \to Y$ or $Y \to X$ is an edge in $\mathcal{G}^*$.

The basic idea is to use independence queries of the form $(X \perp Y \mid U)$ for different sets of variables U. This idea is based on the observation that if X and Y are adjacent in $\mathcal{G}^*$, we cannot separate them with any set of variables.

Lemma 3.1

Let $\mathcal{G}^*$ be a P-map of a distribution $\mathcal{P}$, and let X and Y be two variables such that $X \to Y$ is in $\mathcal{G}^*$. Then, $P \not\models (X \perp Y \mid U)$ for any set U that does not include X and Y.

PROOF Assume that that $X \to Y \in \mathcal{G}^*$, and let U be a set of variables. According to d-separation the trail $X \to Y$ cannot be blocked by the evidence set U. Thus, X and Y are not d-separated by U. Since $\mathcal{G}^*$ is a P-map of P, we have that $P \not\models (X \perp Y \mid U)$. ∎

This lemma implies that if X and Y are adjacent in $\mathcal{G}^*$, all conditional independence queries that involve both of them would fail. Conversely, if X and Y are not adjacent in $\mathcal{G}$, we would hope to be able to find a set of variables that makes these two variables conditionally independent. Indeed, as we now show, we can provide a precise characterization of such a set:

Lemma 3.2

Let $\mathcal{G}^*$ be an I-map of a distribution P, and let X and Y be two variables that are not adjacent in $\mathcal{G}^*$. Then either $P \models (X \perp Y \mid \mathrm{Pa}_X^{\mathcal{G}^*})$ or $P \models (X \perp Y \mid \mathrm{Pa}_Y^{\mathcal{G}^*})$.

The proof is left as an exercise (exercise 3.19).

witness

Thus, if X and Y are not adjacent in $\mathcal{G}^*$, then we can find a set U so that $\mathcal{P} \models (X \perp Y \mid U)$. We call this set U a *witness* of their independence. Moreover, the lemma shows that we can find a witness of bounded size. Thus, if we assume that $\mathcal{G}^*$ has bounded indegree, say less than or equal to d, then we do not need to consider witness sets larger than d.

Algorithm 3.3 Recovering the undirected skeleton for a distribution P that has a P-map

    Procedure Build-PMap-Skeleton (
        $\mathcal{X} = \{X_1, \ldots, X_n\}$,   // Set of random variables
        P,   // Distribution over $\mathcal{X}$
        d // Bound on witness set
)
1   Let $\mathcal{H}$ be the complete undirected graph over $\mathcal{X}$
2   **for** X_i, X_j in $\mathcal{X}$
3        $U_{X_i,X_j} \leftarrow \emptyset$
4   **for** $U \in \text{Witnesses}(X_i, X_j, \mathcal{H}, d)$
5   // Consider U as a witness set for X_i, X_j
6   **if** $P \models (X_i \perp X_j \mid U)$ **then**
7               $U_{X_i,X_j} \leftarrow U$
8   Remove X_i—X_j from $\mathcal{H}$
9   **break**
10   **return** $(\mathcal{H}, \{U_{X_i,X_j} : i, j \in \{1, \ldots, n\}\})$

With these tools in hand, we can now construct an algorithm for building a skeleton of $\mathcal{G}^*$, shown in algorithm 3.3. For each pair of variables, we consider all potential witness sets and test for independence. If we find a witness that separates the two variables, we record it (we will soon see why) and move on to the next pair of variables. If we do not find a witness, then we conclude that the two variables are adjacent in $\mathcal{G}^*$ and add them to the skeleton. The list Witnesses$(X_i, X_j, \mathcal{H}, d)$ in line 4 specifies the set of possible witness sets that we consider for separating X_i and X_j. From our earlier discussion, if we assume a bound d on the indegree, then we can restrict attention to sets U of size at most d. Moreover, using the same analysis, we saw that we have a witness that consists either of the parents of X_i or of the parents of X_j. In the first case, we can restrict attention to sets $U \subseteq \text{Nb}_{X_i}^{\mathcal{H}} - \{X_j\}$, where $\text{Nb}_{X_i}^{\mathcal{H}}$ are the neighbors of X_i in the current graph $\mathcal{H}$; in the second, we can similarly restrict attention to sets $U \subseteq \text{Nb}_{X_j}^{\mathcal{H}} - \{X_i\}$. Finally, we note that if U separates X_i and X_j, then also many of U's supersets will separate X_i and X_j. Thus, we search the set of possible witnesses in order of increasing size.

This algorithm will recover the correct skeleton given that $\mathcal{G}^*$ is a P-map of P and has bounded indegree d. If P does not have a P-map, then the algorithm can fail; see exercise 3.22. This algorithm has complexity of $O(n^{d+2})$ since we consider $O(n^2)$ pairs, and for each we perform $O((n-2)^d)$ independence tests. We greatly reduce the number of independence tests by ordering potential witnesses accordingly, and by aborting the inner loop once we find a witness for a pair (after line 9). However, for pairs of variables that are directly connected in the skeleton, we still need to evaluate all potential witnesses.

Algorithm 3.4 Marking immoralities in the construction of a perfect map

Procedure Mark-Immoralities (
$\quad \mathcal{X} = \{X_1, \ldots, X_n\}$,
$\quad S$ // Skeleton
$\quad \{\boldsymbol{U}_{X_i, X_j} : 1 \leq i, j \leq n\}$ // Witnesses found by Build-PMap-Skeleton
)
1      $\mathcal{K} \leftarrow S$
2   **for** X_i, X_j, X_k such that $X_i{-}X_j{-}X_k \in S$ and $X_i{-}X_k \notin S$
3   // $X_i{-}X_j{-}X_k$ is a potential immorality
4   **if** $X_j \notin \boldsymbol{U}_{X_i, X_k}$ **then**
5   Add the orientations $X_i \to X_j$ and $X_j \leftarrow X_k$ to $\mathcal{K}$
6   **return** $\mathcal{K}$

3.4.3.2   Identifying Immoralities

At this stage we have reconstructed the undirected skeleton S using Build-PMap-Skeleton. Now, we want to reconstruct edge direction. The main cue for learning about edge directions in $\mathcal{G}^*$ are immoralities. As shown in theorem 3.8, all DAGs in the equivalence class of $\mathcal{G}^*$ share the same set of immoralities. Thus, our goal is to consider *potential immoralities* in the skeleton and for each one determine whether it is indeed an immorality. A triplet of variables X, Z, Y is a *potential immorality* if the skeleton contains $X{-}Z{-}Y$ but does not contain an edge between X and Y. If such a triplet is indeed an immorality in $\mathcal{G}^*$, then X and Y cannot be independent given Z. Nor will they be independent given a set $\boldsymbol{U}$ that contains Z. More precisely,

potential
immorality

Proposition 3.1

Let $\mathcal{G}^*$ be a P-map of a distribution P, and let X, Y and Z be variables that form an immorality $X \to Z \leftarrow Y$. Then, $P \not\models (X \perp Y \mid \boldsymbol{U})$ for any set $\boldsymbol{U}$ that contains Z.

PROOF Let $\boldsymbol{U}$ be a set of variables that contains Z. Since Z is observed, the trail $X \to Z \leftarrow Y$ is active, and so X and Y are not d-separated in $\mathcal{G}^*$. Since $\mathcal{G}^*$ is a P-map of P, we have that $P^* \not\models (X \perp Y \mid \boldsymbol{U})$.   ∎

What happens in the complementary situation? Suppose $X{-}Z{-}Y$ in the skeleton, but is not an immorality. This means that one of the following three cases is in $\mathcal{G}^*$: $X \to Z \to Y$, $Y \to Z \to X$, or $X \leftarrow Z \to Y$. In all three cases, X and Y are d-separated only if Z is observed.

Proposition 3.2

Let $\mathcal{G}^*$ be a P-map of a distribution P, and let the triplet X, Y, Z be a potential immorality in the skeleton of $\mathcal{G}^*$, such that $X \to Z \leftarrow Y$ is not in $\mathcal{G}^*$. If $\boldsymbol{U}$ is such that $P \models (X \perp Y \mid \boldsymbol{U})$, then $Z \in \boldsymbol{U}$.

PROOF Consider all three configurations of the trail $X \rightleftharpoons Z \rightleftharpoons Y$. In all three, Z must be observed in order to block the trail. Since $\mathcal{G}^*$ is a P-map of P, we have that if $P \models (X \perp Y \mid \boldsymbol{U})$, then $Z \in \boldsymbol{U}$.   ∎

Combining these two results, we see that a potential immorality $X{-}Z{-}Y$ is an immorality if and only if Z is not in the witness set(s) for X and Y. That is, if $X{-}Z{-}Y$ is an immorality, then proposition 3.1 shows that Z is not in any witness set $\boldsymbol{U}$; conversely, if $X\!-\!Z\!-\!Y$ is not an immorality, the Z must be in every witness set $\boldsymbol{U}$. Thus, we can use the specific witness set $\boldsymbol{U}_{X,Y}$ that we recorded for X, Y in order to determine whether this triplet is an immorality or not: we simply check whether $Z \in \boldsymbol{U}_{X,Y}$. If $Z \notin \boldsymbol{U}_{X,Y}$, then we declare the triplet an immorality. Otherwise, we declare that it is not an immorality. The Mark-Immoralities procedure shown in algorithm 3.4 summarizes this process.

3.4.3.3 Representing Equivalence Classes

Once we have the skeleton and identified the immoralities, we have a specification of the equivalence class of $\mathcal{G}^*$. For example, to test if $\mathcal{G}$ is equivalent to $\mathcal{G}^*$ we can check whether it has the same skeleton as $\mathcal{G}^*$ and whether it agrees on the location of the immoralities.

The description of an equivalence class using only the skeleton and the set of immoralities is somewhat unsatisfying. For example, we might want to know whether the fact that our network is in the equivalence class implies that there is an arc $X \to Y$. Although the definition does tell us whether there is some edge between X and Y, it leaves the direction unresolved. In other cases, however, the direction of an edge is fully determined, for example, by the presence of an immorality. To encode both of these cases, we use a graph that allows both directed and undirected edges, as defined in section 2.2. Indeed, as we show, the chain graph, or PDAG, representation (definition 2.21) provides precisely the right framework.

Definition 3.15

class PDAG

Let $\mathcal{G}$ be a DAG. A chain graph $\mathcal{K}$ is a class PDAG *of the equivalence class of $\mathcal{G}$ if shares the same skeleton as $\mathcal{G}$, and contains a directed edge $X \to Y$ if and only if all $\mathcal{G}'$ that are I-equivalent to $\mathcal{G}$ contain the edge $X \to Y$.*[2] ∎

In other words, a class PDAG represents potential edge orientations in the equivalence classes. If the edge is directed, then all the members of the equivalence class agree on the orientation of the edge. If the edge is undirected, there are two DAGs in the equivalence class that disagree on the orientation of the edge.

For example, the networks in figure 3.5a–(c) are I-equivalent. The class PDAG of this equivalence class is the graph $X\!-\!Z\!-\!Y$, since both edges can be oriented in either direction in some member of the equivalence class. Note that, although both edges in this PDAG are undirected, not all joint orientations of these edges are in the equivalence class. As discussed earlier, setting the orientations $X \to Z \leftarrow Y$ results in the network of figure 3.5d, which does not belong this equivalence class. More generally, if the class PDAG has k undirected edges, the equivalence class can contain at most 2^k networks, but the actual number can be much smaller.

Can we effectively construct the class PDAG $\mathcal{K}$ for $\mathcal{G}^*$ from the reconstructed skeleton and immoralities? Clearly, edges involved in immoralities must be directed in $\mathcal{K}$. The obvious question is whether $\mathcal{K}$ can contain directed edges that are not involved in immoralities. In other words, can there be additional edges whose direction is necessarily the same in every member of the equivalence class? To understand this issue better, consider the following example:

Example 3.9

Consider the DAG of figure 3.11a. This DAG has a single immorality $A \to C \leftarrow B$. This immorality implies that the class PDAG of this DAG must have the arcs $A \to C$ and $B \to C$ directed, as

---

2. For consistency with standard terminology, we use the PDAG terminology when referring to the chain graph representing an I-equivalence class.

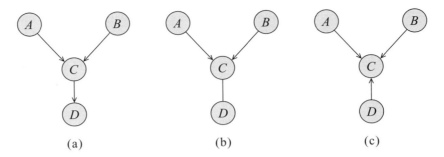

        (a)   (b)   (c)

Figure 3.11   Simple example of compelled edges in the representation of an equivalence class. (a) Original DAG $\mathcal{G}^*$. (b) Skeleton of $\mathcal{G}^*$ annotated with immoralities. (c) a DAG that is not equivalent to $\mathcal{G}^*$.

shown in figure 3.11b. This PDAG representation suggests that the edge $C—D$ can assume either orientation. Note, however, that the DAG of figure 3.11c, where we orient the edge between C and D as $D \to C$, contains additional immoralities (that is, $A \to C \leftarrow D$ and $B \to C \leftarrow D$). Thus, this DAG is not *equivalent to our original DAG.*

In this example, there is only one possible orientation of $C—D$ that is consistent with the finding that $A—C—D$ is not an immorality. Thus, we conclude that the class PDAG for the DAG of figure 3.11a is simply the DAG itself. In other words, the equivalence class of this DAG is a singleton.∎

As this example shows, a negative result in an immorality test also provides information about edge orientation. In particular, in any case where the PDAG $\mathcal{K}$ contains a structure $X \to Y—Z$ and there is no edge from X to Z, then we must orient the edge $Y \to Z$, for otherwise we would create an immorality $X \to Y \leftarrow Z$.

Some thought reveals that there are other local configurations of edges where some ways of orienting edges are inconsistent, forcing a particular direction for an edge. Each such configuration can be viewed as a local constraint on edge orientation, give rise to a rule that can be used to orient more edges in the PDAG. Three such rules are shown in figure 3.12.

Let us understand the intuition behind these rules. Rule R1 is precisely the one we discussed earlier. Rule R2 is derived from the standard acyclicity constraint: If we have the directed path $X \to Y \to Z$, and an undirected edge $X—Z$, we cannot direct the edge $X \leftarrow Z$ without creating a cycle. Hence, we can conclude that the edge must be directed $X \to Z$. The third rule seems a little more complex, but it is also easily motivated. Assume, by contradiction, that we direct the edge $Z \to X$. In this case, we cannot direct the edge $X—Y_1$ as $X \to Y_1$ without creating a cycle; thus, we must have $Y_1 \to X$. Similarly, we must have $Y_2 \to X$. But, in this case, $Y_1 \to X \leftarrow Y_2$ forms an immorality (as there is no edge between Y_1 and Y_2), which contradicts the fact that the edges $X—Y_1$ and $X—Y_2$ are undirected in the original PDAG.

These three rules can be applied constructively in an obvious way: A rule applies to a PDAG whenever the induced subgraph on a subset of variables exactly matches the graph on the left-hand side of the rule. In that case, we modify this subgraph to match the subgraph on the right-hand side of the rule. Note that, by applying one rule and orienting a previously undirected edge, we create a new graph. This might create a subgraph that matches the antecedent of a rule, enforcing the orientation of additional edges. This process, however, must terminate at

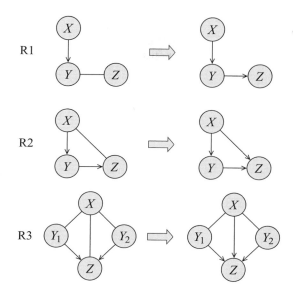

Figure 3.12   Rules for orienting edges in PDAG. Each rule lists a configuration of edges before and after an application of the rule.

constraint
propagation

some point (since we are only adding orientations at each step, and the number of edges is finite). This implies that iterated application of this local constraint to the graph (a process known as *constraint propagation*) is guaranteed to converge.

Algorithm 3.5 Finding the class PDAG characterizing the P-map of a distribution P

    Procedure Build-PDAG (
      $\mathcal{X} = \{X_1, \ldots, X_n\}$ // A specification of the random variables
      P // Distribution of interest
)
1    $S, \{\boldsymbol{U}_{X_i, X_j}\} \leftarrow$ Build-PMap-Skeleton$(\mathcal{X}, P)$
2    $\mathcal{K} \leftarrow$ Find-Immoralities$(\mathcal{X}, S, \{\boldsymbol{U}_{X_i, X_j}\})$
3   **while** not converged
4   Find a subgraph in $\mathcal{K}$ matching the left-hand side of a rule R1–R3
5   Replace the subgraph with the right-hand side of the rule
6   **return** K

Algorithm 3.5 implements this process. It builds an initial graph using Build-PMap-Skeleton and Mark-Immoralities, and then iteratively applies the three rules until convergence, that is, until we cannot find a subgraph that matches a left-hand side of any of the rules.

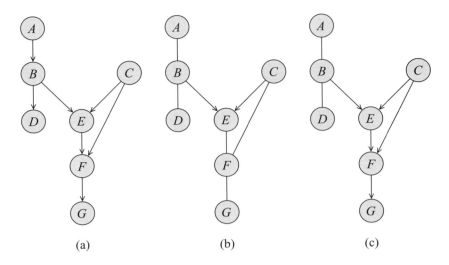

(a)   (b)   (c)

Figure 3.13   More complex example of compelled edges in the representation of an equivalence class. (a) Original DAG $\mathcal{G}^*$. (b) Skeleton of $\mathcal{G}^*$ annotated with immoralities. (c) Complete PDAG representation of the equivalence class of $\mathcal{G}^*$.

Example 3.10   *Consider the DAG shown in figure 3.13a. After checking for immoralities, we find the graph shown in figure 3.13b. Now, we can start applying the preceding rules. For example, consider the variables B, E, and F. They induce a subgraph that matches the left-hand side of rule R1. Thus, we orient the edge between E and F to $E \rightarrow F$. Now, consider the variables C, E, and F. Their induced subgraph matches the left-hand side of rule R2, so we now orient the edge between C and F to $C \rightarrow F$. At this stage, if we consider the variables E, F, G, we can apply the rule R1, and orient the edge $F \rightarrow G$. (Alternatively, we could have arrived at the same orientation using C, F, and G.) The resulting PDAG is shown in figure 3.13c.*   ∎

It seems fairly obvious that this algorithm is guaranteed to be sound: Any edge that is oriented by this procedure is, indeed, directed in exactly the same way in all of the members of the equivalence class. Much more surprising is the fact that it is also complete: Repeated application of these three local rules is guaranteed to capture all edge orientations in the equivalence class, without the need for additional global constraints. More precisely, we can prove that this algorithm produces the correct class PDAG for the distribution P:

Theorem 3.10   *Let P be a distribution that has a P-map $\mathcal{G}^*$, and let $\mathcal{K}$ be the PDAG returned by* Build-PDAG$(\mathcal{X}, P)$. *Then, $\mathcal{K}$ is a class PDAG of $\mathcal{G}^*$.*

The proof of this theorem can be decomposed into several aspects of correctness. We have already established the correctness of the skeleton found by Build-PMap-Skeleton. Thus, it remains to show that the directionality of the edges is correct. Specifically, we need to establish three basic facts:

• **Acyclicity:** The graph returned by Build-PDAG$(\mathcal{X},P)$ is acyclic.

- **Soundness:** If $X \to Y \in \mathcal{K}$, then $X \to Y$ appears in all DAGs in $\mathcal{G}^*$'s I-equivalence class.
- **Completeness:** If $X{-}Y \in \mathcal{K}$, then we can find a DAG $\mathcal{G}$ that is I-equivalent to $\mathcal{G}^*$ such that $X \to Y \in \mathcal{G}$.

The last condition establishes completeness, since there is no constraint on the direction of the arc. In other words, the same condition can be used to prove the existence of a graph with $X \to Y$ and of a graph with $Y \to X$. Hence, it shows that either direction is possible within the equivalence class.

We begin with the soundness of the procedure.

Proposition 3.3

Let P be a distribution that has a P-map $\mathcal{G}^*$, and let $\mathcal{K}$ be the graph returned by Build-PDAG$(\mathcal{X}, P)$. *Then, if $X \to Y \in \mathcal{K}$, then $X \to Y$ appears in all DAGs in the I-equivalence class of $\mathcal{G}^*$.*

The proof is left as an exercise (exercise 3.23).

Next, we consider the acyclicity of the graph. We start by proving a property of graphs returned by the procedure. (Note that, once we prove that the graph returned by the procedure is the correct PDAG, it will follow that this property also holds for class PDAGs in general.)

Proposition 3.4

Let $\mathcal{K}$ be the graph returned by Build-PDAG. *Then, if $X \to Y \in \mathcal{K}$ and $Y{-}Z \in \mathcal{K}$, then $X \to Z \in \mathcal{K}$.*

The proof is left as an exercise (exercise 3.24).

Proposition 3.5

Let $\mathcal{K}$ be the chain graph returned by Build-PDAG. *Then $\mathcal{K}$ is acyclic.*

Proof Suppose, by way of contradiction, that $\mathcal{K}$ contains a cycle. That is, there is a (partially) directed path $X_1 \rightleftharpoons X_2 \rightleftharpoons \ldots \rightleftharpoons X_n \rightleftharpoons X_1$. Without loss of generality, assume that this path is the shortest cycle in $\mathcal{K}$. We claim that the path cannot contain an undirected edge. To see that, suppose that the the path contains the triplet $X_i \to X_{i+1}{-}X_{i+2}$. Then, invoking proposition 3.4, we have that $X_i \to X_{i+2} \in \mathcal{K}$, and thus, we can construct a shorter path without X_{i+1} that contains the edge $X_i \to X_{i+2}$. At this stage, we have a directed cycle $X_1 \to X_2 \to \ldots X_n \to X_1$. Using proposition 3.3, we conclude that this cycle appears in any DAG in the I-equivalence class, and in particular in $\mathcal{G}^*$. This conclusion contradicts the assumption that $\mathcal{G}^*$ is acyclic. It follows that $\mathcal{K}$ is acyclic. ∎

The final step is the completeness proof. Again, we start by examining a property of the graph $\mathcal{K}$.

Proposition 3.6

The PDAG $\mathcal{K}$ returned by Build-PDAG *is necessarily chordal.*

The proof is left as an exercise (exercise 3.25).

This property allows us to characterize the structure of the PDAG $\mathcal{K}$ returned by Build-PDAG. Recall that, since $\mathcal{K}$ is an undirected chain graph, we can partition $\mathcal{X}$ into chain components $\boldsymbol{K}_1, \ldots, \boldsymbol{K}_\ell$, where each chain component contains variables that are connected by undirected edges (see definition 2.21). It turns out that, in an undirected chordal graph, we can orient any edge in any direction without creating an immorality.

Proposition 3.7   *Let $\mathcal{K}$ be a undirected chordal graph over $\mathcal{X}$, and let $X, Y \in \mathcal{X}$. Then, there is a DAG $\mathcal{G}$ such that*

    (a) The skeleton of $\mathcal{G}$ is $\mathcal{K}$.

    (b) $\mathcal{G}$ does not contain immoralities.

    (c) $X \rightarrow Y \in \mathcal{G}$.

The proof of this proposition requires some additional machinery that we introduce in chapter 4, so we defer the proof to that chapter.

Using this proposition, we see that we can orient edges in the chain component $\boldsymbol{K}_j$ without introducing immoralities within the component. We still need to ensure that orienting an edge $X-Y$ within a component cannot introduce an immorality involving edges from outside the component. To see why this situation cannot occur, suppose we orient the edge $X \rightarrow Y$, and suppose that $Z \rightarrow Y \in \mathcal{K}$. This seems like a potential immorality. However, applying proposition 3.4, we see that since $Z \rightarrow Y$ and $Y-X$ are in $\mathcal{K}$, then so must be $Z \rightarrow X$. Since Z is a parent of both X and Y, we have that $X \rightarrow Y \leftarrow Z$ is not an immorality. This argument applies to any edge we orient within an undirected component, and thus no new immoralities are introduced.

With these tools, we can complete the completeness proof of Build-PDAG.

Proposition 3.8   *Let P be a distribution that has a P-map $\mathcal{G}^*$, and let $\mathcal{K}$ be the graph returned by* Build-PDAG$(\mathcal{X}, P)$. *If $X-Y \in \mathcal{K}$, then we can find a DAG $\mathcal{G}$ that is I-equivalent to $\mathcal{G}^*$ such that $X \rightarrow Y \in \mathcal{G}$.*

PROOF Suppose we have an undirected edge $X-Y \in \mathcal{K}$. We want to show that there is a DAG $\mathcal{G}$ that has the same skeleton and immoralities as $\mathcal{K}$ such that $X \rightarrow Y \in \mathcal{G}$. If can build such a graph $\mathcal{G}$, then clearly it is in the I-equivalence class of $\mathcal{G}^*$.

The construction is simple. We start with the chain component that contains $X-Y$, and use proposition 3.7 to orient the edges in the component so that $X \rightarrow Y$ is in the resulting DAG. Then, we use the same construction to orient all other chain components. Since the chain components are ordered and acyclic, and our orientation of each chain component is acyclic, the resulting directed graph is acyclic. Moreover, as shown, the new orientation in each component does not introduce immoralities. Thus, the resulting DAG has exactly the same skeleton and immoralities as $\mathcal{K}$.   ∎

3.5   Summary

In this chapter, we discussed the issue of specifying a high-dimensional joint distribution compactly by exploiting its independence properties. We provided two complementary definitions of a Bayesian network. The first is as a directed graph $\mathcal{G}$, annotated with a set of conditional probability distributions $P(X_i \mid \mathrm{Pa}_{X_i})$. The network together with the CPDs define a distribution via the chain rule for Bayesian networks. In this case, we say that P factorizes over $\mathcal{G}$. We also defined the independence assumptions associated with the graph: the local independencies, the set of basic independence assumptions induced by the network structure; and the larger set of global independencies that are derived from the d-separation criterion. We showed the equivalence of these three fundamental notions: P factorizes over $\mathcal{G}$ if and only if P satisfies the local independencies of $\mathcal{G}$, which holds if and only if P satisfies the global independencies derived from d-separation. This result shows the equivalence of our two views of a Bayesian network: as a scaffolding for factoring a probability distribution P, and as a representation of a set of independence assumptions that hold for P. We also showed that the set of independencies derived from d-separation is a complete characterization of the independence properties that are implied by the graph structure alone, rather than by properties of a specific distribution over $\mathcal{G}$.

We defined a set of basic notions that use the characterization of a graph as a set of independencies. We defined the notion of a *minimal I-map* and showed that almost every distribution has multiple minimal I-maps, but that a minimal I-map for P does not necessarily capture all of the independence properties in P. We then defined a more stringent notion of a *perfect map*, and showed that not every distribution has a perfect map. We defined *I-equivalence*, which captures an independence-equivalence relationship between two graphs, one where they specify precisely the same set of independencies.

Finally, we defined the notion of a *class PDAG*, a partially directed graph that provides a compact representation for an entire I-equivalence class, and we provided an algorithm for constructing this graph.

These definitions and results are fundamental properties of the Bayesian network representation and its semantics. Some of the algorithms that we discussed are never used as is; for example, we never directly use the procedure to find a minimal I-map given an explicit representation of the distribution. However, these results are crucial to understanding the cases where we can construct a Bayesian network that reflects our understanding of a given domain, and what the resulting network means.

3.6 Relevant Literature

The use of a directed graph as a framework for analyzing properties of distributions can be traced back to the path analysis of Wright (1921, 1934).

influence
diagram

The use of a directed acyclic graph to encode a general probability distribution (not within a specific domain) was first proposed within the context of *influence diagrams*, a decision-theoretic framework for making decisions under uncertainty (see chapter 23). Within this setting, Howard and Matheson (1984b) and Smith (1989) both proved the equivalence between the ability to represent a distribution as a DAG and the local independencies (our theorem 3.1 and theorem 3.2).

The notion of Bayesian networks as a qualitative data structure encoding independence relationships was first proposed by Pearl and his colleagues in a series of papers (for example, Verma and Pearl 1988; Geiger and Pearl 1988; Geiger et al. 1989, 1990), and in Pearl's book *Probabilistic Reasoning in Intelligent Systems* (Pearl 1988). Our presentation of I-maps, P-maps, and Bayesian networks largely follows the trajectory laid forth in this body of work.

The definition of d-separation was first set forth by Pearl (1986b), although without formal justification. The soundness of d-separation was shown by Verma (1988), and its completeness for the case of Gaussian distributions by Geiger and Pearl (1993). The measure-theoretic notion of completeness of d-separation, stating that almost all distributions are faithful (theorem 3.5), was shown by Meek (1995b). Several papers have been written exploring the yet stronger notion of completeness for d-separation (faithfulness for all distributions that are minimal I-maps), in various subclasses of models (for example, Becker et al. 2000). The *BayesBall* algorithm, an elegant and efficient algorithm for d-separation and a class of related problems, was proposed by (Shachter 1998).

The notion of I-equivalence was defined by Verma and Pearl (1990, 1992), who also provided and proved the graph-theoretic characterization of theorem 3.8. Chickering (1995) provided the alternative characterization of I-equivalence in terms of covered edge reversal. This definition provides an easy mechanism for proving important properties of I-equivalent networks. As we will see later in the book, the notion of I-equivalence class plays an important role in identifying networks, particularly when learning networks from data. The first algorithm for constructing a perfect map for a distribution, in the form of an I-equivalence class, was proposed by Pearl and Verma (1991); Verma and Pearl (1992). This algorithm was subsequently extended by Spirtes et al. (1993) and by Meek (1995a). Meek also provides an algorithm for finding all of the directed edges that occur in every member of the I-equivalence class.

A notion related to I-equivalence is that of *inclusion*, where the set of independencies $\mathcal{I}(\mathcal{G}')$ is *included* in the set of independencies $\mathcal{I}(\mathcal{G})$ (so that $\mathcal{G}$ is an I-map for any distribution that factorizes over $\mathcal{G}'$). Shachter (1989) showed how to construct a graph $\mathcal{G}'$ that includes a graph $\mathcal{G}$, but with one edge reversed. Meek (1997) conjectured that inclusion holds if and only if one can transform $\mathcal{G}$ to $\mathcal{G}'$ using the operations of edge addition and covered edge reversal. A limited version of this conjecture was subsequently proved by Kočka, Bouckaert, and Studený (2001).

The naive Bayes model, although naturally represented as a graphical model, far predates this view. It was applied with great success within expert systems in the 1960s and 1970s (de Bombal et al. 1972; Gorry and Barnett 1968; Warner et al. 1961). It has also seen significant use as a simple yet highly effective method for classification tasks in machine learning, starting as early as the 1970s (for example, Duda and Hart 1973), and continuing to this day.

The general usefulness of the types of reasoning patterns supported by a Bayesian network, including the very important pattern of intercausal reasoning, was one of the key points raised by Pearl in his book (Pearl 1988). These qualitative patterns were subsequently formalized by Wellman (1990) in his framework of *qualitative probabilistic networks*, which explicitly annotate arcs with the direction of influence of one variable on another. This framework has been used to facilitate knowledge elicitation and knowledge-guided learning (Renooij and van der Gaag 2002; Hartemink et al. 2002) and to provide verbal explanations of probabilistic inference (Druzdzel 1993).

There have been many applications of the Bayesian network framework in the context of real-world problems. The idea of using directed graphs as a model for genetic inheritance appeared as far back as the work on path analysis of Wright (1921, 1934). A presentation much closer to modern-day Bayesian networks was proposed by Elston and colleagues in the 1970s (Elston and Stewart 1971; Lange and Elston 1975). More recent developments include the development of better algorithms for inference using these models (for example, Kong 1991; Becker et al. 1998; Friedman et al. 2000) and the construction of systems for genetic linkage analysis based on this technology (Szolovits and Pauker 1992; Schäffer 1996).

Many of the first applications of the Bayesian network framework were to medical expert systems. The Pathfinder system is largely the work of David Heckerman and his colleagues (Heckerman and Nathwani 1992a; Heckerman et al. 1992; Heckerman and Nathwani 1992b). The success of this system as a diagnostic tool, including its ability to outperform expert physicians, was one of the major factors that led to the rise in popularity of probabilistic methods in the early 1990s. Several other large diagnostic networks were developed around the same period, including MUNIN (Andreassen et al. 1989), a network of over 1000 nodes used for interpreting electromyographic data, and QMR-DT (Shwe et al. 1991; Middleton et al. 1991), a probabilistic reconstruction of the QMR/INTERNIST system (Miller et al. 1982) for general medical diagnosis.

The problem of knowledge acquisition of network models has received some attention. Probability elicitation is a long-standing question in decision analysis; see, for example, Spetzler and von Holstein (1975); Chesley (1978). Unfortunately, elicitation of probabilities from humans is a difficult process, and one subject to numerous biases (Tversky and Kahneman 1974; Daneshkhah 2004). Shachter and Heckerman (1987) propose the "backward elicitation" approach for obtaining both the network structure and the parameters from an expert. *Similarity networks* (Heckerman and Nathwani 1992a; Geiger and Heckerman 1996) generalize this idea by allowing an expert to construct several small networks for differentiating between "competing" diagnoses, and then superimposing them to construct a single large network. Morgan and Henrion (1990) provide an overview of knowledge elicitation methods.

similarity network

The difficulties in eliciting accurate probability estimates from experts are well recognized across a wide range of disciplines. In the specific context of Bayesian networks, this issue has been tackled in several ways. First, there has been both empirical (Pradhan et al. 1996) and theoretical (Chan and Darwiche 2002) analysis of the extent to which the choice of parameters affects the conclusions of the inference. Overall, the results suggest that even fairly significant changes to network parameters cause only small degradations in performance, except when the changes relate to extreme parameters — those very close to 0 and 1. Second, the concept of *sensitivity analysis* (Morgan and Henrion 1990) is used to allow researchers to evaluate the sensitivity of their specific network to variations in parameters. Largely, sensitivity has been measured using the derivative of network queries relative to various parameters (Laskey 1995; Castillo et al. 1997b; Kjærulff and van der Gaag 2000; Chan and Darwiche 2002), with the focus of most of the work being on properties of sensitivity values and on efficient algorithms for estimating them.

sensitivity analysis

As pointed out by Pearl (1988), the notion of a Bayesian network structure as a representation of independence relationships is a fundamental one, which transcends the specifics of probabilistic representations. There have been many proposed variants of Bayesian networks that use a nonprobabilistic "parameterization" of the local dependency models. Examples include various logical calculi (Darwiche 1993), Dempster-Shafer belief functions (Shenoy 1989), possibility values (Dubois and Prade 1990), qualitative (order-of-magnitude) probabilities (known as kappa rankings; Darwiche and Goldszmidt 1994), and interval constraints on probabilities (Fertig and Breese 1989; Cozman 2000).

The acyclicity constraint of Bayesian networks has led to many concerns about its ability to express certain types of interactions. There have been many proposals intended to address this limitation. Markov networks, based on undirected graphs, present a solution for certain types of interactions; this class of probability models are described in chapter 4. Dynamic Bayesian networks "stretch out" the interactions over time, therefore providing an acyclic version of feedback loops; these models are are described in section 6.2.

There has also been some work on directed models that encode cyclic dependencies directly. *Cyclic graphical models* (Richardson 1994; Spirtes 1995; Koster 1996; Pearl and Dechter 1996) are based on distributions over systems of simultaneous linear equations. These models are a

cyclic graphical model

natural generalization of Gaussian Bayesian networks (see chapter 7), and are also associated with notions of d-separation or I-equivalence. Spirtes (1995) shows that this connection breaks down when the system of equations is nonlinear and provides a weaker version for the cyclic case.

dependency
network

Dependency networks (Heckerman et al. 2000) encode a set of local dependency models, representing the conditional distribution of each variable on all of the others (which can be compactly represented by its dependence on its Markov blanket). A dependency network represents a probability distribution only indirectly, and is only guaranteed to be coherent under certain conditions. However, it provides a local model of dependencies that is very naturally interpreted by people.

3.7 Exercises

Exercise 3.1

Provide an example of a distribution $P(X_1, X_2, X_3)$ where for each $i \neq j$, we have that $(X_i \perp X_j) \in \mathcal{I}(P)$, but we also have that $(X_1, X_2 \perp X_3) \notin \mathcal{I}(P)$.

Exercise 3.2

a. Show that the naive Bayes factorization of equation (3.7) follows from the naive Bayes independence assumptions of equation (3.6).

b. Show that equation (3.8) follows from equation (3.7).

c. Show that, if all the variables $C, X_1, \ldots, X_n$ are binary-valued, then $\log \frac{P(C=c^1 | x_1, \ldots, x_n)}{P(C=c^2 | x_1, \ldots, x_n)}$ is a linear function of the value of the finding variables, that is, can be written as $\sum_{i=1}^{n} \alpha_i X_i + \alpha_0$ (where $X_i = 0$ if $X = x^0$ and 1 otherwise).

Exercise 3.3

Consider a simple example (due to Pearl), where a burglar alarm (A) can be set off by either a burglary (B) or an earthquake (E).

a. Define constraints on the CPD of $P(A \mid B, E)$ that imply the *explaining away* property.

b. Show that if our model is such that the alarm always (deterministically) goes off whenever there is a earthquake:

$$P(a^1 \mid b^1, e^1) = P(a^1 \mid b^0, e^1) = 1$$

then $P(b^1 \mid a^1, e^1) = P(b^1)$, that is, observing an earthquake provides a full explanation for the alarm.

Exercise 3.4

We have mentioned that explaining away is one type of intercausal reasoning, but that other type of intercausal interactions are also possible. Provide a realistic example that exhibits the opposite type of interaction. More precisely, consider a v-structure $X \rightarrow Z \leftarrow Y$ over three binary-valued variables. Construct a CPD $P(Z \mid X, Y)$ such that:

- X and Y both increase the probability of the effect, that is, $P(z^1 \mid x^1) > P(z^1)$ and $P(z^1 \mid y^1) > P(z^1)$,

- each of X and Y increases the probability of the other, that is, $P(x^1 \mid z^1) < P(x^1 \mid y^1, z^1)$, and similarly $P(y^1 \mid z^1) < P(y^1 \mid x^1, z^1)$.

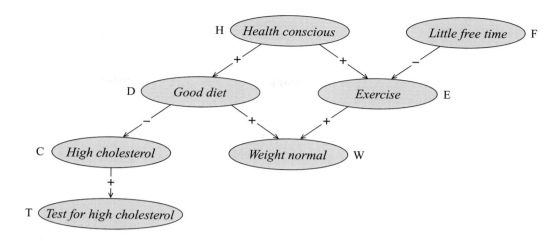

Figure 3.14   A Bayesian network with qualitative influences

Note that strong (rather than weak) inequality must hold in all cases.

Your example should be realistic, that is, X, Y, Z should correspond to real-world variables, and the CPD should be reasonable.

Exercise 3.5

Consider the Bayesian network of figure 3.14.

Assume that all variables are binary-valued. We do not know the CPDs, but do know how each random variable *qualitatively* affects its children. The influences, shown in the figure, have the following interpretation:

- $X \xrightarrow{+} Y$ means $P(y^1 \mid x^1, \boldsymbol{u}) > P(y^1 \mid x^0, \boldsymbol{u})$, for all values $\boldsymbol{u}$ of Y's other parents.
- $X \xrightarrow{-} Y$ means $P(y^1 \mid x^1, \boldsymbol{u}) < P(y^1 \mid x^0, \boldsymbol{u})$, for all values $\boldsymbol{u}$ of Y's other parents.

We also assume explaining away as the interaction for all cases of intercausal reasoning.

For each of the following pairs of conditional probability queries, use the information in the network to determine if one is larger than the other, if they are equal, or if they are incomparable. For each pair of queries, indicate all relevant active trails, and their direction of influence.

(a)	$P(t^1 \mid d^1)$	$P(t^1)$
(b)	$P(d^1 \mid t^0)$	$P(d^1)$
(c)	$P(h^1 \mid e^1, f^1)$	$P(h^1 \mid e^1)$
(d)	$P(c^1 \mid f^0)$	$P(c^1)$
(e)	$P(c^1 \mid h^0)$	$P(c^1)$
(f)	$P(c^1 \mid h^0, f^0)$	$P(c^1 \mid h^0)$
(g)	$P(d^1 \mid h^1, e^0)$	$P(d^1 \mid h^1)$
(h)	$P(d^1 \mid e^1, f^0, w^1)$	$P(d^1 \mid e^1, f^0)$
(i)	$P(t^1 \mid w^1, f^0)$	$P(t^1 \mid w^1)$

Exercise 3.6

Consider a set of variables $X_1, \ldots, X_n$ where each X_i has $|Val(X_i)| = \ell$.

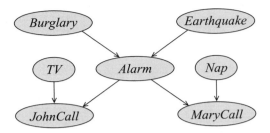

Figure 3.15   A simple network for a burglary alarm **domain**

a. Assume that we have a Bayesian network over $X_1, \ldots, X_n$, such that each node has at most k parents. What is a simple upper bound on the number of independent parameters in the Bayesian network? How many independent parameters are in the full joint distribution over $X_1, \ldots, X_n$?

b. Now, assume that each variable X_i has the parents $X_1, \ldots, X_{i-1}$. How many independent parameters are there in the Bayesian network? What can you conclude about the expressive power of this type of network?

c. Now, consider a naive Bayes model where $X_1, \ldots, X_n$ are evidence variables, and we have an additional class variable C, which has k possible values $c_1, \ldots, c_k$. How many independent parameters are required to specify the naive Bayes model? How many independent parameters are required for an explicit representation of the joint distribution?

Exercise 3.7

Show how you could efficiently compute the distribution over a variable X_i given some assignment to all the other variables in the network: $P(X_i \mid x_1, \ldots, x_{i-1}, x_{i+1}, \ldots, x_n)$. Your procedure should not require the construction of the entire joint distribution $P(X_1, \ldots, X_n)$. Specify the computational complexity of your procedure.

Exercise 3.8

Let $\mathcal{B} = (\mathcal{G}, P)$ be a Bayesian network over some set of variables $\mathcal{X}$. Consider some subset of evidence nodes $\boldsymbol{Z}$, and let $\boldsymbol{X}$ be all of the ancestors of the nodes in $\boldsymbol{Z}$. Let $\mathcal{B}'$ be a network over the induced subgraph over $\boldsymbol{X}$, where the CPD for every node $X \in \boldsymbol{X}$ is the same in $\mathcal{B}'$ as in $\mathcal{B}$. Prove that the joint distribution over $\boldsymbol{X}$ is the same in $\mathcal{B}$ and in $\mathcal{B}'$. The nodes in $\mathcal{X} - \boldsymbol{X}$ are called *barren nodes* relative to $\boldsymbol{X}$, because (when not instantiated) they are irrelevant to computations concerning $\boldsymbol{X}$.

Exercise 3.9★

Prove theorem 3.2 for a general BN structure $\mathcal{G}$. Your proof should not use the soundness of d-separation.

Exercise 3.10

Prove that the global independencies, derived from d-separation, imply the local independencies. In other words, prove that a node is d-separated from its nondescendants given its parents.

Exercise 3.11★

One operation on Bayesian networks that arises in many settings is the marginalization of some node in the network.

a. Consider the Burglary Alarm network $\mathcal{B}$ shown in figure 3.15. Construct a Bayesian network $\mathcal{B}'$ over all of the nodes except for *Alarm* that is a minimal I-map for the marginal distribution $P_{\mathcal{B}}(B, E, T, N, J, M)$. Be sure to get *all* dependencies that remain from the original network.

barren node

b. Generalize the procedure you used to solve the preceding problem into a node elimination algorithm. That is, define an algorithm that transforms the structure of $\mathcal{G}$ into $\mathcal{G}'$ such that one of the nodes X_i of $\mathcal{G}$ is not in $\mathcal{G}'$ and $\mathcal{G}'$ is an I-map of the marginal distribution over the remaining variables as defined by $\mathcal{G}$.

Exercise 3.12⋆⋆

edge reversal

Another operation on Bayesian networks that arises often is *edge reversal*. This involves transforming a Bayesian network $\mathcal{G}$ containing nodes X and Y as well as arc $X \to Y$ into another Bayesian network $\mathcal{G}'$ with reversed arc $Y \to X$. However, we want $\mathcal{G}'$ to represent the same distribution as $\mathcal{G}$; therefore, $\mathcal{G}'$ will need to be an I-map of the original distribution.

a. Consider the Bayesian network structure of figure 3.15. Suppose we wish to reverse the arc $B \to A$. What additional minimal modifications to the structure of the network are needed to ensure that the new network is an I-map of the original distribution? Your network should not reverse any additional edges, and it should differ only minimally from the original in terms of the number of edge additions or deletions. Justify your response.

b. Now consider a general Bayesian network $\mathcal{G}$. For simplicity, assume that the arc $X \to Y$ is the only directed trail from X to Y. Define a general procedure for reversing the arc $X \to Y$, that is, for constructing a graph $\mathcal{G}'$ is an I-map for the original distribution, but that contains an arc $Y \to X$ and otherwise differs minimally from $\mathcal{G}$ (in the same sense as before). Justify your response.

c. Suppose that we use the preceding method to transform $\mathcal{G}$ into a graph $\mathcal{G}'$ with a reversed arc between X and Y. Now, suppose we reverse that arc *back* to its original direction in $\mathcal{G}$ by repeating the preceding method, transforming $\mathcal{G}'$ into $\mathcal{G}''$. Are we guaranteed that the final network structure is equivalent to the original network structure ($\mathcal{G} = \mathcal{G}''$)?

Exercise 3.13⋆

Let $\mathcal{B} = (\mathcal{G}, P)$ be a Bayesian network over $\mathcal{X}$. The Bayesian network is parameterized by a set of CPD parameters of the form $\theta_{x|u}$ for $X \in \mathcal{X}$, $\boldsymbol{U} = \mathrm{Pa}_X^{\mathcal{G}}$, $x \in Val(X)$, $\boldsymbol{u} \in Val(\boldsymbol{U})$. Consider any conditional independence statement of the form $(\boldsymbol{X} \perp \boldsymbol{Y} \mid \boldsymbol{Z})$. Show how this statement translates into a set of polynomial equalities over the set of CPD parameters $\theta_{x|u}$. (Note: A polynomial equality is an assertion of the form $a\theta_1^2 + b\theta_1\theta_2 + c\theta_2^3 + d = 0$.)

Exercise 3.14⋆

Prove theorem 3.6.

Exercise 3.15

Consider the two networks:

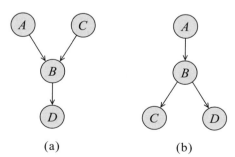

(a)   (b)

For each of them, determine whether there can be any other Bayesian network that is I-equivalent to it.

Exercise 3.16⋆

Prove theorem 3.7.

Exercise 3.17★★

We proved earlier that two networks that have the same skeleton and v-structures imply the same conditional independence assumptions. As shown, this condition is not an if and only if. Two networks can have different v-structures, yet still imply the same conditional independence assumptions. In this problem, you will provide a condition that precisely relates I-equivalence and similarity of network structure.

minimal active trail

a. A key notion in this question is that of a *minimal active trail*. We define an active trail $X_1 \ldots X_m$ to be *minimal* if there is no other active trail from X_1 to X_m that "shortcuts" some of the nodes, that is, there is no active trail $X_1 \rightleftharpoons X_{i_1} \rightleftharpoons \ldots \rightleftharpoons X_{i_k} \rightleftharpoons X_m$ for $1 < i_1 < \ldots < i_k < m$.

Our first goal is to analyze the types of "triangles" that can occur in a minimal active trail, that is, cases where we have $X_{i-1} \rightleftharpoons X_i \rightleftharpoons X_{i+1}$ with a direct edge between X_{i-1} and X_{i+1}. Prove that the only possible triangle in a minimal active trail is one where $X_{i-1} \leftarrow X_i \rightarrow X_{i+1}$, with an edge between X_{i-1} and X_{i+1}, and where either X_{i-1} or X_{i+1} are the center of a v-structure in the trail.

b. Now, consider two networks $\mathcal{G}_1$ and $\mathcal{G}_2$ that have the same skeleton and same immoralities. Prove, using the notion of minimal active trail, that $\mathcal{G}_1$ and $\mathcal{G}_2$ imply precisely the same conditional independence assumptions, that is, that if X and Y are d-separated given $\boldsymbol{Z}$ in $\mathcal{G}_1$, then X and Y are also d-separated given $\boldsymbol{Z}$ in $\mathcal{G}_2$.

c. Finally, prove the other direction. That is, prove that two networks $\mathcal{G}_1$ and $\mathcal{G}_2$ that induce the same conditional independence assumptions must have the same skeleton and the same immoralities.

Exercise 3.18★

In this exercise, you will prove theorem 3.9. This result provides an alternative reformulation of I-equivalence in terms of local operations on the graph structure.

covered edge

a. Let $\mathcal{G}$ be a directed graph with a *covered edge* $X \rightarrow Y$ (as in definition 3.12), and $\mathcal{G}'$ the graph that results by reversing the edge $X \rightarrow Y$ to produce $Y \rightarrow X$, but leaving everything else unchanged. Prove that $\mathcal{G}$ and $\mathcal{G}'$ are I-equivalent.

b. Provide a counterexample to this result in the case where $X \rightarrow Y$ is not a covered edge.

c. Now, prove that for every pair of I-equivalent networks $\mathcal{G}$ and $\mathcal{G}'$, there exists a sequence of covered edge reversal operations that converts $\mathcal{G}$ to $\mathcal{G}'$. Your proof should show how to construct this sequence.

Exercise 3.19★

Prove lemma 3.2.

Exercise 3.20★

requisite CPD

In this question, we will consider the sensitivity of a particular query $P(X \mid \boldsymbol{Y})$ to the CPD of a particular node Z. Let X and Z be nodes, and $\boldsymbol{Y}$ be a set of nodes. We say that Z has a *requisite CPD* for answering the query $P(\boldsymbol{X} \mid \boldsymbol{Y})$ if there are two networks $\mathcal{B}_1$ and $\mathcal{B}_2$ that have identical graph structure $\mathcal{G}$ and identical CPDs everywhere except at the node Z, and where $P_{\mathcal{B}_1}(X \mid \boldsymbol{Y}) \neq P_{\mathcal{B}_2}(X \mid \boldsymbol{Y})$; in other words, the CPD of Z affects the answer to this query.

This type of analysis is useful in various settings, including determining which CPDs we need to acquire for a certain query (and others that we discuss later in the book).

Show that we can test whether Z is a requisite probability node for $P(X \mid \boldsymbol{Y})$ using the following procedure: We modify $\mathcal{G}$ into a graph $\mathcal{G}'$ that contains a new "dummy" parent $\widehat{Z}$, and then test whether $\widehat{Z}$ has an active trail to $\boldsymbol{X}$ given $\boldsymbol{Y}$.

- Show that this is a sound criterion for determining whether Z is a requisite probability node for $P(X \mid \boldsymbol{Y})$ in $\mathcal{G}$, that is, for all pairs of networks $\mathcal{B}_1, \mathcal{B}_2$ as before, $P_{\mathcal{B}_1}(X \mid \boldsymbol{Y}) = P_{\mathcal{B}_2}(X \mid \boldsymbol{Y})$.
- Show that this criterion is weakly complete (like d-separation), in the sense that, if it fails to identify Z as requisite in $\mathcal{G}$, there exists some pair of networks $\mathcal{B}_1, \mathcal{B}_2$ as before, $P_{\mathcal{B}_1}(X \mid \boldsymbol{Y}) \neq P_{\mathcal{B}_2}(X \mid \boldsymbol{Y})$.

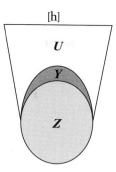

Figure 3.16   Illustration of the concept of a self-contained set

Exercise 3.21★

Define a set Z of nodes to be *self-contained* if, for every pair of nodes $A, B \in Z$, and any *directed* path between A and B, all nodes along the trail are also in Z.

a. Consider a self-contained set Z, and let Y be the set of all nodes that are a parent of some node in Z but are not themselves in Z. Let U be the set of nodes that are an ancestor of some node in Z but that are not already in $Y \cup Z$. (See figure 3.16.)

Prove, based on the d-separation properties of the network, that $(Z \perp U \mid Y)$. Make sure that your proof covers all possible cases.

b. Provide a counterexample to this result if we retract the assumption that Z is self-contained. (Hint: 4 nodes are enough.)

Exercise 3.22★

We showed that the algorithm Build-PMap-Skeleton of algorithm 3.3 constructs the skeleton of the P-map of a distribution P if P has a P-map (and that P-map has indegrees bounded by the parameter d). In this question, we ask you consider what happens if P does not have a P-map.

There are two types of errors we might want to consider:

- **Missing edges:** The edge $X \rightleftharpoons Y$ appears in all the minimal I-maps of P, yet X—Y is not in the skeleton S returned by Build-PMap-Skeleton.
- **Spurious edges:** The edge $X \rightleftharpoons Y$ does not appear in all of the minimal I-maps of P (but may appear in some of them), yet X—Y is in the skeleton S returned by Build-PMap-Skeleton.

For each of these two types of errors, either prove that they cannot happen, or provide a counterexample (that is, a distribution P for which Build-PMap-Skeleton makes that type of an error).

Exercise 3.23★

In this exercise, we prove proposition 3.3. To help us with the proof, we need an auxiliary definition. We say that a partially directed graph $\mathcal{K}$ is a *partial class graph* for a DAG $\mathcal{G}^*$ if

a. $\mathcal{K}$ has the same skeleton as $\mathcal{G}^*$;
b. $\mathcal{K}$ has the same immoralities as $\mathcal{G}^*$;
c. if $X \rightarrow Y \in \mathcal{K}$, then $X \rightarrow Y \in \mathcal{G}$ for any DAG $\mathcal{G}$ that is I-equivalent to $\mathcal{G}^*$.

Clearly, the graph returned by by Mark-Immoralities is a partial class graph of $\mathcal{G}^*$.

Prove that if $\mathcal{K}$ is a partial class graph of $\mathcal{G}^*$, and we apply one of the rules R1–R3 of figure 3.12, then the resulting graph is also a partial class graph $\mathcal{G}^*$. Use this result to prove proposition 3.3 by induction.

Exercise 3.24★

Prove proposition 3.4. Hint: consider the different cases by which the edge $X \rightarrow Y$ was oriented during the procedure.

Exercise 3.25

Prove proposition 3.6. Hint: Show that this property is true of the graph returned by Mark-Immoralities.

Exercise 3.26

Implement an efficient algorithm that takes a Bayesian network over a set of variables $\mathcal{X}$ and a full instantiation ξ to $\mathcal{X}$, and computes the probability of ξ according to the network.

Exercise 3.27

Implement Reachable of algorithm 3.1.

Exercise 3.28★

Implement an efficient algorithm that determines, for a given set $\mathbf{Z}$ of observed variables and all pairs of nodes X and Y, whether X, Y are d-separated in $\mathcal{G}$ given $\mathbf{Z}$. Your algorithm should be significantly more efficient than simply running Reachable of algorithm 3.1 separately for each possible source variable X_i.

4 *Undirected Graphical Models*

So far, we have dealt only with directed graphical models, or Bayesian networks. These models are useful because both the structure and the parameters provide a natural representation for many types of real-world domains. In this chapter, we turn our attention to another important class of graphical models, defined on the basis of undirected graphs.

As we will see, these models are useful in modeling a variety of phenomena where one cannot naturally ascribe a directionality to the interaction between variables. Furthermore, the undirected models also offer a different and often simpler perspective on directed models, in terms of both the independence structure and the inference task. We also introduce a combined framework that allows both directed and undirected edges. We note that, unlike our results in the previous chapter, some of the results in this chapter require that we restrict attention to distributions over discrete state spaces.

4.1 The Misconception Example

To motivate our discussion of an alternative graphical representation, let us reexamine the Misconception example of section 3.4.2 (example 3.8). In this example, we have four students who get together in pairs to work on their homework for a class. The pairs that meet are shown via the edges in the undirected graph of figure 3.10a.

As we discussed, we intuitively want to model a distribution that satisfies $(A \perp C \mid \{B, D\})$ and $(B \perp D \mid \{A, C\})$, but no other independencies. As we showed, these independencies cannot be naturally captured in a Bayesian network: any Bayesian network I-map of such a distribution would necessarily have extraneous edges, and it would not capture at least one of the desired independence statements. More broadly, a Bayesian network requires that we ascribe a directionality to each influence. In this case, the interactions between the variables seem symmetrical, and we would like a model that allows us to represent these correlations without forcing a specific direction to the influence.

A representation that implements this intuition is an undirected graph. As in a Bayesian *Markov network* network, the nodes in the graph of a *Markov network* represent the variables, and the edges correspond to a notion of direct probabilistic interaction between the neighboring variables — an interaction that is not mediated by any other variable in the network. In this case, the graph of figure 3.10, which captures the interacting pairs, is precisely the Markov network structure that captures our intuitions for this example. As we will see, this similarity is not an accident.

$\phi_1(A, B)$			$\phi_2(B, C)$			$\phi_3(C, D)$			$\phi_4(D, A)$		
a^0	b^0	30	b^0	c^0	100	c^0	d^0	1	d^0	a^0	100
a^0	b^1	5	b^0	c^1	1	c^0	d^1	100	d^0	a^1	1
a^1	b^0	1	b^1	c^0	1	c^1	d^0	100	d^1	a^0	1
a^1	b^1	10	b^1	c^1	100	c^1	d^1	1	d^1	a^1	100
(a)			(b)			(c)			(d)		

Figure 4.1   **Factors for the** Misconception **example**

The remaining question is how to parameterize this undirected graph. Because the interaction is not directed, there is no reason to use a standard CPD, where we represent the distribution over one node given others. Rather, we need a more symmetric parameterization. Intuitively, what we want to capture is the *affinities* between related variables. For example, we might want to represent the fact that Alice and Bob are more likely to agree than to disagree. We associate with A, B a general-purpose function, also called a *factor*:

Definition 4.1

factor

scope

Let $\boldsymbol{D}$ be a set of random variables. We define a factor *ϕ to be a function from $Val(\boldsymbol{D})$ to $\mathbb{R}$. A factor is* nonnegative *if all its entries are nonnegative. The set of variables $\boldsymbol{D}$ is called the* scope *of the factor and denoted $Scope[\phi]$.* ∎

Unless stated otherwise, we restrict attention to nonnegative factors.

In our example, we have a factor $\phi_1(A, B) : Val(A, B) \mapsto \mathbb{R}^+$. The value associated with a particular assignment a, b denotes the affinity between these two values: the higher the value $\phi_1(a, b)$, the more compatible these two values are.

Figure 4.1a shows one possible compatibility factor for these variables. Note that this factor is not normalized; indeed, the entries are not even in $[0, 1]$. Roughly speaking, $\phi_1(A, B)$ asserts that it is more likely that Alice and Bob agree. It also adds more weight for the case where they are both right than for the case where they are both wrong. This factor function also has the property that $\phi_1(a^1, b^0) < \phi_1(a^0, b^1)$. Thus, if they disagree, there is less weight for the case where Alice has the misconception but Bob does not than for the converse case.

In a similar way, we define a compatibility factor for each other interacting pair: $\{B, C\}$, $\{C, D\}$, and $\{A, D\}$. Figure 4.1 shows one possible choice of factors for all four pairs. For example, the factor over C, D represents the compatibility of Charles and Debbie. It indicates that Charles and Debbie argue all the time, so that the most likely instantiations are those where they end up disagreeing.

As in a Bayesian network, the parameterization of the Markov network defines the local interactions between directly related variables. To define a global model, we need to combine these interactions. As in Bayesian networks, we combine the local models by multiplying them. Thus, we want $P(a, b, c, d)$ to be $\phi_1(a, b) \cdot \phi_2(b, c) \cdot \phi_3(c, d) \cdot \phi_4(d, a)$. In this case, however, we have no guarantees that the result of this process is a normalized joint distribution. Indeed, in this example, it definitely is not. Thus, we define the distribution by taking the product of

Assignment				Unnormalized	Normalized
a^0	b^0	c^0	d^0	300,000	0.04
a^0	b^0	c^0	d^1	300,000	0.04
a^0	b^0	c^1	d^0	300,000	0.04
a^0	b^0	c^1	d^1	30	$4.1 \cdot 10^{-6}$
a^0	b^1	c^0	d^0	500	$6.9 \cdot 10^{-5}$
a^0	b^1	c^0	d^1	500	$6.9 \cdot 10^{-5}$
a^0	b^1	c^1	d^0	5,000,000	0.69
a^0	b^1	c^1	d^1	500	$6.9 \cdot 10^{-5}$
a^1	b^0	c^0	d^0	100	$1.4 \cdot 10^{-5}$
a^1	b^0	c^0	d^1	1,000,000	0.14
a^1	b^0	c^1	d^0	100	$1.4 \cdot 10^{-5}$
a^1	b^0	c^1	d^1	100	$1.4 \cdot 10^{-5}$
a^1	b^1	c^0	d^0	10	$1.4 \cdot 10^{-6}$
a^1	b^1	c^0	d^1	100,000	0.014
a^1	b^1	c^1	d^0	100,000	0.014
a^1	b^1	c^1	d^1	100,000	0.014

Figure 4.2   Joint distribution for the Misconception **example.** The unnormalized measure and the normalized joint distribution over A, B, C, D, obtained from the parameterization of figure 4.1. The value of the partition function in this example is $7,201,840$.

the local factors, and then normalizing it to define a legal distribution. Specifically, we define

$$P(a, b, c, d) = \frac{1}{Z}\phi_1(a, b) \cdot \phi_2(b, c) \cdot \phi_3(c, d) \cdot \phi_4(d, a),$$

where

$$Z = \sum_{a,b,c,d} \phi_1(a, b) \cdot \phi_2(b, c) \cdot \phi_3(c, d) \cdot \phi_4(d, a)$$

partition function

Markov random field

is a normalizing constant known as the *partition function*. The term "partition" originates from the early history of Markov networks, which originated from the concept of *Markov random field* (or *MRF*) in statistical physics (see box 4.C); the "function" is because the value of Z is a function of the parameters, a dependence that will play a significant role in our discussion of learning.

In our example, the unnormalized measure (the simple product of the four factors) is shown in the next-to-last column in figure 4.2. For example, the entry corresponding to a^1, b^1, c^0, d^1 is obtained by multiplying:

$$\phi_1(a^1, b^1) \cdot \phi_2(b^1, c^0) \cdot \phi_3(c^0, d^1) \cdot \phi_4(d^1, a^1) = 10 \cdot 1 \cdot 100 \cdot 100 = 100,000.$$

The last column shows the normalized distribution.

We can use this joint distribution to answer queries, as usual. For example, by summing out A, C, and D, we obtain $P(b^1) \approx 0.732$ and $P(b^0) \approx 0.268$; that is, Bob is 26 percent likely to have the misconception. On the other hand, if we now observe that Charles does not have the misconception (c^0), we obtain $P(b^1 \mid c^0) \approx 0.06$.

The benefit of this representation is that it allows us great flexibility in representing interactions between variables. For example, if we want to change the nature of the interaction between A and B, we can simply modify the entries in that factor, without having to deal with normalization constraints and the interaction with other factors. The flip side of this flexibility, as we will see later, is that the effects of these changes are not always intuitively understandable.

As in Bayesian networks, there is a tight connection between the factorization of the distribution and its independence properties. The key result here is stated in exercise 2.5: $P \models (\boldsymbol{X} \perp \boldsymbol{Y} \mid \boldsymbol{Z})$ if and only if we can write P in the form $P(\mathcal{X}) = \phi_1(\boldsymbol{X}, \boldsymbol{Z})\phi_2(\boldsymbol{Y}, \boldsymbol{Z})$. In our example, the structure of the factors allows us to decompose the distribution in several ways; for example:

$$P(A, B, C, D) = \left[\frac{1}{Z}\phi_1(A, B)\phi_2(B, C)\right]\phi_3(C, D)\phi_4(A, D).$$

From this decomposition, we can infer that $P \models (B \perp D \mid A, C)$. We can similarly infer that $P \models (A \perp C \mid B, D)$. These are precisely the two independencies that we tried, unsuccessfully, to achieve using a Bayesian network, in example 3.8. Moreover, these properties correspond to our intuition of "paths of influence" in the graph, where we have that B and D are separated given A, C, and that A and C are separated given B, D. Indeed, as in a Bayesian network, independence properties of the distribution P correspond directly to separation properties in the graph over which P factorizes.

4.2   Parameterization

We begin our formal discussion by describing the parameterization used in the class of undirected graphical models that are the focus of this chapter. In the next section, we make the connection to the graph structure and demonstrate how it captures the independence properties of the distribution.

To represent a distribution, we need to associate the graph structure with a set of parameters, in the same way that CPDs were used to parameterize the directed graph structure. However, the parameterization of Markov networks is not as intuitive as that of Bayesian networks, since the factors do not correspond either to probabilities or to conditional probabilities. As a consequence, the parameters are not intuitively understandable, making them hard to elicit from people. As we will see in chapter 20, they are also significantly harder to estimate from data.

4.2.1   Factors

A key issue in parameterizing a Markov network is that the representation is undirected, so that the parameterization cannot be directed in nature. We therefore use factors, as defined in definition 4.1. Note that a factor subsumes both the notion of a joint distribution and the notion of a CPD. A joint distribution over $\boldsymbol{D}$ is a factor over $\boldsymbol{D}$: it specifies a real number for every assignment of values of $\boldsymbol{D}$. A conditional distribution $P(X \mid \boldsymbol{U})$ is a factor over $\{X\} \cup \boldsymbol{U}$. However, both CPDs and joint distributions must satisfy certain normalization constraints (for example, in a joint distribution the numbers must sum to 1), whereas there are no constraints on the parameters in a factor.

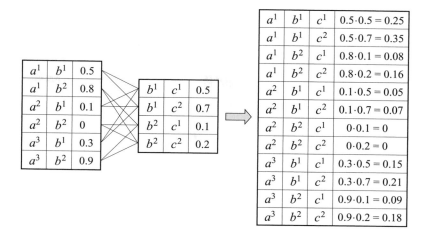

Figure 4.3 **An example of factor product**

As we discussed, we can view a factor as roughly describing the "compatibilities" between different values of the variables in its scope. We can now parameterize the graph by associating a set of a factors with it. One obvious idea might be to associate parameters directly with the edges in the graph. However, a simple calculation will convince us that this approach is insufficient to parameterize a full distribution.

Example 4.1

Consider a fully connected graph over $\mathcal{X}$; in this case, the graph specifies no conditional independence assumptions, so that we should be able to specify an arbitrary joint distribution over $\mathcal{X}$. If all of the variables are binary, each factor over an edge would have 4 parameters, and the total number of parameters in the graph would be $4\binom{n}{2}$. However, the number of parameters required to specify a joint distribution over n binary variables is $2^n - 1$. Thus, pairwise factors simply do not have enough parameters to encompass the space of joint distributions. More intuitively, such factors capture only the pairwise interactions, and not interactions that involve combinations of values of larger subsets of variables. ∎

A more general representation can be obtained by allowing factors over arbitrary subsets of variables. To provide a formal definition, we first introduce the following important operation on factors.

Definition 4.2

factor product

Let $\boldsymbol{X}$, $\boldsymbol{Y}$, and $\boldsymbol{Z}$ be three disjoint sets of variables, and let $\phi_1(\boldsymbol{X}, \boldsymbol{Y})$ and $\phi_2(\boldsymbol{Y}, \boldsymbol{Z})$ be two factors. We define the factor product $\phi_1 \times \phi_2$ to be a factor $\psi : Val(\boldsymbol{X}, \boldsymbol{Y}, \boldsymbol{Z}) \mapsto \mathbb{R}$ as follows:

$$\psi(\boldsymbol{X}, \boldsymbol{Y}, \boldsymbol{Z}) = \phi_1(\boldsymbol{X}, \boldsymbol{Y}) \cdot \phi_2(\boldsymbol{Y}, \boldsymbol{Z}).$$ ∎

The key aspect to note about this definition is the fact that the two factors ϕ_1 and ϕ_2 are multiplied in a way that "matches up" the common part $\boldsymbol{Y}$. Figure 4.3 shows an example of the product of two factors. We have deliberately chosen factors that do not correspond either to probabilities or to conditional probabilities, in order to emphasize the generality of this operation.

As we have already observed, both CPDs and joint distributions are factors. Indeed, the chain rule for Bayesian networks defines the joint distribution factor as the product of the CPD factors. For example, when computing $P(A, B) = P(A)P(B \mid A)$, we always multiply entries in the $P(A)$ and $P(B \mid A)$ tables that have the same value for A. Thus, letting $\phi_{X_i}(X_i, \text{Pa}_{X_i})$ represent $P(X_i \mid \text{Pa}_{X_i})$, we have that

$$P(X_1, \ldots, X_n) = \prod_i \phi_{X_i}.$$

4.2.2   Gibbs Distributions and Markov Networks

We can now use the more general notion of factor product to define an undirected parameterization of a distribution.

Definition 4.3

Gibbs distribution

A distribution P_Φ is a Gibbs distribution *parameterized by a set of factors $\Phi = \{\phi_1(\boldsymbol{D}_1), \ldots, \phi_K(\boldsymbol{D}_K)\}$ if it is defined as follows:*

$$P_\Phi(X_1, \ldots, X_n) = \frac{1}{Z} \tilde{P}_\Phi(X_1, \ldots, X_n),$$

where

$$\tilde{P}_\Phi(X_1, \ldots, X_n) = \phi_1(\boldsymbol{D}_1) \times \phi_2(\boldsymbol{D}_2) \times \cdots \times \phi_m(\boldsymbol{D}_m)$$

is an unnormalized measure and

$$Z = \sum_{X_1, \ldots, X_n} \tilde{P}_\Phi(X_1, \ldots, X_n)$$

partition function

is a normalizing constant called the partition function.   ∎

It is tempting to think of the factors as representing the marginal probabilities of the variables in their scope. Thus, looking at any individual factor, we might be led to believe that the behavior of the distribution defined by the Markov network as a whole corresponds to the behavior defined by the factor. However, this intuition is overly simplistic. **A factor is only one contribution to the overall joint distribution. The distribution as a whole has to take into consideration the contributions from all of the factors involved.**

Example 4.2

Consider the distribution of figure 4.2. The marginal distribution over A, B, is

a^0	b^0	0.13
a^0	b^1	0.69
a^1	b^0	0.14
a^1	b^1	0.04

The most likely configuration is the one where Alice and Bob disagree. By contrast, the highest entry in the factor $\phi_1(A, B)$ in figure 4.1 corresponds to the assignment a^0, b^0. The reason for the discrepancy is the influence of the other factors on the distribution. In particular, $\phi_3(C, D)$ asserts that Charles and Debbie disagree, whereas $\phi_2(B, C)$ and $\phi_4(D, A)$ assert that Bob and Charles agree and that Debbie and Alice agree. Taking just these factors into consideration, we would conclude that Alice and Bob are likely to disagree. In this case, the "strength" of these other factors is much stronger than that of the $\phi_1(A, B)$ factor, so that the influence of the latter is overwhelmed. ∎

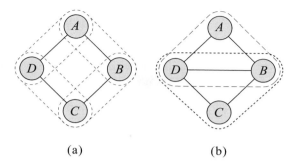

(a)   (b)

Figure 4.4   The cliques in two simple Markov networks. In (a), the cliques are the pairs $\{A, B\}$, $\{B, C\}$, $\{C, D\}$, and $\{D, A\}$. In (b), the cliques are $\{A, B, D\}$ and $\{B, C, D\}$.

We now want to relate the parameterization of a Gibbs distribution to a graph structure. If our parameterization contains a factor whose scope contains both X and Y, we are introducing a direct interaction between them. Intuitively, we would like these direct interactions to be represented in the graph structure. Thus, if our parameterization contains such a factor, we would like the associated Markov network structure $\mathcal{H}$ to contain an edge between X and Y.

Definition 4.4

Markov network
factorization

We say that a distribution P_Φ with $\Phi = \{\phi_1(\boldsymbol{D}_1), \dots, \phi_K(\boldsymbol{D}_K)\}$ factorizes over a Markov network $\mathcal{H}$ if each $\boldsymbol{D}_k$ $(k = 1, \dots, K)$ is a complete subgraph of $\mathcal{H}$. ∎

clique potentials

The factors that parameterize a Markov network are often called *clique potentials*.

As we will see, if we associate factors only with complete subgraphs, as in this definition, we are not violating the independence assumptions induced by the network structure, as defined later in this chapter.

Note that, because every complete subgraph is a subset of some (maximal) clique, we can reduce the number of factors in our parameterization by allowing factors only for maximal cliques. More precisely, let $\boldsymbol{C}_1, \dots, \boldsymbol{C}_k$ be the cliques in $\mathcal{H}$. We can parameterize P using a set of factors $\phi_1(\boldsymbol{C}_1), \dots, \phi_l(\boldsymbol{C}_l)$. Any factorization in terms of complete subgraphs can be converted into this form simply by assigning each factor to a clique that encompasses its scope and multiplying all of the factors assigned to each clique to produce a clique potential. In our Misconception example, we have four cliques: $\{A, B\}$, $\{B, C\}$, $\{C, D\}$, and $\{A, D\}$. Each of these cliques can have its own clique potential. One possible setting of the parameters in these clique potential is shown in figure 4.1. Figure 4.4 shows two examples of a Markov network and the (maximal) cliques in that network.

☞ **Although it can be used without loss of generality, the parameterization using maximal clique potentials generally obscures structure that is present in the original set of factors.** For example, consider the Gibbs distribution described in example 4.1. Here, we have a potential for every pair of variables, so the Markov network associated with this distribution is a single large clique containing all variables. If we associate a factor with this single clique, it would be exponentially large in the number of variables, whereas the original parameterization in terms of edges requires only a quadratic number of parameters. See section 4.4.1.1 for further discussion.

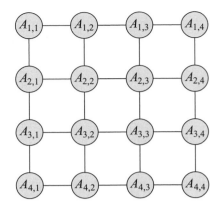

Figure 4.A.1 — A pairwise Markov network (MRF) structured as a grid.

---

pairwise Markov
network

node potential

edge potential

Box 4.A — Concept: Pairwise Markov Networks. *A subclass of Markov networks that arises in many contexts is that of* pairwise Markov networks, *representing distributions where all of the factors are over single variables or pairs of variables. More precisely, a pairwise Markov network over a graph $\mathcal{H}$ is associated with a set of* node potentials $\{\phi(X_i) : i = 1, \ldots, n\}$ *and a set of* edge potentials $\{\phi(X_i, X_j) : (X_i, X_j) \in \mathcal{H}\}$. *The overall distribution is (as always) the normalized product of all of the potentials (both node and edge). Pairwise MRFs are attractive because of their simplicity, and because interactions on edges are an important special case that often arises in practice (see, for example, box 4.C and box 4.B).*

A class of pairwise Markov networks that often arises, and that is commonly used as a benchmark for inference, is the class of networks structured in the form of a grid, as shown in figure 4.A.1. As we discuss in the inference chapters of this book, although these networks have a simple and compact representation, they pose a significant challenge for inference algorithms.

4.2.3   Reduced Markov Networks

We end this section with one final concept that will prove very useful in later sections. Consider the process of conditioning a distribution on some assignment u to some subset of variables U. Conditioning a distribution corresponds to eliminating all entries in the joint distribution that are inconsistent with the event $U = u$, and renormalizing the remaining entries to sum to 1. Now, consider the case where our distribution has the form P_Φ for some set of factors Φ. Each entry in the unnormalized measure $\tilde{P}_\Phi$ is a product of entries from the factors Φ, one entry from each factor. If, in some factor, we have an entry that is inconsistent with $U = u$, it will only contribute to entries in $\tilde{P}_\Phi$ that are also inconsistent with this event. Thus, we can eliminate all such entries from every factor in Φ.

More generally, we can define:

a^1	b^1	c^1	0.25
a^1	b^2	c^1	0.08
a^2	b^1	c^1	0.05
a^2	b^2	c^1	0
a^3	b^1	c^1	0.15
a^3	b^2	c^1	0.09

Figure 4.5 Factor reduction: The factor computed in figure 4.3, reduced to the context $C = c^1$.

Definition 4.5

factor reduction

Let $\phi(Y)$ be a factor, and $U = u$ an assignment for $U \subseteq Y$. We define the reduction of the factor ϕ to the context $U = u$, denoted $\phi[U = u]$ (and abbreviated $\phi[u]$), to be a factor over scope $Y' = Y - U$, such that

$$\phi[u](y') = \phi(y', u).$$

For $U \not\subseteq Y$, we define $\phi[u]$ to be $\phi[U' = u']$, where $U' = U \cap Y$, and $u' = u\langle U'\rangle$, where $u\langle U'\rangle$ denotes the assignment in u to the variables in U'. ∎

Figure 4.5 illustrates this operation, reducing the of figure 4.3 to the context $C = c^1$.

Now, consider a product of factors. An entry in the product is consistent with u if and only if it is a product of entries that are all consistent with u. We can therefore define:

Definition 4.6

reduced Gibbs distribution

Let P_Φ be a Gibbs distribution parameterized by $\Phi = \{\phi_1, \ldots, \phi_K\}$ and let u be a context. The reduced Gibbs distribution $P_\Phi[u]$ is the Gibbs distribution defined by the set of factors $\Phi[u] = \{\phi_1[u], \ldots, \phi_K[u]\}$. ∎

Reducing the set of factors defining P_Φ to some context u corresponds directly to the operation of conditioning P_Φ on the observation u. More formally:

Proposition 4.1

Let $P_\Phi(X)$ be a Gibbs distribution. Then $P_\Phi[u] = P_\Phi(W \mid u)$ where $W = X - U$.

Thus, to condition a Gibbs distribution on a context u, we simply reduce every one of its factors to that context. Intuitively, the renormalization step needed to account for u is simply folded into the standard renormalization of any Gibbs distribution. This result immediately provides us with a construction for the Markov network that we obtain when we condition the associated distribution on some observation u.

Definition 4.7

reduced Markov network

Let $\mathcal{H}$ be a Markov network over X and $U = u$ a context. The reduced Markov network $\mathcal{H}[u]$ is a Markov network over the nodes $W = X - U$, where we have an edge $X - Y$ if there is an edge $X - Y$ in $\mathcal{H}$. ∎

Proposition 4.2

Let $P_\Phi(X)$ be a Gibbs distribution that factorizes over $\mathcal{H}$, and $U = u$ a context. Then $P_\Phi[u]$ factorizes over $\mathcal{H}[u]$.

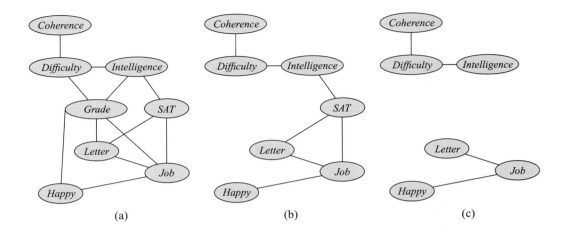

Figure 4.6   Markov networks for the factors in an extended Student **example:** (a) The initial set of factors; (b) Reduced to the context $G = g$; (c) Reduced to the context $G = g, S = s$.

Note the contrast to the effect of conditioning in a Bayesian network: Here, conditioning on a context u only eliminates edges from the graph; in a Bayesian network, conditioning on evidence can activate a v-structure, creating new dependencies. We return to this issue in section 4.5.1.1.

Example 4.3   *Consider, for example, the Markov network shown in figure 4.6a; as we will see, this network is the Markov network required to capture the distribution encoded by an extended version of our* Student *Bayesian network (see figure 9.8). Figure 4.6b shows the same Markov network reduced over a context of the form $G = g$, and (c) shows the network reduced over a context of the form $G = g, S = s$. As we can see, the network structures are considerably simplified.* ∎

Box 4.B — Case Study: Markov Networks for Computer Vision. *One important application area for Markov networks is computer vision. Markov networks, typically called MRFs in this vision community, have been used for a wide variety of visual processing tasks, such as image segmentation, removal of blur or noise, stereo reconstruction, object recognition, and many more.*

In most of these applications, the network takes the structure of a pairwise MRF, where the variables correspond to pixels and the edges (factors) to interactions between adjacent pixels in the grid that represents the image; thus, each (interior) pixel has exactly four neighbors. The value space of the variables and the exact form of factors depend on the task. These models are usually formulated in terms of energies (negative log-potentials), so that values represent "penalties," and a lower value corresponds to a higher-probability configuration.

image denoising   *In* image denoising, *for example, the task is to restore the "true" value of all of the pixels given possibly noisy pixel values. Here, we have a node potential for each pixel X_i that penalizes large discrepancies from the observed pixel value y_i. The edge potential encodes a preference for continuity between adjacent pixel values, penalizing cases where the inferred value for X_i is too* *far from the inferred pixel value for one of its neighbors X_j. However, it is important not to overpenalize true disparities (such as edges between objects or regions), leading to oversmoothing of the image. Thus, we bound the penalty, using, for example, some truncated norm, as described in box 4.D: $\epsilon(x_i, x_j) = \min(c\|x_i - x_j\|_p, \text{dist}_{\max})$ (for $p \in \{1, 2\}$).*

stereo reconstruction

Slight variants of the same model are used in many other applications. For example, in stereo reconstruction, *the goal is to reconstruct the depth disparity of each pixel in the image. Here, the values of the variables represent some discretized version of the depth dimension (usually more finely discretized for distances close to the camera and more coarsely discretized as the distance from the camera increases). The individual node potential for each pixel X_i uses standard techniques from computer vision to estimate, from a pair of stereo images, the individual depth disparity of this pixel. The edge potentials, precisely as before, often use a truncated metric to enforce continuity of the depth estimates, with the truncation avoiding an overpenalization of true depth disparities (for example, when one object is partially in front of the other). Here, it is also quite common to make the penalty inversely proportional to the image gradient between the two pixels, allowing a smaller penalty to be applied in cases where a large image gradient suggests an edge between the pixels, possibly corresponding to an occlusion boundary.*

image segmentation

In image segmentation, *the task is to partition the image pixels into regions corresponding to distinct parts of the scene. There are different variants of the segmentation task, many of which can be formulated as a Markov network. In one formulation, known as multiclass segmentation, each variable X_i has a domain $\{1, \ldots, K\}$, where the value of X_i represents a region assignment for pixel i (for example, grass, water, sky, car). Since classifying every pixel can be computationally expensive, some state-of-the-art methods for image segmentation and other tasks first oversegment the image into superpixels (or small coherent regions) and classify each region — all pixels within a region are assigned the same label. The oversegmented image induces a graph in which there is one node for each superpixel and an edge between two nodes if the superpixels are adjacent (share a boundary) in the underlying image. We can now define our distribution in terms of this graph.*

Features are extracted from the image for each pixel or superpixel. The appearance features depend on the specific task. In image segmentation, for example, features typically include statistics over color, texture, and location. Often the features are clustered or provided as input to local classifiers to reduce dimensionality. The features used in the model are then the soft cluster assignments or local classifier outputs for each superpixel. The node potential for a pixel or superpixel is then a function of these features. We note that the factors used in defining this model depend on the specific values of the pixels in the image, so that each image defines a different probability distribution over the segment labels for the pixels or superpixels. In effect, the model used here is a

conditional random field

conditional random field, *a concept that we define more formally in section 4.6.1.*

The model contains an edge potential between every pair of neighboring superpixels X_i, X_j. Most simply, this potential encodes a contiguity preference, with a penalty of λ whenever $X_i \neq X_j$. Again, we can improve the model by making the penalty depend on the presence of an image gradient between the two pixels. An even better model does more than penalize discontinuities. We can have nondefault values for other class pairs, allowing us to encode the fact that we more often find tigers adjacent to vegetation than adjacent to water; we can even make the model depend on the relative pixel location, allowing us to encode the fact that we usually find water below vegetation, cars over roads, and sky above everything.

Figure 4.B.1 shows segmentation results in a model containing only potentials on single pixels (thereby labeling each of them independently) versus results obtained from a model also containing

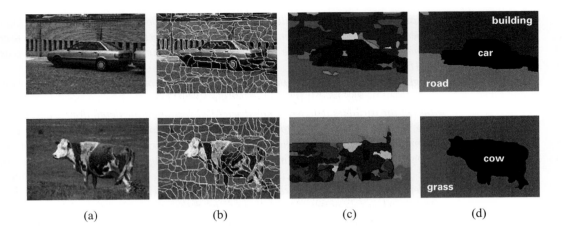

(a)   (b)   (c)   (d)

Figure 4.B.1 — Two examples of image segmentation results (a) The original image. (b) An overseg-
mentation known as superpixels; each superpixel is associated with a random variable that designates its
segment assignment. The use of superpixels reduces the size of the problems. (c) Result of segmentation
using node potentials alone, so that each superpixel is classified independently. (d) Result of segmentation
using a pairwise Markov network encoding interactions between adjacent superpixels.

*pairwise potentials. The difference in the quality of the results clearly illustrates the importance of
modeling the correlations between the superpixels.*

4.3   Markov Network Independencies

In section 4.1, we gave an intuitive justification of why an undirected graph seemed to capture
the types of interactions in the Misconception example. We now provide a formal presentation
of the undirected graph as a representation of independence assertions.

4.3.1   Basic Independencies

As in the case of Bayesian networks, the graph structure in a Markov network can be viewed
as encoding a set of independence assumptions. Intuitively, in Markov networks, probabilistic
influence "flows" along the undirected paths in the graph, but it is blocked if we condition on
the intervening nodes.

Definition 4.8

observed variable

active path

*Let $\mathcal{H}$ be a Markov network structure, and let $X_1 - \ldots - X_k$ be a path in $\mathcal{H}$. Let $\mathbf{Z} \subseteq \mathcal{X}$ be a set
of observed variables. The path $X_1 - \ldots - X_k$ is active given $\mathbf{Z}$ if none of the X_i's, $i = 1, \ldots, k$,
is in $\mathbf{Z}$.*   ∎

Using this notion, we can define a notion of *separation* in the graph.

Definition 4.9

separation

global
independencies

We say that a set of nodes $\boldsymbol{Z}$ separates $\boldsymbol{X}$ and $\boldsymbol{Y}$ in $\mathcal{H}$, denoted $\mathrm{sep}_{\mathcal{H}}(\boldsymbol{X}; \boldsymbol{Y} \mid \boldsymbol{Z})$, if there is no active path between any node $X \in \boldsymbol{X}$ and $Y \in \boldsymbol{Y}$ given $\boldsymbol{Z}$. We define the global independencies associated with $\mathcal{H}$ to be:

$$\mathcal{I}(\mathcal{H}) = \{(\boldsymbol{X} \perp \boldsymbol{Y} \mid \boldsymbol{Z}) \: : \: \mathrm{sep}_{\mathcal{H}}(\boldsymbol{X}; \boldsymbol{Y} \mid \boldsymbol{Z})\}.$$

∎

As we will discuss, the independencies in $\mathcal{I}(\mathcal{H})$ are precisely those that are guaranteed to hold for every distribution P over $\mathcal{H}$. In other words, the separation criterion is sound for detecting independence properties in distributions over $\mathcal{H}$.

Note that the definition of separation is monotonic in $\boldsymbol{Z}$, that is, if $sep_{\mathcal{H}}(\boldsymbol{X}; \boldsymbol{Y} \mid \boldsymbol{Z})$, then $sep_{\mathcal{H}}(\boldsymbol{X}; \boldsymbol{Y} \mid \boldsymbol{Z}')$ for any $\boldsymbol{Z}' \supset \boldsymbol{Z}$. Thus, if we take separation as our definition of the independencies induced by the network structure, we are effectively restricting our ability to encode nonmonotonic independence relations. Recall that in the context of intercausal reasoning in Bayesian networks, nonmonotonic reasoning patterns are quite useful in many situations — for example, when two diseases are independent, but dependent given some common symptom. The nature of the separation property implies that such independence patterns cannot be expressed in the structure of a Markov network. We return to this issue in section 4.5.

As for Bayesian networks, we can show a connection between the independence properties implied by the Markov network structure, and the possibility of factorizing a distribution over the graph. As before, we can now state the analogue to both of our representation theorems for Bayesian networks, which assert the equivalence between the Gibbs factorization of a distribution P over a graph $\mathcal{H}$ and the assertion that $\mathcal{H}$ is an I-map for P, that is, that P satisfies the Markov assumptions $\mathcal{I}(\mathcal{H})$.

4.3.1.1 Soundness

soundness

We first consider the analogue to theorem 3.2, which asserts that a Gibbs distribution satisfies the independencies associated with the graph. In other words, this result states the *soundness* of the separation criterion.

Theorem 4.1

Let P be a distribution over $\mathcal{X}$, and $\mathcal{H}$ a Markov network structure over $\mathcal{X}$. If P is a Gibbs distribution that factorizes over $\mathcal{H}$, then $\mathcal{H}$ is an I-map for P.

PROOF Let $\boldsymbol{X}, \boldsymbol{Y}, \boldsymbol{Z}$ be any three disjoint subsets in $\mathcal{X}$ such that $\boldsymbol{Z}$ separates $\boldsymbol{X}$ and $\boldsymbol{Y}$ in $\mathcal{H}$. We want to show that $P \models (\boldsymbol{X} \perp \boldsymbol{Y} \mid \boldsymbol{Z})$.

We start by considering the case where $\boldsymbol{X} \cup \boldsymbol{Y} \cup \boldsymbol{Z} = \mathcal{X}$. As $\boldsymbol{Z}$ separates $\boldsymbol{X}$ from $\boldsymbol{Y}$, there are no direct edges between $\boldsymbol{X}$ and $\boldsymbol{Y}$. Hence, any clique in $\mathcal{H}$ is fully contained either in $\boldsymbol{X} \cup \boldsymbol{Z}$ or in $\boldsymbol{Y} \cup \boldsymbol{Z}$. Let $\mathcal{I}_{\boldsymbol{X}}$ be the indexes of the set of cliques that are contained in $\boldsymbol{X} \cup \boldsymbol{Z}$, and let $\mathcal{I}_{\boldsymbol{Y}}$ be the indexes of the remaining cliques. We know that

$$P(X_1, \ldots, X_n) = \frac{1}{Z} \prod_{i \in \mathcal{I}_{\boldsymbol{X}}} \phi_i(\boldsymbol{D}_i) \cdot \prod_{i \in \mathcal{I}_{\boldsymbol{Y}}} \phi_i(\boldsymbol{D}_i).$$

As we discussed, none of the factors in the first product involve any variable in $\boldsymbol{Y}$, and none in the second product involve any variable in $\boldsymbol{X}$. Hence, we can rewrite this product in the form:

$$P(X_1, \ldots, X_n) = \frac{1}{Z} f(\boldsymbol{X}, \boldsymbol{Z}) g(\boldsymbol{Y}, \boldsymbol{Z}).$$

From this decomposition, the desired independence follows immediately (exercise 2.5).

Now consider the case where $X \cup Y \cup Z \subset \mathcal{X}$. Let $U = \mathcal{X} - (X \cup Y \cup Z)$. We can partition U into two disjoint sets U_1 and U_2 such that Z separates $X \cup U_1$ from $Y \cup U_2$ in $\mathcal{H}$. Using the preceding argument, we conclude that $P \models (X, U_1 \perp Y, U_2 \mid Z)$. Using the decomposition property (equation (2.8)), we conclude that $P \models (X \perp Y \mid Z)$. ∎

Hammersley-
Clifford
theorem

The other direction (the analogue to theorem 3.1), which goes from the independence properties of a distribution to its factorization, is known as the *Hammersley-Clifford theorem*. Unlike for Bayesian networks, this direction does not hold in general. As we will show, it holds only under the additional assumption that P is a positive distribution (see definition 2.5).

Theorem 4.2

> *Let P be a positive distribution over $\mathcal{X}$, and $\mathcal{H}$ a Markov network graph over $\mathcal{X}$. If $\mathcal{H}$ is an I-map for P, then P is a Gibbs distribution that factorizes over $\mathcal{H}$.*

To prove this result, we would need to use the independence assumptions to construct a set of factors over $\mathcal{H}$ that give rise to the distribution P. In the case of Bayesian networks, these factors were simply CPDs, which we could derive directly from P. As we have discussed, the correspondence between the factors in a Gibbs distribution and the distribution P is much more indirect. The construction required here is therefore significantly more subtle, and relies on concepts that we develop later in this chapter; hence, we defer the proof to section 4.4 (theorem 4.8).

 This result shows that, **for positive distributions, the global independencies imply that the distribution factorizes according the network structure. Thus, for this class of distributions, we have that a distribution P factorizes over a Markov network $\mathcal{H}$ if and only if $\mathcal{H}$ is an I-map of P.** The positivity assumption is necessary for this result to hold:

Example 4.4

> *Consider a distribution P over four binary random variables X_1, X_2, X_3, X_4, which gives probability $1/8$ to each of the following eight configurations, and probability zero to all others:*
>
> | (0,0,0,0) | (1,0,0,0) | (1,1,0,0) | (1,1,1,0) |
> | (0,0,0,1) | (0,0,1,1) | (0,1,1,1) | (1,1,1,1) |
>
> *Let $\mathcal{H}$ be the graph $X_1{-}X_2{-}X_3{-}X_4{-}X_1$. Then P satisfies the global independencies with respect to $\mathcal{H}$. For example, consider the independence $(X_1 \perp X_3 \mid X_2, X_4)$. For the assignment $X_2 = x_2^1, X_4 = x_4^0$, we have that only assignments where $X_1 = x_1^1$ receive positive probability. Thus, $P(x_1^1 \mid x_2^1, x_4^0) = 1$, and X_1 is trivially independent of X_3 in this conditional distribution. A similar analysis applies to all other cases, so that the global independencies hold. However, the distribution P does not factorize according to $\mathcal{H}$. The proof of this fact is left as an exercise (see exercise 4.1).* ∎

4.3.1.2 Completeness

completeness

The preceding discussion shows the soundness of the separation condition as a criterion for detecting independencies in Markov networks: any distribution that factorizes over $\mathcal{G}$ satisfies the independence assertions implied by separation. The next obvious issue is the *completeness* of this criterion.

As for Bayesian networks, the strong version of completeness does not hold in this setting. In other words, it is not the case that every pair of nodes X and Y that are not separated in $\mathcal{H}$ are dependent in every distribution P which factorizes over $\mathcal{H}$. However, as in theorem 3.3, we can use a weaker definition of completeness that does hold:

Theorem 4.3

Let $\mathcal{H}$ be a Markov network structure. If X and Y are not separated given $\mathbf{Z}$ in $\mathcal{H}$, then X and Y are dependent given $\mathbf{Z}$ in some distribution P that factorizes over $\mathcal{H}$.

PROOF The proof is a constructive one: we construct a distribution P that factorizes over $\mathcal{H}$ where X and Y are dependent. We assume, without loss of generality, that all variables are binary-valued. If this is not the case, we can treat them as binary-valued by restricting attention to two distinguished values for each variable.

By assumption, X and Y are not separated given $\mathbf{Z}$ in $\mathcal{H}$; hence, they must be connected by some unblocked trail. Let $X = U_1 - U_2 - \ldots - U_k = Y$ be some minimal trail in the graph such that, for all i, $U_i \notin \mathbf{Z}$, where we define a minimal trail in $\mathcal{H}$ to be a path with no shortcuts: thus, for any i and $j \neq i \pm 1$, there is no edge $U_i - U_j$. We can always find such a path: If we have a nonminimal path where we have $U_i - U_j$ for $j > i + 1$, we can always "shortcut" the original trail, converting it to one that goes directly from U_i to U_j.

For any $i = 1, \ldots, k - 1$, as there is an edge $U_i - U_{i+1}$, it follows that U_i, U_{i+1} must both appear in some clique $\mathbf{C}_i$. We pick some very large weight W, and for each i we define the clique potential $\phi_i(\mathbf{C}_i)$ to assign weight W if $U_i = U_{i+1}$ and weight 1 otherwise, regardless of the values of the other variables in the clique. Note that the cliques $\mathbf{C}_i$ for U_i, U_{i+1} and $\mathbf{C}_j$ for U_j, U_{j+1} must be different cliques: If $\mathbf{C}_i = \mathbf{C}_j$, then U_j is in the same clique as U_i, and we have an edge $U_i - U_j$, contradicting the minimality of the trail. Hence, we can define the clique potential for each clique $\mathbf{C}_i$ separately. We define the clique potential for any other clique to be uniformly 1.

We now consider the distribution P resulting from multiplying all of these clique potentials. Intuitively, the distribution $P(U_1, \ldots, U_k)$ is simply the distribution defined by multiplying the pairwise factors for the pairs U_i, U_{i+1}, regardless of the other variables (including the ones in $\mathbf{Z}$). One can verify that, in $P(U_1, \ldots, U_k)$, we have that $X = U_1$ and $Y = U_k$ are dependent. We leave the conclusion of this argument as an exercise (exercise 4.5). ∎

We can use the same argument as theorem 3.5 to conclude that, for almost all distributions P that factorize over $\mathcal{H}$ (that is, for all distributions except for a set of measure zero in the space of factor parameterizations) we have that $\mathcal{I}(P) = \mathcal{I}(\mathcal{H})$.

Once again, we can view this result as telling us that our definition of $\mathcal{I}(\mathcal{H})$ is the maximal one. For any independence assertion that is not a consequence of separation in $\mathcal{H}$, we can always find a counterexample distribution P that factorizes over $\mathcal{H}$.

4.3.2 Independencies Revisited

When characterizing the independencies in a Bayesian network, we provided two definitions: the local independencies (each node is independent of its nondescendants given its parents), and the global independencies induced by d-separation. As we showed, these two sets of independencies are equivalent, in that one implies the other.

So far, our discussion for Markov networks provides only a global criterion. While the global criterion characterizes the entire set of independencies induced by the network structure, a local criterion is also valuable, since it allows us to focus on a smaller set of properties when examining the distribution, significantly simplifying the process of finding an I-map for a distribution P.

Thus, it is natural to ask whether we can provide a local definition of the independencies induced by a Markov network, analogously to the local independencies of Bayesian networks. Surprisingly, as we now show, in the context of Markov networks, there are three different possible definitions of the independencies associated with the network structure — two local ones and the global one in definition 4.9. While these definitions are related, they are equivalent only for positive distributions. As we will see, nonpositive distributions allow for deterministic dependencies between the variables. Such deterministic interactions can "fool" local independence tests, allowing us to construct networks that are not I-maps of the distribution, yet the local independencies hold.

4.3.2.1   Local Markov Assumptions

The first, and weakest, definition is based on the following intuition: Whenever two variables are directly connected, they have the potential of being directly correlated in a way that is not mediated by other variables. Conversely, when two variables are not directly linked, there must be some way of rendering them conditionally independent. Specifically, we can require that X and Y be independent given all other nodes in the graph.

Definition 4.10

pairwise independencies

Let $\mathcal{H}$ be a Markov network. We define the pairwise independencies *associated with $\mathcal{H}$ to be:*

$$\mathcal{I}_p(\mathcal{H}) = \{(X \perp Y \mid \mathcal{X} - \{X, Y\}) \; : \; X{-}Y \notin \mathcal{H}\}. \qquad \blacksquare$$

Using this definition, we can easily represent the independencies in our Misconception example using a Markov network: We simply connect the nodes up in exactly the same way as the interaction structure between the students.

The second local definition is an undirected analogue to the local independencies associated with a Bayesian network. It is based on the intuition that we can block all influences on a node by conditioning on its immediate neighbors.

Definition 4.11

Markov blanket

local independencies

For a given graph $\mathcal{H}$, we define the Markov blanket *of X in $\mathcal{H}$, denoted $\mathrm{MB}_{\mathcal{H}}(X)$, to be the neighbors of X in $\mathcal{H}$. We define the* local independencies *associated with $\mathcal{H}$ to be:*

$$\mathcal{I}_{\ell}(\mathcal{H}) = \{(X \perp \mathcal{X} - \{X\} - \mathrm{MB}_{\mathcal{H}}(X) \mid \mathrm{MB}_{\mathcal{H}}(X)) \; : \; X \in \mathcal{X}\}. \qquad \blacksquare$$

In other words, the local independencies state that X is independent of the rest of the nodes in the graph given its immediate neighbors. We will show that these local independence assumptions hold for any distribution that factorizes over $\mathcal{H}$, so that X's Markov blanket in $\mathcal{H}$ truly does separate it from all other variables.

4.3.2.2   Relationships between Markov Properties

We have now presented three sets of independence assertions associated with a network structure $\mathcal{H}$. For general distributions, $\mathcal{I}_p(\mathcal{H})$ is strictly weaker than $\mathcal{I}_{\ell}(\mathcal{H})$, which in turn is strictly weaker than $\mathcal{I}(\mathcal{H})$. However, all three definitions are equivalent for positive distributions.

Proposition 4.3

For any Markov network $\mathcal{H}$, and any distribution P, we have that if $P \models \mathcal{I}_\ell(\mathcal{H})$ then $P \models \mathcal{I}_p(\mathcal{H})$.

The proof of this result is left as an exercise (exercise 4.8).

Proposition 4.4

For any Markov network $\mathcal{H}$, and any distribution P, we have that if $P \models \mathcal{I}(\mathcal{H})$ then $P \models \mathcal{I}_\ell(\mathcal{H})$.

The proof of this result follows directly from the fact that if X and Y are not connected by an edge, then they are necessarily separated by all of the remaining nodes in the graph.

The converse of these inclusion results holds only for positive distributions (see definition 2.5). More specifically, if we assume the intersection property (equation (2.11)), all three of the Markov conditions are equivalent.

Theorem 4.4

Let P be a positive distribution. If P satisfies $\mathcal{I}_p(\mathcal{H})$, then P satisfies $\mathcal{I}(\mathcal{H})$.

PROOF We want to prove that for all disjoint sets X, Y, Z:

$$sep_{\mathcal{H}}(X; Y \mid Z) \implies P \models (X \perp Y \mid Z). \tag{4.1}$$

The proof proceeds by descending induction on the size of Z.

The base case is $|Z| = n - 2$; equation (4.1) follows immediately from the definition of $\mathcal{I}_p(\mathcal{H})$.

For the inductive step, assume that equation (4.1) holds for every Z' with size $|Z'| = k$, and let Z be any set such that $|Z| = k - 1$. We distinguish between two cases.

In the first case, $X \cup Z \cup Y = \mathcal{X}$. As $|Z| < n - 2$, we have that either $|X| \geq 2$ or $|Y| \geq 2$. Without loss of generality, assume that the latter holds; let $A \in Y$ and $Y' = Y - \{A\}$. From the fact that $sep_{\mathcal{H}}(X; Y \mid Z)$, we also have that $sep_{\mathcal{H}}(X; Y' \mid Z)$ on one hand and $sep_{\mathcal{H}}(X; A \mid Z)$ on the other hand. As separation is monotonic, we also have that $sep_{\mathcal{H}}(X; Y' \mid Z \cup \{A\})$ and $sep_{\mathcal{H}}(X; A \mid Z \cup Y')$. The separating sets $Z \cup \{A\}$ and $Z \cup Y'$ are each at least size $|Z| + 1 = k$ in size, so that equation (4.1) applies, and we can conclude that P satisfies:

$$(X \perp Y' \mid Z \cup \{A\}) \quad \& \quad (X \perp A \mid Z \cup Y').$$

Because P is positive, we can apply the intersection property (equation (2.11)) and conclude that $P \models (X \perp Y' \cup \{A\} \mid Z)$, that is, $(X \perp Y \mid Z)$.

The second case is where $X \cup Y \cup Z \neq \mathcal{X}$. Here, we might have that both X and Y are singletons. This case requires a similar argument that uses the induction hypothesis and properties of independence. We leave it as an exercise (exercise 4.9). ∎

Our previous results entail that, for positive distributions, the three conditions are equivalent.

Corollary 4.1

The following three statements are equivalent for a positive distribution P:

1. $P \models \mathcal{I}_\ell(\mathcal{H})$.
2. $P \models \mathcal{I}_p(\mathcal{H})$.
3. $P \models \mathcal{I}(\mathcal{H})$.

This equivalence relies on the positivity assumption. In particular, for nonpositive distributions, we can provide examples of a distribution P that satisfies one of these properties, but not the stronger one.

Example 4.5

Let P be any distribution over $\mathcal{X} = \{X_1, \ldots, X_n\}$; let $\mathcal{X}' = \{X_1', \ldots, X_n'\}$. We now construct a distribution $P'(\mathcal{X}, \mathcal{X}')$ whose marginal over $X_1, \ldots, X_n$ is the same as P, and where X_i' is deterministically equal to X_i. Let $\mathcal{H}$ be a Markov network over $\mathcal{X}, \mathcal{X}'$ that contains no edges other than X_i—X_i'. Then, in P', X_i is independent of the rest of the variables in the network given its neighbor X_i', and similarly for X_i'; thus, $\mathcal{H}$ satisfies the local independencies for every node in the network. Yet clearly $\mathcal{H}$ is not an I-map for P', since $\mathcal{H}$ makes many independence assertions regarding the X_i's that do not hold in P (or in P'). ∎

Thus, for nonpositive distributions, the local independencies do not imply the global ones.

A similar construction can be used to show that, for nonpositive distributions, the pairwise independencies do necessarily imply the local independencies.

Example 4.6

Let P be any distribution over $\mathcal{X} = \{X_1, \ldots, X_n\}$, and now consider two auxiliary sets of variables $\mathcal{X}'$ and $\mathcal{X}''$, and define $\mathcal{X}^* = \mathcal{X} \cup \mathcal{X}' \cup \mathcal{X}''$. We now construct a distribution $P'(\mathcal{X}^*)$ whose marginal over $X_1, \ldots, X_n$ is the same as P, and where X_i' and X_i'' are both deterministically equal to X_i. Let $\mathcal{H}$ be the empty Markov network over $\mathcal{X}^*$. We argue that this empty network satisfies the pairwise assumptions for every pair of nodes in the network. For example, X_i and X_i' are rendered independent because $\mathcal{X}^* - \{X_i, X_i'\}$ contains X_i''. Similarly, X_i and X_j are independent given X_i'. Thus, $\mathcal{H}$ satisfies the pairwise independencies, but not the local or global independencies. ∎

4.3.3 From Distributions to Graphs

Based on our deeper understanding of the independence properties associated with a Markov network, we can now turn to the question of encoding the independencies in a given distribution P using a graph structure. As for Bayesian networks, the notion of an I-map is not sufficient by itself: The complete graph implies no independence assumptions and is hence an I-map for any distribution. We therefore return to the notion of a minimal I-map, defined in definition 3.13, which was defined broadly enough to apply to Markov networks as well.

How can we construct a minimal I-map for a distribution P? Our discussion in section 4.3.2 immediately suggests two approaches for constructing a minimal I-map: one based on the pairwise Markov independencies, and the other based on the local independencies.

In the first approach, we consider the pairwise independencies. They assert that, if the edge $\{X, Y\}$ is not in $\mathcal{H}$, then X and Y must be independent given all other nodes in the graph, regardless of which other edges the graph contains. Thus, at the very least, to guarantee that $\mathcal{H}$ is an I-map, we must add direct edges between all pairs of nodes X and Y such that

$$P \not\models (X \perp Y \mid \mathcal{X} - \{X, Y\}). \tag{4.2}$$

We can now define $\mathcal{H}$ to include an edge X—Y for all X, Y for which equation (4.2) holds.

In the second approach, we use the local independencies and the notion of minimality. For each variable X, we define the neighbors of X to be a minimal set of nodes $\boldsymbol{Y}$ that render X independent of the rest of the nodes. More precisely, define:

Definition 4.12
Markov blanket

A set U is a Markov blanket *of X in a distribution P if $X \notin U$ and if U is a minimal set of nodes such that*

$$(X \perp \mathcal{X} - \{X\} - U \mid U) \in \mathcal{I}(P). \tag{4.3}$$

∎

We then define a graph $\mathcal{H}$ by introducing an edge $\{X, Y\}$ for all X and all $Y \in \text{MB}_P(X)$. As defined, this construction is not unique, since there may be several sets U satisfying equation (4.3). However, theorem 4.6 will show that there is only one such minimal set. In fact, we now show that any positive distribution P has a unique minimal I-map, and that both of these constructions produce this I-map.

We begin with the proof for the pairwise definition:

Theorem 4.5

Let P be a positive distribution, and let $\mathcal{H}$ be defined by introducing an edge $\{X, Y\}$ for all X, Y for which equation (4.2) holds. Then the Markov network $\mathcal{H}$ is the unique minimal I-map for P.

PROOF The fact that $\mathcal{H}$ is an I-map for P follows immediately from fact that P, by construction, satisfies $\mathcal{I}_p(\mathcal{H})$, and, therefore, by corollary 4.1, also satisfies $\mathcal{I}(\mathcal{H})$. The fact that it is minimal follows from the fact that if we eliminate some edge $\{X, Y\}$ from $\mathcal{H}$, the graph would imply the pairwise independence $(X \perp Y \mid \mathcal{X} - \{X, Y\})$, which we know to be false for P (otherwise, the edge would have been omitted in the construction of $\mathcal{H}$). The uniqueness of the minimal I-map also follows trivially: By the same argument, any other I-map $\mathcal{H}'$ for P must contain at least the edges in $\mathcal{H}$ and is therefore either equal to $\mathcal{H}$ or contains additional edges and is therefore not minimal.

∎

It remains to show that the second definition results in the same minimal I-map.

Theorem 4.6

Let P be a positive distribution. For each node X, let $\text{MB}_P(X)$ be a minimal set of nodes U satisfying equation (4.3). We define a graph $\mathcal{H}$ by introducing an edge $\{X, Y\}$ for all X and all $Y \in \text{MB}_P(X)$. Then the Markov network $\mathcal{H}$ is the unique minimal I-map for P.

The proof is left as an exercise (exercise 4.11).

Both of the techniques for constructing a minimal I-map make the assumption that the distribution P is positive. As we have shown, for nonpositive distributions, neither the pairwise independencies nor the local independencies imply the global one. Hence, for a nonpositive distribution P, constructing a graph $\mathcal{H}$ such that P satisfies the pairwise assumptions for $\mathcal{H}$ does not guarantee that $\mathcal{H}$ is an I-map for P. Indeed, we can easily demonstrate that both of these constructions break down for nonpositive distributions.

Example 4.7

Consider a nonpositive distribution P over four binary variables A, B, C, D that assigns nonzero probability only to cases where all four variables take on exactly the same value; for example, we might have $P(a^1, b^1, c^1, d^1) = 0.5$ and $P(a^0, b^0, c^0, d^0) = 0.5$. The graph $\mathcal{H}$ shown in figure 4.7 is one possible output of applying the local independence I-map construction algorithm to P: For example, $P \models (A \perp C, D \mid B)$, and hence $\{B\}$ is a legal choice for $\text{MB}_P(A)$. A similar analysis shows that this network satisfies the Markov blanket condition for all nodes. However, it is not an I-map for the distribution.

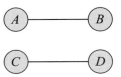

Figure 4.7   An attempt at an I-map for a nonpositive distribution P

If we use the pairwise independence I-map construction algorithm for this distribution, the network constructed is the empty network. For example, the algorithm would not place an edge between A and B, because $P \models (A \perp B \mid C, D)$. Exactly the same analysis shows that no edges will be placed into the graph. However, the resulting network is not an I-map for P.   ∎

Both these examples show that deterministic relations between variables can lead to failure in the construction based on local and pairwise independence. Suppose that A and B are two variables that are identical to each other and that both C and D are variables that correlated to both A and B so that $(C \perp D \mid A, B)$ holds. Since A is identical to B, we have that both $(A, D \perp C \mid B)$ and $(B, D \perp C \mid A)$ hold. In other words, it suffices to observe one of these two variables to capture the relevant information both have about C and separate C from D. In this case the Markov blanket of C is not uniquely defined. This ambiguity leads to the failure of both local and pairwise constructions. Clearly, identical variables are only one way of getting such ambiguities in local independencies. Once we allow nonpositive distribution, other distributions can have similar problems.

Having defined the notion of a minimal I-map for a distribution P, we can now ask to what extent it represents the independencies in P. More formally, we can ask whether every distribution has a perfect map. Clearly, the answer is no, even for positive distributions:

Example 4.8

Consider a distribution arising from a three-node Bayesian network with a v-structure, for example, the distribution induced in the Student *example over the nodes Intelligence, Difficulty, and Grade (figure 3.3). In the Markov network for this distribution, we must clearly have an edge between I and G and between D and G. Can we omit the edge between I and D? No, because we do not have that $(I \perp D \mid G)$ holds for the distribution; rather, we have the opposite: I and D are* dependent *given G. Therefore, the only minimal I-map for this P is the fully connected graph, which does not capture the marginal independence $(I \perp D)$ that holds in P.*   ∎

This example provides another counterexample to the strong version of completeness mentioned earlier. The only distributions for which separation is a sound and complete criterion for determining conditional independence are those for which $\mathcal{H}$ is a perfect map.

4.4   Parameterization Revisited

Now that we understand the semantics and independence properties of Markov networks, we revisit some alternative representations for the parameterization of a Markov network.

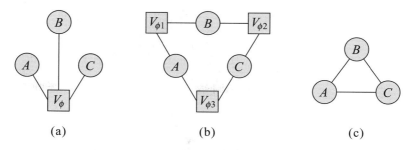

Figure 4.8 Different factor graphs for the same Markov network: (a) One factor graph over A, B, C, with a single factor over all three variables. (b) An alternative factor graph, with three pairwise factors. (c) The induced Markov network for both is a clique over A, B, C.

4.4.1 Finer-Grained Parameterization

4.4.1.1 Factor Graphs

A Markov network structure does not generally reveal all of the structure in a Gibbs parameterization. In particular, one cannot tell from the graph structure whether the factors in the parameterization involve maximal cliques or subsets thereof. Consider, for example, a Gibbs distribution P over a fully connected pairwise Markov network; that is, P is parameterized by a factor for each pair of variables $X, Y \in \mathcal{X}$. The clique potential parameterization would utilize a factor whose scope is the entire graph, and which therefore uses an exponential number of parameters. On the other hand, as we discussed in section 4.2.1, the number of parameters in the pairwise parameterization is quadratic in the number of variables. Note that the complete Markov network is not redundant in terms of conditional independencies — P does not factorize over any smaller network. Thus, although the finer-grained structure does not imply additional independencies in the distribution (see exercise 4.6), it is still very significant.

An alternative representation that makes this structure explicit is a *factor graph*. A factor graph is a graph containing two types of nodes: one type corresponds, as usual, to random variables; the other corresponds to factors over the variables. Formally:

Definition 4.13

factor graph

A factor graph $\mathcal{F}$ is an undirected graph containing two types of nodes: variable nodes *(denoted as ovals) and* factor nodes *(denoted as squares). The graph only contains edges between variable nodes and factor nodes. A factor graph $\mathcal{F}$ is parameterized by a set of factors, where each factor node V_ϕ is associated with precisely one factor ϕ, whose scope is the set of variables that are neighbors of V_ϕ*

factorization

in the graph. A distribution P factorizes over $\mathcal{F}$ if it can be represented as a set of factors of this form. ∎

Factor graphs make explicit the structure of the factors in the network. For example, in a fully connected pairwise Markov network, the factor graph would contain a factor node for each of the $\binom{n}{2}$ pairs of nodes; the factor node for a pair X_i, X_j would be connected to X_i and X_j; by contrast, a factor graph for a distribution with a single factor over $X_1, \ldots, X_n$ would have a single factor node connected to all of $X_1, \ldots, X_n$ (see figure 4.8). Thus, although the Markov networks for these two distributions are identical, their factor graphs make explicit the

$$\epsilon_1(A, B) \qquad\qquad \epsilon_2(B,C) \qquad\qquad \epsilon_3(C,D) \qquad\qquad \epsilon_4(D,A)$$

a^0	b^0	-3.4		b^0	c^0	-4.61		c^0	d^0	0		d^0	a^0	-4.61
a^0	b^1	-1.61		b^0	c^1	0		c^0	d^1	-4.61		d^0	a^1	0
a^1	b^0	0		b^1	c^0	0		c^1	d^0	-4.61		d^1	a^0	0
a^1	b^1	-2.3		b^1	c^1	-4.61		c^1	d^1	0		d^1	a^1	-4.61

$$\qquad(a)\qquad\qquad\qquad\qquad(b)\qquad\qquad\qquad\qquad(c)\qquad\qquad\qquad\qquad(d)$$

Figure 4.9   Energy functions for the Misconception **example**

difference in their factorization.

4.4.1.2   Log-Linear Models

Although factor graphs make certain types of structure more explicit, they still encode factors as complete tables over the scope of the factor. As in Bayesian networks, factors can also exhibit a type of context-specific structure — patterns that involve particular values of the variables. These patterns are often more easily seen in terms of an alternative parameterization of the factors that converts them into log-space.

More precisely, we can rewrite a factor $\phi(\boldsymbol{D})$ as

$$\phi(\boldsymbol{D}) = \exp(-\epsilon(\boldsymbol{D})),$$

energy function
where $\epsilon(\boldsymbol{D}) = -\ln \phi(\boldsymbol{D})$ is often called an *energy function*. The use of the word "energy" derives from statistical physics, where the probability of a physical state (for example, a configuration of a set of electrons), depends inversely on its energy. In this logarithmic representation, we have

$$P(X_1, \ldots, X_n) \propto \exp\left[-\sum_{i=1}^{m} \epsilon_i(\boldsymbol{D}_i) \right].$$

The logarithmic representation ensures that the probability distribution is positive. Moreover, the logarithmic parameters can take any value along the real line.

Any Markov network parameterized using positive factors can be converted to a logarithmic representation.

Example 4.9
Figure 4.9 shows the logarithmic representation of the clique potential parameters in figure 4.1. We can see that the "1" entries in the clique potentials translate into "0" entries in the energy function.∎

This representation makes certain types of structure in the potentials more apparent. For example, we can see that both $\epsilon_2(B,C)$ and $\epsilon_4(D,A)$ are constant multiples of an energy function that ascribes 1 to instantiations where the values of the two variables agree, and 0 to the instantiations where they do not.

We can provide a general framework for capturing such structure using the following notion:

Definition 4.14

feature

Let D be a subset of variables. We define a feature $f(D)$ *to be a function from* $Val(D)$ *to* $\mathbb{R}$. ■

A feature is simply a factor without the nonnegativity requirement. One type of feature of particular interest is the *indicator feature* that takes on value 1 for some values $y \in Val(D)$ and

indicator feature

0 otherwise.

Features provide us with an easy mechanism for specifying certain types of interactions more compactly.

Example 4.10

Consider a situation where A_1 and A_2 each have ℓ values $a^1, \ldots, a^\ell$. Assume that our distribution is such that we prefer situations where A_1 and A_2 take on the same value, but otherwise have no preference. Thus, our energy function might have the following form:

$$\epsilon(A_1, A_2) = \begin{cases} -3 & A_1 = A_2 \\ 0 & otherwise \end{cases}$$

Represented as a full factor, this clique potential requires ℓ^2 values. However, it can also be represented as a log-linear function in terms of a feature $f(A_1, A_2)$ that is an indicator function for the event $A_1 = A_2$. The energy function is then simply a constant multiple -3 of this feature.■

Thus, we can provide a more general definition for our notion of log-linear models:

Definition 4.15

log-linear model

A distribution P is a log-linear model *over a Markov network $\mathcal{H}$ if it is associated with:*

- *a set of features $\mathcal{F} = \{f_1(D_1), \ldots, f_k(D_k)\}$, where each D_i is a complete subgraph in $\mathcal{H}$,*
- *a set of weights $w_1, \ldots, w_k$,*

such that

$$P(X_1, \ldots, X_n) = \frac{1}{Z} \exp\left[-\sum_{i=1}^{k} w_i f_i(D_i) \right].$$

■

Note that we can have several features over the same scope, so that we can, in fact, represent a standard set of table potentials. (See exercise 4.13.)

The log-linear model provides a much more compact representation for many distributions, especially in situations where variables have large domains such as text (such as box 4.E).

4.4.1.3 Discussion

We now have three representations of the parameterization of a Markov network. The Markov network denotes a product over potentials on cliques. A factor graph denotes a product of factors. And a set of features denotes a product over feature weights. Clearly, each representation is finer-grained than the previous one and as rich. A factor graph can describe the Gibbs distribution, and a set of features can describe all the entries in each of the factors of a factor graph.

☞ **Depending on the question of interest, different representations may be more appropriate. For example, a Markov network provides the right level of abstraction for discussing independence queries: The finer-grained representations of factor graphs or log-linear**

models do not change the independence assertions made by the model. On the other hand, as we will see in later chapters, factor graphs are useful when we discuss inference, and features are useful when we discuss parameterizations, both for hand-coded models and for learning.

Ising model

Box 4.C — Concept: Ising Models and Boltzmann Machines. *One of the earliest types of Markov network models is the* Ising *model, which first arose in statistical physics as a model for the energy of a physical system involving a system of interacting atoms. In these systems, each atom is associated with a binary-valued random variable $X_i \in \{+1, -1\}$, whose value defines the direction of the atom's spin. The energy function associated with the edges is defined by a particularly simple parametric form:*

$$\epsilon_{i,j}(x_i, x_j) = w_{i,j} x_i x_j \tag{4.4}$$

This energy is symmetric in X_i, X_j; it makes a contribution of $w_{i,j}$ to the energy function when $X_i = X_j$ (so both atoms have the same spin) and a contribution of $-w_{i,j}$ otherwise. Our model also contains a set of parameters u_i that encode individual node potentials; these bias individual variables to have one spin or another.

As usual, the energy function defines the following distribution:

$$P(\xi) = \frac{1}{Z} \exp\left(-\sum_{i<j} w_{i,j} x_i x_j - \sum_i u_i x_i \right).$$

As we can see, when $w_{i,j} > 0$ the model prefers to align the spins of the two atoms; in this case, the interaction is called ferromagnetic. *When $w_{i,j} < 0$ the interaction is called* antiferromagnetic. *When $w_{i,j} = 0$ the atoms are* non-interacting.

Much work has gone into studying particular types of Ising models, attempting to answer a variety of questions, usually as the number of atoms goes to infinity. For example, we might ask the probability of a configuration in which a majority of the spins are $+1$ or -1, versus the probability of more mixed configurations. The answer to this question depends heavily on the strength of the interaction between the variables; so, we can consider adapting this strength (by multiplying all weights by a temperature parameter*) and asking whether this change causes a phase transition in the probability of skewed versus mixed configurations. These questions, and many others, have been investigated extensively by physicists, and the answers are known (in some cases even analytically) for several cases.*

temperature parameter

Boltzmann distribution

Related to the Ising model is the Boltzmann *distribution; here, the variables are usually taken to have values $\{0, 1\}$, but still with the energy form of equation (4.4). Here, we get a nonzero contribution to the model from an edge (X_i, X_j) only when $X_i = X_j = 1$; however, the resulting energy can still be reformulated in terms of an Ising model (exercise 4.12).*

The popularity of the Boltzmann machine was primarily driven by its similarity to an activation model for neurons. To understand the relationship, we note that the probability distribution over each variable X_i given an assignment to is neighbors is $\text{sigmoid}(z)$ where

$$z = -\left(\sum_j w_{i,j} x_j\right) - w_i.$$

This function is a sigmoid of a weighted combination of X_i's neighbors, weighted by the strength and direction of the connections between them. This is the simplest but also most popular mathematical approximation of the function employed by a neuron in the brain. Thus, if we imagine a process by which the network continuously adapts its assignment by resampling the value of each variable as a stochastic function of its neighbors, then the "activation" probability of each variable resembles a neuron's activity. This model is a very simple variant of a stochastic, recurrent neural network.

---

labeling MRF

Box 4.D — Concept: Metric MRFs. *One important class of MRFs comprises those used for labeling. Here, we have a graph of nodes $X_1, \ldots, X_n$ related by a set of edges $\mathcal{E}$, and we wish to assign to each X_i a label in the space $\mathcal{V} = \{v_1, \ldots, v_K\}$. Each node, taken in isolation, has its preferences among the possible labels. However, we also want to impose a soft "smoothness" constraint over the graph, in that neighboring nodes should take "similar" values.*

We encode the individual node preferences as node potentials in a pairwise MRF and the smoothness preferences as edge potentials. For reasons that will become clear, it is traditional to encode these models in negative log-space, using energy functions. As our objective in these models is inevitably the MAP objective, we can also ignore the partition function, and simply consider the energy function:

$$E(x_1, \ldots, x_n) = \sum_i \epsilon_i(x_i) + \sum_{(i,j) \in \mathcal{E}} \epsilon_{i,j}(x_i, x_j). \tag{4.5}$$

Our goal is then to minimize the energy:

$$\arg \min_{x_1, \ldots, x_n} E(x_1, \ldots, x_n).$$

We now need to provide a formal definition for the intuition of "smoothness" described earlier. There are many different types of conditions that we can impose; different conditions allow different methods to be applied.

Ising model

One of the simplest in this class of models is a slight variant of the Ising model, where we have that, for any i, j:

$$\epsilon_{i,j}(x_i, x_j) = \begin{cases} 0 & x_i = x_j \\ \lambda_{i,j} & x_i \neq x_j, \end{cases} \tag{4.6}$$

for $\lambda_{i,j} \geq 0$. In this model, we obtain the lowest possible pairwise energy (0) when two neighboring nodes X_i, X_j take the same value, and a higher energy $\lambda_{i,j}$ when they do not.

Potts model

This simple model has been generalized in many ways. The Potts model extends it to the setting of more than two labels. An even broader class contains models where we have a distance function on the labels, and where we prefer neighboring nodes to have labels that are a smaller distance apart. More precisely, a function $\mu : \mathcal{V} \times \mathcal{V} \mapsto [0, \infty)$ is a metric if it satisfies:

metric function

- **Reflexivity:** $\mu(v_k, v_l) = 0$ *if and only if $k = l$;*
- **Symmetry:** $\mu(v_k, v_l) = \mu(v_l, v_k)$;

$$\epsilon_1'(A, B) \qquad\qquad \epsilon_2'(B, C)$$

a^0	b^0	-4.4		b^0	c^0	-3.61
a^0	b^1	-1.61		b^0	c^1	$+1$
a^1	b^0	-1		b^1	c^0	0
a^1	b^1	-2.3		b^1	c^1	-4.61

(a)   (b)

Figure 4.10   **Alternative but equivalent energy functions**

- **Triangle Inequality:** $\mu(v_k, v_l) + \mu(v_l, v_m) \geq \mu(v_k, v_m)$.

semimetric

We say that μ is a semimetric *if it satisfies reflexivity and symmetry. We can now define a* metric *MRF (or a semimetric MRF) by defining $\epsilon_{i,j}(v_k, v_l) = \mu(v_k, v_l)$ for all i, j, where μ is a metric (semimetric). We note that, as defined, this model assumes that the distance metric used is the same for all pairs of variables. This assumption is made because it simplifies notation, it often holds in practice, and it reduces the number of parameters that must be acquired. It is not required for the inference algorithms that we present in later chapters. Metric interactions arise in many applications, and play a particularly important role in computer vision (see box 4.B and box 13.B).*

truncated norm

For example, one common metric used is some form of truncated p-norm *(usually $p = 1$ or $p = 2$):*

$$\epsilon(x_i, x_j) = \min(c\|x_i - x_j\|_p, \text{dist}_{\max}). \tag{4.7}$$

4.4.2   Overparameterization

Even if we use finer-grained factors, and in some cases, even features, the Markov network parameterization is generally overparameterized. That is, for any given distribution, there are multiple choices of parameters to describe it in the model. Most obviously, if our graph is a single clique over n binary variables $X_1, \ldots, X_n$, then the network is associated with a clique potential that has 2^n parameters, whereas the joint distribution only has $2^n - 1$ independent parameters.

A more subtle point arises in the context of a nontrivial clique structure. Consider a pair of cliques $\{A, B\}$ and $\{B, C\}$. The energy function $\epsilon_1(A, B)$ (or its corresponding clique potential) contains information not only about the interaction between A and B, but also about the distribution of the individual variables A and B. Similarly, $\epsilon_2(B, C)$ gives us information about the individual variables B and C. The information about B can be placed in either of the two cliques, or its contribution can be split between them in arbitrary ways, resulting in many different ways of specifying the same distribution.

Example 4.11

Consider the energy functions $\epsilon_1(A, B)$ and $\epsilon_2(B, C)$ in figure 4.9. The pair of energy functions shown in figure 4.10 result in an equivalent distribution: Here, we have simply subtracted 1 from $\epsilon_1(A, B)$ and added 1 to $\epsilon_2(B, C)$ for all instantiations where $B = b^0$. It is straightforward to *check that this results in an identical distribution to that of figure 4.9. In instances where $B \neq b^0$ the energy function returns exactly the same value as before. In cases where $B = b^0$, the actual values of the energy functions have changed. However, because the sum of the energy functions on each instance is identical to the original sum, the probability of the instance will not change.* ∎

Intuitively, the standard Markov network representation gives us too many places to account for the influence of variables in shared cliques. Thus, the same distribution can be represented as a Markov network (of a given structure) in infinitely many ways. It is often useful to pick one of this infinite set as our chosen parameterization for the distribution.

4.4.2.1 Canonical Parameterization

canonical parameterization

The *canonical parameterization* provides one very natural approach to avoiding this ambiguity in the parameterization of a Gibbs distribution P. This canonical parameterization requires that the distribution P be positive. It is most convenient to describe this parameterization using energy functions rather then clique potentials. For this reason, it is also useful to consider a log-transform of P: For any assignment ξ to $\mathcal{X}$, we use $\ell(\xi)$ to denote $\ln P(\xi)$. This transformation is well defined because of our positivity assumption.

The canonical parameterization of a Gibbs distribution over $\mathcal{H}$ is defined via a set of energy functions over all cliques. Thus, for example, the Markov network in figure 4.4b would have energy functions for the two cliques $\{A, B, D\}$ and $\{B, C, D\}$, energy functions for all possible pairs of variables except the pair $\{A, C\}$ (a total of five pairs), energy functions for all four singleton sets, and a constant energy function for the empty clique.

At first glance, it appears that we have only increased the number of parameters in the specification. However, as we will see, this approach uniquely associates the interaction parameters for a subset of variables with that subset, avoiding the ambiguity described earlier. As a consequence, many of the parameters in this canonical parameterization are often zero.

The canonical parameterization is defined relative to a particular fixed assignment $\xi^* = (x_1^*, \ldots, x_n^*)$ to the network variables $\mathcal{X}$. This assignment can be chosen arbitrarily. For any subset of variables $\boldsymbol{Z}$, and any assignment $\boldsymbol{x}$ to some subset of $\mathcal{X}$ that contains $\boldsymbol{Z}$, we define the assignment $\boldsymbol{x}_{\boldsymbol{Z}}$ to be $\boldsymbol{x}\langle \boldsymbol{Z} \rangle$, that is, the assignment in $\boldsymbol{x}$ to the variables in $\boldsymbol{Z}$. Conversely, we define $\xi^*_{-\boldsymbol{Z}}$ to be $\xi^*\langle \mathcal{X} - \boldsymbol{Z} \rangle$, that is, the assignment in ξ^* to the variables outside $\boldsymbol{Z}$. We can now construct an assignment $(\boldsymbol{x}_{\boldsymbol{Z}}, \xi^*_{-\boldsymbol{Z}})$ that keeps the assignments to the variables in $\boldsymbol{Z}$ as specified in $\boldsymbol{x}$, and augments it using the default values in ξ^*.

canonical energy function

The *canonical energy function* for a clique $\boldsymbol{D}$ is now defined as follows:

$$\epsilon_{\boldsymbol{D}}^*(\boldsymbol{d}) = \sum_{\boldsymbol{Z} \subseteq \boldsymbol{D}} (-1)^{|\boldsymbol{D} - \boldsymbol{Z}|} \ell(\boldsymbol{d}_{\boldsymbol{Z}}, \xi^*_{-\boldsymbol{Z}}), \tag{4.8}$$

where the sum is over all subsets of $\boldsymbol{D}$, including $\boldsymbol{D}$ itself and the empty set $\emptyset$. Note that all of the terms in the summation have a scope that is contained in $\boldsymbol{D}$, which in turn is part of a clique, so that these energy functions are legal relative to our Markov network structure.

This formula performs an inclusion-exclusion computation. For a set $\{A, B, C\}$, it first subtracts out the influence of all of the pairs: $\{A, B\}$, $\{B, C\}$, and $\{C, A\}$. However, this process oversubtracts the influence of the individual variables. Thus, their influence is added back in, to compensate. More generally, consider any subset of variables $\boldsymbol{Z} \subseteq \boldsymbol{D}$. Intuitively, it

$$\epsilon_1^*(A, B) \qquad\qquad \epsilon_2^*(B, C) \qquad\qquad \epsilon_3^*(C, D) \qquad\qquad \epsilon_4^*(D, A)$$

a^0	b^0	0	b^0	c^0	0	c^0	d^0	0	d^0	a^0	0
a^0	b^1	0	b^0	c^1	0	c^0	d^1	0	d^0	a^1	0
a^1	b^0	0	b^1	c^0	0	c^1	d^0	0	d^1	a^0	0
a^1	b^1	4.09	b^1	c^1	9.21	c^1	d^1	-9.21	d^1	a^1	9.21

$$\epsilon_5^*(A) \qquad\qquad \epsilon_6^*(B) \qquad\quad \epsilon_7^*(C) \qquad\quad \epsilon_8^*(D) \qquad\quad \epsilon_9^*(\emptyset)$$

a^0	0	b^0	0	c^0	0	d^0	0	
a^1	-8.01	b^1	-6.4	c^1	0	d^1	0	-3.18

Figure 4.11   Canonical energy function for the Misconception **example**

makes a "contribution" once for every subset $U \supseteq Z$. Except for $U = D$, the number of times that Z appears is even — there is an even number of subsets $U \supseteq Z$ — and the number of times it appears with a positive sign is equal to the number of times it appears with a negative sign. Thus, we have effectively eliminated the net contribution of the subsets from the canonical energy function.

Let us consider the effect of the canonical transformation on our Misconception network.

Example 4.12

Let us choose (a^0, b^0, c^0, d^0) as our arbitrary assignment on which to base the canonical parameterization. The resulting energy functions are shown in figure 4.11. For example, the energy value $\epsilon_1^*(a^1, b^1)$ was computed as follows:

$$\ell(a^1, b^1, c^0, d^0) - \ell(a^1, b^0, c^0, d^0) - \ell(a^0, b^1, c^0, d^0) + \ell(a^0, b^0, c^0, d^0) =$$
$$- 13.49 - -11.18 - -9.58 + -3.18 = 4.09$$

Note that many of the entries in the energy functions are zero. As discussed earlier, this phenomenon is fairly general, and occurs because we have accounted for the influence of small subsets of variables separately, leaving the larger factors to deal only with higher-order influences. We also note that these canonical parameters are not very intuitive, highlighting yet again the difficulties of constructing a reasonable parameterization of a Markov network by hand. ∎

This canonical parameterization defines the same distribution as our original distribution P:

Theorem 4.7

Let P be a positive Gibbs distribution over $\mathcal{H}$, and let $\epsilon^*(D_i)$ for each clique D_i be defined as specified in equation (4.8). Then

$$P(\xi) = \exp\left[-\sum_i \epsilon_{D_i}^*(\xi\langle D_i\rangle) \right].$$

The proof for the case where $\mathcal{H}$ consists of a single clique is fairly simple, and it is left as an exercise (exercise 4.4). The general case follows from results in the next section.

The canonical parameterization gives us the tools to prove the Hammersley-Clifford theorem, which we restate for convenience.

Theorem 4.8

Let P be a positive distribution over $\mathcal{X}$, and $\mathcal{H}$ a Markov network graph over $\mathcal{X}$. If $\mathcal{H}$ is an I-map for P, then P is a Gibbs distribution over $\mathcal{H}$.

PROOF To prove this result, we need to show the existence of a Gibbs parameterization for any distribution P that satisfies the Markov assumptions associated with $\mathcal{H}$. The proof is constructive, and simply uses the canonical parameterization shown earlier in this section. Given P, we define an energy function for all subsets D of nodes in the graph, regardless of whether they are cliques in the graph. This energy function is defined exactly as in equation (4.8), relative to some specific fixed assignment ξ^* used to define the canonical parameterization. The distribution defined using this set of energy functions is P: the argument is identical to the proof of theorem 4.7, for the case where the graph consists of a single clique (see exercise 4.4).

It remains only to show that the resulting distribution is a Gibbs distribution over $\mathcal{H}$. To show that, we need to show that the factors $\epsilon^*(D)$ are identically 0 whenever D is not a clique in the graph, that is, whenever the nodes in D do not form a fully connected subgraph. Assume that we have $X, Y \in D$ such that there is no edge between X and Y. For this proof, it helps to introduce the notation

$$\sigma_Z[\boldsymbol{x}] = (\boldsymbol{x}_Z, \xi^*_{-Z}).$$

Plugging this notation into equation (4.8), we have that:

$$\epsilon^*_D(\boldsymbol{d}) = \sum_{Z \subseteq D} (-1)^{|D-Z|} \ell(\sigma_Z[\boldsymbol{d}]).$$

We now rearrange the sum over subsets Z into a sum over groups of subsets. Let $W \subseteq D - \{X, Y\}$; then W, $W \cup \{X\}$, $W \cup \{Y\}$, and $W \cup \{X, Y\}$ are all subsets of Z. Hence, we can rewrite the summation over subsets of D as a summation over subsets of $D - \{X, Y\}$:

$$\epsilon^*_D(\boldsymbol{d}) = \sum_{W \subseteq D - \{X,Y\}} (-1)^{|D - \{X,Y\} - W|} \tag{4.9}$$
$$(\ell(\sigma_W[\boldsymbol{d}]) - \ell(\sigma_{W \cup \{X\}}[\boldsymbol{d}]) - \ell(\sigma_{W \cup \{Y\}}[\boldsymbol{d}]) + \ell(\sigma_{W \cup \{X,Y\}}[\boldsymbol{d}])).$$

Now consider a specific subset W in this sum, and let $\boldsymbol{u}^*$ be $\xi^*\langle \mathcal{X} - D \rangle$ — the assignment to $\mathcal{X} - D$ in ξ. We now have that:

$$
\begin{aligned}
\ell(\sigma_{W \cup \{X,Y\}}[\boldsymbol{d}])) - \ell(\sigma_{W \cup \{X\}}[\boldsymbol{d}]) &= \ln \frac{P(x, y, \boldsymbol{w}, \boldsymbol{u}^*)}{P(x, y^*, \boldsymbol{w}, \boldsymbol{u}^*)} \\
&= \ln \frac{P(y \mid x, \boldsymbol{w}, \boldsymbol{u}^*) P(x, \boldsymbol{w}, \boldsymbol{u}^*)}{P(y^* \mid x, \boldsymbol{w}, \boldsymbol{u}^*) P(x, \boldsymbol{w}, \boldsymbol{u}^*)} \\
&= \ln \frac{P(y \mid x^*, \boldsymbol{w}, \boldsymbol{u}^*) P(x, \boldsymbol{w}, \boldsymbol{u}^*)}{P(y^* \mid x^*, \boldsymbol{w}, \boldsymbol{u}^*) P(x, \boldsymbol{w}, \boldsymbol{u}^*)} \\
&= \ln \frac{P(y \mid x^*, \boldsymbol{w}, \boldsymbol{u}^*) P(x^*, \boldsymbol{w}, \boldsymbol{u}^*)}{P(y^* \mid x^*, \boldsymbol{w}, \boldsymbol{u}^*) P(x^*, \boldsymbol{w}, \boldsymbol{u}^*)} \\
&= \ln \frac{P(x^*, y, \boldsymbol{w}, \boldsymbol{u}^*)}{P(x^*, y^*, \boldsymbol{w}, \boldsymbol{u}^*)} \\
&= \ell(\sigma_{W \cup \{Y\}}[\boldsymbol{d}])) - \ell(\sigma_W[\boldsymbol{d}]),
\end{aligned}
$$

where the third equality is a consequence of the fact that X and Y are not connected directly by an edge, and hence we have that $P \models (X \perp Y \mid \mathcal{X} - \{X, Y\})$. Thus, we have that each term in the outside summation in equation (4.9) adds to zero, and hence the summation as a whole is also zero, as required.   ∎

For positive distributions, we have already shown that all three sets of Markov assumptions are equivalent; putting these results together with theorem 4.1 and theorem 4.2, we obtain that, **for positive distributions, all four conditions — factorization and the three types of Markov assumptions — are all equivalent.**

☞

4.4.2.2   Eliminating Redundancy

An alternative approach to the issue of overparameterization is to try to eliminate it entirely. We can do so in the context of a feature-based representation, which is sufficiently fine-grained to allow us to eliminate redundancies without losing expressive power. The tools for detecting and eliminating redundancies come from linear algebra.

linear
dependence

We say that a set of features $f_1, \ldots, f_k$ is *linearly dependent* if there are constants $\alpha_0, \alpha_1, \ldots, \alpha_k$, not all of which are 0, so that for all ξ

$$\alpha_0 + \sum_i \alpha_i f_i(\xi) = 0.$$

This is the usual definition of linear dependencies in linear algebra, where we view each feature as a vector whose entries are the value of the feature in each of the possible instantiations.

Example 4.13

Consider again the Misconception *example. We can encode the log-factors in example 4.9 as a set of features by introducing indicator features of the form:*

$$f_{a,b}(A, B) = \begin{cases} 1 & A = a, B = b \\ 0 & otherwise. \end{cases}$$

Thus, to represent $\epsilon_1(A, B)$, we introduce four features that correspond to the four entries in the energy function. Since A, B take on exactly one of these possible four values, we have that

$$f_{a^0,b^0}(A, B) + f_{a^0,b^1}(A, B) + f_{a^1,b^0}(A, B) + f_{a^1,b^1}(A, B) = 1.$$

Thus, this set of features is linearly dependent.   ∎

Example 4.14

Now consider also the features that capture $\epsilon_2(B, C)$ and their interplay with the features that capture $\epsilon_1(A, B)$. We start by noting that the sum $f_{a^0,b^0}(A, B) + f_{a^1,b^0}(A, B)$ is equal to 1 when $B = b^0$ and 0 otherwise. Similarly, $f_{b^0,c^0}(B, C) + f_{b^0,c^1}(B, C)$ is also an indicator for $B = b^0$. Thus we get that

$$f_{a^0,b^0}(A, B) + f_{a^1,b^0}(A, B) - f_{b^0,c^0}(B, C) - f_{b^0,c^1}(B, C) = 0.$$

And so these four features are linearly dependent.   ∎

As we now show, linear dependencies imply non-unique parameterization.

Proposition 4.5

Let $f_1, \ldots, f_k$ be a set of features with weights $\boldsymbol{w} = \{w_1, \ldots, w_k\}$ that form a log-linear representation of a distribution P. If there are coefficients $\alpha_0, \alpha_1, \ldots, \alpha_k$ such that for all ξ

$$\alpha_0 + \sum_i \alpha_i f_i(\xi) = 0 \tag{4.10}$$

then the log-linear model with weights $\boldsymbol{w}' = \{w_1 + \alpha_1, \ldots, w_k + \alpha_k\}$ also represents P.

PROOF Consider the distribution

$$P_{\boldsymbol{w}'}(\xi) \propto \exp\left\{ -\sum_i (w_i + \alpha_i) f_i(\xi) \right\}.$$

Using equation (4.10) we see that

$$-\sum_i (w_i + \alpha_i) f_i(\xi) = \alpha_0 - \sum_i w_i f_i(\xi).$$

Thus,

$$P_{\boldsymbol{w}'}(\xi) \propto e^{\alpha_0} \exp\left\{ -\sum_i w_i f_i(\xi) \right\} \propto P(\xi).$$

We conclude that $P_{\boldsymbol{w}'}(\xi) = P(\xi)$. ∎

redundant

Motivated by this result, we say that a set of linearly dependent features is *redundant*. A *nonredundant* set of features is one where the features are not linearly dependent on each other. In fact, if the set of features is nonredundant, then each set of weights describes a unique distribution.

Proposition 4.6

Let $f_1, \ldots, f_k$ be a set of nonredundant features, and let $\boldsymbol{w}, \boldsymbol{w}' \in \mathbb{R}^k$. If $\boldsymbol{w} \neq \boldsymbol{w}'$ then $P_{\boldsymbol{w}} \neq P_{\boldsymbol{w}'}$.

Example 4.15

Can we construct a nonredundant set of features for the Misconception *example? We can determine the number of nonredundant features by building the 16×16 matrix of the values of the 16 features (four factors with four features each) in the 16 instances of the joint distribution. This matrix has rank of 9, which implies that a subset of 8 features will be a nonredundant subset. In fact, there are several such subsets. In particular, the canonical parameterization shown in figure 4.11 has nine features of nonzero weight, which form a nonredundant parameterization. The equivalence of the canonical parameterization (theorem 4.7) implies that this set of features has the same expressive power as the original set of features. To verify this, we can show that adding any other feature will lead to a linear dependency. Consider, for example, the feature f_{a^1, b^0}. We can verify that*

$$f_{a^1, b^0} + f_{a^1, b^1} - f_{a^1} = 0.$$

Similarly, consider the feature f_{a^0, b^0}. Again we can find a linear dependency on other features:

$$f_{a^0, b^0} + f_{a^1} + f_{b^1} - f_{a^1, b^1} = 1.$$

Using similar arguments, we can show that adding any of the original features will lead to redundancy. Thus, this set of features can represent any parameterization in the original model. ∎

4.5 Bayesian Networks and Markov Networks

We have now described two graphical representation languages: Bayesian networks and Markov networks. Example 3.8 and example 4.8 show that these two representations are incomparable as a language for representing independencies: each can represent independence constraints that the other cannot. In this section, we strive to provide more insight about the relationship between these two representations.

4.5.1 From Bayesian Networks to Markov Networks

Let us begin by examining how we might take a distribution represented using one of these frameworks, and represent it in the other. One can view this endeavor from two different perspectives: Given a Bayesian network $\mathcal{B}$, we can ask how to represent the distribution $P_{\mathcal{B}}$ as a parameterized Markov network; or, given a graph $\mathcal{G}$, we can ask how to represent the independencies in $\mathcal{G}$ using an undirected graph $\mathcal{H}$. In other words, we might be interested in finding a minimal I-map for a distribution $P_{\mathcal{B}}$, or a minimal I-map for the independencies $\mathcal{I}(\mathcal{G})$. We can see that these two questions are related, but each perspective offers its own insights.

Let us begin by considering a distribution $P_{\mathcal{B}}$, where $\mathcal{B}$ is a parameterized Bayesian network over a graph $\mathcal{G}$. Importantly, the parameterization of $\mathcal{B}$ can also be viewed as a parameterization for a Gibbs distribution: We simply take each CPD $P(X_i \mid \mathrm{Pa}_{X_i})$ and view it as a factor of scope X_i, Pa_{X_i}. This factor satisfies additional normalization properties that are not generally true of all factors, but it is still a legal factor. This set of factors defines a Gibbs distribution, one whose partition function happens to be 1.

What is more important, **a Bayesian network conditioned on evidence $E = e$ also induces a Gibbs distribution: the one defined by the original factors *reduced* to the context $E = e$.**

☞

Proposition 4.7

Let $\mathcal{B}$ be a Bayesian network over $\mathcal{X}$ and $E = e$ an observation. Let $W = \mathcal{X} - E$. Then $P_{\mathcal{B}}(W \mid e)$ is a Gibbs distribution defined by the factors $\Phi = \{\phi_{X_i}\}_{X_i \in \mathcal{X}}$, where

$$\phi_{X_i} = P_{\mathcal{B}}(X_i \mid \mathrm{Pa}_{X_i})[E = e].$$

The partition function for this Gibbs distribution is $P(e)$.

The proof follows directly from the definitions. This result allows us to view any Bayesian network conditioned as evidence as a Gibbs distribution, and to bring to bear techniques developed for analysis of Markov networks.

What is the structure of the undirected graph that can serve as an I-map for a set of factors in a Bayesian network? In other words, what is the I-map for the Bayesian network structure $\mathcal{G}$? Going back to our construction, we see that we have created a factor for each family of X_i, containing all the variables in the family. Thus, in the undirected I-map, we need to have an edge between X_i and each of its parents, as well as between all of the parents of X_i. This observation motivates the following definition:

Definition 4.16

moralized graph

The moral graph $\mathcal{M}[\mathcal{G}]$ of a Bayesian network structure $\mathcal{G}$ over $\mathcal{X}$ is the undirected graph over $\mathcal{X}$ that contains an undirected edge between X and Y if: (a) there is a directed edge between them (in either direction), or (b) X and Y are both parents of the same node.[1] ∎

---

1. The name *moralized graph* originated because of the supposed "morality" of marrying the parents of a node.

For example, figure 4.6a shows the moralized graph for the extended $\mathcal{B}^{student}$ network of figure 9.8.

The preceding discussion shows the following result:

Corollary 4.2

Let $\mathcal{G}$ be a Bayesian network structure. Then for any distribution $P_{\mathcal{B}}$ such that $\mathcal{B}$ is a parameterization of $\mathcal{G}$, we have that $\mathcal{M}[\mathcal{G}]$ is an I-map for $P_{\mathcal{B}}$.

One can also view the moralized graph construction purely from the perspective of the independencies encoded by a graph, avoiding completely the discussion of parameterizations of the network.

Proposition 4.8

Let $\mathcal{G}$ be any Bayesian network graph. The moralized graph $\mathcal{M}[\mathcal{G}]$ is a minimal I-map for $\mathcal{G}$.

PROOF We want to build a Markov network $\mathcal{H}$ such that $\mathcal{I}(\mathcal{H}) \subseteq \mathcal{I}(\mathcal{G})$, that is, that $\mathcal{H}$ is an I-map for $\mathcal{G}$ (see definition 3.3). We use the algorithm for constructing minimal I-maps based on the Markov independencies. Consider a node X in $\mathcal{X}$: our task is to select as X's neighbors the smallest set of nodes U that are needed to render X independent of all other nodes in the

Markov blanket

network. We define the *Markov blanket* of X in a Bayesian network $\mathcal{G}$, denoted $\mathrm{MB}_{\mathcal{G}}(X)$, to be the nodes consisting of X's parents, X's children, and other parents of X's children. We now need to show that $\mathrm{MB}_{\mathcal{G}}(X)$ d-separates X from all other variables in $\mathcal{G}$; and that no subset of $\mathrm{MB}_{\mathcal{G}}(X)$ has that property. The proof uses straightforward graph-theoretic properties of trails, and it is left as an exercise (exercise 4.14). ∎

Now, let us consider how "close" the moralized graph is to the original graph $\mathcal{G}$. Intuitively, **the addition of the moralizing edges to the Markov network $\mathcal{H}$ leads to the loss of independence information implied by the graph structure.** For example, if our Bayesian network $\mathcal{G}$ has the form $X \to Z \leftarrow Y$, with no edge between X and Y, the Markov network $\mathcal{M}[\mathcal{G}]$ loses the information that X and Y are marginally independent (not given Z). However, information is not always lost. Intuitively, moralization causes loss of information about independencies only

moral graph

when it introduces new edges into the graph. We say that a Bayesian network $\mathcal{G}$ is *moral* if it contains no immoralities (as in definition 3.11); that is, for any pair of variables X, Y that share a child, there is a covering edge between X and Y. It is not difficult to show that:

Proposition 4.9

If the directed graph $\mathcal{G}$ is moral, then its moralized graph $\mathcal{M}[\mathcal{G}]$ is a perfect map of $\mathcal{G}$.

PROOF Let $\mathcal{H} = \mathcal{M}[\mathcal{G}]$. We have already shown that $\mathcal{I}(\mathcal{H}) \subseteq \mathcal{I}(\mathcal{G})$, so it remains to show the opposite inclusion. Assume by contradiction that there is an independence $(\boldsymbol{X} \perp \boldsymbol{Y} \mid \boldsymbol{Z}) \in \mathcal{I}(\mathcal{G})$ which is not in $\mathcal{I}(\mathcal{H})$. Thus, there must exist some trail from $\boldsymbol{X}$ to $\boldsymbol{Y}$ in $\mathcal{H}$ which is active given $\boldsymbol{Z}$. Consider some such trail that is minimal, in the sense that it has no shortcuts. As $\mathcal{H}$ and $\mathcal{G}$ have precisely the same edges, the same trail must exist in $\mathcal{G}$. As, by assumption, it cannot be active in $\mathcal{G}$ given $\boldsymbol{Z}$, we conclude that it must contain a v-structure $X_1 \to X_2 \leftarrow X_3$. However, because $\mathcal{G}$ is moralized, we also have some edge between X_1 and X_3, contradicting the assumption that the trail is minimal. ∎

Thus, a moral graph $\mathcal{G}$ can be converted to a Markov network without losing independence assumptions. This conclusion is fairly intuitive, inasmuch as the only independencies in $\mathcal{G}$ that are not present in an undirected graph containing the same edges are those corresponding to v-structures. But if any v-structure can be short-cut, it induces no independencies that are not represented in the undirected graph.

We note, however, that very few directed graphs are moral. For example, assume that we have a v-structure $X \rightarrow Y \leftarrow Z$, which is moral due to the existence of an arc $X \rightarrow Z$. If Z has another parent W, it also has a v-structure $X \rightarrow Z \leftarrow W$, which, to be moral, requires some edge between X and W. We return to this issue in section 4.5.3.

4.5.1.1 Soundness of d-Separation

The connection between Bayesian networks and Markov networks provides us with the tools for proving the soundness of the d-separation criterion in Bayesian networks.

The idea behind the proof is to leverage the soundness of separation in undirected graphs, a result which (as we showed) is much easier to prove. Thus, we want to construct an undirected graph $\mathcal{H}$ such that active paths in $\mathcal{H}$ correspond to active paths in $\mathcal{G}$. A moment of thought shows that the moralized graph is not the right construct, because there are paths in the undirected graph that correspond to v-structures in $\mathcal{G}$ that may or may not be active. For example, if our graph $\mathcal{G}$ is $X \rightarrow Z \leftarrow Y$ and Z is not observed, d-separation tells us that X and Y are independent; but the moralized graph for $\mathcal{G}$ is the complete undirected graph, which does not have the same independence.

barren node

Therefore, to show the result, we first want to eliminate v-structures that are not active, so as to remove such cases. To do so, we first construct a subgraph where remove all *barren nodes* from the graph, thereby also removing all v-structures that do not have an observed descendant. The elimination of the barren nodes does not change the independence properties of the distribution over the remaining variables, but does eliminate paths in the graph involving v-structures that are not active. If we now consider only the subgraph, we can reduce d-separation to separation and utilize the soundness of separation to show the desired result.

upward closure

We first use these intuitions to provide an alternative formulation for d-separation. Recall that in definition 2.14 we defined the *upward closure* of a set of nodes $\boldsymbol{U}$ in a graph to be $\boldsymbol{U} \cup Ancestors_{\boldsymbol{U}}$. Letting $\boldsymbol{U}^*$ be the closure of a set $\boldsymbol{U}$, we can define the network induced over $\boldsymbol{U}^*$; importantly, as all parents of every node in $\boldsymbol{U}^*$ are also in $\boldsymbol{U}^*$, we have all the variables mentioned in every CPD, so that the induced graph defines a coherent probability distribution. We let $\mathcal{G}^+[\boldsymbol{U}]$ be the induced Bayesian network over $\boldsymbol{U}$ and its ancestors.

Proposition 4.10 *Let $\boldsymbol{X}, \boldsymbol{Y}, \boldsymbol{Z}$ be three disjoint sets of nodes in a Bayesian network $\mathcal{G}$. Let $\boldsymbol{U} = \boldsymbol{X} \cup \boldsymbol{Y} \cup \boldsymbol{Z}$, and let $\mathcal{G}' = \mathcal{G}^+[\boldsymbol{U}]$ be the induced Bayesian network over $\boldsymbol{U} \cup \text{Ancestors}_{\boldsymbol{U}}$. Let $\mathcal{H}$ be the moralized graph $\mathcal{M}[\mathcal{G}']$. Then $\text{d-sep}_{\mathcal{G}}(\boldsymbol{X}; \boldsymbol{Y} \mid \boldsymbol{Z})$ if and only if $\text{sep}_{\mathcal{H}}(\boldsymbol{X}; \boldsymbol{Y} \mid \boldsymbol{Z})$.*

Example 4.16 *To gain some intuition for this result, consider the Bayesian network $\mathcal{G}$ of figure 4.12a (which extends our* Student *network). Consider the d-separation query $\text{d-sep}_{\mathcal{G}}(D; I \mid L)$. In this case, $\boldsymbol{U} = \{D, I, L\}$, and hence the moralized graph $\mathcal{M}[\mathcal{G}^+[\boldsymbol{U}]]$ is the graph shown in figure 4.12b, where we have introduced an undirected moralizing edge between D and I. In the resulting graph, D and I are not separated given L, exactly as we would have concluded using the d-separation procedure on the original graph.*

On the other hand, consider the d-separation query $\text{d-sep}_{\mathcal{G}}(D; I \mid S, A)$. In this case, $\boldsymbol{U} = \{D, I, S, A\}$. Because D and I are not spouses in $\mathcal{G}^+[\boldsymbol{U}]$, the moralization process does not add

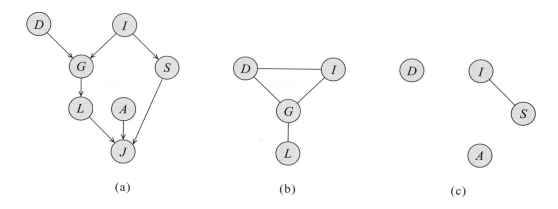

Figure 4.12 **Example of alternative definition of d-separation based on Markov networks.** (a) A Bayesian network $\mathcal{G}$. (b) The Markov network $\mathcal{M}[\mathcal{G}^+[D, I, L]]$. (c) The Markov network $\mathcal{M}[\mathcal{G}^+[D, I, A, S]]$.

an edge between them. The resulting moralized graph is shown in figure 4.12c. As we can see, we have that $\text{sep}_{\mathcal{M}[\mathcal{G}^+[U]]}(D; I \mid S, A)$, *as desired.* ∎

The proof for the general case is similar and is left as an exercise (exercise 4.15).

With this result, the soundness of d-separation follows easily. We repeat the statement of theorem 3.3:

Theorem 4.9

If a distribution $P_\mathcal{B}$ factorizes according to $\mathcal{G}$, then $\mathcal{G}$ is an I-map for P.

PROOF As in proposition 4.10, let $U = X \cup Y \cup Z$, let $U^* = U \cup Ancestors_U$, let $\mathcal{G}_{U^*} = \mathcal{G}^+[U]$ be the induced graph over U^*, and let $\mathcal{H}$ be the moralized graph $\mathcal{M}[\mathcal{G}_{U^*}]$. Let P_{U^*} be the Bayesian network distribution defined over $\mathcal{G}_{U^*}$ in the obvious way: the CPD for any variable in U^* is the same as in $\mathcal{B}$. Because U^* is upwardly closed, all variables used in these CPDs are in U^*.

Now, consider an independence assertion $(X \perp Y \mid Z) \in \mathcal{I}(\mathcal{G})$; we want to prove that $P_\mathcal{B} \models (X \perp Y \mid Z)$. By definition 3.7, if $(X \perp Y \mid Z) \in \mathcal{I}(\mathcal{G})$, we have that $d\text{-}sep_\mathcal{G}(X; Y \mid Z)$. It follows that $sep_\mathcal{H}(X; Y \mid Z)$, and hence that $(X \perp Y \mid Z) \in \mathcal{I}(\mathcal{H})$. P_{U^*} is a Gibbs distribution over $\mathcal{H}$, and hence, from theorem 4.1, $P_{U^*} \models (X \perp Y \mid Z)$. Using exercise 3.8, the distribution $P_{U^*}(U^*)$ is the same as $P_\mathcal{B}(U^*)$. Hence, it follows also that $P_\mathcal{B} \models (X \perp Y \mid Z)$, proving the desired result. ∎

4.5.2 From Markov Networks to Bayesian Networks

The previous section dealt with the conversion from a Bayesian network to a Markov network. We now consider the converse transformation: finding a Bayesian network that is a minimal I-map for a Markov network. It turns out that the transformation in this direction is significantly more difficult, both conceptually and computationally. Indeed, the Bayesian network that is a minimal I-map for a Markov network might be considerably larger than the Markov network.

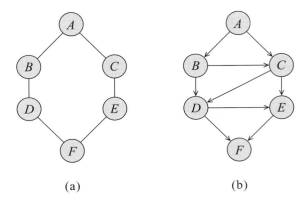

(a)   (b)

Figure 4.13 **Minimal I-map Bayesian networks for a nonchordal Markov network.** (a) A Markov network $\mathcal{H}_\ell$ with a loop. (b) A minimal I-map $\mathcal{G}_\ell$ Bayesian network for $\mathcal{H}$.

Example 4.17
Consider the Markov network structure $\mathcal{H}_\ell$ of figure 4.13a, and assume that we want to find a Bayesian network I-map for $\mathcal{H}_\ell$. As we discussed in section 3.4.1, we can find such an I-map by enumerating the nodes in $\mathcal{X}$ in some ordering, and define the parent set for each one in turn according to the independencies in the distribution. Assume we enumerate the nodes in the order A, B, C, D, E, F. The process for A and B is obvious. Consider what happens when we add C. We must, of course, introduce A as a parent for C. More interestingly, however, C is not independent of B given A; hence, we must also add B as a parent for C. Now, consider the node D. One of its parents must be B. As D is not independent of C given B, we must add C as a parent for B. We do not need to add A, as D is independent of A given B and C. Similarly, E's parents must be C and D. Overall, the minimal Bayesian network I-map according to this ordering has the structure $\mathcal{G}_\ell$ shown in figure 4.13b.   ∎

A quick examination of the structure $\mathcal{G}_\ell$ shows that we have added several edges to the graph, resulting in a set of triangles crisscrossing the loop. In fact, the graph $\mathcal{G}_\ell$ in figure 4.13b is chordal: all loops have been partitioned into triangles.

One might hope that a different ordering might lead to fewer edges being introduced. Unfortunately, this phenomenon is a general one: any Bayesian network I-map for this Markov network must add triangulating edges into the graph, so that the resulting graph is chordal (see definition 2.24). In fact, we can show the following property, which is even stronger:

Theorem 4.10
Let $\mathcal{H}$ be a Markov network structure, and let $\mathcal{G}$ be any Bayesian network minimal I-map for $\mathcal{H}$. Then $\mathcal{G}$ can have no immoralities (see definition 3.11).

PROOF Let $X_1, \ldots, X_n$ be a topological ordering for $\mathcal{G}$. Assume, by contradiction, that there is some immorality $X_i \rightarrow X_j \leftarrow X_k$ in $\mathcal{G}$ such that there is no edge between X_i and X_k; assume (without loss of generality) that $i < k < j$.

Owing to minimality of the I-map $\mathcal{G}$, if X_i is a parent of X_j, then X_i and X_j are not separated by X_j's other parents. Thus, $\mathcal{H}$ necessarily contains one or more paths between X_i

and X_j that are not cut by X_k (or by X_j's other parents). Similarly, $\mathcal{H}$ necessarily contains one or more paths between X_k and X_j that are not cut by X_i (or by X_j's other parents).

Consider the parent set U that was chosen for X_k. By our previous argument, there are one or more paths in $\mathcal{H}$ between X_i and X_k via X_j. As $i < k$, and X_i is not a parent of X_k (by our assumption), we have that U must cut all of those paths. To do so, U must cut either all of the paths between X_i and X_j, or all of the paths between X_j and X_k: As long as there is at least one active path from X_i to X_j and one from X_j to X_k, there is an active path between X_i and X_k that is not cut by U. Assume, without loss of generality, that U cuts all paths between X_j and X_k (the other case is symmetrical). Now, consider the choice of parent set for X_j, and recall that it is the (unique) minimal subset among $X_1, \ldots, X_{j-1}$ that separates X_j from the others. In a Markov network, this set consists of all nodes in $X_1, \ldots, X_{j-1}$ that are the first on some uncut path from X_j. As U separates X_k from X_j, it follows that X_k cannot be the first on any uncut path from X_j, and therefore X_k cannot be a parent of X_j. This result provides the desired contradiction. ∎

Because any nontriangulated loop of length at least 4 in a Bayesian network graph necessarily contains an immorality, we conclude:

Corollary 4.3

Let $\mathcal{H}$ be a Markov network structure, and let $\mathcal{G}$ be any minimal I-map for $\mathcal{H}$. Then $\mathcal{G}$ is necessarily chordal.

triangulation

Thus, the process of turning a Markov network into a Bayesian network requires that we add enough edges to a graph to make it chordal. This process is called *triangulation*. As in the transformation from Bayesian networks to Markov networks, the addition of edges leads to the loss of independence information. For instance, in example 4.17, the Bayesian network $\mathcal{G}_\ell$ in figure 4.13b loses the information that C and D are independent given A and F. In the transformation from directed to undirected models, however, the edges added are only the ones that are, in some sense, implicitly there — the edges required by the fact that each factor in a Bayesian network involves an entire family (a node and its parents). By contrast, the transformation from Markov networks to Bayesian networks can lead to the introduction of a large number of edges, and, in many cases, to the creation of very large families (exercise 4.16).

4.5.3 Chordal Graphs

We have seen that the conversion in either direction between Bayesian networks to Markov networks can lead to the addition of edges to the graph and to the loss of independence information implied by the graph structure. It is interesting to ask when a set of independence assumptions can be represented perfectly by both a Bayesian network and a Markov network. It turns out that this class is precisely the class of undirected chordal graphs.

The proof of one direction is fairly straightforward, based on our earlier results.

Theorem 4.11

Let $\mathcal{H}$ be a nonchordal Markov network. Then there is no Bayesian network $\mathcal{G}$ which is a perfect map for $\mathcal{H}$ (that is, such that $\mathcal{I}(\mathcal{H}) = \mathcal{I}(\mathcal{G})$).

PROOF The proof follows from the fact that the minimal I-map for $\mathcal{G}$ must be chordal. Hence, any I-map $\mathcal{G}$ for $\mathcal{I}(\mathcal{H})$ must include edges that are not present in $\mathcal{H}$. Because any additional edge eliminates independence assumptions, it is not possible for any Bayesian network $\mathcal{G}$ to precisely encode $\mathcal{I}(\mathcal{H})$. ∎

To prove the other direction of this equivalence, we first prove some important properties of chordal graphs. As we will see, chordal graphs and the properties we now show play a central role in the derivation of exact inference algorithms for graphical models. For the remainder of this discussion, we restrict attention to connected graphs; the extension to the general case is straightforward. The basic result we show is that we can decompose any connected chordal graph $\mathcal{H}$ into a *tree of cliques* — a tree whose nodes are the maximal cliques in $\mathcal{H}$ — so that the structure of the tree precisely encodes the independencies in $\mathcal{H}$. (In the case of disconnected graphs, we obtain a forest of cliques, rather than a tree.)

We begin by introducing some notation. Let $\mathcal{H}$ be a connected undirected graph, and let $C_1, \ldots, C_k$ be the set of maximal cliques in $\mathcal{H}$. Let $\mathcal{T}$ be any tree-structured graph whose nodes correspond to the maximal cliques $C_1, \ldots, C_k$. Let C_i, C_j be two cliques in the tree that are directly connected by an edge; we define $S_{i,j} = C_i \cap C_j$ to be a *sepset* between C_i sepset and C_j. Let $W_{<(i,j)}$ ($W_{<(j,i)}$) be all of the variables that appear in any clique on the C_i (C_j) side of the edge. Thus, each edge decomposes $\mathcal{X}$ into three disjoint sets: $W_{<(i,j)} - S_{i,j}$, $W_{<(j,i)} - S_{i,j}$, and $S_{i,j}$.

Definition 4.17

clique tree

We say that a tree $\mathcal{T}$ is a clique tree *for $\mathcal{H}$ if:*

- *each node corresponds to a clique in $\mathcal{H}$, and each maximal clique in $\mathcal{H}$ is a node in $\mathcal{T}$;*
- *each sepset $S_{i,j}$ separates $W_{<(i,j)}$ and $W_{<(j,i)}$ in $\mathcal{H}$.* ∎

Note that this definition implies that each separator $S_{i,j}$ renders its two sides conditionally independent in $\mathcal{H}$.

Example 4.18

Consider the Bayesian network graph $\mathcal{G}_\ell$ in figure 4.13b. Since it contains no immoralities, its moralized graph $\mathcal{H}'_\ell$ is simply the same graph, but where all edges have been made undirected. As $\mathcal{G}_\ell$ is chordal, so is $\mathcal{H}'_\ell$. The clique tree for $\mathcal{H}'_\ell$ is simply a chain $\{A, B, C\} \rightarrow \{B, C, D\} \rightarrow \{C, D, E\} \rightarrow \{D, E, F\}$, which clearly satisfies the separation requirements of the clique tree definition. ∎

Theorem 4.12

Every undirected chordal graph $\mathcal{H}$ has a clique tree $\mathcal{T}$.

PROOF We prove the theorem by induction on the number of nodes in the graph. The base case of a single node is trivial. Now, consider a chordal graph $\mathcal{H}$ of size > 1. If $\mathcal{H}$ consists of a single clique, then the theorem holds trivially. Therefore, consider the case where we have at least two nodes X_1, X_2 that are not connected directly by an edge. Assume that X_1 and X_2

are connected, otherwise the inductive step holds trivially. Let S be a minimal subset of nodes that separates X_1 and X_2.

The removal of the set S breaks up the graph into at least two disconnected components — one containing X_1, another containing X_2, and perhaps additional ones. Let W_1, W_2 be some partition of the variables in $\mathcal{X} - S$ into two disjoint components, such that W_i encompasses the connected component containing X_i. (The other connected components can be assigned to W_1 or W_2 arbitrarily.) We first show that S must be a complete subgraph. Let Z_1, Z_2 be any two variables in S. Due to the minimality of S, each Z_i must lie on a path between X_1 and X_2 that does not go through any other node in S. (Otherwise, we could eliminate Z_i from S while still maintaining separation.) We can therefore construct a minimal path from Z_1 to Z_2 that goes only through nodes in W_1 by constructing a path from Z_1 to X_1 to Z_2 that goes only through W_1, and by eliminating any shortcuts. We can similarly construct a minimal path from Z_1 to Z_2 that goes only through nodes in W_2. The two paths together form a cycle of length ≥ 4. Because of chordality, the cycle must have a chord, which, by construction, must be the edge $Z_1 — Z_2$.

Now consider the induced graph $\mathcal{H}_1 = \mathcal{H}[W_1 \cup S]$. As $X_2 \notin \mathcal{H}_1$, this induced graph is smaller than $\mathcal{H}$. Moreover, $\mathcal{H}_1$ is chordal, so we can apply the inductive hypothesis. Let $\mathcal{T}_1$ be the clique tree for $\mathcal{H}_1$. Because S is a complete connected subgraph, it is either a maximal clique or a subset of some maximal clique in $\mathcal{H}_1$. Let C_1 be some clique in $\mathcal{T}_1$ containing S (there may be more than one such clique). We can similarly define $\mathcal{H}_2$ and C_2 for X_2. If neither C_1 nor C_2 is equal to S, we construct a tree $\mathcal{T}$ that contains the union of the cliques in $\mathcal{T}_1$ and $\mathcal{T}_2$, and connects C_1 and C_2 by an edge. Otherwise, without loss of generality, let $C_1 = S$; we create $\mathcal{T}$ by merging $\mathcal{T}_1$ minus C_1 into $\mathcal{T}_2$, making all of C_1's neighbors adjacent to C_2 instead.

It remains to show that the resulting structure is a clique tree for $\mathcal{H}$. First, we note that there is no clique in $\mathcal{H}$ that intersects both W_1 and W_2; hence, any maximal clique in $\mathcal{H}$ is a maximal clique in either $\mathcal{H}_1$ or $\mathcal{H}_2$ (or both in the possible case of S), so that all maximal cliques in $\mathcal{H}$ appear in $\mathcal{T}$. Thus, the nodes in $\mathcal{T}$ are precisely the maximal cliques in $\mathcal{H}$. Second, we need to show that any $S_{i,j}$ separates $W_{<(i,j)}$ and $W_{<(j,i)}$. Consider two variables $X \in W_{<(i,j)}$ and $Y \in W_{<(j,i)}$. First, assume that $X, Y \in \mathcal{H}_1$; as all the nodes in $\mathcal{H}_1$ are on the $\mathcal{T}_1$ side of the tree, we also have that $S_{i,j} \subset \mathcal{H}_1$. Any path between two nodes in $\mathcal{H}_1$ that goes through W_2 can be shortcut to go only through $\mathcal{H}_1$. Thus, if $S_{i,j}$ separates X, Y in $\mathcal{H}_1$, then it also separates them in $\mathcal{H}$. The same argument applies for $X, Y \in \mathcal{H}_2$. Now, consider $X \in W_1$ and $Y \in W_2$. If $S_{i,j} = S$, the result follows from the fact that S separates W_1 and W_2. Otherwise, assume that $S_{i,j}$ is in $\mathcal{T}_1$, on the path from X to C_1. In this case, we have that $S_{i,j}$ separates X from S, and S separates $S_{i,j}$ from Y. The conclusion now follows from the transitivity of graph separation.

We have therefore constructed a clique tree for $\mathcal{H}$, proving the inductive claim. ∎

Using this result, we can show that the independencies in an undirected graph $\mathcal{H}$ can be captured perfectly in a Bayesian network if and only if $\mathcal{H}$ is chordal.

Theorem 4.13

Let $\mathcal{H}$ be a chordal Markov network. Then there is a Bayesian network $\mathcal{G}$ such that $\mathcal{I}(\mathcal{H}) = \mathcal{I}(\mathcal{G})$.

PROOF Let $\mathcal{T}$ be the clique tree for $\mathcal{H}$, whose existence is guaranteed by theorem 4.12. We can select an ordering over the nodes in the Bayesian network as follows. We select an arbitrary clique C_1 to be the root of the clique tree, and then order the cliques $C_1, \ldots, C_k$ using any topological ordering, that is, where cliques closer to the root are ordered first. We now order the nodes in the network in any ordering consistent with the clique ordering: if X_l first appears in C_i and X_m first appears in C_j, for $i < j$, then X_l must precede X_m in the ordering. We now construct a Bayesian network using the procedure Build-Minimal-I-Map of algorithm 3.2 applied to the resulting node ordering $X_1, \ldots, X_n$ and to $\mathcal{I}(\mathcal{H})$.

Let $\mathcal{G}$ be the resulting network. We first show that, when X_i is added to the graph, then X_i's parents are precisely $\boldsymbol{U}_i = \mathrm{Nb}_{X_i} \cap \{X_1, \ldots, X_{i-1}\}$, where Nb_{X_i} is the set of neighbors of X_i in $\mathcal{H}$. In other words, we want to show that X_i is independent of $\{X_1, \ldots, X_{i-1}\} - \boldsymbol{U}_i$ given $\boldsymbol{U}_i$. Let $\boldsymbol{C}_k$ be the first clique in the clique ordering to which X_i belongs. Then $\boldsymbol{U}_i \subset \boldsymbol{C}_k$. Let $\boldsymbol{C}_l$ be the parent of $\boldsymbol{C}_k$ in the rooted clique tree. According to our selected ordering, all of the variables in $\boldsymbol{C}_l$ are ordered before any variable in $\boldsymbol{C}_k - \boldsymbol{C}_l$. Thus, $\boldsymbol{S}_{l,k} \subset \{X_1, \ldots, X_{i-1}\}$. Moreover, from our choice of ordering, none of $\{X_1, \ldots, X_{i-1}\} - \boldsymbol{U}_i$ are in any descendants of $\boldsymbol{C}_k$ in the clique tree. Thus, they are all in $\boldsymbol{W}_{<(l,k)}$. From theorem 4.12, it follows that $\boldsymbol{S}_{l,k}$ separates X_i from all of $\{X_1, \ldots, X_{i-1}\} - \boldsymbol{U}_i$, and hence that X_i is independent of all of $\{X_1, \ldots, X_{i-1}\} - \boldsymbol{U}_i$ given $\boldsymbol{U}_i$. It follows that $\mathcal{G}$ and $\mathcal{H}$ have the same set of edges. Moreover, we note that all of $\boldsymbol{U}_i$ are in $\boldsymbol{C}_k$, and hence are connected in $\mathcal{G}$. Therefore, $\mathcal{G}$ is moralized. As $\mathcal{H}$ is the moralized undirected graph of $\mathcal{G}$, the result now follows from proposition 4.9.   ∎

For example, the graph $\mathcal{G}_\ell$ of figure 4.13b, and its moralized network $\mathcal{H}'_\ell$ encode precisely the same independencies. By contrast, as we discussed, there exists no Bayesian network that encodes precisely the independencies in the nonchordal network $\mathcal{H}_\ell$ of figure 4.13a.

Thus, we have shown that chordal graphs are precisely the intersection between Markov networks and Bayesian networks, in that the independencies in a graph can be represented exactly in both types of models if and only if the graph is chordal.

4.6 Partially Directed Models

So far, we have presented two distinct types of graphical models, based on directed and undirected graphs. We can unify both representations by allowing models that incorporate both directed and undirected dependencies. We begin by describing the notion of *conditional random field*, a Markov network with a directed dependency on some subset of variables. We then present a generalization of this framework to the class of *chain graphs*, an entire network in which undirected components depend on each other in a directed fashion.

4.6.1 Conditional Random Fields

So far, we have described the Markov network representation as encoding a joint distribution over $\mathcal{X}$. The same undirected graph representation and parameterization can also be used to encode a *conditional distribution* $P(\boldsymbol{Y} \mid \boldsymbol{X})$, where $\boldsymbol{Y}$ is a set of *target variables* and $\boldsymbol{X}$ is a (disjoint) set of *observed variables*. We will also see a directed analogue of this concept in section 5.6. In the case of Markov networks, this representation is generally called a *conditional random field* (CRF).

target variable

observed variable

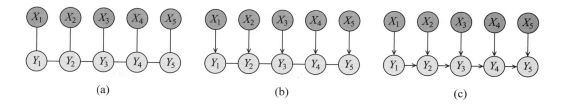

Figure 4.14 **Different linear-chain graphical models:** (a) a linear-chain-structured conditional random field, where the feature variables are denoted using grayed-out ovals; (b) a partially directed variant; (c) a fully directed, non-equivalent model. The X_i's are assumed to be always observed when the network is used, and hence they are shown as darker gray.

4.6.1.1   CRF Representation and Semantics

More formally, a CRF is an undirected graph whose nodes correspond to $Y \cup X$. At a high level, this graph is parameterized in the same way as an ordinary Markov network, as a set of factors $\phi_1(D_1), \dots, \phi_m(D_m)$. (As before, these factors can also be encoded more compactly as a log-linear model; for uniformity of presentation, we view the log-linear model as encoding a set of factors.) However, rather than encoding the distribution $P(Y, X)$, we view it as representing the conditional distribution $P(Y \mid X)$. To have the network structure and parameterization correspond naturally to a conditional distribution, we want to avoid representing a probabilistic model over X. We therefore disallow potentials that involve only variables in X.

Definition 4.18

conditional random field

A *conditional random field is an undirected graph $\mathcal{H}$ whose nodes correspond to $X \cup Y$; the network is annotated with a set of factors $\phi_1(D_1), \dots, \phi_m(D_m)$ such that each $D_i \nsubseteq X$. The network encodes a conditional distribution as follows:*

$$P(Y \mid X) = \frac{1}{Z(X)} \tilde{P}(Y, X)$$

$$\tilde{P}(Y, X) = \prod_{i=1}^{m} \phi_i(D_i) \tag{4.11}$$

$$Z(X) = \sum_{Y} \tilde{P}(Y, X).$$

Two variables in $\mathcal{H}$ are connected by an (undirected) edge whenever they appear together in the scope of some factor. ∎

The only difference between equation (4.11) and the (unconditional) Gibbs distribution of definition 4.3 is the different normalization used in the partition function $Z(X)$. The definition of a CRF induces a different value for the partition function for every assignment x to X. This difference is denoted graphically by having the feature variables grayed out.

Example 4.19

Consider a CRF over $Y = \{Y_1, \dots, Y_k\}$ and $X = \{X_1, \dots, X_k\}$, with an edge $Y_i \!-\! Y_{i+1}$ ($i = 1, \dots, k-1$) and an edge $Y_i \!-\! X_i$ ($i = 1, \dots, k$), as shown in figure 4.14a. The distribution *represented by this network has the form:*

$$P(\boldsymbol{Y} \mid \boldsymbol{X}) \;\; = \;\; \frac{1}{Z(\boldsymbol{X})} \tilde{P}(\boldsymbol{Y}, \boldsymbol{X})$$

$$\tilde{P}(\boldsymbol{Y}, \boldsymbol{X}) \;\; = \;\; \prod_{i=1}^{k-1} \phi(Y_i, Y_{i+1}) \prod_{i=1}^{k} \phi(Y_i, X_i)$$

$$Z(\boldsymbol{X}) \;\; = \;\; \sum_{\boldsymbol{Y}} \tilde{P}(\boldsymbol{Y}, \boldsymbol{X}). \qquad\qquad\blacksquare$$

Note that, unlike the definition of a conditional Bayesian network, the structure of a CRF may still contain edges between variables in $\boldsymbol{X}$, which arise when two such variables appear together in a factor that also contains a target variable. However, these edges do not encode the structure of any distribution over $\boldsymbol{X}$, since the network explicitly does not encode any such distribution.

 The fact that we avoid encoding the distribution over the variables in X is one of the main strengths of the CRF representation. This flexibility allows us to incorporate into the model a rich set of observed variables whose dependencies may be quite complex or even poorly understood. It also allows us to include continuous variables whose distribution may not have a simple parametric form. This flexibility allows us to use domain knowledge in order to define a rich set of features characterizing our domain, without worrying about modeling their joint distribution. For example, returning to the vision MRFs of box 4.B, rather than defining a joint distribution over pixel values and their region assignment, we can define a conditional distribution over segment assignments *given* the pixel values. The use of a conditional distribution here allows us to avoid making a parametric assumption over the (continuous) pixel values. Even more important, we can use image-processing routines to define rich features, such as the presence or direction of an image gradient at a pixel. Such features can be highly informative in determining the region assignment of a pixel. However, the definition of such features usually relies on multiple pixels, and defining a correct joint distribution or a set of independence assumptions over these features is far from trivial. The fact that we can condition on these features and avoid this whole issue allows us the flexibility to include them in the model. See box 4.E for another example.

4.6.1.2   Directed and Undirected Dependencies

A CRF defines a conditional distribution of $\boldsymbol{Y}$ on $\boldsymbol{X}$; thus, it can be viewed as a partially directed graph, where we have an undirected component over Y, which has the variables in $\boldsymbol{X}$ as parents.

Example 4.20

naive Markov

Consider a CRF over the binary-valued variables $\boldsymbol{X} = \{X_1, \ldots, X_k\}$ and $\boldsymbol{Y} = \{Y\}$, and a pairwise potential between Y and each X_i; this model is sometimes known as a naive Markov *model, due to its similarity to the naive Bayes model. Assume that the pairwise potentials defined via the following log-linear model*

$$\phi_i(X_i, Y) = \exp\left\{ w_i \boldsymbol{I}\{X_i = 1, Y = 1\}\right\}.$$

We also introduce a single-node potential $\phi_0(Y) = \exp\left\{w_0 \boldsymbol{I}\{Y = 1\}\right\}$. Following equation (4.11), *we now have:*

$$\tilde{P}(Y = 1 \mid x_1, \ldots, x_k) = \exp\left\{w_0 + \sum_{i=1}^{k} w_i x_i\right\}$$

$$\tilde{P}(Y = 0 \mid x_1, \ldots, x_k) = \exp\{0\} = 1.$$

In this case, we can show (exercise 5.16) that

$$P(Y = 1 \mid x_1, \ldots, x_k) = \text{sigmoid}\left(w_0 + \sum_{i=1}^{k} w_i x_i\right),$$

where

$$\text{sigmoid}(z) = \frac{e^z}{1 + e^z}$$

sigmoid

logistic CPD

is the sigmoid *function. This conditional distribution $P(Y \mid \boldsymbol{X})$ is of great practical interest: It defines a CPD that is not structured as a table, but that is induced by a small set of parameters $w_0, \ldots, w_k$ — parameters whose number is linear, rather than exponential, in the number of parents. This type of CPD, often called a* logistic CPD, *is a natural model for many real-world applications, inasmuch as it naturally aggregates the influence of different parents. We discuss this CPD in greater detail in section 5.4.2 as part of our general presentation of structured CPDs.* ∎

The partially directed model for the CRF of example 4.19 is shown in figure 4.14b. We may be tempted to believe that we can construct an equivalent model that is fully directed, such as the one in figure 4.14c. In particular, conditioned on any assignment x, the posterior distributions over $\boldsymbol{Y}$ in the two models satisfy the same independence assignments (the ones defined by the chain structure). However, the two models are not equivalent: In the Bayesian network, we have that Y_1 is independent of X_2 if we are not given Y_2. By contrast, in the original CRF, the unnormalized marginal measure of $\boldsymbol{Y}$ depends on the entire parameterization of the chain, and specifically the values of all of the variables in $\boldsymbol{X}$. A sound conditional Bayesian network for this distribution would require edges from all of the variables in $\boldsymbol{X}$ to each of the variables Y_i, thereby losing much of the structure in the distribution. See also box 20.A for further discussion.

Box 4.E — Case Study: CRFs for Text Analysis. *One important use for the CRF framework is in the domain of text analysis. Various models have been proposed for different tasks, including part-of-speech labeling, identifying named entities (people, places, organizations, and so forth), and extracting structured information from the text (for example, extracting from a reference list the publication titles, authors, journals, years, and the like). Most of these models share a similar structure: We have a target variable for each word (or perhaps short phrase) in the document, which encodes the possible labels for that word. Each target variable is connected to a set of feature variables that capture properties relevant to the target distinction. These methods are very popular in text analysis, both because the structure of the networks is a good fit for this domain, and because they produce state-of-the-art results for a broad range of natural-language processing problems.*

As a concrete example, consider the named entity recognition task, as described by Sutton and McCallum (2004, 2007). Entities often span multiple words, and the type of an entity may not be *apparent from individual words; for example, "New York" is a location, but "New York Times" is an organization. The problem of extracting entities from a word sequence of length T can be cast as a graphical model by introducing for each word, $X_t, 1 \leq t \leq T$, a target variable, Y_t, which indicates the entity type of the word. The outcomes of Y_t include* B-PERSON, I-PERSON, B-LOCATION, I-LOCATION, B-ORGANIZATION, I-ORGANIZATION, *and* OTHER. *In this so-called "BIO notation,"* OTHER *indicates that the word is not part of an entity, the* B- *outcomes indicate the beginning of a named entity phrase, and the* I- *outcomes indicate the inside or end of the named entity phrase. Having a distinguishing label for the beginning versus inside of an entity phrase allows the model to segment adjacent entities of the same type.*

linear-chain CRF
A common structure for this problem is a linear-chain CRF *often having two factors for each word: one factor* $\phi_t^1(Y_t, Y_{t+1})$ *to represent the dependency between neighboring target variables, and another factor* $\phi_t^2(Y_t, X_1, \ldots, X_T)$ *that represents the dependency between a target and its context in the word sequence. Note that the second factor can depend on arbitrary features of the entire input word sequence. We generally do not encode this model using table factors, but using a log-linear model. Thus, the factors are derived from a number of feature functions, such as* $f_t(Y_t, X_t) = \mathbf{1}\{Y_t = $ B-ORGANIZATION$, X_t = $ *"Times"*$\}$. *We note that, just as logistic CPDs are the conditional analog of the naive Bayes classifier (example 4.20), the linear-chain CRF is the conditional analog of the* hidden Markov model *(HMM) that we present in section 6.2.3.1.*

hidden Markov
model
A large number of features of the word X_t *and neighboring words are relevant to the named entity decision. These include features of the word itself: is it capitalized; does it appear in a list of common person names; does it appear in an atlas of location names; does it end with the character string "ton"; is it exactly the string "York"; is the following word "Times." Also relevant are aggregate features of the entire word sequence, such as whether it contains more than two sports-related words, which might be an indicator that "New York" is an organization (sports team) rather than a location. In addition, including features that are conjunctions of all these features often increases accuracy. The total number of features can be quite large, often in the hundreds of thousands or more if conjunctions of word pairs are used as features. However, the features are sparse, meaning that most features are zero for most words.*

Note that the same feature variable can be connected to multiple target variables, so that Y_t *would typically be dependent on the identity of several words in a window around position* t. *These contextual features are often highly indicative: for example, "Mrs." before a word and "spoke" after a word are both strong indicators that the word is a person. These context words would generally be used as a feature for multiple target variables. Thus, if we were using a simple naive-Bayes-style generative model, where each target variable is a parent of its associated feature, we either would have to deal with the fact that a context word has multiple parents or we would have to duplicate its occurrences (with one copy for each target variable for which it is in the context), and thereby overcount its contribution.*

Linear-chain CRFs frequently provide per-token accuracies in the high 90 percent range on many natural data sets. Per-field precision and recall (where the entire phrase category and boundaries must be correct) are more often around 80–95 percent, depending on the data set.

Although the linear-chain model is often effective, additional information can be incorporated into the model by augmenting the graphical structure. For example, often when a word occurs multiple times in the same document, it often has the same label. This knowledge can be incorporated by including factors that connect identical words, resulting in a skip-chain CRF, *as shown*

skip-chain CRF
in figure 4.E.1a. The first occurrence of the word "Green" has neighboring words that provide strong

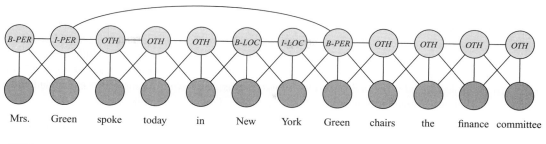

KEY

B-PER	Begin person name	*I-LOC*	Within location name
I-PER	Within person name	*OTH*	Not an entitiy
B-LOC	Begin location name		

(a)

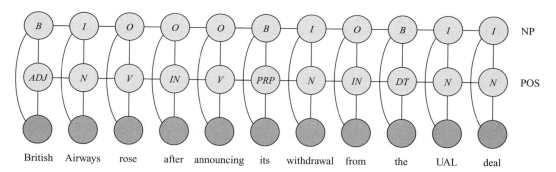

KEY

B	Begin noun phrase	*V*	Verb
I	Within noun phrase	*IN*	Preposition
O	Not a noun phrase	*PRP*	Possesive pronoun
N	Noun	*DT*	Determiner (e.g., a, an, the)
ADJ	Adjective		

(b)

Figure 4.E.1 — Two models for text analysis based on a linear chain CRF Gray nodes indicate X and clear nodes Y. The annotations inside the Y are the true labels. (a) A skip chain CRF for named entity recognition, with connections between adjacent words and long-range connections between multiple occurrences of the same word. (b) A pair of coupled linear-chain CRFs that performs joint part-of-speech labeling and noun-phrase segmentation. Here, B indicates the beginning of a noun phrase, I other words in the noun phrase, and O words not in a noun phrase. The labels for the second chain are parts of speech.

evidence that it is a PERSON*'s name; however, the second occurrence is much more ambiguous. By augmenting the original linear-chain CRF with an additional long-range factor that prefers its connected target variables to have the same value, the model is more likely to predict correctly that the second occurrence is also a* PERSON*. This example demonstrates another flexibility of conditional models, which is that the graphical structure over* $\boldsymbol{Y}$ *can easily depend on the value of the* $\boldsymbol{X}$*'s.*

　　CRFs having a wide variety of model structures have been successfully applied to many different tasks. Joint inference of both part-of-speech labels and noun-phrase segmentation has been coupled HMM *performed with two connected linear chains (somewhat analogous to a* coupled hidden Markov *mode, shown in figure 6.3). This structure is illustrated in figure 4.E.1b.*

4.6.2　Chain Graph Models ⋆

We now present a more general framework that builds on the CRF representation and can be used to provide a general treatment of the independence assumptions made in these partially directed models. Recall from definition 2.21 that, in a *partially directed acyclic graph* (PDAG), partially directed acyclic graph the nodes can be disjointly partitioned into several *chain components*. An edge between two chain component nodes in the same chain component must be undirected, while an edge between two nodes in chain graph different chain components must be directed. Thus, PDAGs are also called *chain graphs*.

4.6.2.1　Factorization

As in our other graphical representations, the structure of a PDAG $\mathcal{K}$ can be used to define a factorization for a probability distribution over $\mathcal{K}$. Intuitively, the factorization for PDAGs represents the distribution as a product of each of the chain components given its parents. chain graph Thus, we call such a representation a *chain graph model*.
model 　　Intuitively, each chain component $\boldsymbol{K}_i$ in the chain graph model is associated with a CRF that defines $\mathcal{P}(\boldsymbol{K}_i \mid \mathrm{Pa}_{\boldsymbol{K}_i})$ — the conditional distribution of $\boldsymbol{K}_i$ given its parents in the graph. More precisely, each is defined via a set of factors that involve the variables in $\boldsymbol{K}_i$ and their parents; the distribution $\mathcal{P}(\boldsymbol{K}_i \mid \mathrm{Pa}_{\boldsymbol{K}_i})$ is defined by using the factors associated with $\boldsymbol{K}_i$ to define a CRF whose target variables are $\boldsymbol{K}_i$ and whose observable variables are $\mathrm{Pa}_{\boldsymbol{K}_i}$.
　　To provide a formal definition, it helps to introduce the concept of a moralized PDAG.

Definition 4.19　*Let* $\mathcal{K}$ *be a PDAG and* $\boldsymbol{K}_1, \ldots, \boldsymbol{K}_\ell$ *be its chain components. We define* $\mathrm{Pa}_{\boldsymbol{K}_i}$ *to be the parents of* moralized graph　*nodes in* $\boldsymbol{K}_i$*. The* moralized graph *of* $\mathcal{K}$ *is an undirected graph* $\mathcal{M}[\mathcal{K}]$ *produced by first connecting, using undirected edges, any pair of nodes* $X, Y \in \mathrm{Pa}_{\boldsymbol{K}_i}$ *for all* $i = 1, \ldots, \ell$*, and then converting all directed edges into undirected edges.*　■

This definition generalizes our earlier notion of a moralized directed graph. In the case of directed graphs, each node is its own chain component, and hence we are simply adding undirected edges between the parents of each node.

Example 4.21　*Figure 4.15 shows a chain graph and its moral graph. We have added the edge between* A *and* B*, since they are both parents of the chain component* $\{C, D, E\}$*, and edges between* C*,* E*, and* H*, because they are parents of the chain component* $\{I\}$*. Note that we did not add an edge between*

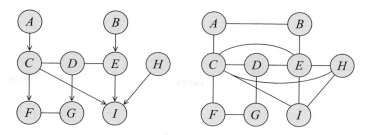

Figure 4.15 **A chain graph $\mathcal{K}$ and its moralized version**

D and H *(even though D and C, E are in the same chain component), since D is not a parent of I.* ∎

We can now define the factorization of a chain graph:

Definition 4.20

chain graph distribution

Let $\mathcal{K}$ be a PDAG, and $\mathbf{K}_1, \ldots, \mathbf{K}_\ell$ be its chain components. A chain graph distribution is defined via a set of factors $\phi_i(\mathbf{D}_i)$ ($i = 1, \ldots, m$), such that each $\mathbf{D}_i$ is a complete subgraph in the moralized graph $\mathcal{M}[\mathcal{K}^+[\mathbf{D}_i]]$. We associate each factor $\phi_i(\mathbf{D}_i)$ with a single chain component $\mathbf{K}_j$, such that $\mathbf{D}_i \subseteq \mathbf{K}_i \cup \mathrm{Pa}_{\mathbf{K}_i}$ and define $P(\mathbf{K}_i \mid \mathrm{Pa}_{\mathbf{K}_i})$ as a CRF with these factors, and with $\mathbf{Y}_i = \mathbf{K}_i$ and $\mathbf{X}_i = \mathrm{Pa}_{\mathbf{K}_i}$. We now define

$$P(\mathcal{X}) = \prod_{i=1}^{\ell} P(\mathbf{K}_i \mid \mathrm{Pa}_{\mathbf{K}_i}).$$

∎

We say that a distribution P factorizes over $\mathcal{K}$ if it can be represented as a chain graph distribution over $\mathcal{K}$.

Example 4.22

In the chain graph model defined by the graph of figure 4.15, we require that the conditional distribution $P(C, D, E \mid A, B)$ factorize according to the graph of figure 4.16a. Specifically, we would have to define the conditional probability as a normalized product of factors:

$$\frac{1}{Z(A, B)} \phi_1(A, C) \phi_2(B, E) \phi_3(C, D) \phi_4(D, E).$$

A similar factorization applies to $P(F, G \mid C, D)$. ∎

4.6.2.2 Independencies in Chain Graphs

As for undirected graphs, there are three distinct interpretations for the independence properties induced by a PDAG. Recall that in a PDAG, we have both the notion of parents of X (variables Y such that $Y \to X$ is in the graph) and neighbors of X (variables Y such that $Y\!-\!X$ is in the graph). Recall that the union of these two sets is the *boundary* of X, denoted $\mathrm{Boundary}_X$. Also

boundary

recall, from definition 2.15, that the descendants of X are those nodes Y that can be reached using any directed path, where a directed path can involve both directed and undirected edges but must contain at least one edge directed from X to Y, and no edges directed from Y to

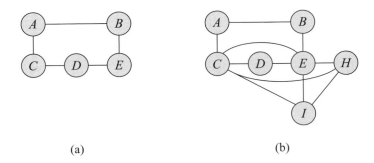

(a)   (b)

Figure 4.16   Example for definition of c-separation in a chain graph. (a) The Markov network $\mathcal{M}[\mathcal{K}^+[C, D, E]]$. (b) The Markov network $\mathcal{M}[\mathcal{K}^+[C, D, E, I]]$.

X. Thus, in the case of PDAGs, it follows that if Y is a descendant of X, then Y must be in a "lower" chain component.

Definition 4.21

pairwise
independencies

For a PDAG $\mathcal{K}$, we define the pairwise independencies *associated with $\mathcal{K}$ to be:*

$$\mathcal{I}_p(\mathcal{K}) = \{(X \perp Y \mid (\text{NonDescendants}_X - \{X, Y\})) :$$
$$X, Y \text{ non-adjacent}, Y \in \text{NonDescendants}_X\}. \qquad \blacksquare$$

This definition generalizes the pairwise independencies for undirected graphs: in an undirected graph, nodes have no descendants, so $\text{NonDescendants}_X = \mathcal{X}$. Similarly, it is not too hard to show that these independencies also hold in a directed graph.

Definition 4.22

local
independencies

For a PDAG $\mathcal{K}$, we define the local independencies *associated with $\mathcal{K}$ to be:*

$$\mathcal{I}_\ell(\mathcal{K}) = \{(X \perp \text{NonDescendants}_X - \text{Boundary}_X \mid \text{Boundary}_X) : X \in \mathcal{X}\}. \qquad \blacksquare$$

This definition generalizes the definition of local independencies for both directed and undirected graphs. For directed graphs, NonDescendants_X is precisely the set of nondescendants, whereas Boundary_X is the set of parents. For undirected graphs, NonDescendants_X is $\mathcal{X}$, whereas $\text{Boundary}_X = \text{Nb}_X$.

We define the global independencies in a PDAG using the definition of moral graph. Our definition follows the lines of proposition 4.10.

Definition 4.23

c-separation

Let $X, Y, Z \subset \mathcal{X}$ be three disjoint sets, and let $U = X \cup Y \cup Z$. We say that X is c-separated from Y given Z if X is separated from Y given Z in the undirected graph $\mathcal{M}[\mathcal{K}^+[X \cup Y \cup Z]]$. $\blacksquare$

Example 4.23

Consider again the PDAG of figure 4.15. Then C is c-separated from E given D, A, because C and E are separated given D, A in the undirected graph $\mathcal{M}[\mathcal{K}^+[\{C, D, E\}]]$, shown in figure 4.16a. However, C is not c-separated from E given only D, since there is a path between C and E via A, B. On the other hand, C is not separated from E given D, A, I. The graph $\mathcal{M}[\mathcal{K}^+[\{C, D, E, I\}]]$ is shown in figure 4.16b. As we can see, the introduction of I into the set U causes us to introduce a direct edge between C and E in order to moralize the graph. Thus, we cannot block the path between C and E using D, A, I. $\blacksquare$

This notion of c-separation clearly generalizes the notion of separation in undirected graphs, since the ancestors of a set U in an undirected graph are simply the entire set of nodes $\mathcal{X}$. It also generalizes the notion of d-separation in directed graphs, using the equivalent definition provided in proposition 4.10. Using the definition of c-separation, we can finally define the notion of global Markov independencies:

Definition 4.24

global
independencies

Let $\mathcal{K}$ be a PDAG. We define the global independencies *associated with $\mathcal{K}$ to be:*

$$\mathcal{I}(\mathcal{K}) = \{(\boldsymbol{X} \perp \boldsymbol{Y} \mid \boldsymbol{Z}) \ : \ \boldsymbol{X}, \boldsymbol{Y}, \boldsymbol{Z} \subset \mathcal{X}, \boldsymbol{X} \text{ is c-separated from } \boldsymbol{Y} \text{ given } \boldsymbol{Z}\}.$$ ∎

As in the case of undirected models, these three criteria for independence are not equivalent for nonpositive distributions. The inclusions are the same: the global independencies imply the local independencies, which in turn imply the pairwise independencies. Because undirected models are a subclass of PDAGs, the same counterexamples used in section 4.3.3 show that the inclusions are strict for nonpositive distributions. For positive distributions, we again have that the three definitions are equivalent.

We note that, as in the case of Bayesian networks, the parents of a chain component are always fully connected in $\mathcal{M}[\mathcal{K}[\boldsymbol{K}_i \cup \mathrm{Pa}_{\boldsymbol{K}_i}]]$. Thus, while the structure over the parents helps factorize the distribution over the chain components containing the parents, it does not give rise to independence assertions in the conditional distribution over the child chain component. Importantly, however, it does give rise to structure in the form of the parameterization of $P(\boldsymbol{K}_i \mid \mathrm{Pa}_{\boldsymbol{K}_i})$, as we saw in example 4.20.

As in the case of directed and undirected models, we have an equivalence between the requirement of factorization of a distribution and the requirement that it satisfy the independencies associated with the graph. Not surprisingly, since PDAGs generalize undirected graphs, this equivalence only holds for positive distributions:

Theorem 4.14

A positive distribution P factorizes over a PDAG $\mathcal{K}$ if and only if $P \models \mathcal{I}(\mathcal{K})$.

We omit the proof.

4.7 Summary and Discussion

In this chapter, we introduced *Markov networks*, an alternative graphical modeling language for probability distributions, based on undirected graphs.

We showed that Markov networks, like Bayesian networks, can be viewed as defining a set of independence assumptions determined by the graph structure. In the case of undirected models, there are several possible definitions for the independence assumptions induced by the graph, which are equivalent for positive distributions. As in the case of Bayesian network, we also showed that the graph can be viewed as a data structure for specifying a probability distribution in a factored form. The factorization is defined as a product of factors (general nonnegative functions) over cliques in the graph. We showed that, for positive distributions, the two characterizations of undirected graphs — as specifying a set of independence assumptions and as defining a factorization — are equivalent.

Markov networks also provide useful insight on Bayesian networks. In particular, we showed how a Bayesian network can be viewed as a Gibbs distribution. More importantly, the unnormalized measure we obtain by introducing evidence into a Bayesian network is also a Gibbs distribution, whose partition function is the probability of the evidence. This observation will play a critical role in providing a unified view of inference in graphical models.

We investigated the relationship between Bayesian networks and Markov networks and showed that the two represent different families of independence assumptions. The difference in these independence assumptions is a key factor in deciding which of the two representations to use in encoding a particular domain. There are domains where interactions have a natural directionality, often derived from causal intuitions. **In this case, the independencies derived from the network structure directly reflect patterns such as intercausal reasoning. Markov networks represent only monotonic independence patterns: observing a variable can only serve to remove dependencies, not to activate them.** Of course, we can encode a distribution with "causal" connections as a Gibbs distribution, and it will exhibit the same nonmonotonic independencies. However, these independencies will not be manifest in the network structure.

In other domains, the interactions are more symmetrical, and attempts to force a directionality give rise to models that are unintuitive and that often are incapable of capturing the independencies in the domain (see, for example, section 6.6). As a consequence, the use of undirected models has increased steadily, most notably in fields such as computer vision and natural language processing, where the acyclicity requirements of directed graphical models are often at odds with the nature of the model. The flexibility of the undirected model also allows the distribution to be decomposed into factors over multiple overlapping "features" without having to worry about defining a single normalized generating distribution for each variable. Conversely, **this very flexibility and the associated lack of clear semantics for the model parameters often make it difficult to elicit models from experts. Therefore, many recent applications use learning techniques to estimate parameters from data, avoiding the need to provide a precise semantic meaning for each of them.**

Finally, the question of which class of models better encodes the properties of the distribution is only one factor in the selection of a representation. There are other important distinctions between these two classes of models, especially when it comes to learning from data. We return to these topics later in the book (see, for example, box 20.A).

4.8   Relevant Literature

contingency table

The representation of a probability distribution as an undirected graph has its roots in the *contingency table* representation that is a staple in statistical modeling. The idea of representing probabilistic interactions in this representation dates back at least as early as the work of Bartlett (1935). This line of work is reviewed in detail by Whittaker (1990) and Lauritzen (1996), and we refer the reader to those sources and the references therein.

A parallel line of work involved the development of the Markov network (or Markov random field) representation. Here, the starting point was a graph object rather than a distribution object (such as a contingency table). Isham (1981) surveys some of the early work along these lines.

The connection between the undirected graph representation and the Gibbs factorization of the distribution was first made in the unpublished work of Hammersley and Clifford (1971). As a consequence, they also showed the equivalence of the different types of (local, pairwise, and global) independence properties for undirected graphs in the case of positive distributions.

Lauritzen (1982) made the connection between MRFs and contingency tables, and proved some of the key results regarding the independence properties arising form the undirected representation. The line of work analyzing independence properties was then significantly extended by Pearl and Paz (1987). The history of these developments and other key references are presented by Pearl (1988) and Lauritzen (1996). The independence properties of chain graphs were studied in detail by Frydenberg (1990); see also Lauritzen (1996). Studený and Bouckaert (1998) also provide an alternative definition of the independence properties in chain graphs, one that is equivalent to c-separation but more directly analogous to the definition of d-separation in directed graphs.

Factor graphs were presented by Kschischang et al. (2001a) and extended by Frey (2003) to encompass both Bayesian networks and Markov networks. The framework of conditional random fields (CRFs) was first proposed by Lafferty, McCallum, and Pereira (2001). They have subsequently been used in a broad range of applications in natural language processing, computer vision, and many more. Skip-chain CRFs were introduced by Sutton and McCallum (2004), and factorial CRFs by Sutton et al. (2007). Sutton and McCallum (2007) also provide an overview of this framework and some of its applications.

Ising models were first proposed by Ising (1925). The literature on this topic is too vast to mention; we refer the reader to any textbook in the area of statistical physics. The connection between Markov networks and these models in statistical physics is the origin of some of the terminology associated with these models, such as partition function or energy. In fact, many of the recent developments in inference for these models arise from approximations that were first proposed in the statistical physics community. Boltzmann machines were first proposed by Hinton and Sejnowski (1983).

Computer vision is another application domain that has motivated much of the work in undirected graphical models. The applications of MRFs to computer vision are too numerous to list; they span problems in low-level vision (such as image denoising, stereo reconstruction, or image segmentation) and in high-level vision (such as object recognition). Li (2001) provides a detailed description of some early applications; Szeliski et al. (2008) describe some applications that are viewed as standard benchmark problems for MRFs in the computer vision field.

4.9   Exercises

Exercise 4.1

Complete the analysis of example 4.4, showing that the distribution P defined in the example does not factorize over $\mathcal{H}$. (Hint: Use a proof by contradiction.)

Exercise 4.2

In this exercise, you will prove that the modified energy functions $\epsilon'_1(A, B)$ and $\epsilon'_2(B, C)$ of figure 4.10 result in precisely the same distribution as our original energy functions. More generally, for any constants λ^1 and λ^0, we can redefine

$$
\begin{aligned}
\epsilon'_1(a, b^i) &:= \epsilon_1(a, b^i) + \lambda^i \\
\epsilon'_2(b^i, c) &:= \epsilon_2(b^i, c) - \lambda^i
\end{aligned}
$$

Show that the resulting energy function is equivalent.

Exercise 4.3★

Provide an example a class of Markov networks $\mathcal{H}_n$ over n such that the size of the largest clique in $\mathcal{H}_n$ is constant, yet *any* Bayesian network I-map for $\mathcal{H}_n$ is exponentially large in n.

Exercise 4.4★

Prove theorem 4.7 for the case where $\mathcal{H}$ consists of a single clique.

Exercise 4.5

Complete the proof of theorem 4.3, by showing that U_1 and U_k are dependent given $\boldsymbol{Z}$ in the distribution P defined by the product of potentials described in the proof.

Exercise 4.6★

Consider a factor graph $\mathcal{F}$, as in definition 4.13. Define the minimal Markov network $\mathcal{H}$ that is guaranteed to be an I-map for any distribution defined over $\mathcal{F}$. Prove that $\mathcal{H}$ is a sound and complete representation of the independencies in $\mathcal{F}$:

a. If $sep_{\mathcal{H}}(\boldsymbol{X};\boldsymbol{Y} \mid \boldsymbol{Z})$ holds, then $(\boldsymbol{X} \perp \boldsymbol{Y} \mid \boldsymbol{Z})$ holds for all distributions over $\mathcal{F}$.
b. If $sep_{\mathcal{H}}(\boldsymbol{X};\boldsymbol{Y} \mid \boldsymbol{Z})$ does not hold, then there is some distribution P that factorizes over $\mathcal{F}$ such that $(\boldsymbol{X} \perp \boldsymbol{Y} \mid \boldsymbol{Z})$ does not hold in P.

Exercise 4.7★

The canonical parameterization in the Hammersley-Clifford theorem is stated in terms of the maximal cliques in a Markov network. In this exercise, you will show that it also captures the finer-grained representation of factor graphs. Specifically, let P be a distribution that factorizes over a factor graph $\mathcal{F}$, as in definition 4.13. Show that the canonical parameterization of P also factorizes over $\mathcal{F}$.

Exercise 4.8

Prove proposition 4.3. More precisely, let P satisfy $\mathcal{I}_\ell(\mathcal{H})$, and assume that X and Y are two nodes in $\mathcal{H}$ that are not connected directly by an edge. Prove that P satisfies $(X \perp Y \mid \mathcal{X} - \{X,Y\})$.

Exercise 4.9★

Complete the proof of theorem 4.4. Assume that equation (4.1) holds for all disjoint sets $\boldsymbol{X},\boldsymbol{Y},\boldsymbol{Z}$, with $|\boldsymbol{Z}| \geq k$. Prove that equation (4.1) also holds for any disjoint $\boldsymbol{X},\boldsymbol{Y},\boldsymbol{Z}$ such that $\boldsymbol{X} \cup \boldsymbol{Y} \cup \boldsymbol{Z} \neq \mathcal{X}$ and $|\boldsymbol{Z}| = k - 1$.

Exercise 4.10

We define the following properties for a set of independencies:

strong union   • **Strong Union:**

$$(\boldsymbol{X} \perp \boldsymbol{Y} \mid \boldsymbol{Z}) \Longrightarrow (\boldsymbol{X} \perp \boldsymbol{Y} \mid \boldsymbol{Z}, \boldsymbol{W}). \tag{4.12}$$

In other words, additional evidence $\boldsymbol{W}$ cannot induce dependence.

transitivity   • **Transitivity:** For all disjoint sets $\boldsymbol{X},\boldsymbol{Y},\boldsymbol{Z}$ and all variables A:

$$\neg(\boldsymbol{X} \perp A \mid \boldsymbol{Z}) \& \neg(A \perp \boldsymbol{Y} \mid \boldsymbol{Z}) \Longrightarrow \neg(\boldsymbol{X} \perp \boldsymbol{Y} \mid \boldsymbol{Z}). \tag{4.13}$$

Intuitively, this statement asserts that if $\boldsymbol{X}$ and $\boldsymbol{Y}$ are both correlated with some A (given $\boldsymbol{Z}$), then they are also correlated with each other (given $\boldsymbol{Z}$). We can also write the contrapositive of this statement, which is less obvious but easier to read. For all $\boldsymbol{X},\boldsymbol{Y},\boldsymbol{Z},A$:

$$(\boldsymbol{X} \perp \boldsymbol{Y} \mid \boldsymbol{Z}) \longrightarrow (\boldsymbol{X} \perp A \mid \boldsymbol{Z}) \vee (A \perp \boldsymbol{Y} \mid \boldsymbol{Z}).$$

Prove that if $\mathcal{I} = \mathcal{I}(\mathcal{H})$ for some Markov network $\mathcal{H}$, then $\mathcal{I}$ satisfies strong union and transitivity.

Exercise 4.11★

In this exercise you will prove theorem 4.6. Consider some specific node X, and let $\mathcal{U}$ be the set of all subsets $\boldsymbol{U}$ satisfying definition 4.12. Define $\boldsymbol{U}^*$ to be the intersection of all $\boldsymbol{U} \in \mathcal{U}$.

a. Prove that $\boldsymbol{U}^* \in \mathcal{U}$. Conclude that $\mathrm{MB}_P(X) = \boldsymbol{U}^*$.

b. Prove that if $P \models (X \perp Y \mid \mathcal{X} - \{X, Y\})$, then $Y \notin \mathrm{MB}_P(X)$.

c. Prove that if $Y \notin \mathrm{MB}_P(X)$, then $P \models (X \perp Y \mid \mathcal{X} - \{X, Y\})$.

d. Conclude that $\mathrm{MB}_P(X)$ is precisely the set of neighbors of X in the graph defined in theorem 4.5, showing that the construction of theorem 4.6 also produces a minimal I-map.

Exercise 4.12

Show that a Boltzmann machine distribution (with variables taking values in $\{0, 1\}$) can be rewritten as an Ising model, where we use the value space $\{-1, +1\}$ (mapping 0 to -1).

Exercise 4.13

Show that we can represent any Gibbs distribution as a log-linear model, as defined in definition 4.15.

Exercise 4.14

Complete the proof of proposition 4.8. In particular, show the following:

a. For any variable X, let $\boldsymbol{W} = \mathcal{X} - \{X\} - \mathrm{MB}_\mathcal{G}(X)$. Then $d\text{-}sep_\mathcal{G}(X; \boldsymbol{W} \mid \mathrm{MB}_\mathcal{G}(X))$.

b. The set $\mathrm{MB}_\mathcal{G}(X)$ is the minimal set for which this property holds.

Exercise 4.15

Prove proposition 4.10.

Exercise 4.16★

Provide an example of a class of Markov networks $\mathcal{H}_n$ over n nodes for arbitrarily large n (not necessarily for every n), where the size of the largest clique is a constant independent of n, yet the size of the largest clique in any chordal graph $\mathcal{H}_n^C$ that contains $\mathcal{H}_n$ is exponential in n. Explain why the size of the largest clique is necessarily exponential in n for all $\mathcal{H}_n^C$.

Exercise 4.17★

In this exercise, you will prove that the chordality requirement for graphs is equivalent to two other conditions of independent interest.

Definition 4.25

Markov network decomposition

Let $\boldsymbol{X}, \boldsymbol{Y}, \boldsymbol{Z}$ be disjoint sets such that $\mathcal{X} = \boldsymbol{X} \cup \boldsymbol{Y} \cup \boldsymbol{Z}$ and $\boldsymbol{X}, \boldsymbol{Y} \neq \emptyset$. We say that $(\boldsymbol{X}, \boldsymbol{Z}, \boldsymbol{Y})$ is a decomposition of a Markov network $\mathcal{H}$ if $\boldsymbol{Z}$ separates $\boldsymbol{X}$ from $\boldsymbol{Y}$ and $\boldsymbol{Z}$ is a complete subgraph in $\mathcal{H}$. ∎

Definition 4.26

We say that a graph $\mathcal{H}$ is decomposable if there is a decomposition $(\boldsymbol{X}, \boldsymbol{Z}, \boldsymbol{Y})$ of $\mathcal{H}$, such that the graphs induced by $\boldsymbol{X} \cup \boldsymbol{Z}$ and $\boldsymbol{Y} \cup \boldsymbol{Z}$ are also decomposable. ∎

Show that, for any undirected graph $\mathcal{H}$, the following conditions are equivalent:

a. $\mathcal{H}$ is decomposable;

b. $\mathcal{H}$ is chordal;

c. for every X, Y, every minimal set $\boldsymbol{Z}$ that separates X and Y is complete.

The proof of equivalence proceeds by induction on the number of vertices in $\mathcal{X}$. Assume that the three conditions are equivalent for all graphs with $|\mathcal{X}| \leq n$, and consider a graph $\mathcal{H}$ with $|\mathcal{X}| = n + 1$.

a. Prove that if $\mathcal{H}$ is decomposable, it is chordal.

b. Prove that if $\mathcal{H}$ is chordal, then for any X, Y, and any minimal set $\boldsymbol{Z}$ that separates X and Y, $\boldsymbol{Z}$ is complete.

c. Prove that for any X, Y, any minimal set $\boldsymbol{Z}$ that separates X and Y is complete, then $\mathcal{H}$ is decomposable.

Exercise 4.18

Let $\mathcal{G}$ be a Bayesian network structure and $\mathcal{H}$ a Markov network structure over $\mathcal{X}$ such that the skeleton of $\mathcal{G}$ is precisely $\mathcal{H}$. Prove that if $\mathcal{G}$ has no immoralities, then $\mathcal{I}(\mathcal{G}) = \mathcal{I}(\mathcal{H})$.

Exercise 4.19

Consider the PDAG of figure 4.15. Write down all c-separation statements that are valid given $\{G\}$; write down all valid statements given $\{G, D\}$; write down all statements that are valid given $\{C, D, E\}$.

5 *Local Probabilistic Models*

In chapter 3 and chapter 4, we discussed the representation of global properties of independence by graphs. These properties of independence allowed us to factorize a high-dimensional joint distribution into a product of lower-dimensional CPDs or factors. So far, we have mostly ignored the representation of these factors. In this chapter, we examine CPDs in more detail. We describe a range of representations and consider their implications in terms of additional regularities we can exploit. We have chosen to phrase our discussion in terms of CPDs, since they are more constrained than factors (because of the local normalization constraints). However, many of the representations we discuss in the context of CPDs can also be applied to factors.

5.1 Tabular CPDs

When dealing with spaces composed solely of discrete-valued random variables, we can always resort to a *tabular* representation of CPDs, where we encode $P(X \mid \mathrm{Pa}_X)$ as a table that contains an entry for each joint assignment to X and Pa_X. For this table to be a proper CPD, we require that all the values are nonnegative, and that, for each value pa_X, we have

$$\sum_{x \in Val(X)} P(x \mid \mathrm{pa}_X) = 1. \tag{5.1}$$

It is clear that this representation is as general as possible. We can represent every possible discrete CPD using such a table. As we will also see, table-CPDs can be used in a natural way in inference algorithms that we discuss in chapter 9. These advantages often lead to the perception

table-CPD

that *table-CPDs*, also known as *conditional probability tables* (CPTs), are an inherent part of the Bayesian network representation.

However, the tabular representation also has several significant disadvantages. First, it is clear that if we consider random variables with infinite domains (for example, random variables with continuous values), we cannot store each possible conditional probability in a table. But even in the discrete setting, we encounter difficulties. The number of parameters needed to describe a table-CPD is the number of joint assignments to X and Pa_X, that is, $|Val(\mathrm{Pa}_X)| \cdot |Val(X)|$.[1] This number grows exponentially in the number of parents. Thus, for example, if we have 5 binary parents of a binary variable X, we need specify $2^5 = 32$ values; if we have 10 parents, we need to specify $2^{10} = 1,024$ values.

---

1. We can save some space by storing only independent parameters, but this saving is not significant.

Clearly, the tabular representation rapidly becomes large and unwieldy as the number of parents grows. This problem is a serious one in many settings. Consider a medical domain where a symptom, *Fever*, depends on 10 diseases. It would be quite tiresome to ask our expert 1,024 questions of the format: "What is the probability of high fever when the patient has disease A, does not have disease B, ...?" Clearly, our expert will lose patience with us at some point!

This example illustrates another problem with the tabular representation: it ignores *structure* within the CPD. If the CPD is such that there are no similarity between the various cases, that is, each combination of disease has drastically different probability of high fever, then the expert might be more patient. However, in this example, like many others, there is some regularity in the parameters for different values of the parents of X. For example, it might be the case that, if the patient suffers from disease A, then she is certain to have high fever and thus $P(X \mid \mathrm{pa}_X)$ is the same for all values pa_X in which A is true. Indeed, many of the representations we consider in this chapter attempt to describe such regularities explicitly and to exploit them in order to reduce the number of parameters needed to specify a CPD.

The key insight that allows us to avoid these problems is the following observation: **A CPD needs to specify a conditional probability $P(x \mid \mathrm{pa}_X)$ for every assignment of values pa_X and x, but it does not have to do so by listing each such value explicitly. We should view CPDs not as tables listing all of the conditional probabilities, but rather as functions that given pa_x and x, return the conditional probability $P(x \mid \mathrm{pa}_X)$.** This implicit representation suffices in order to specify a well-defined joint distribution as a BN. In the remainder of the chapter, we will explore some of the possible representations of such functions.

5.2 Deterministic CPDs

5.2.1 Representation

deterministic
CPD

Perhaps the simplest type of nontabular CPD arises when a variable X is a *deterministic* function of its parents Pa_X. That is, there is a function $f : Val(\mathrm{Pa}_X) \mapsto Val(X)$, such that

$$P(x \mid \mathrm{pa}_X) = \begin{cases} 1 & x = f(\mathrm{pa}_X) \\ 0 & \text{otherwise.} \end{cases}$$

For example, in the case of binary-valued variables, X might be the "or" of its parents. In a continuous domain, we might want to assert in $P(X \mid Y, Z)$ that X is equal to $Y + Z$.

Of course, the extent to which this representation is more compact than a table (that is, takes less space in the computer) depends on the expressive power that our BN modeling language offers us for specifying deterministic functions. For example, some languages might allow a vocabulary that includes only logical OR and AND of the parents, so that all other functions must be specified explicitly as a table.[2] In a domain with continuous variables, a language might choose to allow only linear dependencies of the form $X = 2Y + -3Z + 1$, and not arbitrary functions such as $X = \sin(y + e^z)$.

Deterministic relations are useful in modeling many domains. In some cases, they occur naturally. Most obviously, when modeling constructed artifacts such as machines or electronic circuits, deterministic dependencies are often part of the device specification. For example, the

---

2. Other logical functions, however, can be described by introducing intermediate nodes and composing ORs and ANDs.

behavior of an OR gate in an electronic circuit (in the case of no faults) is that the gate output is a deterministic OR of the gate inputs. However, we can also find deterministic dependencies in "natural" domains.

Example 5.1

Recall that the genotype of a person is determined by two copies of each gene, called alleles. *Each allele can take on one of several values corresponding to different genetic tendencies. The person's phenotype is often a deterministic function of these values. For example, the gene responsible for determining blood type has three values: a, b, and o. Letting G_1 and G_2 be variables representing the two alleles, and T the variable representing the phenotypical blood type, then we have that:*

$$T = \begin{cases} ab & \text{if } G_1 \text{ or } G_2 \text{ is } a \text{ and the other is } b \\ a & \text{if at least one of } G_1 \text{ or } G_2 \text{ is equal to } a \text{ and the other is} \\ & \text{either } a \text{ or } o \\ b & \text{if at least one of } G_1 \text{ or } G_2 \text{ is equal to } b \text{ and the other is} \\ & \text{either } b \text{ or } o \\ o & \text{if } G_1 = o \text{ and } G_2 = o \end{cases}$$

■

Deterministic variables can also help simplify the dependencies in a complex model.

Example 5.2

When modeling a car, we might have four variables $T_1, \ldots, T_4$, each corresponding to a flat in one of the four tires. When one or more of these tires is flat, there are several effects; for example, the steering may be affected, the ride can be rougher, and so forth. Naively, we can make all of the T_i's parents of all of the affected variables — Steering, Ride, and so on. However, it can significantly simplify the model to introduce a new variable Flat-Tire, which is the deterministic OR of $T_1, \ldots, T_4$. We can then replace a complex dependency of Steering and Ride on $T_1, \ldots, T_4$ with a dependency on a single parent Flat-Tire, significantly reducing their indegree. If these variables have other parents, the savings can be considerable.

■

5.2.2 Independencies

Aside from a more compact representation, we get an additional advantage from making the structure explicit. Recall that conditional independence is a numerical property — it is defined using equality of probabilities. However, the graphical structure in a BN makes certain properties of a distribution explicit, allowing us to deduce that some independencies hold without looking at the numbers. By making structure explicit in the CPD, we can do even more of the same.

Example 5.3

Consider the simple network structure in figure 5.1. If C is a deterministic function of A and B, what new conditional independencies do we have? Suppose that we are given the values of A and B. Then, since C is deterministic, we also know the value of C. As a consequence, we have that D and E are independent. Thus, we conclude that $(D \perp E \mid A, B)$ holds in the distribution. Note that, had C not been a deterministic function of A and B, this independence would not necessarily hold. Indeed, d-separation would not deduce that D and E are independent given A and B. ■

Can we augment the d-separation procedure to discover independencies in cases such as this? Consider an independence assertion $(\boldsymbol{X} \perp \boldsymbol{Y} \mid \boldsymbol{Z})$; in our example, we are interested in the case where $\boldsymbol{Z} = \{A, B\}$. The variable C is not in $\boldsymbol{Z}$ and is therefore not considered observed.

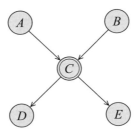

Figure 5.1   Example of a network with a deterministic CPD. The double-line notation represents the fact that C is a deterministic function of A and B.

Algorithm 5.1 Computing d-separation in the presence of deterministic CPDs

---

Procedure DET-SEP(
    $Graph$, // network structure
    D, // set of deterministic variables
    $\boldsymbol{X}, \boldsymbol{Y}, \boldsymbol{Z}$ // query
)
    Let $\boldsymbol{Z}^{+} \leftarrow \boldsymbol{Z}$
    While there is an X_i such that
        (1) $X_i \in \boldsymbol{D}$ // X_i has a deterministic CPD
        (2) $\mathrm{Pa}_{X_i} \subseteq \boldsymbol{Z}^{+}$
        $\boldsymbol{Z}^{+} \leftarrow \boldsymbol{Z}^{+} \cup \{X_i\}$
    return $d\text{-}sep_{\mathcal{G}}(\boldsymbol{X}; \boldsymbol{Y} \mid \boldsymbol{Z}^{+})$

But when A and B are observed, then the value of C is also known with certainty, so we can consider it as part of our observed set $\boldsymbol{Z}$. In our example, this simple modification would suffice for inferring that D and E are independent given A and B.

In other examples, however, we might need to continue this process. For example, if we had another variable F that was a deterministic function of C, then F is also de facto observed when C is observed, and hence when A and B are observed. Thus, F should also be introduced into $\boldsymbol{Z}$. Thus, we have to extend $\boldsymbol{Z}$ iteratively to contain all the variables that are determined by it. This discussion suggests the simple procedure shown in algorithm 5.1.

deterministic
separation

This algorithm provides a procedural definition for *deterministic separation* of $\boldsymbol{X}$ from $\boldsymbol{Y}$ given $\boldsymbol{Z}$. This definition is sound, in the same sense that d-separation is sound.

Theorem 5.1

Let $\mathcal{G}$ be a network structure, and let $\boldsymbol{D}, \boldsymbol{X}, \boldsymbol{Y}, \boldsymbol{Z}$ be sets of variables. If $\boldsymbol{X}$ is deterministically separated from $\boldsymbol{Y}$ given $\boldsymbol{Z}$ (as defined by DET-SEP($\mathcal{G}, \boldsymbol{D}, \boldsymbol{X}, \boldsymbol{Y}, \boldsymbol{Z}$)), then for all distributions P such that $P \models \mathcal{I}_{\ell}(\mathcal{G})$ and where, for each $X \in \boldsymbol{D}$, $P(X \mid \mathrm{Pa}_X)$ is a deterministic CPD, we have that $P \models (\boldsymbol{X} \perp \boldsymbol{Y} \mid \boldsymbol{Z})$.

The proof is straightforward and is left as an exercise (exercise 5.1).

Does this procedure capture all of the independencies implied by the deterministic functions? As with d-separation, the answer must be qualified: Given only the graph structure and the set

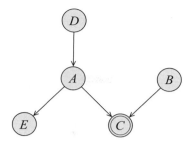

Figure 5.2 **A slightly more complex example with deterministic CPDs**

of deterministic CPDs, we cannot find additional independencies.

Theorem 5.2

Let $\mathcal{G}$ be a network structure, and let $\boldsymbol{D}, \boldsymbol{X}, \boldsymbol{Y}, \boldsymbol{Z}$ be sets of variables. If $\mathrm{DET\text{-}SEP}(\mathcal{G}, \boldsymbol{D}, \boldsymbol{X}, \boldsymbol{Y}, \boldsymbol{Z})$ returns false, then there is a distribution P such that $P \models \mathcal{I}_\ell(\mathcal{G})$ and where, for each $X \in \boldsymbol{D}$, $P(X \mid \mathrm{Pa}_X)$ is deterministic CPD, but we have that $P \not\models (\boldsymbol{X} \perp \boldsymbol{Y} \mid \boldsymbol{Z})$.

Of course, the DET-SEP procedure detects independencies that are derived purely from the fact that a variable is a deterministic function of its parents. However, particular deterministic functions can imply additional independencies.

Example 5.4

Consider the network of figure 5.2, where C is the exclusive or of A and B. What additional independencies do we have here? In the case of XOR (although not for all other deterministic functions), the values of C and B fully determine that of A. Therefore, we have that $(D \perp E \mid B, C)$ holds in the distribution. ∎

Specific deterministic functions can also induce other independencies, ones that are more refined than the variable-level independencies discussed in chapter 3.

Example 5.5

Consider the Bayesian network of figure 5.1, but where we also know that the deterministic function at C is an OR. Assume we are given the evidence $A = a^1$. Because C is an OR of its parents, we immediately know that $C = c^1$, regardless of the value of B. Thus, we can conclude that B and D are now independent: In other words, we have that

$$P(D \mid B, a^1) = P(D \mid a^1).$$

On the other hand, if we are given $A = a^0$, the value of C is not determined, and it does depend on the value of B. Hence, the corresponding statement conditioned on a^0 is false. ∎

Thus, deterministic variables can induce a form of independence that is different from the standard notion on which we have focused so far. Up to now, we have restricted attention to independence properties of the form $(\boldsymbol{X} \perp \boldsymbol{Y} \mid \boldsymbol{Z})$, which represent the assumption that $P(\boldsymbol{X} \mid \boldsymbol{Y}, \boldsymbol{Z}) = P(\boldsymbol{X} \mid \boldsymbol{Z})$ for all values of $\boldsymbol{X}, \boldsymbol{Y}$ and $\boldsymbol{Z}$. Deterministic functions can imply a type of independence that only holds for *particular* values of some variables.

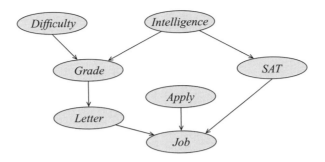

Figure 5.3 The Student **example augmented with a** *Job* **variable**

Definition 5.1	*Let* X, Y, Z *be pairwise disjoint sets of variables, let* C *be a set of variables (that might overlap*
context-specific independence	*with* $X \cup Y \cup Z$*), and let* $c \in Val(C)$*. We say that* X *and* Y *are* contextually independent *given* Z *and the context* c *denoted* $(X \perp_c Y \mid Z, c)$*, if*

$$P(X \mid Y, Z, c) = P(X \mid Z, c) \text{ whenever } P(Y, Z, c) > 0.$$ ∎

Independence statements of this form are called *context-specific independencies* (CSI). They arise in many forms in the context of deterministic dependencies.

Example 5.6	*As we saw in example 5.5, we can have that some value of one parent* A *can be enough to determine the value of the child* C*. Thus, we have that* $(C \perp_c B \mid a^1)$*, and hence also that* $(D \perp_c B \mid a^1)$*. We can make additional conclusions if we use properties of the OR function. For example, if we know that* $C = c^0$*, we can conclude that both* $A = a^0$ *and* $B = b^0$*. Thus, in particular, we can conclude both that* $(A \perp_c B \mid c^0)$ *and that* $(D \perp_c E \mid c^0)$*. Similarly, if we know that* $C = c^1$ *and* $B = b^0$*, we can conclude that* $A = a^1$*, and hence we have that* $(D \perp_c E \mid b^0, c^1)$*.* ∎

 It is important to note that **context-specific independencies can also arise when we have tabular CPDs. However, in the case of tabular CPDs, the independencies would only become apparent if we examine the network parameters. By making the structure of the CPD explicit, we can use qualitative arguments to deduce these independencies.**

5.3 Context-Specific CPDs

5.3.1 Representation

Structure in CPDs does not arise only in the case of deterministic dependencies. A very common type of regularity arises when we have precisely the same effect in several contexts.

Example 5.7	*We augment our* Student *example to model the event that the student will be offered a job at Acme Consulting. Thus, we have a binary-valued variable* J*, whose value is* j^1 *if the student is offered this job, and* j^0 *otherwise. The probability of this event depends on the student's SAT scores and the strength of his recommendation letter. We also have to represent the fact that our student might*

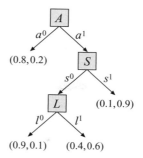

Figure 5.4   A tree-CPD for $P(J \mid A, S, L)$**.** Internal nodes in the tree denote tests on parent variables. Leaves are annotated with the distribution over J.

choose not to apply for a job at Acme Consulting. Thus, we have a binary variable Applied, whose value (a^1 or a^0) indicates whether the student applied or not. The structure of the augmented network is shown in figure 5.3.

   Now, we need to describe the CPD $P(J \mid A, S, L)$. In our domain, even if the student does not apply, there is still a chance that Acme Consulting is sufficiently desperate for employees to offer him a job anyway. (This phenomenon was quite common during the days of the Internet Gold Rush.) In this case, however, the recruiter has no access to the student's SAT scores or recommendation letters, and therefore the decision to make an offer cannot depend on these variables. Thus, among the 8 values of the parents A, S, L, the four that have $A = a^0$ must induce an identical distribution over the variable J.

   We can elaborate this model even further. Assume that our recruiter, knowing that SAT scores are a far more reliable indicator of the student's intelligence than a recommendation letter, first considers the SAT score. If it is high, he generates an offer immediately. (As we said, Acme Consulting is somewhat desperate for employees.) If, on the other hand, the SAT score is low, he goes to the effort of obtaining the professor's letter of recommendation, and makes his decision accordingly. In this case, we have yet more regularity in the CPD: $P(J \mid a^1, s^1, l^1) = P(J \mid a^1, s^1, l^0)$.   ∎

   In this simple example, we have a CPD in which several values of Pa_J specify the same conditional probability over J. In general, we often have CPDs where, for certain partial assignments u to subsets $U \subset \mathrm{Pa}_X$, the values of the remaining parents are not relevant. In such cases, several different distributions $P(X \mid \mathrm{pa}_X)$ are identical. In this section, we discuss how we might capture this regularity in our CPD representation and what implications this structure has on conditional independence. There are many possible approaches for capturing functions over a scope X that are constant over certain subsets of instantiations to X. In this section, we present two common and useful choices: trees and rules.

5.3.1.1   Tree-CPDs

A very natural representation for capturing common elements in a CPD is via a *tree*, where the leaves of the tree represent different possible (conditional) distributions over J, and where the path to each leaf dictates the contexts in which this distribution is used.

Example 5.8

Figure 5.4 shows a tree for the CPD of the variable J in example 5.7. Given this tree, we find $P(J \mid A, S, L)$ by traversing the tree from the root downward. At each internal node, we see a test on one of the attributes. For example, in the root node of our tree we see a test on the value A. We then follow the arc that is labeled with the value a, which is given in the current setting of the parents. Assume, for example, that we are interested in $P(J \mid a^1, s^1, l^0)$. Thus, we have that $A = a^1$, and we would follow the right-hand arc labeled a^1. The next test is over S. We have $S = s^1$, and we would also follow the right-hand arc. We have now reached a leaf, which is annotated with a particular distribution over J: $P(j^1) = 0.9$, and $P(j^0) = 0.1$. This distribution is the one we use for $P(J \mid a^1, s^1, l^0)$. ∎

Formally, we use the following recursive definition of trees.

Definition 5.2

tree-CPD

A tree-CPD representing a CPD for variable X is a rooted tree; each t-node in the tree is either a leaf t-node or an interior t-node. Each leaf is labeled with a distribution $P(X)$. Each interior t-node is labeled with some variable $Z \in \mathrm{Pa}_X$. Each interior t-node has a set of outgoing arcs to its children, each one associated with a unique variable assignment $Z = z_i$ for $z_i \in Val(Z)$.

A branch β through a tree-CPD is a path beginning at the root and proceeding to a leaf node. We assume that no branch contains two interior nodes labeled by the same variable. The parent context induced by branch β is the set of variable assignments $Z = z_i$ encountered on the arcs along the branch. ∎

Note that, to avoid confusion, we use t-nodes and arcs for a tree-CPD, as opposed to our use of nodes and edges as the terminology in a BN.

Example 5.9

Consider again the tree in figure 5.4. There are four branches in this tree. One induces the parent context $\langle a^0 \rangle$, corresponding to the situation where the student did not apply for the job. A second induces the parent context $\langle a^1, s^1 \rangle$, corresponding to an application with a high SAT score. The remaining two branches induce complete assignments to all the parents of J: $\langle a^1, s^0, l^1 \rangle$ and $\langle a^1, s^0, l^0 \rangle$. Thus, this representation breaks down the conditional distribution of J given its parents into four parent contexts by grouping the possible assignments in $Val(\mathrm{Pa}_J)$ into subsets that have the same effect on J. Note that now we need only 4 parameters to describe the behavior of J, instead of 8 in the table representation. ∎

Regularities of this type occur in many domains. Some events can occur only in certain situations. For example, we can have a *Wet* variable, denoting whether we get wet; that variable would depend on the *Raining* variable, but only in the context where we are outside. Another type of example arises in cases where we have a sensor, for example, a thermometer; in general, the thermometer depends on the temperature, but not if it is broken.

This type of regularity is very common in cases where a variable can depend on one of a large set of variables: it depends only on one, but we have uncertainty about the choice of variable on which it depends.

Example 5.10

Let us revisit example 3.7, where George had to decide whether to give the recruiter at Acme Consulting the letter from his professor in Computer Science 101 or his professor in Computer Science 102. George's chances of getting a job can depend on the quality of both letters L_1 and L_2,

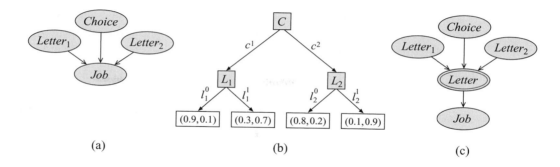

(a)　　　　　　　　　　　　(b)　　　　　　　　　　　　(c)

Figure 5.5 **The** OneLetter **example.** (a) The network fragment. (b) tree-CPD for $P(J \mid C, L_1, L_2)$. (c) Modified network with a new variable L that has a multiplexer CPD.

and hence both are parents. However, depending on which choice C George makes, the dependence will only be on one of the two. Figure 5.5a shows the network fragment, and b shows the tree-CPD for the variable J. (For simplicity, we have eliminated the dependence on S and A that we had in figure 5.4.) ■

More formally, we define the following:

A CPD $P(Y \mid A, Z_1, \ldots, Z_k)$ *is said to be a* multiplexer CPD *if $Val(A) = \{1, \ldots, k\}$, and*

$$P(Y \mid a, Z_1, \ldots, Z_k) = \mathbf{I}\{Y = Z_a\},$$

selector variable

where a is the value of A. The variable A is called the selector variable *for the CPD.* ■

In other words, the value of the multiplexer variable is a copy of the value of one of its parents, $Z_1, \ldots, Z_k$. The role of A is to select the parent who is being copied. Thus, we can think of a multiplexer CPD as a switch.

We can apply this definition to example 5.10 by introducing a new variable L, which is a multiplexer of L_1 and L_2, using C as the selector. The variable J now depends directly only on L. The modified network is shown in figure 5.5c.

This type of model arises in many settings. For example, it can arise when we have different actions in our model; it is often the case that the set of parents for a variables varies considerably based on the action taken. Configuration variables also result in such situations: depending on the specific configuration of a physical system, the interactions between variables might differ (see box 5.A).

Another setting where this type of model is particularly useful is in dealing with uncertainty about *correspondence* between different objects. This problem arises, for example, in *data association*, where we obtain sensor measurements about real-world objects, but we are uncertain about which object gave rise to which sensor measurement. For example, we might get a blip on a radar screen without knowing which of of several airplanes the source of the signal. Such cases also arise in robotics (see box 15.A and box 19.D). Similar situations also arise in other applications, such as *identity resolution*: associating names mentioned in text to the real-

correspondence

data association

identity
resolution

correspondence
variable

world objects to which they refer (see box 6.D). We can model this type of situation using a *correspondence variable U* that associates, with each sensor measurement, the identity u of the object that gave rise to the measurement. The actual sensor measurement is then defined using a multiplexer CPD that depends on the correspondence variable U (which plays the role of the selector variable), and on the value of $A(u)$ for all u from which the measurement could have been derived. The value of the measurement will be the value of $A(u)$ for $U = u$, usually with some added noise due to measurement error. Box 12.D describes this problem in more detail and presents algorithms for dealing with the difficult inference problem it entails.

Trees provide a very natural framework for representing context-specificity in the CPD. In particular, it turns out that people find it very convenient to represent this type of structure using trees. Furthermore, the tree representation lends itself very well to automated learning algorithms that construct a tree automatically from a data set.

---

troubleshooting

Box 5.A — Case Study: Context-Specificity in Diagnostic Networks. *A common setting where context-specific CPDs arise is in* troubleshooting *of physical systems, as described, for example, by Heckerman, Breese, and Rommelse (1995). In such networks, the context specificity is due to the presence of alternative configurations. For example, consider a network for diagnosis of faults in a printer, developed as part of a suite of troubleshooting networks for Microsoft's Windows 95^{TM} operating system. This network, shown in figure 5.A.1a, models the fact that the printer can be hooked up either to the network via an Ethernet cable or to a local computer via a cable, and therefore depends on both the status of the local transport medium and the network transport medium. However, the status of the Ethernet cable only affects the printer's output if the printer is hooked up to the network. The tree-CPD for the variable Printer-Output is shown in figure 5.A.1b. Even in this very simple network, this use of local structure in the CPD reduced the number of parameters required from 145 to 55.*

We return to the topic of Bayesian networks for troubleshooting in box 21.C and box 23.C.

5.3.1.2   Rule CPDs

As we seen, trees are appealing for several reasons. However, trees are a global representation that captures the entire CPD in a single data structure. In many cases, it is easier to reason using a CPD if we break down the dependency structure into finer-grained elements. A finer-grained representation of context-specific dependencies is via *rules*. Roughly speaking, each rule corresponds to a single entry in the CPD of the variable. It specifies a context in which the CPD entry applies and its numerical value.

Definition 5.4

rule

scope

A rule ρ is a pair $\langle c; p \rangle$ where c is an assignment to some subset of variables C, and $p \in [0, 1]$. We define C to be the scope of ρ, denoted $Scope[\rho]$.   ∎

This representation decomposes a tree-CPD into its most basic elements.

(a)

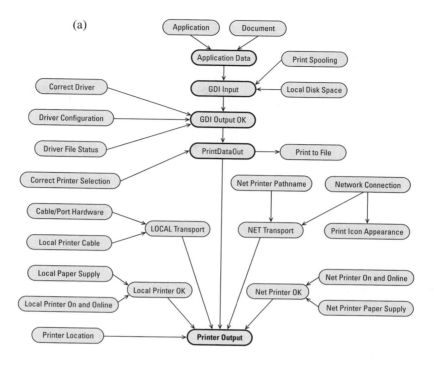

(b)

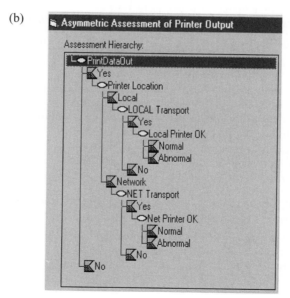

Figure 5.A.1 — Context-specific independencies for diagnostic networks. (a) A real-life Bayesian network that uses context-specific dependencies. The network is used for diagnosing printing problems in Microsoft's online troubleshooting system. (b) The structure of the tree-CPD for the *Printer Output* variable in that network.

Example 5.11

Consider the tree of figure 5.4. There are eight entries in the CPD tree, such that each one corresponds to a branch in the tree and an assignment to the variable J itself. Thus, the CPD defines eight rules:

$$
\left\{
\begin{array}{l}
\rho_1\!:\!\langle a^0, j^0; 0.8\rangle \\
\rho_2\!:\!\langle a^0, j^1; 0.2\rangle \\
\rho_3\!:\!\langle a^1, s^0, l^0, j^0; 0.9\rangle \\
\rho_4\!:\!\langle a^1, s^0, l^0, j^1; 0.1\rangle \\
\rho_5\!:\!\langle a^1, s^0, l^1, j^0; 0.4\rangle \\
\rho_6\!:\!\langle a^1, s^0, l^1, j^1; 0.6\rangle \\
\rho_7\!:\!\langle a^1, s^1, j^0; 0.1\rangle \\
\rho_8\!:\!\langle a^1, s^1, j^1; 0.9\rangle
\end{array}
\right\}
$$

For example, the rule ρ_4 is derived by following the branch a^1, s^0, l^0 and then selecting the probability associated with the assignment $J = j^1$. ∎

Although we can decompose any tree-CPD into its constituent rules, we wish to define rule-based CPDs as an independent notion. To define a coherent CPD from a set of rules, we need to make sure that each conditional distribution of the form $P(X \mid \mathrm{pa}_X)$ is specified by precisely one rule. Thus, the rules in a CPD must be mutually exclusive and exhaustive.

Definition 5.5

rule-based CPD

A rule-based CPD $P(X \mid \mathrm{Pa}_X)$ is a set of rules $\mathcal{R}$ such that:

- *For each rule $\rho \in \mathcal{R}$, we have that $Scope[\rho] \subseteq \{X\} \cup \mathrm{Pa}_X$.*

- *For each assignment $(x, \boldsymbol{u})$ to $\{X\} \cup \mathrm{Pa}_X$, we have precisely one rule $\langle \boldsymbol{c}; p\rangle \in \mathcal{R}$ such that $\boldsymbol{c}$ is compatible with $(x, \boldsymbol{u})$. In this case, we say that $P(X = x \mid \mathrm{Pa}_X = \boldsymbol{u}) = p$.*

- *The resulting CPD $P(X \mid \boldsymbol{U})$ is a legal CPD, in that*

$$
\sum_x P(x \mid \boldsymbol{u}) = 1.
$$

∎

The rule set in example 5.11 satisfies these conditions.

Consider the following, more complex, example.

Example 5.12

Let X be a variable with $\mathrm{Pa}_X = \{A, B, C\}$, and assume that X's CPD is defined via the following set of rules:

$$
\begin{array}{ll}
\rho_1\!:\!\langle a^1, b^1, x^0; 0.1\rangle & \rho_2\!:\!\langle a^1, b^1, x^1; 0.9\rangle \\
\rho_3\!:\!\langle a^0, c^1, x^0; 0.2\rangle & \rho_4\!:\!\langle a^0, c^1, x^1; 0.8\rangle \\
\rho_5\!:\!\langle b^0, c^0, x^0; 0.3\rangle & \rho_6\!:\!\langle b^0, c^0, x^1; 0.7\rangle \\
\rho_7\!:\!\langle a^1, b^0, c^1, x^0; 0.4\rangle & \rho_8\!:\!\langle a^1, b^0, c^1, x^1; 0.6\rangle \\
\rho_9\!:\!\langle a^0, b^1, c^0; 0.5\rangle &
\end{array}
$$

This set of rules defines the following CPD:

X	$a^0 b^0 c^0$	$a^0 b^0 c^1$	$a^0 b^1 c^0$	$a^0 b^1 c^1$	$a^1 b^0 c^0$	$a^1 b^0 c^1$	$a^1 b^1 c^0$	$a^1 b^1 c^1$
x^0	0.3	0.2	0.5	0.2	0.3	0.4	0.1	0.1
x^1	0.7	0.8	0.5	0.8	0.7	0.6	0.9	0.9

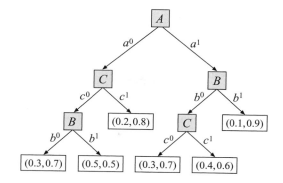

Figure 5.6 **A tree-CPD for the rule-based CPD** $P(X \mid A, B, C)$ **of example 5.12.**

For example, the CPD entry $P(x^0 \mid a^0, b^1, c^1)$ is determined by the rule ρ_3, resulting in the CPD entry 0.2; we can verify that no other rule is compatible with the context a^0, b^1, c^1, x^1. We can also verify that each of the CPD entries is also compatible with precisely one context, and hence that the different contexts are mutually exclusive and exhaustive.

Note that both CPD entries $P(x^1 \mid a^0, b^1, c^0)$ and $P(x^0 \mid a^0, b^1, c^0)$ are determined by a single rule ρ_9. As the probabilities for the different contexts in this case must sum up to 1, this phenomenon is only possible when the rule defines a uniform distribution, as it does in this case. ∎

This perspective views rules as a decomposition of a CPD. We can also view a rule as a finer-grained factorization of an entire distribution.

Proposition 5.1

Let $\mathcal{B}$ be a Bayesian network, and assume that each CPD $P(X \mid \mathrm{Pa}_X)$ in $\mathcal{B}$ is represented as a set of rules $\mathcal{R}_X$. Let $\mathcal{R}$ be the multiset defined as $\uplus_{X \in \mathcal{X}} \mathcal{R}_X$, where $\uplus$ denotes multiset join, which puts together all of the rule instances (including duplicates). Then, the probability of any instantiation ξ to the network variables $\mathcal{X}$ can be computed as

$$P(\xi) = \prod_{\langle c; p \rangle \in \mathcal{R}, \xi \sim c} p.$$

The proof is left as an exercise (exercise 5.3).

The rule representation is more than a simple transformation of tree-CPDs. In particular, although every tree-CPDs can be represented compactly as a set of rules, the converse does not necessarily hold: not every rule-based CPD can be represented compactly as a tree.

Example 5.13

Consider the rule-based CPD of example 5.12. In any rule set that is derived from a tree, one variable — the one at the root — appears in all rules. In the rule set $\mathcal{R}$, none of the parent variables A, B, C appears in all rules, and hence the rule set is not derived from a tree. If we try to represent it as a tree-CPD, we would have to select one of A, B, or C to be the root. Say, for example, that we select A to be the root. In this case, rules that do not contain A would necessarily correspond to more than one branch (one for a^1 and one for a^0). Thus, the transformation would result in more branches than rules. For example, figure 5.6 shows a minimal tree-CPD that represents the rule-based CPD of example 5.12. ∎

5.3.1.3   Other Representations

The tree and rule representations provide two possibilities for representing context-specific structure. We have focused on these two approaches as they have been demonstrated to be useful for representation, for inference, or for learning. However, other representations are also possible, and can also be used for these tasks. In general, if we abstract away from the details of these representations, we see that they both simply induce *partitions* of $Val(\{X\} \cup \mathrm{Pa}_X)$, defined by the branches in the tree on one hand or the rule contexts on the other. Each partition is associated with a different entry in X's CPD.

This perspective allows us to understand the strengths and limitations of the different representations. In both trees and rules, all the partitions are described via an assignment to a subset of the variables. Thus, for example, we cannot represent the partition that contains only a^1, s^1, l^0 and a^1, s^0, l^1, a partition that we might obtain if the recruiter lumped together candidates that had a high SAT score or a strong recommendation letter, but not both. As defined, these representations also require that we either split on a variable (within a branch of the tree or within a rule) or ignore it entirely. In particular, this restriction does not allow us to capture dependencies that utilize a taxonomic hierarchy on some parent attribute, as described in box 5.B

Of course, we can still represent distributions with these properties by simply having multiple tree branches or multiple rules that are associated with the same parameterization. However, this solution both is less compact and fails to capture some aspects of the structure of the CPD. A very flexible representation, that allows these structures, might use general logical formulas to describe partitions. This representation is very flexible, and it can precisely capture any partition we might consider; however, the formulas might get fairly complex. Somewhat more restrictive is decision diagram the use of a *decision diagram*, which allows different t-nodes in a tree to share children, avoiding duplication of subtrees where possible. This representation is more general than trees, in that any structure that can be represented compactly as a tree can be represented compactly as a decision diagram, but the converse does not hold. Decision diagrams are incomparable to rules, in that there are examples where each is more compact than the other. In general, different representations offer different trade-offs and might be appropriate for different applications.

multinet **Box 5.B — Concept: Multinets and Similarity Networks.** *The* multinet *representation provides a more global approach to capturing context-specific independence. In its simple form, a multinet is a network centered on a single distinguished class variable C, which is a root of the network. The multinet defines a separate network $\mathcal{B}_c$ for each value of C, where the structure as well as the parameters can differ for these different networks. In most cases, a multinet defines a single network where every variable X has as its parents C, and all variables Y in any of the networks $\mathcal{B}_c$. However, the CPD of X is such that, in context $C = c$, it depends only on $\mathrm{Pa}_X^{\mathcal{B}_c}$. In some cases, however, a subtlety arises, where Y is a parent of X in $\mathcal{B}_{c^1}$, and X is a parent of Y in $\mathcal{B}_{c^2}$. In this case, the Bayesian network induced by the multinet is cyclic; nevertheless, because of the context-specific independence properties of this network, it specifies a coherent distribution. (See also exercise 5.2.) Although, in most cases, a multinet can be represented as a standard BN with context-specific CPDs, it is nevertheless useful, since it explicitly shows the independencies in a graphical form, making them easier to understand and elicit.*

similarity
network

A related representation, the similarity network, *was developed as part of the Pathfinder system (see box 3.D). In a similarity network, we define a network $\mathcal{B}_S$ for certain subsets of values $S \subset Val(C)$, which contains only those attributes relevant for distinguishing between the values in S. The underlying assumption is that, if a variable X does not appear in the network $\mathcal{B}_S$, then $P(X \mid C = c)$ is the same for all $c \in S$. Moreover, if X does not have Y as a parent in this network, then X is contextually independent of Y given $C \in S$ and X's other parents in this network. A similarity network easily captures structure where the dependence of X on C is defined in terms of a taxonomic hierarchy on C. For example, we might have that our class variable is Disease. While Sore-throat depends on Disease, it does not have a different conditional distribution for every value d of Disease. For example, we might partition diseases into diseases that do not cause sore throat and those that do, and the latter might be further split into diffuse disease (causing soreness throughout the throat) and localized diseases (such as abscesses). Using this partition, we might have only three different conditional distributions for $P(\textit{Sore-Throat} \mid \textit{Disease} = d)$. Multinets facilitate elicitation both by focusing the expert's attention on attributes that matter, and by reducing the number of distinct probabilities that must be elicited.*

5.3.2 Independencies

In many of our preceding examples, we used phrases such as "In the case a^0, where the student does not apply, the recruiter's decision cannot depend on the variables S and L." These phrases suggest that context-specific CPDs induce context-specific independence. In this section, we analyze the independencies induced by context-specific dependency models.

Consider a CPD $P(X \mid \mathrm{Pa}_X)$, where certain distributions over X are shared across different instantiations of Pa_X. The structure of such a CPD allows us to infer certain independencies *locally* without having to consider any global aspects of the network.

Example 5.14

Returning to example 5.7, we can see that $(J \perp_c S, L \mid a^0)$: By the definition of the CPD, $P(J \mid a^0, s, l)$ is the same for all values of s and l. Note that this equality holds regardless of the structure or the parameters of the network in which this CPD is embedded. Similarly, we have that $(J \perp_c L \mid a^1, s^1)$. ∎

In general, if we define c to be the context associated with a branch in the tree-CPD for X, then X is independent of the remaining parents $(\mathrm{Pa}_X - Scope[c])$ given the context c. However, there might be additional CSI statements that we can determine locally, conditioned on contexts that are not induced by complete branches.

Example 5.15

Consider, the tree-CPD of figure 5.5b. Here, once George chooses to request a letter from one professor, his job prospects still depend on the quality of that professor's letter, but not on that of the other. More precisely, we have that $(J \perp_c L_2 \mid c^1)$; note that c^1 is not the full assignment associated with a branch. ∎

Example 5.16

More interestingly, consider again the tree of figure 5.4, and suppose we are given the context s^1. Clearly, we should only consider branches that are consistent with this value. There are two such *branches. One associated with the assignment a^0 and the other with the assignment a^1, s^1. We can immediately see that the choice between these two branches does not depend on the value of L. Thus, we conclude that $(J \perp_c L \mid s^1)$ holds in this case.* ∎

We can generalize this line of reasoning by considering the rules compatible with a particular context c. Intuitively, if none of these rules mentions a particular parent Y of X, then X is conditionally independent of Y given c. More generally, we can define the notion of conditioning a rule on a context:

Definition 5.6

reduced rule

Let $\rho = \langle c'; p \rangle$ be a rule and $C = c$ be a context. If c' is compatible with c, we say that $\rho \sim c$. In this case, let $c'' = c'\langle Scope[c'] - Scope[c]\rangle$ be the assignment in c' to the variables in $Scope[c'] - Scope[c]$. We then define the reduced rule $\rho[c] = \langle c''; p \rangle$. For $\mathcal{R}$ a set of rules, we define the reduced rule set

$$\mathcal{R}[c] = \{\rho[c] \; : \; \rho \in \mathcal{R}, \rho \sim c\}.$$
∎

Example 5.17

In the rule set $\mathcal{R}$ of example 5.12, $\mathcal{R}[a^1]$ is the set

$$
\begin{array}{ll}
\rho_1': \langle b^1, x^0; 0.1 \rangle & \rho_2: \langle b^1, x^1; 0.9 \rangle \\
\rho_5': \langle b^0, c^0, x^0; 0.3 \rangle & \rho_6: \langle b^0, c^0, x^1; 0.7 \rangle \\
\rho_7': \langle b^0, c^1, x^0; 0.4 \rangle & \rho_8': \langle b^0, c^1, x^1; 0.6 \rangle.
\end{array}
$$

Thus, we have left only the rules compatible with a^1, and eliminated a^1 from the context in the rules where it appeared. ∎

Proposition 5.2

Let $\mathcal{R}$ be the rules in the rule-based CPD for a variable X, and let $\mathcal{R}_c$ be the rules in $\mathcal{R}$ that are compatible with c. Let $\boldsymbol{Y} \subseteq \mathrm{Pa}_X$ be some subset of parents of X such that $\boldsymbol{Y} \cap Scope[c] = \emptyset$. If for every $\rho \in \mathcal{R}[c]$, we have that $\boldsymbol{Y} \cap Scope[\rho] = \emptyset$, then $(X \perp_c \boldsymbol{Y} \mid \mathrm{Pa}_X - \boldsymbol{Y}, c)$.

The proof is left as an exercise (exercise 5.4).

This proposition specifies a computational tool for deducing "local" CSI relations from the rule representation. We can check whether a variable Y is being tested in the reduced rule set given a context in linear time in the number of rules. (See also exercise 5.6 for a similar procedure for trees.)

This procedure, however, is incomplete in two ways. First, since the procedure does not examine the actual parameter values, it can miss additional independencies that are true for the specific parameter assignments. However, as in the case of completeness for d-separation in BNs, this violation only occurs in degenerate cases. (See exercise 5.7.)

The more severe limitation of this procedure is that it only tests for independencies between X and some of its parents given a context and the other parents. Are there are other, more global, implications of such CSI relations?

Example 5.18

Consider example 5.7 again. In general, finding out that Gump got the job at Acme will increase our belief that he is intelligent, via evidential reasoning. However, now assume that we know that Gump did not apply. Intuitively, we now learn nothing about his intelligence from the fact that he got the job. ∎

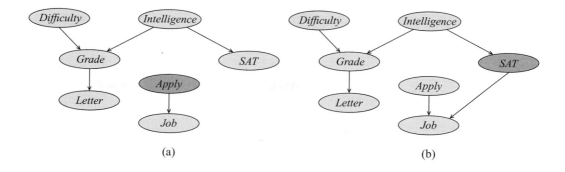

Figure 5.7 **The graph of figure 5.3, after we remove spurious edges:** (a) in the context $A = a^0$; (b) in the context $S = s^1$.

Can we capture this intuition formally? Consider the dependence structure in the context $A = a^0$. Intuitively, in this context, the edges $S \to J$ and $L \to J$ are both redundant, since we know that $(J \perp_c S, L \mid a^0)$. Thus, our intuition is that we should check for d-separation in the graph without this edge. Indeed, we can show that this is a sound check for CSI conditions.

<table>
<tr><td>

Definition 5.7

spurious edge
</td><td>

Let $P(X \mid \mathrm{Pa}_X)$ be a CPD, let $Y \in \mathrm{Pa}_X$, and let c be a context. We say that the edge $Y \to X$ is spurious in the context c if $P(X \mid \mathrm{Pa}_X)$ satisfies $(X \perp_c Y \mid \mathrm{Pa}_X - \{Y\}, c')$, where $c' = c\langle \mathrm{Pa}_X \rangle$ is the restriction of c to variables in Pa_X.   ■
</td></tr>
</table>

If we represent CPDs with rules, then we can determine whether an edge is spurious by examining the reduced rule set. Let $\mathcal{R}$ be the rule-based CPD for $P(X \mid \mathrm{Pa}_X)$, then the edge $Y \to X$ is spurious in context c if Y does not appear in the reduced rule set $\mathcal{R}[c]$.

Algorithm 5.2 Computing d-separation in the presence of context-specific CPDs

---

    Procedure CSI-sep (

      $\mathcal{G}$,   // Bayesian network structure

      c,   // Context

      $\boldsymbol{X}, \boldsymbol{Y}, \boldsymbol{Z}$ // Is $\boldsymbol{X}$ CSI-separated from $\boldsymbol{Y}$ given $\boldsymbol{Z}, c$

)

1      $\mathcal{G}' \leftarrow \mathcal{G}$

2   **for** each edge $Y \to X$ in $\mathcal{G}'$

3   **if** $Y \to X$ is spurious given c in $\mathcal{G}$ **then**

4   Remove $Y \to X$ in $\mathcal{G}'$

5   **return** $d\text{-}sep_{\mathcal{G}'}(\boldsymbol{X}; \boldsymbol{Y} \mid \boldsymbol{Z}, c)$

6

CSI-separation   Now we can define *CSI-separation*, a variant of d-separation that takes CSI into account. This notion, defined procedurally in algorithm 5.2, is straightforward: we use local considerations to remove spurious edges and then apply standard d-separation to the resulting graph. We say that

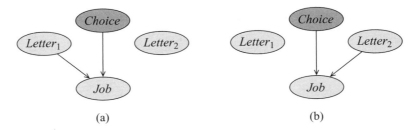

Figure 5.8   **Two reductions of the CPD for the** OneLetter **example:**   (a) in the context $C = c^1$; (b) in the context $C = c^2$.

X is *CSI-separated* from Y given Z in the context c if CSI-SEP$(\mathcal{G}, c, X, Y, Z)$ returns *true*.

As an example, consider the network of example 5.7, in the context $A = a^0$. In this case, we get that the arcs $S \to J$ and $L \to J$ are spurious, leading to the reduced graph in figure 5.7a. As we can see, J and I are d-separated in the reduced graph, as are J and D. Thus, using CSI-SEP, we get that I and J are d-separated given the context a^0. Figure 5.7b shows the reduced graph in the context s^1.

It is not hard to show that CSI-separation provides a sound test for determining context-specific independence.

Theorem 5.3

Let $\mathcal{G}$ be a network structure, let P be a distribution such that $P \models \mathcal{I}_\ell(\mathcal{G})$, let c be a context, and let X, Y, Z be sets of variables. If X is CSI-separated from Y given Z in the context c, then $P \models (X \perp_c Y \mid Z, c)$.

The proof is left as an exercise (exercise 5.8).

Of course, we also want to know if CSI-separation is complete — that is, whether it discovers all the context-specific independencies in the distribution. At best, we can hope for the same type of qualified completeness that we had before: discovering all CSI assertions that are a direct consequence of the structural properties of the model, regardless of the particular choice of parameters. In this case, the structural properties consist of the graph structure (as usual) and the structure of the rule sets or trees. Unfortunately, even this weak notion of completeness does not hold in this case.

Example 5.19

Consider the example of figure 5.5b and the context $C = c^1$. In this context, the arc $L_2 \to J$ is spurious. Thus, there is no path between L_1 and L_2, even given J. Hence, CSI-SEP will report that L_1 and L_2 are d-separated given J and the context $C = c^1$. This case is shown in figure 5.8a. Therefore, we conclude that $(L_1 \perp_c L_2 \mid J, c^1)$. Similarly, in the context $C = c^2$, the arc $L_1 \to J$ is spurious, and we have that L_1 and L_2 are d-separated given J and c^2, and hence that $(L_1 \perp_c L_2 \mid J, c^2)$. Thus, reasoning by cases, we conclude that once we know the value of C, we have that L_1 and L_2 are always d-separated given J, and hence that $(L_1 \perp L_2 \mid J, C)$.

Can we get this conclusion using CSI-separation? Unfortunately, the answer is no. If we invoke CSI-separation with the empty context, then no edges are spurious and CSI-separation reduces to d-separation. Since both L_1 and L_2 are parents of J, we conclude that they are not separated given J and C.   ∎

The problem here is that CSI-separation does not perform reasoning by cases. Of course, if we want to determine whether X and Y are independent given $\mathbf{Z}$ and a context c, we can invoke CSI-separation on the context c, z for each possible value of $\mathbf{Z}$, and see if X and Y are separated in all of these contexts. This procedure, however, is exponential in the number of variables of $\mathbf{Z}$. Thus, it is practical only for small evidence sets. Can we do better than reasoning by cases? The answer is that sometimes we cannot. See exercise 5.10 for a more detailed examination of this issue.

5.4 Independence of Causal Influence

In this section, we describe a very different type of structure in the local probability model. Consider a variable Y whose distribution depends on some set of causes $X_1, \ldots, X_k$. In general, Y can depend on its parents in arbitrary ways — the X_i can interact with each other in complex ways, making the effect of each combination of values unrelated to any other combination. However, in many cases, the combined influence of the X_i's on Y is a simple combination of the influence of each of the X_i's on Y in isolation. In other words, each of the X_i's influences Y independently, and the influence of several of them is simply combined in some way.

We begin by describing two very useful models of this type — the *noisy-or* model, and the class of *generalized linear models*. We then provide a general definition for this type of interaction.

5.4.1 The Noisy-Or Model

Let us begin by considering an example in which a different professor writes a recommendation letter for a student. Unlike our earlier example, this professor teaches a small seminar class, where she gets to know every student. The quality of her letter depends on two things: whether the student participated in class, for example, by asking good questions (Q); and whether he wrote a good final paper (F). Roughly speaking, each of these events is enough to cause the professor to write a good letter. However, the professor might fail to remember the student's participation. On the other hand, she might not have been able to read the student's handwriting, and hence may not appreciate the quality of his final paper. Thus, there is some noise in the process.

Let us consider each of the two causes in isolation. Assume that $P(l^1 \mid q^1, f^0) = 0.8$, that is, the professor is 80 percent likely to remember class participation. On the other hand, $P(l^1 \mid q^0, f^1) = 0.9$, that is, the student's handwriting is readable in 90 percent of the cases. What happens if both occur: the student participates in class and writes a good final paper? The key assumption is that these are two independent *causal mechanisms* for causing a strong letter, and that the letter is weak only if neither of them succeeded. The first causal mechanism — class participation q^1 — fails with probability 0.2. The second mechanism — a good final paper f^1 — fails with probability 0.1. If both q^1 and f^1 occurred, the probability that both mechanisms fail (independently) is $0.2 \cdot 0.1 = 0.02$. Thus, we have that $P(l^0 \mid q^1, f^1) = 0.02$

causal
mechanism

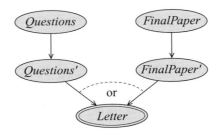

Figure 5.9   Decomposition of the noisy-or model for *Letter*

and $P(l^1 \mid q^1, f^1) = 0.98$. In other words, our CPD for $P(L \mid Q, F)$ is:

Q, F	l^0	l^1
$q^0 f^0$	1	0
$q^0 f^1$	0.1	0.9
$q^1 f^0$	0.2	0.8
$q^1 f^1$	0.02	0.98

noisy-or CPD   This type of interaction between causes is called the *noisy-or* model. Note that we assumed that a student cannot end up with a strong letter if he neither participated in class nor wrote a good final paper. We relax this assumption later on.

An alternative way of understanding this interaction is by assuming that the letter-writing process can be represented by a more elaborate probabilistic model, as shown in figure 5.9. This figure represents the conditional distribution for the *Letter* variable given *Questions* and *FinalPaper*. It also uses two intermediate variables that reveal the associated causal mechanisms. The variable Q' is true if the professor remembers the student's participation; the variable F' is true if the professor could read and appreciate the student's high-quality final paper. The letter is strong if and only if one of these events holds. We can verify that the conditional distribution $P(L \mid Q, F)$ induced by this network is precisely the one shown before.

noise parameter   The probability that Q causes L (0.8 in this example) is called the *noise parameter*, and denoted λ_Q. In the context of our decomposition, $\lambda_Q = P(q'^1 \mid q^1)$. Similarly, we have a noise parameter λ_F, which in this context is $\lambda_F = P(f'^1 \mid f^1)$.

leak probability   We can also incorporate a *leak probability* that represents the probability — say 0.0001 — that the professor would write a good recommendation letter for no good reason, simply because she is having a good day. We simply introduce another variable into the network to represent this event. This variable has no parents, and is true with probability $\lambda_0 = 0.0001$. It is also a parent of the *Letter* variable, which remains a deterministic or.

The decomposition of this CPD clearly shows why this local probability model is called a noisy-or. The basic interaction of the effect with its causes is that of an OR, but there is some noise in the "effective value" of each cause.

We can define this model in the more general setting:

Definition 5.8   *Let Y be a binary-valued random variable with k binary-valued parents $X_1, \ldots, X_k$. The CPD*
noisy-or CPD   *$P(Y \mid X_1, \ldots, X_k)$ is a* noisy-or *if there are $k + 1$ noise parameters $\lambda_0, \lambda_1, \ldots, \lambda_k$ such that*

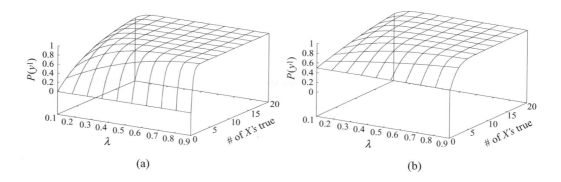

Figure 5.10 **The behavior of the noisy-or model as a function of λ and the number of parents that have value *true*:** (a) with a leak probability of 0; (b) with a leak probability of 0.5.

$$P(y^0 \mid X_1, \ldots, X_k) = (1 - \lambda_0) \prod_{i \,:\, X_i = x_i^1} (1 - \lambda_i) \tag{5.2}$$

$$P(y^1 \mid X_1, \ldots, X_k) = 1 - \left[(1 - \lambda_0) \prod_{i \,:\, X_i = x_i^1} (1 - \lambda_i)\right]$$

We note that, if we interpret x_i^1 as 1 and x_i^0 as 0, we can rewrite equation (5.2) somewhat more compactly as:

$$P(y^0 \mid x_1, \ldots, x_k) = (1 - \lambda_0) \prod_{i=1}^{k} (1 - \lambda_i)^{x_i}. \tag{5.3}$$

Although this transformation might seem cumbersome, it will turn out to be very useful in a variety of settings.

Figure 5.10 shows a graph of the behavior of a special-case noisy or model, where all the variables have the same noise parameter λ. The graph shows the probability of the child Y in terms of the parameter λ and the number of X_i's that have the value *true*.

The noisy-or model is applicable in a wide variety of settings, but perhaps the most obvious is in the medical domain. For example, as we discussed earlier, a symptom variable such as *Fever* usually has a very large number of parents, corresponding to different diseases that can cause the symptom. However, it is often a reasonable approximation to assume that the different diseases use different causal mechanisms, and that if any disease succeeds in activating its mechanism, the symptom is present. Hence, the noisy-or model is a reasonable approximation.

BN2O network

Box 5.C — Concept: BN2O Networks. *A class of networks that has received some attention in the domain of medical diagnosis is the class of BN2O networks.*

A BN2O network, illustrated in figure 5.C.1, is a two-layer Bayesian network, where the top layer corresponds to a set of causes, such as diseases, and the second to findings that might indicate these

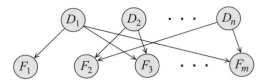

Figure 5.C.1 — A two-layer noisy-or network

causes, such as symptoms or test results. All variables are binary-valued, and the variables in the second layer all have noisy-or models. Specifically, the CPD of F_i is given by:

$$P(f_i^0 \mid \mathrm{Pa}_{F_i}) \;=\; (1 - \lambda_{i,0}) \prod_{D_j \in \mathrm{Pa}_{F_i}} (1 - \lambda_{i,j})^{d_j}.$$

These networks are conceptually very simple and require a small number of easy-to-understand parameters: Each edge denotes a causal association between a cause d_i and a finding f_j; each is associated with a parameter $\lambda_{i,j}$ that encodes the probability that d_i, in isolation, causes f_j to manifest. Thus, these networks resemble a simple set of noisy rules, a similarity that greatly facilitates the knowledge-elicitation task. Although simple, BN2O networks are a reasonable first approximation for a medical diagnosis network.

BN2O networks also have another useful property. In Bayesian networks, observing a variable generally induces a correlation between all of its parents. In medical diagnosis networks, where findings can be caused by a large number of diseases, this phenomenon might lead to significant complexity, both cognitively and in terms of inference. However, in medical diagnosis, most of the findings in any specific case are false — a patient generally only has a small handful of symptoms. As discussed in section 5.4.4, the parents of a noisy-or variable F are conditionally independent given that we observe that F is false. As a consequence, a BN2O network where we observe $F = f^0$ is equivalent to a network where F disappears from the network entirely (see exercise 5.13). This observation can greatly reduce the cost of inference.

5.4.2   Generalized Linear Models

generalized linear
model

An apparently very different class of models that also satisfy independence of causal influence are the *generalized linear models*. Although there are many models of this type, in this section we focus on models that define probability distributions $P(Y \mid X_1, \ldots, X_k)$ where Y takes on values in some discrete finite space. We first discuss the case where Y and all of the X_i's are binary-valued. We then extend the model to deal with the multinomial case.

5.4.2.1   Binary-Valued Variables

Roughly speaking, our models in this case are a soft version of a linear threshold function. As a motivating example, we can think of applying this model in a medical setting: In practice, our body's immune system is constantly fighting off multiple invaders. Each of them adds to the burden, with some adding more than others. We can imagine that when the total burden passes some threshold, we begin to exhibit a fever and other symptoms of infection. That is, as the total burden increases, the probability of fever increases. This requires us to clarify two terms in this discussion. The first is the "total burden" value and how it depends on the particular possible disease causes. The second is a specification of how the probability of fever depends on the total burden.

More generally, we examine a CPD of Y given $X_1, \ldots, X_k$. We assume that the effect of the X_i's on Y can be summarized via a linear function $f(X_1, \ldots, X_k) = \sum_{i=1}^{k} w_i X_i$, where we again interpret x_i^1 as 1 and x_i^0 as 0. In our example, this function will be the total burden on the immune system, and the w_i coefficient describes how much burden is contributed by each disease cause.

The next question is how the probability of $Y = y^1$ depends on $f(X_1, \ldots, X_k)$. In general, this probability undergoes a phase transition around some threshold value τ: when $f(X_1, \ldots, X_k) \geq \tau$, then Y is very likely to be 1; when $f(X_1, \ldots, X_k) < \tau$, then Y is very likely to be 0. It is easier to eliminate τ by simply defining $f(X_1, \ldots, X_k) = w_0 + \sum_{i=1}^{k} w_i X_i$, so that w_0 takes the role of $-\tau$.

To provide a realistic model for immune system example and others, we do not use a hard threshold function to define the probability of Y, but rather a smoother transition function. One common choice (although not the only one) is the *sigmoid* or *logit* function:

sigmoid

$$\text{sigmoid}(z) = \frac{e^z}{1 + e^z}.$$

Figure 5.11a shows the sigmoid function. This function implies that the probability *saturates* to 1 when $f(X_1, \ldots, X_k)$ is large, and saturates to 0 when $f(X_1, \ldots, X_k)$ is small. And so, activation of another disease cause for a sick patient will not change the probability of fever by much, since it is already close to 1. Similarly, if the patient is healthy, a minor burden on the immune system will not increase the probability of fever, since $f(X_1, \ldots, X_k)$ is far from the threshold. In the area of the phase transition, the behavior is close to linear.

We can now define:

Definition 5.9

logistic CPD

Let Y be a binary-valued random variable with k parents $X_1, \ldots, X_k$ that take on numerical values. The CPD $P(Y \mid X_1, \ldots, X_k)$ is a logistic CPD *if there are $k + 1$ weights $w_0, w_1, \ldots, w_k$ such that:*

$$P(y^1 \mid X_1, \ldots, X_k) \quad = \quad \text{sigmoid}\left(w_0 + \sum_{i=1}^{k} w_i X_i\right). \qquad \blacksquare$$

We have already encountered this CPD in example 4.20, where we saw that it can be derived by taking a naive Markov network and reformulating it as a conditional distribution.

log-odds

We can interpret the parameter w_i in terms of its effect on the *log-odds* of Y. In general, the odds ratio for a binary variable is the ratio of the probability of y^1 and the probability of y^0. It is the same concept used when we say that the odds of some event (for example, a sports team winning the Super Bowl) are "2 to 1." Consider the odds ratio for the variable Y, where we use Z to represent $w_0 + \sum_i w_i X_i$:

$$O(\boldsymbol{X}) = \frac{P(y^1 \mid X_1, \ldots, X_k)}{P(y^0 \mid X_1, \ldots, X_k)} = \frac{e^Z/(1 + e^Z)}{1/(1 + e^Z)} = e^Z.$$

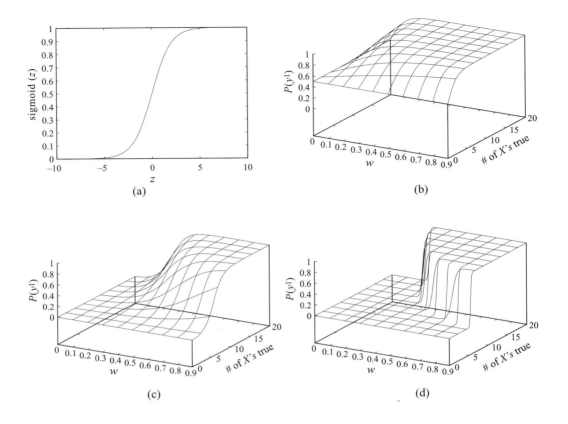

Figure 5.11   The behavior of the sigmoid CPD: (a) The sigmoid function. (b),(c), & (d) The behavior of the linear sigmoid model as a function of w and the number of parents that have value *true*: (b) when the threshold $w_0 = 0$; (c) when $w_0 = -5$; (d) when w and w_0 are multiplied by 10.

Now, consider the effect on this odds ratio as some variable X_j changes its value from *false* to *true*. Let $\boldsymbol{X}_{-j}$ be the variables in $X_1, \ldots, X_k$ except for X_j. Then:

$$\frac{O(\boldsymbol{X}_{-j}, x_j^1)}{O(\boldsymbol{X}_{-j}, x_j^0)} = \frac{\exp(w_0 + \sum_{i \neq j} w_i X_i + w_j)}{\exp(w_0 + \sum_{i \neq j} w_i X_i)} = e^{w_j}.$$

Thus, $X_j = true$ changes the odds ratio by a multiplicative factor of e^{w_j}. A positive coefficient $w_j > 0$ implies that $e^{w_j} > 1$ so that the odds ratio increases, hence making y^1 more likely. Conversely, a negative coefficient $w_j < 0$ implies that $e^{w_j} < 1$ and hence the odds ratio decreases, making y^1 less likely.

Figure 5.11b shows a graph of the behavior of a special case of the logistic CPD model, where all the variables have the same weight w. The graph shows $P(Y \mid X_1, \ldots, X_k)$ as a function of w and the number of X_i's that take the value *true*. The graph shows two cases: one where $w_0 = 0$ and the other where $w_0 = -5$. In the first case, the probability starts out at 0.5

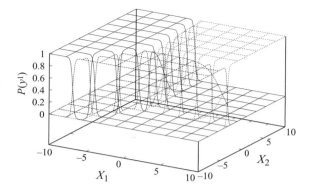

Figure 5.12   A multinomial logistic CPD: The distribution of $P(Y \mid X_1, X_2)$ using the multinomial logistic model for $\ell_1(X_1, X_2) = -3X_1 - 2X_2 + 1$, $\ell_2(X_1, X_2) = 5X_1 - 8X_2 - 4$, and $\ell_3 = x - y + 10$.

when none of the causes are in effect, and rapidly goes up to 1; the rate of increase is, as expected, much higher for high values of w. It is interesting to compare this graph to the graph of figure 5.10b that shows the behavior of the noisy-or model with $\lambda_0 = 0.5$. The graphs exhibit very similar behavior for $\lambda = w$, showing that the incremental effect of a new cause is similar in both. However, the logistic CPD also allows for a negative influence of some X_i on Y by making w_i negative. Furthermore, the parameterization of the logistic model also provides substantially more flexibility in generating qualitatively different distributions. For example, as shown in figure 5.11c, setting w_0 to a different value allows us to obtain the threshold effect discussed earlier. Furthermore, as shown in figure 5.11d, we can adapt the scale of the parameters to obtain a sharper transition. However, the noisy-or model is cognitively very plausible in many settings. Furthermore, as we discuss, it has certain benefits both in reasoning with the models and in learning the models from data.

5.4.2.2   Multivalued Variables

We can extend the logistic CPD to the case where Y takes on multiple values $y^1, \ldots, y^m$. In this case, we can imagine that the different values of Y are each supported in a different way by the X_i's, where the support is again defined via a linear function. The choice of Y can be viewed as a soft version of "winner takes all," where the y^i that has the most support gets probability 1 and the others get probability 0.

More precisely, we have:

Definition 5.10

multinomial
logistic CPD

Let Y be an m-valued random variable with k parents $X_1, \ldots, X_k$ that take on numerical values. The CPD $P(Y \mid X_1, \ldots, X_k)$ is a multinomial logistic *if for each $j = 1, \ldots, m$, there are $k + 1$*

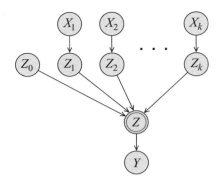

Figure 5.13   Independence of causal influence

weights $w_{j,0}, w_{j,1}, \ldots, w_{j,k}$ such that:

$$\ell_j(X_1, \ldots, X_k) \;=\; w_{j,0} + \sum_{i=1}^{k} w_{j,i} X_i$$

$$P(y^j \mid X_1, \ldots, X_k) \;=\; \frac{\exp(\ell_j(X_1, \ldots, X_k))}{\sum_{j'=1}^{m} \exp(\ell_{j'}(X_1, \ldots, X_k))}. \qquad \blacksquare$$

Figure 5.12 shows one example of this model for the case of two parents and a three-valued child Y. We note that one of the weights $w_{j,1}, \ldots, w_{j,k}$ is redundant, as it can be folded into the bias term $w_{j,0}$.

We can also deal with the case where the parent variables X_i take on more than two values. The approach taken is usually straightforward. If $X_i = x_i^1, \ldots, x_i^m$, we define a new set of binary-valued variables $X_{i,1}, \ldots, X_{i,m}$, where $X_{i,j} = x_{i,j}^1$ precisely when $X_i = j$. Each of these new variables gets its own coefficient (or set of coefficients) in the logistic function. For example, if we have a binary-valued child Y with an m-valued parent X, our logistic function would be parameterized using $m + 1$ weights, $w_0, w_1, \ldots, w_m$, such that

$$P(y^1 \mid X) \;=\; \text{sigmoid}\left(w_0 + \sum_{j=1}^{m} w_j \boldsymbol{I}\{X = x^j\}\right). \qquad (5.4)$$

We note that, for any assignment to X_i, precisely one of the weights $w_1, \ldots, w_m$ will make a contribution to the linear function. As a consequence, one of the weights is redundant, since it can be folded into the bias weight w_0.

We noted before that we can view a binary-valued logistic CPD as a conditional version of a naive Markov model. We can generalize this observation to the nonbinary case, and show that the multinomial logit CPD is also a particular type of pairwise CRF (see exercise 5.16).

5.4.3   The General Formulation

Both of these models are specials case of a general class of local probability models, which satisfy a property called *causal independence* or *independence of causal influence* (ICI). These

causal
independence

models all share the property that the influence of multiple causes can be decomposed into separate influences. We can define the resulting class of models more precisely as follows:

Definition 5.11

Let Y be a random variable with parents $X_1, \ldots, X_k$. The CPD $P(Y \mid X_1, \ldots, X_k)$ exhibits independence of causal influence *if it is described via a network fragment of the structure shown in figure 5.13, where the CPD of Z is a deterministic function f.* ∎

Intuitively, each variable X_i can be transformed separately using its own individual noise model. The resulting variables Z_i are combined using some deterministic combination function. Finally, an additional stochastic choice can be applied to the result Z, so that the final value of Y is not necessarily a deterministic function of the variables Z_i's. The key here is that any stochastic parts of the model are applied independently to each of the X_i's, so that there can be no interactions between them. The only interaction between the X_i's occurs in the context of the function f.

As stated, this definition is not particularly meaningful. Given an arbitrarily complex function f, we can represent any CPD using the representation of figure 5.13. (See exercise 5.15.) It is possible to place various restrictions on the form of the function f that would make the definition more meaningful. For our purposes, we provide a fairly stringent definition that fortunately turns out to capture the standard uses of ICI models.

Definition 5.12

We say that a deterministic binary function $x \diamond y$ is commutative *if $x \diamond y = y \diamond x$, and* associative *if $(x \diamond y) \diamond z = x \diamond (y \diamond z)$. We say that a function $f(x_1, \ldots, x_k)$ is a* symmetric decomposable *function if there is a commutative associative function $x \diamond y$ such that $f(x_1, \ldots, x_k) = x_1 \diamond x_2 \diamond \cdots \diamond x_k$.[3]* ∎

Definition 5.13

symmetric ICI

We say that the CPD $P(Y \mid X_1, \ldots, X_k)$ exhibits symmetric ICI *if it is described via a network fragment of the structure shown in figure 5.13, where the CPD of Z is a deterministic symmetric decomposable function f. The CPD exhibits* fully symmetric ICI *if the CPDs of the different Z_i variables are identical.* ∎

There are many instantiations of the symmetric ICI model, with different noise models — $P(Z_i \mid X_i)$ — and different combination functions. Our noisy-or model uses the combination function OR and a simple noise model with binary variables. The generalized linear models use the Z_i to produce $w_i X_i$, and then summation as the combination function f. The final soft thresholding effect is accomplished in the distribution of Y given Z.

These types of models turn out to be very useful in practice, both because of their cognitive plausibility and because they provide a significant reduction in the number of parameters required to represent the distribution. The number of parameters in the CPD is linear in the number of parents, as opposed to the usual exponential.

Box 5.D — Case Study: Noisy Rule Models for Medical Diagnosis. *As discussed in box 5.C, noisy rule interactions such as noisy-or are a simple yet plausible first approximation of models for*

noisy-max

medical diagnosis. A generalization that is also useful in this setting is the noisy-max *model. Like*

---

3. Because $\diamond$ is associative, the order of application of the operations does not matter.

the application of the noisy-or model for diagnosis, the parents X_i correspond to different diseases that the patient might have. In this case, however, the value space of the symptom variable Y can be more refined than simply $\{present, absent\}$; it can encode the severity of the symptom. Each Z_i corresponds (intuitively) to the effect of the disease X_i on the symptom Y in isolation, that is, the severity of the symptom in case only the disease X_i is present. The value of Z is the maximum of the different Z_i's.

Both noisy-or and noisy-max models have been used in several medical diagnosis networks. Two of the largest are the QMR-DT (Shwe et al. 1991) and CPCS (Pradhan et al. 1994) networks, both based on various versions of a knowledge-based system called QMR (Quick Medical Reference), compiled for diagnosis of internal medicine. QMR-DT is a BN2O network (see box 5.C) that contains more than five hundred significant diseases, about four thousand associated findings, and more than forty thousand disease-finding associations.

CPCS is a somewhat smaller network, containing close to five hundred variables and more than nine hundred edges. Unlike QMR-DT, the network contains not only diseases and findings but also variables for predisposing factors and intermediate physiological states. Thus, CPCS has at least four distinct layers. All variables representing diseases and intermediate states take on one of four values. A specification of the network using full conditional probability tables would require close to 134 million parameters. However, the network is constructed using only noisy-or and noisy-max interactions, so that the number of actual parameters is only 8,254. Furthermore, most of the parameters were generated automatically from "frequency weights" in the original knowledge base. Thus, the number of parameters that were, in fact, elicited during the construction of the network is around 560.

Finally, the symmetric ICI models allow certain decompositions of the CPD that can be exploited by probabilistic inference algorithms for computational gain, when the domain of the variables Z_i and the variable Z are reasonably small.

5.4.4   Independencies

As we have seen, structured CPDs often induce independence properties that go beyond those represented explicitly in the Bayesian network structure. Understanding these independencies can be useful for gaining insight into the properties of our distribution. Also, as we will see, the additional structure can be exploited for improving the performance of various probabilistic inference algorithms.

The additional independence properties that arise in general ICI models $P(Y \mid X_1, \ldots, X_k)$ are more indirect than those we have seen in the context of deterministic CPDs or tree-CPDs. In particular, they do not manifest directly in terms of the original variables, but only if we decompose it by adding auxiliary variables. In particular, as we can easily see from figure 5.13, each X_i is conditionally independent of Y, and of the other X_j's, given Z_i.

We can obtain even more independencies by decomposing the CPD of Z in various ways. For example, assume that $k = 4$, so that our CPD has the form $P(Y \mid X_1, X_2, X_3, X_4)$. We can introduce two new variables W_1 and W_2, such that:

$$
\begin{aligned}
W_1 &= Z_0 \diamond Z_1 \diamond Z_2 \\
W_2 &= Z_3 \diamond Z_4 \\
Z &= W_1 \diamond W_2
\end{aligned}
$$

By the associativity of $\diamond$, the decomposed CPD is precisely equivalent to the original one. In this CPD, we can use the results of section 5.2 to conclude, for example, that X_4 is independent of Y given W_2.

Although these independencies might appear somewhat artificial, it turns out that the associated decomposition of the network can be exploited by inference algorithms (see section 9.6.1). However, as we will see, they are only useful when the domain of the intermediate variables (W_1 and W_2 in our example) are small. This restriction should not be surprising given our earlier observation that any CPD can be decomposed in this way if we allow the Z_i's and Z to be arbitrarily complex.

The independencies that we just saw are derived simply from the fact that the CPD of Z is deterministic and symmetric. As in section 5.2, there are often additional independencies that are associated with the particular choice of deterministic function. The best-known independence of this type is the one arising for noisy-or models:

Proposition 5.3

Let $P(Y \mid X_1, \ldots, X_k)$ be a noisy-or CPD. Then for each $i \neq j$, X_i is independent of X_j given $Y = y^0$.

The proof is left as an exercise (exercise 5.11). Note that this independence is not derived from the network structure via d-separation: Instantiating Y enables the v-structure between X_i and X_j, and hence potentially renders them correlated. Furthermore, this independence is context-specific: it holds only for the specific value $Y = y^0$. Other deterministic functions are associated with other context-specific independencies.

5.5 Continuous Variables

So far, we have restricted attention to discrete variables with finitely many values. In many situations, some variables are best modeled as taking values in some continuous space. Examples include variables such as position, velocity, temperature, and pressure. Clearly, we cannot use a table representation in this case. One common solution is to circumvent the entire issue by discretizing all continuous variables. Unfortunately, this solution can be problematic in many cases. In order to get a reasonably accurate model, we often have to use a fairly fine discretization, with tens or even hundreds of values. For example, when applying probabilistic models to a robot navigation task, a typical discretization granularity might be 15 centimeters for the x and y coordinates of the robot location. For a reasonably sized environment, each of these variables might have more than a thousand values, leading to more than a million discretized values for the robot's position. CPDs of this magnitude are outside the range of most systems.

 Furthermore, **when we discretize a continuous variable we often lose much of the structure that characterizes it. It is not generally the case that each of the million values that defines a robot position can be associated with an arbitrary probability. Basic**

continuity assumptions that hold in almost all domains imply certain relationships that hold between probabilities associated with "nearby" discretized values of a continuous variable. However, such constraints are very hard to capture in a discrete distribution, where there is no notion that two values of the variable are "close" to each other.

Fortunately, nothing in our formulation of a Bayesian network requires that we restrict attention to discrete variables. Our only requirement is that the CPD $P(X \mid \text{Pa}_X)$ represent, for every assignment of values pa_X to Pa_X, a distribution over X. In this case, X might be continuous, in which case the CPD would need to represent distributions over a continuum of values; we might also have some of X's parents be continuous, so that the CPD would also need to represent a continuum of different probability distributions. However, as we now show, we can provide implicit representations for CPDs of this type, allowing us to apply all of the machinery we developed for the continuous case as well as for *hybrid networks* involving both discrete and continuous variables.

hybrid network

In this section, we describe how continuous variables can be integrated into the BN framework. We first describe the purely continuous case, where the CPDs involve only continuous variables, both as parents and as children. We then examine the case of hybrid networks, which involve both discrete and continuous variables.

There are many possible models one could use for any of these cases; we briefly describe only one prototypical example for each of them, focusing on the models that are most commonly used. Of course, there is an unlimited range of representations that we can use: any parametric representation for a CPD is eligible in principle. The only difficulty, as far as representation is concerned, is in creating a language that allows for it. Other tasks, such as inference and learning, are a different issue. As we will see, these tasks can be difficult even for very simple hybrid models.

The most commonly used parametric form for continuous density functions is the Gaussian distribution. We have already described the univariate Gaussian distribution in chapter 2. We now describe how it can be used within the context of a Bayesian network representation.

First, let us consider the problem of representing a dependency of a continuous variable Y on a continuous parent X. One simple solution is to decide to model the distribution of Y as a Gaussian, whose parameters depend on the value of X. In this case, we need to have a set of parameters for every one of the infinitely many values $x \in Val(X)$. A common solution is to decide that the mean of Y is a linear function of X, and that the variance of Y does not depend on X. For example, we might have that

$$p(Y \mid x) = \mathcal{N}\left(-2x + 0.9; 1\right).$$

Example 5.20   *Consider a vehicle (for example, a car) moving over time. For simplicity, assume that the vehicle is moving along a straight line, so that its position (measured in meters) at the t'th second is described using a single variable $X^{(t)}$. Let $V^{(t)}$ represent the velocity of the car at the kth second, measured in meters per second. Then, under ideal motion, we would have that $X^{(t+1)} = X^{(t)} + V^{(t)}$ — if the car is at meter #510 along the road, and its current velocity is 15 meters/second, then we expect its position at the next second to be meter #525. However, there is invariably some stochasticity in the motion. Hence, it is much more realistic to assert that the car's position $X^{(t+1)}$ is described using a Gaussian distribution whose mean is 525 and whose variance is 5 meters.*   ∎

This type of dependence is called a *linear Gaussian* model. It extends to multiple continuous parents in a straightforward way:

Definition 5.14

linear Gaussian
CPD

Let Y be a continuous variable with continuous parents $X_1, \ldots, X_k$. We say that Y has a linear
Gaussian model if there are parameters $\beta_0, \ldots, \beta_k$ and σ^2 such that

$$p(Y \mid x_1, \ldots, x_k) = \mathcal{N}\left(\beta_0 + \beta_1 x_1 + \cdots + \beta_k x_k ; \sigma^2\right).$$

In vector notation,

$$p(Y \mid \boldsymbol{x}) = \mathcal{N}\left(\beta_0 + \boldsymbol{\beta}^T \boldsymbol{x} ; \sigma^2\right).$$

Viewed slightly differently, this formulation says that Y is a linear function of the variables
$X_1, \ldots, X_k$, with the addition of Gaussian noise with mean 0 and variance σ^2:

$$Y = \beta_0 + \beta_1 x_1 + \cdots + \beta_k x_k + \epsilon,$$

where ϵ is a Gaussian random variable with mean 0 and variance σ^2, representing the noise in
the system.

This simple model captures many interesting dependencies. However, there are certain facets
of the situation that it might not capture. For example, the variance of the child variable Y
cannot depend on the actual values of the parents. In example 5.20, we might wish to construct
a model in which there is more variance about a car's future position if it is currently moving
very quickly. The linear Gaussian model cannot capture this type of interaction.

Of course, we can easily extend this model to have the mean and variance of Y depend
on the values of its parents in arbitrary way. For example, we can easily construct a richer
representation where we allow the mean of Y to be $\sin(x_1)^{x_2}$ and its variance to be $(x_3/x_4)^2$.
However, the linear Gaussian model is a very natural one, which is a useful approximation in
many practical applications. Furthermore, as we will see in section 7.2, networks based on the
linear Gaussian model provide us with an alternative representation for multivariate Gaussian
distributions, one that directly reveals more of the underlying structure.

---

robot localization

Box 5.E — Case Study: Robot Motion and Sensors. *One interesting application of hybrid models is in the domain of robot localization. In this application, the robot must keep track of its
location as it moves in an environment, and obtains sensor readings that depend on its location.
This application is an example of a temporal model, a topic that will be discussed in detail in
section 6.2; we also return to the robot example specifically in box 15.A. There are two main local
probability models associated with this application. The first specifies the robot dynamics — the
distribution over its position at the next time step L' given its current position L and the action
taken A; the second specifies the robot sensor model — the distribution over its observed sensor
reading S at the current time given its current location L.*

*We describe one model for this application, as proposed by Fox et al. (1999) and Thrun et al.
(2000). Here, the robot location L is a three-dimensional vector containing its X, Y coordinates
and an angular orientation θ. The action A specifies a distance to travel and a rotation (offset
from the current θ). The model uses the assumption that the errors in both translation and rotation
are normally distributed with zero mean. Specifically, $P(L' \mid L, A)$ is defined as a product of two
independent Gaussians with cut off tails, $P(\theta' \mid \theta, A)$ and $P(X', Y' \mid X, Y, A)$, whose variances*

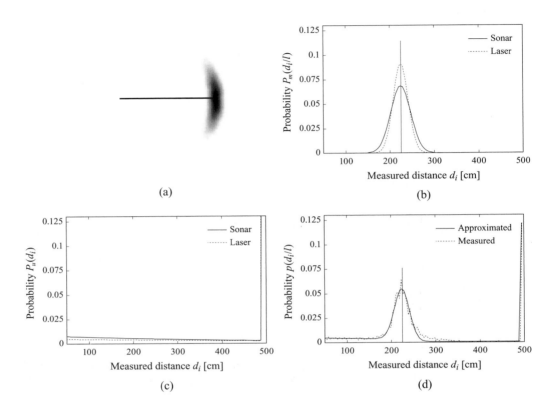

(a)

(b)

(c)

(d)

Figure 5.E.1 — Probabilistic model for robot localization track. (a) A typical "banana-shaped" distribution for the robot motion model. The figure shows the projection of the conditional distribution over $L' = \langle X', Y', \theta' \rangle$ onto the X', Y' space, given the robot's starting position and action shown. (b) Two distributions $P_m(D \mid L)$ for the distance returned by a range sensor given that the distance to the closest obstacle is $o_L = 230cm$. The figure shows a distribution for both an ultrasound sensor and a laser range finder; the laser sensor has a higher accuracy than the ultrasound sensor, as indicated by the smaller variance. (c) Two distributions $P_u(D)$ for the distance returned by a range sensor, for an ultrasound sensor and a laser range finder. The relatively large probability of measuring 500 centimeters owes to the fact that the maximum range of the proximity sensors is set to 500 centimeters. Thus, this distance represents the probability of measuring *at least* 500 centimeters. (d) Overall model distribution (solid line) and empirical distribution (dashed line) of $P(D \mid o_L)$, for $o_L = 230cm$ for a laser sensor.

are proportional to the length of the motion. The robot's conditional distribution over (X', Y') is a banana-shaped cloud (see figure 5.E.1a, where the banana shape is due to the noise in the rotation).

The sensor is generally some type of range sensor, either a sonar or a laser, which provides a reading D of the distance between the robot and the nearest obstacle along the direction of the sensor. There are two distinct cases to consider. If the sensor signal results from an obstacle in the map, then the resulting distribution is modeled by a Gaussian distribution with mean at the distance to this obstacle. Letting o_L be the distance to the closest obstacle to the position L (along the sensor beam), we can define $P_m(D \mid L) = \mathcal{N}\left(o_L; \sigma^2\right)$, where the variance σ^2 represents the uncertainty of the measured distance, based on the accuracy of the world model and the accuracy of the sensor. Figure 5.E.1b shows an example of such a distribution for an ultrasound sensor and a laser range finder. The laser sensor has a higher accuracy than the ultrasound sensor, as indicated by the smaller variance.

The second case arises when the sensor beam is reflected by an obstacle not represented in the world model (for example, a dynamic obstacle, such as a person or a chair, which is not in the robot's map). Assuming that these objects are equally distributed in the environment, the probability $P_u(D)$ of detecting an unknown obstacle at distance D is independent of the location of the robot and can be modeled by an exponential distribution. This distribution results from the observation that a distance d is measured if the sensor is not reflected by an obstacle at a shorter distance and is reflected at distance d. An example exponential distribution is shown in figure 5.E.1c.

Only one of these two cases can hold for a given measurement. Thus, $P(D \mid L)$ is a combination of the two distributions P_m and P_u. The combined probability $P(D \mid L)$ is based on the observation that d is measured in one of two cases:

- *The sensor beam in not reflected by an unknown obstacle before reaching distance d, and is reflected by the known obstacle at distance d (an event that happens only with some probability).*

- *The beam is reflected neither by an unknown obstacle nor by the known obstacle before reaching distance d, and it is reflected by an unknown obstacle at distance d.*

Overall, the probability of sensor measurements is computed incrementally for the different distances starting at 0cm; for each distance, we consider the probability that the sensor beam reaches the corresponding distance and is reflected either by the closest obstacle in the map (along the sensor beam) or by an unknown obstacle. Putting these different cases together, we obtain a single distribution for $P(D \mid L)$. This distribution is shown in figure 5.E.1d, along with an empirical distribution obtained from data pairs consisting of the distance o_L to the closest obstacle on the map and the measured distance d during the typical operation of the robot.

5.5.1 Hybrid Models

We now turn our attention to models incorporating both discrete and continuous variables. We have to address two types of dependencies: a continuous variable with continuous and discrete parents, and a discrete variable with continuous and discrete parents.

Let us first consider the case of a continuous child X. If we ignore the discrete parents of X, we can simply represent the CPD of X as a linear Gaussian of X's continuous parents. The simplest way of making the continuous variable X depend on a discrete variable U is to define a different set of parameters for every value of the discrete parent. More precisely:

Definition 5.15

conditional linear
Gaussian CPD

Let X be a continuous variable, and let $U = \{U_1, \ldots, U_m\}$ be its discrete parents and $Y = \{Y_1, \ldots, Y_k\}$ be its continuous parents. We say that X has a conditional linear Gaussian *(CLG) CPD if, for every value $u \in Val(U)$, we have a set of $k + 1$ coefficients $a_{u,0}, \ldots, a_{u,k}$ and a variance σ_u^2 such that*

$$p(X \mid u, y) = \mathcal{N}\left(a_{u,0} + \sum_{i=1}^{k} a_{u,i} y_i ; \sigma_u^2\right)$$ ∎

If we restrict attention to this type of CPD, we get an interesting class of models. More precisely, we have:

Definition 5.16

CLG network

A Bayesian network is called a CLG *network if every discrete variable has only discrete parents and every continuous variable has a CLG CPD.*   ∎

Gaussian mixture
distribution

Note that the conditional linear Gaussian model does not allow for continuous variables to have discrete children. A CLG model induces a joint distribution that has the form of a *mixture* — a weighted average — of Gaussians. The mixture contains one Gaussian component for each instantiation of the discrete network variables; the weight of the component is the probability of that instantiation. Thus, the number of mixture components is (in the worst case) exponential in the number of discrete network variables.

Finally, we address the case of a discrete child with a continuous parent. The simplest model is a threshold model. Assume we have a binary discrete variable U with a continuous parent Y. We may want to define:

$$P(u^1) = \begin{cases} 0.9 & y \leq 65 \\ 0.05 & \text{otherwise.} \end{cases}$$

Such a model may be appropriate, for example, if Y is the temperature (in Fahrenheit) and U is the thermostat turning the heater on.

The problem with the threshold model is that the change in probability is discontinuous as a function of Y, which is both inconvenient from a mathematical perspective and implausible in many settings. However, we can address this problem by simply using the logistic model or its multinomial extension, as defined in definition 5.9 or definition 5.10.

Figure 5.14 shows how a multinomial CPD can be used to model a simple sensor that has three values: *low*, *medium* and *high*. The probability of each of these values depends on the value of the continuous parent Y. As discussed in section 5.4.2, we can easily accommodate a variety of noise models for the sensor: we can make it less reliable in borderline situations by making the transitions between regions more moderate. It is also fairly straightforward to generalize the model to allow the probabilities of the different values in each of the regions to be values other than 0 or 1.

As for the conditional linear Gaussian CPD, we address the existence of discrete parents for Y by simply introducing a separate set of parameters for each instantiation of the discrete parents.

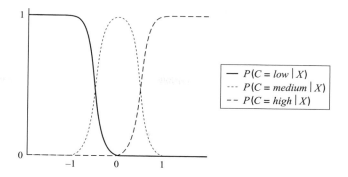

Figure 5.14 **Generalized linear model for a thermostat**

5.6 Conditional Bayesian Networks

The previous sections all describe various compact representations of a CPD. Another very useful way of compactly representing a conditional probability distribution is via a Bayesian network fragment. We have already seen one very simple example of this idea: Our decomposition of the noisy-or CPD for the *Letter* variable, shown in figure 5.9. There, our decomposition used a Bayesian network to represent the internal model of the *Letter* variable. The network included explicit variables for the parents of the variable, as well as auxiliary variables that are not in the original network. This entire network represented the CPD for *Letter*. In this section, we generalize this idea to a much wider setting.

Note that the network fragment in this example is not a full Bayesian network. In particular, it does not specify a probabilistic model — parents and a CPD — for the parent variables *Questions* and *FinalPaper*. This network fragment specifies not a joint distribution over the variables in the fragment, but a *conditional* distribution of *Letter given Questions* and *FinalPaper*. More generally, we can define the following:

Definition 5.17

conditional
Bayesian network

A conditional Bayesian network $\mathcal{B}$ over $\boldsymbol{Y}$ given $\boldsymbol{X}$ is defined as a directed acyclic graph $\mathcal{G}$ whose nodes are $\boldsymbol{X} \cup \boldsymbol{Y} \cup \boldsymbol{Z}$, where $\boldsymbol{X}, \boldsymbol{Y}, \boldsymbol{Z}$ are disjoint. The variables in $\boldsymbol{X}$ are called inputs, the variables in $\boldsymbol{Y}$ outputs, and the variables in $\boldsymbol{Z}$ encapsulated. The variables in $\boldsymbol{X}$ have no parents in $\mathcal{G}$. The variables in $\boldsymbol{Z} \cup \boldsymbol{Y}$ are associated with a conditional probability distribution. The network defines a conditional distribution using a chain rule:

$$P_{\mathcal{B}}(\boldsymbol{Y}, \boldsymbol{Z} \mid \boldsymbol{X}) = \prod_{X \in \boldsymbol{Y} \cup \boldsymbol{Z}} P(X \mid \mathrm{Pa}_X^{\mathcal{G}}).$$

The distribution $P_{\mathcal{B}}(\boldsymbol{Y} \mid \boldsymbol{X})$ is defined as the marginal of $P_{\mathcal{B}}(\boldsymbol{Y}, \boldsymbol{Z} \mid \boldsymbol{X})$:

$$P_{\mathcal{B}}(\boldsymbol{Y} \mid \boldsymbol{X}) = \sum_{\boldsymbol{Z}} P_{\mathcal{B}}(\boldsymbol{Y}, \boldsymbol{Z} \mid \boldsymbol{X}).$$

■

conditional
random field

The *conditional random field* of section 4.6.1 is the undirected analogue of this definition.

The notion of a conditional BN turns out to be useful in many settings. In particular, we can use it to define an encapsulated CPD.

Definition 5.18

encapsulated
CPD

Let Y be a random variable with k parents $X_1, \ldots, X_k$. The CPD $P(Y \mid X_1, \ldots, X_k)$ is an encapsulated CPD if it is represented using a conditional Bayesian network over Y given $X_1, \ldots, X_k$. ∎

At some level, it is clear that the representation of an individual CPD for a variable Y as a conditional Bayesian network $\mathcal{B}_Y$ does not add expressive power to the model. After all, we could simply take the network $\mathcal{B}_Y$ and "substitute it in" for the atomic CPD $P(Y \mid \mathrm{Pa}_Y)$. One key advantage of the encapsulated representation over a more explicit model is that the encapsulation can simplify the model significantly from a cognitive perspective. Consider again our noisy-or model. Externally, to the rest of the network, we can still view *Letter* as a single variable with its two parents: *Questions* and *FinalPaper*. All of the internal structure is encapsulated, so that, to the rest of the network, the variable can be viewed as any other variable. In particular, a knowledge engineer specifying the network does not have to ascribe meaning to the encapsulated variables.

The encapsulation advantage can be even more significant when we want to describe a complex system where components are composed of other, lower-level, subsystems. When specifying a model for such a system, we would like to model each subsystem separately, without having to consider the internal model of its lower level components.

In particular, consider a model for a physical device such as a computer; we might construct such a model for fault diagnosis purposes. When modeling the computer, we would like to avoid thinking about the detailed structure and fault models of its individual components, such as the hard drive, and within the hard drive the disk surfaces, the controller, and more, each of which has yet other components. By using an encapsulated CPD, we can decouple the model of the computer from the detailed model of the hard drive. We need only specify which global aspects of the computer state the hard drive behavior depends on, and which it influences. Furthermore, we can hierarchically compose encapsulated CPDs, modeling, in turn, the hard drive's behavior in terms of its yet-lower-level components.

In figure 5.15 we show a simple hierarchical model for a computer system. This high-level model for a computer, figure 5.15a, uses encapsulated CPDs for *Power-Source*, *Motherboard*, *Hard-Drive*, *Printer*, and more. The *Hard-Drive* CPD has inputs *Temperature*, *Age* and *OS-Status*, and the outputs *Status* and *Full*. Although the hard drive has a rich internal state, the only aspects of its state that influence objects outside the hard drive are whether it is working properly and whether it is full. The *Temperature* input of the hard drive in a computer is outside the probabilistic model and will be mapped to the *Temperature* parent of the *Hard-Drive* variable in the computer model. A similar mapping happens for other inputs.

The *Hard-Drive* encapsulated network, figure 5.15b, in turn uses encapsulated CPDs for *Controller*, *Surface1*, *Drive-Mechanism*, and more. The hierarchy can continue as necessary. In this case, the model for the variable *Motor* (in the *Drive-Mechanism*) is "simple," in that none of its CPDs are encapsulated.

object-oriented
Bayesian network

One obvious observation that can be derived from looking at this example is that an encapsulated CPD is often appropriate for more than one variable in the model. For example, the encapsulated CPD for the variable *Surface1* in the hard drive is almost certainly the same as the CPDs for the variables *Surface2*, *Surface2*, and *Surface4*. Thus, we can imagine creating a *template* of an encapsulated CPD, and reusing it multiple times, for several variables in the model. This idea forms the basis for a framework known as *object-oriented Bayesian networks*.

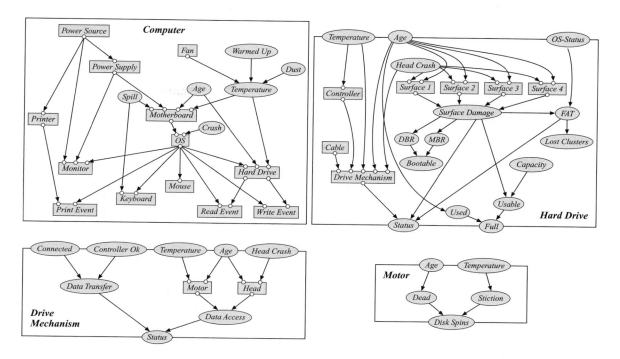

Figure 5.15   Example of encapsulated CPDs for a computer system model: Four levels of a hierarchy of encapsulated CPDs in a model of a computer system. Variables with encapsulated CPDs are shown as rectangles, while nonhierarchical variables are ovals (as usual). Each encapsulated CPD is contained in a box. Input variables intersect the top edge of the box, indicating the fact that their values are received from outside the class, while output variables intersect the bottom. The rectangles representing the complex components also have little bubbles on their borders, showing that variables are passed into and out of those components.

5.7   Summary

In this chapter, we have shown that our ability to represent structure in the distribution does not end at the level of the graph. In many cases, here is important structure within the CPDs that we wish to make explicit. In particular, we discussed several important types of discrete structured CPDs.

- deterministic functions;

- asymmetric, or context specific, dependencies;

- cases where different influences combine independently within the CPD, including noisy-or, logistic functions, and more.

In many cases, we showed that the additional structure provides not only a more compact parameterization, but also additional independencies that are not visible at the level of the original graph.

As we discussed, the idea of structured CPDs is critical in the case of continuous variables, where a table-based representation is clearly irrelevant. We discussed various representations for CPDs in hybrid (discrete/continuous) networks, of which the most common is the linear Gaussian representation. For this case, we showed some important connections between the linear Gaussian representation and multivariate Gaussian distributions.

Finally, we discussed the notion of a conditional Bayesian network, which allows us to decompose a conditional probability distribution recursively, into another Bayesian network.

5.8   Relevant Literature

This chapter addresses the issue of modeling the probabilistic dependence of one variable on some set of others. This issue plays a central role in both statistics and (supervised) machine learning, where much of the work is devoted precisely to the prediction of one variable from others. Indeed, many of the representations described in this chapter are derived from statistical models such as decision trees or regression models. More information on these models can be found in a variety of statistics textbooks (for example, Breiman et al. 1984; Duda et al. 2000; Hastie et al. 2001; McCullagh and Nelder 1989).

The extension of d-separation to the case of Bayesian networks with deterministic CPDs was done by Geiger et al. (1990). Nielsen et al. (2000) proposed a specific representation for deterministic CPDs with particular benefits for inference.

The notion of context-specific independence, which recurs throughout much of this chapter, was first formalized by Shimony (1991). Representations for asymmetric dependencies, which most directly capture CSI, were a key component in Heckerman's similarity networks and the related multinets (Heckerman and Nathwani 1992a; Geiger and Heckerman 1996). Smith, Holtzman, and Matheson (1993) proposed a "conditional" version of influence diagrams, which used many of the same concepts. There have been various proposals for local probabilistic models that encode a mapping between context and parameters. Poole (1993a); Poole and Zhang (2003) used a rule-based representation; Boutilier, Friedman, Goldszmidt, and Koller (1996) proposed both the use of a a fully general partition, and the more specific tree-CPDs; and Chickering, Heckerman, and Meek (1997) suggested the use of general DAG-structured CPDs, which allow "paths" corresponding to different context to "merge." Boutilier *et al.* also define the notion of CSI-separation, which extends the notion of d-separation to networks involving asymmetric dependencies; they provide an efficient algorithm for CSI-separation based on cutting spurious arcs, and they discuss how to check for spurious arcs for various types of local probability models.

The notion of causal independence, and specifically the noisy-or model, was proposed independently by Pearl (1986b) and by Peng and Reggia (1986). This model was subsequently generalized to allow a variety of interactions, such as AND or MAX (Pearl 1988; Heckerman 1993; Srinivas 1993; Pradhan et al. 1994).

Lauritzen and Wermuth (1989) introduced the notion of conditional linear Gaussian Bayesian networks. Shachter and Kenley (1989) used linear Gaussian dependencies in the context of Gaussian influence diagrams, a framework that was later extended to the hybrid case Poland (1994). Lerner et al. (2001) suggest the use of a softmax CPD as a local probability model for discrete variables with continuous parents. Murphy (1998) and Lerner (2002) provide a good introduction to these topics.

Several papers have proposed the use of richer (semiparametric or nonparametric) models for representing dependencies in networks involving continuous variables, including kernel estimators (Hofmann and Tresp 1995), neural networks (Monti and Cooper 1997), and Gaussian processes (Friedman and Nachman 2000). These representations are generally opaque to humans, and therefore are useful only in the context of learning networks from data.

The notion of encapsulated Bayesian networks as a representation of local probability models was introduced by Srinivas (1994) in the context of modeling hierarchically structured physical systems. This idea was later generalized within the framework of object-oriented Bayesian networks by Koller and Pfeffer (1997), which allows the definition of a general object model as a network fragment, in a way that encapsulates it from the rest of the model. A further generalization was proposed by Heckerman and Meek (1997), which allows the dependencies in the encapsulated network fragment to be oriented in the opposite direction than is implied by the parent-child relations of the CPD.

5.9 Exercises

Exercise 5.1★

Prove theorem 5.1.

Exercise 5.2★★

a. Show that a multinet where each Bayesian network $\mathcal{B}_c$ is acyclic always defines a coherent probability distribution — one where all of the probabilities sum to 1. Your proof should apply even when the induced Bayesian network that contains the union of all of the edges in the networks $\mathcal{B}_c$ contains a cycle.

b. Now, consider a more general case, where each variable X is associated with a rule-based CPD $P(X \mid \mathrm{Pa}_X)$ (as in definition 5.5). Provide a general sufficient condition on the set of rule-based CPDs that guarantee that the distribution defined by the chain rule:

$$P(X_1, \ldots, X_n) = \prod_i P(X_i \mid \mathrm{Pa}_{X_i})$$

is coherent. Your condition should also encompass cases (including the case of multinets) where the induced global network — one containing a directed edge $Y \to X$ whenever $Y \in \mathrm{Pa}_X$ — is not necessarily acyclic.

Exercise 5.3

Prove proposition 5.1.

Exercise 5.4★

Prove proposition 5.2.

Exercise 5.5

In this exercise, we consider the use of tree-structured local models in the undirected setting.

a. Show how we can use a structure similar to tree-CPDs to represent a factor in a Markov network. What do the values at the leaves of such a tree represent?

b. Given a context $\boldsymbol{U} = \boldsymbol{u}$, define a simple algorithm that takes a tree factor $\phi(\boldsymbol{Y})$ and returns the reduced factor $\phi[\boldsymbol{U} = \boldsymbol{u}](\boldsymbol{Y} - \boldsymbol{U})$ (see definition 4.5).

c. The preceding expression takes $Y - U$ to be the scope of the reduced factor. In some cases it turns out that we can further reduce the scope. Give an example and specify a general rule for when a variable in $Y - U$ can be eliminated from the scope of the reduced tree-factor.

Exercise 5.6

Provide an algorithm for constructing a tree-CPD that is the reduced tree reduced for a given context c. In other words, assume you are given a tree-CPD for $P(X \mid \text{Pa}_X)$ and a context c such that $Scope[c] \subset \text{Pa}_X$. Provide a linear time algorithm for constructing a tree that represents $P(X \mid \text{Pa}_X, c)$.

Exercise 5.7

Consider a BN $\mathcal{B}$ with a variable X that has a tree-CPD. Assume that all of the distributions at the leaves of X's tree-CPD are different. Let $c \in Val(C)$ for $C \subseteq \text{Pa}_X$ be a context, and let $Z \subseteq \text{Pa}_X$. Show that if $P_{\mathcal{B}} \models (X \perp_c Z \mid \text{Pa}_X - Z, c)$, then T^c does not test any variable in Z.

Exercise 5.8★

Prove theorem 5.3. (Hint: Use exercise 5.5 and follow the lines of the proof of theorem 4.9.)

Exercise 5.9★

Prove the following statement, or disprove it by finding a counterexample: CSI-separation statements are monotonic in the context; that is, if c is an assignment to a set of variables C, and $C \subset C'$, and c' is an assignment to C' that is consistent with c, then if X and Y are CSI-separated given c, they are also CSI-separated given c'.

Exercise 5.10★★

Prove that the problem of determining CSI-separation given a set of variables is $\mathcal{NP}$-complete. More precisely, we define the following decision problem:

> Each instance is a graph $\mathcal{G}$, where some of the variables have tree-structured CPDs of known structure, and a query X, Y, Z. An instance is in the language if there is a $z \in Val(Z)$ such that X and Y are not CSI-separated given z.

Show that this problem is $\mathcal{NP}$-complete.

a. Show that this problem is in $\mathcal{NP}$.
b. Provide a reduction to this problem from 3-SAT. Hint: The set Z corresponds to the propositional variables, and an assignment z to some particular truth assignment. Define a variable Y_i for the ith clause in the 3-SAT formula. Provide a construction that contains an active path from Y_i to Y_{i+1} in a context z iff the assignment z satisfies the ith clause.

Exercise 5.11

In this exercise, we consider the context-specific independencies arising in ICI models.

a. Prove proposition 5.3.
b. What context-specific independencies arise if $P(Y \mid X_1, \ldots, X_k)$ is a noisy-and (analogous to figure 5.9, but where the aggregation function is a deterministic AND)?
c. What context-specific independencies arise if $P(Y \mid X_1, \ldots, X_k)$ is a noisy-max (where the aggregation function is a deterministic max), where we assume that Y and the X_i's take ordinal values $v_1, \ldots, v_l$?

Exercise 5.12★

In this exercise, we study intercausal reasoning in noisy-or models. Consider the v-structure Bayesian network: $X \rightarrow Z \leftarrow Y$, where X, Y, and Z are all binary random variables. Assume that the CPD for Z explaining away   is a noisy-or, as in equation (5.2). Show that this network must satisfy the *explaining away* property:

$$P(x^1 \mid z^1) \geq P(x^1 \mid y^1, z^1).$$

Exercise 5.13

Consider a BN2O network $\mathcal{B}$, as described in box 5.C, and assume we are given a negative observation $F_1 = F_1^0$. Show that the the posterior distribution $P_\mathcal{B}(\cdot \mid F_1 = F_1^0$ can be encoded using another BN2O network $\mathcal{B}'$ that has the same structure as $\mathcal{B}$, except that F_1 is omitted from the network. Specify the parameters of $\mathcal{B}'$ in terms of the parameters of $\mathcal{B}$.

Exercise 5.14★

Consider a BN2O network $\mathcal{B}$, as described in box 5.C, and the task of medical diagnosis: Computing the posterior probability in some set of diseases given evidence concerning some of the findings. However, we are only interested in computing the probability of a particular subset of the diseases, so that we wish (for reasons of computational efficiency) to remove from the network those disease variables that are not of interest at the moment.

a. Begin by considering a particular variable F_i, and assume (without loss of generality) that the parents of F_i are $D_1, \ldots, D_k$ and that we wish to maintain only the parents $D_1, \ldots, D_\ell$ for $\ell < k$. Show how we can construct a new noisy-or CPD for F_i that preserves the correct joint distribution over $D_1, \ldots, D_\ell, F_i$.

b. We now remove some fixed set of disease variables $\mathcal{D}$ from the network, executing this pruning procedure for all the finding variables F_i, removing all parents $D_j \in \mathcal{D}$. Is this transformation exact? In other words, if we compute the posterior probability over some variable $D_i \notin \mathcal{D}$, will we get the correct posterior probability (relative to our original model)? Justify your answer.

Exercise 5.15★

Consider the symmetric ICI model, as defined in definition 5.13. Show that, if we allow the domain of the intermediate variables Z_i to be arbitrarily large, we can represent any CPD $P(Y \mid X_1, \ldots, X_k)$ using this type of model.

Exercise 5.16

a. Consider a naive Markov model over the multivalued variables $\boldsymbol{X} = \{X_1, \ldots, X_k\}$ and $\boldsymbol{Y} = \{Y\}$, with the pairwise potentials defined via the following log-linear model

$$\phi_i(x_i^l, y^m) = \exp\left\{ w_i^{lm} \boldsymbol{I}\{X_i = x_i^l, Y = y^m\} \right\}.$$

Again, we have a single-node potential $\phi_0(y^m) = \exp\{w_0^m \boldsymbol{I}\{Y = y^m\}\}$. Show that the distribution $P(Y \mid X_1, \ldots, X_k)$ defined by this model when viewed as a CRF is equivalent to the multinomial logistic CPD of equation (5.4).

b. Determine the appropriate form of CPD for which this result holds for a CRF defined in terms of a general log-linear model, where features are not necessarily pairwise.

6 *Template-Based Representations*

6.1 Introduction

A probabilistic graphical model (whether a Bayesian network or a Markov network) specifies a joint distribution over a fixed set $\mathcal{X}$ of random variables. This fixed distribution is then used in a variety of different situations. For example, a network for medical diagnosis can be applied to multiple patients, each with different symptoms and diseases. However, in this example, the different situations to which the network is applied all share the same general structure — all patients can be described by the same set of attributes, only the attributes' values differ across patients. We call this type of model *variable-based*, since the focus of the representation is a set of random variables.

In many domains, however, the probabilistic model relates to a much more complex space than can be encoded as a fixed set of variables. In a temporal setting, we wish to represent distributions over systems whose state changes over time. For example, we may be monitoring a patient in an intensive care unit. In this setting, we obtain sensor readings at regular intervals — heart rate, blood pressure, EKG — and are interested in tracking the patient's state over time. As another example, we may be interested in tracking a robot's location as it moves in the world and gathers observations. Here, we want a single model to apply to trajectories of different lengths, or perhaps even infinite trajectories.

An even more complex setting arises in our Genetics example; here, each pedigree (family tree) consists of an entire set of individuals, all with their own properties. Our probabilistic model should encode a joint distribution over the properties of all of the family members. Clearly, we cannot define a single variable-based model that applies universally to this application: each family has a different family tree; the networks that represent the genetic inheritance process within the tree have different random variables, and different connectivities. Yet the mechanism by which genes are transmitted from parent to child is identical both for different individuals within a pedigree and across different pedigrees.

In both of these examples, and in many others, we might hope to construct a single, compact model that provides a *template* for an entire class of distributions from the same type: trajectories of different lengths, or different pedigrees. In this chapter, we define representations that allow us to define distributions over richly structured spaces, consisting of multiple objects, interrelated in a variety of ways. These template-based representations have been used in two main settings. The first is temporal modeling, where the language of *dynamic Bayesian networks* allows us to construct a single compact model that captures the properties of the system dynamics, and to produce distributions over different trajectories. The second involves domains such as the Genetics example, where we have multiple objects that are somehow related to each other. Here, various languages have been proposed that allow us to produce distributions over different worlds, each with its own set of individuals and set of relations between them.

Once we consider higher-level representations that allow us to model objects, relations, and probabilistic statements about those entities, we open the door to very rich and expressive languages and to queries about concepts that are not even within the scope of a variable-based framework. For example, in the Genetics example, our space consists of multiple people with different types of relationships such as *Mother*, *Father-of*, and perhaps *Married*. In this type probability space, we can also express uncertainty about the identity of Michael's father, or how many children Great-aunt Ethel had. Thus, we may wish to construct a probability distribution over a space consisting of distinct pedigree structures, which may even contain a varying set of objects. As we will see, this richer modeling language will allow us both to answer new types of queries, and to provide more informed answers to "traditional" queries.

6.2 Temporal Models

Our focus in this section is on modeling dynamic settings, where we are interested in reasoning about the state of the world as it evolves over time. We can model such settings in terms of a

system state — *system state*, whose value at time t is a snapshot of the relevant attributes (hidden or observed) of the system at time t. We assume that the system state is represented, as usual, as an assignment of values to some set of random variables $\mathcal{X}$. We use $X_i^{(t)}$ to represent the instantiation of the variable X_i at time t. Note that X_i itself is no longer a variable that takes a value; rather, it is a

template variable — *template variable*. This template is instantiated at different points in time t, and each $X_i^{(t)}$ is a variable that takes a value in $Val(X_i)$. For a set of variables $\boldsymbol{X} \subseteq \mathcal{X}$, we use $\boldsymbol{X}^{(t_1:t_2)}$ ($t_1 < t_2$) to denote the set of variables $\{\boldsymbol{X}^{(t)} : t \in [t_1, t_2]\}$. As usual, we use the notation $\boldsymbol{x}^{(t:t')}$ for an assignment of values to this set of variables.

trajectory — Each "possible world" in our probability space is now a *trajectory*: an assignment of values to each variable $X_i^{(t)}$ for each relevant time t. Our goal therefore is to represent a joint distribution over such trajectories. Clearly, the space of possible trajectories is a very complex probability space, so representing such a distribution can be very difficult. We therefore make a series of simplifying assumptions that help make this representational problem more tractable.

Example 6.1
Consider a vehicle localization task, where a moving car tries to track its current location using the data obtained from a, possibly faulty, sensor. The system state can be encoded (very simply) using the: Location — the car's current location, Velocity — the car's current velocity, Weather — the current weather, Failure — the failure status of the sensor, and Obs — the current observation. We have one such set of variables for every point t. A joint probability distribution over all of these sets defines a probability distribution over trajectories of the car. Using this distribution, we can answer a variety queries, such as: Given a sequence of observations about the car, where is it now? Where is it likely to be in ten minutes? Did it stop at the red light? ∎

6.2.1 Basic Assumptions

time slice

Our first simplification is to discretize the timeline into a set of *time slices*: measurements of the system state taken at intervals that are regularly spaced with a predetermined time granularity Δ. Thus, we can now restrict our set of random variables to $\mathcal{X}^{(0)}, \mathcal{X}^{(1)}, \ldots$, where $\mathcal{X}^{(t)}$ are the ground random variables that represent the system state at time $t \cdot \Delta$. For example, in the patient monitoring example, we might be interested in monitoring the patient's state every second, so that $\Delta = 1sec$. This assumption simplifies our problem from representing distributions over a continuum of random variables to representing distributions over countably many random variables, sampled at discrete intervals.

Consider a distribution over trajectories sampled over a prefix of time $t = 0, \ldots, T - P(\mathcal{X}^{(0)}, \mathcal{X}^{(1)}, \ldots, \mathcal{X}^{(T)})$, often abbreviated $P(\mathcal{X}^{(0:T)})$. We can reparameterize the distribution using the chain rule for probabilities, in a direction consistent with time:

$$P(\mathcal{X}^{(0:T)}) = P(\mathcal{X}^{(0)}) \prod_{t=0}^{T-1} P(\mathcal{X}^{(t+1)} \mid \mathcal{X}^{(0:t)}).$$

Thus, the distribution over trajectories is the product of conditional distributions, for the variables in each time slice given the preceding ones. We can considerably simplify this formulation by using our usual tool — conditional independence assumptions. One very natural approach is to assume that the future is conditionally independent of the past given the present:

Definition 6.1

Markov assumption

We say that a dynamic system over the template variables $\mathcal{X}$ satisfies the Markov assumption *if, for all $t \geq 0$,*

$$(\mathcal{X}^{(t+1)} \perp \mathcal{X}^{(0:(t-1))} \mid \mathcal{X}^{(t)}).$$

Markovian system

Such systems are called Markovian.   ∎

The Markov assumptions states that the variables in $\mathcal{X}^{(t+1)}$ cannot depend directly on variables in $\mathcal{X}^{(t')}$ for $t' < t$. If we were to draw our dependency model as an (infinite) Bayesian network, the Markov assumption would correspond to the constraint on the graph that there are no edges into $\mathcal{X}^{(t+1)}$ from variables in time slices $t - 1$ or earlier. Like many other conditional independence assumptions, the Markov assumption allows us to define a more compact representation of the distribution:

$$P(\mathcal{X}^{(0)}, \mathcal{X}^{(1)}, \ldots, \mathcal{X}^{(T)}) = P(\mathcal{X}^{(0)}) \prod_{t=0}^{T-1} P(\mathcal{X}^{(t+1)} \mid \mathcal{X}^{(t)}). \tag{6.1}$$

Like any conditional independence assumption, the Markov assumption may or may not be reasonable in a particular setting.

Example 6.2

Let us return to the setting of example 6.1, but assume we had, instead, selected $\mathcal{X} = \{L, O\}$, where L is the location of the object and O its observed location. At first glance, we might be tempted to make the Markov assumption in this setting: after all, the location at time $t + 1$ does not appear to depend directly on the location at time $t - 1$. However, assuming the object's motion is coherent, the location at time $t + 1$ is not independent of the previous locations given only the location at *time t, because the previous locations give us information about the object's direction of motion and speed. By adding Velocity, we make the Markov assumption closer to being satisfied. If, however, the driver is more likely to accelerate and decelerate sharply in certain types of weather (say heavy winds), then our V, L model does not satisfy the Markov assumption relative to V; we can, again, make the model more Markovian by adding the Weather variable. Finally, in many cases, a sensor failure at one point is usually accompanied with a sensor failure at nearby time points, rendering nearby Obs variables correlated. By adding all of these variables into our state model, we define a state space whereby the Markov assumption is arguably a reasonable approximation.* ∎

Philosophically, one might argue whether, given a sufficiently rich description of the world state, the past is independent of the future given the present. However, that question is not central to the use of the Markov assumption in practice. Rather, **we need only consider whether the Markov assumption is a sufficiently reasonable approximation to the dependencies in our distribution. In most cases, if we use a reasonably rich state description, the approximation is quite reasonable.** In other cases, we can also define models that are *semi-Markov*, where the independence assumption is relaxed (see exercise 6.1).

semi-Markov

Because the process can continue indefinitely, equation (6.1) still leaves us with the task of acquiring an infinite set of conditional distributions, or a very large one, in the case of finite-horizon processes. Therefore, we usually make one last simplifying assumption:

Definition 6.2

We say that a Markovian dynamic system is stationary (*also called* time invariant *or* homogeneous) *if $P(\mathcal{X}^{(t+1)} \mid \mathcal{X}^{(t)})$ is the same for all t. In this case, we can represent the process using a* transition model $P(\mathcal{X}' \mid \mathcal{X})$, *so that, for any $t \geq 0$,*

stationary
dynamical system

transitional
model

$$P(\mathcal{X}^{(t+1)} = \xi' \mid \mathcal{X}^{(t)} = \xi) = P(\mathcal{X}' = \xi' \mid \mathcal{X} = \xi).$$ ∎

6.2.2 Dynamic Bayesian Networks

The Markov and stationarity assumptions described in the previous section allow us to represent the probability distribution over infinite trajectories very compactly: We need only represent the initial state distribution and the transition model $P(\mathcal{X}' \mid \mathcal{X})$. This transition model is a conditional probability distribution, which we can represent using a conditional Bayesian network, as described in section 5.6.

Example 6.3

Let us return to the setting of example 6.1. Here, we might want to represent the system dynamics using the model shown in figure 6.1a, the current observation depends on the car's location (and the map, which is not explicitly modeled) and on the error status of the sensor. Bad weather makes the sensor more likely to fail. And the car's location depends on the previous position and the velocity. All of the variables are interface variables except for Obs, since we assume that the sensor observation is generated at each time point independently given the other variables. ∎

This type of conditional Bayesian network is called a *2-time-slice Bayesian network (2-TBN)*.

Definition 6.3

A 2-time-slice Bayesian network (2-TBN) for a process over $\mathcal{X}$ is a conditional Bayesian network over $\mathcal{X}'$ given $\mathcal{X}_I$, where $\mathcal{X}_I \subseteq \mathcal{X}$ is a set of interface variables. ∎

2-TBN

interface variable

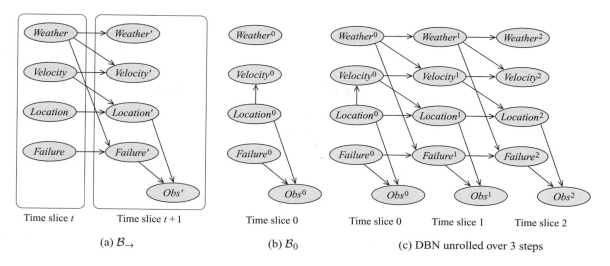

Figure 6.1   A highly simplified DBN for monitoring a vehicle: (a) the 2-TBN; (b) the time 0 network; (c) resulting unrolled DBN over three time slices.

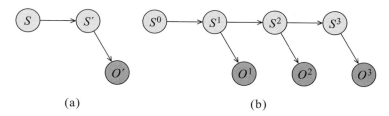

Figure 6.2   HMM as a DBN: (a) The 2-TBN for a generic HMM. (b) The unrolled DBN for four time slices.

As a reminder, in a conditional Bayesian network, only the variables $\mathcal{X}'$ have parents or CPDs. The interface variables $\mathcal{X}_I$ are those variables whose values at time t have a direct effect on the variables at time $t + 1$. Thus, only the variables in $\mathcal{X}_I$ can be parents of variables in $\mathcal{X}'$. In our example, all variables except O are in the interface.

Overall, the 2-TBN represents the conditional distribution:

$$P(\mathcal{X}' \mid \mathcal{X}) = P(\mathcal{X}' \mid \mathcal{X}_I) = \prod_{i=1}^{n} P(X_i' \mid \mathrm{Pa}_{X_i'}). \tag{6.2}$$

template factor   For each template variable X_i, the CPD $P(X_i' \mid \mathrm{Pa}_{X_i'})$ is a *template factor*: it will be instantiated multiple times within the model, for multiple variables $X_i^{(t)}$ (and their parents).

hidden Markov model   Perhaps the simplest nontrivial example of a temporal model of this kind is the *hidden Markov model* (see section 6.2.3.1). It has a single state variable S and a single observation variable O. Viewed as a DBN, an HMM has the structure shown in figure 6.2.

Example 6.4
Consider a robot moving around in a grid. Most simply, the robot is the only aspect of the world that is changing, so that the state of the system S is simply the robot's position. Our transition model $P(S' \mid S)$ then represents the probability that, if the robot is in some state (position) s, it will move to another state s'. Our task is to keep track of the robot's location, using a noisy sensor (for example, a sonar) whose value depends on the robot's location. The observation model $P(O \mid S)$ tells us the probability of making a particular sensor reading o given that the robot's current position is s. (See box 5.E for more details on the state transition and observation models in a real robot localization task.) ∎

inter-time-slice edge

intra-time-slice edge

In a 2-TBN, some of the edges are *inter-time-slice edges*, going between time slices, whereas others are *intra-time-slice edges*, connecting variables in the same time slice. Intuitively, our decision of how to relate two variables depends on how tight the coupling is between them. If the effect of one variable on the other is immediate — much shorter than the time granularity in the model — the influence would manifest (roughly) within a time slice. If the effect is slightly longer-term, the influence manifests from one time slice to the next. In our simple examples, the effect on the observations is almost immediate, and hence is modeled as an intra-time-slice edge, whereas other dependencies are inter-time-slice. In other examples, when time slices have a coarser granularity, more effects might be short relative to the length of the time slice, and so we might have other dependencies that are intra-time-slice.

persistence edge

persistent variable

Many of the inter-time-slice edges are of the form $X \to X'$. Such edges are called *persistence edges*, and they represent the tendency of the variable X (for example, sensor failure) to persist over time with high probability. A variable X for which we have an edge $X \to X'$ in the 2-TBN is called a *persistent variable*.

Based on the stationarity property, a 2-TBN defines the probability distribution $P(\mathcal{X}^{(t+1)} \mid \mathcal{X}^{(t)})$ for any t. Given a distribution over the initial states, we can *unroll* the network over sequences of any length, to define a Bayesian network that induces a distribution over trajectories of that length. In these networks, all the copies of the variable $X_i^{(t)}$ for $t > 0$ have the same dependency structure and the same CPD. Figure 6.1 demonstrates a transition model, initial state network, and a resulting unrolled DBN, for our car example.

Definition 6.4

dynamic Bayesian network

unrolled Bayesian network

A dynamic Bayesian network *(DBN) is a pair $\langle \mathcal{B}_0, \mathcal{B}_\to \rangle$, where $\mathcal{B}_0$ is a Bayesian network over $\mathcal{X}^{(0)}$, representing the initial distribution over states, and $\mathcal{B}_\to$ is a 2-TBN for the process. For any desired time span $T \geq 0$, the distribution over $\mathcal{X}^{(0:T)}$ is defined as a* unrolled Bayesian network, *where, for any $i = 1, \ldots, n$:*

- *the structure and CPDs of $X_i^{(0)}$ are the same as those for X_i in $\mathcal{B}_0$,*

- *the structure and CPD of $X_i^{(t)}$ for $t > 0$ are the same as those for X_i' in $\mathcal{B}_\to$.* ∎

Thus, we can view a DBN as a compact representation from which we can generate an infinite set of Bayesian networks (one for every $T > 0$).

factorial HMM

coupled HMM

Figure 6.3 shows two useful classes of DBNs that are constructed from HMMs. A *factorial HMM*, on the left, is a DBN whose 2-TBN has the structure of a set of chains $X_i \to X_i'$ ($i = 1, \ldots, n$), with a single (always) observed variable Y', which is a child of all the variables X_i'. This type of model is very useful in a variety of applications, for example, when several sources of sound are being heard simultaneously through a single microphone. A *coupled HMM*,

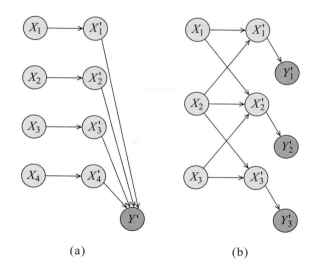

$$(a) \qquad\qquad (b)$$

Figure 6.3 **Two classes of DBNs constructed from HMMs:** (a) A factorial HMM. (b) A coupled HMM.

on the right, is also constructed from a set of chains X_i, but now, each chain is an HMM with its private own observation variable Y_i. The chains now interact directly via their state variables, with each chain affecting its adjacent chains. These models are also useful in a variety of applications. For example, consider monitoring the temperature in a building over time (for example, for fire alarms). Here, X_i might be the true (hidden) state of the ith room, and Y_i the value returned by the room's own temperature sensor. In this case, we would expect to have interactions between the hidden states of adjacent rooms.

In DBNs, it is often the case that our observation pattern is constant over time. That is, we can partition the variables $\mathcal{X}$ into disjoint subsets $\boldsymbol{X}$ and $\boldsymbol{O}$, such that the variables in $\boldsymbol{X}^{(t)}$ are always hidden and $\boldsymbol{O}^{(t)}$ are always observed. For uniformity of presentation, we generally make this assumption; however, the algorithms we present also apply to the more general case.

A DBN can enable fairly sophisticated reasoning patterns.

Example 6.5

By explicitly encoding sensor failure, we allow the agent to reach the conclusion that the sensor has failed. Thus, for example, if we suddenly get a reading that tells us something unexpected, for example, the car is suddenly 15 feet to the left of where we thought it was 0.1 seconds ago, then in addition to considering the option that the car has suddenly teleported, we will also consider the option that the sensor has simply failed. Note that the model only considers options that are built into it. If we had no "sensor failure" variable, and had the sensor reading depend only on the current location, then the different sensor readings would be independent given the car's trajectory, so that there would be no way to explain correlations of unexpected sensor readings except via the trajectory. Similarly, if the system knows (perhaps from a weather report or from prior observations) that it is raining, it will expect the sensor to be less accurate, and therefore be less likely to believe that the car is out of position. ∎

Box 6.A — Case Study: HMMs and Phylo-HMMs for Gene Finding. *HMMs are a primary tool in algorithms that extract information from biological sequences. Key applications (among many) include: modeling families of related proteins within and between organisms, finding genes in DNA sequences, and modeling the correlation structure of the genetic variation between individuals in a population. We describe the second of these applications, as an illustration of the methods used.*

The DNA of an organism is composed of two paired helical strands consisting of a long sequence of nucleotides, each of which can take on one of four values — A,C,G,T; in the double helix structure, A is paired with T and C is paired with G, to form a base pair. The DNA sequence consists of multiple regions that can play different roles. Some regions are genes, whose DNA is transcribed into mRNA, some of which is subsequently translated into protein. In the translation process, triplets of base pairs, known as codons, are converted into amino acids. There are $4^3 = 64$ different codons, but only 20 different amino acids, so that the code is redundant. Not all transcribed regions are necessarily translated. Genes can contain exons, which are translated, and introns, which are spliced out during the translation process. The DNA thus consists of multiple genes that are separated by intergenic regions; and genes are themselves structured, consisting of multiple exons separated by introns. The sequences in each of these regions is characterized by certain statistical properties; for example, a region that produces protein has a very regular codon structure, where the codon triplets exhibit the usage statistics of the amino acids they produce. Moreover, boundaries between these regions are also often demarcated with sequence elements that help the cell determine where transcription should begin and end, and where translation ought to begin and end. Nevertheless, the signals in the sequence are not always clear, and therefore identifying the relevant sequence units (genes, exons, and more) is a difficult task.

HMMs are a critical tool in this analysis. Here, we have a hidden state variable for each base pair, which denotes the type of region to which this base pair belongs. To satisfy the Markov assumption, one generally needs to refine the state space. For example, to capture the codon structure, we generally include different hidden states for the first, second, and third base pairs within a codon. This larger state space allows us to encode the fact that coding regions are sequences of triplets of base pairs, as well as encode the different statistical properties of these three positions. We can further refine the state space to include different statistics for codons in the first exon and in the last exon in the gene, which can exhibit different characteristics than exons in the middle of the gene. The observed state of the HMM naturally includes the base pair itself, with the observation model reflecting the different statistics of the nucleotide composition of the different regions. It can also include other forms of evidence, such as the extent to which measurements of mRNA taken from the cell have suggested that a particular region is transcribed. And, very importantly, it can contain evidence regarding the conservation of a base pair across other species. This last key piece of evidence derives from the fact that base pairs that play a functional role in the cell, such as those that code for protein, are much more likely to be conserved across related species; base pairs that are nonfunctional, such as most of those in the intergenic regions, evolve much more rapidly, since they are not subject to selective pressure. Thus, we can use conservation as evidence regarding the role of a particular base pair.

phylogenetic
HMM

One way of incorporating the evolutionary model more explicitly into the model is via a phylogenetic HMM *(of which we now present a simplified version). Here, we encode not a single DNA sequence, but the sequences of an entire phylogeny (or evolutionary tree) of related species. We* *let $X_{k,i}$ be the ith nucleotide for species s_k. We also introduce a species-independent variable Y_i denoting the functional role of the ith base pair (intergenic, intron, and so on). The base pair $X_{k,i}$ will depend on the corresponding base pair $X_{\ell,i}$ where s_ℓ is the ancestral species from which s_k evolved. The parameters of this dependency will depend on the evolutionary distance between s_k and s_l (the extent to which s_k has diverged) and on the rate at which a base pair playing a particular role evolves. For example, as we mentioned, a base pair in an intergenic region generally evolves much faster than one in a coding region. Moreover, the base pair in the third position in a codon also often evolves more rapidly, since this position encodes most of the redundancy between codons and amino acids, and so allows evolution without changing the amino acid composition. Thus, overall, we define $X_{k,i}$'s parents in the model to be Y_i (the type of region in which $X_{k,i}$ resides), $X_{k,i-1}$ (the previous nucleotide in species s) and $X_{\ell,i}$ (the ith nucleotide in the parent species s_ℓ). This model captures both the correlations in the functional roles (as in the simple gene finding model) and the fact that evolution of a particular base pair can depend on the adjacent base pairs. This model allows us to combine information from multiple species in order to infer which are the regions that are functional, and to suggest a segmentation of the sequence into its constituent units.*

Overall, the structure of this model is roughly a set of trees connected by chains: For each i we have a tree over the variables $\{X_{k,i}\}_{s_k}$, where the structure of the tree is that of the evolutionary tree; in addition, all of the $X_{k,i}$ are connected by chains to $X_{k,i+1}$; finally, we also have the variables Y_i, which also form a chain and are parents of all of the $X_{k,i}$. Unfortunately, the structure of this model is highly intractable for inference, and requires the use of approximate inference methods; see exercise 11.29.

6.2.3 State-Observation Models

state-observation
model

An alternative way of thinking about a temporal process is as a *state-observation model*. In a state-observation model, we view the system as evolving naturally on its own, with our observations of it occurring in a separate process. This view separates out the system dynamics from our observation model, allowing us to consider each of them separately. It is particularly useful when our observations are obtained from a (usually noisy) sensor, so that it makes sense to model separately the dynamics of the system and our ability to sense it.

A state-observation model utilizes two independence assumptions: that the state variables evolve in a Markovian way, so that

$$(\boldsymbol{X}^{(t+1)} \perp \boldsymbol{X}^{(0:(t-1))} \mid \boldsymbol{X}^{(t)});$$

and that the observation variables at time t are conditionally independent of the entire state sequence given the state variables at time t:

$$(\boldsymbol{O}^{(t)} \perp \boldsymbol{X}^{(0:(t-1))}, \boldsymbol{X}^{(t+1:\infty)} \mid \boldsymbol{X}^{(t)}).$$

transition model

observation
model

We now view our probabilistic model as consisting of two components: the *transition model*, $P(\boldsymbol{X}' \mid \boldsymbol{X})$, and the *observation model*, $P(\boldsymbol{O} \mid \boldsymbol{X})$.

From the perspective of DBNs, this type of model corresponds to a 2-TBN structure where the observation variables $\boldsymbol{O}'$ are all leaves, and have parents only in $\boldsymbol{X}'$. This type of situation arises

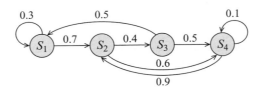

Figure 6.4   A simple 4-state HMM

quite naturally in a variety of real-world systems, where we do not have direct observations of the system state, but only access to a set of (generally noisy) sensors that depend on the state. The sensor observations do not directly effect the system dynamics, and therefore are naturally viewed as leaves.

We note that we can convert any 2-TBN to a state-observation representation as follows: For any observed variable Y that does not already satisfy the structural restrictions, we introduce a new variable $\tilde{Y}$ whose only parent is Y, and that is deterministically equal to Y. Then, we view Y as being hidden, and we interpret our observations of Y as observations on $\tilde{Y}$. In effect, we construct $\tilde{Y}$ to be a perfectly reliable sensor of Y. Note, however, that, while the resulting transformed network is probabilistically equivalent to the original, it does obscure structural independence properties of the network (for example, various independencies given that Y is observed), which are now apparent only if we account for the deterministic dependency between Y and $\tilde{Y}$.

It is often convenient to view a temporal system as a state-observation model, both because it lends a certain uniformity of notation to a range of different systems, and because the state transition and observation models often induce different computational operations, and it is convenient to consider them separately.

State-observation models encompass two important architectures that have been used in a wide variety of applications: hidden Markov models and linear dynamical systems. We now briefly describe each of them.

6.2.3.1   Hidden Markov Models

hidden Markov model

A *hidden Markov model*, which we illustrated in figure 6.2, is the simplest example of a state-observation model. While an HMM is a special case of a simple DBN, it is often used to encode structure that is left implicit in the DBN representation. Specifically, the transition model $P(S' \mid S)$ in an HMM is often assumed to be sparse, with many of the possible transitions having zero probability. In such cases, HMMs are often represented using a different graphical notation, which visualizes this sparse transition model. In this representation, the HMM transition model is encoded using a directed (generally cyclic) graph, whose nodes represent the *different states* of the system, that is, the values in $Val(S)$. We have a directed arc from s to s' if it is possible to transition from s to s' — that is, $P(s' \mid s) > 0$. The edge from s to s' can also be annotated with its associated transition probability $P(s' \mid s)$.

Example 6.6   *Consider an HMM with a state variable S that takes 4 values s_1, s_2, s_3, s_4, and with a transition* *model:*

	s_1	s_2	s_3	s_4
s_1	0.3	0.7	0	0
s_2	0	0	0.4	0.6
s_3	0.5	0	0	0.5
s_4	0	0.9	0	0.1

where the rows correspond to states s and the columns to successor states s' (so that each row must sum to 1). The transition graph for this model is shown in figure 6.4. ∎

Importantly, the transition graph for an HMM is a very different entity from the graph encoding a graphical model. Here, the nodes in the graph are *state*, or possible values of the state variable; the directed edges represent possible transitions between the states, or entries in the CPD that have nonzero probability. Thus, the weights of the edges leaving a node must sum to 1. This graph representation can also be viewed as *probabilistic finite-state automaton*. Note that this graph-based representation does not encode the observation model of the HMM. In some cases, the observation model is deterministic, in that, for each s, there is a single observation o for which $P(o \mid s) = 1$ (although the same observation can arise in multiple states). In this case, the observation is often annotated on the node associated with the state.

probabilistic
finite-state
automaton

It turns out that HMMs, despite their simplicity, are an extremely useful architecture. For example, they are the primary architecture for speech recognition systems (see box 6.B) and for many problems related to analysis of biological sequences (see, for example, box 6.A). Moreover, these applications and others have inspired a variety of valuable generalizations of the basic HMM framework (see, for example, Exercises 6.2–6.5).

---

speech
recognition

language model

Box 6.B — Case Study: HMMs for Speech Recognition. *Hidden Markov models are currently the key technology in all* speech- recognition *systems. The HMM for speech is composed of three distinct layers: the* language model, *which generates sentences as sequences of words; the* word model, *where words are described as a sequence of phonemes; and the* acoustic model, *which shows the progression of the acoustic signal through a phoneme.*

At the highest level, the language model represents a probability distribution over sequences of words in the language. Most simply, one can use a bigram *model, which is a Markov model over words, defined via a probability distribution $P(W_i \mid W_{i-1})$ for each position i in the sentence. We can view this model as a Markov model where the state is the current word in the sequence. (Note that this model does not take into account the actual position in the sentence, so that $P(W_i \mid W_{i-1})$ is the same for all $i > 1$.) A somewhat richer model is the* trigram *model, where the states correspond to pairs of successive words in the sentence, so that our model defines a probability distribution $P(W_i \mid W_{i-1}, W_{i-2})$. Both of these distributions define a ridiculously naive model of language, since they only capture local correlations between neighboring words, with no attempt at modeling global coherence. Nevertheless, these models prove surprisingly hard to beat, probably because they are quite easy to train robustly from the (virtually unlimited amounts of) available training data, without the need for any manual labeling.*

The middle layer describes the composition of individual words in terms of phonemes — *basic phonetic units corresponding to distinct sounds. These units vary not just on the basic sound uttered*

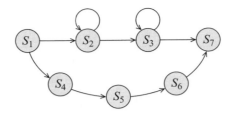

Figure 6.B.1 — A phoneme-level HMM for a fairly complex phoneme.

("p" versus "b"), but also on whether the sound is breathy, aspirated, nasalized, and more. There is an international agreement on an International Phonetic Alphabet, which contains about 100 phonemes. Each word is modeled as a sequence of phonemes. Of course, a word can have multiple different pronunciations, in which case it corresponds to several such sequences.

At the acoustic level, the acoustic signal is segmented into short time frames (around 10–25ms). A given phoneme lasts over a sequence of these partitions. The phoneme is also not homogenous. Different acoustics are associated with its beginning, its middle, and its end. We thus create an HMM for each phoneme, with its hidden variable corresponding to stages in the expression of the phoneme. HMMs for phonemes are usually quite simple, with three states, but can get more complicated, as in figure 6.B.1. The observation represents some set of features extracted from the acoustic signal; the feature vector is generally either discretized into a set of bins or treated as a continuous observation with a Gaussian or mixture of Gaussian distribution.

hierarchical HMM   Given these three models, we can put them all together to form a single huge hierarchical HMM that defines a joint probability distribution over a state space encompassing words, phonemes, and basic acoustic units. In a bigram model, the states in the space have the form (w, i, j), where w is the current word, i is a phoneme within that word, and j is an acoustic position within that phoneme. The sequence of states corresponding to a word w is governed by a word-HMM representing the distribution over pronunciations of w. This word-HMM has a start-state and an end-state. When we exit from the end-state of the HMM for one word w, we branch to the start-state of another word w' with probability $P(w' \mid w)$. Each sequence is thus a trajectory through acoustic HMMs of individual phonemes, transitioning from the end-state of one phoneme's HMM to the start-state of the next phoneme's HMM.

A hierarchical HMM can be converted into a DBN, whose variables represent the states of the different levels of the hierarchy (word, phoneme, and intraphone state), along with some auxiliary variables to capture the "control architecture" of the hierarchical HMM; see exercise 6.5. The DBN formulation has the benefit of being a much more flexible framework in which to introduce extensions to the model. One extension addresses the coarticulation problem, where the proximity of one phoneme changes the pronunciation of another. Thus, for example, the last phoneme in the word "don't" sounds very different if the word after it is "go" or if it is "you." Similarly, we often pronounce "going to" as "gonna." The reason for coarticulation is the fact that a person's speech articulators (such as the tongue or the lips) have some inertia and therefore do not always move all the way to where they are supposed to be. Within the DBN framework, we can easily solve this problem by introducing a dependency of the pronunciation model for one phoneme on the value of the preceding phoneme and the next one. Note that "previous" and "next" need to be interpreted *with care: These are not the values of the phoneme variable at the previous or next states in the HMM, which are generally exactly the same as the current phoneme; rather, these are the values of the variables prior to the previous phoneme change, and following the next phoneme change. This extension gives rise to a non-Markovian model, which is more easily represented as a structured graphical model. Another extension that is facilitated by a DBN structure is the introduction of variables that denote states at which a transition between phonemes occurs. These variables can then be connected to observations that are indicative of such a transition, such as a significant change in the spectrum. Such features can also be incorporated into the standard HMM model, but it is difficult to restrict the model so that these features affect only our beliefs in phoneme transitions.*

Finally, graphical model structure has also been used to model the structure in the Gaussian distribution over the acoustic signal features given the state. Here, two "traditional" models are: a diagonal Gaussian over the features, a model that generally loses many important correlations between the features; and a full covariance Gaussian, a model that requires many parameters and is hard to estimate from data (especially since the Gaussian is different for every state in the HMM). As we discuss in chapter 7, graphical models provide an intermediate point along the spectrum: we can use a Gaussian graphical model that captures the most important of the correlations between the features. The structure of this Gaussian can be learned from data, allowing a flexible trade-off to be determined based on the available data.

6.2.3.2 Linear Dynamical Systems

linear dynamical system

Kaman filter

Another very useful temporal model is a *linear dynamical system*, which represents a system of one or more real-valued variables that evolve linearly over time, with some Gaussian noise. Such systems are also often called *Kalman filters*, after the algorithm used to perform tracking. **A linear dynamical system can be viewed as a dynamic Bayesian network where the variables are all continuous and all of the dependencies are linear Gaussian.**

 Linear dynamical systems are often used to model the dynamics of moving objects and to track their current positions given noisy measurements. (See also box 15.A.)

Example 6.7

Recall example 5.20, where we have a (vector) variable X denoting a vehicle's current position (in each relevant dimension) and a variable V denoting its velocity (also in each dimension). As we discussed earlier, a first level approximation may be a model where $P(X' \mid X, V) = X + V\Delta + \mathcal{N}\left(0; \sigma_X^2\right)$ and $P(V' \mid V) = V + \mathcal{N}\left(0; \sigma_V^2\right)$ (where Δ, as before, is the length of our time slice). The observation — for example, a GPS signal measured from the car — is a noisy Gaussian measurement of X. ∎

These systems and their extensions are at the heart of most target tracking systems, for example, tracking airplanes in an air traffic control system using radar data.

Traditionally, linear dynamical systems have not been viewed from the perspective of factorized representations of the distribution. They are traditionally represented as a state-observation model, where the state and observation are both vector-valued random variables, and the transition and observation models are encoded using matrices. More precisely, the model is generally defined via the following set of equations:

$$P(\boldsymbol{X}^{(t)} \mid \boldsymbol{X}^{(t-1)}) = \mathcal{N}\left(A\boldsymbol{X}^{(t-1)}; Q\right), \tag{6.3}$$

$$P(O^{(t)} \mid \boldsymbol{X}^{(t)}) = \mathcal{N}\left(H\boldsymbol{X}^{(t)}; R\right), \tag{6.4}$$

where: $\boldsymbol{X}$ is an n-vector of state variables, O is an m-vector of observation variables, A is an $n \times n$ matrix defining the linear transition model, Q is an $n \times n$ matrix defining the Gaussian noise associated with the system dynamics, H is an $n \times m$ matrix defining the linear observation model, and R is an $m \times m$ matrix defining the Gaussian noise associated with the observations. This type of model encodes independence structure implicitly, in the parameters of the matrices (see exercise 7.5).

There are many interesting generalizations of the basic linear dynamical system, which can also be placed within the DBN framework. For example, a nonlinear variant, often called an *extended Kalman filter*, is a system where the state and observation variables are still vectors of real numbers, but where the state transition and observation models can be nonlinear functions rather than linear matrix multiplications as in equation (6.3) and equation (6.4). Specifically, we usually write:

extended Kalman filter

$$P(\boldsymbol{X}^{(t)} \mid \boldsymbol{X}^{(t-1)}) = f(\boldsymbol{X}^{(t-1)}, \boldsymbol{U}^{(t-1)})$$
$$P(O^{(t)} \mid \boldsymbol{X}^{(t)}) = g(\boldsymbol{X}^{(t)}, \boldsymbol{W}^{(t)}),$$

where f and g are deterministic nonlinear functions, and $\boldsymbol{U}^{(t)}, \boldsymbol{W}^{(t)}$ are Gaussian random variables that explicitly encode the noise in the transition and observation models, respectively. In other words, rather than model the system in terms of stochastic CPDs, we use an equivalent representation that partitions the model into a deterministic function and a noise component.

Another interesting class of models are systems where the continuous dynamics are linear, but that also include discrete variables. For example, in our tracking example, we might introduce a discrete variable that denotes the driver's target lane in the freeway: the driver can stay in her current lane, or she can switch to a lane on the right or a lane on the left. Each of these discrete settings leads to different dynamics for the vehicle velocity, in both the lateral (across the road) and frontal velocity.

switching linear dynamical system

Systems that model such phenomena are called *switching linear dynamical system (SLDS)*. In such models, we system can switch between a set of discrete *modes*. While within a fixed mode, the system evolves using standard linear (or nonlinear) Gaussian dynamics, but the equations governing the dynamics are different in different modes. We can view this type of system as a DBN by including a discrete variable D that encodes the mode, where $\mathrm{Pa}_{D'} = \{D\}$, and allowing D to be the parent of (some of) the continuous variables in the model, so that they use a conditional linear Gaussian CPDs.

6.3   Template Variables and Template Factors

Having seen one concrete example of a template-based model, we now describe a more general framework that provides the fundamental building blocks for a template-based model. This framework provides a more formal perspective on the temporal models of the previous section, and a sound foundation for the richer models in the remaining sections of this chapter.

Template Attributes   The key concept in the definition of the models we describe in this chapter is that of a *template* that is instantiated many times within the model. A template for a random variable allows us to encode models that contain multiple random variables with the same value space and the same semantics. For example, we can have a *Blood-Type* template, which has a particular value space (say, *A*, *B*, *AB*, or *O*) and is reused for multiple individuals within a pedigree. That is, when reasoning about a pedigree, as in box 3.B, we would want to have multiple instances of blood-type variables, such as *Blood-Type(Bart)* or *Blood-Type(Homer)*.

attribute

template variable

We use the word *attribute* or *template variable* to distinguish a template, such as *Blood-Type*, from a specific random variable, such as *Blood-Type(Bart)*.

In a very different type of example, we can have a template attribute *Location*, which can be instantiated to produce random variables *Location(t)* for a set of different time points *t*. This type of model allows us to represent a joint distribution over a vehicle's position at different points in time, as in the previous section.

In these example, the template was a property of a single object — a person. More broadly, attributes may be properties of entire tuples of objects. For example, a student's grade in a course is associated with a student-course pair; a person's opinion of a book is associated with the person-book pair; the affinity between a regulatory protein in the cell and one of its gene targets is also associated with the pair. More specifically, in our Student example, we want to have a *Grade* template, which we can instantiate for different (student,course) pairs s, c to produce multiple random variables *Grade(s, c)*, such as *Grade(George, CS101)*.

object

class

Because many domains involve heterogeneous objects, such as courses and students, it is convenient to view the world as being composed of a set of *objects*. Most simply, objects can be divided into a set of mutually exclusive and exhaustive *classes* $\mathcal{Q} = Q_1, \ldots, Q_k$. In the Student scenario, we might have a Student class and a Course class.

argument

Attributes have a tuple of *arguments*, each of which is associated with a particular class of objects. This class defines the set of objects that can be used to instantiate the argument in a given domain. For example, in our *Grade* template, we have one argument S that can be instantiated with a "student" object, and another argument C that can be instantiated with a "course" object. Template attributes thus provide us with a "generator" for random variables in a given probability space.

Definition 6.5

attribute

logical variable

argument
signature

An attribute A *is a function* $A(U_1, \ldots, U_k)$, *whose* range *is some set* $Val(A)$ *and where each argument* U_i *is a typed* logical variable *associated with a particular class* $Q[U_i]$. *The tuple* $U_1, \ldots, U_k$ *is called the* argument signature *of the attribute* A, *and denoted* $\alpha(A)$. ∎

From here on, we assume without loss of generality that each logical variable U_i is uniquely associated with a particular class $Q[U_i]$; thus, any mention of a logical variable U_i uniquely specifies the class over which it ranges.

For example, the argument signature of the *Grade* attribute would have two logical variables S, C, where S is of class Student and C is of class Course. We note that the classes associated with an attribute's argument signature are not necessarily distinct. For example, we might have a binary-valued *Cited* attribute with argument signature A_1, A_2, where both are of type Article. We assume, for simplicity of presentation, that attribute names are uniquely defined; thus, for example, the attribute denoting the age for a person will be named differently from the attribute denoting the age for a car.

relation

This last example demonstrates a basic concept in this framework: that of a *relation*. A relation is a property of a tuple of objects, which tells us whether the objects in this tuple satisfy a certain relationship with each other. For example, *Took-Course* is a relation over student-course object pairs s, c, which is true is student s took the course c. As another example *Mother* is a relation between person-person pairs p_1, p_2, which is true if p_1 is the mother of p_2. Relations are not restricted to involving only pairs of objects. For example, we can have a *Go* relation, which takes triples of objects — a person, a source location, and a destination location.

At some level, a relation is simply a binary-valued attribute, as in our *Cited* example. However, this perspective obscures the fundamental property of a relation — that it *relates* a tuple of objects to each other. Thus, when we introduce the notion of probabilistic dependencies that can occur between related objects, the presence or absence of a relation between a pair of objects will play a central role in defining the probabilistic dependency model.

Instantiations   Given a set of template attributes, we can instantiate them in different ways, to produce probability spaces with multiple random variables of the same type. For example, we can consider a particular university, with a set of students and a set of courses, and use the notion of a template attribute to define a probability space that contains a random variable *Grade*(s, c) for different (student,course) pairs s, c. The resulting model encodes a joint distribution over the grades of multiple students in multiple courses. Similarly, in a temporal model, we can have the template attribute *Location*(T); we can then select a set of relevant time points and generate a trajectory with specific random variables *Location*(t)

To instantiate a set of template attributes to a particular setting, we need to define a set of objects for each class in our domain. For example, we may want to take a particular set of students and set of courses and define a model that contains a ground random variable *Intelligence*(s) and *SAT*(s) for every student object s, a ground random variable *Difficulty*(c) for every course object c, and ground random variables *Grade*(s, c) and *Satisfaction*(s, c) for every valid pair of (student,course) objects.

More formally, we now show how a set of template attributes can be used to generate an infinite set of probability spaces, each involving instantiations of the template attributes induced by some set of objects. We begin with a simple definition, deferring discussion of some of the more complicated extensions to section 6.6.

Definition 6.6

object skeleton

Let $\mathcal{Q}$ be a set of classes, and $\aleph$ a set of template attributes over $\mathcal{Q}$. An object skeleton κ specifies a fixed, finite set of objects $\mathcal{O}^{\kappa}[Q]$ for every $Q \in \mathcal{Q}$. We also define

$$\mathcal{O}^{\kappa}[U_1, \ldots, U_k] = \mathcal{O}^{\kappa}[Q[U_1]] \times \ldots \times \mathcal{O}^{\kappa}[Q[U_k]].$$

By default, we define $\Gamma_{\kappa}[A] = \mathcal{O}^{\kappa}[\alpha(A)]$ to be the set of possible assignments to the logical variables in the argument signature of A. However, an object skeleton may also specify a subset of legal assignments. $\Gamma_{\kappa}[A] \subset \mathcal{O}^{\kappa}[\alpha(A)]$. ∎

We can now define the set of instantiations of the attributes:

Student	Intelligence	SAT
George	*low*	*High*
Alice	*high*	*High*

Course	Difficulty
CS101	*high*
Econ101	*low*

Student	Course	Grade	Satisfaction
George	**CS101**	*C*	*low*
George	**Econ101**	*A*	*high*
Alice	**CS101**	*A*	*low*

Figure 6.5 **One possible world for the** University **example.** Here, we have two student objects and two course objects. The attributes *Grade* and *Satisfaction* are restricted to three of their possible four legal assignments.

Definition 6.7

ground random
variable

Let κ be an object skeleton over $\mathcal{Q}, \aleph$. We define sets of ground random variables:

$$
\begin{aligned}
\mathcal{X}_\kappa[A] &= \{A(\gamma) \; : \; \gamma \in \Gamma_\kappa[A]\} \\
\mathcal{X}_\kappa[\aleph] &= \cup_{A \in \aleph} \mathcal{X}_\kappa[A].
\end{aligned}
\tag{6.5}
$$

Note that we are abusing notation here, identifying an assignment $\gamma = \langle U_1 \mapsto u_1, \ldots, U_k \mapsto u_k \rangle$ *with the tuple* $\langle u_1 \ldots, u_k \rangle$; *this abuse of notation is unambiguous in this context due to the ordering of the tuples.* ∎

The ability to specify a subset of $\mathcal{O}^\kappa[\alpha(A)]$ is useful in eliminating the need to consider random variables that do not really appear in the model. For example, in most cases, not every student takes every course, and so we would not want to include a *Grade* variable for every possible (student, course) pair at our university. See figure 6.5 as an example.

Clearly, the set of random variables is different for different skeletons; hence the model is a template for an infinite set of probability distributions, each spanning a different set of objects that induces a different set of random variables. In a sense, this is similar to the situation we had in DBNs, where the same 2TBN could induce a distribution over different numbers of time slices. Here, however, the variation between the different instantiations of the template is significantly greater.

Our discussion so far makes several important simplifying assumptions. First, we portrayed the skeleton as defining a set of objects for each of the classes. As we discuss in later sections, it can be important to allow the skeleton to provide additional background information about the set of possible worlds, such as some relationships that hold between objects (such as the structure of a family tree). Conversely, we may also want the skeleton to provide less information: In particular, the premise underlying equation (6.5) is that the set of objects is predefined by the skeleton. As we briefly discuss in section 6.6.2, we may also want to deal with settings in which we have uncertainty over the number of objects in the domain. In this case, different possible worlds may have different sets of objects, so that a random variable such as $A(u)$ may be defined in some worlds (those that contain the object u) but not in others. Settings like this pose significant challenges, a discussion of which is outside the scope of this book.

Template Factors   The final component in a template-based probabilistic model is one that defines the actual probability distribution over a set of a ground random variables generated from a set of template attributes. Clearly, we want the specification of the model to be defined in a template-based way. Specifically, we would like to take a factor — whether an undirected factor or a CPD — and instantiate it to apply to multiple scopes in the domain. We have already seen one simple example of this notion: in a 2-TBN, we had a template CPD $P(X_i' \mid \mathrm{Pa}_{X_i'})$, which we instantiated to apply to different scopes $X_i^{(t)}, \mathrm{Pa}_{X_i^{(t)}}$, by instantiating any occurrence X_j' to $X_j^{(t)}$, and any occurrence X_j to $X_j^{(t-1)}$. In effect, there we had template variables of the form X_j and X_j' as arguments to the CPD, and we instantiated them in different ways for different time points. We can now generalize this notion by defining a factor with arguments. Recall that a factor ϕ is a function from a tuple of random variables $\boldsymbol{X} = Scope[\phi]$ to the reals; this function returns a number for each assignment $\boldsymbol{x}$ to the variables $\boldsymbol{X}$. We can now define

Definition 6.8

template factor

instantiated factor

A template factor is a function ξ defined over a tuple of template attributes $A_1, \ldots, A_l$, where each A_j has a range $Val(A)$. It defines a mapping from $Val(A_1) \times \ldots \times Val(A_l)$ to $\mathbb{R}$. Given a tuple of random variables $X_1, \ldots, X_l$, such that $Val(X_j) = Val(A_j)$ for all $j = 1, \ldots, l$, we define $\xi(X_1, \ldots, X_l)$ to be the instantiated factor *from $\boldsymbol{X}$ to $\mathbb{R}$.* ∎

In the subsequent sections of this chapter, we use these notions to define various languages for encoding template-based probabilistic models. As we will see, some of these representational frameworks subsume and generalize on the DBN framework defined earlier.

6.4   Directed Probabilistic Models for Object-Relational Domains

Based on the framework described in the previous section, we now describe template-based representation languages that can encode directed probabilistic models.

6.4.1   Plate Models

plate model

We begin our discussion by presenting the *plate model*, the simplest and best-established of the object-relational frameworks. Although restricted in several important ways, the plate modeling framework is perhaps the approach that has been most commonly used in practice, notably for encoding the assumptions made in various learning tasks. This framework also provides an excellent starting point for describing the key ideas of template-based languages and for motivating some of the extensions that have been pursued in richer languages.

In the plate formalism, object types are called *plates*. The fact that multiple objects in the class share the same set of attributes and same probabilistic model is the basis for the use of the term "plate," which suggests a stack of identical objects. We begin with some motivating examples and then describe the formal framework.

6.4.1.1   Examples

Example 6.8

The simplest example of a plate model, shown in figure 6.6, describes multiple random variables generated from the same distribution. In this case, we have a set of random variables $X(d)$ $(d \in \mathcal{D})$

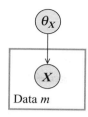

Figure 6.6   Plate model for a set of coin tosses sampled from a single coin

that all have the same domain $Val(X)$ *and are sampled from the same distribution. In a plate representation, we encode the fact that these variables are all generated from the same template by drawing only a single node* $X(d)$ *and enclosing it in a box denoting that d ranges over* $\mathcal{D}$, *so that we know that the box represents an entire "stack" of these identically distributed variables. This box*

plate   *is called a* plate, *with the analogy that it represents a stack of identical plates.*

In many cases, we want to explicitly encode the fact that these variables have an identical distribution. We therefore often explicitly add to the model the variable θ_X, *which denotes the parameterization of the CPD from which the variables* $X(d)$ *are sampled. Most simply, if the X's are coin tosses of a single (possibly biased) coin,* θ_X *would take on values in the range* $[0, 1]$, *and its value would denote the bias of the coin (the probability with which it comes up "Heads").* ∎

The idea of including the CPD parameters directly in the probabilistic model plays a central role in our discussion of learning later in this book (see section 17.3). For the moment, we note only that including the parameters directly in the model allows us to make explicit the fact that all of the variables $X(d)$ are sampled from the same CPD. By contrast, we could also have used a model where a variable $\theta_X(d)$ is included *inside* the plate, allowing us to encode the setting where each of the coin tosses was sampled from a different coin. We note that this transformation is equivalent to adding the coin ID d as a parent to X; however, the explicit placement of θ_X within the plate makes the nature of the dependence more explicit. In this chapter, to reduce clutter, we use the convention that parameters not explicitly included in the model (or in the figure) are outside of all plates.

Example 6.9

Let us return to our Student *example. We can have a* Student *plate that includes the* attributes $I(S), G(S)$. *As shown in figure 6.7a, we can have* $G(S)$ *depend on* $I(s)$. *In this model we have*

ground Bayesian   *a set of (Intelligence,Grade) pairs, one for each student. The figure also shows the* ground Bayesian
network   network *that would result from instantiating this model for two students. As we discussed, this model implicitly makes the assumption that the CPDs for* $P(I(s))$ *and for* $P(G(s) \mid I(s))$ *is the same for all students s. Clearly, we can further enrich the model by introducing additional variables, such as an SAT-score variable for each student.* ∎

Our examples thus far have included only a single type of object, and do not significantly expand the expressive power of our language beyond that of plain graphical models. The key benefit of the plate framework is that it allows for multiple plates that can overlap with each other in various ways.

Example 6.10

nested plate

Assume we want to capture the fact that a course has multiple students in it, each with his or her own grade, and that this grade depends on the difficulty of the course. Thus, we can introduce a second type of plate, labeled Course, *where the Grade attribute is now associated with a (student, course) pair. There are several ways in which we can modify the model to include courses. In figure 6.7b, the* Student *plate is nested within the* Course *plate. The Difficulty variable is enclosed within the* Course *plate, whereas Intelligence and Grade are enclosed within both plates. We thus have that Grade(s, c) for a particular (student, course) pair (s, c) depends on Difficulty(c) and on Intelligence(s, c).* ∎

This formulation ignores an important facet of this problem. As illustrated in figure 6.7b, it induces networks where the *Intelligence* variable is associated not with a student, but rather with a (student, course) pair. Thus, if we have the same student in two different courses, we would have two different variables corresponding to his intelligence, and these could take on different values. This formulation may make sense in some settings, where different notions of "intelligence" may be appropriate to different topics (for example, math versus art); however, it is clearly not a suitable model for all settings. Fortunately, the plate framework allows us to come up with a different formulation.

Example 6.11

plate intersection

Figure 6.7c shows a construction that avoids this limitation. Here, the Student *plate and the* Course *plates intersect, so that the Intelligence attribute is now associated only with the* Student *plate, and Difficulty with the* Course *plate; the Grade attribute is associated with the pair (comprising the intersection between the plates). The interpretation of this dependence is that, for any pair of (student, course) objects s, c, the attribute Grade(s, c) depends on Intelligence(s) and on Difficulty(c). The figure also shows the network that results for two students both taking two courses.* ∎

In these examples, we see that even simple plate representations can induce fairly complex ground Bayesian networks. Such networks model a rich network of interdependencies between different variables, allowing for paths of influence that one may not anticipate.

Example 6.12

Consider the plate model of figure 6.7c, where we know that a student Jane took CS101 and got an A. This fact changes our belief about Jane's intelligence and increases our belief that CS101 is an easy class. If we now observe that Jane got a C in Math 101, it decreases our beliefs that she is intelligent, and therefore should increase our beliefs that CS101 is an easy class. If we now observe that George got a C in CS101, our probability that George has high intelligence is significantly lower. Thus, our beliefs about George's intelligence can be affected by the grades of other students in other classes.

Figure 6.8 shows a ground Bayesian network induced by a more complex skeleton involving fifteen students and four courses. Somewhat surprisingly, the additional pieces of "weak" evidence regarding other students in other courses can accumulate to change our conclusions fairly radically: Considering only the evidence that relates directly to George's grades in the two classes that he took, our posterior probability that George has high intelligence is 0.8. If we consider our entire body of evidence about all students in all classes, this probability decreases from 0.8 to 0.25. When we examine the evidence more closely, this conclusion is quite intuitive. We note, for example, that of the students who took CS101, only George got a C. In fact, even Alice, who got a C in both of her other classes, got an A in CS101. This evidence suggests strongly that CS101 is not a difficult class, so that George's grade of a C in CS101 is a very strong indicator that he does not have high intelligence.

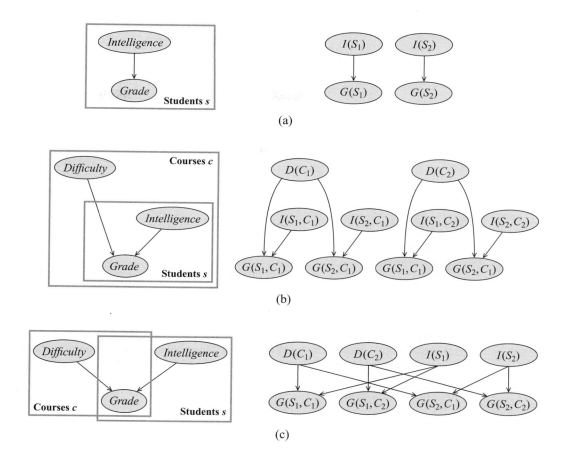

Figure 6.7   Plate models and induced ground Bayesian networks for a simplified Student **example.** (a) Single plate: Multiple independent samples from the same distribution. (b) Nested plates: Multiple courses, each with a separate set of students. (c) Intersecting plates: Multiple courses with overlapping sets of students.

Thus, we obtain much more informed conclusions by defining probabilistic models that encompass all of the relevant evidence.   ∎

As we can see, **a plate model provides a language for encoding models with repeated structure and shared parameters. As in the case of DBNs, the models are represented at the** *template level*; **given a particular set of objects, they can then be instantiated to induce a** *ground Bayesian network* **over the random variables induced by these objects. Because there are infinitely many sets of objects, this template can induce an infinite set of ground networks.**

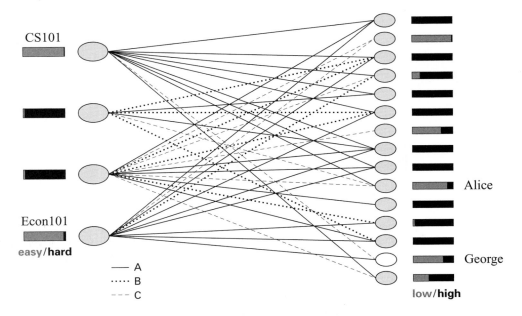

Figure 6.8   Illustration of probabilistic interactions in the University **domain.** The ground network contains random variables for the *Intelligence* of fifteen students (right ovals), including George (denoted by the white oval), and the *Difficulty* for four courses (left ovals), including CS101 and Econ101. There are also observed random variables for some subset of the (student, course) pairs. For clarity, these observed grades are not denoted as variables in the network, but rather as edges relating the relevant (student, course) pairs. Thus, for example George received an A in Econ101 but a C in CS101. Also shown are the final probabilities obtained by running inference over the resulting network.

6.4.1.2   Plate Models: Formal Framework

We now provide a more formal description of the plate modeling language: its representation and its semantics. The plate formalism uses the basic object-relational framework described in section 6.4. As we mentioned earlier, plates correspond to object types.

Each template attribute in the model is embedded in zero, one, or more plates (when plates intersect). If an attribute A is embedded in a set of plates $Q_1, \ldots, Q_k$, we can view it as being associated with the argument signature $U_1, \ldots, U_k$, where each logical variable U_i ranges over the objects in the plate (class) Q_i. Recall that a plate model can also have attributes that are external to any plate; these are attributes for which there is always a single copy in the model. We can view this attribute as being associated with an argument signature of arity zero.

In a plate model, the set of random variables induced by a template attribute A is defined by the complete set of assignments: $\Gamma_\kappa[A] = \mathcal{O}^\kappa[\alpha(A)]$. Thus, for example, we would have a *Grade* random variable for every (student,course) pair, whereas, intuitively, these variables are only defined in cases where the student has taken the course. We can take the values of such variables to be unobserved, and, if the model is well designed, its descendants in the probabilistic dependency graph will also be unobserved. In this case, the resulting random variables in the network will be barren, and can be dropped from the network without affecting the marginal distribution over the others. This solution, however, while providing the right semantics, is not particularly elegant.

We now define the probabilistic dependency structures allowed in the plate framework. To provide a simple, well-specified dependency structure, plate models place strong restrictions on the types of dependencies allowed. For example, in example 6.11, if we define *Intelligence* to be a parent of *Difficulty* (reflecting, say, our intuition that intelligent students may choose to take harder classes), the semantics of the ground model is not clear: for a ground random variable $D(c)$, the model does not specify which specific $I(s)$ is the parent. To avoid this problem, plate models require that an attribute can only depend on attributes in the same plate. This requirement is precisely the intuition behind the notion of plate intersection: Attributes in the intersection of plates can depend on other attributes in any of the plates to which they belong.

Formally, we have the following definition:

Definition 6.9

plate model

A plate model $\mathcal{M}_{Plate}$ *defines, for each template attribute* $A \in \aleph$ *with argument signature* $U_1, \ldots, U_k$:

template parent

- *a set of* template parents

$$\mathrm{Pa}_A = \{B_1(\boldsymbol{U}_1), \ldots, B_l(\boldsymbol{U}_l)\}$$

parent argument signature

such that for each $B_i(\boldsymbol{U}_i)$, *we have that* $\boldsymbol{U}_i \subseteq \{U_1, \ldots, U_k\}$. *The variables* $\boldsymbol{U}_i$ *are the* argument signature *of the parent* B_i.

- *a template CPD* $P(A \mid \mathrm{Pa}_A)$. ∎

This definition allows the *Grade* attribute *Grade*(S, C) to depend on *Intelligence*(S), but not vice versa. Note that, as in Bayesian networks, this definition allows any form of CPD, with or without local structure.

Note that the straightforward graphical representation of plates fails to make certain distinctions that are clear in the symbolic representation.

Example 6.13

Assume that our model contains an attribute Cited(U_1, U_2), *where* U_1, U_2 *are both in the* Paper *class. We might want the dependency model of this attribute to depend on properties of both papers, for example, Topic*(U_1), *Topic*(U_2), *or Review-Paper*(U_1). *To encode this dependency graphically, we first need to have two sets of attributes from the* Paper *class, one for* U_1 *and the other for* U_2. *Moreover, we need to denote somehow which attributes of which of the two arguments are the parents. The symbolic representation makes these distinctions unambiguously.* ∎

To instantiate the template parents and template CPDs, it helps to introduce some shorthand notation. Let $\gamma = \langle U_1 \mapsto u_1, \ldots, U_k \mapsto u_k \rangle$ be some assignment to some set of logical variables, and $B(U_{i_1}, \ldots, U_{i_l})$ be an attribute whose argument signature involves only a subset of these variables. We define $B(\gamma)$ to be the ground random variable $B(u_{i_1}, \ldots, u_{i_l})$.

The template-level plate model, when applied to a particular skeleton, defines a ground probabilistic model, in the form of a Bayesian network:

Definition 6.10

ground Bayesian network

A plate model $\mathcal{M}_{Plate}$ *and object skeleton* κ *define a* ground Bayesian network $\mathcal{B}_\kappa^{\mathcal{M}_{Plate}}$ *as fol-*

lows. Let $A(U_1, \ldots, U_k)$ be any template attribute in $\aleph$. Then, for any assignment $\gamma = \langle U_1 \mapsto u_1, \ldots, U_k \mapsto u_k \rangle \in \Gamma_\kappa[A]$, we have a variable $A(\gamma)$ in the ground network, with parents $B(\gamma)$ for all $B \in \mathrm{Pa}_A$, and the instantiated CPD $P(A(\gamma) \mid \mathrm{Pa}_A(\gamma))$.   ∎

Thus, in our example, we have that the network contains a set of ground random variables *Grade*(s, c), one for every student s and every course c. Each such variable depends on *Intelligence*(s) and on *Difficulty*(c).

The ground network $\mathcal{B}_\kappa^{\mathcal{M}_{Plate}}$ specifies a well-defined joint distribution over $\mathcal{X}_\kappa[\aleph]$, as required. The BN in figure 6.7b is precisely the network structure we would obtain from this definition, using the plate model of figure 6.7 and the object skeleton $\mathcal{O}^\kappa[\mathsf{Student}] = \{s_1, s_2\}$ and $\mathcal{O}^\kappa[\mathsf{Course}] = \{c_1, c_2\}$. In general, despite the compact parameterization (only one local probabilistic model for every attribute in the model), the resulting ground Bayesian network can be quite complex, and models a rich set of interactions. As we saw in example 6.12, the ability to incorporate all of the relevant evidence into the single network shown in the figure can significantly improve our ability to obtain meaningful conclusions even from weak indicators.

☞   **The plate model is simple and easy to understand, but it is also highly limited in several ways. Most important is the first condition of definition 6.9, whereby $A(U_1, \ldots, U_k)$ can only depend on attributes of the form $B(U_{i_1}, \ldots, U_{i_l})$, where $U_{i_1}, \ldots, U_{i_l}$ is a subtuple of $U_1, \ldots, U_k$. This restriction significantly constrains our ability to encode a rich network of probabilistic dependencies between the objects in the domain. For example, in the Genetics domain, we cannot encode a dependence of *Genotype*(U_1) on *Genotype*(U_2), where U_2 is (say) the mother of U_1. Similarly, we cannot encode temporal models such as those described in section 6.2, where the car's position at a point in time depends on its position at the previous time point.** In the next section, we describe a more expressive representation that addresses these limitations.

6.4.2   Probabilistic Relational Models

As we discussed, the greatest limitation of the plate formalism is its restriction on the argument signature of an attribute's parents. In particular, in our genetic inheritance example, we would like to have a model where *Genotype*(u) depends on *Genotype*(u'), where u' is the mother of u. This type of dependency is not encodable within plate models, because it uses a logical variable in the attribute's parent that is not used within the attribute itself. To allow such models, we must relax this restriction on plate models. However, relaxing this assumption without care can lead to nonsensical models. In particular, if we simply allow *Genotype*(U) to depend on *Genotype*(U'), we end up with a dependency model where every ground variable *Genotype*(u) depends on every other such variable. Such models are intractably dense, and (more importantly) cyclic. What we really want is to allow a dependence of *Genotype*(U) on *Genotype*(U'), but only for those assignments to U' that correspond to U's mother. We now describe one representation that allows such dependencies, and then discuss some of the subtleties that arise when we introduce this significant extension to our expressive power.

6.4.2.1   Contingent Dependencies

To capture such situations, we introduce the notion of a *contingent dependency*, which specifies the context in which a particular dependency holds. A contingent dependency is defined in terms of a *guard* — a formula that must hold for the dependency to be applicable.

Example 6.14

Consider again our University *example. As usual, we can define Grade(S, C) for a student S and a course C to have the parents Difficulty(C) and Intelligence(S). Now, however, we can make this dependency contingent on the guard Registered(S, C). Here, the parent's argument signature is the same as the child's. More interestingly, contingent dependencies allow us to model the dependency of the student's satisfaction in a course, Satisfaction(S, C), on the teaching ability of the professor who teaches the course. In this setting, we can make Teaching-Ability(P) the parent of Satisfaction(S, C), where the dependency is contingent on the guard Registered$(S, C) \wedge$ Teaches(P, C). Note that here, we have more logical variables in the parents of Satisfaction(S, C) than in the attribute itself: the attribute's argument signature is S, C, whereas its parent argument signature is the tuple S, C, P.* ∎

We can also represent chains of dependencies within objects in the same class.

Example 6.15

For example, to encode temporal models, we could have Location(U) depend on Location(V), contingent on the guard Precedes(V, U). In our Genetics *example, for the attribute Genotype(U), we would define the template parents Genotype(V) and Genotype(W), the guard Mother$(V, U) \wedge$ Father(W, U), and the parent signature U, V, W.* ∎

We now provide the formal definition underlying these examples:

Definition 6.11

contingent
dependency
model

parent argument
signature

guard

template parent

For a template attribute A, we define a contingent dependency model *as a tuple consisting of:*

- *A* parent argument signature *$\alpha(\mathrm{Pa}_A)$, which is a tuple of typed logical variables U_i such that $\alpha(\mathrm{Pa}_A) \supseteq \alpha(A)$.*

- *A* guard *Γ, which is a binary-valued formula defined in terms of a set of template attributes Pa_A^Γ over the argument signature $\alpha(\mathrm{Pa}_A)$.*

- *a set of* template parents

$$\mathrm{Pa}_A = \{B_1(\boldsymbol{U}_1), \ldots, B_l(\boldsymbol{U}_l)\}$$

such that for each $B_i(\boldsymbol{U}_i)$, we have that $\boldsymbol{U}_i \subseteq \alpha(\mathrm{Pa}_A)$. ∎

probabilistic
relational model

A *probabilistic relational model* (PRM) $\mathcal{M}_{PRM}$ defines, for each $A \in \aleph$ a contingent dependency model, as in definition 6.11, and a template CPD. The structure of the template CPD in this case is more complex, and we discuss it in detail in section 6.4.2.2.

Intuitively, the template parents in a PRM, as in the plate model, define a template for the parent assignments in the ground network, which will correspond to specific assignments of the logical variables to objects of the appropriate type. In this setting, however, the set of logical variables in the parents is not necessarily a subset of the logical variables in the child.

The ability to introduce new logical variables into the specification of an attribute's parents gives us significant expressive power, but introduces some significant challenges. These challenges clearly manifest in the construction of the ground network.

Definition 6.12

ground Bayesian
network

A PRM $\mathcal{M}_{PRM}$ and object skeleton κ define a ground Bayesian network $\mathcal{B}_\kappa^{\mathcal{M}_{PRM}}$ *as follows. Let $A(U_1, \ldots, U_k)$ be any template attribute in $\aleph$. Then, for any assignment $\gamma \in \Gamma_\kappa[A]$, we have a variable $A(\gamma)$ in the ground network. This variable has, for any $B \in \text{Pa}_A^\Gamma \cup \text{Pa}_A$ and any assignment γ' to $\alpha(\text{Pa}_A) - \alpha(A)$, the parent that is the instantiated variable $B(\gamma, \gamma')$.* ∎

An important subtlety in this definition is that the attributes that appear in the guard are also parents of the ground variable. This requirement is necessary, because the values of the guard attributes determine whether there is a dependency on the parents or not, and hence they affect the probabilistic model.

Using this definition for the model of example 6.14, we have that *Satisfaction*(s, c) has the parents: *Teaching-Ability*(p), *Registered*(s, c), and *Teaches*(p, c) for every professor p. The guard in the contingent dependency is intended to encode the fact that the dependency on *Teaching-Ability*(p) is only present for a subset of individuals p, but it is not obvious how that fact affects the construction of our model. The situation is even more complex in example 6.15, where we have as parents of *Genotype*(u) all of the variables of the form *Father*(v, u) and *Genotype*(v), for all person objects v, and similarly for *Mother*(v, u) and *Genotype*(v). In both cases, the resulting ground network is very densely connected. In the Genetics network, it is also obviously cyclic. We will describe how to encode such dependencies correctly within the CPD of the ground network, and how to deal with the issues of potential cyclicity.

6.4.2.2   CPDs in the Ground Network

As we just discussed, the ground network induced by a PRM can introduce a dependency of a variable on a set of parents that is not fixed in advance, and which may be arbitrarily large. How do we encode a probabilistic dependency model for such dependencies?

Exploiting the Guard Structure   The first key observation is that the notion of a contingent dependency is intended to specifically capture context-specific independence: In definition 6.12, if the guard for a parent B of A is false for a particular assignment (γ, γ'), then there is no dependency of $A(\gamma)$ on $B(\gamma, \gamma')$. For example, unless *Mother*(v, u) is true for a particular pair (u, v), we have no dependence of *Genotype*(u) on *Genotype*(v). Similarly, unless *Registered*$(s, c) \wedge$ *Teaches*(p, c) is true, there is no dependence of *Satisfaction*(s, c) on *Teaching-Ability*(p). We can easily capture this type of context-specific independence in the CPD using a variant of the multiplexer CPD of definition 5.3.

While this approach helps us specify the CPD in these networks of potentially unbounded indegree, it does not address the fundamental problem: the dense, and often cyclic, connectivity structure. A common solution to this problem is to assume that the guard predicates are properties of the basic relational structure in the domain, and are often fixed in advance. For example, in the temporal setting, the *Precedes* relation is always fixed: time point $t - 1$ always precedes time point t. Somewhat less obviously, in our Genetics example, it may be reasonable to assume that the pedigree is known in advance.

relational
skeleton

We can encode this assumption by defining a *relational skeleton* κ_r, which defines a certain set of facts (usually relationships between objects) that are given in advance, and are not part of the probabilistic model. In cases where the values of the attributes in the guards are specified as part of the relational skeleton, we can simply use that information to determine the set of parents that are active in the model, usually a very limited set. Thus, for example, if *Registered* and *Teaches* are part of the relational skeleton, then *Satisfaction*(s, c) has the parent *Teaching-Ability*(p) only when *Registered*(s, c) and *Teaches*(p, c) both hold. Similarly, in the Genetics example, if the pedigree structure is given in the skeleton, we would have that *Genotype*(v) is a parent of *Genotype*(u) only if *Mother*(v, u) or *Father*(v, u) are present in the skeleton. Moreover, we see that, assuming a legal pedigree, the resulting ground network in the Genetics domain is guaranteed to be acyclic. Indeed, the resulting model produces ground networks that are precisely of the same type demonstrated in box 3.B. **The use of contingent dependencies allows us to exploit relations that are determined by our skeleton to produce greatly simplified models, and to make explicit the fact that the model is acyclic.**

relational
uncertainty

The situation becomes more complex, however, if the guard predicates are associated with a probabilistic model, and therefore are random variables in the domain. Because the guards are typically associated with relational structure, we refer to this type of uncertainty as *relational uncertainty*. Relational uncertainty introduces significant complexity into our model, as we now cannot use background knowledge (from our skeleton) to simplify the dependency structure in contingent dependencies. In this case, when the family tree is uncertain, we may indeed have that *Genotype*(u) can depend on every other variable *Genotype*(v), a model that is cyclic and ill defined. However, if we restrict the distribution over the *Mother* and *Father* relations so as to ensure that only "reasonable" pedigrees get positive probability, we can still guarantee that our probabilistic model defines a coherent probability distribution. However, defining a probability distribution over the *Mother* and *Father* relations that is guaranteed to have this property is far from trivial; we return to this issue in section 6.6.

Aggregating Dependencies By itself, the use of guards may not fully address the problem of defining a parameterization for the CPDs in a PRM. Consider again a dependency of A on an attribute B that involves some set of logical variables $\boldsymbol{U}'$ that are not in $\alpha(A)$. Even if we assume that we have a relational skeleton that fully determines the values of all the guards, there may be multiple assignments γ' to $\boldsymbol{U}'$ for which the guard holds, and hence multiple different ground parents $B(\gamma, \gamma')$ — one for each distinct assignment γ' to $\boldsymbol{U}'$. Even in our simple University example, there may be multiple instructors for a course c, and therefore multiple ground variables *Teaching-Ability*(p) that are parents of a ground variable *Satisfaction*(s, c). In general, the number of possible instantiations of a given parent B is not known in advance, and may not even be bounded. Thus, we need to define a mechanism for specifying a template-level local dependency model that allows a variable number of parents. Moreover, because the parents corresponding to different instantiations are interchangeable, the local dependency model must be symmetric.

There are many possible ways of specifying such a model. One approach is to use one of the symmetric local probability models that we saw in chapter 5. For example, we can use a noisy-or (section 5.4.1) or logistic model (see section 5.4.2), where all parents have the same parameter. An alternative approach is to define an *aggregator CPD* that uses certain aggregate statistics or summaries of the set of parents of a variable. (See exercise 6.7 for some analysis of the expressive power of such CPDs.)

aggregator CPD

Example 6.16

Let us consider again the dependence of Satisfaction(S, C) on Teaching-Ability(P), with the guard Taking$(S, C) \wedge$ Teaching(C, P). Assuming, for the moment, that both Satisfaction and Teaching-

Ability are binary-valued, we might use a noisy-or model: Given a parameterization for Satisfaction given a single Teaching-Ability, we can use the noisy-or model to define a general CPD for Satisfaction(s, c) given any set of parents Teaching-Ability(p_1), ..., Teaching-Ability(p_m). Alternatively, we can assume that the student's satisfaction depends on the worst instructor and best instructor in the course. In this case, we might aggregate the teaching abilities using the min *and* max *functions, and then use a CPD of our choosing to denote the student's satisfaction as a function of the resulting pair of values. As another example, a student's job prospects can depend on the average grade in all the courses she has taken.* ∎

When designing such a combination rule, it is important to consider any possible boundary cases. On one side, in many settings the set of parents can be empty: a course may have no instructors (if it is a seminar composed entirely of invited lecturers); or a person may not have a parent in the pedigree. On the other side, we may also need to consider cases where the number of parents is large, in which case noisy-or and logistic models often become degenerate (see figure 5.10 and figure 5.11).

The situation becomes even more complex when there are multiple distinct parents at the template level, each of which may result in a set of ground parents. For example, a student's satisfaction in a course may depend both on the teaching ability of multiple instructors and on the quality of the design of the different problem sets in the course. We therefore need to address both the aggregation of each type of parent set (instructors or problem sets) as well as combining them into a single CPD. Thus, we need to define some way of combining a set of CPDs $\{P(X \mid Y_{i,1}, \ldots, Y_{i,j_i}) : i = 1, \ldots, l\}$, to a single joint CPD $\{P(X \mid Y_{1,1}, \ldots, Y_{1,j_1}, \ldots, Y_{l,1}, \ldots, Y_{l,j_l})$. Here, as before, there is no single right answer, and the particular choice is likely to depend heavily on the properties of the application.

We note that the issue of multiple parents is distinct from the multiple parents that arise when we have relational uncertainty. In the case of relational uncertainty, we also have multiple parents for a variable in the ground network; yet, it may well be the case that, in any situation, at most one assignment to the logical variables in the parent signature will satisfy the guard condition. For example, even if we are uncertain about John's paternity, we would like it to be the case that, in any world, there is a unique object v for which *Father*$(v, John)$ holds.

(As we discuss in section 6.6, however, defining a probabilistic model that ensures this type of constraint can be far from trivial.) In this type of situation, the concept of a guard is again useful, since it allows us to avoid defining local dependency models with a variable number of parents in domains (such as genetic inheritance) where such situations do not actually arise.

6.4.2.3 Checking Acyclicity

One important issue in relational dependency models is that the dependency structure of the ground network is, in general, not determined in advance, but rather induced from the model structure and the skeleton. How, then, can we guarantee that we obtain a coherent probability distribution? Most obviously, we can simply check, *post hoc*, that any particular ground network resulting from this process is acyclic. However, this approach is unsatisfying from a model design perspective. When constructing a model, whether by hand or using learning methods, we would like to have some guarantees that it will lead to coherent probability distributions.

Thus, we would like to provide a test that we can execute on a model $\mathcal{M}_{PRM}$ *at the template level*, and which will guarantee that ground distributions induced from $\mathcal{M}_{PRM}$ will be coherent.

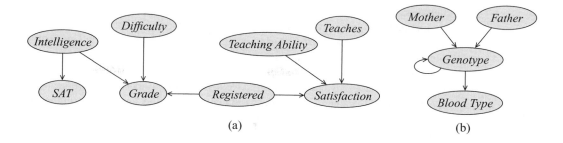

Figure 6.9   **Examples of dependency graphs:** (a) Dependency graph for the University example. (b) Dependency graph for the Genetics example.

One approach for doing so is to construct a template-level graph that encodes a set of potential dependencies that may happen at the ground level. The nodes in the graph are the template-level attributes; there is an edge from B to A if there is any possibility that a ground variable of type B will influence one of type A.

Definition 6.13

template dependency graph

A template dependency graph *for a template dependency model* $\mathcal{M}_{PRM}$ *contains a node for each template-level attribute* A, *and a directed edge from* B *to* A *whenever there is an attribute of type* B *in* $\mathrm{Pa}_A^\Gamma \cup \mathrm{Pa}_A$.   ∎

This graph can easily be constructed from the definition of the dependency model. For example, the template dependency graph of our University model (example 6.14) is shown in figure 6.9a.

It is not difficult to show that if template dependency graph for a model $\mathcal{M}_{PRM}$ is acyclic (as in this case), it is clear that any ground network generated from $\mathcal{M}_{PRM}$ must also be acyclic (see exercise 6.8). However, a cycle in the template dependency graph for $\mathcal{M}_{PRM}$ does not imply that every ground network induced by $\mathcal{M}_{PRM}$ is cyclic. (This is the case, however, for any nondegenerate instantiation of a plate model; see exercise 6.9.) Indeed, there are template models that, although cyclic at the template levels, reflect a natural dependency structure whose ground networks are guaranteed to be acyclic in practice.

Example 6.17

Consider the template dependency graph of our Genetics *domain, shown in figure 6.9b. The template-level self-loop involving Genotype(*Person*) reflects a ground-level dependency of a person's genotype on that of his or her parents. This type of dependency can only lead to cycles in the ground network if the pedigree is cyclic, that is, a person is his/her own ancestor. Because such cases (time travel aside) are impossible, this template model cannot result in cyclic ground networks for the skeletons that arise in practice. Intuitively, in this case, we have an (acyclic) ordering* $\prec$ *over the objects (people) in the domain, which implies that* u' *can be a parent of* u *only when* $u' \prec u$; *therefore, Genotype(*u*) can depend on Genotype(*u'*) only when* $u' \prec u$. *This ordering on objects is acyclic, and therefore so is the resulting dependency structure.*   ∎

The template dependency graph does not account for these constraints on the skeleton, and therefore we cannot conclude by examining the graph whether cycles can occur in ground networks for such ordered skeletons. However, exercise 6.10 discusses a richer form of the template dependency network that explicitly incorporates such constraints, and it is therefore able to determine that our Genetics model results in acyclic ground networks for any skeleton representing an acyclic pedigree.

So far in our discussion of acyclicity, we have largely sidestepped the issue of relational uncertainty. As we discussed, in the case of relational uncertainty, the ground network contains many "potential" edges, only a few of which will ever be "active" simultaneously. In such cases, the resulting ground network may not even be acyclic, even though it may well define a coherent (and sparse) probability distribution for every relational structure. Indeed, as we discussed in example 6.17, there are models that are potentially cyclic but are guaranteed to be acyclic by virtue of specific constraints on the dependency structure. It is thus possible to guarantee the coherence of a cyclic model of this type by ensuring that there is no positive-probability assignment to the guards that actually induces a cyclic dependence between the attributes.

6.5   Undirected Representation

The previous sections describe template-based formalisms that use a directed graphical model as their foundation. One can define similar extensions for undirected graphical models. Many of the ideas underlying this extension are fairly similar to the directed case. However, the greater flexibility of the undirected representation in avoiding local normalization requirements and acyclicity constraints can be particularly helpful in the context of these richer representations. Eliminating these requirements allows us to easily encode a much richer set of patterns about the relationships between objects in the domain; see, for example, box 6.C. In particular, as we discuss in section 6.6, these benefits can be very significant when we wish to define distributions over complex relational structures.

The basic component in a template-based undirected model is some expression, written in terms of template-level attributes with logical variables as arguments, and associated with a template factor. For a given object skeleton, each possible assignment γ to the logical variables in the expression induces a factor in the ground undirected network, all sharing the same parameters. As for variable-based undirected representations, one can parameterize a template-based undirected probabilistic model using full factors, or using features, as in a log-linear model. This decision is largely orthogonal to other issues. We choose to use log-linear features, which are the finest-grained representation and subsume table factors.

Example 6.18

Let us begin by revisiting our Misconception *example in section 4.1. Now, assume that we are interested in defining a probabilistic model over an entire set of students, where some number of pairs study together. We define a binary predicate (relation) Study-Pair(S_1, S_2), which is true when two students S_1, S_2 study together, and a predicate (attribute) Misconception(S) that encodes the*

template feature   *level of understanding of a student S. We can now define a* template feature *f_M, over pairs Misconception(S_1), Misconception(S_2), which takes value 1 whenever*

$$[\textit{Study-Pair}(S_1, S_2) = \textit{true} \land \textit{Misconception}(S_1) = \textit{Misconception}(S_2)] = \textit{true} \qquad (6.6)$$

and has value 0 otherwise.   ■

Definition 6.14

relational Markov
network

feature argument

A relational Markov network $\mathcal{M}_{RMN}$ *is defined in terms of a set* Λ *of* template features, *where each* $\lambda \in \Lambda$ *comprises:*

- *a real-valued* template feature f_λ *whose arguments are* $\aleph(\lambda) = \{A_1(\boldsymbol{U}_1), \ldots, A_l(U_l)\}$;
- *a weight* $w_\lambda \in \mathbb{R}$.

We define $\alpha(\lambda)$ *so that for all* i, $\boldsymbol{U}_i \subseteq \alpha(\lambda)$.   ∎

In example 6.18, we have that $\alpha(\lambda_M) = \{S_1, S_2\}$, both of type Student; and

$$\aleph(\lambda_M) = \{\textit{Study-Pair}(S_1, S_2), \textit{Misconception}(S_1), \textit{Misconception}(S_2)\}.$$

object skeleton

To specify a ground network using an RMN, we must provide an *object skeleton* κ that defines a finite set of objects $\mathcal{O}^\kappa[Q]$ for each class Q. As before, we can also define a restricted set $\Gamma_\kappa[A] \subset \mathcal{O}^\kappa[\alpha(A)]$. Given a skeleton, we can now define a *ground Gibbs distribution* in the natural way:

Example 6.19

Continuing example 6.18, assume we are given a skeleton containing a particular set of students and the set of study pairs within this set. This model induces a Markov network where the ground random variables have the form Misconception(s) for every student s in the domain. In this model, we have a feature f_M for every triple of variables Misconception(s_1), Misconception(s_2), Study-Pair(s_1, s_2). As usual in log-linear models, features can be associated with a weight; in this example, we might choose $w_M = 10$. In this case, the unnormalized measure for a given assignment to the ground variables would be $\exp(10K)$, where K is the number of pairs s_1, s_2 for which equation (6.6) holds.   ∎

More formally:

Definition 6.15

ground Gibbs
distribution

Given an RMN $\mathcal{M}_{RMN}$ and an object skeleton κ, we can define a ground Gibbs distribution $P_\kappa^{\mathcal{M}_{RMN}}$ *as follows:*

- *The variables in the network are $\mathcal{X}_\kappa[\aleph]$ (as in definition 6.7);*
- $P_\kappa^{\mathcal{M}_{RMN}}$ *contains a term*

$$\exp(w_\lambda \cdot f_\lambda(\gamma))$$

for each feature template $\lambda \in \Lambda$ and each assignment $\gamma \in \Gamma_\kappa[\alpha(\lambda)]$.   ∎

As always, a (ground) Gibbs distribution defines a Markov network, where we connect every pair of variables that appear together in some factor.

In the directed setting, the dense connectivity arising in ground networks raised several concerns: acyclicity, aggregation of dependencies, and computational cost of the resulting model. The first of these is obviously not a concern in undirected models. The other two, however, deserve some discussion.

Although better hidden, the issue of aggregating the contribution of multiple assignments of a feature also arises in undirected model. Here, the definition of the Gibbs distribution dictates the form of the aggregation we use. In this case, each grounding of the feature defines a factor in the unnormalized measure, and they are combined by a product operation, or an addition in log-space. In other words, each occurrence of the feature has a log-linear contribution to the unnormalized density. Importantly, however, this type of aggregation may not be appropriate for every application.

Example 6.20   *Consider a model for "viral marketing" — a social network of individuals related by the Friends(P, P') relation, where the attribute of interest Gadget(P) is the purchase of some cool new gadget G. We may want to construct a model where it is more likely that two friends either both own or both do not own G. That is, we have a feature similar to λ_M in example 6.18. In the log-linear model, the unnormalized probability that a person p purchases G grows log-linearly with the number k_p of his friends who own G. However, a more realistic model may involve a saturation effect, where the impact diminishes as k_p grows; that is, the increase in probability of Gadget(P) between $k_p = 0$ and $k_p = 1$ is greater than the increase in probability between $k_p = 20$ and $k_p = 21$.*   ∎

Thus, in concrete applications, we may wish to extend the framework to allow for other forms of combination, or even simply to define auxiliary variables corresponding to relevant aggregates (for example, the value of k_p in our example).

The issue of dense connectivity is as much an issue in the undirected case as in the directed case. The typical solution is similar: If we have background knowledge in the form of a relational skeleton, we can significantly simplify the resulting model. Here, the operation is very simple: we simply reduce every one of the factors in our model using the evidence contained in the relational skeleton, producing a reduced Markov network, as in definition 4.7. In this network, we would eliminate any ground variables whose values are observed in the skeleton and instantiate their (fixed) values in any ground factors containing them. In many cases, this process can greatly simplify the resulting features, often making them degenerate.

Example 6.21   *Returning to example 6.18, assume now that our skeleton specifies the instantiation of the relation Study-Pair, so that we know exactly which pairs of students study together and which do not. Now, consider the reduced Markov network obtained by conditioning on the skeleton. As all the variables Study-Pair(s_1, s_2) are observed, they are all eliminated from the network. Moreover, for any pair of students s_1, s_2 for which Study-Pair$(s_1, s_2) = false$, the feature $\lambda_M(s_1, s_2)$ necessarily takes the value 0, regardless of the values of the other variables in the feature. Because this ground feature is vacuous and has no impact on the distribution, it can be eliminated from the model. The resulting Markov network is much simpler, containing edges only between pairs of students who study together (according to the information in the relational skeleton).*   ∎

We note that we could have introduced the notion of a guarded dependency, as we did for PRMs. However, this component is far less useful here than it was in the directed case, where it also served a role in eliminating the need to aggregate parents that are not actually present in the network and in helping clarify the acyclicity in the model. Neither of these issues arises in the undirected framework, obviating the need for the additional notational complexity.

Finally, we mention one subtlety that is specific to the undirected setting. An undirected model uses nonlocal factors, which can have a dramatic influence on the global probability measure of the model. Thus, the probability distribution defined by an undirected relational model is not modular: Introducing a new object into the domain can drastically change the distribution over the properties of existing objects, even when the newly introduced object seems to have no meaningful interactions with the previous objects.

Example 6.22

Let us return to our example 6.19, and assume that any pair of students study together with some probability p; that is, we have an additional template feature over Study-Pair(S_1, S_2) that takes the value $\log p$ when this binary attribute is true and $\log(1 - p)$ otherwise.

Assume that we have a probability distribution over the properties of some set of students $\mathcal{O}^\kappa[\text{Student}] = \{s_1, \ldots, s_n\}$, and let us study how this distribution changes if we add a new student s_{n+1}. Consider an assignment to the properties of $s_1, \ldots, s_n$ in which m of the n students s_i have Misconception$(s_i) = 1$, whereas the remaining $n - m$ have Misconception$(s_i) = 0$. We can now consider the following situations with respect to s_{n+1}: he studies with k of the m students for whom Misconception$(s_i) = 1$, with ℓ of the $n - m$ students for whom Misconception$(s_i) = 0$, and himself has Misconception$(s_{n+1}) = c$ (for $c \in \{0, 1\}$). The probability of each such event is

$$\binom{m}{k}\binom{(n-m)}{\ell} p^\ell (1-p)^{(n-m-\ell)})(10^{kc} \cdot 10^{\ell(1-c)}),$$

where the first two terms come from the factors over the Study-Pair(S_1, S_2) structure, and the final term comes from the template feature λ_M. We want to compute the marginal distribution over our original variables (not involving s_{n+1}), to see whether introducing s_{n+1} changes this distribution. Thus, we sum out over all of the preceding events, which (using simple algebra) is $(10p + (1 - p))^m + (10p + (1 - p))^{n-m}$.

This analysis shows that the assignments to our original variables are multiplied by very different terms, depending on the value m. In particular, the probability of joint assignments where $m = 0$, so that all students agree, are multiplied by a factor of $(10p + (1 - p))^n$, whereas the probability of joint assignments where the students are equally divided in their opinion are multiplied by $(10p + (1 - p))^{n/2}$, an exponentially smaller factor. Thus, adding a new student, even one about whom we appear to know nothing, can drastically change the properties of our probability distribution. ∎

☞ **Thus, for undirected models, it can be problematic to construct (by learning or by hand) a template-based model for domains of a certain size, and apply it to models of a very different size. The impact of the domain size on the probability distribution varies, and therefore the implications regarding our ability to apply learning in this setting need to be evaluated per application.**

Box 6.C — Case Study: Collective Classification of Web Pages. *One application that calls for interesting models of interobject relationships is the classification of a network of interlinked webpages. One example is that of a university website, where webpages can be associated with students, faculty members, courses, projects, and more. We can associate each webpage w with a hidden variable $T(w)$ whose value denotes the type of the entity to which the webpage belongs. In a standard classification setting, we would use some learned classifier to label each webpage based on its features, such as the words on the page. However, we can obtain more information by also considering the interactions between the entities and the correlations they induce over their labels. For example, an examination of the data reveals that student pages are more likely to link to faculty webpages than to other student pages.*

One can capture this type of interaction both in a directed and in an undirected model. In a directed model, we might have a binary attribute $Links(W_1, W_2)$ that takes the value true if W_1 links to W_2 and false otherwise. We can then have $Links(W_1, W_2)$ depend on $T(W_1)$ and $T(W_2)$, capturing the dependence of the link probability on the classes of the two linked pages. An alternative approach is to use an undirected model, where we directly introduce a pairwise template feature over $T(W_1), T(W_2)$ for pairs W_1, W_2 such that $Links(W_1, W_2)$. Here, we can give higher potentials to pairs of types that tend to link, for example, student-faculty, faculty-course, faculty-project, project-student, and more.

A priori, both models appear to capture the basic structure of the domain. However, the directed model has some significant disadvantages in this setting. First, since the link structure of the webpages is known, the $Links(W_1, W_2)$ is always observed. Thus, we have an active v-structure connecting every pair of webpages, whether they are linked or not. The computational disadvantages of this requirement are obvious. Less obvious but equally important is the fact that there are many more non-links than links, and so the signal from the absent links tends to overwhelm the signal that could derived from the links that are present. In an undirected model, the absent links are simply omitted from the model; we simply introduce a potential that correlates the topics of two webpages only if they are linked. Therefore, an undirected model generally achieves much better performance on this task.

Another important advantage of the undirected model for this task is its flexibility in incorporating a much richer set of interactions. For example, it is often the case that a faculty member has a section in her webpage where she lists courses that she teaches, and another section that lists students whom she advises. Thus, another useful correlation that we may wish to model is one between the types of two webpages that are both linked from a third, and whose links are in close proximity on the page. We can model this type of interaction using features of the form $Close\text{-}Links(W, W_1, W_2) \wedge T(W_1) = t_1 \wedge T(W_2) = t_2$, where $Close\text{-}Links(W, W_1, W_2)$ is derived directly from the structure of the page.

Finally, an extension of the same model can be used to label not only the entities (webpages) but also the links between them. For example, we might want to determine whether a student-professor (s, p) pair with a link from s to p represents an advising relationship, or whether a linked professor-course pair represents an instructor relationship. Once again, a standard classifier would make use of features such as words in the vicinity of the hyperlink. At the next level, we can use an extension of the model described earlier to classify jointly both the types of the entities and the types of the links that relate them. In a more interesting extension, a relational model can also utilize higher-level patterns; for example, using a template feature over triplets of template attributes $T(W)$, we can encode the fact that students and their advisors belong to the same research group, or that students often serve as teaching assistants in courses that their advisors teach.

6.6   Structural Uncertainty ⋆

The object-relational probabilistic models we described allow us to encode a very rich family of distributions over possible worlds. In addition to encoding distributions over the attributes of objects, these approaches can allow us to encode *structural uncertainty* — a probabilistic model

structural
uncertainty

over the actual structure of the worlds, both the set of objects they contain and the relations between them. The different models we presented exhibit significant differences in the types of structural uncertainty that they naturally encompass. In this section, we discuss some of the major issues that arise when representing structural uncertainty, and how these issues are handled by the different models.

relational
uncertainty

object
uncertainty

There are two main types of structural uncertainty that we can consider: *relational uncertainty*, which models a distribution over the presence or absence of relations between objects; and *object uncertainty*, which models a distribution over the existence or number of actual objects in the domain. We discuss each in turn.

6.6.1 Relational Uncertainty

The template representations we have already developed already allow us to encode uncertainty about the relational structure. As in example 6.22, we can simply make the existence of a relationship a stochastic event. What types of probability distributions over relational structure can we encode using these representational tools? In example 6.22, each possible relation *Study-Pair*(s_1, s_2) is selected independently, at random, with probability p. Unfortunately, such graphs are not representative of most relational structures that we observe in real-world settings.

Example 6.23

Let us select an even simpler example, where the graph we are constructing is bipartite. Consider the relation Teaches(P, C), and assume that it takes the value true with probability 0.1. Consider a skeleton that contains 10 professors and 20 courses. Then the expected number of courses per professor is 2, and the expected number of professors per course is 1. So far, everything appears quite reasonable. However, the probability that, in the resulting graph, a particular professor teaches ℓ courses is distributed binomially: $\binom{20}{\ell}0.1^{\ell}0.9^{20-\ell}$. For example, the probability that any single professor teaches 5 or more courses is 4.3 percent, and the probability that at least one of them does is around 29 percent. This is much higher than is realistic in real-world graphs. The situation becomes much worse if we increase the number of professors and courses in our skeleton.

Of course, we can add parents to this attribute. For example, we can let the presence of an edge depend on the research area of the professor and the topic of the course, so that this attribute is more likely to take the value true if the area and topic match. However, this solution does not address the fundamental problem: it is still the case that, given all of the research areas and topics, the relationship status for different pairs of objects (the edges in the relational graph) are chosen independently. ∎

In this example, we wish to model certain global constraints on the distribution over the graph: the fact that each faculty member tends to teach only a small subset of courses. Unfortunately, it is far from clear how to incorporate this constraint into a template-level generative (directed) model over attributes corresponding to the presence of individual relations. Indeed, consider even the simpler case where we wish to encode the prior knowledge that each course has exactly one instructor. This model induces a correlation among all of the binary random variables corresponding to different instantiations *Teaches*(p, c) for different professors p and the same course c: once we have *Teaches*$(p, c) = true$, we must have *Teaches*$(p', c) = false$ for all $p' \neq p$. In order to incorporate this correlation, we would have to define a generative process that "selects" the relation variables *Teaches*(p, c) in some sequence, in a way that allows each *Teaches*(p', c) to depend on all of the preceding variables *Teaches*(p, c). This induces dependency models with dense connectivity, an arbitrary number of parents per variable (in the ground network), and a fairly complex dependency structure.

object-valued
attribute

An alternative approach is to use a different encoding for the course-instructor relationship. In logical languages, an alternative mechanism for relating objects to each other is via functions. A *function*, or *object-valued attribute* takes as argument a tuple of objects from a given set of classes, and returns a set of objects in another class. Thus, for example, rather than having a relation $Mother(P_1, P_2)$, we might use a function $Mother\text{-}of(P_1) \mapsto$ Person that takes, as argument, a person-object p_1, and returns the person object p_2, which is p_1's mother. In this case, the return-value of the function is just a single object, but, in general, we can define functions that return an entire set of objects. In our University example, the relation *Teaches* defines the function *Courses-Of*, which maps from professors to the courses they teach, and the function *Instructor*, which maps from courses to the professors that teach them. We note that these functions are *inverses* of each other: We have that a professor p is in $Instructor(c)$ if and only if c is in $Courses\text{-}Of(p)$.

As we can see, we can easily convert between set-valued functions and relations. Indeed, as long as the relational structure is fixed, the decision on which representation to use is largely a matter of convenience (or convention). However, once we introduce probabilistic models over the relational structure, the two representations lend themselves more naturally to quite different types of model. Thus, for example, if we encode the course-instructor relationship as a function from professors to courses, then rather than select pairwise relations at random, we might select, for any professor p, the set of courses $Courses\text{-}Of(p)$. We can define a distribution over sets in two components: a distribution over the size ℓ of the set, and a distribution that then selects ℓ distinct objects that will make up the set.

Example 6.24

Assume we want to define a probability distribution over the set Courses-Of(p) of courses taught by a professor p. We may first define a distribution over the number ℓ of courses c in Courses-Of(p). This distribution may depend on properties of the professor, such as her department or her level of seniority. Given the size ℓ, we now have to select the actual set of ℓ courses taught by p. We can define a model that selects ℓ courses independently from among the set of courses at the university. This choice can depend on properties of both the professor and the course. For example, if the professor's specialization is in artificial intelligence, she is more likely to teach a course in that area than in operating systems. Thus, the probability of selecting c to be in Courses-Of(p) depends both on Topic(c) and on Research-Area(p). Importantly, since we have already chosen $\ell = |Courses\text{-}Of(p)|$, we need to ensure that we actually select ℓ distinct courses, that is, we must sample from the courses without replacement. Thus, our ℓ sampling events for the different courses cannot be completely independent. ∎

While useful in certain settings, this model does not solve the fundamental problem. For example, although it allows us to enforce that every professor teaches between two and four courses, it still leaves open the possibility that a single course is taught by ten professors. We can, of course, consider a model that reverses the direction of the function, encoding a distribution over the instructors of each course rather than the courses taught by a professor, but this solution would simply raise the converse problem of the possibility that a single professor teaches a large number of classes.

It follows from this discussion that **it is difficult, in generative (directed) representations, to define distributions over relational structures that guarantee (or prefer) certain structural**

properties of the relation. For example, there is no natural way in which we can construct a probabilistic model that exhibits (a preference for) transitivity, that is, one satisfying that if $R(u, v)$ and $R(v, w)$ then (it is more likely that) $R(u, w)$.

These problems have a natural solution within the undirected framework. For example, a preference for transitivity can be encoded simply as a template feature that ascribes a high value to the (template) event

$$R(U, V) = true, R(V, W) = true, R(U, W) = true.$$

A (soft) constraint enforcing at most one instructor per course can be encoded similarly as a (very) low potential on the template event

$$Teaches(P, C) = true, Teaches(P', C) = true.$$

A constraint enforcing at least one instructor per course cannot be encoded in the framework of relational Markov networks, which allow only features with a bounded set of arguments. However, it is not difficult to extend the language to include potentials over unbounded sets of variables, as long as these potentials have a compact, finite-size representation. For example, we could incorporate an aggregator feature that counts the number t_p of objects c such that $Teaches(p, c)$, and introduce a potential over the value of t_p. This extension would allow us to incorporate arbitrary preferences about the number of courses taught by a professor. At the same time, the model could also include potentials over the aggregator i_c that counts the number of instructors p for a course c. Thus, we can simultaneously include global preferences on both sides of the relation between courses and professors.

However, while this approach addresses the issue of expressing such constraints, it leaves unresolved the problem of the complexity of the resulting ground network. In all of these examples, the induced ground network is very densely connected, with a ground variable for every potential edge in the relational graph (for example, $R(u, v)$), and a factor relating every pair or even every triple of these variables. In the latter examples, involving the aggregator, we have potentials whose scope is unbounded, containing all of the ground variables $R(u, v)$.

6.6.2 Object Uncertainty

So far, we have focused our discussion on representing probabilistic models about the presence or absence of certain relations, given a set of base objects. One can also consider settings in which even the set of objects in the world is not predetermined, and so we wish to define a probability distribution over this set.

Perhaps the most common setting in which this type of reasoning arises is in situations where different objects in our domain may be equal to each other. This situation arises quite often. For example, a single person can be student #34 in CS101, student #57 in Econ203, the eldest daughter of John and Mary, the girlfriend of Tom, and so on.

One solution is to allow objects in the domain to correspond to different "names," or ways of referring to an object, but explicitly reason about the probability that some of these names refer to the same object. But how do we model a distribution over equality relationships between the objects playing different roles in the model?

The key insight is to introduce explicitly into the model the notion of a "reference" to an object, where the same object can be referred to in several different ways. That is, we include in the model objects that correspond to the different "references" to the object. Thus, for example, we could have a class of "person objects" and another class for "person reference objects." We can use a relation-based representation in this setting, using a relation *Refers-to*(r, p) that is *true* whenever the reference r refers to a person p. However, we must also introduce uniqueness constraints to ensure that a reference r refers to precisely a single person p. Alternatively, a more natural approach is to use a function, or object-valued attribute, *Referent*(r), which designates the person to whom r refers. This approach automatically enforces the uniqueness constraints, and it is thus perhaps more appropriate to this application.

In either case, the relationship between references and the objects to which they refer is generally probabilistic and interacts probabilistically with other attributes in the domain. In particular, we would generally introduce factors that model the similarity of the properties of a "reference object" r and those of the true object p to which it refers. These *attribute similarity potentials* can be constructed to allow for noise and variation. For example, we can model the fact that a person whose name is "John Franklin Adams" may decide to go by "J.F. Adams" in one setting and "Frank Adams" in another, but is unlikely to go by the name "Peggy Smith." We can also model the fact that a person may decide to "round down" his or her reported age in some settings (for example, social interactions) but not in others (for example, tax forms). The problem of determining the correspondence between references and the entities to which

correspondence   they refer is an instance of the *correspondence* problem, which is described in detail in box 12.D. Box 6.D describes an application of this type of model to the problem of matching bibliographical citations.

In an alternative approach, we might go one step further, we can eliminate any mention of the true underlying objects, and restrict the model only to object references. In this solution, the domain contains only "reference objects" (at least for some classes). Now, rather than mapping references to the object to which they refer, we simply allow for different references to "correspond" to each other. Specifically, we might include a binary predicate *Same-as*(r, r'), which asserts that r and r' both refer to the same underlying object (not included as an object in the domain).

To ensure that *Same-As* is consistent with the semantics of an equality relation, we need to introduce various constraints on its properties. (Because these constraints are standard axioms of equality, we can include them as part of the formalism rather than require each user to specify them.) First, using the ideas described in section 6.6.1, we can introduce undirected (hard) potentials to constrain the relation to satisfy:

- *Reflexivity* — *Same-As*(r, r);

- *Symmetry* — *Same-As*(r, r') if and only if *Same-As*(r', r);

- *Transitivity* — *Same-As*(r, r') and *Same-As*(r', r'') implies *Same-As*(r, r'').

These conditions imply that the *Same-As* relation defines an equivalence relation on reference objects, and thus partitions them into mutually exclusive and exhaustive equivalence classes. Importantly, however, these constraints can only be encoded in an undirected model, and therefore this approach to dealing with equality only applies in that setting. In addition, we include in the model attribute similarity potentials, as before, which indicate the extent to which we expect attributes or predicates for two *Same-As* reference objects r and r' to be similar to each other. This approach, applied to a set of named objects, tends to cluster them together into groups whose attributes are similar and that participate in relations with objects that are also in equivalent groups.

There are, however, several problems with the reference-only solution. First, there is no natural place to put factors that should apply once per underlying entity.

Example 6.25

Suppose we are interested in inferring people's gender from their names. We might have a potential saying that someone named "Alex" is more likely to be male than female. But if we make this a template factor on $\{Name(R), Gender(R)\}$ where R ranges over references, then the factor will apply many times to people with many references. Thus, the probability that a person named "Alex" is male will increase exponentially with the number of references to that person. ∎

A related but more subtle problem is the dependence of the outcome of our inference on the number of references.

Example 6.26

Consider a very simple example where we have only references to one type of object and only the attribute A, which takes values $1, 2, 3$. For each pair of object references r, r' such that Same-As(r, r') holds, we have an attribute similarity potential relating $A(r)$ and $A(r')$: the cases of $A(r) = A(r')$ have the highest weight w; $A(r) = 1$, $A(r') = 3$ has very low weight; and $A(r) = 2$, $A(r') = 1$ and $A(r) = 2$, $A(r') = 3$ both have the same medium potential q. Now, consider the graph of people related by the Same-As relation: since Same-As is an equivalence relation, the graph is a set of mutually exclusive and exhaustive partitions, each corresponding to a set of references that correspond to the same object. Now, assume we have a configuration of evidence where we observe k_i references with $A(r) = i$, for $i = 1, 2, 3$. The most likely assignment relative to this model will have one cluster with all the $A(r) = 1$ references, and another with all the $A(r) = 3$ references. What about the references with $A(r) = 2$?

Somewhat surprisingly, their disposition depends on the relative sizes of the clusters. To understand why, we first note that (assuming $w > 1$) there are only three solutions with reasonably high probability: three separate clusters; a "1+2" and a "3" cluster; and a "1" and a "2+3" cluster. All other solutions have much lower probability, and the discrepancy decays exponentially with the size of the domain. Now, consider the case where $k_2 = 1$, so that there only one r^* with $A(r) = 2$. If we add r^* to the "1" cluster, we introduce an attribute similarity potential between $A(r^*)$ and all of the $A(r)$'s in the "1" cluster. This multiplies the overall probability of the configuration by q^{k_1}. Similarly, if we add r^* to the "3" cluster, the probability of the configuration is multiplied by q^{k_3}. Thus, if $q < 1$, the reference r^* is more likely to be placed in the smaller of the two clusters; if $q > 1$, it is more likely to be placed in the larger cluster. As k_2 grows, the optimal solution may now be one where we put the 2's into their own, separate cluster; the benefit of doing so depends on the relative sizes of the different parameters q, w, k_1, k_2, k_3. ∎

Thus, in this type of model, the resulting posterior is often highly peaked, and the probabilities of the different high-probability outcomes very sensitive to the parameters. By contrast, a model where each equivalence cluster is associated with a single actual object is a lot "smoother," for the number of attribute similarity potentials induced by a cluster of references grows linearly, not quadratically, in the size of the cluster.

Box 6.D — Case Study: Object Uncertainty and Citation Matching. *Being able to browse the network of citations between academic works is a valuable tool for research. For instance, given one citation to a relevant publication, one might want a list of other papers that cite the same work. There are several services that attempt to construct such lists automatically by extracting citations from online papers. This task is difficult because the citations come in a wide variety of formats, and often contain errors — owing both to the original author and to the vagaries of the extraction process. For example, consider the two citations:*

```
Elston R, Stewart A. A General Model for the Genetic Analysis of
    Pedigree Data. Hum. Hered. 1971;21:523-542.

Elston RC, Stewart J (1971): A general model for the analysis of
    pedigree data. Hum Hered 21523-542.
```

These citations refer to the same paper, but the first one gives the wrong first initial for J. Stewart, and the second one omits the word "genetic" in the title. The colon between the journal volume and page numbers has also been lost in the second citation. A citation matching system must handle this kind of variation, but must also avoid lumping together distinct papers that have similar titles and author lists.

Probabilistic object-relational models have proven to be an effective approach to this problem. One way to handle the inherent object uncertainty is to use a directed model with a Citation *class, as well as* Publication *and* Author *classes. The set of observed* Citation *objects can be included in the object skeleton, but the number of* Publication *and* Author *objects is unknown.*

A directed object-relational model for this problem (based roughly on the model of Milch et al. (2004)) is shown in figure 6.D.1a. The model includes random variables for the sizes of the Author *and* Publication *classes. The* Citation *class has an object-valued attribute PubCited(C), whose value is the* Publication *object that the citation refers to. The* Publication *class has a set-valued attribute Authors(P), indicating the set of authors on the publication. These attributes are given very simple CPDs: for PubCited(C), we use a uniform distribution over the set of* Publication *objects, and for Authors(P) we use a prior for the number of contributors along with a uniform selection distribution.*

To complete this model, we include string-valued attributes Name(A) and Title(P), whose CPDs encode prior distributions over name and title strings (for now, we ignore other attributes such as date and journal name). Finally, the Citation *class has an attribute Text(C), containing the observed text of the citation. The citation text attribute depends on the title and author names of the publication it refers to; its CPD encodes the way citation strings are formatted, and the probabilities of various errors and abbreviations.*

Thus, given observed values for all the $Text(c_i)$ attributes, our goal is to infer an assignment of values to the PubCited attributes — which induces a partition of the citations into coreferring groups. To get a sense of how this process works, consider the two preceding citations. One hypothesis, H_1, is that the two citations c_1 and c_2 refer to a single publication p_1, which has "genetic" in its title. An alternative, H_2, is that there is an additional publication p_2 whose title is identical except for the omission of "genetic," and c_2 refers to p_2 instead. H_1 obviously involves an unlikely event — a word being left out of a citation; this is reflected in the probability of $Text(c_2)$ given $Title(p_1)$. But the probability of H_2 involves an additional factor for $Title(p_2)$, reflecting the prior probability of the string "A general model for the analysis of pedigree data" under our model of academic paper titles. Since there are so many possible titles, this probability will be extremely small, allowing H_1 to win out. As this example shows, probabilistic models of this form exhibit

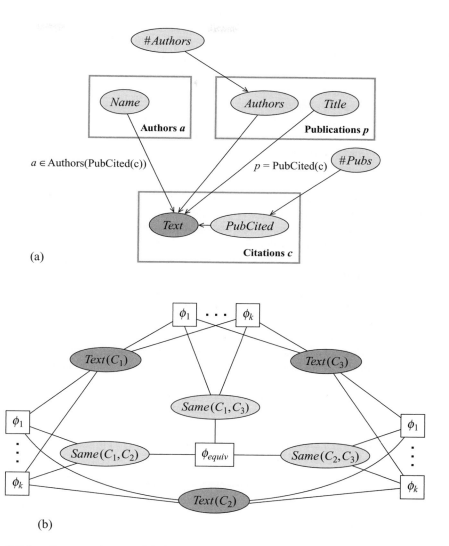

(a)

(b)

Figure 6.D.1 — Two template models for citation-matching (a) A directed model. (b) An undirected model instantiated for three citations.

a built-in Ockham's razor effect: the highest probability goes to hypotheses that do not include any more objects — and hence any more instantiated attributes — than necessary to explain the observed data.

Another line of work (for example, Wellner et al. (2004)) tackle the citation-matching problem using undirected template models, whose ground instantiation is a CRF (as in section 4.6.1). As we saw in the main text, one approach is to eliminate the Author *and* Publication *classes and simply reason about a relation Same(C, C') between citations (constrained to be an equivalence relation). Figure 6.D.1b shows an instantiation of such a model for three citations. For each pair of citations C, C', there is an array of factors $\phi_1, \ldots, \phi_k$ that look at various features of Text(C) and Text(C') — whether they have same surname for the first author, whether their titles are within an edit distance of two, and so on — and relate these features to Same(C_1, C_2). These factors encode preferences for and against coreference more explicitly than the factors in the directed model.*

However, as we have discussed, a reference-only model produces overly peaked posteriors that are very sensitive to parameters and to the number of mentions. Moreover, there are some examples where pairwise compatibility factors are insufficient for finding the right partition. For instance, suppose we have three references to people: "Jane," which is clearly a female's given name; "Smith," which is clearly a surname; and "Stanley," which could be a surname or a male's given name. Any pair of these references could refer to the same person: there could easily be a Jane Smith, a Stanley Smith, or a Jane Stanley. But it is unlikely that all three names corefer. Thus, a reasonable approach uses an undirected model that has explicit (hidden) variables for each entity and its attributes. The same potentials can be used as in the reference-only model. However, due to the use of undirected dependencies, we can allow the use of a much richer feature set, as described in box 4.E.

Systems that use template-based probabilistic models can now achieve accuracies in the high 90s for identifying coreferent citations. Identifying multiple mentions of the same author is harder; accuracies vary considerably depending on the data set, but tend to be around 70 percent. These models are also useful for segmenting citations into fields such as the title, author names, journal, and date. This is done by treating the citation text not as a single attribute but as a sequence of tokens (words and punctuation marks), each of which has an associated variable indicating which field it belongs to. These "field" variables can be thought of as the state variables in a hidden Markov model in the directed setting, or a conditional random field in the undirected setting (as in box 4.E). The resulting model can segment ambiguous citations more accurately than one that treats each citation in isolation, because it prefers for segmentations of coreferring citations to be consistent.

6.7 Summary

The representation languages discussed in earlier chapters — Bayesian networks and Markov networks — allow us to write down a model that encodes a specific probability distribution over a fixed, finite set of random variables. In this chapter, we have provided a general framework for defining *templates* for fragments of the probabilistic model. These templates can be reused both within a single model, and across multiple models of different structures. **Thus, a template-based representation language allows us to encode a potentially infinite set of distributions, over arbitrarily large probability spaces. The rich models that one can**

produce from such a representation can capture complex interactions between many interrelated objects, and thus utilize many pieces of evidence that we may otherwise ignore; as we have seen, these pieces of evidence can provide substantial improvements in the quality of our predictions.

We described several different representation languages: one specialized to temporal representations, and several that allow the specification of models over general object-relational domains. In the latter category, we first described two directed representations: plate models, and probabilistic relational models. The latter allow a considerably richer set of dependencies to be encoded, but at the cost of both conceptual and computational complexity. We also described an undirected representation, which, by avoiding the need to guarantee acyclicity and coherent local probability models, avoids some of the complexities of the directed models. As we discussed, the flexibility of undirected models is particularly valuable when we want to encode a probability distribution over richer representations, such as the structure of the relational graph.

There are, of course, other ways to produce these large, richly structured models. Most obviously, for any given application, we can define a procedural method that can take a skeleton, and produce a concrete model for that specific set of objects (and possibly relations). For example, we can easily build a program that takes a pedigree and produces a Bayesian network for genetic inheritance over that pedigree. The benefit of the template-based representations that we have described here is that they provide a uniform, modular, declarative language for models of this type. Unlike specialized representations, such a language allows the template-based model to be modified easily, whether by hand or as part of an automated learning algorithm. Indeed, learning is perhaps one of the key advantages of the template-based representations. In particular, as we will discuss, the model is learned at the template level, allowing a model to be learned from a domain with one set of objects, and applied seamlessly to a domain with a completely different set of objects (see section 17.5.1.2 and section 18.6.2).

In addition, by making objects and relations first-class citizens in the model, we have laid a foundation for the option of allowing probability distributions over probability spaces that are significantly richer than simply properties of objects. For example, as we saw, we can consider modeling uncertainty about the network of interrelationships between objects, and even about the actual set of objects included in our domain. These extensions raise many important and difficult questions regarding the appropriate type of distribution that one should use for such richly structured probability spaces. These questions become even more complex as we introduce more of the expressive power of relational languages, such as function symbols, quantifiers, and more. These issues are an active area of research.

These representations also raise important questions regarding inference. At first glance, the problem appears straightforward: The semantics for each of our representation languages depends on instantiating the template-based model to produce a specific ground network; clearly, we can simply run standard inference algorithms on the resulting network. This approach is knowledge-based model construction has been called *knowledge-based model construction*, because a knowledge-base (or skeleton) is used to construct a model. However, this approach is problematic, because the models produced by this process can pose a significant challenge to inference algorithms. First, the network produced by this process is often quite large — much larger than models that one can reasonably construct by hand. Second, such models are often quite densely connected, due to the multiple interactions between variables. Finally, structural uncertainty, both about the relations and about the presence of objects, also makes for densely connected models. On the other side, such models often have unique characteristics, such as multiple similar fragments across the network, or large amounts of context-specific independence, which could, perhaps, be exploited by an appropriate choice of inference algorithm. Chapter 15 presents some techniques for addressing the inference problems in temporal models. The question of inference in the models defined by the object-relational frameworks — and specifically of inference algorithms that exploit their special structure — is very much a topic of current work.

6.8   Relevant Literature

Probabilistic models of temporal processes go back many years. Hidden Markov models were discussed as early as Rabiner and Juang (1986), and expanded on in Rabiner (1989). Kalman filters were first described by Kalman (1960). The first temporal extension of probabilistic graphical models is due to Dean and Kanazawa (1989), who also coined the term *dynamic Bayesian network*. Much work has been done on defining various representations that are based on hidden Markov models or on dynamic Bayesian networks; these include generalizations of the basic framework, or special cases that allow more tractable inference. Examples include mixed-memory Markov models (Saul and Jordan 1999); variable-duration HMMs (Rabiner 1989) and their extension segment models (Ostendorf et al. 1996); factorial HMMs (Ghahramani and Jordan 1997); and hierarchical HMMs (Fine et al. 1998; Bui et al. 2001). Smyth, Heckerman, and Jordan (1997) is a review paper that was influential in providing a clear exposition of the connections between HMMs and DBNs. Murphy and Paskin (2001) show how hierarchical HMMs can be reduced to DBNs, a connection that provided a much faster inference algorithm than previously proposed for this representation. Murphy (2002) provides an excellent tutorial on the topics of dynamic Bayesian networks and related representations.

continuous time
Bayesian network

Nodelman et al. (2002, 2003) build on continuous-time Markov processes to define *continuous time Bayesian networks*. As the name suggests, this representation is similar to a dynamic Bayesian network but encodes a probability distribution over trajectories over a continuum of time points.

The topic of integrating object-relational frameworks and probabilistic representations has received much attention over the past few years. Getoor and Taskar (2007) contains reviews of many of the important contributions, and citations to others. Work on this topic goes back to the

knowledge-based
model
construction

idea of *knowledge-based model construction*, which was proposed in the early 1990s; Wellman, Breese, and Goldman (1992) review some of this earlier work. These ideas were then extended and formalized, using logic programming as a foundation (Poole 1993a; Ngo and Haddawy 1996; Kersting and De Raedt 2007).

Plate models were introduced by Gilks, Thomas, and Spiegelhalter (1994) and Buntine (1994) as a language for sharing parameters within and between models. Probabilistic relational models were proposed in Friedman et al. (1999); see Getoor et al. (2007) for a more detailed presentation. Heckerman, Meek, and Koller (2007) define a language that unifies plate models and probabilistic relational models, which was the inspiration for our presentation of PRMs in terms of contingent dependencies.

Undirected probabilistic models for relational domains originated with the framework of relational Markov networks of Taskar et al. (2002, 2007). Richardson and Domingos (2006) provide a particularly elegant representation of features, in terms of logical formulas. In a Markov logic network (MLN), there is no separation between the specification of cliques and the specification of features in the potential. Rather, the model is defined in terms of a collection of logical formulas, each associated with a weight.

Getoor et al. (2002) discuss some strategies for modeling structural uncertainty in a directed setting. Taskar et al. (2002) investigate the same issues in an undirected setting, and demonstrate the advantages of the increased flexibility. Reasoning about object identity has been used in various applications, including data association (Pasula et al. 1999), coreference resolution in natural language text (McCallum and Wellner 2005; Culotta et al. 2007), and the citation matching application discussed in box 6.D (Pasula et al. 2002; Wellner et al. 2004; Milch et al. 2004; Poon and Domingos 2007). Milch et al. (2005, 2007) define BLOG (Bayesian Logic), a directed language explicitly designed to model uncertainty over the number of objects in the domain.

In addition to the logic-based representations we discuss in this chapter, a very different perspective on incorporating template-based structure in probabilistic models utilizes a programming-language framework. Here, we can view a random variable as a stochastic function from its inputs (its parents) to its output. If we explicitly define the stochastic function, one can then reuse it in in multiple places. More importantly, one can define functions that call other functions, or perhaps even functions that recursively call themselves. Important languages

probabilistic context-free grammar

based on this framework include *probabilistic context-free grammars*, which play a key role in statistical models for natural language (see, for example, Manning and Schuetze (1999)) and in modeling RNA secondary structure (see, for example, Durbin et al. 1998), and object-oriented Bayesian networks (Koller and Pfeffer 1997; Pfeffer et al. 1999), which generalizes encapsulated Bayesian networks to allow for repeated elements.

6.9 Exercises

Exercise 6.1

Consider a temporal process where the state variables at time t depend directly not only on the variables at time $t - 1$, but rather on the variables at time $t - 1, \ldots, t - k$ for some fixed k. Such processes are

semi-Markov order k

called *semi-Markov* of order k.

a. Extend definition 6.3 and definition 6.4 to richer notions, that encode such a kth order semi-Markov processes.

b. Show how you can convert a kth order Markov process to a regular (first-order) Markov process representable by a DBN over an extended set of state variables. Describe both the variables and the transition model.

Exercise 6.2⋆

Markov models of different orders are the standard representation of text sequences. For example, in a first-order Markov model, we define our distribution over word sequences in terms of a probability $P(W^{(t)} \mid W^{(t-1)})$. This model is also called a *bigram model*, because it requires that we collected statistics over pairs of words. A second-order Markov model, often called a *trigram model*, defines the distribution is terms of a probability $P(W^{(t)} \mid W^{(t-1)}, W^{(t-2)})$.

Unfortunately, because the set of words in our vocabulary is very large, trigram models define very large CPDs with very many parameters. These are very hard to estimate reliably from data (see section 17.2.3).

shrinkage

One approach for producing more robust estimates while still making use of higher-order dependencies is *shrinkage*. Here, we define our transition model to be a weighted average of transition models of different orders:

$$P(W^{(t)} \mid W^{(t-1)}, W^{(t-2)}) = \alpha_0(W^{(t-1)}, W^{(t-2)})Q_0(W^{(t)}) +$$
$$\alpha_1(W^{(t-1)}, W^{(t-2)})Q_1(W^{(t)} \mid W^{(t-1)}) + \alpha_2(W^{(t-1)}, W^{(t-2)})Q_2(W^{(t)} \mid W^{(t-1)}, W^{(t-2)}),$$

where the Q_i's are different transition models, and the α_i's are nonnegative coefficients such that, for every $W^{(t-1)}, W^{(t-2)}$,

$$\alpha_0(W^{(t-1)}, W^{(t-2)}) + \alpha_1(W^{(t-1)}, W^{(t-2)}) + \alpha_2(W^{(t-1)}, W^{(t-2)}) = 1.$$

Show how we can construct a DBN model that gives rise to equivalent dynamics using standard CPDs, by introducing a new hidden variable $S^{(t)}$. This model is called *mixed-memory HMM*.

mixed-memory
HMM

Exercise 6.3

In this exercise, we construct an HMM model that allows for a richer class of distributions over the duration for which the process stays in a given state.

a. Consider an HMM where the hidden variable has k states, and let $P(s'_j \mid s_i)$ denote the transition model. Assuming that the process is at state s_i at time t, what is the distribution over the number of steps until it first transitions out of state s_i (that is, the smallest number d such that $S^{(t+d)} \neq s_i$)?

b. Construct a DBN model that allows us to incorporate an arbitrary distribution over the duration d_i that a process stays in state s_i after it first transitions to s_i. Your model should allow the distribution over d_i to depend on s_i. Do not worry about parameterizing the distribution over d_i. (Hint: Your model can include variables whose value changes deterministically.) This type of model is called a *duration HMM*.

duration HMM

Exercise 6.4⋆

segment HMM

A *segment HMM* is a Markov chain over the hidden states, but where each state emits not a single symbol as output, but rather a string of unknown length. Thus, at each state $S^{(t)} = s$, the model selects a segment length $L^{(t)}$, using a distribution that can depend on s. The model then emits a segment $Y^{(t,1)}, \ldots, Y^{(t,L^{(t)})}$ of length $L^{(t)}$. In this exercise, we assume that the distribution on the output segment is modeled by a separate HMM $\mathcal{H}_s$. Write down a 2-TBN model that encodes this model. (Hint: Use your answer to exercise 6.3.)

Exercise 6.5⋆

hierarchical HMM

A *hierarchical HMM* is similar to the segment HMM, except that there is no explicit selection of the segment length. Rather, the HMM at a state calls a "subroutine" HMM $\mathcal{H}_s$ that defines the output at the state s; when the "subroutine" HMM enters a finish-state, the control returns to the top-level HMM, which then transitions to its next state. This hierarchical HMM (with three levels) is precisely the framework used as the standard speech recognition architecture.

a. Show how a three-level hierarchical HMM can be represented as a DBN. (Hint: Use "finish variables" — binary variables that are true when a lower-level HMMs finishes its transition.)

b. Explain how you would modify the hierarchical HMM framework to deal with a motion tracking task, where, for example, the higher-level HMM represents motion between floors, the mid-level HMM motion between corridors, and the lowest-level HMM motion between rooms. (Hint: Consider situations where there are multiple staircases between floors.)

Exercise 6.6⋆

data association

Consider the following *data association* problem. We track K moving objects $u_1, \ldots, u_K$, using readings obtained over a trajectory of length T. Each object k has some (unknown) basic appearance A_k, and some position $X_k^{(t)}$ at every time point t. Our sensor provides, at each time point t, a set of L noisy sensor readings, each corresponding to one object: for each $l = 1, \ldots, L$, it returns $B_l^{(t)}$ — the measured object appearance, and $Y_l^{(t)}$ — the measured object position. Unfortunately, our sensor cannot determine the identity of the sensed objects, so sensed object l does not generally correspond to the true object l. In fact, the labeling of the sensed objects is completely arbitrary — all labelings are equally likely.

Write down a DBN that represents the dynamics of this model.

Exercise 6.7

Consider a template-level CPD where $A(U)$ depends on $B(U, V)$, allowing for situations where the ground variable $A(u)$ can depend on unbounded number of ground variables $B(u, v)$. As discussed in the text, we can specify the parameterization for the resulting CPD in various ways: we can use a symmetric noisy-or or sigmoid model, or define a dependency of $A(u)$ on some aggregated statistics of the parent set $\{B(u, v)\}$. Assume that both $A(U)$ and $B(U, V)$ are binary-valued.

aggregator CPD

Show that both a symmetric noisy-or model and a symmetric logistic model can be formulated easily using an *aggregator CPDs*.

Exercise 6.8

Consider the template dependency graph for a model $\mathcal{M}_{PRM}$, as specified in definition 6.13. Show that if the template dependency graph is acyclic, then for any skeleton κ, the ground network $\mathcal{B}_\kappa^{\mathcal{M}_{PRM}}$ is also acyclic.

Exercise 6.9

Let $\mathcal{M}_{Plate}$ be a plate model, and assume that its template dependency graph contains a cycle. Let κ be any skeleton such that $\mathcal{O}^\kappa[\mathbf{Q}] \neq \emptyset$ for every class $\mathbf{Q}$. Show that $\mathcal{B}_\kappa^{\mathcal{M}_{Plate}}$ is necessarily cyclic.

Exercise 6.10⋆⋆

Consider the cyclic dependency graph for the Genetics model shown in figure 6.9b. Clearly, for any valid pedigree — one where a person cannot be his or her own ancestor — the ground network is acyclic. We now describe a refinement of the dependency graph structure that would allow us to detect such acyclicity in this and other similar settings. Here, we assume for simplicity that all attributes in the guards are part of the relational skeleton, and therefore not part of the probabilistic model.

Let γ denote a tuple of objects from our skeleton. Assume that we have some prior knowledge about our domain in the following form: for any skeleton κ, there necessarily exists a partial ordering $\prec$ on tuples of objects γ that is transitive ($\gamma_1 \prec \gamma_2$ and $\gamma_2 \prec \gamma_3$ implies $\gamma_1 \prec \gamma_3$) and irreflexive ($\gamma \not\prec \gamma$). For example, in the Genetics example, we can use ancestry to define our ordering, where $u' \prec u$ whenever u' is an ancestor of u. We further assume that some of the guards used in the probabilistic model imply ordering constraints.

More precisely, let $B(U') \in \mathrm{Pa}_{U(A)}$. We say that a pair of assignments γ to U and γ' to U' is *valid* if they agree on the assignment to the overlapping variables in $U \cap U'$ and if they are consistent with the guard for A. The valid pairs are those that lead to actual edges $B(\gamma') \to A(\gamma)$ in the ground Bayesian network. (The definition here is slightly different than definition 6.12 because there γ' is an assignment to the variables in U' but not in U.) We say that the dependence of A on B is *ordering-consistent* if, for any valid pair of assignments γ to U and γ' to U', we have that $\gamma' \prec \gamma$. Continuing our example, consider the dependence of $Genotype(U)$ on $Genotype(V)$ subject to the guard $Mother(V, U)$. Here, for any pair of assignments u to U and v to V such that the guard $Mother(v, u)$ holds, we have that $v \prec u$. Thus, this dependence is ordering-consistent.

We now define the following extension to our dependency graph. Let $U'(B) \in \mathrm{Pa}_{U(A)}$.

- If $U' = U$, we introduce an edge from B to A whose color is yellow.
- If the dependence is ordering-consistent, we introduce an edge from B to A whose color is green.
- Otherwise, we introduce an edge from B to A whose color is red.

Prove that if every cycle in the colored dependency graph for $\mathcal{M}_{PRM}$ has at least one green edge and no red edges, then for any skeleton satisfying the ordering constraints, the ground BN $\mathcal{B}_\kappa^{\mathcal{M}_{PRM}}$ is acyclic.

7 Gaussian Network Models

Although much of our presentation focuses on discrete variables, we mentioned in chapter 5 that the Bayesian network framework, and the associated results relating independencies to factorization of the distribution, also apply to continuous variables. The same statement holds for Markov networks. However, whereas table CPDs provide a general-purpose mechanism for describing any discrete distribution (albeit potentially not very compactly), the space of possible parameterizations in the case of continuous variables is essentially unbounded. In this chapter, we focus on a type of continuous distribution that is of particular interest: the class of multivariate Gaussian distributions. Gaussians are a particularly simple subclass of distributions that make very strong assumptions, such as the exponential decay of the distribution away from its mean, and the linearity of interactions between variables. While these assumptions are often invalid, Gaussians are nevertheless a surprisingly good approximation for many real-world distributions. Moreover, the Gaussian distribution has been generalized in many ways, to nonlinear interactions, or mixtures of Gaussians; many of the tools developed for Gaussians can be extended to that setting, so that the study of Gaussian provides a good foundation for dealing with a broad class of distributions.

In the remainder of this chapter, we first review the class of multivariate Gaussian distributions and some of its properties. We then discuss how a multivariate Gaussian can be encoded using probabilistic graphical models, both directed and undirected.

7.1 Multivariate Gaussians

7.1.1 Basic Parameterization

We have already described the univariate Gaussian distribution in chapter 2. We now describe its generalization to the multivariate case. As we discuss, there are two different parameterizations for a joint Gaussian density, with quite different properties.

The univariate Gaussian is defined in terms of two parameters: a mean and a variance. In its most common representation, a multivariate Gaussian distribution over $X_1, \ldots, X_n$ is characterized by an n-dimensional *mean vector* $\boldsymbol{\mu}$, and a symmetric $n \times n$ *covariance matrix* Σ; the density function is most often defined as:

$$p(\boldsymbol{x}) = \frac{1}{(2\pi)^{n/2} |\Sigma|^{1/2}} \exp\left[-\frac{1}{2}(\boldsymbol{x} - \boldsymbol{\mu})^T \Sigma^{-1} (\boldsymbol{x} - \boldsymbol{\mu}) \right] \tag{7.1}$$

mean vector
covariance matrix

standard
Gaussian

where $|\Sigma|$ is the determinant of Σ.

We extend the notion of a *standard Gaussian* to the multidimensional case, defining it to be a Gaussian whose mean is the all-zero vector $\mathbf{0}$ and whose covariance matrix is the identity matrix I, which has 1's on the diagonal and zeros elsewhere. The multidimensional standard Gaussian is simply a product of independent standard Gaussians for each of the dimensions.

positive definite

In order for this equation to induce a well-defined density (that integrates to 1), the matrix Σ must be *positive definite*: for any $x \in I\!R^n$ such that $x \neq 0$, we have that $x^T \Sigma x > 0$. Positive definite matrices are guaranteed to be nonsingular, and hence have nonzero determinant, a necessary requirement for the coherence of this definition. A somewhat more complex definition can be used to generalize the multivariate Gaussian to the case of a *positive semi-definite* covariance matrix: for any $x \in I\!R^n$, we have that $x^T \Sigma x \geq 0$. This extension is useful, since it allows for singular covariance matrices, which arise in several applications. For the remainder of our discussion, we focus our attention on Gaussians with positive definite covariance matrices.

positive
semi-definite

Because positive definite matrices are invertible, one can also utilize an alternative parameterization, where the Gaussian is defined in terms of its inverse covariance matrix $J = \Sigma^{-1}$, called *information matrix* (or *precision matrix*). This representation induces an alternative form for the Gaussian density. Consider the expression in the exponent of equation (7.1):

information
matrix

$$-\frac{1}{2}(x - \mu)^T \Sigma^{-1}(x - \mu) \;=\; -\frac{1}{2}(x - \mu)^T J(x - \mu)$$
$$=\; -\frac{1}{2}\left[x^T J x - 2x^T J \mu + \mu^T J \mu\right].$$

The last term is constant, so we obtain:

$$p(x) \propto \exp\left[-\frac{1}{2}x^T J x + (J\mu)^T x\right]. \tag{7.2}$$

information form

This formulation of the Gaussian density is generally called the *information form*, and the vector $h = J\mu$ is called the *potential vector*. The information form defines a valid Gaussian density if and only if the information matrix is symmetric and positive definite, since Σ is positive definite if and only if Σ^{-1} is positive definite. The information form is useful in several settings, some of which are described here.

Intuitively, a multivariate Gaussian distribution specifies a set of ellipsoidal contours around the mean vector μ. The contours are parallel, and each corresponds to some particular value of the density function. The shape of the ellipsoid, as well as the "steepness" of the contours, are determined by the covariance matrix Σ. Figure 7.1 shows two multivariate Gaussians, one where the covariances are zero, and one where they are positive. As in the univariate case, the mean vector and covariance matrix correspond to the first two moments of the normal distribution. In matrix notation, $\mu = E[X]$ and $\Sigma = E[X X^T] - E[X]E[X]^T$. Breaking this expression down to the level of individual variables, we have that μ_i is the mean of X_i, $\Sigma_{i,i}$ is the variance of X_i, and $\Sigma_{i,j} = \Sigma_{j,i}$ (for $i \neq j$) is the *covariance* between X_i and X_j: $Cov[X_i; X_j] = E[X_i X_j] - E[X_i]E[X_j]$.

Example 7.1   *Consider a particular joint distribution $p(X_1, X_2, X_3)$ over three random variables. We can*

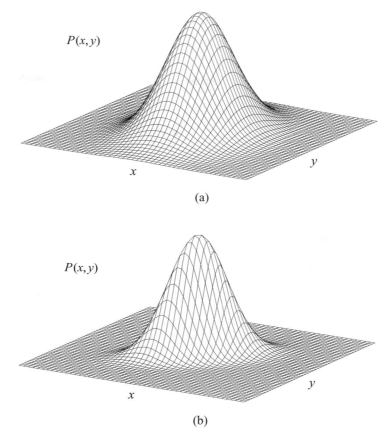

$P(x,y)$

x y

(a)

$P(x,y)$

x y

(b)

Figure 7.1 **Gaussians over two variables** X **and** Y**.** (a) X and Y uncorrelated. (b) X and Y correlated.

parameterize it via a mean vector $\boldsymbol{\mu}$ *and a covariance matrix* Σ:

$$\boldsymbol{\mu} = \begin{pmatrix} 1 \\ -3 \\ 4 \end{pmatrix} \qquad \Sigma = \begin{pmatrix} 4 & 2 & -2 \\ 2 & 5 & -5 \\ -2 & -5 & 8 \end{pmatrix}$$

As we can see, the covariances $\mathbf{Cov}[X_1; X_3]$ *and* $\mathbf{Cov}[X_2; X_3]$ *are both negative. Thus,* X_3 *is negatively correlated with* X_1: *when* X_1 *goes up,* X_3 *goes down (and similarly for* X_3 *and* X_2). ∎

7.1.2 Operations on Gaussians

There are two main operations that we wish to perform on a distribution: compute the marginal distribution over some subset of the variables $\boldsymbol{Y}$, and conditioning the distribution on some assignment of values $\boldsymbol{Z} = \boldsymbol{z}$. It turns out that each of these operations is very easy to perform in one of the two ways of encoding a Gaussian, and not so easy in the other.

Marginalization is trivial to perform in the covariance form. Specifically, the marginal Gaussian distribution over any subset of the variables can simply be read from the mean and covariance matrix. For instance, in example 7.1, we can obtain the marginal Gaussian distribution over X_2 and X_3 by simply considering only the relevant entries in both the mean vector the covariance matrix. More generally, assume that we have a joint normal distribution over $\{X, Y\}$ where $X \in \mathbb{R}^n$ and $Y \in \mathbb{R}^m$. Then we can decompose the mean and covariance of this joint distribution as follows:

$$p(X, Y) = \mathcal{N}\left(\left(\begin{array}{c} \mu_X \\ \mu_Y \end{array}\right); \left[\begin{array}{cc} \Sigma_{XX} & \Sigma_{XY} \\ \Sigma_{YX} & \Sigma_{YY} \end{array}\right]\right) \tag{7.3}$$

where $\mu_X \in \mathbb{R}^n$, $\mu_Y \in \mathbb{R}^m$, Σ_{XX} is a matrix of size $n \times n$, Σ_{XY} is a matrix of size $n \times m$, $\Sigma_{YX} = \Sigma_{XT}^T$ is a matrix of size $m \times n$ and Σ_{YY} is a matrix of size $m \times m$.

Lemma 7.1

> Let $\{X, Y\}$ have a joint normal distribution defined in equation (7.3). Then the marginal distribution over Y is a normal distribution $\mathcal{N}(\mu_Y; \Sigma_{YY})$.

The proof follows directly from the definitions (see exercise 7.1).

On the other hand, conditioning a Gaussian on an observation $Z = z$ is very easy to perform in the information form. We simply assign the values $Z = z$ in equation (7.2). This process turns some of the quadratic terms into linear terms or even constant terms, and some of the linear terms into constant terms. The resulting expression, however, is still in the same form as in equation (7.2), albeit over a smaller subset of variables.

In summary, although the two representations both encode the same information, they have different computational properties. **To marginalize a Gaussian over a subset of the variables, one essentially needs to compute their pairwise covariances, which is precisely generating the distribution in its covariance form. Similarly, to condition a Gaussian on an observation, one essentially needs to invert the covariance matrix to obtain the information form. For small matrices, inverting a matrix may be feasible, but in high-dimensional spaces, matrix inversion may be far too costly.**

7.1.3   Independencies in Gaussians

For multivariate Gaussians, independence is easy to determine directly from the parameters of the distribution.

Theorem 7.1

> Let $X = X_1, ..., X_n$ have a joint normal distribution $\mathcal{N}(\mu; \Sigma)$. Then X_i and X_j are independent if and only if $\Sigma_{i,j} = 0$.

The proof is left as an exercise (exercise 7.2).

Note that this property does not hold in general. In other words, if $p(X, Y)$ is not Gaussian, then it is possible that $Cov[X; Y] = 0$ while X and Y are still dependent in p. (See exercise 7.2.)

At first glance, it seems that conditional independencies are not quite as apparent as marginal independencies. However, it turns out that the independence structure in the distribution is apparent not in the covariance matrix, but in the information matrix.

Theorem 7.2

Consider a Gaussian distribution $p(X_1, \ldots, X_n) = \mathcal{N}(\boldsymbol{\mu}; \Sigma)$, and let $J = \Sigma^{-1}$ be the information matrix. Then $J_{i,j} = 0$ if and only if $p \models (X_i \perp X_j \mid \mathcal{X} - \{X_i, X_j\})$.

The proof is left as an exercise (exercise 7.3).

Example 7.2

Consider the covariance matrix of example 7.1. Simple algebraic operations allow us to compute its inverse:

$$J = \begin{pmatrix} 0.3125 & -0.125 & 0 \\ -0.125 & 0.5833 & 0.3333 \\ 0 & 0.3333 & 0.3333 \end{pmatrix}$$

As we can see, the entry in the matrix corresponding to X_1, X_3 is zero, reflecting the fact that they are conditionally independent given X_2. ∎

Theorem 7.2 asserts the fact that the information matrix captures independencies between pairs of variables, conditioned on all of the remaining variables in the model. These are precisely the same independencies as the pairwise Markov independencies of definition 4.10. Thus, we can view the information matrix J for a Gaussian density p as precisely capturing the pairwise Markov independencies in a Markov network representing p. Because a Gaussian density is a positive distribution, we can now use theorem 4.5 to construct a Markov network that is a unique minimal I-map for p: As stated in this theorem, the construction simply introduces an edge between X_i and X_j whenever $(X_i \perp X_j \mid \mathcal{X} - \{X_i, X_j\})$ does not hold in p. But this latter condition holds precisely when $J_{i,j} \neq 0$. **Thus, we can view the information matrix as directly defining a minimal I-map Markov network for p, whereby nonzero entries correspond to edges in the network.**

7.2 Gaussian Bayesian Networks

We now show how we can define a continuous joint distribution using a Bayesian network. This representation is based on the *linear Gaussian model*, which we defined in definition 5.14. Although this model can be used as a CPD within any network, it turns out that continuous networks defined solely in terms of linear Gaussian CPDs are of particular interest:

Definition 7.1

Gaussian
Bayesian network

We define a Gaussian Bayesian network *to be a Bayesian network all of whose variables are continuous, and where all of the CPDs are linear Gaussians.* ∎

An important and surprising result is that linear Gaussian Bayesian networks are an alternative representation for the class of multivariate Gaussian distributions. This result has two parts. The first is that a linear Gaussian network always defines a joint multivariate Gaussian distribution.

Theorem 7.3

Let Y be a linear Gaussian of its parents $X_1, \ldots, X_k$:

$$p(Y \mid \boldsymbol{x}) = \mathcal{N}\left(\beta_0 + \boldsymbol{\beta}^T \boldsymbol{x}; \sigma^2\right).$$

Assume that $X_1, \ldots, X_k$ are jointly Gaussian with distribution $\mathcal{N}(\boldsymbol{\mu}; \Sigma)$. Then:

- *The distribution of Y is a normal distribution $p(Y) = \mathcal{N}\left(\mu_Y; \sigma_Y^2\right)$ where:*

$$\begin{aligned}\mu_Y &= \beta_0 + \boldsymbol{\beta}^T \boldsymbol{\mu} \\ \sigma_Y^2 &= \sigma^2 + \boldsymbol{\beta}^T \Sigma \boldsymbol{\beta}.\end{aligned}$$

- *The joint distribution over $\{\boldsymbol{X}, Y\}$ is a normal distribution where:*

$$\boldsymbol{Cov}[X_i; Y] = \sum_{j=1}^{k} \beta_j \Sigma_{i,j}.$$

From this theorem, it follows easily by induction that if $\mathcal{B}$ is a linear Gaussian Bayesian network, then it defines a joint distribution that is jointly Gaussian.

Example 7.3   *Consider the linear Gaussian network $X_1 \to X_2 \to X_3$, where*

$$\begin{aligned}p(X_1) &= \mathcal{N}\left(1; 4\right) \\ p(X_2 \mid X_1) &= \mathcal{N}\left(0.5X_1 - 3.5; 4\right) \\ p(X_3 \mid X_2) &= \mathcal{N}\left(-X_2 + 1; 3\right).\end{aligned}$$

Using the equations in theorem 7.3, we can compute the joint Gaussian distribution $p(X_1, X_2, X_3)$. For the mean, we have that:

$$\begin{aligned}\mu_2 &= 0.5\mu_1 - 3.5 = 0.5 \cdot 1 - 3.5 = -3 \\ \mu_3 &= (-1)\mu_2 + 1 = (-1) \cdot (-3) + 1 = 4.\end{aligned}$$

The variance of X_2 and X_3 can be computed as:

$$\begin{aligned}\Sigma_{22} &= 4 + (1/2)^2 \cdot 4 = 5 \\ \Sigma_{33} &= 3 + (-1)^2 \cdot 5 = 8.\end{aligned}$$

We see that the variance of the variable is a sum of two terms: the variance arising from its own Gaussian noise parameter, and the variance of its parent variables weighted by the strength of the dependence. Finally, we can compute the covariances as follows:

$$\begin{aligned}\Sigma_{12} &= (1/2) \cdot 4 = 2 \\ \Sigma_{23} &= (-1) \cdot \Sigma_{22} = -5 \\ \Sigma_{13} &= (-1) \cdot \Sigma_{12} = -2.\end{aligned}$$

The third equation shows that, although X_3 does not depend directly on X_1, they have a nonzero covariance. Intuitively, this is clear: X_3 depends on X_2, which depends on X_1; hence, we expect X_1 and X_3 to be correlated, a fact that is reflected in their covariance. As we can see, the covariance between X_1 and X_3 is the covariance between X_1 and X_2, weighted by the strength of the dependence of X_3 on X_2.

 In general, putting these results together, we can see that the mean and covariance matrix for $p(X_1, X_2, X_3)$ is precisely our covariance matrix of example 7.1.   ∎

The converse to this theorem also holds: the result of conditioning is a normal distribution where there is a linear dependency on the conditioning variables. The expressions for converting a multivariate Gaussian to a linear Gaussian network appear complex, but they are based on simple algebra. They can be derived by taking the linear equations specified in theorem 7.3, and reformulating them as defining the parameters β_i in terms of the means and covariance matrix entries.

Theorem 7.4

Let $\{\boldsymbol{X}, Y\}$ *have a joint normal distribution defined in equation (7.3). Then the conditional density*

$$p(Y \mid \boldsymbol{X}) = \mathcal{N}\left(\beta_0 + \boldsymbol{\beta}^T \boldsymbol{X}; \sigma^2\right),$$

is such that:

$$
\begin{aligned}
\beta_0 &= \mu_Y - \Sigma_{Y\boldsymbol{X}} \Sigma_{\boldsymbol{X}\boldsymbol{X}}^{-1} \mu_{\boldsymbol{X}} \\
\boldsymbol{\beta} &= \Sigma_{\boldsymbol{X}\boldsymbol{X}}^{-1} \Sigma_{Y\boldsymbol{X}} \\
\sigma^2 &= \Sigma_{YY} - \Sigma_{Y\boldsymbol{X}} \Sigma_{\boldsymbol{X}\boldsymbol{X}}^{-1} \Sigma_{\boldsymbol{X}Y}.
\end{aligned}
$$

This result allows us to take a joint Gaussian distribution and produce a Bayesian network, using an identical process to our construction of a minimal I-map in section 3.4.1.

Theorem 7.5

Let $\mathcal{X} = \{X_1, \ldots, X_n\}$, *and let p be a joint Gaussian distribution over $\mathcal{X}$. Given any ordering $X_1, \ldots, X_n$ over $\mathcal{X}$, we can construct a Bayesian network graph $\mathcal{G}$ and a Bayesian network $\mathcal{B}$ over $\mathcal{G}$ such that:*

1. $\mathrm{Pa}_{X_i}^{\mathcal{G}} \subseteq \{X_1, \ldots, X_{i-1}\}$;

2. the CPD of X_i in $\mathcal{B}$ is a linear Gaussian of its parents;

3. $\mathcal{G}$ is a minimal I-map for p.

The proof is left as an exercise (exercise 7.4). As for the case of discrete networks, the minimal I-map is not unique: different choices of orderings over the variables will lead to different network structures. For example, the distribution in figure 7.1b can be represented either as the network where $X \rightarrow Y$ or as the network where $Y \rightarrow X$.

This equivalence between Gaussian distributions and linear Gaussian networks has important practical ramifications. On one hand, we can conclude that, for linear Gaussian networks, the joint distribution has a compact representation (one that is quadratic in the number of variables). Furthermore, the transformations from the network to the joint and back have a fairly simple and efficiently computable closed form. Thus, we can easily convert one representation to another, using whichever is more convenient for the current task. Conversely, **while the two representations are equivalent in their expressive power, there is not a one-to-one correspondence between their parameterizations. In particular, although in the worst case, the linear Gaussian representation and the Gaussian representation have the same number of parameters (exercise 7.6), there are cases where one representation can be significantly more compact than the other.**

Example 7.4

Consider a linear Gaussian network structured as a chain:

$$X_1 \to X_2 \to \cdots \to X_n.$$

Assuming the network parameterization is not degenerate (that is, the network is a minimal I-map of its distribution), we have that each pair of variables X_i, X_j are correlated. In this case, as shown in theorem 7.1, the covariance matrix would be dense — none of the entries would be zero. Thus, the representation of the covariance matrix would require a quadratic number of parameters. In the information matrix, however, for all X_i, X_j that are not neighbors in the chain, we have that X_i and X_j are conditionally independent given the rest of the variables in the network; hence, by theorem 7.2, $J_{i,j} = 0$. Thus, the information matrix has most of the entries being zero; the only nonzero entries are on the tridiagonal (the entries i, j for $j = i - 1, i, i + 1$). ∎

However, not all structure in a linear Gaussian network is represented in the information matrix.

Example 7.5

In a v-structure $X \to Z \leftarrow Y$, we have that X and Y are marginally independent, but not conditionally independent given Z. Thus, according to theorem 7.2, the X, Y entry in the information matrix would not be 0. Conversely, because the variables are marginally independent, the X, Y entry in the covariance entry would be zero.

Complicating the example somewhat, assume that X and Y also have a joint parent W; that is, the network is structured as a diamond. In this case, X and Y are still not independent given the remaining network variables Z, W, and hence the X, Y entry in the information matrix is nonzero. Conversely, they are also not marginally independent, and thus the X, Y entry in the covariance matrix is also nonzero. ∎

These examples simply recapitulate, in the context of Gaussian networks, the fundamental difference in expressive power between Bayesian networks and Markov networks.

7.3 Gaussian Markov Random Fields

We now turn to the representation of multivariate Gaussian distributions via an undirected graphical model. We first show how a Gaussian distribution can be viewed as an MRF. This formulation is derived almost immediately from the information form of the Gaussian. Consider again equation (7.2). We can break up the expression in the exponent into two types of terms: those that involve single variables X_i and those that involve pairs of variables X_i, X_j. The terms that involve only the variable X_i are:

$$-\frac{1}{2} J_{i,i} x_i^2 + h_i x_i, \tag{7.4}$$

where we recall that the potential vector $\boldsymbol{h} = J\boldsymbol{\mu}$. The terms that involve the pair X_i, X_j are:

$$-\frac{1}{2}[J_{i,j} x_i x_j + J_{j,i} x_j x_i] = -J_{i,j} x_i x_j, \tag{7.5}$$

due to the symmetry of the information matrix. Thus, the information form immediately induces a pairwise Markov network, whose node potentials are derived from the potential vector and the diagonal elements of the information matrix, and whose edge potentials are derived from the off-diagonal entries of the information matrix. We also note that, when $J_{i,j} = 0$, there is no edge between X_i and X_j in the model, corresponding directly to the independence assumption of the Markov network.

Gaussian MRF

Thus, any Gaussian distribution can be represented as a pairwise Markov network with quadratic node and edge potentials. This Markov network is generally called a *Gaussian Markov random field (GMRF)*. Conversely, consider any pairwise Markov network with quadratic node and edge potentials. Ignoring constant factors, which can be assimilated into the partition function, we can write the node and edge energy functions (log-potentials) as:

$$\begin{aligned}\epsilon_i(x_i) &= d_0^i + d_1^i x_i + d_2^i x_i^2 \\ \epsilon_{i,j}(x_i, x_j) &= a_{00}^{i,j} + a_{01}^{i,j} x_i + a_{10}^{i,j} x_j + a_{11}^{i,j} x_i x_j + a_{02}^{i,j} x_i^2 + a_{20}^{i,j} x_j^2,\end{aligned} \tag{7.6}$$

where we used the log-linear notation of section 4.4.1.2. By aggregating like terms, we can reformulate any such set of potentials in the log-quadratic form:

$$p'(\boldsymbol{x}) = \exp(-\frac{1}{2}\boldsymbol{x}^T J \boldsymbol{x} + \boldsymbol{h}^T \boldsymbol{x}), \tag{7.7}$$

where we can assume without loss of generality that J is symmetric. This Markov network defines a valid Gaussian density if and only if J is a positive definite matrix. If so, then J is a legal information matrix, and we can take $\boldsymbol{h}$ to be a potential vector, resulting in a distribution in the form of equation (7.2).

However, unlike the case of Gaussian Bayesian networks, it is not the case that every set of quadratic node and edge potentials induces a legal Gaussian distribution. Indeed, the decomposition of equation (7.4) and equation (7.5) can be performed for any quadratic form, including one not corresponding to a positive definite matrix. For such matrices, the resulting function $\exp(\boldsymbol{x}^T A \boldsymbol{x} + \boldsymbol{b}^T \boldsymbol{x})$ will have an infinite integral, and cannot be normalized to produce a valid density. Unfortunately, **other than generating the entire information matrix and testing whether it is positive definite, there is no simple way to check whether the MRF is valid. In particular, there is no local test that can be applied to the network parameters that precisely characterizes valid Gaussian densities.** However, there *are* simple tests that are sufficient to induce a valid density. While these conditions are not necessary, they appear to cover many of the cases that occur in practice.

We first provide one very simple test that can be verified by direct examination of the information matrix.

Definition 7.2

diagonally dominant

A quadratic MRF parameterized by J is said to be diagonally dominant *if, for all* i,

$$\sum_{j \neq i} |J_{i,j}| < J_{i,i}.$$

∎

For example, the information matrix in example 7.2 is diagonally dominant; for instance, for $i = 2$ we have:

$$|-0.125| + 0.3333 < 0.5833.$$

One can now show the following result:

Proposition 7.1

Let $p'(\boldsymbol{x}) = \exp(-\frac{1}{2}\boldsymbol{x}^T J \boldsymbol{x} + \boldsymbol{h}^T \boldsymbol{x})$ be a quadratic pairwise MRF. If J is diagonally dominant, then p' defines a valid Gaussian MRF.

The proof is straightforward algebra and is left as an exercise (exercise 7.8).

The following condition is less easily verified, since it cannot be tested by simple examination of the information matrix. Rather, it checks whether the distribution can be written as a quadratic pairwise MRF whose node and edge potentials satisfy certain conditions. Specifically, recall that a Gaussian MRF consists of a set of node potentials, which are log-quadratic forms in x_i, and a set of edge potentials, which are log-quadratic forms in x_i, x_j. We can state a condition in terms of the coefficients for the nonlinear components of this parameterization:

Definition 7.3

pairwise
normalizable

A quadratic MRF parameterized as in equation (7.6) is said to be pairwise normalizable *if:*

- *for all i, $d_2^i > 0$;*
- *for all i, j, the 2×2 matrix*

$$
\begin{pmatrix}
a_{02}^{i,j} & a_{11}^{i,j}/2 \\
a_{11}^{i,j}/2 & a_{20}^{i,j}
\end{pmatrix}
$$

is positive semidefinite.   ∎

Intuitively, this definition states that each edge potential, considered in isolation, is normalizable (hence the name "pairwise-normalizable").

We can show the following result:

Proposition 7.2

Let $p'(\boldsymbol{x})$ be a quadratic pairwise MRF, parameterized as in equation (7.6). If p' is pairwise normalizable, then it defines a valid Gaussian distribution.

Once again, the proof follows from standard algebraic manipulations, and is left as an exercise (exercise 7.9).

We note that, like the preceding conditions, this condition is sufficient but not necessary:

Example 7.6

Consider the following information matrix:

$$
\begin{pmatrix}
1 & 0.6 & 0.6 \\
0.6 & 1 & 0.6 \\
0.6 & 0.6 & 1
\end{pmatrix}
$$

It is not difficult to show that this information matrix is positive definite, and hence defines a legal Gaussian distribution. However, it turns out that it is not possible to decompose this matrix into a set of three edge potentials, each of which is positive definite.   ∎

Unfortunately, evaluating whether pairwise normalizability holds for a given MRF is not always trivial, since it can be the case that one parameterization is not pairwise normalizable, yet a different parameterization that induces precisely the same density function is pairwise normalizable.

Example 7.7

Consider the information matrix of example 7.2, with a mean vector $\mathbf{0}$. We can define this distribution using an MRF by simply choosing the node potential for X_i to be $J_{i,i}x_i^2$ and the edge potential for X_i, X_j to be $2J_{i,j}x_ix_j$. Clearly, the X_1, X_2 edge does not define a normalizable density over X_1, X_2, and hence this MRF is not pairwise normalizable. However, as we discussed in the context of discrete MRFs, the MRF parameterization is nonunique, and the same density can be induced using a continuum of different parameterizations. In this case, one alternative parameterization of the same density is to define all node potentials as $\epsilon_i(x_i) = 0.05x_i^2$, and the edge potentials to be $\epsilon_{1,2}(x_1, x_2) = 0.2625x_1^2 + 0.0033x_2^2 - 0.25x_1x_2$, and $\epsilon_{2,3}(x_2, x_3) = 0.53x_2^2 + 0.2833x_3^2 + 0.6666x_2x_3$. Straightforward arithmetic shows that this set of potentials induces the information matrix of example 7.2. Moreover, we can show that this formulation is pairwise normalizable: The three node potentials are all positive, and the two edge potentials are both positive definite. (This latter fact can be shown either directly or as a consequence of the fact that each of the edge potentials is diagonally dominant, and hence also positive definite.) ∎

This example illustrates that the pairwise normalizability condition is easily checked for a specific MRF parameterization. However, if our aim is to encode a particular Gaussian density as an MRF, we may have to actively search for a decomposition that satisfies the relevant constraints. If the information matrix is small enough to manipulate directly, this process is not difficult, but if the information matrix is large, finding an appropriate parameterization may incur a nontrivial computational cost.

7.4 Summary

This chapter focused on the representation and independence properties of Gaussian networks.

We showed an equivalence of expressive power between three representational classes: multivariate Gaussians, linear Gaussian Bayesian networks, and Gaussian MRFs. In particular, any distribution that can be represented in one of those forms can also be represented in another. We provided closed-form formulas that allow us convert between the multivariate Gaussian representation and the linear Gaussian Bayesian network. The conversion for Markov networks is simpler in some sense, inasmuch as there is a direct mapping between the entries in the information (inverse covariance) matrix of the Gaussian and the quadratic forms that parameterize the edge potentials in the Markov network. However, unlike the case of Bayesian networks, here we must take care, since not every quadratic parameterization of a pairwise Markov network induces a legal Gaussian distribution: The quadratic form that arises when we combine all the pairwise potentials may not have a finite integral, and therefore may not be normalizable. In general, there is no local way of determining whether a pairwise MRF with quadratic potentials is normalizable; however, we provided some easily checkable sufficient conditions that are often sufficient in practice.

The equivalence between the different representations is analogous to the equivalence of Bayesian networks, Markov networks, and discrete distributions: any discrete distribution can be encoded both as a Bayesian network and as a Markov network, and vice versa. However, as in the discrete case, this equivalence does *not* imply equivalence of expressive power with respect to independence assumptions. In particular, **the expressive power of the directed** and undirected representations in terms of independence assumptions is exactly the same as in the discrete case: Directed models can encode the independencies associated with immoralities, whereas undirected models cannot; conversely, undirected models can encode a symmetric diamond, whereas directed models cannot. As we saw, the undirected models have a particularly elegant connection to the natural representation of the Gaussian distribution in terms of the information matrix; in particular, zeros in the information matrix for p correspond precisely to missing edges in the minimal I-map Markov network for p.

Finally, we note that the class of Gaussian distributions is highly restrictive, making strong assumptions that often do not hold in practice. Nevertheless, it is a very useful class, due to its compact representation and computational tractability (see section 14.2). Thus, in many cases, we may be willing to make the assumption that a distribution is Gaussian even when that is only a rough approximation. This approximation may happen a priori, in encoding a distribution as a Gaussian even when it is not. Or, in many cases, we perform the approximation as part of our inference process, representing intermediate results as a Gaussian, in order to keep the computation tractable. Indeed, as we will see, the Gaussian representation is ubiquitous in methods that perform inference in a broad range of continuous models.

7.5   Relevant Literature

The equivalence between the multivariate and linear Gaussian representations was first derived by Wermuth (1980), who also provided the one-to-one transformations between them. The introduction of linear Gaussian dependencies into a Bayesian network framework was first proposed by Shachter and Kenley (1989), in the context of influence diagrams.

Speed and Kiiveri (1986) were the first to make the connection between the structure of the information matrix and the independence assumptions in the distribution. Building on earlier results for discrete Markov networks, they also made the connection to the undirected graph as a representation. Lauritzen (1996, Chapter 5) and Malioutov et al. (2006) give a good overview of the properties of Gaussian MRFs.

7.6   Exercises

Exercise 7.1

Prove lemma 7.1. Note that you need to show both that the marginal distribution is a Gaussian, and that it is parameterized as $\mathcal{N}(\mu_{\boldsymbol{Y}}; \Sigma_{\boldsymbol{Y}\boldsymbol{Y}})$.

Exercise 7.2

a. Show that, for any joint density function $p(X, Y)$, if we have $(X \perp Y)$ in p, then $\boldsymbol{Cov}[X; Y] = 0$.

b. Show that, if $p(X, Y)$ is Gaussian, and $\boldsymbol{Cov}[X; Y] = 0$, then $(X \perp Y)$ holds in p.

c. Show a counterexample to 2 for non-Gaussian distributions. More precisely, show a construction of a joint density function $p(X, Y)$ such that $\boldsymbol{Cov}[X; Y] = 0$, while $(X \perp Y)$ does not hold in p.

Exercise 7.3

Prove theorem 7.2.

Exercise 7.4

Prove theorem 7.5.

Exercise 7.5

Consider a Kalman filter whose transition model is defined in terms of a pair of matrices A, Q, and whose observation model is defined in terms of a pair of matrices H, R, as specified in equation (6.3) and equation (6.4). Describe how we can extract a 2-TBN structure representing the conditional independencies in this process from these matrices. (Hint: Use theorem 7.2.)

Exercise 7.6

In this question, we compare the number of independent parameters in a multivariate Gaussian distribution and in a linear Gaussian Bayesian network.

a. Show that the number of independent parameters in Gaussian distribution over $X_1, \ldots, X_n$ is the same as the number of independent parameters in a fully connected linear Gaussian Bayesian network over $X_1, \ldots, X_n$.

b. In example 7.4, we showed that the number of parameters in a linear Gaussian network can be substantially smaller than in its multivariate Gaussian representation. Show that the converse phenomenon can also happen. In particular, show an example of a distribution where the multivariate Gaussian representation requires a linear number of nonzero entries in the covariance matrix, while a corresponding linear Gaussian network (one that is a minimal I-map) requires a quadratic number of nonzero parameters. (Hint: The minimal I-map does not have to be the optimal one.)

Exercise 7.7

conditional covariance

Let p be a joint Gaussian density over $\mathcal{X}$ with mean vector $\boldsymbol{\mu}$ and information matrix J. Let $X_i \in \mathcal{X}$, and $\boldsymbol{Z} \subset \mathcal{X} - \{X_i\}$. We define the *conditional covariance* of X_i, X_j given $\boldsymbol{Z}$ as:

$$\boldsymbol{Cov}_p[X_i; X_j \mid \boldsymbol{Z}] = \boldsymbol{E}_p[(X_i - \mu_i)(X_j - \mu_j) \mid \boldsymbol{Z}] = \boldsymbol{E}_{\boldsymbol{z} \sim p(\boldsymbol{Z})}\left[\boldsymbol{E}_{p(X_i, X_j \mid \boldsymbol{z})}[(x_i - \mu_i)(x_j - \mu_j)]\right].$$

partial correlation coefficient

The conditional variance of X_i is defined by setting $j = i$. We now define the *partial correlation coefficient*

$$\rho_{i,j} = \frac{\boldsymbol{Cov}_p[X_i; X_j \mid \mathcal{X} - \{X_i, X_j\}]}{\sqrt{\boldsymbol{Var}_p[X_i \mid \mathcal{X} - \{X_i, X_j\}]\boldsymbol{Var}_p[X_j \mid \mathcal{X} - \{X_i, X_j\}]}}.$$

Show that

$$\rho_{i,j} = -\frac{J_{i,j}}{\sqrt{J_{i,i}J_{j,j}}}.$$

Exercise 7.8

Prove proposition 7.1.

Exercise 7.9

Prove proposition 7.2.

8 *The Exponential Family*

8.1 Introduction

In the previous chapters, we discussed several different representations of complex distributions. These included both representations of global structures (for example, Bayesian networks and Markov networks) and representations of local structures (for example, representations of CPDs and of potentials). In this chapter, we revisit these representations and view them from a different perspective. This view allows us to consider several basic questions and derive generic answers for these questions for a wide variety of representations. As we will see in later chapters, these solutions play a role in both inference and learning for the different representations we consider.

We note, however, that this chapter is somewhat abstract and heavily mathematical. Although the ideas described in this chapter are of central importance to understanding the theoretical foundations of learning and inference, the algorithms themselves can be understood even without the material presented in this chapter. Thus, this chapter can be skipped by readers who are interested primarily in the algorithms themselves.

8.2 Exponential Families

parametric family

Our discussion so far has focused on the representation of a single distribution (using, say, a Bayesian or Markov network). We now consider *families of distributions*. Intuitively, a family is a set of distributions that all share the same parametric form and differ only in choice of particular parameters (for example, the entries in table-CPDs). In general, once we choose the global structure and local structure of the network, we define a family of all distributions that can be attained by different parameters for this specific choice of CPDs.

Example 8.1

Consider the empty graph structure $\mathcal{G}_\emptyset$ over the variables $\mathcal{X} = \{X_1, \ldots, X_n\}$. We can define the family $\mathcal{P}_\emptyset$ to be the set of distributions that are consistent with $\mathcal{G}_\emptyset$. If all the variables in $\mathcal{X}$ are binary, then we can specify a particular distribution in the family by using n parameters, $\boldsymbol{\theta} = \{P(x_i^1) : i = 1, \ldots, n\}$. ∎

We will be interested in families that can be written in a particular form.

Definition 8.1

exponential family

Let $\mathcal{X}$ be a set of variables. An exponential family *$\mathcal{P}$ over $\mathcal{X}$ is specified by four components:*

sufficient statistic
function

parameter space

legal parameter

natural parameter

- *A sufficient statistics function τ from assignments to $\mathcal{X}$ to $\mathcal{R}^K$.*
- *A parameter space that is a convex set $\Theta \subseteq \mathcal{R}^M$ of legal parameters.*
- *A natural parameter function t from $\mathcal{R}^M$ to R^K.*
- *An auxiliary measure A over $\mathcal{X}$.*

Each vector of parameters $\boldsymbol{\theta} \in \Theta$ specifies a distribution $P_{\boldsymbol{\theta}}$ in the family as

$$P_{\boldsymbol{\theta}}(\xi) = \frac{1}{Z(\boldsymbol{\theta})} A(\xi) \exp\left\{ \langle t(\boldsymbol{\theta}), \tau(\xi) \rangle \right\} \tag{8.1}$$

where $\langle t(\boldsymbol{\theta}), \tau(\xi) \rangle$ is the inner product of the vectors $t(\boldsymbol{\theta})$ and $\tau(\xi)$, and

$$Z(\boldsymbol{\theta}) = \sum_{\xi} A(\xi) \exp\left\{ \langle t(\boldsymbol{\theta}), \tau(\xi) \rangle \right\}$$

partition function

is the partition function of $\mathcal{P}$, which must be finite. The parametric family $\mathcal{P}$ is defined as:

$$\mathcal{P} = \{ P_{\boldsymbol{\theta}} : \boldsymbol{\theta} \in \Theta \}. \qquad\blacksquare$$

We see that an exponential family is a concise representation of a class of probability distributions that share a similar functional form. A member of the family is determined by the parameter vector $\boldsymbol{\theta}$ in the set of legal parameters. The sufficient statistic function τ summarizes the aspects of an instance that are relevant for assigning it a probability. The function t maps the parameters to space of the sufficient statistics.

The measure A assigns additional preferences among instances that do not depend on the parameters. However, in most of the examples we consider here A is a constant, and we will mention it explicitly only when it is not a constant.

Although this definition seems quite abstract, many distributions we already have encountered are exponential families.

Example 8.2

Consider a simple Bernoulli distribution. In this case, the distribution over a binary outcome (such as a coin toss) is controlled by a single parameter θ that represents the probability of x^1. To show that this distribution is in the exponential family, we can set

$$\tau(X) = \langle \boldsymbol{I}\{X = x^1\}, \boldsymbol{I}\{X = x^0\} \rangle, \tag{8.2}$$

a numerical vector representation of the value of X, and

$$t(\theta) = \langle \ln \theta, \ln(1 - \theta) \rangle. \tag{8.3}$$

It is easy to see that for $X = x^1$, we have $\tau(X) = \langle 1, 0 \rangle$, and thus

$$\exp\left\{ \langle t(\theta), \tau(X) \rangle \right\} = e^{1 \cdot \ln \theta + 0 \cdot \ln(1-\theta)} = \theta.$$

Similarly, for $X = x^0$, we get that $\exp\left\{ \langle t(\theta), \tau(X) \rangle \right\} = 1 - \theta$. We conclude that, by setting $Z(\theta) = 1$, this representation is identical to the Bernoulli distribution. $\blacksquare$

Example 8.3

Consider a Gaussian distribution over a single variable. Recall that

$$P(x) = \frac{1}{\sqrt{2\pi}\sigma} \exp\left\{-\frac{(x-\mu)^2}{2\sigma^2}\right\}.$$

Define

$$\tau(x) = \langle x, x^2 \rangle \tag{8.4}$$

$$\mathsf{t}(\mu, \sigma^2) = \langle \frac{\mu}{\sigma^2}, -\frac{1}{2\sigma^2} \rangle \tag{8.5}$$

$$Z(\mu, \sigma^2) = \sqrt{2\pi}\sigma \exp\left\{\frac{\mu^2}{2\sigma^2}\right\}. \tag{8.6}$$

We can easily verify that

$$P(x) = \frac{1}{Z(\mu, \sigma^2)} \exp\left\{\langle \mathsf{t}(\theta), \tau(X) \rangle\right\}.$$

■

In fact, most of the parameterized distributions we encounter in probability textbooks can be represented as exponential families. This includes the Poisson distributions, exponential distributions, geometric distributions, Gamma distributions, and many others (see, for example, exercise 8.1).

We can often construct multiple exponential families that encode precisely the same class of distributions. There are, however, desiderata that we want from our representation of a class of distributions as an exponential family. First, we want the parameter space Θ to be "well-behaved," in particular, to be a convex, open subset of $\mathcal{R}^M$. Second, we want the parametric family to be *nonredundant* — to have each choice of parameters represent a unique distribution. More precisely, we want $\theta \neq \theta'$ to imply $P_\theta \neq P_{\theta'}$. It is easy check that a family is nonredundant if and only if the function t is invertible (over the set Θ). Such exponential families are called *invertible*. As we will discuss, these desiderata help us execute certain operations effectively, in particular, finding a distribution Q in some exponential family that is a "good approximation" to some other distribution P.

nonredundant
parameterization

invertible
exponential
family

8.2.1 Linear Exponential Families

A special class of exponential families is made up of families where the function t is the identity function. This implies that the parameters are the same dimension K as the representation of the data. Such parameters are also called the *natural parameters* for the given sufficient statistic function. The name reflects that these parameters do not need to be modified in the exponential form. When using natural parameters, equation (8.1) simplifies to

natural parameter

$$P_\theta(\xi) = \frac{1}{Z(\theta)} \exp\left\{\langle \theta, \tau(\xi) \rangle\right\}.$$

Clearly, for any given sufficient statistics function, we can reparameterize the exponential family using the natural parameters. However, as we discussed earlier, we want the space of parameters Θ to satisfy certain desiderata, which may not hold for the space of natural parameters. In fact, for the case of linear exponential families, we want to strengthen our desiderata, and require that any parameter vector in $\mathcal{R}^K$ defines a distribution in the family. Unfortunately, as stated, this desideratum is not always achievable. To understand why, recall that the definition of a legal parameter space Θ requires that each parameter vector $\boldsymbol{\theta} \in \Theta$ give rise to a legal (normalizable) distribution $P_{\boldsymbol{\theta}}$. These normalization requirements can impose constraints on the space of legal parameters.

Example 8.4
Consider again the Gaussian distribution. Suppose we define a new parameter space using the definition of t. *That is let* $\boldsymbol{\eta} = t(\mu, \sigma^2) = \langle \frac{2\mu}{2\sigma^2}, -\frac{1}{2\sigma^2} \rangle$ *be the natural parameters that corresponds to* $\boldsymbol{\theta} = \langle \mu, \sigma^2 \rangle$. *Clearly, we can now write*

$$P_{\boldsymbol{\eta}}(x) \propto \exp\left\{ \langle \boldsymbol{\eta}, \tau(x) \rangle \right\}.$$

However, not every choice of $\boldsymbol{\eta}$ *would lead to a legal distribution. For the distribution to be normalized, we need to be able to compute*

$$
\begin{aligned}
Z(\boldsymbol{\eta}) &= \int \exp\left\{ \langle \boldsymbol{\eta}, \tau(x) \rangle \right\} dx \\
&= \int_{-\infty}^{\infty} \exp\left\{ \eta_1 x + \eta_2 x^2 \right\} dx.
\end{aligned}
$$

If $\eta_2 \geq 0$ *this integral is undefined, since the function grows when* x *approaches* ∞ *and* $-\infty$. *When* $\eta_2 < 0$, *the integral has a finite value. Fortunately, if we consider* $\boldsymbol{\eta} = t(\mu, \sigma^2)$ *of equation (8.5), we see that the second component is always negative (since* $\sigma^2 > 0$). *In fact, we can see that the image of the original parameter space,* $\langle \mu, \sigma^2 \rangle \in \mathcal{R} \times \mathcal{R}^+$, *through the function* $t(\mu, \sigma^2)$, *is the space* $\mathcal{R} \times \mathcal{R}^-$. *We can verify that, for every* $\boldsymbol{\eta}$ *in that space, the normalization constant is well defined.* ∎

natural parameter space
More generally, when we consider natural parameters for a sufficient statistics function τ, we define the set of allowable natural parameters, the *natural parameter space*, to be the set of natural parameters that can be normalized

$$\Theta = \left\{ \boldsymbol{\theta} \in \mathcal{R}^K : \int \exp\left\{ \langle \boldsymbol{\theta}, \tau(\xi) \rangle \right\} d\xi < \infty \right\}.$$

In the case of distributions over finite discrete spaces, all parameter choices lead to normalizable distributions, and so $\Theta = \mathcal{R}^K$. In other examples, such as the Gaussian distribution, the natural parameter space can be more constrained. An exponential family over the natural parameter space, and for which the natural parameter space is open and convex, is called a

linear exponential family
linear exponential family.
The use of linear exponential families significantly simplifies the definition of a family. To specify such a family, we need to define only the function τ; all other parts of the definition are implicit based on this function. This gives us a tool to describe distributions in a concise manner. As we will see, linear exponential families have several additional attractive properties.

Where do find linear exponential families? The two examples we presented earlier were not phrased as linear exponential families. However, as we saw in example 8.4, we may be able to provide an alternative parameterization of a nonlinear exponential family as a linear exponential family. This example may give rise to the impression that any family can be reparameterized in a trivial manner. However, there are more subtle situations.

Example 8.5

Consider the Bernoulli distribution. Again, we might reparameterize θ by $t(\theta)$. However, the image of the function t of example 8.2 is the curve $\langle \ln \theta, \ln(1 - \theta) \rangle$. This curve is not a convex set, and it is clearly a subspace of the natural parameter space.

Alternatively, we might consider using the entire natural parameter space $\mathcal{R}^2$, corresponding to the sufficient statistic function $\tau(X) = \langle \mathbf{1}\{X = x^1\}, \mathbf{1}\{X = x^0\} \rangle$ of equation (8.2). This gives rise to the parametric form:

$$P_{\boldsymbol{\theta}}(x) \propto \exp\left\{ \langle \boldsymbol{\theta}, \tau(x) \rangle \right\} = \exp\left\{ \theta_1 \mathbf{1}\{X = x^1\} + \theta_2 \mathbf{1}\{X = x^0\} \right\}.$$

Because the probability space is finite, this form does define a distribution for every choice of $\langle \theta_1, \theta_2 \rangle$. However, it is not difficult to verify that this family is redundant: for every constant c, the parameters $\langle \theta_1 + c, \theta_2 + c \rangle$ define the same distribution as $\langle \theta_1, \theta_2 \rangle$.

Thus, a two-dimensional space is overparameterized for this distribution; conversely, the one-dimensional subspace defined by the natural parameter function is not well behaved. The solution is to use an alternative representation of a one-dimensional space. Since we have a redundancy, we may as well clamp θ_2 to be 0. This results in the following representation of the Bernoulli distribution:

$$\tau(x) = \mathbf{1}\{x = x^1\}$$
$$t(\theta) = \ln \frac{\theta}{1 - \theta}.$$

We see that

$$\exp\left\{ \langle t(\theta), \tau(x^1) \rangle \right\} = \frac{\theta}{1 - \theta}$$
$$\exp\left\{ \langle t(\theta), \tau(x^0) \rangle \right\} = 1.$$

Thus,

$$Z(\theta) = 1 + \frac{\theta}{1 - \theta} = \frac{1}{1 - \theta}.$$

Using these, we can verify that

$$P_\theta(x^1) = (1 - \theta) \frac{\theta}{1 - \theta} = \theta.$$

We conclude that this exponential representation captures the Bernoulli distribution. Notice now that, in the new representation, the image of t is the whole real line $\mathcal{R}$. Thus, we can define a linear exponential family with this sufficient statistic function. ∎

Example 8.6

Now, consider a multinomial variable X with k values $x^1, \ldots, x^k$. The situation here is similar to the one we had with the Bernoulli distribution. If we use the simplest exponential representation, we find that the legal natural parameters are on a curved manifold of $\mathcal{R}^k$. Thus, instead we define the sufficient statistic as a function from values of x to $\mathcal{R}^{k-1}$:

$$\tau(x) = \langle \mathbf{1}\{x = x^2\}, \ldots, \mathbf{1}\{x = x^k\} \rangle.$$

Using a similar argument as with the Bernoulli distribution, we see that if we define

$$t(\boldsymbol{\theta}) = \langle \ln \frac{\theta_2}{\theta_1}, \dots, \ln \frac{\theta_k}{\theta_1} \rangle,$$

then we reconstruct the original multinomial distribution. It is also easy to check that the image of t *is* $\mathcal{R}^{k-1}$. *Thus, by reparameterizing, we get a linear exponential family.* ∎

All these examples define linear exponential families. An immediate question is whether there exist families that are not linear. As we will see, there are such cases. However, the examples we present require additional machinery.

8.3   Factored Exponential Families

The two examples of exponential families so far were of univariate distributions. Clearly, we can extend the notion to multivariate distributions as well. In fact, we have already seen one such example. Recall that, in definition 4.15, we defined log-linear models as distributions of the form:

$$P(X_1, \dots, X_n) \propto \exp \left\{ \sum_{i=1}^{k} \theta_i \cdot f_i(\boldsymbol{D}_i) \right\}$$

where each feature f_i is a function whose scope is $\boldsymbol{D}_i$. Such a distribution is clearly a linear exponential family where the sufficient statistics are the vector of features

$$\tau(\xi) = \langle f_1(\boldsymbol{d}_1), \dots, f_k(\boldsymbol{d}_k) \rangle.$$

As we have shown, by choosing the appropriate features, we can devise a log-linear model to represent a given discrete Markov network structure. This suffices to show that discrete Markov networks are linear exponential families.

8.3.1   Product Distributions

What about other distributions with product forms? Initially the issues seem deceptively easy. A product form of terms corresponds to a simple composition of exponential families

Definition 8.2

exponential factor family

An (unnormalized) exponential factor family Φ *is defined by* τ, t, A, *and* Θ *(as in the exponential family). A factor in this family is*

$$\phi_{\boldsymbol{\theta}}(\xi) = A(\xi) \exp \left\{ \langle t(\boldsymbol{\theta}), \tau(\xi) \rangle \right\}.$$ ∎

Definition 8.3

family composition

Let $\Phi_1, \dots, \Phi_k$ *be exponential factor families, where each* Φ_i *is specified by* τ_i, t_i, A_i, *and* Θ_i. *The composition of* $\Phi_1, \dots, \Phi_k$ *is the family* $\Phi_1 \times \Phi_2 \times \dots \times \Phi_k$ *parameterized by* $\boldsymbol{\theta} = \boldsymbol{\theta}_1 \circ \boldsymbol{\theta}_2 \circ \dots \circ \boldsymbol{\theta}_k \in \Theta_1 \times \Theta_2 \times \dots \times \Theta_k$, *defined as*

$$P_{\boldsymbol{\theta}}(\xi) \propto \prod_i \phi_{\boldsymbol{\theta}_i}(\xi) = \left(\prod_i A_i(\xi) \right) \exp \left\{ \sum_i \langle t_i(\boldsymbol{\theta}_i), \tau_i(\xi) \rangle \right\}$$

where $\phi_{\boldsymbol{\theta}_i}$ *is a factor in the i'th factor family.* ∎

It is clear from this definition that the composition of exponential factors is an exponential family with $\tau(\xi) = \tau_1(\xi) \circ \tau_2(\xi) \circ \cdots \circ \tau_k(\xi)$ and natural parameters $t(\theta) = t_1(\theta_1) \circ t_2(\theta_2) \circ \cdots \circ t_k(\theta_k)$.

This simple observation suffices to show that if we have exponential representation for potentials in a Markov network (not necessarily simple potentials), then their product is also an exponential family. Moreover, it follows that the product of linear exponential factor families is a linear exponential family.

8.3.2 Bayesian Networks

Taking the same line of reasoning, we can also show that, if we have a set of CPDs from an exponential family, then their product is also in the exponential family. Thus, we can conclude that a Bayesian network with exponential CPDs defines an exponential family. To show this, we first note that many of the CPDs we saw in previous chapters can be represented as exponential factors.

Example 8.7

We start by examining a simple table-CPD $P(X \mid U)$. Similar to the case of Bernoulli distribution, we can define the sufficient statistics to be indicators for different entries in $P(X \mid U)$. Thus, we set

$$\tau_{P(X|U)}(\mathcal{X}) = \langle I\{X = x, U = u\} : x \in Val(X), u \in Val(U) \rangle.$$

We set the natural parameters to be the corresponding parameters

$$t_{P(X|U)}(\theta) = \langle \ln P(x \mid u) : x \in Val(X), u \in Val(U) \rangle.$$

It is easy to verify that

$$P(x \mid u) = \exp\left\{ \langle t_{P(X|U)}(\theta), \tau_{P(X|U)}(x, u) \rangle \right\},$$

since exactly one entry of $\tau_{P(X|U)}(x, u)$ is 1 and the rest are 0. Note that this representation is not a linear exponential factor. ∎

Clearly, we can use the same representation to capture any CPD for discrete variables. In some cases, however, we can be more efficient. In tree-CPDs, for example, we can have a feature set for each leaf in tree, since all parent assignment that reach the leaf lead to the same parameter over the children.

What happens with continuous CPDs? In this case, not every CPD can be represented by an exponential factor. However, some cases can.

Example 8.8

Consider a linear Gaussian CPD for $P(X \mid U)$ where

$$X = \beta_0 + \beta_1 u_1 + \cdots + \beta_k u_k + \epsilon,$$

where ϵ is a Gaussian random variable with mean 0 and variance σ^2, representing the noise in the system. Stated differently, the conditional density function of X is

$$P(x \mid u) = \frac{1}{\sqrt{2\pi}\sigma} \exp\left\{ -\frac{1}{2\sigma^2}(x - (\beta_0 + \beta_1 u_1 + \cdots + \beta_k u_k))^2 \right\}.$$

By expanding the squared term, we find that the sufficient statistics are the first and second moments of all the variables

$$\tau_{P(X|U)}(\mathcal{X}) = \langle 1, x, u_1, \ldots, u_k, x^2, xu_1, \ldots, xu_k, u_1^2, u_1u_2, \ldots, u_k^2 \rangle,$$

and the natural parameters are the coefficients of each of these terms. ∎

As the product of exponential factors is an exponential family, we conclude that a Bayesian network that is the product of CPDs that have exponential form defines an exponential family.

However, there is one subtlety that arises in the case of Bayesian networks that does not arise for a general product form. When we defined the product of a set of exponential factors in definition 8.3, we ignored the partition functions of the individual factors, allowing the partition function of the overall distribution to ensure global normalization.

However, in both of our examples of exponential factors for CPDs, we were careful to construct a normalized conditional distribution. This allows us to use the chain rule to compose these factors into a joint distribution without the requirement of a partition function. This requirement turns out to be critical: We cannot construct a Bayesian network from a product of unnormalized exponential factors.

Example 8.9

Consider the network structure $A \rightarrow B$, with binary variables. Now, suppose we want to represent the CPD $P(B \mid A)$ using a more concise representation than the one of example 8.7. As suggested by example 8.5, we might consider defining

$$\tau(A, B) = \langle \mathbf{I}\{A = a^1\}, \mathbf{I}\{B = b^1, A = a^1\}, \mathbf{I}\{B = b^1, A = a^0\} \rangle.$$

That is, for each conditional distribution, we have an indicator only for one of the two relevant cases. The representation of example 8.5 suggests that we should define

$$\mathbf{t}(\boldsymbol{\theta}) = \left\langle \ln \frac{\theta_{a^1}}{\theta_{a^0}}, \ln \frac{\theta_{b^1|a^1}}{\theta_{b^0|a^1}}, \ln \frac{\theta_{b^1|a^0}}{\theta_{b^0|a^0}} \right\rangle.$$

Does this construction give us the desired distribution? Under this construction, we would have

$$P_{\boldsymbol{\theta}}(a^1, b^1) = \frac{1}{Z(\boldsymbol{\theta})} \frac{\theta_{a^1} \theta_{b^1|a^1}}{\theta_{a^0} \theta_{b^0|a^1}}.$$

Thus, if this representation was faithful for the intended interpretation of the parameter values, we would have $Z(\boldsymbol{\theta}) = \frac{1}{\theta_{a^0} \theta_{b^0|a^1}}$. On the other hand,

$$P_{\boldsymbol{\theta}}(a^0, b^0) = \frac{1}{Z(\boldsymbol{\theta})},$$

which requires that $Z(\boldsymbol{\theta}) = \frac{1}{\theta_{a^0} \theta_{b^0|a^0}}$ in order to be faithful to the desired distribution. Because these two constants are, in general, not equal, we conclude that this representation cannot be faithful to the original Bayesian network. ∎

The failure in this example is that the global normalization constant cannot play the role of a local normalization constant within each conditional distribution. This implies that to have an exponential representation of a Bayesian network, we need to ensure that each CPD is locally normalized. For every exponential CPD this is easy to do. We simply increase the dimension of τ by adding another dimension that has a constant value, say 1. Then the matching element of $t(\boldsymbol{\theta})$ can be the logarithm of the partition function. This is essentially what we did in example 8.8.

We still might wonder whether a Bayesian network defines a linear exponential family.

Example 8.10

Consider the network structure $A \rightarrow C \leftarrow B$, with binary variables. Assuming a representation that captures general CPDs, our sufficient statistics need to include features that distinguish between the following four assignments:

$$
\begin{aligned}
\xi_1 &= \langle a^1, b^1, c^1 \rangle \\
\xi_2 &= \langle a^1, b^0, c^1 \rangle \\
\xi_3 &= \langle a^0, b^1, c^1 \rangle \\
\xi_4 &= \langle a^0, b^0, c^1 \rangle
\end{aligned}
$$

More precisely, we need to be able to modify the CPD $P(C \mid A, B)$ to change the probability of one of these assignments without modifying the probability of the other three. This implies that $\tau(\xi_1), \ldots, \tau(\xi_4)$ must be linearly independent: otherwise, we could not change the probability of one assignment without changing the others. Because our model is a linear function of the sufficient statistics, we can choose any set of orthogonal basis vectors that we want; in particular, we can assume without loss of generality that the first four coordinates of the sufficient statistics are $\tau_i(\xi) = \mathbf{1}\{\xi = \xi_i\}$, and that any additional coordinates of the sufficient statistics are not linearly dependent on these four. Moreover, since the model is over a finite set of events, any choice of parameters can be normalized. Thus, the space of natural parameters is $\mathcal{R}^K$, where K is dimension of the sufficient statistics vector. The linear family over such features is essentially a Markov network over the clique $\{A, B, C\}$. Thus, the parameterization of this family includes cases where A and B are not independent, violating the independence properties of the Bayesian network. ∎

Thus, this simple Bayesian network cannot be represented by a linear family. **More broadly, although a Bayesian network with suitable CPDs defines an exponential family, this family is not generally a linear one. In particular, any network that contains immoralities does not induce a linear exponential family.**

8.4 Entropy and Relative Entropy

We now explore some of the consequences of representation of models in factored form and of their exponential family representation. These both suggest some implications of these representations and will be useful in developments in subsequent chapters.

8.4.1 Entropy

We start with the notion of entropy. Recall that the entropy of a distribution is a measure of the amount of "stochasticity" or "noise" in the distribution. A low entropy implies that most of the distribution mass is on a few instances, while a larger entropy suggests a more uniform distribution. Another interpretation we discussed in appendix A.1 is the number of bits needed, on average, to encode instances in the distribution.

In various tasks we need to compute the entropy of given distributions. As we will see, we also encounter situations where we want to choose a distribution that maximizes the entropy subject to some constraints. A characterization of entropy will allow us to perform both tasks more efficiently.

8.4.1.1　Entropy of an Exponential Model

We now consider the task of computing the entropy for distributions in an an exponential family defined by τ and t.

Theorem 8.1

Let P_θ be a distribution in an exponential family defined by the functions τ and t. Then

$$\boldsymbol{H}_{P_\theta}(\mathcal{X}) = \ln Z(\boldsymbol{\theta}) - \langle \boldsymbol{E}_{P_\theta}[\tau(\mathcal{X})], \mathsf{t}(\boldsymbol{\theta}) \rangle. \tag{8.7}$$

While this formulation seems fairly abstract, it does provide some insight. The entropy decomposes as a difference of two terms. The first is the partition function $Z(\boldsymbol{\theta})$. The second depends only on the *expected value* of the sufficient statistics $\tau(\mathcal{X})$. Thus, instead of considering each assignment to $\mathcal{X}$, we need to know only the expectations of the statistics under P_θ. As we will see, this is a recurring theme in our discussion of exponential families.

Example 8.11

We now apply this result to a Gaussian distribution $X \sim N(\mu, \sigma^2)$, as formulated in the exponential family in example 8.3. Plugging into equation (8.7) the definitions of τ, t, and Z from equation (8.4), equation (8.5), and equation (8.6), respectively, we get

$$
\begin{aligned}
\boldsymbol{H}_P(X) &= \frac{1}{2}\ln 2\pi\sigma^2 + \frac{\mu^2}{2\sigma^2} - \frac{2\mu}{2\sigma^2}\boldsymbol{E}_P[X] + \frac{1}{2\sigma^2}\boldsymbol{E}_P[X^2] \\
&= \frac{1}{2}\ln 2\pi\sigma^2 + \frac{\mu^2}{2\sigma^2} - \frac{2\mu}{2\sigma^2}\mu + \frac{1}{2\sigma^2}(\sigma^2 + \mu^2) \\
&= \frac{1}{2}\ln 2\pi e\sigma^2
\end{aligned}
$$

where we used the fact that $\boldsymbol{E}_P[X] = \mu$ and $\boldsymbol{E}_P[X^2] = \mu^2 + \sigma^2$. ∎

We can apply the formulation of theorem 8.1 directly to write the entropy of a Markov network.

Proposition 8.1

If $P(\mathcal{X}) = \frac{1}{Z}\prod_k \phi_k(\boldsymbol{D}_k)$ is a Markov network, then

$$\boldsymbol{H}_P(\mathcal{X}) = \ln Z + \sum_k \boldsymbol{E}_P[-\ln \phi_k(\boldsymbol{D}_k)].$$

Example 8.12

Consider a simple Markov network with two potentials $\beta_1(A, B)$ and $\beta_2(B, C)$, so that

		$\beta_1(A, B)$
a^0	b^0	2
a^0	b^1	1
a^1	b^0	1
a^1	b^1	5

		$\beta_2(B, C)$
b^0	c^0	6
b^0	c^1	1
b^1	c^0	1
b^1	c^1	0.5

Simple calculations show that $Z = 30$, *and the marginal distributions are*

A	B	$P(A,B)$
a^0	b^0	0.47
a^0	b^1	0.05
a^1	b^0	0.23
a^1	b^1	0.25

B	C	$P(B,C)$
b^0	c^0	0.6
b^0	c^1	0.1
b^1	c^0	0.2
b^1	c^1	0.1

Using proposition 8.1, we can calculate the entropy:

$$
\begin{aligned}
\boldsymbol{H}_P(A,B,C) &= \ln Z + \boldsymbol{E}_P[-\ln\beta_1(A,B)] + \boldsymbol{E}_P[-\ln\beta_2(B,C)] \\
&= \ln Z \\
&\quad -P(a^0,b^0)\ln\beta_1(a^0,b^0) - P(a^0,b^1)\ln\beta_1(a^0,b^1) \\
&\quad -P(a^1,b^0)\ln\beta_1(a^1,b^0) - P(a^1,b^1)\ln\beta_1(a^1,b^1) \\
&\quad -P(b^0,c^0)\ln\beta_2(b^0,c^0) - P(b^0,c^1)\ln\beta_2(b^0,c^1) \\
&\quad -P(b^1,c^0)\ln\beta_2(b^1,c^0) - P(b^1,c^1)\ln\beta_2(b^1,c^1) \\
&= 3.4012 \\
&\quad -0.47*0.69 - 0.05*0 - 0.23*0 - 0.25*1.60 \\
&\quad -0.6*1.79 - 0.1*0 - 0.2*0 - 0.1*-0.69 \\
&= 1.670.
\end{aligned}
$$

■

In this example, the number of terms we evaluated is the same as what we would have considered using the original formulation of the entropy where we sum over all possible joint assignments. However, if we consider more complex networks, the number of joint assignments is exponentially large while the number of potentials is typically reasonable, and each one involves the joint assignments to only a few variables.

Note, however, that to use the formulation of proposition 8.1 we need to perform a global computation to find the value of the partition function Z as well as the marginal distribution over the scope of each potential $\boldsymbol{D}_k$. As we will see in later chapters, in some network structures, these computations can be done efficiently.

Terms such as $\boldsymbol{E}_P[-\ln\beta_k(\boldsymbol{D}_k)]$ resemble the entropy of $\boldsymbol{D}_k$. However, since the marginal over $\boldsymbol{D}_k$ is usually not identical to the potential β_k, such terms are not entropy terms. In some sense we can think of $\ln Z$ as a correction for this discrepancy. For example, if we multiply all the entries of β_k by a constant c, the corresponding term $\boldsymbol{E}_P[-\ln\beta_k(\boldsymbol{D}_k)]$ will decrease by $\ln c$. However, at the same time $\ln Z$ will increase by the same constant, since it is canceled out in the normalization.

8.4.1.2 Entropy of Bayesian Networks

We now consider the entropy of a Bayesian network. Although we can address this computation using our general result in theorem 8.1, it turns out that the formulation for Bayesian networks is simpler. Intuitively, as we saw, we can represent Bayesian networks as an exponential family where the partition function is 1. This removes the global term from the entropy.

Theorem 8.2 $\overline{\textit{If } P(\mathcal{X}) = \prod_i P(X_i \mid \text{Pa}_i^{\mathcal{G}}) \textit{ is a distribution consistent with a Bayesian network } \mathcal{G}, \textit{ then}}$

$$\boldsymbol{H}_P(\mathcal{X}) = \sum_i \boldsymbol{H}_P(X_i \mid \text{Pa}_i^{\mathcal{G}})$$

PROOF

$$
\begin{aligned}
\boldsymbol{H}_P(\mathcal{X}) &= \boldsymbol{E}_P[-\ln P(\mathcal{X})] \\
&= \boldsymbol{E}_P\left[-\sum_i \ln P(X_i \mid \text{Pa}_i^{\mathcal{G}})\right] \\
&= \sum_i \boldsymbol{E}_P[-\ln P(X_i \mid \text{Pa}_i^{\mathcal{G}})] \\
&= \sum_i \boldsymbol{H}_P(X_i \mid \text{Pa}_i^{\mathcal{G}}),
\end{aligned}
$$

where the first and last steps invoke the definitions of entropy and conditional entropy.   ∎

We see that the entropy of a Bayesian network decomposes as a sum of conditional entropies of the individual conditional distributions. This representation suggests that the entropy of a Bayesian network can be directly "read off" from the CPDs. This impression is misleading. Recall that the conditional entropy term $\boldsymbol{H}_P(X_i \mid \text{Pa}_i^{\mathcal{G}})$ can be written as a weighted average of simpler entropies of conditional distributions

$$\boldsymbol{H}_P(X_i \mid \text{Pa}_i^{\mathcal{G}}) = \sum_{\text{pa}_i^{\mathcal{G}}} P(\text{pa}_i^{\mathcal{G}})\boldsymbol{H}_P(X_i \mid \text{pa}_i^{\mathcal{G}}).$$

While each of the simpler entropy terms in the summation can be computed based on the CPD entries alone, the weighting term $P(\text{pa}_i^{\mathcal{G}})$ is a marginal over $\text{pa}_i^{\mathcal{G}}$ of the joint distribution, and depends on other CPDs upstream of X_i. Thus, computing the entropy of the network requires that we answer probability queries over the network.

However, based on local considerations alone, we can analyze the amount of entropy introduced by each CPD, and thereby provide bounds on the overall entropy:

Proposition 8.2 $\overline{\textit{If } P(\mathcal{X}) = \prod_i P(X_i \mid \text{Pa}_i^{\mathcal{G}}) \textit{ is a distribution consistent with a Bayesian network } \mathcal{G}, \textit{ then}}$

$$\sum_i \min_{\text{pa}_i^{\mathcal{G}}} \boldsymbol{H}_P(X_i \mid \text{pa}_i^{\mathcal{G}}) \leq \boldsymbol{H}_P(\mathcal{X}) \leq \sum_i \max_{\text{pa}_i^{\mathcal{G}}} \boldsymbol{H}_P(X_i \mid \text{pa}_i^{\mathcal{G}}).$$

Thus, if all the CPDs in a Bayesian network are almost deterministic (low conditional entropy given each parent configuration), then the overall entropy of the network is small. Conversely, if all the CPDs are highly stochastic (high conditional entropy) then the overall entropy of the network is high.

8.4.2   Relative Entropy

A related notion is the relative entropy between models. This measure of distance plays an important role in many of the developments of later chapters.

If we consider the relative entropy between an arbitrary distribution Q and a distribution P_θ within an exponential family, we see that the form of P_θ can be exploited to simplify the form of the relative entropy.

Theorem 8.3

Consider a distribution Q and a distribution P_θ in an exponential family defined by τ and t. Then

$$\mathbf{D}(Q\|P_\theta) = -\mathbf{H}_Q(\mathcal{X}) - \langle \mathbf{E}_Q[\tau(\mathcal{X})], t(\boldsymbol{\theta}) \rangle + \ln Z(\boldsymbol{\theta}).$$

The proof is left as an exercise (exercise 8.2).

We see that the quantities of interest are again the expected sufficient statistics and the partition function. Unlike the entropy, in this case we compute the expectation of the sufficient statistics according to Q.

If both distributions are in the same exponential family, then we can further simplify the form of the relative entropy.

Theorem 8.4

Consider two distribution $P_{\boldsymbol{\theta}_1}$ and $P_{\boldsymbol{\theta}_2}$ within the same exponential family. Then

$$\mathbf{D}(P_{\boldsymbol{\theta}_1}\|P_{\boldsymbol{\theta}_2}) = \langle \mathbf{E}_{P_{\boldsymbol{\theta}_1}}[\tau(\mathcal{X})], t(\boldsymbol{\theta}_1) - t(\boldsymbol{\theta}_2) \rangle - \ln \frac{Z(\boldsymbol{\theta}_1)}{Z(\boldsymbol{\theta}_2)}$$

PROOF Combine theorem 8.3 with theorem 8.1. ∎

When we consider Bayesian networks, we can use the fact that the partition function is constant to simplify the terms in both results.

Theorem 8.5

If P is a distribution consistent with a Bayesian network $\mathcal{G}$, then

$$\mathbf{D}(Q\|P) = -\mathbf{H}_Q(\mathcal{X}) - \sum_i \sum_{\mathrm{pa}_i^{\mathcal{G}}} Q(\mathrm{pa}_i^{\mathcal{G}}) \mathbf{E}_{Q(X_i|\mathrm{pa}_i^{\mathcal{G}})}\left[\ln P(X_i \mid \mathrm{pa}_i^{\mathcal{G}})\right];$$

If Q is also consistent with $\mathcal{G}$, then

$$\mathbf{D}(Q\|P) = \sum_i \sum_{\mathrm{pa}_i^{\mathcal{G}}} Q(\mathrm{pa}_i^{\mathcal{G}}) \mathbf{D}(Q(X_i \mid \mathrm{pa}_i^{\mathcal{G}})\|P(X_i \mid \mathrm{pa}_i^{\mathcal{G}})).$$

The second result shows that, analogously to the form of the entropy of Bayesian networks, we can write the relative entropy between two distributions consistent with $\mathcal{G}$ as a weighted sum of the relative entropies between the conditional distributions. These conditional relative entropies can be evaluated directly using the CPDs of the two networks. The weighting of these relative entropies depends on the the joint distribution Q.

8.5 Projections

projection

As we discuss in appendix A.1.3, we can view the relative entropy as a notion of distance between two distributions. We can therefore use it as the basis for an important operation — the *projection* operation — which we will utilize extensively in subsequent chapters. Similar to the geometric concept of projecting a point onto a hyperplane, we consider the problem of finding the distribution, within a given exponential family, that is closest to a given distribution in terms of relative entropy. For example, we want to perform such a projection when we approximate a complex distribution with one with a simple structure. As we will see, this is a crucial strategy for approximate inference in networks where exact inference is infeasible. In such an approximation we would like to find the best (that is, closest) approximation within a family in which we can perform inference. Moreover, the problem of *learning* a graphical model can also be posed as a projection problem of the empirical distribution observed in the data onto a desired family.

Suppose we have a distribution P and we want to approximate it with another distribution Q in a class of distributions $\mathcal{Q}$ (for example, an exponential family). For example, we might want to approximate P with a product of marginal distributions. Because the notion of relative entropy is not symmetric, we can use it to define two types of approximations.

Definition 8.4

Let P be a distribution and let $\mathcal{Q}$ be a convex set of distributions.

I-projection

- *The* I-projection *(information projection) of P onto $\mathcal{Q}$ is the distribution*

$$Q^I = \arg\min_{Q \in \mathcal{Q}} \boldsymbol{D}(Q\|P).$$

M-projection

- *The* M-projection *(moment projection) of P onto $\mathcal{Q}$ is the distribution*

$$Q^M = \arg\min_{Q \in \mathcal{Q}} \boldsymbol{D}(P\|Q). \qquad \blacksquare$$

8.5.1   Comparison

We can think of both Q^I and Q^M as the projection of P into the set $\mathcal{Q}$ in the sense that it is the distribution closest to P. Moreover, if $P \in \mathcal{Q}$, then in both definitions the projection would be P. However, because the relative entropy is not symmetric, these two projections are, in general, different. To understand the differences between these two projections, let us consider a few examples.

Example 8.13

Suppose we have a non-Gaussian distribution P over the reals. We can consider the M-projection and the I-projection on the family of Gaussian distributions. As a concrete example, consider the distribution P of figure 8.1. As we can see, the two projections are different Gaussian distributions. (The M-projection was found using the analytic form that we will discuss, and the I-projection by gradient ascent in the (μ, σ^2) space.) Although the means of the two projected distributions are relatively close, the M-projection has larger variance than the I-projection. $\blacksquare$

We can better understand these differences if we examine the objective function optimized by each projection. Recall that the M-projection Q^M minimizes

$$\boldsymbol{D}(P\|Q) = -\boldsymbol{H}_P(X) + \boldsymbol{E}_P[-\ln Q(X)].$$

We see that, in general, we want Q^M to have high density in regions that are probable according to P, since a small $-\ln Q(X)$ in these regions will lead to a smaller second term. At the same time, there is a high penalty for assigning low density to regions where $P(X)$ is nonnegligible.

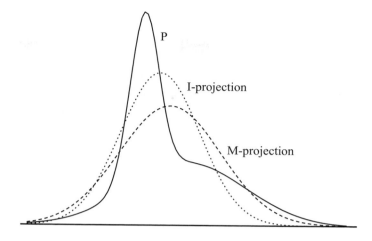

Figure 8.1   Example of M- and I-projections into the family of Gaussian distributions

As a consequence, although the M-projection attempts to match the main mass of P, its high variance is a compromise to ensure that it assigns reasonably high density to all regions that are in the support of P.

On the other hand, the I-projection minimizes

$$\boldsymbol{D}(Q\|P) = -\boldsymbol{H}_Q(X) + \boldsymbol{E}_Q[-\ln P(X)].$$

Thus, the first term incurs a penalty for low entropy, which in the case of a Gaussian Q translates to a penalty on small variance. The second term, $\boldsymbol{E}_Q[-\ln P(X)]$, encodes a preference for assigning higher density to regions where $P(X)$ is large and very low density to regions where $P(X)$ is small. Without the first term, we can minimize the second by putting all of the mass of Q on the most probable point according to P. The compromise between the two terms results in the distribution we see in figure 8.1.

A similar phenomenon occurs in discrete distributions.

Example 8.14

Now consider the projection of a distribution $P(A, B)$ onto the family of factored distributions $Q(A, B) = Q(A)Q(B)$. Suppose $P(A, B)$ is the following distribution:

$$
\begin{aligned}
P(a^0, b^0) &= 0.45 \\
P(a^0, b^1) &= 0.05 \\
P(a^1, b^0) &= 0.05 \\
P(a^1, b^1) &= 0.45.
\end{aligned}
$$

That is, the distribution P puts almost all of the mass on the event $A = B$. This distribution is a particularly difficult one to approximate using a factored distribution, since in P the two variables A and B are highly correlated, a dependency that cannot be captured using a fully factored Q.

Again, it is instructive to compare the M-projection and the I-projection of this distribution (see figure 8.2). It follows from example A.7 (appendix A.5.3) that the M-projection of this distribution is

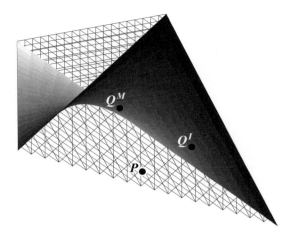

Figure 8.2 **Example of M- and I-projections of a two variable discrete distribution** where $P(a^0 = b^0) = P(a^1 = b^1) = 0.45$ and $P(a^0 = b^1) = P(a^0 = b^1) = 0.05$ onto factorized distribution. Each axis denotes the probability of an instance: $P(a^1, b^1)$, $P(a^1, b^0)$, and $P(a^0, b^1)$. The wire surfaces mark the region of legal distributions. The solid surface shows the distributions where A and independent of B. The points show P and its two projections.

the uniform distribution:

$$
\begin{aligned}
Q^M(a^0, b^0) &= 0.5 * 0.5 = 0.25 \\
Q^M(a^0, b^1) &= 0.5 * 0.5 = 0.25 \\
Q^M(a^1, b^0) &= 0.5 * 0.5 = 0.25 \\
Q^M(a^1, b^1) &= 0.5 * 0.5 = 0.25.
\end{aligned}
$$

In contrast, the I-projection focuses on one of the two "modes" of the distribution, either when both A and B are true or when both are false. Since the distribution is symmetric about these modes, there are two I-projections. One of them is

$$
\begin{aligned}
Q^I(a^0, b^0) &= 0.25 * 0.25 = 0.0625 \\
Q^I(a^0, b^1) &= 0.25 * 0.75 = 0.1875 \\
Q^I(a^1, b^0) &= 0.75 * 0.25 = 0.1875 \\
Q^I(a^1, b^1) &= 0.75 * 0.75 = 0.5625.
\end{aligned}
$$

The second I-projection is symmetric around the opposite mode a^0, b^0.   ∎

As in example 8.13, we can understand these differences by considering the underlying mathematics. **The M-projection attempts to give all assignments reasonably high probability, whereas the I-projection attempts to focus on high-probability assignments in P while maintaining a reasonable entropy.** In this case, this behavior results in a uniform distribution for the M-projection, whereas the I-projection places most of the probability mass on one of the two assignments where P has high probability.

8.5.2 M-Projections

Can we say more about the form of these projections? We start by considering M-projections onto a simple family of distributions.

Proposition 8.3

Let P be a distribution over $X_1, \ldots, X_n$, and let $\mathcal{Q}$ be the family of distributions consistent with $\mathcal{G}_\emptyset$, the empty graph. Then

$$Q^M = \arg \min_{Q \models \mathcal{G}_\emptyset} D(P\|Q)$$

is the distribution:

$$Q^M(X_1, \ldots, X_n) = P(X_1)P(X_2)\cdots P(X_n).$$

PROOF Consider a distribution $Q \models \mathcal{G}_\emptyset$. Since Q factorizes, we can rewrite $D(P\|Q)$:

$$
\begin{aligned}
D(P\|Q) &= E_P[\ln P(X_1, \ldots, X_n) - \ln Q(X_1, \ldots, X_n)] \\
&= E_P[\ln P(X_1, \ldots, X_n)] - \sum_i E_P[\ln Q(X_i)] \\
&= E_P\left[\ln \frac{P(X_1, \ldots, X_n)}{P(X_1)\cdots P(X_n)}\right] + \sum_i E_P\left[\ln \frac{P(X_i)}{Q(X_i)}\right] \\
&= D(P\|Q^M) + \sum_i D(P(X_i)\|Q(X_i)) \\
&\geq D(P\|Q^M).
\end{aligned}
$$

The last step relies on the nonnegativity of the relative entropy. We conclude that $D(P\|Q) \geq D(P\|Q^M)$ with equality only if $Q(X_i) = P(X_i)$ for all i. That is, only when $Q = Q^M$. ∎

Hence, the M-projection of P onto factored distribution is simply the product of marginals of P.

This theorem is an instance of a much more general result. To understand the generalization, we observe that the family $\mathcal{Q}$ of fully factored distributions is characterized by a vector of sufficient statistics that simply counts, for each variable X_i, the number of occurrences of each of its values. The marginal distributions over the X_i's are simply the expectations, relative to P, of these sufficient statistics. We see that, by selecting Q to match these expectations, we obtain the M-projection.

As we now show, this is not an accident. The characterization of a distribution P that is relevant to computing its M-projection into $\mathcal{Q}$ is precisely the expectation, relative to P, of the sufficient statistic function of $\mathcal{Q}$.

Theorem 8.6

Let P be a distribution over $\mathcal{X}$, and let $\mathcal{Q}$ be an exponential family defined by the functions $\tau(\xi)$ and $t(\boldsymbol{\theta})$. If there is a set of parameters $\boldsymbol{\theta}$ such that $\boldsymbol{E}_{Q_{\boldsymbol{\theta}}}[\tau(\mathcal{X})] = \boldsymbol{E}_P[\tau(\mathcal{X})]$, then the M-projection of P is $Q_{\boldsymbol{\theta}}$.

Proof Suppose that $\boldsymbol{E}_P[\tau(\mathcal{X})] = \boldsymbol{E}_{Q_{\boldsymbol{\theta}}}[\tau(\mathcal{X})]$, and let $\boldsymbol{\theta}'$ be some set of parameters. Then,

$$
\begin{aligned}
\boldsymbol{D}(P\|Q_{\boldsymbol{\theta}'}) - \boldsymbol{D}(P\|Q_{\boldsymbol{\theta}}) &= -H_P(\mathcal{X}) - \langle \boldsymbol{E}_P[\tau(\mathcal{X})], t(\boldsymbol{\theta}')\rangle + \ln Z(\boldsymbol{\theta}') \\
&\quad + H_P(\mathcal{X}) + \langle \boldsymbol{E}_P[\tau(\mathcal{X})], t(\boldsymbol{\theta})\rangle - \ln Z(\boldsymbol{\theta}) \\
&= \langle \boldsymbol{E}_P[\tau(\mathcal{X})], t(\boldsymbol{\theta}) - t(\boldsymbol{\theta}')\rangle - \ln \frac{Z(\boldsymbol{\theta})}{Z(\boldsymbol{\theta}')} \\
&= \langle \boldsymbol{E}_{Q_{\boldsymbol{\theta}}}[\tau(\mathcal{X})], t(\boldsymbol{\theta}) - t(\boldsymbol{\theta}')\rangle - \ln \frac{Z(\boldsymbol{\theta})}{Z(\boldsymbol{\theta}')} \\
&= \boldsymbol{D}(Q_{\boldsymbol{\theta}}\|Q_{\boldsymbol{\theta}'}) \geq 0.
\end{aligned}
$$

We conclude that the M-projection of P is $Q_{\boldsymbol{\theta}}$.   ∎

This theorem suggests that we can consider both the distribution P and the distributions in $\mathcal{Q}$ in terms of the expectations of $\tau(\mathcal{X})$. Thus, instead of describing a distribution in the family by the set of parameters, we can describe it in terms of the *expected sufficient statistics*.

To formalize this intuition, we need some additional notation. We define a mapping from legal parameters in Θ to vectors of sufficient statistics

expected sufficient statistics

$$
ess(\boldsymbol{\theta}) = \boldsymbol{E}_{Q_{\boldsymbol{\theta}}}[\tau(\mathcal{X})].
$$

Theorem 8.6 shows that if $\boldsymbol{E}_P[\tau(\mathcal{X})]$ is in the image of *ess*, then the M-projection of P is the distribution $Q_{\boldsymbol{\theta}}$ that matches the expected sufficient statistics of P. In other words,

$$
\boldsymbol{E}_{Q^M}[\tau(\mathcal{X})] = \boldsymbol{E}_P[\tau(\mathcal{X})].
$$

moment matching

This result explains why M-projection is also referred to as *moment matching*. In many exponential families the sufficient statistics are moments (mean, variance, and so forth) of the distribution. In such cases, the M-projection of P is the distribution in the family that matches these moments in P.

We illustrate these concepts in figure 8.3. As we can see, the mapping $ess(\boldsymbol{\theta})$ directly relates parameters to expected sufficient statistics. By comparing the expected sufficient statistics of P to these of distributions in $\mathcal{Q}$, we can find the M-projection.

Moreover, using theorem 8.6, we obtain a general characterization of the M-projection function M-project(s), which maps a vector of expected sufficient statistics to a parameter vector:

Corollary 8.1

Let $\boldsymbol{s}$ be a vector. If $\boldsymbol{s} \in image(ess)$ and ess is invertible, then

$$
\text{M-project}(\boldsymbol{s}) = ess^{-1}(\boldsymbol{s}).
$$

That is, the parameters of the M-projection of P are simply the inverse of the *ess* mapping, applied to the expected sufficient statistic vector of P. This result allows us to describe the M-projection operation in terms of a specific function. This result assumes, of course, that $\boldsymbol{E}_P[\tau]$ is in the image of *ess* and that *ess* is invertible. In many examples that we consider, the image of *ess* includes all possible vectors of expected sufficient statistics we might encounter. Moreover, if the parameterization is nonredundant, then *ess* is invertible.

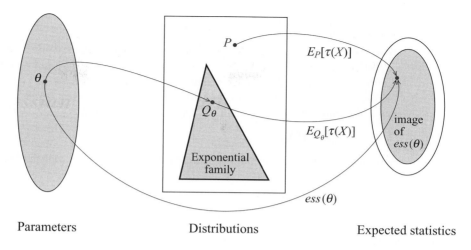

Figure 8.3 **Illustration of the relations between parameters, distributions and expected sufficient statistics.** Each parameter corresponds to a distribution, which in turn corresponds to a value of the expected statistics. The function *ess* maps parameters directly to expected statistics. If the expected statistics of P and Q_θ match, then Q_θ is the M-projection of P.

Example 8.15

Consider the exponential family of Gaussian distributions. Recall that the sufficient statistics function for this family is $\tau(x) = \langle x, x^2 \rangle$. Given parameters $\boldsymbol{\theta} = \langle \mu, \sigma^2 \rangle$, the expected value of τ is

$$ess(\langle \mu, \sigma^2 \rangle) = \boldsymbol{E}_{Q_{\langle \mu, \sigma^2 \rangle}}[\tau(X)] = \langle\ \mu, \sigma^2 + \mu^2\ \rangle.$$

It is not difficult to show that, for any distribution P, $\boldsymbol{E}_P[\tau(X)]$ must be in the image of this function (see exercise 8.4). Thus, for any choice of P, we can apply theorem 8.6.

Finally, we can easily invert this function:

$$\text{M-project}(\langle s_1, s_2 \rangle) = ess^{-1}(\langle s_1, s_2 \rangle) = \langle s_1, s_2 - s_1^2 \rangle.$$

Recall that $s_1 = \boldsymbol{E}_P[X]$ and $s_2 = \boldsymbol{E}_P[X^2]$. Thus, the estimated parameters are the mean and variance of X according to P, as we would expect. ∎

This example shows that the "naive" choice of Gaussian distribution, obtained by matching the mean and variance of a variable X, provides the best Gaussian approximation (in the M-projection sense) to a non-Gaussian distribution over X. We have also provided a solution to the M-projection problem in the case of a factored product of multinomials, in proposition 8.3, which can be viewed as a special case of theorem 8.6. In a more general application of this result, we show in section 11.4.4 a general result on the form of the M-projection for a linear exponential family over discrete state space, including the class of Markov networks.

The analysis for other families of distributions can be subtler.

Example 8.16

We now consider a more complex example of M-projection onto a chain network. Suppose we have a distribution P over variables $X_1, \ldots, X_n$, and want to project it onto the family of distributions Q of the distributions that are consistent with the network structure $X_1 \to X_2 \to \cdots \to X_n$.

What are the sufficient statistics for this network? Based on our previous discussion, we see that each conditional distribution $Q(X_{i+1} \mid X_i)$ requires a statistic of the form

$$\tau_{x_i, x_{i+1}}(\xi) = \boldsymbol{I}\{X_i = x_i, X_{i+1} = x_{i+1}\} \ \forall \langle x_i, x_{i+1} \rangle \in Val(X_i) \times Val(X_{i+1}).$$

These statistics are sufficient but are redundant. To see this, note that the "marginal statistics" must agree. That is,

$$\sum_{x_i} \tau_{x_i, x_{i+1}}(\xi) = \sum_{x_{i+2}} \tau_{x_{i+1}, x_{i+2}}(\xi) \ \forall x_{i+1} \in Val(X_{i+1}). \tag{8.8}$$

Although this representation is redundant, we can still apply the mechanisms discussed earlier and consider the function ess that maps parameters of such a network to the sufficient statistics. The expectation of an indicator function is the marginal probability of that event, so that

$$\boldsymbol{E}_{Q_{\boldsymbol{\theta}}}\left[\tau_{x_i, x_{i+1}}(\mathcal{X})\right] = Q_{\boldsymbol{\theta}}(x_i, x_{i+1}).$$

Thus, the function ess simply maps from $\boldsymbol{\theta}$ to the pairwise marginals of consecutive variables in $Q_{\boldsymbol{\theta}}$. Because these are pairwise marginals of an actual distribution, it follows that these sufficient statistics satisfy the consistency constraints of equation (8.8).

How do we invert this function? Given the statistics from P, we want to find a distribution Q that matches them. We start building Q along the structure of the chain. We choose $Q(X_1)$ and $Q(X_2 \mid X_1)$ so that $Q(x_1, x_2) = \boldsymbol{E}_P[\tau_{x_1, x_2}(\mathcal{X})] = P(x_1, x_2)$. In fact, there is a unique choice that satisfies this equality, where $Q(X_1, X_2) = P(X_1, X_2)$. This choice implies that the marginal distribution $Q(X_2)$ matches the marginal distribution $P(X_2)$. Now, consider our choice of $Q(X_3 \mid X_2)$. We need to ensure that

$$Q(x_3, x_2) = \boldsymbol{E}_P[\tau_{x_2, x_3}(\mathcal{X})] = P(x_2, x_3).$$

We note that, because $Q(x_3, x_2) = Q(x_3 \mid x_2)Q(x_2) = Q(x_3 \mid x_2)P(x_2)$, we can achieve this equality by setting $Q(x_3 \mid x_2) = P(x_3 \mid x_2)$. Moreover, this implies that $Q(x_3) = P(x_3)$. We can continue this construction recursively to set

$$Q(x_{i+1} \mid x_i) = P(x_{i+1} \mid x_i).$$

Using the preceding argument, we can show that this choice will match the sufficient statistics of P. This suffices to show that this Q is the M-projection of P.

Note that, although this choice of Q coincides with P on pairwise marginals of consecutive variables, it does not necessarily agree with P on other marginals. As an extreme example, consider a distribution P where X_1 and X_3 are identical and both are independent of X_2. If we project this distribution onto a distribution Q with the structure $X_1 \to X_2 \to X_3$, then P and Q will not necessarily agree on the joint marginals of X_1, X_3. In Q this distribution will be

$$Q(x_1, x_3) = \sum_{x_2} Q(x_1, x_2)Q(x_3 \mid x_2).$$

Since $Q(x_1, x_2) = P(x_1, x_2) = P(x_1)P(x_2)$ *and* $Q(x_3 \mid x_2) = P(x_3 \mid x_2) = P(x_3)$, *we conclude that* $Q(x_1, x_3) = P(x_1)P(x_3)$, *losing the equality between* X_1 *and* X_3 *in* P. ∎

This analysis used a redundant parameterization; exercise 8.6 shows how we can reparameterize a directed chain within the linear exponential family and thereby obtain an alternative perspective on the M-projection operation.

So far, all of our examples have had the characteristic that the vector of expected sufficient statistics for a distribution P is always in the image of *ess*; thus, our task has only been to invert *ess*. Unfortunately, there are examples where not every vector of expected sufficient statistics can also be derived from a distribution in our exponential family.

Example 8.17

Consider again the family Q *from example 8.10, of distributions parameterized using network structure* $A \to C \leftarrow B$, *with binary variables* A, B, C. *We can show that the sufficient statistics for this distribution are indicators for all the joint assignments to* A, B, *and* C *except one. That is,*

$$\tau(A, B, C) = \langle \; \pmb{I}\{A = a^1, B = b^1, C = c^1\},$$
$$\pmb{I}\{A = a^0, B = b^1, C = c^1\},$$
$$\pmb{I}\{A = a^1, B = b^0, C = c^1\},$$
$$\pmb{I}\{A = a^1, B = b^1, C = c^0\},$$
$$\pmb{I}\{A = a^1, B = b^0, C = c^0\},$$
$$\pmb{I}\{A = a^0, B = b^1, C = c^0\},$$
$$\pmb{I}\{A = a^0, B = b^0, C = c^1\}\rangle.$$

If we look at the expected value of these statistics given some member of the family, we have that, since A *and* B *are independent in* $Q_{\boldsymbol{\theta}}$, $Q_{\boldsymbol{\theta}}(a^1, b^1) = Q_{\boldsymbol{\theta}}(a^1)Q_{\boldsymbol{\theta}}(b^1)$. *Thus, the expected statistics should satisfy*

$$\pmb{E}_{Q_{\boldsymbol{\theta}}}\left[\pmb{I}\{A = a^1, B = b^1, C = c^1\}\right] + \pmb{E}_{Q_{\boldsymbol{\theta}}}\left[\pmb{I}\{A = a^1, B = b^1, C = c^0\}\right] =$$
$$\left(\pmb{E}_{Q_{\boldsymbol{\theta}}}\left[\pmb{I}\{A = a^1, B = b^1, C = c^1\}\right] + \pmb{E}_{Q_{\boldsymbol{\theta}}}\left[\pmb{I}\{A = a^1, B = b^1, C = c^0\}\right]\right.$$
$$\left. + \pmb{E}_{Q_{\boldsymbol{\theta}}}\left[\pmb{I}\{A = a^1, B = b^0, C = c^1\}\right] + \pmb{E}_{Q_{\boldsymbol{\theta}}}\left[\pmb{I}\{A = a^1, B = b^0, C = c^0\}\right]\right)$$
$$\left(\pmb{E}_{Q_{\boldsymbol{\theta}}}\left[\pmb{I}\{A = a^1, B = b^1, C = c^1\}\right] + \pmb{E}_{Q_{\boldsymbol{\theta}}}\left[\pmb{I}\{A = a^1, B = b^1, C = c^0\}\right]\right.$$
$$\left. + \pmb{E}_{Q_{\boldsymbol{\theta}}}\left[\pmb{I}\{A = a^0, B = b^1, C = c^1\}\right] + \pmb{E}_{Q_{\boldsymbol{\theta}}}\left[\pmb{I}\{A = a^0, B = b^1, C = c^0\}\right]\right).$$

This constraint is not typically satisfied by the expected statistics from a general distribution P *we might consider projecting. Thus, in this case, there are expected statistics vectors that do not fall within the image of* ess. ∎

In such cases, and in Bayesian networks in general, the projection procedure is more complex than inverting the *ess* function. Nevertheless, we can show that the projection operation still has an analytic solution.

Theorem 8.7

Let P *be a distribution over* $X_1, \ldots, X_n$, *and let* $\mathcal{G}$ *be a Bayesian network structure. Then the M-projection* Q^M *is:*

$$Q^M(X_1, \ldots, X_n) = \prod_i P(X_i \mid \text{Pa}_{X_i}^{\mathcal{G}}).$$

Because the mapping *ess* for Bayesian networks is not invertible, the proof of this result (see exercise 8.5) does not build on theorem 8.6 but rather directly on theorem 8.5. This result turns out to be central to our derivation of Bayesian network learning in chapter 17.

8.5.3   I-Projections

What about I-projections? Recall that

$$\boldsymbol{D}(Q\|P) = -\boldsymbol{H}_Q(\mathcal{X}) - \boldsymbol{E}_Q[\ln P(\mathcal{X})].$$

If Q is in some exponential family, we can use the derivation of theorem 8.1 to simplify the entropy term. However, the exponential form of Q does not provide insights into the second term. When dealing with the I-projection of a general distribution P, we are left without further simplifications. However, if the distribution P has some structure, we might be able to simplify $\boldsymbol{E}_Q[\ln P(\mathcal{X})]$ into simpler terms, although the projection problem is still a nontrivial one. We discuss this problem in much more detail in chapter 11.

8.6   Summary

In this chapter, we presented some of the basic technical concepts that underlie many of the techniques we explore in depth later in the book. We defined the formalism of exponential families, which provides the fundamental basis for considering families of related distributions. We also defined the subclass of linear exponential families, which are significantly simpler and yet cover a large fraction of the distributions that arise in practice.

We discussed how the types of distributions described so far in this book fit into this framework, showing that Gaussians, linear Gaussians, and multinomials are all in the linear exponential family. Any class of distributions representable by parameterizing a Markov network of some fixed structure is also in the linear exponential family. By contrast, the class of distributions representable by a Bayesian network of some fixed structure is in the exponential family, but is not in the linear exponential family when the network structure includes an immorality.

We showed how we can use the formulation of an exponential family to facilitate computations such as the entropy of a distribution or the relative entropy between two distributions. The latter computation formed the basis for analyzing a basic operation on distributions: that of projecting a general distribution P into some exponential family $\mathcal{Q}$, that is, finding the distribution within $\mathcal{Q}$ that is closest to P. Because the notion of relative entropy is not symmetric, this concept gave rise to two different definitions: I-projection, where we minimize $\boldsymbol{D}(Q\|P)$, and M-projection, where we minimize $\boldsymbol{D}(P\|Q)$. We analyzed the differences between these two definitions and showed that solving the M-projection problem can be viewed in a particularly elegant way, constructing a distribution Q that matches the expected sufficient statistics (or moments) of P.

As we discuss later in the book, both the I-projection and M-projection turn out to play an important role in graphical models. The M-projection is the formal foundation for addressing the learning problem: there, our goal is to find a distribution in a particular class (for example, a Bayesian network or Markov network of a given structure) that is closest (in the M-projection sense) to the *empirical distribution* observed in a data set from which we wish to learn (see equation (16.4)). The I-projection operation is used when we wish to take a given graphical model P and answer probability queries; when P is too complex to allow queries to be answered efficiently, one strategy is to construct a simpler distribution Q, which is a good approximation to P (in the I-projection sense).

8.7 Relevant Literature

The concept of exponential families plays a central role in formal statistic theory. Much of the theory is covered by classic textbooks such as Barndorff-Nielsen (1978). See also Lauritzen (1996). Geiger and Meek (1998) discuss the representation of graphical models as exponential families and show that a Bayesian network usually does not define a linear exponential family.

The notion of I-projections was introduced by Csiszàr (1975), who developed the "information geometry" of such projections and their connection to different estimation procedures. In his terminology, M-projections are called "reverse I-projections." The notion of M-projection is closely related to parameter learning, which we revisit in chapter 17 and chapter 20.

8.8 Exercises

Exercise 8.1⋆

Poisson distribution

A variable X with $Val(X) = 0, 1, 2, \ldots$ is *Poisson-distributed* with parameter $\theta > 0$ if

$$P(X = k) = \frac{1}{k!} \exp -\theta \theta^k.$$

This distribution has the property that $\boldsymbol{E}_P[X] = \theta$.

a. Show how to represent the Poisson distribution as a linear exponential family. (Note that unlike most of our running examples, you need to use the auxiliary measure A in the definition.)

b. Use results developed in this chapter to find the entropy of a Poisson distribution and the relative entropy between two Poisson distributions.

c. What is the function *ess* associated with this family? Is it invertible?

Exercise 8.2

Prove theorem 8.3.

Exercise 8.3

In this exercise, we will provide a characterization of when two distributions P_1 and P_2 will have the same M-projection.

a. Let P_1 and P_2 be two distribution over $\mathcal{X}$, and let $\mathcal{Q}$ be an exponential family defined by the functions $\tau(\xi)$ and $\mathsf{t}(\boldsymbol{\theta})$. If $\boldsymbol{E}_{P_1}[\tau(\mathcal{X})] = \boldsymbol{E}_{P_2}[\tau(\mathcal{X})]$, then the M-projection of P_1 and P_2 onto $\mathcal{Q}$ is identical.

b. Now, show that if the function $ess(\boldsymbol{\theta})$ is invertible, then we can prove the converse, showing that the M-projection of P_1 and P_2 is identical only if $\boldsymbol{E}_{P_1}[\tau(\mathcal{X})] = \boldsymbol{E}_{P_2}[\tau(\mathcal{X})]$. Conclude that this is the case for linear exponential families.

Exercise 8.4

Consider the function *ess* for Gaussian variables as described in example 8.15.

a. What is the image of *ess*?

b. Consider terms of the form $\boldsymbol{E}_P[\tau(X)]$ for the Gaussian sufficient statistics from that example. Show that for any distribution P, the expected sufficient statistics is in the image of *ess*.

Exercise 8.5★

Prove theorem 8.7. (Hint: Use theorem 8.5.)

Exercise 8.6★

Suppose we are given a family $\mathcal{Q}$ of chain distributions of the form $Q(X_1, \ldots, X_n) = Q(X_1)Q(X_2 \mid X_1) \cdots Q(X_n \mid X_{n+1})$. We now show how to reformulate this family as a linear exponential family.

a. Show that the following vector of statistics is sufficient and nonredundant for distributions in the family:

$$\tau(X_1, \ldots, X_n) = \begin{pmatrix} \boldsymbol{I}\{X_1 = x_1^1\}, \\ \cdots \\ \boldsymbol{I}\{X_k = x_n^1\}, \\ \boldsymbol{I}\{X_1 = x_1^1, X_2 = x_2^1\}, \\ \cdots \\ \boldsymbol{I}\{X_{n-1} = x_{n-1}^1, X_n = n_k^1\} \end{pmatrix}.$$

b. Show that you can reconstruct the distributions $Q(X_1)$ and $Q(X_{i+1} \mid X_i)$ from the the expectation $\boldsymbol{E}_Q[\tau(X_1, \ldots, X_n)]$. This shows that given the expected sufficient statistics you can reconstruct $\mathcal{Q}$.

c. Suppose you know $\mathcal{Q}$. Show how to reparameterize it as a linear exponential model

$$Q(X_1, \ldots, X_n) = \frac{1}{Z} \exp \left\{ \sum_i \theta_i \boldsymbol{I}\{X_i = x_i^1\} + \sum_i \theta_{i,i+1} \boldsymbol{I}\{X_i = x_i^1, X_{i+1} = x_{i+1}^1\} \right\}. \quad (8.9)$$

Note that, because the statistics are sufficient, we know that there are some parameters for which we get equality; the question is to determine their values. Specifically, show that if we choose:

$$\theta_i = \ln \frac{Q(x_1^0, \ldots, x_{i-1}^0, x_i^1, x_{i+1}^0, \ldots, x_n^0)}{Q(x_1^0, \ldots, x_n^0)}$$

and

$$\theta_{i,i+1} = \ln \frac{Q(x_1^0, \ldots, x_{i-1}^0, x_i^1, x_{i+1}^1 x_{i+2}^0, \ldots, x_n^0)}{Q(x_1^0, \ldots, x_n^0)} - \theta_i - \theta_{i+1}$$

then we get equality in equation (8.9) for all assignments to $X_1, \ldots, X_n$.

PART II

Inference

9 *Exact Inference: Variable Elimination*

In this chapter, we discuss the problem of performing inference in graphical models. We show that the structure of the network, both the conditional independence assertions it makes and the associated factorization of the joint distribution, is critical to our ability to perform inference effectively, allowing tractable inference even in complex networks.

Our focus in this chapter is on the most common query type: the *conditional probability query*, $P(\boldsymbol{Y} \mid \boldsymbol{E} = \boldsymbol{e})$ (see section 2.1.5). We have already seen several examples of conditional probability queries in chapter 3 and chapter 4; as we saw, such queries allow for many useful reasoning patterns, including explanation, prediction, intercausal reasoning, and many more.

By the definition of conditional probability, we know that

$$P(\boldsymbol{Y} \mid \boldsymbol{E} = \boldsymbol{e}) = \frac{P(\boldsymbol{Y}, \boldsymbol{e})}{P(\boldsymbol{e})}. \tag{9.1}$$

Each of the instantiations of the numerator is a probability expression $P(\boldsymbol{y}, \boldsymbol{e})$, which can be computed by summing out all entries in the joint that correspond to assignments consistent with $\boldsymbol{y}, \boldsymbol{e}$. More precisely, let $\boldsymbol{W} = \mathcal{X} - \boldsymbol{Y} - \boldsymbol{E}$ be the random variables that are neither query nor evidence. Then

$$P(\boldsymbol{y}, \boldsymbol{e}) = \sum_{\boldsymbol{w}} P(\boldsymbol{y}, \boldsymbol{e}, \boldsymbol{w}). \tag{9.2}$$

Because $\boldsymbol{Y}, \boldsymbol{E}, \boldsymbol{W}$ are all of the network variables, each term $P(\boldsymbol{y}, \boldsymbol{e}, \boldsymbol{w})$ in the summation is simply an entry in the joint distribution.

The probability $P(\boldsymbol{e})$ can also be computed directly by summing out the joint. However, it can also be computed as

$$P(\boldsymbol{e}) = \sum_{\boldsymbol{y}} P(\boldsymbol{y}, \boldsymbol{e}), \tag{9.3}$$

which allows us to reuse our computation for equation (9.2). If we compute both equation (9.2) and equation (9.3), we can then divide each $P(\boldsymbol{y}, \boldsymbol{e})$ by $P(\boldsymbol{e})$, to get the desired conditional probability $P(\boldsymbol{y} \mid \boldsymbol{e})$. Note that this process corresponds to taking the vector of marginal probabilities $P(\boldsymbol{y}^1, \boldsymbol{e}), \ldots, P(\boldsymbol{y}^k, \boldsymbol{e})$ (where $k = |Val(\boldsymbol{Y})|$) and *renormalizing* the entries to sum to 1.

9.1   Analysis of Complexity

In principle, a graphical model can be used to answer all of the query types described earlier. We simply generate the joint distribution and exhaustively sum out the joint (in the case of a conditional probability query), search for the most likely entry (in the case of a MAP query), or both (in the case of a marginal MAP query). However, this approach to the inference problem is not very satisfactory, since it returns us to the exponential blowup of the joint distribution that the graphical model representation was precisely designed to avoid.

☞   Unfortunately, we now show that **exponential blowup of the inference task is (almost certainly) unavoidable in the worst case: The problem of inference in graphical models is $\mathcal{NP}$-hard, and therefore it probably requires exponential time in the worst case (except in the unlikely event that $\mathcal{P} = \mathcal{NP}$). Even worse, approximate inference is also $\mathcal{NP}$-hard. Importantly, however, the story does not end with this negative result. In general, we care not about the worst case, but about the cases that we encounter in practice. As we show in the remainder of this part of the book, many real-world applications can be tackled very effectively using exact or approximate inference algorithms for graphical models.**

In our theoretical analysis, we focus our discussion on Bayesian networks. Because any Bayesian network can be encoded as a Markov network with no increase in its representation size, a hardness proof for inference in Bayesian networks immediately implies hardness of inference in Markov networks.

9.1.1   Analysis of Exact Inference

To address the question of the complexity of BN inference, we need to address the question of how we encode a Bayesian network. Without going into too much detail, we can assume that the encoding specifies the DAG structure and the CPDs. For the following results, we assume the worst-case representation of a CPD as a full table of size $|Val(\{X_i\} \cup \mathrm{Pa}_{X_i})|$.

As we discuss in appendix A.3.4, most analyses of complexity are stated in terms of decision problems. We therefore begin with a formulation of the inference problem as a decision problem, and then discuss the numerical version. One natural decision version of the conditional probability task is the problem *BN-Pr-DP*, defined as follows:

> Given a Bayesian network $\mathcal{B}$ over $\mathcal{X}$, a variable $X \in \mathcal{X}$, and a value $x \in Val(X)$, decide whether $P_{\mathcal{B}}(X = x) > 0$.

Theorem 9.1 *The decision problem BN-Pr-DP is $\mathcal{NP}$-complete.*

PROOF   It is straightforward to prove that *BN-Pr-DP* is in $\mathcal{NP}$: In the guessing phase, we guess a full assignment ξ to the network variables. In the verification phase, we check whether $X = x$ in ξ, and whether $P(\xi) > 0$. One of these guesses succeeds if and only if $P(X = x) > 0$. Computing $P(\xi)$ for a full assignment of the network variables requires only that we multiply the relevant entries in the factors, as per the chain rule for Bayesian networks, and hence can be done in linear time.

To prove $\mathcal{NP}$-hardness, we need to show that, if we can answer instances in *BN-Pr-DP*, we can use that as a subroutine to answer questions in a class of problems that is known 3-SAT   to be $\mathcal{NP}$-hard. We will use a reduction from the *3-SAT* problem defined in definition A.8.

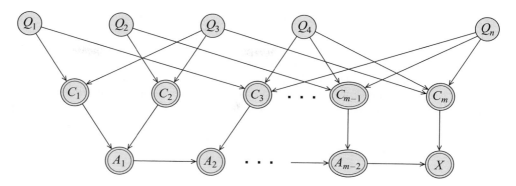

Figure 9.1 **An outline of the network structure used in the reduction of 3-SAT to Bayesian network inference.**

To show the reduction, we show the following: Given any 3-SAT formula ϕ, we can create a Bayesian network $\mathcal{B}_\phi$ with some distinguished variable X, such that ϕ is satisfiable if and only if $P_{\mathcal{B}_\phi}(X = x^1) > 0$. Thus, if we can solve the Bayesian network inference problem in polynomial time, we can also solve the 3-SAT problem in polynomial time. To enable this conclusion, our BN $\mathcal{B}_\phi$ has to be constructible in time that is polynomial in the length of the formula ϕ.

Consider a 3-SAT instance ϕ over the propositional variables $q_1, \ldots, q_n$. Figure 9.1 illustrates the structure of the network constructed in this reduction. Our Bayesian network $\mathcal{B}_\phi$ has a node Q_k for each propositional variable q_k; these variables are roots, with $P(q_k^1) = 0.5$. It also has a node C_i for each clause C_i. There is an edge from Q_k to C_i if q_k or $\neg q_k$ is one of the literals in C_i. The CPD for C_i is deterministic, and chosen such that it exactly duplicates the behavior of the clause. Note that, because C_i contains at most three variables, the CPD has at most eight distributions, and at most sixteen entries.

We want to introduce a variable X that has the value 1 if and only if all the C_i's have the value 1. We can achieve this requirement by having $C_1, \ldots, C_m$ be parents of X. This construction, however, has the property that $P(X \mid C_1, \ldots, C_m)$ is exponentially large when written as a table. To avoid this difficulty, we introduce intermediate "AND" gates $A_1, \ldots, A_{m-2}$, so that A_1 is the "AND" of C_1 and C_2, A_2 is the "AND" of A_1 and C_3, and so on. The last variable X is the "AND" of A_{m-2} and C_m. This construction achieves the desired effect: X has value 1 if and only if all the clauses are satisfied. Furthermore, in this construction, all variables have at most three (binary-valued) parents, so that the size of $\mathcal{B}_\phi$ is polynomial in the size of ϕ.

It follows that $P_{\mathcal{B}_\phi}(x^1 \mid q_1, \ldots, q_n) = 1$ if and only if $q_1, \ldots, q_n$ is a satisfying assignment for ϕ. Because the prior probability of each possible assignment is $1/2^n$, we get that the overall probability $P_{\mathcal{B}_\phi}(x^1)$ is the number of satisfying assignments to ϕ, divided by 2^n. We can therefore test whether ϕ has a satisfying assignment simply by checking whether $P(x^1) > 0$. ∎

This analysis shows that the decision problem associated with Bayesian network inference is $\mathcal{NP}$-complete. However, the problem is originally a numerical problem. Precisely the same construction allows us to provide an analysis for the original problem formulation. We define the problem *BN-Pr* as follows:

Given: a Bayesian network $\mathcal{B}$ over $\mathcal{X}$, a variable $X \in \mathcal{X}$, and a value $x \in Val(X)$, compute $P_{\mathcal{B}}(X = x)$.

Our task here is to compute the total probability of network instantiations that are consistent with $X = x$. Or, in other words, to do a weighted count of instantiations, with the weight being the probability. An appropriate complexity class for counting problems is $\#\mathcal{P}$: Whereas $\mathcal{NP}$ represents problems of deciding "are there any solutions that satisfy certain requirements," $\#\mathcal{P}$ represents problems that ask "how many solutions are there that satisfy certain requirements." It is not surprising that we can relate the complexity of the BN inference problem to the counting class $\#\mathcal{P}$:

Theorem 9.2

The problem BN-Pr is $\#\mathcal{P}$-complete.

We leave the proof as an exercise (exercise 9.1).

9.1.2   Analysis of Approximate Inference

Upon noting the hardness of exact inference, a natural question is whether we can circumvent the difficulties by compromising, to some extent, on the accuracies of our answers. Indeed, in many applications we can tolerate some imprecision in the final probabilities: it is often unlikely that a change in probability from 0.87 to 0.92 will change our course of action. Thus, we now explore the computational complexity of approximate inference.

To analyze the approximate inference task formally, we must first define a metric for evaluating the quality of our approximation. We can consider two perspectives on this issue, depending on how we choose to define our query. Consider first our previous formulation of the conditional probability query task, where our goal is to compute the probability $P(\boldsymbol{Y} \mid \boldsymbol{e})$ for some set of variables $\boldsymbol{Y}$ and evidence $\boldsymbol{e}$. The result of this type of query is a probability distribution over $\boldsymbol{Y}$. Given an approximate answer to this query, we can evaluate its quality using any of the distance metrics we define for probability distributions in appendix A.1.3.3.

There is, however, another way of looking at this task, one that is somewhat simpler and will be very useful for analyzing its complexity. Consider a *specific* query $P(\boldsymbol{y} \mid \boldsymbol{e})$, where we are focusing on one particular assignment $\boldsymbol{y}$. The approximate answer to this query is a number ρ, whose accuracy we wish to evaluate relative to the correct probability. One way of evaluating the accuracy of an estimate is as simple as the difference between the approximate answer and the right one.

Definition 9.1

absolute error

An estimate ρ has absolute error ϵ *for* $P(\boldsymbol{y} \mid \boldsymbol{e})$ *if:*

$$|P(\boldsymbol{y} \mid \boldsymbol{e}) - \rho| \leq \epsilon.$$ ∎

This definition, although plausible, is somewhat weak. Consider, for example, a situation in which we are trying to compute the probability of a really rare disease, one whose true probability is, say, 0.00001. In this case, an absolute error of 0.0001 is unacceptable, even though such an error may be an excellent approximation for an event whose probability is 0.3. A stronger definition of accuracy takes into consideration the value of the probability that we are trying to estimate:

Definition 9.2

relative error

An estimate ρ has relative error ϵ for $P(\boldsymbol{y} \mid \boldsymbol{e})$ if:

$$\frac{\rho}{1 + \epsilon} \leq P(\boldsymbol{y} \mid \boldsymbol{e}) \leq \rho(1 + \epsilon).$$

∎

Note that, unlike absolute error, relative error makes sense even for $\epsilon > 1$. For example, $\epsilon = 4$ means that $P(\boldsymbol{y} \mid \boldsymbol{e})$ is at least 20 percent of ρ and at most 600 percent of ρ. For probabilities, where low values are often very important, relative error appears much more relevant than absolute error.

With these definitions, we can turn to answering the question of whether approximate inference is actually an easier problem. A priori, it seems as if the extra slack provided by the approximation might help. Unfortunately, this hope turns out to be unfounded. As we now show, approximate inference in Bayesian networks is also $\mathcal{NP}$-hard.

This result is straightforward for the case of relative error.

Theorem 9.3

The following problem is $\mathcal{NP}$-hard:

> Given a Bayesian network $\mathcal{B}$ over $\mathcal{X}$, a variable $X \in \mathcal{X}$, and a value $x \in \mathit{Val}(X)$, find a number ρ that has relative error ϵ for $P_{\mathcal{B}}(X = x)$.

PROOF The proof is obvious based on the original $\mathcal{NP}$-hardness proof for exact Bayesian network inference (theorem 9.1). There, we proved that it is $\mathcal{NP}$-hard to decide whether $P_{\mathcal{B}}(x^1) > 0$. Now, assume that we have an algorithm that returns an estimate ρ to the same $P_{\mathcal{B}}(x^1)$, which is guaranteed to have relative error ϵ for some $\epsilon > 0$. Then $\rho > 0$ if and only if $P_{\mathcal{B}}(x^1) > 0$. Thus, achieving this relative error is as $\mathcal{NP}$-hard as the original problem. ∎

We can generalize this result to make $\epsilon(n)$ a function that grows with the input size n. Thus, for example, we can define $\epsilon(n) = 2^{2^n}$ and the theorem still holds. Thus, in a sense, this result is not so interesting as a statement about hardness of approximation. Rather, it tells us that relative error is too strong a notion of approximation to use in this context.

What about absolute error? As we will see in section 12.1.2, the problem of just approximating $P(X = x)$ up to some fixed absolute error ϵ has a randomized polynomial time algorithm. Therefore, the problem cannot be $\mathcal{NP}$-hard unless $\mathcal{NP} = \mathcal{RP}$. This result is an improvement on the exact case, where even the task of computing $P(X = x)$ is $\mathcal{NP}$-hard.

Unfortunately, the good news is very limited in scope, in that it disappears once we introduce evidence. Specifically, it is $\mathcal{NP}$-hard to find an absolute approximation to $P(x \mid \boldsymbol{e})$ for any $\epsilon < 1/2$.

Theorem 9.4

The following problem is $\mathcal{NP}$-hard for any $\epsilon \in (0, 1/2)$:

> Given a Bayesian network $\mathcal{B}$ over $\mathcal{X}$, a variable $X \in \mathcal{X}$, a value $x \in \mathit{Val}(X)$, and an observation $\boldsymbol{E} = \boldsymbol{e}$ for $\boldsymbol{E} \subset \mathcal{X}$ and $\boldsymbol{e} \in \mathit{Val}(\boldsymbol{E})$, find a number ρ that has absolute error ϵ for $P_{\mathcal{B}}(X = x \mid \boldsymbol{e})$.

PROOF The proof uses the same construction that we used before. Consider a formula ϕ, and consider the analogous BN $\mathcal{B}$, as described in theorem 9.1. Recall that our BN had a variable Q_i for each propositional variable q_i in our Boolean formula, a bunch of other intermediate variables, and then a variable X whose value, given any assignment of values q_1^1, q_1^0 to the Q_i's, was the associated truth value of the formula. We now show that, given such an approximation algorithm, we can decide whether the formula is satisfiable. We begin by computing $P(Q_1 \mid x^1)$. We pick the value v_1 for Q_1 that is most likely given x^1, and we instantiate it to this value. That is, we generate a network $\mathcal{B}_2$ that does not contain Q_1, and that represents the distribution $\mathcal{B}$ conditioned on $Q_1 = v_1$. We repeat this process for $Q_2, \ldots, Q_n$. This results in some assignment $v_1, \ldots, v_n$ to the Q_i's. We now prove that this is a satisfying assignment if and only if the original formula ϕ was satisfiable.

We begin with the easy case. If ϕ is not satisfiable, then $v_1, \ldots, v_n$ can hardly be a satisfying assignment for it. Now, assume that ϕ is satisfiable. We show that it also has a satisfying assignment with $Q_1 = v_1$. If ϕ is satisfiable with both $Q_1 = q_1^1$ and $Q_1 = q_1^0$, then this is obvious. Assume, however, that ϕ is satisfiable, but not when $Q_1 = v$. Then necessarily, we will have that $P(Q_1 = v \mid x^1)$ is 0, and the probability of the complementary event is 1. If we have an approximation ρ whose error is guaranteed to be $< 1/2$, then choosing the v that maximizes this probability is guaranteed to pick the v whose probability is 1. Thus, in either case the formula has a satisfying assignment where $Q_1 = v$.

We can continue in this fashion, proving by induction on k that ϕ has a satisfying assignment with $Q_1 = v_1, \ldots, Q_k = v_k$. In the case where ϕ is satisfiable, this process will terminate with a satisfying assignment. In the case where ϕ is not, it clearly will not terminate with a satisfying assignment. We can determine which is the case simply by checking whether the resulting assignment satisfies ϕ. This gives us a polynomial time process for deciding satisfiability. ∎

Because $\epsilon = 1/2$ corresponds to random guessing, this result is quite discouraging. It tells us that, in the case where we have evidence, approximate inference is no easier than exact inference, in the worst case.

9.2   Variable Elimination: The Basic Ideas

We begin our discussion of inference by discussing the principles underlying exact inference in graphical models. As we show, the same graphical structure that allows a compact representation of complex distributions also help support inference. In particular, we can use dynamic programming techniques (as discussed in appendix A.3.3) to perform inference even for certain large and complex networks in a very reasonable time. We now provide the intuition underlying these algorithms, an intuition that is presented more formally in the remainder of this chapter.

We begin by considering the inference task in a very simple network $A \to B \to C \to D$. We first provide a phased computation, which uses results from the previous phase for the computation in the next phase. We then reformulate this process in terms of a global computation on the joint distribution.

Assume that our first goal is to compute the probability $P(B)$, that is, the distribution over values b of B. Basic probabilistic reasoning (with no assumptions) tells us that

$$P(B) = \sum_a P(a) P(B \mid a). \tag{9.4}$$

Fortunately, we have all the required numbers in our Bayesian network representation: each number $P(a)$ is in the CPD for A, and each number $P(b \mid a)$ is in the CPD for B. Note that if A has k values and B has m values, the number of basic arithmetic operations required is $O(k \times m)$: to compute $P(b)$, we must multiply $P(b \mid a)$ with $P(a)$ for each of the k values of A, and then add them up, that is, k multiplications and $k - 1$ additions; this process must be repeated for each of the m values b.

Now, assume we want to compute $P(C)$. Using the same analysis, we have that

$$P(C) = \sum_b P(b)P(C \mid b). \tag{9.5}$$

Again, the conditional probabilities $P(c \mid b)$ are known: they constitute the CPD for C. The probability of B is not specified as part of the network parameters, but equation (9.4) shows us how it can be computed. Thus, we can compute $P(C)$. We can continue the process in an analogous way, in order to compute $P(D)$.

Note that the structure of the network, and its effect on the parameterization of the CPDs, is critical for our ability to perform this computation as described. Specifically, assume that A had been a parent of C. In this case, the CPD for C would have included A, and our computation of $P(B)$ would not have sufficed for equation (9.5).

Also note that this algorithm does not compute single values, but rather sets of values at a time. In particular equation (9.4) computes an entire distribution over all of the possible values of B. All of these are then used in equation (9.5) to compute $P(C)$. This property turns out to be critical for the performance of the general algorithm.

Let us analyze the complexity of this process on a general chain. Assume that we have a chain with n variables $X_1 \to \ldots \to X_n$, where each variable in the chain has k values. As described, the algorithm would compute $P(X_{i+1})$ from $P(X_i)$, for $i = 1, \ldots, n-1$. Each such step would consist of the following computation:

$$P(X_{i+1}) = \sum_{x_i} P(X_{i+1} \mid x_i)P(x_i),$$

where $P(X_i)$ is computed in the previous step. The cost of each such step is $O(k^2)$: The distribution over X_i has k values, and the CPD $P(X_{i+1} \mid X_i)$ has k^2 values; we need to multiply $P(x_i)$, for each value x_i, with each CPD entry $P(x_{i+1} \mid x_i)$ (k^2 multiplications), and then, for each value x_{i+1}, sum up the corresponding entries ($k \times (k - 1)$ additions). We need to perform this process for every variable $X_2, \ldots, X_n$; hence, the total cost is $O(nk^2)$.

By comparison, consider the process of generating the entire joint and summing it out, which requires that we generate k^n probabilities for the different events $x_1, \ldots, x_n$. Hence, we have at least one example where, despite the exponential size of the joint distribution, we can do inference in linear time.

Using this process, we have managed to do inference over the joint distribution without ever generating it explicitly. What is the basic insight that allows us to avoid the exhaustive enumeration? Let us reexamine this process in terms of the joint $P(A, B, C, D)$. By the chain rule for Bayesian networks, the joint decomposes as

$$P(A)P(B \mid A)P(C \mid B)P(D \mid C)$$

To compute $P(D)$, we need to sum together all of the entries where $D = d^1$, and to (separately) sum together all of the entries where $D = d^2$. The exact computation that needs to be

$$
\begin{aligned}
& P(a^1) & P(b^1 \mid a^1) & \quad P(c^1 \mid b^1) & \quad P(d^1 \mid c^1) \\
+\ & P(a^2) & P(b^1 \mid a^2) & \quad P(c^1 \mid b^1) & \quad P(d^1 \mid c^1) \\
+\ & P(a^1) & P(b^2 \mid a^1) & \quad P(c^1 \mid b^2) & \quad P(d^1 \mid c^1) \\
+\ & P(a^2) & P(b^2 \mid a^2) & \quad P(c^1 \mid b^2) & \quad P(d^1 \mid c^1) \\
+\ & P(a^1) & P(b^1 \mid a^1) & \quad P(c^2 \mid b^1) & \quad P(d^1 \mid c^2) \\
+\ & P(a^2) & P(b^1 \mid a^2) & \quad P(c^2 \mid b^1) & \quad P(d^1 \mid c^2) \\
+\ & P(a^1) & P(b^2 \mid a^1) & \quad P(c^2 \mid b^2) & \quad P(d^1 \mid c^2) \\
+\ & P(a^2) & P(b^2 \mid a^2) & \quad P(c^2 \mid b^2) & \quad P(d^1 \mid c^2) \\
\\
& P(a^1) & P(b^1 \mid a^1) & \quad P(c^1 \mid b^1) & \quad P(d^2 \mid c^1) \\
+\ & P(a^2) & P(b^1 \mid a^2) & \quad P(c^1 \mid b^1) & \quad P(d^2 \mid c^1) \\
+\ & P(a^1) & P(b^2 \mid a^1) & \quad P(c^1 \mid b^2) & \quad P(d^2 \mid c^1) \\
+\ & P(a^2) & P(b^2 \mid a^2) & \quad P(c^1 \mid b^2) & \quad P(d^2 \mid c^1) \\
+\ & P(a^1) & P(b^1 \mid a^1) & \quad P(c^2 \mid b^1) & \quad P(d^2 \mid c^2) \\
+\ & P(a^2) & P(b^1 \mid a^2) & \quad P(c^2 \mid b^1) & \quad P(d^2 \mid c^2) \\
+\ & P(a^1) & P(b^2 \mid a^1) & \quad P(c^2 \mid b^2) & \quad P(d^2 \mid c^2) \\
+\ & P(a^2) & P(b^2 \mid a^2) & \quad P(c^2 \mid b^2) & \quad P(d^2 \mid c^2)
\end{aligned}
$$

Figure 9.2   **Computing $P(D)$ by summing over the joint distribution for a chain** $A \to B \to C \to D$; all of the variables are binary valued.

performed, for binary-valued variables A, B, C, D, is shown in figure 9.2.[1]

Examining this summation, we see that it has a lot of structure. For example, the third and fourth terms in the first two entries are both $P(c^1 \mid b^1)P(d^1 \mid c^1)$. We can therefore modify the computation to first compute

$$
P(a^1)P(b^1 \mid a^1) + P(a^2)P(b^1 \mid a^2)
$$

and only then multiply by the common term. The same structure is repeated throughout the table. If we perform the same transformation, we get a new expression, as shown in figure 9.3.

We now observe that certain terms are repeated several times in this expression. Specifically, $P(a^1)P(b^1 \mid a^1) + P(a^2)P(b^1 \mid a^2)$ and $P(a^1)P(b^2 \mid a^1) + P(a^2)P(b^2 \mid a^2)$ are each repeated four times. Thus, it seems clear that we can gain significant computational savings by computing them once and then storing them. There are two such expressions, one for each value of B. Thus, we define a function $\tau_1 : Val(B) \mapsto \mathbb{R}$, where $\tau_1(b^1)$ is the first of these two expressions, and $\tau_1(b^2)$ is the second. Note that $\tau_1(B)$ corresponds exactly to $P(B)$.

The resulting expression, assuming $\tau_1(B)$ has been computed, is shown in figure 9.4. Examining this new expression, we see that we once again can reverse the order of a sum and a product, resulting in the expression of figure 9.5. And, once again, we notice some shared expressions, that are better computed once and used multiple times. We define $\tau_2 : Val(C) \mapsto \mathbb{R}$.

$$
\begin{aligned}
\tau_2(c^1) &= \tau_1(b^1)P(c^1 \mid b^1) + \tau_1(b^2)P(c^1 \mid b^2) \\
\tau_2(c^2) &= \tau_1(b^1)P(c^2 \mid b^1) + \tau_1(b^2)P(c^2 \mid b^2)
\end{aligned}
$$

---

1. When D is binary-valued, we can get away with doing only the first of these computations. However, this trick does not carry over to the case of variables with more than two values or to the case where we have evidence. Therefore, our example will show the computation in its generality.

$$
\begin{array}{llll}
& (P(a^1)P(b^1 \mid a^1) + P(a^2)P(b^1 \mid a^2)) & P(c^1 \mid b^1) & P(d^1 \mid c^1) \\
+ & (P(a^1)P(b^2 \mid a^1) + P(a^2)P(b^2 \mid a^2)) & P(c^1 \mid b^2) & P(d^1 \mid c^1) \\
+ & (P(a^1)P(b^1 \mid a^1) + P(a^2)P(b^1 \mid a^2)) & P(c^2 \mid b^1) & P(d^1 \mid c^2) \\
+ & (P(a^1)P(b^2 \mid a^1) + P(a^2)P(b^2 \mid a^2)) & P(c^2 \mid b^2) & P(d^1 \mid c^2)
\end{array}
$$

$$
\begin{array}{llll}
& (P(a^1)P(b^1 \mid a^1) + P(a^2)P(b^1 \mid a^2)) & P(c^1 \mid b^1) & P(d^2 \mid c^1) \\
+ & (P(a^1)P(b^2 \mid a^1) + P(a^2)P(b^2 \mid a^2)) & P(c^1 \mid b^2) & P(d^2 \mid c^1) \\
+ & (P(a^1)P(b^1 \mid a^1) + P(a^2)P(b^1 \mid a^2)) & P(c^2 \mid b^1) & P(d^2 \mid c^2) \\
+ & (P(a^1)P(b^2 \mid a^1) + P(a^2)P(b^2 \mid a^2)) & P(c^2 \mid b^2) & P(d^2 \mid c^2)
\end{array}
$$

Figure 9.3 **The first transformation on the sum of figure 9.2**

$$
\begin{array}{llll}
& \tau_1(b^1) & P(c^1 \mid b^1) & P(d^1 \mid c^1) \\
+ & \tau_1(b^2) & P(c^1 \mid b^2) & P(d^1 \mid c^1) \\
+ & \tau_1(b^1) & P(c^2 \mid b^1) & P(d^1 \mid c^2) \\
+ & \tau_1(b^2) & P(c^2 \mid b^2) & P(d^1 \mid c^2)
\end{array}
$$

$$
\begin{array}{llll}
& \tau_1(b^1) & P(c^1 \mid b^1) & P(d^2 \mid c^1) \\
+ & \tau_1(b^2) & P(c^1 \mid b^2) & P(d^2 \mid c^1) \\
+ & \tau_1(b^1) & P(c^2 \mid b^1) & P(d^2 \mid c^2) \\
+ & \tau_1(b^2) & P(c^2 \mid b^2) & P(d^2 \mid c^2)
\end{array}
$$

Figure 9.4 **The second transformation on the sum of figure 9.2**

$$
\begin{array}{lll}
& (\tau_1(b^1)P(c^1 \mid b^1) + \tau_1(b^2)P(c^1 \mid b^2)) & P(d^1 \mid c^1) \\
+ & (\tau_1(b^1)P(c^2 \mid b^1) + \tau_1(b^2)P(c^2 \mid b^2)) & P(d^1 \mid c^2)
\end{array}
$$

$$
\begin{array}{lll}
& (\tau_1(b^1)P(c^1 \mid b^1) + \tau_1(b^2)P(c^1 \mid b^2)) & P(d^2 \mid c^1) \\
+ & (\tau_1(b^1)P(c^2 \mid b^1) + \tau_1(b^2)P(c^2 \mid b^2)) & P(d^2 \mid c^2)
\end{array}
$$

Figure 9.5 **The third transformation on the sum of figure 9.2**

$$
\begin{array}{lll}
& \tau_2(c^1) & P(d^1 \mid c^1) \\
+ & \tau_2(c^2) & P(d^1 \mid c^2)
\end{array}
$$

$$
\begin{array}{lll}
& \tau_2(c^1) & P(d^2 \mid c^1) \\
+ & \tau_2(c^2) & P(d^2 \mid c^2)
\end{array}
$$

Figure 9.6 **The fourth transformation on the sum of figure 9.2**

The final expression is shown in figure 9.6.

Summarizing, we begin by computing $\tau_1(B)$, which requires four multiplications and two additions. Using it, we can compute $\tau_2(C)$, which also requires four multiplications and two additions. Finally, we can compute $P(D)$, again, at the same cost. The total number of operations is therefore 18. By comparison, generating the joint distribution requires $16 \cdot 3 = 48$

multiplications (three for each of the 16 entries in the joint), and 14 additions (7 for each of $P(d^1)$ and $P(d^2)$).

Written somewhat more compactly, the transformation we have performed takes the following steps: We want to compute

$$P(D) = \sum_C \sum_B \sum_A P(A)P(B \mid A)P(C \mid B)P(D \mid C).$$

We push in the first summation, resulting in

$$\sum_C P(D \mid C) \sum_B P(C \mid B) \sum_A P(A)P(B \mid A).$$

We compute the product $\psi_1(A, B) = P(A)P(B \mid A)$ and then sum out A to obtain the function $\tau_1(B) = \sum_A \psi_1(A, B)$. Specifically, for each value b, we compute $\tau_1(b) = \sum_A \psi_1(A, b) = \sum_A P(A)P(b \mid A)$. We then continue by computing:

$$\psi_2(B, C) = \tau_1(B)P(C \mid B)$$
$$\tau_2(C) = \sum_B \psi_2(B, C).$$

This computation results in a new vector $\tau_2(C)$, which we then proceed to use in the final phase of computing $P(D)$.

dynamic
programming   This procedure is performing *dynamic programming* (see appendix A.3.3); doing this summation the naive way would have us compute every $P(b) = \sum_A P(A)P(b \mid A)$ many times, once for every value of C and D. In general, in a chain of length n, this internal summation would be computed exponentially many times. Dynamic programming "inverts" the order of computation — performing it inside out instead of outside in. Specifically, we perform the innermost summation first, computing once and for all the values in $\tau_1(B)$; that allows us to compute $\tau_2(C)$ once and for all, and so on.

   To summarize, the two ideas that help us address the exponential blowup of the joint distribution are:

- **Because of the structure of the Bayesian network, some subexpressions in the joint depend only on a small number of variables.**

- **By computing these expressions once and caching the results, we can avoid generating them exponentially many times.**

9.3   Variable Elimination

factor   To formalize the algorithm demonstrated in the previous section, we need to introduce some basic concepts. In chapter 4, we introduced the notion of a *factor* ϕ over a scope *Scope*$[\phi] = \boldsymbol{X}$, which is a function $\phi : Val(\boldsymbol{X}) \mapsto \mathbb{R}$. The main steps in the algorithm described here can be viewed as a manipulation of factors. Importantly, by using the factor-based view, we can define the algorithm in a general form that applies equally to Bayesian networks and Markov networks.

Applying equation (9.6), we can now conclude:

$$
\begin{aligned}
P(D) &= \sum_C \sum_B \sum_A \phi_A \cdot \phi_B \cdot \phi_C \cdot \phi_D \\
&= \sum_C \sum_B \phi_C \cdot \phi_D \cdot \left(\sum_A \phi_A \cdot \phi_B \right) \\
&= \sum_C \phi_D \cdot \left(\sum_B \phi_C \cdot \left(\sum_A \phi_A \cdot \phi_B \right) \right),
\end{aligned}
$$

where the different transformations are justified by the limited scope of the CPD factors; for example, the second equality is justified by the fact that the scope of ϕ_C and ϕ_D does not contain A. In general, any marginal probability computation involves taking the product of all the CPDs, and doing a summation on all the variables except the query variables. We can do these steps in any order we want, as long as we only do a summation on a variable X *after* multiplying in all of the factors that involve X.

In general, we can view the task at hand as that of computing the value of an expression of the form:

$$
\sum_{\boldsymbol{Z}} \prod_{\phi \in \Phi} \phi.
$$

sum-product

We call this task the *sum-product* inference task. The key insight that allows the effective computation of this expression is the fact that the scope of the factors is limited, allowing us to "push in" some of the summations, performing them over the product of only a subset of

variable elimination

factors. One simple instantiation of this algorithm is a procedure called *sum-product variable elimination* (VE), shown in algorithm 9.1. The basic idea in the algorithm is that we sum out variables one at a time. When we sum out any variable, we multiply all the factors that mention that variable, generating a product factor. Now, we sum out the variable from this combined factor, generating a new factor that we enter into our set of factors to be dealt with.

Based on equation (9.6), the following result follows easily:

Theorem 9.5

Let $\boldsymbol{X}$ be some set of variables, and let Φ be a set of factors such that for each $\phi \in \Phi$, $Scope[\phi] \subseteq \boldsymbol{X}$. Let $\boldsymbol{Y} \subset \boldsymbol{X}$ be a set of query variables, and let $\boldsymbol{Z} = \boldsymbol{X} - \boldsymbol{Y}$. Then for any ordering $\prec$ over $\boldsymbol{Z}$, Sum-Product-VE(Φ, $\boldsymbol{Z}$, $\prec$) returns a factor $\phi^*(\boldsymbol{Y})$ such that

$$
\phi^*(\boldsymbol{Y}) = \sum_{\boldsymbol{Z}} \prod_{\phi \in \Phi} \phi.
$$

We can apply this algorithm to the task of computing the probability distribution $P_\mathcal{B}(\boldsymbol{Y})$ for a Bayesian network $\mathcal{B}$. We simply instantiate Φ to consist of all of the CPDs:

$$
\Phi = \{\phi_{X_i}\}_{i=1}^n
$$

where $\phi_{X_i} = P(X_i \mid \mathrm{Pa}_{X_i})$. We then apply the variable elimination algorithm to the set $\{Z_1, \ldots, Z_m\} = \mathcal{X} - \boldsymbol{Y}$ (that is, we eliminate all the nonquery variables).

We can also apply precisely the same algorithm to the task of computing conditional probabilities in a Markov network. We simply initialize the factors to be the clique potentials and

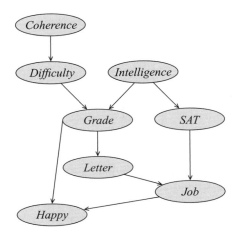

Figure 9.8   **The** Extended-Student **Bayesian network**

run the elimination algorithm. As for Bayesian networks, we then apply the variable elimination algorithm to the set $\boldsymbol{Z} = \mathcal{X} - \boldsymbol{Y}$. The procedure returns an *unnormalized* factor over the query variables $\boldsymbol{Y}$. The distribution over $\boldsymbol{Y}$ can be obtained by normalizing the factor; the partition function is simply the normalizing constant.

Example 9.1   *Let us demonstrate the procedure on a nontrivial example. Consider the network demonstrated in figure 9.8, which is an extension of our* Student *network. The chain rule for this network asserts that*

$$
\begin{aligned}
P(C, D, I, G, S, L, J, H) \;&=\; P(C)P(D \mid C)P(I)P(G \mid I, D)P(S \mid I)\\
&\qquad P(L \mid G)P(J \mid L, S)P(H \mid G, J)\\
&=\; \phi_C(C)\phi_D(D, C)\phi_I(I)\phi_G(G, I, D)\phi_S(S, I)\\
&\qquad \phi_L(L, G)\phi_J(J, L, S)\phi_H(H, G, J).
\end{aligned}
$$

We will now apply the VE algorithm to compute $P(J)$. We will use the elimination ordering: C, D, I, H, G, S, L:

1. *Eliminating C: We compute the factors*

$$
\begin{aligned}
\psi_1(C, D) \;&=\; \phi_C(C) \cdot \phi_D(D, C)\\
\tau_1(D) \;&=\; \sum_C \psi_1.
\end{aligned}
$$

2. *Eliminating D: Note that we have already eliminated one of the original factors that involve D — $\phi_D(D, C) = P(D \mid C)$. On the other hand, we introduced the factor $\tau_1(D)$ that involves*

D. Hence, we now compute:

$$\psi_2(G, I, D) = \phi_G(G, I, D) \cdot \tau_1(D)$$
$$\tau_2(G, I) = \sum_D \psi_2(G, I, D).$$

3. *Eliminating I: We compute the factors*

$$\psi_3(G, I, S) = \phi_I(I) \cdot \phi_S(S, I) \cdot \tau_2(G, I)$$
$$\tau_3(G, S) = \sum_I \psi_3(G, I, S).$$

4. *Eliminating H: We compute the factors*

$$\psi_4(G, J, H) = \phi_H(H, G, J)$$
$$\tau_4(G, J) = \sum_H \psi_4(G, J, H).$$

Note that $\tau_4 \equiv 1$ (all of its entries are exactly 1): we are simply computing $\sum_H P(H \mid G, J)$, which is a probability distribution for every G, J, and hence sums to 1. A naive execution of this algorithm will end up generating this factor, which has no value. Generating it has no impact on the final answer, but it does complicate the algorithm. In particular, the existence of this factor complicates our computation in the next step.

5. *Eliminating G: We compute the factors*

$$\psi_5(G, J, L, S) = \tau_4(G, J) \cdot \tau_3(G, S) \cdot \phi_L(L, G)$$
$$\tau_5(J, L, S) = \sum_G \psi_5(G, J, L, S).$$

Note that, without the factor $\tau_4(G, J)$, the results of this step would not have involved J.

6. *Eliminating S: We compute the factors*

$$\psi_6(J, L, S) = \tau_5(J, L, S) \cdot \phi_J(J, L, S)$$
$$\tau_6(J, L) = \sum_S \psi_6(J, L, S).$$

7. *Eliminating L: We compute the factors*

$$\psi_7(J, L) = \tau_6(J, L)$$
$$\tau_7(J) = \sum_L \psi_7(J, L).$$

We summarize these steps in table 9.1.

Note that we can use any elimination ordering. For example, consider eliminating variables in the order G, I, S, L, H, C, D. We would then get the behavior of table 9.2. The result, as before, is precisely $P(J)$. However, note that this elimination ordering introduces factors with much larger scope. We return to this point later on. ∎

Step	Variable eliminated	Factors used	Variables involved	New factor
1	C	$\phi_C(C), \phi_D(D,C)$	C, D	$\tau_1(D)$
2	D	$\phi_G(G,I,D), \tau_1(D)$	G, I, D	$\tau_2(G,I)$
3	I	$\phi_I(I), \phi_S(S,I), \tau_2(G,I)$	G, S, I	$\tau_3(G,S)$
4	H	$\phi_H(H,G,J)$	H, G, J	$\tau_4(G,J)$
5	G	$\tau_4(G,J), \tau_3(G,S), \phi_L(L,G)$	G, J, L, S	$\tau_5(J,L,S)$
6	S	$\tau_5(J,L,S), \phi_J(J,L,S)$	J, L, S	$\tau_6(J,L)$
7	L	$\tau_6(J,L)$	J, L	$\tau_7(J)$

Table 9.1   A run of variable elimination for the query $P(J)$

Step	Variable eliminated	Factors used	Variables involved	New factor
1	G	$\phi_G(G,I,D), \phi_L(L,G), \phi_H(H,G,J)$	G, I, D, L, J, H	$\tau_1(I,D,L,J,H)$
2	I	$\phi_I(I), \phi_S(S,I), \tau_1(I,D,L,S,J,H)$	S, I, D, L, J, H	$\tau_2(D,L,S,J,H)$
3	S	$\phi_J(J,L,S), \tau_2(D,L,S,J,H)$	D, L, S, J, H	$\tau_3(D,L,J,H)$
4	L	$\tau_3(D,L,J,H)$	D, L, J, H	$\tau_4(D,J,H)$
5	H	$\tau_4(D,J,H)$	D, J, H	$\tau_5(D,J)$
6	C	$\phi_C(C), \phi_D(D,C)$	D, J, C	$\tau_6(D)$
7	D	$\tau_5(D,J), \tau_6(D)$	D, J	$\tau_7(J)$

Table 9.2   A different run of variable elimination for the query $P(J)$

9.3.1.3   Semantics of Factors

It is interesting to consider the semantics of the intermediate factors generated as part of this computation. In many of the examples we have given, they correspond to marginal or conditional probabilities in the network. However, although these factors often correspond to such probabilities, this is not always the case. Consider, for example, the network of figure 9.9a. The result of eliminating the variable X is a factor

$$\tau(A, B, C) = \sum_X P(X) \cdot P(A \mid X) \cdot P(C \mid B, X).$$

This factor does not correspond to any probability or conditional probability in this network. To understand why, consider the various options for the meaning of this factor. Clearly, it cannot be a conditional distribution where B is on the left hand side of the conditioning bar (for example, $P(A, B, C)$), as $P(B \mid A)$ has not yet been multiplied in. The most obvious candidate is $P(A, C \mid B)$. However, this conjecture is also false. The probability $P(A \mid B)$ relies heavily on the properties of the CPD $P(B \mid A)$; for example, if B is deterministically equal to A, $P(A \mid B)$ has a very different form than if B depends only very weakly on A. Since the CPD $P(B \mid A)$ was not taken into consideration when computing $\tau(A, B, C)$, it cannot represent the conditional probability $P(A, C \mid B)$. In general, we can verify that this factor

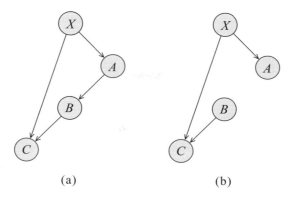

Figure 9.9 **Understanding intermediate factors in variable elimination** as conditional probabilities: (a) A Bayesian network where elimination does not lead to factors that have an interpretation as conditional probabilities. (b) A different Bayesian network where the resulting factor does correspond to a conditional probability.

does not correspond to any conditional probability expression in this network.

It is interesting to note, however, that the resulting factor does, in fact, correspond to a conditional probability $P(A, C \mid B)$, but *in a different network*: the one shown in figure 9.9b, where all CPDs except for B are the same. In fact, this phenomenon is a general one (see exercise 9.2).

9.3.2 Dealing with Evidence

It remains only to consider how we would introduce evidence. For example, assume we observe the value i^1 (the student is intelligent) and h^0 (the student is unhappy). Our goal is to compute $P(J \mid i^1, h^0)$. First, we reduce this problem to computing the unnormalized distribution $P(J, i^1, h^0)$. From this intermediate result, we can compute the conditional probability as in equation (9.1), by renormalizing by the probability of the evidence $P(i^1, h^0)$.

How do we compute $P(J, i^1, h^0)$? The key observation is proposition 4.7, which shows us how to view, as a Gibbs distribution, an unnormalized measure derived from introducing evidence into a Bayesian network. Thus, we can view this computation as summing out all of the entries in the *reduced factor*: $P[i^1 h^0]$ whose scope is $\{C, D, G, L, S, J\}$. This factor is no longer normalized, but it is still a valid factor.

factor reduction

Based on this observation, we can now apply precisely the same sum-product variable elimination algorithm to the task of computing $P(\boldsymbol{Y}, \boldsymbol{e})$. We simply apply the algorithm to the set of factors in the network, reduced by $\boldsymbol{E} = \boldsymbol{e}$, and eliminate the variables in $\mathcal{X} - \boldsymbol{Y} - \boldsymbol{E}$. The returned factor $\phi^*(\boldsymbol{Y})$ is precisely $P(\boldsymbol{Y}, \boldsymbol{e})$. To obtain $P(\boldsymbol{Y} \mid \boldsymbol{e})$ we simply renormalize $\phi^*(\boldsymbol{Y})$ by multiplying it by $\frac{1}{\alpha}$ to obtain a legal distribution, where α is the sum over the entries in our unnormalized distribution, which represents the probability of the evidence. To summarize, the algorithm for computing conditional probabilities in a Bayesian or Markov network is shown in algorithm 9.2.

We demonstrate this process on the example of computing $P(J, i^1, h^0)$. We use the same

---

Algorithm 9.2 Using Sum-Product-VE **for computing conditional probabilities**

    Procedure Cond-Prob-VE (
        $\mathcal{K}$,   // A network over $\mathcal{X}$
        $\boldsymbol{Y}$,   // Set of query variables
        $\boldsymbol{E} = \boldsymbol{e}$ // Evidence
)
1      $\Phi \leftarrow$ Factors parameterizing $\mathcal{K}$
2   Replace each $\phi \in \Phi$ by $\phi[\boldsymbol{E} = \boldsymbol{e}]$
3   Select an elimination ordering $\prec$
4      $\boldsymbol{Z} \leftarrow = \mathcal{X} - \boldsymbol{Y} - \boldsymbol{E}$
5      $\phi^* \leftarrow$ Sum-Product-VE($\Phi, \prec, \boldsymbol{Z}$)
6      $\alpha \leftarrow \sum_{\boldsymbol{y} \in Val(\boldsymbol{Y})} \phi^*(\boldsymbol{y})$
7   **return** α, ϕ^*

Step	Variable eliminated	Factors used	Variables involved	New factor
1'	C	$\phi_C(C), \phi_D(D,C)$	C, D	$\tau_1'(D)$
2'	D	$\phi_G[I = i^1](G,D), \phi_I[I = i^1](), \tau_1'(D)$	G, D	$\tau_2'(G)$
5'	G	$\tau_2'(G), \phi_L(L,G), \phi_H[H = h^0](G,J)$	G, L, J	$\tau_5'(L,J)$
6'	S	$\phi_S[I = i^1](S), \phi_J(J,L,S)$	J, L, S	$\tau_6'(J,L)$
7'	L	$\tau_6'(J,L), \tau_5'(J,L)$	J, L	$\tau_7'(J)$

Table 9.3 **A run of sum-product variable elimination for** $P(J, i^1, h^0)$

elimination ordering that we used in table 9.1. The results are shown in table 9.3; the step numbers correspond to the steps in table 9.1. It is interesting to note the differences between the two runs of the algorithm. First, we notice that steps (3) and (4) disappear in the computation with evidence, since I and H do not need to be eliminated. More interestingly, by not eliminating I, we avoid the step that correlates G and S. In this execution, G and S never appear together in the same factor; they are both eliminated, and only their end results are combined. Intuitively, G and S are conditionally independent given I; hence, observing I renders them independent, so that we do not have to consider their joint distribution explicitly. Finally, we notice that $\phi_I[I = i^1] = P(i^1)$ is a factor over an empty scope, which is simply a number. It can be multiplied into any factor at any point in the computation. We chose arbitrarily to incorporate it into step (2'). Note that if our goal is to compute a conditional probability given the evidence, and not the probability of the evidence itself, we can avoid multiplying in this factor entirely, since its effect will disappear in the renormalization step at the end.

---

network
polynomial

Box 9.A — Concept: The Network Polynomial. *The* network polynomial *provides an interesting and useful alternative view of variable elimination. We begin with describing the concept for the case of a Gibbs distribution parameterized via a set of full table factors* Φ. *The polynomial* f_Φ

is defined over the following set of variables:

- *For each factor $\phi_c \in \Phi$ with scope $\boldsymbol{X}_c$, we have a variable $\theta_{\boldsymbol{x}_c}$ for every $\boldsymbol{x}_c \in Val(\boldsymbol{X}_c)$.*
- *For each variable X_i and every value $x_i \in Val(X_i)$, we have a binary-valued variable λ_{x_i}.*

In other words, the polynomial has one argument for each of the network parameters and for each possible assignment to a network variable. The polynomial f_Φ is now defined as follows:

$$f_\Phi(\boldsymbol{\theta}, \boldsymbol{\lambda}) = \sum_{x_1, \ldots, x_n} \left(\prod_{\phi_c \in \Phi} \theta_{\boldsymbol{x}_c} \cdot \prod_{i=1}^{n} \lambda_{x_i} \right). \tag{9.7}$$

Evaluating the network polynomial is equivalent to the inference task. In particular, let $\boldsymbol{Y} = \boldsymbol{y}$ be an assignment to some subset of network variables; define an assignment $\boldsymbol{\lambda}^{\boldsymbol{y}}$ as follows:

- *for each $Y_i \in \boldsymbol{Y}$, define $\lambda_{y_i}^{\boldsymbol{y}} = 1$ and $\lambda_{y_i'}^{\boldsymbol{y}} = 0$ for all $y_i' \neq y_i$;*
- *for each $Y_i \notin \boldsymbol{Y}$, define $\lambda_{y_i}^{\boldsymbol{y}} = 1$ for all $y_i \in Val(Y_i)$.*

With this definition, we can now show (exercise 9.4a) that:

$$f_\Phi(\boldsymbol{\theta}, \boldsymbol{\lambda}^{\boldsymbol{y}}) = \tilde{P}_\Phi(\boldsymbol{Y} = \boldsymbol{y} \mid \boldsymbol{\theta}). \tag{9.8}$$

The derivatives of the network polynomial are also of significant interest. We can show (exercise 9.4b) that

$$\frac{\partial f_\Phi(\boldsymbol{\theta}, \boldsymbol{\lambda}^{\boldsymbol{y}})}{\partial \lambda_{x_i}} = \tilde{P}_\Phi(x_i, \boldsymbol{y}_{-i} \mid \boldsymbol{\theta}), \tag{9.9}$$

where $\boldsymbol{y}_{-i}$ is the assignment in $\boldsymbol{y}$ to all variables other than X_i. We can also show that

$$\frac{\partial f_\Phi(\boldsymbol{\theta}, \boldsymbol{\lambda}^{\boldsymbol{y}})}{\partial \theta_{\boldsymbol{x}_c}} = \frac{\tilde{P}_\Phi(\boldsymbol{y}, \boldsymbol{x}_c \mid \boldsymbol{\theta})}{\theta_{\boldsymbol{x}_c}} \; ; \tag{9.10}$$

sensitivity analysis

this fact is proved in lemma 19.1. These derivatives can be used for various purposes, including retracting or modifying evidence in the network (exercise 9.4c), and sensitivity analysis — *computing the effect of changes in a network parameter on the answer to a particular probabilistic query (exercise 9.5).*

Of course, as defined, the representation of the network polynomial is exponentially large in the number of variables in the network. However, we can use the algebraic operations performed in a run of variable elimination to define a network polynomial that has precisely the same complexity as the VE run. More interesting, we can also use the same structure to compute efficiently all of the derivatives of the network polynomial, relative both to the λ_i and the $\theta_{\boldsymbol{x}_c}$ (see exercise 9.6).

9.4 Complexity and Graph Structure: Variable Elimination

From the examples we have seen, it is clear that the VE algorithm can be computationally much more efficient than a full enumeration of the joint. In this section, we analyze the complexity of the algorithm, and understand the source of the computational gains.

We also note that, aside from the asymptotic analysis, a careful implementation of this algorithm can have significant ramifications on performance; see box 10.A.

9.4.1   Simple Analysis

Let us begin with a simple analysis of the basic computational operations taken by algorithm 9.1. Assume we have n random variables, and m initial factors; in a Bayesian network, we have $m = n$; in a Markov network, we may have more factors than variables. For simplicity, assume we run the algorithm until all variables are eliminated.

The algorithm consists of a set of elimination steps, where, in each step, the algorithm picks a variable X_i, then multiplies all factors involving that variable. The result is a single large factor ψ_i. The variable then gets summed out of ψ_i, resulting in a new factor τ_i whose scope is the scope of ψ_i minus X_i. Thus, the work revolves around these factors that get created and processed. Let N_i be the number of entries in the factor ψ_i, and let $N_{\max} = \max_i N_i$.

We begin by counting the number of multiplication steps. Here, we note that the total number of factors ever entered into the set of factors Φ is $m + n$: the m initial factors, plus the n factors τ_i. Each of these factors ϕ is multiplied exactly once: when it is multiplied in line 3 of Sum-Product-Eliminate-Var to produce a large factor ψ_i, it is also extracted from Φ. The cost of multiplying ϕ to produce ψ_i is at most N_i, since each entry of ϕ is multiplied into exactly one entry of ψ_i. Thus, the total number of multiplication steps is at most $(n + m)N_i \leq (n + m)N_{\max} = O(mN_{\max})$. To analyze the number of addition steps, we note that the marginalization operation in line 4 touches each entry in ψ_i exactly once. Thus, the cost of this operation is exactly N_i; we execute this operation once for each factor ψ_i, so that the total number of additions is at most $nN_{\max}$. Overall, the total amount of work required is $O(mN_{\max})$.

The source of the inevitable exponential blowup is the potentially exponential size of the factors ψ_i. If each variable has no more than v values, and a factor ψ_i has a scope that contains k_i variables, then $N_i \leq v^{k_i}$. Thus, we see that the computational cost of the VE algorithm is dominated by the sizes of the intermediate factors generated, with an exponential growth in the number of variables in a factor.

9.4.2   Graph-Theoretic Analysis

Although the size of the factors created during the algorithm is clearly the dominant quantity in the complexity of the algorithm, it is not clear how it relates to the properties of our problem instance. In our case, the only aspect of the problem instance that affects the complexity of the algorithm is the structure of the underlying graph that induced the set of factors on which the algorithm was run. In this section, we reformulate our complexity analysis in terms of this graph structure.

9.4.2.1   Factors and Undirected Graphs

We begin with the observation that the algorithm does not care whether the graph that generated the factors is directed, undirected, or partly directed. The algorithm's input is a set of factors Φ, and the only relevant aspect to the computation is the scope of the factors. Thus, it is easiest to view the algorithm as operating on an undirected graph $\mathcal{H}$.

More precisely, we can define the notion of an undirected graph associated with a set of factors:

Definition 9.4

Let Φ be a set of factors. We define

$$Scope[\Phi] = \cup_{\phi \in \Phi} Scope[\phi]$$

to be the set of all variables appearing in any of the factors in Φ. We define $\mathcal{H}_\Phi$ to be the undirected graph whose nodes correspond to the variables in $Scope[\Phi]$ and where we have an edge $X_i - X_j \in \mathcal{H}_\Phi$ if and only if there exists a factor $\phi \in \Phi$ such that $X_i, X_j \in Scope[\phi]$. ∎

In words, the undirected graph $\mathcal{H}_\Phi$ introduces a fully connected subgraph over the scope of each factor $\phi \in \Phi$, and hence is the minimal I-map for the distribution induced by Φ.

We can now show that:

Proposition 9.1

Let P be a distribution defined by multiplying the factors in Φ and normalizing to define a distribution. Letting $\boldsymbol{X} = Scope[\Phi]$,

$$P(\boldsymbol{X}) = \frac{1}{Z} \prod_{\phi \in \Phi} \phi,$$

where $Z = \sum_{\boldsymbol{X}} \prod_{\phi \in \Phi} \phi$. Then $\mathcal{H}_\Phi$ is the minimal Markov network I-map for P, and the factors Φ are a parameterization of this network that defines the distribution P.

The proof is left as an exercise (exercise 9.7).

Note that, for a set of factors Φ defined by a Bayesian network $\mathcal{G}$, in the case without evidence, the undirected graph $\mathcal{H}_\Phi$ is precisely the moralized graph of $\mathcal{G}$. In this case, the product of the factors is a normalized distribution, so the partition function of the resulting Markov network is simply 1. Figure 4.6a shows the initial graph for our Student example.

More interesting is the Markov network induced by a set of factors $\Phi[e]$ defined by the reduction of the factors in a Bayesian network to some context $\boldsymbol{E} = \boldsymbol{e}$. In this case, recall that the variables in $\boldsymbol{E}$ are removed from the factors, so $\boldsymbol{X} = Scope[\Phi_e] = \mathcal{X} - \boldsymbol{E}$. Furthermore, as we discussed, the unnormalized product of the factors is $P(\boldsymbol{X}, \boldsymbol{e})$, and the partition function of the resulting Markov network is precisely $P(\boldsymbol{e})$. Figure 4.6b shows the initial graph for our Student example with evidence $G = g$, and figure 4.6c shows the case with evidence $G = g, S = s$.

9.4.2.2 Elimination as Graph Transformation

Now, consider the effect of a variable elimination step on the set of factors maintained by the algorithm and on the associated Markov network. When a variable X is eliminated, several operations take place. First, we create a single factor ψ that contains X and all of the variables $\boldsymbol{Y}$ with which it appears in factors. Then, we eliminate X from ψ, replacing it with a new factor τ that contains all of the variables $\boldsymbol{Y}$ but does not contain X. Let Φ_X be the resulting set of factors.

How does the graph $\mathcal{H}_{\Phi_X}$ differ from $\mathcal{H}_\Phi$? The step of constructing ψ generates edges between all of the variables $Y \in \boldsymbol{Y}$. Some of them were present in $\mathcal{H}_\Phi$, whereas others are introduced due to the elimination step; edges that are introduced by an elimination step are called *fill edges*. The step of eliminating X from ψ to construct τ has the effect of removing X and all of its incident edges from the graph.

fill edge

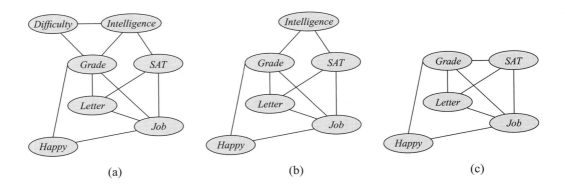

Figure 9.10   Variable elimination as graph transformation in the Student **example,** using the elimination order of table 9.1: (a) after eliminating C; (b) after eliminating D; (c) after eliminating I.

Consider again our Student network, in the case without evidence. As we said, figure 4.6a shows the original Markov network. Figure 9.10a shows the result of eliminating the variable C. Note that there are no fill edges introduced in this step.

After an elimination step, the subsequent elimination steps use the new set of factors. In other words, they can be seen as operations over the new graph. Figure 9.10b and c show the graphs resulting from eliminating first D and then I. Note that the step of eliminating I results in a (new) fill edge G—S, induced by the factor G, I, S.

The computational steps of the algorithm are reflected in this series of graphs. Every factor that appears in one of the steps in the algorithm is reflected in the graph as a clique. In fact, we can summarize the computational cost using a single graph structure.

9.4.2.3   The Induced Graph

We define an undirected graph that is the union of all of the graphs resulting from the different steps of the variable elimination algorithm.

Definition 9.5

induced graph

> *Let Φ be a set of factors over $\mathcal{X} = \{X_1, \ldots, X_n\}$, and $\prec$ be an elimination ordering for some subset $\boldsymbol{X} \subseteq \mathcal{X}$. The* induced graph *$\mathcal{I}_{\Phi,\prec}$ is an undirected graph over $\mathcal{X}$, where X_i and X_j are connected by an edge if they both appear in some intermediate factor ψ generated by the VE algorithm using $\prec$ as an elimination ordering.*   ∎

For a Bayesian network graph $\mathcal{G}$, we use $\mathcal{I}_{\mathcal{G},\prec}$ to denote the induced graph for the factors Φ corresponding to the CPDs in $\mathcal{G}$; similarly, for a Markov network $\mathcal{H}$, we use $\mathcal{I}_{\mathcal{H},\prec}$ to denote the induced graph for the factors Φ corresponding to the potentials in $\mathcal{H}$.

The induced graph $\mathcal{I}_{\mathcal{G},\prec}$ for our Student example is shown in figure 9.11a. We can see that the fill edge G—S, introduced in step (3) when we eliminated I, is the only fill edge introduced.

As we discussed, each factor ψ used in the computation corresponds to a complete subgraph of the graph $\mathcal{I}_{\mathcal{G},\prec}$ and is therefore a clique in the graph. The connection between cliques in $\mathcal{I}_{\mathcal{G},\prec}$ and factors ψ is, in fact, much tighter:

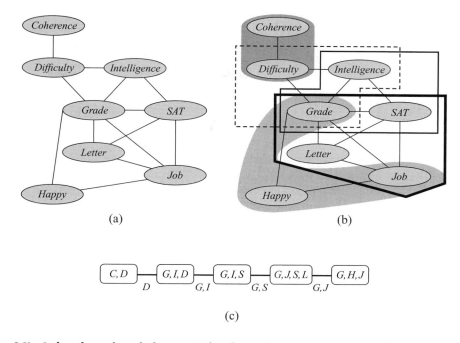

Figure 9.11 **Induced graph and clique tree for the** Student **example.** (a) Induced graph for variable elimination in the Student example, using the elimination order of table 9.1. (b) Cliques in the induced graph: $\{C, D\}$, $\{D, I, G\}$, $\{G, I, S\}$, $\{G, J, S, L\}$, and $\{G, H, J\}$. (c) Clique tree for the induced graph.

Theorem 9.6

Let $\mathcal{I}_{\Phi, \prec}$ be the induced graph for a set of factors Φ and some elimination ordering $\prec$. Then:

1. *The scope of every factor generated during the variable elimination process is a clique in $\mathcal{I}_{\Phi, \prec}$.*

2. *Every maximal clique in $\mathcal{I}_{\Phi, \prec}$ is the scope of some intermediate factor in the computation.*

PROOF We begin with the first statement. Consider a factor $\psi(Y_1, \ldots, Y_k)$ generated during the VE process. By the definition of the induced graph, there must be an edge between each Y_i and Y_j. Hence $Y_1, \ldots, Y_k$ form a clique.

To prove the second statement, consider some maximal clique $\boldsymbol{Y} = \{Y_1, \ldots, Y_k\}$. Assume, without loss of generality, that Y_1 is the first of the variables in $\boldsymbol{Y}$ in the ordering $\prec$, and is therefore the first among this set to be eliminated. Since $\boldsymbol{Y}$ is a clique, there is an edge from Y_1 to each other Y_i. Note that, once Y_1 is eliminated, it can appear in no more factors, so there can be no new edges added to it. Hence, the edges involving Y_1 were added prior to this point in the computation. The existence of an edge between Y_1 and Y_i therefore implies that, at this point, there is a factor containing both Y_1 and Y_i. When Y_1 is eliminated, all these factors must be multiplied. Therefore, the product step results in a factor ψ that contains all of $Y_1, Y_2, \ldots, Y_k$. Note that this factor can contain no other variables; if it did, these variables would also have an edge to all of $Y_1, \ldots, Y_k$, so that $Y_1, \ldots, Y_k$ would not constitute a maximal connected subgraph. ∎

Let us verify that the second property holds for our example. Figure 9.11b shows the maximal cliques in $\mathcal{I}_{\mathcal{G},\prec}$:

$$
\begin{aligned}
\boldsymbol{C}_1 &= \{C, D\} \\
\boldsymbol{C}_2 &= \{D, I, G\} \\
\boldsymbol{C}_3 &= \{I, G, S\} \\
\boldsymbol{C}_4 &= \{G, J, L, S\} \\
\boldsymbol{C}_5 &= \{G, H, J\}.
\end{aligned}
$$

Both these properties hold for this set of cliques. For example, $\boldsymbol{C}_3$ corresponds to the factor ψ generated in step (5).

 Thus, there is a direct correspondence between the maximal factors generated by our algorithm and maximal cliques in the induced graph. Importantly, the induced graph and the size of the maximal cliques within it depend strongly on the elimination ordering. Consider, for example, our other elimination ordering for the Student network. In this case, we can verify that our induced graph has a maximal clique over G, I, D, L, J, H, a second over S, I, D, L, J, H, and a third over C, D, J; indeed, the graph is missing only the edge between S and G, and some edges involving C. In this case, the largest clique contains six variables, as opposed to four in our original ordering. Therefore, the cost of computation here is substantially more expensive.

Definition 9.6

induced width

tree-width

We define the width *of an induced graph to be the number of nodes in the largest clique in the graph minus 1. We define the* induced width $w_{\mathcal{K},\prec}$ *of an ordering $\prec$ relative to a graph $\mathcal{K}$ (directed or undirected) to be the width of the graph $\mathcal{I}_{\mathcal{K},\prec}$ induced by applying VE to $\mathcal{K}$ using the ordering $\prec$. We define the* tree-width *of a graph $\mathcal{K}$ to be its minimal induced width $w_{\mathcal{K}}^* = \min_{\prec} w(\mathcal{I}_{\mathcal{K},\prec})$.* ∎

The minimal induced width of the graph $\mathcal{K}$ provides us a bound on the best performance we can hope for by applying VE to a probabilistic model that factorizes over $\mathcal{K}$.

9.4.3 Finding Elimination Orderings ★

How can we compute the minimal induced width of the graph, and the elimination ordering achieving that width? Unfortunately, there is no easy way to answer this question.

Theorem 9.7

The following decision problem is $\mathcal{NP}$-complete:

> *Given a graph $\mathcal{H}$ and some bound K, determine whether there exists an elimination ordering achieving an induced width $\leq K$.*

It follows directly that finding the optimal elimination ordering is also $\mathcal{NP}$-hard. Thus, we cannot easily tell by looking at a graph how computationally expensive inference on it will be. Note that this $\mathcal{NP}$-completeness result is distinct from the $\mathcal{NP}$-hardness of inference itself. That is, even if some oracle gives us the best elimination ordering, the induced width might still be large, and the inference task using that ordering can still require exponential time.

However, as usual, $\mathcal{NP}$-hardness is not the end of the story. There are several techniques that one can use to find good elimination orderings. The first uses an important graph-theoretic property of induced graphs, and the second uses heuristic ideas.

9.4.3.1 Chordal Graphs

chordal graph

Recall from definition 2.24 that an undirected graph is *chordal* if it contains no cycle of length greater than three that has no "shortcut," that is, every minimal loop in the graph is of length three. As we now show, somewhat surprisingly, the class of induced graphs is equivalent to the class of chordal graphs. We then show that this property can be used to provide one heuristic for constructing an elimination ordering.

Theorem 9.8

Every induced graph is chordal.

PROOF Assume by contradiction that we have such a cycle $X_1 — X_2 — \ldots — X_k — X_1$ for $k > 3$, and assume without loss of generality that X_1 is the first variable to be eliminated. As in the proof of theorem 9.6, no edge incident on X_1 is added after X_1 is eliminated; hence, both edges $X_1 — X_2$ and $X_1 — X_k$ must exist at this point. Therefore, the edge $X_2 — X_k$ will be added at the same time, contradicting our assumption. ∎

Indeed, we can verify that the graph of figure 9.11a is chordal. For example, the loop $H \to G \to L \to J \to H$ is cut by the chord $G \to J$.

The converse of this theorem states that any chordal graph $\mathcal{H}$ is an induced graph for some ordering. One way of showing that is to show that there is an elimination ordering for $\mathcal{H}$ for which $\mathcal{H}$ itself is the induced graph.

Theorem 9.9

Any chordal graph $\mathcal{H}$ admits an elimination ordering that does not introduce any fill edges into the graph.

PROOF We prove this result by induction on the number of nodes in the tree. Let $\mathcal{H}$ be a chordal graph with n nodes. As we showed in theorem 4.12, there is a clique tree $\mathcal{T}$ for $\mathcal{H}$. Let C_k be a clique in the tree that is a leaf, that is, it has only a single other clique as a neighbor. Let X_i be some variable that is in C_k but not in its neighbor. Let $\mathcal{H}'$ be the graph obtained by eliminating X_i. Because X_i belongs only to the clique C_k, its neighbors are precisely $C_k - \{X_i\}$. Because all of them are also in C_k, they are connected to each other. Hence, eliminating X_i introduces no fill edges. Because $\mathcal{H}'$ is also chordal, we can now apply the inductive hypothesis, proving the result. ∎

Algorithm 9.3 Maximum cardinality search for constructing an elimination ordering

    Procedure Max-Cardinality (
        $\mathcal{H}$ // An undirected graph over $\mathcal{X}$
)
1   Initialize all nodes in $\mathcal{X}$ as unmarked
2   **for** $k = |\mathcal{X}| \ldots 1$
3         $X \leftarrow$ unmarked variable in $\mathcal{X}$ with largest number of marked neighbors
4         $\pi(X) \leftarrow k$
5   Mark X
6   **return** π

Example 9.2

We can illustrate this construction on the graph of figure 9.11a. The maximal cliques in the induced graph are shown in b, and a clique tree for this graph is shown in c. One can easily verify that each sepset separates the two sides of the tree; for example, the sepset $\{G, S\}$ separates C, I, D (on the left) from L, J, H (on the right). The elimination ordering C, D, I, H, G, S, L, J, an extension of the elimination in table 9.1 that generated this induced graph, is one ordering that might arise from the construction of theorem 9.9. For example, it first eliminates C, D, which are both in a leaf clique; it then eliminates I, which is in a clique that is now a leaf, following the elimination of C, D. Indeed, it is not hard to see that this ordering introduces no fill edges. By contrast, the ordering in table 9.2 is not consistent with this construction, since it begins by eliminating the variables G, I, S, none of which are in a leaf clique. Indeed, this elimination ordering introduces additional fill edges, for example, the edge $H \rightarrow D$. ∎

An alternative method for constructing an elimination ordering that introduces no fill edges in a chordal graph is the Max-Cardinality algorithm, shown in algorithm 9.3. This method does not use the clique tree as its starting point, but rather operates directly on the graph. When applied to a chordal graph, it constructs an elimination ordering that eliminates cliques one at a time, starting from the leaves of the clique tree; and it does so without ever considering the clique tree structure explicitly.

maximum
cardinality

Example 9.3

Consider applying Max-Cardinality to the chordal graph of figure 9.11. Assume that the first node selected is S. The second node selected must be one of S's neighbors, say J. The node that has the largest number of marked neighbors are now G and L, which are chosen subsequently. Now, the unmarked nodes that have the largest number of marked neighbors (two) are H and I. Assume we select I. Then the next nodes selected are D and H, in any order. The last node to be selected is C. One possible resulting ordering in which nodes are marked is thus S, J, G, L, I, H, D, C. Importantly, the actual elimination ordering proceeds in reverse. Thus, we first eliminate C, D, then H, and so on. We can now see that this ordering always eliminates a variable from a clique that is a leaf clique at the time. For example, we first eliminate C, D from a leaf clique, then H, then G from the clique $\{G, I, D\}$, which is now (following the elimination of C, D) a leaf. ∎

As in this example, Max-Cardinality always produces an elimination ordering that is consistent with the construction of theorem 9.9. As a consequence, it follows that Max-Cardinality, when applied to a chordal graph, introduces no fill edges.

Theorem 9.10

$\overline{\text{Let } \mathcal{H} \text{ be a chordal graph. Let } \pi}$ *be the ranking obtained by running* Max-Cardinality *on* $\mathcal{H}$*. Then* Sum-Product-VE *(algorithm 9.1), eliminating variables in order of increasing* π*, does not introduce any fill edges.*

The proof is left as an exercise (exercise 9.8).

The maximum cardinality search algorithm can also be used to construct an elimination ordering for a nonchordal graph. However, it turns out that the orderings produced by this method are generally not as good as those produced by various other algorithms, such as those described in what follows.

triangulation

To summarize, we have shown that, if we construct a chordal graph that contains the graph $\mathcal{H}_\Phi$ corresponding to our set of factors Φ, we can use it as the basis for inference using Φ. The process of turning a graph $\mathcal{H}$ into a chordal graph is also called *triangulation*, since it ensures that the largest unbroken cycle in the graph is a triangle. Thus, we can reformulate our goal of finding an elimination ordering as that of triangulating a graph $\mathcal{H}$ so that the largest clique in the resulting graph is as small as possible. Of course, this insight only reformulates the problem: Inevitably, the problem of finding such a minimal triangulation is also $\mathcal{NP}$-hard. Nevertheless, there are several graph-theoretic algorithms that address this precise problem and offer different levels of performance guarantee; we discuss this task further in section 10.4.2.

polytree

Box 9.B — Concept: Polytrees. *One particularly simple class of chordal graphs is the class of Bayesian networks whose graph $\mathcal{G}$ is a* polytree. *Recall from definition 2.22 that a polytree is a graph where there is at most one trail between every pair of nodes.*

Polytrees received a lot of attention in the early days of Bayesian networks, because the first widely known inference algorithm for any type of Bayesian network was Pearl's message passing algorithm for polytrees. This algorithm, a special case of the message passing algorithms described in subsequent chapters of this book, is particularly compelling in the case of polytree networks, since it consists of nodes passing messages directly to other nodes along edges in the graph. Moreover, the cost of this computation is linear in the size of the network (where the size of the network is measured as the total sizes of the CPDs in the network, not the number of nodes; see exercise 9.9). From the perspective of the results presented in this section, this simplicity is not surprising: In a polytree, any maximal clique is a family of some variable in the network, and the clique tree structure roughly follows the network topology. (We simply throw out families that do not correspond to a maximal clique, because they are subsumed by another clique.)

Somewhat ironically, the compelling nature of the polytree algorithm gave rise to a long-standing misconception that there was a sharp tractability boundary between polytrees and other networks, in that inference was tractable only in polytrees and NP-hard in other networks. As we discuss in this chapter, this is not the case; rather, there is a continuum of complexity defined by the size of the largest clique in the induced graph.

9.4.3.2 Minimum Fill/Size/Weight Search

An alternative approach for finding elimination orderings is based on a very straightforward intuition. Our goal is to construct an ordering that induces a "small" graph. While we cannot

---

Algorithm 9.4 Greedy search for constructing an elimination ordering

    Procedure Greedy-Ordering (
        $\mathcal{H}$ // An undirected graph over $\mathcal{X}$,
        s // An evaluation metric
)
1   Initialize all nodes in $\mathcal{X}$ as unmarked
2   **for** $k = 1 \ldots |\mathcal{X}|$
3   Select an unmarked variable $X \in \mathcal{X}$ that minimizes $s(\mathcal{H}, X)$
4         $\pi(X) \leftarrow k$
5   Introduce edges in $\mathcal{H}$ between all neighbors of X
6   Mark X
7   **return** π

---

find an ordering that achieves the global minimum, we can eliminate variables one at a time in a greedy way, so that each step tends to lead to a small blowup in size.

The general algorithm is shown in algorithm 9.4. At each point, the algorithm evaluates each of the remaining variables in the network based on its heuristic cost function. Some common cost criteria that have been used for evaluating variables are:

- **Min-neighbors:** The cost of a vertex is the number of neighbors it has in the current graph.

- **Min-weight:** The cost of a vertex is the product of *weights* — domain cardinality — of its neighbors.

- **Min-fill:** - The cost of a vertex is the number of edges that need to be added to the graph due to its elimination.

- **Weighted-min-fill:** The cost of a vertex is the sum of weights of the edges that need to be added to the graph due to its elimination, where a weight of an edge is the product of weights of its constituent vertices.

Intuitively, min-neighbors and min-weight count the size or weight of the largest clique in $\mathcal{H}$ after eliminating X. Min-fill and weighted-min-fill count the number or weight of edges that would be introduced into $\mathcal{H}$ by eliminating X. It can be shown (exercise 9.10) that none of these criteria is universally better than the others.

This type of greedy search can be done either deterministically (as shown in algorithm 9.4), or stochastically. In the stochastic variant, at each step we select some number of low-scoring vertices, and then choose among them using their score (where lower-scoring vertices are selected with higher probability). In the stochastic variants, we run multiple iterations of the algorithm, and then select the ordering that leads to the most efficient elimination — the one where the sum of the sizes of the factors produced is smallest.

Empirical results show that these heuristic algorithms perform surprisingly well in practice. Generally, Min-Fill and Weighted-Min-Fill tend to work better on more problems. Not surprisingly, Weighted-Min-Fill usually has the most significant gains when there is some significant variability in the sizes of the domains of the variables in the network. Box 9.C presents a case study comparing these algorithms on a suite of standard benchmark networks.

Box 9.C — Case Study: Variable Elimination Orderings. *Fishelson and Geiger (2003) performed a comprehensive case study of different heuristics for computing an elimination ordering, testing them on eight standard Bayesian network benchmarks, ranging from 24 nodes to more than 1,000. For each network, they compared both to the best elimination ordering known previously, obtained by an expensive process of simulated annealing search, and to the network obtained by a state-of-the-art Bayesian network package. They compared to stochastic versions of the four heuristics described in the text, running each of them for 1 minute or 10 minutes, and selecting the best network obtained in the different random runs. Maximum cardinality search was not used, since it is known to perform quite poorly in practice.*

The results, shown in figure 9.C.1, suggest several conclusions. First, we see that running the stochastic algorithms for longer improves the quality of the answer obtained, although usually not by a huge amount. We also see that different heuristics can result in orderings whose computational cost can vary in almost an order of magnitude. Overall, Min-Fill and Weighted-Min-Fill achieve the best performance, but they are not universally better. The best answer obtained by the greedy algorithms is generally very good; it is often significantly better than the answer obtained by a deterministic state-of-the-art scheme, and it is usually quite close to the best-known ordering, even when the latter is obtained using much more expensive techniques. Because the computational cost of the heuristic ordering-selection algorithms is usually negligible relative to the running time of the inference itself, we conclude that for large networks it is worthwhile to run several heuristic algorithms in order to find the best ordering obtained by any of them.

9.5 Conditioning ★

conditioning

An alternative approach to inference is based on the idea of *conditioning*. The conditioning algorithm is based on the fact (illustrated in section 9.3.2), that observing the value of certain variables can simplify the variable elimination process. When a variable is not observed, we can use a case analysis to enumerate its possible values, perform the simplified VE computation, and then aggregate the results for the different values. As we will discuss, **in terms of number of operations, the conditioning algorithm offers no benefit over the variable elimination algorithm. However, it offers a continuum of time-space trade-offs, which can be extremely important in cases where the factors created by variable elimination are too big to fit in main memory.**

9.5.1 The Conditioning Algorithm

The conditioning algorithm is easiest to explain in the context of a Markov network. Let Φ be a set of factors over X and P_Φ be the associated distribution. We assume that any observations were already assimilated into Φ, so that our goal is to compute $P_\Phi(Y)$ for some set of query variables Y. For example, if we want to do inference in the Student network given the evidence $G = g$, we would reduce the factors reduced to this context, giving rise to the network structure shown in figure 4.6b.

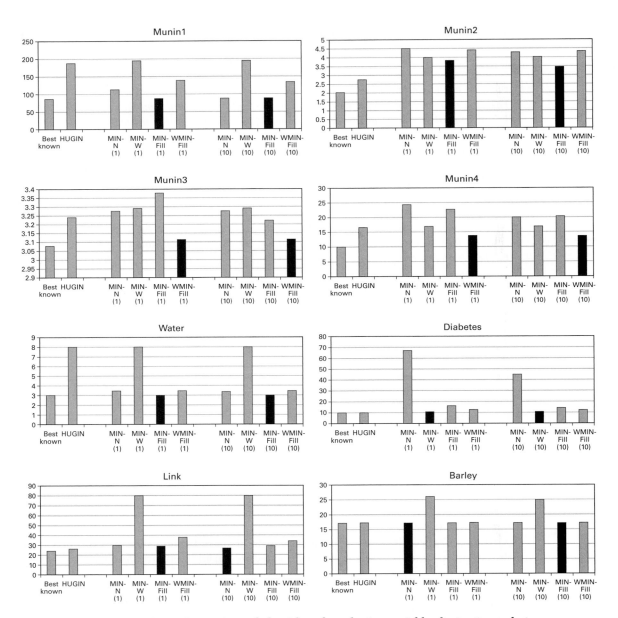

Figure 9.C.1 — Comparison of algorithms for selecting variable elimination ordering.
Computational cost of variable elimination inference in a range of benchmark networks, obtained by various algorithms for selecting an elimination ordering. The cost is measured as the size of the factors generated during the process of variable elimination. For each network, we see the cost of the best-known ordering, the ordering obtained by HUGIN (a state-of-the-art Bayesian network package), and the ordering obtained by stochastic greedy search using four different search heuristics — Min-Neighbors, Min-Weight, Min-Fill, and Weighted-Min-Fill — run for 1 minute and for 10 minutes.

Algorithm 9.5 Conditioning algorithm

Procedure Sum-Product-Conditioning (
    Φ,   // Set of factors, possibly reduced by evidence
    Y,   // Set of query variables
    U // Set of variables on which to condition
)

1   **for** each $u \in Val(U)$
2        $\Phi_u \leftarrow \{\phi[U = u] \; : \; \phi \in \Phi\}$
3   Construct $\mathcal{H}_{\Phi_u}$
4        $(\alpha_u, \phi_u(Y)) \leftarrow$ Cond-Prob-VE$(\mathcal{H}_{\Phi_u}, Y, \emptyset)$
5        $\phi^*(Y) \leftarrow \frac{\sum_u \phi_u(Y)}{\sum_u \alpha_u}$
6   Return $\phi^*(Y)$

The conditioning algorithm is based on the following simple derivation. Let $U \subseteq X$ be any set of variables. Then we have that:

$$\tilde{P}_\Phi(Y) = \sum_{u \in Val(U)} \tilde{P}_\Phi(Y, u). \tag{9.11}$$

The key observation is that each term $\tilde{P}_\Phi(Y, u)$ can be computed by marginalizing out the variables in $X - U - Y$ *in the unnormalized measure* $\tilde{P}_\Phi[u]$ *obtained by reducing* $\tilde{P}_\Phi$ *to the context* u. As we have already discussed, the reduced measure is simply the measure defined by reducing each of the factors to the context u. The reduction process generally produces a simpler structure, with a reduced inference cost.

We can use this formula to compute $P_\Phi(Y)$ as follows: We construct a network $\mathcal{H}_\Phi[u]$ for each assignment u; these networks have identical structures, but different parameters. We run sum-product inference in each of them, to obtain a factor over the desired query set Y. We then simply add up these factors to obtain $\tilde{P}_\Phi(Y)$. We can also derive $P_\Phi(Y)$ by renormalizing this factor to obtain a distribution. As usual, the normalizing constant is the partition function for P_Φ. However, applying equation (9.11) to the case of $Y = \emptyset$, we conclude that

$$Z_\Phi = \sum_u Z_{\Phi[u]}.$$

Thus, we can derive the overall partition function from the partition functions for the different subnetworks $\mathcal{H}_{\Phi[u]}$. The final algorithm is shown in algorithm 9.5. (We note that Cond-Prob-VE was called without evidence, since we assumed for simplicity that our factors Φ have already been reduced with the evidence.)

Example 9.4   *Assume that we want to compute $P(J)$ in the Student network with evidence $G = g^1$, so that our initial graph would be the one shown in figure 4.6b. We can now perform inference by enumerating all of the assignments s to the variable S. For each such assignment, we run inference on a graph structured as in figure 4.6c, with the factors reduced to the assignment g^1, s. In each such network we compute a factor over J, and add them all up. Note that the reduced network contains two disconnected components, and so we might be tempted to run inference only on the component that contains J. However, that procedure would not produce a correct answer: The value we get by summing out the variables in the second component multiplies our final factor. Although this is a constant multiple for each value of s, these values are generally different for the different values of S. Because the factors are added before the final renormalization, this constant influences the weight of one factor in the summation relative to the other. Thus, if we ignore this constant component, the answers we get from the s^1 computation and the s^0 computation would be weighted incorrectly.* ∎*

Historically, owing to the initial popularity of the polytree algorithm, the conditioning approach was mostly used in the case where the transformed network is a polytree. In this case, the algorithm is called *cutset conditioning*.

cutset
conditioning

9.5.2   Conditioning and Variable Elimination

At first glance, it might appear as if this process saves us considerable computational cost over the variable elimination algorithm. After all, we have reduced the computation to one that performs variable elimination in a much simpler network. The cost arises, of course, from the fact that, when we condition on U, we need to perform variable elimination on the conditioned network multiple times, once for each assignment $u \in Val(U)$. The cost of this computation is $O(|Val(U)|)$, which is exponential in the number of variables in U. Thus, we have not avoided the exponential blowup associated with the probabilistic inference process. In this section, we provide a formal complexity analysis of the conditioning algorithm, and compare it to the complexity of elimination. This analysis also reveals various interesting improvements to the basic conditioning algorithm, which can dramatically improve its performance in certain cases.

To understand the operation of the conditioning algorithm, we return to the basic description of the probabilistic inference task. Consider our query J in the Extended Student network. We know that:

$$p(J) = \sum_C \sum_D \sum_I \sum_S \sum_G \sum_L \sum_H P(C, D, I, S, G, L, H, J).$$

Reordering this expression slightly, we have that:

$$p(J) = \sum_g \left[\sum_C \sum_D \sum_I \sum_S \sum_L \sum_H P(C, D, I, S, g, L, H, J) \right].$$

The expression inside the parentheses is precisely the result of computing the probability of J in the network $\mathcal{H}_{\Phi_{G=g}}$, where Φ is the set of CPD factors in $\mathcal{B}$.

In other words, the conditioning algorithm is simply executing parts of the basic summation defining the inference task by case analysis, enumerating the possible values of the conditioning

Step	Variable eliminated	Factors used	Variables involved	New factor
1	C	$\phi_C^+(C,G),\, \phi_D^+(D,C,G)$	C,D,G	$\tau_1(D,G)$
2	D	$\phi_G^+(G,I,D),\, \tau_1(D,G)$	G,I,D	$\tau_2(G,I)$
3	I	$\phi_I^+(I,G),\, \phi_S^+(S,I,G),\, \tau_2(G,I)$	G,S,I	$\tau_3(G,S)$
4	H	$\phi_H^+(H,G,J)$	H,G,J	$\tau_4(G,J)$
5	S	$\tau_3(G,S),\, \phi_J^+(J,L,S,G)$	J,L,S,G	$\tau_5(J,L,G)$
6	L	$\tau_5(J,L,G),\, \phi_L^+(L,G)$	J,L	$\tau_6(J)$
7	—	$\tau_6(J),\, \tau_4(G,J)$	G,J	$\tau_7(G,J)$

Table 9.4   Example of relationship between variable elimination and conditioning. A run of variable elimination for the query $P(J)$ corresponding to conditioning on G.

variables. By contrast, variable elimination performs the same summation from the inside out, using dynamic programming to reuse computation.

Indeed, if we simply did conditioning on all of the variables, the result would be an explicit summation of the entire joint distribution. In conditioning, however, we perform the conditioning step only on some of the variables, and use standard variable elimination — dynamic programming — to perform the rest of the summation, avoiding exponential blowup (at least over that part).

In general, it follows that both algorithms are performing the same set of basic operations (sums and products). However, where the variable elimination algorithm uses the caching of dynamic programming to save redundant computation throughout the summation, conditioning uses a full enumeration of cases for some of the variables, and dynamic programming only at the end.

From this argument, it follows that conditioning always performs no fewer steps than variable elimination. To understand why, consider the network of example 9.4 and assume that we are trying to compute $P(J)$. The conditioned network $\mathcal{H}_{\Phi_{G=g}}$ has a set of factors most of which are identical to those in the original network. The exceptions are the reduced factors: $\phi_L[G = g](L)$ and $\phi_H[G = g](H, J)$. For each of the three values g of G, we are performing variable elimination over these factors, eliminating all variables except for G and J.

We can imagine "lumping" these three computations into one, by augmenting the scope of each factor with the variable G. More precisely, we define a set of augmented factors ϕ^+ as follows: The scope of the factor ϕ_G already contains G, so $\phi_G^+(G, D, I) = \phi_G(G, D, I)$. For the factor ϕ_L^+, we simply combine the three factors $\phi_{L,g}(L)$, so that $\phi_L^+(L, g) = \phi_L[G = g](L)$ for all g. Not surprisingly, the resulting factor $\phi_L^+(L, G)$ is simply our original CPD factor $\phi_L(L, G)$. We define ϕ_H^+ in the same way. The remaining factors are unrelated to G. For each other variable X over scope $\boldsymbol{Y}$, we simply define $\phi_X^+(\boldsymbol{Y}, G) = \phi_X(\boldsymbol{Y})$; that is, the value of the factor does not depend on the value of G.

We can easily verify that, if we run variable elimination over the set of factors $\mathcal{F}_X^+$ for $X \in \{C, D, I, G, S, L, J, H\}$, eliminating all variables except for J and G, we are performing precisely the same computation as the three iterations of variable elimination for the three different conditioned networks $\mathcal{H}_{\Phi_{G=g}}$: Factor entries involving different values g of G never in-

Step	Variable eliminated	Factors used	Variables involved	New factor
1	C	$\phi_C(C), \phi_D(D,C)$	C,D	$\tau_1(D)$
2	D	$\phi_G(G,I,D), \tau_1(D)$	G,I,D	$\tau_2(G,I)$
3	I	$\phi_I(I), \phi_S(S,I), \tau_2(G,I)$	G,S,I	$\tau_3(G,S)$
4	H	$\phi_H(H,G,J)$	H,G,J	$\tau_4(G,J)$
5	S	$\tau_3(G,S), \phi_J(J,L,S)$	J,L,S,G	$\tau_5(J,L,G)$
6	L	$\tau_5(J,L,G), \phi_L(L,G)$	J,L	$\tau_6(J)$
7	G	$\tau_6(J), \tau_4(G,J)$	G,J	$\tau_7(J)$

Table 9.5   A run of variable elimination for the query $P(J)$ **with** G **eliminated last**

teract, and the computation performed for the entries where $G = g$ is precisely the computation performed in the network $\mathcal{H}_{\Phi_{G=g}}$.

Specifically, assume we are using the ordering C, D, I, H, S, L to perform the elimination within each conditioned network $\mathcal{H}_{\Phi_{G=g}}$. The steps of the computation are shown in table 9.4. Step (7) corresponds to the product of all of the remaining factors, which is the last step in variable elimination. The final step in the conditioning algorithm, where we add together the results of the three computations, is precisely the same as eliminating G from the resulting factor $\tau_7(G, J)$.

It is instructive to compare this execution to the one obtained by running variable elimination on the original set of factors, with the elimination ordering C, D, I, H, S, L, G; that is, we follow the ordering used within the conditioned networks for the variables other than G, J, and then eliminate G at the very end. In this process, shown in table 9.5, some of the factors involve G, but others do not. In particular, step (1) in the elimination algorithm involves only C, D, whereas in the conditioning algorithm, we are performing precisely the same computation over C, D three times: once for each value g of G.

In general, we can show:

Theorem 9.11

Let Φ be a set of factors, and $\mathbf{Y}$ be a query. Let $\mathbf{U}$ be a set of conditioning variables, and $\mathbf{Z} = \mathcal{X} - \mathbf{Y} - \mathbf{U}$. Let $\prec$ be the elimination ordering over $\mathbf{Z}$ used by the variable elimination algorithm over the network $\mathcal{H}_{\Phi_u}$ in the conditioning algorithm. Let $\prec^+$ be an ordering that is consistent with $\prec$ over the variables in $\mathbf{Z}$, and where, for each variable $U \in \mathbf{U}$, we have that $\mathbf{Z} \prec^+ U$. Then the number of operations performed by the conditioning is no less than the number of operations performed by variable elimination with the ordering $\prec^+$.

We omit the proof of this theorem, which follows precisely the lines of our example.

Thus, conditioning always requires no fewer computations than variable elimination with some particular ordering (which may or may not be a good one). In our example, the wasted computation from conditioning is negligible. In other cases, however, as we will discuss, we can end up with a large amount of redundant computation. In fact, in some cases, conditioning can be significantly worse:

Example 9.5

Consider the network shown in figure 9.12a, and assume we choose to condition on A_k in order

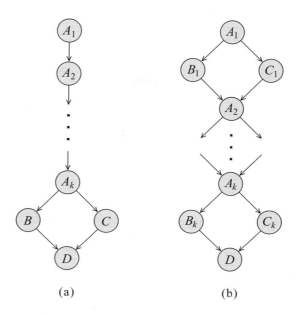

Figure 9.12   Networks where conditioning performs unnecessary computation

to cut the single loop in the network. In this case, we would perform the entire elimination of the chain $A_1 \to \ldots \to A_{k-1}$ multiple times — once for every value of A_k. ∎

Example 9.6

Consider the network shown in figure 9.12b and assume that we wish to use cutset conditioning, where we cut every loop in the network. The most efficient way of doing so is to condition on every other A_i variable, for example, $A_2, A_4, \ldots, A_k$ (assuming for simplicity that k is even). The cost of the conditioning algorithm in this case is exponential in k, whereas the induced width of the network is 2, and the cost of variable elimination is linear in k. ∎

Given this discussion, one might wonder why anyone bothers with the conditioning algorithm. There are two main reasons. First, variable elimination gains its computational savings from caching factors computed as intermediate results. In complex networks, these factors can grow very large. In cases where memory is scarce, it might not be possible to keep these factors in memory, and the variable elimination computation becomes infeasible (or very costly due to constant thrashing to disk). On the other hand, conditioning does not require significant amounts of memory: We run inference separately for each assignment u to U and simply accumulate the results. Overall, the computation requires space that is linear only in the size of the network. Thus, we can view the trade-off of conditioning versus variable elimination as a time-space trade-off. Conditioning saves space by not storing intermediate results in memory, but then it may cost additional time by having to repeat the computation to generate them.

The second reason for using conditioning is that it forms the basis for a useful approximate inference algorithm. In particular, in certain cases, we can get a reasonable approximate solution by enumerating only some of the possible assignment $u \in Val(U)$. We return to this approach in section 12.5

9.5.3 Graph-Theoretic Analysis

As in the case of variable elimination, it helps to reformulate the complexity analysis of the conditioning algorithm in graph-theoretic terms. Assume that we choose to condition on a set U, and perform variable elimination on the remaining variables. We can view each of these steps in terms of its effect on the graph structure.

Let us begin with the step of conditioning the network on some variable U. Once again, it is easiest to view this process in terms of its effect on an undirected graph. As we discussed, this step effectively introduces U into every factor parameterizing the current graph. In graph-theoretic terms, we have introduced U into every clique in the graph, or, more simply, introduced an edge between U and every other node currently in the graph.

When we finish the conditioning process, we perform elimination on the remaining variables. We have already analyzed the effect on the graph of eliminating a variable X: When we eliminate X, we add edges between all of the current neighbors of X in the graph. We then remove X from the graph.

We can now define an induced graph for the conditioning algorithm. Unlike the graph for variable elimination, this graph has two types of fill edges: those induced by conditioning steps, and those induced by the elimination steps for the remaining variables.

Definition 9.7

conditioning
induced graph

Let Φ be a set of factors over $\mathcal{X} = \{X_1, \ldots, X_n\}$, $U \subset \mathcal{X}$ be a set of conditioning variables, and $\prec$ be an elimination ordering for some subset $X \subseteq \mathcal{X} - U$. The induced graph $\mathcal{I}_{\Phi,\prec,U}$ is an undirected graph over $\mathcal{X}$ with the following edges:

- *a* conditioning edge *between every variable $U \in U$ and every other variable $X \in \mathcal{X}$;*
- *a* factor edge *between every pair of variables $X_i, X_j \in X$ that both appear in some intermediate factor ψ generated by the VE algorithm using $\prec$ as an elimination ordering.* ∎

Example 9.7

Consider the Student *example of figure 9.8, where our query is $P(J)$. Assume that (for some reason) we condition on the variable L and perform elimination on the remaining variables using the ordering C, D, I, H, G, S. The graph induced by this conditioning set and this elimination ordering is shown in figure 9.13, with the conditioning edges shown as dashed lines and the factor edges shown, as usual, by complete lines. The step of conditioning on L causes the introduction of the edges between L and all the other variables. The set of factors we have after the conditioning step immediately leads to the introduction of all the factor edges except for the edge G—S; this latter edge results from the elimination of I.* ∎

We can now use this graph to analyze the complexity of the conditioning algorithm.

Theorem 9.12

Consider an application of the conditioning algorithm to a set of factors Φ, where $U \subset \mathcal{X}$ is the set of conditioning variables, and $\prec$ is the elimination ordering used for the eliminated variables $X \subseteq \mathcal{X} - U$. Then the running time of the algorithm is $O(n \cdot v^m)$, where v is a bound on the domain size of any variable, and m is the size of the largest clique in the graph, using both conditioning and factor edges.

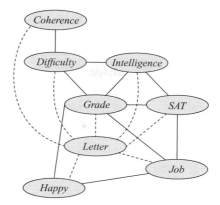

Figure 9.13 **Induced graph for the** Student **example using both conditioning and elimination:** we condition on L and eliminate the remaining variables using the ordering C, D, I, H, G, S.

The proof is left as an exercise (exercise 9.12).

This theorem provides another perspective on the trade-off between conditioning and elimination in terms of their time complexity. Consider, as we did earlier, an algorithm that simply defers the elimination of the conditioning variables U until the end. Consider the effect on the graph of the earlier steps of the elimination algorithm (those preceding the elimination of U). As variables are eliminated, certain edges might be added between the variables in U and other variables (in particular, we add an edge between X and $U \in U$ whenever they are both neighbors of some eliminated variable Y). However, conditioning adds edges between the variables U and *all* other variables X. Thus, conditioning always results in a graph that contains at least as many edges as the induced graph from elimination using this ordering.

However, we can also use the same graph to precisely estimate the time-space trade-off provided by the conditioning algorithm.

Theorem 9.13

Consider an application of the conditioning algorithm to a set of factors Φ, where $U \subset \mathcal{X}$ is the set of conditioning variables, and $\prec$ is the elimination ordering used for the eliminated variables $X \subseteq \mathcal{X} - U$. The space complexity of the algorithm is $O(n \cdot v^{m_f})$, where v is a bound on the domain size of any variable, and m_f is the size of the largest clique in the graph using only factor edges.

The proof is left as an exercise (exercise 9.13).

By comparison, the asymptotic space complexity of variable elimination is the same as its time complexity: exponential in the size of the largest clique containing both types of edges. Thus, we see precisely that conditioning allows us to perform the computation using less space, at the cost (usually) of additional running time.

9.5.4 Improved Conditioning

As we discussed, in terms of the total operations performed, conditioning cannot be better than variable elimination. As we now show, conditioning, naively applied, can be significantly worse.

However, the insights gained from these examples can be used to improve the conditioning algorithm, reducing its cost significantly in many cases.

9.5.4.1 Alternating Conditioning and Elimination

As we discussed, the main problem associated with conditioning is the fact that all computations are repeated for all values of the conditioning variables, even in cases where the different computations are, in fact, identical. This phenomenon arose in the network of example 9.5.

It seems clear, in this example, that we would prefer to eliminate the chain $A_1 \rightarrow \ldots \rightarrow A_{k-1}$ once and for all, before conditioning on A_k. Having eliminated the chain, we would then end up with a much simpler network, involving factors only over A_k, B, C, and D, to which we can then apply conditioning.

The perspective described in section 9.5.3 provides the foundation for implementing this idea. As we discussed, variable elimination works from the inside out, summing out variables in the innermost summation first and caching the results. On the other hand, conditioning works from the outside in, performing the entire internal summation (using elimination) for each value of the conditioning variables, and only then summing the results. However, there is nothing that forces us to split our computation on the outermost summations before considering the inner ones. Specifically, we can eliminate one or more variables on the inside of the summation before conditioning on any variable on the outside.

Example 9.8

Consider again the network of figure 9.12a, and assume that our goal is to compute $P(D)$. We might formulate the expression as:

$$\sum_{A_k} \sum_B \sum_C \sum_{A_1} \ldots \sum_{A_{k-1}} P(A_1, \ldots, A_k, B, C, D).$$

We can first perform the internal summations on $A_{k-1}, \ldots, A_1$, resulting in a set of factors over the scope A_k, B, C, D. We can now condition this network (that is, the Markov network induced by the resulting set of factors) on A_k, resulting in a set of simplified networks over B, C, D (one for each value of A_k). In each such network, we use variable elimination on B and C to compute a factor over D, and aggregate the factors from the different networks, as in standard conditioning. ■

In this example, we first perform some elimination, then condition, and then elimination on the remaining network. Clearly, we can generalize this idea to define an algorithm that alternates the operations of elimination and conditioning arbitrarily. (See exercise 9.14.)

9.5.4.2 Network Decomposition

A second class of examples where we can significantly improve the performance of conditioning arises in networks where conditioning on some subset of variables splits the graph into independent pieces.

Example 9.9

Consider the network of example 9.6, and assume that $k = 16$, and that we begin by conditioning on A_2. After this step, the network is decomposed into two independent pieces. The standard conditioning algorithm would continue by conditioning further, say on A_3. However, there is really no need to condition the top part of the network — the one associated with the variables

A_1, B_1, C_1 on the variable A_3: none of the factors mention A_3, and we would be repeating exactly the same computation for each of its values. ∎

Clearly, having partitioned the network into two completely independent pieces, we can now perform the computation on each of them separately, and then combine the results. In particular, the conditioning variables used on one part would not be used at all to condition the other. More precisely, we can define an algorithm that checks, after each conditioning step, whether the resulting set of factors has been disconnected or not. If it has, it simply partitions them into two or more disjoint sets and calls the algorithm recursively on each subset.

9.6 Inference with Structured CPDs ★

We have seen that BN inference exploits the network structure, in particular the conditional independence and the locality of influence. But when we discussed representation, we also allowed for the representation of finer-grained structure within the CPDs. It turns out that a carefully designed inference algorithm can also exploit certain types of local CPD structure. We focus on two types of structure where this issue has been particularly well studied — independence of causal influence, and asymmetric dependencies — using each of them to illustrate a different type of method for exploiting local structure in variable elimination. We defer the discussion of inference in networks involving continuous variables to chapter 14.

9.6.1 Independence of Causal Influence

The earliest and simplest instance of exploiting local structure was for CPDs that exhibit independence of causal influence, such as noisy-or.

9.6.1.1 Noisy-Or Decompositions

Consider a simple network consisting of a binary variable Y and its four binary parents X_1, X_2, X_3, X_4, where the CPD of Y is a noisy-or. Our goal is to compute the probability of Y. The operations required to execute this process, assuming we use an optimal ordering, is:

- 4 multiplications for $P(X_1) \cdot P(X_2)$
- 8 multiplications for $P(X_1, X_2) \cdot P(X_3)$
- 16 multiplications for $P(X_1, X_2, X_3) \cdot P(X_4)$
- 32 multiplications for $P(X_1, X_2, X_3, X_4) \cdot P(Y \mid X_1, X_2, X_3, X_4)$

The total is 60 multiplications, plus another 30 additions to sum out $X_1, \ldots, X_4$, in order to reduce the resulting factor $P(X_1, X_2, X_3, X_4, Y)$, of size 32, into the factor $P(Y)$ of size 2.

However, we can exploit the structure of the CPD to substantially reduce the amount of computation. As we discussed in section 5.4.1, a noisy-or variable can be decomposed into a deterministic OR of independent noise variables, resulting in the subnetwork shown in figure 9.14a. This transformation, by itself, is not very helpful. The factor $P(Y \mid Z_1, Z_2, Z_3, Z_4)$ is still of size 32 if we represent it as a full factor, so we achieve no gains.

The key idea is that the deterministic OR variable can be decomposed into various cascades of deterministic OR variables, each with a very small indegree. Figure 9.14b shows a simple

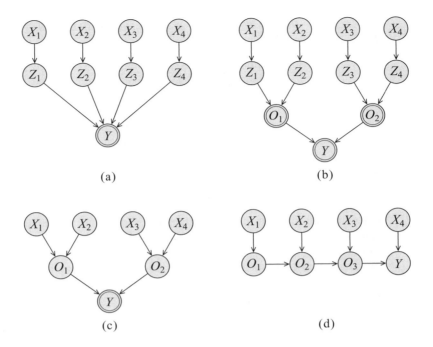

Figure 9.14   Different decompositions for a noisy-or CPD: (a) The standard decomposition of a noisy-or. (b) A tree decomposition of the deterministic-or. (c) A tree-based decomposition of the noisy-or. (d) A chain-based decomposition of the noisy-or.

decomposition of the deterministic OR as a tree. We can simplify this construction by eliminating the intermediate variables Z_i, integrating the "noise" for each X_i into the appropriate O_i. In particular, O_1 would be the noisy-or of X_1 and X_2, with the original noise parameters and a leak parameter of 0. The resulting construction is shown in figure 9.14c.

We can now revisit the inference task in this apparently more complex network. An optimal ordering for variable elimination is $X_1, X_2, X_3, X_4, O_1, O_2$. The cost of performing elimination of X_1, X_2 is:

- 8 multiplications for $\psi_1(X_1, X_2, O_1) = P(X_1) \cdot P(O_1 \mid X_1, X_2)$
- 4 additions to sum out X_1 in $\tau_1(X_2, O_1) = \sum_{X_1} \psi_1(X_1, X_2, O_1)$
- 4 multiplications for $\psi_2(X_2, O_1) = \tau_1(X_2, O_1) \cdot P(X_2)$
- 2 additions for $\tau_2(O_1) = \sum_{X_2} \psi_2(X_2, O_1)$

The cost for eliminating X_3, X_4 is identical, as is the cost for subsequently eliminating O_1, O_2. Thus, the total number of operations is $3 \cdot (8 + 4) = 36$ multiplications and $3 \cdot (4 + 2) = 18$ additions.

A different decomposition of the OR variable is as a simple cascade, where each Z_i is consecutively OR'ed with the previous intermediate result. This decomposition leads to the construction of figure 9.14d. For this construction, an optimal elimination ordering is $X_1, O_1, X_2, O_2, X_3, O_3, X_4$. A simple analysis shows that it takes 4 multiplications and 2 additions to eliminate each of $X_1, \ldots, X_4$, and 8 multiplications and 4 additions to eliminate each of O_1, O_2, O_3. The total cost is $4 \cdot 4 + 3 \cdot 8 = 40$ multiplications and $4 \cdot 2 + 3 \cdot 4 = 20$ additions.

9.6.1.2 The General Decomposition

Clearly, the construction used in the preceding example is a general one that can be applied to more complex networks and other types of CPDs that have independence of causal influence. We take a variable whose CPD has independence of causal influence, and generate its decomposition into a set of independent noise models and a deterministic function, as in figure 5.13.

We then cascade the computation of the deterministic function into a set of smaller steps. Given our assumption about the symmetry and associativity of the deterministic function in the definition of symmetric ICI (definition 5.13), any decomposition of the deterministic function results in the same answer. Specifically, consider a variable Y with parents $X_1, \ldots, X_k$, whose CPD satisfies definition 5.13. We can decompose Y by introducing $k - 1$ intermediate variables $O_1, \ldots, O_{k-1}$, such that:

- the variable Z, and each of the O_i's, has exactly two parents in $Z_1, \ldots, Z_k, O_1, \ldots, O_{i-1}$;
- the CPD of Z and of O_i is the deterministic $\diamond$ of its two parents;
- each Z_l and each O_i is a parent of at most one variable in $O_1, \ldots, O_{k-1}, Z$.

These conditions ensure that $Z = Z_1 \diamond Z_2 \diamond \ldots \diamond Z_k$, but that this function is computed gradually, where the node corresponding to each intermediate result has an indegree of 2.

We note that we can save some extraneous nodes, as in our example, by aggregating the noisy dependence of Z_i on X_i into the CPD where Z_i is used.

After executing this decomposition for every ICI variable in the network, we can simply apply variable elimination to the decomposed network with the smaller factors. As we saw, the complexity of the inference can go down substantially if we have smaller CPDs and thereby smaller factors.

We note that the sizes of the intermediate factors depend not only on the number of variables in their scope, but also on the domains of these variables. For the case of noisy-or variables (as well as noisy-max, noisy-and, and so on), the domain size of these variables is fixed and fairly small. However, in other cases, the domain might be quite large. In particular, in the case of generalized linear models, the domain of the intermediate variable Z generally grows linearly with the number of parents.

Example 9.10

Consider a variable Y with $\mathrm{Pa}_Y = \{X_1, \ldots, X_k\}$, where each X_i is binary. Assume that Y's CPD is a generalized linear model, whose parameters are $w_0 = 0$ and $w_i = w$ for all $i > 1$. Then the domain of the intermediate variable Z is $\{0, 1, \ldots, k\}$. In this case, the decomposition provides a trade-off: The size of the original CPD for $P(Y \mid X_1, \ldots, X_k)$ grows as 2^k; the size of the factors in the decomposed network grow roughly as k^3. In different situations, one approach might be better than the other. ∎

Thus, the decomposition of symmetric ICI variables might not always be beneficial.

9.6.1.3 Global Structure

Our decomposition of the function f that defines the variable Z can be done in many ways, all of which are equivalent in terms of their final result. However, they are not equivalent from the perspective of computational cost. Even in our simple example, we saw that one decomposition can result in fewer operations than the other. The situation is significantly more complicated when we take into consideration other dependencies in the network.

Example 9.11

Consider the network of figure 9.14c, and assume that X_1 and X_2 have a joint parent A. In this case, we eliminate A first, and end up with a factor over X_1, X_2. Aside from the $4 + 8 = 12$ multiplications and 4 additions required to compute this factor $\tau_0(X_1, X_2)$, it now takes 8 multiplications to compute $\psi_1(X_1, X_2, O_1) = \tau_0(X_1, X_2) \cdot P(O_1 \mid X_1, X_2)$, and $4 + 2 = 6$ additions to sum out X_1 and X_2 in ψ_1. The rest of the computation remains unchanged. Thus, the total number of operations required to eliminate all of $X_1, \ldots, X_4$ (after the elimination of A) is $8 + 12 = 20$ multiplications and $6 + 6 = 12$ additions.

Conversely, assume that X_1 and X_3 have the joint parent A. In this case, it still requires 12 multiplications and 4 additions to compute a factor $\tau_0(X_1, X_3)$, but the remaining operations become significantly more complex. In particular, it takes:

- *8 multiplications for $\psi_1(X_1, X_2, X_3) = \tau_0(X_1, X_3) \cdot P(X_2)$*
- *16 multiplications for $\psi_2(X_1, X_2, X_3, O_1) = \psi_1(X_1, X_2, X_3) \cdot P(O_1 \mid X_1, X_2)$*
- *8 additions for $\tau_2(X_3, O_1) = \sum_{X_1, X_2} \psi_2(X_1, X_2, X_3, O_1)$*

The same number of operations is required to eliminate X_3 and X_4. (Once these steps are completed, we can eliminate O_1, O_2 as usual.) Thus, the total number of operations required to eliminate all of $X_1, \ldots, X_4$ (after the elimination of A) is $2 \cdot (8 + 16) = 48$ multiplications and $2 \cdot 8 = 16$ additions, considerably more than our previous case. ∎

Clearly, in the second network structure, had we done the decomposition of the noisy-or variable so as to make X_1 and X_3 parents of O_1 (and X_2, X_4 parents of O_2), we would get the same cost as we did in the first case. However, in order to do that, we need to take into consideration the global structure of the network, and even the order in which other variables are eliminated, at the same time that we are determining how to decompose a particular variable with symmetric ICI. In particular, we should determine the structure of the decomposition at the same time that we are considering the elimination ordering for the network as a whole.

9.6.1.4 Heterogeneous Factorization

An alternative approach that achieves this goal uses a different factorization for a network — one that factorizes the joint distribution for the network into CPDs, as well as the CPDs of symmetric ICI variables into smaller components. This factorization is *heterogeneous*, in that some factors must be combined by product, whereas others need to be combined using the type of operation that corresponds to the symmetric ICI function in the corresponding CPD. One can then define a heterogeneous variable elimination algorithm that combines factors, using whichever operation is appropriate, and that eliminates variables. Using this construction, we can determine a global ordering for the operations that determines the order in which both local

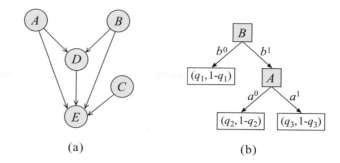

Figure 9.15   A Bayesian network with rule-based structure: (a) the network structure; (b) the CPD for the variable D.

factors and global factors are combined. Thus, in effect, the algorithm determines the order in which the components of an ICI CPD are "recombined" in a way that takes into consideration the structure of the factors created in a variable elimination algorithm.

9.6.2   Context-Specific Independence

A second important type of local CPD structure is the context-specific independence, typically encoded in a CPD as trees or rules. As in the case of ICI, there are two main ways of exploiting this type of structure in the context of a variable elimination algorithm. One approach (exercise 9.15) uses a decomposition of the CPD, which is performed as a preprocessing step on the network structure; standard variable elimination can then be performed on the modified network. The second approach, which we now describe, modifies the variable elimination algorithm itself to conduct its basic operations on structured factors. We can also exploit this structure within the context of a conditioning algorithm.

9.6.2.1   Rule-Based Variable Elimination

An alternative approach is to introduce the structure directly into the factors used in the variable elimination algorithm, allowing it to take advantage of the finer-grained structure. It turns out that this approach is easier to understand and implement for CPDs and factors represented as rules, and hence we present the algorithm in this context.

As specified in section 5.3.1.2, a rule-based CPD is described as a set of mutually exclusive and exhaustive rules, where each rule ρ has the form $\langle c; p \rangle$. As we already discussed, a tree-CPD and a tabular CPD can each be converted into a set of rules in the obvious way.

Example 9.12

Consider the network structure shown in figure 9.15a. Assume that the CPD for the variable D is a tree, whose structure is shown in figure 9.15b. Decomposing this CPD into rules, we get the following *set of rules:*

$$\left\{\begin{array}{cc} \rho_1 & \langle b^0, d^0; 1-q_1 \rangle \\ \rho_2 & \langle b^0, d^1; q_1 \rangle \\ \\ \rho_3 & \langle a^0, b^1, d^0; 1-q_2 \rangle \\ \rho_4 & \langle a^0, b^1, d^1; q_2 \rangle \\ \rho_5 & \langle a^1, b^1, d^0; 1-q_3 \rangle \\ \rho_6 & \langle a^1, b^1, d^0; q_3 \rangle \end{array}\right\}$$

Assume that the CPD $P(E \mid A, B, C, D)$ is also associated with a set of rules. Our discussion will focus on rules involving the variable D, so we show only that part of the rule set:

$$\left\{\begin{array}{cc} \rho_7 & \langle a^0, d^0, e^0; 1-p_1 \rangle \\ \rho_8 & \langle a^0, d^0, e^1; p_1 \rangle \\ \rho_9 & \langle a^0, d^1, e^0; 1-p_2 \rangle \\ \rho_{10} & \langle a^0, d^1, e^1; p_2 \rangle \\ \\ \rho_{11} & \langle a^1, b^0, c^1, d^0, e^0; 1-p_4 \rangle \\ \rho_{12} & \langle a^1, b^0, c^1, d^0, e^1; p_4 \rangle \\ \rho_{13} & \langle a^1, b^0, c^1, d^1, e^0; 1-p_5 \rangle \\ \rho_{14} & \langle a^1, b^0, c^1, d^1, e^1; p_5 \rangle \end{array}\right\}$$

■

Using this type of process, the entire distribution can be factorized into a multiset of rules $\mathcal{R}$, which is the union of all of the rules associated with the CPDs of the different variables in the network. Then, the probability of any instantiation ξ to the network variables $\mathcal{X}$ can be computed as

$$P(\xi) = \prod_{\langle \boldsymbol{c}; p \rangle \in \mathcal{R}, \xi \sim \boldsymbol{c}} p,$$

where we recall that $\xi \sim \boldsymbol{c}$ holds if the assignments ξ and $\boldsymbol{c}$ are compatible, in that they assign the same values to those variables that are assigned values in both.

Thus, as for the tabular CPDs, the distribution is defined in terms of a product of smaller components. In this case, however, we have broken up the tables into their component rows. This definition immediately suggests that we can use similar ideas to those used in the table-based variable elimination algorithm. In particular, we can multiply rules with each other and sum out a variable by adding up rules that give different values to the variables but are the same otherwise.

In general, we define the following two key operations:

Definition 9.8

rule product

Let $\rho_1 = \langle \boldsymbol{c}; p_1 \rangle$ and $\rho_2 = \langle \boldsymbol{c}; p_2 \rangle$ be two rules. Then their product $\rho_1 \cdot \rho_2 = \langle \boldsymbol{c}; p_1 \cdot p_2 \rangle$.   ■

This definition is significantly more restricted than the product of tabular factors, since it requires that the two rules have precisely the same context. We return to this issue in a moment.

Definition 9.9

rule sum

Let Y be a variable with $Val(Y) = \{y^1, \ldots, y^k\}$, and let ρ_i for $i = 1, \ldots, k$ be a rule of the form $\rho_i = \langle \boldsymbol{c}, Y = y^i; p_i \rangle$. Then for $\mathcal{R} = \{\rho_1, \ldots, \rho_k\}$, the sum $\sum_Y \mathcal{R} = \langle \boldsymbol{c}; \sum_{i=1}^{k} p_i \rangle$.   ■

After this operation, Y is summed out in the context c.

Both of these operations can only be applied in very restricted settings, that is, to sets of rules that satisfy certain stringent conditions. In order to make our set of rules amenable to the application of these operations, we might need to refine some of our rules. We therefore define the following final operation:

Definition 9.10

rule split

Let $\rho = \langle c; p \rangle$ be a rule, and let Y be a variable. We define the rule split $Split(\rho \angle Y)$ as follows: If $Y \in Scope[c]$, then $Split(\rho \angle Y) = \{\rho\}$; otherwise,

$$Split(\rho \angle Y) = \{\langle c, Y = y; p \rangle \ : \ y \in Val(Y)\}.$$

∎

In general, the purpose of rule splitting is to make the context of one rule $\rho = \langle c; p \rangle$ compatible with the context c' of another rule ρ'. Naively, we might take all the variables in $Scope[c'] - Scope[c]$ and split ρ recursively on each one of them. However, this process creates unnecessarily many rules.

Example 9.13

Consider ρ_2 and ρ_{14} in example 9.12, and assume we want to multiply them together. To do so, we need to split ρ_2 in order to produce a rule with an identical context. If we naively split ρ_2 on all three variables A, C, E that appear in ρ_{14} and not in ρ_2, the result would be eight rules of the form: $\langle a, b^0, c, d^1, e; q_1 \rangle$, one for each combination of values a, c, e. However, the only rule we really need in order to perform the rule product operation is $\langle a^1, b^0, c^1, d^1, e^1; q_1 \rangle$.

Intuitively, having split ρ_2 on the variable A, it is wasteful to continue splitting the rule whose context is a^0, since this rule (and any derived from it) will not participate in the desired rule product operation with ρ_{14}. Thus, a more parsimonious split of ρ_{14} that still generates this last rule is:

$$\left\{ \begin{array}{l} \langle a^0, b^0, d^1; q_1 \rangle \\ \langle a^1, b^0, c^0, d^1; q_1 \rangle \\ \langle a^1, b^0, c^1, d^1, e^0; q_1 \rangle \\ \langle a^1, b^0, c^1, d^1, e^1; q_1 \rangle \end{array} \right\}$$

This new rule set is still a mutually exclusive and exhaustive partition of the space originally covered by ρ_2, but contains only four rules rather than eight.

∎

In general, we can construct these more parsimonious splits using the recursive procedure shown in algorithm 9.6. This procedure gives precisely the desired result shown in the example.

Rule splitting gives us the tool to take a set of rules and refine them, allowing us to apply either the rule-product operation or the rule-sum operation. The elimination algorithm is shown in algorithm 9.7. Note that the figure only shows the procedure for eliminating a single variable Y. The outer loop, which iteratively eliminates nonquery variables one at a time, is precisely the same as the Sum-Product-VE procedure in algorithm 9.1, except that it takes as input a set of rule factors rather than table factors.

To understand the operation of the algorithm more concretely, consider the following example:

Example 9.14

Consider the network in example 9.12, and assume that we want to eliminate D in this network. Our initial rule set $\mathcal{R}^+$ is the multiset of all of the rules whose scope contains D, which is precisely the set $\{\rho_1, \ldots, \rho_{14}\}$. Initially, none of the rules allows for the direct application of either rule product or rule sum. Hence, we have to split rules.

Algorithm 9.6 Rule splitting algorithm

     Procedure Rule-Split (
        $\rho = \langle c; p \rangle$,   // Rule to be split
        c' // Context to split on
)

1   **if** $c \not\sim c'$ **then return** ρ
2   **if** $Scope[c] \subseteq Scope[c']$ **then return** ρ
3   Select $Y \in Scope[c'] - Scope[c]$
4      $\mathcal{R} \leftarrow Split(\rho \angle Y)$
5      $\mathcal{R}' \leftarrow \cup_{\rho'' \in \mathcal{R}}$ Rule-Split(ρ'', c')
6   **return** $\mathcal{R}'$

The rules ρ_3 on the one hand, and ρ_7, ρ_8 on the other, have compatible contexts, so we can choose to combine them. We begin by splitting ρ_3 and ρ_7 on each other's context, which results in:

$$\left\{ \begin{array}{ll} \rho_{15} & \langle a^0, b^1, d^0, e^0; 1 - q_2 \rangle \\ \rho_{16} & \langle a^0, b^1, d^0, e^1; 1 - q_2 \rangle \\[1em] \rho_{17} & \langle a^0, b^0, d^0, e^0; 1 - p_1 \rangle \\ \rho_{18} & \langle a^0, b^1, d^0, e^0; 1 - p_1 \rangle \end{array} \right\}$$

The contexts of ρ_{15} and $\rho18$ match, so we can now apply rule product, replacing the pair by:

$$\left\{ \begin{array}{ll} \rho_{19} & \langle a^0, b^1, d^0, e^0; (1 - q_2)(1 - p_1) \rangle \end{array} \right\}$$

We can now split ρ_8 using the context of ρ_{16} and multiply the matching rules together, obtaining

$$\left\{ \begin{array}{ll} \rho_{20} & \langle a^0, b^0, d^0, e^1; p_1 \rangle \\ \rho_{21} & \langle a^0, b^1, d^0, e^1; (1 - q_2)p_1 \rangle \end{array} \right\}.$$

The resulting rule set contains $\rho_{17}, \rho_{19}, \rho_{20}, \rho_{21}$ in place of ρ_3, ρ_7, ρ_8.

We can apply a similar process to ρ_4 and ρ_9, ρ_{10}, which leads to their substitution by the rule set:

$$\left\{ \begin{array}{ll} \rho_{22} & \langle a^0, b^0, d^1, e^0; 1 - p_2 \rangle \\ \rho_{23} & \langle a^0, b^1, d^1, e^0; q_2(1 - p_2) \rangle \\ \rho_{24} & \langle a^0, b^0, d^1, e^1; p_2 \rangle \\ \rho_{25} & \langle a^0, b^1, d^1, e^1; q_2 p_2 \rangle \end{array} \right\}.$$

We can now eliminate D in the context a^0, b^1, e^1. The only rules in $\mathcal{R}^+$ compatible with this context are ρ_{21} and ρ_{25}. We extract them from $\mathcal{R}^+$ and sum them; the resulting rule $\langle a^0, b^1, e^1; (1 - q_2)p_1 + q_2 p_2 \rangle$, is then inserted into $\mathcal{R}^-$. We can similarly eliminate D in the context a^0, b^1, e^0.

The process continues, with rules being split and multiplied. When D has been eliminated in a set of mutually exclusive and exhaustive contexts, then we have exhausted all rules involving D; at this point, $\mathcal{R}^+$ is empty, and the process of eliminating D terminates. ∎

Algorithm 9.7 Sum-product variable elimination for sets of rules

Procedure Rule-Sum-Product-Eliminate-Var (
    $\mathcal{R}$,   // Set of rules
    Y // Variable to be eliminated
)

1    $\mathcal{R}^+ \leftarrow \{\rho \in \mathcal{R} \; : \; Scope[\rho] \ni Y\}$
2    $\mathcal{R}^- \leftarrow \mathcal{R} - \mathcal{R}^+$
3   **while** $\mathcal{R}^+ \neq \emptyset$
4   Apply one of the following actions, when applicable
5   **Rule sum:**
6   Select $\mathcal{R}_c \subseteq \mathcal{R}^+$ such that
7        $\mathcal{R}_c = \{\langle c, Y = y^1; p_1\rangle, \ldots, \langle c, Y = y^k; p_k\rangle\}$
8   no other $\rho \in \mathcal{R}^+$ is compatible with c)
9       $\mathcal{R}^- \leftarrow \mathcal{R}^- \cup \sum_Y \mathcal{R}_c$
10      $\mathcal{R}^+ \leftarrow \mathcal{R}^+ - \mathcal{R}_c$
11   **Rule product:**
12   Select $\langle c; p_1\rangle, \langle c; p_2\rangle \in \mathcal{R}^+$
13       $\mathcal{R}^+ \leftarrow \mathcal{R}^+ - \{\langle c; p_1\rangle, \langle c; p_2\rangle\} \cup \{\langle c; p_1 \cdot p_2\rangle\}$
14   **Rule splitting for rule product:**
15   Select $\rho_1, \rho_2 \in \mathcal{R}^+$ such that
16       $\rho_1 = \langle c_1; p_1\rangle$
17       $\rho_2 = \langle c_2; p_2\rangle$
18       $c_1 \sim c_2$
19       $\mathcal{R}^+ \leftarrow \mathcal{R}^+ - \{\rho_1, \rho_2\} \cup \text{Rule-Split}(\rho_1, c_2) \cup \text{Rule-Split}(\rho_2, c_1)$
20   **Rule splitting for rule sum:**
21   Select $\rho_1, \rho_2 \in \mathcal{R}^+$ such that
22       $\rho_1 = \langle c_1, Y = y^i; p_1\rangle$
23       $\rho_2 = \langle c_2, Y = y^j; p_2\rangle$
24       $c_1 \sim c_2$
25       $i \neq j$
26       $\mathcal{R}^+ \leftarrow \mathcal{R}^+ - \{\rho_1, \rho_2\} \cup \text{Rule-Split}(\rho_1, c_2) \cup \text{Rule-Split}(\rho_2, c_1)$
27   **return** $\mathcal{R}^-$

A different way of understanding the algorithm is to consider its application to rule sets that originate from standard table-CPDs. It is not difficult to verify that the algorithm performs exactly the same set of operations as standard variable elimination. For example, the standard operation of factor product is simply the application of rule splitting on all of the rules that constitute the two tables, followed by a sequence of rule product operations on the resulting rule pairs. (See exercise 9.16.)

To prove that the algorithm computes the correct result, we need to show that each operation performed in the context of the algorithm maintains a certain correctness invariant. Let $\mathcal{R}$ be the current set of rules maintained by the algorithm, and $\boldsymbol{W}$ be the variables that have not yet been eliminated. Each operation must maintain the following condition:

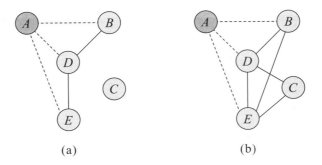

Figure 9.16 Conditioning a Bayesian network whose CPDs have CSI: (a) conditioning on a^0; (b) conditioning on a^1.

The probability of a context c such that $Scope[c] \subseteq W$ can be obtained by multiplying all rules $\langle c'; p \rangle \in \mathcal{R}$ whose context is compatible with c.

It is not difficult to show that the invariant holds initially, and that each step in the algorithm maintains it. Thus, the algorithm as a whole is correct.

9.6.2.2   Conditioning

We can also use other techniques for exploiting CSI in inference. In particular, we can generalize the notion of conditioning to this setting in an interesting way. Consider a network $\mathcal{B}$, and assume that we condition it on a variable U. So far, we have assumed that the structure of the different conditioned networks, for the different values u of U, is the same. When the CPDs are tables, with no extra structure, this assumption generally holds. However, when the CPDs have CSI, we might be able to utilize the additional structure to simplify the conditioned networks considerably.

Example 9.15

Consider the network shown in figure 9.15, as described in example 9.12. Assume we condition this network on the variable A. If we condition on a^0, we see that the reduced CPD for E no longer depends on C. Thus, the conditioned Markov network for this set of factors is the one shown in figure 9.16a. By contrast, when we condition on a^1, the reduced factors do not "lose" any variables aside from A, and we obtain the conditioned Markov network shown in figure 9.16b. Note that the network in figure 9.16a is so simple that there is no point performing any further conditioning on it. Thus, we can continue the conditioning process for only one of the two branches of the computation — the one corresponding to a^1. ∎

In general, we can extend the conditioning algorithm of section 9.5 to account for CSI in the CPDs or in the factors of a Markov network. Consider a single conditioning step on a variable U. As we enumerate the different possible values u of U, we generate a possibly different conditioned network for each one. Depending on the structure of this network, we select which step to take next in the context of this particular network. In different networks, we might choose a different variable to use for the next conditioning step, or we might decide to stop the conditioning process for some networks altogether.

9.6.3 Discussion

We have presented two approaches to variable elimination in the case of local structure in the CPDs: preprocessing followed by standard variable elimination, and specialized variable elimination algorithms that use a factorization of the structured CPD. These approaches offer different trade-offs. On the one hand, the specialized variable elimination approach reveals more of the structure of the CPDs to the inference algorithm, allowing the algorithm more flexibility in exploiting this structure. Thus, this approach can achieve lower computational cost than any fixed decomposition scheme (see box 9.D). By comparison, the preprocessing approach embeds some of the structure within deterministic CPDs, a structure that most variable elimination algorithms do not fully exploit.

On the other hand, specialized variable elimination schemes such as those for rules require the use of special-purpose variable elimination algorithms rather than off-the-shelf packages. Furthermore, the data structures for tables are significantly more efficient than those for other types of factors such as rules. Although this difference seems to be an implementation issue, it turns out to be quite significant in practice. One can somewhat address this limitation by the use of more sophisticated algorithms that exploit efficient table-based operations whenever possible (see exercise 9.18).

 Although the trade-offs between these two approaches is not always clear, it is generally the case that, in networks with significant amounts of local structure, it is valuable to design an inference scheme that exploits this structure for increased computational efficiency.

Box 9.D — Case Study: Inference with Local Structure. *A natural question is the extent to which local structure can actually help speed up inference.*

In one experimental comparison by Zhang and Poole (1996), four algorithms were applied to fragments of the CPCS network (see box 5.D): standard variable elimination (with table representation of factors), the two decompositions illustrated in figure 9.14 for the case of noisy-or, and a special-purpose elimination algorithm that uses a heterogeneous factorization. The results show that in a network such as CPCS, which uses predominantly noisy-or and noisy-max CPDs, significant gains in performance can be obtained. They results also showed that the two decomposition schemes (tree-based and chain-based) are largely equivalent in their performance, and the heterogeneous factorization outperforms both of them, due to its greater flexibility in dynamically determining the elimination ordering during the course of the algorithm.

For rule-based variable elimination, no large networks with extensive rule-based structure had been constructed. So, Poole and Zhang (2003) used a standard benchmark network, with 32 variables and 11,018 entries. Entries that were within 0.05 of each other were collaped, to construct a more compact rule-based representation, with a total of 5,834 distinct entries. As expected, there are a large number of cases where the use of rule-based inference provided significant savings. However, there were also many cases where contextual independence does not provide significant help, in which case the increased overhead of the rule-based inference dominates, and standard VE performs better.

At a high level, the main conclusion is that table-based approaches are amenable to numerous optimizations, such as those described in box 10.A, which can improve the performance by an order of magnitude or even more. Such optimizations are harder to define for more complex data structures. Thus, it is only useful to consider algorithms that exploit local structure either when it is extensively present in the model, or when it has specific structure that can, itself, be exploited using specialized algorithms.

9.7 Summary and Discussion

In this chapter, we described the basic algorithms for exact inference in graphical models. As we saw, probability queries essentially require that we sum out an exponentially large joint distribution. The fundamental idea that allows us to avoid the exponential blowup in this task is the use of dynamic programming, where we perform the summation of the joint distribution from the inside out rather than from the outside in, and cache the intermediate results, thereby avoiding repeated computation.

We presented an algorithm based on this insight, called variable elimination. The algorithm works using two fundamental operations over factors — multiplying factors and summing out variables in factors. We analyzed the computational complexity of this algorithm using the structural properties of the graph, showing that the key computational metric was the induced width of the graph.

We also presented another algorithm, called conditioning, which performs some of the summation operations from the outside in rather than from the inside out, and then uses variable elimination for the rest of the computation. Although the conditioning algorithm is never less expensive than variable elimination in terms of running time, it requires less storage space and hence provides a time-space trade-off for variable elimination.

We showed that both variable elimination and conditioning can take advantage of local structure within the CPDs. Specifically, we presented methods for making use of CPDs with independence of causal influence, and of CPDs with context-specific independence. In both cases, techniques tend to fall into two categories: In one class of methods, we modify the network structure, adding auxiliary variables that reveal some of the structure inside the CPD and break up large factors. In the other, we modify the variable elimination algorithm directly to use structured factors rather than tables.

Although exact inference is tractable for surprisingly many real-world graphical models, it is still limited by its worst-case exponential performance. There are many models that are simply too complex for exact inference. As one example, consider the $n \times n$ grid-structured pairwise Markov networks of box 4.A. It is not difficult to show that the minimal tree-width of this network is n. Because these networks are often used to model pixels in an image, where $n = 1,000$ is quite common, it is clear that exact inference is intractable for such networks. Another example is the family of networks that we obtain from the template model of example 6.11. Here, the moralized network, given the evidence, is a fully connected bipartite graph; if we have n variables on one side and m on the other, the minimal tree-width is $\min(n, m)$, which can be very large for many practical models. Although this example is obviously a toy domain, examples of similar structure arise often in practice. In later chapters, we will see many other examples where exact inference fails to scale up. Therefore, in chapter 11 and chapter 12 we discuss approximate inference methods that trade off the accuracy of the results for the ability to scale up to much larger models.

One class of networks that poses great challenges to inference is the class of networks induced by template-based representations. These languages allow us to specify (or learn) very small, compact models, yet use them to construct arbitrarily large, and often densely connected, networks. Chapter 15 discusses some of the techniques that have been used to deal with dynamic Bayesian networks.

Our focus in this chapter has been on inference in networks involving only discrete variables. The introduction of continuous variables into the network also adds a significant challenge. Although the ideas that we described here are instrumental in constructing algorithms for this richer class of models, many additional ideas are required. We discuss the problems and the solutions in chapter 14.

9.8   Relevant Literature

The first formal analysis of the computational complexity of probabilistic inference in Bayesian networks is due to Cooper (1990).

peeling

forward-backward algorithm

nonserial dynamic programming

Variants of the variable elimination algorithm were invented independently in multiple communities. One early variant is the *peeling* algorithm of Cannings et al. (1976, 1978), formulated for the analysis of genetic pedigrees. Another early variant is the *forward-backward algorithm*, which performs inference in hidden Markov models (Rabiner and Juang 1986). An even earlier variant of this algorithm was proposed as early as 1880, in the context of continuous models (Thiele 1880). Interestingly, the first variable elimination algorithm for fully general models was invented as early as 1972 by Bertelé and Brioschi (1972), under the name *nonserial dynamic programming*. However, they did not present the algorithm in the setting of probabilistic inference in graph-structured models, and therefore it was many years before the connection to their work was recognized. Other early work with similar ideas but a very different application was done in the database community (Beeri et al. 1983).

The general problem of probabilistic inference in graphical models was first tackled by Kim and Pearl (1983), who proposed a local message passing algorithm in polytree-structured Bayesian networks. These ideas motivated the development of a wide variety of more general algorithms. One such trajectory includes the clique tree methods that we discuss at length in the next chapter (see also section 10.6). A second includes a specrum of other methods (for example, Shachter 1988; Shachter et al. 1990), culminating in the variable elimination algorithm, as presented here, first described by Zhang and Poole (1994) and subsequently by Dechter (1999). Huang and Darwiche (1996) provide some useful tips on an efficient implementation of algorithms of this type.

Dechter (1999) presents interesting connections between these algorithms and constraint-satisfaction algorithms, connections that have led to fruitful work in both communities. Other generalizations of the algorithm to settings other than pure probabilistic inference were described by Shenoy and Shafer (1990); Shafer and Shenoy (1990) and by Dawid (1992). The construction of the network polynomial was proposed by Darwiche (2003).

The complexity analysis of the variable elimination algorithm is described by Bertelé and Brioschi (1972); Dechter (1999). The analysis is based on core concepts in graph theory that have been the subject of extensive theoretical analysis; see Golumbic (1980); Tarjan and Yannakakis (1984); Arnborg (1985) for an introduction to some of the key concepts and algorithms.

Much work has been done on the problem of finding low-tree-width triangulations or (equivalently) elimination orderings. One of the earliest algorithms is the maximum cardinality search of Tarjan and Yannakakis (1984). Arnborg, Corneil, and Proskurowski (1987) show that the problem of finding the minimal tree-width elimination ordering is $\mathcal{NP}$-hard. Shoikhet and Geiger (1997) describe a relatively efficient algorithm for finding this optimal elimination ordering — one whose cost is approximately the same as the cost of inference with the resulting ordering. Becker and Geiger (2001) present an algorithm that finds a close-to-optimal ordering. Nevertheless, most implementations use one of the standard heuristics. A good survey of these heuristic methods is presented by Kjærulff (1990), who also provides an extensive empirical comparison. Fishelson and Geiger (2003) suggest the use of stochastic search as a heuristic and provide another set of comprehensive experimental comparisons, focusing on the problem of genetic linkage analysis. Bodlaender, Koster, van den Eijkhof, and van der Gaag (2001) provide a series of simple preprocessing steps that can greatly reduce the cost of triangulation.

The first incarnation of the conditioning algorithm was presented by Pearl (1986a), in the context of cutset conditioning, where the conditioning variables cut all loops in the network, forming a polytree. Becker and Geiger (1994); Becker, Bar-Yehuda, and Geiger (1999) present a variety of algorithms for finding a small loop cutset. The general algorithm, under the name *global conditioning*, was presented by Shachter et al. (1994). They also demonstrated the equivalence of conditioning and variable elimination (or rather, the clique tree algorithm) in terms of the underlying computations, and pointed out the time-space trade-offs between these two approaches. These time-space trade-offs were then placed in a comprehensive computational framework in the *recursive conditioning* method of Darwiche (2001b); Allen and Darwiche (2003a,b). Cutset algorithms have made a significant impact on the application of genetic linkage analysis Schäffer (1996); Becker et al. (1998), which is particularly well suited to this type of method.

The two noisy-or decomposition methods were described by Olesen, Kjærulff, Jensen, Falck, Andreassen, and Andersen (1989) and Heckerman and Breese (1996). An alternative approach that utilizes a heterogeneous factorization was described by Zhang and Poole (1996); this approach is more flexible, but requires the use of a special-purpose inference algorithm. For the case of CPDs with context-specific independence, the decomposition approach was proposed by Boutilier, Friedman, Goldszmidt, and Koller (1996). The rule-based variable elimination algorithm was proposed by Poole and Zhang (2003). The trade-offs here are similar to the case of the noisy-or methods.

9.9   Exercises

Exercise 9.1★

Prove theorem 9.2.

Exercise 9.2★

Consider a factor produced as a product of some of the CPDs in a Bayesian network $\mathcal{B}$:

$$\tau(\boldsymbol{W}) = \prod_{i=1}^{k} P(Y_i \mid \mathrm{Pa}_{Y_i})$$

where $\boldsymbol{W} = \cup_{i=1}^{k}(\{Y_i\} \cup \mathrm{Pa}_{Y_i})$.

a. Show that τ is a conditional probability in some network. More precisely, construct another Bayesian network $\mathcal{B}'$ and a disjoint partition $\boldsymbol{W} = \boldsymbol{Y} \cup \boldsymbol{Z}$ such that $\tau(\boldsymbol{W}) = P_{\mathcal{B}'}(\boldsymbol{Y} \mid \boldsymbol{Z})$.

b. Conclude that all of the intermediate factors produced by the variable elimination algorithm are also conditional probabilities in some network.

Exercise 9.3

Consider a modified variable elimination algorithm that is allowed to multiply all of the entries in a single factor by some arbitrary constant. (For example, it may choose to renormalize a factor to sum to 1.) If we run this algorithm on the factors resulting from a Bayesian network with evidence, which types of queries can we still obtain the right answer to, and which not?

Exercise 9.4★

This exercise shows basic properties of the network polynomial and its derivatives:

a. Prove equation (9.8).

b. Prove equation (9.9).

evidence
retraction

c. Let $\boldsymbol{Y} = \boldsymbol{y}$ be some assignment. For $Y_i \in \boldsymbol{Y}$, we now consider what happens if we *retract* the observation $Y_i = y_i$. More precisely, let $\boldsymbol{y}_{-i}$ be the assignment in $\boldsymbol{y}$ to all variables other than Y_i. Show that

$$P(\boldsymbol{y}_{-i}, Y_i = y_i' \mid \theta) = \frac{\partial f_{\Phi}(\boldsymbol{\theta}, \boldsymbol{\lambda}^{\boldsymbol{y}})}{\lambda_{y_i'}}$$

$$P(\boldsymbol{y}_{-i} \mid \theta) = \sum_{y_i'} \frac{\partial f_{\Phi}(\boldsymbol{\theta}, \boldsymbol{\lambda}^{\boldsymbol{y}})}{\lambda_{y_i'}}.$$

Exercise 9.5★

sensitivity
analysis

In this exercise, you will show how you can use the gradient of the probability of a Bayesian network to perform *sensitivity analysis*, that is, to compute the effect on a probability query of changing the parameters in a single CPD $P(X \mid \boldsymbol{U})$. More precisely, let $\boldsymbol{\theta}$ be one set of parameters for a network $\mathcal{G}$, where we have that $\theta_{x|\boldsymbol{u}}$ is the parameter associated with the conditional probability entry $P(X \mid \boldsymbol{U})$. Let $\boldsymbol{\theta}'$ be another parameter assignment that is the same except that we replace the parameters $\boldsymbol{\theta}_{x|\boldsymbol{u}}$ with $\theta'_{x|\boldsymbol{u}} = \theta_{x|\boldsymbol{u}} + \Delta_{x|\boldsymbol{u}}$.

For an assignment $\boldsymbol{e}$ (which may or may not involve variables in $X, \boldsymbol{U}$, compute the change $P(\boldsymbol{e} : \boldsymbol{\theta}) - P(\boldsymbol{e} : \boldsymbol{\theta}')$ in terms of $\Delta_{x|\boldsymbol{u}}$, and the network derivatives.

Exercise 9.6★

Consider some run of variable elimination over the factors Φ, where all variables are eliminated. This run generates some set of intermediate factors $\tau_i(\boldsymbol{W}_i)$. We can define a set of intermediate (arithmetic, not random) variables v_{ik} corresponding to the different entries $\tau_i(\boldsymbol{w}_i^k)$.

a. Show how, for each variable v_{ij}, we can write down an algebraic expression that defines v_{ij} in terms of: the parameters λ_{x_i}; the parameters θ_{x_c}; and variables v_{jl} for $j < i$.

b. Use your answer to the previous part to define an alternative representation whose complexity is linear in the total size of the intermediate factors in the VE run.

c. Show how the same representation can be used to compute all of the derivatives of the network polynomial; the complexity of your algorithm should be linear in the compact representation of the network polynomial that you derived in the previous part. (Hint: Consider the partial derivatives of the network polynomial relative to each v_{ij}, and use the chain rule for derivatives.)

Exercise 9.7

Prove proposition 9.1.

Exercise 9.8★

Prove theorem 9.10, by showing that any ordering produced by the maximum cardinality search algorithm eliminates cliques one by one, starting from the leaves of the clique tree.

Exercise 9.9

a. Show that variable elimination on polytrees can be performed in linear time, assuming that the local probability models are represented as full tables. Specifically, for any polytree, describe an elimination ordering, and show that the complexity of variable elimination with your ordering is linear in the size of the network. Note that the linear time bound here is in terms of the size of the CPTs in the network, so that the cost of the algorithm grows exponentially with the number of parents of a node.

b. Extend your result from (1) to apply to cases where the CPDs satisfy independence of causal influence. Note that, in this case, the network representation is linear in the number of variables in the network, and the algorithm should be linear in that number.

c. Now extend your result from (1) to apply to cases where the CPDs are tree-structured. In this case, the network representation is the sum of the sizes of the trees in the individual CPDs, and the algorithm should be linear in that number.

Exercise 9.10★

Consider the four criteria described in connection with Greedy-Ordering of algorithm 9.4: Min-Neighbors, Min-Weight, Min-Fill, and Weighted-Min-Fill. Show that none of these criteria dominate the others; that is, for any pair, there is always a graph where the ordering produced by one of them is better than that produced by the other. As our measure of performance, use the computational cost of full variable elimination (that is, for computing the partition function). For each counterexample, define the structure of the graph and the cardinality of the variables, and show the ordering produced by each member of the pair.

Exercise 9.11★

Let $\mathcal{H}$ be an undirected graph, and $\prec$ an elimination ordering. Prove that $X-Y$ is a fill edge in the induced graph if and only if there is a path $X-Z_1-\ldots Z_k-Y$ in $\mathcal{H}$ such that $Z_i \prec X$ and $Z_i \prec Y$ for all $i = 1, \ldots, k$.

Exercise 9.12★

Prove theorem 9.12.

Exercise 9.13★

Prove theorem 9.13.

Exercise 9.14★

The standard conditioning algorithm first conditions the network on the conditioning variables $\boldsymbol{U}$, splitting the computation into a set of computations, one for every instantiation $\boldsymbol{u}$ to $\boldsymbol{U}$; it then performs variable elimination on the remaining network. As we discussed in section 9.5.4.1, we can generalize conditioning so that it alternates conditioning steps and elimination in an arbitrary way. In this question, you will formulate such an algorithm and provide a graph-theoretic analysis of its complexity.

Let Φ be a set of factors over $\mathcal{X}$, and let $\boldsymbol{X}$ be a set of nonquery variables. Define a *summation procedure* σ to be a sequence of operations, each of which is either $elim(X)$ or $cond(X)$ for some $X \in \boldsymbol{X}$, such that each $X \in \boldsymbol{X}$ appears in the sequence σ precisely once. The semantics of this procedure is that, going from left to right, we perform the operation described on the variables in sequence. For example, the summation procedure of example 9.5 would be written as:

$$elim(A_{k-1}), elim(A_{k-2}), \ldots elim(A_1), cond(A_k), elim(C), elim(B).$$

a. Define an algorithm that takes a summation sequence as input and performs the operations in the order stated. Provide precise pseudo-code for the algorithm.

b. Define the notion of an induced graph for this algorithm, and define the time and space complexity of the algorithm in terms of the induced graph.

Exercise 9.15$\star$

In section 9.6.1.1, we described an approach to decomposing noisy-or CPDs, aimed at reducing the cost of variable elimination. In this exercise, we derive a construction for CPD-trees in a similar spirit.

a. Consider a variable Y that has a binary-valued parent A and four additional parents $X_1, \ldots, X_4$. Assume that the CPD of Y is structured as a tree whose first split is A, and where Y depends only on X_1, X_2 in the $A = a^1$ branch, and only on X_3, X_4 in the $A = a^0$ branch. Define two new variables, Y_{a^1} and Y_{a^0}, which represent the value that Y would take if A were to have the value a^1, and the value that Y would take if A were to have the value a^0. Define a new model for Y that is defined in terms of these new variables. Your model should precisely specify the CPDs for Y_{a^1}, Y_{a^0}, and Y in terms of Y's original CPD.

b. Define a general procedure that recursively decomposes a tree-CPD using the same principles.

Exercise 9.16

In this exercise, we show that rule-based variable elimination performs exactly the same operations as table-based variable elimination, when applied to rules generated from table-CPDs. Consider two table factors $\phi(\boldsymbol{X}), \phi'(\boldsymbol{Y})$. Let $\mathcal{R}$ be the set of constituent rules for $\phi(\boldsymbol{X})$ and $\mathcal{R}'$ the set of constituent rules for $\phi(\boldsymbol{Y})$.

a. Show that the operation of multiplying $\phi \cdot \phi'$ can be implemented as a series of rule splits on $\mathcal{R} \cup \mathcal{R}'$, followed by a series of rule products.

b. Show that the operation of summing out $Y \in \boldsymbol{X}$ in ϕ can be implemented as a series of rule sums in $\mathcal{R}$.

Exercise 9.17$\star$

Prove that each step in the algorithm of algorithm 9.7 maintains the program-correctness invariant described in the text: Let $\mathcal{R}$ be the current set of rules maintained by the algorithm, and $\boldsymbol{W}$ be the variables that have not yet been eliminated. The invariant is that:

> The probability of a context $\boldsymbol{c}$ such that $Scope[\boldsymbol{c}] \subseteq \boldsymbol{W}$ can be obtained by multiplying all rules $\langle \boldsymbol{c}'; p \rangle \in \mathcal{R}$ whose context is compatible with $\boldsymbol{c}$.

Exercise 9.18$\star\star$

Consider an alternative factorization of a Bayesian network where each factor is a hybrid between a rule and a table, called a *confactor*. Like a rule, a confactor associated with a context $\boldsymbol{c}$; however, rather than a single number, each confactor contains not a single number, but a standard table-based factor. For example, the CPD of figure 5.4a would have a confactor, associated with the middle branch, whose context is a^1, s^0, and whose associated table is

l^0, j^0	0.9
l^0, j^1	0.1
l^1, j^0	0.4
l^1, j^1	0.6

Extend the rule splitting algorithm of algorithm 9.6 and the rule-based variable elimination algorithm of algorithm 9.7 to operate on confactors rather than rules. Your algorithm should use the efficient table-based data structures and operations when possible, resorting to the explicit partition of tables into rules only when absolutely necessary.

Exercise 9.19★★

We have shown that the sum-product variable elimination algorithm is sound, in that it returns the same answer as first multiplying all the factors, and then summing out the nonquery variables. Exercise 13.3 asks for a similar argument for max-product. One can prove similar results for other pairs of operations, such as max-sum. Rather than prove the same result for each pair of operations we encounter, we now provide a *generalized variable elimination* algorithm from which these special cases, as well as others, follow directly. This general algorithm is based on the following result, which is stated in terms of a pair of abstract operators: generalized combination of two factors, denoted $\phi_1 \otimes \phi_2$; and generalized marginalization of a factor ϕ over a subset W, denoted $\Lambda_W(\phi)$. We define our generalized variable elimination algorithm in direct analogy to the sum-product algorithm of algorithm 9.1, replacing factor product with $\otimes$ and summation for variable elimination with Λ.

generalized
variable
elimination

We now show that if these two operators satisfy certain conditions, the variable elimination algorithm for these two operations is sound:

Commutativity of combination: For any factors ϕ_1, ϕ_2:

$$\phi_1 \otimes \phi_2 = \phi_2 \otimes \phi_1. \tag{9.12}$$

Associativity of combination: For any factors ϕ_1, ϕ_2, ϕ_3:

$$\phi_1 \otimes (\phi_2 \otimes \phi_3) = (\phi_1 \otimes \phi_2) \otimes \phi_3. \tag{9.13}$$

Consonance of marginalization: If ϕ is a factor of scope W, and Y, Z are disjoint subsets of W, then:

$$\Lambda_Y(\Lambda_Z(\phi)) = \Lambda_{(Y \cup Z)}(\phi). \tag{9.14}$$

Marginalization over combination: If ϕ_1 is a factor of scope W and $Y \cap W = \emptyset$, then:

$$\Lambda_Y(\phi_1 \otimes \phi_2) = \phi_1 \otimes \Lambda_Y(\phi_2). \tag{9.15}$$

Show that if $\otimes$ and Λ satisfy the preceding axioms, then we obtain a theorem analogous to theorem 9.5. That is, the algorithm, when applied to a set of factors Φ and a set of variables to be eliminated Z, returns a factor

$$\phi^*(Y) = \Lambda_Z(\bigotimes_{\phi \in \Phi} \phi).$$

Exercise 9.20★★

You are taking the final exam for a course on computational complexity theory. Being somewhat too theoretical, your professor has insidiously sneaked in some unsolvable problems and has told you that exactly K of the N problems have a solution. Out of generosity, the professor has also given you a probability distribution over the solvability of the N problems.

To formalize the scenario, let $\mathcal{X} = \{X_1, \ldots, X_N\}$ be binary-valued random variables corresponding to the N questions in the exam where $Val(X_i) = \{0(\text{unsolvable}), 1(\text{solvable})\}$. Furthermore, let $\mathcal{B}$ be a Bayesian network parameterizing a probability distribution over $\mathcal{X}$ (that is, problem i may be easily used to solve problem j so that the probabilities that i and j are solvable are not independent in general).

a. We begin by describing a method for computing the probability of a question being solvable. That is we want to compute $P(X_i = 1, \text{Possible}(\mathcal{X}) = K)$ where

$$\text{Possible}(\mathcal{X}) = \sum_i \mathbf{1}\{X_i = 1\}$$

is the number of solvable problems assigned by the professor.

To this end, we define an *extended factor* ϕ as a "regular" factor ψ and an index so that it defines a function $\phi(\boldsymbol{X}, L) : Val(\boldsymbol{X}) \times \{0, \ldots, N\} \mapsto I\!\!R$ where $\boldsymbol{X} = Scope[\phi]$. A projection of such a factor $[\phi]_l$ is a regular factor $\psi : Val(\boldsymbol{X}) \mapsto I\!\!R$, such that $\psi(\boldsymbol{X}) = \phi(\boldsymbol{X}, l)$.

Provide a definition of factor combination and factor marginalization for these extended factors such that

$$P(X_i, \text{Possible}(\mathcal{X}) = K) = \left[\sum_{\mathcal{X} - \{X_i\}} \prod_{\phi \in \Phi} \phi \right]_K, \tag{9.16}$$

where each $\phi \in \Phi$ is an extended factor corresponding to some CPD of the Bayesian network, defined as follows:

$$\phi_{X_i}(\{X_i\} \cup \mathbf{Pa}_{X_i}, k) = \begin{cases} P(X_i \mid \text{Pa}_{X_i}) & \text{if } X_i = k \\ 0 & \text{otherwise} \end{cases}$$

b. Show that your operations satisfy the condition of exercise 9.19 so that you can compute equation (9.16) use the generalized variable elimination algorithm.

c. Realistically, you will have time to work on exactly M problems ($1 \le M \le N$). Obviously, your goal is to maximize the expected number of solvable problems that you attempt. (Luckily for you, every solvable problem that you attempt you will solve correctly, and you neither gain nor lose credit for working on an unsolvable problem.) Let $\boldsymbol{Y}$ be a subset of $\mathcal{X}$ indicating exactly M problems you choose to work on, and let

$$\text{Correct}(\mathcal{X}, \boldsymbol{Y}) = \sum_{X_i \in \boldsymbol{Y}} X_i$$

be the number of solvable problems that you attempt. The expected number of problems you solve is

$$\boldsymbol{E}_{P_B}[\text{Correct}(\mathcal{X}, \boldsymbol{Y}) \mid \text{Possible}(\mathcal{X}) = K]. \tag{9.17}$$

Using your generalized variable elimination algorithm, provide an efficient algorithm for computing this expectation.

d. Your goal is to find $\boldsymbol{Y}$ that optimizes equation (9.17). Provide a simple example showing that:

$$\arg \max_{\boldsymbol{Y}:|\boldsymbol{Y}|=M} \boldsymbol{E}_{P_B}[\text{Correct}(\mathcal{X}, \boldsymbol{Y})] \ne \arg \max_{\boldsymbol{Y}:|\boldsymbol{Y}|=M} \boldsymbol{E}_{P_B}[\text{Correct}(\mathcal{X}, \boldsymbol{Y}) \mid \text{Possible}(\mathcal{X}) = K].$$

e. Give an efficient algorithm for finding

$$\arg \max_{\boldsymbol{Y}:|\boldsymbol{Y}|=M} \boldsymbol{E}_{P_B}[\text{Correct}(\mathcal{X}, \boldsymbol{Y}) \mid \text{Possible}(\mathcal{X}) = K].$$

(Hint: Use linearity of expectations.)

10 *Exact Inference: Clique Trees*

In the previous chapter, we showed how we can exploit the structure of a graphical model to perform exact inference effectively. The fundamental insight in this process is that the factorization of the distribution allows us to perform local operations on the factors defining the distribution, rather than simply generate the entire joint distribution. We implemented this insight in the context of the *variable elimination* algorithm, which sums out variables one at a time, multiplying the factors necessary for that operation.

In this chapter, we present an alternative implementation of the same insight. As in the case of variable elimination, the algorithm uses manipulation of factors as its basic computational step. However, the algorithm uses a more global data structure for scheduling these operations, with surprising computational benefits.

Throughout this chapter, we will assume that we are dealing with a set of factors Φ over a set of variables $\mathcal{X}$, where each factor ϕ_i has a scope $\boldsymbol{X}_i$. This set of factors defines a (usually) unnormalized measure

$$\tilde{P}_\Phi(\mathcal{X}) = \prod_{\phi_i \in \Phi} \phi_i(\boldsymbol{X}_i). \tag{10.1}$$

For a Bayesian network without evidence, the factors are simply the CPDs, and the measure $\tilde{P}_\Phi$ is a normalized distribution. For a Bayesian network $\mathcal{B}$ with evidence $\boldsymbol{E} = \boldsymbol{e}$, the factors are the CPDs restricted to $\boldsymbol{e}$, and $\tilde{P}_\Phi(\mathcal{X}) = P_\mathcal{B}(\mathcal{X}, \boldsymbol{e})$. For a Gibbs distribution (with or without evidence), the factors are the (restricted) potentials, and $\tilde{P}_\Phi$ is the unnormalized Gibbs measure.

It is important to note that all of the operations that one can perform on a normalized distribution can also be performed on an unnormalized measure. In particular, we can marginalize $\tilde{P}_\Phi$ on a subset of the variables by summing out the others. We can also consider a conditional measure, $\tilde{P}_\Phi(\boldsymbol{X} \mid \boldsymbol{Y}) = \tilde{P}_\Phi(\boldsymbol{X}, \boldsymbol{Y})/\tilde{P}_\Phi(\boldsymbol{Y})$ (which, in fact, is the same as $P_\Phi(\boldsymbol{X} \mid \boldsymbol{Y})$).

10.1 Variable Elimination and Clique Trees

Recall that the basic operation of the variable elimination algorithm is the manipulation of factors. Each step in the computation creates a factor ψ_i by multiplying existing factors. A variable is then eliminated in ψ_i to generate a new factor τ_i, which is then used to create another factor. In this section, we present another view of this computation. We consider a

message   factor ψ_i to be a computational data structure, which takes *"messages"* τ_j generated by other factors ψ_j, and generates a message τ_i that is used by another factor ψ_l.

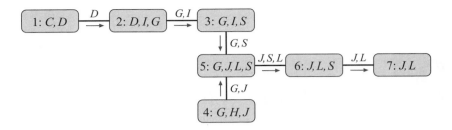

Figure 10.1   Cluster tree for the VE execution in table 9.1

10.1.1   Cluster Graphs

We begin by defining a *cluster graph* — a data structure that provides a graphical flowchart of the factor-manipulation process. Each node in the cluster graph is a *cluster*, which is associated with a subset of variables; the graph contains undirected edges that connect clusters whose scopes have some non-empty intersection. We note that this definition is more general than the data structures we use in this chapter, but this generality will be important in the next chapter, where we significantly extend the algorithms of this chapter.

Definition 10.1

cluster graph

family
preservation

sepset

A cluster graph $\mathcal{U}$ *for a set of factors* Φ *over* $\mathcal{X}$ *is an undirected graph, each of whose nodes i is associated with a subset $C_i \subseteq \mathcal{X}$. A cluster graph must be* family-preserving *— each factor $\phi \in \Phi$ must be associated with a cluster C_i, denoted $\alpha(\phi)$, such that $Scope[\phi] \subseteq C_i$. Each edge between a pair of clusters C_i and C_j is associated with a sepset $S_{i,j} \subseteq C_i \cap C_j$.* ∎

An execution of variable elimination defines a cluster graph: We have a cluster for each factor ψ_i used in the computation, which is associated with the set of variables $C_i = Scope[\psi_i]$. We draw an edge between two clusters C_i and C_j if the message τ_i, produced by eliminating a variable in ψ_i, is used in the computation of τ_j.

Example 10.1

Consider the elimination process of table 9.1. In this case, we have seven factors $\psi_1, \ldots, \psi_7$, whose scope is shown in the table. The message $\tau_1(D)$, generated from $\psi_1(C,D)$, participates in the computation of ψ_2. Thus, we would have an edge from C_1 to C_2. Similarly, the message $\tau_3(G,S)$ is generated from ψ_3 and used in the computation of ψ_5. Hence, we introduce an edge between C_3 and C_5. The entire graph is shown in figure 10.1. The edges in the graph are annotated with directions, indicating the flow of messages between clusters in the execution of the variable elimination algorithm. Each of the factors in the initial set of factors Φ is also associated with a cluster C_i. For example, the cluster $\phi_D(D,C)$ (corresponding to the CPD $P(D \mid C)$) is associated with C_1, and the cluster $\phi_H(H,G,J)$ (corresponding to the CPD $P(H \mid G,J)$) is associated with C_4. ∎

10.1.2   Clique Trees

The cluster graph associated with an execution of variable elimination is guaranteed to have certain properties that turn out to be very important.

First, recall that the variable elimination algorithm uses each intermediate factor τ_i at most once: when ϕ_i is used in Sum-Product-Eliminate-Var to create ψ_j, it is removed from the set of factors Φ, and thus cannot be used again. Hence, the cluster graph induced by an execution of variable elimination is necessarily a tree.

We note that although a cluster graph is defined to be an undirected graph, an execution of variable elimination does define a direction for the edges, as induced by the flow of messages between the clusters. The directed graph induced by the messages is a directed tree, with all the messages flowing toward a single cluster where the final result is computed. This cluster is called the *root* of the directed tree. Using standard conventions in computer science, we assume that the root of the tree is "up," so that messages sent toward the root are sent upward. If C_i is

upstream clique

downstream clique

on the path from C_j to the root we say that C_i is *upstream* from C_j, and C_j is *downstream* from C_i. We note that, for reasons that will become clear later on, the directions of the edges and the root are not part of the definition of a cluster graph.

The cluster tree defined by variable elimination satisfies an important structural constraint:

Definition 10.2

running intersection property

Let $\mathcal{T}$ be a cluster tree over a set of factors Φ. We denote by $\mathcal{V}_\mathcal{T}$ the vertices of $\mathcal{T}$ and by $\mathcal{E}_\mathcal{T}$ its edges. We say that $\mathcal{T}$ has the running intersection property *if, whenever there is a variable X such that $X \in C_i$ and $X \in C_j$, then X is also in every cluster in the (unique) path in $\mathcal{T}$ between C_i and C_j.* ∎

Note that the running intersection property implies that $S_{i,j} = C_i \cap C_j$.

Example 10.2

We can easily check that the running intersection property holds for the cluster tree of figure 10.1. For example, G is present in C_2 and in C_4, so it is also present in the cliques on the path between them: C_3 and C_5. ∎

Intuitively, the running intersection property must hold for cluster trees induced by variable elimination because a variable appears in every factor from the moment it is introduced (by multiplying in a factor that mentions it) until it is summed out. We now prove that this property holds in general.

Theorem 10.1

Let $\mathcal{T}$ be a cluster tree induced by a variable elimination algorithm over some set of factors Φ. Then $\mathcal{T}$ satisfies the running intersection property.

PROOF Let C and C' be two clusters that contain X. Let C_X be the cluster where X is eliminated. (If X is a query variable, we assume that it is eliminated in the last cluster.) We will prove that X must be present in every cluster on the path between C and C_X, and analogously for C', thereby proving the result.

First, we observe that the computation at C_X must take place later in the algorithm's execution than the computation at C: When X is eliminated in C_X, all of the factors involving X are multiplied into C_X; the result of the summation does not have X in its domain. Hence, after this elimination, Φ no longer has any factors containing X, so no factor generated afterward will contain X in its domain.

By assumption, X is in the domain of the factor in C. We also know that X is not eliminated in C. Therefore, the message computed in C must have X in its domain. By definition, the recipient of X's message, which is C's upstream neighbor in the tree, multiplies in the message from C. Hence, it will also have X in its scope. The same argument applies to show that all cliques upstream from C will have X in their scope, until X is eliminated, which happens only in C_X. Thus, X must appear in all cliques between C and C_X, as required.   ∎

A very similar proof can be used to show the following result:

Proposition 10.1

Let $\mathcal{T}$ be a cluster tree induced by a variable elimination algorithm over some set of factors Φ. Let C_i and C_j be two neighboring clusters, such that C_i passes the message τ_i to C_j. Then the scope of the message τ_i is precisely $C_i \cap C_j$.

The proof is left as an exercise (exercise 10.1).

It turns out that a cluster tree that satisfies the running intersection property is an extremely useful data structure for exact inference in graphical models. We therefore define:

Definition 10.3

clique tree

clique

Let Φ be a set of factors over $\mathcal{X}$. A cluster tree over Φ that satisfies the running intersection property is called a clique *tree (sometimes also called a* junction *tree or a* join *tree). In the case of a clique tree, the clusters are also called* cliques.   ∎

Note that we have already defined one notion of a clique tree in definition 4.17. This double definition is not an overload of terminology, because the two definitions are actually equivalent: It follows from the results of this chapter that $\mathcal{T}$ is a clique tree for Φ (in the sense of definition 10.3) if and only if it is a clique tree for a chordal graph containing $\mathcal{H}_\Phi$ (in the sense of definition 4.17), and these properties are true if and only if the clique-tree data structure admits variable elimination by passing messages over the tree.

We first show that the running intersection property implies the independence statement, which is at the heart of our first definition of clique trees. Let $\mathcal{T}$ be a cluster tree over Φ, and let $\mathcal{H}_\Phi$ be the undirected graph associated with this set of factors. For any sepset $S_{i,j}$, let $W_{<(i,j)}$ be the set of all variables in the scope of clusters in the C_i side of the tree, and $W_{<(j,i)}$ be the set of all variables in the scope of clusters in the C_j side of the tree.

Theorem 10.2

$\mathcal{T}$ satisfies the running intersection property if and only if, for every sepset $S_{i,j}$, we have that $W_{<(i,j)}$ and $W_{<(j,i)}$ are separated in $\mathcal{H}_\Phi$ given $S_{i,j}$.

The proof is left as an exercise (exercise 10.2).

To conclude the proof of the equivalence of the two definitions, it remains only to show that the running intersection property for a tree $\mathcal{T}$ implies that each node in $\mathcal{T}$ corresponds to a clique in a chordal graph $\mathcal{H}'$ containing $\mathcal{H}$, and that each maximal clique in $\mathcal{H}'$ is represented in $\mathcal{T}$. This result follows from our ability to use any clique tree satisfying the running intersection property to perform inference, as shown in this chapter.

10.2   Message Passing: Sum Product

In the previous section, we started out with an execution of the variable elimination algorithm, and showed that it induces a clique tree. In this section, we go in the opposite direction. We assume that we are given a clique tree as a starting point, and we will show how this data structure can be used to perform variable elimination. As we will see, the clique tree is a very

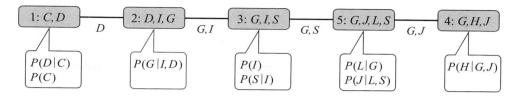

Figure 10.2 **Simplified clique tree** $\mathcal{T}$ **for the** Extended Student **network**

useful and versatile data structure. For one, the same clique tree can be used as the basis for many different executions of variable elimination. More importantly, the clique tree provides a data structure for caching computations, allowing multiple executions of variable elimination to be performed much more efficiently than simply performing each one separately.

Consider some set of factors Φ over $\mathcal{X}$, and assume that we are given a clique tree $\mathcal{T}$ over Φ, as defined in definition 4.17. In particular, $\mathcal{T}$ is guaranteed to satisfy the family preservation and running intersection properties. As we now show, we can use the clique tree in several different ways to perform exact inference in graphical models.

10.2.1 Variable Elimination in a Clique Tree

One way of using a clique tree is simply as guidance for the operations of variable elimination. The factors ψ are computed in the cliques, and messages are sent along the edges. Each clique takes the incoming messages (factors), multiplies them, sums out one or more variables, and sends an outgoing message to another clique. As we will see, the clique-tree data structure dictates the operations that are performed on factors in the clique tree and a partial order over these operations. In particular, if clique C' requires a message from C, then C' must wait with its computation until C performs its computation and sends the appropriate message to C'.

We begin with an example and then describe the general algorithm.

10.2.1.1 An Example

Figure 10.2 shows one possible clique tree $\mathcal{T}$ for the Student network. Note that it is different from the clique tree of figure 10.1, in that nonmaximal cliques (C_6 and C_7) are absent. Nevertheless, it is straightforward to verify that $\mathcal{T}$ satisfies both the family preservation and the running intersection property. The figure also specifies the assignment α of the initial factors (CPDs) to cliques. Note that, in some cases (for example, the CPD $P(I)$), we have more than one possible clique into which the factor can legally be assigned; as we will see, the algorithm applies for any legal choice.

Our first step is to generate a set of *initial potentials* associated with the different cliques. The initial potential $\psi_i(C_i)$ is computed by multiplying the initial factors assigned to the clique C_i. For example, $\psi_5(J, L, G, S) = \phi_L(L, G) \cdot \phi_J(J, L, S)$.

Now, assume that our task is to compute the probability $P(J)$. We want to do the variable elimination process so that J is not eliminated. Thus, we select as our root clique some clique that contains J, for example, C_5. We then execute the following steps:

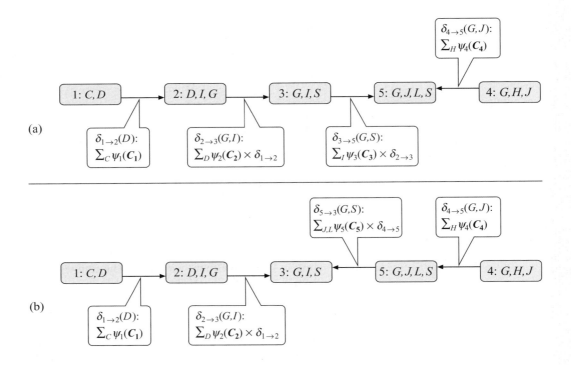

Figure 10.3   Two different message propagations with different root cliques in the Student **clique tree:** (a) C_5 is the root; (b) C_3 is the root.

1. **In C_1:** We eliminate C by performing $\sum_C \psi_1(C, D)$. The resulting factor has scope D. We send it as a message $\delta_{1 \to 2}(D)$ to C_2.

2. **In C_2:** We define $\beta_2(G, I, D) = \delta_{1 \to 2}(D) \cdot \psi_2(G, I, D)$. We then eliminate D to get a factor over G, I. The resulting factor is $\delta_{2 \to 3}(G, I)$, which is sent to C_3.

3. **In C_3:** We define $\beta_3(G, S, I) = \delta_{2 \to 3}(G, I) \cdot \psi_3(G, S, I)$ and eliminate I to get a factor over G, S, which is $\delta_{3 \to 5}(G, S)$.

4. **In C_4:** We eliminate H by performing $\sum_H \psi_4(H, G, J)$ and send out the resulting factor as $\delta_{4 \to 5}(G, J)$ to C_5.

5. **In C_5:** We define $\beta_5(G, J, S, L) = \delta_{3 \to 5}(G, S) \cdot \delta_{4 \to 5}(G, J) \cdot \psi_5(G, J, S, L)$.

The factor β_5 is a factor over G, J, S, L that encodes the joint distribution $P(G, J, L, S)$: all the CPDs have been multiplied in, and all the other variables have been eliminated. If we now want to obtain $P(J)$, we simply sum out G, L, and S.

We note that the operations in the elimination process could also have been done in another order. The only constraint is that a clique get all of its incoming messages from its downstream neighbors before it sends its outgoing message toward its upstream neighbor. We say that a clique is *ready* when it has received all of its incoming messages. Thus, for example, C_4 is ready

ready clique

at the very start of the algorithm, and the computation associated with it can be performed at any point in the execution. However, C_2 is ready only after it receives its message from C_1. Thus, C_1, C_4, C_2, C_3, C_5 is a legal execution ordering for a tree rooted at C_5, whereas C_2, C_1, C_4, C_3, C_5 is not. Overall, the set of messages transmitted throughout the execution of the algorithm is shown in figure 10.3a.

As we mentioned, the choice of root clique is not fully determined. To derive $P(J)$, we could have chosen C_4 as the root. Let us see how the algorithm would have changed in that case:

1. **In C_1:** The computation and message are unchanged.

2. **In C_2:** The computation and message are unchanged.

3. **In C_3:** The computation and message are unchanged.

4. **In C_5:** We define $\beta_5(G, J, S, L) = \delta_{3 \to 5}(G, S) \cdot \psi_5(G, J, S, L)$ and eliminate S and L. We send out the resulting factor as $\delta_{5 \to 4}(G, J)$ to C_4.

5. **In C_4:** We define $\beta_4(H, G, J) = \delta_{5 \to 4}(G, S) \cdot \psi_4(H, G, J)$.

We can now extract $P(J)$ by eliminating H and G from $\beta_4(H, G, J)$.

In a similar way, we can apply exactly the same process to computing the distribution over any other variable. For example, if we want to compute the probability $P(G)$, we could choose any of the cliques where it appears. If we use C_3, for example, the computation in C_1 and C_2 is identical. The computation in C_4 is the same as in the first of our two executions: a message is computed and sent to C_5. In C_5, we compute $\beta_5(G, J, S, L) = \delta_{4 \to 5}(G, J) \cdot \psi_5(G, J, S, L)$, and we eliminate J and L to produce a message $\delta_{5 \to 3}(G, S)$, which can then be sent to C_3 and used in the operation:

$$\beta_3(G, S, I) = \delta_{2 \to 3}(G, I) \cdot \delta_{5 \to 3}(G, S) \cdot \psi_3(G, S, I).$$

Overall, the set of messages transmitted throughout this execution of the algorithm is shown in figure 10.3b.

10.2.1.2 Clique-Tree Message Passing

message passing

We can now specify a general variable elimination algorithm that can be implemented via *message passing* in a clique tree. Let $\mathcal{T}$ be a clique tree with the cliques $C_1, \ldots, C_k$. We begin by multiplying the factors assigned to each clique, resulting in our initial potentials. We then use the clique-tree data structure to pass messages between neighboring cliques, sending all messages toward the root clique. We describe the algorithm in abstract terms; box 10.A provides some important tips for efficient implementation.

initial potential

Recall that each factor $\phi \in \Phi$ is assigned to some clique $\alpha(\phi)$. We define the *initial potential* of C_j to be:

$$\psi_j(C_j) = \prod_{\phi \,:\, \alpha(\phi) = j} \phi.$$

Because each factor is assigned to exactly one clique, we have that

$$\prod_{\phi} \phi = \prod_{j} \psi_j.$$

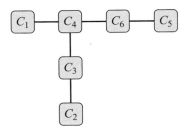

Figure 10.4   An abstract clique tree that is not chain-structured

Let C_r be the selected root clique. We now perform sum-product variable elimination over the cliques, starting from the leaves of the clique tree and moving inward. More precisely, for each clique C_i, we define Nb_i to be the set of indexes of cliques that are neighbors of C_i. Let $p_r(i)$ be the upstream neighbor of i (the one on the path to the root clique r). Each clique C_i, except for the root, performs a message passing computation and sends a message to its upstream neighbor $C_{p_r(i)}$.

sum-product
message passing

The message from C_i to another clique C_j is computed using the following *sum-product message passing* computation:

$$\delta_{i \to j} = \sum_{C_i - S_{i,j}} \psi_i \cdot \prod_{k \in (\mathrm{Nb}_i - \{j\})} \delta_{k \to i}. \tag{10.2}$$

In words, the clique C_i multiplies all incoming messages from its other neighbors with its initial clique potential, resulting in a factor ψ whose scope is the clique. It then sums out all variables except those in the sepset between C_i and C_j, and sends the resulting factor as a message to C_j.

This message passing process proceeds up the tree, culminating at the root clique. When the root clique has received all messages, it multiplies them with its own initial potential. The result is a factor called the *beliefs*, denoted $\beta_r(C_r)$. It represents, as we show,

beliefs

$$\tilde{P}_\Phi(C_r) = \sum_{\mathcal{X} - C_r} \prod_\phi \phi.$$

The complete algorithm is shown in algorithm 10.1.

Example 10.3

Consider the abstract clique tree of figure 10.4, and assume that we have selected C_6 as our root clique. The numbering of the cliques denotes one possible ordering of the operations, with C_1 being the first to compute its message. However, multiple other orderings are legitimate, for example, $2, 5, 1, 3, 4, 6$; in general, any ordering that respects the ordering constraints $\{(2 \prec 3), (3 \prec 4), (1 \prec 4), (4 \prec 6), (5 \prec 6)\}$ is a legal ordering for the message passing process. ∎

We can use this algorithm to compute the marginal probability of any set of query nodes $\boldsymbol{Y}$ which is fully contained in some clique. We select one such clique C_r to be the root, and perform the clique-tree message passing toward that root. We then extract $\tilde{P}_\Phi(\boldsymbol{Y})$ from the final potential at C_r by summing out the other variables $C_r - \boldsymbol{Y}$.

Algorithm 10.1 Upward pass of variable elimination in clique tree

---

Procedure CTree-SP-Upward (
    Φ,   // Set of factors
    $\mathcal{T}$,   // Clique tree over Φ
    α,   // Initial assignment of factors to cliques
    C_r // Some selected root clique
)
1   Initialize-Cliques
2   **while** C_r is not ready
3   Let C_i be a ready clique
4       $\delta_{i \to p_r(i)}(S_{i,p_r(i)}) \leftarrow$ SP-Message$(i, p_r(i))$
5    $\beta_r \leftarrow \psi_r \cdot \prod_{k \in \mathrm{Nb}_{C_r}} \delta_{k \to r}$
6   **return** β_r

Procedure Initialize-Cliques (

)
1   **for** each clique C_i
2       $\psi_i(C_i) \leftarrow \prod_{\phi_j \,:\, \alpha(\phi_j) = i} \phi_j$
3

Procedure SP-Message (
    i,   // sending clique
    j   // receiving clique
)
1    $\psi(C_i) \leftarrow \psi_i \cdot \prod_{k \in (\mathrm{Nb}_i - \{j\})} \delta_{k \to i}$
2    $\tau(S_{i,j}) \leftarrow \sum_{C_i - S_{i,j}} \psi(C_i)$
3   **return** $\tau(S_{i,j})$

10.2.1.3 Correctness

We now prove that this algorithm, when applied to a clique tree that satisfies the family preservation and running intersection property, computes the desired expressions over the messages and the cliques.

In our algorithm, a variable X is eliminated only when a message is sent from C_i to a neighboring C_j such that $X \in C_i$ and $X \notin C_j$. We first prove the following result:

Proposition 10.2

Assume that X is eliminated when a message is sent from C_i to C_j. Then X does not appear anywhere in the tree on the C_j side of the edge $(i\text{–}j)$.

PROOF The proof is a simple consequence of the running intersection property. Assume by contradiction that X appears in some other clique C_k that is on the C_j side of the tree. Then C_j is on the path from C_i to C_k. But we know that X appears in both C_i and C_k but not in C_j, violating the running intersection property. ∎

Based on this result, we can provide a semantic interpretation for the messages used in the clique tree. Let $(i{-}j)$ be some edge in the clique tree. We use $\mathcal{F}_{\prec(i\to j)}$ to denote the set of factors in the cliques on the C_i-side of the edge and $\mathcal{V}_{\prec(i\to j)}$ to denote the set of variables that appear on the C_i-side but are not in the sepset. For example, in the clique tree of figure 10.2, we have that $\mathcal{F}_{\prec(3\to5)} = \{P(C), P(D \mid C), P(G \mid I, D), P(I), P(S \mid I)\}$ and $\mathcal{V}_{\prec(3\to5)} = \{C, D, I\}$. Intuitively, the message passed between the cliques C_i and C_j is the product of all the factors in $\mathcal{F}_{\prec(i\to j)}$, marginalized over the variables in the sepset (that is, summing out all the others).

Theorem 10.3

Let $\delta_{i\to j}$ be a message from C_i to C_j. Then:

$$\delta_{i\to j}(S_{i,j}) = \sum_{\mathcal{V}_{\prec(i\to j)}} \prod_{\phi\in\mathcal{F}_{\prec(i\to j)}} \phi.$$

PROOF The proof proceeds by induction on the length of the path from the leaves. For the base case, the clique C_i is a leaf in the tree. In this case, the result follows from a simple examination of the operations executed at the clique.

Now, consider a clique C_i that is not a leaf, and consider the expression

$$\sum_{\mathcal{V}_{\prec(i\to j)}} \prod_{\phi\in\mathcal{F}_{\prec(i\to j)}} \phi. \tag{10.3}$$

Let $i_1,\ldots,i_m$ be the neighboring cliques of C_i other than C_j. It follows immediately from proposition 10.2 that $\mathcal{V}_{\prec(i\to j)}$ is the disjoint union of $\mathcal{V}_{\prec(i_k\to i)}$ for $k = 1,\ldots,m$ and the variables Y_i eliminated at C_i itself. Similarly, $\mathcal{F}_{\prec(i\to j)}$ is the disjoint union of the $\mathcal{F}_{\prec(i_k\to i)}$ and the factors $\mathcal{F}_i$ from which ψ_i was computed. Thus equation (10.3) is equal to

$$\sum_{Y_i}\sum_{\mathcal{V}_{\prec(i_1\to i)}}\cdots\sum_{\mathcal{V}_{\prec(i_m\to i)}}\left(\prod_{\phi\in\mathcal{F}_{\prec(i_1\to i)}}\phi\right)\cdots\cdot\left(\prod_{\phi\in\mathcal{F}_{\prec(i_m\to i)}}\phi\right)\cdot\left(\prod_{\phi\in\mathcal{F}_i}\phi\right). \tag{10.4}$$

As we just showed, for each k, none of the variables in $\mathcal{V}_{\prec(i_k\to i)}$ appear in any of the other factors. Thus, we can use equation (9.6) and push in the summation over $\mathcal{V}_{\prec(i_k\to i)}$ in equation (10.4), and obtain:

$$\sum_{Y_i}\left(\prod_{\phi\in\mathcal{F}_i}\phi\right)\cdot\sum_{\mathcal{V}_{\prec(i_1\to i)}}\left(\prod_{\phi\in\mathcal{F}_{\prec(i_1\to i)}}\phi\right)\cdots\cdot\sum_{\mathcal{V}_{\prec(i_m\to i)}}\left(\prod_{\phi\in\mathcal{F}_{\prec(i_m\to i)}}\phi\right). \tag{10.5}$$

Using the inductive hypothesis and the definition of ψ_i, this expression is equal to

$$\sum_{Y_i}\psi_i\cdot\delta_{i_1\to i}\cdots\cdot\delta_{i_m\to i}, \tag{10.6}$$

which is precisely the operation used to compute the message $\delta_{i\to j}$.   ∎

This theorem is closely related to theorem 10.2, which tells us that a sepset divides the graph into conditionally independent pieces. It is this conditional independence property that allows the message over the sepset to summarize completely the information in one side of the clique tree that is necessary for the computation in the other.

Based on this analysis, we can show that:

Corollary 10.1

Let C_r be the root clique in a clique tree, and assume that β_r is computed as in the algorithm of algorithm 10.1. Then

$$\beta_r(C_r) = \sum_{\mathcal{X} - C_r} \tilde{P}_\Phi(\mathcal{X}).$$

As we discussed earlier, this algorithm applies both to Bayesian network and Markov network inference. For a Bayesian network $\mathcal{B}$, if Φ consists of the CPDs in $\mathcal{B}$, reduced with some evidence e, then $\beta_r(C_r) = P_{\mathcal{B}}(C_r, e)$. For a Markov network $\mathcal{H}$, if Φ consists of the compatibility functions defining the network, then $\beta_r(C_r) = \tilde{P}_\Phi(C_r)$. In both cases, we can obtain the probability over the variables in C_r as usual, by normalizing the resulting factor to sum to 1. In the Markov network, we can also obtain the value of the partition function simply by summing up all of the entries in the potential of the root clique $\beta_r(C_r)$.

10.2.2 Clique Tree Calibration

We have shown that we can use the same clique tree to compute the probability of any variable in $\mathcal{X}$. In many applications, we often wish to estimate the probability of a large number of variables. For example, in a medical-diagnosis setting, we generally want the probability of several possible diseases. Furthermore, as we will see, when learning Bayesian networks from partially observed data, we always want the probability distributions over each of the unobserved variables in the domain (and their parents).

Therefore, let us consider the task of computing the posterior distribution over every random variable in the network. The most naive approach is to do inference separately for each variable. Letting c be the cost of a single execution of clique tree inference, the total cost of this algorithm is nc. An approach that is slightly less naive is to run the algorithm once for every clique, making it the root. The total cost of this variant is Kc, where K is the number of cliques. However, it turns out that we can do substantially better than either of these approaches.

Let us revisit our clique tree of figure 10.2 and consider the three different executions of the clique tree algorithm that we described: one where C_5 is the root, one where C_4 is the root, and one where C_3 is the root. As we pointed out, the messages sent from C_1 to C_2 and from C_2 to C_3 are the same in all three executions. The message sent from C_4 to C_5 is the same in both of the executions where it appears. In the second of the three executions, there simply is no message from C_4 to C_5 — the message goes the other way, from C_5 to C_4.

More generally, consider two neighboring cliques C_i and C_j in some clique tree. It follows from theorem 10.3 that the value of the message sent from C_i to C_j does not depend on specific choice of root clique: As long as the root clique is on the C_j-side, exactly the same message is sent from C_i to C_j. The same argument applies if the root is on the C_i-side. Thus, in all executions of the clique tree algorithm, whenever a message is sent between two cliques in the same direction, it is necessarily the same. Thus, for any given clique tree, each edge has two messages associated with it: one for each direction of the edge. If we have a total of c cliques, there are $c - 1$ edges in the tree; therefore, we have $2(c - 1)$ messages to compute.

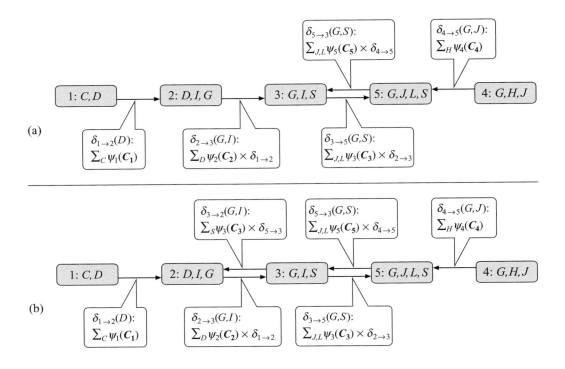

Figure 10.5   Two steps in a downward pass in the Student **network**

We can compute both messages for each edge by the following simple asynchronous algorithm. Recall that a clique can transmit a message upstream toward the root when it has all of the messages from its downstream neighbors. We can generalize this concept as follows:

<table>
<tr><td>

Definition 10.4

ready clique

</td><td>

Let $\mathcal{T}$ be a clique tree. We say that C_i is ready *to transmit to a neighbor C_j when C_i has messages from all of its neighbors except from C_j.*   ∎

</td></tr>
</table>

When C_i is ready to transmit to C_j, it can compute the message $\delta_{i \to j}(S_{i,j})$ by multiplying its initial potential with all of its incoming messages except the one from C_j, and then eliminate the

dynamic
programming

variables in $C_i - S_{i,j}$. In effect, this algorithm uses yet another layer of *dynamic programming* to avoid recomputing the same message multiple times.

sum-product
belief
propagation

Algorithm 10.2 shows the full procedure, often called *sum-product belief propagation*. As written, the algorithm is defined *asynchronously*, with each clique sending a message as soon as it is ready. One might wonder why this process is guaranteed to terminate, that is, why there is always a clique that is ready to transmit to some other clique. In fact, the message passing process performed by the algorithm is equivalent to a much more systematic process that consists of an *upward pass* and a *downward pass*. In the upward pass, we first pick a

upward pass

downward pass

root and send all messages toward the root. When this process is complete, the root has all messages. Therefore, it can now send the appropriate message to all of its children. This

---

Algorithm 10.2 Calibration using sum-product message passing in a clique tree

Procedure CTree-SP-Calibrate (
    Φ,   // Set of factors
    $\mathcal{T}$ // Clique tree over Φ
)
1   Initialize-Cliques
2   **while** exist i, j such that i is ready to transmit to j
3        $\delta_{i \rightarrow j}(\boldsymbol{S}_{i,j}) \leftarrow$ SP-Message(i, j)
4   **for** each clique i
5        $\beta_i \leftarrow \psi_i \cdot \prod_{k \in \mathrm{Nb}_i} \delta_{k \rightarrow i}$
6   **return** $\{\beta_i\}$

---

algorithm continues until the leaves of the tree are reached, at which point no more messages need to be sent. This second phase is called the downward pass. The asynchronous algorithm is equivalent to this systematic algorithm, except that the root is simply the first clique that happens to obtain messages from all of its neighbors. In an actual implementation, we might want to *schedule* this process more explicitly. (At the very least, the algorithm would check in line 2 that a message is not computed more than once.)

message scheduling

Example 10.4

Figure 10.3a shows the upward pass of the clique tree algorithm when C_5 is the root. Figure 10.5a shows a possible first step in a downward pass, where C_5 sends a message to its child C_3, based on the message from C_4 and its initial potential. As soon as a child of the root receives a message, it has all of the information it needs to send a message to its own children. Figure 10.5b shows C_3 sending the downward message to C_2. ∎

beliefs

At the end of this process, we compute the *beliefs* for all cliques in the tree by multiplying the initial potential with each of the incoming messages. The key is to note that the messages used in the computation of β_i are precisely the same messages that would have been used in a standard upward pass of the algorithm with C_i as the root. Thus, we conclude:

Corollary 10.2

Assume that, for each clique i, β_i is computed as in the algorithm of algorithm 10.2. Then

$$\beta_i(\boldsymbol{C}_i) = \sum_{\mathcal{X} - \boldsymbol{C}_i} \tilde{P}_{\Phi}(\mathcal{X}).$$

Note that it is important that C_i compute the message to a neighboring clique C_j based on its *initial potential* ψ_i and not its modified potential β_i. The latter already integrates information from j. If the message were computed based on this latter potential, we would be double-counting the factors assigned to C_j (multiplying them twice into the joint).

When this process concludes, each clique contains the marginal (unnormalized) probability over the variables in its scope. As we discussed, we can compute the marginal probability over a particular variable X by selecting a clique whose scope contains X, and eliminating the redundant variables in the clique. A key point is that the result of this process does not depend on the clique we selected. That is, if X appears in two cliques, they must agree on its marginal.

Definition 10.5

calibrated

Two adjacent cliques C_i and C_j are said to be calibrated *if*

$$\sum_{C_i - S_{i,j}} \beta_i(C_i) = \sum_{C_j - S_{i,j}} \beta_j(C_j).$$

A clique tree $\mathcal{T}$ is calibrated *if all pairs of adjacent cliques are calibrated. For a calibrated clique*

beliefs

tree, we use the term clique beliefs *for $\beta_i(C_i)$ and* sepset beliefs *for*

$$\mu_{i,j}(S_{i,j}) = \sum_{C_i - S_{i,j}} \beta_i(C_i) = \sum_{C_j - S_{i,j}} \beta_j(C_j). \tag{10.7}$$

∎

☞ **The main advantage of the clique tree algorithm is that it computes the posterior probability of all variables in a graphical model using only twice the computation of the upward pass in the same tree.** Letting c once again be the execution cost of message passing in a clique tree to one root, the cost of this algorithm is $2c$. By comparison, recall that the cost of doing a separate computation for each variable is nc and a separate computation for each root clique is Kc, where K is the number of cliques. In most cases, the savings are considerable, making the clique tree algorithm the algorithm of choice in situations where we want to compute the posterior of multiple query variables.

☞ **Box 10.A — Skill: Efficient Implementation of Factor Manipulation Algorithms.** *While simple conceptually, the implementation of algorithms involving manipulation of factors can be surprisingly subtle. In particular,* **different design decisions can lead to orders-of-magnitude differences in performance, as well as differences in the accuracy of the results.** *We now discuss some of the key design decisions in these algorithms. We note that the methods we describe here are equally applicable to the algorithms in many of the other chapters in the book, including the variable elimination algorithm of chapter 9, the exact and approximate sum-product message passing algorithms of chapters 10 and 11, and many of the MAP algorithms of chapter 13.*

The first key decision is the representation of our basic data structure: a factor, or a multidimensional table, with an entry for each possible assignment to the variables. One standard technique for storing multidimensional tables is to flatten them into a single array in computer memory. For

stride

each variable, we also store its cardinality, and its stride, *or step size in the factor. For example, given a factor $\phi(A, B, C)$ over variables A, B, and C, with cardinalities 2, 2, and 3, respectively, we can represent the factor in memory by the array*

$$\texttt{phi}[0\ldots 11] = \left\{ \phi(a^1, b^1, c^1), \phi(a^2, b^1, c^1), \phi(a^1, b^2, c^1), \ldots, \phi(a^2, b^2, c^3) \right\}.$$

Here the stride for variable A is 1, for B is 2 and for C is 4. If we add a fourth variable, D, its stride would be 12, since we would need to step over twelve entries before reaching the next assignment to D. Notice how, using each variable's stride, we can easily go from a variable assignment to a corresponding index into the factor array

$$\texttt{index} = \sum_i \texttt{assignment}[i] \cdot \texttt{phi.stride}[i]$$

Algorithm 10.A.1 — Efficient implementation of a factor product operation.

    Procedure Factor-Product (
      phi1 over scope X_1,
      phi2 over scope X_2
        // Factors represented as a flat array with strides for the variables
)

```
1     j ← 0, k ← 0
2     for l = 0, . . . , |X₁ ∪ X₂|
3       assignment[l] ← 0
4     for i = 0, . . . , |Val(X₁ ∪ X₂)| − 1
5       psi[i] ← phi1[j] · phi2[k]
6       for l = 0, . . . , |X₁ ∪ X₂|
7         assignment[l] ← assignment[l] + 1
8         if assignment[l] = card[l] then
9           assignment[l] ← 0
10          j ← j − (card[l] − 1) · phi1.stride[l]
11          k ← k − (card[l] − 1) · phi2.stride[l]
12        else
13          j ← j + phi1.stride[l]
14          k ← k + phi2.stride[l]
15          break
16    return (psi)
```

---

and vice versa

$$\texttt{assignment}[i] = \lfloor index/\texttt{phi.stride}[i] \rfloor \bmod \texttt{card}[i]$$

With this factor representation, we can now design a library of operations: product, marginalization, maximization, reduction, *and so forth. Since many inference algorithms involve multiple iterations over a series of factor operations, it is important that these be high-performance. One of the key design decisions is indexing the appropriate entries in each factor for the operations that we wish to perform. (In fact, when one uses a naive implementation of index computations, one often discovers that 90 percent of the running time is spent on that task.)*

factor product   *Algorithm 10.A.1 provides an example for the* product *between two arbitrary factors. Here we define* phi.stride$[X] = 0$ *if* $X \notin Scope[\phi]$. *The inner loop (over* l) *advances to the next assignment to the variables in* ψ *and calculates indexes into each other factor array on the fly. It can be understood by considering the equation for computing* index *shown earlier. Similar on-the-fly index calculations can be applied for other factor operations. We leave these as an exercise (exercise 10.3).*

For iterative algorithms or multiple queries, where the same operation (on different data) is performed a large number of times, it may be beneficial to cache these index mappings for later use. Note, however, that the index mappings require the same amount of storage as the factors themselves, that is, are exponential in the number of variables. Thus, this design choice offers a direct trade-off between memory usage and speed, especially in view of the fact that the index computations require approximately the same amount of work as the factor operation itself. Since performance of main memory is orders of magnitudes faster than secondary storage (disk), when memory limitations are an issue, it is better not to cache index mappings for large problems. One exception is template models, where savings can be made by reusing the same indexes for different instantiations of the factor templates.

An additional trick in reducing the computational burden is to preallocate and reuse memory for factor storage. Allocating memory is a relatively expensive procedure, and one does not want to waste time on this task inside a tight loop. To illustrate this point, we consider the example of variable elimination for computing $\psi(A, D)$ as

$$\psi(A, D) = \sum_{B,C} \phi_1(A, B)\phi_2(B, C)\phi_3(C, D) = \sum_{B} \phi_1(A, B) \sum_{C} \phi_2(B, C)\phi_3(C, D).$$

Here we need to compute three intermediate factors: $\tau_1(B, C, D) = \phi_2(B, C)\phi_3(C, D)$; $\tau_2(B, D) = \sum_C \tau_1(B, C, D)$; and $\tau_3(A, B, D) = \phi_1(A, B)\tau_2(B, D)$. Notice that, once $\tau_2(B, D)$ has been calculated, we no longer need the values in $\tau_1(B, C, D)$. By initially allocating memory large enough to hold the larger of $\tau_1(B, C, D)$ and $\tau_3(A, B, D)$, we can use the same memory for both. Because every operation in a variable elimination or message passing algorithm requires the computation of one or more intermediate factors, some of which are much larger than the desired end product, the savings in both time (preallocation) and memory (reusage) can be significant.

We now turn our attention to numerical considerations. Operations such as factor product involve multiplying many small numbers together, which can lead to underflow problems due to finite precision arithmetic. The problem can be alleviated somewhat by renormalizing the factor after each operation (so that the maximum entry in the factor is one); this operation leaves the results to most queries unchanged (see exercise 9.3). However, if each entry in the factor is computed as the product of many terms, underflow can still occur. An alternative solution is to perform the computation in log-space, *replacing multiplications with additions; this transformation allows for greater machine precision to be utilized. Note that* marginalization, *which requires that we sum entries, cannot be performed in log-space; it requires exponentiating each entry in the factor, performing the marginalization, and taking the log of the result. Since moving from log-space to probability-space incurs a significant decrease in dynamic range, factors should be normalized before applying this transform. One standard trick is to shift every entry by the maximum entry*

$$\texttt{phi}[i] \leftarrow \exp\left\{\texttt{logPhi}[i] - c\right\},$$

log-space

factor
marginalization

where $c = \max_i \texttt{logPhi}[i]$; this transformation ensures that the resulting factor has a maximum entry of one and prevents overflow.

We note that there are some caveats to operating in log-space. First, one may incur a performance hit: Floating point multiplication is no slower than floating point addition, but the transformation to and from log-space, as required for marginalization, can take a significant proportion of the total processing time. This caveat does not apply to algorithms such as max-product, where maximization can be performed in log-space; indeed, these algorithms are almost always implemented as max-sum. Moreover, log-space operations require care in handling nonpositive factors (that is, factors with some zero entries).

Finally, at a higher level, as with any software implementation, there is always a trade-off between speed, memory consumption, and reusability of the code. For example, software specialized for the case of pairwise potentials over a grid will almost certainly outperform code written for general graphs with arbitrary potentials. However, the small performance hit in using well designed general purpose code often outweighs the development effort required to reimplement algorithms for each specialized application. However, as always, it is also important not to try to optimize code too early. It is more beneficial to write and profile the code, on real examples, to determine what operations are causing bottlenecks. This allows the development effort to be targeted to areas that can yield the most gain.

10.2.3 A Calibrated Clique Tree as a Distribution

A calibrated clique tree is more than simply a data structure that stores the results of probabilistic inference for all of the cliques in the tree. As we now show, it can also be viewed as an alternative representation of the measure $\tilde{P}_\Phi$.

At calibration, we have that:

$$\beta_i = \psi_i \cdot \prod_{k \in \mathrm{Nb}_i} \delta_{k \to i}. \tag{10.8}$$

We also have that:

$$
\begin{aligned}
\mu_{i,j}(\boldsymbol{S}_{i,j}) &= \sum_{\boldsymbol{C}_i - \boldsymbol{S}_{i,j}} \beta_i(\boldsymbol{C}_i) \\
&= \sum_{\boldsymbol{C}_i - \boldsymbol{S}_{i,j}} \psi_i \cdot \prod_{k \in \mathrm{Nb}_i} \delta_{k \to i} \\
&= \sum_{\boldsymbol{C}_i - \boldsymbol{S}_{i,j}} \psi_i \cdot \delta_{j \to i} \prod_{k \in (\mathrm{Nb}_i - \{j\})} \delta_{k \to i} \\
&= \delta_{j \to i} \sum_{\boldsymbol{C}_i - \boldsymbol{S}_{i,j}} \psi_i \cdot \prod_{k \in (\mathrm{Nb}_i - \{j\})} \delta_{k \to i} \\
&= \delta_{j \to i} \delta_{i \to j},
\end{aligned}
\tag{10.9}
$$

where the fourth equality holds because no variable in the scope of $\delta_{j \to i}$ is involved in the summation.

We can now show the following important result:

Proposition 10.3

At convergence of the clique tree calibration algorithm, we have that:

$$\tilde{P}_\Phi(\mathcal{X}) = \frac{\prod_{i \in \mathcal{V}_\mathcal{T}} \beta_i(\boldsymbol{C}_i)}{\prod_{(i-j) \in \mathcal{E}_\mathcal{T}} \mu_{i,j}(\boldsymbol{S}_{i,j})}. \tag{10.10}$$

PROOF Using equation (10.8), the numerator in the right-hand side of equation (10.10) can be rewritten as:

$$\prod_{i \in \mathcal{V}_\mathcal{T}} \psi_i(\boldsymbol{C}_i) \prod_{k \in \mathrm{Nb}_i} \delta_{k \to i}.$$

Assignment			Marg_C
a^0	b^0	d^0	$600,000$
a^0	b^0	d^1	$300,030$
a^0	b^1	d^0	$5,000,500$
a^0	b^1	d^1	$1,000$
a^1	b^0	d^0	200
a^1	b^0	d^1	$1,000,100$
a^1	b^1	d^0	$100,010$
a^1	b^1	d^1	$200,000$

$\beta_1(A,B,D)$

Assignment		$\text{Marg}_{A,C}$
b^0	d^0	$600,200$
b^0	d^1	$1,300,130$
b^1	d^0	$5,100,510$
b^1	d^1	$201,000$

$\mu_{1,2}(B,D)$

Assignment			Marg_A
b^0	c^0	d^0	$300,100$
b^0	c^0	d^1	$1,300,000$
b^0	c^1	d^0	$300,100$
b^0	c^1	d^1	130
b^1	c^0	d^0	510
b^1	c^0	d^1	$100,500$
b^1	c^1	d^0	$5,100,000$
b^1	c^1	d^1	$100,500$

$\beta_2(B,C,D)$

Figure 10.6   **The clique and sepset beliefs for the** Misconception **example.**

Using equation (10.9), the denominator can be rewritten as:

$$\prod_{(i-j)\in\mathcal{E}_\mathcal{T}} \delta_{i\rightarrow j}\delta_{j\rightarrow i}.$$

Each message $\delta_{i\rightarrow j}$ appears exactly once in the numerator and exactly once in the denominator, so that all messages cancel. The remaining expression is simply:

$$\prod_{i\in\mathcal{V}_\mathcal{T}} \psi_i(\boldsymbol{C}_i) = \tilde{P}_\Phi.$$

$\blacksquare$

reparameteriza-
tion

clique tree
invariant

Thus, via equation (10.10), the clique and sepsets beliefs provide a *reparameterization* of the unnormalized measure. This property is called the *clique tree invariant*, for reasons which will become clear later on in this chapter.

Another intuition for this result can be obtained from the following example:

Example 10.5

Consider a clique tree obtained from Markov network A—B—C—D, with an appropriate set of factors Φ. Our clique tree in this case would have three cliques $\boldsymbol{C}_1 = \{A,B\}$, $\boldsymbol{C}_2 = \{B,C\}$, and $\boldsymbol{C}_3 = \{C,D\}$. When the clique tree is calibrated, we have that $\beta_1(A,B) = \tilde{P}_\Phi(A,B)$ and $\beta_2(B,C) = \tilde{P}_\Phi(B,C)$. From the conditional independence properties of this distribution, we have that

$$\tilde{P}_\Phi(A,B,C) = \tilde{P}_\Phi(A,B)\tilde{P}_\Phi(C \mid B),$$

and

$$\tilde{P}_\Phi(C \mid B) = \frac{\beta_2(B,C)}{\tilde{P}_\Phi(B)}.$$

As $\beta_2(B,C) = \tilde{P}_\Phi(B,C)$, we can obtain $\tilde{P}_\Phi(B)$ by marginalizing $\beta_2(B,C)$. Thus, we can write:

$$\begin{aligned}
\tilde{P}_\Phi(A,B,C) &= \beta_1(A,B)\frac{\beta_2(B,C)}{\sum_C \beta_2(B,C)} \\
&= \frac{\beta_1(A,B)\beta_2(B,C)}{\sum_C \beta_2(B,C)}.
\end{aligned}$$

In fact, when the two cliques are calibrated, they must agree on the marginal of B. Thus, the expression in the denominator can equivalently be replaced by $\sum_A \beta_1(A, B)$. ∎

Based on this analysis, we now formally define the distribution represented by a clique tree:

Definition 10.6

clique tree
measure

We define the measure induced by a calibrated tree $\mathcal{T}$ to be:

$$Q_{\mathcal{T}} = \frac{\prod_{i \in \mathcal{V}_{\mathcal{T}}} \beta_i(\boldsymbol{C}_i)}{\prod_{(i-j) \in \mathcal{E}_{\mathcal{T}}} \mu_{i,j}(\boldsymbol{S}_{i,j})}, \tag{10.11}$$

where

$$\mu_{i,j} = \sum_{\boldsymbol{C}_i - \boldsymbol{S}_{i,j}} \beta_i(\boldsymbol{C}_i) = \sum_{\boldsymbol{C}_j - \boldsymbol{S}_{i,j}} \beta_j(\boldsymbol{C}_j).$$

∎

Example 10.6

Consider, for example, the Markov network of example 3.8, whose joint distribution is shown in figure 4.2. One clique tree for this network consists of the two cliques $\{A, B, D\}$ and $\{B, C, D\}$, with the sepset $\{B, D\}$. The final potentials and sepset for this example are shown in figure 10.6. It is straightforward to confirm that the clique tree is indeed calibrated. One can also verify that this clique tree provides a reparameterization of the original distribution. For example, consider the entry $\tilde{P}_\Phi(a^1, b^0, c^1, d^0) = 100$. According to equation (10.10), the clique tree measure is:

$$\frac{\beta_1(a^1, b^0, d^0)\beta_2(b^0, c^1, d^0)}{\mu_{1,2}(b^0, d^0)} = \frac{200 \cdot 300,100}{600,200} = 100,$$

as required. ∎

Our analysis so far shows that for a set of calibrated potentials derived from clique tree inference, we have two properties: the clique tree measure is $\tilde{P}_\Phi$ and the final beliefs are the marginals of $\tilde{P}_\Phi$. As we now show, these two properties coincide for any calibrated clique tree.

Theorem 10.4

Let $\mathcal{T}$ be a clique tree over Φ, and let $\beta_i(\boldsymbol{C}_i)$ be a set of calibrated potentials for $\mathcal{T}$. Then, $\tilde{P}_\Phi(\mathcal{X}) \propto Q_{\mathcal{T}}$ if and only if, for each $i \in \mathcal{V}_{\mathcal{T}}$, we have that $\beta_i(\boldsymbol{C}_i) \propto \tilde{P}_\Phi(\boldsymbol{C}_i)$.

PROOF Let r be any clique in $\mathcal{T}$, which we choose to be the root. Define the descendant cliques of a clique $\boldsymbol{C}_i$ to be the cliques that are downstream from $\boldsymbol{C}_i$ relative to $\boldsymbol{C}_r$; the nondescendant cliques are then the remaining cliques (other than $\boldsymbol{C}_i$). Let $\boldsymbol{X}$ be the variables in the scope of the nondescendant cliques. It follows immediately from theorem 10.2 that

$$\tilde{P}_\Phi \models (\boldsymbol{C}_i \perp \boldsymbol{X} \mid \boldsymbol{S}_{i,p_r(i)}).$$

From this, we obtain, using the standard chain-rule argument, that:

$$\tilde{P}_\Phi(\mathcal{X}) = \tilde{P}_\Phi(\boldsymbol{C}_r) \cdot \prod_{i \neq r} \tilde{P}_\Phi(\boldsymbol{C}_i \mid \boldsymbol{S}_{i,p_r(i)}).$$

We can rewrite equation (10.11) as a similar product, using the same root:

$$Q_{\mathcal{T}}(\mathcal{X}) = \beta_r(\boldsymbol{C}_r) \cdot \prod_{i \neq r} \beta_i(\boldsymbol{C}_i \mid \boldsymbol{S}_{i,p_r(i)}).$$

The "if" direction now follows from direct substitution of β_i for each $\tilde{P}_\Phi(C_i)$.

To prove the "only if" direction, we note that each of the terms $\beta_i(C_i \mid S_{i,p_r(i)})$ is a conditional distribution; hence, if we marginalize out the variables not in C_r in the distribution Q_T, each of these conditional distributions marginalizes to 1, and so we are left with $Q_T(C_r) = \beta_r(C_r)$. It now follows that if $\tilde{P}_\Phi \propto Q_T$, then $\tilde{P}_\Phi(C_r) \propto Q_T(C_r) = \beta_r(C_r)$. Because this argument applies to any choice of root clique, we have proved that this equality holds for every clique.   ∎

   Thus, we can view the clique tree as an alternative representation of the joint measure, one that directly reveals the clique marginals. As we will see, this view turns out to be very useful, both in the next section and in chapter 11.

10.3   Message Passing: Belief Update

The previous section showed one approach to message passing in clique trees, based on the same ideas of variable elimination that we discussed in chapter 9. In this section, we present a related approach, but one that is based on very different intuitions. We begin by describing an alternative message passing scheme that is different from but mathematically equivalent to that of the previous section. We then show how this new approach can be viewed as operations on the reparameterization of the distribution in terms of the clique and sepset beliefs $\{\beta_i(C_i)\}_{i \in \mathcal{V}_T}$ and $\{\mu_{i,j}(S_{i,j})\}_{(i-j) \in \mathcal{E}_T}$. Each message passing step will change this representation while leaving it a reparameterization of $\tilde{P}_\Phi$.

10.3.1   Message Passing with Division

Consider again the message passing process used in CTree-SP-Calibrate (algorithm 10.2). There, two messages are passed along each link $(i-j)$. Assume, without loss of generality, that the first message is passed from C_j to C_i. A return message from C_i to C_j is passed when C_i has received messages from all of its other neighbors.

At this point, C_i has all of the necessary information to compute its final potential. It multiplies the initial potential with the incoming messages from all of its neighbors:

$$\beta_i = \psi_i \cdot \prod_{k \in \mathrm{Nb}_i} \delta_{k \to i}. \tag{10.12}$$

As we discussed, this final potential is not used in computing the message to C_j: this potential already incorporates the information (message) passed from C_j; if we used it when computing the message to C_j, this information would be double-counted. Thus, the message from C_i to C_j is computed in a way that omits the information obtained from C_j: we multiply the initial potential with all of the messages except for the message from C_i, and then marginalize over the sepset (equation (10.2)).

A different approach to computing the same expression is to multiply in *all* of the messages, and then *divide* the resulting factor by $\delta_{j \to i}$. To make this notion precise, we must define a factor-division operation:

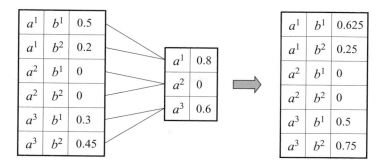

Figure 10.7   An example of factor division

Definition 10.7

factor division

Let X and Y be disjoint sets of variables, and let $\phi_1(X, Y)$ and $\phi_2(Y)$ be two factors. We define the division $\frac{\phi_1}{\phi_2}$ to be a factor ψ of scope X, Y defined as follows:

$$\psi(X, Y) = \frac{\phi_1(X, Y)}{\phi_2(Y)},$$

where we define $0/0 = 0$. ∎

Note that, as in the case of other factor operations, factor division is done component by component. Figure 10.7 shows an example. Also note that the operation is not well defined if the denominator is zero and the numerator is not.

We now see that we can compute the expression of equation (10.2) by computing the beliefs as in equation (10.12), and then dividing by the remaining message:

$$\delta_{i \to j} = \frac{\sum_{C_i - S_{i,j}} \beta_i}{\delta_{j \to i}}. \tag{10.13}$$

Example 10.7

Let us return to the simple clique tree in example 10.5, and assume that C_2 serves as the (de facto) root, so that we first pass messages from C_1 to C_2 and from C_3 to C_2. The message $\delta_{1 \to 2}$ is computed as $\sum_A \psi_1(A, B)$. Using the variable elimination message (CTree-SP-Calibrate), we pass a return message $\delta_{2 \to 1}(B) = \sum_C \psi_2(B, C)\delta_{3 \to 2}(C)$. Alternatively, we can compute $\beta_2(B, C) = \delta_{1 \to 2}(B) \cdot \delta_{3 \to 2}(C) \cdot \psi_2(B, C)$, and then send a message

$$\frac{\sum_C \beta_2(B, C)}{\delta_{1 \to 2}(B)} = \sum_C \frac{\beta_2(B, C)}{\delta_{1 \to 2}(B)} = \sum_C \psi_2(B, C) \cdot \delta_{3 \to 2}(C).$$

Thus, the two approaches are equivalent. ∎

sum-product-divide

beliefs

Based on this insight, we can define the *sum-product-divide* message passing scheme, where each clique C_i maintains its fully updated current beliefs β_i, which are defined as in equation (10.8). Each sepset also maintains its beliefs $\mu_{i,j}$ defined as the product of the messages in both directions, as in equation (10.9). We now show that the entire message passing process can be executed in an equivalent way in terms of the clique and sepset beliefs, without having to remember the initial potentials ψ_i or to compute explicitly the messages $\delta_{i \to j}$.

The message passing process follows the lines of example 10.7. Each clique C_i initializes β_i as ψ_i and then updates it by multiplying with message updates received from its neighbors. Each sepset $S_{i,j}$ maintains $\mu_{i,j}$ as the previous message passed along the edge $(i\text{–}j)$, regardless of the direction. This message is used to ensure that we do not double-count: Whenever a new message is passed along the edge, it is divided by the old message, eliminating the previous message from the update to the clique. Somewhat surprisingly, as we will show, the message passing operation is correct regardless of the clique that sent the last message on the edge. Intuitively, once the message is passed, its information is incorporated into both cliques; thus, each needs to divide by it when passing a message to the other. We can view this algorithm as maintaining a set of belief over the cliques in the tree. The message passing operation takes the beliefs of one clique and uses them to update the beliefs of a neighbor. Thus, we belief-update call this algorithm *belief-update* message passing; it is also known as the *Lauritzen-Spiegelhalter algorithm*.

Example 10.8

Continuing with example 10.7, assume that C_2 initially passes an uninformed message to C_3: $\sigma_{2 \to 3} = \sum_B \psi_2(B, C)$. This message multiplies the beliefs about C_3, so that, at this point:

$$\beta_3(C, D) = \psi_3(C, D) \sum_B \psi_2(B, C).$$

This message is also stored in the sepset as $\mu_{2,3}$. Now, assume that C_3 sends a message to C_2: $\sigma_{3 \to 2}(C) = \sum_D \beta_3(C, D)$. This message is divided by $\mu_{2,3}$, so the actual update for C_2 is:

$$
\begin{aligned}
\frac{\sigma_{3 \to 2}(C)}{\mu_{2,3}(C)} &= \frac{\sum_D \beta_3(C, D)}{\mu_{2,3}(C)} \\
&= \frac{\sum_D \psi_3(C, D)\mu_{2,3}(C)}{\mu_{2,3}(C)} \\
&= \sum_D \psi_3(C, D).
\end{aligned}
$$

This expression is precisely the update that C_2 would have received from C_3 in the case where C_2 does not first send an uninformed message. At this point, the message stored in the sepset is

$$\sum_D \beta_3(C, D) = \sum_D \left(\psi_3(C, D) \cdot \sum_B \psi_2(B, C) \right).$$

Assume that at the next step C_2 receives a message from C_1, containing $\sum_A \psi_1(A, B)$. The sepset $S_{1,2}$ contains a message that is identically 1, so that this message is transmitted as is. At this point, C_2 has received informed messages from both sides and is therefore informed. Indeed, we have shown that:

$$\beta_2(B, C) = \psi_2(B, C) \cdot \sum_A \psi_1(A, B) \cdot \sum_D \psi_3(C, D).$$

as required.   ∎

Algorithm 10.3 Calibration using belief propagation in clique tree

---

Procedure CTree-BU-Calibrate (
    Φ,   // Set of factors
    $\mathcal{T}$ // Clique tree over Φ

)
1   Initialize-CTree
2   **while** exists an uninformed clique in $\mathcal{T}$
3   Select $(i\text{-}j) \in \mathcal{E}_{\mathcal{T}}$
4   BU-Message(i, j)
5   **return** $\{\beta_i\}$

Procedure Initialize-CTree (

)
1   **for** each clique C_i
2        $\beta_i \leftarrow \prod_{\phi \,:\, \alpha(\phi)=i} \phi$
3   **for** each edge $(i\text{-}j) \in \mathcal{E}_{\mathcal{T}}$
4        $\mu_{i,j} \leftarrow \mathbf{1}$

Procedure BU-Message (
    i,   // sending clique
    j   // receiving clique
)
1    $\sigma_{i \rightarrow j} \leftarrow \sum_{C_i - S_{i,j}} \beta_i$
2   // marginalize the clique over the sepset
3    $\beta_j \leftarrow \beta_j \cdot \frac{\sigma_{i \rightarrow j}}{\mu_{i,j}}$
4    $\mu_{i,j} \leftarrow \sigma_{i \rightarrow j}$

The precise algorithm is shown in algorithm 10.3. Note that, as written, the message passing algorithm is underspecified: in line 3, we can select any pair of cliques C_i and C_j between which we will pass a message. Interestingly, we can make this choice arbitrarily, without damaging the correctness of the algorithm. For example, if C_i (for some reason) passes the same message to C_j a second time, the process of dividing out by the stored message reduces the message actually passed to $\mathbf{1}$, so that it has no influence. Furthermore, if C_i passes a message to C_j based on partial information (that is, without taking into consideration all of its incoming messages), and then resends a more updated message later on, the effect is identical to simply sending the updated message once. Moreover, at convergence, regardless of the message passing steps used, we necessarily have a calibrated clique tree. This property follows from the fact that, in order for all message updates to have no effect, we need to have

$\sigma_{i \to j} = \mu_{i,j} = \sigma_{j \to i}$ for all i, j, and so:

$$\sum_{C_i - S_{i,j}} \beta_i = \mu_{i,j} = \sum_{C_j - S_{i,j}} \beta_j.$$

Thus, at convergence, each pair of neighboring cliques i, j must agree on the variables in sepset, and the message $\mu_{i,j}$ is precisely the sepset marginal. These properties also follow from the equivalence between belief-update message passing and sum-product message passing, which we show next.

10.3.2   Equivalence of Sum-Product and Belief Update Messages

So far, although we used sum-product message propagation to motivate the definition of the belief update steps, we have not shown a direct connection between them. We now show a simple and elegant equivalence between the two types of message passing operations. From this result, it immediately follows that belief-update message passing is guaranteed to converge to the correct marginals.

Our proof is based on equation (10.8) and equation (10.9), which provide a mapping between the sum-product and belief-update representations. We consider corresponding runs of the two algorithms in which an identical sequence of message passing steps is executed. We show that these two properties hold as an invariant between the data structures maintained by the two algorithms. The invariant holds initially, and it is maintained throughout the corresponding runs.

Theorem 10.5

Consider a set of sum-product initial potentials $\{\psi_i \; : \; i \in \mathcal{V}_{\mathcal{T}}\}$ and messages $\{\delta_{i \to j}, \delta_{j \to i} \; : \; (i{-}j) \in \mathcal{E}_{\mathcal{T}}\}$, and a set of belief-update beliefs $\{\beta_i \; : \; i \in \mathcal{V}_{\mathcal{T}}\}$ and messages $\{\mu_{i,j} \; : \; (i{-}j) \in \mathcal{E}_{\mathcal{T}}\}$, for which equation (10.8) and equation (10.9) hold. For any pair of neighboring cliques C_i, C_j, let $\{\delta'_{i \to j}, \delta'_{j \to i} \; : \; (i{-}j) \in \mathcal{E}_{\mathcal{T}}\}$ be the set of sum-product messages following an application of SP-Message(i, j), and $\{\beta'_i \; : \; C_i \in \mathcal{T}\}, \{\mu'_{i,j} \; : \; (i{-}j) \in \mathcal{E}_{\mathcal{T}}\}$, be the set of belief-update beliefs following an application of BU-Message(i, j). Then equation (10.8) and equation (10.9) also hold for the new beliefs $\delta'_{i \to j}, \beta'_i, \mu'_{i,j}$.

The proof uses simple algebraic manipulation, and it is left as an exercise (exercise 10.4).

This equivalence implies another result that will prove important in subsequent developments:

Corollary 10.3

In an execution of belief-update message passing, the clique tree invariant equation (10.10) holds initially and after every message passing step.

PROOF The proof of proposition 10.3 relied only on equation (10.8) and equation (10.9). Because these two equalities hold in every step of the belief-update message passing algorithm, we have that the clique tree invariant also holds continuously. ∎

This equivalence also allows us to define a message schedule that guarantees convergence to the correct clique marginals in two passes: We simply follow the same upward-downward-pass schedule used in CTree-SP-Calibrate, using any (arbitrarily chosen) root clique C_r.

10.3.3 Answering Queries

As we have seen, a calibrated clique tree contains the answer to multiple queries at once: the posterior probability of any set of variables that are present together in a single clique. A particular type of query that turns out to be important in this setting is the computation of the posterior for families of variables in a probabilistic network: a node and its parents in the context of Bayesian networks, or a clique in a Markov network. The family preservation property for cluster graphs (and hence for clique trees) implies that a family must be a subset of some cluster in the cluster graph.

In addition to these queries, which we get immediately as a by-product of calibration, we can also use a clique tree for other queries. We describe the algorithm for these queries in terms of a calibrated clique tree that satisfies the clique tree invariant. Due to the equivalence of sum-product and belief-update message passing, we can obtain such a clique tree using either method.

10.3.3.1 Incremental Updates

incremental
update

Consider a situation where, at some point in time, we have a certain set of observations, which we use to condition our distribution and reach conclusions. At some later time, we obtain additional evidence, and want to update our conclusions accordingly. This type of situation, where we want to perform *incremental update* is very common in a wide variety of settings. For example, in a medical setting, we often perform diagnosis on the basis of limited evidence; the initial diagnosis helps us decide which tests to perform, and the results need to be incorporated into our diagnosis.

The most naive approach to dealing with this task is simply to condition the initial factors (for example, the CPDs) on all of the evidence, and then redo the calibration process from the beginning, starting from these factors. A somewhat more efficient approach is based on the view of the clique tree as representing the distribution $\tilde{P}_\Phi$.

Assume that our initial distribution $\tilde{P}_\Phi$ (prior to the new information) is represented via a set of factors Φ, as in equation (10.1). Given some evidence $Z = z$, we can obtain $\tilde{P}_\Phi(\mathcal{X}, Z = z)$ by zeroing out the entries in the unnormalized distribution that are inconsistent with the evidence $Z = z$. We can accomplish this effect by multiplying $\tilde{P}_\Phi$ with an additional factor which is the indicator function $\mathbf{1}\{Z = z\}$. More precisely, assume that our current distribution over $\mathcal{X}$ is defined by a set of factors Φ, so that

$$\tilde{P}_\Phi(\mathcal{X}) = \prod_{\phi \in \Phi} \phi.$$

Then,

$$\tilde{P}_\Phi(\mathcal{X}, Z = z) = \mathbf{1}\{Z = z\} \cdot \prod_{\phi \in \Phi} \phi.$$

Let $\tilde{P}'_\Phi(\mathcal{X}) = \tilde{P}_\Phi(\mathcal{X}, Z = z)$.

Now, assume that we have a clique tree (calibrated or not) that represents this distribution using the clique tree invariant. That is:

$$\tilde{P}_\Phi(\mathcal{X}) = Q_\mathcal{T} = \frac{\prod_{i \in \mathcal{V}_\mathcal{T}} \beta_i(\boldsymbol{C}_i)}{\prod_{(i-j) \in \mathcal{E}_\mathcal{T}} \mu_{i,j}(\boldsymbol{S}_{i,j})}.$$

We can represent the distribution $\tilde{P}'_\Phi(\mathcal{X})$ as

$$\tilde{P}'_\Phi(\mathcal{X}) = \mathbf{1}\{Z = z\} \cdot \frac{\prod_{i \in \mathcal{V}_T} \beta_i(\boldsymbol{C}_i)}{\prod_{(i-j) \in \mathcal{E}_T} \mu_{i,j}(\boldsymbol{S}_{i,j})}.$$

Thus, we obtain a representation of $\tilde{P}'_\Phi$ in the clique tree simply by multiplying in the new factor $\mathbf{1}\{Z = z\}$ into some clique $\boldsymbol{C}_i$ containing the variable Z.

If the clique tree is calibrated before this new factor is introduced, then the clique $\boldsymbol{C}_i$ has already assimilated all of the other information in the graph. Thus, the clique $\boldsymbol{C}_i$ itself is now fully informed, and no additional message passing is required in order to obtain $\tilde{P}'_\Phi(\boldsymbol{C}_i)$. Other cliques, however, still need to be updated with the new information. To obtain $\tilde{P}'_\Phi(\boldsymbol{C}_j)$ for another clique $\boldsymbol{C}_j$, we need only transmit messages from $\boldsymbol{C}_i$ to $\boldsymbol{C}_j$, via the intervening cliques on the path between them. (See exercise 10.10.) As a consequence, the entire tree can be recalibrated to account for the new evidence using a single pass. Note that retracting evidence is not as simple: Once we multiply parts of the distribution by zero, these parts are lost, and they cannot be recovered. Thus, if we want to reserve the ability to retract evidence, we must store the beliefs prior to the conditioning step (see exercise 10.12).

Interestingly, the same incremental-update approach applies to other forms of updating the distribution. In particular, we can multiply the distribution with a factor that is not an indicator function for some variable, an operation that is useful in various applications. The same analysis holds unchanged.

10.3.3.2 Queries Outside a Clique

Consider a query $P(\boldsymbol{Y} \mid e)$ where the variables $\boldsymbol{Y}$ are not present together in a single clique. One naive approach is to construct a clique tree where we force one of the cliques to contain $\boldsymbol{Y}$ (see exercise 10.13). However, this approach forces us to tailor our clique tree to different queries, negating many of its advantages. An alternative approach is to perform variable elimination over a calibrated clique tree.

Example 10.9

Consider the simple clique tree of example 10.7, and assume that we have calibrated the clique tree, so that the beliefs represent the joint distribution as in equation (10.10). Assume that we now want to compute the probability $\tilde{P}_\Phi(B, D)$. If the entire clique tree is calibrated, so is any (connected) subtree $\mathcal{T}'$. Letting $\mathcal{T}'$ consist of the two cliques $\boldsymbol{C}_2$ and $\boldsymbol{C}_3$, it follows from theorem 10.4 that:

$$\tilde{P}_\Phi(B, C, D) = Q_{\mathcal{T}'}.$$

By the clique tree invariant (equation (10.10)), we have that:

$$
\begin{aligned}
\tilde{P}_\Phi(B, D) &= \sum_C \tilde{P}_\Phi(B, C, D) \\
&= \sum_C \frac{\beta_2(B, C)\beta_3(C, D)}{\mu_{2,3}(C)} \\
&= \sum_C \tilde{P}_\Phi(B \mid C)\tilde{P}_\Phi(C, D),
\end{aligned}
$$

where the last equality follows from calibration. Each of these probability expressions corresponds to a set of clique beliefs divided by a message. We can now perform variable elimination, using these factors in the usual way.   ∎

Algorithm 10.4 Out-of-clique inference in clique tree

    Procedure CTree-Query (
        $\mathcal{T}$,   // Clique tree over Φ
        $\{\beta_i\}, \{\mu_{i,j}\}$,   // Calibrated clique and sepset beliefs for $\mathcal{T}$
        $\boldsymbol{Y}$ // A query
)
1   Let $\mathcal{T}'$ be a subtree of $\mathcal{T}$ such that $\boldsymbol{Y} \subseteq Scope[\mathcal{T}']$
2   Select a clique $r \in \mathcal{V}_{\mathcal{T}'}$ to be the root
3      $\Phi \leftarrow \beta_r$
4   **for** each $i \in \mathcal{V}_{\mathcal{T}'} - \{r\}$
5        $\phi \leftarrow \dfrac{\beta_i}{\mu_{i,p_r(i)}}$
6        $\Phi \leftarrow \Phi \cup \{\phi\}$
7      $\boldsymbol{Z} \leftarrow Scope[\mathcal{T}'] - \boldsymbol{Y}$
8   Let $\prec$ be some ordering over $\boldsymbol{Z}$
9   **return** Sum-Product-VE($\Phi, \boldsymbol{Z}, \prec$)

More generally, we can compute the joint probability $\tilde{P}_\Phi(\boldsymbol{Y})$ for an arbitrary subset $\boldsymbol{Y}$ by using the beliefs in a calibrated clique tree to define factors corresponding to conditional probabilities in $\tilde{P}_\Phi$, and then performing variable elimination over the resulting set of factors. The precise algorithm is shown in algorithm 10.4. The savings over simple variable elimination arise because we do not have to perform inference over the entire clique tree, but only over a portion of the tree that contains the variables $\boldsymbol{Y}$ that constitute our query. In cases where we have a very large clique tree, the savings can be significant.

10.3.3.3   Multiple Queries

Now, assume that we want to compute the probabilities of an entire set of queries where the variables are not together in a clique. For example, we might wish to compute $\tilde{P}_\Phi(X, Y)$ for every pair of variables $X, Y \in \mathcal{X} - \boldsymbol{E}$. Clearly, the approach of constructing a clique tree to ensure that our query variables are present in a single clique breaks down in this case: If every pair of variables is present in some clique, there must be some clique that contains all of the variables (see exercise 10.14).

A somewhat less naive approach is simply to run the variable elimination algorithm of algorithm 10.4 $\binom{n}{2}$ times, once for each pair of variables X, Y. However, because pairs of variables, on average, are fairly far from each other in the clique tree, this approach requires fairly substantial running time (see exercise 10.15). An even better approach can be obtained by using *dynamic programming*.

dynamic
programming

Consider a calibrated clique tree $\mathcal{T}$ over Φ, and assume we want to compute the probability $\tilde{P}_\Phi(X, Y)$ for every pair of variables X, Y. We execute this process by gradually constructing a table for each C_i, C_j that contains $\tilde{P}_\Phi(C_i, C_j)$. We construct the table for i, j in order of the distance between C_i and C_j in the tree.

The base case is when i, j are neighboring cliques. In this case, we simply extract $\tilde{P}_\Phi(C_i)$ from its clique beliefs, and compute

$$\tilde{P}_\Phi(C_j \mid C_i) = \frac{\beta_j(C_j)}{\mu_{i,j}(C_i \cap C_j)}.$$

From these, we can compute $\tilde{P}_\Phi(C_i, C_j)$.

Now, consider a pair of cliques C_i, C_j that are not neighbors, and let C_l be the neighbor of C_j that is one step closer in the clique tree to C_i. By construction, we have already computed $\tilde{P}_\Phi(C_i, C_l)$ and $\tilde{P}_\Phi(C_l, C_j)$. The key now, is to observe that

$$\tilde{P}_\Phi \models (C_i \perp C_j \mid C_l).$$

Thus, we can compute

$$\tilde{P}_\Phi(C_i, C_j) = \sum_{C_l - C_j} \tilde{P}_\Phi(C_i, C_l)\tilde{P}_\Phi(C_j \mid C_l),$$

where $\tilde{P}_\Phi(C_j \mid C_l)$ can be easily computed from the marginal $\tilde{P}_\Phi(C_j, C_l)$.

The cost of this computation is significantly lower than that of running variable elimination in the clique tree $\binom{n}{2}$ times (see exercise 10.15).

10.4   Constructing a Clique Tree

So far, we have assumed that a clique tree is given to us. How do we construct a clique tree for a set of factors, or, equivalently, for its underlying undirected graph $\mathcal{H}_\Phi$? There are two basic approaches, the first based on variable elimination and the second on direct graph manipulation.

10.4.1   Clique Trees from Variable Elimination

The first approach is based on variable elimination. As we discussed in section 10.1.1, the execution of a variable elimination algorithm can be associated with a cluster graph: A cluster C_i corresponds to the factor ψ_i generated during the execution of the algorithm, and an undirected edge connects C_i and C_j when τ_i is used (directly) in the computation of ψ_j (or vice versa). As we showed in section 10.1.1, this cluster graph is a tree, and it satisfies the running intersection property; hence, it is a clique tree.

As we showed in theorem 9.6, each factor in an execution of variable elimination with the ordering $\prec$ is a subset of a clique in the induced graph $\mathcal{I}_{\Phi, \prec}$. Furthermore, every maximal clique is a factor in the computation. Based on this result, we can conclude that, in the clique tree $\mathcal{T}$ induced by variable elimination using the ordering $\prec$, each clique is also a clique in the induced graph $\mathcal{I}_{\Phi, \prec}$, and each clique in $\mathcal{I}_{\Phi, \prec}$ is a clique in $\mathcal{T}$. This equivalence is the reason for the use of term *clique* in this context.

In the context of clique tree inference, it is standard to reduce the tree to contain only clusters that are (maximal) cliques in $\mathcal{I}_{\Phi, \prec}$. Specifically, we eliminate from the tree a cluster C_j which is a strict subset of some other cluster C_i:

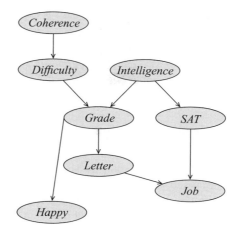

Figure 10.8 **A modified** Student **BN with an unambitious student**

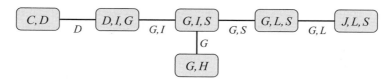

Figure 10.9 **A clique tree for the modified** Student **BN of figure 10.8**

Theorem 10.6

Let $\mathcal{T}$ be a clique tree for a set of factors Φ. Then there exists a clique tree $\mathcal{T}'$ such that:

- *each clique in $\mathcal{T}'$ is also a clique in $\mathcal{T}$;*
- *there is no pair of cliques C_i, C_j in $\mathcal{T}'$ such that $C_j \subset C_i$.*

PROOF The proof is constructive, eliminating redundant cliques one by one. We begin with $\mathcal{T}' = \mathcal{T}$. Let C_j, C_i be a pair of cliques in $\mathcal{T}'$ such that $C_j \subset C_i$. By the running intersection property, C_j is a subset of all cliques on the path between C_j and C_i. Let C_l be some neighbor clique of C_j such that $C_j \subseteq C_l$. We simply remove C_j from the tree, and connect all of the neighbors of C_j, except for C_l itself, directly to C_l. The proof that the resulting structure is a valid clique tree is not difficult, and it is left as an exercise (exercise 10.16).

As each of the clique elimination steps reduces the number of cliques in the tree, the process must terminate. When it does, we have a valid clique tree that does not contain a pair of cliques where one subsumes the other. ∎

The reduction used in this theorem is precisely the one we used to transform the tree in figure 10.1 to the one in figure 10.2. Consider also the following slightly more complex example (one that does not result in a clique tree that has the form of a chain):

Example 10.10   *Assume that our student plans to be a beach bum upon graduation, so his happiness does not depend on getting a job. On the other hand, his happiness still depends on his grade. The network is shown in figure 10.8. Variable elimination with the ordering J, L, S, H, C, D, I, G, followed by a pruning of the nonmaximal clusters from the tree (as in theorem 10.6), we obtain the clique tree shown in figure 10.9.* ∎

10.4.2   Clique Trees from Chordal Graphs

Theorem 10.6 shows that there exists a clique tree for Φ whose cliques are precisely the maximal cliques in $\mathcal{I}_{\Phi,\prec}$. This observation leads us to an alternative approach to constructing a clique tree. As we discussed in section 9.4.3.1, the induced graph $\mathcal{I}_{\prec,\Phi}$ is necessarily a chordal graph. In fact, the converse also holds: any chordal graph can be used as the basis for inference. To see that, recall that theorem 4.12 states that any chordal graph is associated with a clique tree. The algorithms presented in this chapter show that any clique tree can be used for inference. Thus, for any set of factors Φ, we can construct a clique tree for inference over Φ by constructing a chordal graph $\mathcal{H}^*$ that contains the edges in $\mathcal{H}_\Phi$, finding the maximal cliques in it, and connecting them appropriately. We now discuss each of these steps.

triangulation   The process of constructing a chordal graph that subsumes an existing graph $\mathcal{H}$ is called *triangulation*. Not surprisingly, finding a minimum triangulation, that is, one where the largest clique in the resulting chordal graph has minimum size, is $\mathcal{NP}$-hard. There are exact algorithms for finding an optimal triangulation of a graph, which are exponential in the size of the largest clique in the graph. In practice, these algorithms are too expensive, and one typically resorts to heuristic algorithms. Other triangulation methods provide a guarantee that the size of the largest clique is within a certain constant factor of optimal. These algorithms are less expensive, but they are still typically too costly in most applications. In practice, the most common approach to triangulation in graphical models uses the node-elimination techniques that we discussed in section 9.4.3.2.

Given a chordal graph $\mathcal{H}$, we now must find the maximal cliques in the graph. In general, finding the maximal cliques in a graph is also an $\mathcal{NP}$-hard problem. However, for chordal graphs the problem is quite easy. There are several methods. One of the simplest is to run maximum cardinality search on the resulting chordal graph and collect the maximal cliques generated in the process. By theorem 9.10, this process introduces no fill edges. It follows from theorem 9.6 that this process therefore generates all of the maximal cliques in $\mathcal{H}$. Another method is to begin with a family, each member of which is guaranteed to be a clique, and then use a greedy algorithm that adds nodes to the clique until it no longer induces a fully connected subgraph. This algorithm can be performed efficiently, because the number of maximal cliques in a chordal graph is at most n.

Finally, we must determine the edges in the clique tree. Again, one approach for achieving this task is maximum cardinality search, which also dictates which clique transmits information to which other clique. A somewhat more efficient approach that achieves the same effect is maximum   via a *maximum spanning tree* algorithm. More specifically, we build an undirected graph whose spanning tree   nodes are the maximal cliques in $\mathcal{H}$, and where every pair of nodes C_i, C_j is connected by an edge whose *weight* is $|C_i \cap C_j|$. We then use a standard algorithm to find a tree in this graph whose weight — that is, the sum of the weights of the edges in the graph — is maximal.

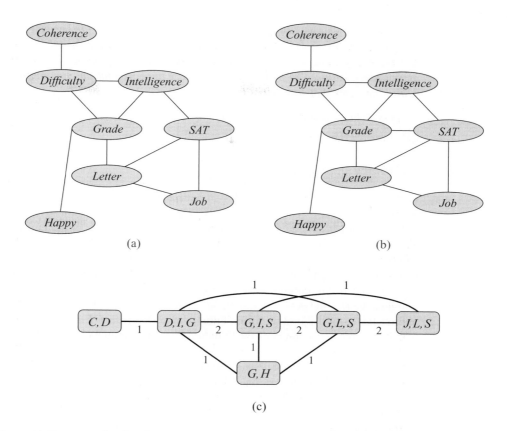

Figure 10.10 Example of a clique-tree construction algorithm: (a) Undirected factor graph (moralized graph for the network of figure 10.8). (b) One possible triangulation for this graph. (c) Cluster graph with edge weights where clusters are the cliques.

One such algorithm is shown in algorithm A.2. We can show that this approach results in the same structure as the one constructed by maximal cardinality search, and therefore it satisfies the running intersection property (see exercise 10.17).

To summarize, we can construct a clique tree as follows:

1. Given a set of factors, construct the undirected graph $\mathcal{H}_\Phi$.

2. Triangulate $\mathcal{H}_\Phi$ to construct a chordal graph $\mathcal{H}^*$.

3. Find cliques in $\mathcal{H}^*$, and make each one a node in a cluster graph.

4. Run the maximum spanning tree algorithm on the cluster graph to construct a tree.

Example 10.11

Consider again the network of figure 10.8. The undirected graph $\mathcal{H}_\Phi$ for this example is simply the moralized graph, shown in figure 10.10a. One possible triangulation for this graph, shown in *figure 10.10b, is obtained by introducing the fill edge G—S. This triangulation is a minimal one, as is the one that introduces the edge L, I. There are also, of course, other nonminimal triangulations. The graph of the maximal cliques for the chordal graph of figure 10.10b, with the associated weights on the edges, is shown in figure 10.10c. It is easy to verify that the maximum weight spanning tree is the clique tree shown in figure 10.9.* ∎

Once a clique tree is constructed, it can be used for inference using either of the two message passing algorithms described earlier.

10.5   Summary

In this chapter, we have described a somewhat different perspective on the basic task of exact inference. This approach uses a preconstructed clique tree as a data structure for exact inference. Messages are passed between the cliques in the clique tree, with the end result that the cliques are calibrated — all cliques agree on the same marginal beliefs of any variable they share. We showed two different approaches to message passing in clique trees. The first uses the same operations as variable elimination, using dynamic programming to cache messages in order to avoid repeated computation. The second uses belief propagation messages, which propagate marginal beliefs between cliques in an attempt to make them agree with each other. Both approaches allow calibration of the entire clique tree within two passes over the tree.

It is instructive to compare the standard variable elimination algorithm of chapter 9 and the algorithm obtained by variable elimination in a clique tree. In principle, they are equivalent, in that they both use the same basic operations of multiplying factors and summing out variables. Furthermore, the cliques in the clique tree are basically the factors in variable elimination. Thus, we can use any variable elimination algorithm to find a clique tree, and any clique tree to define an elimination ordering. It follows that the two approaches have basically the same computational complexity.

In practice, however, the two algorithms offer different trade-offs. On one hand, clique trees have several advantages. Most importantly, **through the use of dynamic programming, the clique tree provides answers to multiple cliques using a single computation. Additional layers of dynamic programming allow the same data structure to answer an even broader range of queries, and to dynamically introduce and retract evidence.** Moreover, the clique tree approach executes a nontrivial number of the required operations in advance, including the construction of the basic data structures, the choice of elimination ordering (which is almost determined), and the product of the CPDs assigned to a single clique.

On the other hand, clique trees, as typically implemented, also have disadvantages. First, clique trees are more expensive in terms of space. In a clique tree, we keep all intermediate factors, whereas in variable elimination we can throw them out. If there are c cliques, the cost of the clique tree algorithm can be as much as $2c$ times as expensive. More importantly, **in a clique tree, the structure of the computation is fixed and predetermined. We therefore have less flexibility to take advantage of computational efficiencies that arise because of specific features of the evidence and query.** For example, in the Student network with evidence i^1, the variable elimination algorithm could avoid introducing a dependence between G and S, resulting in substantially smaller factors. In the clique tree algorithm, the clique structure is usually predetermined, precluding these online optimizations. The difference in cost can be quite dramatic in situations where there is a lot of evidence. This type of situation-specific simplification occurs even more often in networks that exhibit context-specific independence. Finally, in standard implementaitons, the cliques in a clique tree are typically the maximal cliques in a triangulated graph. Furthermore, the operations performed in the clique tree computation are typically implemented in a fairly standard way, where the incoming messages are multiplied with the clique beliefs, and the outgoing message is generated. This approach is not always optimal (see exercise 10.7).

We can modify each of these algorithms to have some of the advantages of the other. For example, we can choose to define a clique tree online, after the evidence is obtained. In this case, the clique tree structure can take advantage of simplifications resulting from the evidence. However, we lose the advantage of precomputing the clique tree offline. As another example, we can store intermediate results in a variable elimination execution, and then do a downward pass to obtain the marginal posteriors of all variables. Here, we gain the advantage of reusing computation, at the cost of additional space. In general, we can view these two algorithms as two examples in a space of variable elimination algorithms. There are many other variants that make somewhat different trade-offs, but, fundamentally, they are performing essentially equivalent computations.

10.6 Relevant Literature

The content of this chapter is closely related to that of chapter 9; hence, many of the citations in section 9.8 are highly relevant to the discussion in this chapter, and the reader is encouraged to explore those as well.

Following the lines of the polytree algorithm, Pearl (1988) proposed a simple approach that clustered nodes so as to produce a polytree; this approach, however, produced very inefficient trees. The sum-product message passing algorithm in clique trees was developed by Shenoy and Shafer (1990); Shafer and Shenoy (1990), who described it in a much broader form that applies to many factored models other than probabilistic graphical models. The sum-product-divide approach was developed in parallel, in a series of papers by Lauritzen and Spiegelhalter (1988) and Jensen, Olesen, and Andersen (1990). This line of work also generated the perspective of message passing operations as performing a reparameterization of the original distribution, an intuition that has been very influential in some of the work of chapter 11. The sum-product-

Hugin

divide algorithm formed the basis for the HUGIN Bayesian network system, described by Andersen, Olesen, Jensen, and Jensen (1989), leading to the common use of the name "HUGIN algorithm" for this method.

Some simple modifications to the clique tree algorithm can greatly improve its efficiency. For example, Kjaerulff (1997) describes a method for improving the in-clique computations by

nested clique tree

using a *nested clique tree* data structure. Jensen (1995) provides a method for efficiently doing incremental update and retraction of evidence in a clique tree. Park and Darwiche (2004b) provide a derivation of the clique tree algorithm in terms of gradients of the network polynomial (see box 9.A). This approach can also be used as the basis for incremental update and evidence retraction.

Clique trees and their variants have been extended in many ways, and used for many tasks other than simple probabilistic inference. A complete survey is outside the scope of this book, but some of these applications and generalizations are mentioned in later chapters.

10.7 Exercises

Exercise 10.1

Prove proposition 10.1.

Exercise 10.2

Prove theorem 10.2.

Exercise 10.3

factor
marginalization

Show how to perform efficient index computations for *factor marginalization* along the lines discussed in box 10.A.

Exercise 10.4

Prove theorem 10.5.

Exercise 10.5★

Let $\mathcal{T}$ be a clique tree and $\boldsymbol{C}_r$ be a root. Let $\boldsymbol{C}_j$ be a clique in the tree and $\boldsymbol{C}_i$ its upward neighbor. Let β_j be the potential at $\boldsymbol{C}_j$ after the upward pass of CTree-SP-Upward (algorithm 10.1). Show that β_j represents the correct *conditional* probability $\tilde{P}_\Phi(\boldsymbol{C}_j \mid \boldsymbol{S}_{i,j})$. In other words, letting $\boldsymbol{X} = \boldsymbol{C}_j - \boldsymbol{S}_{i,j}$ and $\boldsymbol{S} = \boldsymbol{S}_{i,j}$, we have that:

$$\frac{\beta_j(\boldsymbol{X}, \boldsymbol{S})}{\beta_j(\boldsymbol{S})} = \tilde{P}_\Phi(\boldsymbol{X} \mid \boldsymbol{S}).$$

Exercise 10.6★

Assume that we have constructed a clique tree $\mathcal{T}$ for a given Bayesian network graph $\mathcal{G}$, and that each of the cliques in $\mathcal{T}$ contains at most k nodes. Now, the user decides to add a single edge to the Bayesian network, resulting in a network $\mathcal{G}'$. (The edge can be added between any pair of nodes in the network, so long as it maintains acyclicity.) What is the tightest bound you can provide on the maximum clique size in a clique tree $\mathcal{T}'$ for $\mathcal{G}'$? Justify your response by explaining how to construct such a clique tree. (Note: You do not need to provide the optimal clique tree $\mathcal{T}'$. The question asks for the tightest clique tree that you can construct, using only the fact that $\mathcal{T}$ is a clique tree for $\mathcal{G}$.)

Exercise 10.7

The algorithm for performing variable elimination in a clique tree (algorithm 10.1 and algorithm 10.2) specifies a particular approach for sending a variable elimination message: First (in a preprocessing step), the initial clique potentials are generated by multiplying all the factors assigned to a clique. Then, in the message passing operation (SP-Message), the initial potential is multiplied by all of the incoming messages, and the variables not on the sepset are summed out.

Is this the most computationally efficient procedure for executing the message passing step? Either explain why or provide a counterexample.

Exercise 10.8★

network
polynomial

Consider again the *network polynomial* construction of box 9.A and the algorithm of exercise 9.6 for efficiently computing all the polynomial's derivatives. Show that this algorithm provides an alternative derivation of the up/down clique-tree calibration procedure for sum-product clique trees. In other words, find a correspondence between the computation of the partial derivatives in exercise 9.6 and the message passing operations in the clique tree algorithm.

Exercise 10.9★★

Use your answer to exercise 10.8 to come up with a *rule-based clique tree* algorithm, based on the rule-based variable elimination procedure of section 9.6.2.1. Your algorithm should compute in a single upward-downward pass all of the marginal probabilities for all variables in the network; its complexity should be twice the cost of the simple rule-based variable elimination of section 9.6.2.1.

Exercise 10.10★

Let $\mathcal{T}$ be a calibrated clique tree representing the unnormalized distribution $\tilde{P}_\Phi = \prod_{\phi \in \Phi} \phi$. Let ϕ' be a new factor, and let $\tilde{P}'_\Phi = \tilde{P}_\Phi \cdot \phi'$. Let $\boldsymbol{C}_i$ be some clique such that $Scope[\phi] \subseteq \boldsymbol{C}_i$. Show that we can

perform an *incremental update* to obtain $\tilde{P}'_\Phi(\boldsymbol{C}_j)$ for any clique $\boldsymbol{C}_j$ by multiplying ϕ' into β_i and then propagating messages from $\boldsymbol{C}_i$ to $\boldsymbol{C}_j$ along the path between them.

Exercise 10.11★

Consider the problem of eliminating extraneous variables from a clique tree. More precisely, given a calibrated clique tree $\mathcal{T}$ over $\mathcal{X}$, we want to generate a (calibrated) clique tree $\mathcal{T}'$ whose scope is some subset $\boldsymbol{Y} \subset \mathcal{X}$. Clearly, we want to make $\mathcal{T}'$ as small as possible (in terms of the size of the resulting cliques). However, we do not want to construct and calibrate a clique tree from scratch; rather, we want to reuse our previous computation.

a. Suppose $\mathcal{T}$ consists of two cliques $\boldsymbol{C}_1$ and $\boldsymbol{C}_2$ over variables $\{A, B, C\}$ and $\{C, D\}$, respectively. What is the resulting $\mathcal{T}'$ if $\boldsymbol{Y} = \{B, C\}$?

b. For the clique tree $\mathcal{T}$ defined in part 1, what is $\mathcal{T}'$ if $\boldsymbol{Y} = \{B, D\}$?

c. Now consider an arbitrary clique tree $\mathcal{T}$ over $\mathcal{X}$ and an arbitrary $\boldsymbol{Y} \subseteq \mathcal{X}$. Provide an algorithm to transform a calibrated tree $\mathcal{T}$ into a calibrated tree $\mathcal{T}'$ over $\boldsymbol{Y}$. Your algorithm should *not* resort to manipulating the underlying network or factors; all operations should be performed directly on the clique tree.

d. Give an example where the resulting tree $\mathcal{T}'$ is larger than the original clique tree $\mathcal{T}$.

Exercise 10.12★★

Let $\mathcal{T}$ be a clique tree over $\mathcal{X}$, defined via a set of initial factors Φ. Let $\boldsymbol{Y} = \boldsymbol{y}$ be some observed assignment to a subset of the variables. Consider a setting where we might be unsure about a particular observation $Y_i = y_i^j$ for $Y_i \in \boldsymbol{Y}$, and that we want to compute the effect on the unnormalized probability of the other possible values y_i^k. More precisely, let $\boldsymbol{Y}_{-i} = \boldsymbol{Y} - \{Y_i\}$ and $\boldsymbol{y}_{-i}$ be the assignment in $\boldsymbol{y}$ to $\boldsymbol{Y}_{-i}$. We want to compute $\tilde{P}_\Phi(y_i^k, \boldsymbol{Y}_{-i} = \boldsymbol{y}_{-i})$ for every Y_i and every y_i^k. Describe a variant of sum-product message passing (algorithm 10.2) that can perform this task without requiring more messages than the standard two-pass calibration. (Hint: Rather than reducing the factors prior to message passing, consider reducing factors during the message passing process.)

Exercise 10.13

Provide a simple method for constructing a clique tree such that a given set of variables $\boldsymbol{Y}$ is guaranteed to be to together in some clique. Your algorithm should use a standard clique-tree construction algorithm as a black box.

Exercise 10.14

Assume that we have a clique tree $\mathcal{T}$ over $\mathcal{X}$ such that, for every pair of nodes $X, Y \in \mathcal{X}$, there exists a clique that contains both X and Y. Prove that $\mathcal{T}$ must contain a single clique that contains all of $\mathcal{X}$.

Exercise 10.15★

Consider the task of using a calibrated clique tree $\mathcal{T}$ over Φ to compute all of the pairwise marginals of variables, $\tilde{P}_\Phi(X, Y)$ for all X, Y. Assume that our probabilistic network consists of a chain $X_1 - X_2 - \ldots - X_n$, and that our clique tree has the form $1 - \ldots - n - 1$ where $Scope[\boldsymbol{C}_i] = \{X_i, X_{i+1}\}$. Also assume that each variable X_i has $|Val(X_i)| = d$.

a. What is the total cost (number of multiplication steps and number of addition steps) of doing variable elimination over this chain-structured clique tree, as described in the algorithm of algorithm 10.4, for all $\binom{n}{2}$ variable pairs?

b. What is the total cost of the algorithm described in section 10.3.3.3? Provide a precise expression, not merely an asymptotic upper bound.

c. Since we are computing marginals for all variable pairs, we may store any computations done for the previous pairs and use them to save time for the remaining pairs. Construct a dynamic programming algorithm for this specific chain structure that reduces the complexity of this task by a factor of d^2 over the algorithm described in section 10.3.3.3? Explain why this approach idea would not work for a general clique tree.

Exercise 10.16

Complete the proof of theorem 10.6 by showing that, if we have a valid clique tree, then a clique elimination step, as described, results in a valid clique tree. In particular, show that it is a tree, that it satisfies the family preservation property, and that it satisfies the running intersection property.

Exercise 10.17★

Show that the clique tree constructed using the maximum-weight-spanning tree procedure of section 10.4.2 satisfies the running intersection property. (Hint: Show that this tree structure can also be constructed using the maximum-cardinality algorithm.)

11 Inference as Optimization

11.1 Introduction

In the previous chapters we examined exact inference. We have seen that for many networks we can perform exact inference efficiently. As we have seen, the computational and space complexity of the clique tree is exponential in the tree-width of the network. This means that the exact algorithms we examined become infeasible for networks with a large tree-width. In many real-life applications, we encounter such networks. This motivates examination of approximate inference methods that are applicable to networks where exact inference is intractable.

☞ **In this chapter we consider a class of approximate inference methods, where the approximation arises from constructing an approximation to the target distribution P_Φ. This approximation takes a simpler form that allows for inference.** In general, the simpler approximating form exploits a local factorization structure that is similar in nature to the structure exploited by graphical models.

The specific algorithms we consider differ in many details, and yet they share some common conceptual principles. We now review these principles to provide a common framework for the remaining presentation. In each method, we define a target class $\mathcal{Q}$ of "easy" distributions Q and then search for an instance within that class that is the "best" approximation to P_Φ. Queries can then be answered using inference on Q rather than on P_Φ. All of the methods we describe optimize (roughly) the same target function for measuring the similarity between Q and P_Φ.

constrained optimization

This approach reformulates the inference task as one of optimizing an objective function over the class $\mathcal{Q}$. This problem falls into the category of *constrained optimization*. Such problems can be solved using a variety of different methods. Thus, the formulation of inference from this perspective opens the door to the application of a range of techniques developed in the optimization literature. Currently, the technique most often used in the setting of graphical models is one based on the use of *Lagrange multipliers*, which we review in appendix A.5.3. This method produce a set of equations that characterize the optima of the objective. In our setting, this characterization takes the form of a set of fixed-point equations that define each variable in terms of others. A particularly compelling and elegant result is that **the fixed-point equations derived from the constrained energy optimization, for any of the methods we describe, can be viewed as passing messages over a graph object.** Indeed, as we will show, even the standard sum-product algorithm for clique trees (algorithm 10.2) can be rederived from this perspective. Moreover, many other message passing algorithms follow from the same derivation.

Methods in this class fall into three main categories. The first category includes methods that use clique-tree message passing schemes on structures other than trees. This class of methods, which includes the famous *loopy belief propagation* algorithm, can be understood as optimizing approximate versions of the energy functional. The second category includes methods that use message propagation on clique trees with approximate messages. This class of methods, often known as the *expectation propagation* algorithm, maximize the exact energy functional, but with relaxed consistency constraints on the representation Q. Finally, in the third category there are methods that generalize the *mean field* method originating in statistical physics. These methods use the exact energy functional, but they restrict attention to a class Q consisting of distributions Q that have a particular simple factorization. This factorization is chosen to be simple enough to ensure that we can perform inference with Q.

More broadly, each of these algorithms can be described from two perspectives: as a procedural description of a message passing algorithm, or as an optimization problem consisting of an objective and a constraint space. Historically, the message passing algorithm generally originated first, sometimes long before the optimization interpretation was understood. However, the optimization perspective provides a much deeper understanding of these methods, and it shows that message passing is only one way of performing the optimization; it also helps point the way toward useful generalizations. In the ensuing discussion, we usually begin the presentation of each class of methods by describing a simple variant of the algorithm, providing a concrete manifestation to ground the concepts. We then present the optimization perspective on the algorithm, allowing a deeper understanding of the algorithm. Finally, we discuss generalizations of the simple algorithm, often ones that are derived directly from the optimization perspective.

11.1.1   Exact Inference Revisited ★

Before considering approximate inference methods, we start by casting exact inference as an optimization problem. The concepts we introduce here will serve in the discussion of the following approximate inference methods.

Assume we have a factorized distribution of the form

$$P_\Phi(\mathcal{X}) = \frac{1}{Z} \prod_{\phi \in \Phi} \phi(\boldsymbol{U}_\phi), \tag{11.1}$$

where the factors ϕ in Φ comprise the distribution, and the variables $\boldsymbol{U}_\phi = Scope[\phi] \subseteq \mathcal{X}$ are the scope of each factor. For example, the factors might be CPDs in a Bayesian network, generally restricted by an evidence set, or they might be potentials in a Markov network. We are interested in answering queries about the distribution P_Φ. These include queries about marginal probabilities of variables and queries about the partition function Z. As we discussed, if P_Φ is a Bayesian network with instantiated evidence on some variables, then the partition function Z is the probability of the evidence.

Recall that the end product of belief propagation is a calibrated cluster tree. Also recall that a calibrated set of beliefs for the cluster tree represents a distribution. In exact inference we find a set of calibrated beliefs that represent $P_\Phi(\mathcal{X})$. That is, we find beliefs that match the distribution represented by given set of initial potentials. Thus, we can view exact inference as searching over the set of distributions Q that are representable by the cluster tree to find a distribution Q^* that matches P_Φ.

Intuitively, we can rephrase this question as searching for a calibrated distribution that is as close as possible to P_Φ. There are many possible ways of measuring the distance between two distributions, such as the Euclidean distance (L_2), or the L_1 distance and the related variational distance (see appendix A.1.3.3). As we will see, our main challenge, however, is our aim to avoid performing inference with the distribution P_Φ; in particular, we cannot effectively compute marginal distributions in P_Φ. Hence, we need methods that allow us to optimize the distance between Q and P_Φ without answering hard queries about P_Φ. A priori, this requirement may seem impossible to satisfy. However, it turns out that there exists a distance measure — the relative entropy (or KL-divergence) — that allows us to exploit the structure of P_Φ without performing reasoning with it.

Recall that the relative entropy between P_1 and P_2 is defined as[1]

$$\boldsymbol{D}(P_1 \| P_2) = \boldsymbol{E}_{P_1} \left[\ln \frac{P_1(\mathcal{X})}{P_2(\mathcal{X})} \right].$$

Also recall that the relative entropy is always nonnegative, and equal to 0 if and only if $P_1 = P_2$. Thus, we can use it as a distance measure, and choose to find an approximation Q to P_Φ that minimizes the relative entropy.

However, as we discussed, the relative entropy is not symmetric — $\boldsymbol{D}(P_1 \| P_2) \neq \boldsymbol{D}(P_2 \| P_1)$. In section 8.5, we discussed the use of relative entropy for projecting a distribution into a

M-projection   restricted class; this projection can aim to minimize either $\boldsymbol{D}(P_\Phi \| Q)$, via the *M-projection*,

I-projection   or $\boldsymbol{D}(Q \| P_\Phi)$, via the *I-projection*. A priori, it might appear that the M-projection is more appropriate, since one of the main information-theoretic justifications for the relative entropy $\boldsymbol{D}(P_\Phi \| Q)$ is the number of bits lost when coding a true message distribution P_Φ using an (approximate) estimate Q. However, as the discussion of section 8.5.2 shows, computing the M-projection Q — $\arg\min_Q \boldsymbol{D}(P_\Phi \| Q)$ — requires that we compute marginals of P_Φ and is therefore equivalent to running inference in P_Φ. Somewhat surprisingly, as we show in the subsequent discussion, this does not apply to I-projection: we can exploit the structure of P_Φ to optimize $\arg\min_Q \boldsymbol{D}(Q \| P_\Phi)$ efficiently, *without* running inference in P_Φ.

To summarize this discussion, we want to search for a distribution Q that minimizes $\boldsymbol{D}(Q \| P_\Phi)$. To define and analyze this optimization problem formally, we also need to specify the objects we optimize over. Suppose we are given a clique tree structure $\mathcal{T}$ for P_Φ; that is, $\mathcal{T}$ satisfies the running intersection property and the family preservation property. Moreover, suppose we are given a set of beliefs

$$\boldsymbol{Q} = \{\beta_i : i \in \mathcal{V}_\mathcal{T}\} \cup \{\mu_{i,j} : (i\text{-}j) \in \mathcal{E}_\mathcal{T}, \}$$

where $\boldsymbol{C}_i$ denotes clusters in $\mathcal{T}$, β_i denotes beliefs over $\boldsymbol{C}_i$, and $\mu_{i,j}$ denotes beliefs over $\boldsymbol{S}_{i,j}$ of edges in $\mathcal{T}$.

As in definition 10.6, the set of beliefs in $\mathcal{T}$ defines a distribution Q by the formula

$$Q(\mathcal{X}) = \frac{\prod_{i \in \mathcal{V}_\mathcal{T}} \beta_i}{\prod_{(i\text{-}j) \in \mathcal{E}_\mathcal{T}} \mu_{i,j}}. \tag{11.2}$$

---

1. Note that, until now, we defined the relative entropy and other information-theoretic terms, such as mutual information, using logarithms to base 2. As will become apparent, in the context of the discussion in this chapter, the natural logarithm (base e) is more suitable. This change is a simple rescaling of the relevant information-theoretic quantities and does not change their basic properties.

calibration

marginal
consistency

(See section 10.2.3.) Due to the *calibration* requirement, the set of beliefs Q satisfies the *marginal consistency constraints* if, for each $(i\text{–}j) \in \mathcal{E}_\mathcal{T}$, the beliefs on $S_{i,j}$ are the marginal of β_i (and β_j). Recall that theorem 10.4 shows that if Q is a set of calibrated beliefs for $\mathcal{T}$ and Q is the distribution defined by equation (11.2), then

$$\begin{aligned} \beta_i[c_i] &= Q(c_i) \\ \mu_{i,j}[s_{i,j}] &= Q(s_{i,j}). \end{aligned}$$

Thus, the beliefs correspond to marginals of the distribution Q defined by equation (11.2).

Thus, we are now searching over a set of distributions Q that are representable by a set of beliefs Q over the cliques and sepsets in a particular clique tree structure $\mathcal{T}$. Note that when deciding on the representation of Q we are actually making two decisions: We are deciding both on the space of distributions that we are considering (all distributions for which $\mathcal{T}$ is an I-map), and on the representation of these distributions (as a set of calibrated clique beliefs). Both of these decisions are significant components in the specification of our optimization problem.

With these definitions in hand, we can now view exact inference as maximizing $-D(Q \| P_\Phi)$ over the space of calibrated sets Q.

> **CTree-Optimize-KL:**
>
> **Find** $Q = \{\beta_i : i \in \mathcal{V}_\mathcal{T}\} \cup \{\mu_{i,j} : (i\text{–}j) \in \mathcal{E}_\mathcal{T}\}$
>
> **maximizing** $-D(Q \| P_\Phi)$
>
> **subject to**
>
> $$\mu_{i,j}[s_{i,j}] = \sum_{C_i - S_{i,j}} \beta_i(c_i) \quad \forall(i\text{–}j) \in \mathcal{E}_\mathcal{T}, \forall s_{i,j} \in Val(S_{i,j})$$
>
> $$\sum_{c_i} \beta_i(c_i) = 1 \qquad \forall i \in \mathcal{V}_\mathcal{T}.$$

In solving this optimization problem, we conceptually examine different configurations of beliefs that satisfy the marginal consistency constraints, and we select the configuration that maximizes the objective. Such an exhaustive examination, of course, is impossible to perform in practice. However, there are effective solutions to this problem that find the maximum point. We have already seen that, if $\mathcal{T}$ is a proper cluster tree for the set of original potentials Φ, we know that there is a set Q that induces, via equation (11.2), a distribution $Q = P_\Phi$. Because this solution achieves a relative entropy of 0, which is the highest value possible, it is the unique global optimum of this optimization.

Theorem 11.1

If $\mathcal{T}$ is an I-map of P_Φ, then there is a unique solution to CTree-Optimize-KL.

This optimum can be found using the exact inference algorithms we developed in chapter 10.

11.1.2 The Energy Functional

The preceding discussion suggests a strategy for constructing approximations of P_Φ. Instead of searching over the space of all calibrated cluster trees, we can search over a space of "simpler" distributions. In this search we will not find a distribution equivalent to P_Φ, yet we might find one that is reasonably close to P_Φ. Moreover, as part of the design of the target set of distributions, we can ensure that these distributions are ones in which we can perform inference efficiently.

One problem that we will face is that the target of the optimization $D(Q\|P_\Phi)$ is unwieldy for direct optimization. The relative entropy term contains an explicit summation over all possible instantiations of $\mathcal{X}$, an operation that is infeasible in practice. However, since we know the form of $\ln P_\Phi(\xi)$ from equation (11.1), we can exploit its structure to rewrite the relative entropy in a simpler form, as shown in the following theorem.

Theorem 11.2

energy functional

$$D(Q\|P_\Phi) = \ln Z - F[\tilde{P}_\Phi, Q]$$
where $F[\tilde{P}_\Phi, Q]$ is the energy functional

$$F[\tilde{P}_\Phi, Q] = E_Q\left[\ln \tilde{P}(\mathcal{X})\right] + H_Q(\mathcal{X}) = \sum_{\phi \in \Phi} E_Q[\ln \phi] + H_Q(\mathcal{X}). \tag{11.3}$$

PROOF

$$D(Q\|P_\Phi) = E_Q[\ln Q(\mathcal{X})] - E_Q[\ln P_\Phi(\mathcal{X})]. \tag{11.4}$$

Using the product form of P_Φ, we have that

$$\ln P_\Phi(\mathcal{X}) = \sum_{\phi \in \Phi} \ln \phi(U_\phi) - \ln Z.$$

Moreover, recall that $H_Q(\mathcal{X}) = -E_Q[\ln Q(\mathcal{X})]$. Plugging these into equation (11.4), we get

$$\begin{aligned} D(Q\|P_\Phi) &= -H_Q(\mathcal{X}) - E_Q\left[\sum_{\phi \in \Phi} \ln \phi(U_\phi)\right] + E_Q[\ln Z] \\ &= -F[\tilde{P}_\Phi, Q] + \ln Z. \end{aligned}$$ ■

Importantly, the term $\ln Z$ does not depend on Q. Hence, minimizing the relative entropy $D(Q\|P_\Phi)$ is equivalent to maximizing the energy functional $F[\tilde{P}_\Phi, Q]$.

free energy

This latter term relates to concepts from statistical physics, and it is the negative of what is referred to in that field as the *(Helmholtz) free energy*. While explaining the physics-based motivation for this term is out of the scope of this book, we continue to use the standard terminology of energy functional.

energy term

The energy functional contains two terms. The first, called the *energy term*, involves expectations of the logarithms of factors in Φ. Here, each factor in Φ appears as a separate term. Thus, if the factors that comprise Φ are small, each expectation deals with relatively few variables. The difficulties in dealing with these expectations depends on the properties of the distribution Q. Assuming that inference is "easy" in Q, we should be able to evaluate such expectations

entropy term

relatively easily. The second term, called the *entropy term*, is the entropy of Q. Again, the choice of Q determines whether we can evaluate this term. However, we will see that, for the choices we make, this term will also be tractable.

11.1.3   Optimizing the Energy Functional

In the remainder of this chapter, **we pose the problem of finding a good approximation Q as one of maximizing the energy functional, or, equivalently, minimizing the relative entropy.** Importantly, the energy functional involves expectations in Q. As we show, by choosing approximations Q that allow for efficient inference, we can both evaluate the energy functional and optimize it effectively.

Moreover, since $\boldsymbol{D}(Q\|P_\Phi) \geq 0$, we have that

$$\ln Z \geq F[\tilde{P}_\Phi, Q]. \tag{11.5}$$

lower bound   That is, the energy functional is a *lower bound* on the logarithm of the partition function Z, for any choice of Q. Why is this fact significant? Recall that, in directed models, the partition function Z is the probability of the evidence. Computing the partition function is often the hardest part of inference. And so, this theorem shows that if we have a good approximation (that is, $\boldsymbol{D}(Q\|P_\Phi)$ is small), then we can get a good lower-bound approximation to Z. The fact that this approximation is a lower bound will play an important role in later chapters on learning.

variational   In this chapter, we explore inference methods that can be viewed as strategies for optimizing
method   the energy functional. These kinds of methods are often referred to as *variational methods*. The name refers to a general strategy in which we want to solve a problem by introducing new variational parameters that increase the degrees of freedom over which we optimize. Each choice of these parameters gives an approximate answer. We then attempt to optimize the variational parameters to get the best approximation. In our case, the task is to answer queries about P_Φ, and the variational parameters describe the distribution Q. In the methods we consider, we vary these parameters to try to find a good approximation to the target query.

11.2   Exact Inference as Optimization

Before considering approximate inference methods, we illustrate the the use of a variational approach to rederive an exact inference procedure. The concepts we introduce here will serve in discussion of the following approximate inference methods.

As we have already seen, the optimization problem CTree-Optimize-KL has a unique solution. We start by reformulating the optimization problem in terms of the energy functional. As we have seen, maximizing the energy functional is equivalent to minimizing the relative entropy between Q and P_Φ.

Once we restrict attention to calibrated cluster trees, we can further simplify the objective function. More precisely, we can rewrite the energy functional in a factored form as a sum of terms each of which depends directly only on one of the beliefs in $\boldsymbol{Q}$. This form reveals the structure in the distribution, and it is therefore a much better starting point for further analysis. As we will see, this form is also the basis for our approximations in subsequent sections.

Definition 11.1

factored energy
functional

Given a cluster tree $\mathcal{T}$ with a set of beliefs $\boldsymbol{Q}$ and an assignment α that maps factors in P_Φ to clusters in $\mathcal{T}$, we define the factored energy functional*:*

$$\tilde{F}[\tilde{P}_\Phi, \boldsymbol{Q}] = \sum_{i\in\mathcal{V}_\mathcal{T}} \boldsymbol{E}_{\boldsymbol{C}_i \sim \beta_i}[\ln \psi_i] + \sum_{i\in\mathcal{V}_\mathcal{T}} \boldsymbol{H}_{\beta_i}(\boldsymbol{C}_i) - \sum_{(i-j)\in\mathcal{E}_\mathcal{T}} \boldsymbol{H}_{\mu_{i,j}}(\boldsymbol{S}_{i,j}), \tag{11.6}$$

where ψ_i is the initial potential assigned to C_i:

$$\psi_i = \prod_{\phi,\alpha(\phi)=i} \phi,$$

∎

and $E_{C_i \sim \beta_i}[\cdot]$ denotes expectation on the value C_i given the beliefs β_i

Before we prove that the energy functional is equivalent to its factored variant, let us first study its components. The first term is a sum of terms of the form $E_{C_i \sim \beta_i}[\ln \psi_i]$. Recall that ψ_i is a factor (not necessarily a distribution) over the scope C_i, that is, a function from $Val(C_i)$ to $\mathbb{R}^+$. Its logarithm is therefore a function from $Val(C_i)$ to $\mathbb{R}$. The beliefs β_i are a distribution over $Val(C_i)$. We can therefore compute the expectation $\sum_{c_i} \beta_i(c_i) \ln \psi_i$. The last two terms are entropies of the beliefs associated with the clusters and sepsets in the tree. The important benefit of this reformulation is that all the terms are *local*, in the sense that they refer to a specific belief factor. As we will see, this will make our tasks much simpler.

Proposition 11.1

If Q is a set of calibrated beliefs for $\mathcal{T}$, and Q is defined by equation (11.2), then

$$\tilde{F}[\tilde{P}_\Phi, Q] = F[\tilde{P}_\Phi, Q].$$

PROOF Note that $\ln \psi_i = \sum_{\phi,\alpha(\phi)=i} \ln \phi$. Moreover, since $\beta_i(c_i) = Q(c_i)$, we conclude that

$$\sum_i E_{C_i \sim \beta_i}[\ln \psi_i] = \sum_\phi E_{C_i \sim Q}[\ln \phi].$$

It remains to show that

$$H_Q(\mathcal{X}) = \sum_{i \in \mathcal{V}_\mathcal{T}} H_{\beta_i}(C_i) - \sum_{(i-j) \in \mathcal{E}_\mathcal{T}} H_{\mu_{i,j}}(S_{i,j}).$$

This equality follows directly from equation (11.2) and theorem 10.4. ∎

Using this form of the energy, we can now define the optimization problem. We first need to define the space over which we are optimizing. If Q is factorized according to $\mathcal{T}$, we can represent it by a set of calibrated beliefs. Marginal consistency is a constraint on the beliefs that requires neighboring beliefs to agree on the marginal distribution on their joint subset. It is equivalent to requiring that the beliefs be calibrated. Thus, we pose the following constrained optimization procedure:

CTree-Optimize:

Find $Q = \{\beta_i : i \in \mathcal{V}_\mathcal{T}\} \cup \{\mu_{i,j} : (i-j) \in \mathcal{E}_\mathcal{T}\}$
maximizing $\tilde{F}[\tilde{P}_\Phi, Q]$
subject to

$$\mu_{i,j}[s_{i,j}] = \sum_{C_i - S_{i,j}} \beta_i(c_i) \tag{11.7}$$

$$\forall(i-j) \in \mathcal{E}_\mathcal{T}, \forall s_{i,j} \in Val(S_{i,j})$$

$$\sum_{c_i} \beta_i(c_i) = 1 \qquad \forall i \in \mathcal{V}_\mathcal{T} \tag{11.8}$$

$$\beta_i(c_i) \geq 0 \qquad \forall i \in \mathcal{V}_\mathcal{T}, c_i \in Val(C_i). \tag{11.9}$$

The constraints equation (11.7), equation (11.8), and equation (11.9) ensure that the beliefs in Q are calibrated and represent legal distributions (exercise 11.2).

11.2.1 Fixed-Point Characterization

We can now prove that the *stationary points* of this constrained optimization function — the points at which the gradient is orthogonal to all the constraints — can be characterized by a set of *fixed-point equations*. As we show, these equations turn out to be the update equations in the sum-product belief-propagation procedure (CTree-SP-calibrate in algorithm 10.2). Thus, if we turn these equations into an iterative algorithm, as we will describe, we obtain precisely the belief propagation algorithm in clique trees. We note that for this derivation and other similar ones later in the chapter, we restrict attention to models where all of the potentials are strictly positive (contain no zero entries). Although the results generally hold also for the case of deterministic potentials (zero entries), the proofs are considerably more complex and are outside the scope of this book.

Recall that a stationary point of a function is either a local maximum, a local minimum, or a saddle point. In the optimization problem CTree-Optimize, there is a single global maximum (see theorem 11.1). Although we do not show it here, one can show that it is also the only stationary point (see exercise 11.3), and thus once we find a stationary point, we know that we have found the maximum.

We want to *characterize* this stationary point by a set of equations that must hold when the choice of beliefs in Q is at the stationary point. Recall that our aim is to maximize the function
Lagrange
multipliers
$\tilde{F}[\tilde{P}_\Phi, Q]$ under the consistency constraints. The method of *Lagrange multipliers*, reviewed in appendix A.5.3, provides us with tools for dealing with constrained optimization. Because the characterization of the stationary point is of central importance to later developments, we examine how to construct such a characterization using the method of Lagrange multipliers.

When using the method of Lagrange multipliers, we start by defining a Lagrangian with a Lagrange multiplier for each of the constraints on the function we want to optimize. In our case, we have the constraints in equation (11.7) and equation (11.8). We note that, in principle, we also need to introduce a Lagrange multiplier for the inequality constraint that ensures that all beliefs are nonnegative. However, as we will see, the assumption that factors are strictly positive implies that the beliefs we construct in the solution to the optimization problem will be nonnegative, and thus we do not need to enforce these constraints actively. We therefore obtain the following Lagrangian:

$$
\begin{aligned}
\mathcal{J} \;=\; & \tilde{F}[\tilde{P}_\Phi, Q] \\
& - \sum_{i \in \mathcal{V}_\mathcal{T}} \lambda_i \left(\sum_{c_i} \beta_i(c_i) - 1 \right) \\
& - \sum_i \sum_{j \in \mathrm{Nb}_i} \sum_{s_{i,j}} \lambda_{j \to i}[s_{i,j}] \left(\sum_{c_i \sim s_{i,j}} \beta_i(c_i) - \mu_{i,j}[s_{i,j}] \right),
\end{aligned}
$$

where Nb_i is the neighbors of C_i in the clique tree. We introduce Lagrange multipliers λ_i for each beliefs factor β_i to ensure that it sums to 1. We also introduce, for each pair of neighboring cliques i and j and assignment to their sepset $s_{i,j}$, a Lagrange multiplier $\lambda_{j \to i}[s_{i,j}]$ to ensure that the marginal distribution of $s_{i,j}$ in β_j is consistent with its value in the sepset beliefs $\mu_{i,j}$. (Note that we also introduce another Lagrange multiplier for the direction $i \to j$.)

Remember that $\mathcal{J}$ is a function of the clique beliefs $\{\beta_i\}$, the sepset beliefs $\{\mu_{i,j}\}$, and the Lagrange multipliers. To find the maximum of the Lagrangian, we take its partial derivatives with respect to $\beta_i(c_i)$, $\mu_{i,j}[s_{i,j}]$, and the Lagrange multipliers. These last derivatives reconstruct the original constraints. The first two types of derivatives require some work. Differentiating the Lagrangian (see exercise 11.1), we get that

$$\frac{\partial}{\partial \beta_i(c_i)} \mathcal{J} = \ln \psi_i[c_i] - \ln \beta_i(c_i) - 1 - \lambda_i - \sum_{j \in \mathrm{Nb}_i} \lambda_{j \to i}[s_{i,j}]$$

$$\frac{\partial}{\partial \mu_{i,j}[s_{i,j}]} \mathcal{J} = \ln \mu_{i,j}[s_{i,j}] + 1 + \lambda_{i \to j}[s_{i,j}] + \lambda_{j \to i}[s_{i,j}].$$

At the stationary point, these derivatives are zero. Equating each derivative to 0, rearranging terms, and exponentiating, we get

$$\beta_i(c_i) = \exp\{-1 - \lambda_i\} \psi_i[c_i] \prod_{j \in \mathrm{Nb}_i} \exp\{-\lambda_{j \to i}[s_{i,j}]\}$$

$$\mu_{i,j}[s_{i,j}] = \exp\{-1\} \exp\{-\lambda_{i \to j}[s_{i,j}]\} \exp\{-\lambda_{j \to i}[s_{i,j}]\}.$$

These equations describe beliefs as functions of terms of the form $\exp\{-\lambda_{i \to j}[s_{i,j}]\}$. In fact, $\mu_{i,j}$ is a product of two such terms (and a constant). This suggests that these terms play the role of a message $\delta_{i \to j}$. To make this more explicit, we define

$$\delta_{i \to j}[s_{i,j}] = \exp\left\{-\lambda_{i \to j}[s_{i,j}] - \frac{1}{2}\right\}.$$

(We add the term $-\frac{1}{2}$ to deal with the additional $\exp\{-1\}$ term, but since this is a multiplicative constant, it is not that crucial.) We can now rewrite the resulting system of equations as

$$\beta_i(c_i) = \exp\left\{-\lambda_i - 1 + \frac{1}{2}|\mathrm{Nb}_i|\right\} \psi_i(c_i) \prod_{j \in \mathrm{Nb}_i} \delta_{j \to i}[s_{i,j}]$$

$$\mu_{i,j}[s_{i,j}] = \delta_{i \to j}[s_{i,j}] \delta_{j \to i}[s_{i,j}].$$

Combining these equations with equation (11.7), we now rewrite the message $\delta_{i \to j}$ as a function of other messages:

$$\delta_{i \to j}[s_{i,j}] = \frac{\mu_{i,j}[s_{i,j}]}{\delta_{j \to i}[s_{i,j}]}$$

$$= \frac{\sum_{c_i \sim s_{i,j}} \beta_i(c_i)}{\delta_{j \to i}[s_{i,j}]}$$

$$= \exp\left\{-\lambda_i - 1 + \frac{1}{2}|\mathrm{Nb}_i|\right\} \sum_{c_i \sim s_{i,j}} \psi_i(c_i) \prod_{k \in \mathrm{Nb}_i - \{j\}} \delta_{k \to i}[s_{i,k}].$$

Note that the term $\exp\left\{-\lambda_i - 1 + \frac{1}{2}|\mathrm{Nb}_i|\right\}$ is a constant (since it does not depend on c_i), and when we combine these equations with equation (11.8), we can solve for λ_i to ensure that this constant normalizes the clique beliefs β_i. We note that if the original factors define a distribution that sums to 1, then the solution for λ_i that satisfies equation (11.8) will be one where $\lambda_i = \frac{1}{2}(|\mathrm{Nb}_i| - 1)$, that is, the normalizing constant is 1.

This derivation proves the following result.

Theorem 11.3

A set of beliefs $\boldsymbol{Q}$ is a stationary point of CTree-Optimize *if and only if there exists a set of factors* $\{\delta_{i \to j}[\boldsymbol{S}_{i,j}] : (i\text{-}j) \in \mathcal{E}_\mathcal{T}\}$ *such that*

$$\delta_{i \to j} \propto \sum_{\boldsymbol{C}_i - \boldsymbol{S}_{i,j}} \psi_i \left(\prod_{k \in \mathrm{Nb}_i - \{j\}} \delta_{k \to i} \right) \tag{11.10}$$

and moreover, we have that

$$\beta_i \propto \psi_i \left(\prod_{j \in \mathrm{Nb}_i} \delta_{j \to i} \right)$$

$$\mu_{i,j} = \delta_{j \to i} \cdot \delta_{i \to j}.$$

fixed-point equations

This theorem characterizes the solution of the optimization problem in terms of *fixed-point equations* that must hold when we find a maximal Q. These fixed-point equations define the relationships that must hold between the different parameters involved in the optimization problem. Most importantly, equation (11.10) defines each message in terms of *other* messages, allowing an easy iterative approach to solving the fixed point equations. These same themes appear in all the approaches we will discuss later in this chapter.

11.2.2   Inference as Optimization

The fixed-point characterization of theorem 11.3 focuses on the relationships that hold at the maximum point (or points). However, they also hint at a way of achieving these relationships. Intuitively, a change in Q that reduces the differences between the left-hand and right-hand side of these equations will get us closer to a maximum point. The most direct way of reducing such discrepancies is to apply the equations as assignments and iteratively apply equations to the current values of the right-hand side to define a new value for the left-hand side.

More precisely, we initialize all of the $\delta_{i \to j}$'s to 1 and then iteratively apply equation (11.10), computing the left-hand side $\delta_{i \to j}$ of each equality in terms of the right-hand side (essentially converting each equality sign to an assignment). Clearly, a single iteration of this process does not usually suffice to make the equalities hold; however, under certain conditions (which hold in a clique tree), we can guarantee that this process converges to a solution satisfying all of the equations in equation (11.10); the other equations are now easy to satisfy.

Each assignment step defined by a fixed-point equation corresponds to a message passing step, where an outgoing message $\delta_{i \to j}$ is defined in terms of incoming messages $\delta_{k \to i}$. The fact that the process requires multiple assignments to converge corresponds to the fact that inference requires multiple message passing steps. In this specific example, a particular order of applying the fixed-point equation reconstructs the sum-product message passing algorithm in cluster trees shown in algorithm 10.2. As we will see, however, when we consider other variants of the optimization problem, the associated fixed-point equations result in new algorithms.

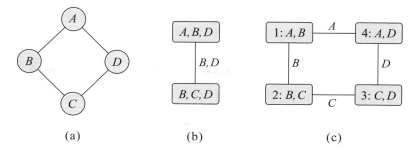

Figure 11.1   An example of a cluster graph. (a) A simple network. (b) A clique tree for the network in (a). (c) A cluster graph for the same network.

11.3   Propagation-Based Approximation

In this section, we consider approximation methods that use exactly the same message propagation as in exact inference. However, these propagation schemes use a general-purpose cluster graph, as in definition 10.1, rather than a clique tree. Since the constraints defining a clique tree were crucial in ensuring exact inference, the message-propagation schemes that use cluster graphs will generally not provide the correct answers.

We begin by defining the general message passing algorithm in a cluster graph. We then show that it can be derived, using the same process as in the previous section, from a set of fixed-point equations induced by the stationary points of an approximate energy functional.

11.3.1   A Simple Example

Consider the simple Markov network of figure 11.1a. Recall that, to perform exact inference within this network, we must first reduce it to a tree, such as the tree of figure 11.1b. Inference in this simple tree involves passing messages over the sepset, which consists of the variables $\{B, D\}$.

Now suppose that, instead, we perform inference as follows. We set up four clusters, which correspond to the four initial potentials: $C_1 = \{A, B\}$, $C_2 = \{B, C\}$, $C_3 = \{C, D\}$, $C_4 = \{A, D\}$. We connect these clusters to each other as shown in the *cluster graph* of figure 11.1c. Note that this cluster graph contains loops (undirected cycles), and is therefore not a tree; such graphs are often called *loopy*. Nevertheless, we can apply the belief-update propagation algorithm CTree-BU-calibrate (algorithm 10.3). Although in our discussion of that algorithm we assumed that the input is a tree, there is nothing in the algorithm itself that relies on that fact. In each step of the algorithm we propagate a message between neighboring clusters. Thus, it is perfectly applicable to a general cluster graph that may not necessarily be a tree.

The clusters in this cluster graph are smaller than those in the clique tree of figure 11.1b; therefore, the message passing steps are less expensive. But what is the result of this procedure? Suppose we propagate messages in the following order $\mu_{1,2}$, $\mu_{2,3}$, $\mu_{3,4}$, and then $\mu_{4,1}$. In the first message, the $\{A, B\}$ cluster passes information to the $\{B, C\}$ cluster through a marginal distribution on B. This information is then propagated to next cluster, and so on. However, in the final message $\mu_{4,1}$, this information reaches the original cluster, but this time as observation

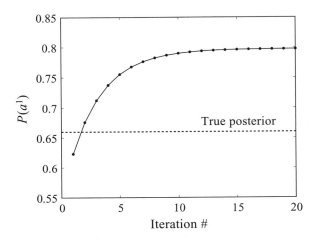

Figure 11.2   An example run of loopy belief propagation in the simple network of figure 11.1a. In this run, all potentials prefer consensus assignments over nonconsensus ones. In each iteration, we perform message passing for all the edges in the cluster graph of figure 11.1b.

about the values of A. As an example, suppose all clusters favor consensus joint assignments; that is, $\beta_1(a^0, b^0)$ and $\beta_1(a^1, b^1)$ are much larger than $\beta_1(a^1, b^0)$ and $\beta_1(a^0, b^1)$, and similarly for the other beliefs. Thus, if the message $\mu_{1,2}$ strengthens the belief that $B = b^1$, then the message $\mu_{2,3}$ will increase the belief in $C = c^1$ and so on. Once we get around the loop, the message $\mu_{4,1}$ will strengthen the support in $A = a^1$. This message will be incorporated into the cluster as though it were independent evidence that did not depend on the initial propagation. Now, if we continue to apply the same sequence of propagations again, we will keep increasing the beliefs in the assignment of $A = a^1$. This behavior is illustrated in figure 11.2. As we can see, in later iterations the procedure overestimates the marginal probability of A. However, the effect of the "feedback" decays until the iterations converge.

This simple experiment already suggests several important issues we need to consider:

- In the case of cluster trees, we described a sequence of message propagations that calibrate the tree in two passes. Once the tree is calibrated, additional message propagations do not change any of the beliefs. Thus, we can say that the propagation process has *converged*. When we consider our example, it seems clear that **the process may not converge in two passes, since information from one pass will circulate and affect the next round. Indeed, it is far from clear that the propagation of beliefs necessarily converges at all.**

- In the case of cluster trees, we saw that, in a calibrated tree, each cluster of beliefs is the joint marginal of the cluster variables. As our example suggests, for cluster graph propagation, the beliefs on A are not necessarily the marginal probability in P_Φ. Thus, the question is the relationship between the calibrated cluster graph and the actual probability distribution.

Before we address these questions, we present the algorithm in more general terms.

convergence

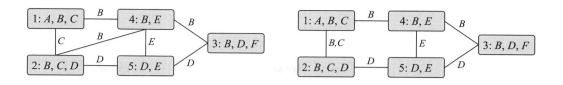

Figure 11.3 **Two examples of generalized cluster graph for an MRF** with potentials over $\{A, B, C\}$, $\{B, C, D\}$, $\{B, D, F\}$, $\{B, E\}$ and $\{D, E\}$.

Box 11.A — Case Study: Turbocodes and loopy belief propagation. *The idea of propagating messages in loopy graphs was first proposed in the early days of the field, in parallel with the introduction of the first exact inference algorithms. As we discussed in box 9.B, one of the first inference algorithms was Pearl's message passing for singly connected Bayesian networks (polytrees). In his 1988 book, Pearl says:*

> *When loops are present, the network is no longer singly connected and local propagation schemes will invariably run into trouble ... If we ignore the existence of loops and permit the nodes to continue communicating with each other as if the network were singly connected, messages may circulate indefinitely around the loops and the process may not converge to a stable equilibrium ... Such oscillations do not normally occur in probabilistic networks ... which tend to bring all messages to some stable equilibrium as time goes on. However, this asymptotic equilibrium is not coherent, in the sense that it does not represent the posterior probabilities of all nodes of the networks.*

loopy belief propagation

As a consequence of these problems, the idea of loopy belief propagation *was largely abandoned for many years.*

 Surprisingly, the revival of loopy belief propagation is due to a seemingly unrelated advance in coding theory. The area of coding addresses the problem of sending messages over a noisy channel, and recovering it from the garbled result. Formally, the coding task can be defined as follows. We wish to send a k-bit message $u_1, \ldots, u_k$. We code the message using a number of bits $x_1, \ldots, x_n$, which are then sent over the noisy channel, resulting in a set of (possibly corrupted) outputs $y_1, \ldots, y_n$, which can be either discrete or continuous. Different channels introduce noise in different ways: In a simple Gaussian noise model, each bit sent is corrupted independently by the addition of some Gaussian noise; another simple model flips each bit independently with some probability; more complex channel models, where noise is added in a correlated way to consecutive bits, are also used. The message decoding *task is to recover an estimate $\hat{u}_1, \ldots, \hat{u}_k$ from $y_1, \ldots, y_n$. The* bit error rate *is the probability that a bit is ultimately decoded incorrectly. This error rate depends on the code and decoding algorithm used and on the amount of noise in the channel. The* rate *of a code is k/n — the ratio between the number of bits in the message and the number of bits used to transmit it.*

message decoding

 For example, a very simple repetition code takes each bit and transmits it three times, then decodes the bit by majority voting on the three (noisy) copies received. If the channel corrupts each bit with probability p, the bit error rate of this algorithm is $p^3 + 3p^2$, which, for reasonable values

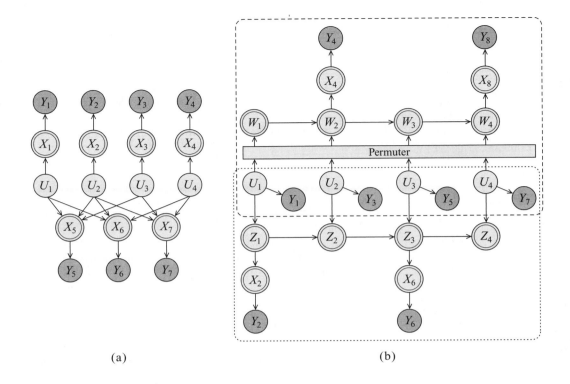

(a)   (b)

Figure 11.A.1 — Two examples of codes (a) A $k = 4, n = 7$ parity check code, where every four message bits are sent along with three bits that encode parity checks. (b) A $k = 4, n = 8$ turbocode. Here, the $\boldsymbol{X}^a$ bits X_1, X_3, X_5, X_7 are simply the original bits U_1, U_2, U_3, U_4 and are omitted for clarity of the diagram; the $\boldsymbol{X}^b$ bits use a *shift register* — a state bit that changes with each bit of the message, where the ith state bit depends on the $(i-1)$st state bit and on the ith message bit. The code uses two shift registers, one applied to the original message bits and one to a set of permuted message bits (using some predetermined permutations). The sent bits contain both the original message bits and some number of the state bits.

of p, is much lower than p. The rate of this code is 1/3, because for every message bit, three bits are transmitted. In general, we can get better bit error rates by increasing the redundancy of the code, so we want to compare the bit error rate of different codes that have the same rate. Repetition codes are some of the least efficient codes designed. Figure 11.A.1a shows a simple rate 4/7 parity check code, where every four message bits are sent along with three bits that encode parity checks (exclusive ORs) of different subsets of the four bits.

In 1948, Claude Shannon provided a theoretical analysis of the coding problem (Shannon 1948). For a given rate, Shannon provided an upper bound on the maximum noise level that can be tolerated while still achieving a certain bit error rate, no matter which code is used. Shannon also showed that there exist channel codes that achieve this limit, but his proof was nonconstructive —

he did not present practical encoders and decoders that achieve this limit.

Since Shannon's landmark result, multiple codes were suggested. However, despite a gradual improvement in the quality of the code (bit-error rate for a given noise level), none of the codes even came close to the Shannon limit. The big breakthrough came in the early 1990s, when Berrou et al. (1993) came up with a new scheme that they called a turbocode, *which, empirically, came much closer to achieving the Shannon limit than any other code proposed up to that point. However, their decoding algorithm had no theoretical justification, and, while it seemed to work well in real examples, could be made to diverge or converge to the wrong answer. The second big breakthrough was the subsequent realization that turbocodes were simply performing belief propagation on a Bayesian network representing the probability model for the code and the channel noise.*

turbocode

To understand this, we first observe that message decoding can easily be reformulated as a probabilistic inference task: We have a prior over the message bits $U = \langle U_1, \ldots, U_k \rangle$, a (usually deterministic) function that defines how a message is converted into a sequence of transmitted bits $X_1, \ldots, X_n$, and another (stochastic) model that defines how the channel randomly corrupts the X_i's to produce Y_i's. The decoding task can then be viewed as finding the most likely joint assignment to U given the observed message bits $y = \langle y_1, \ldots, y_n \rangle$, or (alternatively) as finding the posterior $P(U_i \mid y)$ for each bit U_i. The first task is a MAP inference task, and the second task one of computing posterior probabilities. Unfortunately, the probability distribution is of high dimension, and the network structure of the associated graphical model is quite densely connected and with many loops.

The turbocode approach, as first proposed, comprised both a particular coding scheme, and the use of a message passing algorithm to decode it. The coding scheme transmits two sets of bits: one set comprises the original message bits $X^a = \langle X_1^a, \ldots, X_k^a \rangle = u$, and the second some set $X^b = \langle X_1^b, \ldots, X_k^b \rangle$ of transformed bits (like the parity check bits, but more complicated). The received bits then can also be partitioned into the noisy y^a, y^b. Importantly, the code is designed so that the message can be decoded (albeit with errors) using either y^a or y^b. The turbocoding algorithm then works as follows: It uses the model of X^a (trivial in this case) and of the channel noise to compute a posterior probability over U given y^a. It then uses that posterior $\pi_a(U_1), \ldots, \pi_a(U_k)$ as a prior over U and computes a new posterior over U, using the model for X^b and the channel, and y^b as the evidence, to compute a new posterior $\pi_b(U_1), \ldots, \pi_b(U_k)$. The "new information," which is $\pi_b(U_i)/\pi_a(U_i)$, is then transmitted back to the first decoder, and the process repeats until a stopping criterion is reached. In effect, the turbocoding idea was to use two weak coding schemes, but to "turbocharge" them using a feedback loop. Each decoder is used to decode one subset of received bits, generating a more informed distribution over the message bits to be subsequently updated by the other. The specific method proposed used particular coding scheme for the X^b bits, illustrated in figure 11.A.1b.

This process looked a lot like black magic, and in the beginning, many people did not even believe that the algorithm worked. However, when the empirical success of these properties was demonstrated conclusively, an attempt was made to understand its theoretical properties. McEliece et al. (1998) subsequently showed that the specific message passing procedure proposed by Berrou et al. is precisely an application of belief propagation (with a particular message passing schedule) to the Bayesian network representing the turbocode (as in figure 11.A.1b).

☞ **This revelation had a tremendous impact on both the coding theory community and the graphical models community. For the former, loopy belief propagation provides a general-purpose algorithm for decoding a large family of codes. By separating the** algorithmic question of decoding from the question of the code design, it allowed the development of many new coding schemes with improved properties. These codes have come much, much closer to the Shannon limit than any previous codes, and they have revolutionized both the theory and the practice of coding. For the graphical models community, it was the astounding success of loopy belief propagation for this application that led to the resurgence of interest in these approaches, and subsequently to much of the work described in this chapter.

11.3.2   Cluster-Graph Belief Propagation

cluster graph

The basis for our message passing algorithm is the *cluster graph* of definition 10.1, first defined in section 10.1.1. In that section, we required that cluster graphs be trees and that they respect the running intersection property. Those requirements led us to the definition of a clique tree. Here, we remove the first of these two assumptions, allowing inference to be performed on a loopy cluster graph. However, we still wish to require a variant of the running intersection property that is generalized to this case: for any two clusters containing X, there is precisely one path between them over which information about X can be propagated.

Definition 11.2

running intersection property

We say that $\mathcal{U}$ satisfies the running intersection property *if, whenever there is a variable X such that $X \in \boldsymbol{C}_i$ and $X \in \boldsymbol{C}_j$, then there is a single path between $\boldsymbol{C}_i$ and $\boldsymbol{C}_j$ for which $X \in \boldsymbol{S}_e$ for all edges e in the path.* ■

This generalized running intersection property implies that all edges associated with X form a tree that spans all the clusters that contain X. Thus, intuitively, there is only a single path by which information *that is directly about X* can flow in the graph. Both parts of this assumption are significant. The fact that some path must exist forces information about X to flow between all clusters that contain it, so that, in a calibrated cluster graph, all clusters must agree about the marginal distribution of X. The fact that there is at most one path prevents information about X from cycling endlessly in a loop, making our beliefs more extreme due to "cyclic arguments."

Importantly, however, since the graph is not necessarily a tree, the same pair of clusters might also be connected by other paths. For example, in the cluster graph of figure 11.3a, we see that the edges labeled with B form a subtree that spans all the clusters that contain B. However, there are loops in the graph. For example, there are two paths from $\boldsymbol{C}_3 = \{B, D, F\}$ to $\boldsymbol{C}_2 = \{B, C, D\}$. The first, through $\boldsymbol{C}_4$, propagates information about B, and the second, through $\boldsymbol{C}_5$, propagates information about D. Thus, we can still get circular reasoning, albeit less directly than we would in a graph that did not satisfy the running intersection property; we return to this point in section 11.3.8. Note that while in the case of trees the definition of running intersection implied that $\boldsymbol{S}_{i,j} = \boldsymbol{C}_i \cap \boldsymbol{C}_j$, in a graph this equality is no longer enforced by the running intersection property. For example, cliques $\boldsymbol{C}_1$ and $\boldsymbol{C}_2$ in figure 11.3a have B in common, but $\boldsymbol{S}_{1,2} = \{C\}$.

In clique trees, inference is performed by calibrating beliefs. In a cluster graph, we can also

beliefs

calibrated cluster graph

associate cluster $\boldsymbol{C}_i$ with *beliefs* β_i. We now say that a cluster graph is *calibrated* if for each edge $(i\text{--}j)$, connecting the clusters $\boldsymbol{C}_i$ and $\boldsymbol{C}_j$, we have that

$$\sum_{\boldsymbol{C}_i - \boldsymbol{S}_{i,j}} \beta_i = \sum_{\boldsymbol{C}_j - \boldsymbol{S}_{i,j}} \beta_j;$$

that is, the two clusters agree on the marginal of variables in $\boldsymbol{S}_{i,j}$. Note that this definition is weaker than cluster tree calibration, since the clusters do not necessarily agree on the joint marginal of all the variables they have in common, but only on those variables in the sepset. However, if a calibrated cluster graph satisfies the running intersection property, then the marginal of a variable X is identical in all the clusters that contain it.

Algorithm 11.1 Calibration using sum-product belief propagation in a cluster graph

---

    Procedure CGraph-SP-Calibrate (
      Φ,   // Set of factors
      $\mathcal{U}$ // Generalized cluster graph Φ
)
1   Initialize-CGraph
2   **while** graph is not calibrated
3   Select $(i\text{--}j) \in \mathcal{E}_{\mathcal{U}}$
4        $\delta_{i \to j}(\boldsymbol{S}_{i,j}) \leftarrow$ SP-Message(i,j)
5   **for** each clique i
6        $\beta_i \leftarrow \psi_i \cdot \prod_{k \in \text{Nb}_i} \delta_{k \to i}$
7   **return** $\{\beta_i\}$

    Procedure Initialize-CGraph (
      $\mathcal{U}$
)
1   **for** each cluster $\boldsymbol{C}_i$
2        $\beta_i \leftarrow \prod_{\phi \,:\, \alpha(\phi)=i} \phi$
3   **for** each edge $(i\text{--}j) \in \mathcal{E}_{\mathcal{U}}$
4        $\delta_{i \to j} \leftarrow 1$
5        $\delta_{j \to i} \leftarrow 1$
6

    Procedure SP-Message (
      i,   // sending clique
      j   // receiving clique
)
1      $\psi(\boldsymbol{C}_i) \leftarrow \psi_i \cdot \prod_{k \in (\text{Nb}_i - \{j\})} \delta_{k \to i}$
2      $\tau(\boldsymbol{S}_{i,j}) \leftarrow \sum_{\boldsymbol{C}_i - \boldsymbol{S}_{i,j}} \psi(\boldsymbol{C}_i)$
3   **return** $\tau(\boldsymbol{S}_{i,j})$

How do we calibrate a cluster graph? Because calibration is a local property that relates adjoining clusters, we want to try to ensure that each cluster is sharing information with its

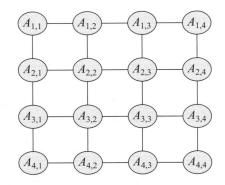

Figure 11.4   **An example of a** 4×4 **two-dimensional grid network**

neighbors. From the perspective of a single cluster C_i, there is not much difference between a cluster graph and a cluster tree. The cluster is related to each neighboring cluster through an edge that conveys information on variables in the sepset. Thus, we can transmit information by simply having one cluster pass a message to the other.

However, a priori, it is not clear how we can execute a message passing algorithm over a loopy clustergraph. In particular, the sum-product calibration of algorithm 10.2 sends a message only when the sending clique is ready to transmit, that is, when all other incoming messages have been received. In the loopy cluster graph, initially, there is no cluster that has received any incoming messages. Thus, no cluster is ready to transmit, and the algorithm is deadlocked. However, in section 10.3, we showed that the two algorithms are actually equivalent; that is, any sequence of sum-product propagation steps can be emulated by the same sequence of belief-update propagation steps and leads to the same beliefs. In this transformation, we have that $\mu_{i,j} = \delta_{i \to j} \delta_{j \to i}$. Thus, we can construct a "deadlock-free" variant of the sum-product message passing algorithm simply by initializing all messages $\delta_{i \to j} = 1$. This initialization of the sum-product algorithm is equivalent to the standard initialization of the belief update algorithm, in which $\mu_{i,j} = 1$. Importantly, in this variant of the sum-product algorithm, each cluster begins with all of the incoming messages initialized, and therefore it can send any of the outgoing messages at any time, without waiting for any other cluster.

Algorithm 11.1 shows the sum-product message passing algorithm for cluster graphs; other than the fact that the algorithm is applied to graphs rather than trees, the algorithm is identical to CTree-SP-Calibrate. In much the same manner, we can adapt CTree-BU-Calibrate to define a procedure CGraph-BU-Calibrate that operates over cluster graphs using belief-update message passing steps. Both of these algorithms are instances of a general class of algorithms called *cluster-graph belief propagation*, which passes messages over cluster graphs.

cluster-graph
belief
propagation

Before we continue, we note that cluster-graph belief propagation can be significantly cheaper than performing exact inference. A canonical example of a class of networks that is compactly representable yet hard for inference is the class of grid-structured Markov networks (such as the ones used in image analysis; see box 4.B). In these networks, each variable $A_{i,j}$ corresponds to a point on a two-dimensional grid. Each edge in this network corresponds to a potential between adjacent points on the grid, with $A_{i,j}$ connected to the four nodes $A_{i-1,j}$, $A_{i+1,j}$,

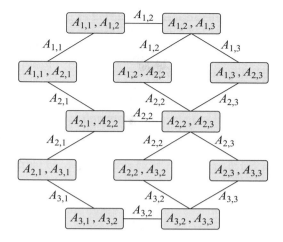

Figure 11.5 **An example of generalized cluster graph for a** 3×3 **grid network**

$A_{i,j-1}$, $A_{i,j+1}$ (except for nodes $A_{i,j}$ on the boundary of the grid); see figure 11.4. Such a network has only pairwise potentials, and hence it is very compactly represented. Yet, exact inference requires separating sets, which are as large as cutsets in the grid. Hence, in an $n \times n$ grid, exact computation is exponential in n.

However, we can easily create a generalized cluster graph for grid networks that directly corresponds to the factors in the network. In this cluster graph, each cluster represents beliefs over two neighboring grid variables, and each cluster has a small number of adjoining edges that connect it to other clusters that share one of the two variables. See figure 11.5 for an example for a small 3×3 grid. (Note that there are several ways of constructing such a cluster graph; this figure represents one reasonable choice.) A round of propagations in the generalized cluster graph is linear in the size of the grid (quadratic in n).

11.3.3 Properties of Cluster-Graph Belief Propagation

What can we say about the properties and guarantees provided by cluster-graph belief propagation? We now consider some of the ramifications of the "mechanical" operation of message passing in the graph. Later, when we discuss cluster-graph belief propagation as an optimization procedure, we will revisit this question from a different perspective.

11.3.3.1 Reparameterization

reparameterization

cluster graph invariant

Recall that in section 10.2.3 we showed that belief propagation maintains an invariant property. This allowed us to show that the convergence point represents a *reparameterization* of the original distribution. We can directly extend this property to cluster graphs, resulting in a *cluster graph invariant*.

Theorem 11.4

Let $\mathcal{U}$ be a generalized cluster graph over a set of factors Φ. Consider the set of beliefs $\{\beta_i\}$ and sepsets $\{\mu_{i,j}\}$ at any iteration of CGraph-BU-Calibrate; then

$$\tilde{P}_\Phi(\mathcal{X}) = \frac{\prod_{i \in \mathcal{V}_\mathcal{U}} \beta_i[\boldsymbol{C}_i]}{\prod_{(i-j) \in \mathcal{E}_\mathcal{U}} \mu_{i,j}[\boldsymbol{S}_{i,j}]}.$$

where $\tilde{P}_\Phi(\mathcal{X}) = \prod_{\phi \in \Phi} \phi$ is the unnormalized distribution defined by Φ.

PROOF Recall that $\beta_i = \psi_i \prod_{j \in \text{Nb}_i} \delta_{j \to i}$ and that $\mu_{i,j} = \delta_{j \to i} \delta_{i \to j}$. We now have

$$\frac{\prod_{i \in \mathcal{V}_\mathcal{U}} \beta_i[\boldsymbol{C}_i]}{\prod_{(i-j) \in \mathcal{E}_\mathcal{U}} \mu_{i,j}[\boldsymbol{S}_{i,j}]} = \frac{\prod_{i \in \mathcal{V}_\mathcal{U}} \psi_i[\boldsymbol{C}_i] \prod_{j \in \text{Nb}_i} \delta_{j \to i}[\boldsymbol{S}_{i,j}]}{\prod_{(i-j) \in \mathcal{E}_\mathcal{U}} \delta_{j \to i}[\boldsymbol{S}_{i,j}] \delta_{i \to j}[\boldsymbol{S}_{i,j}]}$$

$$= \prod_{i \in \mathcal{V}_\mathcal{U}} \psi_i[\boldsymbol{C}_i]$$

$$= \prod_{\phi \in \Phi} \phi(\boldsymbol{U}_\phi) = \tilde{P}_\Phi(\mathcal{X}).$$

Note that the second step is based on the fact that each message $\delta_{i \to j}$ appears exactly once in the numerator and the denominator and thus can be canceled. ∎

☞   **This property shows that cluster-graph belief propagation preserves all of the information about the original distribution. In particular, it does not "dilute" the original factors by performing propagation along loops. Hence, we can view the process as trying to represent the original factors anew in a more useful form.**

11.3.3.2   Tree Consistency

Recall that theorem 10.4 implies that, in a calibrated cluster tree, the belief over a cluster is the marginal of the distribution. Thus, in a calibrated cluster tree, we can "read off" the marginals of P_Φ locally from clusters that contain them. More precisely, by normalizing the beliefs factor β_i (so that it sums to 1), we get the marginal distribution over $\boldsymbol{C}_i$. An obvious question is whether a corresponding property holds for cluster-graph belief propagation. Suppose we manage to calibrate a generalized cluster graph and normalize the resulting beliefs; do we have an interpretation for the beliefs in each cluster?

As we saw in our simple example (figure 11.2), the beliefs we compute by BU-message are not necessarily marginals of P_Φ, but rather an approximation. Can we say anything about the quality of this approximation? To characterize the beliefs we get at the end of the process, we can use the cluster tree invariant property applied to subtrees of a cluster graph.

Consider a subtree $\mathcal{T}$ of $\mathcal{U}$; that is, a subset of clusters and edges that together form a tree that satisfies the running intersection property. For example, consider the cluster graph of figure 11.1c. If we remove one of the clusters and its incident edges, we are left with a proper cluster tree. Note that the running intersection property is not necessarily as easy to achieve in general, since removing some edges from the cluster graph may result in a graph that violates the running intersection property relative to a variable, necessitating the removal of additional edges, and so on.

Once we select a tree $\mathcal{T}$, we can think of it as defining a distribution

$$P_{\mathcal{T}}(\mathcal{X}) = \frac{\prod_{i \in \mathcal{V}_{\mathcal{T}}} \beta_i(\boldsymbol{C}_i)}{\prod_{(i-j) \in \mathcal{E}_{\mathcal{T}}} \mu_{i,j}[\boldsymbol{S}_{i,j}]}.$$

If the cluster graph is calibrated, then by definition so is $\mathcal{T}$. And so, because $\mathcal{T}$ is a tree that satisfies the running intersection property, we can apply theorem 10.4, and we conclude that

$$\beta_i(\boldsymbol{C}_i) = P_{\mathcal{T}}(\boldsymbol{C}_i). \tag{11.11}$$

tree consistency

That is, the beliefs over $\boldsymbol{C}_i$ in the tree are the marginal of $P_{\mathcal{T}}$, a property called *tree consistency*.

As a concrete example, consider the cluster graph of figure 11.1c. Removing the cluster $\boldsymbol{C}_4 = \{A, D\}$, we are left with a proper cluster tree $\mathcal{T}$. The preceding argument implies that once we have calibrated the cluster graph, we have $\beta_1(A, B) = P_{\mathcal{T}}(A, B)$. This result suggests that $\beta_1(A, B) \neq P_{\Phi}(A, B)$; to show this formally, contrast equation (11.11) with theorem 11.4. We see that the tree distribution involves some of the terms that define the joint distribution. Thus, we can conclude that

$$P_{\mathcal{T}}(A, B, C, D) = P_{\Phi}(A, B, C, D) \frac{\mu_{3,4}[D]\mu_{1,4}[A]}{\beta_4(A, D)}.$$

We see that unless $\beta_4(A, D) = \mu_{3,4}[D]\mu_{1,4}[A]$, $P_{\mathcal{T}}$ will be different from P_{Φ}. This conclusion suggests that, in this example, the beliefs $\beta_1(A, B)$ in the calibrated cluster graph are not the marginal $P_{\Phi}(A, B)$.

Clearly, we can apply the same type of reasoning using other subtrees of $\mathcal{U}$. And so we reach the surprising conclusion that equation (11.11) must hold with respect to every cluster tree embedded in $\mathcal{U}$. In our example, we can see that by removing a single cluster, we can construct three different trees that contain $\boldsymbol{C}_1$. The same beliefs $\beta_1(A, B)$ are the marginal of the three distributions defined by each of these trees. While these three distributions agree on the joint marginal of A and B, they can differ on the joint marginal distributions of other pairs of variables.

Moreover, these subtrees allow us to get insight about the quality of the marginal distributions we read from the calibrated cluster graph. Consider our example again: we can use the *residual* term $\frac{\mu_{3,4}[D]\mu_{1,4}[A]}{\beta_4(A,D)}$ to analyze the error in the marginal distribution. In this simple example, this analysis is fairly straightforward (see exercise 11.4).

cluster graph residual

In other cases, the analysis can be more complex. For example, suppose we want to find a subtree in the cluster graph for a grid (e.g., figure 11.5). To construct a tree, we must remove a nontrivial number of clusters. More precisely, because each cluster corresponds to an edge in the grid, a cluster tree corresponds to a subtree of the grid. For an $n \times n$ grid, such a tree will have at most $n^2 - 1$ edges of the $2n(n-1)$ edges in the grid. Thus, each cluster tree contains about half of the clusters in the original cluster graph. In such a situation the residual term is more complex, and we cannot necessarily evaluate it.

11.3.4 Analyzing Convergence ⋆

A key question regarding the belief propagation algorithm is whether and when it converges. Indeed, there are many networks for which belief propagation does not converge; see box 11.C.

Although we cannot hope for convergence in all cases, it is important to understand when this algorithm does converge. We know that if the cluster graph is a tree then the algorithm will converge. Can we find other classes of cluster graphs for which we can prove convergence?

synchronous BP

One method of analyzing convergence is based on the following important perspective on belief propagation. This analysis is easier to perform on a variant of BP called *synchronous BP* that performs all of the message updates simultaneously. Consider the update step that takes all of the messages δ^t at a particular iteration t and produces a new set of messages δ^{t+1} for the next step. Letting Δ be the space of all possible messages in the cluster graph, we can view the belief-propagation update operator as a function $G_{BP} : \Delta \mapsto \Delta$. Consider the standard sum-product message update:

$$\delta'_{i \to j} \propto \sum_{C_i - S_{i,j}} \psi_i \cdot \prod_{k \in (\mathrm{Nb}_i - \{j\})} \delta_{k \to i},$$

BP operator

where we normalize each message to sum to 1; this renormalization step is essential to avoid a degenerate convergence to the $\mathbf{0}$ message. We can now define the *BP operator* as the function that simultaneously takes one set of messages and computes a new one:

$$G_{BP}(\{\delta_{i \to j}\}) = \{\delta'_{i \to j}\}.$$

The question of convergence of the algorithm now reduces to one of asking whether repeated applications of the operator G_{BP} are guaranteed to converge.

One interesting, albeit strong, condition that guarantees convergence is the *contraction property*:

Definition 11.3

contraction

For a number $\alpha \in [0, 1)$, an operator G over a metric space $(\Delta, \mathbf{D}(;))$ is an α-contraction relative to the distance function $\mathbf{D}(;)$ if, for any $\delta, \delta' \in \Delta$, we have that:

$$\mathbf{D}(G(\delta); G(\delta')) \leq \alpha \mathbf{D}(\delta; \delta'). \tag{11.12}$$

∎

In other words, an operator is a contraction if its application to two points in the space is guaranteed to decrease the distance between them by at least some constant factor $\alpha < 1$.

A basic result in analysis shows that, under fairly weak conditions, if an operator G is a contraction, we have that repeated applications of G are guaranteed to converge to a unique fixed point:

Proposition 11.2

fixed-point

Let G be an α-contraction of a complete metric space $(\Delta, \mathbf{D}(;))$. Then there is a unique fixed-point δ^* for which $G(\delta^*) = \delta^*$. Moreover, for any δ, we have that

$$\lim_{n \to \infty} G^n(\delta) = \delta^*.$$

The proof is left as an exercise (exercise 11.5).

Indeed, the *contraction rate* α can be used to provide bounds on the rate of convergence of the algorithm to its unique fixed point: To reach a point that is guaranteed to be within ϵ of δ^*, it suffices to apply G the following number of times:

$$\log_\alpha \frac{\epsilon}{\mathrm{diameter}(\Delta)},$$

where $\text{diameter}(\Delta) = \max_{\boldsymbol{\delta}, \boldsymbol{\delta}' \in \Delta} \boldsymbol{D}(\boldsymbol{\delta}; \boldsymbol{\delta}')$.

Applying this analysis to the operator G induced by the belief-propagation message update is far from trivial. This operator is complex and nonlinear, because it involves both multiplying messages and a renormalization step. A review of these analyses is outside the scope of this book. At a high level, these results show that if the factors in the network are fairly "smooth," one can guarantee that the synchronous BP operator is a contraction and hence converges to a unique fixed point. We describe one of the simplest of these results, in order to give a flavor for this type of analysis.

This analysis applies to synchronous loopy belief propagation over a pairwise Markov network with two-valued random variables $X_i \in \{-1, +1\}$. Specifically, we assume that the network model is parameterized as follows:

$$P(x_1, \ldots, x_n) = \frac{1}{Z} \exp \left(\sum_{(i,j)} \epsilon_{i,j}(x_i, x_j) + \sum_i \epsilon_i(x_i), \right),$$

where we assume for simplicity of notation that $\epsilon_{i,j} = 0$ when X_i and X_j are not neighbors in the network.

hyperbolic tangent

We begin by introducing some notation. The *hyperbolic tangent* function is defined as:

$$\tanh(w) = \frac{e^w - e^{-w}}{e^w + e^{-w}} = \frac{e^{2w} - 1}{e^{2w} + 1}.$$

The hyperbolic tangent has a shape very similar to the sigmoid function of figure 5.11a. The following condition can be shown to suffice for G_{BP} to be a contraction, and hence for the convergence of belief propagation to a unique fixed point:

$$\max_i \max_{j \in \text{Nb}_i} \sum_{k \in \text{Nb}_i - \{j\}} \tanh |\epsilon_{k,i}| < 1. \tag{11.13}$$

Intuitively, this expression measures the total extent to which i's neighbors other than j can influence the message from i to j. The larger the magnitude of the parameters in the network, the larger this sum.

The analysis of the more general case is significantly more complex but shares the same intuitions. At a very high level, if we can place strong bounds on the skew of the parameters in a factor:

$$\max_{\boldsymbol{x}, \boldsymbol{x}'} \phi(\boldsymbol{x}) / \phi(\boldsymbol{x}'),$$

we can guarantee convergence of belief propagation. Intuitively, the lower the skew of the factors in our network, the more each message update "smoothes out" differences between entries in the messages, and therefore also makes different messages more similar to each other.

While the conditions that underlie these theorems are usually too stringent to hold in practice, this analysis does provide useful insight. First, it suggests that **networks with potentials that are closer to deterministic are more likely to have problems with convergence, an observation that certainly holds in practice.** Second, although global contraction throughout the space is a very strong assumption, a contraction property in a region of the space may be plausible, guaranteeing convergence of the algorithm if it winds up (or is initialized) in this region. These results and their ramifications are only now being explored.

11.3.5   Constructing Cluster Graphs

So far, we have taken the cluster graph to be given. However, the choice of cluster graph is generally far from obvious, and it can make a significant difference to the algorithm. Recall that, even in exact inference, more than one clique tree can be used to perform inference for a given distribution. However, while these different trees can vary in their computational cost, they all give rise to the same answers. **In the case of cluster graph approximations, different graphs can lead to very different answers. Thus, when selecting a cluster graph, we have to consider trade-offs between cost and accuracy, since cluster graphs that allow fast propagation might result in a poor approximation.**

☞   It is important to keep in mind that the structure of the cluster graph determines the propagation steps the algorithm can perform, and thus dictate what type of information is passed during the propagations. These choices directly influence the quality of the results.

Example 11.1

Consider, for example, the cluster graphs $\mathcal{U}_1$ and $\mathcal{U}_2$ of figure 11.3a and figure 11.3b. Both are fairly similar, yet in $\mathcal{U}_2$ the edge between C_1 and C_2 involves the marginal distribution over B and C. On the other hand, in $\mathcal{U}_1$, we propagate the marginal only over C. Intuitively, we expect inference in $\mathcal{U}_2$ to better capture the dependencies between B and C. For example, assume that the potential of C_1 introduces strong correlations between B and C (say $B = C$). In $\mathcal{U}_2$, this correlation is conveyed to C_2 directly. In $\mathcal{U}_1$, the marginal on C is conveyed on the edge (1–2), while the marginal on B is conveyed through C_4. In this case, the strong dependency between the two variables is lost. In particular, if the marginal on C is diffuse (close to uniform), then the message C_1 sends to C_4 will also have a uniform distribution on B, and from C_2's perspective the messages on B and C will appear as two independent variables. ∎

On the other hand, if we introduce many messages between clusters or increase the scope of these messages, we run the risk of constructing a tree that violates the running intersection property. And so, we have to worry about methods that ensure that the resulting structure is a proper cluster graph. We now consider several approaches for constructing cluster graphs.

11.3.5.1   Pairwise Markov Networks

pairwise Markov
networks

We start with the class of *pairwise Markov networks*. In these networks, we have a univariate potential $\phi_i[X_i]$ over each variable X_i, and in addition a pairwise potential $\phi_{(i,j)}[X_i, X_j]$ over some pairs of variables. These pairwise potentials correspond to edges in the Markov network. Many problems are naturally formulated as pairwise Markov networks, including the grid networks we discussed earlier and Boltzmann distributions (see box 4.C). Indeed, if we are willing to transform our variables, any distribution can be reformulated as a pairwise Markov network (see exercise 11.10).

One straightforward transformation of such a network into a cluster graph is as follows: For each potential, we introduce a corresponding cluster, and put edges between the clusters that have overlapping scope. In other words, there is an edge between the cluster $C_{(i,j)}$ that corresponds to the edge $X_i—X_j$ and the clusters C_i and C_j that correspond to the univariate factors over X_i and X_j. Figure 11.6 illustrates this construction in the case of a 3 by 3 grid network.

Because there is a direct correspondence between the clusters in the cluster graphs and vari-

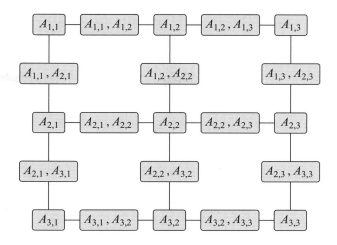

Figure 11.6 **A generalized cluster graph for the 3×3 grid when viewed as pairwise MRF**

ables or edges in the original Markov network, it is often convenient to think of the propagation steps as operations on the original network. Moreover, since each pairwise cluster has only two neighbors, we consider two propagation steps along the path $C_i—C_{(i,j)}—C_j$ as propagating information between X_i and X_j. (See exercise 11.9.) Indeed, early versions of cluster-graph belief propagation were stated in these terms. This algorithm is known as *loopy belief propagation*, since it uses propagation steps used by algorithms for Markov trees, except that it was applied to networks with loops.

loopy belief
propagation

11.3.5.2 Bethe Cluster Graph

A natural question is how we can extend this idea to networks that are more complex than pairwise Markov networks. Once we have larger potentials, they may overlap in ways that result in complex interactions among them.

Bethe cluster
graph

One simple construction, called the *Bethe cluster graph*, uses a bipartite graph. The first layer consists of "large" clusters, with one cluster for each factor ϕ in Φ, whose scope is $Scope[\phi]$. These clusters ensure that we satisfy the family-preservation property. The second layer consists of "small" univariate clusters, one for each random variable. Finally, we place an edge between each univariate cluster X on the second layer and each cluster in the first layer that includes X; the scope of this edge is X itself. For a concrete example, see figure 11.7a.

We can easily verify that this cluster graph is a proper one. First, by construction, it satisfies the family preservation property. Second, the edges that mention a variable X form a star-shaped subgraph with edges from the univariate cluster for X to all the large clusters that contain X. It is also easy to check that, if we apply this procedure to a pairwise Markov network, it results in the "natural" cluster graph for the pairwise network that we discussed. The construction of this cluster graph is simple and can easily be automated.

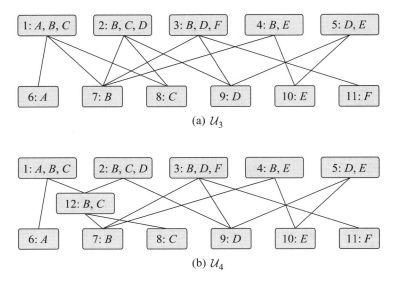

Figure 11.7 **Examples of generalized cluster graphs for network with potentials over** $\{A, B, C\}$, $\{B, C, D\}$, $\{B, D, F\}$, $\{B, E\}$ **and** $\{D, E\}$**.** For visual clarity, sepsets have been omitted — the sepset between any pair of clusters is the intersection of their scopes. (a) Bethe factorization. (b) Capturing interactions between $\{A, B, C\}$ and $\{B, C, D\}$.

11.3.5.3   Beyond Marginal Probabilities

The main limitation of using the Bethe cluster graph is that information between different clusters in the top level is passed through univariate marginal distributions. Thus, interactions between variables are lost during propagations. Consider the example of figure 11.7a. Suppose that C_1 creates a strong dependency between B and C. These two variables are shared with C_2. However, the messages between two clusters are mediated through the univariate factors. And thus, interactions introduced by one cluster are not directly propagated to the other.

One possible solution is to merge some of the large clusters. For example, if we want to capture the interactions between C_1 and C_2 in figure 11.7a, we can replace both of them by a cluster with the score A, B, C, D. This new cluster will allow us to capture the interactions between the factors involved in these two clusters. This modification, however, comes at a price, since the cost of manipulating a cluster grows exponentially with this scope. Moreover, this approach seems excessive in this case, since we can summarize these interactions simply using a distribution over B and C. This intuition suggests the construction of figure 11.7b. Note that this cluster graph is equivalent to figure 11.3b; see exercise 11.6.

Can we generalize this construction? A reasonable goal might be to capture all pairwise interactions. We can try to use a construction similar to the Bethe approximation, but introducing an intermediate level that includes pairwise clusters. In the same manner as we introduced C_{12} in figure 11.7b, we can introduce other pairs that are shared by more than two clusters. As a concrete example, consider the factors $C_1 = \{A, B, C\}$, $C_2 = \{B, C, D\}$, and $C_3 = \{A, C, D\}$. The relevant pairwise factors that capture interactions among these clusters

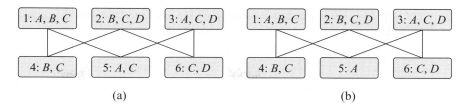

Figure 11.8 **Examples of generalized cluster graph for network with potentials over** $\{A, B, C\}$, $\{B, C, D\}$, **and** $\{A, C, D\}$. For visual clarity, sepsets have been omitted — the sepset between any pair of clusters is the intersection of their scopes. (a) A Bethe-like factorization with pairwise marginals that leads to an illegal cluster graph. (b) One possible way to make this graph legal.

are $\{B, C\} = \boldsymbol{C}_1 \cap \boldsymbol{C}_2$, $\{C, D\} = \boldsymbol{C}_2 \cap \boldsymbol{C}_3$, and $\{A, C\} = \boldsymbol{C}_1 \cap \boldsymbol{C}_3$. The resulting cluster graph appears in figure 11.8a. Unfortunately, a quick check shows that this cluster graph does *not* satisfy the running intersection property — all the edges in this graph are labeled by C, and together they form a loop. As a result, information concerning C can propagate indefinitely around the loop, "overcounting" the effect of C in the result.

How do we avoid this problem? In this specific example, we can consider a weaker approximation by removing C from one of the intersection sets. For example, if we remove C from $\boldsymbol{C}_5$, we get the cluster graph of figure 11.8b. This cluster graph satisfies the running intersection property. An alternative approach tries to "compensate" somehow for the violation of the running intersection property using a more complex message passing algorithm; see section 11.3.7.3.

---

belief
propagation
nonconvergence

Box 11.B — Skill: Making loopy belief propagation work in practice. *One of the main problems with loopy belief propagation is* nonconvergence. *This problem is particularly serious when we build systems that use inference as a subroutine within other tasks, for example, as the inner loop of a learning algorithm (see, for example, section 20.5.1). Several approaches have been used for addressing this nonconvergence issue. Some are fairly simple heuristics. Others are more sophisticated, and typically are based on the characterization of cluster-graph belief propagation as optimizing the approximate free-energy functional.*

A first observation is that, often, nonconvergence is a local problem. In many practical cases, most of the beliefs in the network do converge, and only a small portion of the network remains problematic. In such cases, it is often quite reasonable simply to stop the algorithm at some point (for example, when some predetermined amount of time has elapsed) and use the beliefs at that point, or a running average of the beliefs over some time window. This heuristic is particularly reasonable when we are not interested in individual beliefs, but rather in some aggregate over the entire network, for example, in a learning setting.

A second observation is that nonconvergence is often due to oscillations in the beliefs (see section 11.3.1). This observation suggests that we dampen the oscillations by reducing the difference between two subsequent updates. Consider the belief-propagation update rule in SP-Message(i, j):

$$\delta_{i \to j} \leftarrow \sum_{\boldsymbol{C}_i - \boldsymbol{S}_{i,j}} \psi_i \prod_{k \neq j} \delta_{k \to i}.$$

damping

We can replace this line by a damped *(or smoothed) version that averages the update $\delta_{i \to j}$ with the previous message between the two cliques:*

$$\delta_{i \to j} \leftarrow \lambda \left(\sum_{\boldsymbol{C}_i - \boldsymbol{S}_{i,j}} \psi_i \prod_{k \neq j} \delta_{k \to i} \right) + (1 - \lambda) \delta_{i \to j}^{\text{old}}, \tag{11.14}$$

where λ is the damping weight and $\delta_{i \to j}^{\text{old}}$ is the previous value of the message. When $\lambda = 1$, this update is equivalent to standard belief propagation. For $0 < \lambda < 1$, the update is partial and although it shifts β_j toward agreement with β_i, it leaves some momentum for the old value of the belief, a dampening effect that in turn reduces the fluctuations in the beliefs. It turns out that this damped update rule is "equivalent" to the original update rule, in that a set of beliefs is a convergence point of the damped update if and only if it is a convergence point of standard

stable
convergence
point

updates (see exercise 11.13). Moreover, one can show that, if run from a point close enough to a stable convergence point of the algorithm, with a sufficiently small λ, this damped update rule is guaranteed to converge. Of course, this guarantee is not very useful in practice, but there are indeed many cases where the damped update rule is convergent, whereas the original update rule oscillates indefinitely.

message
scheduling

A broader-spectrum heuristic, which plays an important role not only in ensuring convergence but also in speeding it up considerably, is intelligent message scheduling. *It is tempting to implement BP message passing as a synchronous algorithm, where all messages are updated at once. It turns out that, in most cases, this schedule is far from optimal, both in terms of reaching convergence, and in the number of messages required for convergence. The latter problem is easy to understand: In a cluster graph with m edges, and diameter d, synchronous message passing requires $m(d - 1)$ messages to pass information from one side of the graph to the other. By contrast, asynchronous message passing, appropriately scheduled, can pass information between two clusters at opposite ends of the graph using $d - 1$ messages. Moreover, the fact that, in synchronous message passing, each cluster uses messages from its neighbors that are based on their previous beliefs appears to increase the chances of oscillatory behavior and nonconvergence in general.*

☞

In practice, an *asynchronous* message passing schedule works significantly better than the synchronous approach. Moreover, even greater improvements can be obtained by scheduling messages in a guided way. *One approach, called* tree reparameterization (TRP), *selects a set of trees, each of which spans a large number of the clusters, and whose union covers all of the edges in the network. The TRP algorithm then iteratively selects a tree and does an upward-downward calibration of the tree, keeping all other messages fixed. Of course, calibrating this tree has the effect of "uncalibrating" other trees, and so this process repeats. This approach has the advantage of passing information more globally within the graph. It therefore converges more often, and more quickly, than other asynchronous schedules, particularly if the trees are selected using a careful design that accounts for the properties of the problem.*

asynchronous BP

tree reparameter-
ization

An even more flexible approach attempts to detect dynamically in which parts of the network messages would be most useful. Specifically, as we observed, often some parts of the network converge fairly quickly, whereas others require more messages. We can schedule messages in a way that accounts for their potential usefulness; for example, we can pass a message between clusters where the beliefs disagree most strongly on the sepset. This approach, called residual belief propagation *is convenient, since it is fully general and does not require a deep understanding of the properties of the network. It also works well across a range of different real-world networks.*

residual belief
propagation

An alternative general-purpose approach to avoiding nonconvergence is to directly optimize the energy functional. Here, several methods have been proposed. The simplest is to use standard optimization methods such as gradient ascent to optimize $\tilde{F}[\tilde{P}_\Phi, Q]$ (see appendix A.5.2 and exercise 11.12). Other methods are more specialized to the form of the energy functional, and they often turn out to be more efficient (see section 11.7). Although these methods do improve convergence, they are somewhat complex to implement, and have not (at this time) been used extensively in practice.

It turns out that many of the parameter settings encountered during a learning algorithm are problematic, and cause cluster-graph belief propagation to diverge. Intuitively, in many real-world problems, "appropriate" parameters encode strong constraints that tend to drive the algorithm toward well-behaved regions of the space. However, the parameters encountered during an iterative learning procedure have no such properties, and often allow the algorithm to end up in difficult regions. One approach is to train some parameters of the model separately, using a simpler network. We then use these parameters as our starting point in the general learning procedure. The use of "reasonable" parameters in the model can stabilize BP, allowing it to converge within the context of the general learning algorithm.

local maxima

A final problem with cluster-graph belief propagation is the fact that the energy functional objective is multimodal, and so there are many local maxima *to which a cluster-graph belief propagation algorithm might converge (if it converges). One can, of course, apply any of the standard approaches for addressing optimization of multimodal functions, such as initializing the algorithm heuristically, or using multiple restarts with different initializations. In the setting of BP, initialization must be done with care, so as not to lose the connection to the correct underlying distribution P_Φ, as reflected by the invariant of theorem 11.4. In sum-product belief propagation, we can simply initialize the messages to something other than 1. In belief update propagation, care must be taken to initialize messages and beliefs in a coordinate way, to preserve P_Φ.*

Box 11.C — Case Study: BP in practice. *To convey the behavior of belief propagation in practice, we demonstrate its performance on an 11×11 (121 binary variables) Ising grid (see box 4.C). The potentials of the network were randomly sampled as follows: Each univariate potential was sampled uniformly in the interval $[0, 1]$; for each pair of variables X_i, Z_j, $w_{i,j}$ is sampled uniformly in the range $[-C, C]$ (recall that in an Ising model, we define the negative log potential $\epsilon_{i,j}(x_i, x_j) = -w_{i,j}x_i x_j$). This sampling process creates an energy function where some potentials are attractive ($w_{i,j} > 0$) and some are repulsive ($w_{i,j} < 0$), resulting in a nontrivial inference problem. The magnitude of C (11 in this example) controls the magnitude of the energy forces and higher values correspond, on average, to more challenging inference problems.*

Figure 11.C.1 illustrates the convergence behavior on this problem. Panel (a) shows the percentage of messages converged as a function of time for three variants of the belief propagation algorithm: synchronous BP with damping (dashed line), where only a small fraction of the messages ever converge; asynchronous BP with damping (smoothing) that converges (solid line); asynchronous BP with no damping (dash-dot line) that does not fully converge. The benefit of using asynchronous propagation over synchronous updating is obvious. At first, it appears as if smoothing messages is not beneficial. This is because some percentage of messages can converge quickly when updates are

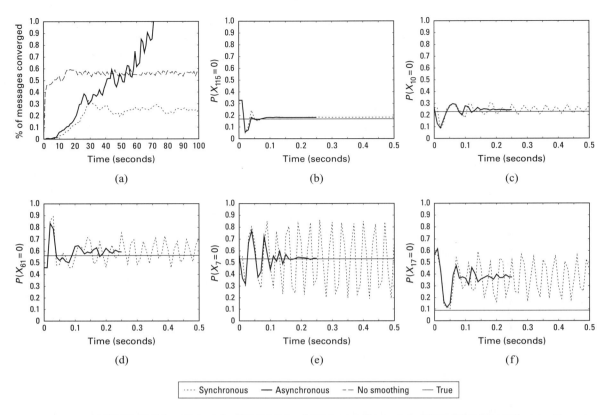

Figure 11.C.1 — Example of behavior of BP in practice on an 11 × 11 Ising grid. (a)   Percentage of messages converged as a function of time for three different BP variants. (b) A marginal where both variants converge rapidly. (c–e) Marginals where the synchronous BP marginals oscillate around the asynchronous BP marginals. (f) A marginal where both variants are inaccurate.

not slowed down by smoothing. However, the overall benefit of damping is evident, and without it the algorithm never converges.

The remaining panels illustrate the progression of the marginal beliefs over the course of the algorithm. (b) shows a marginal where both the synchronous and asynchronous updates converge quite rapidly and are close to the true marginal (thin solid black). Such behavior is atypical, and it comprises only around 10 percent of the marginals in this example. In the vast majority of the cases (almost 80 percent in this example), the synchronous beliefs oscillate around the asynchronous ones ((c)–(e)). In many cases, such as the ones shown in (e), the entropy of the synchronous beliefs is quite significant. For about 10 percent of the marginals (for example (f)), both the asynchronous and synchronous marginals are inaccurate. In these cases, using more informed message schedules can significantly improve the algorithms performance.

These qualitative differences between the BP variants are quite consistent across many random and real-life models. Typically, the more complex the inference problem, the larger the gaps in performance. For very complex real-life networks involving tens of thousands of variables and multiple cycles, even asynchronous BP is not very useful and more elaborate propagation methods or convergent alternatives must be adopted.

11.3.6 Variational Analysis

So far, our discussion of cluster-graph belief propagation has been procedural, motivated purely by similarity to message passing algorithms for cluster trees. Is there any formal justification for this approach? Is there a sense in which we can view this algorithm as providing an approximation to the exact inference task? In this section, we show that cluster-graph belief propagation can be justified using the energy functional formulation of section 11.1. Specifically, the messages passed by cluster-graph belief propagation can be derived from fixed-point equations for the stationary points of an approximate version of the energy functional of equation (11.3). As we will see, this formulation provides significant insight into the generalized belief propagation algorithm. It allows us to understand better the convergence properties of cluster-graph belief propagation and to characterize its convergence points. It also suggests generalizations of the algorithm that have better convergence properties, or that optimize a better approximation to the energy functional.

Our construction will be similar to the one in section 11.2 for exact inference. However, there are important differences that underlie the fact that this algorithm is only an approximate inference algorithm.

First, the exact energy functional $F[\tilde{P}_\Phi, Q]$ has terms involving the entropy of an entire joint distribution; thus, it cannot be tractably optimized. However, the *factored energy functional* $\tilde{F}[\tilde{P}_\Phi, Q]$ is defined in terms of entropies of clusters and sepsets, each of which can be computed efficiently based purely on local information at the clusters. Importantly, however, unlike for clique trees, $\tilde{F}[\tilde{P}_\Phi, Q]$ is no longer simply a reformulation of the energy functional, but rather an approximation of it.

factored energy functional

However, even the factored energy functional cannot be optimized over the space of all marginals Q that correspond to some actual distribution P_Φ. More precisely, consider some cluster graph $\mathcal{U}$; for a distribution P, we define $Q_P = \{P(C_i)\}_{i \in \mathcal{V}_\mathcal{U}} \cup \{P(S_{i,j})\}_{(i-j) \in \mathcal{E}_\mathcal{U}}$. We now define the *marginal polytope* of $\mathcal{U}$ to be

marginal polytope

$$Marg[\mathcal{U}] = \{Q_P : P \text{ is a distribution over } \mathcal{X}\} \tag{11.15}$$

That is, the marginal polytope is the set of all cluster (and sepset) beliefs that can be obtained from marginalizing an actual distribution P. It is called the marginal polytope because it is the set of marginals obtained from the polytope of all probability distributions over $\mathcal{X}$. Unfortunately, not every set of beliefs that correspond to clusters in $\mathcal{U}$ is in the marginal polytope; that is, there are calibrated cluster graph beliefs that do not represent the marginals of any single coherent joint distribution over $\mathcal{X}$ (see exercise 11.2). However, the marginal polytope is a complex object with exponentially many facets. (In fact, the problem of determining whether a set of beliefs is in the marginal polytope can be shown to be $\mathcal{NP}$-hard.) Thus, optimizing a function over the

marginal polytope is a computationally difficult task that is generally as hard as exact inference over the cluster graph. To circumvent these problems, we perform our optimization over the
local consistency polytope
local consistency polytope:

$$Local[\mathcal{U}] = \tag{11.16}$$

$$\left\{ \begin{array}{l} \{\beta_i : i \in \mathcal{V}_\mathcal{U}\} \cup \\ \{\mu_{i,j} : (i\text{-}j) \in \mathcal{E}_\mathcal{U}\} \end{array} \middle| \begin{array}{lll} \mu_{i,j}[\boldsymbol{s}_{i,j}] &= \sum_{\boldsymbol{C}_i - \boldsymbol{S}_{i,j}} \beta_i(\boldsymbol{c}_i) & \forall (i\text{-}j) \in \mathcal{E}_\mathcal{U}, \forall \boldsymbol{s}_{i,j} \in Val(\boldsymbol{S}_{i,j}) \\ 1 &= \sum_{\boldsymbol{c}_i} \beta_i(\boldsymbol{c}_i) & \forall i \in \mathcal{V}_\mathcal{U} \\ \beta_i(\boldsymbol{c}_i) &\geq 0 & \forall i \in \mathcal{V}_\mathcal{U}, \boldsymbol{c}_i \in Val(\boldsymbol{C}_i). \end{array} \right\}$$

pseudo-marginals
We can think of the local consistency polytope as defining a set of *pseudo-marginal distributions*, each one over the variables in one cluster. The constraints imply that these pseudo-marginals must be calibrated and therefore locally consistent with each other. However, they are not necessarily marginals of a single underlying joint distribution.

Overall, we can write down an optimization problem as follows:

CGraph-Optimize:
Find $\boldsymbol{Q}$
maximizing $\tilde{F}[\tilde{P}_\Phi, \boldsymbol{Q}]$
subject to

$$\boldsymbol{Q} \in Local[\mathcal{U}] \tag{11.17}$$

☞   **Thus, our optimization problem contains two approximations: We are using an approximation, rather than an exact, energy functional; and we are optimizing it over the space of pseudo-marginals, which is a relaxation (a superspace) of the space of all coherent probability distributions that factorize over the cluster graph.**

In section 11.1, we noted that the energy functional is a lower bound on the log-partition function; thus, by maximizing it, we get better approximations of P_Φ. Unfortunately, the factored energy functional, which is only an approximation to the true energy functional, is not necessarily also a lower bound. Nonetheless, it is still a reasonable strategy to maximize the approximate energy functional, since it may lead to a good approximation of the log-partition function.

This maximization problem directly generalizes CTree-Optimize to the case of cluster graphs. Not surprisingly, we can derive a similar analogue to theorem 11.3, where we characterize the
fixed-point equations
stationary points of this optimization problem as solutions to a set of *fixed-point equations*.

Theorem 11.5   *A set of beliefs $\boldsymbol{Q}$ is a stationary point of CGraph-Optimize if and only if for every edge $(i\text{-}j) \in \mathcal{E}_\mathcal{U}$ there are auxiliary factors $\delta_{i \to j}(\boldsymbol{S}_{i,j})$ and $\delta_{j \to i}(\boldsymbol{S}_{j,i})$ so that*

$$\delta_{i \to j} \propto \sum_{\boldsymbol{C}_i - \boldsymbol{S}_{i,j}} \psi_i \cdot \prod_{k \in \text{Nb}_i - \{j\}} \delta_{k \to i}. \tag{11.18}$$

and moreover, we have that

$$\begin{aligned} \beta_i &\propto \psi_i \cdot \prod_{j \in \text{Nb}_i} \delta_{j \to i} \\ \mu_{i,j} &= \delta_{j \to i} \cdot \delta_{i \to j}. \end{aligned}$$

The proof is identical to the proof of theorem 11.3.

This theorem shows that we can characterize convergence points of the energy function in terms of the original potentials and messages between clusters. We can, once again, define a procedural variant, in which we initialize $\delta_{i \rightarrow j}$, and then iteratively use equation (11.18) to redefine each $\delta_{i \rightarrow j}$ in terms of the current values of other $\delta_{k \rightarrow i}$. This process is identical (up to a renormalization step) to the update formula we use in CTree-SP-calibrate (algorithm 10.2). Indeed, we defined CGraph-SP-Calibrate, a cluster graph version of CTree-SP-Calibrate, the message passing steps are simply executing this iterative process using the fixed-point equation. Theorem 11.5 shows that convergence points of this procedure are related to stationary points of $\tilde{F}[\tilde{P}_\Phi, Q]$.

Corollary 11.1

Q is the convergence point of applying CGraph-SP-Calibrate($\Phi, \mathcal{U}$) *if and only if Q is a stationary point of $\tilde{F}[\tilde{P}_\Phi, Q]$.*

Due to the equivalence between sum-product and belief update messages, it follows that convergence points of CGraph-BU-Calibrate are also convergence points of CGraph-SP-Calibrate.

Corollary 11.2

At convergence of CGraph-BU-Calibrate, *the set of beliefs is a stationary point of $\tilde{F}[\tilde{P}_\Phi, Q]$.*

It is tempting to interpret this result as stating that the convergence points of belief propagation are maxima of the factored energy functional. However, there are several gaps between the theorem and this idealized interpretation, which it is important to understand. First, we note that maxima of a function are not necessarily fixed points. In this case, we can verify that $\tilde{F}[\tilde{P}_\Phi, Q]$ is bounded from above, and thus must have a maximum. However, if the maximum is a boundary point (where some of the probabilities in Q are 0), it may not be a fixed point. Fortunately, this situation is rare in practice, and it can be guaranteed not to arise under fairly benign assumptions.

Second, we note that maxima are not the only fixed points of the belief propagation algorithm; minima and saddle points are also fixed points. Intuitively, however, such solutions are not likely to be stable, in the sense that slight perturbations to the messages will drive the process away from them. Indeed, it is possible to show (although this result is outside the scope of this book) that *stable convergence points* of belief propagation are always local maxima of the function.

stable convergence point

The most important limitation of this result, however, is that it does not show that we can reach these maxima by applying belief propagation steps. There is no guarantee that the message passing steps of cluster-graph belief propagation necessarily improve the energy functional: a message passing step may increase or decrease the energy functional. Indeed, as we showed, there are examples where the belief propagation procedure oscillates indefinitely and fails to converge. Even more surprisingly, this problem is not simply a matter of the algorithm being unable to "find" the maximum. One can show examples where the global maximum is not a *stable* convergence point of belief propagation. That is, while it is, in principle, a fixed point of the algorithm, it will never be reached in practice, since even a slight perturbation will give rise to oscillatory behavior.

Nevertheless, this result is of significant importance in several ways. First, it provides us with a declarative semantics for cluster-graph belief propagation in terms of optimization of a target functional. The success of the belief propagation algorithm, when it converges, leads us to hope that the development of new, possibly more convergent, methods to solve the optimization problem may give rise to good solutions. Second, the declarative view defines the problem in terms of an objective — the factored energy functional — and a set of constraints — the set of locally consistent pseudo-marginals. Both of these are approximations to the ones used in the optimization problem for exact inference. When we view the task from this perspective, some potential directions for improvements become obvious: We can perhaps achieve a better approximation by making our objective a better approximation to the true energy functional, or by tightening our constraints so as to make the constraint space closer to the exact marginal polytope. We will describe some of the extensions based on these ideas; others are mentioned in section 11.7.

11.3.7   Other Entropy Approximations ⋆

The variational analysis of the previous section provides us with a framework for understanding the properties of this type of approximation, and for providing significant generalizations.

11.3.7.1   Motivation

To understand this general framework, consider first the form of the factored energy functional when our cluster graph $\mathcal{U}$ has the form of the Bethe approximation. Recall that in the Bethe approximation graph there are two layers: one consisting of clusters that correspond to factors in Φ, and the other consisting of univariate clusters. When the cluster graph is calibrated, these univariate clusters have the same distribution as the sepsets between them and the factors in the first layer. As such, we can combine together the entropy terms for all the sepsets labeled by X and the associated univariate cluster and rewrite the energy functional, as follows:

Proposition 11.3

> *If* $Q = \{\beta_\phi : \phi \in \Phi\} \cup \{\beta_i(X_i)\}$ *is a calibrated set of beliefs for a Bethe cluster graph* $\mathcal{U}$ *with clusters* $\{C_\phi : \phi \in \Phi\} \cup \{X_i : X_i \in \mathcal{X}\}$, *then*
>
> $$\tilde{F}[\tilde{P}_\Phi, Q] = \sum_{\phi \in \Phi} E_{Scope[\phi] \sim \beta_\phi}[\ln \phi] + \sum_{\phi \in \Phi} H_{\beta_\phi}(C_\phi) - \sum_i (d_i - 1) H_{\beta_i}(X_i), \qquad (11.19)$$
>
> *where* $d_i = |\{\phi : X_i \in Scope[\phi]\}|$ *is the number of factors that contain* X_i.

Note that equation (11.19) is equivalent to the factored energy functional only when Q is calibrated. However, because we are interested only in such cases, we can freely alternate between the two forms for the purpose of finding fixed points of the factored energy functional. Equa-

Bethe free energy   tion (11.19) is the negation of a term known as the *Bethe free energy* in statistical mechanics. The Bethe cluster graph we discussed earlier is a construction that is designed to match the Bethe free energy functional.

Why is this reformulation useful? Recall that, in our discussion of generalized cluster graphs, we required the running intersection property. This property has two important implications. First is that the set of clusters that contain some variable X are connected; hence, the marginal over X will be the same in all of these clusters at the calibration point. Second is that there is no cycle of clusters and sepsets all of which contain X. We motivated this assumption by noting that it prevents us from allowing information about X to cycle endlessly through a loop. This new formulation provides a more formal justification. As we can see, if the variable X_i

appears in d_i clusters in the cluster graph, then it appears in an entropy term with a positive sign exactly d_i times. Owing to the running intersection property, the number of sepsets that contain X_i is $d_i - 1$ (the number of edges in a tree with k vertices is $k - 1$), so that X_i appears in an entropy term with a negative sign exactly $d_i - 1$ times. In this case, the entropy of X_i appears with positive sign d_i times, and with negative sign $d_i - 1$ times, so overall it is counted exactly once.

This reformulation suggests a much more general class of entropy approximations. We can construct define a set of *regions* $\mathbf{R}$, each with its own scope $\boldsymbol{C}_r$ and its own *counting number* κ_r. We can now define the following *weighted approximate entropy*:

counting number

weighted approximate entropy

$$\tilde{H}^{\kappa}_{\boldsymbol{Q}}(\mathcal{X}) = \sum_r \kappa_r \boldsymbol{H}_{\beta_r}(\boldsymbol{C}_r). \tag{11.20}$$

Example 11.2

Bethe cluster graph

The simple Bethe cluster graph of section 11.3.5.2 fits easily into this new framework. The construction has two levels of regions: a set of "large" $\mathbf{R}^+$, where each $r \in \mathbf{R}^+$ contains multiple variables, and singleton regions containing the individual variables $X_i \in \mathcal{X}$. Both types of regions have counting numbers: κ_r for $r \in \mathbf{R}^+$ and κ_i for $X_i \in \mathcal{X}$. All factors in Φ are assigned only to large regions, so that $\psi_i = 1$ for all i. We use Nb_r to denote the set $\{X_i \in \boldsymbol{C}_r\}$, and Nb_i to denote the set $\{r : X_i \in \boldsymbol{C}_r\}$.

To capture exactly the Bethe free energy, we set each large region to have a counting number of 1, and each singleton region corresponding to X_i to have a counting number of $1 - d_i$ where d_i is the number of large regions that contain X_i in their scope. We see that in this construction the region graph energy functional is identical to the Bethe free energy of equation (11.19). ∎

However, this framework also allows us to capture much richer constructions.

Example 11.3

Consider again the example of figure 11.8a. As we discussed in section 11.3.5.3, this cluster graph has the benefit of maintaining the pairwise correlations between all pairs of variables when passing messages between clusters. Unfortunately, it is not a legal cluster graph, since it does not satisfy the running intersection property. We can obtain another perspective on the problem with this cluster graph by examining the energy functional associated with it:

$$
\begin{aligned}
\tilde{F}[\tilde{P}_\Phi, \boldsymbol{Q}] =\ & \boldsymbol{E}_{\beta_1}[\ln \phi_1(A, B, C)] + \boldsymbol{E}_{\beta_2}[\ln \phi_2(B, C, D)] + \boldsymbol{E}_{\beta_3}[\ln \phi_3(A, C, D)] \\
& + \boldsymbol{H}_{\beta_1}(A, B, C) + \boldsymbol{H}_{\beta_2}(B, C, D) + \boldsymbol{H}_{\beta_3}(A, C, D) \\
& - \boldsymbol{H}_{\beta_4}(B, C) - \boldsymbol{H}_{\beta_5}(A, C) - \boldsymbol{H}_{\beta_6}(C, D).
\end{aligned}
$$

As we can see, the variable C appears in three clusters and three sepsets. As a consequence, the counting number of C in the energy functional is 0. This means that we are undercounting the entropy of C in the approximation. Indeed, as we discussed, this cluster graph does not satisfy the running intersection property. Thus, we considered modifying the graph by removing C from one of the sepsets. However, if we consider this problem from the perspective of the energy functional, we can deal with the problem by adding another factor β_7 that has C as its scope. If we add $\boldsymbol{H}_{\beta_7}(C)$ to the energy functional we solve the undercounting problem. This results in a modified energy functional

$$
\begin{aligned}
\tilde{F}[\tilde{P}_\Phi, \boldsymbol{Q}] =\ & \boldsymbol{E}_{\beta_1}[\ln \phi_1(A, B, C)] + \boldsymbol{E}_{\beta_2}[\ln \phi_2(B, C, D)] + \boldsymbol{E}_{\beta_3}[\ln \phi_3(A, C, D)] \\
& + \boldsymbol{H}_{\beta_1}(A, B, C) + \boldsymbol{H}_{\beta_2}(B, C, D) + \boldsymbol{H}_{\beta_3}(A, C, D) + \boldsymbol{H}_{\beta_7}(C) \\
& - \boldsymbol{H}_{\beta_4}(B, C) - \boldsymbol{H}_{\beta_5}(A, C) - \boldsymbol{H}_{\beta_6}(C, D).
\end{aligned}
$$

This is simply an instance of our weighted entropy approximation, with seven regions: the three triplets, the three pairs, and the singleton C.   ∎

This perspective provides a clean and simple framework for proposing generalizations to the class of approximations defined by the cluster graph framework. Of course, to formulate our optimization problem fully, we need to define the constraints and construct algorithms that solve the resulting optimization problems. We now address these issues in the context of two different classes of weighted entropy approximations.

11.3.7.2 Convex Approximations

One of the biggest problems with the objective used in standard loopy BP is that it gives rise to a nonconvex optimization problem. In fact, the objective often has multiple local optima. These properties make the optimization hard and the answers nonrobust. However, a different choice of counting numbers can lead to a concave optimization objective, and hence to a convex optimization problem. Such problems are much easier to solve using a range of algorithms, and the solutions offer a satisfying guarantee of optimality. We first define the class of convex BP objectives and then describe one solution algorithm.

We focus our discussion on energy functionals whose structure uses the two-layer Bethe cluster graph structure of example 11.2, but where the counting numbers are different. To preserve the desired semantics of the counting numbers, we require:

$$\kappa_i = 1 - \sum_{r \in \mathrm{Nb}_i} \kappa_r, \tag{11.21}$$

ensuring that the total counting number of terms involving the entropy of X_i is precisely 1. When we define $\kappa_r = 1$ for all $r \in \mathbf{R}$, this constraint implies the counting numbers in the Bethe free energy.

We now introduce the following condition on the counting numbers:

Definition 11.4

convex counting
numbers

We say that a vector of counting numbers κ_r is convex *if there exist* nonnegative *numbers ν_r, ν_i, and $\nu_{r,i}$ such that:*

$$\begin{aligned} \kappa_r &= \nu_r + \sum_{i \,:\, X_i \in \mathbf{C}_r} \nu_{r,i} \quad \textit{for all } r \\ \kappa_i &= \nu_i - \sum_{r \,:\, X_i \in \mathbf{C}_r} \nu_{r,i} \quad \textit{for all } i \end{aligned} \tag{11.22}$$
∎

Assuming that we have a set of convex counting numbers, we can rewrite the weighted approximate entropy of equation (11.20) as:

$$\sum_r \kappa_r \boldsymbol{H}_{\beta_r}(\boldsymbol{C}_r) + \sum_i \kappa_i \boldsymbol{H}_{\beta_i}(X_i) = \tag{11.23}$$
$$\sum_r \nu_r \boldsymbol{H}_{\beta_r}(\boldsymbol{C}_r) + \sum_{r, X_i \in \boldsymbol{C}_r} \nu_{r,i}(\boldsymbol{H}_{\beta_r}(\boldsymbol{C}_r) - \boldsymbol{H}_{\beta_i}(X_i)) + \sum_i \nu_i \boldsymbol{H}_{\beta_i}(X_i).$$

Importantly, when the beliefs satisfy the marginal-consistency constraints, the terms in the second summation can be rewritten as conditional entropies:

$$\boldsymbol{H}_{\beta_r}(\boldsymbol{C}_r) - \boldsymbol{H}_{\beta_i}(X_i) = \boldsymbol{H}_{\beta_r}(\boldsymbol{C}_r \mid X_i).$$

Plugging this result back into equation (11.23), we obtain an objective that is a summation of terms each of which is either an entropy or a conditional entropy, all with positive coefficients. Because both entropies and conditional entropies are convex, we obtain the following result:

Proposition 11.4

The function in equation (11.23) is a concave function for any set of beliefs $\mathbf{Q}$ that satisfies the marginal consistency constraints.

concave over constraints

convex entropy

This type of objective function is called *concave over the constraints*, since it is not generally concave, but it is concave over the subspace that satisfies the constraints of our optimization problem. An entropy as in equation (11.23) that uses convex counting numbers is called a *convex entropy*.

Assuming that the potentials are all strictly positive, we can now conclude that the optimization problem CGraph-Optimize with convex counting numbers is a convex optimization problem that has a unique global optimum.

Convex optimization problems can, in principle, be solved by a range of different algorithms, all of which guarantee convergence to the unique global optimum. However, the basic optimization problem can easily get intractably large. Recall that to formulate our optimization space, we need to introduce an optimization variable for each assignment of values to each cluster in our cluster graph, and a constraint for each assignment of values to each sepset in the graph.

Example 11.4

Consider a grid-structured network corresponding to a modestly sized 500×500 image, where each pixel can take 100 values. The structure of the graph is a pairwise network, with approximately $2 \times 250,000$ clusters (pairwise edges), each of which can take $100 \times 100 = 10,000$ values. The total number of variables is therefore $500,000 \times 10,000 = 5 \times 10^9$, an unmanageable number for most optimizers. ∎

Fortunately, due to the convexity of this problem, we have that strong duality holds (see appendix A.5.4), and therefore we can find a solution to this problem by solving its dual. The message passing algorithms that we derive from the Lagrange multipliers are one method that we can use for solving the dual. (For example, exercise 11.17 provides one message passing algorithm for a Bethe cluster graph with general counting numbers.) However, the message passing algorithms are not directly optimizing the objective. Rather, they characterize the optimum using a set of fixed-point equations, and attempt to converge to the optimum by iterating through these equations. This process is generally not guaranteed to achieve the optimum, even when the problem is convex. Again, we can consider using other optimization algorithms over the dual problem. However, a message passing approach has some important advantages, such as modularity and efficiency.

Fortunately, a careful reformulation of the message passing scheme can be shown to guarantee convergence to the global optimum. This reformulation is different for synchronous and asynchronous message passing. We present the asynchronous version, which is simpler and also likely to be more efficient in practice.

The algorithm, shown in algorithm 11.2, uses the following quantities in its computations:

$$\hat{\nu}_i = \nu_i + \sum_{r \in \mathrm{Nb}_i} \nu_r; \qquad\qquad \hat{\nu}_{i,r} = \nu_r + \nu_{i,r}. \qquad\qquad (11.24)$$

In each message passing iteration, it traverses the variables X_i in a round-robin fashion; for each X_i, it computes two sets of messages: incoming messages $\delta_{r \to i}(X_i)$ from regions to variables,

---

Algorithm 11.2 Convergent message passing for Bethe cluster graph with convex counting numbers

       Procedure Convex-BP-Msg (
          $\psi_r(\boldsymbol{C}_r)$ // set of initial potentials
          $\sigma_{i \rightarrow r}(\boldsymbol{C}_r)$ // Current node-to-region messages
)

1   **for** $i = 1, \ldots, n$
2   // Compute incoming messages from neighboring regions to X_i
3   **for** $r \in \mathrm{Nb}_i$
4            $\delta_{r \rightarrow i}(X_i) \leftarrow \sum_{\boldsymbol{C}_r - X_i} \left(\psi_r(\boldsymbol{C}_r) \prod_{j \in \mathrm{Nb}_r - \{i\}} \sigma_{j \rightarrow r}(\boldsymbol{C}_r) \right)^{\frac{1}{\tilde{\nu}_{i,r}}}$
5   // Compute beliefs for X_i, renormalizing to avoid numerical underflows
6          $\beta_i(X_i) \leftarrow \frac{1}{Z_{X_i}} \prod_{r \in \mathrm{Nb}_i} (\delta_{r \rightarrow i}(X_i))^{\hat{\nu}_{i,r}/\hat{\nu}_i}$
7   // Compute outgoing messages from X_i to neighboring regions
8   **for** $r \in \mathrm{Nb}_i$
9            $\sigma_{i \rightarrow r}(\boldsymbol{C}_r) \leftarrow \left(\psi_r(\boldsymbol{C}_r) \prod_{j \in \mathrm{Nb}_r - \{i\}} \sigma_{j \rightarrow r}(\boldsymbol{C}_r) \right)^{-\frac{\nu_{i,r}}{\tilde{\nu}_{i,r}}} \left(\frac{\beta_i(X_i)}{\delta_{r \rightarrow i}(X_i)} \right)^{\nu_r}$
10   **return** $\{\sigma_{i \rightarrow r}(\boldsymbol{C}_r)\}_{i, r \in \mathrm{Nb}_i}$

---

factor graph

and outgoing messages $\sigma_{i \rightarrow r}(\boldsymbol{C}_r)$ from variables to regions (essentially passing messages over the *factor graph*). The overall process is initialized (in the first message passing iteration) by setting $\sigma_{i \rightarrow r} = 1$. This algorithm is guaranteed to converge to the global maximum of our convex energy functional.

This derivation applies to any set of convex counting numbers, leaving open the question of which counting numbers are likely to be give the best approximation. Although there is currently no theoretical analysis answering this question, intuitively, we might argue that we want the counting numbers for different regions to be as close as possible to uniform. This intuition is also supported by the fact that the Bethe approximation, which sets all $\kappa_r = 1$, obtains very high-quality approximations when it converges. Thus, we can try to select nonnegative coefficients ν_i, ν_r, and $\nu_{i,r}$ for which κ_r and κ_i, defined via equation (11.22), satisfy equation (11.21) and minimize

$$\sum_{r \in \mathbf{R}^+} (\kappa_r - 1)^2. \tag{11.25}$$

TRW

Other choices are also possible. For example, the *tree-reweighted belief propagation (TRW)* algorithm computes convex counting numbers for a pairwise Markov network using the following process: We first define a probability distribution ρ over trees $\mathcal{T}$ in the network, such that each edge in the pairwise network is present in at least one tree. This distribution defines a set of weights:

$$\begin{aligned} \kappa_i &= -\sum_{\mathcal{T} \ni X_i} \rho(\mathcal{T}) \\ \kappa_{i,j} &= \sum_{\mathcal{T} \ni (X_i, X_j)} \rho(\mathcal{T}) \end{aligned} \tag{11.26}$$

convex counting
numbers

This computation results in a set of *convex counting numbers* (see exercise 11.18). Preliminary results suggest that the TRW counting numbers and the ones derived from optimizing equation (11.25) appear to achieve similar performance in practice.

 However, the comparison to standard (Bethe-approximation) BP is less clear. **When standard BP converges, it generally tends to produce better results than the convex counterparts, and almost universally it converges much faster. Conversely, when standard BP does not converge, the convex algorithms have an advantage;** but, as we discuss in box 11.B, there are many tricks we can use to improve the convergence of BP, so it is not clear how often nonconvergence is a problem. One setting where a convergent algorithm can have important benefits is in those settings (chapter 19 and chapter 20) where we generally learn the model using iterative, hill-climbing methods that use inference in the inner loop for tasks such as gradient computations. There, the use of a nonconvergent algorithm for computing the gradient can severely destabilize the learning algorithm. In other settings, however, the decision of whether to use standard or convex BP is one of approximate optimization of a pretty good (although still approximate) objective, versus exact optimization of an objective that is generally not as good. The right decision in this trade-off is not clear, and needs to be made specifically for the target application.

11.3.7.3 Region Graph Approximations

As illustrated in example 11.3, a very different motivation for using an objective based on different counting numbers is to improve the quality of the approximation by better capturing interactions between variables. As we showed in this example, we can use the notion of a weighted entropy approximation to define a (hopefully) better approximation to the entropy. Of course, to specify the optimization problem fully, we also need to specify the constraints. In this example, it is fairly straightforward to do so: we want $\beta_7(C)$ to be consistent with the marginal probability of C in one of the other beliefs that mention C. Now, we have an optimization problem that seems to solve the problem we set out to solve: It can compute beliefs on each of the original factors while maintaining consistency at the level of each pairwise marginal shared among these factors.

However, the new optimization problem we defined is not one that corresponds to a cluster graph. To see this, notice that β_7 appears in the role of a cluster. But, if it is a cluster, it would have to be connected to one of the other factors by a sepset with scope C, which would require an additional term in the energy functional associated with this cluster graph. Thus, it is not immediately clear how we would go about optimizing the new modified functional.

We now discuss a general framework that defines the form of the optimization objective and the constraints for constructions that capture higher-level interactions between the variables. We also describe a message passing algorithm that can be used to find fixed points of this optimization problem.

Region Graphs The basic structure we consider is similar to a cluster graph, but unlike cluster graphs we no longer distinguish two types of vertices (clusters and sepsets). Rather, we can have a more deeply nested hierarchy of regions, related by containment.

Definition 11.5

region graph

A region graph $\mathcal{R}$ is a directed graph over a set of vertices $\mathbf{R}$. Each vertex r is called a region *and*

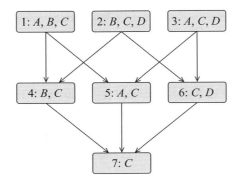

Figure 11.9   An example of simple region graph

is associated with a distinct set of random variables C_r. Whenever there is an arc from a region r_i to a region r_j we require that Scope$[r_i] \supset$ Scope$[r_j]$. Regions that have no incoming edges are called top regions.

counting
numbers

Each region is associated with a counting number $\kappa_r \in \mathbb{R}$. *Note that the counting number may be negative, positive, or zero.*   ∎

Because containment is transitive, we have that if there is a directed path from r to r', then Scope$[r'] \subset$ Scope$[r]$. Thus, a region graph is acyclic.

To define the energy term in the free energy, we must assign our original factors in Φ to regions in the region graph. Here, because different regions are counted to different extents, it is useful to allow a factor to be assigned to more than one region. Thus, for each $\phi \in \Phi$ we have a set of regions $\alpha(\phi) \subset \mathbf{R}$ such that Scope$[\phi] \subseteq C_r$. This property is analogous to the

family
preservation

family-preservation property for cluster graphs. Throughout this book, we assume without loss of generality that any $r \in \alpha(\phi)$ is a top region.

We are now ready to define the energy functional associated with a region graph:

$$\tilde{F}[\tilde{P}_\Phi, \boldsymbol{Q}] = \sum_r \kappa_r \boldsymbol{E}_{\boldsymbol{C}_r \sim \beta_r}[\ln \psi_r] + \tilde{H}_{\boldsymbol{Q}}^\kappa(\mathcal{X}), \tag{11.27}$$

where ψ_r is defined as:

$$\psi_r = \prod_{\phi \in \Phi \,:\, r \in \alpha(\phi)} \phi.$$

As with cluster graphs, a region graph defines a set of beliefs, one per region. We use the notation $\beta_r(\boldsymbol{C}_r)$ to denote the belief associated with the region r over the set of variables $\boldsymbol{C}_r =$ Scope$[r]$.

Figure 11.9 demonstrates the region graph construction for the approximation of example 11.3. This example contains three regions that correspond to the initial factors in the distribution we want to approximate. The lower set of regions are the pairwise intersections between the three factors. The lowest region is associated with the variable C.

Whereas the counting numbers specify the energy functional, the graph structure specifies the constraints over the beliefs that we wish to associate with the regions. In particular, we

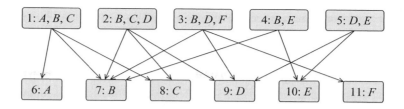

Figure 11.10 **The region graph corresponding to the Bethe cluster graph of figure 11.7a**

want the beliefs to satisfy the calibration constraints that are implied by the edges in the region graph structure.

Definition 11.6

region graph
calibration

Given a region graph, a set of region beliefs is calibrated if whenever $r \rightarrow r'$ appears in the region graph then

$$\sum_{\boldsymbol{C}_r - \boldsymbol{C}_{r'}} \beta_r(\boldsymbol{C}_r) = \beta_{r'}(\boldsymbol{C}'_r) \tag{11.28}$$

∎

The region graph structure provides a very general framework for specifying energy-functional optimization problems. The set of regions and their counting numbers tell us which components appear in the energy functional, and with what weight. The arcs in the region graph tell us which consistency constraints we wish to enforce. We can choose which consistency constraints we want to enforce by adding or removing regions or edges between them; we note that this can be done without affecting the energy functional by simply giving the regions introduced a counting number of 0. The more edges we have, the more constraints we require regarding the calibration of different beliefs.

The region graph framework is very general, and it can encode a broad spectrum of optimization objectives and constraints. However, not all such formulations are equally reasonable as approximations to our true objective, which is the exact energy functional of equation (11.3). The following requirement captures some of the essential properties that make a region graph construction suitable for this purpose.

Definition 11.7

For each variable X_i, let $\mathbf{R}_i = \{r : X_i \in Scope[r]\}$; for each factor $\phi \in \Phi$, let $\mathbf{R}_\phi = \{r : Scope[\phi] \subseteq \boldsymbol{C}_r\}$. We now define the following conditions for a region graph:

- *variable connectedness: for every variable X_i, the set $\mathbf{R}_i$ forms a single connected component;*

- *factor connectedness: for every factor ϕ, the set $\mathbf{R}_\phi$ forms a single connected component;*

- *factor preservation:*

$$\sum_{r \in \alpha(\phi)} \kappa_r = 1.$$

- running intersection: *for every variable X_i,*

$$\sum_{r \in \mathbf{R}_i} \kappa_r = 1.$$

∎

The connectedness requirement for variables ensures that the beliefs about any individual variable X_i in any calibrated region graph will be consistent for all beliefs that contain X_i. The counting number condition for variables ensures that we are not not overcounting or undercounting the entropy of X_i. Together, these conditions extend the running intersection property, ensuring that the subgraph containing X_i is connected and that X_i is "counted" once in total. We note that, like the running intersection property, this requirement does not address the extent to which we count the contribution associated with the interactions between pairs or larger subsets of variables.

The factor-preservation condition for factors ensures that, when we sum up the energy terms of the different regions, each factor is counted exactly once in total. As we will see, this ensures that a calibrated region graph will still encode our original distribution P_Φ. Finally, the connectedness condition for factors ensures that we cannot double-count contributions of our initial factors: that is, a factor cannot "flow" around a loop.

An examination confirms that all parts of the region graph condition hold both for the Bethe region graph and for the region graph of figure 11.9.

How do we construct a valid region graph? One approach is based on the following simple recursive construction for the counting numbers. We first define, for a region r,

$$\mathbf{Up}(r) = \{r' \; : \; (r' \to r) \in \mathcal{E}_\mathcal{R}\}$$

to be the set of regions that are directly upwards of r; similarly, we define

$$\mathbf{Down}(r) = \{r' \; : \; (r \to r') \in \mathcal{E}_\mathcal{R}\}.$$

We also define the *upward closure* of r to be the set $\mathbf{Up}^*(r)$ of all the regions from which there is a directed path to r, and the *downward closure* $\mathbf{Down}^*(r)$ to be all the regions that can be reached by a directed path from r; finally, we define $\mathbf{Down}^+(r) = \{r\} \cup \mathbf{Down}^*(r)$.[2]

We can now define the counting numbers recursively, using the following formula:

$$\kappa_r = 1 - \sum_{r' \in \mathbf{Up}^*(r)} \kappa_{r'}. \tag{11.29}$$

This condition ensures that the sum of the counting numbers of r and all of the regions above it will be 1. Intuitively, we can think of the counting number of the region r as correcting for overcounting or undercounting of the weight of the scope of r by regions above it. Now, assume that our region graph is structured so that, for each variable X_i, there is a unique region r_i such that every other region whose scope contains X_i is an ancestor of r_i. Then, we are guaranteed that both the connectedness and the counting number condition for X_i hold. We can similarly require that for any factor ϕ, there is a unique region r_ϕ such that any other region whose scope contains $Scope[\phi]$ is an ancestor of r_ϕ. This construction guarantees the requirements for factor connectedness and counting numbers.

It is easy to see that the Bethe region graph of example 11.2 satisfies both of these conditions. Moreover, this process of guaranteeing that a unique minimal region exists for each X_i is essentially what we did in example 11.3 to produce a valid region graph.

saturated region graph

These conditions provide us with a simple strategy for constructing a *saturated region graph*.

---

2. Some of the literature on region graphs use the terminology of parents and children of regions. To avoid the confusion with similar terminology in Bayesian networks, we use the terms *up* and *down* here.

Algorithm 11.3 Algorithm to construct a saturated region graph

    Procedure Build-Saturated-Region-Graph (
        R   // a set of initial regions
)

```
1      repeat
2        For any r₁, r₂ ∈ R
3          Z ← Scope[r₁] ∩ Scope[r₂]
4          if Z ≠ ∅
5            and R does not contain a region with scope Z then
6              create region r with Scope[r] = Z
7                R ← R ∪ {r}
8        Until convergence
9        Initialize R as an empty graph with R as vertices
10       for each r₁ ≠ r₂ ∈ R
11         if Scope[r₂] ⊂ Scope[r₁] and
12         not exist r₃ ∈ R such that Scope[r₂] ⊂ Scope[r₃] ⊂ Scope[r₁] then
13             add an arc r₁ → r₂ to the region graph R
14       return R
```

We start with initial set of regions. Often, these regions will be the initial factors in P_Φ, although we can decide to work with bigger regions that capture some more global interactions. We then extend this set of regions into a valid region graph, where our goal is to represent appropriately any subset of variables that is shared by some of the regions. We therefore expand the set of regions to be closed under intersections. We connect these regions so that the upward closure of each region contains all of its supersets. The full procedure is shown in algorithm 11.3. Unlike the Bethe approximation, this region graph maintains the consistency of higher-order marginals. The example of figure 11.9 is an example of running this procedure on the original set of regions $\{A, B, C\}$, $\{B, C, D\}$, and $\{A, C, D\}$. As our previous discussion suggests, this procedure guarantees a region graph that satisfies the region graph condition.

Belief Propagation in Region Graphs Given a region graph, we are faced with the task of optimizing the free energy associated with its structure:

RegionGraph-Optimize:

Find $Q = \{\beta_r : i \in \mathcal{V}_{\mathcal{R}}\}$

maximizing $\tilde{F}[\tilde{P}_\Phi, Q]$

subject to

$$\sum_{C_{r'} - C_r} \beta_{r'}(c_{r'}) \;=\; \beta_r(c_r) \tag{11.30}$$

$$\forall r \in \mathcal{V}_{\mathcal{R}}, \forall r' \in \mathbf{Up}(r), \forall c_r \in Val(C_r)$$

$$\sum_{C_r} \beta_r(c_r) \;=\; 1 \qquad \forall r \in \mathcal{V}_{\mathcal{R}} \tag{11.31}$$

$$\beta_r(c_r) \;\geq\; 0 \qquad \forall r \in \mathcal{V}_{\mathcal{R}}, c_r \in Val(C_r). \tag{11.32}$$

Our strategy for devising algorithms for solving this optimization problem is similar to the approach we took in section 11.3.6. Using the method of Lagrange multipliers, we characterize the stationary points of the target function (given the constraints) as a set of fixed-point equations. We then find an iterative algorithm that attempts to reach such a stationary point.

We first characterize the fixed point via the Lagrange multipliers. As before, we form a Lagrangian by introducing terms for each of the constraints: from equation (11.30), we obtain a Lagrange multiplier $\lambda_{r,r'}(c_r)$ for every pair $r' \in \mathbf{Up}(r)$ and every $c_r \in Val(C_r)$; from equation (11.31), we obtain a Lagrange multiplier λ_r for every r and every $c_r \in Val(C_r)$; as before, we assume that we are dealing with interior fixed points only, and so do not have to worry about the inequality constraint. We differentiate the Lagrangian relative to each of the region beliefs $\beta_r(c_r)$, and obtain the following set of *fixed-point equations*:

fixed-point equations

$$\kappa_r \ln \beta_r(c_r) = \lambda_r + \kappa_r \ln \psi_r(c_r) - \sum_{r' \in \mathbf{Up}(r)} \lambda_{r,r'}(c_r) + \sum_{r' \in \mathbf{Down}(r)} \lambda_{r',r}(c_{r'}) - \kappa_r. \tag{11.33}$$

For regions for which $\kappa_r \neq 0$, we can rewrite this equation to conclude that:

$$\beta_r(c_r) = \frac{1}{Z_r} \psi_r(c_r) \left(\frac{\prod_{r' \in \mathbf{Down}(r)} \delta_{r \to r'}(c_{r'})}{\prod_{r' \in \mathbf{Up}(r)} \delta_{r' \to r}(c_r)} \right)^{1/\kappa_r}. \tag{11.34}$$

From this equality, one can conclude the following result:

Theorem 11.6

Assume that our region graph satisfies the family preservation property. Then, at fixed points of the RegionGraph-Optimize *optimization problem, we have that:*

$$P_\Phi(\mathcal{X}) \propto \prod_r (\beta_r)^{\kappa_r}. \tag{11.35}$$

The proof is derived from a simple cancellation of messages in the different terms (see exercise 11.16).

This result tells us that we can reparameterize the initial distribution P_Φ in terms of the final beliefs obtained as fixed points of the region graph optimization problem. It tells us that we can represent the distribution in terms of a calibrated set of beliefs for the individual regions. This

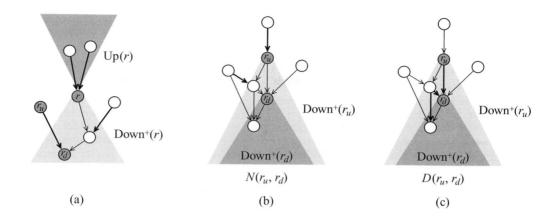

Figure 11.11 The messages participating in different region graph computations. Participating messages are marked with thicker arrows. (a) The computation of the beliefs $\beta_r(\boldsymbol{C}_r)$; (b) The set $N(r_u, r_d)$ participating in the computation of the message $\delta_{r_u \to r_d}$; (c) The set $D(r_u, r_d)$ participating in the computation of the message $\delta_{r_u \to r_d}$.

result is a very powerful one, because it holds for any set of counting numbers that satisfies the family preservation property — a very large class. Of course, this result only shows that any fixed point is a reparameterization of the distribution, but not that such a reparameterization exists. However, under our assumption that all of the initial factors in Φ are strictly positive, one can show that such a fixed point, and hence a corresponding reparameterization, necessarily exists.

As we can see, unlike the case of cluster graphs, the fixed-point equations for region graphs are more involved, and do not lead directly to an elegant *message passing algorithm*. Indeed, the derivation of the update rules from the Lagrange multipliers often involves multiple steps of algebraic manipulation. (Although such derivations are possible in restricted cases; see exercise 11.17 and exercise 11.19.) In the remainder of this section, we present, without derivation, one set of update rules that can be derived from equation (11.34), in the specific case where the counting numbers are as presented in equation (11.29).

The basic idea is similar to the message update used in cluster graphs. There, each sepset carried messages sent by one of the clusters through the sepset to the neighboring cluster. We can think of that as a message from the cluster to the sepset. The analog of this in region graph is a message from a region to a region below it. Thus, for each pair $r_1 \to r_2$ in the region graph we will have a message $\delta_{r_1 \to r_2}$ whose scope is $Scope[r_2]$. All messages are associated with "downward" edges of the form $r_1 \to r_2$, but they are used to define the beliefs and messages of regions that contain them.

The definitions of the messages and the beliefs are somewhat involved. We begin by defining

message passing
algorithm

the beliefs of a region as a function of these messages, which is somewhat simpler:

$$
\beta_r(\boldsymbol{C}_r) \;=\; \psi_r(\boldsymbol{C}_r) \left(\prod_{r_u \in \mathbf{Up}(r)} \delta_{r_u \to r}(\boldsymbol{C}_r) \right) \tag{11.36}
$$

$$
\left(\prod_{r_d \in \mathbf{Down}^*(r)} \; \prod_{r_u \in \mathbf{Up}(r_d)-\mathbf{Down}^+(r)} \delta_{r_u \to r_d}(\boldsymbol{C}_{r_d}) \right).
$$

In other words, the belief of a region is the product of three groups of terms. The first two are very natural: the initial beliefs ψ_r (1 for all regions except the top ones) and the messages from its upward regions. The last group contain all messages sent to regions below the region r from regions other than the region r itself and regions below it. In other words, these are messages from "external" sources to regions below the region r; see figure 11.11a. Thus, the beliefs of the region are not influenced by messages it sends down, but only by messages sent to it or its subsets from other regions.

Again, it is instructive to compare this definition to our definition of beliefs in cluster graphs. In cluster graphs, the belief over a sepset is the product of messages from the neighboring clusters. These messages correspond in our case to messages from upward regions. The belief over a cluster $\boldsymbol{C}$ is the product of its initial potential and messages sent from neighboring clusters to the sepsets adjacent to $\boldsymbol{C}$. The sepsets correspond to the regions in $\mathbf{Down}^+(\boldsymbol{C})$; the message sent by another clique $\boldsymbol{C}'$ to this sepset corresponds to messages sent by an "external" source to a region in $\mathbf{Down}^+(\boldsymbol{C})$.

We now move to defining the computation of a message from r_u to r_d, also illustrated in figure 11.11b,c:

$$
\delta_{r_u \to r_d}(\boldsymbol{C}_{r_d}) = \frac{\sum_{\boldsymbol{C}_{r_u}-\boldsymbol{C}_{r_d}} \psi_{r_u}(\boldsymbol{C}_{r_u}) \prod_{r_1 \to r_2 \in N(r_u,r_d)} \delta_{r_1 \to r_2}(\boldsymbol{C}_{r_2})}{\prod_{r_1 \to r_2 \in D(r_u,r_d)} \delta_{r_1 \to r_2}(\boldsymbol{C}_{r_2})}. \tag{11.37}
$$

The numerator involves the initial factor assigned to the region, and a product of messages associated with the set of edges

$$
N(r_u, r_d) = \{(r_1 \to r_2) \in \mathcal{E}_{\mathcal{R}} : r_1 \notin \mathbf{Down}^+(r_u), r_2 \in \mathbf{Down}^+(r_u) - \mathbf{Down}^+(r_d)\}.
$$

This set contains edges from sources "external" to r_u that are outside the scope of influence of r_d; that is, they either enter r_u directly, or enter regions below r_u that are not below r_d. The denominator involves a product of the messages in the set:

$$
D(r_u, r_d) = \{(r_1 \to r_2) \in \mathcal{E}_{\mathcal{R}} : r_1 \in \mathbf{Down}^+(r_u) - \mathbf{Down}^+(r_d), r_2 \in \mathbf{Down}^+(r_d)\}.
$$

This set counts information that would be passed from r_u to regions below r_d *indirectly* — not through r_d. We want to divide by these messages, since otherwise the same information would be incorporated multiple times into the beliefs and the messages.

Example 11.5 *We now return to the region graph of figure 11.9 that corresponds to the potential function we discussed in example 11.3.*

Applying equation (11.36) to this example, we get a set of equation representing the potentials as function of the initial factors and the messages:

$$\beta_1 = \psi_1 \delta_{2\to4} \delta_{3\to5} \delta_{6\to7}$$
$$\beta_2 = \psi_2 \delta_{1\to4} \delta_{3\to6} \delta_{5\to7}$$
$$\beta_3 = \psi_3 \delta_{1\to5} \delta_{2\to6} \delta_{4\to7}$$
$$\beta_4 = \delta_{1\to4} \delta_{2\to4} \delta_{5\to7} \delta_{6\to7}$$
$$\beta_5 = \delta_{1\to5} \delta_{3\to5} \delta_{4\to7} \delta_{6\to7}$$
$$\beta_6 = \delta_{2\to6} \delta_{3\to6} \delta_{4\to7} \delta_{5\to7}$$
$$\beta_7 = \delta_{4\to7} \delta_{5\to7} \delta_{6\to7}.$$

Applying equation (11.37), we can construct the messages. For example,

$$\delta_{4\to7} = \sum_B \delta_{1\to4} \delta_{2\to4}.$$

One easy way to derive this message directly is to use the marginal consistency constraint:

$$\beta_7 = \sum_B \beta_4.$$

Plugging in the expanded form of the two beliefs, we get

$$\delta_{4\to7} \delta_{5\to7} \delta_{6\to7} = \sum_B \delta_{1\to4} \delta_{2\to4} \delta_{5\to7} \delta_{6\to7}.$$

If we now isolate $\delta_{4\to7}$ we get

$$\delta_{4\to7} = \frac{\sum_b \delta_{1\to4} \delta_{2\to4} \delta_{5\to7} \delta_{6\to7}}{\delta_{5\to7} \delta_{6\to7}}.$$

After we cancel out the common terms $\delta_{5\to7}$ and $\delta_{6\to7}$, we get the desired form.

This message is essentially identical to the message in a cluster graph where we marginalize the other incoming messages in the cluster to send a message to a particular sepset. Here region 4 behaves as a cluster and region 7 as a sepset. The other messages incoming to region 7 have a similar form.

Messages into the middle layer regions have more complex form. For example

$$\delta_{1\to4} = \frac{\sum_A \psi_1 \delta_{3\to5}}{\delta_{5\to7}}.$$

Again, we can use the marginal consistency constraint

$$\beta_4 = \sum_A \beta_1$$

to reconstruct the message. Plugging in the expanded form of the two beliefs, and isolating $\delta_{1\to4}$ we get

$$\delta_{1\to4} = \frac{\sum_A \psi_1 \delta_{2\to4} \delta_{3\to5} \delta_{6\to7}}{\delta_{2\to4} \delta_{5\to7} \delta_{6\to7}}.$$

After we cancel out $\delta_{2\to4}$ and $\delta_{6\to7}$, we get the desired form.   ∎

These definitions set up a message passing algorithm similar to CGraph-SP-Calibrate, except that we use the messages as formulated in equation (11.37). As with belief propagation on cluster graphs, we can prove that convergence points of such propagations are stationary points of the RegionGraph-Optimize optimization problem.

Theorem 11.7

> *A set of beliefs* Q *is a stationary point of* RegionGraph-Optimize *for region graph* $\mathcal{R}$ *if and only if for every edge* $(i\text{--}j) \in \mathcal{E}_\mathcal{R}$ *there are auxiliary factors* $\delta_{u \to d}(C_d)$ *that satisfy equation (11.36) and equation (11.37).*

This result is a direct generalization of theorem 11.5, and is proved in a similar way. We leave the detail as an exercise (see exercise 11.14). Much of the discussion following theorem 11.5 applies here. In particular, we do not have guarantees that iterations of message passing will converge. However, if they do, we have reached a stationary point of the energy functional. In practice, the experience is that when we consider moving from the Bethe approximation to "richer" region graphs that contain intermediate regions with larger subsets, problems of nonconverging runs are less common. For example, a region graph construction for grids is much more convergent than the corresponding cluster graph (see exercise 11.15). However, except for special cases (for example, region graphs that correspond to cluster trees), we do not know how to characterize region graphs where belief propagation converges.

11.3.8   Discussion

Cluster-graph belief propagation methods such as the ones we have described in this chapter provide a general-purpose mechanism for approximate inference in graphical models. In principle, they apply to any network, including networks with high tree-width, for which exact inference is intractable. They have been applied successfully to a large number of dramatically different applications, including (among many others) message decoding in communication over a noisy channel (see box 11.A), predicting protein structure (see box 20.B), and image segmentation (see box 4.B).

☞ However, **it is important to keep in mind that cluster-graph belief propagation is not a global panacea to the problem of inference in graphical models. The algorithm may not converge, and when it does converge, there may be multiple different convergence points.** Although there are currently no conditions characterizing precisely when cluster-graph belief propagation converges, several factors seem to play a role.

The first is the topology of the network: A network containing a large number of short loops is more likely to be nonconvergent. Although this notion has been elusive to characterize in practice, it has been shown that cluster-graph belief propagation is guaranteed to converge on networks with a single loop.

An even more significant factor is the extent to which the factors parameterizing the network are skewed, or close to deterministic. Intuitively, deterministic factors can cause difficulties in several ways. First, they often induce strong correlations between variables, which cluster-graph belief propagation (depending on the approximation chosen) can lose. This error can have an effect not only for the correlated variables, but also for marginals of variables that interact with both. Second, close-to-deterministic factors allow information to be propagated reliably through long paths in the network. Recall that part of our motivation for the running intersection property was to prevent information about some variable to be propagated infinitely through a loop. While the running intersection property prevents such loops from occurring structurally, deterministic potentials allow us to recreate them using an appropriate choice of parameters. For example, if A is deterministically equal to B, then we can have a cycle of clusters where A appears in some of the clusters and B in others. Although this cluster graph may satisfy the running intersection property relative to A, effectively there is a cycle in which the same variable appears in all clusters. Finally, as we discussed in section 11.3.4, factors that are less skewed provide smoothing of the messages, reducing oscillations; indeed, one can even prove that, if the skew of the factors in the network is sufficiently bounded, it can give rise to a contraction property that guarantees convergence.

☞ **In summary, the key factor relating to convergence of belief propagation appears to be the extent to which the network contains strong influences that "pull" a variable in different directions. Owing to its local nature, the algorithm is incapable of reconciling these different constraints, and it can therefore oscillate as different messages arrive that pull it in one direction or another.**

A second problem relates to the quality of the results obtained. Despite the appeal and importance of the energy-based analysis, it does not (except in a few rare cases — see section 11.7) provide any guarantees on the accuracy of the marginals obtained by cluster-graph belief propagation. This is in contrast to the sampling-based methods of chapter 12, where we are at least assured that, if we run the algorithm for long enough, we will obtain accurate estimates of the posteriors. (Of course, the key question of "how long is long enough" does not usually have an answer, so it is not clear how important this distinction is in practice.) Empirical results show that, in the settings where cluster-graph belief propagation convergence is more likely (not too many tight loops, no highly skewed factors), one also often obtains reasonable answers.

Importantly, these answers are often good but overconfident: The value $x \in Val(X)$ to which cluster-graph belief propagation gives the highest probability is often the value for which $P_\Phi(X = x)$ is indeed the highest, but the probability assigned to x by the approximation is often too high. This phenomenon arises (partly) from the fact the cluster-graph belief propagation ignores correlations between messages and can therefore count the same piece of evidence multiple times as it arrives along different paths, leading to overly strong conclusions. In other cases, however, the answers obtained by cluster-graph belief propagation are simply wrong (see section 11.3.1); unfortunately, there is currently no way of determining when the answers returned by a run of cluster-graph belief propagation are reasonable approximations to the true marginals.

The intuitions described previously do help us, however, to design approximations that are more likely to produce good answers. In general, we cannot construct a cluster graph that preserves all of the higher-order interactions among the factors. Hence, we need to decide which factors to include in the cluster graph and how to relate them. As the preceding discussion suggests, we do better if we construct approximations that incorporate tight loops and maintain the strongest factors within clusters as much as possible. While these intuitions provide reasonable rules of thumb on how to construct approximations, it is not obvious how to capture them within a general-purpose automated cluster-graph construction procedure.

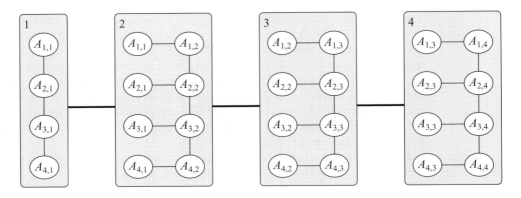

Figure 11.12　**A cluster for a** 4×4 **grid network.** The structure within each cluster represents the arcs whose factors are assigned to that cluster.

11.4　Propagation with Approximate Messages ⋆

The cluster-graph belief propagation methods achieved approximation by "relaxing" the requirement of having a cluster tree. Instead, we used a cluster graph and constructed an approximation by a set of pseudo-marginals. This approximation avoided the need to construct large clusters, which incur an exponential cost in terms of memory and running time. **In this section, we consider an approach in which the simplification occurs within a given cluster structure; rather than simplify this structure, we perform approximate propagation steps within it. This allows us to keep the correct clusters and gain efficiency by using more compact representations and operations on these clusters (at the cost of introducing approximations).** Importantly, this approach is orthogonal to the methods we described in the previous section, in that approximate message passing can occur both within a clique tree or a cluster graph approximation. For ease of presentation, we focus on the case of clique trees in this section, but the ideas easily carry through to the more general setting.

　　The basic concept behind the methods described in this section is the use of approximate messages in clique tree (or cluster graph) propagation. Instead of representing messages in the clique trees as factors, we use more compact representations of approximate factors. There are many different schemes that we can use for approximating messages. To ground the discussion, we begin in section 11.4.1 by describing one important instantiation of this general framework, which is very natural in our setting — message approximation using a factored form (for example, as a product of marginals). In section 11.4.2 and section 11.4.3 we then discuss algorithms that perform message passing using approximate messages. In section 11.4.4 we describe a general algorithm, called *expectation propagation*, that applies to any approximation in the exponential family. Finally, in section 11.4.5, we show that the expectation propagation algorithm for the exponential family can be derived from a variational analysis, similar to the ones we discussed in the previous section.

expectation
propagation

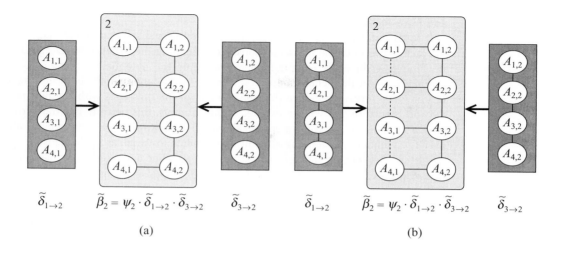

Figure 11.13 **Effect of different message factorizations on the structure of the belief factor** $\tilde{\beta}_2$ **in the example of figure 11.12.**

11.4.1 Factorized Messages

We begin by considering a concrete example where message approximation may be useful. Consider again the 4×4 grid network of figure 11.4. We can construct a cluster tree for this network as shown in figure 11.12. (Note that this cluster tree is not the optimal one for this network.) In our discussion of inference until now, we ignored the inner structure of clusters and treated the cluster as being associated with a single factor. Now, however, we keep track of the structure of the original factors associated with the cluster. We will see shortly why keeping this structure will help us. In figure 11.12 this structure is portrayed by "subnetworks" within each cluster.

Calibration in this cluster tree involves sending messages that are joint measures over four variables. An intuitive idea is to simplify the messages by using a factored representation. For example, consider the message $\delta_{1 \rightarrow 2}$, and suppose that we approximate it by a factored form

$$\tilde{\delta}_{1 \rightarrow 2}[A_{1,1}, A_{2,1}, A_{3,1}, A_{4,1}] = \tilde{\delta}_{1 \rightarrow 2}[A_{1,1}]\tilde{\delta}_{1 \rightarrow 2}[A_{2,1}]\tilde{\delta}_{1 \rightarrow 2}[A_{3,1}]\tilde{\delta}_{1 \rightarrow 2}[A_{4,1}].$$

What can we gain from such an approximation? Clearly, this form provides a more concise representation of the message. Instead of representing the message with values for the sixteen possible joint assignments, we send a message with two values for each variables (leading to a total of eight "parameters"). If we consider bigger grids, then the saving will be more substantial.

Does this saving gain us efficiency? Naively, the answer is no. Recall that the main computational cost of exact inference is generating the message in a cluster. This requires multiplying incoming messages with the original factor of the cluster and marginalization. In our example, C_2 involves eight variables, and so this operation should involve operations over the $2^8 = 256$ joint values of these variables. Thus, one might argue that the saving of eight parameters in representing the message does not deal with the core computational problem we have at hand

and leads to negligible improvement.

It turns out, however, that if we consider the internal structure of the cluster we can exploit the factored form of the message. Consider computing the beliefs over $\boldsymbol{C}_2$ given messages from $\boldsymbol{C}_1$ and $\boldsymbol{C}_3$. In exact computation, we multiply the potential ψ_2 with $\delta_{1\to 2}$ and $\delta_{3\to 2}$ and normalize to get the beliefs about $\boldsymbol{C}_2$. However, if we approximate both messages by a product of univariate terms, we notice that the product of the messages with the factors in $\boldsymbol{C}_2$ forms a network structure that we can exploit. In our example, this network is a tree-structured network shown in figure 11.13a. We can easily calibrate this network and answer queries about the beliefs over $\boldsymbol{C}_2$ without enumerating all joint assignments to variables in this cluster.

We can get similar savings even if we use a richer approximation that can better capture dependencies among variables in the message. Suppose we approximate $\delta_{1\to 2}$ by a factored form that corresponds to the chain structure $A_{1,1}$—$A_{2,1}$—$A_{3,1}$—$A_{4,1}$. This approximation makes conditional independence assumptions about the variables, but it captures some of the dependencies among them. In this example, this representation actually captures the message $\delta_{1\to 2}$. However, we can check that $\delta_{3\to 2}$ does not satisfy the conditional independence of $A_{1,2}$ and $A_{3,2}$ given $A_{2,2}$. Thus, in this case, a chain representation is an approximation. If we multiply these two approximations with the cluster $\boldsymbol{C}_2$, we get a set of factors that has the structure shown in figure 11.13b. Although not a tree, this graph has a tree-width of 2, regardless of the grid size. Thus, once again, we can use exact inference methods on the resulting product of factors.

factor set

We can exploit this intuition by maintaining both the initial potentials and the messages as *factor sets*. For the initial potential, these factors are the parameterization of the original network; for messages, these factors are introduced by the approximation. A factor set $\vec{\phi} = \{\phi_1, \ldots, \phi_k\}$ provides a compact representation for the higher-dimensional factor $\phi_1 \cdot \phi_2 \cdot \cdots \cdot \phi_k$.

factor set product

Recall that belief propagation involves two main operations: product and marginalization. The *product* of factor sets is easy. Suppose we have the factor sets $\vec{\phi}_1$ and $\vec{\phi}_2$. The factor set $\vec{\phi}_1 \cdot \vec{\phi}_2$ is simply the factor set that contains the union of the two factor sets (we assume that the components of the factor sets are distinct).

factor set marginalization

The difficult operation is *marginalization*. Suppose we have a factor set $\vec{\phi} = \{\phi_1, \ldots, \phi_k\}$, and we consider the marginalization $\sum_X \vec{\phi}$. This operation couples all the factors that contain X. In general, for a well-constructed clique tree, the message that results from marginalizing a clique will not satisfy any conditional independence statements and therefore cannot be factorized (see exercise 11.21).

Returning to our example of figure 11.12, one of our inference steps is to compute:

$$\delta_{2\to 3} = \sum_{A_{1,1},A_{2,1},A_{3,1},A_{4,1}} \psi_2 \cdot \delta_{1\to 2}.$$

To achieve the efficient inference steps we discussed earlier, we want to approximate $\delta_{2\to 3}$ by a product of simpler terms. This is an instance of a problem we encountered in section 8.5 — approximating a given distribution by another distribution from a given family of distributions. In our case, the approximating family is the set of distributions that take a particular product form. As we discussed in section 8.5, there are several ways of projecting a distribution P into some constrained class of distributions $\mathcal{Q}$. Indeed, throughout our discussion in this chapter so far, we have been searching for a distribution Q which is the I-projection of P_Φ — the one

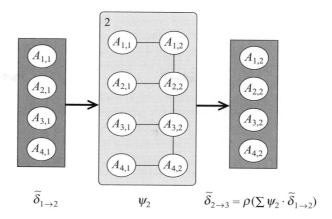

Figure 11.14 **Example of propagation in cluster tree with factorized messages**

M-projection

that minimizes $D(Q\|P_\Phi)$. However, as we discussed in section 8.5, we cannot compute the I-projection of a distribution in closed form. Conversely, the *M-projection* for many classes Q of factored distributions has an easy closed form expression. For example, the M-projection of P onto the class of factored distributions is simply the product of marginals of P (see proposition 8.3). We can compute these marginals by computing the marginals of the individual variables in the message. We note that the message is often not normalized, and is therefore not a distribution; however, we can normalize the message and treat it as a distribution without changing the final posterior obtained by the inference algorithm (see exercise 9.3).

Given the simplicity of M-projections, one might wonder why we used I-projections in our discussion so far. There are two main reasons. First, the theory of energy functionals allows us to use I-projections to lower-bound the partition function. Second, and more importantly, computing marginal probabilities for our distribution P_Φ is generally intractable. Thus, although the M-projection has a simple form, it is infeasible to apply when we have an intractable network structure. In our current setting, the situation is different. Here, our projection task is to approximate the outgoing message from a cluster C. As discussed earlier, we assume that the product of the beliefs and the approximated messages for C is a factor set with tractable structure. This assumption allows us to use exact inference *within the cluster C* to compute the marginal probabilities needed for M-projections.

11.4.2 Approximate Message Computation

We now discuss in greater detail the task of computing factorized messages. To begin, consider the projection of $\delta_{2\to3}$ into a fully factorized representation. According to proposition 8.3, to build the M-projection of $\delta_{2\to3}$ onto a factored distribution, we need to compute the marginals of $A_{1,2}$, $A_{2,2}$, $A_{3,2}$, and $A_{4,2}$ in $\delta_{2\to3}$. Since $\delta_{2\to3}$ is defined to be the marginal of the factor set $\psi_2 \cdot \delta_{1\to2}$, we need to compute terms such as $P_{\psi_2 \cdot \delta_{1\to2}}(A_{1,2})$, where we use $P_{\psi_2 \cdot \delta_{1\to2}}$ to denote the measure represented by the factor set. At an abstract level, this operation involves computing $\delta_{2\to3}$ and then projecting it onto the simpler representation. However, if we examine the structure of this factor set (see figure 11.14), we see that this computation can be done using standard exact inference on the factor set. In this particular case, the factor set is a tree network, and so inference is cheap. Similarly, we can compute the marginals over the variables in $\delta_{2 \to 3}$. Thus, we can use the properties of M-projections and exact inference to compute the resulting projected message without ever explicitly enumerating the joint over the four variables in $\delta_{2 \to 3}$. In an $n \times n$ grid, for example, we can perform these operations in time that is linear in n, whereas the explicit computation would have constructed a factor of exponential size. As we discussed, the more complex approximation of figure 11.13b also has a bounded tree-width of 2, and therefore also allows linear time inference.

Algorithm 11.4 Projecting a factor set to produce a set of marginals over a given set of scopes

       Procedure Factored-Project (
          $\vec{\phi}$,   // an input factor set
          $\mathcal{Y}$,   // A set of desired output scopes
)

1   Build an inference data structure $\mathcal{U}$
2   Initialize $\mathcal{U}$ with factors in $\vec{\phi}$
3   Perform inference in $\mathcal{U}$
4      $\vec{\phi}^o \leftarrow \emptyset$
5   **for** all $\boldsymbol{Y}_j \in \mathcal{Y}$
6   Find a clique $\boldsymbol{C}_j$ in $\mathcal{U}$ so that $\boldsymbol{Y}_j \subseteq \boldsymbol{C}_j$
7         $\psi_j \leftarrow \beta_{\mathcal{U}}(\boldsymbol{Y}_j)$ // marginal of $\boldsymbol{Y}_j$ in $\mathcal{U}$
8         $\vec{\phi}^o \leftarrow \vec{\phi}^o \cup \{\psi_j\}$
9   **return** $\vec{\phi}^o$

The overall structure of the algorithm is shown in algorithm 11.4. At a high level, we can view exact inference in the factor set as a "black box" and not concern ourselves with the exact implementation. However, it is also useful to consider how this type of projected message may be computed efficiently. Most often, the inference data structure $\mathcal{U}$ is a cluster graph or a clique tree, into which the initial factors $\vec{\phi}$ can easily be incorporated, and from which the target factors, of the appropriate scopes, can be easily extracted. To allow for that, we typically design the cluster graph to be family-preserving with respect to both sets of factors. Under that assumption, we can extract a factor ψ_j over the scope $\boldsymbol{Y}_j$ by identifying a cluster $\boldsymbol{C}_j$ in $\mathcal{U}$ whose scope contains $\boldsymbol{Y}_j$, and marginalizing it over $\boldsymbol{Y}_j$. As an alternative approach, we can construct an unconstrained clique tree, and use the out-of-clique inference algorithm of section 10.3.3.2 to extract from the graph the joint marginals of subsets of variables that are not together in a clique. (We note that out-of-clique inference is more challenging in the context of cluster graphs, since the path used to relate the clusters containing the query variables can affect the outcome of the computation; see exercise 11.22.)

If our representation of a message is simply a product of marginals over disjoint subsets of variables, this algorithm suffices to produce the output message: We produce a factor over each (disjoint) scope $\boldsymbol{Y}_j$; the product of these factors is the M-projection of the distribution. But for richer approximations, the required operation is somewhat more complex.

Example 11.6
Consider again our grid example of figure 11.13b. Here, the clique needs to take in messages that involve factors over pairs of variables $A_{i,1}, A_{i+1,1}$ and produce messages over pairs of variables $A_{i,2}, A_{i+1,2}$. We can accomplish this goal by constructing, within this clique, a nested clique tree that has at least one clique containing each of these pairs. For this purpose, we can use any clique tree based on triangulating the structure inside the clique in figure 11.13b. For example, we can have a tree with the cliques $\{A_{1,1}, A_{1,2}, A_{2,2}\}, \{A_{1,1}, A_{2,1}, A_{2,2}\}, \{A_{2,1}, A_{2,2}, A_{3,2}\},$ $\{A_{2,1}, A_{3,1}, A_{3,2}\}, \{A_{3,1}, A_{3,2}, A_{4,2}\}, \{A_{3,1}, A_{4,1}, A_{4,2}\}.$

Our goal now is to produce a set of factors that is the M-projection of the true message onto the chain. Assume that we extract from the clique tree the pairwise marginals $P(A_{1,2}, A_{2,2})$, $P(A_{2,2}, A_{3,2})$, and $P(A_{3,2}, A_{4,2})$. However, we cannot directly encode the distribution using these factors as a factor set, since this approach would double-count the probability of the singletons $A_{2,2}$ and $A_{3,2}$, each of which appears in two factors. To produce the true M-projected distribution, we need to divide by the double-counted marginals $A_{2,2}$ and $A_{3,2}$.

We can achieve this correction easily by computing these node marginals and adding to our factor set representation two factors

$$\frac{1}{\tilde{\delta}_{2\to3}(A_{2,2})}, \frac{1}{\tilde{\delta}_{2\to3}(A_{3,2})}$$

that compensate for the double-counting. This factor set represents the M-projection of the distribution, and it can be used in subsequent message passing steps. Equivalently, we are representing the distribution as a calibrated clique tree over $A_{1,2}, A_{2,2}, A_{3,2}, A_{4,2}$, where the factors derived from the pairwise marginals are the clique beliefs, the factors derived from inverted node marginals are the sepset messages, and the overall distribution is encoded as in equation (10.11). ∎

We can easily generalize this approach to more complex message representations. Most generally, assume that we choose to encode our message between cluster i and cluster j using a representation as in equation (11.35); more precisely, we have some set of factors $\{\beta_r(\boldsymbol{X}_r) : \boldsymbol{X}_r \in \mathbf{R}_{i,j}\}$, each raised to some power:

$$\tilde{\delta}_{i\to j} = \prod_{r\in\mathbf{R}_{i,j}} (\beta_r(\boldsymbol{X}_r))^{\kappa_r}. \tag{11.38}$$

We can compute this approximation in our procedure using the Factored-Project algorithm, using the set $\{\boldsymbol{X}_r \in \mathbf{R}_{i,j}\}$ as $\mathcal{Y}$. The output of this procedure is a set of calibrated marginals over these scopes, and can therefore be plugged into equation (11.38) to produce a message in the appropriate factored form. The same factors, each raised to its appropriate power, can be used as the factorized message passed to cluster j.

Importantly, we note that this approach does not always provide an exact M-projection of the true message. This property is guaranteed only when the set of regions forms a clique tree, allowing the M-projection to be calculated in closed form using equation (11.38); in other cases, we obtain only an approximation to the M-projection. In practice, we often choose some simple clique tree, as in example 11.6, to encode our distribution, allowing the M-projection to be performed easily and exactly. However, in some cases, a clique tree approximation, to stay tractable, is forced to make too many independence assumptions, leading to a poor approximation of the message. Hence, the approach of an approximation M-projection into a richer class of distributions may provide a useful alternative in some cases.

In summary, we have shown how we can perform an approximate message passing step with structured messages. The algorithm maintains messages and beliefs in factored form. Factors are entered into a nested clique tree or cluster graph, and message propagation is used to compute the factors describing the output message. In the fully factored case, this representation is simply a set of factors, and the multiplication operation used in message passing is simply a union of factor sets. In the more complex setting, we may need to postprocess the set of marginals — exponentiating them by appropriate counting numbers — to eliminate double-counting. The resulting set of factors is the compact representation of the outgoing message.

11.4.3   Inference with Approximate Messages

Equipped with these operations on factor sets, we can consider how to use these tools within cluster tree propagation. Our algorithm will maintain the beliefs at each cluster i using a factor set $\vec{\phi}_i$. Initially, we are given a cluster tree $\mathcal{T}$ with an assignment of the original factors to clusters. We have also determined the factorized form for each sepset, in terms of a network structure $\mathcal{G}_{i,j}$ that describes the desired factorization for $\tilde{\delta}_{i \to j}$ and $\tilde{\delta}_{j \to i}$.

There are two main strategies for applying the ideas described in the previous section to define an approximate message passing algorithms in clique trees. One is based on a sum-product message passing scheme, and the other on belief update messages. As we will see, although these two strategies are equivalent in the exact inference setting, they lead to fairly different algorithms when we consider approximations.

For notational simplicity, we introduce the notation M-project-distr$_{i,j}$ to be the combined operation of marginalizing variables that do not appear in $\boldsymbol{S}_{i,j}$ and performing the M-projection onto the set of distributions that can be factorized according to $\mathcal{G}_{i,j}$.

11.4.3.1   Sum-Product Propagation

Consider the application of sum-product propagation (CTree-SP-Calibrate, algorithm 10.2), to our grid example. In this case, we need to perform the following operations:

$$
\begin{aligned}
\delta_{1 \to 2} &= \psi_1 \\
\delta_{2 \to 3} &= \sum_{A_{1,1},\dots,A_{4,1}} \psi_2 \cdot \delta_{1 \to 2} \\
\delta_{3 \to 4} &= \sum_{A_{1,2},\dots,A_{4,2}} \psi_3 \cdot \delta_{2 \to 3} \\
\delta_{4 \to 3} &= \sum_{A_{1,4},\dots,A_{4,4}} \psi_4 \\
\delta_{3 \to 2} &= \sum_{A_{1,3},\dots,A_{4,3}} \psi_3 \cdot \delta_{4 \to 3} \\
\delta_{2 \to 1} &= \sum_{A_{1,2},\dots,A_{4,2}} \psi_2 \cdot \delta_{3 \to 2}.
\end{aligned}
$$

A straightforward application of approximation is to replace each of the messages by the M-projected version:

$$\tilde{\delta}_{1\to2} = \text{M-project-distr}_{1,2}(\psi_1)$$

$$\tilde{\delta}_{2\to3} = \text{M-project-distr}_{2,3}(\psi_2 \cdot \tilde{\delta}_{1\to2})$$

$$\tilde{\delta}_{3\to4} = \text{M-project-distr}_{3,4}(\psi_3 \cdot \tilde{\delta}_{2\to3})$$

$$\tilde{\delta}_{4\to3} = \text{M-project-distr}_{3,4}(\psi_4)$$

$$\tilde{\delta}_{3\to2} = \text{M-project-distr}_{2,3}(\psi_3 \cdot \tilde{\delta}_{4\to3})$$

$$\tilde{\delta}_{2\to1} = \text{M-project-distr}_{1,2}(\psi_2 \cdot \tilde{\delta}_{3\to2}).$$

Each of these projection operations can be performed using the procedure described in the previous section.

sum-product expectation propagation

More generally, our *sum-product expectation propagation* procedure is identical to sum-product propagation of section 10.2, except that we modify the basic propagation procedure SP-Message so that, rather than simply marginalizing the product of factors, it computes their M-projection. Otherwise, the general structure of the propagation procedure is maintained exactly as before. Each cluster collects the messages from its neighbors and when possible sends outgoing messages. As for the original sum-product message passing, this process converges after a single upward-and-downward pass over the clique tree structure. Thus, unlike most of the other approximations we discuss in this chapter, this procedure terminates in a fixed number of steps.

Note that, after performing the propagation, the final beliefs over the individual clusters in the tree are not an explicit representation of a joint distribution over the variables in the cluster. In this case, the final beliefs over a cluster C are represented in a factorized form, as a set of beliefs. In our running example, the computed beliefs over C_2 have the form

$$\tilde{\beta}_2 = \psi_2 \cdot \tilde{\delta}_{1\to2}\tilde{\delta}_{3\to2},$$

and the structure shown in figure 11.13a. Because this structure allows tractable inference, we can answer queries about the posterior distribution of these variables using standard inference methods, such as variable elimination or clique tree inference.

These computational benefits come at a price. The exact beliefs over C_2 are not decomposable (see exercise 11.20). And thus, although forcing a particular independence structure on the marginal distribution has computational advantages, it does lose information. With this caveat in mind, we can still question whether the algorithm finds the best possible approximation within the set of constraints we are considering.

Example 11.7

To get a sense of the issues, consider the simple example shown in figure 11.15a. In this small network, the potential $\phi_1(A, B)$ introduces a strong coupling between A and B, with a strong preference for $A = B$. The potentials $\phi_2(A, C)$, $\phi_3(B, D)$ incur weaker coupling between A and C and B and D. Finally, the potential $\phi_4(C, D)$ has a strong preference for $C = c^1$, with a weaker preference for $C \neq D$.

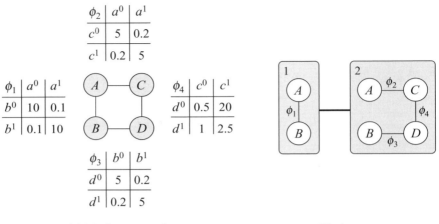

(a) Markov network   (b) cluster tree

Figure 11.15   Markov network used to demonstrate approximate message passing. (a) A simple Markov network with four pairwise potentials. (b) A clique tree for this network.

If we perform exact inference in this network, we find the following marginal posteriors:

$$
\begin{array}{llll}
P(a^0, b^0) & = & 0.274 & \qquad P(c^0, d^0) = 0.102 \\
P(a^0, b^1) & = & 0.002 & \qquad P(c^0, d^1) = 0.018 \\
P(a^1, b^0) & = & 0.041 & \qquad P(c^1, d^0) = 0.368 \\
P(a^1, b^1) & = & 0.682 & \qquad P(c^1, d^1) = 0.512.
\end{array}
$$

We see that the preference for $C = c^1$ is reflected in this distribution (with the marginal distribution $P(c^1) = 0.88$). In addition, the strong coupling between A and B is propagated through the network, which results in making $D = d^1$ more probable when $C = c^1$.

What happens when we perform inference using the cluster tree of figure 11.15b and use approximate messages that are products of marginals? It is easy to see that, because ϕ_1 is symmetric, we get that $\tilde{\delta}_{1\to2}[a^1] = 0.5$, and $\tilde{\delta}_{1\to2}[b^1] = 0.5$. We can compare the exact message and the approximate one

$$
\begin{array}{llllll}
\delta_{1\to2}(a^0, b^0) & = & 0.495 & \qquad \tilde{\delta}_{1\to2}(a^0, b^0) & = & 0.5 * 0.5 = 0.25 \\
\delta_{1\to2}(a^0, b^1) & = & 0.005 & \qquad \tilde{\delta}_{1\to2}(a^0, b^1) & = & 0.5 * 0.5 = 0.25 \\
\delta_{1\to2}(a^1, b^0) & = & 0.005 & \qquad \tilde{\delta}_{1\to2}(a^1, b^0) & = & 0.5 * 0.5 = 0.25 \\
\delta_{1\to2}(a^1, b^1) & = & 0.495 & \qquad \tilde{\delta}_{1\to2}(a^1, b^1) & = & 0.5 * 0.5 = 0.25.
\end{array}
$$

Thus, the approximate message is one where each joint assignment to A and B is equiprobable. This approximation loses the coupling that ϕ_1 introduces between A and B, and therefore it is a poor approximation to the exact message.

Next, we multiply this approximate message into the clique C_2. The initial factor here is $\psi_2 =$

$\phi_2 \cdot \phi_3 \cdot \phi_4$, *and after multiplying it with* $\tilde{\delta}_{1\to2}$ *we get the beliefs*

$$\tilde{\beta}_2(c^0, d^0) = 0.021$$
$$\tilde{\beta}_2(c^0, d^1) = 0.042$$
$$\tilde{\beta}_2(c^1, d^0) = 0.833$$
$$\tilde{\beta}_2(c^1, d^1) = 0.104.$$

Note that this factor is essentially a normalization of ϕ_4, *since the message* $\tilde{\delta}_{1\to2}$ *puts a uniform distribution of* A *and* B, *and since* ϕ_2 *and* ϕ_3 *are symmetric. Because the information about the coupling between* A *and* B *is not propagated into this cluster, we lose the consequent coupling between* C *and* D, *and the resulting approximation to* $P(C, D)$ *is also quite poor.*

By contrast, if we now compute $\tilde{\delta}_{2\to1}$, *we get that*

$$\tilde{\delta}_{2\to1}[a^1] = 0.904$$
$$\tilde{\delta}_{2\to1}[b^1] = 0.173.$$

This message ascribes high probability to a^1 *and a low one to* b^1. *This is quite different from the original coupling introduced by* ϕ_1. *Thus, when we combine the two to get the approximated posterior over* A *and* B, *we get the following beliefs factor:*

$$\tilde{\beta}_1(a^0, b^0) = 0.326$$
$$\tilde{\beta}_1(a^0, b^1) = 0.001$$
$$\tilde{\beta}_1(a^1, b^0) = 0.031$$
$$\tilde{\beta}_1(a^1, b^1) = 0.642.$$

This approximation is fairly close to the exact marginal over A *and* B. ∎

The problem in this example is that the approximation of $\tilde{\delta}_{1\to2}$ is done blindly, not taking into account its effect on approximations on computations in downstream clusters. Thus, the message did not place sufficient emphasis on obtaining a correct approximation for the case $A = a^1$, which roughly corresponds to the "important" (high-probability) case $C = c^1$.

11.4.3.2 Belief Update Propagation

Example 11.7 shows a case where factorizing messages leads to a big error in the approximated beliefs. Let us examine this example in somewhat more detail. Given the update of $\tilde{\delta}_{2\to1}$, the approximation of $P(A, B)$ is more or less on target. Can we use this information to improve our approximation of $P(C, D)$? To do this, we would need to revise $\tilde{\delta}_{1\to2}$. The posterior over A and B informs us that most of the mass of the probability distribution is on a^1, b^1. With this information, we might want to change $\tilde{\delta}_{1\to2}$ to reflect preferences for a^1 and b^1.

A priori, it appears that this idea is inherently problematic; after all, a key constraint for exact inference is to avoid feedback from $\delta_{2\to1}$ to $\delta_{1\to2}$, so as not to double-count evidence. For this reason, we took care, in the sum-product message-passing algorithm, *not* to multiply in the message $\delta_{2\to1}$ when passing messages from C_1 to C_2.

However, recall that in section 10.3, we presented the sum-product-divide update rule and showed its equivalence to the sum-product rule. Briefly recapping, we can take the sum-product update rule:

$$\delta_{i \to j} = \sum_{\boldsymbol{C}_i - \boldsymbol{S}_{i,j}} \psi_i \left(\prod_{k \in \mathrm{Nb}_i - \{j\}} \delta_{k \to i}, \right)$$

and multiply and divide by $\delta_{j \to i}$, resulting in the rule:

$$\delta_{i \to j} \leftarrow \frac{\sum_{\boldsymbol{C}_i - \boldsymbol{S}_{i,j}} \beta_i}{\delta_{j \to i}}.$$

These two rules are therefore equivalent *in the exact case*. However, when we consider approximate inference, the situation is more complex.

belief-update
expectation
propagation

Consider now performing *belief-update expectation propagation* message passing. Assume, as before, that the algorithm is maintaining a set of approximate beliefs $\tilde{\beta}_i$. We now have two possible stages in which to do the project. The first is:

$$\tilde{\delta}_{i \to j} \leftarrow \text{M-project-distr}_{i,j} \left(\frac{\tilde{\beta}_i}{\tilde{\delta}_{j \to i}} \right).$$

This version is identical to the approximate sum-product we discussed before: The beliefs factor β_i accounts for all of the incoming messages; when we divide by the message $\tilde{\delta}_{j \to i}$ before projecting, we are projecting the product of all the other incoming message, which is precisely the sum-product message.

Alternatively, we can do the projection before we divide by $\tilde{\delta}_{j \to i}$. In this second approach, we first project $\tilde{\beta}_i$, and then divide by $\tilde{\delta}_{j \to i}$:

$$
\begin{aligned}
\tilde{\sigma}_{i \to j} &\leftarrow & \text{M-project-distr}_{i,j}(\tilde{\beta}_i) \\
\tilde{\delta}_{i \to j} &\leftarrow & \frac{\tilde{\sigma}_{i \to j}}{\tilde{\delta}_{j \to i}}.
\end{aligned}
\tag{11.39}
$$

In this update, we first collect all messages into $\boldsymbol{C}_i$; we then compute the beliefs about $\boldsymbol{C}_i$ and project this to the required form of the message. As in the exact belief update algorithm, this term accounts for information sent from $\boldsymbol{C}_j$. Note that both $\tilde{\sigma}_{i \to j}$ and $\tilde{\delta}_{j \to i}$ have the same factorization, and hence so does their quotient $\tilde{\delta}_{i \to j}$.

This message, which is subsequently used to update $\boldsymbol{C}_j$, is very different from the sum-product update. This is because the incoming message was used in determining the approximation. **The approximation process is invariably a trade-off, in that a better approximation of some regions of the probability space results in a worse approximation in others. In the belief-update form of message passing, we can take into account the current approximation of the message from the target clique when deciding on our approximation, potentially focusing more of our attention on "more relevant" parts of the space.**

To integrate this update rule into the standard belief-update message passing algorithm, we simply replace BU-Message of algorithm 10.3 with EP-Message, shown in algorithm 11.5. We note that the data structures in this procedure are slightly different from those in the original algorithm. First, we maintain the cluster beliefs implicitly, as a factor set, and consider the product of these factors only as part of the M-projection operation. Second, we do not only

---

Algorithm 11.5 Modified version of BU-Message **that incorporates message projection**

    Procedure EP-Message (
        i,   // sending clique
        j   // receiving clique
)
1       $\tilde{\sigma}_{i \to j} \leftarrow$ M-project-distr$_{\mathcal{Q}_{i,j}}(\vec{\phi}_i)$
2   // marginalize and project the clique over the sepset
3   Remove old $\tilde{\delta}_{i \to j}$ from $\vec{\phi}_j$
4       $\tilde{\delta}_{i \to j} \leftarrow \frac{\tilde{\sigma}_{i \to j}}{\tilde{\delta}_{j \to i}}$
5   // divide by the message from from C_j to C_i
6   Insert new $\tilde{\delta}_{i \to j}$ into $\vec{\phi}_j$

---

keep the previous message sent over the edge, but rather keep the messages $\tilde{\delta}_{i \to j}$ sent in both directions; this more refined bookkeeping is necessary for dividing by the correct message following the approximation. This algorithm is called *expectation propagation* (EP) for reasons that are explained later.

expectation propagation

Example 11.8

> *To understand the behavior of this algorithm, consider the application of belief-update message propagation to example 11.7. Suppose we initialize all messages to 1 and use updates of the form equation (11.39). We start by propagating a message $\tilde{\sigma}_{1 \to 2} =$ M-project-distr$_{1,2}(\tilde{\beta}_1)$ from C_1 to C_2. Because $\tilde{\delta}_{1 \to 2}$ at this stage is 1, the resulting update is exactly the one we discussed in example 11.7. If we now perform propagation from C_2 to C_1, we get the message $\tilde{\delta}_{2 \to 1}$ derived in example 11.7, multiplied by a constant (since $\tilde{\delta}_{1 \to 2}$ is uniform). At this point, as we discussed, the clique beliefs β_1 are a fairly reasonable approximation to the posterior.*
>
> *Using the revised update rule, we now project $\tilde{\beta}_1$, and then divide by $\tilde{\delta}_{2 \to 1}$:*
>
> $$\tilde{\delta}_{1 \to 2} \leftarrow \frac{\text{M-project-distr}_{1,2}(\tilde{\beta}_1)}{\tilde{\delta}_{2 \to 1}}.$$
>
> *This quotient, which is then subsequently used to update C_2, is very different from the previous update $\tilde{\delta}_{2 \to 1}$. Specifically, The marginal $\tilde{\beta}_1$ at this stage puts a posterior of $0.642 + 0.031 = 0.673$ on a^1, and 0.642 on b^1. To avoid double-counting the contribution of $\tilde{\delta}_{2 \to 1}$, we need to divide this marginal by this message. After we normalize messages, we obtain:*
>
> $$\tilde{\delta}_{1 \to 2}[a^1] \leftarrow \frac{\frac{0.673}{0.904}}{\frac{0.673}{0.904} + \frac{0.327}{0.086}} = \frac{0.744}{4.15} = 0.179$$
>
> $$\tilde{\delta}_{1 \to 2}[a^0] \leftarrow \frac{\frac{0.327}{0.086}}{\frac{0.673}{0.904} + \frac{0.327}{0.086}} = \frac{3.406}{4.15} = 0.821$$
>
> $$\tilde{\delta}_{1 \to 2}[b^1] \leftarrow \frac{\frac{0.642}{0.173}}{\frac{0.642}{0.173} + \frac{0.358}{0.827}} = \frac{3.710}{4.413} = 0.895$$
>
> $$\tilde{\delta}_{1 \to 2}[b^0] \leftarrow \frac{\frac{0.358}{0.827}}{\frac{0.642}{0.173} + \frac{0.358}{0.827}} = \frac{0.433}{4.144} = 0.105.$$

Recall that this message can be viewed as a "correction term" for the sepset marginals, relative to the message $\tilde{\delta}_{2\to1}$. Thus, its effect is to reduce the support for a^1 (which was very high in $\tilde{\beta}_2$), and at the same time to increase the support for b^1.

Propagating this message to C_2 and updating the clique beliefs $\tilde{\beta}_2$, we get a normalized factor of:

$$\tilde{\beta}_2(c^0, d^0) \quad \leftarrow \quad 0.031$$
$$\tilde{\beta}_2(c^0, d^1) \quad \leftarrow \quad 0.397$$
$$\tilde{\beta}_2(c^1, d^0) \quad \leftarrow \quad 0.317$$
$$\tilde{\beta}_2(c^1, d^1) \quad \leftarrow \quad 0.254.$$

These beliefs are closer to the exact marginals in that they distribute some of the mass that was previously assigned to c^1, d^0 to two other cases. However, they are still quite far from the exact marginal.   ∎

This example demonstrates several issues. First, there is significant difference between the two update rules. After incorporating $\tilde{\delta}_{2\to1}$, the message from C_1 to C_2 is readjusted to account for the new information, leading to a different approximation and hence a different update $\tilde{\delta}_{1\to2}$. Second, unlike sum-product propagation, belief update propagation does not generally converge within two rounds, even in a clique tree. In fact, an immediate question is whether these iterations converge at all. And if they do, is there anything we can say about the convergence points? As we show in section 11.4.5, the answers to these questions are very similar to the answers we got in the case of cluster-graph belief propagation.

11.4.4   Expectation Propagation

So far, our discussion of approximate message passing has focused on a particular type of approximation: approximating a complex joint distribution as a product of small factors. However, the same ideas are applicable to a broad range of approximations. We now consider this process from a more general perspective, which will allow us to use a wider range of structured approximation in messages. Moreover, as we will see, this generalized form simplifies the variational analysis of this approach.

exponential
family

The framework we use is based on the idea of the *exponential family*, as presented in chapter 8. As we discussed there, the exponential families are a general class of distributions, that contains many of the distributions of interest. Recall that a family of distributions Q is in the exponential family if it can be defined by two functions: the sufficient statistic function $\tau(x)$, and the natural parameter function $t(\theta)$, so that any distribution Q in the family can be written as

$$Q(x) = \frac{1}{Z(\theta)} \exp\left\{\langle \tau(x), t(\theta)\rangle\right\},$$

where θ is a set of parameters that specify the particular member of the family.

To simplify the discussion we will focus on linear exponential families, where $t(\theta) = \theta$. Recall that linear exponential families include Markov networks (and consequently chordal Bayesian networks).

exponential
family messages

As we now show, the approximate message passing approach described earlier applies when

---

Algorithm 11.6 The message passing step in the expectation propagation algorithm. The algorithm performs approximate message propagation by projecting expected sufficient statistics.

       Procedure M-Project-Distr (
           $\mathcal{Q},$ // target exponential family for projection
           $\vec{\phi}$ // Factor set
)
1        $\boldsymbol{X} \leftarrow Scope[\vec{\phi}]$ // Variables in factor set
2        $\bar{\tau} \leftarrow \boldsymbol{E}_{\boldsymbol{x} \sim \prod_{\phi \in \vec{\phi}}}[\tau_{\mathcal{Q}_{i,j}}(\boldsymbol{x})]$
3   // Compute expectation of sufficient statistics relative to distribution defined by product of factors
4        $\theta \leftarrow$ M-project$(\bar{\tau})$
5   **return** (θ)

---

we choose to approximate messages by distributions from (linear) exponential families. Specifically, assume that we restrict each sepset $\boldsymbol{S}_{i,j}$ to be represented within an exponential family $\mathcal{Q}_{i,j}$ defined by a sufficient statistics function $\tau_{i,j}$. When performing message passing from $\boldsymbol{C}_i$ to $\boldsymbol{C}_j$, we compute the marginal of $\tilde{\beta}_i$, usually represented as a factor set $\vec{\phi}_i$, and project it into **M-projection** $\mathcal{Q}_{i,j}$ using the *M-projection* operator M-project-distr$_{i,j}$. This computation is often done using inference procedure that takes into account the structure of $\tilde{\beta}_i$ as a factor set.

It turns out that the entire message passing operation can be formulated cleanly within this framework. If we are using an exponential family to represent our messages, then both the approximate clique marginal $\tilde{\sigma}_{i \rightarrow j}$ and the previous message $\tilde{\delta}_{j \rightarrow i}$ can be represented in the exponential form. Thus, if we ignore normalization factors, we have:

$$
\begin{aligned}
\tilde{\sigma}_{i \rightarrow j} &\propto \exp\left\{\langle \theta_{\tilde{\sigma}_{i \rightarrow j}}, \tau_{i,j}(\boldsymbol{s}_{i,j})\rangle\right\} \\
\tilde{\delta}_{j \rightarrow i} &\propto \exp\left\{\langle \theta_{\tilde{\delta}_{j \rightarrow i}}, \tau_{i,j}(\boldsymbol{s}_{i,j})\rangle\right\} \\
\tilde{\delta}_{i \rightarrow j} &= \frac{\tilde{\sigma}_{i \rightarrow j}}{\tilde{\delta}_{j \rightarrow i}} \propto \exp\left\{\langle (\theta_{\tilde{\sigma}_{i \rightarrow j}} - \theta_{\tilde{\delta}_{j \rightarrow i}}), \tau_{i,j}(\boldsymbol{s}_{i,j})\rangle\right\},
\end{aligned}
$$

where $\theta_{\tilde{\sigma}_{i \rightarrow j}}$ and $\theta_{\tilde{\delta}_{i \rightarrow j}}$ are the parameters of the messages $\tilde{\sigma}_{i \rightarrow j}$ and $\tilde{\delta}_{i \rightarrow j}$ respectively.

Since these messages are in an exponential family, it suffices to represent each of them by the parameters that describe them. We can then view propagation steps as updating these parameters. Specifically, we can rewrite the update step in line 4 of EP-Message as

$$
\theta_{\tilde{\delta}_{i \rightarrow j}} \leftarrow (\theta_{\tilde{\sigma}_{i \rightarrow j}} - \theta_{\tilde{\delta}_{j \rightarrow i}}). \tag{11.40}
$$

Note that, in the case of exact inference in discrete networks, the original update and the one using the exponential family representation are essentially identical, since the exponential family representation of factors is of the same size as the factor. Indeed, the standard update is often performed in a logarithmic representation (for reasons of numerical stability; see box 10.A), which gives rise precisely to the exponential family update.

The final issue we must address is the construction of the exponential-family representation of $\tilde{\sigma}_{i \rightarrow j}$ in line 1 of the algorithm. Recall that this process involves the M-projection of $\tilde{\beta}_i$

onto $\mathcal{Q}_{i,j}$. As we discussed in section 8.5, the M-projection of a distribution P within an exponential family $\mathcal{Q}$ is the distribution $Q \in \mathcal{Q}$, which defines the same expectation over the sufficient statistics as defined by P. In that section, we described a two-phase procedure for computing this approximating distribution: We compute the expected sufficient statistics induced by P, and then find the parameters for a distribution $Q \in \mathcal{Q}$ that induces the same expected sufficient statistics. We can apply this approach to define a general procedure for performing the operation M-project-distr$_{i,j}(\vec{\phi}_i)$ for a general exponential family. We first compute the expected expectation of $\tau_{i,j}$ according to $\tilde{\beta}_i$. We then find the distribution within $\mathcal{Q}_{i,j}$ that gives rise to the same expected sufficient statistics. This step is accomplished by the application of the function M-project that takes a vector of sufficient statistics and returns a parameter vector in the exponential family that induces precisely the expected sufficient statistics $\bar{\tau}_{i,j}$. This function is shown in algorithm 11.6. The use of the expectation step in computing the messages is

expectation propagation

the basis for the name *expectation propagation*, which describes the general procedure for any member of the exponential family.

As we have already discussed, dividing by the previous message corresponds to subtraction of the parameters. Thus, overall, we obtain the following *EP message passing* step:

EP message passing

$$\theta_{\tilde{\delta}_{i \to j}} \leftarrow \text{M-project}_{i,j}(\boldsymbol{E}_{\boldsymbol{S}_{i,j} \sim \tilde{\beta}_i}[\tau_{i,j}(\boldsymbol{S}_{i,j})]) - \theta_{\tilde{\delta}_{j \to i}}. \tag{11.41}$$

How expensive are the two key steps in the expectation-propagation message passing procedure? The first step is computing the expectation $\boldsymbol{E}_{\boldsymbol{S}_{i,j} \sim \tilde{\beta}_i}[\tau_{i,j}]$. In the case of discrete networks, this step might be as expensive as the number of possible values of $\boldsymbol{C}_i$. However, as we saw in the previous section, in many cases we can use the structure of the factors that make up $\tilde{\beta}_i$ to perform this expectation much more efficiently. The second step is computing M-project on these factors. For some exponential families, this step is trivial, and can be done using a plug-in formula. In other families, this is a complex problem by itself. We will return to these issues in much greater detail in later chapters (particularly chapter 19 and chapter 20). Usually, when we design an approximation algorithm, we choose an exponential family for which this second step is easy.

The factored distributions we discussed earlier are perhaps the simplest example of an exponential family that we can use in this algorithm. However, other representations also fall into this class.

Example 11.9

Consider using a chain network to approximate each message, as in figure 11.15b. In example 8.16 and exercise 8.6 we showed that this class of distributions is a linear family and constructed the function M-project for it. Following the derivation, suppose the variables in the sepset are $X_1, \ldots, X_k$ and we want to represent messages using the network structure $X_1 \to X_2 \to \ldots \to X_k$. In example 8.16, we showed that the expected sufficient statistics are summarized in the vector of indicators comprising: $\boldsymbol{1}\{x_i^j\}$ for $i = 1, \ldots, k$ and $x_i^j \in Val(X_i)$; and $\boldsymbol{1}\{x_i^j, x_{i+1}^\ell\}$ for $i = 1, \ldots, k - 1$, $x_i^j \in Val(X_i)$, $x_i^\ell \in Val(X_{i+1})$. Once we have the expected value of these statistics, we can reconstruct the distribution $Q(X_{i+1} \mid X_i)$ as described in exercise 8.6. Given these subprocedures, the remaining propagation steps have been described earlier.

If we consider a chain approximation to grid network, we can use the chains as in figure 11.13b. In this case, the main cost of a propagation step is the projection. The messages incoming to a cluster consists of univariate beliefs and pairwise beliefs along the column. When combined with the clique factors, these result in a ladder-like network (shown in figure 11.13b). As we discussed, we can a build a (nested) clique tree for this network that involves clusters of at most three variables. Thus, we can perform inference efficiently on this nested clique tree to compute the expectations on pairwise beliefs in the outgoing column, and then use those expectations to reconstruct parameters.∎

The view of the expectation propagation algorithm in these terms allows us to understand the scope of this approach. We can apply this approach to any class of distributions in the linear exponential family for which the computation of expected sufficient statistics and the M-projection operation can be implemented effectively. Note that not every class of distributions we have discussed obeys these two restrictions. In particular, Markov networks are in the linear exponential family, but they do not have an effective M-projection procedure. Thus, a general Markov network structure is not often used to represent messages in the expectation propagation algorithm. Conversely, Bayesian networks are not in the linear exponential family (see section 8.3.2). One might argue that Bayesian networks should be usable, since the M-projection operation can be implemented analytically (see theorem 8.7), allowing the expectation propagation algorithm to be applied effectively.

This argument, however, brings up one important caveat regarding the expectation propagation update rule. In the rule, we subtract two sets of parameters: the parameters associated with the message from j to i are subtracted from the parameters obtained from M-projection operation for C_i. For the classes of distributions that we considered earlier, the space of legal parameters Θ was the entire real space $\mathbb{R}^K$; this space is closed under subtraction, guaranteeing that the result of this update rule is a valid distribution in our space. This property does not hold for every exponential family; in particular, it is not the case for Bayesian networks with immoralities. Thus, we may end up in situations where the resulting parameters do not actually define a legal distribution in our space. We return to this issue when we discuss the application of expectation propagation to Gaussians in section 14.3.3, where it can give rise to severe problems. Thus, the most commonly used class of distributions in the expectation propagation algorithm is the class of low tree-width chordal graphs. Because these graphs are both Bayesian networks and Markov networks, they are both in the linear exponential family and admit an effective M-projection operation. The example of the chain distribution we used earlier falls into this category. In more general cases, these distributions are represented as clique trees, allowing them to be represented using a smaller set of factors: clique beliefs and sepset messages.

11.4.5 Variational Analysis

To define a variational principle for expectation propagation, we take an approach similar to the one we discussed in the context of cluster-graph belief propagation. Again, we consider an approximation Q that consists of a set of pseudo marginals. In fact, we use the same energy functional $\tilde{F}[\tilde{P}_\Phi, Q]$.

The main difference from the case of cluster-graph belief propagation is that, in the current approximation, the cluster tree is *not* calibrated. As we project messages into an approximate form, we no longer ensure that beliefs of neighboring clusters agree on the joint distribution of the variables they share. Instead, we maintain a weaker property that depends on the nature of the approximation. For example, in the case where we use messages that are product of marginal distributions, intuition suggests that neighboring clusters will eventually agree on the marginal distributions of the variables in the sepset.

To gain better insight into the constraints we need, we start with reasoning about properties of convergence points of the algorithm. Suppose we iterate expectation-propagation belief update propagations until convergence. Now, consider two neighboring clusters i and j. Since the algorithm has converged, it follows that further updates do not change the cluster beliefs. Thus, the assignment of the expectation-propagation update rule of equation (11.41) becomes an equality for all clusters. We can then reason that

$$\text{M-project}_{i,j}(\boldsymbol{E}_{\boldsymbol{S}_{i,j} \sim \tilde{\beta}_i}[\tau_{i,j}]) = \theta_{\tilde{\delta}_{i \to j}} + \theta_{\tilde{\delta}_{j \to i}}.$$

A similar argument holds for the M-projection of the j cluster beliefs, $\text{M-project}_{i,j}(\boldsymbol{E}_{\boldsymbol{S}_{i,j} \sim \tilde{\beta}_j}[\tau_{i,j}])$. It follows that the projection of the two beliefs onto $\mathcal{Q}_{i,j}$ result in the same distribution.

This discussion suggests that we can pose the problem as optimizing the same objective as CTree-Optimize, except that we now replace the constraint equation (11.7) with an *expectation consistency constraint*:

$$\boldsymbol{E}_{\boldsymbol{S}_{i,j} \sim \mu_{i,j}}[\tau_{i,j}] = \boldsymbol{E}_{\boldsymbol{S}_{i,j} \sim \beta_j}[\tau_{i,j}]. \tag{11.42}$$

In general, $\tau_{i,j}$ is a vector, and so this equation defines a vector of constraints.

We now can define the optimization problem. It is identical to the optimization problems we already encountered, except that we relax the marginal consistency constraints.

EP-Optimize:

Find $\boldsymbol{Q}$
maximizing $\tilde{F}[\tilde{P}_\Phi, \boldsymbol{Q}]$
subject to

$$\boldsymbol{E}_{\boldsymbol{S}_{i,j} \sim \mu_{i,j}}[\tau_{i,j}] = \boldsymbol{E}_{\boldsymbol{S}_{i,j} \sim \beta_j}[\tau_{i,j}] \quad \forall (i\text{–}j) \in \mathcal{E}_\mathcal{T} \tag{11.43}$$

$$\sum_{\boldsymbol{c}_i} \beta_i(\boldsymbol{c}_i) = 1 \quad \forall i \in \mathcal{V}_\mathcal{T} \tag{11.44}$$

$$\sum_{\boldsymbol{s}_{i,j}} \mu_{i,j}[\boldsymbol{s}_{i,j}] = 1 \quad \forall (i\text{–}j) \in \mathcal{E}_\mathcal{T} \tag{11.45}$$

$$\beta_i(\boldsymbol{c}_i) \geq 0 \quad \forall i \in \mathcal{V}_\mathcal{T}, \boldsymbol{c}_i \in \mathit{Val}(\boldsymbol{C}_i). \tag{11.46}$$

Like CGraph-Optimize, this optimization problem is an approximation to the problem of optimizing the energy functional in two ways. First, the space over which we are optimizing is the set of pseudo-marginals $\boldsymbol{Q}$. Because our marginalization constraint only requires that two neighboring clusters agree on their expectations, they will generally not agree on the full marginals. Thus, in general, a solution to this problem will generally *not* correspond to the marginals of an actual distribution Q, even in the context of a clique tree. Second, because we can define the true energy function $F[\tilde{P}_\Phi, Q]$ only for coherent joint distributions, we must resort here to optimizing its factored form $\tilde{F}[\tilde{P}_\Phi, \boldsymbol{Q}]$. Although the two forms are equivalent for distributions over clique trees, they are not equivalent in this setting (since the exact energy functional is not even defined outside the space of coherent distributions Q). Thus, as in the case of the CGBP algorithms in the previous section, we are approximating both the objective and the optimization space.

The generalization of theorem 11.3 for this relaxed optimization problem follows using more or less the same proof structure, where we characterize the stationary points of the constrained objective using a set of *fixed-point equations* that define a message passing algorithm.

fixed-point
equations

Theorem 11.8

Let $\boldsymbol{Q}$ be a set of beliefs such that $\mu_{i,j}$ is in the exponential family $\mathcal{Q}_{i,j}$ for all $(i\text{-}j) \in \mathcal{E}_{\mathcal{T}}$. Let M-project-distr$_{i,j}$ be the M-projection operation into the family $\mathcal{Q}_{i,j}$. Then $\boldsymbol{Q}$ is a stationary point of EP-Optimize if and only if for every edge $(i\text{-}j) \in \mathcal{E}_{\mathcal{T}}$ there are auxiliary beliefs $\delta_{i \to j}(\boldsymbol{S}_{i,j})$ and $\delta_{j \to i}(\boldsymbol{S}_{i,j})$ so that

$$\delta_{i \to j} = \frac{\text{M-project-distr}_{i,j}(\beta_i)}{\delta_{j \to i}} \tag{11.47}$$

$$\beta_i \propto \psi_i \cdot \prod_{j \in \mathrm{Nb}_i} \delta_{j \to i}$$

$$\mu_{i,j} \propto \delta_{j \to i} \cdot \delta_{i \to j}.$$

PROOF As in previous proofs of this kind, we define a Lagrangian that consists of the (approximate) energy functional $\tilde{F}[\tilde{P}_\Phi, \boldsymbol{Q}]$ as well as Lagrange multiplier terms for each of the constraints. In our case, we have a vector of Lagrange multipliers for each of the constraints in equation (11.43).

$$
\begin{aligned}
\mathcal{J} =\ & \sum_{i \in \mathcal{V}_\mathcal{T}} \boldsymbol{E}_{\boldsymbol{C}_i \sim \beta_i}[\ln \psi_i] + \sum_{i \in \mathcal{V}_\mathcal{T}} \boldsymbol{H}_{\beta_i}(\boldsymbol{C}_i) - \sum_{(i\text{-}j) \in \mathcal{E}_\mathcal{T}} \boldsymbol{H}_{\mu_{i,j}}(\boldsymbol{S}_{i,j}) \\
& - \sum_i \sum_{j \in \mathrm{Nb}_i} \vec{\lambda}_{j \to i} \cdot \left(\boldsymbol{E}_{\boldsymbol{S}_{i,j} \sim \mu_{i,j}}[\tau_{i,j}] - \boldsymbol{E}_{\boldsymbol{S}_{i,j} \sim \beta_j}[\tau_{i,j}] \right) \\
& - \sum_{i \in \mathcal{V}_\mathcal{T}} \lambda_i \left(\sum_{\boldsymbol{c}_i} \beta_i(\boldsymbol{c}_i) - 1 \right) - \sum_{(i\text{-}j) \in \mathcal{E}_\mathcal{T}} \lambda_{i,j} \left(\sum_{\boldsymbol{s}_{i,j}} \mu_{i,j}[\boldsymbol{s}_{i,j}] - 1 \right).
\end{aligned}
$$

Taking partial derivatives of J with respect to $\beta_i(\boldsymbol{c}_i)$ and $\mu_{i,j}[\boldsymbol{s}_{i,j}]$ and equating these derivatives to zero, we get the following equalities that must hold at a stationary point:

$$\beta_i(\boldsymbol{c}_i) \propto \psi_i(\boldsymbol{c}_i) \prod_{j \in \mathrm{Nb}_i} \exp\left\{ \vec{\lambda}_{j \to i} \cdot \tau_{i,j}(\boldsymbol{s}_{i,j}) \right\}$$

$$\mu_{i,j}[\boldsymbol{s}_{i,j}] \propto \exp\left\{ (\vec{\lambda}_{j \to i} + \vec{\lambda}_{i \to j}) \cdot \tau_{i,j}(\boldsymbol{s}_{i,j}) \right\}.$$

Note that $(\vec{\lambda}_{j \to i} + \vec{\lambda}_{i \to j})$ serves as the natural parameters of $\mu_{i,j}$ in its exponential form representation.

Moreover, the constraint of equation (11.43) implies that

$$\boldsymbol{E}_{\boldsymbol{S}_{i,j} \sim \mu_{i,j}}[\tau_{i,j}] = \boldsymbol{E}_{\boldsymbol{S}_{i,j} \sim \beta_j}[\tau_{i,j}].$$

Thus, using theorem 8.6, we conclude that $\mu_{i,j} = \text{M-project-distr}_{i,j}(\beta_i)$. By defining

$$\delta_{i \to j}[\boldsymbol{s}_{i,j}] \propto \exp\left\{ \vec{\lambda}_{i \to j} \cdot \tau_{i,}(\boldsymbol{s}_{i,j}) \right\},$$

we can verify that the statement of the theorem is satisfied. ∎

Theorem 11.8 shows that, if we perform EP belief update propagation until convergence, then we reach a stationary point of EP-Optimize. Thus, this result provides an optimization semantics for expectation-propagation message passing.

Our discussion of expectation propagation and the proof were posed in the context of linear exponential families. The same ideas can be extended to nonlinear families but require additional subtleties that we do not discuss.

11.4.6 Discussion

In this section, we presented an alternative approach for inference in large graphical models. Rather than modifying the global structure of the inference object, we modify the structure of the messages and how they are computed. Although we focused on the application of this approximation to clique trees, it is equally applicable in the context of cluster graphs. The message passing algorithms (both the sum-product algorithm and the belief update algorithm) can be used for passing messages between clusters in general graphs. Moreover, the variational analysis of section 11.4.5 also applies, essentially without change, to cluster graphs, using the same derivation as in section 11.3.

Note that the expectation propagation algorithm suffers from the same caveats we discussed in the previous section: **Iterations of EP message propagation are not guaranteed to induce monotonic improvements to the objective function, and the algorithm does not always converge. Moreover, even when the algorithm does converge, the clusters are only approximately calibrated: their marginals agree on the expectations of the sufficient statistics (say the individual marginals), but not on other aspects of the distribution (say marginals over pairs of variables). As a consequence, if we want to answer a query using the network, it may make a difference from which cluster we extract the answer.**

We presented expectation propagation in the context of its application to factored messages. In the simplest case, of fully factored messages, the messages are simply cluster marginals over individual variables. The similarity between this variant of expectation propagation and belief propagation is quite striking; indeed, one can simulate expectation propagation with fully factored messages using cluster-graph belief propagation with a particular factor graph structure (see exercise 11.23). The more general case of messages that are not fully factored (for example, figure 11.13b) is more complex, and they cannot be mapped directly to belief propagation in cluster graphs. However, a mapping does exist between expectation propagation in discrete networks with factorized messages and cluster-graph belief propagation with region graphs.

More important, however, is the fact that expectation propagation provides a general approach for dealing with distributions in the exponential family. It therefore provides message passing algorithms for a broad class of models. For example, we will see an application of expectation propagation to hybrid (continuous/discrete) graphical models in section 14.3.3. Its broad applicability makes expectation propagation an important component in the approximate inference toolbox.

11.5   Structured Variational Approximations

In the previous two sections, we examined approximations based on belief propagation. As we saw, both methods can be viewed as optimizing an approximate energy functional over the class of pseudo-marginals. These pseudo-marginals generally do not correspond to a globally coherent joint distribution Q. **The *structured variational* approach aims to optimize the energy functional over a family $\mathcal{Q}$ of *coherent* distributions Q. This family is chosen to be computationally tractable, and hence it is generally not sufficiently expressive to capture all of the information in P_Φ.**

structured
variational

More precisely, we aim to address the following maximization problem:

Structured-Variational:
Find $\qquad Q \in \mathcal{Q}$
maximizing $\quad F[\tilde{P}_\Phi, Q]$

where $\mathcal{Q}$ is a given family of distributions. In these methods, we are using the exact energy functional $F[\tilde{P}_\Phi, Q]$, which satisfies theorem 11.2. Thus, maximizing the energy functional corresponds directly to obtaining a better approximation to P_Φ (in terms of $D(Q\|P_\Phi)$).

The main parameter in this maximization problem is the choice of family $\mathcal{Q}$. This choice induces a trade-off. On one hand, families that are "simpler," that is, that can be described by a Bayesian network or a Markov network with small tree-width, allow more efficient inference. As we will see, simpler families also allow us to solve the maximization problem efficiently. On the other hand, if the family $\mathcal{Q}$ is too restrictive, then it cannot represent distributions that are good approximations of P_Φ, giving rise to a poor approximation Q. In either case, this family is generally chosen to have enough structure that allows inference to be tractable, giving rise to the name *structured variational* approximation.

structured
variational

As we will see, the methods of this type differ from generalized belief propagation in several ways. They are guaranteed to lower-bound the log-partition function, and they also are guaranteed to converge.

11.5.1 The Mean Field Approximation

mean field

The first approach we consider is called the *mean field* approximation. As we will see, in many respects, it resembles the algorithm obtained using the Bethe approximation to the energy functional. In particular, the resulting algorithm performs message passing where the messages are distributions over single variables. As we will see, however, the form of the updates is somewhat different.

11.5.1.1 The Mean Field Energy

Unlike our presentation in earlier sections, we begin our discussion with the energy functional, and we derive the algorithm directly from analyzing it. The mean field algorithm finds the distribution Q, which is closest to P_Φ in terms of $D(Q\|P_\Phi)$ within the class of distributions representable as a product of independent marginals:

$$Q(\mathcal{X}) = \prod_i Q(X_i).$$ (11.48)

On the one hand, the approximation of P_Φ as a fully factored distribution is likely to lose a lot of information in the distribution. On the other hand, this approximation is computationally attractive, since we can easily evaluate any query on Q by a product over terms that involve the variables in the scope of the query. Moreover, to represent Q, we need only to describe the marginal probabilities of each of the variables.

As in previous sections, the mean field algorithm is derived by considering fixed points of the *energy functional*. We thus begin by considering the form of the energy functional in equation (11.3) when Q has the form of a product distribution as in equation (11.48). We can then characterize its fixed points and thereby derive an iterative algorithm to find such fixed points.

energy functional

The functional contains two terms. The first is a sum of terms of the form $\boldsymbol{E}_{U_\phi \sim Q}[\ln \phi]$, where we need to evaluate

$$\boldsymbol{E}_{U_\phi \sim Q}[\ln \phi] \;\;=\;\; \sum_{\boldsymbol{u}_\phi} Q(\boldsymbol{u}_\phi) \ln \phi(\boldsymbol{u}_\phi)$$

$$=\;\; \sum_{\boldsymbol{u}_\phi} \left(\prod_{X_i \in \boldsymbol{U}_\phi} Q(x_i) \right) \ln \phi(\boldsymbol{u}_\phi).$$

As shown, we can use the form of Q to compute $Q(\boldsymbol{u}_\phi)$ as a product of marginals, allowing the evaluation of this term to be performed in time linear in the number of values of $\boldsymbol{U}_\phi$. Because this cost is linear in the description size of the factors of P_Φ, we cannot expect to do much better.

As we saw in section 8.4.1, the term $\boldsymbol{H}_Q(\mathcal{X})$ also decomposes in this case.

Corollary 11.3

If $Q(\mathcal{X}) = \prod_i Q(X_i)$, then

$$\boldsymbol{H}_Q(\mathcal{X}) = \sum_i \boldsymbol{H}_Q(X_i). \tag{11.49}$$

Thus, the energy functional for a fully factored distribution Q can be rewritten simply as a sum of expectations, each one over a small set of variables. Importantly, the complexity of this expression depends on the size of the factors in P_Φ, and *not* on the topology of the network. Thus, the energy functional in this case can be represented and manipulated effectively, even in networks that would require exponential time for exact inference.

Example 11.10

Continuing our running example, consider the form of the mean field energy for a 4×4 grid network. Based on our discussion, we see that it has the form

$$F[\tilde{P}_\Phi, Q] \;\;=\;\; \sum_{i \in \{1,2,3\}, j \in \{1,2,3,4\}} \boldsymbol{E}_Q[\ln \phi(A_{i,j}, A_{i+1,j})]$$

$$+ \sum_{i \in \{1,2,3,4\}, j \in \{1,2,3\}} \boldsymbol{E}_Q[\ln \phi(A_{i,j}, A_{i,j+1})]$$

$$+ \sum_{i \in \{1,2,3,4\}, j \in \{1,2,3,4\}} \boldsymbol{H}_Q(A_{i,j}).$$

We see that the energy functional involves only expectations over single variables and pairs of neighboring variables. The expression has the same general form for an $n \times n$ grid. Thus, although the tree-width of an $n \times n$ grid is exponential in n, the energy functional can be represented and computed in cost $O(n^2)$; that is, in a time linear in the number of variables. ∎

11.5.1.2 Maximizing the Energy Functional: Fixed-point Characterization

The next step is to consider the task of optimizing the energy function: finding the distribution Q for which this energy functional is maximized:

Mean-Field:

Find $\{Q(X_i)\}$
maximizing $F[\tilde{P}_\Phi, Q]$
subject to

$$Q(\mathcal{X}) = \prod_i Q(X_i) \tag{11.50}$$

$$\sum_{x_i} Q(x_i) = 1 \quad \forall i. \tag{11.51}$$

To simplify notation, from now on we use $\boldsymbol{X}_{-i}$ to denote $\mathcal{X} - \{X_i\}$.

Note that, unlike the cluster-graph belief propagation algorithms of section 11.3 and the expectation propagation algorithm of section 11.4, here we are not approximating the objective. We are approximating only the optimization space by selecting a space of distributions Q that generally does not contain our original distribution P_Φ.

As with the previous optimization problems in this chapter, we use the method of Lagrange multipliers to derive a characterization of the stationary points of $F[\tilde{P}_\Phi, Q]$. However, the structure of Q allows us to consider the optimal value of each component (that is, marginal distribution) given the rest. (This iterative optimization procedure was not feasible in cluster trees and graphs due to constraints that relate different beliefs.)

fixed-point equations

We now provide a set of *fixed-point equations* that characterize the stationary points of the mean field optimization problem:

Theorem 11.9

The distribution $Q(X_i)$ is a local maximum of Mean-Field *given $\{Q(X_j)\}_{j \neq i}$ if and only if*

$$Q(x_i) = \frac{1}{Z_i} \exp\left\{ \sum_{\phi \in \Phi} \boldsymbol{E}_{\mathcal{X} \sim Q}[\ln \phi \mid x_i] \right\}, \tag{11.52}$$

conditional expectation

where Z_i is a local normalizing constant and $\boldsymbol{E}_{\mathcal{X} \sim Q}[\ln \phi \mid x_i]$ is the conditional expectation *given the value x_i*

$$\boldsymbol{E}_{\mathcal{X} \sim Q}[\ln \phi \mid x_i] = \sum_{\boldsymbol{u}_\phi} Q(\boldsymbol{u}_\phi \mid x_i) \ln \phi(\boldsymbol{u}_\phi).$$

PROOF The proof of this theorem relies on proving the fixed-point characterization of the individual marginal $Q(X_i)$ in terms of the other components, $Q(X_1), \ldots, Q(X_{i-1}), Q(X_{i+1}), \ldots, Q(X_n)$, as specified in equation (11.52).

We first consider the restriction of our objective $F[\tilde{P}_\Phi, Q]$ to those terms that involve $Q(X_i)$:

$$F_i[Q] = \sum_{\phi \in \Phi} \boldsymbol{E}_{\boldsymbol{U}_\phi \sim Q}[\ln \phi] + \boldsymbol{H}_Q(X_i). \tag{11.53}$$

To optimize $Q(X_i)$, we define the Lagrangian that consists of all terms in $F[\tilde{P}_\Phi, Q]$ that involve $Q(X_i)$

$$L_i[Q] = \sum_{\phi \in \Phi} \boldsymbol{E}_{\boldsymbol{U}_\phi \sim Q}[\ln \phi] + \boldsymbol{H}_Q(X_i) + \lambda \left(\sum_{x_i} Q(x_i) - 1 \right).$$

The Lagrange multiplier λ corresponds to the constraint that $Q(X_i)$ is a distribution. We now take derivatives with respect to $Q(x_i)$. The following result plays an important role in the remainder of the derivation:

Lemma 11.1

If $Q(\mathcal{X}) = \prod_i Q(X_i)$ then, for any function f with scope $\boldsymbol{U}$,

$$\frac{\partial}{\partial Q(x_i)} \boldsymbol{E}_{\boldsymbol{U} \sim Q}[f(\boldsymbol{U})] = \boldsymbol{E}_{\boldsymbol{U} \sim Q}[f(\boldsymbol{U}) \mid x_i].$$

The proof of this lemma is left as an exercise (see exercise 11.24).

Using this lemma, and standard derivatives of entropies, we see that

$$\frac{\partial}{\partial Q(x_i)} L_i = \sum_{\phi \in \Phi} \boldsymbol{E}_{\mathcal{X} \sim Q}[\ln \phi \mid x_i] - \ln Q(x_i) - 1 + \lambda.$$

Setting the derivative to 0, and rearranging terms, we get that

$$\ln Q(x_i) = \lambda - 1 + \sum_{\phi \in \Phi} \boldsymbol{E}_{\mathcal{X} \sim Q}[\ln \phi \mid x_i].$$

We take exponents of both sides and renormalize; because λ is constant relative to x_i, it drops out in the renormalization, so that we obtain the formula in equation (11.52).
This derivation, by itself, shows only that the solution of equation (11.52) is a stationary point of equation (11.53). To prove that it is a maximum, we note that equation (11.53) is a sum of two terms: $\sum_{\phi \in \Phi} \boldsymbol{E}_{\boldsymbol{U}_\phi \sim Q}[\ln \phi]$ is linear in $Q(X_i)$, given all the other components $Q(X_j)$; $\boldsymbol{H}_Q(X_i)$ is a concave function in $Q(X_i)$. As a whole, given the other components of Q, the function F_i is concave in $Q(X_i)$, and therefore has a unique global optimum, which is easily verified to be equation (11.52) rather than any of the extremal points.   ∎

From this it follows that:

Corollary 11.4

The distribution Q is a stationary point of Mean-Field *if and only if, for each X_i, equation (11.52) holds.*

In contrast to theorem 11.9, this result only provides a characterization of stationary points of the objective, and not necessarily of its optima. The stationary points include local maxima, local minima, and saddle points. The reason for the difference is that, although each "coordinate" $Q(X_i)$ is guaranteed to be locally maximal given the others, the direction that locally improves the objective may require a coordinated change in several components. We return to this point in section 11.5.1.3.
We now move to interpreting this characterization. The key term in equation (11.52) is the argument in the expectation. We can prove the following property.

Corollary 11.5

In the mean field approximation, $Q(X_i)$ is locally optimal only if

$$Q(x_i) = \frac{1}{Z_i} \exp \left\{ \boldsymbol{E}_{\boldsymbol{X}_{-i} \sim Q} [\ln P_\Phi(x_i \mid \boldsymbol{X}_{-i})] \right\} \tag{11.54}$$

where Z_i is a normalizing constant.

PROOF Recall that $\tilde{P}_\Phi = \prod_{\phi \in \Phi} \phi$ is the unnormalized measure defined by Φ. Due to the linearity of expectation:

$$\sum_{\phi \in \Phi} \boldsymbol{E}_{\mathcal{X} \sim Q}[\ln \phi \mid x_i] = \boldsymbol{E}_{\mathcal{X} \sim Q} \left[\ln \tilde{P}_\Phi(X_i, \boldsymbol{X}_{-i}) \mid x_i \right].$$

Because Q is a product of marginals, we can rewrite $Q(\boldsymbol{X}_{-i} \mid x_i) = Q(\boldsymbol{X}_{-i})$, and get that:

$$\boldsymbol{E}_{\mathcal{X} \sim Q} \left[\ln \tilde{P}_\Phi(X_i, \boldsymbol{X}_{-i}) \mid x_i \right] = \boldsymbol{E}_{\boldsymbol{X}_{-i} \sim Q} \left[\ln \tilde{P}_\Phi(x_i, \boldsymbol{X}_{-i}) \right].$$

Using properties of conditional distributions, it follows that:

$$\tilde{P}_\Phi(x_i, \boldsymbol{X}_{-i}) = Z P_\Phi(x_i, \boldsymbol{X}_{-i}) = Z P_\Phi(\boldsymbol{X}_{-i}) P_\Phi(x_i \mid \boldsymbol{X}_{-i}).$$

We conclude that

$$\sum_{\phi \in \Phi} \boldsymbol{E}_{\mathcal{X} \sim Q}[\ln \phi \mid x_i] = \boldsymbol{E}_{\boldsymbol{X}_{-i} \sim Q}[\ln P_\Phi(x_i \mid \boldsymbol{X}_{-i})] + \boldsymbol{E}_{\boldsymbol{X}_{-i} \sim Q}[\ln P_\Phi(\boldsymbol{X}_{-i}) Z].$$

Plugging this equality into the update equation equation (11.52), we get that

$$Q(x_i) = \frac{1}{Z_i} \exp \left\{ \boldsymbol{E}_{\boldsymbol{X}_{-i} \sim Q}[\ln P_\Phi(x_i \mid \boldsymbol{X}_{-i})] \right\} \exp \left\{ \boldsymbol{E}_{\boldsymbol{X}_{-i} \sim Q}[\ln P_\Phi(\boldsymbol{X}_{-i}) Z] \right\}.$$

The term $\ln P_\Phi(\boldsymbol{X}_{-i}) Z$ does not depend on the value of x_i. Recall that when we multiply a belief by a constant factor, it does not change the distribution Q; in fact, as we renormalize the distribution at the end to sum to 1, this constant will be "absorbed" into the normalizing function, to achieve normalization. Thus, we can simply ignore this term, thereby achieving the desired conclusion. We note that this type of algebraic manipulation will prove useful multiple times throughout this section. ∎

This corollary shows that $Q(x_i)$ is the *geometric average* of the conditional probability of x_i given all other variables in the domain. The average is based on the probability that Q assigns to all possible assignments to the variables in the domain. In this sense, the mean field approximation requires that the marginal of X_i be "consistent" with the marginals of other variables.

Note that, in P_Φ, we can also represent the marginal of X_i as an average:

$$P_\Phi(x_i) = \sum_{\boldsymbol{x}_{-i}} P_\Phi(\boldsymbol{x}_{-i}) P_\Phi(x_i \mid \boldsymbol{x}_{-i}) = \boldsymbol{E}_{\boldsymbol{X}_{-i} \sim P_\Phi}[P_\Phi(x_i \mid \boldsymbol{X}_{-i})]. \tag{11.55}$$

This average is an arithmetic average, whereas the one used in the mean field approximation is a geometric average. In general, the latter tends to lead to marginals that are more sharply peaked than the original marginals in P_Φ. More significant, however, is the fact that the expectations in equation (11.55) are taken relative to P_Φ, whereas the ones in equation (11.54) are taken relative to the approximation Q. Thus, this similarity does not imply as a consequence that our approximation in Q to the marginals in P_Φ is a good one.

11.5.1.3   Maximizing the Energy Functional: The Mean Field Algorithm

How do we convert the fixed-point equation of equation (11.52) into an update algorithm? We start by observing that if $X_i \notin Scope[\phi]$ then $\boldsymbol{E}_{\boldsymbol{U}_\phi \sim Q}[\ln \phi \mid x_i] = \boldsymbol{E}_{\boldsymbol{U}_\phi \sim Q}[\ln \phi]$. Thus, expectation terms on such factors are independent of the value of X_i. Consequently, we can absorb them into the normalization constant Z_i and get the following simplification.

In the mean field approximation, $Q(X_i)$ is locally optimal only if

$$Q(x_i) = \frac{1}{Z_i} \exp \left\{ \sum_{\phi : X_i \in Scope[\phi]} \boldsymbol{E}_{(\boldsymbol{U}_\phi - \{X_i\}) \sim Q}[\ln \phi(\boldsymbol{U}_\phi, x_i)] \right\}. \tag{11.56}$$

where Z_i is a normalizing constant.

This representation shows that $Q(X_i)$ has to be consistent with the expectation of the potentials in which it appears. In our grid network example, this characterization implies that $Q(A_{i,j})$ is a product of four terms measuring its interaction with each of its four neighbors:

$$Q(a_{i,j}) = \frac{1}{Z_{i,j}} \exp \left\{ \begin{array}{l} \sum_{a_{i-1,j}} Q(a_{i-1,j}) \ln(\phi(a_{i-1,j}, a_{i,j})) + \\ \sum_{a_{i,j-1}} Q(a_{i,j-1}) \ln(\phi(a_{i,j-1}, a_{i,j})) + \\ \sum_{a_{i+1,j}} Q(a_{i+1,j}) \ln(\phi(a_{i,j}, a_{i+1,j})) + \\ \sum_{a_{i,j+1}} Q(a_{i,j+1}) \ln(\phi(a_{i,j}, a_{i,j+1})) \end{array} \right\}. \tag{11.57}$$

Each term is a (geometric) average of one of the potentials involving $A_{i,j}$. For example, the final term in the exponent represents a geometric average of the potential between $A_{i,j}$ and $A_{i,j+1}$, averaged using the distribution $Q(A_{i,j+1})$.

The characterization of corollary 11.6 provides tools for developing an algorithm to maximize $F[\tilde{P}_\Phi, Q]$. For example, examining equation (11.57), we see that we can easily evaluate the term within the exponential by considering each of $A_{i,j}$'s neighbors and computing the interaction between the values that neighbor can take and possible values of $A_{i,j}$. Moreover, in this example, we see that $Q(A_{i,j})$ does *not* appear on the right-hand side of the update rule. Thus, we can choose $Q(A_{i,j})$, which satisfies the required equality by assigning it to the term denoted by the right-hand side of the equation.

This last observation is true in general. All the terms on the right-hand side of equation (11.56) involve expectations of variables other than X_i, and do not depend on the choice of $Q(X_i)$. We can achieve equality simply by evaluating the exponential terms for each value x_i, normalizing the results to sum to 1, and then assigning them to $Q(X_i)$. As a consequence, we reach the optimal value of $Q(X_i)$ in one easy step.

This last statement must be interpreted with some care. The resulting value for $Q(X_i)$ is its optimal value *given* the choice of all other marginals. Thus, this step optimizes our function relative only to a single coordinate in the space — the marginal of $Q(X_i)$. To optimize the function in its entirety, we need to optimize relative to all of the coordinates. We can embed this step in an iterated *coordinate ascent* algorithm, which repeatedly optimizes a single marginal at a time, given fixed choices to all of the others. The resulting algorithm is shown in algorithm 11.7. Importantly, a single optimization of $Q(X_i)$ does not usually suffice: a subsequent modification

---

Algorithm 11.7 The Mean-Field approximation algorithm

Procedure Mean-Field (
    Φ,   // factors that define P_Φ
    Q_0 // Initial choice of Q
)
1    $Q \leftarrow Q_0$
2   *Unprocessed* $\leftarrow \mathcal{X}$
3   **while** *Unprocessed* $\neq \emptyset$
4   Choose X_i from *Unprocessed*
5        $Q_{old}(X_i) \leftarrow Q(X_i)$
6   **for** $x_i \in Val(X_i)$ **do**
7            $Q(x_i) \leftarrow \exp\left\{\sum_{\phi : X_i \in Scope[\phi]} \boldsymbol{E}_{(\boldsymbol{U}_\phi - \{X_i\}) \sim Q}[\ln \phi[\boldsymbol{U}_\phi, x_i]]\right\}$
8   Normalize $Q(X_i)$ to sum to one
9   **if** $Q_{old}(X_i) \neq Q(X_i)$ **then**
10   *Unprocessed* $\leftarrow$ *Unprocessed* $\cup \left(\cup_{\phi : X_i \in Scope[\phi]} Scope[\phi]\right)$
11   *Unprocessed* $\leftarrow$ *Unprocessed* $- \{X_i\}$
12   **return** Q

---

to another marginal $Q(X_j)$ may result in a different optimal parameterization for $Q(X_i)$. Thus, the algorithm repeats these steps until convergence. Note that, in practice, we do not test for equality in line 9, but rather for equality up to some fixed small-error tolerance.

A key property of the coordinate ascent procedure is that each step leads to an increase in the energy functional. **Thus, each iteration of** Mean-Field **results in a better approximation** Q **to the target density** P_Φ**, guaranteeing convergence.**

Theorem 11.10   *The* Mean-Field *iterations are guaranteed to converge. Moreover, the distribution Q^* returned by* Mean-Field *is a stationary point of $F[\tilde{P}_\Phi, Q]$, subject to the constraint that $Q(\mathcal{X}) = \prod_i Q(X_i)$ is a distribution.*

PROOF We showed earlier that each iteration of Mean-Field is monotonically nondecreasing in $F[\tilde{P}_\Phi, Q]$. Because the energy functional is bounded, the sequence of distributions represented by successive iterations of Mean-Field must converge. At the convergence point the fixed-point equations of theorem 11.9 hold for all the variables in the domain. As a consequence, the convergence point is a stationary point of the energy functional.   ∎

As we discussed, the distribution Q^* returned by Mean-Field is not necessarily a local optimum of the algorithm. However, local minima and saddle points are not stable convergence points of the algorithm, in the sense that a small perturbation of Q followed by optimization will lead to a better convergence point. Because the algorithm is unlikely to accidentally land precisely on the unstable point and get stuck there, in practice, the convergence points of the algorithm are local maxima.

In general, however, the result of the mean field approximation is a local maximum, and not necessarily a global one.

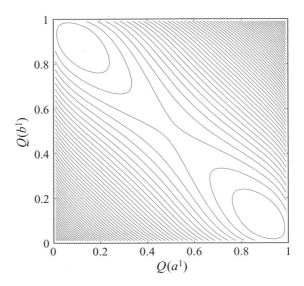

Figure 11.16   An example of a multimodal mean field energy functional landscape. In this network, $P(a, b) = 0.5 - \epsilon$ if $a \neq b$ and ϵ if $a = b$. The axes correspond to the mean field marginal for $Q(a^1)$ and $Q(b^1)$ and the contours show equi-values of the energy functional for different choices of these variational parameters. As we can see, there are two modes to the energy functional, one roughly corresponds to a^1, b^0 and the other one to a^0, b^1. In addition, there is a saddle point at $(0.5, 0.5)$.

Example 11.11   *Consider a distribution P_Φ that is an approximate XOR (exclusive or) of two variables A and B, so that $P_\Phi(a, b) = 0.5 - \epsilon$ if $a \neq b$ and $P_\Phi(a, b) = \epsilon$ if $a = b$. Clearly, we cannot approximate P_Φ by a product of marginals, since such a product cannot capture the relationship between A and B. It turns out that if ϵ is sufficiently small, say 0.01, then the energy potential surface has two local maxima that correspond to the two cases where $a \neq b$. See figure 11.16. (For sufficiently large ϵ, such as 0.1, the mean field approximation has a single maximum point at the uniform distribution.)* ∎

We can use standard strategies, such as multiple random restarts, to try to avoid getting stuck in local maxima. However, these do not overcome the basic shortcoming of the mean field approximation, which is apparent in this example. The approximation cannot describe complex posteriors, such as the XOR posterior we discussed. And thus, we cannot expect it to give satisfactory approximations in these situations. To provide better approximations, we must use a richer class of distributions $\mathcal{Q}$, which has greater expressive power.

11.5.2   Structured Approximations

The mean field algorithm provides an easy approximation method. However, it is limited by forcing Q to be a very simple distribution. As we just saw, the fact that all variables are independent of each other in Q can lead to very poor approximations. Intuitively, if we

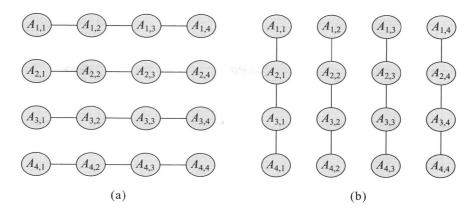

Figure 11.17 **Two structures for variational approximation of a** 4×4 **grid network**

use a distribution Q that can capture some of the dependencies in P_Φ, we can get a better approximation. Thus, we would like to explore the spectrum of approximations between the mean field approximation and exact inference.

A natural approach to get richer approximations that capture some of the dependencies in P_Φ is to use network structures of different complexity. By adding and removing edges from the network we can control the cost of inference in the approximating distribution and how well it captures dependencies in the target distribution. We can achieve this type of flexibility by using either Bayesian networks or Markov networks. Both types of networks lead to similar approximations, and so we focus on the undirected case, parameterized as Gibbs distributions (so that we are not restricted to factors over maximal cliques). Exercise 11.34 develops similar ideas using a Bayesian network approximation.

11.5.2.1 Fixed-Point Characterization

We now consider the form of the variational approximation when we are given a general form of Q as a Gibbs parametric family. Formally, we assume we are given a set of potential scopes $\{C_j \subseteq \mathcal{X} : j = 1, \ldots, J\}$. We can then choose an approximation Q that has the form:

$$Q(\mathcal{X}) = \frac{1}{Z_Q} \prod_{j=1}^{J} \psi_j, \tag{11.58}$$

where ψ_j is a factor with $Scope[\psi_j] = C_j$.

Example 11.12   *Consider again the grid network example. There are many possible approximating network structures we can choose that allow for efficient inference. As a concrete example, we might choose potential scopes $\{A_{1,1}, A_{1,2}\}, \{A_{1,2}, A_{1,3}\}, \ldots, \{A_{2,1}, A_{2,2}\}, \{A_{2,2}, A_{2,3}\}, \ldots$. That is, we preserve the dependencies between variables in the same row, but ignore the ones that relate different columns. Alternatively, we can consider an approximation that preserves dependencies along columns and ignores the dependencies between rows. As we can see in figure 11.17, in both cases,* *the structure we use is a collection of independent chain structures. Exact inference with such structures is linear, and so the cost of inference is not much worse than in the mean field approximation. Clearly, we can also consider many other structures for the approximating distributions. These might introduce additional dependencies and can have higher cost in terms of inference. We will return to the question of what structure to use.* ∎

Assume that we decide on the form of the potentials for the approximating family Q. As before, we consider the form of the energy functional for a distribution Q in this family. We then characterize the stationary points of the functional, and we use those to derive an iterative optimization algorithm.

As before, evaluating the terms that involve $E_{U_\phi \sim Q}[\ln \phi]$ requires performing expectations with respect to the variables in $Scope[\phi]$. Unlike the case of mean field approximation, the complexity of computing this expectation depends on the structure of the approximating distribution. However, we assume that we can solve this problem by exact inference (in the network corresponding to Q), using the methods we discussed in previous chapters.

As discussed in section 8.4.1, the entropy term in the energy functional also reduces to computing a similar set of expectation terms:

Proposition 11.5

If $Q(\mathcal{X}) = \frac{1}{Z_Q} \prod_j \psi_j$, then

$$H_Q(\mathcal{X}) = -\sum_{j=1}^{J} E_{C_j \sim Q}[\ln \psi_j(C_j)] + \ln Z_Q.$$

Overall, we obtain the following form for the energy functional, for distributions Q in the family $\mathcal{Q}$:

$$F[\tilde{P}_\Phi, Q] = \sum_{k=1}^{K} E_Q[\ln \phi_k] - \sum_{j=1}^{J} E_Q[\ln \psi_j] + \ln Z_Q. \tag{11.59}$$

As before, the hard question is how to optimize the potential to get the best approximation. We solve this problem using the same general strategy we discussed in the context of the mean field approximation. First, we derive the fixed-point equations that hold when the approximation is a local maximum (or, more precisely, a stationary point) of the energy functional. We then use these fixed-point equations to help derive an optimization algorithm.

fixed-point equations

We derive the *fixed-point equations* by taking derivatives of $F[\tilde{P}_\Phi, Q]$ with respect to parameters of the distribution Q. In our case, the parameters will be an entry $\psi_j(c_j)$ in each of the factors that define the distribution. We then set those equations to zero, obtaining the following result:

Theorem 11.11

If $Q(\mathcal{X}) = \frac{1}{Z_Q} \prod_j \psi_j$, then the potential ψ_j is a stationary point of the energy functional if and only if

$$\psi_j(c_j) \propto \exp\left\{ E_Q\left[\ln \tilde{P}_\Phi \mid c_j\right] - \sum_{k \neq j} E_Q[\ln \psi_k \mid c_j] - F[\tilde{P}_\Phi, Q] \right\}. \tag{11.60}$$

The proof is straightforward algebraic manipulation and is left as an exercise (exercise 11.26).

This theorem establishes a characterization of the fixed point as the difference between the expected value of logarithm of the original potentials and the expected value of the logarithm of the approximating potentials. The last term in equation (11.60) is the energy functional $F[\tilde{P}_\Phi, Q]$, which is independent of the assignment c_j; thus, as we discussed in the proof of corollary 11.5, we can absorb this term into the normalization constant of the distribution and ignore it.

Corollary 11.7

If $Q(\mathcal{X}) = \frac{1}{Z_Q} \prod_j \psi_j$, then the potential ψ_j is a stationary point of the energy functional if and only if:

$$\psi_j(\boldsymbol{c}_j) \propto \exp\left\{ \boldsymbol{E}_Q\left[\ln \tilde{P}_\Phi \mid \boldsymbol{c}_j\right] - \sum_{k \neq j} \boldsymbol{E}_Q[\ln \psi_k \mid \boldsymbol{c}_j] \right\}. \tag{11.61}$$

As we show in section 11.5.2.3 and section 11.5.2.4, we can often exploit additional structure in Q to reduce further the complexity of the fixed-point equations, and hence of the resulting update steps. The following discussion, which describes the procedure of applying the fixed-point equations to find a stationary point of the energy functional, is orthogonal to these simplifications.

11.5.2.2 Optimization

Given a set of fixed-point equations as in equation (11.61), our task is to find a distribution Q that satisfies them. As in section 11.5.1, our strategy is based on the key observation that the factor ψ_j does not affect the right-hand side of the fixed-point equations defining its value: The first expectation, $\boldsymbol{E}_Q\left[\ln \tilde{P}_\Phi \mid \boldsymbol{c}_j\right]$, is conditioned on $\boldsymbol{c}_j$ and therefore does not depend on the parameterization of ψ_j. The same observation holds for the second expectation, $\boldsymbol{E}_Q[\ln \psi_k \mid \boldsymbol{c}_j]$, for any $k \neq j$. (Importantly, there is no such term for $k = j$ in the right-hand side.) Thus, we can use the same general approach as in Mean-Field: We can optimize each potential ψ_j, *given values for the other potentials*, by simply selecting ψ_j to satisfy the fixed-point equation. As for the case of mean field, this step is guaranteed to increase (or not decrease) the value of the objective; thus, the overall process is guaranteed to converge to a stationary point of the objective.

This last step requires that we perform inference in the approximating distribution Q to compute the requisite expectations. Although this step was also present (implicitly) in the mean field approximation, there the structure of the approximating distribution was trivial, and so the inference step involved only individual marginals. Here, we need to collect the expectation of several factors, and each of these requires that we compute expectations given different assignments to the factor of interest. (See exercise 11.27 for a discussion of how these expectations can be computed efficiently.) For a general distribution Q, even one with tractable structure, running inference in the corresponding network $\mathcal{H}_Q$ can be costly, and we may want to reduce the number of calls to the inference subroutine.

This observation leads to a question of how best to perform updates for several factors in Q. We can consider two strategies. The *sequential update* strategy is similar to our strategy in the mean field algorithm: We choose a factor ψ_j, apply the fixed-point equation to that factor by running inference in $\mathcal{H}_Q$, update the distribution, and then repeat this process with another factor until convergence. The problematic aspect of this approach is that we need to perform inference after each update step. For example, if we are using cluster tree inference in the network $\mathcal{H}_Q$, the network parameterization changes after each update step, so we need to recalibrate the clique tree every time. Some of these steps can be made more efficient by selecting an appropriate order of updates and using dynamic programming (see exercise 11.27), but the process can still be quite expensive.

An alternative approach is the *parallel update* strategy, where we compute the right-hand side of our fixed-point equations (for example, equation (11.61)) simultaneously for each of the factors in Q. If we are using a cluster tree for inference, this process involves multiple queries from the same calibrated cluster tree. Thus, we can perform a single calibration step and use the resulting tree to reestimate all of our potentials. However, the different queries required all have different evidence; hence, it is not easy to obtain significant computational savings, and the algorithms needed are fairly tricky. Nevertheless, this approach might be less costly than recalibrating the clique tree J times.

On the other hand, the guarantees provided by these two update steps are different. For the sequential update strategy, we can prove that each update step is monotonic in the energy functional: each step maximizes the value of one potential given the values of all the others, and therefore is guaranteed not to decrease (and generally to increase) the energy functional. This monotonic improvement implies that iterations of sequential updates necessarily converge, generally to a local maximum. The issue of convergence is more complicated in the parallel update strategy. Because we update all the potentials at once, we have no guarantees that any fixed-point equation holds after the update; a value that was optimal for ψ_j with respect to the values of all other factors before the parallel update step is not necessarily optimal given their new values. As such, it is conceivable that parallel updates will not converge (for example, oscillate between two sets of values for the potentials). Such oscillations can generally be avoided using damped update steps, similar to these we discussed in the case of cluster-graph belief propagation (see box 11.B), but this modified procedure still does not guarantee convergence.

At this point, there is no generally accepted procedure for scheduling updates in variational methods, and different approaches are likely to be best for different applications.

11.5.2.3   Simplifying the Update Equations

Equation (11.61) provides a general characterization of the fixed points of the energy functional, for any approximating class of distributions Q obeying a particular factorization, as in equation (11.58). In many cases, we can exploit additional structure of the approximating class Q and of the distribution P_Φ to simplify significantly the form of these fixed-point equations and thereby make the update step more efficient.

The simplifications we describe take two forms. The first utilizes marginal independencies in Q to simplify the right-hand side of the fixed-point equation, equation (11.61), eliminating irrelevant terms. The second exploits interactions between the form of Q and the form of P_Φ to simplify the factorization of Q, without loss in expressive power. Both simplifications allow the fixed-point updates to be performed more efficiently. We motivate each of the simplifications using an example, and then we present the general result.

Example 11.13

Once again, consider the 4×4 grid network. Assume that we approximate it by a "row" network that has the structure shown in figure 11.17a. This approximating network consists of four independent chains. Now we can apply the general form of the fixed-point equation equation (11.61) for a specific entry in our approximation, say:

$$
\psi_{1,1}(a_{1,1}, a_{1,2}) \propto \exp \left\{ \begin{array}{l} \boldsymbol{E}_Q\left[\ln \tilde{P}_\Phi \mid a_{1,1}, a_{1,2}\right] \\ -\sum_{\substack{i=1,\ldots,4 \\ j=1,\ldots,3 \\ (i,j) \neq (1,1)}} \boldsymbol{E}_Q\left[\ln \psi_{(i,j)}(A_{i,j}, A_{i,j+1}) \mid a_{1,1}, a_{1,2}\right] \end{array} \right\}.
$$

As in the proof of corollary 11.5, the expectation of $\ln \tilde{P}_\Phi$ is the sum of expectations of the logarithm of each of the potentials in Φ. Some of these terms, however, do not depend on the choice of value of $A_{1,1}, A_{1,2}$ we are evaluating. For example, because $A_{2,1}$ and $A_{2,2}$ are independent of $A_{1,1}, A_{1,2}$ in Q, we conclude that

$$
\boldsymbol{E}_{\{A_{2,1}, A_{2,2}\} \sim Q}[\ln \phi(A_{2,1}, A_{2,2}) \mid a_{1,1}, a_{1,2}] = \boldsymbol{E}_{\{A_{2,1}, A_{2,2}\} \sim Q}[\ln \phi(A_{2,1}, A_{2,2})].
$$

Thus, this term will contribute the same value to each of the entries of $\psi(A_{1,1}, A_{1,2})$, and can therefore be absorbed into the corresponding normalizing term. We can continue in this manner and remove all terms that are not dependent on the context of the factor we are interested in. Overall, we can remove any term $\boldsymbol{E}_Q[\ln \phi(A_{i,j}, A_{i,j+1}) \mid a_{1,1}, a_{1,2}]$ and any term $\boldsymbol{E}_Q[\ln \phi(A_{i,j}, A_{i+1,j}) \mid a_{1,1}, a_{1,2}]$ except those where $i = 1$. Similarly, we can remove any term $\boldsymbol{E}_Q[\ln \psi_{(i,j)}(A_{i,j}, A_{i,j+1}) \mid a_{1,1}, a_{1,2}]$ except those where $i = 1$. These simplifications result in the following update rule:

$$
\psi_{1,1}(a_{1,1}, a_{1,2}) \propto
$$
$$
\exp \left\{ \begin{array}{l} \sum_{j=1,\ldots,3} \boldsymbol{E}_{\{A_{1,j}, A_{1,j+1}\} \sim Q}\left[\ln \phi_{(1,j)}(A_{1,j}, A_{1,j+1}) \mid a_{1,1}, a_{1,2}\right] \\ + \sum_{j=1,\ldots,4} \boldsymbol{E}_{\{A_{1,j}, A_{2,j}\} \sim Q}\left[\ln \phi_{(1,j)}(A_{1,j}, A_{2,j}) \mid a_{1,1}, a_{1,2}\right] \\ - \sum_{j=2,3} \boldsymbol{E}_{\{A_{1,j}, A_{1,j+1}\} \sim Q}\left[\ln \psi_{(1,j)}(A_{1,j}, A_{1,j+1}) \mid a_{1,1}, a_{1,2}\right] \end{array} \right\}.
$$

∎

We can generalize this analysis to arbitrary sets of factors:

Theorem 11.12

If $Q(\mathcal{X}) = \frac{1}{Z_Q} \prod_j \psi_j$, then the potential ψ_j is locally optimal only if

$$
\psi_j(\boldsymbol{c}_j) \propto \exp \left\{ \sum_{\phi \in A_j} \boldsymbol{E}_{\mathcal{X} \sim Q}[\ln \phi \mid \boldsymbol{c}_j] - \sum_{\psi_k \in B_j} \boldsymbol{E}_{\mathcal{X} \sim Q}[\ln \psi_k \mid \boldsymbol{c}_j] \right\}, \tag{11.62}
$$

where

$$
A_j = \{\phi \in \Phi : Q \not\models (\boldsymbol{U}_\phi \perp \boldsymbol{C}_j)\}
$$

and

$$
B_j = \{\psi_k : Q \not\models (\boldsymbol{C}_k \perp \boldsymbol{C}_j)\} - \{\boldsymbol{C}_j\}.
$$

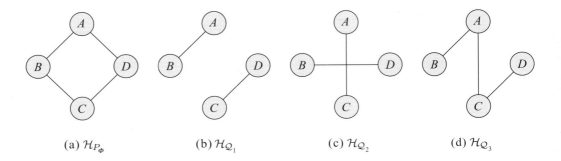

(a) $\mathcal{H}_{P_\Phi}$ (b) $\mathcal{H}_{\mathcal{Q}_1}$ (c) $\mathcal{H}_{\mathcal{Q}_2}$ (d) $\mathcal{H}_{\mathcal{Q}_3}$

Figure 11.18   A diamond network and three possible approximating structures

Stated in words, this result shows that the parameterization of a factor $\psi_j(\boldsymbol{C}_j)$ depends only on factors in P_Φ and in Q whose scopes are not independent of $\boldsymbol{C}_j$ in Q. This result, applied to example 11.13, provides us precisely with the simplification shown: only factors whose scopes intersect with the first row are relevant to $\psi_{1,1}(A_{1,1}, A_{1,2})$. Thus, we can use independence properties of the approximating family Q to simplify the right-hand side of equation (11.61) by removing irrelevant terms.

11.5.2.4   Simplifying the Family Q

It turns out that a similar analysis allows us to simplify the form of the approximating family Q without loss in the quality of the approximation.

We start by considering a simple example.

Example 11.14

Consider again the four-variable pairwise Markov network of figure 11.18a, which is parameterized by the pairwise factors:

$$P_\Phi(A, B, C, D) \propto \phi_{AB}(A, B) \cdot \phi_{BC}(B, C) \cdot \phi_{CD}(C, D) \cdot \phi_{AD}(A, D).$$

Consider applying the variational approximation with the distribution

$$Q(A, B, C, D) = \frac{1}{Z_Q} \psi_1(A, B) \cdot \psi_2(C, D) \tag{11.63}$$

that has the structure shown in figure 11.18b. Using equation (11.62), we conclude that the fixed-point characterization of ψ_1 is

$$\psi_1(a, b) \propto \exp\left\{\boldsymbol{E}_Q[\ln \phi_{AB}(A, B) \mid a, b] + \boldsymbol{E}_Q[\ln \phi_{BC}(B, C) \mid a, b] + \boldsymbol{E}_Q[\ln \phi_{AD}(A, D) \mid a, b]\right\}.$$

Can we further simplify this equation? Consider the first term. Clearly, $\boldsymbol{E}_Q[\ln \phi_{AB}(A, B) \mid a, b] = \ln \phi_{AB}(a, b)$. What about the second term, $\boldsymbol{E}_Q[\ln \phi_{BC}(B, C) \mid a, b]$? To compute this expectation, we need to compute $Q(B, C \mid a, b)$. According to the structure of Q, we can see that

$$Q(B, C \mid a, b) = \begin{cases} Q(C) & \text{If } B = b \\ 0 & \text{otherwise.} \end{cases}$$

Thus, we conclude that

$$\boldsymbol{E}_{A,B,C\sim Q}[\ln\phi_{BC}(B,C)\mid a,b] = \boldsymbol{E}_{C\sim Q}[\ln\phi_{BC}(b,C)].$$

We can simplify the third term in exactly the same way, concluding that:

$$\psi_1(a,b) \propto \exp\left\{\ln\phi_{AB}(a,b) + \boldsymbol{E}_{C\sim Q}[\ln\phi_{BC}(b,C)] + \boldsymbol{E}_{D\sim Q}[\ln\phi_{AD}(a,D)]\right\}.$$

Setting $\psi_1'(a) = \exp\{\boldsymbol{E}_{D\sim Q}[\ln\phi_{AD}(a,D)]\}$ *and* $\psi_1''(b) = \exp\{\boldsymbol{E}_{C\sim Q}[\ln\phi_{BC}(b,C)]\}$, *we conclude that the optimal* ψ_1 *factorizes as a product of three factors:*

$$\psi_1(A,B) = \phi_{AB}(A,B) \cdot \psi_1'(A) \cdot \psi_1''(B).$$

Have we gained anything from this decomposition? First, we see that Q *preserves the original pairwise interaction term* $\phi(A,B)$ *from* P_Φ. *Moreover, the effect of the interactions between these variables and the rest of the network (C and D in this example) is summarized by a univariate factor for each of the variables. Thus,* Q *does not change the interaction between* A *and* B.

Applying the same set of arguments to ψ_2, *we conclude that we can rewrite* Q *as*

$$Q'(A,B,C,D) = \frac{1}{Z_Q}\phi_{AB}(A,B) \cdot \phi_{CD}(C,D) \cdot \psi_1'(A) \cdot \psi_1''(B) \cdot \psi_2'(C) \cdot \psi_2''(D). \quad (11.64)$$

The preceding discussion shows that the best approximation to P_Φ *within* Q *can be rewritten in the form of* Q'. *Thus, there is no point in using the more complicated form of the approximating family of equation (11.63); we may as well use the form of* Q' *in equation (11.64). Note that the form of* Q' *involves a product of a subset of the original factors, which we keep intact without change, and a set of new factors, which we need to optimize. In this example, instead of estimating two pairwise potentials, we estimate four univariate potentials, which utilize a smaller number of parameters.*

Moreover, the update equations for Q' *are simpler. Consider, for example, applying equation (11.62) for* ψ_1':

$$\ln\psi_1'(a) \propto \boldsymbol{E}_{B\sim Q'}[\ln\phi_{AB}(a,B)\mid a] + \boldsymbol{E}_{D\sim Q'}[\ln\phi_{AD}(a,D)\mid a] + \boldsymbol{E}_{B,C\sim Q'}[\ln\phi_{BC}(B,C)\mid a]$$
$$- \boldsymbol{E}_{B\sim Q'}[\ln\phi_{AB}(a,B)\mid a] - \boldsymbol{E}_{B\sim Q'}[\ln\psi_1''(B)\mid a]$$
$$= \boldsymbol{E}_{D\sim Q'}[\ln\phi_{AD}(a,D)\mid a] + \boldsymbol{E}_{B,C\sim Q'}[\ln\phi_{BC}(B,C)\mid a] - \boldsymbol{E}_{B\sim Q'}[\ln\psi_1''(B)\mid a].$$

The terms involving $\boldsymbol{E}_{Q'}[\ln\phi_{AB}\mid a]$ *appear twice, once as a factor in* P_Φ *and once as a factor in* Q'. *These two terms cancel out, and we are left with the simpler update equation. Although this equation does not explicitly mention* ϕ_{AB}, *this factor participates in the computation of* $Q'(B\mid a)$ *that implicitly appears in* $\boldsymbol{E}_{B\sim Q'}[\ln\psi_1''(B)\mid a]$. ∎

Note that this result is somewhat counterintuitive, since it shows that the interactions between A and B are captured by the original potential in P_Φ. Intuitively, we would expect the chain of influence A—D—C—B to introduce additional interactions between A and B that should be represented in Q. This is not the only counterintuitive result.

Example 11.15

Consider another approximating family for the same network, using the network structure shown in figure 11.18c. In this approximation, we have two pairwise factors, $\psi_1(A, C)$, and $\psi_2(B, D)$. Applying the same set of arguments as before, we can show that the update equation can be written as

$$
\begin{aligned}
\ln \psi_1(a, c) \quad \propto \quad & \boldsymbol{E}_{B \sim Q}[\ln \phi_{AB}(a, B)] + \boldsymbol{E}_{D \sim Q}[\ln \phi_{AD}(a, D)] \\
& + \boldsymbol{E}_{B \sim Q}[\ln \phi_{BC}(B, c)] + \boldsymbol{E}_{D \sim Q}[\ln \phi_{CD}(c, D)].
\end{aligned}
$$

Thus, we can factorize ψ_1 into two factors, one with a scope of A and the other with C

$$
\psi_1(A, C) = \psi_1'(A) \cdot \psi_1''(C).
$$

In other words, the approximation in this case is equivalent to the mean field approximation. This result shows that, in some cases, we can remove spurious dependencies in the approximating distribution. However, this result is surprising, since it holds regardless of the actual values of the potentials in P_Φ. And so, we can imagine a network where there are very strong interactions between A and C and between B and D in P_Φ, and yet the variational approximation with a network structure of figure 11.18c will not capture these dependencies. This is a consequence of using I-projections. Had we used an M-projection that minimizes $\boldsymbol{D}(P_\Phi \| Q)$, then we would have represented the dependencies between A and C; see exercise 11.30. ∎

These two examples suggest that we can use the fixed-point characterization to refine an initial approximating network by factorizing its factors into a product of, possibly smaller, factors and potentials from P_Φ. We now consider the general theory of such factorizations and then discuss its implications.

We start with a simple definition and a proposition that form the basis of the simplifications we consider.

Definition 11.8

interface

Let $\mathcal{H}$ be a Markov network structure and let $\boldsymbol{X}, \boldsymbol{Y} \subseteq \mathcal{X}$. We define the $\boldsymbol{Y}$-interface of $\boldsymbol{X}$, denoted $\mathrm{Interface}_{\mathcal{H}}(\boldsymbol{X}; \boldsymbol{Y})$, to be the minimal subset of $\boldsymbol{X}$ such that $\mathrm{sep}_{\mathcal{H}}(\boldsymbol{X}; \boldsymbol{Y} \mid \mathrm{Interface}_{\mathcal{H}}(\boldsymbol{X}; \boldsymbol{Y}))$. ∎

That is, the $\boldsymbol{Y}$-interface of $\boldsymbol{X}$ is the subset of $\boldsymbol{X}$ that suffices to separate it from $\boldsymbol{Y}$.

Example 11.16

The $\{A, D\}$-interface of $\{A, B\}$ in $\mathcal{H}_{P_\Phi}$ of figure 11.18 is $\{A, B\}$, since neither A is separated from $\{A, D\}$ given B, nor is B is separated from $\{A, D\}$ given A. In $\mathcal{H}_{\mathcal{Q}_1}$, we have that B is separated from $\{A, D\}$ given A, so that $\mathrm{Interface}_{\mathcal{H}_{\mathcal{Q}_1}}(\{A, B\}; \{A, D\})$ is $\{A\}$. The same holds in $\mathcal{H}_{\mathcal{Q}_3}$. In $\mathcal{H}_{\mathcal{Q}_2}$, we have that, again, neither A nor B suffices to separate the other from $\{A, D\}$, and hence, $\mathrm{Interface}_{\mathcal{H}_{\mathcal{Q}_2}}(\{A, B\}; \{A, D\}) = \{A, B\}$. ∎

The definition of interface can be used to reduce the scope of the conditional expectations in the fixed-point equations:

Proposition 11.6

If $\mathcal{H}$ is an I-map of $Q(\mathcal{X}) = \frac{1}{Z_Q} \prod_j \psi_j$ and ϕ is a potential with scope $\boldsymbol{U}_\phi$. Then,

$$
\boldsymbol{E}_{\boldsymbol{U}_\phi \sim Q}[\phi \mid \boldsymbol{c}_j] = \boldsymbol{E}_{\boldsymbol{U}_\phi \sim Q}[\phi \mid \boldsymbol{c}_j \langle \mathrm{Interface}_{\mathcal{H}}(\boldsymbol{C}_j; \boldsymbol{U}_\phi) \rangle].
$$

The proof follows immediately from the definition of conditional independence.

This proposition provides a principled approach for reformulating terms on the right-hand side of the fixed-point equation.

☞ **We can use this simplification result to define a two-phase strategy for designing approximation. First, we define a "rough" outline for approximation by defining Q over factors with a fairly large scope. We use this outline to obtain a set of update equations, as implied by equation (11.62) on Q. We then derive a finer-grained representation by factorizing each of these factors using proposition 11.6. This process results in a finer-grained approximation that is provably equivalent to the one with which we started.**

Theorem 11.13

(Factorization) *Let Q be an approximating family defined in terms of factors $\{\psi_j(C_k)\}$, which induce a Markov network structure $\mathcal{H}_Q$. Let $Q \in Q$ be a stationary point of the energy functional $F[\tilde{P}_\Phi, Q]$ subject to the given factorization. Then, factors in Q are factorized as*

$$\psi_j(C_j) = \prod_{\phi \in \Phi_j} \phi \prod_{D_l \in \mathcal{D}_j} \psi_{j,l}(D_l), \tag{11.65}$$

where

$$\Phi_j = \{\phi \in \Phi : Scope[\phi] \subseteq C_j\}$$

and

$$\mathcal{D}_j = \{Interface_{\mathcal{H}_Q}(C_j; X) : X \in \{Scope[\phi] : \phi \in \Phi - \Phi_j\} \cup \{Scope[\psi_k] : k \neq j\}\}.$$

This theorem states that ψ_j can be written as the product of two sets of factors. The first set contains factors in the original distribution P_Φ whose scope is a subset of the scope of ψ_j. The factors in the second set are the interfaces of ψ_j with other factors that appear in the update equation. These include factors in P_Φ that are partially "covered" by the scope of ψ_k, and other factors in Q. The set $\mathcal{D}_k$ defines the set of interfaces between ψ_k and these factors.

To gain a better understanding of this theorem, let us consider various approximations in two concrete examples. The first example serves to demonstrate the ease with which this theorem allows us to determine the form of the factorization of Q.

Example 11.17

Let us return to example 11.14. In example 11.16, we have already shown the interfaces of $\{A, B\}$ with $\{A, D\}$ in $\mathcal{H}_1$. This analysis, together with theorem 11.13, directly imply the reduced factorization of example 11.14 In particular, for $\psi_1(\{A, B\})$, we have that Φ_1 contains only the factor $\phi(\{A, B\})$ in P_Φ, which therefore constitutes the first term in the factorization of equation (11.65). The second set of terms in the equation corresponds to the interfaces of $\{A, B\}$ with other factors in both $\mathcal{H}_{P_\Phi}$ and in $\mathcal{H}_{Q_1}$. We get two such interfaces: one with scope $\{A\}$ from the factor $\phi(\{A, D\})$ in P_Φ, and one with scope $\{B\}$ from the factor $\phi(\{B, C\})$.

Assume that we add the edge A—C, as in figure 11.18d. Now, $Interface_{\mathcal{H}_{Q_3}}(\{A, B\}; \{B, C\})$ is the entire set $\{A, B\}$, since B no longer separates C from A. Thus, in this case, the second set of terms in the factorization of ψ also contains a new pairwise interaction factor $\psi_{1,\{A,B\}}$. As a consequence, the pairwise interaction of A, B is no longer the same in Q and in P_Φ. This result is somewhat counterintuitive: In the simpler network $\mathcal{H}_{Q_1}$, which contained no factors allowing any interaction between the A, B pair and the C, D pair, the A, B interaction was the same in P_Φ and *in Q. But if we enrich our approximation (presumably allowing a better fit via the introduction of the A, C factor), the pairwise interaction term does change.*

Finally, $\mathcal{H}_{Q_2}$ does not contain an $\{A, B\}$ factor. Here, $\Phi_j = \emptyset$ for both factors in $\mathcal{H}_{Q_2}$, and each $\mathcal{D}_j$ consists solely of singleton scopes; for example, $\text{Interface}_{\mathcal{H}_{Q_2}}(\{A, C\}; \{A, D\}) = \{A\}$. ∎

Our second example serves to illustrate the two-phase strategy described earlier, where we first select a "rough" approximation containing a few large factors and then use the theorem to refine them.

Example 11.18

Consider again our running example of the 4×4 grid. Suppose we select an approximation where each factor consists of the variables in a single row in the grid. Thus, for example, $C_1 = \{A_{1,1}, \ldots, A_{1,4}\}$. Note that this approximation is not the one shown in figure 11.17a, since the structure in our approximation here is a full clique over each row. We now apply theorem 11.13. What is the factorization of C_1? First, we search for factors in Φ_1. We see that the factors $\phi(A_{1,1}, A_{1,2})$, $\phi(A_{1,2}, A_{1,3})$, and $\phi(A_{1,3}, A_{1,4})$ have a scope that is a subset of C_1. Next, we consider the interfaces between C_1 and other factors in P_Φ and Q. For example, the interface with $\phi(A_{1,1}, A_{2,1})$ is $\{A_{1,1}\}$. Similarly, $\{A_{1,2}\}$, $\{A_{1,3}\}$, and $\{A_{1,4}\}$ are interfaces with other factors in P_Φ. It is easy to convince ourselves that these are the only non-empty interfaces in $\mathcal{I}_1$. Thus, by applying theorem 11.13, we get the following factorization:

$$\psi_1(A_{1,1}, \ldots, A_{1,4}) = \phi(A_{1,1}, A_{1,2}) \cdot \phi(A_{1,2}, A_{1,3}) \cdot \phi(A_{1,3}, A_{1,4})$$
$$\psi_{1,1}(A_{1,1}) \cdot \psi_{1,2}(A_{1,2}) \cdot \psi_{1,3}(A_{1,3}) \cdot \psi_{1,4}(A_{1,4}).$$

We conclude that, once we decide that the approximation should decouple the rows in the group, we might as well work with an approximation where we keep all original potentials along each row and introduce univariate potentials only to capture interactions along columns. Additional potentials, such as a potential between $A_{1,1}$ and $A_{1,3}$, would not improve the approximation. Thus, while we started with an approximation containing full cliques on each of the rows, we ended up with an approximation whose structure is that of figure 11.17a, and where we have only the original factors and new factors over single variables.

We can work directly with this new factorized form of Q, ignoring our original factorization entirely. More precisely, we define Q' to be

$$Q'(\mathcal{X}) = \phi(A_{1,1}, A_{1,2}) \cdot \phi(A_{1,2}, A_{1,3}) \cdot \phi(A_{1,3}, A_{1,4})$$
$$\cdots$$
$$\phi(A_{4,1}, A_{4,2}) \cdot \phi(A_{4,2}, A_{4,3}) \cdot \phi(A_{4,3}, A_{4,4})$$
$$\psi_{1,1}(A_{1,1}) \cdot \ \cdot \psi_{4,4}(A_{4,4}).$$

In this new form, we fix the value of all the pairwise potentials, and so we have to define an update rule only for the new singleton potentials. For example, consider the fixed-point equation for $\psi_{1,1}(A_{1,1})$. Applying theorem 11.12 we get that

$$\ln \psi_{1,1}(a_{1,1}) \propto$$
$$+ \boldsymbol{E}_{Q'}[\ln \phi(A_{1,1}, A_{2,1}) \mid a_{1,1}] + \boldsymbol{E}_{Q'}[\ln \phi(A_{1,2}, A_{2,2}) \mid a_{1,1}]$$
$$+ \boldsymbol{E}_{Q'}[\ln \phi(A_{1,3}, A_{2,3}) \mid a_{1,1}] + \boldsymbol{E}_{Q'}[\ln \phi(A_{1,4}, A_{2,4}) \mid a_{1,1}]$$
$$- \boldsymbol{E}_{Q'}[\ln \psi_{1,2}(A_{1,2}) \mid a_{1,1}] - \boldsymbol{E}_{Q'}[\ln \psi_{1,3}(A_{1,3}) \mid a_{1,1}] - \boldsymbol{E}_{Q'}[\ln \psi_{1,4}(A_{1,4}) \mid a_{1,1}]$$

where we have exploited the fact that the terms involving factors such as $\phi(A_{1,1}, A_{1,2})$ appear in both P_Φ and Q, and so cancel out of the equation. Note that to compute terms such as $\boldsymbol{E}_{Q'}[\ln \phi(A_{1,2}, A_{2,2}) \mid a_{1,1}]$ we need to evaluate $Q'(A_{1,2}, A_{2,2} \mid a_{1,1}) = Q'(A_{1,2} \mid a_{1,1}) \cdot Q'(A_{2,2})$ (where we used the independencies in Q' to simplify the joint marginal). Note that $Q'(A_{2,2})$ does not change when we update factors in the first row, such as $\psi_{1,1}(A_{1,1})$. Thus, we can cache the computation of this marginal when updating the factors $\psi_{1,1}(A_{1,1}), \ldots, \psi_{1,4}(A_{1,4})$. When performing inference in a large model this can result in dramatic effect. ∎

cluster mean field

This example is a special case of an approximation approach called *cluster mean field*. In this case, our initial approximation has the form

$$Q(\mathcal{X}) = \frac{1}{Z_Q} \prod_j \psi_j(\boldsymbol{C}_j),$$

where the scopes $\boldsymbol{C}_1, \ldots, \boldsymbol{C}_K$ are partition of $\mathcal{X}$. That is, each pair of factors have disjoint scopes, and each variable in $\mathcal{X}$ appears in one factor. This approximation resembles the mean field approximation, except that it is clusters, rather than individual variables, that are marginally independent. We can now apply theorem 11.13 to refine the approximation. Because the factors are all disjoint, there are no chains of influence, and so the interfaces take a particularly simple form:

Proposition 11.7

Let $Q(\mathcal{X}) = \frac{1}{Z_Q} \prod_j \psi_j(\boldsymbol{C}_j)$ be a cluster mean field approximation to a set of factors P_Φ, and let ψ_j be a factor of Q. Then, the set $\mathcal{D}_j$ of theorem 11.13 can be written as

$$\mathcal{D}_j = \{\boldsymbol{C}_j \cap Scope[\phi] : \phi \in \Phi - \Phi_j\} - \{\emptyset\}.$$

The proof follows directly from the independence properties in Q, and is left as an exercise (exercise 11.31).

In words, this result states that the interfaces of a cluster are simply the places where the cluster scope intersects potentials in Φ that are not fully contained in the cluster. In our grid example, when we choose the clusters to be the individual columns, the interfaces are the intersections with the row potentials, which are precisely the singleton variables that we discussed in example 11.18.

We conclude this discussion with a slightly more elaborate example, demonstrating again the strength of this result:

Example 11.19

Consider again our 4×4 grid, and the "comb" approximation whose structure is shown in figure 11.19a. In this structure, we have a fully connected clique over each of the columns, and a "backbone" connecting the columns to each other. Consider again the factorization of the potential over $\boldsymbol{C}_1 = \{A_{1,1}, \ldots, A_{4,1}\}$. As in the previous example, the first term in the new factorization contains the pairwise factors $\phi(A_{1,1}, A_{2,1})$, $\phi(A_{2,1}, A_{3,1})$, and $\phi(A_{3,1}, A_{4,1})$. The second set of terms contains the interfaces with other factors in P_Φ and Q. Due to the structure of the approximation, the Q interfaces introduce only singleton potentials. The factors in P_Φ, however, are more interesting. Consider, for example, the factor $\phi(A_{4,1}, A_{4,2})$. The interface of $\boldsymbol{C}_1$ with $\{A_{4,1}, A_{4,2}\}$ is $A_{1,1}, A_{4,1}$ — the variable $A_{4,1}$ separates $\boldsymbol{C}_1$ from itself, and the variable $A_{1,1}$ from $A_{4,2}$. Now, consider a factor $\phi(A_{2,3}, A_{3,3})$; in this case, the interface is simply $A_{1,1}$, which separates the first column from both of these variables. Continuing this argument, it follows that all other factors in

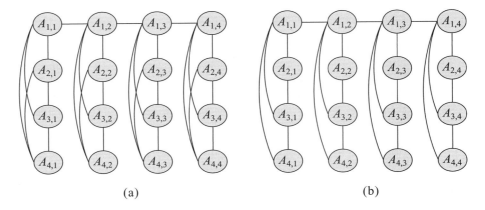

Figure 11.19 Simplification of approximating structure in cluster mean field. (a) Example of an approximating structure we can use in a variational approximation of a 4×4 grid network. (b) Simplification of that network using theorem 11.13.

P_Φ induce an interface containing a variable at the head of the column and (possibly) another variable in the column. Thus, we can eliminate any (new) pairwise interaction terms between any other pair of variables. For a general $n \times n$ grid, this reduces the overall number of (new) pairwise potentials from $n \cdot \binom{n}{2}$ to $n \times (n-1)$.   ■

11.5.2.5 Selecting the Approximation

In general, both the quality and the computational complexity of the variational approximation depend on the structure of P_Φ and the structure of the approximating family $\mathcal{Q}$. There are several guiding intuitions. First, we want to be able to perform efficient inference in the approximating network. In example 11.18, the approximating structure was a chain of variables, where we can perform inference in linear time (as a function of the number of variables in the chain). In general, we often select our network so that the resulting factorization leads to a tractable network (that is, one of low tree-width).

It is important to note, however, that the structure of the original distribution is not the only aspect in determining the complexity of inference in $\mathcal{Q}$. We also need to take into account factors that correspond to the interfaces of the cluster. In our grid example, these interfaces involved a single variable at time, and so they did not add to the network complexity. However, in more complex networks, these factors can have a significant effect.

Another consideration besides computational complexity is the quality of our approximation. Intuitively, we should design $\mathcal{Q}$ so as to preserve the strong dependencies in P_Φ. By preserving such dependencies we maintain the main effects in the distribution we want to apply.

These intuitions provide some guidelines in choosing the approximating distribution. However, these choices are far from an exact science at this stage. The theory we described here allows to automate two parts of the process: defining the form of the approximation given some initial rough set of (disjoint or overlapping) clusters; and defining the fixed-point iterations to

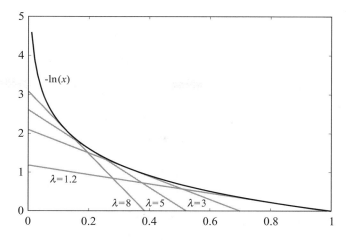

Figure 11.20 **Illustration of the variational bound** $-\ln(x) \geq -\lambda x + \ln(\lambda) + 1$. For any x, the bound holds for all λ, and is tight for some value of λ.

optimize such an approximation. The current tools do not provide for an automated way for determining what are reasonable sets of clusters to achieve a desired degree of approximation.

11.5.3 Local Variational Methods ★

variational method

lower bound

The general method that we used throughout this chapter is an instance of a general class of methods known as *variational methods*. In this class of methods, we take a complex objective function $f_{\mathrm{obj}}(x)$, and lower or upper bound it using a parameterized family of functions $g(x, \lambda)$. Focusing, for concreteness, on the case of a *lower bound*, this family has the property that $f_{\mathrm{obj}}(x) \geq g(x, \lambda)$ for any value of λ, and that, for any x, the bound is tight for some value of λ (a different one for every x).

variational lower bound

As an example, we can show the *variational lower bound*:

Lemma 11.2

For any choice of λ and x

$$-\ln(x) \geq -\lambda x + \ln(\lambda) + 1,$$

and, for any x, this bound is tight for some value of λ.

PROOF Consider the tangent of $\ln(x)$ at the point x_0

$$f_{\mathrm{obj}}(x : x_0) = \ln(x_0) + (x - x_0)\frac{1}{x_0} = \frac{x}{x_0} + \ln(x_0) - 1.$$

Since $\ln(x)$ is a concave function, it is upper bounded by each of its tangents. And so, $-\ln(x) \geq -f_{\mathrm{obj}}(x : x_0)$ for any choice of x and x_0. Setting $x_0 = \lambda^{-1}$ leads to the desired result. ∎

convex duality

This result is illustrated in figure 11.20. It is a special case of a general result in the field of *convex duality*, which guarantees the existence of such bounds for a broad class of functions.

variational parameter

This lower bound allows to approximate a nonlinear function $-\ln(x)$ with a term that is linear in x. This simplification comes at the price of introducing a new *variational parameter* λ, whose value is undetermined. If we optimize λ exactly for each value of x, we obtain a tight lower bound, but a bound is obtained for any value of λ.

The techniques we have used in this chapter so far also fall into this category. Equation (11.5) shows that the energy functional is a lower bound on the log-partition function for any distribution Q. Thus, we can take f_{obj} to be the partition function, $\boldsymbol{x}$ to correspond to the parameters of the true distribution P_Φ, and $\boldsymbol{\lambda}$ to correspond to the parameters of the approximating distribution Q. Although the lower bound is tight when Q precisely represents P_Φ, for reasons of efficiency, we generally optimize Q in a restricted space that provides a bound, but not a tight one, on the log-partition function.

This general approach of introducing auxiliary variational parameters that help in simplifying a complex objective function appears in many other domains. While it is beyond our scope to introduce a general theory of variational methods, we now briefly describe one other application of variational methods that is relevant to probabilistic inference and does not fall directly within the scope of optimizing the energy functional. This application arises in the context of exact inference using an algorithm such as variable elimination. Here, we use variational bounds to avoid creating large factors that can lead to exponential complexity in the algorithm, giving rise to an approximate *variational variable elimination* algorithm. Such simplifications can be achieved in several ways; we describe two.

variational variable elimination

11.5.3.1   Variational Bounds

Consider, for example, the diamond network of figure 11.18a. Assume that we run variable elimination to sum out the variable B, which we assume for convenience is binary-valued. The elimination of B introduces a new factor:

$$\phi_B(A, C) = \sum_b \phi_1(A, b)\phi_2(b, C)$$

Coupling A and C in a single factor may be expensive, for example, if A and C have many values. In more complex networks, this type of coupling can induce complexity if an elimination step couples a larger set of variables, or if the local coupling leads to additional cost later in the computation, when we eliminate A or C.

As we now show, we can use a variational bound to avoid this coupling. Consider the following bound:

Proposition 11.8

If $0 \le \lambda \le 1$, *then*

$$\ln(1 + e^x) \ge \lambda x + \boldsymbol{H}(\lambda),$$

where $\boldsymbol{H}(\lambda) = -\lambda \ln \lambda - (1 - \lambda) \ln(1 - \lambda).$

This bound implies that

$$1 + e^x \ge e^{\lambda x + \boldsymbol{H}(\lambda)}.$$

Why is this useful? Using some algebraic manipulation, we can bound each of the entries in our newly generated factor:

$$
\begin{aligned}
\phi_B(a,c) &= \phi_1(b^0,a)\phi_2(b^0,c) + \phi_1(b^1,a)\phi_2(b^1,c) \\
&= \phi_1(b^0,a)\phi_2(b^0,c)\left[1 + \exp\left\{\ln\frac{\phi_1(b^1,a)\phi_2(b^1,c)}{\phi_1(b^0,a)\phi_2(b^0,c)}\right\}\right] \\
&\geq \phi_1(b^0,a)\phi_2(b^0,c)\exp\left\{\lambda_{a,c}\ln\frac{\phi_1(b^1,a)\phi_2(b^1,c)}{\phi_1(b^0,a)\phi_2(b^0,c)} + \boldsymbol{H}(\lambda_{a,c})\right\} \\
&= \left(\phi_1(b^0,a)^{1-\lambda_{a,c}}\phi_1(b^1,a)^{\lambda_{a,c}}\right) \cdot \\
&\qquad \left(\phi_2(b^0,c)^{1-\lambda_{a,c}}\phi_2(b^1,c)^{\lambda_{a,c}}\right) \cdot e^{\boldsymbol{H}(\lambda_{a,c})}.
\end{aligned}
\tag{11.66}
$$

Thus, we can replace a factor that couples A and C by a product of three terms: an expression involving only factors of A, an expression involving only factors of C, and the final factor $e^{\boldsymbol{H}(\lambda_{a,c})}$. However, all three terms also involve the variational parameter $\lambda_{a,c}$ and therefore also depend on both A and C. At this point, it is unclear what we gain from the transformation.

However, we can choose the same λ for all joint assignments to A, C. In doing so, we replace four variational parameters by a single parameter λ. This operation relaxes the bound, which is no longer tight. On the other hand, it also decouples A and C, leading to a product of terms none of which depends on both variables:

$$
\left(\phi_1(b^0,a)^{1-\lambda}\phi_1(b^1,a)^{\lambda}\right) \cdot \left(\phi_2(b^0,c)^{1-\lambda}\phi_2(b^1,c)^{\lambda}\right) \cdot e^{\boldsymbol{H}(\lambda)} = \tilde{\phi}_1(a,\lambda)\tilde{\phi}_2(c,\lambda)e^{\boldsymbol{H}(\lambda)}.
\tag{11.67}
$$

Thus, if we use this approximation, we have effectively eliminated B without coupling A and C. As we saw in chapter 9, this type of simplification can circumvent the need for coupling yet more variables in later stages in variable elimination, potentially leading to significant savings.

It is interesting to observe how λ decouples the two factors. Each factor is replaced by a geometric average over the values of B. The variational parameter specifies the weight we assign to each of the two cases. Note that the original bound in equation (11.66) is tight; thus, if we pick the "right" variational parameter $\lambda_{a,c}$ for each assignment a, c, we reproduce the correct factor $\phi_B(A,C)$. However, these variational parameters are generally different for each assignment a, c, and hence, a single variational parameter cannot optimize all of the terms. Our choice of λ effectively determines the quality of our approximation for each of the terms $\phi_B(a,c)$. Thus, the overall quality of our approximation for a particular choice of λ depends on the importance of these different terms in the variable elimination computation as a whole.

Other variational approximations exploit specific parametric forms of CPDs in the network. For example, consider networks with sigmoid CPDs (see section 5.4.2). Recall that a logistic CPD $P(X \mid \boldsymbol{U})$ has the parametric form:

$$
P(x^1 \mid \boldsymbol{u}) = \text{sigmoid}\left(\sum_i w_i u_i + w_0\right),
$$

where $\text{sigmoid}(x) = \frac{1}{1+e^{-x}}$. The observation of X couples the parents $\boldsymbol{U}$. Can we decouple these parents using an approximation? Using proposition 11.8 we can find an upper bound:

$$
\ln\text{sigmoid}\left(\sum_i w_i u_i + w_0\right) \leq \lambda\left(\sum_i w_i u_i + w_0\right) - \boldsymbol{H}(\lambda).
\tag{11.68}
$$

Similarly to our earlier example, such an approximation allows us to replace a factor over several variables by a product of smaller factors. In this case, all the parents of X are decoupled by the approximate form.

11.5.3.2 Variational Variable Elimination

How do we use this approximation in the course of inference? Note that equation (11.67) provides a lower bound to $\phi_B(a, c)$ for every value of a, c. Assume that, in the course of running variable elimination, rather than generating $\phi_B(A, C)$, we introduce the expression in equation (11.67) and continue the variable elimination process with these decoupled factors. From a graph-theoretic perspective, the result of this approximation when applied to a variable B is analogous to the effect of conditioning on B, as described in section 9.5.3: it removes from the graph B and all its adjacent edges. However, unlike conditioning, we do not enumerate and perform inference for all values of B (of course, at the cost of obtaining an approximate result).

More generally, in an execution of variable elimination, there may be some set of elimination steps that create large factors that couple many variables. This variational approximation can allow us to avoid this coupling. Like conditioning (see section 9.5.4.1), we can perform such an approximation step not only at the very beginning, but also in a way that is interleaved with variable elimination, allowing us to reuse computation. This class of algorithms is called

variational
variable
elimination

variational variable elimination.

What is the result of this approximation? Each of the entries in the approximated factor is replaced with a lower bound; thus, each entry in every subsequent factor produced by the algorithm is also a lower bound to the original entry. If we proceed to eliminate all variables, either exactly or using additional variational approximation steps for other intermediate factors, the outcome of this process is a lower bound to the partition function. If we do not eliminate all of the variables, the result is an approximate factor in which every entry is a lower bound to the original. Of course, once we renormalize the factor to produce a distribution, we can make no guarantees about the direction of the approximation for any given entry. Nevertheless, the resulting factor might be a reasonable approximation to the original.

The quality of our approximation depends on the choice of variational parameters introduced during the course of the variable elimination algorithm. How do we select them? One approach is simply to select the variational parameter at each step to optimize the quality of our approximation at that step. However, this high-level goal is not fully defined. For example, we can choose λ so as to make $\tilde{\phi}_1(a^1, \lambda)\tilde{\phi}_2(b^1, \lambda)e^{\boldsymbol{H}(\lambda)}$ as close as possible to $\phi_B(a^1, c^1)$; or, we can focus on a^1, c^0. The decision of where to focus our "approximation effort" depends on the impact of these components of the factor on the final outcome of the computation. Thus, a more correct approach is to identify the actual expression that we are trying to estimate — for example, the partition function — and to try to maximize our bound to that expression. In our simple example, we can write down the partition function as a function of the variational parameter λ introduced when eliminating B:

$$\tilde{Z}(\lambda) = e^{\boldsymbol{H}(\lambda)} \sum_a \sum_c \phi_3(a, d)\phi_4(c, d) \sum_d \tilde{\phi}_1(a, \lambda)\tilde{\phi}_2(c, \lambda).$$

This expression is a function of λ; we can then try to identify the best bound by maximizing $\max_\lambda \tilde{Z}(\lambda)$, say, using gradient ascent or another numerical optimization method.

However, as we discussed, in most cases we would use several approximate elimination steps within the variable elimination algorithm. In our example, the elimination of D also couples A and C, a situation we may wish to avoid. Thus, we could apply the same type of variational bound to the internal summation:

$$\phi_D(A, C) = \sum_d \tilde{\phi}_1(a, \lambda)\tilde{\phi}_2(c, \lambda),$$

giving rise to the bound $\tilde{\phi}_3(a, \lambda')\tilde{\phi}_4(c, \lambda')e^{\boldsymbol{H}(\lambda')}$. The resulting approximate partition function now has the form

$$\tilde{Z}(\lambda, \lambda') = e^{\boldsymbol{H}(\lambda)}e^{\boldsymbol{H}(\lambda')} \sum_a \tilde{\phi}_1(a, \lambda)\tilde{\phi}_3(a, \lambda') \sum_c \tilde{\phi}_2(c, \lambda)\tilde{\phi}_4(c, \lambda').$$

We can now maximize $\max_{\lambda, \lambda'} \tilde{Z}(\lambda, \lambda')$; the higher the value we find, the better our approximation to the true partition function.

In general, our approximate partition function will be a function $\tilde{Z}(\boldsymbol{\lambda})$, where $\boldsymbol{\lambda}$ is the vector of all variational parameters λ produced during the different approximation steps in the algorithm. One approach is to reformulate the variable elimination algorithm so that it produces factors that are not purely numerical, but rather symbolic expressions in the variables $\boldsymbol{\lambda}$; see exercise 11.32. Given the multivariate function $\tilde{Z}(\boldsymbol{\lambda})$, we can optimize it numerically to produce the best possible bound. A similar principle applies when we use the variational bound for sigmoid CPDs; see exercise 11.36 for an example. While we only sketch the basic idea here, this approach can be used as the basis for an algorithm that interleaves variational approximation steps with exact elimination steps to form a variational variable elimination algorithm.

11.6 Summary and Discussion

In this chapter, we have described a general class of methods for performing approximate inference in a distribution P_Φ defined by a graphical model. These methods all attempt to construct a representation Q within some approximation class $\mathcal{Q}$ that best approximates P_Φ. The key issue that must be tackled in this task is the construction of such an approximation without performing inference on the (intractable) P_Φ. The methods that we described all take a similar approach of using the I-projection framework, whereby we minimize the KL-divergence $\boldsymbol{D}(Q\|P_\Phi)$; these methods all reformulate this problem as one of maximizing the energy functional.

The different methods that we described all follow a similar template. We optimize the free energy, or an approximation thereof, over a class of representations $\mathcal{Q}$. We provided an optimization-based view for four different methods, each of which makes a particular choice regarding the objective and the constraints. We now recap these choices and their repercussions.

Clique tree calibration optimizes the factored energy functional, which is exact for clique trees. The optimization is performed over the space of calibrated clique potentials. For clique trees, any set of calibrated clique potentials must arise from a real distribution, and so this space is precisely the marginal polytope. As a consequence, both our objective and our constraint space are exact, so our solution represents the exact posterior.

Cluster-graph (loopy) belief propagation optimizes the factored energy functional, which is approximate for loopy graphs. The optimization is performed over the space of locally consistent pseudo-marginals, which is a relaxation of the marginal polytope. Thus, both the objective and the constraint space are approximate.

Expectation propagation over clique trees optimizes the factored energy functional, which is the exact objective in this case. However, the constraints define a space of pseudo-marginals that are not even entirely consistent — only their moments are required to match. Thus, the constraint space here is a relaxation of our constraints, even when the structure is a tree. Expectation propagation over cluster graphs adds, on top of these, all of the approximations induced by belief propagation.

Finally, structured variational methods optimize the exact factored energy functional, for a class of distributions that is (generally) less expressive than necessary to encode P_Φ. As a consequence, the constraint space is actually a tightening of our original constraint space. Because the objective function is exact, the result of this optimization provides a lower bound on the value of the exact optimization problem. As we will see, such bounds can play an important role in the context of learning.

In the approaches we discussed, the optimization method was based on the method of Lagrange multipliers. This derivation gave rise to a series of fixed-point equations for the solution, where one variable is defined in terms of the others. As we showed, an iterative solution for these fixed-point equations gives rise to a suite of message passing algorithms that generalize the clique-tree message passing algorithms of chapter 10.

This general framework opens the door to the development of many other approaches. **Each of the methods that we described here involves a design choice in three dimensions: the objective function that we aim to optimize, the space of (pseudo-)distributions over which we perform our optimization, and the algorithm that we choose to use in order to perform the optimization. Although these three decisions affect each other, these dimensions are sufficiently independent that we can improve each one separately.** Some recent work takes exactly this approach. For example, we have already seen some work that focuses on better approximations to the energy functional; other work (see section 11.7) focuses on identifying constraints over the space of pseudo-marginals that make it a tighter relaxation to the (exact) marginal polytope; yet other work aims to find better (for example, more convergent) algorithms for a general class of optimization problems.

Many of these improvements aimed to address a fundamental problem arising in these algorithms: the possible lack of convergence to a stable solution. The issue of convergence is one on which some progress has been made. Some recently developed methods have better convergence properties in practice, and others are even guaranteed to converge. There are also theoretical analyses that can help determine when the algorithms converge.

A second key question, and one on which relatively little progress has been made, is the quality of the approximation. There is very little work that provides guarantees on the error made in the answers (for example, individual marginals) by any of these methods. As a consequence, there is almost no work that provides help in choosing a low-error approximation for a particular problem. Thus, the problem of applying these methods in practice is still a combination of manual tuning on one hand, and luck on the other. The development of automated methods that can help guide the selection of an approximating class for a particular problem is an important direction for future work.

11.7 Relevant Literature

Variational approximation methods are applicable in a wide variety of settings, including quantum mechanics (Sakurai 1985), statistical mechanics (Parisi 1988), neural networks (Elfadel 1995), and statistics (Rustagi 1976). The literature on the use of these approximation methods for inference problems in graphical models is heavily influenced by ideas in statistical mechanics, although the connections are beyond the scope of this book.

The rules for belief propagation were developed in Pearl's (1986b; 1988) analysis of inference in singly connected networks. In his book, Pearl noted that these propagation rules lead to wrong answers in multiconnected networks. Interest in "loopy" inference on pairwise networks was raised in several publications (Frey 1998; MacKay and Neal 1996; Weiss 1996) that reported good empirical results when running Pearl's algorithm on networks with loops. The most impressive success was in the error-correcting code scheme known as turbocodes (Berrou et al. 1993; McEliece et al. 1995). These decoding algorithms were shown to be equivalent to belief propagation in a network with loops (Kschischang and Frey 1998; McEliece et al. 1998; Frey and MacKay 1997). Researchers using these ad hoc algorithms reported that although they did not compute the correct posteriors, they often did compute the correct MAP assignment.

These empirical successes led to examination of both the reasons for the success of such methods and evaluated their performance in other domains. Weiss (2000) showed that when a network has a single loop, an analytic relationship exists between the belief computed and the true marginals, and it used that relationship to characterize network topologies for which the MAP solution is provably correct. Weiss and Freeman (2001a) then showed that if loopy belief propagation converges in a linear Gaussian network, then it computes correct posterior means; see section 14.7 for more references on results for the Gaussian case. These initial analytic results were supplemented by promising empirical results (Murphy et al. 1999; Horn and McEliece 1997; Weiss 2001).

Several alternative formulations of loopy belief propagation appeared in the literature. These include factor graphs (Kschischang et al. 2001b), tree-based reparameterization (Wainwright et al. 2003a), and algebraic formulations (Aji and McEliece 2000). In parallel, several authors (Yedidia et al. 2000; Dechter et al. 2002) proposed extending the idea of belief propagation to more generalized cluster graphs (or region graphs). Welling (2004) examined methods for automatically choosing the regions for this approximation.

A major advance in our understanding of this general class of approximation algorithms came with the analysis of Yedidia et al. (2000, 2005), showing that these methods can be posed as attempting to maximize an approximate energy functional. This result connected the algorithmic developments in the field with literature on free-energy approximations developed in statistical mechanics (Bethe 1935; Kikuchi 1951). This insight is the basis for our discussion of the Bethe energy functional in section 11.3.6. This result also provided a connection between belief propagation algorithms and other approximation procedures, such as structured variational methods. In addition, it led to the development of new types of approximate inference procedures. These include direct optimization of the Bethe energy functional (Welling and Teh 2001) using gradient-based methods, as well as provably convergent "double loop" variants of belief propagation (Yuille 2002; Heskes et al. 2003). Another class of algorithms combined exact inference on subtrees of the cluster graph within cluster-graph belief propagation (Wainwright et al. 2001, 2002a). This combination leads to faster convergence and introduces new directions for characterizing the errors of the approximation. Wainwright and Jordan (2003) review the connections between belief propagation and optimization of the energy functional.

An active area of research is improving and analyzing the convergence properties of generalized belief propagation algorithms. In practice, the techniques most important for improving convergence are damping (Murphy et al. 1999; Heskes 2002) and message scheduling algorithms, such as the tree-based reparameterization algorithm of (Wainwright et al. 2003a) or the residual belief propagation algorithm of Elidan et al. (2006). On the theoretical side, key directions include the identification of a set of sufficient conditions for the existence of unique local maxima of the Bethe energy functional (Pakzad and Anantharam 2002; Heskes 2002), as well as conditions that ensure a unique convergence point of belief propagation (Tatikonda and Jordan 2002; Ihler et al. 2003; Heskes 2004; Ihler et al. 2005; Mooij and Kappen 2007). Our discussion in section 11.3.4 is based on that of Mooij and Kappen (2007). A related trajectory attempts to estimate the error of the approximate marginal probabilities; see, for example, Leisink and Kappen (2003); Ihler (2007).

A different generalizatiohn of belief-propagation algorithms develops variants of the energy functional that have desired properties. For example, if the energy functional is convex, then it is guaranteed to have a single maximum. An initial convexified free-energy functional was introduced in Wainwright et al. (2002b). This functional has the additional property of providing an upper bound on the partition function. Recently, there has been significant work that provides a more detailed characterization of convex energy functionals (Wainwright and Jordan 2003; Heskes 2006; Wainwright and Jordan 2004; Sontag and Jaakkola 2007). Although the convexity of energy functional implies a unique maximum, it does not generally guarantee that a message passing algorithm will converge. Recent alternative algorithms provide guaranteed convergence for such energy functionals (Heskes 2006; Globerson and Jaakkola 2007a; Hazan and Shashua 2008).

A recent extension is the combination of belief propagation with particle-based methods. The basic idea is to use particle sampling to perform the basic belief propagation steps (Sudderth et al. 2003; Isard 2003). This combination allows us to use cluster-graph belief propagation on networks with continuous non-Gaussian variables, which appear in applications in vision and signal processing; see also section 14.7.

The idea of factored messages first appeared in several contexts (Dechter and Rish 1997; Dechter 1997; Boyen and Koller 1998b; Murphy and Weiss 2001). Similar ideas that involve projection of messages onto "simple" representations (such as Gaussians) are common in the control and tracking literature (Bar-Shalom 1992). The generalization of these ideas in the form of expectation propagation was introduced by Minka (2001b). The connection between expectation propagation and the Bethe energy functional was explored by (Minka 2001b; Heskes and Zoeter 2002; Heskes et al. 2005). The connection between expectation propagation and generalized belief propagation in the case of discrete variables is explored by Welling et al. (2005).

One of the early applications of variational methods to probabilistic methods was the use of mean field approximation for Boltzmann machines (Peterson and Anderson 1987), an approach then widely adopted in computer vision (see, for example, Li (2001)). This methodology was introduced into the field of directed graphical models by Saul et al. (1996), following ideas that appeared in the context of Helmholtz machines (Hinton et al. 1995).

The mean field approximation cannot capture strong dependencies in the posterior distribution. Saul and Jordan (1996) suggested to circumvent this problem by using structured variational methods. These ideas were extended for different forms of approximating distributions and target networks. Ghahramani and Jordan (1997) used independent hidden Markov chains to approximate factorial HMMs, a specific form of a dynamic Bayesian network. Barber and Wiegerinck (1998) use a Boltzmann machine approximation. Wiegerinck (2000) uses Markov networks and cluster trees as approximate distribution. Xing et al. (2003) describe cluster mean field and suggest efficient implementations. Geiger et al. (2006) further extend Wiegerinck's procedure and introduce efficient propagation schemes to maximize parameters in the approximating distribution.

The idea of using a mixture of mean field approximation was developed by Jaakkola and Jordan (1998). These ideas were extended to general networks with auxiliary variables by El-Hay and Friedman (2001).

Jaakkola and Jordan (1996b,a) introduced local variational approximations to compute both upper bounds and lower bounds of the log-likelihood function (that is, of $\log Z$). They demonstrated these methods by a large scale study of inference in the QMR-DT network (Jaakkola and Jordan 1999), where they show that variational methods are more effective than particle based methods. Additional extension on these methods appeared in Ng and Jordan (2000).

Tutorials discussing the use of structured and local methods appear in Jordan et al. (1998); Jaakkola (2001).

11.8 Exercises

Exercise 11.1

entropy

Show that the derivative of the *entropy* is:

$$\frac{\partial}{\partial Q(\boldsymbol{x})} H_Q(\boldsymbol{X}) = -\ln Q(\boldsymbol{x}) - 1.$$

Exercise 11.2⋆

local consistency polytope

Consider the set of locally consistent distributions, as in the *local consistency polytope* of equation (11.16).

marginal polytope

a. For a cluster graph $\mathcal{U}$ that is a clique tree $\mathcal{T}$ (satisfying the running intersection property), show that the set of distributions satisfying these constraints is precisely the *marginal polytope* — the set of legal distributions Q that can be represented over $\mathcal{T}$.

b. Show that, for a cluster graph that is not a tree, there are parameterizations that satisfy these constraints but are not marginals of any legal probability distribution.

Exercise 11.3⋆

In the text, we showed that CTree-Optimize has a unique global optimum. Show that it also has a unique fixed point.

Exercise 11.4⋆

Consider the network of figure 11.1c. We have shown that

$$P_{\mathcal{T}}(A, B, C, D) \propto P_{\Phi}(A, B, C, D) \frac{\mu_{3,4}[D]\mu_{1,4}[A]}{\beta_4(A, D)},$$

cluster graph residual

where $\mathcal{T}$ is the cluster tree we get if we remove $C_4 = \{A, D\}$. Show how to use this result on the *residual* to bound the error in the estimation of marginals in this cluster graph.

Exercise 11.5⋆

Prove proposition 11.2.

Exercise 11.6

Compare the cluster graphs in figure 11.3 and figure 11.7.

a. Show that $\mathcal{U}_2$ and $\mathcal{U}_3$ are equivalent in the following sense: Any set of calibrated potentials of one can be transformed to a calibrated set of potentials of the other.

b. Show similar equivalence between $\mathcal{U}_1$ and $\mathcal{U}_3$.

Exercise 11.7

image
segmentation

As we discussed in box 4.B, Markov networks are commonly used for image-processing tasks. We now consider the application of Markov networks to foreground-background *image segmentation*. Here, we do not have a predetermined set of classes, each with its own model. Rather, for each pixel, we have a binary-valued variable X_i, where x_i^1 means that X_i is a foreground pixel, and x_i^0 a background pixel. In this case, we have no node potentials (say that features of an individual pixel cannot help distinguish foreground from background). The network structure is a grid, where for each pair of adjacent pixels i, j, we have an edge potential $\phi_{i,j}(X_i, X_j) = \alpha_{i,j}$ if $X_i = X_j$ and 1 otherwise. A large value of $\alpha_{i,j}$ makes it more likely that $X_i = X_j$ (that is, pixels i and j are assigned to the same segment), and a small value makes it less likely.

Using the standard Bethe cluster graph, compute the first message sent by loopy belief propagation from any variable to any one of its neighbors. What is wrong with this approach?

Exercise 11.8

Let X be a node with parents $\boldsymbol{U} = \{U_1, \ldots, U_k\}$, where $P(X \mid \boldsymbol{U})$ is a tree-CPD. Assume that we have a cluster consisting of X and $\boldsymbol{U}$. In this question, you will show how to exploit the structure of the tree-CPD to perform message passing more efficiently.

a. Consider a step where our cluster gets incoming messages about $U_1, \ldots, U_k$, and sends a message about X. Show how this step can be executed in time linear in the size of the tree-CPD of $P(X \mid \boldsymbol{U})$.

b. Now, consider the step where our clique gets incoming messages about $U_1, \ldots, U_{k-1}$ and X, and send a message about U_k. Show how this step can also be executed in time linear in the size of the tree-CPD.

Exercise 11.9

Recall that a pairwise Markov network consists of univariate potential $\phi_i[X_i]$ over each variable X_i, and in addition a pairwise potential $\phi_{(i,j)}[X_i, X_j]$ over some pairs of variables. In section 11.3.5.1 we showed a simple transformation of such a network to a cluster graph where we introduce a cluster for each potential. Write the update equations for cluster-graph belief propagation for such a network. Show that we can formulate inference as the direct propagation of messages between variables by implicitly combining messages through pairwise potentials.

Exercise 11.10★

Suppose we are given a set of factors $\Phi = \{\phi_1, \ldots, \phi_K\}$ over $\boldsymbol{X} = \{X_1, \ldots, X_n\}$. Our aim is to convert these factors into a pairwise Markov network by introducing new auxiliary variables $\boldsymbol{Y} = \{Y_1, \ldots, Y_k\}$ so that Y_j denotes a joint assignment to $Scope[\phi_j]$. Show how to construct a set of factors Φ' that is a pairwise Markov network over $\boldsymbol{X} \cup \boldsymbol{Y}$ such that $P_{\Phi'}(\boldsymbol{x}) = P_\Phi(\boldsymbol{x})$ for each assignment to $\boldsymbol{X}$.

Exercise 11.11★

BN cluster graph

In this question, we study how we can define a *cluster graph for a Bayesian network*.

a. Consider the following two schemes for converting a Bayesian network structure $\mathcal{G}$ to a cluster graph $\mathcal{U}$. For each of these two schemes, either show (by proving the necessary properties) that it produces a valid cluster graph for a general Bayesian network, or disprove this result by showing a counterexample.

- **Scheme 1:** For each node X_i in $\mathcal{G}$, define a cluster $\boldsymbol{C}_i$ over X_i's family. Connect $\boldsymbol{C}_i$ and $\boldsymbol{C}_j$ if X_j is a parent of X_i in $\mathcal{G}$, and define the sepset to be the intersection of the clusters.

- **Scheme 2:** For each node X_i in $\mathcal{G}$, define a cluster $\boldsymbol{C}_i$ over X_i's family. Connect $\boldsymbol{C}_i$ and $\boldsymbol{C}_j$ if X_j is a parent of X_i in $\mathcal{G}$, and define the sepset to be the $\{X_j\}$.

b. Construct an alternative scheme to the ones proposed earlier that uses a minimum spanning tree algorithm to transform any Bayesian network into a valid cluster graph.

Exercise 11.12★

Suppose we want to use a gradient method to directly maximize $\tilde{F}[\tilde{P}_\Phi, \boldsymbol{Q}]$ with respect to entries in $\boldsymbol{Q}$. For simplicity, assume that we are dealing with a pairwise network for both P_Φ and Q, and so the entries in $\boldsymbol{Q}$ are all univariate and pairwise potentials.

a. Derive $\frac{\partial \tilde{F}[\tilde{P}_\Phi, \boldsymbol{Q}]}{\partial Q(x_i)}$ and $\frac{\partial \tilde{F}[\tilde{P}_\Phi, \boldsymbol{Q}]}{\partial Q(x_i, y_i)}$ for the two types of potentials we have.

b. Recall that we cannot simply choose $\boldsymbol{Q}$ to maximize $\tilde{F}[\tilde{P}_\Phi, \boldsymbol{Q}]$. We need to ensure that every potential in $\boldsymbol{Q}$ is nonnegative and sums to 1. In addition, we need to maintain the marginal consistency between each pairwise potential and the associated univariate potential marginals in $\boldsymbol{Q}$. A standard solution is to write $\boldsymbol{Q}$ as a function of meta parameters η, so that the transformation from η to $\boldsymbol{Q}$ ensures the consistency. Suggest such a reparameterization, and derive the gradient of $\tilde{F}[\tilde{P}_\Phi, \boldsymbol{Q}]$ with respect to this reparameterization. (Hint: use the chain law of partial derivatives.)

Exercise 11.13★

Prove that if the damped updates of equation (11.14) converge, then they converge to a stationary point of $\tilde{F}[\tilde{P}_\Phi, \boldsymbol{Q}]$.

Exercise 11.14★

In this exercise you will prove theorem 11.7.

a. Start by defining the Lagrangian

$$
\begin{aligned}
\mathcal{J} \;=\; & \tilde{F}[\tilde{P}_\Phi, \boldsymbol{Q}] \\
& - \sum_{i \in \mathcal{V}_\mathcal{R}} \lambda_r \left(\sum_{\boldsymbol{c}_r} \beta_r(\boldsymbol{c}_r) - 1 \right) \\
& - \sum_{(r' \to r) \in \mathcal{E}_\mathcal{R}} \sum_{\boldsymbol{c}_r} \lambda_{r' \to r}[\boldsymbol{c}_r] \left(\sum_{\boldsymbol{c}_{r'} \sim \boldsymbol{c}_r} \beta_{r'}(\boldsymbol{c}_{r'}) - \beta_r(\boldsymbol{c}_r) \right),
\end{aligned}
$$

and show that any maximum point satisfies

$$
\frac{\partial}{\partial \beta_r(\boldsymbol{c}_r)} \mathcal{J} = \kappa_r \ln \psi_r(\boldsymbol{c}_r) - \kappa_r \ln \beta_r(\boldsymbol{c}_r) - \kappa_r - \lambda_r - \sum_{r' \in \mathbf{Up}(r)} \lambda_{r' \to r}[\boldsymbol{c}_r] + \sum_{r' \in \mathbf{Down}(r)} \lambda_{r \to r'}[\boldsymbol{c}_{r'}],
$$

where κ_r is the counting number of the region, as defined in equation (11.20).

b. Define $\delta_{r_u \to r_d}$ in terms of the Lagrange multipliers and show that your solution satisfies equation (11.37) and equation (11.36).

Exercise 11.15

Consider an $n \times n$ grid network. Construct a region-graph approximation for this network that would have a region r for each small square $A_{i,j}, A_{i,j+1}, A_{i+1,j}, A_{i+1,j+1}$. Describe the structure of the graph and the computation of the messages and beliefs.

Exercise 11.16

Prove theorem 11.6.

Exercise 11.17★

In this exercise, we will derive a message passing algorithm, called the *child-parent algorithm*, for Bethe region graphs. Although the messages in this algorithm are somewhat convoluted, they have the advantage of corresponding directly to the Lagrange multipliers for the region graph energy functional. Moreover, although somewhat opaque, the message passing algorithm is a very slight variation on the scheme used in standard belief propagation: the standard messages are simply raised to a power.

Consider the Bethe cluster graph of example 11.2, where we assume that all counting numbers are nonzero. Let $\rho_i = 1/\kappa_i$ and $\rho_r = 1/\kappa_r$. Starting from equation (11.34), derive the correctness of the following update rules for the messages and the beliefs. For all $r \in \mathbf{R}^+$:

$$\beta_r(\boldsymbol{C}_r) \leftarrow \left(\tilde{\psi}_r(\boldsymbol{C}_r) \prod_{X_i \in \boldsymbol{C}_r} \delta_{i \rightarrow r}(X_i) \right)^{\rho_r} . \tag{11.69}$$

$$\delta_{i \rightarrow r}(X_i) \leftarrow \left[\left(\prod_{r' \neg r} \delta_{i \rightarrow r'}(X_i) \right)^{\rho_i} \left(\sum_{\boldsymbol{C}_r - X_i} \psi_r(\boldsymbol{C}_r) \left(\prod_{X_j \in \boldsymbol{C}_r, j \neq i} \delta_{j \rightarrow r} \right)^{\rho_r} \right) \right]^{-\frac{1}{\rho_i + \rho_r}} . \tag{11.70}$$

Exercise 11.18★

Show that the counting numbers defined by equation (11.26) are convex. (Hint: Show first the convexity of counting numbers obtained from this analysis for a tree-structured MRF.)

Exercise 11.19★

In this exercise, we will derive another message passing algorithm, called the *two-way algorithm*, for finding fixed points of region-based energy functionals; this algorithm allows for more general region graphs than in exercise 11.17. It uses two messages along each link $r \rightarrow r'$: one from r to r' and one from r' to r.

Consider a region-based free energy as in equation (11.27). For any region r, let $p_r = |\mathbf{Up}(r)|$ be the number of regions that are directly upward of r. Assume that for any top region (so that $p_r = 0$), we have that $\kappa_r = 1$. We now define $q_r = (1 - \kappa_r)/p_{region}$, taking $q_r = 1$ when $p_r = 0$ (so that $\kappa_r = 1$, as per our assumption). Assume furthermore that $q_r \neq 2$, and define $\beta_r = 1/(2 - q_r)$.

The following equalities define the messages and the potentials in this algorithm:

$$\tilde{\psi}_r(\boldsymbol{C}_r) = (\psi_r(\boldsymbol{C}_r))^{\kappa_r}) \tag{11.71}$$

$$\tilde{\delta}^{up}_{r \rightarrow r'} = \tilde{\psi}_r(\boldsymbol{C}_r) \prod_{r'' \in \mathbf{Up}(r) - \{r'\}} \tilde{\delta}^{down}_{r'' \rightarrow r}(\boldsymbol{C}_r) \prod_{r'' \in \mathbf{Down}(r)} \tilde{\delta}^{up}_{r'' \rightarrow r}(\boldsymbol{C}_{r''}) \tag{11.72}$$

$$\tilde{\delta}^{down}_{r' \rightarrow r} = \sum_{\boldsymbol{C}_{r'} - \boldsymbol{C}_r} \tilde{\psi}_r(\boldsymbol{C}_r) \prod_{r'' \in \mathbf{Up}(r)} \tilde{\delta}^{down}_{r'' \rightarrow r}(\boldsymbol{C}_r) \prod_{r'' \in \mathbf{Down}(r) - \{r'\}} \tilde{\delta}^{up}_{r'' \rightarrow r}(\boldsymbol{C}_{r''}) \tag{11.73}$$

$$\delta^{up}_{r \rightarrow r'} = (\tilde{\delta}^{up}_{r \rightarrow r'}(\boldsymbol{C}_r))^{\beta_r} (\tilde{\delta}^{down}_{r' \rightarrow r}(\boldsymbol{C}_r))^{\beta_r - 1} \tag{11.74}$$

$$\delta^{down}_{r \rightarrow r'} = (\tilde{\delta}^{up}_{r \rightarrow r'}(\boldsymbol{C}_r))^{\beta_r - 1} (\tilde{\delta}^{down}_{r' \rightarrow r}(\boldsymbol{C}_r))^{\beta_r} \tag{11.75}$$

$$\beta_r(\boldsymbol{C}_r) = \tilde{\psi}_r(\boldsymbol{C}_r) \prod_{r' \in \mathbf{Up}(r)} \delta^{down}_{r' \rightarrow r}(\boldsymbol{C}_r) \prod_{r'' \in \mathbf{Down}(r)} \delta^{up}_{r'' \rightarrow r}(\boldsymbol{C}_{r''}). \tag{11.76}$$

Note that the messages $\tilde{\delta}^{up}_{r \rightarrow r'}$ (or $\tilde{\delta}^{down}_{r \rightarrow r'}$) are as we would expect: the message sent from r to r' is simply the product of r's initial potential with all of its incoming messages except the one from r' (and similarly for the message sent from r' to r). However, as we discussed in our derivation of the region graph algorithm, this computation will double-count the information that arrived at r from r' via an indirect path. The final computation of the messages $\delta^{up}_{r \to r'}$ and $\delta^{down}_{r \to r'}$ is intended to correct for that double-counting.

In this exercise, you will show that the fixed points of equation (11.33) are precisely the same as the fixed points of the update equations equation (11.71)–equation (11.76).

a. We begin by defining the messages in terms of the beliefs and the Lagrange multipliers:

$$\tilde{\delta}^{up}_{r \to r'}(\boldsymbol{c}_r) = \exp\{\lambda_{r,r'}(\boldsymbol{c}_r)\} \tag{11.77}$$

$$\tilde{\delta}^{down}_{r \to r'}(\boldsymbol{c}_{r'}) = \beta_{r'}(\boldsymbol{c}_{r'})^{q_{r'}} \exp\{\lambda_{r,r'}(\boldsymbol{c}_{r'})\}. \tag{11.78}$$

Show that these messages satisfy the fixed-point equations in equation (11.33).

b. Show that

$$(\beta_r(\boldsymbol{c}_r))^{\kappa_r} \propto (\beta_r(\boldsymbol{c}_r))^{\kappa_r - 1} \tilde{\psi}_r(\boldsymbol{c}_r) \prod_{r' \in \mathbf{Up}(r)} \delta^{down}_{r' \to r}(\boldsymbol{C}_r) \prod_{r'' \in \mathbf{Down}(r)} \delta^{up}_{r'' \to r}(\boldsymbol{C}_{r''}). \tag{11.79}$$

Conclude that equation (11.76) holds.

c. Show that

$$\tilde{\delta}^{up}_{r \to r'} \delta^{down}_{r' \to r} = \beta_r = \delta^{up}_{r \to r'} \tilde{\delta}^{down}_{r' \to r}, \tag{11.80}$$

and that

$$\delta^{up}_{r \to r'} \delta^{down}_{r' \to r} = (\beta_r)^{q_r}. \tag{11.81}$$

Show that the only solution to these equations is given by equation (11.73) and equation (11.74).

d. Show by direct derivation that theorem 11.6 holds for the potentials defined by equation (11.71)–11.76. (Hint: consider separately those regions that have no parents, and recall our previous assumptions.)

Exercise 11.20

Consider the marginal probability over $\boldsymbol{C}_2 = \{A_{1,2}, A_{2,2}, A_{3,2}, A_{4,2}\}$ in the 4×4 grid network. Show that, if we assume general potentials, this marginal probability does not satisfy any conditional independencies.

Exercise 11.21★

Consider a clique tree $\mathcal{T}$ that is constructed from a run of variable elimination over a Gibbs distribution P'_Φ, as described in section 10.1.2. Let $\boldsymbol{S}_{i,j}$ be a sepset in the clique tree. Show that, for general potentials, P'_Φ does not satisfy any conditional independencies relative to the marginal distribution over $\boldsymbol{S}_{i,j}$. (Hint: consider the actual run of variable elimination that led to the construction of this sepset, and the effect on the graph of eliminating these variables.)

Exercise 11.22★

In section 10.3.3.2 we described an algorithm that uses a calibrated clique tree to compute a joint marginal over a set of variables that is not a subset of any clique in the tree. We can consider using the same procedure in a calibrated cluster graph $\mathcal{U}$. For example, consider a pair of variables X, Y that are not found together in any cluster in $\mathcal{U}$.

a. Define an algorithm that, given clusters $\boldsymbol{C}_1 \ni X$ and $\boldsymbol{C}_2 \ni Y$ and a path between them in the cluster graph computes a joint distribution $P(X, Y)$.

b. A cluster graph can contain more than one path between $\boldsymbol{C}_1$ and $\boldsymbol{C}_2$. Provide an example showing that the answer returned by this algorithm can depend on the choice of path used.

Exercise 11.23★

Suppose we have a cluster $\mathcal{T}$, and we consider two approximation schemes for inference in this cluster tree.

   a. Using expectation propagation with fully factored approximation to the messages (as in figure 11.13).

   b. Using a cluster graph approach on the cluster graph $\mathcal{U}$ constructed in the following manner. Each cluster $\boldsymbol{C}_i \in \mathcal{T}$ appears in $\mathcal{U}$. In addition, for each variable X_k that appears in some sepset $\boldsymbol{S}_{i,j}$ in $\mathcal{U}$ we introduce a new cluster with scope $\{X_k\}$ and connect it to both $\boldsymbol{C}_i$ and $\boldsymbol{C}_j$. (This construction is similar to the Bethe cluster graph, except that we use the clusters in $\mathcal{T}$ as the "big" clusters in the construction.)

Show that both approximations are equivalent in the sense that their respective energy functionals (and the constraints they satisfy) coincide.

Exercise 11.24

Prove lemma 11.1.

Exercise 11.25★

In this exercise, you will provide a simpler proof for a special case of theorem 11.9 and corollary 11.6. Assume that each $X_i \in \mathcal{X}$ is a binary-valued random variable, parameterized with a single parameter $q_i = Q(x_i^1)$.

   a. By considering the derivative of $F[\tilde{P}_\Phi, Q]$ and using lemma 11.1, prove theorem 11.9 without using Lagrange multipliers.

   b. Now, prove corollary 11.6.

Exercise 11.26★

In this exercise, we will prove theorem 11.11. The proof relies on the following proposition, which characterizes the derivatives of an expectation relative to a Gibbs distribution.

Proposition 11.9

$$\overline{\text{If } Q(\mathcal{X}) = \tfrac{1}{Z_Q} \prod_j \psi_j, \text{ then, for any function } f \text{ with scope } \boldsymbol{U},}$$

$$\frac{\partial}{\partial \psi_j(\boldsymbol{c}_j)} \boldsymbol{E}_Q[f(\boldsymbol{U})] = \frac{Q(\boldsymbol{c}_j)}{\psi_j(\boldsymbol{c}_j)} \left(\boldsymbol{E}_Q[f(\boldsymbol{U}) \mid \boldsymbol{c}_j] - \boldsymbol{E}_Q[f(\boldsymbol{U})] \right) + \boldsymbol{E}_Q\left[\frac{\partial}{\partial \psi_j(\boldsymbol{c}_j)} f(\boldsymbol{U}) \right].$$

   a. Prove proposition 11.9.

   b. Apply this proposition to prove theorem 11.11.

Exercise 11.27★

Consider the structured variational approximation of equation (11.61). As we discussed, to execute this update we need to collect the expectation of several factors, and each of these requires that we compute expectations given different assignments to the factor of interest. Assuming we use a clique tree to perform our inference, show how we can use a dynamic programming algorithm to reuse computation so as to evaluate these updates more efficiently.

Exercise 11.28

factorial HMM

Consider the *factorial HMM* model shown in figure 6.3a. Find a variational update rule for the structured variational approximation that factorizes as a set of clusters corresponding to the individual chains, that is, a cluster $\{X_i^{(0)}, \ldots, X_i^{(T)}\}$ for each i. Make sure that you simplify your clusters as much as possible, as in section 11.5.2.3 and section 11.5.2.4.

Exercise 11.29★

Now, consider the DBN whose structure is a set of variables X_i that evolve over time, where, at each time point, the chains are correlated by a tree structure that is the same for all time slices. Let Pa_i be the parents of X_i in the tree. Such structures (or extensions along similar lines) arise in several applications, such as those involving evolution of DNA or protein sequences; here, the chains encode spatial correlations over the sequence, and the tree the evolutionary process of each letter (DNA base pair or protein amino acid), whether across species, as in the *phylogenetic HMM* of box 6.A, or within a family tree (as in box 3.B).

phylogenetic
HMM

Consider an unrolled network over time $0, \ldots, T$, where the initial $X_i^{(0)}$ are all independent. Find update rules for the following cluster variational approximations. In both cases, make sure that you simplify your clusters as much as possible, as in section 11.5.2.3 and section 11.5.2.4.

a. Find an update rule for the cluster variational approximation that has a cluster for each chain; that is, a cluster $\{X_i^{(0)}, \ldots, X_i^{(T)}\}$ for each i.

b. Find an update rule for the cluster variational approximation that has a cluster for each time slice; that is, a cluster $\{X_1^{(t)}, \ldots, X_n^{(t)}\}$ for each t.

Exercise 11.30

Consider a distribution P_Φ that consists of the pairwise Markov network of figure 11.18a, and consider approximating it with distribution Q that is represented by a pairwise Markov network of figure 11.18c. Derive the potentials $\psi_1(A, C)$ and $\psi_1(B, D)$ that maximize $D(P_\Phi \| Q)$. If A and C are not independent in P_Φ, will they be independent in the Q with these potentials?

Exercise 11.31★

Prove proposition 11.7.

Exercise 11.32★★

Describe an algorithm that performs variational variable elimination with optimization of the variational parameters. Your algorithm should take as input a set of factors Φ, an elimination ordering $X_1, \ldots, X_n$, and a subset $X_{i_1}, \ldots, X_{i_k}$ of steps at which variational approximations should be performed. Your algorithm should use the technique of section 11.5.3.2 to avoid coupling the factors in the elimination step of each X_{i_j}, introducing a single variational parameter λ_j for each such step. The result of the variable elimination procedure should be a symbolic expression in the λ_j's. Explain precisely how to construct this symbolic expression and how to optimize it to select the optimal set of variational parameters $\lambda_1, \ldots, \lambda_k$.

Exercise 11.33★

Show that if $Q = \prod_i Q(X_i \mid \boldsymbol{U}_i)$, then

$$\frac{\partial}{\partial Q(x_i \mid \boldsymbol{u}_i)} \boldsymbol{E}_Q[f(\boldsymbol{Y})] = \frac{1}{Q(x_i \mid \boldsymbol{u}_i)} \boldsymbol{E}_Q[f(\boldsymbol{Y}) \mid x_i, \boldsymbol{u}_i].$$

Exercise 11.34★★

Develop a variational approximation using Bayesian networks. Assume that Q is represented by a Bayesian network of a given structure $\mathcal{G}$. Derive the fixed-point characterization of parameters that maximize the energy functional (that is, the analog of theorem 11.11) for this type of approximation. (Hint: use theorem 8.2 and exercise 11.33 in your derivation.)

Exercise 11.35★

Prove proposition 11.8.

Exercise 11.36★

In this exercise we consider inference in two-layered Bayesian networks with logistic CPDs. In these networks, all variables are binary. The variables $X_1, \ldots, X_n$ in the top layer are all independent of each other. The variables $Y_1, \ldots, Y_m$ in the bottom layer depend on the variables in the top layer using a logistic CPD:

$$P(y_j^1 \mid x_1, \ldots, x_n) = \text{sigmoid}(\sum_i w_{i,j} x_i + w_{0,j}).$$

Suppose we observe some of the variables on the bottom layer, and we want to estimate the probability of the evidence.

a. Show that when computing the probability of evidence, we can remove variables Y_i that are not observed.

b. Given evidence $\boldsymbol{y}$, use the bound of equation (11.68) to write a variational upper bound of $P(\boldsymbol{y})$.

c. Develop a mean field approximation to lower-bound the probability of the same evidence.

Exercise 11.37⋆⋆

a. Show that the following function f_{obj} is convex:

$$f_{\text{obj}}(a_1, \ldots, a_n) = \ln \sum_i e^{a_i}.$$

b. Prove the following bound

$$f_{\text{obj}}(a_1, \ldots, a_n) \geq \sum_i \lambda_i a_i - \sum_i \lambda_i \ln \lambda_i,$$

for any choice of $\{\lambda_i\}$ so that $\sum_i \lambda_i = 1$ and $\lambda_i \geq 0$ for all i. (Hint: use the convexity of $f_{\text{obj}}()$.)

c. Write

$$\ln Z = f_{\text{obj}}(\{\ln \tilde{P}_\Phi(\xi)\}),$$

and use the preceding bound to write a lower bound for $\ln Z$. Use this analysis to provide an alternative derivation of the mean field approximation.

Exercise 11.38⋆⋆

mixture
distribution

We now consider a very different class of representations that we can use in a structure variational approximation. Here, our distribution Q is a *mixture distribution*. More precisely, in addition to the variables $\mathcal{X}$ in P_Φ, we introduce a new variable T. We can now define an approximating family:

$$Q(\mathcal{X}) = \sum_t Q(t, \mathcal{X}) = \sum_t Q(t) Q(\mathcal{X} \mid t), \tag{11.82}$$

which is a mixture of different approximations $Q(\mathcal{X} \mid t)$. As one simple instantiation, we might have:

$$Q(\mathcal{X}, T) = Q(T) \prod_i Q(X_i \mid T), \tag{11.83}$$

where each mixture component, $Q(\mathcal{X} \mid t)$, is a mean field distribution.

a. Assuming that Q is structured as in equation (11.82), prove that

$$F[\tilde{P}_\Phi, Q] = \sum_t Q(t) F[\tilde{P}_\Phi, Q(\mathcal{X} \mid t)] + \boldsymbol{I}_Q(T; \mathcal{X}), \tag{11.84}$$

where $\boldsymbol{I}_Q(T; \mathcal{X})$ is the mutual information between T and $\mathcal{X}$.

This result quantifies the gains that we can obtain by using a mixture distribution. The first term is the weighted average of the energy functional of the mixture components; this term is bounded by the best of the components in isolation. The improvement over using specific components is captured by the second term, which measures the extent to which T influences the distribution over $\mathcal{X}$. If the components represent identical distributions, then this term is zero, and, as expected, the value of the energy functional is identical to the one obtained using the component representation. By contrast, if the components are different, then T is informative on $\mathcal{X}$, and the lower bound provided by Q can be better than that of each of the components. We note that the "best scenario" for improvement is when the component terms $F[\tilde{P}_\Phi, Q(\mathcal{X} \mid t)]$ are similar, and yet $I_Q(T; \mathcal{X})$ is large. In other words, the best approximation is obtained when each of the components provides a good approximation, but by using a different distribution.

We now use this formulation of the energy functional to produce an update rule for this variational approximation. This derivation uses a local variational approximation.

b. Let $\{\lambda(\xi, t) : \xi \in \mathit{Val}(\mathcal{X}), t \in \mathit{Val}(T)\}$ be a family of constants. Use lemma 11.2 to show that:

$$I_Q(T; \mathcal{X}) \geq H_Q(T) - \sum_{\xi, t} \lambda(\xi, t) Q(t) + E_Q[\ln \lambda(\mathcal{X}, T)] + 1.$$

Show also that this bound is tight if we choose a different value of $\lambda(\xi, t)$ for each possible combination of ξ, t.

c. Suppose that we use a factorized form for the variational parameters:

$$\lambda(\xi, t) = \lambda(t) \prod_i \lambda(x_i \mid t).$$

Show that

$$F[\tilde{P}_\Phi, Q] \;\geq\; \sum_t Q(t) F[\tilde{P}_\Phi, Q(\mathcal{X} \mid t)] + H_Q(T) - E_Q\left[\tilde{\lambda}(T)\right] + \tag{11.85}$$

$$E_Q[\lambda(T)] + \sum_i E_Q[\lambda(X_i \mid T)], \tag{11.86}$$

where $\tilde{\lambda}(t) = \sum_\xi \lambda(\xi, t)$.

d. Now, assume concretely that our approximation Q has the form of equation (11.83), and that all of the X_i variables are binary-valued. Use your lower bound in equation (11.85) as an approximation to the free energy, and provide an update rule for all of the parameters in Q: $P(t)$ and $P(x_i^1 \mid t)$.

e. Analyze the computational complexity of applying these fixed-point equations.

12 *Particle-Based Approximate Inference*

In the previous chapter, we discussed one class of approximate inference methods. The techniques used in that chapter gave rise to algorithms that were very similar in flavor to the factor-manipulation methods underlying exact inference. In this chapter, we discuss a very different class of methods, ones that can be roughly classified as *particle-based methods*. In these methods, we approximate the joint distribution as a set of instantiations to all or some of the

particle   variables in the network. These instantiations, often called *particles*, are designed to provide a good representation of the overall probability distribution.

Particle-based methods can be roughly characterized along two axes. On one axis, approaches vary on the process by which particles are generated. There is a wide variety of possible processes. At one extreme, we can generate particles using some deterministic process. At another, we can sample particles from some distribution. Within each category, there are many possible variations.

On the other axis, techniques use different notions of particles. Most simply, we can consider

full particle   *full particles* — complete assignments to all of the network variables $\mathcal{X}$. The disadvantage of this approach is that each particle covers only a very small part of the space. A more effective

collapsed particle   notion is that of a *collapsed particle*. A collapsed particle specifies an assignment $\boldsymbol{w}$ only to some subset of the variables $\boldsymbol{W}$, associating with it the conditional distribution $P(\mathcal{X} \mid \boldsymbol{w})$ or some "summary" of it.

The general framework for most of the discussion in this chapter is as follows. Consider some distribution $P(\mathcal{X})$, and assume we want to estimate the probability of some event $\boldsymbol{Y} = \boldsymbol{y}$ relative to P, for some $\boldsymbol{Y} \subseteq \mathcal{X}$ and $\boldsymbol{y} \in Val(\boldsymbol{Y})$. More generally, we might want to estimate the expectation of some function $f(\mathcal{X})$ relative to P; this task is a generalization, since we can choose $f(\xi) = \boldsymbol{I}\{\xi\langle \boldsymbol{Y} \rangle = \boldsymbol{y}\}$, where we recall that $\xi\langle \boldsymbol{Y} \rangle$ is the assignment in ξ to the variables $\boldsymbol{Y}$. We approximate this expectation by generating a set of M particles, estimating the value of the function or its expectation relative to each of the generated particles, and then aggregating the results.

For most of this chapter, we focus on methods that generate particles using random sampling: In section 12.1, we consider the simplest possible method, which simply generates samples from the original network. In section 12.2, we present a significantly improved method that generates samples from a distribution that is closer to the posterior distribution. In section 12.3, we discuss a method based on Markov chains that defines a sampling process that, as it converges, generates samples from distributions arbitrarily close to the posterior. In section 12.5, we consider a very different type of method, one that generates particles deterministically by

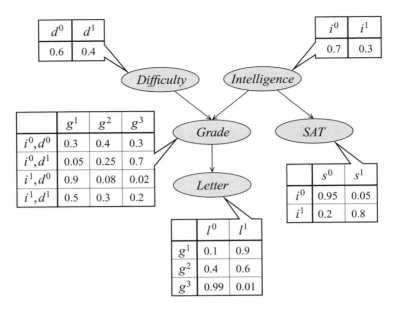

Figure 12.1 **The** Student **network** $\mathcal{B}^{student}$ revisited

searching for high-probability instantiations in the joint distribution. Finally, in section 12.4, we extend these methods to the case of collapsed particles. We note that, unlike our discussion of exact inference, some of the methods presented in this chapter — forward sampling and likelihood weighting — apply (at least in their simple form) only to Bayesian networks, and not to Markov networks or chain graphs.

12.1 Forward Sampling

forward sampling

The simplest approach to the generation of particles is *forward sampling*. Here, we generate random samples $\xi[1], \ldots, \xi[M]$ from the distribution $P(\mathcal{X})$. We first show how we can easily generate particles from $P_{\mathcal{B}}(\mathcal{X})$ by sampling from a Bayesian network. We then analyze the number of particles needed in order to get a good approximation of the expectation of some target function f. We finally discuss the difficulties in generating samples from the posterior $P_{\mathcal{B}}(\mathcal{X} \mid e)$. We note that, in undirected models, even generating a sample from the prior distribution is a difficult task.

12.1.1 Sampling from a Bayesian Network

Sampling from a Bayesian network is a very simple process.

Example 12.1

Consider the Student *network, shown again in figure 12.1. We begin by sampling D with the appropriate (unconditional) distribution; that is, we figuratively toss a coin that lands heads (d^1) 40 percent of the time and tails (d^0) the remaining 60 percent. Let us assume that the coin landed*

Algorithm 12.1 Forward Sampling in a Bayesian network

    Procedure Forward-Sample (
       $\mathcal{B}$ // Bayesian network over $\mathcal{X}$
)
1   Let $X_1, \ldots, X_n$ be a topological ordering of $\mathcal{X}$
2   **for** $i = 1, \ldots, n$
3          $\boldsymbol{u}_i \leftarrow \boldsymbol{x}\langle \mathrm{Pa}_{X_i} \rangle$ // Assignment to Pa_{X_i} in $x_1, \ldots, x_{i-1}$
4   Sample x_i from $P(X_i \mid \boldsymbol{u}_i)$
5   **return** $(x_1, \ldots, x_n)$

---

heads, so that we pick the value d^1 for D. Similarly, we sample I from its distribution; say that the result is i^0. Given those, we know the right distribution from which to sample G: $P(G \mid i^0, d^1)$, as defined by G's CPD; we therefore pick G to be g^1 with probability 0.05, g^2 with probability 0.25, and g^3 with probability 0.7. The process continues similarly for S and L. ∎

As shown in algorithm 12.1, we sample the nodes in some order consistent with the partial order of the BN, so that by the time we sample a node we have values for all of its parents. We can then sample from the distribution defined by the CPD and by the chosen values for the node's parents. Note that the algorithm requires that we have the ability to sample from the distributions underlying our CPD. Such sampling is straightforward in the discrete case (see box 12.A), but subtler when dealing with continuous measures (see section 14.5.1).

Box 12.A — Skill: Sampling from a Discrete Distribution. *How do we generate a sample from a distribution? For a uniform distribution, we can use any pseudo-random number generator on our machine. Other distributions require more thought, and much work has been devoted in statistics to the problem of sampling from a variety of parametric distributions. Most obviously, consider a multinomial distribution $P(X)$ for $Val(X) = \{x^1, \ldots, x^k\}$, which is defined by parameters $\theta_1, \ldots, \theta_k$. This process can be done quite simply as follows: We generate a sample s uniformly from the interval $[0, 1]$. We then partition the interval into k subintervals: $[0, \theta_1), [\theta_1, \theta_1 + \theta_2), \ldots;$ that is, the ith interval is $[\sum_{j=1}^{i-1} \theta_j, \sum_{j=1}^{i} \theta_j)$. If s is in the ith interval, then the sampled value is x^i. We can determine the interval for s using binary search in time $O(\log k)$.*

This approach gives us a general-purpose solution for generating samples from the CPD of any discrete-valued variable: given a parent assignment $\boldsymbol{u}$, we can always generate the full conditional distribution $P(X \mid \boldsymbol{u})$ and sample from it. (Of course, more efficient methods may exist if X has a large value space or a CPD that requires an expensive computation.) As we discuss in section 14.5.1, the problem of sampling from continuous CPDs is considerably more complex.

---

convergence
bound

Using basic *convergence bounds* (see appendix A.2), we know that from a set of particles $\mathcal{D} = \{\xi[1], \ldots, \xi[M]\}$ generated via this sampling process, we can estimate the expectation of any function f as:

$$\hat{\boldsymbol{E}}_{\mathcal{D}}(f) = \frac{1}{M} \sum_{m=1}^{M} f(\xi[m]). \tag{12.1}$$

In the case where our task is to compute $P(\boldsymbol{y})$, this estimate is simply the fraction of particles where we have seen the event $\boldsymbol{y}$:

$$\hat{P}_{\mathcal{D}}(\boldsymbol{y}) = \frac{1}{M} \sum_{m=1}^{M} \boldsymbol{I}\{\boldsymbol{y}[m] = \boldsymbol{y}\}, \tag{12.2}$$

where we use $\boldsymbol{y}[m]$ to denote $\xi[m]\langle\boldsymbol{Y}\rangle$ — the assignment to $\boldsymbol{Y}$ in the particle $\xi[m]$. For example, our estimate for the probability of an event such as i^1, l^0 (a smart student getting a bad letter) is the fraction of particles in which I got the value i^1 and L the value l^0. Note that the same set of particles can be used to estimate the probabilities of multiple events.

This sampling process requires one sampling operation for each random variable in the network. For each variable X, we need to index into the CPD using the current partial instantiation of the parents of X. Using an appropriate data structure, this indexing can be accomplishing in time $O(|\text{Pa}_X|)$. The actual sampling process, as we discussed, requires time $O(\log|Val(X)|)$ (assuming appropriate preprocessing). Letting M be the total number of particles generated, $n = |\mathcal{X}|$, $p = \max_i |\text{Pa}_{X_i}|$, and $d = \max_i |Val(X_i)|$, the overall cost is $O(Mnp \log d)$.

12.1.2   Analysis of Error

Of course, the quality of the estimate obtained depends heavily on the number of particles generated. We now analyze the question of the number of particles required to obtain certain performance guarantees. We focus on the analysis for the case where our goal is to estimate $P(\boldsymbol{y})$.

The techniques of appendix A.2 provide us with the necessary tools for this analysis. Consider the quality of our estimate for a particular event $\boldsymbol{Y} = \boldsymbol{y}$. We define a new random variable over the probability space of P, using the indicator function $\boldsymbol{I}\{\boldsymbol{Y} = \boldsymbol{y}\}$. This is a Bernoulli random variable, and hence our M particles in $\mathcal{D}$ define M independent Bernoulli trials, each with success probability $P(\boldsymbol{y})$.

Hoeffding bound

We can now apply the *Hoeffding bound* (theorem A.3) to show that this estimate is close to the truth with high probability:

$$P_{\mathcal{D}}(\hat{P}_{\mathcal{D}}(\boldsymbol{y}) \notin [P(\boldsymbol{y}) - \epsilon, P(\boldsymbol{y}) + \epsilon]) \leq 2e^{-2M\epsilon^2}. \tag{12.3}$$

This analysis provides us with an estimate of how many samples are required to achieve an estimate whose error is bounded by ϵ, with probability at least $1 - \delta$. Setting

$$2e^{-2M\epsilon^2} \leq \delta$$

sample size

estimator

and doing simple algebraic manipulations, we get that the required *sample size* to get an *estimator* with (ϵ, δ) reliability is:

$$M \geq \frac{\ln(2/\delta)}{2\epsilon^2}.$$

Chernoff bound

relative error
We can similarly apply the *Chernoff bound* (theorem A.4) to conclude that $\hat{P}_\mathcal{D}(\boldsymbol{y})$ is also within a *relative error* ϵ of the true value $P(\boldsymbol{y})$, with high probability. Specifically, we have that:

$$P_\mathcal{D}(\hat{P}_\mathcal{D}(\boldsymbol{y}) \notin P(\boldsymbol{y})(1 \pm \epsilon)) \leq 2e^{-MP(\boldsymbol{y})\epsilon^2/3}. \tag{12.4}$$

Note that in this analysis, unlike the one based on Hoeffding's bound, the error probability (the chance of getting an estimate that is more than ϵ away from the true value) depends on the actual target value $P(\boldsymbol{y})$. This dependence is not surprising for a relative error bound. Assume that we generate M samples, but we generate none where $\boldsymbol{y}[m] = \boldsymbol{y}$. Our estimate $\hat{P}_\mathcal{D}(\boldsymbol{y})$ is simply 0. However, if $P(\boldsymbol{y})$ is very small, it is fairly likely that we simply have not generated any samples where this event holds. In this case, our estimate of 0 is not going to be within any relative error of $P(\boldsymbol{y})$. Thus, for very small values of $P(\boldsymbol{y})$, we need many more samples in order to guarantee that our estimate is close with high probability.

Examining equation (12.4), we can see that, for a given ϵ, the number of samples needed to guarantee a certain error probability δ is:

$$M \geq 3\frac{\ln(2/\delta)}{P(\boldsymbol{y})\epsilon^2}. \tag{12.5}$$

Thus, the number of required samples grows inversely with the probability $P(\boldsymbol{y})$.

In summary, to guarantee an absolute error of ϵ with probability at least $1 - \delta$, we need a number of samples that grows logarithmically in $1/\delta$ and quadratically in $1/\epsilon$. To guarantee a relative error of ϵ with error probability at most δ, we need a number of samples that grows similarly in δ and ϵ, but that also grows linearly with $1/P(\boldsymbol{y})$. A significant problem with using this latter bound is that we do not know $P(\boldsymbol{y})$. (If we did, we would not have to estimate it.) Thus, we cannot determine how many samples we need in order to ensure a good estimate.

12.1.3 Conditional Probability Queries

So far, we have focused on the problem of estimating marginal probabilities, that is, probabilities of events $\boldsymbol{Y} = \boldsymbol{y}$ relative to the original joint distribution. In general, however, we are interested in conditional probabilities of the form $P(\boldsymbol{y} \mid \boldsymbol{E} = \boldsymbol{e})$. Unfortunately, it turns out that this estimation task is significantly harder.

rejection
sampling
One approach to this task is simply to generate samples from the posterior probability $P(\mathcal{X} \mid \boldsymbol{e})$. We can do so by a process called *rejection sampling*: We generate samples $\boldsymbol{x}$ from $P(\boldsymbol{X})$, as in section 12.1.1. We then reject any sample that is not compatible with $\boldsymbol{e}$. The resulting samples are sampled from the posterior $P(\boldsymbol{X} \mid \boldsymbol{e})$. The results of the analysis in section 12.1.2 now apply unchanged.

The problem, of course, is that the number of unrejected particles can be quite small. In general, the expected number of particles that are not rejected from an original sample set of size M is $MP(\boldsymbol{e})$. For example, if $P(\boldsymbol{e}) = 0.001$, then even for $M = 10,000$ samples, the expected number of unrejected particles is 10. Conversely, to obtain at least M^* unrejected particles, we need to generate on average $M = M^*/P(\boldsymbol{e})$ samples from $P(\boldsymbol{X})$.

Unfortunately, in many applications, low-probability evidence is the rule rather than the exception. For example, in medical diagnosis, any set of symptoms typically has low probability. In general, as the number of observed variables $k = |\mathbf{E}|$ grows, the probability of the evidence usually decreases exponentially with k.

An alternative approach to the problem is to use a separate estimator for $P(e)$ and for $P(y, e)$ and then compute the ratio. We can show that if we have estimators of low relative error for both of these quantities, then their ratio will also have a low relative error (exercise 12.2). Unfortunately, this approach only moves the problem from one place to the other. As we said, **the number of samples required to achieve a low relative error also grows linearly with** $1/P(e)$. **The number of samples required to get low *absolute* error does not grow with** $P(e)$. **However, it is not hard to verify (exercise 12.2) that a bound on the absolute error for** $P(e)$ **does not suffice to get any type of bound (relative or absolute) for the ratio** $P(y, e)/P(e)$.

12.2   Likelihood Weighting and Importance Sampling

The rejection sampling process seems very wasteful in the way it handles evidence. We generate multiple samples that are inconsistent with our evidence and that are ultimately rejected without contributing to our estimator. In this section, we consider an approach that makes our samples more relevant to our evidence.

12.2.1   Likelihood Weighting: Intuition

Consider the network in example 12.1, and assume that our evidence is d^1, s^1. Our forward sampling process might begin by generating a value of d^0 for D. No matter how the sampling process proceeds, this sample will always be rejected as being incompatible with the evidence.

It seems much more sensible to simply force the samples to take on the appropriate values at observed nodes. That is, when we come to sampling a node X_i whose value has been observed, we simply set it to its observed value.

In general, however, this simple approach can generate incorrect results:

Example 12.2   *Consider the network of example 12.1, and assume that our evidence is s^1 — a student who received a high SAT score. Using the naive process, we sample D and I from their prior distribution, set $S = s^1$, and then sample G and L appropriately. All of our samples will have $S = s^1$, as desired. However, the expected number of samples that have $I = i^1$ — an intelligent student — is 30 percent, the same as in the prior distribution. Thus, this approach fails to conclude that the posterior probability of i^1 is higher when we observe s^1.*   ∎

The problem with this approach is that it fails to account for the fact that, in the standard forward sampling process, the node S is more likely to take the value s^1 when its parent I has the value i^1 than when I has the value i^0. In particular, consider an imaginary process where we run rejection sampling many times; samples where we generated the value $I = i^1$ would have generated $S = s^1$ in 80 percent of the samples, whereas samples where we generated the value $I = i^0$ would have generated $S = s^1$ in only 5 percent of the samples. To simulate this long-run behavior within a single sample, we should conclude that a sample where we have

---

Algorithm 12.2 Likelihood-weighted particle generation

Procedure LW-Sample (
    $\mathcal{B}$,   // Bayesian network over $\mathcal{X}$
    $\boldsymbol{Z} = \boldsymbol{z}$ // Event in the network
)

1   Let $X_1, \ldots, X_n$ be a topological ordering of $\mathcal{X}$
2    $w \leftarrow 1$
3   **for** $i = 1, \ldots, n$
4        $\boldsymbol{u}_i \leftarrow \boldsymbol{x}\langle \mathrm{Pa}_{X_i} \rangle$ // Assignment to Pa_{X_i} in $x_1, \ldots, x_{i-1}$
5   **if** $X_i \notin \boldsymbol{Z}$ **then**
6   Sample x_i from $P(X_i \mid \boldsymbol{u}_i)$
7   **else**
8            $x_i \leftarrow \boldsymbol{z}\langle X_i \rangle$ // Assignment to X_i in $\boldsymbol{z}$
9            $w \leftarrow w \cdot P(x_i \mid \boldsymbol{u}_i)$ // Multiply weight by probability of desired value
10   **return** $(x_1, \ldots, x_n), w$

$I = i^1$ and force $S = s^1$ should be worth 80 percent of a sample, whereas one where we have $I = i^0$ and force $S = s^1$ should only be worth 5 percent of a sample.

When we have multiple observations and we want our sampling process to set all of them to their observed values, we need to consider the probability that each of the observation nodes, had it been sampled using the standard forward sampling process, would have resulted in the observed values. The sampling events for each node in forward sampling are independent, and hence the weight for each sample should be the product of the weights induced by each evidence node separately.

Example 12.3

Consider the same network, where our evidence set now consists of l^0, s^1. Assume that we sample $D = d^1$, $I = i^0$, set $S = s^1$, sample $G = g^2$, and set $L = l^0$. The probability that, given $I = i^0$, forward sampling would have generated $S = s^1$ is 0.05. The probability that, given $G = g^2$, forward sampling would have generated $L = l^0$ is 0.4. If we consider the standard forward sampling process, each of these events is the result of an independent coin toss. Hence, the probability that both would have occurred is simply the product of their probabilities. Thus, the weight required for this sample to compensate for the setting of the evidence is $0.05 \cdot 0.4 = 0.02$.∎

likelihood
weighting

Generalizing this intuition results in an algorithm called *likelihood weighting (LW)*, shown in algorithm 12.2. The name indicates that the weights of different samples are derived from the likelihood of the evidence accumulated throughout the sampling process.

weighted particle

This process generates a *weighted particle*. We can now estimate a conditional probability $P(\boldsymbol{y} \mid \boldsymbol{e})$ by using LW-Sample M times to generate a set $\mathcal{D}$ of weighted particles $\langle \xi[1], w[1] \rangle, \ldots, \langle \xi[M], w[M] \rangle$. We then estimate:

$$\hat{P}_{\mathcal{D}}(\boldsymbol{y} \mid \boldsymbol{e}) = \frac{\sum_{m=1}^{M} w[m]\boldsymbol{I}\{\boldsymbol{y}[m] = \boldsymbol{y}\}}{\sum_{m=1}^{M} w[m]}. \tag{12.6}$$

estimator

This *estimator* is an obvious generalization of the one we used for unweighted particles in equation (12.2). There, each particle had weight 1; hence, the terms in the numerator were unweighted, and the denominator, which is the sum of all the particle weights, was simply M. It is also important to note that, as in forward sampling, the same set of samples can be used to estimate the probability of any event $\boldsymbol{y}$.

Aside from some intuition, we have provided no formal justification for the correctness of LW as yet. It turns out that LW is a special case of a very general approach called *importance sampling*, which also provides us the basis for an analysis. We begin by providing a general description and analysis of importance sampling, and then reformulate LW as a special case of this framework.

12.2.2   Importance Sampling

importance
sampling

target
distribution

Let $\boldsymbol{X}$ be a variable (or set of variables) that takes on values in some space $Val(\boldsymbol{X})$. *Importance sampling* is a general approach for estimating the expectation of a function $f(\boldsymbol{x})$ relative to some distribution $P(\boldsymbol{X})$, typically called the *target distribution*. As we discussed, we can estimate this expectation by generating samples $\boldsymbol{x}[1], \ldots, \boldsymbol{x}[M]$ from P, and then estimating

$$\boldsymbol{E}_P[f] \approx \frac{1}{M} \sum_{m=1}^{M} f(\boldsymbol{x}[m]).$$

proposal
distribution

In some cases, however, we might prefer to generate samples from a different distribution Q, known as the *proposal distribution* or *sampling distribution*. There are several reasons why we might wish to sample from a different distribution. Most importantly for our purposes, it might be impossible or computationally very expensive to generate samples from P. For example, P might be a posterior distribution for a Bayesian network, or even a prior distribution for a Markov network.

In this section, we discuss how we might obtain estimates of an expectation relative to P by generating samples from a different distribution Q. In general, the proposal distribution Q can be arbitrary; we require only that $Q(\boldsymbol{x}) > 0$ whenever $P(\boldsymbol{x}) > 0$, so that Q does not

support

"ignore" any states that have nonzero probability relative to P. (More formally, the *support* of a distribution P is the set of points $\boldsymbol{x}$ for which $P(\boldsymbol{x}) > 0$; we require that the support of Q contain the support of P.) However, as we will see, the computational performance of this approach does depend strongly on the extent to which Q is similar to P.

12.2.2.1   Unnormalized Importance Sampling

If we generate samples from Q instead of P, we cannot simply average the f-value of the samples generated. We need to adjust our estimator to compensate for the incorrect sampling distribution. The most obvious way of adjusting our estimator is based on the observation that

$$\boldsymbol{E}_{P(\boldsymbol{X})}[f(\boldsymbol{X})] = \boldsymbol{E}_{Q(\boldsymbol{X})}\left[f(\boldsymbol{X}) \frac{P(\boldsymbol{X})}{Q(\boldsymbol{X})} \right]. \tag{12.7}$$

This equality follows directly:[1]

$$
\mathbf{E}_{Q(\boldsymbol{X})}\left[f(\boldsymbol{X})\frac{P(\boldsymbol{X})}{Q(\boldsymbol{X})}\right] = \sum_{\boldsymbol{x}} Q(\boldsymbol{x})f(\boldsymbol{x})\frac{P(\boldsymbol{x})}{Q(\boldsymbol{x})}
$$
$$
= \sum_{\boldsymbol{x}} f(\boldsymbol{x})P(\boldsymbol{x})
$$
$$
= \mathbf{E}_{P(\boldsymbol{X})}[f(\boldsymbol{X})].
$$

Based on this observation, we can use the standard estimator for expectations relative to Q. We generate a set of samples $\mathcal{D} = \{\boldsymbol{x}[1], \ldots, \boldsymbol{x}[M]\}$ from Q, and then estimate:

$$
\hat{\mathbf{E}}_{\mathcal{D}}(f) = \frac{1}{M}\sum_{m=1}^{M} f(\boldsymbol{x}[m])\frac{P(\boldsymbol{x}[m])}{Q(\boldsymbol{x}[m])}. \tag{12.8}
$$

unnormalized importance sampling estimator

We call this estimator the *unnormalized importance sampling* estimator; this method is also often called *unweighted* importance sampling (this terminology is confusing, inasmuch as the particles here are also associated with weights). The factor $P(\boldsymbol{x}[m])/Q(\boldsymbol{x}[m])$ can be viewed as a correction weight to the term $f(\boldsymbol{x}[m])$, which we would have used had Q been our target distribution. We use $w(\boldsymbol{x})$ to denote $P(\boldsymbol{x})/Q(\boldsymbol{x})$.

unbiased estimator

Our analysis immediately implies that this estimator is *unbiased*, that is, its mean for any data set is precisely the desired value:

Proposition 12.1

For data sets $\mathcal{D}$ sampled from Q, we have that:

$$
\mathbf{E}_{\mathcal{D}}\left[\hat{\mathbf{E}}_{\mathcal{D}}(f)\right] = \mathbf{E}_{Q(\boldsymbol{X})}[f(\boldsymbol{X})w(\boldsymbol{X})] = \mathbf{E}_{P(\boldsymbol{X})}[f(\boldsymbol{X})].
$$

We can also estimate the distribution of this estimator around its mean. Letting $\epsilon_{\mathcal{D}} = \hat{\mathbf{E}}_{\mathcal{D}}(f) - \mathbf{E}_{P}[f(\boldsymbol{x})]$, we have that, since $M \to \infty$:

$$
\mathbf{E}_{\mathcal{D}}[\epsilon_{\mathcal{D}}] \sim \mathcal{N}\left(0; \sigma_Q^2/M\right),
$$

where

$$
\sigma_Q^2 = \mathbf{E}_{Q(\boldsymbol{X})}\left[(f(\boldsymbol{X})w(\boldsymbol{X}))^2\right] - \mathbf{E}_{Q(\boldsymbol{X})}[(f(\boldsymbol{X})w(\boldsymbol{X}))]^2
$$
$$
= \mathbf{E}_{Q(\boldsymbol{X})}\left[(f(\boldsymbol{X})w(\boldsymbol{X}))^2\right] - (\mathbf{E}_{P(\boldsymbol{X})}[f(\boldsymbol{X})])^2. \tag{12.9}
$$

estimator variance

As we discussed in appendix A.2, the *variance* of this type of estimator — an average of M independent random samples from a distribution — decreases linearly with the number of samples. This point is important, since it allows us to provide a bound on the number of samples required to obtain a reliable estimate.

To understand the constant term in this expression, consider the (uninteresting) case where the function f is the constant function $f(\xi) \equiv 1$. In this case, equation (12.9) simplifies to:

$$
\mathbf{E}_{Q(\boldsymbol{X})}\left[w(\boldsymbol{X})^2\right] - \mathbf{E}_{P(\boldsymbol{X})}[1] = \mathbf{E}_{Q(\boldsymbol{X})}\left[\left(\frac{P(\boldsymbol{X})}{Q(\boldsymbol{X})}\right)^2\right] - \left(\mathbf{E}_{Q(\boldsymbol{X})}\left[\frac{P(\boldsymbol{X})}{Q(\boldsymbol{X})}\right]\right)^2,
$$

---

1. We present the proof in terms of discrete state spaces, but it holds equally for continuous state spaces.

which is simply the variance of the weighting function $P(x)/Q(x)$. Thus, the more different Q is from P, the higher the variance of this estimator. When f is an indicator function over part of the space, we obtain an identical expression restricted to the relevant subspace. In general, one can show that the lowest variance is achieved when

$$Q(\boldsymbol{X}) \propto |f(\boldsymbol{X})| P(\boldsymbol{X});$$

thus, for example, if f is an indicator function over part of the space, we want our sampling distribution to be P conditioned on the subspace.

Note that we should avoid cases where our sampling probability $Q(\boldsymbol{X}) \ll P(\boldsymbol{X})f(\boldsymbol{X})$ in any part of the space, since these cases can lead to very large or even infinite variance. Thus, care must be taken when using very skewed sampling distributions, to ensure that probabilities in Q are close to zero only when $P(\boldsymbol{X})f(\boldsymbol{X})$ is also very small.

12.2.2.2   Normalized Importance Sampling

One problem with the preceding discussion is that it assumes that P is known. A frequent situation, and one of the most common reasons why we must resort to sampling from a different distribution Q, is that P is known only up to a normalizing constant Z. Specifically, what we have access to is a function $\tilde{P}(\boldsymbol{X})$ such that $\tilde{P}$ is not a normalized distribution, but $\tilde{P}(\boldsymbol{X}) = ZP(\boldsymbol{X})$. For example, in a Bayesian network $\mathcal{B}$, we might have (for $\boldsymbol{X} = \mathcal{X}$) $P(\mathcal{X})$ be our posterior distribution $P_{\mathcal{B}}(\mathcal{X} \mid \boldsymbol{e})$, and $\tilde{P}(\mathcal{X})$ be the unnormalized distribution $P_{\mathcal{B}}(\mathcal{X}, \boldsymbol{e})$. In a Markov network, $P(\mathcal{X})$ might be $P_{\mathcal{H}}(\mathcal{X})$, and $\tilde{P}$ might be the unnormalized distribution obtained by multiplying together the clique potentials, but without normalizing by the partition function.

In this context, we cannot define the weights relative to P, so we define:

$$w(\boldsymbol{X}) = \frac{\tilde{P}(\boldsymbol{X})}{Q(\boldsymbol{X})}. \tag{12.10}$$

Unfortunately, with this definition of weights, the analysis justifying the use of equation (12.8) breaks down. However, we can use a slightly different estimator based on similar intuitions. As before, the weight $w(\boldsymbol{X})$ is a random variable. Its expected value is simply Z:

$$\boldsymbol{E}_{Q(\boldsymbol{X})}[w(\boldsymbol{X})] = \sum_{\boldsymbol{x}} Q(\boldsymbol{x}) \frac{\tilde{P}(\boldsymbol{x})}{Q(\boldsymbol{x})} = \sum_{\boldsymbol{x}} \tilde{P}(\boldsymbol{x}) = Z. \tag{12.11}$$

This quantity is the normalizing constant of the distribution $\tilde{P}$, which is itself often of considerable interest, as we will see in our discussion of learning algorithms.

We can now rewrite equation (12.7):

$$
\begin{aligned}
\boldsymbol{E}_{P(\boldsymbol{X})}[f(\boldsymbol{X})] &= \sum_{\boldsymbol{x}} P(\boldsymbol{x})f(\boldsymbol{x}) \\
&= \sum_{\boldsymbol{x}} Q(\boldsymbol{x})f(\boldsymbol{x})\frac{P(\boldsymbol{x})}{Q(\boldsymbol{x})} \\
&= \frac{1}{Z}\sum_{\boldsymbol{x}} Q(\boldsymbol{x})f(\boldsymbol{x})\frac{\tilde{P}(\boldsymbol{x})}{Q(\boldsymbol{x})} \\
&= \frac{1}{Z}\boldsymbol{E}_{Q(\boldsymbol{X})}[f(\boldsymbol{X})w(\boldsymbol{X})] \\
&= \frac{\boldsymbol{E}_{Q(\boldsymbol{X})}[f(\boldsymbol{X})w(\boldsymbol{X})]}{\boldsymbol{E}_{Q(\boldsymbol{X})}[w(\boldsymbol{X})]}.
\end{aligned}
\tag{12.12}
$$

We can use an empirical estimator for both the numerator and denominator. Given M samples $\mathcal{D} = \{\boldsymbol{x}[1],\ldots,\boldsymbol{x}[M]\}$ from Q, we can estimate:

$$
\hat{\boldsymbol{E}}_{\mathcal{D}}(f) = \frac{\sum_{m=1}^{M} f(\boldsymbol{x}[m])w(\boldsymbol{x}[m])}{\sum_{m=1}^{M} w(\boldsymbol{x}[m])}.
\tag{12.13}
$$

normalized
importance
sampling
estimator

We call this estimator the *normalized importance sampling estimator*; it is also known as the *weighted* importance sampling estimator.

The normalized estimator involves a quotient, and it is therefore much more difficult to analyze theoretically. However, unlike the unnormalized estimator of equation (12.8), the normalized estimator is not unbiased. This bias is particularly immediate in the case $M = 1$. Here, the estimator reduces to:

$$
\frac{f(\boldsymbol{x}[1])w(\boldsymbol{x}[1])}{w(\boldsymbol{x}[1])} = f(\boldsymbol{x}[1]).
$$

Because $\boldsymbol{x}[1]$ is sampled from Q, the mean of the estimator in this case is $\boldsymbol{E}_{Q(\boldsymbol{X})}[f(\boldsymbol{X})]$ rather than the desired $\boldsymbol{E}_{P(\boldsymbol{X})}[f(\boldsymbol{X})]$. Conversely, when M goes to infinity, we have that each of the numerators and denominators converges to the expected value, and our analysis of the expectation applies. In general, for finite M, the estimator is biased, and the bias goes down as $1/M$.

One can show that the variance of the importance sampling estimator with M data instances is approximately:

$$
\boldsymbol{Var}_{P}\left[\hat{\boldsymbol{E}}_{\mathcal{D}}(f(\boldsymbol{X}))\right] \approx \frac{1}{M}\boldsymbol{Var}_{P}[f(\boldsymbol{X})](1 + \boldsymbol{Var}_{Q}[w(\boldsymbol{X})]),
\tag{12.14}
$$

which also goes down as $1/M$. Theoretically, this variance and the variance of the unnormalized estimator (equation (12.8)) are incomparable, and each of them can be larger than the other. Indeed, it is possible to construct examples where each of them performs better than the other. In practice, however, the variance of the normalized estimator is typically lower than that of the unnormalized estimator. This reduction in variance often outweighs the bias term, so that the normalized estimator is often used in place of the unnormalized estimator, even in cases where P is known and we can sample from it effectively.

Note that equation (12.14) can be used to provide a rough estimate on the quality of a set of samples generated using normalized importance sampling. Assume that we were to estimate $\boldsymbol{E}_P[f]$ using a standard sampling method, where we generate M IID samples from $P(\boldsymbol{X})$. (Obviously, this is generally intractable, but it provides a useful benchmark for comparison.) This approach would result in a variance $\boldsymbol{Var}_P[f(\boldsymbol{X})]/M$. The ratio between these two variances is:

$$\frac{1}{1 + \boldsymbol{Var}_Q[w(\boldsymbol{x})]}.$$

Thus, we would expect M weighted samples generated by importance sampling to be "equivalent" to $M/(1 + \boldsymbol{Var}_Q[w(\boldsymbol{x})])$ samples generated by IID sampling from P. We can use this observation to define a rule of thumb for the *effective sample size* of a particular set $\mathcal{D}$ of M samples resulting from a particular run of importance sampling:

effective sample
size

$$M_{\text{eff}} = \frac{M}{1 + \boldsymbol{Var}[\mathcal{D}]} \tag{12.15}$$

$$\boldsymbol{Var}[\mathcal{D}] = \sum_{m=1}^{M} w(\boldsymbol{x}[m])^2 - \left(\sum_{m=1}^{M} w(\boldsymbol{x}[m])\right)^2.$$

This estimate can tell us whether we should continue generating additional samples.

12.2.3   Importance Sampling for Bayesian Networks

With this theoretical foundation, we can now describe the application of importance sampling to Bayesian networks. We begin by providing the proposal distribution most commonly used for Bayesian networks. This distribution Q uses the network structure and its CPDs to focus the sampling process on a particular part of the joint distribution — the one consistent with a particular event $\boldsymbol{Z} = \boldsymbol{z}$. We show several ways in which this construction can be applied to the Bayesian network inference task, dealing with various types of probability queries. Finally, we briefly discuss several other proposal distributions, which are somewhat more complicated to implement but may perform better in practice.

12.2.3.1   The Mutilated Network Proposal Distribution

Assume that we are interested in a particular event $\boldsymbol{Z} = \boldsymbol{z}$, either because we wish to estimate its probability, or because we have observed it as evidence. We wish to focus our sampling process on the parts of the joint that are consistent with this event. In this section, we define an importance sampling process that achieves this goal.

To gain some intuition, consider the network of figure 12.1 and assume that we are interested in a particular event concerning a student's grade: $G = g^2$. We wish to bias our sampling toward parts of the space where this event holds. It is easy to take this event into consideration when sampling L: we simply sample L from $P(L \mid g^2)$. However, it is considerably more difficult to account for G's influence on D, I, and S without doing inference in the network.

Our goal is to define a simple proposal distribution that allows for the efficient generation of particles. We therefore avoid the problem of accounting for the effect of the event on nondescendants; we define a proposal distribution that "sets" the value of a $Z \in \boldsymbol{Z}$ to take the

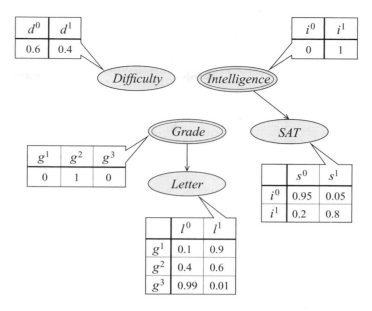

Figure 12.2 **The mutilated network $\mathcal{B}^{student}_{I=i^1,G=g^2}$ used for likelihood weighting**

prespecified value in a way that influences the sampling process for its descendants, but not for the other nodes in the network. The proposal distribution is most easily described in terms of a Bayesian network:

Definition 12.1

mutilated network

Let $\mathcal{B}$ be a network, and $\mathcal{Z}_1 = z_1, \ldots, Z_k = z_k$, abbreviated $\mathbf{Z} = \mathbf{z}$, an instantiation of variables. We define the mutilated network $\mathcal{B}_{\mathbf{Z}=\mathbf{z}}$ as follows:

- *Each node $Z_i \in \mathbf{Z}$ has no parents in $\mathcal{B}_{\mathbf{Z}=\mathbf{z}}$; the CPD of Z_i in $\mathcal{B}_{\mathbf{Z}=\mathbf{z}}$ gives probability 1 to $Z_i = z_i$ and probability 0 to all other values $z_i' \in Val(Z_i)$.*
- *The parents and CPDs of all other nodes $X \notin \mathbf{Z}$ are unchanged.* ∎

For example, the network $\mathcal{B}^{student}_{I=i^1,G=g^2}$ is shown in figure 12.2. As we can see, the node G is decoupled from its parents, eliminating its dependence on them (the node I has no parents in the original network, so its parent set remains empty). Furthermore, both I and G have CPDs that are deterministic, ascribing probability 1 to their (respective) observed values.

Importance sampling with this proposal distribution is precisely equivalent to the LW algorithm shown in algorithm 12.2, with $\tilde{P}(\mathcal{X}) = P_\mathcal{B}(\mathcal{X}, \mathbf{z})$ and the proposal distribution Q induced by the mutilated network $\mathcal{B}_{\mathbf{Z}=\mathbf{z}}$. More formally, we can show the following proposition:

Proposition 12.2

Let ξ be a sample generated by algorithm 12.2 and w be its weight. Then the distribution over ξ is as defined by the network $\mathcal{B}_{\mathbf{Z}=\mathbf{z}}$, and

$$w(\xi) = \frac{P_\mathcal{B}(\xi)}{P_{\mathcal{B}_{\mathbf{Z}=\mathbf{z}}}(\xi)}.$$

The proof is not difficult and is left as an exercise (exercise 12.4). It is important to note, however, that the algorithm does not require the explicit construction of the mutilated network. It simply traverses the original network, using the process shown in algorithm 12.2.

As we now show, this proposal distribution can be used for estimating a variety of Bayesian network queries.

12.2.3.2   Unconditional Probability of an Event ⋆

We begin by considering the simple problem of computing the unconditional probability of an event $\boldsymbol{Z} = \boldsymbol{z}$. Although we can clearly use forward sampling for estimating this probability, we can also use unnormalized importance sampling, where the target distribution P is simply our prior distribution $P_\mathcal{B}(\mathcal{X})$, and the proposal distribution Q is the one defined by the mutilated network $\mathcal{B}_{\boldsymbol{Z}=\boldsymbol{z}}$. Our goal is to estimate the expectation of a function f, which is the indicator function of the query $\boldsymbol{z}$: $f(\xi) = \boldsymbol{I}\{\xi\langle\boldsymbol{Z}\rangle = \boldsymbol{z}\}$.

The unnormalized importance-sampling estimator for this case is simply:

$$
\begin{aligned}
\hat{P}_{\mathcal{D}}(\boldsymbol{z}) &= \frac{1}{M}\sum_{m=1}^{M} \boldsymbol{I}\{\xi[m]\langle\boldsymbol{Z}\rangle = \boldsymbol{z}\}w(\xi[m]) \\
&= \frac{1}{M}\sum_{m=1}^{M} w[m],
\end{aligned}
\tag{12.16}
$$

where the equality follows because, by definition of Q, our sampling process generates samples $\xi[m]$ only where $\boldsymbol{z}$ holds.

When trying to bound the relative error of an estimator, a key quantity is the variance of the estimator *relative to its mean*. In the Chernoff bound, when we are estimating the probability p of a very low-probability event, the variance of the estimator, which is $p(1 - p)$, is very high relative to the mean p. Importance sampling removes some of the variance associated with this sampling process, and it can therefore achieve better performance in certain cases.

In this case, the samples are derived from our proposal distribution Q, and the value of the function whose expectation we are computing is simply the weight. Thus, we need to bound the variance of the function $w(\mathcal{X})$ under our distribution Q. Let us consider the sampling process in the algorithm. As we go through the variables in the network, we encounter the observed variables $Z_1, \ldots, Z_k$. At each point, we multiply our current weight w by some conditional probability number $P_\mathcal{B}(Z_i = z_i \mid \mathrm{Pa}_{Z_i})$.

One situation where we can bound the variance arises in a restricted class of networks, one where the entries in the CPD of the variables Z_i are bounded away from the extremes of 0 and 1. More precisely, we assume that there is some pair of numbers $\ell > 0$ and $u < 1$ such that: for each variable $Z \in \boldsymbol{Z}$, $z \in Val(Z)$, and $\boldsymbol{u} \in Val(\mathrm{Pa}_Z)$, we have that $P_\mathcal{B}(Z = z \mid \mathrm{Pa}_Z = \boldsymbol{u}) \in [\ell, u]$. Next, we assume that $|\boldsymbol{Z}| = k$ for some small k. This assumption is not a trivial one; while queries often involve only a small number of variables, we often have a fairly large number of observations that we wish to incorporate.

Under these assumptions, the weight w generated through the LW process is necessarily in the interval ℓ^k and u^k. We can now redefine our weights by dividing each $w[m]$ by u^k:

$$w'[m] = w[m]/u^k.$$

Each weight $w'[m]$ is now a real-valued random variable in the range $[(\ell/u)^k, 1]$. For a data set $\mathcal{D}$ of weights $w[1], \ldots, w[M]$, we can now define:

$$\hat{p}'_{\mathcal{D}} = \frac{1}{M} \sum_{m=1}^{M} w'[m].$$

The key point is that the mean of this random variable, which is $P_\mathcal{B}(z)/u^k$, is therefore also in the range $[(\ell/u)^k, 1]$, and its variance is, at worst, the variance of a Bernoulli random variable with the same mean. Thus, we now have a random variable whose variance is not that small relative to its mean.

A simple generalization of Chernoff's bound (theorem A.4) to the case of real-valued variables can now be used to show that:

$$
\begin{aligned}
P_\mathcal{D}(\hat{P}_\mathcal{D}(z) \notin P_\mathcal{B}(z)(1 \pm \epsilon)) &= P_\mathcal{D}(\hat{p}'_\mathcal{D} \notin \frac{1}{u^k} P_\mathcal{B}(z)(1 \pm \epsilon)) \\
&\leq 2e^{-M \frac{1}{u^k} P_\mathcal{B}(z)\epsilon^2/3}.
\end{aligned}
$$

sample size We can use this equation, as in the case of Bernoulli random variables, to derive a sufficient condition for the *sample size* that can guarantee that the estimator $\hat{P}_\mathcal{D}(z)$ of equation (12.16) has error at most ϵ with probability at least $1 - \delta$:

$$M \geq \frac{3 \ln(2/\delta) u^k}{P_\mathcal{B}(z)\epsilon^2}. \tag{12.17}$$

Since $P_\mathcal{B}(z) \geq \ell^k$, a (stronger) sufficient condition is that:

$$M \geq \frac{3 \ln(2/\delta)}{\epsilon^2} \left(\frac{u}{\ell}\right)^k. \tag{12.18}$$

Chernoff bound It is instructive to compare this bound to the one we obtain from the *Chernoff bound* in equation (12.5). The bound in equation (12.18) makes a weaker assumption about the probability of the event z. Equation (12.5) requires that $P_\mathcal{B}(z)$ not be too low. By contrast, equation (12.17) assumes only that this probability is in a bounded range ℓ^k, u^k; the actual probability of the event z can still be very low — we have no guarantee on the actual magnitude of ℓ. Thus, for example, if our event z corresponds to a rare medical condition — one that has low probability given any instantiation of its parents — the estimator of equation (12.16) would give us a relative error bound, whereas standard sampling would not.

We can use this bound to determine in advance the number of samples required for a certain desired accuracy. A disadvantage of this approach is that it does not take into consideration the specific samples we happened to generate during our sampling process. Intuitively, not all samples contribute equally to the quality of the estimate. A sample whose weight is high is more compatible with the evidence e, and it arguably provides us with more information. Conversely, a low-weight sample is not as informative, and a data set that contains a large number of low-weight samples might not be representative and might lead to a poor estimate. A somewhat more sophisticated approach is to preselect not the number of particles, but a predefined total weight. We then stop sampling when the total weight of the generated particles reaches our predefined lower bound.

Algorithm 12.3 Likelihood weighting with a data-dependent stopping rule

---

    Procedure Data-Dependent-LW (
        $\mathcal{B}$,   // Bayesian network over $\mathcal{X}$
        $\boldsymbol{Z} = \boldsymbol{z}$,   // Instantiation of interest
        u,   // Upper bound on CPD entries of $\boldsymbol{Z}$
        ϵ,   // Desired error bound
        δ // Desired probability of error
)
1      $\gamma \leftarrow \frac{4(1+\epsilon)}{\epsilon^2} \ln \frac{2}{\delta}$
2      $k \leftarrow |\boldsymbol{Z}|$
3      $W \leftarrow 0$
4      $M \leftarrow 0$
5   **while** $W < \gamma u^k$
6        $\xi, w \leftarrow$ LW-Sample$(\mathcal{B}, \boldsymbol{Z} = \boldsymbol{z})$
7        $W \leftarrow W + w$
8        $M \leftarrow M + 1$
9   **return** W/M

---

data-dependent likelihood weighting

For this algorithm, we can provide a similar theoretical analysis with certain guarantees for this *data-dependent likelihood weighting* approach. Algorithm 12.3 shows an algorithm that uses a data-dependent stopping rule to terminate the sampling process when enough weight has been accumulated. We can show that:

Theorem 12.1

Data-Dependent-LW *returns an estimate $\hat{p}$ for $P_{\mathcal{B}}(\boldsymbol{Z} = \boldsymbol{z})$ which, with probability at least $1 - \delta$, has a relative error of ϵ.*

expected sample size

We can also place an upper bound on the *expected sample size* used by the algorithm:

Theorem 12.2

The expected number of samples used by Data-Dependent-LW *is*

$$\frac{u^k}{P_{\mathcal{B}}(\boldsymbol{z})}\gamma \leq \left(\frac{u}{\ell}\right)^k \gamma,$$

where $\gamma = \frac{4(1+\epsilon)}{\epsilon^2} \ln \frac{2}{\delta}$.

The intuition behind this result is straightforward. The algorithm terminates when $W \geq \gamma u^k$. The expected contribution of each sample is $\boldsymbol{E}_{Q(\mathcal{X})}[w(\xi)] = P_{\mathcal{B}}(\boldsymbol{z})$. Thus, the total number of samples required to achieve a total weight of $W \geq \gamma u^k$ is $M \geq \gamma u^k / P_{\mathcal{B}}(\boldsymbol{z})$. Although this bound on the expected number of samples is no better than our bound in equation (12.17), the data-dependent bound allows us to stop early in cases where we were lucky in our random choice of samples, and to continue sampling in cases where we were unlucky.

12.2.3.3   Ratio Likelihood Weighting

ratio likelihood weighting

We now move to the problem of computing a conditional probability $P(\boldsymbol{y} \mid \boldsymbol{e})$ for a specific event $\boldsymbol{y}$. One obvious approach is *ratio likelihood weighting*: we compute the conditional probability as $P(\boldsymbol{y}, \boldsymbol{e})/P(\boldsymbol{e})$, and use unnormalized importance sampling (equation (12.16)) for both the numerator and denominator.

We can therefore estimate the conditional probability $P(\boldsymbol{y} \mid \boldsymbol{e})$ in two phases: We use the algorithm of algorithm 12.2 M times with the argument $\boldsymbol{Y} = \boldsymbol{y}, \boldsymbol{E} = \boldsymbol{e}$, to generate one set $\mathcal{D}$ of weighted samples $(\xi[1], w[1]), \ldots, (\xi[M], w[M])$. We use the same algorithm M' times with the argument $\boldsymbol{E} = \boldsymbol{e}$, to generate another set $\mathcal{D}'$ of weighted samples $(\xi'[1], w'[1]), \ldots, (\xi'[M'], w'[M'])$. We can then estimate:

$$\hat{P}_{\mathcal{D}}(\boldsymbol{y} \mid \boldsymbol{e}) = \frac{\hat{P}_{\mathcal{D}}(\boldsymbol{y}, \boldsymbol{e})}{\hat{P}_{\mathcal{D}'}(\boldsymbol{e})} = \frac{1/M \sum_{m=1}^{M} w[m]}{1/M' \sum_{m=1}^{M'} w'[m]}. \tag{12.19}$$

In ratio LW, the numerator and denominator are both using unnormalized importance sampling, which admits a rigorous theoretical analysis. Thus, we can now provide bounds on the number of samples M required to obtain a good estimate for both $P(\boldsymbol{y}, \boldsymbol{e})$ and $P(\boldsymbol{e})$.

12.2.3.4 Normalized Likelihood Weighting

Ratio LW allows us to estimate the probability of a single query $P(\boldsymbol{y} \mid \boldsymbol{e})$. In many cases, however, we are interested in estimating an entire joint distribution $P(\boldsymbol{Y} \mid \boldsymbol{e})$ for some variable or subset of variables $\boldsymbol{Y}$. We can answer such a query by running ratio LW for each $\boldsymbol{y} \in Val(\boldsymbol{Y})$, but this approach is typically too computationally expensive to be practical.

normalized
likelihood
weighting

An alternative approach is to use *normalized likelihood weighting*, which is based on the normalized importance sampling estimator of equation (12.13). In this application, our target distribution is $P(\mathcal{X}) = P_{\mathcal{B}}(\mathcal{X} \mid \boldsymbol{e})$. As we mentioned, we do not have access to P directly; rather, we can evaluate $\tilde{P}(\mathcal{X}) = P_{\mathcal{B}}(\mathcal{X}, \boldsymbol{e})$, which is the probability of a full assignment and can be easily computed via the chain rule. In this case, we are trying to estimate the expectation of a function f which is the indicator function of the query $\boldsymbol{y}$: $f(\xi) = \boldsymbol{I}\{\xi\langle \boldsymbol{Y} \rangle = \boldsymbol{y}\}$. Applying the normalized importance sampling estimator of equation (12.13) to this setting, we obtain precisely the estimator of equation (12.6).

☞ **The quality of the importance sampling estimator depends largely on how close the proposal distribution Q is to the target distribution P. We can gain intuition for this question by considering two extreme cases. If all of the evidence in our network is at the roots, the proposal distribution is precisely the posterior, and there is no need to compensate; indeed, no evidence is encountered along the way, and all samples will have the same weight $P(\boldsymbol{e})$. On the other side of the spectrum, if all of the evidence is at the leaves, our proposal distribution $Q(\mathcal{X})$ is the prior distribution $P_{\mathcal{B}}(\mathcal{X})$, leaving the correction purely to the weights. In this situation, LW will work reasonably only if the prior is similar to the posterior. Otherwise, most of our samples will be irrelevant, a fact that will be reflected by their low weight.** For example, consider a medical-diagnosis setting, and assume that our evidence is a very unusual combination of symptoms generated by only one very rare disease. Most samples will not involve this disease and will give only very low probability to this combination of symptoms. Indeed, the combinations sampled are likely to be irrelevant and are not useful at all for understanding what disease the patient has. We return to this issue in section 12.2.4.

To understand the relationship between the prior and the posterior, note that the prior is a weighted average of the posteriors, weighted over different instantiations of the evidence:

$$P(\mathcal{X}) = \sum_e P(e)P(\mathcal{X} \mid e).$$

If the evidence is very likely, then it is a major component in this summation, and it is probably not too far from the prior. For example, in the network $\mathcal{B}^{student}$, the event $S = s^1$ is fairly likely, and the posterior distribution $P_{\mathcal{B}^{student}}(\mathcal{X} \mid s^1)$ is fairly similar to the prior. However, for unlikely evidence, the weight of $P(\mathcal{X} \mid e)$ is negligible, and there is nothing constraining the posterior to be similar to the prior. Indeed, our distribution $P_{\mathcal{B}^{student}}(\mathcal{X} \mid l^0)$ is very different from the prior.

Unfortunately, there is currently no formal analysis for the number of particles required to achieve a certain quality of estimate using normalized importance sampling. In many cases, we simply preselect a number of particles that seems large enough, and we generate that number. Alternatively, we can use a heuristic approach that uses the total weight of the particles generated so far as guidance as to the extent to which they are representative. Thus, for example, we might decide to generate samples until a certain minimum bound on the total weight has been reached, as in Data-Dependent-LW. We note, however, that this approach is entirely heuristic in this case (as in all cases where we do not have bounds $[\ell, u]$ on our CPDs). Furthermore, there are cases where the evidence is simply unlikely in all configurations, and therefore all samples will have low weights.

12.2.3.5   Conditional Probabilities: Comparison

We have seen two variants of likelihood weighting: normalized LW and ratio LW. Ratio LW has two related advantages. The normalized LW process samples an assignment of the variables $\boldsymbol{Y}$ (those not in $\boldsymbol{E}$), whereas ratio LW simply sets the values of these variables. The additional sampling step for $\boldsymbol{Y}$ introduces additional variance into the overall process, leading to a reduction in the robustness of the estimate. Thus, in many cases, the variance of this estimator is lower than that of equation (12.6), leading to more robust estimates.

A second advantage of ratio LW is that it is much easier to analyze, and therefore it is associated with stronger guarantees regarding the number of samples required to get a good estimate. However, these bounds are useful only under very strong conditions: a small number of evidence variables, and a bound on the skew of the CPD entries in the network.

On the other hand, a significant disadvantage of ratio LW is the fact that each query $\boldsymbol{y}$ requires that we generate a new set of samples for the event $\boldsymbol{y}, \boldsymbol{e}$. It is often the case that we want to evaluate the probability of multiple queries relative to the same set of evidence. The normalized LW approach allows these multiple computations to be executed relative to the same set of samples, whereas ratio LW requires a separate sample set for each query $\boldsymbol{y}$. This cost is particularly problematic when we are interested in computing the joint distribution over a subset of variables. Probably due to this last point, normalized LW is used more often in practice.

12.2.4   Importance Sampling Revisited

The likelihood weighting algorithm uses, as its proposal distribution, the very simple distribution obtained from mutilating the network by eliminating edges incoming to observed variables. However, this proposal distribution can be far from optimal. For example, if the CPDs associated with these evidence variables are skewed, the importance weights are likely to be quite large, resulting in estimators with high variance. Indeed, somewhat surprisingly, even in very simple cases, the obvious proposal distribution may not be optimal. For example, if X is not a root node in the network, the optimal proposal distribution for computing $P(X = x)$ may not be the distribution P, even without evidence! (See exercise 12.5.)

backward importance sampling

The importance sampling framework is very general, however, and several other proposal distributions have been utilized. For example, *backward importance sampling* generates samples for parents of evidence variables using the likelihood of their children. Most simply, if X is a variable whose child Y is observed to be $Y = y$, we might generate some samples for X from a renormalized distribution $Q(X) \propto P(Y = y \mid X)$. We can continue this process, sampling X's parents from the likelihood of X's sampled value. We can also propose more complex schemes that sample the value of a variable given a combination of sampled or observed values for some of its parents and/or children. One can also consider hybrid approaches that use some global approximate inference algorithm (such as those in chapter 11) to construct a proposal distribution, which is then used as the basis for sampling. **As long as the importance weights** **are computed correctly, we are guaranteed that this process is correct.** (See exercise 12.7.) This process can lead to significant improvements in theory, and it does lead to improvements in some cases in practice.

12.3 Markov Chain Monte Carlo Methods

One of the limitations of likelihood weighting is that an evidence node affects the sampling only for nodes that are its descendants. The effect on nodes that are nondescendants is accounted for only by the weights. As we discussed, in cases where much of the evidence is at the leaves of the network, we are essentially sampling from the prior distribution, which is often very far from the desired posterior. We now present an alternative sampling approach that generates a *sequence* of samples. This sequence is constructed so that, although the first sample may be generated from the prior, successive samples are generated from distributions that provably get closer and closer to the desired posterior. We note that, unlike forward sampling methods (including likelihood weighting), Markov chain methods apply equally well to directed and to undirected models. Indeed, the algorithm is easier to present in the context of a distribution P_Φ defined in terms of a general set of factors Φ.

12.3.1 Gibbs Sampling Algorithm

One idea for addressing the problem with forward sampling approaches is to try to "fix" the sample we generated by resampling some of the variables we generated early in the process. Perhaps the simplest method for doing this is presented in algorithm 12.4. This method, called

Gibbs sampling

Gibbs sampling, starts out by generating a sample of the unobserved variables from some initial distribution; for example, we may use the mutilated network to generate a sample using forward sampling. Starting from that sample, we then iterate over each of the unobserved variables, sampling a new value for each variable *given* our current sample for all other variables. This process allows information to "flow" across the network as we sample each variable.

To apply this algorithm to a network with evidence, we first reduce all of the factors by the observations e, so that the distribution P_Φ used in the algorithm corresponds to $P(\boldsymbol{X} \mid e)$.

Algorithm 12.4 Generating a Gibbs chain trajectory

    Procedure Gibbs-Sample (
        X // Set of variables to be sampled
        Φ // Set of factors defining P_Φ
        $P^{(0)}(X)$,   // Initial state distribution
        T // Number of time steps
)
1   Sample $x^{(0)}$ from $P^{(0)}(X)$
2   **for** $t = 1, \ldots, T$
3        $x^{(t)} \leftarrow x^{(t-1)}$
4   **for** each $X_i \in X$
5   Sample $x_i^{(t)}$ from $P_\Phi(X_i \mid x_{-i})$
6   // Change X_i in $x^{(t)}$
7   **return** $x^{(0)}, \ldots, x^{(T)}$

Example 12.4

Let us revisit example 12.3, recalling that we have the observations s^1, l^0. In this case, our algorithm will generate samples over the variables D, I, G. The set of reduced factors Φ is therefore: $P(I), P(D), P(G \mid I, D), P(s^1 \mid I), P(l^0 \mid G)$. Our algorithm begins by generating one sample, say by forward sampling. Assume that this sample is $d^{(0)} = d^1, i^{(0)} = i^0, g^{(0)} = g^2$. In the first iteration, it would now resample all of the unobserved variables, one at a time, in some predetermined order, say G, I, D. Thus, we first sample $g^{(1)}$ from the distribution $P_\Phi(G \mid d^1, i^0)$.

Note that because we are computing the distribution over a single variable given all the others, this computation can be performed very efficiently:

$$
\begin{aligned}
P_\Phi(G \mid d^1, i^0) &= \frac{P(i^0)P(d^1)P(G \mid i^0, d^1)P(l^0 \mid G)P(s^1 \mid i^0)}{\sum_g P(i^0)P(d^1)P(g \mid i^0, d^1)P(l^0 \mid g)P(s^1 \mid i^0)} \\
&= \frac{P(G \mid i^0, d^1)P(l^0 \mid G)}{\sum_g P(g \mid i^0, d^1)P(l^0 \mid g)}.
\end{aligned}
$$

Thus, we can compute the distribution simply by multiplying all factors that contain G, with all other variables instantiated, and renormalizing to obtain a distribution over G.

Having sampled $g^{(1)} = g^3$, we now continue to resampling $i^{(1)}$ from the distribution $P_\Phi(I \mid d^1, g^3)$, obtaining, for example, $i^{(1)} = i^1$; note that the distribution for I is conditioned on the newly sampled value $g^{(1)}$. Finally, we sample $d^{(1)}$ from $P_\Phi(D \mid g^3, i^1)$, obtaining d^1. The result of the first iteration of sampling is, then, the sample (i^1, d^1, g^3). The process now repeats. ∎

Note that, unlike forward sampling, the sampling process for G takes into consideration the downstream evidence at its child L. Thus, its sampling distribution is arguably closer to the posterior. Of course, it is not the true posterior, since it still conditions on the originally sampled values for I, D, which were sampled from the prior distribution. However, we now resample I and D from a distribution that conditions on the new value of G, so one can imagine that their sampling distribution may also be closer to the posterior. Thus, perhaps the next sample of G,

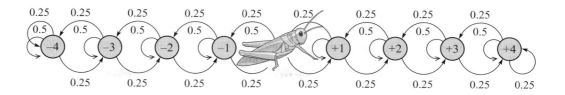

Figure 12.3 The Grasshopper **Markov chain**

which uses these new values for I, D (and conditions on the evidence l^0), will be sampled from a distribution even closer to the posterior. Indeed, this intuition is correct. One can show that, as we repeat this sampling process, the distribution from which we generate each sample gets closer and closer to the posterior $P_\Phi(\boldsymbol{X}) = P(\boldsymbol{X} \mid \boldsymbol{e})$.

Markov chain
Monte Carlo

In the subsequent sections, we formalize this intuitive argument using a framework called *Markov chain Monte Carlo (MCMC)*. This framework provides a general approach for generating samples from the posterior distribution, in cases where we cannot efficiently sample from the posterior directly. In MCMC, we construct an iterative process that gradually samples from distributions that are closer and closer to the posterior. A key question is, of course, how many iterations we should perform before we can collect a sample as being (almost) generated from the posterior. In the following discussion, we provide the formal foundations for MCMC algorithms, and we try to address this and other important questions. We also present several valuable generalizations.

12.3.2 Markov Chains

12.3.2.1 Basic Definition

At a high level, a Markov chain is defined in terms of a graph of states over which the sampling algorithm takes a random walk. In the case of graphical models, this graph is *not* the original graph, but rather a graph whose nodes are the possible assignments to our variables $\boldsymbol{X}$.

Definition 12.2

Markov chain

transition model

A Markov chain *is defined via a state space* $Val(\boldsymbol{X})$ *and a model that defines, for every state* $\boldsymbol{x} \in Val(X)$ *a next-state* distribution over $Val(X)$. *More precisely, the* transition model $\mathcal{T}$ *specifies for each pair of state* $\boldsymbol{x}, \boldsymbol{x}'$ *the probability* $\mathcal{T}(\boldsymbol{x} \rightarrow \boldsymbol{x}')$ *of going from* $\boldsymbol{x}$ *to* $\boldsymbol{x}'$. *This transition probability applies whenever the chain is in state* $\boldsymbol{x}$. ∎

We note that, in this definition and in the subsequent discussion, we restrict attention to *homogeneous*, where the system dynamics do not change over time.

homogeneous
Markov chain

We illustrate this concept with a simple example.

Example 12.5

Consider a Markov chain whose states consist of the nine integers $-4, \ldots, +4$, *arranged as points on a line. Assume that a drunken grasshopper starts out in position 0 on the line. At each point in time, it stays where it is with probability 0.5, or it jumps left and right with equal probability. Thus,* $\mathcal{T}(i \rightarrow i) = 0.5$, $\mathcal{T}(i \rightarrow i+1) = 0.25$, *and* $\mathcal{T}(i \rightarrow i-1) = 0.25$. *However, the two end positions are blocked by walls; hence, if the grasshopper is in position* $+4$ *and tries to jump right, it remains in position* $+4$. *Thus, for example,* $\mathcal{T}(+4 \to +4) = 0.75$. *We can visualize the state space as a graph, with probability-weighted directed edges corresponding to transitions between different states. The graph for our example is shown in figure 12.3.* ∎

We can imagine a random sampling process, that defines a random sequence of states $\boldsymbol{x}^{(0)}, \boldsymbol{x}^{(1)}, \boldsymbol{x}^{(2)}, \dots$. Because the transition model is random, the state of the process at step t can be viewed as a random variable $\boldsymbol{X}^{(t)}$. We assume that the initial state $\boldsymbol{X}^{(0)}$ is distributed according to some initial state distribution $P^{(0)}(\boldsymbol{X}^{(0)})$. We can now define distributions over the subsequent states $P^{(1)}(\boldsymbol{X}^{(1)}), P^{(2)}(\boldsymbol{X}^{(2)}), \dots$ using the chain dynamics:

$$P^{(t+1)}(\boldsymbol{X}^{(t+1)} = \boldsymbol{x}') = \sum_{\boldsymbol{x} \in Val(\boldsymbol{X})} P^{(t)}(\boldsymbol{X}^{(t)} = \boldsymbol{x})\mathcal{T}(\boldsymbol{x} \to \boldsymbol{x}'). \tag{12.20}$$

Intuitively, the probability of being at state $\boldsymbol{x}'$ at time $t+1$ is the sum over all possible states $\boldsymbol{x}$ that the chain could have been in at time t of the probability being in state $\boldsymbol{x}$ times the probability that the chain took a transition from $\boldsymbol{x}$ to $\boldsymbol{x}'$.

12.3.2.2   Asymptotic Behavior

For our purposes, the most important aspect of a Markov chain is its long-term behavior.

Example 12.6

Because the grasshopper's motion is random, we can consider its location at time t to be a random variable, which we denote $X^{(t)}$. Consider the distribution over $X^{(t)}$. Initially, the grasshopper is at 0, so that $P(X^{(0)} = 0) = 1$. At time 1, we have that $X^{(1)}$ is 0 with probability 0.5, and $+1$ or -1, each with probability 0.25. At time 2, we have that $X^{(2)}$ is 0 with probability $0.5^2 + 2 \cdot 0.25^2 = 0.375$, $+1$ and -1 each with probability $2(0.5 \cdot 0.25) = 0.25$, and $+2$ and -2 each with probability $0.25^2 = 0.0625$. As the process continues, the probability gets spread out over more and more of the states. For example, at time $t = 10$, the probabilities of the different states range from 0.1762 for the value 0, and 0.0518 for the values ± 4. At $t = 50$, the distribution is almost uniform, with a range of 0.1107–0.1116. ∎

Thus, one approach for sampling from the uniform distribution over the set $-4, \dots, +4$ is to start off at 0 and then randomly choose the next state from the transition model for this chain. After some number of such steps t, our state $X^{(t)}$ would be sampled from a distribution that is very close to uniform over this space. We note that this approach is not a very good one for sampling from a uniform distribution; indeed, the expected time required for such a chain even to reach the boundaries of the interval $[-K, K]$ is K^2 steps. However, this general approach applies much more broadly, including in cases where our "long-term" distribution is not one from which we can easily sample.

MCMC sampling   *Markov chain Monte carlo (MCMC) sampling* is a process that mirrors the dynamics of the Markov chain; the process of generating an MCMC trajectory is shown in algorithm 12.5. The sample $\boldsymbol{x}^{(t)}$ is drawn from the distribution $P^{(t)}$. We are interested in the limit of this process, that is, whether $P^{(t)}$ converges, and if so, to what limit.

Algorithm 12.5 Generating a Markov chain trajectory

    Procedure MCMC-Sample (
      $P^{(0)}(\boldsymbol{X})$,   // Initial state distribution
      $\mathcal{T}$,   // Markov chain transition model
      T // Number of time steps
)
1   Sample $\boldsymbol{x}^{(0)}$ from $P^{(0)}(\boldsymbol{X})$
2   **for** $t = 1, \ldots, T$
3   Sample $\boldsymbol{x}^{(t)}$ from $\mathcal{T}(\boldsymbol{x}^{(t-1)} \rightarrow \boldsymbol{X})$
4   **return** $\boldsymbol{x}^{(0)}, \ldots, \boldsymbol{x}^{(T)}$

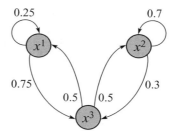

Figure 12.4   A simple Markov chain

12.3.2.3   Stationary Distributions

Intuitively, as the process converges, we would expect $P^{(t+1)}$ to be close to $P^{(t)}$. Using equation (12.20), we obtain:

$$P^{(t)}(\boldsymbol{x}') \approx P^{(t+1)}(\boldsymbol{x}') = \sum_{\boldsymbol{x} \in Val(\boldsymbol{X})} P^{(t)}(\boldsymbol{x})\mathcal{T}(\boldsymbol{x} \rightarrow \boldsymbol{x}').$$

At convergence, we would expect the resulting distribution $\pi(\boldsymbol{X})$ to be an equilibrium relative to the transition model; that is, the probability of being in a state is the same as the probability of transitioning into it from a randomly sampled predecessor. Formally:

Definition 12.3

stationary
distribution

A *distribution* $\pi(\boldsymbol{X})$ *is a* stationary distribution *for a Markov chain* $\mathcal{T}$ *if it satisfies:*

$$\pi(\boldsymbol{X} = \boldsymbol{x}') = \sum_{\boldsymbol{x} \in Val(\boldsymbol{X})} \pi(\boldsymbol{X} = \boldsymbol{x})\mathcal{T}(\boldsymbol{x} \rightarrow \boldsymbol{x}'). \tag{12.21}$$

                                                                             ■

A stationary distribution is also called an *invariant distribution.*[2]

---

2. If we view the transition model as a matrix defined as $A_{i,j} = \mathcal{T}(\boldsymbol{x}_i \rightarrow \boldsymbol{x}_j)$, then a stationary distribution is an eigen-vector of the matrix, corresponding to the eigen-value 1. In general, many aspects of the theory of Markov chains have an algebraic interpretation in terms of matrices and vectors.

As we have already discussed, the uniform distribution is a stationary distribution for the Markov chain of example 12.5. To take a slightly different example:

Example 12.7   *Figure 12.4 shows an example of a different simple Markov chain where the transition probabilities are less uniform. By definition, the stationary distribution π must satisfy the following three equations:*

$$\begin{aligned}
\pi(x^1) &= 0.25\pi(x^1) + 0.5\pi(x^3) \\
\pi(x^2) &= 0.7\pi(x^2) + 0.5\pi(x^3) \\
\pi(x^3) &= 0.75\pi(x^1) + 0.3\pi(x^2),
\end{aligned}$$

as well as the one asserting that it is a legal distribution:

$$\pi(x^1) + \pi(x^2) + \pi(x^3) = 1.$$

It is straightforward to verify that this system has a unique solution: $\pi(x^1) = 0.2$, $\pi(x^2) = 0.5$, $\pi(x^3) = 0.3$. For example, the first equation asserts that

$$0.2 = 0.25 \cdot 0.2 + 0.5 \cdot 0.3,$$

which clearly holds.   ∎

In general, there is no guarantee that our MCMC sampling process converges to a stationary distribution.

Example 12.8   *Consider the Markov chain over two states x^1 and x^2, such that $\mathcal{T}(x^1 \to x^2) = 1$ and $\mathcal{T}(x^2 \to x^1) = 1$. If $P^{(0)}$ is such that $P^{(0)}(x^1) = 1$, then the step t distribution $P^{(t)}$ has $P^{(t)}(x^1) = 1$ if t is even, and $P^{(t)}(x^2) = 1$ if t is odd. Thus, there is no convergence to a stationary distribution.∎*

periodic Markov   Markov chains such as this, which exhibit a fixed cyclic behavior, are called *periodic Markov*
chain   *chains.*

There is also no guarantee that the stationary distribution is unique: In some chains, the stationary distribution reached depends on our starting distribution $P^{(0)}$. Situations like this occur when the chain has several distinct regions that are not reachable from each other. Chains
reducible Markov   such as this are called *reducible Markov chains.*
chain

We wish to restrict attention to Markov chains that have a unique stationary distribution, which is reached from any starting distribution $P^{(0)}$. There are various conditions that suffice to guarantee this property. The condition most commonly used is a fairly technical one: that
ergodic Markov   the chain be *ergodic*. In the context of Markov chains where the state space $Val(\boldsymbol{X})$ is finite,
chain   the following condition is equivalent to this requirement:

Definition 12.4   *A Markov chain is said to be* regular *if there exists some number k such that, for every $\boldsymbol{x}, \boldsymbol{x}' \in$*
regular Markov $Val(\boldsymbol{X})$, *the probability of getting from $\boldsymbol{x}$ to $\boldsymbol{x}'$ in exactly k steps is > 0.*   ∎
chain

In our Markov chain of example 12.5, the probability of getting from any state to any state in exactly 9 steps is greater than 0. Thus, this Markov chain is regular. Similarly, in the Markov chain of example 12.7, we can get from any state to any state in exactly two steps.

The following result can be shown to hold:

Theorem 12.3

If a finite state Markov chain $\mathcal{T}$ is regular, then it has a unique stationary distribution.

Ensuring regularity is usually straightforward. Two simple conditions that together guarantee regularity in finite-state Markov chains are as follows. First, it is possible to get from any state to any state using a positive probability path in the state graph. Second, for each state x, there is a positive probability of transitioning from x to x in one step (a self-loop). These two conditions together are sufficient but not necessary to guarantee regularity (see exercise 12.12). However, they often hold in the chains used in practice.

12.3.2.4 Multiple Transition Models

In the case of graphical models, our state space has a factorized structure — each state is an assignment to several variables. When defining a transition model over this state space, we can consider a fully general case, where a transition can go from any state to any state. However, it is often convenient to decompose the transition model, considering transitions that update only a single component of the state vector at a time, that is, only a value for a single variable.

Example 12.9

Consider an extension to our Grasshopper *chain, where the grasshopper lives, not on a line, but in a two-dimensional plane. In this case, the state of the system is defined via a pair of random variables X, Y. Although we could define a joint transition model over both dimensions simultaneously, it might be easier to have separate transition models for the X and Y coordinate.* ∎

kernel

In this case, as in several other settings, we often define a set of transition models, each with its own dynamics. Each such transition model $\mathcal{T}_i$ is called a *kernel*. In certain cases, the different kernels are necessary, because no single kernel on its own suffices to ensure regularity. This is the case in example 12.9. In other cases, having multiple kernels simply makes the state space more "connected" and therefore speeds the convergence to a stationary distribution.

multi-kernel
Markov chain

There are several ways of constructing a single Markov chain from *multiple kernels*. One common approach is simply to select randomly between them at each step, using any distribution. Thus, for example, at each step, we might select one of $\mathcal{T}_1, \ldots, \mathcal{T}_k$, each with probability $1/k$. Alternatively, we can simply cycle over the different kernels, taking each one in turn. Clearly, this approach does not define a homogeneous chain, since the kernel used in step i is different from the one used in step $i + 1$. However, we can simply view the process as defining a single chain $\mathcal{T}$, each of whose steps is an aggregate step, consisting of first taking $\mathcal{T}_1$, then $\mathcal{T}_2$, ..., through $\mathcal{T}_k$.

In the case of graphical models, one approach is to define a multikernel chain, where we have a kernel $\mathcal{T}_i$ for each variable $X_i \in \boldsymbol{X}$. Let $\boldsymbol{X}_{-i} = \mathcal{X} - \{X_i\}$, and let $\boldsymbol{x}_i$ denote an instantiation to $\boldsymbol{X}_i$. The model $\mathcal{T}_i$ takes a state $(\boldsymbol{x}_{-i}, x_i)$ and transitions to a state of the form $(\boldsymbol{x}_{-i}, x'_i)$. As we discussed, we can combine the different kernels into a single global model in various ways.

Regardless of the structure of the different kernels, **we can prove that a distribution is a stationary distribution for the multiple kernel chain by proving that it is a stationary distribution (satisfies equation (12.21)) for each of individual kernels $\mathcal{T}_i$. Note that each kernel by itself is generally not ergodic; but as long as each kernel satisfies certain conditions (specified in definition 12.5) that imply that it has the desired stationary distribution, we can combine them to produce a coherent chain, which may be ergodic as a whole. This**

ability to add new types of transitions to our chain is an important asset in dealing with the issue of local maxima, as we will discuss.

12.3.3   Gibbs Sampling Revisited

The theory of Markov chains provides a general framework for generating samples from a target distribution π. In this section, we discuss the application of this framework to the sampling tasks encountered in probabilistic graphical models. In this case, we typically wish to generate samples from the posterior distribution $P(\boldsymbol{X} \mid \boldsymbol{E} = \boldsymbol{e})$, where $\boldsymbol{X} = \mathcal{X} - \boldsymbol{E}$. Thus, we wish to define a chain for which $P(\boldsymbol{X} \mid \boldsymbol{e})$ is the stationary distribution. Thus, we define the states of the Markov chain to be instantiations $\boldsymbol{x}$ to $\mathcal{X} - \boldsymbol{E}$. In order to define a Markov chain, we need to define a process that transitions from one state to the other, converging to a stationary distribution $\pi(\boldsymbol{X})$, which is the desired posterior distribution $P(\boldsymbol{X} \mid \boldsymbol{e})$. As in our earlier example, we assume that $P(\boldsymbol{X} \mid \boldsymbol{e}) = P_\Phi$ for some set of factors Φ that are defined by reducing the original factors in our graphical model by the evidence $\boldsymbol{e}$. This reduction allows us to simplify notation and to discuss the methods in a way that applies both to directed and undirected graphical models.

Gibbs chain

Gibbs sampling is based on one yet effective *Markov chain* for factored state spaces, which is particularly efficient for graphical models. We define the kernel $\mathcal{T}_i$ as follows. Intuitively, we simply "forget" the value of X_i in the current state and sample a new value for X_i from its posterior given the rest of the current state. More precisely, let $(\boldsymbol{x}_{-i}, x_i)$ be a state in the chain. We define:

$$\mathcal{T}_i((\boldsymbol{x}_{-i}, x_i) \to (\boldsymbol{x}_{-i}, x_i')) = P(x_i' \mid \boldsymbol{x}_{-i}). \tag{12.22}$$

Note that the transition probability does not depend on the current value x_i of X_i, but only on the remaining state $\boldsymbol{x}_{-i}$. It is not difficult to show that the posterior distribution $P_\Phi(\boldsymbol{X}) = P(\mathcal{X} \mid \boldsymbol{e})$ is a *stationary distribution* of this process. (See exercise 12.13.)

Gibbs stationary distribution

The sampling algorithm for a single trajectory of the Gibbs chain was shown earlier in this section, in algorithm 12.4. Recall that the Gibbs chain is defined via a set of kernels; we use the multistep approach to combine them. Thus, the different local kernels are taken consecutively; having changed the value for a variable X_1, the value for X_2 is sampled based on the new value. Note that a step in the aggregate chain occurs only once we have executed every local transition once.

Gibbs sampling is particularly easy to implement in the many graphical models where we can compute the transition probability $P(X_i \mid \boldsymbol{x}_{-i})$ (in line 5 of the algorithm) very efficiently. In particular, as we now show, this distribution can be done based only on the *Markov blanket* of X_i. We show this analysis for a Markov network; the application to Bayesian networks is straightforward. Recalling definition 4.4, we have that:

Markov blanket

$$
\begin{aligned}
P_\Phi(\boldsymbol{X}) &= \frac{1}{Z} \prod_j \phi_j(\boldsymbol{D}_j) \\
&= \frac{1}{Z} \prod_{j \,:\, X_i \in \boldsymbol{D}_j} \phi_j(\boldsymbol{D}_j) \prod_{j \,:\, X_i \notin \boldsymbol{D}_j} \phi_j(\boldsymbol{D}_j).
\end{aligned}
$$

Let $\boldsymbol{x}_{j,-i}$ denote the assignment in $\boldsymbol{x}_{-i}$ to $\boldsymbol{D}_j - \{X_i\}$, noting that when $X_i \notin \boldsymbol{D}_j$, $\boldsymbol{x}_{j,-i}$ is a full assignment to $\boldsymbol{D}_j$. We can now derive:

$$
\begin{aligned}
P(x_i' \mid \boldsymbol{x}_{-i}) &= \frac{P(x_i', \boldsymbol{x}_{-i})}{\sum_{x_i''} P(x_i'', \boldsymbol{x}_{-i})} \\[2mm]
&= \frac{\frac{1}{Z} \prod_{\boldsymbol{D}_j \ni X_i} \phi_j(x_i', \boldsymbol{x}_{j,-i}) \prod_{\boldsymbol{D}_j \not\ni X_i} \phi_j(x_i', \boldsymbol{x}_{j,-i})}{\frac{1}{Z} \sum_{x_i''} \prod_{\boldsymbol{D}_j \ni X_i} \phi_j(x_i'', \boldsymbol{x}_{j,-i}) \prod_{\boldsymbol{D}_j \not\ni X_i} \phi_j(x_i'', \boldsymbol{x}_{j,-i})} \\[2mm]
&= \frac{\prod_{\boldsymbol{D}_j \ni X_i} \phi_j(x_i', \boldsymbol{x}_{j,-i}) \prod_{\boldsymbol{D}_j \not\ni X_i} \phi_j(\boldsymbol{x}_{j,-i})}{\sum_{x_i''} \prod_{\boldsymbol{D}_j \ni X_i} \phi_j(x_i'', \boldsymbol{x}_{j,-i}) \prod_{\boldsymbol{D}_j \not\ni X_i} \phi_j(\boldsymbol{x}_{j,-i})} \\[2mm]
&= \frac{\prod_{\boldsymbol{D}_j \ni X_i} \phi_j(x_i', \boldsymbol{x}_{j,-i})}{\sum_{x_i''} \prod_{\boldsymbol{D}_j \ni X_i} \phi_j(x_i'', \boldsymbol{x}_{j,-i})}.
\end{aligned} \tag{12.23}
$$

This last expression uses only the factors involving X_i, and depends only on the instantiation in $\boldsymbol{x}_{-i}$ of X_i's Markov blanket. In the case of Bayesian networks, this expression reduces to a formula involving only the CPDs of X_i and its children, and its value, again, depends only on the assignment in $\boldsymbol{x}_{-i}$ to the Markov blanket of X_i.

Example 12.10

Consider again the Student *network of figure 12.1, with the evidence s^1, l^0. The kernel for the variable G is defined as follows. Given a state (i, d, g, s^1, l^0), we define $\mathcal{T}((i, g, d, s^1, l^0) \rightarrow (i, g', d, s^1, l^0)) = P(g' \mid i, d, s^1, l^0)$. This value can be computed locally, using only the CPDs that involve G, that is, the CPDs of G and L:*

$$
P(g' \mid i, d, s^1, l^0) = \frac{P(g' \mid i, d) P(l^0 \mid g')}{\sum_{g''} P(g'' \mid i, d) P(l^0 \mid g'')}.
$$

Similarly, the kernel for the variable I is defined to be $\mathcal{T}((i, g, d, s^1, l^0) \rightarrow (i', g, d, s^1, l^0)) = P(i' \mid g, d, s^1, l^0)$, which simplifies as follows:

$$
P(i' \mid g, d, s^1, l^0) = \frac{P(i') P(g \mid i', d) P(s^1 \mid i')}{\sum_{i''} P(i'') P(g \mid i'', d) P(s^1 \mid i'')}.
$$

∎

As presented, the algorithm is defined via a sequence of local kernels, where each samples a single variable conditioned on all the rest. The reason for this approach is computational. As we showed, we can easily compute the transition model for a single variable given the rest. However, there are cases where we can simultaneously sample several variables efficiently. Specifically, assume we can partition the variables $\boldsymbol{X}$ into several disjoint *blocks* of variables $\boldsymbol{X}_1, \ldots, \boldsymbol{X}_k$, such that we can efficiently sample $\boldsymbol{x}_i$ from $P_\Phi(\boldsymbol{X}_i \mid \boldsymbol{x}_1, \ldots, \boldsymbol{x}_{i-1}, \boldsymbol{x}_{i+1}, \ldots, \boldsymbol{x}_k)$. In this case, we can modify our Gibbs sampling algorithm to iteratively sample blocks of variables, rather than individual variables, thereby taking much "longer-range" transitions in the state space in a single sampling step. Here, like in Gibbs sampling, we define the algorithm to be producing a new sample only once all blocks have been resampled. This algorithm is called *block Gibbs*. Note that standard Gibbs sampling is a special case of block Gibbs sampling, with the blocks corresponding to individual variables.

block Gibbs sampling

Example 12.11

Consider the Bayesian network induced by the plate model of example 6.11. Here, we generally have n students, each with a variable representing his or her intelligence, and m courses, each

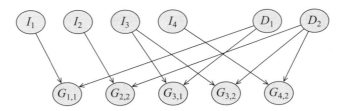

Figure 12.5   A Bayesian network with four students, two courses, and five grades

with a variable representing its difficulty. We also have a set of grades for students in classes (not necessarily a grade for each student in every class). Using an abbreviated notation, we have a set of variables $I_1, \ldots, I_n$ for the students (where each $I_j = I(s_j)$), $\boldsymbol{D} = \{D_1, \ldots, D_\ell\}$ for the courses, and $\boldsymbol{G} = \{G_{j,k}\}$ for the grades, where each variable $G_{j,k}$ has the parents I_j and D_k. See figure 12.5 for an example with $n = 4$ and $\ell = 2$. Let us assume that we observe the grades, so that we have evidence $\boldsymbol{G} = \boldsymbol{g}$. An examination of active paths shows that the different variables I_j are conditionally independent given an assignment $\boldsymbol{d}$ to $\boldsymbol{D}$. Thus, given $\boldsymbol{D} = \boldsymbol{d}, \boldsymbol{G} = \boldsymbol{g}$, we can efficiently sample all of the $\boldsymbol{I}$ variables as a block by sampling each I_j independently of the others. Similarly, we can sample all of the $\boldsymbol{D}$ variables as a block given an assignment $\boldsymbol{I} = \boldsymbol{i}, \boldsymbol{G} = \boldsymbol{g}$. Thus, we can alternate steps where in one we sample $\boldsymbol{i}[m]$ given $\boldsymbol{g}$ and $\boldsymbol{d}[m]$, and in the other we sample $\boldsymbol{d}[m+1]$ given $\boldsymbol{g}$ and $\boldsymbol{i}[m]$. ∎

In this example, we can easily apply block Gibbs because the variables in each block are marginally independent given the variables outside the block. This independence property allows us to compute efficiently the conditional distribution $P_\Phi(\boldsymbol{X}_i \mid \boldsymbol{x}_1, \ldots, \boldsymbol{x}_{i-1}, \boldsymbol{x}_{i+1}, \ldots, \boldsymbol{x}_k)$, and to sample from it. Importantly, however, full independence is not essential: we need only have the property that the block-conditional distribution can be efficiently manipulated. For example, in a grid-structured network, we can easily define our blocks to consist of separate rows or of separate columns. In this case, the structure of each block is a simple chain-structured network; we can easily compute the conditional distribution of one row given all the others, and sample from it (see exercise 12.3).

We note that the Gibbs chain is not necessarily regular, and might not converge to a unique stationary distribution.

Example 12.12   *Consider a simple network that consists of a single v-structure $X \to Z \leftarrow Y$, where the variables are all binary, X and Y are both uniformly distributed, and Z is the deterministic exclusive or of X and Y (that is, $Z = z^1$ iff $X \neq Y$). Consider applying Gibbs sampling to this network with the evidence z^1. The true posterior assigns probability $1/2$ to each of the two states x^1, y^0, z^1 and x^0, y^1, z^1. Assume that we start in the first of these two states. In this case, $P(X \mid y^0, z^1)$ assigns probability 1 to x^1, so that the X transition leaves the value of X unchanged. Similarly, the Y transition leaves the value of Y unchanged. Therefore, the chain will simply stay at the initial state forever, and it will never sample from the other state. The analogous phenomenon occurs for the other starting state. This chain is an example of a reducible Markov chain.* ∎

However, this chain is guaranteed to be regular whenever the distribution is positive, so that every value of X_i has positive probability given an assignment $\boldsymbol{x}_{-i}$ to the remaining variables.

Theorem 12.4

Let $\mathcal{H}$ be a Markov network such that all of the clique potentials are strictly positive. Then the Gibbs-sampling Markov chain is regular.

The proof is not difficult, and is left as an exercise (exercise 12.20).

Positivity is, however, not necessary; there are many examples of nonpositive distributions where the Gibbs chain is regular.

mixing

Importantly, however, even chains that are regular may require a long time to *mix*, that is, get close to the stationary distribution. In this case, instances generated from early in the sampling process will not be representative of the desired stationary distribution.

12.3.4 A Broader Class of Markov Chains ★

As we discussed, the use of MCMC methods relies on the construction of a Markov chain that has the desired properties: regularity, and the target stationary distribution. In the previous section, we described the Gibbs chain, a simple Markov chain that is guaranteed to have these properties under certain assumptions. However, Gibbs sampling is applicable only in certain circumstances; in particular, we must be able to sample from the distribution $P(X_i \mid \boldsymbol{x}_{-i})$. Although this sampling step is easy for discrete graphical models, in continuous models, the conditional distribution may not be one that has a parametric form that allows sampling, so that Gibbs is not applicable.

 Even more important, **the Gibbs chain uses only very local moves over the state space: moves that change one variable at a time. In models where variables are tightly correlated, such moves often lead from states whose probability is high to states whose probability is very low. In this case, the high-probability states will form strong basins of attraction, and the chain will be very unlikely to move away from such a state; that is, the chain will mix very slowly. In this case, we often want to consider chains that allow a broader range of moves, including much larger steps in the space.** The framework we develop in this section allows us to construct a broad family of chains in a way that guarantees the desired stationary distribution.

12.3.4.1 Detailed Balance

Before we address the question of how to construct a Markov chain with a particular stationary distribution, we address the question of how to verify easily that our Markov chain has the desired stationary distribution. Fortunately, we can define a test that is local and easy to check, and that suffices to characterize the stationary distribution. As we will see, this test also provides us with a simple method for constructing an appropriate chain.

Definition 12.5

reversible Markov chain

A finite-state Markov chain $\mathcal{T}$ is reversible *if there exists a unique distribution π such that, for all $\boldsymbol{x}, \boldsymbol{x}' \in Val(\boldsymbol{X})$:*

$$\pi(\boldsymbol{x})\mathcal{T}(\boldsymbol{x} \to \boldsymbol{x}') = \pi(\boldsymbol{x}')\mathcal{T}(\boldsymbol{x}' \to \boldsymbol{x}).$$

(12.24)

detailed balance

This equation is called the detailed balance. ∎

The product $\pi(x)\mathcal{T}(x \rightarrow x')$ represents a process where we pick a starting state at random according to π, and then take a random transition from the chosen state according to the transition model. The detailed balance equation asserts that, using this process, the probability of a transition from x to x' is the same as the probability of a transition for x' to x.

Reversibility implies that π is a stationary distribution of $\mathcal{T}$, but not necessarily that the chain will converge to π (see example 12.8). However, if $\mathcal{T}$ is regular, then convergence is guaranteed, and the reversibility condition provides a simple characterization of its stationary distribution:

Proposition 12.3

If $\mathcal{T}$ is regular and it satisfies the detailed balance equation relative to π, then π is the unique stationary distribution of $\mathcal{T}$.

The proof is left as an exercise (exercise 12.14).

Example 12.13

We can test this proposition on the Markov chain of figure 12.4. Our detailed balance equation for the two states x^1 and x^3 asserts that

$$\pi(x^1)\mathcal{T}(x^1 \rightarrow x^3) = \pi(x^3)\mathcal{T}(x^3 \rightarrow x^1).$$

Testing this equation for the stationary distribution π described in example 12.7, we have:

$$0.2 \cdot 0.75 = 0.3 \cdot 0.5 = 0.15. \qquad \blacksquare$$

The detailed balance equation can also be applied to multiple kernels. If each kernel $\mathcal{T}_i$ satisfies the detailed balance equation relative to some stationary distribution π, then so does the mixture transition model $\mathcal{T}$ (see exercise 12.16). The application to the multistep transition model $\mathcal{T}$ is also possible, but requires some care (see exercise 12.17).

12.3.4.2   Metropolis-Hastings Algorithm

The reversibility condition gives us a condition for verifying that our Markov chain has the desired stationary distribution. However, it does not provide us with a constructive approach for producing such a Markov chain. The *Metropolis-Hastings algorithm* is a general construction that allows us to build a reversible Markov chain with a particular stationary distribution.

Unlike the Gibbs chain, the algorithm does not assume that we can generate next-state samples from a particular target distribution. Rather, it uses the idea of a *proposal distribution* that we have already seen in the case of importance sampling.

As for importance sampling, the proposal distribution in the Metropolis-Hastings algorithm is intended to deal with cases where we cannot sample directly from a desired distribution. In the case of a Markov chain, the target distribution is our next-state sampling distribution at a given state. We would like to deal with cases where we cannot sample directly from this target. Therefore, we sample from a different distribution — the proposal distribution — and then correct for the resulting error. However, unlike importance sampling, we do not want to keep track of importance weights, which are going to decay exponentially with the number of transitions, leading to a whole slew of problems. Therefore, we instead randomly choose whether to accept the proposed transition, with a probability that corrects for the discrepancy between the proposal distribution and the target.

More precisely, our proposal distribution $\mathcal{T}^Q$ defines a transition model over our state space: For each state x, $\mathcal{T}^Q$ defines a distribution over possible successor states in $Val(\boldsymbol{X})$, from

(margin notes) Metropolis-Hastings algorithm

proposal distribution

which we select randomly a candidate next state x'. We can either accept the proposal and transition to x', or reject it and stay at x. Thus, for each pair of states x, x' we have an *acceptance probability* $\mathcal{A}(x \rightarrow x')$. The actual transition model of the Markov chain is then:

acceptance probability

$$\begin{aligned} \mathcal{T}(x \rightarrow x') &= \mathcal{T}^Q(x \rightarrow x')\mathcal{A}(x \rightarrow x') \quad x \neq x' \\ \mathcal{T}(x \rightarrow x) &= \mathcal{T}^Q(x \rightarrow x) + \sum_{x' \neq x} \mathcal{T}^Q(x \rightarrow x')(1 - \mathcal{A}(x \rightarrow x')). \end{aligned} \quad (12.25)$$

By using a proposal distribution, we allow the Metropolis-Hastings algorithm to be applied even in cases where we cannot directly sample from the desired next-state distribution; for example, where the distribution in equation (12.22) is too complex to represent. The choice of proposal distribution can be arbitrary, so long as it induces a regular chain. One simple choice in discrete factored state spaces is to use a multiple transition model, where $\mathcal{T}_i^Q$ is a uniform distribution over the values of the variable X_i.

Given a proposal distribution, we can use the detailed balance equation to select the acceptance probabilities so as to obtain the desired stationary distribution. For this Markov chain, the detailed balance equations assert that, for all $x \neq x'$,

$$\pi(x)\mathcal{T}^Q(x \rightarrow x')\mathcal{A}(x \rightarrow x') = \pi(x')\mathcal{T}^Q(x' \rightarrow x)\mathcal{A}(x' \rightarrow x).$$

We can verify that the following acceptance probabilities satisfy these equations:

$$\mathcal{A}(x \rightarrow x') = \min\left[1, \frac{\pi(x')\mathcal{T}^Q(x' \rightarrow x)}{\pi(x)\mathcal{T}^Q(x \rightarrow x')}\right], \quad (12.26)$$

and hence that the chain has the desired stationary distribution:

Theorem 12.5

Let $\mathcal{T}^Q$ be any proposal distribution, and consider the Markov chain defined by equation (12.25) and equation (12.26). If this Markov chain is regular, then it has the stationary distribution π.

The proof is not difficult, and is left as an exercise (exercise 12.15).
Let us see how this construction process works.

Example 12.14

Assume that our proposal distribution $\mathcal{T}^Q$ is given by the chain of figure 12.4, but that we want to sample from a stationary distribution π' where: $\pi'(x^1) = 0.6$, $\pi'(x^2) = 0.3$, and $\pi'(x^3) = 0.1$. To define the chain, we need to compute the acceptance probabilities. Applying equation (12.26), we obtain, for example, that:

$$\begin{aligned} \mathcal{A}(x^1 \rightarrow x^3) &= \min\left[1, \frac{\pi'(x^3)\mathcal{T}^Q(x^3 \rightarrow x^1)}{\pi'(x^1)\mathcal{T}^Q(x^1 \rightarrow x^3)}\right] = \min\left[1, \frac{0.1 \cdot 0.5}{0.6 \cdot 0.75}\right] = 0.11 \\ \mathcal{A}(x^3 \rightarrow x^1) &= \min\left[1, \frac{\pi'(x^1)\mathcal{T}^Q(x^1 \rightarrow x^3)}{\pi'(x^3)\mathcal{T}^Q(x^3 \rightarrow x^1)}\right] = \min\left[1, \frac{0.6 \cdot 0.75}{0.1 \cdot 0.5}\right] = 1. \end{aligned}$$

We can now easily verify that the stationary distribution of the chain resulting from equation (12.25) and these acceptance probabilities gives the desired stationary distribution π'. ∎

The Metropolis-Hastings algorithm has a particularly natural implementation in the context of graphical models. Each local transition model $\mathcal{T}_i$ is defined via an associated proposal distribution $\mathcal{T}_i^{Q_i}$. The acceptance probability for this chain has the form

$$
\begin{aligned}
\mathcal{A}(\boldsymbol{x}_{-i}, x_i \rightarrow \boldsymbol{x}_{-i}, x_i') &= \min\left[1, \frac{\pi(\boldsymbol{x}_{-i}, x_i')\mathcal{T}_i^{Q_i}(\boldsymbol{x}_{-i}, x_i' \rightarrow \boldsymbol{x}_{-i}, x_i)}{\pi(\boldsymbol{x}_{-i}, x_i)\mathcal{T}_i^{Q_i}(\boldsymbol{x}_{-i}, x_i \rightarrow \boldsymbol{x}_{-i}, x_i')}\right] \\
&= \min\left[1, \frac{P_\Phi(x_i', \boldsymbol{x}_{-i})\,\mathcal{T}_i^{Q_i}(\boldsymbol{x}_{-i}, x_i' \rightarrow \boldsymbol{x}_{-i}, x_i)}{P_\Phi(x_i, \boldsymbol{x}_{-i})\,\mathcal{T}_i^{Q_i}(\boldsymbol{x}_{-i}, x_i \rightarrow \boldsymbol{x}_{-i}, x_i')}\right].
\end{aligned}
$$

The proposal distributions are usually fairly simple, so it is easy to compute their ratios. In the case of graphical models, the first ratio can also be computed easily:

$$
\begin{aligned}
\frac{P_\Phi(x_i', \boldsymbol{x}_{-i})}{P_\Phi(x_i, \boldsymbol{x}_{-i})} &= \frac{P_\Phi(x_i' \mid \boldsymbol{x}_{-i})P_\Phi(\boldsymbol{x}_{-i})}{P_\Phi(x_i \mid \boldsymbol{x}_{-i})P_\Phi(\boldsymbol{x}_{-i})} \\
&= \frac{P_\Phi(x_i' \mid \boldsymbol{x}_{-i})}{P_\Phi(x_i \mid \boldsymbol{x}_{-i})}.
\end{aligned}
$$

As for Gibbs sampling, we can use the observation that each variable X_i is conditionally independent of the remaining variables in the network given its Markov blanket. Letting $\boldsymbol{U}_i$ denote $\mathrm{MB}_\mathcal{K}(X_i)$, and $\boldsymbol{u}_i = (\boldsymbol{x}_{-i})\langle\boldsymbol{U}_i\rangle$, we have that:

$$
\frac{P_\Phi(x_i' \mid \boldsymbol{x}_{-i})}{P_\Phi(x_i \mid \boldsymbol{x}_{-i})} = \frac{P_\Phi(x_i' \mid \boldsymbol{u}_i)}{P_\Phi(x_i \mid \boldsymbol{u}_i)}.
$$

This expression can be computed locally and efficiently, based only on the local parameterization of X_i and its Markov blanket (exercise 12.18).

The similarity to the derivation of Gibbs sampling is not accidental. Indeed, it is not difficult to show that Gibbs sampling is simply a special case of Metropolis-Hastings, one with a particular choice of proposal distribution (exercise 12.19).

The Metropolis-Hastings construction allows us to produce a Markov chain for an arbitrary stationary distribution. Importantly, however, we point out that the key theorem still requires that the constructed chain be regular. This property does not follow directly from the construction. In particular, the exclusive-or network of example 12.12 induces a nonregular Markov chain for any Metropolis-Hastings construction that uses a local proposal distribution — one that proposes changes to only a single variable at a time. In order to obtain a regular chain for this example, we would need a proposal distribution that allows simultaneous changes to both X and Y at a single step.

12.3.5   Using a Markov Chain

So far, we have discussed methods for defining Markov chains that induce the desired stationary distribution. Assume that we have constructed a chain that has a unique stationary distribution π, which is the one from which we wish to sample. How do we use this chain to answer queries? A naive answer is straightforward. We run the chain using the algorithm of algorithm 12.5 until it converges to the stationary distribution (or close to it). We then collect a sample from π. We repeat this process once for each particle we want to collect. The result is a data set $\mathcal{D}$ consisting of independent particles, each of which is sampled (approximately) from the stationary distribution π. The analysis of section 12.1 is applicable to this setting, so we can provide tight bounds on the number of samples required to get estimators of a certain quality. Unfortunately, matters are not so straightforward, as we now discuss.

12.3.5.1 Mixing Time

A critical gap in this description of the MCMC algorithm is a specification of the *burn-in time* T — the number of steps we take until we collect a sample from the chain. Clearly, we want to wait until the state distribution is reasonably close to π. More precisely, we want to find a T that guarantees that, regardless of our starting distribution $P^{(0)}$, $P^{(T)}$ is within some small ϵ of π. In this context, we usually use variational distance (see section A.1.3.3) as our notion of "within ϵ."

Definition 12.6

Let $\mathcal{T}$ be a Markov chain. Let T_ϵ be the minimal T such that, for any starting distribution $P^{(0)}$, we have that:

$$\boldsymbol{D}_{\mathrm{var}}(P^{(T)}; \pi) \leq \epsilon.$$

mixing time

Then T_ϵ is called the ϵ-mixing time of $\mathcal{T}$. ∎

In certain cases, the mixing time can be extremely long. This situation arises in chains where the state space has several distinct regions each of which is well connected, but where transitions between regions are low probability. In particular, we can estimate the extent to which the chain allows mixing using the following quantity:

Definition 12.7

conductance

Let $\mathcal{T}$ be a Markov chain transition model and π its stationary distribution. The conductance of $\mathcal{T}$ is defined as follows:

$$\min_{\substack{\mathcal{S} \subset Val(\boldsymbol{X}) \\ 0 < \pi(\mathcal{S}) \leq 1/2}} \frac{P(\mathcal{S} \rightsquigarrow \mathcal{S}^c)}{\pi(\mathcal{S})},$$

where $\pi(\mathcal{S})$ is the probability assigned by the stationary distribution to the set of states $\mathcal{S}$, $\mathcal{S}^c = Val(\boldsymbol{X}) - \mathcal{S}$, and

$$P(\mathcal{S} \rightsquigarrow \mathcal{S}^c) = \sum_{\boldsymbol{x} \in \mathcal{S}, \boldsymbol{x}' \in \mathcal{S}^c} \mathcal{T}(\boldsymbol{x} \rightarrow \boldsymbol{x}').$$

∎

Intuitively, $P(\mathcal{S} \rightsquigarrow \mathcal{S}^c)$ is the total "bandwidth" for transitioning from $\mathcal{S}$ to its complement. In cases where the conductance is low, there is some set of states $\mathcal{S}$ where, once in $\mathcal{S}$, it is very difficult to transition out of it. Figure 12.6 visualizes this type of situation, where the only transition between $\mathcal{S} = \{x^1, x^2, x^3\}$ and its complement is the dashed transition between x^2 and x^4, which has a very low probability. In cases such as this, if we start in a state within $\mathcal{S}$, the chain is likely to stay in $\mathcal{S}$ and to take a very long time before exploring other regions of the state space. Indeed, it is possible to provide both upper and lower bounds on the mixing rate of a Markov chain in terms of its conductance.

In the context of Markov chains corresponding to graphical models, chains with low conductance are most common in networks that have deterministic or highly skewed parameterization.

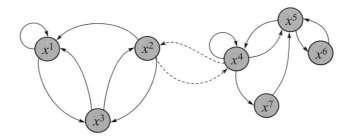

Figure 12.6   Visualization of a Markov chain with low conductance

In fact, as we saw in example 12.12, networks with deterministic CPDs might even lead to reducible chains, where different regions are entirely disconnected. However, even when the distribution is positive, we might still have regions that are connected only by very low-probability transitions. (See exercise 12.21.)

There are methods for providing tight bounds on the ϵ-mixing time of a given Markov chain. These methods are based on an analysis of the transition matrix between the states in the Markov chain.[3] Unfortunately, in the case of graphical models, an exhaustive enumeration of the exponentially many states is precisely what we wish to avoid. (If this enumeration were feasible, we would not have to resort to approximate inference techniques in the first place.) Alternatively, there is a suite of indirect techniques that allow us to provide bounds on the mixing time for some general class of chains. However, the application of these methods to each new class of chains requires a separate and usually quite sophisticated mathematical analysis. As of yet, there is no such analysis for the chains that are useful in the setting of graphical models. A more common approach is to use a variety of heuristics to try to evaluate the extent to which a sample trajectory has "mixed." See box 12.B for some further discussion.

12.3.5.2   Collecting Samples

The burn-in time for a large Markov chain is often quite large. Thus, the naive algorithm described above has to execute a large number of sampling steps for every usable sample. However, a key observation is that, if $x^{(t)}$ is sampled from π, then $x^{(t+1)}$ is also sampled from π. Thus, once we have run the chain long enough that we are sampling from the stationary distribution (or a distribution close to it), we can continue generating samples from the same trajectory and obtain a large number of samples from the stationary distribution.

More formally, assume that we use $x^{(0)}, \ldots, x^{(T)}$ as our burn-in phase, and then collect M samples $\mathcal{D} = \{x[1], \ldots, x[M]\}$ from the stationary distribution. Most simply, we might collect M consecutive samples, so that $x[m] = x^{(T+m)}$, for $m = 1, \ldots, M$. If $x^{(T+1)}$ is sampled from π, then so are all of the samples in $\mathcal{D}$. Thus, if our chain has mixed by the time we collect

---

3. Specifically, they involve computing the second largest eigen-value of the matrix.

our first sample, then for any function f,

$$\hat{\boldsymbol{E}}_{\mathcal{D}}(f) = \frac{1}{M} \sum_{m=1}^{M} f(\boldsymbol{x}[m], \boldsymbol{e})$$

estimator

is an unbiased *estimator* for $\boldsymbol{E}_{\pi(\boldsymbol{X})}[f(\boldsymbol{X}, \boldsymbol{e})]$.

How good is this estimator? As we discussed in appendix A.2.1, the quality of an unbiased estimator is measured by its variance: the lower the variance, the higher the probability that the estimator is close to its mean. In theorem A.2, we showed an analysis of the variance of an estimator obtained from M independent samples. Unfortunately, we cannot apply that analysis in this setting. The key problem, of course, is that consecutive samples from the same trajectory are correlated. Thus, we cannot expect the same performance as we would from M independent samples from π. More formally, the variance of the estimator is significantly higher than that of an estimator generated by M independent samples from π, as discussed before.

Example 12.15

Consider the Gibbs chain for the deterministic exclusive-or network of example 12.12, and assume we compute, for a given run of the chain, the fraction of states in which x^1 holds in the last 100 states traversed by the chain. A chain started in the state x^1, y^0 would have that $100/100$ of the states have x^1, whereas a chain started in the state x^0, y^1 would have that $0/100$ of the states have x^1. Thus, the variance of the estimator is very high in this case. ∎

central limit theorem

One can formalize this intuition by the following generalization of the *central limit theorem* that applies to samples collected from a Markov chain:

Theorem 12.6

Let $\mathcal{T}$ be a Markov chain and $\boldsymbol{X}[1], \ldots, \boldsymbol{X}[M]$ a set of samples collected from $\mathcal{T}$ at its stationary distribution P. Then, since $M \longrightarrow \infty$:

$$\left(\hat{\boldsymbol{E}}_{\mathcal{D}}(f) - \boldsymbol{E}_{\boldsymbol{X} \sim P}[f(\boldsymbol{X})] \right) \longrightarrow \mathcal{N}\left(0; \sigma_f^2\right)$$

where

$$\sigma_f^2 = \boldsymbol{Var}_{\boldsymbol{X} \sim \mathcal{T}}[f(\boldsymbol{X})] + 2 \sum_{\ell=1}^{\infty} \boldsymbol{Cov}_{\mathcal{T}}[f(\boldsymbol{X}[m]); f(\boldsymbol{X}[m+\ell])] < \infty.$$

autocovariance

The terms in the summation are called *autocovariance* terms, since they measure the covariance between samples from the chain, taken at different lags. The stronger the correlations between different samples, the larger the autocovariance terms, the higher the variance of our estimator. This result is consistent with the behavior we discussed in example 12.12.

We want to use theorem 12.6 in order to assess the quality of our estimator. In order to do so, we need to estimate the quantity σ_f^2. We can estimate the variance from our empirical data using the standard estimator:

$$\boldsymbol{Var}_{\boldsymbol{X} \sim \mathcal{T}}[f(\boldsymbol{X})] \approx \frac{1}{M-1} \left[\sum_{m=1}^{M} \left(f(\boldsymbol{X}) - \hat{\boldsymbol{E}}_{\mathcal{D}}(f) \right)^2 \right]. \tag{12.27}$$

To estimate the autocovariance terms from the empirical data, we compute:

$$\boldsymbol{Cov}_{\mathcal{T}}[f(\boldsymbol{X}[m]); f(\boldsymbol{X}[m+\ell])] \approx \frac{1}{M-\ell} \sum_{m=1}^{M-\ell} (f(\boldsymbol{X}[m] - \hat{\boldsymbol{E}}_{\mathcal{D}}(f))(f(\boldsymbol{X}[m+\ell] - \hat{\boldsymbol{E}}_{\mathcal{D}}(f)).$$

$$(12.28)$$

At first glance, theorem 12.6 suggests that the variance of the estimate could be reduced if the chain is allowed a sufficient number of iterations between sample collections. Thus, having collected a particle $x^{(T)}$, we can let the chain run for a while, and collect a second particle $x^{(T+d)}$ for some appropriate choice of d. For d large enough, $x^{(T)}$ and $x^{(T+d)}$ are only slightly correlated, reducing the correlation in the preceding theorem.

However, this approach is suboptimal for various reasons. First, the time d required for "forgetting" the correlation is clearly related to the mixing time of the chain. Thus, chains that are slow to mix initially also require larger d in order to produce close-to-independent particles. Nevertheless, the samples do come from the correct distribution for any value of d, and hence it is often better to compromise and use a shorter d than it is to use a shorter burn-in time T. This method thus allows us to collect a larger number of usable particles with fewer transitions of the Markov chain. Indeed, **although the samples between $x^{(T)}$ and $x^{(T+d)}$ are not independent samples, there is no reason to discard them. That is, one can show that using all of the samples $x^{(T)}, x^{(T+1)}, \ldots, x^{(T+d)}$ produces a provably better estimator than using just the two samples $x^{(T)}$ and $x^{(T+d)}$: our variance is always no higher if we use all of the samples we generated rather than a subset. Thus, the strategy of picking only a subset of the samples is useful primarily in settings where there is a significant cost associated with using each sample (for example, the evaluation of f is costly), so that we might want to reduce the overall number of particles used.**

Box 12.B — Skill: MCMC in Practice. *A key question when using a Markov chain is evaluating the time required for the chain to "mix" — that is, approach the stationary distribution. As we discussed, no general-purpose theoretical analysis exists for the mixing time of graphical models. However, we can still hope to estimate the extent to which a sample trajectory has "forgotten" its origin. Recall that, as we discussed, the most common problem with mixing arises when the state space consists of several regions that are connected only by low-probability transitions. If we start the chain in a state in one of these regions, it is likely to spend some amount of time in that same region before transitioning to another region. Intuitively, the states sampled in the initial phase are clearly not from the stationary distribution, since they are strongly correlated with our initial state, which is arbitrary. However, later in the trajectory, we might reach a state where the current state is as likely to have originated in any initial state. In this case, we might consider the chain to have mixed.*

Diagnosing convergence of a Markov chain Monte Carlo method is a notoriously hard problem. The chain may appear to have converged simply by spending a large number of iterations in a particular mode due to low conductance between modes. However, there are approaches that can tell us if a chain has not converged.

One technique is based directly on theorem 12.6. In particular, we can compute the ratio ρ_ℓ of the estimated autocovariance in equation (12.28) to the estimated variance in equation (12.27). This ratio is known as the autocorrelation *of lag ℓ; it provides a normalized estimate of the extent to which the chain has mixed in ℓ steps. In practice, the autocorrelation should drop off exponentially with the length of the lag, and one way to diagnose a poorly mixing chain is to observe high autocorrelation at distant lags. Note, however, that the number of samples available for computing autocorrelation decreases with lag, leading to large variance in the autocorrelation estimates at large lags.*

A different technique uses the observation that multiple chains sampling the same distribution should, upon convergence, all yield similar estimates. In addition, estimates based on a complete set of samples collected from all of the chains should have variance comparable to variance in each of the chains. More formally, assume that K separate chains are each run for $T + M$ steps starting from a diverse set of starting points. After discarding the first T samples from each chain, let $\boldsymbol{X}_k[m]$ denote a sample from chain k after iteration $T + m$. We can now compute the B (between-chains) and W (within-chain) variances:

$$\bar{f}_k = \frac{1}{M} \sum_{m=1}^{M} f(\boldsymbol{X}_k[m])$$

$$\bar{f} = \frac{1}{K} \sum_{k=1}^{K} \bar{f}_k$$

$$B = \frac{M}{K-1} \sum_{k=1}^{K} (\bar{f}_k - \bar{f})^2$$

$$W = \frac{1}{K} \frac{1}{M-1} \sum_{k=1}^{K} \sum_{m=1}^{M} \left(f(\boldsymbol{X}_k[m]) - \bar{f}_k \right)^2.$$

The expression $V = \frac{M-1}{M} W + \frac{1}{M} B$ can now be shown to overestimate the variance of our estimate of f based on the collected samples. In the limit of $M \longrightarrow \infty$, both W and V converge to the true variance of the estimate. One measure of disagreement between chains is given by $\hat{R} = \sqrt{\frac{V}{W}}$. If the chains have not all converged to the stationary distribution, this estimate will be high. If this value is close to 1, either the chains have all converged to the true distribution, or the starting points were not sufficiently dispersed and all of the chains have converged to the same mode or a set of modes. We can use this strategy with multiple different functions f in order to increase our confidence that our chain has mixed. We can, for example, use indicator functions of various events, as well as more complex functions of multiple variables.

Overall, although the strategy of using only a single chain produces more viable particles using lower computational cost, there are still significant advantages to the multichain approach. First, by starting out in very different regions of the space, we are more likely to explore a more representative subset of states. Second, the use of multiple chains allows us to evaluate the extent to which our chains are mixing. Thus, **to summarize, a good strategy for using a Markov chain in practice is a hybrid approach, where we run a small number of chains in parallel for a reasonably long time, using their behavior to evaluate mixing. After the burn-in phase, we then use the existence of multiple chains to estimate convergence. If mixing appears to have occurred, we can use each of our chains to generate multiple particles, remembering that the particles generated in this fashion are not independent.**

12.3.5.3 Discussion

MCMC methods have many advantages over other methods. Unlike the global approximate inference methods of the previous chapter, they can, at least in principle, get arbitrarily close to the true posterior. Unlike forward sampling methods, these methods do not degrade when the probability of the evidence is low, or when the posterior is very different from the prior. Furthermore, unlike forward sampling, MCMC methods apply to undirected models as well as to directed models. As such, they are an important component in the suite of approximate inference techniques.

However, MCMC methods are not generally an out-of-the-box solution for dealing with inference in complex models. First, the application of MCMC methods leaves many options that need to be specified: the proposal distribution, the number of chains to run, the metrics for evaluating mixing, techniques for determining the delay between samples that would allow them to be considered independent, and more. Unfortunately, at this point, there is little theoretical analysis that can help answer these questions for the chains that are of interest to us. Thus, the application of Markov chains is more of an art than a science, and it often requires significant experimentation and hand-tuning of parameters.

Second, MCMC methods are only viable if the chain we are using mixes reasonably quickly. Unfortunately, many of the chains derived from real-world graphical models frequently have multimodal posterior distributions, with slow mixing between the modes. For such chains, the straightforward MCMC methods described in this chapter are unlikely to work. In such cases, diagnostics such as the ones described in box 12.B can be used to determine that the chain is not mixing, and better methods must then be applied. **The key to improving the convergence of a Markov chain is to introduce transitions that take larger steps in the space, allowing the chain to move more rapidly between modes, and thereby to better explore the space. The best strategy is often to analyze the properties of the posterior landscape of interest, and to construct moves that are tailored for this specific space. (See, for example, exercise 12.23.) Fortunately, the ability to mix different reversible kernels within a single chain (as discussed in section 12.3.4) allows us to introduce a variety of long-range moves while still maintaining the same target posterior.**

In addition to the use of long-range steps that are specifically designed for particular (classes of) chains, there are also some general-purpose methods that try to achieve that goal. The block Gibbs approach (section 12.3.3) is an instance of this general class of methods. Another strategy uses the same ideas in *simulated annealing* to improve convergence of local search to a better optimum. Here, we can define an intermediate distribution parameterized by a *temperature parameter T*: T:

simulated
annealing

temperature
parameter

$$\tilde{P}_T(\boldsymbol{X}) \propto \exp\{-\frac{1}{T} \log \tilde{P}(\boldsymbol{X})\}.$$

This distribution is similar to our original target distribution $\tilde{P}$. At a low temperature of $T = 1$, this equation yields the original target distribution. But as the temperature increases, modes become broader and merge, reducing the multimodality of the distribution and increasing its mixing rate. We can now define various methods that use a combination of related chains running at different temperatures. At a high level, the higher-temperature chain can be viewed as proposing a step, which we can accept or reject using the acceptance probability of our true target distribution. (See section 12.7 for references to some of these more advanced methods.) In effect, these approaches use the higher-temperature chains to define a set of larger steps in the space, thereby providing a general-purpose method for achieving more rapid movement between multiple modes. However, this generality comes at the computational cost of running parallel

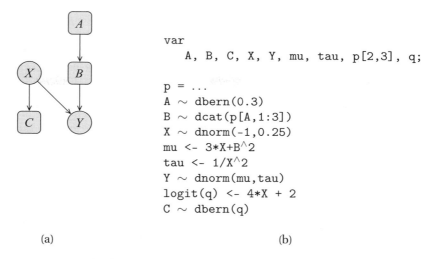

```
var
    A, B, C, X, Y, mu, tau, p[2,3], q;

p = ...
A ~ dbern(0.3)
B ~ dcat(p[A,1:3])
X ~ dnorm(-1,0.25)
mu <- 3*X+B^2
tau <- 1/X^2
Y ~ dnorm(mu,tau)
logit(q) <- 4*X + 2
C ~ dbern(q)
```

(a)   (b)

Figure 12.C.1 — **Example of** BUGS **model specification** (a) A simple hybrid Bayesian network. (b) A BUGS definition of a probabilistic model over this network.

chains; thus, if we can understand our specific posterior well enough to construct specialized operators that move between modes, that often provides a more effective solution.

Box 12.C — Case Study: The BUGS **System.** *One of the main advantages of MCMC methods is their broad applicability to a very general class of networks. Not only do they apply (at least in principle) to any discrete network, regardless of its complexity, they also generalize fairly simply to continuous variables (see section 14.5.3). One very useful system that exploits this generality is the* BUGS *system, developed by Thomas et al. (1992). This system provides a general-purpose language for representing a broad range of probabilistic models and uses MCMC to run inference over these models.*

BUGS system   *The* BUGS *system provides a programming-language-based representation of a probabilistic model. The model defines a joint distribution over a set of random variables. Variables can be defined as functions of each other; these functions can be deterministic functions, or stochastic functions utilizing a rich set of predefined distributions. For example, consider the simple Bayesian network shown in figure 12.C.1a, where A, B, C are discrete and X, Y are continuous. One possible probabilistic model can be written in* BUGS *using the commands shown in figure 12.C.1b. This model defines: A to be a binary-valued variable, with $P(a^1) = 0.3$; B is a 3-valued variable that depends on A, whose CPT is defined in the matrix P; X is a Gaussian random variable with mean -1 and precision (inverse variance) 0.25; Y is a conditional Gaussian whose mean depends on X and B and whose precision also depends on X; and C is a logistic function of $4X + 2$. Even in this very simple example, we can see that the* BUGS *language provides a rich language for encoding different families of functional and stochastic dependencies between variables.*

Given a probabilistic model defined in this way, the BUGS *system can instantiate evidence for some of the variables (for example, by reading their values from a file) and then perform inference over the model by running various MCMC algorithms. The system analyzes the parametric form specifying the distribution of the different variables, and it selects an appropriate sampling algorithm to use. The user specifies the number of sampling iterations to perform, and which variables are to be monitored — their values are to be stored during the MCMC iterations. We can then compute such values as the mean and standard deviation of these monitored variables. The system also provides various methods to help detect convergence of the MCMC runs (see also box 12.B).*

Overall, the BUGS *tool provides a general-purpose and highly flexible framework for specifying and reasoning with probabilistic models. Its ability to provide such a high level of expression power rests on the generality of MCMC as an inference method, and its applicability to a very broad range of distributions (broader than any other inference method currently available).*

12.4   Collapsed Particles

So far, we have restricted our attention to methods that use as their particles only instantiations ξ to all the network variables. Clearly, covering an exponentially large state space with a small number of instantiations is difficult, and it often takes a large number of full particles to obtain reasonable estimates. One approach for improving the performance of particle-based methods is to use as particles *partial* assignments to some subset of the network variables, combined with some type of closed-form representation of a distribution over the rest.

More precisely, assume that we partition $\mathcal{X}$ into two subsets: $\boldsymbol{X}_p$ — the variables whose assignment defines the particle, and $\boldsymbol{X}_d$ — the variables over which we will maintain a closed-form distribution. Then *collapsed particles* consist of an instantiation $\boldsymbol{x}_p \in Val(\boldsymbol{X}_p)$, coupled with some representation of the distribution $P(\boldsymbol{X}_d \mid \boldsymbol{x}_p, \boldsymbol{e})$. The particle is "collapsed" because some of the variables are not assigned but rather summarized using a distribution. Collapsed particles are also known as *Rao-Blackwellized particles*.

collapsed
particles

Assume that we want to estimate an expectation of some function $f(\xi)$ relative to our posterior distribution $P(\boldsymbol{X}_p, \boldsymbol{X}_d \mid \boldsymbol{e})$. We now have:

$$
\begin{aligned}
\boldsymbol{E}_{P(\xi|e)}[f(\xi)] &= \sum_{\boldsymbol{x}_p, \boldsymbol{x}_d} P(\boldsymbol{x}_p, \boldsymbol{x}_d \mid \boldsymbol{e}) f(\boldsymbol{x}_p, \boldsymbol{x}_d, \boldsymbol{e}) \\
&= \sum_{\boldsymbol{x}_p} P(\boldsymbol{x}_p \mid \boldsymbol{e}) \sum_{\boldsymbol{x}_d} P(\boldsymbol{x}_d \mid \boldsymbol{x}_p, \boldsymbol{e}) f(\boldsymbol{x}_p, \boldsymbol{x}_d, \boldsymbol{e}) \\
&= \sum_{\boldsymbol{x}_p} P(\boldsymbol{x}_p \mid \boldsymbol{e}) \left(\boldsymbol{E}_{P(\boldsymbol{X}_d|\boldsymbol{x}_p, \boldsymbol{e})}[f(\boldsymbol{x}_p, \boldsymbol{X}_d, \boldsymbol{e})] \right).
\end{aligned} \tag{12.29}
$$

We can use the samples $\boldsymbol{x}_p[m]$ to approximate any expectation relative to the distribution $P(\boldsymbol{X}_p \mid \boldsymbol{e})$, using the techniques described above. In particular, we can approximate the expectation of the expression in parentheses — an expression that is itself an expectation.

In the case of collapsed particles, we assume that the internal expectation can be computed (or approximated) efficiently. As we will discuss, we can explicitly represent the distribution $P(\boldsymbol{X}_d \mid \boldsymbol{x}_p, \boldsymbol{e})$ as a graphical model, using constructions such as the reduced Markov network of section 4.2.3. We thus have a hybrid approach where we generate samples $\boldsymbol{x}_p$ from $\boldsymbol{X}_p$

and perform exact inference on X_d given x_p. Thus, this approach defines a spectrum: When $X_p = X$, collapsed particles are simply full particles, and we are simply applying the methods defined earlier in this chapter; when $X_p = \emptyset$, we have a single particle whose associated distribution is our original network, so that we are back in the regime of exact inference. We note that, in some cases, the network might not be sufficiently simple for exact inference. However, it might be amenable to some other form of approximation — for example, one of the methods we discuss in chapter 11. **Intuitively, the fewer variables we sample (keep in X_p), the larger the part of the probability mass that we cover using each collapsed particle $x_p[m]$. From an alternative perspective, we are performing an exact computation for the expectation relative to X_d, thereby eliminating any contribution it makes to the bias or the variance of the estimator. Thus, if $|X_p|$ is fairly small, we can obtain much better estimates for the distribution using significantly fewer particles.**

In this section, we describe extensions of the approaches discussed earlier in this chapter to the case of collapsed particles.

12.4.1 Collapsed Likelihood Weighting ★

We begin by describing a collapsed extension to likelihood weighting. We first describe the algorithm generally, from the perspective of normalized importance sampling. We then consider a specific application that is a direct extension to the full-particle version of likelihood weighting.

12.4.1.1 Collapsed Importance Sampling

Recall that, in importance sampling, we generate our samples from an alternative proposal distribution Q, and we compensate for the discrepancy by associating with each particle a weight $w[m]$. In the case of collapsed particles, we are generating specific particles only for the variables in X_p, so that Q would be a distribution over x_p. We thus generate a data set

$$\mathcal{D} = \{(x_p[m], w[m], P(X_d \mid x_p[m], e))\}_{m=1}^{M},$$

where each $x_p[m]$ is sampled from Q. Here we will discuss both the choice of proposal distribution Q and the computation of the weights $w[m]$.

Our goal is to estimate the expectation of equation (12.29). In effect, we are estimating the expectation of a new function g, which represents the internal expectation: $g = E_{P(X_d \mid x_p, e)}[f(x_p, X_d, e)]$. Using normalized importance sampling, we estimate this expectation as:

$$\hat{E}_{\mathcal{D}}(f) = \frac{\sum_{m=1}^{M} w[m] \left(E_{P(X_d \mid x_p[m], e)}[f(x_p[m], X_d, e)] \right)}{\sum_{m=1}^{M} w[m]}. \tag{12.30}$$

Example 12.16

Consider the Extended Student *network, repeated in figure 12.7a, with the evidence d^1, h^0. Assume that we choose to partition the variables as follows: $X_p = \{D, G\}$, and $X_d = \{C, I, S, L, J, H\}$. In this case, each particle defines an assignment (d, g) to the variables D, G. Assuming that our algorithm follows the template of full-particle likelihood weighting, we would ensure that our proposal distribution Q ascribes only positive probability to particles (d^1, g) that are compatible with our evidence d^1. Each such particle is also associated with a distribution $P(C, I, S, L, H \mid g, d^1, h^0)$. The reduced Markov network shown in figure 12.7b represents this distribution.*

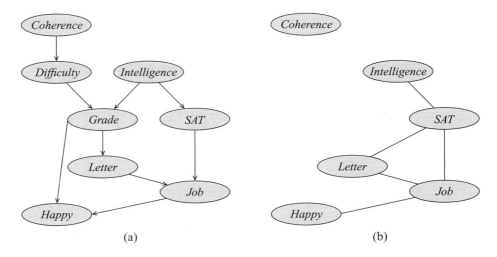

Figure 12.7   Networks illustrating collapsed importance sampling: (a) The Extended-Student Bayesian network $\mathcal{B}^{student}$; (b) The network $\mathcal{B}^{student}_{G=g,D=d}$ reduced by $G=g, D=d$.

Now, assume that our query is $P(j^1 \mid d^1, h^0)$, so that our function f is the indicator function $\boldsymbol{I}\{\xi\langle J \rangle = j^1\}$. For each particle, we evaluate

$$\boldsymbol{E}_{P(C,I,S,L,J,H\mid g,d^1,h^0)} \big[\boldsymbol{I}\{\xi\langle J \rangle = j^1\} \big] = P(j^1 \mid g, d^1, h^0).$$

We then compute an average of these probabilities, weighted by the importance weights (which we will discuss). The computation of the probabilities $P(J \mid g, d^1, h^0)$ can be done using inference in the reduced network, which is simpler than inference in the original network. (Although, in this very simple network, the savings are not significant.)

Note that the extent to which the reduced network allows more effective inference than the original one depends on our choice of variables $\boldsymbol{X}_p$. For example, if (for some reason) we choose $\boldsymbol{X}_p = \{H\}$, the resulting conditioned network is no simpler than the original, and the use of particle-based methods has no computational benefits over the use of exact inference. ∎

12.4.1.2   Formal Description

To specify the algorithm formally, we must define the proposal distribution Q and the associated importance weights. We begin by partitioning our evidence set $\boldsymbol{E}$ into two subsets: $\boldsymbol{E}_p = \boldsymbol{E} \cap \boldsymbol{X}_p$, and $\boldsymbol{E}_d = \boldsymbol{E} \cap \boldsymbol{X}_d$, with $\boldsymbol{e}_p$ and $\boldsymbol{e}_d$ defined accordingly. This partition determines how we handle each of the observed variables: evidence in $\boldsymbol{E}_p$ will be treated as in likelihood weighting, modifying our sampling process and the importance weights; evidence in $\boldsymbol{E}_d$ will be accounted for as part of the exact inference process.

Now, consider an arbitrary proposal distribution Q. We can go through an analysis similar to the one that allowed us to derive equation (12.12):

$$
\begin{aligned}
\boldsymbol{E}_{P(\xi|e)}[f(\xi)] &= \sum_{\boldsymbol{x}_p, \boldsymbol{x}_d} P(\boldsymbol{x}_p, \boldsymbol{x}_d \mid e) f(\boldsymbol{x}_p, \boldsymbol{x}_d, e) \\
&= \sum_{\boldsymbol{x}_p} Q(\boldsymbol{x}_p) \frac{P(\boldsymbol{x}_p \mid e)}{Q(\boldsymbol{x}_p)} \sum_{\boldsymbol{x}_d} P(\boldsymbol{x}_d \mid \boldsymbol{x}_p, e) f(\boldsymbol{x}_p, \boldsymbol{x}_d, e).
\end{aligned}
$$

We can now reformulate the term $P(\boldsymbol{x}_p \mid e)$ as:

$$
\begin{aligned}
P(\boldsymbol{x}_p \mid e) &= \frac{P(\boldsymbol{x}_p, e)}{P(e)} \\
&= \frac{P(\boldsymbol{x}_p, e_p, e_d)}{P(e)} \\
&= \frac{1}{P(e)} P(\boldsymbol{x}_p, e_p) P(e_d \mid \boldsymbol{x}_p, e_p).
\end{aligned}
$$

Plugging this result back into our derivation, we obtain that:

$$
\begin{aligned}
\boldsymbol{E}_{P(\xi|e)}[f(\xi)] &= \frac{1}{P(e)} \sum_{\boldsymbol{x}_p} Q(\boldsymbol{x}_p) \frac{P(\boldsymbol{x}_p, e_p)}{Q(\boldsymbol{x}_p)} P(e_d \mid \boldsymbol{x}_p, e_p) \sum_{\boldsymbol{x}_d} P(\boldsymbol{x}_d \mid \boldsymbol{x}_p, e) f(\boldsymbol{x}_p, \boldsymbol{x}_d, e) \\
&= \frac{1}{P(e)} \boldsymbol{E}_{Q(\boldsymbol{X}_p)} \left[\frac{P(\boldsymbol{x}_p, e_p)}{Q(\boldsymbol{x}_p)} P(e_d \mid \boldsymbol{x}_p, e_p) \boldsymbol{E}_{P(\boldsymbol{x}_d|\boldsymbol{x}_p, e)}[f(\boldsymbol{x}_p, \boldsymbol{x}_d, e)] \right].
\end{aligned} \tag{12.31}
$$

This analysis suggests that the appropriate importance weights should be defined as:

$$
w(\boldsymbol{x}_p) = \frac{P(\boldsymbol{x}_p, e_p)}{Q(\boldsymbol{x}_p)} P(e_d \mid \boldsymbol{x}_p, e_p). \tag{12.32}
$$

Indeed, if we compute the mean of our importance weights, as in equation (12.11), we obtain the following formula for the normalized importance sampling estimator:

$$
\begin{aligned}
\boldsymbol{E}_{Q(\boldsymbol{X}_p)}[w(\boldsymbol{X}_p)] &= \sum_{\boldsymbol{x}_p} Q(\boldsymbol{x}_p) \frac{P(\boldsymbol{x}_p, e_p)}{Q(\boldsymbol{x}_p)} P(e_d \mid \boldsymbol{x}_p, e_p) \\
&= \sum_{\boldsymbol{x}_p} P(\boldsymbol{x}_p, e_p) P(e_d \mid \boldsymbol{x}_p, e_p) \\
&= \sum_{\boldsymbol{x}_p} P(e_d, \boldsymbol{x}_p, e_p) = P(e_d, e_p).
\end{aligned}
$$

Thus, if we select our importance weights as in equation (12.32), we have that:

$$
\boldsymbol{E}_{P(\xi|e)}[f(\xi)] = \frac{\boldsymbol{E}_{Q(\boldsymbol{X}_p)}\left[w(\boldsymbol{X}_p) \boldsymbol{E}_{P(\boldsymbol{x}_d|\boldsymbol{x}_p, e)}[f(\boldsymbol{x}_p, \boldsymbol{x}_d, e)] \right]}{\boldsymbol{E}_{Q(\boldsymbol{X}_p)}[w(\boldsymbol{X}_p)]},
$$

as desired.

12.4.1.3   Proposal Distribution

Our preceding analysis does not place any restrictions on the proposal distribution; we can choose any proposal distribution Q that seems appropriate (as long as it dominates P). However, it is important to remember our two main desiderata for a proposal distribution: easy generation of samples from Q and similarity between Q and our target distribution $P(\boldsymbol{X}_p \mid \boldsymbol{e})$. The proposal distribution we used for the full particle case attempted to address both of these desiderata, at least to some extent. In this section, we describe a generalization of that proposal distribution for the case of collapsed particles.

Our goal is to generate particles from a distribution $Q(\boldsymbol{X}_p)$. Following the template for the full particle case, we would sample each unobserved variable $X \in \boldsymbol{X}_p$ from its CPD. The reason we can execute this process is that we were careful to sample the parents of X_i before that, so that the distribution from which we should sample X is uniquely defined. In the collapsed case, however, Pa_X might not be within the set $\boldsymbol{X}_p$, in which case we would not have values for them when sampling X. For example, returning to the setting of example 12.16, it is not clear how we would define a sampling distribution for the variable G.

Most simply, we can select as our subset $\boldsymbol{X}_p$ an upwardly closed subset of nodes in the network. In our example, we might select $\boldsymbol{X}_p$ to consist of all of the variables C, D, I, G. This variant of the algorithm is very close to full-particle likelihood weighting. Here, we are back in a situation where we can order the variables in such a way that each unobserved variable can be sampled using its original CPD, using the previously sampled assignment to its parents. Each observed variable X_i is "sampled," as in standard likelihood weighting, from a mutilated network that ascribes probability 1 to its observed value $\boldsymbol{e}_p\langle X_i \rangle$. The computation of the importance weights for this case is straightforward: We first compute the part of the importance weight corresponding to $P(\boldsymbol{x}_p, \boldsymbol{e}_p)/Q(\boldsymbol{x}_p)$, using precisely the same incremental computation as in standard likelihood weighting. We then compute $P(\boldsymbol{e}_d \mid \boldsymbol{x}_p, \boldsymbol{e}_p)$ in the network conditioned on $\boldsymbol{x}_p, \boldsymbol{e}_p$, and we multiply the importance weight by this additional factor.

Example 12.17

Continuing example 12.16, assume we choose $\boldsymbol{X}_p$ to be the upwardly closed set C, D, I, G. We would sample C, I, and G from their CPD; D would be sampled from a CPD that has no parents, and ascribes probability 1 to d^1. The importance weight for a particle (c, d^1, i, g) is then computed as $P(d^1 \mid c) \cdot P(h^0 \mid c, d^1, i, g)$. Note that this last term requires inference in the network, specifically, the marginalization of L, J, S. ∎

More generally, we can define a fully general proposal distribution Q by specifying a topological ordering $X_1, \ldots, X_k$ over $\boldsymbol{X}_p$, and a proposal distribution defined in terms of a Bayesian network $\mathcal{G}_Q$ over $\boldsymbol{X}_p$ that specifies, for each variable X_i, $i = 1, \ldots, k$, a parent set $\mathrm{Pa}_{X_i}^{\mathcal{G}_Q} \subseteq \{X_1, \ldots, X_{i-1}\}$, and a CPD $Q(X_i \mid \mathrm{Pa}_{X_i}^{\mathcal{G}_Q})$. This approach allows us to represent an arbitrary proposal distribution.

We can now consider the computation of each of the three terms in the definition of the importance weights in equation (12.32). The term $Q(\boldsymbol{x}_p)$ can be computed very simply via the chain rule for the proposal network $\mathcal{G}_Q$. The terms $P(\boldsymbol{e}_d \mid \boldsymbol{x}_p, \boldsymbol{e}_p)$ can be computed using inference in the conditioned network $\mathcal{B}_{\boldsymbol{E}_p = \boldsymbol{e}_p, \boldsymbol{X}_p = \boldsymbol{x}_p}$ — we compute the probability of the query $P(\boldsymbol{e}_d \mid \boldsymbol{e}_p, \boldsymbol{x}_p)$ in this network. As we stated, the whole approach of collapsed particles is based on the premise that (exact or approximate) inference in this conditioned network is feasible. Similarly, we can compute $P(\boldsymbol{x}_p, \boldsymbol{e}_p)$ in the same network.

12.4.2 Collapsed MCMC

The collapsed MCMC algorithm is also based on equation (12.29); however, as in section 12.3, we simplify our notation by defining Φ to be the set of factors reduced by the evidence e, so that $P_\Phi(X) = P(X \mid e)$ (for $X = \mathcal{X} - E$). As in collapsed likelihood weighting, we approximate the outer expectation by generating particles that are instantiations of X_p. Here, we generate the particles $x_p[m]$ using a Markov chain process; at the limit of the chain, these particles will be sampled from the marginal posterior $P_\Phi(X_p)$. For each such particle x_p, we maintain some representation of the distribution $P_\Phi(X_d \mid x_p)$, and perform (exact or approximate) inference to compute the expectation of f relative to this distribution. For simplicity, we focus our discussion on Gibbs sampling; the extension to a Metropolis-Hastings algorithm with a general proposal distribution is straightforward.

We define the collapsed Gibbs sampling algorithm via a Markov chain whose states are instantiations to X_p. As we discuss, to provide an unbiased estimator for the expectation over $P_\Phi(X_p)$ in equation (12.29), we want the stationary distribution of this Markov chain to be $P_\Phi(X_p)$. We thus modify our Gibbs sampling algorithm as follows: As in standard Gibbs sampling, we define a kernel for each variable $X_i \in X_p$. Let x_{-i} be an assignment to $X_p - \{X_i\}$. The kernel for X_i is defined as follows:

$$\mathcal{T}_i((x_{-i}, x_i) \to (x_{-i}, x_i')) = P_\Phi(x_i' \mid x_{-i}). \tag{12.33}$$

This equation is very similar to equation (12.22). The only difference is that the event x_{-i} on which we condition the distribution over X_i is not a full assignment to all the other variables in the network, but rather only to the remaining variables in X_p. Thus, the efficient computation of equation (12.23), where we simply compute the distribution over X_i given its Markov blanket, may not apply. However, we can compute this probability using inference in the network $\mathcal{H}_{\Phi[x_p]}$ — the Markov network reduced over the assignment x_p — a model that we assume to be tractable. Indeed, the sampling approach can be well integrated with a clique tree over the variables X_p to allow the sampling process to be executed efficiently (see exercise 12.27).

Having defined the chain, we can use it in any of the ways described in section 12.3.5 to collect a data set $\mathcal{D}$ of particles $x_p^{(m)}$, each of which is associated with a distribution over X_d. Using these particles, we can estimate:

$$\hat{E}_{\mathcal{D}}(f) = \frac{1}{M} \sum_{m=1}^{M} \left(E_{P_{\Phi[x_p[m]]}(X_d)}[f(x_p, x_d, e)] \right).$$

Example 12.18

Consider the Markov network defined by the Bayesian network of example 12.11, reduced over the evidence $G = g$. This Markov network is a bipartite graph, where we have two sets of variables I, D such that the only edges in the network are of the form I_j, D_k. Let $\phi_{j,k}(I_j, D_k)$ be the factor associated with each such edge. (For simplicity of notation, we assume that there is a factor for every such edge; the factors corresponding to grades that were not in the model will be vacuous.) We can now apply collapsed Gibbs sampling, selecting to sample the variables D and maintain a distribution over I; this choice will (in most cases) be better, since there are usually more students than courses, and it is generally better to sample in a lower-dimensional space. Thus, our Markov *chain needs to be able to sample D_k from $P_\Phi(D_k \mid \boldsymbol{d}_{-k})$:*

$$P_\Phi(d_k \mid \boldsymbol{d}_{-k}) = \frac{P_\Phi(d_1, d_2, \ldots, d_m)}{\sum_{D_k} P_\Phi(D_k, \boldsymbol{d}_{-k})}.$$

The expression in the numerator, and each term in the sum in the denominator, is the probability of a full assignment to $\boldsymbol{D}$. This type of expression can be computed as the partition function of the reduced Markov network that we obtain by setting $\boldsymbol{D} = \boldsymbol{d}$:

$$P_{\Phi[\boldsymbol{d}]}(I_1, \ldots, I_n) = \frac{1}{Z(\boldsymbol{d})} \prod_{j,k} \phi_{j,k}(I_j, d_k).$$

Each of the factors in the product is a singleton over an individual I_j. Thus, we can compute the partition function or marginals of this distribution in linear time. In particular, we can easily compute the Gibbs transition probability. Using the same analysis, we can easily compute an expression for $P_{\Phi[\boldsymbol{d}]}(\boldsymbol{I})$ in closed form, as a product of individual factors over the I_j's. Thus, we can also easily compute expectations of any function over these variables, such as the expected value of the number of smart students who got a grade of a C in an easy class (see exercise 12.28). ∎

Box 12.D — Concept: Correspondence and Data Association. *One simple but important problem that arises in many settings is the* correspondence problem: *mapping between one set of objects $\mathcal{U} = \{u_1, \ldots, u_k\}$ and another $\mathcal{V} = \{v_1, \ldots, v_m\}$. This problem arises across multiple diverse application domains. Perhaps the most familiar are physical sensing applications, where $\mathcal{U}$ are sensor measurements and $\mathcal{V}$ objects that can generate such measurements; we want to know which object generated which measurement. For example, the objects may be airplanes, and the sensor measurements blips on a radar screen; or the objects may be obstacles in a robot's environments, and the sensor measurements readings from a laser range finder. (See box 15.A.) In this incarnation, this task is known as* data association. *However, there are also many other applications, of which we list only a few examples:*

correspondence problem

data association

- *Matching citations in text to the entities to which they refer (see, for example, box 6.D); this problem has been called* identity resolution *and* record matching.

identity resolution

- *Image registration, or matching features in an image representing one view of an object to features in an image representing a different view; this problem arises in applications such as* stereo reconstruction *or* structure from motion.

image registration

- *Matching words in a sentence in one language to words in the same sentence, translated to a different language; this problem is called* word alignment.

- *Matching genes in a DNA sequence of one organism to* orthologous genes *in the DNA sequence of another organism.*

This problem has been tackled using a range of models and a variety of inference methods. In this case study, we describe some of these approaches and the design decisions they made.

correspondence
variable

mutual exclusion
constraints

Probabilistic Correspondence *To formulate the correspondence problem as a probabilistic model, we can introduce a set of* correspondence variables *that indicate the correspondence between one set of objects and the other. One approach is to have binary-valued variables C_{ij}, such that C_{ij} = true when u_i matches v_j, and false otherwise. While this approach is simple, it places no constraints on the number of matches for each i or for each j. Typically, we want to restrict our model so that, at least on one side, the match is unique. For example, in the radar tracking application, we typically want to assume that each measurement was derived from only a single object. In order to accommodate that in this model, we would need to add (hard)* mutual exclusion constraints *(mutex) that C_{ij} = true implies that $C_{ij'}$ = false for all $j' \neq j$. The resulting model is very densely connected, and it can be challenging for inference algorithms to deal with.*

A more parsimonious representation uses nonbinary variables C_i such that $Val(C_i) = \{1, \ldots, m\}$; here, $C_i = j$ indicates that u_i is matched to v_j. The mutex constraints for the matches to u_i are then forced by the fact that each variable C_i takes on only a single value. Of course, we might also have mutex constraints in the other direction, where we want to assume that each v_j matches only a single u_i. We will return to this setting.

The probabilistic model and the evidence generally combine to produce a set of affinities w_{ij} that specify how likely u_i is to match to v_j. For convenience, we assume that these affinities are represented in log-space. These affinities can be derived from a generative probability — how likely is v_j to have generated u_i, or from an undirected potential that measures the quality of the match. (See, for example, box 6.D for an application where both approaches have been utilized.) For example, in the robot obstacle matching task, $\exp(w_{ij})$ may evaluate how likely it is that an obstacle v_j, at location L_j, generated a measurement u_i at sensed location S_i. In the citation matching problem, we may have a model that tells us how likely it is that an individual with the name "John X. Smyth" generated a citation "J. Smith." The affinities w_{ij} define a node potential over the variables C_i: $\phi_i(C_i = j) = \exp(w_{ij})$. In addition, there may well be other components to the model, as we will discuss.

The general inference task is then to compute the distribution over the correspondence variables or (as a fallback) to find the most likely assignment. We now discuss the challenges posed by this task in different variants of the correspondence problem, and some of the solutions.

The simplest case (which rarely arises in practice) is the one just described: We have only a set of node potentials $\phi_i(C_i = j)$ that specify the affinity of each u_i to each v_j. In this case, we have only a set of unrelated node potentials over the variables C_i, making the model one comprising independent random variables. We can then easily find the most likely assignment $c_i^* = \arg\max_j \phi_i(C_i = j)$, or even compute the full posterior $P(C_i) \propto \phi_i(C_i)$.

The inference task becomes much more challenging when we extend the model in a way that induces correlations over the correspondence variables. We now describe three settings where such complications arise, noting that these are largely orthogonal, so that more than one of these complicating factors may arise in any given application.

Two-Sided Mutex Constraints *The first complication arises if we want to impose mutex constraints not just on the correspondence of the u_i's to the v_j's, but also in the reverse direction; that is, we want each v_j to be assigned to exactly one u_i. (We note that, unless $k = m$, a perfect match is not possible; however, we can circumvent this detail by adding "dummy" objects that are equally good matches to anything.) This requirement induces a model where all of the C_i's are connected to each other with potentials that impose mutex constraints, making it clearly intractable for a* *variable elimination algorithm. Nevertheless, several techniques have been effectively applied to this problem.*

First, we note that we can view the correspondence problem, in this case, as a bipartite graph, where the u_i's are one set of vertices, the v_j's are a second set, and an edge of weight w_{ij} connects u_i to each v_j. An assignment satisfying the mutex constraints is a bipartite matching *in this graph — a subset of edges such that each node has exactly one adjacent edge in the matching. Finding the single highest-probability assignment — the* MAP assignment *— is equivalent to one of finding the bipartite matching whose weight (sum of the edge weights in the matching) is largest. This problem can be solved very efficiently using combinatorial optimization algorithms: If we can match any u_i to any v_j, the total running time is $O(k^3)$; if we have only ℓ possible edges, the running time is $O(k^2 \log(k) + k\ell)$.*

bipartite
matching

MAP assignment

Of course, in many cases, a single assignment is not an adequate solution for our problem, since there may be many distinct solutions that have high probability. For computing posteriors in this setting, two main approaches have been used.

The first is an MCMC sampler over the space of possible matchings. Here, Gibbs sampling is not a viable approach, since any change to the value of a single variable would give rise to an assignment that violates the mutex constraints. A Metropolis-Hastings chain is a good solution, but for good performance, the proposal distribution must be carefully chosen. Most obvious is to pick two variables such that $C_{i_1} = j_1$ and $C_{i_2} = j_2$, and propose a move that flips them, setting $C_{i_1} = j_2$ and $C_{i_2} = j_1$. While this approach is a legal chain, it can be slow to mix: in most local maxima, a single flip of this type will often produce a very poor solution. Thus, some work has been done on constructing proposal distributions with larger steps that flip multiple variables at a time while still maintaining a legal matching.

The second solution uses loopy belief propagation over the Markov network with the C_i variables. As we mentioned, however, the mutex constraints give rise to a fully connected network, which is a very challenging setup for belief propagation. Here, however, we can adopt a simple heuristic: We begin without including the mutex constraints and run inference. We then examine the resulting posterior and identify those mutex constraints that are violated with high probability (those where $P(C_{i_1} = C_{i_2})$ is high). We then add those violated constraints and repeat the inference. In practice, this process usually converges long before all k^2 mutex constraints are added.

Unobserved Attributes *The second type of complication arises when the affinities w_{ij} depend on properties $\boldsymbol{A}_j$ of the objects that are unobserved, and need to be inferred. For example, in the citation matching problem, we generally do not know the correct name of an author or a paper; given the name of an author v_j, we can determine the affinity w_{ij} of any citation u_i. Conversely, given a set of citations u_i that match v_j, we can infer the most likely name to have generated the observed names in these citations. The same situation arises in many other applications. For example, in the airplane tracking application, the plane's location at a given point in time is generally unknown, although we may have a prior on this location based on observations at previous time points. Given the location of airplane v_j, we can determine how likely it was to have generate the blip u_i. Conversely, once we assign blip u_i to v_j, we can update our posterior on v_j's location.*

More formally, here we have a set of observed features $\boldsymbol{B}_i$ for each object u_i, and a set of hidden attributes $\boldsymbol{A}_j$ for each v_j. We have a prior $P(\boldsymbol{A}_j)$, and a set of factors $\phi_i(\boldsymbol{A}_j, \boldsymbol{B}_i, C_i)$ that are vacuous (uniform) if $C_i \neq j$. We want to compute the posterior over $\boldsymbol{A}_j$. For this problem,

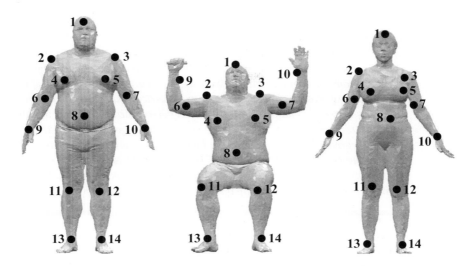

Figure 12.D.1 — Results of a correspondence algorithm for 3D human body scans The model explicitly captures correlations between different correspondence variables.

one obvious extension of the MCMC approach we described is to use a collapsed MCMC approach, where we sample the correspondence variables but maintain a closed-form distribution over A_j; see exercise 12.29. When the features A_k are such that maintaining a closed-form posterior is challenging (for example, they are continuous and do not have a simple parametric form), we often adopt an approach where we pick a single assignment to each A_j; this solution can be

EM *implemented in the framework of an* EM *algorithm (see box 19.D for one example).*

Directly Correlated Correspondences *A final complication arises when we wish to model direct*
correlated
correspondence *correlations between the values of different correspondence variables. This type of correlation would arise naturally, for example, in an application where we match visual features (small patches) in two images of the same real-world object (say a person). Here, we not only want the local appearance of a feature to be reasonably similar in the two images; we also want the relative geometry of the features to be consistent. For example, if a patch containing the eye is next to a patch containing the nose in one image, they should also be close in the other. These (soft) constraints can be represented as potentials on pairs of variables C_{i_1}, C_{i_2} (or even on larger subsets). Similar situations arise in sequence alignment, where we often prefer to match adjacent sequence elements (whether words or genes) on one side to adjacent sequence elements on the other.*

Here, we can often exploit the fact that the spatial structure we are trying to encode is local, so that the resulting Markov network over the C_i variables is often reasonably structured. As a consequence, techniques such as loopy belief propagation, if carefully designed, can be surprisingly effective in this setting.

Figure 12.D.1 demonstrates the results of one model along these lines. The model, due to Anguelov et al. (2004), aims to find a correspondence between a set of (automatically selected) landmarks *on different three-dimensional scans of human bodies. Here, $\phi_i(C_i = j)$ represents the extent to which the local appearance around the ith landmark in one scan is similar to the local appearance of the jth landmark on the other. This task is very challenging for two reasons. First, the local appearance of corresponding patches on two scans can be very different, so that the node potentials are only somewhat informative. Second, the two scans exhibit significant deformation, both of shape and of pose, so that standard parametric models of deformation do not work. In this task, modeling correlations between the different correspondence variables allows us to capture constraints regarding the preservation of the object geometry, specifically, the fact that distances between corresponding points should be roughly preserved. This feature was essential for obtaining reasonable correspondences. Most of the constraints regarding preservation of distances relate to pairs of nearby points, so that the resulting Markov network was not too densely connected, allowing a judicious application of loopy belief propagation to work well.*

12.5   Deterministic Search Methods ★

So far, we have focused on particles generated by random sampling. Random sampling tries to explore the state space of the distribution "uniformly," generating each state proportionately to its probability. **A sampling-based approach can, however, be problematic when we have a highly skewed distribution, where only a small number of states have nonnegligible probability. In this case, sampling methods will tend to sample the same small set of states repeatedly, wasting computational resources to no gain. An alternative approach, designed for settings such as this, is to use a deterministic method that explicitly searches for high-probability states.**

Intuitively, in these search methods, we deterministically generate some set of distinct assignments $\mathcal{D} = \{\xi[1], \ldots, \xi[M]\}$. We then approximate the joint distribution P by considering only these instantiations, ignoring the rest. Intuitively, if our particles account for a large proportion of the probability mass, we have a reasonable approximation to the joint.

Example 12.19

In the Student *network of figure 12.1, the most likely ten instantiations (of 48), in decreasing order, are:*

d^1	i^0	g^3	s^0	l^0	0.184
d^0	i^0	g^3	s^0	l^0	0.119
d^0	i^1	g^1	s^1	l^1	0.117
d^0	i^0	g^1	s^0	l^1	0.108
d^0	i^0	g^2	s^0	l^1	0.096
d^0	i^0	g^2	s^0	l^0	0.064
d^1	i^1	g^1	s^1	l^1	0.043
d^1	i^0	g^2	s^0	l^1	0.04
d^0	i^1	g^1	s^0	l^1	0.029
d^1	i^0	g^2	s^0	l^0	0.027.

Together, they account for about 82.6 percent of the probability mass. To account for 95 percent of the probability mass, we would need twenty-two instantiations, and to account for 99 percent of the mass, we would need thirty-three instantiations. ∎

Intuitively, the quality of our particle-based approximation to the joint distribution improves with the amount of probability mass accounted for in our particles. Thus, our goal is to enumerate instantiations that have high probability in some unnormalized measure $\tilde{P}$. Specifically, in the case of Bayesian networks, we want to enumerate instantiations that are likely given e.

We can formalize this goal as one of finding the highest-probability instantiations in $\tilde{P}$, until we reach some upper bound K on the number of particles. (One might be tempted to use, as a stopping criterion, a lower bound on the total mass accumulated in the enumerated particles; however, because we generally do not know the normalizing constant, the absolute mass accumulated has no real interpretation.) Clearly, this problem encompasses within it the **MAP assignment** task of finding the single most likely instantiation, also known as the *maximum a posteriori (MAP) assignment*. The problem of finding the MAP assignment is the focus of chapter 13. Not surprisingly, many of the successful methods for enumerating high-probability assignments are extensions of methods for finding the MAP assignment. Thus, we largely defer discussion of methods for finding high-probability particles to that chapter. (See, for example, exercise 15.10.) In this section, we focus primarily on the question of using a set of deterministically selected particles to approximate the answers to conditional probability queries.

Consider any event $Z = z$. Approximating the joint using the set of high-probability assignments $\mathcal{D}$, we have that one natural estimate for $\tilde{P}(z)$ is:

$$\sum_{m=1}^{M} \mathbf{1}\{z[m] = z\}\tilde{P}(\xi[m]), \qquad (12.34)$$

where $z[m] = \xi[m]\langle Z \rangle = z$. It is important to note a key difference between this estimate and the one of equation (12.2). There, we merely counted particles, whereas in this case, the particles are weighted by their probability. Intuitively, the difference is due to the fact that, in sampling methods, particles are generated proportionately to their probability. Thus, higher-probability instantiations are generated more often. If we then also weighted the sampled particles by their probability, we would be "double-counting" the probability. Alternatively, we can view this formula as an instance of the importance sampling estimator, equation (12.8). In this case, each particle $\xi[m]$ is generated from a (different) deterministic proposal distribution Q, that ascribes it probability 1.[4] Hence, we must weight the particle by $\tilde{P}(\xi[m])/Q(\xi[m]) = \tilde{P}(\xi[m])$.

Here, however, we have no guarantees about the bias of the estimator. Depending on our search procedure, we might end up with estimates that are arbitrarily bad. At one extreme, for example, we might have a search procedure that avoids (for as many particles as possible) any instantiation ξ where $\xi\langle Z \rangle = z$. Clearly, our estimate will then be biased in favor of low values. A more correct approach is to use our particles to provide both an upper and lower bound to the unnormalized probability of any event z. Some of our particles $\xi[m]$ have $z[m] = z$. The total **lower bound** probability mass of these particles is a *lower bound* on the probability mass of all instantiations ξ where $Z = z$. Similarly, the total probability mass of particles that have $z[m] \neq z$ is a lower bound on the complementary mass, and hence provides an upper bound on the probability

---

4. We cannot, of course, actually apply importance sampling in this way, since this deterministic proposal distribution violates our assumption that Q's support contains that of P. However, this perspective provides intuition for our choice of weights.

mass of the assignments where $\boldsymbol{Z} = \boldsymbol{z}$:

$$\sum_{m=1}^{M} \boldsymbol{I}\{\boldsymbol{z}[m] = \boldsymbol{z}\} \tilde{P}(\xi[m]) \leq \tilde{P}(\boldsymbol{Z} = \boldsymbol{z}) \leq \left(1 - \sum_{m=1}^{M} \boldsymbol{I}\{\boldsymbol{z}[m] \neq \boldsymbol{z}\} \tilde{P}(\xi[m]) \right). \qquad (12.35)$$

Equivalently, we can define $\rho = 1 - \sum_{m=1}^{M} \tilde{P}(\xi[m])$ to be the probability mass not accounted for by our particles. The bound can then be rewritten as:

$$\sum_{m=1}^{M} \boldsymbol{I}\{\boldsymbol{z}[m] = \boldsymbol{z}\} \tilde{P}(\xi[m]) \leq \tilde{P}(\boldsymbol{Z} = \boldsymbol{z}) \leq \rho + \sum_{m=1}^{M} \boldsymbol{I}\{\boldsymbol{z}[m] = \boldsymbol{z}\} \tilde{P}(\xi[m]).$$

This reformulation reflects the fact that the unaccounted probability mass can be associated either with the event $\boldsymbol{Z} = \boldsymbol{z}$ or with its complement. If all of the unaccounted probability mass is associated with $\boldsymbol{z}$, we get the upper bound, and if all of it is associated with the complement, we get the lower bound.

Example 12.20

Consider the Student *network of figure 12.1, and the event l^1 — a strong letter. Within the ten particles of example 12.19, the particles compatible with this event are: $\xi[3]$, $\xi[4]$, $\xi[5]$, $\xi[7]$, $\xi[8]$, and $\xi[9]$. Together, they account for 0.433 of the probability mass in the joint. The complementary event accounts for 0.393 of the mass. The total unaccounted mass is $1 - 0.826 = 0.174$. From these numbers, we obtain the following bounds:*

$$0.433 \leq P(l^1) \leq 1 - 0.393 = 0.433 + 0.174 = 0.607.$$

The true probability of this event is 0.502.

  Now, consider the event i^0, l^1 — a weak student getting a strong letter. In this case, we have three instantiations compatible with this assignment: $\xi[4]$, $\xi[5]$, and $\xi[8]$. Together, they account for 0.244 of the probability mass. The remaining assignments, which are incompatible with the event, account for 0.582 of the mass. Altogether, we obtain the following bounds for the probability of this event:

$$0.244 \leq P(i^0, l^1) \leq (1 - 0.582) = 0.418.$$

The true probability of this event is 0.272.

  When evaluating these results, it is important to note that they are based on the use of only ten particles. The results from any sampling-based method that uses only ten particles would probably be significantly worse. ∎

We can see that the true probability can lie anywhere within the interval specified by equation (12.35), and there is no particular justification for choosing any particular point within the range (for example, the point specified by equation (12.34)). If our search happens to first find instantiations that are compatible with $\boldsymbol{z}$, the lower bound is likely to be a better estimate; if it first finds instantiations that are incompatible with $\boldsymbol{z}$, the upper bound is likely to be closer.

  The fact that we obtain both lower and upper bounds on the mass in the unnormalized measure also allows us to provide bounds on the value of probability (or conditional) probability queries. Assume we are trying to compute $P(\boldsymbol{y} \mid \boldsymbol{e}) = \tilde{P}(\boldsymbol{y}, \boldsymbol{e}) / \tilde{P}(\boldsymbol{e})$. We can now obtain upper and lower bounds for both the numerator and denominator, using equation (12.35). A lower bound for the numerator and an upper bound for the denominator provide a lower bound for the ratio. Similarly, an upper bound for the numerator and a lower bound for the denominator provide an upper bound for the ratio. Analogously, we can obtain bounds on the marginal probability $P(\boldsymbol{y})$ by normalizing $\tilde{P}(\boldsymbol{y})$ by $\tilde{P}(\text{true})$.

More precisely, assume that we have bounds:

$$
\begin{aligned}
\ell_{\boldsymbol{y},e} &\leq P(\boldsymbol{y}, \boldsymbol{e}) \leq u_{\boldsymbol{y},e} \\
\ell_e &\leq P(\boldsymbol{e}) \leq u_e.
\end{aligned}
$$

Then we can bound:

$$
\frac{\ell_{\boldsymbol{y},e}}{u_e} \leq P(\boldsymbol{y} \mid \boldsymbol{e}) \leq \frac{u_{\boldsymbol{y},e}}{\ell_e}. \tag{12.36}
$$

Example 12.21

Assume we want to compute $P(i^0 \mid l^1)$ — the probability that a student with a strong letter is not intelligent — using our ten samples from example 12.19. We have already shown the bounds for both the numerator and the denominator. We obtain:

$$
\begin{aligned}
0.265/0.607 &\leq P(i^0 \mid l^1) \leq 0.439/0.433 \\
0.437 &\leq P(i^0 \mid l^1) \leq 1.014.
\end{aligned}
$$

The true probability of this event is 0.542, which is fairly far away from any of the bounds. ∎

Interestingly, the upper bound in this case is greater than 1. Although clearly valid, this conclusion is not particularly interesting. In general, the deterministic approximations are of value only if we can cover a very large fraction of our probability mass with a small number of particles. While this constraint might seem very restrictive (as it often is), there are nevertheless applications where the probabilities are very skewed, and this assumption is a very good one; we return to this point in section 12.6.

A very similar discussion applies to the extension of deterministic search methods to collapsed particles. In this case, we approximate the outer expectation in equation (12.29) using a set $\mathcal{D}$ of particles $\boldsymbol{x}_p[1], \ldots, \boldsymbol{x}_p[M]$, which are selected using some deterministic search procedure. As usual, each particle is associated with a distribution $P(\boldsymbol{X}_d \mid \boldsymbol{x}_p[m], \boldsymbol{e})$.

Consider the task of computing the probability $P(\boldsymbol{z})$. As in the case of full particles, each of the generated particles $\boldsymbol{x}_p[m]$ accounts for a certain part of the probability mass. However, in this case, we cannot simply test whether the event $\boldsymbol{z}$ is compatible with the particle, since the particle might not specify a full assignment to the variables $\boldsymbol{Z}$. Rather, we have to compute the probability of $\boldsymbol{z}$ relative to the distribution over all of the variables defined by the particle. (See exercise 12.26.)

Thus, in particular, the lower bound on $P(\boldsymbol{z})$ defined by our particles is:

$$
\begin{aligned}
&\sum_{m=1}^{M} P(\boldsymbol{x}_p[m]) \left(\boldsymbol{E}_{P(\boldsymbol{X}_d \mid \boldsymbol{x}_p[m])}[\boldsymbol{I}\{\boldsymbol{x}_p, \boldsymbol{X}_d\langle \boldsymbol{Z}\rangle = \boldsymbol{z}\}] \right) \\
&= \sum_{m=1}^{M} P(\boldsymbol{x}_p[m]) P(\boldsymbol{z} \mid \boldsymbol{x}_p[m]).
\end{aligned}
$$

The lower bound assumes that none of the unaccounted probability mass is compatible with z. Similarly, the upper bound assumes that all of this unaccounted mass is compatible with z, leading to:

$$\left(\sum_{m=1}^{M} P(z, x_p[m]) \right) + \left(1 - \sum_{m=1}^{M} P(x_p[m]) \right).$$

Once again, we can compute both the probability $P(z \mid x_p)$ and the weight $P(x_p)$ by using inference in the conditioned network.

incremental
conditioning

This method is simply an incremental version of the *conditioning* algorithm of section 9.5, using x_p as a conditioning set. However, rather than enumerating all possible assignments to the conditioning set, we enumerate only a subset of them, attempting to cover most of the probability mass. This algorithm is also called *bounded conditioning*.

bounded
conditioning

12.6   Summary

This chapter presents a series of methods that attempt to approximate a joint distribution using a set of particles. We discussed three main classes of techniques for generating particles.

Importance sampling, and specifically likelihood weighting, generates particles by random sampling from a distribution. Because we cannot, in general, sample from the posterior distribution, we generate particles from a different distribution, called the proposal distribution, and then adjust their weights so as to get an unbiased estimator. The proposal distribution is a mutilated Bayesian network, and the samples are generated using forward sampling. Owing to the use of forward sampling, likelihood weighting applies only to directed graphical models. However, if we somehow choose a proposal distribution, the more general framework of importance sampling can also be applied to undirected graphical models (see exercise 12.9).

Markov chain Monte Carlo techniques attempt to generate samples from the posterior distribution. We define a Markov chain, or a stochastic sampling process, whose stationary distribution (the asymptotic result of the sampling process) is the correct posterior distribution. The Metropolis-Hastings algorithm is a general scheme for specifying a Markov chain that induces a particular posterior distribution. We showed how we can use the Metropolis-Hastings algorithm for graphical models. We also showed a particular instantiation, called Gibbs sampling, that is specifically designed for graphical models.

Finally, search methods take a different approach, where particles are generated deterministically, trying to focus on instantiations that have high probability. Unlike random sampling methods, deterministic search methods do not provide an unbiased estimator for the target query. However, they do provide sound upper and lower bounds. When the probability distribution is highly diffuse, so that many particles are necessary to cover most of the probability mass, these bounds will be very loose, and generally without value. However, when a small number of instantiations account for a large fraction of the probability mass, deterministic search techniques can obtain very accurate results with a very small number of particles, often much more accurate results than sampling-based methods with a comparable number of particles. There are several applications that have this property. For example, when performing fault diagnosis (see, for example, box 5.A), where faults are very rare, it can be very efficient to enumerate all hypotheses where the system has up to K faults. Because multiple faults are highly unlikely, even a small value of K (2 or 3) will likely suffice to cover most of the probability mass — even the mass consistent with our evidence. Another example is speech recognition (box 6.B), where only very few trajectories through the HMM are likely. In both of these applications, deterministic search methods have been successfully applied.

From a high level, it appears that sampling methods are the ultimate general-purpose inference algorithm. They are the only method that can be applied to arbitrary probabilistic models and that is guaranteed to achieve the correct results at the large sample limit. Indeed, when faced with a complex probabilistic model that involves continuous variables or a nonparametric model, there are often very few other choices available to us. While optimization-based methods, such as those of chapter 11, can sometimes be applied, the application often requires a nontrivial derivation, specific to the problem at hand. Moreover, these methods provide no accuracy guarantee. Conversely, it seems that sampling-based methods can be applied easily, off-the-shelf, to virtually any model.

This impression, however, is somewhat misleading. **While it is true that sampling methods provide asymptotic guarantees, their performance for reasonable sample sizes is very difficult to predict. In practice, a naive application of sampling methods to a complex probabilistic model often fails dismally, in that the estimates obtained from any reasonable number of samples are highly inaccurate. Thus, the success of these methods depends heavily on the properties of the distribution, and on a careful design of our sampling algorithm. Moreover, there is little theoretic basis for this design, so that the process of getting sampling methods to work is largely a matter of intuition and intensive experimentation.**

Nevertheless, the methods described in this chapter do provide an important component in our arsenal of methods for inference in complex models. Moreover, they are often used very successfully in combination with exact or approximate global inference methods. Standard combinations include the use of global inference for providing more informed proposal distributions, and for manipulating collapsed particles. Such combinations are highly successful in practice, and they often lead to much better results than any of the two types of inference methods in isolation.

Having described these basic methods, we showed how they can be extended to the case of collapsed particles, which consist of an assignment to a subset of network variables, associated with a closed-form distribution over the remaining ones. The answer to a query is then a (possibly weighted) sum over the particles, of the answer to the query within each associated distribution. This approach approximates part of the inference task via particles, while performing exact inference on a subnetwork, which may be simpler than the original network.

12.7   Relevant Literature

The $\mathcal{NP}$-hardness of approximate probabilistic inference in Bayesian networks was shown by Dagum and Luby (1993).

There is a vast literature on the use of Monte Carlo methods for estimating integrals in general, and the expectation of functions in particular. See Robert and Casella (2005) for one good review. Geweke (1989) proves some of the basic results regarding the accuracy of the importance sampling estimates.

Probabilistic logic sampling was first proposed by Henrion (1986). The improvement to likelihood weighting was proposed independently by Fung and Chang (1989) and by Shachter and Peot (1989). Cano et al. (2006) and Dagum and Luby (1997) proposed a variant of likelihood weighting based on unnormalized importance sampling, which separately estimates $P(y, e)$ and $P(e)$, as described in section 12.2.3.2. Dagum and Luby also proposed the use of a data-dependent stopping rule for the case where the CPD entries are bounded away from 0 or 1. For this case, they provide guaranteed bounds on the expected number of samples required to achieve a certain error rate, as discussed in section 12.2.3.2. Pradhan and Dagum (1996) provide some empirical validation of this algorithm, applied to a large medical diagnosis network.

Various heuristic approaches for improving the proposal distribution have been proposed. Fung and del Favero (1994) proposed the backward sampling algorithm, that allows for generating samples from evidence nodes in the direction that is opposite to the topological order of nodes in the network, combining their likelihood function with the CPDs of some previously sampled variables. Other variants use some alternative form of approximate inference over the network to produce an approximation $Q(X)$ to $P(X \mid E = e)$, and then use Q as a proposal distribution for importance sampling. For example, de Freitas et al. (2001) use variational methods (described in chapter 11) in this way.

Shachter and Peot (1989) proposed an adaptive approach, called *self-importance sampling*, which adapts the proposal distribution to the samples obtained, attempting to increase the probability of sampling in higher-probability regions of the posterior distribution $P(X \mid e)$. This approach was subsequently improved by Cheng and Druzdzel (2000) and by Ortiz and Kaelbling (2000), who proposed an *adaptive importance sampling* method that uses the variance of the estimator in a more principled way.

Shwe and Cooper (1991) applied importance sampling to the QMR-DT network for medical diagnosis (Shwe et al. 1991). Their variant of the algorithm combined self-importance sampling and an improved proposal distribution called Markov blanket scoring. This proposal distribution was designed to be computed efficiently in the context of BN2O networks.

Sampling methods based on Markov chains were first proposed for models arising in statistical physics. In particular, the Metropolis-Hastings algorithm was first proposed by Metropolis et al. (1953). Geman and Geman (1984) applied Gibbs sampling to image restoration in a paper that was very influential in popularizing this method within computer vision, and subsequently in related communities. Ackley, Hinton, and Sejnowski (1985) propose the use of Gibbs sampling within Boltzmann machines, for both inference and parameter estimation. Pearl (1987) introduced Gibbs sampling for Bayesian networks. York (1992) continues this work, specifically addressing the problem of networks where some states have low (or even zero) probability.

Over the years, extensive work has been done on the topic of MCMC methods, addressing a broad range of topics including: theoretical analyses, application to new classes of models, improved algorithms that are faster or have better convergence properties, specific applications, and many more. Works reviewing some of these developments include Neal (1993); Smith and Roberts (1993); Gilks et al. (1996); Gamerman and Lopes (2006). Neal (1993), in particular, provides both an excellent tutorial, guidelines for practitioners, and a comprehensive annotated bibliography for relevant papers (up to 1993). Tierney (1994) discusses the conditions under which we can use multiple kernels within a single chain. Nummelin (1984, 2002) shows the central limit theorem for samples from a Markov chain. MacEachern and Berliner (1994) show that subsampling the samples derived from a Markov chain is suboptimal. Gelman and Rubin

adaptive
importance
sampling

(1992) provide a specific strategy for applying MCMC: they specify the number of chains, the burn-in time, and the intervals at which samples should be selected. Their strategy is applicable mostly for problems for which a good initial distribution is available, but provides insights more broadly.

Algorithms that improve convergence are particularly relevant for the high-dimensional, multimodal distributions that often arise in the setting of graphical models. Some methods for addressing this issue use larger, nonlocal steps in the search space, which are helpful in breaking out of local optima; for example, for pairwise MRFs where all variables have a uniform set of values, Swendsen and Wang (1987) and Barbu and Zhu (2005) propose moves that simultaneously flip an entire subgraph from one value to another. Higdon (1998) discusses the general idea of introducing auxiliary variables as a mechanism for taking larger steps in the space. The temperature-based methods draw on the idea of simulated annealing (Kirkpatrick et al. 1983). These methods include simulated tempering (Marinari and Parisi 1992; Geyer and Thompson 1995) in which the state of the model is augmented with a temperature variable for purposes of sampling; parallel tempering (Swendsen and Wang 1986; Geyer 1991) runs multiple chains at different temperatures at the same time and allows chains to exchange datapoints; tempered transitions (Neal 1996) proposes a new sample by moving up and down the temperature schedules; and *annealed importance sampling* Neal (2001) uses a similar approach in combination with an importance sampling reweighting scheme.

annealed importance sampling

The BUGS system is an invaluable tool for the application of MCMC methods, inasmuch as it encompasses a large class of models and implements a number of fairly advanced techniques for improving the mixing rate. The system itself, and some of the ideas in it, are described by Thomas et al. (1992) and Gilks et al. (1994).

The task of finding high-probability assignments in a probabilistic model is very closely related to the problem of finding a MAP assignment. We refer the reader to section 13.9 for many of the relevant references on that topic.

Techniques based on deterministic search were popular in the early days of the Bayesian network community, often motivated by connections with search algorithms and constraint-satisfaction algorithms in AI. As a few examples, Henrion (1991) proposes a search-based algorithm aimed specifically at BN2O networks, whereas Poole (1993b, 1989) describes an algorithm aimed at more general network structures. Horvitz et al. (1989) propose an algorithm that combines conditioning with search, to obtain some of the benefits of exact inference in a search-based algorithm. More recently, Lerner et al. (2000) use a collapsed search-based approach for the problem of fault diagnosis, where probabilities are also highly skewed.

The use of collapsed particles, and specifically the Rao-Blackwell theorem, for sampling-based estimation was proposed by Gelfand and Smith (1990) and further developed by Liu et al. (1994) and Robert and Casella (1996). This idea has since been used in many applications of sampling techniques to graphical models, where sampling some of the variables can often allow us to perform tractable inference over the others. The idea of sampling some of the variables in a Bayesian network and performing exact inference over the others was explored by Bidyuk and Dechter (2007).

We have focused our presentation here on the problem of generating samples from a distribution so as to estimate the posterior. However, another task of significant practical importance is that of computing the *partition function* of an unnormalized measure. As we will see, this task is of particular importance in learning, since the normalizing constant is often used as

partition function

a measure of the overall quality of a learned model. Computation of a partition function by directly sampling the distribution leads to estimates of high variance, since such estimates are usually dominated by a small number of samples with high unnormalized probabilities. In order to avoid this problem, a ratio of partition functions is computed; see exercise 12.8. An equivalent problem of free energy difference, or partition function ratio, has been tackled in computational physics and computation chemistry communities with a range of methods, including free energy perturbation, thermodynamic integration, bridge sampling, umbrella sampling, and Jarzynski equation; these methods are in essence importance-sampling algorithms. See Gelman and Meng (1998); Jarzynski (1997); Neal (2001) for some examples.

correspondence
Extensive work has been done on the *correspondence* problem in its various incarnations: data association, record matching, identity resolution, and more. See, for example, Bar-Shalom and Fortmann (1988); Bar-Shalom (1992) for a review of some of the key ideas. Pasula et al. (1999) first proposed the use of MCMC methods to sample the space of possible associations in target tracking, based on the analysis of Jerrum and Sinclair (1997) for sampling over matchings. Dellaert et al. (2003) also use MCMC for the data-association problem in a structure-from-motion task in computer vision; they propose a more sophisticated proposal distribution that allows more rapid mixing. Anguelov et al. (2004) suggest the use of belief propagation for solving the correspondence problem, in settings where the correspondences are correlated with each other. Their results form part of box 12.D. Some work has also been done on finding the MAP assignment to a data-association problem using variants of belief propagation (Chen et al. 2003; Duchi et al. 2006) or other methods (Lacoste-Julien et al. 2006).

12.8   Exercises

Exercise 12.1⋆

Consider the sequence of variables T_n defined in appendix A.2. Given ϵ and δ, we define $m(\epsilon, \delta) = \min_n P(|T_n - p| \geq \epsilon) \leq \delta$ to be smallest number of trials we need to ensure that the probability of deviating ϵ standard deviations from p is less than δ. Although we cannot compute $m(\epsilon, \delta)$ exactly, we can upper-bound it using the bounds we discuss earlier.

a. Use Chebyshev's bound to give an upper bound on $m(\epsilon, \delta)$. (You can use the fact that $\boldsymbol{Var}[T_n] \leq \frac{1}{4n}$.)
b. Use the Chernoff bounds discussed above to give an upper bound on $m(\epsilon, \delta)$.
c. What is the difference between these two bounds? How does each of these depend on ϵ and δ? What are the implications if we want design a test to estimate p?

Exercise 12.2

Consider the problem of providing a reliable estimate for a ratio p/q, where we have estimators $\hat{p}$ for p and $\hat{q}$ for q.

a. Assume that $\hat{p}$ has ϵ relative error to p and $\hat{q}$ has ϵ relative error to q. More precisely, $\hat{p} \in [(1 - \epsilon)p, (1 + \epsilon)p]$, and $\hat{q} \in [(1 - \epsilon)q, (1 + \epsilon)q]$. Provide a relative error bound on $\hat{p}/\hat{q}$ relative to p/q.
b. Now, assume that we have only an absolute error bound for $\hat{p}$ and $\hat{q}$: $\hat{p} \in [p - \epsilon, p + \epsilon]$, $\hat{q} \in [q - \epsilon, q + \epsilon]$. Show by example that $\hat{p}/\hat{q}$ can be an arbitrarily bad estimate for p/q.
c. What type of guarantee (if any) can you provide if $\hat{p}$ has low absolute error but $\hat{q}$ has low relative error?

Exercise 12.3

Assume we have a calibrated clique tree for a distribution P_Φ. Show how we can use the clique tree to generate samples that are sampled exactly from P_Φ.

Exercise 12.4

Prove proposition 12.2.

Exercise 12.5★

Let $\mathcal{B}$ be a Bayesian network, and X a variable that is not a root of the network. Show that $P_{\mathcal{B}}$ may not be the optimal proposal distribution for estimating $P_{\mathcal{B}}(X = x)$.

Exercise 12.6

edge reversal

In exercise 3.12, we defined the *edge reversal* transformation, which reverses the directionality of an edge $X \to Y$. How would you apply such a transformation in the context of likelihood weighting, and what would be the benefits?

Exercise 12.7★

Let $\mathcal{B}$ be a network, $\boldsymbol{E} = \boldsymbol{e}$ an observation, and $X_1, \ldots, X_k$ be some ordering (not necessarily topological) of the unobserved variables in $\mathcal{B}$. Consider a sampling algorithm that, for each $X_1, \ldots, X_k$ in order, randomly selects a value $x_i[m]$ for X_i using some distribution $Q(X_i \mid x_1[m], \ldots, x_{i-1}[m], \boldsymbol{e})$.

a. Write the formula for the importance weights in normalized importance sampling using this sampling process.

b. Using your answer in part 1, define an improved likelihood weighting algorithm that samples variables in the network in topological order, but, for a variable X_i with parents $\boldsymbol{U}_i$, samples X_i using a distribution that uses both $\boldsymbol{x}_{-i}[m]$ and the evidence in X_i's Markov blanket.

c. Does your approach from part 2 generate samples from the posterior distribution $P(X_1, \ldots, X_k \mid \boldsymbol{e})$? Explain.

Exercise 12.8★

Consider the normalized importance sampling algorithm, but now assume that, in equation (12.10), an unnormalized measure $\tilde{Q}(\boldsymbol{X})$ is used in place of $Q(\boldsymbol{X})$. Show that the average of the weights converges to $\frac{\sum_{\boldsymbol{X}} \tilde{P}(\boldsymbol{x})}{\sum_{\boldsymbol{X}} \tilde{Q}(\boldsymbol{X})}$, which is the ratio of the normalizing constants of the two distributions.

Exercise 12.9

In this question, we consider the application of importance sampling to Markov networks.

a. Explain intuitively why we cannot simply apply likelihood weighting to Markov networks.

b. Show how likelihood weighting can be applied to chordal Markov networks. Is this approach interesting? Explain.

c. Provide a technique by which the more general framework of importance sampling can be applied to Markov networks. Be sure to define both a reasonable proposal distribution and an algorithmic technique for computing the weights.

Exercise 12.10★

Consider the Grasshopper example of figure 12.3

a. Assume that the probability space is the full set of (positive and negative) integers; the transition model is now the same for all i (and not different for $i = \pm 4$). Assuming that the grasshopper starts from 0, use the central limit theorem (theorem A.2) to bound the probability that the grasshopper reaches an integer larger than $\left\lceil \sqrt{2T} \right\rceil$ after T steps.

b. Returning to the setting where the grasshopper moves over the integers in the range $\pm B$, try constructing a chain that reaches every state in time linear in B (Hint: Your chain will be nonreversible, and it will require the addition of another variable to the state description.)

Exercise 12.11

Show an example of a Markov chain where the limiting distribution reached via repeated applications of equation (12.20) depends on the initial distribution $P^{(0)}$.

Exercise 12.12★

Consider the following two conditions on a Markov chain $\mathcal{T}$:

a. It is possible to get from any state to any state using a positive probability path in the state graph.
b. For each state x, there is a positive probability of transitioning directly from x to x (a self-loop).

a. Show that, for a finite-state Markov chain, these two conditions together imply that $\mathcal{T}$ is regular.
b. Show that regularity of the Markov chain implies condition 1.
c. Show an example of a regular Markov chain that does not satisfy the condition 2.
d. Now let us weaken condition 2, requiring only that there exists a state x with a positive probability of transitioning directly from x to x. Show that this weaker condition and condition 1 together still suffice to ensure regularity.

Exercise 12.13

Show directly from equation (12.21) (without using the detailed balance equation) that the posterior distribution $P(\mathcal{X} \mid e)$ is a stationary distribution of the Gibbs chain (equation (12.22)).

Exercise 12.14

Show that any distribution π that satisfies the detailed balance equation, equation (12.24), must be a stationary distribution of $\mathcal{T}$.

Exercise 12.15

Prove theorem 12.5.

Exercise 12.16

Let $Val(\boldsymbol{X})$ be a set of states, and let $\mathcal{T}_1, \ldots, \mathcal{T}_k$ be a set of kernels, each of which satisfies the detailed balance equation relative to some stationary distribution π. Let $p_1, \ldots, p_k$ be any distribution over the indexes $1, \ldots, k$. Prove that the mixture Markov chain $\mathcal{T}$, which at each step takes a step sampled from $\mathcal{T}_i$ with probability p_i, also satisfies the detailed balance equation relative to π.

Exercise 12.17★

Let $\mathcal{T}_1, \ldots, \mathcal{T}_k$ be as in exercise 12.16. Consider the aggregate Markov chain $\mathcal{T}$, where each step consists of a sequence of k steps, with step i being sampled from $\mathcal{T}_i$.

a. Provide an example demonstrating that $\mathcal{T}$ may not satisfy the detailed balance equation relative to π.
b. Show that, nevertheless, $\mathcal{T}$ has the stationary distribution π.
c. Provide a multistep kernel using $\mathcal{T}_1, \ldots, \mathcal{T}_k$ that satisfies the detailed balance equation relative to π.

Exercise 12.18

a. Let X be a node in a Markov network $\mathcal{H}$, and let $\boldsymbol{y}$ be an assignment of values to X's Markov blanket $\boldsymbol{Y} = \mathrm{MB}_{\mathcal{H}}(X)$. Provide an efficient algorithm for computing

$$\frac{P(x' \mid \boldsymbol{y})}{P(x \mid \boldsymbol{y})}$$

for $x, x' \in Val(X)$. Your algorithm should involve only local parameters — potentials for cliques involving X.
b. Show how this algorithm applies to Bayesian networks.

Exercise 12.19

Show that Gibbs sampling is a special case of the Metropolis-Hastings algorithm. More precisely, provide a particular proposal distribution Q_i for each local transition $\mathcal{T}^{Q_i}$ that induces precisely the same distribution over the transitions taken as the associated Gibbs transition distribution $\mathcal{T}_i$.

Exercise 12.20★

Prove theorem 12.4.

Exercise 12.21

Consider the network of example 12.12, but now assume that we make Z a slightly noisy exclusive or of its parents. That is, with some small probability q, Z is chosen uniformly at random regardless of the values of X and Y, and with probability $1 - q$, Z is the exclusive or of X and Y. Analyze the transitions between states in the Gibbs chain for this network, with the evidence z^1, and particularly the expected time required to transition between different regions.

Exercise 12.22

Consider the same situation as in importance sampling, where we have an unnormalized measure $\tilde{P}(\boldsymbol{X})$ from which it is hard to sample, and a proposal distribution Q which is (hopefully) close to the normalized distribution $P \propto \tilde{P}$, from which we can draw independent samples. Consider a Markov chain where we define

$$\mathcal{T}(\boldsymbol{x} \to \boldsymbol{x}') = Q(\boldsymbol{x}') \min\left[1, \frac{w(\boldsymbol{x}')}{w(\boldsymbol{x})}\right]$$

for $\boldsymbol{x}' \neq \boldsymbol{x}$, where $w(\boldsymbol{x})$ is as defined in equation (12.10); we define $\mathcal{T}(\boldsymbol{x} \to \boldsymbol{x}) = 1 - \sum_{\boldsymbol{x}' \neq \boldsymbol{x}} \mathcal{T}(\boldsymbol{x} \to \boldsymbol{x}')$. Intuitively, the transition from $\boldsymbol{x}$ to $\boldsymbol{x}'$ selects an independent sample $\boldsymbol{x}'$ from Q and then moves toward it, depending on whether its importance weight is better than that of our current point $\boldsymbol{x}$.

a. Show that $\mathcal{T}$ defines a legal transition model.

b. Show that P is the stationary distribution of $\mathcal{T}$.

Exercise 12.23★★

Swendson-Wang algorithm

labeling MRF

The *Swendson-Wang algorithm* is a Metropolis-Hastings algorithm for *labeling MRFs*, such as those used in image segmentation (box 4.B), where all variables take values in the same set of labels $l = 1, \ldots, L$. Unlike the standard application of Metropolis-Hastings, this variant takes large steps in the space, which simultaneously flip the values of multiple variables together.

The algorithm proceeds as follows. Let $\mathcal{X}$ be the variables in the network, and $\mathcal{E}$ the set of edges in the pairwise MRF. Let ξ be our current state in the sampling process. We begin by partitioning the variables $\mathcal{X}$ into L disjoint subsets, based on their value in ξ: $\boldsymbol{X}_l = \{X_i \ : \ x_i = l\}$. We now randomly select a subset of the edges in the graph to produce a new set $\mathcal{E}'$: each edge $(i, j) \in \mathcal{E}$ is selected independently, with probability $q_{i,j}$. We now use $\mathcal{E}'$ to partition the graph into connected components. Each set $\boldsymbol{X}_l$ is partitioned into K_l connected components, where a connected component $\boldsymbol{X}_{lk}$ is a maximal subset of nodes $\boldsymbol{X}_{lk} \subseteq \boldsymbol{X}_l$ such that each $X_i, X_j \in \boldsymbol{X}_{lk}$ is connected by a path within $\mathcal{E}'$. The result is a set of connected components:

$$\mathcal{C} = \{\boldsymbol{X}_{lk} \ : \ l = 1, \ldots, L; k = 1, \ldots, K_l\}.$$

We now select a connected component $\boldsymbol{X}_{lk} \in \mathcal{C}$ uniformly at random. Let $q(\boldsymbol{Y} \mid \xi)$ be the probability with which a set $\boldsymbol{Y}$ is selected using this procedure.

We now propose to assign $\boldsymbol{X}_{lk}$ a new label l' with probability $q(l' \mid \xi, \boldsymbol{X}_{lk})$; note that this probability can depend in arbitrary ways on the current state. This proposed move, if accepted, defines a new state ξ', where $X_i' = l'$ for any $X_i \in \boldsymbol{X}_{lk}$, and $X_i' = X_i$ otherwise. Note that this proposed move flips a large number of variables at the same time, and thus it takes much larger steps in the space than a local Gibbs or Metropolis-Hastings sampler for this MRF.

In this exercise, we show how to use this proposal distribution within the Metropolis-Hastings algorithm.

a. Let ξ and ξ' be a pair of states such that:
  - $\boldsymbol{Y}$ forms a connected component in $\mathcal{E}$;
  - the variables in $\boldsymbol{Y}$ all take the value l in ξ;
  - the variables in $\boldsymbol{Y}$ all take the value l' in ξ';
  - all other variables in $\mathcal{X} - \boldsymbol{Y}$ take the same value in ξ and ξ'.

Show that

$$\frac{q(\boldsymbol{Y} \mid \xi)}{q(\boldsymbol{Y} \mid \xi')} = \frac{\prod_{(i,j)\in\mathcal{E}(\boldsymbol{Y},\boldsymbol{X}_l-\boldsymbol{Y})}(1 - q_{i,j})}{\prod_{(i,j)\in\mathcal{E}(\boldsymbol{Y},\boldsymbol{X}'_{l'}-\boldsymbol{Y})}(1 - q_{i,j})}$$

where: $\boldsymbol{X}_l$ is the set of vertices with label l in ξ, $\boldsymbol{X}'_{l'}$ the set of vertices with label l' in ξ'; and where $\mathcal{E}(\boldsymbol{Y},\boldsymbol{Z})$ (between two disjoint sets $\boldsymbol{Y},\boldsymbol{Z}$) is the set of edges connecting nodes in $\boldsymbol{Y}$ to nodes in $\boldsymbol{Z}$.

b. Use this fact to obtain an acceptance probability for the proposed move that respects the detailed-balance equation.

Exercise 12.24★★

Let us return to the setting of exercise 9.20.

a. Suppose now that you want to sample from $P(\mathcal{X} \mid \text{Possible}(\mathcal{X}) = K)$ using a Gibbs sampling strategy. Why is this a bad idea?
b. Can you devise a Metropolis-Hastings MCMC strategy that generates samples from the correct posterior? Describe the proposal distribution and the acceptance probabilities, and prove that your scheme does sample from the correct posterior. Explain intuitively why your chain is likely to work better than the Gibbs chain. You may assume that $1 < K < N$.

Exercise 12.25★★

annealed
importance
sampling

In this exercise we develop the *annealed importance sampling* procedure. Assume that we want to generate samples from a distribution $p(\boldsymbol{x}) \propto f(\boldsymbol{x})$ from which sampling is hard. Assume also that we have a distribution $q(\boldsymbol{x}) \propto g(\boldsymbol{x})$ from which sampling is easy. In principle, we can use q as a proposal distribution for p, and apply importance sampling. However, if q and p are very different, the results are likely to be quite poor. We now construct a sequence of Markov chains that will allow us to incrementally produce a lower-variance importance-sampling estimator.

The technique is as follows. We define a sequence of distributions, $p_0, \ldots, p_k$, where $p_i(\boldsymbol{x}) \propto f_i(\boldsymbol{x})$, and f_i is defined as:

$$f_i(\boldsymbol{x}) = p(\boldsymbol{x})^{\beta_i} q(\boldsymbol{x})^{1-\beta_i},$$

where $1 = \beta_0 > \beta_1 > \ldots > \beta_k = 0$. Note that $p_0 = p$ and $p_k = q$. We assume that we can generate samples from p_k, and that, for each p_i, $i = 1, \ldots, k-1$, we have a Markov chain $\mathcal{T}_i$ whose stationary distribution is p_i. To generate a weighted sample $\boldsymbol{x}, w$ relative to our target distribution p, we follow the following algorithm:

$$\begin{aligned}\boldsymbol{x}_k &\sim p_k(\boldsymbol{X}) \\ \boldsymbol{x}_i &\sim \mathcal{T}_i(\boldsymbol{x}_{i+1} \to \boldsymbol{X}) \quad i = (k-1), \ldots, 1.\end{aligned} \qquad (12.37)$$

Finally, we define our sample to be $\boldsymbol{x} = \boldsymbol{x}_1$, with weight

$$w = \prod_{i=1}^{k} \frac{f_{i-1}(\boldsymbol{x}_i)}{f_i(\boldsymbol{x}_i)}. \qquad (12.38)$$

To prove that these importance weights are correct, we define both a target distribution and a proposal distribution over the larger state space $(\boldsymbol{x}_1, \ldots, \boldsymbol{x}_k)$. We then show that the importance weights defined in equation (12.38) are correct relative to these distributions over the larger space.

a. Let

$$\mathcal{T}_i^{-1}(\boldsymbol{x} \to \boldsymbol{x}') = \mathcal{T}_i(\boldsymbol{x}' \to \boldsymbol{x}) \frac{f_i(\boldsymbol{x}')}{f_i(\boldsymbol{x})}$$

define the reversal of the transition model defined by $\mathcal{T}_i$. Show that $\mathcal{T}_i^{-1}(\boldsymbol{X} \to \boldsymbol{X}')$ is a valid transition model.

b. Define

$$f^*(\boldsymbol{x}_1, \ldots, \boldsymbol{x}_k) = f_0(\boldsymbol{x}_1) \prod_{i=1}^{k-1} \mathcal{T}_i^{-1}(\boldsymbol{x}_i \to \boldsymbol{x}_{i+1}),$$

and define $p^*(\boldsymbol{x}_1, \ldots, \boldsymbol{x}_k) \propto f^*(\boldsymbol{x}_1, \ldots, \boldsymbol{x}_k)$. Use your answer from above to conclude that $p^*(\boldsymbol{x}_1) = p(\boldsymbol{x}_1)$.

c. Let g^* be the function encoding the joint distribution from which $\boldsymbol{x}_1, \ldots, \boldsymbol{x}_k$ are sampled in the annealed importance sampling procedure equation (12.37). Show that the weight in equation (12.38) can be obtained as

$$\frac{f^*(\boldsymbol{x}_1, \ldots, \boldsymbol{x}_k)}{g^*(\boldsymbol{x}_1, \ldots, \boldsymbol{x}_k)}.$$

One can show, under certain assumptions, that the variance of the weights obtained by this procedure grows linearly in the dimension n of the number of variables $\boldsymbol{X}$, whereas the variance in a traditional importance sampling procedure grows exponentially in n.

Exercise 12.26

This exercise explores one heuristic approach for deterministic search in a Bayesian network. It is an intermediate method between full-particle search and collapsed-particle search: It uses partial instantiations as particles but does not perform inference on the resulting conditional distribution.

Assume that our goal is to provide upper and lower bounds on the probability of some event $\boldsymbol{y}$ in a Bayesian network $\mathcal{B}$ over $\mathcal{X}$. Let $X_1, \ldots, X_n$ be some topological ordering of $\mathcal{X}$. We enumerate particles that are partial assignments to $\mathcal{X}$, where each partial assignment instantiates some subset $X_1, \ldots, X_k$; note that the set $X_1, \ldots, X_k$ is not an arbitrary subset of $X_1, \ldots, X_n$, but rather the first k variables in the ordering. Different partial assignments may instantiate different prefixes of the variables. We organize these partial assignments in a tree, where each node is labeled with some partial assignment $(x_1, \ldots, x_k)$. The children of a node labeled $(x_1, \ldots, x_k)$ are $(x_1, \ldots, x_k, x_{k+1})$, for each $x_{k+1} \in Val(X_{k+1})$. We can iteratively grow the tree by choosing some leaf in the tree, corresponding to an assignment $(x_1, \ldots, x_k)$, and expanding the tree to include its children $(x_1, \ldots, x_j, x_{k+1})$ for all possible values x_{k+1}.

Consider a particular tree, with a set of leaves $L = \{\ell[1], \ldots, \ell[M]\}$, where each leaf $\ell[m] \in L$ is associated with the assignment $\boldsymbol{x}[m]$ to some subset of variables $\boldsymbol{X}[m]$.

a. Each leaf $\ell[m]$ in the tree defines a particle. Specify the assignment and probability associated with this particle, and describe how we would compute its probability efficiently.

b. Show how to use your probability estimates from part 1 (a) to provide both a lower and an upper bound for $P(\boldsymbol{y})$.

c. Based on your answer from part 1, provide a simple heuristic for choosing the next leaf to expand in the partial search tree.

Exercise 12.27★★

Consider the application of collapsed Gibbs sampling, where we use a clique tree to manipulate the conditional distribution $\tilde{P}(\boldsymbol{X}_d \mid \boldsymbol{X}_p)$. Develop an algorithm in which, after an initial calibration step, *all* of the variables $X_i \in \boldsymbol{X}_p$ in can be resampled using a single pass over the clique tree. (Hint: Use the algorithm developed in exercise 10.12.)

Exercise 12.28

Consider the setting of example 12.18, where we assume that all grades are observed but none of the I_j or D_k variables are observed. Show how you would use the set of collapsed samples generated in this example to compute the expected value of the number of smart students (i^1) who got a grade of a C (g^3) in an easy class (d^0).

Exercise 12.29★

Consider the data-association problem described in box 12.D: We have two sets of objects $\mathcal{U} = \{u_1, \ldots, u_k\}$ and another $\mathcal{V} = \{v_1, \ldots, v_m\}$, and we wish to map $\mathcal{U}$'s to $\mathcal{V}$'s. We have a set of observed features $\boldsymbol{B}_i$ for each object u_i, and a set of hidden attributes $\boldsymbol{A}_j$ for each v_j. We have a prior $P(\boldsymbol{A}_j)$, and a set of factors $\phi_i(\boldsymbol{A}_j, \boldsymbol{B}_i, C_i)$ such that $\phi_i(\boldsymbol{a}_j, \boldsymbol{b}_i, C_i) = 1$ for all $\boldsymbol{a}_j, \boldsymbol{b}_i$ if $C_i \neq j$. The model contains no other potentials.

We wish to compute the posterior over $\boldsymbol{A}_j$ using collapsed Gibbs sampling, where we sample the C_i's but maintain a closed-form posterior over the $\boldsymbol{A}_j$'s. Provide a sampling scheme for this task, showing clearly both the sampling distribution for the C_i variables and the computation of the closed form over the $\boldsymbol{A}_i$ variables given the assignment to the C_i's.

13 *MAP Inference*

13.1 Overview

So far, we have dealt solely with conditional probability queries. However, MAP queries, which we defined in section 2.1.5, are also very useful in a variety of applications. As a reminder, a MAP query aims to find the most likely assignment to all of the (non-evidence) variables. A marginal MAP query aims to find the most likely assignment to a subset of the variables, marginalizing out over the rest.

MAP queries are often used as a way of "filling in" unknown information. For example, we might be trying to diagnose a complex device, and we want to find a single consistent hypothesis about failures in different components that explains the observed behavior. Another example arises when we are trying to decode messages transmitted over a noisy channel. In such cases, the receiver observes a sequence of bits received over the channel, and then it attempts to find the most likely assignment of input bits that could have generated this observation (taking into account the code used and a model of the channel noise). This type of query is much better viewed as a MAP query than as a standard probability query, because we are not interested in the most likely values for the individual bits sent, but rather in the message whose overall probability is highest. A similar phenomenon arises in speech recognition, where we are trying to decode the most likely utterance given the (noisy) acoustic signal; here also we are not interested in the most likely value of individual phonemes uttered.

13.1.1 Computational Complexity

As for the case of conditional probability queries, it is instructive to analyze the computational complexity of the problem. There are many possible ways of formulating the MAP problem as a decision problem. One that is convenient for our purposes is the problem *BN-MAP-DP*, defined as follows:

> Given a Bayesian network $\mathcal{B}$ over $\mathcal{X}$ and a number τ, decide whether there exists an assignment x to $\mathcal{X}$ such that $P(x) > \tau$.

It turns out that a very similar construction to theorem 9.1 can be used to show that the *BN-MAP-DP* problem is also $\mathcal{NP}$-complete.

Theorem 13.1   *The decision problem BN-MAP-DP is $\mathcal{NP}$-complete*

The proof is left as an exercise (exercise 13.1).

We can also define an analogous decision problem *BN-margMAP-DP* for marginal MAP:

> Given a Bayesian network $\mathcal{B}$ over $\mathcal{X}$, a number τ, and a subset $\mathbf{Y} \subset \mathcal{X}$, decide whether there exists an assignment $\mathbf{y}$ to $\mathbf{Y}$ such that $P(\mathbf{y}) > \tau$.

Because marginal MAP is a generalization of MAP, we immediately conclude the following:

Corollary 13.1

The decision problem BN-margMAP-DP is $\mathcal{NP}$-hard.

However, for the case of marginal MAP, we cannot conclude that *BN-margMAP-DP* is in $\mathcal{NP}$. Intuitively, as we said, the marginal MAP problem involves elements of both maximization and summation, a combination that is significantly harder than either subtask in isolation. In fact, it is possible to show that *BN-margMAP-DP* is complete for a much harder complexity class:

Theorem 13.2

The decision problem BN-margMAP-DP is complete for $\mathcal{NP}^{\mathcal{PP}}$.

Defining the complexity class $\mathcal{NP}^{\mathcal{PP}}$ is outside the scope of this book (see section 9.8), but it is generally considered very hard, since it is known to contain the entire polynomial hierarchy, of which $\mathcal{NP}$ is only the first level.

While the "harder" complexity class of the marginal MAP problem indicates that it is more difficult, the implications of this formulation may be somewhat abstract. A more concrete ramification is the following result, which states that the marginal MAP problem is $\mathcal{NP}$-hard even for polytree networks:

Theorem 13.3

The following decision problem is $\mathcal{NP}$-hard:

polytree

> *Given a* polytree *Bayesian network $\mathcal{B}$ over $\mathcal{X}$, a subset $\mathbf{Y} \subset \mathcal{X}$, and a number τ, decide whether there exists an assignment $\mathbf{y}$ to $\mathbf{Y}$ such that $P(\mathbf{y}) > \tau$.*

We defer the justification for this result to section 13.2.3.

13.1.2   Overview of Solution Methods

As for conditional probability queries, when addressing MAP queries, it is useful to reformulate the joint distribution somewhat more abstractly, as a product of factors. Consider a distribution $P_\Phi(\mathcal{X})$ defined via a set of factors Φ and an unnormalized density $\tilde{P}_\Phi$. We need to compute:

$$\xi^{map} = \arg\max_\xi P_\Phi(\xi) = \arg\max_\xi \frac{1}{Z}\tilde{P}_\Phi(\xi) = \arg\max_\xi \tilde{P}_\Phi(\xi). \tag{13.1}$$

In particular, if $P_\Phi(\mathcal{X}) = P(\mathcal{X} \mid \boldsymbol{e})$, then we aim to maximize $P(\mathcal{X}, \boldsymbol{e})$.

The MAP task goes hand in hand with finding the value of the unnormalized probability of the most likely assignment: $\max_\xi \tilde{P}_\Phi(\xi)$. We note that, given an assignment ξ, we can easily compute its unnormalized probability simply by multiplying all of the factors in Φ, evaluated at ξ. However, we cannot retrieve the *actual* probability of ξ without computing the partition function, a problem that requires that we also solve the sum-product task.

max-product

Because $\tilde{P}_\Phi$ is a product of factors, tasks that involve maximizing $\tilde{P}_\Phi$ are often called *max-*

product inference tasks. Note that we often convert the max-product problem into log-space and maximize $\log \tilde{P}_\Phi$. This logarithm is a sum of factors that correspond to negative energies (see section 4.4.1.2), and hence this version of the problem is often called the *max-sum* problem. It is also common to negate the factors and minimize the sum of the energies for the different potentials; this version is generally called an *energy minimization* problem. The transformation into log-space has several significant advantages. First, it avoids the numerical issues associated with multiplying many small numbers together. More importantly, it transforms the problem into a linear one; as we will see, this transformation allows certain valuable tools to be brought to bear. For consistency with the rest of the book, we mostly use the max-product variant of the problem in the remainder of this chapter. However, all of our discussion carries over with minimal changes to the analogous max-sum (or min-sum) problem: we simply take the logarithm of all factors, and replace factor product steps with factor additions.

Many different algorithms, both exact and approximate, have been proposed for addressing the MAP problem. Most obviously, the goal of the MAP task is find an assignment to a set of variables whose score (unnormalized probability) is maximal. Thus, it is an instance of an *optimization problem* (see appendix A.4.1), a class of problems for which many general-purpose solutions have been developed. These methods include heuristic hill-climbing methods (see appendix A.4.2), as well as more specialized optimization methods. Some of these solutions have also been usefully applied to the MAP problem.

There are also many algorithms that are specifically targeted at the max-product (or min-sum) task, and exploit some of its special structure, most notably the connection to the graph representation. A large subset of algorithms operate by first computing a set of factors that are *max-marginals*. Max-marginals are a general notion that can be defined for any function:

Definition 13.1

max-marginal

The max-marginal *of a function f relative to a set of variables Y is:*

$$MaxMarg_f(\boldsymbol{y}) = \max_{\xi\langle \boldsymbol{Y}\rangle = \boldsymbol{y}} f(\xi), \tag{13.2}$$

for any assignment $\boldsymbol{y} \in Val(\boldsymbol{Y})$. ∎

For example, the max-marginal $MaxMarg_{\tilde{P}_\Phi}(\boldsymbol{Y})$ is a factor that determines a value for each assignment $\boldsymbol{y}$ to $\boldsymbol{Y}$; this value is the unnormalized probability of the most likely joint assignment consistent with $\boldsymbol{y}$.

A large class of MAP algorithms proceed by first computing an exact or approximate set of max-marginals for all of the variables in $\mathcal{X}$, and then attempting to extract an exact or approximate MAP assignment from these max-marginals. The first phase generally uses techniques such as variable elimination or message passing in clique trees or cluster graphs, algorithms similar to those we applied in the context of sum-product inference. Now, assume we have a set of (exact or approximate) max-marginals $\{MaxMarg_f(X_i)\}_{X_i \in \mathcal{X}}$. A key question is how we use those max-marginals to construct an overall assignment. **As we show, the computation of (approximate) max-marginals allows us to solve a global optimization problem as a set of local optimization problems for individual variables. This task, known as *decoding*, is to construct a joint assignment that locally optimizes each of the beliefs. If we can construct such an assignment, we will see that we can provide guarantees on its (strong local or even global) optimality.** One such setting is when the max-marginals are *unambiguous*: For each variable X_i, there is a unique x_i^* that maximizes:

$$x_i^* = \arg \max_{x_i \in Val(X_i)} MaxMarg_f(x_i). \tag{13.3}$$

When the max-marginals are unambiguous, identifying the locally optimizing assignment is easy. When they are ambiguous, the solution is nontrivial even for exact max-marginals, and can require an expensive computational procedure in its own right.

The marginal MAP problem appears deceptively similar to the MAP task. Here, we aim to find the assignment whose (conditional) *marginal* probability is maximal. Here, we partition $\mathcal{X}$ into two disjoint subsets, $\mathcal{X} = \boldsymbol{Y} \cup \boldsymbol{W}$, and aim to compute:

$$\boldsymbol{y}^{m\text{-}map} = \arg \max_{\boldsymbol{y}} P_\Phi(\boldsymbol{y}) = \arg \max_{\boldsymbol{y}} \sum_{\boldsymbol{W}} \tilde{P}_\Phi(\boldsymbol{y}, \boldsymbol{W}). \tag{13.4}$$

☞   **Thus, the marginal MAP problem involves both multiplication and summation, a combination that makes the task much more difficult, both theoretically and in practice. In particular, exact inference methods such as variable elimination can be intractable, even in simple networks.** And many of the approximate methods that have been developed for MAP queries do not extend easily to marginal MAP. So far, the only effective approximation technique for the marginal MAP task uses a heuristic search over the assignments $\boldsymbol{y}$, while employing some (exact or approximate) sum-product inference over $\boldsymbol{W}$ in the inner loop.

13.2   Variable Elimination for (Marginal) MAP

We begin our discussion with the most basic inference algorithm: variable elimination. We first present the simpler case of pure MAP queries, which turns out to be quite straightforward. We then discuss the issues that arise in marginal MAP queries.

13.2.1   Max-Product Variable Elimination

To gain some intuition for the MAP problem, let us begin with a very simple example.

Example 13.1

Consider the Bayesian network $A \to B$. Assume we have no evidence, so that our goal is to compute:

$$\begin{aligned} \max_{a,b} P(a,b) &= \max_{a,b} P(a)P(b \mid a) \\ &= \max_a \max_b P(a)P(b \mid a). \end{aligned}$$

Consider any particular value a of A, and let us consider possible completions of that assignment. Among all possible completions, we want to pick one that maximizes the probability:

$$\max_b P(a)P(b \mid a) = P(a) \max_b P(b \mid a).$$

Thus, a necessary condition for our assignment a, b to have the maximum probability is that B must be chosen so as to maximize $P(b \mid a)$. Note that this condition is not sufficient: we must also choose the value of A appropriately; but for any choice of A, we must choose B as described.

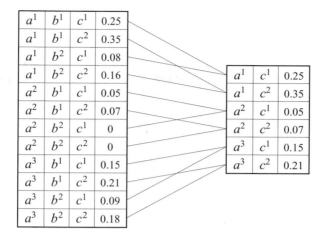

Figure 13.1   Example of the max-marginalization factor operation for variable B

Let $\phi(a)$ denote the internal expression $\max_b P(b \mid a)$. For example, consider the following assignment of parameters:

$$
\begin{array}{cc}
a^0 & a^1 \\
\hline
0.4 & 0.6
\end{array}
\qquad
\begin{array}{c||cc}
A & b^0 & b^1 \\
\hline
a^0 & 0.1 & 0.9 \\
a^1 & 0.55 & 0.45.
\end{array}
\tag{13.5}
$$

In this case, we have that $\phi(a^1) = \max_b P(b \mid a^1) = 0.55$ and $\phi(a^0) = \max_b P(b \mid a^0) = 0.9$. To compute the max-marginal over A, we now compute:

$$
\max_a P(a)\phi(a) = \max\left[0.4 \cdot 0.9, 0.6 \cdot 0.55\right] = 0.36.
$$
∎

As in the case of sum-product queries, we can reinterpret the computation in this example in terms of factors. We define a new operation on factors, as follows:

Definition 13.2

factor
maximization

Let $\boldsymbol{X}$ be a set of variables, and $Y \notin \boldsymbol{X}$ a variable. Let $\phi(\boldsymbol{X}, Y)$ be a factor. We define *the* factor maximization *of Y in ϕ to be factor ψ over $\boldsymbol{X}$ such that:*

$$
\psi(\boldsymbol{X}) = \max_Y \phi(\boldsymbol{X}, Y).
$$
∎

The operation over the factor $P(B \mid A)$ in example 13.1 is performing $\phi(A) = \max_B P(B \mid A)$. Figure 13.1 presents a somewhat larger example.

The key observation is that, like equation (9.6), we can sometimes exchange the order of maximization and product operations: If $X \notin Scope[\phi_1]$, then

$$
\max_X(\phi_1 \cdot \phi_2) = \phi_1 \cdot \max_X \phi_2.
\tag{13.6}
$$

In other words, we can "push in" a maximization operation over factors that do not involve the variable being maximized. A similar property holds for exchanging a maximization with a factor

Step	Variable eliminated	Factors used	Intermediate factor	New factor
1	S	$\phi_S(I,S)$	$\psi_1(I,S)$	$\tau_1(I)$
2	I	$\phi_I(I), \phi_G(G,I,D), \tau_1(I)$	$\psi_2(G,I,D)$	$\tau_2(G,D)$
3	D	$\phi_D(D), \tau_2(G,D)$	$\psi_3(G,D)$	$\tau_3(G)$
4	L	$\phi_L(L,G)$	$\psi_4(L,G)$	$\tau_4(G)$
5	G	$\tau_4(G), \tau_3(G)$	$\psi_5(G)$	$\tau_5(\emptyset)$

Table 13.1   A run of max-product variable elimination

summation operation: If $X \notin Scope[\phi_1]$, then

$$\max_X(\phi_1 + \phi_2) = \phi_1 + \max_X \phi_2. \tag{13.7}$$

max-product
variable
elimination

This insight leads directly to a *max-product variable elimination* algorithm, which is directly analogous to the algorithm in algorithm 9.1. The difference is that in line 4, we replace the expression $\sum_Z \psi$ with the expression $\max_Z \psi$. The algorithm is shown in algorithm 13.1. The same template also covers max-sum, if we replace product of factors with addition of factors.

If X_i is the final variable in this elimination process, we have maximized all variables other than X_i, so that the resulting factor ϕ_{X_i} is the max-marginal over X_i.

Example 13.2

Consider again our very simple Student *network, shown in figure 3.4. Our goal is to compute the most likely instantiation to the entire network, without evidence. We will use the elimination ordering S, I, D, L, G. Note that, unlike the case of sum-product queries, we have no query variables, so that all variables are eliminated.*

The computation generates the factors shown in table 13.1. For example, the first step would compute $\tau_1(I) = \max_s \phi_S(I,s)$. Specifically, we would get $\tau_1(i^0) = 0.95$ and $\tau_1(i^1) = 0.8$. Note, by contrast, that the same factor computed with summation instead of maximization would give $\tau_1(I) \equiv 1$, as we discussed.

The final factor, $\tau_5(\emptyset)$, is simply a number, whose value is

$$\max_{S,I,D,L,G} P(S,I,D,L,G).$$

For this network, we can verify that the value is 0.184.   ∎

 The factors generated by max-product variable elimination have an identical structure to those generated by the sum-product algorithm using the same ordering. **Thus, our entire analysis of the computational complexity of variable elimination, which we performed for sum-product in section 9.4, applies unchanged. In particular, we can use the same algorithms for finding elimination orderings, and the complexity of the execution is precisely the same induced width as in the sum-product case.** We can also use similar ideas to exploit structure in the CPDs; see, for example, exercise 13.2.

13.2.2   Finding the Most Probable Assignment

decoding

We now tackle the original MAP problem: *decoding*, or finding the most likely assignment itself.

Algorithm 13.1 Variable elimination algorithm for MAP. The algorithm can be used both in its max-product form, as shown, or in its max-sum form, replacing factor product with factor addition.

Procedure Max-Product-VE (
    $\Phi,$ // Set of factors over X
    $\prec$ // Ordering on X
)
1   Let $X_1, \ldots, X_k$ be an ordering of X such that
2        $X_i \prec X_j$ iff $i < j$
3   **for** $i = 1, \ldots, k$
4        $(\Phi, \phi_{X_i}) \leftarrow$ Max-Product-Eliminate-Var(Φ, X_i)
5    $x^* \leftarrow$ Traceback-MAP$(\{\phi_{X_i} \ : \ i = 1, \ldots, k\})$
6   **return** x^*, Φ // Φ contains the probability of the MAP

Procedure Max-Product-Eliminate-Var (
    $\Phi,$ // Set of factors
    Z // Variable to be eliminated
)
1    $\Phi' \leftarrow \{\phi \in \Phi \ : \ Z \in Scope[\phi]\}$
2    $\Phi'' \leftarrow \Phi - \Phi'$
3    $\psi \leftarrow \prod_{\phi \in \Phi'} \phi$
4    $\tau \leftarrow \max_Z \psi$
5   **return** $(\Phi'' \cup \{\tau\}, \psi)$

Procedure Traceback-MAP (
    $\{\phi_{X_i} \ : \ i = 1, \ldots, k\}$
)
1   **for** $i = k, \ldots, 1$
2        $u_i \leftarrow (x_{i+1}^*, \ldots, x_k^*)\langle Scope[\phi_{X_i}] - \{X_i\}\rangle$
3   // The maximizing assignment to the variables eliminated after
            X_i
4        $x_i^* \leftarrow \arg\max_{x_i} \phi_{X_i}(x_i, u_i)$
5   // x_i^* is chosen so as to maximize the corresponding entry in
            the factor, relative to the previous choices u_i
6   **return** x^*

As we have discussed, the result of the computation is a max-marginal $MaxMarg_{\tilde{P}_\Phi}(X_i)$ over the final uneliminated variable, X_i. We can now choose the maximizing value x_i^* for X_i. Importantly, from the definition of max-marginals, we are guaranteed that there exists some assignment ξ^* consistent with x_i^*. But how do we construct such an assignment?

We return once again to our simple example:

Example 13.3   *Consider the network of example 13.1, but now assume that we wish to find the actual assignment $a^*, b^* = \arg\max_{A,B} P(A, B)$. As we discussed, we first compute the internal maximization*

$\max_b P(a, b)$. *This computation tells us, for each value of* a, *which value of* b *we must choose to complete the assignment in a way that maximizes the probability. In our example, the maximizing value of* B *for* a^1 *is* b^0, *and the maximizing value of* B *for* a^0 *is* b^1. *However, we cannot actually select the value of* B *at this point, since we do not yet know the correct (maximizing) value of* A. *We therefore proceed with the computation of example 13.1, and compute both the max-marginal over* A, $\max_a P(a)\phi(a)$, *and the value* a *that maximizes this expression. In this case,* $P(a^1)\phi(a^1) = 0.6 \cdot 0.55 = 0.33$, *and* $P(a^0)\phi(a^0) = 0.4 \cdot 0.9 = 0.36$. *The maximizing value* a^* *of* A *is therefore* a^0. *The key insight is that, given this value of* A, *we can now go back and select the corresponding value of* B — *the one that maximizes* $\phi(a^*)$. *Thus, we obtain that our maximizing assignment is* a^0, b^1, *as expected.*   ∎

The key intuition in this computation is that, as we eliminate variables, we cannot determine their maximizing value. However, we can determine a "conditional" maximizing value — their maximizing value given the values of the variables that have not yet been eliminated. When we pick the value of the final variable, we can then go back and pick the values of the other variables accordingly. For the last variable eliminated, say X, the factor for the value x contains the probability of the most likely assignment that contains $X = x$. Thus, we correctly select the most likely assignment to X, and therefore to all the other variables. This process is called

traceback

traceback of the solution.

The algorithm implementing this intuition is shown in algorithm 13.1. Note that the operation in line 2 of Traceback-MAP is well defined, since all of the variables remaining in $Scope[\phi_{X_i}]$ were eliminated after X_i, and hence must be within the set $\{X_{i+1}, \ldots, X_k\}$. We can show that the algorithm returns the MAP:

Theorem 13.4

The algorithm of algorithm 13.1 returns

$$\boldsymbol{x}^* = \arg\max_{\boldsymbol{x}} \prod_{\phi \in \Phi} \phi,$$

and Φ, *which contains a single factor of empty scope whose value is:*

$$\max_{\boldsymbol{x}} \prod_{\phi \in \Phi} \phi.$$

The proof follows in a straightforward way from the preceding intuitions, and we leave it as an exercise (exercise 13.3).

We note that the traceback procedure is not an expensive one, since it simply involves a linear traversal over the factors defined by variable elimination. In each case, when we select a value x_i^* for a variable X_i in line 2, we are guaranteed that x_i^* is, indeed, a part of a jointly coherent MAP assignment. Thus, we will never need to backtrack and revisit this decision, trying a different value for X_i.

Example 13.4

Returning to example 13.2, we now consider the traceback phase. We begin by computing $g^* = \arg\max_g \psi_5(g)$. *It is important to remember that* g^* *is not the value that maximizes* $P(G)$. *It is the value of* G *that participates in the most likely complete assignment to all the network variables* $\mathcal{X} = \{S, I, D, L, G\}$. *Given* g^*, *we can now compute* $l^* = \arg\max_l \psi_4(g^*, l)$. *The value* l^* *is* *the value of L in the most likely complete assignment to $\mathcal{X}$. We use the same procedure for the remaining variables. Thus,*

$$
\begin{aligned}
d^* &= \arg\max_d \psi_3(g^*, d) \\
i^* &= \arg\max_i \psi_2(g^*, i, d^*) \\
s^* &= \arg\max_s \psi_1(i^*, s).
\end{aligned}
$$

It is straightforward (albeit somewhat tedious) to verify that the most likely assignment is d^1, i^0, g^3, s^0, l^0, and its probability is (approximately) the value 0.184 that we obtained in the first part of the computation. ∎

The additional step of computing the actual assignment does not add significant time complexity to the basic max-product task, since it simply does a second pass over the same set of factors computed in the max-product pass. With an appropriate choice of data structures, this cost can be linear in the number n of variables in the network. The cost in terms of space is a little greater, inasmuch as the MAP pass requires that we store the intermediate results in the max-product computation. However, the total cost is at most a factor of n greater than the cost of the computation without this additional storage.

The algorithm of algorithm 13.1 finds the one assignment of highest probability. This assignment gives us the single most likely explanation of the situation. In many cases, however, we want to consider more than one possible explanation. Thus, a common task is to find the set of the K most likely assignments. This computation can also be performed using the output of a run of variable elimination, but the algorithm is significantly more intricate. (See exercise 13.5 for one simpler case.) An alternative approach is to use one of the search-based algorithms that we discuss in section 13.7.

13.2.3 Variable Elimination for Marginal MAP ⋆

We now turn our attention to the application of variable elimination algorithms to the marginal MAP problem. Recall that our marginal MAP problem can be written as $\arg\max_{\boldsymbol{y}} \sum_{\boldsymbol{W}} \tilde{P}_\Phi(\boldsymbol{y}, \boldsymbol{W})$, where $\boldsymbol{y} \cup \boldsymbol{W} = \mathcal{X}$, so that $\tilde{P}_\Phi(\boldsymbol{y}, \boldsymbol{W})$ is a product of factors in some set Φ. Thus, our computation has the following *max-sum-product* form:

max-sum-product

$$
\max_{\boldsymbol{Y}} \sum_{\boldsymbol{W}} \prod_{\phi \in \Phi} \phi. \tag{13.8}
$$

This form immediately suggests a variable elimination algorithm, along the lines of similar algorithms for sum-product and max-product. This algorithm simply puts together the ideas we used for probability queries on one hand and MAP queries on the other. Specifically, the summations and maximizations outside the product can be viewed as operations on factors. Thus, to compute the value of this expression, we simply have to eliminate the variables $\boldsymbol{W}$ by summing them out, and the variables in $\boldsymbol{Y}$ by maximizing them out. When eliminating a variable X, whether by summation or by maximization, we simply multiply all the factors whose scope involves X, and then eliminate X to produce the resulting factor. Our ability to perform this step is justified by the exchangeability of factor summation/maximization and factor product (equation (9.6) and equation (13.6)).

Example 13.5   *Consider again the network of figure 3.4, and assume that we wish to find the probability of the most likely instantiation of SAT result and letter quality:*

$$\max_{S,L} \sum_{G,I,D} P(I, D, G, S, L).$$

We can perform this computation by eliminating the variables one at a time, as appropriate. Specifically, we perform the following operations:

$$
\begin{aligned}
\psi_1(I, G, D) &= \phi_D(D) \cdot \phi_G(G, I, D) \\
\tau_1(I, G) &= \sum_D \psi_1(I, G, D) \\[1em]
\psi_2(S, G, I) &= \phi_I(I) \cdot \phi_S(S, I) \cdot \tau_1(I, G) \\
\tau_2(S, G) &= \sum_I \psi_2(S, G, I) \\[1em]
\psi_3(S, G, L) &= \tau_2(S, G) \cdot \phi_L(L, G) \\
\tau_3(S, L) &= \sum_G \psi_3(S, G, L) \\[1em]
\psi_4(S, L) &= \tau_3(S, L) \\
\tau_4(L) &= \max_S \psi_4(S, L) \\[1em]
\psi_5(L) &= \tau_4(L) \\
\tau_5(\emptyset) &= \max_L \psi_5(L).
\end{aligned}
$$

Note that the first three factors τ_1, τ_2, τ_3 are generated via the operation of summing out, whereas the last two are generated via the operation of maxing out.   ∎

This process computes the unnormalized probability of the marginal MAP assignment. We can find the most likely values to the max-variables exactly as we did in the case of MAP: We simply keep track of the factors associated with them, and then we work our way backward to compute the most likely assignment; see exercise 13.4.

Example 13.6   *Continuing our example, after completing the different elimination steps, we compute the value $l^* = \arg\max_l \psi_5(L)$. We then compute $s^* = \arg\max_s \psi_4(s, l^*)$.*   ∎

The similarity between this algorithm and the previous variable elimination algorithms we described may naturally lead one to conclude that the computational complexity is also similar. Unfortunately, that is not the case: this process is computationally much more expensive than the corresponding variable elimination process for pure sum-product or pure max-product. The difficulty stems from the fact that we are not free to choose an arbitrary elimination ordering. When summing out variables, we can utilize the fact that the operations of summing out different variables commute. Thus, when performing summing-out operations for sum-product variable

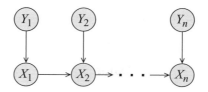

Figure 13.2 **A network where a marginal MAP query requires exponential time**

elimination, we could sum out the variables in any order. Similarly, we could use the same freedom in the case of max-product elimination. Unfortunately, the max and sum operations do not commute (exercise 13.19). Thus, in order to maintain the correct semantics of marginal MAP queries, as specified in equation (13.4), we *must* perform all the variable summations before we can perform any of the variable maximizations.

As we saw in example 9.1, different elimination orderings can induce very different widths. **When we *constrain* the set of legal elimination orderings, we have a smaller range of possibilities, and even the best elimination ordering consistent with the constraint might have significantly larger width than a good unconstrained ordering.**

constrained
elimination
ordering

Example 13.7

Consider the network shown in figure 13.2, and assume that we wish to compute

$$\boldsymbol{y}^{m\text{-}map} = \arg \max_{Y_1,\ldots,Y_n} \sum_{X_1,\ldots,X_n} P(Y_1,\ldots,Y_n,X_1,\ldots,X_n).$$

As we discussed, we must first sum out $X_1,\ldots,X_n$, and only then deal with the maximization over the Y_i's. Unfortunately, the factor generated after summing out all of the X_i's contains all of their neighbors, that is, all of the Y_i's. This factor is exponential in n. By contrast, the minimal induced width of this network is 2, so that any probability query (assuming a small number of query variables) or MAP query can be performed on this network in linear time. ∎

As we can see, **even on very simple polytree networks, elimination algorithms can require exponential time to solve a marginal MAP query.** One might hope that this blowup is a consequence of the algorithm we use, and that perhaps a more clever algorithm would avoid this problem. Unfortunately, theorem 13.3 shows that this difficulty is unavoidable, and unless $\mathcal{P} = \mathcal{NP}$, some exact marginal MAP computation require exponential time, even in very simple networks. Importantly, however, we must keep in mind that this result does not affect *every* marginal MAP query. Depending on the structure of the network and the choice of maximization variables, the additional cost induced by the constrained elimination ordering may or may not be prohibitive.

Putting aside the issue of computational cost, once we have executed a run of variable elimination for the marginal MAP problem, the task of finding the actual marginal MAP assignment can be addressed using a *traceback* procedure that is directly analogous to Traceback-MAP of algorithm 13.1; we leave the details as an exercise (exercise 13.4).

traceback

Algorithm 13.2 Max-product message computation for MAP

    Procedure Max-Message (
      i,   // sending clique
      j   // receiving clique
)
1    $\psi(C_i) \leftarrow \psi_i \cdot \prod_{k \in (\mathrm{Nb}_i - \{j\})} \delta_{k \rightarrow i}$
2    $\tau(S_{i,j}) \leftarrow \max_{C_i - S_{i,j}} \psi(C_i)$
3   **return** $\tau(S_{i,j})$

13.3   Max-Product in Clique Trees

We now extend the ideas used in the MAP variable elimination algorithm to the case of clique trees. As for the case of sum-product, the benefit of the clique tree algorithm is that it uses dynamic programming to compute an entire set of marginals simultaneously. For sum-product, we used clique trees to compute the *sum-marginals* over each of the cliques in our tree. Here, we compute a set of *max-marginals* over each of those cliques.

At this point, one might ask why we want to compute an entire set of max-marginals simultaneously. After all, if our only task is to compute a single MAP assignment, the variable elimination algorithm provides us with a method for doing so. There are two reasons for considering this extension.

First, a set of max-marginals can be a useful indicator for how confident we are in particular components of the MAP assignment. Assume, for example, that our variables are binary-valued, and that the max-marginal for X_1 has $MaxMarg(x_1^1) = 3$ and $MaxMarg(x_1^0) = 2.95$, whereas the max-marginal for X_2 has $MaxMarg(x_2^1) = 3$ and $MaxMarg(x_2^0) = 1$. In this case, we know that there is an alternative joint assignment whose probability is very close to the optimum, in which X_1 takes a different value; by contrast, the best alternative assignment in which X_2 takes a different value has a much lower probability. Note that, without knowing the partition function, we cannot determine the actual magnitude of these differences in terms of probability. But we can determine the relative difference between the change in X_1 and the change in X_2.

Second, in many cases, an exact solution to the MAP problem via a variable elimination procedure is intractable. In this case, we can use message passing procedures in cluster graphs, similar to the clique tree procedure, to compute *approximate* max-marginals. These *pseudo-max-marginals* can be used for selecting an assignment; while this assignment is not generally the MAP assignment, we can nevertheless provide some guarantees in certain cases. As before, our task has two parts: computing the max-marginals and decoding them to extract a MAP assignment. We describe each of those steps in turn.

pseudo-max-marginal

13.3.1   Computing Max-Marginals

In the same way that we used dynamic programming to modify the sum-product variable elimination algorithm to the case of clique trees, we can also modify the max-product algorithm to define a *max-product belief propagation* algorithm in clique trees. The resulting algorithm executes precisely the same initialization and overall message scheduling as in the sum-product

max-product belief propagation

max-product
message passing

belief propagation algorithm of algorithm 10.2; the only difference is the use of *max-product* rather than sum-product message passing, as shown in algorithm 13.2; as for variable elimination, the procedure has both a max-product and a max-sum variant.

As for sum-product message passing, the algorithm will converge after a single upward and downward pass. After those steps, the resulting clique tree $\mathcal{T}$ will contain the appropriate *max-marginal* in every clique.

max-marginal

Proposition 13.1

Consider a run of the max-product clique tree algorithm, where we initialize with a set of factors Φ. Let β_i be a set of beliefs arising from an upward and downward pass of this algorithm. Then for each clique C_i and each assignment c_i to C_i, we have that

$$\beta_i(c_i) = MaxMarg_{\tilde{P}_\Phi}(c_i). \tag{13.9}$$

That is, the clique belief contains, for each assignment c_i to the clique variables, the (unnormalized) measure $\tilde{P}_\Phi(\xi)$ of the most likely assignment ξ consistent with c_i. The proof is exactly the same as the proof of theorem 10.3 and corollary 10.2 for sum-product clique trees, and so we do not repeat the proof. Note that, because the max-product message passing process does not compute the partition function, we *cannot* derive from these max-marginals the actual probability of any assignment; however, because the partition function is a constant, we can still compare the values associated with different assignments, and therefore compute the assignment ξ that maximizes $\tilde{P}_\Phi(\xi)$.

Because max-product message passing over a clique tree produces max-marginals in every clique, and because max-marginals must agree, it follows that any two adjacent cliques must agree on their sepset:

$$\max_{C_i - S_{i,j}} \beta_i = \max_{C_j - S_{i,j}} \beta_j = \mu_{i,j}(S_{i,j}). \tag{13.10}$$

max-calibrated

In this case, the clusters are said to be *max-calibrated*. We say that a clique tree is *max-calibrated* if all pairs of adjacent cliques are max-calibrated.

Corollary 13.2

The beliefs in a clique tree resulting from an upward and downward pass of the max-product clique tree algorithm are max-calibrated.

Example 13.8

Consider, for example, the Markov network of example 3.8, whose joint distribution is shown in figure 4.2. One clique tree for this network consists of the two cliques $\{A, B, D\}$ and $\{B, C, D\}$, with the sepset $\{B, D\}$. The max-marginal beliefs for the clique and sepset for this example are shown in figure 13.3. We can easily confirm that the clique tree is calibrated. ∎

max-product
belief update

We can also define a *max-product belief update* message passing algorithm that is entirely analogous to the belief update variant of sum-product message passing. In particular, in line 1 of algorithm 10.3, we simply replace the summation with the maximization operation:

$$\sigma_{i \to j} \leftarrow \max_{C_i - S_{i,j}} \beta_i.$$

The remainder of the algorithm remains completely unchanged. As in the sum-product case, the max-product belief propagation algorithm and the max-product belief update algorithm

Assignment			$\max_C$
a^0	b^0	d^0	$300,000$
a^0	b^0	d^1	$300,000$
a^0	b^1	d^0	$5,000,000$
a^0	b^1	d^1	500
a^1	b^0	d^0	100
a^1	b^0	d^1	$1,000,000$
a^1	b^1	d^0	$100,000$
a^1	b^1	d^1	$100,000$

$$\beta_1(A,B,D)$$

Assignment		$\max_{A,C}$
b^0	d^0	$300,000$
b^0	d^1	$1,000,000$
b^1	d^0	$5,000,000$
b^1	d^1	$100,000$

$$\mu_{1,2}(B,D)$$

Assignment			$\max_A$
b^0	c^0	d^0	$300,000$
b^0	c^0	d^1	$1,000,000$
b^0	c^1	d^0	$300,000$
b^0	c^1	d^1	100
b^1	c^0	d^0	500
b^1	c^0	d^1	$100,000$
b^1	c^1	d^0	$5,000,000$
b^1	c^1	d^1	$100,000$

$$\beta_2(B,C,D)$$

Figure 13.3 The max-marginals for the Misconception **example.** Listed are the beliefs for the two cliques and the sepset.

are exactly equivalent. Thus, we can show that the analogue to equation (10.9) holds also for max-product:

$$\mu_{i,j}(\boldsymbol{S}_{i,j}) = \delta_{j \to i}(\boldsymbol{S}_{i,j}) \cdot \delta_{i \to j}(\boldsymbol{S}_{i,j}). \tag{13.11}$$

In particular, this equivalence holds at convergence, so that a clique's max-marginal over a sepset can be computed from the max-product messages.

13.3.2 Message Passing as Reparameterization

reparameterization

clique tree measure

Somewhat surprisingly, as for the sum-product case, we can view the max-product message passing steps as *reparameterizing* the original distribution, in a way that leaves the distribution invariant. More precisely, we view a set of beliefs β_i and sepset messages $\mu_{i,j}$ in a max-product clique tree as defining a *measure* using equation (10.11), precisely as for sum-product trees:

$$Q_{\mathcal{T}} = \frac{\prod_{i \in \mathcal{V}_{\mathcal{T}}} \beta_i(\boldsymbol{C}_i)}{\prod_{(i-j) \in \mathcal{E}_{\mathcal{T}}} \mu_{i,j}(\boldsymbol{S}_{i,j})}. \tag{13.12}$$

When we begin a run of max-product belief propagation, the initial potentials are simply the initial potentials in Φ, and the messages are all $\mathbf{1}$, so that $Q_{\mathcal{T}}$ is precisely $\tilde{P}_\Phi$. Examining the proof of corollary 10.3, we can see that it does not depend on the definition of the messages in terms of summing out the beliefs, but only on the way in which the messages are then used to update the receiving beliefs. Therefore, the proof of the theorem holds unchanged for max-product message passing, proving the following result:

Proposition 13.2

In an execution of max-product message passing (whether belief propagation or belief update) in a clique tree, equation (13.12) holds throughout the algorithm.

We can now directly conclude the following result:

Theorem 13.5

Let $\{\beta_i\}$ and $\{\mu_{i,j}\}$ be the max-calibrated set of beliefs obtained from executing max-product message passing, and let $Q_{\mathcal{T}}$ be the distribution induced by these beliefs. Then $Q_{\mathcal{T}}$ is a representation of the distribution $\tilde{P}_\Phi$ that also satisfies the max-product calibration constraints of equation (13.10).

Example 13.9

Continuing with example 13.8, it is straightforward to confirm that the original measure $\tilde{P}_\Phi$ can be reconstructed directly from the max-marginals and the sepset message. For example, consider the entry $\tilde{P}_\Phi(a^1, b^0, c^1, d^0) = 100$. According to equation (10.10), the clique tree measure is:

$$\frac{\beta_1(a^1, b^0, d^0)\beta_2(b^0, c^1, d^0)}{\mu_{1,2}(b^0, d^0)} = \frac{100 \cdot 300,000}{300,000} = 100,$$

as required. The equivalence for other entries can be verified similarly. ∎

Comparing this computation to example 10.6, we see that the sum-product clique tree and the max-product clique tree both induce reparameterizations of the original measure $\tilde{P}_\Phi$, but these two reparameterizations are different, since they must satisfy different constraints.

13.3.3 Decoding Max-Marginals

Given the max-marginals, can we find the actual MAP assignment? In the case of variable elimination, we had the max-marginal only for a single variable X_i (the last to be eliminated). Therefore, although we could identify the assignment for X_i in the MAP assignment, we had to perform a traceback procedure to compute the assignments to the other variables. Now the situation appears different: we have max-marginals for all of the variables in the network. Can we use this property to simplify this process?

One obvious solution is to use the max-marginal for each variable X_i to compute its own optimal assignment, and thereby compose a full joint assignment to all variables. However, this simplistic approach may not always work.

Example 13.10

Consider a simple XOR-like distribution $P(X_1, X_2)$ that gives probability 0.1 to the assignments where $X_1 = X_2$ and 0.4 to the assignments where $X_1 \neq X_2$. In this case, for each assignment to X_1, there is a corresponding assignment to X_2 whose probability is 0.4. Thus, the max-marginal of X_1 is the symmetric factor $(0.4, 0.4)$, and similarly for X_2. Indeed, we can choose either of the two values for X_1 and complete it to a MAP assignment, and similarly for X_2. However, if we choose the values for X_1 and X_2 in an inconsistent way, we may get an assignment whose probability is much lower. Thus, our joint assignment cannot be chosen by separately optimizing the individual max-marginals. ∎

Recall that we defined a set of node beliefs to be unambiguous if each belief has a unique maximal value. This condition prevents symmetric cases like the one in the preceding example. Indeed, it is not difficult to show the following result:

Proposition 13.3

The following two conditions are equivalent:

- *The set of node beliefs $\{MaxMarg_{\tilde{P}_\Phi}(X_i) \; : \; X_i \in \mathcal{X}\}$ is unambiguous, with*

$$x_i^* = \arg\max_{x_i} MaxMarg_{\tilde{P}_\Phi}(X_i)$$

the unique optimizing value for X_i;

- $\tilde{P}_\Phi$ *has a unique MAP assignment* $(x_1^*, \ldots, x_n^*)$.

See exercise 13.8. For generic probability measures, the assumption of unambiguity is not overly stringent, since we can always break ties by introducing a slight random perturbation into all of the factors, making all of the elements in the joint distribution have slightly different probabilities. However, if the distribution has special structure — deterministic relationships or shared parameters — that we want to preserve, this type of ambiguity may be unavoidable.

Thus, if there are no ties in any of the calibrated node beliefs, we can find the unique MAP assignment by locally optimizing the assignment to each variable separately.

If there are ties in the node beliefs, our task can be reformulated as follows:

Definition 13.3

local optimality

Let $\beta_i(C_i)$ be a belief in a max-calibrated clique tree. We say that an assignment ξ^* has the local optimality *property if, for each clique C_i in the tree, we have that*

$$\xi^*\langle C_i \rangle \in \arg\max_{c_i} \beta_i(c_i), \tag{13.13}$$

decoding

that is, the assignment to C_i in ξ^* optimizes the C_i belief. The task of finding a locally optimal assignment ξ^* given a max-calibrated set of beliefs is called the decoding *task.* ∎

traceback

Solving the decoding task in the ambiguous case can be done using a *traceback* procedure as in algorithm 13.1. However, local optimality provides us with a simple, local test for *verifying* whether a given assignment is the MAP assignment:

Theorem 13.6

Let $\beta_i(C_i)$ be a set of max-calibrated beliefs in a clique tree $\mathcal{T}$, with $\mu_{i,j}$ the associated sepset beliefs. Let $Q_\mathcal{T}$ be the clique tree measure defined as in equation (13.12). Then an assignment ξ^* satisfies the local optimality property relative to the beliefs $\{\beta_i(C_i)\}_{i \in \mathcal{V}_\mathcal{T}}$ if and only if it is the global MAP assignment relative to $Q_\mathcal{T}$.

PROOF The proof of the "if" direction follows directly from our previous results. We have that $Q_\mathcal{T}$ is max-calibrated, and hence is a fixed point of the max-product algorithm. (In other words, if we run max-product inference on the distribution defined by $Q_\mathcal{T}$, we would get precisely the beliefs $\beta_i(C_i)$.) Thus, these beliefs are max-marginals of $Q_\mathcal{T}$. If ξ^* is the MAP assignment to $Q_\mathcal{T}$, it must maximize each one of its max-marginals, proving the desired result.

The proof of the only if direction requires the following lemma, which plays an even more significant role in later analyses.

Lemma 13.1

Let ϕ be a factor over scope Y and ψ be a factor over scope $Z \subset Y$ such that ψ is a max-marginal of ϕ over Z; that is, for any z:

$$\psi(z) = \max_{y \sim z} \phi(y).$$

Let $y^* = \arg\max_y \phi(y)$. Then y^* is also an optimal assignment for the factor ϕ/ψ, where, as usual, we take $\psi(y^*) = \psi(y^*\langle Z \rangle)$.

PROOF Recall that, due to the properties of max-marginalization, each entry $\psi(z)$ arises from some entry $\phi(y)$ such that $y \sim z$. Because y^* achieves the optimal value in ϕ, and ψ is the max-marginal of ϕ, we have that z^* achieves the optimal value in ψ. Hence, $\phi(y^*) = \psi(z^*)$, so that $\left(\frac{\phi}{\psi}\right)(y^*) = 1$. Now, consider any other assignment y and the assignment $z = y\langle Z\rangle$. Either the value of z is obtained from y, or it is obtained from some other y' whose value is larger. In the first case, we have that $\phi(y) = \psi(z)$, so that $\left(\frac{\phi}{\psi}\right)(y) = 1$. In the second case, we have that $\phi(y) < \psi(z)$ and $\left(\frac{\phi}{\psi}\right)(y) < 1$. In either case,

$$\left(\frac{\phi}{\psi}\right)(y) \leq \left(\frac{\phi}{\psi}\right)(y^*),$$

as required. ∎

To prove the only-if direction, we first rewrite the clique tree distribution of equation (13.12) in a directed way. We select a root clique C_r; for each clique $i \neq r$, let $\pi(i)$ be the parent clique of i in this rooted tree. We then assign each sepset $S_{i,\pi(i)}$ to the *child* clique i. Note that, because each clique has at most one parent, each clique is assigned at most one sepset. Thus, we obtain the following rewrite of equation (13.12):

$$\beta_r(C_r) \prod_{i \neq r} \frac{\beta_i(C_i)}{\mu_{i,\pi(i)}(S_{i,\pi(i)})}. \tag{13.14}$$

Now, let ξ^* be an assignment that satisfies the local optimality property. By assumption, it optimizes every one of the beliefs. Thus, the conditions of lemma 13.1 hold for each of the ratios in this product, and for the first term involving the root clique. Thus, ξ^* also optimizes each one of the terms in this product, and therefore it optimizes the product as a whole. It must therefore be the MAP assignment. ∎

As we will see, these concepts and related results have important implications in some of our later derivations.

13.4 Max-Product Belief Propagation in Loopy Cluster Graphs

In section 11.3 we applied the sum-product message passing using the clique tree algorithm to a loopy cluster graph, obtaining an approximate inference algorithm. In the same way, we can generalize max-product message passing to the case of cluster graphs. The algorithms that we present in this section are directly analogous to their sum-product counterparts in section 11.3. However, as we discuss, the guarantees that we can provide are much stronger in this case.

13.4.1 Standard Max-Product Message Passing

As for the case of clique trees, the algorithm divides into two phases: computing the beliefs using message passing and using those beliefs to identify a single joint assignment.

13.4.1.1 Message Passing Algorithm

The message passing algorithm is straightforward: it is precisely the same as the algorithm of algorithm 11.1, except that we use the procedure of algorithm 13.2 in place of the SP-Message procedure. As for sum-product, there are no guarantees that this algorithm will converge. Indeed, in practice, it tends to converge somewhat less often than the sum-product algorithm, perhaps because the averaging effect of the summation operation tends to smooth out messages, and reduce oscillations. Many of the same ideas that we discussed in box 11.B can be used to improve convergence in this algorithm as well.

At convergence, the result will be a set of calibrated clusters: As for sum-product, if the clusters are not calibrated, convergence has not been achieved, and the algorithm will continue iterating. However, the resulting beliefs will not generally be the exact max-marginals; these resulting beliefs are often called *pseudo-max-marginals*.

pseudo-max-marginal

As we saw in section 11.3.3.1 for sum-product, the distribution invariance property that holds for clique trees is a consequence only of the message passing procedure, and does not depend on the assumption that the cluster graph is a tree. The same argument holds here; thus, proposition 13.2 can be used to show that max-product message passing in a cluster graph is also simply reparameterizing the distribution:

Corollary 13.3

In an execution of max-product message passing (whether belief propagation or belief update) in a cluster graph, the invariant equation (10.10) holds initially, and after every message passing step.

13.4.1.2   Decoding the Pseudo-Max-Marginals

Given a set of pseudo-max-marginals, we now have to solve the decoding problem in order to identify a joint assignment. In general, we cannot expect this assignment to be the exact MAP, but we can hope for some reasonable approximation. But how do we identify such an assignment? It turns out that our ability to do so depends strongly on whether there exists some assignment that satisfies the local optimality property of definition 13.3 for the max-calibrated beliefs in the cluster graph. Unlike in the case of clique trees, such a joint assignment does not necessarily exist:

Example 13.11

Consider a cluster graph with the three clusters $\{A, B\}, \{B, C\}, \{A, C\}$ and the beliefs

	a^1	a^0
b^1	1	2
b^0	2	1

	b^1	b^0
c^1	1	2
c^0	2	1

	a^1	a^0
c^1	1	2
c^0	2	1

These beliefs are max-calibrated, in that all messages are $(2, 2)$. However, there is no joint assignment that maximizes all of the cluster beliefs simultaneously. For example, if we select a^0, b^1, we maximize the value in the A, B belief. We can now select c^0 to maximize the value in the B, C belief. However, we now have a nonmaximizing assignment a^0, c^0 in the A, C belief. No matter which assignment of values we select in this example, we do not obtain a single joint assignment that maximizes all three beliefs. ■

frustrated loop

Loops such as this are often called *frustrated*.

In other cases, a locally optimal joint assignment does exist. In particular, **when all the node beliefs are all unambiguous, it is not difficult to show that all of the cluster beliefs also have a unique maximizing assignment, and that these local cluster-maximizing assignments are necessarily consistent with each other** (exercise 13.9). However, there are also other cases where the node beliefs are ambiguous, and yet a locally optimal joint assignment exists:

Example 13.12

Consider a cluster graph of the same structure as in example 13.11, but with the beliefs:

	a^1	a^0
b^1	2	1
b^0	1	2

	b^1	b^0
c^1	2	1
c^0	1	2

	a^1	a^0
c^1	2	1
c^0	1	2

In this case, the beliefs are ambiguous, yet a locally optimal joint assignment exists (both a^1, b^1, c^1 and a^0, b^0, c^0 are locally optimal). ∎

In general, the decoding problem in a loopy cluster graph is not a trivial task. Recall that, in clique trees, we could simply choose any of the maximizing assignments for the beliefs at a clique, and be assured that it could be extended into a joint MAP assignment. Here, as illustrated by example 13.11, we may make a choice for one cluster that cannot be extended into a consistent joint assignment. In that example, of course, there is no assignment that works. However, it is not difficult to construct examples where one choice of locally optimal assignments would give rise to a consistent joint assignment, whereas another would not (exercise 13.10).

How do we find a locally optimal joint assignment, if one exists? Recall from the definition that an assignment is locally optimal if and only if it selects one of the optimizing assignments in every single cluster. Thus, we can essentially label the assignments in each cluster as either "legal" if they optimize the belief or "illegal" if they do not. We now must search for an assignment to $\mathcal{X}$ that results in a legal value for each cluster. This problem is precisely a

constraint
satisfaction
problem

constraint satisfaction problem (CSP), where the constraints are derived from the local optimality condition. More precisely, a constraint satisfaction problem can be defined in terms of a Markov network (or factor graph) where all of the entries in the beliefs are either 0 or 1. The CSP problem is now one of finding an assignment whose (unnormalized) measure is 1, if one exists, and otherwise reporting failure. In other words, the CSP problem is simply that of finding the MAP assignment in this model with $\{0, 1\}$-valued beliefs. The field of CSP algorithms is a large one, and a detailed survey is outside the scope of the book; see section 13.9 for some background reading. We note, however, that the CSP problem is itself $\mathcal{NP}$-hard, and therefore we have no guarantees that a locally optimal assignment, even if one exists, can be found efficiently.

Thus, given a max-product calibrated cluster graph, we can convert it to a discrete-valued CSP by simply taking the belief in each cluster, changing each assignment that locally optimizes the belief to 1 and all other assignments to 0. We then run some CSP solution method. If the outcome is an assignment that achieves 1 in every belief, this assignment is guaranteed to be a locally optimal assignment. Otherwise, there is no locally optimal assignment. In this case, we must resort to the use of alternative solution methods. One heuristic in this latter situation is to use information obtained from the max-product propagation to construct a partial assignment. For example, assume that a variable X_i is unambiguous in the calibrated cluster graph, so that the only value that locally optimizes its node marginal is x_i. In this case, we may

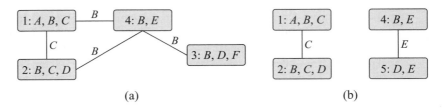

Figure 13.4   Two induced subgraphs derived from figure 11.3a. (a) Graph over $\{B, C\}$; (b) Graph over $\{C, E\}$.

decide to restrict attention only to assignments where $X_i = x_i$. In many real-world problems, a large fraction of the variables in the network are unambiguous in the calibrated max-product cluster graph. Thus, this heuristic can greatly simplify the model, potentially even allowing exact methods (such as clique tree inference) to be used for the resulting restricted model. We note, however, that the resulting assignment would not necessarily satisfy the local optimality condition, and all of the guarantees we will present hold only under that assumption.

13.4.1.3   Strong Local Maximum

What type of guarantee can we provide for a decoded assignment from the pseudo-max-marginals produced by the max-product belief propagation algorithm? It is certainly not the case that this assignment is the MAP assignment; nor is it even the case that we can guarantee that the probability of this assignment is "close" in any sense to that of the true MAP assignment.

strong local
maximum

However, if we can construct a locally optimal assignment ξ^* relative to the beliefs produced by max-product BP, we can prove that ξ^* is a *strong local maximum*, in the following sense: For certain subsets of variables $\mathbf{Y} \subset \mathcal{X}$, there is no assignment ξ' that is higher-scoring than ξ^* and differs from it only in the assignment to $\mathbf{Y}$. These subsets $\mathbf{Y}$ are those that induce any disjoint union of subgraphs each of which contains at most a single loop (including trees, which contain no loops).

Definition 13.4

induced
subgraph

Let $\mathcal{U}$ be a cluster graph over $\mathcal{X}$, and $\mathbf{Y} \subset \mathcal{X}$ be some set of variables. We define the induced subgraph $\mathcal{U}_{\mathbf{Y}}$ to be the subgraph of clusters and sepsets in $\mathcal{U}$ that contain some variable in $\mathbf{Y}$. ∎

This definition is most easily understood in the context of a pairwise Markov network, where the cluster graph is simply the set of edges in the MRF and the sepsets are the individual variables. In this case, the induced subgraph for a set $\mathbf{Y}$ is simply the set of nodes corresponding to $\mathbf{Y}$ and any edges that contain them. In a more general cluster graph, the result is somewhat more complex:

Example 13.13

Consider the cluster graph of figure 11.3a. Figure 13.4a shows the induced subgraph over $\{B, C\}$; this subgraph contains at exactly one loop, which is connected to an additional cluster. Figure 13.4b shows the induced subgraph over $\{C, E\}$; this subgraph is a union of two disjoint trees. ∎

We can now state the following important theorem:

Theorem 13.7

Let $\mathcal{U}$ be a max-product calibrated cluster graph for $\tilde{P}_\Phi$, and let ξ^* be a locally optimal assignment for $\mathcal{U}$. Let $\boldsymbol{Z}$ be any set of variables for which $\mathcal{U}_{\boldsymbol{Z}}$ is a collection of disjoint subgraphs each of which contains at most a single loop. Then for any assignment ξ' which is the same as ξ^* except for the assignment to the variables in $\boldsymbol{Z}$, we have that

$$\tilde{P}_\Phi(\xi') \leq \tilde{P}_\Phi(\xi^*). \tag{13.15}$$

PROOF We prove the theorem under the assumption that $\mathcal{U}_{\boldsymbol{Z}}$ is a single tree, leaving the rest of the proof as an exercise (exercise 13.12). Owing to the recalibration property, we can rewrite the joint probability $\tilde{P}_\Phi$ as in equation (13.12). We can partition the terms in this expression into two groups: those that involve variables in $\boldsymbol{Z}$ and those that do not. Let $\boldsymbol{Y} = \mathcal{X} - \boldsymbol{Z}$ and $\boldsymbol{y}^*$ be the locally optimal assignment to $\boldsymbol{Y}$. We now consider the unnormalized measure obtained over $\boldsymbol{Z}$ when we restrict the distribution to the event $\boldsymbol{Y} = \boldsymbol{y}^*$ (as in definition 4.5). Since we set $\boldsymbol{Y} = \boldsymbol{y}^*$, the terms corresponding to beliefs that do not involve $\boldsymbol{Z}$ are constant, and hence they do not affect the comparison between ξ' and ξ^*.

We can now define $\tilde{P}'_{\boldsymbol{y}^*}(\boldsymbol{Z})$ to be the measure obtained by restricting equation (13.12) only to the terms in the beliefs (at both clusters and sepsets) that involve variables in $\boldsymbol{Z}$. It follows that an assignment $\boldsymbol{z}$ optimizes $\tilde{P}_\Phi(\boldsymbol{z}, \boldsymbol{y}^*)$ if and only if it optimizes $\tilde{P}'_{\boldsymbol{y}^*}$. This measure precisely corresponds to a clique tree whose structure is $\mathcal{U}_{\boldsymbol{Z}}$ and whose beliefs are the beliefs in our original calibrated cluster graph $\mathcal{U}$, but restricted to $\boldsymbol{Y} = \boldsymbol{y}^*$. Let $\mathcal{T}_{\boldsymbol{y}^*}$ represent this clique tree and its associated beliefs.

Because $\mathcal{U}$ is max-product calibrated, so is its subgraph $\mathcal{T}_{\boldsymbol{y}^*}$. Moreover, if an assignment $(\boldsymbol{y}^*, \boldsymbol{z}^*)$ is optimal for some belief β_i, then $\boldsymbol{z}^*$ is also optimal for the restricted belief $\beta_i[\boldsymbol{Y} = \boldsymbol{y}^*]$. We therefore have a max-product calibrated clique tree $\mathcal{T}_{\boldsymbol{y}^*}$ and $\boldsymbol{z}^*$ is a locally optimal assignment for it. Because this is a clique tree, local optimality implies MAP, and so $\boldsymbol{z}^*$ must be a MAP assignment in this clique tree. As a consequence, there is no assignment $\boldsymbol{z}'$ that has a higher probability in $\tilde{P}'_{\boldsymbol{y}^*}$, proving the desired result. ∎

To illustrate the power of this theorem, consider the following example:

Example 13.14

Consider the 4×4 grid network in figure 11.4, and assume that we use the pairwise cluster graph construction of figure 11.6 (shown there for a 3×3 grid). This result implies that the MAP solution found by max-product belief propagation has higher probability than any assignment obtained by changing the assignment to any of the following subsets of variables $\boldsymbol{Y}$:

- a set of variables in any single row, such as $\boldsymbol{Y} = \{A_{1,1}, A_{1,2}, A_{1,3}, A_{1,4}\}$;
- a set of variables in any single column;
- a "comb" structure such as the variables in row 1, column 2 and column 4;
- a single loop, such as $\boldsymbol{Y} = \{A_{1,1}, A_{1,2}, A_{2,2}, A_{2,1}\}$;
- a collection of disconnected subsets of the preceding form, for example: the union of the variables in rows 1 and 3; or the loop above union with the L-structure consisting of the variables in row 4 and the variables in column 4. ∎

☞ This result is a powerful one, inasmuch as it shows that the solution obtained from max-product belief propagation is robust against large perturbations. Thus, although one can construct examples where max-product belief propagation obtains the wrong solutions, these solutions are strong local maxima, and therefore they often have high probability.

13.4.2 Max-Product BP with Counting Numbers ★

The preceding algorithm performs max-product message passing that is analogous to the sum-product message passing with the Bethe free-energy approximation. We can also construct analogues of the various generalizations of sum-product message passing, as defined in section 11.3.7. We can derive max-product variants based both on the region-graph methods, which allow us to introduce larger clusters, and based on the notion of alternative counting numbers. From an algorithmic perspective, the transformation of sum-product to max-product algorithms is straightforward: we simply replace summation with maximization. The key question is the extent to which we can provide a formal justification for these methods.

Recall that, in our discussion of sum-product algorithms, we derived the belief propagation algorithms in two different ways. The first was simply by taking the message passing algorithm on clique trees and running it on loopy cluster graphs, ignoring the presence of loops. The second derivation was obtained by a variational analysis, where the algorithm arose naturally as the fixed points of an approximate energy functional. This view was compelling both because it suggested some theoretical justification for the algorithm and, even more important, because it immediately gave rise to a variety of generalizations, obtained from different approximations to the energy functional, different methods for optimizing the objective, and more.

For the case of max-product, our discussion so far follows the first approach, viewing the message passing algorithm as a simple generalization of the max-product clique tree algorithm. Given the similarity between the sum-product and max-product algorithms presented so far, one may assume that we can analogously provide a variational justification for max-product, for example, as optimizing the same energy functional, but with max-calibration rather than sum-calibration constraints on adjacent clusters. For example, in a variational derivation of the max-product clique tree algorithm, we would replace the sum-calibration constraint of equation (11.7) with the analogous max-calibration constraint of equation (13.10). Although plausible, this analogy turns out to be incorrect. The key problem is that, whereas the sum-marginalization constraint of equation (11.7) is a simple linear equality, the constraint of equation (13.10) is not. Indeed, the max function involved in the constraint is not even smoothly differentiable, so that the framework of Lagrange multipliers cannot be applied.

However, as we now show, we can provide an optimization-based derivation and more formal justification for max-product BP with convex counting numbers. For these variants, we can even show conditions under which these algorithms are guaranteed to produce the correct MAP assignment. We begin this section by describing the basic algorithm, and proving the key optimality result: that any locally optimal assignment for convex max-product BP is guaranteed to be the MAP assignment. Then, in section 13.5, we provide an alternative view of this approach in terms of its relationship to two other classes of algorithms. This perspective will shed additional insight on the properties of this algorithm and on the cases in which it provides a useful guarantee.

Algorithm 13.3 Calibration using max-product BP in a Bethe-structured cluster graph

---

Procedure Generalized-MP-BP (
  Φ, // Set of factors
  $\mathbf{R}$, // Set of regions
  $\{\kappa_r\}_{r \in \mathbf{R}}, \{\kappa_i\}_{X_i \in \mathcal{X}}$ // Counting numbers
)
1 $\rho_i \leftarrow 1/\kappa_i$
2 $\rho_r \leftarrow 1/\kappa_r$
3 Initialize-CGraph
4 **while** region graph is not max-calibrated
5 Select $\boldsymbol{C}_r$ and $X_i \in \boldsymbol{C}_r$
6 $\delta_{i \to r}(X_i) \leftarrow \left[\left(\prod_{r' \neq r} \delta_{i \to r'}(X_i) \right)^{\rho_i} \left(\max_{\boldsymbol{C}_r - X_i} \psi_r(\boldsymbol{C}_r) \left(\prod_{X_j \in \boldsymbol{C}_r, j \neq i} \delta_{j \to r} \right)^{\rho_r} \right) \right]^{-\frac{1}{\rho_i + \rho_r}}$
7 **for** each region $r \in \mathbf{R} \cup \{1, \ldots, n\}$
8   $\beta_r(\boldsymbol{C}_r) \leftarrow \left(\psi_r(\boldsymbol{C}_r) \prod_{X_i \in \boldsymbol{C}_r} \delta_{i \to r}(X_i) \right)^{\rho_r}$
9 **return** $\{\beta_r\}_{r \in \mathbf{R}}$

13.4.2.1 Max-Product with Counting Numbers

We begin with a reminder of the notion of belief propagation with counting numbers. For concreteness, we also provide the max-product variant of a message passing algorithm for this case, although (as we mentioned) the max-product variant can be obtained from the sum-product algorithm using a simple syntactic substitution.

In section 11.3.7, we defined a set of sum-product message passing algorithms; these algorithms were defined in terms of a set of *counting numbers* that specify the extent to which entropy terms for different subsets of variables are counted in the entropy approximation used in the energy functional. For a given set of counting numbers, one can derive a message passing algorithm by using the fixed point equations obtained by differentiating the Lagrangian for the energy functional, with its sum-product calibration constraints. The standard belief propagation algorithm is obtained from the Bethe energy approximation; other sets of counting numbers give rise to other message passing algorithms.

As we discussed, one can take these sum-product message passing algorithms (for example, those in exercise 11.17 and exercise 11.19) and convert them to produce a max-product variant by simply replacing each summation operation as maximization. For concreteness, in algorithm 13.3, we repeat the algorithm of exercise 11.17, instantiated to the max-product setting. Recall that this algorithm applies only to *Bethe cluster graphs*, that is, graphs that have two levels of regions: "large" regions r containing multiple variables with counting numbers κ_r, and singleton regions containing individual variables X_i with counting numbers κ_i; all factors in Φ are assigned only to large regions, so that $\psi_i = 1$ for all i.

reparameteriza-
tion

A critical observation is that, **like the sum-product algorithms, and like the max-product clique tree algorithm (see theorem 13.5), these new message passing algorithms are a *reparameterization* of the original distribution.** In other words, their fixed points are a different representation of the same distribution, in terms of a set of max-calibrated beliefs. This property, which is stated for sum-product in theorem 11.6, asserts that, at fixed points of the message passing algorithm, we have that:

$$\tilde{P}_\Phi(\mathcal{X}) = \prod_r (\beta_r)^{\kappa_r}. \tag{13.16}$$

The proof of this equality (see exercise 11.16 and exercise 11.19) is a consequence only of the way in which we define region beliefs in terms of the messages. Therefore, the reparameterization property applies equally to fixed points of the max-product algorithms. It is this property that will be critical in our derivation.

13.4.2.2 Optimality Guarantee

As in the case of standard max-product belief propagation algorithms, given a set of max-product calibrated beliefs that reparameterize the distribution, we now search for an assignment that is locally optimal for this set of beliefs. However, as we now show, under certain assumptions, such an assignment is guaranteed to be the MAP assignment.

Although more general variants of this theorem exist, we focus on the case of a Bethe-structured region graph, as described. Here, we also assume that our large regions in $\mathbf{R}$ have counting number 1. We assume also that factors in the network are assigned only to large regions, so that $\psi_i = 1$ for all i. Finally, in a property that is critical to the upcoming derivation, we assume that the counting numbers κ_r are convex, as defined in definition 11.4. Recall that

convex counting
numbers

a vector of counting numbers κ_r is *convex* if there exist *nonnegative* numbers ν_r, ν_i, and $\nu_{r,i}$ such that:

$$
\begin{aligned}
\kappa_r &= \nu_r + \sum_{i\,:\,X_i \in \boldsymbol{C}_r} \nu_{r,i} & \text{for all } r \\
\kappa_i &= \nu_i - \sum_{r\,:\,X_i \in \boldsymbol{C}_r} \nu_{r,i} & \text{for all } i.
\end{aligned}
$$

This is the same assumption used to guarantee that the region-graph energy functional in equation (11.27) is a concave function. Although here we have no energy functional, the purpose of this assumption is similar: As we will see, it allows us to redistribute the terms in the reparameterization of the probability distribution, so as to guarantee that all terms have a positive coefficient.

From these assumptions, we can now prove the following theorem:

Theorem 13.8

Let P_Φ be a distribution, and consider a Bethe-structured region graph with large regions and singleton regions, where the counting numbers are convex. Assume that we have a set of max-calibrated beliefs $\beta_r(\boldsymbol{C}_r)$ and $\beta_i(X_i)$ such that equation (13.16) holds. If there exists an assignment ξ^* that is locally optimal relative to each of the beliefs β_r, then ξ^* is the optimal MAP assignment.

PROOF Applying equation (13.16) to our Bethe-structured graph, we have that:

$$\tilde{P}_\Phi(\mathcal{X}) = \prod_{r \in \mathbf{R}} \beta_r \prod_i \beta_i^{\kappa_i}.$$

Owing to the convexity of the counting numbers, we can rewrite the right-hand side as:

$$\prod_r (\beta_r)^{\nu_r} \prod_i (\beta_i)^{\nu_i} \prod_{i,r \,:\, X_i \in \boldsymbol{C}_r} \left(\frac{\beta_r}{\beta_i}\right)^{\nu_{r,i}}.$$

Owing to the nonnegativity of the coefficients ν, we have that:

$$\max_{\xi} \tilde{P}_\Phi(\xi) \;=\; \max_{\xi} \prod_r (\beta_r(\boldsymbol{c}_r))^{\nu_r} \prod_i (\beta_i(x_i))^{\nu_i} \prod_{i,r \,:\, X_i \in \boldsymbol{C}_r} \left(\frac{\beta_r}{\beta_i}(\boldsymbol{c}_r)\right)^{\nu_{r,i}}$$

$$\leq \prod_r (\max_{\boldsymbol{c}_r} \beta_r(\boldsymbol{c}_r))^{\nu_r} \prod_i (\max_{x_i} \beta_i(x_i))^{\nu_i} \prod_{i,r \,:\, X_i \in \boldsymbol{C}_r} \left(\max_{\boldsymbol{c}_r} \frac{\beta_r}{\beta_i}(\boldsymbol{c}_r)\right)^{\nu_{r,i}}.$$

We have now reduced this expression to a product of terms, each raised to the power of a positive exponent. Some of these terms are factors in the max-product calibrated network, and others are ratios of factors and their max-product marginal over an individual variable. The proof now is exactly the same as the proof of theorem 13.6. Let ξ^* be an assignment that satisfies the local optimality property. By assumption, it optimizes every one of the region beliefs. Because the ratios involve a factor and its max-marginal, the conditions of lemma 13.1 hold for each of the ratios in this expression. Thus, ξ^* optimizes each one of the terms in this product, and therefore it optimizes the product as a whole. It therefore optimizes $\tilde{P}_\Phi(\xi)$, and must therefore be the MAP assignment. ∎

We can also derive the following useful corollary, which allows us, in certain cases, to characterize parts of the MAP solution even if the local optimality property does not hold:

Corollary 13.4

Under the setting of theorem 13.8, if a variable X_i takes a particular value x_i^* in all locally optimal assignments ξ^* then $x_i^{map} = x_i^*$ in the MAP assignment. More generally, if there is some set S_i such that, in any locally optimal assignment ξ^* we have that $x_i^* \in S_i$, then $x_i^{map} \in S_i$.

At first glance, the application of this result seems deceptively easy. After all, in order to be locally optimal, an assignment must assign to X_i one of the values that maximizes its individual node marginal. Thus, it appears that we can easily extract, for each X_i, some set S_i (perhaps an overestimate) to which corollary 13.4 applies. Unfortunately, when we use this procedure, we cannot guarantee that x_i^{map} is actually in the set S_i. The corollary applies only if there exists a locally optimal assignment to the entire set of beliefs. If no such assignment exists, the set of locally maximizing values in X_i's node belief may have no relation to the true MAP assignment.

13.4.3 Discussion

 In this section, we have shown that max-product message passing algorithms, if they converge, provide a max-calibrated reparameterization of the distribution $\tilde{P}_\Phi$. This reparameterization essentially converts the global optimization problem of finding a single joint MAP assignment to a local optimization problem: finding a set of locally optimal assignments to the individual cliques that are consistent with each other. Importantly, we can show that this locally consistent assignment, if it exists, satisfies strong optimality properties: In the case of the standard (Bethe-approximation) reparameterization,

the joint assignment satisfies strong local optimality; in the case of reparameterizations based on convex counting numbers, it is actually guaranteed to be the MAP assignment.

Although these guarantees are very satisfying, their usefulness relies on several important questions that we have not yet addressed. The first two relate to the max-product calibrated reparameterization of the distribution: does one exist, and can we find it? First, for a given set of counting numbers, does there always exist a max-calibrated reparameterization of $\tilde{P}_\Phi$ in terms of these counting numbers? Somewhat surprisingly, as we show in section 13.5.3, the answer to that question is yes for convex counting numbers; it turns out to hold also for the Bethe counting numbers, although we do not show this result. Second, we must ask whether we can always find such a reparameterization. We know that if the max-product message passing algorithm converges, it must converge to such a fixed point. But unfortunately, there is no guarantee that the algorithm will converge. In practice, standard max-product message passing often does not converge. For certain specific choices of convex counting numbers (see, for example, box 13.A), one can design algorithms that are guaranteed to be convergent.

However, even if we find an appropriate reparameterization, we are still left with the problem of extracting a joint assignment that satisfies the local optimality property. Indeed, such as assignment may not even exist. In section 13.5.3.3, we present a necessary condition for the existence of such an assignment.

It is currently not known how the choice of counting numbers affects either of these two issues: our ability to find effectively a max-product calibrated reparameterization, and our ability to use the result to find a locally consistent joint assignment. Empirically, preliminary results suggest that nonconvex counting numbers (such as those obtained from the Bethe approximation) converge less often than the the convex variants, but converge more quickly when they do converge. The different convex variants converge at different rates, but tend to converge to fixed points that have a similar set of ambiguities in the beliefs. Moreover, in cases where convex max-product BP converges whereas standard max-product does not, the resulting beliefs often contain many ambiguities (beliefs with equal values), making it difficult to determine whether the local optimality property holds, and to identify such an assignment if it exists.

---

tree-reweighted belief propagation

LP relaxation

Box 13.A — Concept: Tree-Reweighted Belief Propagation. *One algorithm that is worthy of special mention, both because of historical precedence and because of its popularity, is the* tree-reweighted belief propagation *algorithm (often known as* TRW*). This algorithm was the first message passing algorithm to use convex counting numbers; it was also the context in which message passing algorithms were first shown to be related to the* linear program relaxation *of the MAP optimization problem that we discuss in section 13.5. This algorithm, developed in the context of a pairwise Markov network, utilizes the same approach as in the TRW variant of sum-product message passing: It defines a probability distribution over trees* $\mathcal{T}$ *in the network, so that each edge in the pairwise network appears in at least one tree, and it then defines the counting numbers to be the edge and negative node appearance probabilities, as defined in equation (11.26). Note that, unlike the algorithms of section 13.4.2.1, here the factors do not have a counting number of 1, so that the algorithms we presented there require some modification. Briefly, the max-product TRW* *algorithm uses the following update rule:*

$$\delta_{i \to j} = \max_{x_i} \left[\left(\psi_i(x_i) \prod_{k \in \mathrm{Nb}_i} \delta_{k \to i}(x_i) \right)^{\frac{\kappa_{i,j}}{\kappa_i}} \frac{1}{\delta_{j \to i}(x_i)} \psi_{i,j}(x_i, x_j) \right]. \tag{13.17}$$

One particular variant of the TRW algorithm, called TRW-S, provides some particularly satisfying guarantees. Assume that we order the nodes in the network in some fixed ordering $X_1, \ldots, X_n$, and consider a set of trees each of which is a subsequence of this ordering, that is, of the form $X_{i_1}, \ldots, X_{i_k}$ for $i_1 < \ldots i_k$. We now pass messages in the network by repeating two phases, where in one phase we pass messages from X_1 towards X_n, and in the other from X_n towards X_1. With this message passing scheme, it is possible to guarantee that the algorithm continuously increases the dual objective, and hence it is convergent.

13.5 MAP as a Linear Optimization Problem ★

A very important and useful insight on the MAP problem is derived from viewing it directly as an optimization problem. This perspective allows us to draw upon the vast literature on optimization algorithms and apply some of these ideas to the specific case of MAP inference. Somewhat surprisingly, some of the algorithms that we describe elsewhere in this chapter turn out to be related to the optimization perspective; the insights obtained from understanding the connections can provide the basis for a theoretical analysis of these methods, and they can suggest improvements.

For the purposes of this section, we assume that the distribution specified in the MRF is positive, so that all of the entries in all of the factors are positive. This assumption turns out to be critical for some of the derivations in this section, and facilitates many others.

13.5.1 The Integer Program Formulation

integer linear
program

The basic MAP problem can be viewed as an *integer linear program* — an optimization problem (see appendix A.4.1) over a set of integer valued variables, where both the objective and the constraints are linear. To define a linear optimization problem, we must first turn all of our products into summations. This transformation gives rise to the following *max-sum* form:

max-sum

$$\arg\max_{\xi} \log \tilde{P}_\Phi(\xi) = \arg\max_{\xi} \sum_{r \in \mathbf{R}} \log(\phi_r(\mathbf{c}_r)), \tag{13.18}$$

where $\Phi = \{\phi_r \ : \ r \in \mathbf{R}\}$, and $\mathbf{C}_r$ is the scope of ϕ_r.

For $r \in \mathbf{R}$, we define $n_r = |Val(\mathbf{C}_r)|$. For any joint assignment ξ, if $\xi\langle \mathbf{C}_r \rangle = \mathbf{c}_r^j$, the factor $\log(\phi_r)$ makes a contribution to the objective of $\log(\phi_r(\mathbf{c}_r^j))$, a quantity that we denote as η_r^j.

optimization
variables

We introduce *optimization variables* $q(\mathbf{x}_r^j)$, where $r \in \mathbf{R}$ enumerates the different factors, and $j = 1, \ldots, n_r$ enumerates the different possible assignments to the variables $\mathbf{C}_r$ that comprise the factor $\mathbf{C}_r$. These variables take binary values, so that $q(\mathbf{x}_r^j) = 1$ if and only if $\mathbf{C}_r = \mathbf{c}_r^j$, and 0 otherwise. It is important to distinguish the optimization variables from the random variables in our original graphical model; here we have an optimization variable $q(\boldsymbol{x}_r^j)$ for each joint assignment $\boldsymbol{c}_r^j$ to the model variables $\boldsymbol{C}_r$.

Let $\boldsymbol{q}$ denote a vector of the optimization variables $\{q(\boldsymbol{x}_r^j) \; : \; r \in \mathbf{R}; \; j = 1, \dots, n_r\}$, and $\boldsymbol{\eta}$ denote a vector of the coefficient η_r^j sorted in the same order. Both of these are vectors of dimension $N = \sum_{k=1}^K n_r$. With this interpretation, the MAP objective can be rewritten as:

$$\mathbf{maximize}_q \sum_{r \in \mathbf{R}} \sum_{j=1}^{n_r} \eta_r^j q(\boldsymbol{x}_r^j), \tag{13.19}$$

or, in shorthand, $\mathbf{maximize}_q \boldsymbol{\eta}^T \boldsymbol{q}$.

Example 13.15

Assume that we have a pairwise MRF shaped like a triangle A—B—C—A, so that we have three factors over pairs of connected random variables: $\phi_1(A,B), \phi_2(B,C), \phi_3(A,C)$. Assume that A, B are binary-valued, whereas C takes three values. Here, we would have the optimization variables $q(\boldsymbol{x}_1^1), \dots, q(\boldsymbol{x}_1^4), q(\boldsymbol{x}_2^1), \dots, q(\boldsymbol{x}_2^6), q(\boldsymbol{x}_3^1), \dots, q(\boldsymbol{x}_3^6)$. We assume that the values of the variables are enumerated lexicographically, so that $q(\boldsymbol{x}_3^4)$, for example, corresponds to a^2, c^1. ∎

We can view our MAP inference problem as optimizing this linear objective over the space of assignments to $\boldsymbol{q} \in \{0,1\}^N$ that correspond to legal assignments to $\mathcal{X}$. What constraints on $\boldsymbol{q}$ do we need to impose in order to guarantee that it corresponds to some assignment to $\mathcal{X}$? Most obviously, we need to ensure that, in each factor, only a single assignment is selected. Thus, in our example, we cannot have both $q(\boldsymbol{x}_1^1) = 1$ and $q(\boldsymbol{x}_1^2) = 1$. Slightly subtler are the cross-factor consistency constraints: If two factors share a variable, we need to ensure that the assignment to this variable according to $\boldsymbol{q}$ is consistent in those two factors. In our example, for instance, if we have that $q(\boldsymbol{x}_1^1) = 1$, so that $B = b^1$, we would need to have $q(\boldsymbol{x}_2^1) = 1$, $q(\boldsymbol{x}_2^2) = 1$, or $q(\boldsymbol{x}_2^3) = 1$.

There are several ways of encoding these consistency constraints. First, we require that we restrict attention to integer solutions:

$$q(\boldsymbol{x}_r^j) \in \{0,1\} \qquad\qquad \text{For all } r \in \mathbf{R}; \; j \in \{1, \dots, n_r\}. \tag{13.20}$$

We can now utilize two linear equalities to enforce the consistency of these integer solutions. The first constraint enforces the mutual exclusivity within a factor:

$$\sum_{j=1}^{n_r} q(\boldsymbol{x}_r^j) = 1 \qquad\qquad \text{For all } r \in \mathbf{R}. \tag{13.21}$$

The second constraint implies that factors in our MRF agree on the variables in the intersection of their scopes:

$$\sum_{j \; : \; \boldsymbol{c}_r^j \sim \boldsymbol{s}_{r,r'}} q(\boldsymbol{x}_r^j) = \sum_{l \; : \; \boldsymbol{c}_{r'}^l \sim \boldsymbol{s}_{r,r'}} q(\boldsymbol{x}_{r'}^l) \tag{13.22}$$

$$\text{For all } r, r' \in \mathbf{R} \text{ and all } \boldsymbol{s}_{r,r'} \in \mathit{Val}(\boldsymbol{C}_r \cap \boldsymbol{C}_{r'}).$$

Note that this constraint is vacuous for pairs of clusters whose intersection is empty, since there are no assignments $\boldsymbol{s}_{r,r'} \in \mathit{Val}(\boldsymbol{C}_r \cap \boldsymbol{C}_{r'})$.

Example 13.16

Returning to example 13.15, the mutual exclusivity constraints for ϕ_1 would assert that $\sum_{j=1}^{4} q(\boldsymbol{x}_1^j) = 1$. Altogether, we would have three such constraints — one for each factor. The consistency constraints associated with $\phi_1(A, B)$ and $\phi_2(B, C)$ assert that:

$$q(\boldsymbol{x}_1^1) + q(\boldsymbol{x}_1^3) = q(\boldsymbol{x}_2^1) + q(\boldsymbol{x}_2^2) + q(\boldsymbol{x}_2^3)$$
$$q(\boldsymbol{x}_1^2) + q(\boldsymbol{x}_1^4) = q(\boldsymbol{x}_2^4) + q(\boldsymbol{x}_2^5) + q(\boldsymbol{x}_2^6),$$

where the first constraint ensures consistency when $B = b^1$ and the second when $B = b^2$. Overall, we would have three such constraints for $\phi_2(B, C)$, $\phi_3(A, C)$, corresponding to the three values of C, and two constraints for $\phi_1(A, B)$, $\phi_3(A, C)$, corresponding to the two values of A.

Together, these constraints imply that there is a one-to-one mapping between possible assignments to the $q(\boldsymbol{x}_r^j)$ optimization variables and legal assignments to A, B, C. ∎

In general, equation (13.20), equation (13.21), and equation (13.22) together imply that the assignment $q(\boldsymbol{x}_r^j)$'s correspond to a single legal assignment:

Proposition 13.4

Any assignment to the optimization variables $\boldsymbol{q}$ that satisfies equation (13.20), equation (13.21), and equation (13.22) corresponds to a single legal assignment to $X_1, \ldots, X_n$.

The proof is left as an exercise (see exercise 13.13).

Thus, we have now reformulated the MAP task as an integer linear program, where we optimize the linear objective of equation (13.19) subject to the constraints equation (13.20), equation (13.21), and equation (13.22). We note that the problem of solving integer linear programs is itself $\mathcal{NP}$-hard, so that (not surprisingly) we have not avoided the basic hardness of the MAP problem. However, there are several techniques that have been developed for this class of problems, which can be usefully applied to integer programs arising from MAP problems. One of the most useful is described in the next section.

13.5.2 Linear Programming Relaxation

LP relaxation

One of the methods most often used for tackling integer linear programs is the method of *linear program relaxation*. In this approach, we turn a discrete, combinatorial optimization problem into a continuous problem. This problem is a *linear program* (LP), which can be solved in polynomial time, and for which a range of very efficient algorithms exist. One can then use the solutions to this LP to obtain approximate solutions to the MAP problem. To perform this relaxation, we substitute the constraint equation (13.20) with a *relaxed constraint*:

linear program

$$q(\boldsymbol{x}_r^j) \geq 0 \quad \text{For all } r \in \mathbf{R}, \; j \in \{1, \ldots, n_r\}. \tag{13.23}$$

This constraint and equation (13.21) together imply that each $q(\boldsymbol{x}_j^r) \in [0, 1]$; thus, we have relaxed the combinatorial constraint into a continuous one. This relaxation gives rise to the following *linear program* (LP):

linear program

MAP-LP:

Find $\{q(\boldsymbol{x}_r^j)\ :\ r \in \mathbf{R}; j = 1, \ldots, n_r\}$

maximizing $\boldsymbol{\eta}^\top \boldsymbol{q}$

subject to

$$\sum_{j=1}^{n_r} q(\boldsymbol{x}_r^j) \;=\; 1 \qquad\qquad\qquad r \in \mathbf{R}$$

$$\sum_{j\,:\,\boldsymbol{c}_r^j \sim \boldsymbol{s}_{r,r'}} q(\boldsymbol{x}_r^j) \;=\; \sum_{l\,:\,\boldsymbol{c}_{r'}^l \sim \boldsymbol{s}_{r,r'}} q(\boldsymbol{x}_{r'}^l) \qquad \begin{array}{l} r, r' \in \mathbf{R} \\ \boldsymbol{s}_{r,r'} \in Val(\boldsymbol{C}_r \cap \boldsymbol{C}_{r'}) \end{array}$$

$$\boldsymbol{q} \;\geq\; \mathbf{0}$$

This linear program is a relaxation of our original integer program, since every assignment to $\boldsymbol{q}$ that satisfies the constraints of the integer problem also satisfies the constraints of the linear program, but not the other way around. Thus, the optimal value of the objective of the relaxed version will be no less than the value of the (same) objective in the exact version, and it can be greater when the optimal value is achieved at an assignment to $\boldsymbol{q}$ that does not correspond to a legal assignment ξ.

A closer examination shows that the space of assignments to $\boldsymbol{q}$ that satisfies the constraints of MAP-LP corresponds exactly to the locally consistent *pseudo-marginals* for our cluster graph $\mathcal{U}$, which comprise the *local consistency polytope Local*[$\mathcal{U}$], defined in equation (11.16). To see this equivalence, we note that equation (13.23) and equation (13.21) imply that any assignment to $\boldsymbol{q}$ defines a set of locally normalized distributions over the clusters in the cluster graph — nonnegative factors that sum to 1; by equation (13.22), these factors must be sum-calibrated. Thus, there is a one-to-one mapping between consistent pseudo-marginals and possible solutions to the LP.

We can use this observation to answer the following important question: Given a non-integer solution to the relaxed LP, how can we derive a concrete assignment? One obvious approach is a greedy assignment process, which assigns values to the variables X_i one at a time. For each variable, and for each possible assignment x_i, it considers the set of reduced pseudo-marginals that would result by setting $X_i = x_i$. We can now compute the energy term (or, equivalently, the LP objective) for each such assignment, and select the value x_i that gives the maximum value. We then permanently reduce each of the pseudo-marginals with the assignment $X_i = x_i$, and continue. We note that, at the point when we assign X_i, some of the variables have already been assigned, whereas others are still undetermined. At the end of the process, all of the variables have been assigned a specific value, and we have a single joint assignment.

To understand the result obtained by this algorithm, recall that *Local*[$\mathcal{U}$] is a superset of the *marginal polytope Marg*[$\mathcal{U}$] — the space of legal distributions that factorize over $\mathcal{U}$ (see equation (11.15)). Because our objective in equation (13.19) is linear, it has the same optimum over the marginal polytope as over the original space of $\{0, 1\}$ solutions: The value of the objective at a point corresponding to a distribution $P(\mathcal{X})$ is the expectation of its value at the assignments ξ that receive positive probability in P; therefore, one cannot achieve a higher value of the objective with respect to P than with respect to the highest-value assignment ξ. Thus, if we could perform our optimization over the continuous space *Marg*[$\mathcal{U}$], we would find the optimal solution to our MAP objective. However, as we have already discussed, the marginal

pseudo-marginals

local consistency polytope

marginal polytope

polytope is a complex object, which can be specified only using exponentially many constraints. Thus, we cannot feasibly perform this optimization.

By contrast, the optimization problem obtained by this relaxed version has a linear objective with linear constraints, and both involve a number of terms which is linear in the size of the cluster graph. Thus, this linear program admits a range of efficient solutions, including ones with polynomial time guarantees. We can thus apply off-the-shelf methods for solving such problems. Of course, the result is often fractional, in which case it is clearly not an optimal solution to the MAP problem.

The LP formulation has advantages and disadvantages. **By formulating our problem as a linear program, we obtain a very flexible framework for solving it; in particular, we can easily incorporate additional constraints into the LP, which reduce the space of possible assignments to q, eliminating some solutions that do not correspond to actual distributions over $\mathcal{X}$.** (See section 13.9 for some references.) On the other hand, as we discussed in example 11.4, the optimization problems defined over this space of constraints are very large, making standard optimization methods very expensive. Of course, the LP has special structure: For example, when viewed as a matrix, the equality constraints in this LP all have a particular block structure that corresponds to the structure of adjacent clusters; moreover, when the MRF is not densely connected, the constraint matrix is also sparse. However, standard LP solvers may not be ideally suited for exploiting this special structure. Thus, empirical evidence suggests that the more specialized solution methods for the MAP problems are often more effective than using off-the-shelf LP solvers. As we now discuss, the convex message passing algorithms described in section 13.4.2 can be viewed as specialized solution methods to the *dual* of this LP. More recent work explicitly aims to solve this dual using general-purpose optimization techniques that do take advantage of the structure; see section 13.9 for some references.

13.5.3 Low-Temperature Limits

In this section, we show how we can use a limit process to understand the connection between the relaxed MAP-LP and both sum-product and max-product algorithms with convex counting numbers. As we show, this connection provides significant new insight on all three algorithms.

13.5.3.1 LP as Sum-Product Limit

More precisely, recall that the energy functional is defined as:

$$F[P_\Phi, Q] = \sum_{\phi_r \in \Phi} \boldsymbol{E}_{C_r \sim Q}[\log \phi_r(\boldsymbol{C}_r)] + \tilde{\boldsymbol{H}}_Q(\mathcal{X}),$$

where $\tilde{H}_Q(\mathcal{X})$ is some (exact or approximate) version of the entropy of Q. Consider the first term in this expression, also called the energy term. Let $q(\boldsymbol{x}_r^j)$ denote the cluster marginal $\beta_r(\boldsymbol{c}_r^j)$. Then we can rewrite the energy term as:

$$\sum_{r \in \mathbf{R}} \sum_{j=1}^{n_r} q(\boldsymbol{x}_r^j) \log(\phi_r(\boldsymbol{c}_r^j)),$$

which is precisely the objective in our LP relaxation of the MAP problem. Thus, the energy functional is simply a sum of two terms: the LP relaxation objective, and an entropy term. In the energy functional, both of these terms receive equal weight. Now, however, consider an alternative objective, called the *temperature-weighted energy function*. This objective is defined in terms of a *temperature* parameter $T > 0$:

$$\tilde{F}^{(T)}[P_\Phi, Q] = \sum_{\phi_r \in \Phi} \boldsymbol{E}_{\boldsymbol{C}_r \sim Q}[\log \phi_r(\boldsymbol{C}_r)] + T\tilde{\boldsymbol{H}}_Q(\mathcal{X}). \tag{13.24}$$

As usual in our derivations, we consider the task of maximizing this objective subject to the sum-marginalization constraints, that is, that $Q \in Local[\mathcal{U}]$.

The temperature-weighted energy functional reweights the importance of the two terms in the objective. Since $T \longrightarrow 0$, we will place a greater emphasis on the linear energy term (the first term), which is precisely the objective of the relaxed LP. Thus, since $T \longrightarrow 0$, the objective $\tilde{F}^{(T)}[P_\Phi, Q]$ tends to the LP objective. Can we then infer that the fixed points of the objective (say, those obtained from a message passing algorithm) are necessarily optima of the LP? The answer to this question is positive for concave versions of the entropy, and negative otherwise.

In particular, assume that $\tilde{\boldsymbol{H}}_Q(\mathcal{X})$ is a weighted entropy $\tilde{\boldsymbol{H}}_Q^\kappa(\mathcal{X})$, such that κ is a convex set of counting numbers, as in equation (11.20). From the assumption on convexity of the counting numbers and the positivity of the distribution, it follows that the function $\tilde{F}^{(T)}[P_\Phi, Q]$ is strongly convex in the distribution Q. The space $Local[\mathcal{U}]$ is a convex space. Thus, there is a unique global minimum $Q^{*(T)}$ for every T, and that optimum changes continuously in T. Standard results now imply that the limit of $Q^{*(T)}$ is optimal for the limiting problem, which is precisely the LP.

On the other hand, this result does not hold for nonconvex entropies, such as the one obtained by the Bethe approximation, where the objective can have several distinct optima. In this case, there are examples where a sequence of optima obtained for different values of T converges to a point that is not a solution to the LP. Thus, for the remainder of this section, we assume that $\tilde{\boldsymbol{H}}_Q(\mathcal{X})$ is derived from a convex set of counting numbers.

13.5.3.2   Max-Product as Sum-Product Limit

What do we gain from this perspective? It does not appear practical to use this characterization as a constructive solution method. For one thing, we do not want to solve multiple optimization problems, for different values of T. For another, the optimization problem becomes close to degenerate as T grows small, making the problem hard to solve. However, if we consider the dual problem to each of the optimization problems of this sequence, we can analytically characterize the limit of these duals. Surprisingly, this limit turns out to be a fixed point of the max-product belief propagation algorithm.

We first note that the temperature-weighted energy functional is virtually identical in its form to the original functional. Indeed, we can formalize this intuition if we divide the objective through by T; since $T > 0$, this step does not change the optima. The resulting objective has the form:

$$\frac{1}{T} \sum_{\phi_r \in \Phi} \boldsymbol{E}_{\boldsymbol{C}_r \sim Q}[\log \phi_r(\boldsymbol{C}_r)] + \boldsymbol{H}_Q(\mathcal{X}) = \sum_{\phi_r \in \Phi} \boldsymbol{E}_{\boldsymbol{C}_r \sim Q}\left[\frac{1}{T}(\log \phi_r(\boldsymbol{C}_r))\right] + \tilde{\boldsymbol{H}}_Q(\mathcal{X})$$

$$= \sum_{\phi_r \in \Phi} \boldsymbol{E}_{\boldsymbol{C}_r \sim Q}\left[\log \phi_r^{1/T}\right] + \tilde{\boldsymbol{H}}_Q(\mathcal{X}). \tag{13.25}$$

This objective has precisely the same form as the standard approximate energy functional, but for a different set of factors: the original factors, raised to the power of $1/T$. This set of factors defines a new unnormalized density:

$$\tilde{P}_{\Phi}^{(T)}(\mathcal{X}) = (\tilde{P}_{\Phi}(\mathcal{X}))^{1/T}.$$

Because our entropy is concave, and using our assumption that the distribution is positive, the approximate free energy $\tilde{F}[P_{\Phi}, Q]$ is strictly convex and hence has a unique global minimum $Q^{(T)}$ for each temperature T. We can now consider the Lagrangian dual of this new objective, and characterize this unique optimum via its dual parameterization $Q^{(T)}$. In particular, as we have previously shown, $Q^{(T)}$ is a reparameterization of the distribution $\tilde{P}_{\Phi}^{(T)}(\mathcal{X})$:

$$\tilde{P}_{\Phi}^{(T)} = \prod_{r \in \mathbf{R}} \beta_r^{(T)} \prod_i (\beta_i^{(T)})^{\kappa_i} = \prod_{r \in \mathbf{R}^+} (\beta_r^{(T)})^{\kappa_i}, \tag{13.26}$$

where, for simplicity of notation, we define $\mathbf{R}^+ = \mathbf{R} \cup \mathcal{X}$ and $\kappa_r = 1$ for $r \in \mathbf{R}$.

Our goal is now to understand what happens to $Q^{(T)}$ as we take $T \longrightarrow 0$. We first reformulate these beliefs by defining, for every region $r \in \mathbf{R}^+$:

$$\bar{\beta}_r^{(T)} = \max_{\boldsymbol{x}_r'} \beta_r^{(T)}(\boldsymbol{x}_r') \tag{13.27}$$

$$\tilde{\beta}_r^{(T)}(\boldsymbol{c}_r) = \left(\frac{\beta_r^{(T)}(\boldsymbol{c}_r)}{\bar{\beta}_r^{(T)}} \right)^T. \tag{13.28}$$

The entries in the new beliefs $\tilde{\boldsymbol{\beta}}^{(T)} = \{\tilde{\beta}_r^{(T)}(\boldsymbol{C}_r)\}$ take values between 0 and 1, with the maximal entry in each belief always having the value 1.

We now define the limiting value of these beliefs:

$$\tilde{\beta}_r^{(0)}(\boldsymbol{c}_r) = \lim_{T \longrightarrow 0} \tilde{\beta}_r^{(T)}(\boldsymbol{c}_r). \tag{13.29}$$

Because the optimum changes continuously in T, and because the beliefs take values in a convex space (all are in the range $[0, 1]$), the limit is well defined. Our goal is to show that the limit beliefs $\tilde{\boldsymbol{\beta}}^{(0)}$ are a fixed point of the max-product belief propagation algorithm for the model $\tilde{P}_{\Phi}$. We show this result in two parts. We first show that the limit is max-calibrated, and then that it provides a reparameterization of our distribution $\tilde{P}_{\Phi}$.

Proposition 13.5

The limiting beliefs $\tilde{\boldsymbol{\beta}}^{(0)}$ are max-calibrated.

Proof We wish to show that for any region r, any $X_i \in \boldsymbol{C}_r$, and any $x_i \in Val(X_i)$, we have:

$$\max_{\boldsymbol{c}_r \sim x_i} \tilde{\beta}_r^{(0)}(\boldsymbol{c}_r) = \tilde{\beta}_i^{(0)}(x_i). \tag{13.30}$$

Consider the left-hand side of this equality.

$$\max_{c_r \sim x_i} \tilde{\beta}_r^{(0)}(\boldsymbol{c}_r) \;=\; \max_{c_r \sim x_i}\left[\lim_{T \longrightarrow 0} \tilde{\beta}_r^{(T)}(\boldsymbol{c}_r)\right]$$

$$(i) \;=\; \lim_{T \longrightarrow 0}\left[\sum_{c_r \sim x_i}(\tilde{\beta}_r^{(T)}(\boldsymbol{c}_r))^{1/T}\right]^T$$

$$=\; \lim_{T \longrightarrow 0}\left[\sum_{c_r \sim x_i}\left(\frac{\beta_r^{(T)}(\boldsymbol{c}_r)}{\bar{\beta}_r^{(T)}}\right)\right]^T$$

$$=\; \lim_{T \longrightarrow 0}\left[\frac{1}{\bar{\beta}_r^{(T)}}\left(\sum_{c_r \sim x_i}\beta_r^{(T)}(\boldsymbol{c}_r)\right)\right]^T$$

$$(ii) \;=\; \lim_{T \longrightarrow 0}\left[\frac{1}{\bar{\beta}_r^{(T)}}\beta_i^{(T)}(x_i)\right]^T$$

$$(iii) \;=\; \lim_{T \longrightarrow 0}\left[\frac{\bar{\beta}_i^{(T)}}{\bar{\beta}_r^{(T)}}\frac{\beta_i^{(T)}(x_i)}{\bar{\beta}_i^{(T)}}\right]^T$$

$$(iv) \;=\; \lim_{T \longrightarrow 0}\left[\frac{\bar{\beta}_i^{(T)}}{\bar{\beta}_r^{(T)}}(\tilde{\beta}_i^{(T)}(x_i))^{1/T}\right]^T$$

$$=\; \lim_{T \longrightarrow 0}\left[\left(\frac{\bar{\beta}_i^{(T)}}{\bar{\beta}_r^{(T)}}\right)^T \tilde{\beta}_i^{(T)}(x_i)\right]$$

$$(v) \;=\; \lim_{T \longrightarrow 0}\tilde{\beta}_i^{(T)}(x_i)] \;\;=\; \tilde{\beta}_i^{(0)}(x_i),$$

as required. In this derivation, the step marked (i) is a general relationship between maximization and summation; see lemma 13.2. The step marked (ii) is a consequence of the sum-marginalization property of the region beliefs $\beta_r^{(T)}(\boldsymbol{C}_r)$ relative to the individual node belief. The step marked (iii) is simply multiplying and dividing by the same expression. The step marked (iv) is derived directly by substituting the definition of $\tilde{\beta}_i^{(T)}(x_i)$. The step marked (v) is a consequence of the fact that, because of sum-marginalization, $\bar{\beta}_i^{(T)}/\bar{\beta}_r^{(T)}$ (for $X_i \in \boldsymbol{C}_r$) is bounded in the range $[1, |Val(\boldsymbol{C}_r - \{X_i\})|]$ for any $T > 0$, and therefore its limit, since $T \longrightarrow 0$ is 1.

It remains to prove the following lemma:

Lemma 13.2

For $i = 1, \ldots, k$, let $a_i(T)$ be a continuous function of T for $T > 0$. Then

$$\max_i \lim_{T \longrightarrow 0} a_i(T) = \lim_{T \longrightarrow 0}\left(\sum_i(a_i(T))^{1/T}\right)^T. \tag{13.31}$$

PROOF Because the functions are continuous, we have that, for some T_0, there exists some j such that, for any $T < T_0$, $a_j(T) \geq a_i(T)$ for all $i \neq j$; assume, for simplicity, that this j

is unique. (The proof of the more general case is similar.) Let $a_j^* = \lim_{T \longrightarrow 0} a_j(T)$. The left-hand side of equation (13.31) is then clearly a_j^*. The expression on the right-hand side can be rewritten:

$$\lim_{T \longrightarrow 0} a_j(T) \left(\sum_i \left(\frac{a_i(T)}{a_j(T)} \right)^{1/T} \right)^T = a_j^* \left[\lim_{T \longrightarrow 0} \left(\sum_i \left(\frac{a_i(T)}{a_j(T)} \right)^T \right)^{1/T} \right] = a_j^*.$$

The first equality follows from the fact that the $a_j(T)$ sequence is convergent. The second follows from the fact that, because $a_j(T) > a_i(T)$ for all $i \neq j$ and all $T < T_0$, the ratio $a_i(T)/a_j(T)$ is bounded in $[0, 1]$, with $a_j(T)/a_j(T) = 1$; therefore the limit is simply 1. ∎

The proof of this lemma concludes the proof of the theorem. ∎

We now wish to show the second important fact:

Theorem 13.9

The limit $\tilde{\boldsymbol{\beta}}^{(0)}$ is a proportional reparameterization of $\tilde{P}_\Phi$, that is:

$$\tilde{P}_\Phi(\mathcal{X}) \propto \prod_{r \in \mathbf{R}} (\tilde{\beta}_r^{(0)}(\boldsymbol{c}_r))^{\kappa_r}.$$

PROOF Due to equation (13.26), we have that

$$\tilde{P}_\Phi^{(T)}(\mathcal{X}) = \prod_{r \in \mathbf{R}} (\beta_r^{(T)}(\boldsymbol{c}_r))^{\kappa_r}.$$

We can raise each side to the power T, and obtain that:

$$\tilde{P}_\Phi(\mathcal{X}) = \left(\prod_{r \in \mathbf{R}} (\beta_r^{(T)}(\boldsymbol{c}_r))^{\kappa_r} \right)^T.$$

We can divide each side by

$$\left(\prod_{r \in \mathbf{R}^+} (\bar{\beta}_r^{(T)})^{\kappa_r} \right)^T,$$

to obtain the equality

$$\frac{\tilde{P}_\Phi(\mathcal{X})}{\left(\prod_{r \in \mathbf{R}^+} (\bar{\beta}_r^{(T)})^{\kappa_r} \right)^T} = \prod_r (\tilde{\beta}_r^{(T)}(\boldsymbol{c}_r))^{\kappa_r}.$$

This equality holds for every value of $T > 0$. Moreover, as we argued, the right-hand side is bounded in $[0, 1]$, and hence so is the left-hand side. As a consequence, we have an equality of two bounded continuous functions of T, so that they must also be equal at the limit $T \longrightarrow 0$. It follows that the limiting beliefs $\tilde{\boldsymbol{\beta}}^{(0)}$ are proportional to a reparameterization of $\tilde{P}_\Phi$. ∎

13.5.3.3   Discussion

Overall, the analysis in this section reveals interesting connections between three separate algorithms: the linear program relaxation of the MAP problem, the low-temperature limit of sum-product belief propagation with convex counting numbers, and the max-product reparameterization with (the same) convex counting numbers. These connections hold for any set of convex counting numbers and any (positive) distribution $\tilde{P}_\Phi$.

Specifically, we have characterized the solution to the relaxed LP as the limit of a sequence of optimization problems, each defined by a temperature-weighted convex energy functional. Each of these optimization problems can be solved using an algorithm such as convex sum-product BP, which (assuming convergence) produces optimal beliefs for that problem. We have also shown that the beliefs obtained in this sequence can be reformulated to converge to a new set of beliefs that are max-product calibrated. These beliefs are fixed points of the convex max-product BP algorithm. Thus, we can hope to use max-product BP to find these limiting beliefs.

Our earlier results show that the fixed points of convex max-product BP, if they admit a locally optimal assignment, are guaranteed to produce the MAP assignment. We can now make use of the results in this section to shed new light on this analysis.

Theorem 13.10

Assume that we have a set of max-calibrated beliefs $\beta_r(\boldsymbol{C}_r)$ and $\beta_i(X_i)$ such that equation (13.16) holds. Assume furthermore that ξ^* is a locally consistent joint assignment relative to these beliefs. Then the MAP-LP relaxation is tight.

PROOF We first observe that

$$
\max_\xi \log \tilde{P}_\Phi(\xi) = \max_{Q \in Marg[\mathcal{U}]} \boldsymbol{E}_{\xi \sim Q}\left[\log \tilde{P}_\Phi(\xi)\right]
$$

$$
\leq \max_{\boldsymbol{Q} \in Local[\mathcal{U}]} \boldsymbol{E}_{\xi \sim \boldsymbol{Q}}\left[\log \tilde{P}_\Phi(\xi)\right], \tag{13.32}
$$

which is equal to the value of MAP-LP. Note that we are abusing notation in the expectation used in the last expression, since $\boldsymbol{Q} \in Local[\mathcal{U}]$ is not a distribution but a set of pseudo-marginals. However, because $\log \tilde{P}_\Phi(\xi)$ factors according to the structure of the clusters in the pseudo-marginals, we can use a set of pseudo-marginals to compute the expectation.

Next, we note that for any set of functions $f_r(\boldsymbol{C}_r)$ whose scopes align with the clusters $\boldsymbol{C}_r$, we have that:

$$
\max_{Q \in Local[\mathcal{U}]} \boldsymbol{E}_{\boldsymbol{C}_r \sim Q}\left[\sum_r f_r(\boldsymbol{C}_r)\right] = \max_{Q \in Local[\mathcal{U}]} \sum_r \boldsymbol{E}_{\boldsymbol{C}_r \sim Q}[f_r(\boldsymbol{C}_r)]
$$

$$
\leq \sum_r \max_{\boldsymbol{C}_r} f_r(\boldsymbol{C}_r),
$$

because an expectation is smaller than the max.

We can now apply this derivation to the reformulation of $\tilde{P}_\Phi$ that we get from the reparameterization:

$$
\max_{Q \in Local[\mathcal{U}]} \boldsymbol{E}_Q\left[\log \tilde{P}_\Phi(\xi)\right] = \max_{Q \in Local[\mathcal{U}]} \boldsymbol{E}_Q\left[\begin{array}{l} \sum_r \nu_r \log(\beta_r(\boldsymbol{c}_r)) + \sum_i \nu_i \log \beta_i(x_i) \\ + \sum_{i,r\,:\,X_i \in \boldsymbol{C}_r} \nu_{r,i}\left(\log \beta_r(\boldsymbol{c}_r) - \log \beta_i(x_i)\right) \end{array}\right].
$$

From the preceding derivation, it follows that:

$$\leq \sum_r \max_{\boldsymbol{c}_r} \nu_r \log(\beta_r(\boldsymbol{c}_r)) + \sum_i \max_{x_i} \nu_i \log \beta_i(x_i)$$
$$+ \sum_{i,r \,:\, X_i \in \boldsymbol{C}_r} \max_{\boldsymbol{c}_r; x_i = \boldsymbol{c}_r \langle X_i \rangle} \nu_{r,i} \left(\log \beta_r(\boldsymbol{c}_r) - \log \beta_i(x_i)\right).$$

And from the positivity of the counting numbers, we get

$$= \sum_r \nu_r \max_{\boldsymbol{c}_r} \log(\beta_r(\boldsymbol{c}_r)) + \sum_i \nu_i \max_{x_i} \log \beta_i(x_i)$$
$$+ \sum_{i,r \,:\, X_i \in \boldsymbol{C}_r} \nu_{r,i} \max_{\boldsymbol{c}_r; x_i = \boldsymbol{c}_r \langle X_i \rangle} \left(\log \beta_r(\boldsymbol{c}_r) - \log \beta_i(x_i)\right).$$

Now, due to lemma 13.1 (reformulated for log-factors), we have that ξ^* optimizes each of the maximization expressions, so that we conclude:

$$= \sum_r \nu_r \log(\beta_r(\boldsymbol{c}_r^*)) + \sum_i \nu_i \log \beta_i(x_i^*)$$
$$+ \sum_{i,r \,:\, X_i \in \boldsymbol{C}_r} \nu_{r,i} \left(\log \beta_r(\boldsymbol{c}_r^*) - \log \beta_i(x_i^*)\right)$$
$$= \log \tilde{P}_\Phi(\xi^*).$$

Putting this conclusion together with equation (13.32), we obtain:

$$\max_\xi \log \tilde{P}_\Phi(\xi) \leq \max_{Q \in Local[\mathcal{U}]} \boldsymbol{E}_{\xi \sim Q}\left[\log \tilde{P}_\Phi(\xi)\right]$$
$$\leq \log \tilde{P}_\Phi(\xi^*).$$

Because the right-hand side is clearly $\leq$ the left-hand side, the entire inequality holds as an equality, proving that

$$\max_\xi \log \tilde{P}_\Phi(\xi) = \max_{Q \in Local[\mathcal{U}]} \boldsymbol{E}_{\xi \sim Q}\left[\log \tilde{P}_\Phi(\xi)\right],$$

that is, the value of the integer program optimization is the same as that of the relaxed LP. ∎

 This last fact has important repercussions. In particular, it shows that **convex max-product BP can be decoded only if the LP is tight; otherwise, there is no locally optimal joint assignment, and no decoding is possible. It follows that convex max-product BP provides provably useful results only in cases where** MAP-LP **itself provides the optimal answer to the MAP problem. We note that a similar conclusion does *not* hold for nonconvex variants such as those based on the standard Bethe counting numbers; in particular, standard max-product BP is not an upper bound to** MAP-LP**, and therefore it can return solutions in the interior of the polytope of** MAP-LP**. As a consequence, it may be decodable**

even when the LP is not tight; in that case, the returned joint assignment may be the MAP, or it may be a suboptimal assignment.

This result leaves several intriguing open questions. First, we note that this result shows a connection between the results of max-product and the LP only when the LP is tight. It is an open question whether we can show a general connection between the max-product beliefs and the dual of the original LP. A second question is whether we can construct better techniques that directly solve the LP or its dual; indeed, recent work (see section 13.9) explores a range of other techniques for this task. A third question is whether this technique provides a useful heuristic: Even if the reparameterization we derive does not have a locally consistent joint assignment, we can still use it to construct an assignment using various heuristic methods, such as selecting for each variable X_i the assignment $x_i^* = \arg\max_{x_i} \beta_i(x_i)$. While there are no guarantees about this solution, it may still work well in practice.

13.6   Using Graph Cuts for MAP

In this section, we discuss the important class of *metric* and *semi-metric* MRFs, which we defined in box 4.D. This class has received considerable attention, largely owing to its importance in computer- vision applications. We show how this class of networks, although possibly very densely connected, can admit an optimal or close-to-optimal solution, by virtue of structure in the potentials.

13.6.1   Inference Using Graph Cuts

The basic graph construction is defined for pairwise MRFs consisting solely of binary-valued variables ($\mathcal{V} = \{0, 1\}$). Although this case has restricted applicability, it forms the basis for the general case. As we now show, **the MAP problem for a certain class of binary-valued MRFs can be solved *optimally* using a very simple and efficient *graph-cut* algorithm. Perhaps the most surprising aspect of this reduction is that this algorithm is guaranteed to return the optimal solution in polynomial time, regardless of the structural complexity of the underlying graph. This result stands in contrast to most of the other results presented in this book, where polynomial-time solutions were obtainable only for graphs of low tree width. Equally noteworthy is the fact that a similar result does not hold for sum-product computations over this class of graphs; thus, we have an example of a class of networks where sum-product inference and MAP inference have very different computational properties.**

We first define the min-cut problem for a graph, and then show how the MAP problem can be reduced to it. The min-cut problem is defined by a set of vertices $\mathcal{Z}$, plus two distinguished nodes generally known as s and t. We have a set of directed edges $\mathcal{E}$ over $\mathcal{Z} \cup \{s, t\}$, where each edge $(z_1, z_2) \in \mathcal{E}$ is associated with a nonnegative cost $cost(z_1, z_2)$. A *graph cut* is a disjoint partition of $\mathcal{Z}$ into $\mathcal{Z}_s \cup \mathcal{Z}_t$ such that $s \in \mathcal{Z}_s$ and $t \in \mathcal{Z}_t$. The *cost* of the cut is:

graph cut

$$cost(\mathcal{Z}_s, \mathcal{Z}_t) = \sum_{z_1 \in \mathcal{Z}_s, z_2 \in \mathcal{Z}_t} cost(z_1, z_2).$$

In words, the cost is the total sum of the edges that cross from the $\mathcal{Z}_s$ side of the partition to the $\mathcal{Z}_t$ side. The *minimal cut* is the partition $\mathcal{Z}_s, \mathcal{Z}_t$ that achieves the minimal cost. While presenting a min-cut algorithm is outside the scope of this book, such algorithms are standard, have polynomial-time complexity, and are very fast in practice.

How do we reduce the MAP problem to one of computing cuts on a graph? Intuitively, we need to design our graph so that a cut corresponds to an assignment to $\mathcal{X}$, and its cost to the value of the assignment. The construction follows straightforwardly from this intuition. Our vertices (other than s, t) represent the variables in our MRF. We use the s side of the cut to represent the label 0, and the t side to represent the label 1. Thus, we map a cut $\mathcal{C} = (\mathcal{Z}_s, \mathcal{Z}_t)$ to the following assignment $\xi^{\mathcal{C}}$:

$$x_i^{\mathcal{C}} = 0 \quad \text{if and only if} \quad z_i \in \mathcal{Z}_s.$$

We begin by demonstrating the construction on the simple case of the generalized Ising model of equation (4.6). Note that energy functions are invariant to additive changes in all of the components, since these just serve to move all entries in $E(x_1, \ldots, x_n)$ by some additive factor, leaving their relative order invariant. Thus, we can assume, without loss of generality, that all components of the energy function are nonnegative. Moreover, we can assume that, for every node i, either $\epsilon_i(1) = 0$ or $\epsilon_i(0) = 0$. We now construct the graph as follows:

- If $\epsilon_i(1) = 0$, we introduce an edge $z_i \rightarrow t$, with cost $\epsilon_i(0)$.

- If $\epsilon_i(0) = 0$, we introduce an edge $s \rightarrow z_i$, with cost $\epsilon_i(1)$.

- For each pair of variables X_i, X_j that are connected by an edge in the MRF, we introduce both an edge (z_i, z_j) and the edge (z_j, z_i), both with cost $\lambda_{i,j} \geq 0$.

Now, consider the cost of a cut $(\mathcal{Z}_s, \mathcal{Z}_t)$. If $z_i \in \mathcal{Z}_s$, then X_i is assigned a value of 0. In this case, z_i and t are on opposite sides of the cut, and so we will get a contribution of $\epsilon_i(0)$ to the cost of the cut; this contribution is precisely the X_i node energy of the assignment $X_i = 0$, as we would want. The analogous argument applies when $z_i \in \mathcal{Z}_t$. We now consider the edge potential. The edge (z_i, z_j) only makes a contribution to the cut if we place z_i and z_j on opposite sides of the cut; in this case, the contribution is $\lambda_{i,j}$. Conversely, the pair X_i, X_j makes a contribution of $\lambda_{i,j}$ to the energy function if $X_i \neq X_j$, and otherwise it contributes 0. Thus, the contribution of the edge to the cut is precisely the same as the contribution of the node pair to the energy function. Overall, we have shown that the cost of the cut is precisely the same as the energy of the corresponding assignment. Thus, the min-cut algorithm is guaranteed to find the assignment to $\mathcal{X}$ that minimizes the energy function, that is, ξ^{map}.

Example 13.17

Consider a simple example where we have four variables X_1, X_2, X_3, X_4 connected in a loop with the edges X_1—X_2, X_2—X_3, X_3—X_4, X_1—X_4. Assume we have the following energies, where we list only components that are nonzero:

$$\epsilon_1(0) = 7 \quad \epsilon_2(1) = 2 \quad \epsilon_3(1) = 1 \quad \epsilon_4(1) = 6$$
$$\lambda_{1,2} = 6 \quad \lambda_{2,3} = 6 \quad \lambda_{3,4} = 2 \quad \lambda_{1,4} = 1.$$

The graph construction and the minimum cut for this example are shown in figure 13.5.

Going by the node potentials alone, the optimal assignment is $X_1 = 1$, $X_2 = 0$, $X_3 = 0$, $X_4 = 0$. However, we also have interaction potentials that encourage agreement between neighboring nodes. In particular, there are fairly strong potentials that induce $X_1 = X_2$ and $X_2 = X_3$. Thus, the node-optimal assignment achieves a penalty of 7 from the contributions of $\lambda_{1,2}$ and $\lambda_{1,4}$.

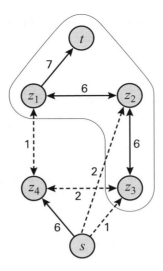

Figure 13.5   Example graph construction for applying min-cut to the binary MAP problem, based on example 13.17. Numbers on the edges represent their weight. The cut is represented by the set of nodes in $\mathcal{Z}_t$. Dashed edges are ones that participate in the cut; note that only one of the two directions of a bidirected edge contributes to the weight of the cut, which is 6 in this example.

Conversely, the assignment where X_2 and X_3 agree with X_1 gets a penalty of only 6 from the X_2 and X_3 node contributions and from the weaker edge potentials $\lambda_{3,4}$ and $\lambda_{1,4}$. Thus, the overall MAP assignment has $X_1 = 1$, $X_2 = 1$, $X_3 = 1$, $X_4 = 0$. ∎

As we mentioned, the MAP problem in such graphs reduces to a minimum cut problem regardless of the network connectivity. Thus, this approach allows us to find MAP solution for a class of MRFs for which probability computations are intractable.

We can easily extend this construction beyond the generalized Ising model:

Definition 13.5

submodular
energy function

A pairwise energy $\epsilon_{i,j}(\cdot, \cdot)$ is said to be submodular *if*

$$\epsilon_{i,j}(1, 1) + \epsilon_{i,j}(0, 0) \leq \epsilon_{i,j}(1, 0) + \epsilon_{i,j}(0, 1). \tag{13.33}$$

∎

The graph construction for submodular energies, which is shown in detail in algorithm 13.4, is a little more elaborate. It first normalizes each edge potential by subtracting $\epsilon_{i,j}(0, 0)$ from all entries; this operation subtracts a constant amount from the energies of all assignments, corresponding to a constant multiple in probability space, which only changes the (in this case irrelevant) partition function. It then moves as much mass as possible to the individual node potentials for i and j. These steps leave a single pairwise term that defines an energy only for the assignment $v_i = 0, v_j = 1$:

$$\epsilon'_{i,j}(0, 1) = \epsilon_{i,j}(1, 0) + \epsilon_{i,j}(0, 1) - \epsilon_{i,j}(0, 0) - \epsilon_{i,j}(1, 1).$$

Algorithm 13.4 Graph-cut algorithm for MAP in pairwise binary MRFs with submodular potentials

---

Procedure MinCut-MAP (
  ϵ // Singleton and pairwise submodular energy factors
)
1   // Define the energy function
2   **for** all i
3       $\epsilon'_i \leftarrow \epsilon_i$
4   Initialize $\epsilon'_{i,j}$ to 0 for all i, j
5   **for** all pairs $i < j$
6       $\epsilon'_i(1) \leftarrow \epsilon'_i(1) + (\epsilon_{i,j}(1,0) - \epsilon_{i,j}(0,0))$
7       $\epsilon'_j(1) \leftarrow \epsilon'_j(1) + (\epsilon_{i,j}(1,1) - \epsilon_{i,j}(1,0))$
8       $\epsilon'_{i,j}(0,1) \leftarrow \epsilon_{i,j}(1,0) + \epsilon_{i,j}(0,1) - \epsilon_{i,j}(0,0) - \epsilon_{i,j}(1,1)$
9
10   // Construct the graph
11   **for** all i
12   **if** $\epsilon'_i(1) > \epsilon'_i(0)$ **then**
13          $\mathcal{E} \leftarrow \mathcal{E} \cup \{(s, z_i)\}$
14          $cost(s, z_i) \leftarrow \epsilon'_i(1) - \epsilon'_i(0)$
15   **else**
16          $\mathcal{E} \leftarrow \mathcal{E} \cup \{(z_i, t)\}$
17          $cost(z_i, t) \leftarrow \epsilon'_i(0) - \epsilon'_i(1)$
18   **for** all pairs $i < j$ such that $\epsilon'_{i,j}(0,1) > 0$
19          $\mathcal{E} \leftarrow \mathcal{E} \cup \{(z_i, z_j)\}$
20          $cost(z_i, z_j) \leftarrow \epsilon'_{i,j}(0,1)$
21
22       $t \leftarrow$ MinCut$(\{z_1, \ldots, z_n\}, \mathcal{E})$
23   // MinCut returns $t_i = 1$ iff $z_i \in \mathcal{Z}_t$
24   **return** t

Because of submodularity, this term satisfies $\epsilon'_{i,j}(0,1) \geq 0$. The algorithm executes this transformation for every pairwise potential i, j. The resulting energy function can easily be converted into a graph using essentially the same construction that we used earlier; the only slight difference is that for our new energy function $\epsilon'_{i,j}(v_i, v_j)$ we need to introduce only the edge (z_i, z_j), with cost $\epsilon'_{i,j}(0,1)$; we do not introduce the opposite edge (z_j, z_i). We now use the same mapping between s-t cuts in the graph and assignment to the variables $X_1, \ldots, X_n$. It is not difficult to verify that the cost of an s-t cut $\mathcal{C}$ in the resulting graph is precisely $E(\xi^{\mathcal{C}}) + \text{Const}$ (see exercise 13.14). Thus, finding the minimum cut in this graph directly gives us the cost-minimizing assignment ξ^{map}.

Note that for pairwise submodular energy, there is an LP relaxation of the MAP integer optimization, which is tight. Thus, this result provides another example where having a tight LP relaxation allows us to find the optimal MAP assignment.

13.6.2   Nonbinary Variables

In the case of nonbinary variables, we can no longer use a graph construction to solve the MRF optimally. Indeed, the problem of optimizing the energy function, even if it is submodular, is $\mathcal{NP}$-hard in this case. Here, a very useful technique is to take greedy hill-climbing steps, but where each step involves a globally optimal solution to a simplified problem. Two types of steps have been utilized extensively: *alpha-expansion* and *alpha-beta swap*. As we will show, under appropriate conditions on the energy function, both the alpha-expansion step and the alpha-beta-swap steps can be performed optimally by applying the min-cut procedure to an appropriately constructed MRF. Thus, the search procedure can take a global step in the space.

alpha-expansion   The *alpha-expansion* considers a particular value v; the step simultaneously considers all of the variables X_i in the MRF, and allows each of them to take one of two values: it can keep its current value x_i, or change its value to v. Thus, the step expands the set of variables that take the label v; the label v is often denoted α in the literature; hence the name alpha-expansion.

The alpha-expansion algorithm is shown in algorithm 13.5. It consists of repeated applications of alpha-expansion steps, for different labels v. Each alpha-expansion step is defined relative to our current assignment $\boldsymbol{x}$ and a target label v. Our goal is to select, for each variable X_i whose current label x_i is other than v, whether in the new assignment $\boldsymbol{x}'$ its new label will remain x_i or move to v. We do so using a new MRF that has binary variables T_i for each variable X_i; we then define a new assignment $\boldsymbol{x}'$ so that $x_i' = x_i$ if $T_i = t_i^0$, and $x_i' = v$ if $T_i = t_i^1$.

restricted energy   We define a new *restricted energy function* E' using the following set of potentials:
function

$$
\begin{aligned}
\epsilon_i'(t_i^0) &= \epsilon_i(x_i) \\
\epsilon_i'(t_i^1) &= \epsilon_i(v) \\
\epsilon_{i,j}'(t_i^0, t_j^0) &= \epsilon_{i,j}(x_i, x_j) \\
\epsilon_{i,j}'(t_i^0, t_j^1) &= \epsilon_{i,j}(x_i, v) \\
\epsilon_{i,j}'(t_i^1, t_j^0) &= \epsilon_{i,j}(v, x_j) \\
\epsilon_{i,j}'(t_i^1, t_j^1) &= \epsilon_{i,j}(v, v)
\end{aligned}
\tag{13.34}
$$

It is straightforward to see that for any assignment $\boldsymbol{t}$, $E'(\boldsymbol{t}) = E(\boldsymbol{x}')$. Thus, finding the optimal $\boldsymbol{t}$ corresponds to finding the optimal $\boldsymbol{x}'$ in the restricted space of v-expansions of $\boldsymbol{x}$.

In order to optimize $\boldsymbol{t}$ using graph cuts, the new energy E' needs to be submodular, as in equation (13.33). Plugging in the definition of the new potentials, we get the following constraint:

$$
\epsilon_{i,j}(x_i, x_j) + \epsilon_{i,j}(v, v) \leq \epsilon_{i,j}(x_i, v) + \epsilon_{i,j}(v, x_j).
$$

Now, if we have an MRF defined by some distance function μ, then $\epsilon_{i,j}(v, v) = 0$ by reflexivity, and the remaining inequality is a direct consequence of the triangle inequality. Thus, we can apply the alpha-expansion procedure to any metric MRF.

alpha-beta swap   The second type of step is the *alpha-beta swap*. Here, we consider two labels: v_1 and v_2. The step allows each variable whose current label is v_1 to keep its value or change it to v_2, and conversely for variables currently labeled v_2. Like the alpha-expansion step, the alpha-beta swap over a given assignment $\boldsymbol{x}$ can be defined easily by constructing a new energy function, over which min-cut can be performed. The details are left as an exercise (exercise 13.15). We note that the alpha-beta-swap operation requires only that the energy function be a semimetric (that is, the triangle inequality is not required).

These two steps allow us to use the min-cut procedure as a subroutine in solving the MAP problem in metric or semimetric MRFs with nonbinary variables.

Algorithm 13.5 Alpha-expansion algorithm

Procedure Alpha-Expansion (
    ϵ,   // Singleton and pairwise energies
    x // Some initial assignment
)
1   **repeat**
2   change $\leftarrow$ *false*
3   **for** $k = 1, \ldots, K$
4        $t \leftarrow$ Alpha-Expand(ϵ, x, v_k)
5   **for** $i = 1, \ldots, n$
6   **if** $t_i = 1$ **then**
7            $x_i \leftarrow v_k$ // If $t_i = 0$, x_i doesn't change
8   change $\leftarrow$ *true*
9   **until** change $= false$
10   **return** (x)

Procedure Alpha-Expand (
    ϵ,
    x // Current assignment
    v // Expansion label
)
1   Define ϵ' as in equation (13.34)
2   **return** MinCut-MAP(ϵ')

Box 13.B — Case Study: Energy Minimization in Computer Vision. *Over the past few years, MRFs have become a standard tool for addressing a range of low-level vision tasks, some of which we reviewed in box 4.B. As we discussed, the pairwise potentials in these models are often aimed at penalizing discrepancies between the values of adjacent pixels, and hence they often naturally satisfy the submodularity assumption that are necessary for the application of graph cut methods. Also very popular is the TRW-S variant of the convex belief propagation algorithms, described in box 13.A. Standard belief propagation has also been used in multiple applications.*

    Vision problems pose some significant challenges. Although the grid structures associated with images are not dense, they are very large, and they contain many tight loops, which can pose difficulties for convergence of the message passing algorithm. Moreover, in some tasks, such as stereo reconstruction, *the value space of the variables is a discretization of a continuous space, and therefore many values are required to get a reasonable approximation. As a consequence, the representation of the pairwise potentials can get very large, leading to memory problems.*

    A number of fairly comprehensive empirical studies have been done comparing the various methods on a suite of computer-vision benchmark problems. By and large, it seems that for the grid-structured networks that we described, graph-cut methods with the alpha-expansion step and TRW-S are fairly comparable, with the graph-cut methods dominating in running time; both significantly

stereo reconstruction

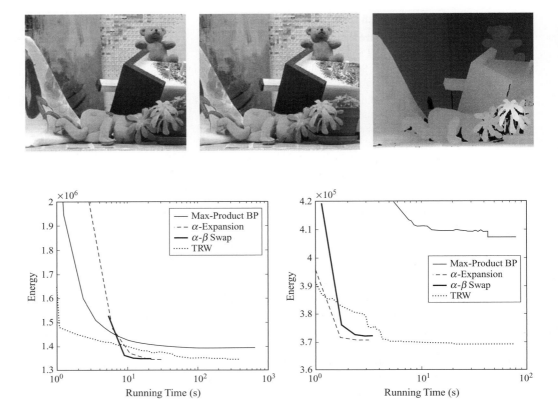

Figure 13.B.1 — MAP inference for stereo reconstruction The top row contains a pair of stereo images for a problem known as Teddy and the target output (darker pixels denote a larger z value); the images are taken from Scharstein and Szeliski (2003). The bottom row shows the best energy obtained as a function of time by several different MAP algorithms:max-product BP, the TRW variant of convex BP, min-cut with alpha-expansion, and min-cut with alpha-beta swap. The left image is for Teddy, and the right is for a different stereo problem called Tsukuba.

outperform the other methods. Figure 13.B.1 shows some sample results on stereo-reconstruction problems; here, the energies are close to submodular, allowing the application of a range of different methods.

The fact that convex BP is solving the dual problem to the relaxed LP allows it to provide a lower bound on the energy of the true MAP assignment. Moreover, as we discussed, it can sometimes provide optimality guarantees on the inferred solution. Thus, it is sometimes possible to compare the results of these methods to the true global optimum of the energy function. Somewhat surprisingly, it appears that both methods come very close to achieving optimal energies on a large fraction of these benchmark problems, suggesting that the problem of energy minimization for these MRFs is essentially solved.

In contrast to this optimistic viewpoint is the observation that the energy minimizing configuration is often significantly worse than the "target" assignment (for example, the true depth disparity in a stereo reconstruction problem). In other words, the ground truth often has a worse energy (lower probability) than the assignment that optimizes the energy function. This finding suggests that a key problem is that of designing better energy functions, which better capture the structure of our target assignments. This topic has been the focus of much recent work. In many cases, the resulting energies involve nonlocal interactions between the pixels, and are therefore significantly more complex. Some evidence suggests that as the graph becomes more dense and less local, belief propagation methods start to degrade. Conversely, as the potentials become less submodular, the graph-cut methods become less applicable. Thus, the design of new energy-minimization methods that are applicable to these richer energy functions is a topic of significant current interest.

13.7 Local Search Algorithms ★

A final class of methods that have been applied to MAP and marginal MAP queries are methods that search over the space of assignments. The task of searching for a high-weight (or low-cost) assignment of values to a set of variables is a central one in many applications, and it has received attention in a number of communities. Methods for addressing this task come in many flavors.

systematic search

Some of those methods are *systematic*: They search the space so as to ensure that assignments that are not considered are not optimal, and thereby guarantee an optimal solution. Such methods generally search over the space of partial assignments, starting with the empty assignment, and assigning variables one at a time. One such method, known as *branch-and-bound*, is described in appendix A.4.3.

branch-and-bound

Other methods are nonsystematic, and they come without performance guarantees. Here, many of the methods search over the space of full assignments, usually by making local changes to the assignment so as to improve its score. These *local search* methods generally provide no guarantees of optimality. Appendix A.4.2 describes some of the techniques that are most commonly applied in practice.

local search

The application of search techniques to the MAP problem is a fairly straightforward process: The *search space* is defined by the possible assignments ξ to $\mathcal{X}$, and $\log \tilde{P}(\xi)$ is the score; we omit details. Although generally less powerful than the methods we described earlier, these methods do have some advantages. For example, the *beam search* method of appendix A.4.2 provides a useful alternative in cases where the complete model is too large to fit into memory; see exercise 15.10. We also note that branch-and-bound does provide a simple method for finding the K most likely assignment; see exercise 13.18. This algorithm requires at least as much computation time as the clique tree–based algorithm, but significantly less space.

search space

beam search

These methods have much greater applicability in the context of *marginal MAP* problem, where most other methods are not (currently) applicable. Here, we search over the space of assignments y to the max-variables Y. Here, we conduct the search so that we can fix some or all of the max-variables to have a concrete assignment. As we show, this allows us to remove the constraint on the variable elimination ordering, allowing an unrestricted ordering to be used.

marginal MAP

Here, we search over the space of assignments $\boldsymbol{y}$ for those that maximize

$$\mathrm{score}(\boldsymbol{y}) = \sum_{\boldsymbol{W}} \tilde{P}_\Phi(\boldsymbol{y}, \boldsymbol{W}). \qquad (13.35)$$

search operator

Several search procedures are appropriate in this setting. In one approach, we use some local search algorithm, as in appendix A.4.2. As usual in local search, the algorithm begins with some complete assignment $\boldsymbol{y}_0$ to $\boldsymbol{Y}$. We then consider applying different *search operators* to $\boldsymbol{y}$; for each such operator o, we produce a new partial assignment $\boldsymbol{y}' = o(\boldsymbol{y})$ as a successor to the current state, which is evaluated by computing $\mathrm{score}(\boldsymbol{y}')$. Importantly, since we now have a complete assignment to the max-variables $\boldsymbol{y}' = o(\boldsymbol{y})$, the resulting score is simply a sum-product expression, and it can be computed by standard sum-product elimination of the variables $\boldsymbol{W}$, with no constraints on the variable ordering. The tree-width in these cases is usually much smaller than in the constrained case; for example, in the network of figure 13.2, the network for a fixed assignment $\boldsymbol{y}'$ is simply a chain, and the computation of the score can therefore be done in time linear in n.

tabu search

While we can consider a variety of search operators, the most straightforward are operators of the form $do(Y_i = y_i^j)$, which set a variable $Y_i \in \boldsymbol{Y}$ to the value y_i^j. We can now apply any greedy local-search algorithm, such as those described in appendix A.4.2. Empirical evidence suggests that greedy hill climbing with *tabu search* performs very well on this task, especially if initialized intelligently. In particular, one simple yet good heuristic is to calibrate the clique tree with no assignment to the max-variables; we then compute, for each Y_i its unnormalized probability $\tilde{P}_\Phi(Y_i)$ (which can be extracted from any clique containing Y_i), and initialize $y_i = \arg\max_{Y_i} \tilde{P}_\Phi(Y_i)$.

While simple in principle, a naive implementation of this algorithm can be quite costly. Let $k = |\boldsymbol{Y}|$ and assume for simplicity that $|Val(Y_i)| = d$ for all $Y_i \in \boldsymbol{Y}$. Each step of the search requires $k \times (d - 1)$ evaluations of score, each of which involves a run of probabilistic inference over the network. Even for simple networks, this cost can often be prohibitive.

dynamic programming

Fortunately, we can greatly improve the computational performance of this algorithm using the same type of *dynamic programming* tricks that we used in other parts of this book. Most important is the observation that **we can compute the score of all of the operators in our search using a single run of clique tree propagation, in the clique tree corresponding to an unconstrained elimination ordering.** Let $\mathcal{T}$ be an *unconstrained* clique tree over $\mathcal{X} = \boldsymbol{Y} \cup \boldsymbol{W}$, initialized with the original potentials of $\tilde{P}_\Phi$. Let $\boldsymbol{y}$ be our current assignment to $\boldsymbol{Y}$. For any Y_i, let $\boldsymbol{Y}_{-i} = \boldsymbol{Y} - \{Y_i\}$ and $\boldsymbol{y}_{-i}$ be the assignment in $\boldsymbol{y}$ to $\boldsymbol{Y}_{-i}$. We can use the algorithm developed in exercise 10.12 to compute $\tilde{P}_\Phi(Y_i, \boldsymbol{y}_{-i})$ for every $Y_i \in \boldsymbol{Y}$. Recall that this algorithm requires only a single clique tree calibration that computes all of the messages; with those messages, each clique that contains a variable Y_i can locally compute $\tilde{P}_\Phi(Y_i, \boldsymbol{y}_{-i})$ in time that is linear in the size of the clique. This idea reduces the cost of each step by a factor of $O(kd)$, an enormous saving. For example, in the network of figure 13.2, we can use a clique tree whose cliques are of the form X_i, Y_{i+1}, X_{i+1}, with sepsets X_i between cliques. Here, the maximum clique size is 3, and the computation requires time linear in k.

We can also use search methods other than local hill climbing. One alternative is to utilize a systematic search procedure that is guaranteed to find the exact solution. Particularly well suited to this task is the branch-and-bound search described in appendix A.4.3. Recall that branch-and-bound systematically explores partial assignments to the variables $\boldsymbol{Y}$; it only discards a partial assignment $\boldsymbol{y}'$ if it already has a complete solution $\boldsymbol{y}$ that is provably better than the best possible solution that one can obtain by extending $\boldsymbol{y}'$ to a complete assignment. This pruning relies on having a way of estimating the upper bound on a partial assignment $\boldsymbol{y}'$. In our setting, such an upper bound can be obtained by using variable elimination, ignoring the constraint on the ordering whereby all summations occur before all maximizations. An algorithm based on these ideas is developed further in exercise 13.20.

13.8 Summary

In this chapter, we have considered the problem of finding the MAP assignment and described a number of methods for addressing it. The MAP problem has a broad range of applications, in computer vision, computational biology, speech recognition, and more. Although the use of MAP inference loses us the ability to measure our confidence (or uncertainty) in our conclusions, there are good reasons nevertheless for using a single MAP assignment rather than using the marginal probabilities of the different variables. One is the preference for obtaining a single coherent joint assignment, whereas a set of individual marginals may not make sense as a whole. The second is that there are inference methods that are applicable to the MAP problem and not to the task of computing probabilities, so that the former may be tractable even when the latter is not.

The methods we discussed fall into several major categories. The variable elimination method is very similar to the approaches we discussed in chapter 9, where we replace summation with maximization. The only slight extension is the traceback procedure, which allows us to identify the MAP assignment once the variable elimination process is complete.

Although one can view the max-product clique tree algorithm as a dynamic programming extension of variable elimination, it is more illuminating to view it as a method for reparameterizing the distribution to produce a max-calibrated set of beliefs. With this reparameterization, we can convert the global optimization problem — finding a coherent joint assignment — to a local optimization problem — finding a set of local assignments each of which optimizes its (calibrated) belief. Importantly, the same view also characterizes the cluster-graph-based belief propagation algorithms. The properties of max-calibrated beliefs allow us to prove strong (local or global) optimality properties for the results of these different message passing algorithms. In particular, for message passing with convex counting numbers we can sometimes construct an assignment that is the true MAP.

A seemingly very different class of methods is based on considering the integer program that directly encodes our optimization problem, and then constructing a relaxation as a linear program. Somewhat surprisingly, there is a deep connection between the convex max-product BP algorithm and the linear program relaxation. In particular, the solution to the dual problem of this LP is a fixed point of any convex max-product BP algorithm; thus, these algorithms can be viewed as a computational method for solving this dual problem. The use of these message passing methods offers a trade-off: they are space-efficient and easy to implement, but they may not converge to the optimum of the dual problem.

Importantly, the fixed point of a convex BP algorithm can be used to provide a MAP assignment only if the MAP LP is a tight relaxation of the integer MAP optimization problem. Thus, it appears that the LP relaxation is the fundamental construct in the application and analysis of the convex BP algorithms. This conclusion motivates two recent lines of work in MAP inference: One line attempts to construct tighter relaxations to the MAP optimization problem; importantly, since the same relaxation is used for both the free energy optimization in section 11.3.6 and for the MAP relaxations, progress made on improved relaxations for one task is directly useful for the other. The second line of work attempts to solve the LP or its dual using techniques other than message passing. While the problems are convex and hence can in principle be solved directly using standard techniques, the size of the problems makes the cost of this simple approach prohibitive in many practical applications. However, the rich and well-developed theory of convex optimization provides a wealth of potential tools, and some are already being adapted to take advantage of the structure of the MAP problem. It is likely that eventually these algorithms will replace convex BP as the method of choice for solving the dual. See section 13.9 for some references along those lines.

A different class of algorithms is based on reducing the MAP problem in pairwise, binary MRFs to one of finding the minimum cut in a graph. Although seemingly restrictive, this procedure forms a basic building block for solving a much broader class of MRFs. These methods provide an effective solution method only for MRFs where the potentials satisfy (or almost satisfy) the submodularity property. Conversely, their complexity depends fairly little on the complexity of the graph (the number of edges); as such, they allow certain MRFs to be solved efficiently that are not tractable to any other method. **Empirically, for energies that are close to submodular, the methods based on graph cuts are significantly faster than those based on message passing.** We note that in this case, also, there is an interesting connection to the linear programming view: The case that admits an optimal solution using minimum cut (pairwise, binary MRFs whose potentials are submodular) are also ones where there is a tight LP relaxation to the MAP problem. Thus, one can view the minimum-cut algorithm as a specialized method for exploiting special structure in the LP for solving it more efficiently.

In contrast to the huge volume of work on the MAP problem, relatively little work has been done on the marginal MAP problem. This lack is, in some sense, not surprising: the intrinsic difficulty of the problem is daunting and eliminates any hope of a general-purpose solution. Nevertheless, it would be interesting to see whether some of the recent algorithmic techniques developed for the MAP problem could be extended to apply to the marginal MAP case, leading to new solutions to the marginal MAP problem for at least a subset of MRFs.

13.9   Relevant Literature

We begin by reminding the reader, before tackling the literature, that there is a conflict of terminologies here: In some papers, the MAP problem is called MPE, whereas the marginal MAP problem is called simply MAP.

Viterbi algorithm

The problem of finding the MAP assignment in a probabilistic model was first addressed by Viterbi (1967), in the context of hidden Markov models; this algorithm came to be called the *Viterbi algorithm*. A generalization to other singly connected Bayesian networks was first proposed by Pearl (1988). The clique tree algorithm for this problem was described by Lauritzen and Spiegelhalter (1988). Shimony (1994) showed that the MAP problem is $\mathcal{NP}$-hard in general networks.

The problem of finding a MAP assignment to an MRF is equivalent (up to a negative-logarithm transformation) to the task of minimizing an energy function that is defined as a sum of terms, each involving a small number of variables. There is a considerably body of literature on the *energy minimization* problem, in both continuous and discrete space. Extensive work on energy minimization in MRFs has been done in the computer-vision community, where the locality of the spatial structure naturally defines a highly structured, often pairwise, MRF.

energy
minimization

iterated
conditional
modes

Early work on the energy minimization task focused on hill-climbing techniques, such as simple coordinate ascent (known under the name *iterated conditional modes* (Besag 1986)) or simulated annealing (Barnard 1989). Many other search methods for the MAP problem have been proposed, including systematic approaches such as branch-and-bound (Santos 1991; Marinescu et al. 2003).

The interest in max-product belief propagation on a loopy graph first arose in the context of turbo-decoding. The first general-purpose theoretical analysis for this approach was provided by Weiss and Freeman (2001b), who showed optimality properties of an assignment derived from an unambiguous set of beliefs reached at convergence of max-product BP. In particular, they showed that the assignment is the global optimum for networks involving only a single loop, and a strong local optimum (robust to changes in the assignments for a disjoint collection of single loops and trees) in general.

Wainwright, Jaakkola, and Willsky (2004) first proposed the view of message passing as reparameterizing the distribution so as to get the local beliefs to correspond to max-marginals. In subsequent work, Wainwright, Jaakkola, and Willsky (2005) developed the first convexified message passing algorithm for the MAP problem. The algorithm, known as TRW, used an approximation of the energy function based on a convex combination of trees. This paper was the first to show lemma 13.1. It also showed that if a fixed point of the TRW algorithm satisfied a stronger property than local optimality, it provided the MAP assignment. However, the TRW algorithm did not monotonically improve its objective, and indeed the algorithm was generally not convergent. Kolmogorov (2006) defined TRW-S, a variant of TRW that passes message asynchronously, in a particular order. TRW-S is guaranteed to increase the objective monotonically, and hence is convergent. However, TRW-S is not guaranteed to converge to the global optimum of the dual objective, since it can get stuck in local optima.

The connections between max-product BP, the lower-temperature limit of sum-product BP, and the linear programming relaxation were studied by Weiss, Yanover, and Meltzer (2007). They also showed results on the optimality of partial assignments extracted from unambiguous beliefs derived from convex BP fixed points, extending earlier results of Kolmogorov and Wainwright (2005) for TRW-S.

Max flow techniques to solve submodular binary problems were originally developed by Boros, Hammer and collaborators (Hammer 1965; Boros and Hammer 2002). These techniques were popularized in the vision-MRF community by Greig, Porteous, and Seheult (1989), who were the first to apply these techniques to images. Ishikawa (2003) extended this work to the nonbinary case, but assuming that the interaction between variables is convex. Boykov, Veksler, and Zabih (2001) were the first to propose the alpha-expansion and alpha-beta swap steps, which allow the application of graph-cut methods to nonbinary problems; they also prove certain guarantees regarding the energy of the assignment found by these global steps, relative to the energy of the optimal MAP assignment. Kolmogorov and Zabih (2004) generalized and analyzed the graph constructions used in these methods, using techniques similar to those described by Boros and Hammer (2002). Recent work extends the scope of the MRFs to which these techniques can be applied, by introducing preprocessing steps that modify factors that do not satisfy the submodularity assumptions. For example, Rother et al. (2005) consider a method that truncates the potentials that do not conform to submodularity, as part of the iterative alpha-expansion algorithm, and they show that this approach, although not making optimal alpha-expansion steps, is still guaranteed to improve the objective at each iteration. We note that, for the case of metric potentials, belief propagation algorithms such as TRW also do well (see box 13.B); moreover, Felzenszwalb and Huttenlocher (2006) show how the computational cost of each message passing step can be reduced from $O(K^2)$ to $O(K)$, where K is the total number of labels, reducing the cost of these algorithms in this setting.

Szeliski et al. (2008) perform an in-depth empirical comparison of the performance of different methods on an ensemble of computer vision benchmark problems. Other empirical comparisons include Meltzer et al. (2005); Kolmogorov and Rother (2006); Yanover et al. (2006).

The LP relaxation for MRFs was first proposed by Schlesinger (1976), and then subsequently rediscovered independently by several researchers. Of these, the most relevant to our presentation is the work of Wainwright, Jaakkola, and Willsky (2005), who also established the first connection between the LP dual and message passing algorithms, and proposed the TRW algorithm. Various extensions were subsequently proposed by various authors, based on different relaxations that require more complex convex optimization algorithms (Muramatsu and Suzuki 2003; Kumar et al. 2006; Ravikumar and Lafferty 2006). Surprisingly, Kumar et al. (2007) subsequently showed that the simple LP relaxation was tighter (that is, better) relaxation than all of those more sophisticated methods.

A spate of recent works (Komodakis et al. 2007; Schlesinger and Giginyak 2007a,b; Sontag and Jaakkola 2007; Globerson and Jaakkola 2007b; Werner 2007; Sontag et al. 2008) make much deeper use of the linear programming relaxation of the MAP problem and of its dual. Globerson and Jaakkola (2007b); Komodakis et al. (2007) both demonstrate a message passing algorithm derived from this dual. The algorithm of Komodakis et al. is based on a dual decomposition algorithm, and is therefore guaranteed to converge to the optimum of the dual objective. Solving the LP relaxation or its dual does not generally give rise to the optimal MAP assignment. The work of Sontag and Jaakkola (2007); Sontag et al. (2008) shows how we can use the LP formulation to gradually add local constraints that hold for any set of pseudo-marginals defined by a real distribution. These constraints make the optimization space a tighter relaxation of the marginal polytope and thereby lead to improved approximations. Sontag et al. present empirical results that show that a small number of constraints often suffice to define the optimal MAP assignment.

Komodakis and colleagues 2005; 2007 also make use of LP duality in the context of graph cut methods, where it corresponds to the well-known duality between min-cut and max-flow. They use this approach to derive primal-dual methods that speed up and extend the alpha-expansion method in several ways.

Santos (1991, 1994) studied the question of finding the M most likely assignments. He presented an exact algorithm that uses the linear programming relaxation of the integer program, augmented with a branch-and-bound search that uses the LP as the bound. Nilsson (1998) provides an alternative algorithm that uses propagation in clique trees. Yanover and Weiss (2003) subsequently generalized this algorithm for the case of loopy BP.

Park and Darwiche extensively studied the marginal MAP problem, providing complexity results (Park 2002; Park and Darwiche 2001), local search algorithms (Park and Darwiche 2004a) (including an efficient clique tree implementation), and a systematic branch-and-bound algorithm (Park and Darwiche 2003) based on the bound obtained by exchanging summation and maximization.

The study of constraint satisfaction problems, and related problems such as Boolean satisfiability (see appendix A.3.4) is the focus of a thriving research community, and much progress has been made. One recent overview can be found in the textbook of Dechter (2003). There has been a growing interest recently in relating CSP methods to belief propagation techniques, most

survey
propagation

notably the *survey propagation* (for example, (Maneva et al. 2007)).

13.10 Exercises

Exercise 13.1⋆⋆

Prove theorem 13.1.

Exercise 13.2⋆

Provide a structured variable elimination algorithm that solves the MAP task for networks with rule-based CPDs.

a. Modify the algorithm Rule-Sum-Product-Eliminate-Var in algorithm 9.7 to deal with the max-product task.

b. Show how we can perform the backward phase that constructs the most likely assignment to $\mathcal{X}$. Make sure you describe which information needs to be stored in the forward phase so as to enable the backward phase.

Exercise 13.3

Prove theorem 13.4.

Exercise 13.4

Show how to adapt Traceback-MAP of algorithm 13.1 to find the marginal MAP assignment, given the factors computed by a run of variable elimination for marginal MAP.

Exercise 13.5⋆

Consider the task of finding the second-most-likely assignment in a graphical model. Assume that we have produced a max-calibrated clique tree.

a. Assume that the probabilistic model is unambiguous. Show how we can find the second-best assignment using a single pass over the clique tree.

b. Now answer the same question in the case where the probabilistic model is ambiguous. Your method should use only the precomputed max-marginals.

Exercise 13.6⋆

Now, consider the task of finding the third-most-likely assignment in a graphical model. Finding the third-most-probable assignment is more complicated, since it cannot be computed from max-marginals alone.

a. We define the notion of *constrained max-marginal*: a max-marginal in a distribution that has some variable X_k constrained to take on only certain values. For $D_k \subset Val(X_k)$, we define the constrained max-marginal of X_i to be:

$$MaxMarg_{\tilde{P}_{X_k \in D_k}}(X_i = x_i) = \max_{\{\boldsymbol{x}: X_i = x_i, X_k \in D_k\}} \tilde{P}(\boldsymbol{x}).$$

Explain how to compute the preceding constrained max-marginals for all i and x_i using max-product message passing.

b. Find the third-most-probable assignment by using two sets of constrained max-marginals.

Exercise 13.7

Prove proposition 13.1.

Exercise 13.8

Prove proposition 13.3.

Exercise 13.9

Assume that max-product belief propagation converges to a set of calibrated beliefs $\beta_i(C_i)$. Assume that each belief is unambiguous, so that it has a unique maximizing assignment c_i^*. Prove that all of these locally optimizing assignments are consistent with each other, in that if $X_k = x_k^*$ in one assignment c_i^*, then $X_k = x_k^*$ in every other assignment c_j^* for which $X_k \in C_j$.

Exercise 13.10

Construct an example of a max-product calibrated cluster graph in which (at least) some beliefs have two locally optimal assignments, such that one local assignment can be extended into a globally consistent joint assignment (across all beliefs), and the other cannot.

Exercise 13.11★

Consider a cluster graph U that contains only a single loop, and assume that we have a set of max-product calibrated beliefs $\{\beta_i\}$ for U and an assignment ξ^* that is locally optimal relative to $\{\beta_i\}$. Prove that ξ^* is the MAP assignment relative to the distribution P_U. (Hint: Use lemma 13.1 and a proof similar to that of theorem 13.6.)

Exercise 13.12

Using exercise 13.11, complete the proof of theorem 13.6. First prove the result for sets Z for which U_Z contains only a single loop. Then prove the result for any Z for which U_Z is a combination of disconnected trees and loops.

Exercise 13.13

Prove proposition 13.4.

Exercise 13.14

Show that the algorithm in algorithm 13.4 returns the correct MAP assignment. First show that for any cut $C = Z_s, Z_t$, we have that

$$cost(C) = E(\xi^C) + \text{Const}.$$

Conclude the desired result.

Exercise 13.15★

Show how the optimal alpha-beta swap step can be found by running min-cut on an appropriately constructed graph. More precisely:

a. Define a set of binary variables $t_1, \ldots, t_n$, such that the value of the t_i's defines an alpha-beta-swap transformation on the x_i's.
b. Define an energy function E' over the T variables such that $E'(t) = E(x')$.
c. Show that the energy function E' is submodular if the original energy function E is a semimetric.

Exercise 13.16★

truncation

As we discussed, many energy functions are not submodular. We now describe a method that allows min-cut methods to be applied to energy functions where most of the terms are submodular, but some small subset is not submodular. This method is based on the *truncation* of the nonsubmodular potentials, so as to make them submodular.

Algorithm 13.6 Efficient min-sum message passing for untruncated 1-norm energies

   Procedure Msg-Truncated-1-Norm (
       c // Parameters defining the pairwise factor
       $h_i(x_i)$ // Single-variable term in equation (13.36)
)
1   **for** $x_j = 1, \ldots, K - 1$
2         $r(x_j) \leftarrow \min[h_i(x_j), r(x_j - 1) + c]$
3   **for** $x_j = K - 2, \ldots, 0$
4         $r(x_j) \leftarrow \min[r(x_j), r(x_j + 1) + c]$
5   **return** (r)

---

a. Let E be an energy function over binary-valued variables that contains some number of pairwise terms $\epsilon_{i,j}(v_i, v_j)$ that do not satisfy equation (13.33). Assume that we replace each such pairwise term $\epsilon_{i,j}$ with a term $\epsilon'_{i,j}$ that satisfies this inequality, by decreasing $\epsilon_{i,j}(0,0)$, by increasing $\epsilon_{i,j}(1,0)$ or $\epsilon_{i,j}(0,1)$, or both. The node energies remain unchanged. Let E' be the resulting energy.
   Show that if ξ^* optimizes E', then $E(\xi^*) \leq E(\mathbf{0})$

b. Describe how, in the multilabel case, this procedure can be used within the alpha-expansion algorithm to find a local optimum of the energy function.

Exercise 13.17★

Consider the task of passing a message over an edge X_i—X_j in a metric MRF; our goal is to make the message passing step more efficient by exploiting the metric structure. As usual in metric MRFs, we consider the problem in terms of energies; thus, our message computation takes the form:

$$\delta_{i \to j}(x_j) = \min_{x_i}(\epsilon_{i,j}(x_i, x_j) + h_i(x_i)), \qquad (13.36)$$

where $h_i(x_i) = \epsilon_i(x_i) + \sum_{k \neq j} \delta_{i \to j}(x_k)$. In general, this computation requires $O(K^2)$ steps. However, we now consider two special cases where this computation can be done in $O(K)$ steps.

a. Assume that $\epsilon_{i,j}(x_i, x_j)$ is an Ising energy function, as in equation (4.6). Show how the message can be computed in $O(K)$ steps.

b. Now assume that both X_i, X_j take on values in $\{0, \ldots, K - 1\}$. Assume that $\epsilon_{i,j}(x_i, x_j)$ is a nontruncated 1-norm, as in equation (4.7) with $p = 1$ and $\text{dist}_{\max} = \infty$. Show that the algorithm in algorithm 13.6 computes the correct message in $O(K)$ steps.

c. Extend the algorithm of algorithm 13.6 to the case of a truncated 1-norm (where $\text{dist}_{\max} < \infty$).

Exercise 13.18★

Consider the use of the branch-and-bound algorithm of appendix A.4.3 for finding the top K highest-probability assignments in an (unnormalized) distribution $\tilde{P}_\Phi$ defined by a set of factors Φ.

a. Consider a partial assignment $\mathbf{y}$ to some set of variables $\mathbf{Y}$. Provide both an upper and a lower bound to $\log \tilde{P}_\Phi(\mathbf{y})$.

b. Describe how to use your bounds in the context of a branch-and-bound algorithm to find the MAP assignment for $\tilde{P}_\Phi$. Can you use both the lower and upper bounds in your search?

c. Extend your algorithm to find the K highest probability joint assignments in $\tilde{P}_\Phi$. Hint: Your algorithm should find the assignments in order of decreasing probability, starting with the MAP. Be sure to reuse as much of your previous computations as possible as you continue the search for the next assignment.

Exercise 13.19

Show that, for any function f,

$$\max_x \sum_y f(x, y) \leq \sum_y \max_x f(x, y), \tag{13.37}$$

and provide necessary and sufficient conditions for when equation (13.37) holds as equality.

Exercise 13.20★

a. Use equation (13.37) to provide an efficient algorithm for computing an upper bound

$$\text{bound}(\boldsymbol{y}_{1\ldots i}) = \max_{y_{i+1}, \ldots, y_n} \text{score}(\boldsymbol{y}_{1\ldots i}, y_{i+1}, \ldots, y_n),$$

where $\text{score}(\boldsymbol{y})$ is defined as in equation (13.35). Your computation of the bound should take no more than a run of variable elimination in an *unconstrained* elimination ordering over all of the network variables.

b. Use this bound to construct a branch-and-bound algorithm for the marginal-MAP problem.

Exercise 13.21★

In this question, we consider the application of conditioning to a marginal MAP query:

$$\arg\max_{\boldsymbol{Y}} \sum_{\boldsymbol{Z}} \prod_{\phi \in \Phi} \phi.$$

Let $\boldsymbol{U}$ be a set of conditioning variables.

a. Consider first the case of a simple MAP query, so that $\boldsymbol{Z} = \emptyset$ and $\boldsymbol{Y} = \mathcal{X}$. Show how you would adapt Conditioning in algorithm 9.5 to deal with the max-product rather than the sum-product task.

b. Now, consider a max-sum-product task. When is $\boldsymbol{U}$ a legal set of conditioning variables for this query? Justify your response. (Hint: Recall that the order of the operations we perform must respect the ordering constraint discussed in section 2.1.5, and that the elimination operations work from the outside in, and the conditioning operations from the inside out.)

c. Now, assuming that $\boldsymbol{U}$ is a legal set of conditioning variables, specify a conditioning algorithm that computes the value of the corresponding max-sum-product query, as in equation (13.8).

d. Extend your max-sum-product algorithm to compute the actual maximizing assignment to $\boldsymbol{Y}$, as in the MAP query. Your algorithm should work for any legal conditioning set $\boldsymbol{U}$.

14 *Inference in Hybrid Networks*

In our discussion of inference so far, we have focused on the case of discrete probabilistic models. However, many interesting domains also contain continuous variables such as temperature, location, or distance. In this chapter, we address the task of inference in graphical models that involve such variables.

For this chapter, let $\mathcal{X} = \Gamma \cup \Delta$, where Γ denotes the continuous variables and Δ the discrete variables. In cases where we wish to distinguish discrete and continuous variables, we use the convention that discrete variables are named with letters near the beginning of the alphabet (A, B, C), whereas continuous ones are named with letters near the end (X, Y, Z).

14.1 Introduction

14.1.1 Challenges

At an abstract level, the introduction of continuous variables in a graphical model is not difficult. As we saw in section 5.5, we can use a range of different representations for the CPDs or factors in our network. We now have a set of factors, over which we can perform the same operations that we utilize for inference in the discrete case: We can multiply factors, which in this case corresponds to multiplying the multidimensional continuous functions representing the factors; and we can marginalize out variables in a factor, which in this case is done using integration rather than summation. It is not difficult to show that, with these operations in hand, the sum-product inference algorithms that we used in the discrete case can be applied without change, and are guaranteed to lead to correct answers.

Unfortunately, a little more thought reveals that the correct implementation of these basic operations poses a range of challenges, whose solution is far from obvious.

The first challenge involves the representation of factors involving continuous variables. Unlike discrete variables, there is no universal representation of a factor over continuous variables, and so we must usually select a parametric family for each CPD or initial factor in our network. Even if we pick the same parametric family for each of our initial factor in the network, it may not be the case that multiplying factors or marginalizing a factor leaves it within the parametric family. If not, then it is not even clear how we would represent the intermediate results in our inference process. The situation becomes even more complex when factors in the original network call for the use of different parametric families. In this case, **it is generally unlikely that we can find a single parametric family that can correctly encode all of the intermediate factors**

in our network. In fact, in some cases — most notably networks involving both discrete and continuous variables — one can show that the intermediate factors cannot be represented using any fixed number of parameters; in fact, the representation size of those factors grows exponentially with the size of the network.

A second challenge involves the marginalization step, which now requires integration rather than summation. Integration introduces a new set of subtleties. First, not all functions are integrable: in some cases, the integral may be infinite or even ill defined. Second, even functions where the integral is well defined may not have a closed-form integral, requiring the use of a numerical integration method, which is usually approximate.

14.1.2   Discretization

An alternative approach to inference in hybrid models is to circumvent the entire problem of dealing with continuous factors: We simply convert all continuous variable to discrete ones by *discretizing* their domain into some finite set of intervals. Once all variables have been discretized, the result is a standard discrete probabilistic model, which we can handle using the standard inference techniques described in the preceding chapters.

discretization

How do we convert a hybrid CPDs into a table? Assume that we have a variable Y with a continuous parent X. Let A be the discrete variable that replaces X and B the discrete variable that replaces Y. Let $a \in Val(A)$ correspond to the interval $[x^1, x^2]$ for X, and $b \in Val(B)$ correspond to the interval $[y^1, y^2]$ for Y.

In principle, one approach for discretization is to define

$$P(b \mid a) = \int_{x^1}^{x^2} p(Y \in [y^1, y^2] \mid X = x) p(X = x \mid X \in [x^1, x^2]) dx.$$

This integral averages out the conditional probability that Y is in the interval $[y^1, y^2]$ given X, aggregating over the different values of x in the relevant interval for X. The distribution used in this formulation is the prior probability $p(X)$, which has the effect of weighting the average more toward more likely values of X. While plausible, this computation is expensive, since it requires that we perform inference in the model. Moreover, even if we perform our estimation relative to the prior $p(X)$, we have no guarantees of a good approximation, since our *posterior* over X may be quite different.

Therefore, for simplicity, we often use simpler approximations. In one, we simply select some particular value $x^* \in [x^1, x^2]$, and estimate $P(b \mid a)$ as the total probability mass of the interval $[y^1, y^2]$ given x^*: :

$$P(Y \in [y^1, y^2] \mid x^*) = \int_{y^1}^{y^2} p(y \mid x^*) dy.$$

For some density functions p, we can compute this interval in closed form. In others, we might have to resort to numerical integration methods. Alternatively, we can average the values over the interval $[x^1, x^2]$ using a predefined distribution, such as the uniform distribution over the interval.

Although discretization is used very often in practice, as we discussed in section 5.5, it has several significant limitations. The discretization is only an approximation of the true probability distribution. In order to get answers that do not lose most of the information, our discretization scheme must have a fine resolution where the posterior probability mass lies. Unfortunately, before we actually perform the inference, we do not know the posterior distribution. Thus, we must often resort to a discretization that is fairly fine-grained over the entire space, leading to a very large domain for the resulting discrete variable.

This problem is particularly serious when we need to approximate distributions over more than a handful of discretized variables. As in any table-based CPD, the size of the resulting factor is exponential in the number of variables. When this number is large and the discretization is anything but trivial, the size of the factor can be huge. For example, if we need to represent a distribution over d continuous variables, each of which is discretized into m values, the total number of parameters required is $O(m^d)$. By contrast, if the distribution is a d-dimensional Gaussian, the number of parameters required is only $O(d^2)$. Thus, not only does the discretization process introduce approximations into our joint probability distribution, but we also often end up converting a polynomial parameterization into an exponential one.

 Overall, discretization provides a trade-off between accuracy of the approximation and cost of computation. In certain cases, acceptable accuracy can be obtained at reasonable cost. However, in many practical applications, the computational cost required to obtain the accuracy needed for the task is prohibitive, making discretization a nonviable option.

14.1.3 Overview

Thus, we see that inference in continuous and hybrid models, although similar in principle to discrete inference, brings forth a new set of challenges. In this chapter, we discuss some of these issues and show how, in certain settings, these challenges can be addressed.

As for inference in discrete networks, the bulk of the work on inference in continuous and hybrid networks falls largely into two categories: Approaches that are based on message passing methods, and approaches that use one of the particle-based methods discussed in chapter 12. However, unlike the discrete case, even the message passing algorithms are rarely exact.

The message passing inference methods have largely revolved around the use of the Gaussian distribution. The easiest case is when the distribution is, in fact, a multivariate Gaussian. In this case, many of the challenges described before disappear. In particular, the intermediate factors in a Gaussian network can be described compactly using a simple parametric representation called the *canonical form*. This representation is closed under the basic operations using in inference: factor product, factor division, factor reduction, and marginalization. Thus, we can define a set of simple data structures that allow the inference process to be performed. Moreover, the integration operation required by marginalization is always well defined, and it is guaranteed to produce a finite integral under certain conditions; when it is well defined, it has a simple analytical solution.

As a consequence, a fairly straightforward modification of the discrete sum-product algorithms (whether variable elimination or clique tree) gives rise to an exact inference algorithm for Gaussian networks. A similar extension results in a Gaussian version of the loopy belief propagation algorithm; here, however, the conditions on integrability impose certain constraints about the form of the distribution. Importantly, under these conditions, loopy belief propagation for Gaussian distributions is guaranteed to return the correct means for the variables in the network, although it can underestimate the variances, leading to overconfident estimates.

There are two main extensions to the purely Gaussian case: non-Gaussian continuous densities, and hybrid models that involve both discrete and continuous variables. The most common method for dealing with these extensions is the same: we approximate intermediate factors in the computation as Gaussians; in effect, these algorithms are an instance of the expectation propagation algorithm discussed in section 11.4.4. As we discuss, Gaussians provide a good basis for the basic operations in these algorithms, including the key operation of approximate marginalization. Interestingly, there is one class of inference tasks where this general algorithm is guaranteed to produce exact answers to a certain subset of queries. This is the class of CLG networks in which we use a particular form of clique tree for the inference. Unfortunately, although of conceptual interest, this "exact" variant is rarely useful except in fairly small problems.

An alternative approach is to use an approximation method that makes no parametric assumptions about the distribution. Specifically, we can approximate the distribution as a set of particles, as described in chapter 12. As we will see, particle-based methods often provide the easiest approach to inference in a hybrid network. They make almost no assumptions about the form of the CPDs, and can approximate an arbitrarily complex posterior. Their primary disadvantage is, as usual, the fact that a very large number of particles might be required for a good approximation.

14.2   Variable Elimination in Gaussian Networks

The first class of networks we consider is the class of Gaussian networks, as described in chapter 7: These are networks where all of the variables are continuous, and all of the local factors encode linear dependencies. In the case of Bayesian networks, the CPDs take the form of linear Gaussians (as in definition 5.15). In the case of Markov networks, they can take the form of general log-quadratic form, as in equation (7.7).

As we showed in chapter 7, both of these representations are simply alternative parameterizations of a joint multivariate Gaussian distributions. This observation immediately suggests one approach to performing exact inference in this class of networks: We simply convert the LG network into the equivalent multivariate Gaussian, and perform the necessary operations — marginalization and conditioning — on that representation. Specifically, as we discussed, if we have a Gaussian distribution $p(X, Y)$ represented as a mean vector and a covariance matrix, we can extract the marginal distribution $p(Y)$ simply by restricting attention to the elements of the mean and the covariance matrix that correspond to the variables in Y. The operation of conditioning a Gaussian on evidence $Y = y$ is also easy: we simply instantiate the variables Y to their observed values y in the joint density function, and renormalize the resulting unnormalized measure over X to obtain a new Gaussian density.

This approach simply generates the joint distribution over the entire set of variables in the network, and then manipulates it directly. However, unlike the case of discrete distributions, the representation size of the joint density in the Gaussian case is quadratic, rather than exponential, in the number of variables. Thus, these operations are often feasible in a Gaussian network in cases that would not be feasible in a comparable discrete network.

Still, even quadratic cost might not be feasible in many cases, for example, when the network is over thousands of variables. Furthermore, this approach does not exploit any of the structure represented in the network distribution. An alternative approach to inference is to adapt the message passing algorithms, such as variable elimination (or clique trees) for exact inference, or belief propagation for approximation inference, to the linear Gaussian case. We now describe this approach. We begin with describing the basic representation used for these message passing schemes, and then present these two classes of algorithms.

14.2.1 Canonical Forms

As we discussed, the key difference between inference in the continuous and the discrete case is that the factors can no longer be represented as tables. Naively, we might think that we can represent factors as Gaussians, but this is not the case. The reason is that linear Gaussian CPDs are generally not Gaussians, but are rather a conditional distribution. Thus, we need to find a more general representation for factors, that accommodates both Gaussian distributions and linear Gaussian models, as well as any combination of these models that might arise during the course of inference.

14.2.1.1 The Canonical Form Representation

The simplest representation used in this setting is the *canonical form*, which represents the intermediate result as a log-quadratic form $\exp(Q(\boldsymbol{x}))$ where Q is some quadratic function. In the inference setting, it is useful to make the components of this representation more explicit:

Definition 14.1

canonical form

A canonical form $\mathcal{C}\,(\boldsymbol{X};K,\boldsymbol{h},g)$ *(or $\mathcal{C}\,(K,\boldsymbol{h},g)$ if we omit the explicit reference to $\boldsymbol{X}$) is defined as:*

$$\mathcal{C}\,(\boldsymbol{X};K,\boldsymbol{h},g) = \exp\left(-\frac{1}{2}\boldsymbol{X}^T K \boldsymbol{X} + \boldsymbol{h}^T \boldsymbol{X} + g\right). \tag{14.1}$$

∎

We can represent every Gaussian as a canonical form. Rewriting equation (7.1), we obtain:

$$\frac{1}{(2\pi)^{n/2}|\Sigma|^{1/2}} \exp\left(-\frac{1}{2}(\boldsymbol{x}-\boldsymbol{\mu})^T \Sigma^{-1}(\boldsymbol{x}-\boldsymbol{\mu})\right)$$

$$= \exp\left(-\frac{1}{2}\boldsymbol{x}^T \Sigma^{-1}\boldsymbol{x} + \boldsymbol{\mu}^T \Sigma^{-1}\boldsymbol{x} - \frac{1}{2}\boldsymbol{\mu}^T \Sigma^{-1}\boldsymbol{\mu} - \log\left((2\pi)^{n/2}|\Sigma|^{1/2}\right)\right).$$

Thus, $\mathcal{N}\,(\boldsymbol{\mu};\Sigma) = \mathcal{C}\,(K,\boldsymbol{h},g)$ where:

$$K = \Sigma^{-1}$$
$$\boldsymbol{h} = \Sigma^{-1}\boldsymbol{\mu}$$
$$g = -\frac{1}{2}\boldsymbol{\mu}^T \Sigma^{-1}\boldsymbol{\mu} - \log\left((2\pi)^{n/2}|\Sigma|^{1/2}\right).$$

However, canonical forms are more general than Gaussians: If K is not invertible, the canonical form is well defined, but it is not the inverse of a legal covariance matrix. In particular, we can easily represent linear Gaussian CPDs as canonical forms (exercise 14.1).

14.2.1.2   Operations on Canonical Forms

canonical form
product

It is possible to perform various operations on canonical forms. The *product* of two canonical form factors over the same scope $\boldsymbol{X}$ is simply:

$$\mathcal{C}\left(K_1, \boldsymbol{h}_1, g_1\right) \cdot \mathcal{C}\left(K_2, \boldsymbol{h}_2, g_2\right) = \mathcal{C}\left(K_1 + K_2, \boldsymbol{h}_1 + \boldsymbol{h}_2, g_1 + g_2\right). \tag{14.2}$$

When we have two canonical factors over different scopes $\boldsymbol{X}$ and $\boldsymbol{Y}$, we simply extend the scope of both to make their scopes match and then perform the operation of equation (14.2). The extension of the scope is performed by simply adding zero entries to both the K matrices and the $\boldsymbol{h}$ vectors.

Example 14.1

Consider the following two canonical forms:

$$\phi_1(X, Y) \;=\; \mathcal{C}\left(X, Y; \begin{bmatrix} 1 & -1 \\ -1 & 1 \end{bmatrix}, \begin{pmatrix} 1 \\ -1 \end{pmatrix}, -3\right)$$

$$\phi_2(Y, Z) \;=\; \mathcal{C}\left(Y, Z; \begin{bmatrix} 3 & -2 \\ -2 & 4 \end{bmatrix}, \begin{pmatrix} 5 \\ -1 \end{pmatrix}, 1\right).$$

We can extend the scope of both of these by simply introducing zeros into the canonical form. For example, we can reformulate:

$$\phi_1(X, Y, Z) = \mathcal{C}\left(X, Y, Z; \begin{bmatrix} 1 & -1 & 0 \\ -1 & 1 & 0 \\ 0 & 0 & 0 \end{bmatrix}, \begin{pmatrix} 1 \\ -1 \\ 0 \end{pmatrix}, -3\right),$$

and similarly for $\phi_2(X, Y, Z)$. The two canonical forms now have the same scope, and can be multiplied using equation (14.2) to produce:

$$\mathcal{C}\left(X, Y, Z; \begin{bmatrix} 1 & -1 & 0 \\ -1 & 4 & -2 \\ 0 & -2 & 4 \end{bmatrix}, \begin{pmatrix} 1 \\ 4 \\ -1 \end{pmatrix}, -2\right).$$ ∎

canonical form
division

The *division* of canonical forms (which is required for message passing in the belief propagation algorithm) is defined analogously:

$$\frac{\mathcal{C}\left(K_1, \boldsymbol{h}_1, g_1\right)}{\mathcal{C}\left(K_2, \boldsymbol{h}_2, g_2\right)} = \mathcal{C}\left(K_1 - K_2, \boldsymbol{h}_1 - \boldsymbol{h}_2, g_1 - g_2\right). \tag{14.3}$$

vacuous
canonical form

Note that the *vacuous canonical form*, which is the analogue of the "all 1" factor in the discrete case, is defined as $K = 0$, $\boldsymbol{h} = \boldsymbol{0}$, $g = 0$. Multiplying or dividing by this factor has no effect.

canonical form
marginalization

Less obviously, we can *marginalize* a canonical form onto a subset of its variables. Let $\mathcal{C}\left(\boldsymbol{X}, \boldsymbol{Y}; K, \boldsymbol{h}, g\right)$ be some canonical form over $\{\boldsymbol{X}, \boldsymbol{Y}\}$ where

$$K = \begin{bmatrix} K_{\boldsymbol{XX}} & K_{\boldsymbol{XY}} \\ K_{\boldsymbol{YX}} & K_{\boldsymbol{YY}} \end{bmatrix} \quad ; \quad h = \begin{pmatrix} h_{\boldsymbol{X}} \\ h_{\boldsymbol{Y}} \end{pmatrix}. \tag{14.4}$$

The marginalization of this function onto the variables $\boldsymbol{X}$ is, as usual, the integral over the variables $\boldsymbol{Y}$:

$$\int \mathcal{C}\left(\boldsymbol{X}, \boldsymbol{Y}; K, \boldsymbol{h}, g\right) d\boldsymbol{Y}.$$

As we discussed, we have to guarantee that all of the integrals resulting from marginalization operations are well defined. In the case of canonical forms, the integral is finite if and only if K_{YY} is positive definite, or equivalently, that it is the inverse of a legal covariance matrix. In this case, the result of the integration operation is a canonical form $\mathcal{C}\left(X; K', h', g'\right)$ given by:

$$
\begin{aligned}
K' &= K_{XX} - K_{XY} K_{YY}^{-1} K_{YX} \\
h' &= h_X - K_{XY} K_{YY}^{-1} h_Y \\
g' &= g + \tfrac{1}{2}\left(\log |2\pi K_{YY}^{-1}| + h_Y^T K_{YY}^{-1} h_Y\right).
\end{aligned}
\tag{14.5}
$$

canonical form reduction   Finally, it is possible to *reduce* a canonical form to a context representing evidence. Assume that the canonical form $\mathcal{C}\left(X, Y; K, h, g\right)$ is given by equation (14.4). Then setting $Y = y$ results in the canonical form $\mathcal{C}\left(X; K', h', g'\right)$ given by:

$$
\begin{aligned}
K' &= K_{XX} \\
h' &= h_X - K_{XY} y \\
g' &= g + h_Y^T y - \frac{1}{2} y^T K_{YY} y.
\end{aligned}
\tag{14.6}
$$

See exercise 14.3.

Importantly, all of the factor operations can be done in time that is polynomial in the scope of the factor. In particular, the product or division of factors requires quadratic time; factor marginalization, which requires matrix inversion, can be done naively in cubic time, and more efficiently using advanced methods.

14.2.2   Sum-Product Algorithms

sum-product   The operations described earlier are the basic building blocks for all of our *sum-product* exact inference algorithms: variable elimination and both types of clique tree algorithms. Thus, we can adapt these algorithms to apply to linear Gaussian networks, using canonical forms as our representation of factors. For example, in the Sum-Product-VE algorithm of algorithm 9.1, we simply implement the factor product operation as in equation (14.2), and replace the summation operation in Sum-Product-Eliminate-Var with an integration operation, implemented as in equation (14.5).

Care must be taken regarding the treatment of evidence. In discrete factors, when instantiating a variable $Z = z$, we could leave the variable Z in the factors involving it, simply zeroing the entries that are not consistent with $Z = z$. In the case of continuous variables, our representation of factors does not allow that option: when we instantiate $Z = z$, the variable Z is no longer part of the canonical form. Thus, it is necessary to reduce all the factors participating in the inference process to a scope that no longer contains Z. This reduction step is already part of the variable elimination algorithm of algorithm 9.2. It is straightforward to ensure that the clique tree algorithms of chapter 10 similarly reduce all clique and sepset potentials with the evidence prior to any message passing steps.

A more important problem that we must consider is that the marginalization operation may not be well defined for an arbitrary canonical form. In order to show the correctness of an inference algorithm, we must show that it executes a marginalization step only on canonical **well-defined marginalization**   forms for which this operation is *well defined*. We prove this result in the context of the sum-product clique tree algorithm; the proof for the other cases follows in a straightforward way, due to the equivalence between the upward pass of the different message passing algorithms.

Proposition 14.1

Whenever SP-Message *is called, within the* CTree-SP-Upward *algorithm (algorithm 10.1) the marginalization operation is well-defined.*

Proof Consider a call SP-Message(i, j), and let $\psi(C_i)$ be the factor constructed in the clique prior to sending the message. Let $\mathcal{C}\left(C_i; K, h, g\right)$ be the canonical form associated with $\psi(C_i)$. Let $\beta_i(C_i) = \mathcal{C}\left(C_i; K', h', g'\right)$ be the final clique potential that would be obtained at C_i in the case where C_i is the root of the clique tree computation. The only difference between these two potentials is that the latter also incorporates the message $\delta_{j \to i}$ from C_j.

Let $Y = C_j - S_{i,j}$ be the variables that are marginalized when the message is computed. By the running intersection property, none of the variables Y appear in the scope of the sepset $S_{i,j}$. Thus, the message $\delta_{j \to i}$ does not mention any of the variables Y. We can verify, by examining equation (14.2), that multiplying a canonical form by a factor that does not mention Y does not change the entries in the matrix K that are associated with the variables in Y. It follows that $K_{YY} = K'_{YY}$, that is, the submatrices for Y in K and K' are the same. Because the final clique potential $\beta_i(C_i)$ is its (unnormalized) marginal posterior, it is a normalizable Gaussian distribution, and hence the matrix K' is positive definite. As a consequence, the submatrix K'_{YY} is also positive definite. It follows that K_{YY} is positive definite, and therefore the marginalization operation is well defined. ∎

 It follows that we can adapt any of our exact inference algorithms to the case of linear Gaussian networks. The algorithms are essentially unchanged; only the representation of factors and the implementation of the basic factor operations are different. In particular, since all factor operations can be done in polynomial time, inference in linear Gaussian networks is linear in the number of cliques, and at most cubic in the size of the largest clique. By comparison, recall that the representation of table factors is, by itself, exponential in the scope, leading to the exponential complexity of inference in discrete networks.

It is interesting to compare the clique tree inference algorithm to the naive approach of simply generating the joint Gaussian distribution and marginalizing it. The exact inference algorithm requires multiple steps, each of which involves matrix product and inversion. By comparison, the joint distribution can be computed, as discussed in theorem 7.3, by a set of vector-matrix products, and the marginalization of a joint Gaussian over any subset of variables is trivial (as in lemma 7.1). Thus, in cases where the Gaussian has sufficiently low dimension, it may be less computationally intensive to use the naive approach for inference. Conversely, in cases where the distribution has high dimension and the network has reasonably low tree-width, the message passing algorithms can offer considerable savings.

14.2.3 Gaussian Belief Propagation

Gaussian belief propagation

The *Gaussian belief propagation* algorithm utilizes the information form, or canonical form, of the Gaussian distribution. As we discussed, a Gaussian network is encoded using a set of local quadratic potentials, as in equation (14.1). Reducing a canonical-form factor on evidence also results in a canonical-form factor (as in equation (14.6)), and so we can focus attention on a representation that consists of a product of canonical-form factors. This product results in an overall quadratic form:

$$p(X_1, \ldots, X_n) \propto \exp\left(-\frac{1}{2}\boldsymbol{X}^T J \boldsymbol{X} + \boldsymbol{h}^T \boldsymbol{X}\right).$$

The measure p is normalizable and defines a legal Gaussian distribution if and only if J is positive definite. Note that we can obtain J by adding together the individual matrices K defined by the various potentials parameterizing the network.

In order to apply the belief propagation algorithm, we must define a cluster graph and assign the components of this parameterization to the different clusters in the graph. As in any belief propagation algorithm, we need the cluster graph to respect the family preservation property. In our setting, the only terms in the quadratic form involve single variables — the $h_i X_i$ terms — and pairs of variables X_i, X_j for which $J_{ij} \neq 0$. Thus, the minimal cluster graph that satisfies the family preservation requirement would contain a cluster for each edge X_i, X_j (pairs for which $J_{ij} \neq 0$). We choose to use a Bethe-structured cluster graph that has a cluster for each variable X_i and a cluster for each edge X_i, X_j. While it is certainly possible to define a belief propagation algorithm on a cluster graph with larger cliques, the standard application runs belief propagation directly on this pairwise network.

We note that the parameterization of the cluster graph is not uniquely defined. In particular, a term of the form $J_{ii}X_i^2$ can be partitioned in infinitely many ways among the node's own cluster and among the edges that contain X_i. Each of these partitions defines a different set of potentials in the cluster graph, and hence will induce a different execution of belief propagation. We describe the algorithm in terms of the simplest partition, where each diagonal term J_{ii} is assigned to the corresponding X_i cluster, and the off-diagonal terms J_{ij} are assigned to the X_i, X_j cluster.

With this decision, the belief propagation algorithm for Gaussian networks is simply derived from the standard message passing operations, implemented with the canonical-form operations for the factor product and marginalization steps. For concreteness, we now provide the precise message passing steps. The message from X_i to X_j has the form

$$\delta_{i \to j}(x_j) = \exp\left(-\frac{1}{2}J_{i \to j}x_j^2 + h_{i \to j}x_j\right). \tag{14.7}$$

We compute the coefficients in this expression via a two-stage process. The first step corresponds to the message sent from the X_i cluster to the X_i, X_j edge; in this step, X_i aggregates all of the information from its own local potential and the messages sent from its other incident edges:

$$\begin{aligned}
\hat{J}_{i \backslash j} &= J_{ii} + \sum_{k \in \mathrm{Nb}_i - \{j\}} J_{k \to i} \\
\hat{h}_{i \backslash j} &= h_i + \sum_{k \in \mathrm{Nb}_i - \{j\}} h_{k \to i}.
\end{aligned} \tag{14.8}$$

In the second step, the X_i, X_j edge takes the message received from X_i and sends the appropriate message to X_j. The form of the message can be computed (with some algebraic manipulation) from the formulas for the conditional mean and conditional variance, that we used in theorem 7.4, giving rise to the following update equations:

$$\begin{aligned}
J_{i \to j} &= -J_{ji}\hat{J}_{i \backslash j}^{-1}J_{ji} \\
h_{i \to j} &= -J_{ji}\hat{J}_{i \backslash j}^{-1}\hat{h}_{i \backslash j}.
\end{aligned} \tag{14.9}$$

These messages can be scheduled in various ways, either synchronously or asynchronously (see box 11.B).

If and when the message passing process has converged, we can compute the X_i-entries of the information form by combining the messages in the usual way:

$$\hat{J}_i = J_{ii} + \sum_{k \in \text{Nb}_i} J_{k \to i}$$

$$\hat{h}_i = h_i + \sum_{k \in \text{Nb}_i} h_{k \to i}.$$

From this information-form representation, X_i's approximate mean $\hat{\mu}_i$ and covariance $\hat{\sigma}_i^2$ can be reconstructed as usual:

$$\hat{\mu}_i = (\hat{J}_i)^{-1} \hat{h}_i$$

$$\hat{\sigma}_i^2 = (\hat{J}_i)^{-1}$$

One can now show the following result, whose proof we omit:

Theorem 14.1

Let $\hat{\mu}_i, \hat{\sigma}_i^2$ be a set of fixed points of the message passing process defined in equation (14.8), equation (14.9). Then $\hat{\mu}_i$ is the correct posterior mean for the variable X_i in the distribution p.

Thus, if the BP message passing process converges, the resulting beliefs encode the correct mean of the joint distribution. The estimated variances $\hat{\sigma}_i^2$ are generally not correct; rather, they are an underestimate of the true variances, so that the resulting posteriors are "overconfident."

This correctness result is predicated on convergence. In general, this message passing process may or may not converge. Moreover, their convergence may depend on the order in which messages are sent. However, unlike the discrete case, one can provide a very detailed characterization of the convergence properties of this process, as well as sufficient conditions for convergence (see section 14.7 for some references). In particular, one can show that the pairwise normalizability condition, as in definition 7.3, suffices to guarantee the convergence of the belief propagation algorithm for any order of messages. Recall that this condition guarantees that each edge can be associated with a potential that is a normalized Gaussian distribution. As a consequence, when the Gaussian parameterizing the edge is multiplied with Gaussians encoding the incoming message, the result is also a well-normalized Gaussian.

We note that pairwise normalizability is sufficient, but not necessary, for the convergence of belief propagation.

Example 14.2

Consider the Gaussian MRF shown in figure 14.1. This model defines a frustrated loop, since three of the edges in the loop are driving X_1, X_2 toward a positive correlation, but the edge between them is driving in the opposite direction. The larger the value of r, the worse the frustration. This model is diagonally dominant for any value of $r < 1/3$. It is pairwise normalizable for any $r < 0.39030$; however, it defines a valid Gaussian distribution for values of r up to 0.5. ∎

In practice, the Gaussian belief propagation algorithm often converges and provides an excellent alternative for reasoning in Gaussian distributions that are too large for exact techniques.

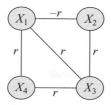

Figure 14.1   A Gaussian MRF used to illustrate convergence properties of Gaussian belief propagation. In this model, $J_{ii} = 1$, and $J_{ij} = r$ for all edges (i, j), except for $J_{12} = -r$.

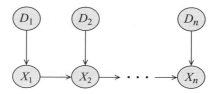

Figure 14.2   Simple CLG network used to demonstrate hardness of inference. $D_1, \ldots, D_n$ are discrete, and $X_1, \ldots, X_n$ are continuous.

14.3   Hybrid Networks

So far, we have dealt with models that involve only continuous variables. We now begin our discussion of hybrid networks — those that include both continuous and discrete variables. We focus the bulk of our discussion on conditional linear Gaussian (CLG) networks (definition 5.16), where there are no discrete variables with continuous parents, and where all the local probability models of continuous variables are conditional linear Gaussian CPDs. In the next section, we discuss inference for non-Gaussian dependencies, which will allow us to deal with non-CLG dependencies.

Even for this restricted class of networks, we can show that inference is very challenging. Indeed, we can show that inference in this class of networks is $\mathcal{NP}$-hard, even when the network structure is a polytree. We then show how the expectation propagation approach described in the previous section can be applied in this setting. Somewhat surprisingly, we show that this approach provides "exact" results in certain cases, albeit mostly ones of theoretical interest.

14.3.1   The Difficulties

As we discussed earlier, at an abstract level, variable elimination algorithms are all the same: They perform operations over factors to produce new factors. In the case of discrete models, factors can be represented as tables, and these operations can be performed effectively as table operations. In the case of Gaussian networks, the factors can be represented as canonical forms. As we now show, in the hybrid case, the representation of the intermediate factors can grow arbitrarily complex.

Consider the simple network shown in figure 14.2, where we assume that each D_i is a discrete binary variable, X_1 is a conditional Gaussian, and each X_i for $i > 1$ is a conditional linear

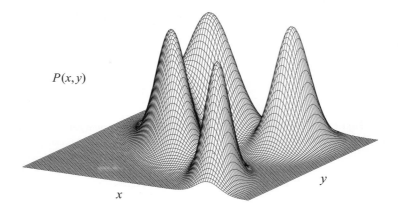

$P(x,y)$

x

y

Figure 14.3   **Joint marginal distribution $p(X_1, X_2)$ for a network as in figure 14.2**

Gaussian (CLG) (see definition 5.15):

$$p(X_i \mid X_{i-1}, D_i) = \mathcal{N}\left(X_i \mid \alpha_{i,d_i} x_{i-1} + \beta_{i,d_i}; \sigma^2_{i,d_i}\right),$$

where for simplicity we take $\alpha_{1,d_1} = 0$, so the same formula applies for all i.

Assume that our goal is to compute $P(X_n)$. To do so, we marginalize the joint distribution:

$$p(X_n) = \sum_{D_1,\ldots,D_n} \int p(D_1,\ldots,D_n, X_1,\ldots,X_n) dX_1 \ldots dX_{n-1}.$$

Using the chain rule, the joint distribution is defined as:

$$p(D_1,\ldots,D_n, X_1,\ldots,X_n) = \prod_{i=1}^{n} P(D_i) p(X_1 \mid D_1) \prod_{i=2}^{n} p(X_i \mid D_i, X_{i-1}).$$

We can reorder the sums and integrals and push each of them in over factors that do not involve the variable to be marginalized. Thus, for example, we have that:

$$p(X_2) = \sum_{D_2} P(D_2) \int p(X_2 \mid X_1, D_2) \sum_{D_1} p(X_1 \mid D_1) P(D_1) dX_1.$$

Using the same variable elimination approach that we used in the discrete case, we first generate a factor over X_1 by multiplying $P(D_1) p(X_1 \mid D_1)$ and summing out D_1. This factor is then multiplied with $p(X_2 \mid X_1, D_2)$ to generate a factor over X_1, X_2, D_2. We can now eliminate X_1 by integrating the function that corresponds to this factor, to generate a factor over X_2, D_2. We can now sum out D_2 to get $p(X_2)$. The process of computing $p(X_n)$ is analogous.

Now, consider the marginal distribution $p(X_i)$ for $i = 1, \ldots, n$. For $i = 1$, this distribution is a mixture of two Gaussians, one corresponding to the value $D_1 = d_1^1$ and the other to $D_1 = d_1^0$. For $i = 2$, let us first consider the distribution $p(X_1, X_2)$. This distribution is a mixture of four

Gaussians, for the four different instantiations of D_1, D_2. For example, assume that we have:

$$
\begin{aligned}
p(X_1 \mid d_1^0) &= \mathcal{N}\left(0; 0.7^2\right) \\
p(X_1 \mid d_1^1) &= \mathcal{N}\left(1.5; 0.6^2\right) \\
p(X_2 \mid X_1, d_2^0) &= \mathcal{N}\left(-1.5X_1; 0.6^2\right) \\
p(X_2 \mid X_1, d_2^1) &= \mathcal{N}\left(1.5; 0.7^2\right).
\end{aligned}
$$

The joint marginal distribution $p(X_1, X_2)$ is shown in figure 14.3. Note that the mixture contains two components where X_1 and X_2 are independent; these components correspond to the instantiations where $D_2 = d_2^1$, in which we have $\alpha_{2,1} = 0$. As shown in lemma 7.1, the marginal distribution of a Gaussian is also a Gaussian, and the same applies to a mixture. Hence the marginal distribution $p(X_2)$ is also a mixture of four Gaussians. We can easily extend this argument, showing that $p(X_i)$ is a mixture of 2^i Gaussians. **In general, even representing the correct marginal distribution in a hybrid network can require space that is exponential in the size of network.**

Indeed, this type of example can be used as the basis for proving a result about the hardness of inference in models of this type. Clearly, as CLG networks subsume standard discrete networks, exact inference in such networks is necessarily $\mathcal{NP}$-hard. More surprising, however, is the fact that this task is $\mathcal{NP}$-hard even in very simple network structures such as polytrees. In fact, the problem of computing the probability of a single discrete variable, or even approximating this probability with any absolute error strictly less than $1/2$, is $\mathcal{NP}$-hard.

To define the problem precisely, assume we are working with finite precision continuous variables. We define the following decision problem *CLG-DP*:

Input: A CLG Bayesian network $\mathcal{B}$ over $\Delta \cup \Gamma$, evidence $\boldsymbol{E} = \boldsymbol{e}$, and a discrete variable $A \in \Delta$.

Output: "Yes" if $P_{\mathcal{B}}(A = a^1 \mid \boldsymbol{E} = \boldsymbol{e}) > 0.5$.

Theorem 14.2

The problem CLG-DP is $\mathcal{NP}$-hard even if $\mathcal{B}$ is a polytree.

The fact that exact inference in polytree CLGs is $\mathcal{NP}$-hard may not be very surprising by itself. After all, the distribution of a continuous variable in a CLG distribution, even in a simple polytree, can be a mixture of exponentially many Gaussians. Therefore, it might be expected that tasks that require that we reason directly with such a distribution are hard. Somewhat more surprisingly, this phenomenon arises even in networks where the prior distribution of every continuous variable is a mixture of at most two Gaussians.

Theorem 14.3

The problem CLG-DP is $\mathcal{NP}$-hard even if $\mathcal{B}$ is a polytree where all of the discrete variables are binary-valued, and where every continuous variable has at most one discrete ancestor.

Intuitively, this proof relies on the use of activated v-structures to introduce, in the posterior, dependencies where a continuous variable can have exponentially many modes.

Overall, these results show that even the easiest approximate inference task — inference over a binary-valued variable that achieves absolute error less than 0.5 — is intractable in CLG networks. This fact implies that one should not expect to find a polynomial-time approximate inference algorithm with a useful error bound without further restrictions on the structure or the parameters of the CLGs.

14.3.2   Factor Operations for Hybrid Gaussian Networks

Despite the discouraging results in the previous section, one can try to produce useful algorithms for hybrid networks in order to construct an approximate inference algorithm that has good performance, at least in practice. We now present the basic factor operations required for message passing or variable elimination in hybrid networks. In subsequent sections, we describe two algorithms that use these operations for inference.

14.3.2.1   Canonical Tables

As we discussed, the key decision in adapting an exact inference algorithm to a class of hybrid networks is the representation of the factors involved in the process. In section 14.2, when doing inference for linear Gaussian networks, we used canonical forms to represent factors. This representation is rich enough to capture both Gaussians and linear Gaussian CPDs, as well as all of the intermediate expressions that arise during the course of inference. In the case of CLG networks, we must contend with discrete variables as well as continuous ones. In particular, a CLG CPD has a linear Gaussian model for each instantiation of the discrete parents.

Extending on the canonical form, we can represent this CPD as a table, with one entry for each instantiation of the discrete variables, each associated with a canonical form over the continuous ones:

Definition 14.2

canonical table

A canonical table ϕ over D, X for $D \subseteq \Delta$ and $X \subseteq \Gamma$ is a table with an entry for each $d \in Val(D)$, where each entry contains a canonical form $\mathcal{C}(X; K_d, h_d, g_d)$ over X. We use $\phi(d)$ to denote the canonical form over X associated with the instantiation d.   ∎

The canonical table representation subsumes both the canonical form and the table factors used in the context of discrete networks. For the former, $D = \emptyset$, so we have only a single canonical form over X. For the latter, $X = \emptyset$, so that the parameters K_d and h_d are vacuous, and we remain with a canonical form $\exp(g_d)$ for each entry $\phi(d)$. Clearly, a standard table factor $\phi(D)$ can be reformulated in this way by simply taking $g_d = \ln(\phi(d))$. Therefore, we can represent any of the original CPDs or Gaussian potentials in a CLG network as a canonical table.

Now, consider the operations on factors used by the various exact inference algorithms. Let us first consider the operations of factor product and factor division. As for the table representation of discrete factors, these operations are performed between corresponding table entries, in the usual way. The product or division operations for the individual entries are performed over the associated canonical forms, as specified in equation (14.2) and equation (14.3).

Example 14.3

Assume we have two factors $\phi_1(A, B, X, Y)$ and $\phi_2(B, C, Y, Z)$, which we want to multiply in order to produce $\tau(A, B, C, X, Y, Z)$. The resulting factor τ has an entry for each instantiation of the discrete variables A, B, C. The entry for a particular instantiation a, b, c is a canonical form, which is derived as the product of the two canonical forms: the one associated with a, b in ϕ_1 and the one associated with b, c in ϕ_2. The product operation for two canonical forms of different scopes is illustrated in example 14.1.   ∎

Similarly, reducing a canonical table with evidence is straightforward. Let $\{d, x\}$ be a set of observations (where d is discrete and x is continuous). We instantiate d in a canonical table by

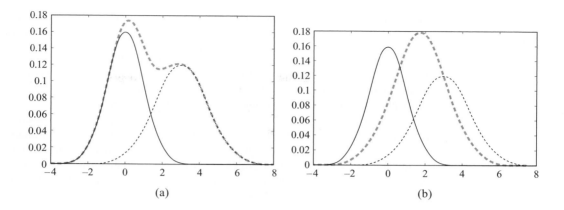

Figure 14.4 Summing and collapsing a Gaussian mixture. (a) Two Gaussian measures and the measure resulting from summing them. (b) The measure resulting from collapsing the same two measures into a single Gaussian.

setting the entries which are not consistent with d to zero. We instantiate x by instantiating every canonical form with this evidence, as in equation (14.6).

Finally, consider the marginalization operation. Here, we have two very different cases: integrating out a continuous variable and summing out a discrete one. The operation of continuous marginalization (integration of of a continuous variable) is straightforward: we simply apply the operation of equation (14.5) to each of the canonical forms in our table. More precisely, assume that our canonical table consists of a set of canonical forms $\mathcal{C}\left(X, Y; K_d, h_d, g_d\right)$, indexed by d. Now, for each d separately, we can integrate out Y in the appropriate canonical form, as in equation (14.5). This results in a new canonical table $\mathcal{C}\left(X; K'_d, h'_d, g'_d\right)$, indexed by d. Clearly, this has the desired effect: Before the operation, we have a mixture, where each mixture is a function over a set of variables X, Y; after the operation, we have a mixture with the same set of components, but now each component only represents the function over the variables X. The only important restriction, as we noted in the derivation of equation (14.5), that each of the matrices $K_{d,YY}$ be positive definite, so that the integral is well defined.

14.3.2.2 Weak Marginalization

The task of discrete marginalization, however, is significantly more complex. To understand the difficulty, consider the following example:

Example 14.4

Assume that we have a canonical form $\phi(A, X)$, for a binary-valued variable A and a continuous variable X. Furthermore, assume that the two canonical forms in the table (associated with a^0 and a^1) are both weighted Gaussians:

$$
\begin{aligned}
\phi(a^0) &= 0.4 \times \mathcal{N}\left(X \mid 0; 1\right) \\
\phi(a^1) &= 0.6 \times \mathcal{N}\left(X \mid 3; 4\right).
\end{aligned}
$$

Figure 14.4a shows the two canonical forms, as well as the marginal distribution over X. Clearly, this distribution is not a Gaussian; in fact, it cannot be represented at all as a canonical form. ■

We see that the family of canonical tables is not closed under discrete marginalization: this operation takes a canonical table and produces something that is not representable in this form.

We now have two alternatives. The first is to enrich the family that we use for our representation of factors. Specifically, we would have to use a table where, for each instantiations of the discrete variables, we have a *mixture* of canonical forms. In this case, the discrete marginalization operation is trivial: In example 14.4, we would have a single entry that contains the marginal distribution shown in figure 14.4a. While this family is closed under discrete marginalization, we have only resurrected our original problem: As discrete variables are "eliminated," they simply induce more components in the mixture of canonical forms; the end result of this process is simply the original exponentially large mixture that we were trying to avoid.

The second alternative is to approximate the result of the discrete marginalization operation. In our example, when marginalizing D, we can approximate the resulting mixture of Gaussians by *collapsing* it into a single Gaussian, as shown in figure 14.4b. An appropriate approximation to use in this setting is the *M-projection* operation introduced in definition 8.4. Here, we select the Gaussian distribution $\hat{p}$ that minimizes $\boldsymbol{D}(p\|\hat{p})$. In example 8.15 we provided a precise characterization of this operation:

mixture
collapsing

M-projection

Proposition 14.2

Let p be an arbitrary distribution over $X_1, \ldots, X_k$. Let $\boldsymbol{\mu}$ be the mean vector of p, and Σ be the matrix of covariances in p:

$$\mu_i = \boldsymbol{E}_p[X_i]$$
$$\Sigma_{i,j} = \boldsymbol{Cov}_p[X_i; X_j].$$

Then the Gaussian distribution $\hat{p} = \mathcal{N}(\boldsymbol{\mu}; \Sigma)$ is the one that minimizes $\boldsymbol{D}(p\|\hat{p})$ among all Gaussian distributions.

Using this result, we have:

Proposition 14.3

Let p be the density function of a mixture of k Gaussians $\{\langle w_i, \mathcal{N}(\boldsymbol{\mu}_i; \Sigma_i)\rangle\}_{i=1}^k$ for $\sum_{i=1}^k w_i = 1$. Let $q = \mathcal{N}(\boldsymbol{\mu}; \Sigma)$ be a Gaussian distribution defined as:

$$\boldsymbol{\mu} = \sum_{i=1}^k w_i \boldsymbol{\mu}_i \tag{14.10}$$

$$\Sigma = \sum_{i=1}^k w_i \Sigma_i + \sum_{i=1}^k w_i (\boldsymbol{\mu}_i - \boldsymbol{\mu})(\boldsymbol{\mu}_i - \boldsymbol{\mu})^T. \tag{14.11}$$

Then q has the same first two moments (means and covariances) as p, and is therefore the Gaussian distribution that minimizes $\boldsymbol{D}(p\|q)$ among all Gaussian distributions.

The proof is left as an exercise (exercise 14.2).

Note that the covariance matrix, as defined by the collapsing operation, has two terms: one term is the weighted average of the covariance matrices of the mixture components; the second corresponds to the distances between the means of the mixture components — the larger these distances, the larger the "space" between the mixture components, and thus the larger the variances in the new covariance matrix.

Example 14.5

Consider again the discrete marginalization problem in example 14.4. Using proposition 14.3, we have that the mean and variance of the optimal Gaussian approximation to the mixture are:

$$\mu = 0.4 \cdot 0 + 0.6 \cdot 3 = 1.8$$
$$\sigma^2 = (0.4 \cdot 1 + 0.6 \cdot 4) + (0.4 \cdot (1.8 - 0)^2 + 0.6 \cdot (3 - 1.8)^2 = 4.96.$$

The resulting Gaussian approximation is shown in figure 14.4b. ∎

Clearly, when approximating a mixture of Gaussians by one Gaussian, the quality of the approximation depends on how close the mixture density is to a single multivariate Gaussian. When the Gaussians are very different, the approximation can be quite bad.

Using these tools, we can define the discrete marginalization operation.

Definition 14.3

weak
marginalization

Assume we have a canonical table defined over $\{A, B, X\}$, where $A, B \subseteq \Delta$ and $X \subseteq \Gamma$. Its weak marginal *is a canonical table over A, X, defined as follows: For every value $a \in Val(A)$, we select the table entries consistent with a and sum them together to obtain a single table entry. The summation operation uses the collapsing operation of proposition 14.3.* ∎

One problem with this definition is that the collapsing operation was defined in proposition 14.3 only for a mixture of Gaussians and not for a mixture of general canonical forms. Indeed, the operation of combining canonical forms is well defined if and only if the canonical forms have finite first two moments, which is the case only if they can be represented as Gaussians. This restriction places a constraint on our inference algorithm: we can marginalize a discrete variable only when the associated canonical forms represent Gaussians. We will return to this point.

14.3.3 EP for CLG Networks

The previous section described the basic data structure that we can use to encode factors in a hybrid Gaussian network, and the basic factor operations needed to manipulate them. Most important was the definition of the weak marginalization operation, which approximates a mixture of Gaussians as a single Gaussian, using the concept of M-projection.

Gaussian EP

With our definition of weak marginalization and other operations on canonical tables, we can now define a message passing algorithm based on the framework of the *expectation propagation* described in section 11.4. As a reminder, to perform a message passing step in the EP algorithm, a cluster multiplies all incoming messages, and then performs an approximate marginalization on the resulting product factor. This last step, which can be viewed abstractly as a two-step process — exact marginalization followed by M-projection — is generally performed in a single approximate marginalization step. For example, in section 11.4.2, the approximate marginals were computed by calibrating a cluster graph (or clique tree) and then extracting from it a set of required marginals.

To apply EP in our setting, we need only to define the implementation of the M-projection operation M-Project-Distr, as needed in line 1 of algorithm 11.5. This operation can be performed using the weak marginalization operation described in section 14.3.2.2, as shown in detail in algorithm 14.1. The marginalization step uses two types of operation. The continuous variables

---

Algorithm 14.1 Expectation propagation message passing for CLG networks

    Procedure CLG-M-Project-Distr (
        $\boldsymbol{Z}$,   // Scope to remain following projection
        $\vec{\phi}$ // Set of canonical tables
)
1   // Compute overall measure using product of canonical tables
2      $\tilde{\beta} \leftarrow \prod_{\phi \in \vec{\phi}} \phi$
3   // Variables to be preserved
4      $\boldsymbol{A} \leftarrow \boldsymbol{Z} \cap \Delta$
5      $\boldsymbol{X} \leftarrow \boldsymbol{Z} \cap \Gamma$
6   // Variables to be eliminated
7      $\boldsymbol{B} \leftarrow (Scope[\vec{\phi}] - \boldsymbol{A}) \cap \Delta$
8      $\boldsymbol{Y} \leftarrow (Scope[\vec{\phi}] - \boldsymbol{X}) \cap \Gamma$
9   **for** each $\boldsymbol{a}, \boldsymbol{b} \in Val(\boldsymbol{A}, \boldsymbol{B})$
10     $\tau(\boldsymbol{a}, \boldsymbol{b}) \leftarrow \int \beta_i(\boldsymbol{a}, \boldsymbol{b}) d\boldsymbol{Y}$ using equation (14.5)
11     $\tilde{\sigma} \leftarrow \sum_{\boldsymbol{B}} \tau$ using definition 14.3
12   **return** $\tilde{\sigma}$

---

are eliminated using the marginalization operation of equation (14.5) over each entry in the canonical table separately. The discrete variables are then summed out. For this last step, there are two cases. If the factor τ contains only discrete variables, then we use standard discrete marginalization. If not, then we use the weak marginalization operation of definition 14.3.

In principle, this application of the EP framework is fairly straightforward. There are, however, two important subtleties that arise in this setting.

14.3.3.1   Ordering Constraints

First, as we discussed, for weak marginalization to be well defined, the canonical form being marginalized needs to be a Gaussian distribution, and not merely a canonical form. In some cases, this requirement is satisfied simply because of the form of the potentials in the original network factorization. For example, in the Gaussian case, recall that our message passing process is well defined if our distribution is pairwise normalizable. In the conditional Gaussian case, we can guarantee normalizability if for each cluster over scope $\boldsymbol{X}$, the initial factor in that cluster is a canonical table such that each canonical form entry $\mathcal{C}(\boldsymbol{X}; K_d, h_d, g_d)$ is normalizable. Because normalizability is closed under factor product (because the sum of two PSD matrices is also PSD) and (both weak and strong) marginalization, this requirement guarantees us that all factors produced by a sum-product algorithm will be normalizable.

However, this requirement is not always easy to satisfy:

Example 14.6   *Consider a CLG network structured as in figure 14.5a, and the clique tree shown in figure 14.5b. Note that the only clique where each canonical form is normalizable is $C_1 = \{A, X\}$; in $C_2 = \{B, X, Y\}$, the canonical forms in the canonical table are all linear Gaussians whose integral is infinite, and hence cannot be collapsed in the operation of proposition 14.3.*

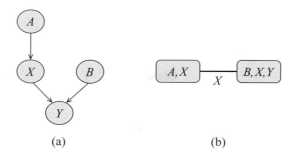

(a)   (b)

Figure 14.5   Example of unnormalizable potentials in a CLG clique tree. (a) A simple CLG (b) A clique tree for it.

In this network, with an appropriate message passing order, one can guarantee that canonical tables are normalizable at the appropriate time point. In particular, we can first pass a message from C_1 to C_2; this message is a Gaussian obtained by weak marginalization of $p(A, X)$ onto X. The resulting potential at C_2 is now a product of a legal Gaussian density over X (derived from the incoming message) multiplied by $P(B)$ and the conditional linear Gaussian $p(Y \mid X, B)$. The resulting distribution is a standard mixture of Gaussians, where each component in the mixture is normalizable. Thus, weak marginalization onto Y can be performed, allowing the message passing process to continue. ∎

This example illustrates that we can sometimes find a legal message passing order even in cases where the initial potentials are not normalizable. However, such a message passing order may not always exist.

Example 14.7   *Consider the network in figure 14.6a. After moralization, the graph, shown in figure 14.6b, is already triangulated. If we now extract the maximum cliques and build the clique tree, we get the tree shown in figure 14.6c. Unfortunately, at this point, neither of the two leaves in this clique tree can send a message. For example, the clique $\{B, X, Y\}$ contains the CPDs for $P(B)$ and $P(Y \mid B, X)$, but not the CPD for X. Hence, the canonical forms over $\{X, Y\}$ represent linear Gaussian CPDs and not Gaussians. It follows that we cannot marginalize out B, and thus this clique cannot send a message. For similar reasons, the clique $\{A, C, Y, Z\}$ cannot send a message. Note, however, that a different clique tree can admit message passing. In particular, both the trees in (d) and (e) admit message passing using weak marginalization.* ∎

As we can see, not every cluster graph allows message passing based on weak marginalization

order-constrained message passing

to take place. More formally, we say that a variable X_i is *order constrained* at cluster C_k if we require a normalizable probability distribution over X_i at C_k in order to send messages. In a cluster graph for a CLG Bayesian network, if C_k requires weak marginalization, then any continuous variable X_i in C_k is order constrained.

If X_i is order constrained in C_k, then we need to ensure that X_i has a well-defined distribution. In order for that to hold, C_k must have obtained a valid probability distribution over X_i's parents, whether from a factor within C_k or from a message sent by another cluster. Now, consider a continuous parent X_j of X_i, and let C_l be the cluster from which C_k obtains the

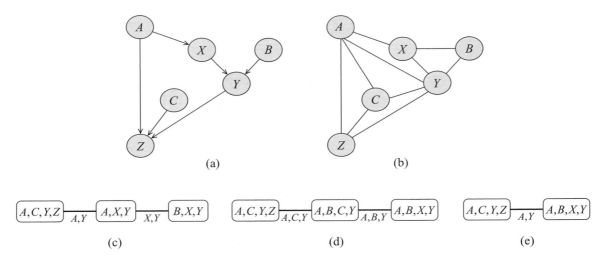

Figure 14.6   A simple CLG and possible clique trees with different correctness properties. (a) A simple CLG. (b) The moralized (and triangulated) graph for the CLG. (c) Clique tree derived the moralized graph in (b). (d) A clique tree with a strong root. (e) A clique tree without a strong root that allows message passing using weak marginalization.

distribution over X_j. (In the first case, we have $k = l$.) In order for X_j to have a well-defined distribution in C_l, we must have that X_j is order constrained in C_l. This process continues until the roots of the networks are reached.

In the context of Markov networks, the situation is somewhat less complex. The use of a global partition function allows us to specify models where individual factors are all normalizable, ensuring a legal measure. In particular, extending definition 7.3, we can require that all of the entries in all of the canonical tables parameterizing the network be *normalizable*. In this case, the factors in all clusters in the network can be normalized to produce valid distributions, avoiding any constraints on message passing. Of course, the factor normalizability constraint is only a sufficient condition, in that there are models that do not satisfy this constraint and yet allow weak marginalization to take place, if messages are carefully ordered.

Gaussian
normalizability

As this discussion shows, **the constraint on normalizability of the Gaussian in the context of weak marginalization can impose significant constraints on the structure of the cluster graph and/or the order of the message passing.**

☞

14.3.3.2   Ill-Defined Messages

belief-update

The second subtlety arises when we apply the *belief-update* message passing algorithm rather than sum product. Recall that the belief update algorithm has important advantages, in that it gradually tunes the approximation to the more relevant regions in the probability space.

Example 14.8

Consider the application of the sum-product belief propagation algorithm to the network of figure 14.5. Assume furthermore that the distribution at C_1 is as described in example 14.5, so that *the result of weak marginalization is as shown in figure 14.4b. Now, assume that $p(Y \mid x, b^1) = \mathcal{N}(Y \mid x; 1)$, and that we observe $b = b^1$ and $Y = -2$. The evidence $Y = -2$ is much more consistent with the left-hand (mean 0) component in the mixture, which is the one derived from $p(X \mid a^0)$. Given this evidence, the exact posterior over X would give much higher probability to the a^0 component in the mixture. However, our Gaussian approximation was the M-projection of a mixture where this component had its prior weight of 0.4. To obtain a better approximation, we would want to construct a new M-projection in $\mathbf{C}_1$ that gives more weight to the $\mathcal{N}(X \mid 0; 1)$ component in the mixture when collapsing the two Gaussians.* ∎

This issue is precisely the motivation that we used in section 11.4.3.2 for the belief update algorithm. **However, the use of division in the Gaussian EP algorithm can create unexpected complications, in that the messages passed can represent unnormalizable densities, with negative variance terms; these can lead, in turn, to unnormalizable densities in the clusters, and thereby to nonsensical results.**

Example 14.9

Consider again the CLG in figure 14.5a, but where we now assume that the CPDs of the discrete variables are uniform and the CPDs of the continuous nodes are defined as follows:

$$p(X \mid A) = \begin{cases} \mathcal{N}(0; 2) & A = a^0 \\ \mathcal{N}(0; 6) & A = a^1 \end{cases}$$

$$p(Y \mid B, X) = \begin{cases} \mathcal{N}(X; 0.01) & B = b^0 \\ \mathcal{N}(-X; 0.01) & B = b^1. \end{cases}$$

Consider the execution of message passing on the clique tree of figure 14.5b, with the evidence $Y = 4$. To make the presentation easier to follow, we present some of the intermediate results in moments form (means and covariances) rather than canonical form.

The message passing algorithm first sends a message from the clique $\{A, X\}$ to the clique $\{B, X, Y\}$. Since the cliques share only the variable X, we collapse the prior distribution of X using proposition 14.3 and get the message $\delta_{1 \to 2} = \mathcal{N}(X \mid 0; 4)$. This message is stored in the sepset at $\mu_{1,2}$. We also multiply the potential of $\{B, X, Y\}$ by this message, getting a mixture of two Gaussians with equal weights:

$$\beta_2(b^0) = \mathcal{N}\left(X, Y \mid \begin{pmatrix} 0 \\ 0 \end{pmatrix}; \begin{bmatrix} 4 & 4 \\ 4 & 4.01 \end{bmatrix}\right)$$

$$\beta_2(b^1) = \mathcal{N}\left(X, Y \mid \begin{pmatrix} 0 \\ 0 \end{pmatrix}; \begin{bmatrix} 4 & -4 \\ -4 & 4.01 \end{bmatrix}\right).$$

After instantiating the evidence $Y = 4$ we get a new mixture of two Gaussians:

$$\beta_2(b^0) = \mathcal{N}(X \mid 3.99; 0.00998)$$
$$\beta_2(b^1) = \mathcal{N}(X \mid -3.99; 0.00998).$$

Note that, in the original mixture, Y has the same marginal — $\mathcal{N}(Y \mid 0; 4.01)$ — for both b^0 and b^1. Therefore, the evidence $Y = 4$ has the same likelihood in both cases, so that the posterior weights of the two cases are still the same as each other.

We now need to send a message back to the clique $\{A, X\}$. To do so, we collapse the two Gaussians, resulting in a message $\delta_{2 \to 1} = \mathcal{N}(X \mid 0; 15.93)$. Note that, in this example, the evidence causes the variance of X to increase.

To incorporate the message $\delta_{2\to1}$ into the clique $\{A, X\}$, we divide it by the message $\mu_{1,2}$, and multiply each entry in $\beta_1(A, X)$ by the resulting quotient $\frac{\delta_{2\to1}}{\mu_{1,2}}$. In particular, for $A = a^1$, we perform the operation:

$$\frac{\mathcal{N}(0; 6) \cdot \mathcal{N}(0; 15.93)}{\mathcal{N}(0; 4)}.$$

This operation can be carried out using the canonical form $\mathcal{C}(K, \boldsymbol{h}, g)$. Consider the operation over the coefficient K, which represents the inverse of the covariance matrix: $K = \Sigma^{-1}$. In our case, the Gaussians are all one-dimensional, so $K = \frac{1}{\sigma^2}$. As in equation (14.2) and equation (14.3), the product and division operation reduces to addition and subtraction of the coefficient K. Thus, the K for the resulting potential is:

$$\frac{1}{6} + \frac{1}{15.93} - \frac{1}{4} = -0.0206 < 0!$$

However, $K < 0$ does not represent a legal Gaussian: it corresponds to $\sigma^2 = \frac{1}{-0.0206}$, which is not a legal variance.   ∎

In practice, this type of situation does occur, but not often. Therefore, despite this complication, the belief-update variant of the EP algorithm is often used in practice.

14.3.4   An "Exact" CLG Algorithm ⋆

There is one case where we can guarantee that the belief update algorithm is not only well defined but even returns "exact" answers. Of course, truly exact answers are not generally possible in a CLG network. Recall that a CLG distribution is a mixture of possibly exponentially many Gaussian hypotheses. The marginal distribution over a single variable can similarly be an exponentially large mixture (as in figure 14.2). Thus, for a query regarding the marginal distribution of a single continuous variable, even representing the answer might be intractable.

Lauritzen's algorithm

Lauritzen's algorithm provides a compromise between correctness and computational efficiency. It is exact for queries involving discrete variables, and it provides the exact first and second moments — means and (co)variances — for the continuous variables. More precisely, for a query $p(\boldsymbol{D}, \boldsymbol{X} \mid e)$ for $\boldsymbol{D} \subseteq \Delta$ and $\boldsymbol{X} \subseteq \Gamma$, Lauritzen's algorithm returns an answer $\hat{p}$ such that $\hat{p}(\boldsymbol{D}) = p(\boldsymbol{D} \mid e)$ is correct, and for each $\boldsymbol{d}$, $\hat{p}(\boldsymbol{X} \mid \boldsymbol{d})$ has the correct first and second moments. For many applications, queries of this type are sufficient.

The algorithm is a modification of the clique tree algorithm for discrete networks. It uses precisely the same type of message passing that we described, but over a highly restricted clique tree data structure. As we will show, these restrictions can (and do) cause significant blowup in the size of the clique tree (much larger than the induced width of the network) and hence do not violate the $\mathcal{NP}$-hardness of inference in these graphs. Indeed, these restrictions restrict the algorithm's usefulness to a fairly narrow class of problems, and they render it primarily of theoretical interest.

14.3.4.1   Strongly Rooted Trees

The ordering constraint we employ on the clique tree, as described in section 14.3.3.1, guarantees that we can calibrate the clique tree without requiring weak marginalization on the upward pass.

That is, the clique tree has a particular clique that is a *strong root*; when we pass messages from the leaves of the clique tree toward this root, no weak marginalization is required.

This property has several implications. First, weak marginalization is the only operation that requires that the clique potential be normalizable. In the upward pass (from the leaves toward the root), no such marginalization is required, and so we need not worry about this constraint. Intuitively, in the downward pass, each clique has already obtained all of the necessary information to define a probability distribution over its scope. This distribution allows weak marginalization to be performed in a well-defined way. Second, because weak marginalization is the only operation that involves approximation, this requirement guarantees that the message passing in the upward pass is exact. As we will discuss, this property suffices to guarantee that the weak marginals in the downward pass are all legal. Indeed, as we will show, this property also allows us to guarantee that these marginals all possess the correct first and second moments (means and covariances).

When does our clique tree structure guarantee that no weak marginalization has to take place? Consider two cliques C_1 and C_2, such that C_2 is the upward neighbor of C_1. There are two cases. Either no marginalization of any discrete variable takes place between C_1 and C_2, or it does. In the first case, only continuous variables are eliminated, so that $C_2 - C_1 \subseteq \Gamma$. In the second case, in order to avoid weak marginalization, we must avoid collapsing any Gaussians. Thus, we must have that the message from C_2 to C_1 contains no continuous variables, so that $C_1 \cap C_2 \subseteq \Delta$. Overall, we have:

Definition 14.4

strong root

A clique C_r in a clique tree $\mathcal{T}$ is a strong root if for every clique C_1 and its upward neighbor C_2, we have that $C_2 - C_1 \subseteq \Gamma$ or that $C_1 \cap C_2 \subseteq \Delta$. A clique tree is called strongly rooted if it has a strong root. ∎

Example 14.10

figure 14.6d shows a strongly rooted clique tree for the network of figure 14.6a. Here, the strong root is the clique $\{A, B, C, Y\}$. For both of the two nonroot cliques, we have that no discrete variables are marginalized between them and their upward neighbor $\{A, B, C, Y\}$, so that the first condition of the definition holds. ∎

If we now apply the message passing procedure of algorithm 14.1, we note that the weak marginalization can only occur in the downward pass. In the downward pass, C_i has received messages from all of its neighbors, and therefore $\beta_i(C_i)$ represents a probability distribution over C_i. Hence, $\tau(A, B, X)$ is a mixture of Gaussians over X, so that the weak marginalization operation is well defined.

14.3.4.2 Strong Roots and Correctness

So far, we have shown that the strongly rooted requirement ensures that the operations in the clique tree are well defined, specifically, that no collapsing is performed unless the canonical forms represent Gaussians. However, as we have already shown, this condition is not necessary for the message passing to be well defined; for example, the clique tree of figure 14.6e is not strongly rooted, but allows weak marginalization. However, as we now show, there are other reasons for using strongly rooted trees for inference in hybrid networks. Specifically, the presence of a strong root ensures not only that message passing is well defined, but also that message passing leads to exact results, in the sense that we described.

Theorem 14.4   *Let $\mathcal{T}$ be a clique tree and C_r be a strong root in $\mathcal{T}$. After instantiating the evidence and running* CTree-BU-Calibrate *using* EP-Message *with* CLG-M-Project-Distr *for message passing, the tree $\mathcal{T}$ is calibrated and every potential contains the correct (weak) marginal. In particular, every clique C contains the correct probability distribution over the discrete variables $C \cap \Delta$ and the correct mean and covariances of the continuous variables $C \cap \Gamma$.*

PROOF   Consider the steps in the algorithm CTree-BU-Calibrate. The clique initialization step is exact: each CPD is multiplied into a clique that contains all of its variables, and the product operation for canonical tables is exact. Similarly, evidence instantiation is also exact. It remains only to show that the message passing phase is exact.

The upward pass is simple — as we discussed, all of the marginalization operations involved are strong marginalizations, and therefore all the operations are exact. Thus, the upward pass is equivalent to running the variable elimination algorithm for the variables in the strong root, and its correctness follows from the correctness of the variable elimination algorithm. The result of the upward pass is the correct (strong) marginal in the strong root C_r.

The downward pass involves weak marginalization, and therefore, it will generally not result in the correct distribution. We wish to show that the resulting distributions in the cliques are the correct *weak* marginals. The proof is by induction on the distance of the clique from the strong root C_r. The base case is the root clique itself, where we have already shown that we have exactly the correct marginal. Now, assume now that we have two cliques C_i and C_j such that C_i is the upward neighbor of C_j. By the inductive hypothesis, after C_i receives the message from its upward neighbor, it has the correct weak marginals. We need to show that, after C_j receives the message from C_i, it also has the correct weak marginal.

Let β_j and β_j' denote the potential of C_j before and after C_i sends the downward message to C_j. Let $\mu_{i,j}$ denote the sepset message before the downward message, and $\delta_{i \rightarrow j}$ denote the actual message sent. Note that $\delta_{i \rightarrow j}$ is the (weak) marginal of the clique potential β_i, and is therefore the correct weak marginal, by the inductive hypothesis.

We first show that, after the message is sent, C_j agrees with C_i on the marginal distribution of the sepset $S_{i,j}$.

$$\sum_{C_j - S_{i,j}} \beta_j' = \sum_{C_j - S_{i,j}} \beta_j \cdot \frac{\delta_{i \rightarrow j}}{\mu_{i,j}} = \frac{\delta_{i \rightarrow j}}{\mu_{i,j}} \cdot \sum_{C_j - S_{i,j}} \beta_j = \frac{\delta_{i \rightarrow j}}{\mu_{i,j}} \cdot \mu_{i,j} = \delta_{i \rightarrow j}, \qquad (14.12)$$

where the marginalization $\sum_{C_j - S_{i,j}}$ also denotes integration, when appropriate. This derivation is correct because this marginalization $\sum_{C_j - S_{i,j}}$ is an exact operation: By the strong root property, all marginalizations toward the strong root (that is, from j to i) are strong marginalizations. Thus, the marginal of C_j after the message is the same as the (weak) marginal of C_i, and the two cliques are calibrated. Because the (weak) marginal of C_i is correct, so is the marginal of C_j. Note that this property does *not* hold in a tree that is not strongly rooted. In general, in such a clique tree, $\mu_{i,j} \neq \sum_{C_j - S_{i,j}} \beta_j$.

It remains to show that β_j' is the correct weak marginal for all the variables in C_j. As shown in exercise 10.5, the premessage potential β_j already encodes the correct posterior conditional distribution $P(C_j \mid S_{i,j})$. (We use P to denote the posterior, after conditioning on the evidence.) In other words, letting $\boldsymbol{X} = \boldsymbol{C}_j - \boldsymbol{S}_{i,j}$ and $\boldsymbol{S} = \boldsymbol{S}_{i,j}$, we have that:

$$\frac{\beta_j(\boldsymbol{X}, \boldsymbol{S})}{\beta_j(\boldsymbol{S})} = P(\boldsymbol{X} \mid \boldsymbol{S}).$$

Now, since the last message along the edge $\boldsymbol{C}_i$—$\boldsymbol{C}_j$ was sent from $\boldsymbol{C}_j$ to $\boldsymbol{C}_i$, we have that

$$\frac{\beta_j(\boldsymbol{X}, \boldsymbol{S})}{\beta_j(\boldsymbol{S})} = \frac{\beta_j(\boldsymbol{X}, \boldsymbol{S})}{\mu_{i,j}(\boldsymbol{S})}.$$

We therefore have that the potential of $\boldsymbol{C}_j$ after the message from $\boldsymbol{C}_i$ is:

$$\beta_j' = \frac{\beta_j \delta_{i \to j}}{\mu_{i,j}} = \delta_{i \to j} P(\boldsymbol{X} \mid \boldsymbol{S}).$$

Thus, every entry in β_j' is a Gaussian, which is derived as a product of two terms: one is a Gaussian over $\boldsymbol{S}$ that has the same first and second moments as P, and the second is the correct $P(\boldsymbol{X} \mid \boldsymbol{S})$. It follows easily that the resulting product $\beta_j(\boldsymbol{X}, \boldsymbol{S})$ is a Gaussian that has the same first and second moments as P (see exercise 14.5). This concludes the proof of the result. ∎

Note that, if the tree is not strongly rooted, the proof breaks down in two places: the upward pass is not exact, and equation (14.12) does not hold. Both of these arise from the fact that $\sum_{\boldsymbol{C}_j - \boldsymbol{S}_{i,j}} \beta_j$ is the exact marginal, whereas, if the tree is not strongly rooted, $\mu_{i,j}$ is computed (in some cliques) using weak marginalization. Thus, the two are not, in general, equal. As a consequence, although the weak marginal of β_j is $\mu_{i,j}$, the second equality fails in the derivation: the weak marginal of a product $\beta_j \cdot \frac{\delta_{i \to j}}{\mu_{i,j}}$ is not generally equal to the product of $\frac{\delta_{i \to j}}{\mu_{i,j}}$ and the weak marginal of β_j. Thus, the strong root property is essential for the strong correctness properties of this algorithm.

14.3.4.3 Strong Triangulation and Complexity

We see that strongly rooted trees are necessary for the correct execution of the clique tree algorithm in CLG networks. Thus, we next consider the task of constructing a strongly rooted tree for a given network. As we discussed, one of our main motivations for the strong root requirement was the ability to perform the upward pass of the clique tree algorithm without weak marginalization. Intuitively, the requirement that discrete variables not be eliminated before their continuous neighbors implies a *constraint on the elimination ordering* within the clique tree. Unfortunately, constrained elimination orderings can give rise to clique trees that are much larger — exponentially larger — than the optimal, unconstrained clique tree for the same network. We can analyze the implications of the strong triangulation constraint in terms of the network structure.

constrained
elimination
ordering

Definition 14.5

Let $\mathcal{G}$ be a hybrid network. A continuous connected component *in $\mathcal{G}$ is a set of variables* $\boldsymbol{X} \subseteq \Gamma$ *such that: if* $X_1, X_2 \in \boldsymbol{X}$ *then there exists a path between* X_1 *and* X_2 *in the moralized graph* $\mathcal{M}[\mathcal{G}]$ *such that all the nodes on the path are in* Γ. *A continuous connected component is* maximal *if it is not a proper subset of any larger continuous connected component. The* discrete neighbors *of a continuous connected component* $\boldsymbol{X}$ *are all the discrete variables that are adjacent to some node* $X \in \boldsymbol{X}$ *in the moralized graph* $\mathcal{M}[\mathcal{G}]$. ∎

For example, in the network figure 14.6a, all of the continuous variables $\{X, Y, Z\}$ are in a single continuous connected component, and all the discrete variables are its neighbors.

We can understand the implications of the strong triangulation requirement in terms of continuous connected components:

Theorem 14.5

> *Let $\mathcal{G}$ be a hybrid network, and let $\mathcal{T}$ be a strongly rooted clique tree for $\mathcal{G}$. Then for any maximal continuous connected component $\boldsymbol{X}$ in $\mathcal{G}$, with $\boldsymbol{D}$ its discrete neighbors, $\mathcal{T}$ includes a clique that contains (at least) all of the nodes in $\boldsymbol{D}$ and some node $X \in \boldsymbol{X}$.*

The proof is left as an exercise (exercise 14.6).

In the CLG of figure 14.6a, all of the continuous variables are in one connected component, and all of the discrete variables are its neighbors. Thus, a strongly rooted tree must have a clique that contains all of the discrete variables and one of the continuous ones. Indeed, the strongly rooted clique tree of figure 14.6d contains such a clique — the clique $\{A, B, C, Y\}$.

This analysis allows us to examine a CLG network, and immediately conclude lower bounds on the computational complexity of clique tree inference in that network. For example, the polytree CLG in figure 14.2 has a continuous connected component containing all of the continuous variables $\{X_1, \ldots, X_n\}$, which has all of the discrete variables as neighbors. Thus, any strongly rooted clique tree for this network necessarily has an exponentially large clique, which is as we would expect, given that this network is the basis for our $\mathcal{NP}$-hardness theorem.

☞ **Because many CLG networks have large continuous connected components that are adjacent to many discrete variables, a strongly rooted clique tree is often far too large to be useful. However, the algorithm presented in this section is of conceptual interest, since it clearly illustrates when the EP message passing can lead to inaccurate or even nonsensical answers.**

14.4 Nonlinear Dependencies

In the previous sections, we dealt with a very narrow class of continuous networks: those where all of the CPDs are parameterized as linear Gaussians. Unfortunately, this class of networks is inadequate as a model for many practical applications, even those involving only continuous variables. For example, as we discussed, in modeling the car's position as a function of its previous position and its velocity (as in example 5.20), rather than assume that the variance in the car's position is constant, it might be more reasonable to assume that the variance of the car is larger if the velocity is large. This dependence is nonlinear, and it cannot be accommodated within the framework of linear Gaussians. In this section, we relax the assumption of linearity and present one approach for dealing with continuous networks that include nonlinear dependencies. We note that the techniques in this section can be combined with the ones described in section 14.3 to allow our algorithms to extend to networks that allow discrete variables to depend on continuous parents.

Once again, the standard solution in this setting can be viewed as an instance of the general expectation propagation framework described in section 11.4.4, and used before in the context of the CLG models. Since we cannot tractably represent and manipulate the exact factors in this setting, we need to use an approximation, by which intermediate factors in the computation are approximated using a compact parametric family. Once again, we choose the family of Gaussian measures as our representation.

At a high level, the algorithm proceeds as follows. As in expectation propagation (EP), each cluster C_i maintains its potentials in a nonexplicit form, as a factor set $\vec{\phi}_i$; some of these factors are from the initial factor set Φ, and others from the incoming messages into C_i. Importantly, due to our use of a Gaussian representation for the EP factors, the messages are all Gaussian measures.

To pass a message from C_i to C_j, C_i approximates the product of the factors in $\vec{\phi}_i$ as a Gaussian distribution, a process called *linearization*, for reasons we will explain. The resulting Gaussian distribution can then be marginalized onto $S_{i,j}$ to produce the approximate cluster marginal $\tilde{\sigma}_{i \to j}$. Essentially, the combination of the linearization step and the marginalization of the resulting Gaussian over the sepset give rise to the weak marginalization operation that we described.

The basic computational step in this algorithm is the linearization operation. We provide several options for performing this operation, and discuss their trade-offs. We then describe in greater detail how this operation can be used within the EP algorithm, and the constraints that it imposes on its detailed implementation.

14.4.1 Linearization

We first consider the basic computational task of approximating a distribution $p(X_1, \ldots, X_d)$ as a Gaussian distribution $\hat{p}(\boldsymbol{X})$. For the purposes of this section, we assume that our distribution is defined in terms of a Gaussian distribution $p_0(Z_1, \ldots, Z_l) = \mathcal{N}(\boldsymbol{Z} \mid \boldsymbol{\mu}; \Sigma)$ and a set of deterministic functions $X_i = f_i(\boldsymbol{Z}_i)$. Intuitively, the auxiliary $\boldsymbol{Z}$ variables encompass all of the stochasticity in the distribution, and the functions f_i serve to convert these auxiliary variables into the variables of interest. For a vector of functions $\vec{f} = (f_1, \ldots, f_d)$ and a Gaussian distribution p_0, we can now define $p(\boldsymbol{X}, \boldsymbol{Z})$ to be the distribution that has $p(\boldsymbol{Z}) = p_0(\boldsymbol{Z})$ and $X_i = f_i(\boldsymbol{Z})$ with probability 1. (Note that this distribution has a point mass at the discrete points where $\boldsymbol{X} = \vec{f}(\boldsymbol{Z})$.) We use $p(\boldsymbol{X}) = p_0(\boldsymbol{Z}) \bigoplus [\boldsymbol{X} = \vec{f}(\boldsymbol{Z})]$ to refer to the marginal of this distribution over $\boldsymbol{X}$.

Our goal is to compute a Gaussian distribution $\hat{p}(\boldsymbol{X})$ that approximates this marginal distribution $p(\boldsymbol{X})$. We call the procedure of determining $\hat{p}$ from p_0 and $\vec{f}$ a *linearization* of $\vec{f}$. We now describe the two main approaches that have been proposed for the linearization operation.

14.4.1.1 Taylor Series Linearization

As we know, if $p_0(\boldsymbol{Z})$ is a Gaussian distribution and $X = f(\boldsymbol{Z})$ is a linear function, then $p(X) = p(f(\boldsymbol{Z}))$ is also a Gaussian distribution. Thus, one very simple and commonly used approach is to approximate f as a linear function $\hat{f}$, and then define $\hat{p}$ in terms of $\hat{f}$.

Taylor series   The most standard linear approximation for $f(\boldsymbol{Z})$ is the *Taylor series* expansion around the mean of $p_0(\boldsymbol{Z})$:

$$\hat{f}(\boldsymbol{Z}) = f(\boldsymbol{\mu}) + \nabla f|_{\boldsymbol{\mu}} \, \boldsymbol{Z}. \tag{14.13}$$

Extended Kalman filter   The Taylor series is used as the basis for the famous *extended Kalman filter* (see section 15.4.1.2).

Although the Taylor series expansion provides us with the optimal linear approximation to *f*, the Gaussian $\hat{p}(X) = p_0(\mathbf{Z}) \oplus \hat{f}(\mathbf{Z})$ may not be the optimal Gaussian approximation to $p(X) = p_0(\mathbf{Z}) \oplus f(\mathbf{Z})$.

Example 14.11

Consider the function $X = Z^2$, and assume that $p(Z) = \mathcal{N}(Z \mid 0; 1)$. The mean of X is simply $E_p[X] = E_p[Z^2] = 1$. The variance of X is

$$\boldsymbol{Var}_p[X] = E_p[X^2] - E_p[X]^2 = E_p[Z^4] - E_p[Z^2]^2 = 3 - 1^2 = 2.$$

On the other hand, the first-order Taylor series approximation of f at the mean value $Z = 0$ is:

$$\hat{f}(Z) = 0^2 + (2Z)_{U=0}Z \equiv 0.$$

Thus, $\hat{p}(X)$ will simply be a delta function where all the mass is located at $X = 0$, a very poor approximation to p.   ∎

This example illustrates a limitation of this simple approach. In general, **the quality of the Taylor series approximation depends on how well $\hat{f}$ approximates f in the neighborhood of the mean of Z, where the size of the neighborhood is determined by the variance of $p_0(\mathbf{Z})$. The approximation is good only if the linear term in the Taylor expansion of f dominates in this neighborhood, and the higher-order terms are small.** In many practical situations, this is not the case, for example, when f changes very rapidly relative to the variance of $p_0(\mathbf{Z})$. In this case, using the simple Taylor series approach can lead to a very poor approximation.

14.4.1.2   M-Projection Using Numerical Integration

The Taylor series approach uses what may be considered an indirect approach to approximating p: we first simplify the nonlinear function f and only then compute the resulting distribution.

M-projection

Alternatively, we can directly approximate p using a Gaussian distribution $\hat{p}$, by using the *M-projection* operation introduced in definition 8.4. Here, we select the Gaussian distribution $\hat{p}$ that minimizes $D(p\|\hat{p})$.

In proposition 14.2 we provided a precise characterization of this operation. In particular, we showed that we can obtain the M-projection of p by evaluating the following set of integrals, corresponding to the moments of p:

$$E_p[X_i] = \int_{-\infty}^{\infty} f_i(\mathbf{z})p_0(\mathbf{z})d\mathbf{z} \tag{14.14}$$

$$E_p[X_iX_j] = \int_{-\infty}^{\infty} f_i(\mathbf{z})f_j(\mathbf{z})p_0(\mathbf{z})d\mathbf{z}. \tag{14.15}$$

From these moments, we can derive the mean and covariance matrix for p, which gives us precisely the M-projection. Thus, the M-projection task reduces to one of computing the expectation of some function f (which may be f_i or a product f_if_j) relative to our distribution p_0. Before we discuss the solution of these integrals, it is important to inject a note of caution. Even if p is a valid density, its moments may be infinite, preventing it from being approximated by a Gaussian.

In some cases, it is possible to solve the integrals in closed form, leading to an efficient and optimal way of computing the best Gaussian approximation. For instance, in the case of example 14.11, equation (14.14) reduces to computing $E_p[Z^2]$, where p is $\mathcal{N}(0; 1)$, an integral that can be easily solved in closed form. Unfortunately, for many functions f, these integrals have no closed-form solutions. However, because our goal is simply to estimate these quantities, we can use *numerical integration* methods. There are many such methods, with various trade-offs. In our setting, we can exploit the fact that our task is to integrate the product of a function and a Gaussian. Two methods that are particularly effective in this setting are described in the following subsections.

numerical integration (margin)

Gaussian Quadrature *Gaussian quadrature* is a method that was developed for the case of one-dimensional integrals. It approximates integrals of the form $\int_a^b W(z) f(z) dz$ where $W(z)$ is a known nonnegative function (in our case a Gaussian). Based on the function W, we choose m points $z_1, \ldots, z_m$ and m weights $w_1, \ldots, w_m$ and approximate the integral as:

Gaussian quadrature (margin)

$$\int_a^b W(z) f(z) dz \approx \sum_{j=1}^m w_j f(z_j). \tag{14.16}$$

The points and weights are chosen such that the integral is exact if f is a polynomial of degree $2m - 1$ or less. Such *rules* are said to have *precision* $2m - 1$.

integration rule (margin)

precision (margin)

To understand this construction, assume that we have chosen m points $z_1, \ldots, z_m$ and m weights $w_1, \ldots, w_m$ so that equation (14.16) holds with equality for any monomial z^i for $i = 1, \ldots, 2m - 1$. Now, consider any polynomial of degree at most $2m - 1$: $f(z) = \sum_{i=0}^{2m-1} \alpha_i z^i$. For such an f, we can show (exercise 14.8) that:

$$\int_a^b W(z) f(z) dz = \sum_{j=1}^m w_j z_j^i.$$

Thus, if our points are exact for any monomial of degree up to $2m - 1$, it is also exact for any polynomial of this degree.

Example 14.12

Consider the case of $m = 2$. In order for the rule to be exact for $f_0, \ldots, f_3$, it must be the case that for $i = 0, \ldots, 3$ we have

$$\int_a^b W(z) f_i(z) dz = w_1 f_i(z_1) + w_2 f_i(z_2).$$

Assuming that $W(z) = \mathcal{N}(0; 1)$, $a = -\infty$, and $b = \infty$, we get the following set of four nonlinear equations

$$w_1 + w_2 = \int_{-\infty}^\infty \mathcal{N}(z \mid 0; 1) \, dz = 1$$

$$w_1 z_1 + w_2 z_2 = \int_{-\infty}^\infty \mathcal{N}(z \mid 0; 1) \, z \, dz = 0$$

$$w_1 z_1^2 + w_2 z_2^2 = \int_{-\infty}^\infty \mathcal{N}(z \mid 0; 1) \, z^2 \, dz = 1$$

$$w_1 z_1^3 + w_2 z_2^3 = \int_{-\infty}^\infty \mathcal{N}(z \mid 0; 1) \, z^3 \, dz = 0$$

The solution for these equations (up to swapping z_1 and z_2) is $w_1 = w_2 = 0.5$, $z_1 = -1$, $z_2 = 1$. This solution gives rise to the following approximation, which we apply to any function f:

$$\int_{-\infty}^{\infty} \mathcal{N}(0;1) f(z) dz \approx 0.5 f(-1) + 0.5 f(1).$$

This approximation is exact for any polynomial of degree 3 or less, and approximate for other functions. The error in the approximation depends on the extent to which f can be well approximated by a polynomial of degree 3. ∎

This basic idea generalizes to larger values of m.

Now, consider the more complex task of integrating a multidimensional function $f(Z_1, \ldots, Z_d)$. One approach is to use the Gaussian quadrature grid in each dimension, giving rise to a d-dimensional grid with m^d points. We can then evaluate the function at each of the grid points, and combine the evaluations together using the appropriate weights. Viewed slightly differently, this approach computes the d-dimensional integral recursively, computing, for each point in dimension i, the Gaussian-quadrature approximation of the integral up to dimension $i - 1$. This integration rule is accurate for any polynomial which is a sum of monomial terms, each of the form $\prod_{i=1}^{d} z_i^{a_i}$, where each $a_i \leq 2m - 1$. Unfortunately, this grid grows exponentially with d, which can be prohibitive in certain applications.

unscented transformation

exact monomials

Unscented Transformation   An alternative approach, called the *unscented transformation*, is based on the integration method of *exact monomials*. This approach uses grids designed specifically for Gaussians over $\mathbb{R}^d$. Intuitively, it uses the symmetry of the Gaussian around its axes to reduce the density of the required grid.

The simplest instance of the exact monomials framework uses $2d + 1$ points, as compared to the 2^d points required for the rule derived from Gaussian quadrature. To apply this transformation, it helps to assume that $p_o(\mathbf{Z})$ is a standard Gaussian $p_0(\mathbf{Z}) = \mathcal{N}(\mathbf{Z} \mid \mathbf{0}; I)$ where I is the identity matrix. In cases where p_0 is not of that form, so that $p_0(\mathbf{Z}) = \mathcal{N}(\mathbf{Z} \mid \boldsymbol{\mu}; \Sigma)$, we can do the following change of variable transformation: Let A be the *matrix square root* of Σ, that is, A is a $d \times d$ matrix such that $\Sigma = A^T A$. We define $\tilde{p}_0(\mathbf{Z}') = \mathcal{N}(\mathbf{Z}' \mid \mathbf{0}; I)$. We can now show that

$$p_0(\mathbf{Z}) = \tilde{p}_0(\mathbf{Z}') \bigoplus [\mathbf{Z} = A\mathbf{Z}' + \boldsymbol{\mu}]. \tag{14.17}$$

We can now perform a change of variables for each of our functions, defining $\tilde{f}_i(\mathbf{Z}) = f_i(A\mathbf{Z} + \boldsymbol{\mu})$, and perform our moment computation relative to the functions $\tilde{f}_i$ rather than f_i.

Now, for $i = 1, \ldots, d$, let $\mathbf{z}_i^+$ be the point in $\mathbb{R}^d$ which has $z_i = +1$ and $z_j = 0$ for all $j \neq i$. Similarly, let $\mathbf{z}_i^- = -\mathbf{z}_i^+$. Let $\lambda \neq 0$ be any number. We then use the following integration rule:

$$\int_{-\infty}^{\infty} W(\mathbf{z}) f(\mathbf{z}) d\mathbf{z} \approx \left(1 - \frac{d}{\lambda^2}\right) f(\mathbf{0}) + \sum_{i=1}^{d} \frac{1}{2\lambda^2} f(\lambda \mathbf{z}_i^+) + \sum_{i=1}^{d} \frac{1}{2\lambda^2} f(\lambda \mathbf{z}_i^-). \tag{14.18}$$

In other words, we evaluate f at the mean of the Gaussian, $\mathbf{0}$, and then at every point which is $\pm \lambda$ away from the mean for one of the variables Z_i. We then take a weighted average of these points, for appropriately chosen weights. Thus, this rule, like Gaussian quadrature, is defined in terms of a set of points $z_0, \ldots, z_{2d}$ and weights $w_0, \ldots, w_{2d}$, so that

$$\int_{-\infty}^{\infty} W(z)f(z)dz = \sum_{i=0}^{2d} w_i f(z_i).$$

This integration rule is used as the basis for the *unscented Kalman filter* (see section 15.4.1.2).

The method of exact monomials can be used to provide exact integration for all polynomials of degree p or less, that is, to all polynomials where each monomial term $\prod_{i=1}^{d} z_i^{a_i}$ has $\sum_{i=1}^{d} a_i \le p$. Therefore, the method of exact monomials has precision p. In particular, equation (14.18) provides us with a rule of precision 3. Similar rules exist that achieve higher precision. For example, we can obtain a method of precision 5 by evaluating f at $\mathbf{0}$, at the $2d$ points that are $\pm \lambda$ away from the mean along one dimension, and at the $2d(d-1)$ points that are $\pm \lambda$ away from the mean along two dimensions. The total number of points is therefore $2d^2 + 1$.

Note that the precision-3 rule is less precise than the one obtained by using Gaussian quadrature separately for each dimension: For example, if we combine one-dimensional Gaussian quadrature rules of precision 2, we will get a rule that is also exact for monomials such as $z_1^2 z_2^2$ (but not for the degree 3 monomial z_1^3). However, the number of grid points used in this method is exponentially lower.

The parameter λ is a free parameter. Every choice of $\lambda \ne 0$ results in a rule of precision 3, but different choices lead to different approximations. Small values of λ lead to more local approximations, which are based on the behavior of f near the mean of the Gaussian and are less affected by the higher order terms of f.

14.4.1.3 Discussion

We have suggested several different methods for approximating q as a Gaussian distribution. What are the trade-offs between them? We begin with two examples.

Example 14.13

Figure 14.7(top) illustrates the two different approximations in comparison to the optimal approximation (the correct mean and covariance) obtained by sampling. We can see that the unscented transformation is almost exact, whereas the linearization method makes significant errors in both mean and covariance.

The bottom row provides a more quantitative analysis for the simple nonlinear function $Y = \sqrt{(\sigma Z_1)^2 + (\sigma Z_2)^2}$. The left panel presents results for $\sigma = 2$, showing the optimal Gaussian M-projection and the approximations using three methods: Taylor series, exact monomials with precision 3, and exact monomials with precision 5. The "optimal" approximation is estimated using a very accurate Gaussian quadrature rule with a grid of 100×100 integration points. We can see that the precision-5 rule is very accurate, but even the precision-3 rule is significantly more accurate than the Taylor series. The right panel shows the KL-divergence between the different approximations and the optimal approximation. We see that the quality of approximation of every method degrades as σ increases. This behavior is to be expected, since all of the methods are accurate for low-order polynomials, and the larger the σ, the larger the contribution of the higher-order terms. For small and medium variances, the Taylor series is the least exact of the three methods. For large variances, the precision 3 rule becomes significantly less accurate. The reason is that for $\sigma > 0.23$, the covariance matrices returned by the numerical integration procedure

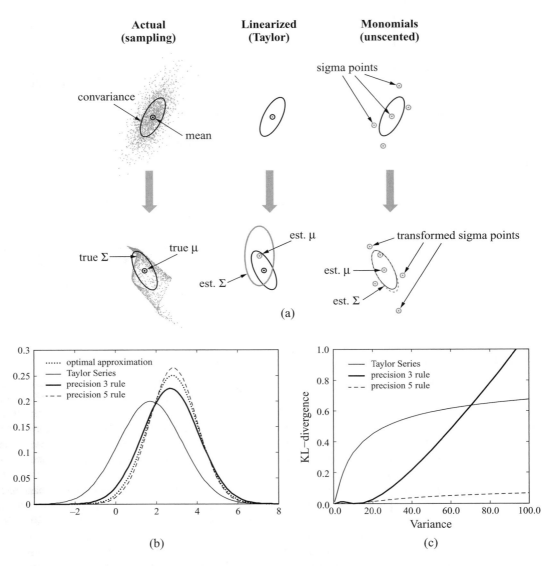

Figure 14.7   Comparisons of different Gaussian approximations for a nonlinear dependency. The top row (adapted with permission from van der Merwe et al. (2000a)) illustrates the process of different approximation methods and the results they obtain; the function being linearized is a feed-forward neural network with random weights. The bottom row shows a more quantitative analysis for the function $f(Z_1, Z_2) = \sqrt{(\sigma Z_1)^2 + (\sigma Z_2)^2}$. The left panel shows the different approximations when $\sigma^2 = 4$, and the right panel the KL-divergence from optimal approximation as a function of σ^2.

are illegal, and must be corrected. The correction produces reasonable answers for values of σ up to $\sigma = 4$, and then degrades. However, it is important to note that, for high variances, the Gaussian approximation is a poor approximation to p in any case, so that the whole approach of using a Gaussian approximation in inference breaks down. For low variances, where the Gaussian approximation is reasonable, even the corrected precision-3 rule significantly dominates the Taylor series approach. ∎

From a computational perspective, the Gaussian quadrature method is the most precise, but also the most expensive. In practice, one would only apply it in cases where precision was of critical important and the dimension was very low. The cost of the other two methods depends on the function f that we are trying to linearize. The linearization method (equation (14.13)) requires that we evaluate f and each of f's d partial derivatives at the point $\mathbf{0}$. In cases where the partial derivative functions can be written in closed form, this process requires only $d + 1$ function evaluations. By contrast, the precision 3 method requires $2d + 1$ evaluations of the function f, and the precision 5 method requires $2d^2 + 1$ function evaluations of f. In addition, to use the numerical integration methods we need to convert our distribution to the form of equation (14.17), which is not always required for the Taylor series linearization. Finally, one subtle problem can arise when using numerical integration to perform the M-projection operation: Here, the quantities of equation (14.14)–(14.15) are computed using an approximate procedure. Thus, although for exact integration, the covariance matrix defined by these equations is guaranteed to be positive definite; this is not the case for the approximate quantities, where the approximation may give rise to a matrix that is not positive definite. See section 14.7 for references to a modified approach that avoids this problem.

☞ Putting computational costs aside, **the requirement in the linearization approach of computing the gradient may be a significant issue in some settings. Some functions may not be differentiable (for example, the $\max$ function), preventing the use of the Taylor series expansion. Furthermore, even if f is differentiable, computing its gradient may still be difficult. In some applications, f might not even be given in a parametric closed form; rather, it might be implemented as a lookup table, or as a function in some programming language. In such cases, there is no simple way to compute the derivatives of the Taylor series expansion, but it is easy to evaluate f on a given point, as required for the numerical integration approach.**

14.4.2 Expectation Propagation with Gaussian Approximation

The preceding section developed the basic tool of approximating a distribution as a Gaussian. We now show how these methods can be used to perform expectation propagation message passing inference in nonlinear graphical models. Roughly speaking, at each step, we take all of the factors that need to be multiplied, and we approximate as a Gaussian the measure derived by multiplying all of these factors. We can then use this Gaussian distribution to form the desired message.

For the application of this method, we assume that each of the factors in our original distribution Φ defines a conditional distribution over a single random variable in terms of others. This assumption certainly holds in the case of standard Bayesian networks, where all CPDs are (by definition) of this form. It is also the case for many Markov networks, especially in the continuous case. We further assume that each of the factors in Φ can be written as a deterministic function:

$$X_i = f_i(\boldsymbol{Y}_i, \boldsymbol{W}_i), \tag{14.19}$$

where $\boldsymbol{Y}_i \subseteq \mathcal{X}$ are the model variables on which X_i's CPD depends, and $\boldsymbol{W}_i$ are "new" standard Gaussian random variables that capture all of the stochasticity in the CPD of X_i. We call these W's *exogenous* variables, since they capture stochasticity that is outside the model. (See also section 21.4.)

Although there are certainly factors that do not satisfy these assumptions, this representational class is quite general, and encompasses many of the factors used in practical applications. Most obviously, using the transformation of equation (14.17), this representation encompasses any Gaussian distribution. However, it also allows us to represent many nonlinear dependencies:

Example 14.14   *Consider the nonlinear CPD $X \sim \mathcal{N}\left(\sqrt{Y_1^2 + Y_2^2}; \sigma^2\right)$. We can reformulate this CPD in terms of a deterministic, nonlinear function, as follows: We introduce a new exogenous variable W that captures the stochasticity in the CPD. We then define $X = f(Y_1, Y_2, W)$ where $f(Y_1, Y_2, W) = \sqrt{Y_1^2 + Y_2^2} + \sigma W$.*   ∎

In this example, as in many real-world CPDs, the dependence of X on the stochastic variable W is linear. However, the same idea also applies in cases where the variance is a more complex function:

Example 14.15   *Returning again to example 5.20, assume we want the variance of the vehicle's position at time $t + 1$ to depend on its time t velocity, so that we have more uncertainty about the vehicle's next position if it is moving faster. Thus, for example, we might want to encode a distribution $\mathcal{N}\left(X' \mid X + V; \rho V^2\right)$. We can do so by introducing an exogenous standard Gaussian variable Z and defining*

$$X' = X + V + \sqrt{\rho}V Z.$$

It is not difficult to verify that X' has the appropriate Gaussian distribution.   ∎

We now apply the EP framework of section 11.4.3.2. As there, we maintain the potential at each cluster $\boldsymbol{C}_i$ as a factor set $\vec{\phi}_i$; some of those factors are initial factors, whereas others are messages sent to cluster $\boldsymbol{C}_i$. Our initial potentials are all of the form equation (14.19), and since we project all messages into the Gaussian parametric family, all of the incoming messages are Gaussians, which we can reformulate as a standard Gaussian and a set of deterministic functions.

In principle, it now appears that we can take all of the incoming messages, along with all of the exogenous Gaussian variables $\boldsymbol{W}_i$, to produce a single Gaussian distribution p_0. We can then apply the linearization procedures of section 14.4.1 to obtain a Gaussian approximation. However, a closer examination reveals several important subtleties that must be addressed.

14.4.2.1   Dealing with Evidence

Our discussion so far has sidestepped the issue of evidence: nowhere did we discuss the place at which evidence is instantiated into the factors. In the context of discrete variables, this issue was resolved in a straightforward way by restricting the factors (as we discussed in section 9.3.2). This restriction could be done at any time during the course of the message passing algorithm (as long as the observed variable is never eliminated).

In the context of continuous variables, the situation is more complex, since any assignment to a variable induces a density over a subspace of measure 0. Thus, when we observe $X = x$, we must a priori restrict all factors to the space where $X = x$. This operation is straightforward in the case of canonical forms, but it is somewhat subtler for nonlinear functions. For example, consider the nonlinear dependence of example 14.14. If we have evidence $Y_1 = y_1$, we can easily redefine our model as $X = g(Y_2, W)$ where $g(Y_2, W) = \sqrt{y_1^2 + Y_2^2} + \sigma W$. However, it is not so clear what to do with evidence on the dependent variable X.

The simplest solution to this problem, and one which is often used in practice, is to instantiate "downstream" the evidence in a cluster after the cluster is linearized. That is, in the preceding example, we would first linearize the function f in its cluster, resulting in a Gaussian distribution $p(X, Y_1, Y_2)$; we would then instantiate the evidence $X = x$ in the canonical form associated with this distribution, to obtain a new canonical form that is proportional to $p(Y_1, Y_2 \mid x)$.

This approach is simple, but it can be very inaccurate. In particular, the linearization operation (no matter how it is executed) depends on the distribution p_0 relative to which the linearization is performed. Our posterior distribution, given the evidence, can be very different from the prior distribution p_0, leading to a very different linearization of f. Better methods for taking the evidence into account during the linearization operation exist, but they are outside the scope of this book.

14.4.2.2 Valid Linearization

A second key issue is that all of the linearization procedures described earlier require that p_o be a Gaussian distribution, and not a general canonical form. This requirement is a significant one, and imposes constraints on one or more of: the structure of the cluster graph, the order in which messages are passed, and even the probabilistic model itself.

Example 14.16

Most simply, consider a chain-structured Bayesian network $X_1 \to X_2 \to X_3$, where all variables have nonlinear CPDs. We thus have a function $X_1 = f_1(W_1)$ and functions $X_2 = f_2(X_1, W_2)$ and $X_3 = f_3(X_2, W_3)$. In the obvious clique tree, we would have a clique C_1 with scope X_1, X_2, containing f_1 and f_2, and a clique C_2 with scope X_2, X_3, containing f_3. Further, assume that we have evidence $X_3 = x_3$.

In the discrete setting, we can simply restrict the factor in C_2 with the observation over X_3, producing a factor over X_2. This factor can then be passed as a message to C_1. In the nonlinear setting, however, we must first linearize f_3, and for that, we must have a distribution over X_2. But we can only obtain this distribution by multiplying into the factor $p(X_1)$, $p(X_2 \mid X_1)$, and $p(X_3 \mid X_2)$. In other words, in order to deal with C_2, we must first pass a message from C_1. ∎

order-constrained
message passing

In general, this requirement on *order constrained* message passing is precisely the same one that we faced for CLG distributions in section 14.3.3.1, with the same consequences. In a Bayesian network, this requirement constrains us to pass messages in a topological order. In other words, before we linearize a function $X_i = f_i(\mathrm{Pa}_{X_i}, \boldsymbol{W}_i)$, we must first linearize and obtain a Gaussian for every $Y_j \in \mathrm{Pa}_{X_i}$.

Example 14.17   *Consider the network of figure 14.6, but where we now assume that all variables (including A, B, C)
are continuous and utilize nonlinear CPDs. As in example 14.7, the clique tree of figure 14.6c does
not allow messages to be passed at all, since none of the cliques respect the ordering constraint. The
clique tree of (e) does allow messages to be passed, but only if the $\{A, B, X, Y\}$ clique is the first
to act, passing a message over A, Y to the clique $\{A, C, Y, Z\}$; this message defines a distribution
over the parents of C and Z, allowing them to be linearized.*   ∎

In the context of Markov networks, we again can partially circumvent the problem if we
assume that the factors in the network are all factor normalizable. However, this requirement
may not always hold in practice.

14.4.2.3   Linearization of Multiple Functions

A final issue, related to the previous one, arises from our assumption in section 14.4.1 that all of
the functions in a cluster depend only on a set of standard Gaussian random variables $\mathbf{Z}$. This
assumption is overly restrictive in almost all cluster graphs.

Example 14.18   *Consider again our chain-structured Bayesian network $X_1 \to X_2 \to X_3$ of example 14.16. Here,
C_1 has scope X_1, X_2, and contains both f_1 and f_2. This structure does not satisfy the preceding
requirements, since X_2 relies also on X_1.*

incremental
linearization

There are two ways to address the issue in this case. The first, which we call incremental
linearization *first linearizes f_1, and subsequently linearize f_2. This approach can be implemented
by making a separate clique containing just X_1. In this clique, we have a Gaussian distribution
$p_1(Z_1)$, and so we can compute a Gaussian approximation to $f_1(Z_1)$, producing a Gaussian
message $\tilde{\delta}_{1\to2}(X_1)$. We can then pass this message to a clique containing X_1, X_2, but only
the $f_2(X_2, Z_2)$ function; we now have a Gaussian distribution $p_2(X_2, Z_2)$, and can use the
techniques of section 14.4.1 to linearize $f_2(X_2, Z_2)$ into this Gaussian, to produce a Gaussian
distribution over X_1, X_2. (Note that $p_2(X_2, Z_2)$ is not a standard Gaussian, but we can use the
trick of equation (14.17) to change the coordinate system; see exercise 14.7.) We can then marginalize
the resulting Gaussian distribution onto X_2, and send a message $\tilde{\delta}_{2\to3}(X_2)$ to C_3, continuing this
process.*

simultaneous
linearization

As a second alternative, called simultaneous linearization, *we can linearize multiple nonlinear
functions into the same cluster together. We can implement this solution by substituting the variable
X_1 with its functional definition; that is, we can define $X_2 = g_2(Z_1, Z_2) = f_2(f_1(Z_1), Z_2)$, and
use g_2 rather than f_2 in the analysis of section 14.4.1.*   ∎

Both of these solutions generalize beyond this simple example. In the incremental approach,
we simply define smaller clusters, each of which contains at most one nonlinear function. We
then linearize one such function at a time, producing each time a Gaussian distribution. This
Gaussian is passed on to another cluster, and used as the basis for linearizing another nonlinear
function. The simultaneous approach linearizes several functions at once, substituting each
variable with the function that defines it, so as to make all the functions depend only on
Gaussian variables.

These two approaches make different approximations: Going back to example 14.16, in the
incremental approach, the approximation of f_2 uses a Gaussian approximation to the distribution

over X_1. If this approximation is poor, it may lead to a poor approximation for the distribution over X_2. Conversely, in the simultaneous approach, we are performing an integral for the function g_2, which may be more complicated than f_2, possibly leading to a poor approximation of its moments. In general, the errors made by each method are going to depend on the specific case at hand, and neither necessarily dominates the other.

14.4.2.4 Iterated Message Passing

The general algorithm we described can be applied within the context of different message passing schemes. Most simply, we define a clique tree, and we then do a standard upward and downward pass. However, as we discussed in the context of the general expectation propagation algorithm, our approximation within a cluster depends on the contents of that factor. In our setting, the approximation of $p \oplus f$ as a Gaussian depends on the distribution p. Depending on the order in which CPDs are linearized and messages are passed, the resulting distribution might be very different.

Example 14.19

Consider a network consisting of a variable X and its two children Y and Z, so that our two cliques are $C_1 = \{X, Y\}$ and $C_2 = \{X, Z\}$. Assume that we observe $Y = y$. If we include the prior $p(X)$ within C_1, then we first compute C_1's message, producing a Gaussian approximation to the joint posterior $\hat{p}(X, Y)$. This posterior is marginalized to produce $\hat{p}(Y)$, which is then used as the basis for linearizing Z. Conversely, if we include $p(X)$ within C_2, then the linearization of Z is done based on an approximation to X's prior distribution, generally leading to a very different linearization for $p(X, Z)$. ∎

In general, the closer the distribution used for the approximation is to the true posterior, the higher the quality of the approximation. Thus, in this example, we would prefer to first linearize Y and condition on its value, and only then to linearize Z. However, we cannot usually linearize all CPDs using the posterior. For example, assume that Z has another child W whose value is also observed; the constraint of using a topological ordering prevents us from linearizing W before linearizing Z, so we are forced to use a distribution over X that does not take into account the evidence on W.

The iterative EP algorithm helps address this limitation. In particular, once we linearize all of the initial clusters (subject to the constraints we mentioned), we can now continue to pass messages between the clusters using the standard belief update algorithm. At any point in the algorithm where cluster C_i must send a message (whether at every message or on a more intermittent basis), it can use the Gaussian defined by the incoming messages and the exogenous Gaussian variables to relinearize the functions assigned to it. The revised messages arising from this new linearization are then sent to adjacent clusters, using the standard EP update rule of equation (11.41) (that is, by subtracting the sufficient statistics of the previously sent message). This generally has the effect of linearizing using a distribution that is closer to the posterior, and hence leading to improved accuracy. However, it is important to remember that, as in section 14.3.3.2, the messages that arise in belief-update message passing may not be positive definite, and hence can give rise to canonical forms that are not legal Gaussians, and for which integration is impossible. To avoid catastrophic failures in the algorithm, it is important to check that any canonical form used for integration is, indeed, a normalizable Gaussian.

14.4.2.5   Augmented CLG Networks

One very important consequence of this algorithm is its ability to address one of the main limitations of CLG networks — namely, that discrete variables cannot have continuous parents. This limitation is a significant one. For example, it prevents us from modeling simple objects such as a thermostat, whose discrete on/off state depends on a continuous temperature variable. As we showed, we can easily model such dependencies, using, for example, a linear sigmoid model or its multinomial extension. The algorithm described earlier, which accommodates nonlinear CPDs, easily deals with this case.

Example 14.20
Consider the network $X \rightarrow A$, where X has a Gaussian distribution given by $p(X) = \mathcal{N}\left(X \mid \mu; \sigma\right)$ and the CPD of A is a softmax given by $P(A = a^1 \mid X = x) = 1/(1 + e^{c_1 x + c_0})$. The clique tree has a single clique (X, A), whose potential should contain the product of these two CPDs. Thus, it should contain two entries, one for a^1 and one for a^0, each of which is a continuous function: $p(x)P(a^1 \mid x)$ and $p(x)P(a^0 \mid x)$. As before, we approximate this potential using a canonical table, with an entry for every assignment to A, and where each entry is a Gaussian distribution. Thus, we would approximate each entry $p(x)P(a \mid x)$ as a Gaussian. Once again, we have a product of a function — $P(a \mid x)$ — with a Gaussian — $p(x)$. We can therefore use any of the methods in section 14.4.1 to approximate the result as a single Gaussian. ∎

This simple extension forms the basis for a general algorithm that deals with hybrid networks involving both continuous and discrete variables; see exercise 14.11.

14.5   Particle-Based Approximation Methods

All of the message passing schemes described before now utilize the Gaussian distribution as a parametric representation for the messages and (in some cases) even the clique potentials. This representation allows for truly exact inference only in the case of pure Gaussian networks, and it is exact in a weak sense for a very restricted class of CLG networks. If our original factors are far from being Gaussian, and, most particularly if they are multimodal, the Gaussian approximation used in these algorithms can be a very poor one. Unfortunately, there is no general-purpose parametric class that allows us to encode arbitrarily continuous densities.

An alternative approach is to use a semiparametric or nonparametric method, which allows us to avoid parametric assumptions that may not be appropriate in our setting. Such approaches are often applied in continuous settings, since there are many settings where the Gaussian approximation is clearly inappropriate. In this section, we discuss how such methods can be used in the context of inference in graphical models.

14.5.1   Sampling in Continuous Spaces

We begin by discussing the basic task of sampling from a continuous univariate measure. As we discussed in box 12.A, sampling from a discrete distribution can be done using straightforward methods. Efficient solutions, albeit somewhat more complex, are also available for other parametric forms, including Gaussians, Gamma distributions, and others; see section 14.7. Unfortunately, such solutions do not exist for all continuous densities. In fact, even distributions that can be characterized analytically may not have established sampling procedures.

A number of general-purpose approaches have been designed for sampling from arbitrary continuous densities. Perhaps the simplest is *rejection sampling* (or sometimes acceptance-rejection sampling). The basic idea is to sample a value from a proxy distribution and then "accept" it with probability that is proportional to the skew introduced by the proxy distribution. Suppose that we want to sample a variable X with density $p(x)$. Now suppose that we have a function $q(x)$ such that $q(x) > 0$ whenever $p(x) > 0$ and $q(x) \geq p(x)$. Moreover, suppose that we can sample from the distribution

$$Q(X) = \frac{1}{Z}q(x),$$

where Z is a normalizing constant. (Note that $Z > 1$, as $q(x)$ is an upper bound on a density function whose integral is 1.) In this case, we repeatedly perform the following steps, until acceptance:

1. Sample $x \sim Q(x)$, $u \sim \text{Unif}([0,1])$.

2. If $u < \frac{p(x)}{q(x)}$ return x.

The probability that this procedure returns the value x is exactly $p(x)$. The proof is analogous to the one we saw for importance sampling (see proposition 12.1). (The difference here is that we need to return a single sample rather than weight multiple samples.) In practice, we can speed up the procedure if we have also a lower bound $b(x)$ for $p(x)$ (that is $b(x) \leq p(x)$). In that case, if $u < \frac{b(x)}{q(x)}$ than we accept the sample x without evaluating $p(x)$. By finding easy to evaluate functions $q(x)$ and $b(x)$ that allow sampling from Q, and provide relatively tight bounds on the density $p(x)$, this method can accept samples within few iterations.

This straightforward approach, however, can be very slow if $q(x)$ is not a good approximation to $p(x)$. Fortunately, more sophisticated methods have also been designed. One example is described in exercise 12.22. See section 14.7 for other references.

14.5.2 Forward Sampling in Bayesian Networks

We now consider the application of forward sampling in a Bayesian network (algorithm 12.1). Assuming that we have a method for generating samples from the continuous density of each variable X_i given an assignment to its parents, we can use this method to generate samples from a broad range of continuous and hybrid Bayesian networks — far greater than the set of networks for which message passing algorithms are applicable. This approach generates random samples from the prior distribution $p(\mathcal{X})$. As we discussed, these samples can be used to estimate the expectation of any function $f(\mathcal{X})$, as in equation (12.1). For example, we can compute the mean $E[X]$ of any variable X by taking $f(\mathcal{X}) = X$, or its variance by taking $f(\mathcal{X}) = X^2$ and subtracting $E[X]^2$.

However, plain forward sampling is generally inapplicable to hybrid networks as soon as we have evidence over some of the network variables. Consider a simple network $X \rightarrow Y$, where we have evidence $Y = y$. Forward sampling generates $(x[m], y[m])$ samples, and rejects those in which $y[m] \neq y$. Because the domain of Y is a continuum, the probability that our sampling process will generate any particular value y is zero. Thus, none of our samples will match the observations, and all will be rejected.

A more relevant approach in the case of hybrid networks is a form of importance sampling, such as the likelihood weighting algorithm of algorithm 12.2. The likelihood weighting algorithm generalizes to the hybrid case in the most obvious way: Unobserved variables are sampled, based on the assignment to their parents, as described earlier. Observed variables are instantiated to their observed values, and the sample weight is adjusted. We can verify, using the same analysis as in section 12.2, that the appropriate adjustment to the importance weight for a continuous variable X_i observed to be x_i is the density $p(x_i \mid \boldsymbol{u}_i)$, where $\boldsymbol{u}_i$ is the assignment, in the sample, to Pa_{X_i}. The only implication of this change is that, unlike in the discrete case, this adjustment may be greater than 1 (because a density function is not bounded in the range $[0, 1]$). Although this difference does not influence the algorithm, it can increase the variance of the sample weights, potentially decreasing the overall quality of our estimator.

14.5.3   MCMC Methods

MCMC

The second type of sampling algorithms is based on *Markov chain Monte Carlo* (MCMC) methods. In general, the theory of Markov chains for continuous spaces is quite complicated, and many of the basic theorems that hold for discrete spaces do not hold, or hold only under certain assumptions. A discussion of these issues is outside the scope of this book. However, simple variants of the MCMC algorithms for graphical models continue to apply in this setting.

Gibbs sampling

The *Gibbs sampling* algorithm (algorithm 12.4) can also be applied to the case of hybrid networks. As in the discrete case, we maintain a sample, which is a complete assignment to the network variables. We then select variables X_i one at a time, and sample a new value for X_i from $p(X_i \mid \boldsymbol{u}_i)$, where $\boldsymbol{u}_i$ is the current assignment to X_i's Markov blanket. The only issue that might arise relates to this sampling step. In equation (12.23), we showed a particularly simple form for the probability of $X_i = x_i'$ given the assignment $\boldsymbol{u}_i$ to X_i's. The same derivation applies to the continuous case, simply replacing the summation with an integral:

$$p(x_i' \mid \boldsymbol{u}_i) = \frac{\prod_{\boldsymbol{C}_j \ni X_i} \beta_j(x_i', \boldsymbol{u}_i)}{\int_{x_i''} \prod_{\boldsymbol{C}_j \ni X_i} \beta_j((x_i'', \boldsymbol{u}_i))},$$

where $\boldsymbol{C}_j$ are the different potentials in the network.

In the discrete case, we could simply multiply the relevant potentials as tables, renormalize the resulting factor, and use it as a sampling distribution. In the continuous case, matters may not be so simple: the product of factors might not have a closed form that allows sampling.

Metropolis-Hastings

In this case, we generally resort to the use of a *Metropolis-Hastings* algorithm. Here, as in the discrete case, changes to a variable X_i are proposed by sampling from a local proposal distribution $\mathcal{T}_i^Q$. The proposed local change is either accepted or rejected using the appropriate acceptance probability, as in equation (12.26):

$$\mathcal{A}(\boldsymbol{u}_i, x_i \rightarrow \boldsymbol{u}_i, x_i') \;\; = \;\; \min\left[1, \frac{p(x_i' \mid \boldsymbol{u}_i)}{p(x_i \mid \boldsymbol{u}_i)} \frac{\mathcal{T}^Q(\boldsymbol{u}_i, x_i' \rightarrow \boldsymbol{u}_i, x_i)}{\mathcal{T}^Q(\boldsymbol{u}_i, x_i \rightarrow \boldsymbol{u}_i, x_i')}\right],$$

where (as usual) probabilities are replaced with densities.

The only question that needs to be addressed is the choice of the proposal distribution. One common choice is a Gaussian or student-t distribution centered on the current value x_i. Another is a uniform distribution over a finite interval also centered on x_i. These proposal distributions define a random walk over the space, and so a Markov chain that uses such a proposal distribution is often called a *random-walk chain*. These choices all guarantee that the Markov chain converges to the correct posterior, as long as the posterior is positive: for all $z \in Val(\mathcal{X} - \boldsymbol{E})$, we have that $p(z \mid e) > 0$. However, the rate of convergence can depend heavily on the window size: the variance of the proposal distribution. Different distributions, and even different regions within the same distribution, can require radically different window sizes. Picking an appropriate window size and adjusting it dynamically are important questions that can greatly impact performance.

14.5.4 Collapsed Particles

As we discussed in section 12.4, it is difficult to cover a large state space using full instantiations to the network variables. Much better estimates — often with much lower variance — can be obtained if we use collapsed particles. Recall that, when using collapsed particles, the variables $\mathcal{X}$ are partitioned into two subsets $\mathcal{X} = \boldsymbol{X}_p \cup \boldsymbol{X}_d$. A collapsed particle then consists of an instantiation $\boldsymbol{x}_p \in Val(\boldsymbol{X}_p)$, coupled with some representation of the distribution $P(\boldsymbol{X}_d \mid \boldsymbol{x}_p, \boldsymbol{e})$. The use of such particles relies on our ability to do two things: to generate samples from $\boldsymbol{X}_p$ effectively, and to represent compactly and reason with the distribution $P(\boldsymbol{X}_d \mid \boldsymbol{x}_p, \boldsymbol{e})$.

The notion of collapsed particles carries over unchanged to the hybrid case, and virtually every algorithm that applied in the discrete case also applies here. Indeed, collapsed particles are often particularly suitable in the setting of continuous or hybrid networks: In many such networks, if we select an assignment to some of the variables, the conditional distribution over the remaining variables can be represented (or well approximated) as a Gaussian. Since we can efficiently manipulate Gaussian distributions, it is generally much better, in terms of our time/accuracy trade-off, to try and maintain a closed-form Gaussian representation for the parts of the distribution for which such an approximation is appropriate.

Although this property can be usefully exploited in a variety of networks, one particularly useful application of collapsed particles is motivated by the observation that inference in a purely continuous network is fairly tractable, whereas inference in the simplest hybrid networks — polytree CLGs — can be very expensive. Thus, if we can "erase" the discrete variables from the network, the result is a much simpler, purely continuous network, which can be manipulated using the methods of section 14.2 in the case of linear Gaussians, and the methods of section 14.4 in the more general case.

 Thus, CLG networks are often effectively tackled using a collapsed particles approach, where the instantiated variables in each particular are the discrete variables, $\boldsymbol{X}_p = \Delta$, and the variables maintained in a closed-form distribution are the continuous variables, $\boldsymbol{X}_d = \Gamma$. We can now apply any of the methods described in section 12.4, often with some useful shortcuts. As one example, likelihood weighting can be easily applied, since discrete variables cannot have continuous parents, so that the set $\boldsymbol{X}_p$ is upwardly closed, allowing for easy sampling (see exercise 14.12). The application of MCMC methods is also compelling in this case, and it can be made more efficient using incremental update methods such as those of exercise 12.27.

14.5.5   Nonparametric Message Passing

Yet another alternative is to use a hybrid approach that combines elements from both particle-based and message passing algorithms. Here, the overall algorithm uses the structure of a message passing (clique tree or cluster graph) approach. However, we use the ideas of particle-based inference to address the limitations of using a parametric representation for the intermediate factors in the computation. Specifically, rather than representing these factors using a single parametric model, we encode them using a nonparametric representation that allows greater flexibility to capture the properties of the distribution. Thus, we are essentially still reparameter-izing the distribution in terms of a product of cluster potentials (divided by messages), but each cluster potential is now encoded using a nonparametric representation.

The advantage of this approach over the pure particle-based methods is that the samples are generated in a much lower-dimensional space: single cluster, rather than the entire joint distribution. This can alleviate many of the issues associated with sampling in a high-dimensional space. On the other side, we might introduce additional sources of error. In particular, if we are using a loopy cluster graph rather than a clique tree as the basis for this algorithm, we have all of the same errors that arise from the representation of a distribution as a set of pseudo-marginals.

One can construct different instantiations of this general approach, which can vary along several axes. The first is the message passing algorithm used: clique tree versus cluster graph, sum-product versus belief update. The second is the form of the representation used for factors and messages: plain particles; a nonparametric density representation; or a semiparametric representation such as a histogram. Finally, we have the approach used to approximate the factor in the chosen representation: importance sampling, MCMC, or a deterministic approximation.

14.6   Summary and Discussion

In this chapter, we have discussed some of the issues arising in applying inference to networks involving continuous variables. While the semantics of such networks is easy to define, they raise considerable challenges for the inference task.

The heart of the problem lies in the fact that we do not have a universal representation for factors involving continuous variables — one that is closed under basic factor operations such as multiplication, marginalization, and restriction with evidence. This limitation makes it very difficult to design algorithms based on variable elimination or message passing. Moreover, the difficulty is not simply a matter of our inability to find good algorithms. Theoretical analysis shows that classes of network structures that are tractable in the discrete case (such as polytrees) give rise to $\mathcal{NP}$-hard inference problems in the hybrid case.

Despite the difficulties, continuous variables are ubiquitous in practice, and so significant work has been done on the inference problem for such models.

Most message passing algorithms developed in the continuous case use Gaussians as a lingua franca for factors. This representation allows for exact inference only in a very limited class of models: clique trees for linear Gaussian networks. However, the exact same factor operations provide a basis for a belief propagation algorithm for Gaussian networks. This algorithm is easy to implement and has several satisfying guarantees, such as guaranteed convergence under fairly weak conditions, producing the exact mean if convergence is achieved. This technique allows us to perform inference in Gaussians where manipulating the full covariance matrix is intractable.

In cases where not all the potentials in the network are Gaussians, the Gaussian representation is generally used as an approximation. In particular, a standard approach uses an instantiation of the expectation propagation algorithm, using M-projection to approximate each non-Gaussian distribution as a Gaussian during message passing. In CLG networks, where factors represent a mixture of (possibly exponentially many) Gaussians derived from different instantiations of discrete variables, the M-projection is used to collapse the mixture components into a single Gaussian. In networks involving nonlinear dependencies between continuous variables, or between continuous and discrete variables, the M-projection involves linearization of the nonlinear dependencies, using either a Taylor expansion or numerical integration. While simple in principle, this application of EP raises several important subtleties. In particular, the M-projection steps can only be done over intermediate factors that are legal distributions. This restriction imposes significant constraints on the structure of the cluster graph and on the order in which messages can be passed. **Finally, we note that the Gaussian approximation is good in some cases, but can be very poor in others. For example, when the distribution is multimodal, the Gaussian M-projection can be a very broad (perhaps even useless) agglomeration of the different peaks.**

This observation often leads to the use of approaches that use a nonparametric approximation. Commonly used approaches include standard sampling methods, such as importance sampling or MCMC. Even more useful in some cases is the use of collapsed particles, which avoid sampling a high-dimensional space in cases where parts of the distribution can be well approximated as a Gaussian. Finally, there are also useful methods that integrate message passing with nonparametric approximations for the messages, allowing us to combine some of the advantages (and some of the disadvantages) of both types of approaches.

Our presentation in this chapter has only briefly surveyed a few of the key ideas related to inference in continuous and hybrid models, focusing mostly on the techniques that are specifically designed for graphical models. Manipulation of continuous densities is a staple of statistical inference, and many of the techniques developed there could be applied in this setting as well. For example, one could easily imagine message passing techniques that use other representations of continuous densities, or the use of other numerical integration techniques. Moreover, the use of different parametric forms in hybrid networks tends to give rise to a host of "special-case" models where specialized techniques can be usefully applied. In particular, even more so than for discrete models, it is likely that a good solution for a given hybrid model will require a combination of different techniques.

14.7 Relevant Literature

Perhaps the earliest variant of inference in Gaussian networks is presented by Thiele (1880), who defined what is the simplest special case of what is now known as the Kalman filtering algorithm (Kalman 1960; Kalman and Bucy 1961). Shachter and Kenley (1989) proposed the general idea of network-based probabilistic models, in the context of Gaussian influence diagrams. The first presentation of the general elimination algorithm for Gaussian networks is due to Normand and Tritchler (1992). However, the task of inference in pure Gaussian networks is highly related to the basic mathematical problem of solving a system of linear equations, and the elimination-based inference algorithms very similar to Gaussian elimination for solving such systems. Indeed, some of the early incarnations of these algorithms were viewed from that perspective; see Parter (1961) and Rose (1970) for some early work along those lines.

Iterative methods from linear algebra (Varga 2000) can also be used for solving systems of linear equations; in effect, these methods employ a form of local message passing to compute the marginal means of a Gaussian distribution. Loopy belief propagation methods were first proposed as a way of also estimating the marginal variances. Over the years, multiple authors (Rusmevichientong and Van Roy 2001; Weiss and Freeman 2001a; Wainwright et al. 2003a; Malioutov et al. 2006) have analyzed the convergence and correctness properties of Gaussian belief propagation, for larger and larger classes of models. All of these papers provide conditions that ensure convergence for the algorithm, and demonstrate that if the algorithm converges, the means are guaranteed to be correct. The recent analysis of Malioutov et al. (2006) is the most comprehensive; they show that their sufficient condition, called *walk-summability*, is equivalent to pairwise normalizability, and encompasses all of the classes of Gaussian models that were previously shown to be solvable via LBP (including attractive, nonfrustrated, and diagonally dominant models). They also show that the variances at convergence are an underestimate of the true variances, so that the LBP results are overconfident; their results point the way to partially correcting these inaccuracies. The results also analyze LBP for non-walksummable models, relating convergence of the variance to validity of the LBP computation tree.

The properties of conditional Gaussian distributions were studied by Lauritzen and Wermuth (1989). Lauritzen (1992) extended the clique tree algorithm to the task of inference in these models, and showed, for strongly rooted clique trees, the correctness of the discrete marginals and of the continuous means and variances. Lauritzen and Jensen (2001) and Cowell (2005) provided alternative variants of this algorithm, somewhat different representations, which are numerically more stable and better able to handle deterministic linear relationships, where the associated covariance matrix is not invertible. Lerner et al. (2001) extend Lauritzen's algorithm to CLG networks where continuous variables can have discrete children, and provide conditions under which this algorithm also has the same correctness guarantees on the discrete marginals and the moments of the continuous variables. The $\mathcal{NP}$-hardness of inference in CLG networks of simple structures (such as polytrees) was shown by Lerner and Parr (2001). Collapsed particles have been proposed by several researchers as a successful alternative to full collapsing of the potentials into a single Gaussian; methods include random sampling over particles (Doucet et al. 2000; Paskin 2003a), and deterministic search over the particle assignment (Lerner et al. 2000; Lerner 2002), a method particularly suitable in applications such as fault diagnosis, when the evidence is likely to be of low probability.

The idea of adaptively modifying the approximation of a continuous cluster potential during the course of message passing was first proposed by Kozlov and Koller (1997), who used a variable-resolution discretization approach (a semiparametric approximation). Koller et al. (1999) generalized this approach to other forms of approximate potentials. The expectation propagation algorithm, which uses a parametric approximation, was first proposed by Minka (2001b), who also made the connection to optimizing the energy functional under expectation matching constraints. Opper and Winther (2005) present an alternative algorithm based on a similar idea. Heskes et al. (2005) provide a unifying view of these two works. Heskes and Zoeter (2003) discuss the use of weak marginalization within the generalized belief propagation in a network involving both discrete and continuous variables.

The use of the Taylor series expansion to deal with nonlinearities in probabilistic models is

walk-summability

a key component of the extended Kalman filter, which extends the Kalman filtering method to nonlinear systems; see Bar-Shalom, Li, and Kirubarajan (2001) for a more in-depth presentation of these methods. The method of exact monomials, under the name unscented filter, was first proposed by Julier and Uhlmann (1997). Julier (2002) shows how this approach can be modified to address the problem of producing approximations that are not positive definite.

Sampling from continuous distributions is a core problem in statistics, on which extensive work has been done. Fishman (1996) provides a good overview of methods for various parametric families. Methods for sampling from other distributions include adaptive rejection sampling (Gilks and Wild 1992; Gilks 1992), adaptive rejection metropolis sampling (Gilks et al. 1995) and slice sampling (Neal 2003).

The BUGS system supports sampling for many continuous families, within their general MCMC framework. Several approaches combine sampling with message passing. Dawid et al. (1995); Hernández and Moral (1997); Kjærulff (1995b) propose methods that sample a continuous factor to turn it into a discrete factor, on which standard message passing can be applied. These methods vary on how the samples are generated, but the sampling is performed only once, in the initial message passing step, so that no adaptation to subsequent information is possible. Sudderth et al. (2003) propose *nonparametric belief propagation*, which uses a nonparametric approximation of the potentials and messages, as a mixture of Gaussians — a set of particles each with a small Gaussian kernel. They use MCMC methods to regenerate the samples multiple times during the course of message passing.

nonparametric
belief
propagation

Many of the ideas and techniques involving inference in hybrid systems were first developed in a temporal setting; we therefore also refer the reader to the relevant references in section 15.6.

14.8 Exercises

Exercise 14.1

Let $\boldsymbol{X}$ and $\boldsymbol{Y}$ be two sets of continuous variables, with $|\boldsymbol{X}| = n$ and $|\boldsymbol{Y}| = m$. Let

$$p(\boldsymbol{Y} \mid \boldsymbol{X}) = \mathcal{N}\left(\boldsymbol{Y} \mid \boldsymbol{a} + B\boldsymbol{X}; C\right)$$

where $\boldsymbol{a}$ is a vector of dimension m, B is an $m \times n$ matrix, and C is an $m \times m$ matrix. This dependence is a multidimensional generalization of a linear Gaussian CPD. Show how $p(\boldsymbol{Y} \mid \boldsymbol{X})$ can be represented as a canonical form.

Exercise 14.2

Prove proposition 14.3.

Exercise 14.3

Prove that setting evidence in a canonical form can be done as shown in equation (14.6).

Exercise 14.4★

Describe a method that efficiently computes the covariance of any pair of variables X, Y in a calibrated Gaussian clique tree.

Exercise 14.5★

Let $\boldsymbol{X}$ and $\boldsymbol{Y}$ be two sets of continuous variables, with $|\boldsymbol{X}| = n$ and $|\boldsymbol{Y}| = m$. Let $p(\boldsymbol{X})$ be an arbitrary density, and let

$$p(\boldsymbol{Y} \mid \boldsymbol{X}) = \mathcal{N}\left(\boldsymbol{Y} \mid \boldsymbol{a} + B\boldsymbol{X}; C\right)$$

where $\boldsymbol{a}$ is a vector of dimension m, B is an $m \times n$ matrix, and C is an $m \times m$ matrix. Show that the first two moments of $p(\boldsymbol{X}, \boldsymbol{Y}) = p(\boldsymbol{X})p(\boldsymbol{Y} \mid \boldsymbol{X})$ depend only on the first two moments of $p(\boldsymbol{X})$ and not on the distribution $p(\boldsymbol{X})$ itself.

Exercise 14.6★

Prove theorem 14.5.

Exercise 14.7

Prove the equality in equation (14.17).

Exercise 14.8

Prove equation (14.17).

Exercise 14.9★

Our derivation in section 14.4.1 assumes that p and $Y = f(\boldsymbol{U})$ have the same scope $\boldsymbol{U}$. Now, assume that $Scope[f] = \boldsymbol{U}$, whereas our distribution p has scope $\boldsymbol{U}, \boldsymbol{Z}$. We can still use the same method if we define $g(\boldsymbol{u}, \boldsymbol{u}') = f(\boldsymbol{u})$, and integrate g. This solution, however, requires that we perform integration in dimension $|\boldsymbol{U} \cup \boldsymbol{Z}|$, which is often much higher than $|\boldsymbol{U}|$. Since the cost of numerical integration grows with the dimension of the integrals, we can gain considerable savings by using only $\boldsymbol{U}$ to compute our approximation.

In this exercise, you will use the interchangeability of the Gaussian and linear Gaussian representations to perform integration in higher dimension.

   a. For $Z \in \boldsymbol{Z}$, show how we can write Z as a linear combination of variables in $\boldsymbol{X}$, with Gaussian noise.
   b. Use this expression to write $\boldsymbol{Cov}[Z; Y]$ as a function of the covariances $\boldsymbol{Cov}[X_i; Y]$.
   c. Put these results together in order to show how we can obtain a Gaussian approximation to $p(Y, \boldsymbol{Z})$.

Exercise 14.10

In some cases, it is possible to decompose a nonlinear dependency $Y = f(\boldsymbol{X})$ into finer-grained dependencies. For example, we may be able to decompose the nonlinear function f as $f(\boldsymbol{X}) = g(g_1(\boldsymbol{X}_1), g_2(\boldsymbol{X}_2))$, where $\boldsymbol{X}_1, \boldsymbol{X}_2 \subset \boldsymbol{X}$ are smaller subsets of variables.

Show how this decomposition can be used in the context of linearizing the function f in several steps rather than in a single step. What are the trade-offs for this approach versus linearizing f directly?

Exercise 14.11★★

Show how to combine the EP-based algorithms for CLGs and for nonlinear CPDs to address CLGs where discrete variables can have continuous parents. Your algorithm should specify any constraints on the message passing derived from the need to allow for valid M-projection.

Exercise 14.12★

Assume we have a CLG network with discrete variables Δ and continuous variables Γ. In this exercise, we consider collapsed methods that sample the discrete variables and perform exact inference over the continuous variables. Let $\boldsymbol{e}_d$ denote the discrete evidence and $\boldsymbol{e}_p$ the continuous evidence.

   a. Given a set of weighted particles such as those described earlier, show how we can estimate the expectation of a function $f(X_i)$ for some $X_i \in \Gamma$. For what functions f do you expect this analysis to give you drastically different answers from the "exact" CLG algorithm of section 14.3.4. (Ignore issues of inaccuracies arising from sampling noise or insufficient number of samples.)
   b. Show how we can efficiently apply collapsed likelihood weighting, and show precisely how the importance weights are computed.
   c. Now, consider a combined algorithm that generates a clique tree over Δ to generate particles $\boldsymbol{x}_p[1]$, ..., $\boldsymbol{x}_p[M]$ sampled exactly from $P(\Delta \mid \boldsymbol{e}_d)$. Show the computation of importance weights in this case. Explain the computational benefit of this approach over doing clique tree inference over the entire network.

15 *Inference in Temporal Models*

In chapter 6, we presented several frameworks that provide a higher-level representation language. We now consider the issue of performing probabilistic inference relative to these representations. The obvious approach is based on the observation that a template-based model can be viewed as a generator of ground graphical models: Given a skeleton, the template-based model defines a distribution over a ground set of random variables induced by the skeleton. We can then use any of our favorite inference algorithms to answer queries over this ground network. This process is called *knowledge-based model construction*, often abbreviated as *KBMC*. However, applying this simple idea is far from straightforward.

First, these models can easily produce models that are very large, or even infinite. Several approaches can be used to reduce the size of the network produced by KBMC; most obviously, given a set of ground query variables Y and evidence $E = e$, we can produce only the part of the network that is needed for answering the query $P(Y \mid e)$. In other words, our ground network is a dynamically generated object, and we can generate only the parts that we need for our current query. While this approach can certainly give rise to considerable savings in certain cases, in many applications the network generated is still very large. Thus, we have to consider whether our various inference algorithms scale to this setting, and what additional approximations we must introduce in order to achieve reasonable performance.

Second, the ground networks induced by a template-based model can often be densely connected; most obviously, both aggregate dependencies on properties of multiple objects and relational uncertainty can give rise to dense connectivity. Again, dense connectivity causes difficulties for all exact and most approximate inference algorithms, requiring algorithmic treatment.

Finally, these models give rise to new types of queries that are not easily expressible as standard probabilistic queries. For example, we may want to determine the probability that every person in our family tree has at least one close relative (for some appropriate definition of "close") with a particular disease. This query involves both universal and existential quantifiers; while it can be translated into a ground-level conjunction (over people in the family tree) of disjunctions (over their close relatives), this translation is awkward and gives rise to a query over a very large number of variables.

The development of methods that address these issues is very much an open area of research. In the context of general template-based models, the great variability of the models expressed in these languages limits our ability to provide general-purpose solutions; the existing approaches offer only partial solutions whose applicability at the moment is somewhat limited. We therefore do not review these methods in this book; see section 15.6 for some references. In the more

knowledge-based model construction

restricted context of temporal models, the networks have a uniform structure, and the set of relevant queries is better established. Thus, more work has been done on this setting. In the remainder of this chapter, we describe some of the exact and approximate methods that have been developed for inference in temporal models.

15.1   Inference Tasks

We now move to the question of inference in a dynamic Bayesian network (DBN). As we discussed, we can view a DBN as a "generator" for Bayesian networks for different time intervals. Thus, one might think that the inference task is solved. Once we generate a specific Bayesian network, we can simply run the inference algorithm of our choice to answer any queries. However, this view is overly simplistic in two different respects.

First, the Bayesian networks generated from a DBN can be arbitrarily large. Second, the type of reasoning we want to perform in a temporal setting is often different from the reasoning applicable in static settings. In particular, many of the reasoning tasks in a temporal domain are executed *online* as the system evolves.

filtering

For example, a common task in temporal settings is *filtering* (also called *tracking*): at any time point t, we compute our most informed beliefs about the current system state, given all of the evidence obtained so far. Formally, let $o^{(t)}$ denote the observation at time t; we want to keep track of $P(\mathcal{X}^{(t)} \mid o^{(1:t)})$ (or of some marginal of this distribution over some subset of variables).

belief state

As a probabilistic query, we can define the *belief state* at time t to be:

$$\sigma^{(t)}(\mathcal{X}^{(t)}) = P(\mathcal{X}^{(t)} \mid o^{(1:t)}).$$

Note that the belief state is exponentially large in the number of unobserved variables in $\mathcal{X}$. We therefore will not, in general, be interested in the belief state in its entirety. Rather, we must find an effective way of encoding and maintaining the belief state, allowing us to query the current probability of various events of interest (for example, marginal distributions over smaller subsets of variables).

prediction

The tracking task is the task of maintaining the belief state over time. A related task is the *prediction* task: at time t, given the observations $o^{(1:t)}$, predict the distribution over (some subset of) the variables at time $t' > t$.

smoothing

A third task, often called *smoothing*, involves computing the posterior probability of $\mathcal{X}^{(t)}$ given all of the evidence $o^{(1:T)}$ in some longer trajectory. The term "smoothing" refers to the fact that, in tracking, the evidence accumulates gradually. In cases where new evidence can have significant impact, the belief state can change drastically from one time slice to the next. By incorporating some future evidence, we reduce these temporary fluctuations. This process is particularly important when the lack of the relevant evidence can lead to temporary "misconceptions" in our belief state.

Example 15.1

In cases of a sensor failure (such as example 6.5), a single anomalous observation may not be enough to cause the system to realize that a failure has occurred. Thus, the first anomalous sensor reading, at time t_1, may cause the system to conclude that the car did, in fact, move in an unexpected direction. It may take several anomalous observations to reach the conclusion that the sensor has failed. By passing these messages backward, we can conclude that the sensor was already broken *at t_1, allowing us to discount its observation and avoid reaching the incorrect conclusion about the vehicle location.* ■

We note that smoothing can be executed with different time horizons of evidence going forward. That is, we may want to use all of the available evidence, or perhaps just the evidence from some window of a few time slices ahead.

A final task is that of finding the most likely trajectory of the system, given the evidence — $\arg\max_{\xi^{(0:T)}} P(\xi^{(0:T)} \mid o^{(1:T)})$. This task is an instance of the MAP problem.

☞ **In all of these tasks, we are trying to compute answers to standard probabilistic queries. Thus, we can simply use one of the standard inference algorithms that we described earlier in this book. However, this type of approach, applied naively, would require us to run inference on larger and larger networks over time and to maintain our entire history of observations indefinitely. Both of these requirements can be prohibitive in practice. Thus, alternative solutions are necessary to avoid this potentially unbounded blowup in the network size.**

In the remainder of our discussion of inference, we focus mainly on the tracking task, which presents us with many of the challenges that arise in other tasks. The solutions that we present for tracking can generally be extended in a fairly straightforward way to other inference tasks.

15.2 Exact Inference

We now consider the problem of exact inference in DBNs. We begin by focusing on the filtering problem, showing how the Markovian independence assumptions underlying our representation provide a simple recursive rule that does not require maintaining an unboundedly large representation. We then show how this recursive rule corresponds directly to the upward pass of inference in the unrolled network.

15.2.1 Filtering in State-Observation Models

We begin by considering the filtering task for state-observation models. Our goal here is to maintain the belief state $\sigma^{(t)}(\boldsymbol{X}^{(t)}) = P(\boldsymbol{X}^{(t)} \mid o^{(1:t)})$. As we now show, we can provide a simple recursive algorithm for propagating these belief states, computing $\sigma^{(t+1)}$ from $\sigma^{(t)}$.

Initially, $P(\boldsymbol{X}^{(0)})$ is precisely $\sigma^{(0)}$. Now, assume that we have already computed $\sigma^{(t)}(\boldsymbol{X}^{(t)})$. To compute $\sigma^{(t+1)}$ based on $\sigma^{(t)}$ and the evidence $o^{(t+1)}$, we first propagate the state forward:

$$
\begin{aligned}
\sigma^{(\cdot t+1)}(\boldsymbol{X}^{(t+1)}) &= P(\boldsymbol{X}^{(t+1)} \mid o^{(1:t)}) \\
&= \sum_{\boldsymbol{X}^{(t)}} P(\boldsymbol{X}^{(t+1)} \mid \boldsymbol{X}^{(t)}, o^{(1:t)}) P(\boldsymbol{X}^{(t)} \mid o^{(1:t)}) \\
&= \sum_{\boldsymbol{X}^{(t)}} P(\boldsymbol{X}^{(t+1)} \mid \boldsymbol{X}^{(t)}) \sigma^{(t)}(\boldsymbol{X}^{(t)}).
\end{aligned}
\tag{15.1}
$$

prior belief state

In words, this expression is the beliefs over the state variables at time $t+1$, given the observations only up to time t (as indicated by the $\cdot$ in the superscript). We can call this expression the *prior belief state* at time $t + 1$. In the next step, we condition this prior belief state to account for the

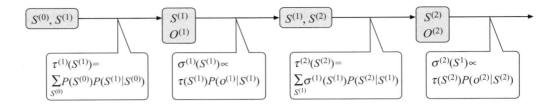

Figure 15.1   Clique tree for HMM

most recent observation $o^{(t+1)}$:

$$
\begin{aligned}
\sigma^{(t+1)}(\boldsymbol{X}^{(t+1)}) &= P(\boldsymbol{X}^{(t+1)} \mid o^{(1:t)}, o^{(t+1)}) \\
&= \frac{P(o^{(t+1)} \mid \boldsymbol{X}^{(t+1)}, o^{(1:t)})P(\boldsymbol{X}^{(t+1)} \mid o^{(1:t)})}{P(o^{(t+1)} \mid o^{(1:t)})} \\
&= \frac{P(o^{(t+1)} \mid \boldsymbol{X}^{(t+1)})\sigma^{(\cdot t+1)}(\boldsymbol{X}^{(t+1)})}{P(o^{(t+1)} \mid o^{(1:t)})}.
\end{aligned}
\tag{15.2}
$$

recursive filter   This simple *recursive filtering* procedure maintains the belief state over time, without keeping track of a network or a sequence of observations of growing length. To analyze the cost of this operation, let N be the number of states at each time point and T the total number of time slices. The belief-state forward-propagation step considers every pair of states s, s', and therefore it has a cost of $O(N^2)$. The conditioning step considers every state s' (multiplying it by the evidence likelihood and then renormalizing), and therefore it takes $O(N)$ time. Thus, the overall time cost of the message-passing algorithm is $O(N^2T)$. The space cost of the algorithm is $O(N^2)$, which is needed to maintain the state transition model.

15.2.2   Filtering as Clique Tree Propagation

The simple recursive filtering process is closely related to message passing in a clique tree. To understand this relationship, we focus on the simplest state-observation model — the HMM. Consider the process of exact clique tree inference, applied to the DBN for an HMM (as shown in figure 6.2). One possible clique tree for this network is shown in figure 15.1. Let us examine the messages passed in this tree in a sum-product clique tree algorithm, taking the last clique in the chain to be the root. In this context, the upward pass is also called the *forward pass*.

forward pass   The first message has the scope $S^{(1)}$ and represents $\sum_{S^{(0)}} P(S^{(0)})P(S^{(1)} \mid S^{(0)}) = P(S^{(1)})$. The next message, sent from the $S^{(1)}, O^{(1)}$ clique, also has the scope $S^{(1)}$, and represents $P(S^{(1)})P(o^{(1)} \mid S^{(1)}) = P(S^{(1)}, o^{(1)})$. Note that, if we renormalize this message to sum to 1, we obtain $P(S^{(1)} \mid o^{(1)})$, which is precisely $\sigma^{(1)}(S^{(1)})$. Continuing, one can verify that the message from the $S^{(1)}, S^{(2)}$ clique is $P(S^{(2)}, o^{(1)})$, and the one from the $S^{(2)}, O^{(2)}$ clique is $P(S^{(2)}, o^{(1)}, o^{(2)})$. Once again, if we renormalize this last message to sum to 1, we obtain exactly $P(S^{(2)} \mid o^{(1)}, o^{(2)}) = \sigma^{(2)}(S^{(2)})$.

     We therefore see that the forward pass of the standard clique-tree message passing algorithm provides us with a solution to the filtering problem. A slight variant of the algorithm gives us precisely the recursive update equations of equation (15.1) and (15.2). Specifically, assume that we normalize the messages from the $S^{(1)}, O^{(1)}$ clique as they are sent, resulting in a probability distribution. (As we saw in exercise 9.3, such a normalization step has no effect on the belief state computed in later stages, and it is beneficial in reducing underflow.)[1]

It is not hard to see that, with this slight modification, the sum-product message passing algorithm results in precisely the recursive update equations shown here: The message passing step executed by the $S^{(t)}, S^{(t+1)}$ clique is precisely equation (15.1), whereas the message passing step executed by the $S^{(t+1)}, O^{(t+1)}$ clique is precisely equation (15.2), with the division corresponding to the renormalization step.

Thus, the upward (forward) pass of the clique tree algorithm provides a solution to the filtering task, with no need for a downward pass. Similarly, for the prediction task, we can use essentially the same message-passing algorithm, but without conditioning on the (unavailable future) evidence. In terms of the clique tree formulation, the unobserved evidence nodes are barren nodes, and they can thus be dropped from the network; thus, the $S^{(t)}, O^{(t)}$ cliques would simply disappear. When viewed in terms of the iterative algorithm, the operation of equation (15.2) would be eliminated.

For the smoothing task, however, we also need to propagate messages backward in time. Once again, this task is clearly an inference task in the unrolled DBN, which can be accomplished using a clique tree algorithm. In this case, messages are passed in both directions in the clique tree. The resulting algorithm is known as the *forward-backward algorithm*. In this algorithm, the backward messages also have semantics. Assume that our clique tree is over the time slices $0, \ldots, T$. If we use the variable-elimination message passing scheme (without renormalization), the backward message sent to the clique $S^{(t)}, S^{(t+1)}$ represents $P(o^{((t+1):T)} \mid S^{(t+1)})$. If we use the belief propagation scheme, the backward message sent to this clique represents the fully informed (smoothed) distribution $P(S^{(t+1)} \mid o^{(1:T)})$. (See exercise 15.1.)

For the smoothing task, we need to keep enough information to reconstruct a full belief state at each time t. Naively, we might maintain the entire clique tree at space cost $O(N^2 T)$. However, by more carefully analyzing the role of cliques and messages, we can reduce this cost considerably. Consider variable-elimination message passing; in this case, the cliques contain only the initial clique potentials, all of which can be read from the 2-TBN template. Thus, we can cache only the messages and the evidence, at a total space cost of $O(NT)$. Unfortunately, space requirements that grow linearly in the length of the sequence can be computationally prohibitive when we are tracking the system for extended periods. (Of course, some linear growth is unavoidable if we want to remember the observation sequence; however, the size of the state space N is usually much larger than the space required to store the observations.) Exercise 15.2 discusses one approach to reducing this computational burden using a time-space trade-off.

15.2.3 Clique Tree Inference in DBNs

The clique tree perspective provides us with a general algorithm for tracking in DBNs. To derive the algorithm, let us consider the clique tree algorithm for HMMs more closely.

---

1. Indeed, in this case, the probability $P(S^{(t)}, o^{(1:t)})$ generally decays exponentially at t grows. Thus, without a renormalization step, numerical underflow would be inevitable.

forward-backward algorithm

Although we can view the filtering algorithm as performing inference on an unboundedly large clique tree, we never need to maintain more than a clique tree over two consecutive time slices. Specifically, we can create a (tiny) clique tree over the variables $S^{(t)}, S^{(t+1)}, O^{(t+1)}$; we then pass in the message $\sigma^{(t)}$ to the $S^{(t)}, S^{(t+1)}$ clique, pass the message to the $S^{(t+1)}, O^{(t+1)}$ clique, and extract the outgoing message $\sigma^{(t+1)}$. We can now forget this time slice's clique tree and move on to the next time slice's.

It is now apparent that the clique trees for all of the time slices are identical — only the messages passed into them differ. Thus, we can perform this propagation using a *template clique tree* Υ over the variables in the 2-TBN. In this setting, Υ would contain the two cliques $\{S, S'\}$ and $\{S', O'\}$, initialized with the potentials $P(S' \mid S)$ and $P(O' \mid S')$ respectively. To propagate the belief state from time t to time $t + 1$, we pass the time t belief state into Υ, by multiplying it into the clique $P(S' \mid S)$, taking S to represent $S^{(t)}$ and S' to represent $S^{(t+1)}$. We then run inference over this clique tree, including conditioning on $O' = o^{(t+1)}$. We can now extract the posterior distribution over S', which is precisely the required belief state $P(S^{(t+1)} \mid o^{(1:(t+1))})$. This belief state can be used as the input message for the next step of propagation.

The generalization to arbitrary DBNs is now fairly straightforward. We maintain a belief state $\sigma^{(t)}(\mathcal{X}^{(t)})$ and propagate it forward from time t to time $t + 1$. We perform this propagation step using clique tree inference as follows: We construct a template clique tree Υ, defined over the variables of the 2-TBN. We then pass a time t message into Υ, and we obtain as the result of clique tree inference in Υ a time $t + 1$ message that can be used as the input message for the next step. Most naively, the messages are full belief states, specifying the distribution over all of the unobserved variables.

In general, however, we can often reduce the scope of the message passed: As can be seen from equation (6.2), only the time t interface variables are relevant to the time $t + 1$ distribution. For example, consider the two generalized HMM structures of figure 6.3. In the factorial HMM structure, all variables but the single observation variable are in the interface, providing little savings. However, in the coupled HMM, all of the private observation variables are not in the interface, leaving a much smaller belief state whose scope is X_1, X_2, X_3. Observation variables are not the only ones that can be omitted from the interface; for example, in the network of figure 15.3a, the nonpersistent variable B is also not in the interface.

The algorithm is shown in algorithm 15.1. It passes messages corresponding to *reduced belief states* $\sigma^{(t)}(\mathcal{X}_I^{(t)})$. At phase t, it passes a time $t - 1$ reduced belief state into the template clique tree, calibrates it, and extracts the time t reduced belief state to be used as input for the next step. Note that the clique-tree calibration step is useful not only for the propagation step. It also provides us with other useful conclusions, such as the marginal beliefs over all individual variables $X^{(t)}$ (and some subsets of variables) given the observations $o^{(1:t)}$.

15.2.4   Entanglement

The use of a clique tree immediately suggests that we are exploiting structure in the algorithm, and so can expect the inference process to be tractable, at least in a wide range of situations. Unfortunately, that does not turn out to be the case. The problem arises from the need to represent and manipulate the (reduced) belief state $\sigma^{(t)}$ (a process not specified in the algorithm). Semantically, this belief state is a joint distribution over $\mathcal{X}_I^{(t)}$; if represented naively,

(left margin notes:)
template clique tree

reduced belief state

Algorithm 15.1 Filtering in a DBN using a template clique tree

---

Procedure CTree-Filter-DBN (
  $\langle \mathcal{B}_0, \mathcal{B}_\rightarrow \rangle,$ // DBN
  $o^{(1)}, o^{(2)}, \ldots$ // Observation sequence
)
1   Construct template clique tree Υ over $\mathcal{X}_I \cup \mathcal{X}'$
2     $\sigma^{(0)} \leftarrow P_{\mathcal{B}_0}(\mathcal{X}_I^{(0)})$
3   **for** $t = 1, 2, \ldots$
4         $\mathcal{T}^{(t)} \leftarrow \Upsilon$
5   Multiply $\sigma^{(t-1)}(\mathcal{X}_I^{(t-1)})$ into $\mathcal{T}^{(t)}$
6   Instantiate $\mathcal{T}^{(t)}$ with $o^{(t)}$
7   Calibrate $\mathcal{T}^{(t)}$ using clique tree inference
8   Extract $\sigma^{(t)}(\mathcal{X}_I^{(t)})$ by marginalization

---

it would require an exponential number of entries in the joint. At first glance, this argument appears specious. After all, one of the key benefits of graphical models is that high-dimensional distributions can be represented compactly by using factorization. It certainly appears plausible that we should be able to find a compact representation for our belief state and use our structured inference algorithms to manipulate it efficiently. As we now show, this very plausible impression turns out to be false.

Example 15.2

Consider our car network of figure 6.1, and consider our belief state at some time t. Intuitively, it seems as if there should be some conditional independence relations that hold in this network. For example, it seems as if Weather$^{(2)}$ and Location$^{(2)}$ should be uncorrelated. Unfortunately, they are not: if we examine the unrolled DBN, we see that there is an active trail between them going through Velocity$^{(1)}$ and Weather$^{(0)}$, Weather$^{(1)}$. This path is not blocked by any of the time 2 variables; in particular, Weather$^{(2)}$ and Location$^{(2)}$ are not conditionally independent given Velocity$^{(2)}$. In general, a similar analysis can be used to show that, for $t \geq 2$, no conditional independence assumptions hold in $\sigma^{(t)}$.   ∎

entanglement

This phenomenon, known as *entanglement*, has significant implications. As we discussed, there is a direct relationship between conditional independence properties of a distribution and our ability to represent it as a product of factors. Thus, a distribution that has no independence properties does not admit a compact representation in a factored form.

Unfortunately, the entanglement phenomenon is not specific to this example. Indeed, it holds for a very broad class of DBNs. We demonstrate it for a large subclass of DBNs that exhibit a very regular structure. We begin by introducing a few useful concepts.

Definition 15.1

persistent independence

For a DBN over $\mathcal{X}$, and $X, Y, Z \subset \mathcal{X}$, we say that the independence $(X \perp Y \mid Z)$ is persistent if $(X^{(t)} \perp Y^{(t)} \mid Z^{(t)})$ holds for every t.   ∎

Persistent independencies are independence properties of the belief state, and are therefore precisely what we need in order to provide a time-invariant factorization of the belief state.

The following concept turns out to be a useful one, in this setting and others.

Definition 15.2

influence graph

Let $\mathcal{B}_\rightarrow$ be a 2-TBN over $\mathcal{X}$. We define the influence graph for $\mathcal{B}_\rightarrow$ to be a directed cyclic graph $\mathcal{I}$ over $\mathcal{X}$ whose nodes correspond to $\mathcal{X}$, and that contains a directed arc $X \rightarrow Y$ if $X \rightarrow Y'$ or $X' \rightarrow Y'$ appear in $\mathcal{B}_\rightarrow$. Note that a persistence arc $X \rightarrow X'$ induces a self-cycle in the influence graph. ∎

The influence graph corresponds to influence in the unrolled DBN:

Proposition 15.1

Let $\mathcal{I}$ be the influence graph for a 2-TBN $\mathcal{B}_\rightarrow$. Then $\mathcal{I}$ contains a directed path from X to Y if and only if, in the unrolled DBN, for every t, there exists a path from $X^{(t)}$ to $Y^{(t')}$ for some $t' \geq t$.

See exercise 15.3.

The following result demonstrates the inevitability of the entanglement phenomenon, by proving that it holds in a broad class of networks. A DBN is called *fully persistent* if it encodes a state-observation model, and, for each state variable $X \in \boldsymbol{X}$, the 2-TBN contains a *persistence edge* $X \rightarrow X'$.

fully persistent

persistence edge

Theorem 15.1

Let $\langle \mathcal{G}_0, \mathcal{G}_\rightarrow \rangle$ be a fully persistent DBN structure over $\mathcal{X} = \boldsymbol{X} \cup \boldsymbol{O}$, where the state variables $\boldsymbol{X}^{(t)}$ are hidden in every time slice, and the observation variables $\boldsymbol{O}^{(t)}$ are observed in every time slice. Furthermore, assume that, in the influence graph for $\mathcal{G}_\rightarrow$:

- there is a trail (not necessarily a directed path) between every pair of nodes, that is, the graph is connected;
- every state variable X has some directed path to some evidence variable in $\boldsymbol{O}$.

Then there is no persistent independence $(\boldsymbol{X} \perp \boldsymbol{Y} \mid \boldsymbol{Z})$ that holds for every DBN $\langle \mathcal{B}_0, \mathcal{B}_\rightarrow \rangle$ over this DBN structure.

The proof is left as an exercise (see exercise 15.4). Note that, as in every other setting, there may be spurious independencies that hold due to specific choices of the parameters. But, for almost all choices of the parameters, there will be no independence that holds persistently.

☞ **In fully persistent DBNs, the tracking problem is precisely one of maintaining a belief state — a distribution over $\boldsymbol{X}^{(t)}$. The entanglement theorem shows that the only exact representation for this belief state is as a full joint distribution, rendering any belief-state-algorithm computationally infeasible except in very small networks.**

More generally, if we want to track the system as it evolves, we need to maintain a representation that summarizes all of our information about the past. Specifically, as we showed in theorem 10.2, any sepset in a clique tree must render the two parts of the tree conditionally independent. In a temporal setting, we cannot allow the sepsets to grow unboundedly with the number of time slices. Therefore, there must exist some sepset over a scope $\mathcal{Y}$ that cuts across the network, in that any path that starts from a time 0 variable and continues to infinity must intersect $\mathcal{Y}$. In fully persistent networks, the set of state variables $\boldsymbol{X}^{(t)}$ is a minimal set satisfying this condition. The entanglement theorem states that this set exhibits no persistent independencies, and therefore the message over this sepset can only be represented as an explicit joint distribution. The resulting sepsets are therefore very large — exponential in the number of state variables. Moreover, as these large messages must be incorporated into some clique in the clique tree, the cliques also become exponentially large.

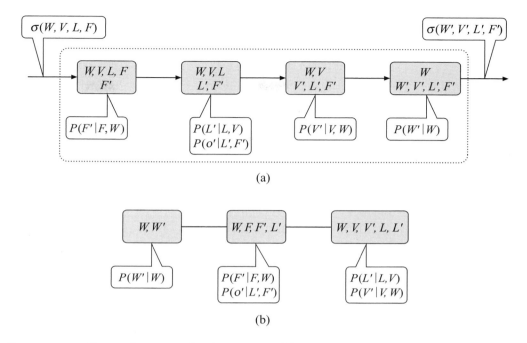

Figure 15.2 **Different clique trees for the Car DBN of figure 6.1.** (a) A template clique tree that allows exact filtering for the Car DBN of figure 6.1. Because the variable O' is always observed, it is not in the scope of any clique (although the factor $P(o' \mid F', L')$ is associated with a clique). (b) A clique tree for the 2-TBN of the Car network, which does not allow exact filtering.

Example 15.3   *Consider again the* Car *network of figure 6.1. To support exact belief state propagation, our template clique tree must include a clique containing W, V, L, F, where we can incorporate the previous belief state $\sigma^{(t)}(W^{(t)}, V^{(t)}, L^{(t)}, F^{(t)})$. It must also contain a clique containing W', V', L', F', from which we can extract $\sigma^{(t+1)}(W^{(t+1)}, V^{(t+1)}, L^{(t+1)}, F^{(t+1)})$. A minimally sized clique tree containing these cliques is shown in figure 15.2a. All of the cliques in the tree are of size 5. By contrast, if we were simply to construct a template clique tree over the dependency structure defined by the 2-TBN, we could obtain a clique tree where the maximal clique size is 4, as illustrated in figure 15.2b.*

A clique size of 4 is the minimum we can hope for: In general, all of the cliques for a fully persistent network over n variables contain at least $n + 1$ variables: one representative (either X or X') of each variable X in $\mathcal{X}$, plus an additional variable that we are currently eliminating. (See exercise 15.5.) In many cases, the minimal induced width would actually be higher. For example, if we introduce an arc $L \rightarrow V'$ into our 2-TBN (for example, because different locations have different speed limits), the smallest template clique tree allowing for exact filtering has a clique size of 6. ∎

Even in networks when not all variables are persistent, entanglement is still an issue: We still need to represent a distribution that cuts across the entire width of the network. In most cases

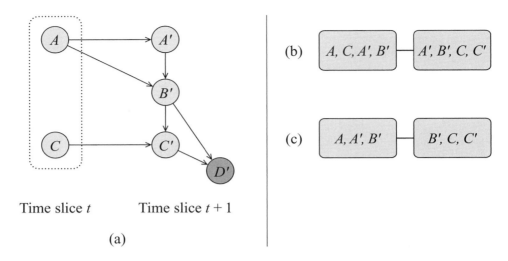

Figure 15.3   Nonpersistent 2-TBN and different possible clique trees: (a) A 2-TBN where not all of the unobserved variables are persistent. (b) A clique tree for this 2-TBN that allows exact filtering; as before, D' is always observed, and hence it is not in the scope of any clique. (c) A clique tree for the 2-TBN, which does not allow exact filtering.

— except for specific structures and observation patterns — these distributions do not exhibit independence structure, and they must therefore be represented as an explicit joint distribution over the interface variables. Because the "width" of the network is often fairly large, this results in large messages and large cliques.

Example 15.4   *Consider the 2-TBN shown in figure 15.3, where not all of the unobserved variables are persistent. In this network, our interface variables are A, C. Thus, we can construct a template clique tree over A, C, A', B', C', D' as the basis for our message passing step. To allow exact filtering, we must have a clique whose scope contains A, C and a clique whose scope contains A', C'. A minimally sized clique tree satisfying these constraints is shown in figure 15.3b. It has a maximum clique size of 4; without these constraints, we can construct a clique tree where the maximal clique size is 3 (figure 15.3c).*   ■

We note that it is sometimes possible to find better clique trees than those constrained to use $\mathcal{X}_I^{(t)}$ as the message. In fact, in some cases, the best sepset actually spans variables in multiple time slices. However, these improvements do not address the fundamental problem, which is that the computational cost of exact inference in DBNs grows exponentially with the "width" of the network. Specifically, we cannot avoid including in our messages at least the set of persistent variables. In many applications, a large fraction of the variables are persistent, rendering this approach intractable.

15.3 Approximate Inference

The computational problems with exact inference force us, in many cases, to fall back on approximate inference. In principle, when viewing the DBN as a large unrolled BN, we can apply any of the approximate inference algorithms that we discussed for BNs. Indeed, there has been significant success, for example, in using a variational approach for reasoning about weakly coupled processes that evolve in parallel. (See exercise 15.6.)

However, several complications arise in this setting. First, as we discussed in section 15.1, the types of tasks that we wish to address in the temporal setting often involve reasoning about arbitrarily large, and possibly unbounded, trajectories. Although we can always address these tasks using inference over the unrolled DBNs, as in the case of exact inference, algorithms that require us to maintain the entire unrolled BN during the inference process may be impractical. In the approximate case, we must also address an additional complication: An approximate inference algorithm that achieves reasonable errors for static networks of bounded size may not work well in arbitrarily large networks. Indeed, the quality of the approximation may degrade with the size of the network.

For the remainder of this section, we focus on the filtering task. As in the case of exact inference, methods developed for filtering extend directly to prediction, and (with a little work) to smoothing with a bounded lookahead. There are many algorithms that have been proposed for approximate tracking in dynamic systems, and one could come up with several others based on the methods described earlier in this book. We begin by providing a high-level overview of a general framework that encompasses these algorithms. We then describe two specific methods that are commonly used in practice, one that uses a message passing approach and the other a sampling approach.

15.3.1 Key Ideas

15.3.1.1 Bounded History Updates

A general approach to addressing the filtering (or prediction) task without maintaining a full history is related to the approach we used for exact filtering. There, we passed messages in the clique tree forward in time, which allowed us to throw away the observations and the cliques in previous time slices once they have been processed.

In principle, the same idea can be applied in the case of approximate inference: We can execute the appropriate inference steps for a time slice and then move forward to the next time slice. However, most approximate inference algorithms require that the same variable in the network be visited multiple times during the course of inference. For example, belief propagation algorithms (as in section 11.3 or section 11.4) send multiple messages through the same cluster. Similarly, structured variational approximation methods (as in section 11.5) are also iterative, running inference on the network multiple times, with different values for the variational parameters. Markov chain Monte Carlo algorithms also require that each node be visited and sampled multiple times. Thus, we cannot just throw away the history and apply our approximate inference to the current time slice alone.

One common solution is to use a form of "limited history" in the inference steps. The various steps associated with the inference algorithm are executed not over the whole network, but over a subnetwork covering only the recent history. Most simply, the subnetwork for time t

is simply a bounded window covering some predetermined number k of previous time slices $t - k, \ldots, t - 1, t$. More generally, the subnetwork can be determined in a dynamic fashion, using a variety of techniques.

We will describe the two methods most commonly used in practice, one based on importance sampling, and the other on approximate message propagation. Both take $k = 1$, using only the current approximate belief state $\hat{\sigma}^{(t)}$ and the current time slice in estimating the next approximate belief state $\hat{\sigma}^{(t+1)}$. In effect, these methods perform a type of approximate message propagation, as in equation (15.1) and equation (15.2). Although this type of approximation is clearly weak in some cases, it turns out to work fairly well in practice.

15.3.1.2   Analysis of Convergence

The idea of running approximate inference with some bounded history over networks of increasing size immediately raises concerns about the quality of the approximation obtained. Consider, for example, the simplest approximate belief-state filtering process. Here, we maintain an approximate belief state $\hat{\sigma}^{(t)}$, which is (hopefully) similar to our true belief state $\sigma^{(t)}$. We use $\hat{\sigma}^{(t)}$ to compute the subsequent belief state $\hat{\sigma}^{(t+1)}$. This step uses approximate inference and therefore introduces some additional error into our approximation. Therefore, as time evolves, our approximation appears to be accumulating more and more errors. In principle, it might be the case that, at some point, our approximate belief state $\hat{\sigma}^{(t)}$ bears no resemblance to the true belief state $\sigma^{(t)}$.

☞   **Although unbounded errors can occur, it turns out that such situations are rare in practice (for algorithms that are carefully designed). The main reason is that the dynamic system itself is typically stochastic. Thus, the effect of approximations that occur far in the past tends to diminish over time, and the overall error (for well-designed algorithms) tends to remain bounded indefinitely.** For several algorithms (including the two described in more detail later), one can prove a formal result along these lines. All of these results make some assumptions about the stochasticity of the system — the rate at which it "forgets" the past. Somewhat more formally, assume that propagating two distributions through the system dynamics (equation (15.1) and 15.2) reduces some notion of distance between them. In this case, discrepancies between $\hat{\sigma}^{(t)}$ and $\sigma^{(t)}$, which result from approximations in previous time slices, decay over time. Of course, new errors are introduced by subsequent approximations, but, in stochastic systems, we can show that they do not accumulate unboundedly.

Formal theorems proving a uniform bound on the distance between the approximate and true belief state — a bound that holds for all time points t — exist for a few algorithms. These theorems are quite technical, and the actual bounds obtained on the error are fairly large. For this reason, we do not present them here. However, in practice, when the underlying system is stochastic, we do see a bounded error for approximate propagation algorithms. Conversely, when the system evolution includes a deterministic component — for example, when the state contains a variable that (once chosen) does not evolve over time — the errors of the approximate inference algorithms often do diverge over time. Thus, while the specific bounds obtain in the theoretical analyses may not be directly useful, they do provide a theoretical explanation for the behavior of the approximate inference algorithms in practice.

15.3.2 Factored Belief State Methods

 The issue underlying the entanglement result is that, over time, all variables in a belief state slice eventually become correlated via active trails through past time slices. In many cases, however, these trails can be fairly long, and, as a consequence, the resulting correlations can be quite weak. This raises the idea of replacing the exact, fully correlated, belief state, with an approximate, factorized belief state that imposes some independence assumptions. For a carefully chosen factorization structure, these independence assumptions may be a reasonable approximation to the structure in the belief state.

This idea gives rise to the following general structure for a filtering algorithm: At each time point t, we have a factored representation $\hat{\sigma}^{(t)}$ of our time t belief state. We then compute the correct update of this time t belief state to produce a new time $t + 1$ belief state $\sigma^{(\cdot t+1)}$. The update step consists of propagating the belief state forward through the system dynamics and conditioning on the time $t + 1$ observations. Owing to the correlations induced by the system dynamics (as in section 15.2.4), $\sigma^{(\cdot t+1)}$ has more correlations than $\hat{\sigma}^{(t)}$, and therefore it requires larger factors to represent correctly. If we continue this process, we rapidly end up with a belief state that has no independence structure and must be represented as a full joint distribution.

belief-state projection

Therefore, we introduce a *projection* step, where we approximation $\sigma^{(\cdot t+1)}$ using a more factored representation, giving rise to a new $\hat{\sigma}^{(t+1)}$, with which we continue the process. This update-project cycle ensures that our approximate belief state remains in a class of distributions that we can tractably maintain and update.

Most simply, we can represent the approximate belief state $\hat{\sigma}^{(t)}$ in terms of a set of factors $\Phi^{(t)} = \{\beta_r^{(t)}(\boldsymbol{X}_r^{(t)})\}$, where we assume (for simplicity) that the factorization of the messages (that is, the choice of scopes $\boldsymbol{X}_r$) does not change over time. Most simply, the scopes of the different factors are disjoint, in which case the belief state is simply a product of marginals over disjoint variables or subsets of variables. As a richer but more complex representation, we can represent $\hat{\sigma}^{(t)}$ using a calibrated cluster tree, or even a calibrated cluster graph $\mathcal{U}$. Indeed, we can even use a general representation that uses overlapping regions and associated counting numbers:

$$\hat{\sigma}^{(t)}(\mathcal{X}^{(t)}) = \prod_r (\beta_r^{(t)}(\boldsymbol{X}_r^{(t)}))^{\kappa_r}.$$

Example 15.5

Consider the task of monitoring a freeway with k cars. As we discussed, after a certain amount of time, the states of the different cars become entangled, so our only option for representing the belief state is as a joint distribution over the states of all *the cars. An obvious approximation is to assume that the correlations between the different cars are not very strong. Thus, although the cars do influence each other, the current state of one car does not tell us too much about the current state of another. Thus, we can choose to approximate the belief state over the entire system using an approximate belief state that ignores or approximates these weak correlations. Specifically, let $\boldsymbol{Y}_i$ be the set of variables representing the state of car i, and let $\boldsymbol{Z}$ be a set of variables that encode global conditions, such as the weather or the current traffic density. Most simply, we can represent the belief state $\hat{\sigma}^{(t)}$ as a product of marginals*

$$\beta_g^{(t)}(\boldsymbol{Z}^{(t)}) \prod_{i=1}^{k} \beta_i^{(t)}(\boldsymbol{Y}_i^{(t)}).$$

In a better approximation, we might preserve the correlations between the state of each individual vehicle and the global system state, by selecting as our factorization

$$\left(\beta_g^{(t)}(\boldsymbol{Z}^{(t)})\right)^{-(k-1)} \prod_{i=1}^{k} \beta_i^{(t)}(\boldsymbol{Z}^{(t)}, \boldsymbol{Y}_i^{(t)}),$$

where the initial term compensates for the multiple counting of the probability of $\boldsymbol{Z}^{(t)}$ in the other factors. Here, the representation of the approximate belief state makes the assumption that the states of the different cars are conditionally independent given the global state variables. ∎

We showed that exact filtering is equivalent to a forward pass of message passing in a clique tree, with the belief states playing the role of messages. Hence, filtering with factored belief states is simply a form of message passing with approximate messages. The use of an approximate belief state in a particular parametric class is also known as *assumed density filtering*. This algorithm is a special case of the more general algorithm that we developed in the context of the *expectation propagation* (EP) algorithm of section 11.4. Viewed abstractly, each *slice-cluster* in our monolithic DBN clique tree (one that captures the entire trajectory) corresponds to a pair of adjacent time slices $t-1, t$ and contains the variables $\mathcal{X}_I^{(t)} \cup \mathcal{X}^{(t+1)}$. As in the general EP algorithm, the potential in a slice-cluster is never represented explicitly, but in a decomposed form, as a product over factors. Each slice-cluster is connected to its predecessor and successor slice-clusters. The messages between these slice-clusters correspond to approximate belief states $\hat{\sigma}^{(t)}(\mathcal{X}_I^{(t)})$, which are represented in a factorized form. For uniformity of exposition, we assume that the initial state distribution — the time 0 belief state — also takes (or is approximated as) the same form. Thus, when propagating messages in this chain, each slice-cluster takes messages in this factorized form and produces messages in this form.

As we discussed in section 11.4.2, the use of factorized messages allows us to perform the operations in each cluster much more efficiently, by using a nested clique tree or cluster graph that exploits the joint structure of the messages and cluster potential. For example, if the belief state is fully factored as a product over the variables in the interface, the message structure imposes no constraints on the nested data structure used for inference within a time slice. In particular, we can use any clique tree over the 2-TBN structure; for instance, in example 15.3, we can use the structure of figure 15.2b. By contrast, for exact filtering, the messages are full belief states over the interface variables, requiring the use of a nested clique tree with very large cliques. Of course, a fully factorized belief state generally provides a fairly poor approximation to the belief state. As we discussed in the context of the EP algorithm, we can also use much more refined approximations, which use a clique tree or even a general region-based approximation to the belief state.

The algorithm used for the message passing is precisely as we described in section 11.4.2, and we do not repeat it here. We make only three important observations. First, unlike a traditional application of EP, when doing filtering, we generally do only a single upward pass of message propagation, starting at time 0 and propagating toward higher time slices. Because we do not have a backward pass, the distinctions between the sum-product algorithm (section 11.4.3.1) and the belief update algorithm (section 11.4.3.2) are irrelevant in this setting, since the difference arises only in the backward pass. Second, without a backward pass, we do not need to keep track of a clique once it has propagated its message forward. Thus, as in exact inference for

assumed density filter

expectation propagation

DBNs, we can keep only a single (factored) message and single (factored) slice-cluster in memory at each point in time and perform the message propagation in space that is constant in the number of time slices.

template cluster graph

If we continue to assume that the belief state representation is the same for every time slice, then the factorization structure used in each of the message passing steps is identical. In this case, we can perform all the message passing steps using the same *template cluster graph* that has a fixed cluster structure and fixed initial factors (those derived from the 2-TBN); at each time t, the factors representing $\hat{\sigma}^{(t)}$ are introduced into the template cluster graph, which is then calibrated and used to produce the factors representing $\hat{\sigma}^{(t+1)}$. The reuse of the template can reduce the cost of the message propagation step. An alternative approach allows the structure used in our approximation to change over time. This flexibility allows us to adapt our structure to reflect the strengths of the interactions between the variables in our domain. For example, in example 15.5, we might expect the variables associated with cars that are directly adjacent to be highly correlated; but the pairs of cars that are close to each other change over time. Section 15.6 describes some methods for dynamically adapting the representation to the current distribution.

15.3.3 Particle Filtering

Of the different particle-based methods that we discussed, forward sampling appears best suited to the temporal setting, since it generates samples incrementally, starting from the root of the network. In the temporal setting, this would correspond to generating trajectories starting from the beginning of time, and going forward. This type of sampling, we might hope, is more amenable to a setting where we do not have to keep sampled trajectories that go indefinitely far back. Obviously, rejection sampling is not an appropriate basis for a temporal sampling algorithm. For an indefinitely long trajectory, all samples will eventually be inconsistent with our observations, so we will end up rejecting all samples. In this section, we present a family of filtering algorithms based on importance sampling and analyze their behavior.

15.3.3.1 Naive Likelihood Weighting

It is fairly straightforward to generalize likelihood weighting to the temporal setting. Recall that LW generates samples by sampling nodes that are not observed from their appropriate distribution, and instantiating nodes that are observed to their observed values. Every node that is instantiated in this way causes the weight of the sample to be changed. However, LW generates samples one at a time, starting from the root and continuing until a full assignment is generated. In an online setting, we can have arbitrarily many variables, so there is no natural end to this sampling process. Moreover, in the filtering problem, we want to be able to answer queries online as the system evolves. Therefore, we first adapt our sampling process to return intermediate answers.

The LW algorithm for the temporal setting maintains a set of samples, each of which is a trajectory up to the current time slice t: $\xi^{(t)}[1], \ldots, \xi^{(t)}[M]$. Each sampled trajectory is associated with a weight $w[m]$. At each time slice, the algorithm takes each of the samples, propagates it forward to sample the variables at time t, and adjusts its weight to suit the new evidence at time t. The algorithm uses a likelihood-weighting algorithm as a subroutine to propagate a time t sample to time $t + 1$. The version of the algorithm for 2-TBNs is almost identical to the algorithm 12.2; it is shown in algorithm 15.2 primarily as a reminder.

Algorithm 15.2 Likelihood-weighted particle generation for a 2-TBN

Procedure LW-2TBN (
  $\mathcal{B}_\rightarrow$ // 2-TBN
  ξ // Instantiation to time $t - 1$ variables
  $O^{(t)} = o^{(t)}$ // time t evidence
)
1   Let $X'_1, \ldots, X'_n$ be a topological ordering of $\mathcal{X}'$ in $\mathcal{B}_\rightarrow$
2   $w \leftarrow 1$
3   **for** $i = 1, \ldots, n$
4     $u_i \leftarrow (\xi, x')\langle \mathrm{Pa}_{X'_i} \rangle$
5   // Assignment to $\mathrm{Pa}_{X'_i}$ in $x_1, \ldots, x_n, x'_1, \ldots, x'_{i-1}$
6   **if** $X'_i \notin O^{(t)}$ **then**
7   Sample x'_i from $P(X'_i \mid u_i)$
8   **else**
9       $x'_i \leftarrow o^{(t)}\langle X'_i \rangle$ // Assignment to X'_i in $o^{(t)}$
10      $w \leftarrow w \cdot P(x'_i \mid u_i)$ // Multiply weight by probability of desired value
11 **return** $(x'_1, \ldots, x'_n), w$

Algorithm 15.3 Likelihood weighting for filtering in DBNs

Procedure LW-DBN (
  $\langle \mathcal{B}_0, \mathcal{B}_\rightarrow \rangle$, // DBN
  M // Number of samples
  $o^{(1)}, o^{(2)}, \ldots$ // Observation sequence
)
1   **for** $m = 1, \ldots, M$
2   Sample $\xi^{(0)}[m]$ from $\mathcal{B}_0$
3     $w[m] \leftarrow 1$
4   **for** $t = 1, 2, \ldots$
5   **for** $m = 1, \ldots, M$
6       $(\xi^{(t)}[m], w) \leftarrow \text{LW-2TBN}(\mathcal{B}_\rightarrow, \xi^{(t-1)}[m], o^{(t)})$
7   // Sample time t variables starting from time $t - 1$ sample
8       $w[m] \leftarrow w[m] \cdot w$
9   // Multiply weight of m'th sample with weight of time t evidence
10    $\hat{\sigma}^{(t)}(\xi) \leftarrow \frac{\sum_{m=1}^M w[m] \boldsymbol{I}\{\xi^{(t)}[m] = \xi\}}{\sum_{m=1}^M w[m]}$

Unfortunately, this extension of the basic LW algorithm is generally a very poor algorithm for DBNs. To understand why, consider the application of this algorithm to any state-observation

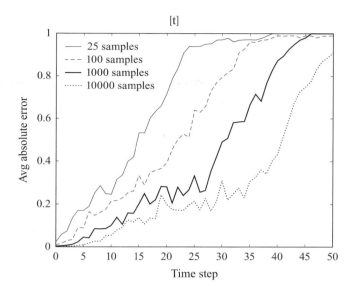

Figure 15.4 **Performance of likelihood weighting over time** with different numbers of samples, for a state-observation model with one state and one observation variable.

model. In this case, we have a very long network, where all of the evidence is at the leaves. Unfortunately, as we discussed, in such networks, LW generates samples according to the prior distribution, with the evidence affecting only the weights. In other words, the algorithm generates completely random state trajectories, which "match" the evidence only by chance. For example, in our Car example, the algorithm would generate completely random trajectories for the car, and check whether one of them happens to match the observed sensor readings for the car's location. Clearly, the probability that such a match occurs — which is precisely the weight of the sample — decreases exponentially (and quite quickly) with time. This problem can also arise in a static BN, but it is particularly severe in this setting, where the network size grows unboundedly. In this case, as time evolves, more and more evidence is ignored in the sample-generation process (affecting only the weight), so that the samples become less and less relevant. Indeed, we can see in figure 15.4 that, in practice, the samples generated get increasingly irrelevant as t grows, so that LW diverges rapidly as time goes by. From a technical perspective, this occurs because, over time, the variance of the weights of the samples grows very quickly, and unboundedly. Thus, the quality of the estimator obtained from this procedure — the probability that it returns an answer within a certain error tolerance — gets increasingly worse.

particle filter

sequential importance sampling

One approach called *particle filtering* (or *sequential importance sampling*) for addressing this problem is based on the key observation that not all samples are equally "good." In particular, **samples that have higher weight explain the evidence observed so far much better, and are likely to be closer to the current state. Thus, rather than propagate all samples forward to the next time step, we should preferentially select "good" samples for propagation, where "good" samples are ones that have high weight.** There are many ways of implementing this basic intuition: We can select samples for propagation deterministically or stochastically.

We can use a fixed number of samples, or vary the number of samples to achieve a certain quality of approximation (estimated heuristically).

15.3.3.2   The Bootstrap Filter

bootstrap filter   The simplest and most common variant of particle filtering is called the *bootstrap filter*. It maintains a set $\mathcal{D}^{(t)}$ of M time t trajectories $\bar{x}^{(0:t)}[m]$, each associated with its own weight $w^{(t)}[m]$. When propagating samples to the next time slice, each sample is chosen randomly for propagation, proportionately to its current weight. The higher the weight of the sample, the more likely it is to be selected for propagation; thus, higher-weight samples may "spawn" multiple copies, whereas lower-weight ones "die off" to make space for the others.

More formally, consider a data set $\mathcal{D}^{(t)}$ consisting of M weighted sample trajectories $(\bar{x}^{(0:t)}[m]$, $w^{(t)}[m])$. We can define the empirical distribution generated by the data set:

$$\hat{P}_{\mathcal{D}^{(t)}}(x^{(0:t)}) \propto \sum_{m=1}^{M} w^{(t)}[m] \boldsymbol{I}\{\bar{x}^{(0:t)}[m] = x^{(0:t)}\}.$$

This distribution is a weighted sum of delta distributions, where the probability of each assignment is its total weight in $\mathcal{D}^{(t)}$, renormalized to sum to 1.

The algorithm then generates M new samples for time $t+1$ as follows: For each sample m, it selects a time t sample for propagation by randomly sampling from $\hat{P}_{\mathcal{D}^{(t)}}$. Each of the M selected samples is used to generate a new time $t+1$ sample using the transition model, which is weighted using the observation model. Note that the weight of the sample $w^{(t)}[m]$ manifests in the relative proportion with which the mth sample is propagated. Thus, we do not need to account for its previous weight when determining the weight of the time $t+1$ sample generated from it. If we did include its weight, we would effectively be double-counting it. The algorithm is shown in algorithm 15.4 and illustrated in figure 15.5.

We can view $\hat{P}_{\mathcal{D}^{(t)}}$ as an approximation to the time t belief state (one where only the sampled states have nonzero probability), and the sampling step as using it to generate an approximate belief state for time $t+1$. Thus, this algorithm can be viewed as performing a stochastic version of the belief-state filtering process.

Note that we view the algorithm as maintaining entire trajectories $\bar{x}^{(0:t)}$, rather than simply the current state. In fact, each sample generated does correspond to an entire trajectory. However, for the purpose of filtering, the earlier parts of the trajectory are not relevant, and we can throw out all but the current state $\bar{x}^{(t)}$.

The bootstrap particle filter works much better than likelihood weighting, as illustrated in figure 15.6a. Indeed, the error seems to remain bounded indefinitely over time (b).

We can generalize the basic bootstrap filter along two dimensions. The first modifies the forward sampling procedure — the process by which we extend a partial trajectory $\bar{x}^{(0:t-1)}$ to include a time t state variable assignment $\bar{x}^{(t)}$. The second modifies the particle selection scheme, by which we take a set of weighted time t samples $\mathcal{D}^{(t)}$ and use their weights to select a new set of time t samples. We will describe these two extensions in more detail.

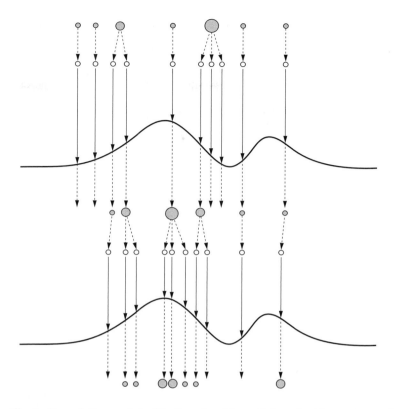

Figure 15.5 **Illustration of the particle filtering algorithm.** (Adapted with permission from van der Merwe et al. (2000a).) At each time slice, we begin with a set of weighted samples (dark circles), we sample from them to generate a set of unweighted samples (light circles). We propagate each sample forward through the system dynamics, and we update the weight of each sample to reflect the likelihood of the evidence (black line), producing a new set of weighted samples (some of which have weight so small as to be invisible). The process then repeats for the next time slice.

15.3.3.3 Sequential Importance Sampling

We can generalize our forward sampling process by viewing it in terms of importance sampling, as in section 12.2.2. Here, however, we are sampling entire trajectories rather than static states. Our goal is to sample a trajectory $\bar{x}^{(0:t)}$ from the distribution $P(x^{(0:t)} \mid o^{(0:t)})$. To use importance sampling, we must construct a proposal distribution α for trajectories and then use importance weights to correct for the difference between our proposal distribution and our target distribution.

To maintain the ability to execute our filtering algorithm in an online fashion, we must construct our proposal distribution so that trajectories are constructed incrementally. Assume that, at time t, we have sampled some set of partial trajectories $\mathcal{D}^{(t)}$, each possibly associated with some weight. If we want to avoid the need to maintain full trajectories and a full observation

---

Algorithm 15.4 Particle filtering for DBNs

    Procedure Particle-Filter-DBN (
        $\langle \mathcal{B}_0, \mathcal{B}_{\rightarrow} \rangle$,   // DBN
        M // Number of samples
        $\boldsymbol{o}^{(1)}, \boldsymbol{o}^{(2)}, \ldots$ // Observation sequence
)

1	**for** $m = 1, \ldots, M$
2	Sample $\bar{\boldsymbol{x}}^{(0)}[m]$ from $\mathcal{B}_0$
3	$w^{(0)}[m] \leftarrow 1/M$
4	**for** $t = 1, 2, \ldots$
5	**for** $m = 1, \ldots, M$
6	Sample $\bar{\boldsymbol{x}}^{(0:t-1)}$ from the distribution $\hat{P}_{\mathcal{D}^{(t-1)}}$.
7	// Select sample for propagation
8	$(\bar{\boldsymbol{x}}^{(t)}[m], w^{(t)}[m]) \leftarrow$ LW-2TBN$(\mathcal{B}_{\rightarrow}, \bar{\boldsymbol{x}}^{(t-1)}, \boldsymbol{o}^{(t)})$
9	// Generate time t sample and weight from selected sample
	$\bar{\boldsymbol{x}}^{(t-1)}$
10	$\mathcal{D}^{(t)} \leftarrow \{(\bar{\boldsymbol{x}}^{(0:t)}[m], w^{(t)}[m]) : m = 1, \ldots, M\}$
11	$\hat{\sigma}^{(t)}(\boldsymbol{x}) \leftarrow \hat{P}_{\mathcal{D}^{(t)}}$

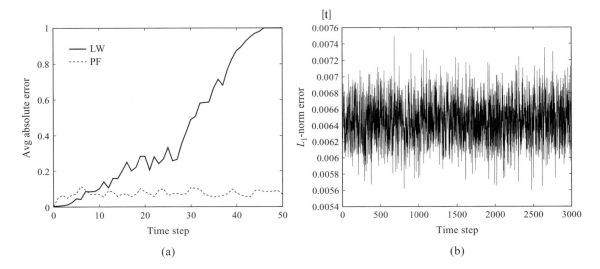

Figure 15.6   Likelihood weighting and particle filtering over time. (a) A comparison for 1,000 time slices. (b) A very long run of particle filtering.

history, each of our proposed sample trajectories for time $t + 1$ must be an extension of one of our time t sample trajectories. More precisely, each proposed trajectory at time $t + 1$ must have the form $\bar{x}^{(0:t)}, x^{(t+1)}$, for some $\bar{x}^{(0:t)}[m] \in \mathcal{D}^{(t)}$.

Note that this requirement, while desirable from a computational perspective, does have disadvantages. In certain cases, our set of time t sample trajectories might be unrepresentative of the true underlying distribution; this might occur simply because of bad luck in sampling, or because our evidence sequence up to time t was misleading, causing us to select for trajectories that turn out to be a bad match to later observations. Thus, it might be desirable to *rejuvenate* our sample trajectories, allowing the states prior to time t to be modified based on evidence observed later on. However, this type of process is difficult to execute efficiently, and is not often done in practice.

If we proceed under the previous assumption, we can compute the appropriate importance weights for our importance sampling process incrementally:

$$
\begin{aligned}
w(\bar{x}^{(0:t)}) &= \frac{P(x^{(0:t)} \mid o^{(0:t)})}{\alpha^{(t)}(x^{(0:t)})} \\
&= \frac{P(x^{(0:t-1)} \mid o^{(0:t)})}{\alpha^{(t-1)}(x^{(0:t-1)})} \frac{P(\bar{x}^{(t)} \mid x^{(0:t-1)}, o^{(0:t)})}{\alpha^{(t)}(\bar{x}^{(t)} \mid x^{(0:t-1)})}.
\end{aligned}
$$

As we have discussed, the quality of an importance sampler is a function of the variance of the weights: the lower the variance, the better the sampler. Thus, we aim to choose our proposal distribution $\alpha^{(t)}(\bar{x}^{(t)} \mid \bar{x}^{(0:t-1)})$ so as to reduce the variance of the preceding expression. Note that only the second of the two terms in this product depends on our time t proposal distribution $\alpha^{(t)}(\bar{x}^{(t)} \mid \bar{x}^{(0:t-1)})$. By assumption, the samples $x^{(0:t-1)}$ are fixed, and hence so is the first term. It is now not difficult to show that the time t proposal distribution that minimizes the overall variance is

$$
\alpha^{(t)}(X^{(t)} \mid x^{(0:t-1)}) = P(X^{(t)} \mid x^{(0:t-1)}, o^{(0:t)}), \tag{15.3}
$$

making the second term uniformly 1. In words, we should sample the time t state variable assignment $\bar{x}^{(t)}$ from its posterior distribution given the chosen sample from the previous state and the time t observations.

Using this proposal, the appropriate importance weight for our time t trajectory is

$$
\frac{P(x^{(0:t-1)} \mid o^{(0:t)})}{\alpha^{(t-1)}(x^{(0:t-1)})}.
$$

What is the proposal distribution we use for the time $t - 1$ trajectories? If we use this idea in combination with resampling, we can make the approximation that our uniformly sampled particles at time $t - 1$ are an approximation to $P(x^{(0:t-1)} \mid o^{(0:t-1)})$. In this case, we have

$$
\begin{aligned}
\frac{P(x^{(0:t-1)} \mid o^{(0:t)})}{\alpha^{(t-1)}(\bar{x}^{(0:t-1)})} &\approx \frac{P(x^{(0:t-1)} \mid o^{(0:t)})}{P(x^{(0:t-1)} \mid o^{(0:t-1)})} \\
&\propto \frac{P(x^{(0:t-1)} \mid o^{(0:t-1)}) P(o^{(t)} \mid x^{(0:t-1)}, o^{(0:t-1)})}{P(x^{(0:t-1)} \mid o^{(0:t-1)})} \\
&= P(o^{(t)} \mid \bar{x}^{(t-1)}),
\end{aligned}
$$

where the last step uses the Markov independence properties. Thus, our importance weights
here are proportional to the probability of the time t observation given the time $t-1$ particle

posterior particle
filter

$x^{(t-1)}$, marginalizing out the time t state variables. We call this approach *posterior particle
filtering*, because the samples are generated using our *posterior* over the time t state, given the
time t observation, rather than using our prior.

However, sampling from the posterior over the state variables given the time t observations
may not be tractable. Indeed, the whole purpose of the (static) likelihood-weighting algorithm is
to address this problem, defining a proposal distribution that is (perhaps) closer to the posterior
while still allowing forward sampling according to the network structure. However, likelihood
weighting is only one of many importance distributions that can be used in this setting. In
many cases, significant advantages can be gained by constructing proposal distributions that are
even closer to this posterior; we describe some ideas in section 15.3.3.5 below.

15.3.3.4   Sample Selection Scheme

We can also generalize the particle selection scheme. A general selection procedure associates
with each particle $\bar{x}^{(0:t)}[m]$ a number of *offspring* $K_m^{(t)}$. Each of the offspring of this particle
is a (possibly weighted) copy of it, which is then propagated independently to the next step, as
discussed in the previous section. Let $\mathcal{D}^{(t)}$ be the original sample set, and $\tilde{\mathcal{D}}^{(t)}$ be the new
sample set.

Assuming that we want to keep the total number of particles constant, we must have
$\sum_{m=1}^{M} K_m^{(t)} = M$. There are many approaches to selecting the number of offspring $K_m^{(t)}$
for each particle $\bar{x}^{(0:t)}[m]$. The bootstrap filter implicitly selects $K_m^{(t)}$ using a multinomial
distribution, where one performs M IID trials, in each of which we obtain the outcome m
with probability $w(\bar{x}^{(0:t)}[m])$ (assuming the weights have been renormalized). This distribution
guarantees that the expectation of $K_m^{(t)}$ is $M \cdot w(\bar{x}^{(0:t)}[m])$. Because each of the particles in
$\tilde{\mathcal{D}}^{(t)}$ is weighed equally, this property guarantees that the expectation (relative to our random
resampling procedure) of $\hat{P}_{\tilde{\mathcal{D}}^{(t)}}$ is our original distribution $\hat{P}_{\mathcal{D}^{(t)}}$. Thus, the resampling procedure does not introduce any bias into our algorithm, in the sense that the expectation of any
estimator relative to $\hat{P}_{\mathcal{D}^{(t)}}$ is the same as its expectation relative to $\hat{P}_{\tilde{\mathcal{D}}^{(t)}}$.

While the multinomial scheme is quite natural, there are many other selection schemes that
also satisfy this constraint. In general, we can use some other method to select the number of
offspring $K_m^{(t)}$ for each sample m, so long as this number satisfies (perhaps approximately) the
constraint on the expectation. Assuming $K_m^{(t)} > 0$, we then assign the weight of each of these
$K_m^{(t)}$ offspring to be:

$$\frac{w(\bar{x}^{(0:t)}[m])}{K_m^{(t)} \Pr(K_m^{(t)} > 0)};$$

intuitively, we divide the original weight of the mth sample between its $K_m^{(t)}$ offspring. The
second term in the denominator accounts for the fact that the sample was not eliminated entirely.
To justify this expression, we observe that the total weight of the sample m offspring *conditioned
on the fact that* $K_m^{(t)} > 0$ is precisely $w(\bar{x}^{(0:t)}[m])$. Thus, the unconditional expectation of the
total of these weights is $w(\bar{x}^{(0:t)}[m])P(K_m^{(t)} > 0)$, causing us to divide by this latter probability

in the new weights.

There are many possible choices for generating the vector of offspring $(K_1^{(t)}, \ldots, K_M^{(t)})$, which tells us how many copies (if any) of each of the M samples in $\mathcal{D}^{(t)}$ we wish to propagate forward. Although the different schemes all have the same expectation, they can differ significantly in terms of their variance. The higher the variance, the greater the probability of obtaining unrepresentative distributions, leading to poor answers. The multinomial sampling scheme induced by the bootstrap filter tends to have a fairly high variance, and other schemes often perform better in practice. Moreover, it is not necessarily optimal to perform a resampling step at every time slice. For example, one can monitor the weights of the samples, and only resample when the variance exceeds a certain threshold, or, equivalently, when the number of effective samples equation (12.15) goes below a certain minimal amount.

Finally, we note that one can also consider methods that vary the number of samples M over time. In certain cases, such as tracking a system in real time, we may be forced to maintain rigid constraints on the amount of time spent in each time slice. In other cases, however, it may be possible to spend more computational resources in some time slices, at the expense of using less in others. For example, if we have the ability to cache our observations a few time slices back (which we may be doing in any case for the purposes of performing smoothing), we can allow more samples in one time slice (perhaps falling a bit behind), and catch up in a subsequent time slice. If so, we can determine whether additional samples are required for the current time slice by using our estimate of the number of effective samples in the current time slice. Empirically, this flexibility in the number of samples used per time slice can also improve the quality of the results, since it helps reduce the variance of the estimator and thereby reduces the harm done by a poor set of samples obtained at one time slice.

15.3.3.5   Other Extensions

As for importance sampling in the static case, there are multiple ways in which we can improve particle filtering by utilizing other inference methods. For example, a key problem in particle filtering is the fact that the diversity of particles often decreases over time, so that we are only generating samples from a relatively small part of our space. In cases where there are multiple reasonably likely hypotheses, this loss of diversity can result in bad situations, where a surprising observation (surprising relative to our current sample population) can suddenly rule out all or most of our samples.

There are several ways of addressing that problem. For example, we can use MCMC methods within a time slice to obtain a more diverse set of samples. While this cannot regenerate hypotheses that are very far away from our current set of samples, it can build up and maintain a broader set of hypotheses that is less likely to become depleted in subsequent steps. A related approach is to generate a clique tree for the time slice in isolation, and then use forward sampling to generate samples from the clique tree (see exercise 12.3). We note that exact inference for a single time slice may be feasible, even if it is infeasible for the DBN as a whole due to the entanglement phenomenon.

Another use for alternative inference methods is to reduce the variance of the generated samples. Here also, multiple approaches are possible. For example, as we discussed in section 15.3.3.3 (in equation (15.3)), we want to generate our time t state variable assignment from its posterior distribution given the chosen sample from the previous state and the time t observations. We can generate this posterior using exact inference on the 2-TBN structure. Again, this approach may be feasible even if exact inference on the DBN is not. If exact inference is infeasible even for the 2-TBN, we may still be able to use some intermediate alternative. We might be able to reverse some edges that point to observed variables (see exercise 3.12), making the time t distribution closer to the optimal sampling distribution at time t. Alternatively, we might use approximate inference method (for example, the EP-based approach of the previous section) to generate a proposal distribution that is closer to the true posterior than the one obtained by the simple likelihood-weighting sampling distribution.

Rao-Blackwellized
particle filter Finally, we can also use collapsed particles rather than fully sampled states in particle filtering. This method is often known as *Rao-Blackwellized particle filtering*, or *RBPF*. As we observed in section 12.4, the use of collapsed particles reduces the bias of the estimator. The procedure is based on the collapsed importance-sampling procedure for static networks, as described in section 12.4.1. As there, we partition the state variables $\boldsymbol{X}$ into two disjoint sets: the sampled variables $\boldsymbol{X}_p$, and the variables $\boldsymbol{X}_d$ whose distribution we maintain in closed form. Each particle now consists of three components: $(\boldsymbol{x}_p^{(t)}[m], w^{(t)}[m], q^{(t)}[m](\boldsymbol{X}_d^{(t)}))$. The particle structure is generally chosen so as to allow $q^{(t)}[m](\boldsymbol{X}_d^{(t)})$ to be represented effectively, for example, in a factorized form.

At a high level, we use importance sampling from some appropriate proposal distribution Q (as described earlier) to sample the variables $\boldsymbol{X}_p^{(t)}$ and exact inference to compute the importance weights and the distribution $q^{(t)}[m](\boldsymbol{X}_d^{(t)})$. This process is described in detail in section 12.4.1. When applying this procedure in the context of particle filtering, we generate the time t particle from a distribution defined by a time $t-1$ particle and the 2-TBN.

More precisely, consider a time $t-1$ particle $\boldsymbol{x}_p^{(t-1)}[m], w^{(t-1)}[m], q^{(t-1)}[m](\boldsymbol{X}_d^{(t-1)})$. We define a joint probability distribution $P_m^{(t)}(\boldsymbol{X}^{(t-1)} \cup \mathcal{X}^{(t)})$ by taking the time $t-1$ particle $\boldsymbol{x}_p^{(t-1)}[m], q^{(t-1)}[m](\boldsymbol{X}_d^{(t-1)})$ as a distribution over $\boldsymbol{X}^{(t-1)}$ (one that gives probability 1 to $\boldsymbol{x}_p^{(t-1)}[m]$), and then using the 2-TBN to define $P(\mathcal{X}^{(t)} \mid \boldsymbol{X}^{(t-1)})$. The distribution $P_m^{(t)}$ is represented in a factored form, which is derived from the factorization of $q^{(t-1)}[m](\boldsymbol{X}_d^{(t-1)})$ and from the structure of the 2-TBN. We can now use $P_m^{(t)}$, and the time t observation $o^{(t)}$, as input to the procedure of section 12.4.1. The output is a new particle and weight w; the particle is added to the time t data set $\mathcal{D}^{(t)}$, with a weight $w \cdot w^{(t-1)}[m]$. The resulting data set, consisting now of collapsed particles, is then utilized as in standard particle filtering. In particular, an additional sample selection step may be used to choose which particles are to be propagated to the next time step.

As defined, however, the collapsed importance-sampling procedure over $P_m^{(t)}$ computes a particle over all of the variables in this model: both time t and time $t-1$ variables. For our purpose, we need to extract a particle involving only time t variables. This process is fairly straightforward. The particle specifies as assignment to $\boldsymbol{X}_p^{(t)}$; the marginal distribution over $\boldsymbol{X}_d^{(t)}$ can be extracted using standard probabilistic inference techniques. We must take care, however: in general, the distribution over $\boldsymbol{X}_d$ can be subject to the same entanglement phenomena as the distribution as a whole. Thus, we must select the factorization (if any) of $q^{(t)}[m](\boldsymbol{X}_d^{(t)})$ so as to be sustainable over time; that is, $q^{(t-1)}[m](\boldsymbol{X}_d^{(t-1)})$ factorizes in a certain way, then so does the marginal distribution $q^{(t)}[m](\boldsymbol{X}_d^{(t)})$ induced by $P_m^{(t)}$. Box 15.A provides an example of such a model for the application of collapsed particle filtering to the task of robot localization and mapping.

15.3.4 Deterministic Search Techniques

Random sampling methods such as particle filtering are not always the best approach for generating particles that search the space of possibilities. In particular, if the transition model is discrete and highly skewed — some successor states have much higher probability than others — then a random sampling of successor states is likely to generate many identical samples. This greatly reduces sample diversity, wastes computational resources, and leads to a poor representation of the space of possibilities. In this case, *search*-based methods may provide a better alternative. Here, we aim to find a set of particles that span the high-probability assignments and that we hope will allow us to keep track of the most likely trajectories through the system.

search

These techniques are commonly used in applications such as *speech recognition* (see box 6.B), where the transitions between phonemes, and even between words, are often highly constrained, with most transitions having probability (close to) 0. Here, we often formulate the problem as that of finding the single highest-probability trajectory through the system. In this case, an exact solution can be found by running a variable elimination algorithm such as that of section 13.2. In the context of HMMs, this algorithm is known as the *Viterbi algorithm*.

speech
recognition

Viterbi algorithm

In many cases, however, the HMM for speech recognition does not fit into memory. Moreover, if our task is continuous speech recognition, there is no natural end to the sequence. In such settings, we often resort to approximate techniques that are more memory efficient. A commonly used technique is beam search, which has the advantage that it can be applied in an online fashion as the data sequence is acquired. See exercise 15.10.

Finally, we note that deterministic search in temporal models is often used within the framework of collapsed particles, combining search over a subset of the variables with marginalization over others. This type of approach provides an approximate solution to the marginal MAP problem, which is often a more appropriate formulation of the problem. For example, in the speech-recognition problem, the MAP solution finds the most likely trajectory through the speech HMM. However, this complete trajectory tells us not only which words were spoken in a given utterance, but also which phones and subphones were traversed; we are rarely interested in the trajectory through these finer-grained states. A more appropriate goal is to find the most likely sequence of words when we marginalize over the possible sequences of phonemes and subphones. Methods such as beam search can also be adapted to the marginal MAP problem, allowing it to be applied in this setting; see exercise 15.10.

15.4 Hybrid DBNs

So far, we have focused on inference in the context of discrete models. However, many (perhaps even most) dynamical systems tend to include continuous, as well as discrete, variables. From a representational perspective, there is no difficulty incorporating continuous variables into the network model, using the techniques described in section 5.5. However, as usual, inference in models incorporating continuous variables poses new challenges. In general, the techniques developed in chapter 14 for the case of static networks also extend to the case of DBNs, in the same way that we extended inference techniques for static discrete networks in earlier sections in this chapter.

We now describe a few of the combinations, focusing on issues that are specific to the combination between DBNs and hybrid models. We emphasize that many of the other techniques described in this chapter and in chapter 14 can be successfully combined. For example, one of the most popular combinations is the application of particle filtering to continuous or hybrid systems; however, the combination does not raise significant new issues, and so we omit a detailed presentation.

15.4.1   Continuous Models

We begin by considering systems composed solely of continuous variables.

15.4.1.1   The Kalman Filter

The simplest such system is the linear dynamical system (see section 6.2.3.2), where the variables are related using linear Gaussian CPDs. These systems can be tracked very efficiently using a set of update equations called the *Kalman filter*.

Recall that the key difficulty with tracking a dynamical system is the entanglement phenomenon, which generally forces us to maintain, as our belief state, a full joint distribution over the state variables at time t. For discrete systems, this distribution has size exponential in the number of variables, which is generally intractably large. By contrast, as a linear Gaussian network defines a joint Gaussian distribution, and Gaussian distributions are closed under conditioning and marginalization, we know that the posterior distribution over any subset of variables given any set of observations is a Gaussian. In particular, the belief state over the state variables $\boldsymbol{X}^{(t)}$ is a multivariate Gaussian. A Gaussian can be represented as a mean vector and covariance matrix, which requires (at most) quadratic space in the number of state variables. Thus, in a Kalman filter, we can represent the belief state fairly compactly.

As we now show, we can also maintain the belief state efficiently, using simple matrix operations over the matrices corresponding to the belief state, the transition model, and the observation model. Specifically, consider a linear Gaussian DBN defined over a set of state variables $\boldsymbol{X}$ with $n = |\boldsymbol{X}|$ and a set of observation variables O with $m = |O|$. Let the probabilistic model be defined as in equation (6.3) and equation (6.4), which we review for convenience:

$$P(\boldsymbol{X}^{(t)} \mid \boldsymbol{X}^{(t-1)}) = \mathcal{N}\left(A\boldsymbol{X}^{(t-1)}; Q\right),$$
$$P(O^{(t)} \mid \boldsymbol{X}^{(t)}) = \mathcal{N}\left(H\boldsymbol{X}^{(t)}; R\right).$$

Kalman filter

state transition update

We now show the *Kalman filter* update equations, which provide an efficient implementation for equation (15.1) and equation (15.2). Assume that the Gaussian distribution encoding $\sigma^{(t)}$ is maintained using a mean vector $\mu^{(t)}$ and a covariance distribution $\Sigma^{(t)}$. The *state transition update* equation is easy to implement:

$$\begin{aligned}
\mu^{(\cdot t+1)} &= A\mu^{(t)} \\
\Sigma^{(\cdot t+1)} &= A\Sigma^{(t)}A^T + Q,
\end{aligned} \tag{15.4}$$

where $\mu^{(\cdot t+1)}$ and $\Sigma^{(\cdot t+1)}$ are the mean and covariance matrix for the prior belief state $\sigma^{(\cdot t+1)}$. Intuitively, the new mean vector is simply the application of the linear transformation A to the mean vector in the previous time step. The new covariance matrix is the transformation of the previous covariance matrix via A, plus the covariance introduced by the noise.

observation
update

The *observation update* is somewhat more elaborate:

$$
\begin{aligned}
K^{(t+1)} &= \Sigma^{(\cdot t+1)} H^T (H \Sigma^{(\cdot t+1)} H^T + R)^{-1} \\
\mu^{(t+1)} &= \mu^{(\cdot t+1)} + K^{(t+1)} (o^{(t+1)} - H \mu^{(\cdot t+1)}) \\
\Sigma^{(t+1)} &= (I - K^{(t+1)} H) \Sigma^{(\cdot t+1)}.
\end{aligned}
\tag{15.5}
$$

This update rule can be obtained using tedious but straightforward algebraic manipulations, by forming the joint Gaussian distribution over $X^{(t+1)}, O^{(t+1)}$ defined by the prior belief state $\sigma^{(\cdot t+1)}$ and the observation model $P(O^{(t+1)} \mid X^{(t+1)})$, and then conditioning the resulting joint Gaussian on the observation $o^{(t+1)}$.

To understand the intuition behind this rule, note first that the mean of $\sigma^{(t+1)}$ is simply the mean of $\sigma^{(\cdot t+1)}$, plus a correction term arising from the evidence. The correction term involves the *observation residual* — the difference between our *expected observation* $H \mu^{(\cdot t+1)}$ and the actual observation $o^{(t+1)}$. This residual is multiplied by a matrix called the *Kalman gain* $K^{(t+1)}$, which dictates the importance that we ascribe to the observation. We can see, for example, that when the measurement error covariance R approaches 0, the Kalman gain approaches H^{-1}; in this case, we are exactly "reverse engineering" the residual in the observation and using it to correct the belief state mean. Thus, the actual observation is trusted more and more, and the predicted observation $H \mu^{(\cdot t+1)}$ is trusted less. We also then have that the covariance of the new belief state approaches 0, corresponding to the fact that the observation tells us the current state with close to certainty. Conversely, when the covariance in our belief state $\Sigma^{(\cdot t+1)}$ tends to 0, the Kalman gain approaches 0 as well. In this case, we trust our predicted distribution, and pay less and less attention to the observation: both the mean and the covariance of the posterior belief state $\sigma^{(t+1)}$ are the same as those of the prior belief state $\sigma^{(\cdot t+1)}$. Finally, it can be shown that the posterior covariance matrix of our estimate approaches some limiting value as $T \longrightarrow \infty$, which reflects our "steady state" uncertainty about the system state. We note that both the time t covariance and its limiting value do not depend on the data. On one hand, this fact offers computational savings, since it allows the covariance matrix to be computed offline. However, it also points to a fundamental weakness of linear-Gaussian models, since we would naturally expect our uncertainty to depend on what we have seen.

The Kalman filtering process maintains the belief state as a mean and covariance matrix. An alternative is to maintain the belief state using information matrices (that is, a canonical form representation, as in equation (14.1)). The resulting set of update equations, called the

information form

information form of the Kalman filter, can be derived in a straightforward way from the basic operations on canonical forms described in section 14.2.1.2; the details are left as an exercise (exercise 15.11). We note that, in the Kalman filter, which maintains covariance matrices, the state transition update (equation (15.4)) is straightforward, and the observation update (equation (15.5)) is complex, requiring the inversion of an $n \times n$ matrix. In the information filter, which maintains information matrices, the situation is precisely the reverse.

15.4.1.2   Nonlinear Systems

The Kalman filter can also be extended to deal with nonlinear continuous dynamics, using the techniques described in section 14.4. In these methods, we maintain all of the intermediate factors arising in the course of inference as multivariate Gaussian distributions. When encountering a nonlinear CPD, which would give rise to a non-Gaussian factor, we simply linearize the result to produce a new Gaussian. We described two main methods for performing the linearization, either by taking the Taylor series expansion of the nonlinear function, or by using one of several numerical integration techniques. The same methods apply without change to the setting of tracking nonlinear continuous DBNs. In this case, the application is particularly straightforward, as the factors that we wish to manipulate in the course of tracking are all distributions; in a general clique tree, some factors do not represent distributions, preventing us from applying these linearization techniques and constraining the order in which messages are passed.

Concretely, assume that our nonlinear system has the model:

$$
\begin{aligned}
P(\boldsymbol{X}^{(t)} \mid \boldsymbol{X}^{(t-1)}) &= f(\boldsymbol{X}^{(t-1)}, \boldsymbol{U}^{(t-1)}) \\
P(O^{(t)} \mid \boldsymbol{X}^{(t)}) &= g(\boldsymbol{X}^{(t)}, \boldsymbol{W}^{(t)}),
\end{aligned}
$$

where f and g are deterministic nonlinear (continuous) functions, and $\boldsymbol{U}^{(t)}, \boldsymbol{W}^{(t)}$ are Gaussian random variables, which explicitly encode the noise in the transition and observation models, respectively. (In other words, rather than modeling the system in terms of stochastic CPDs, we use an equivalent representation that partitions the model into a deterministic function and a noise component.)

To address the filtering task here, we can apply either of the linearization methods described
extended Kalman filter   earlier. The Taylor-series linearization of section 14.4.1.1 gives rise to a method called the *extended Kalman filter*. The unscented transformation of section 14.4.1.2 gives rise to a method called the
unscented Kalman filter   *unscented Kalman filter*. In this latter approach, we maintain our belief state using the same representation as in the Kalman filter: $\sigma^{(t)} = \mathcal{N}\left(\mu^{(t)}; \Sigma^{(t)}\right)$. To perform the transition update, we construct a joint Gaussian distribution $p(\boldsymbol{X}^{(t)}, \boldsymbol{U}^{(t)})$ by multiplying the Gaussians for $\sigma^{(t)}$ and $\boldsymbol{U}^{(t)}$. The result is a Gaussian distribution and a nonlinear function f, to which we can now apply the unscented transformation of section 14.4.1.2. The result is a mean vector $\mu^{(\cdot t+1)}$ and covariance matrix $\Sigma^{(\cdot t+1)}$ for the prior belief state $\sigma^{(\cdot t+1)}$.

To obtain the posterior belief state, we must perform the observation update. We construct a joint Gaussian distribution $p(\boldsymbol{X}^{(t+1)}, \boldsymbol{W}^{(t+1)})$ by multiplying the Gaussians for $\sigma^{(\cdot t+1)}$ and $\boldsymbol{W}^{(t+1)}$. We then estimate a *joint* Gaussian distribution over $\boldsymbol{X}^{(t+1)}, O^{(t+1)}$, using the unscented transformation of equation (14.18) to estimate the integrals required for computing the mean and covariance matrix of this joint distribution. We now have a Gaussian joint distribution over $\boldsymbol{X}^{(t+1)}, O^{(t+1)}$, which we can condition on our observation $o^{(t+1)}$ in the usual way. The resulting posterior over $\boldsymbol{X}^{(t+1)}$ is the new belief state $\sigma^{(t+1)}$. Note that this approach computes a full joint covariance matrix over $\boldsymbol{X}^{(t+1)}, O^{(t+1)}$. When the dependency model of the observation on the state is factored, where we have individual observation variables each of which depends only on a few state variables, we can perform this computation in a more structured way (see exercise 15.12).

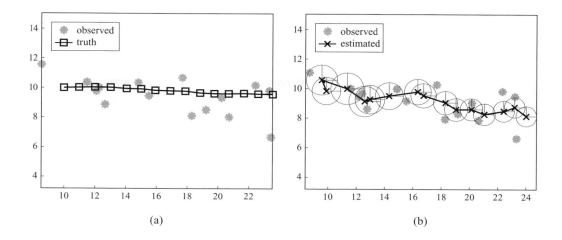

Figure 15.A.1 — Illustration of Kalman filtering for tracking (a) Raw data (dots) generated by an object moving to the right (line). (b) Estimated location of object: crosses are the posterior mean, circles are 95 percent confidence ellipsoids.

Box 15.A — Case Study: Tracking, Localization, and Mapping. *A key application of probabilistic models is to the task of tracking moving objects from noisy measurements. One example is* target tracking, *where we measure the location of an object, such as an airplane, using an external sensor. Another example is* robot localization, *where the moving object itself collects measurements, such as sonar or laser data, that can help it localize itself on a map.*

Kalman filtering was applied to this problem as early as the 1960s. Here, we give a simplified example to illustrate the key ideas. Consider an object moving in a two-dimensional plane. Let $X_1^{(t)}$ and $X_2^{(t)}$ be the horizontal and vertical locations of the object, and $\dot{X}_1^{(t)}$ and $\dot{X}_2^{(t)}$ be the corresponding velocity. We can represent this as a state vector $\boldsymbol{X}^{(t)} \in \mathbb{R}^4$. Let us assume that the object is moving at constant velocity, but is "perturbed" by random Gaussian noise (for example, due to the wind). Thus we can define $X_i' \sim \mathcal{N}\left(X_i + \dot{X}_i; \sigma^2\right)$ for $i = 1, 2$. Assume we can obtain a (noisy) measurement of the location of the object but not its velocity. Let $\boldsymbol{Y}^{(t)} \in \mathbb{R}^2$ represent our observation, where $\boldsymbol{Y}^{(t)} \sim \mathcal{N}\left((X_1^{(t)}, X_2^{(t)}); \Sigma_o\right)$, where Σ_o is the covariance matrix that governs our observation noise model. Here, we do not necessarily assume that noise is added separately to each dimension of the object location. Finally, we need to specify our initial (prior) beliefs about the state of the object, $p(\boldsymbol{X}^{(0)})$. We assume that this distribution is also a Gaussian $p(\boldsymbol{X}^{(0)}) = \mathcal{N}\left(\mu^{(0)}; \Sigma^{(0)}\right)$. We can represent prior ignorance by making $\Sigma^{(0)}$ suitably "broad." These parameters fully specify the model, allowing us to apply the Kalman filter, as described in section 15.4.1.1.

Figure 15.A.1 gives an example. The object moves to the right and generates an observation at each time step (such as a "blip" on a radar screen). We observe these blips and filter out the noise by using the Kalman filter; the resulting posterior distribution is plotted on the right. Our best guess

target tracking

robot localization

about the location of the object is the posterior mean, $\boldsymbol{E}_{\sigma^{(t)}}\left[\boldsymbol{X}_{1:2}^{(t)} \mid \boldsymbol{y}^{(1:t)}\right]$, denoted as a cross. Our uncertainty associated with this estimate is represented as an ellipse that contains 95 percent of the probability mass. We see that our uncertainty goes down over time, as the effects of the initial uncertainty get "washed out." As we discussed, the covariance converges to some steady-state value, which will remain indefinitely.

We have demonstrated this approach in the setting where the measurement is of the object's location, from an external measurement device. It is also applicable when the measurements are collected by the moving object, and estimate, for example, the distance of the robot to various landmarks on a map. If the error in the measured distance to the landmark has Gaussian noise, the Kalman filter approach still applies.

In practice, it is rarely possible to apply the Kalman filter in its simplest form. First, the dynamics and/or observation models are often nonlinear. A common example is when we get to observe the range and bearing to an object, but not its $X_1^{(t)}$ and $X_2^{(t)}$ coordinates. In this case, the observation model contains some trigonometric functions, which are nonlinear. If the Gaussian noise assumption is still reasonable, we apply either the extended Kalman filter or the unscented Kalman filter to linearize the model. Another problem arises when the noise is non-Gaussian, for example, when we have clutter or outliers. In this case, we might use the multivariate T distribution; this solution gains robustness at the cost of computational tractability. An alternative is to assume that each observation comes from a mixture model; one component corresponds to the observation being generated by the object, and another corresponds to the observation being generated by the background. However, this model is now an instance of a conditional linear Gaussian, raising all of the computational issues associated with multiple modes (see section 15.4.2). The same difficulty arises if we are tracking multiple objects, where we are uncertain which observation was generated data association *by which object; this problem is an instance of the* data association *problem (see box 12.D).*

In nonlinear settings, and particularly in those involving multiple modes, another very popular alternative is to use particle filtering. This approach is particularly appealing in an online setting such as robot motion, where computations need to happen in real time, and using limited computational resources. As a simple example, assume that we have a known map, $\mathcal{M}$. The map can be encoded as a occupancy grid — a discretized grid of the environment, where each square is 1 if the environment contains an obstacle at that location, and 0 otherwise. Or we can encode it using a more geometric representation, such as a set of line segments representing walls. We can represent the robot's location in the environment either in continuous coordinates, or in terms of the discretized grid (if we use that representation). In addition, the robot needs to keep track of its pose, or orientation, which we may also choose to discretize into an appropriate number of bins. The measurement $\boldsymbol{y}^{(t)}$ is a vector of measured distances to the nearest obstacles, as described in box 5.E. Our goal is to maintain $P(\boldsymbol{X}^{(t)} \mid \boldsymbol{y}^{(1:t)}, \mathcal{M})$, which is our posterior over the robot location.

Note that our motion model is nonlinear. Moreover, although the error model on the measured distance is a Gaussian around the true distance, the true distance to the nearest obstacle (in any given direction) is not even a continuous function of the robot's position. In fact, belief states can easily become multimodal owing to perceptual aliasing, *that is, when the robot's percepts can match two or more locations. Thus, a Gaussian model is a very poor approximation here.*

Monte Carlo
localization
Thrun et al. (2000) propose the use of particle filtering for localization, giving rise to an algorithm called Monte Carlo localization. *Figure 15.A.2 demonstrates one sample trajectory of the particles over time. We see that, as the robot acquires more measurements, its belief state becomes more*

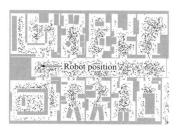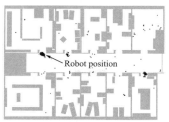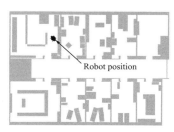

Figure 15.A.2 — Sample trajectory of particle filtering for robot localization

sharply peaked. More importantly, we see that the use of a particle-based belief state makes it easy to model multimodal posteriors.

One important issue when implementing a particle filter is the choice of proposal distribution. The simplest method is to use the standard bootstrap filter, where we propose directly from the dynamics model, and weight the proposals by how closely they match the evidence. However, when the robot is lost, so that our current belief state is very diffuse, this approach will not work very well, since the proposals will literally be all over the map but will then get "killed off" (given essentially zero weight) if they do not match the (high-quality) measurements. As we discussed, the ideal solution is to use posterior particle filtering, where we sample a particle $\boldsymbol{x}^{(t+1)}[m]$ from the posterior distribution $P(\mathcal{X}^{(t+1)} \mid \boldsymbol{x}^{(t)}[m], \boldsymbol{y}^{(t+1)})$. However, this solution requires that we be able to invert the evidence model using Bayes rule, a process that is not always feasible for complex, nonlinear models. One ad hoc fix is to inflate the noise level in the measurement model artificially, giving particles an artificially high chance of surviving until the belief state has the chance to adapt to the evidence. A better approach is to use a proposal that takes the evidence into account; for example, we can compare $\boldsymbol{y}^{(t)}$ with the map and then use a proposal that is a mixture of the bootstrap proposal $P(\mathcal{X}^{(t+1)} \mid \boldsymbol{x}^{(t)}[m])$ and some set of candidate locations that are most consistent with the recent observations.

robot mapping

SLAM

We now turn to the harder problem of localizing a robot in an unknown environment, while mapping *the environment at the same time. This problem is known as* simultaneous localization and mapping (SLAM). *In our previous terminology, this task corresponds to computing $p(\boldsymbol{X}^{(t)}, \mathcal{M} \mid \boldsymbol{y}^{(1:t)})$. Here, again, our task is much easier in the linear setting, where we represent the map in terms of K landmarks whose locations, denoted $L_1, \ldots, L_k$, are now unknown. Assume that we have a Gaussian prior over the location of each landmark, and that our observations $Y_k^{(t)}$ measure the Euclidean distance between the robot position $\boldsymbol{X}^{(t)}$ and the kth landmark location L_k, with some Gaussian noise. It is not difficult to see that $P(Y_k^{(t)} \mid \boldsymbol{X}^{(t)}, L_k)$ is a Gaussian distribution, so that our entire model is now a linear dynamical system. Therefore, we can naturally apply a Kalman filter to this task, where now our belief state represents $P(\boldsymbol{X}^{(t)}, L_1, \ldots, L_k \mid \boldsymbol{y}^{(1:t)})$. Figure 15.A.3a demonstrates this process for a simple two-dimensional map. We can see that the uncertainty of the landmark locations is larger for landmarks encountered later in the process, owing to the accumulation of uncertainty about the robot location. However, when the robot closes the loop and reencounters the first landmark, the uncertainty about its position reduces dramatically; the*

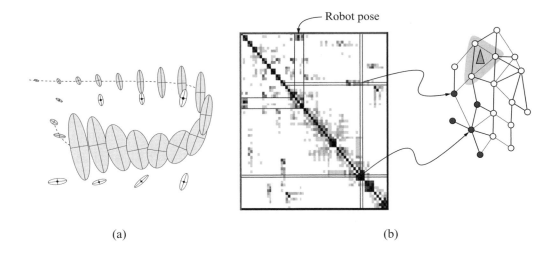

(a)   (b)

Figure 15.A.3 — Kalman filters for the SLAM problem (a) Visualization of marginals in Gaussian SLAM. (b) Visualization of information (inverse covariance) matrix in Gaussian SLAM, and its Markov network structure; very few entries have high values.

landmark uncertainty reduces at the same point. If we were to use a smoothing algorithm, we would also be able to reduce much of the uncertainty about the robot's intermediate locations, and hence also about the intermediate landmarks.

Gaussian inference is attractive in this setting because even representing a posterior over the landmark positions would require exponential space in the discrete case. Here, because of our ability to use Gaussians, the belief state grows quadratically only in the number of landmarks. However, for large maps, even quadratic growth can be too inefficient, particularly in an online setting. To address this computational burden, two classes of methods based on approximate inference have been proposed.

The first is based on factored belief-state idea of section 15.3.2. Here, we utilize the observation that, although the variables representing the different landmark locations become correlated due to entanglement, the correlations between them are often rather weak. In particular, two landmarks that were just observed at the same time become correlated due to uncertainty about the robot position. But, as the robot moves, the direct correlation often decays, and two landmarks often become almost conditionally independent given some subset of other landmarks; this conditional independence manifests as sparsity in the information (inverse covariance) matrix, as illustrated in figure 15.A.3b. Thus, these approaches approximate the belief state by using a sparser representation that maintains only the strong correlations. We note that the set of strongly correlated landmark pairs changes over time, and hence the structure of our approximation must be adaptive. We can consider a range of sparser representations for a Gaussian distribution. One approach is to use a clique tree, which admits exact M-projection operations but grows quadratically in the maximum size of cliques in our clique tree. Another is to use the Markov network representation of the Gaussian (or, equivalently, its inverse covariance). The two main challenges are to determine dynamically the approximation to use at each step, and to perform the (approximate) M-projection in an efficient way, essential in this real-time setting.

A second approach, and one that is more generally applicable, is to use collapsed particles. This approach is based on the important observation that the robot makes independent observations of the different landmarks. Thus, as we can see in figure 15.A.4a, the landmarks are conditionally independent given the robot's trajectory. Importantly, the landmarks are not independent given the robot's current location, owing to entanglement, but if we sample an entire robot trajectory $x^{(1:t)}$, we can now maintain the conditional distribution over the landmark positions in a fully factored form. In this approach, known as FastSLAM, each particle is associated with a (factorized) distribution over maps. In this approach, rather than maintaining a Gaussian of dimension $2K + 2$ (two coordinates for each landmark and two for the robot), we maintain a set of particles, each associated with K two-dimensional Gaussians. Because the Gaussian representation is quadratic in the dimension and the matrix inversion operations are cubic in the dimension, this approach provides considerable savings. (See exercise 15.13.) Experimental results show that, in the settings where Kalman filtering is suitable, this approximation achieves almost the same performance with a small number of particles (as few as ten); see figure 15.A.4b.

This approach is far more general than the sparse Kalman-filtering approach we have described, since it also allows us to deal with other map representations, such as the occupancy-grid map described earlier. Moreover, it also provides an integrated solution in cases where we have uncertainty about data association; here, we can sample over data-association hypotheses and still maintain a factored distribution over the map representation. Overall, by combining the different techniques we have described, we can effectively address most instances of the SLAM problem, so that this problem is now essentially considered solved.

15.4.2 Hybrid Models

We now move to considering a system that includes both discrete and continuous variables. As in the static case, inference in such a model is very challenging, even in cases where the continuous dynamics are simple.

CLG dynamics

switching linear dynamical system

Consider a conditional linear Gaussian DBN, where all of the continuous variables have *CLG dynamics*, and the discrete variables have no continuous parents. This type of model is known as a *switching linear dynamical system*, as it can be viewed as a dynamical system where changes in the discrete state cause the continuous (linear) dynamics to switch. The unrolled DBN is a standard CLG network, and, as in any CLG network, given a full assignment to the discrete variables, the distribution over the continuous variables (or any subset of them) is a multivariate Gaussian. However, if we are not given an assignment to the discrete variables, the marginal distribution over the continuous variables is a mixture of Gaussians, where the number of mixture components is exponential in the number of discrete variables. (This phenomenon is precisely what underlies the $\mathcal{NP}$-hardness proof for inference in CLG networks.)

In the case of a temporal model, this problem is even more severe, since the number of mixture components grows exponentially, and unboundedly, over time.

Example 15.6

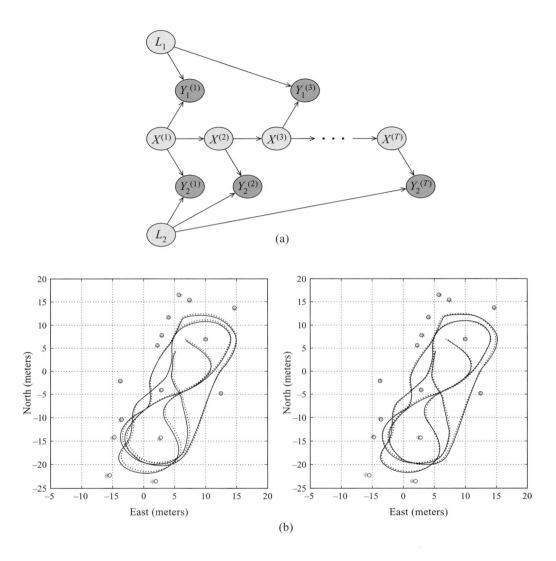

(a)

(b)

Figure 15.A.4 — Collapsed particle filtering for SLAM (a) DBN illustrating the conditional independencies in the SLAM probabilistic model. (b) Sample predictions using a SLAM algorithm (solid path, star
landmarks) in comparison to ground-truth measurements obtained from a GPS sensor, not given as input
to the algorithm (dotted lines, circle landmarks): left — EKF; right — FastSLAM. The results show that both
approaches achieve excellent results, and that there is very little difference in the quality of the estimates
between these two approaches.

Consider a one-dimensional switching linear-dynamical system whose state consists of a single discrete variable D and a single continuous variable X. In the SLDS, both X and D are persistent, and in addition we have $D' \to X'$. The transition model is defined by a CLG model for X', and a standard discrete CPD for D. In this example, the belief state at time t, marginalized over $X^{(t)}$, is a mixture distribution with a number of mixture components that is exponential in t. ∎

In order to make the propagation algorithm tractable, we must make sure that the mixture of Gaussians represented by the belief state does not get too large. The two main techniques to do so are *pruning* and *collapsing*.

mixture pruning

Pruning algorithms reduce the size of the mixture distribution in the belief state by discarding some of its Gaussians. The standard pruning algorithm simply keeps the N Gaussians with the highest probabilities, discards the rest, and then renormalizes the probabilities of the remaining

beam search

Gaussian to sum to 1. This is a form of *beam search* for marginal MAP, as described in exercise 15.10.

mixture collapsing

Collapsing algorithms partition the mixture of Gaussians into subsets, and then they collapse all the Gaussians in each subset into one Gaussian. Thus, if the belief state were partitioned into N subsets, the result would be a belief state with exactly N Gaussians. The different collapsing algorithms can differ in the choice of subsets of Gaussians to be collapsed, and on exactly when the collapsing is performed. The collapsed Gaussian $\hat{p}$ for a mixture p is generally chosen to be its M-projection — the one that minimizes $D(p\|\hat{p})$. Recall from proposition 14.3 that the M-projection can be computed simply and efficiently by matching moments.

We now describe some commonly used collapsing algorithms in this setting and show how they can be viewed within the framework of expectation propagation, as described in section 14.3.3. More precisely, consider a CLG DBN containing the discrete state variables $\boldsymbol{A}$ and the continuous state variables $\boldsymbol{X}$. Let $M = |Val(\boldsymbol{A})|$ — the number of discrete states at any given time slice. Assume, for simplicity, that the state at time 0 is fully observed. We note that, throughout this section, we focus solely on techniques for CLG networks. It is not difficult to combine these techniques with linearization methods (as described earlier), allowing us to accommodate both nonlinear dynamics and discrete children of continuous parents (see section 14.4.2.5).

As we discussed, after t time slices, the total number of Gaussian mixture components in the belief state is M^t — one for every assignment to $\boldsymbol{A}^{(1)}, \ldots, \boldsymbol{A}^{(t)}$. One common approximation

general pseudo-Bayes

is the class of *general pseudo-Bayesian* algorithms, which lump together components whose assignment in the recent past is identical. That is, for a positive integer k, $\sigma^{(t)}$ has M^{k-1} mixture components — one for every assignment to $\boldsymbol{A}^{((t-(k-2)):t)}$ (which for $k = 1$ we take to be a single "vacuous" assignment). If $\sigma^{(t)}$ has M^{k-1} mixture components, then one step of forward propagation results in a distribution with M^k components. Each of these Gaussian components is conditioned on the evidence, and its weight in the mixture is multiplied with the likelihood it gives the evidence. The resulting mixture is now partitioned into M^{k-1} subsets, and each subset is collapsed separately, producing a new belief state.

GPB1

For $k = 1$, the algorithm, known as *GPB1*, maintains exactly one Gaussian in the belief state; it also maintains a distribution over $\boldsymbol{A}^{(t)}$. Thus, $\sigma^{(t)}$ is essentially a product of a discrete distribution $\sigma^{(t)}(\boldsymbol{A}^{(t)})$ and a Gaussian $\sigma^{(t)}(\boldsymbol{X}^{(t)})$. In the forward-propagation step (corresponding to equation (15.1)), we obtain a mixture of M Gaussians: one Gaussian $p_{\boldsymbol{a}^{(t+1)}}^{(t+1)}$

for each assignment $a^{(t+1)}$, whose weight is

$$\sum_{A^{(t)}} P(a^{(t+1)} \mid A^{(t)}) \sigma^{(t)}(a^{(t)}).$$

In the conditioning step, each of these Gaussian components is conditioned on the evidence $o^{(t+1)}$, and its weight is multiplied by the likelihood of the evidence relative to this Gaussian. The resulting weighted mixture is then collapsed into a single Gaussian, using the M-projection operation described in proposition 14.3.

We can view this algorithm as an instance of the *expectation propagation* (EP) algorithm applied to the clique tree for the DBN described in section 15.2.3. The messages, which correspond to the belief states, are approximated using a factorization that decouples the discrete variables $A^{(t)}$ and the continuous variables $X^{(t)}$; the distribution over the continuous variables is maintained as a Gaussian. This approximation is illustrated in figure 15.7a. It is not hard to verify that the GPB1 propagation update is precisely equivalent to the operation of incorporating the time t message into the $(t, t+1)$ clique, performing the in-clique computation, and then doing the M-projection to produce the time $t+1$ message.

The *GPB2* algorithm maintains M Gaussians $\{p^{(t)}_{a^{(t)}}\}$ in the time t belief state, each one corresponding to an assignment $a^{(t)}$. After propagation, we get M^2 Gaussians, each one corresponding to an assignment to both $a^{(t)}$ and $a^{(t+1)}$. We now partition this mixture into M subsets, one for each assignment to $a^{(t+1)}$. Each subset is collapsed, resulting in a new belief state with M Gaussians. Once again, we can view this algorithm as an instance of EP, using the message structure $A^{(t)} \to X^{(t)}$, where the distribution of $X^{(t)}$ given $A^{(t)}$ in the message has the form of a conditional Gaussian. This EP formulation is illustrated in figure 15.7b.

A limitation of the GPB2 algorithm is that, at every propagation step we must generate M^2 Gaussians, a computation that may be too expensive. An alternative approach is called the *interacting multiple model (IMM)* algorithm. Like GPB2, it maintains a belief state with M Gaussians $p^{(t)}_{a^{(t)}}$; but like GPB1, it collapses all of the Gaussians in the mixture into a single Gaussian, prior to incorporating them into the transition model $\mathcal{P}(X^{(t+1)} \mid X^{(t)})$. Importantly, however, it performs the collapsing step after incorporating the discrete transition model $P(A^{(t+1)} \mid A^{(t)})$. This produces a different mixture distribution for each assignment $a^{(t+1)}$ — these distributions are all mixtures of the same set of time t Gaussians, but with different mixture weights, generally producing a better approximation. Each of these components, representing a conditional distribution over $X^{(t)}$ given $A^{(t+1)}$, is then propagated using the continuous dynamics $P(X^{(t+1)} \mid X^{(t)}, A^{(t+1)})$, and conditioned on the evidence, producing the time $t+1$ belief state. The IMM algorithm can also be formulated in the EP framework, as illustrated in figure 15.7c.

Although we collapse our belief state in M different ways when using the IMM algorithm, we only create M Gaussians at time $t+1$ (as opposed to M^2 Gaussians in GPB2). The extra work compared to GPB1 is the computation of the new mixture probabilities and collapsing M mixtures instead of just one, but usually this extra computational cost is small relative to the cost of computing the M Gaussians at time $t+1$. Therefore, the computational cost of the IMM algorithm is only slightly higher than the GPB1 algorithm, since both algorithms generate only M Gaussians at every propagation step, and it is significantly lower than GPB2. In practice, it seems that IMM often performs significantly better than GPB1 and almost as well as GPB2. Thus,

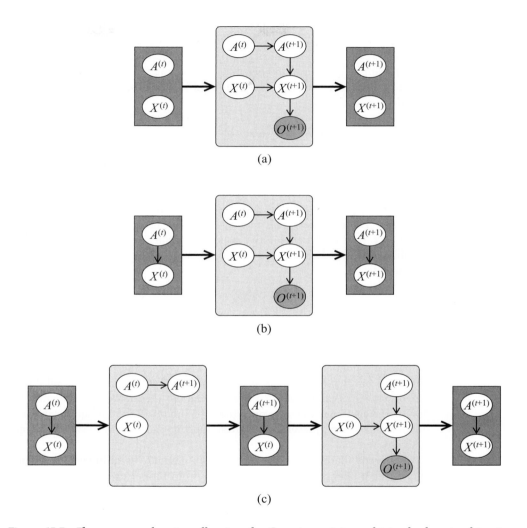

Figure 15.7 Three approaches to collapsing the Gaussian mixture obtained when tracking in a hybrid CLG DBN, viewed as instances of the EP algorithm. The figure illustrates the case where the network contains a single discrete variable A and a single continuous variable X: (a) the GPB1 algorithm; (b) the GPB2 algorithm; (c) the IMM algorithm.

the IMM algorithm appears to be a good compromise between complexity and performance.

We note that all of these clustering methods define the set of mixture components to correspond to assignments of discrete variables in the network. For example, in both GPB1 and IMM, each component in the mixture for $\sigma^{(t)}$ corresponds to an assignment to $\boldsymbol{A}^{(t)}$. In general, this approach may not be optimal, as Gaussians that correspond to different discrete assignments of $\boldsymbol{A}^{(t)}$ may be more similar to each other than Gaussians that correspond to the same assignment. In this case, the collapsed Gaussian would have a variance that is unnecessarily large, leading to a poorer approximation to the belief state. The solution to this problem is to select dynamically a partition of Gaussians in the current belief state, where the Gaussian components in the same partition are collapsed.

Our discussion has focused on cases where the number of discrete states at each time step is tractably small. How do we extend these ideas to more complex systems, where the number of discrete states at each time step is too large to represent explicitly? The EP formulation of these collapsing strategies provides an easy route to generalization. In section 14.3.3, we discussed the application of the EP algorithm to CLG networks, using either a clique tree or a cluster-graph message passing scheme. When faced with a more complex DBN, we can construct a cluster graph or a clique tree for the 2-TBN, where the clusters contain both discrete and continuous variables. These clusters pass messages to each other, using M-projection to the appropriate parametric family chosen for the messages. When a cluster contains one or more discrete variables, the M-projection operation may involve one of the collapsing procedures described. We omit details.

15.5   Summary

In this chapter, we discussed the problem of performing inference in a dynamic Bayesian network. We showed how the most natural inference tasks in this setting map directly onto probabilistic queries in the ground DBN. Thus, at a high level, the inference task here is not different from that of any other Bayesian network: we can simply unroll the network, instantiate our observations, and run inference to compute the answers to our queries. A key challenge lies in the fact that the networks that are produced in this approach can be very (or even unboundedly) large, preventing us from applying many of our exact and approximate inference schemes.

We showed that the tracking problem can naturally be formulated as a single upward pass of clique tree propagation, sending messages from time 0 toward later time slices. The messages naturally represent a *belief state*: our current beliefs about the state of the system. Importantly, this forward-propagation pass can be done in a way that does not require that the clique tree for the entire unrolled network be maintained.

Unfortunately, the entanglement property that arises in all but the most degenerate DBNs typically implies that the belief state has no conditional independence structure, and therefore it does not admit any factorization. This fact generally prevents the use of exact inference, except in DBNs over a fairly small state space.

As for exact inference, approximate inference techniques can be mapped directly to the unrolled DBN. Here also, however, we wish to avoid maintaining the entire DBN in memory during the course of inference. Some algorithms lend themselves more naturally than others to this "online" message passing. Two methods that have been specifically adapted to this setting are the factored message passing that lies at the heart of the expectation propagation algorithm, and the likelihood-weighting (importance-sampling) algorithm.

The application of the factored message passing is straightforward: We represent the message as a product of factors, possibly with counting number corrections to avoid double-counting; we employ a nested clique tree or cluster graph within each time slice to perform approximate message propagation, mapping a time t approximate belief state to a time $t + 1$ approximate belief state. In the case of filtering, this application is even simpler than the original EP, since no backward message passing is used.

The adaptation of the likelihood-weighting algorithm is somewhat more radical. Here, if we simply propagate particles forward, using the evidence to adjust their weights, the weight of the particles will generally rapidly go to zero, leaving us with a set of particles that has little or no relevance to the true state. From a technical perspective, the variance of this estimator rapidly grows as a function of t. We therefore introduce a resampling step, which allows "good" samples to duplicate while removing poor samples. We discussed several approaches for selecting new samples, including the simple (but very common) bootstrap filter, as well as more complex schemes that use MCMC or other approaches to generate a better proposal distribution for new samples. As in static networks, hybrid schemes that combine sampling with some form of (exact or approximate) global inference can be very useful in DBNs. Indeed, we saw examples of practical applications where this type of approach has been used successfully.

Finally, we discussed the task of inference in continuous or hybrid models. The issues here are generally very similar to the ones we already tackled in the case of static networks. In purely Gaussian networks, a straightforward application of the basic Gaussian message propagation algorithm gives rise to the famous Kalman filter. For nonlinear systems, we can apply the techniques of chapter 14 to derive the extended Kalman filter and the unscented filter. More interesting is the case where we have a hybrid system that contains both discrete and continuous variables. Here, in order to avoid an unbounded blowup in the number of components in the Gaussian mixture representing the belief state, we must collapse multiple Gaussians into one. Various standard approaches for performing this collapsing procedure have been developed in the tracking community; these approaches use a window length in the trajectory to determine which Gaussians to collapse. Interestingly, these approaches can be naturally viewed as variants of the EP algorithm, with different message approximation schemes.

Our presentation in this chapter has focused on inference methods that exploit in some way the specific properties of inference in temporal models. However, all of the inference methods that we have discussed earlier in this book (as well as others) can be adapted to this setting.

If we are willing to unroll the network fully, we can use any inference algorithm that has been developed for static networks. But even in the temporal setting, we can adapt other inference algorithms to the task of inference within a pair of consecutive time slices for the purpose of propagating a belief state. Thus, for example, we can consider variational approximation over a pair of time slices, or MCMC sampling, or various combinations of different algorithms. Many such variants have been proposed; see section 15.6 for references to some of them.

All of our analysis in this chapter has assumed that the structure of the different time slices is the same, so that a fixed approximation structure could be reasonably used for all time slices t. In practice, we might have processes that are not homogeneous, whether because the process itself is nonstationary or because of sparse events that can radically change the momentary system dynamics. The problem of dynamically adapting the approximation structure to changing circumstances is an exciting direction of research. Also related is the question of dealing with systems whose very structure changes over time, for example, a road where the number of cars can change dynamically.

This last point is related, naturally, to the problem of inference in other types of template-based models, some of which we described in chapter 6 but whose inference properties we did not tackle here. Of course, one option is to construct the full ground network and perform standard inference. However, this approach can be quite costly and even intractable. An important goal is to develop methods that somehow exploit structure in the template-based model to reduce the computational burden. Ideally, we would run our entire inference at the template level, avoiding the step of generating a ground network. This process is called *lifted inference*, in analogy to terms used in first-order logic theorem proving. As a somewhat less lofty goal, we would like to develop algorithms that use properties of the ground network, for example, the fact that it has repeated substructures due to the use of templates. These directions provide an important trajectory for current and future research.

lifted inference

15.6 Relevant Literature

The earliest instances of inference in temporal models were also some of the first applications of dynamic programming for probabilistic inference in graphical models: the forward-backward algorithm of Rabiner and Juang (1986) for hidden Markov models, and the Kalman filtering algorithm.

Kjærulff (1992, 1995a) provided the first algorithm for probabilistic inference in DBNs, based on a clique tree formulation applied to a moving window in the DBNs. Darwiche (2001a) studies the concept of slice-by-slice triangulation in a DBN, and suggests some new elimination strategies. Bilmes and Bartels (2003) extend on this work by providing a triangulation algorithm specifically designed for DBNs; they also show that it can be beneficial to allow the inference data structure to span a window larger than two time slices. Binder, Murphy, and Russell (1997) show how exact clique tree propagation in a DBN can be performed in a space-efficient manner by using a time-space trade-off.

Simple variants of particle-based filtering were first introduced by Handschin and Mayne (1969); Akashi and Kumamoto (1977). The popularity of these methods dates to the mid-1990s, where the resampling step was first introduced to avoid the degeneracy problems inherent to the naive approaches. This idea was introduced independently in several communities under several names, including: dynamic mixture models (West 1993), bootstrap filters (Gordon et al. 1993), survival of the fittest (Kanazawa et al. 1995), condensation (Isard and Blake 1998a), Monte Carlo filters (Kitagawa 1996), and sequential importance sampling (SIS) with resampling (SIR) (Doucet 1998). Kanazawa et al. (1995) propose the use of arc reversal for generating a better sampling distribution. Particle smoothing was first proposed by Isard and Blake (1998b) and Godsill, Doucet, and West (2000).

The success of these methods on a range of practical applications led to the development of multiple improvements, as well as some significant analysis of the theoretical properties of these methods. Doucet, de Freitas, and Gordon (2001) and Ristic, Arulampalam, and Gordon (2004) provide an excellent overview of some of the key developments.

The Viterbi algorithm for finding the MAP assignment in an HMM was proposed by Viterbi (1967); it is the first incarnation of the variable elimination algorithm for MAP algorithm.

The application of collapsed particles to switching linear systems were proposed by Akashi and Kumamoto (1977); Chen and Liu (2000); Doucet et al. (2000). Lerner et al. (2002) propose deterministic particle methods as an alternative to the sampling based approach, and demonstrate significant advantages in cases (such as fault diagnosis) where the distribution is highly peaked, so that sampling would generate the same particle multiple times. Marthi et al. (2002) describe an alternative sampling algorithm based on an MCMC approach with a decaying time window.

BK algorithm

Boyen and Koller (1998b) were the first to study the entanglement phenomenon explicitly and to propose factorization of belief states as an approximation in DBN inference, describing an algorithm that came to be known as the *BK algorithm*. They also provided a theoretical analysis demonstrating that, under certain conditions, the error accumulated over time remains bounded. Boyen and Koller (1998a) extend these ideas to the smoothing task. Murphy and Weiss (2001) suggested an algorithm that uses belief propagation within a time slice rather than a clique tree algorithm, as well as an iterated variant of the BK algorithm. Boyen and Koller (1999) study the properties of different belief state factorizations and offer some guidance on how to select a factorization that leads to a good approximation; Paskin (2003b); Frogner and Pfeffer (2007) suggest concrete heuristics for this adaptive process. Several methods that combine factorization with particle-based methods have also been proposed (Koller and Fratkina 1998; Murphy 1999; Lerner et al. 2000; Montemerlo et al. 2002; Ng et al. 2002).

The Kalman filtering algorithm was proposed by (Kalman 1960; Kalman and Bucy 1961); a simpler version was proposed as early as the nineteenth century (Thiele 1880). Normand and Tritchler (1992) provide a derivation of the Kalman filtering equations from a Bayesian network perspective. Many extensions and improvements have been developed over the years. Bar-Shalom, Li, and Kirubarajan (2001) provide a good overview of these methods, including the extended Kalman filter, and a variety of methods for collapsing the Gaussian mixture in a switching linear-dynamical system. Kim and Nelson (1998) reviews a range of deterministic and MCMC-based methods for these systems.

Lerner et al. (2000) and Lerner (2002) describe an alternative collapsing algorithm that provides more flexibility in defining the Gaussian mixture; they also show how collapsing can be applied in a factored system, where discrete variables are present in multiple clusters. Heskes and Zoeter (2002) apply the EP algorithm to switching linear systems. Zoeter and Heskes (2006) describe the relationship between the GPB algorithms and expectation propagation and provide an experimental comparison of various collapsing methods. The unscented filter, which we described in chapter 14, was first developed in the context of a filtering task by Julier and Uhlmann (1997). It has also been used as a proposal distribution for particle filtering (van der Merwe et al. 2000b,a), producing a filter of higher accuracy and asymptotic correctness guarantees.

When a temporal model is viewed in terms of the ground network it generates, it is amenable to the application of a range of other approximate inference methods. In particular, the global variational methods of section 11.5 have been applied to various classes of sequence-based models (Ghahramani and Jordan 1997; Ghahramani and Hinton 1998).

Temporal models have been applied to very many real-world problems, too numerous to list. Bar-Shalom, Li, and Kirubarajan (2001) describe the key role of these methods to target tracking. Isard and Blake (1998a) first proposed the use of particle filtering for visual tracking tasks; this approach, often called *condensation*, has been highly influential in the computer-vision community and has led to much follow-on work. The use of probabilistic models has also revolutionized the field of mobile robotics, providing greatly improved solutions to the tasks of navigation and mapping (Fox et al. 1999; Thrun et al. 2000); see Thrun et al. (2005) for a detailed overview of this area. The use of particle filtering, under the name *Monte Carlo localization*, has been particularly influential (Thrun et al. 2000; Montemerlo et al. 2002). However, factored models, both separately and in combination with particle-based methods, have also played a role in these applications (Murphy 1999; Paskin 2003b; Thrun et al. 2004,?), in particular as a representation of complex maps. Dynamic Bayesian models are also playing a role in speech-recognition systems (Zweig and Russell 1998; Bilmes and Bartels 2005) and in fault monitoring and diagnosis of complex systems (Lerner et al. 2002).

Monte Carlo localization

Finally, we also refer the reader to Murphy (2002), who provides an excellent tutorial on inference in DBNs.

15.7   Exercises

Exercise 15.1

Consider clique-tree message passing, described in section 15.2.2, where the messages in the clique tree of figure 15.1 are passed first from the beginning of the clique tree chain toward the end.

  a. Show that if sum-product message passing is used, the backward messages (in the "downward" pass) represent $P(o^{((t+1):T)} \mid S^{(t+1)})$.

  b. Show that if belief-update message passing is used, the backward messages (in the "downward" pass) represent $P(S^{(t+1)} \mid o^{(1:T)})$.

Exercise 15.2⋆

Consider the smoothing task for HMMs, implemented using the clique tree algorithm described in section 15.2.2. As discussed, the $O(NT)$ space requirement may be computationally prohibitive in certain settings. Let K be the space required to store the evidence at each time slice. Show how, by caching a certain subset of messages, we can trade off time for space, resulting in an algorithm whose time requirements are $O(N^2 T \log T)$ and space requirements are $O(N \log T)$.

Exercise 15.3

Prove proposition 15.1.

Exercise 15.4⋆

  a. Prove the entanglement theorem, theorem 15.1.

  b. Is there any 2-TBN (not necessarily fully persistent) where $(X \perp Y \mid \boldsymbol{Z})$ holds persistently but $(X \perp Y \mid \emptyset)$ does not? If so, give an example. If not, explain formally why not.

Exercise 15.5

Consider a fully persistent DBN over n state variables $\boldsymbol{X}$. Show that any clique tree over $\boldsymbol{X}^{(t)}, \boldsymbol{X}^{(t+1)}$ that we can construct for performing the belief-state propagation step has induced width at least $n + 1$.

Exercise 15.6⋆

Recall that a hidden Markov model!factorialfactorial HMMfactorial HMM (see figure 6.3) is a DBN over $X_1, \ldots, X_n, O$ such that the only parent of X_i' in the 2-TBN is X_i, and the parents of O' are $X_1', \ldots, X_n'$. (Typically, some structured model is used to encode this CPD compactly.) Consider the problem of using a structured variational approximation (as in section 11.5) to perform inference over the unrolled network for a fixed number T of time slices.

a. Consider a space of approximate distributions Q composed of disjoint clusters $\{X_i^{(0)}, \ldots, X_i^{(T)}\}$ for $i = 1, \ldots, n$. Show the variational update equations, describe the use of inference to compute the messages, and analyze the computational complexity of the resulting algorithm.

b. Consider a space of approximate distributions Q composed of disjoint clusters $\{X_1^{(1)}, \ldots, X_n^{(t)}\}$ for $t = 1 \ldots, T$. Show the variational update equations, describe the use of inference to compute the messages, and analyze the computational complexity of the resulting algorithm.

c. Discuss the circumstances when you would use one approximation over the other.

Exercise 15.7★

particle smoothing

In this question, you will extend the particle filtering algorithm to address the *smoothing* task, that is, computing $P(\boldsymbol{X}^{(t)} \mid \boldsymbol{o}^{(1:T)})$ where $T > t$ is the "end" of the sequence.

a. Prove using formal probabilistic reasoning that

$$P(\boldsymbol{X}^{(t)} \mid \boldsymbol{o}^{(1:T)}) = P(\boldsymbol{X}^{(t)} \mid \boldsymbol{o}^{(1:t)}) \cdot \sum_{\boldsymbol{X}^{(t+1)}} \frac{P(\boldsymbol{X}^{(t+1)} \mid \boldsymbol{o}^{(1:T)}) P(\boldsymbol{X}^{(t+1)} \mid \boldsymbol{X}^{(t)})}{P(\boldsymbol{X}^{(t+1)} \mid \boldsymbol{o}^{(1:t)})}.$$

b. Based on this formula, construct an extension to the particle filtering algorithm that:

- Does a first pass that is identical to particle filtering, but where it keeps the samples $\bar{\boldsymbol{x}}^{(0:t)}[m], w^{(t)}[m]$ generated for each time slice t.
- Has a second phase that updates the weights of the samples at the different time slices based on the formula in part 1, with the goal of getting an approximation to $P(\boldsymbol{X}^{(t)} \mid o^{(1:T)})$.

Write your algorithm using a few lines of pseudo-code. Provide a brief description of how you would use your algorithm to estimate the probability of $P(\boldsymbol{X}^{(t)} = \boldsymbol{x} \mid o^{(1:T)})$ for some assignment $\boldsymbol{x}$.

Exercise 15.8★

One of the problems with the particle filtering algorithm is that it uses only the observations obtained so far to select the samples that we continue to propagate. In many cases, the "right" choice of samples at time t is not clear based on the evidence up to time t, but it manifests a small number of time slices into the future. By that time, however, the relevant samples may have been eliminated in favor of ones that appeared (based on the limited evidence available) to be better.

In this problem, your task is to extend the particle filtering algorithm to deal with this problem. More precisely, assume that the relevant evidence usually manifests within k time slices (for some small k). Consider performing the particle-filtering algorithm with a lookahead of k time slices rather than a single time slice. Present the algorithm clearly and mathematically, specifying exactly how the weights are computed and how the next-state samples are generated. Briefly explain why your algorithm is sampling from (roughly) the right distribution (right in the same sense that standard particle filtering is sampling from the right distribution).

For simplicity of notation, assume that the process is structured as a state-observation model.

(Hint: Use your answer to exercise 15.7.)

Exercise 15.9★★

collapsed particle filtering

In this chapter, we have discussed only the application of particle filtering that samples all of the network variables. In this exercise, you will construct a collapsed-particle-filtering method. Consider a DBN where we have observed variables $\boldsymbol{O}$, and the unobserved variables are divided into two disjoint groups $\boldsymbol{X}$ and $\boldsymbol{Y}$. Assume that, in the 2-TBN, the parents of $\boldsymbol{X}'$ are only variables in $\boldsymbol{X}$, and that we can efficiently perform exact inference on $P(\boldsymbol{Y}' \mid \boldsymbol{X}, \boldsymbol{X}', \boldsymbol{o}')$. Describe an algorithm that uses *collapsed particle filtering* for this type of DBN, where we represent the approximate belief state $\hat{\sigma}^{(t)}$ using a set of weighted collapsed particles, where the variables $\boldsymbol{X}^{(t)}$ are sampled, and for each sample $\boldsymbol{X}^{(t)}[m]$, we have an exact distribution over the variables $\boldsymbol{Y}^{(t)}$. Specifically, describe how each sample $\boldsymbol{x}^{(t+1)}[m]$ is generated from the time t samples, how to compute the associated distribution over $\boldsymbol{Y}^{(t)}$, and how to compute the appropriate importance weights. Make sure your derivation is consistent with the analysis used in section 12.4.1.

Exercise 15.10

beam search

In this exercise, we apply the *beam-search* methods described in section 13.7 to the task of finding high-probability assignments in an HMM. Assume that our hidden state in the HMM is denoted $X^{(t)}$ and the observations are denoted $O^{(t)}$. Our goal is to find $x^{*(1:T)} = \arg\max_{x^{(1:T)}} P(X^{(1:T)} \mid O^{(1:t)})$. Importantly, however, we want to perform the beam search in an online fashion, adapting our set of candidates as new observations arrive.

   a. Describe how beam search can be applied in the context of an HMM. Specify the search procedure and suggest a number of pruning strategies.

   b. Now, assume we have an additional set of variables $Y^{(t)}$ that are part of our model and whose value we do not care about. That is, our goal has not changed. Assume that our 2-TBN model does not contain an arc $Y \rightarrow X'$. Describe how beam search can be applied in this setting.

Exercise 15.11

Construct an algorithm that performs tracking in linear-Gaussian dynamical systems, maintaining the belief state in terms of the canonical form described in section 14.2.1.2.

Exercise 15.12★

Consider a nonlinear 2-TBN where we have a set of observation variables $O_1, \ldots, O_k$, where each O_i' is a leaf in the 2-TBN and has the parents $\boldsymbol{U}_i \in \boldsymbol{X}'$. Show how we can use the methods of exercise 14.9 to perform the step of conditioning on the observation $\boldsymbol{O}' = \boldsymbol{o}'$, without forming an entire joint distribution over $\boldsymbol{X}' \cup O$. Your method should perform the numerical integration over a space with as small a dimension as possible.

Exercise 15.13★

   a. Write down the probabilistic model for the Gaussian SLAM problem with K landmarks.

   b. Derive the equations for a collapsed bootstrap-sampling particle-filtering algorithm in FastSLAM. Show how the samples are generated, how the importance weights are computed, and how the posterior is maintained.

   c. Derive the equations for the posterior collapsed-particle-filtering algorithm, where $\boldsymbol{x}^{(t+1)}[m]$ is generated from $P(\boldsymbol{X}^{(t+1)} \mid \boldsymbol{x}^{(t)}[m], \boldsymbol{y}^{(t+1)})$. Show how the samples are generated, how the importance weights are computed, and how the posterior is maintained.

   d. Now, consider a different approach of applied collapsed-particle filtering to this problem. Here, we select the landmark positions $\boldsymbol{L} = \{L_1, \ldots, L_k\}$ as our set of sampled variables, and for each particle $\boldsymbol{l}[m]$, we maintain a distribution over the robot's state at time t. Without describing the details of this algorithm, explain qualitatively what will happen to the particles and their weights eventually (as T grows). Are there conditions under which this algorithm will converge to the correct map?

PART III

Learning

16 *Learning Graphical Models: Overview*

16.1 Motivation

In most of our discussions so far, our starting point has been a given graphical model. For example, in our discussions of conditional independencies and of inference, we assumed that the model — structure as well as parameters — was part of the input.

There are two approaches to the task of acquiring a model. The first, as we discussed in box 3.C, is to construct the network by hand, typically with the help of an expert. However, as we saw, knowledge acquisition from experts is a nontrivial task. The construction of even a modestly sized network requires a skilled knowledge engineer who spends several hours (at least) with one or more domain experts. Larger networks can require weeks or even months of work. This process also generally involves significant testing of the model by evaluating results of some "typical" queries in order to see whether the model returns plausible answers.

 Such "manual" network construction is problematic for several reasons. **In some domains, the amount of knowledge required is just too large or the expert's time is too valuable. In others, there are simply no experts who have sufficient understanding of the domain. In many domains, the properties of the distribution change from one application site to another or over time, and we cannot expect an expert to sit and redesign the network every few weeks.** In many settings, however, we may have access to a set of examples generated from the distribution we wish to model. In fact, in the Information Age, it is often easier to obtain even large amounts of data in electronic form than it is to obtain human expertise. For example, in the setting of medical diagnosis (such as box 3.D), we may have access to a large collection of patient records, each listing various attributes such as the patient's history (age, sex, smoking, previous medical complications, and so on), reported symptoms, results of various tests, the physician's initial diagnosis and prescribed treatment, and the treatment's outcome. We may hope to use these data to learn a model of the distribution of patients in our population. In the case of pedigree analysis (box 3.B), we may have some set of family trees where a particular disease (for example, breast cancer) occurs frequently. We can use these family trees to learn the parameters of the genetics of the disease: the transmission model, which describes how often a disease genotype is passed from the parents to a child, and the penetrance model, which defines the probability of different phenotypes given the genotype. In an image segmentation application (box 4.B), we might have a set of images segmented by a person, and we might wish to learn the parameters of the MRF that define the characteristics of different regions, or those that define how strongly we believe that two neighboring pixels should be in the same segment.

It seems clear that such instances can be of use in constructing a good model for the underlying distribution, either in isolation or in conjunction with some prior knowledge acquired from a human. This task of constructing a model from a set of instances is generally called *model learning*. In this part of the book, we focus on methods for addressing different variants of this task. In the remainder of this chapter, we describe some of these variants and some of the issues that they raise.

To make this discussion more concrete, let us assume that the domain is governed by some underlying distribution P^*, which is induced by some (directed or undirected) network model $\mathcal{M}^* = (\mathcal{K}^*, \boldsymbol{\theta}^*)$. We are given a data set $\mathcal{D} = \{\boldsymbol{d}[1], \ldots, \boldsymbol{d}[M]\}$ of M samples from P^*. The standard assumption, whose statistical implications were briefly discussed in appendix A.2, is that the data instances are sampled independently from P^*; as we discussed, such data

IID samples   instances are called *independent and identically distributed (IID)*. We are also given some family of models, and our task is to learn some model $\tilde{\mathcal{M}}$ in this family that defines a distribution $P_{\tilde{\mathcal{M}}}$ (or $\tilde{P}$ when $\tilde{\mathcal{M}}$ is clear from the context). We may want to learn only model parameters for a fixed structure, or some or all of the structure of the model. In some cases, we might wish to present a spectrum of different hypotheses, and so we might return not a single model but rather a probability distribution over models, or perhaps some estimate of our confidence in the model learned.

We first describe the set of goals that we might have when learning a model and the different evaluation metrics to which they give rise. We then discuss how learning can be viewed as an optimization problem and the issues raised by the design of that problem. Finally, we provide a detailed taxonomy of the different types of learning tasks and discuss some of their computational ramifications.

16.2   Goals of Learning

To evaluate the merits of different learning methods, it is important to consider our goal in learning a probabilistic model from data. A priori, it is not clear why the goal of the learning is important. After all, our ideal solution is to return a model $\tilde{\mathcal{M}}$ that precisely captures the distribution P^* from which our data were sampled. Unfortunately, this goal is not generally achievable, because of computational reasons and (more importantly) because a limited data set provides only a rough approximation of the true underlying distribution. **In practice, the amount of data we have is rarely sufficient to obtain an accurate representation of a high-dimensional distribution involving many variables. Thus, we have to select $\tilde{\mathcal{M}}$ so as to construct the "best" approximation to $\mathcal{M}^*$. The notion of "best" depends on our goals. Different models will generally embody different trade-offs.** One approximate model may be better according to one performance metric but worse according to another. Therefore, to guide our development of learning algorithms, we must define the goals of our learning task and the corresponding metrics by which different results will be evaluated.

16.2.1   Density Estimation

The most common reason for learning a network model is to use it for some inference task that we wish to perform. Most obviously, as we have discussed throughout most of this book so far, a graphical model can be used to answer a range of probabilistic inference queries. In this

density
estimation

setting, we can formulate our learning goal as one of *density estimation*: constructing a model $\tilde{\mathcal{M}}$ such that $\tilde{P}$ is "close" to the generating distribution P^*.

How do we evaluate the quality of an approximation $\tilde{\mathcal{M}}$? One commonly used option is to use the relative entropy distance measure defined in definition A.5:

$$\boldsymbol{D}(P^* \| \tilde{P}) = \boldsymbol{E}_{\xi \sim P^*} \left[\log \left(\frac{P^*(\xi)}{\tilde{P}(\xi)} \right) \right].$$

Recall that this measure is zero when $\tilde{P} = P^*$ and positive otherwise. Intuitively, it measures the extent of the compression loss (in bits) of using $\tilde{P}$ rather than P^*.

To evaluate this metric, we need to know P^*. In some cases, we are evaluating a learning algorithm on synthetically generated data, and so P^* may be known. In real-world applications, however, P^* is not known. (If it were, we would not need to learn a model for it from a data set.) However, we can simplify this metric to obtain one that is easier to evaluate:

Proposition 16.1

For any distributions P, P' over $\mathcal{X}$:

$$\boldsymbol{D}(P \| P') = -\boldsymbol{H}_P(\mathcal{X}) - \boldsymbol{E}_{\xi \sim P}[\log P'(\xi)].$$

PROOF

$$\begin{aligned}
\boldsymbol{D}(P \| P') &= \boldsymbol{E}_{\xi \sim P} \left[\log \left(\frac{P(\xi)}{P'(\xi)} \right) \right] \\
&= \boldsymbol{E}_{\xi \sim P}[\log P(\xi) - \log P'(\xi)] \\
&= \boldsymbol{E}_{\xi \sim P}[\log P(\xi)] - \boldsymbol{E}_{\xi \sim P}[\log P'(\xi)] \\
&= -\boldsymbol{H}_P(\mathcal{X}) - \boldsymbol{E}_{\xi \sim P}[\log P'(\xi)].
\end{aligned}$$
∎

Applying this derivation to $P^*, \tilde{P}$, we see that the first of these two terms is the negative entropy of P^*; because it does not depend on $\tilde{P}$, it does not affect the comparison between different approximate models. We can therefore focus our evaluation metric on the second term $\boldsymbol{E}_{\xi \sim P^*} \left[\log \tilde{P}(\xi) \right]$ and prefer models that make this term as large as possible. This term is called

expected
log-likelihood

an *expected log-likelihood*. It encodes our preference for models that assign high probability to instances sampled from P^*. Intuitively, the higher the probability that $\tilde{\mathcal{M}}$ gives to points sampled from the true distribution, the more reflective it is of this distribution. We note that, in moving from the relative entropy to the expected log-likelihood, we have lost our baseline $\boldsymbol{E}_{P^*}[\log P^*]$, an inevitable loss since we do not know P^*. As a consequence, although we can use the log-likelihood as a metric for comparing one learned model to another, we cannot evaluate a particular $\tilde{\mathcal{M}}$ in how close it is to the unknown optimum.

likelihood

log-likelihood

More generally, in our discussion of learning we will be interested in the *likelihood* of the data, given a model $\mathcal{M}$, which is $P(\mathcal{D} : \mathcal{M})$, or for convenience using the *log-likelihood* $\ell(\mathcal{D} : \mathcal{M}) = \log P(\mathcal{D} : \mathcal{M})$.

log-loss

loss function

It is also customary to consider the negated form of the log-likelihood, called the *log-loss*. The log-loss reflects our cost (in bits) per instance of using the model $\tilde{P}$. The log-loss is our first example of a *loss function*, a key component in the statistical machine-learning paradigm. A loss function $loss(\xi : \mathcal{M})$ measures the loss that a model $\mathcal{M}$ makes on a particular instance

risk

ξ. When instances are sampled from some distribution P^*, our goal is to find a model that minimizes the *expected loss*, or the *risk*:

$$\mathbf{E}_{\xi \sim P^*}[loss(\xi \; : \; \mathcal{M})].$$

empirical risk

In general, of course, P^* is unknown. However, we can approximate the expectation using an empirical average and estimate the risk relative to P^* with an *empirical risk* averaged over a set $\mathcal{D}$ of instances sampled from P^*:

$$\mathbf{E}_{\mathcal{D}}[loss(\xi \; : \; \mathcal{M})] = \frac{1}{|\mathcal{D}|} \sum_{\xi \in \mathcal{D}} loss(\xi \; : \; \mathcal{M}). \tag{16.1}$$

In the case of the log-loss, this expression has a very intuitive interpretation. Consider a data set $\mathcal{D} = \{\xi[1], \dots, \xi[M]\}$ composed of IID instances. The probability that $\mathcal{M}$ ascribes to $\mathcal{D}$ is

$$P(\mathcal{D} \; : \; \mathcal{M}) = \prod_{m=1}^{M} P(\xi[m] \; : \; \mathcal{M}).$$

Taking the logarithm, we obtain

$$\log P(\mathcal{D} \; : \; \mathcal{M}) = \sum_{m=1}^{M} \log P(\xi[m] \; : \; \mathcal{M}),$$

which is precisely the negative of the empirical log-loss that appears inside the summation of equation (16.1).

The risk can be used both as a metric for evaluating the quality of a particular model and as a factor for selecting a model among a given class, given a training set $\mathcal{D}$. We return to these ideas in section 16.3.1 and box 16.A.

16.2.2   Specific Prediction Tasks

In the preceding discussion, we assumed that our goal was to use the learning model to perform probabilistic inference. With that assumption, we jumped to the conclusion that we wish to fit the overall distribution P^* as well as possible. However, that objective measures only our ability to evaluate the overall probability of a full instance ξ. In reality, the model can be used for answering a whole range of queries of the form $P(Y \mid X)$. In general, we can devise a test suite of queries for our learned model, which allows us to evaluate its performance on a range of queries. Most attention, however, has been paid to the special case where we have a particular set of variables Y that we are interested in predicting, given a certain set of variables X.

classification task

Most simply, we may want to solve a simple *classification task*, the goal of a large fraction of the work in machine learning. For example, consider the task of document classification, where we have a set X of words and other features characterizing the document, and a variable Y that labels the document topic. In image segmentation, we are interested in predicting the class labels for all of the pixels in the image (Y), *given* the image features (X). There are many other such examples.

A model trained for a prediction task should be able to produce for any instance characterized by x, the probability distribution $\tilde{P}(Y \mid x)$. We might also wish to select the MAP assignment of this conditional distribution to produce a specific prediction:

$$h_{\tilde{P}}(x) = \arg\max_{y} \tilde{P}(y \mid x).$$

What loss function do we want to use for evaluating a model designed for a prediction task? We can, for example, use the *classification error*, also called the *0/1 loss*:

classification
error

0/1 loss

$$E_{(x,y) \sim \tilde{P}}[\mathbf{1}\{h_{\tilde{P}}(x) \neq y\}], \tag{16.2}$$

which is simply the probability, over all (x, y) pairs sampled from $\tilde{P}$, that our classifier selects the wrong label. While this metric is suitable for labeling a single variable, it is not well suited to situations, such as image segmentation, where we simultaneously provide labels to a large number of variables. In this case, we do not want to penalize an entire prediction with an error of 1 if we make a mistake on only a few of the target variables. Thus, in this case, we might consider performance metrics such as the *Hamming loss*, which, instead of using the indicator function $\mathbf{1}\{h_{\tilde{P}}(x) \neq y\}$, counts the number of variables Y in which $h_{\tilde{P}}(x)$ differs from the ground truth y.

Hamming loss

We might also wish to take into account the confidence of the prediction. One such criterion is the *conditional likelihood* or its logarithm (sometimes called the *conditional log-likelihood*):

conditional
likelihood

$$E_{(x,y) \sim P^*}\left[\log \tilde{P}(y \mid x)\right]. \tag{16.3}$$

Like the log-likelihood criterion, this metric evaluates the extent to which our learned model is able to predict data generated from the distribution. However, it requires the model to predict only Y given X, and not the distribution of the X variables. As before, we can negate this expression to define a loss function and compute an empirical estimate by taking the average relative to a data set $\mathcal{D}$.

☞ **If we determine, in advance, that the model will be used only to perform a particular task, we may want to train the model to make trade-offs that make it more suited to that task. In particular, if the model is never evaluated on predictions of the variables X, we may want to design our training regime to optimize the quality of its answers for Y.** We return to this issue in section 16.3.2.

16.2.3 Knowledge Discovery

Finally, a very different motivation for learning a model for a distribution P^* is as a tool for *discovering knowledge* about P^*. We may hope that an examination of the learned model can reveal some important properties of the domain: what are the direct and indirect dependencies, what characterizes the nature of the dependencies (for example, positive or negative correlation), and so forth. For example, in the genetic inheritance domain, it may be of great interest to discover the parameter governing the inheritance of a certain property, since this parameter can provide significant biological insight regarding the inheritance mechanism for the allele(s) governing the disease. In a medical diagnosis domain, we may want to learn the structure of the model to discover which predisposing factors lead to certain diseases and which symptoms

knowledge
discovery

are associated with different diseases. Of course, simpler statistical methods can be used to explore the data, for example, by highlighting the most significant correlations between pairs of variables. However, a learned network model can provide parameters that have direct causal interpretation and can also reveal much finer structure, for example, by distinguishing between direct and indirect dependencies, both of which lead to correlations in the resulting distribution.

The knowledge discovery task calls for a very different evaluation criterion and a different set of compromises from a prediction task. In this setting, we really do care about reconstructing the correct model $\mathcal{M}^*$, rather than some other model $\tilde{\mathcal{M}}$ that induces a distribution similar to $\mathcal{M}^*$. Thus, in contrast to density estimation, where our metric was on the distribution defined by the model (for example, $\boldsymbol{D}(P^*\|\tilde{P})$), here our measure of success is in terms of the model, that is, differences between $\mathcal{M}^*$ and $\tilde{\mathcal{M}}$. Unfortunately, this goal is often not achievable, for several reasons.

identifiability   First, even with large amounts of data, the true model may not be *identifiable*. Consider, for example, the task of recovering aspects of the correct network structure $\mathcal{K}^*$. As one obvious difficulty, recall that a given Bayesian network structure often has several I-equivalent structures. If such is the case for our target distribution $\mathcal{K}^*$, the best we can hope for is to recover an I-equivalent structure. The problems are significantly greater when the data are limited. Here, for example, if X and Y are directly related in $\mathcal{K}^*$ but the parameters relating them induce only a weak relationship, it may be very difficult to detect the correlation in the data and distinguish it from a random fluctuations. This limitation is less of a problem for a density estimation task, where ignoring such weak correlations often has very few repercussions on the quality of our learned density; however, if our task focuses on correct reconstruction of structure, examples such as this reduce our accuracy. Conversely, when the number of variables is large relative to the amount of the training data, there may well be pairs of variables that appear strongly correlated simply by chance. Thus, we are also likely, in such settings, to infer the presence of edges that do not exist in the underlying model. Similar issues arise when attempting to infer other aspects of the model.

The relatively high probability of making model identification errors can be significant if the goal is to discover the correct structure of the underlying distribution. For example, if our goal is to infer which genes regulate which other genes (as in box 21.D), and if we plan to use the results of our analysis for a set of (expensive and time-consuming) wet-lab experiments, we may want to have some confidence in the inferred relationship. **Thus, in a knowledge discovery application, it is far more critical to assess the confidence in a prediction, taking into account the extent to which it can be identified given the available data and the number of hypotheses that would give rise to similar observed behavior.** We return to these issues more concretely later on in the book (see, in particular, section 18.5).

16.3   Learning as Optimization

The previous section discussed different ways in which we can evaluate our learned model. In many of the cases, we defined a numerical criterion — a loss function — that we would like to optimize. This perspective suggests that the learning task should be viewed as an optimization

hypothesis space   problem: we have a *hypothesis space*, that is, a set of candidate models, and an *objective function*,

objective function   a criterion for quantifying our preference for different models. Thus, our learning task can be formulated as one of finding a high-scoring model within our model class. The view of learning as optimization is currently the predominant approach to learning (not only of probabilistic models).

In this section, we discuss different choices of objective functions and their ramification on the results of our learning procedure. This discussion raises important points that will accompany us throughout this part of the book. We note that the formal foundations for this discussion will be established in later chapters, but the discussion is of fundamental importance to our entire discussion of learning, and therefore we introduce these concepts here.

16.3.1 Empirical Risk and Overfitting

Consider the task of constructing a model $\mathcal{M}$ that optimizes the expectation of a particular loss function $\boldsymbol{E}_{\xi \sim P^*}[loss(\xi : \mathcal{M})]$. Of course, we generally do not know P^*, but, as we have discussed, we can use a data set $\mathcal{D}$ sampled from P^* to produce an empirical estimate for this expectation. More formally, we can use the data $\mathcal{D}$ to define an *empirical distribution* $\hat{P}_{\mathcal{D}}$, as follows:

empirical
distribution

$$\hat{P}_{\mathcal{D}}(A) = \frac{1}{M} \sum_m \boldsymbol{I}\{\xi[m] \in A\}. \tag{16.4}$$

That is, the probability of the event A is simply the fraction of training examples that satisfy A. It is clear that $\hat{P}_{\mathcal{D}}(A)$ is a probability distribution. Moreover, as the number of training examples grows, the empirical distribution approaches the true distribution.

Theorem 16.1

Let $\xi[1], \xi[2], \ldots$ be a sequence of IID samples from $P^*(\mathcal{X})$, and let $\mathcal{D}_M = \langle \xi[1], \ldots, \xi[M] \rangle$, then

$$\lim_{M \to \infty} \hat{P}_{\mathcal{D}_M}(A) = P^*(A)$$

almost surely.

Thus, for a sufficiently large training set, $\hat{P}_{\mathcal{D}}$ will be quite close to the original distribution P^* with high probability (one that converges to 1 as $M \to \infty$). Since we do not have direct access to P^*, we can use $\hat{P}_{\mathcal{D}}$ as the best proxy and try to minimize our loss function relative to $\hat{P}_{\mathcal{D}}$. Unfortunately, a naive application of this optimization objective can easily lead to very poor results.

Consider, for example, the use of the empirical log-loss (or log-likelihood) as the objective. It is not difficult to show (see section 17.1) that the distribution that maximizes the likelihood of a data set $\mathcal{D}$ is the empirical distribution $\hat{P}_{\mathcal{D}}$ itself. Now, assume that we have a distribution over a probability space defined by 100 binary random variables, for a total of 2^{100} possible joint assignments. If our data set $\mathcal{D}$ contains 1,000 instances (most likely distinct from each other), the empirical distribution will give probability 0.001 to each of the assignments that appear in $\mathcal{D}$, and probability 0 to all $2^{100} - 1,000$ other assignments. While this example is obviously extreme, the phenomenon is quite general. For example, assume that $\mathcal{M}^*$ is a Bayesian network where some variables, such as *Fever*, have a large number of parents $X_1, \ldots, X_k$. In a table-CPD, the number of parameters grows exponentially with the number of parents k. For large k, we are highly unlikely to encounter, in $\mathcal{D}$, instances that are relevant to all possible parent instantiations, that is, all possible combinations of diseases $X_1, \ldots, X_k$. If we do not have enough data, many of the cases arising in our CPD will have very little (or no) relevant training data, leading to very poor estimates of the conditional probability of *Fever* in this context. In general, as we will see in later chapters, the amount of data required to estimate parameters reliably grows linearly with the number of parameters, so that the amount of data required can grow exponentially with the network connectivity.

As we see in this example, there is a significant problem with using the empirical risk (the loss on our training data) as a surrogate for our true risk. In particular, this type of objective

overfitting
tends to *overfit* the learned model to the training data. However, our goal is to answer queries about examples that were not in our training set. Thus, for example, in our medical diagnosis example, the patients to which the learned network will be applied are new patients, not the ones on whose data the network was trained. In our image-segmentation example, the model will be applied to new (unsegmented) images, not the (segmented) images on which the model

generalization
was trained. Thus, it is critical that the network *generalize* to perform well on unseen data.

The need to generalize to unseen instances and the risk of overfitting to the training set raise an important trade-off that will accompany us throughout our discussion. On one hand, if our hypothesis space is very limited, it might not be able to represent our true target distribution P^*. Thus, even with unlimited data, we may be unable to capture P^*, and thereby remain with a suboptimal model. This type of limitation in a hypothesis space introduces inherent error in the result of the learning procedure, which is called *bias*, since the learning procedure is limited in how close it can approximate the target distribution. Conversely, if we select a hypothesis space that is highly expressive, we are more likely to be able to represent correctly the target distribution P^*. However, given a small data set, we may not have the ability to select the "right" model among the large number of models in the hypothesis space, many of which may provide equal or perhaps even better loss on our limited (and thereby unrepresentative) training set $\mathcal{D}$. Intuitively, when we have a rich hypothesis space and limited number of samples, small random fluctuations in the choice of $\mathcal{D}$ can radically change the properties of the selected model, often resulting in models that have little relationship to P^*. As a result, the learning procedure will suffer from a high *variance* — running it on multiple data sets from the same P^* will lead to highly variable results. Conversely, if we have a more limited hypothesis space, we are less likely to find, by chance, a model that provides a good fit to $\mathcal{D}$. Thus, a high-scoring model within our limited hypothesis space is more likely to be a good fit to P^*, and thereby is more likely to generalize to unseen data.

☞

bias-variance
trade-off
This *bias-variance trade-off* underlies many of our design choices in learning. When selecting a hypothesis space of different models, we must take care not to allow too rich a class of possible models. Indeed, with limited data, the error introduced by variance may be larger than the potential error introduced by bias, and we may choose to restrict our learning to models that are too simple to correctly encode P^*. Although the learned model is guaranteed to be incorrect, our ability to estimate its parameters more reliably may well compensate for the error arising from incorrect structural assumptions. Moreover, when learning structure, although we will not correctly learn all of the edges, this restriction may allow us to more reliably learn the most important edges. In other words, restricting the space of possible models leads us to select models $\tilde{\mathcal{M}}$ whose performance on the training objective is poorer, but whose distance to P^* is better.

Restricting our model class is one way to reduce overfitting. In effect, it imposes a hard constraint that prevents us from selecting a model that precisely captures the training data. A second, more refined approach is to change our training objective so as to incorporate a soft preference for simpler models. Thus, our learning objective will usually incorporate competing components: some components will tend to move us toward models that fit well with our observed data; others will provide *regularization* that prevents us from taking the specifics of the data to extremes. **In many cases, we adopt a combination of the two approaches, utilizing both a hard constraint over the model class and an optimization objective that leads us away from overfitting.**

regularization

The preceding discussion described the phenomenon of overfitting and the importance of ensuring that our learned model generalizes to unseen data. However, we did not discuss how to tell whether our model generalizes, and how to design our hypothesis space and/or objective function so as to reduce this risk. Box 16.A discusses some of the basic experimental protocols that one uses in the design and evaluation of machine learning procedures. Box 16.B discusses a basic theoretical framework that one can use to try and answer questions regarding the appropriate complexity of our model class.

Box 16.A — Skill: Design and Evaluation of Learning Procedures. *A basic question in learning is to evaluate the performance of our learning procedure. We might ask this question in a relative sense, to compare two or more alternatives (for example, different hypothesis spaces, or different training objectives), or in an absolute sense, when we want to test whether the model we have learned "captures" the distribution. Both questions are nontrivial, and there is a large literature on how to address them. We briefly summarize some of the main ideas here.*

In both cases, we would ideally like to compare the learned model to the real underlying distribution that generated the data. This is indeed the strategy we use for evaluating performance when learning from synthetic data sets where we know (by design) the generating distribution (see, for example, box 17.C). Unfortunately, this strategy is infeasible when we learn from real-life data sets where we do not have access to the true generating distribution. And while synthetic studies can help us understand the properties of learning procedures, they are limited in that they are often not representative of the properties of the actual data we are interested in.

Evaluating Generalization Performance   *We start with the first question of evaluating the performance of a given model, or a set of models, on unseen data. One approach is to use* holdout testing. *In this approach, we divide our data set into two disjoint sets: the* training set $\mathcal{D}_{train}$ *and* test set $\mathcal{D}_{test}$*. To avoid artifacts, we usually use a randomized procedure to decide on this partition. We then learn a model using $\mathcal{D}_{train}$ (with some appropriate objective function), and we measure our performance (using some appropriate loss function) on $\mathcal{D}_{test}$. Because $\mathcal{D}_{test}$ is also sampled from P^*, it provides us with an empirical estimate of the risk. Importantly, however, because $\mathcal{D}_{test}$ is disjoint from $\mathcal{D}_{train}$, we are measuring our loss using instances that were unseen during the training, and not on ones for which we optimized our performance. Thus, this approach provides us with an unbiased estimate of the performance on new instances.*

holdout testing

training set

test set

Holdout testing can be used to compare the performance of different learning procedures. It can also be used to obtain insight into the performance of a single learning procedure. In particular, we can compare the performance of the procedure (say the empirical log-loss per instance, or the classification error) on the training set and on the held-out test set. Naturally, the training set performance will be better, but if the difference is very large, we are probably overfitting to the training data and may want to consider a less expressive model class, or some other method for discouraging overfitting.

Holdout testing poses a dilemma. To get better estimates of our performance, we want to increase the size of the test set. Such an increase, however, decreases the size of the training set, which results in degradation of quality of the learned model. When we have ample training data, we can find reasonable compromises between these two considerations. When we have few training samples, there is no good compromise, since decreasing either the training or the test set has large ramifications either on the quality of the learned model or the ability to evaluate it.

An alternative solution is to attempt to use available data for both training and testing. Of course, we cannot test on our training data; the trick is to combine our estimates of performance from repeated rounds of holdout testing. That is, in each iteration we train on some subset of the instances and test on the remaining ones. If we perform multiple iterations, we can use a relatively small test set in each iteration, and pool the performance counts from all of them to estimate performance. The question is how to allocate the training and test data sets. A commonly used

cross-validation

procedure is k-fold cross-validation, where we use each instance once for testing. This is done by partitioning the original data into k equally sized sets, and then in each iteration holding as test data one partition and training from all the remaining instances; see algorithm 16.A.1. An extreme case of cross-validation is leave one out cross-validation, *where we set $k = M$, that is, in each iteration we remove one instance and use it as a testing case. Both cross-validation schemes allow us to estimate not only the average performance of our learning algorithm, but also the extent to which this performance varies across the different folds.*

Both holdout testing and cross-validation are primarily used as methods for evaluating a learning procedure. In particular, a cross-validation procedure constructs k different models, one for each partition of the training set into training/test folds, and therefore does not result in even a single model that we can subsequently use. If we want to write a paper on a new learning procedure, the results of cross-validation provide a good regime for evaluating our procedure and comparing it to others. If we actually want to end up with a real model that we can use in a given domain, we would probably use cross-validation or holdout testing to select an algorithm and ensure that it is working satisfactorily, but then train our model on the entire data set $\mathcal{D}$, thereby making use of the maximum amount of available data to learn a single model.

Selecting a Learning Procedure   *One common use for these evaluation procedures is as a mechanism for selecting a learning algorithm that is likely to perform well on a particular application. That is, we often want to choose among a (possibly large) set of different options for our learning procedure: different learning algorithms, or different algorithmic parameters for the same algorithm (for example, different constraints on the complexity of the learned network structures). At first glance, it is straightforward to use holdout testing or cross-validation for this purpose: we take each option LearnProc$_j$, evaluate its performance, and select the algorithm whose estimated loss is smallest. While this use is legitimate, it is also tempting to use the performance estimate that we obtained using this procedure as a measure for how well our algorithm will generalize to unseen data. This use is likely to lead to misleading and overly optimistic estimates of performance, since we have selected our particular learning procedure to optimize for this particular performance metric. Thus, if we use cross-validation or a holdout set to select a learning procedure, and we want to have an unbiased estimate of how well our selected procedure will perform on unseen data, we must hold back a completely separate test set that is never used in selecting any aspect of the*

Algorithm 16.A.1 — Algorithms for holdout and cross-validation tests.

Procedure Evaluate (
    $\mathcal{M}$,   // parameters to evaluate
    $\mathcal{D}$ // test data set
)
1    $loss \leftarrow 0$
2   **for** $m = 1, \ldots, M$
3      $loss \leftarrow loss + loss(\xi[m] : \mathcal{M})$
4   **return** $\frac{loss}{M}$

Procedure Train-And-Test (
    LearnProc,   // Learning procedure
    $\mathcal{D}_{train}$,   // Training data
    $\mathcal{D}_{test}$,   // Test data
)
1    $\mathcal{M} \leftarrow \text{LearnProc}(\mathcal{D}_{train})$
2   **return** $\text{Evaluate}(\mathcal{M}, \mathcal{D}_{test})$

Procedure Holdout-Test (
    LearnProc,   // Learning procedure
    $\mathcal{D}$,   // Data Set
    p_{test} // Fraction of data for testing
)
1   Randomly reshuffle instances in $\mathcal{D}$
2    $M_{train} \leftarrow \text{round}(M \cdot (1 - p_{test}))$
3    $\mathcal{D}_{train} \leftarrow \{\xi[1], \ldots, \xi[M_{train}]\}$
4    $\mathcal{D}_{test} \leftarrow \{\xi[M_{train} + 1], \ldots, \xi[M]\}$
5   **return** $\text{Train-And-Test}(\text{LearnProc}, \mathcal{D}_{train}, \mathcal{D}_{test})$

Procedure Cross-Validation (
    LearnProc,   // Learning procedure
    $\mathcal{D}$,   // Data Set
    K,   // number of cross-validation folds
)
1   Randomly reshuffle instances in $\mathcal{D}$
2   Partition $\mathcal{D}$ into K disjoint data sets $\mathcal{D}_1, \ldots, \mathcal{D}_K$
3    $loss \leftarrow 0$
4   **for** $k = 1, \ldots, K$
5      $\mathcal{D}_{-k} \leftarrow \mathcal{D} - \mathcal{D}_k$
6      $loss \leftarrow loss + \text{Train-And-Test}(\text{LearnProc}, \mathcal{D}_{-k}, \mathcal{D}_k)$
7   **return** $\frac{loss}{K}$

---

validation set

model, on which our model's final performance will be evaluated. In this setting, we might have: a training set, using which we learn the model; a validation set, *which we use to evaluate different variants of our learning procedure and select among them; and a separate test set, on which our final performance is actually evaluated. This approach, of course, only exacerbates the problem of fragmenting our training data, and so one can develop nested cross-validation schemes that achieve the same goal.*

Goodness of Fit   *Cross-validation and holdout tests allow us to evaluate performance of different learning procedures on unseen data. However, without a "gold standard" for comparison, they do not allow us to evaluate whether our learned model really captures everything there is to capture about the distribution. This question is inherently harder to answer. In statistics, methods for answering*

goodness of fit

such questions fall under the category of goodness of fit *tests. The general idea is the following. After learning the parameters, we have a hypothesis about a distribution that generated the data. Now we can ask whether the data behave as though they were sampled from this distribution. To do this, we compare properties of the training data set to properties of* simulated *data sets of the same size that we generate according to the learned distribution. If the training data behave in a manner that deviates significantly from what we observed in the majority of the simulations, we have reason to believe that the data were not generated from the learned distribution.*

More precisely, we consider some property f of data sets, and evaluate $f(\mathcal{D}_{train})$ for the training set. We then generate new data sets $\mathcal{D}$ from our learned model $\mathcal{M}$ and evaluate $f(\mathcal{D})$ for these randomly generated data sets. If $f(\mathcal{D}_{train})$ deviates significantly from the distribution of the values $f(\mathcal{D})$ among our randomly sampled data sets, we would probably reject the hypothesis that $\mathcal{D}_{train}$ was generated from $\mathcal{M}$. Of course, there are many choices regarding which properties f we should evaluate. One natural choice is to define f as the empirical log-loss in the data set, $E_{\mathcal{D}}[loss(\xi : \mathcal{M})]$, as per equation (16.1). We can then ask whether the empirical log-loss for $\mathcal{D}_{train}$ differs significantly from the expected empirical log-loss for data set $\mathcal{D}$ sampled from $\mathcal{M}$. Note that the expected value of this last expression is simply the entropy of $\mathcal{M}$, and, as we saw in section 8.4, we can compute the entropy of a Bayesian network fairly efficiently. To check for significance, we also need to consider the tail distribution of the log-loss, which is more involved. However, we can approximate that computation by computing the variance of the log-loss as a function of $\mathcal{M}$. Alternatively, because generating samples from a Bayesian network is relatively inexpensive (as in section 12.1.1), we might find it easier to generate a large number of data sets $\mathcal{D}$ of size M sampled from the model and use those to estimate the distribution over $E_{\mathcal{D}}[loss(\xi : \mathcal{M})]$.

Box 16.B — Concept: PAC-bounds. *As we discussed, given a target loss function, we can estimate the empirical risk on our training set $\mathcal{D}_{train}$. However, because of possible overfitting to the training data, the performance of our learned model on the training set might not be representative of its performance on unseen data. One might hope, however, that these two quantities are related, so that a model that achieves low training loss also achieves low expected loss (risk).*

Before we tackle a proof of this type, however, we must realize that we cannot guarantee with certainty the quality of our learned model. Recall that the data set $\mathcal{D}$ is sampled stochastically from P^*, so there is always a chance that we would have "bad luck" and sample a very unrepresentative *data set from P^*. For example, we might sample a data set where we get the same joint assignment in all of the instances. It is clear that we cannot expect to learn useful parameters from such a data set (assuming, of course, that P^* is not degenerate). The probability of getting such a data set is very low, but it is not zero. Thus, our analysis must allow for the chance that our data set will be highly unrepresentative, in which case our learned model (which presumably performed well on the training set) may not perform well on expectation.*

Our goal is then to prove that our learning procedure is probably approximately correct: *that is, for most training sets $\mathcal{D}$, the learning procedure will return a model whose error is low. Making this discussion concrete, assume we use relative entropy to the true distribution as our loss function. Let P^*_M be the distribution over data sets $\mathcal{D}$ of size M sampled IID from P^*. Now, assume that we have a learning procedure L that, given a data set $\mathcal{D}$, returns a model $\mathcal{M}_{L(\mathcal{D})}$. We want to prove results of the form:*

probably approximately correct

> *Let $\epsilon > 0$ be our* approximation parameter *and $\delta > 0$ our* confidence parameter. *Then, for M "large enough," we have that*
> $$P^*_M(\{\mathcal{D} : \boldsymbol{D}(P^* \| P_{\mathcal{M}_{L(\mathcal{D})}}) \leq \epsilon\}) \geq 1 - \delta.$$

That is, for sufficiently large M, we have that, for most data sets $\mathcal{D}$ of size M sampled from P^*, the learning procedure, applied to $\mathcal{D}$, will learn a close approximation to P^*. The number of samples M required to achieve such a bound is called the sample complexity. *This type of result is called a* PAC-bound.

sample complexity

PAC-bound

This type of bound can only be obtained if the hypothesis space contains a model that can correctly represent P^*. In many cases, however, we are learning with a hypothesis space that is not guaranteed to be able to express P^*. In this case, we cannot expect to learn a model whose relative entropy to P^* is guaranteed to be low. In such a setting, the best we can hope for is to get a model whose error is at most ϵ worse than the lowest error found within our hypothesis space. The expected loss beyond the minimal possible error is called the excess risk. *See section 17.6.2.2 for one example of a generalization bound for this case.*

excess risk

16.3.2 Discriminative versus Generative Training

In the previous discussion, we implicitly assumed that our goal is to get the learned model $\tilde{\mathcal{M}}$ to be a good approximation to P^*. However, as we discussed in section 16.2.2, we often know in advance that we want the model to perform well on a particular task, such as predicting $\boldsymbol{Y}$ from $\boldsymbol{X}$. The training regime that we described would aim to get $\tilde{\mathcal{M}}$ close to the overall joint distribution $P^*(\boldsymbol{Y}, \boldsymbol{X})$. This type of objective is known as *generative training*, because we are training the model to *generate* all of the variables, both the ones that we care to predict and the features that we use for the prediction. Alternatively, we can train the model *discriminatively*, where our goal is to get $\tilde{P}(\boldsymbol{Y} \mid \boldsymbol{X})$ to be close to $P^*(\boldsymbol{Y} \mid \boldsymbol{X})$. The same model class can be trained in these two different ways, producing different results.

generative training

discriminative training

Example 16.1

naive Markov

As the simplest example, consider a simple "star" Markov network structure with a single target variable Y connected by edges to each of a set of features $X_1, \ldots, X_n$. If we train the model generatively, we are learning a naive Markov *model, which, because the network is singly connected, is equivalent to a naive Bayes model. On the other hand, we can train the same network structure discriminatively, to obtain a good fit to $P^*(Y \mid X_1, \ldots, X_n)$. In this case, as we showed in example 4.20, we are learning a model that is a logistic regression model for Y given its features.*∎

Note that a model that is trained generatively can still be used for specific prediction tasks. For example, we often train a naive Bayes model generatively but use it for classification. However, a model that is trained for a particular prediction task $P(Y \mid X)$ does not encode a distribution over X, and hence it cannot be used to reach any conclusions about these variables.

Discriminative training can be used for any class of models. However, its application in the context of Bayesian networks is less appealing, since this form of training changes the interpretation of the parameters in the learned model. For example, if we discriminatively train a (directed) naive Bayes model, as in example 16.1, the resulting model would essentially represent the same logistic regression as before, except that the pairwise potentials between Y and each X_i would be locally normalized to look like a CPD. Moreover, most of the computational properties that facilitate Bayesian network learning do not carry through to discriminative training. For this reason, discriminative training is usually performed in the context of undirected models.

conditional
random field

In this setting, we are essentially training a *conditional random field* (CRF), as in section 4.6.1: a model that directly encodes a conditional distribution $P(Y \mid X)$.

There are various trade-offs between generative and discriminative training, both statistical and computational, and the question of which to use has been the topic of heated debates. We now briefly enumerate some of these trade-offs.

bias

Generally speaking, generative models have a higher *bias* — they make more assumptions about the form of the distribution. First, they encode independence assumptions about the feature variables X, whereas discriminative models make independence assumptions only about Y and about their dependence on X. An alternative intuition arises from the following view. A generative model defines $\tilde{P}(Y, X)$, and thereby also induces $\tilde{P}(Y \mid X)$ and $\tilde{P}(X)$, using the same overall model for both. To obtain a good fit to P^*, we must therefore tune our model to get good fits to both $P^*(Y \mid X)$ and $P^*(X)$. Conversely, a discriminative model aims to get a good fit only to $P^*(Y \mid X)$, without constraining the same model to provide a good fit to $P^*(X)$ as well.

☞ **The additional bias in the setting offers a standard trade-off. On one hand, it can help regularize and constrain the learned model, thereby reducing its ability to overfit the data. Therefore, generative training often works better when we are learning from limited amounts of data.** However, imposing constraints can hurt us when the constraints are wrong, by preventing us from learning the correct model. In practice, the class of models we use always imposes some constraints that do not hold in the true generating distribution P^*. For limited amounts of data, the constraints might still help reduce overfitting, giving rise to better generalization. **However, as the amount of data grows, the bias imposed by the constraints**
☞ **starts to dominate the error of our learned model. Because discriminative models make fewer assumptions, they will tend to be less affected by incorrect model assumptions and will often outperform the generatively trained models for larger data sets.**

Example 16.2

Consider the problem of optical character recognition — identifying letters from handwritten images. Here, the target variable Y is the character label (for example, "A"). Most obviously, we can use the individual pixels as our feature variables $X_1, \ldots, X_n$. We can then either generatively train a naive Markov model or discriminatively train a logistic regression model. The naive Bayes (or Markov) model separately learns the distribution over the 256 pixel values given each of the 26 labels; each of these is estimated independently, giving rise to a set of fairly low-dimensional estimation problems. Conversely, the discriminative model is jointly optimizing all of the approximately 26×256 parameters of the multinomial logit distribution, a much higher-dimensional estimation problem. Thus, for sparse data, the naive Bayes model may often perform better.

However, even in this simple setting, the independence assumption made by the naive Bayes model — that pixels are independent given the image label — is clearly false. As a consequence, the naive Bayes model may be counting, as independent, features that are actually correlated, leading to errors in the estimation. The discriminative model is not making these assumptions; by fitting the parameters jointly, it can compensate for redundancy and other correlations between the features. Thus, as we get enough data to fit the logistic model reasonably well, we would expect it to perform better. ∎

A related benefit of discriminative models is that they are able to make use a much richer feature set, where independence assumptions are clearly violated. These richer features can often greatly improve classification accuracy.

Example 16.3

Continuing our example, the raw pixels are fairly poor features to use for the image classification task. Much work has been spent by researchers in computer vision and image processing in developing richer feature sets, such as the direction of the edge at a given image pixel, the value of a certain filter applied to an image patch centered at the pixel, and many other features that are even more refined. In general, we would expect to be able to classify images much better using these features than using the raw pixels directly. However, each of these features depends on the values of multiple pixels, and the same pixels are used in computing the values of many different features. Therefore, these features are certainly not independent, and using them in the context of a naive Bayes classifier is likely to lead to fairly poor answers. However, there is no reason not to include such correlated features within a logistic regression or other discriminative classifier. ∎

 Conversely, generative models have their own advantages. They often offer a more natural interpretation of a domain. And they are better able to deal with missing values and unlabeled data. Thus, the appropriate choice of model is application dependent, and often a combination of different training regimes may be the best choice.

16.4 Learning Tasks

We now discuss in greater detail the different variants of the learning task. As we briefly mentioned, the input of a learning procedure is:

- Some prior knowledge or constraints about $\tilde{\mathcal{M}}$.

- A set $\mathcal{D}$ of data instances $\{d[1], \ldots, d[M]\}$, which are independent and identically distributed (IID) samples from P^*.

The output is a model $\tilde{\mathcal{M}}$, which may include the structure, the parameters, or both.

There are many variants of this fairly abstract learning problem; roughly speaking, they vary along three axes, representing the two types of input and the type of output. First, and most obviously, the problem formulation depends on our output — the type of graphical model we are trying to learn — a Bayesian network or a Markov network. The other two axes summarize the input of the learning procedure. The first of these two characterizes the extent of the constraints that we are given about $\tilde{\mathcal{M}}$, and the second characterizes the extent to which the data in our training set are fully observed. We now discuss each of these in turn. We then present a taxonomy of the different tasks that are defined by these axes, and we review some of their computational implications.

16.4.1   Model Constraints

hypothesis space

The first question is the extent to which our input constrains the *hypothesis space* — the class of models that we are allowed to consider as possible outputs of our learning algorithm. There is an almost unbounded set of options here, since we can place various constraints on the structure or on the parameters of the model. Some of the key points along the spectrum are:

- At one extreme, we may be given a graph structure, and we have to learn only (some of) the parameters; note that we generally do not assume that the given structure is necessarily the correct one $\mathcal{K}^*$.

- We may not know the structure, and we have to learn both parameters and structure from the data.

- Even worse, we may not even know the complete set of variables over which the distribution P^* is defined. In other words, we may only observe some subset of the variables in the domain and possibly be unaware of others.

The less prior knowledge we are given, the larger the hypothesis space, and the more possibilities we need to consider when selecting a model. As we discussed in section 16.3.1, the complexity of the hypothesis space defines several important trade-offs. The first is statistical. If we restrict the hypothesis space too much, it may be unable to represent P^* adequately. Conversely, if we leave it too flexible, our chances increase of finding a model within the hypothesis space that accidentally has high score but is a poor fit to P^*. The second trade-off is computational: in many cases (although not always), the richer the hypothesis space, the more difficult the search to find a high-scoring model.

16.4.2   Data Observability

data observability

Along the second input axis, the problem depends on the extent of the *observability* of our training set. Here, there are several options:

complete data
- The data are *complete*, or *fully observed*, so that each of our training instances $\boldsymbol{d}[m]$ is a full instantiation to all of the variables in $\mathcal{X}^*$.

incomplete data
- The data are *incomplete*, or *partially observed*, so that, in each training instance, some variables are not observed.

hidden variable

- The data contain *hidden variables* whose value is never observed in any training instance. This option is the only one compatible with the case where the set of variables $\mathcal{X}^*$ is unknown, but it may also arise if we know of the existence of a hidden variable but never have the opportunity to observe it directly.

As we move along this axis, more and more aspects of our data are unobserved. When data are unobserved, we must hypothesize possible values for these variables. The greater the extent to which data are missing, the less we are able to hypothesize reliably values for the missing entries.

Dealing with partially observed data is critical in many settings. First, in many settings, observing the values of all variables can be difficult or even impossible. For example, in the case of patient records, we may not perform all tests on all patients, and therefore some of the variables may be unobserved in some records. Other variables, such as the disease the patient had, may never be observed with certainty. The ability to deal with partially observed data cases is also crucial to adapting a Bayesian network using data cases obtained after the network is operational. In such situations, the training instances are the ones provided to the network as queries, and as such, are never fully observed (at least when presented as a query).

A particularly difficult case of missing data occurs when we have hidden variables. Why should we worry about learning such variables? For the task of knowledge discovery, these variables may play an important role in the model, and therefore they may be critical for our understanding of the domain. For example, in medical settings, the genetic susceptibility of a patient to a particular disease might be an important variable. This might be true even if we do not know what the genetic cause is, and thus cannot observe it. As another example, the tendency to be an "impulse shopper" can be an important hidden variable in an application to supermarket data mining. In these cases, our domain expert can find it convenient to specify a model that contains these variables, even if we never expect to observe their values directly.

In other cases, we might care about the hidden variable even when it has no predefined semantic meaning. Consider, for example, a naive Bayes model, such as the one shown in figure 3.2, but where we assume that the X_i's are observed but the class variable C is hidden. In this model, we have a *mixture distribution*: Each value of the hidden variable represents a separate distribution over the X_i's, where each such mixture component distribution is "simple" — all of the X_i's are independent in each of the mixture components. Thus, the population is composed of some number of separate subpopulations, each of which is generated by a distinct distribution. If we could learn this model, we could recover the distinct subpopulations, that is, figure out what types of individuals we have in our population. This type of analysis is very useful from the perspective of knowledge discovery.

mixture distribution

 Finally, we note that **the inclusion of a hidden variable in the network can greatly simplify the structure, reducing the complexity of the network that needs to be learned.** Even a sparse model over some set of variables can induce a large number of dependencies over a subset of its variables; for example, returning to the earlier naive Bayes example, if the class variable C is hidden and therefore is not included in the model, the distribution over the variables $X_1, \ldots, X_n$ has no independencies and requires a fully connected graph to be represented correctly. Figure 16.1 shows another example. (This figure illustrates another visual convention that will accompany us throughout this part of the book: Variables whose values are always hidden are shown as white ovals.) Thus, in many cases, ignoring the hidden variable

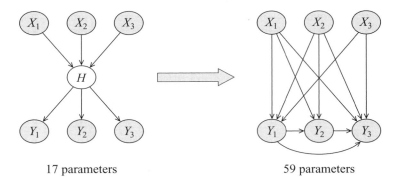

17 parameters   59 parameters

Figure 16.1 The effect of ignoring hidden variables. The model on the right is an I-map for the distribution represented by the model on the left, where the hidden variable is marginalized out. The counts indicate the number of independent parameters, under the assumption that the variables are binary-valued. The variable H is hidden and hence is shown as a white oval.

leads to a significant increase in the complexity of the "true" model (the one that best fits P^*), making it harder to estimate robustly. Conversely, learning a model that usefully incorporates a hidden variable is far from trivial. Thus, the decision of whether to incorporate a hidden variable is far from trivial, and it requires a careful evaluation of the trade-offs.

16.4.3   Taxonomy of Learning Tasks

Based on these three axes, we can provide a taxonomy of different learning tasks and discuss some of the computational issues they raise.

The problem of parameter estimation for a known structure is one of numerical optimization. Although straightforward in principle, this task is an important one, both because numbers are difficult to elicit from people and because parameter estimation forms the basis for the more advanced learning scenarios.

In the case of Bayesian networks, when the data are complete, the parameter estimation problem is generally easily solved, and it often even admits a closed-form solution. Unfortunately, this very convenient property does not hold for Markov networks. Here, the global partition function induces entanglement of the parameters, so that the dependence of the distribution on any single parameter is not straightforward. Nevertheless, for the case of a fixed structure and complete data, the optimization problem is convex and can be solved optimally using simple iterated numerical optimization algorithms. Unfortunately, each step of the optimization algorithm requires inference over the network, which can be expensive for large models.

When the structure is not given, the learning task now incorporates an additional level of complexity: the fact that our hypothesis space now contains an enormous (generally superexponentially large) set of possible structures. In most cases, as we will see, the problem of structure selection is also formulated as an optimization problem, where different network structures are given a score, and we aim to find the network whose score is highest. In the case of Bayesian networks, the same property that allowed a closed-form solution for the parameters also allows the score for a candidate network to be computed in closed form. In the case of Markov network, most natural scores for a network structure cannot be computed in closed form because of the partition function. However, we can define a convex optimization problem that jointly searches over parameter and structure, allowing for a single global optimum.

The problem of dealing with incomplete data is much more significant. Here, the multiple hypotheses regarding the values of the unobserved variables give rise to a combinatorial range of different alternative models, and induce a nonconvex, multimodal optimization problem even in parameter space. The known algorithms generally work by iteratively using the current parameters to fill in values for the missing data, and then using the completion to reestimate the model parameters. This process requires multiple calls to inference as a subroutine, making this process expensive for large networks. The case where the structure is not known is even harder, since we need to combine a discrete search over network structure with nonconvex optimization over parameter space.

16.5 Relevant Literature

Most of the topics reviewed here are discussed in greater technical depth in subsequent chapters, and so we defer the bibliographic references to the appropriate places. Hastie, Tibshirani, and Friedman (2001) and Bishop (2006) provide an excellent overview of basic concepts in machine learning, many of which are relevant to the discussion in this book.

17 *Parameter Estimation*

In this chapter, we discuss the problem of estimating parameters for a Bayesian network. We assume that the network structure is fixed and that our data set $\mathcal{D}$ consists of fully observed instances of the network variables: $\mathcal{D} = \{\xi[1], \ldots, \xi[M]\}$. This problem arises fairly often in practice, since numerical parameters are harder to elicit from human experts than structure is. It also plays a key role as a building block for both structure learning and learning from incomplete data. As we will see, despite the apparent simplicity of our task definition, there is surprisingly much to say about it.

As we will see, there are two main approaches to dealing with the parameter-estimation task: one based on *maximum likelihood estimation*, and the other using Bayesian approaches. For each of these approaches, we first discuss the general principles, demonstrating their application in the simplest context: a Bayesian network with a single random variable. We then show how the structure of the distribution allows the techniques developed in this very simple case to generalize to arbitrary network structures. Finally, we show how to deal with parameter estimation in the context of structured CPDs.

17.1 Maximum Likelihood Estimation

In this section, we describe the basic principles behind maximum likelihood estimation.

17.1.1 The Thumbtack Example

We start with what may be considered the simplest learning problem: parameter learning for a single variable. This is a classical Statistics 101 problem that illustrates some of the issues that we will encounter in more complex learning problems. Surprisingly, this simple problem already contains some interesting issues that we need to tackle.

Imagine that we have a thumbtack, and we conduct an experiment whereby we flip the thumbtack in the air. It comes to land as either heads or tails, as in figure 17.1. We toss the thumbtack several times, obtaining a data set consisting of *heads* or *tails* outcomes. Based on this data set, we want to estimate the probability with which the next flip will land heads or tails. In this description, we already made the implicit assumption that the thumbtack tosses are controlled by an (unknown) *parameter* θ, which describes the frequency of heads in thumbtack tosses. In addition, we also assume that the data instances are independent and identically distributed (IID).

Figure 17.1   A simple thumbtack tossing experiment

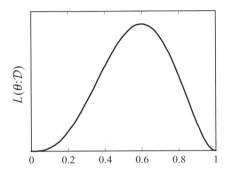

Figure 17.2   The likelihood function for the sequence of tosses H, T, T, H, H

Assume that we toss the thumbtack 100 times, of which 35 come up heads. What is our estimate for θ? Our intuition suggests that the best estimate is 0.35. Had θ been 0.1, for example, our chances of seeing $35/100$ heads would have been much lower. In fact, we examined a similar situation in our discussion of sampling methods in section 12.1, where we used samples from a distribution to estimate the probability of a query. As we discussed, the central limit theorem shows that, as the number of coin tosses grows, it is increasingly unlikely to sample a sequence of IID thumbtack flips where the fraction of tosses that come out heads is very far from θ. Thus, for sufficiently large M, the fraction of heads among the tosses is a good estimate with high probability.

To formalize this intuition, assume that we have a set of thumbtack tosses $x[1], \ldots, x[M]$ that are IID, that is, each is sampled independently from the same distribution in which $X[m]$ is equal to H (heads) or T (tails) with probability θ and $1 - \theta$, respectively. Our task is to find a good value for the parameter θ. As in many formulations of learning tasks, we define a *hypothesis space* Θ — a set of possibilities that we are considering — and an *objective function* that tells us how good different hypotheses in the space are relative to our data set $\mathcal{D}$. In this case, our hypothesis space Θ is the set of all parameters $\theta \in [0, 1]$.

How do we score different possible parameters θ? As we discussed in section 16.3.1, one way of evaluating θ is by how well it predicts the data. In other words, if the data are likely given the parameter, the parameter is a good predictor. For example, suppose we observe the sequence of outcomes H, T, T, H, H. If we know θ, we could assign a probability to observing this particular sequence. The probability of the first toss is $P(X[1] = H) = \theta$. The probability of the second toss is $P(X[2] = T \mid X[1] = H)$, but our assumption that the coin tosses are independent allows us to conclude that this probability is simply $P(X[2] = T) = 1 - \theta$. This is also the probability of the third outcome, and so on. Thus, the probability of the sequence is

$$P(\langle H, T, T, H, H \rangle : \theta) = \theta(1 - \theta)(1 - \theta)\theta\theta = \theta^3(1 - \theta)^2.$$

As expected, this probability depends on the particular value θ. As we consider different values of θ, we get different probabilities for the sequence. Thus, we can examine how the probability of the data changes as a *function* of θ. We thus define the *likelihood function* to be

likelihood
function

$$L(\theta : \langle H, T, T, H, H \rangle) = P(\langle H, T, T, H, H \rangle : \theta) = \theta^3(1 - \theta)^2.$$

Figure 17.2 plots the likelihood function in our example.

Clearly, parameter values with higher likelihood are more likely to generate the observed sequences. Thus, we can use the likelihood function as our measure of quality for different parameter values and select the parameter value that maximizes the likelihood; this value is called the *maximum likelihood estimator (MLE)*. By viewing figure 17.2 we see that $\hat{\theta} = 0.6 = 3/5$ maximizes the likelihood for the sequence H, T, T, H, H.

maximum
likelihood
estimator

Can we find the MLE for the general case? Assume that our data set $\mathcal{D}$ of observations contains $M[1]$ heads and $M[0]$ tails. We want to find the value $\hat{\theta}$ that maximizes the likelihood of θ relative to $\mathcal{D}$. The likelihood function in this case is:

$$L(\theta : \mathcal{D}) = \theta^{M[1]}(1 - \theta)^{M[0]}.$$

It turns out that it is easier to maximize the logarithm of the likelihood function. In our case, the *log-likelihood* function is:

log-likelihood

$$\ell(\theta : \mathcal{D}) = M[1] \log \theta + M[0] \log(1 - \theta).$$

Note that the log-likelihood is monotonically related to the likelihood. Therefore, maximizing the one is equivalent to maximizing the other. However, the log-likelihood is more convenient to work with, since products are converted to summations.

Differentiating the log-likelihood, setting the derivative to 0, and solving for θ, we get that the maximum likelihood parameter, which we denote $\hat{\theta}$, is

$$\hat{\theta} = \frac{M[1]}{M[1] + M[0]}, \tag{17.1}$$

as expected (see exercise 17.1).

As we will see, the maximum likelihood approach has many advantages. However, the approach also has some limitations. For example, if we get 3 heads out of 10 tosses, the MLE estimate is 0.3. We get the same estimate if we get 300 heads out of 1,000 tosses. Clearly, the two experiments are not equivalent. Our intuition is that, in the second experiment, we should be more confident of our estimate. Indeed, statistical estimation theory deals with *confidence intervals*. These are common in news reports, for example, when describing the results of election polls, where we often hear that "61 ± 2 percent" plan to vote for a certain candidate. The 2 percent is a confidence interval — the poll is designed to select enough people so that the MLE estimate will be within 0.02 of the true parameter, with high probability.

confidence
interval

17.1.2 The Maximum Likelihood Principle

We now generalize the discussion of maximum likelihood estimation to a broader range of learning problems. We then consider how to apply it to the task of learning the parameters of a Bayesian network.

We start by describing the setting of the learning problem. Assume that we observe several IID samples of a set of random variables $\mathcal{X}$ from an unknown distribution $P^*(\mathcal{X})$. We assume we know in advance the sample space we are dealing with (that is, which random variables, and what values they can take). However, we do not make any additional assumptions about P^*. We denote the *training set* of samples as $\mathcal{D}$ and assume that it consists of M instances of $\mathcal{X}$: $\xi[1], \ldots \xi[M]$.

training set

Next, we need to consider what exactly we want to learn. We assume that we are given a *parametric model* for which we wish to estimate *parameters*. Formally, a *parametric model* (also known as a parametric family; see section 8.2) is defined by a function $P(\xi : \boldsymbol{\theta})$, specified in terms of a set of *parameters*. Given a particular set of parameter values $\boldsymbol{\theta}$ and an instance ξ of $\mathcal{X}$, the model assigns a probability (or density) to ξ. Of course, we require that for each choice of parameters $\boldsymbol{\theta}$, $P(\xi : \boldsymbol{\theta})$ is a legal distribution; that is, it is nonnegative and

parametric model

parameters

$$\sum_{\xi} P(\xi : \boldsymbol{\theta}) = 1.$$

In general, for each model, not all parameter values are legal. Thus, we need to define the *parameter space* Θ, which is the set of allowable parameters.

parameter space

To get some intuition, we consider concrete examples. The model we examined in section 17.1.1 has parameter space $\Theta_{thumbtack} = [0, 1]$ and is defined as

$$P_{thumbtack}(x : \theta) = \begin{cases} \theta & \text{if } x = H \\ 1 - \theta & \text{if } x = T. \end{cases}$$

There are many additional examples.

Example 17.1

multinomial

Suppose that X is a multinomial variable that can take values $x^1, \ldots, x^K$. The simplest representation of a multinomial distribution is as a vector $\boldsymbol{\theta} \in \mathbb{R}^K$, such that

$$P_{multinomial}(x : \boldsymbol{\theta}) = \theta_k \text{ if } x = x^k.$$

The parameter space of this model is

$$\Theta_{multinomial} = \left\{ \boldsymbol{\theta} \in [0, 1]^K : \sum_i \theta_i = 1 \right\}.$$

∎

Example 17.2

Gaussian

Suppose that X is a continuous variable that can take values in the real line. A Gaussian model for X is

$$P_{Gaussian}(x : \mu, \sigma) = \frac{1}{\sqrt{2\pi}\sigma} e^{-\frac{(x-\mu)^2}{2\sigma^2}},$$

where $\boldsymbol{\theta} = \langle \mu, \sigma \rangle$. The parameter space for this model is $\Theta_{Gaussian} = \mathbb{R} \times \mathbb{R}^+$. That is, we allow any real value of μ and any positive real value for σ.

∎

likelihood function

The next step in maximum likelihood estimation is defining the *likelihood function*. As we saw in our example, the likelihood function for a given choice of parameters $\boldsymbol{\theta}$ is the probability (or density) the model assigns the training data:

$$L(\boldsymbol{\theta} : \mathcal{D}) = \prod_m P(\xi[m] : \boldsymbol{\theta}).$$

In the thumbtack example, we saw that we can write the likelihood function using simpler terms. That is, using the counts $M[1]$ and $M[0]$, we managed to have a compact description of the likelihood. More precisely, once we knew the values of $M[1]$ and $M[0]$, we did not need to consider other aspects of training data (for example, the order of tosses). These are the *sufficient statistics* for the thumbtack learning problem. In a more general setting, a sufficient statistic is a function of the data that summarizes the relevant information for computing the likelihood.

Definition 17.1

sufficient statistics

A function $\tau(\xi)$ from instances of $\mathcal{X}$ to $\mathbb{R}^\ell$ (for some ℓ) is a sufficient statistic *if, for any two data sets $\mathcal{D}$ and $\mathcal{D}'$ and any $\boldsymbol{\theta} \in \Theta$, we have that*

$$\sum_{\xi[m] \in \mathcal{D}} \tau(\xi[m]) = \sum_{\xi'[m] \in \mathcal{D}'} \tau(\xi'[m]) \quad \implies \quad L(\boldsymbol{\theta} : \mathcal{D}) = L(\boldsymbol{\theta} : \mathcal{D}').$$

∎

We often refer to the tuple $\sum_{\xi[m] \in \mathcal{D}} \tau(\xi[m])$ as the sufficient statistics of the data set $\mathcal{D}$.

Example 17.3

Let us reconsider the multinomial model of example 17.1. It is easy to see that a sufficient statistic for the data set is the tuple of counts $\langle M[1], \ldots, M[K] \rangle$, such that $M[k]$ is number of times the value x^k appears in the training data. To obtain these counts by summing instance-level statistics, we define $\tau(x)$ to be a tuple of dimension K, such that $\tau(x)$ has a 0 in every position, except at the position k for which $x = x^k$, where its value is 1:

$$\tau(x^k) = (\overbrace{0, \ldots, 0}^{k-1}, 1, \overbrace{0, \ldots, 0}^{n-k}).$$

Given the vector of counts, we can write the likelihood function as

$$L(\boldsymbol{\theta} : \mathcal{D}) = \prod_k \theta_k^{M[k]}.$$

∎

Example 17.4

Let us reconsider the Gaussian model of example 17.2. In this case, it is less obvious how to construct sufficient statistics. However, if we expand the term $(x - \mu)^2$ in the exponent, we can rewrite the model as

$$P_{Gaussian}(x : \mu, \sigma) = e^{-x^2 \frac{1}{2\sigma^2} + x \frac{\mu}{\sigma^2} - \frac{\mu^2}{2\sigma^2} - \frac{1}{2}\log(2\pi) - \log(\sigma)}.$$

We then see that the function

$$s_{Gaussian}(x) = \langle 1, x, x^2 \rangle$$

is a sufficient statistic for this model. Note that the first element in the sufficient statistics tuple is "1," which does not depend on the value of the data item; it serves, as in the multinomial case, to count the number of data items.

∎

We venture several comments about the likelihood function. First, we stress that the likelihood function measures the effect of the choice of parameters on the training data. Thus, for example, if we have two sets of parameters $\boldsymbol{\theta}$ and $\boldsymbol{\theta}'$, so that $L(\boldsymbol{\theta} : \mathcal{D}) = L(\boldsymbol{\theta}' : \mathcal{D})$, then we *cannot*, given only the data, distinguish between the two choices of parameters. Moreover, if $L(\boldsymbol{\theta} : \mathcal{D}) = L(\boldsymbol{\theta}' : \mathcal{D})$ for all possible choices of $\mathcal{D}$, then the two parameters are *indistinguishable* for any outcome. In such a situation, we can say in advance (that is, before seeing the data) that some distinctions cannot be resolved based on the data alone.

Second, since we are maximizing the likelihood function, we usually want it to be continuous (and preferably smooth) function of $\boldsymbol{\theta}$. To ensure these properties, most of the theory of statistical estimation requires that $P(\xi : \boldsymbol{\theta})$ is a continuous and differentiable function of $\boldsymbol{\theta}$, and moreover that Θ is a continuous set of points (which is often assumed to be convex).

maximum
likelihood
estimation

Once we have defined the likelihood function, we can use *maximum likelihood estimation* to choose the parameter values. Formally, we state this principle as follows.

Maximum Likelihood Estimation: Given a data set $\mathcal{D}$, choose parameters $\hat{\boldsymbol{\theta}}$ that satisfy

$$L(\hat{\boldsymbol{\theta}} : \mathcal{D}) = \max_{\boldsymbol{\theta} \in \Theta} L(\boldsymbol{\theta} : \mathcal{D}).$$

Example 17.5

Consider estimating the parameters of the multinomial distribution of example 17.3. As one might guess, the maximum likelihood is attained when

$$\hat{\theta}_k = \frac{M[k]}{M}$$

(see exercise 17.2). That is, the probability of each value of X corresponds to its frequency in the training data.   ∎

Example 17.6

empirical mean,
variance

Consider estimating the parameters of a Gaussian distribution of example 17.4. It turns out that the maximum is attained when μ and σ correspond to the empirical *mean and variance of the training data:*

$$\hat{\mu} = \frac{1}{M} \sum_m x[m]$$

$$\hat{\sigma} = \sqrt{\frac{1}{M} \sum_m (x[m] - \hat{\mu})^2}$$

(see exercise 17.3).   ∎

17.2   MLE for Bayesian Networks

We now move to the more general problem of estimating parameters for a Bayesian network. It turns out that the structure of the Bayesian network allows us to reduce the parameter estimation problem to a set of unrelated problems, each of which can be addressed using the techniques of the previous section. We begin by considering a simple example to clarify our intuition, and then generalize to more complicated networks.

17.2.1 A Simple Example

The simplest example of a nontrivial network structure is a network consisting of two binary variables, say X and Y, with an arc $X \rightarrow Y$. (A network without such an arc trivially reduces to the cases we already discussed.)

As for a single parameter, our goal in maximum likelihood estimation is to maximize the likelihood (or log-likelihood) function. In this case, our network is parameterized by a parameter vector $\boldsymbol{\theta}$, which defines the set of parameters for all the CPDs in the network. In this example, our parameterization would consist of the following parameters: θ_{x^1}, and θ_{x^0} specify the probability of the two values of X; $\theta_{y^1|x^1}$, and $\theta_{y^0|x^1}$ specify the probability of Y given that $X = x^1$; and $\theta_{y^1|x^0}$, and $\theta_{y^0|x^0}$ describe the probability of Y given that $X = x^0$. For brevity, we also use the shorthand $\boldsymbol{\theta}_{Y|x^0}$ to refer to the set $\{\theta_{y^1|x^0}, \theta_{y^0|x^0}\}$, and $\boldsymbol{\theta}_{Y|X}$ to refer to $\boldsymbol{\theta}_{Y|x^1} \cup \boldsymbol{\theta}_{Y|x^0}$.

In this example, each training instance is a tuple $\langle x[m], y[m] \rangle$ that describes a particular assignment to X and Y. Our likelihood function is:

$$L(\boldsymbol{\theta} : \mathcal{D}) = \prod_{m=1}^{M} P(x[m], y[m] : \boldsymbol{\theta}).$$

Our network model specifies that $P(X, Y : \boldsymbol{\theta})$ has a product form. Thus, we can write

$$L(\boldsymbol{\theta} : \mathcal{D}) = \prod_{m} P(x[m] : \boldsymbol{\theta}) P(y[m] \mid x[m] : \boldsymbol{\theta}).$$

Exchanging the order of multiplication, we can equivalently write this term as

$$L(\boldsymbol{\theta} : \mathcal{D}) = \left(\prod_{m} P(x[m] : \boldsymbol{\theta}) \right) \left(\prod_{m} P(y[m] \mid x[m] : \boldsymbol{\theta}) \right).$$

That is, the likelihood decomposes into two separate terms, one for each variable. Moreover, each of these terms is a *local likelihood* function that measures how well the variable is predicted given its parents.

Now consider the two individual terms. Clearly, each one depends only on the parameters for that variable's CPD. Thus, the first is $\prod_m P(x[m] : \boldsymbol{\theta}_X)$. This term is identical to the multinomial likelihood function we discussed earlier. The second term is more interesting, since we can decompose it even further:

$$\prod_{m} P(y[m] \mid x[m] : \boldsymbol{\theta}_{Y|X})$$

$$= \prod_{m:x[m]=x^0} P(y[m] \mid x[m] : \boldsymbol{\theta}_{Y|X}) \cdot \prod_{m:x[m]=x^1} P(y[m] \mid x[m] : \boldsymbol{\theta}_{Y|X})$$

$$= \prod_{m:x[m]=x^0} P(y[m] \mid x[m] : \boldsymbol{\theta}_{Y|x^0}) \cdot \prod_{m:x[m]=x^1} P(y[m] \mid x[m] : \boldsymbol{\theta}_{Y|x^1}).$$

likelihood decomposability

Thus, in this example, the likelihood function decomposes into a product of terms, one for each group of parameters in $\boldsymbol{\theta}$. This property is called the *decomposability* of the likelihood function.

We can do one more simplification by using the notion of sufficient statistics. Let us consider one term in this expression:

$$\prod_{m:x[m]=x^0} P(y[m] \mid x[m] : \boldsymbol{\theta}_{Y|x^0}). \tag{17.2}$$

Each of the individual terms $P(y[m] \mid x[m] : \boldsymbol{\theta}_{Y|x^0})$ can take one of two values, depending on the value of $y[m]$. If $y[m] = y^1$, it is equal to $\theta_{y^1|x^0}$. If $y[m] = y^0$, it is equal to $\theta_{y^0|x^0}$. How many cases of each type do we get? First, we restrict attention only to those data cases where $x[m] = x^0$. These, in turn, partition into the two categories. Thus, we get $\theta_{y^1|x^0}$ in those data cases where $x[m] = x^0$ and $y[m] = y^1$; we use $M[x^0, y^1]$ to denote their number. We get $\theta_{y^0|x^0}$ in those data cases where $x[m] = x^0$ and $y[m] = y^0$, and use $M[x^0, y^0]$ to denote their number. Thus, the term in equation (17.2) is equal to:

$$\prod_{m:x[m]=x^0} P(y[m] \mid x[m] : \boldsymbol{\theta}_{Y|x^0}) \quad = \quad \theta_{y^1|x^0}^{M[x^0,y^1]} \cdot \theta_{y^0|x^0}^{M[x^0,y^0]}.$$

Based on our discussion of the multinomial likelihood in example 17.5, we know that we maximize $\theta_{Y|x^0}$ by setting:

$$\theta_{y^1|x^0} = \frac{M[x^0, y^1]}{M[x^0, y^1] + M[x^0, y^0]} = \frac{M[x^0, y^1]}{M[x^0]},$$

and similarly for $\theta_{y^0|x^0}$. Thus, we can find the maximum likelihood parameters in this CPD by simply counting how many times each of the possible assignments of X and Y appears in the training data. It turns out that these counts of the various assignments for some set of variables are useful in general. We therefore define:

Definition 17.2   *Let $\boldsymbol{Z}$ be some set of random variables, and $\boldsymbol{z}$ be some instantiation to these random variables. Let $\mathcal{D}$ be a data set. We define $M[\boldsymbol{z}]$ to be the number of entries in $\mathcal{D}$ that have $\boldsymbol{Z}[m] = \boldsymbol{z}$*

$$M[\boldsymbol{z}] = \sum_m \boldsymbol{I}\{\boldsymbol{Z}[m] = \boldsymbol{z}\}. \tag{17.3}$$

∎

17.2.2   Global Likelihood Decomposition

As we can expect, the arguments we used for deriving the MLE of $\theta_{Y|x^0}$ apply for the parameters of other CPDs in that example and indeed for other networks as well. We now develop, in several steps, the formal machinery for proving such properties in Bayesian networks.

We start by examining the likelihood function of a Bayesian network. Suppose we want to learn the parameters for a Bayesian network with structure $\mathcal{G}$ and parameters $\boldsymbol{\theta}$. This means that we agree in advance on the type of CPDs we want to learn (say table-CPDs, or noisy-ors). As we discussed, we are also given a data set $\mathcal{D}$ consisting of samples $\xi[1], \ldots, \xi[M]$. Writing the likelihood, and repeating the steps we performed in our example, we get

$$
\begin{aligned}
L(\boldsymbol{\theta} : \mathcal{D}) &= \prod_m P_{\mathcal{G}}(\xi[m] : \boldsymbol{\theta}) \\
&= \prod_m \prod_i P(x_i[m] \mid \mathrm{pa}_{X_i}[m] : \boldsymbol{\theta}) \\
&= \prod_i \left[\prod_m P(x_i[m] \mid \mathrm{pa}_{X_i}[m] : \boldsymbol{\theta}) \right].
\end{aligned}
$$

conditional
likelihood

Note that each of the terms in the square brackets refers to the *conditional likelihood* of a particular variable given its parents in the network. We use $\boldsymbol{\theta}_{X_i \mid \mathrm{Pa}_{X_i}}$ to denote the subset of parameters that determines $P(X_i \mid \mathrm{Pa}_{X_i})$ in our model. Then, we can write

$$
L(\boldsymbol{\theta} : \mathcal{D}) = \prod_i L_i(\boldsymbol{\theta}_{X_i \mid \mathrm{Pa}_{X_i}} : \mathcal{D}),
$$

local likelihood

where the *local likelihood* function for X_i is:

$$
L_i(\boldsymbol{\theta}_{X_i \mid \mathrm{Pa}_{X_i}} : \mathcal{D}) = \prod_m P(x_i[m] \mid \mathrm{pa}_{X_i}[m] : \boldsymbol{\theta}_{X_i \mid \mathrm{Pa}_{X_i}}).
$$

This form is particularly useful when the parameter sets $\boldsymbol{\theta}_{X_i \mid \mathrm{Pa}_{X_i}}$ are *disjoint*. That is, each CPD is parameterized by a separate set of parameters that do not overlap. This assumption is quite natural in all our examples so far. (Although, as we will see in section 17.5, parameter sharing can be handy in many domains.) **This analysis shows that the likelihood decomposes as a product of independent terms, one for each CPD in the network. This important property is called the *global decomposition* of the likelihood function.**

global
decomposability

We can now immediately derive the following result:

Proposition 17.1   *Let $\mathcal{D}$ be a complete data set for $X_1, \dots, X_n$, let $\mathcal{G}$ be a network structure over these variables, and suppose that the parameters $\boldsymbol{\theta}_{X_i \mid \mathrm{Pa}_{X_i}}$ are disjoint from $\boldsymbol{\theta}_{X_j \mid \mathrm{Pa}_{X_j}}$ for all $j \neq i$. Let $\hat{\boldsymbol{\theta}}_{X_i \mid \mathrm{Pa}_{X_i}}$ be the parameters that maximize $L_i(\boldsymbol{\theta}_{X_i \mid \mathrm{Pa}_{X_i}} : \mathcal{D})$. Then, $\hat{\boldsymbol{\theta}} = \langle \hat{\boldsymbol{\theta}}_{X_1 \mid \mathrm{Pa}_1}, \dots, \hat{\boldsymbol{\theta}}_{X_n \mid \mathrm{Pa}_n} \rangle$ maximizes $L(\boldsymbol{\theta} : \mathcal{D})$.*

In other words, **we can maximize each local likelihood function *independently* of rest of the network, and then combine the solutions to get an MLE solution.** This decomposition of the global problem to independent subproblems allows us to devise efficient solutions to the MLE problem. Moreover, this decomposition is an immediate consequence of the network structure and does not depend on any particular choice of parameterization for the CPDs.

17.2.3   Table-CPDs

Based on the preceding discussion, we know that the likelihood of a Bayesian network decomposes into local terms that depend on the parameterization of CPDs. The choice of parameters determines how we can maximize each of the local likelihood functions. We now consider what

table-CPD

is perhaps the simplest parameterization of the CPD: a *table-CPD*.

Suppose we have a variable X with parents $\boldsymbol{U}$. If we represent that CPD $P(X \mid \boldsymbol{U})$ as a table, then we will have a parameter $\theta_{x|\boldsymbol{u}}$ for each combination of $x \in Val(X)$ and $\boldsymbol{u} \in Val(\boldsymbol{U})$. In this case, we can rewrite the local likelihood function as follows:

$$
\begin{aligned}
L_X(\boldsymbol{\theta}_{X|\boldsymbol{U}} : \mathcal{D}) &= \prod_m \theta_{x[m]|\boldsymbol{u}[m]} \\
&= \prod_{\boldsymbol{u} \in Val(\boldsymbol{U})} \left[\prod_{x \in Val(X)} \theta_{x|\boldsymbol{u}}^{M[\boldsymbol{u},x]} \right],
\end{aligned}
\tag{17.4}
$$

where $M[\boldsymbol{u}, x]$ is the number of times $\xi[m] = x$ and $\boldsymbol{u}[m] = \boldsymbol{u}$ in $\mathcal{D}$. That is, we grouped together all the occurrences of $\theta_{x|\boldsymbol{u}}$ in the product over all instances. This provides a further *local decomposition* of the likelihood function.

local decomposability

We need to maximize this term under the constraints that, for each choice of value for the parents $\boldsymbol{U}$, the conditional probability is legal, that is:

$$
\sum_x \theta_{x|\boldsymbol{u}} = 1 \quad \text{for all } \boldsymbol{u}.
$$

These constraints imply that the choice of value for $\theta_{x|\boldsymbol{u}}$ can impact the choice of value for $\theta_{x'|\boldsymbol{u}}$. However, the choice of parameters given different values $\boldsymbol{u}$ of $\boldsymbol{U}$ are independent of each other. Thus, we can maximize each of the terms in square brackets in equation (17.4) independently.

We can thus further decompose the local likelihood function for a tabular CPD into a product of simple likelihood functions. Each of these likelihood functions is a *multinomial* likelihood, of the type that we examined in example 17.3. The counts in the data for the different outcomes x are simply $\{M[\boldsymbol{u}, x] : x \in Val(X)\}$. We can then immediately use the maximum likelihood estimation for multinomial likelihood of example 17.5 and see that the MLE parameters are

$$
\hat{\theta}_{x|\boldsymbol{u}} = \frac{M[\boldsymbol{u}, x]}{M[\boldsymbol{u}]},
\tag{17.5}
$$

where we use the fact that $M[\boldsymbol{u}] = \sum_x M[\boldsymbol{u}, x]$.

This simple formula reveals a key challenge when estimating parameters for a Bayesian networks. Note that the number of data points used to estimate the parameter $\hat{\theta}_{x|\boldsymbol{u}}$ is $M[\boldsymbol{u}]$. Data points that do not agree with the parent assignment $\boldsymbol{u}$ play no role in this computation. As the number of parents $\boldsymbol{U}$ grows, the number of different parent assignments grows exponentially. Therefore, the number of data instances that we expect to have for a single parent assignment

data fragmentation

shrinks exponentially. This phenomenon is called *data fragmentation*, since the data set is partitioned into a large number of small subsets. Intuitively, when we have a very small number of data instances from which we estimate a parameter, the estimates we get can be very noisy (this intuition is formalized in section 17.6), leading to *overfitting*. We are also more likely to

overfitting

get a large number of zeros in the distribution, which can lead to very poor performance. **Our** **inability to estimate parameters reliably as the dimensionality of the parent set grows is one of the key limiting factors in learning Bayesian networks from data.** This problem is even more severe when the variables can take on a large number of values, for example, in text applications.

Box 17.A — Concept: Naive Bayes Classifier. *One of the basic tasks of learning is* classification. *In this task, our goal is build a classifier — a procedure that assigns instances into two or more categories, for example, deciding whether an email message is junk mail that should be discarded or a relevant message that should be presented to the user. In the usual setting, we are given a training example of instances from each category, where instances are represented by various features. In our email classification example, a message might be analyzed by multiple features: its length, the type of attachments it contains, the domain of the sender, whether that sender appears in the user's address book, whether a particular word appears in the subject, and so on.*

One general approach to this problem, which is referred to as Bayesian classifier, *is to learn a probability distribution of the features of instances of each class. In the language of probabilistic models, we use the random variables* $\boldsymbol{X}$ *to represent the instance, and the random variable* C *to represent the category of the instance. The distribution* $P(\boldsymbol{X} \mid C)$ *is the probability of a particular combination of features given the category. Using Bayes rule, we have that*

$$P(C \mid \boldsymbol{X}) \propto P(C)P(\boldsymbol{X} \mid C).$$

Thus, if we have a good model of how instances of each category behave (that is, of $P(\boldsymbol{X} \mid C)$*), we can combine it with our prior estimate for the frequency of each category (that is,* $P(C)$*) to estimate the posterior probability of each of the categories (that is,* $P(C \mid \boldsymbol{X})$*). We can then decide either to predict the most likely category or to perform a more complex decision based on the strength of likelihood of each option. For example, to reduce the number of erroneously removed messages, a junk-mail filter might remove email messages only when the probability that it is junk mail is higher than a strict threshold.*

This Bayesian classification approach is quite intuitive. Loosely speaking, it states that to classify objects successfully, we need to recognize the characteristics of objects of each category. Then, we can classify a new object by considering whether it matches the characteristic of each of the classes. More formally, we use the language of probability to describe each category, assigning higher probability to objects that are typical for the category and low probability to ones that are not.

The main hurdle in constructing a Bayesian classifier is the question of representation of the multivariate distribution $p(\boldsymbol{X} \mid C)$*. The* naive Bayes *classifier is one where we use the simplest representation we can think of. That is, we assume that each feature* X_i *is independent of all the other features given the class variable* C*. That is,*

$$P(\boldsymbol{X} \mid C) = \prod_i P(X_i \mid C).$$

Learning the distribution $P(C)P(\boldsymbol{X} \mid C)$ *is thus reduced to learning the parameters in the naive Bayes structure, with the category variable* C *rendering all other features as conditionally independent of each other.*

As can be expected, learning this classifier is a straightforward application of the parameter estimation that we consider in this chapter. Moreover, classifying new examples requires simple computation, evaluating $P(c) \prod_i P(x_i \mid c)$ *for each category* c*.*

Although this simple classifier is often dismissed as naive, in practice it is often surprisingly effective. From a training perspective, this classifier is quite robust, since in most applications, even with relatively few training examples, we can learn the parameters of conditional distribution

$P(X_i \mid C)$. *However, one might argue that robust learning does not compensate for oversimplified independence assumption. Indeed, the strong independence assumption usually results in poor representation of the distribution of instances. However,* **errors in estimating the probability of an instance do not necessarily lead to classification errors. For classification, we are interested in the relative size of the conditional distribution of the instances given different categories. The ranking of different labels may not be that sensitive to errors in estimating the actual probability of the instance.** *Empirically, one often finds that the naive Bayes classifier correctly classifies an example to the right category, yet its posterior probability is very skewed and quite far from the correct distribution.*

In practice, the naive Bayes classifier is often a good baseline classifier to try before considering more complex solutions. It is easy to implement, it is robust, and it can handle different choices of descriptions of instances (for example, box 17.E).

17.2.4   Gaussian Bayesian Networks ★

Our discussion until now has focused on learning discrete-state Bayesian networks with multi-nomial parameters. However, the concepts we have developed in this section carry through to a wide variety of other types of Bayesian networks. In particular, the global decomposition properties we proved for a Bayesian network apply, without any change, to any other type of CPD. That is, if the data are complete, the learning problem reduces to a set of local learning problems, one for each variable. The main difference is in applying the maximum likelihood estimation process to a CPD of a different type: how we define the sufficient statistics, and how we compute the maximum likelihood estimate from them. In this section, we demonstrate how MLE principles can be applied in the setting of linear Gaussian Bayesian networks. In section 17.2.5 we provide a general procedure for CPDs in the exponential family.

Consider a variable X with parents $\boldsymbol{U} = \{U_1, \dots, U_k\}$ with a linear Gaussian CPD:

$$P(X \mid \boldsymbol{u}) = \mathcal{N}\left(\beta_0 + \beta_1 u_1 + \dots, \beta_k u_k; \sigma^2\right).$$

Our task is to learn the parameters $\boldsymbol{\theta}_{X \mid \boldsymbol{U}} = \langle \beta_0, \dots, \beta_k, \sigma \rangle$. To find the MLE values of these parameters, we need to differentiate the likelihood and solve the equations that define a stationary point. As usual, it will be easier to work with the log-likelihood function. Using the definition of the Gaussian distribution, we have that

$$\ell_X(\boldsymbol{\theta}_{X \mid \boldsymbol{U}} : \mathcal{D}) = \log L_X(\boldsymbol{\theta}_{X \mid \boldsymbol{U}} : \mathcal{D})$$
$$= \sum_m \left[-\frac{1}{2}\log(2\pi\sigma^2) - \frac{1}{2}\frac{1}{\sigma^2}\left(\beta_0 + \beta_1 u_1[m] + \dots + \beta_k u_k[m] - x[m]\right)^2 \right].$$

We start by considering the gradient of the log-likelihood with respect to β_0:

$$\frac{\partial}{\partial \beta_0}\ell_X(\boldsymbol{\theta}_{X \mid \boldsymbol{U}} : \mathcal{D}) = \sum_m -\frac{1}{\sigma^2}\left(\beta_0 + \beta_1 u_1[m] + \dots + \beta_k u_k[m] - x[m]\right)$$
$$= -\frac{1}{\sigma^2}\left(M\beta_0 + \beta_1 \sum_m u_1[m] + \dots + \beta_k \sum_m u_k[m] - \sum_m x[m]\right).$$

Equating this gradient to 0, and multiplying both sides with $\frac{\sigma^2}{M}$, we get the equation

$$\frac{1}{M}\sum_m x[m] = \beta_0 + \beta_1 \frac{1}{M}\sum_m u_1[m] + \ldots + \beta_k \frac{1}{M}\sum_m u_k[m].$$

Each of the terms is the average value of one of the variables in the data. We use the notation

$$\boldsymbol{E}_{\mathcal{D}}[X] = \frac{1}{M}\sum_m x[m]$$

to denote this expectation. Using this notation, we see that we get the following equation:

$$\boldsymbol{E}_{\mathcal{D}}[X] = \beta_0 + \beta_1 \boldsymbol{E}_{\mathcal{D}}[U_1] + \ldots + \beta_k \boldsymbol{E}_{\mathcal{D}}[U_k]. \tag{17.6}$$

Recall that theorem 7.3 specifies the mean of a linear Gaussian variable X in terms of the means of its parents $U_1, \ldots, U_k$, using an expression that has precisely this form. Thus, equation (17.6) tells us that the MLE parameters should be such that the mean of X in the data is consistent with the predicted mean of X according to the parameters.

Next, consider the gradient with respect to one of the parameters β_i. Using similar arithmetic manipulations, we see that the equation $0 = \frac{\partial}{\partial \beta_i}\ell_X(\boldsymbol{\theta}_{X|U} : \mathcal{D})$ can be formulated as:

$$\boldsymbol{E}_{\mathcal{D}}[X \cdot U_i] = \beta_0 \boldsymbol{E}_{\mathcal{D}}[U_i] + \beta_1 \boldsymbol{E}_{\mathcal{D}}[U_1 \cdot U_i] + \ldots + \beta_k \boldsymbol{E}_{\mathcal{D}}[U_k \cdot U_i]. \tag{17.7}$$

At this stage, we have $k+1$ linear equations with $k+1$ unknowns, and we can use standard linear algebra techniques for solving for the value of $\beta_0, \beta_1, \ldots, \beta_k$. We can get additional intuition, however, by doing additional manipulation of equation (17.7). Recall that the covariance $\boldsymbol{Cov}[X;Y] = \boldsymbol{E}[X \cdot Y] - \boldsymbol{E}[X] \cdot \boldsymbol{E}[Y]$. Thus, if we subtract $\boldsymbol{E}_{\mathcal{D}}[X] \cdot \boldsymbol{E}_{\mathcal{D}}[U_i]$ from the left-hand side of equation (17.7), we would get the empirical covariance of X and U_i. Using equation (17.6), we have that this term can also be written as:

$$\boldsymbol{E}_{\mathcal{D}}[X] \cdot \boldsymbol{E}_{\mathcal{D}}[U_i] = \beta_0 \boldsymbol{E}_{\mathcal{D}}[U_i] + \beta_1 \boldsymbol{E}_{\mathcal{D}}[U_1] \cdot \boldsymbol{E}_{\mathcal{D}}[U_i] + \ldots + \beta_k \boldsymbol{E}_{\mathcal{D}}[U_k] \cdot \boldsymbol{E}_{\mathcal{D}}[U_i].$$

Subtracting this equation from equation (17.7), we get:

$$\begin{aligned}\boldsymbol{E}_{\mathcal{D}}[X \cdot U_i] - \boldsymbol{E}_{\mathcal{D}}[X] \cdot \boldsymbol{E}_{\mathcal{D}}[U_i] &= \beta_1\left(\boldsymbol{E}_{\mathcal{D}}[U_1 \cdot U_i] - \boldsymbol{E}_{\mathcal{D}}[U_1] \cdot \boldsymbol{E}_{\mathcal{D}}[U_i]\right) + \ldots + \\ &\quad \beta_k\left(\boldsymbol{E}_{\mathcal{D}}[U_k \cdot U_i] - \boldsymbol{E}_{\mathcal{D}}[U_k] \cdot \boldsymbol{E}_{\mathcal{D}}[U_i]\right).\end{aligned}$$

Using $\boldsymbol{Cov}_{\mathcal{D}}[X;U_i]$ to denote the observed covariance of X and U_i in the data, we get:

$$\boldsymbol{Cov}_{\mathcal{D}}[X;U_i] = \beta_1 \boldsymbol{Cov}_{\mathcal{D}}[U_1;U_i] + \ldots + \beta_k \boldsymbol{Cov}_{\mathcal{D}}[U_k;U_i].$$

In other words, the observed covariance of X with U_i should be the one predicted by theorem 7.3 given the parameters and the observed covariances between the parents of X.

Finally, we need to find the value of the σ^2 parameter. Taking the derivative of the likelihood and equating to 0, we get an equation that, after suitable reformulation, can be written as

$$\sigma^2 = \boldsymbol{Cov}_{\mathcal{D}}[X;X] - \sum_i \sum_j \beta_i \beta_j \boldsymbol{Cov}_{\mathcal{D}}[U_i;U_j] \tag{17.8}$$

(see exercise 17.4). Again, we see that the MLE estimate has to match the constraints implied by theorem 7.3.

The global picture that emerges is as follows. To estimate $P(X \mid U)$, we estimate the means of X and U and covariance matrix of $\{X\} \cup U$ from the data. The vector of means and covariance matrix defines a joint Gaussian distribution over $\{X\} \cup U$. (In fact, this is the MLE estimate of the joint Gaussian; see exercise 17.5.) We then solve for the (unique) linear Gaussian that matches the joint Gaussian with these parameters. For this purpose, we can use the formulas provided by theorem 7.4. While these equations seem somewhat complex, they are merely describing the solution to a system of linear equations.

This discussion also identifies the sufficient statistics we need to collect to estimate linear Gaussians. These are the univariate terms of the form $\sum_m x[m]$ and $\sum_m u_i[m]$, and the interaction terms of the form $\sum_m x[m] \cdot u_i[m]$ and $\sum_m u_i[m] \cdot u_j[m]$. From these, we can estimate the mean and covariance matrix of the joint distribution.

Box 17.B — Concept: Nonparametric Models. *The discussion in this chapter has focused on estimating parameters for specific parametric models of CPDs: multinomials and linear Gaussians. However, a theory of maximum likelihood and Bayesian estimation exists for a wide variety of other parametric models. Moreover, in recent years, there has been a growing interest in the use of*

nonparametric Bayesian estimation

nonparametric Bayesian estimation methods, where a (conditional) distribution is not defined to be in some particular parametric class with a fixed number of parameters, but rather the complexity of the representation is allowed to grow as we get more data instances. In the case of discrete variables, any CPD can be described as a table, albeit perhaps a very large one; thus a nonparametric method is less essential (although see section 19.5.2.2 for a very useful example of a nonparametric method in the discrete case). In the case of continuous variables, we do not have a "universal" parametric distribution. While Gaussians are often the default, many distributions are not well fit by them, and it is often difficult to determine which parametric family (if any) will be appropriate for a given variable. In such cases, nonparametric methods offer a useful substitute. In such methods, we use the data points themselves as the basis for a probability distribution. Many nonparametric methods have been developed; we describe one simple variant that serves to illustrate this type of approach.

Suppose we want to learn the distribution $P(X \mid U)$ from data. A reasonable assumption is that the CPD is smooth. Thus, if we observe x, u in a training sample, it should increase the probability of seeing similar values of X for similar values of U. More precisely, we increase the density of $p(X = x + \epsilon \mid U = u + \delta)$ for small values of ϵ and δ.

kernel density estimation

One simple approach that captures this intuition is the use of kernel density estimation (also known as Parzen windows). The idea is fairly simple: given the data $\mathcal{D}$, we estimate a "local" joint density $\tilde{p}_X(X, U)$ by spreading out density around each example $x[m], u[m]$. Formally, we write

$$\tilde{p}_X(x, u) = \frac{1}{M} \sum_m K(x, u; x[m], u[m], \alpha),$$

where K is a kernel density function and α is a parameter (or vector of parameters) controlling K. A common choice of kernel is a simple round Gaussian distribution with radius α around $x[m], u[m]$:

$$K(x, u; x[m], u[m], \alpha) = \mathcal{N}\left(\begin{pmatrix} x[m] \\ u[m] \end{pmatrix}; \alpha^2 I \right),$$

where I is the identity matrix and α is the width of the window. Of course, many other choices for kernel function are possible; in fact, if K defines a probability measure (nonnegative and integrates to 1), then $\tilde{p}_X(x, \boldsymbol{u})$ is also a probability measure. Usually we choose kernel functions that are local, in that they put most of the mass in the vicinity of their argument. For such kernels, the resulting density $\tilde{p}_X(x, \boldsymbol{u})$ will have high mass in regions where we have seen many data instances $(x[m], \boldsymbol{u}[m])$ and low mass in regions where we have seen none.

We can now reformulate this local joint distribution to produce a conditional distribution:

$$p(x \mid \boldsymbol{u}) = \frac{\sum_m K(x, \boldsymbol{u}; x[m], \boldsymbol{u}[m], \alpha)}{\sum_m K(\boldsymbol{u}; \boldsymbol{u}[m], \alpha)}$$

where $K(\boldsymbol{u}; \boldsymbol{u}[m], \alpha)$ is $K(x, \boldsymbol{u}; x[m], \boldsymbol{u}[m], \alpha)$ marginalized over x.

Note that this learning procedure estimates virtually no parameters: the CPD is derived directly from the training instances. The only free parameter is α, which is the width of the window. Importantly, this parameter cannot be estimated using maximum likelihood: The α that maximizes the likelihood of the training set is $\alpha = 0$, which gives maximum density to the training instances themselves. This, of course, will simply memorize the training instances without any generalization. Thus, this parameter is generally selected using cross-validation.

The learned CPD here is essentially the list of training instances, which has both advantages and disadvantages. On the positive side, the estimates are very flexible and tailor themselves to the observations; indeed, as we get more training data, we can produce arbitrarily expressive representations of our joint density. On the negative side, there is no "compression" of the original data, which has both computational and statistical ramifications. Computationally, when there are many training samples the learned CPDs can become unwieldy. Statistically, this learning procedure makes no attempt to generalize beyond the data instances that we have seen. In high-dimensional spaces with limited data, most points in the space will be "far" from data instances, and therefore the estimated density will tend to be quite poor in most parts of the space. Thus, this approach is primarily useful in cases where we have a large number of training instances relative to the dimension of the space.

Finally, while these approaches help us avoid parametric assumptions on the learning side, we are left with the question of how to avoid them on the inference side. As we saw, most inference procedures are geared to working with parametric representations, mostly Gaussians. Thus, when performing inference with nonparametric CPDs, we must generally either use parametric approximations, or resort to sampling.

17.2.5 Maximum Likelihood Estimation as M-Projection ★

The MLE principle is a general one, in that it gives a recipe how to construct estimators for different statistical models (for example, multinomials and Gaussians). As we have seen, for simple examples the resulting estimators are quite intuitive. However, the same principle can be applied in a much broader range of parametric models. Indeed, as we now show, we have already discussed the framework that forms the basis for this generalization.

In section 8.5, we defined the notion of *projection*: finding the distribution, within a specified class, that is closest to a given target distribution. Parameter estimation is similar in the sense that we select a distribution from a given class — all of those that can be described by the model — that is "closest" to our data. Indeed, we can show that maximum likelihood estimation aims to find the distribution that is "closest" to the empirical distribution $\hat{P}_{\mathcal{D}}$ (see equation (16.4)).

We start by rewriting the likelihood function in terms of the empirical distribution.

Proposition 17.2

Let $\mathcal{D}$ be a data set, then

$$\log L(\boldsymbol{\theta} : \mathcal{D}) = M \cdot \boldsymbol{E}_{\hat{P}_{\mathcal{D}}}[\log P(\mathcal{X} : \boldsymbol{\theta})].$$

PROOF We rewrite the likelihood by combining all identical instances in our training set and then writing the likelihood in terms of the empirical probability of each entry in our joint distribution:

$$
\begin{aligned}
\log L(\boldsymbol{\theta} : \mathcal{D}) &= \sum_m \log P(\xi[m] : \boldsymbol{\theta}) \\
&= \sum_\xi \left[\sum_m \boldsymbol{I}\{\xi[m] = \xi\} \right] \log P(\xi : \boldsymbol{\theta}) \\
&= \sum_\xi M \cdot \hat{P}_{\mathcal{D}}(\xi) \log P(\xi : \boldsymbol{\theta}) \\
&= M \cdot \boldsymbol{E}_{\hat{P}_{\mathcal{D}}}[\log P(\mathcal{X} : \boldsymbol{\theta})].
\end{aligned}
$$
∎

We can now apply proposition 16.1 to the empirical distribution to conclude that

$$\ell(\boldsymbol{\theta} : \mathcal{D}) = M \left(\boldsymbol{H}_{\hat{P}_{\mathcal{D}}}(\mathcal{X}) - \boldsymbol{D}(\hat{P}_{\mathcal{D}}(\mathcal{X}) \| P(\mathcal{X} : \boldsymbol{\theta})) \right). \tag{17.9}$$

From this result, we immediately derive the following relationship between MLE and M-projections.

Theorem 17.1

The MLE $\boldsymbol{\theta}$ in a parametric family relative to a data set $\mathcal{D}$ is the M-projection of $\hat{P}_{\mathcal{D}}$ onto the parametric family

$$\hat{\boldsymbol{\theta}} = \arg \min_{\boldsymbol{\theta} \in \Theta} \boldsymbol{D}(\hat{P}_{\mathcal{D}} \| P_{\boldsymbol{\theta}}).$$

We see that MLE finds the distribution $P(\mathcal{X} : \boldsymbol{\theta})$ that is the M-projection of $\hat{P}_{\mathcal{D}}$ onto the set of distributions representable in our parametric family.

This result allows us to call upon our detailed analysis of M-projections in order to generalize MLE to other parametric classes in the exponential family. In particular, in section 8.5.2, we discussed the general notion of sufficient statistics and showed that the M-projection of a distribution P into a class of distributions $\mathcal{Q}$ was defined by the parameters $\boldsymbol{\theta}$ such that $\boldsymbol{E}_{Q_{\boldsymbol{\theta}}}[\tau(\mathcal{X})] = \boldsymbol{E}_P[\tau(\mathcal{X})]$. In our setting, we seek the parameters $\boldsymbol{\theta}$ whose expected sufficient statistics match those in $\hat{P}_{\mathcal{D}}$, that is, the sufficient statistics in $\mathcal{D}$.

If our CPDs are in an exponential family where the mapping *ess* from parameters to sufficient statistics is invertible, we can simply take the sufficient statistic vector from $\hat{P}_{\mathcal{D}}$, and invert this mapping to produce the MLE. Indeed, this process is precisely the one that gave rise to our MLE for multinomials and for linear Gaussians, as described earlier. However, the same process can be applied to many other classes of distributions in the exponential family.

This analysis provides us with a notion of sufficient statistics $\tau(\mathcal{X})$ and a clearly defined path to deriving MLE parameters for any distribution in the exponential family. Somewhat more surprisingly, it turns out that a parametric family has a sufficient statistic *only if* it is in the exponential family.

17.3 Bayesian Parameter Estimation

17.3.1 The Thumbtack Example Revisited

Although the MLE approach seems plausible, it can be overly simplistic in many cases. Assume again that we perform the thumbtack experiment and get 3 heads out of 10. It may be quite reasonable to conclude that the parameter θ is 0.3. But what if we do the same experiment with a standard coin, and we also get 3 heads? We would be much less likely to jump to the conclusion that the parameter of the coin is 0.3. Why? Because we have a lot more experience with tossing coins, so we have a lot more *prior knowledge* about their behavior. Note that we do not want our prior knowledge to be an absolute guide, but rather a reasonable starting assumption that allows us to counterbalance our current set of 10 tosses, under the assumption that they may not be typical. However, if we observe 1,000,000 tosses of the coin, of which 300,000 came out heads, then we may be more willing to conclude that this is a trick coin, one whose parameter is closer to 0.3.

Maximum likelihood allows us to make neither of these distinctions: between a thumbtack and a coin, and between 10 tosses and 1,000,000 tosses of the coin. There is, however, another approach, the one recommended by Bayesian statistics.

17.3.1.1 Joint Probabilistic Model

In this approach, we encode our prior knowledge about θ with a probability distribution; this distribution represents how likely we are a priori to believe the different choices of parameters. Once we quantify our knowledge (or lack thereof) about possible values of θ, we can create a joint distribution over the parameter θ and the data cases that we are about to observe $X[1], \ldots, X[M]$. This joint distribution captures our assumptions about the experiment.

Let us reconsider these assumptions. Recall that we assumed that tosses are independent of each other. Note, however, that this assumption was made when θ was fixed. If we do not know θ, then the tosses are not marginally independent: Each toss tells us something about the parameter θ, and thereby about the probability of the next toss. However, once θ is known, we cannot learn about the outcome of one toss from observing the results of others. Thus, we assume that the tosses are *conditionally independent* given θ. We can describe these assumptions using the probabilistic model of figure 17.3.

Having determined the model structure, it remains to specify the local probability models in this network. We begin by considering the probability $P(X[m] \mid \theta)$. Clearly,

$$P(x[m] \mid \theta) = \begin{cases} \theta & \text{if } x[m] = x^1 \\ 1 - \theta & \text{if } x[m] = x^0. \end{cases}$$

Note that since we now treat θ as a random variable, we use the conditioning bar, instead of $P(x[m] : \theta)$.

prior parameter distribution

To finish the description of the joint distribution, we need to describe $P(\theta)$. This is our *prior distribution* over the value of θ. In our case, this is a continuous density over the interval $[0, 1]$. Before we discuss particular choices for this distribution, let us consider how we use it.

The network structure implies that the joint distribution of a particular data set and θ

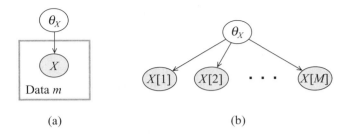

Figure 17.3   Meta-network for IID samples of a random variable X**.** (a) Plate model; (b) Ground Bayesian network.

factorizes as

$$
\begin{aligned}
P(x[1], \ldots, x[M], \theta) &= P(x[1], \ldots, x[M] \mid \theta) P(\theta) \\
&= P(\theta) \prod_{m=1}^{M} P(x[m] \mid \theta) \\
&= P(\theta) \theta^{M[1]} (1 - \theta)^{M[0]},
\end{aligned}
$$

where $M[1]$ is the number of heads in the data, and $M[0]$ is the number of tails. Note that the expression $P(x[1], \ldots, x[M] \mid \theta)$ is simply the likelihood function $L(\theta : \mathcal{D})$.

This network specifies a joint probability model over parameters and data. There are several ways in which we can use this network. Most obviously, we can take an observed data set $\mathcal{D}$ of M outcomes, and use it to instantiate the values of $x[1], \ldots, x[M]$; we can then compute the *posterior distribution* over θ:

posterior
parameter
distribution

$$
P(\theta \mid x[1], \ldots, x[M]) = \frac{P(x[1], \ldots, x[M] \mid \theta) P(\theta)}{P(x[1], \ldots, x[M])}.
$$

In this posterior, the first term in the numerator is the likelihood, the second is the prior over parameters, and the denominator is a normalizing factor that we will not expand on right now. We see that the posterior is (proportional to) a product of the likelihood and the prior. This product is normalized so that it will be a proper density function. In fact, if the prior is a uniform distribution (that is, $P(\theta) = 1$ for all $\theta \in [0, 1]$), then the posterior is just the normalized likelihood function.

17.3.1.2   Prediction

If we do use a uniform prior, what then is the difference between the Bayesian approach and the MLE approach of the previous section? The main philosophical difference is in the use of the posterior. Instead of selecting from the posterior a single value for the parameter θ, we use it, in its entirety, for *predicting* the probability over the next toss.

To derive this prediction in a principled fashion, we introduce the value of the next coin toss $x[M + 1]$ to our network. We can then compute the probability over $x[M + 1]$ given the observations of the first M tosses. Note that, in this model, the parameter θ is unknown, and we are considering all of its possible values. By reasoning over the possible values of θ and using the chain rule, we see that

$$P(x[M+1] \mid x[1], \ldots, x[M]) =$$

$$= \int P(x[M+1] \mid \theta, x[1], \ldots, x[M]) P(\theta \mid x[1], \ldots, x[M]) d\theta$$

$$= \int P(x[M+1] \mid \theta) P(\theta \mid x[1], \ldots, x[M]) d\theta,$$

where we use the conditional independencies implied by the meta-network to rewrite $P(x[M+1] \mid \theta, x[1], \ldots, x[M])$ as $P(x[M+1] \mid \theta)$. In other words, we are *integrating* our posterior over θ to predict the probability of heads for the next toss.

Let us go back to our thumbtack example. Assume that our prior is uniform over θ in the interval $[0, 1]$. Then $P(\theta \mid x[1], \ldots, x[M])$ is proportional to the likelihood $P(x[1], \ldots, x[M] \mid \theta) = \theta^{M[1]}(1-\theta)^{M[0]}$. Plugging this into the integral, we need to compute

$$P(X[M+1] = x^1 \mid x[1], \ldots, x[M]) = \frac{1}{P(x[1], \ldots, x[M])} \int \theta \cdot \theta^{M[1]}(1-\theta)^{M[0]} d\theta.$$

Doing all the math (see exercise 17.6), we get (for uniform priors)

$$P(X[M+1] = x^1 \mid x[1], \ldots, x[M]) = \frac{M[1]+1}{M[1]+M[0]+2}. \tag{17.10}$$

Bayesian
estimator

This prediction, called the *Bayesian estimator*, is quite similar to the MLE prediction of equation (17.1), except that it adds one "imaginary" sample to each count. Clearly, as the number of samples grows, the Bayesian estimator and the MLE estimator converge to the same value. The particular estimator that corresponds to a uniform prior is often referred to as *Laplace's correction*.

Laplace's
correction

17.3.1.3 Priors

Beta distribution

We now want to consider nonuniform priors. The challenge here is to pick a distribution over this continuous space that we can represent compactly (for example, using an analytic formula), and update efficiently as we get new data. For reasons that we will discuss, an appropriate prior in this case is the *Beta distribution*:

Definition 17.3

Beta
hyperparameters

A Beta distribution *is parameterized by two* hyperparameters α_1, α_0, *which are positive reals. The distribution is defined as follows:*

$$\theta \sim \text{Beta}(\alpha_1, \alpha_0) \ \text{if} \ p(\theta) = \gamma \theta^{\alpha_1 - 1}(1-\theta)^{\alpha_0 - 1}.$$

The constant γ is a normalizing constant, defined as follows:

$$\gamma = \frac{\Gamma(\alpha_1 + \alpha_0)}{\Gamma(\alpha_1)\Gamma(\alpha_0)},$$

Gamma function

where $\Gamma(x) = \int_0^\infty t^{x-1} e^{-t} dt$ is the Gamma function. ∎

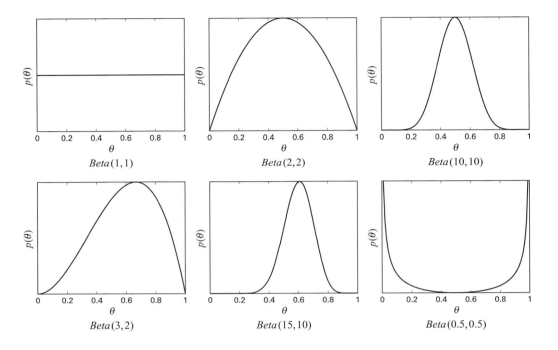

Figure 17.4 Examples of Beta distributions for different choices of hyperparameters

Intuitively, the hyperparameters α_1 and α_0 correspond to the number of imaginary heads and tails that we have "seen" before starting the experiment. Figure 17.4 shows Beta distributions for different values of α.

At first glance, the normalizing constant for the Beta distribution might seem somewhat obscure. However, the Gamma function is actually a very natural one: it is simply a continuous generalization of factorials. More precisely, it satisfies the properties $\Gamma(1) = 1$ and $\Gamma(x + 1) = x\Gamma(x)$. As a consequence, we easily see that $\Gamma(n + 1) = n!$ when n is an integer.

Beta distributions have properties that make them particularly useful for parameter estimation. Assume our distribution $P(\theta)$ is $Beta(\alpha_1, \alpha_0)$, and consider a single coin toss X. Let us compute the *marginal probability* over X, based on $P(\theta)$. To compute the marginal probability, we need to integrate out θ; standard integration techniques can be used to show that:

$$
\begin{aligned}
P(X[1] = x^1) &= \int_0^1 P(X[1] = x^1 \mid \theta) \cdot P(\theta)d\theta \\
&= \int_0^1 \theta \cdot P(\theta)d\theta = \frac{\alpha_1}{\alpha_1 + \alpha_0}.
\end{aligned}
$$

This conclusion supports our intuition that the Beta prior indicates that we have seen α_1 (imaginary) heads α_0 (imaginary) tails.

Now, let us see what happens as we get more observations. Specifically, we observe $M[1]$ heads and $M[0]$ tails. It follows easily that:

$$
\begin{aligned}
P(\theta \mid x[1], \ldots, x[M]) &\propto P(x[1], \ldots, x[M] \mid \theta) P(\theta) \\
&\propto \theta^{M[1]} (1 - \theta)^{M[0]} \cdot \theta^{\alpha_1 - 1} (1 - \theta)^{\alpha_0 - 1} \\
&= \theta^{\alpha_1 + M[1] - 1} (1 - \theta)^{\alpha_0 + M[0] - 1},
\end{aligned}
$$

which is precisely $Beta(\alpha_1 + M[1], \alpha_0 + M[0])$. This result illustrates a key property of the Beta distribution: If the prior is a Beta distribution, then the posterior distribution, that is, the prior conditioned on the evidence, is also a Beta distribution. In this case, we say that the Beta

conjugate prior   distribution is *conjugate* to the Bernoulli likelihood function (see definition 17.4).

An immediate consequence is that we can compute the probabilities over the next toss:

$$
P(X[M + 1] = x^1 \mid x[1], \ldots, x[M]) = \frac{\alpha_1 + M[1]}{\alpha + M},
$$

where $\alpha = \alpha_1 + \alpha_0$. In this case, our posterior Beta distribution tells us that we have seen $\alpha_1 + M[1]$ heads (imaginary and real) and $\alpha_0 + M[0]$ tails.

It is interesting to examine the effect of the prior on the probability over the next coin toss. For example, the prior $Beta(1, 1)$ is very different than $Beta(10, 10)$: Although both predict that the probability of heads in the first toss is 0.5, the second prior is more entrenched, and it requires more observations to deviate from the prediction 0.5. To see this, suppose we observe 3 heads in 10 tosses. Using the first prior, our estimate is $\frac{3+1}{10+2} = \frac{1}{3} \approx 0.33$. On the other hand, using the second prior, our estimate is $\frac{3+10}{10+20} = \frac{13}{30} \approx 0.43$. However, as we obtain more data, the effect of the prior diminishes. If we obtain $1,000$ tosses of which 300 are heads, the first prior gives us an estimate of $\frac{300+1}{1,000+2}$ and the second an estimate of $\frac{300+10}{1,000+20}$, both of which are very close to 0.3. Thus, **the Bayesian framework allows us to capture both of the relevant**

 distinctions. The distinction between the thumbtack and the coin can be captured by the strength of the prior: for a coin, we might use $\alpha_1 = \alpha_0 = 100$, whereas for a thumbtack, we might use $\alpha_1 = \alpha_0 = 1$. The distinction between a few samples and many samples is captured by the peakedness of our posterior, which increases with the amount of data.

17.3.2   Priors and Posteriors

We now turn to examine in more detail the Bayesian approach to dealing with unknown parameters. We start with a discussion of the general principle and deal with the case of Bayesian networks in the next section.

As before, we assume a general learning problem where we observe a training set $\mathcal{D}$ that contains M IID samples of a set of random variable $\mathcal{X}$ from an unknown distribution $P^*(\mathcal{X})$. We also assume that we have a parametric model $P(\xi \mid \boldsymbol{\theta})$ where we can choose parameters from a parameter space Θ.

Recall that the MLE approach attempts to find the parameters $\hat{\boldsymbol{\theta}}$ in Θ that are "best" given the

point estimate   data. The Bayesian approach, on the other hand, does not attempt to find such a *point estimate*. Instead, the underlying principle is that we should keep track of our *beliefs* about $\boldsymbol{\theta}$'s values, and use these beliefs for reaching conclusions. That is, we should quantify the subjective probability we assign to different values of $\boldsymbol{\theta}$ after we have seen the evidence. Note that, in representing such subjective probabilities, we now treat $\boldsymbol{\theta}$ as a random variable. Thus, the Bayesian approach requires that we use probabilities to describe our initial uncertainty about the parameters $\boldsymbol{\theta}$, and then use probabilistic reasoning (that is, Bayes rule) to take into account our observations.

To perform this task, we need to describe a joint distribution $P(\mathcal{D}, \boldsymbol{\theta})$ over the data and the parameters. We can easily write

$$P(\mathcal{D}, \boldsymbol{\theta}) = P(\mathcal{D} \mid \boldsymbol{\theta})P(\boldsymbol{\theta}).$$

parameter prior
The first term is just the likelihood function we discussed earlier. The second term is the *prior distribution* over the possible values in Θ. This prior captures our initial uncertainty about the parameters. It can also capture our previous experience before starting the experiment. For example, if we study coin tossing, we might have prior experience that suggests that most coins are unbiased (or nearly unbiased).

Once we have specified the likelihood function and the prior, we can use the data to derive parameter posterior the *posterior distribution* over the parameters. Since we have specified a joint distribution over all the quantities in question, the posterior is immediately derived by Bayes rule:

$$P(\boldsymbol{\theta} \mid \mathcal{D}) = \frac{P(\mathcal{D} \mid \boldsymbol{\theta})P(\boldsymbol{\theta})}{P(\mathcal{D})}.$$

marginal likelihood
The term $P(\mathcal{D})$ is the *marginal likelihood* of the data

$$P(\mathcal{D}) = \int_{\Theta} P(\mathcal{D} \mid \boldsymbol{\theta})P(\boldsymbol{\theta})d\boldsymbol{\theta},$$

that is, the integration of the likelihood over all possible parameter assignments. This is the a priori probability of seeing this particular data set given our prior beliefs.

As we saw, for some probabilistic models, the likelihood function can be compactly described by using sufficient statistics. Can we also compactly describe the posterior distribution? In general, this depends on the form of the prior. As we saw in the thumbtack example of section 17.1.1, we can sometimes find priors for which we have a description of the posterior.

As another example of the forms of priors and posteriors, let us examine the learning problem of example 17.3. Here we need to describe our uncertainty about the parameters of a multinomial distribution. The parameter space Θ is the space of all nonnegative vectors $\boldsymbol{\theta} = \langle \theta_1, \ldots, \theta_K \rangle$ such that $\sum_k \theta_k = 1$. As we saw in example 17.3, the likelihood function in this model has the form:

$$L(\boldsymbol{\theta} : \mathcal{D}) = \prod_k \theta_k^{M[k]}.$$

Since the posterior is a product of the prior and the likelihood, it seems natural to require that the prior also have a form similar to the likelihood.

Dirichlet distribution
One such prior is the *Dirichlet distribution*, which generalizes the Beta distribution we discussed earlier. A Dirichlet distribution is specified by a set of *hyperparameters* $\alpha_1, \ldots, \alpha_K$, so Dirichlet hyperparameters that

$$\boldsymbol{\theta} \sim Dirichlet(\alpha_1, \ldots, \alpha_K) \quad \text{if } P(\boldsymbol{\theta}) \propto \prod_k \theta_k^{\alpha_k - 1}.$$

Dirichlet posterior
We use α to denote $\sum_j \alpha_j$. If we use a Dirichlet prior, then the *posterior* is also Dirichlet:

Proposition 17.3

If $P(\boldsymbol{\theta})$ is $Dirichlet(\alpha_1, \ldots, \alpha_K)$ then $P(\boldsymbol{\theta} \mid \mathcal{D})$ is $Dirichlet(\alpha_1 + M[1], \ldots, \alpha_K + M[K])$, where $M[k]$ is the number of occurrences of x^k.

Priors such as the Dirichlet are useful, since they ensure that the posterior has a nice compact description. Moreover, this description uses the same representation as the prior. This phenomenon is a general one, and one that we strive to achieve, since it makes our computation and representation much easier.

Definition 17.4

conjugate prior

A family of priors $P(\boldsymbol{\theta} : \boldsymbol{\alpha})$ is conjugate *to a particular model $P(\xi \mid \boldsymbol{\theta})$ if for any possible data set $\mathcal{D}$ of IID samples from $P(\xi \mid \boldsymbol{\theta})$, and any choice of legal hyperparameters $\boldsymbol{\alpha}$ for the prior over $\boldsymbol{\theta}$, there are hyperparameters $\boldsymbol{\alpha}'$ that describe the posterior. That is,*

$$P(\boldsymbol{\theta} : \boldsymbol{\alpha}') \propto P(\mathcal{D} \mid \boldsymbol{\theta})P(\boldsymbol{\theta} : \boldsymbol{\alpha}). \qquad \blacksquare$$

For example, Dirichlet priors are conjugate to the multinomial model. We note that this does not preclude the possibility of other families that are also conjugate to the same model. See exercise 17.7 for an example of such a prior for the multinomial model. We can find conjugate priors for other models as well. See exercise 17.8 and exercise 17.11 for the development of conjugate priors for the Gaussian distribution.

This discussion shows some examples where we can easily update our beliefs about $\boldsymbol{\theta}$ after observing a set of instances $\mathcal{D}$. This update process results in a posterior that combines our prior knowledge and our observations. What can we do with the posterior? We can use the posterior to determine properties of the model at hand. For example, to assess our beliefs that a coin we experimented with is biased toward heads, we might compute the posterior probability that $\theta > t$ for some threshold t, say 0.6.

Another use of the posterior is to predict the probability of future examples. Suppose that we are about to sample a new instance $\xi[M+1]$. Since we already have observations over previous instances, the *Bayesian estimator* is the posterior distribution over a new example:

Bayesian
estimator

$$
\begin{aligned}
P(\xi[M+1] \mid \mathcal{D}) &= \int P(\xi[M+1] \mid \mathcal{D}, \boldsymbol{\theta})P(\boldsymbol{\theta} \mid \mathcal{D})d\boldsymbol{\theta} \\
&= \int P(\xi[M+1] \mid \boldsymbol{\theta})P(\boldsymbol{\theta} \mid \mathcal{D})d\boldsymbol{\theta} \\
&= \boldsymbol{E}_{P(\boldsymbol{\theta}|\mathcal{D})}[P(\xi[M+1] \mid \boldsymbol{\theta})],
\end{aligned}
$$

where, in the second step, we use the fact that instances are independent given $\boldsymbol{\theta}$. Thus, our prediction is the average over all parameters according to the posterior.

Let us examine prediction with the Dirichlet prior. We need to compute

$$P(x[M+1] = x^k \mid \mathcal{D}) = \boldsymbol{E}_{P(\boldsymbol{\theta}|\mathcal{D})}[\theta_k].$$

To compute the prediction on a new data case, we need to compute the expectation of particular parameters with respect for a Dirichlet distribution over $\boldsymbol{\theta}$.

Proposition 17.4

Let $P(\boldsymbol{\theta})$ be a Dirichlet distribution with hyperparameters $\alpha_1, \ldots, \alpha_k$, and $\alpha = \sum_j \alpha_j$, then $E[\theta_k] = \frac{\alpha_k}{\alpha}$.

Recall that our posterior is $Dirichlet(\alpha_1 + M[1], \ldots, \alpha_K + M[K])$ where $M[1], \ldots, M[K]$ are the sufficient statistics from the data. Hence, the prediction with Dirichlet priors is

$$P(x[M + 1] = x^k \mid \mathcal{D}) = \frac{M[k] + \alpha_k}{M + \alpha}.$$

This prediction is similar to prediction with the MLE parameters. The only difference is that we added the hyperparameters to our counts when making the prediction. For this reason the Dirichlet hyperparameters are often called *pseudo-counts*. We can think of these as the number of times we have seen the different outcomes in our prior experience before conducting our current experiment.

The total α of the pseudo-counts reflects how confident we are in our prior, and is often called the *equivalent sample size*. Using α, we can rewrite the hyperparameters as $\alpha_k = \alpha\theta'_k$, where $\boldsymbol{\theta}' = \{\theta'_k : k = 1, \ldots, K\}$ is a distribution describing the *mean prediction* of our prior. We can see that the prior prediction (before observing any data) is simply $\boldsymbol{\theta}'$. Moreover, we can rewrite the prediction given the posterior as:

$$P(x[M + 1] = x^k \mid \mathcal{D}) = \frac{\alpha}{M + \alpha}\theta'_k + \frac{M}{M + \alpha} \cdot \frac{M[k]}{M}. \tag{17.11}$$

That is, the prediction is a weighted average (convex combination) of the prior mean and the MLE estimate. The combination weights are determined by the relative magnitude of α — the confidence of the prior (or total weight of the pseudo-counts) — and M — the number of observed samples. We see that the Bayesian prediction converges to the MLE estimate when $M \to \infty$. Intuitively, when we have a very large training set the contribution of the prior is negligible, and the prediction will be dominated by the frequency of outcomes in the data. We also get convergence to the MLE estimate when $\alpha \to 0$, so that we have only a very weak prior. Note that the case where $\alpha = 0$ is not achievable: the normalization constant for the Dirichlet prior grows to infinity when the hyperparameters are close to 0. Thus, the prior with $\alpha = 0$ (that is, $\alpha_k = 0$ for all k) is not well defined. The prior with $\alpha = 0$ is often called a *improper prior*. The difference between the Bayesian estimate and the MLE estimate arises when M is not too large, and α is not close to 0. In these situations, the Bayesian estimate is "biased" toward the prior probability $\boldsymbol{\theta}'$.

To gain some intuition for the interaction between these different factors, figure 17.5 shows the effect of the strength and means of the prior on our estimates. We can see that, as the amount of real data grows, our estimate converges to the true underlying distribution, regardless of the starting point. The convergence time grows both with the difference between the prior mean and the empirical mean, and with the strength of the prior. We also see that the Bayesian estimate is more stable than the MLE estimate, because with few instances, even single samples will change the MLE estimate dramatically.

Example 17.7

Suppose we are trying to estimate the parameter associated with a coin, and we observe one head and one tail. Our MLE estimate of θ_1 is $1/2 = 0.5$. Now, if the next observation is a head, we will change our estimate to be $2/3 \approx 0.66$. On the other hand, if our next observation is a tail, we will change our estimate to $1/3 \approx 0.33$. In contrast, consider the Bayesian estimate with a Dirichlet prior with $\alpha = 1$ and $\theta'_1 = 0.5$. With this estimator, our original estimate is $1.5/3 = 0.5$. If we observe another head, we revise to $2.5/4 = 0.625$, and if observe another tail, we revise to

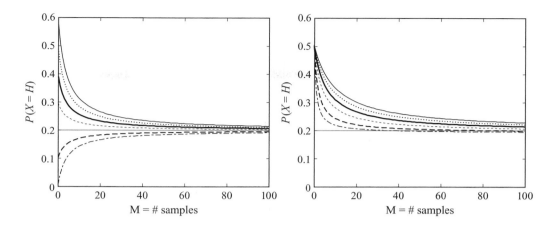

Figure 17.5 **The effect of the strength and means of the Beta prior on our posterior estimates.** Our data set is an idealized version of samples from a biased coin where the frequency of heads is 0.2: for a given data set size M, we assume that $\mathcal{D}$ contains $0.2M$ heads and $0.8M$ tails. The x axis represents the number of samples (M) in our data set $\mathcal{D}$, and the y axis the expected probability of heads according to the Bayesian estimate. (a) shows the effect of varying the prior means θ'_1, θ'_0, for a fixed prior strength α. (b) shows the effect of varying the prior strength for a fixed prior mean $\theta'_1 = \theta'_0 = 0.5$.

$1.5/4 = 0.375$. *We see that the estimate changes by slightly less after the update. If α is larger, then the smoothing is more aggressive. For example, when $\alpha = 5$, our estimate is $4.5/8 = 0.5625$ after observing a head, and $3.5/8 = 0.4375$ after observing a tail. We can also see this effect visually in figure 17.6, which shows our changing estimate for $P(\theta_H)$ as we observe a particular sequence of tosses.* ∎

This smoothing effect results in more robust estimates when we do not have enough data to reach definite conclusions. If we have good prior knowledge, we revert to it. Alternatively, if we do not have prior knowledge, we can use a uniform prior that will keep our estimate from taking extreme values. In general, it is a bad idea to have extreme estimates (ones where some of the parameters are close to 0), since these might assign too small probability to new instances we later observe. In particular, as we already discussed, probability estimates that are actually 0 are dangerous, since no amount of evidence can change them. Thus, if we are unsure about our estimates, it is better to bias them away from extreme estimates. The MLE estimate, on the other hand, often assigns probability 0 to values that were not observed in the training data.

17.4 Bayesian Parameter Estimation in Bayesian Networks

We now turn to Bayesian estimation in the context of a Bayesian network. Recall that the Bayesian framework requires us to specify a joint distribution over the unknown parameters and the data instances. As in the single parameter case, we can understand the joint distribution over parameters and data as a Bayesian network.

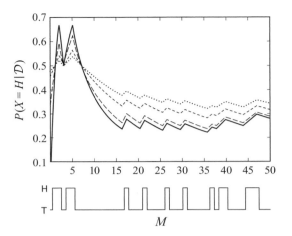

Figure 17.6   The effect of different priors on smoothing our parameter estimates. The graph shows the estimate of $P(X = H|\mathcal{D})$ (y-axis) after seeing different number of samples (x-axis). The graph below the x-axis shows the particular sequence of tosses. The solid line corresponds to the MLE estimate, and the remaining ones to Bayesian estimates with different strengths and uniform prior means. The large-dash line corresponds to *Beta*$(1, 1)$, the small-dash line to *Beta*$(5, 5)$, and the dotted line to *Beta*$(10, 10)$.

17.4.1   Parameter Independence and Global Decomposition

17.4.1.1   A Simple Example

Suppose we want to estimate parameters for a simple network with two variables X and Y so that X is the parent of Y. Our training data consist of observations $X[m], Y[m]$ for $m = 1, \ldots, M$. In addition, we have unknown parameter vectors $\boldsymbol{\theta}_X$ and $\boldsymbol{\theta}_{Y|X}$. The dependencies between these variables are described in the network of figure 17.7. This is the *meta-network* that describes our learning setup.

meta-network

This Bayesian network structure immediately reveals several points. For example, as in our simple thumbtack example, the instances are independent given the unknown parameters. A simple examination of active trails shows that $X[m]$ and $Y[m]$ are d-separated from $X[m']$ and $Y[m']$ once we observe the parameter variables.

In addition, the network structure embodies the assumption that the priors for the individual parameters variables are a priori independent. That is, we believe that knowing the value of one parameter tells us nothing about another. More precisely, we define

---

Definition 17.5

global parameter
independence

Let $\mathcal{G}$ be a Bayesian network structure with parameters $\boldsymbol{\theta} = (\boldsymbol{\theta}_{X_1|\mathrm{Pa}_{X_1}}, \ldots, \boldsymbol{\theta}_{X_n|\mathrm{Pa}_{X_n}})$. A prior $P(\boldsymbol{\theta})$ is said to satisfy global parameter independence *if it has the form:*

$$P(\boldsymbol{\theta}) = \prod_i P(\boldsymbol{\theta}_{X_i|\mathrm{Pa}_{X_i}}).$$ ∎

This assumption may not be suitable for all domains, and it should be considered with care.

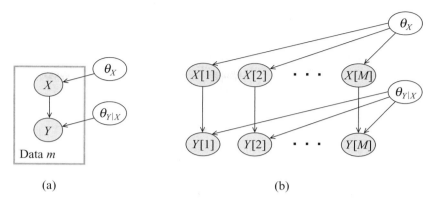

(a) (b)

Figure 17.7 Meta-network for IID samples from a network $X \rightarrow Y$ with global parameter independence. (a) Plate model; (b) Ground Bayesian network.

Example 17.8

Consider an extension of our student example, where our student takes multiple classes. For each class, we want to learn the distribution of Grade given the student's Intelligence and the course Difficulty. For classes taught by the same instructor, we might believe that the grade distribution is the same; for example, if two classes are both difficult, and the student is intelligent, his probability of getting an A is the same in both. However, under the global parameter independence assumption, these are two different random variables, and hence their parameters are independent. ∎

Thus, although we use the global parameter independence in much of our discussion, it is not always appropriate, and we relax it in some of our later discussion (such as section 17.5 and section 18.6.2).

If we accept global parameter independence, we can draw an important conclusion. Complete data d-separates the parameters for different CPDs. For example, if $x[m]$ and $y[m]$ are observed for all m, then $\boldsymbol{\theta}_X$ and $\boldsymbol{\theta}_{Y|X}$ are d-separated. To see this, note that any path between the two has the form

$$\boldsymbol{\theta}_X \rightarrow X[m] \rightarrow Y[m] \leftarrow \boldsymbol{\theta}_{Y|X},$$

so that the observation of $x[m]$ blocks the path. Thus, if these two parameter variables are independent a priori, they are also independent a posteriori. Using the definition of conditional independence, we conclude that

$$P(\boldsymbol{\theta}_X, \boldsymbol{\theta}_{Y|X} \mid \mathcal{D}) = P(\boldsymbol{\theta}_X \mid \mathcal{D})P(\boldsymbol{\theta}_{Y|X} \mid \mathcal{D}).$$

This decomposition has immediate practical ramifications. Given the data set $\mathcal{D}$, we can determine the posterior over $\boldsymbol{\theta}_X$ independently of the posterior over $\boldsymbol{\theta}_{Y|X}$. Once we can solve each problem separately, we can combine the results. This is the analogous result to the likelihood decomposition for MLE estimation of section 17.2.2. In the Bayesian setting this property has additional importance. It tells us the posterior can be represented in a compact factorized form.

17.4.1.2   General Networks

We can generalize this conclusion to the general case of Bayesian network learning. Suppose we are given a network structure $\mathcal{G}$ with parameters $\boldsymbol{\theta}$. In the Bayesian framework, we need to specify a prior $P(\boldsymbol{\theta})$ over all possible parameterizations of the network. The posterior distribution over parameters given the data samples $\mathcal{D}$ is simply

$$P(\boldsymbol{\theta} \mid \mathcal{D}) = \frac{P(\mathcal{D} \mid \boldsymbol{\theta})P(\boldsymbol{\theta})}{P(\mathcal{D})}.$$

marginal
likelihood

The term $P(\boldsymbol{\theta})$ is our prior distribution, $P(\mathcal{D} \mid \boldsymbol{\theta})$ is the probability of the data given a particular parameter settings, which is simply the likelihood function. Finally, $P(\mathcal{D})$ is the normalizing constant. As we discussed, this term is called the *marginal likelihood*; it will play an important role in the next chapter. For now, however, we can ignore it, since it does not depend on $\boldsymbol{\theta}$ and only serves to normalize the posterior.

As we discussed in section 17.2, we can decompose the likelihood into local likelihoods:

$$P(\mathcal{D} \mid \boldsymbol{\theta}) = \prod_i L_i(\boldsymbol{\theta}_{X_i \mid \mathrm{Pa}_{X_i}} : \mathcal{D}).$$

Moreover, if we assume that we have global parameter independence, then

$$P(\boldsymbol{\theta}) = \prod_i P(\boldsymbol{\theta}_{X_i \mid \mathrm{Pa}_{X_i}}).$$

Combining these two decompositions, we see that

$$P(\boldsymbol{\theta} \mid \mathcal{D}) = \frac{1}{P(\mathcal{D})} \prod_i \left[L_i(\boldsymbol{\theta}_{X_i \mid \mathrm{Pa}_{X_i}} : \mathcal{D}) P(\boldsymbol{\theta}_{X_i \mid \mathrm{Pa}_{X_i}}) \right].$$

Now each subset $\boldsymbol{\theta}_{X_i \mid \mathrm{Pa}_{X_i}}$ of $\boldsymbol{\theta}$ appears in just one term in the product. Thus, we have that the posterior can be represented as a product of local terms.

Proposition 17.5

Let $\mathcal{D}$ be a complete data set for $\mathcal{X}$, let $\mathcal{G}$ be a network structure over these variables. If $P(\boldsymbol{\theta})$ satisfies global parameter independence, then

$$P(\boldsymbol{\theta} \mid \mathcal{D}) = \prod_i P(\boldsymbol{\theta}_{X_i \mid \mathrm{Pa}_{X_i}} \mid \mathcal{D}).$$

The proof of this property follows from the steps we discussed. It can also be derived directly from the structure of the meta-Bayesian network (as in the network of figure 17.7).

17.4.1.3   Prediction

This decomposition of the posterior allows us to simplify various tasks. For example, suppose that, in our simple two-variable network, we want to compute the probability of another instance $x[M + 1], y[M + 1]$ based on our previous observations $x[1], y[1], \ldots, x[M], y[M]$. According to the structure of our meta-network, we need to sum out (or more precisely integrate out) the unknown parameter variables

$$P(x[M + 1], y[M + 1] \mid \mathcal{D}) = \int P(x[M + 1], y[M + 1] \mid \mathcal{D}, \boldsymbol{\theta})P(\boldsymbol{\theta} \mid \mathcal{D})d\boldsymbol{\theta},$$

where the integration is over all legal parameter values. Since $\boldsymbol{\theta}$ d-separates instances from each other, we have that

$$
\begin{aligned}
&P(x[M+1], y[M+1] \mid \mathcal{D}, \boldsymbol{\theta}) \\
&= P(x[M+1], y[M+1] \mid \boldsymbol{\theta}) \\
&= P(x[M+1] \mid \boldsymbol{\theta}_X) P(y[M+1] \mid x[M+1], \boldsymbol{\theta}_{Y|X}).
\end{aligned}
$$

Moreover, as we just saw, the posterior probability also decomposes into a product. Thus,

$$
\begin{aligned}
&P(x[M+1], y[M+1] \mid \mathcal{D}) \\
&= \int \int P(x[M+1] \mid \boldsymbol{\theta}_X) P(y[M+1] \mid x[M+1], \boldsymbol{\theta}_{Y|X}) \\
&\qquad P(\boldsymbol{\theta}_X \mid \mathcal{D}) P(\boldsymbol{\theta}_{Y|X} \mid \mathcal{D}) d\boldsymbol{\theta}_X d\boldsymbol{\theta}_{Y|X} \\
&= \left(\int P(x[M+1] \mid \boldsymbol{\theta}_X) P(\boldsymbol{\theta}_X \mid \mathcal{D}) d\boldsymbol{\theta}_X \right) \\
&\qquad \left(\int P(y[M+1] \mid x[M+1], \boldsymbol{\theta}_{Y|X}) P(\boldsymbol{\theta}_{Y|X} \mid \mathcal{D}) d\boldsymbol{\theta}_{Y|X} \right).
\end{aligned}
$$

In the second step, we use the fact that the double integral of two unrelated functions is the product of the integrals. That is:

$$
\int \int f(x) g(y) dx dy = \left(\int f(x) dx \right) \left(\int g(y) dy \right).
$$

Thus, we can solve the prediction problem for the two variables X and Y separately.

The same line of reasoning easily applies to the general case, and thus we can see that, in the setting of proposition 17.5, we have

$$
\begin{aligned}
&P(X_1[M+1], \ldots, X_n[M+1] \mid \mathcal{D}) = \\
&\prod_i \int P(X_i[M+1] \mid \mathrm{Pa}_{X_i}[M+1], \boldsymbol{\theta}_{X_i|\mathrm{Pa}_{X_i}}) P(\boldsymbol{\theta}_{X_i|\mathrm{Pa}_{X_i}} \mid \mathcal{D}) d\boldsymbol{\theta}_{X_i|\mathrm{Pa}_{X_i}}.
\end{aligned} \tag{17.12}
$$

We see that we can solve the prediction problem for each CPD independently and then combine the results.

We stress that the discussion so far was based on the assumption that the priors over parameters for different CPDs are independent. We see that, when learning from complete data, this assumption alone suffices to get a decomposition of the learning problem to several "local" problems, each one involving one CPD.

At this stage it might seem that the Bayesian framework introduces new complications that did not appear in the MLE setup. Note, however, that in deriving the MLE decomposition, we used the property that we can choose parameters for one CPD independently of the others. Thus, we implicitly made a similar assumption to get decomposition. The Bayesian treatment forces us to make such assumptions explicit, allowing us to more carefully evaluate their validity. We view this as a benefit of the Bayesian framework.

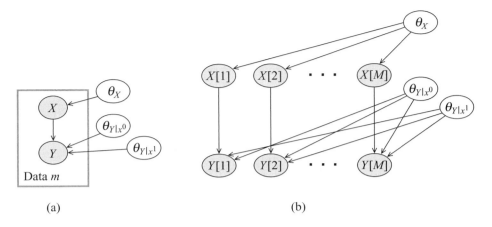

Figure 17.8   **Meta-network for IID samples from a network** $X \to Y$ **with local parameter independence.** (a) Plate model. (b) Ground Bayesian network.

17.4.2   Local Decomposition

Based on the preceding discussion, we now need to solve localized Bayesian estimation problems to get a global Bayesian solution. We now examine this localized estimation task for table-CPDs. The case for tree-CPDs is treated in section 17.5.2.

Consider, for example, the learning setting described in figure 17.7, where we take both X and Y to be binary. As we have seen, we need to represent the posterior $\boldsymbol{\theta}_X$ and $\boldsymbol{\theta}_{Y|X}$ given the data. We already know how to deal with the posterior over $\boldsymbol{\theta}_X$. If we use a Dirichlet prior over $\boldsymbol{\theta}_X$, then the posterior $P(\boldsymbol{\theta}_X \mid x[1], \dots, x[M])$ is also represented as a Dirichlet distribution.

A less obvious question is how to deal with the posterior over $\boldsymbol{\theta}_{Y|X}$. If we are learning table-CPDs, this parameter vector contains four parameters $\theta_{y^0|x^0}, \dots, \theta_{y^1|x^1}$. In our discussion of maximum likelihood estimation, we saw how the local likelihood over these parameters can be further decomposed into two terms, one over the parameters $\boldsymbol{\theta}_{Y|x^0}$ and one over the parameters $\boldsymbol{\theta}_{Y|x^1}$. Do we have a similar phenomenon in the Bayesian setting?

We start with the prior over $\boldsymbol{\theta}_{Y|X}$. One obvious choice is a Dirichlet prior over $\boldsymbol{\theta}_{Y|x^1}$ and another over $\boldsymbol{\theta}_{Y|x^0}$. More precisely, we have

$$P(\boldsymbol{\theta}_{Y|X}) = P(\boldsymbol{\theta}_{Y|x^1})P(\boldsymbol{\theta}_{Y|x^0}),$$

where each of the terms on the right is a Dirichlet prior. Thus, in this case, we assume that the two groups of parameters are independent a priori.

This independence assumption, in effect, allows us to replace the node $\boldsymbol{\theta}_{Y|X}$ in figure 17.7 with two nodes, $\boldsymbol{\theta}_{Y|x^1}$ and $\boldsymbol{\theta}_{Y|x^0}$ that are both roots (see figure 17.8). What can we say about the posterior distribution of these parameter groups? At first, it seems that the two are dependent on each other given the data. Given an observation of $y[m]$, the path

$$\boldsymbol{\theta}_{Y|x^0} \to Y[m] \leftarrow \boldsymbol{\theta}_{Y|x^1}$$

is active (since we observe the sink of a v-structure), and thus the two parameters are not d-separated.

This, however, is not the end of the story. We get more insight if We examine how $y[m]$ depends on the two parameters. Clearly,

$$P(y[m] = y \mid x[m], \boldsymbol{\theta}_{Y|x^0}, \boldsymbol{\theta}_{Y|x^1}) = \begin{cases} \boldsymbol{\theta}_{y|x^0} & \text{if } x[m] = x^0 \\ \boldsymbol{\theta}_{y|x^1} & \text{if } x[m] = x^1. \end{cases}$$

We see that $y[m]$ does not depends on the value of $\boldsymbol{\theta}_{Y|x^0}$ when $x[m] = x^1$. This example is an instance of the same type of context specific independence that we discussed in example 3.7. As discussed in section 5.3, we can perform a more refined form of d-separation test in such a situation by removing arcs that are ruled inactive in particular contexts. For the CPD of $y[m]$, we see that once we observe the value of $x[m]$, one of the two arcs into $y[m]$ is inactive. If $x[m] = x^0$, then the arc $\boldsymbol{\theta}_{Y|x^1} \to y[m]$ is inactive, and if $x[m] = x^1$, then $\boldsymbol{\theta}_{Y|x^0} \to y[m]$ is inactive. In either case, the v-structure $\boldsymbol{\theta}_{Y|x^0} \to y[m] \leftarrow \boldsymbol{\theta}_{Y|x^1}$ is removed. Since this removal occurs for every $m = 1, \ldots, M$, we conclude that no active path exists between $\boldsymbol{\theta}_{Y|x^0}$ and $\boldsymbol{\theta}_{Y|x^1}$ and thus, the two are independent given the observation of the data. In other words, we can write

$$P(\boldsymbol{\theta}_{Y|X} \mid \mathcal{D}) = P(\boldsymbol{\theta}_{Y|x^1} \mid \mathcal{D})P(\boldsymbol{\theta}_{Y|x^0} \mid \mathcal{D}).$$

Suppose that $P(\boldsymbol{\theta}_{Y|x^0})$ is a Dirichlet prior with hyperparameters $\alpha_{y^0|x^0}$ and $\alpha_{y^1|x^0}$. As in our discussion of the local decomposition for the likelihood function in section 17.2.3, we have that the likelihood terms that involve $\boldsymbol{\theta}_{Y|x^0}$ are those that measure the probability of $P(y[m] \mid x[m], \boldsymbol{\theta}_{Y|X})$ when $x[m] = x^0$. Thus, we can decompose the joint distribution over parameters and data as follows:

$$\begin{aligned} P(\boldsymbol{\theta}, \mathcal{D}) \quad = \quad & P(\boldsymbol{\theta}_X)L_X(\boldsymbol{\theta}_X : \mathcal{D}) \\ & P(\boldsymbol{\theta}_{Y|x^1}) \prod_{m:x[m]=x^1} P(y[m] \mid x[m] : \boldsymbol{\theta}_{Y|x^1}) \\ & P(\boldsymbol{\theta}_{Y|x^0}) \prod_{m:x[m]=x^0} P(y[m] \mid x[m] : \boldsymbol{\theta}_{Y|x^0}). \end{aligned}$$

Thus, this joint distribution is a product of three separate joint distributions with a Dirichlet prior for some multinomial parameter and data drawn from this multinomial. Our analysis for updating a single Dirichlet now applies, and we can conclude that the posterior $P(\boldsymbol{\theta}_{Y|x^0} \mid \mathcal{D})$ is Dirichlet with hyperparameters $\alpha_{y^0|x^0} + M[x^0, y^0]$ and $\alpha_{y^1|x^0} + M[x^0, y^1]$.

We can generalize this discussion to arbitrary networks.

Definition 17.6

local parameter independence

Let X be a variable with parents $\boldsymbol{U}$. We say that the prior $P(\boldsymbol{\theta}_{X|U})$ satisfies local parameter independence *if*

$$P(\boldsymbol{\theta}_{X|U}) = \prod_{\boldsymbol{u}} P(\boldsymbol{\theta}_{X|\boldsymbol{u}}).$$

∎

The same pattern of reasoning also applies to the general case.

Proposition 17.6

Let $\mathcal{D}$ be a complete data set for $\mathcal{X}$, let $\mathcal{G}$ be a network structure over these variables with table-CPDs. If the prior $P(\boldsymbol{\theta})$ satisfies global and local parameter independence, then

$$P(\boldsymbol{\theta} \mid \mathcal{D}) = \prod_i \prod_{\mathrm{pa}_{X_i}} P(\boldsymbol{\theta}_{X_i \mid \mathrm{pa}_{X_i}} \mid \mathcal{D}).$$

Moreover, if $P(\boldsymbol{\theta}_{X \mid \boldsymbol{u}})$ is a Dirichlet prior with hyperparameters $\alpha_{x^1 \mid \boldsymbol{u}}, \ldots, \alpha_{x^K \mid \boldsymbol{u}}$, then the posterior $P(\boldsymbol{\theta}_{X \mid \boldsymbol{u}} \mid \mathcal{D})$ is a Dirichlet distribution with hyperparameters $\alpha_{x^1 \mid \boldsymbol{u}} + M[\boldsymbol{u}, x^1], \ldots, \alpha_{x^K \mid \boldsymbol{u}} + M[\boldsymbol{u}, x^K]$.

As in the case of a single multinomial, this result induces a predictive model in which, for the next instance, we have that

$$P(X_i[M+1] = x_i \mid \boldsymbol{U}[M+1] = \boldsymbol{u}, \mathcal{D}) = \frac{\alpha_{x_i \mid \boldsymbol{u}} + M[x_i, \boldsymbol{u}]}{\sum_i \alpha_{x_i \mid \boldsymbol{u}} + M[x_i, \boldsymbol{u}]}. \tag{17.13}$$

Plugging this result into equation (17.12), we see that for computing the probability of a new instance, we can use a single network parameterized as usual, via a set of multinomials, but ones computed as in equation (17.13).

17.4.3 Priors for Bayesian Network Learning

Bayesian network
parameter prior

It remains only to address the question of assessing the set of *parameter priors* required for a Bayesian network. In a general Bayesian network, each node X_i has a set of multinomial distributions $\boldsymbol{\theta}_{X_i \mid \mathrm{pa}_{X_i}}$, one for each instantiation pa_{X_i} of X_i's parents Pa_{X_i}. Each of these parameters will have a separate Dirichlet prior, governed by hyperparameters

$$\boldsymbol{\alpha}_{X_i \mid \mathrm{pa}_{X_i}} = (\alpha_{x_i^1 \mid \mathrm{pa}_{X_i}}, \ldots, \alpha_{x_i^{K_i} \mid \mathrm{pa}_{X_i}}),$$

where K_i is the number of values of X_i.

We can, of course, ask our expert to assign values to each of these hyperparameters based on his or her knowledge. This task, however, is rather unwieldy. Another approach, called the K2 prior, is to use a fixed prior, say $\alpha_{x_i^j \mid \mathrm{pa}_{X_i}} = 1$, for all hyperparameters in the network. As we discuss in the next chapter, this approach has consequences that are conceptually unsatisfying; see exercise 18.10.

A common approach to addressing the specification task uses the intuitions we described in our discussion of Dirichlet priors in section 17.3.1. As we showed, we can think of the hyperparameter α_{x^k} as an imaginary count in our prior experience. This intuition suggests the following representation for a prior over a Bayesian network. Suppose we have an imaginary data set $\mathcal{D}'$ of "prior" examples. Then, we can use counts from this imaginary data set as hyperparameters. More specifically, we set

$$\alpha_{x_i \mid \mathrm{pa}_{X_i}} = \alpha[x_i, \mathrm{pa}_{X_i}],$$

where $\alpha[x_i, \mathrm{pa}_{X_i}]$ is the number of times $X_i = x_i$ and $\mathrm{Pa}_{X_i} = \mathrm{pa}_{X_i}$ in $\mathcal{D}'$. We can easily see that prediction with this setting of hyperparameters is equivalent to MLE prediction from the combined data set that contains instances of both $\mathcal{D}$ and $\mathcal{D}'$.

One problem with this approach is that it requires storing a possibly large data set of pseudo-instances. Instead, we can store the size of the data set α and a representation $P'(X_1, \ldots, X_n)$ of the frequencies of events in this prior data set. If $P'(X_1, \ldots, X_n)$ is the distribution of events in $\mathcal{D}'$, then we get that

$$\alpha_{x_i | \mathrm{pa}_{X_i}} = \alpha \cdot P'(x_i, \mathrm{pa}_{X_i}).$$

How do we represent P'? Clearly, one natural choice is via a Bayesian network. Then, we can use Bayesian network inference to efficiently compute the quantities $P'(x_i, \mathrm{pa}_{X_i})$. Note that P' does not have to be structured in the same way as the network we learn (although it can be). It is, in fact, quite common to define P' as a set of independent marginals over the X_i's. A prior that can be represented in this manner (using α and P') is called a *BDe prior*. Aside from being philosophically pleasing, it has some additional benefits that we will discuss in the next chapter.

BDe prior

Box 17.C — Case Study: Learning the ICU-Alarm **Network.** *To give an example of the techniques described in this chapter, we evaluate them on a synthetic example. Figure 17.C.1 shows the graph structure of the real-world* ICU-Alarm *Bayesian network, hand-constructed by an expert, for monitoring patients in an Intensive Care Unit (ICU). The network has 37 nodes and a total of 504 parameters. We want to evaluate the ability of our parameter estimation algorithms to reconstruct the network parameters from data.*

ICU-Alarm

We generated a training set from the network, by sampling from the distribution specified by the network. We then gave the algorithm only the (correct) network structure, and the generated data, and measured the ability of our algorithms to reconstruct the parameters. We tested the MLE approach, and several Bayesian approaches. All of the approaches used a uniform prior mean, but different prior strengths α.

In performing such an experiment, there are many ways of measuring the quality of the learned network. One possible measure is the difference between the original values of the model parameters and the estimated ones. A related approach measures the distance between the original CPDs and the learned ones (in the case of table-CPDs, these two approaches are the same, but not for general parameterizations). These approaches place equal weights on different parameters, regardless of the extent to which they influence the overall distribution.

The approach we often take is the one described in section 16.2.1, where we measure the relative entropy between the generating distribution P^* and the learned distribution $\tilde{P}$ (see also section 8.4.2). This approach provides a global measure of the extent to which our learned distribution resembles the true distribution. Figure 17.C.2 shows the results for different stages of learning. As we might expect, when more instances are available, the estimation is better. The improvement is drastic in early stages of learning, where additional instances lead to major improvements. When the number of instances in our data set is larger, additional instances lead to improvement, but a smaller one.

More surprisingly, we also see that the MLE achieves the poorest results, a consequence of its extreme sensitivity to the specific training data used. The lowest error is achieved with a very weak prior — $\alpha = 5$ — which is enough to provide smoothing. As the strength of the prior grows, it starts to introduce a bias, not giving the data enough importance. Thus, the error of the estimated probability increases. However, we also note that the effect of the prior, even for $\alpha = 50$, disappears reasonably soon, and all of the approaches converge to the same line. Interestingly, the different

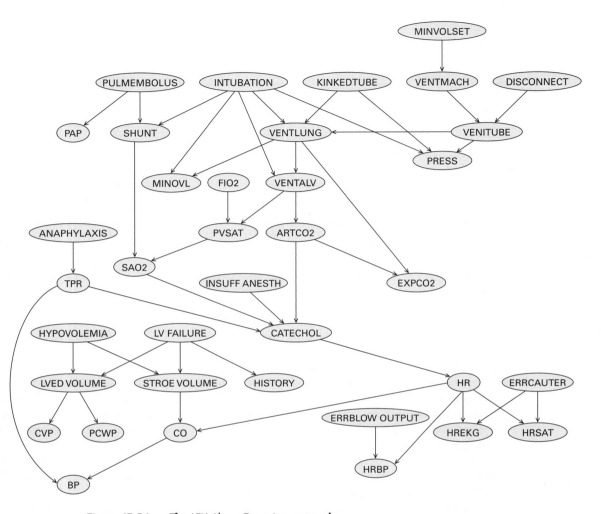

Figure 17.C.1 — The ICU-Alarm **Bayesian network.**

Bayesian approaches converge to this line long before the MLE approach. Thus, at least in this example, an overly strong bias provided by the prior is still a better compromise than the complete lack of smoothing of the MLE approach.

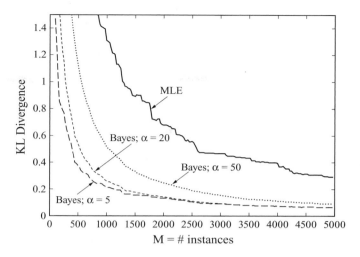

Figure 17.C.2 — Learning curve for parameter estimation for the ICU-Alarm **network** Relative entropy to true model as the amount of training data grows, for different values of the prior strength α.

17.4.4   MAP Estimation ⋆

Our discussion in this chapter has focused solely on Bayesian estimation for multinomial CPDs. Here, we have a closed form solution for the integral required for Bayesian prediction, and thus we can perform it efficiently. In many other representations, the situation is not so simple. In some cases, such as the noisy-or model or the logistic CPDs of section 5.4.2, we do not have a conjugate prior or a closed-form solution for the Bayesian integral. In those cases, Bayesian prediction requires numerical solutions for high-dimensional integrals. In other settings, such normal-Gamma as the linear Gaussian CPD, we do have a conjugate prior (the *normal-Gamma distribution*), but distribution we may prefer other priors that offer other desirable properties (such as the sparsity-inducing Laplacian prior described in section 20.4.1).

MAP estimation   When a full Bayesian solution is impractical, we can resort to using *maximum a posteriori (MAP) estimation*. Here, we search for parameters that maximize the *posterior probability*:

$$\tilde{\boldsymbol{\theta}} = \arg\max_{\boldsymbol{\theta}} \log P(\boldsymbol{\theta} \mid \mathcal{D}).$$

When we have a large amount of data, the posterior is often sharply peaked around its maximum $\tilde{\boldsymbol{\theta}}$. In this case, the integral

$$P(X[M+1] \mid \mathcal{D}) = \int P(X[M+1] \mid \boldsymbol{\theta})P(\boldsymbol{\theta} \mid \mathcal{D})d\boldsymbol{\theta}$$

will be roughly $P(X[M+1] \mid \tilde{\boldsymbol{\theta}})$. More generally, we can view the MAP estimate as a way of regularization   using the prior to provide *regularization* over the likelihood function:

$$\arg\max_{\boldsymbol{\theta}} \log P(\boldsymbol{\theta} \mid \mathcal{D}) = \arg\max_{\boldsymbol{\theta}} \log \left(\frac{P(\boldsymbol{\theta})P(\mathcal{D} \mid \boldsymbol{\theta})}{P(\mathcal{D})} \right)$$

$$= \arg\max_{\boldsymbol{\theta}} \left(\log P(\boldsymbol{\theta}) + \log P(\mathcal{D} \mid \boldsymbol{\theta}) \right). \tag{17.14}$$

That is, $\tilde{\boldsymbol{\theta}}$ is the maximum of a function that sums together the log-likelihood function and $\log P(\boldsymbol{\theta})$. This latter term takes into account the prior on different parameters and therefore biases the parameter estimate away from undesirable parameter values (such as those involving conditional probabilities of 0) when we have few learning instances. When the number of samples is large, the effect of the prior becomes negligible, since $\ell(\boldsymbol{\theta} : \mathcal{D})$ grows linearly with the number of samples whereas the prior does not change.

Because our parameter priors are generally well behaved, MAP estimation is often no harder than maximum likelihood estimation, and is therefore often applicable in practice, even in cases where Bayesian estimation is not. Importantly, however, it does not offer all of the same benefits as a full Bayesian estimation. In particular, it does not attempt to represent the shape of the posterior and thus does not differentiate between a flat posterior and a sharply peaked one. As such, it does not give us a sense of our confidence in different aspects of the parameters, and the predictions do not average over our uncertainty. This approach also suffers from issues regarding representation independence; see box 17.D.

---

representation
independence

Box 17.D — Concept: Representation Independence. *One important property we may want of an estimator is* representation independence. *To understand this concept better, suppose that in our thumbtack example, we choose to use a parameter η, so that $P'(X = H \mid \eta) = \frac{1}{1+e^{-\eta}}$. We have that $\eta = \log \frac{\theta}{1-\theta}$ where θ is the parameter we used earlier. Thus, there is a one-to-one correspondence between a choice θ and a choice η. Although one choice of parameters might seem more natural to us than another, there is no formal reason why we should prefer one over the other, since both can represent exactly the same set of distributions.*

More generally, a reparameterization of a given family is a new set of parameter values η in a space Υ and a mapping from the new parameters to the original one, that is, from η to $\boldsymbol{\theta}(\eta)$ so that $P(\cdot \mid \eta)$ in the new parameterization is equal to $P(\cdot \mid \boldsymbol{\theta}(\eta))$ in the original parameterization. In addition, we require that the reparameterization maintain the same set of distributions, that is, for each choice of $\boldsymbol{\theta}$ there is η such that $P(\cdot \mid \eta) = P(\cdot \mid \boldsymbol{\theta})$.

This concept immediately raises the question as to whether the choice of representation can impact our estimates. While we might prefer a particular way of parameterization because it is more intuitive or interpretable, we may not want this choice to bias our estimated parameters.

Fortunately, it is not difficult to see that maximum likelihood estimation is insensitive to reparameterization. If we have two different ways to represent the same family of the distribution, then the distributions in the family that maximize the likelihood using one parameterization also maximize the likelihood with the other parameterization. More precisely, if $\hat{\eta}$ is MLE, then the matching parameter values $\boldsymbol{\theta}(\hat{\eta})$ are also MLE when we consider the likelihood function in the $\boldsymbol{\theta}$ space. This property is a direct consequence of the fact that the likelihood function is a function of the distribution induced by the parameter values, and not of the actual parameter values.

The situation with Bayesian inference is subtler. Here, instead of identifying the maximum parameter value, we now perform integration over all possible parameter values. Naively, it seems that such an estimation is more sensitive to the parameterization than MLE, which depends only *on the maximum of the likelihood surface. However, a careful choice of prior can account for the representation change and thereby lead to representation independence. Intuitively, if we consider a reparameterization η with a function $\boldsymbol{\theta}(\eta)$ mapping to the original parameter space, then we would like the prior on η to maintain the probability of events. That is,*

$$P(A) = P(\{\boldsymbol{\theta}(\eta) : \eta \in A\}), \ \forall A \subset \Upsilon. \tag{17.15}$$

This constraint implies that the prior over different regions of parameters is maintained. Under this assumption, Bayesian prediction will be identical under the two parameterizations.

We illustrate the notion of a reparameterized prior in the context of a Bernoulli distribution:

Example 17.9

Consider a Beta prior over the parameter θ of a Bernoulli distribution:

$$P(\theta : \alpha_0, \alpha_1) = c\theta^{\alpha_1 - 1}(1 - \theta)^{\alpha_0 - 1}, \tag{17.16}$$

where c is the normalizing constant described in definition 17.3. Recall (example 8.5) that the natural parameter for a Bernoulli distribution is

$$\eta = \log \frac{\theta}{1 - \theta}$$

with the transformation

$$\theta = \frac{1}{1 + e^{-\eta}}, \quad 1 - \theta = \frac{1}{1 + e^{\eta}}.$$

What is the prior distribution on η? To preserve the probability of events, we want to make sure that for every interval $[a, b]$

$$\int_a^b c\theta^{\alpha_1 - 1}(1 - \theta)^{\alpha_0 - 1}d\theta = \int_{\log \frac{a}{1-a}}^{\log \frac{b}{1-b}} P(\eta)d\eta.$$

To do so, we need to perform a change of variables. Using the relation between η and θ, we get

$$d\eta = \frac{1}{\theta(1 - \theta)}d\theta.$$

Plugging this into the equation, we can verify that an appropriate prior is:

$$P(\eta) = c \left(\frac{1}{1 + e^{-\eta}} \right)^{\alpha_1} \left(\frac{1}{1 + e^{\eta}} \right)^{\alpha_0},$$

where c is the same constant as before. This means that the prior on η, when stated in terms of θ, is $\theta^{\alpha_1}(1 - \theta)^{\alpha_0}$, in contrast to equation (17.16). At first this discrepancy seems like a contradiction. However, we have to remember that the transformation from θ to η takes the region $[0, 1]$ and stretches it to the whole real line. Thus, the matching prior cannot be uniform. ∎

This example demonstrates that a uniform prior, which we consider to be unbiased or uninformative, can seem very different when we consider a different parameterization.

Thus, both MLE and Bayesian estimation (when carefully executed) are representation-independent. This property, unfortunately, does not carry through to MAP estimation. Here we are not interested in the integral over all parameters, but rather in the density of the prior at different values of the parameters. This quantity does change when we reparameterize the prior.

Example 17.10   *Consider the setting of example 17.9 and develop the MAP parameters for the priors we considered there. When we use the θ parameterization, we can check that*

$$\tilde{\boldsymbol{\theta}} = \arg\max_{\theta} \log P(\theta) = \frac{\alpha_1 - 1}{\alpha_0 + \alpha_1 - 2}.$$

On the other hand,

$$\tilde{\eta} = \arg\max_{\eta} \log P(\eta) = \log \frac{\alpha_1}{\alpha_0}.$$

To compare the two, we can transform $\tilde{\eta}$ to θ representation and find

$$\boldsymbol{\theta}(\tilde{\eta}) = \frac{\alpha_1}{\alpha_0 + \alpha_1}.$$

In other words, the MAP of the η parameterization gives the same predictions as the mean parameterization if we do the full Bayesian inference.   ∎

☞   **Thus, MAP estimation is more sensitive to choices in formalizing the likelihood and the prior than MLE or full Bayesian inference. This suggests that the MAP parameters involve, to some extent, an arbitrary choice.** *Indeed, we can bias the MAP toward different solutions if we construct a specific reparameterization where the density is particularly large in specific regions of the parameter space. The parameterization dependency of MAP is a serious caveat we should be aware of.*

17.5   Learning Models with Shared Parameters

In the preceding discussion, we focused on parameter estimation for Bayesian networks with table-CPDs. In this discussion, we made the strong assumption that the parameters for each conditional distribution $P(X_i \mid \boldsymbol{u}_i)$ can be estimated separately from parameters of other conditional distributions. In the Bayesian case, we also assumed that the priors on these distributions are independent. This assumption is a very strong one, which often does not

shared
parameters

hold in practice. In real-life systems, we often have *shared parameters*: parameters that occur in multiple places across the network. In this section, we discuss how to perform parameter estimation in networks where the same parameters are used multiple times.

Analogously to our discussion of parameter estimation, we can exploit both global and local structure. Global structure occurs when the same CPD is used across multiple variables in the network. This type of sharing arises naturally from the template-based models of chapter 6. Local structure is finer-grained, allowing parameters to be shared even within a single CPD; it arises naturally in some types of structured CPDs. We discuss each of these scenarios in turn, focusing on the simple case of MLE.

We then discuss the issues arising when we want to use Bayesian estimation. Finally, we discuss the *hierarchical Bayes* framework, a "softer" version of parameter sharing, where parameters are encouraged to be similar but do not have to be identical.

17.5.1 Global Parameter Sharing

Let us begin with a motivating example.

Example 17.11

Let us return to our student, who is now taking two classes c_1, c_2, each of which is associated with a Difficulty variable, D_1, D_2. We assume that the grade G_i of our student in class c_i depends on his intelligence and the class difficulty. Thus, we model G_i as having I and D_i as parents. Moreover, we might assume that these grades share the same conditional distribution. That is, the probability that an intelligent student receives an "A" in an easy class is the same regardless of the identity of the particular class. Stated differently, we assume that the difficulty variable summarizes all the relevant information about the challenge the class presents to the student.

How do we formalize this assumption? A straightforward solution is to require that for all choices of grade g, difficulty d and intelligence i, we have that

$$P(G_1 = g \mid D_1 = d, I = i) = P(G_2 = g \mid D_2 = d, I = i).$$

Importantly, this assumption does not imply that the grades are necessarily the same, but rather that the probability of getting a particular grade is the same if the class has the same difficulty. ∎

This example is simply an instance of a network induced by the simple plate model described in example 6.11 (using D_i to encode $D(c_i)$ and similarly for G_i). Thus, as expected, template models give rise to shared parameters.

17.5.1.1 Likelihood Function with Global Shared Parameters

As usual, the key to parameter estimation lies in understanding the structure of the likelihood function. To analyze this structure, we begin with some notation. Consider a network structure $\mathcal{G}$ over a set of variables $\mathcal{X} = \{X_1, \ldots, X_n\}$, parameterized by a set of parameters $\boldsymbol{\theta}$. Each variable X_i is associated with a CPD $P(X_i \mid \boldsymbol{U}_i, \boldsymbol{\theta})$. Now, rather than assume that each such CPD has its own parameterization $\boldsymbol{\theta}_{X_i \mid \boldsymbol{U}_i}$, we assume that we have a certain set of shared parameters that are used by multiple variables in the network. Thus, the sharing of parameters is *global*, over the entire network.

global parameter sharing

More precisely, we assume that $\boldsymbol{\theta}$ is partitioned into disjoint subsets $\boldsymbol{\theta}^1, \ldots, \boldsymbol{\theta}^K$; with each such subset, we associate a set of variables $\mathcal{V}^k \subset \mathcal{X}$, such that $\mathcal{V}^1, \ldots, \mathcal{V}^K$ is a disjoint partition of $\mathcal{X}$. For $X_i \in \mathcal{V}^k$, we assume that the CPD of X_i depends only on $\boldsymbol{\theta}^k$; that is,

$$P(X_i \mid \boldsymbol{U}_i, \boldsymbol{\theta}) = P(X_i \mid \boldsymbol{U}_i, \boldsymbol{\theta}^k). \tag{17.17}$$

Moreover, we assume that the form of the CPD is the same for all $X_i, X_j \in \mathcal{V}^k$; that is,

$$P(X_i \mid \boldsymbol{U}_i, \boldsymbol{\theta}^k) = P(X_j \mid \boldsymbol{U}_j, \boldsymbol{\theta}^k). \tag{17.18}$$

We note that this last statement makes sense only if $Val(X_i) = Val(X_j)$ and $Val(\boldsymbol{U}_i) = Val(\boldsymbol{U}_j)$. To avoid ambiguous notation, for any variable $X_i \in \mathcal{V}^k$, we use y_k^l to range over possible values of X_i and $\boldsymbol{w}_k^l$ to range over the possible values of its parents.

Consider the decomposition of the probability distribution in this case:

$$
\begin{aligned}
P(X_1, \ldots, X_n \mid \boldsymbol{\theta}) &= \prod_{i=1}^{n} P(X_i \mid \mathrm{Pa}_{X_i}, \boldsymbol{\theta}) \\
&= \prod_{k=1}^{K} \prod_{X_i \in \mathcal{V}^k} P(X_i \mid \mathrm{Pa}_{X_i}, \boldsymbol{\theta}) \\
&= \prod_{k=1}^{K} \prod_{X_i \in \mathcal{V}^k} P(X_i \mid \mathrm{Pa}_{X_i}, \boldsymbol{\theta}^k),
\end{aligned}
$$

where the second equality follows from the fact that $\mathcal{V}^1, \ldots, \mathcal{V}^K$ defines a partition of $\mathcal{X}$, and the third equality follows from equation (17.17).

Now, let $\mathcal{D}$ be some assignment of values to the variables $X_1, \ldots, X_n$; our analysis can easily handle multiple IID instances, as in our earlier discussion, but this extension only clutters the notation. We can now write

$$
L(\boldsymbol{\theta} : \mathcal{D}) = \prod_{k=1}^{K} \prod_{X_i \in \mathcal{V}^k} P(x_i \mid \boldsymbol{u}_i, \boldsymbol{\theta}^k).
$$

This expression is identical to the one we used in section 17.2.2 for the case of IID instances. There, for each set of parameters, we had multiple instances $\{(x_i[m], \boldsymbol{u}_i[m])\}_{m=1}^{M}$, all of which were generated from the same conditional distribution. Here, we have multiple instances $\{(x_i, \boldsymbol{u}_i)\}_{X_i \in \mathcal{V}^k}$, all of which are also generated from the same conditional distribution. Thus, it appears that we can use the same analysis as we did there.

To provide a formal derivation, consider first the case of table-CPDs. Here, our parameterization is a set of multinomial parameters $\theta_{y_k \mid \boldsymbol{w}_k}^k$, where we recall that y_k ranges over the possible values of each of the variables $X_i \in \mathcal{V}^k$ and $\boldsymbol{w}_k$ over the possible value assignments to its parents. Using the same derivation as in section 17.2.2, we can now write:

$$
\begin{aligned}
L(\boldsymbol{\theta} : \mathcal{D}) &= \prod_{k=1}^{K} \prod_{y_k, \boldsymbol{w}_k} \prod_{\substack{X_i \in \mathcal{V}^k \; : \\ x_i = y_k, \boldsymbol{u}_i = \boldsymbol{w}_k}} \theta_{y_k \mid \boldsymbol{w}_k}^k \\
&= \prod_{k=1}^{K} \prod_{y_k, \boldsymbol{w}_k} (\theta_{y_k \mid \boldsymbol{w}_k}^k)^{\check{M}_k[y_k, \boldsymbol{w}_k]},
\end{aligned}
$$

where we now have a new definition of our counts:

$$
\check{M}_k[y_k, \boldsymbol{w}_k] = \sum_{X_i \in \mathcal{V}^k} \boldsymbol{I}\{x_i = y_k, \boldsymbol{u}_i = \boldsymbol{w}_k\}.
$$

aggregate
sufficient
statistics

In other words, we now use *aggregate sufficient statistics*, which combine sufficient statistics from multiple variables across the same network.

Given this formulation of the likelihood, we can now obtain the maximum likelihood solution for each set of shared parameters to get the estimate

$$\hat{\theta}^k_{y_k|\boldsymbol{w}_k} = \frac{\breve{M}_k[y_k, \boldsymbol{w}_k]}{\breve{M}_k[\boldsymbol{w}_k]}.$$

Thus, we use the same estimate as in the case of independent parameters, using our aggregate sufficient statistics. Note that, in cases where our variables X_i have no parents, $\boldsymbol{w}_k$ is the empty tuple ε. In this case, $\breve{M}_k[\varepsilon]$ is the number of variables X_i in $\mathcal{V}^k$.

linear
exponential
family

This aggregation of sufficient statistics applies not only to multinomial distributions. Indeed, for any distribution in the *linear exponential family*, we can perform precisely the same aggregation of sufficient statistics over the variables in $\mathcal{V}^k$. The result is a likelihood function in the same form as we had before, but written in terms of the aggregate sufficient statistics rather than the sufficient statistics for the individual variables. We can then perform precisely the same maximum likelihood estimation process and obtain the same form for the MLE, but using the aggregate sufficient statistics. (See exercise 17.14 for another simple example.)

Does this aggregation of sufficient statistics make sense? Returning to our example, if we treat the grade of the student in each class as independent sample from the same parameters, then each data instance provides us with two independent samples from this distribution. It is important to clarify that, although the grades of the student are dependent on his intelligence, the samples are independent samples from the same distribution. More precisely, if $D_1 = D_2$, then both G_1 and G_2 are governed by the same multinomial distribution, and the student's grades are two independent samples from this distribution.

Thus, when we share parameters, multiple observations from within the same network contribute to the same sufficient statistic, and thereby help estimate the same parameter. Reducing the number of parameters allows us to obtain parameter estimates that are less noisy and closer to the actual generating parameters. This benefit comes at a price, since it requires us to make an assumption about the domain. If the two distributions with shared parameters are actually different, the estimated parameters will be a (weighted) average of the estimate we would have had for each of them separately. When we have a small number of instances, that approximation may still be beneficial, since each of the separate estimates may be far from its generating parameters, owing to sample noise. When we have more data, however, the shared parameters estimate will be worse than the individual ones. We return to this issue in section 17.5.4, where we provide a solution that allows us to gradually move away from the shared parameter assumption as we get more data.

17.5.1.2 Parameter Estimation for Template-Based Models

As we mentioned, the template models of chapter 6 are specifically designed to encode global parameter sharing. Recall that these representations involve a set of template-level parameters, each of which is used multiple times when the ground network is defined. For the purpose of our discussion, we focus mostly on plate models, since they are the simplest of the (directed) template-based representations and serve to illustrate the main points.

As we discussed, it is customary in many plate models to explicitly encode the parameter sharing by including, in the model, random variables that encode the model parameters. This approach allows us to make clear the exact structure of the parameter sharing within the model.

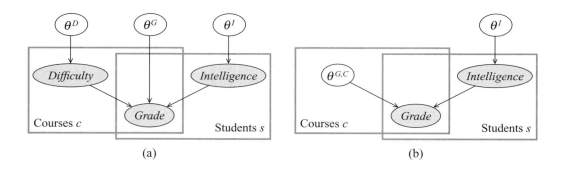

Figure 17.9   Two plate models for the University **example, with explicit parameter variables.** (a) Model where all parameters are global. (b) Model where the difficulty is a course-specific parameter rather than a discrete random variable.

As we mentioned, when parameters are global and shared across over all ground variables derived from a particular template variables, we may choose (purely as a notational convenience) not to include the parameters explicitly in the model. We begin with this simple setting, and then we extend it to the more general case.

Example 17.12

Figure 17.9a is a representation of the plate model of example 6.11, except that we now explicitly encode the parameters as variables within the model. In this representation, we have made it clear that there is only a single parameter $\theta^G_{G|I,D}$, which is the parent of the variables within the plates. Thus, as we can see, the same parameters are used in every CPD $P(G(s,c) \mid I(s), D(c))$ in the ground network. ∎

When all of our parameters are global, the sharing structure is very simple. Let $A(U_1, \ldots, U_k)$ be any attribute in the set of template attributes $\aleph$. Recall from definition 6.10 that this attribute induces a ground random variable $A(\gamma)$ for any assignment $\gamma = \langle U_1 \mapsto u_1, \ldots, U_k \mapsto u_k \rangle \in \Gamma_\kappa[A]$. All of these variables share the same CPD, and hence the same parameters. Let $\boldsymbol{\theta}_A$ be the parameters for the CPD for A in the template-level model. We can now simply define $\mathcal{V}^A$ to be all of the variables of the form $A(\gamma)$ in the ground network. The analysis of section 17.5.1.1 now applies unchanged.

Example 17.13

Continuing example 17.12, the likelihood function for an assignment ξ to a ground network in the University *domain would have the form*

$$\prod_i (\theta^I_i)^{\breve{M}_I[i]} \prod_d (\theta^D_d)^{\breve{M}_D[d]} \prod_{g,i,d} (\theta^G_{g|i,d})^{\breve{M}_G[g,i,d]}.$$

Importantly, the counts are computed as aggregate sufficient statistics, each with its own appropriate set of variables. In particular,

$$\breve{M}_I[i] = \sum_{s \in \mathcal{O}^\kappa [\text{Student}]} \boldsymbol{I}\{I(s) = i\},$$

whereas

$$\check{M}_G[g,i,d] = \sum_{\langle s,c \rangle \in \Gamma_\kappa[Grade]} \boldsymbol{I}\{I(s) = i, D(c) = d, G(s,c) = g\}.$$

For example, the counts for g^1, i^1, d^1 would be the number of all of the student,course pairs s, c such that student s has high intelligence, course c has high difficulty, and the grade of s in c is an A. The MLE for $\theta^G_{g^1 | i^1, d^1}$ is the fraction of those students among all (student,course) pairs. ∎

To provide a concrete formula in the more general case, we focus on table-CPDs. We can now define, for any attribute $A(\mathcal{X})$ with parents $B_1(\boldsymbol{U}_1), \ldots, B_k(\boldsymbol{U}_k)$, the following aggregate sufficient statistics:

$$\check{M}_A[a, b_1, \ldots, b_k] = \sum_{\gamma \in \Gamma_\kappa[A]} \boldsymbol{I}\{A(\gamma) = a, B_1(\gamma[\boldsymbol{U}_1]) = b_1, \ldots, B_k(\gamma[\boldsymbol{U}_k]) = b_k\}, \quad (17.19)$$

where $\gamma[\boldsymbol{U}_j]$ denotes the subtuple of the assignment to $\boldsymbol{U}$ that corresponds to the logical variables in $\boldsymbol{U}_j$. We can now define the template-model log-likelihood for a given skeleton:

$$\ell(\boldsymbol{\theta} : \kappa) = \sum_{A \in \aleph} \sum_{a \in Val(A)} \sum_{\boldsymbol{b} \in Val(\mathrm{Pa}_A)} \check{M}_A[a, \boldsymbol{b}] \log \theta_{A=a, \mathrm{Pa}_A = \boldsymbol{b}}. \quad (17.20)$$

From this formula, the maximum likelihood estimate for each of the model parameters follows easily, using precisely the same analysis as before.

In our derivation so far, we focused on the setting where all of the model parameters are global. However, as we discussed, the plate representation can also encompass more local parameterizations. Fortunately, from a learning perspective, we can reduce this case to the previous one.

Example 17.14

Figure 17.9b represents the setting where each course has its own model for how the course difficulty affects the students' grades. That is, we now have a set of parameters $\boldsymbol{\theta}^{G,c}$, which are used in the CPDs of all of the ground variables $G(c, s)$ for different values of s, but there is no sharing between $G(c, s)$ and $G(c', s)$.

As we discussed, this setting is equivalent to one where we add the course ID c as a parent to the D variable, forcing us to introduce a separate set of parameters for every assignment to c. In this case, the dependence on the specific course ID subsumes the dependence on the difficulty parameter D, which we have (for clarity) dropped from the model. From this perspective, the parameter estimation task can now be handled in exactly the same way as before for the parameters in the (much larger) CPD for G. In effect, this transformation converts global sharing to local sharing, which we will handle. ∎

There is, however, one important subtlety in scenarios such as this one. Recall that, in general, different skeletons will contain different objects. Parameters that are specific to objects in the model do not transfer from one skeleton to another. Thus, we cannot simply transfer the object-specific parameters learned from one skeleton to another. This limitation is important, since a major benefit of the template-based formalisms is the ability to learn models in one instantiation of the template and use it in other instantiations. Nevertheless, the learned models are still useful in many ways. First, the learned model itself often provides significant insight about the objects in the training data. For example, the LDA model of box 17.E tells us, for each of the documents in our training corpus, what the mix of topics in that particular document is. Second, the model parameters that are not object specific can be transferred to other skeletons. For example, the word multinomials associated with different topics that are learned from one document collection can also be used in another, leaving only the new document-specific parameters to be inferred.

Although we have focused our formal presentation on plate models, the same analysis also applies to DBNs, PRMs, and a variety of other template-based languages that share parameters. See exercise 17.16 and exercise 17.17 for two examples.

17.5.2   Local Parameter Sharing

local parameter
sharing

In the first part of this section, we focused on cases where all of the parameter sharing occurred between different CPDs. However, we might also have shared parameters that are shared *locally*, within a single CPD.

Example 17.15

Consider the CPD of figure 5.4 where we model the probability of the student getting a job based on his application, recommendation letter, and SAT scores. As we discussed there, if the student did not formally apply for a position, then the recruiting company does not have access to the recommendation letter or SAT scores. Thus, for example, the conditional probability distribution $P(J \mid a^0, s^0, l^0)$ is equal to $P(J \mid a^0, s^1, l^1)$. ∎

In fact, the representations we considered in section 5.3 can be viewed as encoding parameter sharing for different conditional distributions. That is, each of these representations (for example, tree-CPDs) is a language to specify which of the conditional distributions within a CPD are equal to each other. As we saw, these equality constraints had implications in terms of the independence statements encoded by a model and can also in some cases be exploited in inference. (We note that not all forms of local structure can be reduced to a simple set of equality constraints on conditional distributions within a CPD. For example, noisy-or CPDs or generalized linear models combine their parameters in very different ways and require very different techniques than the ones we discuss in this section.)

Here, we focus on the setting where the CPD defines a set of multinomial distributions, but some of these distributions are shared across multiple contexts. In particular, we assume that our graph $\mathcal{G}$ now defines a set of multinomial distributions that make up CPDs for $\mathcal{G}$: for each variable X_i and each $\boldsymbol{u}_i \in Val(\boldsymbol{U}_i)$ we have a multinomial distribution. We can use the tuple $\langle X_i, \boldsymbol{u}_i \rangle$ to designate this multinomial distribution, and define

$$\mathcal{D} = \cup_{i=1}^n \{ \langle X_i, \boldsymbol{u}_i \rangle : \boldsymbol{u}_i \in Val(\boldsymbol{U}_i) \}$$

to be the set containing all the multinomial distributions in $\mathcal{G}$. We can now define a set of shared parameters $\boldsymbol{\theta}^k$, $k = 1, \ldots, K$, where each $\boldsymbol{\theta}^k$ is associated with a set $\mathcal{D}^k \subseteq \mathcal{D}$ of multinomial distributions. As before, we assume that $\mathcal{D}^1, \ldots, \mathcal{D}^K$ defines a disjoint partition of $\mathcal{D}$. We assume that all conditional distributions within $\mathcal{D}^k$ share the same parameters $\boldsymbol{\theta}^k$. Thus, we have that if $\langle X_i, \boldsymbol{u}_i \rangle \in \mathcal{D}^k$, then

$$P(x_i^j \mid \boldsymbol{u}_i) = \theta_j^k.$$

For this constraint to be coherent, we require that all the multinomial distributions within the same partition have the same set of values: for any $\langle X_i, \boldsymbol{u}_i \rangle, \langle X_j, \boldsymbol{u}_j \rangle \in \mathcal{D}^k$, we have that $Val(X_i) = Val(X_j)$.

Clearly, the case where no parameter is shared can be represented by the trivial partition into singleton sets. However, we can also define more interesting partitions.

Example 17.16

To capture the tree-CPD of figure 5.4, we would define the following partition:

$$
\begin{aligned}
\mathcal{D}^{a^0} &= \{\langle J, (a^0, s^0, l^0) \rangle, \langle J, (a^0, s^0, l^1) \rangle, \langle J, (a^0, s^1, l^0) \rangle, \langle J, (a^0, s^1, l^1) \rangle\} \\
\mathcal{D}^{a^1, s^0, l^0} &= \{\langle J, (a^1, s^0, l^0) \rangle\} \\
\mathcal{D}^{a^1, s^0, l^1} &= \{\langle J, (a^1, s^0, l^1) \rangle\} \\
\mathcal{D}^{a^1, s^1} &= \{\langle J, (a^1, s^1, l^0) \rangle, \langle J, (a^1, s^1, l^1) \rangle\}.
\end{aligned}
$$
∎

More generally, this partition-based model can capture local structure in both tree-CPDs and in rule-based CPDs. In fact, when the network is composed of multinomial CPDs, this finer-grained sharing also generalizes the sharing structure in the global partition models of section 17.5.1.

We can now reformulate the likelihood function in terms of the shared parameters $\boldsymbol{\theta}^1, \ldots, \boldsymbol{\theta}^K$. Recall that we can write

$$
P(\mathcal{D} \mid \boldsymbol{\theta}) = \prod_i \prod_{\boldsymbol{u}_i} \left[\prod_{x_i} P(x_i \mid \boldsymbol{u}_i, \boldsymbol{\theta})^{M[x_i, \boldsymbol{u}_i]} \right].
$$

Each of the terms in square brackets is the local likelihood of a single multinomial distribution. We now rewrite each of the terms in the innermost product in terms of the shared parameters, and we aggregate them according to the partition:

$$
\begin{aligned}
P(\mathcal{D} \mid \boldsymbol{\theta}) &= \prod_i \prod_{\boldsymbol{u}_i} \left[\prod_{x_i} P(x_i \mid \boldsymbol{u}_i, \boldsymbol{\theta})^{M[x_i, \boldsymbol{u}_i]} \right] \\
&= \prod_k \prod_{\langle X_i, \boldsymbol{u}_i \rangle \in \mathcal{D}^k} \prod_j (\theta_j^k)^{M[x_i^j, \boldsymbol{u}_i]} \\
&= \prod_k \left[\prod_j (\theta_j^k)^{\sum_{\langle X_i, \boldsymbol{u}_i \rangle \in \mathcal{D}^k} M[x_i^j, \boldsymbol{u}_i]} \right].
\end{aligned}
\tag{17.21}
$$

This final expression is reminiscent of the likelihood in the case of independent parameters, except that now each of the terms in the square brackets involves the shared parameters. Once again, we can define a notion of aggregate sufficient statistics:

$$
\check{M}_k[j] = \sum_{\langle X_i, \boldsymbol{u}_i \rangle \in \mathcal{D}^k} M[x_i^j, \boldsymbol{u}_i],
$$

and use those aggregate sufficient statistics for parameter estimation, exactly as we used the unaggregated sufficient statistics before.

17.5.3   Bayesian Inference with Shared Parameters

To perform Bayesian estimation, we need to put a prior on the parameters. In the case without parameter sharing, we had a separate (independent) prior for each parameter. This model is clearly in violation of the assumptions made by parameter sharing. If two parameters are shared, we want them to be identical, and thus it is inconsistent to assume they have independent prior. The right approach is to place a prior on the shared parameters.

Consider, in particular, the local analysis of section 17.5.2, we would place a prior on each of the multinomial parameters $\boldsymbol{\theta}^k$. As before, it is very convenient to assume that each of these set of parameters are independent from each other. This assumption corresponds to the local parameter independence we made earlier, but applied in the context where we force the given parameter-sharing strategy. We can use a similar idea in the global analysis of section 17.5.1, introducing a prior over each set of parameters θ^k. If we impose an independence assumption for the priors of the different sets, we obtain a shared-parameter version of the global parameter independence assumption.

One important subtlety relates to the choice of the prior. Given a model with shared parameters, the analysis of section 17.4.3 no longer applies directly. See exercise 17.13 for one possible extension.

As usual, if the prior decomposes as a product and the likelihood decomposes as a product along the same lines, then our posterior also decomposes. For example, returning to equation (17.21), we have that:

$$P(\boldsymbol{\theta} \mid \mathcal{D}) \propto \prod_{k=1}^{K} P(\boldsymbol{\theta}^k) \prod_{j} (\theta_j^k)^{\tilde{M}_k[j]}.$$

The actual form of the posterior depends on the prior. Specifically, if we use multinomial distributions with Dirichlet priors, then the posterior will also be a Dirichlet distribution with the appropriate hyperparameters.

This discussion seems to suggest that the line of reasoning we had in the case of independent parameters is applicable to the case of shared parameters. However, there is one subtle point that can be different. Consider the problem of predicting the probability of the next instance, which can be written as:

$$P(\xi[M + 1] \mid \mathcal{D}) = \int P(\xi[M + 1] \mid \boldsymbol{\theta}) P(\boldsymbol{\theta} \mid \mathcal{D}) d\boldsymbol{\theta}.$$

To compute this formula, we argued that, since $P(\xi[M + 1] \mid \boldsymbol{\theta})$ is a product of parameters $\prod_i \theta_{x_i[M+1]|\boldsymbol{u}_i[M+1]}$, and since the posterior of these parameters are independent, then we can write (for multinomial parameters)

$$P(\xi[M + 1] \mid \mathcal{D}) = \prod_{i} \boldsymbol{E}\big[\theta_{x_i[M+1]|\boldsymbol{u}_i[M+1]} \mid \mathcal{D}\big],$$

where each of the expectations is based on the posterior over $\theta_{x_i[M+1]|\boldsymbol{u}_i[M+1]}$.

When we have shared parameters, we have to be more careful. If we consider the network of example 17.11, then when the $(M + 1)$st instance has two courses of the same difficulty, the likelihood term $P(\xi[M+1] \mid \boldsymbol{\theta})$ involves a product of two parameters that are not independent. More explicitly, the likelihood involves $P(G_1[M+1] \mid I[M+1], D_1[M+1])$ and $P(G_2[M+1] \mid$

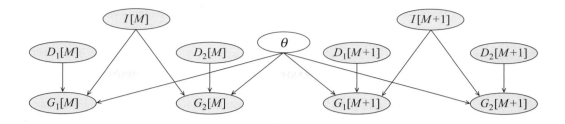

Figure 17.10 **Example meta-network for a model with shared parameters, corresponding to example 17.11.**

$I[M + 1], D_2[M + 1]$); if $D_1[M + 1] = D_2[M + 1]$ then the two parameters are from the same multinomial distribution. Thus, the posterior over these two parameters is not independent, and we cannot write their expectation as a product of expectations.

Another way of understanding this problem is by examining the meta-network for learning in such a situation. The meta-network for Bayesian parameter learning for the network of example 17.11 is shown in figure 17.10. As we can see, in this network $G_1[M + 1]$ is *not* independent of $G_2[M + 1]$ given $I[M + 1]$ because of the trail through the shared parameters. Stated differently, observing $G_1[M + 1]$ will cause us to update the parameters and therefore change our estimate of the probability of $G_2[M + 1]$.

Note that this problem can happen only in particular forms of shared parameters. If the shared parameters are within the same CPD, as in example 17.15, then the $(M + 1)$st instance can involve at most one parameter from each partitions of shared parameters. In such a situation, the problem does not arise and we can use the average of the parameters to compute the probability of the next instance. However, if we have shared parameters across different CPDs (that is, entries in two or more CPDs share parameters), this problem can occur.

How do we solve this problem? The correct Bayesian prediction solution is to compute the average for the product of two (or more) parameters from the same posterior. This is essentially identical to the question of computing the probability of two or more test instances. See exercise 17.18. This solution, however, leads to many complications if we want to use the Bayesian posterior to answer queries about the distribution of the next instance. In particular, this probability no longer factorizes in the form of the original Bayesian network, and thus, we cannot use standard inference procedures to answer queries about future instances. For this reason, a pragmatic approximation is to use the expected parameters for each CPD and ignore the dependencies induced by the shared parameters. When the number of training samples is large, this solution can be quite a good approximation to the true predictive distribution. However, when the number of training examples is small, this assumption can skew the estimate; see exercise 17.18.

17.5.4 Hierarchical Priors ★

In our discussion of Bayesian methods for table-CPDs, we made the strong independence assumption (global and local parameter independence) to decouple the estimation of parameters.

Our discussion of shared parameters relaxed these assumptions by moving to the other end of the spectrum and forcing parameters to be identical. There are many situations where neither solution is appropriate.

Example 17.17

Using our favorite university domain, suppose we learn a model from records of students, classes, teachers, and so on. Now suppose that we have data from several universities. Because of various factors, the data from each university have different properties. These differences can be due to the different population of students in each one (for example, one has more engineering oriented students while the other has more liberal arts students), somewhat different grade scale, or other factors. This leads to a dilemma. We can learn one model over all the data, ignoring the university specific bias. This allows us to pool the data to get a larger and more reliable data set. Alternatively, we can learn a different model for each university, or, equivalently, add a University variable that is the parent of many of the variables in the network. This approach allows us to tailor the parameters to each university. However, this flexibility comes at a price — learning parameters in one university does not help us learn better parameters from the data in the other university. This partitions the data into smaller sets, and in each one we need to learn from scratch that intelligent students tend to get an A in easy classes. ∎

Example 17.18

bigram model

A similar problem might arise when we consider learning dependencies in text domains. As we discussed in box 6.B, a standard model for word sequences is a bigram model, *which views the words as forming a Markov chain, where we have a conditional probability over the next word given the current word. That is, we want to learn $P(W^{(t+1)} \mid W^{(t)})$ where W is a random variable taking values from the dictionary of words. Here again, the context $W^{(t)}$ can definitely change our distribution over $W^{(t+1)}$, but we still want to share some information across these different conditional distributions; for example, the probability of common words, such as "the," should be high in almost all of the conditional distributions we learn.* ∎

In both of these examples, we aim to learn a conditional distribution, say $P(Y \mid X)$. Moreover, we want the different conditional distributions $P(Y \mid x)$ to be similar to each other, yet not identical. Thus, the Bayesian learning problem assuming local independence (figure 17.11a) is not appropriate.

One way to bias the different distributions to be similar to each other is to have the same prior over them. If the prior is very strong, it will bias the different estimates to the same values. In particular, in the domain of example 17.18, we want the prior to bias both distributions toward giving high probability to frequent words. Where do we get such a prior? One simple ad hoc solution is to use the data to set the prior. For example, we can use the frequency of words in training set to construct a prior where more frequent words have a larger hyperparameter. Such an approach ensures that more frequent words have higher posterior in each of the conditional distributions, even if there are few training examples for that particular conditional distribution.

shrinkage

This approach (which is a slightly more Bayesian version of the *shrinkage* of exercise 6.2) works fairly well, and is often used in practice. However, it seems to contradict the whole premise of the Bayesian framework: a prior is a distribution that we formulate over our parameters *prior* to seeing the data. This approach also leaves unaddressed some important questions, such as determining the relative strength of the prior based on the amount of data used to compute it. A more coherent and general approach is to stay within the Bayesian framework, and to

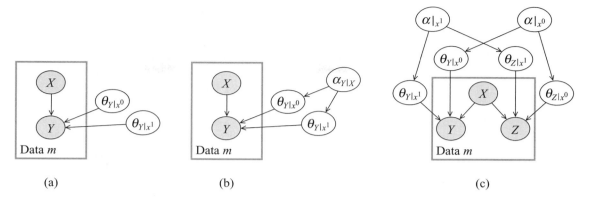

(a)   (b)   (c)

Figure 17.11   Independent and hierarchical priors. (a) A plate model for $P(Y \mid X)$ under assumption of parameter independence. (b) A plate model for a simple hierarchical prior for the same CPD. (c) A plate model for two CPDs $P(Y \mid X)$ and $P(Z \mid X)$ that respond similarly to X.

introduce explicitly our uncertainty about the joint prior as part of the model. Just as we introduced random variables to denote parameters of CPDS, we now take one step further and introduce a random variable to denote the hyperparameters. The resulting model is called a

hierarchical Bayes   *hierarchical Bayesian model.* It uses a factored probabilistic model to describe our prior. This idea, of specifically defining a probabilistic model over our priors, can be used to define rich priors in a broad range of settings. Exercise 17.7 provides another example.

Figure 17.11b shows a simple example, where we have a variable that is the parent of both $\boldsymbol{\theta}_{Y|x^0}$ and $\boldsymbol{\theta}_{Y|x^1}$. As a result, these two parameters are no longer independent in the prior, and consequently in the posterior. Intuitively, the effect of the prior will be to shift both $\boldsymbol{\theta}_{Y|x^0}$ and $\boldsymbol{\theta}_{Y|x^1}$ to be closer to each other. However, as usual when using priors, the effect of the prior diminishes as we get more data. In particular, as we have more data for the different contexts x^1 and x^0, the effect of the prior will gradually decrease. Thus, the hierarchical priors (as for all priors) are particularly useful in the sparse-data regime.

How do we represent the distribution over the hyperparameter $P(\vec{\alpha})$? One option is to create a prior where each component α_y is governed by some distribution, say a Gamma distribution (recall that components are strictly positive). That is,

$$P(\vec{\alpha}) = \prod_y P(\alpha_y),$$

Gamma distribution   where $P(\alpha_y) \sim \text{Gamma}(\mu_y)$ is a *Gamma distribution* with (hyper-)hyperparameter μ_y. The other option is to write $\vec{\alpha}$ as the product of equivalent sample size N_0 with a probability distribution p_0, the first governed by a Gamma distribution and the other by a Dirichlet distribution. (These two representations are actually closely related; see box 19.E.)

Moreover, the same general idea can be used in broader ways. In this example, we used a hierarchical prior to relate two (or more) conditional distributions in the same CPD. We can similarly relax the global parameter independence assumption and introduce dependencies between the parameters for two (or more) CPDs. For example, if we believe that two variables

Y and Z depend on X in a similar (but not identical) way, then we introduce a common prior on $\boldsymbol{\theta}_{Y|x^0}$ and $\boldsymbol{\theta}_{Z|x^0}$, and similarly another common prior for $\boldsymbol{\theta}_{Y|x^1}$ and $\boldsymbol{\theta}_{Z|x^1}$; see figure 17.11c.

The idea of a hierarchical structure can also be extended to additional levels of a hierarchy. For example, in the case of similar CPDs, we might argue that there is a similarity between the distributions of Y and Z given x^0 and x^1. If we believe that this similarity is weaker than the similarity between the distributions of the two variables, we can introduce another hierarchy layer to relate the hyperparameters $\alpha_{\cdot|x^0}$ and $\alpha_{\cdot|x^1}$. To do so, we might introduce hyper-hyperparameters μ that specify a joint prior over $\alpha_{\cdot|x^0}$ and $\alpha_{\cdot|x^1}$.

The notion of hierarchical priors can be readily applied to other types of CPDs. For example, if X is a Gaussian variable without parents, then, as we saw in exercise 17.8, a conjugate prior for the mean of X is simply a Gaussian prior on the mean parameter. This observation suggests that we can easily create a hierarchical prior that relates X and another Gaussian variable Y, by having a common Gaussian over the means of the variables. Since the distribution over the hyperparameter is a Gaussian, we can easily extend the hierarchy upward. Indeed, hierarchical priors over Gaussians are often used to model the dependencies of parameters of related populations. For example, we might have a Gaussian over the SAT score, where one level of the hierarchy corresponds to different classes in the same school, the next one to different schools in the same district, the following to different districts in the same state, and so on.

☞   **The framework of hierarchical priors gives us a flexible language to introduce dependencies in the priors over parameters. Such dependencies are particularly useful when we have small amount of examples relevant to each parameter but many such parameters that we believe are reasonably similar. In such situations, hierarchical priors "spread" the effect of the observations between parameters with shared hyperparameters.**

One question we did not discuss is how to perform learning with hierarchical priors. As with previous discussions of Bayesian learning, we want to compute expectations with respect to the posterior distribution. Since we relaxed the global and local independence assumptions, the posterior distribution no longer decomposes into a product of independent terms. From the perspective of the desired behavior, this lack of independence is precisely what we wanted to achieve. However, it implies that we need to deal with a much harder computational task. In fact, when introducing the hyperparameters as a variable into the model, we transformed our learning problem into one that includes a hidden variable. To address this setting, we therefore need to apply methods for Bayesian learning with hidden variables; see section 19.3 for a discussion of such methods.

---

text classification

bag of words

Box 17.E — Concept: Bag-of-Word Models for Text Classification. *Consider the problem of text classification: classifying a document into one of several categories (or classes). Somewhat surprisingly, some of the most successful techniques for this problem are based on viewing the document as an unordered bag of words, and representing the distribution of this bag in different categories. Most simply, the distribution is encoded using a naive Bayes model. Even in this simple approach, however, there turn out to be design choices that can make important differences to the performance of the model.*

Our first task is to represent a document by a set of random variables. This involves various processing steps. The first removes various characters such as punctuation marks as well as words that are viewed as having no content (such as "the," "and," and so on). In addition, most applica-

tions use a variety of techniques to map words in the document to canonical words in a predefined dictionary $\mathcal{D}$ (for example, replace "apples," "used," and "running" with "apple," "use," and "run" respectively). Once we finish this processing step, there are two common approaches to defining the features that describe the document.

Bernoulli naive
Bayes

In the Bernoulli *naive Bayes model, we define a binary attribute (feature) X_i to denote whether the i'th dictionary word w_i appears in the document. This representation assumes that we care only about the presence of a word, not about how many times it appeared. Moreover, when applying the naive Bayes classifier with this representation we assume that the appearance of one word in the document is independent of the appearance of another (given the document's topic). When learning a naive Bayes classifier with this representation, we learn frequency (over a document) of encountering each dictionary word in documents of specific categories — for example, the probability that the word "ball" appears in a document of category "sports." We learn such a parameter for each pair (dictionary word, category).*

multinomial
naive Bayes

In the multinomial *naive Bayes model, we define attribute to describe the specific sequence of words in the document. The variable X_i denotes which dictionary word appeared the ith word in the document. Thus, each X_i can take on many values, one for each possible word. Here, when we use the naive Bayes classifier, we assume that the choice of word in position i is independent of the choice of word in position j (again, given the document's topic). This model leads to a complication, since the distribution over X_i is over all words in the document, which requires a large number of parameters. Thus, we further assume that the probability that a particular word is used in position i does not depend on i; that is, the probability that $X_i = w$ (given the topic) is the same as the probability that $X_j = w$. In other words, we use parameter sharing between $P(X_i \mid C)$ and $P(X_j \mid C)$. This implies that the total number of parameters is again one for each (dictionary word, category).*

In both models, we might learn that the word "quarterback" is much more likely in documents whose topic is "sports" than in documents whose topic is "economics," the word "bank" is more associated with the latter subjects, and the word "dollar" might appear in both. Nevertheless, the two models give rise to quite different distributions. Most notably, if a word w appears in several different positions in the document, in the Bernoulli model the number of occurrences will be ignored, while in the multinomial model we will multiply the probability $P(w \mid C)$ several times. If this probability is very small in one category, the overall probability of the document given that category will decrease to reflect the number of occurrences. Another difference is in how the document length plays a role here. In the Bernoulli model, each document is described by exactly the same number of variables, while the multinomial model documents of different lengths are associated with a different number of random variables.

The plate model provides a compact and elegant way of making explicit the subtle distinctions between these two models. In both models, we have two different types of objects — documents, and individual words in the documents. Document objects d are associated with the attribute T, representing the document topic. However, the notion of "word objects" is different in the different models.

In the Bernoulli naive Bayes model, our words correspond to words in some dictionary (for example, "cat," "computer," and so on). We then have a binary-valued attribute $A(d, w)$ for each document d and dictionary word w, which takes the value true if the word w appears in the document d. We can model this case using a pair of intersecting plates, one for documents and the other for dictionary words, as shown in figure 17.E.1a.

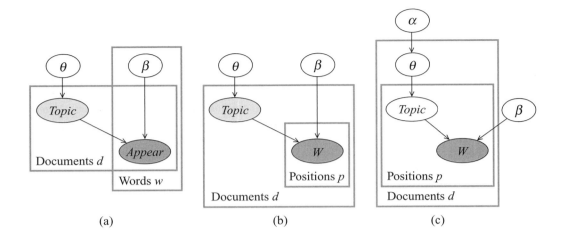

Figure 17.E.1 — Different plate models for text (a) Bernoulli naive Bayes; (b) Multinomial naive Bayes. (c) Latent Dirichlet Allocation.

In the multinomial naive Bayes model, our word objects correspond not to dictionary words, but to word positions P within the document. Thus, we have an attribute W of records representing pairs (D, P), where D is a document and P is a position within it (that is, first word, second word, and so on). This attribute takes on values in the space of dictionary words, so that $W(d, p)$ is the random variable whose value is the actual dictionary word in position p in document d. However, all of these random variables are generated from the same multinomial distribution, which depends on the document topic. The appropriate plate model is shown in figure 17.E.1b.[1]

The plate representation of the two models makes explicit the fact that the Bernoulli parameter $\beta_W[w]$ in the Bernoulli model is different for different words, whereas in the multinomial model, the parameter β_W is the same for all positions within the document. Empirical evidence suggests that the multinomial model is, in general, more successful than the Bernoulli.

In both models, the parameters are estimated from data, and the resulting model used for classifying new documents. The parameters for these models measure the probability of a word given a topic, for example, the probability of "bank" given "economics." For common words, such probabilities can be assessed reasonably well even from a small number of training documents. However, as the number of possible words is enormous, Bayesian parameter estimation is used to avoid overfitting, especially ascribing probability zero to words that do not appear in the training set. With Bayesian estimation, we can learn a naive Bayes model for text from a fairly small corpus, whereas

---

1. Note that, as defined in section 6.4.1.2, a skeleton for a plate model specifies a fixed set of objects for each class. In the multinomial plate model, this assumption implies that we specify a fixed set of word positions, which applies to all documents. In practice, however, documents have different lengths, and so we would want to allow a different set of word positions for each document. Thus, we want the set $\mathcal{O}^\kappa[p]$ to depend on the specific document d. When plates are nested, as they are in this case, we can generalize our notion of skeleton, allowing the set of objects in a nested plate $\mathbf{Q}_1$ to depend on the index of an enclosing plate $\mathbf{Q}_2$.

more realistic models of language are generally much harder to estimate correctly. This ability to define reasonable models with a very small number of parameters, which can be acquired from a small amount of training data, is one of the key advantages of the naive Bayes model.

latent Dirichlet allocation

We can also define much richer representations that capture more fine-grained structure in the distribution. These models are often easily viewed in the plate representation. One such model, shown in figure 17.E.1c, is the latent Dirichlet allocation *(LDA) model, which extends the multinomial naive Bayes model. As in the multinomial naive Bayes model, we have a set of topics associated with a set of multinomial distributions $\boldsymbol{\theta}_W$ over words. However, in the LDA model, we do not assume that an entire document is about a single topic. Rather, we assume that each document d is associated with a continuous mixture of topics, defined using parameters $\boldsymbol{\theta}(d)$. These parameters are selected independently from each document d, from a Dirichlet distribution parameterized by a set of hyperparameters $\boldsymbol{\alpha}$. The word in position p in the document d is then selected by first selecting a topic Topic$(d, p) = t$ from the mixture $\boldsymbol{\theta}(d)$, and then selecting a specific dictionary word from the multinomial $\boldsymbol{\beta}_t$ associated with the chosen topic t. The LDA model generally provides much better results (in measures related to log-likelihood of test data) than other unsupervised clustering approaches to text. In particular, owing to the flexibility of assigning a mixture of topics to a document, there is no problem with words that have low probability relative to a particular topic; thus, this approach largely avoids the overfitting problems with the two naive Bayes models described earlier.*

17.6   Generalization Analysis ⋆

One intuition that permeates our discussion is that more training instances give rise to more accurate parameter estimates. This intuition is supported by our empirical results. In this section, we provide some formal analysis that supports this intuition. This analysis also allows us to quantify the extent to which the error in our estimates decreases as a function of the number of training samples, and increases as a function of the number of parameters we want to learn or the number of variables in our networks.

We begin with studying the asymptotic behavior of our estimator at the large-sample limit. We then provide a more refined analysis that studies the error as a function of the number of samples.

17.6.1   Asymptotic Analysis

We start by considering the asymptotic behavior of the maximum likelihood estimator. In this case, our analysis of section 17.2.5 provides an immediate conclusion: At the large sample limit, $\hat{P}_{\mathcal{D}}$ approaches P^*; thus, as the number of samples grows, $\hat{\boldsymbol{\theta}}$ approaches $\boldsymbol{\theta}^*$ — the projection of P^* onto the parametric family.

A particular case of interest arises when $P^*(\mathcal{X}) = P(\mathcal{X} : \boldsymbol{\theta}^*)$, that is, P^* is representable in the parametric family. Then, we have that $\hat{\boldsymbol{\theta}} \to \boldsymbol{\theta}^*$ as $M \to \infty$. An estimator with this property is called *consistent estimator*. In general, maximum likelihood estimators are consistent.

consistent estimator

We can make this analysis even more precise. Using equation (17.9), we can write

$$\log \frac{P(\mathcal{D} \mid \hat{\boldsymbol{\theta}})}{P(\mathcal{D} \mid \boldsymbol{\theta})} = M \left(\boldsymbol{D}(\hat{P}_{\mathcal{D}} \| P_{\boldsymbol{\theta}}) - \boldsymbol{D}(\hat{P}_{\mathcal{D}} \| P_{\hat{\theta}}) \right).$$

This equality implies that the likelihood function is sharply peaked: the decrease in the likelihood for parameters that are not the MLE is exponential in M. Of course, when we change M the data set $\mathcal{D}$ and hence the distribution $\hat{P}_{\mathcal{D}}$ also change, and thus this result does not guarantee exponential decay in M. However, for sufficiently large M, $\hat{P}_{\mathcal{D}} \to P^*$. Thus, the difference in log-likelihood of different choices of $\boldsymbol{\theta}$ is roughly M times their distance from P^*:

$$\log \frac{P(\mathcal{D} \mid \hat{\boldsymbol{\theta}})}{P(\mathcal{D} \mid \boldsymbol{\theta})} \approx M \left(\boldsymbol{D}(P^* \| P_{\boldsymbol{\theta}}) - \boldsymbol{D}(P^* \| P_{\hat{\theta}}) \right).$$

The terms on the right depend only on M and not on $\hat{P}_{\mathcal{D}}$. Thus, we conclude that for large values of M, the likelihood function approaches a delta function, for which all values are virtually 0 when compared to the maximal value at $\boldsymbol{\theta}^*$, the M-projection of P^*.

To summarize, this argument basically allows us to prove the following result, asserting that the Bayesian estimator is *consistent*:

Theorem 17.2

Let P^* be the generating distribution, let $P(\cdot \mid \theta)$ be a parametric family of distributions, and let $\theta^* = \arg\min_\theta \boldsymbol{D}(P^* \| P(\cdot \mid \theta))$ be the M-projection of P^* on this family. Then

$$\lim_{M \to \infty} P(\cdot \mid \hat{\theta}) = P(\cdot \mid \theta^*)$$

almost surely.

That is, when M grows larger, the estimated parameters describe a distribution that is close to the distribution with the "optimal" parameters in our parametric family.

Is our Bayesian estimator consistent? Recall that equation (17.11) shows that the Bayesian estimate with a Dirichlet prior is an interpolation between the MLE and the prior prediction. The interpolation weight depends on the number of samples M: as M grows, the weight of the prior prediction diminishes and disappears in the limit. Thus, we can conclude that Bayesian learning with Dirichlet priors is also consistent.

17.6.2   PAC-Bounds

This consistency result guarantees that, at the large sample limit, our estimate converges to the true distribution. Though a satisfying result, its practical significance is limited, since in most cases we do not have access to an unlimited number of samples. Hence, it is also important to evaluate the quality of our learned model as a function of the number of samples M. This type of analysis allows us to know how much to trust our learned model as a function of M; or, from the other side, how many samples we need to acquire in order to obtain results of a given quality. Thus, using relative entropy to the true distribution as our notion of solution quality, PAC-bound we use *PAC-bound* analysis (as in box 16.B) to bound $\boldsymbol{D}(P^* \| \hat{P})$ as a function of the number of data samples M.

17.6.2.1   Estimating a Multinomial

We start with the simplest case, which forms the basis for our more extensive analysis. Here, our task is to estimate the multinomial parameters governing the distribution of a single random variable. This task is relevant to many disciplines and has been studied extensively. A basic tool used in this analysis are the *convergence bounds* described in appendix A.2.

convergence bound

Consider a data set $\mathcal{D}$ defined as a set of M IID Bernoulli random variables $\mathcal{D} = \{X[1], \ldots, X[M]\}$, where $P^*(X[m] = x^1) = p^*$ for all m. Note that we are now considering $\mathcal{D}$ itself to be a stochastic event (a random variable), sampled from the distribution $(P^*)^M$. Let

Hoeffding bound

$\hat{p}_{\mathcal{D}} = \frac{1}{M} \sum_m X[m]$. Then an immediate consequence of the *Hoeffding bound* (theorem A.3) is

$$P_M(|\hat{p}_{\mathcal{D}} - p^*| > \epsilon) \leq 2e^{-2M\epsilon^2},$$

where P_M is a shorthand for $P_{\mathcal{D} \sim (P^*)^M}$. Thus, the probability that the MLE $\hat{p}_{\mathcal{D}}$ deviates from the true parameter by more than ϵ is bounded from above by a function that decays exponentially in M.

As an immediate corollary, we can prove a PAC-bound on estimating p^*:

Corollary 17.1

Let $\epsilon, \delta > 0$, and let

$$M > \frac{1}{2\epsilon^2} \log \frac{2}{\delta}.$$

Then $P_M(|\hat{p} - p^*| \leq \epsilon) \geq 1 - \delta$, where P_M, as before, is a probability over data sets $\mathcal{D}$ of size M sampled IID from P^*.

PROOF

$$\begin{aligned} P_M(|\hat{p} - p^*| \leq \epsilon) &= 1 - P_M(\hat{p} - p^* > \epsilon) - P_M(\hat{p} - p^* < -\epsilon) \\ &\geq 1 - 2e^{-2M\epsilon^2} \geq 1 - \delta. \end{aligned} \qquad \blacksquare$$

The number of data instances M required to obtain a PAC-bound grows quadratically in the error $1/\epsilon$, and logarithmically in the confidence $1/\delta$. For example, setting $\epsilon = 0.05$ and $\delta = 0.01$, we get that we need $M \geq 1059.66$. That is, we need a bit more than $1,000$ samples to confidently estimate the probability of an event to within 5 percent error.

This result allows us to bound the absolute value of the error between the parameters. We,

relative entropy

however, are interested in the *relative entropy* between the two distributions. Thus, we want to bound terms of the form $\log \frac{p^*}{\hat{p}}$.

Lemma 17.1

For P_M defined as before:

$$P_M\left(\log \frac{p^*}{\hat{p}} > \epsilon\right) \leq e^{-2Mp^2\epsilon^2 \frac{1}{(1+\epsilon)^2}}.$$

PROOF   The proof relies on the following fact:

If $\epsilon \leq x \leq y \leq 1$ then $\quad (\log y - \log x) \leq \frac{1}{\epsilon}(y - x).$ $\qquad\qquad$ (17.22)

Now, consider some ϵ'. If $p^* - \hat{p} \leq \epsilon'$ then $\hat{p} > p^* - \epsilon'$. Applying equation (17.22), we get that $\log \frac{p^*}{\hat{p}} \leq \frac{\epsilon'}{p^* - \epsilon'}$. Setting $\epsilon' = \frac{\epsilon p^*}{1+\epsilon}$, and taking the contrapositive, we conclude that, if $\log \frac{p^*}{\hat{p}} > \epsilon$ then $p^* - \hat{p} > \frac{\epsilon p^*}{1+\epsilon}$. Using the Hoeffding bound to bound the probability of the latter event, we derive the desired result. ∎

This analysis applies to a binary-valued random variable. We now extend it to the case of a multivalued random variable. The result provides a bound on the relative entropy between $P^*(X)$ and the maximum likelihood estimate for $P(X)$, which is simply its empirical probability $\hat{P}_{\mathcal{D}}(X)$.

Proposition 17.7

Let $P^*(X)$ be a discrete distribution such that $P^*(x) \geq \lambda$ for all $x \in Val(X)$. Let $\mathcal{D} = \{X[1], \ldots, X[M]\}$ consist of M IID samples of X. Then

$$P_M(\boldsymbol{D}(P^*(X)\|\hat{P}_{\mathcal{D}}(X)) > \epsilon) \leq |Val(X)|e^{-2M\lambda^2\epsilon^2 \frac{1}{(1+\epsilon)^2}}.$$

PROOF   We want to bound the error

$$\boldsymbol{D}(P^*(X)\|\hat{P}_{\mathcal{D}}(X)) = \sum_x P^*(x) \log \frac{P^*(x)}{\hat{P}_{\mathcal{D}}(x)}.$$

This expression is a weighted average of log-ratios of the type we bounded in lemma 17.1. If we bound each of the terms in this average by ϵ, we can obtain a bound for the weighted average as a whole. That is, we say that a data set is well behaved if, for each x, $\log \frac{P^*(x)}{\hat{P}_{\mathcal{D}}(x)} \leq \epsilon$. If the data set is well behaved, we have a bound of ϵ on the term for each x, and therefore an overall bound of ϵ for the entire relative entropy.

With what probability is our data set not well behaved? It suffices that there is one x for which $\log \frac{P^*(x)}{\hat{P}_{\mathcal{D}}(x)} > \epsilon$. We can provide an upper bound on this probability using the *union bound*, which bounds the probability of the union of a set of events as the sum of the probabilities of the individual events:

$$P\left(\exists x, \log \frac{P^*(x)}{\hat{P}_{\mathcal{D}}(x)} > \epsilon\right) \leq \sum_x P\left(\log \frac{P^*(x)}{\hat{P}_{\mathcal{D}}(x)} > \epsilon\right).$$

The union bound is an overestimate of the probability, since it essentially represents the case where the different "bad" events are disjoint. However, our focus is on the situation where these events are unlikely, and the error due to such overcounting is not significant.

Formalizing this argument, we obtain:

$$
\begin{aligned}
P_M(\boldsymbol{D}(P^*(X)\|\hat{P}_{\mathcal{D}}(X)) > \epsilon) \quad &\leq \quad P_M\left(\exists x \ : \ \log \frac{P^*(x)}{\hat{P}_{\mathcal{D}}(x)} > \epsilon\right) \\[2mm]
&\leq \quad \sum_x P_M\left(\log \frac{P^*(x)}{\hat{P}_{\mathcal{D}}(x)} > \epsilon\right) \\[2mm]
&\leq \quad \sum_x e^{-2MP^*(x)^2\epsilon^2 \frac{1}{(1+\epsilon)^2}} \\[2mm]
&\leq \quad |Val(X)|e^{-2M\lambda^2\epsilon^2 \frac{1}{(1+\epsilon)^2}},
\end{aligned}
$$

where the second inequality is derived from the union bound, the third inequality from lemma 17.1, and the final inequality from the assumption that $P^*(x) \geq \lambda$. ∎

This result provides us with an error bound for estimating the distribution of a random variable. We can now easily translate this result to a PAC-bound:

Corollary 17.2

Assume that $P^*(x) \geq \lambda$ for all $x \in Val(X)$. For $\epsilon, \delta > 0$, let

$$M \geq \frac{1}{2} \frac{1}{\lambda^2} \frac{(1+\epsilon)^2}{\epsilon^2} \log \frac{|Val(X)|}{\delta}.$$

Then $P_M(\boldsymbol{D}(P^*(X)\|\hat{P}_{\mathcal{D}}(X)) \leq \epsilon) \geq 1 - \delta$.

As with the binary-valued case, the number of samples grows quadratically with $\frac{1}{\epsilon}$ and logarithmically with $\frac{1}{\delta}$. Here, however, we also have a quadratic dependency on $\frac{1}{\lambda}$. The value λ is a measure of the "skewness" of the distribution P^*. This dependence is not that surprising; we expect that, if some values of X have small probability, then we need many more samples to get a good approximation of their probability. Moreover, underestimates of $\hat{P}_{\mathcal{D}}(x)$ for such events can lead to a big error in estimating $\log \frac{P^*(x)}{\hat{P}_{\mathcal{D}}(x)}$. Intuitively, we might suspect that when $P^*(x)$ is small, it is harder to estimate, but at the same time it also less crucial for the total error. Although it is possible to use this intuition to get a better estimation (see exercise 17.19), the asymptotic dependence of M on $\frac{1}{\lambda}$ remains quadratic.

17.6.2.2 Estimating a Bayesian Network

We now consider the problem of learning a Bayesian network. Suppose that P^* is consistent with a Bayesian network $\mathcal{G}$ and that we learned parameters $\boldsymbol{\theta}$ for $\mathcal{G}$ that define a distribution $P_{\boldsymbol{\theta}}$. Using theorem 8.5, we have that

$$\boldsymbol{D}(P^*\|P_{\boldsymbol{\theta}}) = \sum_i \boldsymbol{D}(P^*(X_i \mid \mathrm{Pa}_i^{\mathcal{G}})\|P_{\boldsymbol{\theta}}(X_i \mid \mathrm{Pa}_i^{\mathcal{G}})),$$

where (as shown in appendix A.1.3.2), we have that

$$\boldsymbol{D}(P^*(X \mid \boldsymbol{Y})\|P(X \mid \boldsymbol{Y})) = \sum_{\boldsymbol{y}} P^*(\boldsymbol{y})\boldsymbol{D}(P^*(X \mid \boldsymbol{y})\|P(X \mid \boldsymbol{y})).$$

Thus, as we might expect, the error is a sum of errors in estimating the conditional probabilities.

This error term, however, makes the strong assumption that our generating distribution P^* is consistent with our target class — those distributions representable using the graph $\mathcal{G}$. This assumption is usually not true in practice. When this assumption is false, the given network structure limits the ability of the learning procedure to generalize. For example, if we give a learning procedure a graph where X and Y are independent, then no matter how good our learning procedure, we cannot achieve low generalization error if X and Y are strongly dependent in P^*. More broadly, if the given network structure is inaccurate, then there is inherent error in the learned distribution that the learning procedure cannot overcome.

One approach to deal with this problem is to assume away the cases where P^* does not conform with the given structure $\mathcal{G}$. This solution, however, makes the analysis brittle and of little relevance to real-life scenarios. An alternative solution is to relax our expectations from the learning procedure. Instead of aiming for the error to become very small, we might aim to show that the error is not far away from the inherent error that our procedure must incur due to the limitations in expressive power of the given network structure. In other words, rather than bounding the risk, we provide a bound on the *excess risk* (see box 16.B).

excess risk

More formally, let $\Theta[\mathcal{G}]$ be the set of all possible parameterizations for $\mathcal{G}$. We now define

$$\boldsymbol{\theta}^{\mathrm{opt}} = \arg \min_{\boldsymbol{\theta} \in \Theta[\mathcal{G}]} \boldsymbol{D}(P^* \| P_{\boldsymbol{\theta}}).$$

M-projection

That is, $\boldsymbol{\theta}^{\mathrm{opt}}$ is the best result we might expect the learning procedure to return. (Using the terminology of section 8.5, $\boldsymbol{\theta}^{\mathrm{opt}}$ is the *M-projection* of P^* on the family of distributions defined by $\mathcal{G}, \Theta[\mathcal{G}]$.)

The distance $\boldsymbol{D}(P^* \| P_{\boldsymbol{\theta}^{\mathrm{opt}}})$ reflects the minimal error we might achieve with networks with the structure $\mathcal{G}$. Thus, instead of defining "success" for our learning procedure in terms of obtaining low values for $\boldsymbol{D}(P^* \| P_{\boldsymbol{\theta}})$ (a goal which may not be achievable), we aim to obtain low values for $\boldsymbol{D}(P^* \| P_{\boldsymbol{\theta}}) - \boldsymbol{D}(P^* \| P_{\boldsymbol{\theta}^{\mathrm{opt}}})$.

What is the form of this error term? By solving for $P_{\boldsymbol{\theta}^{\mathrm{opt}}}$ and using basic manipulations we can define it in precise terms.

Theorem 17.3

Let $\mathcal{G}$ be a network structure, let P^* be a distribution, and let $P_{\boldsymbol{\theta}} = (\mathcal{G}, \boldsymbol{\theta})$ be a distribution consistent with $\mathcal{G}$. Then:

$$\boldsymbol{D}(P^* \| P_{\boldsymbol{\theta}}) - \boldsymbol{D}(P^* \| P_{\boldsymbol{\theta}^{\mathrm{opt}}}) = \sum_i \boldsymbol{D}(P^*(X_i \mid \mathrm{Pa}_i^{\mathcal{G}}) \| P_{\boldsymbol{\theta}}(X_i \mid \mathrm{Pa}_i^{\mathcal{G}})).$$

The proof is left as an exercise (exercise 17.20).

This theorem shows that **the error in our learned network decomposes into two components. The first is the error inherent in $\mathcal{G}$, and the second the error due to inaccuracies in estimating the conditional probabilities for parameterizing $\mathcal{G}$.** This theorem also shows that, in terms of error analysis, the treatment of the general case leads to exactly the same terms that we had to bound when we made the assumption that P^* was consistent with $\mathcal{G}$. Thus, in this learning task, the analysis of the inconsistent case is not more difficult than the analysis of the consistent case. As we will see in later chapters, this situation is not usually the case: we can often provide bounds for the consistent case, but not for the inconsistent case.

To continue the analysis, we need to bound the error in estimating conditional probabilities. The preceding treatment showed that we can bound the error in estimating marginal probabilities of a variable or a group of variables. How different is the estimate of conditional probabilities from that of marginal probabilities? It turns out that the two are easily related.

Lemma 17.2

Let P and Q be two distributions on X and Y. Then

$$\boldsymbol{D}(P(X \mid Y) \| Q(X \mid Y)) \le \boldsymbol{D}(P(X, Y) \| Q(X, Y)).$$

See exercise 17.21.

As an immediate corollary, we have that

$$\boldsymbol{D}(P^* \| P) - \boldsymbol{D}(P^* \| P_{\boldsymbol{\theta}^{\mathrm{opt}}}) \le \sum_i \boldsymbol{D}(P^*(X_i, \mathrm{Pa}_i^{\mathcal{G}}) \| P(X_i, \mathrm{Pa}_i^{\mathcal{G}})).$$

Thus, we can use proposition 17.7 to bound the error in estimating $P(X_i, \mathrm{Pa}_i^{\mathcal{G}})$ for each X_i (where we treat $X_i, \mathrm{Pa}_i^{\mathcal{G}}$ as a single variable) and derive a bound on the error in estimating the probability of the whole network.

Theorem 17.4

Let $\mathcal{G}$ be a network structure, and P^* a distribution consistent with some network $\mathcal{G}^*$ such that $P^*(x_i \mid \mathrm{pa}_i^{\mathcal{G}^*}) \geq \lambda$ for all i, x_i, and $\mathrm{pa}_i^{\mathcal{G}^*}$. If P is the distribution learned by maximum likelihood estimate for $\mathcal{G}$, then

$$P\left(\boldsymbol{D}(P^*\|P) - \boldsymbol{D}(P^*\|P_{\boldsymbol{\theta}^{\mathrm{opt}}}) > n\epsilon\right) \leq nK^{d+1} e^{-2M\lambda^{2(d+1)}\epsilon^2 \frac{1}{(1+\epsilon)^2}},$$

where K is the maximal variable cardinality and d is the maximum number of parents in $\mathcal{G}$.

PROOF The proof uses the union bound

$$P\left(\boldsymbol{D}(P^*\|P) - \boldsymbol{D}(P^*\|P_{\boldsymbol{\theta}^{\mathrm{opt}}}) > n\epsilon\right) \leq \sum_i P\left(\boldsymbol{D}(P^*(X_i, \mathrm{Pa}_i^{\mathcal{G}})\|P(X_i, \mathrm{Pa}_i^{\mathcal{G}})) > \epsilon\right)$$

with application of proposition 17.7 to bound the probability of each of these latter events. The only technical detail we need to consider is to show that if conditional probabilities in P^* are always larger than λ, then $P^*(x_i, \mathrm{pa}_i^{\mathcal{G}}) \geq \lambda^{|\mathrm{Pa}_i^{\mathcal{G}}|+1}$; see exercise 17.22. ∎

This theorem shows that we can indeed learn parameters that converge to the optimal ones as the number of samples grows. As with previous bounds, the number of samples we need is

Corollary 17.3

Under the conditions of theorem 17.4, if

$$M \geq \frac{1}{2}\frac{1}{\lambda^{2(d+1)}}\frac{(1+\epsilon)^2}{\epsilon^2}\log\frac{nK^{d+1}}{\delta},$$

then

$$P\left(\boldsymbol{D}(P^*\|P) - \boldsymbol{D}(P^*\|P_{\boldsymbol{\theta}^{\mathrm{opt}}}) < n\epsilon\right) > 1 - \delta.$$

As before, the number of required samples grows quadratically in $\frac{1}{\epsilon}$. Conversely, we expect the error to decrease roughly with $\frac{1}{\sqrt{M}}$, which is commensurate with the behavior we observe in practice (for example, see figure 17.C.2 in box 17.C). We see also that λ and d play a major role in determining M. In practice, we often do not know λ in advance, but such analysis allows us to provides guarantees under the assumption that conditional probabilities are not too small. It also allows us to predict the improvement in error (or at least in our upper bound on it) that we would obtain if we add more samples.

Note that in this analysis we "allow" the error to grow with n as we consider $\boldsymbol{D}(P^*\|P) > n\epsilon$. The argument is that, as we add more variables, we expect to incur some prediction error on each one.

Example 17.19

Consider the network where we have n independent binary-valued variables $X_1, \ldots, X_n$. In this case, we have n independent Bernoulli estimation problems, and would expect a small number of samples to suffice. Indeed, we can obtain an ϵ-close estimate to each of them (with high-probability) using the bound of lemma 17.1. However, the overall relative entropy between P^* and $\hat{P}_\mathcal{D}$ over the joint space $X_1, \ldots, X_n$ will grow as the sum of the relative entropies between the individual marginal distributions $P^*(X_i)$ and $\hat{P}_\mathcal{D}(X_i)$. Thus, even if we perform well at predicting each variable, the total error will scale linearly with n. ∎

Thus, our formulation of the bound in corollary 17.3 is designed to separate out this "inevitable" linear growth in the error from any additional errors that arise from the increase in dimensionality of the distribution to be estimated.

We provided a theoretical analysis for the generalization error of maximum likelihood estimate. A natural question is to carry similar analysis when we use Bayesian estimates. Intuitively, the asymptotic behavior (for $M \to \infty$) will be similar, since the two estimates are asymptotically identical. For small values of M we do expect to see differences, since the Bayesian estimate is smoother and cannot be arbitrarily small and thus the relative entropy is bounded. See exercise 17.23 for an analysis of the Bayesian estimate.

17.7   Summary

In this chapter we examined parameter estimation for Bayesian networks when the data are complete. This is the simplest learning task for Bayesian networks, and it provides the basis for the more challenging learning problems we examine in the next chapters. We discussed two approaches for estimation: MLE and Bayesian prediction. Our primary emphasis here was on table-CPDs, although the ideas generalize to other representations. We touched on a few of these.

As we saw, a central concept in both approaches is the likelihood function, which captures how the probability of the data depends on the choice of parameters. A key property of the likelihood function for Bayesian networks is that it decomposes as a product of local likelihood functions for the individual variables. If we use table-CPDs, the likelihood decomposes even further, as a product of the likelihoods for the individual multinomial distributions $P(X \mid \mathrm{pa}_{X_i})$. This decomposition plays a central role in both maximum likelihood and Bayesian estimation, since it allows us to decompose the estimation problem and treat each of these CPDs or even each of the individual multinomials separately.

When the local likelihood has sufficient statistics, then learning is viewed as mapping values of the statistics to parameters. For networks with discrete variables, these statistics are counts of the form $M[x_i, \mathrm{pa}_{X_i}]$. Thus, learning requires us to collect these for each combination x_i, pa_{X_i} of a value of X_i and an instantiation of values to its parents. We can collect all of these counts in one pass through the data using a data structure whose size is proportional to the representation of the network, since we need one counter for each CPD entry.

Once we collect the sufficient statistics, the estimate of both methods is similar. The MLE estimate for table-CPDs has a simple closed form:

$$\hat{\theta}_{x_i \mid \mathrm{pa}_{X_i}} = \frac{M[x_i, \mathrm{pa}_{X_i}]}{M[\mathrm{pa}_{X_i}]}.$$

The Bayesian estimate is based on the use of a Dirichlet distribution, which is a conjugate prior to the multinomial distribution. In a conjugate prior, the posterior — which is proportional to the prior times the likelihood — has the same form as the prior. This property allows us to maintain posteriors in closed form. In particular, for a discrete Bayesian network with table-CPDs and a Dirichlet prior, the posterior of the local likelihood has the form

$$P(x_i \mid \mathrm{pa}_{X_i}, D) = \frac{M[x_i, \mathrm{pa}_{X_i}] + \alpha_{x_i \mid \mathrm{pa}_{X_i}}}{M[\mathrm{pa}_{X_i}] + \alpha_{\mathrm{pa}_{X_i}}}.$$

Since all we need in order to learn are the sufficient statistics, then we can easily adapt them to learn in an online setting, where additional training examples arrive. We simply store a vector of sufficient statistics, and update it as new instances are obtained.

In the more advanced sections, we saw that the same type of structure applies to other parameterizations in the exponential family. Each family defines the sufficient statistics we need to accumulate and the rules for finding the MLE parameters. We developed these rules for Gaussian CPDs, where again learning can be done using a closed-form analytical formula.

We also discussed networks where some of the parameters are shared, whether between CPDs or within a single CPD. We saw that the same properties described earlier — decomposition and sufficient statistics — allow us to provide an easy analysis for this setting. The likelihood function is now defined in terms of sufficient statistics that aggregate statistics from different parts of the network. Once the sufficient statistics are defined, the estimation procedures, whether MLE or Bayesian, are exactly the same as in the case without shared parameters.

Finally, we examined the theoretical foundations of learning. We saw that parameter estimates are asymptotically correct in the following sense. If the data are actually generated from the given network structure, then, as the number of samples increases, both methods converge to the correct parameter setting. If not, then they converge to the distribution with the given structure that is "closest" to the distribution from which the data were generated. We further analyzed the rate at which the estimates converge. As M grows, we see a *concentration phenomenon*; for most samples, the empirical distribution is in a close neighborhood of the true distribution. Thus, the chances of sampling a data set in which the MLE estimates are far from the true parameters decays exponentially with M. This analysis allowed us to provide a PAC-bound on the number of samples needed to obtain a distribution that is "close" to optimal.

concentration
phenomenon

17.8 Relevant Literature

The foundations of maximum likelihood estimation and Bayesian estimation have a long history; see DeGroot (1989); Schervish (1995); Hastie et al. (2001); Bishop (2006); Bernardo and Smith (1994) for some background material. The thumbtack example is adapted from Howard (1970).

Heckerman (1998) and Buntine (1994, 1996) provide excellent tutorials and reviews on the basic principles of learning Bayesian networks from data, as well as a review of some of the key references.

The most common early application of Bayesian network learning, and perhaps even now, is learning a naive Bayes model for the purpose of classification (see, for example, Duda and Hart 1973). Spiegelhalter and Lauritzen (1990) laid the foundation for the problem of learning general Bayesian networks from data, including the introduction of the global parameter independence assumption, which underlies the decomposability of the likelihood function. This development led to a stream of significant extensions, most notably by Buntine (1991); Spiegelhalter et al. (1993); Cooper and Herskovits (1992). Heckerman et al. (1995) defined the BDe prior and showed its equivalence to a combination of assumptions about the prior.

Many papers (for example, Spiegelhalter and Lauritzen 1990; Neal 1992; Buntine 1991; Diez 1993) have proposed the use of structured CPDs as an approach for reducing the number of parameters that one needs to learn from data. In many cases, the specific learning algorithms are derived from algorithms for learning conditional probability models for probabilistic classification. The probabilistic derivation of tree-CPDs was performed by Buntine (1993) and introduced to Bayesian networks by Friedman and Goldszmidt (1998). The analysis of Bayesian learning of Bayesian networks with linear Gaussian dependencies was performed by Heckerman and Geiger (1995); Geiger and Heckerman (Geiger and Heckerman). Buntine (1994) emphasizes the important connection between the exponential family and the task of learning Bayesian networks. Bernardo and Smith (1994) describe conjugate priors for many distributions in the exponential family. Some material on nonparametric density estimation can be found in Hastie et al. (2001); Bishop (2006). Hofmann and Tresp (1995) use Parzen window to capture conditional distributions in continuous Bayesian networks. Imoto et al. (2003) learn semiparametric spline-regression models as CPDs in Bayesian networks. Rasmussen and Williams (2006) describe

Gaussian
processes

Gaussian processes, a state-of-the-art method for nonparametric estimation, which has also been used for estimating CPDs in Bayesian networks (Friedman and Nachman 2000).

Plate models as a representation for parameter sharing in learning were introduced by Gilks et al. (1994) and Buntine (1994). Hierarchical Bayesian models have a long history in statistics; see, for example, Gelman et al. (1995).

The generalization bounds for parameter estimation in Bayesian networks were first analyzed by Höffgen (1993), and subsequently improved and refined by Friedman and Yakhini (1996) and Dasgupta (1997).

Beinlich et al. (1989) introduced the ICU-Alarm network, which has formed the benchmark for numerous studies of Bayesian network learning.

17.9   Exercises

Exercise 17.1

Show that the estimate of equation (17.1) is the maximum likelihood estimate. Hint: differentiate the log-likelihood with respect to θ.

Exercise 17.2★

Derive the MLE for the multinomial distribution (example 17.5). Hint, maximize the log-likelihood function using a Lagrange coefficient to enforce the constraint $\sum_k \theta_k = 1$.

Exercise 17.3

Derive the MLE for Gaussian distribution (example 17.6). Solve the equations

$$\frac{\partial \log L(\mathcal{D} : \mu, \sigma)}{\partial \mu} = 0$$

$$\frac{\partial \log L(\mathcal{D} : \mu, \sigma)}{\partial \sigma^2} = 0.$$

Exercise 17.4

Derive equation (17.8) by differentiating the log-likelihood function and using equation (17.6) and equation (17.7).

Exercise 17.5★

In this exercise, we examine how to estimate a joint multivariate Gaussian. Consider two continuous variables X and Y, and assume we have a data set consisting of M samples $\mathcal{D} = \{\langle x[m], y[m] \rangle : m = $

$1, \ldots, M\}$. Show that the MLE estimate for a joint Gaussian distribution over X, Y is the Gaussian with mean vector $\langle \boldsymbol{E}_{\mathcal{D}}[X], \boldsymbol{E}_{\mathcal{D}}[Y] \rangle$, and covariance matrix

$$\Sigma_{X,Y} = \begin{pmatrix} \boldsymbol{Cov}_{\mathcal{D}}[X;X] & \boldsymbol{Cov}_{\mathcal{D}}[X;Y] \\ \boldsymbol{Cov}_{\mathcal{D}}[X;Y] & \boldsymbol{Cov}_{\mathcal{D}}[Y;Y] \end{pmatrix}.$$

Exercise 17.6★

Derive equation (17.10) by solving the integral $\int_0^1 \theta^k (1-\theta)^{M-k} d\theta$ for different values of k. (Hint: use integration by parts.)

Exercise 17.7★

mixture of
Dirichlets

In this problem we consider the class of parameter priors defined as a *mixture of Dirichlets*. These comprise a richer class of priors than the single Dirichlet that we discussed in this chapter. A mixture of Dirichlets represents another level of uncertainty, where we are unsure about which Dirichlet distribution is a more appropriate prior for our domain. For example, in a simple coin-tossing situation, we might be uncertain whether the coin whose parameter we are trying to learn is a fair coin, or whether it is a biased one. In this case, our prior might be a mixture of two Dirichlet distributions, representing those two cases.

In this problem, our goal is to show that the family of mixture of Dirichlet priors is conjugate to the multinomial distribution; in other words, if our prior is a mixture of Dirichlets, and our likelihood function is multinomial, then our posterior is also a mixture of Dirichlets.

a. Consider the simple possibly biased coin setting described earlier. Assume that we use a prior that is a mixture of two Dirichlet (Beta) distributions: $P(\theta) = 0.95 \cdot Beta(5000, 5000) + 0.05 \cdot Beta(1, 1)$; the first component represents a fair coin (for which we have seen many imaginary samples), and the second represents a possibly-biased coin, whose parameter we know nothing about. Show that the expected probability of heads given this prior (the probability of heads averaged over the prior) is $1/2$. Suppose that we observe the data sequence $(H, H, T, H, H, H, H, H, H, H)$. Calculate the posterior over θ, $P(\theta \mid \mathcal{D})$. Show that it is also a 2-component mixture of Beta distributions, by writing the posterior in the form $\lambda^1 Beta(\alpha_1^1, \alpha_2^1) + \lambda^2 Beta(\alpha_1^2, \alpha_2^2)$. Provide actual numeric values for the different parameters $\lambda^1, \lambda^2, \alpha_1^1, \alpha_2^1, \alpha_1^2, \alpha_2^2$.

b. Now generalize your calculations from part (1) to the case of a mixture of d Dirichlet priors over a k-valued multinomial parameters. More precisely, assume that the prior has the form

$$P(\boldsymbol{\theta}) = \sum_{i=1}^{d} \lambda^i Dirichlet(\alpha_1^i, \ldots, \alpha_k^i),$$

and prove that the posterior has the same form.

Exercise 17.8★

We now consider a Bayesian approach for learning the mean of a Gaussian distribution. It turns out that in doing Bayesian inference with Gaussians, it is mathematically easier to use the precision $\tau = \frac{1}{\sigma^2}$ rather than the variance. Note that larger the precision, the narrower the distribution around the mean.

Suppose that we have M IID samples $x[1], \ldots, x[M]$ from $X \sim \mathcal{N}\left(\theta; \tau_X^{-1}\right)$. Moreover, assume that we know the value of τ_X. Thus, the unknown parameter θ is the mean. Show that if the prior $P(\theta)$ is $\mathcal{N}\left(\mu; \tau_\theta^{-1}\right)$, then the posterior $P(\theta \mid \mathcal{D})$ is $\mathcal{N}\left(\mu'; (\tau_\theta')^{-1}\right)$ where

$$\begin{aligned} \tau_\theta' &= M\tau_X + \tau_\theta \\ \mu' &= \frac{M\tau_X}{\tau_\theta'} \boldsymbol{E}_{\mathcal{D}}[X] + \frac{\tau_\theta}{\tau_\theta'} \mu_0. \end{aligned}$$

Hint: Start by proving $\sum_m (x[m] - \theta)^2 = M(\theta - \boldsymbol{E}_{\mathcal{D}}[X]) + c$, where c is a constant that does not depend on θ.

Exercise 17.9

We now consider making predictions with the posterior of exercise 17.8. Suppose we now want to compute the probability

$$P(x[M+1] \mid \mathcal{D}) = \int P(x[M+1] \mid \theta) P(\theta \mid \mathcal{D}) d\theta.$$

Show that this distribution is Gaussian. What is the mean and precision of this distribution?

Exercise 17.10★

We now consider the complementary case to exercise 17.8, where we know the mean of X, but do not know the precision. Suppose that $X \sim \mathcal{N}\left(\mu; \theta^{-1}\right)$, where θ is the unknown precision.

We start with a definition. We say that $Y \sim \text{Gamma}\left(\alpha; \beta\right)$ (for $\alpha, \beta > 0$) if

$$P(y : \alpha, \beta) = \frac{\beta^\alpha}{\Gamma(\alpha)} y^\alpha e^{-\beta y}.$$

Gamma
distribution

This distribution is called a *Gamma distribution*. Here, we have that $E[Y] = \frac{\alpha}{\beta}$ and $Var[Y] = \frac{\alpha}{\beta^2}$.

    a. Show that Gamma distributions are a conjugate family for this learning task. More precisely, show that if $P(\theta) \sim \text{Gamma}\left(\alpha; \beta\right)$, then $P(\theta \mid \mathcal{D}) \sim \text{Gamma}\left(\alpha'; \beta'\right)$ where

$$\alpha' = \alpha + \frac{1}{2}M$$

$$\beta' = \beta + \frac{1}{2}\sum_m (x[m] - \mu)^2.$$

       Hint: do not worry about the normalization constant, instead focus on the terms in the posterior that involve θ.

    b. Derive the mean and variance of θ in the posterior. What can we say about beliefs given the data? How do they differ from the MLE estimate?

Exercise 17.11★★

Now consider the case where we know neither the mean nor the precision of X. We examine a family of distributions that are conjugate in this case.

normal-Gamma
distribution

A *normal-Gamma distribution* over μ, τ is of the form:

$$P(\tau)P(\mu \mid \tau)$$

where $P(\tau)$ is $\text{Gamma}\left(\alpha; \beta\right)$ and $P(\mu \mid \tau)$ is $\mathcal{N}\left(\mu_0; \lambda\tau\right)$. That is, the distribution over precisions is a Gamma distribution (as in exercise 17.10), and the distribution over the mean is a Gaussian (as in exercise 17.8), except that the precision of this distribution depends on τ.

Show that if $P(\mu, \tau)$ is normal-Gamma with parameters $\alpha, \beta, \mu_0, \lambda$, then the posterior $P(\mu, \tau \mid \mathcal{D})$ is also a Normal-Gamma distribution.

Exercise 17.12

Suppose that a prior on a parameter vector is $p(\boldsymbol{\theta}) \sim Dirichlet(\alpha_1, \ldots, \alpha_k)$. What is the MAP value of the parameters, that is, $\arg\max_{\boldsymbol{\theta}} p(\boldsymbol{\theta})$?

Exercise 17.13★

In this exercise, we will define a general-purpose prior for models with shared parameters, along the lines of the BDe prior of section 17.4.3.

    a. Following the lines of the derivation of the BDe prior, construct a parameter prior for a network with shared parameters, using the uniform distribution P' as the basis.

    b. Now, extend your analysis to construct a BDe-like parameter prior for a plate model, using, as the basis, a sample size of $\alpha(Q)$ for each Q and a uniform distribution.

Exercise 17.14

Perform the analysis of section 17.5.1.1 in the case where the network is a Gaussian Bayesian network. Derive the form of the likelihood function in terms of the appropriate aggregate sufficient statistics, and show how the MLE is computed from these statistics.

Exercise 17.15

In section 17.5, we discussed sharing at both the global and the local level. We now consider an example where we have both. Consider the following elaboration of our University domain: Each course has an additional attribute *Level*, whose values are *undergrad* (l^0) or *grad* (l^1). The grading curve (distribution for *Grade*) now depends on *Level*: The curve is the same for all undergraduate courses, and depends on *Difficulty* and *Intelligence*, as before. For graduate courses, the distribution is different for each course and, moreover, does not depend on the student's intelligence.

Specify the set of multinomial parameters in this model, and the partition of the multinomial distributions that correctly captures the structure of the parameter sharing.

Exercise 17.16

    a. Using the techniques and notation of section 17.5.1.2, describe the likelihood function for the DBN model of figure 6.2. Your answer should define the set of shared parameters, the partition of the variables in the ground network, the aggregate sufficient statistics, and the MLE.

    b. Now, assume that we want to use Bayesian inference for the parameter estimation in this case. Assuming local parameter independence and a Dirichlet prior, write down the form of the prior and the posterior. How would you use your learned model to compute the probability of a new trajectory?

Exercise 17.17

Consider the application of the global sharing techniques of section 17.5.1.2 to the task of parameter estimation in a PRM. A key difference between PRMs and plate models is that the different instantiations of the same attribute in the ground network may not have the same in-degree. For instance, returning to example 6.16, let $Job(S)$ be an attribute such that S is a logical variable ranging over Student. Assume that $Job(S)$ depends on the average grade of all courses that the student has taken (where we round the grade-point-average to produce a discrete set of values). Show how we can parameterize such a model, and how we can aggregate statistics to learn its parameters.

Exercise 17.18

Suppose we have a single multinomial variable X with K values and we have a prior on the parameters governing X so that $\boldsymbol{\theta} \sim Dirichlet(\alpha_1, \ldots, \alpha_K)$. Assume we have some data set $\mathcal{D} = \{x[1], \ldots, x[M]\}$.

    a. Show how to compute

$$P(X[M+1] = x^i, X[M+2] = x^j \mid \mathcal{D}).$$

    (Hint: use the chain rule for probabilities.)

    b. Suppose we decide to use the approximation

$$P(X[M+1] = x^i, X[M+2] = x^j \mid \mathcal{D}) \approx P(X[M+1] = x^i, \mid \mathcal{D})P(X[M+2] = x^j \mid \mathcal{D}).$$

    That is, we ignore the dependencies between $X[M+1]$ and $X[M+2]$. Analyze the error in this approximation (the ratio between the approximation and the correct probability). What is the quality of this approximation for small M? What is the asymptotic behavior of the approximation when $M \to \infty$. (Hint: deal separately with the case where $i = j$ and the case where $i \neq j$.)

Exercise 17.19★

We now prove a variant on proposition 17.7. Show that in the setting of that proposition 17.7

$$P(\boldsymbol{D}(P\|\hat{P}) > \epsilon) \leq Ke^{-2M\epsilon^2 \frac{1}{K^2} \frac{2}{(1+\frac{\epsilon}{K\lambda})^2}}$$

where $K = |Val(X)|$.

a. Show that

$$P(\boldsymbol{D}(P\|\hat{P}) > \epsilon) \leq \sum_x P(\log \frac{P^*(x)}{\hat{P}(x)} > \frac{\epsilon}{KP(x)}).$$

b. Use this result and lemma 17.1 to prove the stated bound.

c. Show that the stated bound is tighter than the original bound of proposition 17.7. (Hint: examine the case when $\lambda = \frac{1}{K}$.)

Exercise 17.20

Prove theorem 17.3. Specifically, first prove that $P_{\boldsymbol{\theta}^{\mathrm{opt}}} = \prod_i P^*(X_i \mid \mathrm{Pa}_{X_i}^{\mathcal{G}})$ and then use theorem 8.5.

Exercise 17.21

Prove lemma 17.2. Hint: Show that $\boldsymbol{D}(P(X,Y)\|Q(X,Y)) = \boldsymbol{D}(P(X \mid Y)\|Q(X \mid Y)) + \boldsymbol{D}(P(X)\|Q(X))$, and then show that the inequality follows.

Exercise 17.22

Suppose P is a Bayesian network with $P(x_i \mid \mathrm{pa}_i) \geq \lambda$ for all i, x_i and pa_i. Consider a family X_i, Pa_i, show that

$$P(x_i, \mathrm{pa}_i) \geq \lambda^{|\mathrm{Pa}_i|+1}.$$

Exercise 17.23★

We now prove a bound on the error when using Bayesian estimates. Let $\mathcal{D} = \{X[1], \dots, X[M]\}$ consist of M IID samples of a discrete variable X. Let α and P_0 be the equivalent sample size and prior distribution for a Dirichlet prior. The Bayesian estimator will return the distribution

$$\tilde{P}(x) = \frac{M}{M+\alpha}\hat{P}(x) + \frac{\alpha}{M+\alpha}P_0(x).$$

We now want to analyze the error of such an estimate.

a. Prove the analogue of lemma 17.1. Show that

$$P(\log \frac{P^*(x)}{\tilde{P}(x)} > \epsilon) \leq e^{-\frac{2M(\frac{M}{M+\alpha}P^*(x) + \frac{\alpha}{M+\alpha}P_0(x))^2 \epsilon_x^2}{(1+\epsilon_x)^2}}.$$

b. Use the union bound to show that if $P^*(x) \geq \lambda$ and $P_0(x) \geq \lambda_0$ for all $x \in Val(X)$, then

$$P(\boldsymbol{D}(P^*(X)\|\tilde{P}(X)) > \epsilon) \leq |Val(X)|e^{-2M(\frac{M}{M+\alpha}\lambda + \frac{\alpha}{M+\alpha}\lambda_0)^2 \epsilon^2 \frac{1}{(1+\epsilon)^2}}.$$

c. Show that

$$\frac{M}{M+\alpha}\lambda + \frac{\alpha}{M+\alpha}\lambda_0 \geq \max(\lambda, \frac{\alpha}{M+\alpha}\lambda_0).$$

d. Suppose that $\lambda_0 > \lambda$. That is, our prior is less extreme than the real distribution, which is definitely the case if we take P_0 to be the uniform distribution. What can you conclude about a PAC result for the Bayesian estimate?

18 *Structure Learning in Bayesian Networks*

18.1 Introduction

18.1.1 Problem Definition

In the previous chapter, we examined how to learn the parameters of Bayesian networks. We made a strong assumption that we know in advance the network structure, or at least we decide on one regardless of whether it is correct or not. In this chapter, we consider the task of learning in situations where do not know the structure of the Bayesian network in advance. Throughout this chapter, we continue with the (very) strong assumption that our data set is fully observed, deferring the discussion of learning with partially observed data to the next chapter.

As in our discussion so far, we assume that the data $\mathcal{D}$ are generated IID from an underlying distribution $P^*(\mathcal{X})$. Here, we also assume that P^* is induced by some Bayesian network $\mathcal{G}^*$ over $\mathcal{X}$. We begin by considering the extent to which independencies in $\mathcal{G}^*$ manifest in $\mathcal{D}$.

Example 18.1

Consider an experiment where we toss two standard coins X and Y independently. We are given a data set with 100 instances of this experiment. We would like to learn a model for this scenario. A "typical" data set may have 27 head/head, 22 head/tail, 25 tail/head, and 26 tail/tail entries. In the empirical distribution, the two coins are not independent. This may seem reasonable, since the probability of tossing 100 pairs of fair coins and getting exactly 25 outcomes in each category is quite small (approximately $1/1,000$). Thus, even if the two coins are independent, we do not expect the observed empirical distribution to satisfy independence.

Now suppose we get the same results in a very different situation. Say we scan the sports section of our local newspaper for 100 days and choose an article at random each day. We mark $X = x^1$ if the word "rain" appears in the article and $X = x^0$ otherwise. Similarly, Y denotes whether the word "football" appears in the article. Here our intuitions as to whether the two random variables are independent are unclear. If we get the same empirical counts as in the coins described before, we might suspect that there is some weak connection. In other words, it is hard to be sure whether the true underlying model has an edge between X and Y or not. ∎

The importance of correctly reconstructing the network structure depends on our learning goal. As we discussed in chapter 16, there are different reasons for learning the model structure. One is for *knowledge discovery*: by examining the dependencies in the learned network, we can learn the dependency structure relating variables in our domain. Of course, there are other methods that reveal correlations between variables, for example, simple statistical *independence*

knowledge
discovery

independence
test

tests. A Bayesian network structure, however, reveals much finer structure. For instance, it can potentially distinguish between direct and indirect dependencies, both of which lead to correlations in the resulting distribution.

If our goal is to understand the domain structure, then, clearly, the best answer we can aspire to is recovering $\mathcal{G}^*$. Even here, must be careful. Recall that there can be many perfect maps for

I-equivalence a distribution P^*: all of the networks in the same I-equivalence class as $\mathcal{G}^*$. All of these are equally good structures for P^*, and therefore we cannot distinguish between them based only on the data D. In other words, $\mathcal{G}^*$ **is not** *identifiable* **from the data. Thus, the best we can**

 hope for is an algorithm that, asymptotically, recovers $\mathcal{G}^*$'s equivalence class.

identifiability Unfortunately, as our example indicates, the goal of learning $\mathcal{G}^*$ (or an equivalent network) is hard to achieve. The data sampled from P^* are noisy and do not reconstruct this distribution perfectly. We cannot detect with complete reliability which independencies are present in the underlying distribution. Therefore, **we must generally make a decision about our willingness**

 to include in our learned model edges about which we are less sure. If we include more of these edges, we will often learn a model that contains spurious edges. If we include fewer edges, we may miss dependencies. Both compromises lead to inaccurate structures that do not reveal the correct underlying structure. The decision of whether it is better to have spurious correlations or spurious independencies depends on the application.

density estimation The second and more common reason to learn a network structure is in an attempt to perform *density estimation* — that is, to estimate a statistical model of the underlying distribution. As we discussed, our goal is to use this model for reasoning about instances that were not in

generalization our training data. In other words, we want our network model to *generalize* to new instances. It seems intuitively reasonable that because $\mathcal{G}^*$ captures the true dependencies and independencies in the domain, the best generalization will be obtained if we recover the the structure $\mathcal{G}^*$. Moreover, it seems that if we do make mistakes in the structure, it is better to have too many rather than too few edges. With an overly complex structure, we can still capture P^*, and thereby represent the true distribution.

Unfortunately, the situation is somewhat more complex. Let us go back to our coin example and assume that we had 20 data cases with the following frequencies: 3 head/head, 6 head/tail, 5 tail/head, and 6 tail/tail. We can introduce a spurious correlation between X and Y, which would give us, using maximum likelihood estimation, the parameters $P(X = H) = 0.45$, $P(Y = H \mid X = H) = 1/3$, and $P(Y = H \mid X = T) = 5/11$. On the other hand, in the independent structure (with no edge between X and Y), the parameter of Y would be $P(Y = H) = 0.4$. All of these parameter estimates are imperfect, of course, but the ones in the more complex model are significantly more likely to be skewed, because each is estimated from a much smaller data set. In particular, $P(Y = H \mid X = H)$ is estimated from a data set of 9 instances, as opposed to 20 for the estimation of $P(Y = H)$. Recall that the standard deviation of the maximum likelihood estimate behaves as $1/\sqrt{M}$. Thus, if the coins are fair, the standard deviation of the MLE estimate from 20 samples is approximately 0.11, while the standard deviation from 9 samples is approximately 0.17. This example is simply an instance of the *data fragmentation* issue that we discussed in section 17.2.3 in the previous chapter. As

data fragmentation we discussed, when we add more parents to the variable Y, the data used to estimate the CPD fragment into more bins, leaving fewer instances in each bin to estimate the parameters and reducing the quality of the estimated parameters. In a table-CPD, the number of bins grows exponentially with the number of parents, so the (statistical) cost of adding a parent can be very large; moreover, because of the exponential growth, the incremental cost of adding a parent grows with the number of parents already there.

☞ **Thus, when doing density estimation from limited data, it is often better to prefer a sparser structure. The surprising fact is that this observation applies not only to networks that include spurious edges relative to $\mathcal{G}^*$, but also to edges in $\mathcal{G}^*$. That is, we can sometimes learn a better model in term of generalization by learning a structure with fewer edges, even if this structure is incapable of representing the true underlying distribution.**

18.1.2 Overview of Methods

Roughly speaking, there are three approaches to learning without a prespecified structure.

constraint-based structure learning

One approach utilizes *constraint-based structure learning*. These approaches view a Bayesian network as a representation of independencies. They try to test for conditional dependence and independence in the data and then to find a network (or more precisely an equivalence class of networks) that best explains these dependencies and independencies. Constraint-based methods are quite intuitive: they decouple the problem of finding structure from the notion of independence, and they follow more closely the definition of Bayesian network: we have a distribution that satisfies a set of independencies, and our goal is to find an I-map for this distribution. Unfortunately, these methods can be sensitive to failures in individual independence tests. It suffices that one of these tests return a wrong answer to mislead the network construction procedure.

score-based structure learning

model selection

hypothesis space

The second approach is *score-based structure learning*. Score-based methods view a Bayesian network as specifying a statistical model and then address learning as a *model selection* problem. These all operate on the same principle: We define a *hypothesis space* of potential models — the set of possible network structures we are willing to consider — and a scoring function that measures how well the model fits the observed data. Our computational task is then to find the highest-scoring network structure. The space of Bayesian networks is a combinatorial space, consisting of a superexponential number of structures — $2^{O(n^2)}$. Therefore, even with a scoring function, it is not clear how one can find the highest-scoring network. As we will see, there are very special cases where we can find the optimal network. In general, however, the problem is (as usual) $\mathcal{NP}$-hard, and we resort to heuristic search techniques. Score-based methods consider the whole structure at once; they are therefore less sensitive to individual failures and better at making compromises between the extent to which variables are dependent in the data and the "cost" of adding the edge. The disadvantage of the score-based approaches is that they pose a search problem that may not have an elegant and efficient solution.

Bayesian model averaging

Finally, the third approach does not attempt to learn a single structure; instead, it generates an ensemble of possible structures. These *Bayesian model averaging* methods extend the Bayesian reasoning we encountered in the previous chapter and try to average the prediction of all possible structures. Since the number of structures is immense, performing this task seems impossible. For some classes of models this can be done efficiently, and for others we need to resort to approximations.

18.2   Constraint-Based Approaches

18.2.1   General Framework

In constraint-based approaches, we attempt to reconstruct a network structure that best captures the independencies in the domain. In other words, we attempt to find the best minimal I-map for the domain.

Recall that in chapter 3 we discussed algorithms for building I-maps and P-maps that assume that we can test for independence statements in the distribution. The algorithms for constraint-based learning are essentially variants of these algorithms. The main technical question is how to answer independence queries. For now, assume that we have some procedure that can answer such queries. That is, for a given distribution P, the learning algorithm can pose a question, such as "Does P satisfy $(X_1 \perp X_2, X_3 \mid X_4)$?" and receive a yes/no answer. The task of the algorithm is to carry out some algorithm that interacts with this procedure and results in a network structure that is the minimal I-map of P.

We have already seen such an algorithm in chapter 3: Build-Minimal-I-Map constructs a
minimal I-map   *minimal I-map* given a fixed ordering. For each variable X_i, it then searches for the minimal subset of $X_1, \ldots, X_{i-1}$ that render X_i independent of the others. This algorithm was useful in illustrating the definition of an I-map, but it suffers from several drawbacks in the context of learning. First, the input order over variables can have a serious impact on the complexity of the network we find. Second, in learning the parents of X_i, this algorithm poses independence queries of the form $(X_i \perp \{X_1, \ldots, X_{i-1}\} - U \mid U)$. These conditional independence statements involve a large number of variables. Although we do not assume much about the independence testing procedure, we do realize that independence statements with many variables are much more problematic to resolve from empirical data. Finally, Build-Minimal-I-Map performs a large number of queries. For determining the parents of X_i, it must, in principle, examine all the 2^{i-1} possible subsets of $X_1, \ldots, X_{i-1}$.

To avoid these problems, we learn an I-equivalence class rather than a single network, and
class PDAG   we use a *class PDAG* to represent this class. The algorithm that we use is a variant of the Build-PDAG procedure of algorithm 3.5. As we discuss, this algorithm reconstructs the network that best matches the domain without a prespecified order and uses only a polynomial number
independence   of *independence tests* that involve a bounded number of variables.
test

To achieve these performance guarantees, we must make some assumptions:

- The network $\mathcal{G}^*$ has bounded indegree, that is, for all i, $|\mathrm{Pa}_{X_i}^{\mathcal{G}^*}| \leq d$ for some constant d.

- The independence procedure can perfectly answer any independence query that involves up to $2d + 2$ variables.

- The underlying distribution P^* is faithful to $\mathcal{G}^*$, as in definition 3.8.

The first assumption states the boundaries of when we expect the algorithm to work. If the network is simple in this sense, the algorithm will be able to learn it from the data. If the network is more complex, then we cannot hope to learn it with "small" independence queries that involve only a few variables.

The second assumption is stronger, since it requires that the oracle can deal with queries up to a certain size. The learning algorithm does not depend on how the these queries are answered. They might be answered by performing a statistical test for conditional dependence on a training data, or by an active mechanism that gathers more samples until it can reach a significant conclusion about this relations. We discuss how to construct such an oracle in more detail later in this chapter. Note that the oracle can also be a human expert who helps in constructing a model of the network.

The third assumption is the strongest. It is required to ensure that the algorithm is not misled by spurious independencies that are not an artifact of the oracle but rather exist in the domain. By requiring that $\mathcal{G}^*$ is a perfect map of P^*, we rule out quite a few situations, for example, the (noisy) XOR example of example 3.6, and various cases where additional independencies arise from structure in the CPDs.

Once we make these assumptions, the setting is precisely the one we tackled in section 3.4.3. Thus, given an oracle that can answer independence statements perfectly, we can now simply apply Build-PMap-Skeleton. Of course, determining independencies from the data is not a trivial problem, and the answers are rarely guaranteed to be perfect in practice. We will return to these important questions. For the moment, we focus on analyzing the number of independence queries that we need to answer, and thereby the complexity of the algorithm.

Recall that, in the construction of perfect maps, we perform independence queries only in the Build-PMAP-Skeleton procedure, when we search for a witness to the separation between every pair of variables. These witnesses are also used within Mark-Immoralities to determine whether the two parents in a v-structure are conditionally independent. According to lemma 3.2, if X and Y are not adjacent in $\mathcal{G}^*$, then either $\mathrm{Pa}_X^{\mathcal{G}^*}$ or $\mathrm{Pa}_Y^{\mathcal{G}^*}$ is a witness set. If we assume that $\mathcal{G}^*$ has indegree of at most d, we can therefore limit our attention to witness sets of size at most d. Thus, the number of independence queries in this step is polynomial in n, the number of variables. Of course, this number is exponential in d, but we assume that d is a fixed constant throughout the analysis.

Thus, given our assumptions, we can perform a variant of Build-PDAG that performs a polynomial number of independence tests. We can also check all other operations; that is, applying the edge orientation rules, we can also require a polynomial number of steps. Thus, the procedure is polynomial in the number of variables.

18.2.2 Independence Tests

The only remaining question is how to answer queries about conditional independencies between variables in the data. As one might expect, this question has been extensively studied in the statistics literature. We briefly touch on some of the issues and outline one commonly-used methodology to answer this question.

The basic query of this type is to determine whether two variables are independent. As in the example in the introduction to this chapter, we are given joint samples of two variables X and Y, and we want to determine whether X and Y are independent. This basic question is often referred to as *hypothesis testing*.

hypothesis testing

18.2.2.1 Single-Sided Hypothesis Tests

In hypothesis testing, we have a base hypothesis that is usually denoted by H_0 and is referred to as the *null hypothesis*. In the particular case of the independence test, the null hypothesis is "the data were sampled from a distribution $P^*(X, Y) = P^*(X)P^*(Y)$." Note that this

null hypothesis

assumption states that the data were sampled from a *particular* distribution in which X and Y are independent. In real life, we do not have access to $P^*(X)$ and $P^*(Y)$. As a substitute, we use $\hat{P}(X)$ and $\hat{P}(Y)$ as our best approximation for this distribution. Thus, we usually form H_0 as the assumption that $P^*(X, Y) = \hat{P}(X)\hat{P}(Y)$.

decision rule

We want to test whether the data conform to this hypothesis. More precisely, we want to find a procedure that we will call a *decision rule* that will take as input a data set $\mathcal{D}$, and return a verdict, either Accept or Reject. We will denote the function the procedure computes to be $R(\mathcal{D})$. If $R(\mathcal{D}) =$ Accept, then we consider that the data satisfy the hypothesis. In our case, that would mean that we believe that the data were sampled from P^* and that the two variables are independent. Otherwise, we decide to reject the hypothesis, which in our case would imply that the variables are dependent.

The question is then, of course, how to choose a "good" decision rule. A liberal decision rule that accepts many data sets runs the risk of accepting ones that do not satisfy the hypothesis. A conservative rule that rejects many data sets runs the risk of rejecting many that satisfy the hypothesis. The common approach to evaluating a decision rule is analyze the probability of false rejection. Suppose we have access to the distribution $P(\mathcal{D} : H_0, M)$ of data sets of M instances given the null hypothesis. That is, we can evaluate the probability of seeing each particular data set if the hypothesis happens to be correct. In our case, since the hypothesis specifies the distribution P^*, this distribution is just the probability of sampling the particular instances in the data set (we assume that the size of the data set is known in advance).

If we have access to this distribution, we can compute the probability of false rejection:

$$P(\{\mathcal{D} : R(\mathcal{D}) = \text{Reject}\} \mid H_0, M).$$

Then we can say that a decision rule R has a probability of false rejection p. We often refer to $1 - p$ as the confidence in the decision to reject an hypothesis.[1]

At this point we cannot evaluate the probability of false acceptances. Since we are not willing to assume a concrete distribution on data sets that violate H_0, we cannot quantify this probability. For this reason, the decision is not symmetric. That is, rejecting the hypothesis "X and Y are independent" is not the same as accepting the hypothesis "X and Y are dependent." In particular, to define the latter hypothesis we need to specify a distribution over data sets.

18.2.2.2   Deviance Measures

The preceding discussion suggests how to evaluate decision rules. Yet, it leaves open the question of how to design such a rule. A standard framework for this question is to define a measure of

deviance

deviance from the null hypothesis. Such a measure d is a function from possible data sets to the real line. Intuitively, large value of $d(\mathcal{D})$ implies that $\mathcal{D}$ is far away from the null hypothesis.

To consider a concrete example, suppose we have discrete-valued, independent random variables X and Y. Typically, we expect that the counts $M[x, y]$ in the data are close to $M \cdot \hat{P}(x) \cdot \hat{P}(y)$ (where M is the number of samples). This is the expected value of the count, and, as we know, deviances from this value are improbable for large M. Based on this intuition, we can measure the deviance of the data from H_0 in terms of these distances. A common

χ^2 statistic

measure of this type is the χ^2 *statistic*:

---

1. This leads statisticians to state, "We reject the null hypothesis with confidence 95 percent" as a precise statement that can be intuitively interpreted as, "We are quite confident that the variables are correlated."

$$d_{\chi^2}(\mathcal{D}) = \sum_{x,y} \frac{(M[x,y] - M \cdot \hat{P}(x) \cdot \hat{P}(y))^2}{M \cdot \hat{P}(x) \cdot \hat{P}(y)}. \tag{18.1}$$

A data set that perfectly fits the independence assumption has $d_{\chi^2}(\mathcal{D}) = 0$, and a data set where the empirical and expected counts diverge significantly has a larger value.

mutual
information

Another potential deviance measure for the same hypothesis is the *mutual information* $\mathbf{I}_{\hat{P}_{\mathcal{D}}}(X;Y)$ in the empirical distribution defined by the data set $\mathcal{D}$. In terms of counts, this can be written as

$$d_{\mathbf{I}}(\mathcal{D}) = \mathbf{I}_{\hat{P}_{\mathcal{D}}}(X;Y) = \frac{1}{M} \sum_{x,y} M[x,y] \log \frac{M[x,y]}{M[x]M[y]}. \tag{18.2}$$

In fact, these two deviance measures are closely related to each other; see exercise 18.1.

Once we agree on a deviance measure d (say the χ^2 statistic or the empirical mutual information), we can devise a rule for testing whether we want to accept the hypothesis

$$R_{d,t}(\mathcal{D}) = \begin{cases} \text{Accept} & d(\mathcal{D}) \leq t \\ \text{Reject} & d(\mathcal{D}) > t. \end{cases}$$

This rule *accepts* the hypothesis if the deviance is small (less than the predetermined threshold t) and *rejects* the hypothesis if the deviance is large.

The choice of threshold t determines the false rejection probability of the decision rule. The computational problem is to compute the false rejection probability for different values of t. This value is called the *p-value* of t:

p-value

$$\text{p-value}(t) = P(\{\mathcal{D} : d(\mathcal{D}) > t\} \mid H_0, M).$$

18.2.2.3 Testing for Independence

Using the tools we developed so far, we can reexamine the independence test. The basic tool we use is a test to reject the null hypothesis that distribution of X and Y is the one we would estimate if we assume that they are independent. The typical significance level we use is 95 percent. That is, we reject the null hypothesis if the deviance in the observed data has p-value of 0.05 or less.

If we want to test the independence of discrete categorical variables, we usually use the χ^2 statistic or the mutual information. The null hypothesis is that $P^*(X, Y) = \hat{P}(X)\hat{P}(Y)$.

We start by considering how to perform an *exact test*. The definition of p-value requires summing over all possible data sets. In fact, since we care only about the sufficient statistics of X and Y in the data set, we can sum over the smaller space of different sufficient statistics vectors. Suppose we have M samples; we define the space $\mathcal{C}_{X,Y}^M$ to be the set of all empirical counts over X and Y, we might observe in a data set with M samples. Then we write

$$\text{p-value}(t) = \sum_{C[X,Y] \in \mathcal{C}_{X,Y}^M} \mathbf{I}\{d(C[X,Y]) > t\} P(C[X,Y] \mid H_0, M),$$

where $d(C[X,Y])$ is the deviance measure (that is, χ^2 or mutual information) computed with the counts $C[X,Y]$, and

$$P(C[X,Y] \mid H_0, M) = M! \prod_{x,y} \frac{1}{C[x,y]!} P(x,y \mid H_0)^{C[x,y]} \tag{18.3}$$

is the probability of seeing a data set with these counts given H_0; see exercise 18.2.

This exact approach enumerates through all data sets. This is clearly infeasible except for small values of M. A more common approach is to examine the asymptotic distribution of $M[x, y]$ under the null hypothesis. Since this count is a sum of binary indicator variables, its distribution is approximately normal when M is large enough. Statistical theory develops the asymptotic distribution of the deviance measure under the null hypothesis. For the χ^2 statistic,

χ^2 distribution

this distribution is called the χ^2 *distribution*. We can use the tail probability of this distribution to approximate p-values for independence tests. Numerical procedures for such computations are part of most standard statistical packages.

A natural extension of this test exists for testing conditional independence. Suppose we want to test whether X and Y are independent given Z. Then, H_0 is that $P^*(X, Y, Z) = \hat{P}(Z)\hat{P}(X \mid Z)\hat{P}(Y \mid Z)$, and the χ^2 statistic is

$$d_{\chi^2}(\mathcal{D}) = \sum_{x,y,z} \frac{(M[x, y, z] - M \cdot \hat{P}(z)\hat{P}(x \mid z)\hat{P}(y \mid z))^2}{M \cdot \hat{P}(z)\hat{P}(x \mid z)\hat{P}(y \mid z)}.$$

This formula extends easily to conditioning on a set of variables $\mathbf{Z}$.

18.2.2.4 Building Networks

We now return to the problem of learning network structure. With the methods we just discussed, we can evaluate independence queries in the Build-PDAG procedure, so that whenever the test rejects the null hypothesis we treat the variables as dependent. One must realize, however, that these tests are not perfect. Thus, we run the risk of making wrong decisions on some of the queries. In particular, if we use significance level of 95 percent, then we expect that on average 1 in 20 rejections is wrong. When testing a large number of hypotheses, a scenario

multiple
hypothesis testing

called *multiple hypothesis testing*, the number of incorrect conclusions can grow large, reducing our ability to reconstruct the correct network. We can try to reduce this number by taking stricter significance levels (see exercise 18.3). This, however, runs the risk of making more errors of the opposite type.

In conclusion, we have to be aware that some of the independence tests results can be wrong. The procedure Build-PDAG can be sensitive to such errors. In particular, one misleading independence test result can produce multiple errors in the resulting PDAG (see exercise 18.4). When we have relatively few variables and large sample size (and "strong" dependencies among variables), the reconstruction algorithm we described here is efficient and often manages to find a structure that is quite close to the correct structure. When the independence test results are less pronounced, the constraint-based approach can run into trouble.

18.3   Structure Scores

As discussed earlier, score-based methods approach the problem of structure learning as an optimization problem. We define a score function that can score each candidate structure with respect to the training data, and then search for a high-scoring structure. As can be expected, one of the most important decisions we must make in this framework is the choice of scoring function. In this section, we discuss two of the most obvious choices.

18.3.1 Likelihood Scores

18.3.1.1 Maximum Likelihood Parameters

A natural choice for scoring function is the likelihood function, which we used for parameter estimation. Recall that this function measures the probability of the data given a model. Thus, it seems intuitive to find a model that would make the data as probable as possible.

Assume that we want to maximize the likelihood of the model. In this case, our model is a pair $\langle \mathcal{G}, \boldsymbol{\theta}_{\mathcal{G}} \rangle$. Our goal is to find both a graph $\mathcal{G}$ and parameters $\boldsymbol{\theta}_{\mathcal{G}}$ that maximize the likelihood.

In the previous chapter, we determined how to maximize the likelihood for a given structure $\mathcal{G}$. We simply use the maximum likelihood parameters $\hat{\boldsymbol{\theta}}_{\mathcal{G}}$ for that graph. A simple analysis now shows that:

$$
\max_{\mathcal{G}, \boldsymbol{\theta}_{\mathcal{G}}} L(\langle \mathcal{G}, \boldsymbol{\theta}_{\mathcal{G}} \rangle : \mathcal{D}) = \max_{\mathcal{G}} [\max_{\boldsymbol{\theta}_{\mathcal{G}}} L(\langle \mathcal{G}, \boldsymbol{\theta}_{\mathcal{G}} \rangle : \mathcal{D})]
$$
$$
= \max_{\mathcal{G}} [L(\langle \mathcal{G}, \hat{\boldsymbol{\theta}}_{\mathcal{G}} \rangle : \mathcal{D})].
$$

In other words, to find the maximum likelihood $(\mathcal{G}, \boldsymbol{\theta}_{\mathcal{G}})$ pair, we should find the graph structure $\mathcal{G}$ that achieves the highest likelihood *when we use the MLE parameters for* $\mathcal{G}$. We define:

$$
\text{score}_L(\mathcal{G} : \mathcal{D}) = \ell(\hat{\boldsymbol{\theta}}_{\mathcal{G}} : \mathcal{D}),
$$

where $\ell(\hat{\boldsymbol{\theta}}_{\mathcal{G}} : \mathcal{D})$ is the logarithm of the likelihood function and $\hat{\boldsymbol{\theta}}_{\mathcal{G}}$ are the maximum likelihood parameters for $\mathcal{G}$. (As usual, it will be easier to deal with the logarithm of the likelihood.)

18.3.1.2 Information-Theoretic Interpretation

To get a better intuition of the likelihood score, let us consider the scenario of example 18.1. Consider the model $\mathcal{G}_0$ where X and Y are independent. In this case, we get

$$
\text{score}_L(\mathcal{G}_0 : \mathcal{D}) = \sum_m \log \hat{\theta}_{x[m]} + \log \hat{\theta}_{y[m]}.
$$

On the other hand, we can consider the model G_1 where there is an arc $X \to Y$. The log-likelihood for this model is

$$
\text{score}_L(\mathcal{G}_1 : \mathcal{D}) = \sum_m \log \hat{\theta}_{x[m]} + \log \hat{\theta}_{y[m]|x[m]},
$$

where $\hat{\theta}_x$ is again the maximum likelihood estimate for $P(x)$, and $\hat{\theta}_{y|x}$ is the maximum likelihood estimate for $P(y \mid x)$.

We see that the score of two models share a common component (the terms of the form $\log \hat{\theta}_x$). Thus, we can write the difference between the two scores as

$$
\text{score}_L(\mathcal{G}_1 : \mathcal{D}) - \text{score}_L(\mathcal{G}_0 : \mathcal{D}) = \sum_m \log \hat{\theta}_{y[m]|x[m]} - \log \hat{\theta}_{y[m]}.
$$

By counting how many times each conditional probability parameter appears in this term, we can write this sum as:

$$
\text{score}_L(\mathcal{G}_1 : \mathcal{D}) - \text{score}_L(\mathcal{G}_0 : \mathcal{D}) = \sum_{x,y} M[x, y] \log \hat{\theta}_{y|x} - \sum_y M[y] \log \hat{\theta}_y.
$$

Let $\hat{P}$ be the empirical distribution observed in the data; that is, $\hat{P}(x,y)$ is simply the empirical frequency of x, y in D. Then, we can write $M[x,y] = M \cdot \hat{P}(x,y)$, and $M[y] = M\hat{P}(y)$. Moreover, it is easy to check that $\hat{\theta}_{y|x} = \hat{P}(y \mid x)$, and that $\hat{\theta}_y = \hat{P}(y)$. We get:

$$\text{score}_L(\mathcal{G}_1 \; : \; \mathcal{D}) - \text{score}_L(\mathcal{G}_0 \; : \; \mathcal{D}) = M \sum_{x,y} \hat{P}(x,y) \log \frac{\hat{P}(y \mid x)}{\hat{P}(y)} = M \cdot \boldsymbol{I}_{\hat{P}}(X;Y),$$

mutual
information

where $\boldsymbol{I}_{\hat{P}}(X;Y)$ is the *mutual information* between X and Y in the distribution $\hat{P}$.

We see that the likelihood of the model $\mathcal{G}_1$ depends on the mutual information between X and Y. Recall that higher mutual information implies stronger dependency. Thus, stronger dependency implies stronger preference for the model where X and Y depend on each other.

Can we generalize this information-theoretic formulation of the maximum likelihood score to general network structures? Going through a similar arithmetic transformations, we can prove the following result.

Proposition 18.1

decomposable
score

The likelihood score decomposes as follows:

$$\text{score}_L(\mathcal{G} \; : \; \mathcal{D}) = M \sum_{i=1}^n \boldsymbol{I}_{\hat{P}}(X_i; \text{Pa}_{X_i}^{\mathcal{G}}) - M \sum_{i=1}^n \boldsymbol{H}_{\hat{P}}(X_i). \tag{18.4}$$

PROOF We have already seen that by combining all the occurrences of each parameter $\theta_{x_i|\boldsymbol{u}}$, we can rewrite the log-likelihood function as

$$\ell(\hat{\boldsymbol{\theta}}_{\mathcal{G}} : \mathcal{D}) = \sum_{i=1}^n \left[\sum_{\boldsymbol{u}_i \in Val(\text{Pa}_{X_i}^{\mathcal{G}})} \sum_{x_i} M[x_i, \boldsymbol{u}_i] \log \hat{\theta}_{x_i|\boldsymbol{u}_i} \right].$$

Consider one of the terms in the square brackets, and let $\boldsymbol{U}_i = \text{Pa}_{X_i}$.

$$\frac{1}{M} \sum_{\boldsymbol{u}_i} \sum_{x_i} M[x_i, \boldsymbol{u}_i] \log \hat{\theta}_{x_i|\boldsymbol{u}_i}$$

$$= \sum_{\boldsymbol{u}_i} \sum_{x_i} \hat{P}(x_i, \boldsymbol{u}_i) \log \hat{P}(x_i \mid \boldsymbol{u}_i)$$

$$= \sum_{\boldsymbol{u}_i} \sum_{x_i} \hat{P}(x_i, \boldsymbol{u}_i) \log \left(\frac{\hat{P}(x_i, \boldsymbol{u}_i)}{\hat{P}(\boldsymbol{u}_i)} \frac{\hat{P}(x_i)}{\hat{P}(x_i)} \right)$$

$$= \sum_{\boldsymbol{u}_i} \sum_{x_i} \hat{P}(x_i, \boldsymbol{u}_i) \log \frac{\hat{P}(x_i, \boldsymbol{u}_i)}{\hat{P}(\boldsymbol{u}_i)\hat{P}(x_i)} + \sum_{x_i} \left(\sum_{\boldsymbol{u}_i} \hat{P}(x_i, \boldsymbol{u}_i) \right) \log \hat{P}(x_i)$$

$$= \boldsymbol{I}_{\hat{P}}(X_i; \boldsymbol{U}_i) - \sum_{x_i} \hat{P}(x_i) \log \frac{1}{\hat{P}(x_i)}$$

$$= \boldsymbol{I}_{\hat{P}}(X_i; \boldsymbol{U}_i) - \boldsymbol{H}_{\hat{P}}(X_i),$$

where (as implied by the definition) the mutual information $\boldsymbol{I}_{\hat{P}}(X_i; \text{Pa}_{X_i})$ is 0 if $\text{Pa}_{X_i} = \emptyset$. ∎

Note that the second sum in equation (18.4) does not depend on the network structure, and thus we can ignore it when we compare two structures with respect to the same data set.

Recall that we can interpret $I_P(X;Y)$ as the strength of the dependence between X and Y in P. **Thus, the likelihood of a network measures the strength of the dependencies between variables and their parents. In other words, we prefer networks where the parents of each variable are informative about it.**

This result can also be interpreted in a complementary manner.

Corollary 18.1

Let $X_1, \ldots, X_n$ be an ordering of the variables that is consistent with edges in $\mathcal{G}$. Then,

$$\frac{1}{M}\text{score}_L(\mathcal{G} \;:\; \mathcal{D}) = \boldsymbol{H}_{\hat{P}}(X_1, \ldots, X_n) - \sum_{i=1}^{n} \boldsymbol{I}_{\hat{P}}(X_i; \{X_1, \ldots X_{i-1}\} - \text{Pa}_{X_i}^{\mathcal{G}} \mid \text{Pa}_{X_i}^{\mathcal{G}}). \quad (18.5)$$

For proof, see exercise 18.5.

Again, this second reformulation of the likelihood has a term that does not depend on the structure, and one that does. This latter term involves conditional mutual-information expressions of the form $\boldsymbol{I}_{\hat{P}}(X_i; \{X_1, \ldots X_{i-1}\} - \text{Pa}_{X_i}^{\mathcal{G}} \mid \text{Pa}_{X_i}^{\mathcal{G}})$. That is, the information between X_i and the preceding variables in the order given X_i's parents. Smaller conditional mutual-information terms imply higher scores. Recall that conditional independence is equivalent to having zero conditional mutual information. Thus, we can interpret this formulation as measuring to what extent the Markov properties implied by $\mathcal{G}$ are violated in the data. The smaller the violations of the Markov property, the larger the score.

These two interpretations are complementary, one measuring the strength of dependence between X_i and its parents $\text{Pa}_{X_i}^{\mathcal{G}}$, and the other measuring the extent of the independence of X_i from its predecessors given $\text{Pa}_{X_i}^{\mathcal{G}}$.

The process of choosing a network structure is often subject to constraints. Some constraints are a consequence of the acyclicity requirement, others may be due to a preference for simpler structures. Our previous analysis shows that the likelihood score provides valuable guidance in selecting between different candidate networks.

18.3.1.3 Limitations of the Maximum Likelihood Score

Based on the developments in the previous chapter and the preceding analysis, we see that the likelihood score is a good measure of the fit of the estimated Bayesian network and the training data. In learning structure, however, we are also concerned about the performance of the learned network on new instances sampled from the same underlying distribution P^*. Unfortunately, in this respect, the likelihood score can run into problems.

To see this, consider example 18.1. Let $\mathcal{G}_\emptyset$ be the network where X and Y are independent, and $\mathcal{G}_{X \to Y}$ the one where X is the parent of Y. As we have seen, $\text{score}_L(\mathcal{G}_{X \to Y} \;:\; \mathcal{D}) - \text{score}_L(\mathcal{G}_\emptyset \;:\; \mathcal{D}) = M \cdot \boldsymbol{I}_{\hat{P}}(X;Y)$. Recall that the mutual information between two variables is nonnegative. Thus, $\text{score}_L(\mathcal{G}_{X \to Y} \;:\; \mathcal{D}) \geq \text{score}_L(\mathcal{G}_\emptyset \;:\; \mathcal{D})$ for any data set D. This implies that the maximum likelihood score *never* prefers the simpler network over the more complex one. And it assigns both networks the same score only in these rare situations when X and Y are truly independent in the training data.

As explained in the introduction to this chapter, there are situations where we should prefer to learn the simpler network (for example, when X and Y are nearly independent in the training data). We see that the maximum likelihood score would never lead us to make that choice.

This observation applies to more complex networks as well. It is easy to show that adding an edge to a network structure can never decrease the maximum likelihood score. Furthermore, the more complex network will have a higher score in all but a vanishingly small fraction of cases. One approach to proving this follows directly from the notion of likelihood; see exercise 18.6. Another uses the fact that, for any X, Y, Z and any distribution P, we have that:

$$I_P(X; Y \cup Z) \geq I_P(X; Y),$$

with equality holding only if Z is conditionally independent of X given Y, see exercise 2.20. This inequality is fairly intuitive: if Y gives us a certain amount of information about X, adding Z can only give us more information. Thus, the mutual information between a variable and its parents can only go up if we add another parent, and it will go up except in those few cases where we get a conditional independence assertion holding exactly *in the empirical distribution*. It follows that **the maximum likelihood network will exhibit a conditional independence only when that independence happens to hold exactly in the empirical distribution. Due to statistical noise, exact independence almost never occurs, and therefore, in almost all cases, the maximum likelihood network will be a fully connected one. In other words, the likelihood score *overfits* the training data (see section 16.3.1), learning a model that precisely fits the specifics of the empirical distribution in our training set. This model therefore fails to generalize well to new data cases: these are sampled from the underlying distribution, which is not identical to the empirical distribution in our training set.**

overfitting

We note that the discussion of the maximum likelihood score was in the context of networks with table-CPDs. However, the same observations also apply to learning networks with other forms of CPDs (for example, tree-CPDs, noisy-ors, or Gaussians). In these cases, the information-theoretic analysis is somewhat more elaborate, but the general conclusions about the trade-offs between models and about overfitting apply.

Since the likelihood score does not provide us with tools to avoid overfitting, we have to be careful when using it. It is reasonable to use the maximum likelihood score when there are additional mechanisms that disallow overly complicated structures. For example, we will discuss learning networks with a fixed indegree. Such a limitation can constrain the tendency to overfit when using the maximum likelihood score.

18.3.2 Bayesian Score

We now examine an alternative scoring function that is based on a Bayesian perspective; this approach extends ideas that we described in the context of parameter estimation in the previous chapter. We will start by deriving the score from the Bayesian perspective, and then we will try to understand how it avoids overfitting.

Recall that the main principle of the Bayesian approach was that whenever we have uncertainty over anything, we should place a distribution over it. In this case, we have uncertainty both over structure and over parameters. We therefore define a structure prior $P(\mathcal{G})$ that puts a prior probability on different graph structures, and a parameter prior $P(\boldsymbol{\theta}_{\mathcal{G}} \mid \mathcal{G})$, that puts a probability on different choice of parameters once the graph is given. By Bayes rule, we have

$$P(\mathcal{G} \mid \mathcal{D}) = \frac{P(\mathcal{D} \mid \mathcal{G})P(\mathcal{G})}{P(\mathcal{D})},$$

where, as usual, the denominator is simply a normalizing factor that does not help distinguish between different structures. Thus, we define the *Bayesian score* as:

Bayesian score

$$\text{score}_B(\mathcal{G} \; : \; \mathcal{D}) = \log P(\mathcal{D} \mid \mathcal{G}) + \log P(\mathcal{G}). \tag{18.6}$$

The ability to ascribe a prior over structures gives us a way of preferring some structures over others. For example, we can penalize dense structures more than sparse ones. It turns out, however, that the structure-prior term in the score is almost irrelevant compared to the first term. This term, $P(\mathcal{D} \mid \mathcal{G})$, takes into consideration our uncertainty over the parameters:

$$P(\mathcal{D} \mid \mathcal{G}) = \int_{\Theta_{\mathcal{G}}} P(\mathcal{D} \mid \boldsymbol{\theta}_{\mathcal{G}}, \mathcal{G}) P(\boldsymbol{\theta}_{\mathcal{G}} \mid \mathcal{G}) d\boldsymbol{\theta}_{\mathcal{G}}, \tag{18.7}$$

marginal likelihood

where $P(\mathcal{D} \mid \boldsymbol{\theta}_{\mathcal{G}}, \mathcal{G})$ is the likelihood of the data given the network $\langle \mathcal{G}, \boldsymbol{\theta}_{\mathcal{G}} \rangle$ and $P(\boldsymbol{\theta}_{\mathcal{G}} \mid \mathcal{G})$ is our prior distribution over different parameter values for the network $\mathcal{G}$. Recall from section 17.4 that $P(\mathcal{D} \mid \mathcal{G})$ is called the *marginal likelihood* of the data given the structure, since we marginalize out the unknown parameters.

It is important to realize that the marginal likelihood is quite different from the maximum likelihood score. Both terms examine the likelihood of the data given the structure. The maximum likelihood score returns the maximum of this function. In contrast, the marginal likelihood is the average value of this function, where we average based on the prior measure $P(\boldsymbol{\theta}_{\mathcal{G}} \mid \mathcal{G})$. This difference will become apparent when we analyze the marginal likelihood term.

One explanation of why the Bayesian score avoids overfitting examines the sensitivity of the likelihood to the particular choice of parameters. As we discussed, the maximal likelihood is overly "optimistic" in its evaluation of the score: It evaluates the likelihood of the training data using the best parameter values for the given data. This estimate is realistic only if these parameters are also reflective of the data in general, a situation that never occurs.

The Bayesian approach tells us that, although the choice of parameter $\hat{\boldsymbol{\theta}}$ is the most likely given the training set D, it is not the only choice. The posterior over parameters provides us with a range of choices, along with a measure of how likely each of them is. By integrating $P(\mathcal{D} \mid \boldsymbol{\theta}_{\mathcal{G}}, \mathcal{G})$ over the different choices of parameters $\boldsymbol{\theta}_{\mathcal{G}}$, we are measuring the *expected* likelihood, averaged over different possible choices of $\boldsymbol{\theta}_{\mathcal{G}}$. Thus, we are being more conservative in our estimate of the "goodness" of the model.

holdout testing

Another motivation can be derived from the *holdout testing* methods discussed in box 16.A. Here, we consider different network structures, parameterized by the training set, and test their predictiveness (likelihood) on the validation set. When we find a network that best generalizes to the validation set (that is, has the best likelihood on this set), we have some reason to hope that it will also generalize to other unseen instances. As we discussed, the holdout method is sensitive to the particular split into training and test sets, both in terms of the relative sizes of the sets and in terms of which instances fall into which set. Moreover, it does not use all the available data in learning the structure, a potentially serious problem when we have limited amounts of data to learn from.

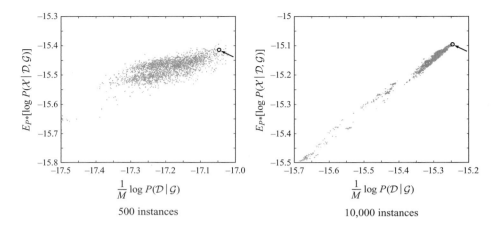

Figure 18.1   Marginal likelihood in training data as predictor of expected likelihood on underlying distribution. Comparison of the average log-marginal-likelihood per sample in training data (x-axis) to the expected log-likelihood of new samples from the underlying distribution (y-axis) for two data sets sampled from the ICU-Alarm network. Each point corresponds to a network structure; the true network structure is marked by a circle.

It turns out that the Bayesian approach can be viewed as performing a similar evaluation without explicitly splitting the data into two parts. Using the chain rule for probabilities, we can rewrite the marginal likelihood as

$$P(\mathcal{D} \mid \mathcal{G}) = \prod_{m=1}^{M} P(\xi[m] \mid \xi[1], \dots, \xi[m-1], \mathcal{G}).$$

Each of the terms in this product — $P(\xi[m] \mid \xi[1], \dots, \xi[m-1], \mathcal{G})$ — is the probability of the m'th instance using the parameters learned from the first $m-1$ instances (using Bayesian estimation). We see that in this term we are using the m'th instance as a test case, since we are computing its probability using what we learned from previous instances. Thus, it provides us with one data point for testing the ability of our model to predict a new data instance, based on the model learned from the previous ones. This type of analysis is called a *prequential analysis*.

prequential
analysis

However, unlike the holdout approach, we are not holding out any data. Each instance is evaluated in incremental order, and contributes both to our evaluation of the model and to our final model score. Moreover, the Bayesian score does not depend on the order of instances. Using the chain law of probabilities, we can generate a similar expansion for any ordering of the instances. Each one of these will give the same result (since these are different ways of expanding the term $P(\mathcal{D} \mid \mathcal{G})$).

This intuition suggests that

$$\frac{1}{M} \log P(\mathcal{D} \mid \mathcal{G}) \approx \boldsymbol{E}_{P^*}[\log P(\mathcal{X} \mid \mathcal{G}, \mathcal{D})] \tag{18.8}$$

is an estimator for the average log-likelihood of a new sample from the distribution P^*. In

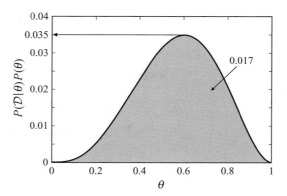

Figure 18.2 **Maximal likelihood score versus marginal likelihood for the data** $\langle H, T, T, H, H \rangle$.

practice, it turns out that for reasonable sample sizes this is indeed a fairly good estimator of the ability of a model to generalize to unseen data. Figure 18.1 demonstrates this property empirically for data sets sampled from the ICU-Alarm network. We generated a collection of network structures by sampling from the posterior distribution over structures given different data sets (see section 18.5). For each structure we evaluated the two sides of the preceding approximation: the average log-likelihood per sample, and the expected likelihood of new samples from the underlying distribution. As we can see, there is a general agreement between the estimate using the training data and the actual generalization error of each network structure. In particular, the difference in scores of two structures correlates with the differences in generalization error. This phenomenon is particularly noticeable in the larger training set.

We note that the Bayesian score is not the only way of providing "test set" performance using each instance. See exercise 18.12 for an alternative score with similar properties.

18.3.3 Marginal Likelihood for a Single Variable

We now examine how to compute the marginal likelihood for simple cases, and then in the next section treat the case of Bayesian networks.

Consider a single binary random variable X, and assume that we have a prior distribution $Dirichlet(\alpha_1, \alpha_0)$ over X. Consider a data set $\mathcal{D}$ that has $M[1]$ heads and $M[0]$ tails. Then, the maximum likelihood value given D is

$$P(\mathcal{D} \mid \hat{\theta}) = \left(\frac{M[1]}{M} \right)^{M[1]} \cdot \left(\frac{M[0]}{M} \right)^{M[0]}.$$

Now, consider the marginal likelihood. Here, we are not conditioning on the parameter. Instead, we need to compute the probability $P(X[1], \ldots, X[M])$ of the data given our prior. One approach to computing this term is to evaluate the integral equation (18.7). An alternative approach uses the chain rule

$$P(x[1], \ldots, x[M]) = P(x[1]) \cdot P(x[2] \mid x[1]) \cdot \ldots \cdot P(x[M] \mid x[1], \ldots, x[M-1]).$$

Recall that if we use a Beta prior, then

$$P(x[m+1] = H \mid x[1], \ldots, x[m]) = \frac{M^m[1] + \alpha_1}{m + \alpha},$$

where $M^m[1]$ is the number of heads in the first m examples. For example, if $\mathcal{D} = \langle H, T, T, H, H \rangle$,

$$
\begin{aligned}
P(x[1], \ldots, x[5]) &= \frac{\alpha_1}{\alpha} \cdot \frac{\alpha_0}{\alpha + 1} \cdot \frac{\alpha_0 + 1}{\alpha + 2} \cdot \frac{\alpha_1 + 1}{\alpha + 3} \cdot \frac{\alpha_1 + 2}{\alpha + 4} \\
&= \frac{[\alpha_1(\alpha_1 + 1)(\alpha_1 + 2)][\alpha_0(\alpha_0 + 1)]}{\alpha \cdots (\alpha + 4)}.
\end{aligned}
$$

Picking $\alpha_1 = \alpha_0 = 1$, so that $\alpha = \alpha_1 + \alpha_0 = 2$, we get

$$\frac{[1 \cdot 2 \cdot 3] \cdot [1 \cdot 2]}{2 \cdot 3 \cdot 4 \cdot 5 \cdot 6} = \frac{12}{720} = 0.017$$

(see figure 18.2), which is significantly lower than the likelihood

$$\left(\frac{3}{5}\right)^3 \cdot \left(\frac{2}{5}\right)^2 = \frac{108}{3125} \approx 0.035.$$

Thus, a model using maximum-likelihood parameters ascribes a much higher probability to this sequence than does the marginal likelihood. The reason is that the log-likelihood is making an overly optimistic assessment, based on a parameter that was designed with full retrospective knowledge to be an optimal fit to the entire sequence.

In general, for a binomial distribution with a Beta prior, we have

$$P(x[1], \ldots, x[M]) = \frac{[\alpha_1 \cdots (\alpha_1 + M[1] - 1)][\alpha_0 \cdots (\alpha_0 + M[0] - 1)]}{\alpha \cdots (\alpha + M - 1)}.$$

Each of the terms in square brackets is a product of a sequence of numbers such as $\alpha \cdot (\alpha + 1) \cdots (\alpha + M - 1)$. If α is an integer, we can write this product as $\frac{(\alpha + M - 1)!}{(\alpha - 1)!}$. However, we do not necessarily know that α is an integer. It turns out that we can use a generalization of the

Gamma function   factorial function for this purpose. Recall that the *Gamma function* is such that $\Gamma(m) = (m-1)!$ and $\Gamma(x + 1) = x \cdot \Gamma(x)$. Using the latter property, we can rewrite

$$\alpha(\alpha + 1) \cdots (\alpha + M - 1) = \frac{\Gamma(\alpha + M)}{\Gamma(\alpha)}.$$

Hence,

$$P(x[1], \ldots, x[M]) = \frac{\Gamma(\alpha)}{\Gamma(\alpha + M)} \cdot \frac{\Gamma(\alpha_1 + M[1])}{\Gamma(\alpha_1)} \cdot \frac{\Gamma(\alpha_0 + M[0])}{\Gamma(\alpha_0)}.$$

A similar formula holds for a multinomial distribution over the space $x^1, \ldots, x^k$, with a Dirichlet prior with hyperparameters $\alpha_1, \ldots, \alpha_k$:

$$P(x[1], \ldots, x[M]) = \frac{\Gamma(\alpha)}{\Gamma(\alpha + M)} \cdot \prod_{i=1}^{k} \frac{\Gamma(\alpha_i + M[x^i])}{\Gamma(\alpha_i)}. \tag{18.9}$$

Note that the final expression for the marginal likelihood is invariant to the order we selected in the expansion via the chain rule. In particular, any other order results in exactly the same final expression. This property is reassuring, because the IID assumption tells us that the specific order in which we get data cases is insignificant. Also note that the marginal likelihood can be computed directly from the same sufficient statistics used in the computation of the likelihood function — the counts of the different values of the variable in the data. This observation will continue to hold in the general case of Bayesian networks.

18.3.4 Bayesian Score for Bayesian Networks

We now generalize the discussion of the Bayesian score to more general Bayesian networks. Consider two possible structures over two binary random variables X and Y. $\mathcal{G}_\emptyset$ is the graph with no edges. Here, we have:

$$P(\mathcal{D} \mid \mathcal{G}_\emptyset) = \int_{\Theta_X \times \Theta_Y} P(\theta_X, \theta_Y \mid \mathcal{G}_\emptyset) P(\mathcal{D} \mid \theta_X, \theta_Y, \mathcal{G}_\emptyset) d[\theta_X, \theta_Y].$$

We know that the likelihood term $P(\mathcal{D} \mid \theta_X, \theta_Y, \mathcal{G}_\emptyset)$ can be written as a product of terms, one involving θ_X and the observations of X in the data, and the other involving θ_Y and the observations of Y in the data. If we also assume *parameter independence*, that is, that $P(\theta_X, \theta_Y \mid \mathcal{G}_\emptyset)$ decomposes as a product $P(\theta_X \mid \mathcal{G}_\emptyset) P(\theta_Y \mid \mathcal{G}_\emptyset)$, then we can simplify the integral

parameter independence

$$P(\mathcal{D} \mid \mathcal{G}_\emptyset) = \left(\int_{\Theta_X} P(\theta_X \mid \mathcal{G}_\emptyset) \prod_m P(x[m] \mid \theta_X, \mathcal{G}_\emptyset) d\theta_X \right)$$
$$\left(\int_{\Theta_Y} P(\theta_Y \mid \mathcal{G}_\emptyset) \prod_m P(y[m] \mid \theta_Y, \mathcal{G}_\emptyset) d\theta_Y \right),$$

where we used the fact that the integral of a product of independent functions is the product of integrals. Now notice that each of the two integrals is the marginal likelihood of a single variable. Thus, if X and Y are multinomials, and each has a Dirichlet prior, then we can write each integral using the closed form of equation (18.9).

Now consider the network $\mathcal{G}_{X \to Y} = (X \to Y)$. Once again, if we assume parameter independence, we can decompose this integral into a product of three integrals, each over a single parameter family.

$$P(\mathcal{D} \mid \mathcal{G}_{X \to Y}) = \left(\int_{\Theta_X} P(\theta_X \mid \mathcal{G}_{X \to Y}) \prod_m P(x[m] \mid \theta_X, \mathcal{G}_{X \to Y}) d\theta_X \right)$$
$$\left(\int_{\Theta_{Y \mid x^0}} P(\theta_{Y \mid x^0} \mid \mathcal{G}_{X \to Y}) \prod_{m:x[m]=x^0} P(y[m] \mid \theta_{Y \mid x^0}, \mathcal{G}_{X \to Y}) d\theta_{Y \mid x^0} \right)$$
$$\left(\int_{\Theta_{Y \mid x^1}} P(\theta_{Y \mid x^1} \mid \mathcal{G}_{X \to Y}) \prod_{m:x[m]=x^1} P(y[m] \mid \theta_{Y \mid x^1}, \mathcal{G}_{X \to Y}) d\theta_{Y \mid x^1} \right).$$

Again, each of these can be written using the closed form solution of equation (18.9).

Comparing the marginal likelihood of the two structures, we see that the term that corresponds to X is similar in both. In fact, the terms $P(x[m] \mid \theta_X, \mathcal{G}_\emptyset)$ and $P(x[m] \mid \theta_X, \mathcal{G}_{X \to Y})$ are identical (both make the same predictions given the parameter values). Thus, if we choose the prior $P(\Theta_X \mid \mathcal{G}_\emptyset)$ to be the same as $P(\Theta_X \mid \mathcal{G}_{X \to Y})$, we have that the first term in the marginal likelihood of both structures is identical.

Thus, given this assumption about the prior, the difference between the marginal likelihood of $\mathcal{G}_\emptyset$ and $\mathcal{G}_{X \to Y}$ is due to the difference between the marginal likelihood of all the observations of Y and the marginal likelihoods of the observations of Y when we partition our examples based on the observed value of X. Intuitively, if Y has a different distribution in these two cases, then the latter term will have better marginal likelihood. On the other hand, if Y is distributed in roughly the same manner in both subsets, then the simpler network will have better marginal likelihood.

To see this behavior, we consider an idealized experiment where the empirical distribution is such that $P(x^1) = 0.5$, and $P(y^1 \mid x^1) = 0.5 + p$ and $P(y^1 \mid x^0) = 0.5 - p$, where p is a free parameter. Larger values of p imply stronger dependence between X and Y. Note, however, that the marginal distributions of X and Y are the same regardless of the value of p. Thus, the score of the empty structure $\mathcal{G}_\emptyset$ does not depend on p. On the other hand, the score of the structure $\mathcal{G}_{X \to Y}$ depends on p. Figure 18.3 illustrates how these scores change as functions of the number of training samples. The graph compares the average score per instance (of equation (18.8)) for both structures for different values of p.

We can see that, as we get more data, the Bayesian score prefers the structure $\mathcal{G}_{X \to Y}$ where X and Y are dependent. When the dependency between them is strong, this preference arises very quickly. But as the dependency becomes weaker, more data are required in order to justify this selection. Thus, if the two variables are independent, small fluctuations in the data, due to sampling noise, are unlikely to cause a preference for the more complex structure. By contrast, any fluctuation from pure independence in the empirical distribution will cause the likelihood score to select the more complex structure.

We now return to consider the general case. As we can expect, the same arguments we applied to the two-variable networks apply to any network structure.

Proposition 18.2

Let $\mathcal{G}$ be a network structure, and let $P(\theta_\mathcal{G} \mid \mathcal{G})$ be a parameter prior satisfying global parameter independence. Then,

$$P(\mathcal{D} \mid \mathcal{G}) = \prod_i \int_{\Theta_{X_i \mid \mathrm{Pa}_{X_i}}} \prod_m P(x_i[m] \mid \mathrm{pa}_{X_i}[m], \theta_{X_i \mid \mathrm{Pa}_{X_i}}, \mathcal{G}) P(\theta_{X_i \mid \mathrm{Pa}_{X_i}} \mid \mathcal{G}) d\theta_{X_i \mid \mathrm{Pa}_{X_i}}.$$

Moreover, if $P(\theta_\mathcal{G})$ also satisfies local parameter independence, then

$$P(\mathcal{D} \mid \mathcal{G}) = \prod_i \prod_{\boldsymbol{u}_i \in Val(\mathrm{Pa}^\mathcal{G}_{X_i})} \int_{\Theta_{X_i \mid \boldsymbol{u}_i}} \prod_{m, \boldsymbol{u}_i[m] = \boldsymbol{u}_i} P(X_i[m] \mid \boldsymbol{u}_i, \theta_{X_i \mid \boldsymbol{u}_i}, \mathcal{G}) P(\theta_{X_i \mid \boldsymbol{u}_i} \mid \mathcal{G}) d\theta_{X_i \mid \boldsymbol{u}_i}.$$

Using this proposition and the results about the marginal likelihood of Dirichlet priors, we conclude the following result: If we consider a network with Dirichlet priors where $P(\theta_{X_i \mid \mathrm{pa}_{X_i}} \mid$

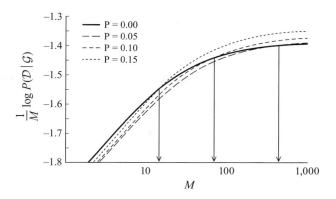

Figure 18.3 The effect of correlation on the Bayesian score. The solid line indicates the score of the independent model $\mathcal{G}_\emptyset$. The remaining lines indicate the score of the more complex structure $\mathcal{G}_{X \to Y}$, for different sampling distributions parameterized by p.

$\mathcal{G}$) has hyperparameters $\{\alpha^{\mathcal{G}}_{x^j_i | \boldsymbol{u}_i} : j = 1, \ldots, |X_i|\}$ then

$$P(\mathcal{D} \mid \mathcal{G}) = \prod_i \prod_{\boldsymbol{u}_i \in Val(\mathrm{Pa}^{\mathcal{G}}_{X_i})} \frac{\Gamma(\alpha^{\mathcal{G}}_{X_i | \boldsymbol{u}_i})}{\Gamma(\alpha^{\mathcal{G}}_{X_i | \boldsymbol{u}_i} + M[\boldsymbol{u}_i])} \prod_{x^j_i \in Val(X_i)} \left[\frac{\Gamma(\alpha^{\mathcal{G}}_{x^j_i | \boldsymbol{u}_i} + M[x^j_i, \boldsymbol{u}_i])}{\Gamma(\alpha^{\mathcal{G}}_{x^j_i | \boldsymbol{u}_i})} \right],$$

where $\alpha^{\mathcal{G}}_{X_i | \boldsymbol{u}_i} = \sum_j \alpha^{\mathcal{G}}_{x^j_i | \boldsymbol{u}_i}$. In practice, we use the logarithm of this formula, which is more manageable to compute numerically.[2]

18.3.5 Understanding the Bayesian Score

overfitting

As we have just seen, **the Bayesian score seems to be biased toward simpler structures, but as it gets more data, it is willing to recognize that a more complex structure is necessary. In other words, it appears to trade off fit to data with model complexity, thereby reducing the extent of *overfitting*.** To understand this behavior, it is useful to consider an approximation to the Bayesian score that better exposes its fundamental properties.

Theorem 18.1

If we use a Dirichlet parameter prior for all parameters in our network, then, when $M \to \infty$, we have that:

$$\log P(\mathcal{D} \mid \mathcal{G}) = \ell(\hat{\boldsymbol{\theta}}_{\mathcal{G}} : \mathcal{D}) - \frac{\log M}{2} \mathrm{Dim}[\mathcal{G}] + O(1),$$

model dimension

where $\mathrm{Dim}[\mathcal{G}]$ *is the* model dimension, *or the number of* independent parameters *in* $\mathcal{G}$.

independent
parameters

See exercise 18.7 for the proof.

2. Most scientific computation libraries have efficient numerical implementation of the function $\log \Gamma(x)$, which enables us to compute this score efficiently.

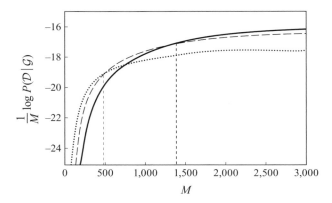

Figure 18.4   The Bayesian score of three structures, evaluated on synthetic data generated from the ICU-Alarm network. The solid line is the original structure, which has 509 parameters. The dashed line is a simplification that has 359 parameters. The dotted line is a tree-structure and has 214 parameters.

Thus, we see that the Bayesian score tends to trade off the likelihood — fit to data — on one hand and some notion of model complexity on the other hand. This approximation is called the *BIC score* (for Bayesian information criterion):

BIC score

$$\text{score}_{BIC}(\mathcal{G} \; : \; \mathcal{D}) = \ell(\hat{\boldsymbol{\theta}}_\mathcal{G} : \mathcal{D}) - \frac{\log M}{2} \text{Dim}[\mathcal{G}].$$

We note that the negation of this quantity can be viewed as the number of bits required to encode both the model ($\log M/2$ bits per model parameter, a derivation whose details we omit) and the data given the model (as per our discussion in section A.1.3). Thus, this objective is also known as *minimum description length*.

minimum
description
length

We can decompose this score even further using our analysis from equation (18.4):

$$\text{score}_{BIC}(\mathcal{G} \; : \; \mathcal{D}) = M \sum_{i=1}^{n} \boldsymbol{I}_{\hat{P}}(X_i; \text{Pa}_{X_i}) - M \sum_{i=1}^{n} \boldsymbol{H}_{\hat{P}}(X_i) - \frac{\log M}{2} \text{Dim}[\mathcal{G}].$$

We can observe several things about the behavior of this score function. First, the entropy terms do not depend on the graph, so they do not influence the choice of structure and can be ignored. The score exhibits a trade-off between fit to data and model complexity: the stronger the dependence of a variable on its parents, the higher the score; the more complex the network, the lower the score. However, the mutual information term grows linearly in M, whereas the complexity term grows logarithmically. Therefore, the larger M is, the more emphasis will be given to the fit to data.

Figure 18.4 illustrates this theorem empirically. It shows the Bayesian score of three structures on a data set generated by the ICU-Alarm network. One of these structures is the correct one, and the other two are simplifications of it. We can see that, for small M, the simpler structures have the highest scores. This is compatible with our analysis: for small data sets, the penalty term outweighs the likelihood term. But as M grows, the score begins to exhibit an increasing preference for the more complex structures. With enough data, the true model is preferred.

This last statement is a general observation about the BIC and Bayesian scores: Asymptotically, these scores will prefer a structure that exactly fits the dependencies in the data. To make this statement precise, we introduce the following definition:

Definition 18.1

consistent score

Assume that our data are generated by some distribution P^* for which the network $\mathcal{G}^*$ is a perfect map. We say that a scoring function is consistent *if the following properties hold as the amount of data $M \to \infty$, with probability that approaches 1 (over possible choices of data set D):*

- *The structure $\mathcal{G}^*$ will maximize the score.*
- *All structures $\mathcal{G}$ that are not I-equivalent to $\mathcal{G}^*$ will have strictly lower score.* ∎

Theorem 18.2

The BIC score is consistent.

PROOF Our goal is to prove that for sufficiently large M, if the graph that maximizes the BIC score is $\mathcal{G}$, then $\mathcal{G}$ is I-equivalent to $\mathcal{G}^*$. We briefly sketch this proof.

Consider some graph $\mathcal{G}$ that implies an independence assumption that $\mathcal{G}^*$ does not support. Then $\mathcal{G}$ cannot be an I-map of the true underlying distribution P. Hence, $\mathcal{G}$ cannot be a maximum likelihood model with respect to the true distribution P^*, so that we must have:

$$\sum_i \boldsymbol{I}_{P^*}(X_i; \mathrm{Pa}_{X_i}^{\mathcal{G}^*}) > \sum_i \boldsymbol{I}_{P^*}(X_i; \mathrm{Pa}_{X_i}^{\mathcal{G}}).$$

As $M \to \infty$, our empirical distribution $\hat{P}$ will converge to P^* with probability 1. Therefore, for large M,

$$\mathrm{score}_L(\mathcal{G}^* : \mathcal{D}) - \mathrm{score}_L(\mathcal{G} : \mathcal{D}) \approx \Delta \cdot M,$$

where $\Delta = \sum_i \boldsymbol{I}_{P^*}(X_i; \mathrm{Pa}_{X_i}^{\mathcal{G}^*}) - \sum_i \boldsymbol{I}_{P^*}(X_i; \mathrm{Pa}_{X_i}^{\mathcal{G}})$. Therefore, asymptotically we have that

$$\mathrm{score}_{BIC}(\mathcal{G}^* : \mathcal{D}) - \mathrm{score}_{BIC}(\mathcal{G} : \mathcal{D}) \approx \Delta M + \frac{1}{2}(\mathrm{Dim}[\mathcal{G}] - \mathrm{Dim}[\mathcal{G}^*]) \log M.$$

The first term grows much faster than the second, so that eventually its effect will dominate, and the score of $\mathcal{G}^*$ will be better.

Now, assume that $\mathcal{G}$ implies all the independence assumptions in $\mathcal{G}^*$, but that $\mathcal{G}^*$ implies an independence assumption that $\mathcal{G}$ does not. (In other words, $\mathcal{G}$ is a superset of $\mathcal{G}^*$.) In this case, $\mathcal{G}$ can represent any distribution that $\mathcal{G}^*$ can. In particular, it can represent P^*. As $\hat{P}$ converges to P^*, we will have that:

$$\mathrm{score}_L(\mathcal{G}^* : \mathcal{D}) - \mathrm{score}_L(\mathcal{G} : \mathcal{D}) \to 0.$$

Therefore, asymptotically we have that for

$$\mathrm{score}_{BIC}(\mathcal{G}^* : \mathcal{D}) - \mathrm{score}_{BIC}(\mathcal{G} : \mathcal{D}) \approx \frac{1}{2}(\mathrm{Dim}[\mathcal{G}] - \mathrm{Dim}[\mathcal{G}^*]) \log M.$$

Now, since $\mathcal{G}$ makes fewer independence assumptions than $\mathcal{G}^*$, it must be parameterized by a larger set of parameters. Thus, $\mathrm{Dim}[\mathcal{G}] > \mathrm{Dim}[\mathcal{G}^*]$, so that $\mathcal{G}^*$ will be preferred to $\mathcal{G}$. ∎

As the Bayesian score is asymptotically identical to BIC (the remaining $O(1)$ terms do not grow with M), we get:

Corollary 18.2

The Bayesian score is consistent.

Note that consistency is an asymptotic property, and thus it does not imply much about the properties of networks learned with limited amounts of data. Nonetheless, the proof illustrates the trade-offs that are playing a role in the definition of score.

18.3.6   Priors

Until now we did not specify the actual choice of priors we use. We now discuss possible choices of priors and their effect on the score.

18.3.6.1   Structure Priors

structure prior

We begin with the *prior* over network structures, $P(\mathcal{G})$. Note that although this term seems to describe our bias for certain structure, in fact, it plays a relatively minor role. As we can see in theorem 18.1, the logarithm of the marginal likelihood grows linearly with the number of examples, while the prior over structures remains constant. Thus, the structure prior does not play an important role in asymptotic analysis as long as it does not rule out (that is, assign probability 0) any structure.

For this reason, we often use a uniform prior over structures. Nonetheless, the structure prior can make some difference when we consider small samples. Thus, we might want to encode some of our preferences in this prior. For example, we might penalize edges in the graph, and use a prior $P(\mathcal{G}) \propto c^{|\mathcal{G}|}$, where c is some constant smaller than 1, and $|\mathcal{G}|$ is the number of edges in the graph.

Note that in both these choices (the uniform and the penalty per edge) it suffices to use a value that is proportional to the prior, since the normalizing constant is the same for all choice of $\mathcal{G}$ and hence can be ignored. For this reason, we do not need to worry about the exact number of possible network structures in order to use these priors.

As we will immediately see, it will be mathematically convenient to assume that the structure prior satisfies *structure modularity*. This condition requires that the prior $P(\mathcal{G})$ be proportional to a product of terms, where each term relates to one family. Formally,

structure modularity

$$P(\mathcal{G}) \propto \prod_i P(\mathrm{Pa}_{X_i} = \mathrm{Pa}_{X_i}^{\mathcal{G}}),$$

where $P(\mathrm{Pa}_{X_i} = \mathrm{Pa}_{X_i}^{\mathcal{G}})$ denotes the prior probability we assign to choosing the specific set of parents for X_i. Structure priors that satisfy this property do not penalize for global properties of the graph (such as its depth) but only for local properties (such as the indegrees of variables). This is clearly the case for both priors we discuss here.

In addition, it also seems reasonable to require that I-equivalent network structures are assigned the same prior. Again, this means that when two networks are equivalent, we do not distinguish between them by subjective preferences.

18.3.6.2 Parameter Priors and Score Decomposability

parameter prior

In order to use Bayesian scores, we also need to have *parameter priors* for the parameterization corresponding to every possible structure. Before we discuss how to represent such priors, we consider the desired properties from these priors.

decomposable score

global parameter independence

Proposition 18.2 shows that the Bayesian score of a network structure $\mathcal{G}$ *decomposes* into a product of terms, one for each family. This is a consequence of the *global parameter independence* assumption. In the case of parameter learning, this assumption was crucial for decomposing the learning problem into independent subproblems. Can we exploit a similar phenomenon in the case of structure learning?

In the simple example we considered in the previous section, we compared the score of two networks $\mathcal{G}_\emptyset$ and $\mathcal{G}_{X \to Y}$. We saw that if we choose the priors $P(\Theta_X \mid \mathcal{G}_\emptyset)$ and $P(\Theta_X \mid \mathcal{G}_{X \to Y})$ to be identical, the score associated with X is the same in both graphs. Thus, not only does the score of both structures have a product form, but in the case where the same variable has the same parents in both structures, the term associated with it also has the same value in both scores.

Considering more general structures, if $\mathrm{Pa}_{X_i}^{\mathcal{G}} = \mathrm{Pa}_{X_i}^{\mathcal{G}'}$ then it would seem natural that the term that measures the score of X_i given its parents in $\mathcal{G}$ would be identical to the one in $\mathcal{G}'$. This seems reasonable. Recall that the score associated with X_i measures how well it can be predicted given its parents. Thus, if X_i has the same set of parents in both structures, this term should have the same value.

Definition 18.2

decomposable score

A *structure score* $\mathrm{score}(\mathcal{G} : \mathcal{D})$ *is* decomposable *if the score of a structure $\mathcal{G}$ can be written as*

$$\mathrm{score}(\mathcal{G} : \mathcal{D}) = \sum_i \mathrm{FamScore}(X_i \mid \mathrm{Pa}_{X_i}^{\mathcal{G}} : \mathcal{D}),$$

family score

where the family score $\mathrm{FamScore}(X \mid \boldsymbol{U} : \mathcal{D})$ *is a score measuring how well a set of variables $\boldsymbol{U}$ serves as parents of X in the data set $\mathcal{D}$.* ∎

As an example, the likelihood score is decomposable. Using proposition 18.1, we see that in this decomposition

$$\mathrm{FamScore}_L(X \mid \boldsymbol{U} : \mathcal{D}) = M \cdot \left[\boldsymbol{I}_{\hat{P}}(X; \boldsymbol{U}) - \boldsymbol{H}_{\hat{P}}(X) \right].$$

☞ **Score decomposability has important ramifications when we search for structures that maximize the scores. The high-level intuition is that if we have a decomposable score, then a local change in the structure (such as adding an edge) does not change the score of other parts of the structure that remained the same. As we will see, the search algorithms we consider can exploit decomposability to reduce dramatically the computational overhead of evaluating different structures during search.**

Under what conditions is the Bayesian score decomposable? It turns out that a natural restriction on the prior suffices.

Definition 18.3

parameter modularity

Let $\{P(\boldsymbol{\theta}_{\mathcal{G}} \mid \mathcal{G}) : \mathcal{G} \in \boldsymbol{\mathcal{G}}\}$ *be a set of parameter priors that satisfy global parameter independence. The prior satisfies* parameter modularity *if for each $\mathcal{G}, \mathcal{G}'$ such that $\mathrm{Pa}_{X_i}^{\mathcal{G}} = \mathrm{Pa}_{X_i}^{\mathcal{G}'} = \boldsymbol{U}$, then $P(\boldsymbol{\theta}_{X_i \mid \boldsymbol{U}} \mid \mathcal{G}) = P(\boldsymbol{\theta}_{X_i \mid \boldsymbol{U}} \mid \mathcal{G}')$.* ∎

Parameter modularity states that the prior over the CPD of X_i depends only on the *local* structure of the network (that is, the set of parents of X_i), and not on other parts of the network. It is straightforward to see that parameter modularity implies that the score is decomposable.

Proposition 18.3   *Let $\mathcal{G}$ be a network structure, let $P(\mathcal{G})$ be a structure prior satisfying structure modularity, and let $P(\boldsymbol{\theta}_{\mathcal{G}} \mid \mathcal{G})$ be a parameter prior satisfying global parameter independence and parameter modularity. Then, the Bayesian score over network structures is decomposable.*

18.3.6.3   Representing Parameter Priors

How do we represent our parameter priors? The number of possible structures is superexponential, which makes it difficult to elicit separate parameters for each one. How do we elicit priors for all these networks? If we require parameter modularity, the number of different priors we need is somewhat smaller, since we need a prior for each choice of parents for each variable. This number, however, is still exponential.

A simpleminded approach is simply to take some fixed Dirichlet distribution, for example, $Dirichlet(\alpha, \ldots, \alpha)$, for every parameter, where α is a predetermined constant. A typical K2 prior   choice is $\alpha = 1$. This prior is often referred to as the *K2 prior*, referring to the name of the software system where it was first used.

The K2 prior is simple to represent and efficient to use. However, it is somewhat inconsistent. Consider a structure where the binary variable Y has no parents. If we take $Dirichlet(1, 1)$ for θ_Y, we are in effect stating that our imaginary sample size is two. But now, consider a different structure where Y has the parent X, which has 4 values. If we take $Dirichlet(1, 1)$ as our prior for all parameters $\theta_{Y \mid x^i}$, we are effectively stating that we have seen two imaginary samples in each context x^i, for a total of eight. It seems that the number of imaginary samples we have seen for different events is a basic concept that should not vary with different candidate structures.

A more elegant approach is one we already saw in the context of parameter estimation: the BDe prior   *BDe prior*. We elicit a prior distribution P' over the entire probability space and an equivalent sample size α for the set of imaginary samples. We then set the parameters as follows:

$$\alpha_{x_i \mid \mathrm{pa}_{X_i}} = \alpha \cdot P'(x_i, \mathrm{pa}_{X_i}).$$

This choice will avoid the inconsistencies we just discussed. If we consider the prior over $\theta_{Y \mid x^i}$ in our example, then

$$\alpha_y = \alpha \cdot P'(y) = \sum_{x^i} \alpha \cdot P'(y, x^i) = \sum_{x^i} \alpha_{y \mid x^i}.$$

Thus, the number of imaginary samples for the different choices of parents for Y will be identical.

As we discussed, we can represent P' as a Bayesian network whose structure can represent our prior about the domain structure. Most simply, when we have no prior knowledge, we set P' to be the uniform distribution, that is, the empty Bayesian network with a uniform marginal distribution for each variable. In any case, it is important to note that the network structure is used *only* to provide parameter priors. It is not used to guide the structure search directly.

18.3.7 Score Equivalence ⋆

The BDe score turns out to satisfy an important property. Recall that two networks are I-equivalent if they encode the same set of independence statements. Hence, based on observed independencies, we cannot distinguish between I-equivalent networks. This suggests that based on observing data cases, we do not expect to distinguish between equivalent networks.

Definition 18.4

score equivalence

Let $\text{score}(\mathcal{G} : \mathcal{D})$ *be some scoring rule. We say that it satisfies* score equivalence *if for all I-equivalent networks* $\mathcal{G}$ *and* $\mathcal{G}'$ *we have* $\text{score}(\mathcal{G} : \mathcal{D}) = \text{score}(\mathcal{G}' : \mathcal{D})$ *for all data sets* $\mathcal{D}$. ■

In other words, score equivalence implies that all networks in the same equivalence class have the same score. In general, if we view I-equivalent networks as equally good at describing the same probability distributions, then we want to have score equivalence. We do not want the score to introduce artificial distinctions when we choose networks.

Do the scores discussed so far satisfy this condition?

Theorem 18.3

The likelihood score and the BIC score satisfy score equivalence.

For a proof, see exercise 18.8 and exercise 18.9

What about the Bayesian score? It turns out that the simpleminded K2 prior we discussed is *not* score-equivalent; see exercise 18.10. The BDe score, on the other hand, is score-equivalent. In fact, something stronger can be said.

Theorem 18.4

Let $P(\mathcal{G})$ *be a structure prior that assigns I-equivalent networks identical prior. Let* $P(\boldsymbol{\theta}_\mathcal{G} \mid \mathcal{G})$ *be a prior over parameters for networks with table-CPDs that satisfies global and local parameter independence and where for each* X_i *and* $\boldsymbol{u}_i \in Val(\text{Pa}_{X_i}^\mathcal{G})$, *we have that* $P(\boldsymbol{\theta}_{X_i \mid \boldsymbol{u}_i} \mid \mathcal{G})$ *is a Dirichlet prior. The Bayesian score with this prior satisfies score equivalence if and only if the prior is a BDe prior for some choice of* α *and* P'.

We do not prove this theorem here. See exercise 18.11 for a proof that the BDe score in this case satisfies score equivalence.

In other words, if we insist on using Dirichlet priors and also want the decomposition property, then to satisfy score equivalence, we must use a BDe prior.

18.4 Structure Search

In the previous section, we discussed scores for evaluating the quality of different candidate Bayesian network structures. These included the likelihood score, the Bayesian score, and the BIC score (which is an asymptotic approximation of the Bayesian score). We now examine how to find a structure with a high score.

We now have a well-defined optimization problem. Our input is:

- training set $\mathcal{D}$;
- scoring function (including priors, if needed);
- a set $\boldsymbol{\mathcal{G}}$ of possible network structures (incorporating any prior knowledge).

Our desired output is a network structure (from the set of possible structures) that maximizes the score.

It turns out that, for this discussion, we can ignore the specific choice of score. Our search algorithms will apply unchanged to all three of these scores.

score decomposability   As we will discuss, the main property of the scores that affect the search is their *decomposability*. That is, we assume we can write the score of a network structure $\mathcal{G}$:

$$\text{score}(\mathcal{G} \; : \; \mathcal{D}) = \sum_i \text{FamScore}(X_i \mid \text{Pa}_{X_i}^{\mathcal{G}} \; : \; \mathcal{D}).$$

score equivalence   Another property that is shared by all these scores is *score equivalence*: if $\mathcal{G}$ is I-equivalent to $\mathcal{G}'$ then $\text{score}(\mathcal{G} \; : \; \mathcal{D}) = \text{score}(\mathcal{G}' \; : \; \mathcal{D})$. This property is less crucial for search, but, as we will see, it can simplify several points.

18.4.1   Learning Tree-Structured Networks

We begin with the simplest variant of the structure learning task — the task of learning a *tree-structured network*. More precisely:

Definition 18.5

tree network

A network structure $\mathcal{G}$ is called tree-structured *if each variable X has at most one parent in $\mathcal{G}$, that is, $\mid \text{Pa}_X^{\mathcal{G}} \mid \leq 1$.*   ∎

Strictly speaking, the notion of tree-structured networks covers a broader class of graphs than those comprising a single tree; it also covers graphs composed of a set of disconnected trees, that is, a forest. In particular, the network of independent variables (no edges) also satisfies this definition. However, as the basic structure of these networks is still a collection of trees, we continue to use the term tree-structure.

Note that the class of trees is narrower than the class of polytrees that we discussed in chapter 9. A polytree can have variables with multiple parents, whereas a tree cannot. In other words, a tree-structured network cannot have v-structures. In fact, the problem of learning polytree-structured networks has very different computational properties than that of learning trees (see section 18.8).

Why do we care about learning trees? Most importantly, because unlike richer classes of structures, they can be learned efficiently — in polynomial time. But learning trees can also be useful in themselves. They are sparse, and therefore they avoid most of the overfitting problems associated with more complex structures. They also capture the most important dependencies in the distribution, and they can therefore provide some insight into the domain. They can also provide a better baseline for approximating the distribution than the set of independent marginals of the different variables (another commonly used simple approximation). They are thus often used as a starting point for learning a more complex structure, or even on their own in cases where we cannot afford significant computational resources.

The key properties we are going to use for learning trees are the decomposability of the score on one hand and the restriction on the number of parents on the other hand. We start by examining the score of a network and performing slight manipulations. Instead of maximizing the score of a tree structure $\mathcal{G}$, we will try to maximize the difference between its score and the score of the empty structure $\mathcal{G}_\emptyset$. We define

$$\Delta(\mathcal{G}) = \text{score}(\mathcal{G} \; : \; \mathcal{D}) - \text{score}(\mathcal{G}_\emptyset \; : \; \mathcal{D}).$$

We know that $\text{score}(\mathcal{G}_\emptyset : \mathcal{D})$ is simply a sum of terms $\text{FamScore}(X_i : \mathcal{D})$ for each X_i. That is the score of X_i if it does not have any parents. The score $\text{score}(\mathcal{G} : \mathcal{D})$ consists of terms $\text{FamScore}(X_i \mid \text{Pa}_{X_i}^{\mathcal{G}} : \mathcal{D})$. Now, there are two cases. If $\text{Pa}_{X_i}^{\mathcal{G}} = \emptyset$, then the term for X_i in both scores cancel out. If $\text{Pa}_{X_i}^{\mathcal{G}} = X_j$, then we are left with the difference between the two terms. Thus, we conclude that

$$\Delta(\mathcal{G}) = \sum_{i, \text{Pa}_{X_i}^{\mathcal{G}} \neq \emptyset} \left(\text{FamScore}(X_i \mid \text{Pa}_{X_i}^{\mathcal{G}} : \mathcal{D}) - \text{FamScore}(X_i : \mathcal{D}) \right).$$

If we define the weight

$$w_{j \to i} = \text{FamScore}(X_i \mid X_j : \mathcal{D}) - \text{FamScore}(X_i : \mathcal{D}),$$

then we see that $\Delta(\mathcal{G})$ is the sum of weights on pairs X_i, X_j such that $X_j \to X_i$ in $\mathcal{G}$

$$\Delta(\mathcal{G}) = \sum_{X_j \to X_i \in \mathcal{G}} w_{j \to i}.$$

maximum weight spanning forest

We have transformed our problem to one of finding a *maximum weight spanning forest* in a directed weighted graph. Define a fully connected directed graph, where each vertex is labeled by a random variable in $\mathcal{X}$, and the weight of the edge from vertex X_j to vertex X_i is $w_{j \to i}$, and then search for a maximum-weight-spanning forest. Clearly, the sum of edge weights in a forest is exactly $\Delta(\mathcal{G})$ of the structure $\mathcal{G}$ with the corresponding set of edges. The graph structure that corresponds to that maximum-weight forest maximizes $\Delta(\mathcal{G})$.

How hard is the problem of finding a maximal-weighted directed spanning tree? It turns out that this problem has a polynomial-time algorithm. This algorithm is efficient but not simple.

The task becomes simpler if the score satisfies score equivalence. In this case, we can show (see exercise 18.13) that $w_{i \to j} = w_{j \to i}$. Thus, we can examine an undirected spanning tree (forest) problem, where we choose which edges participate in the forest, and only afterward determine their direction. (This can be done by choosing an arbitrary root and directing all edges away from it.) Finding a maximum spanning tree in undirected graph is an easy problem. One algorithm for solving it is shown in algorithm A.2; an efficient implementation of this algorithm requires time complexity of $O(n^2 \log n)$, where n is the number of vertices in the graph.

Using this reduction, we end up with an algorithm whose complexity is $O(n^2 \cdot M + n^2 \log n)$ where n is the number of variables in $\mathcal{X}$ and M is the number of data cases. This complexity is a result of two stages. In the first stage we perform a pass over the data to collect the sufficient statistics of each of the $O(n^2)$ edges. This step takes $O(n^2 \cdot M)$ time. The spanning tree computation requires $O(n^2 \log n)$ using standard data structures, but it can be reduced to $O(n^2 + n \log n) = O(n^2)$ using more sophisticated approaches. We see that the first stage dominates the complexity of the algorithm.

18.4.2 Known Order

variable ordering

We now consider a special case that also turns out to be easier than the general case. Suppose we restrict attention to structures that are consistent with some predetermined *variable ordering* $\prec$ over $\mathcal{X}$. In other words, we restrict attention to structures $\mathcal{G}$ where, if $X_i \in \text{Pa}_{X_j}^{\mathcal{G}}$, then $X_i \prec X_j$.

This assumption was a standard one in the early work on learning Bayesian networks from data. In some domains the ordering is indeed known in advance. For example, if there is a clear temporal order by which the variables are assigned values, then it is natural to try to learn a network that is consistent with the temporal flow.

Before we proceed, we stress that choosing an ordering in advance may be problematic. As we have seen in the discussion of minimal I-maps in section 3.4, a wrong choice of order can result in unnecessarily complicated I-map. Although learning does not recover an exact I-map, the same reasoning applies. Thus, a bad choice of order can result in poor learning result.

With this caveat in mind, assume that we select an ordering $\prec$; without loss of generality, assume that our ordering is $X_1 \prec X_2 \prec \ldots \prec X_n$. We want to learn a structure that maximizes the score, but so that $\mathrm{Pa}_{X_i} \subseteq \{X_1, \ldots, X_{i-1}\}$.

The first observation we make is the following. We need to find the network that maximizes the score. This score is a sum of local scores, one per variable. Note that the choice of parents for one variable, say X_i, does not restrict the choice of parents of another variable, say X_j. Since we obey the ordering, none of our choices can create a cycle. Thus, in this scenario, learning the parents of each variable is independent of the other variables. Stated more formally:

Proposition 18.4

Let $\prec$ be an ordering over $\mathcal{X}$, and let $\mathrm{score}(\mathcal{G} : \mathcal{D})$ be a decomposable score. If we choose $\mathcal{G}$ to be the network where

$$\mathrm{Pa}_{X_i}^{\mathcal{G}} = \arg \max_{U_i \subseteq \{X_j : X_j \prec X_i\}} \mathrm{FamScore}(X_i \mid U_i : \mathcal{D})$$

for each i, then $\mathcal{G}$ maximizes the score among the structures consistent with $\prec$.

Based on this observation, we can learn the parents for each variable independently from the parents of other variables. In other words, we now face n small learning problems.

Let us consider these learning problems. Clearly, we are forced to make X_1 a root. In the case of X_2 we have a choice. We can either have the edge $X_1 \to X_2$ or not. In this case, we can evaluate the difference in score between these two options, and choose the best one. Note that this difference is exactly the weight $w_{1 \to 2}$ we defined when learned tree networks. If $w_{1 \to 2} > 0$, we add the edge $X_1 \to X_2$; otherwise we do not.

Now consider X_3. Now we have four options, corresponding to whether we add the edge $X_1 \to X_3$, and whether we add the edge $X_2 \to X_3$. A naive approach to making these choices is to decouple the decision whether to add the edge $X_1 \to X_3$ from the decision about the edge $X_2 \to X_3$. Thus, we might evaluate $w_{1 \to 3}$ and $w_{2 \to 3}$ and based on these two numbers try to decide what is the best choice of parents.

Unfortunately, this approach is flawed. In general, the score $\mathrm{FamScore}(X_3 \mid X_1, X_2 : \mathcal{D})$ is *not* a function of $\mathrm{FamScore}(X_3 \mid X_1 : \mathcal{D})$ and $\mathrm{FamScore}(X_3 \mid X_2 : \mathcal{D})$. An extreme example is an XOR-like CPD where X_3 is a probabilistic function of the XOR of X_1 and X_2. In this case, $\mathrm{FamScore}(X_3 \mid X_1 : \mathcal{D})$ will be small (and potentially smaller than $\mathrm{FamScore}(X_3 \mid : \mathcal{D})$ since the two variables are independent), yet $\mathrm{FamScore}(X_3 \mid X_1, X_2 : \mathcal{D})$ will be large. By choosing the particular dependence of X_3 on the XOR of X_1 and X_2, we can change the magnitude of the latter term.

We conclude that we need to consider all four possible parent sets before we choose the parents of X_3. This does not seem that bad. However, when we examine X_4 we need to consider eight parent sets, and so on. For learning the parents of X_n, we need to consider 2^{n-1} parent sets, which is clearly too expensive for any realistic number of variables.

In practice, we do not want to learn networks with a large number of parents. Such networks are expensive to represent, most often are inefficient to perform inference with, and, most important, are prone to overfitting. So, we may find it reasonable to restrict our attention to networks the indegree of each variable is at most d.

If we make this restriction, our situation is somewhat more reasonable. The number of possible parent sets for X_n is $1 + \binom{n-1}{1} + \ldots + \binom{n-1}{d} = O(d\binom{n-1}{d})$ (when $d < n/2$). Since the number of choices for all other variables is less than the number of choices for X_n, the procedure has to evaluate $O(dn\binom{n-1}{d}) = O(d\binom{n}{d})$ candidate parent sets. This number is polynomial in n (for a fixed d).

We conclude that learning given a fixed order and a bound on the indegree is computationally tractable. However, the computational cost is exponential in d. Hence, the exhaustive algorithm that checks all parent sets of size $\le d$ is impractical for values of d larger than 3 or 4. When a larger d is required, we can use heuristic methods such as those described in the next section.

18.4.3 General Graphs

What happens when we consider the most general problem, where we do not have an ordering over the variables? Even if we restrict our attention to networks with small indegree, other problems arise. Suppose that adding the edge $X_1 \rightarrow X_2$ is beneficial, for example, if the score of X_1 as a parent of X_2 is higher than all other alternatives. If we decide to add this edge, we cannot add other edges — for example, $X_2 \rightarrow X_1$ — since this would introduce a cycle. The restriction on the immediate reverse of the edge we add might not seem so problematic. However, adding this edge also forbids us from adding together pairs of edges, such as $X_2 \rightarrow X_3$ and $X_3 \rightarrow X_1$. Thus, the decision on whether to add $X_1 \rightarrow X_2$ is not simple, since it has ramifications for other choices we make for parents of all the other variables.

This discussion suggests that the problem of finding the maximum-score network might be more complex than in the two cases we examined. If fact, we can make this statement more precise. Let d be an integer, we define $\boldsymbol{G}_d = \{\mathcal{G} : \forall i, |\mathrm{Pa}_{X_i}^{\mathcal{G}}| \le d\}$.

Theorem 18.5

The following problem is $\mathcal{NP}$-hard for any $d \ge 2$:

Given a data set $\mathcal{D}$ and a decomposable score function score, *find*

$$\mathcal{G}^* = \arg\max_{\mathcal{G} \in \boldsymbol{G}_d} \mathrm{score}(\mathcal{G} : \mathcal{D}).$$

The proof of this theorem is quite elaborate, and so we do not provide it here.

Given this result, we realize that it is unlikely that there is an efficient algorithm that constructs the highest-scoring network structure for all input data sets. Unlike the situation in inference, for example, the known intermediate situations where the problem is easier are not the ones we usually encounter in practice; see exercise 18.14.

As with many intractable problems, this is not the end of the story. Instead of aiming for an algorithm that will always find the highest-scoring network, we resort to heuristic algorithms that attempt to find the best network but are not guaranteed to do so. In our case, we are

local search

faced with a combinatorial optimization problem; we need to search the space of graphs (with bounded indegree) and return a high-scoring one. We solve this problem using a *local search* approach. To do so, we define three components: a search space, which defines the set of candidate network structures; a scoring function that we aim to maximize (for example, the BDe score given the data and priors); and the search procedure that explores the search space without necessarily seeing all of it (since it is superexponential in size).

18.4.3.1   The Search Space

search space

We start by considering the *search space*. As discussed in appendix A.4.2, we can think of a search space as a graph over candidate solutions, connected by possible operators that the search procedure can perform to move between different solutions. In the simplest setting, we consider the search space where each search state denotes a complete network structure $\mathcal{G}$ over $\mathcal{X}$. This is the search space we discuss for most of this chapter. However, we will see other formulations for search spaces.

A crucial design choice that has large impact on the success of heuristic search is how the space is interconnected. If each state has few neighbors, then the search procedure has to consider only a few options at each point of the search. Thus, it can afford to evaluate each of these options. However, this comes at a price. Paths from the initial solution to a good one might be long and complex. On the other hand, if each state has many neighbors, we may be able to move quickly from the initial state to a good state, but it may be difficult to determine which step to take at each point in the search. A good trade-off for this problem chooses reasonably few neighbors for each state but ensures that the "diameter" of the search space remains small. A natural choice for the neighbors of a state representing a network structure is a set of structures that are identical to it except for small "local" modifications. Thus, we define

search operators

the connectivity of our search space in terms of *operators* such as:

edge addition

- *edge addition*;

edge deletion

- *edge deletion*;

edge reversal

- *edge reversal*.

In other words, the states adjacent to a state $\mathcal{G}$ are those where we change one edge, either by adding one, deleting one, or reversing the orientation of one. Note that we only consider operations that result in legal networks. That is, acyclic networks that satisfy the constraints we put in advance (such as indegree constraints).

This definition of search space is quite natural and has several desirable properties. First, notice that the diameter of the search space is at most n^2. That is, there is a relatively short path between any two networks we choose. To see this, note that if we consider traversing a path from $\mathcal{G}_1$ to $\mathcal{G}_2$, we can start by deleting all edges in $\mathcal{G}_1$ that do not appear in $\mathcal{G}_2$, and then we can add the edges that are in $\mathcal{G}_2$ and not in $\mathcal{G}_1$. Clearly, the number of steps we take is bounded by the total number of edges we can have, n^2.

Second, recall that the score of a network $\mathcal{G}$ is a sum of local scores. The operations we consider result in changing only one local score term (in the case of addition or deletion of an edge) or two (in the case of edge reversal). Thus, they result in a local change in the score; most components in the score remain the same. This implies that there is some sense of "continuity" in the score of neighboring networks.

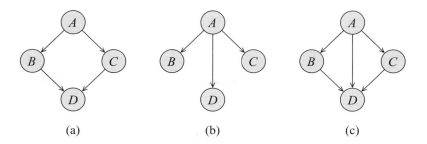

Figure 18.5   **Example of a search problem requiring edge deletion.** (a) original network that generated the data. (b) and (c) intermediate networks encountered during the search.

The choice of the three particular operations we consider also needs some justification. For example, if we always start the search from the empty graph $\mathcal{G}_\emptyset$, we may wonder why we include the option to delete an edge. We can reach every network by adding the appropriate arcs to the empty network. In general, however, we want the search space to allow us to reverse our choices. As we will see, this is an important property in escaping local maxima (see appendix A.4.2).

However, the ability to delete edges is important even if we perform only "greedy" operations that lead to improvement. To see this, consider the following example. Suppose the original network is the one shown in figure 18.5a, and that A is highly informative about both C and B. Starting from an empty network and adding edges greedily, we might add the edges $A \to B$ and $A \to C$. However, in some data sets, we might add the edge $A \to D$. To see why, we need to realize that A is informative about both B and C, and since these are the two parents of D, also about D. Now B and C are also informative about D. However, each of them provides part of the information, and thus neither B nor C by itself is the best parent of D. At this stage, we thus end up with the network figure 18.5b. Continuing the search, we consider different operators, adding the edge $B \to D$ and $C \to D$. Since B is a parent of D in the original network, it will improve the prediction of D when combined with A. Thus, the score of A and B together as parents of D can be larger than the score of A alone. Similarly, if there are enough data to support adding parameters, we will also add the edge $C \to D$, and reach the structure shown in figure 18.5c. This is the correct structure, except for the redundant edge $A \to D$. Now the ability to delete edges comes in handy. Since in the original distribution B and C together separate A from D, we expect that choosing B, C as the parents of D will have higher score than choosing A, B, C. To see this, note that A cannot provide additional information on top of what B and C convey, and having it as an additional parent results in a penalty. After we delete the edge $A \to D$ we get the original structure.

A similar question can be raised about the edge reversal operator. Clearly, we can achieve the effect of reversing an edge $X \to Y$ in two steps, first deleting the edge $X \to Y$ and then adding the edge $Y \to X$. The problem is that when we delete the edge $X \to Y$, we usually reduce the score (assuming that there is some dependency between X and Y). Thus, these two operations require us to go "downhill" in the first step in order to get to a better structure in the next step. The reverse operation allows us to realize the trade-off between a worse parent set for Y and a better one for X.

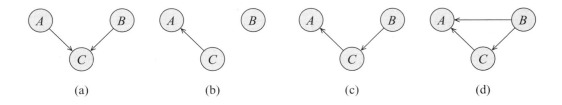

Figure 18.6   Example of a search problem requiring edge reversal. (a) original network that generated the data. (b) and (c) intermediate networks encountered during the search. (d) an undesirable outcome.

To see the utility of the edge reversal operator, consider the following simple example. Suppose the real network generating the data has the v-structure shown in figure 18.6a. Suppose that the dependency between A and C is stronger than that between B and C. Thus, a first step in a greedy-search procedure would add an edge between A and C. Note, however, that score equivalence implies that the network with the edge $A \rightarrow C$ has *exactly* the same score as the network with the edge $C \rightarrow A$. At this stage, we cannot distinguish between the two choices. Thus, the decision between them is arbitrary (or, in some implementations, randomized). It is thus conceivable that at this stage we have the network shown in figure 18.6b. The greedy procedure proceeds, and it decides to add the edge $B \rightarrow C$, resulting in the network of figure 18.6c. Now we are in the position to realize that reversing the edge $C \rightarrow A$ can improve the score (since A and B together should make the best predictions of C). However, if we do not have a reverse operator, a greedy procedure would not delete the edge $C \rightarrow A$, since that would definitely hurt the score.

In this example, note that when we do not perform the edge reversal, we might end up with the network shown in figure 18.6d. To realize why, recall that although A and B are marginally independent, they are dependent given C. Thus, B and C together make better predictions of A than C alone.

18.4.3.2   The Search Procedure

local search

Once we define the search space, we need to design a procedure to explore it and search for high-scoring states. There is a wide literature on heuristic search. The vast majority of the search methods used in structure learning are *local search* procedures such as greedy hill climbing, as described in appendix A.4.2. In the structure-learning setting, we pick an initial network structure $\mathcal{G}$ as a starting point; this network can be the empty one, a random choice, the best tree, or a network obtained from some prior knowledge. We compute its score. We then consider all of the neighbors of $\mathcal{G}$ in the space — all of the legal networks obtained by applying a single operator to $\mathcal{G}$ — and compute the score for each of them. We then apply the change that leads to the best improvement in the score. We continue this process until no modification improves the score.

There are two questions we can ask. First, how expensive is this process, and second, what can we say about the final network it returns?

Computational Cost   We start by briefly considering the time complexity of the procedure. At each iteration, the procedure applies $|\mathcal{O}|$ operators and evaluates the resulting network. Recall that the space of operators we consider is quadratic in the number of variables. Thus, if we perform K steps before convergence, then we perform $O(K \cdot n^2)$ operator applications. Each operator application involves two steps. First, we need to check that the network is acyclic. This check can be done in time linear in the number of edges. If we are considering networks with indegree bounded by d, then there are at most nd edges. Second, if the network is legal, we need to evaluate it. For this, we need to collect sufficient statistics from the data. These might be different for each network and require $O(M)$ steps, and so our rough time estimate is $O(K \cdot n^2 \cdot (M + nd))$. The number of iterations, K, varies and depends on the starting network and on how different the final network is. However, we expect it not to be much larger than n^2 (since this is the diameter of the search space). We emphasize that this is a rough estimate, and not a formal statement. As we will show, we can make this process faster by using properties of the score that allow for smart caching.

When n is large, considering $O(n^2)$ neighbors at each iteration may be too costly. However, most operators attempt to perform a rather bad change to the network. So can we skip evaluating them? One way of avoiding this cost is to use search procedures that replace the exhaustive enumeration in line 5 of Greedy-Local-Search (algorithm A.5) by a randomized choice of operators. This *first-ascent hill climbing* procedure samples operators from $\mathcal{O}$ and evaluates them one by one. Once it finds one that leads to a better-scoring network, it applies it without considering other operators. In the initial stages of the search, this procedure requires relatively few random trials before it finds such an operator. As we get closer to the local maximum, most operators hurt the score, and more trials are needed before an upward step is found (if any).

first-ascent hill climbing

Local Maxima   What can we say about the network returned by a greedy hill-climbing search procedure? Clearly, the resulting network cannot be improved by applying a single operator (that is, changing one edge). This implies that we are in one of two situations. We might have reached a *local maximum* from which all changes are score-reducing. The other option is that we have reached a *plateau*: a large set of neighboring networks that have the same score. By design, the greedy hill-climbing procedure cannot "navigate" through a plateau, since it relies on improvement in score to guide it to better structures.

local maximum

plateau

Upon reflection, we realize that greedy hill climbing will encounter plateaus quite often. Recall that we consider scores that satisfy score equivalence. Thus, all networks in an I-equivalence class will have the same score. Moreover, as shown in theorem 3.9, the set of I-equivalent networks forms a contiguous region in the space, which we can traverse using a set of covered edge-reversal operations. Thus, any I-equivalence class necessarily forms a plateau in the search space.

I-equivalence

Recall that equivalence classes can potentially be exponentially large. Ongoing work studies the average size of an equivalence class (when considering all networks) and the actual distributions of sizes encountered in realistic situations, such as during structure search.

It is clear, however, that most networks we encounter have at least a few equivalent networks. Thus, we conclude that most often, greedy hill climbing will converge to an equivalence class. There are two possible situations: Either there is another network in this equivalence class from which we can continue the upward climb, or the whole equivalence class is a local maximum. Greedy hill climbing cannot deal with either situation, since it cannot explore without upward indications.

As we discussed in appendix A.4.2, there are several strategies to improve on the network $\mathcal{G}$ returned by a greedy search algorithm. One approach that deals with plateaus induced by equivalence classes is to enumerate explicitly all the network structures that are I-equivalent to $\mathcal{G}$, and for each one to examine whether it has neighbors with higher score. This enumeration, however, can be expensive when the equivalence class is large. An alternative solution, described in section 18.4.4, is to search directly over the space of equivalence classes. However, both of these approaches save us from only some of the plateaus, and not from local maxima.

Appendix A.4.2 describes other methods that help address problem of local maxima. For
basin flooding example, *basin flooding* keeps track of all previous networks and considers any operator leading from one of them to a structure that we have not yet visited. A key problem with this approach is that storing the list of networks we visited in the recent past can be expensive (recall that greedy hill climbing stores just one copy of the network). Moreover, we do not necessarily want to explore the whole region surrounding a local maximum, since it contains many variants of the same network. To see why, suppose that three different edges in a local maximum network can be removed with very little change in score. This means that all seven networks that contain at least one deletion will be explored before a more interesting change will be considered.

tabu search A method that solves both problems is the *tabu search* of algorithm A.6. Recall that this procedure keeps a list of recent operators we applied, and in each step we do not consider operators that reverse the effect of recently applied operators. Thus, once the search decides to add an edge, say $X \rightarrow Y$, it cannot delete this edge in the next L steps (for some prechosen L). Similarly, once an arc is reversed, it cannot be reversed again. As for the basin-flooding approach, tabu search cannot use the termination criteria of greedy hill climbing. Since we want the search to proceed after reaching the local maxima, we do not want to stop when the score of the current candidate is smaller than the previous one. Instead, we continue the search with the hope of reaching a better structure. If this does not happen after a prespecified number of steps, we decide to abandon the search and select the best network encountered at any time during the search.

Finally, as we discussed, one can also use randomization to increase our chances of escaping local maxima. In the case of structure learning, these methods do help. In particular, simulated annealing was reported to outperform greedy hill-climbing search. However, in typical example domains (such as the ICU-Alarm domain) it appears that simple methods such as tabu search with random restarts find higher-scoring networks much faster.

Data Perturbation Methods   So far, we have discussed only the application of general-purpose local search methods to the specific problem of structure search. We now discuss one class of
data perturbation methods — *data-perturbation* methods — that are more specific to the learning task. The idea is similar to random restarts: We want to perturb the search in a way that will allow it to overcome local obstacles and make progress toward the global maxima. Random restart methods achieve this perturbation by changing the network. Data perturbation methods, on the other hand, change the training data.

To understand the idea, consider a perturbation that duplicates some instances (say by random choice) and removes others (again randomly). If we do a reasonable number of these modifications, the resulting data set $\mathcal{D}'$ has most of the characteristics of the original data set $\mathcal{D}$. For example, the value of sufficient statistics in the perturbed data are close

---

Algorithm 18.1 Data perturbation search

Procedure Search-with-Data-Perturbation (
    $\mathcal{G}_\emptyset,$ // initial network structure
    $\mathcal{D}$ // Fully observed data set
    score,   // Score
    $\mathcal{O},$ // A set of search operator
    Search,   // Search procedure
    $t_0,$ // Initial perturbation size
    $\gamma,$ // Reduction in perturbation size

)

1    $\mathcal{G} \leftarrow \text{Search}(\mathcal{G}_\emptyset, \mathcal{D}, \text{score}, \mathcal{O})$
2    $\mathcal{G}_{\text{best}} \leftarrow \mathcal{G}$
3    $t \leftarrow t_0$
4   **for** $i = 1, \ldots$ until convergence
5      $\mathcal{D}' \leftarrow \text{Perturb}(\mathcal{D}, t)$
6      $\mathcal{G} \leftarrow \text{Search}(\mathcal{G}, \mathcal{D}', \text{score}, \mathcal{O})$
7   **if** $\text{score}(\mathcal{G} : \mathcal{D}) > \text{score}(\mathcal{G}_{\text{best}} : \mathcal{D})$ **then**
8        $\mathcal{G}_{\text{best}} \leftarrow \mathcal{G}$
9      $t \leftarrow \gamma \cdot t$
10
11   **return** $\mathcal{G}_{\text{best}}$

---

to the values in the original data. Thus, we expect that big differences between networks are preserved. That is, if $\text{score}(\mathcal{G}_{X \to Y} : \mathcal{D}) \gg \text{score}(\mathcal{G}_2 : \mathcal{D})$, then we expect that $\text{score}(\mathcal{G}_{X \to Y} : \mathcal{D}') \gg \text{score}(\mathcal{G}_2 : \mathcal{D}')$. On the other hand, the perturbation does change the comparison between networks that are similar. The basic intuition is that the score using $\mathcal{D}'$ has the same broad outline as the score using $\mathcal{D}$, yet might have different fine-grained topology. This suggests that a structure $\mathcal{G}$ that is a local maximum when using the score on $\mathcal{D}$ is no longer a local maximum when using $\mathcal{D}'$. The magnitude of perturbation determines the level of details that are preserved after the perturbation.

    We note that instead of duplicating and removing instances, we can achieve perturbation by *weighting* data instances. Much of the discussion on scoring networks and related topics applies without change if we assign weight to each instance. Formally, the only difference is the computation of sufficient statistics. If we have weights $w[m]$ for the m'th instance, then the sufficient statistics are redefined as:

weighted data instances

$$M[z] = \sum_m \mathbf{I}\{\mathbf{Z}[m] = z\} \cdot w[m].$$

Note that when $w[m] = 1$, this reduces to the standard definition of sufficient statistics. Instance duplication and deletion lead to integer weights. However, we can easily consider perturbation that results in fractional weights. This leads to a continuous spectrum of data perturbations that range from small changes to weights to drastic ones.

The actual search procedure is shown in algorithm 18.1. The heart of the procedure is the Perturb function. This procedure can implemented in different ways. A simple approach is to sample each $w[m]$ for a distribution whose variance is dictated by t, for example, using a Gamma distribution, with mean 1 and variance t. (Note that we need to use a distribution that attains nonnegative values. Thus, the Gamma distribution is more suitable than a Gaussian distribution.)

18.4.3.3   Score Decomposition and Search

The discussion so far has examined how generic ideas in heuristic search apply to structure learning. We now examine how the particulars of the problem impact the search.

The dominant factor in the cost of the search algorithm is the evaluation of neighboring networks at each stage. As discussed earlier, the number of such networks is approximately n^2. To evaluate each of these network structures, we need to score them. This process requires that we traverse all the different data cases, computing sufficient statistics relative to our new structure. This computation can get quite expensive, and it is the dominant cost in any structure learning algorithm.

score
decomposability

This key task is where the *score decomposability* property turns out to be useful. Recall that the scores we examine decompose into a sum of terms, one for each variable X_i. Each of these family scores is computed relative only to the variables in the family of X_i. A local change — adding, deleting, or reversing an edge — leaves almost all of the families in the network unchanged. (Adding and deleting changes one family, and reversing changes two.) For families whose composition does not change, the associated component of the score also does not change. To understand the importance of this observation, assume that our current candidate network is $\mathcal{G}$. For each operator, we compute the improvement in the score that would result in making that change. We define the *delta score*

delta score

$$\delta(\mathcal{G} : o) = \text{score}(o(\mathcal{G}) : \mathcal{D}) - \text{score}(\mathcal{G} : \mathcal{D})$$

to be the change of score associated with applying o on $\mathcal{G}$. Using score decomposition, we can compute this quantity relatively efficiently.

Proposition 18.5

Let $\mathcal{G}$ be a network structure and score be a decomposable score.

- If o is "Add $X \to Y$," and $X \to Y \notin \mathcal{G}$, then

$$\delta(\mathcal{G} : o) = \text{FamScore}(Y, \text{Pa}_Y^{\mathcal{G}} \cup \{X\} : \mathcal{D}) - \text{FamScore}(Y, \text{Pa}_Y^{\mathcal{G}} : \mathcal{D}).$$

- If o is "Delete $X \to Y$" and $X \to Y \in \mathcal{G}$, then

$$\delta(\mathcal{G} : o) = \text{FamScore}(Y, \text{Pa}_Y^{\mathcal{G}} - \{X\} : \mathcal{D}) - \text{FamScore}(Y, \text{Pa}_Y^{\mathcal{G}} : \mathcal{D}).$$

- If o is "Reverse $X \to Y$" and $X \to Y \in \mathcal{G}$, then

$$\delta(\mathcal{G} : o) = \text{FamScore}(X, \text{Pa}_X^{\mathcal{G}} \cup \{Y\} : \mathcal{D}) + \text{FamScore}(Y, \text{Pa}_Y^{\mathcal{G}} - \{X\} : \mathcal{D})$$
$$-\text{FamScore}(X, \text{Pa}_X^{\mathcal{G}} : \mathcal{D}) - \text{FamScore}(Y, \text{Pa}_Y^{\mathcal{G}} : \mathcal{D}).$$

See exercise 18.18.

Note that these computations involve only the sufficient statistics for the particular family that changed. This requires a pass over only the appropriate columns in the table describing the training data.

Now, assume that we have an operator o, say "Add $X \rightarrow Y$," and instead of applying this edge addition, we have decided to apply another operator o' that changes the family of some variable Z (for $Z \neq Y$), producing a new graph $\mathcal{G}'$. The key observation is that $\delta(\mathcal{G}' : o)$ remains unchanged — we do not need to recompute it. We need only to recompute $\delta(\mathcal{G}' : o')$ for operators o' that involve Y.

Proposition 18.6

> Let $\mathcal{G}$ and $\mathcal{G}'$ be two network structures and score be a decomposable score.
>
> - If o is either "Add $X \rightarrow Y$" or "Delete $X \rightarrow Y$" and $\mathrm{Pa}_Y^{\mathcal{G}} = \mathrm{Pa}_Y^{\mathcal{G}'}$, then $\delta(\mathcal{G} : o) = \delta(\mathcal{G}' : o)$.
>
> - If o is "Reverse $X \rightarrow Y$," $\mathrm{Pa}_Y^{\mathcal{G}} = \mathrm{Pa}_Y^{\mathcal{G}'}$, and $\mathrm{Pa}_X^{\mathcal{G}} = \mathrm{Pa}_X^{\mathcal{G}'}$, then $\delta(\mathcal{G} : o) = \delta(\mathcal{G}' : o)$.

See exercise 18.19.

This shows that we can cache the computed $\delta(\mathcal{G} : o)$ for different operators and then reuse most of them in later search steps. The basic idea is to maintain a data structure that records for each operator o the value of $\delta(\mathcal{G} : o)$ with respect to the current network $\mathcal{G}$. After we apply a step in the search, we have a new current network, and we need to update this data structure. Using proposition 18.6 we see that most of the computed values do not need to be changed. We need to recompute $\delta(\mathcal{G}' : o)$ only for operators that modify one of the families that we modified in the recent step. By careful data-structure design, this cache can save us a lot of computational time; see box 18.A for details. **Overall, the decomposability of the scoring function provides significant reduction in the amount of computation that we need to perform during the search. This observation is critical to making structure search feasible for high-dimensional spaces.**

Box 18.A — Skill: Practical Collection of Sufficient Statistics. *The passes over the training data required to compute sufficient statistics generally turn out to be the most computationally intensive part of structure learning. It is therefore crucial to take advantage of properties of the score in order to ensure efficient computations as well as use straightforward organizational tricks.*

One important source of computational savings derives from proposition 18.6. As we discussed, this proposition allows us to avoid recomputing many of the delta-scores after taking a step in the search. We can exploit this observation in a variety of ways. For example, if we are performing greedy hill climbing, we know that the search will necessarily examine all operators. Thus, after each step we can update the evaluation of all the operators that were "damaged" by the last move. The number of such operators is $O(n)$, and so this requires $O(n \cdot M)$ time (since we need to collect sufficient statistics from data). Moreover, if we keep the score of different operators in a heap, we spend $O(n \log n)$ steps to update the heap but then can retrieve the best operator in constant time. Thus, although the cost of a single step in the greedy hill-climbing procedure seems to involve quadratic number of operations of $O(n^2 \cdot M)$, we can perform it in time $O(n \cdot M + n \log n)$.

We can further reduce the time consumed by the collection of sufficient statistics by considering additional levels of caching. For example, if we use table-CPDs, then the counts needed for evalu-

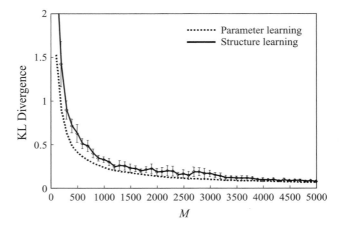

Figure 18.7 Performance of structure and parameter learning for instances generated from the ICU-Alarm **network.** The graph shows the KL-divergence of the learned to the true network, and compares two learning tasks: learning the parameters only, using a correct network structure, and learning both parameters and structure. The curves shows average performance over 10 data sets of the same size, with error bars showing +/- one standard deviation. The error bars for parameter learning are much smaller and are not shown.

ating X as a parent of Y and the counts needed to evaluate Y as a parent of X are the same. Thus, we can save time by caching previously computed counts, and also by marginalizing counts such as $M[x, y]$ to compute $M[x]$. A more elaborate but potentially very effective approach is one where we plan the collection of the entire set of sufficient statistics needed. In this case, we can use efficient algorithms for the set cover problem to choose a smaller set of sufficient statistics that covers all the needed computations. There are also efficient data structures (such as AD-trees for discrete spaces and KD-trees or metric trees for continuous data) that are designed explicitly for maintaining and retrieving sufficient statistics; these data structures can significantly improve the performance of the algorithm, particularly when we are willing to approximate *sufficient statistics in favor of dramatic speed improvements.*

One cannot overemphasize the importance of these seemingly trivial caching tricks. In practice, learning the structure of a network without making use of such tricks is infeasible even for a modest number of variables.

18.4.3.4   Empirical Evaluation

In practice, relatively cheap and simple algorithms, such as tabu search, work quite well. Figure 18.7 shows the results of learning a network from data generated from the ICU-Alarm network. The graph shows the KL-divergence to the true network and compares two learning tasks: learning the parameters only, using a correct network structure, and learning both parameters and structure. Although the graph does show that it is harder to recover both the structure and the parameters, the difference in the performance achieved on the two tasks is surprisingly small. We see that structure learning is not necessarily a harder task than parameter estimation, although computationally, of course, it is more expensive. We note, however, that even the computational cost is not prohibitive. Using simple optimization techniques (such as tabu search with random restarts), learning a network with a hundred variables takes a few minutes on a standard machine.

We stress that the networks learned for different sample sizes in figure 18.7 are not the same as the original networks. They are usually simpler (with fewer edges). As the graph shows, they perform quite similarly to the real network. This means that for the given data, these networks seem to provide a better score, which means a good trade-off between complexity and fit to the data. As the graph suggests, this estimate (based on training data) is quite reasonable.

18.4.4 Learning with Equivalence Classes ⋆

The preceding discussion examined different search procedures that attempt to escape local maxima and plateaus in the search space. An alternative approach to avoid some of these pitfalls is to change the search space. In particular, as discussed, many of the plateaus we encounter during the search are a consequence of score equivalence — equivalent networks have equivalent scores. This observation suggests that we can avoid these plateaus if we consider searching over equivalence classes of networks.

To carry out this idea, we need to examine carefully how to construct the search space. Recall that an equivalence class of networks can be exponential in size. Thus, we need a compact representation of states (equivalence classes) in our search space. Fortunately, we already encountered such a representation. Recall that a *class PDAG* is a partially directed graph that corresponds to an equivalence class of networks. This representation is relatively compact, and thus, we can consider the search space over all possible class PDAGs.

class PDAG

Next, we need to answer the question how to score a given class PDAG. The scores we discussed are defined for over network structures (DAGs) and not over PDAGs. Thus, to score a class PDAG $\mathcal{K}$, we need to build a network $\mathcal{G}$ in the equivalence class represented by $\mathcal{K}$ and then score it. As we saw in section 3.4.3.3, this is a fairly straightforward procedure.

Finally, we need to decide on our search algorithm. Once again, we generally resort to a hill-climbing search using local graph operations. Here, we need to define appropriate search operations on the space of PDAGs. One approach is to use operations at the level of PDAGs. In this case, we need operations that add, remove, and reverse edges; moreover, since PDAGs contain both directed edges and undirected ones, we may wish to consider operations such as adding an undirected edge, orienting an undirected edge, and replacing a directed edge by an undirected one. An alternative approach is to use operators in DAG space that are guaranteed to change the equivalence class. In particular, consider an equivalence class $\mathcal{E}$ (represented as a class PDAG). We can define as our operators any step that takes a DAG $\mathcal{G} \in \mathcal{E}$, adds or deletes an edge from $\mathcal{G}$ to produce a new DAG $\mathcal{G}'$, and then constructs the equivalence class $\mathcal{E}'$ for $\mathcal{G}'$ (represented again as a class PDAG). Since both edge addition and edge deletion change the skeleton, we are guaranteed that $\mathcal{E}$ and $\mathcal{E}'$ are distinct equivalence classes.

GES algorithm

One algorithm based on this last approach is called the *GES algorithm*, for *greedy equivalence search*. GES starts out with the equivalence class for the empty graph and then takes greedy edge-addition steps until no additional edge-addition steps improve the score. It then executes the reverse procedure, removing edges one at a time until no additional edge-removal steps improve the score.

consistent score   When used with a *consistent score* (as in definition 18.1), this simple two-pass algorithm has some satisfying guarantees. Assume that our distribution P^* is faithful for the graph $\mathcal{G}^*$ over $\mathcal{X}$; thus, as in section 18.2, there are no spurious independencies in P^*. Moreover, assume that we have (essentially) infinite data. Under these assumptions, our (consistent) scoring function gives only the correct equivalence class — the equivalence class of $\mathcal{G}^*$ — the highest score. For this setting, one can show that GES is guaranteed to produce the equivalence class of $\mathcal{G}^*$ as its output.

Although the assumptions here are fairly strong, this result is still important and satisfying. Moreover, empirical results suggest that GES works reasonably well even when some of its assumptions are violated (to an extent). Thus, it also provides a reasonable alternative in practice.

Although simple in principle, there are two significant computational issues associated with GES and other algorithms that work in the space of equivalence classes. The first is the cost of generating the equivalence classes that result from the local search operators discussed before. The second is the cost of evaluating their scores while reusing (to the extent possible) the sufficient statistics from our current graph. Although nontrivial (and outside the scope of this book), local operations that address both of these tasks have been constructed, making this algorithm a computationally feasible alternative to search over DAG space.

Box 18.B — Concept: Dependency Networks. *An alternative formalism for parameterizing a Markov network is by associating with each variable X_i a conditional probability distribution (CPD) $P_i(X_i \mid \mathcal{X} - \{X_i\}) = P_i(X_i \mid \mathrm{MB}_{\mathcal{H}}(X_i))$. Networks parameterized in this way are sometimes called* dependency networks *and are drawn as a cyclic directed graph, with edges to each variable from all of the variables in its Markov blanket.*

dependency networks   *This representation offers certain trade-offs over other representations. In terms of semantics, a key limitation of this parameterization is that a set of CPDs $\{P_i(X_i \mid \mathrm{MB}_{\mathcal{H}}(X_i)) : X_i \in \mathcal{X}\}$ may not be consistent with any probability distribution P; that is, there may not be a distribution P such that $P_i(X_i \mid \mathrm{MB}_{\mathcal{H}}(X_i)) = P(X_i \mid \mathrm{MB}_{\mathcal{H}}(X_i))$ for all i (hence the use of the subscript i on P_i). Moreover, determining whether such a set of CPDs is consistent with some distribution P is a computationally difficult problem. Thus, eliciting or learning a consistent dependency network can be quite difficult, and the semantics of an inconsistent network is unclear.*

However, in a noncausal domain, dependency networks arguably provide a more appropriate representation of the dependencies in the distribution than a Bayesian network. Certainly, for a lay user, understanding the notion of a Markov blanket in a Bayesian network is not trivial. On the other hand, in comparison to Markov networks, the CPD parameterization is much more natural and easy to understand. (As we discussed, there is no natural interpretation for a Markov network factor in isolation.)

From the perspective of inference, dependency networks provide a very easy mechanism for answering queries where all the variables except for a single query variable are observed (see box 18.C). However, answering other queries is not as obvious. The representation lends itself very nicely to Gibbs sampling, which requires precisely the distribution of individual variables given their Markov blanket. However, exact inference requires that we transform the network to a standard

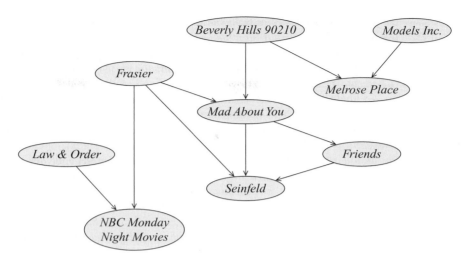

Figure 18.C.1 — Learned Bayesian network for collaborative filtering. A fragment of a Bayesian network for collaborative filtering, learned from Nielsen TV rating data, capturing the viewing record of sample viewers. The variables denote whether a TV program was watched.

parameterization, a task that requires a numerical optimization process.

The biggest advantage arises in the learning setting. If we are willing to relax the consistency requirement, the problem of learning such networks from data becomes quite simple: we simply have to learn a CPD independently for each variable, a task to which we can apply a wide variety of standard supervised learning algorithms. In this case, however, it is arguable whether the resulting network can be considered a unified probabilistic model, rather than a set of stand-alone predictors for individual variables.

Box 18.C — Case Study: Bayesian Networks for Collaborative Filtering. *In many marketing settings, we want to provide to a user a recommendation of an item that he might like, based on previous items that he has bought or liked. For example, a bookseller might want to recommend books that John might like to buy, using John's previous book purchases. Because we rarely have enough data for any single user to determine his or her preferences, the standard solution is an approach called* collaborative filtering, *which uses the observed preferences of other users to try to determine the preferences for any other user. There are many possible approaches to this problem, including ones that explicitly try to infer key aspects of a user's preference model.*

collaborative
filtering

One approach is to learn the dependency structure between different purchases, as observed in the population. We treat each item i as a variable X_i in a joint distribution, and each user as an instance. Most simply, we view a purchase of an item i (or some other indication of preference) as one value for the variable X_i, and the lack of a purchase as a different value. (In certain settings, *we may get explicit ratings from the user, which can be used instead.) We can then use structure learning to obtain a Bayesian network model over this set of random variables. Dependency networks (see box 18.B) have also been used for this task; these arguably provide a more intuitive visualization of the dependency model to a lay user.*

Both models can be used to address the collaborative filtering task. Given a set of purchases for a set of items S, we can compute the probability that the user would like a new item i. In general, this question is reduced to a probabilistic inference task where all purchases other than S and i are set to false; thus, all variables other than the query variable X_i are taken to be observed. In a Bayesian network, this query can be computed easily by simply looking at the Markov blanket of X_i. In a dependency network, the process is even simpler, since we need only consider the CPD for X_i. Bayesian networks and Markov networks offer different trade-offs. For example, the learning and prediction process for dependency networks is somewhat easier, and the models are arguably more understandable. However, Bayesian networks allow answering a broader range of queries — for example, queries where we distinguish between items that the user has viewed and chosen not to purchase and items that the user simply has not viewed (whose variables arguably should be taken to be unobserved).

Heckerman et al. (2000) applied this approach to a range of different collaborative filtering data sets. For example, figure 18.C.1 shows a fragment of a Bayesian network for TV-watching habits learned from Nielsen viewing data. They show that both the Bayesian network and the dependency network methods performed significantly better than previous approaches proposed for this task. The performance of the two methods in terms of predictive accuracy is roughly comparable, and both were fielded successfully as part of Microsoft's E-Commerce software system.

18.5   Bayesian Model Averaging $\star$

18.5.1   Basic Theory

We now reexamine the basic principles of the learning problem. Recall that the Bayesian methodology suggests that we treat unknown parameters as random variables and should consider all possible values when making predictions. When we do not know the structure, the Bayesian methodology suggests that we should consider all possible graph structures. Thus, according to Bayesian theory, given a data set $\mathcal{D} = \{\xi[1], \ldots, \xi[M]\}$ we should make predictions according Bayesian to the *Bayesian estimation* rule:

estimation

$$P(\xi[M+1] \mid \mathcal{D}) = \sum_{\mathcal{G}} P(\xi[M+1] \mid \mathcal{D}, \mathcal{G}) P(\mathcal{G} \mid \mathcal{D}), \qquad (18.10)$$

where $P(\mathcal{G} \mid \mathcal{D})$ is posterior probability of different networks given the data

$$P(\mathcal{G} \mid \mathcal{D}) = \frac{P(\mathcal{G}) P(\mathcal{D} \mid \mathcal{G})}{P(\mathcal{D})}.$$

In our discussion so far, we searched for a single structure $\mathcal{G}$ that maximized the Bayesian score, and thus also the posterior probability. When is the focus on a single structure justified?

Recall that the logarithm of the marginal likelihood $\log P(\mathcal{D} \mid \mathcal{G})$ grows linearly with the number of samples. Thus, when M is large, there will be large differences between the top-scoring structure and all the rest. We used this property in the proof of theorem 18.2 to show that for asymptotically large M, the best-scoring structure is the true one. Even when M is not that large, the posterior probability of this particular equivalence class of structures will be exponentially larger than all other structures and dominate most of the mass of the posterior. In such a case, the posterior mass is dominated by this single equivalence class, and we can approximate equation (18.10) with

$$P(\xi[M+1] \mid \mathcal{D}) \approx P(\xi[M+1] \mid \mathcal{D}, \tilde{\mathcal{G}}),$$

where $\tilde{\mathcal{G}} = \arg\max_{\mathcal{G}} P(\mathcal{G} \mid \mathcal{D})$. The intuition is that $P(\tilde{\mathcal{G}} \mid \mathcal{D}) \approx 1$, and $P(\mathcal{G} \mid \mathcal{D}) \approx 0$ for any other structure.

What happens when we consider learning with smaller number of samples? In such a situation, the posterior mass might be distributed among many structures (which may not include the true structure). If we are interested only in density estimation, this might not be a serious problem. If $P(\xi[M+1] \mid \mathcal{D}, \mathcal{G})$ is similar for the different structure with high posterior, then picking one of them will give us reasonable performance. Thus, if we are doing density estimation, then we might get away with learning a single structure and using it. However, we need to remember that the theory suggests that we consider the whole set of structures when making predictions.

If we are interested in *structure discovery*, we need to be more careful. The fact that several networks have similar scores suggests that one or several of them might be close to the "true" structure. However, we cannot really distinguish between them given the data. If this is the situation, then we should not be satisfied with picking one of these structures (say, even the one with the best score) and drawing conclusions about the domain. Instead, we want to be more cautious and quantify our confidence about the

conclusions we make. Such *confidence estimates* are crucial in many domains where we use Bayesian network learning to learn about the structure of processes that generated data. This is particularly true when the available data is limited.

To consider this problem, we need to specify more precisely what would help us understand the posterior over structures. In most cases, we do not want to quantify the posterior explicitly. Moreover, the set of networks with "high" posterior probability is usually large. To deal with

this issue, there are various approaches we can take. A fairly general one is to consider *network feature* queries. Such a query can ask what is the probability that an edge, say $X \to Y$, appears in the "true" network. Another possible query might be about separation in the "true" network — for example, whether $d\text{-}sep(\boldsymbol{X}; \boldsymbol{Y} \mid \boldsymbol{Z})$ holds in this network. In general, we can formulate such a query as a function $f(\mathcal{G})$. For a binary feature, $f(\mathcal{G})$ can return 1 when the network structure contains the feature, and 0 otherwise. For other features, $f(\mathcal{G})$ may return a numerical quantity that is relevant to the graph structure (such as the length of the shortest trail from X to Y, or the number of nodes whose outdegree is greater than some threshold k).

Given a numerical feature $f(\mathcal{G})$, we can compute its expectation over all possible network structures $\mathcal{G}$:

$$\boldsymbol{E}_{P(\mathcal{G}\mid\mathcal{D})}[f(\mathcal{G})] = \sum_{\mathcal{G}} f(\mathcal{G}) P(\mathcal{G} \mid \mathcal{D}). \tag{18.11}$$

In particular, for a binary feature f, this quantity is simply the posterior probability $P(f \mid \mathcal{D})$.

The problem in computing either equation (18.10) or equation (18.11), of course, is that the number of possible structures is superexponential; see exercise 18.20.

We can reduce this number by restricting attention to structures $\mathcal{G}$ where there is a bound d on the number of parents per variable. This assumption, which we will make throughout this section, is a fairly innocuous one. There are few applications in which very large families are called for, and there are rarely enough data to support robust parameter estimation for such families. From a more formal perspective, networks with very large families tend to have low score. Let $\boldsymbol{\mathcal{G}}_d$ be the set of all graphs with indegree bounded by some constant d. Note that the number of structures in $\boldsymbol{\mathcal{G}}_d$ is still superexponential; see exercise 18.20.

Thus, an exhaustive enumeration over the set of possible network structures is feasible only for tiny domains (4–5 variables). In the next sections we consider several approaches to address this problem. In the following discussion, we assume that we are using a Bayesian score that decomposes according to the assumptions we discussed in section 18.3. Recall that the Bayesian score, as defined earlier, is equal to $\log P(\mathcal{D}, \mathcal{G})$. Thus, we have that

$$P(\mathcal{D}, \mathcal{G}) = \prod_i \exp\{\text{FamScore}_B(X_i \mid \text{Pa}_{X_i}^{\mathcal{G}} \; : \; \mathcal{D})\}. \tag{18.12}$$

18.5.2   Model Averaging Given an Order

In this section, we temporarily turn our attention to a somewhat easier problem. Rather than perform model averaging over the space of *all* structures, we restrict attention to structures that are consistent with some predetermined *variable ordering* $\prec$. As in section 18.4.2, we restrict attention to structures $\mathcal{G}$ such that there is an edge $X_i \to X_j$, only if $X_i \prec X_j$.

variable ordering

18.5.2.1   Computing the marginal likelihood

marginal likelihood

We first consider the problem of computing the *marginal likelihood* of the data given the order:

$$P(\mathcal{D} \mid \prec) = \sum_{\mathcal{G} \in \boldsymbol{\mathcal{G}}_d} P(\mathcal{G} \mid \prec) P(\mathcal{D} \mid \mathcal{G}). \tag{18.13}$$

Note that this summation, although restricted to networks with bounded indegree and consistent with $\prec$, is still exponentially large; see exercise 18.20.

Before we compute the marginal likelihood, we note that computing the marginal likelihood is equivalent to making predictions with equation (18.10). To see this, we can use the definition of probability to see that

$$P(\xi[M+1] \mid \mathcal{D}, \prec) = \frac{P(\xi[M+1], \mathcal{D} \mid \prec)}{P(\mathcal{D} \mid \prec)}.$$

Now both terms on the right are marginal likelihood terms, one for the original data and the other for original data extended by the new instance.

We now return to the computation of the marginal likelihood. The key insight is that when we restrict attention to structures consistent with a given order $\prec$, the choice of family for one variable places no additional constraints on the choice of family for another. Note that this property does not hold without the restriction on the order; for example, if we pick X_i to be a parent of X_j, then X_j cannot in turn be a parent of X_i.

Therefore, we can choose a structure $\mathcal{G}$ consistent with $\prec$ by choosing, independently, a family $\boldsymbol{U}_i$ for each variable X_i. Global parameter modularity assumption states that the choice of parameters for the family of X_i is independent of the choice of family for another family in the network. Hence, summing over possible graphs consistent with $\prec$ is equivalent to summing over possible choices of family for each variable, each with its parameter prior. Given our constraint on the size of the family, the possible parent sets for the variable X_i are

$$\mathcal{U}_{i,\prec} = \{\boldsymbol{U} \ : \ \boldsymbol{U} \prec X_i, |\boldsymbol{U}| \leq d\},$$

where $\boldsymbol{U} \prec X_i$ is defined to hold when all variables in $\boldsymbol{U}$ precede X_i in $\prec$. Let $\boldsymbol{\mathcal{G}}_{d,\prec}$ be the set of structures in $\boldsymbol{\mathcal{G}}_d$ consistent with $\prec$. Using equation (18.12), we have that

$$
\begin{aligned}
P(\mathcal{D} \,|\, \prec) &= \sum_{\mathcal{G} \in \boldsymbol{\mathcal{G}}_{d,\prec}} \prod_i \exp\{\mathrm{FamScore}_B(X_i \mid \mathrm{Pa}^{\mathcal{G}}_{X_i} \ : \ \mathcal{D})\} \\
&= \prod_i \sum_{\boldsymbol{U}_i \in \mathcal{U}_{i,\prec}} \exp\{\mathrm{FamScore}_B(X_i \mid \boldsymbol{U}_i \ : \ \mathcal{D})\}.
\end{aligned}
\tag{18.14}
$$

Intuitively, the equality states that we can sum over all network structures consistent with $\prec$ by summing over the set of possible families for each variable, and then multiplying the results for the different variables. This transformation allows us to compute $P(\mathcal{D} \,|\, \prec)$ efficiently. The expression on the right-hand side consists of a product with a term for each variable X_i, each of which is a summation over all possible families for X_i. Given a bound d over the number of parents, the number of possible families for a variable X_i is at most $\binom{n}{d} \leq n^d$. Hence, the total cost of computing equation (18.14) is at most $n \cdot n^d = n^{d+1}$.

18.5.2.2 Probabilities of features

For certain types of features f, we can use the technique of the previous section to compute, in closed form, the probability $P(f \,|\, \prec, \mathcal{D})$ that f holds in a structure given the order and the data.

In general, if f is a feature. We want to compute

$$P(f \,|\, \prec, \mathcal{D}) = \frac{P(f, \mathcal{D} \,|\, \prec)}{P(\mathcal{D} \,|\, \prec)}.$$

We have just shown how to compute the denominator. The numerator is a sum over all structures that contain the feature and are consistent with the order:

$$P(f, \mathcal{D} \,|\, \prec) = \sum_{\mathcal{G} \in \boldsymbol{\mathcal{G}}_{d,\prec}} f(\mathcal{G}) P(\mathcal{G} \,|\, \prec) P(\mathcal{D} \mid \mathcal{G}). \tag{18.15}$$

The computation of this term depends on the specific type of feature f.

The simplest situation is when we want to compute the posterior probability of a particular choice of parents $\boldsymbol{U}$. This in effect requires us to sum over all graphs where $\mathrm{Pa}^{\mathcal{G}}_{X_i} = \boldsymbol{U}$. In this case, we can apply the same closed-form analysis to (18.15). The only difference is that we restrict $\mathcal{U}_{i,\prec}$ to be the singleton $\{\boldsymbol{U}\}$. Since the terms that sum over the parents of X_j for $i \neq j$ are not disturbed by this constraint, they cancel out from the equation.

Proposition 18.7

$$P(\text{Pa}_{X_i}^{\mathcal{G}} = \boldsymbol{U} \mid D, \prec) \quad = \quad \frac{\exp\{\text{FamScore}_B(X_i \mid \boldsymbol{U} : \mathcal{D})\}}{\sum_{\boldsymbol{U}' \in \mathcal{U}_{i,\prec}} \exp\{\text{FamScore}_B(X_i \mid \boldsymbol{U}' : \mathcal{D})\}}.$$

A slightly more complex situation is when we want to compute the posterior probability of the *edge feature* $X_i \to X_j$. Again, we can apply the same closed-form analysis to (18.15). The only difference is that we restrict $\mathcal{U}_{j,\prec}$ to consist only of subsets that contain X_i.

Proposition 18.8

$$P(X_j \in \text{Pa}_{X_i}^{\mathcal{G}} \mid \prec, \mathcal{D}) = \frac{\sum_{\{\boldsymbol{U} \in \mathcal{U}_{i,\prec} \,:\, X_j \in \boldsymbol{U}\}} \exp\{\text{FamScore}_B(X_i \mid \boldsymbol{U} : \mathcal{D})\}}{\sum_{\boldsymbol{U} \in \mathcal{U}_{i,\prec}} \exp\{\text{FamScore}_B(X_i \mid \boldsymbol{U} : \mathcal{D})\}}.$$

Unfortunately, this approach cannot be used to compute the probability of arbitrary structural features. For example, we cannot compute the probability that there exists some directed path from X_i to X_j, as we would have to consider all possible ways in which a path from X_i to X_j could manifest itself through exponentially many structures.

We can overcome this difficulty using a simple sampling approach. Proposition 18.7 provides us with a closed-form expression for the exact posterior probability of the different possible families of the variable X_i. We can therefore easily sample entire networks from the posterior distribution given the order: we simply sample a family for each variable, according to the distribution specified in proposition 18.7. We can then use the sampled networks to evaluate any feature, such as X_i is ancestor of X_j.

18.5.3   The General Case

In the previous section, we made the simplifying assumption that we were given a predetermined order. Although this assumption might be reasonable in certain cases, it is clearly too restrictive in domains where we have very little prior knowledge. We therefore want to consider structures consistent with all possible orders. Here, unfortunately, we have no elegant tricks that allow an efficient, closed-form solution. As with search problems we discussed, the choices of parents for one variable can interfere with the choices of parents for another.

A general approach is try to approximate the exhaustive summation over structures in quantities of interest (that is, equation (18.10) and equation (18.11)) with approximate sums. For this purpose we utilize ideas that are similar to the ones we discuss in chapter 12 in the context of approximate inference.

A first-cut approach for approximating the sums in our case is to find a set $\boldsymbol{\mathcal{G}}'$ of high scoring structures, and then estimate the relative mass of the structures in $\boldsymbol{\mathcal{G}}'$. Thus, for example, we would approximate equation (18.11) with

$$P(f \mid \mathcal{D}) \approx \frac{\sum_{\mathcal{G} \in \boldsymbol{\mathcal{G}}'} P(\mathcal{G} \mid \mathcal{D}) f(\mathcal{G})}{\sum_{\mathcal{G} \in \boldsymbol{\mathcal{G}}'} P(\mathcal{G} \mid \mathcal{D})}. \tag{18.16}$$

This approach leaves open the question of how we construct $\boldsymbol{\mathcal{G}'}$. The simplest approach is to use model selection to pick a single high-scoring structure and then use that as our approximation. If the amount of data is large relative to the size of the model, then the posterior will be sharply peaked around a single model, and this approximation is a reasonable one. However, as we discussed, when the amount of data is small relative to the size of the model, there is usually a large number of high-scoring models, so that using a single model as our set $\boldsymbol{\mathcal{G}'}$ is a very poor approximation.

We can find a larger set of structures by recording all the structures examined during the search and returning the high-scoring ones. However, the set of structures found in this manner is quite sensitive to the search procedure we use. For example, if we use greedy hill climbing, then the set of structures we collect will all be quite similar. Such a restricted set of candidates also shows up when we consider multiple restarts of greedy hill climbing. This is a serious problem, since we run the risk of getting estimates of confidence that are based on a biased sample of structures.

18.5.3.1 MCMC Over Structures

An alternative approach is based on the use of sampling. As in chapter 12, we aim to approximate the expectation over graph structures in equation (18.10) and equation (18.16) by an empirical average. Thus, if we manage to sample graphs $\mathcal{G}_1, \ldots, \mathcal{G}_K$ from $P(\mathcal{G} \mid \mathcal{D})$, we can approximate equation (18.16) as

$$P(f \mid \mathcal{D}) \approx \frac{1}{K} \sum_k f(\mathcal{G}_k).$$

The question is how to sample from the posterior distribution. One possible answer is to use the general tool of Markov chain Monte Carlo (MCMC) simulation; see section 12.3. In this case, we define a Markov chain over the space of possible structures whose stationary distribution is the posterior distribution $P(G \mid \mathcal{D})$. We then generate a set of possible structures by doing a random walk in this Markov chain. Assuming that we continue this process until the chain converges to the stationary distribution, we can hope to get a set of structures that is representative of the posterior.

How do we construct a Markov chain over the space of structures? The idea is a fairly straightforward application of the principles discussed in section 12.3. The states of the Markov chain are graphs in the set $\boldsymbol{\mathcal{G}}$ of graphs we want to consider. We consider local operations (for example, add edge, delete edge, reverse edge) that transform one structure to another. We assume we have a proposal distribution $\mathcal{T}^Q$ over such operations. We then apply the Metropolis-Hastings acceptance rule. Suppose that the current state is $\mathcal{G}$, and we sample the transition to $\mathcal{G}'$ from the proposal distribution, then we accept this transition with probability

$$\min\left[1, \frac{P(\mathcal{G}', \mathcal{D})\mathcal{T}^Q(\mathcal{G}' \to \mathcal{G})}{P(\mathcal{G}, \mathcal{D})\mathcal{T}^Q(\mathcal{G} \to \mathcal{G}')}\right].$$

As discussed in section 12.3, this strategy ensures that we satisfy the detailed balance condition. To ensure that we have a regular Markov chain, we also need to verify that the space $\boldsymbol{\mathcal{G}}$ is connected, that is, that we can reach each structure in $\boldsymbol{\mathcal{G}}$ from any other structure in $\boldsymbol{\mathcal{G}}$ through a sequence of operations. This is usually easy to ensure with the set of operators we discussed.

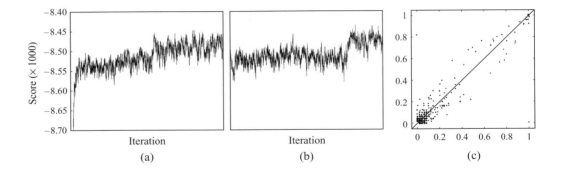

Figure 18.8   MCMC structure search using 500 instances from ICU-Alarm **network.** (a) & (b) Plots of the progression of the MCMC process for two runs for 500 instances sampled from the ICU-Alarm network. The x-axis denotes the iteration number, and the y-axis denotes the score of the current structure. (a) A run initialized from the empty structure, and (b) a run initialized from the structure found by structure search. (c) Comparison of the estimates of the posterior probability of edges using 100 networks sampled every 2,000 iterations from the each of the runs (after initial burn-in of 20,000 iterations). Each point denotes an edge, the x-coordinate is the estimate using networks from run (a) and the y-coordinate is the estimate using networks from run (b).

It is important to stress that if the operations we apply are local (such as edge addition, deletion, and reversal), then we can efficiently compute the ratio $\frac{P(\mathcal{G}',\mathcal{D})}{P(\mathcal{G},\mathcal{D})}$. To see this, note that the logarithm of this ratio is the difference in score between the two graphs. As we discussed in section 18.4.3.3, this difference involves only terms that relate to the families of the variables that are involved in the local move. Thus, performing MCMC over structures can use the same caching schemes discussed in section 18.4.3.3, and thus it can be executed efficiently.

The final detail we need to consider is the choice the proposal distribution $\mathcal{T}^Q$. Many choices are reasonable. The simplest one to use, and one that is often used in practice, is the uniform distribution over all possible operations (excluding ones that violate acyclicity or other constraints we want to impose).

To demonstrate the use of MCMC over structures, we sampled 500 instances from the ICU-Alarm network. This is a fairly small training set, and so we do not hope to recover one network. Instead, we run the MCMC sampling to collect 100 networks and use these to estimate the posterior of different edges. One of the hard problems in using such an approach is checking whether the MCMC simulation converged to the stationary distribution. One ad-hoc test we can perform is to compare the results of running several independent simulations. For example, in figure 18.8a and 18.8b we plot the progression of two runs, one starting from the empty structure and the other from the structure found by structure search on the same data set. We see that initially the two runs are very different; they sample networks with rather different scores. In later stages of the simulation, however, the two runs are in roughly the same range of scores. This suggests that they might be exploring the same region. Another way of testing this is to compare the estimate we computed based on each of the two runs. As we see in figure 18.8c,

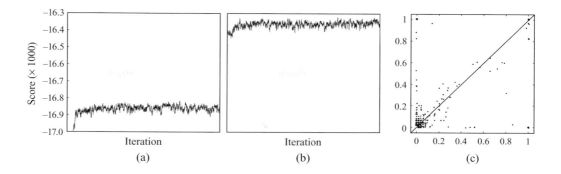

Figure 18.9 **MCMC structure search using 1,000 instances from** ICU-Alarm **network.** The protocol is the same as in figure 18.8. (a) & (b) Plots two MCMC runs. (c) A comparison of the estimates of the posterior probability of edges.

although the estimates are definitely not identical, they are mostly in agreement with each other.

Variants of MCMC simulation have been applied successfully to this task for a variety of small domains, typically with 4–14 variables. However, there are several issues that potentially limit its effectiveness for large domains involving many variables. As we discussed, the space of network structures grows superexponentially with the number of variables. Therefore, the domain of the MCMC traversal is enormous for all but the tiniest domains. More importantly, the posterior distribution over structures is often quite peaked, with neighboring structures having very different scores. The reason is that even small perturbations to the structure — a removal of a single edge — can cause a huge reduction in score. Thus, the "posterior landscape" can be quite jagged, with high "peaks" separated by low "valleys." In such situations, MCMC is known to be slow to mix, requiring many samples to reach the posterior distribution.

To see this effect, consider figure 18.9, where we repeated the same experiment we performed before, but this time for a data set of 1000 instances. As we can see, although we considered a large number of iterations, different MCMC runs converge to quite different ranges of scores. This suggests that the different runs are sampling from different regions of the search space. Indeed, when we compare estimates in figure 18.9c, we see that some edges have estimated posterior of almost 1 in one run and of 0 in another. We conclude that each of the runs became stuck in a local "hill" of the search space and explored networks only in that region.

18.5.3.2 MCMC Over Orders

collapsed MCMC

To avoid some of the problems with MCMC over structures, we consider an approach that utilizes *collapsed particles*, as described in section 12.4. Recall that a collapsed particle consists of two components, one an assignment to a set of random variables that we sampled, and the other a distribution over the remaining variables.

In our case, we utilize the notion of collapsed particle as follows. Instead of working in the space of graphs $\mathcal{G}$, we work in the space of pairs $\langle \prec, \mathcal{G} \rangle$ so that $\mathcal{G}$ is consistent with the ordering $\prec$. As we have seen, we can use closed-form equations to deal the distribution over $\mathcal{G}$ given an ordering $\prec$. Thus, each ordering can represent a collapsed particle.

We now construct a Markov chain over the state of all $n!$ orders. Our construction will guarantee that this chain has the stationary distribution $P(\prec \mid \mathcal{D})$. We can then simulate this Markov chain, obtaining a sequence of samples $\prec_1, \ldots, \prec_T$. We can now approximate the expected value of any function f as

$$P(f \mid \mathcal{D}) \approx \frac{1}{T} \sum_{t=1}^{T} P(f \mid \mathcal{D}, \prec_t),$$

where we compute $P(f \mid \prec_t, \mathcal{D})$ as described in section 18.5.2.2.

It remains only to discuss the construction of the Markov chain. Again, we use a standard *Metropolis-Hastings* algorithm. For each order $\prec$, we define a *proposal probability* $\mathcal{T}^Q(\prec \to \prec')$, which defines the probability that the algorithm will "propose" a move from $\prec$ to $\prec'$. The algorithm then *accepts* this move with probability

Metropolis-
Hastings

$$\min\left[1, \frac{P(\prec', \mathcal{D})\mathcal{T}^Q(\prec' \to \prec)}{P(\prec, \mathcal{D})\mathcal{T}^Q(\prec \to \prec')}\cdot\right]$$

As we discussed in section 12.3, this suffices to ensure the detailed balance condition, and thus the resulting chain is reversible and has the desired stationary distribution.

We can consider several specific constructions for the proposal distribution, based on different neighborhoods in the space of orders. In one very simple construction, we consider only operators that flip two variables in the order (leaving all others unchanged):

$$(X_{i_1} \ldots X_{i_j} \ldots X_{i_d} \ldots X_{i_n}) \mapsto (X_{i_1} \ldots X_{i_d} \ldots X_{i_j} \ldots X_{i_n}).$$

Clearly, such operators allow to get from any ordering to another in relatively few moves.

We note that, again, we can use decomposition to avoid repeated computation during the evaluation of candidate operators. Let $\prec$ be an order and let $\prec'$ be the order obtained by flipping X_{i_j} and X_{i_k}. Now, consider the terms in equation (18.14); those terms corresponding to variables X_{i_ℓ} in the order $\prec$ that precede X_{i_j} or succeed X_{i_k} do not change, since the set of potential parent sets $\mathcal{U}_{i_l, \prec}$ is the same.

Performing MCMC on the space of orderings is much more expensive than MCMC on the space of networks. Each proposed move requires performing summation over a fairly large set of possible parents for each variable. On the other hand, since each ordering corresponds to a large number of networks, a few moves in the space of orderings correspond to a much larger number of moves in the space of networks.

Empirical results show that using MCMC over orders alleviate some of the problems we discussed with MCMC over network structures. For example, figure 18.10 shows two runs of this variant of MCMC for the same data set as in figure 18.9. Although these two runs involve much fewer iterations, they quickly converge to the same "area" of the search space and agree on the estimation of the posterior. This is an empirical indication that these are reasonably close to converging on the posterior.

18.6   Learning Models with Additional Structure

So far, our discussion of structure learning has defined the task as one of finding the best graph structure $\mathcal{G}$, where a graph structure is simply a specification of the parents to each of the

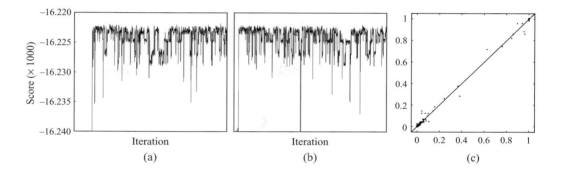

Figure 18.10 MCMC order search using 1,000 instances from ICU-Alarm **network.** The protocol is the same as in figure 18.8. (a) & (b) Plots two MCMC runs. (c) A comparison of the estimates of the posterior probability of edges.

random variables in the network. However, some classes of models have additional forms of structure that we also need to identify. For example, when learning networks with structured CPDs, we may need to select the structure for the CPDs. And when learning template-based models, our model is not even a graph over $\mathcal{X}$, but rather a set of dependencies over a set of template attributes.

Although quite different, the approach in both of these settings is analogous to the one we used for learning a simple graph structure: we define a hypothesis space of potential models and a scoring function that allows us to evaluate different models. We then devise a search procedure that attempts to find a high-scoring model. Even the choice of scoring functions is essentially the same: we generally use either a penalized likelihood (such as the BIC score) or a Bayesian score based on the marginal likelihood. Both of these scoring functions can be computed using the likelihood function for models with shared parameters that we developed in section 17.5. As we now discuss, the key difference in this new setting is the structure of the search space, and thereby of the search procedure. In particular, our new settings require that we make decisions in a space that is not the standard one of deciding on the set of parents for a variable in the network.

18.6.1 Learning with Local Structure

As we discussed, we can often get better performance in learning if we use more compact models of CPDs rather than ones that require a full table-based parameterization. More compact models decrease the number of parameters and thereby reduce overfitting. Some of the techniques that we described earlier in this chapter are applicable to various forms of structured CPDs, including table-CPDs, noisy-or CPDs, and linear Gaussian CPDs (although the closed-form Bayesian score is not applicable to some of those representations). In this section, we discuss the problem of learning CPDs that explicitly encode local parameter sharing within a CPD, focusing on CPD trees as a prototypical (and much-studied) example. Here, we need to make decisions about the local structure of the CPDs as well as about the network structure. Thus, our search space is now much larger: it consists both of "global" decisions — the assignment of parents for each variable — and of "local" decisions — the structure of the CPD tree for each variable.

18.6.1.1 Scoring Networks

We first consider how the development of the score of a network changes when we consider tree-CPDs instead of table-CPDs. Assume that we are given a network structure $\mathcal{G}$; moreover, for each variable X_i, we are given a description of the CPD tree $\mathcal{T}_i$ for $P(X_i \mid \mathrm{Pa}_{X_i}^{\mathcal{G}})$. We view the choice of the trees as part of the specification of the model. Thus, we consider the score of $\mathcal{G}$ with $\mathcal{T}_1, \ldots, \mathcal{T}_n$

$$\mathrm{score}_B(\mathcal{G}, \mathcal{T}_1, \ldots, \mathcal{T}_n \ : \ \mathcal{D}) = \log P(\mathcal{D} \mid \mathcal{G}, \mathcal{T}_1, \ldots, \mathcal{T}_n) + \log P(\mathcal{G}) + \sum_i \log P(\mathcal{T}_i \mid \mathcal{G}),$$

where $P(\mathcal{D} \mid \mathcal{G}, \mathcal{T}_1, \ldots, \mathcal{T}_n)$ is the marginal likelihood when we integrate out the parameters in all the CPDs, and $P(\mathcal{T}_i \mid \mathcal{G})$ is the prior probability over the tree structure.

structure prior   We usually assume that the *structure prior* over trees does not depend on the graph, but only on the choice of parents. Possible priors include the uniform prior

$$P(\mathcal{T}_i \mid \mathcal{G}) \propto 1$$

and a prior that penalizes larger trees

$$P(\mathcal{T}_i \mid \mathcal{G}) \propto c^{|\mathcal{T}_i|}$$

(for some constant $c < 1$).

We now turn to the marginal likelihood term. As in our previous discussion, we assume global and local parameter independence. This means that we have an independent prior over the parameters in each leaf of each tree. For each X_i and each assignment pa_{X_i} to X_i's parents, let $\lambda(\mathrm{pa}_{X_i})$ be the leaf in $\mathcal{T}_i$ to which pa_{X_i} is assigned.

Using an argument that parallels our development of proposition 18.2, we have:

Proposition 18.9

Let $\mathcal{G}$ be a network structure, $\mathcal{T}_1, \ldots, \mathcal{T}_n$ be CPD trees, and $P(\boldsymbol{\theta}_{\mathcal{G}} \mid \mathcal{G}, \mathcal{T}_1, \ldots, \mathcal{T}_n)$ be a parameter prior satisfying global and local parameter independence. Then,

$$P(\mathcal{D} \mid \mathcal{G}, \mathcal{T}_1, \ldots, \mathcal{T}_n) =$$
$$\prod_i \prod_{\ell \in \mathrm{Leaves}(\mathcal{T}_i)} \int \prod_{m \ : \ \lambda(\mathrm{pa}_{X_i}[m]) = \ell} P(X_i[m] \mid \ell, \boldsymbol{\theta}_{X_i \mid \ell}, \mathcal{G}, \mathcal{T}_i) P(\boldsymbol{\theta}_{X_i \mid \ell} \mid \mathcal{G}, \mathcal{T}_i) d\boldsymbol{\theta}_{X_i \mid \ell}.$$

Thus, the score over a tree-CPD decomposes according to the structure of each of the trees. Each of the terms that corresponds to a leaf is the marginal likelihood of a single variable distribution and can be solved using standard techniques. Usually, we use Dirichlet priors at the leaves, and so the term for each leaf will be a term of the form of equation (18.9).

Similar to the developments in table-CPDs, we can extend the notion of score decomposability to networks with tree-CPDs. Recall that score decomposability implies that identical substructures receive the same score (regardless of structure of other parts of the network). Suppose we have two possible tree-CPDs for X (not necessarily with the same set of parents in each), and suppose that there are two leaves ℓ_1 and ℓ_2 in the two trees such that $\boldsymbol{c}_{\ell_1} = \boldsymbol{c}_{\ell_2}$; that is, the same conditions on the parents of X in each of the two CPDs lead to the respective leaves. Thus, the two leaves represent a similar situation (even though the tree structures can differ elsewhere), and intuitively they should receive the same score.

To ensure this property, we need to extend the notion of parameter modularity:

Definition 18.6

tree parameter modularity

Let $\{P(\boldsymbol{\theta}_{\mathcal{G}} \mid \mathcal{G}, \mathcal{T}_1, \ldots, \mathcal{T}_n)\}$ *be a set of parameter priors over networks with tree-CPDs that satisfy global and local parameter independence. The prior satisfies* tree parameter modularity *if for each* $\mathcal{G}, \mathcal{T}_i$ *and* $\mathcal{G}', \mathcal{T}_i'$ *and* $\ell \in \mathrm{Leaves}(\mathcal{T}_i)$, $\ell' \in \mathrm{Leaves}(\mathcal{T}_i')$ *such that* $\boldsymbol{c}_\ell = \boldsymbol{c}_{\ell'}$, *then* $P(\boldsymbol{\theta}_{X_i|\ell} \mid \mathcal{G}, \mathcal{T}_i) = P(\boldsymbol{\theta}_{X_i|\ell'} \mid \mathcal{G}', \mathcal{T}_i')$. ■

BDe prior

A natural extension of the *BDe prior* satisfies this condition. Suppose we choose a prior distribution P' and an equivalent sample size α; then we choose hyperparameters for $P(\boldsymbol{\theta}_{X_i|\ell} \mid \mathcal{G}, \mathcal{T}_i)$ to be $\alpha_{x_i|\ell} = \alpha \cdot P'(x_i, \boldsymbol{c}_\ell)$. Using this assumption, we can now decompose the Bayesian score for a tree-CPD; see exercise 18.17.

18.6.1.2 Search with Local Structure

Our next task is to describe how to search our space of possible hypotheses for one that has high score. Recall that we now have a much richer hypothesis space over which we need to search.

The key question in the case of learning tree-CPDs is that of choosing a tree structure for $P(X \mid \boldsymbol{U})$ for some set of possible parents $\boldsymbol{U}$. (We will discuss different schemes for choosing $\boldsymbol{U}$.) There are two natural operators in this space.

- Split – replace a leaf in the tree by an internal variable that leads to a leaf. This step increases the tree height by 1 on a selected branch.

- Prune – replace an internal variable by a leaf.

Starting from the "empty" tree (that consists of single leaf), we can reach any other tree by a sequence of split operations.

The prune operations allow searches to retract some of the moves. This capability is critical, since there are many local maxima in growing trees. For example, it is often the case that a sequence of two splits leads to a high-scoring tree, but the intermediate step (after performing only the first split) has low score. Thus, greedy hill-climbing search can get stuck in a local maximum early on. As a consequence, the search algorithm for tree-CPDs is often not a straightforward hill climbing. In particular, one common strategy is to choose the best-scoring split at each point, even if that split actually decreases the score. Once we have finished constructing the tree, we go back and evaluate earlier splitting decisions that we made.

The search space over trees can explored using a "divide and conquer" approach. Once we decide to split our tree on one variable, say Y, the choices we make in one subtree, say the one corresponding to $Y = y^0$, are independent of the choices we make in other subtrees. Thus, we can explore for the best structure for each one of the subsets independently of the other. Thus, we can devise recursive search procedures for trees, which works roughly as follows. The procedure receives a set of instances to learn from, a target variable X, and a set $\boldsymbol{U}$ of possible parents. It evaluates each $Y \in \boldsymbol{U}$ as a potential question by scoring the tree-CPD that has Y as a root and has no further question. This *myopic* evaluation can miss pairs or longer sequences of questions that lead to high accuracy predictions and hence to a high scoring model. However, it can be performed very efficiently (see exercise 18.17), and so it is often used, under the hope that other approaches (such as random restarts or tabu search) can help us deal with local optima.

After choosing the root, the procedure divides the data set into smaller data sets, each one corresponding to one value of the variable Y. Using the decomposition property we just discussed, the procedure is recursively invoked to learn a subtree in each D_y. These subtrees are put together into a tree-CPD that has Y as a root. The final test compares the score of the constructed subtree to that of the empty subtree. This test is necessary, since our construction algorithm selected the best split at this point, but without checking that this split actually improves the score. This strategy helps avoid local maxima arising from the myopic nature of the search, but it does potentially lead to unnecessary splits that actually hurt our score. This final check helps avoid those. In effect, this procedure evaluates, as it climbs back up the tree in the recursion, all of the possible merge steps relative to our constructed tree.

There are two standard strategies for applying this procedure. One is an *encapsulated search*. Here, we perform one of the network-structure search procedures we described in earlier sections of this chapter. Whenever we perform a search operation (such as adding or removing an edge), we use a local procedure to find a representation for the newly created CPD. The second option is to use a *unified search*, where we do not distinguish between operators that modify the network and operators that modify the local structure. Instead, we simply apply a local search that modifies the joint representation of the network and local structure for each CPD. Here, each state in the search space consists of a collection of n trees $\langle \mathcal{T}_1, \ldots, \mathcal{T}_n \rangle$, which define both the structure and the parameters of the network. We can structure the operations in the search in many ways: we can update an entire CPD tree for one of the variables, or we can evaluate the delta-score of the various split and merge operations in tree-CPDs all across the network and choose the best one. In either case, we must remember that not every collection of trees defines an acyclic network structure, and so we must construct our search carefully to ensure acyclic structures, as well as any other constraints that we want to enforce (such as bounded indegree).

Each of these two options for learning with local structure has its benefits and drawbacks. Encapsulated search spaces decouple the problem of network structure learning from that of learning CPD structures. This modularity allows us to "plug in" any CPD structure learning procedure within generic search procedure for network learning. Moreover, since we consider the structure of each CPD as a separate problem, we can exploit additional structure in the CPD representation. A shortcoming of encapsulated search spaces is that they can easily cause us to repeat a lot of effort. In particular, we redo the local structure search for X's CPD every time we consider a new parent for X. This new parent is often irrelevant, and so we end up discarding the local structure we just learned. Unified search spaces alleviate this problem. In this search formulation, adding a new parent Y to X is coupled with a proposal as to the specific position in the tree-CPD of X where Y is used. This flexibility comes at a cost: the number of possible operators at a state in the search is very large — we can add a split on any variable at each leaf of the n trees. Moreover, key operations such as edge reversal or even edge deletion can require many steps in the finer-grained search space, and therefore they are susceptible to forming local optima. Overall, there is no clear winner between these two methods, and the best choice is likely to be application-specific.

18.6.2 Learning Template Models

We now briefly discuss the task of structure learning for template-based models. Here, once we define the hypothesis space, the solution becomes straightforward. Importantly, when learning template-based models, our hypothesis space is the space of structures in the template representation, not in the ground network from which we are learning.

For DBNs, we are learning a representation as in definition 6.4: a time 0 network $\mathcal{B}_0$, and a transition network $\mathcal{B}_\rightarrow$. Importantly, the latter network is specified in terms of template attributes; thus, the learned structure will be used for all time slices in the unrolled DBN. The learned structure must satisfy the constraints on our template-based representation. For example, $\mathcal{B}_\rightarrow$ must be a valid conditional Bayesian network for $\mathcal{X}'$ given $\mathcal{X}$: it must be acyclic, and there must be no incoming edges into any variables in $\mathcal{X}$.

For object-relational models, each hypothesis in our space specifies an assignment Pa_A for every $A \in \aleph$. Here also, the learned structure is at the template level and is applied to all instantiations of A in the ground network. The learned structure must satisfy the constraints of the template-based language with which we are dealing. For example, in a plate model, the structure must satisfy the constraints of definition 6.9, for example, the fact that for any template parent $B_i(U_i) \in \mathrm{Pa}_A$, we have that the parent's arguments U_i must be a subset of the attribute's argument signature $\alpha(A)$.

Importantly, in both plate models and DBNs, the set of possible structures is finite. This fact is obvious for DBNs. For plate models, note that the set of template attributes is finite; the possible argument signatures of the parents is also finite, owing to the constraints that $U_i \subseteq \alpha(A)$. However, this constraint does not necessarily hold for richer languages. We will return to this point.

Given a hypothesis space, we need to determine a scoring function and a search algorithm. Based on our analysis in section 17.5.1.2 and section 17.5.3, we can easily generalize any of our standard scoring functions (say the BIC score or the marginal likelihood) to the case of template-based representations. The analysis in section 17.5.1.2 provides us with a formula for a decomposed likelihood function, with terms corresponding to each of the model parameters at the template level. From this formula, the derivation of the BIC score is immediate. We use the decomposed likelihood function and penalize with the number of parameters *in the template model*. For a Bayesian score, we can follow the lines of section 17.5.3 and assume global parameter independence at the template level. The posterior now decomposes in the same way as the likelihood, giving us, once again, a decomposable score.

With a decomposable score, we can now search over the set of legal structures in our hypothesis space, as defined earlier. In this case, the operators that we typically apply are the natural variants of those used in standard structure search: edge addition, deletion, and (possibly) reversal. The only slight subtlety is that we must check, at each step, that the model resulting from the operator satisfies the constraints of our template-based representation. In particular, in many cases for both DBNs and PRMs, edge reversal will lead to an illegal structure. For example, in DBNs, we cannot reverse an edge $A \rightarrow A'$. In plate models, we cannot reverse an edge $B(U) \rightarrow A(U, U')$. Indeed, the notion of edge reversal only makes sense when $\alpha(A) = \alpha(B)$. Thus, a more natural search space for plate models (and other object-relational template models) is one that simply searches for a parent set for each $A \in \aleph$, using operators such as parent addition and deletion.

We conclude this discussion by commenting briefly on the problem of structure learning for more expressive object-relational languages. Here, the complexity of our hypothesis space can be significantly greater. Consider, for example, a PRM. Most obviously, we see that the specification here involves not only possible parents but also possibly a guard and an aggregation function. Thus, our model specification requires many more components. Subtler, but also much more important, is the fact that the space of possible structures is potentially infinite. The key issue here is that the parent signature $\alpha(\mathrm{Pa}_A)$ can be a superset of $\alpha(A)$, and therefore the set of possible parents we can define is unboundedly large.

Example 18.2

Returning to the Genetics *domain, we can, if we choose, allow the genotype of a person depending on the genotype of his or her mother, or of his or her grandmother, or of his or her paternal uncles, or on any type of relative arbitrarily far away in the family tree (as long as acyclicity is maintained). For example, the dependence on paternal uncle (not by marriage) can be written as a dependence of Genotype(U) on Genotype(U') with the guard*

$$Father(V, U) \wedge Brother(U', V),$$

assuming Brother has already been defined.   ∎

In general, by increasing the number of logical variables in the attribute's parent argument signature, we can introduce dependencies over objects that are arbitrarily far away. Thus, when learning PRMs, we generally want to introduce a structure prior that penalizes models that are more complex, for example, as an exponential penalty on the number of logical variables in $\alpha(\mathrm{Pa}_A) - \alpha(A)$, or on the number of clauses in the guard.

 In summary, **the key aspect to learning structure of template-based models is that both the structure and the parameters are specified at the template level, and therefore that is the space over which our structure search is conducted. It is this property that allows us to learn a model from one skeleton (for example, one pedigree, or one university) and apply that same model to a very different skeleton.**

18.7   Summary and Discussion

In this chapter, we considered the problem of structure learning from data. As we have seen, there are two main issues that we need to deal with: the statistical principles that guide the choice between network structures, and the computational problem of applying these principles.

The statistical problem is easy to state. Not all dependencies we can see in the data are real. Some of them are artifacts of the finite sample we have at our disposal. Thus, to learn (that is, generalize to new examples), we must apply caution in deciding which dependencies to model in the learned network.

We discussed two approaches to address this problem. The first is the constraint-based approach. This approach performs statistical tests of independence to collect a set of dependencies that are strongly supported by the data. Then it searches for the network structure that "explains" these dependencies and no other dependencies. The second is the score-based approach. This approach scores whole network structures against the data and searches for a network structure that maximize the score.

What is the difference between these two approaches? Although at the outset they seem quite different, there are some similarities. In particular, if we consider a choice between the two possible networks over two variables, then both approaches use a similar decision rule to make the choice; see exercise 18.27.

When we consider more than two variables, the comparison is less direct. **At some level, we can view the score-based approach as performing a test that is similar to a hypothesis test. However, instead of testing each pair of variables locally, it evaluates a function that is somewhat like testing the complete network structure against the null hypothesis of the empty network. Thus, the score-based approach takes a more global perspective, which allows it to trade off approximations in different part of the network.**

The second issue we considered was the computational issue. Here there are clear differences. In the constraint-based approach, once we collect the independence tests, the construction of the network is an efficient (low-order polynomial) procedure. On the other hand, we saw that the optimization problem in the score-based approach is NP-hard. Thus, we discussed various approaches for heuristic search.

When discussing this computation issue, one has to remember how to interpret the theoretical results. In particular, the NP-hardness of score-based optimization does not mean that the problem is hopeless. When we have a lot of data, the problem actually becomes *easier*, since one structure stands out from the rest. In fact, recent results indicates that there might be search procedures that when applied to sufficiently large data sets are guaranteed to reach the global optimum. This suggests that the hard cases might be the ones where the differences between the maximal scoring network and that other local maxima might not be that dramatic. This is a rough intuition, and it is an open problem to characterize formally the trade-off between quality of solution and hardness of the score-based learning problem.

Another open direction of research attempts to combine the best of both worlds. Can we use the efficient procedures developed for constraint-based learning to find high-scoring network structure? The high-level motivation that the Build-PDAG we discussed uses knowledge about Bayesian networks to direct its actions. On the other hand, the search procedures we discussed so far are fairly uninformed about the problem. A simpleminded combination of these two approaches uses a constraint-based method to find starting point for the heuristic search. More elaborate strategies attempt to use the insight from constraint-based learning to reformulate the search space — for example, to avoid exploring structures that are clearly not going to score well, or to consider global operators.

Another issue that we touched on is estimating the confidence in the structures we learned. We discussed MCMC approaches for answering questions about the posterior. This gives us a measure of our confidence in the structures we learned. In particular, we can see whether a part of the learned network is "crucial" in the sense that it has high posterior probability, or closer to arbitrary when it has low posterior probability. Such an evaluation, however, compares structures only within the class of models we are willing to learn. It is possible that the data do not match any of these structures. In such situations, the posterior may not be informative about goodness of fit the problem. The statistical literature addresses such questions under the name of *goodness of fit* tests, which we briefly described in box 16.A. These tests attempt to evaluate whether a given model would have data such as the one we observed. This topic is still underdeveloped for models such as Bayesian networks.

18.8   Relevant Literature

We begin by noting that many of the works on learning Bayesian networks involve both parameter estimation and structure learning; hence most of the references discussed in section 17.8 are still relevant to the discussion in this chapter.

The constraint-based approaches to learning Bayesian networks were already discussed in chapter 3, under the guise of algorithms for constructing a perfect map for a given distribution. Section 3.6 provides the references relevant to that work; some of the key algorithms of this type include those of Pearl and Verma (1991); Verma and Pearl (1992); Spirtes, Glymour, and Scheines (1993); Meek (1995a); Cheng, Greiner, Kelly, Bell, and Liu (2002). The application of these methods to the task of learning from an empirical distribution requires the use of statistical independence tests (see, for example, Lehmann and Romano (2008)). However, little work has been devoted to analyzing the performance of these algorithms in that setting, when some of the tests may fail.

Much work has been done on the development and analysis of different scoring functions for probabilistic models, including the BIC/MDL score (Schwarz 1978; Rissanen 1987; Barron et al. 1998) and the Bayesian score (Dawid 1984; Kass and Raftery 1995), as well as other scores, such as the AIC (Akaike 1974). These papers also establish the basic properties of these scores, such as the consistency of the BIC/MDL and Bayesian scores.

The Bayesian score for discrete Bayesian networks, using a Dirichlet prior, was first proposed by Buntine (1991) and Cooper and Herskovits (1992) and subsequently generalized by Spiegelhalter, Dawid, Lauritzen, and Cowell (1993); Heckerman, Geiger, and Chickering (1995). In particular, Heckerman et al.   propose the BDe score, and they show that the BDe prior is the only one satisfying certain natural assumptions, including global and local parameter independence and score-equivalence. Geiger and Heckerman (1994) perform a similar analysis for Gaussian networks, resulting in a formal justification for a Normal-Wishart prior that they call the *BGe prior*. The application of MDL principles to Bayesian network learning was developed in parallel (Bouckaert 1993; Lam and Bacchus 1993; Suzuki 1993). These papers also defined the relationship between the maximum likelihood score and information theoretic scores. These connections in a more general setting were explored in early works in information theory Kullback (1959), as well as in early work on decomposable models Chow and Liu (1968).

BGe prior

Buntine (1991) first explored the use of nonuniform priors over Bayesian network structures, utilizing a prior over a fixed node ordering, where all edges were included independently. Heckerman, Mamdani, and Wellman (1995) suggest an alternative approach that uses the extent of deviation between a candidate structure and a "prior network."

Perhaps the earliest application of structure search for learning in Bayesian networks was the work of Chow and Liu (1968) on learning tree-structured networks. The first practical algorithm for learning general Bayesian network structure was proposed by Cooper and Herskovits (1992). Their algorithm, known as the *K2 algorithm*, was limited to the case where an ordering on the variables is given, allowing families for different variables to be selected independently. The approach of using local search over the space of general network structures was proposed and studied in depth by Chickering, Geiger, and Heckerman (1995) (see also Heckerman et al. 1995), although initial ideas along those lines were outlined by Buntine (1991). Chickering et al. compare different search algorithms, including K2 (with different orderings), local search, and simulated annealing. Their results suggest that unless a good ordering is known, local search offers the best time-accuracy trade-off. Tabu search is discussed by Glover and Laguna (1993). Several works considered the combinatorial problem of searching over all network structures (Singh and Moore 2005; Silander and Myllymaki 2006) based on ideas of Koivisto and Sood (2004).

Another line of research proposes local search over a different space, or using different operators. Best known in this category are algorithms, such as the GES algorithm, that search over the space of I-equivalence classes. The foundations for this type of search were developed by Chickering (1996b, 2002a). Chickering shows that GES is guaranteed to learn the optimal Bayesian network structure at the large sample limit, if the data is sampled from a graphical model (directed or undirected). Other algorithms that guarantee identification of the correct structure at the large-sample limit include the constraint-based SGS method of Spirtes, Glymour, and Scheines (1993) and the KES algorithm of Nielsen, Kočka, and Peña (2003).

Other search methods that use alternative search operators or search spaces include the *optimal reinsertion* algorithm of Moore and Wong (2003), which takes a variable and moves it to a new position in the network, and the *ordering-based search* of Teyssier and Koller (2005), which searches over the space of orderings, selecting, for each ordering the optimal (bounded-indegree) network consistent with it. Both of these methods take much larger steps in the space than search over the space of network structures; although each step is also more expensive, empirical results show that these algorithms are nevertheless faster. More importantly, they are significantly less susceptible to local optima. A very different approach to avoiding local optima is taken by the data perturbation method of Elidan et al. (2002), in which the sufficient statistics are perturbed to move the algorithm out of local optima.

Much work has been done on efficient caching of sufficient statistics for machine learning tasks in general, and for Bayesian networks in particular. Moore and Lee (1997); Komarek and Moore (2000) present AD-trees and show their efficacy for learning Bayesian networks in high dimension. Deng and Moore (1989); Moore (2000); Indyk (2004) present some data structures for continuous spaces.

Several papers also study the theoretical properties of the Bayesian network structure learning task. Some of these papers involve the computational feasibility of the task. In particular, Chickering (1996a) showed that the problem of finding the network structure with indegree $\leq d$ that optimizes the Bayesian score for a given data set is $\mathcal{NP}$-hard, for any $d \geq 2$. $\mathcal{NP}$-hardness is also shown for finding the maximum likelihood structures within the class of polytree networks (Dasgupta 1999) and path-structured networks (Meek 2001). Chickering et al. (2003) show that the problem of finding the optimal structure is also $\mathcal{NP}$-hard at the large-sample limit, even when: the generating distribution is perfect with respect to some DAG containing hidden variables, we are given an independence oracle, we are given an inference oracle, and we restrict potential solutions to structures in which each node has at most d parents (for any $d \geq 3$). Importantly, all of these $\mathcal{NP}$-hardness results hold only in the inconsistent case, that is, where the generating distribution is not perfect for some DAG. In the case where the generating distribution is perfect for some DAG over the observed variables, the problem is significantly easier. As we discussed, several algorithms can be guaranteed to identify the correct structure. In fact, the constraint-based algorithms of Spirtes et al. (1993); Cheng et al. (2002) can be shown to have polynomial-time performance if we assume a bounded indegree (in both the generating and the learned network), providing a sharp contrast to the $\mathcal{NP}$-hardness result in the inconsistent setting.

Little work has been done on analyzing the PAC-learnability of Bayesian network structures, that is, their learnability as a function of the number of samples. A notable exception is the work of Höffgen (1993), who analyzes the problem of PAC-learning the structure of Bayesian networks with bounded indegree. He focuses on the maximum likelihood network subject to the indegree constraints and shows that this network has, with high probability, low KL-divergence to the true distribution, if we learn with a number of samples M that grows logarithmically with the number of variables in the network. Friedman and Yakhini (1996) extend this analysis for search with penalized likelihood — for example, MDL/BIC scores. We note that, currently, no efficient algorithm is known for finding the maximum-likelihood Bayesian network of bounded indegree, although the work of Abbeel, Koller, and Ng (2006) does provide a polynomial-time algorithm for learning a low-degree factor graph under these assumptions.

Bayesian model averaging has been used in the context of density estimation, as a way of gaining more robust predictions over those obtained from a single model. However, most often it is used in the context of knowledge discovery, to obtain a measure of confidence in predictions relating to structural properties. In a limited set of cases, the space of legal networks is small enough to allow a full enumeration of the set of possible structures (Heckerman et al. 1999). Buntine (1991) first observed that in the case of a fixed ordering, the exponentially large summation over model structures can be reformulated compactly. Meila and Jaakkola (2000) show that one can efficiently infer and manipulate the full Bayesian posterior over all the superexponentially many tree-structured networks. Koivisto and Sood (2004) suggest an exact method for summing over all network structure. Although this method is exponential in the number of variables, it can deal with domains of reasonable size.

Other approaches attempt to approximate the superexponentially large summation by considering only a subset of possible structures. Madigan and Raftery (1994) propose a heuristic approximation, but most authors use an MCMC approach over the space of Bayesian network structure (Madigan and York 1995; Madigan et al. 1996; Giudici and Green 1999). Friedman and Koller (2003) propose the use of MCMC over orderings, and show that it achieves much faster mixing and therefore more robust estimates than MCMC over network space. Ellis and Wong (2008) improve on this algorithm, both in removing bias in its prior distribution and in improving its computational efficiency.

The idea of using local learning of tree-structured CPDs for learning global network structure was proposed by Friedman and Goldszmidt (1996, 1998), who also observed that reducing the number of parameters in the CPD can help improve the global structure reconstruction. Chickering et al. (1997) extended these ideas to CPDs structured as a decision graph (a more compact generalization of a decision tree). Structure learning in dynamic Bayesian networks was first proposed by Friedman, Murphy, and Russell (1998). Structure learning in object-relational models was first proposed by Friedman, Getoor, Koller, and Pfeffer (1999); Getoor, Friedman, Koller, and Taskar (2002). Segal et al. (2005) present the module network framework, which combines clustering with structure learning.

tree-augmented naive Bayes

Learned Bayesian networks have also been used for specific prediction and classification tasks. Friedman et al. (1997) define a *tree-augmented naive Bayes* structure, which extends the traditional naive Bayes classifier by allowing a tree-structure over the features. They demonstrate that this enriched model provides significant improvements in classification accuracy. Dependency networks were introduced by Heckerman et al. (2000) and applied to a variety of settings, including collaborative filtering. This latter application extended the earlier work of Breese et al. (1998), which demonstrated the success of Bayesian network learning (with tree-structured CPDs) to this task.

18.9 Exercises

Exercise 18.1

Show that the $\chi^2(\mathcal{D})$ statistic of equation (18.1) is approximately twice of $d_{\mathbf{I}}(\mathcal{D})$ of equation (18.2). Hint: examine the first-order Taylor expansion of $\frac{z}{t} \approx 1 + \frac{z-t}{t}$ in z around t.

Exercise 18.2

Derive equation (18.3). Show that this value is the sum of the probability of all possible data sets that have the given empirical counts.

Exercise 18.3★

multiple hypothesis testing

Suppose we are testing *multiple hypotheses* $1, \ldots, N$ for a large value. Each hypothesis has an observed deviance measure D_i, and we computed the associated p-value p_i. Recall that under the null hypothesis, p_i has a uniform distribution between 0 and 1. Thus, $P(p_i < t \mid H_0) = t$ for every $t \in [0, 1]$.

We are worried that one of our tests received a small p-value by chance. Thus, we want to consider the distribution of the *best* p-value out of all of our tests under the assumption that the null hypothesis is true in all of the tests. More formally, we want to examine the behavior of $\min_i p_i$ under H_0.

  a. Show that

  $$P(\min_i p_i < t \mid H_0) \leq t \cdot N.$$

  (Hint: Do *not* assume that the variables p_i are independent of each other.)
  b. Suppose we want to ensure that the probability of a random rejection is below 0.05, what p-value should we use in individual hypothesis tests?
  c. Suppose we assume that the tests (that is, the variables p_i) are independent of each other. Derive the bound in this case. Does this bound give better (higher) decision p-value when we use $N = 100$ and global rejection rate below 0.05 ? How about $N = 1,000$?

Bonferroni correction

The bound you derive in [b] is called *Bonferroni bound*, or more often *Bonferroni correction* for a multiple-hypothesis testing scenario.

Exercise 18.4★

Consider again the Build-PDAG procedure of algorithm 3.5, but now assume that we apply it in a setting where the independence tests might return incorrect answers owing to limited and noisy data.

  a. Provide an example where Build-PDAG can fail to reconstruct the true underlying graph $\mathcal{G}^*$ even in the presence of a single incorrect answer to an independence question.
  b. Now, assume that the algorithm constructs the correct skeleton but can encounter a single incorrect answer when extracting the immoralities.

Exercise 18.5

Prove corollary 18.1. Hint: Start with the result of proposition 18.1, and use the chain rule of entropies and the chain rule mutual information.

Exercise 18.6

Show that adding edges to a network increases the likelihood.

Exercise 18.7★

Stirling's approximation

Prove theorem 18.1. Hint: You can use *Stirling's approximation* for the Gamma function

$$\Gamma(x) \approx \sqrt{2\pi}\, x^{x - \frac{1}{2}} e^{-x}$$

or

$$\log \Gamma(x) \approx \frac{1}{2} \log(2\pi)x \log(x) - \frac{1}{2} \log(x) - x$$

Exercise 18.8

Show that if $\mathcal{G}$ is I-equivalent to $\mathcal{G}'$, then if we use table-CPDs, we have that $score_L(\mathcal{G} : \mathcal{D}) = score_L(\mathcal{G}' : \mathcal{D})$ for any choice of $\mathcal{D}$.

Hint: Consider the set of distributions that can be represented by parameterization each network structure.

Exercise 18.9

Show that if $\mathcal{G}$ is I-equivalent to $\mathcal{G}'$, then if we use table-CPDs, we have that $score_{BIC}(\mathcal{G} : \mathcal{D}) = score_{BIC}(\mathcal{G}' : \mathcal{D})$ for any choice of $\mathcal{D}$. You can use the results of exercise 3.18 and exercise 18.8 in your proof.

Exercise 18.10

Show that the Bayesian score with a K2 prior in which we have a Dirichlet prior $Dirichlet(1, 1, \ldots, 1)$ for each set of multinomial parameters is not score-equivalent.

Hint: Construct a data set for which the score of the network $X \rightarrow Y$ differs from the score of the network $X \leftarrow Y$.

Exercise 18.11⋆

We now examine how to prove score equivalence for the BDe score. Assume that we have a prior specified by an equivalent sample size α and prior distribution P'. Prove the following:

a. Consider networks over the variables X and Y. Show that the BDe score of $X \rightarrow Y$ is equal to that of $X \leftarrow Y$.

b. Show that if $\mathcal{G}$ and $\mathcal{G}'$ are identical except for a covered edge reversal of $X \rightarrow Y$, then the BDe score of both networks is equal.

c. Show that the proof of score equivalence follows from the last result and theorem 3.9.

Exercise 18.12⋆

In section 18.3.2, we have seen that the Bayesian score can be posed a sequential procedure that estimates the performance on new unseen examples. In this example, we consider another score that is based on this motivation.

Recall that leave-one-out cross-validation (LOOCV), described in box 16.A, is a procedure for estimating the performance of a learning method on new samples. In our context, this defines the following score:

$$LOOCV(\mathcal{D} \mid \mathcal{G}) = \prod_{m=1}^{M} P(\xi[m] \mid \mathcal{G}, \mathcal{D}_{-m}),$$

where $\mathcal{D}_{-m}$ is the data with the m'th instance removed.

a. Consider the setting of section 18.3.3 where we observe a series of values of a binary variable. Develop a closed-form equation for $LOOCV(\mathcal{D} \mid \mathcal{G})$ as a function of the number of heads and the number of tails.

b. Now consider the network $\mathcal{G}_{X \rightarrow Y}$ and a data set $\mathcal{D}$ that consists of observations of two binary variables X and Y. Develop a closed-form equation for $LOOCV(\mathcal{D} \mid \mathcal{G}_{X \rightarrow Y})$ as a function of the sufficient statistics of X and Y in $\mathcal{D}$.

c. Based on these two examples, what are the properties of the LOOCV score? Is it decomposable?

Exercise 18.13

Consider the algorithm for learning tree-structured networks in section 18.4.1. Show that the weight $w_{i \to j} = w_{j \to i}$ if the score satisfies score equivalence.

Exercise 18.14⋆⋆

We now consider a situation where we can find the high-scoring structure in a polynomial time. Suppose we are given a directed graph C that imposes constraints on the possible parent-child relationships in the learned network: an edge $X{-}Y$ in C implies that X might be considered as a parent of Y. We define $\boldsymbol{G}_C = \{\mathcal{G} : \mathcal{G} \subseteq C\}$ to be the set of graphs that are consistent with the constraints imposed by C.

Describe an algorithm that, given a decomposable score score and a data set $\mathcal{D}$, finds the maximal scoring network in $\boldsymbol{G}_C$. Show that your algorithm is exponential in the tree-width of the graph C.

Exercise 18.15⋆

Consider the problem of learning a Bayesian network structure over two random variables X and Y.

a. Show a data set — an empirical distribution and a number of samples M — where the optimal network structure according to the BIC scoring function is different from the optimal network structure according to the ML scoring function.

b. Assume that we continue to get more samples that exhibit precisely the same empirical distribution. (For simplicity, we restrict attention to values of M that allow that empirical distribution to be achieved; for example, an empirical distribution of 50 percent heads and 50 percent tails can be achieved only for an even number of samples.) At what value of M will the network that optimizes the BIC score be the same as the network that optimizes the likelihood score?

Exercise 18.16

This problem considers the performance of various types of structure search algorithms. Suppose we have a general network structure search algorithm, $\boldsymbol{A}$, that takes a set of basic operators on network structures as a parameter. This set of operators defines the search space for $\boldsymbol{A}$, since it defines the candidate network structures that are the "immediate successors" of any current candidate network structure—that is, the successor states of any state reached in the search. Thus, for example, if the set of operators is {add an edge not currently in the network}, then the successor states of any candidate network $\mathcal{G}$ is the set of structures obtained by adding a single edge anywhere in $\mathcal{G}$ (so long as acyclicity is maintained).

Given a set of operators, $\boldsymbol{A}$ does a simple greedy search over the set of network structures, starting from the empty network (no edges), using the BIC scoring function. Now, consider two sets of operators we can use in $\boldsymbol{A}$. Let $\boldsymbol{A}_{[add]}$ be $\boldsymbol{A}$ using the set of operations {add an edge not currently in the network}, and let $\boldsymbol{A}_{[add,delete]}$ be $\boldsymbol{A}$ using the set of operations {add an edge not currently in the network, delete an edge currently in the network}.

a. Show a distribution where, regardless of the amount of data in our training set (that is, even with infinitely many samples), the answer produced by $\boldsymbol{A}_{[add]}$ is worse (that is, has a lower BIC score) than the answer produced by $\boldsymbol{A}_{[add,delete]}$. (It is easiest to represent your true distribution in the form of a Bayesian network; that is, a network from which the sample data are generated.)

b. Show a distribution where, regardless of the amount of data in our training set, $\boldsymbol{A}_{[add,delete]}$ will converge to a local maximum. In other words, the answer returned by the algorithm has a lower score than the optimal (highest-scoring) network. What can we conclude about the ability of our algorithm to find the optimal structure?

Exercise 18.17⋆

This problem considers the problem of learning a CPD tree structure for a variable in a Bayesian network, using the Bayesian score. Assume that the network structure $\mathcal{G}$ includes a description of the CPD trees in it; that is, for each variable X_i, we have a CPD tree $\mathcal{T}_i$ for $P(X_i \mid \mathrm{Pa}_{X_i}^{\mathcal{G}})$. We view the choice of the trees as part of the specification of the model. Thus, we consider the score of $\mathcal{G}$ with $\mathcal{T}_1, \ldots, \mathcal{T}_n$

$$\mathrm{score}_B(\mathcal{G}, \mathcal{T}_1, \ldots, \mathcal{T}_n \ : \ \mathcal{D}) = \log P(\mathcal{D} \mid \mathcal{G}, \mathcal{T}_1, \ldots, \mathcal{T}_n) + \log P(\mathcal{G}) + \sum_i \log P(\mathcal{T}_i \mid \mathcal{G}).$$

Here, we assume for simplicity that the two structure priors are uniform, so that we focus on the marginal likelihood term $P(\mathcal{D} \mid \mathcal{G}, \mathcal{T}_1, \ldots, \mathcal{T}_n)$. Assume we have selected a fixed choice of parents $\boldsymbol{U}_i$ for each variable X_i. We would like to find a set of trees $\mathcal{T}_1, \ldots, \mathcal{T}_n$ that together maximizes the Bayesian score.

a. Show how you can decompose the Bayesian score in this case as a sum of simpler terms; make sure you state the assumptions necessary to allow this decomposition.

b. Assume that we consider only a single type of operator in our search, a $split(X_i, \ell, Y)$ operator, where ℓ is a leaf in the current CPD tree for X_i, and $Y \in \boldsymbol{U}_i$ is a possible parent of X_i. This operator replaces the leaf ℓ by an internal variable that splits on the values of Y. Derive a simple formula for the delta-score $\delta(\mathcal{T} \ : \ o)$ of such an operator $o = split(X_i, \ell, Y)$. (Hint: Use the representation of the decomposed score to simplify the formula.)

c. Suppose our greedy search keeps track of the delta-score for all the operators. After we take a step in the search space by applying operator $o = split(X, \ell, Y)$, how should we update the delta score for another operator $o' = split(X', \ell', Y')$? (Hint: Use the representation of the delta-score in terms of decomposed score in the previous question.)

d. Now, consider applying the same process using the likelihood score rather than the Bayesian score. What will the resulting CPD trees look like in the general case? You can make any assumption you want about the behavior of the algorithm in case of ties in the score.

Exercise 18.18

Prove proposition 18.5.

Exercise 18.19

Prove proposition 18.6.

Exercise 18.20★

Recall that the $\Theta(f(n))$ denotes both an asymptotic lower bound and an asymptotic upper bound (up to a constant factor).

a. Show that the number of DAGs with n vertices is $2^{\Theta(n^2)}$.

b. Show that the number of DAGs with n vertices and indegree bounded by d is $2^{\Theta(dn \log n)}$.

c. Show that the number of DAGs with n vertices and indegree bounded by d that are consistent with a given order is $2^{\Theta(dn \log n)}$.

Exercise 18.21

Consider the problem of learning the structure of a 2-TBN over $\mathcal{X} = \{X_1, \ldots, X_n\}$. Assume that we are learning a model with bounded indegree k. Explain, using the argument of asymptotic complexity, why the problem of learning the 2-TBN structure is considerably easier if we assume that there are no intra-time-slice edges in the 2-TBN.

Exercise 18.22★

module network

In this problem, we will consider the task of learning a generalized type of Bayesian networks that involves shared structure and parameters. Let $\mathcal{X}$ be a set of variables, which we assume are all binary-valued. A *module network* over $\mathcal{X}$ partitions the variables $\mathcal{X}$ into K disjoint clusters, for $K \ll n = |\mathcal{X}|$. All of the variables assigned to the same cluster have precisely the same parents and CPD. More precisely, such a network defines:

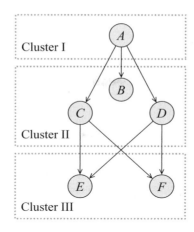

Figure 18.11 A simple module network

- An assignment function $\mathcal{A}$, which defines for each variable X, a cluster assignment $\mathcal{A}(X) \in \{C_1, \ldots, C_K\}$.
- For each cluster C_k ($k = 1, \ldots, K$), a graph $\mathcal{G}$ that defines a set of parents $\mathrm{Pa}_{C_k} = \boldsymbol{U}_k \subset \mathcal{X}$ and a CPD $P_k(X \mid \boldsymbol{U}_k)$.

The module network structure defines a ground Bayesian network where, for each variable X, we have the parents $\boldsymbol{U}_k$ for $k = \mathcal{A}(X)$ and the CPD $P_k(X \mid \boldsymbol{U}_k)$. Figure 18.11 shows an example of such a network.

Assume that our goal is to learn a module network that maximizes the Bayesian score given a data set $\mathcal{D}$, where we need to learn both the assignment of variables to clusters and the graph structure.

a. Define an appropriate set of parameters and an appropriate notion of sufficient statistics for this class of models, and write down a precise formula for the likelihood function of a pair $(\mathcal{A}, \mathcal{G})$ in terms of the parameters and sufficient statistics.

b. Draw the meta-network for the module network shown in figure 18.11. Assuming a uniform prior over each parameter, write down exactly (normalizing constants included) the appropriate parameter posterior given a data set $\mathcal{D}$.

c. We now turn to the problem of learning the structure of the cluster network. We will use local search, using the following types of operators:

- **Add** operators that add a parent for a cluster;
- **Delete** operators that delete a parent for a cluster;
- **Node-Move** operators $o_{k \rightarrow k'}(X)$ that change from $\mathcal{A}(X) = k$ to $\mathcal{A}(X) = k'$.

Describe an efficient implementation of the **Node-Move** operator.

d. For each type of operator, specify (precisely) which other operators need to be reevaluated once the operator has been taken? Briefly justify your response.

e. Why did we not include edge reversal in our set of operators?

Exercise 18.23★

reinsertion
operator

It is often useful when learning the structure of a Bayesian network to consider more global search operations. In this problem we will consider an operator called *reinsertion*, which works as follows: For the current structure $\mathcal{G}$, we choose a variable X_i to be our *target variable*. The first step is to remove the variable from the network by severing all connections to its children and parents. We then select the optimal set of at most K_p parents and at most K_c children for X and reinsert it into the network with edges from the selected parents and to the selected children. Throughout this problem, assume the use of the BIC score for structure evaluation.

a. Let X_i be our current target variable, and assume for the moment that we have somehow chosen U_i to be optimal parents of X_i. Consider the case of $K_c = 1$, where we want to choose the *single* optimal child for X_i. Candidate children — those that do not introduce a cycle in the graph — are $Y_1, \ldots, Y_\ell$. Write an argmax expression for finding the optimal child C. Explain your answer.

b. Now consider the case of $K_c = 2$. How do we find the optimal pair of children? Assuming that our family score for any $\{X_k, U_k\}$ can be computed in a constant time f, what is the best asymptotic computational complexity of finding the optimal pair of children? Explain. Extend your analysis to larger values of K_c. What is the computational complexity of this task?

c. We now consider the choice of parents for X_i. We now assume that we have already somehow chosen the optimal set of children and will hold them fixed. Can we do the same trick when choosing the parents? If so, show how. If not, argue why not.

Exercise 18.24

Prove proposition 18.7.

Exercise 18.25

Prove proposition 18.8.

Exercise 18.26★

Consider the idea of searching for a high-scoring network by searching over the space of orderings $\prec$ over the variables. Our task is to search for a high-scoring network that has bounded indegree k. For simplicity, we focus on the likelihood score. For a given order $\prec$, let $\mathcal{G}_\prec^*$ be the highest-likelihood network consistent with the ordering $\prec$ of bounded indegree k. We define $\text{score}_L(\prec : \mathcal{D}) = \text{score}_L(\mathcal{G}_\prec^* : \mathcal{D})$. We now search over the space of orderings using operators o that swap two adjacent nodes in the ordering, that is:

$$X_1, \ldots, X_{i-1}, X_i, X_{i+1}, \ldots, X_n \mapsto X_1, \ldots, X_{i-1}, X_{i+1}, X_i, \ldots, X_n.$$

Show how to use score decomposition properties to search this space efficiently.

Exercise 18.27★

Consider the choice between $\mathcal{G}_\emptyset$ and $\mathcal{G}_{X \to Y}$ given a data set of joint observations of binary variables X and Y.

a. Show that $\text{score}_{BIC}(\mathcal{G}_{X \to Y} : \mathcal{D}) > \text{score}_{BIC}(\mathcal{G}_\emptyset : \mathcal{D})$ if and only if $\boldsymbol{I}_{\hat{P}_\mathcal{D}}(X;Y) > c$. What is this constant c?

b. Suppose that we have BDe prior with uniform P' and $\alpha = 0$. Write the condition on the counts when $\text{score}_{BDe}(\mathcal{G}_{X \to Y} : \mathcal{D}) > \text{score}_{BDe}(\mathcal{G}_\emptyset : \mathcal{D})$.

c. Consider empirical distributions of the form discussed in figure 18.3. That is, $\hat{P}(x, y) = 0.25 + 0.5 \cdot p$ if x and y are equal, and $\hat{P}(x, y) = 0.25 - 0.5 \cdot p$ otherwise. For different values of p, plot as a function of M both the χ^2 deviance measure and the mutual information. What do you conclude about these different functions?

d. Implement the exact p-value test described in section 18.2.2.3 for the χ^2 and the mutual information deviance measures.

e. Using the same empirical distribution, plot for different values of M the *decision boundary* between $\mathcal{G}_\emptyset$ and $\mathcal{G}_{X \to Y}$ for each of the three methods we considered in this exercise. That is, find the value of p at which the two alternatives have (approximately) equal score, or at which the p-value of rejecting the null hypothesis is (approximately) 0.05.

What can you conclude about the differences between these structure selection methods?

19 *Partially Observed Data*

Until now, our discussion of learning assumed that the training data are *fully observed*: each instance assigns values to all the variables in our domain. This assumption was crucial for some of the technical developments in the previous two chapters. Unfortunately, this assumption is clearly unrealistic in many settings. In some cases, data are missing by accident; for example, some fields in the data may have been omitted in the data collection process. In other cases, certain observations were simply not made; in a medical-diagnosis setting, for example, one never performs all possible tests or asks all of the possible questions. Finally, some variables are *hidden*, in that their values are never observed. For example, some diseases are not observed directly, but only via their symptoms.

In fact, in many real-life applications of learning, the available data contain missing values. Hence, we must address the learning problem in the presence of *incomplete data*. As we will see, incomplete data pose both foundational problems and computational problems. The foundational problems are in formulating an appropriate learning task and determining when we can expect to learn from such data. The computational problems arise from the complications incurred by incomplete data and the construction of algorithms that address these complications.

In the first section, we discuss some of the subtleties encountered in learning from incomplete data and in formulating an appropriate learning problem. In subsequent sections, we examine techniques for addressing various aspects of this task. We focus initially on the parameter-learning task, assuming first that the network structure is given, and then treat the more complex structure-learning question at the end of the chapter.

19.1 Foundations

19.1.1 Likelihood of Data and Observation Models

A central concept in our discussion of learning so far was the likelihood function that measures the probability of the data induced by different choices of models and parameters. The likelihood function plays a central role both in maximum likelihood estimation and in Bayesian learning. In these developments, the likelihood function was determined by the probabilistic model we are learning. Given a choice of parameters, the model defined the probability of each instance. In the case of fully observed data, we assumed that each instance $\xi[m]$ in our training set $\mathcal{D}$ is simply a random sample from the model.

It seems straightforward to extend this idea to incomplete data. Suppose our domain consists of two random variables X and Y, and in one particular instance we observed only the value of X to be $x[m]$, but not the value of Y. Then, it seems natural to assign the instance the probability $P(x[m])$. More generally, the likelihood of an incomplete instance is simply the marginal probability given our model. Indeed, the most common approach to define the likelihood of an incomplete data set is to simply marginalize over the unobserved variables.

This approach, however, embodies some rather strong assumptions about the nature of our data. To learn from incomplete data, we need to understand these assumptions and examine the situation much more carefully. Recall that when learning parameters for a model, we assume that the data were generated according to the model, so that each instance is a sample from the model. **When we have missing data, the data-generation process actually involves two steps. In the first step, data are generated by sampling from the model. In this step, values of all the variables are selected. The next step determines which values we get to observe and which ones are hidden from us.** In some cases, this process is simple; for example, some particular variable may always be hidden. In other situations, this process might be much more complex.

To analyze the probabilistic model of the observed training set, we must consider not only the data-generation mechanism, but also the mechanism by which data are hidden. Consider the following two examples.

Example 19.1

We flip a thumbtack onto a table, and every now and then it rolls off the table. Since a fall from the table to the floor is quite different from our desired experimental setup, we do not use results from these flips (they are missing). How would that change our estimation? The simple solution is to ignore the missing values and simply use the counts from the flips that we did get to observe. That is, we pretend that missing flips never happened. As we will see, this strategy can be shown to be the correct one to use in this case.   ∎

Example 19.2

Now, assume that the experiment is performed by a person who does not like "tails" (because the point that sticks up might be dangerous). So, in some cases when the thumbtack lands "tails," the experimenter throws the thumbtack on the floor and reports a missing value. However, if the thumbtack lands "heads," he will faithfully report it. In this case, the solution is also clear. We can use our knowledge that every missing value is "tails" and count it as such. Note that this leads to very different likelihood function (and hence estimated parameters) from the strategy that we used in the previous case.

While this example may seem contrived, many real-life scenarios have very similar properties. For example, consider a medical trial evaluating the efficacy of a drug, but one where patients can drop out of the trial, in which case their results are not recorded. If patients drop out at random, we are in the situation of example 19.1; on the other hand, if patients tend to drop out only when the drug is not effective for them, the situation is essentially analogous to the one in this example. ∎

Note that in both examples, we observe sequences of the form $H, T, H, ?, T, ?, \ldots$, but nevertheless we treat them differently. The difference between these two examples is our knowledge about the observation mechanism. As we discussed, each observation is derived as a combination of two mechanisms: the one that determines the outcome of the flip, and the one that determines whether we observe the flip. Thus, our training set actually consists of two variables for each flip: the flip outcome X, and the observation variable O_X, which tells us whether we observed the value of X.

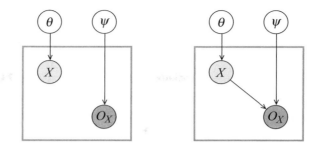

(a) Random missing values   (b) Deliberate missing values

Figure 19.1   **Observation models in two variants of the thumbtack example**

Definition 19.1

observability
variable

observability
model

Let $X = \{X_1, \ldots, X_n\}$ be some set of random variables, and let $O_X = \{O_{X_1}, \ldots, O_{X_n}\}$ be their observability variable. The observability model is a joint distribution $P_{missing}(X, O_X) = P(X) \cdot P_{missing}(O_X \mid X)$, so that $P(X)$ is parameterized by parameters $\boldsymbol{\theta}$, and $P_{missing}(O_X \mid X)$ is parameterized by parameters $\boldsymbol{\psi}$.

We define a new set of random variables $Y = \{Y_1, \ldots, Y_n\}$, where $Val(Y_i) = Val(X_i) \cup \{?\}$. The actual observation is Y, which is a deterministic function of X and O_X,

$$Y_i = \begin{cases} X_i & O_{X_i} = o^1 \\ ? & O_{X_i} = o^0. \end{cases}$$

The variables $Y_1, \ldots, Y_n$ represent the values we actually observe, either an actual value or a ? that represents a missing value. ∎

Thus, we observe the Y variable. This observation always implies that we know the value of the O_X variables, and whenever $Y_i \neq ?$, we also observe the value of X_i. To illustrate the definition of this concept, we consider the probability of the observed value Y in the two preceding examples.

Example 19.3

In the scenario of example 19.1, we have a parameter θ that describes the probability of $X = 1$ (Heads), and another parameter ψ that describes the probability of $O_X = o^1$. Since we assume that the hiding mechanism is random, we can describe this scenario by the meta-network of figure 19.1a. This network describes how the probability of different instances (shown as plates) depend on the parameters. As we can see, this network consists of two independent subnetworks. The first relates the values of X in the different examples to the parameter θ, and the second relates the values of O_X to ψ.

Recall from our earlier discussion that if we can show that θ and ψ are independent given the evidence, then the likelihood decomposes into a product. We can derive this decomposition as follows. Consider the three values of Y and how they could be attained. We see that

$$\begin{aligned} P(Y = 1) &= \theta\psi \\ P(Y = 0) &= (1 - \theta)\psi \\ P(Y = ?) &= (1 - \psi). \end{aligned}$$

Thus, if we see a data set $\mathcal{D}$ of tosses with $M[1]$, $M[0]$, and $M[?]$ instances that are Heads, Tails, and ?, respectively, then the likelihood is

$$L(\theta, \psi : \mathcal{D}) = \theta^{M[1]}(1 - \theta)^{M[0]}\psi^{M[1]+M[0]}(1 - \psi)^{M[?]}.$$

As we expect, the likelihood function in this example is a product of two functions: a function of θ, and a function of ψ. We can easily see that the maximum likelihood estimate of θ is $\frac{M[1]}{M[1]+M[0]}$, while the maximum likelihood estimate of ψ is $\frac{M[1]+M[0]}{M[1]+M[0]+M[?]}$.

We can also reach the conclusion regarding independence using a more qualitative analysis. At first glance, it appears that observing Y activates the v-structure between X and O_X, rendering them dependent. However, the CPD of Y has a particular structure, which induces context-specific independence. In particular, we see that X and O_X are conditionally independent given both values of Y: when $Y = ?$, then O_X is necessarily o^0, in which case the edge $X \to Y$ is spurious (as in definition 5.7); if $Y \neq ?$, then Y deterministically establishes the values of both X and O_X, in which case they are independent.   ∎

Example 19.4

Now consider the scenario of example 19.2. Recall that in this example, the missing values are a consequence of an action of the experimenter after he sees the outcome of the toss. Thus, the probability of missing values depends on the value of X. To define the likelihood function, suppose θ is the probability of $X = 1$. The observation parameters ψ consist of two values: $\psi_{O_X|x^1}$ is probability $O_X = o^1$ when $X = 1$, and $\psi_{O_X|x^0}$ is the probability of $O_X = o^1$ when $X = 0$.

We can describe this scenario by the meta-network of figure 19.1b. In this network, O_X depends directly on X. When we get an observation $Y = ?$, we essentially observe the value of O_X but not of X. In this case, due to the direct edge between X and O_X, the context-specific independence properties of Y do not help: X and O_X are correlated, and therefore so are their associated parameters. Thus, we cannot conclude that the likelihood decomposes.

Indeed, when we examine the form of the likelihood, this becomes apparent. Consider the three values of Y and how they could be attained. We see that

$$
\begin{aligned}
P(Y = 1) &= \theta\psi_{O_X|x^1} \\
P(Y = 0) &= (1 - \theta)\psi_{O_X|x^0} \\
P(Y = ?) &= \theta(1 - \psi_{O_X|x^1}) + (1 - \theta)(1 - \psi_{O_X|x^0}).
\end{aligned}
$$

And so, if we see a data set $\mathcal{D}$ of tosses with $M[1]$, $M[0]$, and $M[?]$ instances that are Heads, Tails, and ?, respectively, then the likelihood is

$$
\begin{aligned}
&L(\theta, \psi_{O_X|x^1}, \psi_{O_X|x^0} : \mathcal{D}) \\
&= \theta^{M[1]}(1 - \theta)^{M[0]}\psi_{O_X|x^1}^{M[1]}\psi_{O_X|x^0}^{M[0]} \\
&\quad (\theta(1 - \psi_{O_X|x^1}) + (1 - \theta)(1 - \psi_{O_X|x^0}))^{M[?]}.
\end{aligned}
$$

As we can see, the likelihood function in this example is more complex than the one in the previous example. In particular, there is no easy way of decoupling the likelihood of θ from the likelihood of $\psi_{O_X|x^1}$ and $\psi_{O_X|x^0}$. This makes sense, since different values of these parameters imply different possible values of X when we see a missing value and so affect our estimate of θ; see exercise 19.1.∎

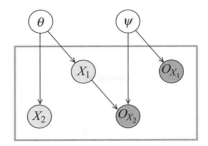

Figure 19.2   An example satisfying MAR but not MCAR. Here, the observability pattern depends on the value of underlying variables.

19.1.2   Decoupling of Observation Mechanism

As we saw in the last two examples, modeling the observation variables, that is, the process that generated the missing values, can result in nontrivial modeling choices, which in some cases result in complex likelihood functions. Ideally, we would hope to avoid dealing with these issues and instead focus on the likelihood of the process that we are interested in (the actual random variables). When can we ignore the observation variables? In the simplest case, the observation mechanism is completely independent of the domain variables. This case is precisely the one we encountered in example 19.1.

Definition 19.2

missing completely at random

A missing data model $P_{missing}$ is missing completely at random *(MCAR) if $P_{missing} \models (\boldsymbol{X} \perp O_{\boldsymbol{X}})$.* ∎

In this case, the likelihood of X and O_X decomposes as a product, and we can maximize each part separately. We have seen this decomposition in the likelihood function of example 19.3. The implications of the decoupling is that we can maximize the likelihood of the parameters of the distribution of $\boldsymbol{X}$ without considering the values of the parameters governing the distribution of $O_{\boldsymbol{X}}$. Since we are usually only interested in the former parameters, we can simply ignore the later parameters.

The MCAR assumption is a very strong one, but it holds in certain settings. For example, momentary sensor failures in medical/scientific imaging (for example, flecks of dust) are typically uncorrelated with the relevant variables being measured, and they induce MCAR observation models. Unfortunately, in many other domains the MCAR simply does not hold. For example, in medical records, the pattern of missing values owes to the tests the patient underwent. These, however, are determined by some of the relevant variables, such as the patient's symptoms, the initial diagnosis, and so on.

As it turns out, MCAR is sufficient but not necessary for the decomposition of the likelihood function. We can provide a more general condition where, rather than assuming marginal independence between $O_{\boldsymbol{X}}$ and the values of $\boldsymbol{X}$, we assume only that the observation mechanism is *conditionally independent* of the underlying variables given other observations.

Example 19.5

Consider a scenario where we flip two coins in sequence. We always observe the first toss X_1, and based on its outcome, we decide whether to hide the outcome of the second toss X_2. See figure 19.2 *for the corresponding model. In this case, $P_{missing} \models (O_{X_2} \perp X_2 \mid X_1)$. In other words, the true values of both coins are independent of whether they are hidden or not, given our observations.*

To understand the issue better, let us write the model and likelihood explicitly. Because we assume that the two coins are independent, we have two parameters θ_{X_1} and θ_{X_2} for the probability of the two coins. In this example, the first coin is always observed, and the observation of the second one depends on the value of the first. Thus, we have parameters $\psi_{O_{X_2} \mid x_1^1}$ and $\psi_{O_{X_2} \mid x_1^0}$ that represent the probability of observing X_2 given that X_1 is heads or tails, respectively.

To derive the likelihood function, we need to consider the probability of all possible observations. There are six possible cases, which fall in two categories.

In the first category are the four cases where we observe both coins. By way of example, consider the observation $Y_1 = y_1^1$ and $Y_2 = y_2^0$. The probability of this observation is clearly $P(X_1 = x_1^1, X_2 = x_2^0, O_{X_1} = o^1, O_{X_2} = o^1)$. Using our modeling assumption, we see that this is simply the product $\theta_{X_1}(1 - \theta_{X_2})\psi_{O_{X_2} \mid x_1^1}$.

In the second category are the two cases where we observe only the first coin. By way of example, consider the observation $Y_1 = y_1^1, Y_2 = ?$. The probability of this observation is $P(X_1 = x_1^1, O_{X_1} = o^1, O_{X_2} = o^0)$. Note that the value of X_2 does not play a role here. This probability is simply the product $\theta_{X_1}(1 - \psi_{O_{X_2} \mid x_1^1})$.

If we write all six possible cases and then rearrange the products, we see that we can write the likelihood function as

$$
\begin{aligned}
L(\boldsymbol{\theta} : \mathcal{D}) \;=\; & \theta_{X_1}^{M[y_1^1]}(1 - \theta_{X_1})^{M[y_1^0]} \\
& \theta_{X_2}^{M[y_2^1]}(1 - \theta_{X_2})^{M[y_2^0]} \\
& \psi_{O_{X_2} \mid x_1^1}^{M[y_1^1, y_2^1] + M[y_1^1, y_2^0]}(1 - \psi_{O_{X_2} \mid x_1^1})^{M[y_1^1, y_2^?]} \\
& \psi_{O_{X_2} \mid x_1^0}^{M[y_1^0, y_2^1] + M[y_1^0, y_2^0]}(1 - \psi_{O_{X_2} \mid x_1^0})^{M[y_1^0, y_2^?]}.
\end{aligned}
$$

This likelihood is a product of four different functions, each involving just one parameter. Thus, we can estimate each parameter independently of the rest. ∎

As we saw in the last example, conditional independence can help us decouple the estimate of parameters of $P(\boldsymbol{X})$ from these of $P(O_{\boldsymbol{X}} \mid \boldsymbol{X})$. Is this a general phenomenon? To answer this question, we start with a definition.

Definition 19.3

Let $\boldsymbol{y}$ be a tuple of observations. These observations partition the variables $\boldsymbol{X}$ into two sets, the observed variables $\boldsymbol{X}_{obs}^{\boldsymbol{y}} = \{X_i : y_i \neq ?\}$ and the hidden ones $\boldsymbol{X}_{hidden}^{\boldsymbol{y}} = \{X_i : y_i = ?\}$. The values of the observed variables are determined by $\boldsymbol{y}$, while the values of the hidden variables are not.

missing at random

We say that a missing data model $P_{missing}$ is missing at random *(MAR) if for all observations $\boldsymbol{y}$ with $P_{missing}(\boldsymbol{y}) > 0$, and for all $\boldsymbol{x}_{hidden}^{\boldsymbol{y}} \in Val(\boldsymbol{X}_{hidden}^{\boldsymbol{y}})$, we have that*

$$
P_{missing} \models (o_{\boldsymbol{X}} \perp \boldsymbol{x}_{hidden}^{\boldsymbol{y}} \mid \boldsymbol{x}_{obs}^{\boldsymbol{y}})
$$

where $o_{\boldsymbol{X}}$ are the specific values of the observation variables given $\boldsymbol{Y}$. ∎

In words, the MAR assumption requires independence between the events $o_{\boldsymbol{X}}$ and $\boldsymbol{x}_{hidden}^{\boldsymbol{y}}$ given $\boldsymbol{x}_{obs}^{\boldsymbol{y}}$. Note that this statement is written in terms of event-level conditional independence rather than conditional independence between random variables. This generality is necessary since every instance might have a different pattern of observed variables; however, if the set of observed variables is known in advance, we can state MAR as conditional independence between random variables.

This statement implies that the observation pattern gives us no additional information about the hidden variables *given the observed variables*:

$$P_{missing}(\boldsymbol{x}_{hidden}^{y} \mid \boldsymbol{x}_{obs}^{y}, o_{\boldsymbol{X}}) = P_{missing}(\boldsymbol{x}_{hidden}^{y} \mid \boldsymbol{x}_{obs}^{y}).$$

Why should we require the MAR assumption? If $P_{missing}$ satisfies this assumption, then we can write

$$
\begin{aligned}
P_{missing}(\boldsymbol{y}) &= \sum_{\boldsymbol{x}_{hidden}^{y}} \left[P(\boldsymbol{x}_{obs}^{y}, \boldsymbol{x}_{hidden}^{y}) P_{missing}(o_{\boldsymbol{X}} \mid \boldsymbol{x}_{hidden}^{y}, \boldsymbol{x}_{obs}^{y}) \right] \\
&= \sum_{\boldsymbol{x}_{hidden}^{y}} \left[P(\boldsymbol{x}_{obs}^{y}, \boldsymbol{x}_{hidden}^{y}) P_{missing}(o_{\boldsymbol{X}} \mid \boldsymbol{x}_{obs}^{y}) \right] \\
&= P_{missing}(o_{\boldsymbol{X}} \mid \boldsymbol{x}_{obs}^{y}) \sum_{\boldsymbol{x}_{hidden}^{y}} P(\boldsymbol{x}_{obs}^{y}, \boldsymbol{x}_{hidden}^{y}) \\
&= P_{missing}(o_{\boldsymbol{X}} \mid \boldsymbol{x}_{obs}^{y}) P(\boldsymbol{x}_{obs}^{y}).
\end{aligned}
$$

The first term depends only on the parameters ψ, and the second term depends only on the parameters $\boldsymbol{\theta}$. Since we write this product for every observed instance, we can write the likelihood as a product of two likelihoods, one for the observation process and the other for the underlying distribution.

Theorem 19.1

If $P_{missing}$ satisfies MAR, then $L(\boldsymbol{\theta}, \psi : \mathcal{D})$ can be written as a product of two likelihood functions $L(\boldsymbol{\theta} : \mathcal{D})$ and $L(\psi : \mathcal{D})$.

This theorem implies that we can optimize the likelihood function in the parameters $\boldsymbol{\theta}$ of the distribution $P(\boldsymbol{X})$ independently of the exact value the observation model parameters. In other words, the MAR assumption is a license to ignore the observation model while learning parameters.

The MAR assumption is applicable in a broad range of settings, but it must be considered with care. For example, consider a sensor that measures blood pressure B but can fail to record a measurement when the patient is overweight. Obesity is a very relevant factor for blood pressure, so that the sensor failure itself is informative about the variable of interest. However, if we always have observations W of the patient's body weight and H of the height, then O_B is conditionally independent of B given W and H. As another example, consider the patient description in hospital records. If the patient does not have an X-ray result X, he probably does not suffer from broken bones. Thus, O_X gives us information about the underlying domain variables. However, assume that the patient's chart also contains a "primary complaint" variable, which was the factor used by the physician in deciding which tests to perform; in this case, the MAR assumption does hold.

In both of these cases, we see that the MAR assumption does not hold given a limited set of observed attributes, but if we expand our set of observations, we can get the MAR assumption to hold. In fact, one can show that we can always extend our model to produce one where the

MAR assumption holds (exercise 19.2). Thus, from this point onward we assume that the data satisfy the MAR assumption, and so our focus is only on the likelihood of the observed data. **However, before applying the methods described later in this chapter, we always need to consider the possible correlations between the variables and the observation variables, and possibly to expand the model so as to guarantee the MAR assumption.**

19.1.3   The Likelihood Function

Throughout our discussion of learning, the likelihood function has played a major role, either on its own, or with the prior in the context of Bayesian estimation. Under the assumption of MAR, we can continue to use the likelihood function in the same roles. From now on, assume we have a network $\mathcal{G}$ over a set of variables $\boldsymbol{X}$. In general, each instance has a different set of observed variables. We will denote by $\boldsymbol{O}[m]$ and $\boldsymbol{o}[m]$ the observed variables and their values in the m'th instance, and by $\boldsymbol{H}[m]$ the missing (or hidden) variables in the m'th instance. We use $L(\boldsymbol{\theta} : \mathcal{D})$ to denote the probability of the observed variables in the data, marginalizing out the hidden variables, and ignoring the observability model:

$$L(\boldsymbol{\theta} : \mathcal{D}) = \prod_{m=1}^{M} P(\boldsymbol{o}[m] \mid \boldsymbol{\theta}).$$

As usual, we use $\ell(\boldsymbol{\theta} : \mathcal{D})$ to denote the logarithm of this function.

With this definition, it might appear that the problem of learning with missing data does not differ substantially from the problem of learning with complete data. We simply use the likelihood function in exactly the same way. Although this intuition is true to some extent, the computational issues associated with the likelihood function are substantially more complex in this case.

To understand the complications, we consider a simple example on the network $\mathcal{G}_{X \rightarrow Y}$ with the edge $X \rightarrow Y$. When we have complete data, the likelihood function for this network has the following form:

$$L(\boldsymbol{\theta}_X, \boldsymbol{\theta}_{Y|x^0}, \boldsymbol{\theta}_{Y|x^1} : \mathcal{D}) =$$
$$\theta_{x^1}^{M[x^1]} \theta_{x^0}^{M[x^0]} \cdot \theta_{y^1|x^0}^{M[x^0,y^1]} \theta_{y^0|x^0}^{M[x^0,y^0]} \cdot \theta_{y^1|x^1}^{M[x^1,y^1]} \theta_{y^0|x^1}^{M[x^1,y^0]}.$$

In the binary case, we can use the constraints to rewrite $\theta_{x^0} = 1 - \theta_{x^1}$, $\theta_{y^0|x^0} = 1 - \theta_{y^1|x^0}$, and $\theta_{y^0|x^1} = 1 - \theta_{y^1|x^1}$. Thus, this is a function of three parameters. For example, if we have a data set with the following sufficient statistics:

x^1, y^1: 13
x^1, y^0: 16
x^0, y^1: 10
x^0, y^0: 4,

then our likelihood function has the form:

$$\theta_{x^1}^{29}(1 - \theta_{x^1})^{14} \cdot \theta_{y^1|x^0}^{10}(1 - \theta_{y^1|x^0})^4 \cdot \theta_{y^1|x^1}^{13}(1 - \theta_{y^1|x^1})^{16}. \tag{19.1}$$

This function is well-behaved: it is log-concave, and it has a unique global maximum that has a simple analytic closed form.

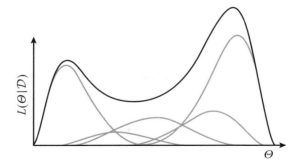

Figure 19.3 **A visualization of a multimodal likelihood function with incomplete data.** The data likelihood is the sum of complete data likelihoods (shown in gray lines). Each of these is unimodal, yet their sum is multimodal.

Assume that the first instance in the data set was $X[1] = x^0, Y[1] = y^1$. Now, consider a situation where, rather than observing this instance, we observed only $Y[1] = y^1$. We now have to reason that this particular data instance could have arisen in two cases: one where $X[1] = x^0$ and one where $X[1] = x^1$. In the former case, our likelihood function is as before. In the second case, we have

$$\theta_{x^1}^{30}(1 - \theta_{x^1})^{13} \cdot \theta_{y^1|x^0}^{9}(1 - \theta_{y^1|x^0})^{4} \cdot \theta_{y^1|x^1}^{14}(1 - \theta_{y^1|x^1})^{16}. \tag{19.2}$$

When we do not observe $X[1]$, the likelihood is the marginal probability of the observations. That is, we need to sum over possible assignments to the unobserved variables. This implies that the likelihood function is the *sum* of the two complete likelihood functions of equation (19.1) and equation (19.2). Since both likelihood functions are quite similar, we can rewrite this sum as

$$\theta_{x^1}^{29}(1 - \theta_{x^1})^{13} \cdot \theta_{y^1|x^0}^{9}(1 - \theta_{y^1|x^0})^{4} \cdot \theta_{y^1|x^1}^{13}(1 - \theta_{y^1|x^1})^{16} \left[\theta_{x^1}\theta_{y^1|x^1} + (1 - \theta_{x^1})\theta_{y^1|x^0} \right].$$

This form seems quite nice, except for the last sum, which couples the parameter θ_{x^1} with $\theta_{y^1|x^1}$ and $\theta_{y^1|x^0}$.

If we have more missing values, there are other cases we have to worry about. For example, if $X[2]$ is also unobserved, we have to consider all possible combinations for $X[1]$ *and* $X[2]$. This results in a sum over four terms similar to equation (19.1), each one with different counts. In general, the likelihood function with incomplete data is the sum of likelihood functions, one for each possible *joint* assignment of the missing values. Note that the number of possible assignments is exponential in the total number of missing values.

We can think of the situation using a geometric intuition. Each one of the complete data likelihood defines a unimodal function. Their sum, however, can be multimodal. In the worst case, the likelihood of each of the possible assignments to the missing values contributes to a different peak in the likelihood function. The total likelihood function can therefore be quite complex. It takes the form of a "mixture of peaks," as illustrated pictorially in figure 19.3.

parameter
independence

likelihood
decomposability

To make matters even more complicated, we lose the property of *parameter independence*, and thereby the *decomposability* of the likelihood function. Again, we can understand this phenomenon either qualitatively, from the perspective of graphical models, or quantitatively, by

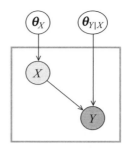

Figure 19.4   The meta-network for parameter estimation for $X \to Y$. When $X[m]$ is hidden but $Y[m]$ is observed, the trail $\theta_X \to X[m] \to Y[m] \leftarrow \theta_{Y|X}$ is active. Thus, the parameters are not independent in the posterior distribution.

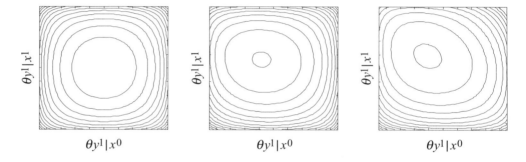

Figure 19.5   Contour plots for the likelihood function for the network $X \to Y$, over the parameters $\theta_{y^1|x^0}$ **and** $\theta_{y^1|x^1}$**.** The total number of data points is 8. (a) No X values are missing. (b) Two X values missing. (c) Three X values missing.

looking at the likelihood function. Qualitatively, recall from section 17.4.2 that, in the complete data case, $\boldsymbol{\theta}_{Y|x^1}$ and $\boldsymbol{\theta}_{Y|x^0}$ are independent given the data, because they are independent given $Y[m]$ and $X[m]$. But if $X[m]$ is unobserved, they are clearly dependent. This fact is clearly illustrated by the meta-network (as in figure 17.7) that represents the learning problem. For example, in a simple network over two variables $X \to Y$, we see that missing data can couple the two parameters' variables; see figure 19.4.

We can also see this phenomenon numerically. Assume for simplicity that $\boldsymbol{\theta}_X$ is known. Then, our likelihood is a function of two parameters $\boldsymbol{\theta}_{y^1|x^1}$ and $\boldsymbol{\theta}_{y^1|x^0}$. Intuitively, if our missing $X[1]$ is H, then it cannot be T. Thus, the likelihood functions of the two parameters are correlated; the more missing data we have, the stronger the correlation. This phenomenon is shown in figure 19.5.

local
decomposability

global
decomposability

This example shows that we have lost the *local decomposability* property in estimating the CPD $P(Y \mid X)$. What about *global decomposability* between different CPDs? Consider a simple model where there is one hidden variable H, and two observed variables X and Y, and edges $H \to X$ and $H \to Y$. Thus, the probability of observing the values x and y is

$$P(x, y) = \sum_h P(h)P(x \mid h)P(y \mid h).$$

The likelihood function is a product of such terms, one for each observed instance $x[m], y[m]$, and thus has the form

$$L(\boldsymbol{\theta} : \mathcal{D}) = \prod_{x,y} \left(\sum_h P(h)P(x \mid h)P(y \mid h) \right)^{M[x,y]}.$$

When we had complete data, we rewrote the likelihood function as a product of local likelihood functions, one for each CPD. This decomposition was crucial for estimating each CPD independently of the others. In this example, we see that the likelihood is a product of sum of products of terms involving different CPDs. The interleaving of products and sums means that we cannot write the likelihood as a product of local likelihood functions. Again, this result is intuitive: Because we do not observe the variable H, we cannot decouple the estimation of $P(X \mid H)$ from that of $P(Y \mid H)$. Roughly speaking, both estimates depend on how we "reconstruct" H in each instance.

We now consider the general case. Assume we have a network $\mathcal{G}$ over a set of variables $\boldsymbol{X}$. In general, each instance has a different set of observed variables. We use $\mathcal{D}$ to denote, as before, the actual observed data values; we use $\mathcal{H} = \cup_m \boldsymbol{h}[m]$ to denote a possible assignment to all of the missing values in the data set. Thus, the pair $(\mathcal{D}, \mathcal{H})$ defines an assignment to all of the variables in all of our instances.

The likelihood function is

$$L(\boldsymbol{\theta} : \mathcal{D}) = P(\mathcal{D} \mid \boldsymbol{\theta}) = \sum_{\mathcal{H}} P(\mathcal{D}, \mathcal{H} \mid \boldsymbol{\theta}).$$

Unfortunately, the number of possible assignments in this sum is exponential in the number of missing values in the entire data set. Thus, although each of the terms $P(\mathcal{D}, \mathcal{H} \mid \boldsymbol{\theta})$ is a unimodal distribution, the sum can have, in the worst case, an exponential number of modes.

However, unimodality is not the only property we lose. Recall that our likelihood function in the complete data case was compactly represented as a product of local terms. This property was important both for the analysis of the likelihood function and for the task of evaluating the likelihood function. What about the incomplete data likelihood? If we use a straightforward representation, we get an exponential sum of terms, which is clearly not useful. Can we use additional properties of the data to help in representing the likelihood? Recall that we assume that different instances are independent of each other. This allows us to write the likelihood function as a product over the probability of each partial instance.

Proposition 19.1   *Assuming IID data, the likelihood can be written as*

$$L(\boldsymbol{\theta} : \mathcal{D}) = \prod_m P(\boldsymbol{o}[m] \mid \boldsymbol{\theta}) = \prod_m \sum_{\boldsymbol{h}[m]} P(\boldsymbol{o}[m], \boldsymbol{h}[m] \mid \boldsymbol{\theta}).$$

This proposition shows that, to compute the likelihood function, we need to perform inference for each instance, computing the probability of the observations. As we discussed in section 9.1, this problem can be intractable, depending on the network structure and the pattern of missing values. Thus, for some learning problems, even the task of evaluating likelihood function for a particular choice of parameters is a difficult computational problem. This observation suggests that optimizing the choice of parameters for such networks can be computationally challenging.

To conclude, **in the presence of partially observed data, we have lost all of the important properties of our likelihood function: its unimodality, its closed-form representation, and the decomposition as a product of likelihoods for the different parameters. Without these properties, the learning problem becomes substantially more complex.**

19.1.4   Identifiability

Another issue that arises in the context of missing data is our ability to identify uniquely a model from the data.

Example 19.6   *Consider again our thumbtack tossing experiments. Suppose the experimenter can randomly choose to toss one of two thumbtacks (say from two different brands). Due to a miscommunication between the statistician and the experimenter, only the toss outcomes were recorded, but not the brand of thumbtack used.*

To model the experiment, we assume that there is a hidden variable H, so that if $H = h^1$, the experimenter tossed the first thumbtack, and if $H = h^2$, the experimenter tossed the second thumbtack. The parameters of our model are θ_H, $\theta_{X|h^1}$, and $\theta_{X|h^2}$, denoting the probability of choosing the first thumbtack, and the probability of heads in each thumbtack. This setting satisfies MCAR (since H is hidden). It is straightforward to write the likelihood function:

$$L(\boldsymbol{\theta} : \mathcal{D}) = P(x^1)^{M[1]}(1 - P(x^1))^{M[0]},$$

where

$$P(x^1) = \theta_H \theta_{X|h^1} + (1 - \theta_H)\theta_{X|h^2}.$$

If we examine this term, we see that $P(x^1)$ is the weighted average of $\theta_{X|h^1}$ and $\theta_{X|h^2}$. There are multiple choices of these two parameters and θ_H that achieve the same value of $P(x^1)$. For example, $\theta_H = 0.5, \theta_{X|h^1} = 0.5, \theta_{X|h^2} = 0.5$ leads to the same behavior as $\theta_H = 0.5, \theta_{X|h^1} = 0.8, \theta_{X|h^2} = 0.2$. Because the likelihood of the data is a function only of $P(x^1)$, we conclude that there is a continuum of parameter choices that achieve the maximum likelihood. ∎

This example illustrates a situation where the learning problem is underconstrained: Given the observations, we cannot hope to recover a unique set of parameters. Recall that in previous sections, we showed that our estimates are consistent and thus will approach the true parameters when sufficient data are available. In this example, we cannot hope that more data will let us recover the true parameters.

Before formally treating the issue, let us examine another example that does not involve hidden variables.

Example 19.7

Suppose we conduct an experiment where we toss two coins X and Y that may be correlated with each other. After each toss, one of the coins is hidden from us using a mechanism that is totally unrelated to the outcome of the coins. Clearly, if we have sufficient observations (that is, the mechanism does not hide one of the coins consistently), then we can estimate the marginal probability of each of the coins. Can we, however, learn anything about how they depend on each other? Consider some pair of marginal probabilities $P(X)$ and $P(Y)$; because we never get to observe both coins together, any joint distribution that has these marginals has the same likelihood. In particular, a model where the two coins are independent achieves maximum likelihood but is not the unique point. In fact, in some cases a model where one is a deterministic function of the other also achieves the same likelihood (for example, if we have the same frequency of observed X heads as of observed Y heads). ∎

These two examples show that in some learning situations we cannot resolve all aspects of the model by learning from data. This issue has been examined extensively in statistics, and is known as *identifiability*, and we briefly review the relevant notions here.

identifiability

Definition 19.4

identifiability

Suppose we have a parametric model with parameters $\boldsymbol{\theta} \in \Theta$ that defines a distribution $P(\boldsymbol{X} \mid \boldsymbol{\theta})$ over a set $\boldsymbol{X}$ of measurable variables. A choice of parameters $\boldsymbol{\theta}$ is identifiable if there is no $\boldsymbol{\theta}' \neq \boldsymbol{\theta}$ such that $P(\boldsymbol{X} \mid \boldsymbol{\theta}) = P(\boldsymbol{X} \mid \boldsymbol{\theta}')$. A model is identifiable if all parameter choices $\boldsymbol{\theta} \in \Theta$ are identifiable. ∎

In other words, a model is identifiable if each choice of parameters implies a different distribution over the observed variables. Nonidentifiability implies that there are parameter settings that are indistinguishable given the data, and therefore cannot be identified from the data. Usually this is a sign that the parameterization is redundant with respect to the actual observations. For example, the model we discuss in example 19.6 is unidentifiable, since there are regions in the parameters space that induce the same probability on the observations.

Another source of nonidentifiability is hidden variables.

Example 19.8

Consider now a different experiment where we toss two thumbtacks from two different brands: Acme (A) and Bond (B). In each round, both thumbtacks are tossed and the entries are recorded. Unfortunately, due to scheduling constraints, two different experimenters participated in the experiment; each used a slightly different hand motion, changing the probability of heads and tails. Unfortunately, the experimenter name was not recorded, and thus we only have measurements of the outcome in each experiment.

To model this situation, we have three random variables to describe each round. Suppose A denotes the outcome of the toss of the Acme thumbtack and B the outcome of the toss of the Bond thumbtack. Because these outcomes depend on the experimenter, we add another (hidden) variable H that denotes the name of the experimenter. We assume that the model is such that A and B are independent given H. Thus,

$$P(A, B) = \sum_h P(h) P(A \mid h) P(B \mid h).$$

Because we never observe H, the parameters of this model can be reshuffled by "renaming" the values of the hidden variable. If we exchange the roles of h^0 and h^1, and change the corresponding entries in the CPDs, we get a model with exactly the same likelihood, but with different parameters. In this case, the likelihood surface is duplicated. For each parameterization, there is an equivalent parameterization by exchanging the names of the hidden variable. We conclude that this model is not identifiable. ∎

This type of unidentifiability exists in any model where we have hidden variables we never observe. When we have several hidden variables, the problem is even worse, and the number of equivalent "reflections" of each solution is exponential in the number of hidden variables.

Although such a model is not identifiable due to "renaming" transformations, it is in some sense better than the model of example 19.6, where we had an entire region of equivalent parameterizations. To capture this distinction, we can define a weaker version of identifiability.

Definition 19.5

locally
identifiable

Suppose we have a parametric model with parameters $\boldsymbol{\theta} \in \Theta$ that defines a distribution $P(\boldsymbol{X} \mid \boldsymbol{\theta})$ over a set $\boldsymbol{X}$ of measurable variables. A choice of parameters $\boldsymbol{\theta}$ is locally identifiable *if there is a constant $\epsilon > 0$ such that there is no $\boldsymbol{\theta}' \neq \boldsymbol{\theta}$ such that $\|\boldsymbol{\theta} - \boldsymbol{\theta}'\|_2 < \epsilon$ and $P(\boldsymbol{X} \mid \boldsymbol{\theta}) = P(\boldsymbol{X} \mid \boldsymbol{\theta}')$. A model is* locally identifiable *if all parameter choices $\boldsymbol{\theta} \in \Theta$ are locally identifiable.* ∎

In other words, a model is locally identifiable if each choice of parameters defines a distribution that is different than the distribution of neighboring parameterization in a sufficiently small neighborhood. This definition implies that, from a local perspective, the model is identifiable. The model of example 19.8 is locally identifiable, while the model of example 19.6 is not.

It is interesting to note that we have encountered similar issues before: As we discussed in chapter 18, our data do not allow us to distinguish between structures in the same I-equivalence class. This limitation did not prevent us from trying to learn a model from data, but we needed to avoid ascribing meaning to directionality of edges that are not consistent throughout the I-equivalence class. The same approach holds for unidentifiability due to missing data: **A nonidentifiable model does not mean that we should not attempt to learn models from data. But it does mean that we should be careful not to read into the learned model more than what can be distinguished given the available data.**

19.2   Parameter Estimation

As for the fully observable case, we first consider the parameter estimation task. As with complete data, we consider two approaches to estimation, maximum likelihood estimation (MLE), and Bayesian estimation. We start with a discussion of methods for MLE estimation, and then consider the Bayesian estimation problem in the next section.

More precisely, suppose we are given a network structure $\mathcal{G}$ and the form of the CPDs. Thus, we only need to set the parameters $\boldsymbol{\theta}$ to define a distribution $P(\mathcal{X} \mid \boldsymbol{\theta})$. We are also given a data set $\mathcal{D}$ that consists of M partial instances to $\mathcal{X}$. We want to find the values $\hat{\boldsymbol{\theta}}$ that maximize the log-likelihood function: $\hat{\boldsymbol{\theta}} = \arg\max_{\boldsymbol{\theta}} \ell(\boldsymbol{\theta} : \mathcal{D})$. As we discussed, in the presence of incomplete data, the likelihood does not decompose. And so the problem requires optimizing a highly nonlinear and multimodal function over a high-dimensional space (one consisting of parameter assignments to all CPDs). There are two main classes of methods for performing this optimization: a generic nonconvex optimization algorithm, such as gradient ascent; and *expectation maximization,* a more specialized approach for optimizing likelihood functions.

19.2.1 Gradient Ascent

gradient ascent

One approach to handle this optimization task is to apply some variant of *gradient ascent,* a standard function-optimization technique applied to the likelihood function (see appendix A.5.2). These algorithms are generic and can be applied if we can evaluate the gradient function at different parameter choices.

19.2.1.1 Computing the Gradient

The main technical question we need to tackle is how to compute the gradient. We begin with considering the derivative relative to a single CPD entry $P(x \mid u)$. We can then use this result as the basis for computing derivatives relative to other parameters, which arise when we have structured CPDs.

Lemma 19.1

Let $\mathcal{B}$ be a Bayesian network with structure $\mathcal{G}$ over $\mathcal{X}$ that induces a probability distribution P, let o be a tuple of obserations for some of the variables, and let $X \in \mathcal{X}$ be some random variable. Then

$$\frac{\partial}{\partial P(x \mid u)} P(o) = \frac{1}{P(x \mid u)} P(x, u, o)$$

if $P(x \mid u) > 0$, where $x \in Val(X)$, $u \in Val(\mathrm{Pa}_X)$.

PROOF We start by considering the case where the evidence is a full assignment ξ to all variables. The probability of such an assignment is a product of the relevant CPD entries. Thus, the gradient of this product with respect to the parameter $P(x \mid u)$ is simply

$$\frac{\partial}{\partial P(x \mid u)} P(\xi) = \begin{cases} \frac{1}{P(x \mid u)} P(\xi) & \text{if } \xi\langle X, \mathrm{Pa}_X \rangle = \langle x, u \rangle \\ 0 & \text{otherwise.} \end{cases}$$

We now consider the general case where the evidence is a partial assignment. As usual, we can write $P(o)$ as a sum over all full assignments consistent with $P(o)$

$$P(o) = \sum_{\xi : \xi\langle O \rangle = o} P(\xi).$$

Applying the differentiation formula to each of these full assignments, we get

$$\begin{aligned}
\frac{\partial}{\partial P(x \mid u)} P(o) &= \sum_{\xi : \xi\langle O \rangle = o} \frac{\partial}{\partial P(x \mid u)} P(\xi) \\
&= \sum_{\xi : \xi\langle O \rangle = o, \xi\langle X, \mathrm{Pa}_X \rangle = \langle x, u \rangle} \frac{1}{P(x \mid u)} P(\xi) \\
&= \frac{1}{P(x \mid u)} P(x, u, o).
\end{aligned}$$

∎

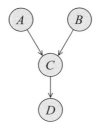

Figure 19.6   A simple network used to illustrate learning algorithms for missing data

When $\boldsymbol{o}$ is inconsistent with x or $\boldsymbol{u}$, then the gradient is 0, since the probability $P(x, \boldsymbol{u}, \boldsymbol{o})$ is 0 in this case. When $\boldsymbol{o}$ is consistent with x and $\boldsymbol{u}$, the gradient is the ratio between the probability $P(x, \boldsymbol{u}, \boldsymbol{o})$ and the parameter $P(x \mid \boldsymbol{u})$. Intuitively, this ratio takes into account the weight of the cases where $P(x \mid \boldsymbol{u})$ is "used" in the computation of $P(\boldsymbol{o})$. Increasing $P(x \mid \boldsymbol{u})$ by a small amount will increase the probability of these cases by a multiplicative factor.

The lemma does not deal with the case where $P(x \mid \boldsymbol{u}) = 0$, since we cannot divide by 0. Note, however, that the proof shows that this division is mainly a neat manner of writing the product of all terms *except* $P(x \mid \boldsymbol{u})$. Thus, even in this extreme case we can use a similar proof to compute the gradient, although writing the term explicitly is less elegant. Since in learning we usually try to avoid extreme parameter assignments, we will continue our discussion with the assumption that $P(x \mid \boldsymbol{u}) > 0$.

An immediate consequence of lemma 19.1 is the form of the gradient of the log-likelihood function.

Theorem 19.2

Let $\mathcal{G}$ be a Bayesian network structure over $\mathcal{X}$, and let $\mathcal{D} = \{\boldsymbol{o}[1], \dots, \boldsymbol{o}[M]\}$ be a partially observable data set. Let X be a variable and $\boldsymbol{U}$ its parents in $\mathcal{G}$. Then

$$\frac{\partial \ell(\boldsymbol{\theta} : \mathcal{D})}{\partial P(x \mid \boldsymbol{u})} = \frac{1}{P(x \mid \boldsymbol{u})} \sum_{m=1}^{M} P(x, \boldsymbol{u} \mid \boldsymbol{o}[m], \boldsymbol{\theta}).$$

The proof is left as an exercise (exercise 19.5).

This theorem provides the form of the gradient for table-CPDs. For other CPDs, such as noisy-or CPDs, we can use the *chain rule of derivatives* to compute the gradient. Suppose that the CPD entries of $P(X \mid \boldsymbol{U})$ are written as functions of some set of parameters $\boldsymbol{\theta}$. Then, for a specific parameter $\theta \in \boldsymbol{\theta}$, we have

$$\frac{\partial \ell(\boldsymbol{\theta} : \mathcal{D})}{\partial \theta} = \sum_{x, \boldsymbol{u}} \frac{\partial \ell(\boldsymbol{\theta} : \mathcal{D})}{\partial P(x \mid \boldsymbol{u})} \frac{\partial P(x \mid \boldsymbol{u})}{\partial \theta},$$

where the first term is the derivative of the log-likelihood function when parameterized in terms of the table-CPDs induced by $\boldsymbol{\theta}$. For structured CPDs, we can use this formula to compute the gradient with respect to the CPD parameters. For some CPDs, however, this may not be the most efficient way of computing these gradients; see exercise 19.4.

19.2.1.2 An Example

We consider a simple example to clarify the concept. Consider the network shown in figure 19.6, and a partially specified data case $o = \langle a^1, ?, ?, d^0 \rangle$.

We want to compute the gradient of one family of parameters $P(D \mid c^0)$ given the observation o. Using theorem 19.2, we know that

$$\frac{\partial \log P(o)}{\partial P(d^0 \mid c^0)} = \frac{P(d^0, c^0 \mid o)}{P(d^0 \mid c^0)},$$

and similarly for other values of D and C.

Assume that our current θ is:

$$
\begin{aligned}
\theta_{a^1} &= 0.3 \\
\theta_{b^1} &= 0.9 \\
\theta_{c^1 \mid a^0, b^0} &= 0.83 \\
\theta_{c^1 \mid a^0, b^1} &= 0.09 \\
\theta_{c^1 \mid a^1, b^0} &= 0.6 \\
\theta_{c^1 \mid a^1, b^1} &= 0.2 \\
\theta_{d^1 \mid c^0} &= 0.1 \\
\theta_{d^1 \mid c^1} &= 0.8.
\end{aligned}
$$

In this case, the probabilities of the four data cases that are consistent with o are

$$
\begin{aligned}
P(\langle a^1, b^1, c^1, d^0 \rangle) &= 0.3 \cdot 0.9 \cdot 0.2 \cdot 0.2 = 0.0108 \\
P(\langle a^1, b^1, c^0, d^0 \rangle) &= 0.3 \cdot 0.9 \cdot 0.8 \cdot 0.9 = 0.1944 \\
P(\langle a^1, b^0, c^1, d^0 \rangle) &= 0.3 \cdot 0.1 \cdot 0.6 \cdot 0.2 = 0.0036 \\
P(\langle a^1, b^0, c^0, d^0 \rangle) &= 0.3 \cdot 0.1 \cdot 0.4 \cdot 0.9 = 0.0108.
\end{aligned}
$$

To compute the *posterior* probability of these instances given the partial observation o, we divide the probability of each instance with the total probability, which is 0.2196, that is,

$$
\begin{aligned}
P(\langle a^1, b^1, c^1, d^0 \rangle \mid o) &= 0.0492 \\
P(\langle a^1, b^1, c^0, d^0 \rangle \mid o) &= 0.8852 \\
P(\langle a^1, b^0, c^1, d^0 \rangle \mid o) &= 0.0164 \\
P(\langle a^1, b^0, c^0, d^0 \rangle \mid o) &= 0.0492.
\end{aligned}
$$

Using these computations, we see that

$$
\begin{aligned}
\frac{\partial \log P(o)}{\partial P(d^1 \mid c^0)} &= \frac{P(d^1, c^0 \mid o)}{P(d^1 \mid c^0)} = \frac{0}{0.1} = 0 \\[6pt]
\frac{\partial \log P(o)}{\partial P(d^0 \mid c^0)} &= \frac{P(d^0, c^0 \mid o)}{P(d^0 \mid c^0)} = \frac{0.8852 + 0.0492}{0.9} = 1.0382 \\[6pt]
\frac{\partial \log P(o)}{\partial P(d^1 \mid c^1)} &= \frac{P(d^1, c^1 \mid o)}{P(d^1 \mid c^1)} = \frac{0}{0.8} = 0 \\[6pt]
\frac{\partial \log P(o)}{\partial P(d^0 \mid c^1)} &= \frac{P(d^0, c^1 \mid o)}{P(d^0 \mid c^1)} = \frac{0.0492 + 0.0164}{0.2} = 0.328.
\end{aligned}
$$

These computations show that we can increase the probability of the observations o by either increasing $P(d^0 \mid c^0)$ or $P(d^0 \mid c^1)$. Moreover, increasing the former parameter will lead to a bigger change in the probability of o than a similar increase in the latter parameter.

Now suppose we have an observation $o' = \langle a^0, ?, ?, d^1 \rangle$. We can repeat the same computation as before and see that

$$\frac{\partial \log P(o')}{\partial P(d^1 \mid c^0)} = \frac{P(d^1, c^0 \mid o')}{P(d^1 \mid c^0)} = \frac{0.2836}{0.1} = 2.8358$$

$$\frac{\partial \log P(o')}{\partial P(d^0 \mid c^0)} = \frac{P(d^0, c^0 \mid o')}{P(d^0 \mid c^0)} = \frac{0}{0.9} = 0$$

$$\frac{\partial \log P(o')}{\partial P(d^1 \mid c^1)} = \frac{P(d^1, c^1 \mid o')}{P(d^1 \mid c^1)} = \frac{0.7164}{0.8} = 0.8955$$

$$\frac{\partial \log P(o')}{\partial P(d^0 \mid c^1)} = \frac{P(d^0, c^1 \mid o')}{P(d^0 \mid c^1)} = \frac{0}{0.2} = 0.$$

Suppose our data set consists only of these two instances. The gradient of the log-likelihood function is the sum of the gradient with respect to the two instances. We get that

$$\frac{\partial \ell(\boldsymbol{\theta} : \mathcal{D})}{\partial P(d^1 \mid c^0)} = 2.8358$$

$$\frac{\partial \ell(\boldsymbol{\theta} : \mathcal{D})}{\partial P(d^0 \mid c^0)} = 1.0382$$

$$\frac{\partial \ell(\boldsymbol{\theta} : \mathcal{D})}{\partial P(d^1 \mid c^1)} = 0.8955$$

$$\frac{\partial \ell(\boldsymbol{\theta} : \mathcal{D})}{\partial P(d^0 \mid c^1)} = 0.328.$$

Note that all the gradients are nonnegative. Thus, increasing any of the parameters in the CPD $P(D \mid C)$ will increase the likelihood of the data. It is clear, however, that we cannot increase both $P(d^1 \mid c^0)$ and $P(d^0 \mid c^0)$ at the same time, since this will lead to an illegal conditional probability. One way of solving this is to use a single parameter $\theta_{d^1 \mid c^0}$ and write

$$P(d^1 \mid c^0) = \theta_{d^1 \mid c^0} \quad P(d^0 \mid c^0) = 1 - \theta_{d^1 \mid c^0}.$$

Using the chain rule of conditional probabilities, we have that

$$\frac{\partial \ell(\boldsymbol{\theta} : \mathcal{D})}{\partial \theta_{d^1 \mid c^0}} = \frac{\partial P(d^1 \mid c^0)}{\partial \theta_{d^1 \mid c^0}} \frac{\partial \ell(\boldsymbol{\theta} : \mathcal{D})}{\partial P(d^1 \mid c^0)} + \frac{\partial P(d^0 \mid c^0)}{\partial \theta_{d^1 \mid c^0}} \frac{\partial \ell(\boldsymbol{\theta} : \mathcal{D})}{\partial P(d^0 \mid c^0)}$$

$$= \frac{\partial \ell(\boldsymbol{\theta} : \mathcal{D})}{\partial P(d^1 \mid c^0)} - \frac{\partial \ell(\boldsymbol{\theta} : \mathcal{D})}{\partial P(d^0 \mid c^0)}$$

$$= 2.8358 - 1.0382 = 1.7976.$$

Thus, in this case, we prefer to increase $P(d^1 \mid c^0)$ and decrease $P(d^0 \mid c^0)$, since the resulting increase in the probability of o' will be larger than the decrease in the probability of o.

Algorithm 19.1 Computing the gradient in a network with table-CPDs

Procedure Compute-Gradient (
  $\mathcal{G}$, // Bayesian network structure over $X_1, \ldots, X_n$
  $\boldsymbol{\theta}$, // Set of parameters for $\mathcal{G}$
  $\mathcal{D}$ // Partially observed data set

)

1 // Initialize data structures
2 **for** each $i = 1, \ldots, n$
3   **for** each $x_i, \boldsymbol{u}_i \in Val(X_i, \mathrm{Pa}_{X_i}^{\mathcal{G}})$
4    $\bar{M}[x_i, \boldsymbol{u}_i] \leftarrow 0$
5 // Collect probabilities from all instances
6 **for** each $m = 1 \ldots M$
7   Run clique tree calibration on $\langle \mathcal{G}, \boldsymbol{\theta} \rangle$ using evidence $\boldsymbol{o}[m]$
8   **for** each $i = 1, \ldots, n$
9   **for** each $x_i, \boldsymbol{u}_i \in Val(X_i, \mathrm{Pa}_{X_i}^{\mathcal{G}})$
10     $\bar{M}[x_i, \boldsymbol{u}_i] \leftarrow \bar{M}[x_i, \boldsymbol{u}_i] + P(x_i, \boldsymbol{u}_i \mid \boldsymbol{o}[m])$
11 // Compute components of the gradient vector
12 **for** each $i = 1, \ldots, n$
13   **for** each $x_i, \boldsymbol{u}_i \in Val(X_i, \mathrm{Pa}_{X_i}^{\mathcal{G}})$
14    $\delta_{x_i \mid \boldsymbol{u}_i} \leftarrow \frac{1}{\theta_{x_i \mid \boldsymbol{u}_i}} \bar{M}[x_i, \boldsymbol{u}_i]$
15 **return** $\{\delta_{x_i, \mid \boldsymbol{u}_i} : \forall i = 1, \ldots, n, \forall (x_i, \boldsymbol{u}_i) \in Val(X_i, \mathrm{Pa}_{X_i}^{\mathcal{G}})\}$

19.2.1.3 Gradient Ascent Algorithm

We now generalize these ideas to case of an arbitrary network. For now we focus on the case of table-CPDs. In this case, the gradient is given by theorem 19.2. To compute the gradient for the CPD $P(X \mid \boldsymbol{U})$, we need to compute the joint probability of x and $\boldsymbol{u}$ relative to our current parameter setting $\boldsymbol{\theta}$ and each observed instance $\boldsymbol{o}[m]$. In other words, we need to compute the joint distribution $P(X[m], \boldsymbol{U}[m] \mid \boldsymbol{o}[m], \boldsymbol{\theta})$ for each m. We can do this by running an inference procedure for each data case. Importantly, **we can do all of the required inference for** **each data case using one clique tree calibration, since the family preservation property guarantees that X and its parents U will be together in some clique in the tree.** Procedure Compute-Gradient, shown in algorithm 19.1, performs these computations.

Once we have a procedure for computing the gradient, it seems that we can simply plug it into a standard package for gradient ascent and optimize the parameters. As we have illustrated, however, there is one issue that we need to deal with. It is not hard to confirm that all components of the gradient vector are nonnegative. This is natural, since increasing each of the parameters will lead to higher likelihood. Thus, a step in the gradient direction will *increase* all the parameters. Remember, however, that we want to ensure that our parameters describe a legal probability distribution. That is, the parameters for each conditional probability are nonnegative and sum to one.

In the preceding example, we saw one approach that works well when we have binary variables. In general networks, there are two common approaches to deal with this issue. The first approach is to modify the gradient ascent procedure we use (for example, conjugate gradient) to respect these constraints. First, we must project each gradient vector onto the hyperplane that satisfies the linear constraints on the parameters; this step is fairly straightforward (see exercise 19.6). Second, we must ensure that parameters are nonnegative; this requires restricting possible steps to avoid stepping out of the allowed bounds.

reparameterization

The second approach is to *reparameterize* the problem. Suppose we introduce new parameters $\lambda_{x|u}$, and define

$$P(x \mid \boldsymbol{u}) = \frac{e^{\lambda_{x|u}}}{\sum_{x' \in Val(X)} e^{\lambda_{x'|u}}}, \tag{19.3}$$

for each X and its parents $\boldsymbol{U}$. Now, any choice of values for the λ parameters will lead to legal conditional probabilities. We can compute the gradient of the log-likelihood with respect to the λ parameters using the chain rule of partial derivatives, and then use standard (unmodified) conjugate gradient ascent procedure. See exercise 19.7.

Lagrange multipliers

Another way of dealing with the constraints implied by conditional probabilities is to use the method of *Lagrange multipliers*, reviewed in appendix A.5.3. Applying this method to the optimization of the log-likelihood leads to the method we discuss in the next section, and we defer this discussion; see also exercise 19.8.

Having dealt with this subtlety, we can now apply any gradient ascent procedure to find a local maximum of the likelihood function. As discussed, in most missing value problems, the likelihood function has many local maxima. Unfortunately, gradient ascent procedures are guaranteed to achieve only a local maximum of the function. Many of the techniques we discussed earlier in the book can be used to avoid local maxima and increase our chances of finding a global maximum, or at least a better local maximum: the general-purpose methods of appendix A.4.2, such as multiple random starting points, or applying random perturbations to convergence points; and the more specialized data perturbation methods of algorithm 18.1.

19.2.2 Expectation Maximization (EM)

expectation maximization

An alternative algorithm for optimizing a likelihood function is the *expectation maximization* algorithm. Unlike gradient ascent, EM is not a general-purpose algorithm for nonlinear function optimization. Rather, it is tailored specifically to optimizing likelihood functions, attempting to build on the tools we had for solving the problem with complete data.

19.2.2.1 Intuition

Recall that when learning from complete data, we can collect sufficient statistics for each CPD. We can then estimate parameters that maximize the likelihood with respect to these statistics. As we saw, in the case of missing data, we do not have access to the full sufficient statistics. Thus, we cannot use the same strategy for our problem. For example, in a simple $X \rightarrow Y$ network, if we see the training instance $\langle ?, y^1 \rangle$, then we do not know whether to count this instance toward the count $M[x^1, y^1]$ or toward the count $M[x^0, y^1]$.

A simple approach is to "fill in" the missing values arbitrarily. For example, there are strategies that fill in missing values with "default values" (say *false*) or by randomly choosing a value. Once we fill in all the missing values, we can use standard, complete data learning procedure. Such approaches are called called *data imputation* methods in statistics.

data imputation

The problem with such an approach is that the procedure we use for filling in the missing values introduces a bias that will be reflected in the parameters we learn. For example, if we fill all missing values with *false*, then our estimate will be skewed toward higher (conditional) probability of *false*. Similarly, if we use a randomized procedure for filling in values, then the probabilities we estimate will be skewed toward the distribution from which we sample missing values. This might be better than a skew toward one value, but it still presents a problem. Moreover, when we consider learning with hidden variables, it is clear that an imputation procedure will not help us. The values we fill in for the hidden variable are conditionally independent from the values of the other variables, and thus, using the imputed values, we will not learn any dependencies between the hidden variable and the other variables in the network.

A different approach to filling in data takes the perspective that, when learning with missing data, we are actually trying to solve two problems at once: learning the parameters, and hypothesizing values for the unobserved variables in each of the data cases. Each of these tasks is fairly easy when we have the solution to the other. Given complete data, we have the statistics, and we can estimate parameters using the MLE formulas we discussed in chapter 17. Conversely, given a choice of parameters, we can use probabilistic inference to hypothesize the likely values (or the distribution over possible values) for unobserved variables. Unfortunately, because we have neither, the problem is difficult.

The EM algorithm solves this "chicken and egg" problem using a bootstrap approach. We start out with some arbitrary starting point. This can be either a choice of parameters, or some initial assignment to the hidden variables; these assignments can be either random, or selected using some heuristic approach. Assuming, for concreteness, that we begin with a parameter assignment, the algorithm then repeats two steps. First, we use our current parameters to *complete* the data, using probabilistic inference. We then treat the completed data as if it were observed and learn a new set of parameters.

data completion

More precisely, suppose we have a guess θ^0 about the parameters of the network. The resulting model defines a joint distribution over all the variables in the domain. Given a partial instance, we can compute the posterior (using our putative parameters) over all possible assignments to the missing values in that instance. The EM algorithm uses this probabilistic completion of the different data instances to estimate the *expected* value of the sufficient statistics. It then finds the parameters θ^1 that maximize the likelihood with respect to these statistics.

Somewhat surprisingly, this sequence of steps provably improves our parameters. In fact, as we will prove formally, unless our parameters have not changed due to these steps (such that $\theta^0 = \theta^1$), our new parameters θ^1 necessarily have a higher likelihood than θ^0. But now we can iteratively repeat this process, using θ^1 as our new starting point. Each of these operations can be thought of as taking an "uphill" step in our search space. More precisely, we will show (under very benign assumptions) that: each iteration is guaranteed to improve the log-likelihood function; that this process is guaranteed to converge; and that the convergence point is a fixed point of the likelihood function, which is essentially always a local maximum. Thus, the guarantees of the EM algorithm are similar to those of gradient ascent.

19.2.2.2   An Example

We start with a simple example to clarify the concepts. Consider the simple network shown in figure 19.6. In the fully observable case, our maximum likelihood parameter estimate for the parameter $\hat{\theta}_{d^1|c^0}$ is:

$$\hat{\theta}_{d^1|c^0} = \frac{M[d^1,c^0]}{M[c^0]} = \frac{\sum_{m=1}^{M} \mathbf{I}\{\xi[m]\langle D,C\rangle = \langle d^1,c^0\rangle\}}{\sum_{m=1}^{M} \mathbf{I}\{\xi[m]\langle C\rangle = c^0\}},$$

where $\xi[m]$ is the m'th training example. In the fully observable case, we knew exactly whether the indicator variables were 0 or 1. Now, however, we do not have complete data cases, so we no longer know the value of the indicator variables.

Consider a partially specified data case $\boldsymbol{o} = \langle a^1, ?, ?, d^0 \rangle$. There are four possible instantiations to the missing variables B, C which could have given rise to this partial data case: $\langle b^1, c^1 \rangle$, $\langle b^1, c^0 \rangle$, $\langle b^0, c^1 \rangle$, $\langle b^0, c^0 \rangle$. We do not know which of them is true, or even which of them is more likely.

However, assume that we have some estimate $\boldsymbol{\theta}$ of the values of the parameters in the model. In this case, we can compute how likely each of these completions is, given our distribution. That is, we can define a distribution $Q(B, C) = P(B, C \mid \boldsymbol{o}, \boldsymbol{\theta})$ that induces a distribution over the four data cases. For example, if our parameters $\boldsymbol{\theta}$ are:

$$
\begin{aligned}
\boldsymbol{\theta}_{a^1} &= 0.3 & \boldsymbol{\theta}_{b^1} &= 0.9 \\
\boldsymbol{\theta}_{d^1|c^0} &= 0.1 & \boldsymbol{\theta}_{d^1|c^1} &= 0.8 \\
\boldsymbol{\theta}_{c^1|a^0,b^0} &= 0.83 & \boldsymbol{\theta}_{c^1|a^1,b^0} &= 0.6 \\
\boldsymbol{\theta}_{c^1|a^0,b^1} &= 0.09 & \boldsymbol{\theta}_{c^1|a^1,b^1} &= 0.2,
\end{aligned}
$$

then $Q(B, C) = P(B, C \mid a^1, d^0, \boldsymbol{\theta})$ is defined as:

$$
\begin{aligned}
Q(\langle b^1, c^1 \rangle) &= 0.3 \cdot 0.9 \cdot 0.2 \cdot 0.2/0.2196 = 0.0492 \\
Q(\langle b^1, c^0 \rangle) &= 0.3 \cdot 0.9 \cdot 0.8 \cdot 0.9/0.2196 = 0.8852 \\
Q(\langle b^0, c^1 \rangle) &= 0.3 \cdot 0.1 \cdot 0.6 \cdot 0.2/0.2196 = 0.0164 \\
Q(\langle b^0, c^0 \rangle) &= 0.3 \cdot 0.1 \cdot 0.4 \cdot 0.9/0.2196 = 0.0492,
\end{aligned}
$$

where 0.2196 is a normalizing factor, equal to $P(a^1, d^0 \mid \boldsymbol{\theta})$.

If we have another example $\boldsymbol{o}' = \langle ?, b^1, ?, d^1 \rangle$. Then $Q'(A, C) = P(A, C \mid b^1, d^1, \boldsymbol{\theta})$ is defined as:

$$
\begin{aligned}
Q'(\langle a^1, c^1 \rangle) &= 0.3 \cdot 0.9 \cdot 0.2 \cdot 0.8/0.1675 = 0.2579 \\
Q'(\langle a^1, c^0 \rangle) &= 0.3 \cdot 0.9 \cdot 0.8 \cdot 0.1/0.1675 = 0.1290 \\
Q'(\langle a^0, c^1 \rangle) &= 0.7 \cdot 0.9 \cdot 0.09 \cdot 0.8/0.1675 = 0.2708 \\
Q'(\langle a^0, c^0 \rangle) &= 0.7 \cdot 0.9 \cdot 0.91 \cdot 0.1/0.1675 = 0.3423.
\end{aligned}
$$

Intuitively, now that we have estimates for how likely each of the cases is, we can treat these estimates as truth. That is, we view our partially observed data case $\langle a^1, ?, ?, d^0 \rangle$ as consisting of four complete data cases, each of which has some *weight* lower than 1. The weights correspond to our estimate, based on our current parameters, on how likely is this particular completion of the partial instance. (This approach is somewhat reminiscent of the weighted particles in the likelihood weighting algorithm.) Importantly, as we will discuss, we do

weighted data
instances

not usually explicitly generate these completed data cases; however, this perspective is the basis for the more sophisticated methods.

More generally, let $\boldsymbol{H}[m]$ denote the variables whose values are missing in the data instance $\boldsymbol{o}[m]$. We now have a data set $\mathcal{D}^+$ consisting of

$$\cup_m \{\langle \boldsymbol{o}[m], \boldsymbol{h}[m] \rangle \ : \ \boldsymbol{h}[m] \in Val(\boldsymbol{H}[m])\},$$

where each data case $\langle \boldsymbol{o}[m], \boldsymbol{h}[m] \rangle$ has weight $Q(\boldsymbol{h}[m]) = P(\boldsymbol{h}[m] \mid \boldsymbol{o}[m], \boldsymbol{\theta})$.

We can now do standard maximum likelihood estimation using these completed data cases. We compute the *expected sufficient statistics*:

expected
sufficient
statistics

$$\bar{M}_{\boldsymbol{\theta}}[\boldsymbol{y}] = \sum_{m=1}^{M} \sum_{\boldsymbol{h}[m] \in Val(\boldsymbol{H}[m])} Q(\boldsymbol{h}[m]) \boldsymbol{I}\{\xi[m]\langle \boldsymbol{Y} \rangle = \boldsymbol{y}\}.$$

We then use these expected sufficient statistics as if they were real in the MLE formula. For example:

$$\tilde{\boldsymbol{\theta}}_{d^1 \mid c^0} = \frac{\bar{M}_{\boldsymbol{\theta}}[d^1, c^0]}{\bar{M}_{\boldsymbol{\theta}}[c^0]}.$$

In our example, suppose the data consist of the two instances $\boldsymbol{o} = \langle a^1, ?, ?, d^0 \rangle$ and $\boldsymbol{o'} = \langle ?, b^1, ?, d^1 \rangle$. Then, using the calculated Q and Q' from above, we have that

$$
\begin{aligned}
\bar{M}_{\boldsymbol{\theta}}[d^1, c^0] &= Q'(\langle a^1, c^0 \rangle) + Q'(\langle a^0, c^0 \rangle) \\
&= 0.1290 + 0.3423 = 0.4713 \\
\bar{M}_{\boldsymbol{\theta}}[c^0] &= Q(\langle b^1, c^0 \rangle) + Q(\langle b^0, c^0 \rangle) + Q'(\langle a^1, c^0 \rangle) + Q'(\langle a^0, c^0 \rangle) \\
&= 0.8852 + 0.0492 + 0.1290 + 0.3423 = 1.4057.
\end{aligned}
$$

Thus, in this example, using these particular parameters to compute expected sufficient statistics, we get

$$\tilde{\boldsymbol{\theta}}_{d^1 \mid c^0} = \frac{0.4713}{1.4057} = 0.3353.$$

Note that this estimate is quite different from the parameter $\theta_{d^1 \mid c^0} = 0.1$ that we used in our estimate of the expected counts. The initial parameter and the estimate are different due to the incorporation of the observations in the data.

This intuition seems nice. However, it may require an unreasonable amount of computation. To compute the expected sufficient statistics, we must sum over all the completed data cases. The number of these completed data cases is much larger than the original data set. For each $\boldsymbol{o}[m]$, the number of completions is exponential in the number of missing values. Thus, if we have more than few missing values in an instances, an implementation of this approach will not be able to finish computing the expected sufficient statistics.

Fortunately, it turns out that there is a better approach to computing the expected sufficient statistic than simply summing over all possible completions. Let us reexamine the formula for an expected sufficient statistic, for example, $\bar{M}_{\boldsymbol{\theta}}[c^1]$. We have that

$$\bar{M}_{\boldsymbol{\theta}}[c^1] = \sum_{m=1}^{M} \sum_{\boldsymbol{h}[m] \in Val(\boldsymbol{H}[m])} Q(\boldsymbol{h}[m]) \boldsymbol{I}\{\xi[m]\langle C \rangle = c^1\}.$$

Let us consider the internal summation, say for a data case $\boldsymbol{o} = \langle a^1, ?, ?, d^0 \rangle$. We have four possible completions, as before, but we are only summing over the two that are consistent with c^1, that is, $Q(b^1, c^1) + Q(b^0, c^1)$. This expression is equal to $Q(c^1) = P(c^1 \mid a^1, d^0, \boldsymbol{\theta}) = P(c^1 \mid \boldsymbol{o}[1], \boldsymbol{\theta})$. This idea clearly generalizes to our other data cases. Thus, we have that

$$\bar{M}_{\boldsymbol{\theta}}[c^1] = \sum_{m=1}^{M} P(c^1 \mid \boldsymbol{o}[m], \boldsymbol{\theta}).$$

Now, recall our formula for sufficient statistics in the fully observable case:

$$M[c^1] = \sum_{m=1}^{M} \boldsymbol{I}\{\xi[m]\langle C \rangle = c^1\}.$$

Our new formula is identical, except that we have substituted our indicator variable — either 0 or 1 — with a probability that is somewhere between 0 and 1. Clearly, if in a certain data case we get to observe C, the indicator variable and the probability are the same. Thus, we can view the expected sufficient statistics as filling in soft estimates for hard data when the hard data are not available.

We stress that we use *posterior* probabilities in computing expected sufficient statistics. Thus, although our choice of $\boldsymbol{\theta}$ clearly influences the result, the data also play a central role. This is in contrast to the probabilistic completion we discussed earlier that used a prior probability to fill in values, regardless of the evidence on the other variables in the same instances.

19.2.2.3 The EM Algorithm for Bayesian Networks

We now present the basic EM algorithm and describe the guarantees that it provides.

Networks with Table-CPDs   Consider the application of the EM algorithm to a general Bayesian network with table-CPDs. Assume that the algorithm begins with some initial parameter assignment $\boldsymbol{\theta}^0$, which can be chosen either randomly or using some other approach. (The case where we begin with some assignment to the missing data is analogous.) The algorithm then repeatedly executes the following phases, for $t = 0, 1, \ldots$

expected
sufficient
statistics

Expectation (E-step): The algorithm uses the current parameters $\boldsymbol{\theta}^t$ to compute the *expected sufficient statistics*.

- For each data case $\boldsymbol{o}[m]$ and each family $X, \boldsymbol{U}$, compute the joint distribution $P(X, \boldsymbol{U} \mid \boldsymbol{o}[m], \boldsymbol{\theta}^t)$.

- Compute the expected sufficient statistics for each $x, \boldsymbol{u}$ as:

$$\bar{M}_{\boldsymbol{\theta}^t}[x, \boldsymbol{u}] = \sum_{m} P(x, \boldsymbol{u} \mid \boldsymbol{o}[m], \boldsymbol{\theta}^t).$$

E-step

This phase is called the *E-step (expectation step)* because the counts used in the formula are the expected sufficient statistics, where the expectation is with respect to the current set of parameters.

Algorithm 19.2 Expectation-maximization algorithm for BN with table-CPDs

---

Procedure Compute-ESS (

    $\mathcal{G}$,   // Bayesian network structure over $X_1, \ldots, X_n$

    $\boldsymbol{\theta}$,   // Set of parameters for $\mathcal{G}$

    $\mathcal{D}$ // Partially observed data set

)

1   // Initialize data structures

2   **for** each $i = 1, \ldots, n$

3   **for** each $x_i, \boldsymbol{u}_i \in Val(X_i, \mathrm{Pa}_{X_i}^{\mathcal{G}})$

4        $\bar{M}[x_i, \boldsymbol{u}_i] \leftarrow 0$

5   // Collect probabilities from all instances

6   **for** each $m = 1 \ldots M$

7   Run inference on $\langle \mathcal{G}, \boldsymbol{\theta} \rangle$ using evidence $\boldsymbol{o}[m]$

8   **for** each $i = 1, \ldots, n$

9   **for** each $x_i, \boldsymbol{u}_i \in Val(X_i, \mathrm{Pa}_{X_i}^{\mathcal{G}})$

10         $\bar{M}[x_i, \boldsymbol{u}_i] \leftarrow \bar{M}[x_i, \boldsymbol{u}_i] + P(x_i, \boldsymbol{u}_i \mid \boldsymbol{o}[m])$

11   **return** $\{\bar{M}[x_i, \boldsymbol{u}_i] : \forall i = 1, \ldots, n, \forall x_i, \boldsymbol{u}_i \in Val(X_i, \mathrm{Pa}_{X_i}^{\mathcal{G}})\}$

Procedure Expectation-Maximization (

    $\mathcal{G}$,   // Bayesian network structure over $X_1, \ldots, X_n$

    $\boldsymbol{\theta}^0$,   // Initial set of parameters for $\mathcal{G}$

    $\mathcal{D}$ // Partially observed data set

)

1   **for** each $t = 0, 1 \ldots$, until convergence

2   // E-step

3      $\{\bar{M}_t[x_i, \boldsymbol{u}_i]\} \leftarrow$ Compute-ESS$(\mathcal{G}, \boldsymbol{\theta}^t, \mathcal{D})$

4   // M-step

5   **for** each $i = 1, \ldots, n$

6   **for** each $x_i, \boldsymbol{u}_i \in Val(X_i, \mathrm{Pa}_{X_i}^{\mathcal{G}})$

7         $\theta_{x_i|\boldsymbol{u}_i}^{t+1} \leftarrow \dfrac{\bar{M}_t[x_i, \boldsymbol{u}_i]}{\bar{M}_t[\boldsymbol{u}_i]}$

8   **return** $\boldsymbol{\theta}^t$

Maximization (M-step): Treat the expected sufficient statistics as observed, and perform maximum likelihood estimation, with respect to them, to derive a new set of parameters. In other words, set

$$\boldsymbol{\theta}_{x|\boldsymbol{u}}^{t+1} = \frac{\bar{M}_{\boldsymbol{\theta}^t}[x, \boldsymbol{u}]}{\bar{M}_{\boldsymbol{\theta}^t}[\boldsymbol{u}]}.$$

M-step   This phase is called the *M-step* (*maximization step*), because we are maximizing the likelihood relative to the expected sufficient statistics.

    A formal version of the algorithm is shown fully in algorithm 19.2.

The maximization step is straightforward. The more difficult step is the expectation step. How do we compute expected sufficient statistics? We must resort to Bayesian network inference over the network $\langle \mathcal{G}, \boldsymbol{\theta}^t \rangle$. Note that, as in the case of gradient ascent, the only expected sufficient statistics that we need involve a variable and its parents. Although one can use a variety of different inference methods to perform the inference task required for the E-step, we can, as in the case of gradient ascent, use the clique tree or cluster graph algorithm. Recall that the family-preservation property guarantees that X and its parents U will be together in some cluster in the tree or graph. **Thus, once again, we can do all of the required inference for each data case using one run of message-passing calibration.**

☞

exponential
family

General Exponential Family ⋆   The same idea generalizes to other distributions where the likelihood has sufficient statistics, in particular, all models in the *exponential family* (see definition 8.1). Recall that such families have a sufficient statistic function $\tau(\xi)$ that maps a complete instance to a vector of sufficient statistics. When learning parameters of such a model, we can summarize the data using the sufficient statistic function τ. For a complete data set $\mathcal{D}^+$, we define

$$\tau(\mathcal{D}^+) = \sum_m \tau(\boldsymbol{o}[m], \boldsymbol{h}[m]).$$

We can now define the same E and M-steps described earlier for this more general case.

E-step

Expectation (*E-step*): For each data case $\boldsymbol{o}[m]$, the algorithm uses the current parameters $\boldsymbol{\theta}^t$ to define a model, and a posterior distribution:

$$Q(\boldsymbol{H}[m]) = P(\boldsymbol{H}[m] \mid \boldsymbol{o}[m], \boldsymbol{\theta}^t).$$

expected
sufficient
statistics

It then uses inference in this distribution to compute the *expected sufficient statistics*:

$$\boldsymbol{E}_Q[\tau(\langle \mathcal{D}, \mathcal{H} \rangle)] = \sum_m \boldsymbol{E}_Q[\tau(\boldsymbol{o}[m], \boldsymbol{h}[m])]. \tag{19.4}$$

M-step

Maximization (*M-step*): As in the case of table-CPDs, once we have the expected sufficient statistics, the algorithm treats them as if they were real and uses them as the basis for maximum likelihood estimation, using the appropriate form of the ML estimator for this family.

Convergence Results   Somewhat surprisingly, this simple algorithm can be shown to have several important properties. We now state somewhat simplified versions of the relevant results, deferring a more precise statement to the next section.

The first result states that each iteration is guaranteed to improve the log-likelihood of the current set of parameters.

Theorem 19.3

During iterations of the EM procedure of algorithm 19.2, we have

$$\ell(\boldsymbol{\theta}^t : \mathcal{D}) \leq \ell(\boldsymbol{\theta}^{t+1} : \mathcal{D}).$$

Thus, the EM procedure is constantly increasing the log-likelihood objective function. Because the objective function can be shown to be bounded (under mild assumptions), this procedure is guaranteed to converge. By itself, this result does not imply that we converge to a maximum of the objective function. Indeed, this result is only "almost true":

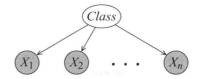

Figure 19.7 **The naive Bayes clustering model.** In this model each observed variables X_i is independent of the other observed variables given the value of the (unobserved) cluster variable C.

Theorem 19.4
Suppose that $\boldsymbol{\theta}^t$ is such that $\boldsymbol{\theta}^{t+1} = \boldsymbol{\theta}^t$ during EM, and $\boldsymbol{\theta}^t$ is also an interior point of the allowed parameter space. Then $\boldsymbol{\theta}^t$ is a stationary point of the log-likelihood function.

This result shows that EM converges to a stationary point of the likelihood function. Recall that a stationary point can be a local maximum, local minimum, or a saddle point. Although it seems counterintuitive that by taking upward steps we reach a local minimum, it is possible to construct examples where EM converges to such a point. **However, nonmaximal convergence points can only be reached from very specific starting points, and are moreover not stable, since even small perturbations to the parameters are likely to move the algorithm away from this point. Thus, in practice, EM generally converges to a local maximum of the likelihood function.**

19.2.2.4 Bayesian Clustering Using EM

clustering

One important application of learning with incomplete data, and EM in particular, is to the problem of *clustering*. Here, we have a set of data points in some feature space $\boldsymbol{X}$. Let us even assume that they are fully observable. We want to classify these data points into coherent categories, that is, categories of points that seem to share similar statistical properties.

Bayesian
clustering

The *Bayesian clustering* paradigm views this task as a learning problem with a single hidden variable C that denotes the category or class from which an instance comes. Each class is associated with a probability distribution over the features of the instances in the class. In most cases, we assume that the instances in each class c come from some coherent, fairly simple, distribution. In other words, we postulate a particular form for the *class-conditional distribution* $P(\boldsymbol{x} \mid c)$. For example, in the case of real-valued data, we typically assume that the class-conditional distribution is a multivariate Gaussian (see section 7.1). In discrete settings, we typically assume that the class-conditional distribution is a naive Bayes structure (section 3.1.3), where each feature is independent of the rest given the class variable. Overall, this approach

mixture
distribution

views the data as coming from a *mixture distribution* and attempts to use the hidden variable to separate out the mixture into its components.

naive Bayes

Suppose we consider the case of a *naive Bayes* model (figure 19.7) where the hidden class variable is the single parent of all the observed feature. In this particular learning scenario, the E-step involves computing the probability of different values of the class variables for each instance. Thus, we can think of EM as performing a soft classification of the instances, that is, each data instance belongs, to some degree, to multiple classes.

In the M-step we compute the parameters for the CPDs in the form $P(X \mid C)$ and the prior

$P(C)$ over the classes. These estimates depends on our expected sufficient statistics. These are:

$$\bar{M}_{\boldsymbol{\theta}}[c] \quad \leftarrow \quad \sum_m P(c \mid x_1[m], \dots, x_n[m], \boldsymbol{\theta}^t)$$

$$\bar{M}_{\boldsymbol{\theta}}[x_i, c] \quad \leftarrow \quad \sum_m P(c, x_i \mid x_1[m], \dots, x_n[m], \boldsymbol{\theta}^t).$$

We see that an instance helps determine the parameters for all of the classes that it participates in (that is, ones where $P(c \mid \boldsymbol{x}[m])$ is bigger than 0). Stated a bit differently, each instance "votes" about the parameters of each cluster by contributing to the statistics of the conditional distribution given that value of the cluster variable. However, the weight of this vote depends on the probability with which we assign the instance to the particular cluster.

Once we have computed the expected sufficient statistics, the M-step is, as usual, simple. The parameters for the class variable CPD are

$$\theta_c^{t+1} \leftarrow \frac{\bar{M}_{\boldsymbol{\theta}}[c]}{M},$$

and for the conditional CPD are

$$\theta_{x_i \mid c}^{t+1} \leftarrow \frac{\bar{M}_{\boldsymbol{\theta}}[x_i, c]}{\bar{M}_{\boldsymbol{\theta}}[c]}.$$

We can develop similar formulas for the case where some of the observed variables are continuous, and we use a conditional Gaussian distribution (a special case of definition 5.15) to model the CPD $P(X_i \mid C)$. The application of EM to this specific model results in a simple and efficient algorithm.

We can think of the clustering problem with continuous observations from a geometrical perspective, where each observed variable X_i represents one coordinate, and instances correspond to points. The parameters in this case represent the distribution of coordinate values in each of the classes. Thus, each class corresponds to a *cloud* of points in the input data. In each iteration, we reestimate the location of these clouds. In general, depending on the particular starting point, EM will proceed to assign each class to a dense cloud.

The EM algorithm for clustering uses a "soft" cluster assignment, allowing each instance to contribute part of its weight to multiple clusters, proportionately to its probability of belonging to each of them. As another alternative, we can consider "hard clustering," where each instance contributes all of its weight to the cluster to which it is most likely to belong. This variant, called

hard-assignment
EM

hard-assigment EM proceeds by performing the following steps.

- Given parameters $\boldsymbol{\theta}^t$, we assign $c[m] = \arg\max_c P(c \mid \boldsymbol{x}[m], \boldsymbol{\theta}^t)$ for each instance m. If we let $\mathcal{H}^t$ comprise all of the assignments $c[m]$, this results in a complete data set $(\mathcal{D}^+)^t = \langle \mathcal{D}, \mathcal{H}^t \rangle$.

- Set $\boldsymbol{\theta}^{t+1} = \arg\max_{\boldsymbol{\theta}} \ell(\boldsymbol{\theta} : (\mathcal{D}^+)^t)$. This step requires collecting sufficient statistics from $(\mathcal{D}^+)^t$, and then choosing MLE parameters based on these.

This approach is often used where the class-conditional distributions $P(\boldsymbol{X} \mid c)$ are all "round" Gaussian distributions with unit variance. Thus, each class c has its own mean vector $\boldsymbol{\mu}_c$, but a unit covariance matrix. In this case, the most likely class for an instance $\boldsymbol{x}$ is simply the

k-means

class c such that the Euclidean distance between x and μ_c is smallest. In other words, each point gravitates to the class to which it is "closest." The reestimation step is also simple. It simply selects the mean of the class to be at the center of the cloud of points that have aligned themselves with it. This process iterates until convergence. This algorithm is called *k-means*.

Although hard-assignment EM is often used for clustering, it can be defined more broadly; we return to it in greater detail in section 19.2.2.6.

---

collaborative
filtering

Box 19.A — Case Study: Discovering User Clusters. *In box 18.C, we discussed the collaborative filtering problem, and the use of Bayesian network structure learning to address it. A different application of Bayesian network learning to the collaborative filtering data task, proposed by Breese et al. (1998), utilized a Bayesian clustering approach. Here, one can introduce a cluster variable C denoting subpopulations of customers. In a simple model, the individual purchases X_i of each user are taken to be conditionally independent given the user's cluster assignment C. Thus, we have a naive Bayes clustering model, to which we can apply the EM algorithm. (As in box 18.C, items i that the user did not purchase are assigned $X_i = x_i^0$.)*

This learned model can be used in several ways. Most obviously, we can use inference to compute the probability that the user will purchase item i, given a set of purchases S. Empirical studies show that this approach achieves lower performance than the structure learning approach of box 18.C, probably because the "user cluster" variable simply cannot capture the complex preference patterns over a large number of items. However, this model can provide significant insight into the types of users present in a population, allowing, for example, a more informed design of advertising campaigns.

As one example, Bayesian clustering was applied to a data set of people browsing the MSNBC website. Each article was associated with a binary random variable X_i, which took the value x_i^1 if the user followed the link to the article. Figure 19.A.1 shows the four largest clusters produced by Bayesian clustering applied to this data set. Cluster 1 appears to represent readers of commerce and technology news (a large segment of the reader population at that period, when Internet news was in its early stages). Cluster 2 are people who mostly read the top-promoted stories in the main page. Cluster 3 are readers of sports news. In all three of these cases, the user population was known in advance, and the website contained a page targeting these readers, from which the articles shown in the table were all linked. The fourth cluster was more surprising. It appears to contain readers interested in "softer" news. The articles read by this population were scattered all over the website, and users often browsed several pages to find them. Thus, the clustering algorithm revealed an unexpected pattern in the data, one that may be useful for redesigning the website.

19.2.2.5 Theoretical Foundations ⋆

So far, we used an intuitive argument to derive the details of the EM algorithm. We now formally analyze this algorithm and prove the results regarding its convergence properties.

At each iteration, EM maintains the "current" set of parameters. Thus, we can view it as a local learning algorithm. Each iteration amounts to taking a step in the parameter space from

Cluster 1 (36 percent)

E-mail delivery isn't exactly guaranteed

Should you buy a DVD player?

Price low, demand high for Nintendo

Cluster 3 (19 percent)

Umps refusing to work is the right thing

Cowboys are reborn in win over eagles

Did Orioles spend money wisely?

Cluster 2 (29 percent)

757 Crashes at sea

Israel, Palestinians agree to direct talks

Fuhrman pleads innocent to perjury

Cluster 4 (12 percent)

The truth about what things cost

Fuhrman pleads innocent to perjury

Real astrology

Figure 19.A.1 — Application of Bayesian clustering to collaborative filtering. Four largest clusters found by Bayesian clustering applied to MSNBC news browsing data. For each cluster, the table shows the three news articles whose probability of being browsed is highest.

θ^t to θ^{t+1}. This is similar to gradient-based algorithms, except that in those algorithms we have good understanding of the nature of the step, since each step attempts to go uphill in the steepest direction. Can we find a similar justification for the EM iterations?

The basic outline of the analysis proceeds as follows. We will show that each iteration can be viewed as maximizing an *auxiliary function*, rather than the actual likelihood function. The choice of auxiliary function depends on the current parameters at the beginning of the iteration. The auxiliary function is nice in the sense that it is similar to the likelihood function in complete data problems. The crucial part of the analysis is to show how the auxiliary function relates to the likelihood function we are trying to maximize. As we will show, the relation is such that we can show that the parameters that maximize the auxiliary function in an iteration also have better likelihood than the parameters with which we started the iteration.

The Expected Log-Likelihood Function   Assume we are given a data set $\mathcal{D}$ that consists of partial observations. Recall that $\mathcal{H}$ denotes a possible assignment to all the missing values in our data set. The combination of $\mathcal{D}, \mathcal{H}$ defines a complete data set $\mathcal{D}^+ = \langle \mathcal{D}, \mathcal{H} \rangle = \{\boldsymbol{o}[m], \boldsymbol{h}[m]\}_m$, where in each instance we now have a full assignment to all the variables. We denote by $\ell(\boldsymbol{\theta} : \langle \mathcal{D}, \mathcal{H} \rangle)$ the log-likelihood of the parameters $\boldsymbol{\theta}$ with respect to this completed data set.

Suppose we are not sure about the true value of $\mathcal{H}$. Rather, we have a probabilistic estimate that we denote by a distribution Q that assigns a probability to each possible value of $\mathcal{H}$. Note that Q is a joint distribution over full assignments to all of the missing values in the entire data set. Thus, for example, in our earlier network, if $\mathcal{D}$ contains two instances $\boldsymbol{o}[1] = \langle a^1, ?, ?, d^0 \rangle$ and $\boldsymbol{o}[2] = \langle ?, b^1, ?, d^1 \rangle$, then Q is a joint distribution over $B[1], C[1], A[2], C[2]$.

In the fully observed case, our score for a set of parameters was the log-likelihood. In this case, given Q, we can use it to define an average score, which takes into account the different possible completions of the data and their probabilities. Specifically, we define the *expected log-likelihood* as:

expected
log-likelihood

$$\boldsymbol{E}_Q[\ell(\boldsymbol{\theta} : \langle \mathcal{D}, \mathcal{H} \rangle)] = \sum_{\mathcal{H}} Q(\mathcal{H})\ell(\boldsymbol{\theta} : \langle \mathcal{D}, \mathcal{H} \rangle)$$

This function has appealing characteristics that are important in the derivation of EM.

The first key property is a consequence of the linearity of expectation. Recall that when learning table-CPDs, we showed that

$$\ell(\boldsymbol{\theta} : \langle \mathcal{D}, \mathcal{H} \rangle) = \sum_{i=1}^{n} \sum_{(x_i, \boldsymbol{u}_i) \in Val(X_i, \mathrm{Pa}_{X_i})} M_{\langle \mathcal{D}, \mathcal{H} \rangle}[x_i, \boldsymbol{u}_i] \log \theta_{x_i | \boldsymbol{u}_i}.$$

Because the only terms in this sum that depend on $\langle \mathcal{D}, \mathcal{H} \rangle$ are the counts $M_{\langle \mathcal{D}, \mathcal{H} \rangle}[x_i, \boldsymbol{u}_i]$, and these appear within a linear function, we can use linearity of expectations to show that

$$\boldsymbol{E}_Q[\ell(\boldsymbol{\theta} : \langle \mathcal{D}, \mathcal{H} \rangle)] = \sum_{i=1}^{n} \sum_{(x_i, \boldsymbol{u}_i) \in Val(X_i, \mathrm{Pa}_{X_i})} \boldsymbol{E}_Q\left[M_{\langle \mathcal{D}, \mathcal{H} \rangle}[x_i, \boldsymbol{u}_i] \right] \log \theta_{x_i | \boldsymbol{u}_i}.$$

If we now generalize our notation to define

$$\bar{M}_Q[x_i, \boldsymbol{u}_i] = \boldsymbol{E}_{\mathcal{H} \sim Q}\left[M_{\langle \mathcal{D}, \mathcal{H} \rangle}[x_i, \boldsymbol{u}_i] \right] \tag{19.5}$$

we obtain

$$\boldsymbol{E}_Q[\ell(\boldsymbol{\theta} : \langle \mathcal{D}, \mathcal{H} \rangle)] = \sum_{i=1}^{n} \sum_{(x_i, \boldsymbol{u}_i) \in Val(X_i, \mathrm{Pa}_{X_i})} \bar{M}_Q[x_i, \boldsymbol{u}_i] \log \theta_{x_i | \boldsymbol{u}_i}.$$

This expression has precisely the same form as the log-likelihood function in the complete data case, but using the expected counts rather than the exact full-data counts. The implication is that instead of summing over all possible completions of the data, we can evaluate the expected log-likelihood based on the expected counts.

The crucial point here is that the log-likelihood function of complete data is *linear* in the counts. This allows us to use linearity of expectations to write the expected likelihood as a function of the expected counts.

The same idea generalizes to any model in the exponential family, which we defined in chapter 8. Recall that a model is in the exponential family if we can write:

$$P(\xi \mid \boldsymbol{\theta}) = \frac{1}{Z(\boldsymbol{\theta})} A(\xi) \exp\left\{ \langle \mathrm{t}(\boldsymbol{\theta}), \tau(\xi) \rangle \right\},$$

where $\langle \cdot, \cdot \rangle$ is the inner product, $A(\xi)$, $\mathrm{t}(\boldsymbol{\theta})$, and $Z(\boldsymbol{\theta})$ are functions that define the family, and $\tau(\xi)$ is the sufficient statistics function that maps a complete instance to a vector of sufficient statistics.

As discussed in section 17.2.5, when learning parameters of such a model, we can summarize the data using the sufficient statistic function τ. We define

$$\tau(\langle \mathcal{D}, \mathcal{H} \rangle) = \sum_m \tau(\boldsymbol{o}[m], \boldsymbol{h}[m]).$$

Because the model is in the exponential family, we can write the log-likelihood $\ell(\boldsymbol{\theta} : \langle \mathcal{D}, \mathcal{H} \rangle)$ as a *linear function* of $\tau(\langle \mathcal{D}, \mathcal{H} \rangle)$

$$\ell(\boldsymbol{\theta} : \langle \mathcal{D}, \mathcal{H} \rangle) = \langle \mathrm{t}(\boldsymbol{\theta}), \tau(\langle \mathcal{D}, \mathcal{H} \rangle) \rangle + \sum_m \log A(\boldsymbol{o}[m], \boldsymbol{h}[m]) - M \log Z(\boldsymbol{\theta}).$$

Using the linearity of expectation, we see that

$$\boldsymbol{E}_Q[\ell(\boldsymbol{\theta} : \langle \mathcal{D}, \mathcal{H} \rangle)] = \langle \mathsf{t}(\boldsymbol{\theta}), \boldsymbol{E}_Q[\tau(\langle \mathcal{D}, \mathcal{H} \rangle)] \rangle + \sum_m \boldsymbol{E}_Q[\log A(\boldsymbol{o}[m], \boldsymbol{h}[m])] - M \log Z(\boldsymbol{\theta}).$$

Because $A(\boldsymbol{o}[m], \boldsymbol{h}[m])$ does not depend on the choice of $\boldsymbol{\theta}$, we can ignore it. We are left with maximizing the function:

$$\boldsymbol{E}_Q[\ell(\boldsymbol{\theta} : \langle \mathcal{D}, \mathcal{H} \rangle)] = \langle \mathsf{t}(\boldsymbol{\theta}), \boldsymbol{E}_Q[\tau(\langle \mathcal{D}, \mathcal{H} \rangle)] \rangle - M \log Z(\boldsymbol{\theta}) + \text{const}. \tag{19.6}$$

In summary, the derivation here is directly analogous to the one for table-CPDs. The expected log-likelihood is a linear function of the expected sufficient statistics $\boldsymbol{E}_Q[\tau(\langle \mathcal{D}, \mathcal{H} \rangle)]$. We can compute these as in equation (19.4), by aggregating their expectation in each instance in the training data. Now, maximizing the right-hand side of equation (19.6) is equivalent to maximum likelihood estimation in a *complete* data set where the sum of the sufficient statistics coincides with the expected sufficient statistics $\boldsymbol{E}_Q[\tau(\langle \mathcal{D}, \mathcal{H} \rangle)]$. These two steps are exactly the E-step and M-step we take in each iteration of the EM procedure shown in algorithm 19.2. In the procedure, the distribution Q that we are using is $P(\mathcal{H} \mid \mathcal{D}, \boldsymbol{\theta}^t)$. Because instances are assumed to be independent given the parameters, it follows that

$$P(\mathcal{H} \mid \mathcal{D}, \boldsymbol{\theta}^t) = \prod_m P(\boldsymbol{h}[m] \mid \boldsymbol{o}[m], \boldsymbol{\theta}^t),$$

where $\boldsymbol{h}[m]$ are the missing variables in the m'th data instance, and $\boldsymbol{o}[m]$ are the observations in the m'th instance. Thus, we see that in the t'th iteration of the EM procedure, we choose $\boldsymbol{\theta}^{t+1}$ to be the ones that maximize $\boldsymbol{E}_Q[\ell(\boldsymbol{\theta} : \langle \mathcal{D}, \mathcal{H} \rangle)]$ with $Q(\mathcal{H}) = P(\mathcal{H} \mid \mathcal{D}, \boldsymbol{\theta}^t)$. This discussion allows us to understand a single iteration as an (implicit) optimization step of a well-defined target function.

Choosing Q   The discussion so far has showed that we can use properties of exponential models to efficiently maximize the expected log-likelihood function. Moreover, we have seen that the t'th EM iteration can be viewed as maximizing $\boldsymbol{E}_Q[\ell(\boldsymbol{\theta} : \langle \mathcal{D}, \mathcal{H} \rangle)]$ where Q is the conditional probability $P(\mathcal{H} \mid \mathcal{D}, \boldsymbol{\theta}^t)$. This discussion, however, does not provide us with guidance as to why we choose this particular auxiliary distribution Q. Note that each iteration uses a different Q distribution, and thus we cannot relate the optimization taken in one iteration to the ones made in the subsequent one. We now show why the choice $Q(\mathcal{H}) = P(\mathcal{H} \mid \mathcal{D}, \boldsymbol{\theta}^t)$ allows us to prove that each EM iteration improves the likelihood function.

To do this, we will define a new function that will be the target of our optimization. Recall that our ultimate goal is to maximize the log-likelihood function. The log-likelihood is a function only of $\boldsymbol{\theta}$; however, in intermediate steps, we also have the current choice of Q. Therefore, we will define a new function that accounts for both $\boldsymbol{\theta}$ and Q, and view each step in the algorithm as maximizing this function.

We already encountered a similar problem in our discussion of approximate inference in chapter 11. Recall that in that setting we had a known distribution P and attempted to find an approximating distribution Q. This problem is similar to the one we face, except that in learning we also change the parameters of target distribution P to maximize the probability of the data.

Let us briefly summarize the main idea that we used in chapter 11. Suppose that $P = \tilde{P}/Z$ is some distribution, where $\tilde{P}$ is an unnormalized part of the distribution, specified by a product of factors, and Z is the partition function that ensures that P sums up to one. We defined the *energy functional* as

$$F[P, Q] = \boldsymbol{E}_Q \left[\log \tilde{P} \right] + \boldsymbol{H}_Q(\mathcal{X}).$$

We then showed that the logarithm of the partition function can be rewritten as:

$$\log Z = F[P, Q] + \boldsymbol{D}(Q \| P).$$

How does this apply to the case of learning from missing data? We can choose

$$P(\mathcal{H} \mid \mathcal{D}, \boldsymbol{\theta}) = P(\mathcal{H}, \mathcal{D} \mid \boldsymbol{\theta}) / P(\mathcal{D} \mid \boldsymbol{\theta})$$

as our distribution over $\mathcal{H}$ (we hold $\mathcal{D}$ and $\boldsymbol{\theta}$ fixed for now). With this choice, the partition function $Z(\boldsymbol{\theta})$ is the data likelihood $P(\mathcal{D} \mid \boldsymbol{\theta})$ and $\tilde{P}$ is the joint probability $P(\mathcal{H}, \mathcal{D} \mid \boldsymbol{\theta})$, so that $\log \tilde{P} = \ell(\boldsymbol{\theta} : \langle \mathcal{D}, \mathcal{H} \rangle)$. Rewriting the energy functional for this new setting, we obtain:

$$F_{\mathcal{D}}[\boldsymbol{\theta}, Q] = \boldsymbol{E}_Q[\ell(\boldsymbol{\theta} : \langle \mathcal{D}, \mathcal{H} \rangle)] + \boldsymbol{H}_Q(\mathcal{H}).$$

Note that the first term is precisely the *expected log-likelihood* relative to Q. Applying our earlier analysis, we now can prove

Corollary 19.1

For any Q,

$$
\begin{aligned}
\ell(\boldsymbol{\theta} : \mathcal{D}) \;&=\; F_{\mathcal{D}}[\boldsymbol{\theta}, Q] + \boldsymbol{D}(Q(\mathcal{H}) \| P(\mathcal{H} \mid \mathcal{D}, \boldsymbol{\theta})) \\
&=\; \boldsymbol{E}_Q[\ell(\boldsymbol{\theta} : \langle \mathcal{D}, \mathcal{H} \rangle)] + \boldsymbol{H}_Q(\mathcal{H}) + \boldsymbol{D}(Q(\mathcal{H}) \| P(\mathcal{H} \mid \mathcal{D}, \boldsymbol{\theta})).
\end{aligned}
$$

Both equalities have important ramifications. Starting from the second equality, since both the entropy $\boldsymbol{H}_Q(\mathcal{H})$ and the relative entropy $\boldsymbol{D}(Q(\mathcal{H}) \| P(\mathcal{H} \mid \mathcal{D}, \boldsymbol{\theta}))$ are nonnegative, we conclude that the expected log-likelihood $\boldsymbol{E}_Q[\ell(\boldsymbol{\theta} : \langle \mathcal{D}, \mathcal{H} \rangle)]$ is a lower bound on $\ell(\boldsymbol{\theta} : \mathcal{D})$. This result is true for any choice of distribution Q. If we select $Q(\mathcal{H})$ to be the *data completion distribution* $P(\mathcal{H} \mid \mathcal{D}, \boldsymbol{\theta})$, the relative entropy term becomes zero. In this case, the remaining term $\boldsymbol{H}_Q(\mathcal{H})$ captures to a certain extent the difference between the expected log-likelihood and the real log-likelihood. Intuitively, when Q is close to being deterministic, the expected value is close to the actual value.

The first equality, for the same reasons, shows that, for any distribution Q, the F function is a lower bound on the log-likelihood. Moreover, this lower bound is tight for every choice of $\boldsymbol{\theta}$: if we choose $Q = P(\mathcal{H} \mid \mathcal{D}, \boldsymbol{\theta})$, the two functions have the same value. Thus, if we maximize the F function, we are bound to maximize the log-likelihood.

There many possible ways to optimize this target function. We now show that the EM procedure we described can be viewed as *implicitly* optimizing the EM functional F using a particular optimization strategy. The strategy we are going to utilize is a *coordinate ascent* optimization. We start with some choice $\boldsymbol{\theta}$ of parameters. We then search for Q that maximizes $F_{\mathcal{D}}[\boldsymbol{\theta}, Q]$ while keeping $\boldsymbol{\theta}$ fixed. Next, we fix Q and search for parameters that maximize $F_{\mathcal{D}}[\boldsymbol{\theta}, Q]$. We continue in this manner until convergence.

We now consider each of these steps.

- **Optimizing Q.** Suppose that $\boldsymbol{\theta}$ are fixed, and we are searching for $\arg\max_Q F_{\mathcal{D}}[\boldsymbol{\theta}, Q]$. Using corollary 19.1, we know that, if $Q^* = P(\mathcal{H} \mid \mathcal{D}, \boldsymbol{\theta})$, then

$$F_{\mathcal{D}}[\boldsymbol{\theta}, Q^*] = \ell(\boldsymbol{\theta} : \mathcal{D}) \geq F_{\mathcal{D}}[\boldsymbol{\theta}, Q].$$

Margin notes:

energy functional

expected log-likelihood

data completion

coordinate ascent

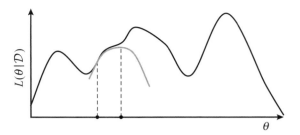

Figure 19.8   An illustration of the hill-climbing process performed by the EM algorithm. The black line represents the log-likelihood function; the point on the left represents $\boldsymbol{\theta}^t$; the gray line represents the expected log-likelihood derived from $\boldsymbol{\theta}^t$; and the point on the right represents the parameters $\boldsymbol{\theta}^{t+1}$ that maximize this expected log-likelihood.

Thus, we maximize the EM functional by choosing the auxiliary distribution Q^*. In other words, we can view the E-step as implicitly optimizing Q by using $P(\mathcal{H} \mid \mathcal{D}, \boldsymbol{\theta}^t)$ in computing the expected sufficient statistics.

- **Optimizing $\boldsymbol{\theta}$.** Suppose Q is fixed, and that we wish to find $\arg\max_{\boldsymbol{\theta}} F_{\mathcal{D}}[\boldsymbol{\theta}, Q]$. Because the only term in F that involves $\boldsymbol{\theta}$ is $\boldsymbol{E}_Q[\ell(\boldsymbol{\theta} : \langle \mathcal{D}, \mathcal{H} \rangle)]$, the maximization is equivalent to maximizing the expected log-likelihood. As we saw, we can find the maximum by computing expected sufficient statistics and then solving the MLE given these expected sufficient statistics.

Convergence of EM   The discussion so far shows that the EM procedure can be viewed as maximizing an objective function; because the objective function can be shown to be bounded, this procedure is guaranteed to converge. However, it is not clear what can be said about the convergence points of this procedure. We now analyze the convergence points of this procedure in terms of our true objective: the log-likelihood function. Intuitively, as our procedure is optimizing the energy functional, which is a tight lower bound of the log-likelihood function, each step of this optimization also improves the log-likelihood. This intuition is illustrated in figure 19.8. In more detail, the E-step is selecting, at the current set of parameters, the distribution Q^t for which the energy functional is a tight lower bound to $\ell(\boldsymbol{\theta} : \mathcal{D})$. The energy functional, which is a well-behaved concave function in $\boldsymbol{\theta}$, can be maximized effectively via the M-step, taking us to the parameters $\boldsymbol{\theta}^{t+1}$. Since the energy functional is guaranteed to remain below the log-likelihood function, this step is guaranteed to improve the log-likelihood. Moreover, the improvement is guaranteed to be at least as large as the improvement in the energy functional. More formally, using corollary 19.1, we can now prove the following generalization of theorem 19.3:

Theorem 19.5

During iterations of the EM procedure of algorithm 19.2, we have that

$$\ell(\boldsymbol{\theta}^{t+1} : \mathcal{D}) - \ell(\boldsymbol{\theta}^t : \mathcal{D}) \geq \boldsymbol{E}_{P(\mathcal{H}|\mathcal{D}, \boldsymbol{\theta}^t)}\big[\ell(\boldsymbol{\theta}^{t+1} : \mathcal{D}, \mathcal{H})\big] - \boldsymbol{E}_{P(\mathcal{H}|\mathcal{D}, \boldsymbol{\theta}^t)}\big[\ell(\boldsymbol{\theta}^t : \mathcal{D}, \mathcal{H})\big].$$

As a consequence, we obtain that:

$$\ell(\boldsymbol{\theta}^t : \mathcal{D}) \leq \ell(\boldsymbol{\theta}^{t+1} : \mathcal{D}).$$

PROOF We begin with the first statement. Using corollary 19.1, with the distribution $Q^t(\mathcal{H}) = P(\mathcal{H} \mid \mathcal{D}, \boldsymbol{\theta}^t)$ we have that

$$
\begin{aligned}
\ell(\boldsymbol{\theta}^{t+1} : \mathcal{D}) &= \boldsymbol{E}_{Q^t}\left[\ell(\boldsymbol{\theta}^{t+1} : \langle \mathcal{D}, \mathcal{H} \rangle)\right] + \boldsymbol{H}_{Q^t}(\mathcal{H}) + \boldsymbol{D}(Q^t(\mathcal{H}) \| P(\mathcal{H} \mid \mathcal{D}, \boldsymbol{\theta}^{t+1})) \\
\ell(\boldsymbol{\theta}^t : \mathcal{D}) &= \boldsymbol{E}_{Q^t}\left[\ell(\boldsymbol{\theta}^t : \langle \mathcal{D}, \mathcal{H} \rangle)\right] + \boldsymbol{H}_{Q^t}(\mathcal{H}) + \boldsymbol{D}(Q^t(\mathcal{H}) \| P(\mathcal{H} \mid \mathcal{D}, \boldsymbol{\theta}^t)) \\
&= \boldsymbol{E}_{Q^t}\left[\ell(\boldsymbol{\theta}^t : \langle \mathcal{D}, \mathcal{H} \rangle)\right] + \boldsymbol{H}_{Q^t}(\mathcal{H}).
\end{aligned}
$$

The last step is justified by our choice of $Q^t(\mathcal{H}) = P(\mathcal{H} \mid \mathcal{D}, \boldsymbol{\theta}^t)$. Subtracting these two terms, we have that

$$
\begin{aligned}
\ell(\boldsymbol{\theta}^{t+1} : \mathcal{D}) - \ell(\boldsymbol{\theta}^t : \mathcal{D}) = \\
\boldsymbol{E}_{Q^t}\left[\ell(\boldsymbol{\theta}^{t+1} : \mathcal{D}, \mathcal{H})\right] - \boldsymbol{E}_{Q^t}\left[\ell(\boldsymbol{\theta}^t : \mathcal{D}, \mathcal{H})\right] + \boldsymbol{D}(Q^t(\mathcal{H}) \| P(\mathcal{H} \mid \mathcal{D}, \boldsymbol{\theta}^{t+1})).
\end{aligned}
$$

Because the last term is nonnegative, we get the desired inequality.

To prove the second statement of the theorem, we note that $\boldsymbol{\theta}^{t+1}$ is the value of $\boldsymbol{\theta}$ that maximizes $\boldsymbol{E}_{P(\mathcal{H} \mid \mathcal{D}, \boldsymbol{\theta}^t)}[\ell(\boldsymbol{\theta} : \mathcal{D}, \mathcal{H})]$. Hence the value obtained for this expression for $\boldsymbol{\theta}^{t+1}$ is at least at large as the value obtained for any other set of parameters, including $\boldsymbol{\theta}^t$. It follows that the right-hand side of the inequality is nonnegative, which implies the second statement. ∎

 We conclude that EM performs a variant of hill climbing, in the sense that it improves the log-likelihood at each step. Moreover, the M-step can be understood as maximizing a lower-bound on the improvement in the likelihood. Thus, in a sense we can view the algorithm as searching for the largest possible improvement, when using the expected log-likelihood as a proxy for the actual log-likelihood.

For most learning problems, we know that the log-likelihood is upper bounded. For example, if we have discrete data, then the maximal likelihood we can assign to the data is 1. Thus, the log-likelihood is bounded by 0. If we have a continuous model, we can construct examples where the likelihood can grow unboundedly; however, we can often introduce constraints on the parameters that guarantee a bound on the likelihood (see exercise 19.10). If the log-likelihood is bounded, and the EM iterations are nondecreasing in the log-likelihood, then the sequence of log-likelihoods at successive iterations must converge.

The question is what can be said about this convergence point. Ideally, we would like to guarantee convergence to the maximum value of our log-likelihood function. Unfortunately, as we mentioned earlier, we cannot provide this guarantee; however, we can now prove theorem 19.4, which shows convergence to a fixed point of the log-likelihood function, that is, one where the gradient is zero. We restate the theorem for convenience:

Theorem 19.6

Suppose that $\boldsymbol{\theta}^t$ is such that $\boldsymbol{\theta}^{t+1} = \boldsymbol{\theta}^t$ during EM, and $\boldsymbol{\theta}^t$ is also an interior point of the allowed parameter space. Then $\boldsymbol{\theta}^t$ is a stationary point of the log-likelihood function.

PROOF We start by rewriting the log-likelihood function using corollary 19.1.

$$\ell(\boldsymbol{\theta} : \mathcal{D}) = \boldsymbol{E}_Q[\ell(\boldsymbol{\theta} : \langle \mathcal{D}, \mathcal{H} \rangle)] + \boldsymbol{H}_Q(\mathcal{H}) + \boldsymbol{D}(Q(\mathcal{H}) \| P(\mathcal{H} \mid \mathcal{D}, \boldsymbol{\theta})).$$

We now consider the gradient of $\ell(\boldsymbol{\theta} : \mathcal{D})$ with respect to $\boldsymbol{\theta}$. Since the term $\boldsymbol{H}_Q(\mathcal{H})$ does not depend on $\boldsymbol{\theta}$, we get that

$$\nabla_{\boldsymbol{\theta}} \ell(\boldsymbol{\theta} : \mathcal{D}) = \nabla_{\boldsymbol{\theta}} \boldsymbol{E}_Q[\ell(\boldsymbol{\theta} : \langle \mathcal{D}, \mathcal{H} \rangle)] + \nabla_{\boldsymbol{\theta}} \boldsymbol{D}(Q(\mathcal{H}) \| P(\mathcal{H} \mid \mathcal{D}, \boldsymbol{\theta})).$$

This observation is true for any choice of Q. Now suppose we are in an EM iteration. In this case, we set $Q = P(\mathcal{H} \mid \mathcal{D}, \boldsymbol{\theta}^t)$ and evaluate the gradient at $\boldsymbol{\theta}^t$.

A somewhat simplified proof runs as follows. Because $\boldsymbol{\theta} = \boldsymbol{\theta}^t$ is a minimum of the KL-divergence term, we know that $\nabla_{\boldsymbol{\theta}} \boldsymbol{D}(Q(\mathcal{H}) \| P(\mathcal{H} \mid \mathcal{D}, \boldsymbol{\theta}^t))$ is 0. This implies that

$$\nabla_{\boldsymbol{\theta}} \ell(\boldsymbol{\theta}^t : \mathcal{D}) = \nabla_{\boldsymbol{\theta}} \boldsymbol{E}_Q\left[\ell(\boldsymbol{\theta}^t : \langle \mathcal{D}, \mathcal{H} \rangle)\right].$$

Or, in other words, $\nabla_{\boldsymbol{\theta}} \ell(\boldsymbol{\theta}^t : \mathcal{D}) = 0$ if and only if $\nabla_{\boldsymbol{\theta}} \boldsymbol{E}_Q\left[\ell(\boldsymbol{\theta}^t : \langle \mathcal{D}, \mathcal{H} \rangle)\right] = 0$.

Recall that $\boldsymbol{\theta}^{t+1} = \arg\max_{\boldsymbol{\theta}} \boldsymbol{E}_Q\left[\ell(\boldsymbol{\theta}^t : \langle \mathcal{D}, \mathcal{H} \rangle)\right]$. Hence the gradient of the expected likelihood at $\boldsymbol{\theta}^{t+1}$ is 0. Thus, we conclude that $\boldsymbol{\theta}^{t+1} = \boldsymbol{\theta}^t$ only if $\nabla_{\boldsymbol{\theta}} \boldsymbol{E}_Q\left[\ell(\boldsymbol{\theta}^t : \langle \mathcal{D}, \mathcal{H} \rangle)\right] = 0$. And so, at this point, $\nabla_{\boldsymbol{\theta}} \ell(\boldsymbol{\theta}^t : \mathcal{D}) = 0$. This implies that this set of parameters is a stationary point of the log-likelihood function.

The actual argument has to be somewhat more careful. Recall that the parameters must lie within some allowable set. For example, the parameters of a discrete random variable must sum up to one. Thus, we are searching within a constrained space of parameters. When we have constraints, we often do not have zero gradient. Instead, we get to a stationary point when the gradient is orthogonal to the constraints (that is, local changes within the allowed space do not improve the likelihood). The arguments we have stated apply equally well when we replace statements about equality to 0 with orthogonality to the constraints on the parameter space. ∎

19.2.2.6   Hard-Assignment EM

In section 19.2.2.4, we briefly mentioned the idea of using a hard assignment to the hidden variables, in the context of applying EM to Bayesian clustering. We now generalize this simple idea to the case of arbitrary Bayesian networks.

hard-assignment EM

This algorithm, called *hard-assignment EM*, also iterates over two steps: one in which it completes the data given the current parameters $\boldsymbol{\theta}^t$, and the other in which it uses the completion to estimate new parameters $\boldsymbol{\theta}^{t+1}$. However, rather than using a soft completion of the data, as in standard EM, it selects for each data instance $\boldsymbol{o}[m]$ the single assignment $\boldsymbol{h}[m]$ that maximizes $P(\boldsymbol{h} \mid \boldsymbol{o}[m], \boldsymbol{\theta}^t)$.

Although hard-assignment EM is similar in outline to EM, there are important differences. In fact, hard-assignment EM can be described as optimizing a different objective function, one that involves both the learned parameters and the learned assignment to the hidden variables. This objective is to maximize the likelihood of the complete data $\langle \mathcal{D}, \mathcal{H} \rangle$, given the parameters:

$$\max_{\boldsymbol{\theta}, \mathcal{H}} \ell(\boldsymbol{\theta} : \mathcal{H}, \mathcal{D}).$$

See exercise 19.14. Compare this objective to the EM objective, which attempts to maximize $\ell(\boldsymbol{\theta} : \mathcal{D})$, averaging over all possible completions of the data.

Does this observation provide us insight on these two learning procedures? The intuition is that these two objectives are similar if $P(\mathcal{H} \mid \mathcal{D}, \boldsymbol{\theta})$ assigns most of the probability mass to

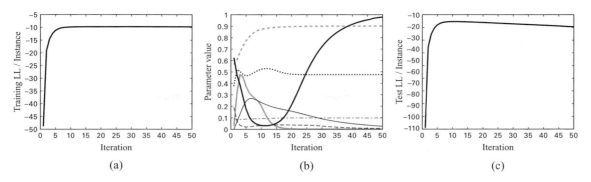

Figure 19.B.1 — Convergence of EM run on the ICU Alarm network. (a) Training log-likelihood. (b) Progress of several sample parameters. (c) Test data log-likelihood.

one completion of the data. In such a case, EM will effectively perform hard assignment during the E-step. However, if $P(\mathcal{H} \mid \mathcal{D}, \boldsymbol{\theta})$ is diffuse, the two algorithms will lead to very different solutions. In clustering, the hard-assignment version tends to increase the contrast between different classes, since assignments have to choose between them. In contrast, EM can learn classes that are overlapping, by having many instances contributing to two or more classes.

Another difference between the two EM variants is in the way they progress during the learning. Note that for a given data set, at the end of an iteration, the hard-assignment EM can be in one of a finite number of parameter values. Namely, there is only one parameter assignment for each possible assignment to $\mathcal{H}$. Thus, hard-assignment EM traverses a path in the combinatorial space of assignments to $\mathcal{H}$. The soft-assignment EM, on the other hand, traverses the continuous space of parameter assignments. The intuition is that hard-assignment EM converges faster, since it makes discrete steps. In contrast, soft-assignment EM can converge very slowly to a local maximum, since close to the maximum, each iteration makes only small changes to the parameters. The flip side of this argument is that soft-assignment EM can traverse paths that are infeasible to the hard-assignment EM. For example, if two clusters need to shift their means in a coordinated fashion, soft-assignment EM can progressively change their means. On the other hand, hard-assignment EM needs to make a "jump," since it cannot simultaneously reassign multiple instances and change the class means.

Box 19.B — Case Study: EM in Practice. *The EM algorithm is guaranteed to monotonically improve the training log-likelihood at each iteration. However, there are no guarantees as to the speed of convergence or the quality of the local maxima attained. To gain a better perspective of how the algorithm behaves in practice, we consider here the application of the method to the* ICU-Alarm *network discussed in earlier learning chapters.*

We start by considering the progress of the training data likelihood during the algorithm's iterations. In this example, $1,000$ samples were generated from the ICU-Alarm *network. For each instance, we then independently and randomly hid 50 percent of the variables. As can be seen in figure 19.B.1a, much of the improvement over the performance of the random starting point is in the*

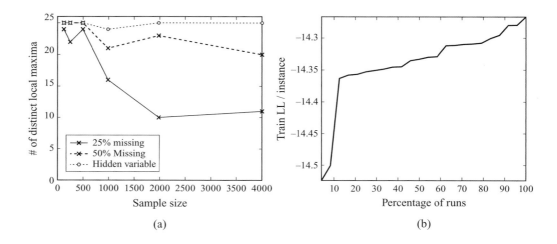

Figure 19.B.2 — Local maxima in likelihood surface. (a) Number of unique local maxima (in 25 runs) for different sample sizes and missing value configurations. (b) Distribution of training likelihood of local maxima attained for 25 random starting points with 1,000 samples and one hidden variable.

first few iterations. However, examining the convergence of different parameters in (b), we see that some parameters change significantly after the fifth iteration, even though changes to the likelihood are relatively small. In practice, any nontrivial model will display a wide range of sensitivity to the network parameters. Given more training data, the sensitivity will, typically, overall decrease. Owing to these changes in parameters, the training likelihood continues to improve after the initial iterations, but very slowly. This behavior of fast initial improvement, followed by slow convergence, is typical of EM.

We next consider the behavior of the learned model on unseen test data. As we can see in (c), early in the process, test-data improvement correlates with training-data performance. However, after the 10th iterations, training performance continues to improve, but test performance decreases. This overfitting *phenomenon is an instance of* overfitting *to the training data. With more data or fewer unobserved values, this phenomenon will be less pronounced. With less data or hidden variables, on the other hand, explicit techniques for coping with the problem may be needed (see box 19.C).*

A second key issue any type of optimization of the likelihood in the case of missing data is that of local maxima. To study this phenomenon, we consider the number of local maxima for 25 random starting points under different settings. As the sample size (x-axis) grows, the number of local maxima diminishes. In addition, the number of local maxima when more values are missing (dashed line) is consistently greater than the number of local maxima in the setting where more data is available (solid line). Importantly, in the case where just a single variable is hidden, the number of local maxima is large, and remains large even when the amount of training data is quite large. To see that this is not just an artifact of possible permutations of the values of the hidden variable, and to demonstrate the importance of achieving a superior local maxima, in (b) we show the training set log-likelihood of the 25 different local maxima attained. The difference between the *best and worst local maxima is over* 0.2 *bit-per-instance. While this may not seem significant, for a training set of* $1,000$ *instances, this corresponds to a factor of* $2^{0.2*1,000} \approx 10^{60}$ *in the training set likelihood. We also note that the spread of the quality in different local maxima is quite uniform, so that it is not easy to attain a good local maximum with a small number of random trials.*

19.2.3 Comparison: Gradient Ascent versus EM

So far we have discussed two algorithms for parameter learning with incomplete data: gradient ascent and EM. As we will discuss (see box 19.C), there are many issues involved in the actual implementation of these algorithms: the choice of initial parameters, the stopping criteria, and so forth. However, before discussing these general points, it is worth comparing the two algorithms.

There are several points of similarity in the overall strategy of both algorithms. Both algorithms are *local* in nature. At each iteration they maintain a "current" set of parameters, and use these to find the next set. Moreover, both perform some version of greedy optimization based on the current point. Gradient ascent attempts to progress in the steepest direction from the current point. EM performs a greedy step in improving its target function given the local parameters. Finally, both algorithms provide a guarantee to converge to local maxima (or, more precisely, to stationary points where the gradient is 0). On one hand, this is an important guarantee, in the sense that both are at least locally maximal. On the other hand, this is a weak guarantee, since many real-world problems have multimodal likelihood functions, and thus we do not know how far the learned parameters are from the global maximum (or maxima).

In terms of the actual computational steps, the two algorithms are also quite similar. For table-CPDs, the main component of either an EM iteration or a gradient step is computing the expected sufficient statistics of the data given the current parameters. This involves performing inference on each instance. Thus, both algorithms can exploit dynamic programming procedures (for example, clique tree inference) to compute all the expected sufficient statistics in an instance efficiently.

In term of implementation details, the algorithms provide different benefits. On one hand, gradient ascent allows to use "black box" nonlinear optimization techniques, such as conjugate gradient ascent (see appendix A.5.2). This allows the implementation to build on a rich set of existing tools. Moreover, gradient ascent can be easily applied to various CPDs by using the chain rule of derivatives. On the other hand, EM relies on maximization from complete data. Thus, it allows for a fairly straightforward use of learning procedure for complete data in the case of incomplete data. The only change is replacing the part that accumulates sufficient statistics by a procedure that computes expected sufficient statistics. As such, most people find EM easier to implement.

A final aspect for consideration is the convergence rate of the algorithm. Although we cannot predict in advance how many iterations are needed to learn parameters, analysis can show the general behavior of the algorithm in terms of how fast it approaches the convergence point. Suppose we denote by $\ell_t = \ell(\theta^t : \mathcal{D})$ the likelihood of the solution found in the t'th iteration (of either EM or gradient ascent). The algorithm converges toward $\ell^* = \lim_{t \to \infty} \ell_t$. The error at the t'th iteration is

$$\epsilon_t = \ell^* - \ell_t.$$

convergence rate

Although we do not go through the proof, one can show that EM has *linear convergence rate.* This means that for each domain there exists a t_0 and $\alpha < 1$ such that for all $t \geq t_0$

$$\epsilon_{t+1} \leq \alpha \epsilon_t.$$

On the face of it, this is good news, since it shows that the error decreases at each iteration. Such a convergence rate means that $\ell_{t+1} - \ell_t = \epsilon_t - \epsilon_{t+1} \geq \epsilon_t(1 - \alpha)$. In other words, if we know α, we can bound the error

$$\epsilon_t \leq \frac{\ell_{t+1} - \ell_t}{1 - \alpha}.$$

While this result provides a bound on the error (and also suggests a way of estimating it), it is not always a useful one. In particular, if α is relatively close to 1, then even when the difference is likelihood between successive iterations is small, the error can be much larger. Moreover, the number of iterations to convergence can be very large. In practice we see this behavior quite often. **The first iterations of EM show huge improvement in the likelihood. These are then** **followed by many iterations that slowly increase the likelihood; see box 19.B. Conjugate gradient often has opposite behavior. The initial iterations (which are far away from the local maxima) often take longer to improve the likelihood. However, once the algorithm is in the vicinity of maximum, where the log-likelihood function is approximately quadratic, this method is much more efficient in zooming on the maximum.** Finally, it is important to keep in mind that these arguments are asymptotic in the number of iterations; the actual number of iterations required for convergence may not be in the asymptotic regime. Thus, the rate of convergence of different algorithms may not be the best indicator as to which of them is likely to work most efficiently in practice.

Box 19.C — Skill: Practical Considerations in Parameter Learning. *There are a few practical considerations in implementing both gradient-based methods and EM for learning parameters with missing data. We now consider a few of these. We present these points mostly in the context of the EM algorithm, but most of our points apply equally to both classes of algorithms.*

In a practical implementation of EM, there are two key issues that one needs to address. The first is the presence of local maxima. As demonstrated in box 19.B, the likelihood of even relatively simple networks can have a large number of local maxima that significantly differ in terms of their quality. There are several adaptations of these local search algorithms that aim to consistently reach beneficial local maxima. These adaptations include a judicious selection of initialization, and methods for modifying the search so as to achieve a better local maximum. The second key issue involves the convergence of the algorithm: determining convergence, and improving the rate of convergence.

Local Maxima   *One of the main limitations of both the EM and the gradient ascent procedures is that they are only guaranteed to reach a stationary point, which is usually a local maximum. How do we improve the odds of finding a global — or at least a good local — maximum?*

The first place where we can try to address the issue of local maxima is in the initialization of the algorithm. EM and gradient ascent, as well as most other "local" algorithms, require a starting point — a set of initial parameters that the algorithm proceeds to improve. Since both algorithms are deterministic, this starting point (implicitly) determines which local maximum is found. **In practice, different initializations can result in radically different convergence points, sometimes with very different likelihood values. Even when the likelihood values are similar, different convergence points may represent semantically different conclusions about the data.** *This issue is particularly severe when hidden variables are involved, where we can easily obtain very different clusterings of the data. For example, when clustering text documents by similarity (for example, using a version of the model in box 17.E where the document cluster variable is hidden), we can learn one model where the clusters correspond to document topics, or another where they correspond to the style of the publication in which the document appeared (for example, newspaper, webpage, or blog). Thus, initialization should generally be considered very seriously in these situations, especially when the amount of missing data is large or hidden variables are involved.*

In general, we can initialize the algorithm either in the E-step, by picking an initial set of parameters, or in the M-step, by picking an initial assignment to the unobserved variables. In the first type of approach, the simplest choices for starting points are either a set of parameters fixed in advance or randomly chosen parameters. If we use predetermined initial parameters, we should exercise care in choosing them, since a misguided choice can lead to very poor outcomes. In particular, for some learning problems, the seemingly natural choice of uniform parameters can lead to disastrous results; see exercise 19.11. Another easy choice is applicable for parts of the network where we have only a moderate amount of missing data. Here, we can sometimes estimate parameters using only the observed data, and then use those to initialize the E-step. Of course, this approach is not always feasible, and it is inapplicable when we have a hidden variable. A different natural choice is to use the mean of our prior over parameters. On one hand, if we have good prior information, this might serve as a good starting position. Note that, although this choice does bias the learning algorithm to prefer the prior's view of the data, the learned parameters can be drastically different in the end. On the other hand, if the prior is not too informative, this choice suffers from the same drawbacks we mentioned earlier. Finally, a common choice is to use a randomized starting point, an approach that avoids any intrinsic bias. However, there is also no reason to expect that a random choice will give rise to a good solution. For this reason, often one tries multiple random starting points, and the convergence point of highest likelihood is chosen.

The second class of methods initializes the procedure at the M-step by completing the missing data. Again, there are many choices for completing the data. For example, we can use a uniform or a random imputation method to assign values to missing observations. This procedure is particularly useful when we have different patterns of missing observations in each sample. Then, the counts from the imputed data consist of actual counts combined with imputed ones. The real data thus bias the estimated parameters to be reasonable. Another alternative is to use a simplified learning procedure to learn initial assignment to missing values. This procedure can be, for example, hard-assignment EM. As we discussed, such a procedure usually converges faster and therefore can serve as a good initialization. However, hard-assignment EM also requires a starting point, or a selection among multiple random starting points.

When learning with hidden variables, such procedures can be more problematic. For example, if we consider a naive Bayes clustering model and use random imputation, the result would be that we randomly assign instances to clusters. With a sufficiently large data set, these clusters will be very similar (since they all sample from the same population). In a smaller data set the sampling noise might distinguish the initial clusters, but nonetheless, this is not a very informed starting point. We discuss some methods for initializing a hidden variable in section 19.5.3.

Other than initialization, we can also consider modifying our search so as to reduce the risk of getting stuck at a poor local optimum. The problem of avoiding local maxima is a standard one, and we describe some of the more common solutions in appendix A.4.2. Many of these solutions are applicable in this setting as well. As we mentioned, the approach of using multiple random restarts is commonly used, often with a beam search *modification to quickly prune poor starting points. In particular, in this beam search variant, K EM runs are carried out in parallel and every few iterations only the most promising ones are retained. A variant of this approach is to generate K EM threads at each step by slightly perturbing the most beneficial $k < K$ threads from the previous iteration. While such adaptations have no formal guarantees, they are extremely useful in practice in terms trading off quality of solution and computational requirements.*

beam search

Annealing methods (appendix A.4.2) have also been used successfully in the context of the EM algorithm. In such methods, we gradually transform from an easy objective with a single local maximum to the desired EM objective, and thereby we potentially avoid many local maxima that are far away from the central basin of attraction. Such an approach can be carried out by directly smoothing the log-likelihood function and gradually reducing the level to which it is smoothed, or implicitly by gradually altering the weights of training instances.

Finally, we note that we can never determine with certainty whether the EM convergence point is truly the global maximum. In some applications this limitation is acceptable — for example, if we care only about fitting the probability distribution over the training examples (say for detecting instances from a particular subpopulation). In this case, if we manage to learn parameters that assign high probability for samples in the target population, then we might be content even if these parameters are not the best ones possible. On the other hand, if we want to use the learned parameters to reveal insight about the domain, then we might care about whether the parameters are truly the optimal ones or not. In addition, if the learning procedure does not perform well, we have to decide whether the problem stems from getting trapped in a poor local maximum, or from the fact that the model is not well suited to the distribution in our particular domain.

Stopping Criteria   *Both algorithms we discussed have the property that they will reach a fixed point once they converged on a stationary point of the likelihood surface. In practice, we never really reach the stationary point, although we can get quite close to it. This raises the question of when we stop the procedure.*

The basic idea is that when solutions at successive iterations are similar to each other, additional iterations will not change the solution by much. The question is how to measure similarity of solutions. There are two main approaches. The first is to compare the parameters from successive iterations. The second is to compare the likelihood of these choices of parameters. Somewhat surprisingly, these two criteria are quite different. In some situations small changes in parameters lead to dramatic changes in likelihood, and in others large changes in parameters lead to small changes in the likelihood.

To understand how there can be a discrepancy between changes in parameters and changes in likelihood, consider the properties of the gradient as shown in theorem 19.2. Using a Taylor expansion of the likelihood, this gradient provides us with an estimate how the likelihood will change when we change the parameters. We see that if $P(x, \boldsymbol{u} \mid \boldsymbol{o}[m], \boldsymbol{\theta})$ is small in most data *instances, then the gradient $\frac{\partial \ell(\boldsymbol{\theta}:\mathcal{D})}{\partial P(x|\boldsymbol{u})}$ will be small. This implies that relatively large changes in $P(x \mid \boldsymbol{u})$ will not change the likelihood by much. This can happen for example if the event $\boldsymbol{u}$ is uncommon in the training data, and the value of $P(x \mid \boldsymbol{u})$ is involved in the likelihood only in a few instances. On the flip side, if the event $x, \boldsymbol{u}$ has a large posterior in all samples, then the gradient $\frac{\partial \ell(\boldsymbol{\theta}:\mathcal{D})}{\partial P(x|\boldsymbol{u})}$ will be of size proportional to M. In such a situation a small change in the parameter can result in a large change in the likelihood.*

In general, since we are aiming to maximize the likelihood, large changes in the parameters that have negligible effect on the likelihood are of less interest. Moreover, measuring the magnitude of changes in parameters is highly dependent on our parameterization. For example, if we use the reparameterization of equation (19.3), the difference (say in Euclidean distance) between two sets of parameters can change dramatically. Using the likelihood for tracking convergence is thus less sensitive to these choices and more directly related to the goal of the optimization.

Even once we decide what to measure, we still need to determine when we should stop the process. Some gradient-based methods, such as conjugate gradient ascent, build an estimate of the second-order derivative of the function. Using these derivatives, they estimate the improvement we expect to have. We can then decide to stop when the expected improvement is not more than a fixed amount of log-likelihood units. We can apply similar stoping criteria to EM, where again, if the change in likelihood in the last iteration is smaller than a predetermined threshold we stop the iterations.

☞ **Importantly, although the training set log-likelihood is guaranteed to increase monotonically until convergence, there is no guarantee that the generalization performance of the model — the expected log-likelihood relative to the underlying distribution — also increases monotonically. (See section 16.3.1.) Indeed, it is often the case that, as we ap-**
overfitting **proach convergence, the generalization performance starts to decrease, due to *overfitting* of the parameters to the specifics of the training data.**

Thus, an alternative approach is to measure directly when additional improvement to the training set likelihood does not contribute to generalization. To do so we need to separate the available data into a training set and a validation set (see box 16.A). We run learning on the training set, but
validation set *at the end of each iteration we evaluate the log-likelihood of the* validation set *(which is not seen during learning). We stop the procedure when the likelihood of the validation set does not improve. (As usual, the actual performance of the model would then need to be evaluated on a separate test set.) This method allows us to judge when the procedure stops learning the interesting phenomena and begins to overfit the training data. On the flip side, such a procedure is both slower (since we need to evaluate likelihood on an additional data set at the end of iteration) and forces us to train on a smaller subset of data, increasing the risk of overfitting. Moreover, if the validation set is small, then the estimate of the generalization ability by the likelihood on this set is noisy. This noise can influence the stopping time.*

Finally, we note that in EM much of the improvement is typically observed in the first few iterations, but the final convergence can be quite slow. Thus, in practice, it is often useful to limit the number of EM iterations or use a lenient convergence threshold. This is particularly important when EM is used as part of a higher-level algorithm (for example, structure learning) and where, in the intermediate stages of the overall learning algorithm, approximate parameter estimates are often sufficient. Moreover, early stopping can help reduce overfitting, as we discussed.

Accelerating Convergence   *There are also several strategies that can help improve the rate of convergence of EM to its local optimum. We briefly list a few of them.*

The first idea is to use hybrid algorithms *that mix EM and gradient methods. The basic intuition is that EM is good at rapidly moving to the general neighborhood of a local maximum in few iterations but bad at pinpointing the actual maximum. Advanced gradient methods, on the other hand, quickly converge once we are close to a maximum. This observation suggests that we should run EM for few iterations and then switch over to using a method such as conjugate gradient. Such hybrid algorithms are often much more efficient. Another alternative is to use* accelerated EM *methods that take even larger steps in the search than standard EM (see section 19.7).*

accelerated EM

incremental EM

Another class of variations comprises incremental methods. *In these methods we do not perform a full E-step or a full M-step. Again, the high-level intuition is that, since we view our procedure as maximizing the energy functional $F_{\mathcal{D}}[\boldsymbol{\theta}, Q]$, we can consider steps that increase this functional but do not necessarily find the maximum value parameters or Q. For example, recall that $\boldsymbol{\theta}$ consists of several subcomponents, one per CPD. Rather than maximizing all the parameters at once, we can consider a partial update where we maximize the energy functional with respect to one of the subcomponents while freezing the others; see exercise 19.16. Another type of partial update is based on writing Q as a product of independent distributions — each one over the missing values in a particular instance. Again, we can optimize the energy functional with respect to one of these while freezing the other; see exercise 19.17. These partial updates can provide two types of benefit: they can require less computation than a full EM update, and they can propagate changes between the statistics and the parameters much faster, reducing the total number of iterations.*

Box 19.D — Case Study: EM for Robot Mapping.   *One interesting application of the EM algorithm is to robotic mapping. Many variants of this applications have been proposed; we focus on one by Thrun et al. (2004) that explicitly tries to use the probabilistic model to capture the structure in the environment.*

The data in this application are a point cloud representation of an indoor environment. The point cloud can be obtained by collecting a sequence of point clouds, measured along a robot's motion trajectory. One can use a robot localization procedure to (approximately) assess the robot's pose (position and heading) along each point in the trajectory, which allows the different measurements to be put on a common frame of reference. Although the localization process is not fully accurate, the estimates are usually reasonable for short trajectories. One can then take the points obtained over the trajectory, and fit the points using polygons, to derive a 3D map of the surfaces in the robot's environment. However, the noise in the laser measurements, combined with the errors in localization, leads adjacent polygons to have slightly different surface normals, giving rise to a very jagged representation of the environment.

The EM algorithm can be used to fit a more compact representation of the environment to the data, reducing the noise and providing a smoother, more realistic output. In particular, in this example, the model consists of a set of 3D planes $p_1, \ldots, p_K$, each characterized by two parameters $\boldsymbol{\alpha}_k, \beta_k$, where $\boldsymbol{\alpha}_k$ is a unit-length vector in $\mathbb{R}^3$ that encodes the plane's surface normal vector, and β_k is a scalar that denotes its distance to the origin of the global coordinate system. Thus, the distance of any point $\boldsymbol{x}$ to the plane is $d(\boldsymbol{x}, p_k) = |\boldsymbol{\alpha}_k \boldsymbol{x} - \beta_k|$.

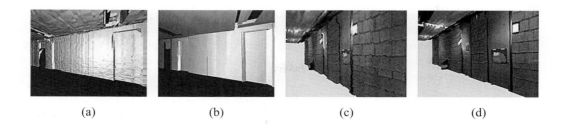

(a)   (b)   (c)   (d)

Figure 19.D.1 — Sample results from EM-based 3D plane mapping (a) Raw data map obtained from range finder. (b) Planes extracted from map using EM. (c) Fragment of reconstructed surface using raw data. (d) The same fragment reconstructed using planes.

correspondence
variable

data association

The probabilistic model also needs to specify, for each point x_m in the point cloud, to which plane x_m belongs. This assignment can be modeled via a set of correspondence *variables C_m such that $C_m = k$ if the measurement point x_m was generated by the kth plane. Each assignment to the correspondence variables, which are unobserved, encodes a possible solution to the* data association *problem. (See box 12.D for more details.) We define $P(X_m \mid C_m = k : \theta_k)$ to be $\propto \mathcal{N}\big(d(x, p_k) \mid 0; \sigma^2\big)$. In addition, we also allow an additional value $C_m = 0$ that encodes points that are not generated by any of the planes; the distribution $P(X_m \mid C_m = 0)$ is taken to be uniform over the (finite) space.*

Given a probabilistic model, the EM algorithm can be applied to find the assignment of points to planes — the correspondence variables, which are taken to be hidden; and the parameters α_k, β_k that characterize the planes. Intuitively, the E-step computes the assignment to the correspondence variables by assigning the weight of each point proportionately to its distance to each of them. The M-step then recomputes the parameters of each plane to fit the points assigned to it. See exercise 19.18 and exercise 19.19. The algorithm also contains an additional outer loop that heuristically suggests new surfaces to be added to the model, and removes surfaces that do not have enough support in the data (for example, one possible criterion can depend on the total weight that different data points assign to the surface).

The results of this algorithm are shown in figure 19.D.1. One can see that the resulting map is considerably smoother and more realistic than the results derived directly from the raw data.

19.2.4   Approximate Inference ⋆

The main computational cost in both gradient ascent and EM is in the computation of expected sufficient statistics. This step requires running probabilistic inference on each instance in the training data. These inference steps are needed both for computing the likelihood and for computing the posterior probability over events of the form x, pa_X for each variable and its parents. For some models, such as the naive Bayes clustering model, this inference step is almost trivial. For other models, this inference step can be extremely costly. In practice, we often want to learn parameters for models where exact inference is impractical. Formally, this happens when the tree-width of the unobserved parts of the model is large. (Note the contrast to learning from complete data, where the cost of learning the model did not depend on the complexity of inference.) In such situations the cost of inference becomes the limiting factor in our ability to learn from data.

Example 19.9

Recall the network discussed in example 6.11 and example 19.9, where we have n students taking m classes, and the grade for each student in each class depends on both the difficulty of the class and his or her intelligence. In the ground network for this example, we have a set of variables $\boldsymbol{I} = \{I(s)\}$ for the n students (denoting the intelligence level of each student s), $\boldsymbol{D} = \{D(c)\}$ for the m courses (denoting the difficulty level of each course c), and $\boldsymbol{G} = \{G(s,c)\}$ for the grades, where each variable $G(s,c)$ has as parents $I(s)$ and $D(c)$. Since this network is derived from a plate model, the CPDs are all shared, so we only have three CPDs that must be learned: $P(I(S))$, $P(D(C))$, $P(G(S,C) \mid I(S), D(C))$.

Suppose we only observe the grades of the students but not their intelligence or the difficulty of courses, and we want to learn this model. First, we note that there is no way to force the model to respect our desired semantics for the (hidden) variables I and D; for example, a model in which we flip the two values for I is equally good. Nevertheless, we can hope that some value for I will correspond to "high intelligence" and the other to "low intelligence," and similarly for D.

To perform EM in this model, we need to infer the expected counts of assignments to triplets of variables of the form $I(s), D(c), G(s,c)$. Since we have parameter sharing, we will aggregate these counts and then estimate the CPD $P(G(S,C) \mid I(S), D(C))$ from the aggregate counts. The problem is that observing a variable $G(s,c)$ couples its two parents. Thus, this network induces a Markov network that has a pairwise potential between any pair of $I(s)$ and $D(c)$ variables that share an observed child. If enough grade variables are observed, this network will be close to a full bipartite graph, and exact inference about the posterior probability becomes intractable. This creates a serious problem in applying either EM or gradient ascent for learning this seemingly simple model from data. ∎

An obvious solution to this problem is to use approximate inference procedures. A simple approach is to view inference as a "black box." Rather than invoking exact inference in the learning procedures shown in algorithm 19.1 and algorithm 19.2, we can simply invoke one of the approximate inference procedures we discussed in earlier chapters. This view is elegant because it decouples the choices made in the design of the learning procedure from the choices made in the approximate inference procedures.

However, this decoupling can obscure important effects of the approximation on our learning procedure. For example, suppose we use approximate inference for computing the gradient in a gradient ascent approach. In this case, our estimate of the gradient is generally somewhat wrong, and the errors in successive iterations are generally not consistent with each other. Such inaccuracies can confuse the gradient ascent procedure, a problem that is particularly significant when the procedure is closer to the convergence point and the gradient is close to 0, so that the errors can easily dominate. **A key question is whether learning with approximate inference results in an approximate learning procedure; that is, whether we are guaranteed to find a local maximum of an approximation of the likelihood function. In general, there are very few cases where we can provide any types of guarantees on the interaction between approximate inference and learning. Nevertheless, in practice, the use of approximate**

inference is often unavoidable, and so many applications use some form of approximate inference despite the lack of theoretical guarantees.

One class of approximation algorithms for which a unifying perspective is useful is in the combination of EM with the global approximate inference methods of chapter 11. Let us consider first the *structured variational* methods of section 11.5, where the integration is easiest to understand. In these methods, we are attempting to find an approximate distribution Q that is close to an unnormalized distribution $\tilde{P}$ in which we are interested. We saw that algorithms in this class can be viewed as finding a distribution Q in a suitable family of distributions that maximizes the energy functional:

$$F[\tilde{P}, Q] = \boldsymbol{E}_Q\left[\log \tilde{P}\right] + \boldsymbol{H}_Q(\mathcal{X}).$$

Thus, in these approximate inference procedures, we search for a distribution Q that maximizes

$$\max_{Q \in \mathcal{Q}} F[\tilde{P}, Q].$$

We saw that we can view EM as an attempt to maximize the same energy functional, with the difference that we are also optimizing over the parameterization $\boldsymbol{\theta}$ of $\tilde{P}$. We can combine both goals into a single objective by requiring that the distribution Q used in the EM functional come from a particular family $\mathcal{Q}$. Thus, we obtain the following *variational EM* problem:

$$\max_{\boldsymbol{\theta}} \max_{Q \in \mathcal{Q}} F_{\mathcal{D}}[\boldsymbol{\theta}, Q], \tag{19.7}$$

where $\mathcal{Q}$ is a family of approximate distributions we are considering for representing the distribution over the unobserved variables.

To apply the variational EM framework, we need to choose the family of distributions $\mathcal{Q}$ that will be used to approximate the distribution $P(\mathcal{H} \mid \mathcal{D}, \boldsymbol{\theta})$. Importantly, because this posterior distribution is a product of the posteriors for the different training instance, our approximation Q can take the same form without incurring any error. Thus, we need only to decide how to represent the posterior $P(\boldsymbol{h}[m] \mid \boldsymbol{o}[m], \boldsymbol{\theta})$ for each instance m. We therefore define a class $\mathcal{Q}$ that we will use to approximate $P(\boldsymbol{h}[m] \mid \boldsymbol{o}[m], \boldsymbol{\theta})$. Importantly, since the evidence $\boldsymbol{o}[m]$ is different for each data instance m, the posterior distribution for each instance is also different, and hence we need to use a different distribution $Q[m] \in \mathcal{Q}$ to approximate the posterior for each data instance. In principle, using the techniques of section 11.5, we can use any class $\mathcal{Q}$ that allows tractable inference. In practice, a common solution is to use the mean field approximation, where we assume that Q is a product of marginal distributions (one per each unobserved value).

Example 19.10

Consider using the mean field approximation (see section 11.5.1) for the learning problem of example 19.9. Recall that in the mean field approximation we approximate the target posterior distribution by a product of marginals. More precisely, we approximate $P(I(s_1), \ldots, I(s_n), D(c_1), \ldots, D(c_m) \mid o, \boldsymbol{\theta})$ by a distribution

$$Q(I(s_1), \ldots, I(s_n), D(c_1), \ldots, D(c_m)) = Q(I(s_1)) \cdots Q(I(s_n))Q(D(c_1)) \cdots Q(D(c_m)).$$

Importantly, although the prior over the variables $I(s_1), \ldots, I(s_n)$ is identical, their posterior is generally different. Thus, the marginal of each of the variable has different parameters in Q (and similarly for the $D(c)$ variables).

In our approximate E-step, given a set of parameters $\boldsymbol{\theta}$ for the model, we need to compute approximate expected sufficient statistics. We do so in two steps. First, we use iterations of the mean field update equation equation (11.54) to find the best choice of marginals in Q to approximate $P(I(s_1), \ldots, I(s_n), D(c_1), \ldots, D(c_m) \mid \boldsymbol{o}, \boldsymbol{\theta})$. We then use the distribution Q to compute approximate expected sufficient statistics by finding:

$$
\begin{aligned}
\bar{M}_Q[g_{(i,j)}, I(s_i), D(c_j)] &= Q(I(s_i), D(c_j))\mathbf{I}\{G(s_i, c_j) = g_{(i,j)}\} \\
&= Q(I(s_i))Q(D(c_j))\mathbf{I}\{G(s_i, c_j) = g_{(i,j)}\}. \qquad \blacksquare
\end{aligned}
$$

Given our choice of $\mathcal{Q}$, we can optimize the variational EM objective very similarly to the optimization of the exact EM objective, by iterating over two steps:

variational E-step • **Variational E-step** For each m, find

$$
Q^t[m] = \arg \max_{Q \in \mathcal{Q}} F_{\boldsymbol{o}[m]}[\boldsymbol{\theta}, Q].
$$

This step is identical to our definition of variational inference in chapter 11, and it can be implemented using the algorithms we discussed there, usually involving iterations of local updates until convergence.

At the end of this step, we have an approximate distribution $Q^t = \prod_m Q^t[m]$ and can collect the expected sufficient statistics. To compute the expected sufficient statistics, we combine the observed values in the data with expected counts from the distribution Q. This process requires answering queries about events in the distribution Q. For some approximations, such as the mean field approximation, we can answer such queries efficiently (that is, by multiplying the marginal probabilities over each variables); see example 19.10. If we use a richer class of approximate distributions, we must perform a more elaborate inference process. Note that, because the approximation Q is simpler than the original distribution P, we have no guarantee that a clique tree for Q will respect the family-preservation property relative to families in P. Thus, in some cases, we may need to perform queries that are outside the clique tree used to perform the E-step (see section 10.3.3.2).

• **M-step** We find a new set of parameters

$$
\boldsymbol{\theta}^{t+1} = \arg \max_{\theta} F_{\mathcal{D}}[\boldsymbol{\theta}, Q^t];
$$

this step is identical to the M-step in standard EM.

The preceding algorithm is essentially performing coordinate-wise ascent alternating between optimization of Q and $\boldsymbol{\theta}$. It opens up the way to alternative ways of maximizing the same objective function. For example, we can limit the number of iterations in the variational E-step. Since each such iteration improves the energy functional, we do not need to reach a maximum in the Q dimension before making an improvement to the parameters.

Importantly, regardless of the method used to optimize the variational EM functional of equation (19.7), we can provide some guarantee regarding the properties of our optimum. Recall that we showed that

$$
\ell(\boldsymbol{\theta} : \mathcal{D}) = \max_{Q} F_{\mathcal{D}}[\boldsymbol{\theta}, Q] \geq \max_{Q \in \mathcal{Q}} F_{\mathcal{D}}[\boldsymbol{\theta}, Q].
$$

lower bound

Thus, maximizing the objective of equation (19.7) maximizes a *lower bound* of the likelihood. When we limit the choice of Q to be in a particular family, we cannot necessarily get a tight bound on the likelihood. However, since we are maximizing a lower bound, we know that we do not overestimate the likelihood of parameters we are considering. If the lower bound is relatively good, this property implies that we distinguish high-likelihood regions in the parameter space from very low ones. Of course, if the lower bound is loose, this guarantee is not very meaningful.

We can try to extend these ideas to other approximation methods. For example, generalized belief propagation section 11.3 is an attractive algorithm in this context, since it can be fairly efficient. Moreover, because the cluster graph satisfies the family preservation property, computation of an expected sufficient statistic can be done locally within a single cluster in the graph. The question is whether such an approximation can be understood as maximizing a clear objective. Recall that cluster-graph belief propagation can be viewed as attempting to maximize an approximation of the energy functional where we replace the term $H_Q(\mathcal{X})$ by approximate entropy terms. Using exactly the same arguments as before, we can then show that, if we use generalized belief propagation for computing expected sufficient statistics in the E-step, then we are effectively attempting to maximize the approximate version of the energy functional. In this case, we cannot prove that this approximation is a lower bound to the correct likelihood. Moreover, if we use a standard message passing algorithm to compute the fixed points of the energy functional, we have no guarantees of convergence, and we may get oscillations both within an E-step and over several steps, which can cause significant problems in practice. Of course, we can use other approximations of the energy functional, including ones that are guaranteed to be lower bounds of the likelihood, and algorithms that are guaranteed to be convergent. These approaches, although less commonly used at the moment, share the same benefits of the structured variational approximation.

 More broadly, the ability to characterize the approximate algorithm as attempting to optimize a clear objective function is important. For example, an immediate consequence is that, to monitor the progress of the algorithm, we should evaluate the approximate energy functional, since we know that, at least when all goes well, this quantity should increase until the convergence point.

19.3 Bayesian Learning with Incomplete Data ⋆

19.3.1 Overview

In our discussion of parameter learning from complete data, we discussed the limitations of maximum likelihood estimation, many of which can be addressed by the Bayesian approach. In the Bayesian approach, we view the parameters as unobserved variables that influence the probability of all training instances. Learning then amounts to computing the probability of new examples based on the observation, which can be performed by computing the posterior probability over the parameters, and using it for prediction.

More precisely, in Bayesian reasoning, we introduce a prior $P(\boldsymbol{\theta})$ over the parameters, and are interested in computing the posterior $P(\boldsymbol{\theta} \mid \mathcal{D})$ given the data. In the case of complete data, we saw that if the prior has some properties (for example, the priors over the parameters of different CPDs are independent, and the prior is conjugate), then the posterior has a nice form and is representable in a compact manner. Because the posterior is a product of the prior and the likelihood, it follows from our discussion in section 19.1.3 that these useful properties are lost in the case of incomplete data. In particular, as we can see from figure 19.4 and figure 19.5, the parameter variables are generally correlated in the posterior. Thus, we can no longer represent the posterior as a product of posteriors over each set of parameters. Moreover, the posterior will generally be highly complex and even multimodal. Bayesian inference over this posterior would generally require a complex integration procedure, which generally has no analytic solution.

MAP estimation   One approach, once we realize that incomplete data makes the prospects of exact Bayesian reasoning unlikely, is to focus on the more modest goal of *MAP estimation*, which we first discussed in section 17.4.4. In this approach, rather than integrating over the entire posterior $P(\mathcal{D}, \boldsymbol{\theta})$, we search for a maximum of this distribution:

$$\tilde{\boldsymbol{\theta}} = \arg\max_{\boldsymbol{\theta}} P(\boldsymbol{\theta} \mid \mathcal{D}) = \arg\max_{\boldsymbol{\theta}} \frac{P(\boldsymbol{\theta})P(\mathcal{D} \mid \boldsymbol{\theta})}{P(\mathcal{D})}.$$

Ideally, the neighborhood of the MAP parameters is the center of mass of the posterior, and therefore, using them might be a reasonable approximation for averaging over parameters in their neighborhood. Using the same transformations as in equation (17.14), the problem reduces to one of computing the optimum:

$$\text{score}_{\text{MAP}}(\boldsymbol{\theta} \ : \ \mathcal{D}) = \ell(\boldsymbol{\theta} : \mathcal{D}) + \log P(\boldsymbol{\theta}).$$

MAP-EM   This function is simply the log-likelihood function with an additional prior term. Because this prior term is usually well behaved, we can generally easily extend both gradient-based methods and the EM algorithm to this case; see, for example, exercise 19.20 and exercise 19.21. Thus, finding MAP parameters is essentially as hard or as easy as finding MLE parameters. As such, it is often applicable in practice. Of course, the same caveats that we discussed in section 17.4.4 — the sensitivity to parameterization, and the insensitivity to the form of the posterior — also apply here.

A second approach is to try to address the task of full Bayesian learning using an approximate method. Recall from section 17.3 that we can cast Bayesian learning as inference in the meta-network that includes all the variables in all the instances as well as the parameters. Computing the probability of future events amounts to performing queries about the posterior probability of the $(M+1)$st instance given the observations about the first M instances. In the case of complete data, we could derive closed-form solutions to this inference problem. In the case of incomplete data, these solutions do not exist, and so we need to resort to approximate inference procedures.

In theory, we can apply any approximate inference procedure for Bayesian networks to this problem. Thus, all the procedures we discussed in the inference chapter can potentially be used for performing Bayesian inference with incomplete data. Of course, some are more suitable than others.

For example, we can conceivably perform likelihood weighting, as described in section 12.2: we first sample parameters from the prior, and then the unobserved variables. Each such sample will be weighted by the probability of the observed data given the sampled parameter and hidden variables. Such a procedure is relatively easy to implement and does not require running complex inference procedure. However, since the parameter space is a high-dimensional continuous region, the chance of sampling high-posterior parameters is exceedingly small. As a result, virtually all the samples will be assigned negligible weight. Unless the learning problem is relatively easy and the prior is quite informative, this inference procedure would provide a poor approximation and would require a huge number of samples.

In the next two sections, we consider two approximate inference procedures that can be applied to this problem with some degree of success.

19.3.2 MCMC Sampling

A common strategy for dealing with hard Bayesian learning problems is to perform MCMC simulation (see section 12.3). Recall that, in these methods, we construct a Markov chain whose state is the assignment to all unobserved variables, such that the stationary distribution of the chain is posterior probability over these variables. In our case, the state of the chain consists of $\boldsymbol{\theta}$ and $\mathcal{H}$, and we need to ensure that the stationary distribution of the chain is the desired posterior distribution.

19.3.2.1 Gibbs Sampling

Gibbs sampling

One of the simplest MCMC strategies for complex multivariable chains is *Gibbs sampling*. In Gibbs sampling, we choose one of the variables and sample its value given the value of all the other variables. In our setting, there are two types of variables: those in $\mathcal{H}$ and those in $\boldsymbol{\theta}$; we deal with each separately.

Suppose $X[m]$ is one of the variables in $\mathcal{H}$. The current state of the MCMC sampler has a value for all other variables in $\mathcal{H}$ and for $\boldsymbol{\theta}$. Since the parameters are known, selecting a value for $X[m]$ requires a sampling step that is essentially the same as the one we did when we performed Gibbs sampling for inference in the m'th instance. This step can be performed using the same sampling procedure as in section 12.3.

meta-network

Now suppose that $\boldsymbol{\theta}_{X|U}$ are the parameters for a particular CPD. Again, the current state of the sampler assigns value for all of the variables in $\mathcal{H}$. Since the structure of the *meta-network* is such that $\boldsymbol{\theta}_{X|U}$ are independent of the parameters of all other CPDs given $\mathcal{D}$ and $\mathcal{H}$, then we need to sample from $P(\boldsymbol{\theta}_{X|U} \mid \mathcal{D}, \mathcal{H})$ — the posterior distribution over the parameters given the complete data $\mathcal{D}, \mathcal{H}$. In section 17.4 we showed that, if the prior is of a particular form (for example, a product of Dirichlet priors), then the posterior based on complete data also has a compact form. Now, we can use these properties to sample from the posterior. To be concrete, if we consider table-CPDs with Dirichlet priors, then the posterior is a product of Dirichlet distributions, one for each assignment of values for U. Thus, if we know how to sample from a Dirichlet distribution, then we can sample from this posterior. It turns out that sampling from a Dirichlet distribution can be done using a reasonably efficient procedure; see box 19.E.

Thus, we can apply Gibbs sampling to the meta-network. If we simulate a sufficiently long run, the samples we generate will be from the joint posterior probability of the parameters and the hidden variables. We can then use these samples to make predictions about new samples and to estimate the marginal posterior of parameters or hidden variables. The BUGS system (see box 12.C) provides a simple, general-purpose tool for Gibbs-sampling-based Bayesian learning.

Box 19.E — Skill: Sampling from a Dirichlet distribution. *Suppose we have a parameter vector random variable* $\boldsymbol{\theta} = \langle \theta_1, \ldots, \theta_k \rangle$ *that is distributed according to a Dirichlet distribution* $\boldsymbol{\theta} \sim Dirichlet(\alpha_1, \ldots, \alpha_K)$. *How do we sample a parameter vector from this distribution?*

An effective procedure for sampling such a vector relies on an alternative definition of the Dirichlet distribution. We need to start with some definitions.

Definition 19.6

Gamma
distribution

A continuous random variable X *has a* Gamma *distribution* $\mathrm{Gamma}(\alpha, \beta)$ *if it has the density*

$$p(x) = \frac{\beta^\alpha}{\Gamma(\alpha)} x^{\alpha-1} e^{-\beta x}.$$
∎

We can see that the $x^{\alpha-1}$ *term is reminiscent of components of the Dirichlet distribution. Thus, it might not be too surprising that there is a connection between the two distributions.*

Theorem 19.7

Let $X_1, \ldots, X_k$ *be independent continuous random variables such that* $X_i \sim \mathrm{Gamma}(\alpha_i, 1)$. *Define the random vector*

$$\boldsymbol{\theta} = \left\langle \frac{X_1}{X_1 + \cdots + X_k}, \ldots, \frac{X_k}{X_1 + \cdots + X_k} \right\rangle.$$

Then, $\boldsymbol{\theta} \sim Dirichlet(\alpha_1, \ldots, \alpha_k)$.

Thus, we can think of a Dirichlet distribution as a two-step process. First, we sample k independent values, each from a separate Gamma distribution. Then we normalize these values to get a distribution. The normalization creates the dependency between the components of the vector $\boldsymbol{\theta}$.

This theorem suggests a natural way to sample from a Dirichlet distribution. If we can sample from Gamma distributions, we can sample values $X_1, \ldots, X_k$ from the appropriate Gamma distribution and then normalize these values.

The only remaining question is how to sample from a Gamma distribution. We start with a special case. If we consider a variable $X \sim \mathrm{Gamma}(1, 1)$, then the density function is

$$p(X = x) = e^{-x}.$$

In this case, we can solve the cumulative distribution using simple integration and get that

$$P(X < x) = 1 - e^{-x}.$$

From this, it is not hard to show the following result:

Lemma 19.2

If $U \sim \mathrm{Unif}([0:1])$, *then* $-\ln U \sim \mathrm{Gamma}(1, 1)$.

In particular, if we want to sample parameter vectors from $Dirichlet(1, \ldots, 1)$, which is the uniform distribution over parameter vectors, we need to sample k values from the uniform distribution, take their negative logarithm, and normalize. Since a sample from a uniform distribution can be readily obtained from a pseudo-random number generator, we get a simple procedure for sampling from the uniform distribution over multinomial distributions. Note that this procedure is not the one we intuitively would consider for this problem. When $\alpha \neq 1$ the sampling problem is harder, and requires more sophisticated methods, often based on the rejection sampling *approach described in section 14.5.1; these methods are outside the scope of this book.*

19.3.2.2   Collapsed MCMC

Recall that, in many situations, we can make MCMC sampling more efficient by using collapsed particles that represent a partial state of the system. If we can perform exact inference over the remaining state, then we can use MCMC sampling over the smaller state space and thereby get more efficient sampling.

We can apply this idea in the context of Bayesian inference in two different ways. In one approach, we have *parameter collapsed particles*, where each particle is an assignment to the parameters $\boldsymbol{\theta}$, associated with a distribution over $\mathcal{H}$; in the other, we have *data completion distribution particles*, where each particle is an assignment to the unobserved variables $\mathcal{H}$, associated with a distribution over $\boldsymbol{\theta}$. We now discuss each of these approaches in turn.

Parameter Collapsed Particles   Suppose we choose the collapsed particles to contain assignments to the parameters $\boldsymbol{\theta}$, accompanied by distributions over the hidden variables. Thus, we need to be able to deal with queries about $P(\mathcal{H} \mid \boldsymbol{\theta}, \mathcal{D})$ and $P(\mathcal{D}, \boldsymbol{\theta})$. First note that given $\boldsymbol{\theta}$, the different training instances are conditionally independent. Thus, we can perform inference in each instance separately. The question now is whether we can perform this instance-level inference efficiently. This depends on the structure of the network we are learning.

Consider, for example, the task of learning the naive Bayes clustering model of section 19.2.2.4. In this case, each instance has a single hidden variable, denoting the cluster of the instance. Given the value of the parameters, inference over the hidden variable involves summing over all the values of the hidden variable, and computing the probability of the observation variables given each value. These operations are linear in the size of the network, and thus can be done efficiently. This means that evaluating the likelihood of a proposed particle is quite fast. On the other hand, if we are learning parameters for the network of example 19.9, then the network structure is such that we cannot perform efficient exact inference. In this case the cost of evaluating the likelihood of a proposed particle is nontrivial and requires additional approximations. Thus, the ease of operations with this type of collapsed particle depends on the network structure.

In addition to evaluating the previous queries, we need to be able to perform the sampling steps. In particular, for Gibbs sampling, we need to be able to sample from the distribution:

$$P(\boldsymbol{\theta}_{X_i|\mathrm{Pa}_{X_i}} \mid \{\boldsymbol{\theta}_{X_j|\mathrm{Pa}_{X_j}}\}_{j \neq i}, \mathcal{D}).$$

Unfortunately, sampling from this conditional distribution is unwieldy. Even though we assume the value of all other parameters, since we do not have complete data, we are not guaranteed to have a simple form for this conditional distribution (see example 19.11). As an alternative to Gibbs sampling, we can use Metropolis-Hastings. Here, in each proposal step, we suggest new parameter values and evaluate the likelihood of these new parameters relative to the old ones. This step requires that we evaluate $P(\mathcal{D}, \boldsymbol{\theta})$, which is as costly as computing the likelihood. This fact makes it critical that we construct a good proposal distribution, since a poor one can lead to many (expensive) rejected proposals.

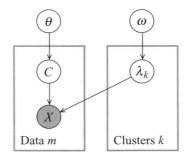

Figure 19.9   Plate model for Bayesian clustering

Example 19.11

Bayesian
clustering

Let us return to the setting of Bayesian clustering described in section 19.2.2.4. In this model, which is illustrated in figure 19.9, we have a set of data points $\mathcal{D} = \{\boldsymbol{x}[1], \ldots, \boldsymbol{x}[M]\}$, which are taken from one of a set of K clusters. We assume that each cluster is characterized by a distribution $Q(\boldsymbol{X} \mid \boldsymbol{\lambda})$, which has the same form for each cluster, but different parameters. As we discussed, the form of the class-conditional distribution depends on the data; typical models include naive Bayes for discrete data, or Gaussian distributions for continuous data. This decision is orthogonal to our discussion here. Thus, we have a set of K parameter vectors $\boldsymbol{\lambda}_1, \ldots, \boldsymbol{\lambda}_K$, each sampled from a distribution $P(\boldsymbol{\lambda}_k \mid \boldsymbol{\omega})$. We use a hidden variable $C[m]$ to represent the cluster from which the m'th data point was sampled. Thus, the class-conditional distribution $P(\boldsymbol{X} \mid C = c^k, \boldsymbol{\lambda}_{1,\ldots,K}) = Q(\boldsymbol{X} \mid \boldsymbol{\lambda}_k)$. We assume that the cluster variable C is sampled from a multinomial with parameters $\boldsymbol{\theta}$, sampled from a Dirichlet distribution $\boldsymbol{\theta} \sim Dirichlet(\alpha_0/K, \ldots, \alpha_0/K)$. The symmetry of the model relative to clusters reflects the fact that cluster identifiers are meaningless placeholders; it is only the partition of instances to clusters that is significant.

To consider the use of parameter collapsed particles, let us begin by writing down the data likelihood given the parameters.

$$P(\mathcal{D} \mid \boldsymbol{\lambda}_1, \ldots, \boldsymbol{\lambda}_K, \boldsymbol{\theta}) = \prod_{m=1}^{M} \left(\sum_{k=1}^{K} P(C[m] = c^k \mid \boldsymbol{\theta}) P(\boldsymbol{x}[m] \mid C[m] = c^k, \boldsymbol{\lambda}_k) \right). \quad (19.8)$$

In this case, because of the simple structure of the graphical model, this expression is easy to evaluate for any fixed set of parameters.

Now, let us consider the task of sampling $\boldsymbol{\lambda}_k$ given $\boldsymbol{\theta}$ and $\boldsymbol{\lambda}_{-k}$, where we use a subscript of $-k$ to denote the set consisting of all of the values $k' \in \{1, \ldots, K\} - \{k\}$. The distribution with which we wish to sample $\boldsymbol{\lambda}_k$ is

$$P(\boldsymbol{\lambda}_k \mid \boldsymbol{\lambda}_{-k}, \mathcal{D}, \boldsymbol{\theta}, \boldsymbol{\omega}) \propto P(\mathcal{D} \mid \boldsymbol{\lambda}_1, \ldots, \boldsymbol{\lambda}_K, \boldsymbol{\theta}) P(\boldsymbol{\lambda}_k \mid \boldsymbol{\omega}).$$

Examining equation (19.8), we see that, given the data and the parameters other than $\boldsymbol{\lambda}_k$, all of the terms $P(\boldsymbol{x}[m] \mid C[m] = c^{k'}, \boldsymbol{\lambda}_{k'})$ for $k' \neq k$ can now be treated as a constant, and aggregated into a single number; similarly, the terms $P(C[m] = c^k \mid \boldsymbol{\theta})$ are also constant. Hence, each of the terms inside the outermost product can be written as a linear function $a_m P(\boldsymbol{x}[m] \mid C[m] =$

$c^k) + b_m$. *Unfortunately, the entire expression is a product of these linear functions, making the sampling distribution for $\boldsymbol{\lambda}_k$ proportional to a degree M polynomial in its likelihood function (multiplied by $P(\boldsymbol{\lambda}_k \mid \boldsymbol{\omega})$). This distribution is rarely one from which we can easily sample.*

The Metropolis-Hastings approach is more feasible in this case. Here, as discussed in section 14.5.3,

random-walk chain

we can use a random-walk chain *as a proposal distribution and use the data likelihood to compute the acceptance probabilities. In this case, the computation is fairly straightforward, since it involves the ratio of two expressions of the form of equation (19.8), which are the same except for the values of $\boldsymbol{\lambda}_k$. Unfortunately, although most of the terms in the numerator and denominator are identical, they appear within the scope of a summation over k and therefore do not cancel. Thus, to compute the acceptance probability, we need to compute the full data likelihood for both the current and proposed parameter choices; in this particular network, this computation can be performed fairly efficiently.* ∎

Data Completion Collapsed Particles An alternative choice is to use collapsed particles that assign a value to $\mathcal{H}$. In this case, each particle represents a complete data set. Recall that if the prior distribution satisfies certain properties, then we can use closed-form formulas to compute parameter posteriors from $P(\boldsymbol{\lambda} \mid \mathcal{D}, \mathcal{H})$ and to evaluate the marginal likelihood $P(\mathcal{D}, \mathcal{H})$. This implies that if we are using a well-behaved prior, we can evaluate the likelihood of particles in time that does not depend on the network structure.

For concreteness, consider the case where we are learning a Bayesian network with table-CPDs where we have an independent Dirichlet prior over each distribution $P(X_i \mid \mathrm{pa}_{X_i})$. In this case, if we have a particle that represents a complete data instance, we can summarize it by the sufficient statistics $M[x_i, \mathrm{pa}_{x_i}]$, and using these we can compute both the posterior over parameters and the marginal likelihood using closed-form formulas; see section 17.4 and section 18.3.4.

Example 19.12

Let us return to the Bayesian clustering task, but now consider the setting where each particle $\boldsymbol{c}$ is an assignment to the hidden variables $C[1], \ldots, C[M]$. Given an assignment to these variables, we are now in the regime of complete data, in which the different parameters are independent in the posterior. In particular, let $I_k(\boldsymbol{c})$ be the set of indexes $\{m : c[m] = k\}$. We can now compute the distribution associated with our particle $\boldsymbol{c}$ as a Dirichlet posterior

$$P(\boldsymbol{\theta} \mid \boldsymbol{c}) = Dirichlet(\alpha_0/K + |I_1(\boldsymbol{c})|, \ldots, \alpha_0/K + |I_K(\boldsymbol{c})|).$$

We also have that:

$$P(\boldsymbol{\lambda}_k \mid \boldsymbol{c}, \mathcal{D}, \boldsymbol{\omega}) = P(\boldsymbol{\lambda}_k \mid \mathcal{D}_{I_k(\boldsymbol{c})}, \boldsymbol{\omega}) \propto P(\boldsymbol{\lambda}_k \mid \boldsymbol{\omega}) \prod_{m \in I_k(\boldsymbol{c})} P(\boldsymbol{x}[m] \mid \boldsymbol{\lambda}_k),$$

that is, the posterior over $\boldsymbol{\lambda}_k$ starting from the prior defined by $\boldsymbol{\omega}$ and conditioning on the data instances in $I_k(\boldsymbol{c})$. If we now further assume that $P(\boldsymbol{\lambda} \mid \boldsymbol{\omega})$ is a conjugate prior to $Q(\boldsymbol{X} \mid \boldsymbol{\lambda})$, this posterior can be computed in closed form.

To apply Gibbs sampling, we also need to specify a distribution for sampling a new value for $C[m']$ given $\boldsymbol{c}_{-m'}$, where again, we use the notation $-m'$ to indicate all values $\{1, \ldots, M\} - \{m'\}$. Similarly, let $I_k(\boldsymbol{c}_{-m'})$ denote the set of indexes $\{m \neq m' : c[m] = k\}$. Due to the *independencies represented by the model structure, we have:*

$$P(C[m'] = k \mid \boldsymbol{c}_{-m'}, \mathcal{D}, \boldsymbol{\omega}) \propto \tag{19.9}$$
$$P(C[m'] = k \mid \boldsymbol{c}_{-m'})P(\boldsymbol{x}[m'] \mid C[m'] = k, \boldsymbol{x}[I_k(\boldsymbol{c}_{-m'})], \boldsymbol{\omega}).$$

The second term on the right-hand side is simply a Bayesian prediction over $\boldsymbol{X}$ from the parameter posterior $P(\boldsymbol{\lambda}_k \mid \mathcal{D}_{I_k(\boldsymbol{c}_{-m'})}, \boldsymbol{\omega})$, as defined. Because of the symmetry of the parameters for the different clusters, the term does not depend on m' or on k, but only on the data set on which we condition. We can rewrite this term as $Q(\boldsymbol{X} \mid \mathcal{D}_{I_k(\boldsymbol{c}_{-m'})}, \boldsymbol{\omega})$. The first term on the right-hand side is the prior on cluster assignment for the instance m', as determined by the Dirichlet prior and the assignments of the other instances. Some algebra allows us to simplify this expression, resulting in:

$$P(C[m'] = k \mid \boldsymbol{c}_{-m'}, \mathcal{D}, \boldsymbol{\omega}) \propto (|I_k(\boldsymbol{c}_{-m'})| + \alpha_0/K)Q(\boldsymbol{X} \mid \mathcal{D}_{I_k(\boldsymbol{c}_{-m'})}, \boldsymbol{\omega}). \tag{19.10}$$

Assuming we have a conjugate prior, this expression can be easily computed. Overall, for most conjugate priors, the cost of computing the sampling distribution for $C[m]$ in this model is $O(MK)$. ■

It turns out that efficient sampling is also possible in more general models; see exercise 19.22. Alternatively we can use a Metropolis-Hastings approach where the proposal distribution can propose to modify several hidden values at once; see exercise 19.23.

Comparison Both types of collapsed particles can be useful for learning in practice, but they have quite different characteristics.

As we discussed, when we use parameter collapsed particles, the cost of evaluating a particle (for example, in a Metropolis-Hastings iteration) is determined by cost of inference in the network. In the worst case, this cost can be exponential, but in many examples it can be efficient. In contrast, the cost of evaluating data collapsed particles depends on properties of the prior. If the prior is properly chosen, the cost is linear in the size of the network.

Another aspect is the space in which we perform MCMC. In the case of parameter collapsed particles, the MCMC procedure is performing integration over a high-dimensional continuous space. The simple Metropolis-Hastings procedures we discussed in this book are usually quite poor for addressing this type of task. However, there is an extensive literature of more efficient MCMC procedures for this task (these are beyond the scope of this book). In the case of data collapsed particles, we perform the integration over parameters in closed form and use MCMC to explore the discrete (but exponential) space of assignments to the unobserved variables. In this problem, relatively simple MCMC methods, such as Gibbs sampling, can be fairly efficient.

To summarize, there is no clear choice between these two options. Both types of collapsed particles can speed up the convergence of the sampling procedure and the accuracy of the estimates of the parameters.

19.3.3   Variational Bayesian Learning

Another class of approximate inference procedures that we can apply to perform Bayesian inference in the case of incomplete data are variational approximations. Here, we can use the methods we developed in chapter 11 to the inference problem posed by Bayesian learning paradigm, resulting in an approach called *variational Bayes*. Recall that, in a variational approximation, we

variational Bayes

aim to find a distribution Q, from a predetermined family of distributions $\mathcal{Q}$, that is close to the real posterior distribution. In our case, we attempt to approximate $P(\mathcal{H}, \boldsymbol{\theta} \mid \mathcal{D})$; thus, the unnormalized measure $\tilde{P}$ in equation (11.3) is $P(\mathcal{H}, \boldsymbol{\theta}, \mathcal{D})$, and our approximating distribution Q is over the parameters and the hidden variables.

Plugging in the variational principle for our problem, and using the fact that $P(\mathcal{H}, \boldsymbol{\theta}, \mathcal{D}) = P(\boldsymbol{\theta})P(\mathcal{H}, \mathcal{D} \mid \boldsymbol{\theta})$, we have that the energy functional takes the form:

$$F[P, Q] = \boldsymbol{E}_Q[\log P(\boldsymbol{\theta})] + \boldsymbol{E}_Q[\log P(\mathcal{H}, \mathcal{D} \mid \boldsymbol{\theta})] + \boldsymbol{H}_Q(\boldsymbol{\theta}, \mathcal{H}).$$

The development of such an approximation requires that we decide on the class of approximate distributions we want to consider. While there are many choices here, a natural one is to decouple the posterior over the parameters from the posterior over the missing data. That is, assume that

$$Q(\boldsymbol{\theta}, \mathcal{H}) = Q(\boldsymbol{\theta})Q(\mathcal{H}). \tag{19.11}$$

This is clearly a nontrivial assumption, since our previous discussion shows that these two posteriors are coupled by the data. Nonetheless, we can hope that an approximation that decouples the two distributions will be more tractable.

Recall that, in our discussion of structured variational methods, we saw that the interactions between the structure of the approximation Q and the true distribution P can lead to further structural simplifications in Q (see section 11.5.2.4). Using these tools, we can find the following simplification.

Theorem 19.8

Let $P(\boldsymbol{\theta})$ be a parameter prior satisfying global parameter independence, $P(\boldsymbol{\theta}) = \prod_i P(\boldsymbol{\theta}_{X_i \mid U_i})$. Let $\mathcal{D}$ be a partially observable IID data set. If we consider a variational approximation with distributions satisfying $Q(\boldsymbol{\theta}, \mathcal{H}) = Q(\boldsymbol{\theta})Q(\mathcal{H})$, then Q can be decomposed as

$$Q(\boldsymbol{\theta}, \mathcal{H}) = \prod_i Q(\boldsymbol{\theta}_{X_i \mid U_i}) \prod_m Q(\boldsymbol{h}[m]).$$

The proof is by direct application of proposition 11.7 and is left as an exercise (exercise 19.24).

This theorem shows that, once we decouple the posteriors over the parameters and missing data, we also lose the coupling between components of the two distributions (that is, different parameters or different instances). Thus, we can further decompose each of the two posteriors into a product of independent terms. This result matches our intuition, since the coupling between the parameters and missing data was the source of dependence between components of the two distributions. That is, the posteriors of two parameters were dependent due to incomplete data, and the posterior of missing data in two instances were dependent due to uncertainty about the parameters.

This theorem does not necessarily justify the (strong) assumption of equation (19.11), but it does suggest that it provides significant computational gains. In this case, we see that we can assume that the approximate posterior also satisfies global parameter independence, and similarly the approximate distribution over $\mathcal{H}$ consists of independent posteriors, one per instance. This simplification already makes the representation of Q much more tractable. Other simplifications, following the same logic, are also possible.

The variational Bayes approach often gives rise to very natural update rules.

Example 19.13

Consider again the Bayesian clustering model of section 19.2.2.4. In this case, we aim to represent the posterior over the parameters $\boldsymbol{\theta}_H, \boldsymbol{\theta}_{X_1|H}, \ldots, \boldsymbol{\theta}_{X_n|H}$ and over the hidden variables $H[1], \ldots, H[M]$. The decomposition of theorem 19.8 allows us write Q as a product distribution, with a term for each of these variables. Thus, we have that

$$Q = Q(\boldsymbol{\theta}_H)\left[\prod_i Q(\boldsymbol{\theta}_{X_i|H})\right]\left[\prod_m Q(H[m])\right].$$

mean field

This factorization is essentially a mean field *approximation. Using the results of section 11.5.1, we see that the fixed-point equations for this approximation are of the form*

$$Q(\boldsymbol{\theta}_H) \quad \propto \quad \exp\left\{\ln P(\boldsymbol{\theta}_H) + \sum_m \boldsymbol{E}_{Q(H[m])}[\ln P(H[m] \mid \boldsymbol{\theta}_H)]\right\}$$

$$Q(\boldsymbol{\theta}_{X_i|H}) \quad \propto \quad \exp\left\{\ln P(\boldsymbol{\theta}_{X_i|H}) + \sum_m \boldsymbol{E}_{Q(H[m])}\big[\ln P(x_i[m] \mid H[m], \boldsymbol{\theta}_{X_i|H})\big]\right\}$$

$$Q(H[m]) \quad \propto \quad \exp\Big\{\boldsymbol{E}_{Q(\boldsymbol{\theta}_H)}[\ln P(H[m] \mid \boldsymbol{\theta}_H)]$$
$$+ \sum_i \boldsymbol{E}_{Q(\boldsymbol{\theta}_{X_i|H})}\big[\ln P(x_i[m] \mid H[m], \boldsymbol{\theta}_{X_i|H})\big]\Big\}.$$

The application of the mean-field theory allows us to identify the structure of the update equation. To provide a constructive solution, we also need to determine how to evaluate the expectations in these update equations. We now examine these expectations in the case where all the variables are binary and the priors over parameters are simple Dirichlet distributions (Beta distributions, in fact).

We start with the first fixed-point equation. A value for $\boldsymbol{\theta}_H$ is a pair $\langle \theta_{h^0}, \theta_{h^1} \rangle$. Using the definition of the Dirichlet prior, we have that

$$\ln P(\boldsymbol{\theta}_H = \langle \theta_{h^0}, \theta_{h^1} \rangle) = \ln c + (\alpha_{h^0} - 1)\ln \theta_{h^0} + (\alpha_{h^1} - 1)\ln \theta_{h^1},$$

where α_{h^0} and α_{h^1} are the hyperparameters of the prior $P(\boldsymbol{\theta}_H)$, and c is the normalizing constant of the prior (which we can ignore). Similarly, we can see that

$$\boldsymbol{E}_{Q(H[m])}[\ln P(H[m] \mid \boldsymbol{\theta}_H = \langle \theta_{h^0}, \theta_{h^1} \rangle)] = Q(H[m] = h^0)\ln \theta_{h^0} + Q(H[m] = h^1)\ln \theta_{h^1}.$$

Combining these results, we get that

$$Q(\boldsymbol{\theta}_H = \langle \theta_{h^0}, \theta_{h^1} \rangle) \quad \propto \quad \exp\Bigg\{\left(\alpha_{h^0} + \sum_m Q(H[m] = h^0) - 1\right)\ln \theta_{h^0} +$$
$$\left(\alpha_{h^1} + \sum_m Q(H[m] = h^1) - 1\right)\ln \theta_{h^1}\Bigg\}$$
$$= \quad \theta_{h^0}^{\alpha_{h^0} + \sum_m Q(H[m]=h^0) - 1}\theta_{h^1}^{\alpha_{h^1} + \sum_m Q(H[m]=h^1) - 1}.$$

In other words, $Q(\theta_H)$ is a Beta distribution with hyperparameters

$$
\alpha'_{h^0} = \alpha_{h^0} + \sum_m Q(H[m] = h^0)
$$

$$
\alpha'_{h^1} = \alpha_{h^1} + \sum_m Q(H[m] = h^1).
$$

Note that this is exactly the Bayesian update for θ_H with the expected sufficient statistics given $Q(\mathcal{H})$.

A similar derivation shows that $Q(\theta_{X_i \mid H})$ is also a pair of independent Beta distributions (one for each value of H) that are updated with the expected sufficient statistics given $Q(\mathcal{H})$.

M-step

E-step

These updates are reminiscent of the EM-update (M-step), since we use expected sufficient statistics to update the posterior. In the EM M-step, we update the MLE using the expected sufficient statistics. If we carry the analogy further, the last fixed-point equation, which updates $Q(H[m])$, corresponds to the E-step, since it updates the expectations over the missing values. Recall that, in the E-step of EM, we use the current parameters to compute

$$
Q(H[m]) = P(H[m] \mid x_1[m], \ldots x_n[m]) \propto P(H[m] \mid \theta_H) \prod_i P(x_i[m] \mid H[m], \theta_{X_i \mid H}).
$$

If we were doing a Bayesian approach, we would not simply take our current values for the parameters $\theta_H, \theta_{X_i \mid H}$; rather, we would average over their posteriors. Examining this last fixed-point equation, we see that we indeed average over the (approximate) posteriors $Q(\theta_H)$ and $Q(\theta_{X_i \mid H})$. However, unlike standard Bayesian averaging, where we compute the average value of the parameter itself, here we average its logarithm; that is, we evaluate terms of the form

$$
\boldsymbol{E}_{Q(\theta_{X_i \mid H})}\left[\ln P(x_i \mid H[m], \theta_{X_i \mid H})\right] = \int_0^1 Q(\theta_{x_i \mid H[m]}) \ln \theta_{x_i \mid H[m]} d\theta_{x_i \mid H[m]}.
$$

Using methods that are beyond the scope of this book, one can show that this integral has a closed-form solution:

$$
\boldsymbol{E}_{Q(\theta_{X_i \mid H})}\left[\ln P(x_i \mid H[m], \theta_{X_i \mid H})\right] = \varphi(\alpha'_{x_i \mid h}) - \varphi(\sum_{x'_i} \alpha'_{x'_i \mid h}),
$$

digamma
function

where α' are the hyperparameters of the posterior approximation in $Q(\theta_{X_i \mid H})$ and $\varphi(z) = (\ln \Gamma(z))' = \frac{\Gamma'(z)}{\Gamma(z)}$ is the digamma function, which is equal to $\ln(z)$ plus a polynomial function of $\frac{1}{z}$. And so, for $z \gg 1$, $\varphi(z) \approx \ln(z)$. Using this approximation, we see that

$$
\boldsymbol{E}_{Q(\theta_{X_i \mid H})}\left[\ln P(x_i \mid H[m], \theta_{X_i \mid H})\right] \approx \ln \frac{\alpha'_{x_i \mid h}}{\sum_{x'_i} \alpha'_{x'_i \mid h}},
$$

that is, the logarithm of the expected conditional probability according to the posterior $Q(\theta_{X_i \mid H})$. This shows that if the posterior hyperparameters are large the variational update is almost identical to EM's E-step.

To wrap up, we applied the structured variational approximation to the Bayesian learning problem. Using the tools we developed in previous chapters, we defined tractable fixed-point equations.

As with the mean field approximation we discussed in section 11.5.1, we can find a fixed-point solution for Q by iteratively applying these equations.

The resulting algorithm is very similar to applications of EM. Applications of the update equations for the parameters are almost identical to standard EM of section 19.2.2.4 in the sense that we use expected sufficient statistics. However, instead of finding the MLE parameters given these expected sufficient statistics, we compute the posterior assuming these were observed. The update for $Q(H[m])$ is reminiscent to the computation of $P(H[m])$ when we know the parameters. However, instead of using parameter values we use expectations of their logarithm and then take the exponent.   ∎

This example shows that we can find a variational approximation to the Bayesian posterior using an EM-like algorithm in which we iterate between updates to the parameter posteriors and updates to the missing data posterior. These ideas generalize to other network structures in a fairly straightforward way. The update for the posterior over parameter is similar to Bayesian update with expected sufficient statistics, and the update of the posterior over hidden variable is similar to a computation with the expected parameters (with the differences discussed earlier). In more complex examples we might need to make further assumptions about the distribution Q in order to get a tractable approximation. For example, if there are multiple missing values per instance, then we might not be able to afford to represent their distribution by the joint distribution and would instead need to introduce structure into Q. The basic ideas are similar to ones we explored before, and so we do not elaborate them. See exercise 15.6 for one example.

Of course, this method has some clear drawbacks. Because we are representing the parameter posterior by a factored distribution, we cannot expect to represent a multimodal posterior. Unfortunately, we know that the posterior is often multimodal. For example, in the clustering problem, we know that change in names of values of H would not change the prediction. Thus, the posterior in this example should be symmetric under such renaming. This implies that a unimodal distribution can only be a partial approximation to the true posterior. In multimodal cases, the effect of the variational approximation cannot be predicted. It may select one of the peaks and try to approximate it using Q, or it may choose a "broad" distribution that averages over some or all of the peaks.

19.4   Structure Learning

We now move to discuss the more complex task of learning the network structure as well as the parameters, again in the presence of incomplete data. Recall that in the case of complete data, we started by defining a score for evaluating different network structures and then examined search procedures that can maximize this score. As we will see, **both components of structure learning — the scoring function and the search procedure — are considerably more complicated in the case of incomplete data. Moreover, in the presence of hidden variables, even our search space becomes significantly more complex, since we now have to select the value space for the hidden variables, and even the number of hidden variables that the model contains.**

19.4.1 Scoring Structures

In section 18.3, we defined three scores: the likelihood score, the BIC score, and the Bayesian score. As we discussed, the likelihood score does not penalize more complex models, and it is therefore not useful when we want to compare between models of different complexity. Both the BIC and Bayesian score have built-in penalization for complex models and thus trade off the model complexity with its fit to the data. Therefore, they are far less likely to overfit.

We now consider how to extend these scores to the case when some of the data are missing. On the face of it, the score we want to evaluate is the same Bayesian score we considered in the case of complete data:

$$\text{score}_B(\mathcal{G} : \mathcal{D}) = \log P(\mathcal{D} \mid \mathcal{G}) + \log P(\mathcal{G})$$

where $P(\mathcal{D} \mid \mathcal{G})$ is the marginal likelihood of the data:

$$P(\mathcal{D} \mid \mathcal{G}) = \int_{\Theta_{\mathcal{G}}} P(\mathcal{D} \mid \boldsymbol{\theta}_{\mathcal{G}}, \mathcal{G}) P(\boldsymbol{\theta}_{\mathcal{G}} \mid \mathcal{G}) d\boldsymbol{\theta}_{\mathcal{G}}.$$

In the complete data case, the likelihood term inside the integral had a multiplicative factorization, and thus we could simplify it. In the case of incomplete data, the likelihood involves summing out over the unobserved variables, and thus it does not decompose.

As we discussed, we can view the computation of the marginal likelihood as an inference problem. For most learning problems with incomplete data, this inference problem is a difficult one. We now consider different strategies for dealing with this issue.

19.4.1.1 Laplace Approximation

One approach for approximating an integral in a high-dimensional space is to provide a simpler approximation to it, which we can then integrate in closed form. One such method is the *Laplace approximation*, described in box 19.F.

Laplace
approximation

Box 19.F — Concept: Laplace Approximation. *The Laplace approximation can be applied to any function of the form $f(\boldsymbol{w}) = e^{g(\boldsymbol{w})}$ for some vector $\boldsymbol{w}$. Our task is to compute the integral*

$$F = \int f(\boldsymbol{w}) d\boldsymbol{w}.$$

Using Taylor's expansion, we can expand an approximation of g around a point $\boldsymbol{w}_0$

$$g(\boldsymbol{w}) \approx g(\boldsymbol{w}_0) + \left[\frac{\partial g(\boldsymbol{w})}{\partial x_i}\right]\Bigg|_{\boldsymbol{w}=\boldsymbol{w}_0} (\boldsymbol{w} - \boldsymbol{w}_0) + \frac{1}{2}(\boldsymbol{w} - \boldsymbol{w}_0)^T \left[\frac{\partial \partial g(\boldsymbol{w})}{\partial x_i \partial x_j}\right]\Bigg|_{\boldsymbol{w}=\boldsymbol{w}_0} (\boldsymbol{w} - \boldsymbol{w}_0),$$

where $\left[\frac{\partial g(\boldsymbol{w})}{\partial x_i}\right]\Big|_{\boldsymbol{w}=\boldsymbol{w}_0}$ denotes the vector of first derivatives and $\left[\frac{\partial \partial g(\boldsymbol{w})}{\partial x_i \partial x_j}\right]\Big|_{\boldsymbol{w}=\boldsymbol{w}_0}$ denotes the

Hessian

Hessian — the matrix of second derivatives.

If $\boldsymbol{w}_0$ is the maximum of $g(\boldsymbol{w})$, then the second term disappears. We now set

$$C = -\left[\frac{\partial^2 g(\boldsymbol{w})}{\partial x_i \partial x_j}\right]\Bigg|_{\boldsymbol{w}=\boldsymbol{w}_0}$$

to be the negative of the matrix of second derivatives of $g(\boldsymbol{w})$ at $\boldsymbol{w}_0$. Since $\boldsymbol{w}_0$ is a maximum, this matrix is positive semi-definitive. Thus, we get the approximation

$$g(\boldsymbol{w}) \approx g(\boldsymbol{w}_0) - \frac{1}{2}(\boldsymbol{w} - \boldsymbol{w}_0)^T \boldsymbol{C}(\boldsymbol{w} - \boldsymbol{w}_0).$$

Plugging this approximation into the definition of $f(x)$, we can write

$$\int f(\boldsymbol{w})d\boldsymbol{w} \approx f(\boldsymbol{w}_0) \int e^{-\frac{1}{2}(\boldsymbol{w} - \boldsymbol{w}_0)^T \boldsymbol{C}(\boldsymbol{w} - \boldsymbol{w}_0)} d\boldsymbol{w}.$$

The integral is identical to the integral of an unnormalized Gaussian distribution with covariance matrix $\Sigma = C^{-1}$. We can therefore solve this integral analytically and obtain:

$$\int f(\boldsymbol{w})d\boldsymbol{w} \approx f(\boldsymbol{w}_0)|\boldsymbol{C}|^{-\frac{1}{2}}(2\pi)^{\frac{1}{2}\dim(\boldsymbol{C})},$$

where $\dim(C)$ is the dimension of the matrix $\boldsymbol{C}$.

At a high level, the Laplace approximation uses the value at the maximum and the curvature (the matrix of second derivatives) to approximate the integral of the function. This approximation works well when the function f is dominated by a single peak that has roughly a Gaussian shape.

How do we use the Laplace approximation in our setting? Taking g to be the log-likelihood function combined with the prior $\log P(\mathcal{D} \mid \boldsymbol{\theta}, \mathcal{G}) + \log P(\boldsymbol{\theta}|\mathcal{G})$, we get that $\log P(\mathcal{D}, \mathcal{G})$ can

Laplace score

be approximated by the *Laplace score*:

$$\text{score}_{Laplace}(\mathcal{G} \; : \; \mathcal{D}) = \log P(\mathcal{G}) + \log P(\mathcal{D} \mid \tilde{\boldsymbol{\theta}}_{\mathcal{G}}, \mathcal{G}) + \frac{\dim(\boldsymbol{C})}{2}\log 2\pi - \frac{1}{2}\log |\boldsymbol{C}|,$$

where $\tilde{\boldsymbol{\theta}}_{\mathcal{G}}$ are the MAP parameters and $\boldsymbol{C}$ is the negative of the Hessian matrix of the log-likelihood function. More precisely, the entries of $\boldsymbol{C}$ are of the form

$$-\left.\frac{\partial^2 \log P(\mathcal{D} \mid \boldsymbol{\theta}, \mathcal{G})}{\partial \theta_{x_i|\boldsymbol{u}_i} \partial \theta_{x_j|\boldsymbol{u}_j}}\right|_{\tilde{\boldsymbol{\theta}}_{\mathcal{G}}} = -\sum_m \left.\frac{\partial^2 \log P(\boldsymbol{o}[m] \mid \boldsymbol{\theta}, \mathcal{G})}{\partial \theta_{x_i|\boldsymbol{u}_i} \partial \theta_{x_j|\boldsymbol{u}_j}}\right|_{\tilde{\boldsymbol{\theta}}_{\mathcal{G}}},$$

where $\theta_{x_i|\boldsymbol{u}_i}$ and $\theta_{x_j|\boldsymbol{u}_j}$ are two parameters (not necessarily from the same CPD) in the parameterization of the network.

The Laplace score takes into account not only the number of free parameters but also the curvature of the posterior distribution in each direction. Although the form of this expression arises directly by approximating the posterior marginal likelihood, it is also consistent with our intuitions about the desired behavior. Recall that the parameter posterior is a concave function, and hence has a negative definitive Hessian. Thus, the negative Hessian C is positive definite and therefore has a positive determinant. A large determinant implies that the curvature at the MAP point is sharp; that is, the peak is relatively narrow and most of its mass is at the maximum. In this case, the model is probably overfitting to the training data, and we incur a large penalty. Conversely, if the curvature is small, the peak is wider, and the mass of the posterior is distributed over a larger set of parameters. In this case, overfitting is less likely, and, indeed, the Laplace score imposes a smaller penalty on the model.

To compute the Laplace score, we first need to use one of the methods we discussed earlier to find the MAP parameters of the distribution, and then compute the Hessian matrix. The computation of the Hessian is somewhat involved. To compute the entry for the derivative relative to $\theta_{x_i|u_i}$ and $\theta_{x_j|u_j}$, we need to compute the joint distribution over x_i, x_j, u_i, u_j given the observation; see exercise 19.9. Because these variables are not necessarily together in a clique (or cluster), the cost of doing such computations can be much higher than computing the likelihood. Thus, this approximation, while tractable, is still expensive in practice.

19.4.1.2 Asymptotic Approximations

One way of avoiding the high cost of the Laplace approximation is to approximate the term $|C|^{-\frac{1}{2}}$. Recall that the likelihood is the sum of the likelihood of each instance. Thus, the Hessian matrix is the sum of many Hessian matrixes, one per instance. We can consider asymptotic approximations that work well when the number of instances grows ($M \to \infty$). For this analysis, we assume that all data instances have the same observation pattern; that is, the set of variables $O[m] = O$ for all m.

Consider the matrix C. As we just argued, this matrix has the form

$$C = \sum_{m=1}^{M} C_m,$$

where C_m is the negative of the hessian of $\log P(o[m] \mid \theta, \mathcal{G})$. We can view each C_m as a sample from a distribution that is induced by the (random) choice of assignment o to O; each assignment o induces a different matrix C_o. We can now rewrite:

$$C = M \frac{1}{M} \sum_{m=1}^{M} C_m.$$

As M grows, the term $\frac{1}{M} \sum_{m=1}^{M} C_m$ approaches the expectation $E_{P^*}[C_o]$.

Taking the determinant of both sides, and recalling that $\det(\alpha A) = \alpha^{\dim(A)} \det(A)$, we get

$$\det(C) = M^{\dim(C)} \det\left(\frac{1}{M} \sum_{m=1}^{M} C_m\right) \approx M^{\dim(C)} \det(E_{P^*}[C_o]).$$

Taking logarithms of both sides, we get that

$$\log \det(C) \approx \dim(C) \log M + \log \det(E_{P^*}[C_o]).$$

Notice that the last term does not grow with M. Thus, when we consider the asymptotic behavior of the score, we can ignore it. This rough argument is the outline of the proof for the following result.

Theorem 19.9

As $M \to \infty$, we have that:

$$\text{score}_{Laplace}(\mathcal{G} : \mathcal{D}) = \text{score}_{BIC}(\mathcal{G} : \mathcal{D}) + O(1)$$

BIC score

where $\text{score}_{BIC}(\mathcal{G} : \mathcal{D})$ is the BIC score

$$\text{score}_{BIC}(\mathcal{G} : \mathcal{D}) = \log P(\mathcal{D} \mid \tilde{\theta}_{\mathcal{G}}, \mathcal{G}) - \frac{\log M}{2} \text{Dim}[\mathcal{G}] + \log P(\mathcal{G}) + \log P(\tilde{\theta}_{\mathcal{G}} \mid \mathcal{G}).$$

This result shows that the BIC score is an asymptotic approximation to the Laplace score, a conclusion that is interesting for several important reasons. First, it shows that the intuition we had for the case of complete data, where the score trades off the likelihood of the data with a structure penalty, still holds. Second, as in the complete data case, the asymptotic behavior of this penalty is logarithmic in the number of samples; this relationship implies the rate at which more instances can lead us to introduce new parameters.

independent
parameters

An important subtlety in this analysis is hidden in the use of the notation $\text{Dim}[\mathcal{G}]$. In the case of complete data, this notation stood for the number of *independent parameters* in the network, a quantity that we could easily compute. Here, it turns out that for some models, the actual number of degrees of freedom is smaller than the space of parameters. This implies that the matrix C is not of full rank, and so its determinant is 0. In such cases, we need to perform a variant of the Laplace approximation in the appropriate subspace, which leads to a determinant of a smaller matrix. The question of how to determine the right number of degrees of freedom (and thus the magnitude of $\text{Dim}[\mathcal{G}]$) is still an open problem.

19.4.1.3   Cheeseman-Stutz Approximation

We can use the Laplace/BIC approximations to derive an even tighter approximation to the Bayesian score. The intuition is that, in the case of complete data, the full Bayesian score was more precise than the BIC score since it took into account the extent to which each parameter was used and how its range of values influenced the likelihood. These considerations are explicit in the integral form of the likelihood and implicit in the closed-form solution of the integral. When we use the BIC score on incomplete data, we lose these fine-grained distinctions in evaluating the score.

Recall that the closed-form solution of the Bayesian score is a function of the sufficient statistics of the data. An ad hoc approach for constructing a similar (approximate) score when we have incomplete data is to apply the closed-form solution of the Bayesian score on some approximation of the statistics of the data. A natural choice would be the expected sufficient statistics given the MAP parameters. These expected sufficient statistics represent the completion of the data given our most likely estimate of the parameters.

More formally, for a network $\mathcal{G}$ and a set of parameters $\boldsymbol{\theta}$, we define $\mathcal{D}^*_{\mathcal{G},\boldsymbol{\theta}}$ to be a fictitious "complete" data set whose actual counts are the same as the *fractional* expected counts relative to this network; that is, for every event $\boldsymbol{x}$:

$$M_{\mathcal{D}^*_{\mathcal{G},\boldsymbol{\theta}}}[\boldsymbol{x}] = \bar{M}_{P(\mathcal{H}|\mathcal{D},\boldsymbol{\theta},\mathcal{G})}[\boldsymbol{x}]. \tag{19.12}$$

Because the expected counts are based on a coherent distribution, there can be such a data set (although it might have instances with fractional weights). To evaluate a particular network $\mathcal{G}$, we define the data set $\mathcal{D}^*_{\mathcal{G},\tilde{\boldsymbol{\theta}}_\mathcal{G}}$ induced by our network $\mathcal{G}$ and its MAP parameters $\tilde{\boldsymbol{\theta}}_\mathcal{G}$, and approximate the Bayesian score $P(\mathcal{D} \mid \mathcal{G})$ by $P(\mathcal{D}^*_{\mathcal{G},\tilde{\boldsymbol{\theta}}_\mathcal{G}} \mid \mathcal{G})$, using the standard integration over the parameters.

While straightforward in principle, a closer look suggests that this approximation cannot be a very good one. The first term,

$$P(\mathcal{D} \mid \mathcal{G}) = \int \sum_{\mathcal{H}} p(\mathcal{D}, \mathcal{H} \mid \boldsymbol{\theta}, \mathcal{G}) P(\boldsymbol{\theta} \mid \mathcal{G}) d\boldsymbol{\theta} = \sum_{\mathcal{H}} \int p(\mathcal{D}, \mathcal{H} \mid \boldsymbol{\theta}, \mathcal{G}) P(\boldsymbol{\theta} \mid \mathcal{G}) d\boldsymbol{\theta},$$

involves a summation of exponentially many integrals over the parameter space — one for each assignment to the hidden variables $\mathcal{H}$. On the other hand, the approximating term

$$P(\mathcal{D}^*_{\mathcal{G},\tilde{\theta}_{\mathcal{G}}} \mid \mathcal{G}) = \int p(\mathcal{D}^*_{\mathcal{G},\tilde{\theta}_{\mathcal{G}}} \mid \boldsymbol{\theta}, \mathcal{G}) P(\boldsymbol{\theta} \mid \mathcal{G}) d\boldsymbol{\theta}$$

is only a single such integral. In both terms, the integrals are over a "complete" data set, so that one of these sums is on a scale that is exponentially larger than the other.

One ad hoc solution is to simply correct for this discrepancy by estimating the difference:

$$\log P(\mathcal{D} \mid \mathcal{G}) - \log P(\mathcal{D}^*_{\mathcal{G},\tilde{\theta}_{\mathcal{G}}} \mid \mathcal{G}).$$

We use the asymptotic Laplace approximation to write each of these terms, to get:

$$
\begin{aligned}
\log P(\mathcal{D} \mid \mathcal{G}) - \log P(\mathcal{D}^*_{\mathcal{G},\tilde{\theta}_{\mathcal{G}}} \mid \mathcal{G}) &\approx \left[\log P(\mathcal{D} \mid \tilde{\boldsymbol{\theta}}_{\mathcal{G}}, \mathcal{G}) - \frac{1}{2} \mathrm{Dim}[\mathcal{G}] \log M \right] \\
&\quad - \left[\log P(\mathcal{D}^*_{\mathcal{G},\tilde{\theta}_{\mathcal{G}}} \mid \tilde{\boldsymbol{\theta}}_{\mathcal{G}}, \mathcal{G}) - \frac{1}{2} \mathrm{Dim}[\mathcal{G}] \log M \right] \\
&= \log P(\mathcal{D} \mid \tilde{\boldsymbol{\theta}}_{\mathcal{G}}, \mathcal{G}) - \log P(\mathcal{D}^*_{\mathcal{G},\tilde{\theta}_{\mathcal{G}}} \mid \tilde{\boldsymbol{\theta}}_{\mathcal{G}}, \mathcal{G}).
\end{aligned}
$$

The first of these terms is the log-likelihood achieved by the MAP parameters on the observed data. The second is the log-likelihood on the fictional data set, a term that can be computed in closed form based on the statistics of the fictional data set. We see that the first term is, again, a summation of an exponential number of terms representing different assignments to $\mathcal{H}$. We note that the Laplace approximation is valid only at the large sample limit, but more careful arguments can show that this construction is actually fairly accurate for a large class of situations.

Putting these arguments together, we can write:

$$
\begin{aligned}
\log P(\mathcal{D} \mid \mathcal{G}) &= \log P(\mathcal{D}^*_{\mathcal{G},\tilde{\theta}_{\mathcal{G}}} \mid \mathcal{G}) + \log P(\mathcal{D} \mid \mathcal{G}) - \log P(\mathcal{D}^*_{\mathcal{G},\tilde{\theta}_{\mathcal{G}}} \mid \mathcal{G}) \\
&\approx \log P(\mathcal{D}^*_{\mathcal{G},\tilde{\theta}_{\mathcal{G}}} \mid \mathcal{G}) + \log P(\mathcal{D} \mid \tilde{\boldsymbol{\theta}}_{\mathcal{G}}, \mathcal{G}) - \log P(\mathcal{D}^*_{\mathcal{G},\tilde{\theta}_{\mathcal{G}}} \mid \tilde{\boldsymbol{\theta}}_{\mathcal{G}}, \mathcal{G}).
\end{aligned}
$$

Cheeseman-Stutz score

This approximation is the basis for the *Cheeseman-Stutz score*:

$$\mathrm{score}_{CS}(\mathcal{G} : \mathcal{D}) = \log P(\mathcal{G}) + \log P(\mathcal{D}^*_{\mathcal{G},\tilde{\theta}_{\mathcal{G}}} \mid \mathcal{G}) + \log P(\mathcal{D} \mid \tilde{\boldsymbol{\theta}}_{\mathcal{G}}, \mathcal{G}) - \log P(\mathcal{D}^*_{\mathcal{G},\tilde{\theta}_{\mathcal{G}}} \mid \tilde{\boldsymbol{\theta}}_{\mathcal{G}}, \mathcal{G})$$

The appealing property of the Cheeseman-Stutz score is that, unlike the BIC score, it uses the closed-form solution of the complete data marginal likelihood in the context of incomplete data. Experiments in practice (see box 19.G) show that this score is much more accurate than the BIC score and much cheaper to evaluate than the Laplace score.

19.4.1.4 Candidate Method

candidate method

Another strategy for approximating the score is the *candidate method*; it uses a particular choice of parameters (the *candidate*) to evaluate the marginal likelihood. Consider any set of parameters $\boldsymbol{\theta}$. Using the chain law of probability, we can write $P(\mathcal{D}, \boldsymbol{\theta} \mid \mathcal{G})$ in two different ways:

$$
\begin{aligned}
P(\mathcal{D}, \boldsymbol{\theta} \mid \mathcal{G}) &= P(\mathcal{D} \mid \boldsymbol{\theta}, \mathcal{G}) P(\boldsymbol{\theta} \mid \mathcal{G}) \\
P(\mathcal{D}, \boldsymbol{\theta} \mid \mathcal{G}) &= P(\boldsymbol{\theta} \mid \mathcal{D}, \mathcal{G}) P(\mathcal{D} \mid \mathcal{G}).
\end{aligned}
$$

Equating the two right-hand terms, we can write

$$P(\mathcal{D} \mid \mathcal{G}) = \frac{P(\mathcal{D} \mid \boldsymbol{\theta}, \mathcal{G}) P(\boldsymbol{\theta} \mid \mathcal{G})}{P(\boldsymbol{\theta} \mid \mathcal{D}, \mathcal{G})}. \tag{19.13}$$

The first term in the numerator is the likelihood of the observed data given $\boldsymbol{\theta}$, which we should be able to evaluate using inference (exact or approximate). The second term in the numerator is the prior over the parameters, which is usually given. The denominator is the posterior over the parameters, the term most difficult to approximate.

The candidate method reduces the problem of computing the marginal likelihood to the problem of generating a reasonable approximation to the parameter posterior $P(\boldsymbol{\theta} \mid \mathcal{D}, \mathcal{G})$. It lets us estimate the likelihood when using methods such as MCMC sampling to approximate the posterior distribution. Of course, the quality of our approximation depends heavily on the design of the MCMC sampler. If we use a simple sampler, then the precision of our estimate of $P(\boldsymbol{\theta} \mid \mathcal{D}, \mathcal{G})$ will be determined by the number of sampled particles (since each particle either has these parameters or not). If, instead, we use collapsed particles, then each particle will have a distribution over the parameters, providing a better and smoother estimate for the posterior.

The quality of our estimate also depends on the particular choice of candidate $\boldsymbol{\theta}$. We can obtain a more robust estimate by averaging the estimates from several choices of candidate parameters (say several likely parameter assignments based on our simulations). However, each of these requires inference for computing the numerator in equation (19.13), increasing the cost.

An important property of the candidate method is that equation (19.13), on which the method is based, is not an approximation. Thus, if we could compute the denominator exactly, we would have an exact estimate for the marginal likelihood. This gives us the option of using more computational resources in our MCMC approximation to the denominator, to obtain increasingly accurate estimates. By contrast, the other methods all rely on an asymptotic approximation to the score and therefore do not offer a similar trade-off of accuracy to computational cost.

19.4.1.5   Variational Marginal Likelihood

A different approach to estimating the marginal likelihood is using the variational approximations we discussed in section 19.3.3. Recall from corollary 19.1 that, for any distribution Q,

$$\ell(\boldsymbol{\theta} : \mathcal{D}) = F_{\mathcal{D}}[\boldsymbol{\theta}, Q] + \boldsymbol{D}(Q(\mathcal{H}) \| P(\mathcal{H} \mid \mathcal{D}, \boldsymbol{\theta})).$$

It follows that the energy functional is a lower bound of the marginal likelihood, and the difference between them is the relative entropy between the approximate posterior distribution and the true one. Thus, if we find a good approximation Q of the posterior $P(\mathcal{H} \mid \mathcal{D}, \boldsymbol{\theta})$, then the relative entropy term is small, so that that energy functional is a good approximation of the marginal likelihood. As we discussed, the energy functional itself has the form:

$$F_{\mathcal{D}}[\boldsymbol{\theta}, Q] = \boldsymbol{E}_{\mathcal{H} \sim Q}[\ell(\boldsymbol{\theta} : \mathcal{D}, \mathcal{H})] + \boldsymbol{H}_Q(\mathcal{H}).$$

Both of these terms can be computed using inference relative to the distribution Q. Because this distribution was chosen to allow tractable inference, this provides a feasible approach for approximating the marginal likelihood.

Box 19.G — Case Study: Evaluating Structure Scores. *To study the different approximations to the Bayesian score in a restricted setting, Chickering and Heckerman (1997) consider learning a naive Bayes* mixture distribution, *as in section 19.2.2.4, where the cardinality K of the hidden variable (the number of mixture components) is unknown. Adding more values to the class variables increases the representational power of the model, but also introduces new parameters and thus increases the ability of the model to overfit the data. Since the class of distributions that are representable with a cardinality of K is contained within those that are representable with a cardinality of $K' > K$, the likelihood score increases monotonically with the cardinality of the class variable. The question is whether the different structure scores can pinpoint a good cardinality for the hidden variable. To do so, they perform MAP parameter learning on structures of different cardinality and then evaluate the different scores. Since the structure learning problem is one-dimensional (in the sense that the only parameter to learn is the cardinality of the class variable), there is no need to consider a specific search strategy in the evaluation.*

It is instructive to evaluate performance on both real data, and on synthetic data where the true number of clusters is known. However, even in synthetic data cases, where the true cardinality of the hidden variable is known, using this true cardinality as the "gold standard" for evaluating methods may not be appropriate, as with few data instances, the "optimal" model may be one with fewer parameters. Thus, Chickering and Heckerman instead compare all methods to the candidate method, using MCMC to evaluate the denominator; with enough computation, one can use this method to obtain high-quality approximations to the correct marginal likelihood.

The data in the synthetic experiments were generated from a variety of models, which varied along several axes: the true cardinality of the hidden variable (d); the number of observed variables (n); and the number of instances (M). The first round of experiments revealed few differences between the different scores. An analysis showed that this was because synthetic data sets with random parameter choices are too easy. Because of the relatively large number of observed variables, such random models always had distinguished clusters. That is, using the true parameters, the posterior probability $P(c \mid x_1, \ldots, x_n)$ is close to 1 for the true cluster value and 0 for all others. Thus, the instances belonging to different clusters are easily separable, making the learning problem too easy.

To overcome this problem, they considered sampling networks where the values of the parameters for $P(X_i \mid c)$ are correlated for different values of c. If the correlation is absolute, then the clusters overlap. For intermediate correlation the clusters were overlapping but not identical. By tuning the degree of correlation in sampling the distribution, they managed to generate networks with different degree of separation between the clusters. On data sets where the generating distribution did not have any separation between clusters, all the scores preferred to set the cardinality of the cluster variable to 1, as expected. When they examined data sets where the generating distribution had partial overlap between clusters they saw differentiating behavior between scoring methods. They also performed this same analysis on several real-world data sets. Figure 19.G.1 demonstrates the results for two data sets and summarizes the results for many of the synthetic data sets, evaluating the ability of the different methods to come close to the "optimal" cardinality, as determined by the candidate method.

Overall, the results suggest that BIC tends to underfit badly, almost always selecting models with an overly low cardinality for the hidden variable; moreover, its score estimate for models of higher (and more appropriate) cardinality tended to decrease very sharply, making it very unlikely that

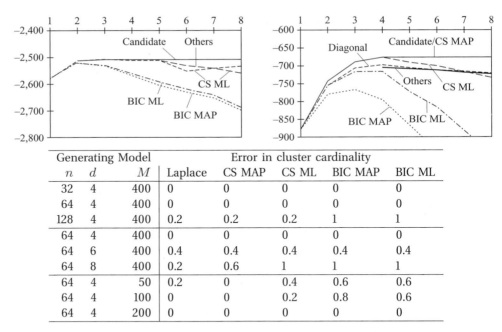

Figure 19.G.1 — Evaluation of structure scores for a naive Bayes clustering model In the graphs in the top row, the x axis denotes different cluster cardinalities, and the y axis the marginal likelihood estimated by the method. The graph on the left represents synthetic data with $d = 4$, $n = 128$, and $M = 400$. The graph on the right represents a real-world data set, with $d = 4$, $n = 35$ and $M = 47$. The table at bottom shows errors in model selection for the number of values of a hidden variable, as made by different approximations to the marginal likelihood. The errors are computed as differences between the cardinality selected by the method and the "optimal" cardinality selected by the "gold standard" candidate method. The errors are averaged over five data sets. The blocks of lines correspond to experiments where one of the three parameters defining the synthetic network varied while the others were held constant. (Adapted from Chickering and Heckerman (1997), with permission.)

they would be chosen. The other asymptotic approximations were all reasonably good, although all of them tended to underestimate the marginal likelihood as the number of clusters grows. A likely reason is that many of the clusters tend to become empty in this setting, giving rise to a "ridge-shaped" likelihood surface, where many parameters have no bearing on the likelihood. In this case, the "peak"-shaped estimate of the likelihood used by the asymptotic approximations tends to underestimate the true value of the integral. Among the different asymptotic approximations, the Cheeseman-Stutz approximation using the MAP configuration of the parameters had a slight edge over the other methods in its accuracy, and was more robust when dealing with parameters that are close to 0 or 1. It was also among the most efficient of the methods (other than the highly inaccurate BIC approach).

19.4.2 Structure Search

19.4.2.1 A Naive Approach

Given a definition of the score, we can now consider the structure learning task. In the most general terms, we want to explore the set of graph structures involving the variables of interest, score each one of these, and select the highest-scoring one. For some learning problems, such as the one discussed in box 19.G, the number of structures we consider is relatively small, and thus we can simply systematically score each structure and find the best one.

This strategy, however, is infeasible for most learning problems. Usually the number of structures we want to consider is very large — exponential or even superexponential in the number of variables — and we do not want to score all them. In section 18.4, we discussed various optimization procedures that can be used to identify a high-scoring structures. As we showed, for certain types of constraints on the network structure — tree-structured networks or a given node ordering (and bounded indegree) — we can actually find the optimal structure efficiently. In the more general case, we apply a hill-climbing procedure, using search operators that consist of local network modifications, such as edge addition, deletion, or reversal.

Unfortunately, the extension of these methods to the case of learning with missing data quickly hits a wall, since all of these methods relied on the decomposition of the score into a sum of individual family scores. This requirement is obvious in the case of learning tree-structured networks and in the case of learning with a fixed ordering: in both cases, the algorithm relied explicitly on the decomposition of the score as a sum of family scores.

The difficulty is a little more subtle in the case of the hill-climbing search. There, in each iteration, we consider applying $O(n^2)$ possible search operators (approximately 1–2 operators for each possible edge in the network). This number is generally quite large, so that the evaluation of the different possible steps in the search is a significant computational bottleneck. Although the same issue arises in the complete data case, there we could significantly reduce the computational burden due to two ideas. First, since the score is based on sufficient statistics, we could cache sufficient statistics and reuse them. Second, since the score is decomposable, the change in score of many of the operators is oblivious to modifications in another part of the network. Thus, as we showed in section 18.4.3.3, once we compute the delta-score of an operator o relative to a candidate solution $\mathcal{G}$:

$$\delta(\mathcal{G} \; : \; o) = \text{score}(o(\mathcal{G}) \; : \; \mathcal{D}) - \text{score}(\mathcal{G} \; : \; \mathcal{D}),$$

the same quantity is also the delta-score $\delta(\mathcal{G}' \; : \; o)$ for any other $\mathcal{G}'$ that is similar to $\mathcal{G}$ in the local topology that is relevant for o. For example, if o adds $X \to Y$, then the delta-score remains unchanged for any graph $\mathcal{G}'$ for which the family of Y is the same as in $\mathcal{G}$. The decomposition property implied that the search procedure in the case of complete data could maintain a priority queue of the effect of different search operators from previous iterations of the algorithm and avoid repeated computations.

When learning from incomplete data, the situation is quite different. As we discussed, local changes in the structure can result in global changes in the likelihood function. Thus, after a local structure change, the parameters of all the CPDs might change. As a consequence, the score is not decomposable; that is, the delta-score of one local modification (for example, adding an arc) can change after we modify a remote part of the network.

X_1	X_2	X_3	X_4	Count
0	0	0	1	2
0	0	1	0	1
0	1	0	0	2
0	1	0	1	1
0	1	1	0	8
0	1	1	1	1
1	0	0	0	3
1	0	0	1	42
1	0	1	0	10
1	0	1	1	11
1	1	0	1	15
1	1	1	0	3
1	1	1	1	1

(a)

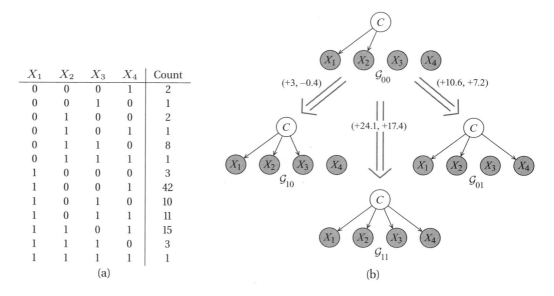

(b)

Figure 19.10 **Nondecomposability of structure scores in the case of missing data.** (a) A training set over variables $X_1, \ldots, X_4$. (b) Four possible networks over $X_1, \ldots, X_4$ and a hidden variable C. Arrows from the top network to the other three are labeled with the change in log-likelihood (LL) and Cheeseman-Stutz (CS) score, respectively. The baseline score (for $\mathcal{G}_{00}$) is: -336.5 for the LL score, and -360 for the CS score. We can see that the contribution of adding the arc $C \to X_3$ is radically different when X_4 is added as a child of C. This example shows that both the log-likelihood and the Cheeseman-Stutz score are not decomposable.

Example 19.14

Consider a task of learning a network structure for clustering, where we are also trying to determine whether different features are relevant. More precisely, assume we have a hidden variable C, and four possibly related variables $X_1, \ldots, X_4$. Assume that we have already decided that both X_1 and X_2 are children of C, and are trying to decide which (if any) of the edges $C \to X_3$ and $C \to X_4$ to include in the model, thereby giving rise to the four possible models in figure 19.10a. Our training set is as shown in figure 19.10b, and the resulting delta-scores relative to the baseline network $\mathcal{G}_{00}$ are shown in (c). As we can see, adding the edge $C \to X_3$ to the original structure $\mathcal{G}_{00}$ leads only to a small improvement in the likelihood, and a slight decrease in the Cheeseman-Stutz score. However, adding the edge $C \to X_3$ to the structure $\mathcal{G}_{01}$, where we also have $C \to X_4$, leads to a substantial improvement.

This example demonstrates a situation where the score is not decomposable. The intuition here is simple. In the structure $\mathcal{G}_{00}$, the hidden variable C is "tuned" to capture the dependency between X_1 and X_2. In this network structure, there is a weak dependency between these two variables and X_3. In $\mathcal{G}_{10}$, the hidden variable has more or less the same role, and therefore there is little explanatory benefit for X_3 in adding the edge to the hidden variable. However, when we add X_3 and X_4 together, the hidden variable shifts to capture the strong dependency between X_3 and X_4 while still capturing some of the dependencies between X_1 and X_2. Thus, the score improves *dramatically, and in a nonadditive manner.* ■

As a consequence of these problems, a search procedure that uses one of the scores we discussed has to evaluate the score of each candidate structure it considers, and it cannot rely on cached computations. In all of the scores we considered, this evaluation involves nontrivial computations (for example, running EM or an MCMC procedure) that are much more expensive than the cost of scoring a structure in the case of complete data. The actual cost of computation in these steps depends on the network structure (that is, the cost of inference in the network) and the number of iterations to convergence. Even in simple networks (for example, ones with a single hidden variable) this computation is an order of magnitude longer than evaluation of the score in complete data.

The main problem is that, in this type of search, most of the computation results are discarded. To understand why, recall that to select a move o from our current graph $\mathcal{G}$, we first evaluate all candidate successors $o(\mathcal{G})$. To evaluate each candidate structure $o(\mathcal{G})$, we compute the MLE or MAP parameters for $o(\mathcal{G})$, score it, and then compare it to the score of other candidates we consider at this point. Since we select to apply only one of the proposed search operators o at each iteration, the parameters learned for other structures $o'(\mathcal{G})$ are not needed. In practice, search using the modify-score-discard strategy is rarely feasible; it is useful only for learning in small domains, or when we have many constraints on the network structure, and so do not have many choices at each decision point.

19.4.2.2 Heuristic Solutions

There are several heuristics for avoiding this significant computational cost. We list a few of them here. We note that nondecomposability is a major issue in the context of Markov network learning, and so we return to these ideas in much greater length in section 20.7.

One approach is to construct "quick and dirty" estimates of the change in score. We can employ such estimates in a variety of ways. In one approach, we can simply use them as our estimates of the delta-score in any of the search algorithms used earlier. Alternatively, we can use them as a pruning mechanism, focusing our attention on the moves whose estimated change in score is highest and evaluating them more carefully. This approach uses the estimates to prioritize our computational resources and invest computation on careful evaluation of the real change in score for fewer modifications.

There are a variety of different approaches we can use to estimate the change in score. One approach is to use computations of delta-scores acquired in previous steps. More precisely, suppose we are at a structure $\mathcal{G}_0$, and evaluate a search operator whose delta-score is $\delta(\mathcal{G}_0 : o)$. We can then assume for the next rounds of search that the delta-score for this operator has not changed, even though the network structure has changed. This approach allows us to cache the results computation for at least some number of subsequent iterations (as though we are learning from complete data). In effect, this approach approximates the score as decomposable, at least for the duration of a few iterations. Clearly, this approach is only an approximation, but one that may be quite reasonable: even if the delta-scores themselves change, it is not unreasonable to assume that a step that was good relative to one structure is often probably good also relative to a closely related structure. Of course, this assumption can also break down: as the score is not decomposable, applying a set of "beneficial" search operators together can lead to structures with worse score.

The implementation of such a scheme requires us to make various decisions. How long do we maintain the estimate $\delta(\mathcal{G} : o)$? Which other search operators invalidate this estimate after they are applied? There is no clear right answer here, and the actual details of implementations of this heuristic approach differ on these counts.

Another approach is to compute the score of the modified network, but assume that only the parameters of the changed CPD can be optimized. That is, we freeze all of the parameters except those of the single CPD $P(X \mid \boldsymbol{U})$ whose family composition has changed, and optimize the parameters of $P(X \mid \boldsymbol{U})$ using gradient ascent or EM. When we optimize only the parameters of a single CPD, EM or gradient ascent should be faster for two reasons. First, because we have only a few parameters to learn, the convergence is faster. Second, because we modify only the parameters of a single CPD, we can cache intermediate computations; see exercise 19.15.

The set of parameterizations where only the CPD $P(X \mid \boldsymbol{U})$ is allowed to change is a subset of the set of all possible parameterizations for our network. Hence, any likelihood that we can achieve in this case would also be achievable if we ran a full optimization. As a consequence, the estimate of the likelihood in the modified network is a lower bound of the actual likelihood we can achieve if we can optimize all the parameters. If we are using a score such as the BIC score, this estimate is a proven lower bound on the score. If we are using a score such as the Cheeseman-Stutz score, this argument is not valid, but appears generally to hold in practice. That is, the score of the network with frozen parameters will be usually either smaller or very close to that of the one were we can optimize all parameters.

More generally, if a heuristic estimate is a proven lower bound or upper bound, we can improve the search procedure in a way that is guaranteed not to lose the optimal candidates. In the case of a lower bound, an estimated value that is higher than moves that we have already evaluated allows us to prune those other moves as guaranteed to be suboptimal. Conversely, if we have a move with a guaranteed upper bound that is lower than previously evaluated candidates, we can safely eliminate it. In practice, however, such bounds are hard to come by.

19.4.3   Structural EM

We now consider a different approach to constructing a heuristic that identifies helpful moves during the search. This approach shares some of the ideas that we discussed. However, by putting them together in a particular way, it provides significant computational savings, as well as certain guarantees.

19.4.3.1   Approximating the Delta-score

One efficient approach for approximating the score of network is to construct some complete data set $\mathcal{D}^*$, and then apply a score based on the complete data set. This was precisely the intuition that motivated the Cheeseman-Stutz approximation. However, the Cheeseman-Stutz approximation is computationally expensive. The data set we use — $\mathcal{D}^*_{\mathcal{G}, \tilde{\theta}_{\mathcal{G}}}$ — is constructed by finding the MAP parameters for our current candidate $\mathcal{G}$. We also needed to introduce a correction term that would improve the approximation to the marginal likelihood; this correction term required that we run inference over $\mathcal{G}$. Because these steps must be executed for each candidate network, this approach quickly becomes infeasible for large search spaces.

However, what if we do not want to obtain an accurate approximation to the marginal likelihood? Rather, we want only a heuristic that would help identify useful moves in the space. In this case, one simple heuristic is to construct a single *completed data set* $\mathcal{D}^*$ and use it to evaluate multiple different search steps. That is, to evaluate a search operator o, we define

completed data

$$\hat{\delta}_{\mathcal{D}^*}(\mathcal{G} : o) = \text{score}(o(\mathcal{G}) : \mathcal{D}^*) - \text{score}(\mathcal{G} : \mathcal{D}^*),$$

where we can use any complete-data scoring function for the two terms on the right-hand side. The key observation is that this expression is simply a delta-score relative to a complete data set, and it can therefore be evaluated very efficiently. We will return to this point.

Clearly, the results of applying this heuristic depend on our choice of completed data set $\mathcal{D}^*$, an observation that immediately raises some important questions: How do we define our completed data set $\mathcal{D}^*$? Can we provide any guarantees on the accuracy of this heuristic? One compelling answer to these questions is obtained from the following result:

Theorem 19.10

Let $\mathcal{G}_0$ be a graph structure and $\tilde{\theta}_0$ be the MAP parameters for $\mathcal{G}_0$ given a data set $\mathcal{D}$. Then for any graph structure $\mathcal{G}$:

$$\text{score}_{BIC}(\mathcal{G} : \mathcal{D}^*_{\mathcal{G}_0, \tilde{\theta}_0}) - \text{score}_{BIC}(\mathcal{G}_0 : \mathcal{D}^*_{\mathcal{G}_0, \tilde{\theta}_0}) \leq \text{score}_{BIC}(\mathcal{G} : \mathcal{D}) - \text{score}_{BIC}(\mathcal{G}_0 : \mathcal{D}).$$

This theorem states that the true improvement in the BIC score of network $\mathcal{G}$, relative to the network $\mathcal{G}_0$ that we used to construct our completed data $\mathcal{D}^*_{\mathcal{G}_0, \tilde{\theta}_0}$, is at least as large as the estimated improvement of the score using the completed data $\mathcal{D}^*_{\mathcal{G}_0, \tilde{\theta}_0}$.

The proof of this theorem is essentially the same as the proof of theorem 19.5; see exercise 19.25. Although the analogous result for the Bayesian score is not true (due to the nonlinearity of the Γ function used in the score), it is approximately true, especially when we have a reasonably large sample size; thus, we often apply the same ideas in the context of the Bayesian score, albeit without the same level of theoretical guarantees.

This result suggests the following scheme. Consider a graph structure $\mathcal{G}_0$. We compute its MAP parameters $\tilde{\theta}_0$, and construct a complete (fractional) data set $\mathcal{D}^*_{\mathcal{G}_0, \tilde{\theta}_0}$. We can now use the BIC score relative to this completed data set to evaluate the delta-score for any modification o to $\mathcal{G}$. We can thus define

$$\hat{\delta}_{\mathcal{D}^*_{\mathcal{G}_0, \tilde{\theta}_0}}(\mathcal{G} : o) = \text{score}_{BIC}(o(\mathcal{G}) : \mathcal{D}^*_{\mathcal{G}_0, \tilde{\theta}_0}) - \text{score}_{BIC}(\mathcal{G} : \mathcal{D}^*_{\mathcal{G}_0, \tilde{\theta}_0}).$$

The theorem guarantees that our heuristic estimate for the delta-score is a lower bound on the true change in the BIC score. The fact that this estimate is a lower bound is significant, since it guarantees that any change that we make that improves the estimated score will also improve the true score.

19.4.3.2 The Structural EM Algorithm

Importantly, the preceding guarantee holds not just for the application of a single operator, but also for any series of changes that modify $\mathcal{G}_0$. Thus, we can use our completed data set $\mathcal{D}^*_{\mathcal{G}_0, \tilde{\theta}_0}$ to estimate and apply an arbitrarily long sequence of operators to $\mathcal{G}_0$; as long as we have that

$$\text{score}_{BIC}(\mathcal{G} : \mathcal{D}^*_{\mathcal{G}_0, \tilde{\theta}_0}) > \text{score}_{BIC}(\mathcal{G}_0 : \mathcal{D}^*_{\mathcal{G}_0, \tilde{\theta}_0})$$

for our new graph $\mathcal{G}$, we are guaranteed that the true score of $\mathcal{G}$ is also better.

Algorithm 19.3 The structural EM algorithm for structure learning

    Procedure Structural-EM (
        $\mathcal{G}^0,$ // Initial bayesian network structure over $X_1, \ldots, X_n$
        $\boldsymbol{\theta}^0,$ // Initial set of parameters for $\mathcal{G}^0$
        $\mathcal{D}$ // Partially observed data set
)
1   **for** each $t = 0, 1 \ldots$, until convergence
2   // Optional parameter learning step
3           $\boldsymbol{\theta}^{t'} \leftarrow$ Expectation-Maximization$(\mathcal{G}^t, \boldsymbol{\theta}^t, \mathcal{D})$
4   // Run EM to generate expected sufficient statistics for $\mathcal{D}^*_{\mathcal{G}^t, \boldsymbol{\theta}^{t'}}$
5           $\mathcal{G}^{t+1} \leftarrow$ Structure-Learn$(\mathcal{D}^*_{\mathcal{G}^t, \boldsymbol{\theta}^{t'}})$
6           $\boldsymbol{\theta}^{t+1} \leftarrow$ Estimate-Parameters$(\mathcal{D}^*_{\mathcal{G}^t, \boldsymbol{\theta}^{t'}}, \mathcal{G}^{t+1})$
7   **return** $\mathcal{G}^t, \boldsymbol{\theta}^t$

However, we must take care in interpreting this guarantee. Assume that we have already modified $\mathcal{G}_0$ in several ways, to obtain a new graph $\mathcal{G}$. Now, we are considering a new operator o, and are interested in determining whether that operator is an improvement; that is, we wish to estimate the delta-score: $\text{score}_{BIC}(o(\mathcal{G}) : \mathcal{D}) - \text{score}_{BIC}(\mathcal{G} : \mathcal{D})$. The theorem tells us that if $o(\mathcal{G})$ satisfies $\text{score}_{BIC}(o(\mathcal{G}) : \mathcal{D}^*_{\mathcal{G}_0, \tilde{\boldsymbol{\theta}}_0}) > \text{score}_{BIC}(\mathcal{G}_0 : \mathcal{D}^*_{\mathcal{G}_0, \tilde{\boldsymbol{\theta}}_0})$, then it is necessarily better than our original graph $\mathcal{G}_0$. However, it does not follow that if $\hat{\delta}_{\mathcal{D}^*_{\mathcal{G}_0, \tilde{\boldsymbol{\theta}}_0}}(\mathcal{G} : o) > 0$, then $o(\mathcal{G})$ is necessarily better than $\mathcal{G}$. In other words, we can verify that each of the graphs we construct improves over the graph used to construct the completed data set, but not that each operator improves over the previous graph in the sequence. Note that we are guaranteed that our estimate is a true lower bound for any operator applied directly to $\mathcal{G}_0$. Intuitively, we believe that our estimates are likely to be reasonable for graphs that are "similar" to $\mathcal{G}_0$. (This intuition was also the basis for some of the heuristics described in section 19.4.2.2.) However, as we move farther away, our estimates are likely to degrade. Thus, at some point during our search, we probably want to select a new graph and construct a more relevant complete data set.

structural EM   This observation suggests an EM-like algorithm, called *structural EM*, shown in algorithm 19.3. In structural EM, we iterate over a pair of steps. In the E-step, we use our current model to generate (perhaps implicitly) a completed data set, based on which we compute expected sufficient statistics. In the M-step, we use these expected sufficient statistics to improve our model. The biggest difference is that now our M-step can improve not only the parameters, but also the structure. (Note that the structure-learning step also reestimates the parameters.) The structure learning procedure in the M-step can be any of the procedures we discussed in section 18.4, whether a general-purpose heuristic search or an exact search procedure for a specialized subset of networks for which we have an exact solution (for example, a maximum weighted spanning tree procedure for learning trees). If we use the BIC score, theorem 19.10 guarantees that, if this search procedure finds a structure that is better than the one we used in the previous iteration, then the structural EM procedure will monotonically improve the score.

Since the scores are upper-bounded, the algorithm must converge. Unlike the case of EM, we cannot, however, prove that the structure it finds is a local maximum.

19.4.3.3 Structural EM for Selective Clustering

We now illustrate the structural EM algorithm on a particular class of networks. Consider the task of learning structures that generalize our example of example 19.14; these networks are similar to the naive Bayes clustering of section 19.2.2.4, except that some observed variables may be independent of the cluster variable. Thus, in our structure, the class variable C is a root, and each observed attribute X_i is either a child of C or a root by itself. This limited set of structures contains 2^n choices.

Before discussing how to learn these structures using the ideas we just explored, let us consider why this problem is an interesting one. One might claim that, instead of structure learning, we can simply run parameter learning within the full structure (where each X_i is a child of C); after all, if X_i is independent of C, then we can capture this independence within the parameters of the CPD $P(X_i \mid C)$. However, as we discussed, statistical noise in the sampling process guarantees that we will never have true independence in the empirical distribution. Learning a more restricted model with fewer edges is likely to result in more robust clustering. Moreover, this approach allows us to detect irrelevant attributes during the clustering, providing insight into the domain.

If we have a complete data set, learning in this class of models is trivial. Since this class of structures is such that we cannot have cycles, if the score is decomposable, the choice of family for X_i does not impact the choice of parents for X_j. Thus, we can simply select the optimal family for each X_i separately: either C is its only parent, or it has no parents. We can thus select the optimal structure using $2n$ local score evaluations.

The structural EM algorithm applies very well in this setting. We initialize each iteration with our current structure $\mathcal{G}_t$. We then perform the following steps:

- Run parameter estimation (such as EM or gradient ascent) to learn parameters $\tilde{\boldsymbol{\theta}}_t$ for $\mathcal{G}_t$.

- Construct a new structure $\mathcal{G}_{t+1}$ so that $\mathcal{G}_{t+1}$ contains the edge $C \to X_i$ if

$$\mathrm{FamScore}(X_i \mid \{C\} \; : \; \mathcal{D}^*_{\mathcal{G}_t, \tilde{\boldsymbol{\theta}}_t}) > \mathrm{FamScore}(X_i \mid \emptyset \; : \; \mathcal{D}^*_{\mathcal{G}_t, \tilde{\boldsymbol{\theta}}_t}).$$

We continue this procedure until convergence, that is, until an iteration that makes no changes to the structure.

According to theorem 19.10, if we use the BIC score in this procedure, then any improvement to our expected score based on $\mathcal{D}^*_{\mathcal{G}_t, \tilde{\boldsymbol{\theta}}_t}$ is guaranteed to give rise to an improvement in the true BIC score; that is, $\mathrm{score}_{BIC}(\mathcal{G}_{t+1} \; : \; \mathcal{D}) \geq \mathrm{score}_{BIC}(\mathcal{G}_t \; : \; \mathcal{D})$. Thus, each iteration (until convergence) improves the score of the model.

One issue in implementing this procedure is how to evaluate the family scores in each iteration: $\mathrm{FamScore}(X_i \mid \emptyset \; : \; \mathcal{D}^*_{\mathcal{G}_t, \tilde{\boldsymbol{\theta}}_t})$ and $\mathrm{FamScore}(X_i \mid \{C\} \; : \; \mathcal{D}^*_{\mathcal{G}_t, \tilde{\boldsymbol{\theta}}_t})$. The first term depends on sufficient statistics for X_i in the data set; as X_i is fully observed, these can be collected once and reused in each iteration. The second term requires sufficient statistics of X_i

and C in $\mathcal{D}^*_{\mathcal{G}_t, \tilde{\boldsymbol{\theta}}_t}$; here, we need to compute:

$$\bar{M}_{\mathcal{D}^*_{\mathcal{G}_t, \tilde{\boldsymbol{\theta}}_t}}[x_i, c] = \sum_m P(C[m] = c, X_i[m] = x_i \mid \boldsymbol{o}[m], \mathcal{G}_t, \tilde{\boldsymbol{\theta}}_t)$$

$$= \sum_{m, X_i[m] = x_i} P(C[m] = c \mid \boldsymbol{o}[m], \mathcal{G}_t, \tilde{\boldsymbol{\theta}}_t).$$

We can collect all of these statistics with a singe pass over the data, where we compute the posterior over C in each instance. Note that these are the statistics we need for parameter learning in the full naive Bayes network, where each X_i is connected to C. In some of the iterations of the algorithm, we will compute these statistics even though X_i and C are independent in $\mathcal{G}_t$. Somewhat surprisingly, even when the joint counts of X_i and C are obtained from a model where these two variables are independent, the expected counts can show a dependency between them; see exercise 19.26.

Note that this algorithm can take very large steps in the space. Specifically, the choice of edges in each iteration is made from scratch, independently of the choice in the previous structure; thus, $\mathcal{G}_{t+1}$ can be quite different from $\mathcal{G}_t$. Of course, this observation is true only up to a point, since the use of the distribution based on $(\mathcal{G}_t, \tilde{\boldsymbol{\theta}}_t)$ does bias the reconstruction to favor some aspects of the previous iteration. This point goes back to the inherent nondecomposability of the score in this case, which we saw in example 19.14. To understand the limitation, consider the convergence point of EM for a particular graph structure where C has a particular set of children $\boldsymbol{X}$. At this point, the learned model is optimized so that C captures (as much as possible) the dependencies between its children in $\boldsymbol{X}$, to allow the variables in $\boldsymbol{X}$ to be conditionally independent given C. Thus, different choices of $\boldsymbol{X}$ will give rise to very different models. When we change the set of children, we change the information that C represents, and thus change the score in a global way. As a consequence, the choice of $\mathcal{G}_t$ that we used to construct the completed data does affect our ability to add certain edges into the graph.

This issue brings up the important question of how we can initialize this search procedure. A simple initialization point is to use the full network, which is essentially the naive Bayes clustering network, and let the search procedure prune edges. An alternative is to start with a random subset of edges. Such a randomized starting point can allow us to discover "local maxima" that are not accessible from the full network. One might also tempted to use the empty network as a starting point, and then consider adding edges. It is not hard to show, however, that the empty network is a bad starting point: structural EM will never add a new edge if we initialize the algorithm with the empty network; see exercise 19.27.

19.4.3.4 An Effective Implementation of Structural EM

Our description of the structural EM procedure is at a somewhat abstract level, and it lends itself to different types of implementations. The big unresolved issue in this description is how to represent and manage the completed data set created in the E-step. Recall that the number of completions of each instance is exponential in the number of missing values in that instance. If we have a single hidden variable, as in the selective naive Bayes classifier of section 19.4.3.3, then storing all completions (and their relative weights) might be a feasible implementation. However, if we have several unobserved variables in each instance, then this solution rapidly becomes impractical.

We can, however, exploit the fact that procedures that learn from complete data sets do not need to access all the instances; they require only sufficient statistics computed from the data set. Thus, we do not need to maintain all the instances of the completed data set; we need only to compute the relevant sufficient statistics in the completed data set. These sufficient statistics are, by definition, the expected sufficient statistics based on the current model $(\mathcal{G}_t, \boldsymbol{\theta}_t)$ and the observed data. This is precisely the same idea that we utilized in the E-step of standard EM for parameter estimation. However, there is one big difference. In parameter estimation, we know in advance the sufficient statistics we need. When we perform structure learning, this is no longer true. When we change the structure, we need a new set of sufficient statistics for the parts of the model we have changed. For example, if in the original network X is a root, then, for parameter estimation, we need only sufficient statistics of X alone. Now, if we consider adding Y as a parent of X, we need the joint statistics of X and Y together. If we do add the edge $Y \rightarrow X$, and now consider Z as an additional parent of X, we now need the joint statistics of X, Y, and Z.

This suggests that the number of sufficient statistics we may need can be quite large. One strategy is to compute in advance the set of sufficient statistics we might need. For specialized classes of structures, we may know this set exactly. For example, in the clustering scenario that we examined in section 19.4.3.3, we know the precise sufficient statistics that are needed for the M-step. Similarly, if we restrict ourselves to trees, we know that we are interested only in pairwise statistics and can collect all of them in advance. If we are willing to assume that our network has a bounded indegree of at most k, then we can also decide to precompute all sufficient statistics involving $k + 1$ or fewer variables; this approach, however, can be expensive for k greater than two or three. An alternative strategy is to compute sufficient statistics "on demand" as the search progresses through the space of different structures. This approach allows us to compute only the sufficient statistics that the search procedure requires. However, it requires that we revisit the data and perform new inference queries on the instances; moreover, this inference generally involves variables that are not together in a family and therefore may require out-of-clique inference, such as the one described in section 10.3.3.2.

Importantly, however, once we compute sufficient statistics, all of the decomposability properties for complete data that we discussed in section 18.4.3.3 hold for the resulting delta-scores. Thus, we can apply our caching-based optimizations in this setting, greatly increasing the computational efficiency of the algorithm. This property is key to allowing the structural EM algorithm to scale up to large domains with many variables.

19.5 Learning Models with Hidden Variables

In the previous section we examined searching for structures when the data are incomplete. In that discussion, we confined ourselves to structures involving a given set of variables. Although this set can include hidden variables, we implicitly assumed that we knew of the existence of these variables, and could simply treat them as an extreme case of missing data. Of course, it is important to remember that hidden variables introduce important subtleties, such as our inability to identify the model. Nevertheless, as we discussed, in section 16.4.2, hidden variables are useful for a variety of reasons.

In some cases, prior knowledge may tell us that a hidden variable belongs in the model, and perhaps even where we should place it relative to the other variables. In other cases (such as the naive Bayes clustering), the placement of the hidden variable is dictated by the goals of our learning (clustering of the instances into coherent groups). In still other cases, however, we may want to infer automatically that it would be beneficial to introduce a hidden variable into the model. This opportunity raises a whole range of new questions: When should we consider introducing a hidden variable? Where in the network should we connect it? How many values should we allow it to have?

In this section, we first present some results that provide intuition regarding the role that a hidden variable can play in the model. We then describe a few heuristics for dealing with some of the computational questions described before.

19.5.1   Information Content of Hidden Variables

One can view the role of a hidden variable as a mechanism for capturing information about the interaction between other variables in the network. In our example of the network of figure 16.1, we saw that the hidden variable "conveyed" information from the parents X_1, X_2, X_3 to the children Y_1, Y_2, Y_3. Similarly, in the naive Bayes clustering network of figure 3.2, the hidden variable captures information between its children. These examples suggest that, in

information
content

learning a model for the hidden variable, we want to maximize the *information* that the hidden variable captures about its children. We now show that learning indeed maximizes a notion of information between the hidden variable and its children. We analyze a specific example, the naive Bayes network of figure 3.2, but the ideas can be generalized to other network structures.

Suppose we observe M samples of $X_1, \ldots, X_n$ and use maximum likelihood to learn the parameters of the network. Any choice of parameter set $\boldsymbol{\theta}$ defines a distribution $\hat{Q}_{\boldsymbol{\theta}}$ over $X_1, \ldots, X_n, H$ so that

$$\hat{Q}_{\boldsymbol{\theta}}(h, x_1, \ldots, x_n) = \hat{P}_{\mathcal{D}}(x_1, \ldots, x_n) P(h \mid x_1, \ldots, x_n, \boldsymbol{\theta}), \tag{19.14}$$

where $\hat{P}_{\mathcal{D}}$ is the empirical distribution of the observed variables in the data. This is essentially the augmentation of the empirical distribution by our stochastic "reconstruction" of the hidden variable.

Consider for a moment a complete data set $\langle \mathcal{D}, \mathcal{H} \rangle$, where H is also observed. Proposition 18.1 shows that

$$\max_{\boldsymbol{\theta}} \frac{1}{M} \ell(\boldsymbol{\theta} : \langle \mathcal{D}, \mathcal{H} \rangle) = \sum_i \boldsymbol{I}_{\hat{P}_{\langle \mathcal{D}, \mathcal{H} \rangle}}(X_i; H) - \boldsymbol{H}_{\hat{P}_{\langle \mathcal{D}, \mathcal{H} \rangle}}(H) - \sum_i \boldsymbol{H}_{\hat{P}_{\langle \mathcal{D}, \mathcal{H} \rangle}}(X_i). \tag{19.15}$$

We now show that a similar relationship holds in the case of incomplete data; in fact, this relationship holds not only at the maximum likelihood point but also in other points of the parameter space:

Proposition 19.2

Let $\mathcal{D}$ be a data set where $X_1, \ldots, X_n$ are observed and $\boldsymbol{\theta}^0$ be a choice of parameters for the network of $figure$ 3.2. Define $\boldsymbol{\theta}^1$ to be the result of an EM-iteration if we start with $\boldsymbol{\theta}^0$ (that is, the result of an M-step if we use sufficient statistics from $\hat{Q}_{\boldsymbol{\theta}^0}$). Then

$$\frac{1}{M} \ell(\boldsymbol{\theta}^0 : \mathcal{D}) \leq \sum_i \boldsymbol{I}_{\hat{Q}_{\boldsymbol{\theta}^0}}(X_i; H) - \sum_i \boldsymbol{H}_{\hat{P}_{\mathcal{D}}}(X_i) \leq \frac{1}{M} \ell(\boldsymbol{\theta}^1 : \mathcal{D}). \tag{19.16}$$

Roughly speaking, this result states that the information-theoretic term is approximately equal to the likelihood. When $\boldsymbol{\theta}^0$ is a local maxima of the likelihood, we have that $\boldsymbol{\theta}^1 = \boldsymbol{\theta}^0$, and so we have equality in the left-hand and right-hand sides of equation (19.16). For other parameter choices, the information-theoretic term can be larger than the likelihood, but not by "too much," since it is bounded above by the next iteration of EM. Both the likelihood and the information-theoretic term have the same maxima.

Because the entropy terms $\boldsymbol{H}_{\hat{P}_{\mathcal{D}}}(X_i)$ do not depend on $\boldsymbol{\theta}$, this result implies that maximizing the likelihood is equivalent to finding a hidden variable H that maximizes the information about each of the observed variables. Note that the information here is defined in terms of the distribution $\hat{Q}_{\boldsymbol{\theta}^0}$, as in equation (19.14). This information measures what H conveys about each of the observed variables in the *posterior distribution* given the observations. This is quite intuitive: For example, assume we learn a model, and after observing $x_1, \ldots, x_n$, our posterior over H has $H = h^1$ with high probability. In this case, we are fairly sure about the cluster assignment of the cluster, so if the clustering is informative, we can conclude quite a bit of information about the value of each of the attributes.

Finally, it is useful to compare this result to the complete data case of equation (19.15); there, we had an additional $-\boldsymbol{H}(H)$ term, which accounts for the observations of H. In the case of incomplete data, we do not observe H and thus do not need to account for it. Intuitively, since we sum over all the possible values of H, we are not penalized for more complex (higher entropy) hidden variables. This difference also shows that adding more values to the hidden variable will always improve the likelihood. As we add more values, the hidden variable can only become more informative about the observed variables. Since our likelihood function does not include a penalty term for the entropy of H, this score does not penalize for the increased number of values of H.

We now turn to the proof of this proposition.

PROOF Define $Q(\mathcal{H}) = P(\mathcal{H} \mid \mathcal{D}, \boldsymbol{\theta}^0)$, then, by corollary 19.1 we have that

$$\ell(\boldsymbol{\theta}^0 : \mathcal{D}) = \boldsymbol{E}_Q\left[\ell(\boldsymbol{\theta}^0 : \langle \mathcal{D}, \mathcal{H}\rangle)\right] + \boldsymbol{H}_Q(\mathcal{H}).$$

Moreover, if $\boldsymbol{\theta}^1 = \arg\max_{\boldsymbol{\theta}} \boldsymbol{E}_Q[\ell(\boldsymbol{\theta} : \langle \mathcal{D}, \mathcal{H}\rangle)]$, then

$$\boldsymbol{E}_Q\left[\ell(\boldsymbol{\theta}^0 : \langle \mathcal{D}, \mathcal{H}\rangle)\right] \leq \boldsymbol{E}_Q\left[\ell(\boldsymbol{\theta}^1 : \langle \mathcal{D}, \mathcal{H}\rangle)\right].$$

Finally, we can use corollary 19.1 again and get that

$$\boldsymbol{E}_Q\left[\ell(\boldsymbol{\theta}^1 : \langle \mathcal{D}, \mathcal{H}\rangle)\right] + \boldsymbol{H}_Q(\mathcal{H}) \leq \ell(\boldsymbol{\theta}^1 : \mathcal{D}).$$

Combining these three inequalities, we conclude that

$$\ell(\boldsymbol{\theta}^0 : \mathcal{D}) \leq \boldsymbol{E}_Q\left[\ell(\boldsymbol{\theta}^1 : \langle \mathcal{D}, \mathcal{H}\rangle)\right] + \boldsymbol{H}_Q(\mathcal{H}) \leq \ell(\boldsymbol{\theta}^1 : \mathcal{D}).$$

Since $\boldsymbol{\theta}^1$ maximize the expected log-likelihood, we can apply equation (19.15) for the completed data set $\langle \mathcal{D}, \mathcal{H}\rangle$, and conclude that

$$\boldsymbol{E}_Q\left[\ell(\boldsymbol{\theta}^1 : \langle \mathcal{D}, \mathcal{H}\rangle)\right] = M\left[\sum_i \boldsymbol{E}_Q\left[\boldsymbol{I}_{\hat{P}_{\langle \mathcal{D}, \mathcal{H}\rangle}}(X_i; H)\right] - \boldsymbol{E}_Q\left[\boldsymbol{H}_{\hat{P}_{\langle \mathcal{D}, \mathcal{H}\rangle}}(H)\right] - \sum_i \boldsymbol{H}_{\hat{P}_{\mathcal{D}}}(X_i)\right].$$

Using basic rewriting, we have that

$$M\boldsymbol{E}_Q\left[\boldsymbol{H}_{\hat{P}_{\langle\mathcal{D},\mathcal{H}\rangle}}(H)\right] = \boldsymbol{H}_Q(\mathcal{H})$$

and that

$$\boldsymbol{E}_Q\left[\boldsymbol{I}_{\hat{P}_{\langle\mathcal{D},\mathcal{H}\rangle}}(X_i;H)\right] = \boldsymbol{I}_{\hat{Q}_{\theta^0}}(X_i;H),$$

which proves the result.   ∎

19.5.2   Determining the Cardinality

One of the key questions that we need to address for a hidden variable is that of its cardinality.

19.5.2.1   Model Selection for Cardinality

model selection
The simplest approach is to use *model selection*, where we consider a number of different cardinalities for H, and then select the best one. For our evaluation criterion, we can use a Bayesian technique, utilizing one of the approximate scores presented in section 19.4.1; box 19.G provides a comparative study of the different scores in precisely this setting. As another alternative, we can measure test generalization performance on a holdout set or using cross-validation. Both of these methods are quite expensive, even for a single hidden variable, since they both require that we learn a full model for each of the different cardinalities that we are considering; for multiple hidden variables, it is generally intractable.

A cheaper approach is to consider a more focused problem of using H to represent a clustering problem, where we cluster instances based only on the features $\boldsymbol{X}$ in H's (putative) Markov blanket. Here, the assumption is that if we give H enough expressive power to capture the distinctions between different classes of instances, we have captured much of the information in $\boldsymbol{X}$. We can now use any clustering algorithm to construct different clusterings and to evaluate their explanatory power. Commonly used variants are EM with a naive Bayes model, the simpler k-means algorithm, or any other of many existing clustering algorithms. We can now evaluate different cardinalities for H at much lower cost, using a score that measures only the quality of the local clustering. An even simpler approach is to introduce H with a low cardinality (say binary-valued), and then use subsequent learning stages to tell us whether there is still information in the vicinity of H. If there is, we can either increase the cardinality of H, or add another hidden variable.

19.5.2.2   Dirichlet Processes

Bayesian model
averaging
A very different alternative is a *Bayesian model averaging* approach, where we do not select a cardinality, but rather average over different possible cardinalities. Here, we use a prior over the set of possible cardinalities of the hidden variable, and use the data to define a posterior. **The Bayesian model averaging approach allows us to circumvent the difficult question of selecting the cardinality of the hidden variable. On the other side, because it fails to make a definitive decision on the set of clusters and on the assignment of instances to clusters, the results of the algorithm may be harder to interpret. Moreover, techniques**

that use Bayesian model averaging are generally computationally even more expensive than approaches that use model selection.

One particularly elegant solution is provided by the *Dirichlet process* approach. We provide an intuitive, bottom-up derivation for this approach, which also has extensive mathematical foundations; see section 19.7 for some references.

To understand the basic idea, consider what happens if we apply the approach of example 19.12 but allow the number of possible clusters K to grow very large, much larger than the number of data instances. In this case, the bound K does not limit the expressive power of model, since we can (in principle) put each instance in its own cluster.

Our natural concern about this solution is the possibility of overfitting: after all, we certainly do not want to put each point in its own cluster. However, recall that we are using a Bayesian approach and not maximum likelihood. To understand the difference, consider the posterior distribution over the $(M + 1)$'st instance given a cluster assignment for the previous instances $1, \ldots, M$. This formula is also proportional to equation (19.10), with $M + 1$ playing the role of m'. (The normalizing constant is different, because here we are also interested in modeling the distribution over $\boldsymbol{X}[M + 1]$, whereas there we took $\boldsymbol{x}[m']$ as given.) The first term in the equation, $|I_k(\boldsymbol{c})| + \alpha_0/K$, captures the relative prior probability that the $(M + 1)$'st instance selects to join cluster k. Note that the more instances are in the k'th cluster, the higher the probability that the new instance will select to join. Thus, the Dirichlet prior naturally causes instances to prefer to cluster together and thereby helps avoid overfitting.

A second concern is the computational burden of maintaining a very large number of clusters. Recall that if we use the collapsed Gibbs sampling approach of example 19.12, the cost per sampling step grows linearly with K. Moreover, most of this computation seems like a waste: with such a large K, many of the clusters are likely to remain empty, so why should we waste our time considering them?

The solution is to abstract the notion of a cluster assignment. Because clusters are completely symmetrical, we do not care about the specific assignment to the variables $C[m]$, but only about the *partition* of the instances into groups. Moreover, we can collapse all of the empty partitions, treating them all as equivalent. We therefore define a particle σ in our collapsed Gibbs process to encode a *partition* of the the data instances $\{1, \ldots, M\}$: an unordered set of non-empty subsets $\{I_1, \ldots, I_l\}$. Each $I \in \sigma$ is associated with a distribution over the parameters $\Theta_\sigma = \{\boldsymbol{\theta}_I\}_{I \in \sigma}$ and over the multinomial $\boldsymbol{\theta}$. As usual, we define $\sigma_{-m'}$ to denote the partition induced when we remove the instance m'.

To define the sampling process, let $C[m']$ be the variable to be sampled. Let L be the number of (non-empty) clusters in the partition $\sigma_{-m'}$. Introducing $C[m']$ (while keeping the other instances fixed) induces $L + 1$ possible partitions: joining one of the L existing clusters, or opening a new one. We can compute the conditional probabilities of each of these outcomes. Let $I \in \sigma$.

$$P(I \leftarrow I \cup \{m'\} \mid \sigma_{-m'}, \mathcal{D}, \boldsymbol{\omega}) \quad \propto \quad \left(|I| + \frac{\alpha_0}{K}\right) Q(\boldsymbol{x}[m'] \mid \mathcal{D}_I, \boldsymbol{\omega}) \tag{19.17}$$

$$P(\sigma \leftarrow \sigma \cup \{\{m'\}\} \mid \sigma_{-m'}, \mathcal{D}, \boldsymbol{\omega}) \quad \propto \quad (K - L)\frac{\alpha_0}{K} Q(\boldsymbol{x}[m'] \mid \boldsymbol{\omega}), \tag{19.18}$$

where the first line denotes the event where m' joins an existing cluster I, and the second the event where it forms a new singleton cluster (containing only m') that is added to σ. To compute these transition probabilities, we needed to sum over all possible concrete cluster assignments that are consistent with σ, but this computation is greatly facilitated by the symmetry of our prior (see exercise 19.29). Using abstract partitions as our particles provides significant computational savings: we need only to compute $L+1$ values for computing the transition distribution, rather than K, reducing the complexity of each Gibbs iteration to $O(NL)$, independent of the number of classes K.

As long as K is larger than the amount of data, it appears to play no real role in the model. Therefore, a more elegant approach is to remove it, allowing the number of clusters to grow to infinity. At the limit, the sampling equation for σ is now even simpler:

$$P(I \leftarrow I \cup \{m'\} \mid \sigma_{-m'}, \mathcal{D}, \boldsymbol{\omega}) \quad \propto \quad |I| \cdot Q(\boldsymbol{x}[m'] \mid \mathcal{D}_I, \boldsymbol{\omega}) \tag{19.19}$$

$$P(\sigma \leftarrow \sigma \cup \{\{m'\}\} \mid \sigma_{-m'}, \mathcal{D}, \boldsymbol{\omega}) \quad \propto \quad \alpha_0 \cdot Q(\boldsymbol{x}[m'] \mid \boldsymbol{\omega}). \tag{19.20}$$

This scheme removes the bound on the number of clusters and induces a prior that allows any possible partition of the samples. Given the data, we obtain a posterior over the space of possible partitions. This posterior gives positive probability to partitions with different numbers of clusters, thereby averaging over models with different complexity. In general, the number of clusters tends to grow logarithmically with the size of the data. This type of model is called a

nonparametric Bayesian model; see also box 17.B.

Of course, with K at infinity, our Dirichlet prior over $\boldsymbol{\theta}$ is not a legal prior. Fortunately, it turns out that one can define a generalization of the Dirichlet prior that induces these conditional probabilities. One simple derivation comes from a sampling process known as the *Chinese restaurant process*. This process generates a random partition as follows: The guests (instances) enter a restaurant one by one, and each guest chooses between joining one of the non-empty tables (clusters) and opening a new table. The probability that a customer chooses to join a table l at which n_l customers are already sitting is $\propto n_l$; the probability that he opens a new tables is $\propto \alpha_0$. The instances assigned to the same table all use the parameters of that table. It is not hard to show (exercise 19.30) that this prior induces precisely the update equations in equation (19.19) and equation (19.20).

A second derivation is called the *stick-breaking prior*; it is parameterized by α_0, and defines an infinite sequence of random variables $\beta_i \sim Beta(1, \alpha_0)$. We can now define an infinite-dimensional vector defined as:

$$\lambda_k = \beta_k \prod_{l=1}^{k-1} (1 - \beta_l).$$

This prior is called a stick-breaking prior because it can be viewed as defining a process of breaking a stick into pieces: We first break a piece of fraction β_1, then the second piece is a fraction β_2 of the remainder, and so on. It is not difficult to see that $\sum_k \lambda_k = 1$. It is also possible to show that, under the appropriate definitions, the limit of the distributions $Dirichlet(\alpha_0/K, \ldots, \alpha_0/K)$ as $K \longrightarrow \infty$ induces the stick-breaking prior.

19.5.3   Introducing Hidden Variables

Finally, we consider the question of determining when and where to introduce a hidden variable. The analysis of section 19.5.1 tells us that a hidden variable in a naive Bayes clustering network is optimized to capture information about the variables to which it is connected. Intuitively,

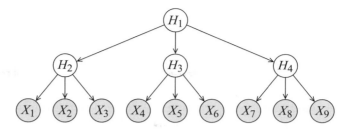

Figure 19.11   An example of a network with a hierarchy of hidden variables

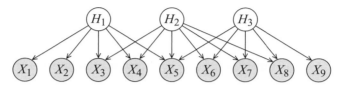

Figure 19.12   An example of a network with overlapping hidden variables

this requirement imposes a significant bias on the parameterization of the hidden variable. This bias helps constrain our search and allows us to learn a hidden variable that plays a meaningful role in the model. Conversely, if we place a hidden variable where the search is not similarly constrained, we run the risk of learning hidden variables that are meaningless, and that only capture the noise in the training data. As a rough rule of thumb, we want the model with the hidden variables to have a lot fewer independent parameters than the number of degrees of freedom of the empirical distribution.

Thus, when selecting network topologies involving hidden variables, we must exercise care. One useful example of such a class of topologies are organized *hierarchically* (for example, figure 19.11), where the hidden variables form a treelike hierarchy. Since each hidden variable is a parent of several other variables (either observed or hidden), it serves to mediate the dependencies among its children and between these children and other variables in the network (through its parent). This constraint implies that the hidden variable can improve the likelihood by capturing such dependencies when they exist. The general topology leaves much freedom in determining what is the best hierarchy structure. Intuitively, distance in the hierarchy should roughly correspond to the degree of dependencies between the variables, so that strongly dependent variables would be closer in the hierarchy. This rule is not exact, of course, since the nature of the dependencies influences whether the hidden variable can capture them.

Another useful class of networks are those with *overlapping* hidden variables; see figure 19.12. In this network, each hidden variable is the parent of several observed variables. The justification is that each hidden variable captures aspects of the instance that several of the observed variables depend on. Such a topology encodes multiple "flavors" of dependencies between the different variables, breaking up the dependency between them as some combination of independent axes. This approach often provides useful information about the structure of the domain. However, once we have an observed variable depending on multiple hidden ones, we might

hierarchical
organization

overlapping
organization

need to introduce many parameters, or restrict attention to some compact parameterization of the CPDs. Moreover, while the tree structure of hierarchical networks ensures efficient inference, overlapping hidden variables can result in a highly intractable structure.

In both of these approaches, as in others, we need to determine the placements of the hidden variables. As we discussed, once we introduce a hidden variable somewhere within our structure and localize it correctly in the model by connecting it to its correct neighbors, we can estimate parameters for it using EM. Even if we locate it approximately correctly, we can use the structural EM algorithm to adapt both the structure and the parameters.

☞   **However, we cannot simply place a hidden variable arbitrarily in the model and expect our learning procedures to learn a reasonable model.** Since these methods are based on iterative improvements, running structural EM with a bad initialization usually leads either to a trivial structure (where the hidden variable has few neighbors or disconnected from the rest of the variables) or to a structure that is very similar to the initial network structure. One extreme example of a bad initialization is to introduce a hidden variable that is disconnected from the rest of the variables; here, we can show that the variable will never be connected to the rest of the model (see exercise 19.27).

This discussion raises the important question of how to induce the existence of a hidden variable, and how to assign it a putative position within the network. One approach is based on finding "signatures" that the hidden variable might leave. As we discussed, a hidden variable captures dependencies between the variables to which it is connected. Indeed, if we assume that a hidden variable truly exists in the underlying distribution, we expect its neighbors in the graph to be dependent. For example, in figure 16.1, marginalizing the hidden variable induces correlations among its children and between its parents and its children (see also exercise 3.11). Thus, a useful heuristic is to look for subsets of variables that seem to be highly interdependent.

There are several approaches that one can use to find such subsets. Most obviously, we can learn a structure over the observed variables and then search for subsets that are connected by many edges. An obvious problem with this approach is that most learning methods are biased against learning networks with large indegrees, especially given limited data. Thus, these methods may return sparser structures even when dependencies may exist, preventing us from using the learned structure to infer the existence of a hidden variable. Nevertheless, these methods can be used successfully given a reasonably large number of samples. Another approach is to avoid the structure learning phase and directly consider the dependencies in the empirical distribution. For example, a quick-and-dirty method is to compute a measure of dependency, such as mutual information, between all pairs of variables. This approach avoids the need to examine marginals over larger sets of variables, and hence it is applicable in the case of limited data. However, we note that children of an observed variable will also be highly correlated, so that this approach does not distinguish between dependencies that can be explained by the observed variables and ones that require introducing hidden variables. Nevertheless, we can use this approach as a heuristic for introducing a hidden variable, and potentially employ a subsequent pruning phase to eliminate the variable if it is not helpful, given the observed variables.

19.6   Summary

In this chapter, we considered the problem of learning in the presence of incomplete data. We saw that learning from such data introduces several significant challenges.

One set of challenges involves the statistical interpretation of the learning problem in this setting. As we saw, we need to be aware of the process that generated the missing data and the effect of nonrandom observation mechanisms on the interpretation of the data. Moreover, we also need to be mindful of the possibility of unidentifiability in the models we learn, and as a consequence, to take care when interpreting the results.

A second challenge involves computational considerations. **Most of the key properties that helped make learning feasible in the fully observable case vanish in the partially observed setting. In particular, the likelihood function no longer decomposes, and is even multimodal. As a consequence, the learning task requires global optimization over a high-dimensional space, with an objective that is highly susceptible to local optima.**

We presented two classes of approaches for performing parameter estimation in this setting: a generic gradient-based process, and the EM algorithm, which is specifically designed for maximizing likelihood functions. Both of these methods perform hill climbing over the parameter space, and are therefore guaranteed only to find a local optimum (or rather, a stationary point) of the likelihood function. Moreover, each iteration in these algorithms requires that we solve an inference problem for each (partially observed) instance in our data set, a requirement that introduces a major computational burden.

In some cases, we want not only a single parameter estimate, but also some evaluation of our confidence in those estimates, as would be obtained from Bayesian learning. Clearly, given the challenges we mentioned, closed-form solutions to the integration are generally impossible. However, several useful approximations have been developed and used in practice; most commonly used are the methods based on MCMC methods, and on variational approximations.

We discussed the problem of structure learning in the partially-observed setting. **Most commonly used are the score-based approaches, where we define the problem as one of finding a high-scoring structure. We presented several approximations to the Bayesian score; most of these are based on an asymptotic approximation, and hence should be treated with care given only a small number of samples.** We then discussed the challenges of searching over the space of networks when the score is not decomposable, a setting that (in principle) forces us to apply a highly expensive evaluation procedure to every candidate that we are considering in the search. The structural EM algorithm provides one approach to reduce this cost. It uses an approximation to the score that is based on some completion of the data, allowing us to use the same efficient algorithms that we applied in the complete data case.

Finally, we briefly discussed some of the important questions that arise when we consider hidden variables: Where in the model should we introduce a hidden variable? What should we select as the cardinality of such a variables? And how do we initialize a variable so as to guide the learning algorithm toward "good" regions of the space? While we briefly described some ideas here, the methods are generally heuristic, and there are no guarantees.

Overall, owing to the challenges of this learning setting, the methods we discussed in this chapter are more heuristic and provide weaker guarantees than methods that we encountered in previous learning chapters. For this reason, the application of these methods is more of an art than a science, and there are often variations and alternatives that can be more effective for particular learning scenarios. This is an active area of study, and even for the simple clustering problem there is still much active research. Thus, we did not attempt to give a complete coverage and rather focused on the core methods and ideas.

       However, while these complications mean that learning from incomplete data is often challenging or even impossible, there are still many real-life applications where the methods we discussed here are highly effective. Indeed, the methods that we described here are some of the most commonly used of any in the book. They simply require that we take care in their application, and generally that we employ a fair amount of hand-tuned engineering.

19.7   Relevant Literature

The problem of statistical estimation from missing data has received a thorough treatment in the field of statistics. The distinction between the data generating mechanism and the observation mechanism was introduced by Rubin (1976) and Little (1976). Follow-on work defined the notion of MAR and MCAR Little and Rubin (1987). Similarly, the question of identifiability is also a central in statistical inference Casella and Berger (1990); Tanner (1993). Treatment of the subject for Bayesian networks appears in Settimi and Smith (1998a); Garcia (2004).

An early discussion that touches on the gradient of likelihood appears in Buntine (1994). Binder et al. (1997); Thiesson (1995) applied gradient methods for learning with missing values in Bayesian networks. They derived the gradient form and suggested how to compute it efficiently using clique tree calibration. Gradient methods are often more flexible in using models that do not have a closed-form MLE estimate even from complete data (see also chapter 20) or when using alternative objectives. For example, Greiner and Zhou (2002) suggest training using gradient ascent for optimizing conditional likelihood.

The framework of expectation maximization was introduced by Dempster et al. (1977), who generalized ideas that were developed independently in several related fields (for example, the Baum-Welch algorithm in hidden Markov models (Rabiner and Juang 1986)). The use of expectation maximization for maximizing the posterior was introduced by Green (1990). There is a wide literature of extensions of expectation maximization, analysis of convergence rates, and speedup methods; see McLachlan and Krishnan (1997) for a survey. Our presentation of the theoretical foundations of expectation maximization follows the discussion by Neal and Hinton (1998).

The use of expectation maximization in specific graphical models first appeared in various forms (Cheeseman, Self et al. 1988b; Cheeseman, Kelly et al. 1988; Ghahramani and Jordan 1993). Its adaptation for parameter estimation in general graphical models is due to Lauritzen (1995). Several approaches for accelerating EM convergence in graphical models were examined by Bauer et al. (1997) and Ortiz and Kaelbling (1999). The idea of incremental updates within expectation maximization was formulated by Neal and Hinton (1998). The application of expectation maximization for learning the parameters of noisy-or CPDs (or more generally CPDs with causal independence) was suggested by Meek and Heckerman (1997). The relationship between expectation maximization and hard-assignment EM was discussed by Kearns et al. (1997).

There are numerous applications of expectation maximization to a wide variety of problems. The collaborative filtering application of box 19.A is based on Breese et al. (1998). The application to robot mapping of box 19.D is due to Thrun et al. (2004).

There is a rich literature combining expectation maximization with different types of approximate inference procedures. Variational EM was introduced by Ghahramani (1994) and further elaborated by Ghahramani and Jordan (1997). The combination of expectation maximization with various types of belief propagation algorithms has been used in many current applications (see, for example, Frey and Kannan (2000); Heskes et al. (2003); Segal et al. (2001)). Similarly, other combinations have been examined in the literature, such as Monte Carlo EM (Caffo et al. 2005).

Applying Bayesian approaches with incomplete data requires approximate inference. A common solution is to use MCMC sampling, such as Gibbs sampling, using data-completion particles (Gilks et al. 1994). Our discussion of sampling the Dirichlet distribution is based on (Ripley 1987). More advanced sampling is based on the method of Fishman (1976) for sampling from Gamma distributions. Bayesian Variational methods were introduced by MacKay (1997); Jaakkola and Jordan (1997); Bishop et al. (1997) and further elaborated by Attias (1999); Ghahramani and Beal (2000). Minka and Lafferty (2002) suggest a Bayesian method based on expectation propagation.

The development of Laplace-approximation structure scores is based mostly on the presentation in Chickering and Heckerman (1997); this work is also the basis for the analysis of box 19.G. The BIC score was originally suggested by Schwarz (1978). Geiger et al. (1996, 2001) developed the foundations for the BIC score for Bayesian networks with hidden variables. This line of work was extended by several works (D. Rusakov 2005; Settimi and Smith 2000, 1998b). The Cheeseman-Stutz approximation was initially introduced for clustering models by Cheeseman and Stutz (1995) and later adapted for graphical models by Chickering and Heckerman (1997). Variational scores were suggested by Attias (1999) and further elaborated by Beal and Ghahramani (2006).

Search based on structural expectation maximization was introduced by Friedman (1997, 1998) and further discussed in Meila and Jordan (2000); Thiesson et al. (1998). The selective clustering example of section 19.4.3.3 is based on Barash and Friedman (2002). Myers et al. (1999) suggested a method based on stochastic search. An alternative approach uses *reversible jump MCMC* methods that perform Monte Carlo search through both parameter space and structure space (Green 1995). More recent proposals use Dirichlet processes to integrate over potential structures (Rasmussen 1999; Wood et al. 2006).

Introduction of hidden variables is a classic problem. Pearl (1988) suggested a method based on algebraic constraints in the distribution. The idea of using algebraic signatures of hidden variables has been proposed in several works (Spirtes et al. 1993; Geiger and Meek 1998; Robins and Wasserman 1997; Tian and Pearl 2002; Kearns and Mansour 1998). Using the structural signature was suggested by Martin and VanLehn (1995) and developed more formally by Elidan et al. (2000). Additional methods include hierarchical methods (Zhang 2004; Elidan and Friedman 2005), the introduction of variables to capture temporal correlations (Boyen et al. 1999), and introduction of variables in networks of continuous variables (Elidan et al. 2007).

reversible jump MCMC

19.8 Exercises

Exercise 19.1
Consider the estimation problem in example 19.4.

a. Provide upper and lower bounds on the maximum likelihood estimate of θ.
b. Prove that your bounds are tight; that is, there are values of $\psi_{O_X|x^1}$ and $\psi_{O_X|x^0}$ for which these estimates are equal to the maximum likelihood.

Exercise 19.2★

Suppose we have a given model $P(\boldsymbol{X} \mid \boldsymbol{\theta})$ on a set of variable $\boldsymbol{X} = \{X_1, \ldots, X_n\}$, and some incomplete data. Suppose we introduce additional variables $\boldsymbol{Y} = \{Y_1, \ldots, Y_n\}$ so that Y_i has the value 1 if X_i is observed and 0 otherwise. We can extend the data, in the obvious way, to include complete observations of the variables $\boldsymbol{Y}$. Show how to augment the model to build a model $P(\boldsymbol{X}, \boldsymbol{Y} \mid \boldsymbol{\theta}, \boldsymbol{\theta}') = P(\boldsymbol{X} \mid \boldsymbol{\theta})P(\boldsymbol{Y} \mid \boldsymbol{X}, \boldsymbol{\theta}')$ so that it satisfies the *missing at random* assumption.

missing at
random

Exercise 19.3

Consider the problem of applying EM to parameter estimation for a variable X whose local probabilistic model is a tree-CPD. We assume that the network structure $\mathcal{G}$ includes the structure of the tree-CPDs in it, so that we have a structure $\mathcal{T}$ for X. We are given a data set $\mathcal{D}$ with some missing values, and we want to run EM to estimate the parameters of $\mathcal{T}$. Explain how we can adapt the EM algorithm in order to accomplish this task. Describe what expected sufficient statistics are computed in the E-step, and how parameters are updated in the M-step.

Exercise 19.4

Consider the problem of applying EM to parameter estimation for a variable X whose local probabilistic model is a noisy-or. Assume that X has parents $Y_1, \ldots, Y_k$, so that our task for X is to estimate the noise parameters $\lambda_0, \ldots, \lambda_k$. Explain how we can use the EM algorithm to accomplish this task. (Hint: Utilize the structural decomposition of the noisy-or node.)

Exercise 19.5

Prove theorem 19.2. (Hint: use lemma 19.1.)

Exercise 19.6

Suppose we are using a gradient method to learn parameters for a network with table-CPDs. Let X be one of the variables in the network with parents $\boldsymbol{U}$. One of the constraints we need to maintain is that

$$\sum_x \theta_{x|\boldsymbol{u}} = 1$$

for every assignment $\boldsymbol{u}$ for $\boldsymbol{U}$. Given the gradient $\frac{\partial}{\partial \theta_{x|\boldsymbol{u}}} \ell(\boldsymbol{\theta} : \mathcal{D})$, show how to project it to null space of this constraint. That is, show how to find a gradient direction that maximizes the likelihood while preserving this constraint.

Exercise 19.7

Suppose we consider reparameterizing table-CPDs using the representation of equation (19.3). Use the chain law of partial derivatives to find the form of $\frac{\partial}{\partial \lambda_{x|\boldsymbol{u}}} \ell(\boldsymbol{\theta} : \mathcal{D})$.

Exercise 19.8★

Suppose we have a Bayesian network with table-CPDs. Apply the method of Lagrange multipliers to characterize the maximum likelihood solution under the constraint that each conditional probability sums to one. How does your characterization relate to EM?

Exercise 19.9★

We now examine how to compute the Hessian of the likelihood function. Recall that the Hessian of the log-likelihood is the matrix of second derivatives. Assume that our model is a Bayesian network with table-CPDs.

a. Prove that the second derivative of the likelihood of an observation $\boldsymbol{o}$ is of the form:

$$\frac{\partial^2 \log P(\boldsymbol{o})}{\partial \theta_{x_i|\boldsymbol{u}_i} \partial \theta_{x_j|\boldsymbol{u}_j}} = \frac{1}{\theta_{x_i|\boldsymbol{u}_i} \theta_{x_j|\boldsymbol{u}_j}} \left[P(x_i, \boldsymbol{u}_i, x_j, \boldsymbol{u}_j \mid \boldsymbol{o}) - P(x_i, \boldsymbol{u}_i \mid \boldsymbol{o})P(x_j, \boldsymbol{u}_j \mid \boldsymbol{o}) \right].$$

b. What is the cost of computing the full Hessian matrix of $\log P(\boldsymbol{o})$ if we use clique tree propagation?

c. What is the computational cost if we are only interested in entries of the form

$$\frac{\partial^2}{\partial \theta_{x_i \mid \boldsymbol{u}_i} \partial \theta_{x_i' \mid \boldsymbol{u}_i'}} \log P(\boldsymbol{o});$$

that is, we are interested in the "diagonal band" that involves only second derivatives of entries from the same family?

Exercise 19.10★

a. Consider the task of estimating the parameters of a univariate Gaussian distribution $\mathcal{N}\left(\mu; \sigma^2\right)$ from a data set $\mathcal{D}$. Show that if we maximize likelihood subject to the constraint $\sigma^2 \geq \epsilon$ for some $\epsilon > 0$, then the likelihood $L(\mu, \sigma^2 : \mathcal{D})$ is guaranteed to remain bounded.

b. Now, consider estimating the parameters of a multivariate Gaussian $\mathcal{N}\left(\boldsymbol{\mu}; \Sigma\right)$ from a data set $\mathcal{D}$. Provide constraints on Σ that achieve the same guarantee.

Exercise 19.11★

Consider learning the parameters of the network $H \rightarrow X, H \rightarrow Y$, where H is a hidden variable. Show that the distribution where $P(H), P(X \mid H), P(Y \mid H)$ are uniform is a stationary point of the likelihood (gradient is 0). What does that imply about gradient ascent and EM starting from this point?

Exercise 19.12

Prove theorem 19.5. Hint, note that $\ell(\boldsymbol{\theta}^t : \mathcal{D}) = F_{\mathcal{D}}[\boldsymbol{\theta}^t, P(\mathcal{H} \mid \mathcal{D}, \boldsymbol{\theta}^t)]$, and use corollary 19.1.

Exercise 19.13

Consider the task of learning the parameters of a DBN with table-CPDs from a data set with missing data. In particular, assume that our data set consists of a sequence of observations $\boldsymbol{o}_0^{(0)}, \boldsymbol{o}_1^{(1)}, \ldots, \boldsymbol{o}_t^{(T)}$. (Note that we do not assume that the same variables are observed in every time-slice.)

a. Describe precisely how you would run EM in this setting to estimate the model parameters; your algorithm should specify exactly how we run the E-step, which sufficient statistics we compute and how, and how the sufficient statistics are used within the M-step.

b. Given a single trajectory, as before, which of the network parameters might you be able to estimate?

Exercise 19.14★

Show that, until convergence, each iteration of hard-assignment EM increases $\ell(\boldsymbol{\theta} : \langle \mathcal{D}, \mathcal{H} \rangle)$.

Exercise 19.15★

Suppose that we have an incomplete data set $\mathcal{D}$, and network structure $\mathcal{G}$ and matching parameters. Moreover, suppose that we are interested in learning the parameters of a single CPD $P(X_i \mid \boldsymbol{U}_i)$. That is, we assume that the parameters we were given for all other families are frozen and do not change during the learning. This scenario can arise for several reasons: we might have good prior knowledge about these parameters; or we might be using an incremental approach, as mentioned in box 19.C (see also exercise 19.16).

We now consider how this scenario can change the computational cost of the EM algorithm.

a. Assume we have a clique tree for the network $\mathcal{G}$ and that the CPD $P(X_i \mid \boldsymbol{U}_i)$ was assigned to clique $\boldsymbol{C}_j$. Analyze which messages change after we update the parameters for $P(X_i \mid \boldsymbol{U}_i)$. Use this analysis to show how, after an initial precomputation step, we can perform iterations of this single-family EM procedure with a computational cost that depends only on the size of $\boldsymbol{C}_j$ and not the size of the rest of the cluster tree.

b. Would this conclusion change if we update the parameters of several families that are all assigned to the same cluster in the cluster tree?

Exercise 19.16★

We can build on the idea of the single-family EM procedure, as described in exercise 19.15, to define an *incremental EM* procedure for learning all the parameters in the network. In this approach, at each step we optimize the parameters of a single CPD (or several CPDs) while freezing the others. We then iterate between these local EM runs until all families have converged.

Is this modified version of EM still guaranteed to converge? In other words, does

$$\ell(\boldsymbol{\theta}^{t+1} : \mathcal{D}) \geq \ell(\boldsymbol{\theta}^t : \mathcal{D})$$

still hold? If so, prove the result. If not, explain why not.

Exercise 19.17★

We now consider how to use the interpretation of the EM as maximizing an energy functional to allow partial or incremental updates over the instances. Consider the EM algorithm of algorithm 19.2. In the Compute-ESS we collect the statistics from all the instances. This requires running inference on all the instances.

We now consider a procedure that performs partial updates where it update the expected sufficient statistics for some, but not all, of the instances. In particular, suppose we replace this procedure by one that runs inference on a single instance and uses the update to replace the old contribution of the instance with a new one; see algorithm 19.4. This procedure, instead of computing all the expected sufficient statistics in each E-step, caches the contribution of each instance to the sufficient statistics, and then updates only a single one in each iteration.

a. Show that the incremental EM algorithm converges to a fixed point of the log-likelihood function. To do so, show that each iteration improves the EM energy functional. Hint: you need to define what is the effect of the partial E-step on the energy functional.

b. How would that analysis generalize if in each iteration the algorithm performs a partial update for k instances (instead of 1)?

c. Assume that the computations in the M-step are relatively negligible compared to the inference in the E-step. Would you expect the incremental EM to be more efficient than standard EM? If so, why?

Exercise 19.18★

Consider the model described in box 19.D.

a. Assume we perform the E-step for each step $\boldsymbol{x}_m$ by defining

$$\tilde{P}(\boldsymbol{x}_m \mid C_m = k : \boldsymbol{\theta}_k) = \mathcal{N}\left(d(\boldsymbol{x}, p_k) \mid 0; \sigma^2\right)$$

and $\tilde{P}(\boldsymbol{x}_m \mid C_m = 0 : \boldsymbol{\theta}_k) = C$ for some constant C. Why is this formula not a correct application of EM? (Hint: Consider the normalizing constants.)

We note that although this approach is mathematically not quite right, it seems to be a reasonable approximation that works in practice.

b. Given a solution to the E-step, show how to perform maximum likelihood estimation of the model parameters $\boldsymbol{\alpha}_k, \beta_k$, subject to the constraint that $\boldsymbol{\alpha}_k$ be a unit-vector, that is, that $\boldsymbol{\alpha}_k \cdot \boldsymbol{\alpha}_k = 1$. (Hint: Use Lagrange multipliers.)

Algorithm 19.4 The incremental EM algorithm for network with table-CPDs

---

    Procedure Incremental-E-Step (
      $\boldsymbol{\theta}$,   // Parameters for update
      m,   // instance to update
)
1   Run inference on $\langle \mathcal{G}, \boldsymbol{\theta} \rangle$ using evidence $\boldsymbol{o}[m]$
2   **for** each $i = 1, \ldots, n$
3   **for** each $x_i, \boldsymbol{u}_i \in Val(X_i, \mathrm{Pa}_{X_i}^{\mathcal{G}})$
4   // Remove old contribution
5        $\bar{M}[x_i, \boldsymbol{u}_i] \leftarrow \bar{M}[x_i, \boldsymbol{u}_i] - \bar{M}_m[x_i, \boldsymbol{u}_i]$
6   // Compute new contribution
7        $\bar{M}_m[x_i, \boldsymbol{u}_i] \leftarrow P(x_i, \boldsymbol{u}_i \mid \boldsymbol{o}[m])$
8        $\bar{M}[x_i, \boldsymbol{u}_i] \leftarrow \bar{M}[x_i, \boldsymbol{u}_i] + \bar{M}_m[x_i, \boldsymbol{u}_i]$

    Procedure Incremental-EM (
      $\mathcal{G}$,   // Bayesian network structure over $X_1, \ldots, X_n$
      $\boldsymbol{\theta}^0$,   // Initial set of parameters for $\mathcal{G}$
      $\mathcal{D}$ // Partially observed data set
)
1   **for** each $i = 1, \ldots, n$
2   **for** each $x_i, \boldsymbol{u}_i \in Val(X_i, \mathrm{Pa}_{X_i}^{\mathcal{G}})$
3        $\bar{M}[x_i, \boldsymbol{u}_i] \leftarrow 0$
4   **for** each $m = 1 \ldots M$
5          $\bar{M}_m[x_i, \boldsymbol{u}_i] \leftarrow 0$
6   // Initialize the expected sufficient statistics
7   **for** each $m = 1 \ldots M$
8   Incremental-E-Step$(\mathcal{G}, \boldsymbol{\theta}^0, \mathcal{D}, m)$
9    $m \leftarrow 1$
10   **for** each $t = 0, 1 \ldots$, until convergence
11   // E-step
12   Incremental-E-Step$(\mathcal{G}, \boldsymbol{\theta}^t, \mathcal{D}, m)$
13     $m \leftarrow (m \bmod M) + 1$
14   // M-step
15   **for** each $i = 1, \ldots, n$
16   **for** each $x_i, \boldsymbol{u}_i \in Val(X_i, \mathrm{Pa}_{X_i}^{\mathcal{G}})$
17       $\theta_{x_i \mid \boldsymbol{u}_i}^{t+1} \leftarrow \frac{\bar{M}[x_i, \boldsymbol{u}_i]}{\bar{M}[\boldsymbol{u}_i]}$
18   **return** $\boldsymbol{\theta}^t$

Exercise 19.19★

Consider the setting of exercise 12.29, but now assume that we cannot (or do not wish to) maintain a distribution over the A_j's. Rather, we want to find the assignment $a_1^*, \ldots, a_m^*$ for which $P(a_1, \ldots, a_m)$ is maximized.

In this exercise, we address this problem using the EM algorithm, treating the values $a_1, \ldots, a_m$ as parameters. In the E-step, we compute the expected value of the C_i variables; in the M-step, we maximize the value of the a_j's given the distribution over the C_j's.

a. Describe how one can implement this EM procedure exactly, that is, with no need for approximate inference.
b. Why is approximate inference necessary in exercise 12.29 but not here? Give a precise answer in terms of the properties of the probabilistic model.

Exercise 19.20

Suppose that a prior on a parameter vector is $p(\boldsymbol{\theta}) \sim Dirichlet(\alpha_1, \ldots, \alpha_k)$. Derive $\frac{\partial}{\partial \theta_i} \log p(\boldsymbol{\theta})$.

Exercise 19.21

Consider the generalization of the EM procedure to the task of finding the MAP parameters. Let

$$\tilde{F}_{\mathcal{D}}[\boldsymbol{\theta}, Q] = F_{\mathcal{D}}[\boldsymbol{\theta}, Q] + \log P(\boldsymbol{\theta}).$$

a. Prove the following result:

Corollary 19.2

$\overline{\textit{For a distribution } Q, \text{score}_{\text{MAP}}(\boldsymbol{\theta} \; : \; \mathcal{D}) \geq \tilde{F}_{\mathcal{D}}[\boldsymbol{\theta}, Q] \textit{ with equality if only if } Q(\mathcal{H}) = P(\mathcal{H} \mid \mathcal{D}, \boldsymbol{\theta}).}$

b. Show that a coordinate ascent approach on $\tilde{F}_{\mathcal{D}}[\boldsymbol{\theta}, Q]$ requires only changing the M-step to perform MAP rather than ML estimation, that is, to maximize:

$$\boldsymbol{E}_Q[\ell(\boldsymbol{\theta} : \langle \mathcal{D}, \mathcal{H} \rangle)] + \log P(\boldsymbol{\theta}).$$

c. Using exercise 17.12, provide a specific formula for the M-step in a network with table-CPDs.

Exercise 19.22

In this case, we analyze the use of collapsed Gibbs with data completion particles for the purpose of sampling from a posterior in the case of incomplete data.

a. Consider first the simple case of example 19.12. Assuming that the data instances $\boldsymbol{x}$ are sampled from a discrete naive Bayes model with a Dirichlet prior, derive a closed form for equation (19.9).
b. Now, consider the general case of sampling from $P(\mathcal{H} \mid \mathcal{D})$. Here, the key step would involve sampling from the distribution

$$P(X_i[m] \mid \langle \mathcal{D}, \mathcal{H} \rangle_{-X_i[m]}) \propto P(X_i[m], \langle \mathcal{D}, \mathcal{H} \rangle_{-X_i[m]}),$$

where $\langle \mathcal{D}, \mathcal{H} \rangle_{-X_i[m]}$ is a complete data set from which the observation of $X_i[m]$ is removed.
Assuming we have table-CPD and independent Dirichlet priors over the parameters, derive this conditional probability from the form of the marginal likelihood of the data. Show how to use sufficient statistics of the particle to perform this sampling efficiently.

Exercise 19.23★

We now consider a Metropolis-Hastings sampler for the same setting as exercise 19.22. For simplicity, we assume that the same variables are hidden in each instance. Consider the proposal distribution for variable X_i specified in algorithm 19.5. (We are using a multiple-transition chain, as in section 12.3.2.4, where each variable has its own kernel.) In this proposal distribution, we resample a value for X_i in all of the instances, based on the current parameters and the completion for all the other variables.

Derive the form of the acceptance probability for this proposal distribution. Show how to use sufficient statistics of the completed data to evaluate this acceptance probability efficiently.

Algorithm 19.5 Proposal distribution for collapsed Metropolis-Hastings over data completions

---

       Procedure Proposal-Distribution (
          $\mathcal{G}$,   // Bayesian network structure over $X_1, \ldots, X_n$
          $\mathcal{D}$ // completed data set
          X_i // A variable to sample
)
1       $\boldsymbol{\theta} \leftarrow$ Estimate-Parameters$(\mathcal{D}, \mathcal{G})$
2       $\mathcal{D}' \leftarrow \mathcal{D}$
3   **for** each $m = 1 \ldots M$
4   Sample $x_i'[m]$ from $P(X_i[m] \mid \boldsymbol{x}_{-i}[m], \boldsymbol{\theta})$
5   **return** $\mathcal{D}'$

Exercise 19.24

Prove theorem 19.8.

Exercise 19.25★

Prove theorem 19.10. Hint: Use the proof of theorem 19.5.

Exercise 19.26

Consider learning structure in the setting discussed in section 19.4.3.3. Describe a data set $\mathcal{D}$ and parameters for a network where X_1 and C are independent, yet the expected sufficient statistics $\bar{M}[X_1, C]$ show dependency between X_1 and C.

Exercise 19.27

Consider using the structural EM algorithm to learn the structure associated with a hidden variable H; all other variables are fully observed. Assume that we start our learning process by performing an E-step in a network where H is not connected to any of $X_1, \ldots, X_n$. Show that, for any initial parameter assignment to $P(H)$, the SEM algorithm will not connect H to the rest of the variables in the network.

Exercise 19.28

Consider the task of learning a model involving a binary-valued hidden variable H using the EM algorithm. Assume that we initialize the EM algorithm using parameters that are symmetric in the two values of H; that is, for any variable X_i that has H has a parent, we have $P(X_i \mid \boldsymbol{U}_i, h^0) = P(X_i \mid \boldsymbol{U}_i, h^1)$. Show that, with this initialization, the model will remain symmetric in the two values of H, over all EM iterations.

Exercise 19.29

Derive the sampling update equations for the partition-based Gibbs sampling of equation (19.17) and equation (19.18) from the corresponding update equations over particles defined as ground assignments (equation (19.10)). Your update rules must sum over all assignments consistent with the partition.

Exercise 19.30

Consider the distribution over partitions induced by the Chinese restaurant process.

a. Find a closed-form formula for the probability induced by this process for any partition σ of the guests. Show that this probability is invariant to the order the guests enter the restaurant.

b. Show that a Gibbs sampling process over the partitions generated by this algorithm satisfies equation (19.19) and equation (19.20).

Algorithm 19.6 Proposal distribution over partitions in the Dirichlet process priof

---

    Procedure DP-Merge-Split-Proposal (
       σ // A partition
)
1   Uniformly choose two different instances m, l
2   **if** m, l are assigned to two different clusters I, I' **then**
3   // Propose partition that merges the two clusters
4          $\sigma' \leftarrow \sigma - \{I, I'\} \cup \{I \cup I'\}$
5   **else**
6   Let I be the cluster to which m, l are both assigned
7   // Propose to randomly split I so as to separate them
8          $I_1 \leftarrow \{m\}$
9          $I_2 \leftarrow \{l\}$
10   **for** $n \in I$
11   Add n to I_1 with probability 0.5 and to I_2 with probability 0.5
12        $\sigma' \leftarrow \sigma - \{I\} \cup \{I_1, I_2\}$
13   **return** (σ')

Exercise 19.31⋆

Algorithm 19.6 presents a Metropolis-Hastings proposal distribution over partitions in the Dirichlet process prior. Compute the acceptance probability of the proposed move.

20 *Learning Undirected Models*

20.1 Overview

In previous chapters, we developed the theory and algorithms for learning Bayesian networks from data. In this chapter, we consider the task of learning Markov networks. Although many of the same concepts and principles arise, the issues and solutions turn out to be quite different.

☞ **Perhaps the most important reason for the differences is a key distinction between Markov networks and Bayesian networks: the use of a global normalization constant (the partition function) rather than local normalization within each CPD. This global factor couples all of the parameters across the network, preventing us from decomposing the problem and estimating local groups of parameters separately.** This global parameter coupling has significant computational ramifications. As we will explain, in contrast to the situation for Bayesian networks, even simple (maximum-likelihood) parameter estimation with complete data cannot be solved in closed form (except for chordal Markov networks, which are therefore also Bayesian networks). Rather, we generally have to resort to iterative methods, such as gradient ascent, for optimizing over the parameter space. The good news is that the likelihood objective is concave, and so these methods are guaranteed to converge to the global optimum. The bad news is that each of the steps in the iterative algorithm requires that we run inference on the network, making even simple parameter estimation a fairly expensive, or even intractable, process. Bayesian estimation, which requires integration over the space of parameters, is even harder, since there is no closed-form expression for the parameter posterior. Thus, the integration associated with Bayesian estimation must be performed using approximate inference (such as variational methods or MCMC), a burden that is often infeasible in practice.

As a consequence of these computational issues, much of the work in this area has gone into the formulation of alternative, more tractable, objectives for this estimation problem. Other work has been focused on the use of approximate inference algorithms for this learning problem and on the development of new algorithms suited to this task.

The same issues have significant impact on structure learning. In particular, because a Bayesian parameter posterior is intractable to compute, the use of exact Bayesian scoring for model selection is generally infeasible. In fact, scoring any model (computing the likelihood) requires that we run inference to compute the partition function, greatly increasing the cost of search over model space. Thus, here also, the focus has been on approximations and heuristics that can reduce the computational cost of this task. Here, however, there is some good news, arising from another key distinction between Bayesian and Markov networks: the lack of a global acyclicity constraint in undirected models. Recall (see theorem 18.5) that the acyclicity constraint couples decisions regarding the family of different variables, thereby making the structure selection problem much harder. The lack of such a global constraint in the undirected case eliminates these interactions, allowing us to choose the local structure locally in different parts of the network. In particular, it turns out that a particular variant of the structure learning task can be formulated as a continuous, convex optimization problem, a class of problems generally viewed as tractable. Thus, elimination of global acyclicity removes the main reason for the $\mathcal{NP}$-hardness of structure learning that we saw in Bayesian networks. However, this does *not* make structure learning of Markov networks efficient; the convex optimization process (as for parameter estimation) still requires multiple executions of inference over the network.

A final important issue that arises in the context of Markov networks is the overwhelmingly common use of these networks for settings, such as image segmentation and others, where we have a particular inference task in mind. In these settings, we often want to train a network *discriminatively* (see section 16.3.2), so as to provide good performance for our particular prediction task. Indeed, much of Markov network learning is currently performed for CRFs.

The remainder of this chapter is structured as follows. We begin with the analysis of the properties of the likelihood function, which, as always, forms the basis for all of our discussion of learning. We then discuss how the likelihood function can be optimized to find the maximum likelihood parameter estimates. The ensuing sections discuss various important extensions to these basic ideas: conditional training, parameter priors for MAP estimation, structure learning, learning with missing data, and approximate learning methods that avoid the computational bottleneck of multiple iterations of network inference. These extensions are usually described as building on top of standard maximum-likelihood parameter estimation. However, it is important to keep in mind that they are largely orthogonal to each other and can be combined. Thus, for example, we can also use the approximate learning methods in the case of structure learning or of learning with missing data. Similarly, all of the methods we described can be used with maximum conditional likelihood training. We return to this issue in section 20.8.

We note that, for convenience and consistency with standard usage, we use natural logarithms throughout this chapter, including in our definitions of entropy or KL-divergence.

20.2   The Likelihood Function

As we saw in earlier chapters, the key component in most learning tasks is the likelihood function. In this section, we discuss the form of the likelihood function for Markov networks, its properties, and their computational implications.

20.2.1   An Example

As we suggested, the existence of a global partition function couples the different parameters in a Markov network, greatly complicating our estimation problem. To understand this issue, consider the very simple network A—B—C, parameterized by two potentials $\phi_1(A, B)$ and $\phi_2(B, C)$. Recall that the log-likelihood of an instance $\langle a, b, c \rangle$ is

$$\ln P(a, b, c) = \ln \phi_1(a, b) + \ln \phi_2(b, c) - \ln Z,$$

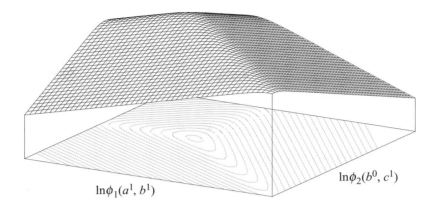

Figure 20.1 **Log-likelihood surface for the Markov network** $A\!-\!B\!-\!C$, **as a function of** $\ln \phi_1(a^1, b^1)$ (*x*-**axis**) **and** $\ln \phi_2(b^0, c^1)$ (*y*-**axis**); all other parameters in both potentials are set to 1. Surface is viewed from the $(+\infty, +\infty)$ point toward the $(-, -)$ quadrant. The data set $\mathcal{D}$ has $M = 100$ instances, for which $M[a^1, b^1] = 40$ and $M[b^0, c^1] = 40$. (The other sufficient statistics are irrelevant, since all of the other log-parameters are 0.)

where Z is the partition function that ensures that the distribution sums up to one. Now, consider the log-likelihood function for a data set $\mathcal{D}$ containing M instances:

$$\ell(\boldsymbol{\theta} : \mathcal{D}) = \sum_m \left(\ln \phi_1(a[m], b[m]) + \ln \phi_2(b[m], c[m]) - \ln Z(\boldsymbol{\theta}) \right)$$

$$= \sum_{a,b} M[a, b] \ln \phi_1(a, b) + \sum_{b,c} M[b, c] \ln \phi_2(b, c) - M \ln Z(\boldsymbol{\theta}).$$

Thus, we have sufficient statistics that summarize the data: the joint counts of variables that appear in each potential. This is analogous to the situation in learning Bayesian networks, where we needed the joint counts of variables that appear within the same family. This likelihood consists of three terms. The first term involves ϕ_1 alone, and the second term involves ϕ_2 alone. The third term, however, is the log-partition function $\ln Z$, where:

$$Z(\boldsymbol{\theta}) = \sum_{a,b,c} \phi_1(a, b)\phi_2(b, c).$$

Thus, $\ln Z(\boldsymbol{\theta})$ is a function of both ϕ_1 and ϕ_2. As a consequence, it *couples* the two potentials in the likelihood function.

Specifically, consider maximum likelihood estimation, where we aim to find parameters that maximize the log-likelihood function. In the case of Bayesian networks, we could estimate each conditional distribution independently of the other ones. Here, however, when we change one of the potentials, say ϕ_1, the partition function changes, possibly changing the value of ϕ_2 that maximizes $-\ln Z(\boldsymbol{\theta})$. Indeed, as illustrated in figure 20.1, the log-likelihood function in our simple example shows clear dependencies between the two potentials.

In this particular example, we can avoid this problem by noting that the network $A\!-\!B\!-\!C$ is equivalent to a Bayesian network, say $A \to B \to C$. Therefore, we can learn the parameters of this BN, and then define $\phi_1(A, B) = P(A)P(B \mid A)$ and $\phi_2(B, C) = P(C \mid B)$. Because the two representations have equivalent expressive power, the same maximum likelihood is achievable in both, and so the resulting parameterization for the Markov network will also be a maximum-likelihood solution. In general, however, there are Markov networks that do not have an equivalent BN structure, for example, the diamond-structured network of figure 4.13 (see section 4.5.2). In such cases, we generally cannot convert a learned BN parameterization into an equivalent MN; indeed, the optimal likelihood achievable in the two representations is generally not the same.

20.2.2   Form of the Likelihood Function

To provide a more general description of the likelihood function, it first helps to provide a more convenient notational basis for the parameterization of these models. For this purpose,

log-linear model   we use the framework of *log-linear models*, as defined in section 4.4.1.2. Given a set of features $\mathcal{F} = \{f_i(\boldsymbol{D}_i)\}_{i=1}^{k}$, where $f_i(\boldsymbol{D}_i)$ is a feature function defined over the variables in $\boldsymbol{D}_i$, we have:

$$P(X_1, \ldots, X_n : \boldsymbol{\theta}) = \frac{1}{Z(\boldsymbol{\theta})} \exp\left\{ \sum_{i=1}^{k} \theta_i f_i(\boldsymbol{D}_i) \right\}. \tag{20.1}$$

As usual, we use $f_i(\xi)$ as shorthand for $f_i(\xi\langle \boldsymbol{D}_i \rangle)$. The parameters of this distribution correspond to the weight we put on each feature. When $\theta_i = 0$, the feature is ignored, and it has no effect on the distribution.

As discussed in chapter 4, this representation is very generic and can capture Markov networks with global structure and local structure. A special case of particular interest is when $f_i(\boldsymbol{D}_i)$ is a binary indicator function that returns the value 0 or 1. With such features, we can encode a "standard" Markov network by simply having one feature per potential entry. In more general, however, we can consider arbitrary valued features.

Example 20.1

As a specific example, consider the simple diamond network of figure 3.10a, where we take all four variables to be binary-valued. The features that correspond to this network are sixteen indicator functions: four for each assignment of variables to each of our four clusters. For example, one such feature would be:

$$f_{a^0, b^0}(a, b) = \boldsymbol{I}\{a = a^0\}\boldsymbol{I}\{b = b^0\}.$$

With this representation, the weight of each indicator feature is simply the natural logarithm of the corresponding potential entry. For example, $\theta_{a^0, b^0} = \ln \phi_1(a^0, b^0)$. ∎

Given a model in this form, the log-likelihood function has a simple form.

Proposition 20.1

Let $\mathcal{D}$ be a data set of M examples, and let $\mathcal{F} = \{f_i : i = 1, \ldots, k\}$ be a set of features that define a model. Then the log-likelihood is

$$\ell(\boldsymbol{\theta} : \mathcal{D}) = \sum_i \theta_i \left(\sum_m f_i(\xi[m]) \right) - M \ln Z(\boldsymbol{\theta}). \tag{20.2}$$

sufficient statistics

The *sufficient statistics* of this likelihood function are the sums of the feature values in the instances in $\mathcal{D}$. We can derive a more elegant formulation if we divide the log-likelihood by the number of samples M.

$$\frac{1}{M}\ell(\boldsymbol{\theta}:\mathcal{D}) = \sum_i \theta_i \boldsymbol{E}_{\mathcal{D}}[f_i(\boldsymbol{d}_i)] - \ln Z(\boldsymbol{\theta}), \tag{20.3}$$

where $\boldsymbol{E}_{\mathcal{D}}[f_i(\boldsymbol{d}_i)]$ is the empirical expectation of f_i, that is, its average in the data set.

20.2.3 Properties of the Likelihood Function

The formulation of proposition 20.1 describes the likelihood function as a sum of two functions. The first function is linear in the parameters; increasing the parameters directly increases this linear term. Clearly, because the log-likelihood function (for a fixed data set) is upper-bounded (the probability of an event is at most 1), the second term $\ln Z(\boldsymbol{\theta})$ balances the first term.

Let us examine this second term in more detail. Recall that the partition function is defined as

$$\ln Z(\boldsymbol{\theta}) = \ln \sum_{\xi} \exp\left\{\sum_i \theta_i f_i(\xi)\right\}.$$

convex partition function

One important property of the partition function is that it is *convex* in the parameters $\boldsymbol{\theta}$. Recall that a function $f(\vec{x})$ is convex if for every $0 \le \alpha \le 1$,

$$f(\alpha\vec{x} + (1-\alpha)\vec{y}) \le \alpha f(\vec{x}) + (1-\alpha)f(\vec{y}).$$

Hessian

In other words, the function is bowl-like, and every interpolation between the images of two points is larger than the image of their interpolation. One way to prove formally that the function f is convex is to show that the *Hessian* — the matrix of the function's second derivatives — is positive semidefinite. Therefore, we now compute the derivatives of $Z(\boldsymbol{\theta})$.

Proposition 20.2

Let $\mathcal{F}$ be a set of features. Then,

$$\frac{\partial}{\partial\theta_i}\ln Z(\boldsymbol{\theta}) = \boldsymbol{E}_{\boldsymbol{\theta}}[f_i]$$

$$\frac{\partial^2}{\partial\theta_i\partial\theta_j}\ln Z(\boldsymbol{\theta}) = \boldsymbol{Cov}_{\boldsymbol{\theta}}[f_i; f_j],$$

where $\boldsymbol{E}_{\boldsymbol{\theta}}[f_i]$ is a shorthand for $\boldsymbol{E}_{P(\mathcal{X}:\boldsymbol{\theta})}[f_i]$.

PROOF The first derivatives are computed as:

$$\frac{\partial}{\partial\theta_i}\ln Z(\boldsymbol{\theta}) = \frac{1}{Z(\boldsymbol{\theta})}\sum_{\xi}\frac{\partial}{\partial\theta_i}\exp\left\{\sum_j \theta_j f_j(\xi)\right\}$$

$$= \frac{1}{Z(\boldsymbol{\theta})}\sum_{\xi} f_i(\xi)\exp\left\{\sum_j \theta_j f_j(\xi)\right\}$$

$$= \boldsymbol{E}_{\boldsymbol{\theta}}[f_i].$$

We now consider the second derivative:

$$
\begin{aligned}
\frac{\partial^2}{\partial \theta_j \partial \theta_i} \ln Z(\boldsymbol{\theta}) &= \frac{\partial}{\partial \theta_j} \left[\frac{1}{Z(\boldsymbol{\theta})} \sum_{\xi} f_i(\xi) \exp \left\{ \sum_k \theta_k f_k(\xi) \right\} \right] \\
&= -\frac{1}{Z(\boldsymbol{\theta})^2} \left(\frac{\partial}{\partial \theta_j} Z(\boldsymbol{\theta}) \right) \sum_{\xi} f_i(\xi) \exp \left\{ \sum_k \theta_k f_k(\xi) \right\} \\
&\quad + \frac{1}{Z(\boldsymbol{\theta})} \sum_{\xi} f_i(\xi) f_j(\xi) \exp \left\{ \sum_k \theta_k f_k(\xi) \right\} \\
&= -\frac{1}{Z(\boldsymbol{\theta})^2} Z(\boldsymbol{\theta}) \boldsymbol{E}_{\boldsymbol{\theta}}[f_j] \sum_{\xi} f_i(\xi) \tilde{P}(\xi : \boldsymbol{\theta}) \\
&\quad + \frac{1}{Z(\boldsymbol{\theta})} \sum_{\xi} f_i(\xi) f_j(\xi) \tilde{P}(\xi : \boldsymbol{\theta}) \\
&= -\boldsymbol{E}_{\boldsymbol{\theta}}[f_j] \sum_{\xi} f_i(\xi) P(\xi : \boldsymbol{\theta}) \\
&\quad + \sum_{\xi} f_i(\xi) f_j(\xi) P(\xi : \boldsymbol{\theta}) \\
&= \boldsymbol{E}_{\boldsymbol{\theta}}[f_i f_j] - \boldsymbol{E}_{\boldsymbol{\theta}}[f_i] \boldsymbol{E}_{\boldsymbol{\theta}}[f_j] \\
&= \boldsymbol{Cov}_{\boldsymbol{\theta}}[f_i; f_j].
\end{aligned}
$$

Thus, the Hessian of $\ln Z(\boldsymbol{\theta})$ is the covariance matrix of the features, viewed as random variables distributed according to distribution defined by $\boldsymbol{\theta}$. Because a covariance matrix is always positive semidefinite, it follows that the Hessian is positive semidefinite, and hence that $\ln Z(\boldsymbol{\theta})$ is a convex function of $\boldsymbol{\theta}$. ∎

Because $\ln Z(\boldsymbol{\theta})$ is convex, its complement $(-\ln Z(\boldsymbol{\theta}))$ is concave. The sum of a linear function and a concave function is concave, implying the following important result:

Corollary 20.1

The log-likelihood function is concave.

redundant
parameterization

This result implies that the log-likelihood is unimodal and therefore has no local optima. It does not, however, imply the uniqueness of the global optimum: Recall that a parameterization of the Markov network can be *redundant*, giving rise to multiple representations of the same distribution. The standard parameterization of a set of table factors for a Markov network — a feature for every entry in the table — is always redundant. In our simple example, for instance, we have:

$$
f_{a^0,b^0} = 1 - f_{a^0,b^1} - f_{a^1,b^0} - f_{a^1,b^1}.
$$

We thus have a continuum of parameterizations that all encode the same distribution, and (necessarily) give rise to the same log-likelihood. Thus, there is a unique globally optimal value for the log-likelihood function, but not necessarily a unique solution. In general, because the function is concave, we are guaranteed that there is a convex region of continuous global optima.

It is possible to eliminate the redundancy by removing some of the features. However, as we discuss in section 20.4, that turns out to be unnecessary, and even harmful, in practice.

We note that we have defined the likelihood function in terms of a standard log-linear parameterization, but the exact same derivation also holds for networks that use shared parameters, as in section 6.5; see exercise 20.1 and exercise 20.2.

20.3 Maximum (Conditional) Likelihood Parameter Estimation

We now move to the question of estimating the parameters of a Markov network with a fixed structure, given a fully observable data set $\mathcal{D}$. We focus in this section on the simplest variant of this task — maximum-likelihood parameter estimation, where we select parameters that maximize the log-likelihood function of equation (20.2). In later sections, we discuss alternative objectives for the parameter estimation task.

20.3.1 Maximum Likelihood Estimation

As for any function, the gradient of the log-likelihood must be zero at its maximum points. For a concave function, the maxima are precisely the points at which the gradient is zero. Using proposition 20.2, we can compute the gradient of the average log-likelihood as follows:

$$\frac{\partial}{\partial \theta_i} \frac{1}{M} \ell(\boldsymbol{\theta} : \mathcal{D}) = \boldsymbol{E}_{\mathcal{D}}[f_i(\mathcal{X})] - \boldsymbol{E}_{\boldsymbol{\theta}}[f_i]. \tag{20.4}$$

This analysis provides us with a precise characterization of the maximum likelihood parameters $\hat{\boldsymbol{\theta}}$:

Theorem 20.1

Let $\mathcal{F}$ be a set of features. Then, $\boldsymbol{\theta}$ is a maximum-likelihood parameter assignment if and only if $\boldsymbol{E}_{\mathcal{D}}[f_i(\mathcal{X})] = \boldsymbol{E}_{\hat{\boldsymbol{\theta}}}[f_i]$ for all i.

In other words, at the maximal likelihood parameters $\hat{\boldsymbol{\theta}}$, the expected value of each feature relative to $P_{\hat{\boldsymbol{\theta}}}$ matches its empirical expectation in $\mathcal{D}$. In other words, we want the *expected sufficient statistics* in the learned distribution to match the empirical expectations. This type of equality constraint is also called *moment matching*. This theorem easily implies that maximum likelihood estimation is *consistent* in the same sense as definition 18.1: if the model is sufficiently expressive to capture the data-generating distribution, then, at the large sample limit, the optimum of the likelihood objective is the true model; see exercise 20.3.

expected
sufficient
statistics

moment
matching

MLE consistency

By itself, this criterion does not provide a constructive definition of the maximum likelihood parameters. **Unfortunately, although the function is concave, there is no analytical form for its maximum. Thus, we must resort to iterative methods that search for the global optimum. Most commonly used are the gradient ascent methods reviewed in appendix A.5.2, which iteratively take steps in parameter space to improve the objective.** At each iteration, they compute the gradient, and possibly the Hessian, at the current point $\boldsymbol{\theta}$, and use those estimates to approximate the function at the current neighborhood. They then take a step in the right direction (as dictated by the approximation) and repeat the process. Due to the convexity of the problem, this process is guaranteed to converge to a global optimum, regardless of our starting point.

To apply these gradient-based methods, we need to compute the gradient. Fortunately, equation (20.4) provides us with an exact formula for the gradient: the difference between the feature's empirical count in the data and its expected count relative to our current parameterization $\boldsymbol{\theta}$. For example, consider again the fully parameterized network of example 20.1. Here, the features are simply indicator functions; the empirical count for a feature such as $f_{a^0,b^0}(a,b) = \boldsymbol{I}\{a = a^0\}\boldsymbol{I}\{b = b^0\}$ is simply the empirical frequency, in the data set $\mathcal{D}$, of the event a^0, b^0. At a particular parameterization $\boldsymbol{\theta}$, the expected count is simply $P_{\boldsymbol{\theta}}(a^0, b^0)$. Very naturally, the gradient for the parameter associated with this feature is the difference between these two numbers.

However, this discussion ignores one important aspect: the computation of the expected counts. In our example, for instance, we must compute the different probabilities of the form $P_{\boldsymbol{\theta}^t}(a, b)$. Clearly, this computation requires that we run inference over the network. As for the case of EM in Bayesian networks, a feature is necessarily part of a factor in the original network, and hence, due to family preservation, all of the variables involved in a feature must occur together in a cluster in a clique tree or cluster graph. Thus, a single inference pass that calibrates an entire cluster graph or tree suffices to compute all of the expected counts. Nevertheless, **a full inference step is required at every iteration of the gradient ascent procedure. Because inference is almost always costly in time and space, the computational cost of parameter estimation in Markov networks is usually high, sometimes prohibitively so.** In section 20.5 we return to this issue, considering the use of approximate methods that reduce the computational burden.

Our discussion does not make a specific choice of algorithm to use for the optimization. In practice, standard gradient ascent is not a particularly good algorithm, both because of its slow convergence rate and because of its sensitivity to the step size. Much faster convergence is obtained with second-order methods, which utilize the Hessian to provide a quadratic approx-

log-likelihood
Hessian

imation to the function. However, from proposition 20.2 we can conclude that the *Hessian* of the log-likelihood function has the form:

$$\frac{\partial}{\partial \theta_i \partial \theta_j} \ell(\boldsymbol{\theta} : \mathcal{D}) = -M \, \boldsymbol{Cov}_{\boldsymbol{\theta}}[f_i; f_j]. \tag{20.5}$$

L-BFGS algorithm

To compute the Hessian, we must compute the joint expectation of two features, a task that is often computationally infeasible. Currently, one commonly used solution is the *L-BFGS algorithm*, a gradient-based algorithm that uses line search to avoid computing the Hessian (see appendix A.5.2 for some background).

20.3.2   Conditionally Trained Models

As we discussed in section 16.3.2, we often want to use a Markov network to perform a particular inference task, where we have a known set of observed variables, or features, $\boldsymbol{X}$, and a predetermined set of variables, $\boldsymbol{Y}$, that we want to query. In this case, we may prefer to

discriminative
training

use *discriminative training*, where we train the network as a *conditional random field* (CRF) that encodes a conditional distribution $P(\boldsymbol{Y} \mid \boldsymbol{X})$.

conditional
random field

conditional
likelihood

More formally, in this setting, our training set consists of pairs $\mathcal{D} = \{(\boldsymbol{y}[m], \boldsymbol{x}[m])\}_{m=1}^{M}$, specifying assignments to $\boldsymbol{Y}, \boldsymbol{X}$. An appropriate objective function to use in this situation is the *conditional likelihood* or its logarithm, defined in equation (16.3). In our setting, the log-conditional-likelihood has the form:

$$\ell_{\boldsymbol{Y}|\boldsymbol{X}}(\boldsymbol{\theta} : \mathcal{D}) = \ln P(\boldsymbol{y}[1,\ldots,M] \mid \boldsymbol{x}[1,\ldots,M], \boldsymbol{\theta}) = \sum_{m=1}^{M} \ln P(\boldsymbol{y}[m] \mid \boldsymbol{x}[m], \boldsymbol{\theta}). \quad (20.6)$$

In this objective, we are optimizing the likelihood of each observed assignment $\boldsymbol{y}[m]$ given the corresponding observed assignment $\boldsymbol{x}[m]$. Each of the terms $\ln P(\boldsymbol{y}[1,\ldots,M] \mid \boldsymbol{x}[1,\ldots,M], \boldsymbol{\theta})$ is a log-likelihood of a Markov network model with a different set of factors — the factors in the original network, reduced by the observation $\boldsymbol{x}[1,\ldots,M]$ — and its own partition function. Each term is thereby a concave function, and because the sum of concave functions is concave, we conclude:

Corollary 20.2

The log conditional likelihood of equation (20.6) is a concave function.

As for corollary 20.1, this result implies that the function has a global optimum and no local optima, but not that the global optimum is unique. Here also, redundancy in the parameterization may give rise to a convex region of contiguous global optima.

The approaches for optimizing this objective are similar to those used for optimizing the likelihood objective in the unconditional case. The objective function is a concave function, and so a gradient ascent process is guaranteed to give rise to the unique global optimum. The form of the gradient here can be derived directly from equation (20.4). We first observe that the gradient of a sum is the sum of the gradients of the individual terms. Here, each term is, in fact, a log-likelihood — the log-likelihood of a single data case $\boldsymbol{y}[m]$ in the Markov network obtained by reducing our original model to the context $\boldsymbol{x}[m]$. A reduced Markov network is itself a Markov network, and so we can apply equation (20.4) and conclude that:

$$\frac{\partial}{\partial \theta_i} \ell_{\boldsymbol{Y}|\boldsymbol{X}}(\boldsymbol{\theta} : \mathcal{D}) = \sum_{m=1}^{M} \left(f_i(\boldsymbol{y}[m], \boldsymbol{x}[m]) - \boldsymbol{E}_{\boldsymbol{\theta}}[f_i \mid \boldsymbol{x}[m]] \right). \quad (20.7)$$

This solution looks deceptively similar to equation (20.4). Indeed, if we aggregate the first component in each of the summands, we obtain precisely the empirical count of f_i in the data set $\mathcal{D}$. There is, however, one key difference. In the unreduced Markov network, the expected feature counts are computed relative to a single model; in the case of the conditional Markov network, these expected counts are computed as the summation of counts in an ensemble of models, defined by the different values of the conditioning variables $\boldsymbol{x}[m]$. This difference has significant computational consequences. Recall that computing these expectations involves running inference over the model. **Whereas in the unconditional case, each gradient step required only a single execution of inference, when training a CRF, we must (in general) execute inference *for every single data case*, conditioning on $\boldsymbol{x}[m]$. On the other hand, the inference is executed on a simpler model, since conditioning on evidence in a Markov network can only reduce the computational cost.** For example, the network of figure 20.2 is very densely connected, whereas the reduced network over $\boldsymbol{Y}$ alone (conditioned on $\boldsymbol{X}$) is a simple chain, allowing linear-time inference.

Discriminative training can be particularly beneficial in cases where the domain of $\boldsymbol{X}$ is very large or even infinite. For example, in our image classification task, the partition function in the

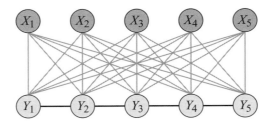

Figure 20.2 A highly connected CRF that allows simple inference when conditioned: The edges that disappear in the reduced Markov network after conditioning on X are marked in gray; the remaining edges form a simple linear chain.

generative setting involves summation (or integration) over the space of all possible images; if we have an $N \times N$ image where each pixel can take 256 values, the resulting space has 256^{N^2} values, giving rise to a highly intractable inference problem (even using approximate inference methods).

Box 20.A — Concept: Generative and Discriminative Models for Sequence Labeling. *One of the main tasks to which probabilistic graphical models have been applied is that of taking a set of interrelated instances and jointly labeling them, a process sometimes called* collective clas-

collective classification

sification. We have already seen examples of this task in box 4.B and in box 4.E; many other examples exist. Here, we discuss some of the trade-offs between different models that one can apply to this task. We focus on the context of labeling instances organized in a sequence, since it is simpler and allows us to illustrate another important point.

sequence labeling

In the sequence labeling *task, we get as input a sequence of observations X and need to label them with some joint label Y. For example, in text analysis (box 4.E), we might have a sequence of words each of which we want to label with some label. In a task of* activity *recognition, we might*

activity recognition

obtain a sequence of images and want to label each frame with the activity taking place in it (for example, running, jumping, walking). We assume that we want to construct a model for this task and to train it using fully labeled training data, where both Y and X are observed.

Figure 20.A.1 illustrates three different types of models that have been proposed and used for sequence labeling, all of which we have seen earlier in this book (see figure 6.2 and figure 4.14). The

hidden Markov model

first model is a hidden Markov model *(or HMM), which is a purely generative model: the model generates both the labels Y and the observations X. The second is called a* maximum entropy

maximum entropy Markov model

Markov model (or MEMM). This model is also directed, but it represents a conditional distribution $P(Y \mid X)$; hence, there is no attempt to model a distribution over the X's. The final model

conditional random field

is the conditional random field *(or CRF) of section 4.6.1. This model also encodes a conditional distribution; hence the arrows from X to Y. However, here the interactions between the Y are modeled as undirected edges.*

These different models present interesting trade-offs in terms of their expressive power and learnability. First, from a computational perspective, HMMs and MEMMs are much more easily learned. As purely directed models, their parameters can be computed in closed form using either maximum-likelihood or Bayesian estimation (see chapter 17); conversely, the CRF requires that we use an

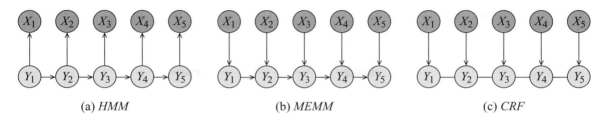

(a) *HMM*   (b) *MEMM*   (c) *CRF*

Figure 20.A.1 — Different models for sequence labeling: HMM, MEMM, and CRF

iterative gradient-based approach, which is considerably more expensive (particularly here, when inference must be run separately for every training sequence; see section 20.3.2).

A second important issue relates to our ability to use a rich feature set. As we discussed in example 16.3 and in box 4.E, our success in a classification task often depends strongly on the quality of our features. In an HMM, we must explicitly model the distribution over the features, including the interactions between them. This type of model is very hard, and often impossible, to construct correctly. The MEMM and the CRF are both discriminative models, and therefore they avoid this challenge entirely.

The third and perhaps subtler issue relates to the independence assumptions made by the model. As we discussed in section 4.6.1.2, the MEMM makes the independence assumption that $(Y_i \perp X_j \mid \boldsymbol{X}_{-j})$ for any $j > i$. Thus, an observation from later in the sequence has absolutely no effect on the posterior probability of the current state; or, in other words, the model does not allow for any smoothing. The implications of this can be severe in many settings. For example, consider the task of activity recognition from a video sequence; here, we generally assume that activities are highly persistent: if a person is walking in one frame, she is also extremely likely to be walking in the next frame. Now, imagine that the person starts running, but our first few observations in the sequence are ambiguous and consistent with both running and walking. The model will pick one — the one whose probability given that one frame is highest — which may well be walking. Assuming that activities are persistent, this choice of activity is likely to stay high for a large number of steps; the posterior of the initial activity will never change. In other words, the best we can expect is a prediction where the initial activity is walking, and then (perhaps) transitions to running. The model is incapable of going back and changing its prediction about the first few frames. This

label bias problem

problem has been called the label bias problem.

 To summarize, the trade-offs between these different models are subtle and non-definitive. In cases where we have many correlated features, discriminative models are probably better; but, if only limited data are available, the stronger bias of the generative model may dominate and allow learning with fewer samples. Among the discriminative models, MEMMs should probably be avoided in cases where many transitions are close to deterministic. In many cases, CRFs are likely to be a safer choice, but the computational cost may be prohibitive for large data sets.

20.3.3 Learning with Missing Data

We now turn to the problem of parameter estimation in the context of missing data. As we saw in section 19.1, the introduction of missing data introduces both conceptual and technical difficulties. In certain settings, we may need to model explicitly the process by which data are observed. Parameters may not be identifiable from the data. And the likelihood function becomes significantly more complex: there is coupling between the likelihood's dependence on different parameters; worse, the function is no longer concave and generally has multiple local maxima.

The same issues regarding observation processes (ones that are not missing at random) and identifiability arise equally in the context of Markov network learning. The issue regarding the complexity of the likelihood function is analogous, although not quite the same. In the case of Markov networks, of course, we have coupling between the parameters even in the likelihood function for complete data. However, as we discuss, in the complete data case, the log-likelihood function is concave and easily optimized using gradient methods. Once we have missing data, we lose the concavity of the function and can have multiple local maxima. Indeed, the example we used was in the context of a Bayesian network of the form $X \rightarrow Y$, which can also be represented as a Markov network. Of course, the parameterization of the two models is not the same, and so the form of the function may differ. However, one can verify that a function that is multimodal in one parameterization will also be multimodal in the other.

20.3.3.1 Gradient Ascent

As in the case of Bayesian networks, if we assume our data is missing at random, we can perform maximum-likelihood parameter estimation by using some form of gradient ascent process to optimize the likelihood function. Let us therefore begin by analyzing the form of the gradient in the case of missing data. Let $\mathcal{D}$ be a data set where some entries are missing; let $o[m]$ be the observed entries in the mth data instance and $\mathcal{H}[m]$ be the random variables that are the missing entries in that instance, so that for any $h[m] \in Val(\mathcal{H}[m])$, $(o[m], h[m])$ is a complete assignment to $\mathcal{X}$.

As usual, the average log-likelihood function has the form:

$$
\begin{aligned}
\frac{1}{M} \ln P(\mathcal{D} \mid \boldsymbol{\theta}) \quad &= \quad \frac{1}{M} \sum_{m=1}^{M} \ln \left(\sum_{h[m]} P(o[m], h[m] \mid \boldsymbol{\theta}) \right) \\
&= \frac{1}{M} \sum_{m=1}^{M} \ln \left(\sum_{h[m]} \tilde{P}(o[m], h[m] \mid \boldsymbol{\theta}) \right) - \ln Z.
\end{aligned}
\tag{20.8}
$$

Now, consider a single term within the summation, $\sum_{h[m]} \tilde{P}(o[m], h[m] \mid \boldsymbol{\theta})$. This expression has the same form as a partition function; indeed, it is precisely the partition function for the Markov network that we would obtain by reducing our original Markov network with the observation $o[m]$, to obtain a Markov network representing the conditional distribution $\tilde{P}(\mathcal{H}[m] \mid o[m])$. Therefore, we can apply proposition 20.2 and conclude that:

$$
\frac{\partial}{\partial \theta_i} \ln \sum_{h[m]} \tilde{P}(o[m], h[m] \mid \boldsymbol{\theta}) \quad = \quad \boldsymbol{E}_{h[m] \sim P(\mathcal{H}[m] \mid o[m], \boldsymbol{\theta})} [f_i],
$$

that is, the gradient of this term is simply the *conditional* expectation of the feature, given the observations in this instance.

Putting this together with previous computations, we obtain the following:

Proposition 20.3

For a data set $\mathcal{D}$

$$\frac{\partial}{\partial \theta_i} \frac{1}{M} \ell(\boldsymbol{\theta} : \mathcal{D}) = \frac{1}{M} \left[\sum_{m=1}^{M} \boldsymbol{E}_{\boldsymbol{h}[m] \sim P(\mathcal{H}[m]|\boldsymbol{o}[m],\boldsymbol{\theta})}[f_i] \right] - \boldsymbol{E}_{\boldsymbol{\theta}}[f_i]. \tag{20.9}$$

In other words, the gradient for feature f_i in the case of missing data is the difference between two expectations — the feature expectation *over the data and the hidden variables* minus the feature expectation over all of the variables.

It is instructive to compare the cost of this computation to that of computing the gradient in equation (20.4). For the latter, to compute the second term in the derivative, we need to run inference once, to compute the expected feature counts relative to our current distribution $P(\mathcal{X} \mid \boldsymbol{\theta})$. The first term is computed by simply aggregating the feature over the data. By comparison, to compute the derivative here, we actually need to run inference separately for every instance m, conditioning on $\boldsymbol{o}[m]$. Although inference in the reduced network may be simpler (since reduced factors are simpler), the cost of this computation is still much higher than learning without missing data. Indeed, not surprisingly, the cost here is comparable to the cost of a single iteration of gradient descent or EM in Bayesian network learning.

20.3.3.2 Expectation Maximization

As for any other probabilistic model, an alternative method for parameter estimation in context of missing data is via the expectation maximization algorithm. In the case of Bayesian network learning, EM seemed to have significant advantages. Can we define a variant of EM for Markov networks? And does it have the same benefits?

The answer to the first question is clearly yes. We can perform an E-step by using our current parameters $\boldsymbol{\theta}^{(t)}$ to compute the expected sufficient statistics, in this case, the expected feature counts. That is, at iteration t of the EM algorithm, we compute, for each feature f_i, the expected sufficient statistic:

$$\bar{M}_{\boldsymbol{\theta}^{(t)}}[f_i] = \frac{1}{M} \left[\sum_{m=1}^{M} \boldsymbol{E}_{\boldsymbol{h}[m] \sim P(\mathcal{H}[m]|\boldsymbol{o}[m],\boldsymbol{\theta})}[f_i] \right].$$

With these expected feature counts, we can perform an M-step by doing maximum likelihood parameter estimation. The proofs of convergence and other properties of the algorithm go through unchanged.

Here, however, there is one critical difference. Recall that, in the case of directed models, given the expected sufficient statistics, we can perform the M-step efficiently, in closed form. By contrast, the M-step for Markov networks requires that we run inference multiple times, once for each iteration of whatever gradient ascent procedure we are using. At step k of this "inner-loop" optimization, we now have a gradient of the form:

$$\bar{M}_{\boldsymbol{\theta}^{(t)}}[f_i] - \boldsymbol{E}_{\boldsymbol{\theta}^{(t,k)}}[f_i].$$

The trade-offs between the two algorithms are now more subtle than in the case of Bayesian networks. For the joint gradient ascent procedure of the previous section, we need to run inference $M + 1$ times in each gradient step: once without evidence, and once for each data case. If we use EM, we run inference M times to compute the expected sufficient statistics in the E-step, and then once for each gradient step, to compute the second term in the gradient. Clearly, there is a computational savings here. However, each of these gradient steps now uses an "out-of-date" set of expected sufficient statistics, making it increasingly less relevant as our optimization proceeds.

In fact, we can view the EM algorithm, in this case, as a form of caching of the first term in the derivative: Rather than compute the expected counts in each iteration, we compute them every few iterations, take a number of gradient steps, and then recompute the expected counts. There is no need to run the "inner-loop" optimization until convergence; indeed, that strategy is often not optimal in practice.

20.3.4   Maximum Entropy and Maximum Likelihood ⋆

We now return to the case of basic maximum likelihood estimation, in order to derive an alternative formulation that provides significant insight. In particular, we now use theorem 20.1 to relate maximum likelihood estimation in log-linear models to another important class or problems examined in statistics: the problem of finding the distribution of maximum entropy subject to a set of constraints.

To motivate this alternative formulation, consider a situation where we are given some summary statistics of an empirical distribution, such as those that may be published in a census report. These statistics may include the marginal distributions of single variables, of certain pairs, and perhaps of other events that the researcher summarizing the data happened to consider of interest. As another example, we might know the average final grade of students in the class and the correlation of their final grade with their homework scores. However, we do not have access to the full data set. While these two numbers constrain the space of possible distributions over the domain, they do not specify it uniquely. Nevertheless, we might want to construct a "typical" distribution that satisfies the constraints and use it to answer other queries.

One compelling intuition is that we should select a distribution that satisfies the given constraints but has no additional "structure" or "information." There are many ways of making this intuition precise. One that has received quite a bit of attention is based on the intuition that entropy is the inverse of information, so that we should search for the distribution of highest entropy. (There are more formal justifications for this intuition, but these are beyond the scope *maximum* of this book.) More formally, in *maximum entropy* estimation, we solve the following problem:
entropy

> Maximum-Entropy:
>
> **Find** $Q(\mathcal{X})$
> **maximizing** $H_Q(\mathcal{X})$
> **subject to**
>
> $$\boldsymbol{E}_Q[f_i] = \boldsymbol{E}_\mathcal{D}[f_i] \quad i = 1, \ldots, k. \tag{20.10}$$

expectation   The constraints of equation (20.10) are called *expectation constraints*, since they constrain us
constraints   to the set of distributions that have a particular set of expectations. We know that this set is

non-empty, since we have one example of a distribution that satisfies these constraints — the empirical distribution.

Somewhat surprisingly, the solution to this problem is a Gibbs distribution over the features $\mathcal{F}$ that matches the given expectations.

Theorem 20.2

The distribution Q^* is the maximum entropy distribution satisfying equation (20.10) if and only if $Q^* = P_{\hat{\theta}}$, where

$$P_{\hat{\theta}}(\mathcal{X}) = \frac{1}{Z(\hat{\theta})} \exp\left\{ \sum_i \hat{\theta}_i f_i(\mathcal{X}) \right\}$$

and $\hat{\theta}$ is the maximum likelihood parameterization relative to $\mathcal{D}$.

PROOF For notational simplicity, let $P = P_{\hat{\theta}}$. From theorem 20.1, it follows that $E_P[f_i] = E_{\mathcal{D}}[f_i(\mathcal{X})]$ for $i = 1, \ldots, k$, and hence that P satisfies the constraints of equation (20.10). Therefore, to prove that $P = Q^*$, we need only show that $H_P(\mathcal{X}) \geq H_Q(\mathcal{X})$ for all other distributions Q that satisfy these constraints. Consider any such distribution Q.

From proposition 8.1, it follows that:

$$H_P(\mathcal{X}) = -\sum_i \hat{\theta}_i E_P[f_i] + \ln Z(\theta). \tag{20.11}$$

Thus,

$$
\begin{aligned}
H_P(\mathcal{X}) - H_Q(\mathcal{X}) &= -\left[\sum_i \hat{\theta}_i E_P[f_i(\mathcal{X})] \right] + \ln Z_P - E_Q[-\ln Q(\mathcal{X})] \\
(i) &= -\left[\sum_i \hat{\theta}_i E_Q[f_i(\mathcal{X})] \right] + \ln Z_P + E_Q[\ln Q(\mathcal{X})] \\
&= E_Q[-\ln P(\mathcal{X})] + E_Q[\ln Q(\mathcal{X})] \\
&= D(Q\|P) \geq 0,
\end{aligned}
$$

where (i) follows from the fact that both $P_{\hat{\theta}}$ and Q satisfy the constraints, so that $E_{P_{\hat{\theta}}}[f_i] = E_Q[f_i]$ for all i.

We conclude that $H_{P_{\hat{\theta}}}(\mathcal{X}) \geq H_Q(\mathcal{X})$ with equality if and only if $P_{\hat{\theta}} = Q$. Thus, the maximum entropy distribution Q^* is necessarily equal to $P_{\hat{\theta}}$, proving the result. ∎

duality

One can also provide an alternative proof of this result based on the concept of *duality* discussed in appendix A.5.4. Using this alternative derivation, one can show that the two problems, maximizing the entropy given expectation constraints and maximizing the likelihood given structural constraints on the distribution, are *convex duals* of each other. (See exercise 20.5.)

Both derivations show that these objective functions provide bounds on each other, and are identical at their convergence point. That is, for the maximum likelihood parameters $\hat{\theta}$,

$$H_{P_{\hat{\theta}}}(\mathcal{X}) = -\frac{1}{M} \ell(\hat{\theta} : \mathcal{D}).$$

As a consequence, we see that for any set of parameters $\boldsymbol{\theta}$ and for any distribution Q that satisfy the expectation constraints equation (20.10), we have that

$$\boldsymbol{H}_Q(\mathcal{X}) \leq \boldsymbol{H}_{P_{\hat{\boldsymbol{\theta}}}}(\mathcal{X}) = -\frac{1}{M}\ell(\hat{\boldsymbol{\theta}} : \mathcal{D}) \leq -\frac{1}{M}\ell(\boldsymbol{\theta} : \mathcal{D})$$

with equality if and only if $Q = \mathcal{P}_{\boldsymbol{\theta}}$. We note that, while we provided a proof for this result from first principles, it also follows directly from the theory of convex duality.

Our discussion has shown an entropy dual only for likelihood. A similar connection can be shown between conditional likelihood and conditional entropy; see exercise 20.6.

20.4 Parameter Priors and Regularization

So far, we have focused on maximum likelihood estimation for selecting parameters in a Markov network. However, as we discussed in chapter 17, maximum likelihood estimation (MLE) is prone to overfitting to the training data. Although the effects are not as transparent in this case (due to the lack of direct correspondence between empirical counts and parameters), overfitting of the maximum likelihood estimator is as much of a problem here.

As for Bayesian networks, we can reduce the effect of overfitting by introducing a prior distribution $P(\boldsymbol{\theta})$ over the model parameters. Note that, because we do not have a decomposable closed form for the likelihood function, we do not obtain a decomposable closed form for the posterior in this case. Thus, a fully Bayesian approach, where we integrate out the parameters to compute the next prediction, is not generally feasible in Markov networks. However, we can

MAP estimation   aim to perform *MAP estimation* — to find the parameters that maximize $P(\boldsymbol{\theta})P(\mathcal{D} \mid \boldsymbol{\theta})$.

Given that we have no constraints on the conjugacy of the prior and the likelihood, we can consider virtually any reasonable distribution as a possible prior. However, only a few priors have been applied in practice.

20.4.1 Local Priors

Most commonly used is a Gaussian prior on the log-linear parameters $\boldsymbol{\theta}$. The most standard form of this prior is simply a zero-mean diagonal Gaussian, usually with equal variances for each of the weights:

$$P(\boldsymbol{\theta} \mid \sigma^2) = \prod_{i=1}^{k} \frac{1}{\sqrt{2\pi}\sigma} \exp\left\{-\frac{\theta_i^2}{2\sigma^2}\right\},$$

hyperparameter   for some choice of the variance σ^2. This variance is a *hyperparameter*, as were the α_i's in the Dirichlet distribution (section 17.3.2). Converting to log-space (in which the optimization is typically done), this prior gives rise to a term of the form:

$$-\frac{1}{2\sigma^2}\sum_{i=1}^{k}\theta_i^2,$$

L_2-regularization   This term places a quadratic penalty on the magnitude of the weights, where the penalty is measured in Euclidean, or L_2-norm, generally called an L_2-*regularization* term. This term is

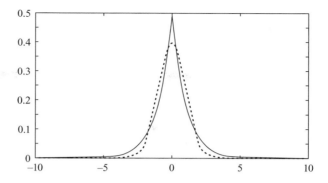

Figure 20.3 **Laplacian distribution ($\beta = 1$) and Gaussian distribution ($\sigma^2 = 1$)**

concave, and therefore it gives rise to a concave objective, which can be optimized using the same set of methods as standard MLE.

Laplacian distribution

A different prior that has been used in practice uses the zero-mean *Laplacian distribution*, which, for a single parameter, has the form

$$P_{Laplacian}(\theta \mid \beta) = \frac{1}{2\beta} \exp\left\{ -\frac{|\theta|}{\beta} \right\}. \tag{20.12}$$

One example of the Laplacian distribution is shown in figure 20.3; it has a nondifferentiable point at $\theta = 0$, arising from the use of the absolute value in the exponent. As for the Gaussian case, one generally assumes that the different parameters θ_i are independent, and often (but not always) that they are identically distributed with the same hyperparameter β. Taking the logarithm, we obtain a term

$$-\frac{1}{\beta} \sum_{i=1}^{k} |\theta_i|$$

L_1-regularization

that also penalizes weights of high magnitude, measured using the L_1-norm. Thus, this approach is generally called L_1-*regularization*.

Both forms of regularization penalize parameters whose magnitude (positive or negative) is large. Why is a bias in favor of parameters of low magnitude a reasonable one? Recall from our discussion in section 17.3 that a prior often serves to pull the distribution toward an "uninformed" one, smoothing out fluctuations in the data. Intuitively, a distribution is "smooth" if the probabilities assigned to different assignments are not radically different. Consider two assignments ξ and ξ'; their relative probability is

$$\frac{P(\xi)}{P(\xi')} = \frac{\tilde{P}(\xi)/Z_{\boldsymbol{\theta}}}{\tilde{P}(\xi')/Z_{\boldsymbol{\theta}}} = \frac{\tilde{P}(\xi)}{\tilde{P}(\xi')}.$$

Moving to log-space and expanding the unnormalized measure $\tilde{P}$, we obtain:

$$
\ln \frac{P(\xi)}{P(\xi')} = \sum_{i=1}^{k} \theta_i f_i(\xi) - \sum_{i=1}^{k} \theta_i f_i(\xi')
$$
$$
= \sum_{i=1}^{k} \theta_i (f_i(\xi) - f_i(\xi')).
$$

When all of the θ_i's have small magnitude, this log-ratio is also bounded, resulting in a smooth distribution. Conversely, when the parameters can be large, we can obtain a "spiky" distribution with arbitrarily large differences between the probabilities of different assignments.

In both the L_2 and the L_1 case, we penalize the magnitude of the parameters. In the Gaussian case, the penalty grows quadratically with the parameter magnitude, implying that an increase in magnitude in a large parameter is penalized more than a similar increase in a small parameter. For example, an increase in θ_i from 0 to 0.1 is penalized less than an increase from 3 to 3.1. In the Laplacian case, the penalty is linear in the parameter magnitude, so that the penalty growth is invariant over the entire range of parameter values. This property has important ramifications. In the quadratic case, as the parameters get close to 0, the effect of the penalty diminishes. Hence, the models that optimize the penalized likelihood tend to have many small weights. Although the resulting models are smooth, as desired, they are structurally quite dense. By comparison, in the L_1 case, the penalty is linear all the way until the parameter value is 0. This penalty provides a continued incentive for parameters to shrink until they actually hit 0. As a consequence, **the models learned with an L_1 penalty tend to be much sparser than those learned with an L_2 penalty, with many parameter weights achieving a value of 0. From a structural perspective, this effect gives rise to models with fewer edges and sparser potentials, which are potentially much more tractable.** We return to this issue in section 20.7.

Importantly, both the L_1 and L_2 regularization terms are concave. Because the log-likelihood is also concave, the resulting posterior is concave, and can therefore be optimized efficiently using the gradient-based methods we described for the likelihood case. Moreover, the introduction of these penalty terms serves to reduce or even eliminate multiple (equivalent) optima that arise when the parameterization of the network is redundant. For example, consider the trivial example where we have no data. In this case, the maximum likelihood solution is (as desired) the uniform distribution. However, due to redundancy, there is a continuum of parameterizations that give rise to the uniform distribution. However, when we introduce either of the earlier prior distributions, the penalty term drives the parameters toward zero, giving rise to the unique optimum $\boldsymbol{\theta} = \mathbf{0}$. Although one can still construct examples where multiple optima occur, they are very rare in practice. Conversely, methods that eliminate redundancies by reexpressing some of the parameters in terms of others can produce undesirable interactions with the regularization terms, giving rise to priors where some parameters are penalized more than others.

The regularization hyperparameters — σ^2 in the L_2 case, and β in the L_1 case — encode the strength in our belief that the model weights should be close to 0. The larger these parameters (both in the denominator), the broader our parameter prior, and the less strong our bias toward 0. In principle, any choice of hyperparameter is legitimate, since a prior is simply a reflection of our beliefs. **In practice, however, the choice of prior can have a significant effect on the quality of our learned model. A standard method for selecting this parameter is via a**

cross-validation procedure, as described in box 16.A: We repeatedly partition the training set, learn a model over one part with some choice of hyperparameter, and measure the performance of the learned model (for example, log-likelihood) on the held-out fragment.

20.4.2 Global Priors

conjugate prior

An alternative approach for defining priors is to search for a *conjugate prior*. Examining the likelihood function, we see that the posterior over parameters has the following general form:

$$
\begin{aligned}
P(\boldsymbol{\theta} \mid \mathcal{D}) \quad &\propto \quad P(\boldsymbol{\theta}) P(\mathcal{D} \mid \boldsymbol{\theta}) \\
&= \quad P(\boldsymbol{\theta}) \exp \left\{ \sum_i M \boldsymbol{E}_{\mathcal{D}}[f_i]\theta_i - M \ln Z(\boldsymbol{\theta}) \right\}.
\end{aligned}
$$

This expression suggests that we use a family of prior distributions of the form:

$$
P(\boldsymbol{\theta}) \propto \exp \left\{ \sum_i M_0 \alpha_i \theta_i - M_0 \ln Z(\boldsymbol{\theta}) \right\}.
$$

This form defines a family of priors with hyperparameters $\{\alpha_i\}$. It is easy to see that the posterior is from the same family with $\alpha_i' = \alpha_i + \boldsymbol{E}_{\mathcal{D}}[f_i]$ and $M_0' = M_0 + M$, so that this prior is conjugate to the log-linear model likelihood function.

We can think of the hyperparameters $\{\alpha_i\}$ as specifying the sufficient statistics from prior observations and of M_0 as specifying the number of these prior observations. This formulation is quite similar to the use of pseudocounts in the BDe priors for directed models (see section 17.4.3). The main difference from directed models is that this conjugate family (both the prior and the likelihood) does not decompose into independent priors for the different features.

20.5 Learning with Approximate Inference

The methods we have discussed here assume that we are able to compute the partition function $Z(\boldsymbol{\theta})$ and expectations such as $\boldsymbol{E}_{P_{\boldsymbol{\theta}}}[f_i]$. In many real-life applications the structure of the network does not allow for exact computation of these terms. For example, in applications to image segmentation (box 4.B), we generally use a grid-structured network, which requires exponential size clusters for exact inference.

The simplest approach for learning in intractable networks is to apply the learning procedure (say, conjugate gradient ascent) using an approximate inference procedure to compute the required queries about the distribution $P_{\boldsymbol{\theta}}$. This view decouples the question of inference from learning and treats the inference procedure as a black box during learning. The success of such an approach depends on whether the approximation method interferes with the learning. In particular, **nonconvergence of the inference method, or convergence to approximate** **answers, can lead to inaccurate and even oscillating estimates of the gradient, potentially harming convergence of the overall learning algorithm.** This type of situation can arise both in particle-based methods (say MCMC sampling) and in global algorithms such as belief propagation. In this section, we describe several methods that better integrate the inference into the learning outer loop in order to reduce problems such as this.

A second approach for dealing with inference-induced costs is to come up with alternative (possibly approximate) objective functions whose optimization does not require (as much) inference. Some of these techniques are reviewed in the next section. However, one of the main messages of this section is that the boundary between these two classes of methods is surprisingly ambiguous. **Approximately optimizing the likelihood objective by using an approximate inference algorithm to compute the gradient can often be reformulated as exactly optimizing an approximate objective.** When applicable, this view is often more insightful and also more usable. First, it provides more insight about the outcome of the optimization. Second, it may allow us to bound the error in the optimum in terms of the distance between the two functions being optimized. Finally, by formulating a clear objective to be optimized, we can apply any applicable optimization algorithm, such as conjugate gradient or Newton's method.

Importantly, while we describe the methods in this section relative to the plain likelihood objective, they apply almost without change to the generalizations and extensions we describe in this chapter: conditional Markov networks; parameter priors and regularization; structure learning; and learning with missing data.

20.5.1 Belief Propagation

belief
propagation

A fairly popular approach for approximate inference is the *belief propagation* algorithm and its variants. Indeed, in many cases, an algorithm in this family would be used for inference in the model resulting from the learning procedure. In this case, it can be shown that **we should learn the model using the same inference algorithm that will be used for querying it. Indeed, it can be shown that using a model trained with the same approximate inference algorithm is better than using a model trained with exact inference.**

At first glance, the use of belief propgation for learning appears straightforward. We can simply run BP within every iteration of gradient ascent to compute the expected feature counts used in the gradient computation. Due to the family preservation property, each feature f_i must be a subset of a cluster C_i in the cluster graph. Hence, to compute the expected feature count $E_\theta[f_i]$, we can compute the BP marginals over C_i, and then compute the expectation. In practice, however, this approach can be highly problematic. As we have seen, BP often does not converge. The marginals that we derive from the algorithm therefore oscillate, and the final results depend on the point at which we choose to stop the algorithm. As a result, the gradient

unstable gradient

computed from these expected counts is also *unstable*. This instability can be a significant problem in a gradient-based procedure, since it can gravely hurt the convergence properties of the algorithm. This problem is even more severe in the context of line-search methods, where the function evaluations can be inconsistent at different points in the line search.

There are several solutions to this problem: One can use one of the convergent alternatives to the BP algorithm that still optimizes the same Bethe energy objective; one can use a convex energy approximation, such as those of section 11.3.7.2; or, as we now show, one can reformulate the task of learning with approximate inference as optimizing an alternative objective, allowing the use of a range of optimization methods with better convergence properties.

20.5.1.1 Pseudo-moment Matching

Let us begin by a simple analysis of the fixed points of the learning algorithm. At convergence, the approximate expectations must satisfy the condition of theorem 20.1; in particular, the converged BP beliefs for C_i must satisfy

$$E_{\beta_i(C_i)}[f_{C_i}] = E_D[f_i(C_i)].$$

Now, let us consider the special case where our feature model defines a set of fully parameterized potentials that precisely match the clusters used in the BP cluster graph. That is, for every cluster C_i in the cluster graph, and every assignment c_i^j to C_i, we have a feature which is an indicator function $I\{c_i^j\}$, that is, it is 1 when $C_i = c_i^j$ and 0 otherwise. In this case, the preceding set of equalities imply that, for every assignment c_i^j to C_i, we have that

$$\beta_i(c_i^j) = \hat{P}(c_i^j). \tag{20.13}$$

That is, at convergence of the gradient ascent algorithm, the convergence point of the underlying belief propagation must be to a set of beliefs that exactly matches the empirical marginals in the data. But if we already know the outcome of our convergence, there is no point to running the algorithm!

This derivation gives us a closed form for the BP potentials at the point when both algorithms — BP inference and parameter gradient ascent — have converged. As we have already discussed, the full-table parameterization of Markov network potentials is redundant, and therefore there are multiple solutions that can give rise to this set of beliefs. One of these solutions can be obtained by dividing each sepset in the calibrated cluster graph into one of the adjacent clique potentials. More precisely, for each sepset $S_{i,j}$ between C_i and C_j, we select the endpoint for which $i < j$ (in some arbitrary ordering), and we then define:

$$\phi_i \leftarrow \frac{\beta_i}{\mu_{i,j}}.$$

We perform this transformation for each sepset. We use the final set of potentials as the parameterization for our Markov network. We can show that a single pass of message passing in a particular order gives rise to a calibrated cluster graph whose potentials are precisely the ones in equation (20.13). Thus, in this particular special case, we can provide a closed-form solution to both the inference and learning problem. This approach is called *pseudo-moment matching*.

pseudo-moment matching

While it is satisfying that we can find a solution so effectively, the form of the solution should be considered with care. In particular, we note that the clique potentials are simply empirical cluster marginals divided by empirical sepset marginals. These quantities depend only on the local structure of the factor and not on any global aspect of the cluster graph, including its structure. For example, the BC factor is estimated in exactly the same way within the diamond network of figure 11.1a and within the chain network $A—B—C—D$. Of course, potentials are also estimated locally in a Bayesian network, but there the local calibration ensures that the distribution can be factorized using purely local computations. As we have already seen, this is not the case for Markov networks, and so we expect different potentials to adjust to fit each other; however, the estimation using loopy BP does not accommodate that. In a sense, this observation is not surprising, since the BP approach also ignores the more global information.

We note, however, that this purely local estimation of the parameters only holds under the very restrictive conditions described earlier. It does not hold when we have parameter priors (regularization), general features rather than table factors, any type of shared parameters (as in section 6.5), or conditional random fields. We discuss this more general case in the next section.

20.5.1.2   Belief Propagation and Entropy Approximations ⋆

We now provide a more general derivation that allows us to reformulate maximum-likelihood learning with belief propagation as a unified optimization problem with an approximate objective. This perspective opens the door to the use of better approximation algorithms.

maximum
entropy

Our analysis starts from the *maximum-entropy* dual of the maximum-likelihood problem.

Maximum-Entropy:

Find $Q(\mathcal{X})$
maximizing $H_Q(\mathcal{X})$
subject to

$$\boldsymbol{E}_Q[f_i] = \boldsymbol{E}_{\mathcal{D}}[f_i] \quad i = 1, \ldots, k.$$

We can obtain a tractable approximation to this problem by applying the same sequence of transformations that we used in section 11.3.6 to derive belief propagation from the energy optimization problem. More precisely, assume we have a cluster graph $\mathcal{U}$ consisting of a set of clusters $\{\boldsymbol{C}_i\}$ connected by sepsets $\boldsymbol{S}_{i,j}$. Now, rather than optimize Maximum-Entropy over the space of distributions Q, we optimize over the set of possible pseudo-marginals in the *local consistency polytope* $Local[\mathcal{U}]$, as defined in equation (11.16). Continuing as in the BP derivation, we also approximate the entropy as in its *factored* form (definition 11.1):

local consistency
polytope

factored entropy

$$\boldsymbol{H}_Q(\mathcal{X}) \approx \sum_{\boldsymbol{C}_i \in \mathcal{U}} \boldsymbol{H}_{\beta_i}(\boldsymbol{C}_i) - \sum_{(\boldsymbol{C}_i - \boldsymbol{C}_j) \in \mathcal{U}} \boldsymbol{H}_{\mu_{i,j}}(\boldsymbol{S}_{i,j}). \tag{20.14}$$

As before, this reformulation is exact when the cluster graph is a tree but is approximate otherwise.

Putting these approximations together, we obtain the following approximation to the maximum-entropy optimization problem:

Approx-Maximum-Entropy:

Find Q
maximizing $\sum_{\boldsymbol{C}_i \in \mathcal{U}} \boldsymbol{H}_{\beta_i}(\boldsymbol{C}_i) - \sum_{(\boldsymbol{C}_i - \boldsymbol{C}_j) \in \mathcal{U}} \boldsymbol{H}_{\mu_{i,j}}(\boldsymbol{S}_{i,j})$
subject to

$$\begin{aligned}
\boldsymbol{E}_{\beta_i}[f_i] &= \boldsymbol{E}_{\mathcal{D}}[f_i] \quad i = 1, \ldots, k \\
Q &\in Local[\mathcal{U}].
\end{aligned} \tag{20.15}$$

CAMEL

This approach is called *CAMEL*, for constrained approximate maximum enropy learning.

Example 20.2

To illustrate this reformulation, consider a simple pairwise Markov network over the binary variables A, B, C, with three clusters: $C_1 = \{A, B\}, C_2 = \{B, C\}, C_3 = \{A, C\}$. We assume that the log-linear model is defined by the following two features, both of which are shared over all clusters: $f_{00}(x, y) = 1$ if $x = 0$ and $y = 0$, and 0 otherwise; and $f_{11}(x, y) = 1$ if $x = 1$ and $y = 1$. Assume we have 3 data instances $[0, 0, 0], [0, 1, 0], [1, 0, 0]$. The unnormalized empirical counts of each feature, pooled over all clusters, is then $E_{\hat{P}}[f_{00}] = (3 + 1 + 1)/3 = 5/3$, $E_{\hat{P}}[f_{11}] = 0$. In this case, the optimization of equation (20.15) would take the following form:

Find $Q = \{\beta_1, \beta_2, \beta_3, \mu_{1,2}, \mu_{2,3}, \mu_{1,3}\}$

maximizing $H_{\beta_1}(A, B) + H_{\beta_2}(B, C) + H_{\beta_3}(A, C)$
$- H_{\mu_{1,2}}(B) - H_{\mu_{2,3}}(C) - H_{\mu_{1,3}}(A)$

subject to

$$\sum_i E_{\beta_i}[f_{00}] = 5/3$$

$$\sum_i E_{\beta_i}[f_{11}] = 0$$

$$\sum_a \beta_1(a, b) - \sum_c \beta_2(b, c) = 0 \quad \forall b$$

$$\sum_b \beta_2(b, c) - \sum_a \beta_3(a, c) = 0 \quad \forall c$$

$$\sum_c \beta_3(a, c) - \sum_b \beta_1(a, b) = 0 \quad \forall a$$

$$\sum_{c_i} \beta_i(c_i) = 1 \quad i = 1, 2, 3$$

$$\beta_i \geq 0 \quad i = 1, 2, 3.$$

The CAMEL optimization problem of equation (20.15) is a constrained maximization problem with linear constraints and a nonconcave objective. The problem actually has two distinct sets of constraints: the first set encodes the moment-matching constraints and comes from the learning problem; and the second set encodes the constraint that Q be in the marginal polytope and arises from the cluster-graph approximation. It thus forms a unified optimization problem that encompasses both the learning task — moment matching — and the inference task — obtaining a set of consistent pseudo-marginals over a cluster graph. Analogously, if we introduce Lagrange multipliers for these constraints (as in appendix A.5.3), they would have very different interpretations. The multipliers for the first set of constraints would correspond to weights θ in the log-linear model, as in the max-likelihood / max-entropy duality ; those in the second set would correspond to messages $\delta_{i \to j}$ in the cluster graph, as in the BP algorithm.

This observation leads to several solution algorithms for this problem. In one class of methods, we could introduce Lagrange multipliers for all of the constraints and then optimize the resulting problem over these new variables. If we perform the optimization by a double-loop algorithm where the outer loop optimizes over θ (say using gradient ascent) and the inner loops "optimizes" the $\delta_{i \to j}$ by iterating their fixed point equations, the result would be precisely gradient ascent over parameters with BP in the inner loop for inference.

20.5.1.3   Sampling-Based Learning ★

The partition function $Z(\theta)$ is a summation over an exponentially large space. One approach to approximating this summation is to reformulate it as an expectation with respect to some distribution $Q(\mathcal{X})$:

$$
\begin{aligned}
Z(\boldsymbol{\theta}) &= \sum_{\xi} \exp\left\{ \sum_i \theta_i f_i(\xi) \right\} \\
&= \sum_{\xi} \frac{Q(\xi)}{Q(\xi)} \exp\left\{ \sum_i \theta_i f_i(\xi) \right\} \\
&= \boldsymbol{E}_Q\left[\frac{1}{Q(\mathcal{X})} \exp\left\{ \sum_i \theta_i f_i(\mathcal{X}) \right\} \right].
\end{aligned}
$$

importance sampling

This is precisely the form of the *importance sampling* estimator described in section 12.2.2. Thus, we can approximate it by generating samples from Q, and correcting appropriately via weights. We can simplify this expression if we choose Q to be $P_{\boldsymbol{\theta}^0}$ for some set of parameters $\boldsymbol{\theta}^0$:

$$
\begin{aligned}
Z(\boldsymbol{\theta}) &= \boldsymbol{E}_{P_{\boldsymbol{\theta}^0}}\left[\frac{Z(\boldsymbol{\theta}^0)\exp\left\{ \sum_i \theta_i f_i(\mathcal{X}) \right\}}{\exp\left\{ \sum_i \theta_i^0 f_i(\mathcal{X}) \right\}} \right] \\
&= Z(\boldsymbol{\theta}^0)\boldsymbol{E}_{P_{\boldsymbol{\theta}^0}}\left[\exp\left\{ \sum_i (\theta_i - \theta_i^0) f_i(\mathcal{X}) \right\} \right].
\end{aligned}
$$

If we can sample instances $\xi^1, \ldots, \xi^K$ from $P_{\boldsymbol{\theta}_0}$, we can approximate the log-partition function as:

$$
\ln Z(\boldsymbol{\theta}) \approx \ln\left(\frac{1}{K}\sum_{k=1}^K \exp\left\{ \sum_i (\theta_i - \theta_i^0) f_i(\xi^k) \right\} \right) + \ln Z(\boldsymbol{\theta}^0). \tag{20.16}
$$

We can plug this approximation of $\ln Z(\boldsymbol{\theta})$ into the log-likelihood of equation (20.3) and optimize it. Note that $\ln Z(\boldsymbol{\theta}^0)$ is a constant that we can ignore in the optimization, and the resulting expression is therefore a simple function of $\boldsymbol{\theta}$, which can be optimized using methods such as gradient ascent or one of its extensions. Interestingly, gradient ascent over $\boldsymbol{\theta}$ relative to equation (20.16) is equivalent to utilizing an importance sampling estimator directly to approximate the expected counts in the gradient of equation (20.4) (see exercise 20.12). However, as we discussed, it is generally more instructive and useful to view such methods as exactly optimizing an approximate objective rather than approximately optimizing the exact likelihood.

Of course, as we discussed in section 12.2.2, the quality of an importance sampling estimator depends on the difference between $\boldsymbol{\theta}$ and $\boldsymbol{\theta}^0$: the greater the difference, the larger the variance of the importance weights. Thus, this type of approximation is reasonable only in a neighborhood surrounding $\boldsymbol{\theta}^0$.

MCMC

How do we use this approximation? One possible strategy is to iterate between two steps. In one we run a sampling procedure, such as *MCMC*, to generate samples from the current parameter set $\boldsymbol{\theta}^t$. Then in the second iteration we use some gradient procedure to find $\boldsymbol{\theta}^{t+1}$

that improve the approximate log-likelihood based on these samples. We can then regenerate samples and repeat the process. As the samples are regenerated from a new distribution, we can hope that they are generated from a distribution not too far from the one we are currently optimizing, maintaining a reasonable approximation.

20.5.2 MAP-Based Learning ★

As another approximation to the inference step in the learning algorithm, we can consider approximating the expected feature counts with their counts in the single *MAP assignment* to the current Markov network. As we discussed in chapter 13, in many classes of models, computing a single MAP assignment is a much easier computational task, making this a very appealing approach in many settings.

MAP assignment

More precisely, to approximate the gradient at a given parameter assignment $\boldsymbol{\theta}$, we compute

$$\boldsymbol{E}_{\mathcal{D}}[f_i(\mathcal{X})] - f_i(\xi^{\mathrm{MAP}}(\boldsymbol{\theta})), \tag{20.17}$$

where $\xi^{\mathrm{MAP}}(\boldsymbol{\theta}) = \arg\max_{\xi} P(\xi \mid \boldsymbol{\theta})$ is the MAP assignment given the current set of parameters $\boldsymbol{\theta}$. This approach is also called *Viterbi training*.

Viterbi training

Once again, we can gain considerable intuition by reformulating this approximate inference step as an exact optimization of an approximate objective. Some straightforward algebra shows that this gradient corresponds exactly to the approximate objective

$$\frac{1}{M}\ell(\boldsymbol{\theta} : \mathcal{D}) - \ln P(\xi^{\mathrm{MAP}}(\boldsymbol{\theta}) \mid \boldsymbol{\theta}), \tag{20.18}$$

or, due to the cancellation of the partition function:

$$\frac{1}{M}\sum_{m=1}^{M} \ln \tilde{P}(\xi[m] \mid \boldsymbol{\theta}) - \ln \tilde{P}(\xi^{\mathrm{MAP}}(\boldsymbol{\theta}) \mid \boldsymbol{\theta}). \tag{20.19}$$

To see this, consider a single data instance $\xi[m]$:

$$\begin{aligned}
\ln P(\xi[m] \mid \boldsymbol{\theta}) &- \ln P(\xi^{\mathrm{MAP}}(\boldsymbol{\theta}) \mid \boldsymbol{\theta}) \\
&= \; [\ln \tilde{P}(\xi[m] \mid \boldsymbol{\theta}) - \ln Z(\boldsymbol{\theta})] - [\ln \tilde{P}(\xi^{\mathrm{MAP}}(\boldsymbol{\theta}) \mid \boldsymbol{\theta}) - \ln Z(\boldsymbol{\theta})] \\
&= \; \ln \tilde{P}(\xi[m] \mid \boldsymbol{\theta}) - \ln \tilde{P}(\xi^{\mathrm{MAP}}(\boldsymbol{\theta}) \mid \boldsymbol{\theta}) \\
&= \; \sum_i \theta_i[f_i(\xi[m]) - f_i(\xi^{\mathrm{MAP}}(\boldsymbol{\theta}))].
\end{aligned}$$

If we average this expression over all data instances and take the partial derivative relative to θ_i, we obtain an expression whose gradient is precisely equation (20.17).

The first term in equation (20.19) is an average of expressions of the form $\ln \tilde{P}(\xi \mid \boldsymbol{\theta})$. Each such expression is a linear function in $\boldsymbol{\theta}$, and hence their average is also linear in $\boldsymbol{\theta}$. The second term, $\tilde{P}(\xi^{\mathrm{MAP}}(\boldsymbol{\theta}) \mid \boldsymbol{\theta})$, may appear to be the log-probability of an instance. However, as indicated by the notation, $\xi^{\mathrm{MAP}}(\boldsymbol{\theta})$ is itself a function of $\boldsymbol{\theta}$: in different regions of the parameter space, the MAP assignment changes. In fact, this term is equal to:

$$\ln P(\xi^{\mathrm{MAP}}(\boldsymbol{\theta}) \mid \boldsymbol{\theta}) = \max_{\xi} \ln P(\xi \mid \boldsymbol{\theta}).$$

This is a maximum of linear functions, which is a convex, piecewise-linear function. Therefore, its negation is concave, and so the entire objective of equation (20.19) is also concave and hence has a global optimum.

Although reasonable at first glance, a closer examination reveals some important issues with this objective. Consider again a single data instance $\xi[m]$. Because $\xi^{\mathrm{MAP}}(\boldsymbol{\theta})$ is the MAP assignment, it follows that $\ln P(\xi[m] \mid \boldsymbol{\theta}) \leq \ln P(\xi^{\mathrm{MAP}}(\boldsymbol{\theta}) \mid \boldsymbol{\theta})$, and therefore the objective is always nonpositive. The maximal value of 0 can be achieved in two ways. The first is if we manage to find a setting of $\boldsymbol{\theta}$ in which the empirical feature counts match the feature counts in $\xi^{\mathrm{MAP}}(\boldsymbol{\theta})$. This optimum may be hard to achieve: Because the counts in $\xi^{\mathrm{MAP}}(\boldsymbol{\theta})$ are discrete, they take on only a finite set of values; for example, if we have a feature that is an indicator function for the event $X_i = x_i$, its count can take on only the values 0 or 1, depending on whether the MAP assignment has $X_i = x_i$ or not. Thus, we may never be able to match the feature counts exactly. The second way of achieving the optimal value of 0 is to set all of the parameters θ_i to 0. In this case, we obtain the uniform distribution over assignments, and the objective achieves its maximum value of 0. This possible behavior may not be obvious when we consider the gradient, but it becomes apparent when we consider the objective we are trying to optimize.

That said, we note that in the early stages of the optimization, when the expected counts are far from the MAP counts, the gradient still makes progress in the general direction of increasing the relative log-probability of the data instances. This approach can therefore work fairly well in practice, especially if not optimized to convergence.

Box 20.B — Case Study: CRFs for Protein Structure Prediction. *One interesting application of CRFs is to the task of predicting the three-dimensional* structure *of proteins. Proteins are constructed as chains of residues, each containing one of twenty possible amino acids. The amino acids are linked together into a common backbone structure onto which amino-specific side-chains are attached. An important computational problem is that of predicting the side-chain conformations given the backbone. The full configuration for a side-chain consists of up to four angles, each of which takes on a continuous value. However, in practice, angles tend to cluster into bins of very similar angles, so that the common practice is to discretize the value space of each angle into a small number (usually up to three) bins, called* rotamers.

With this transformation, side-chain prediction can be formulated as a discrete optimization problem, where the objective is an energy over this discrete set of possible side-chain conformations. Several energy functions have been proposed, all of which include various repulsive and attractive terms between the side-chain angles of nearby residues, as well as terms that represent a prior and internal constraints within the side chain for an individual residue. Rosetta, a state-of-the-art system, uses a combination of eight energy terms, and uses simulated annealing to search for the minimal energy configuration. However, even this highly engineered system still achieves only moderate accuracies (around 72 percent of the discretized angles predicted correctly). An obvious question is whether the errors are due to suboptimal answers returned by the optimization algorithm, or to the design of the energy function, which may not correctly capture the true energy "preferences" of protein structures.

Yanover, Schueler-Furman, and Weiss (2007) propose to address this optimization problem using MAP inference techniques. The energy functions used in this type of model can also be viewed as the log-potentials of a Markov network, where the variables represent the different angles to be inferred, and their values the discretized rotamers. The problem of finding the optimal configuration is then simply the MAP inference problem, and can be tackled using some of the algorithms described in chapter 13. Yanover et al. show that the TRW algorithm of box 13.A finds the provably global optimum of the Rosetta energy function for approximately 85 percent of the proteins in a standard benchmark set; this computation took only a few minutes per protein on a standard workstation. They also tackled the problem by directly solving the LP relaxation of the MAP problem using a commercial LP solver; this approach found the global optimum of the energy function for all proteins in the test set, but at a higher computational cost. However, finding the global minimum gave only negligible improvements on the actual accuracy of the predicted angles, suggesting that the primary source of inaccuracy in these models is in the energy function, not the optimization.

Thus, this problem seems like a natural candidate for the application of learning methods. The task was encoded as a CRF, whose input is a list of amino acids that make up the protein as well as the three-dimensional shape of the backbone. Yanover et al. encoded this distribution as a log-linear model whose features were the (eight) different components of the Rosetta energy function, and whose parameters were the weights of these features. Because exact inference for this model is intractable, it was trained by using a TRW variant for sum-product algorithms (see section 11.3.7.2). This variant uses a set of convex counting numbers to provide a convex approximation, and a lower bound, to the log-partition function. These properties guarantee that the learning process is stable and is continually improving a lower bound on the true objective. This new energy function improves performance from 72 percent to 78 percent, demonstrating that learning can significantly improve models, even those that are carefully engineered and optimized by a human expert. Notably, for the learned energy function, and for other (yet more sophisticated) energy functions, the use of globally optimal inference does lead to improvements in accuracy. Overall, a combination of these techniques gave rise to an accuracy of 82.6 percent, a significant improvement.

20.6 Alternative Objectives

Another class of approximations can be obtained directly by replacing the objective that we aim to optimize with one that is more tractable. To motivate the alternative objectives we present in this chapter, let us consider again the form of the log-likelihood objective, focusing, for simplicity, on the case of a single data instance ξ:

$$
\ell(\boldsymbol{\theta} : \xi) = \ln \tilde{P}(\xi \mid \boldsymbol{\theta}) - \ln Z(\boldsymbol{\theta})
$$

$$
= \ln \tilde{P}(\xi \mid \boldsymbol{\theta}) - \ln \left(\sum_{\xi'} \tilde{P}(\xi' \mid \boldsymbol{\theta}) \right).
$$

Considering the first term, this objective aims to increase the log-measure (logarithm of the unnormalized probability) of the observed data instance ξ. Of course, because the log-measure is a linear function of the parameters in our log-linear representation, that goal can be achieved simply by increasing all of the parameters associated with positive empirical expectations in ξ, and decreasing all of the parameters associated with negative empirical expectations. Indeed, we can increase the first term unboundedly using this approach. The second term, however, balances the first, since it is the logarithm of a sum of the unnormalized measures of instances, in this case, all possible instances in $Val(\mathcal{X})$. In a sense, then, we can view the log-likelihood objective as aiming to increasing the distance between the log-measure of ξ and the aggregate of the measures of all instances. We can thus view it as contrasting two terms. The key difficulty with this formulation, of course, is that the second term involves a summation over the exponentially many instances in $Val(\mathcal{X})$, and therefore requires inference in the network.

This formulation does, however, suggest one approach to approximating this objective: **perhaps we can still move our parameters in the right direction if we aim to increase the difference between the log-measure of the data instances and a *more tractable set* of other instances, one that does not require summation over an exponential space.** The *contrastive objectives* that we describe in this section all take that form.

contrastive
objective

20.6.1   Pseudolikelihood and Its Generalizations

Perhaps the earliest method for circumventing the intractability of network inference is the pseudolikelihood objective. As one motivation for this approximation, consider the likelihood of a single instance ξ. Using the chain rule, we can write

$$P(\xi) = \prod_{j=1}^{n} P(x_j \mid x_1, \ldots, x_{j-1}).$$

We can approximate this formulation by replacing each term $P(x_i \mid x_1, \ldots, x_{i-1})$ by the conditional probability of x_i given all other variables:

$$P(\xi) \approx \prod_{j} P(x_j \mid x_1, \ldots, x_{j-1}, x_{j+1}, \ldots, x_n).$$

pseudolikelihood

This approximation leads to the *pseudolikelihood* objective:

$$\ell_{\mathrm{PL}}(\boldsymbol{\theta} : \mathcal{D}) = \frac{1}{M} \sum_{m} \sum_{j} \ln P(x_j[m] \mid \boldsymbol{x}_{-j}[m], \boldsymbol{\theta}), \qquad (20.20)$$

where $\boldsymbol{x}_{-j}$ stands for $x_1, \ldots, x_{j-1}, x_{j+1}, \ldots, x_n$. Intuitively, this objective measures our ability to predict each variable in the model given a full observation over all other variables. The predictive model takes a form that generalizes the *multinomial logistic CPD* of definition 5.10 and is identical to it in the case where the network contains only pairwise features — factors over edges in the network. As usual, we can use the conditional independence properties in the network to simplify this expression, removing from the right-hand side of $P(X_j \mid \boldsymbol{X}_{-j})$ any variable that is not a neighbor of X_j.

multinomial
logistic CPD

At first glance, this objective may appear to be more complex than the likelihood objective. However, a closer examination shows that we have eliminated the exponential summation over instances with several summations, each of which is far more tractable. In particular:

$$P(x_j \mid \boldsymbol{x}_{-j}) = \frac{P(x_j, \boldsymbol{x}_{-j})}{P(\boldsymbol{x}_{-j})} \;\;=\;\; \frac{\tilde{P}(x_j, \boldsymbol{x}_{-j})}{\tilde{P}(\boldsymbol{x}_{-j})}$$

$$= \frac{\tilde{P}(x_j, \boldsymbol{x}_{-j})}{\sum_{x'_j} \tilde{P}(x'_j, \boldsymbol{x}_{-j})}.$$

The critical feature of this expression is that the global partition function has disappeared, and instead we have a local partition function that requires summing only over the values of X_j.

The contrastive perspective that we described earlier provides an alternative insight on this derivation. Consider the pseudolikelihood objective applied to a single data instance ξ:

$$
\begin{aligned}
\sum_j \ln P(x_j \mid \boldsymbol{x}_{-j}) &= \sum_j \left(\ln \tilde{P}(x_j, \boldsymbol{x}_{-j}) - \ln \sum_{x'_j} \tilde{P}(x'_j, \boldsymbol{x}_{-j}) \right) \\
&= \sum_j \left(\ln \tilde{P}(\xi) - \ln \sum_{x'_j} \tilde{P}(x'_j, \boldsymbol{x}_{-j}) \right).
\end{aligned}
$$

Each of the terms in this final summation is a contrastive term, where we aim to increase the difference between the log-measure of our training instance ξ and an aggregate of the log-measures of instances that differ from ξ in the assignment to precisely one variable. In other words, we are increasing the contrast between our training instance ξ and the instances in a local neighborhood around it.

We can further simplify each of the summands in this expression, obtaining:

$$
\ln P(x_j \mid \boldsymbol{x}_{-j}) =
\left(\sum_{i \,:\, Scope[f_i] \ni X_j} \theta_i f_i(x_j, \boldsymbol{x}_{-j}) \right) - \ln \left(\sum_{x'_j} \exp \left\{ \sum_{i \,:\, Scope[f_i] \ni X_j} \theta_i f_i(x'_j, \boldsymbol{x}_{-j}) \right\} \right).
$$
(20.21)

Each of these terms is precisely a log-conditional-likelihood term for a Markov network over a single variable X_j, conditioned on all the remaining variables. Thus, it follows from corollary 20.2 that the function is concave in the parameters $\boldsymbol{\theta}$. Since a sum of concave functions is also concave, we have that the pseudolikelihood objective of equation (20.20) is concave. Thus, we are guaranteed that gradient ascent over this objective will converge to the global maximum.

To compute the gradient, we use equation (20.21), to obtain:

$$
\frac{\partial}{\partial \theta_i} \ln P(x_j \mid \boldsymbol{x}_{-j}) = f_i(x_j, \boldsymbol{x}_{-j}) - \boldsymbol{E}_{x'_j \sim P_{\boldsymbol{\theta}}(X_j \mid \boldsymbol{x}_{-j})} \left[f_i(x'_j, \boldsymbol{x}_{-j}) \right].
$$
(20.22)

If X_j is not in the scope of f_i, then $f_i(x_j, \boldsymbol{x}_{-j}) = f_i(x'_j, \boldsymbol{x}_{-j})$ for any x'_j, and the two terms are identical, making the derivative 0. Inserting this expression into equation (20.20), we obtain:

Proposition 20.4

$$
\frac{\partial}{\partial \theta_i} \ell_{\mathrm{PL}}(\boldsymbol{\theta} : \mathcal{D}) = \sum_{j : X_j \in Scope[f_i]} \left(\frac{1}{M} \sum_m f_i(\xi[m]) - \boldsymbol{E}_{x'_j \sim P_{\boldsymbol{\theta}}(X_j \mid \boldsymbol{x}_{-j}[m])} \left[f_i(x'_j, \boldsymbol{x}_{-j}[m]) \right] \right).
$$
(20.23)

While this term looks somewhat more involved than the gradient of the likelihood in equation (20.4), it is much easier to compute: each of the expectation terms requires a summation over only a single random variable X_j, conditioned on all of its neighbors, a computation that can generally be performed very efficiently.

What is the relationship between maximum likelihood estimation and maximum pseudo-likelihood? In one specific situation, the two estimators return the same set of parameters.

Theorem 20.3

Assume that our data are generated by a log-linear model P_{θ^*} that is of the form of equation (20.1). Then, as the number of data instances M goes to infinity, with probability that approaches 1, θ^* is a global optimum of the pseudolikelihood objective of equation (20.20).

PROOF To prove the result, we need to show that because the size of the data set tends to infinity, the gradient of the pseudolikelihood objective at θ^* tends to zero. Owing to the concavity of the objective, this equality implies that θ^* is necessarily an optimum of the pseudolikelihood objective. We provide a somewhat informal sketch of the gradient argument, but one that contains all the essential ideas.

Because $M \longrightarrow \infty$, the empirical distribution $\hat{P}$ gets arbitrarily close to P_{θ^*}. Thus, the statistics in the data are precisely representative of their expectations relative to P_{θ^*}. Now, consider one of the summands in equation (20.23), associated with a feature f_i. Due to the convergence of the sufficient statistics,

$$\frac{1}{M} \sum_m f_i(\xi[m]) \longrightarrow E_{\xi \sim P_{\theta^*}(\mathcal{X})}[f_i(\xi)].$$

Conversely,

$$\frac{1}{M} \sum_m E_{x'_j \sim P_{\theta^*}(X_j | \boldsymbol{x}_{-j}[m])} \left[f_i(x'_j, \boldsymbol{x}_{-j}[m]) \right]$$

$$= \sum_{\boldsymbol{x}_{-j}} P_{\mathcal{D}}(\boldsymbol{x}_{-j}) \sum_{x'_j} P_{\theta^*}(x'_j \mid \boldsymbol{x}_{-j}) f_i(x'_j, \boldsymbol{x}_{-j})$$

$$\longrightarrow \sum_{\boldsymbol{x}_{-j}} P_{\theta^*}(\boldsymbol{x}_{-j}) \sum_{x'_j} P_{\theta^*}(x'_j \mid \boldsymbol{x}_{-j}) f_i(x'_j, \boldsymbol{x}_{-j})$$

$$= E_{\xi \sim P_{\theta^*}}[f_i(\xi)].$$

Thus, at the limit, the empirical and expected counts are equal, so that the gradient is zero. ∎

This theorem states that, like the likelihood objective, the pseudolikelihood objective is also

consistent

consistent. If we assume that the models are nondegenerate so that the two objectives are strongly concave, the maxima are unique, and hence the two objectives have the same maximum.

While this result is an important one, it is important to be cognizant of its limitations. In particular, we note that the two assumptions are central to this argument. First, in order for the empirical and expected counts to match, the model being learned needs to be sufficiently expressive to represent the generating distribution. Second, the data distribution needs to be close enough to the generating distribution to be well captured within the model, a situation that is only guaranteed to happen at the large-sample limit. Without these assumptions, the two objectives can have quite different optima that lead to different results.

In practice, these assumptions rarely hold: our model is never a perfect representation of the true underlying distribution, and we often do not have enough data to be close to the large sample limit. Therefore, one must consider the question of how good this objective is in practice. The answer to this question depends partly on the types of queries for which we intend to use the model. **If we plan to run queries where we condition on most of the variables and query the values of only a few, the pseudolikelihood objective is a very close match to the type of predictions we would like to make, and therefore pseudolikelihood may well provide a better training objective than likelihood.** For example, if we are trying to learn a Markov network for collaborative filtering (box 18.C), we generally take the user's preference for all items except the query item to be observed. **Conversely, if a typical query involves most or all of the variables in the model, the likelihood objective is more appropriate.** For example, if we are trying to learn a model for image segmentation (box 4.B), the segment value of all of the pixels is unobserved. (We note that this last application is a CRF, where we would generally use a conditional likelihood objective, conditioned on the actual pixel values.) In this case, a (conditional) likelihood is a more appropriate objective than the (conditional) pseudolikelihood.

However, even in cases where the likelihood is the more appropriate objective, we may have to resort to pseudolikelihood for computational reasons. In many cases, this objective performs surprisingly well. However, in others, it can provide a fairly poor approximation.

Example 20.3

Consider a Markov network over three variables X_1, X_2, Y, where each pair is connected by an edge. Assume that X_1, X_2 are very highly correlated (almost identical) and both are somewhat (but not as strongly) correlated with Y. In this case, the best predictor for X_1 is X_2, and vice versa, so the pseudolikelihood objective is likely to overestimate significantly the parameters on the X_1—X_2, and almost entirely dismiss the X_1—Y and X_2—Y edges. The resulting model would be an excellent predictor for X_2 when X_1 is observed, but virtually useless when only Y and not X_1 is observed. ∎

This example is typical of a general phenomenon: Pseudolikelihood, by assuming that each variable's local neighborhood is fully observed, is less able to exploit information obtained from weaker or longer-range dependencies in the distribution.

generalized pseudolikelihood

This limitation also suggests a spectrum of approaches known as *generalized pseudolikelihood*, which can reduce the extent of this problem. In particular, in the objective of equation (20.20), rather than using a product of terms over individual variables, we can consider terms where the left-hand side consists of several variables, conditioned on the rest. More precisely, we can define a set of subsets of variables $\{ \boldsymbol{X}_s \ : \ s \in \mathcal{S} \}$, and then define an objective:

$$\ell_{\mathrm{GPL}}(\boldsymbol{\theta} : \mathcal{D}) = \frac{1}{M} \sum_m \sum_s \ln P(\boldsymbol{x}_s[m] \mid \boldsymbol{x}_{-s}[m], \boldsymbol{\theta}), \tag{20.24}$$

where $\boldsymbol{X}_{-s} = \mathcal{X} - \boldsymbol{X}_s$.

Clearly, there are many possible choices of subsets $\{ \boldsymbol{X}_s \}$. For different such choices, this expression generalizes several objectives: the likelihood, the pseudolikelihood, and even the conditional likelihood. When variables are together in the same subset $\boldsymbol{X}_s$, the relationship between them is subject (at least in part) to a likelihood-like objective, which tends to induce a more correct model of the joint distribution over them. However, as for the likelihood, this objective requires that we compute expected counts over the variables in each $\boldsymbol{X}_s$ given an assignment to $\boldsymbol{X}_{-s}$. Thus, the choice of $\boldsymbol{X}_s$ offers a trade-off between "accuracy" and computational cost. One common choice of subsets is the set of all cliques in the Markov networks, which guarantees that the factor associated with each clique is optimized in at least one likelihood-like term in the objective.

20.6.2 Contrastive Optimization Criteria

As we discussed, both likelihood and pseudolikelihood can be viewed as attempting to increase the "log-probability gap" between the log-probability of the observed instances in $\mathcal{D}$ and the logarithm of the aggregate probability of a set of instances. Building on this perspective, one can construct a range of methods that aim to increase the log-probability gap between $\mathcal{D}$ and some other instances. The intuition is that, by driving the probability of the observed data higher relative to other instances, we are tuning our parameters to predict the data better.

More precisely, consider again the case of a single training instance ξ. We can define a "contrastive" objective where we aim to maximize the log-probability gap:

$$\left(\ln \tilde{P}(\xi \mid \boldsymbol{\theta}) - \ln \tilde{P}(\xi' \mid \boldsymbol{\theta}) \right),$$

where ξ' is some other instance, whose selection we discuss shortly. Importantly, this expression takes a very simple form:

$$\left(\ln \tilde{P}(\xi \mid \boldsymbol{\theta}) - \ln \tilde{P}(\xi' \mid \boldsymbol{\theta}) \right) = \boldsymbol{\theta}^T [\boldsymbol{f}(\xi) - \boldsymbol{f}(\xi')]. \tag{20.25}$$

Note that, for a fixed instantiation ξ', this expression is a linear function of $\boldsymbol{\theta}$ and hence is unbounded. Thus, in order for this type of function to provide a coherent optimization objective, the choice of ξ' will generally have to change throughout the course of the optimization. Even then, we must take care to prevent the parameters from growing unboundedly, an easy way of arbitrarily increasing the objective.

One can construct many variants of this type of method. Here, we briefly survey two that have been particularly useful in practice.

20.6.2.1 Contrastive Divergence

contrastive
divergence

One approach whose popularity has recently grown is the *contrastive divergence* method. In this method, we "contrast" our data instances $\mathcal{D}$ with a set of randomly perturbed "neighbors" $\mathcal{D}^-$. In particular, we aim to maximize:

$$\ell_{\mathrm{CD}}(\boldsymbol{\theta} : \mathcal{D} \| \mathcal{D}^-) = \left[\boldsymbol{E}_{\xi \sim \hat{P}_{\mathcal{D}}} \left[\ln \tilde{P}_{\boldsymbol{\theta}}(\xi) \right] - \boldsymbol{E}_{\xi \sim \hat{P}_{\mathcal{D}^-}} \left[\ln \tilde{P}_{\boldsymbol{\theta}}(\xi) \right] \right], \tag{20.26}$$

where $\hat{P}_{\mathcal{D}}$ and $\hat{P}_{\mathcal{D}^-}$ are the empirical distributions relative to $\mathcal{D}$ and $\mathcal{D}^-$, respectively.

As we discussed, the set of "contrasted" instances $\mathcal{D}^-$ will necessarily differ at different stages in the search. Given a current parameterization $\boldsymbol{\theta}$, what is a good choice of instances to which we want to contrast our data instances $\mathcal{D}$? One intuition is that we want to move our parameters $\boldsymbol{\theta}$ in a direction that increases the probability of instances in $\mathcal{D}$ relative to "typical" instances in our current distribution; that is, we want to increase the probability gap between instances

$\xi \in \mathcal{D}$ and instances ξ sampled randomly from P_θ. Thus, we can generate a contrastive set $\mathcal{D}^-$ by sampling from P_θ, and then maximizing the objective in equation (20.26).

How do we sample from P_θ? As in section 12.3, we can run a Markov chain defined by the Markov network P_θ, using, for example, Gibbs sampling, and initializing from the instances in $\mathcal{D}$; once the chain mixes, we can collect samples from the distribution P_θ. Unfortunately, sampling from the chain for long enough to achieve mixing usually takes far too long to be feasible as the inner loop of a learning algorithm. However, there is an alternative approach, which is both less expensive and more robust. Rather than run the chain defined by P_θ to convergence, we initialize from the instances in $\mathcal{D}$, and run the chain only for a few steps; we then use the instances generated by these short sampling runs to define $\mathcal{D}^-$.

Intuitively, this approach has significant appeal: We want our model to give high probability to the instances in $\mathcal{D}$; our current parameters, initialized at $\mathcal{D}$, are causing us to move away from the instances in $\mathcal{D}$. Thus, we want to move our parameters in a direction that increases the probability of the instances in $\mathcal{D}$ relative to the "perturbed" instances in $\mathcal{D}^-$.

The gradient of this objective is also very intuitive, and easy to compute:

$$\frac{\partial}{\partial \theta_i} \ell_{CD}(\theta : \mathcal{D} \| \mathcal{D}^-) = E_{\hat{P}_{\mathcal{D}}}[f_i(\mathcal{X})] - E_{\hat{P}_{\mathcal{D}^-}}[f_i(\mathcal{X})]. \tag{20.27}$$

Note that, if we run the Markov chain to the limit, the samples in $\mathcal{D}^-$ are generated from P_θ; in this case, the second term in this difference converges to $E_{P_\theta}[f_i]$, which is precisely the second term in the gradient of the log-likelihood objective in equation (20.4). Thus, at the limit of the Markov chain, this learning procedure is equivalent (on expectation) to maximizing the log-likelihood objective. However, in practice, the approximation that we get by taking only a few steps in the Markov chain provides a good direction for the search, at far lower computational cost. In fact, empirically it appears that, because we are taking fewer sampling steps, there is less variance in our estimation of the gradient, leading to more robust convergence.

20.6.2.2 Margin-Based Training ⋆

A very different intuition arises in settings where our goal is to use the learned network for predicting a MAP assignment. For example in our image segmentation application of box 4.B, we want to use the learned network to predict a single high-probability assignment to the pixels that will encode our final segmentation output. This type of reasoning only arises in the context of conditional queries, since otherwise there is only a single MAP assignment (in the unconditioned network). Thus, we describe the objective in this section in the context of conditional Markov networks.

Recall that, in this setting, our training set consists of a set of pairs $\mathcal{D} = \{(\boldsymbol{y}[m], \boldsymbol{x}[m])\}_{m=1}^M$. Given an observation $\boldsymbol{x}[m]$, we would like our learned model to give the highest probability to $\boldsymbol{y}[m]$. In other words, we would like the probability $P_\theta(\boldsymbol{y}[m] \mid \boldsymbol{x}[m])$ to be higher than any other probability $P_\theta(\boldsymbol{y} \mid \boldsymbol{x}[m])$ for $\boldsymbol{y} \neq \boldsymbol{y}[m]$. In fact, to increase our confidence in this prediction, we would like to increase the log-probability gap as much as possible, by increasing:

$$\ln P_\theta(\boldsymbol{y}[m] \mid \boldsymbol{x}[m]) - \left[\max_{\boldsymbol{y} \neq \boldsymbol{y}[m]} \ln P_\theta(\boldsymbol{y} \mid \boldsymbol{x}[m])\right].$$

margin This difference between the log-probability of the target assignment $\boldsymbol{y}[m]$ and that of the "next best" assignment is called the *margin*. The higher the margin, the more confident the model is

margin-based
estimation
in selecting $\boldsymbol{y}[m]$. Roughly speaking, *margin-based estimation* methods usually aim to maximize the margin.

One way of formulating this of *max-margin* objective as an optimization problem is as follows:

> **Find** $\qquad \gamma, \boldsymbol{\theta}$
>
> **maximizing** $\quad \gamma$
>
> **subject to**
>
> $$\ln P_{\boldsymbol{\theta}}(\boldsymbol{y}[m] \mid \boldsymbol{x}[m]) - \ln P_{\boldsymbol{\theta}}(\boldsymbol{y} \mid \boldsymbol{x}[m]) \;\geq\; \gamma \qquad \forall m, \boldsymbol{y} \neq \boldsymbol{y}[m].$$

The objective here is to maximize a single parameter γ, which encodes the worst-case margin over all data instances, by virtue of the constraints, which impose that the log-probability gap between $\boldsymbol{y}[m]$ and any other assignment $\boldsymbol{y}$ (given $\boldsymbol{x}[m]$) is at least γ. Importantly, due to equation (20.25), the first set of constraints can be rewritten in a simple linear form:

$$\boldsymbol{\theta}^T (\boldsymbol{f}(\boldsymbol{y}[m], \boldsymbol{x}[m]) - \boldsymbol{f}(\boldsymbol{y}, \boldsymbol{x}[m])) \geq \gamma.$$

With this reformulation of the constraints, it becomes clear that, if we find any solution that achieves a positive margin, we can increase the margin unboundedly simply by multiplying all the parameters through by a positive constant factor. To make the objective coherent, we can bound the magnitude of the parameters by constraining their L_2-norm: $\|\boldsymbol{\theta}\|_2^2 = \boldsymbol{\theta}^T \boldsymbol{\theta} = \sum_i \theta_i^2 = 1$; or, equivalently, we can decide on a fixed margin and try to reduce the magnitude of the parameters as much as possible. With the latter approach, we obtain the following optimization problem:

> Simple-Max-Margin:
>
> **Find** $\qquad \boldsymbol{\theta}$
>
> **minimizing** $\quad \|\boldsymbol{\theta}\|_2^2$
>
> **subject to**
>
> $$\boldsymbol{\theta}^T (\boldsymbol{f}(\boldsymbol{y}[m], \boldsymbol{x}[m]) - \boldsymbol{f}(\boldsymbol{y}, \boldsymbol{x}[m])) \geq 1 \qquad \forall m, \boldsymbol{y} \neq \boldsymbol{y}[m]$$

quadratic
program

convex
optimization
At some level, this objective is simple: it is a *quadratic program* (QP) with linear constraints, and hence is a convex problem that can be solved using a variety of *convex optimization* methods. However, a more careful examination reveals that the problem contains a constraint for every m, and (more importantly) for every assignment $\boldsymbol{y} \neq \boldsymbol{y}[m]$. Thus, the number of constraints is exponential in the number of variables $\boldsymbol{Y}$, generally an intractable number.

However, these are not arbitrary constraints: the structure of the underlying Markov network is reflected in the form of the constraints, opening the way toward efficient solution algorithms.
constraint
generation
One simple approach uses *constraint generation*, a general-purpose method for solving optimization problems with a large number of constraints. Constraint generation is an iterative method, which repeatedly solves for $\boldsymbol{\theta}$, each time using a larger set of constraints. Assume we have some algorithm for performing constrained optimization. We initially run this algorithm using none of the margin constraints, and obtain the optimal solution $\boldsymbol{\theta}^0$. In most cases, this solution will not satisfy many of the margin constraints, and it is thus not a feasible solution to our original QP. We add one or more constraints that are violated by $\boldsymbol{\theta}^0$ into a set of *active constraints*. We now repeat the constrained optimization process to obtain a new solution $\boldsymbol{\theta}^1$, which is guaranteed

to satisfy the active constraints. We again examine the constraints, find ones that are violated, and add them to our active constraints. This process repeats until no constraints are violated by our solution. Clearly, since we only add constraints, this procedure is guaranteed to terminate: eventually there will be no more constraints to add. Moreover, when it terminates, the solution is guaranteed to be optimal: At any iteration, the optimization procedure is solving a relaxed problem, whose value is at least as good as that of the fully constrained problem. If the optimal solution to this relaxed problem happens to satisfy all of the constraints, no better solution can be found to the fully constrained problem.

This description leaves unanswered two important questions. First, how many constraints we will have to add before this process terminates? Fortunately, it can be shown that, under reasonable assumptions, at most a polynomial number of constraints will need to be added prior to termination. Second, how do we find violated constraints without exhaustively enumerating and checking every one? As we now show, we can perform this computation by running MAP inference in the Markov network induced by our current parameterization $\boldsymbol{\theta}$. To see how, recall that we either want to show that

$$\ln \tilde{P}(\boldsymbol{y}[m], \boldsymbol{x}[m]) \geq \ln \tilde{P}(\boldsymbol{y}, \boldsymbol{x}[m]) + 1$$

for every $\boldsymbol{y} \in Val(\boldsymbol{Y})$ except $\boldsymbol{y}[m]$, or we want to find an assignment $\boldsymbol{y}$ that violates this inequality constraint. Let

$$\boldsymbol{y}^{map} = \arg \max_{\boldsymbol{y} \neq \boldsymbol{y}[m]} \tilde{P}(\boldsymbol{y}, \boldsymbol{x}[m]).$$

There are now two cases: If $\ln \tilde{P}(\boldsymbol{y}[m], \boldsymbol{x}[m]) < \ln \tilde{P}(\boldsymbol{y}^{map}, \boldsymbol{x}[m]) + 1$, then this is a violated constraint, which can be added to our constraint set. Alternatively, if $\ln \tilde{P}(\boldsymbol{y}[m], \boldsymbol{x}[m]) > \ln \tilde{P}(\boldsymbol{y}^{map}, \boldsymbol{x}[m]) + 1$, then, due to the selection of $\boldsymbol{y}^{map}$, we are guaranteed that

$$\ln \tilde{P}(\boldsymbol{y}[m], \boldsymbol{x}[m]) > \ln \tilde{P}(\boldsymbol{y}^{map}, \boldsymbol{x}[m]) + 1 \geq \ln \tilde{P}(\boldsymbol{y}, \boldsymbol{x}[m]) + 1,$$

for every $\boldsymbol{y} \neq \boldsymbol{y}[m]$. That is, in this second case, all of the exponentially many constraints for the m'th data instance are guaranteed to be satisfied. As written, the task of finding $\boldsymbol{y}^{map}$ is not a simple MAP computation, due to the constraint that $\boldsymbol{y}^{map} \neq \boldsymbol{y}[m]$. However, this difficulty arises only in the case where the MAP assignment is $\boldsymbol{y}[m]$, in which case we need only find the second-best assignment. Fortunately, it is not difficult to adapt most MAP solution methods to the task of finding the second-best assignment (see, for example, exercise 13.5).

The use of MAP rather than sum-product as the inference algorithm used in the inner loop of the learning algorithm can be of significance. As we discussed, MAP inference admits the use of more efficient optimization algorithms that are not applicable to sum-product. In fact, as we discussed in section 13.6, there are even cases where sum-product is intractable, whereas MAP can be solved in polynomial time.

However, the margin constraints we use here fail to address two important issues. First, we are not guaranteed that there exists a model that can correctly select $\boldsymbol{y}[m]$ as the MAP assignment for every data instance m: First, our training data may be noisy, in which case $\boldsymbol{y}[m]$ may not be the actual desired assignment. More importantly, our model may not be expressive enough to always pick the desired target assignment (and the "simple" solution of increasing its expressive power may lead to overfitting). Because of the worst-case nature of our optimization objective, when we cannot achieve a positive margin for every data instance, there is no longer any incentive in getting a better margin for those instances where a positive margin can be achieved. Thus, the solution we obtain becomes meaningless. To address this problem, we must allow for instances to have a nonpositive margin and simply penalize such exceptions in the objective; the penalization takes the form of *slack variables* η_m that measure the extent of the violation for the m'th data instances. This approach allows the optimization to trade off errors in the labels of a few instances for a better solution overall.

A second, related problem arises from our requirement that our model achieve a uniform margin for all $\boldsymbol{y} \neq \boldsymbol{y}[m]$. To see why this requirement can be problematic, consider again our image segmentation problem. Here, $\boldsymbol{x}[m]$ are features derived from the image, $\boldsymbol{y}[m]$ is our "ground truth" segmentation, and other assignments $\boldsymbol{y}$ are other candidate segmentations. Some of these candidate segmentations differ from $\boldsymbol{y}[m]$ only in very limited ways (perhaps a few pixels are assigned a different label). In this case, we expect that a reasonable model $P_{\boldsymbol{\theta}}$ will ascribe a probability to these "almost-correct" candidates that is very close to the probability of the ground truth. If so, it will be difficult to find a good model that achieves a high margin. Again, due to the worst-case nature of the objective, this can lead to inferior models. We address this concern by allowing the required margin $\ln P(\boldsymbol{y}[m] \mid \boldsymbol{x}[m]) - \ln P(\boldsymbol{y} \mid \boldsymbol{x}[m])$ to vary with the "distance" between $\boldsymbol{y}[m]$ and $\boldsymbol{y}$, with assignments $\boldsymbol{y}$ that are more similar to $\boldsymbol{y}[m]$ requiring **Hamming loss** a smaller margin. In particular, using the ideas of the *Hamming loss*, we can define $\Delta_m(\boldsymbol{y})$ to be the number of variables $Y_i \in \boldsymbol{Y}$ such that $y_i \neq y_i[m]$, and require that the margin increase linearly in this discrepancy.

Putting these two modifications together, we obtain our final optimization problem:

Max-Margin:

> **Find** $\boldsymbol{\theta}$
> **maximizing** $\|\boldsymbol{\theta}\|_2^2 + C \sum_m \eta_m$
> **subject to**
>
> $$\boldsymbol{\theta}^T \left(\boldsymbol{f}(\boldsymbol{y}[m], \boldsymbol{x}[m]) - \boldsymbol{f}(\boldsymbol{y}, \boldsymbol{x}[m]) \right) \geq \Delta_m(\boldsymbol{y}) - \eta_m \quad \forall m, \boldsymbol{y} \neq \boldsymbol{y}[m].$$

Here, C is a constant that determines the balance between the two parts of the objective: how much we choose to penalize mistakes (negative margins) for some instances, versus achieving a higher margin overall.

Fortunately, the same constraint generation approach that we discussed can also be applied in this case (see exercise 20.14).

20.7   Structure Learning

model selection We now move to the problem of *model selection*: learning a network structure from data. As usual, there are two types of solution to this problem: the constraint-based approaches, which search for a graph structure satisfying the independence assumptions that we observe in the empirical distribution; and the score-based approaches, which define an objective function for different models, and then search for a high-scoring model.

From one perspective, the constraint-based approaches appear relatively more advantageous here than they did in the case of Bayesian network learning. First, the independencies associated with separation in a Markov network are much simpler than those associated with d-separation in a Bayesian network; therefore, the algorithms for inferring the structure are much simpler here. Second, recall that all of our scoring functions were based on the likelihood function; here, unlike in the case of Bayesian networks, even evaluating the likelihood function is a computationally expensive procedure, and often an intractable one.

On the other side, the disadvantage of the constraint-based approaches remains: their lack of robustness to statistical noise in the empirical distribution, which can give rise to incorrect independence assumptions. We also note that the constraint based approaches produce only a structure, and not a fully specified model of a distribution. To obtain such a distribution, we need to perform parameter estimation, so that we eventually encounter the computational costs associated with the likelihood function. Finally, in the context of Markov network learning, it is not clear that learning the global independence structure is necessarily the appropriate problem. In the context of learning Bayesian networks we distinguished between learning the global structure (the directed graph) and local structure (the form of each CPD). In learning undirected models we can similarly consider both the problem of learning the undirected graph structure and the particular set of factors or features that represent the parameterization of the graph. Here, however, it is quite common to find distributions that have a compact factorization yet have a complex graph structure. One extreme example is the fully connected network with pairwise potentials. Thus, in many domains we want to learn the factorization of the joint distribution, which often cannot be deduced from the global independence assumptions.

We will review both types of approach, but we will focus most of the discussion on score-based approaches, since these have received more attention.

20.7.1   Structure Learning Using Independence Tests

constraint-based
structure learning

independence
tests

We first consider the idea of *constraint-based structure learning*. Recall that the structure of a Markov network specifies a set of independence assertions. We now show how we can use *independence tests* to reconstruct the Markov network structure. For this discussion, assume that the generating distribution P^* is positive and can be represented as a Markov network $\mathcal{H}^*$ that is a perfect map of P^*. Thus, we want to perform a set of independence tests on P^* and recover $\mathcal{H}^*$. To make the problem tractable, we further assume that the degree of nodes in $\mathcal{H}^*$ is at most d^*.

Recall that in section 4.3.2 we considered three set sets of independencies that characterize a Markov network: global independencies that include all consequences of separation in the graph; Markov independencies that describe the independence of each variable X from the rest of the variables given its Markov blanket; and pairwise independencies that describe the independence of each nonadjacent pair of variables X, Y given all other variables. We showed there that these three definitions are equivalent in positive distributions.

Can we use any of these concepts to recover the structure of $\mathcal{H}^*$? Intuitively, we would prefer to examine a smaller set of independencies, since they would require fewer independence tests. Thus, we should focus either on the local Markov independencies or pairwise independencies.

local Markov
independencies

Recall that *local Markov independencies* are of the form

$$(X \perp \mathcal{X} - \{X\} - \mathrm{MB}_{\mathcal{H}^*}(X) \mid \mathrm{MB}_{\mathcal{H}^*}(X)) \quad \forall X$$

pairwise
independencies

and *pairwise independencies* are of the form

$$(X \perp Y \mid \mathcal{X} - \{X, Y\}) \quad \forall (X\!-\!Y) \notin \mathcal{H}.$$

Unfortunately, as written, neither of these sets of independencies can be checked tractably, since both involve the entire set of variables $\mathcal{X}$ and hence require measuring the probability of exponentially many events. The computational infeasibility of this requirement is obvious. But equally problematic are the statistical issues: these independence assertions are evaluated not on the true distribution, but on the empirical distribution. Independencies that involve many variables lead to fragmentation of the data, and are much harder to evaluate without error. To estimate the distribution sufficiently well as to evaluate these independencies reliably, we would need exponentially many data points.

Thus, we need to consider alternative sets of independencies that involve only smaller subsets of variables. Several such approaches have been proposed; we review only one, as an example. Consider the network $\mathcal{H}^*$. Clearly, if X and Y are not neighbors in $\mathcal{H}^*$, then they are separated by the *Markov blanket* $\mathrm{MB}_{\mathcal{H}^*}(X)$ and also by $\mathrm{MB}_{\mathcal{H}^*}(Y)$. Thus, we can find a set $\mathbf{Z}$ with $|\mathbf{Z}| \leq \min(|\mathrm{MB}_{\mathcal{H}^*}(X)|, |\mathrm{MB}_{\mathcal{H}^*}(Y)|)$ so that $sep_{\mathcal{H}^*}(X; Y \mid \mathbf{Z})$ holds. On the other hand, if X and Y are neighbors in $\mathcal{H}^*$, then we cannot find such a set $\mathbf{Z}$. Because $\mathcal{H}^*$ is a perfect map of P^*, we can show that

$$X \!-\! Y \notin \mathcal{H}^* \quad \text{if and only if} \quad \exists \mathbf{Z}, |\mathbf{z}| \leq d^* \& P^* \models (X \perp Y \mid \mathbf{Z}).$$

Thus, we can determine whether $X \!-\! Y$ is in $\mathcal{H}^*$ using $\sum_{k=0}^{d^*} \binom{n-2}{k}$ independence tests. Each of these independence tests involves only $d^* + 2$ variables, which, for low values of d^*, can be tractable. We have already encountered this test in section 3.4.3.1, as part of our Bayesian network construction procedure. If fact, it is not hard to show that, given our assumptions and perfect independence tests, the Build-PMap-Skeleton procedure of algorithm 3.3 reconstructs the correct Markov structure $\mathcal{H}^*$ (exercise 20.15).

This procedure uses a polynomial number of tests. Thus, the procedure runs in polynomial time. Moreover, if the probability of a false answer in any single independence test is at most ϵ, then the probability that any one of the independence tests fails is at most $\sum_{k=0}^{d^*} \binom{n-2}{k} \epsilon$. Therefore, for sufficiently small ϵ, we can use this analysis to prove that we can reconstruct the correct network structure $\mathcal{H}^*$ with high probability.

While this result is satisfying at some level, there are significant limitations. **First, the number of samples required to obtain correct answers for all of the independence tests can be very large in practice. Second, the correctness of the algorithm is based on several important assumptions: that there is a Markov network that is a perfect map of P^*; that this network has a bounded degree; and that we have enough data to obtain reliable answers to the independence tests. When these assumptions are violated, this algorithm can learn incorrect network structures.**

Example 20.4

Assume that the underlying distribution P^* is a Bayesian network with a v-structure $X \to Z \leftarrow Y$. We showed in section 3.4.3 that, assuming perfect independence tests, Build-PMap-Skeleton *learns the skeleton of $\mathcal{G}^*$. However, the Markov network $\mathcal{H}^*$ that is an I-map for P^* is the moralized network, which contains, in addition to the skeleton edges, edges between parents of a joint child. These edges will not be learned correctly by this procedure. In particular, we have that $(X \perp Y \mid \emptyset)$ holds, and so the algorithm will allow us to remove the edge between X and Y, even though it exists in the true network $\mathcal{H}^*$.* ∎

The failure in this example results from the fact that the distribution P^* does not have a perfect

Markov blanket

map that is a Markov network. Because many real-life distributions do not have a perfect map that is a compact graph, the applicability of this approach can be limited.

Moreover, as we discussed, this approach focuses solely on reconstructing the network structure and does not attempt to learn the the structure of the factorization, or to estimate the parameters. In particular, we may not have enough data to reliably estimate parameters for the structure learned by this procedure, limiting its usability in practice. Nevertheless, as in the case of Bayesian network structure learning, constraint-based approaches can be a useful tool for obtaining qualitative insight into the global structure of the distribution, and as a starting point for the search in the score-based methods.

20.7.2 Score-Based Learning: Hypothesis Spaces

hypothesis space

We now move to the score-based structure learning approach. As we discussed earlier, this approach formulates structure learning as an optimization problem: We define a *hypothesis space* consisting of a set of possible networks; we also define an objective function, which is used to score different candidate networks; and then we construct a search algorithm that attempts to identify a high-scoring network in the hypothesis space. We begin in this section by discussing the choice of hypothesis space for learning Markov networks. We discuss objective functions and the search strategy in subsequent sections.

There are several ways of formulating the search space for Markov networks, which vary in terms of the granularity at which they consider the network parameterization. At the coarsest-grained, we can pose the hypothesis space as the space of different structures of the Markov network itself and measure the model complexity in terms of the size of the cliques in the network. At the next level, we can consider parameterizations at the level of the factor graph, and measure complexity in terms of the sizes of the factors in this graph. At the finest level of granularity, we can consider a search space at the level of individual features in a log-linear model, and measure sparsity at the level of features included in the model.

The more fine-grained our hypothesis space, the better it allows us to select a parameterization that matches the properties of our distribution without overfitting. For example, the factor-graph approach allows us to distinguish between a single large factor over k variables and a set of $\binom{k}{2}$ pairwise factors over the same variables, requiring far fewer parameters. The feature-based approach also allows us to distinguish between a full factor over k variables and a single log-linear feature over the same set of variables.

Conversely, the finer-grained spaces can obscure the connection to the network structure, in that sparsity in the space of features selected does not correspond directly to sparsity in the model structure. For example, introducing even a single feature $f(d)$ into the model has the structural effect of introducing edges between all of the variables in d. Thus, even models with a fairly small number of features can give rise to dense connectivity in the induced network. While this is not a problem from the statistical perspective of reliably estimating the model parameters from limited data, it can give rise to significant problems from the perspective of performing inference in the model. Moreover, a finer-grained hypothesis space also means that search algorithms take smaller steps in the space, potentially increasing the cost of our learning procedure. We will return to some of these issues.

We focus our presentation on the formulation of the search space in terms of log-linear models. Here, we have a set of features Ω, which are those that can potentially have nonzero weight. Our task is to select a log-linear model structure $\mathcal{M}$, which is defined by some subset $\Phi[\mathcal{M}] \subseteq \Omega$. Let $\Theta[\mathcal{M}]$ be the set of parameterizations $\boldsymbol{\theta}$ that are compatible with the model structure: that is, those where $\theta_i \neq 0$ only if $f_i \in \Phi[\mathcal{M}]$. A structure and a compatible parameterization define a log-linear distribution via:

$$P(\mathcal{X} \mid \mathcal{M}, \boldsymbol{\theta}) = \frac{1}{Z} \exp \left\{ \sum_{i \in \Phi[\mathcal{M}]} \theta_i f_i(\xi) \right\} = \frac{1}{Z} \exp \left\{ \boldsymbol{f}^T \boldsymbol{\theta} \right\},$$

where, because of the compatibility of $\boldsymbol{\theta}$ with $\mathcal{M}$, a feature not in $\Phi[\mathcal{M}]$ does not influence in the final vector product, since it is multiplied by a parameter that is 0.

Regardless of the formulation chosen, we may sometimes wish to impose structural constraints that restrict the set of graph structures that can be selected, in order to ensure that we learn a network with certain sparsity properties. In particular, one choice that has received some attention is to restrict the class of networks learned to those that have a certain bound on the *tree-width*. By placing a tight bound on the tree-width, we prevent an overly dense network from being selected, and thereby reduce the chance of overfitting. Moreover, because models of low tree-width allow exact inference to be performed efficiently (to some extent), this restriction also allows the computational steps required for evaluating the objective during the search to be performed efficiently. However, this approach also has limitations. First, it turns out to be nontrivial to implement, since computing the tree-width of a graph is itself an intractable problem (see theorem 9.7); even keeping the graph under the required width is not simple. Moreover, many of the distributions that arise in real-world applications cannot be well represented by networks of low tree-width.

bounded
tree-width

20.7.3   Objective Functions

We now move to considering the objective function that we aim to optimize in the score-based approach. We note that our discussion in this section uses the likelihood function as the basis for the objectives we consider; however, we can also consider similar objectives based on various approximations to the likelihood (see section 20.6); most notably, the pseudolikelihood has been used effectively as a substitute for the likelihood in the context of structure learning, and most of our discussion carries over without change to that setting.

20.7.3.1   Likelihood Function

The most straightforward objective function is the likelihood of the training data. As before, we take the score to be the log-likelihood, defining:

$$\text{score}_L(\mathcal{M} \; : \; \mathcal{D}) = \max_{\boldsymbol{\theta} \in \Theta[\mathcal{M}]} \ln P(\mathcal{D} \mid \mathcal{M}, \boldsymbol{\theta}) = \ell(\langle \mathcal{M}, \hat{\boldsymbol{\theta}}_{\mathcal{M}} \rangle : \mathcal{D}),$$

where $\hat{\boldsymbol{\theta}}_{\mathcal{M}}$ are the maximum likelihood parameters compatible with $\mathcal{M}$.

The likelihood score measures the fitness of the model to the data. However, for the same reason discussed in chapter 18, it prefers more complex models. In particular, if $\Phi[\mathcal{M}_1] \subset \Phi[\mathcal{M}_2]$ then $\text{score}_L(\mathcal{M}_1 \; : \; \mathcal{D}) \leq \text{score}_L(\mathcal{M}_2 \; : \; \mathcal{D})$. Typically, this inequality is strict, due to the ability of the richer model to capture noise in the data.

Therefore, the likelihood score can be used only with very strict constraints on the expressive model of the model class that we are considering. Examples include bounds on the structure of the Markov network (for example, networks with low tree-width) or on the number of features used. A second option, which also provides some regularization of parameter values, is to use an alternative objective that penalizes the likelihood in order to avoid overfitting.

20.7.3.2 Bayesian Scores

Bayesian score

Recall that, for Bayesian networks, we used a *Bayesian score*, whose primary term is a marginal likelihood that integrates the likelihood over all possible network parameterizations: $\int P(\mathcal{D} \mid \mathcal{M}, \boldsymbol{\theta})P(\boldsymbol{\theta} \mid \mathcal{M})d\boldsymbol{\theta}$. This score accounts for our uncertainty over parameters using a Bayesian prior; it avoided overfitting by preventing overly optimistic assessments of the model fit to the training data. In the case of Bayesian networks, we could efficiently evaluate the marginal likelihood. In contrast, in the case of undirected models, this quantity is difficult to evaluate, even using approximate inference methods.

Instead, we can use asymptotic approximations of the marginal likelihood. The simplest approximation is the *BIC score*:

BIC score

$$\text{score}_{BIC}(\mathcal{M} \; : \; \mathcal{D}) = \ell(\langle\mathcal{M}, \hat{\boldsymbol{\theta}}_{\mathcal{M}}\rangle : \mathcal{D}) - \frac{\dim(\mathcal{M})}{2} \ln M,$$

model dimension

where $\dim(\mathcal{M})$ is the *dimension* of the model and M the number of instances in $\mathcal{D}$. This quantity measures the degrees of freedom of our parameter space. When the model has nonredundant features, $\dim(\mathcal{M})$ is exactly the number of features. When there is redundancy, the dimension is smaller than the number of features. Formally, it is the rank of the matrix whose rows are complete assignments ξ_i to $\mathcal{X}$, whose columns are features f_j, and whose entries are $f_j(\xi_i)$. This matrix, however, is exponential in the number of variables, and therefore its rank cannot be computed efficiently. Nonetheless, we can often estimate the number of nonredundant parameters in the model. As a very coarse upper bound, we note that the number of nonredundant features is always upper-bounded by the size of the full table representation of the Markov network, which is the total number of entries in the factors.

The BIC approximation penalizes each degree of freedom (that is, free parameter) by a fixed amount, which may not be the most appropriate penalty. Several more refined alternatives have been proposed. One common choice is the *Laplace approximation*, which provides a more explicit approximation to the marginal likelihood:

Laplace approximation

$$\text{score}_{Laplace}(\mathcal{M} \; : \; \mathcal{D}) = \ell(\langle\mathcal{M}, \tilde{\boldsymbol{\theta}}_{\mathcal{M}}\rangle : \mathcal{D}) + \ln P(\tilde{\boldsymbol{\theta}}_{\mathcal{M}} \mid \mathcal{M}) + \frac{\dim(\mathcal{M})}{2} \ln(2\pi) - \frac{1}{2} \ln |A|,$$

MAP estimation

where $\tilde{\boldsymbol{\theta}}_{\mathcal{M}}$ are the parameters for $\mathcal{M}$ obtained from *MAP estimation*:

$$\tilde{\boldsymbol{\theta}}_{\mathcal{M}} = \arg\max_{\boldsymbol{\theta}} P(\mathcal{D} \mid \boldsymbol{\theta}, \mathcal{M})P(\boldsymbol{\theta} \mid \mathcal{M}), \tag{20.28}$$

Hessian

and A is the negative *Hessian* matrix:

$$A_{i,j} = -\frac{\partial}{\partial\theta_i\partial\theta_j} \left(\ell(\langle\mathcal{M}, \boldsymbol{\theta}\rangle : \mathcal{D}) + \ln P(\boldsymbol{\theta} \mid \mathcal{M})\right),$$

evaluated at the point $\tilde{\boldsymbol{\theta}}_{\mathcal{M}}$.

As we discussed in section 19.4.1.1, the Laplace score also takes into account the local shape of the posterior distribution around the MAP parameters. It therefore provides a better approximation than the BIC score. However, as we saw in equation (20.5), to compute the Hessian, we need to evaluate the pairwise covariance of every feature pair given the model, a computation that may be intractable in many cases.

20.7.3.3   Parameter Penalty Scores

An alternative to approximations of the marginal likelihood are methods that simply evaluate the maximum posterior probability

$$\text{score}_{MAP}(\mathcal{M} \; : \; \mathcal{D}) = \max_{\boldsymbol{\theta} \in \Theta[\mathcal{M}]} \ell(\langle \mathcal{M}, \tilde{\boldsymbol{\theta}}_{\mathcal{M}} \rangle : \mathcal{D}) + \ln P(\tilde{\boldsymbol{\theta}}_{\mathcal{M}} \mid \mathcal{M}), \qquad (20.29)$$

MAP score

where $\tilde{\boldsymbol{\theta}}_{\mathcal{M}}$ are the MAP parameters for $\mathcal{M}$, as defined in equation (20.28). One intuition for this type of *MAP score* is that the prior "regularizes" the likelihood, moving it away from the maximum likelihood values. If the likelihood of these parameters is still high, it implies that the model is not too sensitive to particular choice of maximum likelihood parameters, and thus it is more likely to generalize.

Although the regularized parameters may achieve generalization, this approach achieves model selection only for certain types of prior. To understand why, note that the MAP score is based on a distribution not over structures, but over parameters. We can view any parameterization $\boldsymbol{\theta}_{\mathcal{M}}$ as a parameterization to the "universal" model defined over our entire set of features Ω: one where features not in $\Phi[\mathcal{M}]$ receive weight 0. Assuming that our parameter prior simply ignores zero weights, we can view our score as simply evaluating different choices of parameterizations $\boldsymbol{\theta}_{\Omega}$ to this universal model.

We have already discussed several parameter priors and their effect on the learned parameters. Most parameter priors are associated with the magnitude of the parameters, rather than the complexity of the graph as a discrete data structure. In particular, as we discussed, although
L_2-regularization L_2-*regularization* will tend to drive the parameters toward zero, few will actually hit zero, and so structural sparsity will not be achieved. Thus, like the likelihood score, the L_2-regularized MAP objective will generally give rise to a fully connected structure. Therefore, this approach is generally not used in the context of model selection (at least not in isolation).
L_1-regularization   A more appropriate approach for this task is L_1-*regularization*, which does have the effect of driving model parameters toward zero, and thus can give rise to a sparse set of features. In other
L_1-MAP score   words, the structure that optimizes the L_1-*MAP score* is not, in general, the universal structure Ω. Indeed, as we will discuss, an L_1 prior has other useful properties when used as the basis for a structure selection objective.

However, as we have discussed, feature-level sparsity does not necessarily induce sparsity in
block-L_1-regularization   the network. An alternative that does tend to have this property is the *block-L_1-regularization*. Here, we partition all the parameters into groups $\boldsymbol{\theta}_i = \{\theta_{i,1}, \dots, \theta_{i,k_i}\}$ (for $i = 1, \dots, l$). We now define a variant of the L_1 penalty that tends to make each parameter *group* either go to zero together, or not:

$$-\sum_{i=1}^{l} \left| \sqrt{\sum_{j=1}^{k_i} \theta_{i,j}^2} \right|. \qquad (20.30)$$

To understand the behavior of this penalty term, let us consider its derivative for the simple case where we have two parameters in the same group, so that our expression takes the form $\sqrt{\theta_1^2 + \theta_2^2}$. We now have that:

$$\frac{\partial}{\partial \theta_1} \left[-\sqrt{\theta_1^2 + \theta_2^2} \right] = -\frac{\theta_1}{\sqrt{\theta_1^2 + \theta_2^2}}.$$

We therefore see that, when θ_2 is large, the derivative relative to θ_1 is fairly small, so that there is no pressure on θ_1 to go to 0. Conversely, when θ_2 is small, the derivative relative to θ_1 tends to -1, which essentially gives the same behavior as L_1 regularization. Thus, this prior tends to have the following behavior: if the overall magnitude of the parameters in the group is small, all of them will be forced toward zero; if the overall magnitude is large, there is little downward pressure on any of them.

In our setting, we can naturally apply this prior to give rise to sparsity in network structure. Assume that we are willing to consider, within our network, factors over scopes $\boldsymbol{Y}_1, \ldots, \boldsymbol{Y}_l$. For each $\boldsymbol{Y}_i$, let $f_{i,j}$, for $j = 1, \ldots, k_i$, be all of the features whose scope is $\boldsymbol{Y}_i$. We now define a block-L_1 prior where we have a block for each set of parameters $\theta_{i1}, \ldots, \theta_{ik_i}$. The result of this prior would be to select together nonzero parameters for an entire set of features associated with a particular scope.

Finally, we note that one can also use multiple penalty terms on the likelihood function. For example, a combination of a parameter penalty and a structure penalty can often provide both regularization of the parameters and a greater bias toward sparse structures.

20.7.4 Optimization Task

Having selected an objective function for our model structure, it remains to address the optimization problem.

20.7.4.1 Greedy Structure Search

local search

As in the approach used for structure learning of general Bayesian networks (section 18.4.3), the external search over structures is generally implemented by a form of *local search*. Indeed, the general form of the algorithms in appendix A.4.2 applies to our feature-based view of Markov network learning. The general template is shown in algorithm 20.1. Roughly speaking, the algorithm maintains a current structure, defined in terms of a set of features $\mathcal{F}$ in our log-linear model. At each point in the search, the algorithm optimizes the model parameters relative to the current feature set and the structure score. Using the current structure and parameters, it estimates the improvement of different structure modification steps. It then selects some subset of modifications to implement, and returns to the parameter optimization step, initializing from the current parameter setting. This process is repeated until a termination condition is reached. This general template can be instantiated in many ways, including the use of different hypothesis spaces (as in section 20.7.2) and different scoring functions (as described in section 20.7.3).

20.7.4.2 Successor Evaluation

Although this approach is straightforward at a high level, there are significant issues with its implementation. Importantly, the reasons that made this approach efficient in the context of

---

Algorithm 20.1 Greedy score-based structure search algorithm for log-linear models

---

Procedure Greedy-MN-Structure-Search (
    Ω,   // All possible features
    $\mathcal{F}_0$,   // initial set of features
    $\text{score}(\cdot \ : \ \mathcal{D})$,   // Score
)
1        $\mathcal{F}' \leftarrow \mathcal{F}_0$ // New feature set
2        $\boldsymbol{\theta} \leftarrow \mathbf{0}$
3   **do**
4            $\mathcal{F} \leftarrow \mathcal{F}'$
5            $\boldsymbol{\theta} \leftarrow$ Parameter-Optimize$(\mathcal{F}, \boldsymbol{\theta}, \text{score}(\cdot \ : \ \mathcal{D}))$
6   // Find parameters that optimize the score objective, relative to
                   current feature set, initializing from the current parameters
7   **for** each $f_k \in \mathcal{F}$ such that $\theta_k = 0$
8                $\mathcal{F} \leftarrow \mathcal{F} - f_k$
9   // Remove inactive features
10   **for** each operator o applicable to $\mathcal{F}$
11   Let $\hat{\Delta}_o$ be the approximate improvement for o
12   Choose some subset $\mathcal{O}$ of operators based on $\hat{\Delta}$
13                $\mathcal{F}' \leftarrow \mathcal{O}(\mathcal{F})$ // Apply selected operators to $\mathcal{F}$
14   **while** termination condition not reached
15   **return** $(\mathcal{F}, \boldsymbol{\theta})$

Bayesian networks do not apply here. In the case of Bayesian networks, evaluating the score of a candidate structure is a very easy task, which can be executed in closed form, at very low computation cost. Moreover, the Bayesian network score satisfies an important property:

score
decomposability

it *decomposes* according to the structure of the network. As we discussed, this property has two major implications. First, a local modification to the structure involves changing only a single term in the score (proposition 18.5); second, the change in score incurred by a particular change (for example, adding an edge) remains unchanged after modifications to other parts of the network (proposition 18.6). These properties allowed us to design efficient search procedure that does not need to reevaluate all possible candidates after every step, and that can cache intermediate computations to evaluate candidates in the search space quickly.

Unfortunately, none of these properties hold for Markov networks. For concreteness, consider the likelihood score, which is comparable across both network classes. First, as we discussed, even computing the likelihood of a fully specified model — structure as well as parameters — requires that we run inference for every instance in our training set. Second, to score a structure, we need to estimate the parameters for it, a problem for which there is no closed-form solution. Finally, none of the decomposition properties hold in the case of undirected models. By adding a new feature (or a set of features, for example, a factor), we change the weight $(\sum_i \theta_i f_i(\xi))$ associated with different instances. This change can be decomposed, since it is a linear function of the different features. However, this change also affects the partition function, and, as we saw in the context of parameter estimation, the partition function couples the effects of changes in one parameter on the other. We can clearly see this phenomenon in figure 20.1, where the effect on the likelihood of modifying $f_1(b^1, c^1)$ clearly depends on the current value of the parameter for $f_2(a^1, b^1)$.

As a consequence, a local search procedure is considerably more expensive in the context of Markov networks. At each stage of the search, we need to evaluate the score for all of the candidates we wish to examine at that point in the search. This evaluation requires that we estimate the parameters for the structure, a process that itself requires multiple iterations of a numerical optimization algorithm, each involving inference over all of the instances in our training set. We can reduce somewhat the computational cost of the algorithm by using the observation that a single change to the structure of the network often does not result in drastic changes to the model. Thus, if we begin our optimization process from the current set of parameters, a reasonably small number of iterations often suffices to achieve convergence to the new set of parameters. Importantly, because all of the parameter objectives described are convex (when we have fully observable data), the initialization has no effect, and convergence to the global optimum remains guaranteed. Thus, this approach simply provides a way of speeding up the convergence to the optimal answer. (We note, however, that this statement holds only when we use exact inference; the choice of initialization can affect the accuracy of some approximate inference algorithms, and therefore the answers that we get.)

 Unfortunately, although this observation does reduce the cost, the number of candidate hypotheses at each step is generally quite large. **The cost of running inference on each of the candidate successors is prohibitive, especially in cases where, to fit our target distribution well, we need to consider nontrivial structures. Thus, much of the work on the problem of structure learning for Markov networks has been devoted to reducing the computational cost of evaluating the score of different candidates during the search.** In particular, when evaluating different structure-modification operators in line 11, most algorithms use some heuristic to rank different candidates, rather than computing the exact delta-score of each operator. These heuristic estimates can be used either as the basis for the final selection, or as a way of pruning the set of possible successors, where the high-ranking candidates are then evaluated exactly. This design decision is a trade-off between the quality of our operator selection and the computational cost.

Even with the use of heuristics, the cost of taking a step in the search can be prohibitive, since it requires a reestimation of the network parameters and a reevaluation of the (approximate) delta-score for all of the operators. This suggests that it may be beneficial to select, at each structure modification step (line 12), not a single operator but a subset $\mathcal{O}$ of operators. This approach can greatly reduce the computational cost, but at a cost: our (heuristic) estimate of each operator can deteriorate significantly if we fail to take into account interactions between the effects of different operators. Again, this is a trade-off between the quality and cost of the operator selection.

20.7.4.3 Choice of Scoring Function

As we mentioned, the rough template of algorithm 20.1 can be applied to any objective function. However, the choice of objective function has significant implications on our ability to effectively optimize it. Let us consider several of the choices discussed earlier.

We first recall that both the log-likelihood objective and the L_2-regularized log-likelihood generally give nonzero values to all parameters. In other words, if we allow the model to consider a set of features $\mathcal{F}$, an optimal model (maximum-likelihood or maximum L_2-regularized likelihood) over $\mathcal{F}$ will give a nonzero value to the parameters for all of these features. In other words, we cannot rely on these objectives to induce sparsity in the model structure. Thus, if we simply want to optimize these objectives, we should simply choose the richest model available in our hypothesis space and then optimize its parameters relative to the chosen objective.

One approach for deriving more compact models is to restrict the class of models to ones with a certain bound on the complexity (for example, networks of bounded tree-width, or with a bound on the number of edges or features allowed). However, these constraints generally introduce nontrivial combinatorial trade-offs between features, giving rise to a search space with multiple local optima, and making it generally intractable to find a globally optimal solution. A second approach is simply to halt the search when the improvement in score (or an approximation to it) obtained by a single step does not exceed a certain threshold. This heuristic is not unreasonable, since good features are generally introduced earlier, and so there is a general trend of diminishing returns. However, there is no guarantee that the solution we obtain is even close to the optimum, since there is no bound on how much the score would improve if we continue to optimize beyond the current step.

Scoring functions that explicitly penalize structure complexity — such as the BIC score or Laplace approximation — also avoid this degeneracy. Here, as in the case of Bayesian networks, we can consider a large hypothesis space and attempt to find the model in this space that optimizes the score. However, due to the discrete nature of the structure penalty, the score is discontinuous and therefore nonconcave. Thus, there is generally no guarantee of convergence to the global optimum. Of course, this limitation was also the case when learning Bayesian networks; as there, it can be somewhat alleviated by methods that avoid local maxima (such as tabu search, random restarts, or data perturbation).

However, in the case of Markov networks, we have another solution available to us, one that avoids the prospect of combinatorial search spaces and the ensuing problem of local optima. This solution is the use of L_1-regularized likelihood. As we discussed, the L_1-regularized likelihood is a concave function that has a unique global optimum. Moreover, this objective function naturally gives rise to sparse models, in that, at the optimum, many parameters have value 0, corresponding to the elimination of features from the model. We discuss this approach in more detail in the next section.

20.7.4.4 L_1-Regularization for Structure Learning

Recall that the L_1-regularized likelihood is simply the instantiation of equation (20.29) to the case of an L_1-prior:

$$\text{score}_{L_1}(\boldsymbol{\theta} \; : \; \mathcal{D}) = \ell(\langle \mathcal{M}, \boldsymbol{\theta} \rangle : \mathcal{D}) - \|\boldsymbol{\theta}\|_1. \tag{20.31}$$

Somewhat surprisingly, the L_1-regularized likelihood can be optimized in a way that guarantees convergence to the globally optimal solution. To understand why, recall that the task of optimizing the L_1-regularized log-likelihood is a convex optimization problem that has no local optima.[1] Indeed, in theory, we can entirely avoid the combinatorial search component when

---

1. There might be multiple global optima due to redundancy in the parameter space, but these global optima all form a single convex region. Therefore, we use the term "the global optimum" to refer to any point in this optimal region.

using this objective. We can simply introduce all of the possible features into the model and optimize the resulting parameter vector $\boldsymbol{\theta}$ relative to our objective. The sparsifying effect of the L_1 penalty will drive some of the parameters to zero. The parameters that, at convergence, have zero values correspond to features that are absent from the log-linear model. In this approach, we are effectively making a structure selection decision as part of our parameter optimization procedure. Although appealing, this approach is not generally feasible. In most cases, the number of potential features we may consider for inclusion in the model is quite large. Including all of them in the model simultaneously gives rise to an intractable structure for the inference that we use as part of the computation of the gradient.

Therefore, even in the context of the L_1-regularized likelihood, we generally implement the optimization as a double-loop algorithm where we separately consider the structure and parameters. However, there are several benefits to the L_1-regularized objective:

- We do not need to consider feature deletion steps in our combinatorial search.
- We can consider feature introduction steps in any (reasonable) order, and yet achieve convergence to the global optimum.
- We have a simple and efficient test for determining convergence.
- We can prove a PAC-learnability generalization bound for this type of learning.

We now discuss each of these points.

For the purpose of this discussion, assume that we currently have a model over a set of features $\mathcal{F}$, and assume that $\boldsymbol{\theta}^l$ optimizes our L_1-regularized objective, subject to the constraint that θ_k^l can be nonzero only if $f_k \in \mathcal{F}$. At this convergence point, any feature deletion step cannot improve the score: Consider any $f_k \in \mathcal{F}$; the case where f_k is deleted is already in the class of models that was considered when we optimized the choice of $\boldsymbol{\theta}^l$ — it is simply the model where $\theta_k^l = 0$. Indeed, the algorithm already discards features whose parameter was zeroed by the continuous optimization procedure (line 7 of algorithm 20.1). If our current optimized model $\boldsymbol{\theta}^l$ has $\theta_k^l \neq 0$, it follows that setting θ_k to 0 is suboptimal, and so deleting f_k can only reduce the score. Thus, there is no value to considering discrete feature deletion steps: features that should be deleted will have their parameters set to 0 by the continuous optimization procedure. We note that this property also holds, in principle, for other smooth objectives, such as the likelihood or the L_2-regularized likelihood; the difference is that for those objectives, parameters will generally not be set to 0, whereas the L_1 objective does tend to induce sparsity.

The second benefit arises directly from the fact that optimizing the L_1-regularized objective is a convex optimization problem. In such problems, any sequence of steps that continues to improve the objective (when possible) is guaranteed to converge to the global optimum. The restriction imposed by the set $\mathcal{F}$ induces a coordinate ascent approach: at each step, we are optimizing only the features in $\mathcal{F}$, leaving at 0 those parameters θ_k for $f_k \notin \mathcal{F}$. As long as each step continues to improve the objective, we are making progress toward the global optimum. At each point in the search, we consider the steps that we can take. If some step leads to an improvement in the score, we can take that step and continue with our search. If none of the steps lead to an improvement in the score, we are guaranteed that we have reached convergence to the global optimum. **Thus, the decision on which operators to consider at each point in the algorithm (line 12 of algorithm 20.1) is not relevant to the convergence of the**

algorithm to the true global optimum: As long as we repeatedly consider each operator until convergence, we are guaranteed that the global optimum is reached regardless of the order in which the operators are applied.

While this guarantee is an important one, we should interpret it with care. First, when we add features to the model, the underlying network becomes more complex, raising the cost of inference. Because inference is executed many times during the algorithm, adding many irrelevant features, even if they were eventually eliminated, can greatly degrade the computational performance time of the algorithm. Even more problematic is the effect when we utilize approximate inference, as is often the case. As we discussed, for many approximate inference algorithms, not only the running time but also the accuracy tend to degrade as the network becomes more complex. Because inference is used to compute the gradient for the continuous optimization, the degradation of inference quality can lead to models that are suboptimal. Moreover, because the resulting model is generally also used to estimate the benefit of adding new features, any inaccuracy can propagate further, causing yet more suboptimal features to be introduced into the model. **Hence, especially when the quality of approximate inference is a concern, it is worthwhile to select with care the features to be introduced into the model rather than blithely relying on the "guaranteed" convergence to a global optimum.**

Another important issue to be addressed is the problem of determining convergence in line 14 of Greedy-MN-Structure-Search. In other words, how do we test that none of the search operators we currently have available can improve the score? A priori, this task appears daunting, since we certainly do not want to try all possible feature addition/deletion steps, reoptimize the parameters for each of them, and then check whether the score has improved. Fortunately, there is a much more tractable solution. Specifically, we can show the following proposition:

Proposition 20.5

Let $\Delta_L^{\mathrm{grad}}(\theta_k \ : \ \boldsymbol{\theta}^l, \mathcal{D})$ denote the gradient of the likelihood relative to θ_k, evaluated at $\boldsymbol{\theta}^l$. Let β be the hyperparameter defining the L_1 prior. Let $\boldsymbol{\theta}^l$ be a parameter assignment for which the following conditions hold:

- *For any k for which $\theta_k^l \neq 0$ we have that*

$$\Delta_L^{\mathrm{grad}}(\theta_k \ : \ \boldsymbol{\theta}^l, \mathcal{D}) - \frac{1}{\beta}\mathrm{sign}(\theta_k^l) = 0.$$

- *For any k for which $\theta_k^l = 0$ we have that*

$$|\Delta_L^{\mathrm{grad}}(\theta_k \ : \ \boldsymbol{\theta}^l, \mathcal{D})| < \frac{1}{2\beta}.$$

Then $\boldsymbol{\theta}^l$ is a global optimum of the L_1-regularized log-likelihood function:

$$\frac{1}{M}\ell(\boldsymbol{\theta} : \mathcal{D}) - \frac{1}{\beta}\sum_{i=1}^{k}|\theta_i|.$$

PROOF We provide a rough sketch of the proof. The first condition guarantees that the gradient relative to any parameter for which $\theta_k^l \neq 0$ is zero, and hence the objective function cannot be improved by changing its value. The second condition deals with parameters $\theta_k^l = 0$, for which the gradient is discontinuous at the convergence point. However, consider a point $\boldsymbol{\theta}'$ in the nearby vicinity of $\boldsymbol{\theta}$, so that $\theta'_k \neq 0$. At $\boldsymbol{\theta}'$, the gradient of the function relative to θ_k is very close to

$$\Delta_L^{\mathrm{grad}}(\theta_k \;:\; \boldsymbol{\theta}^l, \mathcal{D}) - \frac{1}{\beta}\mathrm{sign}(\theta'_k).$$

The value of this expression is positive if $\theta'_k < 0$ and negative if $\theta'_k > 0$. Thus, $\boldsymbol{\theta}^l$ is a local optimum of the L_1-regularized objective function. Because the function has only global optima, $\boldsymbol{\theta}^l$ must be a global optimum. ∎

Thus, we can test convergence easily as a direct by-product of the continuous parameter optimization procedure executed at each step. We note that we still have to consider every feature that is not included in the model and compute the relevant gradient; but we do not have to go through the (much more expensive) process of trying to introduce the feature, optimizing the resulting model, and evaluating its score.

So far, we have avoided the discussion of optimizing this objective. As we mentioned in
L-BFGS algorithm section 20.3.1, a commonly used method for optimizing the likelihood is the *L-BFGS algorithm*, which uses gradient descent combined with line search (see appendix A.5.2). The problem with applying this method to the L_1-regularized likelihood is that the regularization term is not continuously differentiable: the gradient relative to any parameter θ_i changes at $\theta_i = 0$ from $+1$ to -1. Perhaps the simplest solution to this problem is to adjust the line-search procedure to avoid changing the sign of any parameter θ_i: If, during our line search, θ_i crosses from positive to negative (or vice versa), we simply fix it to be 0, and continue with the line search for the remaining parameters. Note that this decision corresponds to taking f_i out of the set of active features in this iteration. If the optimal parameter assignment has a nonzero value for θ_i, we are guaranteed that f_i will be introduced again in a later stage in the search, as we have discussed.

Finally, as we mentioned, we can prove a useful theoretical guarantee for the results of
PAC-bound L_1-regularized Markov network learning. Specifically, we can show the following *PAC-bound*:

Theorem 20.4

Let $\mathcal{X}$ be a set of variables such that $|Val(X_i)| \leq d$ for all i. Let P^* be a distribution, and $\delta, \epsilon, B > 0$. Let $\mathcal{F}$ be a set of all indicator features over all subsets of variables $\boldsymbol{X} \subset \mathcal{X}$ such that $|\boldsymbol{X}| \leq c$, and let

$$\Theta_{c,B} = \{\boldsymbol{\theta} \in \Theta[\mathcal{F}] \;:\; \|\boldsymbol{\theta}\|_1 \leq B\}$$

be all parameterizations of $\mathcal{F}$ whose L_1-norm is at most B. Let $\beta = \sqrt{c\ln(2nd/\delta)/(2M)}$. Let

$$\boldsymbol{\theta}^*_{c,B} = \arg\max_{\boldsymbol{\theta}\in\Theta_{c,B}} \boldsymbol{D}(P^*\|P_{\boldsymbol{\theta}})$$

be the best parameterization achievable within the class $\Theta_{c,B}$. For any data set $\mathcal{D}$, let

$$\hat{\boldsymbol{\theta}} = \arg\max_{\boldsymbol{\theta}\in\Theta[\mathcal{F}]} \mathrm{score}_{L_1}(\boldsymbol{\theta} \;:\; \mathcal{D}).$$

Then, for

$$M \geq \frac{2cB^2}{\epsilon^2}\ln\left(\frac{2nd}{\delta}\right),$$

with probability at least $1 - \delta$,

$$\boldsymbol{D}(P^* \| P_{\hat{\boldsymbol{\theta}}}) \leq \boldsymbol{D}(P^* \| P_{\boldsymbol{\theta}^* c, B}) + \epsilon.$$

In other words, this theorem states that, with high probability over data sets $\mathcal{D}$, the relative entropy to P^* achieved by the best L_1-regularized model is at most ϵ worse than the relative entropy achieved by the best model within the class of limited-degree Markov networks. This guarantee is achievable with a number of samples that is polynomial in ϵ, c, and B, and logarithmic in δ and d. The logarithmic dependence on n may feel promising, but we note that B is a sum of the absolute values of all network parameters; assuming we bound the magnitude of individual parameters, this terms grows linearly with the total number of network parameters. Thus, L_1-regularized learning provides us with a model that is close to optimal (within the class $\Theta_{c,B}$), using a polynomial number of samples.

20.7.5   Evaluating Changes to the Model

We now consider in more detail the candidate evaluation step that takes place in line 11 of Greedy-MN-Structure-Search. As we discussed, the standard way to reduce the cost of the candidate evaluation step is simply to avoid computing the exact score of each of the candidate successors, and rather to select among them using simpler heuristics. Many approximations are possible, ranging from ones that are very simple and heuristic to ones that are much more elaborate and provide certain guarantees.

Most simply, we can examine statistics of the data to determine features that may be worth including. For example, if two variables X_i and X_j are strongly correlated in the data, it may be worthwhile to consider introducing a factor over X_i, X_j (or a pairwise feature over one or more combinations of their values). The limitation of this approach is that it does not take into account the features that have already been introduced into the model and the extent to which they already explain the observed correlation.

grafting

gradient heuristic

A somewhat more refined approach, called *grafting*, estimates the benefit of introducing a feature f_k by compute the gradient of the likelihood relative to θ_k, evaluated at the current model. More precisely, assume that our current model is $(\mathcal{F}, \boldsymbol{\theta}^0)$. The *gradient heuristic* estimate to the delta-score (for score X) obtained by adding $f_k \notin \mathcal{F}$ is defined as:

$$\Delta_X^{\mathrm{grad}}(\theta_k \; : \; \boldsymbol{\theta}^0, \mathcal{D}) = \frac{\partial}{\partial \theta_k} \mathrm{score}_X(\boldsymbol{\theta} \; : \; \mathcal{D}), \tag{20.32}$$

evaluated at the current parameters $\boldsymbol{\theta}^0$.

The gradient heuristic does account for the parameters already selected; thus, for example, it can avoid introducing features that are not relevant given the parameters already introduced. Intuitively, features that have a high gradient can induce a significant immediate improvement in the score, and therefore they are good candidates for introduction into the model. Indeed, we are guaranteed that, if θ_k has a positive gradient, introducing f_k into $\mathcal{F}$ will result in some improvement to the score. The problem with this approach is that it does not attempt to evaluate how large this improvement can be. Perhaps we can increase θ_k only by a small amount before further changes stop improving the score.

An even more precise approximation is to evaluate a change to the model by computing the score obtained in a model where we keep all other parameters fixed. Consider a step where we introduce or delete a single feature f_k in our model. We can obtain an approximation to the score by evaluating the change in score when we change only the parameter θ_k associated with f_k, keeping all other parameters unchanged. To formalize this idea, let $(\mathcal{F}, \boldsymbol{\theta}^0)$ be our

gain heuristic   current model, and consider changes involving f_k. We define the *gain heuristic* estimate to be the change in the score of the model for different values for θ_k, assuming the other parameters are kept fixed:

$$\Delta_X^{\mathrm{gain}}(\theta_k \; : \; \boldsymbol{\theta}^0, \mathcal{D}) = \mathrm{score}_X((\theta_k, \boldsymbol{\theta}_{-k}^0) \; : \; \mathcal{D}) - \mathrm{score}_X(\boldsymbol{\theta}^0 \; : \; \mathcal{D)}, \tag{20.33}$$

where $\boldsymbol{\theta}_{-k}^0$ is the vector of all parameters other than θ_k^0. As we discussed, due to the nonde-composability of the likelihood, when we change θ_k, the current assignment $\boldsymbol{\theta}_{-k}^0$ to the other parameters is generally no longer optimal. However, it is still reasonable to use this function as a heuristic to rank different steps: Parameters that give rise to a larger improvement in the objective by themselves often also induce a larger improvement when other parameters are optimized. Indeed, changing those other parameters to optimize the score can only improve it further. Thus, the change in score that we obtain when we "freeze" the other parameters and change only θ_k is a lower bound on the change in the score.

The gain function can be used to provide a lower bound on the improvement in score derived from the deletion of a feature f_k currently in the model: we simply evaluate the gain function setting $\theta_k = 0$. We can also obtain a lower bound on the value of a step where we introduce into the model a new feature f_k (that is, one for which the current parameter $\theta_k^0 = 0$). The improvement we can get if we freeze all parameters but one is clearly a lower bound on the improvement we can get if we optimize over all of the parameters. Thus, the value of $\Delta_X^{\mathrm{gain}}(\theta_k \; : \; \boldsymbol{\theta}^0, \mathcal{D})$ is a lower bound on the improvement in the objective that can be gained by setting θ_k to its chosen value and optimizing all other parameters. To compute the best lower bound, we must maximize the function relative to different possible values of θ_k, giving us the score of the best possible model when all parameters other than θ_k are frozen. In particular, we can define:

$$\mathrm{Gain}_X(\boldsymbol{\theta}^0 \; : \; f_k, \mathcal{D}) = \max_{\theta_k} \Delta_X^{\mathrm{gain}}(\theta_k \; : \; \boldsymbol{\theta}^0, \mathcal{D}).$$

This is a lower bound on the change in the objective obtained from introducing a $f_k \notin \mathcal{F}$.

In principle, lower bounds are more useful than simple approximations. If our lower bound of the candidate's score is higher than that of our current best model, then we definitely want to evaluate that candidate; this will result in a better current candidate and allow us to prune additional candidates and focus on the ones that seem more promising for evaluation. Upper bounds are useful as well. If we have a candidate model and obtain an upper bound on its score, then we can remove it from consideration once we evaluate another candidate with higher score; thus, upper bounds help us prune models for which we would never want to evaluate the true score. In practice, however, fully evaluating the score for all but a tiny handful of candidate structures is usually too expensive a proposition. Thus, the gain is generally used simply as an approximation rather than a lower bound.

How do we evaluate the gain function efficiently, or find its optimal value? The gain function is a univariate function of θ_k, which is a projection of the score function onto this single dimension. Importantly, all of our scoring functions — including any of the penalized likelihood functions we described — are concave (for a given set of active features). The projection of a concave function onto a single dimension is also concave, so that this single-parameter delta-score is also concave and therefore has a global optimum.

Nevertheless, given the complexity of the likelihood function, it is not clear how this global optimum can be efficiently found. We now show how the difference between two log-likelihood terms can be considerably simplified, even allowing a closed-form solution in certain important cases. Recall from equation (20.3) that

$$\frac{1}{M}\ell(\boldsymbol{\theta} : \mathcal{D}) = \sum_k \theta_k \boldsymbol{E}_{\mathcal{D}}[f_k] - \ln Z(\boldsymbol{\theta}).$$

Because parameters other than θ_k are the same in the two models, we have that:

$$\frac{1}{M}[\ell((\theta_k, \boldsymbol{\theta}^0_{-k}) : \mathcal{D}) - \ell(\boldsymbol{\theta}^0 : \mathcal{D})] = (\theta_k - \theta_k^0)\boldsymbol{E}_{\mathcal{D}}[f_k] - \left[\ln Z(\theta_k, \boldsymbol{\theta}^0_{-k}) - \ln Z(\boldsymbol{\theta}^0)\right].$$

The first term is a linear function in θ_k, whose coefficient is the empirical expectation of f_k in the data. For the second term, we have:

$$
\begin{aligned}
\ln \frac{Z(\theta_k, \boldsymbol{\theta}^0_{-k})}{Z(\boldsymbol{\theta}^0)} &= \ln \left[\frac{1}{Z(\boldsymbol{\theta}^0)} \sum_\xi \exp \left\{ \sum_j \theta_j^0 f_j(\xi) + (\theta_k - \theta_k^0)f_k(\xi) \right\} \right] \\
&= \ln \sum_\xi \frac{\tilde{P}_{\boldsymbol{\theta}^0}(\xi)}{Z(\boldsymbol{\theta}^0)} \left[\exp \left\{ (\theta_k - \theta_k^0)f_k(\xi) \right\} \right] \\
&= \ln \boldsymbol{E}_{\boldsymbol{\theta}^0} \left[\exp \left\{ (\theta_k - \theta_k^0)f_k(\boldsymbol{d}_k) \right\} \right].
\end{aligned}
$$

Thus, the difference of these two log-partition functions can be rewritten as a log-expectation relative to our original distribution. We can convert this expression into a univariate function of θ_k by computing (via inference in our current model $\boldsymbol{\theta}^0$) the marginal distribution over the variables $\boldsymbol{d}_k$. Altogether, we obtain that:

$$
\begin{aligned}
\frac{1}{M}[\ell((\theta_k, \boldsymbol{\theta}^0_{-k}) : \mathcal{D}) - \ell(\boldsymbol{\theta}^0 : \mathcal{D})] = & \\
(\theta_k - \theta_k^0)\boldsymbol{E}_{\mathcal{D}}[f_k] - \ln \sum_{\boldsymbol{d}_k} P_{\boldsymbol{\theta}^0}(\boldsymbol{d}_k) & \left[\exp \left\{ (\theta_k - \theta_k^0)f_k(\boldsymbol{d}_k) \right\} \right].
\end{aligned}
\tag{20.34}
$$

We can incorporate this simplified form into equation (20.33) for any penalized-likelihood scoring function. We can now easily provide our lower-bound estimates for feature deletions. For introducing a feature f_k, the optimal lower bound can be computed by optimizing the univariate function defined by equation (20.33) over the parameter θ_k. Because this function is concave, it can be optimized using a variety of univariate numerical optimization algorithms. For example, to compute the lower bound for an L_2-regularized likelihood, we would compute:

$$
\max_{\theta_k} \left\{ \theta_k \boldsymbol{E}_{\mathcal{D}}[f_k] - \ln \sum_{\boldsymbol{d}_k} P_{\boldsymbol{\theta}^0}(\boldsymbol{d}_k) \left[\exp \left\{ (\theta_k - \theta_k^0)f_k(\boldsymbol{d}_k) \right\} \right] - \frac{\theta_k^2}{2\sigma^2} \right\}.
$$

However, in certain special cases, we can actually provide a closed-form solution for this optimization problem. We note that this derivation applies only in restricted cases: only in the case of generative training (that is, not for CRFs); only for the likelihood or L_1-penalized objective; and only for binary-valued features.

Proposition 20.6

Let f_k be a binary-valued feature and let $\boldsymbol{\theta}^0$ be a current setting of parameters for a log-linear model. Let $\hat{p}_k = \boldsymbol{E}_\mathcal{D}[f_k]$ be the empirical probability of f_k in $\mathcal{D}$, and $p_k^0 = P_\theta(f_k)$ be its probability relative to the current model. Then:

$$\max_{\theta_k} \left[\text{score}_L((\theta_k, \boldsymbol{\theta}^0_{-k}) \; : \; \mathcal{D}) - \text{score}_L(\boldsymbol{\theta}^0 \; : \; \mathcal{D}) \right] = \boldsymbol{D}(\hat{p}_k \| p_k^0),$$

where the KL-divergence is the relative entropy between the two Bernoulli distributions parameterized by $\hat{p}_k$ and p_k^0 respectively.

The proof is left as an exercise (exercise 20.16).

To see why this result is intuitive, recall that when we maximize the likelihood relative to some log-linear model, we obtain a model where the expected counts match the empirical counts. In the case of a binary-valued feature, in the optimized model we have that the final probability of f_k would be the same as the empirical probability $\hat{p}_k$. Thus, it is reasonable that the amount of improvement we obtain from this optimization is a function of the discrepancy between the empirical probability of the feature and its probability given the current model. The bigger the discrepancy, the bigger the improvement in the likelihood.

A similar analysis applies when we consider several binary-valued features $f_1, \ldots, f_k$, as long as they are mutually exclusive and exhaustive; that is, as long as there are no assignments for which both $f_{i\cdot}(\xi) = 1$ and $f_j(\xi) = 1$. In particular, we can show the following:

Proposition 20.7

Let $\boldsymbol{\theta}^0$ be a current setting of parameters for a log-linear model, and consider introducing into the model a complete factor ϕ over scope $\boldsymbol{d}$, parameterized with $\boldsymbol{\theta_k}$ that correspond to the different assignments to $\boldsymbol{d}$. Then

$$\max_{\boldsymbol{\theta_k}} \left[\text{score}_L((\boldsymbol{\theta_k}, \boldsymbol{\theta}^0_{-\boldsymbol{k}}) \; : \; \mathcal{D}) - \text{score}_L(\boldsymbol{\theta}^0 \; : \; \mathcal{D}) \right] = \boldsymbol{D}(\hat{P}(\boldsymbol{d}) \| P_\theta(\boldsymbol{d})).$$

The proof is left as an exercise (exercise 20.17).

Although the derivations here were performed for the likelihood function, a similar closed-form solution, in the same class of cases, can also be performed for the L_1-regularized likelihood (see exercise 20.18), but not for the L_2-regularized likelihood. Intuitively, the penalty in the L_1-regularized likelihood is a linear function in each θ_k, and therefore it does not complicate the form of equation (20.34), which already contains such a term. However, the L_2 penalty is quadratic, and introducing a quadratic term into the function prevents an analytic solution.

One issue that we did not address is the task of computing the expressions in equation (20.34), or even the closed-form expressions in proposition 20.6 and proposition 20.7. All of these expressions involve expectations over the scope $\boldsymbol{D}_k$ of f_k, where f_k is the feature that we want to eliminate from or introduce into the model. Let us consider first the case where f_k is already in the model. In this case, if we use a *belief propagation* algorithm (whether a clique tree or a loopy cluster graph), the family preservation property guarantees that the feature's scope $\boldsymbol{D}_k$ is necessarily a subset of some cluster in our inference data structure. Thus, we can easily compute the necessary expectations. However, for a feature not currently in the model, we would not generally expect its scope to be included in any cluster. If not, we must somehow compute expectations of sets of variables that are not together in the same cluster. In the case of clique trees, we can use the out-of-clique inference methods described in section 10.3.3.2. For the case of loopy cluster graphs, this problem is more challenging (see exercise 11.22).

20.8   Summary

In this chapter, we discussed the problem of learning undirected graphical models from data. The key challenge in learning these models is that the global partition function couples the parameters, with significant consequences: There is no closed-form solution for the optimal parameters; moreover, we can no longer optimize each of the parameters independently of the others. Thus, even simple maximum-likelihood parameter estimation is no longer trivial. For the same reason, full Bayesian estimation is computationally intractable, and even approximations are expensive and not often used in practice.

Following these pieces of bad news, there are some good ones: the likelihood function is concave, and hence it has no local optima and can be optimized using efficient gradient-based methods over the space of possible parameterizations. We can also extend this method to MAP estimation when we are given a prior over the parameters, which allows us to reduce the overfitting to which maximum likelihood is prone.

The gradient of the likelihood at a point $\boldsymbol{\theta}$ has a particularly compelling form: the gradient relative to the parameter θ_i corresponding to the feature f_i is the difference between the empirical expectation of f_i in the data and its expectation relative to the distribution $P_{\boldsymbol{\theta}}$. While very intuitive and simple in principle, the form of the gradient immediately gives rise to some bad news: to compute the gradient at the point $\boldsymbol{\theta}$, we need to run inference over the model $P_{\boldsymbol{\theta}}$, a costly procedure to execute at every gradient step.

This complexity motivates the use of myriad alternative approaches: ones involving the use of approximate inference for computing the gradient; and ones that utilize a different objective than the likelihood. Methods in the first class included using message passing algorithms such as belief propagation, and methods based on sampling. We also showed that many of the methods that use approximate inference for optimizing the likelihood can be reformulated as exactly optimizing an approximate objective. This perspective can offer significant insight. For example, we showed that learning with belief propagation can be reformulated as optimizing a joint objective that involves both inference and learning; this alternative formulation is more general and allows the use of alternative optimization methods that are more stable and convergent than using BP to estimate the gradient.

Methods that use an approximate objective include pseudolikelihood, contrastive divergence, and maximum-margin (which is specifically geared for discriminative training of conditional models). Importantly, both likelihood and these objectives can be viewed as trying to increase the distance between the log-probability of assignments in our data and those of some set other assignments. This "contrastive" view provides a different view of these objectives, and it suggests that they are only representatives of a much more general class of approximations.

The same analysis that we performed for optimizing the likelihood can also be extended to other cases. In particular, we showed a very similar derivation for conditional training, where the objective is to maximize the likelihood of a set of target variables $\boldsymbol{Y}$ given some set of observed feature variables $\boldsymbol{X}$.

We also showed that similar approaches can be applied to learning with missing data. Here, the optimization task is no longer convex, but the gradient has a very similar form and can be optimized using the same gradient-ascent methods. However, as in the case of Bayesian network learning with missing data, the likelihood function is generally multimodal, and so the gradient ascent algorithm can get stuck in local optima. Thus, we may need to resort to techniques such as data perturbation or random restarts.

We also discussed the problem of structure learning of undirected models. Here again, we can use both constraint-based and score-based methods. Owing to the difficulties arising from the form of the likelihood function, full Bayesian scoring, where we score a model by integrating over all of the parameters, is intractable, and even approximations are generally impractical. Thus, we generally use a simpler scoring function, which combines a likelihood term (measuring fit to data) with some penalty term. We then search over some space of structures for ones that optimize this objective. For most objectives, the resulting optimization problem is combinatorial with multiple local optima, so that we must resort to heuristic search. One notable exception is the use of an L_1-regularized likelihood, where the penalty on the absolute value of the parameters tends to drive many of the parameters to zero, and hence often results in sparse models. This objective allows the structure learning task to be formulated as a convex optimization problem over the space of parameters, allowing the optimization to be performed efficiently and with guaranteed convergence to a global optimum. Of course, even here inference is still an unavoidable component in the inner loop of the learning algorithm, with all of the ensuing difficulties.

As we mentioned, the case of discriminative training is a setting where undirected models are particularly suited, and are very commonly used. However, it is important to carefully weigh the trade-offs of generative versus discriminative training. As we discussed, there are significant differences in the computational cost of the different forms of training, and the trade-off can go either way. More importantly, as we discussed in section 16.3.2, **generative models incorporate** **a higher bias by making assumptions — ones that are often only approximately correct — about the underlying distribution. Discriminative models make fewer assumptions, and therefore tend to require more data to train; generative models, due to the stronger bias, often perform better in the sparse-data regime. But incorrect modeling assumptions also hurt performance; therefore, as the amount of training data grows, the discriminative model, which makes fewer assumptions, often performs better. This difference between the two classes of models is particularly significant when we have complex features whose correlations are hard to model.** However, it is important to remember that models trained discriminatively to predict Y given X will perform well primarily in this setting, and even slight changes may lead to a degradation in performance. For example, a model for predicting $P(Y \mid X_1, X_2)$ would not be useful for predicting $P(Y \mid X_1)$ in situations where X_2 is not observed. In general, discriminative models are much less flexible in their ability to handle missing data.

We focused most of our discussion of learning on the problem of learning log-linear models defined in terms of a set of features. Log-linear models are a finer-grained representation than a Markov network structure or a set of factors. Thus, they can make better trade-offs between model complexity and fit to data. However, sparse log-linear models (with few features) do not directly correspond to sparse Markov network structures, so that we might easily end up learning a model that does not lend itself to tractable inference. It would be useful to consider the development of Markov network structure learning algorithms that more easily support efficient inference. Indeed, some work has been done on learning Markov networks of bounded tree-width, but networks of low tree-width are often poor approximations to the target distribution. Thus, it would be interesting to explore alternative approaches that aim at structures that support approximate inference.

This chapter is structured as a core idea with a set of distinct extensions that build on it: The core idea is the use of the likelihood function and the analysis of its properties. The extensions include conditional likelihood, learning with missing data, the use of parameter priors, approximate inference and/or approximate objectives, and even structure learning. In many cases, these extensions are orthogonal, and we can easily combine them in various useful ways. For example, we can use parameter priors with conditional likelihood or in the case of missing data; we can also use them with approximate methods such as pseudolikelihood, contrastive divergence or in the objective of equation (20.15). Perhaps more surprising is that we can easily perform structure learning with missing data by adding an L_1-regularization term to the likelihood function of equation (20.8) and then using the same ideas as in section 20.7.4.4. In other cases, the combination of the different extensions is more involved. For example, as we discussed, structure learning requires that we be able to evaluate the expected counts for variables that are not in the same family; this task is not so easy if we use an approximate algorithm such as belief propagation. As another example, it is not immediately obvious how we can extend the pseudolikelihood objective to deal with missing data. These combinations provide useful directions for future work.

20.9   Relevant Literature

Log-linear models and contingency tables have been used pervasively in a variety of communities, and so key ideas have often been discovered multiple times, making a complete history too long to include. Early attempts for learning log-linear models were based on the *iterative proportional scaling* algorithm and its extension, *iterative proportional fitting*. These methods were first developed for contingency tables by Deming and Stephan (1940) and applied to log-linear models by Darroch and Ratcliff (1972). The convex duality between the maximum likelihood and maximum entropy problems appears to have been proved independently in several papers in diverse communities, including (at least) Ben-Tal and Charnes (1979); Dykstra and Lemke (1988); Berger, Della-Pietra, and Della-Pietra (1996). It appears that the first application of gradient algorithms to maximum likelihood estimation in graphical models is due to Ackley, Hinton, and Sejnowski (1985) in the context of Boltzmann machines. The importance of the method used to optimize the likelihood was highlighted in the comparative study of Minka (2001a); this study focused on learning for logistic regression, but many of the conclusions hold more broadly. Since then, several better methods have been developed for optimizing likelihood. Successful methods include conjugate gradient, L-BFGS (Liu and Nocedal 1989), and stochastic meta-descent (Vishwanathan et al. 2006).

Conditional random fields were first proposed by Lafferty et al. (2001). They have since been applied in a broad range of applications, such as labeling multiple webpage on a website (Taskar et al. 2002), image segmentation (Shental et al. 2003), or information extraction from text (Sutton and McCallum 2005). The application to protein-structure prediction in box 20.B is due to Yanover et al. (2007).

The use of approximate inference in learning is an inevitable consequence of the intractability of the inference problem. Several papers have studied the interaction between belief propagation and Markov network learning. Teh and Welling (2001) and Wainwright et al. (2003b) present methods for certain special cases; in particular, Wainwright, Jaakkola, and Willsky (2003b) derive

iterative proportional scaling

iterative proportional fitting

the pseudo-moment matching argument. Inspired by the moment-matching behavior of learning with belief propagation, Sutton and McCallum (2005); Sutton and Minka (2006) define the piecewise training objective that directly performs moment matching on all network potentials. Wainwright (2006) provides a strong argument, both theoretical and empirical, for using the same approximate inference method in training as will be used in performing the prediction using the learned model. Indeed, he shows that, if an approximate method is used for inference, then we get better performance guarantees if we use that same method to train the model than if we train a model using exact inference. He also shows that it is detrimental to use an unstable inference algorithm (such as sum-product BP) in the inner loop of the learning algorithm. Ganapathi et al. (2008) define the unified CAMEL formulation that encompasses learning and inference in a single joint objective, allowing the nonconvexity of the BP objective to be taken out of the inner loop of learning.

Although maximum (conditional) likelihood is the most commonly used objective for learning Markov networks, several other objectives have been proposed. The earliest is pseudolikelihood, proposed by Besag (1977b), of which several extensions have been proposed (Huang and Ogata 2002; McCallum et al. 2006). The asymptotic consistency of both the likelihood and the pseudolikelihood objectives is shown by Gidas (1988). The statistical efficiency (convergence as a function of the number of samples) of the pseudolikelihood estimator has also been analyzed (for example, (Besag 1977a; Geyer and Thompson 1992; Guyon and Künsch 1992; Liang and Jordan 2008)).

The use of margin-based estimation methods for probabilistic models was first proposed by Collins (2002) in the context of parsing and sequence modeling, building on the voted-perceptron algorithm (Freund and Schapire 1998). The methods described in this chapter build on a class of large-margin methods called *support vector machines* (Shawe-Taylor and Cristianini 2000; Hastie et al. 2001; Bishop 2006), which have the important benefit of allowing a large or even infinite feature space to be used and trained very efficiently. This formulation was first proposed by Altun, Tsochantaridis, and Hofmann (2003); Taskar, Guestrin, and Koller (2003), who proposed two different approaches for addressing the exponential number of constraints. Altun et al. use a constraint-generation scheme, which was subsequently proven to require at most a polynomial number of steps (Tsochantaridis et al. 2004). Taskar et al. use a closed-form polynomial-size reformulation of the optimization problem that uses a clique tree-like data structure. Taskar, Chatalbashev, and Koller (2004) also show that this formulation also allows tractable training for networks where conditional probability products are intractable, but the MAP assignment can be found efficiently. The contrastive divergence approach was introduced by Hinton (2002); Teh, Welling, Osindero, and Hinton (2003), and was shown to work well in practice in various studies (for example, (Carreira-Perpignan and Hinton 2005)). This work forms part of a larger trend of training using a range of alternative, often contrastive, objectives. LeCun et al. (2007) provide an excellent overview of this area.

Much discussion has taken place in the machine learning community on the relative merits of discriminative versus generative training. Some insightful papers of particular relevance to graphical models include the work of Minka (2005) and LeCun et al. (2007). Also of interest are the theoretical analyses of Ng and Jordan (2002) and Liang and Jordan (2008) that discuss the statistical efficiency of discriminative versus generative training, and provide theoretical support for the empirical observation that generative models, even if not consistent with the true underlying distribution, often work better in the sparse data case, but discriminative models

support vector
machine

tend to work better as the amount of data grows.

The work of learning Markov networks with hidden variables goes back to the seminal paper of Ackley, Hinton, and Sejnowski (1985), who used gradient ascent to train Boltzmann machines with hidden variables. This line of work, largely dormant for many years, has seen a resurgence in the work on *deep belief networks* (Hinton et al. 2006; Hinton and Salakhutdinov 2006), a training regime for a multilayer restricted Boltzmann machine that iteratively tries to learn deeper and deeper hidden structure in the data.

deep belief
networks

Parameter priors and regularization methods for log-linear models originate in statistics, where they have long been applied to a range of statistical models. Many of the techniques described here were first developed for traditional statistical models such as linear or logistic regression, and then extended to the general case of Markov networks and CRFs. See Hastie et al. (2001) for some background on this extensive literature.

The problem of learning the structure of Markov networks has not received as much attention as the task of Bayesian network structure learning. One line of work has focused on the problem of learning a Markov network of bounded tree-width, so as to allow tractable inference. The work of Chow and Liu (1968) shows that the maximum-likelihood tree-structured network can be found in quadratic time.

A tree is a network of tree-width 1. Thus, the obvious generalization of is to learning the class of Markov networks whose tree-width is at most k. Unfortunately, there is a sharp threshold phenomenon, since Srebro (2001) proves that for any tree-width k greater than 1, finding the maximum likelihood tree-width-k network is $\mathcal{NP}$-hard. Interestingly, Narasimhan and Bilmes (2004) provide a constraint-based algorithm for PAC-learning Markov networks of tree-width at most k: Their algorithm is guaranteed to find, with probability $1 - \delta$, a network whose relative entropy is within ϵ of optimal, in polynomial time, and using a polynomial number of samples. Importantly, their result does not contradict the hardness result of Srebro, since their analysis applies only in the consistent case, where the data is derived from a k-width network. This discrepancy highlights again the significant difference between learnability in the consistent and the inconsistent case. Several search-based heuristic algorithms for learning models with small tree-width have been proposed (Bach and Jordan 2001; Deshpande et al. 2001); so far, none of these algorithms have been widely adopted, perhaps because of the limited usefulness of bounded tree-width networks.

Abbeel, Koller, and Ng (2006) provide a different PAC-learnability result in the consistent case, for networks of bounded connectivity. Their constraint-based algorithm is guaranteed to learn, with high probability, a network $\tilde{\mathcal{M}}$ whose (symmetric) relative entropy to the true distribution $(\boldsymbol{D}(\tilde{P}\|P^*) + \boldsymbol{D}(P^*\|\tilde{P}))$ is at most ϵ. The complexity, both in time and in number of samples, grows exponentially in the maximum number of assignments to any local neighborhood (a factor and its Markov blanket). This result is somewhat surprising, since it shows that the class of low-connectivity Markov networks (such as grids) is PAC-learnable, even though inference (including computing the partition function) can be intractable.

Of highest impact has been the work on using local search to optimize a (regularized) likelihood. This line of work originated with the seminal paper of Della Pietra, Della Pietra, and Lafferty (1997), who defined the single-feature gain, and proposed the gain as an effective heuristic for feature selection in learning Markov network structure. McCallum (2003) describes some heuristic approximations that allow this heuristic to be applied to CRFs. The use of L_1-regularization for feature selection originates from the Lasso model proposed for linear regression by Tibshirani (1996). It was first proposed for logistic regression by Perkins et al. (2003); Goodman (2004). Perkins et al. also suggested the gradient heuristic for feature selection and the L_1-based stopping rule. L_1-regularized priors were first proposed for learning log-linear distributions by Riezler and Vasserman (2004); Dudík, Phillips, and Schapire (2004). The use of L_1-regularized objectives for learning the structure of general Markov networks was proposed by Lee et al. (2006). Building on the results of Dudík et al., Lee et al. also showed that the number of samples required to achieve close-to-optimal relative entropy (within the target class) grows only polynomially in the size of the network. Importantly, unlike the PAC-learnability results mentioned earlier, this result also holds in the inconsistent case.

Pseudolikelihood has also been used as a criterion for model selection. Ji and Seymour (1996) define a pseudolikelihood-based objective and show that it is asymptotically consistent, in that the probability of selecting an incorrect model goes to zero as the number of training examples goes to infinity. However, they did not provide a tractable algorithm for finding the highest scoring model in the superexponentially large set of structures. Wainwright et al. (2006) suggested the use of an L_1-regularized pseudolikelihood for model selection, and also proved a theorem that provides guarantees on the near-optimality of the learned model, using a polynomial number of samples. Like the result of Lee et al. (2006), this result applies also in the inconsistent case.

This chapter has largely omitted discussion of the Bayesian learning approach for Markov networks, for both parameter estimation and structure learning. Although an exact approach is computationally intractable, some interesting work has been done on approximate methods. Some of this work uses MCMC methods to sample from the parameter posterior. Murray and Ghahramani (2004) propose and study several diverse methods; owing to the intractability of the posterior, all of these methods are approximate, in that their stationary distribution is only an approximation to the desired parameter posterior. Of these, the most successful methods appear to be a method based on Langevin sampling with approximate gradients given by contrastive divergence, and a method where the acceptance probability is approximated by replacing the log partition function with the Bethe free energy. Two more restricted methods (Møller et al. 2006; Murray et al. 2006) use an approach called "perfect sampling" to avoid the need for estimating the partition function; these methods are elegant but of limited applicability. Other approaches approximate the parameter posterior by a Gaussian distribution, using either expectation propagation (Qi et al. 2005) or a combination of a Bethe and a Laplace approximation (Welling and Parise 2006a). The latter approach was also used to approximate the Bayesian score in order to perform structure learning (Welling and Parise 2006b). Because of the fundamental intractability of the problem, all of these methods are somewhat complex and computationally expensive, and they have therefore not yet made their way into practical applications.

20.10 Exercises

Exercise 20.1

Consider the network of figure 20.2, where we assume that some of the factors share parameters. Let $\boldsymbol{\theta}_i^y$ be the parameter vector associated with all of the features whose scope is Y_i, Y_{i+1}. Let $\boldsymbol{\theta}_{i,j}^{xy}$ be the parameter vector associated with all of the features whose scope is Y_i, X_j.

a. Assume that, for all i, i', $\boldsymbol{\theta}_i^y = \boldsymbol{\theta}_{i'}^y$, and that for all i, i' and j, $\boldsymbol{\theta}_{i,j}^{xy} = \boldsymbol{\theta}_{i',j}^{xy}$. Derive the gradient update for this model.

b. Now (without the previous assumptions) assume for all i and j, j', $\boldsymbol{\theta}_{i,j}^{xy} = \boldsymbol{\theta}_{i,j'}^{xy}$. Derive the gradient update for this model.

Exercise 20.2

In this exercise, we show how to learn Markov networks with shared parameters, such as a *relational Markov network* (RMN).

a. Consider the log-linear model of example 6.18, where we assume that the *Study-Pair* relationship is determined in the relational skeleton. Thus, we have a single template feature, with a single weight, which is applied to all study pairs. Derive the likelihood function for this model, and the gradient.

b. Now provide a formula for the likelihood function and the gradient for a general RMN, as in definition 6.14.

Exercise 20.3

Assume that our data are generated by a log-linear model $P_{\boldsymbol{\theta}^*}$ that is of the form of equation (20.1). Show that, as the number of data instances M goes to infinity, with probability that approaches 1, $\boldsymbol{\theta}^*$ is a global optimum of the likelihood objective of equation (20.3). (Hint: Use the characterization of theorem 20.1.)

Exercise 20.4

Use the techniques described in this chapter to provide a method for performing maximum likelihood estimation for a CPD whose parameterization is a generalized linear model, as in definition 5.10.

Exercise 20.5★

Show using Lagrange multipliers and the definitions of appendix A.5.4 that the problem of maximizing $H_Q(\mathcal{X})$ subject to equation (20.10) is dual to the problem of maximizing the log likelihood $\max \ell(\boldsymbol{\theta} : \mathcal{D})$.

Exercise 20.6★

In this problem, we will show an analogue to theorem 20.2 for the problems of maximizing conditional likelihood and maximizing conditional entropy.

Consider a data set $\mathcal{D} = \{(\boldsymbol{y}[m], \boldsymbol{x}[m])\}_{m=1}^{M}$ as in section 20.3.2, and define the following conditional entropy maximization problem:

Maximum-Conditional-Entropy:

Find $\quad Q(\boldsymbol{Y} \mid \boldsymbol{X})$

maximizing $\quad \sum_{m=1}^{M} H_Q(\boldsymbol{Y} \mid \boldsymbol{x}[m])$

subject to

$$\sum_{m=1}^{M} \boldsymbol{E}_{Q(\boldsymbol{Y}|\boldsymbol{x}[m])}[f_k] = \sum_{m=1}^{M} f_k(\boldsymbol{y}[m], \boldsymbol{x}[m]) \quad i = 1, \ldots, k. \qquad (20.35)$$

Show that $Q^*(\boldsymbol{Y} \mid \boldsymbol{X})$ optimizes this objective if and only if $Q^* = P_{\hat{\boldsymbol{\theta}}}$ where $P_{\hat{\boldsymbol{\theta}}}$ maximizes $\ell_{\boldsymbol{Y}|\boldsymbol{X}}(\boldsymbol{\theta} : \mathcal{D})$ as in equation (20.6).

Exercise 20.7★

One of the earliest approaches for finding maximum likelihood parameters is called *iterative proportional scaling* (IPS). The idea is essentially to use coordinate ascent to improve the match between the empirical feature counts and the expected feature counts. In other words, we change θ_k so as to make $\boldsymbol{E}_{P_{\boldsymbol{\theta}}}[f_k]$ closer to $\boldsymbol{E}_{\mathcal{D}}[f_k]$. Because our model is multiplicative, it seems natural to multiply the weight of instances where $f_k(\xi) = 1$ by the ratio between the two expectations. This intuition leads to the following update rule:

$$\theta_k' \leftarrow \theta_k + \ln \frac{\boldsymbol{E}_{\mathcal{D}}[f_k]}{\boldsymbol{E}_{P_{\boldsymbol{\theta}}}[f_k]}. \qquad (20.36)$$

The IPS algorithm iterates over the different parameters and updates each of them in turn, using this update rule.

Somewhat surprisingly, one can show that each iteration increases the likelihood until it reaches a maximum point. Because the likelihood function is concave, there is a single maximum, and the algorithm is guaranteed to find it.

Theorem 20.5

Let $\boldsymbol{\theta}$ be a parameter vector, and $\boldsymbol{\theta}'$ the vector that results from it after an application of equation (20.36). Then $\ell(\boldsymbol{\theta}' : \mathcal{D}) \geq \ell(\boldsymbol{\theta} : \mathcal{D})$ with equality if only if $\frac{\partial}{\partial \theta_k} \ell(\boldsymbol{\theta}^t : \mathcal{D}) = 0$.

In this exercise, you will prove this theorem for the special case where f_k is binary-valued. More precisely, let $\Delta(\theta_k)$ denote the change in likelihood obtained from modifying a single parameter θ_k, keeping the others fixed. This expression was computed in equation (20.34). You will now show that the IPS update step for θ_k maximizes a lower bound on this single parameter gain.

Define

$$
\begin{aligned}
\tilde{\Delta}(\theta_k) &= (\theta'_k - \theta_k)\boldsymbol{E}_\mathcal{D}[f_k] - \frac{Z(\boldsymbol{\theta}')}{Z(\boldsymbol{\theta})} + 1 \\
&= (\theta'_k - \theta_k)\boldsymbol{E}_\mathcal{D}[f_k] - \boldsymbol{E}_{P_\theta}[1 - f_k] - e^{\theta'_k - \theta_k}\boldsymbol{E}_{P_{\theta'}}[f_k] + 1.
\end{aligned}
$$

a. Show that $\Delta(\theta'_k) \geq \tilde{\Delta}(\theta'_k)$. (Hint: use the bound $\ln(x) \leq x - 1$.)

b. Show that $\theta_k + \ln \frac{\boldsymbol{E}_\mathcal{D}[f_k]}{\boldsymbol{E}_{P_\theta}[f_k]} = \arg\max_{\theta'_k} \tilde{\Delta}(\theta'_k)$.

c. Use these two facts to conclude that IPS steps are monotonically nondecreasing in the likelihood, and that convergence is achieved only when the log-likelihood is maximized.

d. This result shows that we can view IPS as performing coordinatewise ascent on the likelihood surface. At each iteration we make progress along on dimension (one parameter) while freezing the others. Why is coordinate ascent a wasteful procedure in the context of optimizing the likelihood?

Exercise 20.8

hyperbolic prior

Consider the following *hyperbolic prior* for parameters in log-linear models.

$$
P(\theta) = \frac{1}{\left(e^\theta + e^{-\theta}\right)/2}.
$$

a. Derive a gradient-based update rule for this parameter prior.

b. Qualitatively describe the expected behavior of this parameter prior, and compare it to those of the L_2 or L_1 priors discussed in section 20.4. In particular, would you expect this prior to induce sparsity?

Exercise 20.9

piecewise training

We now consider an alternative local training method for Markov networks, known as *piecewise training*. For simplicity, we focus on Markov networks parameterized via full table factors. Thus, we have a set of factors $\phi_c(\boldsymbol{X}_c)$, where $\boldsymbol{X}_c$ is the scope of factor c, and $\phi_c(\boldsymbol{x}_c^j) = \exp(\theta_{cj})$. For a particular parameter assignment $\boldsymbol{\theta}$, we define $Z_c(\boldsymbol{\theta})$ to be the local partition function for this factor in isolation:

$$
Z_c(\boldsymbol{\theta}_c) = \sum_{\boldsymbol{x}_c} \phi_c(\boldsymbol{x}_c),
$$

where $\boldsymbol{\theta}_c$ is the parameter vector associated with the factor $\phi_c(\boldsymbol{X}_c)$. We can approximate the global partition function in the log-likelihood objective of equation (20.3) as a product of the local partition functions, replacing $Z(\theta)$ with $\prod_c Z_c(\boldsymbol{\theta}_c)$.

a. Write down the form of the resulting objective, simplify it, and derive the assignment of parameters that optimizes it.
b. Compare the result of this optimization to the result of the pseudo-moment matching approach described in section 20.5.1.

Exercise 20.10★

CAMEL

In this exercise, we analyze the following simplification of the *CAMEL* optimization problem of equation (20.15):

Simple-Approx-Maximum-Entropy:

Find Q
maximizing $\sum_{C_i \in \mathcal{U}} H_{\beta_i}(C_i)$
subject to

$$\begin{aligned} E_{\beta_i}[f_i] &= E_{\mathcal{D}}[f_i] \quad i = 1, \dots, k \\ \sum_{c_i} \beta_i(c_i) &= 1 \quad i = 1, \dots, k \\ Q &\geq 0 \end{aligned}$$

Here, we approximate both the objective and the constraints. The objective is approximated by the removal of all of the negative entropy terms for the sepsets. The constraints are relaxed by removing the requirement that the potentials in Q be locally consistent (sum-calibrated) — we now require only that they be legal probability distributions.

Show that this optimization problem is the Lagrangian dual of the piecewise training objective in exercise 20.9.

Exercise 20.11★

multiconditional training

Consider a setting, as in section 20.3.2, where we have two sets of variables Y and X. *Multiconditional training* provides a spectrum between pure generative and pure discriminative training by maximizing the following objective:

$$\alpha \ell_{Y|X}(\theta : \mathcal{D}) + (1 - \alpha)\ell_{X|Y}(\theta : \mathcal{D}). \tag{20.37}$$

Consider the model structure shown in figure 20.2, and a partially labeled data set $\mathcal{D}$, where in each instance m we observe all of the feature variables $x[m]$, but only the target variables in $O[m]$.

Write down the objective of equation (20.37) for this case and compute its derivative.

Exercise 20.12★

Consider the problem of maximizing the approximate log-likelihood shown in equation (20.16).

a. Derive the gradient of the approximate likelihood, and show that it is equivalent to utilizing an importance sampling estimator directly to approximate the expected counts in the gradient of equation (20.4).
b. Characterize properties of the maximum point (when the gradient is 0). Is such a maximum always attainable? Prove or suggest a counterexample.

Exercise 20.13★★

One approach to providing a lower bound to the log-likelihood is by upper-bounding the partition function. Assume that we can decompose our model as a convex combination of (hopefully) simpler models, each with a weight α_k and a set of parameters ψ^k. We define these submodels as follows: $\psi^k(\theta) = w^k \bullet \theta$, where we require that, for any feature i,

$$\sum_k \alpha_k w_i^k = 1. \tag{20.38}$$

a. Under this assumption, prove that

$$\ln Z(\boldsymbol{\theta}) \leq \sum_k \alpha_k \ln Z(\boldsymbol{w}^k \bullet \boldsymbol{\theta}). \tag{20.39}$$

This result allows us to define an approximate log-likelihood function:

$$\frac{1}{M} \ell(\boldsymbol{\theta} : \mathcal{D}) \geq \ell_{\text{convex}}(\boldsymbol{\theta} : \mathcal{D}) = \sum_i \theta_i \boldsymbol{E}_{\mathcal{D}}[f_i] - \sum_k \alpha_k \ln Z(\boldsymbol{w}^k \bullet \boldsymbol{\theta}).$$

b. Assuming that the submodels are more tractable, we can efficiently evaluate this lower bound, and also compute its derivatives to be used during optimization. Show that

$$\frac{\partial}{\partial \theta_i} \ell_{\text{convex}}(\boldsymbol{\theta} : \mathcal{D}) = \boldsymbol{E}_{\mathcal{D}}[f_i] - \sum_k \alpha_k w_i^k \boldsymbol{E}_{P_{\boldsymbol{w}^k \bullet \boldsymbol{\theta}}}[f_i]. \tag{20.40}$$

c. We can provide a bound on the error of this approximation. Specifically, show that:

$$\frac{1}{M} \ell(\boldsymbol{\theta} : \mathcal{D}) - \ell_{\text{convex}}(\boldsymbol{\theta} : \mathcal{D}) = \sum_k \alpha_k \boldsymbol{D}(P_{\boldsymbol{\theta}} \| P_{\boldsymbol{w}^k \bullet \boldsymbol{\theta}}),$$

where the KL-divergence measures are defined in terms of the natural logarithm. Thus, we see that the error is an average of the divergence between the true distribution and each of the approximating submodels.

d. The justification for this approach is that we can make the submodels simpler than the original model by having some parameters be equal to 0, thereby eliminating the resulting feature from the model structure. Other than this constraint, however, we still have considerable freedom in choosing the submodel weight vectors $\boldsymbol{w}^k$. Assume that each weight vector $\{\boldsymbol{w}_k\}$ maximizes $\ell_{\text{convex}}(\boldsymbol{\theta} : \mathcal{D})$ subject to the constraint of equation (20.38) plus additional constraints requiring that certain entries w_i^k be equal to 0. Show that if i, k and l are such that $\theta_i \neq 0$ and neither w_i^k nor w_i^l is constrained to be zero, then

$$\boldsymbol{E}_{P_{\boldsymbol{w}^k \bullet \boldsymbol{\theta}}}[f_i] = \boldsymbol{E}_{P_{\boldsymbol{w}^l \bullet \boldsymbol{\theta}}}[f_i].$$

Conclude from this result that for each such i and k, we have that

$$\boldsymbol{E}_{P_{\boldsymbol{w}^k \bullet \boldsymbol{\theta}}}[f_i] = \boldsymbol{E}_{\mathcal{D}}[f_i].$$

Exercise 20.14

Consider a particular parameterization $(\boldsymbol{\theta}, \eta)$ to Max-margin. Show how we can use second-best MAP inference to either find a violated constraint or guarantee that all constraints are satisfied.

Exercise 20.15

Let $\mathcal{H}^*$ be a Markov network where the maximum degree of a node is d^*. Show that if we have an infinitely large data set $\mathcal{D}$ generated from $\mathcal{H}^*$ (so that independence tests are evaluated perfectly), then the Build-PMap-Skeleton procedure of algorithm 3.3 reconstructs the correct Markov structure $\mathcal{H}^*$.

Exercise 20.16★

Prove proposition 20.6. (Hint: Take the derivative of equation (20.34) and set it to zero.)

Exercise 20.17

In this exercise, you will prove proposition 20.7, which allows us to find a closed-form optimum to multiple features in a log-linear model.

a. Prove the following proposition.

Proposition 20.8

Let $\boldsymbol{\theta}^0$ be a current setting of parameters for a log-linear model, and suppose that $f_1, \ldots, f_l$ are mutually exclusive binary features, that is, there is no ξ and $i \neq j$, so that $f_i(\xi) = 1$ and $f_j(\xi) = 1$. Then,

$$\max_{\theta_1, \ldots, \theta_l} \left[\text{score}_L((\theta_1, \ldots, \theta_l, \boldsymbol{\theta}^0_{-\{1, \ldots, l\}}) \; : \; \mathcal{D}) - \text{score}_L(\boldsymbol{\theta}^0 \; : \; \mathcal{D}) \right] = \boldsymbol{D}(\hat{\boldsymbol{p}} \| \boldsymbol{p}^0),$$

where $\hat{\boldsymbol{p}}$ is a distribution over $l+1$ values with $\hat{p}_i = \boldsymbol{E}_{\mathcal{D}}[f_i]$, and $\boldsymbol{p}^0$ is a distribution with $p^0(i) = P_{\boldsymbol{\theta}}(f_i)$.

b. Use this proposition to prove proposition 20.7.

Exercise 20.18

Derive an analog to proposition 20.6 for the case of the L_1 regularized log-likelihood objective.

PART IV

Actions and Decisions

21 *Causality*

21.1 Motivation and Overview

So far, we have been somewhat ambivalent about the relation between Bayesian networks and causality. On one hand, from a formal perspective, all of the definitions refer only to probabilistic properties such as conditional independence. The BN structure may be directed, but the directions of the arrows do not have to be meaningful. They can even be antitemporal. Indeed, we saw in our discussion of I-maps that we can take any ordering on the nodes and create a BN for any distribution. On the other hand, it is common wisdom that a "good" BN structure should correspond to causality, in that an edge $X \rightarrow Y$ often suggests that X "causes" Y, either directly or indirectly. The motivation for this statement is pragmatic: Bayesian networks with a causal structure tend to be sparser and more natural. However, as long as the network structure is capable of representing the underlying joint distribution correctly, the answers that we obtain to probabilistic queries are the same, regardless of whether the network structure corresponds to some notion of causal influence.

Given this observation, is there any deeper value to imposing a causal semantics on a Bayesian network? In this chapter, we discuss a type of reasoning for which a causal interpretation of the network is critical — reasoning about situations where we *intervene* in the world, thereby interfering in the natural course of events. For example, we may wish to know if an intervention where we prevent smoking in all public places is likely to decrease the frequency of lung cancer. To answer such queries, we need to understand the causal relationships between the variables in our model.

In this chapter, we provide a framework for interpreting a Bayesian network as a *causal model* whose edges have causal significance. Not surprisingly, this interpretation distinguishes between models that are equivalent in their ability to represent probabilistic correlations. Thus, although the two networks $X \rightarrow Y$ and $Y \rightarrow X$ are equivalent *as probabilistic models*, they will turn out to be very different *as causal models*.

21.1.1 Conditioning and Intervention

As we discussed, for standard probabilistic queries it does not matter whether our model is causal or not. It matters only that it encode the "right" distribution. The difference between causal models and probabilistic models arise when we care about *interventions* in the model — situations where we do not simply observe the values that variables take but can take actions that can manipulate these values.

In general, actions can affect the world in a variety of ways, and even a single action can have multiple effects. Indeed, in chapter 23, we discuss models that directly incorporate agent actions and allow for a range of effects. In this chapter, however, our goal is to isolate the specific issue of understanding causal relationships between variables. **One approach to modeling causal relationships is using the notion of** *ideal interventions* — **interventions of the form** $do(Z := z)$**, which force the variable** Z **to take the value** z**, and have no other** *immediate* **effect.** An ideal intervention is equivalent to a dedicated action whose only effect is setting Z to z. However, we can consider such an ideal intervention even when such an action does not exist in the world. For example, consider the question of whether a particular mutation in a person's DNA causes a particular disease. This causal question can be formulated as the question of whether an ideal intervention, whose only effect is to generate this mutation in a person's DNA, would lead to the disease. Note that, even if such a process were ethical, current technology does not permit an action whose only effect is to mutate the DNA in all cells of a human organism. However, understanding the causal connection between the mutation and the disease can be a critical step toward finding a cure; the ideal intervention provides us with a way of formalizing this question and trying to provide an answer.

More formally, we consider a new type of "conditioning" on an event of the form $do(Z := z)$, often abbreviated $do(z)$; this information corresponds to settings where an agent directly manipulated the world, to set the variable Z to take the value z with probability 1. We are now interested in answering queries of the form $P(Y \mid do(z))$, or, more generally, $P(Y \mid do(z), X = x)$. These queries are called *intervention queries*. They correspond to settings where we set the variables in Z to take the value z, observe the values x for the variables in X, and wish to find the distribution over the variables Y. Such queries arise naturally in a variety of settings:

- **Diagnosis and Treatment:** "If we get a patient to take this medication, what are her chances of getting well?" This query can be formulated as $P(H \mid do(M := m^1))$, where H is the patient's health, and $M = m^1$ corresponds to her taking the medication. Note that this query is not the same as $P(H \mid m^1)$. For example, if patients who take the medication on their own are more likely to be health-conscious, and therefore healthier in general, the chances of $P(H \mid m^1)$ may be higher than is warranted for the patient in question.

- **Marketing:** "If we lower the price of hamburgers, will people buy more ketchup?" Once again, this query is not a standard observational query, but rather one in which we intervene in the model, and thereby possibly change its behavior.

- **Policy Making:** "If we lower the interest rates, will that give rise to inflation?"

- **Scientific Discovery:** "Does smoking cause cancer?" When we formalize it, this query is an intervention query, meaning: "If we were to force someone to smoke, would they be more likely to get cancer?"

A different type of causal query arises in situations where we *already* have some information about the true state of the world, and want to inquire about the state the world *would be* in had we been able to intervene and set the values of certain variables. For example, we might want to know "Would the U.S. have joined World War II had it not been for the attack on Pearl Harbor?" Such queries are called *counterfactual queries*, because they refer to a world that we

☞

ideal intervention

intervention
query

counterfactual
query

know did not happen. Intuitively, our interpretation for such a query is that it refers to a world that differs only in this one respect. Thus, in this *counterfactual* world, Hitler would still have come into power in Germany, Poland would still have been invaded, and more. On the other hand, events that are direct causal consequences of the variable we are changing are clearly going to be different. For example, in the counterfactual world, the USS *Arizona* (which sank in the attack) would not (with high probability) currently be at the bottom of Pearl Harbor.

At first glance, counterfactual analysis might seem somewhat pointless and convoluted (who cares about what would have happened?). However, such queries actually arise naturally in several settings:

- **Legal liability cases:** "Did the driver's intoxicated state cause the accident?" In other words, would the accident have happened had the driver not been drunk? Here, we may want to preserve many other aspects of the world, for example, that it was a rainy night (so the road was slippery).

- **Treatment and Diagnosis:** "We are faced with a car that does not start, but where the lights work; will replacing the battery make the car start?" Note that this is not an intervention query; it is a counterfactual query: we are actually asking whether the car would be working now had we replaced the battery. As in the previous example, we want to preserve as much of our scenario as possible. For example, given our observation that the lights work, the problem probably is not with the battery; we need to account for this conclusion when reasoning about the situation where the battery has been replaced.

Even without a formal semantics for a causal model, we can see that the answer to an intervention query $P(Y \mid do(z), X = x)$ is generally quite different from the answer to its corresponding probabilistic query $P(Y \mid Z = z, X = x)$.

Example 21.1

Let us revisit our simple Student *example of section 3.1.3.1, and consider a particular student Gump. As we have already discussed, conditioning on an observation that Gump receives an A in the class increases the probability that he has high intelligence, his probability of getting a high SAT score, and his probability of getting a good job.*

By contrast, consider a situation where Gump is lazy, and rather than working hard to get an A in the class, he pays someone to hack into the university registrar's database and change his grade in the course to an A. In this case, what is his probability of getting a good job? Intuitively, the company where Gump is applying only has access to Gump's transcript; thus, the company's response to a manipulated grade would be the same as the response to an authentic grade. Therefore, we would expect $P(J \mid do(g^1)) = P(J \mid g^1)$. What about the other two probabilities? Intuitively, we feel that the manipulation to Gump's grade should not affect our beliefs about his intelligence, nor about his SAT score. Thus, we would expect $P(i^1 \mid do(g^1)) = P(i^1)$ and $P(s^1 \mid do(g^1)) = P(s^1)$. ∎

Why is our response to these queries different? In all three cases, there is a strong correlation between Gump's grade and the variable of interest. However, we perceive the correlation between Gump's grade and his job prospects as being *causal*. Thus, changes to his grade will directly affect his chances of being hired. The correlation between intelligence and grade arises because of an opposite causal connection: intelligence is a causal factor in grade. The correlation between Gump's SAT score and grade arises due to a third mechanism — their joint dependence on Gump's intelligence. Manipulating Gump's grade does not change his intelligence or his chances of doing well in the class. In this chapter, we describe a formal framework of causal models that provides a rigorous basis for answering such queries and allows us to distinguish between these different cases. As we will see, this framework can be used to answer both intervention and counterfactual queries. However, the latter require much finer-grained information, which may be difficult to acquire in practice.

21.1.2 Correlation and Causation

As example 21.1 illustrates, a correlation between two variables X and Y can arise in multiple settings: when X causes Y, when Y causes X, or when X and Y are both effects of a single cause. This observation immediately gives rise to the question of identifiability of causal models: If we observe two variables X, Y to be probabilistically correlated in some observed distribution, what can we infer about the causal relationship between them. As we saw, different relationships give rise to very different answers to causal queries.

This problem is greatly complicated by the broad range of reasons that may lead to an observed correlation between two variables X and Y. As we saw in example 21.1, when some variable W causally affects both X and Y, we generally observe a correlation between them. If we know about the existence of W and can observe it, we can disentangle the correlation between X and Y that is induced by W and compute the residual correlation between X and Y that may be attributed to a direct causal relationship. **In practice, however, there is a huge set of possible *latent variables*, representing factors that exist in the world but that we cannot observe and often are not even aware of. A latent variable may induce correlations between the observed variables that do not correspond to causal relations between them, and hence forms a *confounding factor* in our goal of determining causal interactions.**

latent variable

confounding
factor

As we discussed in section 19.5, when our task is pure probabilistic reasoning, latent variables need not be modeled explicitly, since we can always represent the joint distribution over the observable variables only using a probabilistic graphical model. Of course, this marginalization process can lead to more complicated models (see, for example, figure 16.1), and may therefore be undesirable. We may therefore choose to model certain latent variables explicitly, in order to simplify the resulting network structure. Importantly, however, for the purpose of answering probabilistic queries, we do not need to model all latent variables. As long as our model $\mathcal{B}_{obs}$ over the observable variables allows us to capture exactly the correct marginal distribution over the observed variables, we can answer any query as accurately with $\mathcal{B}_{obs}$ as with the true network, where the latent variables are included explicitly.

However, as we saw, the answer to a causal query over X, Y is quite different when a correlation between them is due to a causal relationship and when it is induced by a latent variable. Thus, for the purposes of causal inference, it is critical to disentangle the component in the correlation between X and Y that is due to causal relationships and the component due to these confounding factors. Unfortunately, this requirement poses a major challenge, since it is virtually impossible, in complex real-world settings, to identify all of the relevant latent variables and quantify their effects.

Example 21.2   *Consider a situation where we observe a significant positive correlation in our patient population between taking PeptAid, an antacid medication (T), and the event of subsequently developing a* *stomach ulcer (O). Because taking PeptAid precedes the ulcer, we might be tempted to conclude that PeptAid causes stomach ulcers. However, an alternative explanation is that the correlation can be attributed to a latent common cause — preulcer discomfort: individuals suffering from preulcer discomfort were more likely to take PeptAid and ultimately more likely to develop ulcers. Even if we account for this latent variable, there are many others that can have a similar effect. For example, some patients who live a more stressful lifestyle may be more inclined to eat irregular meals and therefore more likely to require antacid medication; the same patients may also be more susceptible to stomach ulcers.* ■

selection bias   Latent variables are only one type of mechanism that induces a noncausal correlation between variables. Another important class of confounding factors involves *selection bias*. Selection bias arises when the population that the distribution represents is a segment of the population that exhibits atypical behavior.

Example 21.3   *Consider a university that sends out a survey to its alumni, asking them about their history at the institution. Assume that the observed distribution reveals a negative correlation between students who participated in athletic activities (A) and students whose GPA was high (G). Can we conclude from this finding that participating in athletic activities reduces one's GPA? Or that students with a high GPA tend not to participate in athletic activities? An alternative explanation is that the respondents to the survey ($S = s^1$) are not a representative segment population: Students who did well in courses tended to respond, as did students who participated in athletic activities (and therefore perhaps enjoyed their time at school more); students who did neither tended not to respond. In other words, we have a causal link from A to S and from G to S. In this case, even if A and G are independent in the overall distribution over the student population, we may have a correlation in the subpopulation of respondents. This is an instance of standard intercausal reasoning, where $P(a^1 \mid s^1) > P(a^1 \mid g^1, s^1)$. But without accounting for the possible bias in selecting our population, we may falsely explain the correlation using a causal relationship.* ■

There are many other examples where correlations might arise due to noncausal reasons. One reason involves a mixture of different populations.

Example 21.4   *It is commonly accepted that young girls develop verbal ability at an earlier age than boys. Conversely, boys tend to be taller and heavier than girls. There is certainly no (known) correlation between height and verbal ability in either girls or boys separately. However, if we simply measure height and verbal ability across all children (of the same age), then we may well see a negative correlation between verbal ability and height.* ■

This type of situation is a special case of a latent variable, denoting the class to which the instance belongs (gender, in this case). However, it deserves special mention both because it is quite common, and because these class membership variables are often not perceived as "causes" and may therefore be ignored when looking for a confounding common cause.

A similar situation arises when the distribution we obtain arises from two time series, each of which has a particular trend.

Example 21.5   *Consider data obtained by measuring, in each year over the past century, the average height of the adult population in the world in that year (H), and the total size of the polar caps in that year (S).*

Because average population height has been increasing (due to improved nutrition), and the total size of the polar caps has been decreasing (due to global warming), we would observe a negative correlation between H and S in these data. However, we would not want to conclude that the size of the polar caps causally influences average population height. ∎

In a sense, this situation is also an instance of a latent variable, which in this case is time.

☞ **Thus, we see that the correlation between a pair of variables X and Y may be a consequence of multiple mechanisms, where some are causal and others are not. To answer a causal query regarding an intervention at X, we need to disentangle these different mechanisms, and to isolate the component of the correlation that is due to the**
causal effect **causal effect of X on Y.** A large part of this chapter is devoted to addressing this challenge.

21.2 Causal Models

causal model

causal mechanism

We begin by providing a formal framework for viewing a Bayesian network as a causal model. A *causal model* has the same form as a probabilistic Bayesian network. It consists of a directed acyclic graph over the random variables in the domain. The model asserts that each variable X is governed by a *causal mechanism* that (stochastically) determines its value based on the values of its parents. That is, the value of X is a (stochastic) function of the values of its parents.

A causal mechanism takes the same form as a standard CPD. For a node X and its parents U, the causal model has a stochastic function from the values of U to the values of X. In other words, for each value u of U, it specifies a distribution over the values of X. The difference is in the interpretation of the edges. In a causal model, we assume that X's parents are its direct causes (relative to the variables represented in the model). In other words, we assume that causality flows in the direction of the edges, so that X's value is actually determined via the stochastic function implied by X's CPD.

The assumption that CPDs correspond to causal mechanisms forms the basis for the treatment of intervention queries. When we intervene at a variable X, setting its value to x, we *replace* its original causal mechanism with one that dictates that it take the value x. This manipulation corresponds to replacing X's CPD with a different one, where $X = x$ with probability 1, regardless of anything else.

Example 21.6

For instance, in example 21.1, if Gump changes his grade to an A by hacking into the registrar's database, the result is a model where his grade is no longer determined by his performance in the class, but rather set to the value A, regardless of any other aspects of the situation. An appropriate graphical model for the postintervention situation is shown in figure 21.1a. In this network, the Grade variable no longer depends on Intelligence or Difficulty, nor on anything else. It is simply set to take the value A with probability 1. ∎

mutilated network

The model in figure 21.1a is an instance of the *mutilated network*, a concept introduced in definition 12.1. Recall that, in the mutilated network $\mathcal{B}_{Z=z}$, we eliminate all incoming edges into each variable $Z_i \in Z$, and set its value to be z_i with probability 1.

Based on this intuition, we can now define a causal model as a model that can answer intervention queries using the appropriate mutilated network.

Definition 21.1

causal model

A causal model $\mathcal{C}$ over $\mathcal{X}$ is a Bayesian network over $\mathcal{X}$, which, in addition to answering proba-

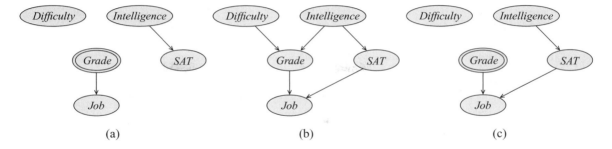

Figure 21.1 **Mutilated** Student **networks representing interventions** (a) Mutilated Student network with an intervention at G. (b) An expanded Student network, with an additional arc $S \to J$. (c) A mutilated network from (b), with an intervention at G.

intervention
query

bilistic queries, can also answer intervention queries $P(\boldsymbol{Y} \mid \mathrm{do}(\boldsymbol{z}), \boldsymbol{x})$, *as follows:*

$$P_{\mathcal{C}}(\boldsymbol{Y} \mid \mathrm{do}(\boldsymbol{z}), \boldsymbol{x}) = P_{\mathcal{C}_{\boldsymbol{Z}=\boldsymbol{z}}}(\boldsymbol{Y} \mid \boldsymbol{x}).$$ ∎

Example 21.7

It is easy to see that this approach deals appropriately with example 21.1. Let $\mathcal{C}^{\mathrm{student}}$ be the appropriate causal model. When we intervene in this model by setting Gump's grade to an A, we obtain the mutilated network shown in figure 21.1a. The distribution induced by this network over Gump's SAT score is the same as the prior distribution over his SAT score in the original network. Thus,

$$P(S \mid \mathrm{do}(G := g^1)) = P_{\mathcal{C}^{\mathrm{student}}_{G=g^1}}(S) = P_{\mathcal{C}^{\mathrm{student}}}(S),$$

as we would expect. Conversely, the distribution induced by this network on Gump's job prospects is $P_{\mathcal{C}^{\mathrm{student}}}(J \mid G = g^1)$. ∎

Note that, in general, the answer to an intervention query does not necessarily reduce to the answer to some observational query.

Example 21.8

Assume that we start out with a somewhat different Student network, as shown in figure 21.1b, which contains an edge from the student's SAT score to his job prospects (for example, because the recruiter can also base her hiring decision on the student's SAT scores). Now, the query $P_{\mathcal{C}^{\mathrm{student}}}(J \mid \mathrm{do}(g^1))$ is answered by the mutilated network of figure 21.1c. In this case, the answer to the query is clearly not $P_{\mathcal{C}^{\mathrm{student}}}(J)$, due to the direct causal influence of his grade on his job prospects. On the other hand, it is also not equal to $P_{\mathcal{C}^{\mathrm{student}}}(J \mid g^1)$, because this last expression also includes the influence via the evidential trail $G \leftarrow I \to S \to J$, which does not apply in the mutilated model. ∎

The ability to provide a formal distinction between observational and causal queries can help resolve some apparent paradoxes that have been the cause of significant debate. One striking example is *Simpson's paradox*, a variant of which is the following:

Simpson's
paradox

Example 21.9

Consider the problem of trying to determine whether a drug is beneficial in curing a particular disease within some population of patients. Statistics show that, within the population, 57.5 percent

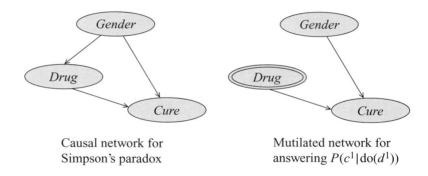

Causal network for   Mutilated network for
Simpson's paradox   answering $P(c^1|\text{do}(d^1))$

Figure 21.2   Causal network for Simpson's paradox

of patients who took the drug (D) are cured (C), whereas only 50 percent of the patients who did not take the drug are cured. Given these statistics, we might be inclined to believe that the drug is beneficial. However, more refined statistics show that within the subpopulation of male patients, 70 percent who took the drug are cured, whereas 80 percent of those who did not take the drug are cured. Moreover, within the subpopulation of female patients, 20 percent of who took the drug are cured, whereas 40 percent of those who did not take the drug are cured. Thus, despite the apparently beneficial effect of the drug on the overall population, the drug appears to be detrimental to both men and women! More precisely, we have that:

$$P(c^1 \mid d^1) \;>\; P(c^1 \mid d^0)$$
$$P(c^1 \mid d^1, G = \textit{male}) \;<\; P(c^1 \mid d^0, G = \textit{male})$$
$$P(c^1 \mid d^1, G = \textit{female}) \;<\; P(c^1 \mid d^0, G = \textit{female}).$$

How is this possible?

This case can occur because taking the drug is correlated with gender: men are much more likely to take the drug than women. In this particular example, 75 percent of men take the drug, whereas only 25 percent of women do. With these parameters, even if men and women are equally represented in the population of patients, we obtain the surprising behavior described earlier.

The conceptual difficulty behind this paradox is that it is not clear which statistics one should use when deciding whether to prescribe the drug to a patient: the general ones or the ones conditioned on gender. In particular, it is not difficult to construct examples where this reversal continues, in that conditioning on yet another variable leads one to the conclusion that the drug is beneficial after all, and conditioning on one more reverses the conclusion yet again. So how can we decide which variables we should condition on?

The causal framework provides an answer to this question. The appropriate query we need to answer in determining whether to prescribe the drug is not $P(c^1 \mid d^1)$, but rather $P(c^1 \mid \text{do}(d^1))$. Figure 21.2 shows the causal network corresponding to this situation, and the mutilated network required for answering the query. We will show how we can use the structure of the causal network to answer queries such as $P(c^1 \mid \text{do}(d^1))$. As we will see in example 21.14, the answer to this query shows that the drug is not beneficial, as expected. ∎

21.3 Structural Causal Identifiability

The framework of the previous section provides us with a mechanism for answering intervention queries, given a fully specified causal model. However, fully specifying a causal model is often impossible. As we discussed, there are often a multitude of (generally unknown) factors that are latent, and that can induce correlations between the variables. In most cases, we cannot fully specify the causal connection between the latent variables and the variables in our model.

In general, the only thing that we can reasonably hope to obtain is the marginal distribution over the observed variables in the model. As we discussed, for probabilistic queries, this marginal distribution suffices (assuming it can be acquired with reasonable accuracy). However, for intervention queries, we must disentangle the causal influence of X and Y from other factors leading to correlations between them. It is far from clear how we could ever accomplish this goal.

Example 21.10

Consider a pair of variables X, Y with an observed correlation between them, and imagine that our task is to determine $P(Y \mid \text{do}(X))$. Let us even assume that X temporally precedes Y, and therefore we know that Y cannot cause X. However, if we consider the possibility that at least some of the correlation between X and Y is due to a hidden common cause, we have no way of determining how much effect perturbing X would have on Y. If all of the correlation is due to a causal link, then $P(Y \mid \text{do}(X)) = P(Y \mid X)$; conversely, if all of the correlation is due to the hidden common cause, then $P(Y \mid \text{do}(X)) = P(Y)$. And, in general, any value between those two distributions is possible. ∎

Thus, given that latent variables are inevitable in many settings, it appears that the situation regarding causal queries is hopeless. Fortunately, as we now show, **we can sometimes answer causal questions in models involving latent variables using observed correlations alone**. More precisely, in this section, we attempt to address to question of which intervention queries are *identifiable*, that is, can be answered using only conditional probabilities involving observable variables. Probabilities over observed variables can be estimated from data or elicited from an expert. Thus, if we can reduce the answer to a query to an expression involving only such probabilities, we may be able to provide a robust and accurate answer to it.

identifiability

21.3.1 Query Simplification Rules

The key observation in this section is that the structure of a causal model give rise to certain equivalence rules over interventional queries, which allow one query to be replaced by an equivalent one that may have a simpler form. By applying one or more of these simplification steps, we may be able to convert one causal query to another query that involves no interventions, and can therefore be answered using observational data alone.

augmented causal model

decision variable

These rules can be defined in terms of an *augmented causal model* that encodes the possible effect of interventions explicitly within the graph structure. More precisely, we view the process of an intervention in terms of a new *decision variable* (see chapter 23) $\widehat{Z}$ that determines whether we intervene at Z, and if so, what its value is. The variable $\widehat{Z}$ takes on values in $\{\epsilon\} \cup Val(Z)$. If $\widehat{Z} = \epsilon$, then Z behaves as a random variable whose distribution is determined by its usual CPD $P(Z \mid \text{Pa}_Z)$; if $\widehat{Z} = z$, then it deterministically sets the value of Z to be z with probability 1. Let $\widehat{\boldsymbol{Z}}$ denote the set $\{\widehat{Z} \; : \; Z \in \boldsymbol{Z}\}$.

Note that, in those cases where Z's value is deterministically set by one parent, all of Z's other parents U become irrelevant, and so we can effectively remove all edges from U to Z. Let $\mathcal{G}^{\dagger}$ be the augmented model for $\mathcal{G}$. Let $\mathcal{G}^{\dagger}_{\underline{Z}}$ be the graph obtained from $\mathcal{G}^{\dagger}$ except that every $Z \in \boldsymbol{Z}$ has only the single parent $\widehat{Z}$. Note that $\mathcal{G}^{\dagger}_{\underline{Z}}$ is similar to the mutilated network used in definition 21.1 to define the semantics of intervention queries; the only difference is that we now make the interventions themselves explicit. As we will now see, this difference allows us to study the effect of one intervention within a model that contain others.

intervention
query
simplification

Based on this construction, we now define three *query simplification* rules. The first simply allows us to insert or delete observations into a query.

Proposition 21.1

Let $\mathcal{C}$ be a causal model over the graph structure $\mathcal{G}$. Then:

$$P(\boldsymbol{Y} \mid \mathrm{do}(\boldsymbol{Z} := \boldsymbol{z}), \boldsymbol{X} = \boldsymbol{x}, \boldsymbol{W} = \boldsymbol{w}) = P(\boldsymbol{Y} \mid \mathrm{do}(\boldsymbol{Z} := \boldsymbol{z}), \boldsymbol{X} = \boldsymbol{x}),$$

if $\boldsymbol{W}$ is d-separated from $\boldsymbol{Y}$ given $\boldsymbol{Z}, \boldsymbol{X}$ in the graph $\mathcal{G}^{\dagger}_{\underline{Z}}$.

This rule is a simple consequence of the fact that probabilities of intervention queries are defined relative to the graph $\mathcal{G}^{\dagger}_{\underline{Z}}$, and the correspondence between independence and d-separation in this graph.

The second rule is subtler, and it allows us to replace an intervention with the corresponding observation.

Proposition 21.2

Let $\mathcal{C}$ be a causal model over the graph structure $\mathcal{G}$. Then:

$$P(\boldsymbol{Y} \mid \mathrm{do}(\boldsymbol{Z} := \boldsymbol{z}), \mathrm{do}(\boldsymbol{X} := \boldsymbol{x}), \boldsymbol{W} = \boldsymbol{w}) = P(\boldsymbol{Y} \mid \mathrm{do}(\boldsymbol{Z} := \boldsymbol{z}), \boldsymbol{X} = \boldsymbol{x}, \boldsymbol{W} = \boldsymbol{w}),$$

if $\boldsymbol{Y}$ is d-separated from $\widehat{\boldsymbol{X}}$ given $\boldsymbol{X}, \boldsymbol{Z}, \boldsymbol{W}$ in the graph $\mathcal{G}^{\dagger}_{\underline{Z}}$.

Intuitively, this rule holds because it tells us that we get no more information regarding $\boldsymbol{Y}$ from the fact that an intervention took place at $\boldsymbol{X}$ than the values $\boldsymbol{x}$ themselves. In other words, knowing that $\boldsymbol{X} = \boldsymbol{x}$, we do not care whether these values were obtained as a result of an intervention or not. This criterion is also equivalent to asking whether $\boldsymbol{X}$ have *requisite CPD* for the query $P(\boldsymbol{Y} \mid do(\boldsymbol{Z} := \boldsymbol{z}), \boldsymbol{X} = \boldsymbol{x}, \boldsymbol{W} = \boldsymbol{w})$, as defined in exercise 3.20. The relationship is not surprising, because an intervention at a variable $X \in \boldsymbol{X}$ corresponds exactly to changing its CPD; if our query is oblivious to changes in the CPD (given X), then we should not care whether there was an intervention at X or not.

requisite CPD

As a simple example, consider the case where $\boldsymbol{Z} = \emptyset$ and $\boldsymbol{W} = \emptyset$, and where we have single variables X, Y. In this case, the rule reduces to the assertion that

$$P(Y \mid do(X := x)) = P(Y \mid X = x),$$

if Y is d-separated from $\widehat{X}$ given X in the graph $\mathcal{G}^{\dagger}$. The separation property holds if the only trails between X and Y in $\mathcal{G}$ emanate causally from X (that is, go through its children). Indeed, in this case, intervening at X has the same effect on Y as observing X. Conversely, if we have an active trail from $\widehat{X}$ to Y given X, then it must go through a v-structure activated by X. In this case, Y is not a descendant of X, and an observation of X has a very different effect than an intervention at X. The proof of this theorem is left as an exercise (exercise 21.1).

Example 21.11

Consider again the network of figure 21.1b, and assume that the student somehow manages to cheat on the SAT exam and get a higher SAT score. We are interested in the effect on the student's grade, that is, in evaluating a query of the form $P(G \mid \mathrm{do}(S), J, I)$. Consider the augmented model $\mathcal{G}^\dagger$, which contains a new decision parent $\widehat{S}$ for the node S. In this graph, we would have that G is d-separated from $\widehat{S}$ given S, J and I. Thus, the theorem would allow us to conclude that $P(G \mid \mathrm{do}(S), J, I) = P(G \mid S, J, I)$. To provide another intuition for this equality, note that, in the original graph, there are two possible trails between G and S: $G \rightarrow J \leftarrow S$ and $G \leftarrow I \rightarrow S$. The effects of the first trail remain unchanged in the mutilated network: there is no difference between an intervention at S and an observation at S, relative to the trail $G \rightarrow J \leftarrow S$. There is a difference between these two cases relative to the trail $G \leftarrow I \rightarrow S$, but that trail is blocked by the observation at I, and hence there is no effect to changing the intervention at S to an observation. Note that this last argument would not apply to the query $P(G \mid \mathrm{do}(S), J)$, and indeed, the theorem would not allow us to conclude that $P(G \mid \mathrm{do}(S), J) = P(G \mid S, J)$. ∎

The final rule allows us to introduce or delete interventions, in the same way that proposition 21.1 allows us to introduce or delete observations.

Proposition 21.3

Let $\mathcal{C}$ be a causal model over the graph structure $\mathcal{G}$. Then:

$$P(\boldsymbol{Y} \mid \mathrm{do}(\boldsymbol{Z} := \boldsymbol{z}), \mathrm{do}(\boldsymbol{X} := \boldsymbol{x}), \boldsymbol{W} = \boldsymbol{w}) = P(\boldsymbol{Y} \mid \mathrm{do}(\boldsymbol{Z} := \boldsymbol{z}), \boldsymbol{W} = \boldsymbol{w}),$$

if $\boldsymbol{Y}$ is d-separated from $\widehat{\boldsymbol{X}}$ given $\boldsymbol{Z}, \boldsymbol{W}$ in the graph $\mathcal{G}^\dagger_{\overline{\boldsymbol{Z}}}$.

This analysis can also be interpreted in terms of requisite CPDs. Here, the premise is equivalent to stating that the CPDs of the variables in $\boldsymbol{X}$ are not requisite for the query even when their values are not observed. In this case, we can ignore both the knowledge of the intervention and the knowledge regarding the values imposed by the intervention.

Again, we can obtain intuition by considering the simpler case where $\boldsymbol{Z} = \emptyset$, $\boldsymbol{W} = \emptyset$, and $\boldsymbol{X}$ is a single variable X. Here, the rule reduces to:

$$P(\boldsymbol{Y} \mid \mathrm{do}(X := x)) = P(\boldsymbol{Y}),$$

if $\boldsymbol{Y}$ is d-separated from $\widehat{X}$ in the graph $\mathcal{G}^\dagger$. The intuition behind this rule is fairly straightforward. Conditioning on $\widehat{X}$ corresponds to changing the causal mechanism for X. If the d-separation condition holds, this operation provably has no effect on $\boldsymbol{Y}$. From a different perspective, changing the causal mechanism for X can only affect $\boldsymbol{Y}$ causally, via X's children. If there are no trails from $\widehat{X}$ to $\boldsymbol{Y}$ *without* conditioning on X, then there are no causal paths from X to $\boldsymbol{Y}$. Not surprisingly, this condition is equivalent to the graphical criterion for identifying requisite probability nodes (see exercise 21.2), which also test whether the CPD of a variable X can affect the outcome of a given query; in this case, the CPDs of the variable X is determined by whether there is an intervention at X.

Example 21.12

Let us revisit figure 21.1b, and a query of the form $P(S \mid \mathrm{do}(G))$ (taking $\boldsymbol{Y} = S$, $\boldsymbol{X} = G$, and $\boldsymbol{Z}, \boldsymbol{W} = \emptyset$). Consider the augmented model $\mathcal{G}^\dagger$, which contains a new decision parent $\widehat{G}$ for the node S. The node $\widehat{G}$ is d-separated from S in this graph, so that we can conclude $P(S \mid \mathrm{do}(G)) = P(S)$, as we would expect. On the other hand, if our query is $P(S \mid \mathrm{do}(G), J)$

(so that now $\boldsymbol{W} = J$), then G itself is an ancestor of our evidence J. In this case, there is an active trail from $\widehat{G}$ to S in the network; hence, the rule does not apply, and we cannot conclude $P(S \mid \text{do}(G), J) = P(S \mid J)$. Again, this is as we would expect, because when we observe J, the fact that we intervened at G is clearly relevant to S, due to standard intercausal reasoning. ∎

21.3.2   Iterated Query Simplification

The rules in the previous section allow us to simplify a query in certain cases. But their applicability appears limited, since there are many queries where none of the rules apply directly. A key insight, however, is that we can also perform other transformations on the query, allowing the rules to be applied.

Example 21.13

To illustrate this approach, consider again the example of example 21.8, which involves the query $P(J \mid \text{do}(G))$ in the network of figure 21.1b. As we discussed, none of our rules apply directly to this query: Obviously we cannot eliminate the intervention — $P(J \mid \text{do}(G)) \neq P(J)$. We also cannot turn the intervention into an observation — $P(J \mid \text{do}(G)) \neq P(J \mid G)$; intuitively, the reason is that intervening at G only affects J via the single edge $G \to J$, whereas conditioning G also influences J by the indirect trail $G \leftarrow I \to S \to J$. This trail is called a back-door *trail, since it leaves G by the "back door."*

However, we can use standard probabilistic reasoning to conclude that:

$$P(J \mid \text{do}(G)) = \sum_S P(J \mid \text{do}(G), S) P(S \mid \text{do}(G)).$$

Both of the terms in the summation can *be further simplified. For the first term, we have that the only active trail from G to J is now the direct edge $G \to J$. More formally, J is d-separated from G given S in the graph where outgoing arcs from G have been deleted. Thus, we can apply proposition 21.2, and conclude:*

$$P(J \mid \text{do}(G), S) = P(J \mid G, S).$$

For the second term, we have already shown in example 21.12 that $P(S \mid \text{do}(G)) = P(S)$. Putting the two together, we obtain that:

$$P(J \mid \text{do}(G)) = \sum_S P(J \mid G, S) P(S).$$

∎

This example illustrates a process whereby we introduce conditioning on some set of variables:

$$P(\boldsymbol{Y} \mid do(\boldsymbol{X}), \boldsymbol{Z}) = \sum_{\boldsymbol{W}} P(\boldsymbol{Y} \mid do(\boldsymbol{X}), \boldsymbol{Z}, \boldsymbol{W}) P(\boldsymbol{W} \mid do(\boldsymbol{X}), \boldsymbol{Z}).$$

Even when none of the transformation rule apply to the query $P(\boldsymbol{Y} \mid do(\boldsymbol{X}), \boldsymbol{Z})$, they may apply to each of the two terms in the summation of the transformed expression.

back-door trail

back-door criterion

The example illustrates one special case of this transformation. A *back-door trail* from X to Y is an active trail that leaves X via a parent of X. For a query $P(\boldsymbol{Y} \mid do(\boldsymbol{X}))$, a set $\boldsymbol{W}$ satisfies the *back-door criterion* if no node in $\boldsymbol{W}$ is a descendant of $\boldsymbol{X}$, and $\boldsymbol{W}$ blocks all back-door paths from X to Y. Using an argument identical to the one in the example, we can show that if a set W satisfies the back-door criterion for a query $P(Y \mid do(X))$, then

$$P(Y \mid do(X)) = \sum_W P(Y \mid X, W)P(W). \tag{21.1}$$

The back-door criterion can be used to address Simpson's paradox, as described in example 21.9.

Example 21.14

Consider again the query $P(c^1 \mid do(d^1))$. The variable G (Gender) introduces a back-door trail between C and D. We can account for its influence using equation (21.1):

$$P(c^1 \mid do(d^1)) = \sum_g P(c^1 \mid d^1, g)P(g).$$

We therefore obtain that:

$$
\begin{aligned}
P(c^1 \mid do(d^1)) &= 0.7 \cdot 0.5 + 0.2 \cdot 0.5 = 0.45 \\
P(c^1 \mid do(d^0)) &= 0.8 \cdot 0.5 + 0.4 \cdot 0.5 = 0.6.
\end{aligned}
$$

And therefore, we should not prescribe the drug.   ∎

More generally, by repeated application of these rules, we can sometimes simplify fairly complex queries, obtaining answers in cases that are far from obvious at first glance.

Box 21.A — Case Study: Identifying the Effect of Smoking on Cancer. *In the early 1960s, following a significant increase in the number of smokers that occurred around World War II, people began to notice a substantial increase in the number of cases of lung cancer. After a great many studies, a correlation was noticed between smoking and lung cancer. This correlation was noticed in both directions: the frequency of smokers among lung cancer patients was substantially higher than in the general population, and the frequency of lung cancer patients within the population of smokers was substantially higher than within the population of nonsmokers.*

These results, together with some experiments of injecting tobacco products into rats, led the Surgeon General, in 1964, to issue a report linking cigarette smoking to cancer and, most particularly, lung cancer. His claim was that the correlation found is causal, namely: If we ban smoking, the rate of cancer cases will be roughly the same as the one we find among nonsmokers in the population.

These studies came under severe attacks from the tobacco industry, backed by some very prominent statisticians. The claim was that the observed correlations can also be explained by a model in which there is no causal connection between smoking and lung cancer. Instead, an unobserved genotype might exist that simultaneously causes cancer and produces an inborn craving for nicotine. In other words, there were two hypothesized models, shown in figure 21.A.1a,b.

The two models can express precisely the same set of distributions over the observable variables S, C. Thus, they can do an equally good job of representing the empirical distribution over these variables, and there is no way to distinguish between them based on observational data alone. Moreover, both models will provide the same answer to standard probabilistic queries such as $P(c^1 \mid$

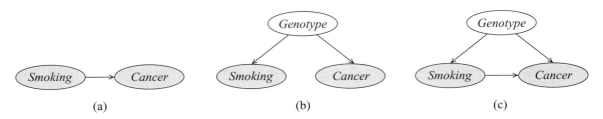

Figure 21.A.1 — Three candidate models for smoking and cancer. (a) a direct causal influence; (b) indirect influence via a latent common parent *Genotype*; (c) incorporating both types of influence.

s^1). *However, relative to interventional queries, these models have very different consequences. According to the Surgeon General's model, we would have:*

$$P(c^1 \mid \mathrm{do}(S := s^1)) = P(c^1 \mid s).$$

In other words, if we force people to smoke, their probability of getting cancer is the same as the probability conditioned on smoking, which is much higher than the prior probability. On the other hand, according to the tobacco industry model, we have that

$$P(c^1 \mid \mathrm{do}(S := s^1)) = P(c^1).$$

In other words, making the population smoke or stop smoking would have no effect on the rate of cancer cases.

Pearl (1995) proposes a formal analysis of this dilemma, which we now present. He proposes that we combine these two models into a single joint model, which accommodates for both possible types of interactions between smoking and cancer, as shown in figure 21.A.1c. We now need to assess, from the marginal distribution over the observed variables alone, the parameterization of the various links. Unfortunately, it is impossible to determine the parameters of these links from observational data alone, since both of the original two models (in figure 21.A.1a,b) can explain the data perfectly.

However, if we refine the model somewhat, introducing an additional assumption, we can provide such estimates. Assume that we determine that the effect of smoking on cancer is not a direct one, but occurs through the accumulation of tar deposits in the lungs, as shown in figure 21.A.2a. Note that this model makes the assumption that the accumulation of tar in the lungs is not directly affected by the latent Genotype variable. As we now show, if we can measure the amount of tar deposits in the lungs of various individuals (for example, by X-ray or in autopsies), we can determine the probability of the intervention query $P(c^1 \mid \mathrm{do}(s^1))$ using observed correlations alone.

We are interested in $P(c^1 \mid \mathrm{do}(s^1))$, which is an intervention query whose mutilated network is $\mathcal{G}_{\overline{S}}^{\dagger}$ in figure 21.A.2b. Standard probabilistic reasoning shows that:

$$P(C \mid \mathrm{do}(s^1)) \;=\; \sum_t P(C \mid \mathrm{do}(s^1), t) P(t \mid \mathrm{do}(s^1)).$$

We now consider and simplify each term in the summation separately.

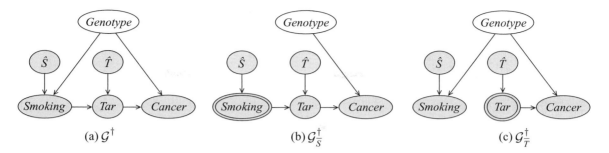

Figure 21.A.2 — Determining causality between smoking and cancer. Augmented network for a model in which the effect of smoking on cancer is indirect, mediated via tar accumulation in the lungs, and mutilated variants for two possible interventions.

The second term, which measures the effect of Smoking on Tar, can be simplified directly using our rule for converting interventions to observations, stated in proposition 21.2. Here, $\widehat{S}$ is d-separated from T given S in the graph $\mathcal{G}^{\dagger}$, shown in figure 21.A.2a. It follows that:

$$P(t \mid \mathrm{do}(s)) \quad = \quad P(t \mid s).$$

Intuitively, the only active trail from $\widehat{S}$ to T goes via S, and the effect of that trail is identical regardless of whether we condition on S or intervene at S.

Now, let us examine the first term, $P(C \mid \mathrm{do}(s^1), t)$, which measures the effect of Tar on Cancer in the presence of our intervention on S. Unfortunately, we cannot directly convert the intervention at S to an observation, since C is not d-separated from $\widehat{S}$ given S, T in $\mathcal{G}^{\dagger}$. However, we can convert the observation at T to an intervention, because C is d-separated from $\widehat{T}$ given S, T in the graph $\mathcal{G}^{\dagger}_{\overline{S}}$.

$$P(C \mid \mathrm{do}(s^1), t) = P(C \mid \mathrm{do}(s^1), \mathrm{do}(t)).$$

We can now eliminate the intervention at S from this expression using proposition 21.3, which applies because C is d-separated from $\widehat{S}$ given T in the graph $\mathcal{G}^{\dagger}_{\overline{T}}$ (figure 21.A.2c), obtaining:

$$P(C \mid \mathrm{do}(s^1), \mathrm{do}(t)) = P(C \mid \mathrm{do}(t)).$$

Considering this last expression, we can apply standard probabilistic reasoning and introduce conditioning on S:

$$
\begin{aligned}
P(C \mid \mathrm{do}(t)) \quad &= \quad \sum_{s'} P(C \mid \mathrm{do}(t), s') P(s' \mid \mathrm{do}(t)) \\
&= \quad \sum_{s'} P(C \mid t, s') P(s' \mid \mathrm{do}(t)) \\
&= \quad \sum_{s'} P(C \mid t, s') P(s').
\end{aligned}
$$

The second equality is an application of proposition 21.2, which applies because C is d-separated from $\widehat{T}$ given T, S in $\mathcal{G}^{\dagger}$. The final equality is a consequence of proposition 21.3, which holds because S is d-separated from $\widehat{T}$ in $\mathcal{G}^{\dagger}$.

Putting everything together, we get:

$$
\begin{aligned}
P(c \mid do(s^1)) &= \sum_t P(c \mid do(s^1), t) P(t \mid do(s^1)) \\
&= \sum_t P(c \mid do(s^1), t) P(t \mid s^1) \\
&= \sum_t P(t \mid s^1) \sum_{s'} P(c \mid t, s') P(s').
\end{aligned}
$$

Thus, if we agree that tar in the lungs is the intermediary between smoking and lung cancer, we can uniquely determine the extent to which smoking causes lung cancer even in the presence of a confounding latent variable!

Of course, this statement is more useful as a thought experiment than as a practical computational tool, both because it is unlikely that smoking affects cancer only via tar accumulation in the lungs and because it would be very difficult in practice to measure this intermediate variable. Nevertheless, this type of analysis provides insight on the extent to which understanding of the underlying causal model allows us to identify the value of intervention queries even in the presence of latent variables.

These rules provide a powerful mechanism for answering intervention queries, even when the causal model involves latent variables (see, for example, box 21.A). More generally, using these rules, we can show that the query $P(Y \mid do(x))$ is identifiable in each of the models shown in figure 21.3 (see exercise 21.4). In these figures, we have used a bidirected dashed arrow, known as a *bow pattern*, to denote the existence of a latent common cause between two variables. For example, the model in (d) has the same structure as in our Smoking model of figure 21.A.2a. (box 21.A). This notation simplifies the diagrams considerably, and it is therefore quite commonly used. Note that none of the diagrams contain a bow pattern between X and one of its children. In general, a necessary condition for identifiability is that the graph contain no bow pattern between X and a child of X that is an ancestor of Y. The reason is that, if such a bow pattern exists between X and one of its children W, we have no mechanism for distinguishing X's direct influence on W and the indirect correlation induced by the latent variable, which is their common parent.

It is interesting to note that, in the models shown in (a), (b), and (e), Y has a parent Z whose effect on Y is not identifiable, yet the effect of X on Y, including its effect via Z, is identifiable. For example, in (a), $P(Y \mid X) = P(Y \mid do(X))$, and so there is no need to disentangle the influence that flows through the $Z \rightarrow Y$ edge, and the influence that flows through the bow pattern. These examples demonstrate that, to identify the influence of one variable on another, it is not necessary to identify every edge involved in the interactions between them.

Figure 21.4 shows a few examples of models where the influence of X on Y is not identifiable. Interestingly, the model in (g) illustrates the converse to the observation we just stated: identification of every edge involved in the interaction between X and Y does not suffice to

bow pattern

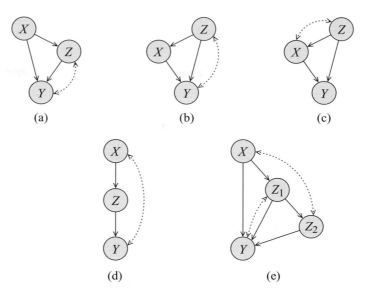

Figure 21.3 **Examples of models where** $P(Y \mid do(X))$ **is identifiable.** The bidirected dashed arrows denote cases where a latent variable affects both of the linked variables.

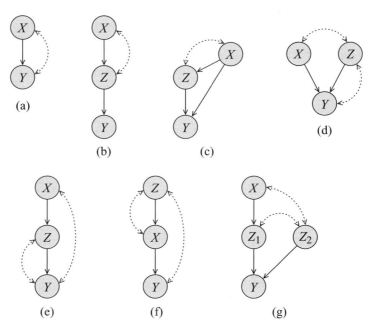

Figure 21.4 **Examples of models where** $P(Y \mid do(X))$ **is not identifiable.** The bidirected dashed arrows denote cases where a latent variable affects both of the linked variables.

identify their interaction. In particular, in this model, we can identify all of $P(Z_1 \mid do(X))$, $P(Z_2 \mid do(X))$, $P(Y \mid do(Z_1))$, and $P(Y \mid do(Z_2))$, but we cannot identify $P(Y \mid do(X))$ (see exercise 21.5).

Note that the removal of any edge (directed or bidirected) from a causal diagram of this type can only help in the identifiability of causal effects, since it can only deactivate active trails. Conversely, adding edges can only reduce identifiability. Hence, any subgraph of the graphs in figure 21.3 is also identifiable, and any extension of the graphs in figure 21.4 is nonidentifiable. Moreover, it is possible to show that the graphs in figure 21.3 are maximal, in the sense that adding any edge renders $P(Y \mid do(X))$ unidentifiable; similarly, the graphs in figure 21.4 are minimal, in that the query is identifiable in any of their (strict) subgraphs. Similarly, the introduction of mediating observed variables onto any edge in a causal graph — transforming an arc $A \rightarrow B$ into a path $A \rightarrow Z \rightarrow B$ for a new observed variable Z — can only increase our ability to identify causal effects.

Finally, we note that the techniques in this section provide us with methods for answering only queries that are identifiable. Unfortunately, unidentifiable queries are quite common in practice. In section 21.5, we describe methods that allow us to provide partial answers for queries that are not identifiable.

21.4   Mechanisms and Response Variables ⋆

The underlying intuition for our definition of a causal model is that the graph structure defines a causal progression, where the value of each variable is selected, in order, based on the values of its parents, via some causal mechanism. So far, the details of this causal mechanism have remained implicit. We simply assumed that it induces, for each variable X, a conditional probability distribution $P(X \mid \mathrm{Pa}_X)$.

This approach suffices in cases, as in the previous section, where we can compute the value of a causal query in terms of standard probabilistic queries. In general, however, this identification may not be possible. In such cases, we have to reason explicitly about the mechanism governing the behavior of the variables in the model and the ways in which the latent variables may influence the observed variables. By obtaining a finer-grained analysis of these mechanisms, we may be able to provide some analysis for intervention queries that are not identifiable from probabilities over the observable variables alone.

A second reason for understanding the mechanism in more detail is to answer additional types of queries: counterfactual queries, and certain types of diagnostic queries. As we discussed, our intuition for a counterfactual query is that we want to keep as much as we can between the real and the counterfactual world. By understanding the exact mechanism that governs each variable, we can obtain more reliable inferences about what took place in the true world, so as to preserve as much as possible when reasoning about the counterfactual events.

Example 21.15   *Consider a significantly simplified causal model for Gump's job prospects, where we have only the edge $G \rightarrow J$ — Gump's job prospects depend only on his grade. Let us also assume that grades are binary-valued — high and low. Our counterfactual query is as follows: "Gump received a low grade. He applied to Acme Consulting, and he was not hired. Would Acme have hired him had he managed to hack into the registrar's computer system and change his grade to a high one?"*

Consider two very different models of the world. In the first, there are two (equally likely) populations of companies. Those in the first population are not in hiring mode, and they always reject Gump without looking at his grade. Companies in the second population are desperate for employees and will take anyone, regardless of their grade. The basic intuition of counterfactual queries is that, when we move to the counterfactual world, Gump does not get another "random draw" of company. We want to answer the counterfactual query assuming that Acme's recruiting response model is the same in both cases. Under this assumption, we can conclude that Acme was not in hiring mode, and that Gump's outcome would have been no different had he managed to change his grade.

In a second model, there are also two (equally likely) populations of companies. In the first, the company is highly selective, and it hires a student if and only if his grades are high. In the second population, the recruiter likes to hire underdogs because he can pay them less, and he hires Gump if and only if his grades are low. In this setting, if we again preserve the hiring response of the company, we would conclude that Acme must use a selective recruiting policy, and so Gump would have been hired had he managed to change his grade to a high one.

Note that these two models are identical from a probabilistic perspective. In both cases, $P(J \mid G) = (0.5, 0.5)$. But they are very different with respect to answering counterfactual queries. ∎

The same type of situation applies in a setting where we act in the world, and where we wish to reason about the probabilities of different events before and after our action. For example, consider the problem of *troubleshooting* a broken device with multiple components, where a fault in each can lead to the observed failure mode. Assume that we replace one of the components, and we wish to reason about the probabilities in the system after the replacement action. Intuitively, if the problem was not with the component we replaced, then the replacement action should have no effect: the probability that the device remains broken should be 1. The mechanism by which different combinations of faults can lead to the observed behavior is quite complex, and we are generally uncertain about which one is currently the case. However, we wish to have this mechanism persist between the prerepair and postrepair state. Box 21.C describes the use of this approach in a real-world diagnostic setting.

As these examples illustrate, to answer certain types of queries, we need to obtain a detailed specification of the causal mechanism that determines the values of different variables in the model. In general, we can argue that (quantum effects aside) any randomness in the model is the cumulative effect of latent variables that are simply unmodeled. For example, consider even a seemingly random event such as the outcome of a coin toss. We can imagine this event depending on a large number of exogenous variables, such as the rotation of the coin when it was tossed, the forces applied to it, any air movement in the room, and more. Given all of these factors, the outcome of the coin toss is arguably deterministic. As another example, a company's decision on whether to hire Gump can depend on the company's current goals and funding level, on the existence of other prospective applicants, and even on whether the recruiter likes the color of Gump's tie. Based on this intuition, we can divide the variables into two groups. The *endogenous variables* are those that we choose to include in the model; the exogenous variables are unmodeled, latent variables. We can now argue, as a hypothetical thought experiment, that the exogenous variables encompass all of the stochasticity in the world; therefore, given all of the exogenous variables, each endogenous variable (say the company's hiring decision) is fully determined by its endogenous parents (Gump's grade in our example).

troubleshooting

endogenous
variable

Clearly, we cannot possibly encode a complete specification of a causal model with all of the relevant exogenous variables. In most cases, we are not even aware of the entire set of relevant exogenous variables, far less able to specify their influence on the model. However, for our purposes, we can abstract away from specific exogenous variables and simply focus on their effect on the endogenous variables. To understand this transformation, consider the following example.

Example 21.16

Consider again the simple setting of example 21.15. Let U be the entire set of exogenous variables affecting the outcome of the variable J in the causal model $G \to J$. We can now assume that the value of J is a deterministic function of G, U — for each assignment u to U and g to G, the variable J deterministically takes the value $f_J(g, u)$. We are interested in modeling only the interaction between G and J. Thus, we can partition the set of assignments u into four classes, based on the mapping μ that they induce between G and J:

- *$\mu^J_{1 \mapsto 1, 0 \mapsto 1}$ — those where $f_J(u, g^1) = j^1$ and $f_J(u, g^0) = j^1$; these are the "always hire" cases, where Gump is hired regardless of his grade.*

- *$\mu^J_{1 \mapsto 1, 0 \mapsto 0}$ — those where $f_J(u, g^1) = j^1$ and $f_J(u, g^0) = j^0$; these are the "rational recruiting" cases, where Gump is hired if and only if his grade is high.*

- *$\mu^J_{1 \mapsto 0, 0 \mapsto 1}$ — those where $f_J(u, g^1) = j^0$ and $f_J(u, g^0) = j^1$; these are the "underdog" cases, where Gump is hired if and only if his grade is low.*

- *$\mu^J_{1 \mapsto 0, 0 \mapsto 0}$ — those where $f_J(u, g^1) = j^0$ and $f_J(u, g^0) = j^0$; these are the "never hire" cases, where Gump is not hired regardless of his grade.*

Each assignment u induces precisely one of these four different mappings between G and J. For example, we might have a situation where, if Gump wears a green tie with pink elephants, he is never hired regardless of his grade, which is the last case in the previous list.

For the purposes of reasoning about the interaction between G and J, we can abstract away from modeling specific exogenous variables U, and simply reason about how likely we are to encounter each of the four categories of functions induced by assignments u. ∎

Generalizing from this example, we define as follows:

Definition 21.2

response variable

Let X be a variable and Y be a set of parents. A response variable *for X given Y is a variable U^X whose domain is the set of all possible functions $\mu(Y)$ from $Val(Y)$ to $Val(X)$. That is, let $y_1, \ldots, y_m$ be an enumeration of $Val(Y)$ (for $m = |Val(Y)|$). For a tuple of (generally not distinct) values $x_1, \ldots, x_m \in Val(X)$, we use $\mu^X_{(y_1, \ldots, y_m) \mapsto (x_1, \ldots, x_m)}$ to denote the function that assigns x_i to y_i, for $i = 1, \ldots, m$. The domain of U^X contains one such function for each such tuple, giving a total of k^m functions of this form (for $|Val(X)| = k$).* ∎

Example 21.17

In example 21.16, we introduce two response variables, U^G and U^J. Because G has no endogenous parents, the variable U^G is degenerate: its values are simply g^1 and g^0. The response variable U^J takes one of the four values $\mu^J_{1 \mapsto 1, 1 \mapsto 1}, \mu^J_{1 \mapsto 1, 1 \mapsto 0}, \mu^J_{1 \mapsto 0, 1 \mapsto 1}, \mu^J_{1 \mapsto 0, 1 \mapsto 0}$, each of which defines a function from G to J. Thus, given U^J and G, the value of J is fully determined. For example, if $U^J = \mu^J_{1 \mapsto 1, 1 \mapsto 0}$, and $G = g^1$, then $J = j^1$. ∎

Definition 21.3

functional causal model

A functional causal model $\mathcal{C}$ over a set of endogenous variables $\mathcal{X}$ is a causal model defined over two sets of variables: the variables $\mathcal{X}$, and a set of response variables $\mathcal{U} = \{U^X : X \in \mathcal{X}\}$. Each variable $X \in \mathcal{X}$ has a set of parents $\mathrm{Pa}_X \subset \mathcal{X}$, and a response variable parent U^X for X given Pa_X. The model for each variable $X \in \mathcal{X}$ is deterministic: When $U^X = \mu$ and $\mathrm{Pa}_X = \boldsymbol{y}$, then $X = \mu(\boldsymbol{y})$ with probability 1.

The model $\mathcal{C}$ also defines a joint probability distribution over the response variables, defined as a Bayesian network over $\mathcal{U}$. Thus, each response variable U has a set of parents $\mathrm{Pa}_U \subset \mathcal{U}$, and a CPD $P(U \mid \mathrm{Pa}_U)$. ∎

Functional causal models provide a much finer-grained specification of the underlying causal mechanisms than do standard causal models, where we only specify a CPD for each endogenous variable. In our $G \to J$ example, rather than specifying a CPD for J, we must specify a distribution over U^J. A table representation of the CPD has two independent parameters — one for each assignment to G. The distribution $P(U^J)$ has three independent parameters — there are four possible values for U^J, and their probabilities must sum to 1. While, in this case, the blowup in the representation may not appear too onerous, the general case is far worse. In general, consider a variable X with parents $\boldsymbol{Y}$, and let $k = |Val(X)|$ and $m = |Val(\boldsymbol{Y})|$. The total number of possible mappings from $Val(\boldsymbol{Y})$ to $Val(\boldsymbol{X})$ is k^m — each function selects one of X's k values for each of the m assignments to $\boldsymbol{Y}$. Thus, the total number of independent parameters in (an explicit representation of) $P(U^X)$ is $k^m - 1$. By comparison, a table-CPD $P(X \mid \boldsymbol{Y})$ requires $m(k-1)$ independent parameters only.

Example 21.18

Consider the network in figure 21.5a, representing a causal model for a randomized clinical trial, where some patients are randomly assigned a medication and others a placebo. The model contains three binary-valued endogenous variables: A indicates the treatment assigned to the patient; T indicates the treatment actually received; and O indicates the observed outcome (positive or negative). The model contains three response variables: U^A, U^T, and U^O. Intuitively, U^A (which has a uniform distribution) represents the stochastic event determining the assignment to the two groups (medication and placebo); U^T determines the patient's model for complying with the prescribed course of treatment; and U^O encodes the form of the patient's response to different treatments.

The domain of U^T consists of the following four functions:

- *$\mu^T_{1 \mapsto 1, 0 \mapsto 1}$ — always taker;*
- *$\mu^T_{1 \mapsto 1, 0 \mapsto 0}$ — complier (takes medicine if and only if prescribed);*
- *$\mu^T_{1 \mapsto 0, 0 \mapsto 1}$ — defier (takes medicine if and only if not prescribed);*
- *$\mu^T_{1 \mapsto 0, 0 \mapsto 0}$ — never taker.*

Similarly, the domain of U^O consists of the following four functions:

- *$\mu^O_{1 \mapsto 1, 0 \mapsto 1}$ — always well;*
- *$\mu^O_{1 \mapsto 1, 0 \mapsto 0}$ — helped (recovers if and only if takes medicine);*
- *$\mu^O_{1 \mapsto 0, 0 \mapsto 1}$ — hurt (recovers if and only if does not take medicine);*
- *$\mu^O_{1 \mapsto 0, 0 \mapsto 0}$ — never well.*

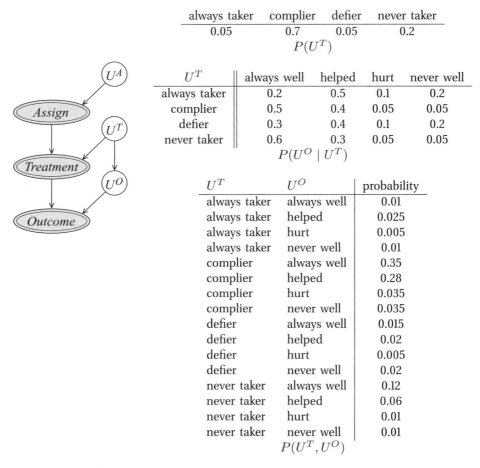

always taker	complier	defier	never taker
0.05	0.7	0.05	0.2

$$P(U^T)$$

U^T	always well	helped	hurt	never well
always taker	0.2	0.5	0.1	0.2
complier	0.5	0.4	0.05	0.05
defier	0.3	0.4	0.1	0.2
never taker	0.6	0.3	0.05	0.05

$$P(U^O \mid U^T)$$

U^T	U^O	probability
always taker	always well	0.01
always taker	helped	0.025
always taker	hurt	0.005
always taker	never well	0.01
complier	always well	0.35
complier	helped	0.28
complier	hurt	0.035
complier	never well	0.035
defier	always well	0.015
defier	helped	0.02
defier	hurt	0.005
defier	never well	0.02
never taker	always well	0.12
never taker	helped	0.06
never taker	hurt	0.01
never taker	never well	0.01

$$P(U^T, U^O)$$

Figure 21.5   A simple functional causal model for a clinical trial. (a) Network structure. (b) Sample CPDs and resulting joint distribution for the response variables U^T and U^O.

The model makes explicit two important assumptions.   First, the assigned treatment A does not influence the response O directly, but only through the actual treatment T; this conditional independence is achieved by the use of a placebo in the control group, so that the patients do not know to which of the two groups they were assigned. The second assumption is that A is marginally independent from $\{U^T, U^O\}$, which is ensured via the randomization of the assignment to the two groups. Note, however, that U^T and U^O are not independent, reflecting the fact that factors determining a patient's decision to comply with the treatment can also affect his final outcome. (For example, patients who are not very ill may neglect to take their medication, and may be more likely to get well regardless.) Thus, to specify this model fully, we need to define a joint probability distribution over U^T, U^O — a total of fifteen independent parameters.   ∎

We see that a full specification of even a simple functional causal model is quite complex.

21.5   Partial Identifiability in Functional Causal Models ★

In section 21.3, we considered the task of answering causal queries involving interventions. As we discussed, unlike probability queries, to answer intervention queries we generally need to consider the effect of latent variables on the observable variables. A functional causal model $\mathcal{C}$ provides a complete specification of a causal model, including the effect of the latent variables, and can thus be used to answer any intervention query. As in definition 21.1, we mutilate the network by eliminating all incoming arcs to intervened variables, and then do inference on the resulting network.

However, this "solution" does not address the fundamental problem. As we discussed in section 21.3, our accumulated statistics are generally only over the observable variables. We will rarely have any direct observations of the finer-grained distributions over the response variables. Thus, it is unrealistic to assume that we can construct a fully specified functional causal model. So, what is the point in defining these constructs if we can never obtain them? As we show in this section and the next one, these networks, even if never fully elicited, can help us provide at least partial answers to both intervention and counterfactual queries.

The basis for this approach rests on the fact that the distributions of response variables are related to the conditional probabilities for the endogenous variables. Thus, we can use our information about the latter to *constrain* the former. To understand this relationship, consider the following example:

response variable
constraints

Example 21.19

Let us revisit example 21.16, and consider the observed probability $P(j^1 \mid g^1)$ — the probability that Gump gets a job if he gets a high grade. Given $G = g^1$, we get j^1 in two cases: in the "always hire" case — $U^J = \mu_{1 \mapsto 1, 0 \mapsto 1}^J$, and in the "rational recruiting" case — $U^J = \mu_{1 \mapsto 1, 0 \mapsto 0}^J$. These two choices for U^J are indistinguishable in the context g^1, but would have led to different outcomes had Gump had a low grade. As a consequence, we have that:

$$
\begin{aligned}
P(j^1 \mid g^1) &= P(\mu_{1 \mapsto 1, 0 \mapsto 1}^J) + P(\mu_{1 \mapsto 1, 0 \mapsto 0}^J) \\
P(j^1 \mid g^0) &= P(\mu_{1 \mapsto 1, 0 \mapsto 1}^J) + P(\mu_{1 \mapsto 0, 0 \mapsto 1}^J) \\
P(j^0 \mid g^1) &= P(\mu_{1 \mapsto 0, 0 \mapsto 1}^J) + P(\mu_{1 \mapsto 0, 0 \mapsto 0}^J) \\
P(j^0 \mid g^0) &= P(\mu_{1 \mapsto 1, 0 \mapsto 0}^J) + P(\mu_{1 \mapsto 0, 0 \mapsto 0}^J).
\end{aligned}
$$

We see that each conditional probability is a sum of some subset of the probabilities associated with the response variable.　■

In the case where the response variable U^X is marginally independent of other response variables, we can generalize this example to provide a full formulation of the constraints that relate the observed conditional probabilities $P(X \mid \boldsymbol{Y})$ and the distributions over U^X:

$$
P(x_i \mid \boldsymbol{y}_i) = \sum_{\bar{x}_{-i} \in Val(X)^{m-1}} P(\mu_{\boldsymbol{y}_i \to x_i, \bar{\boldsymbol{y}}_{-i} \mapsto \bar{x}_{-i}}^X), \tag{21.2}
$$

where $\bar{\boldsymbol{y}}_{-i}$ is the tuple of all assignments $\boldsymbol{y}_j$ except for $j = i$, and similarly for $\bar{x}_{-i}$. The more general case, where response variables may be correlated, is a little more complex, due to the richer parameterization of the model.

Example 21.20

Consider again example 21.18. In this case, our functional causal model has sixteen unknown response parameters, specifying the joint distribution $P(U^T, U^O)$. By observing the statistics over the endogenous variables, we can evaluate the distribution $P(T, O \mid A)$, which can be used to constrain $P(U^T, U^O)$. Let $\nu_{(1,0)\mapsto(i,j),(1,0)\mapsto(k,l)}$ (for $i, j, k, l \in \{0,1\}$ denote $P(\mu^T_{1\mapsto i, 0 \mapsto j}, \mu^O_{1 \mapsto k, 0 \mapsto l})$. Using reasoning similar to that of example 21.19, we can verify that these two sets of probabilities are related by the following linear equalities:

$$P(t^i, o^j \mid a^k) = \sum_{i', j' \in \{0,1\}} \nu_{(k, 1-k) \mapsto (i, i'), (i, 1-i) \mapsto (j, j')} \qquad \forall i, j, k. \tag{21.3}$$

For example,

$$P(t^1, o^0 \mid a^1) = \sum_{i', j' \in \{0,1\}} \nu_{(1,0) \mapsto (1, i'), (1,0) \mapsto (0, j')};$$

that is, to be consistent with the assignment a^1, t^1, o^0, the U^T function should map a^1 to t^1, but it can map a^0 arbitrarily, and the U^O function should map t^1 to o^0, but it can map t^0 arbitrarily.∎

Thus, we see that the observed probabilities over endogenous variables impose constraints on the possible distributions of the response variables. These constraints, in turn, impose constraints over the possible values of various queries of interest. By reasoning about the possible values for the distributions over the response variables, we can often obtain reasonably precise bounds over the values of queries of interest.

Example 21.21

Continuing our clinical trial example, assume that we are interested in determining the extent to which taking the medication improves a patient's chances of a cure. Naively, we might think to answer this question by evaluating $P(o^1 \mid t^1) - P(o^1 \mid t^0)$. This approach would be incorrect, since it does not account for the correlation between compliance and cure. For example, if patients who choose to comply with the treatment are generally sicker and therefore less likely to get well regardless, then the cure rate in the population of patients for which $T = t^1$ will be misleadingly low, giving a pessimistic estimate of the efficacy of treatment.

A correct answer is obtained by the corresponding intervention query,

$$P(o^1 \mid \mathrm{do}(t^1)) - P(o^1 \mid \mathrm{do}(t^0)),$$

average causal
effect

which measures the increase in cure probability between a patient who was forced not to take the treatment and one who was forced to take it. This query is also known as the average causal effect *of T on O and is denoted $\mathrm{ACE}(T \to O)$. Unfortunately, the causal model for this situation contains a bidirected arc between T and O, due to the correlation between their responses. Therefore, the influence of T on O is not identifiable in this model, and, indeed, none of the simplification rules of section 21.3 apply in this case. However, it turns out that we can still obtain surprisingly meaningful bounds over the value of this query.*

We begin by noting that

$$\begin{aligned} P(o^1 \mid \mathrm{do}(t^1)) &= P(\mu^O_{1 \mapsto 1, 0 \mapsto 1}) + P(\mu^O_{1 \mapsto 1, 0 \mapsto 0}) \\ P(o^1 \mid \mathrm{do}(t^0)) &= P(\mu^O_{1 \mapsto 1, 0 \mapsto 1}) + P(\mu^O_{1 \mapsto 0, 0 \mapsto 1}), \end{aligned}$$

so that

$$\mathrm{ACE}(T \to O) = P(o^1 \mid \mathrm{do}(t^1)) - P(o^1 \mid \mathrm{do}(t^0)) = P(\mu^O_{1 \mapsto 1, 0 \mapsto 0}) - P(\mu^O_{1 \mapsto 0, 0 \mapsto 1}). \quad (21.4)$$

These are just marginal probabilities of the distribution $P(U^T, U^O)$, and they can therefore be written as a linear combination of the sixteen ν parameters representing the entries in this joint distribution.

The set of possible values for ν is determined by the constraints equation (21.3), which defines eight linear equations constraining various subsets of these parameters to sum up to probabilities in the observed distribution. We also need to require that ν be nonnegative.

Altogether, we have a linear formulation of $\mathrm{ACE}(T \to O)$ in terms of the ν parameters, and a set of linear constraints on these parameters — the equalities of equation (21.3) and the nonnegativity inequalities. We can now use linear programming techniques to obtain both the maximum and the minimum value of the function representing $\mathrm{ACE}(T \to O)$ subject to the constraints. This gives us bounds over the possible value that $\mathrm{ACE}(T \to O)$ can take in any functional causal model consistent with our observed probabilities.

In this fairly simple problem, we can even provide closed-form expressions for these bounds. Somewhat simpler bounds that are correct but not tight are:

$$\mathrm{ACE}(T \to O) \;\geq\; P(o^1 \mid a^1) - P(o^1 \mid a^0) - P(o^1, t^0 \mid a^1) - P(o^0, t^1 \mid a^0).$$
$$(21.5)$$

$$\mathrm{ACE}(T \to O) \;\leq\; P(o^1 \mid a^1) - P(o^1 \mid a^0) + P(o^0, t^0 \mid a^1) + P(o^1, t^1 \mid a^0).$$
$$(21.6)$$

Both bounds have a base quantity of $P(o^1 \mid a^1) - P(o^1 \mid a^0)$; this quantity, sometimes called the encouragement, *represents the difference in cure rate between the group that was prescribed the medication and the group that was not, ignoring the issue of how many in each group actually took the prescribed treatment. In a regime of full compliance with the treatment, the encouragement would be equivalent to $\mathrm{ACE}(T \to O)$. The correction factors in each of these equations provide a bound on the extent to which noncompliance can affect this estimate. Note that the total difference between the upper and lower bounds is*

$$P(o^0, t^0 \mid a^1) + P(o^1, t^1 \mid a^0) + P(o^1, t^0 \mid a^1) + P(o^0, t^1 \mid a^0) = P(t^0 \mid a^1) + P(t^1 \mid a^0),$$

natural bounds

which is precisely the total rate of noncompliance. These bounds are known as the natural bounds. *They are generally not as tight as the bounds we would obtain from the linear program, but they are tight in cases where no patient is a defier, that is, $P(\mu^T_{1 \mapsto 0, 0 \mapsto 1}) = 0$. In many cases, they provide surprisingly informative bounds on the result of the intervention query, as shown in box 21.B.* ■

Box 21.B — Case Study: The Effect of Cholestyramine. *A study conducted as part of the Lipid Research Clinics Coronary Primary Prevention Trial produced data about a drug trial relating to a drug called cholestyramine. In a portion of this data set (337 subjects), subjects were randomized into two treatment groups of roughly equal size; in one group, all subjects were prescribed the drug (a^1), while in the other group, all subjects were prescribed a placebo (a^0). Patients who were in the* *control group did not have access to the drug. The cholesterol level of each patient was measured several times over the years of treatment, and the average was computed. In this case, both the actual consumption of the drug and the resulting cholesterol level are continuous-valued.*

Balke and Pearl (1994a) provide a formal analysis of this data set. In their analysis, a patient was said to have received treatment (t^1) if his consumption was above the midpoint between minimal and maximal consumption among patients in the study. A patient was said to have responded to treatment (o^1) if his average cholesterol level throughout the treatment was at least 28 points lower than his measurement prior to treatment.

The resulting data exhibited the following statistics:

$$P(t^1, o^1 \mid a^1) = 0.473 \qquad P(t^1, o^1 \mid a^0) = 0$$
$$P(t^1, o^0 \mid a^1) = 0.139 \qquad P(t^1, o^0 \mid a^0) = 0$$
$$P(t^0, o^1 \mid a^1) = 0.073 \qquad P(t^0, o^1 \mid a^0) = 0.081$$
$$P(t^0, o^0 \mid a^1) = 0.315 \qquad P(t^0, o^0 \mid a^0) = 0.919.$$

The encouragement in this case is $P(o^1 \mid a^1) - P(o^1 \mid a^0) = 0.0473 + 0.073 - 0.081 = 0.465$. According to equation (21.5) and 21.6, we obtain:

$$\mathrm{ACE}(T \to O) \quad \geq \quad 0.465 - 0.073 - 0 = 0.392$$
$$\mathrm{ACE}(T \to O) \quad \leq \quad 0.465 + 0.315 + 0 = 0.78.$$

The difference between these bounds represents the noncompliance rate of $P(t^0 \mid a^1) = 0.388$. Thus, despite the fact that 38.8 percent of the subjects deviated from their treatment protocol, we can still assert (ignoring possible errors resulting from using statistics over a limited population) that, when applied uniformly to the population, cholestyramine increases by at least 39.2 percent the probability of reducing a patient's cholesterol level by 28 points or more.

21.6   Counterfactual Queries ⋆

Part of our motivation for moving to functional causal models derived from the issue of answering counterfactual queries. We now turn our attention to such queries, and we show how functional causal models can be used to address them. As we discussed, similar issues arise in the context of diagnostic reasoning, where we wish to reason about the probabilities of various events after taking a repair action; see box 21.C.

21.6.1   Twinned Networks

Recall that a counterfactual query considers a scenario that actually took place in the real world and a corresponding scenario in a counterfactual world, where we modify the chain of events via some causal intervention. Thus, we are actually reasoning in parallel about two different worlds: the real one, and the counterfactual one. A variable X may take on one value in the real world, and a different value in the counterfactual world. To distinguish between these two values, we use X to denote the random variable X in the true world, and X' to denote X in the *counterfactual world*. Intuitively, both the real and the counterfactual worlds are governed

counterfactual
world

by the same causal model, except that one involves an intervention.

However, a critical property of the desired output for a counterfactual query was that it involves minimal modification to the true world. As a degenerate case, consider the following counterfactual query relating to example 21.18: "A patient in the study took the prescribed treatment and did not get well. Would he have gotten well had we forced him to comply with the treatment?" Formally, we can write this query as $P(O' = o^1 \mid T = t^1, O = o^0, do(T' := t^1))$. Our intuition in this case says that we change nothing in the counterfactual world, and so the outcome should be exactly the same. But even this obvious assertion has significant consequences. Clearly, we cannot simply generate a new random assignment to the variables in the counterfactual world. For example, we want the patient's prescribed course of treatment to be the same in both worlds. We also want his response to the treatment to be the same.

The notion of response variables allows us to make this intuition very precise. A response variable U_X in a functional causal model summarizes the effect of all of the stochastic choices that can influence the choice of value for the endogenous variable X. When moving to the counterfactual world, all of these stochastic choices should remain the same. Thus, we assume that the values of the response variables are selected at random in the real world and preserved unchanged in the counterfactual world. Note that, when the counterfactual query includes a nondegenerate intervention $do(X := x')$ — one where x' is different from the value of x in the true world — the values of endogenous variables can change in the counterfactual world. However, the mechanism by which they choose these values remains constant.

Example 21.22

Consider the counterfactual query $P(O' = o^1 \mid T = t^0, O = o^0, do(T' := t^1))$. This query represents a situation where the patient did not take the prescribed treatment and did not get well, and we wish to determine the probability that the patient would have gotten well had he complied with the treatment. Assume that the true world is such that the patient was assigned to the treatment group, so that $U^A = a^1$. Assume also that he is in the "helped" category, so that the response variable $U^O = \mu^O_{1 \mapsto 1, 0 \mapsto 0}$. As we assumed, both of these are conserved in the counterfactual world. Thus, given $T' = t^1$, and applying the deterministic functions specified by the response variables, we obtain that $A' = a^1$, $T' = t^1$, and $O' = o^1$, so that the patient would have recovered had he complied. ∎

In general, of course, we do not know the value of the response variables in the real world. However, a functional causal model specifies a prior distribution over their values. Some values are not consistent with our observations in the real scenario, so the distribution over the response variables is conditioned accordingly. The resulting posterior can be used for inference in the counterfactual world.

Example 21.23

Consider again our clinical trial of example 21.18. Assume that the trial is randomized, so that $P(U^A = a^1) = P(U^A = a^0) = 0.5$. Also, assume that the medication is unavailable outside the treatment group, so that $P(U^T = \mu^T_{1 \mapsto 1, 0 \mapsto 1}) = 0$. One possible joint distribution for $P(U^T, U^O)$ is shown in figure 21.5b. Assume that we have a patient who, in the real world, was assigned to the treatment group, did not comply with the treatment, and did not get well; we are interested in the probability that he would have gotten well had he complied. Thus, our query is:

$$P(O' = o^1 \mid A = a^1, T = t^0, O = o^0, do(T' := t^1)).$$

Our observations in the real world are consistent only with the following values for the response

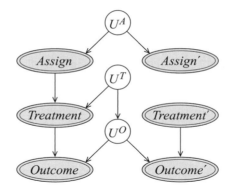

Figure 21.6   Twinned counterfactual network for figure 21.5, with an intervention at T'.

variables: U^T can be "defier" or "never taker"; U^O can be "helped" or "never well." Conditioning our joint distribution, we obtain the following joint posterior:

U^T	U^O	probability
defier	*helped*	$4/13$
defier	*never well*	$2/13$
never taker	*helped*	$6/13$
never taker	*never well*	$1/13$

In the counterfactual world, we intervene by forcing the patient to comply with the treatment. Therefore, his outcome is determined by the value that U^O defines for t^1. This value is o^1 when the patient is of category "helped," and o^0 when he is of category "never well." Using the posterior we computed, we conclude that the answer to the query is $10/13$. ∎

This type of reasoning can be obtained as a direct consequence of inference in a joint causal network that incorporates both the real and the counterfactual world. Specifically, consider the following construction:

Definition 21.4

counterfactual
twinned network

Let C be a functional causal model over the endogenous variables $\mathcal{X}$ and the corresponding response variables $\mathcal{U} = \{U^X \ : \ X \in \mathcal{X}\}$. We define the counterfactual twinned network to be a functional causal model over the variables $\mathcal{X} \cup \mathcal{U} \cup \{X' \ : \ X \in \mathcal{X}\}$, such that:

- *if $\mathrm{Pa}_X = \mathbf{Y}$, then $\mathrm{Pa}_{X'} = \mathbf{Y}'$,*
- *X' also has the response variable U^X as a parent,*
- *the deterministic function governing X' is identical to that of X.* ∎

We can answer counterfactual queries for our original causal model by constructing the twinned causal model and then answering the counterfactual query as an intervention query in that network, in exactly the same way as we would for any functional causal network.

Example 21.24   *Returning to the query of example 21.23, we first construct the twinned network for figure 21.5, and then use it to answer the intervention query $P(O' = o^1 \mid A = a^1, T = t^0, O = o^0, \mathrm{do}(T' := t^1))$ as a simple intervention query in the resulting network. Specifically, we mutilate the counterfactual network so as to eliminate incoming arcs into T', condition on our evidence $A = a^1, T = t^0, O = o^0$, and simply run probabilistic inference. The network is shown in figure 21.6. It is straightforward to verify that the answer is precisely what we obtained in the preceding analysis.*   ■

Note that this approach is very general, and it allows us to answer a broad range of queries. We can account for interventions in the real network as well as the counterfactual one and allow for evidence in both. For example, we can ask, "Assume that we had a patient in the control (placebo) group who did not get well; if I had assigned that patient to the treatment group and observed that he did not get well, what is the probability that he complied with treatment (in the counterfactual world)?" This query is formally written as $P(T' = t^1 \mid A = a^0, O = o^0, \mathrm{do}(A' := a^1), O' = o^0)$, and it can be answered using inference in the twinned network.

Box 21.C — Case Study: Persistence Networks for Diagnosis. *One real-world application of the twinned-network analysis arises in the setting of* troubleshooting *(see also box 5.A). In this application, described in Breese and Heckerman (1996), we have a network with multiple (unobserved) variables $X_1, \ldots, X_k$ that denote the possible failure of various components in a system. Our task is to repair the system. In an ideal setting, our repair actions can correspond to ideal interventions: we take one of the failure variables X_k and set it to a value denoting no failure: $\mathrm{do}(X_i := x_i^0)$.*

troubleshooting

Now, assume that we have some set of observations e about the current state of the system, and want to evaluate the benefit of some repair action $\mathrm{do}(X_i := x_i^0)$. How do we compute the probability that the repair action fixes the problem, or the distribution over the remaining failure variables following the repair action? It is not difficult to see that an intervention query is not appropriate in this setting: The evidence that we had about the failure symptoms prior to the repair is highly relevant for determining these probabilities, and it must be taken into account when computing the posterior following the intervention. For example, if the probability of x_i^0 (i not broken) was very high given e, chances are the system is still broken following the repair. The right query, to determine the postintervention distribution for a variable Y, is a counterfactual one:

$$P(Y' \mid e, \mathrm{do}(X_i' := x_i^0)).$$

Under certain (fairly strong) assumptions — noisy-or CPDs and a single fault only — one can avoid constructing the twinned network and significantly reduce the cost of this computation. (See exercise 21.6.) We return to the issue of troubleshooting networks in box 23.C.

21.6.2   Bounds on Counterfactual Queries

Although a functional causal model gives us enormous ability to answer sophisticated counterfactual queries, such a model is rarely available, as we discussed. However, we can apply the same idea as in section 21.5 to provide bounds on the probabilities of the response variables, and hence on the answers to counterfactual queries. We demonstrate this approach on a ficti-tious example, which also serves to illustrate the differences between different types of causal analysis.

Example 21.25

The marketer of PeptAid, an antacid medication, randomly mailed out product samples to 10 percent of the population. A follow-on market survey determined, for each individual, whether he or she received the sample (A), whether he or she took it (T), and whether he or she subsequently developed an ulcer (O). The accumulated data exhibited the following statistics:

$$P(t^1, o^1 \mid a^1) = 0.14 \qquad P(t^1, o^1 \mid a^0) = 0.32$$
$$P(t^1, o^0 \mid a^1) = 0.17 \qquad P(t^1, o^0 \mid a^0) = 0.32$$
$$P(t^0, o^1 \mid a^1) = 0.67 \qquad P(t^0, o^1 \mid a^0) = 0.04$$
$$P(t^0, o^0 \mid a^1) = 0.02 \qquad P(t^0, o^0 \mid a^0) = 0.32.$$

The functional causal model for this situation is identical to that of figure 21.5a.

Examining these numbers, we see a strong correlation between individuals who consumed Pep-tAid and those who developed ulcers: $P(o^1 \mid t^1) = 0.5$, whereas $P(o^1 \mid t^0) = 0.26$. Moreover, the probability of developing ulcers was 45 percent greater in individuals who received the PeptAid samples than in those who did not: $P(o^1 \mid a^1) = 0.81$, whereas $P(o^1 \mid a^0) = 0.36$. Thus, using the observational data alone, we might conclude that PeptAid causes ulcers, and that its manufacturer is legally liable for damages to the affected population.

As we discussed in example 21.2, an immediate counterargument is that the high positive corre-lation is due to some latent common cause such as preulcer discomfort. Indeed, one can show that, in this case, PeptAid actually helps reduce the risk of ulcers: a causal analysis along the lines of example 21.21 shows that

$$-0.23 \leq \text{ACE}(T \rightarrow O) \leq -0.15.$$

That is, the average causal effect of PeptAid consumption is to reduce an individual's chance of getting an ulcer by at least 15 percent.

However, now consider a particular patient George who received the sample, consumed it, and subsequently developed an ulcer. We would like to know whether the patient would not have devel-oped the ulcer had he not received the sample. In this case, the relevant query is a counterfactual, which has the form:

$$P(O' = o^0 \mid A = a^1, T = t^1, O = o^1, \text{do}(A' := a^0)).$$

Given our evidence $A = a^1, T = t^1, O = o^1$, only the responses "complier" and "always taker" are possible for U^T, and only the responses "hurt" and "never well" for U^O (where here "never well" means "always ulcer", or $\mu^O_{1 \mapsto 1, 0 \mapsto 1}$). Of these, only the combination ("complier," "hurt") is consistent with the query assertion $O' = o^0$: if George is a "never well," he would have developed ulcers regardless; similarly, if George is an "always taker," he would have taken PeptAid regardless, with the same outcome. We conclude that the probability of interest is equal to:

$$\frac{P(U^T = complier, U^O = hurt)}{P(T = t^1, O = o^1 \mid A = a^1)}.$$

Because the numerator is fixed, this expression is linear in the ν parameters, and so we can compute bounds using linear programming. The resulting bounds show that:

$$P(O' = o^0 \mid A = a^1, T = t^1, O = o^1, \text{do}(A' := a^0)) \geq 0.93;$$

thus, at least 93 percent of patients in George's category — those who received PeptAid, consumed it, and developed an ulcer — would not have developed an ulcer had they not received the sample!∎

This example illustrates the subtleties of causal analysis, and the huge differences between the answers to apparently similar queries. Thus, care must be taken when applying causal analysis to understand precisely which query it is that we really wish to answer.

21.7   Learning Causal Models

So far, we have focused on the problem of using a given causal model to answer causal queries such as intervention or counterfactual queries. In this section, we discuss the problem of learning a causal model from data.

As usual, there are several axes along which we can partition the space of learning tasks.

Perhaps the most fundamental axis is the notion of what we mean by a causal model. Most obvious is a causal network with standard CPDs. Here, we are essentially learning a standard Bayesian network, except that we are willing to ascribe to it causal semantics and use it to answer interventional queries. However, we can also consider other choices. For example, at one extreme, we can consider a functional causal model, where our network has response variables and a fully specified parameterization for them; clearly, this problem is far more challenging, and such rich models are much more difficult to identify from available data. At the other extreme, we can simplify our problem considerably by abstracting away all details of the parameterization and focus solely on determining the causal structure.

Second, we need to determine whether we are given the structure and so have to deal only with parameter estimation, or whether we also have to learn the structure.

A third axis is the type of data from which we learn the model. In the case of probabilistic models, we assumed that our data consist of instances sampled randomly from some generating distribution P^*. Such data are called *observational*. In the context of causal models, we may also have access to *interventional data*, where (some or all of) our data instances correspond to cases where we intervene in the model by setting the values of some variables and observe the values of the others. As we will discuss, interventional data provide significant power in disambiguating different causal models that can lead to identical observational patterns. Unfortunately, in many cases, interventions are difficult to perform, and sometimes are even illegal (or immoral), whereas observational data are usually much more plentiful.

A final axis involves the assumptions that we are willing to make regarding the presence of factors that can confound causal effects. If we assume that there are no confounding factors — latent variables or selection bias — the problem of identifying causal structure becomes significantly simpler. In particular, under fairly benign assumptions, we can fully delineate the cases in which we can infer causal direction from observational data. When we allow these confounding factors, the problem becomes significantly harder, and the set of cases where we can reach nontrivial conclusions becomes much smaller.

Of course, not all of the entries in this many-dimensional grid are interesting, and not all have been explored. To structure our discussion, we begin by focusing on the task of learning a causal model, which allows us to infer the causal direction for the interactions between the variables and to answer interventional queries. We consider first the case where there are no confounding factors, so that all relevant variables are observed in the data. We discuss both the case of learning only from observational data, then introduce the use of interventional data. We then discuss the challenges associated with latent variables and approaches for dealing with

observational
data

interventional
data

them, albeit in a limited way. Finally, we move to the task of learning a functional causal model, where even determining the parameterization for a fixed structure is far from trivial.

21.7.1 Learning Causal Models without Confounding Factors

We now consider the problem of learning a causal model from data. At some level, it appears that there is very little to say here that we have not already said. After all, a causal model is essentially a Bayesian network, and we have already devoted three chapters in this book to the problem of learning the structure and the parameters of such networks from data. Although many of the techniques we developed in these chapters will be useful to us here, they do not provide a full solution to the problem of learning causal models. To understand why, recall that our objective in the task of learning probabilistic models was simply to provide a good fit to the true underlying distribution. Any model that fit that distribution (reasonably well) was an adequate solution.

As we discussed in section 21.1.2, there are many different probabilistic models that can give rise to the same marginal distribution over the observed variables. While these models are (in some sense) equivalent with respect to probabilistic queries, as causal models, they generally give rise to very different conclusions for causal queries. Unfortunately, distinguishing between these structures is often impossible. Thus, our task in this section is to obtain the strongest possible conclusion about the causal models that could have given rise to the observed data.

21.7.1.1 Key Assumptions

There are two key assumptions that underlie methods that learn causal models from observational data. They relate the independence assumptions that hold in the true underlying distribution P^* with the causal relationships between the variables in the domain. Assume that P^* is derived from a causal network whose structure is the graph $\mathcal{G}^*$.

causal Markov assumption

The first, known as the *causal Markov assumption*, asserts that, in P^*, each variable is conditionally independent of its non-effects (direct or indirect) given its direct causes. Thus, each variable is conditionally independent of its nondescendants given its parents in $\mathcal{G}^*$. This assumption is precisely the same as the local Markov assumptions of definition 3.1, except that arcs are given a causal interpretation. We can thus restate this assumption as asserting that the causal network $\mathcal{G}^*$ is an I-map for the distribution P^*.

While the difference between the local Markov and causal Markov assumptions might appear purely syntactic, it is fundamental from a philosophical perspective. The local Markov assumptions for Bayesian networks are simply phenomenological: they state properties that a particular distribution has. The causal Markov assumption makes a statement about the world: If we relate variables by the "causes" relationship, these independence assumptions will hold in the empirical distribution we observe in the world.

The justification for this assumption is that causality is local in time and space, so that the direct causes of a variable (stochastically) determine its value. Current quantum theory and experiments show that this assumption does not hold at the quantum level, where there are nonlocal variables that appear to have a direct causal effect on others. While these cases do not imply that the causal Markov assumption does not hold, they do suggest that we may see more violations of this assumption at the quantum level. However, in practice, the causal Markov assumption appears to be a reasonable working assumption in most macroscopic systems.

The second assumption is the *faithfulness assumption*, which states that the only conditional independencies in P^* are those that arise from d-separation in the corresponding causal graph $\mathcal{G}^*$. When combined with the causal Markov assumption, the consequence is that $\mathcal{G}^*$ is a perfect map of P^*. As we discussed in section 3.4.2, there are many examples where the faithfulness assumption is violated, so that there are independencies in P^* not implied by the structure of $\mathcal{G}^*$; however, these correspond to particular parameter values, which (as stated in theorem 3.5) are a set of measure zero within the space of all possible parameterizations. As we discussed, there are still cases where one of these parameterizations arises naturally. However, for the purposes of learning causality, we generally need to make both of these assumptions.

21.7.1.2   Identifying the Model

With these assumptions in hand, we can now assume that our data set consists of samples from P^*, and our task is simply to identify a perfect map for P^*. Of course, as we observed in section 3.4.2, the perfect map for a distribution P^* is not unique. **Because we might have several different structures that are equivalent in the independence assumptions they impose, we cannot, based on observational data alone, distinguish between them. Thus, we cannot, in general, determine a unique causal structure for our domain, even given infinite data.** For the purpose of answering probabilistic queries, this limitation is irrelevant: any of these structures will perform equally well. However, for causal queries, determining the correct direction of the edges in the network is critical. Thus, at best, we can hope to identify the equivalence class of $\mathcal{G}^*$.

The class of *constraint-based structure learning* methods is suitable to this task. Here, we take the independencies observed in the empirical distribution and consider them as representative of P^*. Given enough data instances, the empirical distribution $\hat{P}$ will reflect exactly the independencies that hold in P^*, so that an independence oracle will provide accurate answers about independencies in P^*. The task now reduces to that of identifying an I-equivalence class that is consistent with these observed independencies. For the case without confounding factors, we have already discussed such a method: the Build-PDAG procedure described in section 18.2.

Recall that Build-PDAG constructs a *class PDAG*, which represents an entire I-equivalence class of network structures. In the class PDAG, an edge is oriented $X \to Y$ if and only if it is oriented in that way in every graph that is a member of the I-equivalence class. Thus, the algorithm does not make unwarranted conclusions about directionality of edges.

Even with this conservative approach, we can often infer the directionality of certain edges.

Example 21.26

Assume that our underlying distribution P^* is represented by the causal structure of the Student *network shown in figure 3.3. Assuming that the empirical distribution $\hat{P}$ reflects the independencies in P^*, the* Build-PDAG *will return the PDAG shown in figure 21.7a. Intuitively, we can infer the causal directions for the arcs $D \to G$ and $I \to G$ because of the v-structure at G; we can infer the causal direction for $G \to J$ because the opposite orientation $J \to G$ would create a v-structure involving J, which induces a different set of independencies. However, we are unable to infer a causal direction for the edge $I—S$, because both orientations are consistent with the observed independencies.* ∎

In some sense, the constraint-based approaches are ideally suited to inferring causal direction;

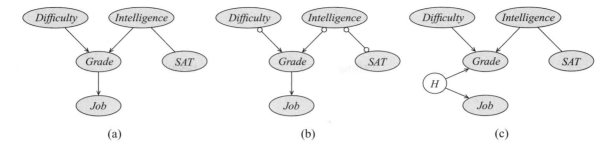

(a)   (b)   (c)

Figure 21.7 **Models corresponding to the equivalence class of the** Student **network.** (a) A PDAG, representing its equivalence class when all relevant variables are observed. (b) A PAG, representing its equivalence class when latent variables are a possibility. (c) An unsuccessful attempt to "undirect" the $G \to J$ arc.

in fact, these algorithms were mostly developed for the purpose of causal discovery. However, as we discussed in section 18.2, these approaches are fairly sensitive to mistakes in the independence tests over the distribution. This property can make them somewhat brittle, especially when the data set size is small relative to the number of variables.

Score-based methods allow us to factor in our confidence in different independencies, finding a solution that is more globally consistent. Thus, an alternative approach is to apply model selection with some appropriate score and then construct the I-equivalence class of the network $\mathcal{G}$ produced by the structure learning algorithm. The I-equivalence class can be represented by a PDAG, which can be constructed simply by applying the procedure Build-PDAG to $\mathcal{G}$.

A better solution, however, accounts for the fact that the highest-scoring network is not generally the only viable hypothesis. **In many cases, especially when data are scarce, there can be several *non-equivalent* network structures that all have a reasonably high posterior probability. Thus, a better approach for causal discovery is to use *Bayesian model averaging*,** as described in section 18.5. The result is a distribution over different network structures that allows us to encode our uncertainty not only about the orientation of an edge, but also about the presence or absence of an edge. Furthermore, we obtain numerical confidence measures in these network features. Thus, whereas the PDAG representation prevents us from orienting an edge $X \to Y$ even if there is only a single member in the equivalence class that has the opposite orientation, the Bayesian model averaging approach allows us to quantify the overall probability mass of structures in which an edge exists and is oriented in a particular direction.

Although constraint-based methods and Bayesian methods are often viewed as competing solutions, one successful approach is to combine them, using a constraint-based method to initialize a graph structure, and then using Bayesian methods to refine it. This approach exploits the strengths of both methods. It uses the global nature of the constraint-based methods to avoid local maxima, and it uses the ability of the Bayesian methods to avoid making irreversible decisions about independencies that may be incorrect due to noise in the data.

Bayesian model
averaging

21.7.2   Learning from Interventional Data

So far, we have focused on the task of learning causal models from observational data alone. In the causal setting, a natural question is to consider data that are obtained (at least partly) from interventional queries. That is, some of our data cases are obtained in a situation where we intervene in the model, sampled from the mutilated network corresponding to an intervention. One important situation where such data are often available is scientific discovery, where the data we obtain can be either a measurement of an existing system or a system that was subjected to perturbations.

Although both constraint-based and score-based approaches can be applied in this setting, constraint-based approaches are not as commonly used. Although it is straightforward to define the independencies associated with the mutilated network, we generally do not have enough data instances for any given type of intervention to measure reliably whether these independencies hold. Score-based approaches are much more flexible at combining data from diverse interventions. We therefore focus our discussion on that setting.

To formalize our analysis, we assume that each data instance in $\mathcal{D}$ is specified by an intervention $do(\boldsymbol{Z}[m] := \boldsymbol{z}[m])$; for each such data case, we have a fully observed data instance $\mathcal{X}[m] = \xi[m]$. As usual in a score-based setting, our first task is to define the likelihood function. Consider first the probability of a single instance $P(\xi \mid do(\boldsymbol{Z} := \boldsymbol{Z}), \mathcal{C})$. This term is defined in terms of the mutilated network $\mathcal{C}_{\boldsymbol{Z}=\boldsymbol{z}}$. In this network, the distribution of each variable $Z \in \boldsymbol{Z}$ is defined by the intervention, so that $Z = z$ with probability 1 and therefore is also the value we necessarily see in ξ. The variables not in $\boldsymbol{Z}$ are subject to their normal probabilistic model, as specified in $\mathcal{C}$. Letting $\boldsymbol{u}_i$ be the assignment to Pa_{X_i} in ξ, we obtain:

$$P(\xi \mid do(\boldsymbol{Z} := \boldsymbol{z}), \mathcal{C}) = \prod_{X_i \notin \boldsymbol{Z}} P(x_i \mid \boldsymbol{u}_i).$$

sufficient
statistics

In the case of table-CPDs (see exercise 21.8 for another example), it follows that the *sufficient statistics* for this type of likelihood function are:

$$M[x_i; \boldsymbol{u}_i] = \sum_{m \ : \ X_i \notin \boldsymbol{Z}[m]} \boldsymbol{I}\{X_i[m] = x_i, \mathrm{Pa}_{X_i}[m] = \boldsymbol{u}_i\}, \tag{21.7}$$

for an assignment x_i to X_i and $\boldsymbol{u}_i$ to Pa_{X_i}. This sufficient statistic counts the number of occurrences of this event, *in data instances where there is no intervention at X_i*. Unlike our original sufficient statistic $M[x_i, \boldsymbol{u}_i]$, this definition of the sufficient statistic treats X_i differently from its parents, hence the change in notation.

It now follows that:

$$L(\mathcal{C} : \mathcal{D}) = \prod_{i=1}^{n} \prod_{x_i \in Val(X_i), \boldsymbol{u}_i \in Val(\mathrm{Pa}_{X_i})} \theta_{x_i \mid \boldsymbol{u}_i}^{M[x_i; \boldsymbol{u}_i]}. \tag{21.8}$$

Because this likelihood function has the same functional form as our original likelihood function, we can proceed to apply any of the likelihood-based approaches described in earlier chapters. We can perform maximum likelihood estimation or incorporate a prior to define a Bayesian posterior over the parameters. We can also define a Bayesian (or other likelihood-based) score in order to perform model selection or model averaging. The formulas and derivations are exactly the same, only with the new sufficient statistics.

 Importantly, **in this framework, we can now distinguish between I-equivalent models, which are indistinguishable given observational data alone.** In particular, consider a network over two variables X, Y, and assume that we have interventional data at either X, Y, or both. As we just mentioned, the sufficient statistics are asymmetrical in the parent and child, so that $M[X; Y]$, for the network $Y \to X$, is different from $M[Y; X]$, for the network $X \to Y$. Therefore, although the two networks are I-equivalent, the likelihood function can be different for the two networks, allowing us to select one over the other.

Example 21.27

Consider the task of learning a causal model over the variables X, Y. Assume that we have the samples with the sufficient statistics shown in the following table:

Intervention	x^1, y^1	x^1, y^0	x^0, y^1	x^0, y^0
None	4	1	1	4
$\mathrm{do}(X := x^1)$	2	0	0	0
$\mathrm{do}(Y := y^1)$	1	0	1	0

The observational data suggest that each of X and Y are (roughly) uniformly distributed, but that they are correlated with each other. The interventional data, although limited, suggest that, when we intervene at X, Y tends to follow, but when we intervene at Y, X is unaffected. These intuitions suggest that the causal model $X \to Y$ is more plausible. Indeed, computing the sufficient statistics for the model $X \to Y$ and these data instances, we obtain:

$$
\begin{aligned}
M[x^1] &= M[x^1 y^1 \mid \text{None}] + M[x^1 y^0 \mid \text{None}] + M[x^1 y^1 \mid \mathrm{do}(y^1)] = 4 + 1 + 1 \\
M[x^0] &= M[x^0 y^1 \mid \text{None}] + M[x^0 y^0 \mid \text{None}] + M[x^0 y^1 \mid \mathrm{do}(y^1)] = 1 + 4 + 1 \\
M[y^1; x^1] &= M[x^1 y^1 \mid \text{None}] + M[x^1 y^1 \mid \mathrm{do}(x^1)] = 4 + 2 \\
M[y^0; x^1] &= M[x^1 y^0 \mid \text{None}] = 1 \\
M[y^1; x^0] &= M[x^0 y^1 \mid \text{None}] = 1 \\
M[y^0; x^0] &= M[x^0 y^0 \mid \text{None}] = 4.
\end{aligned}
$$

Importantly, we note that, unlike the purely observational case, $M[y^1; x^1] + M[y^0; x^1] \neq M[x^1]$; this is because different data instances contribute to the different counts, depending on the variable at which the intervention takes place.

We can now compute the maximum likelihood parameters for this model as $\theta_{x^1} = 0.5$, $\theta_{y^1 \mid x^1} = 6/7$, $\theta_{y^1 \mid x^0} = 1/5$. The log-likelihood of the data is then:

$$
\begin{aligned}
&M[x^1] \log \theta_{x^1} + M[x^0] \log \theta_{x^0} + \\
&M[y^1; x^1] \log \theta_{y^1 \mid x^1} + M[y^0; x^1] \log \theta_{y^0 \mid x^1} + M[y^1; x^0] \log \theta_{y^1 \mid x^0} + M[y^0; x^0] \log \theta_{y^0 \mid x^0},
\end{aligned}
$$

which equals -19.75.

We can analogously execute the same steps for the causal model $Y \to X$, where our sufficient statistics would have the form $M[y]$ and $M[x; y]$, each utilizing a different set of instances. Overall, we would obtain a log-likelihood of -21.41, which is lower than for the causal model $X \to Y$. Thus, the log-likelihood of the two causal models is different, even though they are I-equivalent as probabilistic models. Moreover, the causal model that is consistent with our intuitions is the one that obtains the highest score. ∎

We also note that **an intervention at X can help disambiguate parts of the network not directly adjacent to X**. For example, assume that the true network is a chain $X_1 \rightarrow X_2 \rightarrow \ldots \rightarrow X_n$. There are n I-equivalent directed graphs (where we root the graph at X_i for $i = 1, \ldots, n$). Interventions at any X_i can reveal that $X_{i+1}, \ldots, X_n$ all respond to an intervention at X_i, whereas $X_1, \ldots, X_{i-1}$ do not. Although these experiments would not fully disambiguate the causal structure, they would help direct all of the edges from X_i toward any of its descendants. Perhaps less intuitive is that they also help direct edges that are *not* downstream of our intervention. For example, if X_{i-1} does not respond to an intervention at X_i, but we are convinced that they are directly correlated, we now have more confidence that we can direct the edge $X_{i-1} \rightarrow X_i$. Importantly, directing some edges can have repercussions on others, as we saw in section 3.4.3. Indeed, in practice, a series of interventions at some subset of variables can significantly help disambiguate the directionality of many of the edges; see box 21.D.

Box 21.D — Case Study: Learning Cellular Networks from Intervention Data. *As we mentioned earlier, one of the important settings where interventional data arise naturally is scientific discovery. One application where causal network discovery has been applied is to the task of* cellular network reconstruction. *In the central paradigm of molecular biology, a gene in a cell (as encoded in DNA) is* expressed *to produce mRNA, which in turn is translated to produce protein, which performs its cellular function. The different steps of this process are carefully regulated. There are proteins whose task it is to regulate the expression of their* target *genes; others change the activity level of proteins by a physical change to the protein itself. Unraveling the structure of these networks is a key problem in cell biology. Causal network learning has been successfully applied to this task in a variety of ways.*

cellular network reconstruction

One important type of cellular network is a signaling *network, where a signaling protein physically modifies the structure of a target in a process called* phosphorylation, *thereby changing its activity level. Fluorescence microscopy can be used to measure the level of a* phosphoprotein *— a particular protein in a particular phosphorylation state. The phosphoprotein is fused to a fluorescent marker of a particular color. The fluorescence level of a cell, for a given color channel, indicates the level of the phosphoprotein fused with that color marker. Current technology allows us to measure simultaneously the levels of a small number of phosphoproteins, at the level of single cells. These data provide a unique opportunity to measure the activity levels of several proteins within individual cells, and thereby, we hope, to determine the causal network that underlies their interactions.*

Sachs et al. (2005) measured eleven phosphoproteins in a signaling pathway in human T-cells under nine different perturbation conditions. Of these conditions, two were general perturbations, but the remaining seven activated or inhibited particular phosphoproteins, and hence could be viewed as ideal interventions. Overall, 5,400 measurements were obtained over these nine conditions. The continuous measurements for each gene were discretized using a k-means algorithm, and the system was modeled as a Bayesian network where the variables are the levels of the phosphoproteins and the edges are the causal connections between them. The network was learned using standard score-based search, using a Bayesian score based on the interventional likelihood of equation (21.8). To obtain a measure of confidence in the edges of the learned structure, confidence estimation was bootstrap *performed using a* bootstrap *method, where the same learning procedure was applied to different training sets, each sampled randomly, with replacement, from the original data set. This procedure gave rise to an ensemble of networks, each with its own structure. The confidence associated with an edge was then estimated as the fraction of the learned networks that contained the edge.*

The result of this procedure gave rise to a network with seventeen high-confidence causal arcs between various components. A comparison to the known literature for this pathway revealed that fifteen of the seventeen edges were well established in the literature; the other two were novel but had supporting evidence in at least one literature citation. Only three well-established connections were missed by this analysis. Moreover, in all but one case, the direction of causal influence was correctly inferred. One of the two novel predictions was subsequently tested and validated in a wet-lab experiment. This finding suggested a new interaction between two pathways.

The use of a single global model for inferring the causal connections played an important role in the quality of the results. For example, because causal directions of arcs are often compelled by their interaction with other arcs within the overall structure, the learning algorithm was able to detect correctly causal influences from proteins that were not perturbed in the assay. In other cases, strong correlations did not lead to the inclusion of direct arcs in the model, since the correlations were well explained by indirect pathways. Importantly, although the data were modeled as fully observed, many relevant proteins were not actually measured. In such cases, the resulting indirect paths gave rise to direct edges in the learned network.

Various characteristics of this data set played an important role in the quality of the results. For example, the application of learning to a curtailed data set consisting solely of 1,200 observational data points gave rise to only ten arcs, all undirected, of which eight were expected or reported; ten of the established arcs were missing. Thus, the availability of interventional data played an important role in the accurate reconstruction of the network, especially the directionality of the edges. Another experiment showed the value of single-cell measurements, as compared to an equal-size data set each of whose instances is an average over a population of cells. This result suggests that the cell population is heterogeneous, so that averaging destroys much of the signal present in the data.

Nevertheless, the same techniques were also applied, with some success, to a data set that lacks these enabling properties. These data consist of gene expression measurements — measurements of mRNA levels for different genes, each collected from a population of cells. Here, data were collected from approximately 300 experiments, each acquired from a strain with a different gene deleted. Such perturbations are well modeled as ideal interventions, and hence they allow the use of the same techniques. This data set poses significant challenges: there are only 300 measurements (one for each perturbation experiment), but close to 6,000 variables (genes); the population averaging of each sample obscures much of the signal; and mRNA levels are a much weaker surrogate for activity level than direct protein measurements. Nevertheless, by focusing attention on subgraphs where many edges had high confidence, it was possible to reconstruct correctly some known pathways.

The main limitation of these techniques is the assumption of acyclicity, which does not hold for cellular networks. Nevertheless, these results suggest that causal-network learning, combined with appropriate data, can provide a viable approach for uncovering cellular pathways of different types.

21.7.3   Dealing with Latent Variables ⋆

So far, we have discussed the learning task in the setting where we have no confounding factors. The situation is more complicated if we have confounding effects such as latent variables or selection bias. In this case, the samples in our empirical distribution are generated from a "partial view" of P^*, where some variables have been marginalized out, and others perhaps instantiated to particular values. Because we do not know the set of confounding variables, there is an infinite set of networks that could have given rise to exactly the same set of dependencies over the observable variables. For example, $X \to Y$, $X \to H \to Y$, $X \to H \to H' \to Y$, and so on are completely indistinguishable in terms of their effect on X, Y (assuming that the hidden variables H, H' do not affect other variables). The task of determining a causal model in this case seems unreasonably daunting.

In the remainder of this section, we describe solutions to the problem of discovering causal structure in presence of confounding effects that induce noncausal correlations between the observed variables. These confounding effects include latent variables and selection bias. Although there are methods that cover both of these problems, the treatment of the latter is significantly more complex. Moreover, although selection bias clearly occurs, latent variables are almost ubiquitous, and they have therefore been the focus of more work. We therefore restrict our discussion to the case of learning models with latent variables.

21.7.3.1   Score-Based Approaches

A first thought is to try to learn a model with hidden variables using score-based methods such as those we discussed in chapter 19. The implementation of this idea, however, requires significant care. First, we generally do not know where in the model we need to introduce hidden variables. In fact, we can have an unbounded number of latent variables in the model. Moreover, as we saw, causal conclusions can be quite sensitive to local maxima or to design decisions such as the number of values of the hidden variable. We note, again, that probabilistic conclusions are also somewhat sensitive to these issues, but significantly less so, since models that achieve the same marginals over the observed variables are equivalent with respect to probabilistic queries, but not with respect to causal queries. Nevertheless, when we have some strong prior knowledge about the possible number and placement of hidden variables, and we apply our analysis with care, we can learn informative causal models.

21.7.3.2   Constraint-Based Approaches

An alternative solution is to use constraint-based approaches that try to use the independence properties of the distribution to learn the structure of the causal model, *including* the placement of the hidden variables. Here, as for the fully observable case, the independencies in a distribution only determine a structure up to equivalence. However, in the case of latent variables, we observe independence relationships only between the observable variables. We therefore need to introduce a notion of the independencies induced over the observable variables. We say that a directed acyclic graph $\mathcal{G}$ is a *latent variable network* over $\mathcal{X}$ if it is a causal network structure over $\mathcal{X} \cup \mathcal{H}$, where $\mathcal{H}$ is some arbitrarily large set of (latent) variables disjoint from $\mathcal{X}$.

latent variable
network

Definition 21.5

Let $\mathcal{G}$ be a latent variable network over $\mathcal{X}$. We define $\mathcal{I}_{\mathcal{X}}(\mathcal{G})$ to be the set of independencies $(\boldsymbol{X} \perp \boldsymbol{Y} \mid \boldsymbol{Z}) \in \mathcal{I}(\mathcal{G})$ for $\boldsymbol{X}, \boldsymbol{Y}, \boldsymbol{Z} \subset \mathcal{X}$. ∎

We can now define a notion of I-equivalence over the observable variables.

Definition 21.6

$I_{\mathcal{X}}$-equivalence

Let $\mathcal{G}_1, \mathcal{G}_2$ be two latent variable networks over $\mathcal{X}$ (not necessarily over the same set of latent variables). We say that $\mathcal{G}_1$ and $\mathcal{G}_2$ are $I_{\mathcal{X}}$-equivalent if $\mathcal{I}_{\mathcal{X}}(\mathcal{G}_1) = \mathcal{I}_{\mathcal{X}}(\mathcal{G}_2)$. ∎

Clearly, we cannot explicitly enumerate the $I_{\mathcal{X}}$-equivalence class of a graph. Even in the fully observable case, this equivalence class can be quite large; when we allow latent variables, the equivalence class is generally infinite, since we can have an unbounded number of latent variables placed in a variety of configurations. The constraint-based methods sidestep this difficulty by searching over a space of graphs over the observable variables alone. Like PDAGs, an edge in these graphs can represent different types of relationships between its endpoints.

Definition 21.7

partial ancestral graph

Let $\boldsymbol{G}$ be an $I_{\mathcal{X}}$-equivalence class of latent variable networks over $\mathcal{X}$. A partial ancestral graph (PAG) $\mathcal{P}$ over $\mathcal{X}$ is a graph whose nodes correspond to $\mathcal{X}$, and whose edges represent the dependencies in $\boldsymbol{G}$. The presence of an edge between X and Y in $\mathcal{P}$ corresponds to the existence, in each $\mathcal{G} \in \boldsymbol{G}$, of an active trail between X and Y that utilizes only latent variables. Edges have three types of endpoints: $-$, $>$, and $\circ$; these endpoints on the Y end of an edge between X and Y have the following meanings:

- An arrowhead $>$ implies that Y is not an ancestor of X in any graph in $\boldsymbol{G}$.
- A straight end $-$ implies that Y is an ancestor of X in all graphs in $\boldsymbol{G}$.
- A circle $\circ$ implies that neither of the two previous cases holds. ∎

The interpretation of the different edge types is as follows: An edge $X \to Y$ has (almost) the standard meaning: X is an ancestor of Y in all graphs in $\boldsymbol{G}$, and Y is not an ancestor of X in any graph. Thus, each graph in $\boldsymbol{G}$ contains a directed path from X to Y. However, some graphs may also contain trail where a latent variable is an ancestor of both. Thus, for example, the edge $S \to C$ would represent both of the networks in figure 21.A.1a,b when both G and T are latent.

An edge $X \leftrightarrow Y$ means that X is never an ancestor of Y, and Y is never an ancestor of X; thus, the edge must be due to the presence of a latent common cause. Note that an undirected edge $X - Y$ is illegal relative to this definition, since it is inconsistent with the acyclicity of the graphs in $\boldsymbol{G}$. An edge $X \circ\!\!\to Y$ means that Y is not an ancestor of X in any graph, but X is an ancestor of Y in some, but not all, graphs.

Figure 21.8 shows an example PAG, along with several members of the (infinite) equivalence class that it represents. All of the graphs in the equivalence class have one or more active trails between X and Y, none of which are directed from Y to X.

At first glance, it might appear that the presence of latent variables completely eliminates our ability to infer causal direction. After all, any edge can be ascribed to an indirect correlation via a latent variable. However, somewhat surprisingly, there are configurations where we can infer a causal orientation to an edge.

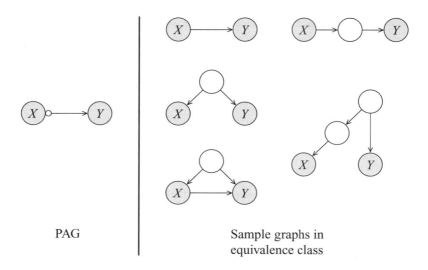

PAG   Sample graphs in
                                             equivalence class

Figure 21.8   Example PAG (left), along with several members of the (infinite) equivalence class that it represents. All of the graphs in the equivalence class have one or more active trails between X and Y, none of which are directed from Y to X.

Example 21.28   *Consider again the learning problem in example 21.26, but where we now allow for the presence of latent variables. Figure 21.7b shows the PAG reflecting the equivalence class of our original network of figure 3.3. Not surprisingly, we cannot reach any conclusions about the edge between I and S. The edge between D and G can arise both from a directed path from D to G and from the presence of a latent variable that is the parent of both. However, a directed path from G to D would not result in a marginal independence between D and I, and a dependence given G. Thus, we have an arrowhead $>$ on the G side of this edge. The same analysis holds for the edge between I and G.*

Most interesting, however, is the directed edge $G \to J$, which asserts that, in any graph in $\mathcal{G}$, there is a directed path from G to J. To understand why, let us try to explain the correlation between G and J by introducing a common latent parent (see figure 21.7c). This model is not $I_{\mathcal{X}}$-equivalent to our original network, because it implies that J is marginally independent of I and D. More generally, we can conclude that J must be a descendant of I and D, because observing J renders them dependent. Because G renders J independent of I, it must block any directed path from I to J. It follows that there is a directed path from G to J in every member of $\mathcal{G}$. In fact, in this case, we can reach the stronger conclusions that all the trails between these two variables are directed paths from G to J. Thus, in this particular case, the causal influence of G on J is simply $P(J \mid G)$, which we can obtain directly from observational data alone.   ∎

This example gives intuition for how we might determine a PAG structure for a given distribution. The algorithm for constructing a PAG for a distribution P proceeds along similar lines to the algorithm for constructing PDAGs, described in section 3.4.3 and 18.2. The full algorithm for learning PAGs is quite intricate, and we do not provide a full description of it, but only give some high-level intuition. The algorithm has two main phases. The first phase constructs an undirected graph over the observed variables, representing direct probabilistic interactions between them in P. In general, we want to connect X and Y with a direct edge if and only if there is no subset $\boldsymbol{Z}$ of $\mathcal{X} - \{X, Y\}$ such that $P \models (X \perp Y \mid \boldsymbol{Z})$. Of course, we cannot actually enumerate over the exponentially many possible subsets $\boldsymbol{Z} \subset \mathcal{X}$. As in Build-PMap-Skeleton, we both bound the size of the possible separating set and prune sets that cannot be separating sets given our current knowledge about the adjacency structure of the graph. The second phase of the algorithm orients as many edges as possible, using reasoning similar to the ideas used for PDAGs, but extended to deal with the confounding effect of latent variables.

The PAG-learning algorithm offers similar (albeit somewhat weaker) guarantees than the PDAG construction algorithm. In particular, one cannot show that the edge orientation rules are complete, that is, produce the strongest possible conclusion about edge orientation that is consistent with the equivalence class. However, one can show that all of the latent variable networks over $\mathcal{X}$ that are consistent with a PAG produced by this algorithm are $\mathrm{I}_{\mathcal{X}}$-equivalent.

Importantly, we note that a PAG is only a partial graph structure, and not a full model; thus, it cannot be used directly for answering causal queries. One possible solution is to use the score-based techniques we described to parameterize the causal model. This approach, however, is fraught with difficulties: First, we have the standard difficulties of using EM to learn parameters for hidden variables; an even bigger problem is that the PAG provides no guidance about the number of latent variables, their domain, or the edges between them.

Another alternative is to use the methods of section 21.3 and 21.5, which use a learned causal structure with latent variables, in conjunction with statistics over the observable data, to answer causal queries. We note, however, that these methods require a known connectivity structure among the hidden variables, whereas the learned PAG does not specify this structure. Nevertheless, if we are willing to introduce some assumptions about this structure, these algorithms may be usable. We return to this option in section 21.7.4, where we discuss more robust ways of estimating the answers to such queries.

21.7.4 Learning Functional Causal Models ⋆

Finally, we turn to the question of learning a much richer class of models: functional causal models, where we have a set of response variables with their associated parameters. As we discussed, these models have two distinct uses. The first is to answer a broader range of causal queries, such as counterfactual queries or queries regarding the average causal effect. The second is to avoid, to some extent, the infinite space of possible configurations of latent variables. As we discussed in section 21.4, a fully specified functional causal model summarizes the effect of all of the exogenous variables on the variables in our model, and thereby, within a finite description, specifies the causal behavior of our endogenous variables. Thus, rather than select a set of concrete latent variables with a particular domain and parameterization for each one, we use response variables to summarize all of the possibilities. Our conclusions in this case are robust, and they apply for any true underlying model of the latent variables.

The difficulty, of course, is that a functional causal model is a very complex object. The parameterization of a response variable is generally exponentially larger than the parameterization of a CPD for the corresponding endogenous variable. Moreover, the data we are given provide the outcome in only one of the exponentially many counterfactual cases given by the response variable. In this section, we describe one approach for learning with these issues.

Recall that a functional causal model is parameterized by a joint distribution $P(\mathcal{U})$ over the response variables. The local models of the endogenous variables are, by definition, deterministic functions of the response variables. A response variable U^X for a variable X with parents $\boldsymbol{Y}$ is a discrete random variable, whose domain is the space of all functions $\mu(\boldsymbol{Y})$ from $Val(\boldsymbol{Y})$ to $Val(X)$. The joint distribution $P(\mathcal{U})$ is encoded by a Bayesian network. We first focus on the case where the structure of the network is known, and our task is only to learn the parameterization. We then briefly discuss the issue of structure learning.

Consider first the simple case, where U^X has no parents. In this case, we can parameterize U^X using a multinomial distribution $\boldsymbol{\nu}^X = (\nu_1^X, \ldots, \nu_m^X)$, where $m = |Val(U^X)|$. In the Bayesian approach, we would take this parameter to be itself a random variable and introduce an appropriate prior, such as a Dirichlet distribution, over $\boldsymbol{\nu}^X$. More generally, we can use the techniques of section 17.3 to parameterize the entire Bayesian network over $\mathcal{U}$.

Our goal is then to compute the posterior distribution $P(\mathcal{U} \mid \mathcal{D})$; this posterior defines an answer to both intervention queries and counterfactual queries. Consider a general causal query $P(\phi \mid \psi)$, where ϕ and ψ may contain both real and counterfactual variables, and ψ may also contain interventions. We have that:

$$P(\phi \mid \psi, \mathcal{D}) = \int P(\phi \mid \psi, \boldsymbol{\nu}) P(\boldsymbol{\nu} \mid \psi, \mathcal{D}) d\boldsymbol{\nu}.$$

Assuming that $\mathcal{D}$ is reasonably large, we can approximate $P(\boldsymbol{\nu} \mid \psi, \mathcal{D})$ as $P(\boldsymbol{\nu} \mid \mathcal{D})$. Thus, to answer a causal query, we simply take the expectation of the answer to the query over the posterior parameter distribution $P(\boldsymbol{\nu} \mid \mathcal{D})$.

The main difficulty in this procedure is that the data set $\mathcal{D}$ is only partly observable: even if we fully observe the endogenous variables $\mathcal{X}$, the response variables $\mathcal{U}$ are not directly observed. As we saw, an observed assignment ξ to $\mathcal{X}$ limits the set of possible values to $\mathcal{U}$ to the subset of functions consistent with ξ. In particular, if $x, \boldsymbol{y}$ is the assignment to $X, \boldsymbol{Y}$ in ξ, then U^X is restricted to the set of possible functions μ for which $\mu(\boldsymbol{y}) = x$, a set that is exponentially large. Thus, to apply Bayesian learning, we must use techniques that approximate the posterior parameter distribution $P(\boldsymbol{\nu} \mid \mathcal{D})$. In section 19.3, we discussed several approaches to approximating this posterior, including variational Bayesian learning and MCMC methods. Both can be applied in this setting as well.

Thus, in principle, this approach is a straightforward application of techniques we have already discussed. However, because of the size of the space, the use of functional causal models in general and in this case in particular is feasible only for fairly small models.

When we also need to learn the structure of the functional causal model, the situation becomes even more complex, since the problem is one of structure learning in the presence of hidden variables. One approach is to use the constraint-based approach of section 21.7.3.2 to learn a structure involving latent variables, and then the approach described here for filling in the parameters. A second approach is to use one of the methods of section 19.4. However, there is an important issue that arises in this approach: Recall that a response variable for a variable X specifies the value of X for each configuration of its endogenous parents $\boldsymbol{U}$. Thus, as our structure learning algorithm adapts the structure of the network, the *domain* of the response variables changes; for example, if our search adds a parent to X, the domain of U^X changes. Thus, when performing the search, we would need to recompute the posterior parameter dis-

tribution and thereby the score after every structure change to the model. However, under
certain independence assumptions, we can use score decomposability to reduce significantly
the amount of recomputation required; see exercise 21.10.

21.8   Summary

In this chapter, we addressed the issue of ascribing a causal interpretation to a Bayesian network.
While a causal interpretation does not provide any additional capabilities in terms of answering
standard probabilistic queries, it provides the basic framework for answering *causal queries* —
queries involving interventions in the world. We provided semantics for causal models in terms
of the causal mechanism by which a variable's value is generated. An intervention query can
then be viewed as a substitution of the existing causal mechanism with one that simply forces
the intervened variable to take on a particular value.

We discussed the greater sensitivity of causal queries to the specifics of the model, including
the specific orientations of the arcs and the presence of latent variables. Latent variables are
particularly tricky, since they can induce correlations between the variables in the model that
are hard to distinguish from causal relationships. These issues make the identification of a
causal model much more difficult than the selection of an adequate probabilistic model.

We presented a class of situations in which a causal query can be answered exactly, using only
a distribution over the observable variables, even when the model as a whole is not identifiable.
In other cases, even if the query is not fully identifiable, we can often provide surprisingly strong
bounds over the answer to a causal query.

Besides intervention queries, causal models can also be used to answer counterfactual queries
— queries about a sequence of events that we know to be different from the sequence that
actually took place in the world. To answer such queries, we need to make explicit the random
choices made in selecting the values of variables in the model; these random choices need to be
preserved between the real and counterfactual worlds in order to maintain the correct semantics
for the idea of a counterfactual. Functional causal models allow us to represent these random
choices in a finite way, regardless of the (potentially unbounded) number of latent variables in
the domain. We showed how to use functional causal models to answer counterfactual queries.
While these models are even harder to identify than standard causal models, the techniques for
partially identifying causal queries can also be used in this case.

Finally, we discussed the controversial and challenging problem of learning causal models
from data. Much of the work in this area has been devoted to the problem of inferring causal
models from observational data alone. This problem is very challenging, especially when we
allow for the possible presence of latent variables. We described both constraint-based and
Bayesian methods for learning causal models from data, and we discussed their advantages and
disadvantages.

Causality is a fundamental concept when reasoning about many topics, ranging from specific
scientific applications to commonsense reasoning. Causal networks provide a framework for
performing this type of reasoning in a systematic and principled way. On the other side, the
learning algorithms we described, by combining prior knowledge about domain structure with
empirical data, can help us identify a more accurate causal structure, and perhaps obtain a
better understanding of the domain. There are many possible applications of this framework in

the realm of scientific discovery, both in the physical and life sciences and in the social sciences.

21.9   Relevant Literature

The use of functional equations to encode causal processes dates back at least as far as the work of Wright (1921), who used them to model genetic inheritance. Wright (1934) also used directed graphs to represent causal structures.

The view of Bayesian networks as encoding causal processes was present throughout much of their history, and certainly played a significant role in early work on constraint-based methods for learning network structure from data (Verma and Pearl 1990; Spirtes et al. 1991, 1993). The formal framework for viewing a Bayesian network as a causal graph was developed in the early and mid 1990s, primarily by two groups: by Spirtes, Glymour, and Scheines, and by Pearl and his students Balke and Galles. Much of this work is summarized in two seminal books: the early book of Spirtes, Glymour, and Scheines (1993) and the more recent book by Pearl (2000), on which much of the content of this chapter is based. The edited collection of Glymour and Cooper (1999) also reviews other important developments.

The use of a causal model for analyzing the effect of interventions was introduced by Pearl and Verma (1991) and Spirtes, Glymour, and Scheines (1993). The formalization of the causal calculus, which allows the simplification of intervention queries and their reformulation in terms of purely observable queries, was first presented in detail in Pearl (1995). The example on smoking and cancer was also presented there. Based on these ideas, Galles and Pearl (1995) provide an algorithm for determining the identifiability of an intervention query. Dawid (2002, 2007) provides an alternative formulation of causal intervention that makes explicit use of decision variables. This perspective, which we used in section 21.3, significantly simplifies certain aspects of causal reasoning.

The idea of making mechanisms explicit via response variables is based on ideas proposed in Rubin's theory of counterfactuals (Rubin 1974). It was introduced into the framework of causal networks by Balke and Pearl (1994b,a), and in parallel by Heckerman and Shachter (1994), who use a somewhat different framework based on influence diagrams. Balke and Pearl (1994a) describe a method that uses the distribution over the observed variables to constrain the distribution of the response variables. The PeptAid example (example 21.25) is due to Balke and Pearl (1994a), who also performed the analysis of the cholesterolymine example (box 21.B). Chickering and Pearl (1997) present a Gibbs sampling approach to Bayesian parameter estimation in causal settings.

The work on constraint-based structure learning (described in section 18.2) was first presented as an approach for learning causal networks. It was proposed and developed in the work of Verma and Pearl (1990) and in the work of Spirtes et al. (1993). Even this very early work was able to deal with latent variables. Since then, there has been significant work on extending and improving these early algorithms. Spirtes, Meek, and Richardson (1999) present a state-of-the-art algorithm for identifying a PAG from data and show that it can accommodate both latent variables and selection bias.

Heckerman, Meek, and Cooper (1999) proposed a Bayesian approach to causal discovery and the use of a Markov chain Monte Carlo algorithm for sampling structures in order to obtain probabilities of causal features.

The extension of Bayesian structure learning to a combination of observational and interventional data was first developed by Cooper and Yoo (1999). These ideas were extended and applied by Pe'er et al. (2001) to the problem of identifying regulatory networks from gene expression data, and by Sachs et al. (2005) to the problem of identifying signaling networks from fluorescent microscopy data, as described in box 21.D. Tong and Koller (2001a,b) build on these ideas in addressing the problem of *active learning* — choosing a set of interventions so as best to learn a causal network model.

active learning

21.10 Exercises

Exercise 21.1★

a. Prove proposition 21.2, which allows us to convert causal interventions in a query into observations.
b. An alternative condition for this proposition works in terms of the original graph $\mathcal{G}$ rather than the graph $\mathcal{G}^\dagger$. Let $\mathcal{G}_{\underline{\boldsymbol{X}}}$ denote the graph $\mathcal{G}$, minus all edges going out of nodes in $\boldsymbol{X}$. Show that the d-separation criterion used in the proposition is equivalent to requiring that $\boldsymbol{Y}$ is d-separated from $\boldsymbol{X}$ given $\boldsymbol{Z}, \boldsymbol{W}$ in the graph $\mathcal{G}_{\overline{\boldsymbol{Z}}\,\underline{\boldsymbol{X}}}$.

Exercise 21.2★

Prove proposition 21.3, which allows us to drop causal interventions from a query entirely.

Exercise 21.3★

For probabilistic queries, we have that

$$\min_x P(y \mid x) \leq P(y) \leq \max_x P(y \mid x).$$

Show that the same property does not hold for intervention queries. Specifically, provide an example where it is not the case that:

$$\min_x P(y \mid do(x)) \leq P(y) \leq \max_x P(y \mid do(x)).$$

Exercise 21.4★★

Show that every one of the diagrams in figure 21.3 is identifiable via the repeated application of proposition 21.1, 21.2, and 21.3.

Exercise 21.5★

a. Show that, in the causal model of figure 21.4g, each of the queries $P(Z_1 \mid do(X))$, $P(Z_2 \mid do(X))$, $P(Y \mid do(Z_1))$, and $P(Y \mid do(Z_2))$ are identifiable.
b. Explain why the effect of X on Y cannot be identifiable in this model.
c. Show that we can identify both $P(Y \mid do(X), do(Z_1))$ and $P(Y \mid do(X), do(Z_2))$. This example illustrates that the effect of a joint intervention may be more easily identified than the effect of each of its components.

Exercise 21.6

As we discussed in box 21.C, under certain assumptions, we can reduce the cost of performing counterfactual inference to that of a standard probabilistic query. In particular, assume that we have a system status variable X that is a noisy-or of the failure variables $X_1, \ldots, X_k$, and that there is no leak probability, so that $X = x^0$ when all $X_i = x_i^0$ (that is, X is normal when all its components are normal). Furthermore, assume that only a single X_i is in the failure mode ($X_i = x_i^1$). Show that

$$P(x'^0 \mid x^1, do(x_i^0), \boldsymbol{e}) = P(d_i^1 \mid x^1, \boldsymbol{e}),$$

where Z_i is the noisy version of X_i, as in definition 5.11.

Exercise 21.7

This exercise demonstrates computation of sufficient statistics with interventional data. The following table shows counts for different interventions.

Intervention	$x^0 y^0 z^0$	$x^0 y^0 z^1$	$x^0 y^1 z^0$	$x^0 y^1 z^1$	$x^1 y^0 z^0$	$x^1 y^0 z^1$	$x^1 y^1 z^0$	$x^1 y^1 z^1$
None	4	2	1	0	3	2	1	4
$do(X := x^0)$	3	1	2	1	0	0	0	0
$do(Y := y^0)$	7	1	0	0	2	1	0	0
$do(Z := z^0)$	1	0	1	0	1	0	1	0

Calculate $M[x^0; y^0 z^0]$, $M[y^0; x^0 z^0]$, and $M[x^0]$.

Exercise 21.8

Consider the problem of learning a Gaussian Bayesian network from interventional data $\mathcal{D}$. As in section 21.7.2, assume that each data instance in $\mathcal{D}$ is specified by an intervention $do(\mathbf{Z}[m] := \mathbf{z}[m])$; for each such data case, we have a fully observed data instance $\mathcal{X}[m] = \xi[m]$. Write down the sufficient statistics that would be used to score a network structure $\mathcal{G}$ from this data set.

Exercise 21.9★

Consider the problem of Bayesian learning for a functional causal model $\mathcal{C}$ over a set of endogenous variables $\mathcal{X}$. Assume we have a data set $\mathcal{D}$ where the endogenous variables $\mathcal{X}$ are fully observed. Describe a way for approximating the parameter posterior $P(\boldsymbol{\nu} \mid \mathcal{X})$ using collapsed Gibbs sampling. Specifically, your algorithm should sample the response variables $\mathcal{U}$ and compute a closed-form distribution over the parameters $\boldsymbol{\nu}$.

Exercise 21.10★★

Consider the problem of learning the structure of a functional causal model $\mathcal{C}$ over a set of endogenous variables $\mathcal{X}$.

a. Using your answer from exercise 21.9, construct an algorithm for learning the structure of a causal model. Describe precisely the key steps used in the algorithm, including the search steps and the use of the Gibbs sampling algorithm to evaluate the score at each step.

b. Now, assume that we are willing to stipulate that the response variables U^X for each variable X are independent. (This assumption is a very strong one, but it may be a reasonable approximation in some cases.) How can you significantly improve the learning algorithm in this case? Provide a new pseudo-code description of the algorithm, and quantify the computational gains.

Exercise 21.11★

causal
independence

As for probabilistic independence, we can define a notion of *causal independence*: $(\mathbf{X} \perp_C \mathbf{Y} \mid \mathbf{Z})$ if, for any values $\mathbf{x}, \mathbf{x}' \in Val(\mathbf{X})$, we have that $P(\mathbf{Y} \mid do(\mathbf{Z}), do(\mathbf{x})) = P(\mathbf{Y} \mid do(\mathbf{Z}), do(\mathbf{x}'))$. (Note that, unlike probabilistic independence — $(\mathbf{X} \perp \mathbf{Y} \mid \mathbf{Z})$ — causal independence is not symmetric over $\mathbf{X}, \mathbf{Y}$.)

a. Is causal independence equivalent to the statement: "For any value $\mathbf{x} \in Val(\mathbf{X})$, we have that $P(\mathbf{Y} \mid do(\mathbf{Z}), do(\mathbf{x})) = P(\mathbf{Y} \mid do(\mathbf{Z}))$." (Hint: Use your result from exercise 21.3.)

b. Prove that $(\mathbf{X} \perp_C \mathbf{Y} \mid \mathbf{Z}, \mathbf{W})$ and $(\mathbf{W} \perp_C \mathbf{Y} \mid \mathbf{X}, \mathbf{Z})$ implies that $(\mathbf{X}, \mathbf{W} \perp_C \mathbf{Y} \mid \mathbf{Z})$. Intuitively, this property states that if changing $\mathbf{X}$ cannot affect $P(\mathbf{Y})$ when $\mathbf{W}$ is fixed, and changing $\mathbf{W}$ cannot affect $P(\mathbf{Y})$ when $\mathbf{X}$ is fixed, then changing $\mathbf{X}$ and $\mathbf{Y}$ together cannot affect $P(\mathbf{Y})$.

Exercise 21.12★

We discussed the issue of trying to use data to extract causal knowledge, that is, the directionality of an influence. In this problem, we will consider the interaction between this problem and both hidden variables and selection bias.

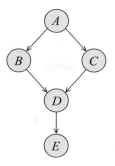

Figure 21.9   Learned causal network for exercise 21.12

Assume that our learning algorithm came up with the network in figure 21.9, which we are willing to assume is a perfect map for the distribution over the variables A, B, C, D, E. Under this assumption, among which pairs of variables between which a causal path exists in this model does there also necessarily exist a causal path ...

a. ... if we assume there are no hidden variables?

b. ... if we allow the possibility of one or more hidden variables?

c. ... if we allow for the possibility of selection bias?

For each of these options, specify the pairs for which a causal path exists, and explain why it exists in every I_X-equivalent structure. For the other pairs, provide an example of an I_X-equivalent structure for which no causal path exists.

22 *Utilities and Decisions*

We now move from the task of simply reasoning under uncertainty — reaching conclusions about the current situation from partial evidence — to the task of deciding how to act in the world. In a decision-making setting, an agent has a set of possible actions and has to choose between them. Each action can lead to one of several outcomes, which the agent can prefer to different degrees.

Most simply, the outcome of each action is known with certainty. In this case, the agent must simply select the action that leads to the outcome that is most preferred. Even this problem is far from trivial, since the set of outcomes can be large and complex and the agent must weigh different factors in determining which of the possible outcomes is most preferred. For example, when deciding which computer to buy, the agent must take into consideration the CPU speed, the amount of memory, the cost, the screen size, and many other factors. Deciding which of the possible configurations he most prefers can be quite difficult.

Even more difficult is the decision-making task in situations where the outcome of an action is not fully determined. In this case, we must take into account both the probabilities of various outcomes and the preferences of the agent between these outcomes. Here, it is not enough to determine a preference ordering between the different outcomes. We must be able to ascribe preferences to complex scenarios involving probability distributions over possible outcomes. The framework of *decision theory* provides a formal foundation for this type of reasoning. This framework requires that we assign numerical *utilities* to the various possible outcome, encoding the agent's preferences. In this chapter, we focus on a discussion of utilities functions and the principle of *maximum expected utility*, which is the foundation for decision making under uncertainty. In the next chapter, we discuss computationally tractable representations of an agent's decision problem and the algorithmic task of finding an optimal strategy.

decision theory

22.1 Foundations: Maximizing Expected Utility

In this section, we formally describe the basic decision-making task and define the principle of maximum expected utility. We also provide a formal justification for this principle from basic axioms of rationality.

22.1.1 Decision Making Under Uncertainty

We begin with a simple motivating example.

Example 22.1

Consider a decision maker who encounters the following situation. She can invest in a high-tech company (A), where she can make a profit of $4 million with 20 percent probability and $0 with 80 percent probability; or she can invest in pork belly futures (B), where she can make $3 million with 25 percent probability and $0 with 75 percent probability. (That is, the pork belly investment is less profitable but also less risky.) In order to choose between these two investment opportunities, the investor must compare her preferences between two scenarios, each of which encodes a probability distribution over outcomes: the first scenario, which we denote π_A, can be written as [$4 million : 0.2; $0 : 0.8]; the second scenario, denoted π_B, has the form [$3 million : 0.25; $0 : 0.75]. ∎*

In order to ascertain which of these scenarios we prefer, it is not enough to determine that we prefer $4 million to $3 million to $0. We need some way to aggregate our preferences for these outcomes with the probabilities with which we will get each of them. One approach for doing this aggregation is to assign each outcome a numerical *utility*, where a higher utility value associated with an outcome indicates that this outcome is more preferred. Importantly, however, utility values indicate more than just an ordinal preference ranking between outcomes; their numerical value is significant by itself, so that the relative values of different states tells us the strength of our preferences between them. This property allows us to combine the utility values of different states, allowing us to ascribe an *expected utility* to situations where we are uncertain about the outcome of an action. Thus, we can compare two possible actions using their expected utility, an ability critical for decision making under uncertainty.

utility

expected utility

We now formalize these intuitions.

Definition 22.1

lottery

preference over lotteries

A lottery π over an outcome space $\mathcal{O}$ is a set $[\pi_1 : \alpha_1; \ldots; \pi_k : \alpha_k]$ such that $\alpha_1, \ldots, \alpha_k \in [0, 1]$, $\sum_i \alpha_i = 1$, and each π_i is an outcome in $\mathcal{O}$. For two lotteries π_1, π_2, if the agent prefers π_1, we say that $\pi_1 \succ \pi_2$. If the agent is indifferent between the two lotteries, we say that $\pi_1 \sim \pi_2$. ∎*

A comparison between two different scenarios involving uncertainty over the outcomes is quite difficult for most people. At first glance, one might think that the "right" decision is the one that optimizes a person's monetary gain. However, that approach rarely reflects the preferences of the decision maker.

Example 22.2

Consider a slightly different decision-making situation. Here, the investor must decide between company C, where she earns $3 million with certainty, and company D, where she can earn $4 million with probability 0.8 and $0 with probability 0.2. In other words, she is now comparing two lotteries $\pi_C = [$3 million : 1]$ and $\pi_D = [$4 million : 0.8; $0 : 0.2]$. The expected profit of lottery D is $3.2 million, which is larger than the profit of $3 million from lottery C. However, a vast majority of people prefer the option of lottery C to that of lottery D. ∎*

The problem becomes far more complicated when one accounts for the fact that many decision-making situations involve aspects other than financial gain.

☞   **A general framework that allows us to make decisions such as these ascribes a numerical *utility* to different outcomes. An agent's utilities describe her overall preferences, which can depend not only on monetary gains and losses, but also on all other relevant aspects.** Each outcome o is associated with a numerical value $U(o)$, which is a numerical encoding of the agent's "happiness" for this outcome. Importantly, utilities are not just ordinal values, denoting the agent's preferences between the outcomes, but are actual numbers whose magnitude is meaningful. Thus, we can probabilistically aggregate utilities and compute their expectations over the different possible outcomes.

We now make these intuitions more formal.

Definition 22.2

decision-making situation

outcome

action

utility function

A decision-making situation $\mathcal{D}$ *is defined by the following elements:*

- *a set of* outcomes $\mathcal{O} = \{o_1, \ldots, o_N\}$;
- *a set of possible* actions *that the agent can take,* $\mathcal{A} = \{a_1, \ldots, a_K\}$;
- *a probabilistic outcome model* $P : \mathcal{A} \mapsto \Delta_{\mathcal{O}}$, *which defines a lottery* π_a, *which specifies a probability distribution over outcomes given that the action a was taken;*
- *a* utility function $U : \mathcal{O} \mapsto \mathbb{R}$, *where $U(o)$ is the agent's preferences for the outcome o.* ∎

Note that the definition of an outcome can also include the action taken; outcomes that involve one action a would then get probability 0 in the lottery induced by another action a'.

Definition 22.3

MEU principle

expected utility

The principle of maximum expected utility (MEU principle) *asserts that, in a decision-making situation $\mathcal{D}$, we should choose the action a that maximizes the* expected utility:

$$\text{EU}[\mathcal{D}[a]] = \sum_{o \in \mathcal{O}} \pi_a(o)U(o).$$

∎

Example 22.3

Consider a decision situation $\mathcal{I}_F$ where a college graduate is trying to decide whether to start up a company that builds widgets. The potential entrepreneur does not know how large the market demand for widgets really is, but he has a distribution: the demand is either m^0—nonexistent, m^1—low, or m^2—high, with probabilities 0.5, 0.3, and 0.2 respectively. The entrepreneur's profit, if he founds the startup, depends on the situation. If the demand is nonexistent, he loses a significant amount of money (outcome o_1); if it is low, he sells the company and makes a small profit (outcome o_2); if it is high, he goes public and makes a fortune (outcome o_3). If he does not found the startup, he loses nothing and earns nothing (outcome o_0). These outcomes might involve attributes other than money. For example, if he loses a significant amount of money, he also loses his credibility and his ability to start another company later on. Let us assume that the agent's utilities for the four outcomes are: $U(o_0) = 0$; $U(o_1) = -7$; $U(o_2) = 5$; $U(o_3) = 20$. The agent's expected utility for the action of founding the company (denoted f^1) is

$$\text{EU}[\mathcal{D}[f^1]] = 0.5 \cdot (-7) + 0.3 \cdot 5 + 0.2 \cdot 20 = 2.$$

His expected utility for the action of not founding the company (denoted f^0) is 0. The action choice maximizing the expected utility is therefore f^1. ∎

Our definition of a decision-making situation is very abstract, resulting in the impression that the setting is one where an agent takes a single simple action, resulting in a single simple outcome. In fact, both actions and outcomes can be quite complex. Actions can be complete strategies involving sequences of decisions, and outcomes (as in box 22.A) can also involve multiple aspects. We will return to these issues later on.

22.1.2   Theoretical Justification ★

What justifies the principle of maximizing expected utility, with its associated assumption regarding the existence of a numerical utility function, as a definition of rational behavior? It turns out that there are several theoretical analyses that can be used to *prove* the existence of such a function. At a high level, these analyses postulate some set of axioms that characterize the behavior of a rational decision maker. They then show that, for any agent whose decisions abide by these postulates, there exists some utility function U such that the agent's decisions are equivalent to maximizing the expected utility relative to U.

The analysis in this chapter is based on the premise that a decision maker under uncertainty must be able to decide between different lotteries. We then make a set of assumptions about the nature of the agent's preferences over lotteries; these assumptions arguably should hold for the preferences of any rational agent. **For an agent whose preferences satisfy these axioms, we prove that there exists a utility function U such that the agent's preferences are equivalent to those obtained by maximizing the expected utility relative to U.**

We first extend the concept of a lottery.

Definition 22.4

compound lottery

A compound lottery π *over an outcome space* $\mathcal{O}$ *is a set* $[\pi_1 : \alpha_1; \ldots; \pi_k : \alpha_k]$ *such that* $\alpha_1, \ldots, \alpha_k \in [0, 1]$, $\sum_i \alpha_i = 1$, *and each* π_i *is either an outcome in* $\mathcal{O}$ *or another lottery.*   ∎

Example 22.4

One example of a compound lottery is a game where we first toss a coin; if it comes up heads, we get $3 ($o_1$); if it comes up tails, we participate in another subgame where we draw a random card from a deck, and if it comes out spades, we get $50 ($o_2$); otherwise we get nothing (o_3). This lottery would be represented as $[o_1 : 0.5; [o_2 : 0.25; o_3 : 0.75] : 0.5]$.   ∎

rationality postulates

We can now state the *postulates of rationality* regarding the agent's preferences over lotteries. At first glance, each these postulates seems fairly reasonable, but each of them has been subject to significant criticism and discussion in the literature.

- **(A1) Orderability:** For all lotteries π_1, π_2, either

$$(\pi_1 \prec \pi_2) \text{ or } (\pi_1 \succ \pi_2) \text{ or } (\pi_1 \sim \pi_2) \tag{22.1}$$

This postulate asserts that an agent must know what he wants; that is, for any pair of lotteries, he must prefer one, prefer the other, or consider them to be equivalent. Note that this assumption is not a trivial one; as we discussed, it is hard for people to come up with preferences over lotteries.

- **(A2) Transitivity:** For all lotteries π_1, π_2, π_3, we have that:

$$\text{If } (\pi_1 \prec \pi_2) \text{ and } (\pi_2 \prec \pi_3) \text{ then } (\pi_1 \prec \pi_3). \tag{22.2}$$

The transitivity postulate asserts that preferences are transitive, so that if the agent prefers lottery 1 to lottery 2 and lottery 2 to lottery 3, he also prefers lottery 1 to lottery 3. Although transitivity seems very compelling on normative grounds, it is the most frequently violated axiom in practice. One hypothesis is that these "mistakes" arise when a person is forced to make choices between inherently incomparable alternatives. The idea is that each pairwise comparison invokes a preference response on a different "attribute" (for instance, money, time, health). Although each scale itself may be transitive, their combination need not be. A similar situation arises when the overall preference arises as an aggregate of the preferences of several individuals.

- **(A3) Continuity:** For all lotteries π_1, π_2, π_3,

$$\text{If } (\pi_1 \prec \pi_2 \prec \pi_3) \text{ then there exists } \alpha \in (0,1) \text{ such that } (\pi_2 \sim [\pi_1 : \alpha; \pi_3 : (1-\alpha)]). \quad (22.3)$$

This postulate asserts that if π_2 is somewhere between π_1 and π_3, then there should be some lottery between π_1 and π_3, which is equivalent to π_2. For our simple Entrepreneur example, we might have that $o_0 \sim [o_1 : 0.8; o_3 : 0.2]$. This axiom excludes the possibility that one alternative is "infinitely better" than another one, in the sense that any probability mixture involving the former is preferable to the latter. It therefore captures the relationship between probabilities and preferences and the form in which they compensate for each other.

- **(A4) Monotonicity:** For all lotteries π_1, π_2, and probabilities α, β,

$$(\pi_1 \succ \pi_2), (\alpha \geq \beta) \Rightarrow ([\pi_1 : \alpha; \pi_2 : (1-\alpha)] \succ [\pi_1 : \beta; \pi_2 : (1-\beta)]). \quad (22.4)$$

This postulate asserts that an agent prefers that better things happen with higher probability. Again, although this attribute seems unobjectionable, it has been argued that risky behavior such as Russian roulette violates this axiom. People who choose to engage in such behavior seem to prefer a probability mixture of "life" and "death" to "life," even though they (presumably) prefer "life" to "death." This argument can be resolved by revising the outcome descriptions, incorporating the aspect of the thrill obtained by playing the game.

- **(A5) Substitutability:** For all lotteries π_1, π_2, π_3, and probabilities α,

$$(\pi_1 \sim \pi_2) \Rightarrow ([\pi_1 : \alpha; \pi_3 : (1-\alpha)] \sim [\pi_2 : \alpha; \pi_3 : (1-\alpha)]). \quad (22.5)$$

This axiom states that if π_1 and π_2 are equally preferred, we can substitute one for the other without changing our preferences.

- **(A6) Decomposability:** For all lotteries π_1, π_2, and probabilities α, β,

$$[\pi_1 : \alpha, [\pi_2 : \beta, \pi_3 : (1-\beta)] : (1-\alpha)] \sim [\pi_1 : \alpha, \pi_2 : (1-\alpha)\beta, \pi_3 : (1-\alpha)(1-\beta)]. \quad (22.6)$$

This postulate says that compound lotteries are equivalent to flat ones. For example, our lottery in example 22.4 would be equivalent to the lottery

$$[o_1 : 0.5; o_2 : 0.125; o_3 : 0.375].$$

Intuitively, this axiom implies that the preferences depend only on outcomes, not the process in which they are obtained. It implies that a person does not derive any additional pleasure (or displeasure) from suspense or participation in the game.

If we are willing to accept these postulates, we can derive the following result:

Theorem 22.1

Assume that we have an agent whose preferences over lotteries satisfy the axioms (A1)–(A6). Then there exists a function $U : \mathcal{O} \mapsto \mathbb{R}$, such that, for any pair of lotteries π, π', we have that $\pi \prec \pi'$ if and only if $U(\pi) < U(\pi')$, where we define (recursively) the expected utility of any lottery as:

$$U([\pi_1 : \alpha_1, \ldots, \pi_k : \alpha_k]) = \sum_{i=1}^{k} \alpha_i U(\pi_i).$$

That is, the utility of a lottery is simply the expectation of the utilities of its components.

PROOF Our goal is to take a preference relation $\prec$ that satisfies these axioms, and to construct a utility function U over consequences such that $\prec$ is equivalent to implementing the MEU principle over the utility function U. We take the least and most preferred outcomes $o_{\min}$ and $o_{\max}$; these outcomes are typically known as *anchor outcomes*. By orderability (A1) and transitivity (A2), such outcomes must exist. We assign $U(o_{\min}) := 0$ and $U(o_{\max}) := 1$. By orderability, we have that for any other outcome o:

anchor outcome

$$o_{\min} \preceq o \preceq o_{\max}.$$

By continuity (A3), there must exist a probability α such that

$$[o : 1] \sim [o_{\min} : (1 - \alpha); o_{\max} : \alpha] \tag{22.7}$$

We assign $U(o) := \alpha$. The axioms can then be used to show that the assignment of utilities to lotteries resulting from applying the expected utility-principle results in an ordering that is consistent with our preferences. We leave the completion of this proof as an exercise (exercise 22.1). ∎

From an operational perspective, this discussion gives us a formal justification for the principle of maximum expected utility. When we have a set of outcomes, we ascribe a numerical *utility* to each one. If we have a set of actions that induce different lotteries over outcomes, we should choose the action whose *expected utility* is largest; as shown by theorem 22.1, this choice is equivalent to choosing the action that induces the lottery we most prefer.

22.2 Utility Curves

The preceding analysis shows that, under certain assumptions, a utility function must exist. However, it does not provide us with an understanding of utility functions. In this section, we take a more detailed look at the form of a utility functions and its connection to the utility function properties.

A utility function assigns numeric values to various possible outcomes. These outcomes can vary along multiple dimensions. Most obvious is monetary gain, but most settings involve other attributes as well. We begin in this section by considering the utility of simple outcomes, involving only a single attribute. We discuss the form of a utility function over a single attribute and the effects of the utility function on the agent's behavior. We focus on monetary outcomes, which are the most common and easy to understand. However, many of the issues we discuss in this section — those relating to risk attitudes and rationality — are general in their scope, and they apply also to other types of outcomes.

22.2.1　Utility of Money

Consider a decision-making situation where the outcomes are simply monetary gains or losses. In this simple setting, it is tempting to assume that the utility of an outcome is simply the amount of money gained in that outcome (with losses corresponding to negative utilities). However, as we discussed in example 22.2, most people do not always choose the outcome that maximizes their expected monetary gain. Making such a decision is not irrational; it simply implies that, for most people, their utility for an outcome is not simply the amount of money they have in that outcome.

utility curve

Consider a graph whose X-axis is the monetary gain a person obtains in an outcome (with losses corresponding to negative amounts), and whose Y-axis is the person's utility for that outcome. In general, most people's utility is monotonic in money, so that they prefer outcomes with more money to outcomes with less. However, if we draw a *curve* representing a person's utility as a function of the amount of money he or she gains in an outcome, that curve is rarely a straight line. This nonlinearity is the "justification" for the rationality of the preferences we observe in practice in example 22.2.

Example 22.5

Let c_0 represent the agent's current financial status, and assume for simplicity that he assigns a utility of 0 to c_0. If he assigns a utility of 10 to the consequence $c_0 + 3$million and 12 to the consequence $c_0 + 4$million, then the expected utility of the gamble in example 22.2 is $0.2 \cdot 0 + 0.8 \cdot 12 = 9.6 < 10$. Therefore, with this utility function, the agent's decision is completely rational. ∎

Saint Petersburg paradox

A famous example of the nonlinearity of the utility of money is the *Saint Petersburg paradox*:

Example 22.6

Suppose you are offered a chance to play a game where a fair coin is tossed repeatedly until it comes up heads. If the first head appears on the nth toss, you get $\$2^n$. How much would you be willing to pay in order to play this game?

The probability of the event H_n—the first head showing up on the nth toss—is $1/2^n$. Therefore, the expected winnings from playing this game are:

$$\sum_{n=1}^{\infty} P(H_n)\text{Payoff}(H_n) = \sum_{n=1}^{\infty} \frac{1}{2^n} 2^n = 1 + 1 + 1 + \ldots = \infty.$$

Therefore, you should be willing to pay any amount to play this game. However, most people are willing to pay only about $2. ∎

Empirical psychological studies show that people's utility functions in a certain range often grow logarithmically in the amount of monetary gain. That is, the utility of the outcome c_k, corresponding to an agent's current financial status plus $\$k$, looks like $\alpha + \beta \log(k + \gamma)$. In the Saint Petersburg example, if we take $U(c_k) = \log_2 k$, we get:

$$\sum_{n=1}^{\infty} P(H_n)U(\textit{Payoff}(H_n)) = \sum_{n=1}^{\infty} \frac{1}{2^n} U(c_{2^n}) = \sum_{n=1}^{\infty} \frac{n}{2^n} = 2,$$

which is precisely the amount that most people are willing to pay in order to play this game.

In general, most people's utility function tends to be concave for positive amount of money, so that the incremental value of additional money decreases as the amount of wealth grows.

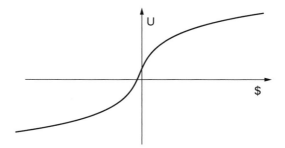

Figure 22.1   Example curve for the utility of money

Conversely, for negative amounts of money (debts), the shape of the curve often has the opposite shape, as shown in figure 22.1. Thus, for many people, going into debt of $1 million has significant negative utility, but the additional negative utility incurred by an extra $1 million of debt is a lot lower. Formally, $|U(-\$2,000,000) - U(-\$1,000,000)|$ is often significantly less than $|U(-\$1,000,000) - U(\$0)|$.

22.2.2   Attitudes Toward Risk

 There is a tight connection between the form of a person's utility curve and his behavior in different decision-making situations. In particular, the shape of this curve determines the person's attitude toward *risk*. A concave function, as in figure 22.2, indicates that the agent is *risk-averse*: he prefers a sure thing to a gamble with the same payoff. Consider in more detail the risk-averse curve of figure 22.2. We see that the utility of a lottery such as $\pi = [\$1000 : 0.5, \$0 : 0.5]$ is lower than the utility of getting $500 with certainty. Indeed, risk-averse preferences are characteristic of most people, especially when large sums of money are involved. In particular, recall example 22.2, where we compared a lottery where we win $3 million with certainty to one where we win $4 million with probability 0.8. As we discussed, most people prefer the first lottery to the second, despite the fact that the expected monetary gain in the first lottery is lower. This behavior can be explained by a risk-averse utility function in that region.

Returning to the lottery π, empirical research shows that many people are indifferent between playing π and the outcome where they get (around) $400 with certainty; that is, the utilities of the lottery and the outcome are similar. The amount $400 is called the *certainty equivalent* of the lottery. It is the amount of "sure thing" money that people are willing to trade for a lottery. The difference between the expected monetary reward of $500 and the certainty equivalent of $500 is called the *insurance premium*, and for good reason. The premium people pay to the insurance company is precisely to guarantee a sure thing (a sure small loss) as opposed to a lottery where one of the consequences involves a large negative utility (for example, the price of rebuilding the house if it burns down).

As we discussed, people are typically risk-averse. However, they often seek risk when the certain loss is small (relative to their financial situation). Indeed, lotteries and other forms of gambling exploit precisely this phenomenon. When the agent prefers the lottery to the certainty

risk

risk-averse

certainty
equivalent

insurance
premium

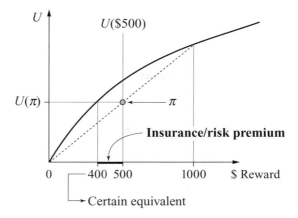

Figure 22.2   Utility curve and its consequences to an agent's attitude toward risk

risk-seeking

risk-neutral

equivalent, he is said to be *risk-seeking*, a behavior that corresponds to a curve whose shape is convex. Finally, if the agent's utility curve is linear, he is said to be *risk-neutral*. Most utility curves are locally linear, which means we can assume risk neutrality for small risks and rewards. Finally, as we noted, people are rarely consistent about risk throughout the entire monetary range: They are often risk-averse for positive gains, but can be risk-seeking for large negative amounts (going into debt). Thus, in our example, someone who is already $10 million in debt might choose to accept a gamble on a fair coin with $10 million payoff on heads and a $20 million loss on tails.

22.2.3   Rationality

The framework of utility curves provides a rich language for describing complex behaviors, including risk-averse, risk-seeking, or risk-neutral behaviors. They can even change their risk preferences over the range. One may thus be tempted to conclude that, for any behavior profile, there is some utility function for which that behavior is rational. However, that conclusion turns out to be false; indeed, empirical evidence shows that people's preferences are rarely rational under our definitions.

Example 22.7

Consider again the two simple lotteries in example 22.2 and example 22.1. In the two examples, we had four lotteries:

$$\pi_A \;:\; [\$4million : 0.2; \$0 : 0.8]$$
$$\pi_B \;:\; [\$3million : 0.25; \$0 : 0.75]$$
$$\pi_C \;:\; [\$3million : 1]$$
$$\pi_D \;:\; [\$4million : 0.8; \$0 : 0.2].$$

Most people, by an overwhelming majority, prefer π_C to π_D. The opinions on π_A versus π_B are more divided, but quite a number of people prefer π_A to π_B. Each of these two preferences — $\pi_D \succ \pi_C$ and $\pi_A \succ \pi_B$ — is rational relative to some utility functions. However, their combination is not — there is no utility function that is consistent with both of these preferences. To understand why, assume (purely to simplify the presentation) that $U(\$0) = 0$. In this case, preferring π_C to π_D is equivalent to saying that

$$U(c_{3,000,000}) > 0.8 \cdot U(c_{4,000,000}).$$

On the other hand, preferring π_A to π_B is equivalent to:

$$
\begin{aligned}
0.2 \cdot U(c_{4,000,000}) &> 0.25 \cdot U(c_{3,000,000}) \\
0.8 \cdot U(c_{4,000,000}) &> U(c_{3,000,000}).
\end{aligned}
$$

Multiplying both sides of the first inequality by 4, we see that these two statements are directly contradictory, so that these preferences are inconsistent with decision-theoretic foundations, for any utility function.   ∎

Thus, people are often irrational, in that their choices do not satisfy the principle of maximum expected utility relative to any utility function. When confronted with their "irrationality," the responses of people vary. Some feel that they have learned an important lesson, which often affects other decisions that they make. For example, some subjects have been observed to cancel their automobile collision insurance and take out more life insurance. In other cases, people stick to their preferences even after seeing the expected utility analysis. These latter cases indicate that the principle of maximizing expected utility is not, in general, an adequate *descriptive model* of human behavior. As a consequence, there have been many proposals for alternative definitions of rationality that attempt to provide a better fit to the behavior of human decision makers. Although of great interest from a psychological perspective, there is no reason to believe that these frameworks will provide a better basis for building automated decision-making systems. Alternatively, we can view decision theory as a *normative model* that provides the "right" formal basis for rational behavior, regardless of human behavior. One can then argue that we should design automated decision-making systems based on these foundations; indeed, so far, most such systems have been based on the precepts of decision theory.

22.3   Utility Elicitation

22.3.1   Utility Elicitation Procedures

How do we acquire an appropriate utility function to use in a given setting? In many ways, this problem is much harder than acquiring a probabilistic model. In general, we can reasonably assume that the probabilities of chance events apply to an entire population and acquire a single probabilistic model for the whole population. For example, when constructing a medical diagnosis network, the probabilities will usually be learned from data or acquired from a human expert who understands the statistics of the domain. By contrast, utilities are inherently personal, and people often have very different preference orderings in the same situation. Thus, the utility function we use needs to be acquired for the individual person or entity for whom the decision is being made. Moreover, as we discussed, probability values can be learned from data by observing empirical frequencies in the population. The individuality of utility values, and the fact that they are never observed directly, makes it difficult to apply similar learning methods to the utility acquisition task.

utility elicitation

standard gamble

There have been several methods proposed for *eliciting* utilities from people. The most classical method is the *standard gamble* method, which is based directly on the axioms of utility theory. In the proof of theorem 22.1, we selected two anchor states — our least preferred and most preferred states $s_\perp$ and $s_\top$. We then used the continuity axiom (equation (22.3)) to place each state on a continuous spectrum between these two anchor states, by finding the

indifference point

indifference point α — a probability value $\alpha \in [0, 1]$ such that $s \sim [s_\perp : (1 - \alpha); s_\top : \alpha]$.

We can convert this idea to a utility elicitation procedure as follows. We select a pair of anchor states. In most cases, these are determined in advance, independently of the user. For example, in a medical decision-making situation, $s_\perp$ is often "death," whereas $s_\top$ is an immediate and complete cure. For any outcome s, we can now try to find the indifference point. It is generally assumed that we cannot ask a user to assess the value of α directly. We therefore use some procedure that searches over the space of possible α's. If $s \prec [s_\perp : (1 - \alpha); s_\top : \alpha]$, we consider lower values of α, and if $s \succ [s_\perp : (1 - \alpha); s_\top : \alpha]$, we consider higher values, until we find the indifference point. Taking $U(s_\perp) = 0$ and $U(s_\top) = 1$, we simply take $U(s) = \alpha$.

The standard gamble procedure is satisfying because of its sound theoretical foundations. However, it is very difficult for people to apply in practice, especially in situations involving large numbers of outcomes. Moreover, many independent studies have shown that the final values obtained in the process of standard gamble elicitation are sensitive to the choice of anchors and to the choice of the search procedure.

time trade-off

Several other methods for utility elicitation have been proposed to address these limitations. For example, *time trade-off* tries to compare two outcomes: (1) t years (where t is the patient's life expectancy) in the current state of health (state s), and (2) t' years (where $t' < t$) in perfect health (the outcome $s_\top$). As in standard gamble, t' is varied until the indifferent point is reached, and the utility of the state s is taken to be proportional to t' at that point. Another

visual-analog scale

method, the *visual-analog scale*, simply asks users to point out their utilities on some scale.

Overall, each of the methods proposed has significant limitations in practice. Moreover, the results obtained for the same individual using different methods are usually quite different, putting into question the results obtained by any method. Indeed, one might wonder whether there even exists such an object as a person's "true utility value." Nevertheless, one can still argue that decisions made for an individual using his or her own utility function (even with the imprecisions involved in the process) are generally better for that individual than decisions made using some "global" utility function determined for the entire population.

22.3.2 Utility of Human Life

Attributes whose utility function is particularly difficult to acquire are those involving human life. Clearly, such factors play a key role in medical decision-making situations. However, they also appear in a wide variety of other settings. For example, even a simple decision such as whether to replace worn-out tires for a car involves the reduced risk of death or serious injury in a car with new tires.

Because utility theory requires that we reduce all outcomes to a single numerical value, we are forced to place a utility value on human life, placing it on the same scale as other factors, such as money. Many people find this notion morally repugnant, and some simply refuse to do so. However, the fact of the matter is that, in making decisions, one makes these trade-offs, whether consciously or unconsciously. For example, airplanes are not overhauled after each trip, even though that would clearly improve safety. Not all cars are made with airbags, even though they are known to save lives. Many people accept an extra stopover on a flight in order to save money, even though most airplane accidents happen on takeoff and landing.

Placing a utility on human life raises severe psychological and philosophical difficulties. One such difficulty relates to actions involving some probability of death. The naive approach would be to elicit the utility of the outcome *death* and then estimate the utility of an outcome involving some probability p of death as $p \cdot U(death)$. However, this approach implies that people's utility is linear in their probability of death, an assumption which is generally false. In other words, even if a person is willing to accept \$50 for an outcome involving a one-in-a-million chance of death, it does not mean that he would be willing to accept \$50 million for the outcome of *death* with certainty. Note that this example shows that, at least for this case, people violate the basic assumption of decision theory: that a person's preference for an uncertain outcome can be evaluated using expected utility, which is linear in the probabilities.

A more appropriate approach is to encode explicitly the chance of death. Thus, a key metric used to measure utilities for outcomes involving risk to human life is the *micromort* — a one-in-a-million chance of death. Several studies across a range of people have shown that a micromort is worth about \$20 in 1980 dollars, or under \$50 in today's dollars. We can consider a utility curve whose X-axis is micromorts. As for monetary utility, this curve behaves differently for positive and negative values. For example, many people are not willing to pay very much to remove a risk of death, but require significant payment in order to assume additional risk.

Micromorts are useful for evaluating situations where the primary consideration is the probability that death will occur. However, in many situations, particularly in medical decision making, the issue is not the chance of immediate death, but rather the amount of life that a person has remaining. In one approach, we can evaluate outcomes using life expectancy, where we would construct a utility curve whose X axis was the number of expected years of life. However, our preferences for outcomes are generally much more complex, since they involve not only the quantity but also the *quality of life*. In some cases, a person may prefer to live for fewer years, but in better health, than to live longer in a state where he is in pain or is unable to perform certain activities. The trade-offs here are quite complex, and highly personal.

One approach to simplifying this complex problem is by measuring outcomes using units called *QALYs* — *quality-adjusted life years*. A year of life in good health with no infirmities is worth 1 QALY. A year of life in poor health is discounted, and it is worth some fraction (less than 1) of a QALY. (In fact, some health states — for example, those involving significant pain and loss of function — may even be worth negative QALYs.) Using QALYs, we can assign a single numerical score to complex outcomes, where a person's state of health can evolve over time. QALYs are much more widely used than micromorts as a measure of utility in medical and social-policy decision making.

micromort

QALY

22.4 Utilities of Complex Outcomes

So far, we have largely focused on outcomes involving only a single attribute. In this case, we can write down our utility function as a simple table in the case of discrete outcomes, or as a curve in the case of continuous-valued outcomes (such as money). In practice, however, outcomes often involve multiple facets. In a medical decision-making situation, outcomes might involve pain and suffering, long-term quality of life, risk of death, financial burdens, and more. Even in a much "simpler" setting such as travel planning, outcomes involve money, comfort of accommodations, availability of desired activities, and more. A utility function must incorporate the importance of these different attributes, and the preferences for various values that they might take, in order to produce a single numeric value for each outcome.

Our utility function in domains such as this has to construct a single number for each outcome that depends on the values of all of the relevant variables. More precisely, assume that an outcome is described by an assignment of values to some set of variables $V = \{V_1, \ldots, V_k\}$; we then have to define a utility function $U : Val(V) \mapsto \mathbb{R}$. As usual, the size of this representation is exponential in k.

In the case of probabilities, we addressed the issue of exponential blowup by exploiting structure in the distribution. We showed a direct connection between independence properties of the distribution and our ability to represent it compactly as a product of smaller factors. As we now discuss, very similar ideas apply in the setting of utility functions.

☞ **Specifically, we can show a correspondence between "independence properties" among utility attributes of an agent and our ability to factor his utility function into a combination of *subutility functions*, each defined over a subset of utility attributes.** A *subutility function* is a function $f : Val(Y) \mapsto \mathbb{R}$, for some $Y \subseteq V$, where Y is the scope of f.

subutility
function

However, the notion of independence in this setting is somewhat subtle. A utility function on its own does not induce behavior; it is a meaningful entity only in the context of a decision-making setting. Thus, our independence properties must be defined in that context as well. As we will see, there is not a single definition of independence that is obviously the right choice; several definitions are plausible, each with its own properties.

22.4.1 Preference and Utility Independence ⋆

To understand the notion of independence in decision making, we begin by considering the simpler setting, where we are making decisions in the absence of uncertainty. Here, we need only consider preferences on outcomes. Let X, Y be a disjoint partition of our set of utility attributes V. We thus have a preference ordering $\prec$ over pairs (x, y).

When can we say that X is "independent" of Y? Intuitively, if we are given $Y = y$, we can now consider our preferences over the possible values x, given that y holds. Thus, we have an induced preference ordering $\prec_y$ over values $x \in Val(X)$, where we write $x_1 \prec_y x_2$ if $(x_1, y) \prec (x_2, y)$. In general, the ordering induced by one value y is different from the ordering induced by another. We say that X is *preferentially independent* of Y if all values y induce the same ordering over $Val(X)$. More precisely:

Definition 22.5

preference
independence

The set of attributes X is preferentially independent *of $Y = V - X$ in $\prec$ if, for all $y, y' \in$*

$Val(\boldsymbol{Y})$, and for all $\boldsymbol{x}_1, \boldsymbol{x}_2 \in Val(\boldsymbol{X})$, we have that

$$\boldsymbol{x}_1 \prec_{\boldsymbol{y}} \boldsymbol{x}_2 \quad \Leftrightarrow \quad \boldsymbol{x}_1 \prec_{\boldsymbol{y}'} \boldsymbol{x}_2. \qquad\blacksquare$$

Note that preferential independence is not a symmetric relation:

Example 22.8
Consider an entrepreneur whose utility function $U(S, F)$ involves two binary-valued attributes: the success of his company (S) and the fame he gets (F). One reasonable preference ordering over outcomes might be:

$$(s^0, f^1) \prec (s^0, f^0) \prec (s^1, f^0) \prec (s^1, f^1).$$

That is, the most-preferred state is where he is successful and famous; the next-preferred state is where he is successful but not famous; then the second-next-preferred state is where he is unsuccessful but unknown; and the least-preferred state is where he is unsuccessful and (in)famous. In this preference ordering, we have that S is preferentially independent of F, since the entrepreneur prefers to be successful whether he is famous or not ($(s^0, f^1) \prec (s^1, f^1)$ and $(s^0, f^0) \prec (s^1, f^0)$). On the other hand, F is not preferentially independent of S, since the entrepreneur prefers to be famous if successful but unknown if he is unsuccessful. $\blacksquare$

When we move to the more complex case of reasoning under uncertainty, we compare decisions that induce lotteries over outcomes. Thus, our notion of independence must be defined relative to this more complex setting. From now on, let $\prec, U$ be a pair, where U is a utility function over $Val(\boldsymbol{V})$, and $\prec$ is the associated preference ordering for lotteries over $Val(\boldsymbol{V})$. We define independence properties for U in terms of $\prec$.

Our first task is to define the notion of a conditional preference structure, where we "fix" the value of some subset of variables $\boldsymbol{Y}$. This structure defines a preference ordering $\prec_{\boldsymbol{y}}$ for lotteries over $Val(\boldsymbol{X})$, given some particular instantiation $\boldsymbol{y}$ to $\boldsymbol{Y}$. The definition is a straightforward generalization of the one we used for preferences over outcomes:

Definition 22.6

conditional
preference
structure

Let $\pi_1^{\boldsymbol{X}}$ and $\pi_2^{\boldsymbol{X}}$ be two distributions over $Val(\boldsymbol{X})$. We define the conditional preference structure $\prec_{\boldsymbol{y}}$ as follows:

$$\pi_1^{\boldsymbol{X}} \prec_{\boldsymbol{y}} \pi_2^{\boldsymbol{X}} \quad \text{if} \quad (\pi_1^{\boldsymbol{X}}, \mathbf{1}_{\boldsymbol{y}}) \prec (\pi_2^{\boldsymbol{X}}, \mathbf{1}_{\boldsymbol{y}}),$$

where $(\pi^{\boldsymbol{X}}, \mathbf{1}_{\boldsymbol{y}})$ assigns probability $\pi^{\boldsymbol{X}}(\boldsymbol{x})$ to any assignment $(\boldsymbol{x}, \boldsymbol{y})$ and probability 0 to any assignment $(\boldsymbol{x}, \boldsymbol{y}')$ for $\boldsymbol{y}' \neq \boldsymbol{y}$. $\blacksquare$

In other words, the preference ordering $\prec_{\boldsymbol{y}}$ "expands" lotteries over $Val(\boldsymbol{X})$ by having $\boldsymbol{Y} = \boldsymbol{y}$ with probability 1, and then using $\prec$.

With this definition, we can now generalize preferential independence in the obvious way: $\boldsymbol{X}$ is utility independent of $\boldsymbol{Y} = \boldsymbol{V} - \boldsymbol{X}$ when conditional preferences for lotteries over $\boldsymbol{X}$ do not depend on the particular value $\boldsymbol{y}$ given to $\boldsymbol{Y}$.

Definition 22.7

utility
independence

We say that $\boldsymbol{X}$ is utility independent of $\boldsymbol{Y} = \boldsymbol{V} - \boldsymbol{X}$ if, for all $\boldsymbol{y}, \boldsymbol{y}' \in Val(\boldsymbol{Y})$, and for any pair of lotteries $\pi_1^{\boldsymbol{X}}, \pi_2^{\boldsymbol{X}}$ over $Val(\boldsymbol{X})$, we have that:

$$\pi_1^{\boldsymbol{X}} \prec_{\boldsymbol{y}} \pi_2^{\boldsymbol{X}} \quad \Leftrightarrow \quad \pi_1^{\boldsymbol{X}} \prec_{\boldsymbol{y}'} \pi_2^{\boldsymbol{X}}. \qquad\blacksquare$$

Because utility independence is a straight generalization of preference independence, it, too, is not symmetric.

Note that utility independence is only defined for a set of variables and its complement. This limitation is inevitable in the context of decision making, since we can define preferences only over entire outcomes, and therefore every variable must be assigned a value somehow.

Different sets of utility independence assumptions give rise to different decompositions of the utility function. Most basically, for a pair $(\prec, U)$ as before, we have that:

Proposition 22.1

A set $\boldsymbol{X}$ is utility independent of $\boldsymbol{Y} = \boldsymbol{V} - \boldsymbol{X}$ in $\prec$ if and only if U has the form:

$$U(\boldsymbol{V}) = f(\boldsymbol{Y}) + g(\boldsymbol{Y})h(\boldsymbol{X}).$$

Note that each of the functions f, g, h has a smaller scope than our original U, and hence this representation requires (in general) fewer parameters.

From this basic theorem, we can obtain two conclusions.

Proposition 22.2

Every subset of variables $\boldsymbol{X} \subset \boldsymbol{V}$ is utility independent of its complement if and only if there exist k functions $U_i(V_i)$ and a constant c such that

$$U(\boldsymbol{V}) = \prod_{i=1}^{k} U_i(V_i) + c,$$

or k functions $U_i(V_i)$ such that

$$U(\boldsymbol{V}) = \sum_{i=1}^{k} U_i(V_i).$$

---

utility
decomposition

In other words, when every subset is utility independent of its complement, the utility function *decomposes* either as a sum or as a product of subutility functions over individual variables. In this case, we need only elicit a linear number of parameters, exponentially fewer than in the general case.

If we weaken our assumption, requiring only that each variable in isolation is utility independent of its complement, we obtain a much weaker result:

Proposition 22.3

If, for every variable $V_i \in \boldsymbol{V}$, V_i is utility independent of $\boldsymbol{V} - \{V_i\}$, then there exist k functions $U_i(V_i)$ $(i = 1, \ldots, k)$ such that U is a multilinear function (a sum of products) of the U_i's.

For example, if $\boldsymbol{V} = \{V_1, V_2, V_3\}$, then this theorem would imply only that $U(V_1, V_2, V_3)$ can be written as

$$c_1 U_1(V_1)U_2(V_2)U_3(V_3) + c_2 U_1(V_1)U_2(V_2) + c_3 U_1(V_1)U_3(V_3) + c_4 U_2(V_2)U_3(V_3) +$$
$$c_5 U_1(V_1) + c_6 U_2(V_2) + c_7 U_3(V_3).$$

In this case, the number of subutility functions is linear, but we must elicit (in the worst case) exponentially many coefficients. Note that, if the domains of the variables are large, this might still result in an overall savings in the number of parameters.

22.4.2   Additive Independence Properties

Utility independence is an elegant assumption, but the resulting decomposition of the utility function can be difficult to work with. The case of a purely additive or purely multiplicative decomposition is generally too limited, since it does not allow us to express preferences that relate to combinations of values for the variables. For example, a person might prefer to take a vacation at a beach destination, but only if the weather is good; such a preference does not easily decompose as a sum or a product of subutilities involving only individual variables.

In this section, we explore progressively richer families of utility factorizations, where the utility is encoded as a sum of subutility functions:

$$U(\boldsymbol{V}) = \sum_{i=1}^{k} U_i(\boldsymbol{Z}_i).   \tag{22.8}$$

We also study how these decompositions correspond to a form of independence assumption about the utility function.

22.4.2.1   Additive Independence

In our first decomposition, we restrict attention to decomposition as in equation (22.8), where $\boldsymbol{Z}_1, \ldots, \boldsymbol{Z}_k$ is a *disjoint* partition of $\boldsymbol{V}$. This decomposition is more restrictive than the one allowed by utility independence, since we allow a decomposition only as a sum, and not as a product. This decomposition turns out to be equivalent to a notion called *additive independence*, which has much closer ties to probabilistic independence. Roughly speaking, $\boldsymbol{X}$ and $\boldsymbol{Y}$ are additively independent if our preference function for lotteries over $\boldsymbol{V}$ depends only on the marginals over $\boldsymbol{X}$ and $\boldsymbol{Y}$. More generally, we define:

additive
independence

Definition 22.8

> *Let $\boldsymbol{Z}_1, \ldots, \boldsymbol{Z}_k$ be a disjoint partition of $\boldsymbol{V}$. We say that $\boldsymbol{Z}_1, \ldots, \boldsymbol{Z}_k$ are additively independent in $\prec$ if, for any lotteries π_1, π_2 that have the same marginals on all $\boldsymbol{Z}_i$, we have that π_1 and π_2 are indifferent under $\prec$.* ∎

Additive independence is strictly stronger than utility independence: For two subsets $\boldsymbol{X} \cup \boldsymbol{Y} = \boldsymbol{V}$ that are additively independent, we have both that $\boldsymbol{X}$ is utility independent of $\boldsymbol{Y}$ and $\boldsymbol{Y}$ is utility independent of $\boldsymbol{X}$. It then follows, for example, that the preference ordering in example 22.8 does not have a corresponding additively independent utility function. Additive independence is equivalent to the decomposition of U as a sum of subutilities over the $\boldsymbol{Z}_i$'s:

Theorem 22.2

> *Let $\boldsymbol{Z}_1, \ldots, \boldsymbol{Z}_k$ be a disjoint partition of $\boldsymbol{V}$, and let $\prec, U$ be a corresponding pair of a preference ordering and a utility function. Then $\boldsymbol{Z}_1, \ldots, \boldsymbol{Z}_k$ are additively independent in $\prec$ if and only if U can be written as: $U(\boldsymbol{V}) = \sum_{i=1}^{k} U_i(\boldsymbol{Z}_i)$.*

PROOF The "if" direction is straightforward. For the "only if" direction, consider first the case where $\boldsymbol{X}, \boldsymbol{Y}$ is a disjoint partition of $\boldsymbol{V}$, and $\boldsymbol{X}, \boldsymbol{Y}$ are additively independent in $\prec$. Let $\boldsymbol{x}, \boldsymbol{y}$ be some arbitrary fixed assignment to $\boldsymbol{X}, \boldsymbol{Y}$. Let $\boldsymbol{x}', \boldsymbol{y}'$ be any other assignment to $\boldsymbol{X}, \boldsymbol{Y}$. Let π_1 be the distribution that assigns probability 0.5 to each of $\boldsymbol{x}, \boldsymbol{y}$ and $\boldsymbol{x}', \boldsymbol{y}'$, and π_2 be the distribution that assigns probability 0.5 to each of $\boldsymbol{x}, \boldsymbol{y}'$ and $\boldsymbol{x}', \boldsymbol{y}$. These two distributions have the same marginals over X and Y. Therefore, by the assumption of additive independence, $\pi_1 \sim \pi_2$, so that

$$
\begin{aligned}
0.5 U(\boldsymbol{x}, \boldsymbol{y}) + 0.5 U(\boldsymbol{x}', \boldsymbol{y}') &= 0.5 U(\boldsymbol{x}, \boldsymbol{y}') + 0.5 U(\boldsymbol{x}', \boldsymbol{y}) \\
U(\boldsymbol{x}', \boldsymbol{y}') &= U(\boldsymbol{x}, \boldsymbol{y}') - U(\boldsymbol{x}, \boldsymbol{y}) + U(\boldsymbol{x}', \boldsymbol{y}).
\end{aligned}
\tag{22.9}
$$

Now, define $U_1(\boldsymbol{X}) = U(\boldsymbol{X}, \boldsymbol{y})$ and $U_2(\boldsymbol{Y}) = U(\boldsymbol{x}, \boldsymbol{Y}) - U(\boldsymbol{x}, \boldsymbol{y})$. It follows directly from equation (22.9) that for any $\boldsymbol{x}', \boldsymbol{y}'$, $U(\boldsymbol{x}', \boldsymbol{y}') = U_1(\boldsymbol{x}') + U_2(\boldsymbol{y}')$, as desired. The case of a decomposition $\boldsymbol{Z}_1, \ldots, \boldsymbol{Z}_k$ follows by a simple induction on k. ∎

Example 22.9

Consider a student who is deciding whether to take a difficult course. Taking the course will require a significant time investment during the semester, so it has a cost. On the other hand, taking the course will result in a more impressive résumé, making the student more likely to get a good job with a high salary after she graduates. The student's utility might depend on the two attributes T (taking the course) and J (the quality of the job obtained). The two attributes are plausibly additively independent, so that we can express the student's utility as $U_1(T) + U_2(J)$. Note that this independence of the utility function is completely unrelated to any possible probabilistic (in)dependencies. For example, taking the class is definitely correlated probabilistically with the student's job prospects, so T and J are dependent as probabilistic attributes but additively independent as utility attributes. ∎

In general, however, additive independence is a strong notion that rarely holds in practice.

Example 22.10

Consider a student planning his course load for the next semester. His utility might depend on two attributes — how interesting the courses are (I), and how much time he has to devote to class work versus social activities (T). It is quite plausible that these two attributes are not utility independent, because the student might be more willing to spend significant time on class work if the material is interesting. ∎

Example 22.11

Consider the task of making travel reservations, and the two attributes H — the quality of one's hotel — and W — the weather. Even these two seemingly unrelated attributes might not be additively independent, because the pleasantness of one's hotel room is (perhaps) more important when one has to spend more time in it on account of bad weather. ∎

22.4.2.2 Conditional Additive Independence

The preceding discussion provides a strong argument for extending additive independence to the case of nondisjoint subsets. For this extension, we turn to probability distributions for intuition: In a sense, additive independence is analogous to marginal independence. We therefore wish to construct a notion analogous to conditional independence:

Definition 22.9

CA-independence

Let X, Y, Z be a disjoint partition of V. We say that X and Y are conditionally additively independent (CA-independent) given Z in $\prec$ if, for every assignment z to Z, X and Y are additively independent in the conditional preference structure $\prec_z$. ∎

The CA-independence condition is equivalent to an assumption that the utility decomposes with overlapping subsets:

Proposition 22.4

Let X, Y, Z be a disjoint partition of V, and let $\prec, U$ be a corresponding pair of a preference ordering and a utility function. Then X and Y are CA-independent given Z in $\prec$ if and only if U can be written as:

$$U(X, Y, Z) = U_1(X, Z) + U_2(Y, Z).$$

The proof is straightforward and is left as an exercise (exercise 22.2).

Example 22.12

Consider again example 22.10, but now we add an attribute F representing how much fun the student has in his free time (for example, does he have a lot of friends and hobbies that he enjoys?). Given an assignment to T, which determines how much time the student has to devote to work versus social activities, it is quite reasonable to assume that I and F are additively independent. Thus, we can write $U(I, T, F)$ as $U_1(I, T) + U_2(T, F)$. ∎

Based on this result, we can prove an important theorem that allows us to view a utility function in terms of a graphical model. **Specifically, we associate a utility function with an undirected graph, like a Markov network. As in probabilistic graphical models, the separation properties in the graph encode the CA-independencies in the utility function. Conversely, the utility function decomposes additively along the maximal cliques in the network.** Formally, we define the two types of relationships between a pair $(\prec, U)$ and an undirected graph:

Definition 22.10

CAI-map

We say that $\mathcal{H}$ is an CAI-map for $\prec$ if, for any disjoint partition X, Y, Z of V, if X and Y are separated in $\mathcal{H}$ given Z, we have that X and Y are CA-independent in $\prec$ given Z. ∎

Definition 22.11

utility
factorization

We say that a utility function U factorizes according to $\mathcal{H}$ if we can write U as a sum

$$U(V) = \sum_{c=1}^{k} U_c(C_c),$$

where $C_1, \ldots, C_k$ are the maximal cliques in $\mathcal{H}$. ∎

We can now show the same type of equivalence between these two definitions as we did for probability distributions. The first theorem goes from factorization to independencies, showing that a factorization of the utility function according to a network $\mathcal{H}$ implies that it satisfies the independence properties implied by the network. It is analogous to theorem 3.2 for Bayesian networks and theorem 4.1 for Markov networks.

Theorem 22.3

Let $(\prec, U)$ be a corresponding pair of a preference function and a utility function. If U factorizes according to $\mathcal{H}$, then $\mathcal{H}$ is a CAI-map for $\prec$.

PROOF The proof of this result follows immediately from proposition 22.4. Assume that U factorizes according to $\mathcal{H}$, so that $U = \sum_c U_c(C_c)$. Any C_c cannot involve variables from both X and Y. Thus, we can divide the cliques into two subsets: $\mathcal{C}_1$, which involve only variables in

$\boldsymbol{X}, \boldsymbol{Z}$, and $\mathcal{C}_2$, which involve only variables in $\boldsymbol{Y}, \boldsymbol{Z}$. Letting $U_i = \sum_{c \in \mathcal{C}_i} U_c(\boldsymbol{C}_c)$, for $i = 1, 2$, we have that $U(\boldsymbol{V}) = U_1(\boldsymbol{X}, \boldsymbol{Z}) + U_2(\boldsymbol{Y}, \boldsymbol{Z})$, precisely the condition in proposition 22.4. The desired CA-independence follows. ∎

The converse result asserts that any utility function that satisfies the CA-independence properties associated with the network can be factorized over the network's cliques. It is analogous to theorem 3.1 for Bayesian networks and theorem 4.2 for Markov networks.

Theorem 22.4

Let $(\prec, U)$ be a corresponding pair of a preference function and a utility function. If $\mathcal{H}$ is a CAI-map for $\prec$, then U factorizes according to $\mathcal{H}$.

Although it is possible to prove this result directly, it also follows from the analogous result for probability distributions (the *Hammersley-Clifford theorem* — theorem 4.2). The basic idea is to construct a probability distribution by exponentiating U and then show that CA-independence properties for U imply corresponding probabilistic conditional independence properties for P:

Hammersley-Clifford theorem

Lemma 22.1

Let U be a utility function, and define $P(\boldsymbol{V}) \propto \exp(U(\boldsymbol{V}))$. For a disjoint partition $\boldsymbol{X}, \boldsymbol{Y}, \boldsymbol{Z}$ of $\boldsymbol{V}$, we have that $\boldsymbol{X}$ and $\boldsymbol{Y}$ are CA-independent given $\boldsymbol{Z}$ in U if and only if $\boldsymbol{X}$ and $\boldsymbol{Y}$ are conditionally independent given $\boldsymbol{Z}$ in P.

The proof is left as an exercise (exercise 22.3).

Based on this correspondence, many of the results and algorithms of chapter 4 now apply without change to utility functions. In particular, the proof of theorem 22.4 follows immediately (see exercise 22.4).

As for probabilistic models, we can consider the task of constructing a graphical model that reflects the independencies that hold for a utility function. Specifically, we define $\mathcal{H}$ to be a *minimal CAI-map* if it is a CAI-map from which no further edges can be removed without rendering it not a CAI-map. Our goal is the construction of an undirected graph which is a minimal CAI-map for a utility function U. We addressed exactly the same problem in the context of probability functions in section 4.3.3 and provided two algorithms. One was based on checking pairwise independencies of the form $(X \perp Y \mid \mathcal{X} - \{X, Y\})$. The other was based on checking local (Markov blanket) independencies of the form $(X \perp \mathcal{X} - \{X\} - \boldsymbol{U} \mid \boldsymbol{U})$. Importantly, both of these types of independencies involve a disjoint and exhaustive partition of the set of variables into three subsets. Thus, we can apply these procedures without change using CA-independencies.

minimal CAI-map

Because of the equivalence of lemma 22.1, and because $P \propto \exp(U)$ is a positive distribution, all of the results in section 4.3.3 hold without change. In particular, we can show that either of the two procedures described produces the unique minimal CAI-map for U. Indeed, we can prove an even stronger result: The unique minimal CAI-map $\mathcal{H}$ for U is a *perfect CAI-map* for U, in the sense that any CA-independence that holds for U is implied by separation in $\mathcal{H}$:

perfect CAI-map

Theorem 22.5

Let $\mathcal{H}$ be any minimal CAI-map for U, and let $\boldsymbol{X}, \boldsymbol{Y}, \boldsymbol{Z}$ be a disjoint partition of $\boldsymbol{V}$. Then if $\boldsymbol{X}$ is CA-independent of $\boldsymbol{Y}$ given $\boldsymbol{Z}$ in U, then $\boldsymbol{X}$ is separated from $\boldsymbol{Y}$ by $\boldsymbol{Z}$ in $\mathcal{H}$.

The proof is left as an exercise (exercise 22.5).

completeness

Note that this result is a strong *completeness* property, allowing us to read any CA-independence that holds in the utility function from a graph. One might wonder why a similar result was so elusive for probability distributions. The reason is not that utility functions are better expressed as graphical models than are probability distributions. Rather, the language of CA-independencies is substantially more limited than that of conditional independencies in the probabilistic case: For probability distributions, we can evaluate any statement of the form "X is independent of Y given Z," whereas for utility functions, the corresponding (CA-independence) statement is well defined only when X, Y, Z form a disjoint partition of V. In other words, although any CA-independence statement that holds in the utility function can be read from the graph, the set of such statements is significantly more restricted. In fact, a similar weak completeness statement can also be shown for probability distributions.

22.4.2.3   Generalized Additive Independence

Because of its limited expressivity, the notion of conditional additive independence allows us only to make fairly coarse assertions regarding independence — independence of two subsets of variables given all of the rest. As a consequence, the associated factorization is also quite coarse. In particular, we can only use CA-independencies to derive a factorization of the utility function over the maximal cliques in the Markov network. As was the case in probabilistic models (see section 4.4.1.1), this type of factorization can obscure the finer-grained structure in the function, and its parameterization may be exponentially larger.

In this section, we present the most general additive decomposition: the decomposition of equation (22.8), but with arbitrarily overlapping subsets. This type of decomposition is the utility analogue to a Gibbs distribution (definition 4.3), with the factors here combining additively rather than multiplicatively.

Once again, we can provide an independence-based formulation for this decomposition:

Definition 22.12

GA-independence

Let $Z_1, \ldots, Z_k$ be (not necessarily disjoint) subsets of V. We say that $Z_1, \ldots, Z_k$ are generalized additively independent (GA-independent) in $\prec$ if, for any lotteries π_1, π_2 that have the same marginals on all Z_i, we have that π_1 and π_2 are indifferent under $\prec$. ∎

This definition is identical to that of additive independence (definition 22.8), with the exception that the subsets $Z_1, \ldots, Z_k$ are not necessarily mutually exclusive nor exhaustive. Thus, this definition allows us to consider cases where our preferences between two distributions depend only on some arbitrary set of marginals. It is also not hard to show that GA-independence subsumes CA-independence (see exercise 22.6).

Satisfyingly, a factorization theorem analogous to theorem 22.2 holds for GA-independence:

Theorem 22.6

Let $Z_1, \ldots, Z_k$ be (not necessarily disjoint) subsets of V, and let $\prec, U$ be a corresponding pair of a preference ordering and a utility function. Then $Z_1, \ldots, Z_k$ are GA-independent in $\prec$ if and only if U can be written as:

$$U(V) = \sum_{i=1}^{k} U_i(Z_i). \tag{22.10}$$

Thus, the set of possible factorizations associated with GA-independence strictly subsumes the set of factorizations associated with CA-independence. For example, using GA-independence, we can obtain a factorization $U(X, Y, Z) = U_1(X, Y) + U_2(Y, Z) + U_3(X, Z)$. The Markov network associated with this factorization is a full clique over X, Y, Z, and therefore no CA-independencies hold for this utility function. Overall, GA-independence provides a rich and natural language for encoding complex utility functions (see, for example, box 22.A).

Box 22.A — Case Study: Prenatal Diagnosis. *An important problem involving utilities arises in the domain of* prenatal diagnosis, *where the goal is to detect chromosomal abnormalities present in a fetus in the early stages of the pregnancy. There are several tests available to diagnose these diseases. These tests have different rates of false negatives and false positives, costs, and health risks. The task is to decide which tests to conduct on a particular patient. This task is quite difficult. The patient's risk for having a child with a serious disease depends on the mother's age, child's sex and race, and the family history. Some tests are not very accurate; others carry a significant risk of inducing miscarriages. Both a miscarriage (spontaneous abortion or SAB) and an elective termination of the pregnancy (induced abortion or IAB) can affect the woman's chances of conceiving again.*

Box 23.A describes a decision-making system called PANDA (Norman et al. 1998) for assisting the parents in deciding on a course of action for prenatal testing. *The PANDA system requires that we have a utility model for the different outcomes that can arise as part of this process. Note that, unlike for probabilistic models, we cannot simply construct a single utility model that applies to all patients. Different patients will typically have very different preferences regarding these outcomes, and certainly regarding lotteries over them. Interestingly, the standard protocol (and the one followed by many health insurance companies), which recommends prenatal diagnosis (under normal circumstances) only for women over the age of thirty-five, was selected so that the risk (probability) of miscarriage is equal to that of having a Down syndrome baby. Thus, this recommendation essentially assumes not only that all women have the same utility function, but also that they have equal utility for these two events.*

The outcomes in this domain have many attributes, such as the inconvenience and expense of fairly invasive testing, the disease status of the fetus, the possibility of test-induced miscarriage, knowledge of the status of the fetus, and future successful pregnancy. Specifically, the utility could be viewed as a function of five attributes: pregnancy loss L, with domain {no loss, miscarriage, elective termination}; Down status D of the fetus, with domain: {normal, Down}; mother's knowledge K, with domain {none, accurate, inaccurate}; future pregnancy F, with domain {yes, no}; type of test T with domain {none, CVS, amnio}. An outcome is an assignment of values to all the attributes. For example, $\langle$no loss, normal, none, yes, none$\rangle$ is one possible outcomes. It represents the situation in which the fetus is not affected by Down syndrome, the patient decides not to take any tests (as a consequence, she is unaware of the Down status of the fetus until the end of the pregnancy), the pregnancy results in normal birth, and there is a future pregnancy. Another outcome, $\langle$miscarriage, normal, accurate, no, CVS$\rangle$ represents a situation where the patient decides to undergo the CVS test. The test result correctly asserts that the fetus is not affected by Down syndrome. However, a miscarriage occurs as a side effect of the procedure, and there is no future pregnancy. Our decision-making situation involves comparing lotteries involving complex (and emotionally difficult) outcomes such as these.

In this domain, we have three ternary attributes and two binary ones, so the total number of outcomes is 108. Even if we remove outcomes that have probability zero (or are very unlikely), a

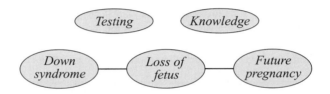

Figure 22.A.1 — Typical utility function decomposition for prenatal diagnosis

large number of outcomes remain. In order to perform utility elicitation, we must assign a numerical utility to each one. A standard utility elicitation process such as standard gamble involves a fairly large number of comparisons for each outcome. Such a process is clearly infeasible in this case.

However, in this domain, many of the utility functions elicited from patients decompose additively in natural ways. For example, as shown by Chajewska and Koller (2000), many patients have utility functions where invasive testing (T) and knowledge (K) are additively independent of other attributes; pregnancy loss (L) is correlated both with Down syndrome (D) and with future pregnancy (F), but there is no direct interaction between Down syndrome and future pregnancy. Thus, for example, one common decomposition is $U_1(T) + U_2(K) + U_3(D, L) + U_4(L, F)$, as encoded in the Markov network of figure 22.A.1.

Box 22.B — Case Study: Utility Elicitation in Medical Diagnosis. *In box 3.D we described the Pathfinder system, designed to assist a pathologist in diagnosing lymph-node diseases. The Pathfinder system was purely probabilistic — it produced only a probability distribution over possible diagnoses. However, the performance evaluation of the Pathfinder system accounted for the implications of a correct or incorrect diagnosis on the patient's utility. For example, if the patient has a viral infection and is diagnosed as having a bacterial infection, the consequences are not so severe: the patient may take antibiotics unnecessarily for a few days or weeks. On the other hand, if the patient has Hodgkin's disease and is incorrectly diagnosed as having a viral infection, the consequences — such as delaying chemotherapy — may be lethal. Thus, to evaluate more accurately the implications of Pathfinder's performance, a utility model was constructed that assigned, for every pair of diseases d, d', a utility value $u_{d,d'}$, which denotes the patient's utility for having the disease d and being diagnosed with disease d'.*

We might be tempted to evaluate the system's performance on a particular case by computing $u_{d^*,d} - u_{d^*,d^*}$, where d^* is the true diagnosis, and d is the most likely diagnosis produced by the system. However, this metric ignores the important distinction between the quality of the decision *and the quality of the* outcome: *A bad decision is one that is not optimal relative to the agent's state of knowledge, whereas a bad outcome can arise simply because the agent is unlucky. In this case, the set of observations may suggest (even to the most knowledgeable expert) that one disease is the most likely, even when another is actually the case. Thus, a better metric is the* inferential

inferential loss

loss — the difference in the expected utility between the gold standard *distribution produced by an expert and the distribution produced by the system, given exactly the same set of observations.*

Estimating the utility values $u_{d,d'}$ is a nontrivial task. One complication arises from the fact that this situation involves outcomes whose consequences are fairly mild, and others that involve a significant risk of morbidity or mortality. Putting these on a single scale is quite challenging. The approach taken in Pathfinder is to convert all utilities to the micromort scale — a one-in-a-million chance of death. For severe outcomes (such as Hodgkins disease), one can ask the patient what probability of immediate, painless death he would be willing to accept in order to avoid both the disease d and the (possibly incorrect) diagnosis d'. For mild outcomes, where the micromort equivalent may be too low to evaluate reliably, utilities were elicited in terms of monetary equivalents — for example, how much the patient would be willing to pay to avoid taking antibiotics for two weeks. At this end of the spectrum, the "conversion" between micromorts and dollars is fairly linear (see section 22.3.2), and so the resulting dollar amounts can be converted into micromorts, putting these utilities on the same scale as that of severe outcomes.

The number of distinct utilities that need to be elicited even in this simple setting is impractically large: with sixty diseases, the number of utilities is $60^2 = 3,600$. Even aggregating diseases that have similar treatments and prognoses, the number of utilities is $36^2 = 1,296$. However, utility independence can be used to decompose the outcomes into independent factors, such as the disutility of a disease d when correctly treated, the disutility of delaying the appropriate treatment, and the disutility of undergoing an unnecessary treatment. This decomposition reduced the number of assessments by 80 percent, allowing the entire process of utility assessment to be performed in approximately sixty hours.

The Pathfinder IV system (based on a full Bayesian network) resulted in a mean inferential loss of 16 micromorts, as compared to 340 micromorts for the Pathfinder III system (based on a naive Bayes model). At a rate of $20/micromort (the rate elicited in the 1980s), the improvement in the expected utility of Pathfinder IV over Pathfinder III is equivalent to around $6,000 per case.

22.5 Summary

In this chapter, we discussed the use of probabilistic models within the context of a decision task. The key new element one needs to introduce in this context is some representation of the preferences of the agent (the decision maker). Under certain assumptions about these preferences, one can show that the agent, whether implicitly or explicitly, must be following the principle of maximum expected utility, relative to some utility function. It is important to note that the assumptions required for this result are controversial, and that they do not necessarily hold for human decision makers. Indeed, there are many examples where human decision makers do not obey the principle of maximum expected utility. Much work has been devoted to developing other precepts of rational decision making that better match human decision making. Most of this study has taken place in fields such as economics, psychology, or philosophy. Much work remains to be done on evaluating the usefulness of these ideas in the context of automated decision-making systems and on developing computational methods that allow them to be used in complex scenarios such as the ones we discuss in this book.

In general, an agent's utility function can involve multiple attributes of the state of the world. In principle, the complexity of the utility function representation grows exponentially with the number of attributes on which the utility function depends. Several representations have been developed that assume some structure in the utility function and exploit it for reducing the number of parameters required to represent the utility function.

Perhaps the biggest challenge in this setting is that of acquiring an agent's utility functions. Unlike the case of probability distributions, where an expert can generally provide a single model that applies to an entire population, an agent's utility function is often highly personal and even idiosyncratic. Moreover, the introspection required for a user to understand her preferences and quantify them is a time-consuming and even distressing process. Thus, there is significant value in developing methods that speed up the process of utility elicitation.

A key step in that direction is in learning better models of utility functions. Here, we can attempt to learn a model for the structure of the utility function (for example, its decomposition), or for the parameters that characterize it. We can also attempt to learn richer models that capture the dependence of the utility function on the user's background and perhaps even its evolution over time. Another useful direction is to try to learn aspects of the agent's utility function by observing previous decisions that he made.

These models can be viewed as narrowing down the space of possibilities for the agent's utility function, allowing it to be elicited using fewer questions. One can also try to develop algorithms that intelligently reduce the number of utility elicitation questions that one needs to ask in order to make good decisions. It may be hoped that a combination of these techniques will make utility-based decision making a usable component in our toolbox.

22.6   Relevant Literature

Pascal's wager

The principle of maximum expected utility dates back at least to the seventeenth century, where it played a role in the famous *Pascal's wager* (Arnauld and Nicole 1662), a decision-theoretic analysis concerning the existence of God. Bernoulli (1738), analyzing the *St. Petersburg paradox*, made the distinction between monetary rewards and utilities. Bentham (1789) first proposed the idea that all outcomes should be reduced to numerical utilities.

Ramsey (1931) was the first to provide a formal derivation of numerical utilities from preferences. The axiomatic derivation described in this chapter is due to von Neumann and Morgenstern (1944). A Bayesian approach to decision theory was developed by Ramsey (1931); de Finetti (1937); Good (1950); Savage (1954). Ramsey and Savage both defined axioms that provide a simultaneous justification for both probabilities and utilities, in contrast to the axioms of von Neumann and Morgenstern, that take probabilities as given. These axioms motivate the Bayesian approach to probabilities in a decision-theoretic setting. The book by Kreps (1988) provides a good review of the topic.

The principle of maximum expected utility has also been the topic of significant criticism, both on normative and descriptive grounds. For example, Kahneman and Tversky, in a long series of papers, demonstrate that human behavior is often inconsistent with the principles of rational decision making under uncertainty for any utility function (see, for example, Kahneman, Slovic, and Tversky 1982). Among other things, Tversky and Kahneman show that people commonly use heuristics in their probability assessments that simplify the problem but often lead to serious errors. Specifically, they often pay disproportionate attention to low-probability events and treat high-probability events as though they were less likely than they actually are.

Motivated both by normative limitations in the MEU principle and by apparent inconsistencies between the MEU principle and human behavior, several researchers have proposed alternative criteria for optimal decision making. For example, Savage (1951) proposed the *minimax risk* criterion, which asserts that we should associate with each outcome not only some utility value but also a regret value. This approach was later refined by Loomes and Sugden (1982) and Bell (1982), who show how regret theory can be used to explain such apparently irrational behaviors as gambling on negative-expected-value lotteries or buying costly insurance.

A discussion of utility functions exhibited by human decision makers can be found in von Winterfeldt and Edwards (1986). Howard (1977) provides an extensive discussion of attitudes toward risk and defines the notions of risk-averse, risk-seeking, and risk-neutral behaviors. Howard (1989) proposes the notion of micromorts for eliciting utilities regarding human life. The Pathfinder system is one of the first automated medical diagnosis systems to use a carefully constructed utility function; a description of the system, and of the process used to construct the utility function, can be found in Heckerman (1990) and Heckerman, Horvitz, and Nathwani (1992).

The basic framework of multiattribute utility theory is presented in detail in the seminal book of Keeney and Raiffa (1976). These ideas were introduced into the AI literature by Wellman (1985). The notion of generalized additive independence (GAI), under the name *interdependent value additivity*, was proposed by Fishburn (1967, 1970), who also provided the conditions under which a GAI model provides an accurate representation of a utility function. The idea of using a graphical model to represent utility decomposition properties was introduced by Wellman and Doyle (1992). The rigorous development was performed by Bacchus and Grove (1995), who also proposed the idea of GAI-networks. These ideas were subsequently extended in various ways by La Mura and Shoham (1999) and Boutilier, Bacchus, and Brafman (2001).

Much work has been done on the problem of utility elicitation. The standard gamble method was first proposed by von Neumann and Morgenstern (1947), based directly on their axioms for utility theory. Time trade-off was proposed by Torrance, Thomas, and Sackett (1972). The visual analog scale dates back at least to the 1970s (Patrick et al. 1973); see Drummond et al. (1997) for a detailed presentation. Chajewska (2002) reviews these different methods and some of the documented difficulties with them. She also provides a fairly recent overview of different approaches to utility elicitation.

There has been some work on eliciting the structure of decomposed utility functions from users (Keeney and Raiffa 1976; Anderson 1974, 1976). However, most often a simple (for example, fully additive) structure is selected, and the parameters are estimated using least-squares regression from elicited utilities of full outcomes. Chajewska and Koller (2000) show how the problem of inferring the decomposition of a utility function can be viewed as a Bayesian model selection problem and solved using techniques along the lines of chapter 18. Their work is based on the idea of explicitly representing an explicit probabilistic model over utility parameters, as proposed by Jimison et al. (1992). The prenatal diagnosis example is taken from Chajewska and Koller (2000), based on data from Kuppermann et al. (1997).

Several authors (for example, Heckerman and Jimison 1989; Poh and Horvitz 2003; Ha and Haddawy 1997, 1999; Chajewska et al. 2000; Boutilier 2002; Braziunas and Boutilier 2005) have proposed that model refinement, including refinement of utility assessments, should be viewed in terms of optimizing expected value of information (see section 23.7). In general, it can be shown that a full utility function need not be elicited to make optimal decisions, and that close-to-optimal decisions can be made after a small number of utility elicitation queries.

22.7   Exercises

Exercise 22.1

Complete the proof of theorem 22.1. In particular, let $U(s) := p$, as defined in equation (22.7), be our utility assignment for outcomes. Show that, if we use the MEU principle for selecting between two lotteries, the resulting preference over lotteries is equivalent to $\prec$. (Hint: Do not forget to address the case of compound lotteries.)

Exercise 22.2

Prove proposition 22.4.

Exercise 22.3

Prove lemma 22.1.

Exercise 22.4

Complete the proof of theorem 22.4.

Exercise 22.5★

Prove theorem 22.5. (Hint: Use exercise 4.11.)

Exercise 22.6★

Prove the following result without using the factorization properties of U. Let $\boldsymbol{X}, \boldsymbol{Y}, \boldsymbol{Z}$ be a disjoint partition of $\boldsymbol{V}$. Then $\boldsymbol{X}, \boldsymbol{Z}$ and $\boldsymbol{Y}, \boldsymbol{Z}$ are GA-independent in $\prec$ if and only if $\boldsymbol{X}$ and $\boldsymbol{Y}$ are CA-independent given $\boldsymbol{Z}$ in $\prec$.

This result shows that CA-independence and GA-independence are equivalent *over the scope of independence assertions to which CA-independence applies* (those involving disjoint partitions of $\boldsymbol{V}$).

Exercise 22.7

Consider the problem of computing the optimal action for an agent whose utility function we are uncertain about. In particular, assume that, rather than a known utility function over outcomes $\mathcal{O}$, we have a probability density function $P(U)$, which assigns a density for each possible utility function $U : \mathcal{O} \mapsto \mathbb{R}$.

a. What is the expected utility for a given action a, taking an expectation both over the outcomes of π_a, and over the possible utility functions that the agent might have?

b. Use your answer to provide an efficient computation of the optimal action for the agent.

Exercise 22.8★

As we discussed, different people have different utility functions. Consider the problem of learning a probability distribution over the utility functions found in a population. Assume that we have a set of samples $U[1], \ldots, U[M]$ of users from a population, where for each user m we have elicited a utility function $U[m] : Val(\boldsymbol{V}) \mapsto \mathbb{R}$.

a. Assume that we want to model our utility function as in equation (22.10). We want to use the same factorization for all users, but where different users have different subutility functions; that is, $U_i[m]$ and $U_i[m']$ are not the same. Moreover, the elicited values $U(\boldsymbol{v})[m]$ are noisy, so that they may not decompose exactly as in equation (22.10). We can model the actual elicited values for a given user as the sum in equation (22.10) plus Gaussian noise.

  Formulate the distribution over the utility functions in the population as a linear Gaussian graphical model. Using the techniques we learned earlier in this book, provide a learning algorithm for the parameters in this model.

b. Now, assume that we allow different users in the population to have one of several different factorizations of their utility functions. Show how you can extend your graphical model and learning algorithm accordingly. Show how your model allows you to infer which factorization a user is likely to have.

23 *Structured Decision Problems*

In the previous chapter, we described the basic principle of decision making under uncertainty — maximizing expected utility. However, our definition for a decision-making problem was completely abstract; it defined a decision problem in terms of a set of abstract states and a set of abstract actions. Yet, our overarching theme in this book has been the observation that the world is structured, and that we can obtain both representational and computational efficiency by exploiting this structure. In this chapter, we discuss structured representations for decision-making problems and algorithms that exploit this structure when addressing the computational task of finding the decision that maximizes the expected utility.

We begin by describing *decision trees* — a simple yet intuitive representation that describes a decision-making situation in terms of the scenarios that the decision maker might encounter. This representation, unfortunately, scales up only to fairly small decision tasks; still, it provides a useful basis for much of the later development. We describe *influence diagrams*, which extend Bayesian networks by introducing decisions and utilities. We then discuss algorithmic techniques for solving and simplifying influence diagrams. Finally, we discuss the concept of *value of information*, which is very naturally encoded within the influence diagram framework.

23.1 Decision Trees

23.1.1 Representation

A decision tree is a representation of the different scenarios that might be encountered by the decision maker in the context of a particular decision problem. A decision tree has two types of internal nodes (denoted t-nodes to distinguish them from nodes in a graphical model) — one set encoding decision points of the agent, and the other set encoding decisions of nature. The outgoing edges at an agent's t-node correspond to different decisions that the agent might make. The outgoing edges at one of nature's t-nodes correspond to random choices that are made by nature. The leaves of the tree are associated with outcomes, and they are annotated with the agent's utility for that outcome.

Definition 23.1

decision tree

t-node

A decision tree $\mathcal{T}$ is a rooted tree with a set of internal t-nodes $\mathcal{V}$ and leaves $\mathcal{V}_L$. The set $\mathcal{V}$ is partitioned into two disjoint sets — agent t-nodes $\mathcal{V}_A$ and nature t-nodes $\mathcal{V}_N$. Each t-node has some set of choices $\mathcal{C}[v]$, associated with its outgoing edges. We let $succ(v, c)$ denote the child of v reached via the edge labeled with c. Each of nature's t-nodes v is associated with a probability

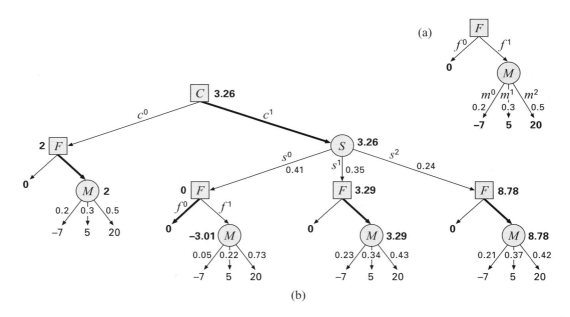

(b)

Figure 23.1 **Decision trees for the** Entrepreneur **example.** (a) one-stage scenario; (b) two-stage scenario, with the solution (optimal strategy) denoted using thicker edges.

distribution P_v over $\mathcal{C}[v]$. Each leaf $v \in \mathcal{V}_L$ in the tree is annotated with a numerical value $U(v)$ corresponding to the agent's utility for reaching that leaf. ∎

Most simply, in our basic decision-making scenario of definition 22.2, a lottery ℓ induces a two-layer tree. The root is an agent t-node v, and it has an outgoing edge for each possible action $a \in \mathcal{A}$, leading to some child $succ(v, a)$. Each node $succ(v, a)$ is a nature t-node; its children are leaves in the tree, with one leaf for each outcome in $\mathcal{O}$ for which $\ell_a(o) > 0$; the corresponding edge is labeled with the probability $\ell_a(o)$. The leaf associated with some outcome o is annotated with $U(o)$. Most simply, in our basic Entrepreneur scenario of example 22.3, the corresponding decision tree would be as shown in figure 23.1a. Note that if the agent decides not to found the company, there is no dependence of the outcome on the market demand, and the agent simply gets a utility of 0.

 The decision-tree representation allows us to encode decision scenarios in a way that reveals much more of their internal structure than the abstract setting of outcomes and utilities. In particular, it allows us to encode explicitly sequential decision settings, where the agent makes several decisions; it also allows us to encode information that is available to the agent at the time a decision must be made.

Example 23.1

Consider an extension of our basic Entrepreneur *example where the entrepreneur has the opportunity to conduct a survey on the demand for widgets before deciding whether to found the company. Thus, the agent now has a sequence of two decisions: the first is whether to conduct the survey, and the second is whether to found the company. If the agent conducts the survey, he obtains informa-*

tion about its outcome, which can be one of three values: a negative reaction, s^0, indicating almost no hope of widget sales; a neutral reaction, s^1, indicating some hope of sales; and an enthusiastic reaction, s^2, indicating a lot of potential demand. The probability distribution over the market demand is different for the different outcomes of the survey. If the agent conducts the survey, his decision on whether to found the company can depend on the outcome.

The decision tree is shown in figure 23.1b. At the root, the agent decides whether to conduct the survey (c^1) or not (c^0). If he does not conduct the survey, the next t-node is another decision by the agent, where he decides whether to found the company (f^1) or not (f^0). If the agent decides to found the company, nature decides on the market demand for widgets, which determines the final outcome. The situation if the agent decides to conduct the survey is more complex. Nature then probabilistically chooses the value of the survey. For each choice, the agent has a t-node where he gets to decide whether he founds the company or not. If he does, nature gets to decide on the distribution of market demand for widgets, which is different for different outcomes of the survey.∎

strategy

We can encode the agent's overall behavior in a decision problem encoded as a decision tree as a *strategy*. There are several possible definitions of a strategy. One that is simple and suitable for our purposes is a mapping from agent t-nodes to possible choices at that t-node.

Definition 23.2

decision-tree
strategy

A decision-tree strategy σ specifies, for each $v \in \mathcal{V}_A$, one of the choices labeling its outgoing edges. ∎

For example, in the decision tree of figure 23.1b, a strategy has to designate an action for the agent t-node, labeled C, and the four agent t-nodes, labeled F. One possible strategy is illustrated by the thick lines in the figures.

Decision trees provide a structured representation for complex decision problems, potentially involving multiple decisions, taken in sequence, and interleaved with choices of nature. However, they are still instances of the abstract framework defined in definition 22.2. Specifically, the outcomes are the leaves in the tree, each of which is annotated with a utility; the set of agent actions is the set of all strategies; and the probabilistic outcome model is the distribution over leaves induced by nature's random choices given a strategy (action) for the agent.

23.1.2 Backward Induction Algorithm

As in the abstract decision-making setting, our goal is to select the strategy that maximizes the agent's expected utility. This computational task, for the decision-tree representation, can be solved using a straightforward tree-traversal algorithm. This approach is an instance of an approach called *backward induction* in the game-theoretic and economic literature, and the *Expectimax* algorithm in the artificial intelligence literature.

backward
induction

Expectimax

The algorithm proceeds from the leaves of the tree upward, computing the maximum expected utility MEU_v achievable by the agent at each t-node v in the tree — his expected utility if he plays the optimal strategy from that point on. At a leaf v, MEU_v is simply the utility $U(v)$ associated with that leaf's outcome. Now, consider an internal t-node v for whose children we have already computed the MEU. If v belongs to nature, the expected utility accruing to the agent if v is reached is simply the weighted average of the expected utilities at each of v's children, where the weighted average is taken relative to the distribution defined by nature over v's children. If v belongs to the agent, the agent has the ability to select the action at v.

Algorithm 23.1 Finding the MEU strategy in a decision tree

     Procedure MEU-for-Decision-Trees (
        $\mathcal{T}$ // Decision tree
)
1    $L \leftarrow Leaves(\mathcal{T})$
2   **for** each node $v \in L$
3   Remove v from L
4   Add v's parents to L
5   **if** v is a leaf **then**
6          $\text{MEU}_v \leftarrow U(v)$
7   **else if** v belongs to nature **then**
8          $\text{MEU}_v \leftarrow \sum_{c \in \mathcal{C}[v]} P_v(c)\text{MEU}_{succ(v,c)}$
9   **else**   // v belongs to the Agent
10       $\sigma(v) \leftarrow \arg\max_{c \in \mathcal{C}[v]} \text{MEU}_{succ(v,c)}$
11      $\text{MEU}_v \leftarrow \text{MEU}_{succ(v,succ(v,))}$
12   **return** (σ)

The optimal action for the agent is the one leading to the child whose MEU is largest, and the MEU accruing to the agent is the MEU associated with that child. The algorithm is shown in algorithm 23.1.

23.2 Influence Diagrams

 The decision-tree representation is a significant improvement over representing the problem as a set of abstract outcomes; however, much of the structure of the problem is still not made explicit. For example, in our simple Entrepreneur scenario, the agent's utility if he founds the company depends only on the market demand M, and not on the results of the survey S. In the decision tree, however, the utility values appear in four separate subtrees: one for each value of the S variable, and one for the subtree where the survey is not performed. An examination of the utility values shows that they are, indeed, identical, but this is not apparent from the structure of the tree.

The tree also loses a subtler structure, which cannot be easily discerned by an examination of the parameters. The tree contains four nodes that encode a probability distribution over the values of the market demand M. These four distributions are different. We can presume that neither the survey nor the agent's decision has an effect on the market demand itself. The reason for the change in the distribution presumably arises from the effect of conditioning the distribution on different observations (or no observation) on the survey variable S. In other words, these distributions represent $P(M \mid s^0)$, $P(M \mid s^1)$, $P(M \mid s^2)$, and $P(M)$ (in the branch where the survey was not performed). These interactions between these different parameters are obscured by the decision-tree representation.

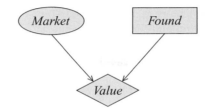

Figure 23.2 **Influence diagram** $\mathcal{I}_F$ **for the basic** Entrepreneur **example**

23.2.1 Basic Representation

influence
diagram

An alternative representation is the *influence diagram* (sometimes also called a *decision network*), a natural extension of the Bayesian network framework. It encodes the decision scenario via a set of variables, each of which takes on values in some space. Some of the variables are random variables, as we have seen so far, and their values are selected by nature using some probabilistic model. Others are under the control of the agent, and their value reflects a choice made by him. Finally, we also have numerically valued variables encoding the agent's utility.

This type of model can be encoded graphically, using a directed acyclic graph containing three types of nodes — corresponding to *chance variables*, *decision variables*, and *utility variables*. These different node types are represented as ovals, rectangles, and diamonds, respectively. An influence diagram $\mathcal{I}$ is a directed acyclic graph over these nodes, such that the utility nodes have no children.

Example 23.2

The influence diagram $\mathcal{I}_F$ for our entrepreneur example is shown in figure 23.2. The utility variable V_E encodes the utility of the entrepreneur's earnings, which are a deterministic function of the utility variable's parents. This function specifies the agent's real-valued utility for each combination of the parent nodes; in this case, the utility is a function from $Val(M) \times Val(F)$ to $\mathbb{R}$. We can represent this function as a table:

	m^0	m^1	m^2
f^1	-7	5	20
f^0	0	0	$0,$

where f^1 represents the decision to found the company and f^0 the decision not to do so. The CPD for the M node is:

m^0	m^1	m^2
0.5	0.3	$0.2.$

∎

chance variable

decision variable

More formally, in an influence diagram, the world in which the agent acts is represented by the set $\mathcal{X}$ of *chance variables*, and by a set $\mathcal{D}$ of *decision variables*. Chance variables are those whose values are chosen by nature. The decision variables are variables whose values the agent gets to choose. Each variable $V \in \mathcal{X} \cup \mathcal{D}$ has a finite domain $Val(V)$ of possible values. We can place this representation within the context of the abstract framework of definition 22.2: The possible actions $\mathcal{A}$ are all of the possible assignments $Val(\mathcal{D})$; the possible outcomes are all of the joint assignments in $Val(\mathcal{X} \cup \mathcal{D})$. Thus, this framework provides a factored representation of both the action and the outcome space.

We can also decompose the agent's utility function. A standard decomposition (see discussion in section 22.4) is as a linear sum of terms, each of which represents a certain component of the

utility variable   agent's utility. More precisely, we have a set of *utility variables* $\mathcal{U}$, which take on real numbers as values. The agent's final utility is the sum of the value of V for all $V \in \mathcal{U}$.

Let $\mathcal{Z}$ be the set of all variables in the network — chance, decision, and utility variables. We

outcome   expand the notion of *outcome* to encompass a full assignment to $\mathcal{Z}$, which we denote as ζ.

The parents of a chance variable X represent, as usual, the direct influences on the choice of X's value. Note that the parents of X can be both other chance variables as well as decision variables, but they cannot be utility variables, since we assumed that utility nodes have no children. Each chance node X is associated with a CPD, which represents $P(X \mid \mathrm{Pa}_X)$.

The parents of a utility variable V represent the set of variables on which the utility V depends. The value of a utility variable V is a deterministic function of the values of Pa_V; we use $V(\boldsymbol{w})$ to denote the value that node V takes when $\mathrm{Pa}_V = \boldsymbol{w}$. Note that, as for any deterministic function, we can also view V as defining a CPD, where for each parent assignment, some value gets probability 1. When convenient, we will abuse notation and interpret a utility node as defining a factor.

Summarizing, we have the following definition:

Definition 23.3   *An influence diagram $\mathcal{I}$ over $\mathcal{Z}$ is a directed acyclic graph whose nodes correspond to $\mathcal{Z}$, and*

influence   *where nodes corresponding to utility variables have no children. Each chance variable $X \in \mathcal{X}$ is*

diagram   *associated with a CPD $P(X \mid \mathrm{Pa}_X)$. Each utility variable $V \in \mathcal{U}$ is associated with a deterministic function $V(\mathrm{Pa}_V)$.* ∎

23.2.2   Decision Rules

So far, we have not discussed the semantics of the decision node. For a decision variable $D \in \mathcal{D}$, Pa_D is the set of variables whose values the agent knows when he chooses a value for

information edge D. The edges incoming into a decision variable are often called *information edges*.

Example 23.3   *Let us return to the setting of example 23.1. Here, we have the chance variable M that represents the market demand, and the chance variable S that represents the results of the survey. The variable S has the values s^0, s^1, s^2, and an additional value $s^\perp$, denoting that the survey was not taken. This additional value is needed in this case, because we allow the agent's decision to depend on the value of S, and therefore we need to allow some value for this variable when the survey is not taken. The variable S has two parents, C and M. We have that $P(s^\perp \mid c^0, m) = 1$, for any value of m. In the case c^1, the probabilities over values of S are:*

	s^0	s^1	s^2
m^0	0.6	0.3	0.1
m^1	0.3	0.4	0.3
m^2	0.1	0.4	0.5

The entrepreneur knows the result of the survey before making his decision whether to found the company. Thus, there is an edge between S and his decision F. We also assume that conducting

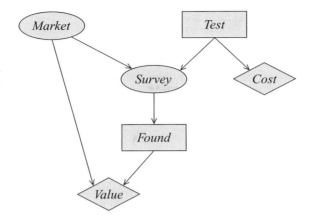

Figure 23.3 **Influence diagram** $\mathcal{I}_{F,C}$ **for** Entrepreneur **example with market survey**

the survey has some cost, so that we have an additional utility node V_S, with the parent C; V_S takes on the value -1 if $C = c^1$ and 0 otherwise. The resulting influence diagram $\mathcal{I}_{F,C}$ is shown in figure 23.3. ∎

The influence diagram representation captures the causal structure of the problem and its parameterization in a much more natural way than the decision tree. It is clear in the influence diagram that S depends on M, that M is parameterized via a simple (unconditional) prior distribution, and so on.

The choice that the agent makes for a decision variable D can be contingent only on the values of its parents. More precisely, in any trajectory through the decision scenario, the agent will encounter D in some particular *information states*, where each information state is an assignment of values to Pa_D. An agent's strategy for D must tell the agent how to act at D, at each of these information states.

information state

Example 23.4

In example 23.3, for instance, the agent's strategy must tell him whether to found the company or not in each possible scenario he may encounter; the agent's information state at this decision is defined by the possible values of the decision variable C and the survey variable S. The agent must therefore decide whether to found the company in four different information states: if he chose not to conduct the survey, and in each of the three different possible outcomes of the survey. ∎

A *decision rule* tells the agent how to act in each possible information state. Thus, the agent is choosing a local conditional model for the decision variable D. In effect, the agent has the ability to choose a CPD for D.

Definition 23.4

decision rule

deterministic
decision rule

complete strategy

A decision rule δ_D for a decision variable D is a conditional probability $P(D \mid \text{Pa}_D)$ — a function that maps each instantiation pa_D of Pa_D to a probability distribution δ_D over $Val(D)$. A decision rule is deterministic *if each probability distribution $\delta_D(D \mid \text{pa}_D)$ assigns nonzero probability to exactly one value of D. A complete assignment σ of decision rules to every decision $D \in \mathcal{D}$ is called a* complete strategy; *we use σ_D to denote the decision rule at D.* ∎

Example 23.5

A decision rule for C is simply a distribution over its two values. A decision rule for F must define, for every value of C, and for every value $s^0, s^1, s^2, s^\perp$ of S, a probability distribution over values of F. Note, however, that there is a deterministic relationship between C and S, so that many of the combinations are inconsistent (for example, c^1 and $s^\perp$, or c^0 and s^1). For example, in the case c^1, s^1, one possible decision rule for the agent is f^0 with probability 0.7 and f^1 with probability 0.3. ■

As we will see, in the case of single-agent decision making, one can always choose an optimal deterministic strategy for the agent. However, it is useful to view a strategy as an assignment of CPDs to the decision variables. Indeed, in this case, the parents of a decision node have the same semantics as the parents of a chance node: the agent's strategy can depend only on the values of the parent variables. Moreover, randomized decision rules will turn out to be a useful concept in some of our constructions that follow. In the common case of deterministic decision rules, which pick a single action $d \in Val(D)$ for each assignment $\boldsymbol{w} \in Val(\mathrm{Pa}_D)$, we sometimes abuse notation and use δ_D to refer to the decision-rule function, in which case $\delta_D(\boldsymbol{w})$ denotes the single action d that has probability 1 given the parent assignment $\boldsymbol{w}$.

23.2.3 Time and Recall

intervention

Unlike a Bayesian network, an influence diagram has an implicit causal semantics. One assumes that the agent can *intervene* at a decision variable D by selecting its value. This intervention will affect the values of variables downstream from D. By choosing a decision rule, the agent determines how he will intervene in the system in different situations.

The acyclicity assumption for influence diagrams, combined with the use of information edges, ensures that an agent cannot observe a variable that his action affects. Thus, acyclicity implies that the network respects some basic causal constraints. In the case of multiple decisions, we often want to impose additional constraints on the network structure. In many cases, one assumes that the decisions in the network are all made by a single agent in some sequence over time; in this case, we have a total ordering $\prec$ on $\mathcal{D}$. An additional assumption that is often made in this case is that the agent does not forget his previous decisions or information it once had. This assumption is typically called the *perfect recall* assumption (or sometime the *no forgetting* assumption), formally defined as follows:

Definition 23.5

temporal ordering

perfect recall

recall edge

An influence diagram $\mathcal{I}$ is said to have a temporal ordering *if there is some total ordering $\prec$ over $\mathcal{D}$, which is consistent with partial ordering imposed by the edges in $\mathcal{I}$. The influence diagram $\mathcal{I}$ satisfies the* perfect recall *assumption relative to $\prec$ if, whenever $D_i \prec D_j$, $\mathrm{Pa}_{D_j} \supset (\mathrm{Pa}_{D_i} \cup \{D_i\})$. The edges from $\mathrm{Pa}_{D_i} \cup \{D_i\}$ to D_j are called* recall edges. ■

Intuitively, a recall edge is an edge from a variable X (chance or decision) to a decision variable D whose presence is implied by the perfect recall assumption. In particular, if D' is a decision that precedes D in the temporal ordering, then we have recall edges $D' \rightarrow D$ and $X \rightarrow D$ for $X \in \mathrm{Pa}_{D'}$. To reduce visual clutter, we often omit recall edges in an influence diagram when the temporal ordering is known. For example, in figure 23.3, we omitted the edge from C to F.

Although the perfect recall assumption appears quite plausible at first glance, there are several arguments against it. First, it is not a suitable model for situations where the "agent" is actually a compound entity, with individual decisions made by different "subagents." For example, our agent might be a large organization, with different members responsible for various decisions. It is also not suitable for cases where an agent might not have the resources (or the desire) to remember an entire history of all previous actions and observations.

☞ **The perfect recall assumption also has significant representational and computational ramifications. The size of the decision rule at a decision node is, in general, exponential in the number of parents of the decision node. In the case of perfect recall, the number of parents grows with every decision, resulting in a very high-dimensional space of possible decision rules for decision variables later in the temporal ordering. This blowup makes computations involving large influence diagrams with perfect recall intractable in many cases.** The computational burden of perfect recall leads us to consider also influence diagrams in which the perfect recall assumption does not hold, also known as *limited memory influence diagrams* (or LIMIDs). In these networks, all information edges must be represented explicitly, since perfect recall is no longer universally true. We return to this topic in section 23.6.

limited memory influence diagram

23.2.4 Semantics and Optimality Criterion

A choice of a decision rule δ_D effectively turns D from a decision variable into a chance variable. Let σ_D be any partial strategy that specifies a decision rule for the decision variables $D \in \boldsymbol{D}$. We can replace each decision variable in $\boldsymbol{D}$ with the CPD defined by its decision rule in σ, resulting in an influence diagram $\mathcal{I}[\sigma]$ whose chance variables are $\mathcal{X} \cup \boldsymbol{D}$ and whose decision variables are $\mathcal{D} - \boldsymbol{D}$. In particular, when σ is a complete strategy, $\mathcal{I}[\sigma]$ is simply a Bayesian network, which we denote by $\mathcal{B}_{\mathcal{I}[\sigma]}$. This Bayesian network defines a probability distribution over possible outcomes ζ.

expected utility

The agent's *expected utility* in this setting is simply:

$$\mathrm{EU}[\mathcal{I}[\sigma]] = \sum_{\zeta} P_{\mathcal{B}_{\mathcal{I}[\sigma]}}(\zeta)U(\zeta) \tag{23.1}$$

where the utility of an outcome is the sum of the individual utility variables in that outcome:

$$U(\zeta) = \sum_{V \in \mathcal{U}} \zeta\langle V \rangle.$$

The linearity of expectation allows us to simplify equation (23.1) by considering each utility variable separately, to obtain:

$$\begin{aligned}
\mathrm{EU}[\mathcal{I}[\sigma]] &= \sum_{V \in \mathcal{U}} \boldsymbol{E}_{\mathcal{B}_{\mathcal{I}[\sigma]}}[V] \\
&= \sum_{V \in \mathcal{U}} \sum_{v \in Val(V)} P_{\mathcal{B}_{\mathcal{I}[\sigma]}}(V = v)v.
\end{aligned}$$

We often drop the subscript $\mathcal{B}_{\mathcal{I}[\sigma]}$ where it is clear from context.

An alternative useful formulation for this expected utility makes explicit the dependence on the factors parameterizing the network:

$$\mathrm{EU}[\mathcal{I}[\sigma]] = \sum_{\mathcal{X} \cup \mathcal{D}} \left[\left(\prod_{X \in \mathcal{X}} P(X \mid \mathrm{Pa}_X) \right) \left(\prod_{D \in \mathcal{D}} \delta_D \right) \left(\sum_{i \,:\, V_i \in \mathcal{U}} V_i \right) \right]. \tag{23.2}$$

The expression inside the summation is constructed as a product of three components. The first is a product of all of the CPD factors in the network; the second is a product of all of the factors corresponding to the decision rules (also viewed as CPDs); and the third is a factor that captures the agent's utility function as a sum of the subutility functions V_i. As a whole, the expression inside the summation is a single factor whose scope is $\mathcal{X} \cup \mathcal{D}$. The value of the entry in the factor corresponding to an assignment o to $\mathcal{X} \cup \mathcal{D}$ is a product of the probability of this outcome (using the decision rules specified by σ) and the utility of this outcome. The summation over this factor is simply the overall expected utility $\mathrm{EU}[\mathcal{I}[\sigma]]$.

Example 23.6

Returning to example 23.3, our outcomes are complete assignments m, c, s, f, u_s, u_f. The agent's utility in such an outcome is $u_s + u_f$. The agent's expected utility given a strategy σ is

$$P_{\mathcal{B}}(V_S = -1) \cdot -1 + P_{\mathcal{B}}(V_S = 0) \cdot 0 +$$
$$P_{\mathcal{B}}(V_E = -7) \cdot -7 + P_{\mathcal{B}}(V_E = 5) \cdot 5 + P_{\mathcal{B}}(V_E = 20) \cdot 20 + P_{\mathcal{B}}(V_E = 0) \cdot 0),$$

where $\mathcal{B} = \mathcal{B}_{\mathcal{I}_{F,C}[\sigma]}$. It is straightforward to verify that the strategy that optimizes the expected utility is: $\delta_C = c^1$; $\delta_F(c^1, s^0) = f^0$, $\delta_F(c^1, s^1) = f^1$, $\delta_F(c^1, s^2) = f^1$. Because the event $C = c^0$ has probability 0 in this strategy, any choice of probability distributions for $\delta_F(c^0, S)$ is optimal. By following the definition, we can compute the overall expected utility for this strategy, which is 3.22, so that $\mathrm{MEU}[\mathcal{I}_{F,C}] = 3.22$. ∎

According to the basic postulate of statistical decision theory, the agent's goal is to maximize his expected utility for a given decision setting. Thus, he should choose the strategy σ that maximizes $\mathrm{EU}[\mathcal{I}[\sigma]]$.

Definition 23.6

MEU strategy

MEU value

An MEU strategy σ^* for an influence diagram $\mathcal{I}$ is one that maximizes $\mathrm{EU}[\mathcal{I}[\sigma]]$. The MEU value $\mathrm{MEU}[\mathcal{I}]$ is $\mathrm{EU}[\mathcal{I}[\sigma^*]]$. ∎

In general, there may be more than one MEU strategy for a given influence diagram, but they all have the same expected utility.

This definition lays out the basic computational task associated with influence diagrams: Given an influence diagram $\mathcal{I}$, our goal is to find the MEU strategy $\mathrm{MEU}[\mathcal{I}]$. Recall that a strategy is an assignment of decision rules to all the decision variables in the network; thus, our goal is to find:

$$\arg \max_{\delta_{D_1}, \ldots, \delta_{D_k}} \mathrm{EU}[\mathcal{I}[\delta_{D_1}, \ldots, \delta_{D_k}]]. \tag{23.3}$$

Each decision rule is itself a complex function, assigning an action (or even a distribution over actions) to each information state. This complex optimization task appears quite daunting at first. Here we present two different ways of tackling it.

---

prenatal
diagnosis

Box 23.A — Case Study: Decision Making for Prenatal Testing. *As we discussed in box 22.A, prenatal diagnosis offers a challenging domain for decision making. It incorporates a sequence of interrelated decisions, each of which has significant effects on variables that determine the patient's preferences. Norman et al. (1998) construct a system called PANDA (which roughly stands* for "Prenatal Testing Decision Analysis"). PANDA uses an influence diagram to model the sequential decision process, the relevant random variables, and the patient's utility. The influence diagram contains a sequence of six decisions: four types of diagnostic test (CVS, triple marker screening, ultrasound, and amniocentesis), as well as early and late termination of the pregnancy. The model focuses on five diseases that are serious, relatively common, diagnosable using prenatal testing, and not readily correctable: Down syndrome, neural-tube defects, cystic fibrosis, sickle-cell anemia, and fragile X mental retardation. The probabilistic component of the network (43 variables) includes predisposing factors that affect the probability of these diseases, and it models the errors in the diagnostic ability of the tests (both false positive and false negative). Utilities were elicited for every patient and placed on a scale of 0–100, where a utility of 100 corresponds to the outcome of a healthy baby with perfect knowledge throughout the course of the pregnancy, and a utility of 0 corresponds to the outcome of both maternal and fetal death.

The strategy space in this model is very complex, since any decision (including a decision to take a test) can depend on the outcome of one or more earlier tests, As a consequence, there are about 1.62×10^{272} different strategies, of which 3.85×10^{38} are "reasonable" relative to a set of constraints. This enormous space of options highlights the importance of using automated methods to guide the decision process.

The system can be applied to different patients who vary both on their predisposing factors and on their utilities. Both the predisposing factors and the utilities give rise to very different strategies. However, a more relevant question is the extent to which the different strategy choices make a difference to the patient's final utility. To provide a reasonable scale for answering this question, the algorithm was applied to select for each patient their best and worst strategy. As an example, for one such patient (a young woman with predisposing factors for sickle-cell anemia), the optimal strategy achieved an expected utility of 98.77 and the worst strategy an expected utility of 87.85, for a difference of 10.92 utility points. Other strategies were then evaluated in terms of the percentage of these 10.92 points that they provided to the patient. For many patients, most of the reasonable strategies performed fairly well, achieving over 99 percent of the utility gap for that patient. However, for some patients, even reasonable strategies gave very poor results. For example, for the patient with sickle-cell anemia, strategies that were selected as optimal for other patients in the study provided her only 65–70 percent of the utility gap. Notably, the "recommended" strategy for women under the age of thirty-five, which is to perform no tests, performed even worse, achieving only 64.7 percent of the utility gap.

Overall, this study demonstrates the importance of personalizing medical decision making to the information and the utility for individual patients.

23.3 Backward Induction in Influence Diagrams

We now turn to the problem of selecting the optimal strategy in an influence diagram. Our first approach to addressing this problem is a fairly simple algorithm that mirrors the backward induction algorithm for decision trees described in section 23.1.2. As we will show, this algorithm can be implemented effectively using the techniques of variable elimination of chapter 9. This algorithm applies only to influence diagrams satisfying the perfect recall assumption, a restriction that has significant computational ramifications.

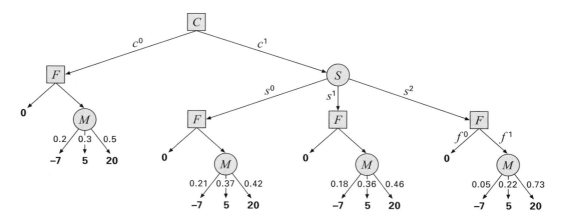

Figure 23.4   Decision tree for the influence diagram $\mathcal{I}_{F,C}$ **in the** Entrepreneur **example.** For clarity, probability zero events are omitted, and edges are labeled only at representative nodes.

23.3.1   Decision Trees for Influence Diagrams

Our starting point for the backward induction algorithm is to view an influence diagram as defining a set of possible trajectories, defined from the perspective of the agent. A trajectory includes both the observations made by the agent and the decisions he makes. The set of possible trajectories can be organized into a decision tree, with a split for every chance variable and every decision variable. We note that this construction gives rise to an exponentially large decision tree. Importantly, we never have to construct this tree explicitly. As we will show, we can use this tree as a conceptual construct, which forms the basis for defining a variable elimination algorithm. The VE algorithm works directly on the influence diagram, never constructing the exponentially large tree.

We begin by illustrating the decision tree construction on a simple example.

Example 23.7

Consider the possible trajectories that might be encountered by our entrepreneur of example 23.3. Initially, he has to decide whether to conduct the survey or not (C). He then gets to observe the value of the survey (S). He then has to decide whether to found the company or not (F). The variable M influences his utility, but he never observes it (at least not in a way that influences any of his decisions). Finally, the utility is selected based on the entire trajectory. We can organize this set of trajectories into a tree, where the first split is on the agent's decision C, the second split (on every branch) is nature's decision regarding the value of S, the third split is on the agent's decision F, and the final split is on M. At each leaf, we place the utility value corresponding to the scenario. Thus, for example, the agent's utility at the leaf of the trajectory c^1, s^1, f^1, m^1 is $V_S(c^1) + V_E(f^1, m^1) = -1 + 5$. The decision tree for this example is the same one shown in figure 23.4.   ∎

Note that the ordering of the nodes in the tree is defined by the agent's observations, not by the topological ordering of the underlying influence diagram. Thus, in this example, S precedes M, despite the fact that, viewed from the perspective of generative causal model, M is "determined by nature" before S.

More generally, we assume (as stated earlier) that the influence diagram satisfies perfect recall relative to some temporal ordering $\prec$ on decisions. Without loss of generality, assume that $D_1 \prec \ldots \prec D_k$. We extend $\prec$ to a partial ordering over $\mathcal{X} \cup \mathcal{D}$ which is consistent with the information edges in the influence diagrams; that is, whenever W is a parent of D for some $D \in \mathcal{D}$ and $W \in \mathcal{X} \cup \mathcal{D}$, we have that $W \prec D$. This ordering is guaranteed to extend the total ordering $\prec$ over $\mathcal{D}$ postulated in definition 23.5, allowing us to abuse notation and use $\prec$ for both.

This partial ordering constrains the orderings that we can use to define the decision tree. Let $\boldsymbol{X}_1$ be the set of variables X such that $X \prec D_1$; these variables are the ones that the agent observes for the first time at decision D_1. More generally, let $\boldsymbol{X}_i$ be those variables X such that $X \prec D_i$ but not $X \prec D_{i-1}$. These variables are the ones that the agent observes for the first time at decision D_i. With the perfect recall assumption, the agent's decision rule at D_i can depend on all of $\boldsymbol{X}_1 \cup \ldots \cup \boldsymbol{X}_i \cup \{D_1, \ldots, D_{i-1}\}$. Let $\boldsymbol{Y}$ be the variables that are not observed prior to any decision. The sets $\boldsymbol{X}_1, \ldots, \boldsymbol{X}_k, \boldsymbol{Y}$ form a disjoint partition of $\mathcal{X}$. We can then define a tree where the first split is on the set of possible assignments $\boldsymbol{x}_1$ to $\boldsymbol{X}_1$, the second is on possible decisions in $Val(D_1)$, and so on, and where the final split is on possible assignments $\boldsymbol{y}$ to $\boldsymbol{Y}$.

The choices at nature's chance moves are associated with probabilities. These probabilities are not the same as the generative probabilities (as reflected in the CPDs in the influence diagrams), but reflect the agent's subjective beliefs in nature's choices given the evidence observed so far.

Example 23.8

Consider the decision tree of figure 23.4, and consider nature's choice for the branch $S = s^1$ at the node corresponding to the trajectory $C = c^1$. The probability that the survey returns s^1 is the marginal probability $\sum_M P(M) \cdot P(s^1 \mid M, c^1)$. Continuing down the same branch, and assuming $F = f^1$, the branching probability for $M = m^1$ is the conditional probability of $M = m^1$ given s^1 (and the two decision variables, although these are irrelevant to this probability).∎

In general, consider a branch down the tree associated with the choices $\boldsymbol{x}_1, d_1, \ldots, \boldsymbol{x}_{i-1}, d_{i-1}$. At this vertex, we have a decision of nature, splitting on possible instantiations $\boldsymbol{x}_i$ to $\boldsymbol{X}_i$. We associate with this vertex a distribution $P(\boldsymbol{x}_i \mid \boldsymbol{x}_1, d_1, \ldots, \boldsymbol{x}_{i-1}, d_{i-1})$.

As written, this probability expression is not well defined, since we have not specified a distribution relative to which it is computed. Specifically, because we do not have a decision rule for the different decision variables in the influence diagram, we do not yet have a fully specified Bayesian network. We can ascribe semantics to this term using the following lemma:

Lemma 23.1

Let $\boldsymbol{x}_1, \ldots, \boldsymbol{x}_{i-1}, d_1, \ldots, d_{i-1}$ be an assignment to $\boldsymbol{X}_1, \ldots, \boldsymbol{X}_{i-1}, D_1, \ldots, D_{i-1}$ respectively. Let σ_1, σ_2 be any two strategies in which $P_{\mathcal{B}_{\mathcal{I}[\sigma_i]}}(\boldsymbol{x}_1, \ldots, \boldsymbol{x}_{i-1}, d_1, \ldots, d_{i-1}) \neq 0$ $(i = 1, 2)$. Then

$$P_{\mathcal{B}_{\mathcal{I}[\sigma_1]}}(\boldsymbol{X}_i \mid \boldsymbol{x}_1, \ldots, \boldsymbol{x}_{i-1}, d_1, \ldots, d_{i-1}) = P_{\mathcal{B}_{\mathcal{I}[\sigma_2]}}(\boldsymbol{X}_i \mid \boldsymbol{x}_1, \ldots, \boldsymbol{x}_{i-1}, d_1, \ldots, d_{i-1}).$$

The proof is left as an exercise (exercise 23.4). Thus, the probability of $\boldsymbol{X}_i$ given $\boldsymbol{x}_1, \ldots, \boldsymbol{x}_{i-1}, d_1, \ldots, d_{i-1}$ does not depend on the choice of strategy σ, and so we can define a probability for this event without defining a particular strategy σ. We use $P(\boldsymbol{X}_i \mid \boldsymbol{x}_1, d_1, \ldots, \boldsymbol{x}_{i-1}, d_{i-1})$ as shorthand for this uniquely defined probability distribution.

23.3.2   Sum-Max-Sum Rule

Given an influence diagram $\mathcal{I}$, we can construct the decision tree using the previous procedure and then simply run MEU-for-Decision-Trees (algorithm 23.1) over the resulting tree. The algorithm computes both the MEU value of the tree and the optimal strategy. We now show that this MEU value and strategy are also the optimal value and strategy for the influence diagram.

Our first key observation is that, in the decision tree we constructed, we can choose a different action at each t-node in the layer for a decision variable D. In other words, the decision tree strategy allows us to take a different action at D for each assignment to the decision and observation variables preceding D in $\prec$. The perfect-recall assumption asserts that these variables are precisely the parents of D in the influence diagram $\mathcal{I}$. Thus, a decision rule at D is precisely as expressive as the set of individual decisions at the t-nodes corresponding to D, and the decision tree algorithm is simply selecting a set of decision rules for all of the decision variables in $\mathcal{I}$ — that is, a complete strategy for $\mathcal{I}$.

Example 23.9

In the F layer (the third layer) of the decision tree in figure 23.4, we maximize over different possible values of the decision variable F. Importantly, this layer is not selecting a single decision, but a (possibly) different action at each node in the layer. Each of these nodes corresponds to an information state — an assignment to C and S. Altogether, the set of decisions at this layer selects the entire decision rule δ_F.   ∎

Note that the perfect recall assumption is critical here. The decision tree semantics (as we defined it) makes the implicit assumption that we can make an independent decision at each t-node in the tree. Hence, if D' follows D in the decision tree, then every variable on the path to D t-nodes also appears on the path to D' t-nodes. Thus, the decision tree semantics can be consistent with the influence diagram semantics only when the influence diagram satisfies the perfect recall assumption.

We now need to show that the strategy selected by this algorithm is the one that maximizes the expected utility for $\mathcal{I}$. To do so, let us examine more closely the expression computed by MEU-for-Decision-Trees when applied to the decision tree constructed before.

Example 23.10

In example 23.3, our computation for the value of the entire tree can be written using the following expression:

$$\max_C \sum_S P(S \mid C) \max_F \sum_M P(M \mid S, F, C)[V_S(C) + V_E(M, F)].$$

Note that we can simplify some of the conditional probability terms in this expression using the conditional independence properties of the network (which are also invariant under any choice of decision rules). For example, M is independent of F, C given S, so that $P(M \mid S, F, C) = P(M \mid S)$.   ∎

More generally, consider an influence diagram where, as before, the sequence of chance and decision variables is: $\boldsymbol{X}_1, D_1, \ldots, \boldsymbol{X}_k, D_k, \boldsymbol{Y}$. We can write the value of the decision-making situation using the following expression, known as the *sum-max-sum rule*:

sum-max-sum
rule

$$\mathrm{MEU}[\mathcal{I}] = \sum_{\boldsymbol{X}_1} P(\boldsymbol{X}_1) \max_{D_1} \sum_{\boldsymbol{X}_2} P(\boldsymbol{X}_2 \mid \boldsymbol{X}_1, D_1) \max_{D_2} \ldots$$
$$\sum_{\boldsymbol{X}_k} P(\boldsymbol{X}_k \mid \boldsymbol{X}_1, \ldots, \boldsymbol{X}_{k-1}, D_1, \ldots, D_{k-1})$$
$$\max_{D_k} \sum_{\boldsymbol{Y}} P(\boldsymbol{Y} \mid \boldsymbol{X}_1, \ldots, \boldsymbol{X}_k, D_1, \ldots, D_k) U(\boldsymbol{Y}, \boldsymbol{X}_1, \ldots, \boldsymbol{X}_k, D_1, \ldots, D_k).$$

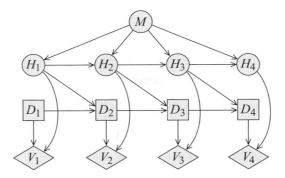

Figure 23.5 **Iterated optimization versus variable elimination.** An influence diagram that allows an efficient solution using iterated optimization, but where variable elimination techniques are considerably less efficient.

This expression is effectively performing the same type of backward induction that we used in decision trees.

We can now push in the conditional probabilities into the summations or maximizations. This operation is the inverse to the one we have used so often earlier in the book, where we move probability factors out of a summation or maximization; the same equivalence is used to justify both. Once all the probabilities are pushed in, all of the conditional probability expressions cancel each other, so that we obtain simply:

$$\text{MEU}[\mathcal{I}] = \sum_{\boldsymbol{X}_1} \max_{D_1} \sum_{\boldsymbol{X}_2} \max_{D_2} \ldots \sum_{\boldsymbol{X}_k} \max_{D_k}$$
$$\sum_{\boldsymbol{Y}} P(\boldsymbol{X}_1, \ldots, \boldsymbol{X}_k, \boldsymbol{Y} \mid D_1, \ldots, D_k) U(\boldsymbol{X}_1, \ldots, \boldsymbol{X}_k, \boldsymbol{Y}, D_1, \ldots, D_k). \quad (23.4)$$

If we view this expression in terms of factors (as in equation (23.2)), we can decompose the joint probability $P(\boldsymbol{X}_1, \ldots, \boldsymbol{X}_k, \boldsymbol{Y} \mid D_1, \ldots, D_k) = P(\mathcal{X} \mid \mathcal{D})$ as the product of all of the factors corresponding to the CPDs of the variables $\mathcal{X}$ in the influence diagram. The joint utility $U(\boldsymbol{X}_1, \ldots, \boldsymbol{X}_k, \boldsymbol{Y}, D_1, \ldots, D_k) = U(\mathcal{X}, \mathcal{D})$ is the sum of all of the utility variables in the network.

Now, consider a strategy σ — an assignment of actions to all of the agent t-nodes in the tree. Given a fixed strategy, the maximizations become vacuous, and we are simply left with a set of summations over the different chance variables in the network. It follows directly from the definitions that the result of this summation is simply the expected utility of σ in the influence diagram, as in equation (23.2). The fact that the sum-max-sum computation results in the MEU strategy now follows directly from the optimality of the strategy produced by the decision tree algorithm.

The form of equation (23.4) suggests an alternative method for computing the MEU value and strategy, one that does not require that we explicitly form a decision tree. Rather, we can apply a *variable elimination* algorithm that directly computes the the sum-max-sum expression: We eliminate both the chance and decision variables, one at a time, using the $\sum$ or max

variable elimination

operations, as appropriate. At first glance, this approach appears straightforward, but the details are somewhat subtle. Unlike most of our applications of the variable elimination algorithm, which involve only two operations (either sum-product or max-product), this expression involves four — sum-marginalization, max-marginalization, factor product (for probabilities and utilities), and factor addition (for utilities). The interactions between these different operations require careful treatment, and the machinery required to handle them correctly has a significant effect on the design and efficiency of the variable elimination algorithm. The biggest complication arises from the fact that sum-marginalization and max-marginalization do not commute, and therefore elimination operations can be executed only in an order satisfying certain constraints; as we showed in section 13.2.3 and section 14.3.1, such constraints can cause inference even in simple networks to become intractable. The same issues arise here:

Example 23.11

Consider a setting where a student must take a series of exams in a course. The hardness H_i of each exam i is not known in advance, but one can assume that it depends on the hardness of the previous exam H_{i-1} (if the class performs well on one exam, the next one tends to be harder, and if the class performs poorly, the next one is often easier). It also depends on the overall meanness of the instructor. The student needs to decide how much to study for each exam (D_i); studying more makes her more likely to succeed in the exam, but it also reduces her quality of life. At the time the student needs to decide on D_i, she knows the difficulty of the previous one and whether she studied for it, but she does not remember farther back than that. The meanness of the instructor is never observed. The influence diagram is shown in figure 23.5.

If we apply a straightforward variable elimination algorithm based on equation (23.4), we would have to work from the inside out in an order that is consistent with the operations in the equation. Thus, we would first have to eliminate M, which is never observed. This step has the effect of creating a single factor over all of the H_i variables, whose size is exponential in k. ∎

Fortunately, as we discuss in section 23.5, there are better solution methods for influence diagrams, which are not based on variable elimination and hence avoid some of these difficulties.

23.4   Computing Expected Utilities

In constructing a more efficient algorithm for finding the optimal decision in an influence diagram, we first consider the special case of an influence diagram with no decision variables. This problem is of interest in its own right, since it allows us to evaluate the expected utility of a given strategy. More importantly, it is also a key subroutine in the algorithm for finding an optimal strategy.

We begin our discussion with the even more restricted setting, where there is a single utility variable, and then discuss how it can be extended to the case of several utility variables. As we will see, although there are straightforward generalizations, an efficient implementation for this extension can involve some subtlety.

23.4.1   Simple Variable Elimination

Assume we have a single utility factor U. In this case, the expected utility is simply a product of factors: the CPDs of the chance variables, the decision rules, and the utility function of the

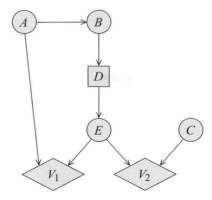

Figure 23.6 **An influence diagram with multiple utility variables**

single utility factor U, summed out over all of the variables in the network. **Thus, in the setting of a single utility variable, we can apply our standard variable elimination algorithm in a straightforward way, to the set of factors defining the expected utility.** Because variable elimination is well defined for any set of factors (whether derived from probabilities or not), there is no obstacle to applying it in this setting.

Example 23.12

Consider the influence diagram in figure 23.6. The influence diagram is drawn with two utility variables, but (proceeding with our assumption of a single utility variable) we analyze the computation for each of them in isolation, assuming it is the only utility variable in the network.

We begin with the utility variable V_1, and use the elimination ordering C, A, E, B, D. Note that C is barren relative to V_1 (that is, it has no effect on the utility V_1) and can therefore be ignored. (Eliminating C would simply produce the all 1's factor.) Eliminating A, we obtain

$$\mu_1^1(B, E) = \sum_A V_1(A, E) P(B \mid A) P(A).$$

Eliminating E, we obtain

$$\mu_2^1(B, D) = \sum_E P(E \mid D) \mu_1^1(B, E).$$

We can now proceed to eliminate D and B, to compute the final expected utility value.

Now, consider the same variable elimination algorithm, with the same ordering, applied to the utility variable V_2. In this case, C is not barren, so we compute:

$$\mu_1^2(E) = \sum_C V_2(C, E) P(C).$$

The variable A does not appear in the scope of μ_1^2, and hence we do not use the utility factor in this step. Rather, we obtain a standard probability factor:

$$\phi_1^2(B) = \sum_A P(A) P(B \mid A).$$

Eliminating E, we obtain:

$$\mu_2^2(D) = \sum_E P(E \mid D)\mu_1^2(E).$$

To eliminate B, we multiply $P(D \mid B)$ (which is a decision rule, and hence simply a CPD) with $\phi_1^2(B)$, and then marginalize out B from the resulting probability factor. Finally, we multiply the result with $\mu_2^2(D)$ to obtain the expected utility of the influence diagram, given the decision rule for D. ∎

23.4.2   Multiple Utility Variables: Simple Approaches

An efficient extension to multiple utility variables is surprisingly subtle. One obvious solution is to collapse all of the utility factors into a single large factor $U = \sum_{V \in \mathcal{U}} V$. We are now back to the same situation as above, and we can run the variable elimination algorithm unchanged. Unfortunately, this solution can lead to unnecessary computational costs:

Example 23.13   *Let us return to the influence diagram of example 23.12, but where we now assume that we have both V_1 and V_2. In this simple solution, we add both together to obtain $U(A, E, C)$. If we now run our variable elimination process (with the same ordering), it produces the following factors: $\mu_1^U(A, E) = \sum_C P(C)U(A, C, E)$; $\mu_2^U(B, E) = \sum_A P(A)P(B \mid A)\mu_1^U(A, E)$; and $\mu_3^U(B, D) = \sum_E P(E \mid D)\mu_2^U(B, E)$. Thus, this process produces a factor over the scope A, C, E, which is not created by either of the two preceding subcomputations; if, for example, both A and C have a large domain, this factor might result in high computational costs.* ∎

Thus, this simple solution requires that we sum up the individual subutility functions to construct a single utility factor whose scope is the union of the scopes of the subutilities. As a consequence, this transformation loses the structure of the utility function and creates a factor that may be exponentially larger. In addition to the immediate costs of creating this larger factor, factors involving more variables can also greatly increase the cost of the variable elimination algorithm by forcing us to multiply in more factors as variables are eliminated.

A second simple solution is based on the linearity of expectations:

$$\sum_{\mathcal{X} - \mathrm{Pa}_D} \prod_{X \in \mathcal{X}} P(X \mid \mathrm{Pa}_X)\Big(\sum_{i \,:\, V_i \in \mathcal{U}} V_i\Big) = \sum_{i \,:\, V_i \in \mathcal{U}} \Big(\sum_{\mathcal{X} - \mathrm{Pa}_D} \prod_{X \in \mathcal{X}} P(X \mid \mathrm{Pa}_X)V_i\Big).$$

Thus, we can run multiple separate executions of variable elimination, one for each utility factor V_i, computing for each of them an expected utility factor μ_{-D}^i; we then sum up these expected utility factors and optimize the decision rule relative to the resulting aggregated utility factor. The limitation of this solution is that, in some cases, it forces us to replicate work that arises for multiple utility factors.

Example 23.14   *Returning to example 23.12, assume that we replace the single variable E between D and the two utility variables with a long chain $D \to E_1 \to \ldots \to E_k$, where E_k is the parent of V_1 and V_2. If we do a separate variable elimination computation for each of V_1 and V_2, we would be executing twice the steps involved in eliminating $E_1, \ldots, E_k$, rather than reusing the computation for both utility variables.* ∎

23.4.3 Generalized Variable Elimination ⋆

A solution that addresses both the limitations described before is to perform variable elimination with multiple utility factors simultaneously, but allow the algorithm to add utility factors to each other, as called for by the variable elimination algorithm. In other words, just as we multiply factors together when we eliminate a variable that they have in common, we would combine two utility factors together in the same situation.

Example 23.15

Let us return to example 23.12 using the same elimination ordering C, A, E. The first steps, of eliminating C and A, are exactly those we took in that example, as applied to each of the two utility factors separately. In other words, the elimination of C does not involve V_1, and hence produces precisely the same factor $\mu_1^1(B, E)$ as before; similarly, the elimination of A does not involve V_2, and produces $\mu_1^2(E), \phi_1^2(B)$. However, when we now eliminate E, we must somehow combine the two utility factors in which E appears. At first glance, it appears as if we can simply add these two factors together. However, a close examination reveals an important subtlety. The utility factor $\mu_1^2(E) = \sum_C P(C)V_2(C, E)$ is a function that defines, for each assignment to E, an expected utility given E. However, the entries in the other utility factor,

$$\mu_1^1(B, E) = \sum_A P(A)P(B \mid A)V_1(A, E)$$
$$= \sum_A P(B)P(A \mid B)V_1(A, E) = P(B)\sum_A P(A \mid B)V_1(A, E),$$

do not represent an expected utility; rather, they are a product of an expected utility with the probability $P(B)$. Thus, the two utility factors are on "different scales," so to speak, and cannot simply be added together. To remedy this problem, we must convert both utility factors into the "utility scale" before adding them together. To do so, we must keep track of $P(B)$ as we do the elimination and divide $\mu_1^1(B, E)$ by $P(B)$ to rescale it appropriately, before adding it to μ_1^2. ∎

Thus, in order to perform the variable elimination computation correctly with multiple utility variables, we must keep track not only of utility factors, but also of the probability factors necessary to normalize them. This intuition suggests an algorithm where our basic data structures — our factors — are actually pairs of factors $\gamma = (\phi, \mu)$, where ϕ is a probability factor, and μ is a utility factor. Intuitively, ϕ is the probability factor that can bring μ into the "expected utility scale." More precisely, assume for simplicity that the probability and utility factors in a joint factor have the same scope; we can make this assumption without loss of generality by simply increasing the scope of either or both factors, duplicating entries as required. Intuitively, the probability factor is maintained as an auxiliary factor, used to normalize the utility factor when necessary, so as to bring it back to the standard "utility scale." Thus, if we have a joint factor $(\phi(\boldsymbol{Y}), \mu(\boldsymbol{Y}))$, then $\mu(\boldsymbol{Y})/\phi(\boldsymbol{Y})$ is a factor whose entries are expected utilities associated with the different assignments $\boldsymbol{y}$.

Our goal is to define a variable-elimination-style algorithm using these joint factors. As for any variable elimination algorithm, we must define operations that combine factors, and operations that marginalize — sum out — variables out of a factor. We now define both of these steps. We consider, for the moment, factors associated only with probability variables and utility variables; we will discuss later how to handle decision variables.

Initially, each variable W that is associated with a CPD induces a probability factor ϕ_W; such variables include both chance variables in $\mathcal{X}$ and decision variables associated with a decision rule. (As we discussed, a decision rule for a decision variable D essentially turns D into a chance variable.) We convert ϕ_W into a joint factor γ_W by attaching to it an all-zeros utility factor over the same scope, $\mathbf{0}_{Scope[\phi_W]}$. Similarly, each utility variable $V \in \mathcal{U}$ is associated with a utility factor μ_V, which we convert to a joint factor by attaching to it an all-ones probability factor: $\gamma_V = (\mathbf{1}_{\mathrm{Pa}_V}, V)$ for $V \in \mathcal{U}$.

Intuitively, we want to multiply probability components (as usual) and add utility components. Thus, we define the *joint factor combination* operation as follows:

- For two joint factors $\gamma_1 = (\phi_1, \mu_1)$, $\gamma_2 = (\phi_2, \mu_2)$, we define the *joint factor combination* operation:

$$\gamma_1 \bigoplus \gamma_2 = (\phi_1 \cdot \phi_2, \mu_1 + \mu_2). \tag{23.5}$$

We see that, if all of the joint factors in the influence diagram are combined, we obtain a single (exponentially large) probability factor that defines the joint distribution over outcomes, and a single (exponentially large) utility factor that defines the utilities of the outcomes. Of course, this procedure is not one that we would ever execute; rather, as in variable elimination, we want to interleave combination steps and marginalization steps in a way that preserves the correct semantics.

The definition of the marginalization operation is subtler. Intuitively, we want the probability of an outcome to be multiplied with its utility. However, as suggested by example 23.15, we must take care that the utility factors derived as intermediate results all maintain the same scale, so that they can be correctly added in the factor combination operation. Thus, when marginalizing a variable W, we divide the utility factor by the associated probability factor, ensuring that it maintains its expected utility interpretation:

- For a joint factor $\gamma = (\phi, \mu)$ over scope $\boldsymbol{W}$, we define the *joint factor marginalization* operation for $\boldsymbol{W}' \subset \boldsymbol{W}$ as follows:

$$marg_{\boldsymbol{W}'}(\gamma) = \left(\sum_{\boldsymbol{W}'} \phi, \frac{\sum_{\boldsymbol{W}'} \phi \cdot \mu}{\sum_{\boldsymbol{W}'} \phi} \right). \tag{23.6}$$

Intuitively, this operation marginalizes out (that is, eliminates) the variables in $\boldsymbol{W}'$, handling both utility and probability factors correctly.

Finally, at the end of the process, we can combine the probability and utility factors to obtain a single factor that corresponds to the overall expected utility:

- For a joint factor $\gamma = (\phi, \mu)$, we define the *joint factor contraction* operation as the factor product of the two components:

$$\mathrm{cont}(\gamma) = \phi \cdot \mu. \tag{23.7}$$

To understand these definitions, consider again the problem of computing the expected utility for some (complete) strategy σ for the influence diagram $\mathcal{I}$. Thus, we now have a probability

---

Algorithm 23.2 Generalized variable elimination for joint factors in influence diagrams

Procedure Generalized-VE-for-IDs (
 Φ, // Set of joint (probability,utility) factors
 $W_1, \ldots, W_k$ // List of variables to be eliminated
)
1 **for** $i = 1, \ldots, k$
2 $\Phi' \leftarrow \{\phi \in \Phi \;:\; W_i \in Scope[\phi]\}$
3 $\psi \leftarrow \bigoplus_{\phi \in \Phi'} \phi$
4 $\tau \leftarrow marg_{W_i}(\psi)$
5 $\Phi \leftarrow \Phi - \Phi' \cup \{\tau\}$
6 $\phi^* \leftarrow \bigoplus_{\phi \in \Phi} \phi$
7 **return** ϕ^*

---

factor for each decision variable. Recall that our expected utility is defined as:

$$\mathrm{EU}[\mathcal{I}[\sigma]] = \sum_{W \in \mathcal{X} \cup \mathcal{D}} \prod_{W \in \mathcal{X} \cup \mathcal{D}} \phi_W \cdot \left(\sum_{V \in \mathcal{U}} \mu_V \right).$$

Let γ^* be the marginalization over all variables of the combination of all of the joint factors:

$$\gamma^* = (\phi^*, \mu^*) = marg_\emptyset \Big(\bigoplus_{(W \in \mathcal{X} \cup \mathcal{U})} [\gamma_W] \Big). \tag{23.8}$$

Note that the factor has empty scope and is therefore simply a pair of numbers. We can now show the following simple result:

Proposition 23.1

For γ^* defined in equation (23.8), we have: $\gamma^* = (1, \mathrm{EU}[\mathcal{I}[\sigma]])$.

The proof follows directly from the definitions and is left as an exercise (exercise 23.2).

 Of course, as we discussed, we want to interleave the marginalization and combination steps. An algorithm implementing this idea is shown in algorithm 23.2. The algorithm returns a single joint factor (ϕ, μ).

Example 23.16

Let us consider the behavior of this algorithm on the influence diagram of example 23.12, assuming again that we have a decision rule for D, so that we have only chance variables and utility variables. Thus, we initially have five joint factors derived from the probability factors for A, B, C, D, E; for example, we have $\gamma_B = (P(B \mid A), \mathbf{0}_{A,B})$. We have two joint factors γ_1, γ_2 derived from the utility variables V_1, V_2; for example, we have $\gamma_2 = (\mathbf{1}_{C,E}, V_2(C, E))$.

 Now, consider running our generalized variable elimination algorithm, using the elimination ordering C, A, E, B, D. Eliminating C, we first combine γ_C, γ_2 to obtain:

$$\gamma_C \bigoplus \gamma_2 = (P(C), V_2(C, E)),$$

where the scope of both components is taken to be C, E. We then marginalize C to obtain:

$$\gamma_3(E) = \left(\mathbf{1}_E, \frac{\sum_C (P(C) V_2(C, E))}{\mathbf{1}_E} \right)$$
$$= (\mathbf{1}_E, \boldsymbol{E}_{P(C)}[V_2(C, E)]).$$

Continuing to eliminate A, we combine γ_A, γ_B, and γ_1 and marginalize A to obtain:

$$
\begin{aligned}
\gamma_4(B, E) &= \left(\sum_A P(A)P(B \mid A), \frac{\sum_A P(A)P(B \mid A)V_1(A, E)}{\sum_A P(A)P(B \mid A)} \right) \\
&= (P(B), \sum_A P(A \mid B)V_1(A, E)) \\
&= (P(B), \boldsymbol{E}_{P(A|B)}[V_1(A, E)]).
\end{aligned}
$$

Importantly, the utility factor here can be interpreted as the expected utility over V_1 given B, where the expectation is taken over values of A. It therefore keeps this utility factor on the same scale as the others, avoiding the problem of incomparable utility factors that we had in example 23.15.

We next eliminate E. We first combine γ_E, γ_3, and γ_4 to obtain:

$$
(P(E \mid D)P(B), \boldsymbol{E}_{P(C)}[V_2(C, E)] + \boldsymbol{E}_{P(A|B)}[V_1(A, E)]).
$$

Marginalizing E, we obtain:

$$
\gamma_5(B, D) \quad = \quad (P(B), \boldsymbol{E}_{P(C,E|D)}[V_2(C, E)] + \boldsymbol{E}_{P(A,E|B,D)}[V_1(A, E)]).
$$

To eliminate B, we first combine γ_5 and γ_D, to obtain:

$$
(P(D \mid B)P(B), \boldsymbol{E}_{P(C,E|D)}[V_2(C, E)] + \boldsymbol{E}_{P(A,E|D)}[V_1(A, E)]).
$$

We then marginalize B, obtaining:

$$
\begin{aligned}
\gamma_6(D) &= \left(P(D), \frac{\sum_B (\boldsymbol{E}_{P(C,E,D,B)}[V_2(C, E)] + \boldsymbol{E}_{P(A,E,D,B)}[V_1(A, E)])}{P(D)} \right) \\
&= \left(P(D), \boldsymbol{E}_{P(C,E|D)}[V_2(C, E)] + \boldsymbol{E}_{P(A,E|B,D)}[V_1(A, E)] \right).
\end{aligned}
$$

Finally, we have only to marginalize D, obtaining:

$$
\gamma_7(\emptyset) \quad = \quad \left(1, \boldsymbol{E}_{P(C,E)}[V_2(C, E)] + \boldsymbol{E}_{P(A,E)}[V_1(A, E)] \right),
$$

as desired.   ∎

How do we show that it is legitimate to reorder these marginalization and combination operators? In exercise 9.19, we defined the notion of generalized marginalize-combine factor operators and stated a result showing that, for any pair of operators satisfying certain conditions, any legal reordering of the operators led to the same result. In particular, this result implied, as special cases, correctness of sum-product, max-product, and max-sum variable elimination. The same analysis can be used for the operators defined here, showing the following result:

Theorem 23.1

Let Φ be a set of joint factors over $\boldsymbol{Z}$. Generalized-VE-for-IDs($\Phi, \boldsymbol{W}$) returns the joint factor

$$
marg_{\boldsymbol{W}} \left(\bigoplus_{(\gamma \in \Phi)} \gamma \right).
$$

The proof is left as an exercise (exercise 23.3).

Note that the complexity of the algorithm is the same (up to a constant factor) as that of a standard VE algorithm, applied to an analogous set of factors — with the same scope — as our initial probability and utility factors. In other words, for a given elimination ordering, the cost of the algorithm grows as the induced tree-width of the graph generated by this initial set of factors.

So far, we have discussed the problem of computing the expected utility of a complete strategy. How can we apply these ideas to our original task, of optimizing a single decision rule? The idea is essentially the same as in section 23.4.1. As there, we apply Generalized-VE-for-IDs to eliminate all of the variables other than $\mathrm{Family}_D = \{D\} \cup \mathrm{Pa}_D$. In this process, the probability factor induced by the decision rule for D is only combined with the other factors at the final step of the algorithm, when the remaining factors are all combined. It thus has no effect on the factors produced up to that point. We can therefore omit ϕ_D from our computation, and produce a joint factor $\gamma_{-D} = (\phi_{-D}, \mu_{-D})$ over Family_D based only on the other factors in the network.

For any decision rule δ_D, if we run Generalized-VE-for-IDs on the factors in the original influence diagram plus a joint factor $\gamma_D = (\delta_D, \mathbf{0}_{\mathrm{Family}_D})$, we would obtain the factor

$$\gamma_{\delta_D} = \gamma_{-D} \bigoplus \gamma_D.$$

Rewriting this expression, we see that the overall expected utility for the influence diagram given the decision rule δ_D is then:

$$\sum_{\boldsymbol{w} \in Val(\mathrm{Pa}_D), d \in Val(D)} \mathrm{cont}(\gamma_{-D})(\boldsymbol{w}, d)\delta_D(\boldsymbol{w}).$$

Based on this observation, and on the fact that we can always select an optimal decision rule that is a deterministic function from $Val(\mathrm{Pa}_D)$ to $Val(D)$, we can easily optimize δ_D. For each assignment $\boldsymbol{w}$ to Pa_D, we select

$$\delta_D(\boldsymbol{w}) = \arg \max_{d \in Val(D)} \mathrm{cont}(\gamma_{-D})(\boldsymbol{w}, d).$$

As before, the problem of optimizing a single decision rule can be solved using a standard variable elimination algorithm, followed by a simple optimization. In this case, we must use a generalized variable elimination algorithm, involving both probability and utility factors.

23.5 Optimization in Influence Diagrams

We now turn to the problem of selecting an optimal strategy in an influence diagram. We begin with the simple case, where we have only a single decision variable. We then show how to extend these ideas to the more general case.

23.5.1 Optimizing a Single Decision Rule

We first make the important observation that, for the case of a single decision variable, the task of finding an optimal decision rule can be reduced to that of computing a single utility factor.

We begin by rewriting the expected utility of the influence diagram in a different order:

$$\text{EU}[\mathcal{I}[\sigma]] = \sum_{D, \text{Pa}_D} \delta_D \sum_{\mathcal{X} - \text{Pa}_D} \prod_{X \in \mathcal{X}} P(X \mid \text{Pa}_X)(\sum_{V \in \mathcal{U}} V). \tag{23.9}$$

Our task is to select δ_D.

expected utility
factor

We now define the *expected utility factor* to be the value of the internal summation in equation (23.9):

$$\mu_{-D} = \sum_{\mathcal{X} - \text{Pa}_D} \prod_{X \in \mathcal{X}} P(X \mid \text{Pa}_X)(\sum_{V \in \mathcal{U}} V). \tag{23.10}$$

This expression is the marginalization of this product onto the variables $D \cup \text{Pa}_D$; importantly, it does not depend on our choice of decision rule for D. Given μ_{-D}, we can compute the expected utility for any decision rule δ_D as:

$$\sum_{D, \text{Pa}_D} \delta_D \mu_{-D}(D, \text{Pa}_D).$$

Our goal is to find δ_D that maximizes this expression.

Proposition 23.2

Consider an influence diagram $\mathcal{I}$ with a single decision variable D. Letting μ_{-D} be as in equation (23.10), the optimal decision rule for D in $\mathcal{I}$ is defined as:

$$\delta_D(\boldsymbol{w}) = \arg \max_{d \in Val(D)} \mu_{-D}(d, \boldsymbol{w}) \qquad\qquad \forall \boldsymbol{w} \in Val(\text{Pa}_D). \tag{23.11}$$

The proof is left as an exercise (see exercise 23.1).

Thus, we have shown how the problem of optimizing a single decision rule can be solved very simply, once we have computed the utility factor $\mu_{-D}(D, \text{Pa}_D)$.

Importantly, any of the algorithms described before, whether the simpler ones in section 23.4.1 and 23.4.2, or the more elaborate generalized variable elimination algorithm of section 23.4.3, can be used to compute this expected utility factor. We simply structure our elimination ordering to eliminate only the variables other than D, Pa_D; we then combine all of the factors that are computed via this process, to produce a single integrated factor $\mu_{-D}(D, \text{Pa}_D)$. We can then use this factor as in proposition 23.2 to find the optimal decision rule for D, and thereby solve the influence diagram.

How do we generalize this approach to the case of an influence diagram with multiple decision rules $D_1, \ldots, D_k$? In principle, we could generate an expected utility factor where we eliminated all variables other than the union $\boldsymbol{Y} = \cup_i(\{D_i\} \cup \text{Pa}_{D_i})$ of all of the decision variables and all of their parents. Intuitively, this factor would specify the expected utility of the influence diagram given an assignment to $\boldsymbol{Y}$. However, in this case, the optimization problem is much more complex, in that it requires that we consider simultaneously the decisions at all of the decision variables in the network. Fortunately, as we show in the next section, we can perform this multivariable optimization using localized optimization steps over single variables.

23.5.2   Iterated Optimization Algorithm

In this section, we describe an iterated approach that breaks up the problem into a series of simpler ones. Rather than optimize all of the decision rules at the same time, we fix all of the decision rules but one, and then optimize the remaining one. The problem of optimizing a single decision rule is significantly simpler, and admits very efficient algorithms, as shown in section 23.5.1. This algorithm is very similar in its structure to the local optimization approach for marginal MAP problems, presented in section 13.7. Both algorithms are intended to deal with the same computational bottleneck: the exponentially large factors generated by a constrained elimination ordering. They both do so by optimizing one variable at a time, keeping the others fixed. The difference is that here we are optimizing an entire decision rule for the decision variable, whereas there we are simply picking a single value for the MAP variable.

We will show that, under certain assumptions, this iterative approach is guaranteed to converge to the optimal strategy. Importantly, this approach also applies to influence diagrams with imperfect recall, and can therefore be considerably more efficient.

The basic idea behind this algorithm is as follows. The algorithm proceeds by sequentially optimizing individual decision rules. We begin with some (almost) arbitrary strategy σ, which assigns a decision rule to all decision variables in the network. We then optimize a single decision rule relative to our current assignment to the others. This decision rule is used to update σ, and another decision rule is now optimized relative to the new strategy. More precisely, let σ_{-D} denote the decision rules in a strategy σ other than the one for D. We say
locally optimal that a decision rule δ_D is *locally optimal* for a strategy σ if, for any other decision rule δ'_D,
decision rule

$$\mathrm{EU}[\mathcal{I}[(\sigma_{-D}, \delta_D)]] \geq \mathrm{EU}[\mathcal{I}[(\sigma_{-D}, \delta'_D)]].$$

Our algorithm starts with some strategy σ, and then iterates over different decision variables D. It then selects a locally optimal decision rule δ_D for σ, and updates σ by replacing σ_D with the new δ_D. Note that the influence diagram $\mathcal{I}[\sigma_{-D}]$ is an influence diagram with the single decision variable D, which can be solved using a variety of methods, as described earlier. The algorithm terminates when no decision rule can be improved by this process.

Perhaps the most important property of this algorithm is its ability to deal with the main computational limitation of the simple variable elimination strategy described in section 23.3.2: the fact that the constrained variable elimination ordering can require creation of large factors even when the network structure does not force them.

Example 23.17

Consider again example 23.11; here, we would begin with some set of decision rules for all of $D_1, \ldots, D_k$. We would then iteratively compute the expected utility factor μ_{-D_i} for one of the D_i variables, using the (current) decision rules for the others. We could then optimize the decision rule for D_i, and continue the process. Importantly, the only constraint on the variable elimination ordering is that D_i and its parents be eliminated last. With these constraints, the largest factor induced in any of these variable elimination procedures has size 4, avoiding the exponential blowup in k that we saw in example 23.11. ∎

In a naive implementation, the algorithm runs variable elimination multiple times — once for each iteration — in order to compute μ_{-D_i}. However, using the approach of joint (probability, utility) factors, as described in section 23.4.3, we can provide a very efficient implementation as a clique tree. See exercise 23.10.

So far, we have ignored several key questions that affect both the algorithm's complexity and its correctness. Most obviously, we can ask whether this iterative algorithm even converges. When we optimize D, we either improve the agent's overall expected utility or we leave the decision rule unchanged. Because the expected utility is bounded from above and the total number of strategies is discrete, the algorithm cannot improve the expected utility indefinitely. Thus, at some point, no additional improvements are possible, and the algorithm will terminate. A second question relates to the quality of the solution obtained. Clearly, this solution is locally optimal, in that no change to a single decision rule can improve the agent's expected utility. However, local optimality does not, in general, imply that the strategy is globally optimal.

Example 23.18

Consider an influence diagram containing only two decision variables D_1 and D_2, and a utility variable $V(D_1, D_2)$ defined as follows:

$$V(d_1, d_2) = \begin{cases} 2 & d_1 = d_2 = 1 \\ 1 & d_1 = d_2 = 0 \\ 0 & d_1 \neq d_2. \end{cases}$$

The strategy $(0, 0)$ is locally optimal for both decision variables, since the unique optimal decision for D_i when $D_j = 0$ $(j \neq i)$ is $D_i = 0$. On the other hand, the globally optimal strategy is $(1, 1)$. ∎

☞ **However, under certain conditions, local optimality does imply global optimality, so that the iterated optimization process is guaranteed to converge to a globally optimal solution. These conditions are more general than perfect recall, so that this algorithm works in every case where the algorithm of the previous section applies. In this case, we can provide an ordering for applying the local optimization steps that guarantees that this process converges to the globally optimal strategy after modifying each decision rule exactly once. However, this algorithm also applies to networks that do not satisfy the perfect recall assumption, and in certain such cases it is even guaranteed to find an optimal solution.** By relaxing the perfect recall assumption, we can avoid some of the exponential blowup of the decision rules in terms of the number of decisions in the network.

23.5.3 Strategic Relevance and Global Optimality ⋆

The algorithm described before iteratively changes the decision rule associated with individual decision variables. In general, changing the decision rule for one variable D' can cause a decision rule previously optimal for another variable D to become suboptimal. Therefore, the algorithm must revisit D and possibly select a new decision rule for it. In this section, we provide conditions under which we can guarantee that changing the decision rule for D' will not necessitate a change in the decision rule for D. In other words, we define conditions under which the decision rule for D' may not be relevant for optimizing the decision rule for D. Thus, if we choose a decision rule for D and later select one for D', we do not have to revisit the selection made for D. As we show, under certain conditions, this criterion allows us to optimize all of the decision rules using a single iteration through them.

23.5.3.1 Strategic Relevance

Intuitively, we would like to define a decision variable D' as *strategically relevant* to D if, to optimize the decision rule at D, the decision maker needs to consider the decision rule at D'. That is, we want to say that D' is is relevant to D if there is a partial strategy profile σ over

$\mathcal{D} - \{D, D'\}$, two decision rules $\delta_{D'}$ and $\delta'_{D'}$, and a decision rule δ_D, such that $(\sigma, \delta_D, \delta_{D'})$ is optimal, but $(\sigma, \delta_D, \delta'_{D'})$ is not.

Example 23.19

Consider a simple influence diagram where we have two decision variables $D_1 \rightarrow D_2$, and a utility $V(D_1, D_2)$ that is the same as the one used in example 23.18. Pick an arbitrary decision rule δ_{D_1} (not necessarily deterministic), and consider the problem of optimizing δ_{D_2} relative to δ_{D_1}. The overall expected utility for the agent is

$$\sum_{d_1} \delta_{D_1}(d_1) \sum_{d_2} \delta_{D_2}(d_2 \mid d_1) V(d_1, d_2).$$

An optimal decision for D_2 given the information state d_1 is $\arg\max_{d_2} V(d_1, d_2)$, regardless of the choice of decision rule for D_1. Thus, in this setting, we can pick an arbitrary decision rule δ_{D_1} and optimize δ_{D_2} relative to it; our selected decision rule will then be locally optimal relative to any decision rule for D_1. However, there is a subtlety that makes the previous statement false in certain settings. Let $\delta'_{D_1} = d_1^0$. Then one optimal decision rule for D_2 is $\delta'_{D_2}(d_1^0) = \delta'_{D_2}(d_1^1) = d_2^0$. Clearly, d_2^0 is the right choice when $D_1 = d_1^0$, but it is suboptimal when $D_1 = d_1^1$. However, because δ_{D_1} gives this latter event probability 0, this choice for δ_{D_2} is locally optimal relative to δ_{D_1}. ∎

As this example shows, a decision rule can make arbitrary choices in information states that have probability zero without loss in utility. In particular, because δ'_{D_1} assigns probability zero to d_1^1, the "suboptimal" δ'_{D_2} is locally optimal relative to δ'_{D_1}; however, δ'_{D_2} is not locally optimal relative to other decision rules for D_1. Thus, if we use the previous definition, D_1 appears relevant to D_2 despite our intuition to the contrary. We therefore want to avoid probability-zero events, which allow situations such as this. We say that a decision rule is *fully mixed* if each probability distribution $\delta_D(D \mid \text{pa}_D)$ assigns nonzero probability to all values of D. We can now formally define strategic relevance.

fully mixed decision rule

Definition 23.7

strategic relevance

Let D and D' be decision nodes in an influence diagram $\mathcal{I}$. We say that D' is strategically relevant to D (or that D strategically relies on D') if there exist:

- *a partial strategy profile σ over $\mathcal{D} - \{D, D'\}$;*
- *two decision rules $\delta_{D'}$ and $\delta'_{D'}$ such that $\delta_{D'}$ is fully mixed;*
- *a decision rule δ_D that is optimal for $(\sigma, \delta_{D'})$ but not for $(\sigma, \delta'_{D'})$.* ∎

This definition does not provide us with an operative procedure for determining relevance. We can obtain such a procedure by considering an alternative mathematical characterization of the notion of local optimality.

Proposition 23.3

Let δ_D be a decision rule for a decision variable D in $\mathcal{I}$, and let σ be a strategy for $\mathcal{I}$. Then δ_D is locally optimal for σ if and only if for every instantiation $\boldsymbol{w}$ of Pa_D where $P_{\mathcal{B}_{\mathcal{I}[\sigma]}}(\boldsymbol{w}) > 0$, the probability distribution $\delta_D(D \mid \boldsymbol{w})$ is a solution to

$$\arg\max_{q(D)} \sum_{d \in Val(D)} q(d) \sum_{V \in \mathcal{U}_{\succ D}} \sum_{v \in Val(V)} P_{\mathcal{B}_{\mathcal{I}[\sigma]}}(v \mid d, \boldsymbol{w}) \cdot v, \tag{23.12}$$

where $\mathcal{U}_{\succ D}$ is the set of utility nodes in $\mathcal{U}$ that are descendants of D in $\mathcal{I}$.

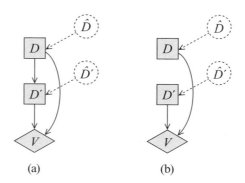

Figure 23.7   **Influence diagrams, augmented to test for s-reachability**

The proof is left as an exercise (exercise 23.6).

The significance of this result arises from two key points. First, the only probability expressions appearing in the optimization criterion are of the form $P_{\mathcal{B}_{\mathcal{I}[\sigma]}}(V \mid \text{Family}_D)$ for some utility variable V and decision variable D. Thus, we care about a decision rule $\delta_{D'}$ only if the CPD induced by this decision rule affects the value of one of these probability expressions. Second, the only utility variables that participate in these expressions are those that are descendants of D in the network.

23.5.3.2   S-Reachability

requisite CPD

We have reduced our problem to one of determining which decision rule CPDs might affect the value of some expression $P_{\mathcal{B}_{\mathcal{I}[\sigma]}}(V \mid \text{Family}_D)$, for $V \in \mathcal{U}_{\succ D}$. In other words, we need to determine whether the decision variable is a *requisite CPD* for this query. We also encountered this question in a very similar context in section 21.3.1, when we wanted to determine whether an intervention (that is, a decision) was relevant to a query. As described in exercise 3.20, we can determine whether the CPD for a variable Z is requisite for answering a query $P(\boldsymbol{X} \mid \boldsymbol{Y})$ with a simple graphical criterion: We introduce a new "dummy" parent $\widehat{Z}$ whose values correspond to different choices for the CPD of Z. Then Z is a requisite probability node for $P(\boldsymbol{X} \mid \boldsymbol{Y})$ if and only if $\widehat{Z}$ has an active trail to $\boldsymbol{X}$ given $\boldsymbol{Y}$.

Based on this concept and equation (23.12), we can define *s-reachability* — a graphical criterion for detecting strategic relevance.

Definition 23.8

s-reachable

A decision variable D' is s-reachable *from a decision variable D in an ID $\mathcal{I}$ if there is some utility node $V \in \mathcal{U}_{\succ D}$ such that if a new parent $\widehat{D'}$ were added to D', there would be an active path in $\mathcal{I}$ from $\widehat{D'}$ to V given* Family_D, *where a path is active in an ID if it is active in the same graph, viewed as a BN.*   ∎

Note that unlike d-separation, s-reachability is not necessarily a symmetric relation.

Example 23.20

Consider the simple influence diagrams in figure 23.7, representing example 23.19 and example 23.18 respectively. In (a), we have a perfect-recall setting. Because the agent can observe D when deciding *on the decision rule for D', he does not need to know the decision rule for D in order to evaluate his options at D'. Thus, D' does not strategically rely on D. Indeed, if we add a dummy parent $\widehat{D}$ to D, we have that V is d-separated from $\widehat{D}$ given $\mathrm{Family}_{D'} = \{D, D'\}$. Thus, D is not s-reachable from D'. Conversely, the agent's decision rule at D' does influence his payoff at D, and so D' is relevant to D. Indeed, if we add a dummy parent $\widehat{D'}$ to D', we have that V is not d-separated from $\widehat{D'}$ given D, Pa_D.*

By contrast, in (b), the agent forgets his action at D when observing D'; as his utility node is influenced by both decisions, we have that each decision is relevant to the other. The s-reachability analysis using d-separation from the dummy parents supports this intuition. ∎

The notion of s-reachability is sound and complete for strategic relevance (almost) in the same sense that d-separation is sound and complete for independence in Bayesian networks. As for d-separation, the soundness result is very strong: without s-reachability, one decision cannot be relevant to another.

Theorem 23.2

If D and D' are two decision nodes in an ID $\mathcal{I}$ and D' is not s-reachable from D in $\mathcal{I}$, then D does not strategically rely on D'.

PROOF Let σ be a strategy profile for $\mathcal{I}$, and let δ_D be a decision rule for D that is optimal for σ. Let $\mathcal{B} = \mathcal{B}_{\mathcal{I}[\sigma]}$. By proposition 23.3, for every $\boldsymbol{w} \in \mathit{Val}(\mathrm{Pa}_D)$ such that $P_{\mathcal{B}}(\boldsymbol{w}) > 0$, the distribution $\delta_D(D \mid \boldsymbol{w})$ must be a solution of the maximization problem:

$$\arg \max_{P(D)} \sum_{d \in \mathit{Val}(D)} P(d) \sum_{V \in \mathcal{U}_{\succ D}} \sum_{v \in \mathit{Val}(V)} P_{\mathcal{B}}(v \mid d, \boldsymbol{w}) \cdot v. \tag{23.13}$$

Now, let σ' be any strategy profile for $\mathcal{I}$ that differs from σ only at D', and let $\mathcal{B}' = \mathcal{B}_{\mathcal{I}[\sigma']}$. We must construct a decision rule δ'_D for D that agrees with δ_D on all $\boldsymbol{w}$ where $P_{\mathcal{B}}(\boldsymbol{w}) > 0$, and that is optimal for σ'. By proposition 23.3, it suffices to show that for every $\boldsymbol{w}$ where $P_{\mathcal{B}'}(\boldsymbol{w}) > 0$, $\delta'_D(D \mid \boldsymbol{w})$ is a solution of:

$$\arg \max_{P(D)} \sum_{d \in \mathit{Val}(D)} P(d) \sum_{V \in \mathcal{U}_{\succ D}} \sum_{v \in \mathit{Val}(V)} P_{\mathcal{B}'}(v \mid d, \boldsymbol{w}) \cdot v. \tag{23.14}$$

If $P_{\mathcal{B}}(\boldsymbol{w}) = 0$, then our choice of $\delta'_D(D \mid \boldsymbol{w})$ is unconstrained; we can simply select a distribution that satisfies equation (23.14). For other $\boldsymbol{w}$, we must let $\delta'_D(D \mid \boldsymbol{w}) = \delta_D(D \mid \boldsymbol{w})$. We know that $\delta_D(D \mid \boldsymbol{w})$ is a solution of equation (23.13), and the two expressions are different only in that equation (23.13) uses $P_{\mathcal{B}}(v \mid d, \boldsymbol{w})$ and equation (23.14) uses $P_{\mathcal{B}'}(v \mid d, \boldsymbol{w})$. The two networks $\mathcal{B}$ and $\mathcal{B}'$ differ only in the CPD for D'. Because D' is not a requisite probability node for any $V \in \mathcal{U}_{\succ D}$ given D, Pa_D, we have that $P_{\mathcal{B}}(v \mid d, \boldsymbol{w}) = P_{\mathcal{B}'}(v \mid d, \boldsymbol{w})$, and that $\delta'_D(D \mid \boldsymbol{w}) = \delta_D(D \mid \boldsymbol{w})$ is a solution of equation (23.14), as required. ∎

Thus, s-reachability provides us with a sound criterion for determining which decision variables D' are strategically relevant for D. As for d-separation, the completeness result is not as strong: s-reachability does not imply relevance in *every* ID. We can choose the probabilities and utilities in the ID in such a way that the influence of one decision rule on another does not manifest itself. However, s-reachability is the most precise graphical criterion we can use: it will not identify a strategic relevance unless that relevance actually exists in some ID that has the given graph structure.

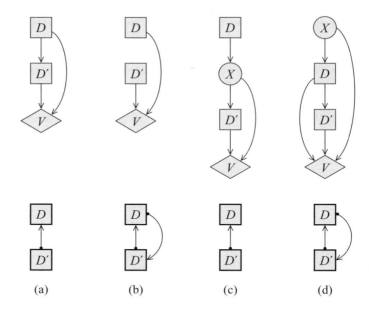

Figure 23.8   Four simple influence diagrams (top), and their relevance graphs (bottom).

Theorem 23.3

If a node D' is s-reachable from a node D in an ID, then there is some ID with the same graph structure in which D strategically relies on D'.

This result is roughly analogous to theorem 3.4, which states that there exists some parameterization that manifests the dependencies not induced by d-separation. A result analogous to the strong completeness result of theorem 3.5 is not known for this case.

23.5.3.3   The Relevance Graph

We can get a global view of the strategic dependencies between different decision variables in an influence diagram by putting them within a single graphical data structure.

Definition 23.9

relevance graph

The relevance graph *for an influence diagram $\mathcal{I}$ is a directed (possibly cyclic) graph whose nodes correspond to the decision variables $\mathcal{D}$ in $\mathcal{I}$, and where there is a directed edge $D' \to D$ if D' is strategically relevant to D.* ∎

To construct the graph for a given ID, we need to determine, for each decision node D, the set of nodes D' that are s-reachable from D. Using standard methods from chapter 3, we can find this set for any given D in time linear in the number of chance and decision variables in the ID. By repeating the algorithm for each D, we can derive the relevance graph in time $O((n + k)k)$ where $n = |\mathcal{X}|$ and $k = |\mathcal{D}|$.

Recall our original statement that a decision node D strategically relies on a decision node D' if one needs to know the decision rule for D' in order to evaluate possible decision rules for

D. Intuitively, if the relevance graph is acyclic, we have a decision variable that has no parents in the graph, and hence relies on no other decisions. We can optimize the decision rule at this variable relative to some arbitrary strategy for the other decision rules. Having optimized that decision rule, we can fix its strategy and proceed to optimize the next one. Conversely, if we have a cycle in the relevance graph, then we have some set of decisions all of which rely on each other, and their decision rules need to be optimized together. In this case, the simple iterative approach we described no longer applies.

However, before we describe this iterative algorithm formally and prove its correctness, it is instructive to examine some simple IDs and see when one decision node relies on another.

Example 23.21

Consider the four examples shown in figure 23.8, all of which relate to a setting where the agent first makes decision D and then D'. Examples (a) and (b) are the ones we previously saw in example 23.20, showing the resulting relevance graphs. As we saw, in (a), we have that D relies on D' but not vice versa, leading to the structure shown on the bottom. In (b) we have that each decision relies on the other, leading to a cyclic relevance graph. Example (c) represents a situation where the agent does not remember D when making the decision D'. However, the agent knows everything he needs to about D: his utility does not depend on D directly, but only on the chance node, which he can observe. Hence D' does not rely on D.

One might conclude that a decision node D' never relies on another D when D is observed by D', but the situation is subtler. Consider example (d), which represents a simple card game: the agent observes a card and decides whether to bet (D); at a later stage, the agent remembers only his bet but not the card, and decides whether to raise his bet (D'); the utility of both depends on the total bet and the value of the card. Even though the agent does remember the actual decision at D, he needs to know the decision rule for D in order to know what the value of D tells him about the value of the card. Thus, D' relies on D; indeed, when D is observed, there is an active trail from a hypothetical parent $\hat{D}$ that runs through the chance node to the utility node. ∎

However, it is the case that perfect recall — remembering both the previous decisions and the previous observations, does imply that the underlying relevance graph is acyclic.

Theorem 23.4

Let $\mathcal{I}$ be an influence diagram satisfying the perfect recall assumption. Then the relevance graph for $\mathcal{I}$ is acyclic.

The proof follows directly from properties of d-separation, and it is left as an exercise (exercise 23.7). We note that the ordering of the decisions in the relevance graph will be the opposite of the ordering in the original ID, as in figure 23.8a.

23.5.3.4 Global Optimality

Using the notion of a relevance graph, we can now provide an algorithm that, under certain conditions, is guaranteed to find an MEU strategy for the influence diagram. In particular, consider an influence diagram $\mathcal{I}$ whose relevance graph is acyclic, and let $D_1, \ldots, D_k$ be a topological ordering of the decision variables according to the relevance graph. We now simply execute the algorithm of section 23.5.2 in the order $D_1, \ldots, D_k$.

Why does this algorithm guarantee global optimality of the inferred strategy? When selecting the decision rule for D_i, we have two cases: for $j < i$, by induction, the decision rules for D_j

---

Algorithm 23.3 Iterated optimization for influence diagrams with acyclic relevance graphs

---

Procedure Iterated-Optimization-for-IDs (
    $\mathcal{I}$,   // Influence diagram
    $\mathcal{G}$ // Acyclic relevance graph for $\mathcal{I}$
)
1   Let $D_1, \ldots, D_k$ be an ordering of $\mathcal{D}$ that is a topological ordering for $\mathcal{G}$
2   Let σ^0 be some fully mixed strategy for $\mathcal{I}$
3   **for** $i = 1, \ldots, k$
4   Choose δ_{D_i} to be locally optimal for σ^{i-1}
5        $\sigma^i \leftarrow (\sigma^{i-1}_{-D_i}, \delta_{D_i})$
6   **return** σ^k

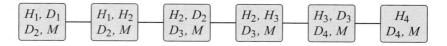

Figure 23.9 **Clique tree for the imperfect-recall influence diagram of figure 23.5.** Although the network has many cascaded decisions, our ability to "forget" previous decisions allows us to solve the problem using a bounded tree-width clique tree.

are already stable, and so will never need to change; for $j > i$, the decision rules for D_j are irrelevant, so that changing them will not require revisiting D_i.

One subtlety with this argument relates, once again, to the issue of probability-zero events. If our arbitrary starting strategy σ assigns probability zero to a certain decision $d \in Val(D)$ (in some setting), then the local optimization of another decision rule D' might end up selecting a suboptimal decision for the zero probability cases. If subsequently, when optimizing the decision rule for D, we ascribe nonzero probability to $D = d$, our overall strategy will not be optimal. To avoid this problem, we can use as our starting point any fully mixed strategy σ. One obvious choice is simply the strategy that, at each decision D and for each assignment to Pa_D, selects uniformly at random between all of the possible values of D.

The overall algorithm is shown in algorithm 23.3.

Theorem 23.5

Applying Iterated-Optimization-for-IDs *on an influence diagram $\mathcal{I}$ whose relevance graph is acyclic, returns a globally optimal strategy for $\mathcal{I}$.*

The proof is not difficult and is left as an exercise (exercise 23.8).

Thus, this algorithm, by iteratively optimizing individual decision rules, finds a globally optimal solution. The algorithm applies to any influence diagram whose relevance graph is acyclic, and hence to any influence diagrams satisfying the perfect recall assumption. Hence, it is at least as general as the variable elimination algorithm of section 23.3. However, as we saw, some influence diagrams that violate the perfect recall assumption have acyclic relevance graphs nonetheless; this algorithm also applies to such cases.

Example 23.22

Consider again the influence diagram of example 23.11. Despite the lack of perfect recall in this *network, an s-reachability analysis shows that each decision variable D_i strategically relies only on D_j for $j > i$. For example, if we add a dummy parent $\widehat{D_1}$ to D_1, we can verify that it is d-separated from V_3 and V_4 given $\mathrm{Pa}_{D_3} = \{H_2, D_2\}$, so that the resulting relevance graph is acyclic.* ∎

The ability to deal with problems where the agent does not have to remember his entire history can provide very large computational savings in large problems. Specifically, we can solve this influence diagram using the clique tree of figure 23.9, at a cost that grows linearly rather than exponentially in the number of decision variables.

This algorithm is guaranteed to find a globally optimal solution only in cases where the relevance graph is acyclic. However, we can extend this algorithm to find a globally optimal solution in more general cases, albeit at some computational cost. In this extension, we simultaneously optimize the rules for subsets of interdependent decision variables. Thus, for example, in example 23.18, we would optimize the decision rules for D_1 and D_2 together, rather than each in isolation. (See exercise 23.9.) This approach is guaranteed to find the globally optimal strategy, but it can be computationally expensive, depending on the number of interdependent decisions that must be considered together.

Box 23.B — Case Study: Coordination Graphs for Robot Soccer. *One subclass of problem in decision making is that of making a joint decision for a team of agents with a shared utility function. Let the world state be defined by a set of variables $\boldsymbol{X} = \{X_1, \ldots, X_n\}$. We now have a team of m agents, each with a decision variable A_i. The team's utility function is described by a function $U(\boldsymbol{X}, \boldsymbol{A})$ (for $\boldsymbol{A} = \{A_1, \ldots, A_m\}$). Given an assignment $\boldsymbol{x}$ to $X_1, \ldots, X_n$, our goal is*

joint action

to find the optimal joint action $\arg\max_{\boldsymbol{a}} U(\boldsymbol{x}, \boldsymbol{a})$.

In a naive encoding, the representation of the utility function grows exponentially both in the number of state variables and in the number of agents. However, we can come up with more efficient algorithms by exploiting the same type of factorization that we have utilized so far. In particular, we assume that we can decompose U as a sum of subutility functions, each of which depends only on the actions of some subset of the agents. More precisely,

$$U(X_1, \ldots, X_n, A_1, \ldots, A_m) = \sum_i V_i(\boldsymbol{X}_i, \boldsymbol{A}_i),$$

where V_i is some subutility function with scope $\boldsymbol{X}_i, \boldsymbol{A}_i$.

This optimization problem is simply a max-sum problem over a factored function, a problem that is precisely equivalent to the MAP problem that we addressed in chapter 13. Thus, we can apply any of the algorithms we described there. In particular, max-sum variable elimination can be used to produce optimal joint actions, whereas max-sum belief propagation can be used to construct approximate max-marginals, which we can decode to produce approximate solutions. The application of these message passing algorithms in this type of distributed setting is satisfying, since the decomposition of the utility function translates to a limited set of interactions between agents who need to coordinate their choice of actions. Thus, this approach has been called a

coordination graph

coordination graph.

RoboSoccer

Kok, Spaan, and Vlassis (2003), in their UvA (Universiteit van Amsterdam) Trilearn team, applied coordination graphs to the RoboSoccer domain, a particularly challenging application of decision making under uncertainty. RoboSoccer is an annual event where teams of real or simulated robotic agents participate in a soccer competition. This application requires rapid decision making under uncertainty and partial observability, along with coordination between the different team members. The simulation league allows teams to compete purely on the quality of their software, eliminating the component of hardware design and maintenance. However, key challenges are faithfully simulated in this environment. For example, each agent can sense its environment via only three sensors: a visual sensor, a body sensor, and an aural sensor. The visual sensor measures relative distance, direction, and velocity of the objects in the player's current field of view. Noise is added to the true quantities and is larger for objects that are farther away. The agent has only a partial view of the world and needs to take viewing actions (such as turning its neck) deliberately in order to view other parts of the field. Players in the simulator have different abilities; for example, some can be faster than others, but they will also tire more easily. Overall, this tournament provides a challenge for real-time, multiagent decision-making architectures.

Kok et al. hand-coded a set of utility rules, each of which represents the incremental gain or loss to the team from a particular combination of joint actions. At every time point t they instantiate the variables representing the current state and solve the resulting coordination graph. Note that there is no attempt to address the problem of sequential decision making, where our choice of action at time t should consider its effect on actions at subsequent time points. The myopic nature of the decision making is based on the assumption that the rules summarize the long-term benefit to the team from a particular joint action.

To apply this framework in this highly dynamic, continuous setting, several adaptations are required. First, to reduce the set of possible actions that need to be considered, each agent is assigned a role: interceptor, passer, receiver, or passive. The assignment of roles is computed directly from the current state information. For example, the fastest player close to the ball will be assigned the passer role when he is able to kick the ball, and the interceptor role otherwise. The assignment of roles defines the structure of the coordination graph: interceptors, passers, and receivers are connected, whereas passive agents do not need to be considered in the joint-action selection process. The roles also determine the possible actions for each agent, which are discrete, high-level actions such as passing a ball to another agent in a given direction. The state variables are also defined as a high-level abstraction of the continuous game state; for example, there is a variable pass-blocked(i, j, d) that indicates whether a pass from agent i to agent j in direction d is blocked by an opponent. With this symbolic representation, one can write value rules that summarize the value gained by a particular combination of actions. For example, one rule says:

$$[\textit{has-role-receiver}(j) \land \neg \textit{isPassBlocked}(i, j, d) \land A_i = \textit{passTo}(j, d) \land A_j = \textit{moveTo}(d) : V(j, d)]$$

where $V(j, d)$ depends on the position where the receiving agent j receive the pass — the closer to the opponent goal the better.

A representation of a utility function as a set of rules is equivalent to a feature-based representation of a Markov network. To perform the optimization efficiently using this representation, we can easily adapt the rule-based variable-elimination scheme described in section 9.6.2.1. Note that the actual rules used in the inference are considerably simpler, since they are conditioned on the state variables, which include the role assignment of the agents and the other aspects of the state (such as isPassBlocked). However, this requirement introduces other complications: because of the limited communication bandwidth, each agent needs to solve the coordination graph on its own. Moreover, the state of the world is not generally fully observed to the agent; thus, one needs to ensure that the agents take the necessary observation actions (such as turning the neck) to obtain enough information to condition the relevant state variables. Depending on the number of agents and their action space, one can now solve this problem using either variable elimination or belief propagation.

The coordination graph framework allows the different agents in the team to conduct complex maneuvers, an agent j would move to receive a pass from agent i even before agent i was in position to kick the ball; by contrast, previous methods required j to observe the trajectory of the ball before being able to act accordingly. This approach greatly increased the capabilities of the UVA Trilearn team. Whereas their entry took fourth place in the RoboSoccer 2002 competition, in 2003 it took first place among the forty-six qualifying team, with a total goal count of 177–7.

23.6   Ignoring Irrelevant Information ⋆

As we saw, there are several significant advantages to reducing the amount of information that the agent considers at each decision. Eliminating an information edge from a variable W into a decision variable D reduces the complexity of its decision rule, and hence the cognitive load on the decision maker. Computationally, it decreases the cost of manipulating its factor and of computing the decision rule. In this section, we consider a procedure for removing information edges from an influence diagram.

Of course, removing information edges reduces the agent's strategy space, and therefore can potentially significantly decrease his maximum expected utility value. If we want to preserve the agent's MEU value, we need to remove information edges with care. We focus here on removing only information edges that do not reduce the agent's MEU. We therefore study when a variable $W \in \mathrm{Pa}_D$ is irrelevant to making the optimal decision at D. In this section, we provide a graphical criterion for guaranteeing that W is irrelevant and can be dropped without penalty from the set Pa_D.

Intuitively, W is not relevant when it has no effect on utility nodes that participate in determining the decision at D.

Example 23.23

Consider the influence diagram $\mathcal{I}_S$ of figure 23.10. Intuitively, the edge from Difficulty (D) to Apply (A) is irrelevant. To understand why, consider its effect on the different utility variables in the network. On one hand, it influences V_S; however, given the variable Grade, which is also observed at A, D is irrelevant to V_S. On the other hand, it influences V_Q; however, V_Q cannot be influenced by the decision at A, and hence is not considered by the decision maker when determining the strategy at A. Overall, D is irrelevant to A given A's other parents.   ∎

We can make this intuition precise as follows:

Definition 23.10

irrelevant
information edge

An information edge $W \to D$ from a (chance or decision) variable W is irrelevant for a decision variable D if there is no active trail from W to $\mathcal{U}_{\succ D}$ given $\mathrm{Pa}_D - \{W\}$.   ∎

According to this criterion, D is irrelevant for A, supporting our intuitive argument. We note that certain recall edges can also be irrelevant according to this definition. For example, assume that we add an edge from *Difficulty* to the decision variable *Take*. The *Difficulty* $\to$ *Take* edge

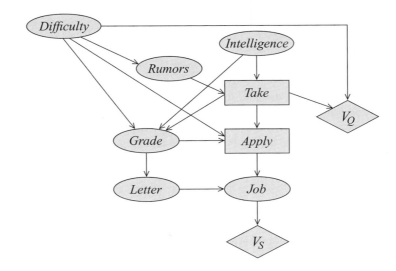

Figure 23.10   More complex influence diagram $\mathcal{I}_S$ **for the** Student **scenario.** Recall edges that follow from the definition are omitted for clarity.

is not irrelevant, but the *Difficulty* → *Apply* edge, which would be implied by perfect recall, is irrelevant.

We can show that irrelevant edges can be removed without penalty from the network.

Proposition 23.4

Let $\mathcal{I}$ be an influence diagram, and $W \rightarrow D$ an irrelevant edge in $\mathcal{I}$. Let $\mathcal{I}'$ be the influence diagram obtained by removing the edge $W \rightarrow D$. Then for any strategy σ in $\mathcal{I}$, there exists a strategy σ' in $\mathcal{I}'$ such that $\mathrm{EU}[\mathcal{I}'[\sigma']] \geq \mathrm{EU}[\mathcal{I}[\sigma]]$.

The proof follows from proposition 23.3 and is left as an exercise (exercise 23.11).

An influence diagram $\mathcal{I}'$ that is obtained from $\mathcal{I}$ via the removal of irrelevant edges is called
reduction   a *reduction* of $\mathcal{I}$. An immediate consequence of proposition 23.4 is the following result:

Theorem 23.6

If $\mathcal{I}'$ is a reduction of $\mathcal{I}$, then any strategy σ that is optimal for $\mathcal{I}'$ is also optimal for $\mathcal{I}$.

The more edges we remove from $\mathcal{I}$, the simpler our computational problem. We would thus like to find a reduction that has the fewest possible edges. One simple method for obtaining a minimal reduction — one that does not admit the removal of any additional edges — is to remove irrelevant edges iteratively from the network one at a time until no further edges can be removed. An obvious question is whether the order in which edges are removed makes a difference to the final result. Fortunately, the following result implies otherwise:

Theorem 23.7

Let $\mathcal{I}$ be an influence diagram and $\mathcal{I}'$ be any reduction of it. An arc $W \rightarrow D$ in $\mathcal{I}'$ is irrelevant in $\mathcal{I}'$ if and only if it is irrelevant in $\mathcal{I}$.

The proof follows from properties of d-separation, and it is left as an exercise (exercise 23.12).

This theorem implies that we can examine each edge independently, and test whether it is irrelevant in $\mathcal{I}$. All such edges can then be removed at once. Thus, we can find all irrelevant edges using a single global computation of d-separation on the original ID.

The removal of irrelevant edges has several important computational benefits. First, it decreases the size of the strategy representation in the ID. Second, by removing edges in the network, it can reduce the complexity of the variable-elimination-based algorithms described in section 23.5. Finally, as we now show, it also has the effect of removing edges from the relevance graph associated with the ID. By breaking cycles in the relevance graph, it allows more decision rules to be optimized in sequence, reducing the need for iterations or for jointly optimizing the decision rules at multiple variables.

Proposition 23.5

If $\mathcal{I}'$ is a reduction of $\mathcal{I}$, then the relevance graph of $\mathcal{I}'$ is a subset (not necessarily strict) of the relevance graph of $\mathcal{I}$.

PROOF It suffices to show the result for the case where $\mathcal{I}'$ is a reduction of $\mathcal{I}$ by a single irrelevant edge. We will show that if D' is not s-reachable from D in $\mathcal{I}$, then it is also not s-reachable from D in $\mathcal{I}'$. If D' is s-reachable from D in $\mathcal{I}'$, then for a dummy parent $\widehat{D'}$, we have that there is some $V \in \mathcal{U}_{\succ D}$ and an active trail in $\mathcal{I}'$ from $\widehat{D}$ to V given $D, \text{Pa}_D^{\mathcal{I}'}$. By assumption, that same trail is not active in $\mathcal{I}$. Since removal of edges cannot make a trail active, this situation can occur only if $\text{Pa}_D^{\mathcal{I}'} = \text{Pa}_D^{\mathcal{I}} - \{W\}$, and W blocks the trail from $\widehat{D'}$ to V in $\mathcal{I}$. Because observing W blocks the trail, it must be part of the trail, in which case there is a subtrail from W to V in $\mathcal{I}$. This subtrail is active given $(\text{Pa}_D^{\mathcal{I}} - \{W\}), D$. However, observing D cannot activate a trail where we condition on D's parents (because then v-structures involving D are blocked). Thus, this subtrail must form an active trail from W to V given $\text{Pa}_D^{\mathcal{I}} - \{W\}$, violating the assumption that $W \to D$ is an irrelevant edge. ∎

23.7 Value of Information

So far, we have focused on the problem of decision making. Influence diagrams provide us with a representation for structured decision problems, and a basis for efficient decision-making algorithms.

One particularly useful type of task, which arises in a broad range of applications, is that of determining which variables we want to observe. Most obviously, in any diagnostic task, we usually have a choice of different tests we can perform. Because tests usually come at a cost (whether monetary or otherwise), we want to select the tests that are most useful in our particular setting. For example, in a medical setting, a diagnostic test such as a biopsy may involve significant pain to the patient and risk of serious injury, as well as high monetary costs. In other settings, we may be interested in determining if and where it is worthwhile to place sensors — such as a thermostat or a smoke alarm — so as to provide the most useful information in case of a fire.

 The decision-theoretic framework provides us with a simple and elegant measure for the value of making a particular observation. Moreover, the influence diagram representation allows us to formulate this measure using a simple, graph-based criterion, which also provides considerable intuition.

23.7.1   Single Observations

We begin with the question of evaluating the benefit of a single observation. In the setting of influence diagrams, we can model this question as one of computing the value of observing the value of some variable. Our *Survey* variable in the Entrepreneur example is precisely such a situation. Although we could (and did) analyze this type of decision using our general framework, it is useful to consider such decisions as a separate (and simpler) class. By doing so, we can gain insight into questions such as these.

The key idea is that the benefit of making an observation is the utility the agent can gain by observing the associated variable, assuming he acts optimally in both settings.

Example 23.24
Let us revisit the Entrepreneur *example, and consider the value to the entrepreneur of conducting the survey, that is, of observing the value of the Survey variable. In effect, we are comparing two scenarios and the utility to the entrepreneur in each of them: One where he conducts the survey, and one where he does not. If the agent does not observe the S variable, that node is barren in the network, and it can therefore be simply eliminated. This would result precisely in the influence diagram of figure 23.2. In example 22.3 we analyzed the agent's optimal action in this setting and showed that his MEU is 2. The second case is one in which the agent conducts the survey. This situation is equivalent to the influence diagram of figure 23.3, where we restrict to strategies where $C = c^1$. As we have already discussed, $C = c^1$ is the optimal strategy in this setting, so that the optimal utility obtainable by the agent in this situation is 3.22, as computed in example 23.6. Hence, the improvement in the entrepreneur's utility, assuming he acts optimally in both cases, is* 1.22. ∎

More generally, we define:

Definition 23.11

value of perfect
information

Let $\mathcal{I}$ be an influence diagram, X a chance variable, and D a decision variable such that there is no (causal) path from D to X. Let $\mathcal{I}'$ be the same as $\mathcal{I}$, except that we add an information edge from X to D, and to all decisions that follow D (that is, we have perfect information about X from D onwards). The value of perfect information *for X at D, denoted $\mathrm{VPI}_{\mathcal{I}}(D \mid X)$, is the difference between the MEU of $\mathcal{I}'$ and the MEU of $\mathcal{I}$.* ∎

Let us analyze the concept of value of perfect information. First, it is not difficult to see that it cannot be negative; if the information is free, it cannot hurt to have it.

Proposition 23.6
Let $\mathcal{I}$ be an influence diagram, D a decision variable in $\mathcal{I}$, and X a chance variable that is a nondescendant of D. Let σ^* be the optimal strategy in $\mathcal{I}$. Then $\mathrm{VPI}_{\mathcal{I}}(D \mid X) \geq 0$, and equality holds if and only if σ^* is still optimal in the new influence diagram with X as a parent of D.

The proof is left as an exercise (exercise 23.13).

Does information always help? What if the numbers had been such that the entrepreneur would have founded the company regardless of the survey? In that case, the expected utility with the survey and without it would have been identical; that is, the VPI of S would have been zero. This property is an important one: **there is no value to information if it does not change the selected action(s) in the optimal strategy.**

Let us analyze more generally when information helps. To do that, consider a different decision problem.

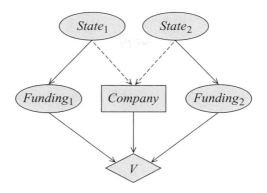

Figure 23.11 Influence diagram for VPI computation in example 23.25. We can compute the value of information for each of the two *State* variables by comparing the value with/without the dashed information edges.

Example 23.25

Our budding entrepreneur has decided that founding a startup is not for him. He is now choosing between two job opportunities at existing companies. Both positions are offering a similar starting salary, so his utility depends on his salary a year down the road, which depends on whether the company is still doing well at that point. The agent has the option of obtaining some information about the current state of the companies.

More formally, the entrepreneur has a decision variable C, whose value c_i is accepting a job with company i ($i = 1, 2$). For each company, we have a variable S_i that represents the current state of the company (quality of the management, the engineering team, and so on); this value takes three values, with s_i^3 being a very high-quality company and s_i^1 a poor company. We also have a binary-valued variable F_i, which represents the funding status of the company in the future, with f_i^1 representing the state of having funding. We assume that the utility of the agent is 1 if he takes a job with a company for which $F_i = f_i^1$ and 0 otherwise. We want to evaluate the value of information of observing S_1 (the case of observing S_2 is essentially the same). The structure of the influence diagram is shown in figure 23.11; the edges that would be added to compute the value of information are shown as dashed.

We now consider three different scenarios and compute the value of information in each of them.

Scenario 1: *Company 1 is well established, whereas Company 2 is a small startup. Thus, $P(S_1) = (0.1, 0.2, 0.7)$ (that is, $P(s_1^1) = 0.1$), and $P(S_2) = (0.4, 0.5, 0.1)$. The economic climate is poor, so the chances of getting funding are not great. Thus, for both companies, $P(f_i^1 \mid S_i) = (0.1, 0.4, 0.9)$ (that is, $P(f_i^1 \mid s_i^1) = 0.1$). Without additional information, the optimal strategy is c_1, with MEU value 0.72. Intuitively, in this case, the information obtained by observing S_1 does not have high value. Although it is possible that c_1 will prove less reliable than c_2, this outcome is very unlikely; with very high probability, c_1 will turn out to be the better choice even with the information. Thus, the probability that the information changes our decision is low, and the value of the information is also low. More formally, a simple calculation shows that the optimal strategy changes to c_2 only if we observe s_1^1, which happens with probability 0.1. The MEU value in this scenario is 0.743, which is not a significant improvement over our original MEU value. If observing*

S_1 *costs more than* 0.023 *utility points, the agent should not make the observation.*

Scenario 2: *The economic climate is still bad, but now c_1 and c_2 are both small startups. In this case, we might have $P(S_2)$ as in Scenario 1, and $P(S_1) = (0.3, 0.4, 0.3)$; $P(F_i \mid S_i)$ is also as in Scenario 1. Intuitively, our value of information in this case is quite high. There is a reasonably high probability that the observation will change our decision, and therefore a high probability that we would gain a lot of utility by finding out more information and making a better decision. Indeed, if we go through the calculation, the MEU strategy in the case without the additional observation is c_1, and the MEU value is 0.546. However, with the observation, we change our decision to c_2 both when $S_1 = s_1^1$ and when $S_1 = s_1^2$, events that are fairly probable. The MEU value in this case is 0.6882, a significant increase over the uninformed MEU.*

Scenario 3: *In this case, c_1 and c_2 are still both small startups, but the time is the middle of the Internet boom, so both companies are likely to be funded by investors desperate to get into this area. Formally, $P(S_1)$ and $P(S_2)$ are as above, but $P(f_i^1 \mid S_i) = (0.6, 0.8, 0.99)$. In this case, the probability that the observation changes the agent's decision is reasonably high, but the change to the agent's expected utility when the decision changes is low. Specifically, the uninformed optimal strategy is c_1, with MEU value 0.816. Observing s_1^1 changes the decision to c_2; but, while this observation occurs with probability 0.3, the difference in the expected utility between the two decisions in this case is less than 0.2. Overall, the MEU of the informed case is 0.8751, which is not much greater than the uninformed MEU value.*   ■

Overall, we see that our definition of VPI allows us to make fairly subtle trade-offs.

The value of information is critical in many applications. For example, in medical or fault diagnosis, it often serves to tell us which diagnostic tests to perform (see box 23.C). Note that its behavior is exactly appropriate in this case. We do not want to perform a test just because it will help us narrow down the probability of the problem. We want to perform tests that will change our diagnosis. For example, if we have an invasive, painful test that will tell us which type of flu a patient has, but knowing that does not change our treatment plan (lie in bed and drink a lot of fluids), there is no point in performing the test.

23.7.2   Multiple Observations

We now turn to the more complex setting where we can make multiple simultaneous observations. In this case, we must decide which subset of the m potentially observable variables we choose to observe. For each such subset, we can evaluate the MEU value with the observations, as in the single variable case, and select the subset whose MEU value is highest. However, this approach is overly simplistic in several ways. First, the number of possible subsets of observations is exponentially large (2^m). A doctor, for example, might have available a large number of tests that she can perform, so the number of possible subsets of tests that she might select is huge. Even if we place a bound on the number of observations that can be performed or on the total cost of these observations, the number of possibilities can be very large.

More importantly, in practice, we often do not select in advance a set of observations to be performed, and then perform all of them at once. Rather, observations are typically made in sequence, so that the choice of which variable to observe next can be made with knowledge about the outcome of the previous observations. In general, the value of an observation can depend strongly on the outcome of a previous one. For example, in example 23.25, if we observe that the current state of Company 1 is excellent — $S_1 = s_1^3$, observing the state of Company 2 is significantly less useful than in a situation where we observe that $S_1 = s_1^2$. Thus, the optimal choice of variable to observe generally depends on the outcomes of the previous observations.

Therefore, when we have the ability to select a sequence of observations, the optimal selection has the form of a *conditional plan*: Start by observing X_1; if we observe $X_1 = x_1^1$, observe X_2; if we observe $X_1 = x_1^2$, observe X_3; and so on. Each such plan is exponentially large in the number k of possible observations that we are allowed to perform. The total number of such plans is therefore doubly exponential. Selecting an optimal observation plan is computationally a very difficult task, for which no good algorithms exist in general.

myopic value of
information

The most common solution to this problem is to approximate the solution using *myopic value of information*, where we incrementally select at each stage the optimal single observation, ignoring its effect on the later choices that we will have to make. The optimal single observation can be selected easily using the methods described in the previous section. This myopic approximation can be highly suboptimal. For example, we might have an observation that, by itself, provides very little information useful for our decision, but does tell us which of two other observations is the most useful one to make.

In situations where the myopic approximation is complex, we can try to generate a conditional plan, as described before. One approach for solving such a problem is to formulate it as an influence diagram, with explicit decisions for which variable to observe. This type of transformation is essentially the one underlying the very simple case of example 23.3, where the variable C represents our decision on whether to observe S or not. The optimal strategy for this extended influence diagram also specifies the optimal observation plan. However, the resulting influence diagram can be quite complex, and finding an optimal strategy for it can be very expensive and often infeasible.

Box 23.C — Case Study: Decision Making for Troubleshooting. *One of the most commonly used applications of Bayesian network technology is to the task of fault diagnosis and repair. Here, we construct a probabilistic model of the device in question, where random variables correspond to different faults and different types of observations about the device state. Actions in this type of domain correspond both to diagnostic tests that can help indicate where the problem lies, and to actions that repair or replace a broken component. Both types of actions have a cost. One can now apply decision-theoretic techniques to help select a sequence of observation and repair actions.*

One of the earliest and largest fielded applications of this type was the decision-theoretic troubleshooting system incorporated into the Microsoft's Windows 95™ operating system. The system, described in Heckerman, Breese, and Rommelse (1995) and Breese and Heckerman (1996), included hundreds of Bayesian networks, each aimed at troubleshooting a type of fault that commonly arises in the system (for example, a failure in printing, or an application that does not launch). Each fault had its own Bayesian network model, ranging in size from a few dozen to a few hundred variables. To compute the probabilities required for an analysis involving repair actions, which intervene in the model, one must take into account the fact that the system state from before the repair also persists afterward (except for the component that was repaired). For this computation, a

counterfactual
twinned network

counterfactual twinned network model was used, as described in box 21.C.

The probabilistic models were augmented with utility models for observing the state of a component in the system (that is, whether it is faulty) and for replacing it. Under carefully crafted assumptions (such as a single fault hypothesis), it was possible to define an optimal series of re-

pair/observation actions, given a current state of information e, *and thereby compute an exact formula for the* expected cost of repair *ECR(e). (See exercise 23.15.) This formula could then be used to compute exactly the benefit of any diagnostic test* D, *using a standard value of information computation:*

$$\sum_{d \in Val(D)} P(D = d \mid e) ECR(e, D = d).$$

One can then add the cost of the observation of D *to choose the optimal diagnostic test. Note that the computation of ECR(e, $D = d$) estimates the cost of the full trajectory of repair actions following the observation, a trajectory that is generally different for different values of the observation* d. *Thus, although this analysis is still myopic in considering only a single observation action* D *at a time, it is nonmyopic in evaluating the cost of the plan of action following the observation.*

Empirical results showed that this technique was very valuable. One test, for example, was applied to the printer diagnosis network of box 5.A. Here, the cost was measured in terms of minutes to repair. In synthetic cases with known failures, sampled from the network, the system saved about 20 percent of the time over the best predetermined plan. Interestingly, the system also performed well, providing equal or even better savings, in cases where there were multiple faults, violating the assumptions of the model.

At a higher level, decision-theoretic techniques are particularly valuable in this setting for several reasons. The standard system used up to that point was a standard static flowchart where the answer to a question would lead to different places in the flowchart. From the user side, the experience was significantly improved in the decision-theoretic system, since there was considerably greater flexibility: diagnostic tests are simply treated as observed variables in the network, so if a user chooses not to answer a question at a particular point, the system can still proceed with other questions or tests. Users also felt that the questions they were asked were intuitive and made sense in context. Finally, there was also significant benefit for the designer of the system, because the decision-theoretic system allowed modular and easily adaptable design. For example, if the system design changes slightly, the changes to the corresponding probabilistic models are usually small (a few CPDs may change, or maybe some variables are added/deleted); but the changes to the "optimal" flowchart are generally quite drastic. Thus, from a software engineering perspective, this approach was also very beneficial.

This application is one of the best-known examples of decision-theoretic troubleshooting, but similar techniques have been successfully used in a large number of applications, including in a decision-support system for car repair shops, in tools for printer and copier repair, and many others.

23.8   Summary

In this chapter, we placed the task of decision making using decision-theoretic principles within the graphical modeling framework that underlies this entire book. Whereas a purely proba-bilistic graphical model provides a factorized description of the probability distribution over possible states of the world, an influence diagram provides such a factorized representation for the agent's actions and utility function as well. The influence diagram clearly encodes the breakdown of these three components of the decision-making situation into variables, as well as the interactions between these variables. These interactions are both probabilistic, where one variable affects the distribution of another, and informational, where observing a variable allows an agent to actively change his action (or his decision rule).

We showed that dynamic programming algorithms, similar to the ones used for pure probabilistic inference, can be used to find an optimal strategy for the agent in an influence diagram. However, as we saw, inference in an influence diagram is more complex than in a Bayesian network, both conceptually and computationally. This complexity is due to the interactions between the different operations involved: products for defining the probability distribution; summation for aggregating utility variables; and maximization for determining the agent's optimal actions.

The influence diagram representation provides a compact encoding of rich and complex decision problems involving multiple interrelated factors. It provides an elegant framework for considering such important issues as which observations are required to make optimal decisions, the definition of recall and the value of the perfect recall assumption, the dependence of a particular decision on particular observations or components of the agent's utility function, and the like. Value of information — a concept that plays a key role in many practical applications — is particularly easy to capture in the influence diagram framework.

However, there are several factors that can cause the complexity of the influence diagram to grow unreasonably large, and significantly reduce its usability in many real-world settings. One such limitation is the perfect recall assumption, which can lead the decision rules to grow exponentially large in the number of actions and observations the agent makes. We note that this limitation is not one of the representation, but rather of the requirements imposed by the notion of optimality and by the algorithms we use to find solutions. A second source of blowup arises when the scenario that arises following one decision by the agent is very different from the scenario following another. For example, imagine that the agent has to decide whether to go from San Francisco to Los Angeles by air or by car. The subsequent decisions he has to make and the variables he may observe in these two cases are likely to be very different. This example is an instance of context-specificity, as described in section 5.2.2; however, the simple solution of modifying our CPD structure to account for context-specificity is usually insufficient to capture compactly these very broad changes in the model structure. The decision-tree structure is better able to capture this type of structure, but it too has its limitations; several works have tried to combine the benefits of both representations (see section 23.9).

Finally, the basic formalism for sequential decision making under uncertainty is only a first step toward a more general formalism for planning and acting under uncertainty in many settings: single-agent, multiagent distributed decision making, and multiagent strategic (game-theoretic) interactions. A complete discussion of the ideas and methods in any of these areas is a book in itself; we encourage the reader who is interested in these topics to pursue some additional readings, some of which are mentioned in section 23.9.

23.9 Relevant Literature

The influence diagram representation was introduced by Howard and Matheson (1984a), albeit more as a guide to formulating a decision problem than as a formal language with well-defined semantics. See also Oliver and Smith (1990) for an overview.

Olmsted (1983) and Shachter (1986, 1988) provided the first algorithm for decision making in influence diagrams, using local network transformations such as edge reversal. This algorithm was gradually improved and refined over the years in a series of papers (Tatman and Shachter 1990; Shenoy 1992; Shachter and Ndilikilikesha 1993; Ndilikilikesha 1994). The most recent algorithm of this type is due to Jensen, Jensen, and Dittmer (1994); their algorithm utilizes the clique tree data structure for addressing this task. All of these solutions use a constrained elimination ordering, and they are therefore generally feasible only for fairly small influence diagrams.

A somewhat different approach is based on reducing the problem of solving an influence diagram to inference in a standard Bayesian network. The first algorithm along these lines is due to Cooper (1988), whose approach applied only to a single decision variable. This idea was subsequently extended and improved considerably by Shachter and Peot (1992) and Zhang (1998).

Nilsson and Lauritzen (2000) and Lauritzen and Nilsson (2001) provide an algorithm based on the concept of limited memory influence diagrams, which relaxes the perfect recall assumption made in almost all previous work. This relaxation allows them to avoid the constraints on the elimination ordering, and thereby leads to a much more efficient clique tree algorithm. Similar ideas were also developed independently by Koller and Milch (2001). The clique-tree approach was further improved by Madsen and Nilsson (2001).

The simple influence diagram framework poses many restrictions on the type of decision-making situation that can be expressed naturally. Key restrictions include the perfect recall assumption (also called "no forgetting"), and the assumption of the uniformity of the paths that traverse the influence diagram.

Regarding this second point, a key limitation of the basic influence diagram representation is that it is designed for encoding situations where all trajectories through the system go through the same set of decisions in the same fixed order. Several authors (Qi et al. 1994; Covaliu and Oliver 1995; Smith et al. 1993; Shenoy 2000; Nielsen and Jensen 2000) propose extensions that deal with asymmetric decision settings, where a choice taken at one decision variable can lead to different decision being encountered later on. Some approaches (Smith et al. 1993; Shenoy 2000) use an approach based on context-specific independence, along the lines of the tree-CPDs of section 5.3. These approaches are restricted to cases where the sequence of observations and decisions is fixed in all trajectories of the system. The approach of Nielsen and Jensen (1999, 2000) circumvents this limitation, allowing for a partial ordering over observations and decisions. The partial ordering allows them to reduce the set of constraints on the elimination ordering in a variable elimination algorithm, resulting in computational savings. This approach was later extended by Jensen and Vomlelová (2003).

In a somewhat related trajectory, Shachter (1998, 1999) notes that some parents of a decision node may be irrelevant for constructing the optimal decision rule, and provided a graphical procedure, based on his BayesBall algorithm, for identifying such irrelevant chance nodes. The LIMID framework of Nilsson and Lauritzen (2000); Lauritzen and Nilsson (2001) makes these notions more explicit by specifically encoding in the influence diagram representation the subset of potentially observable variables relevant to each decision. This allows a relaxation of the ordering constraints induced by the perfect recall assumption. They also define a graphical procedure for identifying which decision rules depend on which others. This approach forms the basis for the recursive algorithm presented in this chapter, and for its efficient implementation using clique trees.

The concept of value of information was first defined by Howard (1966). Over the years, various algorithms (Zhang et al. 1993; Chávez and Henrion 1994; Ezawa 1994) have been proposed for performing value of information computations efficiently in an influence diagram, culminating in the work of Dittmer and Jensen (1997) and Shachter (1999). All of these papers focus on the myopic case and provide an algorithm for computing the value of information only for all single variables in this network (allowing the decision maker to decide which one is best to observe). Recent work of Krause and Guestrin (2005a,b) addresses the nonmyopic problem of selecting an entire sequence of observations to make, within the context of a particular class of utility functions.

There have been several fielded systems that use the decision-theoretic approach described in this chapter, although many use a Bayesian network and a simple utility function rather than a full-fledged influence diagram. Examples of this latter type include the Pathfinder system of Heckerman (1990); Heckerman et al. (1992), and Microsoft's system for decision-theoretic troubleshooting (Heckerman et al. 1995; Breese and Heckerman 1996) that was described in box 23.C. The Vista system of Horvitz and Barry (1995) used an influence diagram to make decisions on display of information at NASA Mission Control Center. Norman et al. (1998) present an influence-diagram system for prenatal testing, as described in box 23.A. Meyer et al. (2004) present a fielded application of an influence diagram for selecting radiation therapy plans for prostate cancer.

Markov decision process

A framework closely related to influence diagrams is that of *Markov decision processes* (MDPs) and its extension to the *partially observable* case (partially observable Markov decision processes, or POMDPs). The formal foundations for this framework were set forth by Bellman (1957); Bertsekas and Tsitsiklis (1996) and Puterman (1994) provide an excellent modern introduction to this topic. Although both an MDP and an influence diagram encode decision problems, the focus of influence diagrams has been on richly spaces that involve rich structure in terms of the state description (and sometimes the utility function), but only a few decisions; conversely, much of the focus of MDPs has been on state spaces that are fairly unstructured (encoded simply as a set of states), but on complex decision settings with long (often infinite) sequences of decisions.

Several groups have worked on the synthesis of these two fields, tackling the problem of sequential decision making in large, richly structured state spaces. Boutilier et al. (1989, 2000) were the first to explore this extension; they used a DBN representation of the MDP, and relied on the use of context-specific structure both in the system dynamics (tree-CPDs) and in the form of the value function. Boutilier, Dean, and Hanks (1999) provide a comprehensive survey of the representational issues and of some of the earlier algorithms in this area. Koller and Parr (1999); Guestrin et al. (2003) were the first to propose the use of factored value functions, which decompose additively as a sum of subutility functions with small scope. Building on the rule-based variable elimination approach described in section 9.6.2.1, they also show how to make use of both context-specific structure and factorization.

Another interesting extension that we did not discuss is the problem of decision making in multiagent systems. At a high level, one can consider two different types of multiagent systems: ones where the agents share a utility function, and need to cooperate in a decentralized setting with limited communication; and ones where different agents have different utility functions, and must optimize their own utility while accounting for the other agents' actions. Guestrin et al. (2003) present some results for the cooperative case, and introduce the notion of coordination graph; they focus on issues that arise within the context of MDPs, but some of their ideas can also be applied to influence diagrams. The coordination graph structure was the basis for the

game theory

RoboSoccer application of Kok et al. (2003); Kok and Vlassis (2005), described in box 23.B.

The problem of optimal decision making in the presence of strategic interactions is the focus of most of the work in the field of *game theory*. In this setting, the notion of a "rational strategy" is somewhat more murky, since what is optimal for one player depends on the actions taken by others. Fudenberg and Tirole (1991) and Osborne and Rubinstein (1994) provide a good introduction to the field of game theory, and to the standard solution concepts used. Generally, work in game theory has represented multiagent interactions in a highly unstructured way: either in the *normal form*, which lists a large matrix indexed by all possible strategies of all agents, or in the *extensive form* — a game tree, a multiplayer version of a decision tree. More recently, there have been several proposals for game representations that build on ideas in graphical models. These proposals include graphical games (Kearns et al. 2001), multiagent influence diagrams (Koller and Milch 2003), and game networks (La Mura 2000). Subsequent work (Vickrey and Koller 2002; Blum et al. 2006) has shown that ideas similar to those used for inference in graphical models and influence diagrams can be used to provide efficient algorithms for finding Nash equilibria (or approximate Nash equilibria) in these structured game representations.

23.10   Exercises

Exercise 23.1

Show that the decision rule δ_D that maximizes: $\sum_{D,\mathrm{Pa}_D} \delta_D \mu_{-D}(D, \mathrm{Pa}_D)$ is defined as:

$$\delta_D(\boldsymbol{w}) = \arg \max_{d \in Val(D)} \mu_{-D}(d, \boldsymbol{w}) \quad \text{for all } \boldsymbol{w} \in Val(\mathrm{Pa}_D).$$

Exercise 23.2

Prove proposition 23.1. In particular:

a. Show that for γ^* defined as in equation (23.8), we have that $\phi^* = \prod_{W \in \mathcal{X} \cup \mathcal{D}} \phi_W$, $\mu^* = \prod_{V \in \mathcal{U}} \mu_V$.

b. For $\boldsymbol{W}' \subset \boldsymbol{W}$, show that

$$\mathrm{cont}(marg_{\boldsymbol{W}'}(\gamma)) = \sum_{\boldsymbol{W} - \boldsymbol{W}'} \mathrm{cont}(\gamma),$$

that is, that contraction and marginalization interchange appropriately.

c. Use your previous results to prove proposition 23.1.

Exercise 23.3★

Prove theorem 23.1 by showing that the combination and marginalization operations defined in equation (23.5) and equation (23.6) satisfy the axioms of exercise 9.19:

a. Commutativity and associativity of combination:

$$\gamma_1 \bigoplus \gamma_2 = \gamma_2 \bigoplus \gamma_1$$
$$\gamma_1 \bigoplus (\gamma_2 \bigoplus \gamma_3) = (\gamma_1 \bigoplus \gamma_2) \bigoplus \gamma_3.$$

b. Consonance of marginalization: Let γ be a factor over scope $\boldsymbol{W}$ and let $\boldsymbol{W}_2 \subseteq \boldsymbol{W}_1 \subseteq \boldsymbol{W}$. Then:

$$marg_{\boldsymbol{W}_2}(marg_{\boldsymbol{W}_1}(\gamma)) = marg_{\boldsymbol{W}_2}(\gamma).$$

c. Interchanging marginalization and combination: Let γ_1 and γ_2 be potentials over $\boldsymbol{W}_1$ and $\boldsymbol{W}_2$ respectively. Then:

$$marg_{\boldsymbol{W}_1}((\gamma_1 \bigoplus \gamma_2)) = \gamma_1 \bigoplus marg_{\boldsymbol{W}_1}(\gamma_2).$$

Exercise 23.4★

Prove lemma 23.1.

Exercise 23.5★★

Extend the variable elimination algorithm of section 23.3 to the case of multiple utility variables, using the mechanism of joint factors used in section 23.4.3. (Hint: Define an operation of max-marginalization, as required for optimizing a decision variable, for a joint factor.)

Exercise 23.6★

Prove proposition 23.3. (Hint: The proof is based on algebraic manipulation of the expected utility $EU[\mathcal{I}[(\sigma_{-D}, \delta_D)]]$.)

Exercise 23.7

Prove theorem 23.4, as follows:

a. Show that if D_i and D_j are two decisions such that $D_i, \mathrm{Pa}_{D_i} \subseteq \mathrm{Pa}_{D_j}$, then D_i is not s-reachable from D_j.
b. Use this result to conclude the theorem.
c. Show that the nodes in the relevance graph in this case will be totally ordered, in the opposite order to the temporal ordering $\prec$ over the decisions in the influence diagram.

Exercise 23.8★

In this exercise, you will prove theorem 23.5, using two steps.

a. We first need to prove a result analogous to theorem 23.2, but showing that a decision rule δ_D remains optimal even if the decision rules at several decisions D' change.
   Let σ be a fully mixed strategy, and δ_D a decision rule for D that is locally optimal for σ. Let σ' be another strategy such that, whenever $\sigma'(D') \neq \sigma(D')$, then D' is not s-reachable from D. Prove that δ_D is also optimal for σ'.
b. Now, let σ^k be the strategy returned by Iterated-Optimization-for-IDs, and σ' be some other strategy for the agent. Let $D_1, \ldots, D_k$ be the ordering on decisions used by the algorithm. Show that $EU[\mathcal{I}[\sigma^n]] \geq EU[\mathcal{I}[\sigma']]$. (Hint: Use induction on the number of variables l at which σ^k and σ' differ.)

Exercise 23.9★★

Extend the algorithm of algorithm 23.3 to find a globally optimal solution even in influence diagrams with cyclic relevance graphs. Your algorithm will have to optimize several decision rules simultaneously, but it should not always optimize all decision rules simultaneously. Explain precisely how you jointly optimize multiple decision rules, and how you select the order in which decision rules are optimized.

Exercise 23.10★★

In this exercise, we will define an efficient clique tree implementation of the algorithm of algorithm 23.3.

a. Describe a clique tree algorithm for a setting where cliques and sepsets are each parameterized with a joint (probability, utility) potential, as described in section 23.4.3. Define: (i) the clique tree initialization in terms of the network parameterization and a complete strategy σ, and (ii) the message passing operations.

b. Show how we can use the clique-tree data structure to reuse computation between different steps of the iterated optimization algorithm. In particular, show how we can easily retract the current decision rule δ_D from the calibrated clique tree, compute a new optimal decision rule for D, and then update the clique tree accordingly. (Hint: Use the ideas of section 10.3.3.1.)

Exercise 23.11★

Prove proposition 23.4.

Exercise 23.12

Prove theorem 23.7: Let $\mathcal{I}$ be an influence diagram and $\mathcal{I}'$ be any reduction of it, and let $W \to D$ be some arc in $\mathcal{I}'$.

a. (easy) Prove that if $W \to D$ is irrelevant in $\mathcal{I}$, then it is also irrelevant in $\mathcal{I}'$.
b. (hard) Prove that if $W \to D$ is irrelevant in $\mathcal{I}'$, then it is also irrelevant in $\mathcal{I}$.

Exercise 23.13

a. Prove proposition 23.6.
b. Is the value of learning the values of two variables equal to the sum of the values of learning each of them? That is to say, is

$$\mathrm{VPI}(\mathcal{I}, D, \{X, Y\}) = \mathrm{VPI}(\mathcal{I}, D, X) + \mathrm{VPI}(\mathcal{I}, D, Y)?$$

Exercise 23.14★

Consider an influence diagram $\mathcal{I}$, and assume that we have computed the optimal strategy for $\mathcal{I}$ using the clique tree algorithm of section 23.5.2. Let D be some decision in D, and X some variable not observed at D in $\mathcal{I}$. Show how we can efficiently compute $\mathrm{VPI}_{\mathcal{I}}(D \mid X)$, using the results of our original clique tree computation, when:

a. D is the only decision variable in $\mathcal{I}$.
b. The influence diagram contains additional decision variables, but the relevance graph is acyclic.

Exercise 23.15★

Consider a setting where we have a faulty device. Assume that the failure can be caused by a failure in one of n components, exactly one of which is faulty. The probability that repairing component c_i will repair the device is p_i. By the single-fault hypothesis, we have that $\sum_{i=1}^{n} p_i = 1$. Further assume that each component c_i can be examined with cost C_i^o and then repaired (if faulty) with cost C_i^r. Finally, assume that the costs of observing and repairing any component do not depend on any previous actions taken.

a. Show that if we observe and repair components in the order $c_1, \ldots, c_n$, then the expected cost until the device is repaired is:

$$\sum_{i=1}^{n} \left[\left(1 - \sum_{j=1}^{i-1} p_j \right) C_i^o + p_i C_i^r \right].$$

b. Use that to show that the optimal sequence of actions is the one in which we repair components in order of their p_i / C_i^o ratio.
c. Extend your analysis to the case where some components can be replaced, but not observed; that is, we cannot determine whether they are broken or not.

Exercise 23.16

value of control

The value of perfect information measures the change in our MEU if we allow observing a variable that was not observed before. In the same spirit, define a notion of a *value of control*, which is the gain to the agent if she is allowed to intervene at a chance variable X and set its value. Make reasonable assumptions about the space of strategies available to the agent, but state your assumptions explicitly.

24 *Epilogue*

Why Probabilistic Graphical Models?

In this book, we have presented a framework of structured probabilistic models. This framework rests on two foundations:

- the use of a probabilistic model — a joint probability distribution — as a representation of our domain knowledge;
- the use of expressive data structures (such as graphs or trees) to encode structural properties of these distributions.

declarative representation

The first of these ideas has several important ramifications. First, **our domain knowledge is encoded *declaratively*, using a representation that has its own inherent semantics. Thus, the conclusions induced by the model are intrinsic to it, and not dependent on a specific implementation or algorithm. This property gives us the flexibility to develop a range of inference algorithms, which may be appropriate in different settings. As long as each algorithm remains faithful to the underlying model semantics, we know it to be correct.**

Moreover, because the basic operations of the calculus of probabilities (conditioning, marginalization) are generally well accepted as being sound reasoning patterns, we obtain an important guarantee: If we obtain surprising or undesirable conclusions from our probabilistic model, the problem is with our model, not with our basic formalism. Of course, this conclusion relies on the assumption that we are using exact probabilistic inference, which implements (albeit efficiently) the operations of this calculus; when we use approximate inference, errors induced by the algorithm may yield undesirable conclusions. Nevertheless, **the existence of a declarative representation allows us to separate out the two sources of error — modeling error and algorithmic error — and consider each separately.** We can ask separately whether our model is a correct reflection of our domain knowledge, and whether, for the model we have, approximate inference is introducing overly large errors. Although the answer to each of these questions may not be trivial to determine, each is more easily considered in isolation. For example, to test the model, we might try different queries or perform sensitivity analysis. To test an approximate inference algorithm, we might try the algorithm on fragments of the network, try a different (approximate or exact) inference algorithm, or compare the probability of the answer obtained to that of an answer we may expect.

A third benefit to the use of a declarative probabilistic representation is the fact that the same representation naturally and seamlessly supports multiple types of reasoning. We can compute the posterior probability of any subset of variables given observations about any others, subsuming reasoning tasks such as prediction, explanation, and more. We can compute the most likely joint assignment to all of the variables in the domain, providing a solution to a problem

abduction   known as *abduction*. With a few extensions to the basic model, we can also answer causal queries and make optimal decisions under uncertainty.

The second of the two ideas is the key to making probabilistic inference practical. **The ability to exploit structure in the distribution is the basis for providing a compact representation of high-dimensional (or even infinite-dimensional) probability spaces. This compact representation is highly modular, allowing a flexible representation of domain knowledge that can easily be adapted, whether by a human expert or by an automated algorithm.** This property is one of the key reasons for the use of probabilistic models. For example, as we discussed in box 23.C, a diagnostic system designed by a human expert to go through a certain set of menus asking questions is very brittle: even small changes to the domain knowledge can lead to a complete reconstruction of the menu system. By contrast, a system that uses inference relative to an underlying probabilistic model can easily be modified simply by revising the model (or small parts of it); these changes automatically give rise to a new interaction with the user.

The compact representation is also the key for the construction of effective reasoning algorithms. All of the inference algorithms we discussed exploit the structure of the graph in fundamental ways to make the inference feasible. Finally, the graphical representation also provides the basis for learning these models from data. First, the smaller parameter space utilized by these models allows parameter estimation even of high-dimensional distributions from a reasonable amount of data. Second, the space of sparse graph structures defines an effective and natural bias for structure learning, owing to the ubiquity of (approximate) conditional independence properties in distributions arising in the real world.

The Modeling Pipeline

The framework of probabilistic graphical models provides support for natural representation, effective inference, and feasible model acquisition. Thus, it naturally leads to an integrated methodology for tackling a new application domain — a methodology that relies on all three of these components.

Consider a new task that we wish to address. We first define a class of models that encode the key properties of the domain that are critical to the task. We then use learning to fill in the missing details of the model. The learned model can be used as the basis for knowledge discovery, with the learned structure and parameters providing important insights about properties of the domain; it can also be used for a variety of reasoning tasks: diagnosis, prediction, or decision making.

Many important design decisions must be made during this process. One is the form of the graphical model. We have described multiple representations throughout this book — directed and undirected, static and temporal, fixed or template-based, with a variety of models for local interactions, and so forth. These should not be considered as mutually exclusive options, but rather as useful building blocks. Thus, a model does not have to be either a Bayesian network or a Markov network — perhaps it should have elements of both. A model may be neither a full dynamic Bayesian network nor a static one: perhaps some parts of the system can be modeled as static, and others as dynamic.

In another decision, when designing our class of models, we can provide a fairly specific description of the models we wish to consider, or one that is more abstract, specifying only high-level properties such as the set of observed variables. Our prior knowledge can be incorporated in a variety of ways: as hard constraints on the learned model, as a prior, or perhaps even only as an initialization for the learning algorithm. Different combinations will be appropriate for different applications.

These decisions, of course, influence the selection of our learning algorithm. In some cases, we will need to fill in only (some) parameters; in others, we can learn significant aspects of the model structure. In some cases, all of the variables will be known in advance; in others, we will need to infer the existence and role of hidden variables.

 When designing a class of models, it is critical to keep in mind the basic trade-off between faithfulness — accurately modeling the variables and interactions in the domain — and identifiability — the ability to reliably determine the details of the model. Given the richness of the representations one can encode in the framework of probabilistic models, it is often very tempting to select a highly expressive representation, which really captures everything that we think is going on in the domain. Unfortunately, such models are often hard to identify from training data, owing both to the potential for overfitting and to the large number of local maxima that can make it difficult to find the optimal model (even when enough training data are available). Thus, one should always keep in mind Einstein's maxim:

Everything should be made as simple as possible, but not simpler.

There are many other design decisions that influence our learning algorithm. Most obviously, there are often multiple learning algorithms that are applicable to the same class of models. Other decisions include what priors to use, when and how to introduce hidden variables, which features to construct, how to initialize the model, and more. Finally, if our goal is to use the model for knowledge discovery, we must consider issues such as methods for evaluating our confidence in the learned model and its sensitivity to various choices that we made in the design. Currently, these decisions are primarily made using individual judgment and experience.

Finally, if we use the model for inference, we also have various decisions to make. For any class of models, there are multiple algorithms — both exact and approximate — that one can apply. Each of these algorithms works well in certain cases and not others. It is important to remember that here, too, we are not restricted to using only a pure version of one of the inference algorithms we described. We have already presented hybrid methods such as collapsed sampling methods, which combine exact inference and sampling. However, many other hybrids are possible and useful. For example, we might use collapsed particle methods combined with belief propagation rather than exact inference, or use a variational approximation to provide a better proposal distribution for MCMC methods.

Overall, it is important to realize that what we have provided is a set of ideas and tools. One can be flexible and combine them in different ways. Indeed, one can also extend these ideas, constructing new representations and algorithms that are based on these concepts. This is precisely the research endeavor in this field.

Some Current and Future Directions

Despite the rapid advances in this field, there are many directions in which significant open problems remain. Clearly, one cannot provide a comprehensive list of all of the interesting open problems; indeed, identifying an open problem is often the first step in a research project. However, we describe here some broad categories of problems where there is clearly much work that needs to be done.

On the pragmatic side, probabilistic models have been used as a key component in addressing some very challenging applications involving automated reasoning and decision making, data analysis, pattern recognition, and knowledge discovery. We have mentioned some of these applications in the case studies provided in this book, but there are many others to which this technology is being applied, and still many more to which it could be applied. There is much work to be done in further developing these methods in order to allow their effective application to an increasing range of real-world problems.

However, our ability to easily apply graphical models to solve a range of problems is limited by the fact that many aspects of their application are more of an art than a science. As we discussed, there are many important design decisions in the selection of the representation, the learning procedure, and the inference algorithm used. Unfortunately, there is no systematic procedure that one can apply in navigating these design spaces. Indeed, there is not even a comprehensive set of guidelines that tell us, for a particular application, which combination of ideas are likely to be useful. At the moment, the design process is more the result of trial-and-error experimentation, combined with some rough intuitions that practitioners learn by experience. It would be an important achievement to turn this process from a black art into a science.

At a higher level, one can ask whether the language of probabilistic graphical models is adequate for the range of problems that we eventually wish to address. Thus, a different direction is to extend the expressive power of probabilistic models to incorporate a richer range of concepts, such as multiple levels of abstractions, complex events and processes, groups of objects with a rich set of interactions between them, and more. If we wish to construct a representation of general world knowledge, and perhaps to solve truly hard problems such as perception, natural language understanding, or commonsense reasoning, we may need a representation that accommodates concepts such as these, as well as associated inference and learning algorithms. Notably, many of these issues were tackled, with varying degrees of success, within the disciplines of philosophy, psychology, linguistics, and traditional knowledge representation within artificial intelligence. Perhaps some of the ideas developed in this long-term effort can be integrated into a probabilistic framework, which also supports reasoning from limited observations and learning from data, providing an alternative starting point for this very long-term endeavor.

The possibility that these models can be used as the basis for solving problems that lie at the heart of human intelligence raises an entirely new and different question: Can we use models such as these as a tool for understanding human cognition? In other words, can these structured models, with their natural information flow over a network of concepts, and their ability to integrate intelligently multiple pieces of weak evidence, provide a good model for human cognitive processes? Some preliminary evidence on this question is promising, and it suggests that this direction is worthy of further study.

A *Background Material*

A.1 Information Theory

Information theory deals with questions involving efficient coding and transmission of information. To address these issues, one must consider how to encode information so as to maximize the amount of data that can sent on a given channel, and how to deal with noisy channels. We briefly touch on some technical definitions that arise in information theory, and use compression as our main motivation. Cover and Thomas (1991) provides an excellent introduction to information theory, including historical perspective on the development and applications of these notions.

A.1.1 Compression and Entropy

Suppose that one plans to transmit a large corpus of say English text over a digital line. One option is to send the text using standard (for example, ASCII) encoding that uses a fixed number of bits per character. A somewhat more efficient approach is to use a code that is tailored to the task of transmitting English text. For example, if we construct a dictionary of all words, we can use binary encoding to describe each word; using 16 bits per word, we can encode a dictionary of up to 65,536 words, which covers most English text.

compression We can gain an additional boost in *compression* by building a *variable-length code*, which encodes different words in bit strings of different length. The intuition is that words that are frequent in English should be encoded by shorter code words, and rare words should be encoded by longer ones. To be unambiguously decodable, a variable-length code must be *prefix free*: no codeword can be a strict prefix of another. Without this property, we would not be able to tell (at least not using a simple scan of the data) when one code word ends and the next begins.

It turns out that variable-length codes can significantly improve our compression rate:

Example A.1

Assume that our dictionary contains four words — w_1, w_2, w_3, w_4 — with frequencies $P(w_1) = 1/2$, $P(w_2) = 1/4$, $P(w_3) = 1/8$, and $P(w_4) = 1/8$. One prefix-free encoding for this dictionary is to encode w_1 using a single bit codeword, say "0"; we would then encode w_2 using the 2-bit sequence "10", and w_3 and w_4 using three bits each "110" and "111".

Now, consider the expected number of bits that we would need for a message sent with this frequency distribution. We must encode the word w_1 on average half the time, and it costs us 1 bit. We must encode the word w_2 a quarter of the time, and it costs us 2 bits. Overall, we get that the *expected number of bits used is:*

$$\frac{1}{2} \cdot 1 + \frac{1}{4} \cdot 2 + \frac{1}{8} \cdot 3 + \frac{1}{8} \cdot 3 = 1.75.$$ ∎

One might ask whether a different encoding would give us better compression performance in this example. It turns out that this encoding is the best we can do, relative to the word-frequency distribution. To provide a formal analysis for this statement, suppose we have a random variable X that denotes the next item we need to encode (for example, a word). In order to analyze the performance of a compression scheme, we need to know the distribution over different values of X. So we assume that we have a distribution $P(X)$ (for example, frequencies of different words in a large corpus of English documents).

The notion of the entropy of a distribution provides us with a precise lower bound for the expected number of bits required to encode instances sampled from $P(X)$.

Definition A.1

entropy

Let $P(X)$ be a distribution over a random variable X. The entropy of X is defined as

$$\boldsymbol{H}_P(X) = \boldsymbol{E}_P\left[\log\frac{1}{P(x)}\right] = \sum_x P(x)\log\frac{1}{P(x)},$$

where we treat $0\log 1/0 = 0$.[1] ∎

When discussing entropies (and other information-theoretic measures) we use logarithms of base 2. We can then interpret the entropy in terms of bits.

The central result in information theory is a theorem by Shannon showing that the entropy of X is the lower bound on the average number of bits that are needed to encode values of X. That is, if we consider a proper codebook for values of X (one that can be decoded unambiguously), then the expected code length, relative to the distribution $P(X)$, cannot be less than $\boldsymbol{H}_P(X)$ bits.

Going back to our example, we see that the average number of bits for this code is precisely the entropy. Thus, the lower bound is tight in this case, in that we can construct a code that achieves precisely that bound. As another example, consider a uniform distribution $P(X)$. In this case, the optimal encoding is to represent each word using the same number of bits, $\log|Val(X)|$. Indeed, it is easy to verify that $\boldsymbol{H}_P(X) = \log|Val(X)|$, so again the bound is tight (at least for cases where $|Val(X)|$ is a power of 2.) Somewhat surprisingly, the entropy bound is tight in general, in that there are codes that come very close to the "optimum" of assigning the value x a code of length $-\log P(x)$.[2]

 Another way of viewing the entropy is as a measure of our uncertainty about the value of X. Consider a game where we are allowed to ask yes/no questions until we pinpoint the value X. Then the entropy of X is average number of questions we need to ask to get to the answer (if we have a good strategy for asking them). If we have little uncertainty about X, then we get to the value with few questions. An extreme case is when $\boldsymbol{H}_P(X) = 0$. It is easy to verify that this can happen only when one value of X has probability 1 and the rest probability

---

1. To justify this, note that $\lim_{\epsilon\to 0}\epsilon\log\frac{1}{\epsilon} = 0$.
2. This value is not generally an integer, so one cannot directly map x to a code word with $-\log P(x)$ bits. However, by coding longer sequences rather than individual values, we can come arbitrarily close to this bound.

0. In this case, we do not need to ask any questions to get to the value of X. On the other hand, if the value of X is very uncertain, then we need to ask many questions.

This discussion in fact identifies the two boundary cases for $H_P(X)$.

Proposition A.1

$$0 \leq H_P(X) \leq \log |Val(X)|$$

The definition of entropy naturally extends to multiple variables.

Definition A.2

joint entropy

Suppose we have a joint distribution over random variables $X_1, \ldots, X_n$. Then the joint entropy of $X_1, \ldots, X_n$ is

$$H_P(X_1, \ldots, X_n) = E_P \left[\log \frac{1}{P(X_1, \ldots, X_n)} \right].$$

∎

The joint entropy captures how many bits are needed (on average) to encode joint instances of the variables.

A.1.2 Conditional Entropy and Information

Suppose we are encoding the values of X and Y. A natural question is what is the cost of encoding X if we are already encoding Y. Formally, we can examine the difference between $H_P(X, Y)$ — the number of bits needed (on average) to encode of both variables, and $H_P(Y)$ — the number of bits needed to encode Y alone.

Definition A.3

conditional entropy

The conditional entropy of X given Y is

$$H_P(X \mid Y) = H_P(X, Y) - H_P(Y) = E_P \left[\log \frac{1}{P(X \mid Y)} \right].$$

∎

This quantity captures the additional cost (in terms of bits) of encoding X when we are already encoding Y. The definition gives rise to the *chain rule of entropy*:

entropy chain rule

Proposition A.2

For any distribution $P(X_1, \ldots, X_n)$, we have that

$$H_P(X_1, \ldots, X_n) = H_P(X_1) + H_P(X_2 \mid X_1) + \ldots + H_P(X_n \mid X_1, \ldots, X_{n-1}).$$

That is, to encode a joint value of $X_1, \ldots, X_n$, we first need to encode X_1, then encode X_2 given that we know the value of X_1, then encode X_3 given the first two, and so on. Note that, similarly to the chain rule of probabilities, we can expand the chain rule in any order we prefer; that is, all orders result in precisely the same value.

Intuitively, we would expect $H_P(X \mid Y)$, the additional cost of encoding X when we already encode Y, to be at least as small as the cost of encoding X alone. To motivate that, we see that the worst case scenario is where we encode X as though we did not know the value of Y. Indeed, one can formally show

Proposition A.3

$$H_P(X \mid Y) \leq H_P(X).$$

The difference between these two quantities is of special interest.

Definition A.4   *The* mutual information *between* X *and* Y *is*

mutual
information
$$\boldsymbol{I}_P(X;Y) = \boldsymbol{H}_P(X) - \boldsymbol{H}_P(X \mid Y) = \boldsymbol{E}_P \left[\log \frac{P(X \mid Y)}{P(X)} \right].$$ ∎

The mutual information captures how many bits we save (on average) in the encoding of X if
we know the value of Y. Put in other words, it represents the extent to which the knowledge of
Y reduces our uncertainty about X.

The mutual information satisfies several nice properties.

Proposition A.4   • $0 \leq \boldsymbol{I}_P(X;Y) \leq \boldsymbol{H}_P(X)$.

• $\boldsymbol{I}_P(X;Y) = \boldsymbol{I}_P(Y;X)$.

• $\boldsymbol{I}_P(X;Y) = 0$ *if and only if* X *and* Y *are independent.*

Thus, the mutual information is nonnegative, and equal to 0 if and only if the two variables
are independent of each other. This is fairly intuitive, since if X and Y are independent, then
learning the value of Y does not tell us any thing new about the value of X. In fact, we can view
the mutual information as a quantitative measure of the strength of the dependency between
X and Y. The bigger the mutual information, the stronger the dependency. The extreme upper
value of the mutual information is when X is a deterministic function of Y (or vice versa). In
this case, once we know Y we are certain about the value of X, and so $\boldsymbol{I}_P(X;Y) = \boldsymbol{H}_P(X)$.
That is, Y supplies the maximal amount of information about X.

A.1.3 Relative Entropy and Distances Between Distributions

In many situations when doing probabilistic reasoning, we want to compare two distributions.
For example, we might want to approximate a distribution by one with desired qualities (say,
simpler representation, more efficient to reason with, and so on) and want to evaluate the quality
of a candidate approximation. Another example is in the context of learning a distribution from
data, where we want to compare the learned distribution to the "true" distribution from which
the data was generated.

distance measure   Thus, we want to construct a *distance measure* d that evaluates the distance between two
distributions. There are some properties that we might wish for in such a distance measure:

Positivity: $d(P, Q)$ is always nonnegative, and is zero if and only if $P = Q$;

Symmetry: $d(P, Q) = d(Q, P)$.

Triangle inequality: for any three distributions P, Q, R, we have that

$$d(P, R) \leq d(P, Q) + d(Q, R).$$

distance metric   When a distance measure d satisfies these criteria, it is called a *distance metric*.

We now review several common approaches used to compare distributions. We begin by
describing one important measure that is motivated by information-theoretic considerations. It
also turns out to arise very naturally in a wide variety of probabilistic settings.

A.1.3.1   Relative Entropy

Consider the preceding discussion of compression. As we discussed, the entropy measures the performance of "optimal" code that assigns the value x a code of length $-\log P(x)$. However, in many cases in practice, we do not have access to the true distribution P that generates the data we plan to compress. Thus, instead of using P we use another distribution Q (say one we estimated from prior data, or supplied by a domain expert), which is our best guess for P.

Suppose we build a code using Q. Treating Q as a proxy to the real distribution, we use $-\log Q(x)$ bits to encode the value x. Thus, the expected number of bits we use on data generated from P is

$$\boldsymbol{E}_P\left[\log \frac{1}{Q(x)}\right].$$

A natural question is how much we lost, due to the inaccuracy of using Q. Thus, we can examine the difference between this encoding and the best achievable one, $H_P(X)$. This difference is called the relative entropy.

Definition A.5

relative entropy

Let P and Q be two distributions over random variables $X_1, \ldots, X_n$. The relative entropy *of P and Q is*

$$\boldsymbol{D}(P(X_1,\ldots,X_n)\|Q(X_1,\ldots,X_n)) = \boldsymbol{E}_P\left[\log \frac{P(X_1,\ldots,X_n)}{Q(X_1,\ldots,X_n)}\right].$$ ∎

When the set of variables in question is clear from the context, we use the shorthand notation $\boldsymbol{D}(P\|Q)$. This measure is also often known as the *Kullback-Liebler divergence* (or *KL-divergence*).

This discussion suggests that the relative entropy measures the additional cost imposed by using a wrong distribution Q instead of P. Thus, Q is close, in the sense of relative entropy, to P if this cost is small. As we expect, the additional cost of using the wrong distribution is always positive. Moreover, the relative entropy is 0 if and only if the two distributions are identical:

Proposition A.5

$\boldsymbol{D}(P\|Q) \geq 0$, *and is equal to zero if and only if $P = Q$.*

It is also natural to ask whether the relative entropy is also bounded from above. As we can quickly convince ourselves, if there is a value x such that $P(x) > 0$ and $Q(x) = 0$, then the relative entropy $\boldsymbol{D}(P\|Q)$ is infinite. More precisely, if we consider a sequence of distributions Q_ϵ such that $Q_\epsilon(x) = \epsilon$, then $\lim_{\epsilon \to 0} \boldsymbol{D}(P\|Q_\epsilon) = \infty$.

It is natural ask whether the relative entropy defines a distance measure over distributions. Proposition A.5 shows that the relative entropy satisfies the positivity property specified above. Unfortunately, **positivity is the only property of distances that relative entropy satisfies; it satisfies neither symmetry nor the triangle inequality.** Given how natural these properties are, one might wonder why relative entropy is used at all. Aside from the fact that it arises very naturally in many settings, it also has a variety of other useful properties, that often make up for the lack of symmetry and the triangle inequality.

A.1.3.2   Conditional Relative Entropy

As with entropies, we can define a notion of conditional relative entropy.

Definition A.6

conditional
relative entropy

Let P and Q be two distributions over random variables X, Y. The conditional relative entropy of P and Q, is

$$\boldsymbol{D}(P(X \mid Y) \| Q(X \mid Y)) = \boldsymbol{E}_P\left[\log \frac{P(X \mid Y)}{Q(X \mid Y)}\right].$$ ∎

We can think of the conditional relative entropy $\boldsymbol{D}(P(X \mid Y) \| Q(X \mid Y))$ as the weighted sum of the relative entropies between the conditional distributions given different values of y

$$\boldsymbol{D}(P(X \mid Y) \| Q(X \mid Y)) = \sum_y P(y) \boldsymbol{D}(P(X \mid y) \| Q(X \mid y)).$$

relative entropy
chain rule

Using the conditional relative entropy, we can write the *chain rule of relative entropy*:

Proposition A.6

Let P and Q be distributions over $X_1, \ldots, X_n$, then

$$\begin{aligned}
\boldsymbol{D}(P\|Q) \quad = \quad & \boldsymbol{D}(P(X_1)\|Q(X_1)) + \\
& \boldsymbol{D}(P(X_2 \mid X_1)\|Q(X_2 \mid X_1)) + \ldots + \\
& \boldsymbol{D}(P(X_n \mid X_1, \ldots, X_{n-1})\|Q(X_n \mid X_1, \ldots, X_{n-1})).
\end{aligned}$$

Using the chain rule, we can prove additional properties of the relative entropy. First, using the chain rule and the fact that $\boldsymbol{D}(P(Y \mid X)\|Q(Y \mid X)) \geq 0$, we can get the following property.

Proposition A.7

$$\boldsymbol{D}(P(X)\|Q(X)) \leq \boldsymbol{D}(P(X, Y)\|Q(X, Y)).$$

That is, the relative entropy of a marginal distributions is upper-bounded by the relative entropy of the joint distributions. This observation generalizes to situations where we consider sets of variables. That is,

$$\boldsymbol{D}(P(X_1, \ldots, X_k)\|Q(X_1, \ldots, X_k)) \leq \boldsymbol{D}(P(X_1, \ldots, X_n)\|Q(X_1, \ldots, X_n))$$

for $k \leq n$.

Suppose that X and Y are independent in both P and Q. Then, we have that $P(Y \mid X) = P(Y)$, and similarly, $Q(Y \mid X) = Q(Y)$. Thus, we conclude that $\boldsymbol{D}(P(Y \mid X)\|Q(Y \mid X)) = \boldsymbol{D}(P(Y)\|Q(Y))$. Combining this observation with the chain rule, we can prove an additional property.

Proposition A.8

If both P and Q satisfy $(X \perp Y)$, then

$$\boldsymbol{D}(P(X, Y)\|Q(X, Y)) = \boldsymbol{D}(P(X)\|Q(X)) + \boldsymbol{D}(P(Y)\|Q(Y)).$$

A.1.3.3 Other Distance Measures

There are several different metric distances between distributions that we may consider. Several simply treat a probability distribution as a vector in $\mathbb{R}^N$ (where N is the dimension of our probability space), and use standard distance metrics for Euclidean spaces. More precisely, let P and Q be two distributions over $X_1, \ldots, X_n$. The three most commonly used distance metrics of this type are:

- The L_1 distance: $\|P - Q\|_1 = \sum_{x_1,\ldots,x_n} |P(x_1, \ldots, x_n) - Q(x_1, \ldots, x_n)|$.

- The L_2 distance: $\|P - Q\|_2 = \left(\sum_{x_1,\ldots,x_n} (P(x_1, \ldots, x_n) - Q(x_1, \ldots, x_n))^2 \right)^{\frac{1}{2}}$.

- The L_∞ distance: $\|P - Q\|_\infty = \max_{x_1,\ldots,x_n} |P(x_1, \ldots, x_n) - Q(x_1, \ldots, x_n)|$.

variational distance An apparently different distance measure is the *variational distance*, which seems more specifically tailored to probability distributions, rather than to general real-valued vectors. It is defined as the maximal difference in the probability that two distributions assign to *any* event that can be described by the distribution. For two distributions P, Q over an event space $\mathcal{S}$, we define:

$$D_{var}(P; Q) = \max_{\alpha \in \mathcal{S}} |P(\alpha) - Q(\alpha)|. \tag{A.1}$$

Interestingly, this distance turns out to be exactly half the L_1 distance:

Proposition A.9

Let P and Q be two distributions over $\mathcal{S}$. Then

$$D_{var}(P; Q) = \frac{1}{2}\|P - Q\|_1.$$

These distance metrics are all useful in the analysis of approximations, but, unlike the relative entropy, they do not decompose by a chain-rule-like construction, often making the analytical analysis of such distances harder. However, we can often use an analysis in terms of relative entropy to provide bounds on the L_1 distance, and hence also on the variational distance:

Theorem A.1

For any two distribution P and Q, we have that

$$\|P - Q\|_1 \leq ((2\ln 2)D(P\|Q))^{1/2}.$$

A.2 Convergence Bounds

In many situations that we cover in this book, we are given a set of samples generated from a distribution, and we wish to estimate certain properties of the generating distribution from the samples. We now review some properties of random variables that are useful for this task. The derivation of these convergence bounds is central to many aspects of probability theory, statistics, and randomized algorithms. Motwani and Raghavan (1995) provide one good introduction on this topic and its applications to the analysis of randomized algorithms.

Specifically, suppose we have a biased coin that has an unknown probability p of landing heads. We can estimate the value of p by tossing the coin several times and counting the frequency of heads. More precisely, assume we have a data set $\mathcal{D}$ consisting of M coin tosses, that is, M trials from a Bernoulli distribution. The m'th coin toss is represented by a binary variable $X[m]$ that has value 1 if the coin lands heads, and 0 otherwise. Since each toss is separate from the previous one, we are assuming that all these random variables are

IID
independent. Thus, these variables are *independence and identically distribution*, or *IID*. It is easy to compute the expectation and variance of each $X[m]$:

- $E[X[m]] = p$.
- $Var[X[m]] = p(1-p)$.

A.2.1 Central Limit Theorem

We are interested in the sum of all the variables $S_{\mathcal{D}} = X[1] + \ldots + X[M]$ and in the fraction of successful trials $T_{\mathcal{D}} = \frac{1}{M} S_{\mathcal{D}}$. Note that $S_{\mathcal{D}}$ and $T_{\mathcal{D}}$ are functions of the data set $\mathcal{D}$. As $\mathcal{D}$ is chosen randomly, they can be viewed as random variables over the probability space defined by different possible data sets $\mathcal{D}$. Using properties of expectation and variance, we can analyze the properties of these random variables.

- $E[S_{\mathcal{D}}] = M \cdot p$, by linearity of expectation.
- $Var[S_{\mathcal{D}}] = M \cdot p(1-p)$, since all the all the $X[i]$'s are independent.
- $E[T_{\mathcal{D}}] = p$.
- $Var[T_{\mathcal{D}}] = \frac{1}{M} p(1-p)$, since $Var\left[\frac{1}{M} S_{\mathcal{D}}\right] = \frac{1}{M^2} Var[S_{\mathcal{D}}]$.

The fact that $Var[T_{\mathcal{D}}] \to 0$ as $M \to \infty$ suggests that for sufficiently large M the distribution of $T_{\mathcal{D}}$ is concentrated around p. In fact, a general result in probability theory allows us to conclude that this distribution has a particular form:

Theorem A.2

central limit
theorem

(Central Limit Theorem) *Let* $X[1], X[2], \ldots$ *be a series of IID random variables, where each* $X[m]$ *is sampled from a distribution such that* $E[X[m]] = \mu$, *and variance* $Var[X[m]] = \sigma^2$ *$(0 < \sigma < \infty)$. Then*

$$\lim_{M \to \infty} P\left(\frac{\sum_m (X[m] - \mu)}{\sqrt{M}\sigma} < r \right) = \Phi(r),$$

where $\Phi(r) = P(Z < r)$ *for a Gaussian variable* Z *with distribution* $\mathcal{N}(0; 1)$.

Gaussian
Thus, if we collect a large number of repeated samples from the same distribution, then the distribution of the random variable $(S_{\mathcal{D}} - E[S_{\mathcal{D}}])/\sqrt{Var[S_{\mathcal{D}}]}$ is roughly *Gaussian*. In other words, the distribution of $S_{\mathcal{D}}$ is, at the limit, close to a Gaussian with the appropriate expectation and variance: $\mathcal{N}(E[S_{\mathcal{D}}]; Var[S_{\mathcal{D}}])$.

There are variants of the central limit theorem for the case where each $X[m]$ has a different distribution. These require additional technical conditions that we do not go into here. However, the general conclusion is similar — the sum of many independent random variables has a distribution that is approximately Gaussian. This is often a justification for using a Gaussian distribution in modeling quantities that are the cumulative effect of many independent (or almost independent) factors.

estimator

The quantity $T_{\mathcal{D}}$ is an *estimator* for the mean μ: a statistical function that we can use to estimate the value of μ. The mean and variance of an estimator are the two key quantities for evaluating it. The mean of the estimator tells us the value around which its values are going to be concentrated. When the mean of the estimator is the target value μ, it is called an *unbiased*

unbiased
estimator

estimator for the quantity μ — an estimator whose mean is precisely the desired value. In general, lack of bias is a desirable property in an estimator: it tells us that, although they are noisy, at least the values obtained by the estimator are centered around the right value. The variance of the estimator tells us the "spread" of values we obtain from it. Estimators with high variance are not very reliable, as their value is likely to be far away from their mean.

Applying the central limit theorem to our problem, we see that, for sufficiently large M, the variable $T_{\mathcal{D}}$ has a roughly Gaussian distribution with mean p and variance $\frac{p(1-p)}{M}$.

A.2.2 Convergence Bounds

In many situations, we are interested not only in the asymptotic distribution of $T_{\mathcal{D}}$, but also in the probability that $T_{\mathcal{D}}$ is close to p for a concrete choice of M. We can bound this probability in several ways. One of the simplest is by using Chebyshev's inequality; see exercise 12.1. This bound, however, is quite loose, as it assumes quadratic decay in the distance $|T_{\mathcal{D}} - p|$. Other, more refined bounds, can be used to prove an exponential rate of decay in this distance. There are many variants of these bounds, of which we describe two.

Hoeffding bound

The first, called *Hoeffding bound*, measures error in terms of the absolute distance $|T_{\mathcal{D}} - p|$.

Theorem A.3

Let $\mathcal{D} = \{X[1], \ldots, X[M]\}$ be a sequence of M independent Bernoulli trials with probability of success p. Let $T_{\mathcal{D}} = \frac{1}{M} \sum_m X[m]$. Then

$$P_{\mathcal{D}}(T_{\mathcal{D}} > p + \epsilon) \leq e^{-2M\epsilon^2}$$
$$P_{\mathcal{D}}(T_{\mathcal{D}} < p - \epsilon) \leq e^{-2M\epsilon^2}.$$

The bound asserts that, with very high probability, $T_{\mathcal{D}}$ is within an additive error ϵ of the true probability p. The probability here is taken relative to possible data sets $\mathcal{D}$. Intuitively, we might end up with really unlikely choices of $\mathcal{D}$, for example, ones where we get the same value all the time; these choices will clearly give wrong results, but they are very unlikely to arise as a result of a random sampling process. Thus, the bound tells us that, **for most data sets $\mathcal{D}$ that we generate at random, we obtain a good estimate. Furthermore, the fraction of "bad" sample sets $\mathcal{D}$, those for which the estimate is more than ϵ from the true value, diminishes exponentially as the number of samples M grows.**

Chernoff bound

The second bound, called the *Chernoff bound*, measures error in terms of the relative size of this distance to the size of p.

Theorem A.4

Let $\mathcal{D} = \{X[1], \ldots, X[M]\}$ be a sequence of M independent Bernoulli trials with probability of success p. Let $T_{\mathcal{D}} = \frac{1}{M} \sum_m X[m]$, then

$$P_{\mathcal{D}}(T_{\mathcal{D}} > p(1 + \epsilon)) \leq e^{-Mp\epsilon^2/3}$$
$$P_{\mathcal{D}}(T_{\mathcal{D}} < p(1 - \epsilon)) \leq e^{-Mp\epsilon^2/2}.$$

Let $\sigma_M = \sqrt{Var[T_\mathcal{D}]}$ be the standard deviation of $T_\mathcal{D}$ for $\mathcal{D}$ of size M. Using the multiplicative Chernoff bound, we can show that

$$P_\mathcal{D}(|T_\mathcal{D} - p| \geq k\sigma) \leq 2e^{-k^2/6}. \tag{A.2}$$

This inequality should be contrasted with the Chebyshev inequality. The big difference owes to the fact that the Chernoff bound exploits the particular properties of the distribution of $T_\mathcal{D}$.

A.3 Algorithms and Algorithmic Complexity

In this section, we briefly review relevant algorithms and notions from algorithmic complexity. Cormen et al. (2001) is a good source for learning about algorithms, data structures, graph algorithms, and algorithmic complexity; Papadimitriou (1993) and Sipser (2005) provide a good introduction to the key concepts in computational complexity.

A.3.1 Basic Graph Algorithms

Given a graph structure, there are many useful operations that we might want to perform. For example, we might want to determine whether there is a certain type of path between two nodes. In this section, we survey algorithms for performing two key tasks that will be of use in several places throughout this book. Additional algorithms, for more specific tasks, are presented as they become relevant.

Algorithm A.1 Topological sort of a graph

    Procedure Topological-Sort (
       $\mathcal{G} = (\mathcal{X}, \mathcal{E})$ // A directed graph
)
1   Set all nodes to be unmarked
2   **for** $i = 1, \ldots, n$
3   Select any unmarked node X all of whose parents are marked
4        $d(X) \leftarrow i$
5   Mark X
6   **return** $(\vec{d})$

---

topological ordering   One algorithm, shown in algorithm A.1, finds a *topological ordering* of the nodes in the graph, as defined in definition 2.19.

maximum weight spanning tree   Another useful algorithm is one that finds, in a weighted undirected graph $\mathcal{H}$ with nonnegative edge weights, a *maximum weight spanning tree*. More precisely, a subgraph is said to be a *spanning tree* if it is a tree and it spans all vertices in the graph. Similarly, a *spanning forest* is a forest that spans all vertices in the graph. A maximum weight spanning tree (or forest) is the tree (forest) whose edge-weight sum is largest among all spanning trees (forests).

Algorithm A.2 Maximum weight spanning tree in an undirected graph

Procedure Max-Weight-Spanning-Tree (
$\quad \mathcal{H} = (\mathcal{N}, \mathcal{E})$
$\quad \{w_{ij} \; : \; (X_i, X_j) \in \mathcal{E}\}$
)
1 $\quad \mathcal{N}_T \leftarrow \{X_1\}$
2 $\quad \mathcal{E}_T \leftarrow \emptyset$
3 $\quad$ **while** $\mathcal{N}_T \neq \mathcal{X}$
4 $\quad\quad \mathcal{E}' \leftarrow \{(i, j) \in \mathcal{E} \; : \; X_i \in \mathcal{N}_T, X_j \notin \mathcal{N}_T\}$
5 $\quad\quad (X_i, X_j) \leftarrow \arg\max_{(X_i, X_j) \in \mathcal{E}'} w_{ij}$
6 $\quad\quad$ // (X_i, X_j) is the highest-weight edge between a node in T
$\quad\quad\quad$ and a node out of T
7 $\quad\quad \mathcal{N}_T \leftarrow \mathcal{N}_T \cup \{X_j\}$
8 $\quad\quad \mathcal{E}_T \leftarrow \mathcal{E}_T \cup \{(X_i, X_j)\}$
9 $\quad$ **return** ($\mathcal{E}_T$)

A.3.2 Analysis of Algorithmic Complexity

A key step in evaluating the usefulness of an algorithm is to analyze its computational cost: the amount of time it takes to complete the computation and the amount of space (memory) required. To evaluate the algorithm, we are usually not interested in the cost for a particular input, but rather in the algorithm's performance over a set of inputs. Of course, we would expect most algorithms to run longer when applied to larger problems. Thus, the complexity of an algorithm is usually measured in terms of its performance, as a function of the size of the input given to it. Of course, to determine the precise cost of the algorithm, we need to know exactly how it is implemented and even which machine it will be run on. However, we can often determine the scalability of an algorithm at a more abstract level, without worrying about the details of its implementation. We now provide a high-level overview of some of the basic concepts underlying such analysis.

Consider an algorithm that takes a list of n numbers and adds them together to compute their sum. Assuming the algorithm simply traverses the list and computes the sum as it goes along, it has to perform some fixed number of basic operations for each element in the list. The precise operations depend on the implementation: we might follow a pointer in a linked list, or simply increment a counter in an array. Thus, the precise cost might vary based on the implementation. But, the total number of operations per list element is some fixed constant factor. Thus, for any reasonable implementation, the running time of the algorithm will be bounded by $C \cdot n$ for some
asymptotic
complexity
constant C. In this case, we say that the *asymptotic complexity* of the algorithm is $O(n)$, where the $O()$ notation makes implicit the precise nature of the constant factor, which can vary from one implementation to another. This idea only makes sense if we consider the running time as a function of n. For any fixed problem size, say up to 100, we can always find a constant C (for instance, a million years) such that the algorithm takes time no more than C. However, even if we are not interested in problems of unbounded size, evaluating the way in which the running time varies as a function of the problem size is the first step to understanding how well it will

scale to large problems.

To take a more relevant example, consider the maximum weight spanning tree procedure of algorithm A.2. A (very) naive implementation of this algorithm traverses all of the edges in the graph every time a node is added to the spanning tree; the resulting cost is $O(mn)$ where m is the number of edges and n the number of nodes. A more careful implementation of the data structures, however, maintains the edges in a sorted data structure known as a heap, and the list of edges adjacent to a node in an *adjacency list*. In this case, the complexity of the algorithm can be $O(m \log n)$ or (with a yet more sophisticated data structure) $O(m + n \log n)$. Surprisingly, even more sophisticated implementations exist whose complexity is very close to linear time in m.

More generally, we can provide the following definition:

Definition A.7

Consider an algorithm $\mathcal{A}$ that takes as input problems Π from a particular class, and returns an output. Assume that the size of each possible input problem Π is measured using some set of parameters $n_1, \ldots, n_k$. We say that the running time of $\mathcal{A}$ is $O(f(n_1, \ldots, n_k))$ for some function f (called "big O of f"), if, for $n_1, \ldots, n_k$ sufficiently large, there exists a constant C such that, for any possible input problem Π, the running time of $\mathcal{A}$ on Π is at most $C \cdot f(n_1, \ldots, n_k)$. ∎

In our example, each problem Π is a graph, and its size is defined by two parameters: the number of nodes n and the number of edges m. The function $f(n, m)$ is simply $n + m$.

running time

polynomial time

exponential time

When the function f is linear in each of the input size parameters, we say that the *running time* of the algorithm is linear, or that the algorithm has *linear time*. We can similarly define notions of *polynomial time* and *exponential time*. It may be useful to distinguish different rates of growth in the different parameters. For example, if we have a function that has the form $f(n, m) = n^2 + 2^m$, we might say that the function is polynomial in n but exponential in m.

Although one can find algorithms at various levels of complexity, the key cutoff between feasible and infeasible computations is typically set between algorithms whose complexity is polynomial and those whose complexity is exponential. Intuitively, an algorithm whose complexity is exponential allows virtually no useful scalability to larger problems. For example, assume we have an algorithm whose complexity is $O(2^n)$, and that we can now solve instances whose size is N. If we wait a few years and get a computer that is twice as fast as the one we have now, we will be able to solve only instances whose size is $N + 1$, a negligible improvement.

We can also see this phenomenon by comparing the growth curves for various cost functions, as in figure A.1. We see that the constant factors in front of the polynomial functions have some impact on very small problem sizes, but even for moderate problem sizes, such as 20, the exponential function quickly dominates and grows to the point of infeasibility. Thus, a major distinction is made between algorithms that run in polynomial time and those whose running time is exponential. **While the exponential-polynomial distinction is a critical one, there is also a tendency to view polynomial-time algorithms as tractable. This view, unfortunately, is overly simplified: an algorithm whose running time is $O(n^3)$ is not generally tractable for problems where n is in the thousands.**

Algorithmic theory offers a suite of tools for constructing efficient algorithms for certain types of problems. One such tool, which we shall use many times throughout the book, is *dynamic programming*, which we describe in more detail in appendix A.3.3. Unfortunately, not all problems are not amenable to these techniques, and a broad class of highly important problems fall into a category for which polynomial-time algorithms are extremely unlikely to

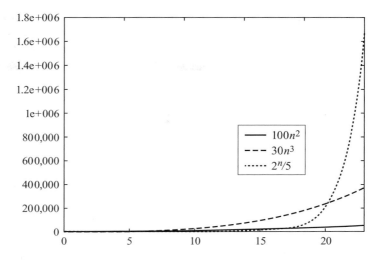

Figure A.1 Illustration of asymptotic complexity. The growth curve of three functions: The solid line is $100n^2$, the dashed line is $30n^3$, and the dotted line is $2^n/5$.

exist; see appendix A.3.4.

A.3.3 Dynamic Programming

As we discussed earlier, several techniques can be used to provide efficient solutions to apparently challenging computational problems. **One important tool is *dynamic programming*, a general method that we can apply when the solution to a problem requires that we solve many smaller subproblems that recur many times. In this case, we are often better off precomputing the solution to the subproblems, storing them, and using them to compute the values to larger problems.**

Perhaps the simplest application of dynamic programming is the problem of computing *Fibonacci numbers*, defined via the recursive equations:

$$F_0 = 1$$
$$F_1 = 1$$
$$F_n = F_{n-1} + F_{n-2}.$$

Thus, we have that $F_2 = 2$, $F_3 = 3$, $F_4 = 5$, $F_5 = 8$, and so on.

One simple algorithm to compute *Fibonacci(n)* is to use the recursive definition directly, as shown in algorithm A.3. Unrolling the computation, we see that the first of these recursive calls, Fibonacci($n-1$), calls Fibonacci($n-2$) and Fibonacci($n-3$). Thus, we are already have two calls to Fibonacci($n-2$). Similarly, Fibonacci($n-2$) also calls Fibonacci($n-3$), another redundant computation. If we carry through the entire recursive analysis, we can show that the running time of the algorithm is exponential in n.

On the other hand, we can compute "bottom up", as in algorithm A.4. Here, we start with F_0

dynamic programming

Algorithm A.3 Recursive algorithm for computing Fibonacci numbers

    Procedure Fibonacci (
       n
)
1   **if** ($n = 0$ or $n = 1$) **then**
2   **return** (1)
3   **return** (Fibonacci($n - 1$) + Fibonacci($n - 2$))

Algorithm A.4 Dynamic programming algorithm for computing Fibonacci numbers

    Procedure Fibonacci (
       n
)
1      $F_0 \leftarrow 1$
2      $F_1 \leftarrow 1$
3   **for** $i = 2, \ldots, n$
4        $F_i \leftarrow F_{i-1} + F_{i-2}$
5   **return** (F_n)

---

and F_1, compute F_2 from F_0 and F_1, compute F_3 from F_1 and F_2, and so forth. Clearly, this process computes F_n in time $O(n)$. We can view this alternative algorithm as precomputing and then caching (or storing) the results of the intermediate computations performed on the way to each F_i, so that each only has to be performed once.

More generally, if we can define the set of intermediate computations required and how they depend on each other, we can often use this caching idea to avoid redundant computation and provide significant savings. This idea underlies most of the exact inference algorithms for graphical models.

A.3.4   Complexity Theory

In appendix A.3.3, we saw how the same problem might be solvable by two algorithms that have radically different complexities. Examples like this raise an important issue regarding the algorithm design process: If we come up with an algorithm for a problem, how do we know whether its computational complexity is the best we can achieve? In general, unfortunately, we cannot tell. There are very few classes of problems for which we can give nontrivial lower bounds on the amount of computation required for solving them.

complexity
theory

However, there are certain types of problems for which we can provide, not a guarantee, but at least a certain expectation regarding the best achievable performance. *Complexity theory* has defined classes of problems that are, in a sense, equivalent to each other in terms of their computational cost. In other words, we can show that an algorithm for solving one problem can be converted into an algorithm that solves another problem. Thus, if we have an efficient algorithm for solving the first problem, it can also be used to solve the second efficiently.

The most prominent such class of problems is that of $\mathcal{NP}$-complete problems; this class contains many problems for which researchers have unsuccessfully tried, for decades, to find efficient algorithms. Thus, by proving that a problem is $\mathcal{NP}$-complete, we are essentially showing that it is "as easy" as all other $\mathcal{NP}$-complete problems. Finding an efficient (polynomial time) algorithm for this problem would therefore give rise to efficient algorithms for all $\mathcal{NP}$-complete problems, an extremely unlikely event. In other words, by showing that a problem is $\mathcal{NP}$-complete, we are essentially showing that it is extremely unlikely to have an efficient solution. We now provide some of the formal basis for this type of discussion.

A.3.4.1 Decision Problems

A *decision problem* Π is a task that has the following form: The program must accept an input ω and decide whether it satisfies a certain condition or not. A prototypical decision problem is the *SAT* problem, which is defined as the problem of taking as input a formula in propositional logic, and returning *true* if the formula has a satisfying assignment and *false* if it does not. For example, an algorithm for the SAT problem should return *true* for the formula

$$(q_1 \vee \neg q_2 \vee q_3) \wedge (\neg q_1 \vee q_2 \vee \neg q_3),$$ (A.3)

which has (among others) the satisfying assignment $q_1 = true$; $q_2 = true$; $q_3 = true$. It would return *false* for the formula

$$(\neg q_1 \vee \neg q_2) \wedge (q_2 \vee q_3) \wedge (\neg q_1 \vee \neg q_3),$$ (A.4)

which has no satisfying assignments.

We often use a somewhat restricted version of the SAT problem, called *3-SAT*.

Definition A.8

3-SAT

A formula ϕ is said to be a 3-SAT *formula over the Boolean (binary-valued) variables $q_1, \ldots, q_n$ if it has the following form: ϕ is a conjunction: $\phi = C_1 \wedge \ldots \wedge C_m$. Each C_i is a clause of the form $\ell_{i,1} \vee \ell_{i,2} \vee \ell_{i,3}$. Each $\ell_{i,j}$ ($i = 1, \ldots, m$; $j = 1, 2, 3$) is a* literal, *which is either q_k or $\neg q_k$ for some $k = 1, \ldots, n$.* ∎

A decision problem Π is associated with a language $\mathcal{L}_\Pi$ that defines the precise set of instances for which a correct algorithm for Π must return *true*. In the case of 3-SAT, $\mathcal{L}_{3SAT}$ is the set of all correct encodings of propositional 3-SAT formulas that are satisfiable.

A.3.4.2 $\mathcal{P}$ and $\mathcal{NP}$

A decision problem is said to be in the class $\mathcal{P}$ if there exists a deterministic algorithm that takes an instance ω and determines whether or not $\omega \in \mathcal{L}_\Pi$, in polynomial time in the size of the input ω. In SAT, for example, the input is the formula, and its size is simply its length.

We can also define a significantly more powerful type of computation that allows us to provide a formal foundation for a very rich class of problems. Consider again our SAT algorithm. The naive algorithm for determining whether a formula is satisfiable enumerates all of the assignments, and returns *true* if one of them satisfies the formula. Imagine that we allow the algorithm a notion of a "lucky guess": the algorithm is allowed to guess an assignment, and then verify whether it satisfies the formula. The algorithm can determine if the formula is satisfiable simply by having one guess that works out. In other words, we assume that the algorithm asserts that the formula is in $\mathcal{L}_{3SAT}$ if there is some guess that works out. This type of computation is called a *nondeterministic computation*. A fully formal definition requires that we introduce a range of concepts (such as Turing Machines) that are outside the scope of this book. Roughly speaking, a *nondeterministic decision algorithm* has the following form. The first stage is a guessing stage, where the algorithm nondeterministically produces some guess γ. The second stage is a deterministic verifying stage that either accepts its input ω based on γ or not. The algorithm as a whole is said to accept ω if it accepts γ using any one of its guesses. A decision problem Π is in the class $\mathcal{NP}$ if there exists a nondeterministic algorithm that accepts an instance ω if and only if $\omega \in \mathcal{L}_\Pi$, and if the verification stage can be executed in polynomial time in the length of ω. Clearly, SAT is in $\mathcal{NP}$: the guesses γ are possible assignments, and they are verified in polynomial time simply by testing whether the assignment γ satisfies the input formula ϕ.

Because deterministic computations are a special case of nondeterministic ones, we have that $\mathcal{P} \subseteq \mathcal{NP}$. The converse of this inclusion is the biggest open problem in computational complexity. In other words, can every problem that can be solved in polynomial time using a lucky guess also be solved in polynomial time without guessing?

As stated, it seems impossible to get a handle on this problem: The number of problems in $\mathcal{NP}$ is potentially unlimited, and even if we find an efficient algorithm for one problem, what does that tell us about the class in general? The notion of $\mathcal{NP}$-complete problems gives us a tool for reducing this unmanageable question into a much more compact one. Roughly speaking, the class $\mathcal{NP}$ has a set of problems that are the "hardest problems in $\mathcal{NP}$": if we can solve them in polynomial time, we can provably solve any problem in $\mathcal{NP}$ in polynomial time. These problems are known as $\mathcal{NP}$-complete problems.

$\mathcal{NP}$-hard

reduction

More formally, we say that a decision problem Π is $\mathcal{NP}$-*hard* if for every decision problem Π' in $\mathcal{NP}$, there is a polynomial-time transformation of inputs such that an input for Π' belongs to $\mathcal{L}_{\Pi'}$ if and only if the transformed instance belongs to $\mathcal{L}_\Pi$. This type of transformation is called a *reduction* of one problem to another. When we have such a reduction, any algorithm $\mathcal{A}$ that solves the decision problem Π can be used to solve Π': We simply convert each instance of Π' to the corresponding instance of Π, and apply $\mathcal{A}$. An $\mathcal{NP}$-hard problem can be used in this way for any problem in $\mathcal{NP}$. Thus, it provides a universal solution for any $\mathcal{NP}$-problem. It is possible to show that the SAT problem is $\mathcal{NP}$-hard. A problem Π is said to be $\mathcal{NP}$-complete if it is both $\mathcal{NP}$-hard and in $\mathcal{NP}$. The 3-SAT problem is $\mathcal{NP}$-complete, as are many other important problems. For example, the *Max-Clique Problem* of deciding whether an undirected graph has a clique of size at least K (where K is a parameter to the algorithm) is also $\mathcal{NP}$-hard.

Max-Clique
Problem

At the moment, it is not yet known whether $\mathcal{P} = \mathcal{NP}$. Much work has been devoted to investigating both sides of this conjecture. In particular, decades of research have been spent on failed attempts to find polynomial-time algorithms for many $\mathcal{NP}$-complete problems, such as SAT or Max-Clique. The lack of success suggests that probably no such algorithm exists for any $\mathcal{NP}$-hard problem, and that $\mathcal{P} \neq \mathcal{NP}$. Thus, a standard way of showing that a particular problem Π probably is unlikely to have a polynomial time algorithm is to show that it is $\mathcal{NP}$-hard. In other words, we try to find a reduction from some known $\mathcal{NP}$-hard problem, such as SAT, to the problem of interest. If we construct such a reduction, then we have shown the following: If we find a polynomial-time algorithm for Π, we have also provided a polynomial-time algorithm for all $\mathcal{NP}$-complete problems, and shown that $\mathcal{NP} = \mathcal{P}$. Although this is not impossible, it is currently believed to be highly unlikely.

☞ **Thus, if we show that a problem is $\mathcal{NP}$-hard, we should probably resign ourselves to algorithms that are exponential-time in the worst case. However, as we will see, there are many cases where algorithms can be exponential-time in the worst case, yet achieve significantly better performance in practice. Because many of the problems we encounter are $\mathcal{NP}$-hard, finding tractable cases and providing algorithms for them is where most of the interesting work takes place.**

A.3.4.3 Other Complexity Classes

The classes $\mathcal{P}$ and $\mathcal{NP}$ are the most important and commonly used classes used to describe the computational complexity of problems, but they are only part of a rich framework used for classifying problems based on their time or space complexity. In particular, the class $\mathcal{NP}$ is only the first level in an infinite hierarchy of increasingly larger classes. Classes higher in the hierarchy might or might not be harder than the lower classes; this problem also is a major open problem in complexity theory.

A different dimension along which complexity can vary relates to the existential nature of the definition of the class $\mathcal{NP}$. A problem is in $\mathcal{NP}$ if there is some guess on which a polynomial time computation succeeds (returns *true*). In our SAT example, the guesses were different assignments that could satisfy the formula ϕ defined in the problem instance. However, we might want to know what fraction of the computations succeed. In our SAT example, we may want to compute the exact number (or fraction) of assignments satisfying ϕ. This problem is no longer a decision problem, but rather a counting problem that returns a numeric output.

$\#\mathcal{P}$ The class $\#\mathcal{P}$ is defined precisely for problems that return a numerical value. Such a problem is in $\#\mathcal{P}$ if the number can be computed as the number of accepting guesses of a nondeterministic polynomial time algorithm. The problem of counting the number of satisfying assignments to a 3-SAT formula is clearly in $\#\mathcal{P}$. Like the class of $\mathcal{NP}$-hard problems, there are problems that are at least as hard as any problem in $\#\mathcal{P}$. The problem of counting satisfying assignments is the canonical $\#\mathcal{P}$-hard problem. This problem is clearly $\mathcal{NP}$-hard: if we can solve it, we can immediately solve the 3-SAT decision problem. For trivial reasons, it is not in $\mathcal{NP}$, because it is a counting problem, not a decision problem. However, it is generally believed that the counting version of the 3-SAT problem is inherently more difficult than the original decision problem, in that we can use them to solve problems that are "harder" than $\mathcal{NP}$.

$\mathcal{RP}$ Finally, another, quite different, complexity class is the class of *randomized polynomial time algorithms* — those that can be solved using a polynomial time algorithm that makes random guesses. There are several ways of defining when a randomized algorithm accepts a particular input; we provide one of them. A decision problem Π is in the class $\mathcal{RP}$ if there exists a randomized algorithm that makes a guess probabilistically, and then processes it in polynomial time, such that the following holds: The algorithm always returns *false* for an input not in $\mathcal{L}_\Pi$; for an input in $\mathcal{L}_\Pi$, the algorithm returns *true* with probability greater than $1/2$. Thus, the algorithm only has to get the "right" answer in half of its guesses; this requirement is much more stringent than that of nondeterministic polynomial time, where the algorithm only had to get one guess right. Thus, many problems are known to be in $\mathcal{NP}$ but are not known to be in $\mathcal{RP}$. Whether $\mathcal{NP} = \mathcal{RP}$ is another important open question, where the common belief is also that the answer is no.

A.4   Combinatorial Optimization and Search

A.4.1   Optimization Problems

Many of the problems we address in this book and in other settings can be formulated as an
optimization problem. Here, we are given a *solution space* Σ of possible solutions σ, and an
objective function $f_{\mathrm{obj}} : \Sigma \mapsto I\!\!R$ that allows us to evaluate the "quality" of each candidate
solution. Our aim is then to find the solution that achieves the maximum score:

optimization
problem

objective
function

$$\sigma^* = \arg\max_{\sigma \in \Sigma} \boldsymbol{f}(\sigma).$$

This optimization task is a maximization problem; we can similarly define a minimization
problem, where our goal is to minimize a *loss function*. One can easily convert one problem to
another (by negating the objective), and so, without loss of generality, we focus on maximization
problems.

Optimization problems can be discrete, where the solution space Σ consists of a certain
(finite) number of discrete hypotheses. In most such cases, this space is (at least) exponentially
large in the size of the problem, and hence, for reasonably sized problems, it cannot simply be
enumerated to find the optimal solution. In other problems the solution space is continuous, so
that enumeration is not even an option.

The available tools for solving an optimization problem depend both on the form of the
solution space Σ and on the form of the objective. For some classes of problems, we can
identify the optimum in terms of a closed-form expression; for others, there exist algorithms
that can provably find the optimum efficiently (in polynomial time), even when the solution
space is large (or infinite); others are $\mathcal{NP}$-hard; and yet others do not (yet) have any theoretical
analysis of their complexity. Throughout this book, multiple optimization problems arise, and
we will see examples of all of these cases.

A.4.2   Local Search

Many optimization problems do not appear to admit tractable solution algorithms exist, and
we are forced to fall back on heuristic methods that have no guarantees of actually finding the
optimal solution. One such class of methods that are in common use is the class of *local search*
methods. Such search procedures operate over a *search space*. A search space is a collection
of candidate solutions, often called *search states*. Each search state is associated with a score
and a set of neighboring states. A search procedure is a procedure that, starting from one state,
explores search space in attempt to find a high-scoring state.

local search

search space

search state

Local search algorithms keep track of a "current" state. At each iteration they consider several
states that are "similar" to the current one, and therefore are viewed as adjacent to it in the
search space. These states are often generated by a set of *search operators*, each of which takes a
state and makes a small modification to it. They select one of these neighboring states and make
it the current candidate. These iterations are repeated until some termination condition. These
local search procedures can be thought of as moving around in the solution space by taking
small steps. Generally, these steps are taken in a direction that tends to improve the objective.
If we assume that "similar" solutions tend to have similar values, this approach is likely to move
toward better regions of the space.

search operators

MAP assignment

structure search

This approach can be applied to a broad range of problems. For example, we can use it to find a *MAP assignment* relative to a distribution P: the space of solutions is the set of assignments ξ to a set of random variables $\mathcal{X}$; the objective function is $P(\xi)$; and the search operators take one assignment x and change the value of one variable X_i from x_i to x_i'. As we discuss in section 18.4, it can also be used to perform *structure search* over the space of Bayesian network structures to find one that optimizes a certain "goodness" function: the search space is the set of network structures, and the search operators make small changes to the current structure, such as adding or deleting an edge.

Algorithm A.5 Greedy local search algorithm with search operators

---

    Procedure Greedy-Local-Search (
        σ_0,   // initial candidate solution
        score,   // Score function
        $\mathcal{O}$,   // Set of search operators
)

```
1       σ_best ← σ_0
2       do
3          σ ← σ_best
4          Progress ← false
5          for each operator o ∈ O
6             σ_o ← o(σ)     // Result of applying o on σ
7             if σ_o is legal solution then
8                if score(σ_o) > score(σ_best) then
9                   σ_best ← σ_o
10                  Progress ← true
11         while Progress
12
13         return σ_best
```

A.4.2.1   Local Hill Climbing

greedy
hill-climbing

One of the simplest, and often used, search procedures is the *greedy hill-climbing* procedure. As the name suggests, at each step we take the step that leads to the largest improvement in the score. This is the search analogue of a continuous gradient-ascent method; see appendix A.5.2. The actual details of the procedure are shown in algorithm A.5. We initialize the search with some solution σ_0. Then we repeatedly execute the following steps: We consider all of the solutions that are neighbors of the current one, and we compute their score. We then select the neighbor that leads to the best improvement in the score. We continue this process until no modification improves the score. One issue with this algorithm is that the number of operators that can be applied may be quite large. A slight variant of this algorithm, called *first-ascent hill*

first-ascent hill
climbing

climbing, samples operators from $\mathcal{O}$ and evaluates them one at a time. Once it finds one that leads to better scoring network, it applies it without considering other operators. In the initial stages of the search, this procedure requires relatively few random trials before it finds such an operator. As we get closer to the local maximum, most operators hurt the score, and more trials are needed before an upward step is found (if any).

What can we say about the solution returned by Greedy-Local-Search? From our stopping criterion, it follows that the score of this solution is no lower than that of its neighbors. This implies that we are in one of two situations. We might have reached a *local maximum* from which all changes are score-reducing. **Except in rare cases, there is no guarantee that the local maximum we find via local search is actually the global optimum σ^*. Indeed, it may be a very poor solution.** The other option is that we have reached a *plateau*: a large set of neighboring solutions that have the same score. By design, greedy hill-climbing procedure cannot "navigate" through a plateau, since it relies on improvement in score to guide it to better solutions. Once again, we have no guarantee that this plateau achieves the highest possible score.

There are many modifications to this basic algorithm, mostly intended to address this problem. We now discuss some basic ideas that are applicable to all local search algorithms. We defer to the main text any detailed discussion of algorithms specific to problems of interest to us.

A.4.2.2   Forcing Exploration in New Directions

One common approach is to try to escape a suboptimal convergence point by systematically exploring the region around that point with the hope of finding an "outlet" that leads to a new direction to climb up. This can be done if we are willing to record all networks we "visited" during the search. Then, instead of choosing the best operator in Line 7 of Greedy-Local-Search, we select the best operator that leads to a solution we have not visited. We then allow the search to continue even when the score does not improve (by changing the termination condition). This variant can take steps that explore new territories even if they do not improve the score. Since it is greedy in nature, it will try to choose the best network that was not visited before. To understand the behavior of this method, visualize it climbing to the hilltop. Once there, the procedure starts pacing parts of the hill that were not visited before. As a result, it will start circling the hilltop in circles that grow wider and wider until it finds a ridge that leads to a new hill. (This procedure is often called *basin flooding* in the context of minimization problems.)

In this variant, even when no further progress can be made, the algorithm keeps moving, trying to find new directions. One possible termination condition is to stop when no progress has been made for some number of steps. Clearly, the final solution produced should not necessarily be the one at which the algorithm stops, but rather the best solution found anywhere during the search. Unfortunately, the computational cost of this algorithm can be quite high, since it needs to keep track of all solutions that have been visited in the past.

Tabu search is a much improved variant of this general idea utilizes the fact that the steps in our search space take the form of local modifications to the current solution. In tabu search, we keep a list not of solutions that have been found, but rather of operators that we have recently applied. In each step, we do not consider operators that reverse the effect of operators applied within a history window of some predetermined length L. Thus, if we flip a variable X_i from x_i to x_i', we cannot flip it back in the next L steps. These restrictions force the search procedure to explore new directions in the search space, instead of tweaking with the same parts of the solution. The size L determines the amount of memory retained by the search.

The tabu search procedure is shown in algorithm A.6. The "tabu list" is the list of operators

local maximum

☞

plateau

basin flooding

tabu search

Algorithm A.6 Local search with tabu list

     Procedure LegalOp (

        $o,$ // Search operator to check

        TABU   // List of recently applied operators

)

1   **if** exists $o' \in TABU$ such that o reverses o' **then return** *false*

2   **else return** *true*

3

     Procedure Tabu-Structure-Search (

        $\sigma_0,$ // initial candidate solution

        score,   // Score

        $\mathcal{O},$ // A set of search operators

        $L,$ // Size of tabu list

        $N,$ // Stopping criterion

)

1     $\sigma_{\text{best}} \leftarrow \sigma_0$

2     $\sigma \leftarrow \sigma_{\text{best}}$

3     $t \leftarrow 1$

4   *LastImprovement* $\leftarrow 0$

5   **while** *LastImprovement* $< N$

6        $o^{(t)} \leftarrow \epsilon$ // Set current operator to be uninitialized

7   **for** each operator $o \in \mathcal{O}$ // Search for best allowed operator

8   **if** LegalOp$(o, \{o^{(t-L)}, \ldots, o^{(t-1)}\})$ **then**

9              $\sigma_o \leftarrow o(\sigma)$

10   **if** σ_o is legal solution **then**

11   **if** $o^{(t)} = \epsilon$ or score$(\sigma_o) >$ score(σ_{o^t}) **then**

12                 $o^{(t)} \leftarrow o$

13        $\sigma \leftarrow \sigma_{o^t}$

14   **if** score$(\sigma) >$ score(σ_{best}) **then**

15           $\sigma_{\text{best}} \leftarrow \sigma_o$

16   *LastImprovement* $\leftarrow 0$

17   **else**

18   *LastImprovement* $\leftarrow$ *LastImprovement* $+ 1$

19        $t \leftarrow t + 1$

20

21   **return** σ_{best}

---

applied in the last L steps. The procedure LegalOp checks if a new operator is legal given the current tabu list. The implementation of this procedure depends on the exact nature of operators we use. As in the basin-flooding approach, tabu search does not stop when it reaches a solution that cannot be improved, but rather continues the search with the hope of reaching a better structure. If this does not happen after a prespecified number of steps, we decide to abandon the search.

Algorithm A.7 Beam search

Procedure Beam-Search (
    σ_0,   // initial candidate solution
    score,   // Score
    $\mathcal{O}$,   // A set of search operators
    K,   // Beam width
)
1    $Beam \leftarrow \{\sigma_0\}$
2   **while** not terminated
3      $H \leftarrow \emptyset$ // Current successors
4   **for** each $\sigma \in L$ and each $o \in \mathcal{O}$
5   Add $o(\sigma)$ to H
6      $Beam \leftarrow$ K-Best(score, H, K)
7    $\sigma_{\text{best}} \leftarrow$ K-Best(score, $H, 1$)
8   **return** (σ_{best})

---

beam search

    Another variant that forces a more systematic search of the space is *beam search*. In beam search, we conduct a hill-climbing search, but we keep track of a certain fixed number K of states. The value K is called the *beam width*. At each step in the search, we take all of the current states and generate and evaluate all of their successors. The best K are kept, and the algorithm repeats. The algorithm is shown in algorithm A.7. Note that with a beam width of 1, beam search reduces to greedy hill-climbing search, and with an infinite beam width, it reduces to breadth-first search. Note that this version of beam search assumes that the (best) steps taken during the search always improve the score. If that is not the case, we would also have to compare the current states in our beam *Beam* to the new candidates in H in order to determine the next set of states to put in the beam. The termination condition can be an upper bound on the number of steps or on the improvement achieved in the last iteration.

A.4.2.3   Randomization in Search

randomization

    Another approach that can help in reducing the impact of local maxima is *randomization*. Here, multiple approaches exist. We note that most randomization procedures can be applied as a wrapper to a variety of local search algorithm, including both hill climbing and tabu search. Most simply, we can initialize the algorithm at different random starting points, and then use a hill-climbing algorithm from each one. Another strategy is to interleave random steps and hill-climbing steps. Here, many strategies are possible. In one approach, we can "revitalize" the search by taking the best network found so far and applying several randomly chosen operators

---

Algorithm A.8 Greedy hill-climbing search with random restarts

---

    Procedure Search-with-Restarts (

      σ_0,   // initial candidate solution

      score,   // Score

      $\mathcal{O}$,   // A set of search operators

      Search,   // Search procedure

      l,   // random restart length

      k // number of random restarts

)

1    $\sigma_{\text{best}} \leftarrow \text{Search}(\sigma_0, \text{score}, \mathcal{O})$

2   **for** $i = 1, \ldots, k$

3       $\sigma \leftarrow \sigma_{\text{best}}$

4   // Perform random walk

5       $j \leftarrow 1$

6   **while** $j < l$

7   **sample** o from $\mathcal{O}$

8   **if** $o(\sigma)$ is a legal network **then**

9             $\sigma \leftarrow o(\sigma)$

10            $j \leftarrow j + 1$

11       $\sigma \leftarrow \text{Search}(\sigma, \text{score}, \mathcal{O})$

12   **if** $\text{score}(\sigma) > \text{score}(\sigma_{\text{best}})$ **then**

13          $\sigma_{\text{best}} \leftarrow \sigma$

14

15   **return** σ_{best}

---

random restart

to get a network that is fairly similar, yet perturbed. We then restart our search procedure from the new network. If we are lucky, this *random restart* step moves us to a network that belongs to a better "basin of attraction," and thus the search will converge to a better structure. A simple random restart procedure is shown in algorithm A.8; it can be applied as wrapper to plain hill climbing, tabu search, or any other search algorithm.

This approach can be effective in escaping from fairly local maxima (which can be thought of as small bumps on the slope of a larger hill). However, it is unlikely to move from one wide hill to another. There are different choices in applying random restart, the most important one is how many random "steps" to take. If we take too few, we are unlikely to escape the local maxima. If we take too many, than we move too far off from the region of high scoring network. One possible strategy is to applying random restarts of growing magnitude. That is, each successive random restart applies more random operations.

To make this method concrete, we need a way of determining how to apply random restarts, and how to interleave hill-climbing steps and randomized moves. A general framework for doing is *simulated annealing*. The basic idea of simulated annealing is similar to Metropolis-Hastings MCMC methods that we discuss in section 12.3, and so we only briefly touch it.

simulated annealing

In broad outline, the simulated annealing procedure attempts to mix hill-climbing steps with moves that can decrease the score. This mixture is controlled by a so-called *temperature parameter*. When the temperature is "hot," the search tries many moves that decrease the score. As the search is annealed (the temperature is slowly reduced) it starts to focus only on moves that improve the score. The intuition is that during the "hot" phase the search explores the space and eventually gets trapped in a region of high scores. As the temperature reduces it is able to distinguish between finer details of the score "landscape" and eventually converge to a good maximum.

To carry out this intuition, a simulated annealing procedure uses a *proposal distribution* over operators to propose candidate operators to apply in the search. At each step, the algorithm selects an operator o using this distribution, and evaluates $\delta(o)$ — the change in score incurred by applying o at the current state. The search accepts this move with probability $\min(1, e^{\frac{\delta(o)}{\tau}})$, where τ is the current temperature. Note that, if $\delta(o) > 0$, the move is automatically accepted. If $\delta(o) < 0$, the move is accepted with probability that depends both on the decrease in score and on the temperature τ. For large value of τ (hot) all moves are applied with probability close to 1. For small values of τ (cold), all moves that decrease the score are applied with small probability. The search procedure anneals τ every fixed number of move attempts. There are various strategies for annealing; the simplest one is simply to have τ decay exponentially. One can actually show that, if the temperature is annealed sufficiently slowly, simulated annealing converges to the globally optimal solution with high probability. However, in practice, this "guaranteed" annealing schedule is both unknown and much too slow to be useful in practice. **In practice, the success of simulated annealing depends heavily on the design of the proposal distribution and annealing schedule.**

A.4.3   Branch and Bound Search

Here we discussed one class of solutions to discrete optimization problems: the class of local hill-climbing search. Those methods are very broadly useful, since they apply to any discrete optimization problem for which we can define a set of search operators. In some cases, however, we may know some additional structure within the problem, allowing more informed methods to be applied. One useful type of information is a mechanism that allows us to evaluate a partial assignment $\boldsymbol{y}_{1\ldots i}$, and to place a bound $\text{bound}(\boldsymbol{y}_{1\ldots i})$ on the best score of any complete assignment that extends $\boldsymbol{y}_{1\ldots i}$. In this case, we can use an algorithm called *branch and bound* *search*, shown in algorithm A.9 for the case of a maximization problem.

Roughly speaking, branch and bound searches the space of partial assignments, beginning with the empty assignment, and assigning the variables $X_1, \ldots, X_n$, one at a time (in some order), using depth-first search. At each point, when considering the current partial assignment $\boldsymbol{y}_{1\ldots i}$, the algorithm evaluates it using $\text{bound}(\boldsymbol{y}_{1\ldots i})$ and compares it to the best full assignment ξ found so far. If $\text{score}(\xi)$ is better than the best score that can possibly be achieved starting from $\boldsymbol{y}_{1\ldots i}$, then there is no point continuing to explore any of those assignments, and the algorithm backtracks to try a different partial assignment. Because the bound is correct, it is not difficult to show that the assignments that were *pruned* without being searched cannot possibly be optimal. When the bound is reasonably tight, this algorithm can be very effective, pruning large parts of the space without searching it.

The algorithm shows the simplest variant of the branch-and-bound procedure, but many extensions exist. One heuristic is to perform the search so as to try and find good assignments

---

Algorithm A.9 Branch and bound algorithm

    Procedure Branch-and-Bound (

      score,   // Score function

      bound,   // Upper bound function

      σ_{best},   // Best full assignment so far

      $\text{score}_{\text{best}}$,   // Best score so far

      i,   // Variable to be assigned next

      $\boldsymbol{y}_{1\ldots i-1}$,   // Current partial assignment

)   // Recursive algorithm, called initially with the following arguments: some arbitrary full assignment σ_{best}, $\text{score}_{\text{best}} = \text{score}(\sigma_{\text{best}})$, $i = 1$, and the empty assignment.

1   **for** each $x_i \in Val(X_i)$

2       $\boldsymbol{y}_{1\ldots i} \leftarrow (\boldsymbol{y}_{1\ldots i-1}, x_i)$ // Extend the assignment

3   **if** $i = n$ and $\text{score}(\boldsymbol{y}_{1\ldots n}) > \text{score}_{\text{best}}$ **then**

4         $(\sigma_{\text{best}}, \text{score}_{\text{best}}) \leftarrow (\boldsymbol{y}_{1\ldots n}, \text{score}(\boldsymbol{y}_{1\ldots n}))$

5   // Found a better full assignment

6   **else if** $\text{bound}(\boldsymbol{y}_{1\ldots i}) > \text{score}_{\text{best}}$ **then**

7         $(\sigma_{\text{best}}, \text{score}_{\text{best}}) \leftarrow$ Branch-and-Bound(score, bound, $\sigma_{\text{best}}, \text{score}_{\text{best}}, i+1, \boldsymbol{y}_{1\ldots i}$)

8   // If bound is better than current solution, try current partial assignment; otherwise, prune and move on

9   **return** $(\sigma_{\text{best}}, \text{score}_{\text{best}})$

---

early. The better the current assignment σ_{best}, the better we can prune suboptimal trajectories. Other heuristics intelligently select, at each point in the search, which variable to assign next, allowing this choice to vary across different points in the search. When available, one can also use a lower bound as well as an upper bound, allowing pruning to take place based on partial (not just full) trajectories. Many other extensions exist, but are outside the scope of this book.

A.5   Continuous Optimization

In the preceding section, we discussed the problem of optimizing an objective over a discrete space. In this section we briefly review methods for solving optimization problems over a *continuous* space. See Avriel (2003); Bertsekas (1999) for more thorough discussion of nonlinear optimization, and see Boyd and Vandenberghe (2004) for an excellent overview of convex optimization methods.

A.5.1   Characterizing Optima of a Continuous Function

At several points in this book we deal with maximization (or minimization) problems. In these problems, we have a function $f_{\text{obj}}(\theta_1, \ldots, \theta_n)$ for several *parameters*, and we wish to find joint values of the parameters that maximizes the value of f_{obj}.

Formally, we face the following problem:

Find values $\theta_1, \ldots, \theta_n$ such that $f_{\text{obj}}(\theta_1, \ldots, \theta_n) = \max_{\theta'_1, \ldots, \theta'_n} f_{\text{obj}}(\theta'_1, \ldots, \theta'_n)$.

Example A.2

Assume we are given a set of points $(x[1], y[1]), \ldots, (x[m], y[m])$. Our goal is to find the "centroid" of these points, defined as a point (θ_x, θ_y) that minimizes the square distance to all of the points. We can formulate this problem into a maximization problem by considering the negative of the sum of squared distances:

$$f_{\mathrm{obj}}(\theta_x, \theta_y) = -\sum_i \left((x[i] - \theta_x)^2 + (y[i] - \theta_y)^2\right).$$

∎

gradient

One way of finding the maximum of a function is to use the fact that, at the maximum, the *gradient* of the function is 0. Recall that the gradient of a function $f_{\mathrm{obj}}(\theta_1, \ldots, \theta_n)$ is the vector of partial derivatives

$$\nabla f = \left\langle \frac{\partial f}{\partial \theta_1}, \ldots \frac{\partial f}{\partial \theta_n} \right\rangle.$$

Theorem A.5

If $\langle \theta_1, \ldots, \theta_n \rangle$ is an interior maximum point of f_{obj}, then

$$\nabla f(\theta_1, \ldots, \theta_n) = 0.$$

stationary point

This property, however, does not characterize maximum points. Formally, a point $\langle \theta_1, \ldots, \theta_n \rangle$ where $\nabla f_{\mathrm{obj}}(\theta_1, \ldots, \theta_n) = 0$ is a *stationary point* of f_{obj}. Such a point can be either a local maximum, a local minimum, or a saddle point. However, finding such a point can often be the first step toward finding a maximum.

To satisfy the requirement that $\nabla f = 0$ we need to solve the set of equations

$$\frac{\partial}{\partial \theta_k} f_{\mathrm{obj}}(\theta_1, \ldots, \theta_n) = 0 \qquad k = 1, \ldots, n$$

Example A.3

Consider the task of example A.2. We can easily verify that:

$$\frac{\partial}{\partial \theta_x} f_{\mathrm{obj}}(\theta_x, \theta_y) = 2\sum_i (x[i] - \theta_x).$$

Equating this term to 0 and performing simple arithmetic manipulations, we get the equation:

$$\theta_x = \frac{1}{m}\sum_i x[i].$$

The exact same reasoning allows us to solve for θ_y.

∎

In this example, we conclude that f_{obj} has a unique stationary point. We next need to verify that this point is a maximum point (rather than a minimum or a saddle point). In our example, we can check that, for any sequence that extends from the origin to infinity (that is, $\theta_x^2 + \theta_y^2 \to \infty$), we have $f_{\mathrm{obj}} \to -\infty$. Thus, the single stationary point is a maximum.

In general, to verify that the stationary point is a maximum, we can check that the second derivative is negative. To see this, recall the multivariate Taylor expansion of degree two:

$$f_{\mathrm{obj}}(\vec{\theta}) = f_{\mathrm{obj}}(\vec{\theta}_0) + (\vec{\theta} - \vec{\theta}_0)^T \nabla f_{\mathrm{obj}}(\vec{\theta}_0) + \frac{1}{2}\left[\vec{\theta} - \vec{\theta}_0\right]^T A(\vec{\theta}_0)\left[\vec{\theta} - \vec{\theta}_0\right],$$

Hessian

where $A(\vec{\theta}_0)$ is the *Hessian* — the matrix of second derivatives at $\vec{\theta}_0$. If we use this expansion around a stationary point, then the gradient is 0, and we only need to examine the term $[\vec{\theta} - \vec{\theta}_0]^T A(\vec{\theta}_0)[\vec{\theta} - \vec{\theta}_0]$. In the univariate case, we can verify that a point is a local maximum by testing that the second derivative is negative. The analogue to this condition in the multivariate

negative definite

case is that $A(\vec{\theta}_0)$ is *negative definite* at the point $\vec{\theta}_0$, that is, that $\vec{\theta}^T A(\vec{\theta}_0)\vec{\theta} < 0$ for all $\vec{\theta}$.

Theorem A.6

Suppose $\vec{\theta} = \langle \theta_1, \ldots, \theta_n \rangle$ is an interior point of f_{obj} with $\nabla \vec{\theta} = 0$. Let $A(\vec{\theta}) = \{ \frac{\partial^2}{\partial \theta_i \partial \theta_j} f_{\text{obj}}(\vec{\theta}) \}$ be the Hessian matrix of second derivatives of f_{obj} at $\vec{\theta}$. $A(\vec{\theta})$ is negative definite at θ if and only if $\vec{\theta}$ is a local maximum of f_{obj}.

Example A.4

Continuing example A.2, we can verify that

$$\frac{\partial^2}{\partial_{\theta_x}^2} f_{\text{obj}}(\theta_x, \theta_y) = -2m$$

$$\frac{\partial^2}{\partial_{\theta_y}^2} f_{\text{obj}}(\theta_x, \theta_y) = -2m$$

$$\frac{\partial^2}{\partial_{\theta_y} \partial_{\theta_x}} f_{\text{obj}}(\theta_x, \theta_y) = 0$$

$$\frac{\partial^2}{\partial_{\theta_x} \partial_{\theta_y}} f_{\text{obj}}(\theta_x, \theta_y) = 0.$$

Thus, the Hessian matrix is simply

$$A = \begin{pmatrix} -2m & 0 \\ 0 & -2m \end{pmatrix}.$$

It is easy to verify that this matrix is negative definite. ∎

A.5.2 Gradient Ascent Methods

The characterization of appendix A.5.1 allows us to provide closed-form solutions for certain continuous optimization problems. However, there are many problems for which such solutions cannot be found. In these problems, the equations posed by $\nabla f_{\text{obj}}(\boldsymbol{\theta}) = 0$ do not have an analytical solution. Moreover, in many practical problems, there are multiple local maxima, and then this set of equations does not even have a unique solution.

One approach for dealing with problems that do not yield analytical solutions is to *search* for a (local) maximum. The idea is very analogous to the discrete local search of appendix A.4.2: We begin with an initial point $\boldsymbol{\theta}^0$, which can be an arbitrary choice, a random guess, or an approximation of the solution based on other considerations. From this starting point, we want to "climb" to a maximum. A great many techniques roughly follow along these lines. In this section, we survey some of the most common ones.

A.5.2.1 Gradient Ascent

gradient ascent

The simplest approach is *gradient ascent*, an approach directly analogous to the hill-climbing

---

Algorithm A.10 Simple gradient ascent algorithm

     Procedure Gradient-Ascent (
       θ^1,   // Initial starting point
       f_{obj},   // Function to be optimized
       δ // Convergence threshold
)
1     $t \leftarrow 1$
2   **do**
3       $\theta^{t+1} \leftarrow \theta^t + \eta \nabla f_{\text{obj}}(\theta^t)$
4       $t \leftarrow t + 1$
5   **while** $\|\theta^t - \theta^{t-1}\| > \delta$
6   **return** (θ^t)

---

search of algorithm A.5 (see appendix A.4.2). Using the Taylor expansion of a function, we know that, in the neighborhood of θ^0, the function can be approximated by the linear equation

$$f_{\text{obj}}(\theta) \approx f_{\text{obj}}(\theta^0) + (\theta - \theta^0)^T \nabla f_{\text{obj}}(\theta^0).$$

Using basic properties of linear algebra, we can check that the slope of this linear function, that is, $\nabla f_{\text{obj}}(\theta^0)$, points to the direction of the steepest ascent. This observation suggests that, if we take a step in the direction of the gradient, we increase the value of f_{obj}. This reasoning leads to the simple gradient ascent algorithm shown in algorithm A.10. Here, η is a constant that determines the *rate* of ascent at each iteration. Since the gradient ∇f_{obj} approaches 0 as we approach a maximum point, the procedure will converge if η is sufficiently small.

Note that, in order to apply gradient ascent, we need to be able to evaluate the function f_{obj} at different points, and also to evaluate its gradient. In several examples we encounter in this book, we can perform these calculations, although in some cases these are costly. Thus, a major objective is to reduce the number of points at which we evaluate f_{obj} or ∇f_{obj}.

The performance of gradient ascent depends on the choice of η. If η is too large, then the algorithm can "overshoot" the maximum in each iteration. For sufficiently small value of η, the gradient ascent algorithm will converge, but if η is too small, we will need many iterations to converge. Thus, one of the difficult points in applying this algorithm is deciding on the value of η. Indeed, in practice, one typically needs to begin with a large η, and decrease it over time; this approach leaves us with the problem of choosing an appropriate schedule for shrinking η.

A.5.2.2   Line Search

An alternative approach is to adaptively choose the step size η at each step. The intuition is that we choose a direction to climb and continue in that direction until we reach a point where we start to descend. In this procedure, at each point θ^t in the search, we define a "line" in the direction of the gradient:

$$g(\eta) = \vec{\theta}^t + \eta \nabla f_{\text{obj}}(\theta^t).$$

line search   We now use a *line search* procedure to find the value of η that defines a (local) maximum of

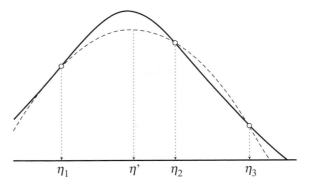

Figure A.2 Illustration of line search with Brent's method. The solid line shows a one-dimensional function. The three points, η_1, η_2, and η_3, bracket the maximum of this function. The dashed line shows the quadratic fit to these three points and the choice of η' proposed by Brent's method.

f_{obj} along the line; that is, we find:

$$\eta^t = \arg\max_\eta g(\eta).$$

We now take an η^t-sized step in the direction of the gradient; that is, we define:

$$\boldsymbol{\theta}^{t+1} \leftarrow \boldsymbol{\theta}^t + \eta^t \nabla f_{\mathrm{obj}}(\boldsymbol{\theta}^t).$$

And the process repeats.

There are several methods for performing the line search. The basic idea is to find three points $\eta_1 < \eta_2 < \eta_3$ so that $f_{\mathrm{obj}}(g(\eta_2))$ is larger than both $f_{\mathrm{obj}}(g(\eta_1))$ and $f_{\mathrm{obj}}(g(\eta_3))$. In this case, we know that there is at least one local maximum between η_1 and η_3, and we say that η_1, η_2 and η_3 *bracket* a maximum; see figure A.2 for an illustration. Once we have a method for finding a bracket, we can zoom in on the maximum. If we choose a point η' so that $\eta_1 < \eta' < \eta_2$ we can find a new, tighter, bracket. To see this, we consider the two possible cases. If $f_{\mathrm{obj}}(g(\eta')) > f_{\mathrm{obj}}(g(\eta_2))$, then η_1, η', η_2 bracket a maximum. Alternatively, if $f_{\mathrm{obj}}(g(\eta')) \leq f_{\mathrm{obj}}(g(\eta_2))$, then η', η_2, η_3 bracket a maximum. In both cases, the new bracket is smaller than the original one. Similar reasoning applies if we choose η' between η_2 and η_3.

The question is how to choose η'. One approach is to perform a binary search and choose $\eta' = (\eta_1 + \eta_3)/2$. This ensures that the size of the new bracket is half of the old one. A faster approach, known as *Brent's method*, fits a quadratic function based on the values of f_{obj} at the three points η_1, η_2, and η_3. We then choose η' to be the maximum point of this quadratic approximation. See figure A.2 for an illustration of this method.

A.5.2.3 Conjugate Gradient Ascent

Line search attempts to maximize the improvement along the direction defined by $\nabla f_{\mathrm{obj}}(\boldsymbol{\theta}^t)$. This approach, however, often has undesired consequences on the convergence of the search. To understand the problem, we start by observing that $\nabla f_{\mathrm{obj}}(\boldsymbol{\theta}^{t+1})$ must be *orthogonal* to

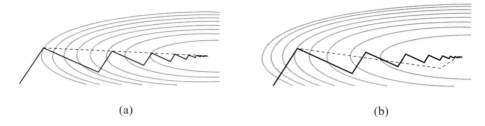

(a)   (b)

Figure A.3   Two examples of the convergence problem with line search. The solid line shows the progression of gradient ascent with line search. The dashed line shows the progression of the conjugate gradient method: (a) a quadratic function $f_{\mathrm{obj}}(x, y) = -(x^2 + 10y^2)$; (b) its exponential $f_{\mathrm{obj}}(x, y) = \exp\{-(x^2 + 10y^2)\}$. In both cases, the two search procedures start from the same initial point (bottom left of the figure), and diverge after the first line search.

$\nabla f_{\mathrm{obj}}(\boldsymbol{\theta}^t)$. To see why, observe that $\boldsymbol{\theta}^{t+1}$ was chosen to be a local maximum along the $\nabla f_{\mathrm{obj}}(\boldsymbol{\theta}^t)$ direction. Thus, the gradient of f_{obj} at $\boldsymbol{\theta}^{t+1}$ must be 0 in this direction. This implies that the two consecutive gradient vectors are orthogonal. As a consequence, the progress of the gradient ascent will be in a zigzag line. As the procedure approaches a maximum point, the size of each step becomes smaller, and the progress slows down. See figure A.3 for an illustration of this phenomenon.

A possible solution is to "remember" past directions of search and to bias the new direction to be a combination of the gradient at the current point and the direction implied by previous steps. This intuitive idea can be developed into a variety of algorithms. It turns out, however, that one variant of this algorithm can be shown to be optimal for finding the maximum of quadratic functions. Since, by the Taylor expansion, all functions are approximately quadratic in the neighborhood of a maximum, it follows that the final steps of the algorithm will converge to a maximum relatively quickly.

conjugate gradient ascent
The algorithm, known as *conjugate gradient ascent*, is shown in algorithm A.11. The vector $\boldsymbol{h}^t$ is the "corrected" direction for search. It combines the gradient $\boldsymbol{g}^t$ with the previous direction of search $\boldsymbol{h}^{t-1}$. The effect of previous search directions on the new one depends on the relative sizes of the gradients.

If our function f_{obj} is a quadratic function, the conjugate gradient ascent procedure is guaranteed to converge in n steps, where n is the dimension of the space. Indeed, in figure A.3a we see that the conjugate method converges in two steps. When the function is not quadratic, conjugate gradient ascent might require more steps, but is still much faster than standard gradient ascent. For example, in figure A.3b, it converges in four steps (the last step is too small to be visible in the figure).

Finallly, we note that **gradient ascent is the continuous analogue of the local hill-climbing approaches described in section A.4.2. As such, it is susceptible to the same issues of local maxima and plateaus. The approaches used to address these issues in this setting are similar to those outlined in the discrete case.**

Algorithm A.11 Conjugate gradient ascent

Procedure Conjugate-Gradient-Ascent (
    $\boldsymbol{\theta}^1$,   // Initial starting point
    f_{obj},   // Function to be optimized
    δ // Convergence threshold
)

1      $t \leftarrow 1$
2      $\boldsymbol{g}^0 \leftarrow \boldsymbol{1}$
3      $\boldsymbol{h}^0 \leftarrow \boldsymbol{0}$
4   **do**
5         $\boldsymbol{g}^t \leftarrow \nabla f_{\text{obj}}(\boldsymbol{\theta}^t)$
6         $\gamma^t \leftarrow \frac{(\boldsymbol{g}^t - \boldsymbol{g}^{t-1})^T \boldsymbol{g}^t}{(\boldsymbol{g}^{t-1})^T \boldsymbol{g}^{t-1}}$
7         $\boldsymbol{h}^t \leftarrow \boldsymbol{g}^t + \gamma^t \boldsymbol{h}^{t-1}$
8   Choose η^t by line search along the line $\boldsymbol{\theta}_t + \eta \boldsymbol{h}^t$
9         $\boldsymbol{\theta}^{t+1} \leftarrow \boldsymbol{\theta}^t + \eta^t \boldsymbol{h}^t$
10        $t \leftarrow t + 1$
11   **while** $\|\boldsymbol{\theta}^t - \boldsymbol{\theta}^{t-1}\| > \delta$
12   **return** $(\boldsymbol{\theta}^t)$

A.5.3 Constrained Optimization

In appendix A.5.1, we considered the problem of optimizing a continuous function over its entire domain (see also appendix A.5.2). In many cases, however, we have certain constraints that the desired solution must satisfy. Thus, we have to optimize the function within a constrained space. We now review some basic methods that address this problem of *constrained optimization*.

constrained
optimization

Example A.5

Suppose we want to find the maximum entropy distribution over a variable X, with $\text{Val}(X) = \{x^1, \ldots, x^K\}$. Consider the entropy of X:

$$\boldsymbol{H}(X) = -\sum_{k=1}^{K} P(x^k) \log P(x^k).$$

We can maximize this function using the gradient method by treating each $P(x^k)$ as a separate parameter θ_k. We compute the gradient of $\boldsymbol{H}_P(X)$ with respect to each of these parameters:

$$\frac{\partial}{\partial \theta_k} \boldsymbol{H}(X) = -\log(\theta_k) - 1.$$

Setting this partial derivative to 0, we get that $\log(\theta_k) = -1$, and thus $\theta_k = 1/2$. This solution seems fine until we realize that the numbers do not sum up to 1, and hence our solution does not *define a probability distribution!*

The flaw in our analysis is that we want to maximize the entropy subject to a constraint *on the parameters, namely, $\sum_k \theta_k = 1$. In addition, we also remember that we need to require that $\theta_k \geq 0$. In this case we see that the gradient drives the solution away from from 0 ($-\log(\theta_k) \to \infty$ as $\theta_k \to 0$), and thus we do not need to enforce this constraint actively.* ∎

Problems of this type appear in many settings, where we are interested in maximizing a function f under a set of *equality constraints*. This problem is posed as follows:

> **Find** θ
> **maximizing** $f(\theta)$
> **subject to**
> $$
> \begin{aligned}
> c_1(\boldsymbol{\theta}) &= 0 \\
> &\cdots \\
> c_m(\boldsymbol{\theta}) &= 0.
> \end{aligned}
> \tag{A.5}
> $$

Note that any equality constraint (such as the one in our example above) can be rephrased as constraining a function c to 0. Formally, we are interested in the behavior of f in the region of points that satisfies all the constraints

$$
\mathcal{C} = \{\boldsymbol{\theta} : \forall j = 1, \ldots, n, \; c_j(\boldsymbol{\theta}) = 0\}.
$$

To define our goal, remember that we want to find a maxima point within $\mathcal{C}$. Since $\mathcal{C}$ is a constrained "surface" we need to adopt the basic definition of maxima (and similarly minima, stationary point, etc.) to this situation. We can define local maxima in two ways. The first definition is in term of neighborhood. We define the ϵ-neighborhood of θ in $\mathcal{C}$ to be all the points $\theta' \in \mathcal{C}$ such that $\|\theta - \theta'\|_2 < \epsilon$. We then say that θ is a local maxima in $\mathcal{C}$ if there is an $\epsilon > 0$ such that $f(\theta) > f(\theta')$ for all θ' in its ϵ-neighborhood. An alternative definition that will be easier for the following is in terms of derivatives. Recall that a stationary point (local maximum, local minimum, or a saddle point) of a function if the derivative is 0. In the constraint case we have a similar definition, but we must ensure that the derivatives are ones that do not take us outside the constrained surface. Stated differently, if we consider a derivative in the direction δ, we want to ensure that the constraints remain 0 if we take a small step in direction δ. Formally, this means that the derivative has to be *tangent* to each constraint c_i, that is $\delta^T \nabla c_i(\boldsymbol{\theta}) = 0$.

A general approach to solving such constrained optimization problems is the method of *Lagrange multipliers*. We define a new function, called the *Lagrangian*, of θ and of a new vector of parameters $\boldsymbol{\lambda} = \langle \lambda_1, \ldots, \lambda_m \rangle$

$$
\mathcal{J}(\boldsymbol{\theta}, \boldsymbol{\lambda}) = f(\boldsymbol{\theta}) - \sum_{j=1}^{m} \lambda_j c_j(\boldsymbol{\theta}).
$$

Theorem A.7

If $\langle \boldsymbol{\theta}, \boldsymbol{\lambda} \rangle$ is a stationary point of the Lagrangian $\mathcal{J}$, then $\boldsymbol{\theta}$ is a stationary point of f subject to the constraints $c_1(\boldsymbol{\theta}) = 0, \ldots, c_m(\boldsymbol{\theta}) = 0$.

PROOF We briefly outline the proof. A formal proof requires the use of more careful tools from functional analysis.

We start by showing that θ satisfies the constraints. Since $\langle \boldsymbol{\theta}, \boldsymbol{\lambda} \rangle$ is a stationary point of $\mathcal{J}$, we have that for each j

$$
\frac{\partial}{\partial \lambda_j} \mathcal{J}(\boldsymbol{\theta}, \boldsymbol{\lambda}) = -c_j(\boldsymbol{\theta}).
$$

Thus, at stationary points of $\mathcal{J}$, the constraint $c_j(\boldsymbol{\theta}) = 0$ must be satisfied.

Now consider $\nabla \boldsymbol{f}(\boldsymbol{\theta})$. For each component θ_i of $\boldsymbol{\theta}$, we have that

$$0 = \frac{\partial}{\partial \theta_i} \mathcal{J}(\boldsymbol{\theta}, \boldsymbol{\lambda}) = \frac{\partial}{\partial \theta_i} \boldsymbol{f}(\boldsymbol{\theta}) - \sum_j \lambda_j \frac{\partial}{\partial \theta_i} c_j(\boldsymbol{\theta}).$$

Thus,

$$\nabla \boldsymbol{f}(\boldsymbol{\theta}) = \sum_j \lambda_j \nabla c_j(\boldsymbol{\theta}). \tag{A.6}$$

In other words, the gradient of $\boldsymbol{f}$ is a linear combination of the gradients of c_j.

We now use this property to prove that $\boldsymbol{\theta}$ is a stationary point of $\boldsymbol{f}$ when constrained to region $\mathcal{C}$. Consider a direction $\boldsymbol{\delta}$ that is tangent to the region $\mathcal{C}$ at $\boldsymbol{\theta}$. As $\boldsymbol{\delta}$ is tangent to $\mathcal{C}$, we expect that moving infinitesimally in this direction will maintain the constraint that c_j is 0; that is, c_j should not change its value when we move in this direction. More formally, the derivative of c_j in the direction $\boldsymbol{\delta}$ is 0. The derivative of c_j in a direction $\boldsymbol{\delta}$ is $\boldsymbol{\delta}^T \nabla c_j$. Thus, if $\boldsymbol{\delta}$ is tangent to $\mathcal{C}$, we have

$$\boldsymbol{\delta}^T \nabla c_j(\boldsymbol{\theta}) = 0$$

for all j. Using equation (A.6), we get

$$\boldsymbol{\delta}^T \nabla \boldsymbol{f}(\boldsymbol{\theta}) = \sum_j \lambda_j \boldsymbol{\delta}^T \nabla c_j(\boldsymbol{\theta}) = 0.$$

Thus, the derivative of $\boldsymbol{f}$ in a direction that is tangent to $\mathcal{C}$ is 0. This implies that when moving away from $\boldsymbol{\theta}$ within the allowed region $\mathcal{C}$ the value of $\boldsymbol{f}$ has 0 derivative. Thus, $\boldsymbol{\theta}$ is a stationary point of $\boldsymbol{f}$ when restricted to $\mathcal{C}$. ∎

We also have the converse property: If $\boldsymbol{f}$ satisfies some regularity conditions, then for every stationary point of $\boldsymbol{f}$ in $\mathcal{C}$ there is a choice of $\boldsymbol{\lambda}$ so that $\langle \boldsymbol{\theta}, \boldsymbol{\lambda} \rangle$ is a stationary point of $\mathcal{J}$.

 We see that the Lagrangian construction allows us to solve constrained optimization problems using tools for unconstrained optimization. We note that a local maximum of $\boldsymbol{f}$ always corresponds to a stationary point of $\mathcal{J}$, but this stationary point is not necessarily a local maximum of $\mathcal{J}$. If, however, we restrict attention to nonnegative constraint functions c, then a local maximum of $\boldsymbol{f}$ must correspond to a local maximum of $\mathcal{J}$.

We now consider two examples of using this technique.

Example A.6

Let us return to example A.5. In order to find the maximum entropy distribution over X, we need to solve the Lagrangian

$$\mathcal{J} = -\sum_k \theta_k \log \theta_k - \lambda \left(\sum_k \theta_k - 1 \right).$$

Setting $\nabla \mathcal{J} = 0$ implies the following system of equations:

$$
\begin{aligned}
0 &= -\log \theta_1 - 1 - \lambda \\
&\cdots \\
0 &= -\log \theta_K - 1 - \lambda \\
0 &= \sum_k \theta_k - 1.
\end{aligned}
$$

Each of the first K equations can be rewritten as $\theta_k = 2^{-1-\lambda}$. Plugging this term into the last equation, we get that $\lambda = \log(K) - 1$, and thus $P(x^k) = 1/K$. We conclude that we achieve maximum entropy with the uniform distribution. ∎

To see an example with more than one constraint, consider the following problem.

Example A.7

M-projection

Suppose we have a distribution $P(X, Y)$ over two random variables, and we want to find the closest distribution $Q(X, Y)$ in which X is independent of Y. As we discussed in section 8.5, this process is called M-projection *(see definition 8.4). Since X and Y are independent in Q, we must have that $Q(X, Y) = Q(X)Q(Y)$. Thus, we are searching for parameters $\theta_x = Q(x)$ and $\theta_y = Q(y)$ for different values $x \in \text{Val}(X)$ and $y \in \text{Val}(Y)$.*

Formally, we want to solve the following problem:

Find $\{\theta_x : x \in \text{Val}(X)\}$ and $\{\theta_y : y \in \text{Val}(y)\}$ that minimize

$$
\boldsymbol{D}(P(X, Y) \| Q(X)Q(Y)) = \sum_x \sum_y P(x, y) \log \frac{P(x, y)}{\theta_x \theta_y},
$$

subject to the constraints

$$
\begin{aligned}
0 &= \sum_x \theta_x - 1 \\
0 &= \sum_y \theta_y - 1.
\end{aligned}
$$

We define the Lagrangian

$$
\mathcal{J} = \sum_x \sum_y P(x, y) \log \frac{P(x, y)}{\theta_x \theta_y} - \lambda_x \left(\sum_x \theta_x - 1 \right) - \lambda_y \left(\sum_y \theta_y - 1 \right).
$$

To simplify the computation of derivatives, we notice that

$$
\log \frac{P(x, y)}{\theta_x \theta_y} = \log P(x, y) - \log \theta_x - \log \theta_y.
$$

Using this simplification, we can compute the derivative with respect to the probability of a particular value of X, say θ_{x^k}. We note that this parameter appears only when the value of x in the summation equals x^k. Thus,

$$
\frac{\partial}{\partial \theta_{x^k}} \mathcal{J} = -\sum_y \frac{P(x^k, y)}{\theta_{x^k}} - \lambda_x.
$$

Equating this derivative to 0, we get

$$\theta_{x^k} = -\frac{\sum_y P(x^k, y)}{\lambda_x} = -\frac{P(x^k)}{\lambda_x}.$$

To solve for the value of λ_x, we use the first constraint, and get that

$$1 = \sum_x \theta_x = -\sum_x \frac{P(x)}{\lambda_x}.$$

∎

Thus, we get that $\lambda_x = -\sum_x P(x)$. Thus, we can conclude that $\lambda_x = -1$, and consequently that $\theta_x = P(x)$. An analogous reasoning shows that $\theta_y = P(y)$.

This solution is very natural. The closest distribution to $P(X, Y)$ in which X and Y are independent is $Q(X, Y) = P(X)P(Y)$. This distribution preserves the marginal distributions of both X and Y, but loses all information about their joint behavior.

A.5.4 Convex Duality

convex duality

The concept of *convex duality* plays a central role in optimization theory. We briefly review the main results here for equality-constrained optimization problems with nonnegativity constraints (although the theory extends quite naturally to the case of general inequality constraints).

In appendix A.5.3, we considered an optimization problem of maximizing $f(\theta)$ subject to certain constraints, which we now call the *primal problem*. We showed how to formulate a Lagrangian $\mathcal{J}(\theta, \lambda)$, and proved that if $\langle \theta, \lambda \rangle$ is a stationary point of $\mathcal{J}$ then θ is a stationary point of the objective function f that we are trying to maximize.

We can extend this idea further and define the *dual function* $g(\lambda)$ as

$$g(\lambda) = \sup_{\theta \geq 0} \mathcal{J}(\theta, \lambda).$$

That is, the dual function $g(\lambda)$, is the *supremum*, or maximum, over the parameters θ for a given λ. In general, we allow the dual function to take the value ∞ when $\mathcal{J}$ is unbounded above (which can occur when the primal constraints are unsatisfied), and refer to the points λ at which this happens as *dual infeasible*.

Example A.8

Let us return to example A.6, where our task is to find the distribution $P(X)$ of maximum entropy. Now, however, we also want the distribution to satisfy the constraint that $E_P[X] = \mu$. Treating each $P(X = k)$ as a separate parameter θ_k, we can write our problem formally as:

Constrained-Entropy:

Find P
maximizing $H_P(X)$
subject to

$$\sum_{k=1}^{K} k\theta_k = \mu$$
$$\sum_{k=1}^{K} \theta_k = 1$$
$$\theta_k \geq 0 \qquad \forall k = 1, \dots, K$$

(A.7)

Introducing Lagrange multipliers *for each of the constraints we can write*

$$\mathcal{J}(\boldsymbol{\theta}, \lambda, \nu) = -\sum_{k=1}^{K} \theta_k \log \theta_k - \lambda \left(\sum_{k=1}^{K} k\theta_k - \mu \right) - \nu \left(\sum_{k=1}^{K} \theta_k - 1 \right).$$

Maximizing over $\boldsymbol{\theta}$ for each $\langle \lambda, \nu \rangle$ we get the dual function

$$\boldsymbol{g}(\lambda, \nu) = \sup_{\boldsymbol{\theta} \geq 0} \mathcal{J}(\boldsymbol{\theta}, \lambda, \nu)$$

$$= \lambda\mu + \nu + e^{-\nu-1} \sum_{k} e^{-k\lambda}.$$

Thus, the convex dual (to be minimized) is $\lambda\mu + \nu + e^{-\nu-1} \sum_{k} e^{-k\lambda}$. We can minimize over ν analytically by taking derivatives and setting them equal to zero, giving $\nu = \log \boldsymbol{g}(\sum_{k} e^{-k\lambda}) - 1$. Substituting into $\boldsymbol{g}$, we arrive at the dual optimization problem

$$minimize \quad \lambda\mu + \log \left(\sum_{k=1}^{K} e^{-k\lambda} \right).$$

This form of optimization problem is known as a geometric program. *The convexity of the objective function can be easily verified by taking second derivatives. Taking the first derivative and setting it to zero provides some insight into the solution to the problem:*

$$\frac{\sum_{k=1}^{K} ke^{-k\lambda}}{\sum_{k=1}^{K} e^{-k\lambda}} = \mu,$$

indicating that the solution has $\theta_k \propto \alpha^k$ for some fixed α. ∎

Importantly, as we can see in this example, the dual function is a pointwise maximization over a family of linear functions (of the dual variables). Thus, the dual function is always convex even when the primal objective function $\boldsymbol{f}$ is not.

One of the most important results in optimization theory is that the dual function gives an upper bound on the optimal value of the optimization problem; that is, for any primal feasible point $\boldsymbol{\theta}$ and any dual feasible point $\boldsymbol{\lambda}$, we have $\boldsymbol{g}(\boldsymbol{\lambda}) \geq f_{\text{obj}}(\boldsymbol{\theta})$. This leads directly to the property of *weak duality*, which states that the minimum value of the dual function is at least as large as the maximum value of the primal problem; that is,

$$\boldsymbol{g}(\boldsymbol{\lambda}^\star) = \inf_{\boldsymbol{\lambda}} \boldsymbol{g}(\boldsymbol{\lambda}) \geq \boldsymbol{f}(\boldsymbol{\theta}^\star).$$

The difference $\boldsymbol{f}(\boldsymbol{\theta}^\star) - \boldsymbol{g}(\boldsymbol{\lambda}^\star)$ is known as the *duality gap*. Under certain conditions the duality gap is zero, that is, $\boldsymbol{f}(\boldsymbol{\theta}^\star) = \boldsymbol{g}(\boldsymbol{\lambda}^\star)$, in which case we have *strong duality*. Thus, duality can be used to provide a *certificate* of optimality. That is, if we can show that $\boldsymbol{g}(\boldsymbol{\lambda}) = \boldsymbol{f}(\boldsymbol{\theta})$ for some value of $\langle \boldsymbol{\theta}, \boldsymbol{\lambda} \rangle$, then we know that $\boldsymbol{f}(\boldsymbol{\theta})$ is optimal.

The concept of a dual function plays an important role in optimization. In a number of situations, the dual objective function is easier to optimize than the primal. Moreover, there are methods that solve the primal and dual together, using the fact that each bounds the other to improve the search for an optimal solution.

Bibliography

Abbeel, P., D. Koller, and A. Ng (2006, August). Learning factor graphs in polynomial time & sample complexity. *Journal of Machine Learning Research 7*, 1743–1788.

Ackley, D., G. Hinton, and T. Sejnowski (1985). A learning algorithm for Boltzmann machines. *Cognitive Science 9*, 147–169.

Aji, S. M. and R. J. McEliece (2000). The generalized distributive law. *IEEE Trans. Information Theory 46*, 325–343.

Akaike, H. (1974). A new look at the statistical identification model. *IEEE Transactions on Automatic Control 19*, 716–723.

Akashi, H. and H. Kumamoto (1977). Random sampling approach to state estimation in switching environments. *Automatica 13*, 429–434.

Allen, D. and A. Darwiche (2003a). New advances in inference by recursive conditioning. In *Proc. 19th Conference on Uncertainty in Artificial Intelligence (UAI)*, pp. 2–10.

Allen, D. and A. Darwiche (2003b). Optimal time–space tradeoff in probabilistic inference. In *Proc. 18th International Joint Conference on Artificial Intelligence (IJCAI)*, pp. 969–975.

Altun, Y., I. Tsochantaridis, and T. Hofmann (2003). Hidden Markov support vector machines. In *Proc. 20th International Conference on Machine Learning (ICML)*.

Andersen, S., K. Olesen, F. Jensen, and F. Jensen (1989). HUGIN—a shell for building Bayesian belief universes for expert systems. In *Proc. 11th International Joint Conference on Artificial Intelligence (IJCAI)*, pp. 1080–1085.

Anderson, N. (1974). Information integration theory: A brief survey. In *Contemporary developments in Mathematical Psychology*, Volume 2, pp. 236–305. San Francisco, California: W.H. Freeman and Company.

Anderson, N. (1976). How functional measurement can yield validated interval scales of mental quantities. *Journal of Applied Psychology 61*(6), 677–692.

Andreassen, S., F. Jensen, S. Andersen, B. Falck, U. Kjærulff, M. Woldbye, A. R. Sørensen, A. Rosenfalck, and F. Jensen (1989). MUNIN — an expert EMG assistant. In J. E. Desmedt (Ed.), *Computer-Aided Electromyography and Expert Systems*, Chapter 21. Amsterdam: Elsevier Science Publishers.

Anguelov, D., D. Koller, P. Srinivasan, S. Thrun, H.-C. Pang, and J. Davis (2004). The correlated correspondence algorithm for unsupervised registration of nonrigid surfaces. In *Proc. 18th Conference on Neural Information Processing Systems (NIPS)*.

Arnauld, A. and P. Nicole (1662). Port-royal logic.

Arnborg, S. (1985). Efficient algorithms for combinatorial problems on graphs with bounded, decomposability—a survey. *BIT 25*(1), 2–23.

Arnborg, S., D. Corneil, and A. Proskurowski (1987). Complexity of finding embeddings in a k-tree. *SIAM J. Algebraic Discrete Methods 8*(2), 277–284.

Attias, H. (1999). Inferring parameters and structure of latent variable models by variational Bayes. In *Proc. 15th Conference on Uncertainty in Artificial Intelligence (UAI)*, pp. 21–30.

Avriel, M. (2003). *Nonlinear Programming: Analysis and Methods*. Dover Publishing.

Bacchus, F. and A. Grove (1995). Graphical models for preference and utility. In *Proc. UAI-95*, pp. 3–10.

Bach, F. and M. Jordan (2001). Thin junction trees. In *Proc. 15th Conference on Neural Information Processing Systems (NIPS)*.

Balke, A. and J. Pearl (1994a). Counterfactual probabilities: Computational methods, bounds and applications. In *Proc. 10th Conference on Uncertainty in Artificial Intelligence (UAI)*, pp. 46–54.

Balke, A. and J. Pearl (1994b). Probabilistic evaluation of counterfactual queries. In *Proc. 10th Conference on Artificial Intelligence (AAAI)*, pp. 230–237.

Bar-Shalom, Y. (Ed.) (1992). *Multitarget multisensor tracking: Advanced applications*. Norwood, Massachusetts: Artech House.

Bar-Shalom, Y. and T. Fortmann (1988). *Tracking and Data Association*. New York: Academic Press.

Bar-Shalom, Y., X. Li, and T. Kirubarajan (2001). *Estimation with Application to Tracking and Navigation*. John Wiley and Sons.

Barash, Y. and N. Friedman (2002). Context-specific Bayesian clustering for gene expression data. *Journal of Computational Biology 9*, 169–191.

Barber, D. and W. Wiegerinck (1998). Tractable variational structures for approximating graphical models. In *Proc. 12th Conference on Neural Information Processing Systems (NIPS)*, pp. 183–189.

Barbu, A. and S. Zhu (2005). Generalizing Swendsen-Wang to sampling arbitrary posterior probabilities. *IEEE Trans. on Pattern Analysis and Machine Intelligence 27*(8), 1239–1253.

Barnard, S. (1989). Stochastic stero matching over scale. *International Journal of Computer Vision 3*, 17–32.

Barndorff-Nielsen, O. (1978). *Information and Exponential Families in Statistical Theory*. Wiley.

Barron, A., J. Rissanen, and B. Yu (1998). The minimum description length principle in coding and modeling. *IEEE Transactions on Information Theory 44*(6), 2743–2760.

Bartlett, M. (1935). Contingency table interactions. *Journal of the Royal Statistical Society, Series B 2*, 248–252.

Bauer, E., D. Koller, and Y. Singer (1997). Update rules for parameter estimation in Bayesian networks. In *Proc. 13th Conference on Uncertainty in Artificial Intelligence (UAI)*, pp. 3–13.

Bayes, T. (1763). An essay towards solving a problem in the doctrine of chances. *Philosophical Transactions of the Royal Society of London 53*, 370–418.

Beal, M. and Z. Ghahramani (2006). Variational Bayesian learning of directed graphical models with hidden variables. *Bayesian Analysis 1*, 793–832.

Becker, A., R. Bar-Yehuda, and D. Geiger (1999). Random algorithms for the loop cutset problem. In *Proc. 15th Conference on Uncertainty in Artificial Intelligence (UAI)*, pp. 49–56.

Becker, A. and D. Geiger (1994). The loop cutset problem. In *Proc. 10th Conference on Uncertainty in Artificial Intelligence (UAI)*, pp. 60–68.

Becker, A. and D. Geiger (2001). A sufficiently fast algorithm for finding close to optimal clique trees. *Artificial Intelligence 125*(1–2), 3–17.

Becker, A., D. Geiger, and C. Meek (2000). Perfect tree-like Markovian distributions. In *Proc. 16th Conference on Uncertainty in Artificial Intelligence (UAI)*, pp. 19–23.

Becker, A., D. Geiger, and A. Schäffer (1998). Automatic selection of loop breakers for genetic linkage analysis. *Human Heredity 48*, 49–60.

Beeri, C., R. Fagin, D. Maier, and M. Yannakakis (1983). On the desirability of acyclic database schemes. *Journal of the Association for Computing Machinery 30*(3), 479–513.

Beinlich, L., H. Suermondt, R. Chavez, and G. Cooper (1989). The ALARM monitoring system: A case study with two probabilistic inference techniques for belief networks. In *Proceedings of the Second European Conference on Artificial Intelligence in Medicine*, pp. 247–256. Springer Verlag.

Bell, D. (1982). egret in decision making under uncertainty. *Operations Research 30*, 961–981.

Bellman, R. E. (1957). *Dynamic Programming*. Princeton, New Jersey: Princeton University Press.

Ben-Tal, A. and A. Charnes (1979). A dual optimization framework for some problems of information theory and statistics. *Problems of Control and Information Theory 8*, 387–401.

Bentham, J. (1789). An introduction to the principles of morals and legislation.

Berger, A., S. Della-Pietra, and V. Della-Pietra (1996). A maximum entropy approach to natural language processing. *Computational Linguistics 16*(2).

Bernardo, J. and A. Smith (1994). *Bayesian Theory*. New York: John Wiley and Sons.

Bernoulli, D. (1738). Specimen theoriae novae de mensura sortis (exposition of a new theory on the measurement of risk). English Translation by L. Sommer, *Econometrica*, 22:23–36, 1954.

Berrou, C., A. Glavieux, and P. Thitimajshima (1993). Near Shannon limit error-correcting coding: Turbo codes. In *Proc. International Conference on Communications*, pp. 1064–1070.

Bertelé, U. and F. Brioschi (1972). *Nonserial Dynamic Programming*. New York: Academic Press.

Bertsekas, D. (1999). *Nonlinear Programming* (2nd ed.). Athena Scientific.

Bertsekas, D. P. and J. N. Tsitsiklis (1996). *Neuro-Dynamic Programming*. Athena Scientific.

Besag, J. (1977a). Efficiency of pseudo-likelihood estimation for simple Gaussian fields. *Biometrika 64*(3), 616–618.

Besag, J. (1977b). Spatial interaction and the statistical analysis of lattice systems. *Journal of the Royal Statistical Society, Series B 36*, 192–236.

Besag, J. (1986). On the statistical analysis of dirty pictures (with discussion). *Journal of the Royal Statistical Society, Series B 48*, 259–302.

Bethe, H. A. (1935). Statistical theory of superlattices. *in Proceedings of the Royal Society of London A*, 552.

Bidyuk, B. and R. Dechter (2007). Cutset sampling for bayesian networks. *Journal of Artificial Intelligence Research 28*, 1–48.

Bilmes, J. and C. Bartels (2003). On triangulating dynamic graphical models. In *Proc. 19th Conference on Uncertainty in Artificial Intelligence (UAI)*.

Bilmes, J. and C. Bartels (2005, September). Graphical model architectures for speech recognition. *IEEE Signal Processing Magazine 22*(5), 89–100.

Binder, J., D. Koller, S. Russell, and K. Kanazawa (1997). Adaptive probabilistic networks with hidden variables. *Machine Learning 29*, 213–244.

Binder, J., K. Murphy, and S. Russell (1997). Space-efficient inference in dynamic probabilistic networks. In *Proc. 15th International Joint Conference on Artificial Intelligence (IJCAI)*.

Bishop, C. (2006). *Pattern Recognition and Machine Learning*. Information Science and Statistics

(M. Jordan, J. Kleinberg, and B. Schökopf, editors). New York: Springer-Verlag.

Bishop, C., N. Lawrence, T. Jaakkola, and M. Jordan (1997). Approximating posterior distributions in belief networks using mixtures. In *Proc. 11th Conference on Neural Information Processing Systems (NIPS)*.

Blalock, Jr., H. (1971). *Causal Models in the Social Sciences*. Chicago, Illinois: Aldine-Atheson.

Blum, B., C. Shelton, and D. Koller (2006). A continuation method for nash equilibria in structured games. *Journal of Artificial Intelligence Resarch 25*, 457–502.

Bodlaender, H., A. Koster, F. van den Eijkhof, and L. van der Gaag (2001). Pre-processing for triangulation of probabilistic networks. In *Proc. 17th Conference on Uncertainty in Artificial Intelligence (UAI)*, pp. 32–39.

Boros, E. and P. Hammer (2002). Pseudo-Boolean optimization. *Discrete Applied Mathematics 123*(1-3).

Bouckaert, R. (1993). Probabilistic network construction using the minimum description length principle. In *Proc. European Conference on Symbolic and Quantitative Approaches to Reasoning with Uncertainty*, pp. 41–48.

Boutilier, C. (2002). A POMDP formulation of preference elicitation problems. In *Proc. 18th Conference on Artificial Intelligence (AAAI)*, pp. 239–46.

Boutilier, C., F. Bacchus, and R. Brafman (2001). UCP-Networks: A directed graphical representation of conditional utilities. In *Proc. 17th Conference on Uncertainty in Artificial Intelligence (UAI)*, pp. 56–64.

Boutilier, C., T. Dean, and S. Hanks (1999). Decision theoretic planning: Structural assumptions and computational leverage. *Journal of Artificial Intelligence Research 11*, 1 – 94.

Boutilier, C., R. Dearden, and M. Goldszmidt (1989). Exploiting structure in policy construction. In *Proc. 14th International Joint Conference on Artificial Intelligence (IJCAI)*, pp. 1104–1111.

Boutilier, C., R. Dearden, and M. Goldszmidt (2000). Stochastic dynamic programming with factored representations. *Artificial Intelligence 121*(1), 49–107.

Boutilier, C., N. Friedman, M. Goldszmidt, and D. Koller (1996). Context-specific independence in Bayesian networks. In *Proc. 12th Conference on Uncertainty in Artificial Intelligence (UAI)*, pp. 115–123.

Boyd, S. and L. Vandenberghe (2004). *Convex Optimization*. Cambridge University Press.

Boyen, X., N. Friedman, and D. Koller (1999). Discovering the hidden structure of complex dynamic systems. In *Proc. 15th Conference on Uncertainty in Artificial Intelligence (UAI)*, pp. 91–100.

Boyen, X. and D. Koller (1998a). Approximate learning of dynamic models. In *Proc. 12th Conference on Neural Information Processing Systems (NIPS)*.

Boyen, X. and D. Koller (1998b). Tractable inference for complex stochastic processes. In *Proc. 14th Conference on Uncertainty in Artificial Intelligence (UAI)*, pp. 33–42.

Boyen, X. and D. Koller (1999). Exploiting the architecture of dynamic systems. In *Proc. 15th Conference on Artificial Intelligence (AAAI)*.

Boykov, Y., O. Veksler, and R. Zabih (2001). Fast approximate energy minimization via graph cuts. *IEEE Transactions on Pattern Analysis and Machine Intelligence 23*(11), 1222–1239.

Braziunas, D. and C. Boutilier (2005). Local utility elicitation in GAI models. In *Proc. 21st Conference on Uncertainty in Artificial Intelligence (UAI)*, pp. 42–49.

Breese, J. and D. Heckerman (1996). Decision-theoretic troubleshooting: A framework for repair and experiment. In *Proc. 12th Conference on Uncertainty in Artificial Intelligence (UAI)*, pp.

124–132.

Breese, J., D. Heckerman, and C. Kadie (1998). Empirical analysis of predictive algorithms for collaborative filtering. In *Proc. 14th Conference on Uncertainty in Artificial Intelligence (UAI)*, pp. 43–52.

Breiman, L., J. Friedman, R. Olshen, and C. Stone (1984). *Classification and Regression Trees*. Monterey,CA: Wadsworth & Brooks.

Buchanan, B. and E. Shortliffe (Eds.) (1984). *Rule-Based Expert Systems: The MYCIN Experiments of the Stanford Heuristic Programming Project*. Reading, MA: Addison-Wesley.

Bui, H., S. Venkatesh, and G. West (2001). Tracking and surveillance in wide-area spatial environments using the Abstract Hidden Markov Model. *International Journal of Pattern Recognition and Artificial Intelligence*.

Buntine, W. (1991). Theory refinement on Bayesian networks. In *Proc. 7th Conference on Uncertainty in Artificial Intelligence (UAI)*, pp. 52–60.

Buntine, W. (1993). Learning classification trees. In D. J. Hand (Ed.), *Artificial Intelligence Frontiers in Statistics*, Number III in AI and Statistics. Chapman & Hall.

Buntine, W. (1994). Operations for learning with graphical models. *Journal of Artificial Intelligence Research 2*, 159–225.

Buntine, W. (1996). A guide to the literature on learning probabilistic networks from data. *IEEE Transactions on Knowledge and Data Engineering 8*, 195–210.

Caffo, B., W. Jank, and G. Jones (2005). Ascent-based Monte Carlo Expectation-Maximization. *Journal of the Royal Statistical Society, Series B*.

Cannings, C., E. A. Thompson, and H. H. Skolnick (1976). The recursive derivation of likelihoods on complex pedigrees. *Advances in Applied Probability 8*(4), 622–625.

Cannings, C., E. A. Thompson, and M. H. Skolnick (1978). Probability functions on complex pedigrees. *Advances in Applied Probability 10*(1), 26–61.

Cano, J., L.D., Hernández, and S. Moral (2006). Importance sampling algorithms for the propagation of probabilities in belief networks. *International Journal of Approximate Reasoning 15*(1), 77–92.

Carreira-Perpignan, M. and G. Hinton (2005). On contrastive divergence learning. In *Proc. 11thWorkshop on Artificial Intelligence and Statistics*.

Casella, G. and R. Berger (1990). *Statistical Inference*. Wadsworth.

Castillo, E., J. Gutiérrez, and A. Hadi (1997a). *Expert Systems and Probabilistic Network Models*. New York: Springer-Verlag.

Castillo, E., J. Gutiérrez, and A. Hadi (1997b). Sensitivity analysis in discrete Bayesian networks. *IEEE Transactions on Systems, Man and Cybernetics 27*, 412–23.

Chajewska, U. (2002). *Acting Rationally with Incomplete Utility Information*. Ph.D. thesis, Stanford University.

Chajewska, U. and D. Koller (2000). Utilities as random variables: Density estimation and structure discovery. In *Proc. 16th Conference on Uncertainty in Artificial Intelligence (UAI)*, pp. 63–71.

Chajewska, U., D. Koller, and R. Parr (2000). Making rational decisions using adaptive utility elicitation. In *Proc. 16th Conference on Artificial Intelligence (AAAI)*, pp. 363–369.

Chan, H. and A. Darwiche (2002). When do numbers really matter? *Journal of Artificial Intelligence Research 17*, 265–287.

Chávez, T. and M. Henrion (1994). Efficient estimation of the value of information in Monte Carlo models. In *Proc. 10th Conference on Uncertainty in Artificial Intelligence (UAI)*, pp. 119–127.

Cheeseman, P., J. Kelly, M. Self, J. Stutz, W. Taylor, and D. Freeman (1988). Autoclass: a Bayesian classification system. In *Proc. 5th International Conference on Machine Learning (ICML)*.

Cheeseman, P., M. Self, J. Kelly, and J. Stutz (1988). Bayesian classification. In *Proc. 4th Conference on Artificial Intelligence (AAAI)*, Volume 2, pp. 607–611.

Cheeseman, P. and J. Stutz (1995). Bayesian classification (AutoClass): Theory and results. In *Proceedings of the First International Conference on Knowledge Discovery and Data Mining (KDD-95)*. AAAI Press.

Chen, L., M. Wainwright, M. Cetin, and A. Willsky (2003). Multitarget-multisensor data association using the tree-reweighted max-product algorithm. In *Proceedings SPIE Aerosense Conference*, Orlando, Florida.

Chen, R. and S. Liu (2000). Mixture Kalman filters. *Journal of the Royal Statistical Society, Series B*.

Cheng, J. and M. Druzdzel (2000). AIS-BN: An adaptive importance sampling algorithm for evidential reasoning in large Bayesian networks. *Journal of Artificial Intelligence Research 13*, 155–188.

Cheng, J., R. Greiner, J. Kelly, D. Bell, and W. Liu (2002). Learning bayesian networks from data: An information-theory based approach. *Artificial Intelligence*.

Chesley, G. (1978). Subjective probability elicitation techniques: A performance comparison. *Journal of Accounting Research 16*(2), 225–241.

Chickering, D. (1996a). Learning Bayesian networks is NP-Complete. In D. Fisher and H. Lenz (Eds.), *Learning from Data: Artificial Intelligence and Statistics V*, pp. 121–130. Springer-Verlag.

Chickering, D. (2002a, February). Learning equivalence classes of Bayesian-network structures. *Journal of Machine Learning Research 2*, 445–498.

Chickering, D., D. Geiger, and D. Heckerman (1995, January). Learning Bayesian networks: Search methods and experimental results. In *Proceedings of the Fifth International Workshop on Artificial Intelligence and Statistics*, pp. 112–128.

Chickering, D., C. Meek, and D. Heckerman (2003). Large-sample learning of Bayesian networks is hard. In *Proc. 19th Conference on Uncertainty in Artificial Intelligence (UAI)*, pp. 124–133.

Chickering, D. and J. Pearl (1997). A clinician's tool for analyzing non-compliance. *Computing Science and Statistics 29*, 424–31.

Chickering, D. M. (1995). A transformational characterization of equivalent Bayesian network structures. In *Proc. 11th Conference on Uncertainty in Artificial Intelligence (UAI)*, pp. 87–98.

Chickering, D. M. (1996b). Learning equivalence classes of Bayesian network structures. In *Proc. 12th Conference on Uncertainty in Artificial Intelligence (UAI)*, pp. 150–157.

Chickering, D. M. (2002b, November). Optimal structure identification with greedy search. *Journal of Machine Learning Research 3*, 507–554.

Chickering, D. M. and D. Heckerman (1997). Efficient approximations for the marginal likelihood of Bayesian networks with hidden variables. *Machine Learning 29*, 181–212.

Chickering, D. M., D. Heckerman, and C. Meek (1997). A Bayesian approach to learning Bayesian networks with local structure. In *Proc. 13th Conference on Uncertainty in Artificial Intelligence (UAI)*, pp. 80–89.

Chow, C. K. and C. N. Liu (1968). Approximating discrete probability distributions with dependence trees. *IEEE Transactions on Information Theory 14*, 462–467.

Collins, M. (2002). Discriminative training methods for hidden Markov models: Theory and experiments with perceptron algorithms. In *Proc. Conference on Empirical Methods in Natural*

Language Processing (EMNLP).

Cooper, G. (1990). Probabilistic inference using belief networks is NP-hard. *Artificial Intelligence 42*, 393–405.

Cooper, G. and E. Herskovits (1992). A Bayesian method for the induction of probabilistic networks from data. *Machine Learning 9*, 309–347.

Cooper, G. and C. Yoo (1999). Causal discovery from a mixture of experimental and observational data. In *Proc. 15th Conference on Uncertainty in Artificial Intelligence (UAI)*, pp. 116–125.

Cooper, G. F. (1988). A method for using belief networks as influence diagrams. In *Proceedings of the Fourth Workshop on Uncertainty in Artificial Intelligence (UAI)*, pp. 55–63.

Cormen, T. H., C. E. Leiserson, R. L. Rivest, and C. Stein (2001). *Introduction to Algorithms*. Cambridge, Massachusetts: MIT Press. 2nd Edition.

Covaliu, Z. and R. Oliver (1995). Representation and solution of decision problems using sequential decision diagrams. *Management Science 41*(12), 1860–81.

Cover, T. M. and J. A. Thomas (1991). *Elements of Information Theory*. John Wiley & Sons.

Cowell, R. (2005). Local propagation in conditional gaussian Bayesian networks. *Journal of Machine Learning Research 6*, 1517–1550.

Cowell, R. G., A. P. Dawid, S. L. Lauritzen, and D. J. Spiegelhalter (1999). *Probabilistic Networks and Expert Systems*. New York: Springer-Verlag.

Cox, R. (2001). *Algebra of Probable Inference*. The Johns Hopkins University Press.

Cozman, F. (2000). Credal networks. *Artificial Intelligence 120*, 199–233.

Csiszàr, I. (1975). I-divergence geometry of probability distributions and minimization problems. *The Annals of Probability 3*(1), 146–158.

Culotta, A., M. Wick, R. Hall, and A. McCallum (2007). First-order probabilistic models for coreference resolution. In *Proc. Conference of the North American Association for Computational Linguistics*.

D. Rusakov, D. G. (2005). Asymptotic model selection for naive Bayesian networks. *Journal of Machine Learning Research 6*, 1–35.

Dagum, P. and M. Luby (1993). Appoximating probabilistic inference in Bayesian belief networks in NP-hard. *Artificial Intelligence 60*(1), 141–153.

Dagum, P. and M. Luby (1997). An optimal approximation algorithm for Baysian inference. *Artificial Intelligence 93*(1–2), 1–27.

Daneshkhah, A. (2004). Psychological aspects influencing elicitation of subjective probability. Technical report, University of Sheffield.

Darroch, J. and D. Ratcliff (1972). Generalized iterative scaling for log-linear models. *Annals of Mathematical Statistics 43*, 1470–1480.

Darwiche, A. (1993). Argument calculus and networks. In *Proc. 9th Conference on Uncertainty in Artificial Intelligence (UAI)*, pp. 420–27.

Darwiche, A. (2001a). Constant space reasoning in dynamic Bayesian networks. *International Journal of Approximate Reasoning 26*, 161–178.

Darwiche, A. (2001b). Recursive conditioning. *Artificial Intelligence 125*(1–2), 5–41.

Darwiche, A. (2003). A differential approach to inference in Bayesian networks. *Journal of the ACM 50*(3), 280–305.

Darwiche, A. and M. Goldszmidt (1994). On the relation between Kappa calculus and probabilistic reasoning. In *Proc. 10th Conference on Uncertainty in Artificial Intelligence (UAI)*.

Dasgupta, S. (1997). The sample complexity of learning fixed-structure Bayesian networks. *Ma-*

chine Learning 29, 165–180.

Dasgupta, S. (1999). Learning polytrees. In *Proc. 15th Conference on Uncertainty in Artificial Intelligence (UAI)*, pp. 134–141.

Dawid, A. (1979). Conditional independence in statistical theory (with discussion). *Journal of the Royal Statistical Society, Series B 41*, 1–31.

Dawid, A. (1980). Conditional independence for statistical operations. *Annals of Statistics 8*, 598–617.

Dawid, A. (1984). Statistical theory: The prequential approach. *Journal of the Royal Statistical Society, Series A 147*(2), 278–292.

Dawid, A. (1992). Applications of a general propagation algorithm for probabilistic expert system. *Statistics and Computing 2*, 25–36.

Dawid, A. (2002). Influence diagrams for causal modelling and inference. *International Statistical Review 70*, 161–189. Corrections p437.

Dawid, A. (2007, September). Fundamentals of statistical causality. Technical Report 279, RSS/EPSRC Graduate Training Programme, University of Sheffield.

Dawid, A., U. Kjærulff, and S. Lauritzen (1995). Hybrid propagation in junction trees. In *Advances in Intelligent Computing*, Volume 945. Springer-Verlag.

de Bombal, F., D. Leaper, J. Staniland, A. McCann, and J. Harrocks (1972). Computer-aided diagnosis of acute abdominal pain. *British Medical Journal 2*, 9–13.

de Finetti, B. (1937). Foresight: Its logical laws, its subjective sources. *Annals Institute H. Poincaré 7*, 1–68. Translated by H. Kyburg in Kyburg et al. (1980).

de Freitas, N., P. Højen-Sørensen, M. Jordan, and S. Russell (2001). Variational MCMC. In *Proc. 17th Conference on Uncertainty in Artificial Intelligence (UAI)*, pp. 120–127.

Dean, T. and K. Kanazawa (1989). A model for reasoning about persistence and causation. *Computational Intelligence 5*(3), 142–150.

Dechter, R. (1997). Mini-Buckets: A general scheme for generating approximations in automated reasoning. In *Proc. 15th International Joint Conference on Artificial Intelligence (IJCAI)*, pp. 1297–1303.

Dechter, R. (1999). Bucket elimination: A unifying framework for reasoning. *Artificial Intelligence 113*(1–2), 41–85.

Dechter, R. (2003). *Constraint Processing*. Morgan Kaufmann.

Dechter, R., K. Kask, and R. Mateescu (2002). Iterative join-graph propagation. In *Proc. 18th Conference on Uncertainty in Artificial Intelligence (UAI)*, pp. 128–136.

Dechter, R. and I. Rish (1997). A scheme for approximating probabilistic inference. In *Proc. 13th Conference on Uncertainty in Artificial Intelligence (UAI)*.

DeGroot, M. H. (1989). *Probability and Statistics*. Reading, MA: Addison Wesley.

Della Pietra, S., V. Della Pietra, and J. Lafferty (1997). Inducing features of random fields. *IEEE Trans. on Pattern Analysis and Machine Intelligence 19*(4), 380–393.

Dellaert, F., S. Seitz, C. Thorpe, and S. Thrun (2003). EM, MCMC, and chain flipping for structure from motion with unknown correspondence. *Machine Learning 50*(1–2), 45–71.

Deming, W. and F. Stephan (1940). On a least squares adjustment of a sampled frequency table when the expected marginal totals are known. *Annals of Mathematical Statistics 11*, 427–444.

Dempster, A., N. M. Laird, and D. Rubin (1977). Maximum likelihood from incomplete data via the EM algorithm. *Journal of the Royal Statistical Society, Series B 39*(1), 1–22.

Deng, K. and A. Moore (1989). Multiresolution instance-based learning. In *Proc. 14th International*

Joint Conference on Artificial Intelligence (IJCAI), pp. 1233–1239.

Deshpande, A., M. Garofalakis, and M. Jordan (2001). Efficient stepwise selection in decomposable models. In *Proc. 17th Conference on Uncertainty in Artificial Intelligence (UAI)*, pp. 128–135.

Diez, F. (1993). Parameter adjustment in Bayes networks: The generalized noisy OR-gate. In *Proc. 9th Conference on Uncertainty in Artificial Intelligence (UAI)*, pp. 99–105.

Dittmer, S. L. and F. V. Jensen (1997). Myopic value of information in influence diagrams. In *Proc. 13th Conference on Uncertainty in Artificial Intelligence (UAI)*, pp. 142–149.

Doucet, A. (1998). On sequential simulation-based methods for Bayesian filtering. Technical Report CUED/FINFENG/TR 310, Department of Engineering, Cambridge University.

Doucet, A., N. de Freitas, and N. Gordon (Eds.) (2001). *Sequential Monte Carlo Methods in Practice*. New York: Springer-Verlag.

Doucet, A., N. de Freitas, K. Murphy, and S. Russell (2000). Rao-Blackwellised particle filtering for dynamic Bayesian networks. In *Proc. 16th Conference on Uncertainty in Artificial Intelligence (UAI)*.

Doucet, A., S. Godsill, and C. Andrieu (2000). On sequential Monte Carlo sampling methods for Bayesian filtering. *Statistics and Computing 10*(3), 197–208.

Drummond, M., B. O'Brien, G. Stoddart, and G. Torrance (1997). *Methods for the Economic Evaluation of Health Care Programmes, 2nd Edition*. Oxford, UK: Oxford University Press.

Druzdzel, M. (1993). *Probabilistic Reasoning in Decision Support Systems: From Computation to Common Sense*. Ph.D. thesis, Carnegie Mellon University.

Dubois, D. and H. Prade (1990). Inference i possibilistic hypergraphs. In *Proc. of the 6th Conference on Information Processing and Management of Uncertainty in Knowledge-Based Systems*.

Duchi, J., D. Tarlow, G. Elidan, and D. Koller (2006). Using combinatorial optimization within max-product belief propagation. In *Proc. 20th Conference on Neural Information Processing Systems (NIPS)*.

Duda, R., J. Gaschnig, and P. Hart (1979). Model design in the PROSPECTOR consultant system for mineral exploration. In D. Michie (Ed.), *Expert Systems in the Microelectronic Age*, pp. 153–167. Edinburgh, Scotland: Edinburgh University Press.

Duda, R. and P. Hart (1973). *Pattern Classification and Scene Analysis*. New York: John Wiley & Sons.

Duda, R., P. Hart, and D. Stork (2000). *Pattern Classification, Second Edition*. Wiley.

Dudík, M., S. Phillips, and R. Schapire (2004). Performance guarantees for regularized maximum entropy density estimation. In *Proc. Conference on Computational Learning Theory (COLT)*.

Durbin, R., S. Eddy, A. Krogh, and G. Mitchison (1998). *Biological Sequence Analysis: Probabilistic Models of Proteins and Nucleic Acids*. Cambridge: Cambridge University Press.

Dykstra, R. and J. Lemke (1988). Duality of I projections and maximum likelihood estimation for log-linear models under cone constraints. *Journal of the American Statistical Association 83*(402), 546–554.

El-Hay, T. and N. Friedman (2001). Incorporating expressive graphical models in variational approximations: Chain-graphs and hidden variables. In *Proc. 17th Conference on Uncertainty in Artificial Intelligence (UAI)*, pp. 136–143.

Elfadel, I. (1995). Convex potentials and their conjugates in analog mean-field optimization. *Neural Computation 7*, 1079–1104.

Elidan, G. and N. Friedman (2005). Learning hidden variable networks: The information bottleneck approach. *Journal of Machine Learning Research 6*, 81–127.

Elidan, G., I. McGraw, and D. Koller (2006). Residual belief propagation: Informed scheduling for asynchronous message passing. In *Proc. 22nd Conf. on Uncertainty in Artificial Intelligence*.

Elidan, G., N. Lotner, N. Friedman, and D. Koller (2000). Discovering hidden variables: A structure-based approach. In *Proc. 14th Conf. on Neural Information Processing Systems (NIPS)*.

Elidan, G., I. Nachman, and N. Friedman (2007). "Ideal Parent" structure learning for continuous variable networks. *Journal of Machine Learning Research 8*, 1799–1833.

Elidan, G., M. Ninio, N. Friedman, and D. Schuurmans (2002). Data perturbation for escaping local maxima in learning. In *Proc. 18th National Conference on Artificial Intelligence (AAAI)*.

Ellis, B. and W. Wong (2008). Learning causal Bayesian network structures from experimental data. *Journal of the American Statistical Association 103*, 778–789.

Elston, R. C. and J. Stewart (1971). A general model for the analysis of pedigree data. *Human Heredity 21*, 523–542.

Ezawa, K. (1994). Value of evidence on influence diagrams. In *Proc. 10th Conference on Uncertainty in Artificial Intelligence (UAI)*, pp. 212–220.

Feller, W. (1970). *An Introduction to Probability Theory and Its Applications* (third ed.), Volume I. New York: John Wiley & Sons.

Felzenszwalb, P. and D. Huttenlocher (2006, October). Efficient belief propagation for early vision. *International Journal of Computer Vision 70*(1).

Fertig, K. and J. Breese (1989). Interval influence diagrams. In *Proc. 5th Conference on Uncertainty in Artificial Intelligence (UAI)*.

Fine, S., Y. Singer, and N. Tishby (1998). The hierarchical Hidden Markov Model: Analysis and applications. *Machine Learning 32*, 41–62.

Fishburn, P. (1967). Interdependence and additivity in multivariate, unidimensional expected utility theory. *International Economic Review 8*, 335–42.

Fishburn, P. (1970). *Utility Theory for Decision Making*. New York: Wiley.

Fishelson, M. and D. Geiger (2003). Optimizing exact genetic linkage computations. In *Proc. International Conf. on Research in Computational Molecular Biology (RECOMB)*, pp. 114–121.

Fishman, G. (1976, July). Sampling from the gamma distribution on a computer. *Communications of the ACM 19*(7), 407–409.

Fishman, G. (1996). *Monte Carlo — Concept, Algorithms, and Applications*. Series in Operations Research. Springer.

Fox, D., W. Burgard, and S. Thrun (1999). Markov localization for mobile robots in dynamic environments. *Journal of Artificial Intelligence Research 11*, 391–427.

Freund, Y. and R. Schapire (1998). Large margin classification using the perceptron algorithm. In *Proc. Conference on Computational Learning Theory (COLT)*.

Frey, B. (2003). Extending factor graphs so as to unify directed and undirected graphical models. In *Proc. 19th Conference on Uncertainty in Artificial Intelligence (UAI)*, pp. 257–264.

Frey, B. and A. Kannan (2000). Accumulator networks: suitors of local probability propagation. In *Proc. 14th Conference on Neural Information Processing Systems (NIPS)*.

Frey, B. and D. MacKay (1997). A revolution: Belief propagation in graphs with cycles. In *Proc. 11th Conference on Neural Information Processing Systems (NIPS)*.

Frey, B. J. (1998). *Graphical Models for Machine Learning and Digital Communication*. Cambridge, Massachusetts: MIT Press.

Friedman, N. (1997). Learning belief networks in the presence of missing values and hidden variables. In *Proc. 14th International Conference on Machine Learning (ICML)*, pp. 125–133.

Friedman, N. (1998). The Bayesian structural em algorithm. In *Proc. 14th Conference on Uncertainty in Artificial Intelligence (UAI)*, pp. 129–138.

Friedman, N., D. Geiger, and M. Goldszmidt (1997). Bayesian network classifiers. *Machine Learning 29*, 131–163.

Friedman, N., D. Geiger, and N. Lotner (2000). Likelihood computations using value abstraction. In *Proc. 16th Conference on Uncertainty in Artificial Intelligence (UAI)*.

Friedman, N., L. Getoor, D. Koller, and A. Pfeffer (1999). Learning probabilistic relational models. In *Proc. 16th International Joint Conference on Artificial Intelligence (IJCAI)*, pp. 1300–1307.

Friedman, N. and M. Goldszmidt (1996). Learning Bayesian networks with local structure. In *Proc. 12th Conference on Uncertainty in Artificial Intelligence (UAI)*, pp. 252–262.

Friedman, N. and M. Goldszmidt (1998). Learning Bayesian networks with local structure. See Jordan (1998), pp. 421–460.

Friedman, N. and D. Koller (2003). Being Bayesian about Bayesian network structure: A Bayesian approach to structure discovery in Bayesian networks. *Machine Learning 50*(1–2), 95–126.

Friedman, N., K. Murphy, and S. Russell (1998). Learning the structure of dynamic probabilistic networks. In *Proc. 14th Conference on Uncertainty in Artificial Intelligence (UAI)*.

Friedman, N. and I. Nachman (2000). Gaussian process networks. In *Proc. 16th Conference on Uncertainty in Artificial Intelligence (UAI)*, pp. 211–219.

Friedman, N. and Z. Yakhini (1996). On the sample complexity of learning Bayesian networks. In *Proc. 12th Conference on Uncertainty in Artificial Intelligence (UAI)*.

Frogner, C. and A. Pfeffer (2007). Discovering weakly-interacting factors in a complex stochastic process. In *Proc. 21st Conference on Neural Information Processing Systems (NIPS)*.

Frydenberg, J. (1990). The chain graph Markov property. *Scandinavian Journal of Statistics 17*, 790–805.

Fudenberg, D. and J. Tirole (1991). *Game Theory*. MIT Press.

Fung, R. and K. C. Chang (1989). Weighting and integrating evidence for stochastic simulation in Bayesian networks. In *Proc. 5th Conference on Uncertainty in Artificial Intelligence (UAI)*, San Mateo, California. Morgan Kaufmann.

Fung, R. and B. del Favero (1994). Backward simulation in Bayesian networks. In *Proc. 10th Conference on Uncertainty in Artificial Intelligence (UAI)*, pp. 227–234.

Galles, D. and J. Pearl (1995). Testing identifiability of causal models. In *Proc. 11th Conference on Uncertainty in Artificial Intelligence (UAI)*, pp. 185–95.

Gamerman, D. and H. Lopes (2006). *Markov Chain Monte Carlo: Stochastic Simulation for Bayesian Inference*. Chapman & Hall, CRC.

Ganapathi, V., D. Vickrey, J. Duchi, and D. Koller (2008). Constrained approximate maximum entropy learning. In *Proc. 24th Conference on Uncertainty in Artificial Intelligence (UAI)*.

Garcia, L. D. (2004). Algebraic statistics in model selection. In *Proc. 20th Conference on Uncertainty in Artificial Intelligence (UAI)*, pp. 177–18.

Geiger, D. and D. Heckerman. A characterization of the bivariate normal-Wishart distribution. *Probability and Mathematical Statistics 18*, 119–131.

Geiger, D. and D. Heckerman (1994). Learning gaussian networks. In *Proc. 10th Conference on Uncertainty in Artificial Intelligence (UAI)*, pp. 235–243.

Geiger, D. and D. Heckerman (1996). Knowledge representation and inference in similarity networks and Bayesian multinets. *Artificial Intelligence 82*(1–2), 45–74.

Geiger, D., D. Heckerman, H. King, and C. Meek (2001). Stratified exponential families: Graphical models and model selection. *Annals of Statistics 29*, 505–529.

Geiger, D., D. Heckerman, and C. Meek (1996). Asymptotic model selection for directed networks with hidden variables. In *Proc. 12th Conference on Uncertainty in Artificial Intelligence (UAI)*, pp. 283–290.

Geiger, D. and C. Meek (1998). Graphical models and exponential families. In *Proc. 14th Conference on Uncertainty in Artificial Intelligence (UAI)*, pp. 156–165.

Geiger, D., C. Meek, and Y. Wexler (2006). A variational inference procedure allowing internal structure for overlapping clusters and deterministic constraints. *Journal of Artificial Intelligence Research 27*, 1–23.

Geiger, D. and J. Pearl (1988). On the logic of causal models. In *Proc. 4th Conference on Uncertainty in Artificial Intelligence (UAI)*, pp. 3–14.

Geiger, D. and J. Pearl (1993). Logical and algorithmic properties of conditional independence and graphical models. *Annals of Statistics 21*(4), 2001–21.

Geiger, D., T. Verma, and J. Pearl (1989). d-separation: From theorems to algorithms. In *Proc. 5th Conference on Uncertainty in Artificial Intelligence (UAI)*, pp. 139–148.

Geiger, D., T. Verma, and J. Pearl (1990). Identifying independence in Bayesian networks. *Networks 20*, 507–534.

Gelfand, A. and A. Smith (1990). Sampling based approaches to calculating marginal densities. *Journal of the American Statistical Association 85*, 398–409.

Gelman, A., J. B. Carlin, H. S. Stern, and D. B. Rubin (1995). *Bayesian Data Analysis*. London: Chapman & Hall.

Gelman, A. and X.-L. Meng (1998). Simulating normalizing constants: From importance sampling to bridge sampling to path sampling. *Statistical Science 13*(2), 163–185.

Gelman, A. and D. Rubin (1992). Inference from iterative simulation using multiple sequences. *Statistical Science 7*, 457–511.

Geman, S. and D. Geman (1984, November). Stochastic relaxation, Gibbs distributions, and the Bayesian restoration of images. *IEEE Trans. on Pattern Analysis and Machine Intelligence 6*(6), 721–741.

Getoor, L., N. Friedman, D. Koller, A. Pfeffer, and B. Taskar (2007). Probabilistic relational models. See Getoor and Taskar (2007).

Getoor, L., N. Friedman, D. Koller, and B. Taskar (2002). Learning probabilistic models of link structure. *Journal of Machine Learning Research 3*(December), 679–707.

Getoor, L. and B. Taskar (Eds.) (2007). *Introduction to Statistical Relational Learning*. MIT Press.

Geweke, J. (1989). Bayesian inference in econometric models using Monte Carlo integration. *Econometrica 57*, 1317–1339.

Geyer, C. and E. Thompson (1992). Constrained Monte Carlo maximum likelihood for dependent data. *Journal of the Royal Statistical Society, Series B.*

Geyer, C. and E. Thompson (1995). Annealing Markov chain Monte Carlo with applications to ancestral inference. *Journal of the American Statistical Association 90*(431), 909–920.

Geyer, C. J. (1991). Markov chain Monte Carlo maximum likelihood. In *Computing Science and Statistics: Proceedings of 23rd Symposium on the Interface Interface Foundation*, pp. 156–163. Fairfax Station.

Ghahramani, Z. (1994). Factorial learning and the em algorithm. In *Proc. 8th Conference on Neural Information Processing Systems (NIPS)*, pp. 617–624.

Ghahramani, Z. and M. Beal (2000). Propagation algorithms for variational Bayesian learning. In

Proc. 14th Conference on Neural Information Processing Systems (NIPS).

Ghahramani, Z. and G. Hinton (1998). Variational learning for switching state-space models. *Neural Computation 12*(4), 963–996.

Ghahramani, Z. and M. Jordan (1993). Supervised learning from incomplete data via an EM approach. In *Proc. 7th Conference on Neural Information Processing Systems (NIPS)*.

Ghahramani, Z. and M. Jordan (1997). Factorial hidden Markov models. *Machine Learning 29*, 245–273.

Gibbs, J. (1902). *Elementary Principles of Statistical Mechanics*. New Haven, Connecticut: Yale University Press.

Gidas, B. (1988). Consistency of maximum likelihood and pseudo-likelihood estimators for Gibbsian distributions. In W. Fleming and P.-L. Lions (Eds.), *Stochastic differential systems, stochastic control theory and applications*. Springer, New York.

Gilks, W. (1992). Derivative-free adaptive rejection sampling for Gibbs sampling. In J. Bernardo, J. Berger, A. Dawid, and A. Smith (Eds.), *Bayesian Statistics 4*, pp. 641–649. Oxford, UK: Clarendon Press.

Gilks, W., N. Best, and K. Tan (1995). Adaptive rejection Metropolis sampling within Gibbs sampling. *Annals of Statistics 44*, 455–472.

Gilks, W., S. Richardson, and D. Spiegelhalter (Eds.) (1996). *Markov Chain Monte Carlo in Practice*. Chapman & Hall, London.

Gilks, W., A. Thomas, and D. Spiegelhalter (1994). A language and program for complex Bayesian modeling. *The Statistician 43*, 169–177.

Gilks, W. and P. Wild (1992). Adaptive rejection sampling for Gibbs sampling. *Annals of Statistics 41*, 337–348.

Giudici, P. and P. Green (1999, December). Decomposable graphical Gaussian model determination. *Biometrika 86*(4), 785–801.

Globerson, A. and T. Jaakkola (2007a). Convergent propagation algorithms via oriented trees. In *Proc. 23rd Conference on Uncertainty in Artificial Intelligence (UAI)*.

Globerson, A. and T. Jaakkola (2007b). Fixing max-product: Convergent message passing algorithms for MAP LP-relaxations. In *Proc. 21st Conference on Neural Information Processing Systems (NIPS)*.

Glover, F. and M. Laguna (1993). Tabu search. In C. Reeves (Ed.), *Modern Heuristic Techniques for Combinatorial Problems*, Oxford, England. Blackwell Scientific Publishing.

Glymour, C. and G. F. Cooper (Eds.) (1999). *Computation, Causation, Discovery*. Cambridge: MIT Press.

Godsill, S., A. Doucet, and M. West (2000). Methodology for Monte Carlo smoothing with application to time-varying autoregressions. In *Proc. International Symposium on Frontiers of Time Series Modelling*.

Golumbic, M. (1980). *Algorithmic Graph Theory and Perfect Graphs*. London: Academic Press.

Good, I. (1950). *Probability and the Weighing of Evidence*. London: Griffin.

Goodman, J. (2004). Exponential priors for maximum entropy models. In *Proc. Conference of the North American Association for Computational Linguistics*.

Goodman, L. (1970). The multivariate analysis of qualitative data: Interaction among multiple classification. *Journal of the American Statistical Association 65*, 226–56.

Gordon, N., D. Salmond, and A. Smith (1993). Novel approach to nonlinear/non-Gaussian Bayesian state estimation. *IEE Proceedings-F 140*(2), 107–113.

Gorry, G. and G. Barnett (1968). Experience with a model of sequential diagnosis. *Computers and Biomedical Research 1*, 490–507.

Green, P. (1995). Reversible jump Markov chain Monte Carlo computation and Bayesian model determination. *Biometrika 82*, 711–732.

Green, P. J. (1990). On use of the EM algorithm for penalized likelihood estimation. *Journal of the Royal Statistical Society, Series B 52*(3), 443–452.

Greig, D., B. Porteous, and A. Seheult (1989). Exact maximum a posteriori estimation for binary images. *Journal of the Royal Statistical Society, Series B 51*(2), 271–279.

Greiner, R. and W. Zhou (2002). Structural extension to logistic regression: Discriminant parameter learning of belief net classifiers. In *Proc. 18th Conference on Artificial Intelligence (AAAI)*.

Guestrin, C. E., D. Koller, R. Parr, and S. Venkataraman (2003). Efficient solution algorithms for factored MDPs. *Journal of Artificial Intelligence Research 19*, 399–468.

Guyon, X. and H. R. Künsch (1992). Asymptotic comparison of estimators in the Ising model. In *Stochastic Models, Statistical Methods, and Algorithms in Image Analysis, Lecture Notes in Statistics*, Volume 74, pp. 177–198. Springer, Berlin.

Ha, V. and P. Haddawy (1997). Problem-focused incremental elicitation of multi-attribute utility models. In *Proc. 13th Conference on Uncertainty in Artificial Intelligence (UAI)*, pp. 215–222.

Ha, V. and P. Haddawy (1999). A hybrid approach to reasoning with partially elicited preference models. In *Proc. 15th Conference on Uncertainty in Artificial Intelligence (UAI)*, pp. 263–270.

Haberman, S. (1974). *The General Log-Linear Model*. Ph.D. thesis, Department of Statistics, University of Chicago.

Halpern, J. Y. (2003). *Reasoning about Uncertainty*. MIT Press.

Hammer, P. (1965). Some network flow problems solved with pseudo-Boolean programming. *Operations Research 13*, 388–399.

Hammersley, J. and P. Clifford (1971). Markov fields on finite graphs and lattices. Unpublished manuscript.

Handschin, J. and D. Mayne (1969). Monte Carlo techniques to estimate the conditional expectation in multi-stage non-linear filtering. *International Journal of Control 9*(5), 547–559.

Hartemink, A., D. Gifford, T. Jaakkola, and R. Young (2002, March/April). Bayesian methods for elucidating genetic regulatory networks. *IEEE Intelligent Systems 17*, 37–43. special issue on Intelligent Systems in Biology.

Hastie, T., R. Tibshirani, and J. Friedman (2001). *The Elements of Statistical Learning*. Springer Series in Statistics.

Hazan, T. and A. Shashua (2008). Convergent message-passing algorithms for inference over general graphs with convex free energies. In *Proc. 24th Conference on Uncertainty in Artificial Intelligence (UAI)*.

Heckerman, D. (1990). *Probabilistic Similarity Networks*. MIT Press.

Heckerman, D. (1993). Causal independence for knowledge acquisition and inference. In *Proc. 9th Conference on Uncertainty in Artificial Intelligence (UAI)*, pp. 122–127.

Heckerman, D. (1998). A tutorial on learning with Bayesian networks. See Jordan (1998).

Heckerman, D. and J. Breese (1996). Causal independence for probability assessment and inference using Bayesian networks. *IEEE Transactions on Systems, Man, and Cybernetics 26*, 826–831.

Heckerman, D., J. Breese, and K. Rommelse (1995, March). Decision-theoretic troubleshooting.

Communications of the ACM 38(3), 49–57.

Heckerman, D., D. M. Chickering, C. Meek, R. Rounthwaite, and C. Kadie (2000). Dependency networks for inference, collaborative filtering, and data visualization. *jmlr 1*, 49–75.

Heckerman, D. and D. Geiger (1995). Learning Bayesian networks: a unification for discrete and Gaussian domains. In *Proc. 11th Conference on Uncertainty in Artificial Intelligence (UAI)*, pp. 274–284.

Heckerman, D., D. Geiger, and D. M. Chickering (1995). Learning Bayesian networks: The combination of knowledge and statistical data. *Machine Learning 20*, 197–243.

Heckerman, D., E. Horvitz, and B. Nathwani (1992). Toward normative expert systems: Part I. The Pathfinder project. *Methods of Information in Medicine 31*, 90–105.

Heckerman, D. and H. Jimison (1989). A Bayesian perspective on confidence. In *Proc. 5th Conference on Uncertainty in Artificial Intelligence (UAI)*, pp. 149–160.

Heckerman, D., A. Mamdani, and M. Wellman (1995). Real-world applications of Bayesian networks. *Communications of the ACM 38*.

Heckerman, D. and C. Meek (1997). Embedded Bayesian network classifiers. Technical Report MSR-TR-97-06, Microsoft Research, Redmond, WA.

Heckerman, D., C. Meek, and G. Cooper (1999). A Bayesian approach to causal discovery. See Glymour and Cooper (1999), pp. 141–166.

Heckerman, D., C. Meek, and D. Koller (2007). Probabilistic entity-relationship models, PRMs, and plate models. See Getoor and Taskar (2007).

Heckerman, D. and B. Nathwani (1992a). An evaluation of the diagnostic accuracy of Pathfinder. *Computers and Biomedical Research 25(1)*, 56–74.

Heckerman, D. and B. Nathwani (1992b). Toward normative expert systems. II. Probability-based representations for efficient knowledge acquisition and inference. *Methods of Information in Medicine 31*, 106–16.

Heckerman, D. and R. Shachter (1994). A decision-based view of causality. In *Proc. 10th Conference on Uncertainty in Artificial Intelligence (UAI)*, pp. 302–310. Morgan Kaufmann.

Henrion, M. (1986). Propagation of uncertainty in Bayesian networks by probabilistic logic sampling. In *Proc. 2nd Conference on Uncertainty in Artificial Intelligence (UAI)*, pp. 149–163.

Henrion, M. (1991). Search-based algorithms to bound diagnostic probabilities in very large belief networks. In *Proc. 7th Conference on Uncertainty in Artificial Intelligence (UAI)*, pp. 142–150.

Hernández, L. and S. Moral (1997). Mixing exact and importance sampling propagation algorithms in dependence graphs. *International Journal of Intelligent Systems 12*, 553–576.

Heskes, T. (2002). Stable fixed points of loopy belief propagation are minima of the Bethe free energy. In *Proc. 16th Conference on Neural Information Processing Systems (NIPS)*, pp. 359–366.

Heskes, T. (2004). On the uniqueness of loopy belief propagation fixed points. *Neural Computation 16*, 2379–2413.

Heskes, T. (2006). Convexity arguments for efficient minimization of the Bethe and Kikuchi free energies. *Journal of Machine Learning Research 26*, 153–190.

Heskes, T., K. Albers, and B. Kappen (2003). Approximate inference and constrained optimization. In *Proc. 19th Conference on Uncertainty in Artificial Intelligence (UAI)*, pp. 313–320.

Heskes, T., M. Opper, W. Wiegerinck, O. Winther, and O. Zoeter (2005). Approximate inference techniques with expectation constraints. *Journal of Statistical Mechanics: Theory and Experiment*.

Heskes, T. and O. Zoeter (2002). Expectation propagation for approximate inference in dynamic

Bayesian networks. In *Proc. 18th Conference on Uncertainty in Artificial Intelligence (UAI)*.

Heskes, T. and O. Zoeter (2003). Generalized belief propagation for approximate inference in hybrid Bayesian networks. In *Proceedings of the Ninth International Workshop on Artificial Intelligence and Statistics*.

Heskes, T., O. Zoeter, and W. Wiegerinck (2003). Approximate expectation maximization. In *Proc. 17th Conference on Neural Information Processing Systems (NIPS)*, pp. 353–360.

Higdon, D. M. (1998). Auxiliary variable methods for Markov chain Monte Carlo with applications. *Journal of the American Statistical Association 93*, 585–595.

Hinton, G. (2002). Training products of experts by minimizing contrastive divergence. *Neural Computation 14*, 1771–1800.

Hinton, G., S. Osindero, and Y. Teh (2006). A fast learning algorithm for deep belief nets. *Neural Computation 18*, 1527–1554.

Hinton, G. and R. Salakhutdinov (2006). Reducing the dimensionality of data with neural networks. *Science 313*, 504–507.

Hinton, G. and T. Sejnowski (1983). Optimal perceptual inference. In *Proceedings of the IEEE conference on Computer Vision and Pattern Recognition*, pp. 448–453.

Hinton, G. E., P. Dayan, B. Frey, and R. M. Neal (1995). The wake-sleep algorithm for unsupervised neural networks. *Science 268*, 1158–1161.

Höffgen, K. (1993). Learning and robust learning of product distributions. In *Proc. Conference on Computational Learning Theory (COLT)*, pp. 77–83.

Hofmann, R. and V. Tresp (1995). Discovering structure in continuous variables using bayesian networks. In *Proc. 9th Conference on Neural Information Processing Systems (NIPS)*.

Horn, G. and R. McEliece (1997). Belief propagation in loopy bayesian networks: experimental results. In *Proceedings if IEEE International Symposium on Information Theory*, pp. 232.

Horvitz, E. and M. Barry (1995). Display of information for time-critical decision making. In *Proc. 11th Conference on Uncertainty in Artificial Intelligence (UAI)*, pp. 296–305.

Horvitz, E., J. Breese, and M. Henrion (1988). Decision theory in expert systems and artificial intelligence. *International Journal of Approximate Reasoning 2*, 247–302. Special Issue on Uncertainty in Artificial Intelligence.

Horvitz, E., H. Suermondt, and G. Cooper (1989). Bounded conditioning: Flexible inference for decisions under scarce resources. In *Proc. 5th Conference on Uncertainty in Artificial Intelligence (UAI)*, pp. 182–193.

Howard, R. (1970). Decision analysis: Perspectives on inference, decision, and experimentation. *Proceedings of the IEEE 58*, 632–643.

Howard, R. (1977). Risk preference. In R. Howard and J. Matheson (Eds.), *Readings in Decision Analysis*, pp. 429–465. Menlo Park, California: Decision Analysis Group, SRI International.

Howard, R. and J. Matheson (1984a). Influence diagrams. See Howard and Matheson (1984b), pp. 721–762.

Howard, R. and J. Matheson (Eds.) (1984b). *The Principle and Applications of Decision Analysis*. Menlo Park, CA, USA: Strategic Decisions Group.

Howard, R. A. (1966). Information value theory. *IEEE Transactions on Systems Science and Cybernetics SSC-2*, 22–26.

Howard, R. A. (1989). Microrisks for medical decision analysis. *International Journal of Technology Assessment in Health Care 5*, 357–370.

Huang, C. and A. Darwiche (1996). Inference in belief networks: A procedural guide. *International

Journal of Approximate Reasoning 15(3), 225–263.

Huang, F. and Y. Ogata (2002). Generalized pseudo-likelihood estimates for Markov random fields on lattice. *Annals of the Institute of Statistical Mathematics 54*, 1–18.

Ihler, A. (2007). Accuracy bounds for belief propagation. In *Proc. 23rd Conference on Uncertainty in Artificial Intelligence (UAI)*.

Ihler, A. T., J. W. Fisher, and A. S. Willsky (2003). Message errors in belief propagation. In *Proc. 17th Conference on Neural Information Processing Systems (NIPS)*.

Ihler, A. T., J. W. Fisher, and A. S. Willsky (2005). Loopy belief propagation: Convergence and effects of message errors. *Journal of Machine Learning Research 6*, 905–936.

Imoto, S., S. Kim, T. Goto, S. Aburatani, K. Tashiro, S. Kuhara, and S. Miyano (2003). Bayesian network and nonparametric heteroscedastic regression for nonlinear modeling of genetic network. *Journal of Bioinformatics and Computational Biology 1*, 231–252.

Indyk, P. (2004). Nearest neighbors in high-dimensional spaces. In J. Goodman and J. O'Rourke (Eds.), *Handbook of Discrete and Computational Geometry* (2nd ed.). CRC Press.

Isard, M. (2003). PAMPAS: Real-valued graphical models for computer vision. In *Proc. Conference on Computer Vision and Pattern Recognition (CVPR)*, pp. 613–620.

Isard, M. and A. Blake (1998a). Condensation — conditional density propagation for visual tracking. *International Journal of Computer Vision 29*(1), 5–28.

Isard, M. and A. Blake (1998b). A smoothing filter for condensation. In *Proc. European Conference on Computer Vision (ECCV)*, Volume 1, pp. 767–781.

Isham, V. (1981). An introduction to spatial point processes and Markov random fields. *International Statistical Review 49*, 21–43.

Ishikawa, H. (2003). Exact optimization for Markov random fields with convex priors. *IEEE Trans. on Pattern Analysis and Machine Intelligence 25*(10), 1333–1336.

Ising, E. (1925). Beitrag zur theorie des ferromagnetismus. *Z. Phys. 31*, 253–258.

Jaakkola, T. (2001). Tutorial on variational approximation methods. In M. Opper and D. Saad (Eds.), *Advanced mean field methods*, pp. 129–160. Cambridge, Massachusetts: MIT Press.

Jaakkola, T. and M. Jordan (1996a). Computing upper and lower bounds on likelihoods in intractable networks. In *Proc. 12th Conference on Uncertainty in Artificial Intelligence (UAI)*, pp. 340–348.

Jaakkola, T. and M. Jordan (1996b). Recursive algorithms for approximating probabilities in graphical models. In *Proc. 10th Conference on Neural Information Processing Systems (NIPS)*, pp. 487–93.

Jaakkola, T. and M. Jordan (1997). A variational approach to bayesian logistic regression models and their extensions. In *Proc. 6thWorkshop on Artificial Intelligence and Statistics*.

Jaakkola, T. and M. Jordan (1998). Improving the mean field approximation via the use of mixture models. See Jordan (1998).

Jaakkola, T. and M. Jordan (1999). Variational probabilistic inference and the QMR-DT network. *Journal of Artificial Intelligence Research 10*, 291–322.

Jarzynski, C. (1997, Apr). Nonequilibrium equality for free energy differences. *Physical Review Letters 78*(14), 2690–2693.

Jaynes, E. (2003). *Probability Theory: The Logic of Science*. Cambridge University Press.

Jensen, F., F. V. Jensen, and S. L. Dittmer (1994). From influence diagrams to junction trees. In *Proc. 10th Conference on Uncertainty in Artificial Intelligence (UAI)*, pp. 367–73.

Jensen, F. and M. Vomlelová (2003). Unconstrained influence diagrams. In *Proc. 19th Conference on Uncertainty in Artificial Intelligence (UAI)*, pp. 234–41.

Jensen, F. V. (1995). Cautious propagation in Bayesian networks. In *Proc. 11th Conference on Uncertainty in Artificial Intelligence (UAI)*, pp. 323–328.

Jensen, F. V. (1996). *An introduction to Bayesian Networks*. London: University College London Press.

Jensen, F. V., K. G. Olesen, and S. K. Andersen (1990, August). An algebra of Bayesian belief universes for knowledge-based systems. *Networks 20*(5), 637–659.

Jerrum, M. and A. Sinclair (1997). The Markov chain Monte Carlo method. In D. Hochbaum (Ed.), *Approximation Algorithms for NP-hard Problems*. Boston: PWS Publishing.

Ji, C. and L. Seymour (1996). A consistent model selection procedure for Markov random fields based on penalized pseudolikelihood. *Annals of Applied Probability*.

Jimison, H., L. Fagan, R. Shachter, and E. Shortliffe (1992). Patient-specific explanation in models of chronic disease. *AI in Medicine 4*, 191–205.

Jordan, M., Z. Ghahramani, T. Jaakkola, and L. K. Saul (1998). An introduction to variational approximations methods for graphical models. See Jordan (1998).

Jordan, M. I. (Ed.) (1998). *Learning in Graphics Models*. Cambridge, MA: The MIT Press.

Julier, S. (2002). The scaled unscented transformation. In *Proceedings of the American Control Conference*, Volume 6, pp. 4555–4559.

Julier, S. and J. Uhlmann (1997). A new extension of the Kalman filter to nonlinear systems. In *Proc. of AeroSense: The 11th International Symposium on Aerospace/Defence Sensing, Simulation and Controls*.

Kahneman, D., P. Slovic, and A. Tversky (Eds.) (1982). *Judgment under Uncertainty: Heuristics and Biases*. Cambridge: Cambridge University Press.

Kalman, R. and R. Bucy (1961). New results in linear filtering and prediction theory. *Trans. ASME, Series D, Journal of Basic Engineering*.

Kalman, R. E. (1960). A new approach to linear filtering and prediction problems. *Transactions of the ASME–Journal of Basic Engineering 82*(Series D), 35–45.

Kanazawa, K., D. Koller, and S. Russell (1995). Stochastic simulation algorithms for dynamic probabilistic networks. In *Proc. 11th Conference on Uncertainty in Artificial Intelligence (UAI)*, pp. 346–351.

Kass, R. and A. Raftery (1995). Bayes factors. *Journal of the American Statistical Association 90*(430), 773–795.

Kearns, M., M. L. Littman, and S. Singh (2001). Graphical models for game theory. In *Proc. 17th Conference on Uncertainty in Artificial Intelligence (UAI)*, pp. 253–260.

Kearns, M. and Y. Mansour (1998). Exact inference of hidden structure from sample data in noisy-or networks. In *Proc. 14th Conference on Uncertainty in Artificial Intelligence (UAI)*, pp. 304–31.

Kearns, M., Y. Mansour, and A. Ng (1997). An information-theoretic analysis of hard and soft assignment methods for clustering. In *Proc. 13th Conference on Uncertainty in Artificial Intelligence (UAI)*, pp. 282–293.

Keeney, R. L. and H. Raiffa (1976). *Decisions with Multiple Objectives: Preferences and Value Tradeoffs*. John Wiley & Sons, Inc.

Kersting, K. and L. De Raedt (2007). Bayesian logic programming: Theory and tool. See Getoor and Taskar (2007).

Kikuchi, R. (1951). A theory of cooperative phenomena. *Physical Review Letters 81*, 988–1003.

Kim, C.-J. and C. Nelson (1998). *State-Space Models with Regime-Switching: Classical and Gibbs-Sampling Approaches with Applications*. MIT Press.

Kim, J. and J. Pearl (1983). A computational model for combined causal and diagnostic reasoning in inference systems. In *Proc. 7th International Joint Conference on Artificial Intelligence (IJCAI)*, pp. 190–193.

Kirkpatrick, S., C. Gelatt, and M. Vecchi (1983). Optimization by simulated annealing. *Science 220*, 671–680.

Kitagawa, G. (1996). Monte Carlo filter and smoother for non-Gaussian nonlinear state space models. *Journal of Computational and Graphical Statistics 5*(1), 1–25.

Kjærulff, U. (1990, March). Triangulation of graph — Algorithms giving small total state space. Technical Report R90-09, Aalborg University, Denmark.

Kjærulff, U. (1992). A computational scheme for reasoning in dynamic probabilistic networks. In *Proc. 8th Conference on Uncertainty in Artificial Intelligence (UAI)*, pp. 121–129.

Kjærulff, U. (1995a). dHugin: A computational system for dynamic time-sliced Bayesian networks. *International Journal of Forecasting 11*, 89–111.

Kjærulff, U. (1995b). HUGS: Combining exact inference and Gibbs sampling in junction trees. In *Proc. 11th Conference on Uncertainty in Artificial Intelligence (UAI)*, pp. 368–375.

Kjaerulff, U. (1997). Nested junction trees. In *Proc. 13th Conference on Uncertainty in Artificial Intelligence (UAI)*, pp. 294–301.

Kjærulff, U. and L. van der Gaag (2000). Making sensitivity analysis computationally efficient. In *Proc. 16th Conference on Uncertainty in Artificial Intelligence (UAI)*, pp. 317–325.

Koivisto, M. and K. Sood (2004). Exact Bayesian structure discovery in Bayesian networks. *Journal of Machine Learning Research 5*, 549–573.

Kok, J., M. Spaan, and N. Vlassis (2003). Multi-robot decision making using coordination graphs. In *Proc. International Conference on Advanced Robotics (ICAR)*, pp. 1124–1129.

Kok, J. and N. Vlassis (2005). Using the max-plus algorithm for multiagent decision making in coordination graphs. In *RoboCup-2005: Robot Soccer World Cup IX*, Osaka, Japan.

Koller, D. and R. Fratkina (1998). Using learning for approximation in stochastic processes. In *Proc. 15th International Conference on Machine Learning (ICML)*, pp. 287–295.

Koller, D., U. Lerner, and D. Anguelov (1999). A general algorithm for approximate inference and its application to hybrid Bayes nets. In *Proc. 15th Conference on Uncertainty in Artificial Intelligence (UAI)*, pp. 324–333.

Koller, D. and B. Milch (2001). Multi-agent influence diagrams for representing and solving games. In *Proc. 17th International Joint Conference on Artificial Intelligence (IJCAI)*, pp. 1027–1034.

Koller, D. and B. Milch (2003). Multi-agent influence diagrams for representing and solving games. *Games and Economic Behavior 45*(1), 181–221. Full version of paper in IJCAI '03.

Koller, D. and R. Parr (1999). Computing factored value functions for policies in structured MDPs. In *Proc. 16th International Joint Conference on Artificial Intelligence (IJCAI)*, pp. 1332–1339.

Koller, D. and A. Pfeffer (1997). Object-oriented Bayesian networks. In *Proc. 13th Conference on Uncertainty in Artificial Intelligence (UAI)*, pp. 302–313.

Kolmogorov, V. (2006). Convergent tree-reweighted message passing for energy minimization. *IEEE Transactions on Pattern Analysis and Machine Intelligence*.

Kolmogorov, V. and C. Rother (2006). Comparison of energy minimization algorithms for highly connected graphs. In *Proc. European Conference on Computer Vision (ECCV)*.

Kolmogorov, V. and M. Wainwright (2005). On the optimality of tree reweighted max-product message passing. In *Proc. 21st Conference on Uncertainty in Artificial Intelligence (UAI)*.

Kolmogorov, V. and R. Zabih (2004). What energy functions can be minimized via graph cuts? *IEEE Transactions on Pattern Analysis and Machine Intelligence 26*(2).

Komarek, P. and A. Moore (2000). A dynamic adaptation of AD-trees for efficient machine learning on large data sets. In *Proc. 17th International Conference on Machine Learning (ICML)*, pp. 495–502.

Komodakis, N., N. Paragios, and G. Tziritas (2007). MRF optimization via dual decomposition: Message-passing revisited. In *Proc. International Conference on Computer Vision (ICCV)*.

Komodakis, N. and G. Tziritas (2005). A new framework for approximate labeling via graph-cuts. In *Proc. International Conference on Computer Vision (ICCV)*.

Komodakis, N., G. Tziritas, and N. Paragios (2007). Fast, approximately optimal solutions for single and dynamic MRFs. In *Proc. Conference on Computer Vision and Pattern Recognition (CVPR)*.

Kong, A. (1991). Efficient methods for computing linkage likelihoods of recessive diseases in inbred pedigrees. *Genetic Epidemiology 8*, 81–103.

Korb, K. and A. Nicholson (2003). *Bayesian Artificial Intelligence*. CRC Press.

Koster, J. (1996). Markov properties of non-recursive causal models. *The Annals of Statistics 24*(5), 2148–77.

Kočka, T., R. Bouckaert, and M. Studený (2001). On characterizing inclusion of Bayesian networks. In *Proc. 17th Conference on Uncertainty in Artificial Intelligence (UAI)*, pp. 261–68.

Kozlov, A. and D. Koller (1997). Nonuniform dynamic discretization in hybrid networks. In *Proc. 13th Conference on Uncertainty in Artificial Intelligence (UAI)*, pp. 314–325.

Krause, A. and C. Guestrin (2005a). Near-optimal nonmyopic value of information in graphical models. In *Proc. 21st Conference on Uncertainty in Artificial Intelligence (UAI)*.

Krause, A. and C. Guestrin (2005b). Optimal nonmyopic value of information in graphical models: Efficient algorithms and theoretical limits. In *Proc. 19th International Joint Conference on Artificial Intelligence (IJCAI)*.

Kreps, D. (1988). *Notes on the Theory of Choice*. Boulder, Colorado: Westview Press.

Kschischang, F. and B. Frey (1998). Iterative decoding of compound codes by probability propagation in graphical models. *IEEE Journal on Selected Areas in Communications 16*, 219–230.

Kschischang, F., B. Frey, and H.-A. Loeliger (2001a). Factor graphs and the sum-product algorithm. *IEEE Transactions on Information Theory 47*, 498–519.

Kschischang, F., B. Frey, and H.-A. Loeliger (2001b). Factor graphs and the sum-product algorithm. *IEEE Trans. Information Theory 47*, 498–519.

Kullback, S. (1959). *Information Theory and Statistics*. New York: John Wiley & Sons.

Kumar, M., V. Kolmogorov, and P. Torr (2007). An analysis of convex relaxations for MAP estimation. In *Proc. 21st Conference on Neural Information Processing Systems (NIPS)*.

Kumar, M., P. Torr, and A. Zisserman (2006). Solving Markov random fields using second order cone programming relaxations. In *Proc. Conference on Computer Vision and Pattern Recognition (CVPR)*, pp. 1045–1052.

Kuppermann, M., S. Shiboski, D. Feeny, E. Elkin, and A. Washington (1997, Jan–Mar). Can preference scores for discrete states be used to derive preference scores for an entire path of events? An application to prenatal diagnosis. *Medical Decision Making 17*(1), 42–55.

Kyburg, H., , and H. Smokler (Eds.) (1980). *Studies in Subjective Probability*. New York: Krieger.

La Mura, P. (2000). Game networks. In *Proc. 16th Conference on Uncertainty in Artificial Intelligence*

(UAI), pp. 335–342.

La Mura, P. and Y. Shoham (1999). Expected utility networks. In *Proc. 15th Conference on Uncertainty in Artificial Intelligence (UAI)*, pp. 366–73.

Lacoste-Julien, S., B. Taskar, D. Klein, and M. Jordan (2006, June). Word alignment via quadratic assignment. In *Proceedings of the Human Language Technology Conference of the NAACL, Main Conference*, pp. 112–119.

Lafferty, J., A. McCallum, and F. Pereira (2001). Conditional random fields: Probabilistic models for segmenting and labeling sequence data. In *Proc. 18th International Conference on Machine Learning (ICML)*.

Lam, W. and F. Bacchus (1993). Using causal information and local measures to learn Bayesian networks. In *Proc. 9th Conference on Uncertainty in Artificial Intelligence (UAI)*, pp. 243–250.

Lange, K. and R. C. Elston (1975). Extensions to pedigree analysis. I. Likelihood calculations for simple and complex pedigrees. *Human Heredity 25*, 95–105.

Laskey, K. (1995). Sensitivity analysis for probability assessments in Bayesian networks. *IEEE Transactions on Systems, Man, and Cybernetics 25*(6), 901 – 909.

Lauritzen, S. (1982). *Lectures on contingency tables* (2 ed.). Aalborg: Denmark: University of Aalborg Press.

Lauritzen, S. (1992). Propagation of probabilities, means, and variances in mixed graphical association models. *Journal of the American Statistical Association 87*(420), 1089–1108.

Lauritzen, S. (1996). *Graphical Models*. New York: Oxford University Press.

Lauritzen, S. and D. Nilsson (2001). Representing and solving decision problems with limited information. *Management Science 47*(9), 1235–51.

Lauritzen, S. L. (1995). The EM algorithm for graphical association models with missing data. *Computational Statistics and Data Analysis 19*, 191–201.

Lauritzen, S. L. and F. Jensen (2001). Stable local computation with conditional Gaussian distributions. *Statistics and Computing 11*, 191–203.

Lauritzen, S. L. and D. J. Spiegelhalter (1988). Local computations with probabilities on graphical structures and their application to expert systems. *Journal of the Royal Statistical Society, Series B B 50*(2), 157–224.

Lauritzen, S. L. and N. Wermuth (1989). Graphical models for associations between variables, some of which are qualitative and some quantitative. *Annals of Statistics 17*, 31–57.

LeCun, Y., S. Chopra, R. Hadsell, R. Marc'Aurelio, and F.-J. Huang (2007). A tutorial on energy-based learning. In G. Bakir, T. Hofmann, B. Schölkopf, A. Smola, B. Taskar, and S. Vishwanathan (Eds.), *Predicting Structured Data*. MIT Press.

Lee, S.-I., V. Ganapathi, and D. Koller (2006). Efficient structure learning of Markov networks using L1-regularization. In *Proc. 20th Conference on Neural Information Processing Systems (NIPS)*.

Lehmann, E. and J. Romano (2008). *Testing Statistical Hypotheses*. Springer Texts in Statistics.

Leisink, M. A. R. and H. J. Kappen (2003). Bound propagation. *Journal of Artificial Intelligence Research 19*, 139–154.

Lerner, U. (2002). *Hybrid Bayesian Networks for Reasoning about Complex Systems*. Ph.D. thesis, Stanford University.

Lerner, U., B. Moses, M. Scott, S. McIlraith, and D. Koller (2002). Monitoring a complex physical system using a hybrid dynamic Bayes net. In *Proc. 18th Conference on Uncertainty in Artificial Intelligence (UAI)*, pp. 301–310.

Lerner, U. and R. Parr (2001). Inference in hybrid networks: Theoretical limits and practical algorithms. In *Proc. 17th Conference on Uncertainty in Artificial Intelligence (UAI)*, pp. 310–318.

Lerner, U., R. Parr, D. Koller, and G. Biswas (2000). Bayesian fault detection and diagnosis in dynamic systems. In *Proc. 16th Conference on Artificial Intelligence (AAAI)*, pp. 531–537.

Lerner, U., E. Segal, and D. Koller (2001). Exact inference in networks with discrete children of continuous parents. In *Proc. 17th Conference on Uncertainty in Artificial Intelligence (UAI)*, pp. 319–328.

Li, S. (2001). *Markov Random Field Modeling in Image Analysis*. Springer.

Liang, P. and M. Jordan (2008). An asymptotic analysis of generative, discriminative, and pseudolikelihood estimators. In *Proc. 25th International Conference on Machine Learning (ICML)*.

Little, R. J. A. (1976). Inference about means for incomplete multivariate data. *Biometrika 63*, 593–604.

Little, R. J. A. and D. B. Rubin (1987). *Statistical Analysis with Missing Data*. New York: John Wiley & Sons.

Liu, D. and J. Nocedal (1989). On the limited memory method for large scale optimization. *Mathematical Programming 45*(3), 503–528.

Liu, J., W. Wong, and A. Kong (1994). Covariance structure of the Gibbs sampler with applications to the comparisons of estimators and sampling schemes. *Biometrika 81*, 27–40.

Loomes, G. and R. Sugden (1982). Regret theory: An alternative theory of rational choice under uncertainty. *The Economic Journal 92*, 805–824.

MacEachern, S. and L. Berliner (1994, August). Subsampling the Gibbs sampler. *The American Statistician 48*(3), 188–190.

MacKay, D. J. C. (1997). Ensemble learning for hidden markov models. Unpublished manuscripts, http://wol.ra.phy.cam.ac.uk/mackay.

MacKay, D. J. C. and R. M. Neal (1996). Near shannon limit performance of low density parity check codes. *Electronics Letters 32*, 1645–1646.

Madigan, D., S. Andersson, M. Perlman, and C. Volinsky (1996). Bayesian model averaging and model selection for Markov equivalence classes of acyclic graphs. *Communications in Statistics: Theory and Methods 25*, 2493–2519.

Madigan, D. and E. Raftery (1994). Model selection and accounting for model uncertainty in graphical models using Occam's window. *Journal of the American Statistical Association 89*, 1535–1546.

Madigan, D. and J. York (1995). Bayesian graphical models for discrete data. *International statistical Review 63*, 215–232.

Madsen, A. and D. Nilsson (2001). Solving influence diagrams using HUGIN, Shafer-Shenoy and lazy propagation. In *Proc. 17th Conference on Uncertainty in Artificial Intelligence (UAI)*, pp. 337–45.

Malioutov, D., J. Johnson, and A. Willsky (2006). Walk-sums and belief propagation in Gaussian graphical models. *Journal of Machine Learning Research 7*, 2031–64.

Maneva, E., E. Mossel, and M. Wainwright (2007, July). A new look at survey propagation and its generalizations. *Journal of the ACM 54*(4), 2–41.

Manning, C. and H. Schuetze (1999). *Foundations of Statistical Natural Language Processing*. MIT Press.

Marinari, E. and G. Parisi (1992). Simulated tempering: A new Monte Carlo scheme. *Europhysics Letters 19*, 451.

Marinescu, R., K. Kask, and R. Dechter (2003). Systematic vs. non-systematic algorithms for solving the MPE task. In *Proc. 19th Conference on Uncertainty in Artificial Intelligence (UAI)*.

Marthi, B., H. Pasula, S. Russell, and Y. Peres (2002). Decayed MCMC filtering. In *Proc. 18th Conference on Uncertainty in Artificial Intelligence (UAI)*.

Martin, J. and K. VanLehn (1995). Discrete factor analysis: Learning hidden variables in Bayesian networks. Technical report, Department of Computer Science, University of Pittsburgh.

McCallum, A. (2003). Efficiently inducing features of conditional random fields. In *Proc. 19th Conference on Uncertainty in Artificial Intelligence (UAI)*, pp. 403–10.

McCallum, A., C. Pal, G. Druck, and X. Wang (2006). Multi-conditional learning: Generative/discriminative training for clustering and classification. In *Proc. 22nd Conference on Artificial Intelligence (AAAI)*.

McCallum, A. and B. Wellner (2005). Conditional models of identity uncertainty with application to noun coreference. In *Proc. 19th Conference on Neural Information Processing Systems (NIPS)*, pp. 905–912.

McCullagh, P. and J. Nelder (1989). *Generalized Linear Models*. London: Chapman & Hall.

McEliece, R., D. MacKay, and J.-F. Cheng (1998, February). Turbo decoding as an instance of Pearl's "belief propagation" algorithm. *IEEE Journal on Selected Areas in Communications 16*(2).

McEliece, R. J., E. R. Rodemich, and J.-F. Cheng (1995). The turbo decision algorithm. In *Proc. 33rd Allerton Conference on Communication Control and Computing*, pp. 366–379.

McLachlan, G. J. and T. Krishnan (1997). *The EM Algorithm and Extensions*. Wiley Interscience.

Meek, C. (1995a). Causal inference and causal explanation with background knowledge. In *Proc. 11th Conference on Uncertainty in Artificial Intelligence (UAI)*, pp. 403–418.

Meek, C. (1995b). Strong completeness and faithfulness in Bayesian networks. In *Proc. 11th Conference on Uncertainty in Artificial Intelligence (UAI)*, pp. 411–418.

Meek, C. (1997). *Graphical Models: Selecting causal and statistical models*. Ph.D. thesis, Carnegie Mellon University.

Meek, C. (2001). Finding a path is harder than finding a tree. *Journal of Artificial Intelligence Research 15*, 383–389.

Meek, C. and D. Heckerman (1997). Structure and parameter learning for causal independence and causal interaction models. In *Proc. 13th Conference on Uncertainty in Artificial Intelligence (UAI)*, pp. 366–375.

Meila, M. and T. Jaakkola (2000). Tractable Bayesian learning of tree belief networks. In *Proc. 16th Conference on Uncertainty in Artificial Intelligence (UAI)*.

Meila, M. and M. Jordan (2000). Learning with mixtures of trees. *Journal of Machine Learning Research 1*, 1–48.

Meltzer, T., C. Yanover, and Y. Weiss (2005). Globally optimal solutions for energy minimization in stereo vision using reweighted belief propagation. In *Proc. International Conference on Computer Vision (ICCV)*, pp. 428–435.

Metropolis, N., A. Rosenbluth, M. Rosenbluth, A. Teller, and E. Teller (1953). Equation of state calculation by fast computing machines. *Journal of Chemical Physics 21*, 1087–1092.

Meyer, J., M. Phillips, P. Cho, I. Kalet, and J. Doctor (2004). Application of influence diagrams to prostate intensity-modulated radiation therapy plan selection. *Physics in Medicine and Biology 49*, 1637–53.

Middleton, B., M. Shwe, D. Heckerman, M. Henrion, E. Horvitz, H. Lehmann, and G. Cooper (1991). Probabilistic diagnosis using a reformulation of the INTERNIST-1/QMR knowledge base.

II. Evaluation of diagnostic performance. *Methods of Information in Medicine 30*, 256–67.

Milch, B., B. Marthi, and S. Russell (2004). BLOG: Relational modeling with unknown objects. In *ICML 2004 Workshop on Statistical Relational Learning and Its Connections to Other Fields*.

Milch, B., B. Marthi, S. Russell, D. Sontag, D. Ong, and A. Kolobov (2005). BLOG: Probabilistic models with unknown objects. In *Proc. 19th International Joint Conference on Artificial Intelligence (IJCAI)*, pp. 1352–1359.

Milch, B., B. Marthi, S. Russell, D. Sontag, D. Ong, and A. Kolobov (2007). BLOG: Probabilistic models with unknown objects. See Getoor and Taskar (2007).

Miller, R., H. Pople, and J. Myers (1982). Internist-1, an experimental computer-based diagnostic consultant for general internal medicine. *New England Journal of Medicine 307*, 468–76.

Minka, T. (2005). Discriminative models, not discriminative training. Technical Report MSR-TR-2005-144, Microsoft Research.

Minka, T. and J. Lafferty (2002). Expectation propagation for the generative aspect model. In *Proc. 18th Conference on Uncertainty in Artificial Intelligence (UAI)*.

Minka, T. P. (2001a). Algorithms for maximum-likelihood logistic regression. Available from `http://www.stat.cmu.edu/~minka/papers/logreg.html`.

Minka, T. P. (2001b). Expectation propagation for approximate Bayesian inference. In *Proc. 17th Conference on Uncertainty in Artificial Intelligence (UAI)*, pp. 362–369.

Møller, J. M., A. Pettitt, K. Berthelsen, and R. Reeves (2006). An efficient Markov chain Monte Carlo method for distributions with intractable normalisation constants. *Biometrika 93*(2), 451–458.

Montemerlo, M., S. Thrun, D. Koller, and B. Wegbreit (2002). FastSLAM: A factored solution to the simultaneous localization and mapping problem. In *Proc. 18th Conference on Artificial Intelligence (AAAI)*, pp. 593–598.

Monti, S. and G. F. Cooper (1997). Learning Bayesian belief networks with neural network estimators. In *Proc. 11th Conference on Neural Information Processing Systems (NIPS)*, pp. 579–584.

Mooij, J. M. and H. J. Kappen (2007). Sufficient conditions for convergence of the sum-product algorithm. *IEEE Trans. Information Theory 53*, 4422–4437.

Moore, A. (2000). The anchors hierarchy: Using the triangle inequality to survive high-dimensional data. In *Proc. 16th Conference on Uncertainty in Artificial Intelligence (UAI)*, pp. 397–405.

Moore, A. and W.-K. Wong (2003). Optimal reinsertion: A new search operator for accelerated and more accurate bayesian network structure learning. In *Proc. 20th International Conference on Machine Learning (ICML)*, pp. 552–559.

Moore, A. W. and M. S. Lee (1997). Cached sufficient statistics for efficient machine learning with large datasets. *Journal of Artificial Intelligence Research 8*, 67–91.

Morgan, M. and M. Henrion (Eds.) (1990). *Uncertainty: A Guide to Dealing with Uncertainty in Quantitative Risk and Policy Analysis*. Cambridge University Press.

Motwani, R. and P. Raghavan (1995). *Randomized Algorithnms*. Cambridge University Press.

Muramatsu, M. and T. Suzuki (2003). A new second-order cone programming relaxation for max-cut problems. *Journal of Operations Research of Japan 43*, 164–177.

Murphy, K. (1999). Bayesian map learning in dynamic environments. In *Proc. 13th Conference on Neural Information Processing Systems (NIPS)*.

Murphy, K. (2002). Dynamic Bayesian Networks: A tutorial. Technical report, Massachussetts Institute of Technology. Available from `http://www.cs.ubc.ca/~murphyk/Papers/dbnchapter.pdf`.

Murphy, K. and M. Paskin (2001). Linear time inference in hierarchical HMMs. In *Proc. 15th Conference on Neural Information Processing Systems (NIPS)*.

Murphy, K. and Y. Weiss (2001). The factored frontier algorithm for approximate inference in DBNs. In *Proc. 17th Conference on Uncertainty in Artificial Intelligence (UAI)*.

Murphy, K. P. (1998). Inference and learning in hybrid Bayesian networks. Technical Report UCB/CSD-98-990, University of California, Berkeley.

Murphy, K. P., Y. Weiss, and M. Jordan (1999). Loopy belief propagation for approximate inference: an empirical study. In *Proc. 15th Conference on Uncertainty in Artificial Intelligence (UAI)*, pp. 467–475.

Murray, I. and Z. Ghahramani (2004). Bayesian learning in undirected graphical models: Approximate MCMC algorithms. In *Proc. 20th Conference on Uncertainty in Artificial Intelligence (UAI)*.

Murray, I., Z. Ghahramani, and D. MacKay (2006). MCMC for doubly-intractable distributions. In *Proc. 22nd Conference on Uncertainty in Artificial Intelligence (UAI)*.

Myers, J., K. Laskey, and T. Levitt (1999). Learning Bayesian networks from incomplete data with stochastic search algorithms. In *Proc. 15th Conference on Uncertainty in Artificial Intelligence (UAI)*, pp. 476–485.

Narasimhan, M. and J. Bilmes (2004). PAC-learning bounded tree-width graphical models. In *Proc. 20th Conference on Uncertainty in Artificial Intelligence (UAI)*.

Ndilikilikesha, P. (1994). Potential influence diagrams. *International Journal of Approximate Reasoning 10*, 251–85.

Neal, R. (1996). Sampling from multimodal distributions using tempered transitions. *Statistics and Computing 6*, 353–366.

Neal, R. (2001). Annealed importance sampling. *Statistics and Computing 11*(2), 25–139.

Neal, R. (2003). Slice sampling. *Annals of Statistics 31*(3), 705–767.

Neal, R. M. (1992). Asymmetric parallel Boltzmann machines are belief networks. *Neural Computation 4*(6), 832–834.

Neal, R. M. (1993). Probabilistic inference using Markov chain Monte Carlo methods. Technical Report CRG-TR-93-1, University of Toronto.

Neal, R. M. and G. E. Hinton (1998). A new view of the EM algorithm that justifies incremental and other variants. See Jordan (1998).

Neapolitan, R. E. (2003). *Learning Bayesian Networks*. Prentice Hall.

Ng, A. and M. Jordan (2000). Approximate inference algorithms for two-layer Bayesian networks. In *Proc. 14th Conference on Neural Information Processing Systems (NIPS)*.

Ng, A. and M. Jordan (2002). On discriminative vs. generative classifiers: A comparison of logistic regression and naive Bayes. In *Proc. 16th Conference on Neural Information Processing Systems (NIPS)*.

Ng, B., L. Peshkin, and A. Pfeffer (2002). Factored particles for scalable monitoring. In *Proc. 18th Conference on Uncertainty in Artificial Intelligence (UAI)*, pp. 370–377.

Ngo, L. and P. Haddawy (1996). Answering queries from context-sensitive probabilistic knowledge bases. *Theoretical Computer Science*.

Nielsen, J., T. Kočka, and J. M. Peña (2003). On local optima in learning Bayesian networks. In *Proc. 19th Conference on Uncertainty in Artificial Intelligence (UAI)*, pp. 435–442.

Nielsen, T. and F. Jensen (1999). Welldefined decision scenarios. In *Proc. 15th Conference on Uncertainty in Artificial Intelligence (UAI)*, pp. 502–11.

Nielsen, T. and F. Jensen (2000). Representing and solving asymmetric Bayesian decision problems. In *Proc. 16th Conference on Uncertainty in Artificial Intelligence (UAI)*, pp. 416–25.

Nielsen, T., P.-H. Wuillemin, F. Jensen, and U. Kjærulff (2000). Using robdds for inference in Bayesian networks with troubleshooting as an example. In *Proc. 16th Conference on Uncertainty in Artificial Intelligence (UAI)*, pp. 426–35.

Nilsson, D. (1998). An efficient algorithm for finding the M most probable configurations in probabilistic expert systems. *Statistics and Computing 8*(2), 159–173.

Nilsson, D. and S. Lauritzen (2000). Evaluating influence diagrams with LIMIDs. In *Proc. 16th Conference on Uncertainty in Artificial Intelligence (UAI)*, pp. 436–445.

Nodelman, U., C. R. Shelton, and D. Koller (2002). Continuous time Bayesian networks. In *Proc. 18th Conference on Uncertainty in Artificial Intelligence (UAI)*, pp. 378–387.

Nodelman, U., C. R. Shelton, and D. Koller (2003). Learning continuous time Bayesian networks. In *Proc. 19th Conference on Uncertainty in Artificial Intelligence (UAI)*.

Norman, J., Y. Shahar, M. Kuppermann, and B. Gold (1998). Decision-theoretic analysis of prenatal testing strategies. Technical Report SMI-98-0711, Stanford University, Section on Medical Informatics.

Normand, S.-L. and D. Tritchler (1992). Parameter updating in a Bayes network. *Journal of the American Statistical Association 87*, 1109–1115.

Nummelin, E. (1984). *General Irreducible Markov Chains and Non-Negative Operators*. Cambridge University Press.

Nummelin, E. (2002). Mc's for mcmc'ists. *International Statistical Review 70*(2), 215–240.

Olesen, K. G., U. Kjærulff, F. Jensen, B. Falck, S. Andreassen, and S. Andersen (1989). A Munin network for the median nerve — A case study on loops. *Applied Artificial Intelligence 3*, 384–403.

Oliver, R. M. and J. Q. Smith (Eds.) (1990). *Influence Diagrams, Belief Nets and Decision Analysis*. New York: John Wiley & Sons.

Olmsted, S. (1983). *On Representing and Solving Influence Diagrams*. Ph.D. thesis, Stanford University.

Opper, M. and O. Winther (2005). Expectation consistent free energies for approximate inference. In *Proc. 19th Conference on Neural Information Processing Systems (NIPS)*.

Ortiz, L. and L. Kaelbling (1999). Accelerating em: An empirical study. In *Proc. 15th Conference on Uncertainty in Artificial Intelligence (UAI)*, pp. 512–521.

Ortiz, L. E. and L. P. Kaelbling (2000). Adaptive importance sampling for estimation in structured domains. In *Proc. 16th Conference on Uncertainty in Artificial Intelligence (UAI)*, pp. 446–454.

Osborne, M. and A. Rubinstein (1994). *A Course in Game Theory*. The MIT Press.

Ostendorf, M., V. Digalakis, and O. Kimball (1996). From HMMs to segment models: A unified view of stochastic modeling for speech recognition. *IEEE Transactions on Speech and Audio Processing 4*(5), 360–378.

Pakzad, P. and V. Anantharam (2002). Minimal graphical representation of Kikuchi regions. In *Proc. 40th Allerton Conference on Communication Control and Computing*, pp. 1585–1594.

Papadimitriou, C. (1993). *Computational Complexity*. Addison Wesley.

Parisi, G. (1988). *Statistical Field Theory*. Reading, Massachusetts: Addison-Wesley.

Park, J. (2002). MAP complexity results and approximation methods. In *Proc. 18th Conference on*

Uncertainty in Artificial Intelligence (UAI), pp. 388–396.

Park, J. and A. Darwiche (2001). Approximating MAP using local search. In *Proc. 17th Conference on Uncertainty in Artificial Intelligence (UAI)*, pp. 403–âÅŞ410.

Park, J. and A. Darwiche (2003). Solving MAP exactly using systematic search. In *Proc. 19th Conference on Uncertainty in Artificial Intelligence (UAI)*.

Park, J. and A. Darwiche (2004a). Complexity results and approximation strategies for MAP explanations. *Journal of Artificial Intelligence Research 21*, 101–133.

Park, J. and A. Darwiche (2004b). A differential semantics for jointree algorithms. *Artificial Intelligence 156*, 197–216.

Parter, S. (1961). The user of linear graphs in Gauss elimination. *SIAM Review 3*, 119–130.

Paskin, M. (2003a). Sample propagation. In *Proc. 17th Conference on Neural Information Processing Systems (NIPS)*.

Paskin, M. (2003b). Thin junction tree filters for simultaneous localization and mapping. In *Proc. 18th International Joint Conference on Artificial Intelligence (IJCAI)*, pp. 1157–1164.

Pasula, H., B. Marthi, B. Milch, S. Russell, and I. Shpitser (2002). Identity uncertainty and citation matching. In *Proc. 16th Conference on Neural Information Processing Systems (NIPS)*, pp. 1401–1408.

Pasula, H., S. Russell, M. Ostland, and Y. Ritov (1999). Tracking many objects with many sensors. In *Proc. 16th International Joint Conference on Artificial Intelligence (IJCAI)*.

Patrick, D., J. Bush, and M. Chen (1973). Methods for measuring levels of well-being for a health status index. *Health Services Research 8*, 228–45.

Pearl, J. (1986a). A constraint-propagation approach to probabilistic reasoning. In *Proc. 2nd Conference on Uncertainty in Artificial Intelligence (UAI)*, pp. 357–370.

Pearl, J. (1986b). Fusion, propagation and structuring in belief networks. *Artificial Intelligence 29*(3), 241–88.

Pearl, J. (1987). Evidential reasoning using stochastic simulation of causal models. *Artificial Intelligence 32*, 245–257.

Pearl, J. (1988). *Probabilistic Reasoning in Intelligent Systems*. San Mateo, California: Morgan Kaufmann.

Pearl, J. (1995). Causal diagrams for empirical research. *Biometrika 82*, 669–710.

Pearl, J. (2000). *Causality: Models, Reasoning, and Inference*. Cambridge Univ. Press.

Pearl, J. and R. Dechter (1996). Identifying independencies in causal graphs with feedback. In *Proc. 12th Conference on Uncertainty in Artificial Intelligence (UAI)*, pp. 420–26.

Pearl, J. and A. Paz (1987). GRAPHOIDS: A graph-based logic for reasoning about relevance relations. In B. Du Boulay, D. Hogg, and L. Steels (Eds.), *Advances in Artificial Intelligence*, Volume 2, pp. 357–363. Amsterdam: North Holland.

Pearl, J. and T. S. Verma (1991). A theory of inferred causation. In *Proc. Conference on Knowledge Representation and Reasoning (KR)*, pp. 441–452.

Pe'er, D., A. Regev, G. Elidan, and N. Friedman (2001). Inferring subnetworks from preturbed expression profiles. *Bioinformatics 17*, S215–S224.

Peng, Y. and J. Reggia (1986). Plausibility of diagnostic hypotheses. In *Proc. 2nd Conference on Artificial Intelligence (AAAI)*, pp. 140–45.

Perkins, S., K. Lacker, and J. Theiler (2003, March). Grafting: Fast, incremental feature selection by gradient descent in function space. *Journal of Machine Learning Research 3*, 1333–1356.

Peterson, C. and J. R. Anderson (1987). A mean field theory learning algorithm for neural networks. *Complex Systems 1*, 995–1019.

Pfeffer, A., D. Koller, B. Milch, and K. Takusagawa (1999). SPOOK: A system for probabilistic object-oriented knowledge representation. In *Proc. 15th Conference on Uncertainty in Artificial Intelligence (UAI)*, pp. 541–550.

Poh, K. and E. Horvitz (2003). Reasoning about the value of decision-model refinement: Methods and application. In *Proc. 19th Conference on Uncertainty in Artificial Intelligence (UAI)*, pp. 174–182.

Poland, W. (1994). *Decision Analysis with Continuous and Discrete Variables: A Mixture Distribution Approach*. Ph.D. thesis, Department of Engineering-Economic Systems, Stanford University.

Poole, D. (1989). Average-case analysis of a search algorithm for estimating prior and posterior probabilities in Bayesian networks with extreme probabilities. In *Proc. 13th International Joint Conference on Artificial Intelligence (IJCAI)*, pp. 606–612.

Poole, D. (1993a). Probabilistic Horn abduction and Bayesian networks. *Artificial Intelligence 64*(1), 81–129.

Poole, D. (1993b). The use of conflicts in searching Bayesian networks. In *Proc. 9th Conference on Uncertainty in Artificial Intelligence (UAI)*, pp. 359–367.

Poole, D. and N. Zhang (2003). Exploiting causal independence in Bayesian network inference. *Journal of Artificial Intelligence Research 18*, 263–313.

Poon, H. and P. Domingos (2007). Joint inference in information extraction. In *Proc. 23rd Conference on Artificial Intelligence (AAAI)*, pp. 913–918.

Pradhan, M. and P. Dagum (1996). Optimal Monte Carlo estimation of belief network inference. In *Proc. 12th Conference on Uncertainty in Artificial Intelligence (UAI)*, pp. 446–453.

Pradhan, M., M. Henrion, G. Provan, B. Del Favero, and K. Huang (1996). The sensitivity of belief networks to imprecise probabilities: An experimental investigation. *Artificial Intelligence 85*, 363–97.

Pradhan, M., G. M. Provan, B. Middleton, and M. Henrion (1994). Knowledge engineering for large belief networks. In *Proc. 10th Conference on Uncertainty in Artificial Intelligence (UAI)*, pp. 484–490.

Puterman, M. L. (1994). *Markov Decision Processes: Discrete Stochastic Dynamic Programming*. John Wiley and Sons, New York.

Qi, R., N. Zhang, and D. Poole (1994). Solving asymmetric decision problems with influence diagrams. In *Proc. 10th Conference on Uncertainty in Artificial Intelligence (UAI)*, pp. 491–497.

Qi, Y., M. Szummer, and T. Minka (2005). Bayesian conditional random fields. In *Proc. 11th Workshop on Artificial Intelligence and Statistics*.

Rabiner, L. R. (1989). A tutorial on Hidden Markov Models and selected applications in speech recognition. *Proceedings of the IEEE 77*(2), 257–286.

Rabiner, L. R. and B. H. Juang (1986, January). An introduction to hidden Markov models. *IEEE ASSP Magazine*, 4–15.

Ramsey, F. (1931). *The Foundations of Mathematics and other Logical Essays*. London: Kegan, Paul, Trench, Trubner & Co., New York: Harcourt, Brace and Company. edited by R.B. Braithwaite.

Rasmussen, C. and C. Williams (2006). *Gaussian Processes for Machine Learning*. MIT Press.

Rasmussen, C. E. (1999). The infinite gaussian mixture model. In *Proc. 13th Conference on Neural Information Processing Systems (NIPS)*, pp. 554–560.

Ravikumar, P. and J. Lafferty (2006). Quadratic programming relaxations for metric labelling and Markov random field MAP estimation. In *Proc. 23rd International Conference on Machine*

Learning (ICML).

Renooij, S. and L. van der Gaag (2002). From qualitative to quantitative probabilistic networks. In *Proc. 18th Conference on Uncertainty in Artificial Intelligence (UAI)*, pp. 422–429.

Richardson, M. and P. Domingos (2006). Markov logic networks. *Machine Learning 62*, 107–136.

Richardson, T. (1994). Properties of cyclic graphical models. Master's thesis, Carnegie Mellon University.

Riezler, S. and A. Vasserman (2004). Incremental feature selection and l1 regularization for relaxed maximum-entropy modeling. In *Proc. Conference on Empirical Methods in Natural Language Processing (EMNLP)*.

Ripley, B. D. (1987). *Stochastic Simulation*. New York: John Wiley & Sons.

Rissanen, J. (1987). Stochastic complexity (with discussion). *Journal of the Royal Statistical Society, Series B 49*, 223–265.

Ristic, B., S. Arulampalam, and N. Gordon (2004). *Beyond the Kalman Filter: Particle Filters for Tracking Applications*. Artech House Publishers.

Robert, C. and G. Casella (1996). Rao-Blackwellisation of sampling schemes. *Biometrika 83*(1), 81–94.

Robert, C. and G. Casella (2005). *Monte Carlo Statistical Methods* (2nd ed.). Springer Texts in Statistics.

Robins, J. M. and L. A. Wasserman (1997). Estimation of effects of sequential treatments by reparameterizing directed acyclic graphs. In *Proc. 13th Conference on Uncertainty in Artificial Intelligence (UAI)*, pp. 409–420.

Rose, D. (1970). Triangulated graphs and the elimination process. *Journal of Mathematical Analysis and Applications 32*, 597–609.

Ross, S. M. (1988). *A First Course in Probability* (third ed.). London: Macmillan.

Rother, C., S. Kumar, V. Kolmogorov, and A. Blake (2005). Digital tapestry. In *Proc. Conference on Computer Vision and Pattern Recognition (CVPR)*.

Rubin, D. (1974). Estimating causal effects of treatments in randomized and nonrandomized studies. *Journal of Educational Psychology 66*(5), 688–701.

Rubin, D. R. (1976). Inference and missing data. *Biometrika 63*, 581–592.

Rusmevichientong, P. and B. Van Roy (2001). An analysis of belief propagation on the turbo decoding graph with Gaussian densities. *IEEE Transactions on Information Theory 48*(2).

Russell, S. and P. Norvig (2003). *Artificial Intelligence: A Modern Approach* (2 ed.). Prentice Hall.

Rustagi, J. (1976). *Variational Methods in Statistics*. New York: Academic Press.

Sachs, K., O. Perez, D. Pe'er, D. Lauffenburger, and G. Nolan (2005, April). Causal protein-signaling networks derived from multiparameter single-cell data. *Science 308*(5721), 523–529.

Sakurai, J. J. (1985). *Modern Quantum Mechanics*. Reading, Massachusetts: Addison-Wesley.

Santos, A. (1994). A linear constraint satisfaction approach to cost-based abduction. *Artificial Intelligence 65*(1), 1–28.

Santos, E. (1991). On the generation of alternative explanations with implications for belief revision. In *Proc. 7th Conference on Uncertainty in Artificial Intelligence (UAI)*, pp. 339–347.

Saul, L., T. Jaakkola, and M. Jordan (1996). Mean field theory for sigmoid belief networks. *Journal of Artificial Intelligence Research 4*, 61–76.

Saul, L. and M. Jordan (1999). Mixed memory Markov models: Decomposing complex stochastic processes as mixture of simpler ones. *Machine Learning 37*(1), 75–87.

Saul, L. K. and M. I. Jordan (1996). Exploiting tractable substructures in intractable networks. In

Proc. 10th Conference on Neural Information Processing Systems (NIPS).

Savage, L. (1951). The theory of statistical decision. *Journal of the American Statistical Association 46*, 55–67.

Savage, L. J. (1954). *Foundations of Statistics*. New York: John Wiley & Sons.

Schäffer, A. (1996). Faster linkage analysis computations for pedigrees with loops or unused alleles. *Human Heredity*, 226–235.

Scharstein, D. and R. Szeliski (2003). High-accuracy stereo depth maps using structured light. In *Proc. Conference on Computer Vision and Pattern Recognition (CVPR)*, Volume 1, pp. 195–202.

Schervish, M. (1995). *Theory of Statistics*. Springer-Verlag.

Schlesinger, M. (1976). Sintaksicheskiy analiz dvumernykh zritelnikh singnalov v usloviyakh pomekh (syntactic analysis of two-dimensional visual signals in noisy conditions). *Kibernetika 4*, 113–130.

Schlesinger, M. and V. Giginyak (2007a). Solution to structural recognition (max,+)-problems by their equivalent transformations (part 1). *Control Systems and Computers 1*, 3–15.

Schlesinger, M. and V. Giginyak (2007b). Solution to structural recognition (max,+)-problems by their equivalent transformations (part 2). *Control Systems and Computers 2*, 3–18.

Schwarz, G. (1978). Estimating the dimension of a model. *The Annals of Statistics 6*(2), 461–464.

Segal, E., D. Pe'er, A. Regev, D. Koller, and N. Friedman (2005, April). Learning module networks. *Journal of Machine Learning Research 6*, 557–588.

Segal, E., B. Taskar, A. Gasch, N. Friedman, and D. Koller (2001). Rich probabilistic models for gene expression. *Bioinformatics 17*(Suppl 1), S243–52.

Settimi, R. and J. Smith (2000). Geometry, moments and conditional independence trees with hidden variables. *Annals of Statistics*.

Settimi, R. and J. Q. Smith (1998a). On the geometry of Bayesian graphical models with hidden variables. In *Proc. 14th Conference on Uncertainty in Artificial Intelligence (UAI)*, pp. 472–479.

Settimi, R. and J. Q. Smith (1998b). On the geometry of Bayesian graphical models with hidden variables. In *Proc. 14th Conference on Uncertainty in Artificial Intelligence (UAI)*, pp. 472–479.

Shachter, R. (1988, July–August). Probabilistic inference and influence diagrams. *Operations Research 36*, 589–605.

Shachter, R. (1999). Efficient value of information computation. In *Proc. 15th Conference on Uncertainty in Artificial Intelligence (UAI)*, pp. 594–601.

Shachter, R., S. K. Andersen, and P. Szolovits (1994). Global conditioning for probabilistic inference in belief networks. In *Proc. 10th Conference on Uncertainty in Artificial Intelligence (UAI)*, pp. 514–522.

Shachter, R. and D. Heckerman (1987). Thinking backwards for knowledge acquisition. *Artificial Intelligence Magazine 8*, 55 – 61.

Shachter, R. and C. Kenley (1989). Gaussian influence diagrams. *Management Science 35*, 527–550.

Shachter, R. and P. Ndilikilikesha (1993). Using influence diagrams for probabilistic inference and decision making. In *Proc. 9th Conference on Uncertainty in Artificial Intelligence (UAI)*, pp. 276–83.

Shachter, R. D. (1986). Evaluating influence diagrams. *Operations Research 34*, 871–882.

Shachter, R. D. (1989). Evidence absorption and propagation through evidence reversals. In *Proc. 5th Conference on Uncertainty in Artificial Intelligence (UAI)*, pp. 173–190.

Shachter, R. D. (1998). Bayes-ball: The rational pastime. In *Proc. 14th Conference on Uncertainty in Artificial Intelligence (UAI)*, pp. 480–487.

Shachter, R. D., B. D'Ambrosio, and B. A. Del Favero (1990). Symbolic probabilistic inference in belief networks. In *Proc. 6th Conference on Artificial Intelligence (AAAI)*, pp. 126–131.

Shachter, R. D. and M. A. Peot (1989). Simulation approaches to general probabilistic inference on belief networks. In *Proc. 5th Conference on Uncertainty in Artificial Intelligence (UAI)*, pp. 221–230.

Shachter, R. D. and M. A. Peot (1992). Decision making using probabilistic inference methods. In *Proc. 8th Conference on Uncertainty in Artificial Intelligence (UAI)*, pp. 276–83.

Shafer, G. and J. Pearl (Eds.) (1990). *Readings in Uncertain Reasoning*. Representation and Reasoning. San Mateo, California: Morgan Kaufmann.

Shafer, G. and P. Shenoy (1990). Probability propagation. *Annals of Mathematics and Artificial Intelligence 2*, 327–352.

Shannon, C. (1948). A mathematical theory of communication. *Bell System Technical Journal 27*, 379–423; 623–656.

Shawe-Taylor, J. and N. Cristianini (2000). *Support Vector Machines and other kernel-based learning methods*. Cambridge University Press.

Shenoy, P. (1989). A valuation-based language for expert systems. *International Journal of Approximate Reasoning 3*, 383–411.

Shenoy, P. (2000). Valuation network representation and solution of asymmetric decision problems. *European Journal of Operational Research 121*(3), 579–608.

Shenoy, P. and G. Shafer (1990). Axioms for probability and belief-function propagation. In *Proc. 6th Conference on Uncertainty in Artificial Intelligence (UAI)*, pp. 169–198.

Shenoy, P. P. (1992). Valuation-based systems for Bayesian decision analysis. *Operations Research 40*, 463–484.

Shental, N., A. Zomet, T. Hertz, and Y. Weiss (2003). Learning and inferring image segmentations using the GBP typical cut algorithm. In *Proc. International Conference on Computer Vision*.

Shimony, S. (1991). Explanation, irrelevance and statistical independence. In *Proc. 7th Conference on Artificial Intelligence (AAAI)*.

Shimony, S. (1994). Finding MAPs for belief networks in NP-hard. *Artificial Intelligence 68*(2), 399–410.

Shoikhet, K. and D. Geiger (1997). A practical algorithm for finding optimal triangulations. In *Proc. 13th Conference on Artificial Intelligence (AAAI)*, pp. 185–190.

Shwe, M. and G. Cooper (1991). An empirical analysis of likelihood-weighting simulation on a large, multiply connected medical belief network. *Computers and Biomedical Research 24*, 453–475.

Shwe, M., B. Middleton, D. Heckerman, M. Henrion, E. Horvitz, H. Lehmann, and G. Cooper (1991). Probabilistic diagnosis using a reformulation of the INTERNIST-1/QMR knowledge base. I. The probabilistic model and inference algorithms. *Methods of Information in Medicine 30*, 241–55.

Silander, T. and P. Myllymaki (2006). A simple approach for finding the globally optimal Bayesian network structure. In *Proc. 22nd Conference on Uncertainty in Artificial Intelligence (UAI)*.

Singh, A. and A. Moore (2005). Finding optimal bayesian networks by dynamic programming. Technical report, Carnegie Mellon University.

Sipser, M. (2005). *Introduction to the Theory of Computation* (Second ed.). Course Technology.

Smith, A. and G. Roberts (1993). Bayesian computation via the Gibbs sampler and related Markov chain Monte Carlo methods. *Journal of the Royal Statistical Society, Series B 55*, 3–23.

Smith, J. (1989). Influence diagrams for statistical modeling. *Annals of Statistics 17*(2), 654–72.

Smith, J., S. Holtzman, and J. Matheson (1993). Structuring conditional relationships in influence diagrams. *Operations Research 41*(2), 280–297.

Smyth, P., D. Heckerman, and M. Jordan (1997). Probabilistic independence networks for hidden Markov probability models. *Neural Computation 9*(2), 227–269.

Sontag, D. and T. Jaakkola (2007). New outer bounds on the marginal polytope. In *Proc. 21st Conference on Neural Information Processing Systems (NIPS)*.

Sontag, D., T. Meltzer, A. Globerson, T. Jaakkola, and Y. Weiss (2008). Tightening LP relaxations for MAP using message passing. In *Proc. 24th Conference on Uncertainty in Artificial Intelligence (UAI)*.

Speed, T. and H. Kiiveri (1986). Gaussian Markov distributions over finite graphs. *The Annals of Statistics 14*(1), 138–150.

Spetzler, C. and C.-A. von Holstein (1975). Probabilistic encoding in decision analysis. *Management Science*, 340–358.

Spiegelhalter, D. and S. Lauritzen (1990). Sequential updating of conditional probabilities on directed graphical structures. *Networks 20*, 579–605.

Spiegelhalter, D. J., A. P. Dawid, S. L. Lauritzen, and R. G. Cowell (1993). Bayesian analysis in expert systems. *Statistical Science 8*, 219–283.

Spirtes, P. (1995). Directed cyclic graphical representations of feedback models. In *Proc. 11th Conference on Uncertainty in Artificial Intelligence (UAI)*, pp. 491–98.

Spirtes, P., C. Glymour, and R. Scheines (1991). An algorithm for fast recovery of sparse causal graphs. *Social Science Computer Review 9*, 62–72.

Spirtes, P., C. Glymour, and R. Scheines (1993). *Causation, Prediction and Search*. Number 81 in Lecture Notes in Statistics. New York: Springer-Verlag.

Spirtes, P., C. Meek, and T. Richardson (1999). An algorithm for causal inference in the presence of latent variables and selection bias. See Glymour and Cooper (1999), pp. 211–52.

Srebro, N. (2001). Maximum likelihood bounded tree-width Markov networks. In *Proc. 17th Conference on Uncertainty in Artificial Intelligence (UAI)*.

Srinivas, S. (1993). A generalization of the noisy-or model. In *Proc. 9th Conference on Uncertainty in Artificial Intelligence (UAI)*, pp. 208–215.

Srinivas, S. (1994). A probabilistic approach to hierarchical model-based diagnosis. In *Proc. 10th Conference on Uncertainty in Artificial Intelligence (UAI)*.

Studený, M. and R. Bouckaert (1998). On chain graph models for description of conditional independence structures. *Annals of Statistics 26*.

Sudderth, E., A. Ihler, W. Freeman, and A. Willsky (2003). Nonparametric belief propagation. In *Proc. Conference on Computer Vision and Pattern Recognition (CVPR)*, pp. 605–612.

Sutton, C. and T. Minka (2006). Local training and belief propagation. Technical Report MSR-TR-2006-121, Microsoft Research.

Sutton, C. and A. McCallum (2004). Collective segmentation and labeling of distant entities in information extraction. In *ICML Workshop on Statistical Relational Learning and Its Connections to Other Fields*.

Sutton, C. and A. McCallum (2005). Piecewise training of undirected models. In *Proc. 21st Conference on Uncertainty in Artificial Intelligence (UAI)*.

Sutton, C. and A. McCallum (2007). An introduction to conditional random fields for relational learning. In L. Getoor and B. Taskar (Eds.), *Introduction to Statistical Relational Learning*. MIT

Press.

Sutton, C., A. McCallum, and K. Rohanimanesh (2007, March). Dynamic conditional random fields: Factorized probabilistic models for labeling and segmenting sequence data. *Journal of Machine Learning Research 8*, 693–723.

Suzuki, J. (1993). A construction of Bayesian networks from databases based on an MDL scheme. In *Proc. 9th Conference on Uncertainty in Artificial Intelligence (UAI)*, pp. 266–273.

Swendsen, R. and J. Wang (1987). Nonuniversal critical dynamics in Monte Carlo simulations. *Physical Review Letters 58*(2), 86–88.

Swendsen, R. H. and J.-S. Wang (1986, Nov). Replica Monte Carlo simulation of spin-glasses. *Physical Review Letters 57*(21), 2607–2609.

Szeliski, R., R. Zabih, D. Scharstein, O. Veksler, V. Kolmogorov, A. Agarwala, M. Tappen, and C. Rother (2008, June). A comparative study of energy minimization methods for Markov random fields with smoothness-based priors. *IEEE Trans. on Pattern Analysis and Machine Intelligence 30*(6), 1068–1080. See http://vision.middlebury.edu/MRF for more detailed results.

Szolovits, P. and S. Pauker (1992). Pedigree analysis for genetic counseling. In *Proceedings of the Seventh World Congress on Medical Informatics (MEDINFO '92)*, pp. 679–683. North-Holland.

Tanner, M. A. (1993). *Tools for Statistical Inference*. New York: Springer-Verlag.

Tarjan, R. and M. Yannakakis (1984). Simple linear-time algorithms to test chordality of graphs, test acyclicity of hypergraphs, and selectively reduce acyclic hypergraphs. *SIAM Journal of Computing 13*(3), 566–579.

Taskar, B., P. Abbeel, and D. Koller (2002). Discriminative probabilistic models for relational data. In *Proc. 18th Conference on Uncertainty in Artificial Intelligence (UAI)*, pp. 485–492.

Taskar, B., P. Abbeel, M.-F. Wong, and D. Koller (2007). Relational Markov networks. See Getoor and Taskar (2007).

Taskar, B., V. Chatalbashev, and D. Koller (2004). Learning associative Markov networks. In *Proc. 21st International Conference on Machine Learning (ICML)*.

Taskar, B., C. Guestrin, and D. Koller (2003). Max margin Markov networks. In *Proc. 17th Conference on Neural Information Processing Systems (NIPS)*.

Tatikonda, S. and M. Jordan (2002). Loopy belief propagation and Gibbs measures. In *Proc. 18th Conference on Uncertainty in Artificial Intelligence (UAI)*.

Tatman, J. A. and R. D. Shachter (1990). Dynamic programming and influence diagrams. *IEEE Transactions on Systems, Man and Cybernetics 20*(2), 365–379.

Teh, Y. and M. Welling (2001). The unified propagation and scaling algorithm. In *Proc. 15th Conference on Neural Information Processing Systems (NIPS)*.

Teh, Y., M. Welling, S. Osindero, and G. Hinton (2003). Energy-based models for sparse over-complete representations. *Journal of Machine Learning Research 4*, 1235–1260. Special Issue on ICA.

Teyssier, M. and D. Koller (2005). Ordering-based search: A simple and effective algorithm for learning bayesian networks. In *Proc. 21st Conference on Uncertainty in Artificial Intelligence (UAI)*, pp. 584–590.

Thiele, T. (1880). *Sur la compensation de quelques erreurs quasisystematiques par la methode des moindres carrees*. Copenhagen: Reitzel.

Thiesson, B. (1995). Accelerated quantification of Bayesian networks with incomplete data. In *Proceedings of the First International Conference on Knowledge Discovery and Data Mining (KDD-*

95), pp. 306–311. AAAI Press.

Thiesson, B., C. Meek, D. M. Chickering, and D. Heckerman (1998). Learning mixtures of Bayesian networks. In *Proc. 14th Conference on Uncertainty in Artificial Intelligence (UAI)*.

Thomas, A., D. Spiegelhalter, and W. Gilks (1992). BUGS: A program to perform Bayesian inference using Gibbs sampling. In J. Bernardo, J. Berger, A. Dawid, and A. Smith (Eds.), *Bayesian Statistics 4*, pp. 837–842. Oxford, UK: Clarendon Press.

Thrun, S., W. Burgard, and D. Fox (2005). *Probabilistic Robotics*. Cambridge, MA: MIT Press.

Thrun, S., D. Fox, W. Burgard, and F. Dellaert (2000). Robust Monte Carlo localization for mobile robots. *Artificial Intelligence 128*(1–2), 99–141.

Thrun, S., Y. Liu, D. Koller, A. Ng, Z. Ghahramani, and H. Durrant-Whyte (2004). Simultaneous localization and mapping with sparse extended information filters. *International Journal of Robotics Research 23*(7/8).

Thrun, S., C. Martin, Y. Liu, D. Hähnel, R. Emery-Montemerlo, D. Chakrabarti, and W. Burgard (2004). A real-time expectation maximization algorithm for acquiring multi-planar maps of indoor environments with mobile robots. *IEEE Transactions on Robotics 20*(3), 433–443.

Thrun, S., M. Montemerlo, D. Koller, B. Wegbreit, J. Nieto, and E. Nebot (2004). FastSLAM: An efficient solution to the simultaneous localization and mapping problem with unknown data association. *Journal of Machine Learning Research*.

Tian, J. and J. Pearl (2002). On the testable implications of causal models with hidden variables. In *Proc. 18th Conference on Uncertainty in Artificial Intelligence (UAI)*, pp. 519–527.

Tibshirani, R. (1996). Regression shrinkage and selection via the lasso. *Journal of the Royal Statistical Society, Series B 58*(1), 267–288.

Tierney, L. (1994). Markov chains for exploring posterior distributions. *Annals of Statistics 22*(4), 1701–1728.

Tong, S. and D. Koller (2001a). Active learning for parameter estimation in Bayesian networks. In *Proc. 15th Conference on Neural Information Processing Systems (NIPS)*, pp. 647–653.

Tong, S. and D. Koller (2001b). Active learning for structure in Bayesian networks. In *Proc. 17th International Joint Conference on Artificial Intelligence (IJCAI)*, pp. 863–869.

Torrance, G., W. Thomas, and D. Sackett (1972). A utility maximization model for evaluation of health care programs. *Health Services Research 7*, 118–133.

Tsochantaridis, I., T. Hofmann, T. Joachims, and Y. Altun (2004). Support vector machine learning for interdependent and structured output spaces. In *Proc. 21st International Conference on Machine Learning (ICML)*.

Tversky, A. and D. Kahneman (1974). Judgment under uncertainty: Heuristics and biases. *Science 185*, 1124–1131.

van der Merwe, R., A. Doucet, N. de Freitas, and E. Wan (2000a, Aug.). The unscented particle filter. Technical Report CUED/F-INFENG/TR 380, Cambridge University Engineering Department.

van der Merwe, R., A. Doucet, N. de Freitas, and E. Wan (2000b). The unscented particle filter. In *Proc. 14th Conference on Neural Information Processing Systems (NIPS)*.

Varga, R. (2000). *Matrix Iterative Analysis*. Springer-Verlag.

Verma, T. (1988). Causal networks: Semantics and expressiveness. In *Proc. 4th Conference on Uncertainty in Artificial Intelligence (UAI)*, pp. 352–359.

Verma, T. and J. Pearl (1988). Causal networks: Semantics and expressiveness. In *Proc. 4th Conference on Uncertainty in Artificial Intelligence (UAI)*, pp. 69–76.

Verma, T. and J. Pearl (1990). Equivalence and synthesis of causal models. In *Proc. 6th Conference on Uncertainty in Artificial Intelligence (UAI)*, pp. 255 –269.

Verma, T. and J. Pearl (1992). An algorithm for deciding if a set of observed independencies has a causal explanation. In *Proc. 8th Conference on Uncertainty in Artificial Intelligence (UAI)*, pp. 323–330.

Vickrey, D. and D. Koller (2002). Multi-agent algorithms for solving graphical games. In *Proceedings of the Eighteenth National Conference on Artificial Intelligence (AAAI-02)*, pp. 345–351.

Vishwanathan, S., N. Schraudolph, M. Schmidt, and K. Murphy (2006). Accelerated training of conditional random fields with stochastic gradient methods. In *Proc. 23rd International Conference on Machine Learning (ICML)*, pp. 969–976.

Viterbi, A. (1967, April). Error bounds for convolutional codes and an asymptotically optimum decoding algorithm. *IEEE Transactions on Information Theory 13*(2), 260–269.

von Neumann, J. and O. Morgenstern (1944). *Theory of games and economic behavior* (first ed.). Princeton, NJ: Princeton Univ. Press.

von Neumann, J. and O. Morgenstern (1947). *Theory of games and economic behavior* (second ed.). Princeton, NJ: Princeton Univ. Press.

von Winterfeldt, D. and W. Edwards (1986). *Decision Analysis and Behavioral Research*. Cambridge, UK: Cambridge University Press.

Vorobev, N. (1962). Consistent families of measures and their extensions. *Theory of Probability and Applications 7*, 147–63.

Wainwright, M. (2006). Estimating the "wrong" graphical model: Benefits in the computation-limited setting. *Journal of Machine Learning Research 7*, 1829–1859.

Wainwright, M., T. Jaakkola, and A. Willsky (2003a). Tree-based reparameterization framework for analysis of sum-product and related algorithms. *IEEE Transactions on Information Theory 49*(5).

Wainwright, M., T. Jaakkola, and A. Willsky (2003b). Tree-reweighted belief propagation and approximate ML estimation by pseudo-moment matching. In *Proc. 9thWorkshop on Artificial Intelligence and Statistics*.

Wainwright, M., T. Jaakkola, and A. Willsky (2004, April). Tree consistency and bounds on the performance of the max-product algorithm and its generalizations. *Statistics and Computing 14*, 143–166.

Wainwright, M., T. Jaakkola, and A. Willsky (2005). MAP estimation via agreement on trees: Message-passing and linear programming. *IEEE Transactions on Information Theory*.

Wainwright, M., T. Jaakkola, and A. S. Willsky (2001). Tree-based reparameterization for approximate estimation on loopy graphs. In *Proc. 15th Conference on Neural Information Processing Systems (NIPS)*.

Wainwright, M., T. Jaakkola, and A. S. Willsky (2002a). Exact map estimates by (hyper)tree agreement. In *Proc. 16th Conference on Neural Information Processing Systems (NIPS)*.

Wainwright, M., T. Jaakkola, and A. S. Willsky (2002b). A new class of upper bounds on the log partition function. In *Proc. 18th Conference on Uncertainty in Artificial Intelligence (UAI)*.

Wainwright, M. and M. Jordan (2003). Graphical models, exponential families, and variational inference. Technical Report 649, Department of Statistics, University of California, Berkeley.

Wainwright, M. and M. Jordan (2004). Semidefinite relaxations for approximate inference on graphs with cycles. In *Proc. 18th Conference on Neural Information Processing Systems (NIPS)*.

Wainwright, M., P. Ravikumar, and J. Lafferty (2006). High-dimensional graphical model selection using ℓ_1-regularized logistic regression. In *Proc. 20th Conference on Neural Information*

Processing Systems (NIPS).

Warner, H., A. Toronto, L. Veasey, and R. Stephenson (1961). A mathematical approach to medical diagnosis — application to congenital heart disease. *Journal of the American Madical Association 177*, 177–184.

Weiss, Y. (1996). Interpreting images by propagating bayesian beliefs. In *Proc. 10th Conference on Neural Information Processing Systems (NIPS)*, pp. 908–914.

Weiss, Y. (2000). Correctness of local probability propagation in graphical models with loops. *Neural Computation 12*, 1–41.

Weiss, Y. (2001). Comparing the mean field method and belief propagation for approximate inference in MRFs. In M. Opper and D. Saad (Eds.), *Advanced mean field methods*, pp. 229–240. Cambridge, Massachusetts: MIT Press.

Weiss, Y. and W. Freeman (2001a). Correctness of belief propagation in Gaussian graphical models of arbitrary topology. *Neural Computation 13*.

Weiss, Y. and W. Freeman (2001b). On the optimality of solutions of the max-product belief propagation algorithm in arbitrary graphs. *IEEE Transactions on Information Theory 47*(2), 723–735.

Weiss, Y., C. Yanover, and T. Meltzer (2007). MAP estimation, linear programming and belief propagation with convex free energies. In *Proc. 23rd Conference on Uncertainty in Artificial Intelligence (UAI)*.

Welling, M. (2004). On the choice of regions for generalized belief propagation. In *Proc. 20th Conference on Uncertainty in Artificial Intelligence (UAI)*.

Welling, M., T. Minka, and Y. Teh (2005). Structured region graphs: Morphing EP into GBP. In *Proc. 21st Conference on Uncertainty in Artificial Intelligence (UAI)*.

Welling, M. and S. Parise (2006a). Bayesian random fields: The Bethe-Laplace approximation. In *Proc. 22nd Conference on Uncertainty in Artificial Intelligence (UAI)*.

Welling, M. and S. Parise (2006b). Structure learning in Markov random fields. In *Proc. 20th Conference on Neural Information Processing Systems (NIPS)*.

Welling, M. and Y.-W. Teh (2001). Belief optimization for binary networks: a stable alternative to loopy belief propagation. In *Proc. 17th Conference on Uncertainty in Artificial Intelligence (UAI)*.

Wellman, M. (1985). Reasoning about preference models. Technical Report MIT/LCS/TR-340, Laboratory for Computer Science, MIT.

Wellman, M., J. Breese, and R. Goldman (1992). From knowledge bases to decision models. *Knowledge Engineering Review 7*(1), 35–53.

Wellman, M. and J. Doyle (1992). Modular utility representation for decision-theoretic planning. In *Procec. First International Conference on AI Planning Systems*, pp. 236–42. Morgan Kaufmann.

Wellman, M. P. (1990). Foundamental concepts of qualitative probabilistic networks. *Artificial Intelligence 44*, 257–303.

Wellner, B., A. McCallum, F. Peng, and M. Hay (2004). An integrated, conditional model of information extraction and coreference with application to citation matching. In *Proc. 20th Conference on Uncertainty in Artificial Intelligence (UAI)*, pp. 593–601.

Wermuth, N. (1980). Linear recursive equations, covariance selection and path analysis. *Journal of the American Statistical Association 75*, 963–975.

Werner, T. (2007). A linear programming approach to max-sum problem: A review. *IEEE Trans. on Pattern Analysis and Machine Intelligence 29*(7), 1165–1179.

West, M. (1993). Mixture models, Monte Carlo, Bayesian updating and dynamic models. *Comput-*

ing Science and Statistics 24, 325–333.

Whittaker, J. (1990). *Graphical Models in Applied Multivariate Statistics.* Chichester, United Kingdom: John Wiley and Sons.

Wiegerinck, W. (2000). Variational approximations between mean field theory and the junction tree algorithm. In *Proc. 16th Conference on Uncertainty in Artificial Intelligence (UAI)*, pp. 626–636.

Wold, H. (1954). Causality and econometrics. *Econometrica 22*, 162–177.

Wood, F., T. Griffiths, and Z. Ghahramani (2006). A non-parametric bayesian method for inferring hidden causes. In *Proc. 22nd Conference on Uncertainty in Artificial Intelligence (UAI)*, pp. 536–543.

Wright, S. (1921). Correlation and causation. *Journal of Agricultural Research 20*, 557–85.

Wright, S. (1934). The method of path coefficients. *Annals of Mathematical Statistics 5*, 161–215.

Xing, E., M. Jordan, and S. Russell (2003). A generalized mean field algorithm for variational inference in exponential families. In *Proc. 19th Conference on Uncertainty in Artificial Intelligence (UAI)*, pp. 583–591.

Yanover, C., T. Meltzer, and Y. Weiss (2006, September). Linear programming relaxations and belief propagation — an empirical study. *Journal of Machine Learning Research 7*, 1887–1907.

Yanover, C., O. Schueler-Furman, and Y. Weiss (2007). Minimizing and learning energy functions for side-chain prediction. In *Proc. International Conference on Research in Computational Molecular Biology (RECOMB)*, pp. 381–395.

Yanover, C. and Y. Weiss (2003). Finding the M most probable configurations using loopy belief propagation. In *Proc. 17th Conference on Neural Information Processing Systems (NIPS)*.

Yedidia, J., W. Freeman, and Y. Weiss (2005). Constructing free-energy approximations and generalized belief propagation algorithms. *IEEE Trans. Information Theory 51*, 2282–2312.

Yedidia, J. S., W. T. Freeman, and Y. Weiss (2000). Generalized belief propagation. In *Proc. 14th Conference on Neural Information Processing Systems (NIPS)*, pp. 689–695.

York, J. (1992). Use of the Gibbs sampler in expert systems. *Artificial Intelligence 56*, 115–130.

Yuille, A. L. (2002). CCCP algorithms to minimize the Bethe and Kikuchi free energies: Convergent alternatives to belief propagation. *Neural Computation 14*, 1691–1722.

Zhang, N. (1998). Probabilistic inference in influence diagrams. In *Proc. 14th Conference on Uncertainty in Artificial Intelligence (UAI)*, pp. 514–522.

Zhang, N. and D. Poole (1994). A simple approach to Bayesian network computations. In *Proceedings of the 10th Biennial Canadian Artificial Intelligence Conference*, pp. 171–178.

Zhang, N. and D. Poole (1996). Exploiting contextual independence in probabilistic inference. *Journal of Artificial Intelligence Research 5*, 301–328.

Zhang, N., R. Qi, and D. Poole (1993). Incremental computation of the value of perfect information in stepwise-decomposable influence diagrams. In *Proc. 9th Conference on Uncertainty in Artificial Intelligence (UAI)*, pp. 400–407.

Zhang, N. L. (2004). Hierarchical latent class models for cluster analysis. *Journal of Machine Learning Research 5*, 697–723.

Zoeter, O. and T. Heskes (2006). Deterministic approximate inference techniques for conditionally Gaussian state space models. *Statistical Computing 16*, 279–292.

Zweig, G. and S. J. Russell (1998). Speech recognition with dynamic Bayesian networks. In *Proc. 14th Conference on Artificial Intelligence (AAAI)*, pp. 173–180.

Notation Index

Subject Index